TECHNIQUES OF CHEMISTRY

VOLUME II

ORGANIC SOLVENTS

PHYSICAL PROPERTIES AND METHODS OF PURIFICATION

Fourth Edition

by

JOHN A. RIDDICK

Baton Rouge, Louisiana

WILLIAM B. BUNGER

Indiana State University
Terre Haute, Indiana

and

THEODORE K. SAKANO

Rose-Hulman Institute of Technology
Terre Haute, Indiana

On the basis of the First Edition
by ARNOLD WEISSBERGER and ERIC S. PROSKAUER
the completely revised Second Edition
by JOHN A. RIDDICK and EMORY E. TOOPS, JR.
  and
the completely revised Third Edition
by JOHN A. RIDDICK and WILLIAM B. BUNGER

A Wiley-Interscience Publication
JOHN WILEY & SONS
New York · Chichester · Brisbane · Toronto · Singapore

Copyright 1934 by Interscience Publishers, Inc.
Copyright 1955 by Interscience Publishers, Inc.
Copyright 1970 by John Wiley & Sons, Inc.
Copyright © 1986 by John Wiley & Sons, Inc.

Library of Congress Cataloging in Publication Data:

Riddick, John A. (John Allen), 1903–
    Organic solvents.
    (Techniques of chemistry; v. 2)
    "On the basis of the first edition by Arnold
Weissberger and Eric S. Proskauer, the completely
revised second edition by John A. Riddick and Emory E.
Toops, Jr., and the completely revised third edition
by John A. Riddick and William B. Bunger."
    "A Wiley-Interscience publication."
    Bibliography: p.
    Includes index.
    1. Nonaqueous solvents. I. Bunger, William B.,
1917–   . II. Sakano, Theodore. III. Weissberger,
Arnold, 1898–   . Organic solvents. IV. Title.
V. Series.
QD61.T4 vol. 2, 1986   542 s [547.1'3423] 86-15698
[QD544.5]
ISBN 0-471-08467-0

Printed in the United States of America

10 9 8 7 6 5 4 3 2 1

TECHNIQUES OF CHEMISTRY

ARNOLD WEISSBERGER, *Editor*

VOLUME II

ORGANIC SOLVENTS

Fourth Edition

In Memoriam
WILLIAM B. BUNGER
1917–1984
teacher, scholar, friend

INTRODUCTION TO THE SERIES

Techniques of Chemistry is the successor to the Technique of Organic Chemistry Series and its companion—Technique of Inorganic Chemistry. Because many of the methods are employed in all branches of chemical science, the division into techniques for organic and inorganic chemistry has become increasingly artificial. Accordingly, the new series reflects the wider application of techniques, and the component volumes for the most part provide complete treatments of the methods covered. Volumes in which limited areas of application are discussed can be easily recognized by their titles.

Like its predecessors, the series is devoted to a comprehensive presentation of the respective techniques. The authors give the theoretical background for an understanding of the various methods and operations and describe the techniques and tools, their modifications, their merits and limitations, and their handling. The series should contribute to a better understanding and a more rational and effective application of the respective techniques.

Authors and editors hope that readers will find the volumes in this series useful and will communicate to them any criticisms and suggestions for improvements.

ARNOLD WEISSBERGER

Research Laboratories
Eastman Kodak Company
Rochester, New York

PREFACE

The Fourth Edition of *Organic Solvents* is considered a continuation of the Third Edition. Besides the updating of physical properties and purification and preparation techniques of Third Edition solvents, 150 solvents have been added, bringing the number of compounds to 504. Many of the new solvents were selected to complete groups of isomers and to expand homologous series of the Third Edition.

Several isomeric and homologous series were initiated in the Fourth Edition, including the xylenols, chlorinated toluenes, xylidenes, picolines, lutidines, and silanes.

The use and importance of aerosol propellants is steadily increasing. Over a dozen propellants are represented among the lighter hydrocarbons, aliphatic fluorinated hydrocarbons, and mixed halogenated hydrocarbons.

The major change in the tables in Chapter III is the listing of numerical values in both cgs and SI units, whereas previous editions used only cgs units. A complete revision of the explanatory material (properties, symbols, unit definitions) is given in the introduction.

The number of tabulated properties has been increased by six: pressure coefficient dp/dt, isothermal compressibility κ_T, adiabatic compressibility κ_S, thermal conductivity λ, solubility parameter δ, and evaporation rates. The most significant addition is the solubility parameter, which is used as a primary factor in the designing of solutions.

Finally, the method of citing journal references in the Bibliography has been changed to conform more nearly to the style in *Chemical Abstracts*. Thus, the year of publication appears first, followed by the volume in italics and the first page of the reference.

We are grateful to the assistance from the staffs of the Chemistry Library of Louisiana State University and the Science Library of Indiana State University, and thank the Waters Computing Center of Rose-Hulman Institute of Technology for the use of computer facilities in the compilation of the bibliography.

For assistance in the preparation of Chapter III tables and the bibliography we are grateful to Troy and Kelly Mazely, Darby Andrew Keeney, Gary L. Wease, Daniel R. Crane, Evan R. Kokoska, Robert J. Tiller, and others.

From John Wiley & Sons, Margery Carazzone provided valuable information for the production of the manuscript and Dr. Stanley F. Kudzin gave encouragement and support throughout the project.

THEODORE K. SAKANO
JOHN A. RIDDICK

Terre Haute, Indiana
Baton Rouge, Louisiana
March 1986

PREFACE TO THE THIRD EDITION

Since the appearance of the Second Edition of *Organic Solvents* the role of the solvent has become much more significant and better understood. Many more solvents are available for industrial use, generally in a good to high grade of purity, and numerous additional solvents have become available for research.

One hundred solvents have been added to the Third Edition; we could not find justification for deleting any that were listed in the Second Edition. Readers of the Second Edition suggested many compounds and properties to be added in the Third Edition. Of these suggestions those were accepted that were in keeping with the aim of the book, provided sufficient information was available. Primary consideration was given to the uniqueness of the solvent property and to present and possible future uses of the solvent, particularly in the newer fields of application such as electrochemical reactions, thin layer chromatography, and preparative solvent extraction. Several solvents were added to complete groups of isomers and to expand homologous series to include a reasonable number of compounds, for instance in the C_nH_{2n+2} and C_nH_{2n} series. The pentane isomers are complete. The *n*-alkyl alcohols through hexanol are given, as are the isomers of the C_3, C_4, and C_5 alcohols.

In most of the functional group classes unsaturated compounds have been added. The amides have been increased from two to nine in keeping with the expanding interest in research and in industrial use. The number of sulfur compounds has been doubled and includes dimethyl sulfoxide and sulfolane. Several heterocyclic solvents have been added. Newly included unique compounds of limited availability are 1,1,2,2-tetramethylurea and hexamethylphosphoric triamide.

Water is a solvent of reference and a standard for calibration and comparison; the physical properties of water have been listed at the beginning of the tables in Chapter III as Tables 0a, 0b, and 0c. We know of no other place where so many of the properties of water are collected in easy-to-use tables.

The number of properties tabulated has been increased by twelve, not including the references to spectra and some properties that have synonymous methods of expression such as the ionization constant K_i, and the negative logarithm of the acid dissociation constant pK_a. The newly added properties were selected from a study of the scientific literature beginning in 1960 to determine which properties were used most frequently. Several properties that were of limited academic interest have become useful in newer areas such as electronics and space science.

Eight Index Tables appear in Chapter III, including four new ones—Ebullioscopic Constants, Cryoscopic Constants K_f and A for Equation 2.72 in Chapter II, and Density.

The format of the book remains the same as in the First and Second Editions, as does the aim, "to make readily accessible the abundant material" that has accumulated in the literature on organic solvents.

Chapter II, "Discussion of Properties and Criteria for their Selection," has been completely revised to facilitate the use of the tables in Chapter III. The newer concepts relating to purity have been included in Chapter IV, and the more recent methods for characterizing purity are discussed.

We have calculated many values to make the tables more useful. Some of the calculated values, such as the temperature coefficients of density and of the refractive index, were used during the evaluation of the data prior to inclusion in the tables.

We express our gratitude for their suggestions for the Third Edition to Dr. Paul D. Bartlett, Harvard University; Professor G. Cauquis, Centre d'Etudes Nucleaires de Grenoble; Dr. G. L. Covert, Eastman Kodak Company; Dr. Lyman C. Craig, Rockefeller Institute; Mr. P. B. Dalton, General Aniline and Film Corporation; the late Dr. Henry de Laszlo, Koch-Light Laboratories; the late Dr. J. B. Dickey, Tennessee Eastman Company; Mr. Hugh J. Hagemeyer, Texas Eastman Company; Dr. Dan H. Moore, South Jersey Medical Research Foundation; Dr. L. J. Roll and Dr. I. F. Salminen, Eastman Kodak Company; Dr. Dietmar Seyferth, Massachusetts Institute of Technology; Dr. Howard Steinberg, U.S. Borax Research Corporation; Dr. W. West and Mr. C. C. Unruh, Eastman Kodak Company; Dr. D. S. Young, Tennessee Eastman Company; and Dr. C. W. Zuehlke, Eastman Kodak Company.

Our sincere thanks also go to Professor Thomas F. Fagley and Dr. Ray Oglukian of Tulane University, Dr. T. C. Wehman of Michigan State University, and Dr. Robert West of Lehigh University for unpublished information from their research, and to Dr. Arthur and Elizabeth Rose of Applied Science Laboratories for their many helpful suggestions and enlightening discussions. We are grateful to Dr. Arthur and Wanda L. Campbell for their assistance during the first two years of the preparation of this edition. We are especially grateful to Mr. Gene Seidel of Crown Zellerbach Corporation for his generous help in supplying extensive information on dimethyl sulfoxide and dimethyl sulfide and his offer to furnish information on other solvents, and to Commercial Solvents Corporation for providing facilities for our work and for freely allowing the use of previously unpublished information.

To our wives we express our appreciation for their patience during the five years of preparation of this book and for their help in tabulation and proofreading.

JOHN A. RIDDICK
WILLIAM B. BUNGER

Baton Rouge, Louisiana
Terre Haute, Indiana
January 1970

FROM THE PREFACE TO
THE FIRST EDITION

In recent years improvements in the methods of organic chemistry and the enhanced interest in the chemical physics of non-aqueous solutions have led to an ever-increasing demand for variety and purity in purification.

For this reason it has seemed desirable to make a collection of physical constants and of methods of purification for these solvents. The aim of this book is to make readily accessible the abundant material which has been accumulated by chemists and physicists in recent years. To this end an organic chemist and a physical chemist have collaborated.

The success of physical and chemical work, both preparation and measurement, is often decided by choice of solvent, and this choice is sometimes not at all simple. We hope that a study of the systematic list of solvents and of the numerical data, classified systematically, contained in this book will make this task easier. The selection of solvents treated is inevitably somewhat arbitrary although we have tried to cater for as many different requirements as possible: we shall be glad of any suggestions for extension or limitation of the list or other amendments. We should be particularly grateful for copies of papers which contain measurements of physical constants of solvents or methods of their organic solvents.

A. WEISSBERGER
E. S. PROSKAUER

Oxford and Leipzig

CONTENTS

TECHNIQUES OF CHEMISTRY

ARNOLD WEISSBERGER, *Editor*

VOLUME II

ORGANIC SOLVENTS

Fourth Edition

Chapter I

CLASSIFICATION OF THE SOLVENTS

The choice of compounds includes not only the common solvents but also other compounds whose properties might render them suitable as solvents in special cases. The solvents are arranged in the order of increasing chemical complexity under the following major classes:

Hydrocarbons
Compounds with one type of characteristic atom or group (hydroxy compounds, esters, halogenateds, etc.)
Compounds with more than one type of characteristic atom or group (ether alcohols, amino alcohols, esters of keto acids, etc.)

Subdivisions are made according to the nature of the hydrocarbon residue (saturated, aromatic, unsaturated); the number of times the characteristic atom or group occurs; heterocyclic or open chain structure; and the primary, secondary, or tertiary nature of amines.

Code numbers have been assigned to compounds for easy identification and reference. The same number appears in boldface type or is underscored with each compound in every tabulation of the solvents and each relevant use of the name throughout the book.

The primary names of the solvents included in this book are, in general, those used by *Chemical Abstracts* over the years and are probably the most familiar ones used by chemists. The exceptions are mainly those compounds whose *Chemical Abstracts* name is not known well enough to be recognized readily: for example, **463.** Sulfolane and **481.** Triethanolamine.

Donaldson and coauthors [2037] present and discuss the nomenclature changes for *Chemical Abstracts* index names beginning with Volume 76, 1972. This is the first volume of the Ninth Collective period. Some of these names have been used as the primary solvent names in this book.

The current *Chemical Abstracts* index name is followed by an asterisk. Previous index names are also followed by an asterisk and the last year of use in parentheses: for example, **39.** Toluene* (1971), methylbenzene*.

WATER

0. Water

HYDROCARBONS

Saturated Aliphatic Hydrocarbons

1. Propane*
2. Butane*

3. 2-Methylpropane*, isobutane* (1916)
4. Cyclopentane*, pentamethylene
5. Pentane*, n-pentane
6. 2-Methylbutane*, ethyldimethylmethane, isopentane
7. 2,2-Dimethylpropane*, neopentane, tetramethylmethane
8. Methylcyclopentane*
9. Cyclohexane*, hexahydrobenzene, hexamethylene
10. Hexane*, n-hexane
11. 2-Methylpentane*, dimethylpropylmethane
12. 3-Methylpentane*, diethylmethylmethane
13. 2,2-Dimethylbutane*, ethyltrimethylmethane, neohexane
14. 2,3-Dimethylbutane*, diisopropyl, isopropyldimethylmethane
15. Methylcyclohexane*, hexahydrotoluene, cyclohexylmethane
16. Heptane*, n-heptane
17. 2-Methylhexane*, ethylisobutylmethane
18. 3-Methylhexane*, ethylmethylpropylmethane
19. 2,3-Dimethylpentane*, ethylisopropylmethylmethane
20. 2,4-Dimethylpentane*, diisopropylmethane
21. 1,2-Dimethylcyclohexane* (mixed isomers), o-xylolhexahydride
22. cis-1,2-Dimethylcyclohexane*
23. trans-1,2-Dimethylcyclohexane*
24. Ethylcyclohexane*
25. Octane*, n-octane
26. 2,2,3-Trimethylpentane*, tert-butylethylmethylmethane
27. 2,2,4-Trimethylpentane*, isooctane, isobutyltrimethylmethane
28. 2,2,3,3-Tetramethylbutane*, hexamethylethane, di-tert-butylethane
29. Nonane*, n-nonane
30. 2,2,5-Trimethylhexane*
31. Decahydronaphthalene* (mixed isomers), decalin, bicyclo[4.4.0]decane, naphthalane, naphthane
32. cis-Decahydronaphthalene*, cis-decalin, cis-bicyclo[4.4.0]decane
33. trans-Decahydronaphthalene*, trans-decalin, trans-bicyclo[4.4.0]decane
34. Decane*, n-decane
35. 1,1'-Bicyclohexyl*, bicyclohexyl* (1971), cyclohexylcyclohexane
36. Dodecane*, duodecane, bihexyl
37. Tridecane*, n-tridecane

Aromatic Hydrocarbons

38. Benzene*, benzol, benzole, phene
39. Toluene* (1971), methylbenzene*, phenylmethane
40. o-Xylene* (1971), 1,2-dimethylbenzene*, 2-xylene
41. m-Xylene* (1971), 1,3-dimethylbenzene*, 3-xylene
42. p-Xylene* (1971), 1,4-dimethylbenzene*, 4-xylene
43. Ethylbenzene*, phenylethane
44. Isopropylbenzene, cumene* (1971), 1-methylethylbenzene*, cumol, 2-phenylpropane

45. Mesitylene* (1971), 1,3,5-trimethylbenzene*, *sym*-trimethylbenzene
46. Naphthalene*, naphthene
47. 1-Methylnaphthalene*, α-methylnaphthalene
48. 1,2-Dimethylnaphthalene*
49. 1,6-Dimethylnaphthalene*
50. 1,2,3,4-Tetrahydronaphthalene*, tetralin, tetrahydronaphthalene, naphthalene-1,2,3,4-tetrahydride
51. Butylbenzene*
52. *sec*-Butylbenzene* (1971), 1-methylpropylbenzene*, 2-phenylbutane
53. Isobutylbenzene* (1971), 2-methylpropylbenzene*, 1-phenyl-2-methylpropane
54. *tert*-Butylbenzene* (1971), 1,1-dimethylethylbenzene*, 2-methyl-2-phenylpropane, pseudobutylbenzene, trimethylphenylmethane
55. *p*-Cymene* (1971), 1-methyl-4-(1-methylethyl)benzene*, cymene, 4-isopropyl-1-methylbenzene, *p*-isopropyltoluene
56. Cyclohexylbenzene*, phenylcyclohexane* (1950)

Unsaturated Hydrocarbons

57. 1-Pentene*, amylene, α-*n*-amylene, propylethylene
58. 2-Pentene* (mixed isomers), β-*n*-amylene, *sym*-ethylmethylethylene
59. *cis*-2-Pentene* (1966), (*Z*)-2-pentene*
60. *trans*-2-Pentene* (1966), (*E*)-2-pentene*
61. 1-Hexene*, α-hexene
62. 1-Heptene*, α-heptene
63. 1-Octene*, α-octene, caprylene, α-octylene
64. 2-Octene* (mixed isomers)*, β-octene
65. *cis*-2-Octene* (1966), (*Z*)-2-octene*
66. *trans*-2-Octene* (1966), (*E*)-2-octene*
67. 1-Nonene*, α-nonene, *n*-heptylethylene, 1-nonylene
68. 1-Decene*, α-decene
69. 1-Dodecene*
70. 1-Tridecene*
71. Cyclohexene*, 1,2,3,4-tetrahydrobenzene, tetrahydrobenzene
72. Dipentene* and *dl*-Limonene* (1956), *dl-p*-mentha-1,8-diene* (1961), (±)-*p*-mentha-1,8-diene* (1971), (±)-1-methyl-4-(1-methylethenyl)cyclohexene*, *p*-menthadiene, 1,8(9)-*p*-menthadiene, 4-isopropenyl-1-methyl-1-cyclohexene
73. *d*-Limonene* (1956), *d-p*-mentha-1,8-diene* (1961), (*R*)-(+)-*p*-mentha-1,8-diene* (1971), (+)-1-methyl-4-(1-methylethenyl)cyclohexene*, carvene, citrene
74. *l*-Limonene* (1956), *l-p*-mentha-1,8-diene* (1961), (*S*)-(−)-*p*-mentha-1,8-diene* (1971), (−)-1-methyl-4-(1-methylethenyl)cyclohexene*
75. Styrene* (1971), ethenylbenzene*, styrol, vinylbenzene, phenylethylene
76. 2-Pinene* (1971), pinene* (1956), α-pinene, 2,6,6-trimethylbicyclo-[3.1.1]hept-2-ene*

77. 2(10)-Pinene* (1971), nopinene* (1956), β-pinene, 6,6-dimethyl-2-methy
lene bicyclo[3.1.1]heptane*

COMPOUNDS WITH ONE TYPE OF
CHARACTERISTIC ATOM OR GROUP

HYDROXY COMPOUNDS

MONOHYDRIC ALCOHOLS

Aliphatic Alcohols

78. Methanol*, methyl alcohol, wood alcohol, carbinol
79. Ethanol*, ethyl alcohol* (1971), methylcarbinol, alcohol
80. 1-Propanol*, propyl alcohol* (1971), n-propyl alcohol, ethylcarbinol
81. 2-Propanol*, isopropyl alcohol* (1971), dimethylcarbinol, sec-propyl alco
hol, isopropanol

Butyl Alcohols

82. 1-Butanol*, butyl alcohol* (1971), n-butyl alcohol, butanol, butan-1-ol, n
propylcarbinol
83. 2-Butanol*, sec-butyl alcohol* (1971), methylethylcarbinol
84. 2-Methyl-1-propanol*, isobutyl alcohol* (1971), isobutanol, isopropylcarbi
nol
85. 2-Methyl-2-propanol*, tert-butyl alcohol* (1971), trimethylcarbinol

Pentyl Alcohols

86. 1-Pentanol*, pentyl alcohol* (1971), amyl alcohol (1961), pri-n-amyl alco
hol, n-amyl alcohol, n-butylcarbinol
87. 2-Pentanol*, sec-amyl alcohol, sec-n-amyl alcohol, 1-methyl-1-butanol,
methylpropylcarbinol
88. 3-Pentanol*, diethylcarbinol, 1-ethyl-1-propanol
89. 2-Methyl-1-butanol*, act-amyl alcohol, d-pri-act-amyl alcohol, d-act-butyl
carbinol
90. 3-Methyl-1-butanol*, isopentyl alcohol* (1971), isoamyl alcohol* (1951), pri
isoamyl alcohol, isobutylcarbinol
91. 2-Methyl-2-butanol* (current and 1952–56), tert-pentyl alcohol* (1971), tert
amyl alcohol* (1951), dimethylethylcarbinol
92. 3-Methyl-2-butanol*, sec-isoamyl alcohol, methylisopropylcarbinol
93. 2,2-Dimethyl-1-propanol*, neopentyl alcohol, neopentanol, tert-butylcarbi
nol

Other Aliphatic Alcohols

94. Cyclohexanol*, cyclohexyl alcohol, hexahydrophenol, Hexalin
95. 1-Hexanol*, hexyl alcohol* (1971), n-hexyl alcohol, amylcarbinol, pentyl
carbinol

96. 2-Methyl-1-pentanol*
97. 2-Methyl-2-pentanol*, dimethylpropylcarbinol
98. 4-Methyl-2-pentanol*, methylamylalcohol, methylisobutylcarbinol
99. 2-Ethyl-1-butanol*, sec-hexyl alcohol, 2-ethylbutyl alcohol, pseudo-hexyl alcohol

Methylcyclohexanols

100. 1-Methylcyclohexanol*
101. 2-Methylcyclohexanol* (mixed isomers), o-methylcyclohexyl alcohol
102. cis-2-Methylcyclohexanol*
103. trans-2-Methylcyclohexanol*
104. 3-Methylcyclohexanol* (mixed isomers), m-methylcyclohexyl alcohol
105. cis-3-Methylcyclohexanol*
106. trans-3-Methylcyclohexanol*
107. 4-Methycyclohexanol* (mixed isomers), p-methylcyclohexyl alcohol
108. cis-4-Methylcyclohexanol*
109. trans-4-Methylcyclohexanol*

Other Aliphatic Alcohols

110. 2-Heptanol*, n-sec-heptyl alcohol, n-heptan-2-ol, n-heptanol-2, methylamylcarbinol
111. 3-Heptanol*, ethylbutylcarbinol
112. 1-Octanol*, octyl alcohol* (1971), caprylic alcohol, n-octyl alcohol, heptylcarbinol
113. 2-Octanol*, capryl alcohol, methylhexylcarbinol, sec-octyl alcohol
114. 2-Ethyl-1-hexanol*, 2-ethylhexyl alcohol, 1-ethyl-n-amylcarbinol

Aromatic Alcohols

115. Benzyl alcohol* (1971), benzenemethanol*, phenylmethyl alcohol, α-hydroxytoluene, phenylcarbinol

Phenols

116. Phenol*, carbolic acid, hydroxybenzene, phenic acid

Cresols

117. o-Cresol* (1971), 2-methylphenol*, o-cresylic acid, o-hydroxytoluene, o-methylphenol, 1-hydroxy-2-methylbenzene
118. m-Cresol* (1971), 3-methylphenol*, m-cresylic acid, m-hydroxytoluene, m-methylphenol, 1-hydroxy-3-methylbenzene
119. p-Cresol* (1971), 4-methylphenol*, p-cresylic acid, p-hydroxytoluene, p-methylphenol, 1-hydroxy-4-methylbenzene

Xylenols

120. 2,4-Xylenol* (1971), 2,4-dimethylphenol*
121. 2,5-Xylenol* (1971), 2,5-dimethylphenol*

122. 2,6-Xylenol* (1971), 2,6-dimethylphenol*
123. 3,4-Xylenol* (1971), 3,4-dimethylphenol*
124. 3,5-Xylenol* (1971), 3,5-dimethylphenol*

Unsaturated Monohydric Alcohols

125. 2-Propen-1-ol*, allyl alcohol* (1971), propene(1)-ol(3), Δ^3-1-propenol, vinylcarbinol
126. *cis*-2-Buten-1-ol* (1966), (Z)-2-buten-1-ol*, *cis*-crotonyl alcohol* (1926), Δ^2-1-butenol* (1916), crotyl alcohol, γ-methylallyl alcohol, propenylcarbinol
127. *trans*-2-Buten-1-ol* (1966), (E)-2-buten-1-ol*
128. 2-Propyn-1-ol*, 2-propin-1-ol* (1936), propargylic alcohol* (1934), propargyl alcohol, acetylenecarbinol
129. α-Terpineol* (1956), *dl*-α-terpineol, *p*-menth-1-en-8-ol* (1971), (±)-*p*-menth-1-en-8-ol* (1971), α,α,4-trimethyl-3-cyclohexene-1-methanol*
130. *d*-α-Terpineol* (1956), (+)-*p*-menth-1-en-8-ol* (1966), (R)-(+)-*p*-menth-1-en-8-ol* (1971), (R)-(+)-α,α,4-trimethyl-3-cyclohexene-1-methanol*
131. *l*-α-Terpineol* (1956), (−)-*p*-menth-1-en-8-ol* (1966), (S)-(−)-*p*-menth-1-en-8-ol* (1971), (S)-(−)-α,α,4-trimethyl-3-cyclohexene-1-methanol*

POLYHYDRIC ALCOHOLS

132. *cis*-2-Butene-1,4-diol* (1966), (Z)-2-butene-1,4-diol*
133. *trans*-2-Butene-1,4-diol* (1966), (E)-2-butene-1,4-diol*
134. 1,2-Ethanediol*, ethylene glycol* (1971), glycol, ethylene alcohol, 1,2-dihydroxyethane
135. 1,2-Propanediol*, propylene glycol, α-propylene glycol, 1,2-dihydroxypropane
136. 1,3-Propanediol*, trimethylene glycol, isopropylene glycol
137. 1,3-Butanediol*, 1,3-butylene glycol
138. 1,4-Butanediol*, tetramethylene glycol, 1,4-butylene glycol, 1,4-dihydroxybutane
139. 2,3-Butanediol* (mixed isomers), 2,3-butylene glycol, *sym*-dimethylene glycol
140. *l*-2,3-Butanediol* (1966), D(−)-2,3-butanediol* (1971), (2R,3R)-(−)-2,3-butanediol* (1971), R-(R*,R*)-2,3-butanediol*, D(−)-2,3-butylene glycol
141. *d*-2,3-Butanediol* (1966), L(+)-2,3-butanediol* (1971), S-(R*,R*)-2,3-butanediol*, L(+)-2,3-butylene glycol
142. *dl*-2,3-Butanediol* (1966), *racemic*-2,3-butanediol, (±)-2,3-butanediol* (1971), (R*,R*)-(±)-2,3-butanediol*, (±)-2,3-butylene glycol
143. *meso*-2,3-butanediol* (1971), (R*,S*)-2,3-butanediol*
144. 1,5-Pentanediol*, pentamethylene glycol
145. Hexylene glycol, 2-methyl-2,4-pentanediol*, 1,1,3-trimethyltrimethylene glycol
146. 2-Ethyl-1,3-hexanediol*, 3-hydroxymethyl-4-heptanol, ethohexadiol
147. Glycerol* (1971), 1,2,3-propanetriol*, glycerin, 1,2,3-trihydroxypropane
148. 1,2,6-Hexanetriol*, 1,2,6-trihydroxyhexane

ETHERS

Aliphatic Open-Chain Ethers

149. Methyl ether* (1914, 1971), dimethyl ether* (1915–16), oxybismethane*
150. Ethyl vinyl ether* (1971), ethoxyethene*, vinyl ethyl ether
151. Ethyl ether* (1971), 1,1'-oxybisethane*, diethyl ether, ethoxyethane, ethyl oxide, sulfuric ether, 3-oxapentane, ether
152. Propyl ether* (1971), 1,1'-oxybispropane*, di-*n*-propyl ether, 1-propoxypropane
153. Isopropyl ether* (1971), 2,2'-oxybispropane*, diisopropyl ether, 2-isopropoxypropane
154. Butyl vinyl ether* (1971), 1-(ethenyloxy)butane*, vinyl butyl ether
155. Butyl ethyl ether* (1971), 1-ethoxybutane*, *n*-butyl ethyl ether
156. Butyl ether* (1971), 1,1'-oxybisbutane*, dibutyl ether, di-*n*-butylether, 1-butoxybutane
157. Pentyl ether* (1971), 1,1'-oxybispentane*, amyl ether* (1951), *n*-amyl ether, di-*n*-amyl ether, 1-pentoxypentane
158. Isopentyl ether* (1971), 1,1'-oxybis(3-methylbutane)*, isoamyl ether* (1951), diisoamyl ether
159. 1,2-Dimethoxyethane*, *sym*-dimethoxyethane* (1941), monoglyme, dimethyl Cellosolve, ethylene glycol dimethyl ether
160. 1,2-Diethoxyethane*, ethylene glycol diethyl ether, ethyl glyme, diethyl Cellosolve
161. 1,2-Dibutoxyethane* (1971), 1,1'-[1,2-ethanediylbis(oxy)]bisbutane*, ethylene glycol dibutyl ether, dibutyl Cellosolve
162. Bis(2-methoxyethyl) ether* (1971), 1,1'-oxybis[2-methoxyethane]*, diglyme, diethylene glycol dimethyl ether
163. Bis(2-ethoxyethyl) ether* (1971), bis(β-ethoxyethyl) ether* (1946), 1,1'-oxybis[2-ethoxyethane]*, diethyl carbitol, diethylene glycol diethyl ether, diethoxydiethylene glycol, ethyl diglyme
164. Bis(2-butoxyethyl) ether* (1971), bis(β-butoxyethyl) ether* (1946), 1,1'-[oxybis(2,1-ethanediyloxy)]bis-butane*, dibutyl carbitol, diethylene glycol dibutyl ether, butyl diglyme, dibutoxydiethylene glycol, butex

Aliphatic Cyclic Ethers

165. Propylene oxide* (1971), methyloxirane*, propene oxide, 1,2-epoxypropane, 2-methyloxacyclopropane
166. 1,3-Dioxolane*, glycol methylene ether, ethylene glycol methylene ether, ethylenemethylenedioxide
167. 1,2-Epoxybutane* (1971), ethyloxirane*, ethylethylene oxide* (1931)
168. Cineole* (1961), 1,8-epoxy-*p*-menthane* (1971), cajeputole, 1,3,3-trimethyl-2-oxabicyclo[2.2.2]octane*, eucalyptole
169. Furan*, furfuran, divinylene oxide, oxole, tetrole
170. Tetrahydrofuran*, oxacyclopentane, diethylene oxide, tetramethylene oxide
171. 2-Methyltetrahydrofuran*, 1,4-oxidopentane, α-methyltetramethylene oxide

172. *p*-Dioxane* (1971), 1,4-dioxane*, diethylene dioxide
173. Tetrahydropyran*, pentamethylene oxide, oxacyclohexane

Aromatic Ethers

174. Benzylethyl ether* (1971), (ethoxymethyl)benzene*, α-ethoxytoluene
175. Anisole* (1971), methoxybenzene*, methyl phenyl ether
176. Phenetole* (1971), ethoxybenzene*, ethyl phenyl ether
177. Benzyl ether* (1971), 1,1'-[oxybis(methylene)]bisbenzene*, dibenzyl ether
178. Phenyl ether* (1971), 1,1'-oxybisbenzene*, diphenyl ether, diphenyl oxide phenoxybenzene
179. Veratrole* (1961), *o*-dimethoxybenzene* (1971), 1,2-dimethoxybenzene*

ACETALS

180. Dimethoxymethane*, methylal* (1961)
181. Acetal* (1961), acetaldehyde diethyl acetal* (1971), 1,1-diethoxyethane* ethylidine diethyl ether

CARBONYLS

ALDEHYDES

Saturated Aliphatic Aldehydes

182. Acetaldehyde*, ethanal, acetic aldehyde
183. Propionaldehyde* (1971), propanal*, methylacetaldehyde
184. Butyraldehyde* (1971), butanal*, butaldehyde, butyral, butyric aldehyde
185. Isobutyraldehyde* (1971), 2-methylpropanal*, isobutylaldehyde

Aromatic Aldehydes

186. Benzaldehyde*, benzenecarbonal, oil of bitter almond

Unsaturated Aldehydes

187. Acrolein* (1971), 2-propenal*, acryl aldehyde, acrylic aldehyde, allyl aldehyde, ethylene aldehyde
188. Crotonaldehyde* (1966), (*E*)-crotonaldehyde* (1971), (*E*)-2-butenal*, *trans*-2-butenal, crotonic aldehyde, β-methylacrolein, propenaldehyde, propylene aldehyde

KETONES

Aliphatic Ketones

189. Acetone* (1971), 2-propanone*, dimethyl ketone, methyl ketone
190. 2-Butanone*, MEK, ethyl methyl ketone, methyl ethyl ketone
191. Cyclopentanone*
192. 2-Pentanone*, methyl propyl ketone* (1916), ethyl acetone

193. 3-Pentanone*, diethyl ketone, ethyl ketone, *sym*-dimethylacetone
194. Cyclohexanone*, cyclohexyl ketone, hexanon
195. 2-Hexanone*, methyl butyl ketone, methyl *n*-butyl ketone
196. 4-Methyl-2-pentanone*, MIBK, methyl isobutyl ketone, isobutyl methyl ketone
197. 2-Heptanone*, methyl *n*-amyl ketone* (1916), methyl pentyl ketone
198. 3-Heptanone*, butyl ethyl ketone, ethyl *n*-butyl ketone
199. 2,4-Dimethyl-3-pentanone*, diisopropyl ketone* (1916)
200. 2-Octanone*, methyl hexyl ketone
201. 2,6-Dimethyl-4-heptanone*, isovalerone* (1956), diisobutyl ketone* (1916)
202. Camphor* (1966), (1*R*,4*R*)-(+)-camphor* (1971), 1,7,7-trimethyl-(1*R*)-bicyclo[2.2.1]heptan-2-one*, 2-camphanone, 2-bornanone, *d*-2-keto-1,7,7-trimethylnorcamphane

Aromatic Ketones

203. Acetophenone* (1971), 1-phenylethanone*, methyl phenyl ketone, acetylbenzene, hypnone

Unsaturated Ketones

204. 4-Methyl-3-penten-2-one*, mesityl oxide* (1956), *conjugated* mesityl oxide
205. 4-Methyl-4-penten-2-one*, isomesityl oxide

Diketones

206. 2,4-Pentanedione*, acetylacetone* (1916), diacetylmethane

ACIDS

Saturated Acids

207. Formic acid*, methanoic acid, aminic acid
208. Acetic acid*, ethanoic acid, ethylic acid, methylformic acid
209. Propionic acid* (1971), propanoic acid*, methylacetic acid
210. Butyric acid* (1971), butanoic acid*, *n*-butyric acid, ethylacetic acid
211. Isobutyric acid* (1971), 2-methylpropanoic acid*, 2-methylpropionic acid, dimethylacetic acid, α-methylpropionic acid, isopropylformic acid
212. Valeric acid* (1971), pentanoic acid*, *n*-valeric acid
213. Isovaleric acid* (1971), 3-methylbutanoic acid*, isopropylacetic acid, 3-methylbutyric acid, β-methylbutyric acid
214. Hexanoic acid*, caproic acid* (1950), capronic acid, *n*-hexoic acid, *n*-caproic acid, butylacetic acid
215. Octanoic acid*, caprylic acid* (1951)
216. Nonanoic acid*, pelargonic acid* (1946), *n*-nonoic acid, nonylic acid

Unsaturated Acids

217. Acrylic acid* (1971), 2-propenoic acid*, propenoic acid, acroleic acid, eth ylenecarboxylic acid

218. Crotonic acid* (1966), (*E*)-crotonic acid* (1971), (*E*)-2-butenoic acid*, *β* methylacrylic acid, *trans*-butenoic acid, α-crotonic acid

219. Methacrylic acid* (1971), 2-methyl-2-propenoic acid*, α-methylacrylic aci

220. Oleic acid* (1971), (*Z*)-9-octadecenoic acid*, *cis*-octadecen(9)oic acid, *cis* 9-octadecenoic acid

ACID ANHYDRIDES

221. Acetic anhydride* (1971), acetic acid anhydride*, ethanoic anhydride

222. Propionic anhydride* (1971), propanoic acid anhydride*

223. Butyric anhydride* (1971), butanoic acid anhydride*

ESTERS

Esters of Saturated Aliphatic Monocarboxylic Acids

224. Methyl formate*, methyl methanoate

225. Ethyl formate*, ethyl methanoate

226. Propyl formate*, propyl methanoate

227. Butyl formate*, butyl methanoate

228. Isobutyl formate* (1971), 2-methylpropyl formate*, isobutyl methanoate, 2 methyl-1-propyl methanoate

229. Methyl acetate*, methyl ethanoate

230. Vinyl acetate* (1971), ethenyl acetate*, ethenyl ethanoate

231. Ethylene glycol diacetate* (1971), 1,2-ethanediol diacetate*, glycol diace tate, ethylene acetate, ethylene diacetate

232. Ethyl acetate*, ethyl ethanoate, acetic ester, acetic ether

233. Propargyl acetate, 2-propyn-1-ol acetate*, propiolic acetate* (1936), 2-pro pyn acetate

234. Allyl acetate* (1971), 2-propenyl acetate*

235. Propyl acetate*, propyl ethanoate, *n*-propyl acetate

236. Isopropyl acetate* (1971), 1-methylethyl acetate*, isopropyl ethanoate, 2 propyl ethanoate

237. Triacetin* (1971), 1,2,3-propanetriol triacetate*, glyceryl triacetate

238. Butyl acetate*, butyl ethanoate, butyl acetic ether

239. Isobutyl acetate* (1971), 2-methylpropyl acetate*, isobutyl ethanoate, 2- methyl-1-propyl ethanoate, *β*-methylpropyl acetate

240. *sec*-Butyl acetate* (1971), 1-methylpropyl acetate*, 2-butyl acetate, α-meth ylpropyl acetate

241. Pentyl acetate*, amyl acetate* (1960), pentyl ethanoate, amyl acetic ether, banana oil

242. Isopentyl acetate* (1971), isoamyl acetate* (1951), 3-methyl-1-butyl ace tate*, 3-methylbutyl ethanoate, γ-methylbutyl ethanoate, amyl acetic ether

243. Hexyl acetate*
244. 4-Methyl-2-pentyl acetate*, methylamyl acetate
245. 2-Ethylhexyl acetate*, 2-ethylhexyl ethanoate
246. Benzyl acetate* (1971), phenylmethyl acetate*, benzyl ethanoate
247. Ethyl propionate* (1971), ethyl propanoate*
248. Ethyl butyrate* (1971), ethyl butanoate*, butyric ether
249. Isobutyl isobutyrate* (1971), 2-methylpropyl-2-methylpropanoate*, isobutyl 2-methylpropanoate
250. Ethyl isovalerate* (1971), ethyl-3-methylbutanoate*, ethyl isopropylacetate
251. Isopentyl isovalerate* (1971), 3-methylbutyl-3-methylbutanoate*, isoamyl isovalerate* (1951), 3-methyl-1-butyl-isovalerate
252. Butyl stearate* (1971), butyl octadecanoate*

Esters of Unsaturated Aliphatic Monocarboxylic Acids

253. Methyl acrylate* (1971), methyl 2-propenoate*, methyl propenoate
254. Ethyl acrylate* (1971), ethyl 2-propenoate*, ethyl propenoate
255. Methyl methacrylate* (1971), methyl 2-methyl-2-propenoate*, methylmethylacrylate, methacrylate monomer
256. Methyl oleate* (1971), methyl (Z)-9-octadecenoate*, methyl cis-9-octadecenoate
257. Butyl oleate* (1971), butyl (Z)-9-octadecenoate*

Esters of Aromatic Monocarboxylic Acids

258. Methyl benzoate*, methyl benzenecarboxylate, Niobe oil
259. Ethyl benzoate*, ethyl benzenecarboxylate
260. Propyl benzoate*, propyl benzenecarboxylate
261. Benzyl benzoate* (1971), phenylmethyl benzoate*, benzyl benzenecarboxylate, benzyl phenylformate, benzenoate
262. Ethyl cinnamate* (1971), ethyl 3-phenyl-2-propenoate*, ethyl trans-3-phenylpropenoate, ethyl trans-3-benzenepropenoate

Lactones

263. γ-Butyrolactone, 2(3H)-dihydrofuranone*, 4-hydroxybutyric acid γ-lactone* (1966), 4-hydroxybutanoic acid lactone, butyrolactone

Esters of Dicarboxylic Acids

264. Ethylene carbonate, cyclic ethylene carbonate* (1971), 1,3-dioxolan-2-one*, vinylene carbonate, glycol carbonate
265. Propylene carbonate* (1971), 4-methyl-1,3-dioxolan-2-one*, propanediol-1,2-carbonate
266. Ethyl carbonate* (1966), diethyl carbonate*
267. Ethyl oxalate* (1966), diethyl oxalate* (1971), diethyl ethanedioate*, ethyl ethanedioate, oxalic ester
268. Ethyl malonate* (1966), diethyl malonate* (1971), diethyl propanedioate*, ethyl propanedioate, malonic ester

269. Methyl maleate* (1966), dimethyl maleate* (1971), (Z)-dimethyl 2-butene dioate*, methyl cis-butenedioate, methyl cis-1,2-ethylenedicarboxylate
270. Ethyl maleate* (1966), diethyl maleate* (1971), (Z)-diethyl 2-butenedioate* ethyl cis-butenedioate, ethyl cis-1,2-ethylenedicarboxylate
271. Butyl maleate* (1966), dibutyl maleate* (1971), (Z)-dibutyl 2-butenedioate* butyl cis-butenedioate, butyl cis-1,2-ethylenedicarboxylate
272. Butyl phthalate* (1966), dibutyl phthalate* (1971), dibutyl 1,2-benzenedi carboxylate*
273. Bis(2-ethylhexyl) phthalate* (1971), bis(2-ethylhexyl) 1,2-benzenedicarbox ylate*, 2-ethylhexyl phthalate, DOP, di-2-ethylhexyl phthalate
274. Butyl sebacate* (1966), dibutyl sebacate* (1971), dibutyl decanedioate*, bu tyl decanedioate

Esters of Polybasic Acids

275. Butyl borate* (1966), tributyl borate*
276. Methyl phosphate* (1966), trimethyl phosphate*
277. Ethyl phosphate* (1966), triethyl phosphate*
278. Butyl phosphate* (1966), tributyl phosphate*
279. Tricresyl phosphate (mixed isomers), tris(methylphenyl) phosphate*, tritoly phosphate* (1971), tolyl phosphate* (1966)
280. Tri-o-cresyl phosphate, tris(2-methylphenyl) phosphate*, tri-o-tolyl phos phate* (1971), o-tolyl phosphate* (1966)
281. Tri-m-cresyl phosphate, tris(3-methylphenyl) phosphate*, tri-m-tolyl phos phate* (1971), m-tolyl phosphate* (1966)
282. Tri-p-cresyl phosphate, tris(4-methylphenyl) phosphate*, tri-p-tolyl phos phate* (1971), p-tolyl phosphate* (1966)

HALOGENATED HYDROCARBONS

FLUORINATED HYDROCARBONS

Aliphatic Fluorinated Hydrocarbons

283. 1,1-Difluoroethane*, ethylidene fluoride, Propellant 152a
284. 1,1,1-Trifluoroethane, Freon-143
285. 1,1,1,2,2-Pentafluoropropane*, Refrigerant 245
286. Octafluorocyclobutane*, perfluorocyclobutane, Freon-C318, Propellant C318
287. Decafluorobutane*, perfluorobutane, n-butforane

Aromatic Fluorinated Hydrocarbons

288. Fluorobenzene*, phenyl fluoride
289. o-Fluorotoluene* (1971), 1-fluoro-2-methylbenzene*
290. m-Fluorotoluene* (1971), 1-fluoro-3-methylbenzene*
291. p-Fluorotoluene* (1971), 1-fluoro-4-methylbenzene*
292. α,α,α-Trifluorotoluene* (1971), trifluoromethylbenzene*, benzotrifluoride trifluorophenylmethane
293. Hexafluorobenzene*

Unsaturated Fluorinated Hydrocarbons

294. 1,1-Difluoroethylene* (1971), 1,1-difluoroethene*, vinylidene fluoride

MONOCHLORINATED HYDROCARBONS

Aliphatic Chlorinated Hydrocarbons

295. Chloromethane*, methyl chloride
296. Chloroethane*, ethyl chloride
297. 1-Chloropropane*, *n*-propyl chloride
298. 2-Chloropropane*, isopropyl chloride
299. 1-Chlorobutane*, *n*-butyl chloride
300. 2-Chlorobutane*, *sec*-butyl chloride, methylethylchloromethane
301. 1-Chloro-2-methylpropane*, isobutyl chloride, isopropylchloromethane
302. 2-Chloro-2-methylpropane*, *tert*-butyl chloride, trimethylchloromethane
303. 1-Chloropentane*, *n*-amyl chloride, pentyl chloride
304. 1-Chloro-3-methylbutane*, isoamyl chloride, isopentyl chloride, 1-chloro-isopentane
305. 1-Chlorooctane*, *n*-octyl chloride
306. 3-Chloromethylheptane*, 2-ethylhexyl chloride

Aromatic Chlorinated Hydrocarbons

307. Chlorobenzene*, phenyl chloride, benzene chloride, monochlorobenzene
308. α-Chlorotoluene* (1971), (chloromethyl)benzene*, benzyl chloride, 1'-chloro-1-methylbenzene
309. *o*-Chlorotoluene* (1971), 1-chloro-2-methylbenzene*, 2-chloro-1-methyl-benzene
310. *m*-Chlorotoluene* (1971), 1-chloro-3-methylbenzene*, 3-chloro-1-methyl-benzene
311. *p*-Chlorotoluene* (1971), 1-chloro-4-methylbenzene*, 4-chloro-1-methyl-benzene
312. 1-Chloronaphthalene*, α-chloronaphthalene

Unsaturated Chlorinated Hydrocarbons

313. Chloroethylene* (1971), chloroethene*, vinyl chloride
314. 3-Chloropropene*, allyl chloride, chloroallylene, 3-chloropropylene, 3-chloroprene

POLYCHLORINATED HYDROCARBONS

Aliphatic Chlorinated Hydrocarbons

315. Dichloromethane*, methylene chloride, methylene dichloride
316. Chloroform* (1971), trichloromethane*, methylene trichloride
317. Carbon tetrachloride* (1971), tetrachloromethane*, perchloromethane
318. 1,1-Dichloroethane*, *asym*-dichloroethane, ethylidene chloride, ethylidene dichloride

319. 1,2-Dichloroethane*, *sym*-dichloroethane, ethylene chloride, ethylene di chloride, EDC
320. 1,2-Dichloropropane*, propylene chloride, propylene dichloride
321. 1,1,1-Trichloroethane*, methylchloroform, trichloroethane, Chlorothene
322. 1,1,2-Trichloroethane*
323. 1,2,3-Trichloropropane*, glycerin trichlorohydrin, trichlorohydrin
324. 1,1,2,2-Tetrachloroethane*, acetylene tetrachloride, tetrachloroethane
325. Pentachloroethane*, 1,1,1,2,2-pentachloroethane, pentalin

Aromatic Chlorinated Hydrocarbons

326. *o*-Dichlorobenzene* (1971), 1,2-dichlorobenzene*
327. *m*-Dichlorobenzene* (1971), 1,3-dichlorobenzene*
328. *p*-Dichlorobenzene* (1971), 1,4-dichlorobenzene*, Dichloricide
329. α,α-Dichlorotoluene* (1971), (dichloromethyl)benzene*, benzal chloride benzylidene chloride
330. 2,4-Dichlorotoluene* (1971), 2,4-dichloro-1-methylbenzene*
331. 3,4-Dichlorotoluene* (1971), 1,2-dichloro-4-methylbenzene*
332. α,α,α-Trichlorotoluene* (1971), (trichloromethyl)benzene*, benzotrichlo ride, benzenyl chloride, phenyl chloroform

Unsaturated Chlorinated Hydrocarbons

333. 1,1-Dichloroethylene* (1971), 1,1-dichloroethene*, vinylidene chloride
334. *cis*-1,2-Dichloroethylene* (1966), (Z)-1,2-dichloroethylene* (1971), (Z)-1,2 dichloroethene*, *cis*-acetylene dichloride
335. *trans*-1,2-Dichloroethylene* (1966), (E)-1,2-dichloroethylene* (1971), (E) 1,2-dichloroethene*, *trans*-acetylene dichloride
336. Trichloroethylene* (1971), trichloroethene*, ethinyl trichloride
337. Tetrachloroethylene* (1971), tetrachloroethene*, ethylene tetrachloride, per chloroethylene, Perclene

BROMINATED HYDROCARBONS

Aliphatic Monobrominated Hydrocarbons

338. Bromomethane*, methyl bromide
339. Bromoethane*, ethyl bromide, monobromoethane
340. 1-Bromopropane*, *n*-propyl bromide
341. 2-Bromopropane*, isopropyl bromide
342. 1-Bromobutane*, *n*-butyl bromide
343. 2-Bromobutane*, *sec*-butyl bromide
344. 2-Bromo-2-methylpropane*, *tert*-butyl bromide, 3-bromo-isobutane
345. 1-Bromopentane*, *n*-amyl bromide, pentyl bromide
346. 1-Bromodecane*, *n*-decyl bromide

Aromatic Monobrominated Hydrocarbons

347. Bromobenzene*, phenyl bromide, monobromobenzene
348. 1-Bromonaphthalene*, α-bromonaphthalene

Unsaturated Monobrominated Hydrocarbons

349. Bromoethylene* (1971), bromoethene*, vinyl bromide
350. 2-Bromopropene*, β-bromopropylene

Aliphatic Polybrominated Hydrocarbons

351. Bromoform* (1961), tribromomethane*
352. 1,2-Dibromoethane*, *sym*-dibromoethane, ethylene bromide, ethylene dibromide
353. 1,2-Dibromopropane*, propylene bromide, propylene dibromide
354. 1,1,2,2-Tetrabromoethane*, acetylene tetrabromide, tetrabromoethane, *sym*-tetrabromoethane

IODINATED HYDROCARBONS

Aliphatic Iodinated Hydrocarbons

355. Iodomethane*, methyl iodide
356. Iodoethane*, ethyl iodide
357. 1-Iodopropane*, *n*-propyl iodide
358. 2-Iodopropane*, isopropyl iodide
359. 1-Iodobutane*, *n*-butyl iodide
360. 1-Iodo-2-methylpropane*, isobutyl iodide
361. Diiodomethane*, methylene iodide

Aromatic Iodinated Hydrocarbons

362. Iodobenzene*

MIXED HALOGENATED HYDROCARBONS

363. Chlorotrifluoromethane*, Freon-13, trifluorochloromethane
364. Dichlorodifluoromethane*, Freon-12, difluorodichloromethane
365. Trichlorofluoromethane*, Freon-11, fluorotrichloromethane
366. Chlorodifluoromethane*, Freon-22, difluorochloromethane
367. Dichlorofluoromethane*, Freon-21, fluorodichloromethane
368. Dibromofluoromethane*, fluorodibromomethane
369. Bromochloromethane*, chlorobromomethane
370. Bromochlorofluoromethane*, fluorochlorobromomethane
371. Chloropentafluoroethane*, Freon-115, pentafluorochloroethane
372. 1,2-Dichlorotetrafluoroethane* (1971), 1,2-dichloro-1,1,2,2-tetrafluoroethane* (current and 1927–46), Freon-114
373. 1,1,2-Trichloro-1,2,2-trifluoroethane*, 1,1,2-trichlorotrifluoroethane* (1966), Freon-113, F-113, Genetron-113
374. 1,1,2,2-Tetrachlorodifluoroethane* (1966), 1,1,2,2-tetrachloro-1,2-difluoroethane*, Freon-112, F-112
375. 1-Chloro-1,1-difluoroethane*, Propellant 142b, 1,1-difluoro-1-chloroethane
376. 1,2-Dibromotetrafluoroethane* (1971), 1,2-dibromo-1,1,2,2-tetrafluoroethane*, Freon-114B2, F-114B2

377. 1,2-Dibromo-1,1-difluoroethane*, 1,1-difluoro-1,2-dibromoethane
378. 1-Bromo-2-chloroethane*, 2-chloro-1-bromoethane, ethylene chlorobromide
379. 1,1-Dichloro-2,2-difluoroethylene* (1971), 1,1-dichloro-2,2-difluoroethene*, 2,2-difluoro-1,1-dichloroethylene

NITROGEN COMPOUNDS

NITRO COMPOUNDS

380. Nitromethane*, NM, nitrocarbol
381. Nitroethane*, NE
382. 1-Nitropropane*, 1-NP
383. 2-Nitropropane*, 2-NP
384. Nitrobenzene*, nitrobenzol, Oil of Mirbane

NITRILES

Aliphatic Nitriles

385. Acetonitrile*, ethanenitrile, methyl cyanide, cyanomethane, ethyl nitrile
386. Propionitrile* (1971), propanenitrile*, ethyl cyanide
387. Succinonitrile* (1971), butanedinitrile*, 1,2-dicyanoethane, ethylene cyanide, ethylene dicyanide
388. Butyronitrile* (1971), butanenitrile*, 1-cyanopropane, n-propyl cyanide
389. Isobutyronitrile* (1971), 2-methylpropanenitrile*, isopropyl cyanide, 2-cyanopropane, 2-methylpropionitrile
390. Valeronitrile* (1971), pentanenitrile*, n-butyl cyanide
391. Hexanenitrile*, capronitrile* (1951), caproic nitrile, n-amyl cyanide, pentyl cyanide
392. 4-Methylvaleronitrile* (1971), 4-methylpentanenitrile*, isocapronitrile* (1952), isoamyl cyanide
393. Octanenitrile*, caprylonitrile* (1950), n-heptyl cyanide, octyl nitrile
394. α-Tolunitrile* (1946), phenylacetonitrile* (1971), benzeneacetonitrile*, benzyl cyanide

Aromatic Nitriles

395. Benzonitrile*, phenyl cyanide, cyanobenzene

Unsaturated Nitriles

396. Acrylonitrile* (1971), 2-propenenitrile*, vinyl cyanide, cyanoethylene
397. Methacrylonitrile* (1971), 2-methyl-2-propenenitrile*, 2-cyanopropene, α-methacrylonitrile

AMINES

Primary Monoamines

Primary Aliphatic Amines

398. Methylamine* (1971), methanamine*
399. Propylamine* (1971), 1-propanamine*, n-propylamine, monopropylamine, 1-aminopropane

400. Isopropylamine* (1971), 2-propanamine*, 2-propylamine, 2-aminopropane, monoisopropylamine

401. Butylamine* (1971), 1-butanamine*, *n*-butylamine, 1-aminobutane, monobutylamine

402. Isobutylamine*(1971), 2-methyl-1-propanamine*, 1-amino-2-methylpropane, monoisobutylamine

403. *sec*-Butylamine* (1971), 2-butanamine*, 2-aminobutane, α-methylpropylamine

404. *tert*-Butylamine* (1971), 2-methyl-2-propanamine*, 2-amino-2-methylpropane, 2-methylisopropylamine, 2-aminoisobutane

405. Pentylamine* (1971), 1-pentanamine*, amylamine* (1946), 1-aminopentane

406. Cyclohexylamine* (1971), cyclohexanamine*, aminocyclohexane, hexahydroaniline

407. 2-Ethylhexylamine* (1971), 2-ethyl-1-hexanamine*

Primary Aromatic Amines

408. Aniline*, aminobenzene, phenylamine

409. *o*-Toluidine* (1971), 2-methylbenzenamine*, 2-toluidine, *o*-aminotoluene, *o*-methylaniline, 2-aminomethylbenzene

410. *m*-Toluidine* (1971), 3-methylbenzenamine*, 3-toluidine, *m*-aminotoluene, *m*-methylaniline, 3-aminomethylbenzene

411. *p*-Toluidine* (1971), 4-methylbenzenamine*, 4-toluidine, *p*-aminotoluene, *p*-methylaniline, 4-aminomethylbenzene, tolylamine

412. 2,4-Xylidene* (1971), 2,4-dimethylbenzenamine*, 2,4-dimethylaniline, 4-amino-*m*-xylene, 4-amino-1,3-dimethylbenzene

413. 2,6-Xylidene* (1971), 2,6-dimethylbenzenamine*, 2,6-dimethylaniline, 2-amino-*m*-xylene, 2-amino-1,3-dimethylbenzene

Primary Unsaturated Amines

414. Allylamine* (1971), 2-propen-1-amine*, 2-propenylamine, 3-aminopropylene

Primary Polyamines

415. Ethylenediamine* (1971), 1,2-ethanediamine*, 1,2-diaminoethane

416. Diethylenetriamine* (1971), *N*-(2-aminoethyl)-1,2-ethanediamine*, 2,2'-diaminodiethylamine

Secondary Monoamines

417. Ethylenimine* (1971), aziridine*, dimethylenimine, vinylimine

418. Dimethylamine* (1971), *N*-methylmethanamine*

419. Diethylamine* (1971), *N*-ethylethanamine*

420. Dipropylamine* (1971), *N*-propyl-1-propanamine*, di-*n*-propylamine

421. Diisopropylamine* (1971), *N*-(1-methylethyl)-2-propanamine*, 2,4-dimethyl-3-azapentane

422. Dibutylamine* (1971), *N*-butyl-1-butanamine*, di-*n*-butylamine

423. Pyrrole* (1971), 1*H*-pyrrole*, azole

424. Pyrrolidine*, tetramethylenimine, tetrahydropyrrole
425. Piperidine*, hexahydropyridine, pentamethylenimine
426. N-Methylaniline* (1971), N-methylbenzenamine*

Tertiary Monoamines

427. Trimethylamine* (1971), N,N-dimethylmethanamine*
428. Triethylamine* (1971), N,N-diethylethanamine*
429. Tributylamine* (1971), N,N-dibutyl-1-butanamine*
430. N,N-Dimethylaniline* (1971), N,N-dimethylbenzenamine*
431. Pyridine*, azine
432. 2-Picoline* (1971), 2-methylpyridine*, α-picoline, α-methylpyridine
433. 3-Picoline* (1971), 3-methylpyridine*, β-picoline, β-methylpyridine
434. 4-Picoline* (1971), 4-methylpyridine*, γ-picoline, γ-methylpyridine
435. 2,4-Lutidine* (1971), 2,4-dimethylpyridine*, α,γ-dimethylpyridine, α,γ-lutidine
436. 2,6-Lutidine* (1971), 2,6-dimethylpyridine*, α,α'-dimethylpyridine, α,α' lutidine
437. 2,4,6-Trimethylpyridine*, s-collidine* (1971)
438. Quinoline*
439. Isoquinoline*, leucoline, 2-benzazine

AMIDES

440. Formamide*, methanamide
441. N-Methylformamide*
442. N,N-Dimethylformamide*, dimethylformamide, DMF
443. Acetamide*
444. N-Methylacetamide*, monomethylacetamide
445. N,N-Dimethylacetamide*, DMAC
446. N-Methylpropionamide* (1971), N-methylpropanamide*, monomethylpropionamide
447. 1,1,3,3-Tetramethylurea* (1966), tetramethylurea*, temur
448. 2-Pyrrolidinone*, butyrolactam, pyrrolidone
449. 1-Methyl-2-pyrrolidinone*, N-methylpyrrolidone, methylpyrrolidone
450. ε-Caprolactam, caprolactam, hexahydro-2H-azepin-2-one*, 2-oxahexamethylenimine* (1956), 2-ketohexamethylenimine

SULFUR COMPOUNDS

Sulfides

451. Carbon disulfide*, carbon bisulfide
452. Dimethyl disulfide*, methyl disulfide* (1971)

Thiols

453. 1-Butanethiol*, n-butyl mercaptan, thiobutyl alcohol
454. Benzenethiol*, thiophenol, phenyl mercaptan

Thioethers

455. Methyl sulfide* (1971), thiobismethane*, dimethyl sulfide, 2-thiapropane
456. Ethyl sulfide* (1971), 1,1'-thiobisethane*, diethyl sulfide, ethyl thioethane, 3-thiapentane
457. Butyl sulfide* (1971), 1,1'-thiobisbutane*, *n*-butyl sulfide, di-*n*-butyl sulfide
458. Thiophene*, thiofuran, thiofurfuran, thiole, thiotetrole
459. Tetrahydrothiophene*, thiacyclopentane, diethylene sulfide, thiophane
460. 2-Methylthiophene*, α-methylthiophene
461. 3-Methylthiophene*, β-methylthiophene

Oxo-Sulfur Compounds

462. Dimethyl sulfoxide, methyl sulfoxide* (1971), DMSO, sulfinylbismethane*, methylsulfinylmethane
463. Sulfolane, tetrahydrothiophene 1,1-dioxide*, 2,3,4,5-tetrahydrothiophene 1,1-dioxide

COMPOUNDS WITH MORE THAN ONE TYPE OF CHARACTERISTIC ATOM OR GROUP

Ether Alcohols

464. 2-Methoxyethanol*, methyl Cellosolve, methyl glycol, ethylene glycol monomethyl ether
465. 2-Ethoxyethanol*, Cellosolve, hydroxy ether, glycol ethyl ether, glycol monoethyl ether, ethylene glycol monoethyl ether
466. 2-Butoxyethanol*, butyl Cellosolve, butyl glycol, ethylene glycol monobutyl ether
467. Furfuryl alcohol* (1971), 2-furanmethanol*, furfurol, 2-furancarbinol, 2-hydroxymethylfuran
468. Tetrahydrofurfuryl alcohol* (1971), tetrahydro-2-furanmethanol*, tetrahydro-2-furancarbinol
469. Diethylene glycol* (1971), 2,2'-oxybisethanol*, diglycol, di-2-hydroxyethyl ether, 2,2'-dihydroxyethyl ether, ethylene diglycol, 2,2'-oxydiethanol, 3-oxa-1,5-pentanediol
470. Triethylene glycol* (1971), 2,2'-[1,2-ethanediylbis(oxy)]bisethanol*, triglycol, 2,2'-ethylene-dioxydiethanol, 3,6-dioxa-1,8-octanediol, glycol bis(hydroxyethyl) ether, ethylene glycol dihydroxydiethyl ether
471. 2-(2-Methoxyethoxy)ethanol*, methyl Carbitol, methyl diglycol, methoxydiethylene glycol, methoxy diglycol, diethylene glycol monomethyl ether
472. 2-(2-Ethoxyethoxy)ethanol*, Carbitol, ethoxy diglycol, ethyl Carbitol, 2-(β-ethoxyethoxy)ethanol, diethylene glycol monoethyl ether

Carbonyl Alcohols

473. Salicylaldehyde* (1971), 2-hydroxybenzaldehyde*, salicylic aldehyde, *o*-hydroxybenzaldehyde

474. 4-Hydroxy-4-methyl-2-pentanone*, diacetone alcohol, diacetone, 4-hydroxy 2-keto-4-methylpentane

Halogenated Alcohols

475. 2,2,2-Trifluoroethanol*
476. 2-Chloroethanol*, ethylene chlorohydrin, 2-chloroethyl alcohol, β-chloro ethyl alcohol

Cyano Alcohols

477. 2-Cyanoethanol, hydracrylonitrile* (1971), ethylene cyanohydrin, 3-hydroxy propanenitrile*, 2-cyanoethyl alcohol, β-cyanoethyl alcohol, 3-hydroxypro pionitrile, glycol cyanohydrin

Amino Alcohols

478. 2-Aminoethanol*, monoethanolamine, ethanolamine, 2-hydroxyethylamine β-hydroxyethylamine, β-aminoethyl alcohol, ethylolamine
479. 2-Amino-2-methyl-1-propanol*, AMP, β-amino-isobutanol
480. Diethanolamine, 2,2'-iminodiethanol* (1971), 2,2'-iminobisethanol*, bis(hy droxyethyl)amine, 2,2'-dihydroxydiethylamine, diethylolamine, 3-aza-1,5 pentanediol
481. Triethanolamine, 2,2',2"-nitrilotriethanol* (1971), 2,2',2"-nitrilotrisetha nol*, tris(hydroxyethyl)amine, trihydroxytriethylamine, triethylolamine, N (2-hydroxyethyl)-3-aza-1,5-pentanediol

Thioether Alcohols

482. 2,2'-Thiodiethanol* (1971), 2,2'-thiobisethanol*, bis(2-hydroxyethyl) sul fide, thiodiglycol, thiodiethylene glycol

Aldehyde Ethers

483. 2-Furaldehyde* (1971), 2-furancarboxaldehyde*, furfural, furfuraldehyde, fural, furole, 2-furancarbinal

Chloro Ethers

484. Bis(2-chloroethyl) ether* (1971), 1,1'-oxybis[2-chloroethane]*, 2,2'-dichlo rodiethyl ether, β,β'-dichlorodiethyl ether, dichlorodiethyl ether, sym-dichlo rodiethyl ether
485. Epichlorohydrin* (1961), 1-chloro-2,3-epoxypropane* (1971), (chlorometh yl)oxirane*, α-epichlorohydrin, γ-chloropropylene oxide

Nitro Ethers

486. o-Nitroanisole* (1971), 1-methoxy-2-nitrobenzene*

Amino Ethers

487. Morpholine*, tetrahydro-2*H*-1,4-oxazine, diethylene oximide, diethylene-imide oxide, tetrahydro-*p*-oxazine

Esters of Hydroxy Acids

488. Ethyl lactate* (1971), ethyl 2-hydroxypropionate*
489. Methyl salicylate* (1971), methyl 2-hydroxybenzoate*, methyl *o*-hydroxy-benzoate, oil of wintergreen

Ether Esters

490. 2-Methoxyethyl acetate* (1966), 2-methoxyethanol acetate*, methyl Cello-solve acetate, methoxyethyl ethanoate, methylethylene glycol acetate, gly-colmonomethyl ether acetate
491. 2-Ethoxyethyl acetate* (1966), 2-ethoxyethanol acetate*, Cellosolve acetate, ethylethylene glycol acetate, ethylene glycol monoethyl ether acetate
492. 2-(2-Ethoxyethoxy)ethyl acetate* (1966), 2-(2-ethoxyethoxy)ethanol ace-tate*, Carbitol acetate, diglycol monoethyl ether acetate, diethylene glycol monoethyl ether acetate

Esters of Keto Acids

493. Methyl acetoacetate* (1971), methyl 3-oxobutanoate*
494. Ethyl acetoacetate* (1971), ethyl 3-oxobutanoate*, diacetic ether, acetoacetic ester

Esters of Cyano Acids

495. Methyl cyanoacetate*
496. Ethyl cyanoacetate*, ethyl cyanoethanoate

Fluoro Acids

497. Trifluoroacetic acid*, perfluoroacetic acid, trifluoroethanoic acid

Halogenated Amines

498. Heptacosafluorotributylamine* (1971), 1,1,2,2,3,3,4,4,4-nonafluoro-*N*,*N*-bis(nonafluorobutyl)-1-butanamine*
499. *o*-Chloroaniline* (1971), 2-chlorobenzenamine*, 2-chlorophenylamine

Phosphoramides

500. Hexamethylphosphoric triamide*, HMPA, hexamethylphosphoramide

Silicon Compounds

501. Tetramethylsilane*
502. Trichloromethylsilane*
503. Trichloroethylsilane*
504. Tetraethylsilane*

Chapter II

PHYSICAL PROPERTIES: DISCUSSION OF PROPERTIES AND CRITERIA FOR THEIR SELECTION

The physical properties and numerical values included in the following tables can be grouped into four main types:

1. Values useful for the identification of pure substances and as criteria of purity, such as boiling point, melting point, and refractive index.

2. Values useful for the execution of physicochemical measurements and determinations, such as ebullioscopic constants.

3. Properties useful for the study of solute systems that make it possible to differentiate between the properties of the solvent and the properties of the solute or the solution, such as electrical conductivity and spectra.

4. Properties important for the handling and storing of solvents, such as toxicity and flash point.

With respect to the reliability and accuracy of a particular numerical value, the literature has been used in the following sequence:

1. Papers devoted to precision measurements of physical constants or groups of physical constants of one compound or of a group of compounds. These furnish, as a rule, the best values.

2. Papers devoted to physicochemical measurements in nonaqueous systems in which purification methods for the particular solvents are mentioned. Although papers of this kind often show great precision in the determination of the constants, they may be doubtful with respect to the degree of purity of the solvent.

3. Papers devoted to the synthesis of the solvent or employing the solvent as a medium, for example, for the study of reaction mechanisms. Although these papers are usually reliable with respect to the purity of the solvent, they must be used with care as far as the physical measurements are concerned.

In the absence of such papers, the data given in the literature were evaluated either mathematically or graphically, or simply by inspection, the original sources having been consulted wherever possible. Extensive use has been made of the *Annual Tables of Physical Constants* and the *International Critical Tables*. Data from these sources have, for the most part, been checked back to the original articles, and care has been taken to refer in every case to the original source.

Six groups of workers have been particularly responsible for many reliable physical data and for the development or adaptation of some of the techniques of measurements: Timmermans and coworkers at the Bureau International des Etalons; Swietoslawski and coworkers of the Institute of Technology, Warsaw; groups at the Bureau of Mines laboratories in Bartlesville and Laramie; the group at the National Chemical Laboratory in Teddington, Middlesex, England; Rossini and coworkers at the National Bureau of Standards, Carnegie Institute of Technology, and cooperating institutions in the American Petroleum Institute Project 44; and Zwolinski and coworkers at the Thermodynamics Research Center at Texas A & M University.

For relatively few compounds, however, have the properties been studied extensively, and often a particular property has been determined at one or a few selected temperatures only. Therefore, in addition to the definitions of the properties listed in the tables and brief references to their determination, in this chapter we emphasize those mathematical relationships that allow the calculation of certain physical properties from values of other properties and the interpolation or extrapolation of data for temperatures not given in the tables.

In some cases, a simpler, although less accurate, relationship has been discussed in preference to a more complex one. For example, the critical temperature has been related to the normal boiling point by several empirical equations. Although these equations are, in general, less accurate than those that involve the parachor as well as the boiling point, they are given in detail because the data necessary for the use of the simpler relations are much more accessible.

While the literature on the correlation and calculation of physical properties and on physical properties–temperature relationships has by no means been completely covered, enough references have been included to guide the reader to the original literature for further information.

Most physical properties are temperature dependent; a clear concept of temperature and its measurement is needed for the optimum use of tables of physical data or the determination of physical properties. A complete treatment of temperature and its measurement has been presented by the American Institute of Physics [151]. Sturtevant and Weissberger [7958, Chapter 6] and Reilly and Rae [6114] have discussed temperature measurement and control and thermometry.

The International Union of Pure and Applied Chemistry has conformed with the Conférence Générale des Poids et Measures and adopted the International System of Units (SI) [3622]. This has resulted in changes in values and names and expressions of units. Many of these changes have been incorporated in this book; some of them have not been included because it is felt that a book of this type should be progressive but not use terms that will not come into general use for some years.

BOILING POINT

The *boiling point* (bp) is defined as that temperature at which the vapor pressure of the liquid is equal to the opposing pressure. Values listed in the tables refer to an opposing pressure of 760 Torr unless otherwise stated. Most routine boiling points for the characterization of substances are measured at "atmospheric pressure"; for comparative purposes these must be corrected to 760 Torr. This may be done by using the rule of Crafts [1706].

$$\Delta t = C(273 + t)(760 - p) \tag{2.1}$$

where t is the observed boiling point at p Torr pressure, Δt is the correction to be applied to the observed boiling point, and C is a constant that is practically the same for all substances. For alcohols, water, and carboxylic acids, $C = 0.00010$; for very low-boiling substances, nitrogen, ammonia, and so forth, $C = 0.00014$; for all other substances, C can be taken as 0.00012.

The *pressure coefficient of temperature* (dt/dp) at 760 Torr has been a property used mainly to correct boiling points taken at ambient pressure to standard atmospheric pressure. It can be found in many tables of physical properties. The reciprocal coefficient dp/dt is a convenient property and is the one usually given in thermodynamic studies involving vapor pressure.

The boiling point is one of the most useful properties for the characterization of organic compounds. Swietoslawski and Anderson [7958] have presented an excellent discussion of boiling and condensing temperatures and have described several reliable methods for their determination.

An empirical method has been devised by Kinney [4018–4020] for calculating the boiling points of aliphatic and aromatic hydrocarbons and correlating boiling points with molecular structure. The method is useful for predicting the boiling points of new compounds as well as for checking literature values that appear doubtful. Within certain limits, it may also be used to determine the structure of unknown hydrocarbons.

Kinney calculates the boiling point from the boiling point number (bpn) by the equation

$$bp = 230.14 \, (bpn)^{1/3} - 543 \qquad (2.2)$$

The boiling point numbers, although empirical, reflect the atomic and structural forces that affect the boiling points of organic compounds. The normal paraffin hydrocarbons and hydrogen were taken as the basis for calculating the boiling point numbers.

Kinney has calculated the boiling point numbers and boiling points for a larger number of compounds and has compared the calculated and experimental values. The average deviation for 323 aromatic hydrocarbons was 5°C. The agreement was poorest in the olefin series where the effect of *cis-trans* isomerism could not be established with certainty.

VAPOR PRESSURE

The *vapor pressure* (p) of a solvent is the pressure exerted by the vapor when in equilibrium with the liquid or solid phase. Vapor pressure, being a specific property of a compound, is widely used for practical calculations in physical chemistry and chemical engineering. It is also helpful in the characterization of organic compounds.

The error in vapor pressure measurements usually is in the pressure rather than the temperature determination. A discussion of vapor pressure measurements, in which pressure measurements are treated in some detail, has been presented by Thomson [7958].

Experimental vapor pressures usually do not cover the full $p-t$ range; to obtain full usefulness, some type of interpolation formula is necessary. Many different equations have been proposed for representing the vapor pressure–temperature relationship.

One of the most familiar and useful vapor pressure formulas can be obtained by integration of the Clausius–Clapeyron equation

$$\frac{(d \ln p)}{dT} = \frac{\Delta H_v}{RT^2} \qquad (2.3)$$

The general integral of (2.3) is

$$\ln p = -\frac{\Delta H_v}{RT} + C$$

$$\log p = A - \frac{B}{T} \tag{2.4}$$

The use of (2.4) implies the following conditions: (1) that ΔH_v, the heat of vaporization, is constant over the temperature range involved; (2) that the volume of the liquid is negligible compared to the volume of vapor; and (3) that the vapor is a perfect gas. Unless the temperature range is very small, these conditions are not fulfilled and the use of (2.4) leads to approximate values.

Antoine [391, 392] first proposed the equation bearing his name for water but later applied it to other liquids. He substituted $(t + C)$ for T in (2.4) to obtain

$$\log p = A - \frac{B}{(t + C)} \tag{2.5}$$

Gutmann and Simmons [3039] have derived the Antoine equation in the form

$$\log p = \frac{(A - B)}{[T - (3\theta/8)]} \tag{2.6}$$

from the theoretical van der Waals and Debye equations. The symbol θ is a characteristic temperature for each compound, and C in (2.5) is equal to $273.15 - (3\theta/8)$. An accurate method for computing θ has been discussed by Dreisbach [2086].

Thomson [7449] has given a comprehensive discussion of the Antoine equation and its use and applicability. He presents several methods for calculating C but states that, for the practical interpretation of data, C may be taken as equal to 230. In studies of vapor pressures of anesthetic substances, Rodgers and Hill [6222] reported a rapid and highly accurate method for obtaining the three Antoine constants from experimental data with a hand-held programmable calculator.

The Antoine equation has been found to represent experimental data over a wider range of pressures than the Clausius–Clapeyron equation and is probably the most reliable three-constant equation in the region between the triple point temperature and a reduced temperature of about 0.85. The Antoine equation fails as the temperature approaches the critical temperature. A four-constant equation has been derived by Frost and Kalkwarf [2587] by integrating the Clapeyron equation, $dp/dT = \Delta H_v/T \Delta V$, after first expressing ΔH_v as a linear function of T and applying the van der Waals' equation to the solution of the volume of the vapor. The resulting vapor pressure equation (2.7) has been applied by Bond and Thodos [956] and by Perry and Thodos [5778] to data for more than 50 aliphatic and aromatic hydrocarbons. It was found that

$$\log p = A + \frac{B}{T} + C \log T + \frac{Dp}{T^2} \tag{2.7}$$

accurately described the vapor pressure behavior over the entire liquid range from triple point to critical temperature.

The Antoine equation is applicable not only to the variation of vapor pressure with temperature but also to the effect of temperature on certain other physical properties.

Ahn [51] has found that the Antoine equation is useful in calculating viscosities at different temperatures. Novikov [5526] has shown that reasonably good results are obtained when the equation is applied to temperature variations of density, refractive index, pK_a, and surface tension.

Polynomials in t (or p) have been employed [6823] to represent accurate vapor pressure measurements. These equations usually are of the form

$$p = 760 + A(t - bp) + B(t - bp)^2 + C(t - bp)^3 \qquad (2.8)$$

where bp is the normal boiling point and A, B, and C are constants, and

$$t = bp + A(p - 760) + B(p - 760)^2 + C(p - 760)^3 \qquad (2.9)$$

Equations (2.8) and (2.9) give a very close fit over a limited range of pressures. For example, the average deviation in t for 1,2-dichloroethane [6823] is only 0.001° over a pressure range of 660–880 Torr and the greatest deviation is 0.002°C. The average deviation in p is 0.01 Torr, and the greatest deviation is 0.04 Torr from the experimental values.

Many empirical and semitheoretical vapor pressure equations have been proposed. They usually attempt to correct for the bending in the curve that always results when data fitted to (2.4) are plotted. The more common of these are

$$\log p = A - \frac{B}{T} + CT \qquad (2.10)$$

$$\log p = A - \frac{B}{T} + \frac{C}{T^2} + \frac{D}{T^3} \qquad (2.11)$$

$$\log p = A - \frac{B}{T} - CT + DT^2 \qquad (2.12)$$

$$\log p = A - \frac{B}{T} - C \log T \qquad (2.13)$$

$$\ln p = A - \frac{B}{T} + C \ln T \qquad (2.14)$$

$$\log p = A - \frac{B}{T} + 1.75 \log T + CT \qquad (2.15)$$

$$\log p = A - \frac{B}{(T - C)} \qquad (2.16)$$

The reader is referred to Thomson [7449, 7958] for well-integrated discussions of vapor pressure–temperature relationships.

In addition to the standard reference works, many articles summarizing and evaluating vapor pressure data have been published. Among these are those of Stull [7058], Dreisbach and Shrader [2091], Dreisbach and Martin [2090], and Zwolinski and Wilhoit [8317]. The most extensive are those of Stull and of Jordan [3779, 3780]. The vapor pressure values for many of the compounds in Chapter III have been taken from Stull, who lists over 1200 compounds. Enough values have been listed to enable the reader to calculate a vapor pressure–temperature relationship. Thomson [7958] advises that the critical investigator smooth these data using the Antoine equation with $C = 230$.

In order to give vapor pressure data of maximum usefulness, values for the constants of (2.4) or (2.5) have been included in the tables. When insufficient data were available to permit evaluation of the constants, values of the vapor pressures at several temperatures have been included. It was desirable to give the vapor pressure at 25°C for each solvent. If this value was not available, the vapor pressure nearest 25°C was listed.

DENSITY

The *density* (d) of a substance is defined as the mass per unit of volume and is expressed in grams per cubic centimeter in the cgs system or in kilograms per cubic meter in SI units. It is one of the most useful of the physical properties and is widely employed in the characterization of material, both single and multicomponent. It is also a factor in the calculation of many other properties. Bauer [7958] discusses the determination and uses of density.

Prior to the mid-1960s, chemists had been accustomed to expressing densities of liquids in grams per milliliter, 1 mL being equivalent to 1.000028 cm^3. In 1964, the Conférence Générale des Poids et Measures redefined the liter so that it is now a specific name for cubic decimeter; thus the milliliter becomes identical with the cubic centimeter. This change reduces the milliliter by 28 parts in 1,000,000. Weighings are made on the routine analytical balance to 0.0001 gram. The weight of 4 mL of water will be less, based on the redefined liter, by 0.00011 gram which is within the sensitivity of the balance. Therefore, the change may be significant to accurate determinations of density and other properties.

The *specific gravity* of a liquid is the ratio of the weight of any volume of the liquid to the weight of an equal volume of some standard, usually water. A dimensionless quantity, it is denoted by the symbol $d_{t_2}^{t_1}$, where t_1 is the temperature of the liquid and t_2 is that of the water. Water attains its maximum density of 0.999973 g/cm^3 at a temperature of 3.98°C, so that for nearly all practical purposes the specific gravity of a liquid referred to water at this temperature may be considered numerically equal to its density; for this reason, the density is commonly designated by d_4^t.

Specific gravities are often expressed with reference to the standard at some temperature other than 3.98°C. For example, the specific gravity of ethanol may be given as $d_{25}^{25} = 0.787$, which is the ratio of some volume of ethanol at 25°C to that of an equal volume of water at 25°. The density of the alcohol may be obtained by multiplying its specific gravity by the density of water at 25°:

$$d_{4(\text{alcohol})}^{25} = d_{25(\text{alcohol})}^{25} \times d_{4(\text{water})}^{25}$$

$$= 0.787 \times 0.997 \text{ g/cm}^3 = 0.785 \text{ g/cm}^3 \qquad (2.17)$$

Often the value of the density of specific gravity of a substance is not available at the desired temperature. If densities at two temperatures are known, the *coefficient of density* (dd/dt) may be used to calculate the density at a particular temperature.

$$\frac{dd}{dt} = \frac{(d_4^{t_1} - d_4^{t_2})}{(t_2 - t_1)} \qquad (2.18)$$

where $t_1 < t_2$. The coefficient varies widely for various liquids and, to some extent with temperature for the same liquid. If great accuracy is not desired, dd/dt can be assumed to be constant from about 0 to 40°, unless there is a phase change within this range.

The *coefficient of thermal expansion* (α) sometimes called the coefficient of cubical expansion, is defined as $\alpha = (1/v)(\partial v/\partial t)_p$. For temperatures reasonably close together α is expressed as

$$\alpha = \frac{1}{v}\frac{\Delta v}{\Delta t} \qquad (2.19)$$

The coefficient α may be calculated from two densities, d_1 and d_2 $(d_1 > d_2)$, at temperatures t_1 and t_2 $(t_2 > t_1)$ as follows: Assume that the liquid has a volume V_1 equal to unity at temperature t_1. Then the volume V_2 at t_2 is greater than V_1, and the increase in volume is

$$\Delta v = \frac{d_1}{d_2} - 1 \qquad (2.20)$$

and

$$\Delta t = t_2 - t_1 \qquad (2.21)$$

Substituting (2.20) and (2.21) in (2.19) gives

$$\alpha = \frac{(d_1/d_2) - 1}{\Delta t} \qquad (2.22)$$

The compressibility of a liquid may be carried out either isothermally or adiabatically resulting in two coefficients of compressibility. The *isothermal compressibility* (κ_T) is defined as $\kappa_T = -(1/v)(\partial v/\partial p)_T$. It is determined by the measurement of the volume of the liquid under various different pressures while the temperature is held constant. κ_T varies with both pressure and temperature, decreasing with increasing pressure and increasing with increasing temperature.

The *adiabatic compressibility* (κ_s) is defined as $\kappa_s = -(1/v)(\partial v/\partial p)_s$. Adiabatic compressibilities are often obtained from measurements of the velocity of sound through liquids. Isothermal and adiabatic compressibilities are related by the equation,

$$\kappa_T = \kappa_s(C_p/C_v) \qquad (2.23)$$

Coefficients of thermal expansion and compressibilities are useful in determining certain thermodynamic properties of liquids [816].

Dreisbach [2084] has developed a method for calculating the density at any temperature between 0 and 40° from the specific gravity at 25°/25° and the coefficient of thermal expansion. Dreisbach and Spencer [2093] have also shown that densities of liquids may be calculated from the molecular weight and the boiling point at any pressure.

Eremenko and Korolev [2281] reported that density and refractive index for a series of *n*-alkanols are related by a linear function to a high degree of accuracy. Nauruzov [5427] showed a similar equation for *n*-alkanes. The application of the Eykman equation for calculating densities from refractive index data is discussed in the following section.

REFRACTIVE INDEX

The *refractive index* (*n*) is the ratio of the velocity of light in a particular substance to the velocity of light in vacuum. Values usually reported refer to the ratio of the velocity in air to that in the substance saturated with air. The refractive index is useful for characterizing compounds and also for calculating other physical properties. Bauer and Fajans [7958] have discussed the refractive index, including some of its uses and methods of determination. The principles and uses of the several refractometers have been treated by Reilly and Rae [6114].

The refractive index is a function of both temperature and the wavelength of the incident light. Usually the sodium D_1 and D_2 lines are used. They have a weighted mean of 5892.6 Ångström units (Å) and are symbolized by D. Ordinary changes in atmospheric pressure do not produce a change in refractive index detectable with the usual instruments.

The most reliable refractive index data have been determined in the range of 15 to 30°. Between these limits, interpolated values will usually be correct to the fourth decimal place. The calculated temperature coefficient dn/dt decreases about 1% per degree of increase in temperature. If dn/dt is calculated from reliable data, the refractive index may be calculated over a small temperature range with an accuracy approaching that of the determined values. Extrapolation or interpolation is not reliable if a transition point lies within the temperature range. When only a single refractive index is available, approximate values at other temperatures may be calculated, the mean value 0.00045 per degree used for dn/dt.

The empirical equation of Eykman [2353],

$$\frac{(n^2 - 1)}{(n + 0.4)} \frac{1}{d} = C \tag{2.24}$$

offers an accurate means not only for checking the accuracy of experimental densities and refractive indices, but also for calculating one from the other. In (2.24), n is the refractive index, d the density at the same temperature, and C a constant. Eykman stated that the constant C is independent of temperature.

In spite of its empirical nature, Eykman's equation is to be preferred to the more familiar Lorenz–Lorentz expression [2784, p. 518]

$$\frac{(n^2 - 1)}{(n^2 + 2)} \cdot \frac{M}{d} = R \tag{2.25}$$

even though the latter has a theoretical basis.

Kurtz, Amon, and Sankin [4334] have emphasized the usefulness of the Eykman equation over a wide range of temperatures. Their summary of the available data, including the original data of Eykman, shows that the latter's equation accurately represents the experimental data. Their conclusions agree with those previously reached by Dreisbach [2085], Bauer and Fajans [7958], Kurtz and Lipkin [4335], Gibson and Kincaid [2738], and Ward and Kurtz [7904].

Taking Eykman's original data for 42 hydrocarbons, Kurtz and coworkers showed that the observed change in refractive index agreed with the calculated value within an average deviation of $\pm 3 \times 10^{-6}$ per degree. The Lorenz–Lorentz equation showed an average deviation of $+38 \times 10^{-6}$ per degree change in temperature. The minimum temperature interval in the group was 31.0° and the maximum 122.3°. They also stated, "...consideration of the currently available data for the change in refractive index and density with temperature leads one to conclude that changes in refractive index calculated from the Eykman equation from experimental data for density at two or more temperatures are likely to be more accurate than experimentally determined changes in refractive index unless the experimental work is very carefully done."

Other equations beside those of Eykman and Lorenz–Lorentz have been proposed for correlating density and refractive index. Since they are, for the most part, of historical interest only, they will not be discussed here.

Relationships have been proposed for specific classes or groups of compounds. Of particular interest are those of Ward and Kurtz [7904]. A careful consideration of existing data led these authors to conclude that the classic Newtonian specific refraction equation

$$\frac{(n^2 - 1)}{d} = r \tag{2.26}$$

accurately represents the relationship between refractive index and density for hydrocarbon isomers.

The equation

$$\Delta n = 0.60 \, \Delta d \tag{2.27}$$

was originally proposed by Ward and Kurtz [7904] to correct refractive indexes of hydrocarbons to 20° with reasonable accuracy. This equation was based on both the Eykman equation and empirical observation. In a later paper Kurtz and coworkers [4334] showed that, for hydrocarbons, this simple equation has a deviation of $\pm 8 \times 10^{-6}$ per degree change in temperature.

VISCOSITY

The *coefficient of viscosity* or *dynamic viscosity* (η) is defined as the force per unit area necessary to maintain a unit velocity gradient between two parallel planes a unit distance apart. The cgs unit of viscosity is the poise (P) (dyne · s/cm^2 = g/s · cm), but numerical values are usually given in smaller subsidiary units such as the millipoise (mP) (10^{-3} poise) or the centipoise (cP) (10^{-2} poise). The SI unit of viscosity is Newton · s/m^2 (N · s/m^2 = kg/s · m).

The *kinematic viscosity* (ν) is defined as the ratio of the dynamic viscosity to the density of a fluid. It is used in fluid mechanics and is the only parameter characteristic of the fluid itself to appear in the equations of hydrodynamics. The kinematic viscosity is directly proportional to the time required for the liquid to flow through a capillary-tube viscometer under its own hydrostatic head. The cgs units are called stokes (cm^2/s) and may be converted to poises by means of the equation poises = stokes × density. The SI units are m^2/s. cgs viscosities in this book will be expressed in centipoises, unless otherwise noted.

The absolute viscosity of only a few substances has been measured directly. It is easier to measure the viscosity of a liquid by comparison with a reference standard of known viscosity. The many kinds of viscometers available and the variables and conditions of viscosity measurements have been described by McGoury and Mark [7958].

The primary reference liquid for viscosity measurements is water. In 1931, Bingham [876, 877] evaluated the viscosity data of many workers and concluded that the most probable value for water at 20° was 1.005 cP. The *International Critical Tables* (3621) give the viscosity of water at 20° as 1.0087 cP. Cannon and Fenske [1340–1342] used Bingham's evaluation in developing a practical viscometer for routine work.

Coe and Godfrey, whose work at the National Bureau of Standards on the determination of the absolute viscosity of water was interrupted during World War II, published a preliminary report in 1944 [1550]. The final value reported by Swindells, Coe, and Godfrey [7142] is 0.010019 ± 0.000003 P at 20°. This figure will be rounded off to 0.01002 P for viscometer calibrations, and the general recognition and use of this value are being encouraged. The values for water at other temperatures between 15 and 60° are best represented by Cragoe's equation [1550]:

$$\log \frac{\eta}{\eta_{20}} = \frac{1.2348(20 - t) - 0.001467(t - 20)^2}{t + 96} \tag{2.28}$$

The lack of consistency in the value used for the viscosity of water prior to the general use of the Swindells, Coe, and Godfrey value has made it difficult to evaluate viscosity data. Few investigators specify the value used for their calibrations. When calibration data are known, viscosities can be corrected to correspond to the accepted value for water by the following simple proportion:

$$\eta = \frac{\eta_1 \eta_2}{\eta_3} \tag{2.29}$$

where η is the corrected viscosity at a particular temperature, η_1 the viscosity of water used in calibration, η_3 the viscosity of water calculated from (2.28), and η_2 the viscosity determined using η_1. Viscosities of water at several temperatures between 0 and 100°C are listed in Table 0b, Chapter III.

Many attempts have been made to correlate viscosity and temperature. One of the earliest, proposed by Arrhenius [457], was of the exponential type:

$$\Phi = Ae^{-E/RT} \tag{2.30}$$

where A and E are constants for a given liquid and $\Phi = 1/\eta$. Values of A and E have been calculated for 44 hydrocarbons by Kierstead and Turkevich [3999] from viscosity data in the literature. It was found that log Φ did not vary by more than 0.027, an amount that corresponds to an error in Φ of 6.5%. The usual difference between the calculated and observed value was about 1%. Qualitative observations on the effect of structure on the values of A and E were also made.

The values of A and E for the hydrocarbons listed in Chapter III, as far as available, are given in Table 2.1 [3999].

Bingham [876] developed an equation of the form

$$T = A\Phi + C - \frac{B}{\Phi} \tag{2.31}$$

where A, B, and C are constants, T is the absolute temperature, and Φ is the fluidity or reciprocal of the viscosity. Using this equation, Bingham determined the mean percentage difference between the observed and calculated values for 87 substances to be 0.17.

Table 2.1 Viscosity–Temperature Relationship: $\Phi = Ae^{-E/RT}$

Number	Compound	A (poise^{-1})	E (kcal mol^{-1}
4	Cyclopentane	5,554	1.878
5	Pentane	5,520	1.504
6	2-Methylbutane	6,099	1.549
8	Methylcyclopentane	6,603	2.055
9	Cyclohexane	15,210	2.905
10	Hexane	5,598	1.690
11	2-Methylpentane	5,818	1.676
12	3-Methylpentane	5,194	1.638
13	2,2-Dimethylbutane	8,002	1.978
14	2,3-Dimethylbutane	8,563	2.035
15	Methylcyclohexane	2,352	1.678
16	Heptane	5,842	1.858
17	2-Methylhexane	9,122	2.070
25	Octane	6,641	2.088
29	Nonane	8,485	2.394
31	Decahydronaphthalene	7,252	3.034
34	Decane	8,936	2.565
36	Dodecane	11,510	3.010
38	Benzene	9,803	2.428
39	Toluene	5,777	2.059
40	o-Xylene	6,141	2.277
43	Ethylbenzene	5,336	2.084
46	Naphthalene	580,300	5.977
50	1,2,3,4-Tetrahydronaphthalene	9,800	3.112

The alcohols had the largest mean percentage difference; if they are omitted, the mean difference for the remaining 70 compounds is 0.09. Values of A, B, and C are given in Table 2.2 [876]; the unit of fluidity is poise^{-1}.

When the fluidity curves vary widely from the simple linear type, (2.31) may be improved by introducing a fourth constant and rewriting it as

$$T = A\Phi + C - \frac{B}{(\Phi + D)} \tag{2.32}$$

Bingham found that, for eight compounds that gave a percentage difference of 0.77 between the observed and calculated value with (2.31), the difference was only 0.07 with (2.32). For example, octane had a difference of 0.16% when the simpler formula was used; with (2.32), this difference was reduced to 0.02%. The constants for (2.32) have been collected in Table 2.3 [876]; the unit of fluidity is poise^{-1}.

Many other equations relating viscosity to temperature have been proposed. Among them are

$$\log \eta = \frac{A}{T} - B \tag{2.33}$$

$$\eta = ae^{b\theta} \qquad \text{(where } \theta = T_{bp}/T) \tag{2.34}$$

Table 2.2 Viscosity–Temperature Relationship: $T = A\Phi + C - (B/\Phi)$; $\Phi = 1/\eta$

Number	Compound	A	B	C
5	Pentane	0.16544	14,937.0	256.61
6	2-Methylbutane	0.18327	10,454.0	234.17
10	Hexane	0.21892	9,137.8	254.03
11	2-Methylpentane	0.20708	9,562.8	252.85
16	Heptane	0.23203	8,681.2	272.70
25	Octane	0.27055	6,332.6	277.25
38	Benzene	0.32052	2,633.1	260.82
39	Toluene	0.32688	4,193.5	262.66
40	o-Xylene	0.37738	3,009.3	271.96
41	m-Xylene	0.34134	4,542.9	266.82
42	p-Xylene	0.32087	5,127.4	277.17
43	Ethylbenzene	0.34180	4,540.8	273.54
78	Methanol	0.24316	4,498.9	279.01
79	Ethanol	0.28395	2,398.6	298.39
80	1-Propanol	0.31496	1,211.7	308.41
81	2-Propanol	0.29810	738.16	300.17
82	1-Butanol	0.33610	877.08	311.94
84	2-Methyl-1-propanol	0.33648	512.18	309.72
85	2-Methyl-2-propanol	0.29657	260.86	305.73
86	1-Pentanol	0.36060	513.18	312.40
89	2-Methyl-1-butanol	0.35020	376.48	311.50
91	2-Methyl-2-butanol	0.29578	259.87	307.20
125	2-Propene-1-ol	0.28815	1,935.7	299.53
151	Ethyl ether	0.16574	14,674.0	256.72
152	Propyl ether	0.23579	6,858.5	266.34
182	Acetaldehyde	0.15205	18,364.0	265.13
189	Acetone	0.23871	8,905.0	247.64
190	2-Butanone	0.27275	5,572.6	252.32
193	3-Pentanone	0.28145	6,179.3	262.41
207	Formic acid	0.52465	963.4	281.57
208	Acetic acid	0.42437	2,716.8	291.81
209	Propionic acid	0.43533	2,908.9	287.53
210	Butyric acid	0.45359	1,951.9	296.43
221	Acetic anhydride	0.40620	2,747.8	273.61
222	Propionic anhydride	0.39705	2,444.6	287.45
224	Methyl formate	0.34444	1,292.0	198.31
225	Ethyl formate	0.29418	4,858.1	239.32
226	Propyl formate	0.29797	4,800.6	260.44
229	Methyl acetate	0.26047	6,475.7	249.48
232	Ethyl acetate	0.27056	5,361.2	257.20
235	Propyl acetate	0.29534	4,262.6	267.50
247	Ethyl propionate	0.29125	4,846.4	264.51
297	1-Chloropropane	0.25540	7,465.2	246.74
298	2-Chloropropane	0.24993	5,881.9	234.22
301	1-Chloro-2-methylpropane	0.28656	4,973.0	253.00
314	3-Chloropropene	0.26292	6,377.2	234.10
315	Dichloromethane	0.39806	2,666.5	212.97

Table 2.2 (Continued)

Number	Compound	A	B	C
316	Chloroform	0.40697	4,400.0	245.73
317	Carbon tetrachloride	0.47337	1,807.5	262.15
318	1,1-Dichloroethane	0.33277	4,651.6	247.99
319	1,2-Dichloroethane	0.44121	2,219.3	258.83
339	Bromoethane	0.35853	4,073.2	217.36
340	1-Bromopropane	0.36084	4,950.6	248.95
341	2-Bromopropane	0.33069	5,009.9	248.57
352	1,2-Dibromoethane	0.68897	1,421.6	277.92
355	Iodomethane	0.44916	3,107.1	215.85
356	Iodoethane	0.43828	4,348.2	243.29
357	1-Iodopropane	0.46571	3,474.4	255.85
358	2-Iodopropane	0.41279	4,108.8	262.07
451	Carbon disulfide	0.26901	16,751.0	282.23
456	Ethyl sulfide	0.28517	6,447.9	258.00
458	Thiophene	0.38204	2,967.2	254.95

$$\eta = A - B \log t \qquad \text{(where } t = \text{temperature above mp)} \qquad (2.35)$$

$$\eta v^{1/2} = A e^{b/T} \qquad \text{(where } v = \text{specific volume)} \qquad (2.36)$$

$$\frac{1}{\eta} = \frac{A}{(T_c - T)^{3/10}} \qquad \text{(where } T_c = \text{critical temperature)} \qquad (2.37)$$

Thomson [7449] has stated that viscosity–temperature data for some liquids can be fitted to the Antoine equation (2.5).

The constants in (2.35) for several liquids between their melting and boiling temperatures have been determined by Stuart [7054]. These are tabulated in Table 2.4.

Lederer [4479] proposed the equation

$$\log \eta = \frac{\overset{\circ}{q}}{4.571} T - 2.75 \log T + ET + C \qquad (2.38)$$

for the variation of viscosity of fatty acids with temperature, where $\eta =$ coefficient of

Table 2.3 Viscosity–Temperature Relationship: $T = A\Phi + C - \{B/(\Phi + D)\}$; $\Phi = 1/\eta$

Number	Compound	A	B	C	D
25	Octane	0.14507	100,745.0	438.10	400
82	1-Butanol	0.23695	4,802.0	349.71	40
84	2-Methyl-1-propanol	0.23700	2,993.7	340.66	30
86	1-Pentanol	0.24191	3,908.7	354.17	35
89	2-Methyl-1-butanol	0.24650	2,942.8	346.82	30
91	2-Methyl-2-butanol	0.22988	2,124.0	328.84	30
210	Butyric acid	0.16154	70,630.0	504.44	250
222	Propionic anhydride	0.23619	52,294.0	425.82	250

Table 2.4 Viscosity–Temperature Relationship:
$\eta = A - B \log t$ (cP)

Number	Compound	A	B
10	Hexane	2.03	0.83
16	Heptane	2.64	1.08
25	Octane	2.70	1.14
79	Ethanol	4.20	1.16
151	Ethyl ether	2.85	1.25
408	Aniline	14.2	7.10

Table 2.5 Viscosity–Temperature Relationship: $\log \eta = \overset{\circ}{q}/4.571T - 2.75 \log T + ET + C$

Number	Compound	$\overset{\circ}{q}$	E	C
210	Butyric acid	1960	0.001770	2.9952
212	Valeric acid	2330	0.001663	2.9050
214	Hexanoic acid	2740	0.001401	2.8200
215	Octanoic acid	3475	0.001576	2.4883

Table 2.6 Viscosity–Temperature Relationship: $\log \eta$ (poise) $= A/T - B$

Number	Compound	A	B	Range (K)	Reference
159	1,2-Dimethoxyethane	425	3.773	198–298	1379
170	Tetrahydrofuran	393	3.655	198–298	1379
173	Tetrahydropyran	591	4.10	228–298	5477
175	Anisole	636.8	4.139	283–353	5742
176	Phenetole	763.5	4.451	283–353	5742
208	Acetic acid	584.4	3.9053		4357
209	Propionic acid	526.4	3.755		4357
210	Butyric acid	634.8	3.9529		4357
212	Valeric acid	676.6	3.9780		4357
214	Hexanoic acid	771.5	4.1417		4357
311	p-Chlorotoluene	445.0	3.5627		2396
330	2,4-Dichlorotoluene	541.0	3.6296		2396
463	Sulfolane	1031.3	2.3892	298–323	7694
484	Bis(2-chloroethyl)ether	882.2	4.621	283–353	5742

viscosity in poises, $\overset{\circ}{q}$ = heat of association at 0°C in cal/mol and E and C are constants. Values of the constants for four fatty acids are given in Table 2.5.

For a discussion of other equations and a general discussion of viscosity–temperature relationships see Højendahl [3464], Thomas [7415], Kimura [4008, 4009], Telang [7207], Lucatu [4679], Bingham [876], Hatschek [3226], Marschalks and Barna [4901], and Williamson [8088]. See also Table 2.6.

Certain organic solvents form glasses at low temperatures and are suitable for use as trapping matrices in studies of electrons, ions, and free radicals produced by radiation. Viscosity data for these compounds are included in the tables of Chapter III. For a discussion of experimental techniques at low temperatures see Ling and Willard [4613, 4614] and Hutzler, Colton, and Ling [3605].

SURFACE TENSION

The *surface tension* (γ) is defined as the force in dynes acting at right angles to any line 1 cm in length on the surface of a liquid. It has the dimensions of force per unit length. The cgs unit of surface tension is dyne/cm and the SI unit is N/m. The energy required to extend the area of a surface by a unit area (1 cm^2 or 1 m^2) is called the surface energy and is numerically equal to the surface tension that opposes the increase. Although the surface energy is usually regarded as the fundamental property of a surface, for purposes of calculation the surface tension is more convenient to use. This is justified by the numerical equivalency of these two quantities.

Surface tensions find their widest use in the determination of parachors, though the importance of surface tensions as engineering data is increasing [5046]. Harkins has described the various methods of measurement, and Thomson has summarized the work done on the parachor [7958]. An interpretation and catalog of parachors of organic compounds have been presented by Quayle [6007].

Meissner and Michaels [5046] have presented a method for estimating the surface tension of multicomponent liquids based in part on the parachor and molar refraction. The method is also applicable to pure liquids. Empirical relationships that permit the surface tensions of aqueous solutions to be computed have also been derived.

The application of this method to compounds with various functional groups showed an average deviation from the experimentally determined value of 2.8%. The largest deviations were for polyhydric alcohols and halogen and nitro compounds.

Like most physical properties, the surface tension can vary widely with temperature. Vogel [7757–7772], in the course of his work on parachors, has determined the surface tension of many compounds at various temperatures.

The earliest attempt to correlate surface tension and temperature was made by Eötvös [2277] in 1886; it was based on the concept of corresponding states. The integrated form of his equation is

$$\frac{\gamma_1(MV_1)^{2/3} - \gamma_2(MV_2)^{2/3}}{t_2 - t_1} = k \tag{2.39}$$

where γ is the surface tension, M the molecular weight, V the specific volume of liquid, and k the temperature coefficient of molar surface energy.

The original idea was that k is a constant. Ramsay and Shields [6040] found k to be constant and equal to about 2.1 for liquids such as carbon disulfide, carbon tetrachloride, ethyl ether, and ethyl acetate. However, certain liquids such as water, alcohols, and carboxylic acids not only had a temperature coefficient of less than 2.1, but the actual value also varied with temperature. Ramsay and Shields argued that this was a measure of molecular association, though Walden and Swinne [7864] later showed that, even for compounds that were neither associated nor dissociated as liquids, the temperature coefficient could vary from about 0.6 to 6.0.

The relationship of Eötvös and other relationships, either derived from it or proposed independently, have not been very useful for the calculation of surface tensions at various temperatures from values at a particular temperature. Frequently, if the temperature range is not large, surface tension–temperature data may be quite accurately fitted to empirical equations of the types

$$\gamma = \gamma_0(1 - Bt) \tag{2.40}$$

and

$$\gamma = A - Bt \tag{2.41}$$

Data for several compounds based on these equations are given in Tables 2.7 and 2.8. Jasper [3684] has summarized and evaluated surface tension data for many organic compounds.

HEAT OF VAPORIZATION

The *heat of vaporization* (ΔH_v) is the amount of heat required to vaporize a definite quantity, for example, 1 gram or 1 mol of a liquid. It is often referred to as the *latent heat of vaporization*.

The heat of vaporization decreases with an increase in temperature. It is most often calculated from vapor pressure measurements rather than experimentally determined. Sturtevant [7958] has given a brief discussion of this property and some methods of determination.

Trouton's equation [2784] may be used for a rough approximation of the heat of vaporization of nonpolar substances. This equation is

$$\Delta H_v = 21 T_{bp} \text{ cal mol}^{-1} \tag{2.42}$$

where T_{bp} is the absolute temperature of the boiling point. A modification of (2.42) that gives more reliable estimates has been proposed by Kistiakowsky [3507]:

$$\Delta H_v = 8.75 T_{bp} + 4.571 \log T_{bp}^{2} \tag{2.43}$$

More exact values of heats of vaporization may be obtained from vapor pressure data; this is advantageous, since the heats of vaporization may be calculated over as wide a range of temperature as covered by the vapor pressure data.

The simplest calculation of the heat of vaporization can be made from the Clapeyron equation:

$$\Delta H_v = \frac{dp}{dt} T(V_g - V_l) \tag{2.44}$$

where V_g is the molar volume of the vapor and V_l the molar volume of the liquid. The values calculated from this equation will agree with experimental values provided that dp/dt is determined over a very small interval. For practical calculations, the molar volume of the liquid may be neglected, since it is small compared to that of the vapor.

The accuracy of (2.44) may be increased by using Haggenmacher's correction [3066]

$$\left\{ 1 - \left(\frac{T_c^3 p}{p_c T^3} \right) \right\}^{1/2} \tag{2.45}$$

Table 2.7 Surface Tension–Temperature Relationship: $\gamma = \gamma_0(1 - Bt)$ (dyne cm^{-1})

Number	Compound	γ_0	B	Range (°C)	Ref.
38	Benzene	30.64	0.00416	0–100	7848
45	Mesitylene	28.98	0.00314	25–165	7848
80	1-Propanol	24.577	0.00318	10–90	4636
134	1,2-Ethanediol	48.48	0.00205	0–132	7848
151	Ethyl ether	18.50	0.00603	0–80	7848
175	Anisole	38.25	0.00323	20–80	5741
		36.43	0.00306	0–153	7848
176	Phenetole	33.79	0.00303	0–152	7848
189	Acetone	25.330	0.00484	0–80	4636
190	2-Butanone	25.964	0.00451	0–45	4636
193	3-Pentanone	25.970	0.00389	0–45	4636
203	Acetophenone	40.34	0.00260	0–171	7848
207	Formic acid	39.35	0.00273	0–80	7848
224	Methyl formate	27.68	0.00553	0–100	7848
232	Ethyl acetate	26.00	0.00463	0–120	7848
307	Chlorobenzene	34.10	0.00320	0–127	7848
317	Carbon tetrachloride	28.17	0.00418	0–100	7848
356	Iodoethane	32.55	0.00411	0–78	7848
384	Nitrobenzene	44.945	0.00262	0–153	7848
385	Acetonitrile	30.62	0.00400	0–73	7848
386	Propionitrile	28.57	0.00385	0–90	7848
395	Benzonitrile	39.45	0.00269	0–150	7848
408	Aniline	44.00	0.00261	0–165	7848
425	Piperidine	31.559	0.00380	0–30	4636

Table 2.8 Surface Tension–Temperature Relationship: $\gamma = A - Bt$ (dyne cm^{-1})

Number	Compound	A	B	Range (°C)	Ref.
5	Pentane	18.25	0.11021	0–100	3687
10	Hexane	20.44	0.1022	0–100	3687
		20.695	0.1001	0–30	4638
16	Heptane	22.10	0.0980	0–100	3687
25	Octane	23.52	0.09509	0–100	3687
29	Nonane	24.72	0.09347	0–100	3687
34	Decane	25.67	0.09197	0–100	3687
36	Dodecane	27.12	0.08843	0–100	3687
61	1-Hexene	20.47	0.10271	0–100	3687
62	1-Heptene	22.28	0.09908	0–100	3687
63	1-Octene	23.68	0.09581	0–100	3687
67	1-Nonene	24.90	0.09379	0–100	3687
68	1-Decene	25.84	0.09190	0–100	3687
78	Methanol	24.23	0.09254	−40–40	7045
79	Ethanol	23.88	0.08807	−40–40	7045
82	1-Butanol	25.348	0.0788	0–40	4637
84	2-Methyl-1-propanol	23.569	0.025	0–40	4637
86	1-Pentanol	26.78	0.08147	−40–40	7045
95	1-Hexanol	27.84	0.08381	−40–40	7045
112	1-Octanol	28.89	0.07782	−40–40	7045
125	2-Propen-1-ol	26.499	0.0864	0–40	4637
134	1,2-Ethanediol	50.21	0.089	25–150	2638
136	1,3-Propanediol	47.43	0.0903	25–140	2638
162	Bis(2-methoxyethyl)ether	32.47	0.1164	25–80	2638
172	p-Dioxane	36.23	0.1391	15–60	2638
		36.1	0.14	−45–70	3464
203	Acetophenone	40.805	0.1162	30–50	4637
207	Formic acid	38.368	0.1025	0–30	4637
209	Propionic acid	27.797 + 0.000183t^2	0.1067	10–60	4637
210	Butyric acid	27.762 + 0.000104t^2	0.0981	0–60	4637
211	Isobutyric acid	26.208	0.0935	0–45	4637
213	Isovaleric acid	26.253	0.0844	20–40	4637
262	Ethyl cinnamate	38.516	0.1042	30–50	4638
267	Ethyl oxalate	33.135	0.1076	10–50	4638
268	Ethyl malonate	33.030	0.1026	10–50	4638
317	Carbon tetrachloride	29.44	0.1406	−29–76	5618
319	1,2-Dichloroethane	33.21	0.1023	25–90	2638
		36.3	0.14	−45–70	3464
380	Nitromethane	40.72	0.1678	0–60	6886
		38.534	0.1423	0–45	4637
381	Nitroethane	35.27	0.1255	0–60	6886
382	1-Nitropropane	32.62	0.1009	0–60	6886
383	2-Nitropropane	32.18	0.1158	0–60	6886
384	Nitrobenzene	46.09	0.1258	0–50	6314
388	Butyronitrile	28.350	0.1004	30–50	4638

Table 2.8 (*Continued*)

Number	Compound	A	B	Range (°C)	Ref.
390	Valeronitrile	26.909 $+ 0.000177t^2$	0.101	10–50	4638
395	Benzonitrile	39.469	0.1218	30–40	4638
409	o-Toluidine	41.867	0.1129	0–30	4637
411	p-Toluidine	38.277	0.0924	50–60	4637
431	Pyridine	39.710	0.138	20–85	5526
440	Formamide	58.306	0.7927	0–40	4637
454	Benzenethiol	40.465	0.1188	10–45	4638
464	2-Methoxyethanol	35.36	0.139	25–80	2638
469	Diethylene glycol	46.97	0.088	20–120	2638
470	Triethylene glycol	47.33	0.088	20–120	2638
483	2-Furaldehyde	44.34	0.1211	0–30	4575
		44.341	0.1221	20–40	4637
484	Bis(2-chloroethyl))ether	40.57	0.1306	25–80	2638
494	Ethyl acetoacetate	33.645	0.1082	30–50	4638
497	Trifluoroacetic acid	15.638	0.08444	24–68	3688

where T_c and p_c are the critical temperature and pressure. Although this factor contains critical constants, which frequently are unknown, Dreisbach and Spencer [2092] have shown that this equation can be applied without knowledge of the values for these constants. They have calculated values for this factor for 10 compounds and have found that it is constant for families of compounds and varies only slightly for different families. For detailed information about the application of this correction factor, the reader is referred to the original article [2092].

A method for calculating the heat of vaporization and also the external work of vaporization has been presented by Haggenmacher [3065]. Equations were developed that are applicable over the entire range of the liquid phase, that is, from the triple point to the critical point. The only limit to their application is the availability of exact vapor pressures and the critical temperature and pressure. When exact vapor pressure data are used, the calculated values show a deviation of only a few tenths of a percent from precise experimental measurements. The author includes tables of calculated values of heat and external work of vaporization at various temperatures for 22 hydrocarbons.

The simple relationship

$$\Delta H_v T = k_1 \qquad (2.46)$$

where k_1 is a constant and T the absolute temperature, has been proposed by Bowden and Jones [1025] for nonassociated liquids. This relationship is applicable over a range to 50° above the boiling point. For compounds that do not deviate greatly from Trouton's rule, the relationship has an accuracy of better than 0.5%.

Bowden and Jones [1025] also studied the variation of the heat of vaporization with density and found that nonassociated substances conformed to

$$\frac{\Delta H_v}{(D - d) D^{1/3}} = k_2 \qquad (2.47)$$

where D is the density of the liquid and d that of the saturated vapor. This equation holds fairly well to within about 50° of the critical temperature. The authors had previously shown that normal liquids followed the relationship

$$\frac{\Delta H_v^{4/5}}{D - d} = k_3 \tag{2.48}$$

If (2.47) and (2.48) are combined and $D - d$ is eliminated,

$$\frac{\Delta H_v^{3/5}}{D} = k_4 \tag{2.49}$$

where k_4 is a constant. The heats of vaporization calculated from (2.49) agree well with the observed values over a range of temperature to about 100° above the boiling point of the liquid.

Equation (2.50) is frequently used to express the heat of vaporization as a function of temperature. The constants for several solvents

$$\Delta H_v = A + BT + CT^2 \text{ cal mol}^{-1} \tag{2.50}$$

are given in Table 2.9.

The *heat of sublimation* (ΔH_s) is the amount of heat required to transform a definite quantity of a solid, usually 1 mol, at a specified temperature directly to the vapor phase at the equilibrium vapor pressure.

HEATS OF FORMATION AND COMBUSTION

The *heat of formation* (ΔH_f) is the heat evolved (or absorbed) in the formation of a compound from its elements. When the product and all of the reactants are in the standard state of 1 atm, the function is called the *standard heat of formation* and has the notation ΔH_f°. The tabulated data in Chapter III are standard values given at 25°C. In this edition, values for both liquid and gaseous states are listed for many compounds.

The *heat of combustion* is the heat evolved in the complete combustion of a compound in oxygen. When the reaction occurs at *constant volume*, the thermodynamic function is called the *internal energy of combustion* with the notation ΔE_c. When the reaction occurs at *constant pressure*, the function is called the *enthalpy of combustion* with the notation ΔH_c. Almost all of the tabulated data in Chapter III are standard enthalpies of combustion ΔH_c° at 25°C.

There are numerous compilations of heats of formation and combustion. References frequently used in this edition for these functions are those of Zwolinski and Wilhoit [8316], Domalski [2036], and Cox and Pilcher [1703]. Excellent sources of heats of formation for C_1 and C_2 compounds are those of Wagman and coworkers [7826] and Wagman and Evans [7827].

CRITICAL CONSTANTS

The *critical temperature* (t_c) is defined as the temperature above which a gas cannot be liquefied. The minimum pressure required for liquefaction at this temperature is called the *critical pressure* (p_c). The *critical volume* (v_c) is the volume occupied by 1 mol of

Table 2.9 Heat of Vaporization–Temperature Relationship: $\Delta H_v = A + BT + CT^2$ (cal mol^{-1})

Number	Compound	A	B	$C \times 10^2$	Range (K)	Ref.
4	Cyclopentane	7,556	6.439	-2.990	290-320	4752
8	Methylcyclo-pentane	9,701	-2.100	-1.704	305-345	4752
82	1-Butanol	12,347.7	20.159	-6.491	355-390	1683
169	Furan	8,854	-1.882	-1.949	279-303	3037
175	Anisole	13,214.6	-1.2941	-1.8504	350-450	3083
193	3-Pentanone	13,363.7	-12.3666	-0.5172	335-375	3083
238	Butyl acetate	16,200	-18.16			6450
275	Butyl borate	12,324	2.1588		390-490	1476
288	Fluorobenzene	10,689	-3.850	-1.448	323-373	6527
293	Hexafluoro-benzene	12,587.5	-10.3365	-1.0917	315-353	1682
380	Nitromethane	11,730	-4.9977	-1.2400	318-373	4754
409	o-Toluidine	7,670.3	-2.557		323-473	796
410	m-Toluidine	6,996.5	-1.673		323-473	796
411	p-Toluidine	7,145.8	-1.951		323-473	796
423	Pyrrole	12,734	-0.97662	1.8971	362-403	6519
451	Carbon disulfide	7,868	-0.096	-1.419	283-323	7818
453	1-Butanethiol	10,655	-0.4640	-2.0136	330-372	6523
454	Benzenethiol	15,808.5	-13.620	-0.1586	373-418	6528
455	Methyl sulfide	9,087	-4.1794	-1.3860	273-300	4751
456	Ethyl sulfide	11,298	-5.0897	-1.3849	325-365	6522

he substance at its critical temperature and critical pressure. F. Brescia [1083] presents a good introductory discussion of critical temperature in his proposal for a more accurate concept. A comprehensive review of critical properties has been given by Kobe and Lynn [4088].

The experimental determination of critical temperature and pressure is difficult, and sometimes impossible, because of decomposition at temperatures below the critical points. Many empirical methods for predicting these critical constants have been devised. It is not our purpose to cover the literature completely, but only to point out those methods that are of interest because of their simplicity, or because they require little auxiliary data. An excellent summary of methods for calculating critical constants has been made by Thomson [7958].

The simplest means of estimating the critical temperature is provided by a relationship proposed by Guldberg [3018]. He showed that the following was approximately true:

$$\frac{T_{bp}}{T_c} = k \tag{2.51}$$

where T_{bp} is the absolute temperature of the boiling point and T_c the absolute critical temperature. The values for the constant k for 64 compounds ranged from 0.566 to 0.708 and had an average value of 0.651. The average value of k for the various types of aliphatic compounds is given in Table 2.10 [3018].

A number of later investigators have inverted Guldberg's original formula and have given the average value of k as about 1.5 with variations of 1.3–1.8. Although this relationship should be used cautiously, it has the advantage that only the boiling point is required to estimate the critical temperature.

Many other physical properties have been correlated with critical constants. The equation

$$T_c = K_c \log P - \beta \tag{2.52}$$

has been proposed by Lewis [4561], where P is the parachor and K_c and β are constants for a particular homologous series. Good agreement between experimental and calculated values of T_c are obtained for normal paraffin hydrocarbons with

$$T_c = 551 \log P - 832 \tag{2.53}$$

For aromatic hydrocarbons $K_c = 551$ and $\beta = 716$.

Many other relations have been proposed, some based on surface tensions, parachors, liquid densities, or densities at the boiling point.

Table 2.10 Boiling Point–Critical Temperature Relationship: $T_{bp}/T_c = k$

Compounds	k
Alcohols	0.686
Acids	0.676
Esters	0.677
Amines	0.658
Halides	0.621

Several useful equations have been proposed by Meissner and Redding [5047] which like Guldberg's formula, require only a knowledge of the boiling point. They are appli cable to both polar and nonpolar compounds. For compounds boiling below 235 K, th equation has the form

$$T_c = 1.70 \, T_{bp} - 2 \qquad (2.54)$$

Three equations are applicable for compounds boiling above 235 K. For compound containing sulfur or halogens

$$T_c = 1.41 \, T_{bp} + 66 - 11F \qquad (2.55)$$

where F is the number of fluorine atoms in the molecule. The equation

$$T_c = 1.41 \, T_{bp} + 66 - r(0.383 \, T_{bp} - 93) \qquad (2.56)$$

applies to halogen- and sulfur-free aromatics or naphthenes. The factor r in (2.56) is th ratio of noncyclic carbon atoms to the total number of carbon atoms in the molecule All halogen- and sulfur-free compounds other than the aromatics or naphthenes follov the relationship

$$T_c = 1.027 \, T_{bp} + 159 \qquad (2.57)$$

When this relationship was tested on compounds boiling below 600 K (with the excep tion of water), the average deviation between the calculated and observed values wa ±5%.

Herzog [3352] has examined various correlations of the critical constants and ha evaluated their accuracy. He has proposed four equations involving the parachors and normal boiling points as auxiliary data. Either the experimental or the calculated para chor may be used.

Fewer relations have been proposed for estimating the critical pressure than for th critical temperature. Attempts have been made to correlate the critical pressure with viscosity [5696]. However, the simplest and most useful formulas are based on the crit ical temperature or critical density.

The relationships

$$p_c = \frac{20.8 T_c}{V_c - 8} \qquad (2.58)$$

and

$$p_c = \frac{20.8 T_c}{(M/D_c - 8)} \qquad (2.59)$$

have been proposed by Meissner and Redding [5047], where V_c is the critical volume, M the molecular weight, and D_c the critical density. These authors found that V_c could be evaluated from

$$V_c = (0.377P + 11.0)^{1.25} \qquad (2.60)$$

where P is the parachor. The authors fully discuss their equations and recommend a procedure for calculation.

Equations developed by Herzog [3352] involving the parachor give somewhat more reliable results in calculating the critical pressure and critical volume than those of Meissner and Redding. Because of the essentially empirical nature of all critical constant equations, careful checks must be made against experimental data whenever possible. Many equations are satisfactory for specific groups of compounds but cannot be applied indiscriminately to all compounds.

Critical densities may be determined from orthobaric densities of liquid and vapor by making use of the law of rectilinear diameter first proposed by Cailletet and Mathias 1305]. This law, (2.61), states that the average density of liquid and vapor is a linear function of the temperature.

$$\tfrac{1}{2}(D + d) = a - bt \tag{2.61}$$

Plots of the density of liquid (D) and density of vapor (d) are made versus temperature, and extrapolation of the average density line to the point of junction of these two curves gives the critical density.

Hakala [3076] solved simultaneously the Katayama–Eötvös [3878] and McCleod 4741] equations for surface tension with the Cailletet–Mathias equation to obtain (2.62), from which critical densities can be estimated with good accuracy from data taken far below the critical temperature.

$$D + d = 2d_c - A(D - d)^{10/3} \tag{2.62}$$

The critical density is determined from the intercept of the plot of $D + d$ versus $D - d)^{10/3}$. Data for 29 compounds gave results with an average accuracy of 0.5%; the average extrapolation was over a range of 100°C.

Kudchadker and coworkers [4299] have compiled the critical constants of several hundred organic compounds.

HEAT CAPACITY

The *specific heat capacity* or *specific heat* of a substance is the amount of heat required to raise the temperature of 1 gram of the substance 1°C at constant pressure and without change of phase. Although the specific heat is a function of temperature, average values are frequently quoted for a specified temperature range. The equation

$$c_p = A + Bt + Ct^2 \text{ cal deg}^{-1} \text{ gram}^{-1} \tag{2.63}$$

has been used to express the variation of the specific heat with temperature, and values of the constants for several solvents are given in Table 2.11.

Molar heat capacities are also temperature dependent and may be precisely defined as $C = dq/dt$, where dq is the differential flow of the heat accompanying the infinitesimal temperature change dT. Conditions of constant pressure or constant volume must be specified and are designated with the symbols C_p and C_v, respectively. The cgs unit of heat capacity is expressed as calorie per degree per mol and the SI unit is joule per degree per mol.

Water has been the standard substance for the intercomparison of heat capacity calorimeters because it has characteristics such as high heat capacity and ease of purification

Table 2.11 Specific Heat-Temperature Relationship: $c_p = A + Bt + Ct^2$ (cal deg^{-1} gram^{-1})

Number	Compound	A	B × 10⁴	C × 10⁵	Range (°C)	Ref.
82	1-Butanol	0.6053	9.76	0.6		5664
83	2-Butanol	0.655	9.20	1.837		5664
86	1-Pentanol	0.5210	15		40–70	4486
87	2-Pentanol	0.592	22.6	0.126		5664
88	3-Pentanol	0.6658	10		40–70	4486
89	2-Methyl-1-butanol	0.5657	13		40–70	4486
90	3-Methyl-1-butanol	0.5277	20		40–70	4486
91	2-Methyl-2-butanol	0.6236	22		40–70	4486
134	1,2-Ethanediol	0.55	12.5		-10–195	7191
156	Butyl ether	0.217	93.2	1.98		5664
190	2-Butanone	0.4242	16.2	1.038		5664
207	Formic acid	0.4966	7.09			6439
250	Ethyl isovalerate	0.4539	11		40–70	4486
267	Ethyl oxalate	0.4336	6.5		40–70	4486
296	Chloroethane	0.3941	7.088		-67–15	4327
297	1-Chloropropane	0.3820	7.954		-38–43	4327
301	1-Chloro-2-methylpropane	0.3915	7.334		14–58	4327
304	1-Chloro-3-methylbutane	0.3734	7.732		13–100	4327
315	Dichloromethane	0.452	7.448		-76–12	4327
316	Chloroform	0.2232	4.4656		-50–50	4327
317	Carbon tetrachloride	0.1933	2.688		-20–72	4327
318	1,1-Dichloroethane	0.2953	4.94		-51–55	4327
319	1,2-Dichloroethane	0.2812	6.1994		25–83	4327

320	1,2-Dichloropropane	0.3087	6.8052	16–156	4327
323	1,2,3-Trichloropropane	0.2662	5.451	17–150	4327
324	1,1,2,2-Tetrachloroethane	0.2286	2.957	15–145	4327
325	Pentachloroethane	0.2208	4.082	16–154	4327
334	cis-1,2-Dichloroethylene	0.2618	5.4886	–31–54	4327
337	Tetrachloroethylene	0.1937	3.799	16–119	4327
338	Bromomethane	0.2841	4.2308	–60–10	4327
339	Bromoethane	0.2088	4.884	–49–37	4327
340	1-Bromopropane	0.2589	5.2502	–30–67	4327
342	1-Bromobutane	0.2794	5.566		4327
351	Bromoform	0.1201	1.38322	9–147	4327
352	1,2-Dibromoethane	0.1638	3.0994	15–127	4327
353	1,2-Dibromopropane	0.1801	3.7124	10–130	4327
354	1,1,2,2-Tetrabromo-ethane	0.1084	2.436	12–132	4327
355	Iodomethane	0.2079	2.50	–56–20	4327
356	Iodoethane	0.1677	3.42	–37–71	4327
361	Diiodomethane	0.1146	1.6424	15–164	4327
365	Trichlorofluoromethane	0.2104	1.51	–30–63	774
366	Chlorodifluoromethane	0.2819	7.84	–30–63	774
367	Dichlorofluoromethane	0.2471	1.89	–30–63	774
373	1,1,2-Trichloro-1,2,2-trifluoromethane	0.2148	2.07	–30–63	774
409	o-Toluidine	0.4706	7.0		6440
419	Diethylamine	0.5604	8.607		3902
428	Triethylamine	0.4987	12.58		3902
483	2-Furaldehyde	0.392	–2.63	1.08	15–150	5587

47

to a high degree of purity, which make it suitable for this purpose. However, measurements below 0°C and above 100°C present problems with water and in some calorimeters it is impossible to use water outside of its normal liquid range. The Fourth Congress on Calorimetry recommended that three additional substances be used as standards for the intercomparison of heat capacity calorimeters: heptane, benzoic acid, and aluminum oxide. Ginnings and Furukawa [2763] determined the heat capacities of the three proposed substances. They also reported later [2615] on diphenyl ether, which was suggested as another additional calorimetric standard (see Chapter V). Cruickshank and coworkers [1739] have given an extensive review for determining heat capacities of liquids and solutions near room temperature.

Since organic solvents are generally used in the liquid state, the heat capacities given in Chapter III are given for the liquid at the stated temperature unless otherwise indicated.

Constants for the equation

$$C_p = A + BT + CT^2 + DT^3 \text{ cal K}^{-1} \text{ mol}^{-1} \tag{2.64}$$

are given in Table 2.12 for a number of solvents. Kurbatow [4326] has expressed the molar heat capacities of several organic solvents as a linear function of the Celsius temperature (see Table 2.13).

The heat capacity of the ideal gas reference state (C_p°) is a useful thermodynamic property. Values of C_p° may be obtained by a linear extrapolation to zero pressure of the plots of the heat capacities of the real gas against pressure at constant temperature. For gases obeying the Berthelot equation of state, values of C_p° may be calculated from the critical constants and the measured heat capacity:

$$C_p^\circ = C_{p(g)} + \frac{81RT_c^3 p}{32T^3 p_c} \tag{2.65}$$

Constants for the equation

$$C_p^\circ = A + BT + CT^2 + DT^3 \text{ cal K}^{-1} \text{ mol}^{-1} \tag{2.66}$$

are given in Table 2.14.

THERMAL CONDUCTIVITY

The *thermal conductivity coefficient* λ is used in the basic law of heat transfer by conduction. The rate of heat flow is equal to the *thermal conductivity coefficient* λ multiplied by the area of heat transfer and the temperature gradient [4027]. The cgs unit of the coefficient is cal s^{-1} cm^{-1} deg^{-1} and the SI unit is J s^{-1} m^{-1} deg^{-1} (W m^{-1} deg^{-1}).

Theoretical discussions of thermal conductivity may be found in chemical engineering texts on heat transfer and references on properties of liquids such as that of Dreisbach [2083].

Sakiadis and Coates [6352] have described in detail the experimental techniques of determining the thermal conductivity coefficients of 53 organic liquids.

Table 2.12 Heat Capacity–Temperature Relationship: $C_P = A + BT + CT^2 + DT^3$ (cal K⁻¹ mol⁻¹)

Number	Compound	A	B	C × 10⁴	D × 10⁷	Range (K)	Ref.
12	3-Methylpentane	33.476	-0.007651	1.6314	2.4051	119-327	2436
16	Heptane	56.582	-0.14490	5.7813	-4.1667	240-370	3557
17	2-Methylhexane	41.850	-0.026750	2.1531	0.10417	160-300	3557
20	2,4-Dimethylpentane	37.649	-0.042300	4.4188	-4.0364	160-310	3557
32	cis-Decahydronaphthalene	12.68	0.1434			300-350	5184
33	trans-Decahydronaphthalene	10.95	0.1464			300-350	5184
35	1,1'-Bicyclohexyl	13.55	0.1814			277-475	5597
46	Naphthalene	19.212	0.092572			357-371	4749
47	1-Methylnaphthalene	67.196	-0.31472	13.098	-13.642	245-360	4749
50	1,2,3,4-Tetrahydro-naphthalene	44.376	-0.12879	7.4003	-7.467	248-319	7146
51	Butylbenzene	65.158	-0.24196	10.343	-10.108	195-380	5071
56	Cyclohexylbenzene	19.2	0.1465			280-475	5597
69	1-Dodecene	350.670	-2.8766	99.525	110.00	240-320	4748
147	Glycerol	32.9	0.0761	-0.269		273-523	5586
169	Furan	35.17	-0.14860	5.695	-5.322	191-299	3037
208	Acetic acid	15.77	0.03511	0.359		295-400	4919
209	Propionic acid	12.26	0.08023	0.0385		305-445	4919
210	Butyric acid	33.91	-0.02796	1.90		275-370	4919
288	Fluorobenzene	39.496	-0.13777	5.7113	-5.3644	235-350	6527
387	Succinonitrile	28.583	0.02922			335-350	8179
396	Acrylonitrile	31.664	-0.088005	3.1604	-2.8358	193-352	2437
399	Propylamine	12.552	0.26979	-9.8973	12.7641	185-340	2437
400	Isopropylamine	-8.896	0.48638	-17.3193	21.4835	179-323	2437
404	tert-Butylamine	-27.785	0.74527	-26.2156	31.8612	207-337	2437
415	Ethylenediamine	46.521	-0.05994	1.6738	-0.8597	293-334	5067
423	Pyrrole	12.699	0.059825			250-365	6519
454	Benzenethiol	34.985	0.002175	0.6500		258-375	6528

Table 2.13 Heat Capacity–Temperature Relationship: $C_p = A + Bt$ (cal deg^{-1} mol^{-1})

Number	Compound	A	B	Range (°C)
38	Benzene	31.08	0.06342	8–80
39	Toluene	35.66	0.08491	−76–+60
40	o-Xylene	42.31	0.095348	15–132
41	m-Xylene	41.37	0.11055	15–132
42	p-Xylene	41.43	0.11065	15–132
43	Ethylbenzene	41.44	0.1172	15–184
44	Isopropylbenzene	48.42	0.10098	16–153
45	Mesitylene	48.35	0.10102	14–155
55	p-Cymene	55.21	0.1092	10–155

OPTICAL ACTIVITY

The optical rotatory power of a substance may be expressed as the *specific rotation* ($[\alpha]_\lambda^t$), which is defined by

$$[\alpha]_\lambda^t = \frac{100\alpha}{lc}$$

where α is the observed angle of rotation, l the length of the light path in decimeters, c the concentration in grams solute per 100 mL solution, and λ the wavelength of light. Among liquid organic solvents, optical activity is a result of the presence of one or more asymmetric carbon atoms in the molecule. Optical activity may be used in the identification of substances, the elucidation of structure, and the determination of the concentration of solutions.

FREEZING POINT CONSTANT

The *freezing point constant* or *cryoscopic constant* (K_f), is defined by

$$K_f = \frac{M_1 R T_0^2}{1000 \Delta H_m} \tag{2.67}$$

where R is 1.9872 cal K^{-1} mol^{-1}, T_0 the freezing point of the solvent in Kelvin, M_1 the molecular weight of the solvent, and ΔH_m the molar heat of fusion of the solvent.

The numerical value of K_f is usually expressed as degrees per mol per 1000 g of solvent but is sometimes expressed as degrees per mol per 100 g of solvent. The former unit has been used in Chapter III.

Equation (2.67) is accurate, provided that the solution is dilute enough to obey Raoult's law and ΔH_m is constant between the freezing point of solvent and solution.

The cryoscopic constant is often used in (2.68) for molecular weight determinations.

$$M_2 = \frac{1000 w_2 K_f}{w_1 \Delta T} \tag{2.68}$$

In this equation w_2 is the weight of the solute, w_1 the weight of the solvent, and ΔT the freezing point depression. Molecular weights so determined may vary somewhat with

Table 2.14 Heat Capacity (Ideal Gas State)–Temperature Relationship: $C_p^o = A + BT + CT^2 + DT^3$ (cal K^{-1} mol^{-1})

Number	Compound	A	$B \times 10^3$	$C \times 10^7$	Range (K)	Ref.
0	Water	7.256	2.298	2.83	300–1500	6926
		7.219	2.374	2.67	300–1500	6927
5	Pentane	5.780	88.843	−273.26	300–1500	6926
		3.140	100.532	−355.60	300–1500	6927
6	2-Methylbutane	2.801	102.820	−367.41	300–1500	6927
7	2,2-Dimethylpropane	6.076	98.954	−353.69	300–1500	6927
8	Methylcyclopentane	−8.096	127.43	−421.1	335–470	4752
10	Hexane	4.298	118.661	−421.30	300–1500	6927
		7.313	104.906	−323.97	300–1500	6926
16	Heptane	8.850	120.974	−374.78	300–1500	6926
		5.401	136.930	−487.71	300–1500	6927
25	Octane	6.231	155.942	−558.57	300–1500	6927
		10.381	137.054	−425.64	300–1500	6926
31	Decahydronaphthalene	−10.48	209	−823	300–1500	7806
38	Benzene	−0.283	77.936	−262.96	300–1500	6926
39	Toluene	0.436	94.254	−312.58	300–1500	6926
40	o-Xylene	4.603	104.476	−336.16	300–1500	6926
41	m-Xylene	1.956	109.147	−355.83	300–1500	6926
42	p-Xylene	1.846	108.594	−352.00	300–1500	6926
45	Mesitylene	3.042	124.059	−397.43	300–1500	6926
46	Naphthalene	−7.66	147.11	−620.8	300–1500	7806
50	1,2,3,4,-Tetrahydro-naphthalene	−8.90	172.5	−707	300–1500	7806
79	Ethanol	3.578	49.847	−169.91	300–1000	6927
82	1-Butanol	−0.780	102.9	−456	400–450	1683
85	2-Methyl-2-propanol	10.836	49.648	240.70	365–435	836
169	Furan	−7.55	93.52	−528.7	320–485	3037
175	Anisole	−2.218	131.15	−655.5	360–500	3083
189	Acetone	5.371	49.227	−151.82	300–1500	6926
193	3-Pentanone	8.316	80.510	−203.0	360–500	3083

51

Table 2.14 (Continued)

Number	Compound	A	$B \times 10^3$	$C \times 10^7$	Range (K)	Ref.
229	Methyl acetate	5.0378	54.768	-113	300-450	1629
232	Ethyl acetate	9.5098	52.653	124	350-450	1629
235	Propyl acetate	-8.7079	154.48	-889	375-450	1629
288	Fluorobenzene	-6.581	115.89	-612.2	345-500	6527
293	Hexafluorobenzene	11.0175	104.91	-584	340-470	1682
315	Dichloromethane	4.309	31.673	-163.51	250-600	6927
316	Chloroform	7.052	35.598	-216.86	275-770	6927
317	Carbon tetrachloride	22.675	3.274	-3.264^a	275-770	6927
321	1,1,1-Trichloroethane	7.114	60.778	-352.39	329-450	353
351	Bromoform	9.356	32.319	-212.72	300-600	6927
354	1,1,2,2-Tetrabromoethane	8.5059	78.346	$-789.76 + 3.0734 \times 10^{-8} T^3$	250-750	1288
355	Iodomethane	4.105	24.487	-97.33	300-600	6927
361	Diiodomethane	5.839	32.571	-195.28	250-600	6927
380	Nitromethane	2.352	42.882	-169.4	365-520	4754
385	Acetonitrile	5.018	27.935	-93.02	295-1200	6927
431	Pyridine	-3.016	88.083	-386.65	295-1000	6926
451	Carbon disulfide	7.692	13.426	-91.16	300-1800	6927
453	1-Butanethiol	2.628	97.293	-410.15	360-500	6523
454	Benzenethiol	-5.541	122.12	-657.7	430-500	6528
455	Methylsulfide	5.628	44.990	-155.35	320-500	4751
456	Ethyl sulfide	2.76	94.965	-379.4	350-485	6522
462	Dimethyl sulfoxide	6.94	56	-227	300-1000	4811

$^a 10^{-7} C/T^2$.

the solute concentration; best values are often obtained by extrapolating a plot of M_2 versus concentration to zero concentration where the behavior of the solution is most nearly ideal. The determination of molecular weights by (2.68) is accurate to about $\pm 5\%$ under favorable conditions. The usual procedures rarely attain this accuracy but are adequate for checking the correctness of postulated empirical formulas.

Brancker and coworkers [1056] have proposed a modification of (2.68) that overcomes the difficulties that arise from variations in K_f and molecular weights with concentration. This modified form is

$$\Delta t = K_f m^b \tag{2.69}$$

where b is a constant and m is the molality of the solution. For benzene as the solvent, the modified equation was found to be

$$\Delta t = 5.229 m^{0.9394} \tag{2.70}$$

The value of K_f in (2.70) was found to vary with different samples of the same cryoscopic solvent. The value of b was constant for all pure solutes but was extremely sensitive to impurities. Table 2.15 illustrates some results.

Using the results of Meldrum, Saxer, and Jones [5050], Brancker and coworkers [1056] found the equation for camphor to be

$$\Delta t = 38.73 m^{0.9580} \tag{2.71}$$

Skau and Wakeham [7958] have discussed the theory and determination of molecular weights by the freezing point method, and also the sources of error. The familiar Beckmann procedure [717] and its sources of error have been considered by many investigators, among them Skau and Wakeham [7958], Rall and Smith [6026], and Daniels, Mathews, and Williams [1808].

The unusually high freezing point constant of camphor, about 40, has been utilized by Rast [6072] to develop a simple, rapid procedure for determining the molecular weights of compounds soluble in camphor. This method is described in detail by Kamm [3841], Meyer [5080], and Durand [2152].

A modification of the Rast procedure has been made by Keller and Halban [3931]. Data are given for cyclopentadecanone, diethyldiphenyl, urea, tetrabromomethane, and borneol as solvents.

Rossini [6274]; Glasgow, Krouskop, et al. [2788]; Streiff and Rossini [7039]; and

Table 2.15 Molecular Weights of 1,2,3,4-Tetrahydronaphthalene in Benzene; mw = 132.196

Concentration (%w)	Molecular Weight [Eq. (2.68)]	Molecular Weight [Eq. (2.70)]
0.68	111.6	133.2
1.98	118.0	132.4
3.98	127.1	137.0
7.94	131.2	135.5
11.13	134.9	136.6

Glasgow, Streiff, and Rossini [2781] have described a method, apparatus, and procedure for the determination of purity by freezing point measurements. This method is adaptable to molecular weight determinations of high accuracy, provided that the fundamental thermodynamic requirements are met by the solvent–solute system.

The basic equation is

$$-\ln N_1 = A(T_0 - T)\{1 + B(T_0 - T) + \cdots\} \tag{2.72}$$

where N_1 is the mole fraction of the solvent, T the freezing point of the solution, and T_0 the freezing point of the pure solvent; $A = \Delta H_m/RT_0^2$ and $B = (1/T_0) - (\Delta C_p/2\Delta H_m)$. Equation (2.72) may be modified to calculate the mole percent purity P.

$$\log P = 2.00000 - \frac{A}{2.30259}(T_0 - T)\{1 + B(T_0 - T) + \cdots\} \tag{2.73}$$

It will be noted that the constant A is related to K_f through the equation

$$K_f = \frac{M_1}{1000A} \tag{2.74}$$

Values of A or of A and B for many hydrocarbons have been given in Chapter III. The terms in brackets should be dropped when only the constant A is given.

$$-\ln N_1 = A(T_0 - T) \tag{2.75}$$

For dilute solutions, (2.75) becomes

$$N_2 = A(T_0 - T) = \frac{w_2 M_1}{w_1 M_2}$$

from whence a convenient equation for the calculation of molecular weights from freezing point depressions is obtained:

$$M_2 = \frac{w_2 M_1}{w_1 A(T_0 - T)} \tag{2.76}$$

Witschonke [8121] has tabulated the freezing point of a number of compounds determined by the equilibrium freezing and melting curves. A cryoscopic constant K', which is described as the mole percent impurity which lowers the freezing point 1°C, was calculated from the equation

$$K' = \frac{100\Delta H_m}{RT_0^2} \tag{2.77}$$

I, the mole percent impurity, is then calculated from the equation

$$I = K'(T_0 - T_f) \tag{2.78}$$

where T_0 is the freezing point of the pure solvent and T_f is the freezing point of the sample. Equation (2.78) can be used to calculate the purity of a sample within the range 95 to 100%. The freezing point of the 100% compound must be known. An example of this calculation may be found in Chapter V, under Criteria of Purity for benzonitrile. Some of Witschonke's values for K' and T_0 are given in Table 2.16.

Table 2.16 Cryoscopic Constants K' (mol%/°C)

Number	Compound	fp 100% m (°C)	K'
118	m-Cresol	12.22	0.749
189	Acetone	−94.6	2.14
326	o-Dichlorobenzene	−17.01	2.4
384	Nitrobenzene	5.74	1.87
387	Succinonitrile	58.1	0.42
390	Valeronitrile	−96.20	3.50
394	α-Tolunitrile	−22.0	1.9
395	Benzonitrile	−12.75	1.93
396	Acrylonitrile	−88.55	2.7
408	Aniline	−6.00	1.38
431	Pyridine	−41.6	1.86
438	Quinoline	−14.85	1.95

BOILING POINT CONSTANT

The *boiling point constant* (K_B) or *ebullioscopic constant* is defined as

$$K_B = \frac{M_1 R T_0^2}{1000 \Delta H_v} \tag{2.79}$$

where R is 1.9872 cal K^{-1} mol^{-1}, T_0 the boiling point of the solvent, and ΔH_v the molar heat of vaporization. The numerical values for K_B in Chapter III are expressed in degrees per mol per 1000 g solvent.

Like the similar equation for K_f (2.67), Eq. (2.79) is valid only for dilute solution for which Raoult's law is at least approximately applicable. Since the boiling point and the heat of vaporization of a solvent vary with pressure, K_B will also vary. In general, the effect is within experimental error for a pressure variation of about 10 Torr around atmospheric pressure.

Molecular weights can be determined with the relation

$$M = K_B \frac{1000 w_2}{w_1 \Delta T_b} \tag{2.80}$$

where ΔT_b is the elevation of the boiling point brought about by the addition of w_2 grams of solute to w_1 grams of solvent. For the experimental details in determining boiling points, see Swietoslawski and Anderson [7958].

The advantages of the ebullioscopic determination of molecular weights over the cryoscopic method were discussed by Menzies and Wright [5060] in 1921. They also considered the errors encountered in ebullioscopic molecular weight determinations and described the application of a differential thermometric method. The errors that can arise when the gas constant R in (2.79) is not corrected by means of the Berthelot equation [2784] have been considered by Hoyt and Fink [3529]. For additional information on the ebullioscopic method, see Rosanoff and Dunphy [6253].

Hill and Brown [3406] have described a stable ebulliometer with highly sensitive differential thermometers to give a wide range of operating temperatures. These authors state that their method represents a 10-fold extension of the range and sensitivity of the Menzies and Wright [5060] ebulliometer. This is made possible by the coupling of solvents boiling above 130° with a conventional water thermometer, or by the use of low boiling thermometric liquids with conventional solvents. A precision of 2% was obtained for solutes whose molecular weight ranged from 300 to 13,400.

A thermoelectric procedure for determining the molecular weights of solutes in organic solvents has been described by Taylor and Hall [7193]. It is a modification of the procedure of Hill [3404] for observing the temperature difference between two different solutions or between the solution and the solvent when opposing thermocouples are in an atmosphere saturated with the solvent. Hill's proposal had been used exclusively for water, but Taylor and Hall applied it to organic solvents. For molecular weight determinations, the underlying principles are those of ebulliometry.

ACIDITY AND BASICITY

The characteristics of acids and bases and the mechanisms of their reactions are often explained on the basis of one or both of the two leading theories proposed in 1923. The Lewis theory [4566, first proposed; 4567, amplified] defines a base as an electron-pair donor and an acid as an electron-pair acceptor in the formation of a coordinate covalent bond. Brønsted [1119, 1120] and Lowry [4675] simultaneously and independently published the proton theory of acids and bases. This theory will be hereinafter referred to as the Brønsted theory. It regards an acid as a proton donor and a base as a proton acceptor. While Brønsted and Lewis bases are the same, Brønsted acids are limited to proton donors; for practical purposes they may be considered as a special, but large, class of Lewis acids.

The acidic or basic character of a solvent often serves as the criterion of its usefulness as a reaction medium, and solvents are sometimes classified on the basis of their acidic or basic properties. Solvents that tend to provide protons are called *protogenic*, while those that act primarily as proton acceptors are said to be *protophilic*. *Amphiprotic* solvents are those that may act either as proton donors or proton acceptors, depending on the nature of the solute. Water and the alcohols are well-known examples. Still other solvents are relatively inert with respect to proton activity and are classified as *aprotic* solvents. Most hydrocarbons under conditions usually encountered fall in this class; however, the concept of the aprotic solvent should be employed with discretion because there is no known organic solvent that will not ionize to some extent under some conditions. For example, benzene, usually found to be inert, is a base in hydrogen fluoride having a pK_{BH^+} value of 9.4 [4814]. The influence of the acidic or basic character of the solvent on chemical reactions has been ably discussed by Luder and Zuffanti [4680]. Davis [1859] has lucidly presented the role of the aprotic solvent.

The strengths of acids and bases have generally been determined from measurements made on aqueous solutions of the compounds. Albert and Serjeant [68] discussed the strengths of acids and bases in water using a minimum of mathematics. More extensive and sophisticated presentations including strengths in water and in nonaqueous solvents are given by Bates [669, 4158, Chapter 10], Kolthoff [4158, Chapter 11], Bruckenstein and Kolthoff [4158, Chapter 12], Kolthoff and Bruckenstein [4158, Chapter 13], Perrin [5774], and Arnett and coworkers [432].

There is no acidity function that expresses the strengths of acids and bases in all different solvent systems. Generally, a substance that is an acid or a base in one solvent is an acid or base in most other solvents, and the measurable strength will vary from solvent to solvent. In the tables of Chapter III, the strengths of acids and bases are expressed as they appeared in the original literature, as equilibrium constants or as functions derived therefrom. Symbols are briefly defined in "Abbreviations and Symbols" and are fully discussed by Bates [669]. Whenever possible, the temperature and the solvent (if different from water) are noted.

Acidity and basicity are chemical properties. Their true measure is their effect on a chemical reaction such as the hydrolysis of sucrose, and so forth. Since this is not a convenient measure, physical means are used such as the electromotive force of cells, conductivity, or freezing point depressions. The physical methods are reliable within a single solvent system for which they are applicable, and between certain solvent systems. Ionization as a criterion of strengths of acids and bases must be used with caution, particularly from solvent to solvent. Perchloric acid is considered completely ionized in water; it has a relatively low ionization in acetic acid. If ionization were the criterion of strength, sucrose should be hydrolyzed more rapidly by perchloric acid in water than in acetic acid. The reverse is true. Hydrochloric acid is also considered completely ionized in water. It shows no ionization in benzene, yet it is an acid in this solvent, according to the Brønsted and the Lewis concepts. Since it reacts with ammonia in benzene, there is some chemical evidence that it is at least a moderately strong acid. Davis has carried out extensive investigations concerning the effect of the solvent and the nature of acid and base reactions in aprotic solvents. She has presented the behavior of acids and bases in aprotic solvents in a recently published monograph [1859].

The *autoprotolysis constant* (K_s) is the product of the ion activity of the self-ionized ions of a solvent. It need not be limited to ionization involving protons, but its name is so limiting. The general case is

$$2\ HR \rightleftarrows H_2R^+ + R^-$$ (2.81)

hence,

$$K_s = \gamma[H_2R^+] \times \gamma[R^-]$$ (2.82)

where brackets indicate the molarity of the enclosed entity. Concentrations of the ionic species are generally so small that the activity coefficients may be considered to approach unity and (2.82) becomes

$$K_s = [H_2R^+] \times [R^-]$$ (2.83)

The autoprotolysis constant for water is also known as the ion product and is derived from equilibrium considerations thus:

$$H_2O + H_2O \rightleftarrows H_3O^+ + OH^-$$ (2.84)

$$K_{aq} = \frac{[H_3O^+] \times [OH^-]}{[H_2O]}$$ (2.85)

The relative change in $[H_2O]$ is negligible in comparison to a large change in $[H_3O^+]$ or $[OH^-]$; therefore $[H_2O]$ is considered a constant and (2.85) becomes

$$K_w = K_s = [H_3O^+] \times [OH^-]$$ (2.86)

The autoprotolysis constants for only a few amphiprotic solvents have been determined. The specific constants are expressed as indicated in (2.81) and (2.82). Examples from three functional group types are as follows.

Methanol:

$$\underset{\text{acid}}{CH_3OH} + \underset{\text{base}}{CH_3OH} \rightleftarrows \underset{\text{acid}}{CH_3OH_2^+} + \underset{\text{base}}{CH_3O^-} \tag{2.87}$$

$$K_s = [CH_3OH_2^+] \times [CH_3O^-] \tag{2.88}$$

Acetonitrile:

$$\underset{\text{acid}}{CH_3CN} + \underset{\text{base}}{CH_3CN} \rightleftarrows \underset{\text{acid}}{CH_3CNH^+} + \underset{\text{base}}{CH_2CN^-} \tag{2.89}$$

$$K_s = [CH_3CNH^+] \times [CH_2CN^-] \tag{2.90}$$

Acetic acid:

$$\underset{\text{acid}}{HC_2H_3O_2} + \underset{\text{base}}{HC_2H_3O_2} \rightleftarrows \underset{\text{acid}}{H_2C_2H_3O_2^+} + \underset{\text{base}}{C_2H_3O_2^-} \tag{2.91}$$

$$K_s = [H_2C_2H_3O_2^+] \times [C_2H_3O_2^-] \tag{2.92}$$

The H_3O^+, $CH_3OH_2^+$, CH_3CNH^+, and $H_2C_2H_3O_2^+$ and all like protonated ions are *onium* ions, for example, hydronium, methanonium, and so forth. The onium ion is more frequently referred to as the *hydrogen ion*. The composition of the onium ion as H_2R^+ is not universally accepted.

The autoprotolysis constant and the dielectric constant are believed to be the principal influencing factors that determine the nature and extent of the effect of solvates, particularly the acidic or basic nature.

The *ionization constant* (K_i) of an organic solvent is a measure of its strength as an acid or base when it is dissolved in some other solvent, usually water. Acid ionization constants are commonly designated as K_a and base ionization constants as K_b. These constants are the equilibrium constants for the ionization reaction that have been modified to include the concentration of the water that is always present in large excess. For example, for an acid,

$$\underset{\text{acid}}{HC_2H_3O_2} + \underset{\text{base}}{H_2O} \rightleftarrows \underset{\text{acid}}{H_3O^+} + \underset{\text{base}}{C_2H_3O_2^-}$$

$$K_a = \frac{[H_3O^+] \times [C_2H_3O_2^-]}{[HC_2H_3O_2]} \tag{2.93}$$

and for a base (C_4H_9, *n*-butyl radical = Bu),

$$\underset{\text{acid}}{H_2O} + \underset{\text{base}}{BuNH_2} \rightleftarrows \underset{\text{base}}{BuNH_3^+} + \underset{\text{base}}{OH^-}$$

$$K_b = \frac{[BuNH_3^+] \times [OH^-]}{[BuNH_2]} \tag{2.94}$$

The ionization constant K_a for acetic acid is 1.754×10^{-5} and K_b for butyl amine is 4.09×10^{-4}. In 1909, Sørensen proposed a more convenient means of expressing the effective hydrogen ion concentration as

$$pH = -\log [H^+]$$

"p" represents the word "potenz," translated into English as power, the exponent of 10 in the arithmetical expression of the concentration of the hydrogen ion. This reasoning was extended to other expressions such as hydroxyl ion concentration, ionization constants, and autoprotolysis constants.

The autoprotolysis constant for water, $K_w = K_s = [H^+] \times [OH^-]$. Numerically, at 25°C, $K_w = 10^{-14} = 10^{-7} \times 10^{-7}$.

$$pK_w = 14 \quad \text{and} \quad pH = pOH = 7 \tag{2.95}$$

The strengths of bases are more generally expressed in terms of the protonated base BH^+. This is derived from K_b by converting to pK_b; for example, for butyl amine in water, $K_b = 4.09 \times 10^{-4}$, $pK_b = 5.389$, and pK_a or pK_{BH^+} is $14.000 - 5.389 = 8.611$.

The subject of acidity and basicity as related to the solvents in this book, of necessity, is brief. The references given herein should be consulted for a more complete presentation of the subject.

ELECTRICAL PROPERTIES

The electrical properties tabulated in Chapter III are the *specific conductance* or *conductivity* (κ), the *dielectric constant* (ϵ), and the *dipole moment* (μ).

The specific conductance is the reciprocal of the specific resistance, which is the resistance that would be offered by the solvent when placed between plane parallel electrodes 1 cm^2 in area and 1 cm apart. The cgs unit of specific conductance is $ohm^{-1}\ cm^{-1}$ or mho cm^{-1} and the SI unit is $ohm^{-1}\ m^{-1}$ or mho m^{-1}.

In general, pure solvents have a low conductivity. For example, the purest water has a specific resistance at 25° of about 20 million ohm cm. Ordinary water, distilled in the presence of air, has a specific resistance of only about 100,000 ohm cm due to the absorption of carbon dioxide and other gases from the air.

The conductivity of a solvent can be used as a criterion of purity in limited cases. It is an excellent method for determining the purity of water and can be used to determine the completeness of removal of dissociable material from other solvents of high dielectric constant. It is sometimes useful for determining the dryness of solvents. Mixtures of two liquids, each of which is practically nonconducting, may have a conductance of the same magnitude as the individual solvents or they may have a higher conductance.

The dielectric constant is a measure of the relative effect a solvent has on the force with which two oppositely charged plates attract each other. The dielectric constant of a vacuum is defined as unity, but for practical purposes measurements are made in air; it is a unitless number.

Qualitatively, substances of high dielectric constant are good ionizing media. This fact may be used for estimating solubilities. If the solubility of a substance is known in a solvent of known dielectric constant, its solubility in other solvents can often be roughly estimated from their dielectric constants.

The most important use of the dielectric constant is the calculation of *molar polarizations* and from them the dipole moments. The molar polarization P is defined by the

Clausius–Mosotti equation [2784] as

$$P = \frac{\epsilon - 1}{\epsilon + 2} \frac{M}{d}$$
(2.96)

where ϵ = dielectric constant, M = molecular weight, and d = density.

The dipole moment is a measure of the displacement of what might be termed the centers of gravity of positive and negative charges of a molecule. It is the product of a charge and a distance. The cgs unit of the dipole moment is esu-cm and the SI unit is C-m. The values tabulated in Chapter III are given in *Debye* units where 1 Debye (D) is equal to 10^{-18} esu-cm or 3.336×10^{-30} C-m.

The dipole moment is important in studying the structure of compounds because it provides a means for measuring the polarity, or electrical dissymmetry, of a molecule. Physical properties, such as solubility, the properties of solutions, the influence of solvents on reaction rates, and deviations from the simple gas laws can be quantitatively interpreted more readily if the dipole moments are known.

Many qualitative conclusions about solvents can be drawn from their dipole moments. Perfectly symmetrical molecules are nonpolar. Unsymmetrical compounds, however, are almost invariably polar, particularly when oxygen, nitrogen, or a halogen atom is present. In general, nonpolar substances are volatile and soluble in other nonpolar compounds, whereas polar compounds are less volatile and are soluble only in other polar compounds.

Numerous dielectric constants and dipole moments of compounds in the gaseous state have been compiled by Maryott and Buckley [4934] and McAlpine and Smyth [4730]. Dielectric constants of liquids have been tabulated by Maryott and Smith [4937] and dipole moments have been tabulated by McClellan [4739].

SOLUBILITY PARAMETER

The *solubility parameter* (δ) is a physical constant used to describe the relationship between the physical properties of the solvent and its effectiveness in dissolving specific solutes. The overall concepts require an extensive theoretical treatment in areas such as thermodynamics, structure of liquids, and molecular forces. The scope of this section will be limited to some fundamental defining terms.

J.H. Hildebrand developed many theories on solubility and solutions and eventually proposed the term "solubility parameter" designated by the symbol δ [3392, 3395]. The thermodynamic term is the cohesive energy density, introduced by Scatchard [6406] and later incorporated by Hildebrand into the solubility parameter concept. The cohesive energy density, sometimes referred as the internal pressure, is the isothermal internal energy of vaporization of the liquid to the ideal gas state divided by the volume of the liquid. In the simplest terms, the cohesive energy density is a measure of the energy required to overcome all of the intermolecular forces holding a liquid together [1271]. The notation for this term is $\Delta E_v/V$.

The solubility parameter is defined as

$$\delta = (\Delta E_v/V)^{1/2}$$
(2.97)

The cgs unit is given as $cal^{1/2}$ $cm^{-3/2}$. Barton [652, 653] has suggested an SI unit of

$MPa^{1/2}$, which equals $J^{1/2} cm^{-3/2}$. Typical magnitudes of solubility parameters (in cgs units) at room temperature are: *n*-pentane, 7.0; bromobenzene, 9.9; and ethanol, 12.7 [652].

In thermodynamics the mixing process or mutual solubility between two liquids is favored when the heat of mixing is low. On a molecular basis this means that the attractive forces holding each liquid together and the forces between the two liquids are all similar in magnitude. Consequently solubility is expected to be the highest when two substances have comparable solubility parameters.

The determination of numerical values of δ is based on the values of ΔE_v and V. The latter is calculated from the molecular weight and density of the liquid. At temperatures below the normal boiling point, Eq. (2.98) is quite valid.

$$\Delta E_v = \Delta H_v - RT \qquad (2.98)$$

With heats of vaporization compiled in many tables, this calculation of ΔE_v is the one most frequently used. Burrell [1271] has summarized numerical evaluation from other physical constants such as surface tension and critical constants.

Extensive reviews of mixing processes, molecular forces, solution theories, mixed solvents, and the solubility parameter concept are given in Hildebrand and Scott [3395], Hildebrand, Prausnitz, et al. [3394], Hildebrand [3393], Barton [652, 653], Burrell [1271], and Godfrey [2804].

Solubility parameters are applied in the selection of the appropriate solvent for dissolving a particular substance. Burrell [1270, 1271] has analyzed the solubility behavior of many polymers. Tortorello and coworkers [7533] used solubility parameters to select acrylic polymers in the design of high-performance coating such as those for aircraft. Sanders [6371] has studied the solubility parameters of various halogenated hydrocarbons for use in propellants and aerosols. Other major industrial applications include paint technology [2611] and formulation of liquid inks [6913].

Selection of solvents for gas and liquid chromatography can be based on solubility parameters [4030, 6240]. Eisenbach and coworkers [2232] have applied the concept of the solubility parameter to the systematic search for suitable nonaqueous media for biological systems. Barton [652] has listed over 30 applications of solubility parameters. The most comprehensive compilations of solubility parameters are those of Barton [652, 653]. Other listings are given in Hildebrand and Scott [3395], Lieberman [4593], Fowler and Katritzky [2500], Takamatsu [7159], and Burrell [1271].

Industrial solvent charts that tabulate solubility parameters include those of Shell Chemical [6628, 6631, 6632], Solvents and Chemical Companies [6905], Eastman Chemicals [2190], Celanese [1399], and Union Carbide [7635].

EVAPORATION RATE

The *evaporation rate* is one of the important parameters of design components for finishes, printing inks, industrial solvent mixtures, and so forth, where the evaporation time of the solvent is a significant consideration. A usable relationship has not been found between boiling point, vapor pressure, and other parameters of a liquid.

A number of methods of expressing the evaporation rate have been advanced. Three are used extensively at the present time; namely, the ratio of the time of evaporation of a specified amount of the solvent under given conditions relative to the time of evaporation of (1) ether or (2) butyl acetate and (3) the time of evaporation of a thin film under specified conditions. The ratio of time of evaporation relative to ether is widely used in Europe whereas the ratio to butyl acetate is preferred in the United States.

The "thin film" or Shell Method [6628, 6704] is becoming the preferred method. The evaporation rate for n-butyl acetate (purity 90–92%) is determined on the Shell Automatic Thin Film Evaporator [6628]. The time for 90% of this solvent to evaporate is determined and this is taken as BuOAc = 1.0 on a relative scale. Then the 90% evaporation times for other solvents are determined on the same instrument under the same conditions. The Shell Chart of Solvent Properties [6628, 6631] lists the time in seconds for 90% of the solvent to evaporate and the rate relative to BuOAc = 1.0. The values tabulated in Chapter III are based on BuOAc = 1.0 and the 90% evaporation time.

Other tabulated values of evaporation rates are given in Hofmann [3455] and Bent and Wik [778] as well as the solvent charts of Union Carbide [7635], Celanese [1399], Eastman Chemicals [2190], and Solvents and Chemical Companies [6905].

FLASH POINT

The *flash point* (fl pt) of a liquid is the minimum temperature at which its equilibrium vapor in admixture with air at a total pressure of 760 Torr will be ignited by an external ignition source applied in the manner prescribed in the test to form a flame that propagates itself over the surface of the liquid. Although not usually considered a common physical property, flash points are included in Chapter III because of their widespread use in classifying solvents for storage and shipping. The National Fire Protection Association defines a flammable liquid as one having a flash point below 100°F and a vapor pressure not exceeding 40 psi (absolute) at 100°F [5416].

Several procedures for measuring flash points are currently in use in the United States. The American Society for Testing and Materials describes the following standard methods [324]:

> *Tagliabue Open Cup* (TOC), ASTM **D 1310.**   Determination of open cup flash points of volatile liquids in the range of 0 to 475°F and having fire points up to 500°F.
>
> *Tagliabue Closed Cup* (TCC), ASTM **D 56.**   Determination of the flash points of all mobile liquids flashing below 200°F.
>
> *Cleveland Open Cup* (COC), ASTM **D 92.**   Determination of flash and fire points of all petroleum products (used for other organic liquids also) except fuel oil and liquids having an open cup flash point below 175°F.
>
> *Pensky–Martens Closed Cup* (PMCC), ASTM **D 93.**   Determination of flash points of such liquids as fuel oil, lubricating oils, viscous liquids, and liquids that tend to form surface films. (Should not be used for common solvents.)

These several types of apparatus give somewhat different values that are not convertible. Reliable results within the limits claimed are obtained when the standardized apparatus is used in the manner prescribed [4782].

SPECTROSCOPY

In spite of the large volume of work published in spectroscopy, suitable data for some organic solvents are surprisingly scarce. An abundance of papers have appeared for certain other solvents; in these instances, primary consideration has been given to those references that deal with the application of spectra to problems of identification and analysis. Spectroscopic data are not given in the tables of Chapter III but references to *ultraviolet* (uv), *infrared* (ir), *Raman, mass,* and *nuclear magnetic resonance* (nmr) spectroscopy are included.

There are available many excellent books on the possibilities and limitations of the several branches of spectroscopy. Among those of a nonmathematical character are Sawyer [6401], Brode [1112], West [7959], Harrison, Lord, and Loofbourow [3191], and Gutowsky [7960].

The ultraviolet is the oldest of the nonvisual spectra that has been used experimentally. The most comprehensive collection of ultraviolet spectra in the region 2100–3200 Å has been compiled by the American Petroleum Institute, Research Project 44 [155]. New spectra are added as they become available.

Many analysts consider infrared spectra the most useful of the absorption spectra. With the commonly used spectrometers, many substances do not show a characteristic absorption in the ultraviolet region, but all substances show characteristic absorption in the infrared. Herzberg [3349] has presented a comprehensive discussion on the theory and interpretation of infrared and Raman spectra. The book by Randall, Fowler, Fuson, and Dangl [6045] can be used easily without an extensive background in mathematics.

The application of infrared spectra to qualitative and quantitative analysis has been discussed by Barnes, Gore, et al. [630, 631]. One such application is the evaluation of hydrogen bonding forces; further discussion is given in Crowley et al. [1730], Nelson et al. [5450, 5451], and Burrell [1272].

Raman spectroscopy has been largely used in the study of molecular structure, but it also has application to problems of analytical chemistry. Braun and Fenske [1068] have outlined some of its analytical possibilities and have presented a bibliography on the subject. Several instruments employing continuous gas lasers of the He/Ne type, instead of conventional low-pressure mercury arc sources, are now commercially available. These more powerful sources permit the extension of Raman spectroscopy to smaller samples and to dilute solutions. Fluorescence and absorption frequently associated with the mercury arc source are avoided and the narrower excitation line width of the laser source is advantageous in measurements made of the closely spaced rotational spectra of gases.

The mass spectrometer has become an extremely important tool for the analysis of organic compounds and for the elucidation of their structures. The analytical applications encompass hydrocarbons, alcohols, esters, ketones, and aldehydes as well as materials containing appreciable quantities of water. A recent companion to electron bombardment mass spectroscopy that results in fragmentation of the molecules is field-ionization mass spectrometry, which gives largely molecular ions. High-resolution mass spectrometers ($M/\Delta M = 10,000$) are becoming common research equipment, and instruments giving resolutions as high as 1,200,000 have been reported. A short, comprehensive survey of the analytical applications and instrumentation has been prepared by Hipple and Shepherd [3414]. They have included an extensive bibliography. The American Petroleum Institute, Research Project 44, and the National Bureau of Standards [154] have issued a catalog of mass spectra data. For recent developments in instrumentation the reader is referred to the review published annually in *Analytical Chemistry*.

Chamberlain [1410] and Stehling [6986] present accurate and rapid methods for the determination of the important features of the molecular structures of organic compounds by nuclear magnetic resonance spectroscopy. These methods also may assist in the identification of impurities in solvents.

There are numerous collections of spectral data used for the tables in Chapter III including those of the American Petroleum Institute Research Project 44 [153–155], the Thermodynamics Research Center Data Project [7225–7229], Aldrich Chemical [5932, 5933], and Sadtler Research [6748, 6749].

SAFETY

The brief statements on toxicology in Chapter V are intended only as guides for safe handling and are not meant to give a complete summary of available data. The present maximum allowable concentrations are taken from lists published by the American Conference of Governmental Industrial Hygienists [139–142]. These are the maximum average atmospheric concentrations to which workers may be exposed for an 8-hr day without injury to health. They are expressed on a volume/volume basis.

Chapter III

PHYSICAL PROPERTIES: TABULATIONS

EXPLANATION OF THE TABLES

Each table is headed by the generally recognized name of the compound. The current *Chemical Abstracts* index name, if different from the more common name, is listed on the second line. The *Chemical Abstract Service Registry Number* and the structural (left) and empirical (right) formulas are on the third line.

Each property is designated by a symbol or abbreviation in the first column. These conform to the recommendations of the International System of Units (SI) so far as possible and practical at the present time. The symbols and units are listed in Table 3.1 and defined in the next two sections. The second column lists the conditions, such as temperature and pressure, at which the property is given. All temperatures in this column are Celsius unless otherwise designated. The degree sign is not used to indicate numerical temperature values. The third and fourth columns list the cgs and SI values, respectively, from the primary references or as calculated therefrom directly by conversion factors. Some common conversion factors are listed after the definitions of symbols.

References to the literature are listed in the fifth column. Sources of the property values used in the table (primary references) are the numbers before the semicolon. Values calculated by the authors are designated by underscoring the references from which the basic data were taken. Additional references are not confined to the condition for the value of the property listed in the second column. Some references have been included because the discussion is of interest, and others are included because the property was determined at other than the usual conditions, such as high or low temperatures. The references constitute a good key for a literature survey.

Beilstein references in the fifth column are given with the series abbreviation, volume, and page number; e.g., Beil EIII4, 527.

At the bottom of each table are the *Beilstein* and spectral references. The latter have not been separated into primary and secondary sources. Data such as wavelengths and absorption coefficients are not given here.

SYMBOLS AND ABBREVIATIONS

The first part of *Organic Solvents* in its first three editions was slanted primarily for those using the metric system. The fourth edition has increased its scope by including additional properties relevant to such areas as solution design, refrigeration, aerosol propellants and solvents, and solvents used at low temperatures (below 0°C). Data for the tables in this chapter are reported in the metric system in both cgs and SI units. This section lists the symbols and names for units in both metric systems and the foot-pound-second system for units pertinent to and used throughout this book (see pages 68–71).

Table 3.1 Sequence of Listing: Properties, Symbols, Unit Definitions

Property	Symbol	cgs	SI
1. Molecular weight	mw	g mol^{-1}	g mol^{-1}
2. Boiling point	bp	°C	K
3. Temperature coefficient of pressure	dt/dp	°C Torr^{-1}	K kPa^{-1}
4. Pressure coefficient	dp/dt	Torr °C^{-1}	kPa K^{-1}
5. Vapor pressure	p	Torr	kPa
6. Constants for Chapter II vapor pressure–temperature equations	A, B, C		
7. Freezing point	fp	°C	K
8. Triple point	tp	°C	K
9. Density	d	g cm^{-3}	kg m^{-3}
10. Coefficient of density	dd/dt	g cm^{-3} °C^{-1}	kg m^{-3} K^{-1}
11. Thermal expansion coefficient	α	°C^{-1}	K^{-1}
12. Isothermal compressibility	κ_T	Torr^{-1}	kPa^{-1}
13. Adiabatic compressibility	κ_S	Torr^{-1}	kPa^{-1}
14. Refractive index	n_D	Dimensionless number	
15. Refractive index coefficient	dn/dt	°C^{-1}	K^{-1}
16. Viscosity coefficient	η	cP / cSt	kg m^{-1} s^{-1} / m^2 s^{-1}
17. Surface tension	γ	dyne cm^{-1}	N m^{-1}
18. Heat of vaporization	ΔH_v	kcal mol^{-1}	kJ mol^{-1}
19. Heat of sublimation	ΔH_s	kcal mol^{-1}	kJ mol^{-1}
20. Heat of fusion	ΔH_m	kcal mol^{-1}	kJ mol^{-1}
21. Heat of transition	ΔH_t	kcal mol^{-1}	kJ mol^{-1}
22. Standard heat of formation	ΔH_f°	kcal mol^{-1}	kJ mol^{-1}
23. Heat of combustion	ΔH_c°	kcal mol^{-1}	kJ mol^{-1}
Standard state	ΔH_c	kcal mol^{-1}	kJ mol^{-1}
Constant pressure			
Constant volume	ΔE_c		
24. Heat of polymerization	ΔH_p	kcal mol^{-1}	kJ mol^{-1}
25. Heat capacity, ideal gas state	C_p°	cal K^{-1} mol^{-1}	J K^{-1} mol^{-1}
26. Heat capacity	C_p	cal K^{-1} mol^{-1}	J K^{-1} mol^{-1}
27. Thermal conductivity coefficient	λ	cal s^{-1} cm^{-1} K^{-1}	J s^{-1} m^{-1} K^{-1}

			Degrees of rotation
32. Specific rotation	$[\alpha]_D$		Degrees of rotation
33. Cryoscopic constant	A	eq 2.72	eq 2.72
34. Cryoscopic constant	K_f	eq 2.67	eq 2.67
35. Ebullioscopic constant	K_B	eq 2.79	eq 2.79
36. Ionization constant in water	K_i	Dimensionless number	Dimensionless number
37. Acid dissociation constant in water	K_a	Dimensionless number	Dimensionless number
38. Base dissociation constant in water	K_b	Dimensionless number	Dimensionless number
39. Negative logarithm of K_a	pK_a	Dimensionless number	Dimensionless number
40. Negative logarithm of K_b	pK_b	Dimensionless number	Dimensionless number
41. Negative logarithm of ionization constant of protonated base	pK_{BH^+}		Dimensionless number
42. Negative logarithm of autoprotolysis constant	pK_s		Dimensionless number
43. Specific conductance	κ	$ohm^{-1}\ cm^{-1}$	$ohm^{-1}\ m^{-1}$
44. Dielectric constant	ϵ	Dimensionless number	Dimensionless number
45. Dipole moment	μ	Debye units (see Chap. II)	
46. Solubility parameter	δ	$cal^{1/2}\ cm^{-3/2}$	$J^{1/2}\ cm^{-3/2}$
47. Evaporation rate BuOAc = 1 90% evap	ER	s	s
48. Solubility with water, mutual in aq aq in	soly	Dimensionless number	Dimensionless number
49. Aqueous azeotrope bp, °C, in column 2	aq az	%w, solute / %w, water / %w, compound	%w, solute / %w, water / %w, compound
50. Flash point	fl pt	°C	K
51. *Beilstein* reference[a]	Beil	References only	
52. Ultraviolet spectra	uv	References only	
53. Infrared spectra	ir	References only	
54. Raman spectra	Raman	References only	
55. Mass spectra	ms	References only	
56. Nuclear magnetic resonance	nmr	References only	

[a]System number is underscored and the page number of the volume in which the system number appeared in Basic Series H follows. When the first citation does not occur in Series H, the abbreviation for the Supplementary Series precedes the page number.

A   Cryoscopic constant defined by Eq. (2.72)
Å   Angstrom unit, 10^{-10} m
A, B, C   Constants for vapor pressure–temperature equations
ACS   American Chemical Society
aq   Water
AR   Analytical Reagent (grade)
ASTM   American Society for Testing and Materials
atm   Atmosphere
av   Average value
av   Avoirdupois
az   Azeotrope
B   Cryoscopic constant defined by Eq. (2.72)
Beil   *Beilstein* reference
bp   Boiling point
Btu   British thermal unit
BuOAc   Butyl acetate for evaporation rate scale
C   Coulomb
°C   Celsius scale of temperature
ca   About
cal   Calorie, thermochemical, also called gram calorie
cal_{IT} Calorie, international table
cgs   Centimeter-gram-second system of units
C_p Heat capacity at constant pressure
C_p° Heat capacity at constant pressure for the ideal gas state
CC   Closed cup flash point tester, kind not known
cm   Centimeter
COC   Cleveland open cup flash point tester, ASTM **D 92**
cP   Centipoise
cSt   Centistoke
D   Debye unit
d   Density
d   Decomposes, when following a temperature
d_c Critical density
dd/dt   Coefficient of density
dn/dt   Refractive index coefficient at average D line of sodium
dp/dt   Pressure coefficient
DR   Distillation range; initial boiling point "ibp" to dry point, "dry" unless otherwise indicated
dt/dp   Temperature coefficient of pressure
eq   Equation
ER   Evaporation rate
est   Estimated value
esu   cgs unit of charge
°F   Fahrenheit scale of temperature
fl pt   Flash point
fp   Freezing point

fps	Foot-pound-second system of units
g	Gram
(g)	Gaseous or vapor state
gal	Gallon
gl	Glass or vitreous condition
hr	Hour
id	Inside diameter
in	Inch
inf	Infinite
ir	Infrared
IUPAC	International Union of Pure and Applied Chemistry
J	Joule, absolute
K	Thermodynamic scale of temperature, Kelvin units
K_a	Ionization constant of acid
K_b	Ionization constant of base
K_B	Ebullioscopic constant defined by Eq. (2.79)
K_f	Cryoscopic constant defined by Eq. (2.67)
K_f'	Cryoscopic constant defined by Eq. (2.77)
K_i	Ionization constant
K_s	Autoprotolysis constant
kcal	Kilocalorie
kg	Kilogram
kJ	Kilojoule
kPa	Kilopascal
(l)	Liquid state
L	Liter
M	Molecular weight in equations only
min	Minute
mL	Milliliter
mmHg	Millimeters of mercury
mol	Mole
ms	Mass spectra
(ms)	Metastable
mw	Molecular weight
MPa	Megapascal
N	Newton, kg m s^{-2}, SI unit of force
na	Information not available
n_D	Refractive index at average sodium D line
NBS	National Bureau of Standards
nm	Nanometer (millicron)
nmr	Nuclear magnetic resonance
OC	Open cup flash point tester, kind not known
od	Outside diameter
oz	Ounce
P	Poise
p	Vapor pressure

Pa	Pascal, N m^{-2}, SI unit of pressure
p_c	Critical pressure
pH	Negative logarithm of hydrogen ion concentration
pK_a	Negative logarithm of acid dissociation constant
pK_b	Negative logarithm of base dissociation constant
pK_{BH^+}	Negative logarithm of ionization constant of protonated base
pK_s	Negative logarithm of autoprotolysis constant
pK_w	Negative logarithm of ion product of water
ppb	Parts per billion
ppm	Parts per million
psi	Pounds per square inch
psia	Pounds per square inch absolute
psig	Pounds per square inch gauge
R	Molar gas constant
s	Second
(s)	Solid state
SI	International System of Units
sl sol	Slightly soluble
soly	Solubility
sp gr	Specific gravity
(st)	Stable
t	Temperature, Celsius unless otherwise specified
T	Thermodynamic temperature, Kelvin units
t_c	Critical temperature
Torr	Pressure unit, 760 Torr = 1 atm
TCC	Tag closed cup flash point tester, ASTM **D 56**
TLV	Threshold limit value
TOC	Tag open cup flash point tester, ASTM **D 1310**
tp	Triple point
USP	United States Pharmacopeia
uv	Ultraviolet
(u)	Unstable
V_m	Molar volume
v_c	Critical volume
W	Watt
α	Thermal expansion coefficient
$[\alpha]$	Specific rotation
γ	Surface tension
δ	Solubility parameter
ΔE_c	Heat of combustion at constant volume
ΔH_c	Heat of combustion at constant pressure
ΔH_c°	Heat of combustion at constant pressure, standard state
ΔH_f°	Standard heat of formation
ΔH_m	Heat of fusion
ΔH_p	Heat of polymerization
ΔH_s	Heat of sublimation

ΔH_t	Heat of transition
ΔH_v	Heat of vaporization
λ	Thermal conductivity coefficient
ϵ	Dielectric constant
η	Viscosity coefficient
κ	Specific conductance
κ_S	Adiabatic compressibility
κ_T	Isothermal compressibility
μ	Dipole moment
μm	Micrometer (micron)
°	Degrees Celsius (not used in Chapter III)
°	Degrees of rotation
°	Indicates the property of a substance in its standard state
%m	Per cent mole
%v	Per cent volume
%w	Per cent weight
%w/v	Per cent weight/volume
(+)	Dextrorotary
(−)	Levorotary
(±)	Racemic mixture

CONVERSION FACTORS

The conversion factors used to place values in both cgs and SI units in the tables are listed in Table 3.2. Only the conversion from cgs to SI is given. Certain fundamental constants such as the molar gas constant (R) and the elementary charge were taken from the compilation of Cohen and Taylor [1570]. Atomic weights were taken from the 1979 report by Holden [3468].

Obviously, conversion factor listing could go on indefinitely. For an extensive set of factors between the two metric systems and the foot-pound system, the *Handbook of Chemistry and Physics* [7939] is a convenient source.

The major addition-type conversions are required for temperature and the constant A in Eq. (2.4) and Eq. (2.5):

$$K = °C + 273.15 \tag{3.1}$$

$$A_{SI} = A_{cgs} - 0.875097 \tag{3.2}$$

Table 3.2 Conversion Factors

Property	To convert from	To	Multiply by
Pressure	Torr	kPa	0.133322
	Torr	Pa	133.322
	atm	kPa	101.325
Temp. coeff. of pressure	$^\circ$C Torr^{-1}	K kPa^{-1}	7.50062
Pressure coefficient	Torr $^\circ$C^{-1}	kPa K^{-1}	0.133322
Density	g cm^{-3}	kg m^{-3}	1000
Coefficient of density	g cm^{-3} $^\circ$C^{-1}	kg m^{-3} K^{-1}	1000
Compressibility	Torr^{-1}	kPa^{-1}	7.50062
	atm^{-1}	kPa^{-1}	0.0098692
Viscosity coefficient	cP	kg m^{-1} s^{-1}	10^{-3}
	cSt	m^2 s^{-1}	10^{-6}
Surface tension	dyne cm^{-1}	N m^{-1}	10^{-3}
Energy	cal	J	4.184
	kcal	kJ	4.184
Heat capacity	cal K^{-1} mol^{-1}	J K^{-1} mol^{-1}	4.184
Thermal conductivity coeff.	cal s^{-1} cm^{-1} K^{-1}	J s^{-1} m^{-1} K^{-1}	418.4
Critical pressure	atm	MPa	0.101325
Critical volume	L mol^{-1}	m^3 mol^{-1}	10^{-3}
Specific conductance	ohm^{-1} cm^{-1}	ohm^{-1} m^{-1}	100
Solubility parameter	cal$^{1/2}$ cm$^{-3/2}$	J$^{1/2}$ cm$^{-3/2}$	2.045

TABLES OF PHYSICAL PROPERTIES

0a. Water

		cgs	SI	
mw		18.0152	18.0152	3468
bp	1 atm	100.00	373.15	definition
dt/dp	1 atm	0.03687	0.2765	7230
dp/dt	1 atm	27.12	3.616	7230
p		see Table 0c	(cgs only)	
eq 2.5	Range	A	B	C
cgs only Ref. 158	0–29	8.184254	1791.3	238.1
	30–39	8.1393986	1767.262	236.29
	40–49	8.0886767	1739.351	234.10
	50–59	8.0464202	1715.429	232.14
	60–69	8.0116295	1695.167	230.41
	70–79	7.9845588	1678.948	228.97
	80–89	7.9634288	1665.924	227.77
	90–99	7.9483960	1656.390	226.86
	100–150	7.9186968	1636.909	224.92
fp	1 atm	0.00	273.15	definition
tp	4.58 Torr	0.0100	273.16	127,832
d		see Table 0b	(cgs only)	
α	20	0.00020661	0.00020661	3928
	25	0.00025705	0.00025705	
	30	0.00030314	0.00030314	
κ_T	20	6.121×10^{-8}	4.591×10^{-7}	3928;3866,4198
	25	6.032×10^{-8}	4.524×10^{-7}	
	30	5.966×10^{-8}	4.475×10^{-7}	
n_D		see Table 0b		
η		see Table 0b	(cgs only)	
γ		see Table 0b	(cgs only)	
ΔH_v		see Table 0c	(cgs only)	
ΔH_m		1.4359	6.0078	2800
ΔH_s		12.185	50.982	2800
ΔH_f°	25(1)	−68.315	−285.830	8316
	25(g)	−57.7979	−241.826	7828
C_p°	0	8.001	33.476	156
	25	8.025	33.577	
	100	8.134	34.033	
	Table 2.14			
C_p		see Table 0c	(cgs only)	
λ	30	0.00146	0.611	4434;668,1388
t_c		373.9	647.1	6375;7231,4299,1089
p_c		217.72	22.06	6375;7231,4299,1089
d_c		0.315	315	4299;1089,7231
v_c		0.0572	5.72×10^{-5}	4299;7231

0a. Water

(Continued)

		cgs	SI	
K_f	eq 2.67	1.853	1.853	2800
K_B	eq 2.79	0.515	0.515	2800
pK_w	5	14.734	14.734	3175
	10	14.535	14.535	
	20	14.169	14.169	
	25	14.000	14.000	
	30	13.837	13.837	
	40	13.542	13.542	
	50	13.272	13.272	
κ,calcd	25	5.49×10^{-8}	5.49×10^{-6}	3099
κ,exp	15	3.30×10^{-8}	3.30×10^{-6}	
	18	3.90×10^{-8}	3.90×10^{-6}	
	20	4.40×10^{-8}	4.40×10^{-6}	
	25	5.89×10^{-8}	5.89×10^{-6}	
	30	7.75×10^{-8}	7.75×10^{-6}	
μ	25(1)	1.82	1.82	4740
ε		see Table 0c		
δ	25	23.53	48.13	6240;6723

0b. Water

t	γ	η	n_D	d
Ref. No.	2800	2800,7142	7232,7468	157;3929,3928
0	75.626	1.7702	1.3339492	0.9998396
1	75.470		1.3339475	0.9998987
5	74.860	1.5108	1.3338835	0.9999641
10	74.113	1.3039	1.3336902	0.9997000
15	73.350	1.1374	1.3333873	0.9991005
15.56			1.3333426	0.9990003
20	72.583	1.0019	1.3329880	0.9982058
21	72.427	0.9764	1.3328976	0.9979944
22	72.270	0.9532	1.3328037	0.9977726
23	72.113	0.9310	1.3327067	0.9975407
24	71.957	0.9100	1.3326064	0.9972989
25	71.810	0.89025	1.3325029	0.9970474
26	71.652	0.8703	1.3323964	0.9967864
27	71.495	0.8512	1.3322869	0.9965160

0b.   Water

(Continued)

t	γ	η	n_D	d
28	71.339	0.8328	1.3321743	0.9962364
29	71.192	0.8145	1.3320589	0.9959479
30	71.035	0.79726	1.3319405	0.9956504
35	70.230	0.71903	1.3313076	0.9940313
40	69.416	0.65263	1.3306096	0.9922191
45	68.592	0.59716	1.3298513	0.9902161
50	67.799	0.54675	1.3290364	0.9880382
55	66.894	0.50415	1.3281683	0.9856959
60	66.040	0.46688	1.3272495	0.9831980
65	65.167	0.43407		0.9805524
70	64.274	0.40503		0.9777657
75	63.393	0.37918		0.9748437
80	62.500	0.35604		0.971791
85	61.587	0.33524		0.968613
90	60.684	0.31647		0.965313
95	59.763	0.29945		0.961893
100	58.802	0.28395		0.958357
105				0.954706

0c.   Water

t	p	ΔH_v	C_p	ε	
Ref. No.	158;832,127	5616	2763	3624	4858
0	4.581	10.7669	18.161	87.87	87.740
1	4.925				
5	6.542	10.7165	18.094		85.763
10	9.208	10.6660	18.051	83.91	83.832
15	12.788	10.6154	18.023		81.946
15.56	13.26				
20	17.536	10.5647	18.006	80.16	80.103
21	18.651		18.004		79.73
22	19.828		18.002		79.38
23	21.070		18.000		79.02
24	22.379		17.998		78.65

0c. Water

(Continued)

t	p	ΔH_v	C_p	ε	
25	23.758	10.5138	17.997	78.36	78.304
26	25.211		17.996		77.94
27	26.741		17.995		77.60
28	28.351		17.994		77.24
29	30.045		17.993		76.90
30	31.827	10.4629	17.992	76.57	76.546
35	42.180	10.4117	17.991		74.828
40	55.33	10.3604	17.992	73.16	73.151
45	71.89	10.3087	17.996		71.512
50	92.55	10.2569	18.001	69.90	69.910
55	118.09	10.2048	18.009		68.345
60	149.44	10.1523	18.017	66.79	66.815
65	187.62	10.0995	18.028		65.319
70	233.77	10.0463	18.040	63.82	63.857
75	289.17	9.99271	18.054		62.427
80	355.26	9.93862	18.069	61.03	61.027
85	433.58	9.88406	18.087		59.659
90	525.87	9.82889	18.106	58.32	58.319
95	634.00	9.77317	18.129		57.007
100	760.00	9.71671	18.153	55.72	55.720
105	906.07				

1. Propane

CH$_3$CH$_2$CH$_3$ 74-98-6   C$_3$H$_8$

		cgs	SI	
mw		44.096	44.096	3468
bp	1 atm	-42.070	231.080	169;3939,3377,1268,4727 2891,3196
dt/dp	1 atm	0.02985	0.2239	4382;159
dp/dt	1 atm	33.50	4.466	4382
p	-94.50	34.3	4.57	1371;7459,1927
	-52.89	459.13	61.21	3939;3196
	25	7111	948.1	169;2088,2891,2511
eq 2.5	A	6.80338	5.92828	176
	B	803.997	803.997	
	C	247.04	247.04	
fp		-187.70	85.45	3939;3377,4227,3196
tp		-187.68	85.47	169
d	-160	0.70498	704.98	4742;3247,5263,164
	-80	0.5915	591.5	2989;164,4857,4727
	20	0.50039	500.39	164;6549,5426,3830,7204
	25	0.4927	492.7	164;7204
dd/dt	-80 to -40	0.00115	1.15	2989
κ_T	-40 to 75	no values in abstract		4804,3831
n_D	-42.2	1.3392	1.3392	2989;2988
	20	1.28624	1.28624	161;2510,2088
	25	1.28143	1.28143	
dn/dt	-80 to -40	0.000698	0.000698	2989
η	-40	0.204	0.000204	163;4618
	20	0.105	0.000105	7141;4618
	40	0.0819	8.19 x 10^{-5}	7141;6962
γ	-40	15.15	0.01515	165;4727
	1.0	9.31	0.00931	3899
	39.5	5.84	0.00584	
ΔH_v	25	3.542	14.820	169
	bp	4.487	18.774	3939,169;3196
ΔH_m		0.8422	3.524	3939,169
ΔH_f°	25(1)	-28.789	-120.453	170;5969
	25(g)	-25.02	-104.68	7380;5969,2036,1419,5873
ΔH_c°	25(1)	-526.63	-2203.42	179;5969
	25(g)	-530.605	-2220.051	5969;2036,5868
C_p°	25	17.59	73.60	1419,7375;2286
C_p	-172.82	20.57	86.05	2866;3939
	23.59	28.33	118.54	2866
t_c		96.70	369.85	7420;5298,5300,166,1964,707 4727,2891
p_c		41.91	4.247	7420;5298,5300,166,1964 707,2891

1. Propane

(Continued)

		cgs	SI	
d_c		0.2185	218.5	7420;5298,5300,166,1964,707
v_c		0.2018	0.0002018	7420;166,1964,707
A	eq 2.72	0.05802	0.05802	169
ε	−158(1)	2.0427	2.0427	5661
	25(1)	1.2898	1.2898	2088;2795
μ	(g)	0.083	0.083	4588,7803
δ	25	6.4	13.1	948;3395,4593,6610,6723
soly	in aq,25	0.00624%w	0.00624%w	4734;524
	aq in,25(g)	3.94%v	3.94%v	1514
fl pt		−104	169	5825
Beil	10,103			
uv	6083			
ir	4796,5849,6068,2692,153			
Raman	836,4105,845			
ms	6199,4097,5173,7914,814,3477			
nmr	5399,5400,6930,2902			

2. Butane

$CH_3CH_2CH_2CH_3$ 106-97-8 C_4H_{10}

		cgs	SI	
mw		58.123	58.123	3468
bp	1 atm	−0.50	272.65	169,480;1561,7814,2891,4382
dt/dp	1 atm	0.03465	0.2599	2088;4382,7814
dp/dt	1 atm	28.86	3.848	2088
p	−78.50	9.31	1.24	1371;480,7459,1927
	−26.64	249.53	33.27	480;7814
	25	1823	243.0	1319;169,2891

2. Butane

(Continued)

		cgs	SI	
eq 2.5	A	6.80776	5.93266	176;480
	B	935.773	935.773	
	C	238.789	238.789	
fp		−138.350	134.800	2778;480,161,1371,1561,3560
d	−130	0.72736	727.36	4742;3247,164,3246
	−50	0.6519	651.9	164;3247,3246
	20	0.57861	578.61	164;3247,1561,5426,3830
	25	0.57287	572.87	3246,7204
n_D	−24.3	1.3613	1.3613	7814;2988
	20	1.32943	1.32943	161;2510
	25	1.32594	1.32594	
dn/dt	−10 to −50	0.00061	0.00061	7814
η	−40	0.314	0.000314	163;4618
	20	0.164	0.000164	7141;4618
	40	0.136	0.000136	
γ	−50	20.88	0.02088	165
	0	14.84	0.01484	165;3899,1561
	45.4	10.6	0.0106	3899
ΔH_v	25	5.035	21.066	169
	bp	5.352	22.393	169,480;7814
ΔH_m		1.114	4.661	169,480;3560
ΔH_t	−165.59	0.4940	2.067	169,480;6335,3560
ΔH_f°	25(l)	−35.34	−147.86	170;5969
	25(g)	−30.03	−125.65	7380;2036,5969,5966,5873
ΔH_c°	25(l)	−682.45	−2855.37	179;5969
	25(g)	−687.68	−2877.25	2036;5868,5966,5969,3560
C_p°	25	23.54	98.49	7375;1443,7211
C_p	−3.16	31.65	132.42	480;3560
t_c		151.99	425.14	4244;166,708,2891,1561,6555
P_c		37.343	3.7838	4244;166,708,2891,6555
d_c		0.228	228	166;2891,708
v_c		0.255	0.000255	166;708
A	eq 2.72	0.03084	0.03084	2778;169,2088
ε	25(l)	1.776	1.776	2088;3246
	25(g)	1.002536	1.002536	7930
μ	(g)	0	0	7929
δ	25	6.8	13.9	948;3395,6747,4593,6723
soly	in aq,25	0.00614%w	0.00614%w	4734
	in aq,20(g)	3.27%v	3.27%v	1514;6149,4455
	aq in,21	0.00753%w	0.00753%w	906
fl pt	COC	−60	213	2088

2. Butane

(Continued)

Beil	10,118
uv	3943,6083
ir	633,5849,6068,520,2158,5076,632
Raman	336,6018,5386,845,4105,5347,3149,6727,6646,5193,2158
ms	814,7914,5173,4097,7913,3477,6199,7008
nmr	2902,6930,63

3. 2-Methylpropane

$(CH_3)_2CHCH_3$ 75-28-5 C_4H_{10}

		cgs	SI	
mw		58.123	58.123	3468
bp	1 atm	−11.730	261.420	171;475,7814,1561,1269,4441
dt/dp	1 atm	0.03368	0.2526	4382;171,7814
dp/dt	1 atm	29.69	3.958	4382
p	−85.08	11.37	1.516	475;1927,7459
	−27.57	391.02	52.132	475;7814,176,1269
	25	2611	348.1	2088,177
eq 2.5	A	6.87782	6.00272	176;475
	B	947.54	947.54	
	C	248.87	248.87	
fp		−159.600	113.550	2779,171;475,1561
d	−130	0.71259	712.59	4742;3247
	−29.5	0.61415	614.15	7814;3247,1561
	20	0.55711	557.11	173;3247,1561,5426,3830,7204
	25	0.55092	550.92	
dd/dt	0	0.001057	1.057	7814
n_D	−24.3	1.3524	1.3524	7814
	20	1.3209	1.3209	2510
	25	1.3175	1.3175	

3. 2-Methylpropane

(Continued)

		cgs	SI	
η	-70 -10	0.528 0.238	0.000528 0.000238	163;4618
γ	-50 -10	18.90 14.1	0.01890 0.0141	174;1561
ΔH_v	25 bp	4.570 5.090	19.121 21.297	177 177;7814,475
ΔH_m		1.085	4.540	475,177
ΔH_f°	25(l) 25(g)	-36.95 -32.07	-154.60 -134.18	181;5969 7380;5966,5969,5873
ΔH_c°	25(l) 25(g)	-680.84 -685.71	-2846.63 -2869.01	179;5969 5868;5966,5969,6280
C_p°	25	23.10	96.65	7377;2286,1443,7813,6337
C_p	-13.15	31.00	129.70	475
t_c		134.69	407.84	6576;704,175,1561,6555,4441
P_c		35.82	3.629	6576;704,175,6555
d_c		0.2255	225.5	6576;704,175
v_c		0.2578	0.0002578	6576;704,175
A	eq 2.72	0.04234	0.04234	2778,177
ϵ	-153(l) 25(l) 25(g)	2.0950 1.734 1.002564	2.0950 1.734 1.002564	5661 2088 7930
μ	(g)	0.132	0.132	4589;7803,7929
δ	25	6.25	12.78	3395;906,4593,6723
soly	in aq,25 aq in,21.8	0.00489%w 0.00835%w	0.00489%w 0.00835%w	4734;6099 906
fl pt	COC	-87	186	2088
Beil		10,124		
uv		6083		
ir		2324,6068,5849,7957,633,632,153		
Raman		335,2324,4105,845,6018		
ms		814,7914,5173,4097,6199,7913,7008		
nmr		2902,6930		

4. Cyclopentane

$\underline{CH_2CH_2CH_2CH_2CH_2}$ 287-92-37 C_5H_{10}

		cgs	SI	
mw		70.134	70.134	3468
bp	1 atm	49.262	322.412	221;2494,3661,8039,8091,7491
dt/dp	1 atm	0.04003	0.3002	221;8091,2494
dp/dt	1 atm	24.98	3.331	221
p	25	318	42.4	231;7058,8091
eq 2.5	A	6.92094	6.04584	224;8091,3913
	B	1142.30	1142.30	
	C	233.463	233.463	
fp		-93.879	179.271	221;2494,3661,2780,8039,474
d	20	0.74538	745.38	221;4381,697,3913,7758,2700
	25	0.74045	740.45	2494,7491,8039
dd/dt	25	0.000986	0.986	$\underline{221}$
α	25	0.001347	0.001347	2340
κ_T	25	1.811×10^{-7}	1.358×10^{-6}	6151;2340
n_D	20	1.40645	1.40645	221;4381,697,7491,8039,2494
	25	1.40363	1.40363	7758,8040
dn/dt	25	0.000564	0.000564	$\underline{221}$
η	10	0.492	0.000492	229;2700,7491
	20	0.439	0.000439	
	25	0.416	0.000416	
	Table 2.1			
γ	10	24.07	0.02407	222;7491,7758,3547,8039
	20	22.61	0.02261	
	25	21.88	0.02188	222;4381
ΔH_v	25	6.818	28.527	225,4752,4670;474,6934,7754
	bp	6.524	27.296	
	Table 2.9			
ΔH_m		0.1455	0.6088	225;3661,474
ΔH_t	-150.76	1.1674	4.8844	225;474
	-135.08	0.08233	0.3445	
ΔH_f°	25(1)	-25.28	-105.77	2036,227;6933,3738,8316
ΔH_c°	25(1)	-786.55	-3290.93	2036,226;6933,3975,3738,8316
C_p°	25	19.82	82.93	227;4752,6934
C_p	25	31.22	130.62	227;7145,3661,6934
t_c		238.46	511.61	223;1451,2770,3913,119
p_c		44.43	4.502	223;3913,2770,758
d_c		0.270	270	223;4299,2770
V_c		0.260	0.000260	223;4299
A	eq 2.72	0.00228	0.00228	225

4.   Cyclopentane

(Continued)

		cgs	SI	
ε	20	1.96875	1.96875	7019;7497,4937
μ	25	0.00	0.00	7660
δ	25	8.10	16.57	5334
soly	in aq, 25 aq in, 20	0.0159%w 0.0142%w	0.0159%w 0.0142%w	4733 2268
fl pt		−37	236	5822

Beil	452,19
uv	5842,6083
ir	153,4384,5892,7901,602,5898,4660,5139,5932
Raman	678,2400,6683,2567,697,6049,80,6546,5139
ms	154,3430,6199,5133,2113
nmr	1263,4806,2420,5933,6748

5.   Pentane

$CH_3CH_2CH_2CH_2CH_3$ 109-66-0 C_5H_{12}

		cgs	SI	
mw		72.150	72.150	3468
bp	1 atm	36.065	309.215	161;5609,1823,2494,6639,8128 3520,7472
dt/dp	1 atm	0.03871	0.2904	1823;159,8091,8128,2494,3520
dp/dt	1 atm	25.83	3.444	1823
p	25	512.5	68.33	169;5070
eq 2.5	A	6.85296	5.97786	176;706,8091,5070,5609,1371
	B	1064.84	1064.84	
	C	232.012	232.012	
fp		−129.730	143.420	161;1371,3048,5679,7479,5068 5070,7472

5. Pentane

(Continued)

		cgs	SI	
tp		-129.721	143.429	1823;1371
d	20	0.62624	626.24	159;2051,3520,8039,2700
	25	0.62139	621.39	2494,7490
dd/dt	0 - 40	0.000975	0.975	1882;3520
α	25	0.001610	0.001610	3131
κ_T	25	2.906×10^{-7}	2.180×10^{-6}	6152
κ_s	25	2.1272×10^{-7}	1.5955×10^{-6}	3131;1437
n_D	20	1.35748	1.35748	161;2051,3520,8039,2494,4834
	25	1.35472	1.35472	
dn/dt	20 - 25	0.00053	0.00053	3520
η	15	0.245	0.000245	172;6639,2751,2700,7490
	20	0.235	0.000235	
	25	0.225	0.000225	
	Table 2.1			
	Table 2.2			
γ	25	15.48	0.01548	174;4791,3686,6008,8039
	Table 2.8			7490,7760
ΔH_v	25	6.316	26.427	169;3505;5070,5613,5872,5153
				6669,1046
	bp	6.163	25.786	3505;169,1823
ΔH_m		2.006	8.393	169;5070,5068,5679
ΔH_f°	25(1)	-41.46	-173.47	2848;180,5854,2908,8316
	25(g)	-35.12	-146.94	7380;2036,2848
ΔH_c°	25(1)	-838.69	-3509.08	2848;5967,5854,178,6276,3975
				2908,8316
	25(g)	-845.16	-3536.15	2036;2848
C_p°	25	28.69	120.04	7375;3505,5780
	Table 2.14			
C_p	25	39.96	167.19	5068;3692,5872,5070,5679
t_c		223.50	496.65	1823;166,105,5153,7400,1046
				706,1671
p_c		33.25	3.369	166,1823;8227,5153,1046,706
d_c		0.237	237	166,1823;8227,1046,706
v_c		0.304	0.000304	166,1823;706,7400,8227
				706,8233
A	eq 2.72	0.04910	0.04910	169;5068
κ	19.5	2×10^{-10}	2×10^{-8}	2685
ε	20	1.841	1.841	6405;4937,2051
μ		0.00	0.00	7660;3463

5. Pentane

(Continued)

		cgs	SI	
δ	25	7.02	14.36	5334;4593,3394,1271,6723
soly	in aq,25	0.0038%w	0.0038%w	4733;6556,1630,7219
	aq in,24.8	0.0120%w	0.0120%w	906
aq az	34.6	98.6%w	98.6%w	3498
fl pt	TOC	−40	233	5712;5713
Beil	10,130			
uv	6083			
ir	153,4384,7077,606,6068,633,5932			
Raman	2400,3435,6642,3337,6049,1790,3149			
ms	154,5133,5523,6887			
nmr	5933,6748			

6. 2-Methylbutane

$CH_3CH(CH_3)CH_2CH_3$ 78-78-4 C_5H_{12}

		cgs	SI	
mw		72.150	72.150	3468
bp	1 atm	27.875	301.025	1824;177,8091,2494,7490 7499,8039
dt/dp	1 atm	0.03778	0.2834	1824;171,8091,2494,7490
dp/dt	1 atm	26.47	3.529	1824
p	25	688	91.7	177;6241,6504,6503
eq 2.5	A	6.79533	5.92023	176;6504,8091
	B	1022.88	1022.88	
	C	233.460	233.460	
fp		−159.900	113.250	177;2780,4079,5680,7480,7581
tp		−159.90	113.25	1824
d	20	0.6193	619.3	173;2494,7470,8039,2700
	25	0.6142	614.2	3520,7499
dd/dt	25	0.00102	1.02	_173_

6. 2-Methylbutane

(Continued)

		cgs	SI	
κ_T	25	3.266×10^{-7}	2.450×10^{-6}	6151
n_D	20	1.35373	1.35373	171;1258,2494,8040,2108,8039
	25	1.35088	1.35088	
dn/dt	20-25	0.00058	0.00058	3520
η	15	0.237	0.000237	172;7494,2700
	20	0.225	0.000225	
	25	0.215	0.000215	
	Table 2.1 Table 2.2			
γ	20	15.00	0.01500	174,3283;8039
	25	14.46	0.01446	
ΔH_v	25	5.937	24.84	177;5153,6504,6503,6526
	bp	5.968	24.97	6151;177
ΔH_m		1.231	5.151	177;3036,6503,5680,6505
ΔH_f°	25(1)	-42.58	-178.15	2848;180,5854,2908,8316
	25(g)	-36.70	-153.55	7380;2848
ΔH_c°	25(1)	-837.57	-3504.39	2848;178,5854,4079,3975 2908,8316
	25(g)	-843.60	-3529.62	2848
C_p°	25	28.41	118.87	7377;6526
	Table 2.14			
C_p	26.84	39.55	165.48	3036;5680,6504,6526,6503
t_c		187.24	460.39	175,1824;8232,5153,8233,7088 119,7776
P_c		33.37	3.381	175,1824;8232,7088,8231 5153,7776
d_c		0.236	236	175,1824;8232,8231,7088,7776
v_c		0.306	0.000306	175,1824;8232,8231,8233
A	eq 2.72	0.04830	0.04830	177
ε	25	1.8275	1.8275	5240;4937
δ	25	6.75	13.81	5334,3395;4593
soly	in aq,25	0.0048%w	0.0048%w	4733;5724
	aq in,21.8	0.0097%w	0.0097%w	906;2268,5724
fl pt		-57	216	5822

Beil	<u>10</u>,134
uv	5929,6083
ir	153,3761,6068,606,4384,5932
Raman	3827,6049,8302,8221,5021,4183
ms	154,2194,6887
nmr	5933,6748

7.   2,2-Dimethylpropane

$C(CH_3)_4$ 463-82-1 C_5H_{12}

		cgs	SI	
mw		72.150	72.150	3468
bp	1 atm	9.500	282.650	5609,176;1825,479,8023,7581
dt/dp	1 atm	0.03642	0.2732	1825;171
dp/dt	1 atm	27.46	3.661	1825
p	25	1285	171.3	177;179,8023
eq 2.5	A	7.71426	6.83916	176;702,5609,1875
	B	938.234	938.234	
	C	235.249	235.249	
fp		-16.58	256.57	177;479,8023,7581
tp		-16.6	256.6	1825;2274
d	0	0.6136	613.6	173;8023
	20	0.5910	591.0	173
	25	0.5852	585.2	
dd/dt	0-25	0.00114	1.14	173
n_D	0	1.3513	1.3513	8023
	6	1.3476	1.3476	
	20	1.342	1.342	171
η	-10	0.391	0.000391	172
	0	0.328	0.000328	
	5	0.303	0.000303	
γ	20	12.05	0.01205	2088
	30	10.98	0.01098	
ΔH_v^*	25	5.205	21.778	177;479
	bp	5.438	22.753	177,3505;1825
ΔH_m		0.752	3.146	177;479,2274
ΔH_t	-132.65	0.6287	2.630	2274
ΔH_f°	25(1)	-45.44	-190.12	2848;180,5854,2908,8316
	25(g)	-40.18	-168.11	7380;2848
ΔH_c°	25(1)	-834.71	-3492.43	2848;178,5854,4079,3975
				2908,8316
	25(g)	-840.06	-3514.81	2848
C_p°	25	28.88	120.83	7379,3505
	Table 2.14			
t_c		160.60	433.75	175,1875;1825,702,479,7400
P_c		31.545	3.1963	1875;1825,175,702,479,7400
d_c		0.2319	231.9	1875;1825,175,702
v_c		0.3111	0.0003111	1875;1825,175,702,7400
A	eq 2.72	0.00575	0.00575	177
ε	20	1.801	1.801	2088;2438

7. 2,2-Dimethylpropane

(Continued)

		cgs	SI	
δ	25	6.2	12.7	3394,4593
soly	in aq,25	0.0033%w	0.0033%w	4733
fl pt		-65	208	5822
* at saturation pressure				

Beil	10,141
uv	6083
ir	153,6068,3761,7957
Raman	3827,6047,4105,5021,6642,4996,6941
ms	1993,5223,6887,2194,154,3279
nmr	4852

8. Methylcyclopentane

$CH_2CH_2CH_2CH_2CHCH_3$ 96-37-7 C_6H_{12}

		cgs	SI	
mw		84.161	84.161	3468
bp	1 atm	71.812	344.962	231,8091,2494;7491,3378,8039
dt/dp	1 atm	0.04274	0.3206	228;2494,8091
dp/dt	1 atm	23.40	3.119	228
p	25	138	18.4	231;8091
eq 2.5	A	6.86283	5.98773	230;4385,8091
	B	1186.059	1186.059	
	C	226.042	226.042	
fp		-142.434	130.716	231;2494,3378,8039,2780,7491
d	20	0.74864	748.64	228;2494,5238,8039,2700
	25	0.74394	743.94	3913,7491
dd/dt	25	0.000940	0.940	228
n_D	20	1.40970	1.40970	228;2494,5238,8039,3378
	25	1.40700	1.40700	7491,8040

8. Methylcyclopentane

(Continued)

		cgs	SI	
dn/dt	25	0.000540	0.000540	<u>228</u>
η	20	0.507	0.000507	229;7491,2700
	25	0.478	0.000478	
	Table 2.1			
γ	20	22.30	0.02230	233;8039
	25	21.72	0.02172	
ΔH_v	25	7.550	31.59	231;4752,5613,7754
	bp	6.950	29.08	
	Table 2.9			
ΔH_m		1.656	6.929	231
ΔH_f°	25(1)	-32.92	-137.74	8316,2857;3738
ΔH_c°	25(1)	-941.28	-3938.32	8316,3738,2857;5237,3975,232
C_p°	25	26.24	109.79	235,6518
	Table 2.14			
t_c		259.61	532.76	234,2770;119,4299
p_c		37.36	3.786	234,2770;4299
d_c		0.264	264.	234,2770,3913,4299
v_c		0.319	0.000319	234,4299
A	eq 2.72	0.04877	0.04877	231
soly	in aq,25	0.00426%w	0.00426%w	4734
	aq in,20	0.0131%w	0.0131%w	2268
aq az	63.3	82%w	82%w	6165
fl pt	TCC	-27	246	5822

Beil	<u>452</u>,27
ir	153,4384,5898,602,5891,633,2161,5932
Raman	697,6049,2400,6683,2161
ms	154,6199,2420,7165
nmr	4619,5933,6748

9. Cyclohexane

$$\underline{CH_2CH_2CH_2CH_2CH_2CH_2}$$ 110–82–7 C_6H_{12}

		cgs	SI	
m̄w		84.161	84.161	3468
bp	1 atm	80.730	353.880	224;2494,6286,2776,8091,7494
dt/dp	1 atm	0.04376	0.3282	221;2494,8091,2970
dp/dt	1 atm	22.85	3.047	221
p	25	97.81	13.04	1740;238
eq 2.5	A	6.83917	5.96407	224;1740,8091,1220
	B	1200.31	1200.31	
	C	222.504	222.504	
fp		6.72	279.87	7020;1740,221,5084,6306,4676
				6454,2780
tp		6.68	279.83	7020
d	20	0.77855	778.55	221;7169,52,3702,4381,4650
	25	0.77389	773.89	2494,5084,6409,2700,4951
				6286,6454,8039,7494,8156
	30	0.76845	768.45	5370
α	25	0.001220	0.001220	2429;8156,8064
κ_T	25	1.520×10^{-7}	1.140×10^{-6}	6151;1262,6152,7169,8156
				54,52
κ_S	25	1.096×10^{-7}	8.220×10^{-7}	7169;5411,5945
n_D	20	1.42623	1.42623	221;4381,4650,6716,1729,6588
	25	1.42354	1.42354	8039,6374,7494,8040
	30	1.4214	1.4214	1982
dn/dt		0.00054	0.00054	2970
η	20	0.9751	0.0009751	326;237,7494,7404,2700
	25	0.898	0.000898	237;7494
	30	0.820	0.000820	5412;6029
	Table 2.1			
γ	20	25.24	0.02524	222;4195,7087,3547,8039,7758
	25	24.65	0.02465	222;4381,3668
	30	23.74	0.02374	5370
ΔH_v	25	7.861	32.89	6331;1740,225,4969,5613,4340
				1046,5153,6934,4670,7754
	bp	7.183	30.05	7110;225
ΔH_m		0.6399	2.6773	225;5680,6306
ΔH_t	−87.06	1.6109	6.7400	225;444
ΔH_f°	25(1)	−37.33	−156.19	227,2036;2857,3738,6933
				6306,8316
ΔH_c°	25(1)	−936.87	−3919.86	2036,2857;226,6933,3975,6156
				5701,5237,3738
C_p°	25	25.40	106.27	227;4624,1088,6934
C_p	25	37.29	156.01	7178;2490,6377,2977,7738,6306
				5817,6934,5680

9. Cyclohexane

(Continued)

		cgs	SI	
t_c		280.49	553.64	3568;223,8233,1671,1046,923 5153,1451,4088
p_c		40.22	4.075	3568;223,125,8231,1046,8234 2776,5156,4088
d_c		0.2718	271.8	223;4088,8231,1046,8234
v_c		0.3097	0.0003097	223;1671,8234,1449
A	eq 2.72	0.004196	0.004196	7020;238
K_f	eq 2.67	20.0	20.0	4940;4941
K_B	eq 2.79	2.75	2.75	4938;4940
κ	25	$\sim 7 \times 10^{-18}$	$\sim 7 \times 10^{-16}$	2486
ε	20	2.02431	2.02431	7019;2438,1983,4650,4937,5023 8075,1729,6588,7497,1584
δ	25	8.203	16.78	8112;4024,4593,1271,652,6240
soly	aq in,20 in aq,25	0.010%w 0.0055%w	0.010%w 0.0055%w	906;794,7186,2268,3734,6219 4733;4302,2823
aq az	68.95	91%w	91%w	3497;3498
fl pt	TCC	−17	256	5712;152,5822
Beil	452,20			
uv	3198,1883,7016,6428,1884,5842,6983			
ir	153,5892,7077,3794,1521,5898,633,4660,4316,4932			
Raman	678,2886,80,2400,6049,3711,1092,6546			
ms	154,6199,2223,3728,3430,2420			
nmr	3435,3794,4806,1263,2464,5933,6748			

10. Hexane

$CH_3(CH_2)_4CH_3$ 110–54–3 C_6H_{14}

		cgs	SI	
mw		86.177	86.177	3468
bp	1 atm	68.736	341.886	161;2494,6462,7495,7521 8128,2776
dt/dp	1 atm	0.04191	0.3144	159;6462,6639,7495,8128
dp/dt	1 atm	23.86	3.181	159
p	25	151.3	20.17	169;6462
eq 2.5	A B C	6.87601 1171.17 224.408	6.00091 1171.17 224.408	186;4536,3912,8091,1371
fp		−95.322	177.828	161;1882,2494,2780,3560,7495
tp		−95.32	177.83	1371
d	20 25	0.65933 0.65484	659.33 654.84	161;2429,6271,2176,1147,1072 1711,2051,2494,3912,4521 2700,6008,6639,7495,7521 7848,5239,3183,3317
	30	0.65018	650.18	6945;52,1980
dd/dt	0–40	0.000891	0.891	1882
α	25	0.001391	0.001391	2429;3131
κ_T	25	2.274×10^{-7}	1.706×10^{-6}	6152;1262,52,1990,1988,3865
κ_S	25	1.7571×10^{-7}	1.3180×10^{-6}	7171;3131,6945,1980
n_D	20 25	1.37486 1.37226	1.37486 1.37226	161;1539,4650,1147,6584,2051 2494,4834,7521,8039,8040 1711,3317
dn/dt	0–40	0.000542	0.000542	1882
η	15 20 25	0.3275 0.3126 0.2942	0.0003275 0.0003126 0.0002942	162;2129,6639,7495,7705,2751 2225,2700 464;162,1999,3317,3724
	Table 2.1 Table 2.2 Table 2.4			
γ	20 25 Table 2.8	18.42 17.94	0.01842 0.01794	512;6314,6271 6465;184,7760,7848,7850,8039 3686,3668
ΔH_v	25	7.541	31.552	169;4521,5153,5613,7592,7816 4340,1046,3572
	bp	6.896	28.853	
ΔH_m		3.126	13.079	169
ΔH_f°	25(1) 25(g)	−47.45 −39.90	−198.53 −166.94	189;2036,2857,2908,8316 7376
ΔH_c°	25(1)	−995.06	−4163.33	188;8316,2857,2036,3975,2908 3717,5967
c_p°	25 Table 2.14	34.08	142.59	7375;5780,7816

10. Hexane

(Continued)

		cgs	SI	
c_p	25	46.72	195.48	191,5068;3560,5680,5817 7059,2061
λ	37.8	0.000245	0.123	6352
t_c		234.53	507.68	5298,5300;166,119,923,3912 8233,1671,2776
p_c		29.951	3.0348	5298,5300;166,105,3912,5153 4086,7400,8235
d_c		0.233	233	166;8231,7410,4086,1046
v_c		0.370	0.000370	166;7410,1671,4086,7400,8233
A	eq 2.72	0.04974	0.04974	169;7466
κ	25	$<10^{-16}$	$<10^{-14}$	2486
ε	20 25	1.8863 1.8799	1.8863 1.8799	7019;4650,6405 5239;4937,2051,8084,3199,6760
μ	25(1)	0.085	0.085	5239
δ	25	7.27	14.87	5334,6240;652,4593,6723
ER	BuOAc 90%	8.9 53	8.9 53	6631 6631
soly	in aq,25 aq in,20	0.00123%w 0.0111%w	0.00123%w 0.0111%w	7219,400;4733,6556,2601 906;2268,6219
aq az	61.6	94.4%w	94.4%w	3498;3497
fl pt	TCC	-26	247	5712;5822,4902,2088

Beil	10,142
uv	1367,6083
ir	153,602,1660,7077,6104,3769,8188,5932
Raman	1533,2400,2886,2889,5494,6261,7162,3149
ms	154,6199,6887,2194,714,4582
nmr	5933,6748

11.   2-Methylpentane

$(CH_3)_2CHCH_2CH_2CH_3$ 107–83–5 C_6H_{14}

		cgs	SI	
mw		86.177	86.177	3468
bp	1 atm	60.271	333.421	187,8091;1210,2494,7059,8039
dt/dp	1 atm	0.04141	0.3106	183;2494,3520
dp/dt	1 atm	24.15	3.220	183
p	25	212	28.3	187;2612
eq 2.5	A	6.86360	5.98850	186;8091
	B	1148.74	1148.74	
	C	228.166	228.166	
fp		−153.670	119.480	187;1210,2494,7059
d	20	0.65315	653.15	183;1210,2494,3520,4521,6188
	25	0.64852	648.52	8039,2700
dd/dt	25	0.000926	0.926	<u>183</u>
α	25	0.001426	0.001426	4317
κ_S	25	1.8948×10^{-7}	1.4212×10^{-6}	7171;4317
n_D	20	1.37145	1.37145	183;1210,2494,6188,7520
	25	1.36873	1.36873	8039,8040
dn/dt	20–25	0.00046	0.00046	7482;3520
η	−195.7(gl)	2.4×10^{15}	2.4×10^{12}	4614,4613
	20.176	0.2868	0.0002868	2225;7482,2129,2700
	25.083	0.2736	0.0002736	
	Table 2.1			
	Table 2.2			
γ	20	17.38	0.01738	184;8039
	25	16.87	0.01687	
	30	16.37	0.01637	
ΔH_v	25	7.138	29.87	187;4521,5613
	bp	6.643	27.79	
ΔH_m		1.498	6.268	187;2061,7059
ΔH_f°	25(1)	−48.91	−204.64	189;8316
	25(g)	−41.75	−174.68	7378
ΔH_c°	25(1)	−993.60	−4157.22	188;8316
C_p°	25	33.99	142.21	190
C_p	25	46.35	193.91	191;2061,5613,7059,4317
λ	37.8	0.000255	0.107	6352
t_c		224.30	497.45	185,119;3912,6352
p_c		29.71	3.010	185;3912
d_c		0.235	235	185
v_c		0.367	0.000367	185
A	eq 2.72	0.05287	0.05287	187
ε		1.881	1.881	2088

11.   2-Methylpentane

(Continued)

		cgs	SI	
δ	25	7.02	14.36	5334
soly	in aq,25	0.0014%w	0.0014%w	4733;1630
aq az	55.1	97 %w	97 %w	6165
fl pt	TCC	<-29	<244	5712;5822
Beil	10,148			
uv	6083			
ir	153,606,4384,4587,5932			
Raman	696,2400,5021,6261,6049,6683			
ms	154,6887,2194			
nmr	5933,6748			

12.   3-Methylpentane

$CH_3CH_2CH(CH_3)CH_2CH_3$ 96–14–0 C_6H_{14}

		cgs	SI	
mw		86.177	86.177	3468
bp	1 atm	63.282	336.432	187,2494,8091;1210,3912 8039,8132
dt/dp	1 atm	0.04182	0.3137	183,2494,8091;3520,8132
dp/dt	1 atm	23.91	3.188	183,2494,8091
p	25	190	25.3	187;2612
eq 2.5	A	6.81366	5.93856	186;8091,1876
	B	1133.52	1133.52	
	C	224.944	224.944	
fp		(gl)	(gl)	2061;2494,2780
tp		−162.898	110.252	2436
d	20	0.66431	664.31	183;1210,1711,2494,3520,3912
	25	0.65976	659.76	6008,6188,7059,2700,8039
dd/dt	25	0.000910	0.910	$\underline{183}$
α	25	0.001396	0.001396	4317
κ_S	25	1.7609×10^{-7}	1.3208×10^{-6}	7171;4317
n_D	20	1.37652	1.37652	183,2494;1711,6188,8039,8040
	25	1.37386	1.37386	
dn/dt	20 – 25	0.00054	0.00054	3520
η	−195.7(gl)	2.2×10^{14}	2.2×10^{11}	4614,4613
	19.986	0.3031	0.0003031	2225
	25	0.307	0.000307	7482;2700
	30	0.292	0.000292	
	Table 2.1			
γ	20	18.12	0.01812	184;8039,6008
	25	17.60	0.01760	
	30	17.08	0.01708	
ΔH_v	25	7.236	30.28	187;5613,7758
	bp	6.711	28.08	
ΔH_m		1.2675	5.3032	2436
ΔH_f°	25(1)	−48.37	−202.38	189;8316
ΔH_c°	25(1)	−994.14	−4159.48	188;8316
C_p°	25	33.49	140.12	217;7816
C_p	25	45.572	190.67	2436,191;4317,2061,5613
	Table 2.12			7059,7816
λ	37.8	0.000260	0.109	6352
t_c		231.2	504.4	185;1876,3912
p_c		30.83	3.124	185;1876,3912

12.   3-Methylpentane

(Continued)

		cgs	SI	
d_c		0.235	235	185
v_c		0.367	0.000367	185
A	eq 2.72	0.05247	0.05247	2436
ε		1.895	1.895	2088
δ	25	7.13	14.58	5334
soly	in aq, 25	0.0013%w	0.0013%w	4733
fl pt	TCC	-31	242	5822

Beil	10,149
uv	6083
ir	153,606,4384,4587,5932
Raman	696,2400,5021,6049,6261,6642,6683
ms	154,6887,2194
nmr	5933,6748

13.   2,2-Dimethylbutane

$CH_3C(CH_3)_2CH_2CH_3$ 75-83-2 C_6H_{14}

		cgs	SI	
mw		86.177	86.177	3468
bp	1 atm	49.741	322.891	187,2494,8091;1125,3912 8039,8132
dt/dp	1 atm	0.04117	0.3088	183,2494,8091;1125,3520,8132
dp/dt	1 atm	24.29	3.238	183,2494,8091
p	25	320	42.7	187;1125,2612
eq 2.5	A	6.75473	5.87963	186;8091
	B	1081.14	1081.14	
	C	229.349	229.349	
fp		-99.865	173.285	187;1125,2494,2780,7059,8039

13.   2,2-Dimethylbutane

(Continued)

d	20	0.64916	649.16	183;1125,1711,2494,3520,3912
	25	0.64446	644.46	6188,7059,8039
dd/dt	25	0.000940	0.940	$\underline{183}$
α	25	0.001468	0.001468	4317
κ_S	25	2.0744×10^{-7}	1.5559×10^{-6}	7171;4317
n_D	20	1.36876	1.36876	183;1282,1125,1711,6188
	25	1.36595	1.36595	8039,8040
dn/dt	20 - 25	0.00058	0.00058	3520
η	25	0.351	0.000351	7482;2225
	30	0.330	0.000330	
	Table 2.1			
γ	20	16.30	0.01630	184;8039
	25	15.81	0.01581	
	30	15.31	0.01531	
ΔH_v	25	6.618	27.69	187;5613,7816,5872
	bp	6.287	26.30	
ΔH_m		0.1384	0.5791	187;2061,7059
ΔH_f°	25(1)	−50.62	−211.79	189;8316
ΔH_c°	25(1)	−991.89	−4150.07	188;8316
C_p°	25	33.81	141.46	218;7816
C_p	25	45.10	188.7	191;4317,2061,7059,6352,5872
λ	37.8	0.000231	0.0967	6352
t_c		215.68	488.83	185,119;7400,6352,3912
P_c		30.40	3.080	185;7400,3912
d_c		0.240	240	185
v_c		0.359	0.000359	185;7397
A	eq 2.72	0.002319	0.002319	187
ϵ		1.873	1.873	2088
δ	25	6.71	13.73	5334
soly	in aq,25	0.0018%w	0.0018%w	4733
fl pt	TCC	−31	242	5822
Beil	$\underline{10}$,150			
uv	4166,6083			
ir	153,4384,7425,5932			
Raman	2400,5021,6049,6261,7162,6683,8302			
ms	154,6887,2194			
nmr	5933			

14. 2,3-Dimethylbutane

$(CH_3)_2CHCH(CH_3)_2$ 79-29-8 C_6H_{14}

		cgs	SI	
mw		86.177	86.177	3468
bp	1 atm	57.988	331.138	187,2494,8091;1125,1210 3912,7059
dt/dp	1 atm	0.04173	0.3130	183,2494,8091;1125,8132
dp/dt	1 atm	23.96	3.195	183,2494,8091
p	25	235	31.3	187;1125,2612
eq 2.5	A	6.81451	5.93941	186;8091
	B	1129.73	1129.73	
	C	229.215	229.215	
fp		-128.538	144.612	187;1125,1210,2494,8039,8132
d	20	0.66164	661.64	183,2494;1125,1210,1711,3912
	25	0.65702	657.02	4521,6188,7059,8039,2700
dd/dt	25	0.000924	0.924	<u>183</u>
α	25	0.001391	0.001391	4317
κ_S	25	1.8521×10^{-7}	1.3892×10^{-6}	7171;4317
n_D	20	1.37495	1.37495	183,2494;1125,1210,1711,3520
	25	1.37231	1.37231	6188,8039,8040
dn/dt	20 - 25	0.00049	0.00049	3520
η	25	0.361	0.000361	7482;2700
	30	0.342	0.000342	
	Table 2.1			
γ	20	17.37	0.01737	184;8039
	25	16.87	0.01687	
	30	16.37	0.01637	
ΔH_v	25	6.961	29.125	187;4521,5153,5613
	bp	6.519	27.276	
ΔH_m		0.191	0.799	187;2061
ΔH_f°	25(1)	-49.02	-205.10	189;8316
ΔH_c°	25(1)	-993.49	-4156.76	188;8316,3975,7446
C_p°	25	33.32	139.41	219
C_p	25	45.10	188.7	191;2061,6526,7059
λ	37.8	0.000245	0.103	6352
t_c		226.78	499.93	185,119;8231,4088,4018 560,8233
P_c		30.86	3.127	185;4018,8231,4088,560 3912,5153
d_c		0.241	241	185;4018,8231,4088,560
v_c		0.358	0.000358	185;560,8233
A	eq 2.72	0.00460	0.00460	187

14. 2,3-Dimethylbutane

(Continued)

ε		1.890	1.890	2088
δ	25	6.97	14.26	5334
soly	aq in,20	0.0110%w	0.0110%w	2268
fl pt	TOC	<-29	<244	5712;5822

Beil	10,151
uv	6083
ir	153,4384,7425,4587,5932
Raman	2400,5021,6049,6261,6683,8302,8221
ms	154,6887,2194
nmr	5933

15. Methylcyclohexane

$CH_2CH_2CH_2CH_2CH_2CHCH_3$ 108-87-2 C_7H_{14}

		cgs	SI	
mw		98.188	98.188	3468
bp	1 atm	100.934	374.084	238,8091;1117,1257,3378 8039,5379
dt/dp	1 atm	0.04671	0.3504	236;2494,8091
dp/dt	1 atm	21.41	2.854	236
p	25	46	6.1	238;2762,8091,3667,2612
eq 2.5	A	6.82300	5.94790	244;7057,5379,3913
	B	1270.763	1270.763	
	C	221.416	221.416	
fp		-126.593	146.557	238;2494,2780,3378,7480,8039
d	20	0.76939	769.39	236;1117,2494,3378,4951,5238
	25	0.76506	765.06	7494,7499,8039,2700
	30	0.76031	760.31	1475;5370

15. Methylcyclohexane

(Continued)

		cgs	SI	
n_D	20	1.42312	1.42312	236;506,1117,2494,5238,6049
	25	1.42058	1.42058	7494,8039,8040
dn/dt	25	0.000508	0.000508	236
η	-195.7(gl)	1.2×10^{20}	1.2×10^{17}	4614
	20	0.734	0.000734	237;7499,2700
	25	0.685	0.000685	
	Table 2.1			
γ	20	23.85	0.02385	242;3283,1969,3547,7758,8039
	25	23.29	0.02329	
	30	22.76	0.02276	5370
ΔH_v	25	8.451	35.359	238,4670;4969,5613,6934,7754
	bp	7.44	31.13	
ΔH_m		1.6135	6.7509	238;5679
ΔH_f°	25(1)	-45.45	-190.16	240,3738;8316
ΔH_c°	25(1)	-1091.13	-4565.29	239,3738,8316;3975,5701
C_p°	25	32.27	135.02	248;6934
C_p	25	44.29	185.29	2977;5679,3692,6934
t_c		298.97	572.12	119;3913,2770,5379
P_c		34.32	3.477	3913,2770
d_c		0.266	266	3913,2770
v_c		0.3445	0.0003445	2087
A	eq 2.72	0.03780	0.03780	238
K_f	eq 2.67	14.13	14.13	3550
κ	25	$<10^{-16}$	$<10^{-14}$	2486
ε	20	2.020	2.020	4937;8075
μ		0	0	8075
δ	25	7.82	15.99	5334;4593,1271,413
soly	in aq,25	0.0014%w	0.0014%w	4733
	aq in,20	0.0116%w	0.0116%w	2268
aq az	80.0	75.9%w	75.9%w	6165
fl pt	TOC	-6	267	5822

Beil	452,29
uv	5929
ir	153,602,5891,5890,5892,5898,633,5932
Raman	697,2400,2886,6049,5466,6683,80
ms	154,3728,6199,2420
nmr	367,5933,6748

16. Heptane

$CH_3(CH_2)_5CH_3$ 142–82–5 C_7H_{16}

		cgs	SI	
mw		100.203	100.203	3468
bp	1 atm	98.424	371.574	161;1125,2569,2197,2494 6821,8091
dt/dp	1 atm	0.04479	0.3360	159;1124,1125,2494,6823 5204,5234
dp/dt	1 atm	22.33	2.977	159
p	25	45.7	6.09	196;2382,6821,6823,6864,8091 1371,2612,7185
eq 2.5	A B C	6.89677 1264.90 216.544	6.02167 1264.90 216.544	195
fp		−90.582	182.568	161;1117,2780,4834,5680 2058,3557
tp		−90.60	182.55	1371
d	20	0.68375	683.75	161,4650,7170;8047,90,1124 1125,1212,1892,2569,2198 3378,6639,6875,8039,2700
	25	0.67946	679.46	161;7170,8047,3183,52 1475,1072
	30	0.67519	675.19	7170,6945;1980,8047
dd/dt	0 – 50	0.000840	0.840	1882
κ_T	25	1.920×10^{-7}	1.440×10^{-6}	52;1990,1988,6152,7170,3865
κ_S	30	1.6055×10^{-7}	1.2043×10^{-6}	7170;6945,1980
n_D	20 25	1.38764 1.38511	1.38764 1.38511	161,2492,4089;1117,1125,1212 2051,1124,2197,2382,6868 6980,7150,8040,6584
dn/dt	0 – 40	0.000508	0.000508	1882
η	−40.01	1.321cSt	1.321×10^{-6}	3732;2129,2569,2382,4570 6639,6875
	15	0.4412	0.0004412	162;7490,7705,2751,2700
	20	0.4181	0.0004181	162;6664
	25	0.3967	0.0003967	162;6664,5037
	35	0.3592	0.0003592	162
	Table 2.1 Table 2.2 Table 2.4			
γ	20	20.21	0.02021	4790;7490,7760,193,2569,6008 512,28,8039,3686
	25	19.70	0.01970	4790;193,3668
	30	19.17	0.01917	4790;193
	Table 2.8			
ΔH_v	25	8.736	36.55	196;4969,5153,5613,7819 5871,4776
	bp	7.576	31.70	196;7784,1046,5871

16. Heptane

(Continued)

		cgs	SI	
ΔH_m		3.355	14.04	196,5871
ΔH_f°	25(1)	−53.59	−224.22	198;2908,1843,8316
	25(g)	−44.85	−187.65	7376
ΔH_c°	25(1)	−1151.29	−4817.00	197;8316,7396,3975,2908,3717 1843,5967
C_p°	25 Table 2.14	39.48	165.18	7375;2763,7819,2058,5780,3557
C_p	25 Table 2.12	53.77	224.98	200;2490,6377,2978,3557,5068 5680,5817,7819,2058,5613
γ	37.8	0.000300	0.126	6352
t_c		267.53	540.68	8267;166,4089,8231,8228,923 5153,8233,119
P_c		27.39	2.775	8267;166,4089,8231,8228,2199 5153,631,4018
d_c		0.2356	235.6	8267;166,4089,8231,8228,4018 705,1046
v_c		0.4253	0.0004253	8267;166,8231,8228,1671,705 631,8233
A	eq 2.72	0.05065	0.05065	196;2199,7466,2058,3557
K_B	eq 2.79	3.43	3.43	2798
κ	25	$<10^{-16}$	$<10^{-14}$	2486
ε	20	1.9246	1.9246	4650;4937,6405,2051,2199,2515
μ		0.0	0.0	3463;6875
δ	25	7.43	15.20	6425,8112,5334;3421,3395
soly	in aq,25	0.000357%w	0.000357%w	7219;4733,6556,1630,2601 906,2268
	aq in,25	0.0091%w	0.0091%w	6425;2722
aq az	79.2	87.1%w	87.1%w	3498
fl pt	TOC	−1	272	5822;5712,5713
Beil	10,154			
uv	5929,6083			
ir	153,602,3313,4384,7077,633,3769,5932			
Raman	696,965,1533,3337,5402,6049,6642			
ms	154,6199,2194,5223,714			
nmr	5933,6748			

17. 2-Methylhexane

$(CH_3)_2CHCH_2CH_2CH_2CH_3$ 591–76–4 C_7H_{16}

		cgs	SI	
mw		100.203	100.203	3468
bp	1 atm	90.052	363.202	196;2199,2198,7058,8039
dt/dp	1 atm	0.04431	0.3324	192
dp/dt	1 atm	22.57	3.009	192
p	25	66	8.8	196,3557;7058
eq 2.5	A	6.87122	5.99612	195;3557
	B	1235.10	1235.10	
	C	219.469	219.469	
fp		-118.276	154.876	196;2199,2198,5680,7036
				8039,3557
d	20	0.67859	678.59	192;2199,2198,6875,8039
	25	0.67439	674.39	
dd/dt	25	0.000840	0.840	192
n_D	20	1.38485	1.38485	192,2488;2199,2198,6875
	25	1.38227	1.38227	8039,8040
dn/dt	25	0.000516	0.000516	192
η	20	0.378	0.000378	6875;2199
	Table 2.1			
γ	20	19.29	0.01929	193;2199,8039,2395,2088
	25	18.80	0.01880	
ΔH_v	25	8.319	34.807	196
	bp	7.330	30.669	
ΔH_m		2.195	9.184	196,3557;5680
ΔH_f°	25(1)	-54.86	-229.53	198;1843,8316
	25(g)	-46.54	-194.72	7378
ΔH_c°	25(1)	-1150.02	-4811.68	197;8316,1843,3975
C_p°	25	39.32	164.51	199
C_p	25	53.28	222.92	200;3557,5680
	Table 2.12			
t_c		257.16	530.31	194;2199,119,4299
P_c		26.98	2.734	194;2199,4299
d_c		0.238	238	194;4299
v_c		0.421	0.000421	194;4299
A	eq 2.72	0.04605	0.04605	196;3557
ε	20	1.919	1.919	4937;2199,6101
μ	20(1)	0	0	4739
δ	25	7.24	14.8	948
soly	aq in,20	0.0103%w	0.0103%w	2268

17.   2-Methylhexane

(Continued)

Beil	10,156
ir	153,4587,5932
Raman	696,965,2400,2991,6261,6642,6683
ms	154,2194,3326
nmr	5933

18.   3-Methylhexane

$CH_3CH_2CH(CH_3)CH_2CH_2CH_3$ C_7H_{16}

		cgs	SI	
mw		100.203	100.203	3468
bp	1 atm	91.850	365.000	196;2199,2198,2777,7058 4412,3274
dt/dp	1 atm	0.04459	0.3345	192
dp/dt	1 atm	22.43	2.990	192
p	25	62	8.3	196;7058,2612
eq 2.5	A B C	6.86503 1238.88 219.100	5.98993 1238.88 219.100	195
fp		−119.4	153.8	7058
d	20 25	0.68713 0.68295	687.13 682.95	192;2199,2198,2777,6875,2191 4412,2869,3274
dd/dt	25	0.000836	0.836	192
n_D	20 25	1.38864 1.38609	1.38864 1.38609	192;2492,2199,2198,2777,6875 2191,4412,2869,3274
dn/dt	25	0.000510	0.000510	192

18. 3-Methylhexane

(Continued)

		cgs	SI	
η	-195.7(gl)	3.2×10^{20}	3.2×10^{17}	4614
	20	0.372	0.000372	7482;2199,6875
	30	0.329	0.000329	
γ	20	19.79	0.01979	193;2199
	25	19.30	0.01930	
ΔH_v	25	8.386	35.087	196;7754
	bp	7.359	30.790	
ΔH_f°	25(1)	-54.12	-226.44	198;1843,8316
ΔH_c°	25(1)	-1150.76	-4814.78	197;8316,1843,3975
C_p°	25	39.10	163.59	217
C_p	16.0	51.2	214.2	3559
t_c		262.04	535.19	194;2199,7400
P_c		27.77	2.814	194;2199,7400
d_c		0.248	248	194
v_c		0.404	0.000404	194;7400
$[\alpha]_D$	25	+9.12	+9.12	2869;2191,4412,3274
ε	20	1.927	1.927	4937;2199,6875
μ	20(1)	0	0	4739
δ	25	7.29	14.9	948
fl pt		-4	269	5822
Beil	10,157			
ir	153,4587,5932			
Raman	2400,2991,6261,6642,6683			
ms	154,2194			
nmr	4933			

19. 2,3-Dimethylpentane

$CH_3CH(CH_3)CH(CH_3)CH_2CH_3$ 565-59-3 C_7H_{16}

		cgs	SI	
mw		100.203	100.203	3468
bp	1 atm	89.784	362.934	196;2199,2198,7058,7150 8039,490
dt/dp	1 atm	0.04482	0.3362	192
dp/dt	1 atm	22.31	2.975	192
p	25	69	9.2	196;7058
eq 2.5	A	6.85069	5.97559	195;5609
	B	1236.40	1236.40	
	C	221.655	221.655	
fp		(gl)	(gl)	8039
d	20	0.69508	695.08	192;2199,2198,6875,7150,8039
	25	0.69091	690.91	490,519,7573
dd/dt	25	0.000834	0.834	192
κ_S	25	1.512×10^{-7}	1.134×10^{-6}	519
n_D	20	1.39196	1.39196	192;1710,2199,2198,6875,7150
	25	1.38945	1.38945	8039,8040,2492,490,7573
dn/dt	25	0.000502	0.000502	192
η	-195.7(gl)	3.6×10^{18}	3.6×10^{15}	4614;2199,6875
	15	0.436	0.000436	7482
	30	0.412	0.000412	
γ	20	19.96	0.01996	193;2199,8039
	25	19.47	0.01947	
ΔH_v	25	8.185	34.246	196;5613
	bp	7.263	30.388	
ΔH_f°	25(1)	-54.58	-228.36	198;1843,7960
ΔH_c°	25(1)	-1150.30	-4812.86	197;7960,1843,3975
C_p°	25	38.44	160.83	219
C_p	25	52.17	218.28	200
t_c		264.14	537.29	194;2199,4298
p_c		28.70	2.908	194;2199,4298
d_c		0.255	255	194;4298
v_c		0.393	0.000393	194;4298
$[\alpha]_D$	25	-10.39	-10.39	4411;7573
ε	20	1.939	1.939	4937;2199
μ	20(1)	0	0	4739
δ	25	7.24	14.8	948

19. 2,3-Dimethylpentane

(Continued)

Beil	10,157
ir	153,4587,5932
Raman	965,2400,2991,6261,6683
ms	154,2194
nmr	4933

20. 2,4-Dimethylpentane

$(CH_3)_2CHCH_2CH(CH_3)_2$ 108–08–7 C_7H_{16}

		cgs	SI	
mw		100.203	100.203	3468
bp	1 atm	80.500	353.650	196;490,2198,2382,7058 7476,8039
dt/dp	1 atm	0.04376	0.3282	192
dp/dt	1 atm	22.85	3.047	192
p	25	98	13	196;2382,7058
eq 2.5	A	6.82427	5.94917	195
	B	1191.06	1191.06	
	C	221.540	221.540	
fp		−119.242	153.908	196;2199,2198,7036,7476 8039,3557
d	20	0.67270	672.70	192;2199,2198,2382,6875,8039
	25	0.66832	668.32	490,519
dd/dt	25	0.000876	0.876	192
κ_S	25	1.748×10^{-7}	1.311×10^{-6}	519
n_D	20	1.38145	1.38145	192,2492;2199,6875,8039
	25	1.37882	1.37882	8040,490
dn/dt	25	0.000526	0.000526	192

20. 2,4-Dimethylpentane

(Continued)

		cgs	SI	
η	15	0.375	0.000375	7482;2199,2382,6875
	20	0.360	0.000360	
γ	20	18.15	0.01815	193;2199,8039
	25	17.66	0.01766	
ΔH_v	25	7.861	32.890	196;5613,7754
	bp	7.051	29.501	
ΔH_m		1.636	6.845	196,3557
ΔH_f°	25(1)	−56.07	−234.60	198;1843,8316
ΔH_c°	25(1)	−1148.81	−4806.62	197;8316,1843,3975
C_p°	25	40.81	170.75	220
C_p	25	53.61	224.3	200;3557
	Table 2.12			
t_c		246.58	519.73	194;2199,4298
p_c		27.01	2.737	194;2199,4298
d_c		0.240	240	194;4298
v_c		0.418	0.000418	194;4298
A	eq 2.72	0.03476	0.03476	196;3557
ϵ	20	1.914	1.914	4937;2199,6875,490
μ	20(1)	0	0	4739
δ	25	6.96	14.24	5334
soly	in aq,25	0.0041%w	0.0041%w	4733
fl pt		−12	261	5822
Beil	10,158			
uv	6083			
ir	153,4587,5932			
Raman	965,2400,2991,6261,6683			
ms	154,2194			
nmr	4933,6748			

21.   1,2-Dimethylcyclohexane

(Mixed isomers)

583–57–3 C_8H_{16}

		cgs	SI	
mw		112.214	112.214	3468
bp	1 atm	127	400	5823;8272,499,1435,6323
p	37.8	31	4.1	5823
d	15	0.798	798	1435
	20	0.7893	789.3	8272;5823,499
n_D	20	1.4321	1.4321	8272;5823,499
ΔH_v	25	9.05	37.87	1703
ΔH_f°	25(1)	−52.28	−218.74	1703
	25(g)	−43.23	−180.87	
ΔH_c°	25(1)	−1246.65	−5215.98	1703
μ		0	0	5662
fl pt	TCC	13	286	5823
Beil	452,36			
ir	5932			
Raman	697			
nmr	5933			

22.   *cis*-1,2-Dimethylcyclohexane

2207–01–4 C_8H_{16}

		cgs	SI	
mw		112.214	112.214	3468
bp	1 atm	129.728	402.878	8091;241,5826,5145,6591
dt/dp	1 atm	0.04988	0.3741	8091;241,5145
dp/dt	1 atm	20.05	2.673	8091

22.   *cis*-1,2-Dimethylcyclohexane

(Continued)

		cgs	SI	
p	25	14.47	1.929	2087
eq 2.5	A	6.84164	5.96654	8091;244
	B	1369.525	1369.525	
	C	216.040	216.040	
fp		-50.023	223.127	241;7037,2494,5145,3561,5826
d	20	0.79627	796.27	2494;241,5145,6591,2234
	25	0.79222	792.22	
dd/dt		0.00075	0.75	5145
	15.56	0.00099	0.00099	5826
n_D	20	1.43596	1.43596	2494;241,6796,5826
	25	1.43358	1.43358	
dn/dt	25	0.000476	0.000476	2494
η	20	1.114	0.001114	2700;5145,6591
	40	0.818	0.000818	
γ	20	25.72	0.02572	242;2087
	25	25.19	0.02519	
ΔH_v	25	9.492	39.715	245;5613
	bp	8.173	34.196	5826;245,4753
ΔH_m		0.3932	1.6451	3561;245
ΔH_t	-100.6	1.9734	8.2567	3561;245
ΔH_f°	25(l)	-50.64	-211.88	247;1703
	25(g)	-41.13	-172.09	1703
ΔH_c°	25(l)	-1248.31	-5222.93	3740;246,5826
C_p°	25	37.4	156.5	713;248,4753
C_p	25	50.24	210.20	3561
t_c		333	606	243;5826
P_c		29	2.9	243;5826
d_c		0.24	240	243
v_c		0.46	0.00046	243
A	eq 2.72	0.00397	0.00397	245;7037
ϵ	25	2.062	2.062	2087
soly	in aq	0.0006%w	0.0006%w	4733
fl pt	TCC	16	289	5826

22. *cis*-1,2-Dimethylcyclohexane

(Continued)

Beil	452,21
ir	4476,5892,781,8285,153
Raman	5146,4476,8303
ms	5407,154
nmr	5362

23. *trans*-1,2-Dimethylcyclohexane

6876–23–9 C_8H_{16}

		cgs	SI	
mw		112.214	112.214	3468
bp	1 atm	123.419	396.569	8091;241,5826,5145,6591
dt/dp	1 atm	0.04951	0.3714	8091;241,5145
dp/dt	1 atm	20.20	2.693	8091
p	25	19.40	2.586	2087
eq 2.5	A	6.83722	5.96212	8091;244
	B	1356.100	1356.100	
	C	219.342	219.342	
fp		−88.194	184.956	241;7037,2494,5145,3561,5826
d	20	0.77601	776.01	2494;241,5145,6591,2234
	25	0.77204	772.04	
dd/dt		0.00079	0.79	5145
α	15.56	0.00104	0.00104	5826
n_D	20	1.42697	1.42697	2494;241,6796,5826
	25	1.42471	1.42471	
dn/dt	25	0.000452	0.000452	2494

23. *trans*-1,2-Dimethylcyclohexane

(Continued)

		cgs	SI	
η	20	0.817	0.000817	2700;5145,6591
	40	0.627	0.000627	
γ	20	24.05	0.02405	242;2087
	25	23.57	0.02357	
ΔH_v	25	9.168	38.359	245;5613
	bp	7.973	33.361	5826;245,4753
ΔH_m		2.5076	10.492	3561;245
ΔH_f°	25(1)	-52.19	-218.36	247;1703
	25(g)	-42.99	-179.87	1703
ΔH_c°	25(1)	-1246.77	-5216.49	3740;246,5826
C_p°	25	38.0	159.0	713;248,4753
C_p	25	50.05	209.41	3561
t_c		323	596	243;5826
P_c		29	2.9	243;5826
d_c		0.24	240	243
v_c		0.46	0.00046	243
A	eq 2.72	0.03664	0.03664	245;7037
ε	25	2.036	2.036	2087
fl pt	TCC	11	284	5826
Beil	452,21			
ir	4476,5892,781,8285,153			
Raman	5146,4476,8303			
ms	5407,2420,154			
nmr	5362			

24. Ethylcyclohexane

CH$_2$CH$_2$CH$_2$CH$_2$CH$_2$CHCH$_2$CH$_3$ 1678-91-7   C$_8$H$_{16}$

		cgs	SI	
mw		112.214	112.214	3468
bp	1 atm	131.795	404.945	244;2494,8091,697,7476
dt/dp	1 atm	0.04969	0.3727	236;2494,8091
dp/dt	1 atm	20.12	2.683	236
p	25	13	1.7	238;8091
eq 2.5	A	6.86728	5.99218	244;8091
	B	1382.466	1382.466	
	C	214.995	214.995	
fp		-111.323	161.827	238;2494,2780 7476,3561,7479
d	20	0.78792	787.92	236,2494;697,2700
	25	0.78390	783.90	
dd/dt	25	0.000804	0.804	236
n$_D$	20	1.43304	1.43304	236,2494;697
	25	1.43073	1.43073	
dn/dt	25	0.000462	0.000462	236
η	-195.7(gl)	4.4 x 10^{24}	4.4 x 10^{21}	4614,2700
	20	0.843	0.000843	237
	25	0.787	0.000787	
γ	20	25.67	0.02567	242
	25	25.14	0.02514	
ΔH$_v$	25	9.720	40.67	7106;238,5613
	bp	8.109	33.93	
ΔH$_m$		1.9919	8.3341	238;3561
ΔH$_f^\circ$	25(1)	-50.72	-212.21	240,3740,3738;8316
ΔH$_c^\circ$	25(1)	-1248.23	-5222.59	239,3740,3738,8316
C$_p^\circ$	25	37.96	158.82	248
C$_p$	25	50.62	211.79	3561
t$_c$		336	609	243;2087
P$_c$		30	3.0	243
d$_c$		0.25	250	243;2087
v$_c$		0.45	0.00045	243;2087
A	eq 2.72	0.03828	0.03828	238
ε		2.054	2.054	2087
μ	100(g)	0	0	4739
δ	25	7.97	16.3	948
aq az	90.6	60.2%w	60.2%w	6165
fl pt		22	295	2087

24. Ethylcyclohexane

(Continued)

Beil	452,35
ir	153,5891,5892,6260,633,5932
Raman	697,2400,6049
ms	154,6199,2420
nmr	5933,6748

25. Octane

$CH_3(CH_2)_6CH_3$ 111–65–9 C_8H_{18}

		cgs	SI	
mw		114.230	114.230	3468
bp	1 atm	125.673	398.823	161;5910,8091,1117,1084 1436,6856
dt/dp	1 atm	0.04738	0.3554	201;2494,8091,8128
dp/dt	1 atm	21.11	2.814	201
p	25	14.0	1.87	169;4611,8091
eq 2.5	A B C	6.91868 1351.99 209.155	6.04358 1351.99 209.155	204;1371
fp		−56.764	216.386	161;5910,2045,3560,6856,6860 2435,1371
tp		−56.78	216.37	1371
d	20 25	0.70267 0.69862	702.67 698.62	161;5910,1424,3183,2273,52 2495,1117,1212,2045,6639 7495,696,2700
dd/dt	0 – 40	0.000803	0.803	1882;2273
α	25	0.001164	0.001164	3131
κ_T	25	1.707×10^{-7}	1.280×10^{-6}	52;1988,1990,6152,3865

25. Octane

(Continued)

		cgs	SI	
κ_S	25	1.3899×10^{-7}	1.0425×10^{-6}	3131;6945,1980
n_D	20	1.39743	1.39743	161;5910,1212,1258,2045,2051
	25	1.39505	1.39505	4834,6639,6856,8039,8040
dn/dt	0 – 40	0.000483	0.000483	1882
η	20	0.5466	0.0005466	162,6639;2129,7495,6459,3642
	25	0.5151	0.0005151	2751,2700,5037
	Table 2.1			
	Table 2.2			
	Table 2.3			
	Table 2.4			
γ	20	21.68	0.02168	512
	25	21.18	0.02118	3668;202,3283,6008,7760
	Table 2.8			8039,3686
ΔH_v	25	9.916	41.49	169;5153,5613,7754,2435,1046
	bp	8.2292	34.431	3505;169
ΔH_m		4.957	20.74	169;3560,5680
ΔH_f°	25(l)	−59.82	−250.29	2849,207;2908,8316
	25(g)	−49.91	−208.82	7376
ΔH_c°	25(l)	−1307.43	−5470.25	2849,206;8316,3975,2908
				3717,5967
C_p°	25	44.88	187.78	7375
	Table 2.14			
C_p	25	60.74	254.15	242;3131,2435,3560,5613
				5680,5068
λ	37.8	3.10×10^{-4}	0.130	6352
t_c		295.61	568.76	166,4299;8233,1671,119,7400
				1631,4018
p_c		24.54	2.487	166,4299,8231;7400,1631,8235
				5778,4018
d_c		0.232	232	166,4299,8231;4018,1046
v_c		0.492	0.000492	166,4299,8231;1671,7400,8233
A	eq 2.72	0.05328	0.05328	169;7466,2435
ε	20	1.948	1.948	4937,6405;2051,7850
μ	20(l)	0	0	2051
δ	25	7.57	15.48	5334;652,1271,4593
soly	in aq,25	6.6×10^{-7}%w	6.6×10^{-7}%w	4733;2601,7219
	aq in,20	0.0095%w	0.0095%w	2268
aq az	89.6	74.5%w	74.5%w	3498
fl pt	TOC	22	295	5628

25. Octane

(Continued)

Beil	<u>10</u>,159
uv	6083
ir	153,606,4384,633,5711,4383,3769,5932
Raman	965,2400,3337,5248,6049,6642,695
ms	154,6199,2421,714
nmr	5933,6748

26. 2,2,3-Trimethylpentane

$(CH_3)_3CCH(CH_3)CH_2CH_3$ 564-02-3 C_8H_{18}

		cgs	SI	
mw		114.230	114.230	3468
bp	1 atm	109.844	382.994	204;2494,8091,7058,8039
dt/dp	1 atm	0.04755	0.3567	201,2494,8091
dp/dt	1 atm	21.03	2.804	201,2494,8091
p	25	32	4.3	205;7058,8091
eq 2.5	A	6.82336	5.94826	204;8091
	B	1293.94	1293.94	
	C	218.355	218.355	
fp		-112.27	160.88	205;2494,5402,7058,8039
d	20	0.71602	716.02	201;2494,4426,2700
	25	0.71207	712.07	
dd/dt	25	0.000790	0.790	<u>201</u>
n_D	20	1.40295	1.40295	201,2494;4426,8039,8040
	25	1.40066	1.40066	
dn/dt	20 - 25	0.00046	0.00046	3520
η	20	0.598	0.000598	2700
	40	0.471	0.000471	

26.   2,2,3-Trimethylpentane

(Continued)

		cgs	SI	
γ	20	20.67	0.02067	202;8039
	25	20.22	0.02022	
ΔH_v	25	8.824	36.920	205;5613
	bp	7.65	32.01	
ΔH_m		2.06	8.62	205
ΔH_f°	25(1)	-61.11	-255.68	207;8316
ΔH_c°	25(1)	-1305.83	-5463.63	206,8316
C_p°	25	44.70	187.02	207
C_p	25	58.69	245.55	208
t_c		290.28	563.43	203,4299
P_c		26.94	2.730	203,4299
d_c		0.262	262	203,4299
v_c		0.436	0.000436	203,4299
A	eq 2.72	0.0401	0.0401	205
ε	20	1.962	1.962	6159;4937
δ	25	7.19	14.7	948
Beil	10,164			
ir	153,4384			
Raman	965,2400			
ms	154,6683			
nmr	5933,6748			

27. 2,2,4-Trimethylpentane

$(CH_3)_3CCH_2CH(CH_3)_2$ 540-84-1 C_8H_{18}

		cgs	SI	
mw		114.230	114.230	3468
bp	1 atm	99.238	372.388	205,2494,3915;1117,1124 1125,5871
dt/dp	1 atm	0.04651	0.3489	201,2494,8091;776,1125,6823
dp/dt	1 atm	21.50	2.867	201,2494,8091
p	25	49	6.5	205;703,4611,6821,8091,2612
eq 2.5	A B C	6.80395 1253.36 220.241	5.92885 1253.36 220.241	204;703,3915,6823
fp		-107.388	165.762	205;1124,2494,2780,3378 5680,7479
d	20 25	0.69193 0.68781	691.93 687.81	201;978,1117,1124,1125,3915 4426,6875,7150,8039,2700
dd/dt	25	0.000824	0.824	201
n_D	20 25	1.39145 1.38898	1.39145 1.38898	201,2494;3623,1117,1124,1125 978,4426,6875,7150,8040
dn/dt	20 - 25	0.00047	0.00047	3520
η	20 30 40	0.504 0.447 0.403	0.000504 0.000447 0.000403	2700;2197,6875
γ	20 25 30	18.77 18.32 17.88	0.01877 0.01832 0.01788	202;8039,978
ΔH_v	25 bp	8.401 7.357	35.15 30.78	7107;205,5613,5871
ΔH_m		2.198	9.196	205;5680,5871
ΔH_f°	25(1)	-61.94	-259.16	207;8316
ΔH_c°		-1305.29	-5461.33	206,8316;3975
C_p°	25	45.03	188.41	207
C_p	25	57.015	238.551	207;5613,5680
λ	37.8	0.000231	0.0967	6352
t_c		270.74	543.89	203,4299;703,3915,7400,119
p_c		25.34	2.568	203,4299;703,3915,7400
d_c		0.244	244	203,4299
v_c		0.468	0.000468	203,4299;7400
A	eq 2.72	0.04026	0.04026	205
ε	20	1.940	1.940	4937;6875
μ	(1)	0	0	6875
δ	25	6.85	14.01	5334;3421,4593

27.   2,2,4-Trimethylpentane

(Continued)

		cgs	SI	
soly	in aq,25	0.00024%w	0.00024%w	4733
	aq in,20	0.0055%w	0.0055%w	794;2268
aq az	79.4	76.7%w	76.7%w	6165
fl pt	TCC	-12	261	5712;5822
Beil	10,164			
uv	3760,6083			
ir	153,344,606,3313,633,6368,5932			
Raman	696,965,2400,6047,6049,7162			
ms	154,6199			
nmr	6748			

28.   2,2,3,3-Tetramethylbutane

$(CH_3)_3CC(CH_3)_3$ 594-82-1 C_8H_{18}

		cgs	SI	
mw		114.230	114.230	3468
bp	1 atm	106.29	379.44	204;5826,1320,2088
dt/dp	1 atm	0.0476	0.357	201
dp/dt	1 atm	21.0	2.80	201
p	25	20.854	2.780	2088;4611,6520
eq 2.5	A	6.77930	5.90420	204;1320,6520
	B	1270.1	1270.1	
	C	219.50	219.50	
fp		100.69	373.84	201,1320;6520,5826,6856
d	25(s)	0.8215	821.5	6589;2088,1320
	30(s)	0.8188	818.8	
	100.71(1)	0.6568	656.8	6589
n_D	20(s)	1.4695	1.4695	1320,2088
η	20	0.6340	0.0006340	4862,5826

28.   2,2,3,3-Tetramethylbutane

(Continued)

		cgs	SI	
γ	20	21.14	0.02114	2088
	30	20.22	0.02022	
ΔH_v	bp	7.530	31.506	1320;2088,5826
ΔH_s	25	10.365	43.367	6520;1320,5613
ΔH_m		1.8022	7.5404	6520;1320,5688
ΔH_t	-120.7	0.478	2.000	6520;5680
ΔH_f°	25(s)	-62.36	-260.91	207;2849,5969,2088
	25(g)	-53.99	-225.89	5969
ΔH_c°	25(s)	-1304.88	-5459.62	206;2849,5969
	25(g)	-1313.28	-5494.76	5969
C_p°	25	44.74	187.19	208
C_p	102.26	66.94	280.08	6520;208
t_c		294.7	567.9	203;2397,5826
P_c		28.3	2.87	203;2397,5826
d_c		0.248	248	203
v_c		0.461	0.000461	203;5826
A	eq 2.72	0.00613	0.00613	2088
Beil	10,165			
ir	4384,5559,1534,5932			
Raman	1534			
ms	154			
nmr	1160,6614,5933			

29. Nonane

$CH_3(CH_2)_7CH_3$ 111–84–2 C_9H_{20}

		cgs	SI	
mw		128.257	128.257	3468
bp	1 atm	150.818	423.968	161;1436,4834,8009,8091,4235
dt/dp	1 atm	0.04967	0.3726	209;2494,8091,5016
dp/dt	1 atm	20.13	2.684	
p	25	4.3	0.57	212;8091,1371
eq 2.5	A	6.93893	6.06383	168;8091
	B	1431.82	1431.82	
	C	202.011	202.011	
fp		–53.489	219.661	161;2045,2494,3560,5016,6856
				2435,1371
tp		–53.500	219.650	1371
d	20	0.71772	717.72	161;2045,2051,2494,6008,6639
	25	0.71375	713.75	8039,4235,5016,52
dd/dt	0 – 40	0.000774	0.774	1882;5016
κ_T	25	1.569×10^{-7}	1.177×10^{-6}	52;1988,6152,3865
κ_S	30	1.33×10^{-7}	9.98×10^{-7}	1980
n_D	20	1.40542	1.40542	161,1058;2045,2051,2494,6639
	25	1.40311	1.40311	8039,5016,6856
dn/dt	0 – 40	0.000463	0.000463	1882;5016
η	20	0.7160	0.0007160	162;6639
	25	0.6696	0.0006696	
	Table 2.1			
γ	25	22.38	0.02238	3687,165;6008,8039,3686
	Table 2.8			
ΔH_v	25	11.100	46.442	212,4360;5613,4186,1046
	bp	8.823	36.915	
ΔH_m		3.697	15.468	212,24353514,6856
ΔH_f°	25(l)	–65.80	–275.30	7381;4360,2846,8316
	25(g)	–54.70	–228.86	7376
ΔH_c°	25(l)	–1463.80	–6124.58	213,3739,4360;2846,5967
				3717,8316
C_p°	25	50.29	210.41	7375
C_p	25	68.01	284.55	215;2435,5068,5613,5680
				6352,3560
λ	37.8	0.000321	0.134	6352
t_c		321.41	594.56	166,119,135;3914,7684,7400
				6854,6352,5778
p_c		22.58	2.288	166,7400,135;5778,7683
				1046,3914
d_c		0.234	234	166;1046

29. Nonane

(Continued)

		cgs	SI	
v_c		0.548	0.000548	166;7400
A	eq 2.72	0.03857	0.03857	212;7466,2435
ε	20	1.970	1.970	6405;4937,2051,6159
δ	25	7.65	15.65	6425;3395,4593
soly	in aq aq in,25	220ppb 0.0079%w	220ppb 0.0079%w	4735 6425
aq az	95	60.2%w	60.2%w	3498
fl pt	TCC	31	304	716;5822

Beil	10,165
uv	6083
ir	153,606,633,4383,5932
Raman	698,2400,3337,5573,6642,4705
ms	154,5204,2421,714
nmr	4105,5933,6748

30. 2,2,5-Trimethylhexane

$(CH_3)_3CCH_2CH_2CH(CH_3)_2$ 3522-94-9 C_9H_{20}

		cgs	SI	
mw		128.257	128.257	3468
bp	1 atm	124.093	397.243	5609;212,3520
dt/dp	1 atm	0.04838	0.3629	209
dp/dt	1 atm	20.67	2.756	209
p	25	16.625	2.216	5609;212
eq 2.5	A B C	6.83531 1324.049 210.737	5.96021 1324.049 210.737	5609
fp		−105.780	167.370	212

30. 2,2,5-Trimethylhexane

(Continued)

		cgs	SI	
d	20	0.70721	707.21	209
	25	0.70322	703.22	
dd/dt	25	0.000798	0.798	209
n_D	20	1.39972	1.39972	209,2492;3520
	25	1.39728	1.39728	
dn/dt	25	0.000488	0.000488	209
γ	20	20.04	0.02004	210
	25	19.60	0.01960	
ΔH_v	25	9.602	40.175	212;4360
	bp	8.07	33.76	
ΔH_m		1.48	6.19	212
ΔH_f°	25(1)	−70.11	−293.34	2846,7381;4360,8316
ΔH_c°	25(1)	−1459.50	−6106.55	2846,213;4360,8316
C_p°	25	49.68	207.86	214
λ	37.8	0.000258	0.108	6352
t_c		294.8	568.0	211,4299
P_c		23.0	2.33	211
d_c		0.247	247	211
v_c		0.519	0.000519	211
A	eq 2.72	0.0266	0.0266	212
δ	25	7.09	14.5	948
fl pt	TOC	13	286	5822

Beil	10,167
ir	153
Raman	2400
ms	154,5204,2421

31. Decahydronaphthalene

(Mixed isomers)

91–17–8 $C_{10}H_{18}$

		cgs	SI	
mw		138.252	138.252	3468
bp	1 atm	191.7	464.9	3348;2659,5251,2776
p	23.3	1.0	0.13	2659;3348,4611
	66.1	9.7	1.3	
	86.8	26.4	3.5	
	118.8	87.4	11.7	
	150.0	240.0	32.0	
fp		−124±2	149±2	3348
d	20	0.8865	886.5	7967;2659,3120,3011
	25	0.8789	878.9	3348
n_D	20	1.4758	1.4758	2659;3120
η	25	2.415	0.002415	3348;3011
	Table 2.1			
γ	33.6	29.36	0.02936	3348
ΔH_v	bp	9.82	41.09	3348
C_p°	25	44.52	186.27	7806
	Table 2.14			
C_p		53.56	224.10	4902;3348,7967,3011
t_c		445	718	3348
P_c		∼29	∼2.9	3348
K_B	eq 2.79	5.762	5.762	3348
ε	25	2.1542	2.1542	7094;5251
μ	25(1)	0	0	7094;2363
δ	25	8.80	18.0	948
soly	in aq,25	<0.02%w	<0.02%w	987
	aq in,20	0.0063%w	0.0063%w	2268
fl pt	TCC	58	331	5712
Beil	453,92			
ir	4385,6311,4383,5932			
Raman	5303,4991			

32. *cis*-Decahydronaphthalene

493–01–6 $C_{10}H_{18}$

		cgs	SI	
mw		138.252	138.252	3468
bp	1 atm	195.774	468.924	1324;2401,2558,2565,7058
dt/dp	1 atm	0.05710	0.4283	1324
dp/dt	1 atm	17.51	2.335	
p	25	0.78	0.10	1324;2401,6592,6878,8297
eq 2.5	A	6.87529	6.00019	1324;6592
	B	1594.460	1594.460	
	C	203.392	203.392	
fp		–43.01	230.14	7041;4749,5677,6588,6593
				7058,6972
d	20	0.89671	896.71	298;6593,6642,3120,3011
	25	0.89288	892.88	4381;1324
dd/dt		0.000760	0.760	1324
α		0.000867	0.000867	4195
n_D	20	1.48098	1.48098	298;4381
	25	1.47878	1.47878	1324;2401,6593,5186,3120
dn/dt		0.000440	0.000440	1324
η	20	3.381	0.003381	6591;3011
	30	2.723	0.002723	
γ	20	32.18	0.03218	6590;4195
	30	31.01	0.03101	6590
ΔH_v	25	12.271	51.342	2087
	bp	9.799	40.999	6592
ΔH_m		2.268	9.489	4749;5677
ΔH_f°	25(1)	–52.75	–220.71	1842;6928,8316
	25(g)	–40.45	–169.24	6928
ΔH_c°	25(1)	–1502.4	–6286.0	1842;6928,3975,8316
C_p°	25	39.84	166.69	5184
C_p	25	55.45	232.00	4749;5677,6587,3011
	Table 2.12			
t_c		429.0	702.2	4299;5184,6592,1451
p_c		27	2.7	5184
A	eq 2.72	0.02154	0.02154	4749;7041
B		0.00299	0.00299	
K_f	eq 2.67	19.47	19.47	3550
ε	20	2.197	2.197	4937;6588,6972

32. *cis*-Decahydronaphthalene

(Continued)

		cgs	SI	
μ	20 – 100 in 38	~0	~0	6588;2363
δ	25	9.19	18.8	948
soly	aq	ins	ins	4026
aq az	99.1	22.4%w	22.4%w	6165

Beil	453,92
ir	153,6458,5585,5932
Raman	6049,5766
ms	154
nmr	5364,3435,5226,5933

33. *trans*-Decahydronaphthalene

493–02–7 $C_{10}H_{18}$

		cgs	SI	
mw		138.252	138.252	3468
bp	1 atm	187.273	460.423	1324;2401,6592,6593,7058
dt/dp	1 atm	0.05656	0.4242	1324
dp/dt	1 atm	17.68	2.357	1324
p	25	1.23	0.164	1324;2401,6592,7058,8297
eq 2.5	A	6.85681	5.98171	1324;6592
	B	1564.683	1564.683	
	C	206.259	206.259	
fp		−30.400	242.750	7041;4749,5677,6588,6593 7058,6972
d	20	0.86971	869.71	298;2031,6590,6593,3120,3011
	25	0.86596	865.96	4381;1324
dd/dt		0.000749	0.749	1324
α		0.000865	0.000865	4195
n_D	20	1.46949	1.46949	4381;298,3623
	25	1.46715	1.46715	1324,3623;6593,3120,2401,5186
dn/dt		0.000434	0.000434	1324

33. *trans*-**Decahydronaphthalene**

(Continued)

		cgs	SI	
η	20 30	2.128 1.774	0.002128 0.001774	6591;2031,3011
γ	20 30	30.15 28.87	0.03015 0.02887	4195;6590,2031 6590
ΔH_v	25 bp	11.92 9.615	49.87 40.229	$\underline{2087}$ 6592
ΔH_m		3.445	14.414	4749,5677
ΔH_f°	25(1) 25(g)	−54.87 −43.54	−229.58 −182.17	1842;6928,8316 6928
ΔH_c°	25(1)	−1502.4	−6286.0	1842;6928,3975,8316
C_p°	25	40.04	167.53	5184
C_p	25	54.61	228.49	4749;5677,6587,3011
	Table 2.12			
t_c		413.8	687.0	4299;6592,1451,5184
P_c		27	2.7	5184
A	eq 2.72 B	0.02941 0.00292	0.02941 0.00292	4749;7041
K_f	eq 2.67	20.81	20.81	3550
ε	20	2.172	2.172	4937;6588,6972
μ	20 − 100 in $\underline{38}$	∿0	∿0	6588;2363
δ		8.8	18.0	652
soly	aq	ins	ins	4026
aq az	98.5	26.5%w	26.5%w	6165
Beil	$\underline{453}$,92			
ir	153,6458,5585,5932			
Raman	6049,5766			
ms	154			
nmr	5364,3435,5933,6748			

34. Decane

$CH_3(CH_2)_8CH_3$ 124-18-5 $C_{10}H_{22}$

		cgs	SI	
mw		142.284	142.284	3468
bp	1 atm	174.155	447.305	161;2494,1436,4834,5251,6639
dt/dp	1 atm	0.05172	0.3879	159,2494,8091
dp/dt	1 atm	19.33	2.578	159,2494,8091
p	25	1.3	0.17	169;8091,5947,1371
eq 2.5	A	6.94363	6.06853	168;8091
	B	1495.17	1495.17	
	C	193.858	193.858	
fp		-29.644	243.506	161;1371,1882,2045,5016,3560
				7038,2435
tp		-29.65	243.50	1371
d	20	0.73012	730.12	161;2494,2051,5329,6008,6639
	25	0.72635	726.35	7760,5016,1072
dd/dt	0 - 40	0.000751	0.751	1882;5016
α	25	0.001051	0.001051	3131
κ_T	25	1.459×10^{-7}	1.094×10^{-6}	1990,1988;6152,3865
κ_S	25	1.2063×10^{-7}	9.0482×10^{-7}	3131
n_D	20	1.41189	1.41189	161;1882,2045,2051,5329,6639
	25	1.40967	1.40967	7760,5016,6584
dn/dt	0 - 40	0.000447	0.000447	1882;5016
η	20	0.9284	0.0009284	162;6639,5329,2751,5037,1355
	25	0.8614	0.0008614	
	Table 2.1			
γ	20	23.74	0.02374	4195;512
	25	23.37	0.02337	3274;165,3283,5239,6008
	Table 2.8			7760,3686
ΔH_v	25	12.277	51.367	169,4360;5613,1046
	bp	9.388	39.279	
ΔH_m		6.864	28.719	169;2435,3560
ΔH_f°	25(l)	-71.92	-300.91	7382;4360,8316
	25(g)	-59.64	-249.53	7376
ΔH_c°	25(l)	-1620.05	-6778.29	216;4360,8316,5967,37173975
C_p°	25	55.70	233.05	7375;5680
C_p	25	75.18	314.54	7383;2435,5068,5613,3560
				2978,3131
λ	37.8	0.000318	0.133	6352
t_c		344.4	617.6	166,119,135;7684,7400,923
				5778,1046,3914
p_c		20.76	2.104	166,135;7400,5778,7683
				1046,3914

34. Decane

(Continued)

		cgs	SI	
d_c		0.236	236	166;1046
v_c		0.603	0.000603	166;7400
A	eq 2.72	0.05826	0.05826	169;7466,2435
ε	20	1.989	1.989	6405;4937,2051
μ	-30-170(1)	~0	~0	2051
δ	25	7.72	15.79	6425,5334,6240
soly	in aq,25	52ppb	72ppb	4735
	aq in,25	72ppm	72ppm	6425
aq az	98	31.5%w	31.5%w	6165
fl pt	TCC	46	319	5712;5822

Beil	10,168
ir	153,606,633,6695,4383,4587,3769,5932
Raman	2400,2886,5573,4105,4705
ms	154,6199,6146,2421,714,6605
nmr	5933,6748

35. 1,1'-Bicyclohexyl

92-51-3 $C_{12}H_{22}$

		cgs	SI	
mw		166.306	166.306	3468
bp	1 atm	239.04	512.19	5017;4942,3796,7767
dt/dp	1 atm	0.555	4.16	5017;8048
dp/dt	1 atm	1.80	0.240	5017
p	25	0.108	0.0144	6299

35. 1,1′-Bicyclohexyl

(Continued)

		cgs	SI	
eq 2.4	A	8.21	7.33	2028
	B	2715	2715	
fp		3.63	276.78	5017;5597,3796,7040,6299,7767
d	20	0.88619	886.19	5017;3796,3011,6299
	25	0.88249	882.49	
dd/dt		0.000740	0.740	5017;6299
n_D	20	1.47995	1.47995	5017;2028,6299
	25	1.47768	1.47768	
dn/dt		0.000454	0.000454	5017
η	20	3.75	0.00375	6299;3796,3011,7767
	25	3.40	0.00340	
γ	20.5	32.68	0.03268	7767;6299
	41.2	30.56	0.03056	
ΔH_v	25	13.86	57.98	5232
	bp	10.4	43.5	8049;2028
ΔH_f°	25(1)	−66.05	−276.35	5232;2854
ΔH_m		1.620	6.780	5597
ΔH_t	17.1	0.368	1.540	5597
	5.7	0.177	0.740	
	0.3	1.692	7.080	
ΔH_c°	25(1)	−1814.03	−7589.90	5232;2854,3975
C_p	22	62.5	261.3	5597
A	eq 2.72	0.0110	0.0110	7040
K_f	eq 2.67	14.52	14.52	4942;3550
μ	25	<0.4	<0.4	6299
δ	25	8.51	17.4	948
fl pt	TOC	101	374	4902

Beil	453,108
ir	153,2656,6580,2654,5932
Raman	4991,5756
nmr	5933

36. Dodecane

$CH_3(CH_2)_{10}CH_3$ 112–40–3 $C_{12}H_{26}$

		cgs	SI	
mw		170.337	170.337	3468
bp	1 atm	216.323	489.473	160;8091,6639,1882,5017
dt/dp	1 atm	0.05528	0.4146	159,8091;5017
dp/dt	1 atm	18.09	2.412	159,8091
p	25	0.12	0.016	169;8091,5947
eq 2.5	A	6.99795	6.12285	167;8091
	B	1639.27	1639.27	
	C	181.835	181.835	
fp		-9.582	263.568	160;6639,1882,5017,2435,3560
d	20	0.74875	748.75	160;2429,6639,1882,5017,6008
	25	0.74518	745.18	7760,1093,6465
dd/dt		0.000719	0.719	1882
α	25	0.000974	0.000974	2429
κ_T	25	1.32×10^{-7}	9.88×10^{-7}	1991,1988;6152,207
n_D	20	1.42167	1.42167	160;6639,1882,4839,5017
	25	1.41952	1.41952	
dn/dt		0.000425	0.000425	1882
η	20	1.508	0.001508	162;6639,1093,2751,5037
	25	1.378	0.001378	
	Table 2.1			
γ	20	25.58	0.02558	4195;512
	25	24.90	0.02490	3668;6465,165,6008,3686
	Table 2.8			
ΔH_v	25	14.648	61.287	169;5242,1046
	bp	10.43	43.64	169
ΔH_m		8.57	35.86	169;2435,3560
ΔH_f°	25(1)	-84.16	-352.13	170
	25(g)	-69.50	-290.79	7376
ΔH_c°	25(1)	-1932.55	-8085.80	7385;5967,3717
C_p°	25	66.52	278.33	7375
C_p	25	89.86	375.97	2435,5068;3560,7386
t_c		385.1	658.3	166,135;5778,7684,119,1046
P_c		18.00	1.824	135;166,5778,7683,1046
d_c		0.239	239	166;1046
v_c		0.713	0.000713	166
A	eq 2.72	0.06208	0.06208	7384;7466,2435,7035
ϵ	30	2.002	2.002	2051
μ	-10-210(1)	0	0	2051

36. Dodecane

(Continued)

		cgs	SI	
δ	25	7.84	16.04	6425,8112;652
soly	in aq,25	3.7ppb	3.7ppb	7091
	aq in,25	65ppm	65ppm	6425
aq az	99.6	11.1%	11.1%	6165
fl pt	TCC	73	346	5712
Beil	10,171			
uv	4062			
ir	153,633,4383,4587,2655,3769,602,5932			
Raman	4105,2889,5572,2400,4705			
ms	5904			
nmr	5933,6748			

37. Tridecane

$CH_3(CH_2)_{11}CH_3$ 629–50–5 $C_{13}H_{28}$

		cgs	SI	
mw		184.364	184.364	3468
bp	1 atm	235.466	508.616	160;1324,1882,5826
dt/dp	1 atm	0.0567	0.425	160;1324
dp/dt	1 atm	17.6	2.35	160
p	25	0.03972	0.005296	2088
eq 2.5	A	7.00756	6.13246	167;1324
	B	1690.67	1690.67	
	C	174.220	174.220	
fp		−5.390	267.760	160;7035,2435,1882,2045,5826
d	20	0.75610	756.10	160;1324,1882,1213,2045,6459
	25	0.75271	752.71	160;1324,1213,3597
dd/dt	25	0.000715	0.715	1324;1882
κ_T	25	1.26×10^{-7}	9.48×10^{-7}	1988;6152

37. Tridecane

(Continued)

		cgs	SI	
n_D	20	1.42560	1.42560	1324,160,3627;6437,7760,1213
	25	1.42346	1.42346	1882,7464
dn/dt	0 – 40	0.000417	0.000417	1882
η	20	1.880	0.001880	163;2045,3038,6437,6459
	25	1.706	0.001706	
γ	20	25.99	0.02599	165,3687;3686,4195,7760
	25	25.55	0.02555	
ΔH_v	25	15.83	66.23	169;5242
	bp	10.91	45.65	169;5826
ΔH_m		6.812	28.501	2435;169
ΔH_t	-18.15	1.831	7.661	2435,169
ΔH_f°	25(1)	-90.27	-377.69	5969;170
	25(g)	-74.45	-311.50	5969;7376
ΔH_c°	25(1)	-2088.85	-8739.75	5969
	25(g)	-2104.67	-8805.94	
C_p°	25	71.93	300.97	7375
C_p	25	97.25	406.89	2435;5068
t_c		402.6	675.8	166;5826
P_c		17	1.7	166;5826
d_c		0.24	240	166
v_c		0.77	0.00077	166;5826
A	eq 2.72	0.04780	0.04780	2435;7035,169,7466
ε	25	2.03	2.03	2088
δ	25	7.89	16.14	6425
soly	aq in,25	0.0060%w	0.0060%w	6425
fl pt	TCC	79	352	5826,5823
Beil	10,171			
ir	4469,3769,6893,153			
Raman	4705			
ms	3425,1562,154			
nmr	2761,6748			

38. Benzene

C_6H_6 71–43–2 C_6H_6

		cgs	SI	
mw		78.113	78.113	3468
bp	1 atm	80.094	353.244	279;676,2729,6822,6823 8131,8304
dt/dp	1 atm	0.04271	0.3204	274,2494,8091;6823
dp/dt	1 atm	23.41	3.122	274,2494,8091
p	25	95.2	12.7	280;2439,2759,3514,6462,8304 2612,115,3834
eq 2.5	A B C	6.89742 1206.53 220.91	6.02232 1206.53 220.91	279;6409,6822,6823,8091 59,5703
fp		5.533	278.683	280;4978,2729,2780,5578,7494
d	10	0.88962	889.62	3634;276,887,1057,1568,2038 2494,4414
	20	0.87900	879.00	7169,3634,978;3084,6121,3079 7170,2431,276,4951,6409 7494,7848,676,8304
	25	0.87360	873.60	3084;4381,3079,7170,7169,1147 3183,2489,6121,3317
	30	0.86829	868.29	3084;3634,7169,7170,5370,6121
dd/dt	20	0.001051	1.051	6121
α	25	0.001213	0.001213	1965;2431,4902,2700,8155
κ_T	25	1.288×10^{-7}	9.660×10^{-7}	54;6151,1965,7169,6487,2431 1880,3866
κ_S	25	9.0427×10^{-8}	6.7826×10^{-7}	7169;1965,802,7158,1966,5411
n_D	20 25	1.50112 1.49792	1.50112 1.49792	274;6121,4089,7081,978,4381 1147,6398,2489,2492,601 1057,1568,1892,2494,2729 4414,7494,8131
dn/dt	20	0.0006337	0.0006337	6121
η	20 25 30 40	0.6487 0.6028 0.5621 0.4923	0.0006487 0.0006028 0.0005621 0.0004923	275;3634,5089,1984,464,2201 2125,2677,4570,5996,7494 4987,2700,3317
	Table 2.1 Table 2.2			
γ	20 25 30	28.85 28.20 27.56	0.02885 0.02820 0.02756	278;5495,4381,5339,6271,978 4195,6465,46,5370,28 1165,2038,3283,7848,7851 7864,7794
	Table 2.7			
ΔH_v	25 bp	8.0887 7.3437	33.843 30.726	7512;7110,280,2439,4969,5613 7592,2355,3759,1046
ΔH_s		10.653	44.572	5130
ΔH_m		2.358	9.866	280;5578,6856,7167
ΔH_f°	25(1) 25(g)	11.718 19.81	49.028 82.89	282;2857,7203,8316,2036 2036

38. Benzene

(Continued)

		cgs	SI	
ΔH°_c	25(1)	-780.97	-3267.58	2851;281,2857,7446,3975,6156 8316,2036
	25(g)	-789.06	-3301.43	2036
C°_p	25	19.52	81.67	273,1088,7203;7512
	Table 2.14			
C_p	25	32.447	135.760	2489;7391,7738,7178,2977,6377 60,5578,1568,3692,4153
	Table 2.13			4326,5817,6439
t_c		289.01	562.16	3084,7390;4299,1328,4089,923 5153,8233,1671,2878,1631
p_c		48.34	4.898	7390,4299;1328,4089,105,5153 4018,125,956,2878
d_c		0.3044	304.4	3084,7390;4299,4018,2878,1046
v_c		0.2566	0.0002566	3084,7390;1671,761,1449,8233
A	eq 2.72	0.01528	0.01528	280;4249
K_f	eq 2.67	5.12	5.12	5704;734,2385
K_B	eq 2.79	2.53	2.53	3406,726,2798;1370,5060 7845,730
$pK_{BH}{}^+$	0 in HF	9.4	9.4	4814
κ	25	4.43×10^{-17}	4.43×10^{-15}	4902;2484
ε	25	2.27401	2.27401	7019;5758,5411,2438,1983,3911 5062,1129,2515,4819,3624 5050,3199,3718
μ	20-60(1)	0	0	5798;3463,4703,8084,106,107
δ	25	9.166	18.74	4024,6240;3759,3421,1271,6723
ER	BuOAc 90%	5.1 90	5.1 90	6631;6905 6631,6634;6905
soly	in aq,25	0.1791%w	0.1791%w	5006;4733,5725,1593,794,946 5496,3320,591,896,6556
	aq in,25	0.0635%w	0.0635%w	8111;7186,3781,1532,794,906
	in aq,(%w)=0.1806-0.001095t+3.179 x $10^{-5}t^2$ 440			
aq az	69.25	91.17%w	91.17%w	3497
fl pt	TCC	-11	262	5712
Beil	463,179			
uv	155,827,1337,6475,6209,3760,5497,6686,6749			
ir	153,1521,1660,3088,633,5849,5479,5932			
Raman	2886,6046,6049,7719,6599,1783,5480			
ms	154,3728,6199,6887,5223			
nmr	811,3435,7021,1022,2464,5933,6748			

39. Toluene

Methylbenzene

$C_6H_5CH_3$ 108–88–3 C_7H_8

		cgs	SI	
mw		92.140	92.140	3468
bp	1 atm	110.630	383.780	279;2494,2729,7470,7499 8091,8304
dt/dp	1 atm	0.04630	0.3473	274;2494,7470,8091
dp/dt	1 atm	21.60	2.880	274
p	25	28.529	3.8036	833;280,4242,6462,6485,8091 2612,5332
eq 2.5	A B C	6.96050 1348.77 219.976	6.08540 1348.77 219.976	279;8091,4011
fp		-94.991	178.159	280;2494,2729,2780,6917 7059,7480
d	10	0.8762	876.2	276;563,879,1265,2038,2494 2729,4951
	20 25 30	0.86683 0.86219 0.85754	866.83 862.19 857.54	3084;276,4381,7470,7494,7499 7522,7758,2700,3900,6095 3084;5370,276,563,3900
dd/dt	25	0.000929	0.929	3084
α	25	0.001067	0.001067	6096;4902,1965
κ_T	25	1.215×10^{-7}	9.115×10^{-7}	53;1987,6151,1965,6096,3866 6487,1880
κ_S	25	9.28×10^{-8}	6.96×10^{-7}	6096;1966,1965,5411
n_D	20 25	1.49693 1.49413	1.49693 1.49413	274;4381,7081,2492,1057,1117 2494,2729,7494,7758,8081
dn/dt	25	0.000560	0.000560	274
η	20 25 35	0.5859 0.5525 0.4928	0.0005859 0.0005525 0.0004928	3062;275,2025,4570,5996,3352 563,2700,6664,464 3062;1984,3642
	Table 2.1 Table 2.2			
γ	20 25 30	28.52 27.92 27.33	0.02852 0.02792 0.02733	278;2038,5495,4195,4381,3283 7522,7758,7864,7794 278;46,5376,6762
ΔH_v	25 bp	9.080 7.931	37.990 33.183	280;4969,5613,7848,3759,1046
ΔH_m		1.586	6.636	280;6856,6917,7059
ΔH_f°	25(1)	2.867	11.996	282;2857,7203,8316
ΔH_c°	25(1)	-934.53	-3910.07	2851;281,7446,2268,6156,8316
c_p°	25	24.80	103.76	273;7203
	Table 2.14			
c_p	25	37.59	157.29	7391;4011,2977,6096,2489,8081 4326,6377,6439,6440 6917,8080
	Table 2.13			

39. Toluene

(Continued)

		cgs	SI	
λ	25	0.0003083	0.1296	3847;1855
t_c		318.65	591.80	7390,3084;4299,105,923,1671 119,956,5970
p_c		40.52	4.106	7390;105,956,5970,8235,1046
d_c		0.2913	291.3	7390;3084,4902,1046
v_c		0.3163	0.0003163	7390;1671
A	eq 2.72	0.02515	0.02515	280
K_B	eq 2.79	3.29	3.29	3406;5595
pK_{BH}^+	0 in HF	6.3	6.3	4814
κ	25	8.0×10^{-16}	8.0×10^{-14}	2485;1998,3965
ε	25	2.3807	2.3807	5240;1983,6760,5411,4937,4251 7850,107,5798,2100,6791
μ	20-60(1)	0.31	0.31	5798;7803,3381,4251,6411,8076 7461,107
δ	25	8.91	18.23	4024,3759;6240,8111,652,8112
ER	BuOAc 90%	1.90 236	1.90 236	6628,1399,2190;3455,6631 6628;1399,6631
soly	in aq,25	0.0515%w	0.0515%w	4733;946,3320,1630,2601,2986 896,7219,6372
	aq in,25	0.0334%w	0.0334%w	8111;7186,3759,2268,2782,3734
aq az	85	79.8%w	79.8%w	3498;3497
fl pt	TCC	4	277	5712

Beil	466,280
uv	155,827,1054,4040,6475,3760,549,6749
ir	153,5479,3888,633,4283,8085,8099,5932
Raman	2886,3825,4101,6049,8122,5797,8099
ms	154,3728,6199,6887,7199,99
nmr	1396,1429,7021,5319,4852,1022,2464,5933,6748

40. o-Xylene

1,2-Dimethylbenzene

$o\text{-}C_6H_4(CH_3)_2$ 95–47–6 C_8H_{10}

		cgs	SI	
mw		106.167	106.167	3468
bp	1 atm	144.429	417.579	279;1257,2494,2729,8091 8132,2776
dt/dp	1 atm	0.04969	0.3727	274;2494,8091,8160
dp/dt	1 atm	20.12	2.683	274
p	25	6.6	0.88	280;3874,4611,7058,2776,8160
eq 2.5	A	7.00582	6.13072	279;3874,7057,2220,2776,4865
	B	1479.82	1479.82	
	C	214.315	214.315	
fp		−25.182	247.968	280;1256,2474,2729,2780 7113,6275
d	20	0.88014	880.14	3084;276,4381,2489,6095,1254
	25	0.87594	875.94	2038,2494,2729,4951,5144 2700,6791
	30	0.87174	871.74	3084;276,6074,6054
dd/dt	25	0.000840	0.840	3084
α	25	0.000952	0.000952	6096
κ_T	25	1.081×10^{-7}	8.105×10^{-7}	6096;6151,3866,6487
κ_S	25	8.48×10^{-8}	6.36×10^{-7}	6096;6054
n_D	20	1.50545	1.50545	274;2492,1258,2494,2729,2489
	25	1.50295	1.50295	
dn/dt	8 – 25	0.00054941	0.00054941	2220
η	20	0.809	0.000809	275;1255,2700,6095
	25	0.756	0.000756	
	Table 2.1			
	Table 2.2			
γ	20	30.04	0.03004	278;2038,4381
	25	29.49	0.02949	
ΔH_v	25	10.381	43.434	280,4670;5613,4969,1155,1046
	bp	8.80	36.82	
ΔH_m		3.250	13.598	280;3558,6275
ΔH_f°	25(1)	−5.841	−24.439	282;7203,8316
ΔH_c°	25(1)	−1088.16	−4552.86	281,8316,2851;6156,3975
C_p°	25	31.85	133.26	283,7203;5211
	Table 2.14			
C_p	25	44.95	188.07	7391;2489,6096,8080,3558,4326
t_c		357.18	630.33	3084,7390;105,2220,956,125 2776,1046
P_c		36.85	3.734	7390;956,125,2776,1046
d_c		0.2877	287.7	3084;7390,1046,4299
v_c		0.369	0.000369	7390,4299

40.   *o*-Xylene

(Continued)

		cgs	SI	
A	eq 2.72	0.02660	0.02660	280
κ	25	6.7 x 10^{-16}	6.7 x 10^{-14}	2485
ε	20	2.568	2.568	4937;1983,107,3177,2100,6791
μ	20-60(1)	0.45	0.45	5798;3593,8076,106,107 4490,7461
δ		8.8	18.0	652;4593
soly	in aq,25	0.0175%w	0.0175%w	4733;6372,7219,7092,8062
	aq in,25	0.043%w	0.043%w	3458
aq az	93.5	50.1%w	50.1%w	6165;3498
fl pt	TCC	17	290	5712
Beil	467,362			
uv	155,582,5276,6475,4388,3760,2643,6749			
ir	153,627,630,7077,2643,3639,3887,2932,5932			
Raman	2400,3336,6599,1772,7721,1782,1789			
ms	154,3728,6199,99			
nmr	7021,6362,6091,2730,5933,6748			

41.   *m*-Xylene

1,3-Dimethylbenzene

m-C$_6$H$_4$(CH$_3$)$_2$ 108-38-3   C$_8$H$_{10}$

		cgs	SI	
mw		106.167	106.167	3468
bp	1 atm	139.120	412.270	279;2494,7488,8010,8091 8132,2776
dt/dp	1 atm	0.04903	0.3678	274;2673,8091,8160
dp/dt	1 atm	20.40	2.719	274
p	25	8.3	1.1	280;3874,4611,7058,2776,8160

41. *m*-Xylene

(Continued)

		cgs	SI	
eq 2.5	A	7.01295	6.13785	279;3874,7057,2220,2776,4864
	B	1465.39	1465.39	
	C	215.512	215.512	
t_p		−47.872	225.278	280;2494,2780,7113,7484 8010,7479
d	20	0.86436	864.36	3084;276,2489,6095,2038,2494
	25	0.86009	860.09	4951,7488,8010,2700
	30	0.85581	855.81	3084;7156,6074,6054
dd/dt	25	0.000855	0.855	3084
α	25	0.000981	0.000981	6096
κ_T	25	1.149×10^{-7}	8.621×10^{-7}	6096;6151,3866,6487
κ_S	25	8.89×10^{-8}	6.67×10^{-7}	6096;6054,7156
n_D	20	1.49722	1.49722	274,2492;2489,2220,2492
	25	1.49464	1.49464	8010,8080
dn/dt	5 − 25	0.00051262	0.00051262	2220
η	20	0.617	0.000617	275;7488,2700,6095
	25	0.581	0.000581	
	Table 2.2			
γ	20	28.66	0.02866	278;2038,3283,7488,7864,28
	25	28.10	0.02810	
ΔH_v	25	10.195	42.656	280;4969,5613,7848,4670,1046
	bp	8.69	36.36	
ΔH_m		2.765	11.569	280;3558,6275
ΔH_f°	25(1)	−6.075	−25.418	282;7203,8316
ΔH_c°	25(1)	−1087.92	−4551.86	281,8316,2851;6155,3975
C_p°	25	30.49	127.57	283,7203
	Table 2.14			
C_p	25	43.84	183.44	7391;2489,6377,8080,3558,4326
	Table 2.13			6439,6440
t_c		343.90	617.05	7390,3084;105,2220,7848,956 125,2776
P_c		34.90	3.536	7390;4299,105,2220,125,2776 956,1046
d_c		0.2822	282.2	3084;7390,4299,1046
v_c		0.376	0.000376	7390,4299
A	eq 2.72	0.02742	0.02742	280
$pK_{BH}{}^+$	0 in HF	3.2	3.2	4814
κ	25	8.6×10^{-16}	8.6×10^{-14}	2485
ε	20	2.3742	2.3742	1983;4937,107,3177,2100,6074
μ	20−60(1)	0.30	0.30	5798;6411,7461,106,107,4490

41. *m*-Xylene

(Continued)

		cgs	SI	
δ		8.8	18.0	1271
soly	in aq,25	0.0146%w	0.0146%w	7092;959,7219,6372,3320,5974
	aq in,20	0.0402%w	0.0402%w	2268;3458
aq az	92.5	54.5%w	54.5%w	6165,3498,3497
fl pt	TCC	25	298	5712

Beil	<u>467</u>,370
uv	155,582,5276,6475,3760,2643,7172,6749
ir	153,7537,7011,3639,8085,3890,8099,2931,5932
Raman	2400,1772,8099,7721,1782,1789,1790
ms	154
nmr	7021,6091,2730,1022,6748,5933

42. *p*-Xylene

1,4-Dimethylbenzene

p-$C_6H_4(CH_3)_2$ 106–42–3 C_8H_{10}

		cgs	SI	
mw		106.167	106.167	3468
bp	1 atm	138.359	411.509	279;5609,2729,7494,8010,8091 8132,2776
dt/dp	1 atm	0.04917	0.3688	274;2494,7494,8091,8160
dp/dt	1 atm	20.34	2.711	274
p	25	8.7	1.2	280;3874,4611,7058,2776,8160
eq 2.5	A	6.98650	6.11140	279;3874,7057,2220,2776,5584
	B	1451.39	1451.39	5609,4865
	C	215.148	215.148	
fp		13.263	286.413	280;2729,2780,3558,7113,7494
d	20	0.86098	860.98	3684;276,2494,2038,2729,4951
	25	0.85661	856.61	7494,4987,2700,6791,6096
	30	0.85225	852.25	6095,2489,6054
dd/dt	25	0.000873	0.873	<u>3084</u>

42. *p*-Xylene

(Continued)

		cgs	SI	
α	25	0.000956	0.000956	6096
κ_T	25	1.145×10^{-7}	8.588×10^{-7}	6096;6151,3866,6487
κ_S	25	8.97×10^{-8}	6.73×10^{-7}	6096;7156,5411,6054
n_D	20	1.49582	1.49582	274,2492;2494,2729,7494,8010
	25	1.49325	1.49325	8080,2489,1414
dn/dt	6 - 25	0.00052667	0.00052667	2220
η	20	0.644	0.000644	275;7494,2700,4987
	25	0.605	0.000605	275;6095
	30	0.566	0.000566	5410,5412
	Table 2.2			
γ	20	28.31	0.02831	278,2038;28,3283,5369
	25	27.76	0.02776	
ΔH_v	25	10.128	42.376	280,4670;5613,1155,4969,1046
	bp	8.60	35.98	5584,3506
ΔH_m		4.090	17.113	280;1668,3558,7167,5705
ΔH_f°	25(1)	-5.838	-24.426	282;7203,8316
	25(g)	4.30	18.0	2076
ΔH_c°	25(1)	-1088.16	-4552.86	281,2851;3975,6156,8316
C_p°	25	30.02	125.6	2076;283,7203,3506
	Table 2.14			
C_p	25	43.42	181.66	7391;6096,2489,5627,8080,2137
	Table 2.13			4326,6439,6440
t_c		343.08	616.23	7390,3084;105,2220,956,125
				2776,1046
P_c		34.65	3.511	7390,4299;105,2220,956,125
				2776,1046
d_c		0.2807	280.7	3084;7390,1046
v_c		0.378	0.000378	7390
A	eq 2.72	0.02509	0.02509	280
K_f	eq 2.67	4.3	4.3	734;5705
pK_{BH}^+	0 in HF	5.7	5.7	4814
κ	25	7.6×10^{-16}	7.6×10^{-14}	2485
ε	20	2.2699	2.2699	5798,4937;1983,2438,5758,5251
				7497,167,6195,2100
μ	20-60(1)	0.02	0.02	5798;7660,2363,3593,6411,8078
				4490,1311
δ	25	8.77	17.94	4024;6240,4593
soly	in aq,25	0.0156%w	0.0156%w	7092;6372,8062,946,3320,5974
	aq in,25	0.0456%w	0.0456%w	4024;3458
aq az	92.6	54.9%w	54.9%w	6165
fl pt	TCC	25	298	5712

42.   *p*-Xylene

(Continued)

Beil	<u>467</u>,382
uv	155,582,5276,6475,3760,2643,7172,6749
ir	153,626,5670,7011,8085,3639,3889,2930,5932
Raman	2400,5720,1772,7721,1782,1789,1790,2930
ms	154,99
nmr	7021,6091,2730,1795,5933,6748

43.   **Ethylbenzene**

$C_6H_5CH_2CH_3$ 100–41–4 C_8H_{10}

		cgs	SI	
mw		106.167	106.167	3468
bp	1 atm	136.193	409.343	5610;279,2494,2729,4969 7494,8091
dt/dp	1 atm	0.04898	0.3674	274;2494,7494,8091,8160
dp/dt	1 atm	20.42	2.722	274
p	25	9.6	1.3	280;3076,6533,7058,8091,8160
eq 2.5	A	6.96790	6.09280	279;7057,2220,6533,8091
	B	1431.71	1431.71	58,4011
	C	214.099	214.099	
fp		−94.975	178.175	280;2494,2729,2780
d	20	0.86696	866.96	3084;276,887,2038,2494,2677
	25	0.86253	862.53	2729,4951,7494,7761,2700 2489,4376
dd/dt	25	0.000881	0.881	<u>3084</u>
κ_T	25	1.15×10^{-7}	8.65×10^{-7}	4376;6151
n_D	20	1.49588	1.49588	274,2492;2494,2729,4969,7494
	25	1.49320	1.49320	7761,8080,2489
dn/dt	7 – 25	0.00051701	0.00051701	2220

43. Ethylbenzene

(Continued)

		cgs	SI	
η	20	0.6783	0.0006783	275;2125,2677,7494,2700
	25	0.6373	0.0006373	
	Table 2.1			
	Table 2.2			
γ	20	29.05	0.02905	278;2038,7761,3283
	25	28.48	0.02848	
ΔH_v	25	10.098	42.25	280,3506,7107;4969,5613,6533
				7848,4670,3759,7754,4164
	bp	8.413	35.20	7107;280
ΔH_m		2.195	9.184	280;3558,6533
ΔH_f°	25(1)	-2.977	-12.456	282;7203,8316
ΔH_c°	25(1)	-1091.03	-4564.87	281,2851;3975,6156,3716,8316
C_p°	25	30.69	128.41	273,7203;3506,5134
C_p	25	44.353	185.572	2489;2490,7391,8080,3558,4153
	Table 2.13			6439,6440,6533,6534
t_c		344.00	617.15	3084;7390,105,923,119,956,5398
P_c		35.59	3.606	7390;105,956,8235,5398
d_c		0.2835	283.5	3084,7390;1046
v_c		0.3745	0.0003745	7390
A	eq 2.72	0.03479	0.03479	280
ε	20	2.4042	2.4042	5798;4937,6159,107
μ	20-60(1)	0.37	0.37	5798;7660,107,106,1761,7803
δ	25	8.79	17.98	3759;898,3395,6151,4593
ER	BuOAc	0.89	0.89	778
soly	in aq,25	0.0152%w	0.0152%w	4733;4902,946,3320,4061,2601
				7219,6372
	aq in,25	0.043%w	0.043%w	3759,3497
aq az	92	67%w	67%w	3498;3497
fl pt	TCC	18	291	5712

Beil	467,351
uv	155,6230,6475,2724,4986,7640,7172,6749
ir	153,3942,7077,3888,2929,3639,8085,5932
Raman	2400,1783,3973,3971,2135
ms	154,3728,6199,99
nmr	3805,102,1429,4852,1022,5933,6748

44. Isopropylbenzene

1-Methylethylbenzene

$C_6H_5CH(CH_3)_2$ 98–82–8 C_9H_{12}

		cgs	SI	
mw		120.194	120.194	3468
bp	1 atm	152.411	425.561	279;2494,2729,3601,4090 8091,2776
dt/dp	1 atm	0.05074	0.3806	274;7476,2494,8091,8160
dp/dt	1 atm	19.71	2.628	274
p	25	4.6	0.61	280;4611,8091,8160,2612
eq 2.5	A	6.94098	6.06588	279;4090,8091
	B	1464.17	1464.17	
	C	208.207	208.207	
fp		-96.033	177.117	280;7476,2494,2729,2780 3426,2783
d	20	0.86168	861.68	3084;276,2494,2038,2729,3426
	25	0.85743	857.43	3601,7761,2700,1082
dd/dt	25	0.000853	0.853	3084
κ_T	25	1.19×10^{-7}	8.93×10^{-7}	6151
n_D	20	1.49145	1.49145	274,2492;2494,2729,3426,3601
	25	1.48890	1.48890	6856,7761
dn/dt	5 - 26	0.0005186	0.0005186	2220
η	-195.7(gl)	1.7×10^{49}	1.7×10^{46}	4614;2700
	20	0.791	0.000791	275
	25	0.739	0.000739	
γ	20	28.20	0.02820	277,2038;7761,7794,3601
	25	27.68	0.02768	
ΔH_v	25	10.789	45.141	280;4090,5613,4670
	bp	8.97	37.53	
ΔH_m		1.861	7.786	280
ΔH_f°	25(1)	-9.848	-41.204	282;7203,8316
ΔH_c°	25(1)	-1246.52	-5215.44	281,8316;2851,6156,3975
C_p°	25	36.26	151.71	283,7203
C_p	15 - 30	47.54	198.91	4153;2220,4326
	Table 2.13			
t_c		357.98	631.13	3084;7390,105,2220,956 125,2776
p_c		31.67	3.209	7390;4299,105,956,125,2776
d_c		0.2790	279.0	3084;7390
v_c		0.429	0.000429	7390
A	eq 2.72	0.02985	0.02985	280
ε	20	2.3833	2.3833	5798;6159,107,4937,2100,6791

44. Isopropylbenzene

(Continued)

		cgs	SI	
μ	20–60(1)	0.39	0.39	5988;107,1761,7803
δ	25	8.61	17.6	653
ER	90%	1075	1075	6704
soly	in aq,25	0.00653%w	0.00653%w	619;4733,6372
	aq in,20	0.0303%w	0.0303%w	2268
aq az	95	56.2%	56.2%	3498
fl pt	TCC	39	312	5712;152
Beil	468,393			
uv	155,6209,4986,7172,6749			
ir	153,6260,633,5711,5801,3639,8085,5932			
Raman	2400,5480,79			
ms	154,6199			
nmr	1396,4852,1022,5933,6748			

45. Mesitylene

1,3,5-Trimethylbenzene

sym-$C_6H_3(CH_3)_3$ 108–67–8 C_9H_{12}

		cgs	SI	
mw		120.194	120.194	3468
bp	1 atm	164.743	437.893	279;2445,7057
dt/dp	1 atm	0.05100	0.3825	274
dp/dt	1 atm	19.61	2.614	274
p	25	2.5	0.33	280;4611,3874,3487
eq 2.5	A	7.08527	6.21017	279;3874,7057
	B	1577.80	1577.80	
	C	210.526	210.526	
fp		−44.720	228.430	280;6275,2780,3426,7036
d	20	0.86518	865.18	274;3426
	25	0.86111	861.11	274;8063
dd/dt	25	0.000814	0.814	274

45. Mesitylene

(Continued)

		cgs	SI	
n_D	20	1.49937	1.49937	274;2492,2445,3426,5758
	25	1.49684	1.49684	
dn/dt	25	0.000506	0.000506	274
η	20	1.154	0.001154	2087
	30	0.936	0.000936	
γ	20	28.84	0.02884	278
	25	28.32	0.02832	
	Table 2.7			
ΔH_v	25	11.348	47.480	280;5613,3235
	bp	9.33	39.04	
ΔH_m		2.274	9.514	280;6275
ΔH_f°	25(1)	−15.184	−63.530	282;7203,8316
ΔH_c°	25(1)	−1241.19	−5193.14	281,8316,2851;3975,6156,7446
C_p°	25	35.91	150.25	283,7203
	Table 2.14			
C_p	25	49.98	209.10	7391;8063
	Table 2.13			
t_c		364.10	637.25	7390;119,956,3914
P_c		30.86	3.127	7390;956,3914
d_c		0.28	280	7390
v_c		0.429	0.000429	7390
A	eq 2.72	0.02193	0.02193	280;7036
pK_{BH^+}	0 in HF	0.4	0.4	4814
κ	25	$<1 \times 10^{-16}$	$<1 \times 10^{-14}$	2485
ε	25	2.273	2.273	2438;107,6159,2974,5758
μ	20−30(1)	0	0	107;7461,4489,106,4490
δ	25	8.80	18.00	3395;4593,652
soly	in aq,25	0.00482%w	0.00482%w	7092;6372
	aq in,20	0.0291%w	0.0291%w	2268
aq az	96.5			3497

Beil	468,406
uv	6911,2724,6935,5881,6937,2394,3760,6749
ir	2252,623,624,4587,5889,8085,3639,2381,2944,5932
Raman	7717,2135,3339,2381,2944
ms	6199
nmr	2002,4436,7021,5933,6748

46. Naphthalene

91–20–3 $C_{10}H_8$

		cgs	SI	
mw		128.173	128.173	3468
bp	1 atm	217.942	491.092	294;1324,1273,4839
dt/dp	1 atm	0.05840	0.4380	291,1324
dp/dt	1 atm	17.12	2.283	291,1324
p	25(s)	0.0820	0.0109	6764
	80.29	7.48	0.997	2501 ;6540
eq 2.5	A	10.0896	9.2145	2501;4813,7058,6797,5141,1048
16 – 80	B	2926.61	2926.61	2747,6016,1597
	C	237.332	237.332	
80 –218	A	7.01065	6.13555	2501,1324,294;8296
	B	1733.71	1733.71	
	C	201.859	201.859	
fp		80.290	353.440	291;4749,1889,2513,4839
				7084,3322
tp		80.273	353.423	1576;2501,1921,128
d	80	0.9782*	978.2*	289;1273,4839,7848,8288,5062
				3282,3545,2893,291
	81.85	0.97672	976.72	3084;289
	96.85	0.96473	964.73	
dd/dt	90 – 210	0.0007865	0.7865	3029
n_D	85	1.5898	1.5898	291,4839
η	99.8	0.7802	0.0007802	6474,1273,3029
	Table 2.1			
γ	100	31.8	0.0318	290
	127.2	27.98	0.02798	7848,1273;290
	bp	18.69	0.01869	
ΔH_v	100	12.009	50.246	1921
	bp	10.32	43.18	7755;7754,5141,2355,643
				1709,4330
ΔH_s	25	17.331	72.513	1921;3635,1504,5243,1281,1597
				3872,6083,6928,5141,6016
				2747,8145,6797
ΔH_m	25	4.27	17.87	6928;4749,7906,4968,7167
				6918,4959
ΔH_f°	25(s)	18.77	78.53	296;2036,1586,6928,8316
	25(g)	35.6	149.0	2036
ΔH_c°	25(s)	-1232.54	-5156.95	295;2036,1586,3975,6928,8316
	25(g)	-1249.4	-5227.5	2036

46. Naphthalene

(Continued)

		cgs	SI	
C_p°	25	31.53	131.92	297;643,6764
	Table 2.14			
C_p	25(s)	39.60	165.69	4749;296,6764,643,3532,3558
	Table 2.12			
t_c		475.28	748.43	3084;293,119,1671,1449,956
				1451,8288,3128,3914
P_c		39.98	4.051	293;6496,956,3050,3914
d_c		0.3155	315.5	3084;293
v_c		0.408	0.000408	1449;1671,293
A	eq 2.72	0.01827	0.01827	4749;7635,4959
	B	0.00261	0.00261	
K_f	eq 2.67	6.94	6.94	2354,503;4968,719,734
K_B	eq 2.79	5.80	5.80	732;716,5595
$pK_{BH}{}^+$	0 in HF	4.0	4.0	4814
κ	81.8	4.35×10^{-10}	4.35×10^{-8}	6028
ε	80	2.54	2.54	4937;1394,3216,8084,5062
μ	25 in 38	0	0	8082;4688,8084
δ	25	9.9*	20.3*	652,4593;4787
soly	in aq,25	0.003169%w	0.003169%w	5006;4807,7933,2233,946,4061
aq az	98.8	16%w	16%w	3497
fl pt	TCC	80	353	5712
* supercooled liquid				

Beil	476,531
uv	155,827,370,2408,2379,6686,6749
ir	153,2631,6311,4385,428,633,8085,5932
Raman	336,5722,1069,2417,1782,1790,4172
ms	154,837,5915
nmr	811,812,3742,7021,5914,5933,6748

47. 1-Methylnaphthalene

90–12–0

$C_{11}H_{10}$

		cgs	SI	
mw		142.200	142.200	3468
bp	1 atm	244.685	517.835	1324,291;4839,4708,3415
dt/dp	1 atm	0.06047	0.4536	1324,291
dp/dt	1 atm	16.54	2.205	1324,291
p	25	0.0671	0.00895	2087
eq 2.5	A	7.03592	6.16082	1324,294
	B	1826.948	1826.948	
	C	195.002	195.002	
fp		−30.480	242.670	7035,291;4839,4708,3415
tp		−30.45	242.70	4749
d	20	1.02028	1020.28	291;3415,1324,458,4708,4839
	25	1.01676	1016.76	6039,3005
dd/dt	25	0.000727	0.727	1324
n_D	20	1.61755	1.61755	1324,291,3623;1862,458,4708
	25	1.61512	1.61512	3415,3605
dn/dt	25	0.000477	0.000477	3623
η	20	3.10	0.00310	6039;4708,3415
	40	1.97	0.00197	6039;3415
γ	20	39.80	0.03980	458;290,2087
	40	38.02	0.03802	
ΔH_v	25	14.357	60.070	2087
	bp	10.871	45.485	2087;8049
ΔH_m		1.660	6.945	4749
ΔH_t	−32.36	1.190	4.979	
ΔH_f°	25(1)	13.43	56.19	6928;1703
	25(g)	27.93	116.86	6928
ΔH_c°	25(1)	−1389.59	−5814.05	6928;3415,1703
C_p	23.11	53.429	223.55	4749
	Table 2.12			
t_c		499	772	113;293,4299,2087
p_c		35.22	3.568	2087;341
d_c		0.319	319	2087;341
v_c		0.446	0.000446	
A	eq 2.72	0.01418	0.01418	7035;4749
pK_{BH}^+	0 in HF	1.67	1.67	4814

47. 1-Methylnaphthalene

(Continued)

		cgs	SI	
ε	20	2.915	2.915	6039
μ	25 in 38	0.37	0.37	1751;4509,6342,4706
soly	in aq,25	0.00285%w	0.00285%w	4807;65
	aq in,20	0.0377%w	0.0377%w	2268
aq az	99.8	6%w	6%w	3499

Beil	478,566
uv	2571,4424,8137,4338,590,5123,4884
ir	4383,1402,2575,3549,5932,153
Raman	8292,2801,4704,1069
ms	5915
nmr	2261,2155,2606,5933

48. 1,2-Dimethylnaphthalene

573-98-8 $C_{12}H_{12}$

		cgs	SI	
mw		156.227	156.227	3468
bp	1 atm	266.3	539.5	291
dt/dp	1 atm	0.0606	0.455	291
dp/dt	1 atm	16.5	2.20	291
p	25	0.0127	0.00169	2087
eq 2.5	A	7.00605	6.13095	294;2087
	B	1805.31	1805.31	
	C	171.37	171.37	
fp		-1.0	272.2	291
d	20	1.0178	1017.8	291;458,505
	25	1.0143	1014.3	
dd/dt	25	0.00070	0.70	291
n_D	20	1.6165	1.6165	291;5008,6498,505,458
	25	1.6143	1.6143	
dn/dt	25	0.00044	0.00044	291
γ	20	39.2	0.0392	265;458
	25	38.7	0.0387	

48. 1,2-Dimethylnaphthalene

(Continued)

		cgs	SI	
ΔH_v	25 bp	16.74 11.70	70.05 48.95	2087
λ		0.000320	0.134	1096
t_c		511	784	2087
P_c		32.28	3.271	2087
d_c		0.330	330	2087
v_c		0.473	0.000473	2087
ε	25	2.613	2.613	2087
μ		0.68	0.68	4706

Beil	478a,EI 267
uv	2571,2691,155
ir	7655,2575,7653,3549,153
Raman	4707
ms	5915
nmr	8215

49. 1,6-Dimethylnaphthalene

575–43–9 $C_{12}H_{12}$

		cgs	SI	
mw		156.227	156.227	3468
bp	1 atm	265.6	538.8	2327;291,2087
dt/dp	1 atm	0.0607	0.455	2087;291
dp/dt	1 atm	16.5	2.20	2087
p	25	0.0146	0.00195	2087
eq 2.5	A B C	7.0478 1846 180	6.1727 1846 180	2087

49.   1,6-Dimethylnaphthalene

(Continued)

		cgs	SI	
fp		−16.9	256.3	2327;291,2087
d	20	1.0019	1001.9	291;2327,458,505,7969
	25	0.9984	998.4	
dd/dt	25	0.00070	0.70	291
n_D	20	1.6072	1.6072	291,2327;458,505
	25	1.6050	1.6050	291
dn/dt	25	0.00044	0.00044	291
γ	20	37.42	0.03742	291;2087
	40	35.40	0.03540	
ΔH_v	25	16.64	69.61	2087
	bp	11.66	48.78	
λ		0.000320	0.134	1096
t_c		505	778	2087
p_c		31.80	3.222	2087
d_c		0.330	330	2087
v_c		0.473	0.000473	2087
ε	25	2.583	2.583	2087
μ		0.32	0.32	4706

Beil	478a,EI 268
uv	2327,3264,1425,2691,2571,155
ir	4383,2327,7653,3549,2575,7653,153
Raman	1069,4707
ms	5915

50.   1,2,3,4-Tetrahydronaphthalene

119-64-2 $C_{10}H_{12}$

		cgs	SI	
mw		132.205	132.205	3468
bp	1 atm	207.65	480.80	292;4840,2659,3348,7058,3120
dt/dp	1 atm	0.0575	0.431	292
dp/dt	1 atm	17.4	2.32	292
p	25	0.40	0.053	4611;2659,3348
eq 2.13	A	11.954	11.079	164;5405
	B	2797.9	2797.9	
	C	1.187	1.187	
fp		-35.749	237.401	292;3435,4840,4749,3348,7479
d	20	0.9695	969.5	292;4840,2031,3348,3120
	25	0.9660	966.0	
dd/dt	25	0.000700	0.700	292
n_D	20	1.54135	1.54135	292;4840,2659,3120
	25	1.53919	1.53919	
dn/dt	25	0.000432	0.000432	292
η	20	2.349	0.002349	2151;2031,7404
	25	2.14	0.00214	2151;3348
	Table 2.1			
γ	21.5	35.46	0.03546	2031;3348,3547,59
	33.0	32.44	0.03244	3547;59
ΔH_v	25	13.20	55.23	164
	bp	10.48	43.85	3348
ΔH_m		2.975	12.447	4749
ΔH_f°	25(1)	-6.83	-28.58	2854,7393;164
	25(g)	5.29	22.13	164
ΔH_c°	25(1)	-1343.57	-5621.50	2854,7392;164,3975
	25(g)	-1355.81	-5672.71	164
C_p°	25	36.41	152.34	7394
	Table 2.14			
C_p	25	51.97	217.46	7393;4749,3348
	Table 2.12			
λ		0.00042	0.18	2151
t_c		481	754	2151
P_c		37.4	3.79	2151
d_c		0.3097	309.7	2151;2087

50.   1,2,3,4-Tetrahydronaphthalene

(Continued)

		cgs	SI	
v_c		0.4269	0.0004269	2151
A	eq 2.72	0.02657	0.02657	4749
	B	0.00248	0.00248	
K_B	eq 2.79	5.582	5.582	3348
ε	20	2.773	2.773	107
μ	20 - 30(1)	0.60	0.60	107
δ		9.5	19.4	652,1271,4593
aq az	99.4	19.6%w	19.6%w	6165
fl pt	TCC	78	351	4902
Beil	473,491			
uv	155,3766,595,5275			
ir	153,6311,6696,4383,8085,1886,5932			
Raman	5303,4991,1886			
ms	154,3922,8109			
nmr	3805,811,7080,3435,5933			

51.   Butylbenzene

$C_6H_5C_4H_9$ 104-51-8 $C_{10}H_{14}$

		cgs	SI	
mw		134.221	134.221	3468
bp	1 atm	183.270	456.420	288;3560,6734,7058,7495
dt/dp	1 atm	0.05358	0.4019	284
dp/dt	1 atm	18.66	2.488	284
p	25	1.1	0.15	288
eq 2.5	A	6.98317	6.10807	287;5071
	B	1577.965	1577.965	
	C	201.378	201.378	
fp	(st)	-87.85	185.30	5071,288;2780,5822,7058
				7495,3560
	(ms)	-88.01	185.14	5071
d	20	0.86013	860.13	284,2038;2220,4951,7495,7758
	25	0.85607	856.07	

51. Butylbenzene

(Continued)

		cgs	SI	
dd/dt	0 - 40	0.0008130	0.8130	2220
n_D	20	1.48979	1.48979	284;2220,7495,7761
	25	1.48742	1.48742	
dn/dt	25	0.000474	0.000474	284
η	-195.7(gl)	3.5×10^{45}	3.5×10^{42}	4614;7404
	20	1.035	0.001035	270
	25	0.960	0.000960	
γ	20	29.23	0.02923	285;2038,3283,7761,7794
	30	28.22	0.02822	
ΔH_v	25	12.28	51.36	7107;5071
	bp	9.29	38.87	7107;288
ΔH_m	(st)	2.682	11.221	5071,288;3560
	(ms)	2.691	11.259	5071
ΔH_f°	25(1)	-15.16	-63.43	2850;8316,7203,272
ΔH_c°	25(1)	-1403.56	-5872.50	2850;271,8316,2851
C_p°	25	41.85	175.10	273;7203
C_p	27	58.357	244.166	5071;3560
	Table 2.12			
t_c		387.3	660.5	286,4299;1451,3011
p_c		28.49	2.887	286,4299
d_c		0.270	270	286,4299;7482
v_c		0.497	0.000497	286,4299
A	eq 2.72(st)	0.03931	0.03931	5071;288
B		0.00289	0.00289	
A	eq 2.72(ms)	0.03951	0.03951	5071
B		0.00264	0.00264	
ϵ	20	2.359	2.359	107
μ	20 - 30(1)	0.36	0.36	107;1761
δ	25	8.59	17.57	3759
soly	in aq,25	0.00118%w	0.00118%w	7092;7219,4061
	aq in,25	0.041%w	0.041%w	3759;2268
aq az	98.2	30%w	30%w	6165
fl pt	TOC	71	344	5712
Beil	469,413			
uv	155,2724,5881,6749			
ir	153,5892,5928,8085,633,2574,5932			
Raman	2400,79			
ms	154,6199,6294			
nmr	5933,6749			

52. *sec*-Butylbenzene

1-Methylpropylbenzene

$C_6H_5CH(CH_3)CH_2CH_3$ 135–98–8 $C_{10}H_{14}$

		cgs	SI	
mw		134.221	134.221	3468
bp	1 atm	173.305	446.455	288;7476,1241,2220,3601 6743,7058
dt/dp	1 atm	0.05313	0.3985	284
dp/dt	1 atm	18.82	2.509	284
p	25	1.7	0.23	288;4611,7058
eq 2.5	A	6.95097	6.07587	287
	B	1540.174	1540.174	
	C	205.101	205.101	
fp		−75.470	197.680	288;7476,1241,2780,7058
d	20	0.86207	862.07	284,2038;1241,1771,2220,3601
	25	0.85797	857.97	
dd/dt	25	0.000820	0.820	284
n_D	20	1.49020	1.49020	284,2492;1241,1771,3601,6743
	25	1.48779	1.48779	
dn/dt	16 – 25	0.0004846	0.0004846	2220
γ	20	28.52	0.02852	285;2038,3601,7794
	25	28.03	0.02803	
	30	27.54	0.02754	
ΔH_v	25	11.83	49.50	288
	bp	9.07	37.95	
ΔH_m		2.35	9.83	288
t_c		383.7	656.9	956
P_c		29.4	2.98	956
d_c		0.263	263	2087
v_c		0.511	0.000511	2087
$[\alpha]_D$	23	−17.9	−17.9	Beil. EIII5, 932
A	eq 2.72	0.0303	0.0303	288
ε	20	2.364	2.364	107
μ	20 – 30(1)	0.37	0.37	107;1761
soly	in aq,25	0.00176%w	0.00176%w	7092
	aq in,20	0.0317%w	0.0317%w	2268
fl pt	TCC	52	325	5712

Beil	469,414
uv	155,5881,6749
ir	153,5892,633,5928,8085,2574,5932
Raman	2400,79
ms	154,6294
nmr	5933,6748

53.   Isobutylbenzene

2-Methylpropylbenzene

$C_6H_5CH_2CH(CH_3)_2$ 538–93–2 $C_{10}H_{14}$

		cgs	SI	
mw		134.221	134.221	3468
bp	1 atm	172.759	445.909	2496,2497,284;5826,884
dt/dp	1 atm	0.05319	0.3990	2497,284
dp/dt	1 atm	18.80	2.507	2497,284
p	25	1.930	0.2573	2087;4611
eq 2.5	A	6.92804	6.05294	287;2496
	B	1525.446	1525.446	
	C	204.122	204.122	
fp		−51.48	221.67	7037,284;341,2780,884,5826
d	20	0.85321	853.21	2497,284;2038,4392,5826
	25	0.84907	849.07	
dd/dt	25	0.000809	0.809	2497
n_D	20	1.48646	1.48646	2497,284;884,3295,5826
	25	1.48400	1.48400	
dn/dt	25	0.000489	0.000489	2497
γ	20	27.47	0.02747	285,2038
	25	26.99	0.02699	285;7794
ΔH_v	25	11.82	49.45	288;5826,341
	bp	9.04	37.82	
ΔH_m		2.99	12.51	288
ΔH_f°	25(l)	−16.70	−69.87	5964;1703
	25(g)	−5.15	−21.55	
ΔH_c°	25(l)	−1402.04	−5866.14	5964;1703
C_p°	25	42	176	323
C_p	25	57.5	240.6	323
t_c		384.72	657.87	119;286,341,2087
p_c		31.10	3.151	5826;286,2087
d_c		0.274	274	2087
v_c		0.490	0.000490	2087
A	eq 2.72	0.0306	0.0306	7037,288
ε	20	2.319	2.319	107;2087,6457
μ		0.44	0.44	4392
fl pt	TOC	60	333	5923

Beil	469,414
uv	6956,155,6749
ir	4383,5889,8085,4384,1664,2574,153,5932
Raman	1775,81
ms	154,6294
nmr	5933,6748

54. *tert*-Butylbenzene

1,1-Dimethylethylbenzene

$C_6H_5C(CH_3)_3$ 98–06–6 $C_{10}H_{14}$

		cgs	SI	
mw		134.221	134.221	3468
bp	1 atm	169.119	442.269	288,1241,2220,3600,7058
dt/dp	1 atm	0.05269	0.3952	284
dp/dt	1 atm	18.98	2.530	284
p	25	2.1	0.28	288;4611,7058
eq 2.5	A	6.92050	6.04540	287
	B	1504.572	1504.572	
	C	203.328	203.328	
fp		-57.850	215.300	288;1241,2220,2780,7058
d	20	0.86650	866.50	284,2038;1241,3600,7150
	25	0.86240	862.40	
dd/dt	25	0.000820	0.820	284
n_D	20	1.49266	1.49266	284,2492;1241,3600,6743,7150
	25	1.49024	1.49024	
dn/dt	25	0.000484	0.000484	284
γ	20	28.13	0.02813	2038,285;3600,7794
	25	27.64	0.02764	285
	30	27.14	0.02714	2038,285
ΔH_v	25	11.73	49.08	288
	bp	8.99	37.61	
ΔH_m		2.006	8.393	288
ΔH_c°	25(1)	-1403.2	-5871.0	3975;6156
t_c		374.1	647.3	956
p_c		29.2	2.96	956
d_c		0.274	274	2087
v_c		0.490	0.000490	2087
A	eq 2.72	0.02173	0.02173	288
ε	20	2.366	2.366	107;6159,4492,4937
μ	20 - 30(1)	0.36	0.36	107;6231,106,1761,4492,7803
soly	in aq,25	0.00295%w	0.00295%w	7092
	aq in,20	0.0292%w	0.0292%w	2268
fl pt	TOC	60	333	5822

Beil	469,415
uv	155,4986,6749
ir	153,5892,633,5928,8085,2574,5932
Raman	2400,5649,640
ms	154,6294
nmr	4852,2730,1022,5933,6748

55. *p*-Cymene

1-Methyl-4-(1-methylethyl)benzene

p-CH$_3$C$_6$H$_4$CH(CH$_3$)$_2$ 99-87-6 C$_{10}$H$_{14}$

		cgs	SI	
mw		134.221	134.221	3468
bp	1 atm	177.10	450.25	288;2220,2400,3560,4090 7058,4760
dt/dp	1 atm	0.0528	0.396	284;8160
dp/dt	1 atm	18.9	2.53	284
p	25	1.5	0.20	288;4611,7058,8160
eq 2.5	A	6.9260	6.0509	287;4090,4760
	B	1538.00	1538.00	
	C	203.10	203.10	
fp		−67.935	205.215	280,7035;2729,3560,4760 7479,2220
d	20	0.8573	857.3	284;2220,2729,4492,7150
	25	0.8533	853.3	284
dd/dt	25	0.00080	0.80	
n$_D$	20	1.4909	1.4909	284;2220,2400,2729,4492,7150
	25	1.4885	1.4885	
dn/dt	7 − 25	0.0004830	0.0004830	2220
η	20	3.402	0.003402	7483;616,2201,2087
	30	1.600	0.001600	
γ	20	28.81	0.02881	2087;7794,160
	30	27.74	0.02774	
ΔH$_v$	25	12.02	50.29	288;4090,7848,1155
	bp	9.12	38.16	
ΔH$_m$		2.309	9.661	288,3560
ΔH$_f^\circ$	25(l)	−18.65	−78.03	2850
ΔH$_c^\circ$	25(l)	−1400.07	−5857.89	2850
C$_p$	−23.9	56.50	236.40	288;6439
	Table 2.13			
t$_c$		378.6	651.8	105;2220,956,286
P$_c$		28.6	2.90	105;956,286
d$_c$		0.266	266	2087
v$_c$		0.505	0.000505	2087
A	eq 2.72	0.02759	0.02759	288;7035
K$_B$	eq 2.79	5.52	5.52	726
κ	25	2 x 10^{-8}	2 x 10^{-6}	5708
ε	20	2.253	2.253	107;4492,4937
μ	20 − 30(l)	0	0	107;4492,8099
soly	in aq,25	0.00234%w	0.00234%w	591
fl pt	TCC	47	320	5712

55.   *p*-Cymene

(Continued)

Beil	469,420
uv	155,6749
ir	153,4384,8085,6665,5181,5269,5932
Raman	2400,2135
ms	154
nmr	6748

56.   Cyclohexylbenzene

827-52-1 $C_{12}H_{16}$

		cgs	SI	
mw		160.258	160.258	3468
bp	1 atm	240.12	513.27	5017;934,1374
dt/dp	1 atm	0.0525	0.394	5017
dp/dt	1 atm	19.0	2.54	5017
p	12.4	0.017	0.0023	4611
fp		7.35	280.50	5597;5017,6739,912,7040,1374
d	20	0.94272	942.72	5017;934,2312,1374
	25	0.93874	938.74	
dd/dt	25	0.000796	0.796	5017
n_D	20	1.52633	1.52633	5017;934,2312,1374
	25	1.52393	1.52393	
dn/dt	25	0.000480	0.000480	5017
η	0	3.681	0.003681	5393
ΔH_v	25	14.33	59.94	5232
ΔH_m		3.66	15.3	5597
ΔH_f°	25(1)	-18.92	-79.18	5232;2854

56. Cyclohexylbenzene

(Continued)

		cgs	SI	
ΔH_c°	25(1)	−1656.21	−6929.58	5232;2854
C_p	21.85	62.45	261.3	5597
	Table 2.12			
A	eq 2.72	0.0295	0.0295	7040
κ		8.85 x 10^{-10}	8.85 x 10^{-8}	4902
fl pt	TOC	99	372	4902
Beil	473,503			
uv	2643			
ir	153,2656,5928,6084,5932			
Raman	4991,594			
ms	74			
nmr	5933,6748			

57. 1-Pentene

$CH_3CH_2CH_2CH{=}CH_2$ 109-67-1 C_5H_{10}

		cgs	SI	
mw		70.134	70.134	3468
bp	1 atm	29.962	303.112	254;2493,1877,2701,3919,8067
dt/dp	1 atm	0.03801	0.2851	249,2493
dp/dt	1 atm	26.31	3.508	
p	25	638	85.1	255;1877,7058,2493
eq 2.5	A	6.84424	5.96914	254;1877,2493
	B	1044.01	1044.01	
	C	233.49	233.49	
fp		−165.219	107.931	7387;1416,7042,7036,4687,7513
d	20	0.64050	640.50	249,2493;2701,3919,1014,8067
	25	0.63533	635.33	8038,1368,6653,4033
dd/dt	22.5	0.001034	1.034	2493,1877

57. 1-Pentene

(Continued)

		cgs	SI	
n_D	20	1.37148	1.37148	249,2493;2701,3919,1014,8067
	25	1.36835	1.36835	8038,108,1368,4033
dn/dt	25	0.000626	0.000626	249
η	-5	0.25	0.00025	250
	0	0.24	0.00024	
γ	0	18.20	0.01820	252
	10	17.10	0.01710	
	20	16.00	0.01600	
	25	15.45	0.01545	
ΔH_v	25	6.088	25.47	255;1877
	bp	6.022	25.20	7387
ΔH_m		1.388	5.807	1416;7387,7513
ΔH_f°	25(1)	-11.23	-46.97	7389,2858;8316
	25(g)	-5.086	-21.28	7389
ΔH_c°	25(1)	-800.60	-3349.72	2858,7388;8316,7134
	25(g)	-806.7	-3375.4	7388
C_p°	25	25.94	108.52	7389;4001
C_p	25	37.69	157.69	7389;1416,7513
t_c		191.63	464.78	258;5300,4044,119
P_c		34.83	3.529	258;5300,4044
d_c		0.237	237	258;2088
v_c		0.296	0.000296	258
A	eq 2.72	0.05996	0.05996	255;7042
ϵ	20	2.017	2.017	108;6651
μ	20(1)	0.34	0.34	108;2654,7660
soly	in aq,25	0.0148%w	0.0148%w	4733
fl pt	estm	-51	222	5822
Beil	11,210			
uv	1366,1368,6653,2679,3760			
ir	153,6652,7439,6645,5445			
Raman	2400,6049,2119,6683,6084			
ms	154,6199,2119,714			

58.   2-Pentene

(Mixed isomers)

CH₃CH₂CH=CHCH₃   109–68–2   C₅H₁₀

		cgs	SI	
mw		70.134	70.134	3468
bp	1 atm	36.7	309.9	5910;2757,3699,5679,1368
				1014,8169
dt/dp	1 atm	0.056	0.42	3779
dp/dt	1 atm	18	2.4	3779
p	25	528	70.4	3779
eq 2.4	A	7.7945	6.9194	3779
–80 to 37	B	1512.5	1512.5	
fp		–138	135	2757
d	15	0.6555	655.5	2757;3699,5513,1014
	20	0.6545	654.5	5910
n_D	20	1.3798	1.3798	8169;3699,5910,1368,1014
	25	1.3839	1.3839	2757
η	20	0.214	0.000214	8169
	25	0.201	0.000201	
γ	14.8	17.68	0.01768	3699
	23.3	16.79	0.01679	
ΔH_v	25	8.42	35.23	3779
C_p	15.9	36.1	151.0	5679
t_c		202.4	475.6	4088
P_c		40.4	4.09	4088,758
soly	in aq,25	0.0203%w	0.0203%w	4733
fl pt	estm	–45	228	5822

Beil	11,210
uv	1368,1366
ir	6652,7439,5932
ms	6199,5133
nmr	6748

59. *cis*-2-Pentene

(Z)-2-Pentene

627–20–3 C_5H_{10}

		cgs	SI	
mw		70.134	70.134	3468
bp	1 atm	36.922	310.072	259
dt/dp	1 atm	0.03830	0.2873	256
dp/dt	1 atm	26.11	3.481	256
p	25	495	66.0	260
eq 2.5	A	6.84308	5.96798	259;5822
	B	1052.44	1052.44	
	C	228.693	228.693	
fp		−151.390	121.760	260;1416,7036,4687,7513
d	20	0.6556	655.6	256
	25	0.6504	650.4	
dd/dt	25	0.00104	1.04	256
n_D	20	1.3830	1.3830	256
	25	1.3798	1.3798	
dn/dt	25	0.00064	0.00064	256
γ	0	19.7	0.0197	257
	10	18.6	0.0186	
	20	17.38	0.01738	
	25	16.8	0.0168	
ΔH_v	25	6.41	26.82	260
	bp	6.24	26.11	
ΔH_m		1.6998	7.112	260,1416;7513
ΔH_f°	25(1)	−12.78	−53.49	2858;8316,262,5968
ΔH_c°	25(1)	−799.05	−3343.21	2858;261,8316,1650,5968
C_p°	25	24.32	101.75	263,4001
C_p	25	36.26	151.71	7513;1416
t_c		203	476	258,4299
P_c		36	3.6	258,4299
d_c		0.24	240	258
v_c		0.292	0.000292	258
A	eq 2.72	0.05770	0.05770	260
fl pt	estm	−45	228	5822
Beil	11,210			
uv	2679,3760			
ir	153,5932			
Raman	2400,6049,6645,6084			
ms	154			

60.   *trans*-2-Pentene

(*E*)-2-Pentene

CH₃ ＼   ＼ ／ H
　　　C＝C
H ／   ＼ CH₂CH₃   646-04-8   C₅H₁₀

		cgs	SI	
mw		70.134	70.134	3468
bp	1 atm	36.344	309.494	259
dt/dp	1 atm	0.03824	0.2868	256
dp/dt	1 atm	26.15	3.486	256
p	25	505	67.3	260;5822
eq 2.5	A	6.89983	6.02473	259
	B	1080.76	1080.76	
	C	232.567	232.567	
fp		-140.244	132.906	260;1416,7036,7513
d	20	0.6482	648.2	256
	25	0.6431	643.1	
dd/dt	25	0.00102	1.02	256
n_D	20	1.3793	1.3793	256
	25	1.3761	1.3761	
dn/dt	25	0.00064	0.00064	256
γ	0	18.9	0.0189	257
	10	17.9	0.0179	
	20	16.90	0.01690	
	25	16.42	0.01642	
ΔH_v	25	6.38	26.69	260
	bp	6.23	26.07	
ΔH_m		1.9961	8.352	260,1416;7513
ΔH_f°	25(1)	-13.86	-57.98	2858;8316,262
ΔH_c°	25(1)	-797.97	-3338.71	2858;261,8316,1650
C_p°	25	25.92	108.45	263;4001
C_p	25	37.52	156.98	7513;1416
t_c		202	475	258,4299
p_c		36	3.6	258,4299
d_c		0.24	240	258
v_c		0.292	0.000292	258
A	eq 2.72	0.05687	0.05687	260

Beil	11,210
uv	2679,3760
ir	153,8297,5445,5932
Raman	2400,6049,6645,6084
ms	154
nmr	5933

61. 1-Hexene

$CH_3(CH_2)_3CH{=}CH_2$ 592–41–6 C_6H_{12}

		cgs	SI	
mw		84.161	84.161	3468
bp	1 atm	63.478	336.628	254;2493,1325,1014,8067 2701,7476
dt/dp	1 atm	0.04149	0.3112	249,2493,1325
dp/dt	1 atm	24.10	3.213	149,2493,1325
p	25	186	24.8	255,1325;2493
eq 2.5	A	6.85770	5.98260	254;2493,1325
	B	1148.62	1148.62	
	C	225.346	225.346	
fp		−139.813	133.337	7387;7476,4748,7677,7042,5016
d	20	0.67317	673.17	249,2493,1325;1014,8067,8038
	25	0.66848	668.48	2701,7677,4166,5016
dd/dt	25	0.000943	0.943	2493,1325;5016
κ_T	25	2.281×10^{-7}	1.711×10^{-6}	1262
n_D	20	1.38788	1.38788	249,2493,1325;1014,8067,8038
	25	1.38502	1.38502	108,7677,4166,5016
dn/dt		0.00057	0.00057	5016
η	20	0.26	0.00026	250;8169
	25	0.25	0.00025	
γ	25	17.90	0.01790	3687;252,3685
	Table 2.8			
ΔH_v	25	7.32	30.63	255
	bp	6.76	28.28	1325;7387
ΔH_m		2.2341	9.3475	255,4748
ΔH_f°	25(1)	−17.35	−72.60	7389;8316
	25(g)	−10.03	−41.95	7389
ΔH_c°	25(1)	−956.84	−4003.40	7388;8316
	25(g)	−964.2	−4034.1	7388
C_p°	25	31.33	131.08	7389
C_p	25	43.81	183.30	7389;4748,4001
t_c		230.88	504.03	253;119
P_c		31.0	3.14	253
d_c		0.238	238	253
v_c		0.354	0.000354	253
A	eq 2.72	0.0509	0.0509	7387;4748,7042
ε	20	2.051	2.051	108
μ	20(1)	0.34	0.34	108;106,2698,4687
soly	in aq,25	0.00697%w	0.00697%w	7219;4733,5408
	aq in,30	0.0477%w	0.0477%w	2268
aq az	57.7	94.3%w	94.3%w	6165

61. 1-Hexene

(Continued)

		cgs	SI	
fl pt	estm	−26	247	5822
Beil	11,215			
uv	1366,2679,4895,3760			
ir	153,3962,607,5932			
Raman	1069,4166,7542,1015,7604,6683,742			
ms	2223,714			
nmr	1006,2457,5933,6748			

62. 1-Heptene

$CH_3(CH_2)_4CH{=}CH_2$ 592-76-7 C_7H_{14}

		cgs	SI	
mw		98.188	98.188	3468
bp	1 atm	93.639	366.789	254;2493,1014,8067,2701 4033,1368
dt/dp	1 atm	0.04447	0.3336	249,2493
dp/dt	1 atm	22.49	2.998	
p	25	56	7.5	255;2493,27,4629
eq 2.5	A	6.90187	6.02677	254;2493
	B	1258.34	1258.34	
	C	219.299	219.299	
fp		−118.856	154.294	7387;4748,5690,7042
d	20	0.69698	696.98	249,2493;1014,8067,8038,2701
	25	0.69267	692.67	4033,1368
dd/dt	25	0.000883	0.883	2493
n_D	20	1.39980	1.39980	249;2493,1014,8067,108,8038
	25	1.39713	1.39713	4033,1368
dn/dt	25	0.000534	0.000534	249
η	20	0.35	0.00035	250
	25	0.34	0.00034	

62. 1-Heptene

(Continued)

		cgs	SI	
γ	25	19.80	0.01980	<u>3687</u>,252;3685
	Table 2.8			
ΔH_v	25	8.52	35.65	255
	bp	7.43	31.09	
ΔH_m		2.964	12.401	255,4748;5690
ΔH_t	-136.26	0.060	0.25	7387
ΔH_f°	25(1)	-23.51	-98.37	2852,7389;8316
	25(g)	-15.00	-62.76	7389
ΔH_c°	25(1)	-1113.05	-4657.00	2852,7388;6218,8316
	25(g)	-1121.6	-4692.6	7388
C_p°	25	36.73	153.67	7389;4001
C_p	25	50.62	211.79	7389,4748;4895
t_c		264.14	537.29	253;119
p_c		27.9	2.83	253
d_c		0.238	238	253
v_c		0.413	0.000413	253
A	eq 2.72	0.06264	0.06264	7387;4748,7042
ε	20	2.071	2.071	108
μ	20(1)	0.34	0.34	108;2698
soly	in aq,25	0.00182%w	0.00182%w	<u>7219</u>
	aq in,20.1	0.1126%w	0.1126%w	906;2268
aq az	77.0	85.2%w	85.2%w	6165

Beil	<u>11</u>,219
uv	1368,1366,1367
ir	153,6652,4384,6645,5932
Raman	1017,7542,1015,6645,7604,6683,6084
ms	714
nmr	5933,6748

63. 1-Octene

$CH_3(CH_2)_5CH{=}CH_2$ 111–66–0 C_8H_{16}

		cgs	SI	
mw		112.214	112.214	3468
bp	1 atm	121.286	394.436	254;2493,2701,5910,8067 1014,8169
dt/dp	1 atm	0.04711	0.3534	249,2493
dp/dt	1 atm	21.23	2.830	249,2493
p	25	17	2.3	255;2493
eq 2.5	A	6.93495	6.05985	254;2493
	B	1355.46	1355.46	
	C	213.054	213.054	
fp		-101.690	171.460	7387;5910,4748,7038
d	20	0.71492	714.92	249,2493;2701,5910,8038,8067
	25	0.71085	710.85	1014,4033,4166
dd/dt	25	0.000834	0.834	2493;2231
n_D	20	1.40870	1.40870	249,2493;108,5910,8038,8067
	25	1.40620	1.40620	1014,8169,4033,4166
dn/dt	25	0.00050	0.00050	249;2231
η	20	0.470	0.000470	250;8169
	25	0.447	0.000447	
γ	25	21.28	0.02128	3687,252;3685
	Table 2.8			
ΔH_v	25	9.69	40.55	7107;4878
	bp	8.11	33.95	7107;7387
ΔH_m		3.721	15.57	7387;4748
ΔH_f°	25(1)	-29.60	-123.86	7389;8316
	25(g)	-19.98	-83.59	7389
ΔH_c°	25(1)	-1269.34	-5310.90	7388;6218,8316
	25(g)	-1278.9	-5351.1	7388
C_p°	25	42.13	176.28	7389;4001
C_p	25	57.60	241.00	7389;4748
λ	37.8	0.000306	0.128	6352
t_c		293.4	566.6	253,4299;119
p_c		25.2	2.55	253
d_c		0.238	238	253
v_c		0.472	0.000472	253
A	eq 2.72	0.06268	0.06268	255;4748,7038
ϵ	20	2.084	2.084	108
μ	20(1)	0.34	0.34	108;2698
δ	25	7.6	15.5	653
soly	in aq,25	0.000410%w	0.000410%w	7219;4733,5408

63. 1-Octene

(Continued)

	cgs		SI	
aq az	88.0	71.3%w	71.3%w	6165
fl pt	TOC	21	294	5822

Beil	11,221
uv	4062,5882,7995
ir	153,5711,4383,3962,7439,6070,6645,5932
Raman	1017,1532,4166,7542,1015,6645,6683
ms	714
nmr	5933,6748

64. 2-Octene

(Mixed isomers)

$$CH_3(CH_2)_4CH{=}CHCH_3 \qquad\qquad 111\text{--}67\text{--}1 \qquad\qquad C_8H_{16}$$

		cgs	SI	
mw		112.214	112.214	3468
bp	1 atm	124.25	397.40	7473;1014,1971,4058,8169
fp		−106.8	166.4	7473
d	20	0.7196	719.6	8169;4058,1014
	25	0.7154	715.4	
dd/dt	25	0.000851	0.851	8169
n_D	20	1.4152	1.4152	8169;3581,5823,1014,1971
	28.5	1.4145	1.4145	4058
η	20	0.506	0.000506	8169
	30	0.451	0.000451	
λ	37.8	0.000318	0.133	6352
fl pt	TOC	21	294	5823

64. 2-Octene

(Continued)

Beil	11,EII 200
ir	7439,6070,5932
Raman	1532
ms	6199
nmr	6748

65. *cis*-2-Octene

(Z)-2-Octene

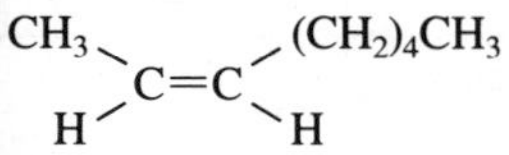

$$CH_3 \diagdown \diagup (CH_2)_4CH_3$$
$$C{=}C$$
$$H \diagup \diagdown H$$

7642–04–8 C_8H_{16}

		cgs	SI	
mw		112.214	112.214	3468
bp	1 atm	125.64	398.79	264;1336,5826
dt/dp	1 atm	0.04870	0.3653	2088;264
dp/dt	1 atm	20.53	2.738	2088
p	25	16.07	2.142	2088
eq 2.5	A	6.87711	6.00201	2088
	B	1361.3	1361.3	
	C	215	215	
fp		−100.2	173.0	264;1336,5826
d	20	0.7243	724.3	264;1336
	25	0.7201	720.1	264
dd/dt	20	0.00085	0.85	2231
α	15.56	0.00115	0.00115	5826
n_D	20	1.4150	1.4150	264;1336
	25	1.4125	1.4125	264;5826
dn/dt	20	0.00048	0.00048	2231

65. *cis*-2-Octene

(Continued)

		cgs	SI	
γ	20	21.91	0.02191	2088
	25	20.90	0.02090	
ΔH_v	25	8.552	35.781	2088
	bp	8.064	33.740	
ΔH_f°	26.0(1)	21.60	90.37	3857
ΔH_c°	25(1)	−1271.2	−5318.7	323
C_p°	25	43	180	323
C_p	25	57	238	323
t_c		308	581	323;5826,2088
P_c		27.40	2.776	323;5826,2088
d_c		0.250	250	2088
v_c		0.449	0.000449	2088
ϵ	25	2.062	2.062	4937

Beil	11,EIII 839
uv	5882
ir	6645,5932
Raman	1016,2919,2918,6645
ms	5908

66. *trans*-2-Octene

(*E*)-2-Octene

13389–42–9 C_8H_{16}

		cgs	SI	
mw		112.214	112.214	3468
bp	1 atm	125.0	398.2	264;1336,5826
dt/dp	1 atm	0.0487	0.365	2088;264
dp/dt	1 atm	20.5	2.74	2088
p	25	16.52	2.202	2088
eq 2.5	A	6.87364	5.99854	2088
	B	1357.6	1357.6	
	C	215	215	
fp		−87.7	185.5	264;1336,5826
d	20	0.7199	719.9	264;1336
	25	0.7157	715.7	264
dd/dt	20	0.00085	0.85	2231
α	15.56	0.00113	0.00113	5826
n_D	20	1.4132	1.4132	264;1336,5826
	25	1.4107	1.4107	264;5826
dn/dt	20	0.00048	0.00048	2231
γ	20	21.38	0.02138	2088
	30	20.39	0.02039	
ΔH_v	25	9.527	39.861	2088
	bp	8.045	33.660	
ΔH_c°	25(1)	−1270.5	−5315.8	323
C_p°	25	45	188	323
C_p	25	58.3	243.9	323
t_c		307	580	323;5826,2088
P_c		27.50	2.786	323;5826,2088
d_c		0.225	225	2088
v_c		0.499	0.000499	2088
ε	25	2.002	2.002	4937

Beil	11,EIII839
uv	5882
ir	6645,5932
Raman	1016,2919,6645
ms	5908
nmr	5933

67. 1-Nonene

$CH_3(CH_2)_6CH{=}CH_2$ 　　　　　　　124–11–8 　　　　　　　　　C_9H_{18}

		cgs	SI	
mw		126.241	126.241	3468
bp	1 atm	146.883	420.033	254;2493,2701,1014,8067
dt/dp	1 atm	0.04944	0.3708	249,2493
dp/dt	1 atm	20.23	2.697	249,2493
p	25	5.3	0.71	255;2493,6459
eq 2.5	A	6.95430	6.07920	254;2493
	B	1436.20	1436.20	
	C	205.690	205.690	
fp		–81.338	191.812	7387;7041,6459
d	20	0.72922	729.22	249,2493;8038,2701,1014
	25	0.72531	725.31	8067,6459
dd/dt	25	0.000788	0.788	2493
n_D	20	1.41572	1.41572	249,2493;8038,1014,8067
	25	1.41333	1.41333	6459,4863
dn/dt	25	0.000478	0.000478	249
η	20	0.620	0.000620	250
	25	0.586	0.000586	
γ	25	22.56	0.02256	3687;252,3685
	Table 2.8			
ΔH_v	25	10.88	45.52	255
	bp	8.68	36.31	
ΔH_m		4.32	18.08	7387
ΔH_f°	25(l)	–35.73	–149.49	7389;8316
	25(g)	–24.85	–103.96	7389
ΔH_c°	25(l)	–1425.57	–5964.60	7388;8316,1650
	25(g)	–1436.4	–6010.1	7388
C_p°	25	47.54	198.90	7389;4001
C_p	25	64.50	269.85	7389
t_c		320.1	593.3	253
P_c		23.0	2.33	253
d_c		0.239	239	253
v_c		0.528	0.000528	253
A	eq 2.72	0.0588	0.0588	255,7041
μ	20 in 38	0.59	0.59	2698
δ	25	7.7	15.8	653
soly	in aq,25	9.93×10^{-5}%w	9.93×10^{-5}%w	7219
aq az	94.5	53.7%w	53.7%w	6165

67. 1-Nonene

(Continued)

Beil	11,223
ir	153,4383,3126,5932
Raman	1017,7542,6084
nmr	5933

68. 1-Decene

$$CH_3(CH_2)_7CH=CH_2 \qquad\qquad 872-05-9 \qquad\qquad C_{10}H_{20}$$

		cgs	SI	
mw		140.268	140.268	3468
bp	1 atm	170.597	443.747	254;2493,2701,7465
dt/dp	1 atm	0.05157	0.3868	249,2493
dp/dt	1 atm	19.39	2.585	249,2493
p	25	1.6	0.21	255,6459
eq 2.5	A	6.93477	6.05967	254;2493
	B	1484.98	1484.98	
	C	195.707	195.707	
fp		-66.307	206.843	7387;4081,4748,6459,7465 5016,7041
d	20	0.74081	740.81	249,2493;2701,8038,6459
	25	0.73693	736.93	7465,2698
dd/dt	25	0.000777	0.777	2493;5016
n_D	20	1.42146	1.42146	249,2493;8038,6459,7465,5016
	25	1.41913	1.41913	2698,8169
dn/dt		0.00051	0.00051	5016
η	20	0.805	0.000805	250;6459,8169
	25	0.756	0.000756	
γ	25	23.54	0.02354	3687;252,3700,3685
	Table 2.8			

68. 1-Decene

(Continued)

		cgs	SI	
ΔH_v	25	12.05	50.43	4878
	bp	9.24	38.66	7387
ΔH_m		5.05	21.13	7387;4748
ΔH_t	−74.826	1.90	7.95	7387
ΔH_f°	25(l)	−41.85	−175.12	7389;8316
	25(g)	−29.80	−124.69	7389
ΔH_c°	25(l)	−1581.81	−6618.30	7388;6218,8316
	25(g)	−1593.9	−6668.7	7388
C_p°	25	52.95	221.53	7389;4001
C_p	25	71.78	300.32	7389,4748
t_c		343.9	617.1	253
P_c		21.1	2.14	253
d_c		0.240	240	253
v_c		0.585	0.000585	253
A	eq 2.72	0.0594	0.0594	7387;7041,4748
μ	20 in 38	0.42	0.42	2698
δ	25	7.8	16.0	653
aq az	96.7	35.8%w	35.8%w	6165
Beil	11,223			
uv	153,2562,4383,606,607,5830			
ir	607,5932			
Raman	7542,3260,6084			
nmr	5933,6748			

69. 1-Dodecene

$CH_3(CH_2)_9CH{=}CH_2$ 　　　　　　　　112–41–47 　　　　　　　　　　$C_{12}H_{24}$

		cgs	SI	
mw		168.322	168.322	3468
bp	1 atm	213.401	486.551	254;2493,3581
dt/dp	1 atm	0.05522	0.4142	2493,249
dp/dt	1 atm	18.11	2.414	2493,249
p	25	0.1446	0.01928	2088
eq 2.5	A	6.97607	6.10097	254,2493
	B	1621.11	1621.11	
	C	182.449	182.449	
fp		−35.209	237.941	7387;7035,6459,3581
tp		−35.22	237.93	4748
d	20	0.75836	758.36	2493,249;2701,3699,6459,4236
	25	0.75474	754.74	
dd/dt		0.000733	0.733	2493
n_D	20	1.43002	1.43002	2493,249;2701,6459,2698,4863
	25	1.42782	1.42782	
dn/dt	25	0.000440	0.000440	2493
η	20	1.30	0.00130	251;6459
	25	1.203	0.001203	251;1725
γ	20	25.60	0.02560	252;3687,3685,3699
	25	25.15	0.02515	
ΔH_v	25	14.53	60.78	4878,7046;255
	bp	10.27	42.97	255
ΔH_m		4.163	17.42	7387;4748
ΔH_t	−60.23	1.09	4.56	7387,4748
ΔH_f°	25(1)	−54.06	−226.20	7389,7046
	25(g)	−39.54	−165.42	7389,7046;4001
ΔH_c°	25(1)	−1894.3	−7925.9	7388,7046
	25(g)	−1908.9	−7986.7	7388
C_p°	25	63.77	266.81	7389;4001
C_p	25	86.20	360.66	7389,4748
t_c		384	657	2088;7399
p_c		18.30	1.855	2088;7399
d_c		0.2475	247.5	7399
v_c		0.6802	0.0006802	7399
A	eq 2.72	0.0370	0.0370	7035,7387;4748
μ	25(1)	0.41	0.41	1725;2698
fl pt		79	352	5823

69. 1-Dodecene

(Continued)

Beil	11,225
ir	3628,4383,5932
Raman	4962
ms	1563
nmr	6011,5933,6748

70. 1-Tridecene

$CH_3(CH_2)_{10}CH{=}CH_2$ 　　　　2437–56–1 　　　　$C_{13}H_{26}$

		cgs	SI	
mw		182.348	182.348	3468
bp	1 atm	232.837	505.987	254;1324,3581
dt/dp	1 atm	0.05680	0.4260	1324,249
dp/dt	1 atm	17.61	2.347	1324,249
p	25	0.05477	0.007302	2088
eq 2.5	A	6.98102	6.10592	254;1324
	B	1672.00	1672.00	
	C	174.947	174.947	
fp		−23.056	250.094	7387;7035,6437,6459,3581
d	20	0.76527	765.27	1324,249;4372,6437,6459,4229
	25	0.76168	761.68	
dd/dt	25	0.000726	0.726	1324
n_D	20	1.43336	1.43336	1324,249;4372,6437,6459,4229
	25	1.43118	1.43118	
dn/dt	25	0.000436	0.000436	1324

70. 1-Tridecene

(Continued)

		cgs	SI	
η	20	1.63	0.00163	251;6459,6437
	25	1.50	0.00150	
γ	20	26.24	0.02624	252;2088
	25	25.80	0.02580	
ΔH_v	25	15.60	65.27	255
	bp	10.75	44.98	
ΔH_m		5.46	22.83	7387
ΔH_f°	25(1)	-60.23	-252.01	7389
	25(g)	-44.61	-186.64	7389;4001
ΔH_c°	25(1)	-2050.5	-8579.4	7388
	25(g)	-2066.2	-8644.8	
C_p°	25	69.18	289.45	7389;4001
C_p	25	93.65	391.84	7389;341
t_c		401	674	2088
P_c		16.81	1.703	2088
A	eq 2.72	0.0439	0.0439	7035,7387
fl pt		79	352	5823
Beil	11,225			
ir	5932			
ms	1563			
nmr	5933			

71. Cyclohexene

$$H_2C-C(H)=CH$$

110–83–8 C_6H_{10}

		cgs	SI	
mw		82.145	82.145	3468
bp	1 atm	82.979	356.129	267,2493;506,3547,4969 7473,7758
dt/dp	1 atm	0.04381	0.3286	267,2493
dp/dt	1 atm	22.83	3.043	267,2493
p	25	88.8	11.8	2493;4611
eq 2.5	A	6.88617	6.01107	2493;5078,4536
	B	1229.973	1229.973	
	C	224.104	224.104	
fp		−103.512	169.638	267;5679,7473,7042
d	20	0.81096	810.96	267,2493;506,3547,5392,7758
	25	0.80609	806.09	2700,3183,4536
dd/dt	25	0.000955	0.0955	2493
κ_T	25	1.388×10^{-7}	1.041×10^{-6}	1262
n_D	20	1.44654	1.44654	267,2493;108,506,3547,4969
	25	1.44377	1.44377	7758,2400,8169
dn/dt	25	0.000554	0.000554	267
η	20	0.650	0.000650	8169;2700
	30	0.608	0.000608	
γ	20	26.54	0.02654	7482
	30	25.22	0.02522	
ΔH_v	25	7.922	33.144	4536;7110
	bp	7.286	30.485	4969
ΔH_m		0.786	3.29	5679
ΔH_f°	25(1)	−9.13	−38.20	8316,2857;269
ΔH_c°	25(1)	−896.75	−3752.00	2857;266,2268,5701,8316
C_p°	25	25.10	105.02	268
C_p	20	34.7	145.2	5679
t_c		287.27	560.42	119;1451,4299
d_c		0.288	288	2087
v_c		0.285	0.000285	2087
A	eq 2.72	0.01376	0.01376	7042
κ	25	1.5×10^{-15}	1.5×10^{-13}	2486
ε	20	2.220	2.220	108,4937;7497
μ	20(1)	0.28	0.28	108;4687

71.   Cyclohexene

(Continued)

		cgs	SI	
soly	in aq,25 aq in,20	0.0213%w 0.0317%w	0.0213%w 0.0317%w	4733;5408 2268
aq az	70.8	90.0%w	90.0%w	3497
fl pt	estm	−23	250	5822

Beil	453,63
uv	3198,4423,5882,5930,3760,7995,7585
ir	153,4385,5892,633,6696,3962,4316,5932
Raman	2400,5466,2135,4895
ms	154.6199,3430
nmr	4623,5933,6748,346

72.   Dipentene

(±)-1-Methyl-4-(1-methylethenyl)cyclohexene

$CH_2{=}\overset{\underset{\displaystyle CH_3}{|}}{C}CHCH_2CH{=}\overset{\underset{\displaystyle CH_3}{|}}{C}CH_2CH_2$ 7705–14–8 $C_{10}H_{16}$

		cgs	SI	
mw		136.236	136.236	3468
bp	1 atm	174.6	447.8	7058;3012,6162,1190,508,3013
p	68.2 128.2	20.0 200.0	2.67 26.66	7058;6162,5843,3012 7058;5843,3012
eq 2.4	A B	7.984 2289	7.109 2289	6301
fp		<−40	<233	3316
d	20.85 25	0.8402 0.8370	840.2 837.0	1445;3013,508,5767 7409;5170,2128,3910
n_D	20 25	1.4735 1.4701	1.4735 1.4701	7741;5843,421,1190,6162,3013 7409
η	25	1.223	0.001223	2128
γ	10.8 24.2	28.22 26.90	0.02822 0.02690	5170
ΔH_v	25	11.5	48.1	1703
ΔH_f°	25(1) 25(g)	−12.1 −0.6	−50.6 −2.5	1703

72. Dipentene

(Continued)

		cgs	SI	
ΔH°_c	25(1)	−1474.7	−6170.1	3236;1703,508
C_p	20.20	59.62	249.45	4153
ϵ	25	2.3810	2.3810	7409;7009
δ		8.5	17.4	4593,2611,1271
fl pt	TCC	49	322	3316;421

Beil	457,137
uv	3144
ir	5887,5181,7741
Raman	3693,1531
ms	7407

73. *d*-Limonene

(+)-1-Methyl-4-(1-methylethenyl)cyclohexene

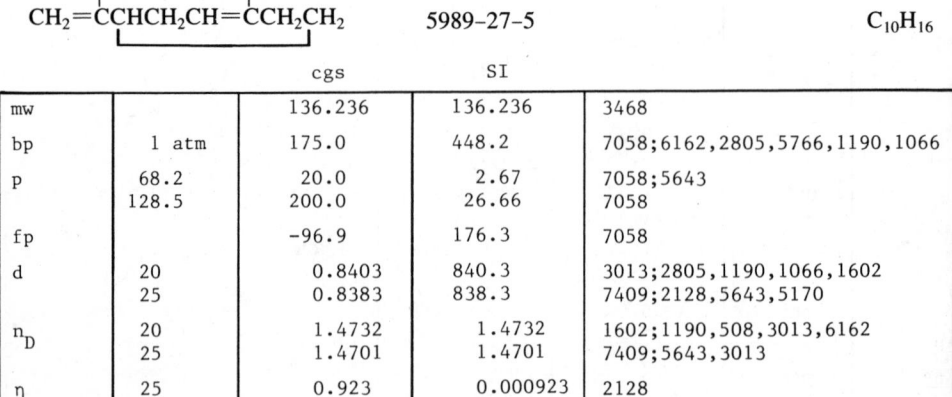

5989–27–5 $C_{10}H_{16}$

		cgs	SI	
mw		136.236	136.236	3468
bp	1 atm	175.0	448.2	7058;6162,2805,5766,1190,1066
p	68.2	20.0	2.67	7058;5643
	128.5	200.0	26.66	7058
fp		−96.9	176.3	7058
d	20	0.8403	840.3	3013;2805,1190,1066,1602
	25	0.8383	838.3	7409;2128,5643,5170
n_D	20	1.4732	1.4732	1602;1190,508,3013,6162
	25	1.4701	1.4701	7409;5643,3013
η	25	0.923	0.000923	2128

73. *d*-Limonene

(Continued)

		cgs	SI	
γ	10.9	28.53	0.02853	5170
	23.1	27.31	0.02731	
ΔH_v	25	11.5	48.1	1703
ΔH_f°	25(1)	-13.0	-54.4	1703
	25(g)	-1.5	-6.3	
ΔH_c°	25(1)	-1473.9	-6166.8	3236;1703,508
$[\alpha]_D$	20	126.1	126.1	3013,1066;1190,2805,6162,1602
				5766,5643,7409,7103
ε	25	2.3746	2.3746	7409;7009
μ	25 in 38	0.61	0.61	3106;7103
soly	in aq,25	0.00138%w	0.00138%w	4950
Beil	457,133			
uv	3241,3144,1742			
ir	5269,5887,5932			
Raman	3693			
ms	6199,6347,7785			
nmr	5933			

74. *l*-Limonene

(−)-1-Methyl-4-(1-methylethenyl)cyclohexene

$$CH_2=\overset{\overset{\textstyle CH_3}{|}}{C}CHCH_2CH=\overset{\overset{\textstyle CH_3}{|}}{C}CH_2CH_2$$ 5989-54-8 $C_{10}H_{16}$

		cgs	SI	
mw		136.236	136.236	3468
bp	755 Torr	177.6-177.8	450.8-451.0	6162;3013,1190,5766,1066
p	64.4	15.0	2.00	6162
d	20.6	0.8417	841.7	6162;1066,1190,3013
	25	0.8384	838.4	7409;5170,2128

74. *l*-Limonene

(Continued)

		cgs	SI	
n_D	20.5	1.47468	1.47468	1190;3013,6162
	25	1.4701	1.4701	7409
η	25	1.806	0.001806	2128
γ	23.2	26.87	0.02687	5170
	38.4	25.73	0.02573	
$[\alpha]_D$	20	−122.1	−122.1	1066,3013;1190,2132,5766
				7409,6162
ε	25	2.3738	2.3738	7409
Beil	457,134			
uv	3144			
ir	5932			
Raman	2134,963,581			

75. Styrene

Ethenylbenzene

$C_6H_5CH{=}CH_2$ 100–42–5 C_8H_8

		cgs	SI	
mw		104.151	104.151	3468
bp	1 atm	145.14	418.29	299;2090,2220,7058,2779
dt/dp	1 atm	0.049	0.37	299,2090
dp/dt	1 atm	20.4	2.7	299,2090
p	25	6.31	0.841	5874;7058,1303,4611,1224
eq 2.5	A	7.22302	6.34792	2090;2220,5874,1224
	B	1629.2	1629.2	
	C	230	230	
fp		−30.628	242.522	299;2090,7058,5874
d	20	0.90600	906.00	299,2090;2220
	25	0.90122	901.22	

75. Styrene

(Continued)

		cgs	SI	
dd/dt	0 - 40	0.0008739	0.8739	2086
n_D	20	1.54682	1.54682	299;2090,2220,1224
	25	1.54395	1.54395	
dn/dt	13 - 25	0.0005186	0.0005186	2220
η	20	0.751	0.000751	2211
	25	0.696	0.000696	
	30	0.650	0.000650	
γ	20	32.3	0.0323	2087;7794
	30	30.98	0.03098	
ΔH_v	25	10.500	43.932	5874;1303
	bp	9.25	38.70	1224
ΔH_m		2.617	10.950	5874
ΔH_f°	25(1)	24.83	103.89	301;5968,8316
ΔH_c°	25(1)	-1050.58	-4395.63	6194;3975,300,509,5968,8316
ΔH_p		16.68	69.79	6194
C_p°	25	29.18	122.09	302;712,6534
C_p	25.39	43.64	182.59	2211
t_c		363.7	636.9	2087;6534
P_c		36.3	3.68	6534
ϵ	20	2.4257	2.4257	5798;4937
μ	20 - 60(1)	0.13	0.13	5798;5878,2330,3339,4485
	25 in 38	0.43	0.43	3765;4137
δ		9.3	19.0	4593
soly	in aq,25	0.031%w	0.031%w	4395;591
	aq in,25	0.066%w	0.066%w	4395
aq az	93.9	59.1%w	59.1%w	3498
fl pt	TCC	30	303	5712
Beil	473,474			
uv	155,827,6230,6475,6209,6569,1799,6749			
ir	153,3942,633,5711,7439,5874,1624,5932			
Raman	3971,742,1001,5874,1624			
ms	154,3728,6199,7199			
nmr	1187,4697,4992,5933,6748			

76. 2-Pinene

2,6,6-Trimethylbicyclo[3.1.1]hept-2-ene
(Mixed isomers)

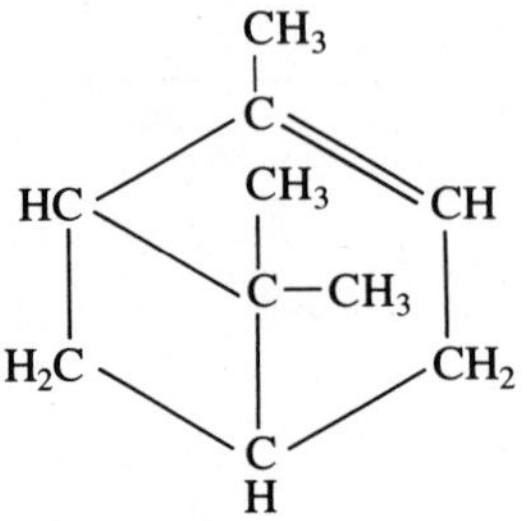

CH₃ structure (chemical structure diagram)

CH3
C
HC   CH3   CH
C—CH3
H2C   CH2
C
H

80-56-8 $C_{10}H_{16}$

		cgs	SI	
mw		136.236	136.236	3468
bp	1 atm	155.9	429.1	3235;495,7925,7457,508 6480,3013
p	27.25	4.91	0.655	3235;3012,2600,4611
eq 2.13	A B C	26.40174 3134.525 6.16045	25.52664 3134.525 6.16045	3235;6301
fp		−64	209	6269;495,3013,3236
d	20 25	0.8582 0.8539	858.2 853.9	7925 3235;2600,495,5843,7457,7409 6480,508,963,1620,2128
n_D	20 25	1.4658 1.4632	1.4658 1.4632	7925;7578,2600,495,5843 7457,7409 3235;508,6480,963,3013 617,3236
dn/dt		0.00045	0.00045	2600
η	20	1.40	0.00140	616;2128
γ	30	25.8	0.0258	Beil. EIII5, 368
ΔH_v	3-15Torr 215-760	11.140 9.482	46.610 39.673	3235;7578 3235
ΔH_c°	25(1)	−1465.3	−6130.8	3975;508,509,3236
$[\alpha]_D$	20 d 20 l	+51.14 −51.28	+51.14 −51.28	7457;508,6480,4722,963,1620 617,3013
K_f	eq 2.67	16.50	16.50	6269
ε	30	2.2588	2.2588	1861;7105,7409,4937
μ	30 in 38	0.36	0.36	1861;7105,5316,3106,7103
δ	25	7.9	16.2	653
fl pt	CC	35	308	419

Beil	458,144
uv	7995,7585
ir	153,633,5887,5181,8105
Raman	5467,846,963,7716,962,8105
ms	154,7407,2215
nmr	407,6748

77. 2(10)-Pinene

6,6-Dimethyl-2-methylene bicyclo[3.1.1]heptane

127–91–3 $C_{10}H_{16}$

		cgs	SI	
mw		136.236	136.236	3468
bp	1 atm	166.0	439.2	3235;7579,5843,3236
p	25	4.6	0.61	5843;2600
eq 2.13	A	28.77768	27.90258	3235
	B	3318.845	3318.845	
	C	6.94243	6.94243	
fp		−61.54	211.61	3236
d	25	0.8667	866.7	3235;2600,5843,7409,495
n_D	15	1.4813	1.4813	2600
	25	1.4768	1.4768	3235;7409,7578,5843,495,3236
dn/dt		0.00045	0.00045	2600
η	20	1.70	0.00170	616
ΔH_v	2–23 Torr	10.390	43.471	3235;7578
	175–760	9.610	40.208	3235
ΔH_c°	25(1)	−1485.1	−6213.7	3236;509
$[\alpha]_D$	d	+12.6	+12.6	Beil. EII5, 102
	1	−21.49	−21.49	2600
ε		2.4970	2.4970	7409
fl pt	CC	38	311	420
Beil	458,154			
uv	7995			
ir	153,8105			
Raman	3693,8105			
ms	7407,2215			
nmr	407,6748			

78. Methanol

CH_3OH 67-56-1 CH_4O

		cgs	SI	
mw		32.042	32.042	3468
bp	1 atm	64.546	337.696	129;2728,7236,4778,4933,7059 8130,7230,7239
dt/dp	1 atm	0.03344	0.2508	129;7239,7488,8130,7230
dp/dt	1 atm	29.90	3.987	129
p	25	127.04	16.937	129;2728,7236,3532,5782,7058 7483,7233,1972,1009,6298
eq 2.5	A B C	8.08029 1581.993 239.711	7.20519 1581.993 239.711	129;1009,7235,7230,3270,2728
fp		-97.68	175.47	7236;3793,3932,4778,6250 6976,7059
tp		-97.56	175.59	1354
d	15	0.79609	796.09	7488;6121,29,2677,3508,3752 3924,3931
	20 25	0.79104 0.78637	791.04 786.37	3080;6121,3634,6357,1989,6298 4698,5782,7241,7470,7768 7230,2963,7239
dd/dt	20	0.0009321	0.9321	6121
α	25	0.001196	0.001196	6357;1965,3481,3080
κ_T	25	1.664×10^{-7}	1.248×10^{-6}	1989;1965,3897
κ_S	25	1.371×10^{-7}	1.028×10^{-6}	4315;1965
n_D	15	1.33034	1.33034	7242;6121,3621,29,3508,3924 4969,5782,7488
	20 25	1.32840 1.32652	1.32840 1.32652	7242;6121,7768,7230
dn/dt	20	0.000383	0.000383	6121;5598
η	20	0.5929	0.0005929	7240;3634,557,2677,7488,7522 7705,7838
	25	0.5513	0.0005513	7240;6080,3752,4315
	Table 2.2			
γ	20	22.55	0.02255	3283,7488;4195,6958,7045,28 3621,3931,5782,7522,7768
	25 30	22.30 21.69	0.02230 0.02169	513 3283,7488;6762
ΔH_v	25 bp	8.946 8.433	37.430 35.284	7236;660,2439,3618,5153,6976 7978,2926,2923,6277,7821 5475,5902,7110
ΔH_m		0.7685	3.215	1354;7236,5674,6976
ΔH_t	-115.81	0.1520	0.6360	1354
ΔH_f°	25(1) 25(g)	-57.04 -47.96	-238.66 -200.66	7826,2036;1418,7237,2923,8316 7826,2036;1445

78. Methanol

(Continued)

		cgs	SI	
ΔH°_c	25(1)	−173.64	−726.51	2036;1418,2923,6277,7446 3975,8316
	25(g)	−182.72	−764.50	2036
C°_p	25	10.53	44.06	1445;7237,7826,6835,7978 2926,7051
C_p	25	19.47	81.47	8269;1004,1354,2439,3508,3932 5817,6976,7978,7745
λ	20	0.00051	0.21	1612,6352
t_c		239.49	512.64	7234;5153,8231,8233,1403,3270 1671,2214
P_c		79.9	8.10	7234;5153,8231,4088,3270,2214
d_c		0.272	272	7234;2214,8233,1403
v_c		0.118	0.000118	7234;2214,8233,1671
A	eq 2.72	0.0125	0.0125	7236
K_B	eq 2.79	0.785	0.785	1370;726,3756,7845,5638,734
pK_a	25	15.5	15.5	578;5887,4766,8115
$pK_{BH}+$	25 in aq H_2SO_4	−2.05	−2.05	5755
pK_s	25	16.91	16.91	6858;1181,4259,1300
κ	25	1.5×10^{-9}	1.5×10^{-7}	2334;3586,4698,5782,6463,7842
ϵ	25	32.66	32.66	3624;6416,2814,3719,1811,2882 4414,74,5905,3199
μ	20(1)	2.87	2.87	4739;776,4830,4933,7032 8141,4511
δ	25	14.5	29.7	652,6240,7159;6723,7635
ER	BuOAc 90%	2.10 221	2.10 221	6628;6631,1399,778 6628,6631;1399
soly	in aq,25	inf	inf	2042
aq az		none	none	3497
fl pt	TCC TOC	12 15	285 288	5712,1612 1612;4193

Beil	<u>19</u>,273
uv	155,842,5864,7229,3573,3189
ir	634,990,1689,1845,3452,4792,6368,3107,5932
Raman	1069,2206,3086,4100,4393,7720,3921,3107
ms	154,6199,2745,2947,1669
nmr	4926,4923,5938,4924,7227,4852,1423,2464,5933,6748

79. Ethanol

CH_3CH_2OH			64–17–5	C_2H_6O
		cgs	SI	
mw		46.069	46.069	3468
bp	1 atm	78.293	351.443	129;1681,7236,1255,2754,4933 8130,2091,1198
dt/dp	1 atm	0.03330	0.2498	129;7239,8130,2090,1198 4380,4665
dp/dt	1 atm	30.03	4.004	129
p	25	59.03	7.870	129;7236,7058,2410,6485,2091 6407,6864,4828
eq 2.5	A	8.04389	7.16879	129;7235,3270,2090
	B	1552.601	1552.601	
	C	222.419	222.419	
fp		−114.49	158.66	2352;7236,7484,3933,2090,7471
d	15	0.79360	793.60	5414;29,2288,2754,3933 4417,7768
	20	0.78920	789.20	3080;7241,7239,6494,1198,2677
	25	0.78493	784.93	2882,4380,4698,8231,8279 6357,1989,5370
dd/dt	25	0.000856	0.856	3080
α	25	0.001096	0.001096	6357;3080,3481
κ_T	25	1.537×10^{-7}	1.153×10^{-6}	1989;3867
κ_S	25	1.261×10^{-7}	9.460×10^{-7}	4315
n_D	20	1.36143	1.36143	7242;6876,1198,3057,6407,29
	25	1.35941	1.35941	2288,2090,4380,7768
dn/dt		0.00040	0.00040	6849;5598
η	25	1.0826	0.0010826	7240;3566,557,2201,7522,598 2677,7838,6080
	30	0.987	0.000987	7240;7505,2020,6664
	Table 2.2 Table 2.4			
γ	20	22.32	0.02232	3283,4195;28,4380,7522,2288 4417,7551,7045,6271,5339
	30	21.48	0.02148	3283,5370
	Table 2.8			
ΔH_v	25	10.112	42.309	7236;7821,4756,768,660,7848 5687,4969,5668,2923,7754 3618,6277,5902,5475
	bp	9.249	38.698	1681;7236
ΔH_m		1.20	5.02	7236;3933,5674
ΔH_f°	25(1)	−66.37	−277.69	2036,7826;7237,1418,2923,2933
	25(g)	−56.19	−235.10	2036,7826
ΔH_c	25(1)	−326.68	−1366.83	2036;1418,7446,3975,2923,6277
	25(g)	−336.86	−1409.42	2036
C_p°	25	15.64	65.44	7237,7826;2926,1088,2933
	Table 2.14			1681,7051

79. Ethanol

(Continued)

		cgs	SI	
Cp	25	26.85	112.34	2490;8269,7826,7738,7745,3933 5668,2288,5674,7846 2439,7805
λ	37.8	0.000406	0.170	6352;6834,1611
t_c		240.77	513.92	116;7234,3270,8231,1671,8233 5153,2214
P_c		60.68	6.148	116;7234,2214,3270,8231,5153
d_c		0.276	276	7234;4088,8231
v_c		0.167	0.000167	7234;1671,8233
A	eq 2.72	0.0024	0.0024	7236
K_B	eq 2.79	1.160	1.160	4437;730,7845,3432,6889 1370,720
pK_a	25	15.9	15.9	578;4766,649,5099
$pK_{BH}{+}$	25,in aq H_2SO_4	-1.94	-1.94	5755
pK_s		19.1	19.1	1181
κ	25	1.35×10^{-9}	1.35×10^{-7}	1810;4698,6463,3586,4703 7842,3210
ε	25	24.55	24.55	3719,1811;6876,3613,2953,4251 8184,6416
μ	20(1)	1.66	1.66	4739;4512,3381,4579,3382 6862,4494
δ		12.78	26.14	6240;7159,652,4593,1271
ER	BuOAc 90%	1.60 278	1.60 278	6628,6631;1399 6628,6631;1399
soly	in aq	inf	inf	2042
aq az	78.174	96.0%w	96.0%w	3497
fl pt	CC TOC	13 18	286 291	152 2352;7618,4193

Beil	<u>20</u>,292
uv	155,3189,5053,841,1883,7229
ir	6448,4586,6384,8291,4388,5165,6368,5932
Raman	1069,4861,7082,6067,1095,7832,7586
ms	6199,7199,5222,2578,1592,5054,2745
nmr	3609,4926,5938,6601,7586,3805,7227,2464,5933,6748

80. 1-Propanol

$CH_3CH_2CH_2OH$		71-23-8		C_3H_8O
		cgs	SI	
mw		60.096	60.096	3468
bp	1 atm	97.151	370.301	129;4294,7236,7768,5329,8130 7485,7239,853
dt/dp	1 atm	0.03476	0.2607	129;7239,7485,1197,853,1198 8130,4665
dp/dt	1 atm	28.77	3.836	129;4294
p	25	20.99	2.798	129;5332,7239,7236,7058,5975 3800,853
eq 2.5	A	7.75123	6.87613	129;3938,7235,134,1151,3270
	B	1441.705	1441.705	4967,853,1220,1655
	C	198.859	198.859	
fp		-126.2	147.0	7236;7239,5678,7058
tp		-124.40	148.75	1686
d	0	0.81930	819.30	4267;393,29,1197,1198 2677,4267
	15	0.80749	807.49	7485;3774,4417,4698,5329 5678,7549
	20	0.80361	803.61	3080;7241,7772,8231,7485,7239
	25	0.79960	799.60	3080;7241,1147,1989,6357
dd/dt	10 - 40	0.00079	0.79	7618
α	25	0.001004	0.001004	6357;3080,3481
κ_T	25	1.368×10^{-7}	1.026×10^{-6}	6152;1989,3867
κ_S	25	1.13×10^{-7}	8.49×10^{-7}	4315;3868
n_D	20	1.38556	1.38556	7242;7772,1198,3057,5329,7549
	25	1.38370	1.38370	1197,29,1147
dn/dt	25	0.000372	0.000372	7242;5598
η	-195.7(gl)	3.0×10^{20}	3.0×10^{17}	4614;557,2129,2677,5329 7522,7549
	15	2.492	0.002492	7240;7485,6080
	25	1.9430	0.0019430	7240;6080,2309,4315,3724
	30	1.725	0.001725	7240;5853,2020,7482,7705,8019
	Table 2.2			
γ	20	23.45	0.02345	4790;7045,3283,1165,4417,5329
	30	22.76	0.02276	7485,7522,7772
	Table 2.7			
ΔH_v	25	11.310	47.321	4294,7236,7821,2926;5902,1670 4665,2923,7754,8089,853 3618,660,7784,6277
	bp	9.850	41.211	4294;7236,7821,2926,7110
ΔH_m		1.284	5.372	1686;5678
ΔH_f°	25(1)	-72.66	-304.01	2036;1418,5283,1915,7237,2923
ΔH_c°	25(1)	-482.75	-2019.83	2036;1418,5283,1915,6888,7446 3975,2923,6277

80. 1-Propanol

(Continued)

		cgs	SI	
C_p°	25	20.82	87.11	7247;7051,2926,4967
C_p	25	34.39	143.87	2490;7738,8269,7745,768,8284 7549,3692,5817,4665
λ	37.8	0.000377	0.158	6352
t_c		264.17	537.32	8267;7234,2214,134,5153,3270 8233,8231,1686
P_c		51.14	5.182	8267;7234,2214,134,8231 3270,5153
d_c		0.2787	278.7	8267;7234,2214,134
v_c		0.2156	0.0002156	8267;7234,2214,1686,8233
A	eq 2.72	0.0289	0.0289	7236
K_B	eq 2.79	1.59	1.59	726
pK_s	25	19.4	19.4	1426
κ	18	9.17×10^{-9}	9.17×10^{-7}	3586;4698,8019
ε	25	20.45	20.45	2309;3719,1811,6119,8184,3586 5251,3613,2814,3724
μ	20(1)	3.09	3.09	4739;8141,4830,6860,3057 4511,4494
δ		12.18	24.91	6240,7635;652,4593,1271
ER	BuOAc 90%	0.86 530	0.86 530	6628,6631;1399,2190,6905 6628,6631;1399
soly	in aq	inf	inf	2042
aq az	87.65	71.7%w	71.7%w	3498
fl pt	CC TOC	25 27	298 300	152 7618;4193
Beil	24,350			
uv	841,5883,5918,3189			
ir	2874,4270,5192,2760,8312,5479,4505,5932			
Raman	7949,1069,6018,4861,4279			
ms	2745,154,2578,2572,5054,2947,7408			
nmr	7227,4926,4340,2464,5933,6748			

81. 2-Propanol

CH$_3$CHOHCH$_3$ 67–63–0   C$_3$H$_8$O

		cgs	SI	
mw		60.096	60.096	3468
bp	1 atm	82.242	355.392	129;7245,7239,7485,7132 4933,1286
dt/dp	1 atm	0.03476	0.2607	129;7239,853,7485,1197,1198
dp/dt	1 atm	28.77	3.836	129
p	25	43.32	5.775	129;7245
eq 2.5	A B C	7.74128 1360.131 197.592	6.86618 1360.131 197.592	129;7244,134,853,4268
fp		−88.0	185.2	7245;357,7485,5682,7484 5678,7476
d	20 25	0.78545 0.78126	785.45 781.26	7241;1333,1197,2677,4417,5581 5676,3934,7239,4969,5687 7522,7768,3080
dd/dt	10 – 40	0.00082	0.82	7618
α	25	0.001064	0.001064	3080;2272
κ_T	40	1.776 x 10^{-7}	1.332 x 10^{-6}	3897
κ_S	35	1.421 x 10^{-7}	1.066 x 10^{-6}	3868
n$_D$	20 25	1.3772 1.3752	1.3772 1.3752	7242;1197,7768,8080,1286,7971
dn/dt	25	0.00041	0.00041	7242
η	15 25 30	2.773 2.0436 1.767	0.002773 0.0020436 0.001767	7242;7485,6808,2129,7522,8019 2677,7705,5853
	Table 2.2			
γ	15 30	21.79 20.96	0.02179 0.02096	7485;3283,7768,4417
ΔH$_v$	25 bp	10.88 9.53	45.52 39.87	7245;8089,660,768,5676,4756 4969,7821,3078,798,7754 5902,5475
ΔH$_m$		1.292	5.406	7245;357,5682,3934
ΔH$_f^\circ$	25(1)	−75.97	−317.86	2036;1418,7246,1232,5686
ΔH$_c^\circ$	25(1)	−479.44	−2005.98	2036;1418,6888,7446,3975 5686,5685
C$_p^\circ$	25	21.21	88.74	7247;798,2934,3078,7051
C$_p$	25.07	36.950	154.599	357;798,8080,3692,5817,5682
λ	37.8	0.000336	0.141	6352,2272
t$_c$		235.15	508.30	116,118;5617,6082,1156
p$_c$		46.993	4.7616	118;5617,134
d$_c$		0.27317	273.17	118;5617,134
v$_c$		0.2230	0.0002230	118;5617,1156

81. 2-Propanol

(Continued)

		cgs	SI	
A	eq 2.72	0.0190	0.0190	7245
pK_a	25,in aq H_2SO_4	-3.2	-3.2	649;4766,1023
pK_s	25	21.08	21.08	6858;1300
κ	25	5.8×10^{-8}	5.8×10^{-6}	3210;6802,8019
ε	25	19.92	19.92	3719,1811;4134,6951,5117,4417 1400,2814
μ	30 in 38	1.66	1.66	4933;3383,7031,4825,6862,6190
δ	25	11.5	23.5	7159,652,4593;6240
ER	BuOAc 90%	2.30 319	2.30 319	2272,7635,6905;6631,6628 6631,6628
soly	in aq	inf	inf	2272
aq aq	80.10	88.0%w	88.0%w	3498
fl pt	TCC TOC	12 13	285 286	2272 152,2271,2272

Beil	24,360
uv	3189,5918,7235
ir	6844,5479,3427,2934,642,2428,4554,5932
Raman	2934,7175,7720,466,8291,153,6384
ms	5054,2578,2572,6018,154,1069.2745
nmr	7227,4926,3450,5938,4852,1423,2464,5933,6748

82. 1-Butanol

$CH_3CH_2CH_2CH_2OH$ 71–36–3 $C_4H_{10}O$

		cgs	SI	
mw		74.122	74.122	3468
bp	1 atm	117.725	390.875	129;7831,7473,7236,7239,1198 3751,5329
dt/dp	1 atm	0.03712	0.2784	129;8130,1286,7239,853,1198 2090,7495
dp/dt	1 atm	26.94	3.592	129
p	25	6.83	0.910	129;7236,5975,2091,1286,7058
eq 2.5	A B C	7.42253 1338.769 177.042	6.54743 1338.769 177.042	129;7235,134,2090,1286,3270 853,3938,1151
fp		−88.62	184.53	7236;2090,5674,7480,7239,1090
tp		−88.64	184.51	1683
d	20	0.80956	809.56	3080;978,7241,7495,4394,1198 1286,3774,6876
	25	0.80575	805.75	3080;3900,1989,6357
dd/dt	10 – 40	0.00076	0.76	7618
α	25	0.000948	0.000948	6357;3080,3481
κ_T	25	1.26×10^{-7}	9.42×10^{-7}	1989;6152,3867
κ_S	35	1.15×10^{-7}	8.66×10^{-7}	3868;5945
n_D	20 25	1.39929 1.39741	1.39929 1.39741	6175;7242,563,5598,978,4615 29,1204,5329,7768,6876 7495,1151,86,4969,7549
dn/dt		0.00039	0.00039	7495;5598
η	15	3.159	0.003159	6080;7495,1198,2129,3774,7549 2201,3751,5329
	25 30 Table 2.2 Table 2.3	2.5710 2.271	0.0025710 0.002271	7240;6080,6175,3900 7495,5853;7240,6175,563 1798,3900
γ	20 30 Table 2.8	24.67 23.78	0.02467 0.02378	978;7045,3283,4203,5329 4417,7768
ΔH_v	25	12.51	52.34	5902;7236,2923,768,3618,2926 660,1286
	bp Table 2.9	10.310	43.137	1683;853,4969,7821,4665,4756 6277,7110
ΔH_m		2.240	9.372	1683,7236;5674
ΔH_f°	25(1)	−78.25	−327.40	2036;5282,5283,6332,6790,7237 1418,3022,2923
ΔH_c°	25(1)	−639.53	−2675.79	2036;6790,5282,5283,3975,1418 3022,2923,6277
C_p°	25 Table 2.14	26.29	110.00	7237;2926,1683,7051

82. 1-Butanol

(Continued)

		cgs	SI	
C_p	25 Table 2.11	42.32	177.08	8269;1683,768,7549,5817 5674,8080
λ	37.8	0.000366	0.153	6352
t_c		289.85	563.00	7234;3270,134,6756,3345,6082 5726,2214
p_c		43.55	4.413	7234;134,6756,3270,3345,2214
d_c		0.270	270	7234;6756,134,2214
v_c		0.274	0.000274	7234;2214
A	eq 2.72	0.0331	0.0331	7236
pK_s	25	20.89	20.89	6858
κ		9.12×10^{-9}	9.12×10^{-7}	3586;1813,3965
ε	25	17.51	17.51	1811;4937,4417,2953,2882,6876 8184,2814
μ	25 in 317	1.75	1.75	4511;3057,6862,2882,6860,6876 8141,6191
δ		11.60	23.73	6240;652,4593,1271
ER	BuOAc 90%	0.43 1076	0.43 1076	6628,6631;3455,778,7635 6628,6631;1399
soly	in aq,25	7.45%w	7.45%w	7018;794,3745,3403,2601,987 2042,7219
	aq in,25	20.5%w	20.5%w	7018;3403,3745
aq az	92.7	57.5%w	57.5%w	3497;7018
fl pt	CC TOC	29 36	302 309	5712 7618;4193

Beil	24,367
uv	5883,4388,841,7229,5918
ir	1801,6398,4279,5192,2655,8291,6384,6770,5932
Raman	1069,7720,4861,4103,7719,4279
ms	5054,2572,154,2745,7572,7408
nmr	7227,5003,3805,2464,5933,6748

83. 2-Butanol

$CH_3CH_2CHOHCH_3$		78–92–2		$C_4H_{10}O$
		cgs	SI	
mw		74.122	74.122	3468
bp	1 atm	99.512	372.662	129;7245,7239,1197,799 86,3509
dt/dp	1 atm	0.03542	0.2657	129;7239,853,1198,1197,7495
dp/dt	1 atm	28.23	3.764	129
p	25	17.38	2.317	129;7245,853,1286,6241,6636 1653,7058
eq 2.5	A	7.22967	6.35457	129;7244,853,134,799,6636
	B	1171.891	1171.891	1286,1147
	C	169.955	169.955	
fp		−114.7	158.5	7058,7239,1549
d	20	0.80652	806.52	8047;3080,7241,6175,6176,7239
	25	0.80241	802.41	4417,4543
dd/dt	10 – 40	0.00080	0.80	7618
α	25	0.001024	0.001024	3080;4195,3481
n_D	15	1.39946	1.39946	7495;86,1197,1069,1286,4543
	20	1.39706	1.39706	6176,6175;7242,1147
	25	1.39530	1.39530	
dn/dt	25	0.000364	0.000364	6176
η	20	3.632	0.003632	7240;7495,4417,4349,7416,7419
	25	2.998	0.002998	
	30	2.4989	0.0024989	
γ	20	23.37	0.02337	4195;3283
	30	22.62	0.02262	3283
ΔH_v	25	11.87	49.66	7245,7821;1286,4756,4969,5902
	bp	9.75	40.79	
ΔH_f°	25(1)	−81.88	−342.59	2036;1418,6332,7246,1232,6790
	25(g)	−70.100	−293.30	6223
ΔH_c°	25(1)	−635.90	−2660.61	2036;1418,6790
C_p°	25	26.878	112.458	6223;7247,799,7051
C_p	25	51.1	213.8	5664
	Table 2.11			
t_c		262.80	535.95	7243,4299;134
p_c		41.39	4.194	7243,4299;134
d_c		0.276	276	7243,4299;134
v_c		0.268	0.000268	7243,4299
$[\alpha]_D$	20	+13.87	+13.87	5833,5835;4550
κ		$<1.0 \times 10^{-7}$	$<1.0 \times 10^{-5}$	1813
ε	25	16.56	16.56	1811;4417,2100,7497,1813 4937,2814
δ		10.8	22.1	652,4593,1271;7159

83. 2-Butanol

(Continued)

		cgs	SI	
R	BuOAc 90%	0.81 563	0.81 563	6628;6631,1399,778 6628,6631;1399
soly	in aq,20 aq in,25	12.5%w 44.1%w	12.5%w 44.1%w	2042;3403,3745,5262
aq az	87.0	73.2%w	73.2%w	3498;110
l pt	TOC	23	296	7618;152
Beil	24,371			
uv	5918,5883			
ir	642,7579,1719,1801,1171,7055,6368,6770,5932			
Raman	1069,799,4183			
ms	5054,2572,154,2745,99,7408			
nmr	7227,5003,5933,6748			

84. 2-Methyl-1-propanol

$(CH_3)_2CHCH_2OH$ 78–83–1 $C_4H_{10}O$

		cgs	SI	
mw		74.122	74.122	3468
bp	1 atm	107.886	381.036	129;7245,1681,86,1301,5329 7239,4969,3427
dt/dp	1 atm	0.03611	0.2709	129;7239,1198,4929,1286
dp/dt	1 atm	27.69	3.692	129
p	25	11.45	1.527	129;7245,7058,3427
eq 2.5	A B C	7.37601 1275.197 175.787	6.50091 1295.197 175.787	129;7244,1286,134,3427
fp		–108	165	7058;7453
t_p		–101.97	171.18	1686

84.   2-Methyl-1-propanol

(Continued)

		cgs	SI	
d	20	0.8016	801.6	7241;6175,7495,1198,1301,4417
	25	0.7978	797.8	7549,7239,86,1286,2677
				5329,7768,1147
dd/dt	10 – 40	0.00076	0.76	7618
α		0.00095	0.00095	3481
κ_S	35	1.27×10^{-7}	9.50×10^{-7}	3868
n_D	15	1.39768	1.39768	7495;86,4656,5329,7768
	20	1.39591	1.39591	6175;7242,1286,4969,7549
	25	1.39389	1.39389	6175;1147,1813,1198,7242
dn/dt		0.00039	0.00039	7495
η	15	4.6556	0.0046556	7240;7495,1198,5329,7549,7852
				2677,7522
	25	3.3330	0.0033330	7240;6175,2129,4349,7705
	Table 2.2			
	Table 2.3			
γ	20	22.98	0.02298	3283
	30	22.11	0.02211	
	Table 2.8			
ΔH_v	25	12.14	50.79	5902;7246,768,660,7754,7821
				1286,853,4969
	bp	10.013	41.893	1681;7246
ΔH_m		1.511	6.322	1686
ΔH_f°	25(1)	−79.85	−334.09	2036;6790,6332,1418,7246
ΔH_c°	25(1)	−637.93	−2669.10	2036;6790,7446,3975,1418
C_p°	25	27	113	7246;1681,7051
C_p	25	43.26	181.00	1686;768,8080,7549,8284
λ	37.8	0.000332	0.139	6352
t_c		274.63	547.78	116;7243,1156,6854,134,6352
P_c		42.39	4.295	7243,4299
d_c		0.272	272	7243,4299
v_c		0.273	0.000273	7243,4299
K_B	eq 2.79	2.166	2.166	4437;726
pK_s	25	21.08	21.08	6858
κ	25	1.6×10^{-8}	1.6×10^{-8}	1813
ε	25	17.93	17.93	1811;4134,1813,7850,2100,4848
				4417,6416,2814
μ		1.79	1.79	4825;3381,3383,6190
δ		10.5	21.5	2190,1399,6905;7635
ER	BuOAc	0.62	0.62	6628,6631,1399;2190,6905
	90%	740	740	6628,6631;1399

84.   2-Methyl-1-propanol

(Continued)

		cgs	SI	
soly	in aq,25	10%w	10%w	2042;3403,2601,987
	aq in,25	16.9%w	16.9%w	
aq az	89.8	67%w	67%w	3497;7018
fl pt	CC	28	301	5712
	TOC	35	308	7618

Beil	24,373
uv	5918
ir	5285,153,6384,466,8291,2275,6770,5932
Raman	5649,4861
ms	2745,154,7408
nmr	7227,6748

85.   2-Methyl-2-propanol

$(CH_3)_3COH$ 75-65-0 $C_4H_{10}O$

		cgs	SI	
mw		74.122	74.122	3468
bp	1 atm	82.347	355.497	129;7245,2090,1286,7485,86
				7121,853
dt/dp	1 atm	0.03343	0.2508	129;7239,2090,7485,1286
dp/dt	1 atm	29.91	3.988	129
p	25	42.28	5.637	129;7245,2091,5676,7058,836
eq 2.5	A	7.23158	6.35648	129;7244,134,2090,1286,836
	B	1107.060	1107.060	
	C	172.102	172.102	
fp	(s)	25.62	298.77	3674;7245,3983,1975,5675,1900
				7239,6750
	(ms)	25.00	298.15	3674

85. 2-Methyl-2-propanol

(Continued)

		cgs	SI	
d	25(a)	0.7812	781.2	7241;86,2090,5631,6925,7121 7485,7239
	30	0.77545	775.45	3081;5853,1147,7241
	40	0.7649	764.9	7241
dd/dt	30	0.001032	1.032	3081
α	30	0.001325	0.001325	3081
n_D	20(a)	1.3877	1.3877	7242;2090,1069,86
	25(a)	1.3852	1.3852	
	27	1.3840	1.3840	1147
η	20	5.942	0.005942	7240
	25	4.438	0.004438	
	30	3.378	0.003378	5853;7240,7485,6925
	Table 2.2			
γ	26	20.02	0.02002	3283;7482,7485,5631
	35	19.10	0.01910	
ΔH_v	25	11.19(a)	46.82	7245;1286,7821,4756,836,853? 5676,1903,5902
	bp	9.33	39.04	
ΔH_s	25	12.73	53.26	7245;8143
ΔH_m	(s)	1.587	6.640	3674;7245,5562,7167,5675,6790 2720,1903
	(ms)	1.101	4.607	3674
ΔH_f°	25(s)	-87.5	-366.1	2036
	25(1)	-85.86	-359.24	2036,7246,8316;6790
ΔH_c°	25(s)	-630.3	-2637.2	2036
	25(1)	-631.92	-2643.95	2036,6790;3975,7446,8316
C_p°	25	27.10	113.39	7246;7051,836
	Table 2.14			
C_p	25.81(s)	35.04	146.61	5562;1903,5675
	25.81(1)	52.66	220.33	5562;6798
λ	37.8	0.000276	0.115	6352
t_c		233.0	506.2	7243;4299,5726,134
p_c		39.20	3.972	7243,4299;134
d_c		0.270	270	7243,4299;134
v_c		0.275	0.000275	4299;7243
A	eq 2.72	0.0090	0.0090	7245
K_f	eq 2.67	8.28	8.28	3674;2720,5691,482,864
K_B	eq 2.79	1.745	1.745	482
pK		19	19	4766;1023,1324,1955
pK_s		26.8	26.8	1300
κ	27	2.66×10^{-8}	2.66×10^{-6}	6870;3965
ε	25	12.47	12.47	1811;1400,4134,6870,7497 6951,5251

85. 2-Methyl-2-propanol

(Continued)

		cgs	SI	
μ	30 in 38	1.66	1.66	4930;5251,6191
δ		10.6	21.7	4593;6631
ER	BuOAc	1.30	1.30	6628;6631,778
	90%	352	352	6628;6631
soly	25	inf	inf	6631
aq az	79.9	88.24%w	88.24%w	3497
fl pt	CC	11	284	5712

(a) Undercooled liquid below normal freezing point

Beil	24,379
uv	5918,5883
ir	3452,153,6384,8291,6696,6826,633,6770,5932
Raman	2886,1069,5649
ms	2745,154,861,6717
nmr	3764,6606,4852,2933,2464,5933,6748

86. 1-Pentanol

$CH_3(CH_2)_3CH_2OH$ 71-41-0 $C_5H_{12}O$

		cgs	SI	
mw		88.149	88.149	3468
bp	1 atm	137.983	411.133	129;7236,1681,29,1286,6733
				7728,5329,7489
dt/dp	1 atm	0.03982	0.2987	129;7239,8130,7489,1286
dp/dt	1 atm	25.11	3.348	129
p	25	2.20	0.293	129;7236,7058,5332
eq 2.5	A	7.17816	6.30306	129;3938,7235,1286
	B	1286.333	1286.333	
	C	161.307	161.307	

86. 1-Pentanol

(Continued)

		cgs	SI	
fp		−78.2	195.0	7236;5681,7479,6733,7489
tp		−77.59	195.56	1686
d	20	0.81445	814.45	3080;7241,5598,7489,1286,4417
	25	0.81080	810.80	5579,29,7239,2764,5329
				6733,7768,1989,6357
dd/dt	10 − 40	0.00072	0.72	7618
α	25	0.000905	0.000905	6357;3080,4026
κ_T	25	1.18×10^{-7}	8.84×10^{-7}	1989;6152,3867
κ_S	35	1.08×10^{-7}	8.12×10^{-7}	3868
n_D	20	1.4100	1.4100	7242;5598,7768,1286,6733,29
	25	1.4080	1.4080	5329,7489
dn/dt	25	0.00042	0.00042	7242;5598
η	−195.7(gl)	2.1×10^{36}	2.1×10^{33}	4614;7416,6733,5329
	15	4.719	0.004719	7240;7489
	25	3.5128	0.0035128	7240;3774
	Table 2.2			
	Table 2.3			
γ	20	25.60	0.02560	3283;4417,7489,5329,7768,7045
	30	24.72	0.02472	
	Table 2.8			
ΔH_v	25	13.61	56.94	7236,7821;2926,1286,6277
				2923,4026
	bp	10.604	44.367	1681;7236,7821
ΔH_m		2.510	10.502	1686,5681;7236
ΔH_f°	25(l)	−84.27	−352.57	3021,2036;5283,1418,7237,2023
	25(g)	−70.66	−295.63	3021
ΔH_c°	25(l)	−795.88	−3329.96	3021,2036;5283,1418,7134
				2923,6277
C_p°	25	31.75	132.84	7247;7807,6352,2926,1681,7051
C_p	25	49.95	208.98	8269;6798,1686,5681,5817,7246
	Table 2.11			
λ	37.8	0.000357	0.149	6352
t_c		315.00	588.15	116;7234,3270,2214
p_c		38.58	3.909	116;2214,3270
d_c		0.270	270	4299,7234,2214
v_c		0.326	0.000326	4299,2214
A	eq 2.72	0.0311	0.0311	7236
pK_s	25	20.81	20.81	6858
ε	25	13.9	13.9	4937;4134,7497,8184,2665,4848
				4132,2814

86.   1-Pentanol

(Continued)

		cgs	SI	
μ	25 in 317	1.7	1.7	4511,4512;3381,4687,4494 2291,4830
δ	25	10.9	22.3	653
soly	in aq,25 aq in,25	2.19%w 7.46%w	2.19%w 7.46%w	2764;29,1287,987,7219
aq az	95.8	45.6%w	45.6%w	3497
fl pt	CC TOC	38 48	311 321	5712 2350;7618
Beil	24,383			
uv	5918			
ir	7225,4461,7055,8270,4021,3243,6844,5932			
Raman	2886,7720,1793			
ms	154,2572,7408			
nmr	7227,5933,6748			

87.   2-Pentanol

$CH_3CH_2CH_2CHOHCH_3$ 6032-29-7 $C_5H_{12}O$

		cgs	SI	
mw		88.149	88.149	3468
bp	1 atm	119.0	392.2	7236;1071,2052,8270,3599 1197,7489
dt/dp	1 atm	0.03762	0.2822	7239;7489
dp/dt	1 atm	26.58	3.544	7239
p	25	5.83	0.777	7236;1071,1286,7058,7418
eq 2.5	A B C	7.27575 1271.92 170.37	6.40065 1271.92 170.37	7244;1286
fp		(gl)	(gl)	7489;7479

87. 2-Pentanol

(Continued)

		cgs	SI	
d	20	0.8094	809.4	7241;7720,4026,1197,1071,7489
	25	0.8054	805.4	7239,5864
dd/dt		0.000799	0.799	7482
α		0.00097	0.00097	7489
n_D	20	1.4064	1.4064	7242;7720,1137,1071,4026,1197
	25	1.4044	1.4044	1286,8270,1818
dn/dt	25	0.00039	0.00039	7242
η	15	5.307	0.005307	7240;1818,7419,7489,7416,2129
	30	2.884	0.002884	
γ	20	23.98	0.02398	5864;7489
	30	22.96	0.02296	
ΔH_v	25	12.7	53.1	7236;1286
	bp	10.6	44.4	
ΔH_f°	25(1)	-87.75	-367.15	1418;7246,8418,6333
ΔH_c°	25(1)	-792.41	-3315.44	1418;8138
C_p	25	57.2	239.4	5664
	Table 2.11			
$[\alpha]_D$	20	+13.901	+13.901	1071;5835,5833
ε	25	13.71	13.71	1818;2291,7497
μ	22 in 38	1.66	1.66	2291
δ	25	10.8	22.0	653
ER	BuOAc	43	43	3455;778
soly	in aq,25	4.46%w	4.46%w	2764;2042
	aq in,25	11.79%w	11.79%w	
aq az	91.7	63.5%w	63.5%w	3497
fl pt	OC	42	315	4026;4193

Beil	24,384
uv	5918
ir	1108,1801,8270,153,5932
Raman	4102
ms	4971,2572,715
nmr	5933

88. 3-Pentanol

$(C_2H_5)_2CHOH$ 584–02–1 $C_5H_{12}O$

		cgs	SI	
mw		88.149	88.149	3468
bp	1 atm	115.3	388.5	7245;2764,7472,1197,5639 8032,6189
dt/dp	1 atm	0.03765	0.2824	7239;7482
dp/dt	1 atm	26.56	3.541	7239
p	25	8.24	1.10	7245;7418
eq 2.5	A	7.41493	6.53983	7244
	B	1354.42	1354.42	
	C	183.41	183.41	
fp		(gl)	(gl)	7472
d	20	0.8203	820.3	7241;7482,8032,2964,1197,7971
	25	0.8160	816.0	7239,5515,1818
dd/dt	25	0.00086	0.86	7241
α		0.00088	0.00088	3481
n_D	20	1.4104	1.4104	7242;7239,5639,1069,5835,8032
	25	1.4079	1.4079	8270,1137
dn/dt	25	0.00049	0.00049	7242
η	15	8.139	0.008139	7240;1818,7482,7416,7419
	30	3.852	0.003852	
γ	20	24.60	0.02460	7482;7489
	30	23.76	0.02376	
ΔH_v	25	12.7	53.1	7245;4756
	bp	10.4	43.5	
ΔH_f°	25(1)	−88.51	−370.33	1418;7246,1232,8316,6333
ΔH_c°	25(1)	−791.65	−3312.26	1418;8316
C_p	25	60.0	251.0	7246
	Table 2.11			
ϵ	25	13.35	13.35	1818;2291,4610,7497
μ	22 in 38	1.64	1.64	2291;4610
δ	25	10.2	20.8	653
soly	in aq,25	5.15%w	5.15%w	2764
	aq in,25	8.32%w	8.32%w	
aq az	91.7	64%w	64%w	3497;523
Beil	24,385			
ir	5932			
nmr	7227,5933,6748			

89. 2-Methyl-1-butanol

$CH_3CH_2CH(CH_3)CH_2OH$ 137–32–6 $C_5H_{12}O$

		cgs	SI	
mw		88.149	88.149	3468
bp	1 atm	128.7	401.9	7245;7216,2052,4801,1070 3510,538
dt/dp	1 atm	0.03898	0.2924	7239
dp/dt	1 atm	25.65	3.420	7239
p	25	3.12	0.416	7245;1070,2192,7418
eq 2.5	A	7.06730	6.19220	7244
	B	1195.26	1195.26	
	C	156.83	156.83	
fp		<−70	<203	2885
d	20	0.8191	819.1	7241;7416,7216,539,7239,8026
	25	0.8150	815.0	1070,2264,538,4543,2192
dd/dt	10 − 40	0.00076	0.76	7618
n_D	20	1.4107	1.4107	7242;7416,1069,8026,1070,3510
	25	1.4086	1.4086	4543,1818
dn/dt	25	0.00045	0.00045	7247
η	15	6.468	0.006468	1818;7240,7416,2885,7419
	25	4.797	0.004797	
	Table 2.2			
	Table 2.3			
γ	25	25.1	0.0251	6830
ΔH_v	25	13.0	54.4	7245;2885,4756
	bp	10.8	45.2	
ΔH_f°	25(1)	−85.24	−356.64	1418;8316,7246
ΔH_c°	25(1)	−794.92	−3325.95	1418;8316
C_p	25	52.6	220.1	7246
	Table 2.11			
$[\alpha]_D$	20.7	−5.900	−5.900	2192;7216,538,8026,2224 1070,5131
ε	25	15.63	15.63	1818;4937
μ	20 in 38	1.88	1.88	1818
soly	in aq,25	2.97%w	2.97%w	2764
	aq in,25	9.19%w	9.19%w	
aq az	93.8	58.5%w	58.5%w	523
fl pt	TOC	46	319	7618;2885

89.   2-Methyl-1-butanol

(Continued)

Beil	24,385
ir	1719,1801,7055,153,633,8270,5932
Raman	1069,4102
ms	2572
nmr	5933,6748

90.   3-Methyl-1-butanol

$(CH_3)_2CHCH_2CH_2OH$ 123–51–3 $C_5H_{12}O$

		cgs	SI	
mw		88.149	88.149	3468
bp	1 atm	130.5	403.7	7245;3806,4969,7476,29 2052,8080
dt/dp	1 atm	0.03886	0.2915	7239;7487
dp/dt	1 atm	25.73	3.431	7239
p	25	2.37	0.316	7245;7058,7596,1286
eq 2.5	A	6.95361	6.07851	7244;1286
	B	1128.19	1128.19	
	C	146.47	146.47	
fp		–117.2	156.0	7058;4026
d	20	0.8104	810.4	7241;7487,1286,4251,7768,7942
	25	0.8071	807.1	7239,29,2764,4969,13
dd/dt	10 – 40	0.00075	0.75	7618
α		0.00092	0.00092	7487
κ_S	35	1.16×10^{-7}	8.67×10^{-7}	3868
n_D	20	1.4072	1.4072	7247;1818,7487,8080,29
	25	1.4052	1.4052	7768,1286
dn/dt		0.00037	0.00037	7487

90.   3-Methyl-1-butanol

(Continued)

		cgs	SI	
η	-195.7(g1)	1.6×10^{35}	1.6×10^{32}	4614
	15	5.054	0.005054	1818;7487,7240
	25	3.738	0.003738	
γ	15	24.77	0.02477	7487;6830
	30	23.44	0.02344	
ΔH_v	25	13.3	55.6	7245;4756,1286,4665,1154,4969
	bp	10.6	44.4	
ΔH_f°	25(1)	-85.18	-356.39	1418;8316
ΔH_c°	25(1)	-794.98	-3326.20	1418;7446,8316
C_p°	178.5	45.80	191.63	7351
C_p	25	50.5	211.3	7246;6440,8080
λ	30	0.000354	0.148	6834
t_c		307	580	7243;6854,1156
K_B	eq 2.79	2.65	2.65	7590;726
κ	25	1.4×10^{-9}	1.4×10^{-7}	4246;3965
ε	25	15.19	15.19	1818;2814,4937,5117,6119,1394 4251,4134,3613
μ		1.82	1.82	4825
δ	25	11.0	22.5	7159;6240,4593
soly	in aq,25	2.67%w	2.67%w	3806;2764,987,2601
	aq in,25	9.61%w	9.61%w	2764
aq az	95.15	50.4%w	50.4%w	3497;523
fl pt	CC	43	316	5712;4026
	TOC	51	324	7618;4193

Beil	24,392
uv	5918
ir	153,1801,466,8270,2428,6384,8291,5932
Raman	1793,7720
ms	154,2572,7408
nmr	5933,6748

91. 2-Methyl-2-butanol

$(CH_3)_2COHCH_2CH_3$ 75–85–4 $C_5H_{12}O$

		cgs	SI	
mw		88.149	88.149	3468
bp	1 atm	102.0	375.2	7245;5681,2764,7472,483 7058,7489
dt/dp	1 atm	0.03727	0.2795	7239;7482
dp/dt	1 atm	26.83	3.577	7239
p	25	16.7	2.23	6166;7058
eq 2.5	A B C	6.5193 863.4 135.3	5.6442 863.4 135.3	7244;1286
fp		−8.8	264.4	7245;5681,7472,7058,7489
d	20 25	0.8096 0.8050	809.6 805.0	7241;7416,1286,5631,7239,2764 7489,1818
dd/dt	25	0.00092	0.92	7241
α		0.00133	0.00133	4026;7489
n_D	20 25	1.4050 1.4020	1.4050 1.4020	7247;7482,1069,1286,4026,1818
dn/dt	25	0.00056	0.00056	7247
η	15 25	5.641 3.548	0.005641 0.003548	1818;7240,7489
	Table 2.2 Table 2.3			
γ	20 30	22.77 21.89	0.02277 0.02189	3283;7482,7489,5631 5370;3283
ΔH_v	25 bp	12.0 9.7	50.2 40.6	7245;1286,4756,7167,1155,18
ΔH_m		1.065	4.456	7245,5681;7167
ΔH_f°	25(1)	−90.71	−379.53	1418;8316
ΔH_c°	25(1)	−789.45	−3303.06	1418;7446,3975,8316
C_p°	108.2	45.22	189.20	7351
C_p	21.2	58.4	244.3	5681;4026,7246
	Table 2.11			
t_c		272	545	7243;6854,1156
A	eq 2.72	0.0077	0.0077	7245
K_f	eq 2.67	10.4	10.4	4043
K_B	eq 2.79	2.255	2.255	7590;726
pK		19	19	4766∓1023
ε	25	5.78	5.78	1818;2814,4937,7497,6657,1394
μ		1.7	1.7	6657
δ		10.0	20.5	6631

91.   2-Methyl-2-butanol

(Continued)

		cgs	SI	
ER	BuOAc	0.91	0.91	6628,6631;778
	90%	505	505	6628,6631
soly	in aq,25	11.00%w	11.00%w	2764
	aq in,25	23.47%w	23.47%w	
aq az	87.35	72.5%w	72.5%w	3497;523
fl ,pt	OC	21	294	4026

Beil	24,388
uv	5918
ir	1719,1801,2275,153,8291,7055,5932
Raman	1069,7720,4861
ms	154,2572
nmr	5575,1423,5933,6748

92.   3-Methyl-2-butanol

$(CH_3)_2CHCHOHCH_3$ 598–75–4 $C_5H_{12}O$

		cgs	SI	
mw		88.149	88.149	3468
bp	1 atm	111.5	384.7	7245;8270,4026,2764
dt/dp	1 atm	0.03781	0.2836	7239
dp/dt	1 atm	26.45	3.526	7239
p	25	8.97	1.20	7245
eq 2.5	A	6.94210	6.06700	7244
	B	1090.93	1090.93	
	C	157.15	157.15	
d	20	0.8184	818.4	7241;5834,4026,7239,2764
	25	0.8138	813.8	
dd/dt	25	0.00092	0.92	7241
n_D	20	1.4096	1.4096	7242;8270,4026,5834
	25	1.4075	1.4075	

92. 3-Methyl-2-butanol

(Continued)

		cgs	SI	
dn/dt	25	0.00042	0.00042	<u>7242</u>
η	25	3.51	0.00351	2129
	30	3.163	0.003163	7240;7416
γ	25	23.0	0.0230	6830
ΔH_v	25	12.4	51.9	7245;4756
	bp	10.0	41.8	
ΔH_f°	25(1)	-87.63	-366.64	1418;8316
ΔH_c°	25(1)	-792.53	-3315.95	1418;8316
$[\alpha]_D$	20	+4.85	+4.85	5834
δ	25	10.0	20.5	653
soly	in aq, 25	5.55%w	5.55%w	2764
	aq in, 25	11.93%w	11.93%w	
aq az	91.0	67%w	67%w	3497
fl pt		38	311	4193
Beil	24,391			
ir	8270,1801,7579			
Raman	4102			
ms	2572			
nmr	7670,5933			

93. 2,2-Dimethyl-1-propanol

(CH₃)₃CCH₂OH   75–84–3   C₅H₁₂O

		cgs	SI	
mw		88.149	88.149	3468
bp	1 atm	113.1	386.3	7239;7704,2764,6410,5285,5918
dt/dp	1 atm	0.03676	0.2757	7239
dp/dt	1 atm	27.20	3.627	7239
p	25(1)*	10	1.3	7244
eq 2.5	A B C	7.8753 1604.7 208.2	7.0002 1604.7 208.2	7244
fp		52	325	7239;6410,7504,8027,2764,690
ΔH_v	bp	2.3	9.6	7244
K_f	eq 2.67	11.0	11.0	8027
ε	60	8.35	8.35	6880
soly	in aq,25 aq in,25	3.50%w 8.36%w	3.50%w 8.36%w	2764

*Undercooled liquid

Beil	24,406
uv	5883
ir	5285,8270,5932
Raman	4102
ms	2572
nmr	5933

94. Cyclohexanol

$CH_2CH_2CH_2CH_2CH_2CHOH$ 108–93–0 $C_6H_{12}O$

		cgs	SI	
mw		100.160	100.160	3468
bp	1 atm	161.10	434.25	7491;1904,2657,2776,1729
				3547,7608
dt/dp	1 atm	0.045	0.34	7491
dp/dt	1 atm	22	3.0	7491
p	56.0	10	1.3	7058;2659
eq 2.5	A	6.80369	5.92859	5379;5524
107 - 160	B	1199.1	1199.1	
	C	145.0	145.0	
fp		25.15	298.30	7491;506,3560,1729,4397,6714
d	25	0.9684	968.4	6293;1904,3547,506,7758,7971
	30	0.94155	941.55	7491
	45	0.92994	929.94	
α		0.00077	0.00077	3481
κ_S	35	6.855×10^{-8}	5.142×10^{-7}	5945
n_D	25	1.46477	1.46477	7482;7608,2659,7758
	30	1.4629	1.4629	7971
	37	1.46055	1.46055	506
η	30	41.067	0.041067	7482;7491
	45	17.194	0.017194	
γ	25.5	33.91	0.03391	7491;3161,3547
	30	33.47	0.03347	
ΔH_v	25	14.82	62.01	7821;1905
	158.7	10.87	45.48	5777
ΔH_m		0.406	1.70	3935;1905
ΔH_f°	25(1)	–83.45	–349.15	6912;6013,8316
ΔH_c°	25(1)	–890.77	–3726.98	6912;6013,6563,3975,8316
C_p	31.9	48.38	202.42	5817
t_c		352	625	2776,4299;5379
p_c		37	3.7	2776,4299
K_f	eq 2.67	39.3	39.3	5124;1905,8103
ε	25	15.0	15.0	4937;4134,1729,8075,7497,5251
μ	25 in <u>317</u>	1.86	1.86	4511;828,5251,8075,3105,2602
δ		11.4	23.3	1271,4593,652
ER	BuOAc	0.05	0.05	6628,6631;1399,6905,7635
	90%	9160	9160	6628,6631;5621
soly	in aq,24.6	3.75%w	3.75%w	6714;8295,1904
	aq in,20	11.78%w	11.78%w	2042
aq az	97.8	30.5%w	30.5%w	3498;2879
fl pt	CC	68	341	152

94. Cyclohexanol

(Continued)

Beil	<u>502</u>,5
uv	4388
ir	4505,1171,5024,1942,8270,5837,633,4960,5932
Raman	7720,5434,6355,5466,7228,1338,5797
ms	2572,5909
nmr	7068,5437,1423,4477,1795,5933,6748

95. 1-Hexanol

$CH_3(CH_2)_4CH_2OH$ 111–27–3 $C_6H_{14}O$

		cgs	SI	
mw		102.176	102.176	3468
bp	1 atm	157.0	430.2	7236;29,1286,1287,5329 7475,7768
dt/dp	1 atm	0.0406	0.305	7233;2770
dp/dt	1 atm	24.6	3.28	7233
p	25	0.82	0.11	7236;1286,3518,7058
eq 2.5	A	7.28781	6.41271	3938;7235,6259,1286,3518,3270
	B	1422.031	1422.031	
	C	165.444	165.444	
fp		−44.6	228.6	7248;2770,3933,7058,7475,7479
d	20	0.81875	818.75	3702;2430,5598,7249,2770,1286
	25	0.81534	815.34	3774,5579,7233,29,1287 7768,6357,1989,1072
dd/dt	10 – 40	0.00070	0.70	7618
α	25	0.000878	0.000878	6357;4317,2430,3702,4195,3481
κ_T	25	1.10×10^{-7}	8.24×10^{-7}	1989;6152
κ_S	25	9.6725×10^{-8}	7.2255×10^{-7}	3131;4317

95. 1-Hexanol

(Continued)

		cgs	SI	
n_D	20	1.4172	1.4172	5598;7233,3542,1286,5329,7768
	25	1.4157	1.4157	29,1287,6753
dn/dt		0.0004	0.0004	5598
η	15	6.293	0.006293	3518;2770,5310,3774,5329
	25	4.592	0.004592	
	30	3.765	0.003765	6759
γ	20.12	26.14	0.02614	7045;4195,3283,3518,5329,6830
	25.20	25.73	0.02573	5310,7768
	Table 2.8			
ΔH_v	25	14.78	61.85	4876;7236,2923,2926,7821,1286
				6277,3518
	bp	11.8	49.4	7236
ΔH_m		3.68	15.40	7236;3933
ΔH_f°	25(1)	-90.86	-380.16	2036;5283,1418,2923,7237,8316
ΔH_c°	25(1)	-951.86	-3982.58	1418,2036;5283,2923,6277,8316
C_p°	25	37.2	155.6	7237;7807,2926
C_p	25	57.68	241.32	8269;1072,7826,3933
λ	37.8	0.000363	0.152	6352
t_c		337	610	3270;7234,2214,3518,6352
P_c		34.1	3.46	2214;3270
d_c		0.268	268	2214;7234
v_c		0.380	0.000380	2214
A	eq 2.72	0.0361	0.0361	7236
ϵ	25	13.3	13.3	4937;6760,6119,4848,2665,5556
μ	20 in 38	1.55	1.55	4273;4511,4494,4687,4512
				4604,4590
δ	25	10.7	21.9	7159,1271,4593
ER	BuOAc	0.05	0.05	6905,7635
soly	in aq,20	0.706%w	0.706%w	29;1287,2601,5725,7219
	aq in,20	7.42%w	7.42%w	4027
aq az	97.8	32.8%w	32.8%w	3498;3497
fl pt	TOC	62	335	2350;7618,4193
Beil	24,407			
uv	5918			
ir	153,6959,4021,8291,5932			
Raman	1719,2428,8270,4861			
ms	2572			
nmr	5933,6748			

96.   2-Methyl-1-pentanol

$CH_3CH_2CH_2CH(CH_3)CH_2OH$ 105–30–6 $C_6H_{14}O$

		cgs	SI	
mw		102.176	102.176	3468
bp	1 atm	147.93	421.08	3518;7248
dt/dp	1 atm	0.0408	0.306	7248
dp/dt	1 atm	24.5	3.27	7248
p	25	2.6	0.35	3518;7618,7418
eq 2.5	A	7.86701	6.99191	7250;3518
	B	1775.12	1775.12	
	C	208.00	208.00	
d	20	0.8242	824.2	7248;7249,3518
	25	0.8206	820.6	
dd/dt	10 – 40	0.00075	0.75	7618
n_D	20	1.4190	1.4190	7248,3518
	25	1.4172	1.4172	
dn/dt	25	0.00039	0.00039	3518
η	15	8.407	0.008407	3518;7419
	25	5.553	0.005553	
γ	15	25.66	0.02566	3518
	25	24.94	0.02494	
ΔH_v	bp	12.000	50.208	3518
t_c		348.9	622.1	3518
δ	25	10.2	20.8	653
soly	in aq,20	0.31%w	0.31%w	7618
	aq in,20	5.4%w	5.4%w	
aq az	97.2	40%w	40%w	3498
fl pt	TOC	57	330	7618
Beil	24,409			
ir	5932			

97. 2-Methyl-2-pentanol

$(CH_3)_2COHCH_2CH_2CH_3$ 590-36-3 $C_6H_{14}O$

		cgs	SI	
mw		102.176	102.176	3468
bp	1 atm	121.4	394.6	7248;8065,7058,3517,1963
dt/dp	1 atm	0.0433	0.325	7248
dp/dt	1 atm	23.1	3.08	7248
p	25	8.6	1.1	3517
eq 2.5	A	6.15175	5.27665	8065;3517
	B	811.05	811.05	
	C	126.60	126.60	
fp		-102	171	7248;8065,3517,7058,1963
d	20	0.8136	813.6	7249;8065,3517,5631
	25	0.8095	809.5	
dd/dt	25	0.00083	0.83	7249
n_D	20	1.4113	1.4113	7248;8065,3517
	25	1.4089	1.4089	
dn/dt	25	0.00048	0.00048	7248
η	25	2.194	0.002194	3517
	35	1.631	0.001631	
γ	25	22.90	0.02290	3517;5631
	35	22.05	0.02205	
ΔH_v	25	14.3	59.8	8065
	bp	9.10	38.07	8065;3517
t_c		286.3	559.5	4432
δ	25	9.6	19.6	653
soly	in aq,25	3.24%w	3.24%w	2766;29
	aq in,25	10.05%w	10.05%w	2766
Beil	24,EII 439			
ms	2572			
nmr	5933,6748			

98. 4-Methyl-2-pentanol

$(CH_3)_2CHCH_2CHOHCH_3$ 108–11–2 $C_6H_{14}O$

		cgs	SI	
mw		102.176	102.176	3468
bp	1 atm	131.7	404.9	7248;4549,6736,1137,2766
				1197,7058
dt/dp	1 atm	0.0413	0.310	7248;1198
dp/dt	1 atm	24.2	3.23	7248
p	25	8.2	1.1	3518;7058,7418
eq 2.5	A	6.15173	5.27663	7250;3518,363
	B	811.05	811.05	
	C	126.60	126.60	
fp		−90	183	7618,6905
d	20	0.8080	808.0	7249;1197,2766,3518,7248
	25	0.8036	803.6	1198,7971
dd/dt	10 − 40	0.00082	0.82	7618;6905
α		0.00103	0.00103	6905;3481,7618
n_D	20	1.4112	1.4112	7248;6736,3518,1198,1197
	25	1.4090	1.4090	7971,1137
dn/dt	25	0.00044	0.00044	7248
η	25	4.074	0.004074	3518;7419
	35	2.715	0.002715	
γ	25	22.63	0.02263	3518
	35	21.87	0.02187	
ΔH_v	25	14.03	58.70	363
	bp	10.57	44.22	363;3518
ΔH_f°	25(1)	−94.34	−394.72	6333
t_c		301.2	574.4	4432
$[\alpha]_D$	19.5	+21.19	+21.19	3952;4549
	17	−20.80	−20.80	
μ		1.7	1.7	2190
δ		10.0	20.5	4593,1271,6905,1399,6631
ER	BuOAc	0.27	0.27	6628,6631;1399,2190
	90%	1711	1711	6628,6631;1399
soly	in aq,25	1.64%w	1.64%w	2766
	aq in,25	6.35%w	6.35%w	
aq az	94.3	56.7%w	56.7%w	3498
fl pt	TOC	41	314	2349,7618
	TCC	46	319	3481

98. 4-Methyl-2-pentanol

(Continued)

Beil	24,410
ir	8270,7579,5932
ms	2572
nmr	7670,5933

99. 2-Ethyl-1-butanol

CH$_3$CH$_2$CH(C$_2$H$_5$)CH$_2$OH   97-95-0   C$_6$H$_{14}$O

		cgs	SI	
mw		102.176	102.176	3468
bp	1 atm	146.5	419.7	7248;8025
dt/dp	1 atm	0.0433	0.325	7248;3519
dp/dt	1 atm	23.1	3.08	3519
p	25	3.8	0.51	3519;7418
eq 2.5	A	6.84055	5.96545	7250;3519
	B	1188.69	1188.69	
	C	153.70	153.70	
fp		-114.4	158.8	7248;4026
d	20	0.8333	833.3	7248;3519,7971
	25	0.8295	829.5	
dd/dt	10 - 40	0.00078	0.78	7618
α		0.00092	0.00092	6905;3481
κ$_S$	25	1.736 x 10^{-7}	1.302 x 10^{-6}	7171
n$_D$	20	1.4224	1.4224	7248;8025,3519,7971
	25	1.4205	1.4205	
dn/dt	25	0.00038	0.00038	7248

99. 2-Ethyl-1-butanol

(Continued)

		cgs	SI	
η	15	8.021	0.008021	3519;7419
	25	5.892	0.005892	
γ	15	25.06	0.02506	3519
	25	24.32	0.02432	
ΔH_v	bp	10.322	43.187	3519
t_c		145.7	418.9	3519
ϵ	90	6.19	6.19	5556
δ		10.5	21.5	652,1271
ER	BuOAc	0.08	0.08	6905
soly	in aq,20	0.63%w	0.63%w	2042;4026
	aq in,20	4.56%w	4.56%w	
aq az	96.7	41.3%w	41.3%w	3497
fl pt	TOC	53	326	7618;2042
Beil	24,408			
ir	6959,5932			
nmr	6748			

100.   1-Methylcyclohexanol

```
    HOCCH₃
  H₂C    CH₂
   |      |
  H₂C    CH₂
     \   /
      C
      H₂
```

590–67–0   C₇H₁₄O

		cgs	SI	
mw		114.187	114.187	3468
bp	1 atm	156–158	429–431	8273;6321
p	70	25	3.3	5278;5059,4896,506,6322 5127,1225
eq 2.4	A	12.08	11.20	6166
50–100	B	3677	3677	
fp		26	299	5059;4896,506,5127
d	0	0.953	953	6321;8273,3237,6322,5059,6721
	24.65	0.9251	925.1	506
n_D	24.65	1.45874	1.45874	506;6721,8273,5278
ΔH_v	25	19	79	6166
	bp	8.4	35.1	
Beil	502,11			
ir	5932			
ms	5909			
nmr	5933			

101.   2-Methylcyclohexanol

(Mixed isomers)

```
    HCOH
  H₂C    CHCH₃
   |      |
  H₂C    CH₂
     \   /
      C
      H₂
```

583–59–5   C₇H₁₄O

		cgs	SI	
mw		114.187	114.187	3468
bp	1 atm	167.6	440.8	506;5430,6793,6321,7758
dt/dp	1 atm	0.0621	0.466	6166
dp/dt	1 atm	16.1	2.15	6166
p	79	29	3.9	5430
eq 2.4	A	8.690	7.815	6166
.	B	2544	2544	

101.   2-Methylcyclohexanol

(Continued)

		cgs	SI	
d	0	0.9452	945.2	6321;506,6721,6793
	20	0.9254	925.4	7758
	24.7	0.9215	921.5	
n_D	13.4	1.46585	1.46585	506;7758,6793
	20	1.4610	1.4610	5430
γ	20	30.75	0.03075	7758
	24.7	20.42	0.02042	
ΔH_v	av	11.6	48.5	6166
ΔH_c°	25(1)	-1048.5	-4386.9	3975
ε	20	13.3	13.3	4937;5251
μ		1.95	1.95	8075;5251
aq az	98.4	20%w	20%w	3497
fl pt	CC	68	341	5712
Beil	502,11			
ir	1041,5932			
Raman	5466			
ms	5909			
nmr	5933			

102. *cis*-2-Methylcyclohexanol

7443-70-1 $C_7H_{14}O$

		cgs	SI	
mw		114.187	114.187	3468
bp	1 atm	165	438	6793;2237,2892,3546,2238
dt/dp	1 atm	0.0529	0.397	6166
dp/dt	1 atm	18.9	2.52	6166
p	65	16	2.1	2238;6793,2892,3548
eq 2.4	A	8.6543	7.7792	6166
60–165	B	2528.8	2528.8	
fp		6.8–7.3	280.0–280.5	3548;7479,3546
d	20	0.93600	936.00	3548;3657,3546,6793,2892
	25	0.93181	931.81	
	30	0.92750	927.50	
dd/dt	25	0.000850	0.850	<u>3548</u>
n_D	20	1.46536	1.46536	3548;6793,2892
	24	1.46325	1.46325	
	30	1.46074	1.46074	
dn/dt	20 – 30	0.000462	0.000462	<u>3548</u>
η	25	18.08	0.01808	3548;2892,3546
	30	13.60	0.01360	
ΔH_v	60 – 165	11.6	48.5	6166
ΔH_f°	25(1)	−86.63	−362.46	7743;8316
ΔH_c°	25(1)	−1045.7	−4375.2	8316;1667,6793
$[\alpha]_D$	20	−6.23	−6.23	2892
Beil	<u>502</u>,11			
ir	1468,1577,5932			

103.   *trans*-2-Methylcyclohexanol

7443-52-9 $C_7H_{14}O$

		cgs	SI	
mw		114.187	114.187	3468
bp	1 atm	166.5	439.7	6793;2237,2238,3657,2892,3546
	1 atm(±)	163-164	436-437	7608
p	72	20	2.7	2238;2892,3657,3548,6793,2237
eq 2.4	A	9.197	8.322	6166
80-165	B	2776.4	2776.4	
fp		-4.3 to -3.7	268.9-269.5	3548;7479,3546
d	20	0.92472	924.72	3548;3546,6793,3657,2892
	25	0.92085	920.85	
	30	0.91682	916.82	
dd/dt	25	0.000790	0.790	3548
n_D	20	1.46164	1.46164	3548;3657,6793,2892
	20(±)	1.4602	1.4602	7608
	25	1.45972	1.45972	3548
	30	1.45762	1.45762	
dn/dt	25	0.000402	0.000402	3548
η	25	37.13	0.03713	3548;2892,3546
	30	25.14	0.02514	
ΔH_v	80 - 165	12.7	53.1	6166
ΔH_f°	25(1)	-97.0	-405.8	8316
ΔH_c°	25(1)	-1039.6	-4349.7	8316;1667,6793
$[\alpha]_D$	20	(+)17.19	(+)17.19	2892
		(-)17.81	(-)17.81	
Beil		502,11		
ir		1468,1577,5932		
nmr		5933,6748		

104. 3-Methylcyclohexanol

(Mixed isomers)

```
      HCOH
H₂C       CH₂
  |         |
H₂C       CHCH₃
    \   C  /
        H₂
```

591–23–1 $C_7H_{14}O$

		cgs	SI	
mw		114.187	114.187	3468
bp	763 Torr	172	445	7758;6321,6793,3047
p	77	14	1.9	506
d	0	0.9336	933.6	6321;3047,6721
	20	0.9168	916.8	7758;506,6793
n_D	20	1.45757	1.45757	7758;6793,3047
	25.5	1.45444	1.45444	506
γ	20	27.75	0.02775	7758
	23.1	27.62	0.02762	
[α]_D		−3.68	−3.68	3047
ε	20	12.3	12.3	4937;5251
μ		1.9	1.9	8075;5251,6794
Beil	502,12			
ir	5932			
Raman	5466			
ms	5909			
nmr	5933			

105. *cis*-3-Methylcyclohexanol

OH   OH

(structure: cyclohexanol with CH₃)   (structure: cyclohexanol with CH₃)   5454–79–5 $C_7H_{14}O$

		cgs	SI	
mw		114.187	114.187	3468
bp	12 Torr	94	367	4800;2803,2238,2237,5753,2809
	1 atm(±)	168	441	6793
p	65	5.6	0.75	2752;4800
	60(±)	2	0.3	4799
fp		−6 to −5	267–268	3551
d	16	0.9201	920.1	2803;4799,6793,2892
	20	0.9155	915.5	3551
	30	0.9065	906.5	
	30(±)	0.9072	907.2	4799
n_D	16	1.4589	1.4589	2803;4799,2892,2809,5753,4800
	20	1.4572	1.4572	6793,3551;2752
η	16	37.74	0.03774	2803
	25	19.7	0.0197	2892
	30	17.23	0.01723	3551
γ	30	29.2	0.0292	2752
ΔH_f°	25(1)	−91.9	−384.5	8316
ΔH_c°	25(1)	−1044.7	−4371.0	8316;6793
$[\alpha]_D$	20(−)	3.91	3.91	3551,2803,2892,2752
ε	20	16.47	16.47	Beil. EIII6, 67
μ	(±)	1.91	1.91	6794
Beil	502,12			
ir	1468,1577			

106. *trans*-3-Methylcyclohexanol

7443-55-2 $C_7H_{14}O$

		cgs	SI	
mw		114.187	114.187	3468
bp	13 Torr	84	357	4800;2237,2803,2238,5753,6793
	1 atm(±)	166–168	439–441	5528;4799,2809,7608,1666
p	61	5.2	0.69	2752
eq 2.4	A	6.890	6.015	6166
	B	2062	2062	
fp		−1 to 0	272–273	3551
d	20	0.9214	921.4	3551;2803,6793,2892,4799
	30	0.9138	913.8	4800,1666
dd/dt	25	0.00076	0.76	3551
n_D	16	1.4590	1.4590	2803;2892,6793,3551,5753
				2808,4799
	20	1.4580	1.4580	2752;2809,5528,4800,1666
	20(±)	1.4545	1.4545	7608
η	16	26.52	0.02652	2803
	25(±)	25.1	0.0251	2892
	30	15.60	0.01560	3551
γ	30	28.80	0.02880	2752
ΔH_v	av	9.44	39.50	6166
ΔH_f°	25(l)	−97.1	−406.3	8316
ΔH_c°	25(l)	−1039.5	−4349.3	8316;6793
$[\alpha]_D$	20	(−)7.34	(−)7.34	3551;2803
	21	(+)6.7	(+)6.7	4800
ε	20	8.05	8.05	Beil. EIII6, 69
μ	(±)	1.75	1.75	6794
Beil	502,12			
ir	1468,1577			

107.   4-Methylcyclohexanol

(Mixed isomers)

HCOH
H$_2$C   CH$_2$
H$_2$C   CH$_2$
HCCH$_3$ 589–91–3   C$_7$H$_{14}$O

		cgs	SI	
mw		114.187	114.187	3468
bp	763 Torr	172	445	7758;6715,6793,6321
	12 Torr	75	348	506
d	0	0.9328	932.8	6321;6721
	20	0.9122	912.2	7758
	22.5	0.9183	918.3	506
n$_D$	20	1.45647	1.45647	7758
	22.5	1.45594	1.45594	506
γ	20	27.63	0.02763	7758
ε	20	13.3	13.3	4937;5251
μ		1.9	1.9	8075;5251
Beil	502,14			
ir	633,5932			
Raman	5466			
nmr	5933			

108.   *cis*-4-Methylcyclohexanol

OH

S

CH$_3$ 7731–28–4   C$_7$H$_{14}$O

		cgs	SI	
mw		114.187	114.187	3468
bp	1 atm	171	444	1666;6793,2237,3657,2238,2892
p	52	2	0.3	3657
d	21.5	0.9129	912.9	6793;2892,1666
	30	0.9173	917.3	3657
n$_D$	20	1.4614	1.4614	3657;6793,1666
	25	1.4584	1.4584	2892

108. *cis*-4-Methylcyclohexanol

(Continued)

		cgs	SI	
η	25	0.247	0.000247	2892
ΔE_c	18(1)	-1036.9	-4338.4	<u>6793</u>

Beil	<u>502</u>,14
ir	2238,1468,1577,5932
nmr	4559,5933,6748

109. *trans*-4-Methylcyclohexanol

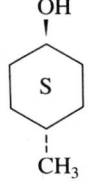

7731-29-5 $C_7H_{14}O$

		cgs	SI	
mw		114.187	114.187	3468
bp	1 atm	171	444	1666;7608,2237,6793,2892,2238
p	54	3	0.4	3657
d	20.7	0.9118	911.8	6793;1666
	25	0.9080	908.0	2892
	30	0.9040	904.0	3657
n_D	20	1.4559	1.4559	1666;3657,6793,7608
	25	1.4544	1.4544	2892
η	25	0.385	0.000385	2892
ΔE_c	18(1)	-1032.1	-4318.3	<u>6793</u>

Beil	<u>502</u>,14
ir	2238,1468,1577,5932
nmr	4559,5933,6748

110.   2-Heptanol

$CH_3(CH_2)_4CHOHCH_3$ 543–49–7 $C_7H_{16}O$

		cgs	SI	
mw		116.203	116.203	3468
bp	1 atm(±)	159.7	432.9	7251;2253,2052,2103,4907,3599 6650,5833,5970
dt/dp	1 atm(±)	0.043	0.32	7251
dp/dt	1 atm(±)	23	3.1	7251
p	20(±)	0.9	0.1	4902;2253,6650
eq 2.5	A B C	6.33006 977.8 124.3	5.45496 977.8 124.3	7252
fp		(gl)	(gl)	6650
d	20(±) 25(±) 20(−) 35(+)	0.8171 0.8134 0.8184 0.8050	817.1 813.4 818.4 805.0	7251;4907,5970,7971,6650 5833
α	ca. 20(±)	0.00094	0.00094	4902
n_D	20(±) 25(±) 20(+)	1.4210 1.4190 1.4209	1.4210 1.4190 1.4209	7251;6650,5970,3599,4907 7971,8270 5833
η	20(±) 30	6.53 3.321	0.00653 0.003321	4902;2129 7416
ΔH_v	65 −160(±)	11.9	49.8	6166
$[\alpha]_D$	20(+)in 79 20(+)in 38 20(+) 25(+) 17(−)	11.45 13.71 10.32 10.21 10.48	11.45 13.71 10.32 10.21 10.48	5835 5833;5970
ε	22	9.21	9.21	2291,4937;5110
μ	22 in 38	1.71	1.71	2291
δ	25	9.8	20.1	653
aq az	98.7	17%w	17%w	3497
Beil	24,415			
ir	6682,8270,5932			
ms	7226			
nmr	5933,6748			

111. 3-Heptanol

$CH_3(CH_2)_3CHOHCH_2CH_3$ 589-82-2 $C_7H_{16}O$

		cgs	SI	
mw		116.203	116.203	3468
bp	1 atm	156.7	429.9	7251;8065,7416,5629,2008,6650
dt/dp	1 atm	0.045	0.34	7251;8065
dp/dt	1 atm	22	3.0	
p	21.5	0.74	0.099	7418
	69.3	20.0	2.67	7252;6650,4540
eq 2.5	A	6.30332	5.42822	8065
	B	948.1	948.1	
	C	120.2	120.2	
fp		-70	203	5629,7251
d	20	0.8212	821.2	8065;7251,7416,6650,2008
	25	0.8170	817.0	
dd/dt	25	0.00084	0.84	8065
n_D	20	1.4201	1.4201	2008;8065,7251,7416,6650,4540
	25	1.4175	1.4175	
dn/dt		0.00040	0.00040	7416
η	38.4	2.986	0.002986	7416;7419
	53.9	1.688	0.001688	
ΔH_v	bp	10.15	42.47	8065
$[\alpha]_D$	25	5.12	5.12	4545;4540,5835
ε	22	6.86	6.86	2291
μ	22 in 38	1.71	1.71	2291
δ	25	9.9	20.2	653
fl pt	COC	60	333	6402,5415
Beil	24,EI 205			
ir	6682,5932			
ms	2572,8205			
nmr	5933			

112. 1-Octanol

$CH_3(CH_2)_6CH_2OH$ 111-87-5 $C_8H_{18}O$

		cgs	SI	
mw		130.230	130.230	3468
bp	1 atm	195.156	468.306	129;7233,1236,1287,1896,2052 5329,7475
dt/dp	1 atm	0.04810	0.3608	129;7233,2090
dp/dt	1 atm	20.79	2.772	129
p	25	0.075	0.010	7236,129;1286,7728,2695,7058 2026,120
eq 2.5	A	6.76021	5.88511	129;7235,3270,2090,1286
	B	1264.322	1264.322	6257,3938
	C	130.73	130.73	
fp		−14.97	258.18	2090;1896,7058,484,2052 7475,1847
d	20	0.82499	824.99	3080;2090,1286,1896,3774,6876
	25	0.82157	821.57	1287,2677,5329,7768,6357 2430,1989,3702
dd/dt	−60 to 120	0.00081	0.81	1812
α	25	0.000827	0.000827	3080;6357,2430,2052,3702
κ_T	25	1.02×10^{-7}	7.64×10^{-7}	6152,1989
n_D	20	1.4296	1.4296	7253;7233,5598,2090,1287,5329
	25	1.4276	1.4276	29,2052,7768
dn/dt		0.00040	0.00040	6166;5598
η	15	10.662	0.010662	6080;1896,2677,5310,3774,5329
	25	7.363	0.007363	
	30	6.125	0.006125	1896
γ	19.96	27.28	0.02728	7045;3283,2052,5329,6830
	24.73	26.92	0.02692	5310,7768
	Table 2.8			
ΔH_v	25	16.96	70.98	4878;7236,1286,1847,5192,5326 6277,2052,1154,7755
	bp	11.2	46.9	7236
ΔH_s		25.2	105.4	1847
ΔH_m		10.1	42.3	1847
ΔH_f°	25(l)	−102.30	−428.02	3021;5283,1418,2923,8316
	25(g)	−85.29	−356.85	3021
ΔH_c°	25(l)	−1264.94	−5292.51	3021;5283,1418,3975,2923,6277
C_p°	25	48.2	201.7	7237;7807,2926
C_p	13	68.0	284.5	2052
	25	73.03	305.55	8269
λ	37.8	0.000383	0.160	6352
t_c		379.3	652.5	116;2214,7234,1156,6854
P_c		28.23	2.86	116;2214

112.   1-Octanol

(Continued)

		cgs	SI	
d_c		0.266	266	2214,7234
v_c		0.489	0.000489	2214
A	eq 2.72	0.077	0.077	7236
κ	23.1	1.39×10^{-7}	1.39×10^{-5}	6870
ε	20	10.34	10.34	4937,5251;6870,3749,6876 5556,2814
μ	25 in 317	1.76	1.76	4511;6876,4687,2882,2697
δ		10.3	21.1	652,6240,6905,4593,1271;7159
ER	BuOAc	<0.01	<0.01	6905
soly	in aq,25	0.0538%w	0.0538%w	2052;29,1287,6677
aq az	99.4	10%w	10%w	3497
fl pt	TOC	85	358	4902;4193
Beil	24,418			
uv	5918			
ir	5165,153,4021,2695,466,8270,5932			
Raman	4861			
ms	2572,7226			
nmr	5933,6748			

113.   2-Octanol

$CH_3(CH_2)_5CHOHCH_3$			123-96-6		$C_8H_{18}O$
		cgs	SI		
mw		130.230	130.230	3468	
bp	1 atm	179.8	453.0	7253;2052,2253,7473,7058	
dt/dp	1 atm	0.0492	0.369	7253	
dp/dt	1 atm	20.3	2.71	7253	

113. 2-Octanol

(Continued)

		cgs	SI	
p	40.0	0.84	0.11	2695
	80.0	13.55	1.81	2695;7254
eq 2.5	A	6.3888	5.5137	7254;2695
	B	1060.4	1060.4	
	C	122.5	122.5	
fp		-32	241	7253;2052,7058,7473
d	20	0.8207	820.7	7253;3725,2052,2253,6813,1656
	25	0.8171	817.1	
dd/dt		0.000766	0.766	3725;6813
α	0 - 25	0.000897	0.000897	2052;2253
n_D	20	1.4261	1.4261	7253;2052,2253,6180
	25	1.4241	1.4241	
dn/dt	25	0.00040	0.00040	6180
η	20	8.190	0.008190	5310;3726
	25	6.490	0.006490	7402
γ	20	26.28	0.02628	3162;5310
	30	35.51	0.03551	
ΔH_v	35 - 24	16.3	68.2	2695
	bp	10.60	44.35	8065
ΔH°_c	25(1)	-1262.2	-5281.0	8065
C_p	25	78.9	330.1	1541
t_c		364	637	7238;4299
ε	20	8.173	8.173	1656;3725,6876,2052
μ	20	1.707	1.707	2697;2882,6191
δ	25	9.9	20.2	653
soly	in aq,25	0.1281	0.1281	5175;29,2052
aq az	98	27%w	27%w	3497
fl pt		74	347	7609

Beil	24,419
uv	5883
ir	2695,5932
Raman	5465
ms	2572
nmr	6748,5933

114.   2-Ethyl-1-hexanol

$CH_3(CH_2)_3CH(C_2H_5)CH_2OH$ 104-76-7 $C_8H_{18}O$

		cgs	SI	
mw		130.230	130.230	3468
bp	1 atm	184.34	457.49	2174;5731,4544,2032,4743 4725,8034
dt/dp	1 atm	0.0479	0.359	7253;7618
dp/dt	1 atm	20.9	2.78	7253
p	25	0.143	0.0191	2174
eq 2.5	A	6.67138	5.79628	7254;2174
	B	1204.50	1204.50	
	C	133.14	133.14	
fp	(gl)	-76	197	4902;7618
d	15	0.8435	843.5	8034;4544,7618,4725,4543
	20	0.8327	832.7	7507
	25	0.8290	829.0	
dd/dt		0.00073	0.73	7618;1812
α	20	0.000875	0.000875	4902
n_D	20	1.4316	1.4316	7253;4725,4543
	25	1.4290	1.4290	7253;4544
dn/dt	25	0.00052	0.00052	7253
η	20	9.8	0.0098	7618
ΔH_v	bp	10.8	54.2	7618
ΔH_f°	25(1)	-103.46	-432.88	7507
ΔH_c°	25(1)	-1263.81	-5287.78	7507
C_p	25	73.45	307.3	7618
t_c		367.0	640.2	4432
$[\alpha]_D$	25(-)	1.53	1.53	4544
	25(+)	1.533	1.533	4543
ε	90	4.41	4.41	5556
μ	25 in 38	1.74	1.74	5731
δ		9.5	19.4	4593,1271,652
ER	BuOAc	0.02	0.02	6628;6631,1399
	90%	25730	25730	6628;6631,1399
soly	in aq,20	0.07%w	0.07%w	7618;2042
	aq in,20	2.6%w	2.6%w	7618
aq az	99.1	20%w	20%w	3498
fl pt	TOC	82	355	2350

114.   2-Ethyl-1-hexanol

(Continued)

Beil	24,421
uv	4725
ir	7055,5932
ms	2572
nmr	5933

115.   Benzyl alcohol

Benzenemethanol

$C_6H_5CH_2OH$ 100–51–6 C_7H_8O

		cgs	SI	
mw		108.140	108.140	3468
bp	1 atm	205.45	478.60	7490;7058,7817,2090,7475,7986
dt/dp	1 atm	0.050	0.38	7490;2090,4665
dp/dt	1 atm	20	2.7	7490
p	25	0.11	0.015	6582
eq 2.4	A	9.838	8.963	6582
	B	3214	3214	
fp		−15.3	257.9	7490,7475;7479,2090,5689,7058
d	15	1.04927	1049.27	7490;2677,4969,7768,7971
				3253,6161
	25	1.04127	1041.27	4913;7848,13
	30	1.03765	1037.65	7490;13
dd/dt	25 – 40	0.00074	0.74	13
α		0.00075	0.00075	7490
n_D	15	1.5426	1.5426	7482;2659,3248,7490,7768
				7986,7971
	20	1.54035	1.54035	2090;1407
	25	1.53837	1.53837	

115. Benzyl alcohol

(Continued)

		cgs	SI	
dn/dt	25	0.000396	0.000396	<u>2090</u>
η	15	7.760	0.007760	7490;2125,5310,2677,7401
	30	4.650	0.004650	7490;6759
γ	15	40.41	0.04041	7490;5310,7768
	20	39.96	0.03996	4518
	30	38.94	0.03894	
ΔH_v	bp	12.066	50.484	4969;4683,4665,7755
ΔH_s		4.95	20.7	8050
ΔH_m		4.074	17.046	5689
ΔH_f°	25(1)	−38.49	−161.04	5684;8316
ΔH_c°	25(1)	−893.15	−3736.94	5684;3975,8316
C_p	25	51.611	215.94	5478;5689,4665,4683
pK_a		18	18	4766
ε	20	13.1	13.1	4937;4134,2100,5428,7850
	30	11.916	11.916	6761
μ		1.66	1.66	8077;4913,2363
δ		12.1	24.8	652,7159,6240,4593
soly	in aq,20	0.08%w	0.08%w	2042
	aq in,20	8.37%w	8.37%w	
aq az	99.9	9%w	9%w	3497
<u>fl pt</u>	CC	101	374	152

Beil	528,428
uv	1337,5010,5918,2609,4388,6686,6749
ir	1285,1486,1849,1942,753,3888,6368,5932
Raman	3340,6778,7711,6774,1783
ms	16
nmr	4923,7227,1423,4924,6606,6748,5933

116. Phenol

C_6H_5OH 108–95–2 C_6H_6O

		cgs	SI	
mw		94.113	94.113	3468
bp	1 atm	181.839	454.989	349,7301;6500,6797,7297 2033,8084
dt/dp	1 atm	0.04788	0.3591	349;2090,732,1897,7297
dp/dt	1 atm	20.89	2.785	349
p	25 41	0.41 1.39	0.055 0.185	<u>7301</u>;2091,732,2816,7058 <u>349</u>
eq 2.5 9 – 40	A B C	11.5638 3586.36 273	10.6887 3586.36 273	349,855;2816,579,7300 6797,2090
100–200	A B C	7.13457 1516.072 174.569	6.25947 1516.072 174.569	349,855;7300
fp		40.90	314.05	349;3523,533,7297,3403 3749,4959
	1.37 Torr	40.88	314.03	7301
d	20	1.0767(a)	1076.7(a)	7297;7491,3523,2090,6161 7590,1057
	25(s) 40(s) 40 45(1) 70(1)	1.132 1.132(b) 1.0587(a) 1.05446 1.0325	1132 1132(b) 1058.7(a) 1054.46 1032.5	349 1579 7298;533 7482 5775
dd/dt	50 – 70	0.00086	0.86	5775
n_D	40.6 45 60	1.54274 1.54027 1.5321	1.54274 1.54027 1.5321	7482;2090,7491,1057,7297 5775
dn/dt	50 – 70	0.00045	0.00045	5775
η	45 60	4.076 2.578	0.004076 0.002578	7491;2125,7401
γ	50 55 60	37.77 37.26 36.69	0.03777 0.03726 0.03669	7491;7482,533
ΔH_v	25 bp	13.82(a) 10.920	57.82(a) 45.689	7301
ΔH_s	25	16.410	68.659	349;958,7754,819,1837,2355 6797,855
	40.88	16.36	68.45	7301
ΔH_m	tp	2.752	11.514	7301;7167,6991,2356,4959,356
ΔH_f°	25(s)	−39.45	−165.06	2036,819;2927,3528,349 7302,1700
	25(g)	−23.040	−96.399	4303,819
ΔH_c°	25(s)	−729.80	−3053.48	349;542,1700,2036,819 3975,7446
C_p°	25	24.671	103.223	4303;7302,2927,356

116. Phenol

(Continued)

		cgs	SI	
C_p	25(s)	30.40	127.21	5478;356,4715,5681
	mp(1)	48.14	201.42	356
λ	30(s)	0.000357	0.149	6834
	75	0.000338	0.141	
t_c		421.1	694.3	7299;4299,6854,3345,113
P_c		60.5	6.13	7299;4299,3345
d_c		0.411	411	7299
v_c		0.229	0.000229	7299
A	eq 2.72	0.01404	0.01404	7301
K_f	eq 2.67	7.40	7.40	7590;6161,4959,6706,2356 719,1404
K_B	eq 2.79	3.60	3.60	732;720,727
K_a	20	2.06×10^{-10}	2.06×10^{-10}	7735
pK_a	20	9.686	9.686	7735;1481,8266,2164
	30	9.654	9.654	7735
κ	50	2.68×10^{-8}	2.68×10^{-6}	5775;7854
ε	40	11.60	11.60	5775
	60	10.00	10.00	5775;8084,1837,2100,4937 4713,5251
μ	30 in 38	1.59	1.59	569;6505,4913,4513,2907,3875 2859,4579,2039
	40	2.20	2.20	5775
δ	25	11.3	23.1	4187;4593
soly	in aq,25	8.66%w	8.66%w	3403;3523,3745,6708
	aq in,25	28.72%w	28.72%w	
aq az	99.6	9.2%w	9.2%w	3497
fl pt	CC	79	352	5712

(a) Supercooled liquid   (b) Air free

Beil	512,110
uv	462,945,3882,4888,4985,5785,6208,6686,6749
ir	8208,2927,5026,7179,7225,2505,4472,4303,3112
Raman	4713,4831,2927,7082,8178,4126,8122,8106
ms	16,837
nmr	4926,4924,3562,1423,5706,2321,4974,2788,5933,6748

117. *o*-Cresol

2-Methylphenol

o-CH$_3$C$_6$H$_4$OH   95–48–7   C$_7$H$_8$O

		cgs	SI	
mw		108.140	108.140	3468
bp	1 atm	191.004	464.154	7297;855,2090,7058,349 5671,2776
dt/dp	1 atm	0.05021	0.3766	7297,349;2090,855
dp/dt	1 atm	19.92	2.655	7297,349
p	25	0.31	0.041	7301;855,2091,7058,5671,2816
eq 2.5	A	12.7778	11.9027	349;855,2816,2090,7300
0 – 30	B	3970.17	3970.17	
	C	273	273	
110 –200	A	7.07055	6.19545	
	B	1542.299	1542.299	
	C	177.110	177.110	
fp		30.944	304.094	7297;3805,2783,1357,5671 1898,3945
	0.52 Torr	30.92	304.07	7301;358
d	20	1.0460(a)	1046.0(a)	7297;2090,6160,5671
	25(s)	1.135	1135	
	fp	1.04	1040	4987
	40	1.0282	1028.2	7298
	46	1.0229	1022.9	5775
dd/dt	40 – 60	0.00088	0.88	5775
n$_D$	20	1.5467(a)	1.5467(a)	7297;2090
	25	1.5442(a)	1.5442(a)	
	50	1.5310	1.5310	5775
dn/dt	40 – 60	0.00050	0.00050	5775
η	25	7.608	0.007608	3945
	35	6.6370	0.0066370	1984
	45	3.506	0.003506	7401
	80	1.47cSt	1.47 x 10^{-6}	5671
γ	40.3	34.8	0.0348	3663
	176	21.5	0.0215	
ΔH$_v$	25	15.39(a)	64.39(a)	7301;855,349,958,6713
	bp	10.801	45.191	349;7301,5671
ΔH$_s$	25	18.17	76.02	819
	fp	18.18	76.07	7301
ΔH$_m$		3.781	15.820	358;7301
ΔH$_f^°$	25(s)	-48.90	-204.60	819,2036;358,7301
	25(g)	-30.730	-128.574	4303,819
ΔH$_c^°$	25(s)	-882.72	-3693.30	349;2036,819,3975,1700,2025
C$_p^°$	25	30.426	127.301	4303;7302,2925
C$_p$	25(s)	36.94	154.56	358;5671,1054
	31.05	55.67	232.92	
t$_c$		424.4	697.6	7299,113;3345,2776,4299
P$_c$		49.4	5.01	7299;3345,2776,4299

117. *o*-Cresol

(Continued)

		cgs	SI	
d_c		0.384	384	7299
v_c		0.282	0.000282	7299
A	eq 2.72	0.01684	0.01684	7301
K_f	eq 2.67	5.60	5.60	6161;349,6713
pK_a	25	10.287	10.287	868,869;3705,1441,795,3324
κ	25	1.27×10^{-9}	1.27×10^{-7}	3945
	50	4.3×10^{-10}	4.3×10^{-8}	5775
ε	25	11.5	11.5	4937;4713,5796
	40	6.31	6.31	5775
μ	25 in 38	1.45	1.45	4692;2906,6871,2039,569,5775
	30	1.64	1.64	5775
δ	25	10.7	21.9	4187
soly	in aq,40	3.08%w	3.08%w	6713;6708,4026
fl pt	CC	81	354	152

(a) Supercooled liquid

Beil	525,349
uv	6059,3646,1884,1564,4040,827,6204,6479
ir	2569,6360,7179,4060,8000,4927,7011,4303,2942,5932
Raman	4713,3336,2925,4112,2942
ms	16,2569,4927,4060,8000
nmr	4926,3562,4923,7227,5933,6748

118. *m*-Cresol

3-Methylphenol

m-CH$_3$C$_6$H$_4$OH   108–39–4   C$_7$H$_8$O

		cgs	SI	
mw		108.140	108.140	3468
bp	1 atm	202.232	475.382	7297,349;7491,855,2816 7054,5671
dt/dp	1 atm	0.05007	0.3756	349,7472;7491,855
dp/dt	1 atm	19.97	2.663	349,7472
p	25	0.143	0.0191	7301;7058,5671,2816,855
eq 2.5	A	9.9653	9.0902	349;2816
11 – 40	B	3223.45	3223.45	
	C	273	273	
110–200	A	7.15904	6.28394	349,7300;855,5405
	B	1603.811	1603.811	
	C	172.646	172.646	
fp	100%	12.22	285.37	349,7297;3805,1867,1357 1898,3945
	tp	12.25	285.40	358
d	fp	1.05	1050	4987;3253,7297,6161,2677 5671,6293
	20	1.03410	1034.10	349;7491
	25	1.03019	1030.19	
	30	1.02628	1026.28	
dd/dt	25 – 60	0.00080	0.80	5775
α	25	0.000759	0.000759	349
n$_D$	20	1.5414	1.5414	7297;5671
	25	1.5396	1.5396	
	50	1.5271	1.5271	5775
dn/dt	25 – 60	0.00045	0.00045	5775
η	15	24.666	0.024666	7491;4987,5671,3045,7401,2677
	30	9.807	0.009807	
	35	8.7224	0.0087224	1084
γ	15	38.01	0.03801	7491
	30	36.54	0.03654	
ΔH$_v$	25	14.75	61.71	349,7301;4682,855,6713 958,5671
	bp	11.329	47.401	
ΔH$_m$		2.559	10.707	358;6293,7301
ΔH$_f^\circ$	25(1)	−46.37	−194.01	819,2036;349,1700,7302
	25(g)	−31.620	−132.298	4303,819
ΔH$_c^\circ$	25(1)	−885.25	−3703.89	349;4303,542,3975,819 1700,2925
C$_p^\circ$	25	29.798	124.675	4303;2925,7302
C$_p$	25	53.76	224.93	358;5671,4682,1054
t$_c$		432.6	705.8	7299,113;4299,3051,2776
P$_c$		45	4.6	7299;4299,3051,2776
d$_c$		0.346	346	7299;4299

118. *m*-Cresol

(Continued)

		cgs	SI	
v_c		0.312	0.000312	7299;4299
A	eq 2.72	0.01340	0.01340	7301
K_f	1%m*	0.749	0.749	349;6713
pK_a	25	10.09	10.09	868,3324,869;1441,993 1481,795
k	25	1.397×10^{-8}	1.397×10^{-6}	3945
	50	6.34×10^{-9}	6.34×10^{-7}	5775
ε	25	12.44	12.44	5775;4937,746
μ	20 in 38	1.48	1.48	5775;2039,6871,569,2905
	20	2.48	2.48	5775
S	25	10.8	22.1	4187;652
soly	in aq,40	2.51%w	2.51%w	6713;7219,6708,4026
fl pt	CC	86	359	5712
*1%m impurity				

Beil	526,373
uv	1883,1884,1357,3440,3460,7742,6204,6749
ir	7011,6360,7179,4927,2569,8000,6368,2942,4303,5932
Raman	4713,3335,2925,4112,2942
ms	16
nmr	7586,4923,2005,4926,7227,6748,5933

119. *p*-Cresol

4-Methylphenol

p-CH$_3$C$_6$H$_4$OH   106–44–5   C$_7$H$_8$O

		cgs	SI	
mw		108.140	108.140	3468
bp	1 atm	201.940	475.090	7297;349,5671,2816,855 1433,2732
dt/dp	1 atm	0.04985	0.3739	349,7297;2090,855
dp/dt	1 atm	20.06	2.674	349,7297
p	25	0.13	0.017	7301,2091;855,5671,2816,7058
eq 2.5	A	12.0298	11.1547	349;2090,7301,2825
0 – 34	B	3861.98	3861.98	
	C	273	273	
110–200	A	7.11767	6.24257	349;855,7300
	B	1566.029	1566.029	
	C	167.680	167.680	
sublm	0.30 Torr	34.70	307.85	7301
fp	100%	34.739	307.889	7301;349,7297,1826,2090,3945
tp		34.79	307.94	358
d	20	1.0348(a)	1034.8(a)	7297;2090,6161,5671
	25(s)	1.154	1154	349;7297
	fp	1.02	1020	4987
	40	1.0185	1018.5	7298
dd/dt	40 – 60	0.00077	0.77	5775
n$_D$	25	1.5391	1.5391	7297;4987
	41	1.53115	1.53115	2090
	46	1.52870	1.52870	
dn/dt	40 – 60	0.00042	0.00042	5775
η	35	9.4022	0.0094022	1984
	45	5.607	0.005607	7401;3945
	80	2.00cSt	2.00 x 10^{-6}	5671
γ	25.6	34.5	0.0345	Beil. EI6,197
	194.5	19.2	0.0192	
ΔH$_v$	25	16.06(a)	67.20(a)	7301
	bp	11.34	47.45	
ΔH$_s$	25	17.67	73.93	349,7301;5671,349,819,855 958,7754
	fp	17.69	74.01	7301
ΔH$_c^°$	25(s)	-883.99	-3698.61	349;2036,819,3975,1700,2925
ΔH$_f^°$	25(s)	-47.63	-199.28	2036,819;349,1700,7302
	25(g)	-29.960	-125.353	4303,819
C$_p^°$	25	29.869	124.972	4303;2925,7302
C$_p$	25(s)	35.91	150.25	358;5671,1199
	25	52.83(a)	221.03(a)	5478
	34.79	54.36	227.44	358
t$_c$		431.4	704.6	113;2776,7299,3345,4299
P$_c$		50.8	5.15	7299;4299,2781,3345

119. *p*-Cresol

(Continued)

		cgs	SI	
d_c		0.391	391	7299
v_c		0.277	0.000277	7229
A	eq 2.72	0.01097	0.01097	7301
K_f	eq 2.67	6.96	6.96	2356;349,6161,6713
pK_a	25	10.26	10.26	868,869,4336;1481,2164,3324 1441,993
κ	25	1.378×10^{-8}	1.378×10^{-6}	3945
	50	6.02×10^{-9}	6.02×10^{-7}	5775
ε	40	11.07	11.07	5775
	58	9.91	9.91	4937;5795
μ	20 in 38	1.48	1.48	5775;6871,2039,2282,569 2859,2906
	40	2.36	2.36	5775
δ	25	10.8	22.1	4187
soly	in aq,40	2.26%w	2.26%w	6713;4026,6708
fl pt	CC	94	367	152

(a) supercooled liquid

Beil	527,389
uv	6059,3646,1884,827,4040,6051,1357,6749
ir	7011,6360,7179,4060,8000,2569,4927,2942,4303,5932
Raman	4713,5720,5248,4112,2942
ms	16
nmr	7227,6748,5933

120. 2,4-Xylenol

2,4-Dimethylphenol

2,4-(CH$_3$)$_2$C$_6$H$_3$OH			105–67–9	C$_8$H$_{10}$O
		cgs	SI	
mw		122.166	122.166	3468
bp	1 atm	210.931	484.081	349;5671,2087,2776,7058
dt/dp	1 atm	0.05206	0.3905	349;2087
dp/dt	1 atm	19.21	2.561	349
p	25	0.1634	0.02178	2087
eq 2.5	A	7.04694	6.17184	349;2087
	B	1581.391	1581.391	
	C	168.652	168.652	
fp		24.54	297.69	349;534,5671,2087,7058
d	20	1.02017	1020.17	349
	25	1.01601	1016.01	
	40	1.0033	1003.3	5775;3552
dd/dt	40	0.00086	0.86	5775
α		0.000818	0.000818	349
n$_D$	50	1.5254	1.5254	5775
dn/dt	50	0.00048	0.00048	5775
γ	40	31.23	0.03123	3552
ΔH_v	25	15.740	65.856	349;2087,819
	bp	11.268	47.145	
ΔH_f°	25(1)	-54.69	-228.82	349;819
	25(g)	-38.95	-162.97	
ΔH_c°	25(1)	-1039.31	-4348.47	349;819
t$_c$		434.4	707.6	113;2776
P$_c$		65	6.6	2776
pK$_a$	25	10.63	10.63	3324;6147,6942,6214
κ	50	8.0 x 10^{-10}	8.0 x 10^{-8}	5775
ε	30	6.16	6.16	5775
μ	20(1)	1.70	1.70	5775
	20 in 9	1.40	1.40	5775;7067
δ	20	10.4	21.3	4187
soly	in aq,25	0.787%w	0.787%w	591

Beil	529,486
uv	6648,1633,8144
ir	4060,8000,2943,2569,5932
Raman	2943,3341
ms	16
nmr	3388,6898,5933

121. 2,5-Xylenol

2,5-Dimethylphenol

$2,5\text{-}(CH_3)_2C_6H_3OH$ 95-87-4 $C_8H_{10}O$

		cgs	SI	
mw		122.166	122.166	3468
bp	1 atm	211.132	484.282	349;5671,2087,7058
dt/dp	1 atm	0.05233	0.3925	349
dp/dt	1 atm	19.11	2.548	349
p	25	0.1634	0.02178	2087
eq 2.5	A	7.03684	6.16174	349;2087
	B	1581.906	1581.906	
	C	169.497	169.497	
fp		74.85	348.00	349;7058,2087,534,5671
d	25(s)	1.189	1189	349
	85	0.9582	958.2	5775;3552,5671
dd/dt	85	0.00090	0.90	5775
n_D	80	1.5092	1.5092	5775
dn/dt	80	0.00049	0.00049	5775
η	80	1.55	0.00155	5671
γ	80	30.20	0.03020	3552
ΔH_v	25	14.33	59.96	2087
	bp	11.219	46.940	349;2087
ΔH_s	25	20.31	84.98	349;819
ΔH_f°	25(s)	-58.96	-246.69	349;819
	25(g)	-38.65	-161.71	
ΔH_c°	25(s)	-1035.04	-4330.61	349;819
t_c		433.7	706.9	116
pK_a	25	10.41	10.41	3324;6147,6214
κ	50	5.0×10^{-10}	5.0×10^{-8}	5775
ϵ	65	6.36	6.36	5775
μ	70(1)	1.66	1.66	5775
	20 in 38	1.45	1.45	5775;7067,3552
δ	20	10.4	21.3	4187

Beil	529,494
uv	8144,1633,6648
ir	2569,2943,8000,4060,5932
Raman	3341,2943
ms	16
nmr	3388,5933

122.   2,6-Xylenol

2,6-Dimethylphenol

2,6-$(CH_3)_2C_6H_3OH$ 576–26–1 $C_8H_{10}O$

		cgs	SI	
mw		122.166	122.166	3468
bp	1 atm	201.030	474.180	349;5671,2087
dt/dp	1 atm	0.05302	0.3977	349;2087
dp/dt	1 atm	18.86	2.515	349
p	25	0.1432	0.01909	2087
eq 2.5	A	7.05753	6.18243	349;2087
	B	1618.528	1618.528	
	C	186.482	186.482	
fp		45.62	318.77	349;534,2087,5671
d	25(s)	1.132	1132	349
	65	0.9777	977.7	5775
dd/dt	65	0.00093	0.93	5775
n_D	60	1.5171	1.5171	5775
dn/dt	60	0.00051	0.00051	5775
ΔH_v	25	14.439	60.412	2087
	bp	10.641	44.522	349;2087
ΔH_s	25	18.07	75.60	349;819
ΔH_f°	25(s)	−56.75	−237.44	349;819
	25(g)	−38.68	−161.84	
ΔH_c°	25(s)	−1037.25	−4339.85	349;819
t_c		427.8	701.0	113
pK_a	25	10.63	10.63	3324;2440,6147,6214
κ	50	1.0×10^{-10}	1.0×10^{-8}	5775
ϵ	40	4.90	4.90	5775
μ	40(l)	1.45	1.45	5775
	20 in 9	1.40	1.40	5775;2471,7067
δ	20	10.5	21.5	4187

Beil	529,485
uv	6648,1633,8144,1883,6749
ir	8000,2943,2569,4060,5932
Raman	2943,3341
ms	8060,16
nmr	3388,2527,6748

123. 3,4-Xylenol

3,4-Dimethylphenol

3,4-$(CH_3)_2C_6H_3OH$		95–65–8		$C_8H_{10}O$
		cgs	SI	
mw		122.166	122.166	3468
bp	1 atm	226.947	500.097	349;5671,2087,2776,7058
dt/dp	1 atm	0.05257	0.3943	349;2087
dp/dt	1 atm	19.02	2.536	349
p	25	0.0387	0.00516	2087
eq 2.5	A	7.07343	6.19833	349;2087
	B	1617.202	1617.202	
	C	158.778	158.778	
fp		65.11	338.26	349;534,2087,5671,7058
d	25(s)	1.138	1138	349
	80	0.9813	981.3	5775;5671,3552
dd/dt	80	0.00080	0.80	5775
n_D	60	1.5268	1.5268	5775
dn/dt	60	0.00044	0.00044	5775
η	80	3.00	0.00300	5775
γ	75	29.02	0.02902	3552;2087
ΔH_v	25	16.11	67.40	2087
	bp	11.871	49.668	349;2087
ΔH_s	25	20.49	85.73	349;819
ΔH_f°	25(s)	−57.93	−242.38	349;819
	25(g)	−37.44	−156.65	
ΔH_c°	25(s)	−1036.07	−4334.92	349;819
t_c		456.7	729.9	113;2776
P_c		100	10.1	2776
pK_a	25	10.36	10.36	3324;6147,6214,3995
κ	50	1.40×10^{-9}	1.40×10^{-7}	5775
ε	60	9.02	9.02	5775
μ	60(1)	2.32	2.32	5775
	20 in 38	1.56	1.56	5775;7067
δ	20	10.5	21.5	4187
Beil	529,480			
uv	8144,1633,6648,6749			
ir	2569,4060,2943,8000,5932			
Raman	3341,2943			
ms	16			
nmr	6748			

124. 3,5-Xylenol

3,5-Dimethylphenol

3,5-$(CH_3)_2C_6H_3OH$ 108–68–9 $C_8H_{10}O$

		cgs	SI	
mw		122.166	122.166	3468
bp	1 atm	221.692	494.842	349;7058,2087,2776,5671
dt/dp	1 atm	0.05190	0.3893	349;2087
dp/dt	1 atm	19.27	2.569	349
p	25	0.0543	0.00724	2087
eq 2.5	A	7.11745	6.24235	349;2087
	B	1630.124	1630.124	
	C	163.076	163.076	
fp		63.24	336.39	349;4959,7058,534,5671
d	25(s)	1.115	1115	349
	75	0.9694	969.4	5775;3552,5671,2087
dd/dt	75	0.00083	0.83	5775
n_D	70	1.5150	1.5150	5775
dn/dt	70	0.00045	0.00045	5775
η	80	2.42	0.00242	5671
γ	74	28.42	0.02842	3552;2087
ΔH_v	25	15.767	65.968	2087
	bp	11.785	49.308	349;2087
ΔH_s	25	19.80	82.84	349;819
ΔH_m		4.164	17.42	4959
ΔH_f°	25(s)	−58.43	−244.47	349;819
	25(g)	−38.63	−161.63	
ΔH_c°	25(g)	−1035.57	−4332.83	349;819
t_c		442.4	715.6	113;2776
p_c		55	5.6	2776
A	eq 2.72	0.01850	0.01850	4959
	B	0.002	0.002	
pK_a	25	10.18	10.18	3324;3995,6147,6214
κ	50	6.00×10^{-9}	6.00×10^{-7}	5775
ε	50	9.06	9.06	5775
μ	50(1)	2.29	2.29	5775
	20 in 38	1.55	1.55	5775;7067,2471
δ	20	10.5	21.5	4187
soly	in aq,28.8	0.62%w	0.62%w	5039
	aq in,51.4	10.00%w	10.00%w	

124.   3,5-Xylenol

(Continued)

Beil	529,492
uv	6648,1633,8144,6749
ir	8000,2943,2569,4060,5932
Raman	2943,3341
ms	16
nmr	3388,6748

125.   2-Propen-1-ol

$CH_2=CHCH_2OH$ 107–18–6 C_3H_6O

		cgs	SI	
mw		58.080	58.080	3468
bp	1 atm	97.08	370.23	2338,7879;7878,4665,2035
dt/dp	1 atm	0.040	0.30	7489;4665
dp/dt	1 atm	25	3.3	7489
p	25	28.1	3.75	7482
eq 2.13	A	32.62580	31.75070	4027
	B	3451.8	3451.8	
	C	7.94975	7.94975	
fp	(gl)	−129	144	4027;7479
d	0	0.86814	868.14	7482;6624,7879,7878,7971
	15	0.85511	855.11	
	30	0.84209	842.09	
dd/dt	0 – 30	0.000868	0.868	7482
n_D	20	1.4135	1.4135	2035
	30	1.4090	1.4090	7971
η	15	1.486	0.001486	7489;2035,7404
	30	1.072	0.001072	
	Table 2.2			

125.   2-Propen-1-ol

(Continued)

		cgs		SI	
γ	15	26.15	0.02615	7489	
	20	25.68	0.02568		
	30	24.92	0.02492		
	Table 2.8				
ΔH_v	bp	9.550	39.957	4027;4665	
ΔH_f°	25(1)	−41.60	−174.05	8316	
ΔH_c°	25(1)	−445.50	−1863.97	8316;3975,7446	
C_p°	20	33.7	141.0	Beil. EIII1, 1874	
C_p	20.5–95.5	38.59	161.46	4665	
t_c		271.9	545.1	5372	
pK_a	25	15.5	15.5	578	
ε	15	21.6	21.6	4937;2100,7458,3840	
μ	in 38	1.77	1.77	2905;4295	
δ		11.8	24.1	652	
soly	in aq	inf	inf	4027	
aq az	88.89	72.3%w	72.3%w	3497;7531,6319,7878	
fl pt	CC	21	294	152	
Beil	25,436				
uv	3575,7174				
ir	633,1924,5576,6724,5932				
Raman	3973,4034,5358,1017,6724				
ms	7750				
nmr	7227,3753,6748,5933				

126. *cis*-2-Buten-1-ol

(*Z*)-2-Buten-1-ol

HC—CH₃
‖
CHCH₂OH   4088–60–2   C₄H₈O

		cgs	SI	
mw		72.107	72.107	3468
bp	1 atm	123.6	396.8	885;3224
dt/dp	1 atm	0.0527	0.395	6166
dp/dt	1 atm	19.0	2.53	6166
p	63	60	8.0	3224
eq 2.4	A	8.998	8.123	6166
60 - 120	B	2427	2427	
fp		−89.44	183.71	885
d	0	0.8726	872.6	3439;4026,3224
	20	0.854	854	
n_D	20	1.4342	1.4342	885;4026
ΔH_v	60 - 120	11.1	46.4	6166
soly	in aq	16.6%w	16.6%w	3439
Beil	25,442			
Raman	5358,2921			
nmr	5933			

127. *trans*-2-Buten-1-ol

(*E*)-2-Buten-1-ol

CH₃CH
‖
CHCH₂OH   504–61–0   C₄H₈O

		cgs	SI	
mw		72.107	72.107	3468
bp	1 atm	121.2	394.4	885,3224;8238,8239
d	25	0.8454	845.4	3224
n_D	20	1.4289	1.4289	885;8238
	25	1.4262	1.4262	3224
Beil	25,442			
Raman	5358,2921			
nmr	5933			

128. 2-Propyn-1-ol

CH≡CCH₂OH   107–19–7   C₃H₄O

		cgs	SI	
mw		56.064	56.064	3468
bp	1 atm	113.6	386.8	3288;6736,3303,6135,3302
				3223,7878
dt/dp	1 atm	0.0502	0.377	389
dp/dt	1 atm	19.9	2.66	389
p	25	15.5	2.07	389
eq 2.4	A	8.656	7.781	389
	B	2226.4	2226.4	
fp		−51.80	221.35	3288;6135
d	20	0.9478	947.8	3288;6736,3223,3297,7878
	25	0.9450	945.0	6233
n_D	20	1.4320	1.4320	3288;6736,3223,3297
	25	1.4300	1.4300	6233
η	20	1.68	0.00168	389;6736

128. 2-Propyn-1-ol

(Continued)

		cgs	SI	
γ	28.5	35.2	0.0352	6736
ΔH_v	112	10.06	42.09	389
ΔH_c°	25(1)	−423.8	−1773.2	3975,7446
C_p	20	34.5	144.3	389
pK_a	25	13.55	13.55	578
ε		24.5	24.5	389
μ	25 in 38	1.78	1.78	6233;6736,2905
soly	in aq	inf	inf	389
aq az	97	45.5%w	45.5%w	3498;6736,7878
fl pt	TOC	36	309	389
Beil	26,454			
uv	7181			
ir	3653,8164,5539			
Raman	682,5539			
nmr	3227,6748			

129. *dl*-α-Terpineol

α,α,4-Trimethyl-3-cyclohexene-1-methanol

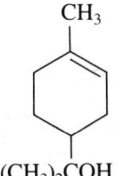

(CH$_3$)$_2$COH   98–55–5   C$_{10}$H$_{18}$O

		cgs	SI	
mw		154.252	154.252	3468
bp	1 atm	217.5	490.7	7058;5843
p	94.3	10.0	1.33	7058
	171.2	200	26.7	7058;5843
fp		35	308	7058

129. *dl-α*-Terpineol

(Continued)

		cgs	SI	
d	20	0.9337	933.7	7939;3910,1601
n_D	20	1.4831	1.4831	6981,7939;3316,1601
η	20	58.57	0.05857	5310
	25	26	0.026	3316;616
γ	20	28.65	0.02865	5310
ΔH_v		19.2	80.3	6581
δ	25	9.3	19.0	653
fl pt	COC	90	363	3316
Beil	507,58			
uv	6395			
ir	5269,5180,5932			
Raman	960,4117			
ms	7143,2193			
nmr	6748			

130. *d-α*-Terpineol

(*R*)-(+)-α,α,4-Trimethyl-3-cyclohexene-1-methanol

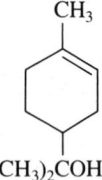

(CH₃)₂COH   7785-53-7 $C_{10}H_{18}O$

		cgs	SI	
mw		154.252	154.252	3468
bp	731 Torr	206-207	479-480	6981
p	104	15	2.0	2610
fp		36.9	310.1	2610
d	11.5	0.9475	947.5	2610
	45.5	0.9179	917.9	
n_D	20	1.4819	1.4819	2610
η	20	59.43	0.05943	5310

130. *d-α*-Terpineol

(Continued)

		cgs	SI	
γ	20	28.63	0.02863	5310
$[\alpha]_D$	20	+92.45	+92.45	6981;2610,5666,2448
Beil	507,56			

131. *l-α*-Terpineol

(*S*)-(−)-α,α,4-Trimethyl-3-cyclohexene-1-methanol

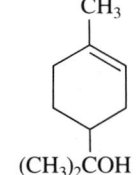

CH₃

(CH₃)₂COH   10482–56–1   C₁₀H₁₈O

		cgs	SI	
mw		154.252	154.252	3468
p	80 – 81	5	0.7	6981
fp		37.0	310.2	2610
d	19.5	0.9402	940.2	2610
	25	0.9364	936.4	
n_D	20	1.4820	1.4820	6981
$[\alpha]_D$	20	−100.5	−100.5	2610;5666,2448
Beil	507,57			

132.   *cis*-2-Butene-1,4-diol

(*Z*)-2-Butene-1,4-diol

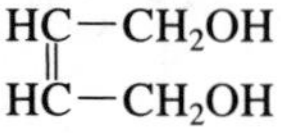

HC—CH$_2$OH
‖
HC—CH$_2$OH   6117–80–2   C$_4$H$_8$O$_2$

		cgs	SI	
mw		88.106	88.106	3468
bp	1 atm	235	508	7657
dt/dp	1 atm	0.045	0.34	6166
dp/dt	1 atm	22	3.0	6166
p	97	2.3	0.31	6853;5949,2446
eq 2.4	A	9.6679	8.7928	6166
	B	3448.7	3448.7	
fp		11.0	284.2	6853
d	15	1.082	1082	5949;7657
	20	1.0740	1074.0	6853
n$_D$	20	1.4793	1.4793	6853;5949,7657
ΔH$_v$	100–235	15.8	66.1	6166
K$_B$	eq 2.79	2.86	2.86	6166
μ		2.48	2.48	7554
Beil	31,499			
ir	6853			

133.   *trans*-2-Butene-1,4-diol

(*E*)-2-Butene-1,4-diol

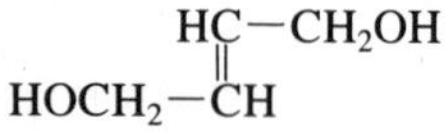

HC—CH$_2$OH
‖
HOCH$_2$—CH   821–11–4   C$_4$H$_8$O$_2$

		cgs	SI	
mw		88.106	88.106	3468
bp	13 Torr	132	405	7657
dt/dp	13 Torr	1.4	10.5	6166
dp/dt	13 Torr	0.71	0.095	6166
p	108	3.8	0.51	6853;2446,5948,5976,7657

133. *trans*-2-Butene-1,4-diol

(Continued)

		cgs	SI	
eq 2.4 100–130	A B	9.9895 3597.3	9.1144 3597.3	6166
fp		27.3	300.5	6853;5948,5976
d	20	1.0685	1068.5	6853;5948,5976,7657
n_D	20	1.4779	1.4779	6853;2446,7657,5948,5976
ΔH_v	100–130	16.5	69.0	6166
μ		2.45	2.45	7554;4678
Beil	31,499			
ir	6853			

134. 1,2-Ethanediol

$(CH_2OH)_2$ 107–21–1 $C_2H_6O_2$

		cgs	SI	
mw		62.068	62.068	3468
bp	1 atm	197.54	470.69	126;7255,6436,1012,2639,7490 7191,4665
dt/dp	1 atm	0.0444	0.333	126;7255,4665,7490
dp/dt	1 atm	22.5	3.00	126
p	25	0.0878	0.0117	126;5053,7058,5978,3832,7553
eq 2.5	A B C	7.71505 1818.591 178.651	6.83995 1818.591 178.651	126,7256,2639,7553,5567
eq 2.4 130–197	A B	9.2477 2994.4	8.3726 2994.4	6436;2661
fp*		–12.6	260.6	5501;7630,7255,7191,2637,1012 7490,7479

* The freezing point of 1,2-ethanediol is in
  doubt because of the tendency to supercool
  and form a glass.

134. 1,2-Ethanediol

(Continued)

		cgs	SI	
d	0	1.12763	1127.63	7490;2469,7191,2737,7845
	15	1.11710	1117.10	6436,7971
	20	1.1135	1113.5	7255;6179,2638,5978
	25	1.1100	1110.0	
	30	1.10635	1106.35	6416
dd/dt	10 - 40	0.00070	0.70	7630
α	20	0.000626	0.000626	4195
n_D	15	1.43312	1.43312	7490;2469,7971,5978,5251,6436
	20	1.4318	1.4318	7255
	25	1.4306	1.4306	
η	15	26.09	0.02609	7490,2478,873,7419
	30	13.759	0.013759	2020
γ	20	48.49	0.04849	4195;7843,7850,2638
	30	46.24	0.04624	5388
	Table 2.7			
	Table 2.8			
ΔH_v	25	16.2	67.8	2661
	bp	12.06	50.46	1902;5613,2639,4683,4665,958
ΔH_m		2.380	9.958	5501,5682,7191,4902,7167,1901
ΔH_f°	25(1)	-108.70	-454.80	7826,2036;8316,2661,5654
	25(g)	-92.7	-387.9	2661
ΔH_c°	25(1)	-284.35	-1189.72	2036;3975,7446,5297,2661,5654
C_p	19.8	35.7	149.4	5682,7191,4665,6352,7000
	27	36.1	151.0	5501
	Table 2.11			
λ	37.8	0.000624	0.261	6352;1766
K_i		5.7×10^{-15}	5.7×10^{-15}	5099;578,2876,4654
pK_a		14.24	14.24	5099
pK_s	25	15.84	15.84	4322;8294,77
κ	25	1.16×10^{-6}	0.000116	5325;5053,7843,7851,4135,4902
ε	25	37.7	37.7	4937;5053,3840,6951,4135,4610
	30	39.4	39.41	6416
μ	25 in 172	2.31	2.31	7534;5251,7535,4135,4610
δ		14.2	29.1	4593,1271;6905,652,6240
ER	BuOAc	<0.01	<0.01	6631
soly	in aq	inf	inf	5053
aq az		none	none	3497
fl pt	COC	116	389	5053

134. 1,2-Ethanediol

(Continued)

Beil	30,465
ir	4332,1234,4287,633,4860,6399,5932
Raman	3159,4103,1234,7228,6399
ms	910
nmr	6601,8222,6400,7227,2464,5933,6748

135. 1,2-Propanediol

CH₃CHOHCH₂OH   57–55–6   C₃H₈O₂

		cgs	SI	
mw		76.095	76.095	3468
bp	1 atm	187.6	460.8	7255;1529,5975,6442,7058
dt/dp	1 atm	0.0415	0.311	7255;5053
dp/dt	1 atm	24.1	3.21	7255
p	25	0.133	0.0177	5975;5978
eq 2.5	A	8.9171	8.0420	7256;6436
	B	2645.7	2645.7	
	C	250.7	250.7	
fp		−60	213	7255;1766
d	15	1.0399	1039.9	1766;1529,2677,5978,5975
	20	1.0362	1036.2	7255
	25	1.0328	1032.8	
	30	1.0290	1029.0	1766
dd/dt	10 − 40	0.00072	0.72	7630
α	20	0.000695	0.000695	6631
n_D	20	1.4329	1.4329	7255;1529,5251,1766,5978
	25	1.4314	1.4314	
dn/dt	20 − 30	0.00030	0.00030	1766

135. 1,2-Propanediol

(Continued)

		cgs	SI	
η	−195.7(gl)	1.45×10^{43}	1.45×10^{40}	3605
	0	243	0.243	1766;4631,2677,7419
	20	56.0	0.0560	
	40	18.0	0.0180	
γ	25	36.51	0.03651	1766;6173
	30	35.46	0.03546	5388
ΔH_v	25	15.4	64.4	6436;958
	bp	12.51	52.35	7630
ΔH_f°	25(l)	−116.1	−485.8	2661;2036,8316
	25(g)	−100.7	−421.3	2661
ΔH_c°	25(l)	−439.31	−1838.07	2661;2036,3975,5297,8316
C_p	20	45.1	188.7	1766
λ	37.8	0.000502	0.210	6352;1766
$[\alpha]_D$	20	−15.37	−15.37	4547
pK_a	25	14.8	14.8	8158
pK_s	25	17.21	17.21	4322
ε	20	32.0	32.0	4937;4165,4610,5251
μ		2.25	2.25	5251;4610
δ		12.6	25.8	652,6631
ER	BuOAc	0.01	0.01	6631
soly	in aq	inf	inf	5053
aq az		none	none	3497
fl pt	COC	99	372	3518

Beil	30,472
ir	1766,5932
Raman	7228
nmr	5003,8222,7227,1423,2464,6748

136. 1,3-Propanediol

HOCH$_2$CH$_2$CH$_2$OH   504-63-2   C$_3$H$_8$O$_2$

		cgs	SI	
mw		76.095	76.095	3468
bp	1 atm	214.4	487.6	7255;7058,1529,2639,6436
dt/dp	1 atm	0.0423	0.317	7255
dp/dt	1 atm	23.6	3.15	7255
p	106.2	10	1.3	7256
eq 2.5	A	9.4710	8.5959	7256;6436,2639
	B	3212.0	3212.0	
	C	273.0	273.0	
fp		-26.7	246.5	7255;2637
d	20	1.0538	1053.8	6436;2638,1529,5267
	25	1.050	1050	7255
α	20 - 40	0.00061	0.00061	5053
n$_D$	20	1.4396	1.4396	7255,2637;5251,6436,1529
	25	1.4386	1.4386	
	40	1.4332	1.4332	5053
η	20	46.6	0.0466	4631;7419
γ	25	45.17	0.04517	2638
	30	46.95	0.04695	5388
	Table 2.8			
ΔH$_v$	25	17.4	72.8	6436
	bp	13.83	57.86	2639
ΔH$_f^\circ$	25(1)	-111.1	-464.8	2661;2036
	25(g)	-93.7	-392.0	2661
ΔH$_c^\circ$	25(1)	-444.30	-1858.95	2661;2036,3975
ε	20	35.0	35.0	4937,5251;3840
μ	25 in 172	2.55	2.55	7534;5251
δ	25	16.1	33.0	653
soly	in aq	inf	inf	5053
aq az		none	none	3498

Beil	30,472
uv	3575
ir	7876,6399,5932
Raman	6399
ms	910
nmr	8222,1423,6748

137.   1,3-Butanediol

$HOCH_2CH_2CHOHCH_3$ 107–88–0 $C_4H_{10}O_2$

		cgs	SI	
mw		90.122	90.122	3468
bp	1 atm	207.5	480.7	6436,5053;1529,792,4548
dt/dp	1 atm	0.0415	0.311	7255
dp/dt	1 atm	24.1	3.21	7255
p	25	0.06	0.008	6905
	74	0.5	0.07	2872
eq 2.5	A	9.8035	8.9284	7256;6436
	B	3470.5	3470.5	
	C	294.8	294.8	
fp		<-50	<223	5053
d	20	1.0053	1005.3	7255;6436,1529,5053
	25	1.0000	1000.0	4763
dd/dt	0 - 40	0.000666	0.666	4763
α	20	0.000664	0.000664	4763
κ_T	20	5.69 x 10^{-8}	4.27 x 10^{-7}	4763
κ_S	20	3.37 x 10^{-8}	2.53 x 10^{-7}	8286
n_D	20	1.441	1.441	7255;792,1529,2872
	25	1.439	1.439	
η	20	130.3	0.1303	4631
	25	98.3	0.0983	5053
	35	89	0.089	
γ	25	37.8	0.0378	5053
	30	37.04	0.03704	5388
ΔH_v	25	16.2	67.8	6436
	bp	13.97	58.45	958
ΔH_f°	25(1)	-119.7	-500.8	2661
	25(g)	-103.5	-433.0	2661
ΔH_c°	25(1)	-598.04	-2502.20	2661;5297
$[\alpha]_D$	22(+)	18.5	18.5	4539;4550
	25(-)	18.8	18.8	4548
pK_a	25	15.1	15.1	8158
δ		11.6	23.7	652,6905
soly	in aq	inf	inf	4026
fl pt	TOC	121	394	5053

Beil	30,477
ir	4860,7225,5932
ms	7804
nmr	2380,4980,5933

138. 1,4-Butanediol

$(CH_2CH_2OH)_2$ 110–63–4 $C_4H_{10}O_2$

		cgs	SI	
mw		90.122	90.122	3468
bp	1 atm	228	501	1529;2627,1237,3123,6471
p	122.9	10.0	1.33	7417;2627
eq 2.4	A	9.39	8.51	2661
	B	3264	3264	
fp		19.65	292.80	2661;6471,1237,2326,2627
d	20	1.0185	1018.5	1529;6471
	25	1.01289	1012.89	5387;1237,7417
α	20 - 40	0.000646	0.000646	1529
κ_S	20	3.85×10^{-8}	2.89×10^{-7}	8287
n_D	20	1.4460	1.4460	1529;7417,1237
	30	1.4425	1.4425	1529
dn/dt	30	0.00070	0.00070	1529
η	20	89.24	0.08924	529;4631
	25	71.5	0.0715	2627,1237;7419,7417
γ	20	44.6	0.0446	6501;2121,6511
	30	43.79	0.04379	5388
ΔH_v	25	18.3	76.6	2661
	182	14.9	62.3	
ΔH_f°	25(1)	−120.3	−503.3	2661
	25(g)	−102.0	−426.8	
ΔH_c°	25(1)	−597.50	−2499.94	2661
pK_a	25	14.5	14.5	8158
ε	30	30.2	30.2	4937
μ	25 in 172	2.58	2.58	7534;7554,7903
δ	25	12.1	24.7	653
soly	in aq	inf	inf	2627
fl pt	OC	>121	>394	2627,6402
Beil	30,478			
ir	5932			
ms	910,7804			
nmr	5933			

139. 2,3-Butanediol

(Mixed isomers)

CH₃CHOHCHOHCH₃   513–85–9 $C_4H_{10}O_2$

		cgs	SI	
mw		90.122	90.122	3468
bp	1 atm	182.0	455.2	7058,1766;6436,665
p	44.0	1.0	0.13	7058
	107.8	40.0	5.33	
eq 2.4	A	9.5521	8.6770	6436
80 - 130	B	3023.9	3023.9	
	A	9.2616	8.3865	
130 - 182	B	2907.1	2907.1	
fp		22.5	295.7	7058
d	20	1.0033	1003.3	6436;1766
	110	0.9285	928.5	7417
n_D	20	1.4377	1.4377	1766
	25	1.4366	1.4366	7417;6436
η	20	107.9	0.1079	4631
	25	121	0.121	1766
ΔH°_c	20	-588.61	-2462.74	5297
C_p	30	54.15	226.58	3987
K_b	25,aq HCl	0.33	0.33	2748
pK_a	25	14.9	14.9	8158
δ	25	11.1	22.7	653
soly	in aq,20	inf	inf	1237
fl pt	TOC	85	358	1766,6402

Beil	30,479
ir	5932
ms	7804
nmr	5933

140. *l*-2,3-Butanediol

R-(*R**,*R**)-2,3-Butanediol

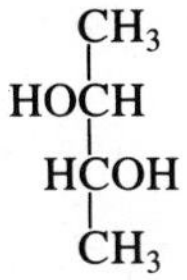

$$
\begin{array}{c}
CH_3 \\
HOCH \\
HCOH \\
CH_3
\end{array}
$$

24347–58–8

$C_4H_{10}O_2$

		cgs	SI	
mw		90.122	90.122	3468
bp	1 atm	179.0	452.2	4080,5624;5246
dt/dp	1 atm	0.085	0.64	4080
dp/dt	1 atm	12	1.6	4080
p	25	0.55	0.073	4080
eq 2.4	A	9.14412	8.26902	4080
	B	2852.58	2852.58	
fp		19	292	7905
d	20	0.991	991	1766
	25	0.9869	986.9	4080,5440;5246,7931
n_D	20	1.4328	1.4328	1527;5246
	25	1.4308	1.4308	1527,4080;7905,5624,5440,7931
dn/dt		0.00036	0.00036	7931
η	20	87.5	0.0875	1528
	25	41.1	0.0411	4080;7905
γ	25	30.61	0.03061	4080;1528
$[\alpha]_D$	25	−13.16	−13.16	7931;4080,7905,5246,5440
				7221,4529
soly	in aq,20	inf	inf	1237
aq az		none	none	3499
Beil	30,EI 250			
ir	2450			

141.   d-2,3-Butanediol

S-(R*,R*)-2,3-Butanediol

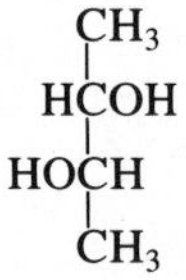

CH$_3$
HCOH
HOCH
CH$_3$

19132–06–0

C$_4$H$_{10}$O$_2$

		cgs	SI	
mw		90.122	90.122	3468
bp	1 atm	179–182	452–455	6981;5246
d	25	0.9872	987.2	4080;6981,1766
	30	0.996	996	5246
n$_D$	25	1.4306	1.4306	4080;6981,1766,5246
γ	25	30.67	0.03067	4080
[α]$_D$		13.19	13.19	4080;2973,7221
soly	in aq,20	inf	inf	1237
Beil	30,EIII 2183			
ir	2450			

142.   dl-2,3-Butanediol

R*,R*-(±)-2,3-Butanediol

CH$_3$CHOHCHOHCH$_3$

6982–25–8

C$_4$H$_{10}$O$_2$

		cgs	SI	
mw		90.122	90.122	3468
bp	742 Torr	176.7	449.9	8101;1766
p	89	19.5	2.60	7189;8101
fp		7.6	280.8	8101;1766,6981
d		0.993	993	5406
n$_D$	25	1.4310	1.4310	7931
	30	1.4293	1.4293	
dn/dt		0.00036	0.00036	7931
ΔH$_v$	25	14.16	59.25	1703
ΔH$_f^\circ$	25(l)	−129.43	−541.54	1703
	25(g)	−115.27	−482.29	
ΔH$_c^\circ$	25(l)	−588.35	−2461.66	1703
soly	in aq,20	inf	inf	1237

142. *dl*-2,3-Butanediol

(Continued)

Beil	30,EII 546
ir	2450
Raman	7189

143. *meso*-2,3-Butanediol

(*R**,*S**)-2,3-Butanediol

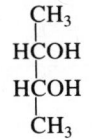

5341-95-7 $C_4H_{10}O_2$

		cgs	SI	
mw		90.122	90.122	3468
bp	1 atm	182.3	455.5	4080;5624,8101,7189
dt/dp	1 atm	0.083	0.62	4080
dp/dt	1 atm	12	1.6	4080
p	25	0.38	0.051	4080
eq 2.4	A	8.95845	8.08335	4080
	B	274.81	274.81	
fp		34.4	307.6	8101;1766,5406
d	20	1.004	1004	5406
	25	0.9939	993.9	4080;1527,8101
n_D	25	1.4372	1.4372	7931;1527,8101,5624
	35	1.4339	1.4339	7931;4080,1766
dn/dt		0.00034	0.00034	7931
η	35	65.8	0.0658	4080;1766
soly	in aq,20	inf	inf	1237
aq az		none	none	3499

143.   *meso*-2,3-Butanediol

(Continued)

Beil	30, EII 546
ir	2450,1446
Raman	7189
ms	3203

144.   1,5-Pentanediol

$HOCH_2(CH_2)_3CH_2OH$ 111-29-5 $C_5H_{12}O_2$

		cgs	SI	
mw		104.149	104.149	3468
bp	1 atm	242.4	515.6	7630;1237,1529,3124
dt/dp	1 atm	0.047	0.35	7630
dp/dt	1 atm	21	2.8	7630
p	133.6	10.0	1.33	7417
eq 2.4	A	9.48	8.60	2661
	B	3390	3390	
fp		−15.6	257.6	7630,1237;3097
d	20	0.9914	991.4	1529;7630,1237,4763,3905
	25	0.9858	985.8	4763
dd/dt	-30 − 40	0.000608	0.608	4763
α	20	0.000615	0.000615	4763;1529
κ_T	20	5.35 x 10^{-8}	4.01 x 10^{-7}	4763
n_D	20	1.4500	1.4500	1529;7417,7630,1237,3905
	25	1.4484	1.4484	
dn/dt	25	0.00032	0.00032	1529
η	20	114.66	0.11466	1529;7630,1237,7417
γ	20	43.4	0.0434	6501;1237

144. 1,5-Pentanediol

(Continued)

		cgs	SI	
ΔH_v	25	19.7	82.4	2661
	bp	14.5	60.7	7630
ΔH_f°	25(1)	-127.0	-531.4	2661
	25(g)	-107.3	-448.9	
ΔH_c°	25(1)	-753.12	-3151.05	2661
μ	25 in 172	2.51	2.51	7534
δ	25	11.5	23.5	653
soly	in aq	inf	inf	7630
aq az		none	none	3498
fl pt	COC	129	402	7630,1237,5415
Beil		30,481		
ir		6615,5932		
ms		910,7804		
nmr		5933,6748		

145. Hexylene glycol

2-Methyl-2,4-pentanediol

$(CH_3)_2COHCH_2CHOHCH_3$ 107–41–5 $C_6H_{14}O_2$

		cgs	SI	
mw		118.175	118.175	3468
bp	1 atm	197.5	470.7	7257;1766,6627,1237,7630,6471
dt/dp	1 atm	0.049	0.37	7257;1766,7630
dp/dt	1 atm	20.4	2.7	7257
p	94	10.0	1.33	1766
	125	50.0	6.67	
fp		-50(g1)	223(g1)	7257,1766,6627,1237,7630
d	20	0.92109	921.09	3610;7257,1463,4763,1766,5558
	25	0.9181	918.1	1463;7257

145. Hexylene glycol

(Continued)

		cgs	SI	
dd/dt	-50 - 40	0.000749	0.749	4763;1766
α	20	0.000813	0.000813	4763;1766,6627
κ_T	20	8.56×10^{-8}	6.42×10^{-7}	4763
n_D	20	1.4275	1.4275	1463;3610,7257,7417,1766,5558
	25	1.4257	1.4257	
dn/dt	25	0.00035	0.00035	3610
η	20	34.4	0.0344	1766;7417,7419,7630,6627
γ	20	33.1	0.0331	6627
ΔH_v	bp	13.7	57.3	1766;6627,7630,1237
ε	20	25.86	25.86	3610;5251
μ	in 172	2.9	2.9	8005
δ		9.7	19.8	6631;6905
ER	BuOAc	<0.01	<0.01	6631
soly	in aq	inf	inf	1766
fl pt	COC	101	374	1766,7630,6402;6627
Beil	30,486			
ir	6615,8197,5932			
Raman	2133			
nmr	5933,6748			

146. 2-Ethyl-1,3-hexanediol

$CH_3CH_2CH_2CHOHCH(C_2H_5)CH_2OH$ 94-96-2 $C_8H_{18}O_2$

		cgs	SI	
mw		146.229	146.229	3468
bp	1 atm	244.2	517.4	1766,4743;7630,1237,2899
dt/dp	1 atm	0.049	0.37	1766,7630
dp/dt	1 atm	20.4	2.7	1766,7630
p	129	10.0	1.33	1766;2967
	163	50.0	6.67	
fp		-40(gl)	233(gl)	1766,7630,1237,2899
d	20	0.93971	939.71	3610;4743,2899,1766
	30	0.93253	932.53	
dd/dt		0.00074	0.74	1766;7630
α	20	0.00080	0.00080	1766
n_D	20	1.45076	1.45076	3610;2899,1766,7630
	25	1.4490	1.4490	7105
dn/dt		0.00037	0.00037	1766
η	20	323	0.323	2899;1766,1237
γ	35	29.02	0.02902	7105
ΔH_v	bp	14.9	62.3	1766;7630,1237
ε	20	18.73	18.73	3610
δ	25	9.4	19.2	653
soly	in aq,20	4.2%w	4.2%w	4743,2899;1766,7219
	aq in,20	11.7%w	11.7%w	4743,2899;1766
fl pt	COC	129	402	1766;5415,6402,4743,2899

Beil	30,EII 556
ir	5932
nmr	5933,6748

147. Glycerol

1,2,3-Propanetriol

$CH_2OHCHOHCH_2OH$ 56-81-5 $C_3H_8O_3$

		cgs	SI	
mw		92.094	92.094	3468
bp	1 atm	290.0	563.2	7058,4563;4433
p	50	0.0025	0.00033	5159;73,7058,4433
eq 2.4	A	11.27423	10.39913	6265;8296,1326
20 - 70	B	4480.5	4480.5	
fp		18.18	291.33	73;2736,5994,7058,4433,7783
d	15	1.26443	1264.43	7490;73,2288,6663,1003,4433
	20	1.26134	1261.34	
	25	1.2559	1255.9	4763
	30	1.25512	1255.12	7490;6416
dd/dt		0.000615	0.615	6905
α	20	0.000520	0.000520	4763
κ_T	20	2.93×10^{-8}	2.19×10^{-7}	4763
n_D	15	1.47547	1.47547	7482;4433,2288
	20	1.4746	1.4746	7033
	25	1.4730	1.4730	
η	-195.7	2.64×10^{59}	2.64×10^{56}	3605
	20	1412	1.412	6552;2478,4433,7490,2288 5048,7788
	25	945	0.945	2241
	30	612	0.612	
	50	142	0.142	6552
γ	20	63.3	0.0633	5053;4433,2288
	90	58.6	0.0586	
	150	51.9	0.0519	
ΔH_v	55	21.06	88.12	5159;726,4433,958
	bp	14.59	61.04	726
ΔH_m		4.370	18.284	2736;1901,7167
ΔH_f°	25(1)	-159.76	-668.44	2036;5692,8316
ΔH_c°	25(1)	-395.65	-1655.40	2036;5692,3975,8316
C_p	26.2	53.37	223.30	2736;2288,6352,4433
	Table 2.12			
λ	37.8	0.000740	0.310	6352;668
K_f	eq 2.67	3.27-3.69	3.27-3.69	5994
K_B	eq 2.79	6.52	6.52	726
pK_a		14.40	14.40	4654;5099
κ	25	ca 6×10^{-8}	ca 6×10^{-6}	6463;6537
ϵ	25	42.5	42.5	4937;1805,5251,73,56 4165,2100
	30	48.70	48.70	6416
μ	in 172	2.56	2.56	5290;3840

147. Glycerol

(Continued)

		cgs	SI	
δ		16.5	33.8	652,4593,1271,6905
soly	in aq	inf	inf	5159
aq az		none	none	5159,3497
fl pt	COC, 99%	177	450	5159

Beil	38,502
uv	6262,7174
ir	1378,1278,8291,8073,5932
Raman	1968,5403,6353,6051,7228,7665,7714
ms	4783
nmr	2380,6400,5933

148. 1,2,6-Hexanetriol

$HOCH_2(CH_2)_3CHOHCH_2OH$ 106-69-4 $C_6H_{14}O_3$

		cgs	SI	
mw		134.175	134.175	3468
bp	5 Torr	178	451	1237
p	120	0.10	0.013	7217
eq 2.5	A	5.4749	4.5998	7217
120–160	B	1154.2	1154.2	
	C	60.1	60.1	
fp		−20(gl)	253(gl)	1237
d	20	1.1030	1103.0	7217;1237
	30	1.0972	1097.2	
dd/dt	25	0.00058	0.58	7217
n_D	20	1.4766	1.4766	7217;1237
	25	1.4752	1.4752	7217;8271
dn/dt	25	0.00028	0.00028	7217

148.   1,2,6-Hexanetriol

(Continued)

		cgs	SI	
η	20 30	2628 1095	2.628 1.095	7217;1237
γ	20	49.9	0.0499	7217
soly	in aq,20	inf	inf	7217
fl pt	COC	193	466	6402;4247,5415

Beil	41, EIV 2784
ir	6615,5932
nmr	5933

149. Methyl ether

Oxybismethane

$(CH_3)_2O$ | 115–10–6 | C_2H_6O

		cgs	SI	
mw		46.068	46.068	3468
bp	1 atm	−24.84	248.31	7289;122,1609,7623,3950,4726
dt/dp	1 atm	0.03012	0.2259	122;7289
dp/dt	1 atm	33.20	4.426	122
p	25	4450	593.3	122;1609,7623
eq 2.5	A	7.31646	6.44136	7293;3950
	B	1025.56	1025.56	
	C	256.05	256.05	
fp		−141.49	131.66	7289,3950;7623,1609,4726
d	−40	0.7538	753.8	2992;4856,4726
	−20	0.7174	717.4	1609;4856,2992,4726
	20(1)	0.6689	668.9	7289;5652,1609,7623
	25(1)	0.6612	661.2	
dd/dt	−60 − bp	0.00132	1.32	2992
n_D	−40	1.3438	1.3438	2992;1609
	20	1.3018	1.3018	2510;7289
	25	1.2984	1.2984	
dn/dt	−60 − bp	0.00050	0.00050	2992
γ	−42.0	21.3	0.0213	4762;1609
	−10.6	16.5	0.0165	
ΔH_v	20	4.443	18.59	1609
	bp	5.143	21.52	1609;3950,4762
ΔH_m		1.181	4.941	7290;3950,1609
ΔH_f°	25(g)	−43.99	−184.05	5856;7826,2036
ΔH_c°	25(g)	−349.04	−1460.38	5856;2036
C_p°	25	15.39	64.39	7826;7291,6553,3059,4042
C_p	−33	24.45	102.30	3950
	25	26.3	110.0	
t_c		126.9	400.1	122,7292;689,1349,4762
P_c		51.7	5.24	122;7292,1349,689
d_c		0.2714	271.4	1349;1609,7292
v_c		0.170	0.000170	7292
A	eq 2.72	0.03425	0.03425	7290
pK_a	aq H_2SO_4	−3.83	−3.83	434
pK_{BH^+}	25	−2.48	−2.48	5755;4556
	aq H_2SO_4			
ε	25(1)	5.02	5.02	1609;4904
$\mu\prime$	25 in 38	1.35	1.35	7734;454,613
	(g)	1.28	1.28	3000;6373
δ		8.8	18.0	1271,4593

149.   Methyl ether

(Continued)

		cgs	SI	
soly	in aq,24 aq in,24	35.3%w 7.0%w	35.3%w 7.0%w	1609
fl pt	TOC	−41	232	1609;6402

Beil	19,281
uv	6428,3189,7430
ir	5754,5845,6894,3845,4943,5849,1712,3059
Raman	3845,4943,7200,6772,4100
ms	4784,7199
nmr	98

150.   Ethyl vinyl ether

Ethoxyethene

$CH_2{=}CHOC_2H_5$ 109-92-2 C_4H_8O

		cgs	SI	
mw		72.107	72.107	3468
bp	1 atm	35.72	308.87	2035,8161,7795;4091,8116,122
dt/dp	1 atm	0.0372	0.279	122;7615
dp/dt	1 atm	26.9	3.58	122
p	25	515	68.7	122;7615,4091
eq 2.4	A B	7.41783 1400.44	6.5427 1400.44	7615
fp		−115.8	157.4	2035;7615
d	9.2 20	0.7723 0.7531	772.3 753.1	1408;8116 8161;7795
dd/dt		0.00117	1.17	7615
n_D	20 25	1.37542 1.37288	1.37542 1.37288	4091;2035,8161,7795,7615
dn/dt	25	0.00051	0.00051	4091
η	20	0.2	0.0002	7615

150.   Ethyl vinyl ether

(Continued)

		cgs	SI	
γ	20	19.00	0.01900	8161;7795
ΔH$_v$	25	6.57	27.5	122;5857
	bp	6.26	26.2	122;7615
ΔH$_f^\circ$	25(g)	−33.50	−140.2	5857
ΔH$_c^\circ$	25(g)	−615.98	−2577.3	5857
t$_c$		202	475	4091;4299
P$_c$		40.2	4.07	4091;4299
μ	20 in 38	1.26	1.26	2094;5634
δ	25	7.8	16.0	653
soly	in aq,20	0.9%w	0.9%w	7615
	aq in,20	0.2%w	0.2%w	
aq az	34.6	98.5%w	98.5%w	3498

Beil	25,433
ir	5113,153,1866,2160,6349,5932
Raman	683,4034,3973,2160,6349
nmr	1187,3219,5933,6748

151.   Ethyl ether

1,1'-Oxybisethane

$(C_2H_5)_2O$ 60-29-7 $C_4H_{10}O$

		cgs	SI	
mw		74.122	74.122	3468
bp	1 atm	34.431	307.581	133;2754,5258,7495,7289,8129 4969,7470
dt/dp	1 atm	0.03711	0.2783	133;7289,8129,7470,7495
dp/dt	1 atm	26.95	3.593	133
p	25	537.21	71.622	133;7290,4663

151. Ethyl ether

(Continued)

		cgs	SI	
eq 2.5	A	6.92625	6.05115	133;7293
	B	1062.409	1062.409	
	C	228.183	228.183	
fp	(st)	−116.3	156.9	7289,5678,7499;4797,7480,5104
	(ms)	−123.3	149.9	4797,7482
tp	(st)	−116.23	156.92	1684
	(ms)	−123.29	149.86	1684
d	15	0.71925	719.25	7495;2677,4969,6980,6791
				2754,6461
	20	0.71361	713.61	3081;7499,7763,4380,7289
	25	0.70782	707.82	
	30	0.70205	702.05	7495
dd/dt	25	0.001154	1.154	7289
α	25	0.001654	0.001654	3081
n_D	15	1.35555	1.35555	7495,4091;1069,7763
	20	1.35243	1.35243	7289
	25	1.34954	1.34954	
	30	1.34682	1.34682	4091
dn/dt		0.00056	0.00056	7495
η	15	0.247	0.000247	7495;557,4570,7705,2129,7522
	20	0.242	0.000242	2677
	Table 2.2			
	Table 2.4			
γ	20	17.06	0.01706	3283;28,7763,5258,7848
	25	16.50	0.01650	4069
	30	15.95	0.01595	3283
	Table 2.7			
ΔH_v	25	6.51	27.2	7290;5678,5153,7592,5851
	bp	6.3356	26.508	1684;7290,4969,6951,7754,7784
ΔH_m	(st)	1.6300	6.820	1684;7290,735,7167
	(ms)	1.718	7.190	1684
ΔH_f°	25(l)	−66.83	−279.62	7291;1384,5857,1703,7826
	25(g)	−60.26	−252.13	7826;1705,5851
ΔH_c°	25(l)	−651.60	−2726.3	5851;7446,7134,5857,3975
	25(g)	−657.52	−2751.1	1703;5851
C_p°	25	28.45	119.05	1684;5869,6469
C_p	25	41.23	172.5	1684
t_c		193.59	466.74	133;4299,4091,7761,8231,5153
				7848,1671
P_c		35.94	3.642	133;4299,5153,7761,8231,2302
				4091,7850
d_c		0.265	265	4299
v_c		0.280	0.000280	4299;1671,8233,4091
A	eq 2.72	0.03569	0.03569	7290

151.   Ethyl ether

(Continued)

		cgs	SI	
K_f	eq 2.67	1.79	1.79	735
K_B	eq 2.79	1.824	1.824	1370;726,718,3432,720,7845
pK_a	aq H_2SO_4	−2.39	−2.39	5755;4484,4556,434,437 7874,8115
pK_b		5.2	5.2	2880
κ		<3 x 10^{-16}	<3 x 10^{-14}	7787;3094,2209,2208
ε	20	4.335	4.335	4937;1518,7497,6657,4251 5251,8184
	25	4.197	4.197	5005
μ	20	1.15	1.15	5251;3000,2094,4579,3827 454,7734
δ		7.4	15.1	1271;4593,3394,6641,7635,3565 652,6723
ER	BuOAc	33	33	7635
soly	in aq,25	6.04%w	6.04%w	3806;3401,3745,5496
	aq in,25	1.468%w	1.468%w	6291;3806,5496
aq az	34.15	98.74%w	98.74%w	3497
fl pt	CC	−45	228	5712

Beil	<u>21</u>,314
uv	1883,3330,4423,6428,1884,4062,5864
ir	5849,633,2250,5888,5906,1942,7077,1497,6894,5932
Raman	1783,1069,1536,7228,4944,1785,1095,1497
ms	4784,154,7199,3851,6199,7408,3423
nmr	7227,5532,1396,4694,3003,2464,1795,5933

152.   Propyl ether

1,1'-Oxybispropane

$(CH_3CH_2CH_2)_2O$ 111-43-3 $C_6H_{14}O$

		cgs	SI	
mw		102.176	102.176	3468
bp	1 atm	90.08	363.23	122;7289,8183,1487
dt/dp	1 atm	0.04353	0.3265	122;7289,8183
dp/dt	1 atm	22.97	3.063	122
p	25	62.51	8.334	122;7290
eq 2.5	A	6.90585	6.03075	122;1487,5078
	B	1233.748	1233.748	
	C	216.442	216.442	
fp		-123.2	150.0	7289;7479
tp		-114.79	158.36	355
d	20	0.7466	746.6	7289
	25	0.7419	741.9	
	30	0.737	737	6791
dd/dt	25	0.00094	0.94	7289
n_D	20	1.3805	1.3805	7289
	25	1.3780	1.3780	
dn/dt		0.00049	0.00049	7488
η	15	0.448	0.000448	7488;2129,2677
	30	0.376	0.000376	
	Table 2.2			
γ	20	20.53	0.02053	3283;7488,7763
	30	19.35	0.01935	
ΔH_v	25	8.53	35.7	122;7290,1599
	bp	7.475	31.274	355;122,3779
ΔH_m	(st)	2.574	10.77	355
	(ms)	2.280	9.540	355
ΔH_f°	25(1)	-78.58	-328.8	8316;7291,1599
ΔH_c°	25(1)	-963.93	-4033.1	1599,8316
C_p°	25	38	159	7291
	106.88	45.521	190.46	355
C_p	25	52.96	221.6	355
t_c		257.45	530.60	116;355,122
p_c		29.88	3.028	116;122
pK_a	in aq	-4.40	-4.40	437
	H_2SO_4			
	25	-1.99	-1.99	4556
ε	26	3.39	3.39	4937;3662,6791
μ	25 in 317	1.32	1.32	454;7734,3662,3000,6374
soly	in aq,25	0.49%w	0.49%w	763
	aq in,25	0.45%w	0.45%w	
aq az	72.5	75%w	75%w	6165;3497

152.   Propyl ether

(Continued)

		cgs	SI	
fl pt	OC	−20.6	252.6	4902

Beil	24,354
ir	2250,7077,3387,1497,5932
Raman	1069,1536,1497
ms	4784
nmr	3805,1396,3003,5933,6748

153.   Isopropyl ether

2,2′-Oxybispropane

$[(CH_3)_2CH]_2O$ 108-20-3 $C_6H_{14}O$

		cgs	SI	
mw		102.176	102.176	3468
bp	1 atm	68.51	341.66	122;7294,2090,3891,4091,2425 7058,1487
dt/dp	1 atm	0.04143	0.3108	122;7294,2090
dp/dt	1 atm	24.14	3.218	122
p	25	149.1	19.88	122;1109,5480,7637,7295
eq 2.5	A	6.85188	5.97678	122;1487,2090
	B	1143.073	1143.073	
	C	219.340	219.340	
fp		−85.5	187.7	7294;2090,5681,7395
tp		−85.38	187.77	354
d	20	0.7239	723.9	563;7294,2090,2425,3891,794 2677,7763
	25	0.71854	718.54	978;7294
	30	0.7135	713.5	563;6483
dd/dt	25	0.00104	1.04	563
α	55	0.00156	0.00156	7615

153. Isopropyl ether

(Continued)

		cgs	SI	
n_D	20	1.3680	1.3680	2821;7294,2090,3891,4091,2425
				7763,978
	25	1.3655	1.3655	7294
	30	1.3630	1.3630	2821
dn/dt	25	0.00050	0.00050	563
η	20	0.3347	0.0003347	563,2677
	25	0.379	0.000379	2425;4902
	30	0.3010	0.0003010	563
γ	20.0	17.73	0.01773	978
	24.5	17.34	0.01734	7763;2425,1109
	30	16.63	0.01663	978
ΔH_v	25	7.65	32.0	122;7295,1599,354
	bp	6.98	29.2	122;7295,2425
ΔH_m		2.876	12.035	354;7295
ΔH_f°	25(1)	−84.01	−351.5	1599;8316
ΔH_c°	25(1)	−958.51	−4010.4	1599;8316
C_p°	86.83	43.621	182.51	354
C_p	25	51.65	216.1	354;5681,2425
t_c		227.17	500.32	116,354;122,4091,2302
P_c		27.95	2.832	116;122
d_c		0.265	265	4091;4299
v_c		0.386	0.000386	4091;4299
A	eq 2.72	0.0377	0.0377	7295
pK_a	in aq H_2SO_4	−4.30	−4.30	437
	25	−1.90	−1.90	4556;4484
ε	25	3.88	3.88	4937;7447,3118
	30	3.805	3.805	6483
μ	20	1.22	1.22	2882;2302,7447
	25 in 38	1.38	1.38	7734
δ		7.06	14.4	7635;6240,7159,6631
ER	BuOAc	8.42	8.42	7635;6631
	90%	57	57	6631
soly	in aq,20	1.2%w	1.2%w	7637;794
	aq in,20	0.57%w	0.57%w	
aq az	62.2	95.5%w	95.5%w	3497
fl pt	TOC	−9.4	263.8	5713;2425
	CC	−27.8	245.4	

153.   Isopropyl ether

(Continued)

Beil	24,362
ir	153,2560,1942,633,6894,1497,5932
Raman	1536,3387,6894,1497
ms	154,4784,2700,7408,3423
nmr	3003,1396,5532,1039,5933,6748

154.   Butyl vinyl ether

1-(Ethenyloxy)butane

$CH_2{=}CHOC_4H_9$ 111–34–2 $C_6H_{12}O$

		cgs	SI	
mw		100.160	100.160	3468
bp	1 atm	93.82	366.97	7795,8161;6692,5329,7615
				7475,7763
dt/dp	1 atm	0.044	0.33	7615
dp/dt	1 atm	23	3.0	7615
p	25	51	6.8	6442
fp		−92	181	6442
d	20	0.7792	779.2	7795,8161;1408
	25	0.7727	772.7	6442
n_D	20	1.4007	1.4007	7615;7795,8161,1408
	25	1.3997	1.3997	6442
η	20	0.5	0.0005	7615
γ	20	21.99	0.02199	7795;8161
ΔH_v	bp	7.71	32.3	6442
C_p	25	55.4	232	6442
pK_b		8.8	8.8	2880
μ	25 in 38	1.25	1.25	6229

154.   Butyl vinyl ether

(Continued)

		cgs	SI	
δ	25	7.9	16.1	653
soly	in aq,20 aq in,20	0.30%w 0.09%w	0.30%w 0.09%w	7615
aq az	77.5	88.4%w	88.4%w	3498;6692
fl pt	COC	−1.1	272.1	6442,5713

Beil	25,433
uv	2586,5962
ir	5113,1866,6349,5932
Raman	683,3973,2586,6349
nmr	1187,2392,3219,5933

155.   Butyl ethyl ether

1-Ethoxybutane

$C_4H_9OC_2H_5$ 628–81–9 $C_6H_{14}O$

		cgs	SI	
mw		102.176	102.176	3468
bp	1 atm	92.24	365.39	122;7294,7475,5517,5329 7763,1487
dt/dp	1 atm	0.04319	0.3239	122;7294
dp/dt	1 atm	23.15	3.087	122
p	25	55.96	7.461	122
eq 2.5	A B C	6.937672 1252.485 216.465	6.06257 1252.485 216.465	1487,122
fp		−103	170	7294;7475
d	20 25	0.7495 0.7448	749.5 744.8	7294,5329;7763,5517
dd/dt	25	0.00094	0.94	7294

155.   Butyl ethyl ether

(Continued)

		cgs	SI	
n_D	20	1.3818	1.3818	7294,7763;5329,5517
	25	1.3793	1.3793	
dn/dt	25	0.00050	0.00050	7294
η	20	0.421	0.000421	5329
	25	0.397	0.000397	
γ	20	20.75	0.02075	5329,7763
	25	20.25	0.02025	
ΔH_v	25	8.80	36.8	122
	bp	7.67	32.1	122;7295
C_p°	25	38	159	7296
t_c		227.17	500.32	122
P_c		27.95	2.832	122
pK_a	in aq H_2SO_4	−4.12	−4.12	437
μ		1.24	1.24	4580;4579
aq az	76.6	88.1%w	88.1%w	3498

Beil	24,369
ir	633
ms	4748,1364
nmr	6748

156. Butyl ether

1,1'-Oxybisbutane

$(C_4H_9)_2O$ 142-96-1 $C_8H_{18}O$

		cgs	SI	
mw		130.230	130.230	3468
bp	1 atm	140.29	413.44	1487;122,7289,4091,4298,7488 2090,2568,4971
dt/dp	1 atm	0.04848	0.3636	122;7289,7488,2090
dp/dt	1 atm	20.63	2.750	122
p	25	6.74	0.898	122;7290,3402,2091
eq 2.5	A	6.805282	5.930185	1487,122;2090,4298
	B	1302.768	1302.768	
	C	191.669	191.669	
fp		-95.2	178	7289;7488,415,2090
d	20	0.7684	768.4	7289;2568,3000,7763,2090,7488
	25	0.7641	764.1	
dd/dt	25	0.00086	0.86	7289
n_D	20	1.3992	1.3992	7289;4091,7763,2568,2090
	25	1.3968	1.3968	
dn/dt		0.00045	0.00045	7488
η	15	0.741	0.000741	7488;881
	30	0.602	0.000602	7488
γ	15	23.40	0.02340	7411;3283,7763
	30	21.99	0.02199	
ΔH_v	25	10.6	44.4	122;7290,1599
	bp	8.70	36.4	122;7290,4971
ΔH_f°	25(1)	-90.34	-377.98	1599;7291,8316
ΔH_c°	25(1)	-1276.92	-5342.63	1599;6801,8316
C_p°	25	48.8	204.2	7291
C_p	Table 2.11			
pK_a	in aq H_2SO_4	-5.40	-5.40	437
	25	-1.94	-1.94	4556
ε	20	3.083	3.083	5027;5023,7447,3480,4937
μ	20	1.18	1.18	5027;7734,454,3000,7447,4580
δ		7.76	15.9	6240;4593
soly	in aq,20	0.03%w	0.03%w	7637;763
	aq in,20	0.19%w	0.19%w	
aq az	92.9	67%w	67%w	3497
fl pt	CC	37.8	311.0	5713,4902

156.   Butyl ether

(Continued)

	cgs	SI
Beil	24,369	
uv	5883	
ir	153,633,2250,5932	
Raman	1535,1536	
ms	4784	
nmr	3805,5532,3003,6748,5933	

157.   Pentyl ether

1,1'-Oxybispentane

$(C_5H_{11})_2O$ 693–65–2 $C_{10}H_{22}O$

		cgs	SI	
mw		158.283	158.283	3468
bp	1 atm	186.8	460.0	7289;122,7476,2090,7495
dt/dp	1 atm	0.0521	0.391	122;7289,2090
dp/dt	1 atm	19.2	2.56	122
p	25	0.6	0.08	122
	105.46	57.04	7.605	2091
eq 2.5	A	7.45597	6.58087	2090
	B	1906.7	1906.7	
	C	230	230	
fp		−69.4	203.8	7289;7475,7495,2090,7476
d	20	0.7830	783.0	7289;7495,7763,2090
	25	0.7790	779.0	
dd/dt	25	0.00080	0.80	7289
n_D	20	1.4119	1.4119	7289;7495,7763,2090
	25	1.4098	1.4098	
dn/dt	25	0.00042	0.00042	7289
η	15	1.188	0.001188	7495;881
	30	0.922	0.000922	

157. Pentyl ether

(Continued)

		cgs	SI	
γ	20	24.76	0.02476	3283;7763
	30	23.78	0.02378	
ΔH_v	25	13.1	54.8	122
	bp	9.99	41.8	122
ΔH_f°	25(1)	−110.0	−460.3	3975
ΔH_c°	25(1)	−1582.0	−6619.1	3975;7134
C_p°	25	59.8	250.2	7291
$pK_{BH}+$	25	−1.92	−1.92	4556
ε	25	2.77	2.77	4937;7447,2100
μ	25 in 317	1.20	1.20	454
fl pt	TCC	57.2	330.4	5712,152

Beil	24,384
ir	633,5932
Raman	1536,3387
ms	4784

158. Isopentyl ether

1,1′-Oxybis(3-methylbutane)

$[(CH_3)_2CHCH_2CH_2]_2O$ 544–01–4 $C_{10}H_{22}O$

		cgs	SI	
mw		158.283	158.283	3468
bp	1 atm	173.4	446.6	7058;7763,4458
dt/dp	1 atm	0.050	3.8	7058
dp/dt	1 atm	20	2.7	7058
p	25	1.4	0.19	7058;6081
eq 2.5	A	7.56926	6.69416	7058;6081
	B	1891.32	1891.32	
	C	230	230	

158.   Isopentyl ether

(Continued)

		cgs	SI	
d	20	0.7777	777.7	7763
	28	0.7713	771.3	
dd/dt	25	0.00080	0.80	7763
n_D	20	1.40850	1.40850	7763
η	11	1.40	0.00140	4271
	20	1.012	0.001012	881
	30	0.866	0.000866	
γ	21.7	22.85	0.02285	4027
	28.0	22.31	0.02231	7763
ΔH_v	20	12.10	50.63	6081
	bp	8.40	35.15	4330
C_p	24 - 170	90.56	378.90	4330
ε	20	2.82	2.82	4937;2295,7845
μ		1.23	1.23	4580;454,2295
soly	in aq,20	0.02%w	0.02%w	4027;987
aq az	97.4	46%w	46%w	3497

Beil	24,401
ir	633
Raman	1536,3387
ms	3424,4784
nmr	5532,6748

159.   1,2-Dimethoxyethane

$CH_3OCH_2CH_2OCH_3$ 110–71–4 $C_4H_{10}O_2$

		cgs	SI	
mw		90.122	90.122	3468
bp	1 atm	84.50	357.65	4339;7058,7881,4091,5656
dt/dp	1 atm	0.049	0.37	7058
dp/dt	1 atm	20	2.7	7058
p	20	48.0	6.40	386
eq 2.5	A	6.6487	5.7736	7058
	B	1217.03	1217.03	
	C	230	230	
fp		−69	204	386;560
d	20	0.8691	869.1	4195;5656,7881,1346,1379
	25	0.86370	863.70	4339;1259
α		0.00119	0.00119	4195
n_D	20	1.37963	1.37963	4091;7881,5656,1259,1379,1346
	25	1.37811	1.37811	
dn/dt	25	0.000304	0.000304	4091
η	25	0.455	0.000455	1379;1259
	Table 2.6			
γ	20	24.61	0.02461	4195
	80	17.71	0.01771	
ΔH_v	25	8.697	36.39	4342
ΔH_m		3.00	12.6	2904
ΔH_f°	25(1)	−90.02	−376.64	2036
ΔH_c°	25(1)	−627.76	−2626.55	2036
C_p	25	46.20	193.3	
t_c		263	536	4299;4091
p_c		38.2	3.87	4299;4091
d_c		0.333	333	4299;4091
v_c		0.271	0.000271	4299;4091
pK_a	in aq H_2SO_4	−2.97	−2.97	438
ε	25	7.20	7.20	1379
μ	25 in 38	1.71	1.71	4005;2833
δ	25	8.6	17.6	653
ER	BuOAc	4.99	4.99	2904
soly	in aq	inf	inf	6165,1259
aq az	77.4	89.9%w	89.9%w	3498
fl pt	OC	1.11	274.26	386
	CC	−6	267	2904

159. 1,2-Dimethoxyethane

(Continued)

Beil	<u>30</u>,467
ir	1097,5183,5533,6894,3654,5932
ms	3424,3423
nmr	11,3003

160. 1,2-Diethoxyethane

$(C_2H_5OCH_2)_2$ 629-14-1 $C_6H_{14}O_2$

		cgs	SI	
mw		118.175	118.175	3468
bp	1 atm	121.2	394.4	4339;1766,2904,4630
dt/dp	1 atm	0.044	0.33	1766
dp/dt	1 atm	23	3.0	1766
p	20	9.40	1.25	2904;1766
t_p		-74.0	199.2	1766;2904,7992
d	20	0.8402	840.2	4339
	25	0.83510	835.10	4339
dd/dt		0.00102	1.02	1766
α	20	0.00121	0.00121	1766
n_D	20	1.3922	1.3922	1766;2904
	25	1.3898	1.3898	4339;2771
η	20	0.7	0.0007	1766;2904
ΔH_v	25	10.33	43.20	4342
C_p	25	62.00	259.4	4341
ε		5.10	5.10	2074

160. 1,2-Diethoxyethane

(Continued)

		cgs	SI	
δ	25	8.3	17.0	653
ER	BuOAc	1.05	1.05	2904;5455
soly	in aq,20	21.0%w	21.0%w	1766;2904
	aq in,20	3.4%w	3.4%w	
aq az	89.4	75%w	75%w	3497
fl pt	CC	27	300	2904

Beil	30,468
ir	3654,5932
ms	3423
nmr	6748

161. 1,2-Dibutyoxyethane

1,1'-[1,2-Ethanediylbis(oxy)]bisbutane

$(C_4H_9OCH_2)_2$ 112-48-1 $C_{10}H_{22}O_2$

		cgs	SI	
mw		174.283	174.283	3468
bp	1 atm	203.3	476.5	1766;7615,4339,2419
dt/dp	1 atm	0.052	0.39	1766;7615
dp/dt	1 atm	19	2.6	1766
p	84	10	1.3	1766;7615
fp		−69.1	204.1	1766;7615
d	20	0.8311	831.1	2419;1766,7615
	25	0.83189	831.89	4339
dd/dt		0.00085	0.85	1766;7615
α	20	0.00100	0.00100	1766
n_D	20	1.4131	1.4131	1766;7615
	25	1.4112	1.4112	4339;2771

161.   1,2-Dibutyoxyethane

(Continued)

		cgs	SI	
dn/dt		0.00043	0.00043	1766
η	20	1.3	0.0013	1766,7615
ΔH$_v$	25	14.04	58.76	4342
	bp	11.4	47.8	7615
C$_p$	20	83.7	350.2	7615
δ	25	8.2	16.8	653
soly	in aq,20	0.2%w	0.2%w	1766,7615
	aq in,20	0.6%w	0.6%w	
aq az	99.1	23.2%w	23.2%w	3498
fl pt	OC	85	358	1766,7615
Beil	30,EIII 2083			
nmr	6748			

162.   Bis(2-methoxyethyl) ether

1,1'-Oxybis(2-methoxyethane)

(CH$_3$OCH$_2$CH$_2$)$_2$O   111–96–6   C$_6$H$_{14}$O$_3$

		cgs	SI	
mw		134.175	134.175	3468
bp	1 atm	159.76(d)	432.91(d)	2639;2637
dt/dp	1 atm	0.066	0.50	2639
dp/dt	1 atm	15	2.0	2639
p	25	3.4	0.45	2639
eq 2.4	A	8.0837	7.2086	2639
	B	2251.5	2251.5	
fp		−64.0	209.2	2904;388
d	20	0.9434	943.4	2904,388
	25	0.9384	938.4	7880;1717,3125*
dd/dt		0.00106	1.06	1688

162. Bis(2-methoxyethyl) ether

(Continued)

		cgs	SI	
n_D	20	1.4078	1.4078	2904,388;2637
	25	1.40576	1.40576	7880;3125*
η	25	0.989	0.000989	7880;3125*
γ	20	27.0	0.0270	2904
	25	29.5	0.0295	2638
	bp	14.44	0.01444	2638
	Table 2.8			
ΔH_v	bp	10.312	43.145	2639
ΔH_m		3.25	13.6	2904
ΔH_f°	25(l)	−14.3	−598	2904
ΔH_c°	25(l)	−902	−3770	2904
C_p		54.1	226	2904
μ	25 in 35	1.97	1.97	4005
δ	25	9.4	19.3	653
ER	BuOAc	0.36	0.36	2904
soly	in aq	inf	inf	6165
aq az	99.8	~25%w	~25%w	6165
fl pt	TCC	62.8	336.0	3125*
	OC	70.0	343.2	388
* Best commercial grade				
Beil	30,468			
ir	3654,5932			
nmr	2464,5933,6748			

163. Bis(2-ethoxyethyl) ether

1,1'-Oxybis(2-ethoxyethane)

$(CH_3CH_2OCH_2CH_2)_2O$ 112–36–7 $C_8H_{18}O_3$

		cgs	SI	
mw		162.228	162.228	3468
bp	1 atm	188.9	462.1	1766;7615,385,2904,1717
dt/dp	1 atm	0.050	0.38	1766;7615
dp/dt	1 atm	20	2.7	1766
p	20	0.38	0.051	7615;1766,2904
fp		-44.3	228.9	1766;7615,385,2904
d	20	0.9063	906.3	7763;7615,385,2904,1766
	42	0.8863	886.3	7763
dd/dt	5 - 40	0.000966	0.966	1766;7615
α	55	0.00111	0.00111	7615
n_D	20	1.41147	1.41147	7763;1766,7615,385,2904
η	20	1.40	0.00140	1766;7615,385,2904
γ	22.1	26.99	0.02699	7763;2904
	26.1	26.68	0.02668	
ΔH_v	bp	11.7	49.0	7615
ΔH_f°		-152	-636	2904
$\Delta H_c^{\circ'}$		-1199	-5017	2904
C_p	15	81.6	341.4	7615
ε		5.70	5.70	2074
δ		8.67	17.73	7635
ER	BuOAc	0.04	0.04	7635,2904
soly	in aq,20	inf	inf	1766;7615,385,2904
aq az	99.4	31%w	31%w	3498
fl pt	TOC	82	355	1766,7615
Beil	30,EII 520			
ir	1097,5932			
nmr	5532,5933,6748			

164.   Bis(2-butoxyethyl) ether

1,1'-[Oxybis(2,1-ethanediyloxy)]bis-butane

$(CH_3CH_2CH_2CH_2OCH_2CH_2)_2O$ 112–73–2 $C_{12}H_{26}O_3$

		cgs	SI	
mw		218.336	218.336	3498
bp	1 atm	254.6	527.8	1766;7615,2904,387
dt/dp	1 atm	0.056	0.42	1766;7615
dp/dt	1 atm	18	2.4	1766
p	122	10	1.3	1766
fp		−60.2	213.0	1766;7615,2904,387
d	20	0.8837	883.7	1766;7615,2904,387
dd/dt	5 – 40	0.000844	0.844	1766;7615
α	55	0.00097	0.00097	7615
n_D	20	1.4233	1.4233	1766;7615,2904,387
η	20	2.39	0.00239	1766;7615,2904,387
γ	20	27.0	0.0270	2904,387
ΔH_v	bp	13.3	55.6	1766,7615;2904
ΔH_f°		−175	−732	2904
ΔH_c°		−1823	−7627	2904
C_p	20	108	452	2904
δ	25	8.3	17.0	653
ER	BuOAc	<0.001	<0.001	2904
soly	in aq,20	0.3%w	0.3%w	1766;7615,2904,387
	aq in,20	1.4%w	1.4%w	
aq az	99.8	5.3%w	5.3%w	3498
fl pt	TOC	118	391	1766,7615
Beil	30,EII 521			
ir	5932			
nmr	6748			

165.   Propylene oxide (Mixed isomers*)

Methyl oxirane

d-form   *l*-form   75–56–9 C_3H_6O

		cgs	SI	
mw		58.080	58.080	3468
bp	1 atm	33.9	307.1	1766,4091;7615,7058,6230
bp(+)	1 atm	36.5–38	309.7–311	7
dt/dp	1 atm	0.036	0.27	1766;7615
dp/dt	1 atm	28	3.7	1766
p	20	445	59.3	1766;7058,1007
	25	569	75.9	2065
eq 2.5	A	6.96997	6.09487	4760;5233,1007
	B	1065.27	1065.27	
	C	226.283	226.283	
fp		−104.4	168.8	7615;3697,5561,7058,4760,1766
d	0	0.8598	859.8	8299;6230
	20	0.8287	828.7	8299
d(+)	20/20	0.8412	841.2	7
dd/dt	0 – 20	0.00156	1.56	8299;7613,7615
α	20	0.00151	0.00151	1766
n_D	5	1.3712	1.3712	8299;1766,6230
	20	1.36603	1.36603	4091
	25	1.36322	1.36322	
dn/dt	25	0.000562	0.000562	4091
η	0	0.410	0.000410	8299
	20	0.327	0.000327	
	25	0.28	0.00028	2068
ΔH_v	25	6.667	27.89	6766,7615
	bp	6.872	28.75	1766
ΔH_m		1.561	6.531	5561
ΔH_f°	25(1)	−28.84	−120.67	6766;2924,8316
ΔH_c°	25(1)	−458.28	−1917.44	6766;3975,5297,8316
C_p°	25	17.36	72.63	2924
C_p	25(1)	28.77	120.4	5561;1766
t_c		209.1	482.3	4091;4299,1766
P_c		48.6	4.92	4091;4299
d_c		0.312	312	4091;4299
v_c		0.186	0.000186	4091;4299

165. Propylene oxide (Mixed isomers*)

(Continued)

		cgs	SI	
$[\alpha]_D$	(+),18	12.72	12.72	7;4547,4549,6658
	(-),18	8.26	8.26	7
pK_b		7.0	7.0	2880
μ	25 in 38	2.00	2.00	6230;3372,7111,7998
δ		9.2	18.8	4593
soly	in aq,20	40.5%w	40.5%w	7615;1766
	aq in,20	12.8%w	12.8%w	
aq az		none	none	1766,3497
fl pt	CC	-35	238	152,5713
	OC	-37	236	3697,7613

* Only properties of the d and l isomers are indicated
as such; mixed dl forms not indicated.

Beil	2362,6
ir	4023,2335,5932
Raman	575,7228,4023,2335
ms	2640
nmr	5363,4862,5933,6748

166. 1,3-Dioxolane

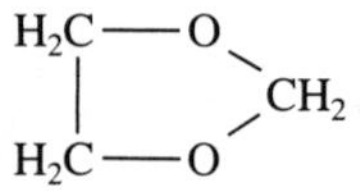

H₂C —— O
| ＼
| CH₂
H₂C —— O ／

646-06-0

$C_3H_6O_2$

		cgs	SI	
mw		74.079	74.079	3468
bp	1 atm	75.6	348.8	7623,2903
p	20	79	11	7623;2903
fp		-97.22	175.93	1522;7623,560
d	20	1.0647	1064.7	7623;1507,5852,2903
n_D	20	1.3992	1.3992	5852;1098,2903
η	20	0.6	0.0006	2903
ΔH_v	25	8.5	35.6	2460,1703
ΔH_m		6.567	27.476	1522

166.   1,3-Dioxolane

(Continued)

		cgs	SI	
ΔH_t	−130.8	2.677	11.201	1522
ΔH_f°	25(1)	−80.60	−337.23	5852;1703
	25(g)	−72.10	−301.67	
ΔH_c°	25(1)	−406.50	−1700.80	5852;2460,1703
pK_a	25	−3.8	−3.8	3850
pK_b		7.5	7.5	2880
μ	in 38	1.47	1.47	1757
δ		10.2	20.9	4593
soly	in aq	inf	inf	7623,2903
aq az	71.9	93%w	93%w	3498;2903
fl pt	TOC	1	274	6402,5415
	CC	−6	267	2903

Beil	2668,2
uv	6857
ir	7187,791,621,5932
Raman	7187,621
ms	2573
nmr	6647,5933,6748

167.   1,2-Epoxybutane

Ethyloxirane

$CH_3CH_2CHCH_2O$		106-88-7		C_4H_8O
		cgs	SI	
mw		72.107	72.107	3468
bp	1 atm	63.424	336.574	5610;4027,2566,7613
dt/dp	740–760 Torr	0.040	0.30	7613
dp/dt	740–760 Torr	25	3.3	7613

167. 1,2-Epoxybutane

(Continued)

		cgs	SI	
p	20 25	141 207	18.7 27.6	4027;7613 2065;5610
eq 2.4	A B	7.8461 1670.3	6.9710 1670.3	4027
fp		−150	123	4027
d	20 25	0.8297 0.824	829.7 824	7613;4027,2566 3101
dd/dt	20 − 30	0.00110	1.10	7613
n_D	20 25	1.3840 1.381	1.3840 1.381	4027 2072
η	20 25	0.41 0.40	0.00041 0.00040	4027 2072
ΔH_v	20 − 63 bp	7.64 7.25	32.0 30.3	4027 7613
ΔH_c°	25(1)	−587.47	−2457.98	2072,7613;5297
μ	20 in 38	2.01	2.01	5161
δ	25	9.0	18.4	653
soly	in aq,20 aq in,20	5.91%w 2.65%w	5.91%w 2.65%w	4027
aq az	60.1	92%w	92%w	6165
fl pt	TOC	−32	241	2072

Beil	2362,11
ir	1369,4023,5932
Raman	4023
nmr	5933,6748

168.   Cineole

1,3,3-Trimethyl-2-oxabicyclo[2.2.2]octane

470–82–6 $C_{10}H_{18}O$

		cgs	SI	
mw		154.252	154.252	3468
bp	1 atm	176.0	449.2	7058;3012,3013
dt/dp	1 atm	0.053	0.40	7058
dp/dt	1 atm	19	2.5	7058
p	25	1.9	0.25	7058;6643,3012
eq 2.5	A	7.27446	6.39936	7058
	B	1783.82	1783.82	
	C	230	230	
fp		1.3	274.5	3013;7058
d	15.5	0.9284	928.4	3013,4026
	20	0.9237	923.7	3125
	25	0.9192	919.2	
dd/dt	25	0.00090	0.90	3125
n_D	20	1.4575	1.4575	3013;7058,7883
	25	1.4555	1.4555	3125
dn/dt	25	0.00040	0.00040	3013
γ	20	32.1	0.0321	3125
	25	31.1	0.0311	
K_f		6.7	6.7	2385
ϵ	23.5	4.57	4.57	5428
soly	in aq,21	0.35%w	0.35%w	2180
aq az	97.6	35.9%w	35.9%w	6165;3497
fl pt	TCC	48	321	3125
Beil	2363,24			
uv	5991			
ir	5269,5932			

169. Furan

OCH=CHCH=CH   110-00-9 C_4H_4O

		cgs	SI	
mw		68.075	68.075	3468
bp	1 atm	31.36	304.51	3037;4091,4971,2035,1451,4026
dt/dp	1 atm	0.04131	0.3099	2087
dp/dt	1 atm	24.21	3.227	2087
p	26.469	634.00	84.526	3037
eq 2.5	A	6.97523	6.10013	3037;5669
	B	1060.851	1060.851	
	C	227.740	227.740	
fp		−85.61	187.54	3037,4026;2035
d	10	0.95144	951.44	3037;498
	15	0.94467	944.67	
	20	0.93781	937.81	4026
n_D	20	1.42140	1.42140	3037;4026
	25	1.41871	1.41871	4091
dn/dt		0.000538	0.000538	4091
η	20	0.380	0.000380	7483
	25	0.361	0.000361	
γ	20	24.10	0.02410	7483
	25	23.38	0.02338	
ΔH_v	20	6.628	27.73	3037;4971,4026,5211
	bp	6.474	27.09	
	Table 2.9			
ΔH_m		0.909	3.80	3037
ΔH_f°	25(1)	−14.903	−62.354	3037;8316
ΔH_c°	25(1)	−497.97	−2083.51	3037;2908,8316
C_p°	25	15.64	65.44	3037;555
	Table 2.14			
C_p	25.93	27.440	114.81	3037
	Table 2.12			
t_c		217.0	490.2	4299;1451,3037
p_c		54.3	5.50	4299;3037
d_c		0.312	312	3037;4299
v_c		0.218	0.000218	3037;4299
A	eq 2.72	0.0130	0.0130	3037
ε	25	2.942	2.942	3177
μ	25 in 38	0.71	0.71	3177;6212,4493
δ		9.4	19.2	652
soly	in aq,25	1%w	1%w	6643,4026
	aq in,25	0.3%w	0.3%w	
fl pt	TCC	−36	237	6005

169. Furan

(Continued)

Beil	2364,27
uv	5056,6022,5953,4808,6086,5841,3503
ir	3037,555,4462,5839,7436,4875,4883,6163,5932
Raman	3037,62,10,967,4121,6126,6125,6163
ms	3430,154,3368
nmr	2981,6419,9,2901,6092,4436,6748

170. Tetrahydrofuran

$OCH_2CH_2CH_2CH_2$ 109-99-9 C_4H_8O

		cgs	SI	
mw		72.107	72.107	3468
bp	1 atm	65.965	339.115	6515;2148,6004,944,5841,4091 6779,1451
dt/dp	1 atm	0.044	0.33	6004
dp/dt	1 atm	23	3.0	6004
p	23.139	149.41	19.920	6515
	25	162.05	21.60	4136;4091
	30	201.54	26.870	3667
eq 2.5	A	7.67206	6.79696	4136;1384,995,5211
	B	1557.06	1557.06	
	C	260.05	260.05	
fp		−108.39	164.76	4442;6004,4902,2148,6779
d	20	0.8892	889.2	944;1379,5074,1018,4195
	30	0.8730	873.0	6483;3667
dd/dt		0.00101	1.01	2148
α		0.001138	0.001138	4195;995
n_D	20	1.40716	1.40716	4091;944,5841,5074,1018
	25	1.40496	1.40496	
dn/dt	25	0.00044	0.00044	4091

170.   Tetrahydrofuran

(Continued)

		cgs	SI	
η	20	0.575	0.000575	2636;4902,1379,4349,5074
	25	0.460	0.000460	849
	Table 2.6			
γ	20	27.31	0.02731	4195;2636
	25	26.4	0.0264	2148
ΔH_v	25	7.65	32.0	5746;1384,933,5211
	bp	7.126	29.815	3504;5211,995
ΔH_m		2.04	8.54	3850
ΔH_f°	25(g)	−44.03	−184.22	5746;1384,6515,4259
	25(l)	−51.68	−216.23	3850
ΔH_c°	25(l)	−752.8	−3149.7	1384;5746
C_p°	25	18.22	76.23	6515;3504
C_p	25	29.61	123.9	3850;5746
λ	20	0.000336	0.141	2636
	30	0.000324	0.136	
t_c		267.0	540.2	4299;1451,4091
P_c		51.2	5.19	4091;4299
d_c		0.322	322	4091;4299
v_c		0.224	0.000224	4091;4299
A	eq 2.72	0.03780	0.03780	4442
pK_a	in aq H_2SO_4	−2.08	−2.08	434,438;436
	25	−1.97	−1.97	4556;4484
pK_b	25	5	5	2880
κ		4.5×10^{-5} to 9.3×10^{-8}	0.0045 to 9.3×10^{-6}	6006
ε	25	7.58	7.58	4902;2148,1379,1757,849
	30	7.261	7.261	6483
μ	25 in 38	1.75	1.75	1757,3372;3689,2710,6212,6880
δ		9.9	20.3	6632,6631,6905;7635,2190,1399
ER	BuOAc 90%	4.72 97	4.72 97	6631;6628,7635,6905,2190,1399 6628,6631;1399
soly	in aq	inf	inf	6004
aq az	63.4	93.3%w	93.3%w	6165;6133
fl pt	CC	−14.4	258.8	6004,2150
	COC	−20	253	6006

170. Tetrahydrofuran

(Continued)

Beil	2362,10
uv	5841
ir	2551,4286,7561,4569,7901,6695,153,4660,5932
Raman	4118,4119,547,62
ms	2641,3430,2113,6803
nmr	7227,5532,3909,4806,2464,5933,6748

171. 2-Methyltetrahydrofuran

H₂C————CH₂
H₂C‖O‖CHCH₃

96–47–9 $C_5H_{10}O$

		cgs	SI	
mw		86.133	86.133	3468
bp	1 atm	79.9	353.1	8016;4091,764,5717
fp		-137.2	136.0	8016
d	0	0.8732	873.2	764;4621
	20	0.8540	854.0	764;8016
n_D	20	1.40751	1.40751	4091;8016
	25	1.40508	1.40508	
dn/dt	25	0.000481	0.000481	4091
ΔH_v	bp	7.73	32.34	5211
t_c		264	537	4091;4299
p_c		37.1	3.76	4091;4299
d_c		0.322	322	4091;4299
v_c		0.267	0.000267	4299
pK_a	aq H_2SO_4	-2.65	-2.65	438
pK_b		4.8	4.8	2880
ε	25	6.97	6.97	2617;2074

171.   2-Methyltetrahydrofuran

(Continued)

		cgs	SI	
soly	in aq,25 aq in,25	13.87%w 6.08%w	13.87%w 6.08%w	764
Beil	2362,12			
ir	645,2161,5932			
Raman	4717			
nmr	5532,5933,6748			

172.   p-Dioxane

1,4-Dioxane

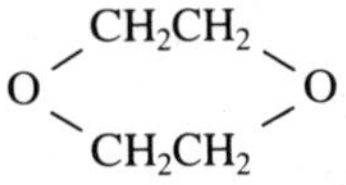

$$O \overset{CH_2CH_2}{\underset{CH_2CH_2}{\diagup}} O$$

123–91–1 $C_4H_8O_2$

		cgs	SI	
mw		88.106	88.106	3468
bp	1 atm	101.320	347.470	6824;4249,676,3354,5155 7015,7491
dt/dp	1 atm	0.0432	0.324	6824;7491
dp/dt	1 atm	23.1	3.09	
p	25 30	37.1 46.02	4.95 6.135	7748;7058,3384,1109, 3515 3667
eq 2.4	A B	7.8642 1866.7	6.9891 1866.7	2639;7748,3464
fp		11.80	284.95	3354,7491;3173,6508,2219 4249,6484
d	20	1.03361	1033.61	7482;2963,676,2414,5155 2555,3344
	25	1.02797	1027.97	3120;6284,7491,3354,6877 6484,1109,13
dd/dt		0.001128	1.128	7482
α		0.001115	0.001115	4195;7624,1965

172. *p*-Dioxane

(Continued)

		cgs	SI	
κ_T	25	9.84×10^{-8}	7.38×10^{-7}	1965
κ_S	25	7.18×10^{-8}	5.39×10^{-7}	1966,1965
n_D	15	1.42436	1.42436	7491;6484,2414,5155,7015
				3344,6877
	20	1.42241	1.42241	3354;4091,7624
	25	1.42025	1.42025	2555;3797
dn/dt	20 – 40	0.00046	0.00046	7624
η	15	1.439	0.001439	7491;3896,3344,2638,3464
	30	1.087	0.001087	
γ	15	34.45	0.03445	7491;2638,3464,3344,1109
	20	33.75	0.03375	7087;4195,3515
	25	32.80	0.03280	3797
	Table 2.8			
ΔH_v	bp	8.505	35.585	7748;2639,3344
ΔH_m		2.978	12.460	6284;3661,4249
ΔH_f°	25(1)	-84.50	-353.55	8316
ΔH_c°	25(1)	-564.99	-2363.92	6888;8316
C_p°	26.84	22.56	94.39	4851
C_p	25	36.01	150.65	2978;6284,3344,3661
t_c		314	587	4299;2776,3344,2639,3464,4091
p_c		51.4	5.21	4091;4299,2776,3344,3464
d_c		0.370	370	4091;4299
v_c		0.238	0.000238	4091;4299
K_f	eq 2.67	4.63	4.63	4249;5635,3344,6284,3354,1404
K_B	eq 2.79	3.270	3.270	3344;3406
pK_a	in aq H_2SO_4	-2.92	-2.92	438;4519
	25	-1.83	-1.83	4556
κ	25	5×10^{-15}	5×10^{-13}	4248
ε	25	2.209	2.209	4937,4320;676,522,4248
				828,1400
μ		0.45	0.45	6376;4296,7692,6877,8075
δ		10.13	20.72	6240;4593
ER	BuOAc	2.42	2.42	3455
soly	in aq	inf	inf	4027
aq az	87.82	82%w	82%w	3497
fl pt	CC	12	285	5712
	TOC	23	296	7624

172. *p*-Dioxane

(Continued)

Beil	2668,3
uv	3331,5841,4026
ir	4781,7077,3281,7901,6311,6844,633,6894,2244,5932
Raman	5029,6355,62,6043,6067,4851,2244
ms	154,3430,4314
nmr	5532,4694,3764,2464,5933,6748

173. Tetrahydropyran

$$\underbrace{CH_2CH_2CH_2CH_2CH_2O}_{}$$
142–68–7
$C_5H_{10}O$

		cgs	SI	
mw		86.133	86.133	3468
bp	1 atm	88	361	1082;90,5841,383,764
eq 2.4	A	7.99	7.12	1384
	B	1660.5	1660.5	
fp		−45	228	5477;6779
d	20	0.8814	881.4	764;383,5477
	25	0.8772	877.2	90
α		0.000104	0.000104	5477
n_D	20	1.42084	1.42084	4089;383,5841
	25	1.41862	1.41862	4089;90
dn/dt	25	0.000444	0.000444	4089
η	20	0.826	0.000826	5477
	25	0.764	0.000764	
	Table 2.6			
ΔH_v	25	8.35	34.9	5746;1384
ΔH_f°	25(g)	−53.50	−223.84	5746
ΔH_c°	25(1)	−750.52	−3140.18	6800;5746,6888,1384,6801

173.   Tetrahydropyran

(Continued)

		cgs	SI	
C_p	25(1)	37.4	156.5	5746
t_c		299.0	572.2	4089
P_c		47.1	4.77	4089
d_c		0.328	328	4089
v_c		0.263	0.000263	4089
pK_a	in aq H_2SO_4	-2.79	-2.79	434,438
$pK_{BH}+$	25	-1.89	-1.89	4556
ε	25	5.61	5.61	5477
μ	25 in 38	1.63	1.63	5747;1757,90
soly	in aq,25	8.02%w	8.02%w	764;383
	aq in,25	3.14%w	3.14%w	
aq az	71	91.5%w	91.5%w	3498;383

Beil	2362,12
ir	7901,6695,547,7702,6894,5932
Raman	547,1882,3822,7702,6894
ms	6803
nmr	4806,5933

174.   Benzylethyl ether

(Ethoxymethyl)benzene

$C_6H_5CH_2OC_2H_5$ 　　　　　　　　 539-30-0 　　　　　　　　 $C_9H_{12}O$

		cgs	SI	
mw		136.193	136.193	3468
bp	1 atm	185.0	458.2	7058
dt/dp	1 atm	0.052	0.39	7058
dp/dt	1 atm	19	2.6	7058

174.   Benzylethyl ether

(Continued)

		cgs	SI	
p	25	0.9	0.1	<u>7058</u>
eq 2.5	A	7.5247	6.6496	<u>7058</u>
	B	1927.21	1927.21	
	C	230	230	
d	20	0.9478	947.8	6340;7563
	25	0.9446	944.6	
	40	0.9343	934.3	3539
dd/dt	25	0.00064	0.64	<u>6340</u>
n_D	20	1.4958	1.4958	6340;7563
	25	1.4934	1.4934	
dn/dt	25	0.00048	0.00048	<u>6340</u>
γ	20	32.82	0.03282	3539
	25	32.18	0.03218	
	40	29.97	0.02997	
pK_{BH}^{+}	25	−2.47	−2.47	4556
ε	20	3.9	3.9	4937;2026
Beil	<u>528</u>,431			
Raman	5348			

175.   Anisole

Methoxybenzene

$C_6H_5OCH_3$ 100–66–3 C_7H_8O

		cgs	SI	
mw		108.140	108.140	3468
bp	1 atm	153.60	426.75	122;3012,3049,7490,7470,7471
				4684,5831
dt/dp	1 atm	0.04894	0.3671	122;2090,4684,7490,7470
dp/dt	1 atm	20.43	2.724	122
p	25	3.54	0.472	122;2090,2091,7058,4912,3049

175. Anisole

(Continued)

		cgs	SI	
eq 2.5	A	7.05105	6.17595	122;2090,5669
	B	1489.502	1489.502	
	C	203.573	203.573	
fp		-37.5	235.7	7490,7479;2090,7471,1057
				3049,5582
d	15	0.99858	998.58	7490;557,2677,3253,3049
				1057,4912
	20	0.99402	994.02	2090;5582,7848,7470,6270,6791
	25	0.98932	989.32	
dd/dt		0.000932	0.932	5742
α		0.000951	0.000951	4195
n_D	20	1.51700	1.51700	2090;1057,5582,7479,3049,7490
	25	1.51430	1.51430	
dn/dt		0.00050	0.00050	7490
η	15	1.52	0.001152	7490;557,2694,7401,2677,3049
	30	0.789	0.000789	
	Table 2.6			
γ	15	36.18	0.03618	7490;7479,7864,7848
	20	35.00	0.03500	4195
	30	34.15	0.03415	7490
	Table 2.7			
ΔH_V	25	11.20	46.84	2404;4450,122,3083
	bp	9.32	39.0	122;4684,7848
	Table 2.9			
ΔH_f°	25(1)	-27.43	-114.77	2404;4450
	25(g)	-16.24	-67.93	
ΔH_c°	25(1)	-904.19	-3783.12	2404;3975,7446,542,4450
	Table 21.4			
C_p	31.6	49.85	208.57	5817;4715,6440,4684
t_c		372.4	645.6	116;122,3049,3051,4299
P_c		41.9	4.25	116;122,4299,3051,3049,758
K_B	eq 2.79	4.502	4.502	2181;5595
pK_a	in aq	-6.54	-6.54	434,437
	H_2SO_4			
	25	-4.07	-4.07	4556
κ	25	1×10^{-13}	1×10^{-11}	794
ε	25	4.33	4.33	4937;5818,1837,2850,3662,5251
μ	25 in 317	1.245	1.245	452;3691,6231,4579,2295,3662
δ	25	9.5	19.5	653
soly	in aq	1.04%w	1.04%w	987
aq az	95.5	59.5%w	59.5%w	3497

175. Anisole

(Continued)

Beil	514,318
uv	2040,2054,3882,8313,1251,6825,7606,6749
ir	4676,2929,3646,6844,1097,3886,633,5633,5932
Raman	4101,3341,3971,742,5720,3973,4113,5633
ms	5079,99
nmr	7026,8314,6042,2788,5933,6748

176. Phenetole

Ethoxybenzene

$C_6H_5OC_2H_5$ 103–73–1 $C_8H_{10}O$

		cgs	SI	
mw		122.166	122.166	3468
bp	1 atm	169.84	442.99	122;2090,3049,3948,5582 7473,7483
dt/dp	1 atm	0.05031	0.3774	122;2090,7483
dp/dt	1 atm	19.88	2.650	122
p	25	1.53	0.204	122;2090
eq 2.5	A	7.02168	6.14658	122;2090
	B	1509.276	1509.276	
	C	194.648	194.648	
fp		−29.52	243.63	2090;5582,7058,7473
d	20	0.96514	965.14	2090;557,2677,3049,3948,5582
	25	0.96049	960.49	7763,7848,7483
dd/dt	25	0.00093	0.93	2090
n_D	20	1.50735	1.50735	2090;3049,7763,5582,7483
	25	1.50485	1.50485	
dn/dt	25	0.00050	0.00050	2090

176. Phenetole

(Continued)

		cgs	SI	
η	15	1.364	0.001364	7483;3948,2677,7401,557,3049
	30	1.040	0.001040	
	Table 2.6			
γ	20	32.88	0.03288	7483;7763
	Table 2.7			
ΔH_v	25	12.20	51.04	2404;122
	160	10.70	44.77	4709
	bp	9.73	40.7	122;7755
ΔH_f°	25(1)	-36.48	-152.64	2404
	25(g)	-24.28	-101.60	
ΔH_c°	25(1)	-1057.50	-4424.59	2404;542,3975
C_p	20	54.48	227.94	6440
t_c'		374.0	647.2	3049;4299,3051,7848
P_c		33.8	3.42	3049;3051,4299,758
K_f	eq 2.67	7.15	7.15	2354;5595
K_B	eq 2.79	5.0	5.0	4709
pK_{BH}^+	25	3.96	3.96	4556
κ	25	$<1.7 \times 10^{-8}$	$<1.7 \times 10^{-6}$	6367
ϵ	20	4.22	4.22	4937,3662;7483
μ	25 in 317	1.36	1.36	452;4108,3662,5858,2295,7734
soly	in aq	0.12%w	0.12%w	987
aq az	97.3	41%w	41%w	3497
Beil	514,140			
uv	582,1885,6210,6060,6749			
ir	1097,633,6359,3888,3886,2929,5943			
Raman	3973,926,3971			
nmr	6748			

177. Benzyl ether

1,1'-[Oxybis(methylene)]bisbenzene

$(C_6H_5CH_2)_2O$ 103–50–4 $C_{14}H_{14}O$

		cgs	SI	
mw		198.264	198.264	3468
bp	1 atm	288.30	561.45	2090;765,7105
dt/dp	1 atm	0.06122	0.4592	2090
dp/dt	1 atm	16.33	2.178	2090
p	25	0.023	0.0031	7515
eq 2.5	A	7.71832	6.84322	2090
	B	2507.3	2507.3	
	C	230	230	
fp		3.60	276.75	765;2090
d	20	1.0428	1042.8	765;7105,2090,8260
	35	1.0341	1034.1	
dd/dt	20 - 35	0.00058	0.58	765
n_D	20	1.54057	1.54057	2090;7105,8260
	25	1.53851	1.53851	
dn/dt	25	0.000412	0.000412	
η	20	5.333	0.005333	881
	30	4.062	0.004062	
	35	3.711	0.003711	7105
γ	35	38.2	0.0382	7105
ΔH_m		4.83	20.21	765
K_f	eq 2.67	6.27	6.27	765
$pK_{BH}{}^+$	25	2.82	2.82	4556
δ	25	9.4	19.2	653
soly	in aq,35	0.0040%w	0.0040%w	7105
Beil	528,434			
uv	5010,4052,6749			
ir	6348,633,5932			
nmr	8282,6748,5933			

178. Phenyl ether

1,1'-Oxybisbenzene

$(C_6H_5)_2O$ 101-84-8 $C_{12}H_{10}O$

		cgs	SI	
mw		170.210	170.210	3468
bp	1 atm	258.06	531.21	122;2090,2087,2776,1637
dt/dp	1 atm	0.06033	0.4525	122;2090,5885
dp/dt	1 atm	16.58	2.210	122
p	25	0.0213	0.00284	2087;122
eq 2.5	A	7.01423	6.13913	122;2087,1384
145 - 325	B	1802.984	1802.984	
	C	178.137	178.137	
fp		26.87	300.02	2615;7920,5885,5234,1637,2090
tp		26.869	300.019	7691;5885
d	30	1.06611	1066.11	2090;1637,7555,5234,7920
	35	1.06117	1061.17	
dd/dt	30 - 170	0.00086	0.86	8288
n_D	30	1.57625	1.57625	2090;7555
	35	1.57387	1.57387	
	50	1.56681	1.56681	2087
dn/dt	30	0.000476	0.000476	2090
η	40	2.4594cSt	2.4594×10^{-6}	2087;7555
	60	1.7065cSt	1.7065×10^{-6}	
	80	1.2716cSt	1.2716×10^{-6}	
γ	30	38.82	0.03882	2087
	40	37.73	0.03773	
ΔH_v	25	15.99	66.90	5243,122;2087,1384
	bp	11.5	48.2	122
ΔH_m		4.115	17.217	2615;5885,2087,2153
ΔH_f°	25(1)	-3.583	-14.991	2615
ΔH_c°	25(1)	-1466.63	-6136.38	2615;1384
C_p	25(s)	51.76	216.56	2615;2763
	26.87(1)	64.15	268.40	
t_c		493.6	766.8	116;2087,3974,2776,8288
P_c		30.4	3.08	116;2776
d_c		0.312	312	2087
v_c		0.545	0.000545	2087
A	eq 2.72	0.02177	0.02177	2087;5885
K_f	eq 2.67	7.88	7.88	2153
$pK_{BH}{}^+$	25	5.79	5.79	4556
ε	20(s)	2.68	2.68	2295,3542;7555
	20(1)	3.686	3.686	

178.   Phenyl ether

(Continued)

		cgs	SI	
ε	25(1)	3.60	3.60	7920
μ	25 in 38	1.16	1.16	2843,4526;6195,4579,3542,2295
δ		10.10	20.66	6240
soly	in aq	0.39%w	0.39%w	987;2068
	in aq	106µM	106µM	591
aq az	99.33	3.25%w	3.25%w	3497
fl pt		96	369	2068

Beil	514,146
uv	1794,6206,7606,7229,3619,6210,7607,6749
ir	633,1794,4699,3885,4869,5269,5932
Raman	3972,3885,3973,3971,6841,926
ms	837,8107
nmr	6748,5933

179.   Veratrole

1,2-Dimethoxybenzene

1,2-$C_6H_4(OCH_3)_2$ 91-16-7 $C_8H_{10}O_2$

		cgs	SI	
mw		138.166	138.166	3468
bp	1 atm	206.25	479.40	7482;5765,6193
dt/dp	1 atm	0.054	0.41	6166;7482,2581
dp/dt	1 atm	19	2.5	
p	25	0.47	0.063	6166
eq 2.4	A	8.1532	7.2781	6166;7482,2581,1767
	B	2527.58	2527.58	
fp	(st)	22.5	295.7	1899;1394,916,1200,1767
	(ms)	20.9	294.1	
d	25	1.0819	1081.9	6193,2581;5765,7482,1767,4711
	40	1.0667	1066.7	
	60	1.0470	1047.0	
dd/dt	25 - 40	0.00101	1.01	6193

179. Veratrole

(Continued)

		cgs	SI	
n_D	25	1.53232	1.53232	6193;4026,1767,2951
η	25	3.281	0.003281	6193;6954,4711,2581
	40	2.184	0.002184	
	60	1.457	0.001457	
γ	0	42.5	0.0425	Beil. EI6, 383
	131	26.0	0.0260	4711
	184	22.6	0.0226	
ΔH_v	bp	11.52	48.20	6166;6954,2581,1767
ΔH_m		3.834	16.04	4668
K_f	eq 2.67	6.38	6.38	5080;4668
ϵ	25	4.09	4.09	6193;2811,1394
μ	25 in 38	1.29	1.29	2342;1997,3689,1767,452,6193
soly	in aq	sl sol	sl sol	7033
aq az	99.8	11%w	11%w	6165;3497
Beil	553,771			
uv	1884,2824,2709,8313,7796,5409,6749			
ir	1097,4473,5932			
Raman	3336,4113			
ms	1705			
nmr	4918,8314,2811,2901,5933,6748			

180. Dimethoxymethane

$CH_2(OCH_3)_2$ 109-87-5 $C_3H_8O_2$

		cgs	SI	
mw		76.095	76.095	3468
bp	1 atm	42.30	315.45	7495;7470,7471,5486,5487
dt/dp	1 atm	0.040	0.30	7470
dp/dt	1 atm	25	3.3	7470
p	25	398.7	53.16	4764;5487,5486
eq 2.5	A	7.10401	6.22891	4764
	B	1162.58	1162.58	
	C	−40	−40	
fp		−105.15	168.00	4764;7495
d	15	0.86645	866.45	7495;7470,5656,7763
	30	0.84745	847.45	
dd/dt		0.001267	1.27	7495
n_D	15	1.35626	1.35626	7495;5486
	20	1.35373	1.35373	4089;7763,5656
	25	1.35132	1.35132	4089
dn/dt		0.00049	0.00049	7495
η	15	0.340	0.000340	7495
	30	0.325	0.000325	
γ	20	21.12	0.02112	3283,7763,7762
	30	19.76	0.01976	
ΔH_v	25	6.904	28.89	4764;5487,5486,820
ΔH_m		1.991	8.330	4764;7762,7761
ΔH_c°	25(1)	−455.8	−1907.07	3975;7446
	25(g)	−472.20	−1975.68	5855
ΔH_f°	25(g)	−83.21	−348.15	
C_p	25	38.58	161.42	4764;3621
t_c		207.4	480.6	4089;4764
P_c		39.0	3.95	4089
d_c		0.357	357	4089
v_c		0.213	0.000213	4089
K_B	eq 2.79	2.125	2.125	726;723,734
pK_a	25	−4.57	−4.57	3849
ε	20	2.645	2.645	7497;4937,7842
μ	34(g)	0.74	0.74	4297
δ		8.4	17.2	1399
ER	BuOAc	8.07	8.07	1399
	90%	58	58	1399
soly	in aq,16	24.4%w	24.4%w	6556
	aq in,16	4.1%w	4.1%w	

180. Dimethoxymethane

(Continued)

		cgs	SI	
aq az	42.05	98.6%w	98.6%w	3497;3498
fl pt	CC	−18	255	5712

Beil	75,574
uv	6506
ir	7077,5534,8097,7147,1097,791,5932
Raman	1069,8097,5188,5535
ms	2573,4453,837,3423
nmr	5536,2522,3003,5532,4129,5933

181. Acetal

1,1-Diethoxyethane

$CH_3CH(OC_2H_5)_2$ 105–57–77 $C_6H_{14}O_2$

		cgs	SI	
mw		118.175	118.175	3468
bp	1 atm	103.6	376.8	5486;7058,2719,6691,4665
				5655,7763
dt/dp	1 atm	0.046	0.35	4665
dp/dt	1 atm	22	2.9	4665
p	25	34.1	4.55	5486;36,3012,7058
eq 2.5	A	7.22012	6.34502	5486
	B	1447.59	1447.59	
	C	230	230	
fp		(gl)	(gl)	7478
d	20	0.8264	826.4	7763;2719,36,6691,7848
	25	0.8213	821.3	
dd/dt		0.00102	1.02	7763
n_D	20	1.38054	1.38054	7763;7764,1069,6691,2719
	25	1.3682	1.3682	36

181. Acetal

(Continued)

		cgs	SI	
γ	15	21.91	0.02191	3684;7763,7793,7762
	20	21.40	0.02140	
ΔH_v	25	9.04	37.82	5851;5486,7848,7850
	bp	7.819	32.71	4665
ΔH_f°	25(g)	−108.41	−453.59	5851
	25(1)	−117.45	−491.41	5851
ΔH_c°	25(g)	−934.11	−3908.32	5851
	25(1)	−925.07	−3870.49	5851;3975
C_p	19.1-99	61.40	256.90	4665
t_c		254	527	4299;7848
ε	25	3.80	3.80	4937;7850,2100
μ	in 38	1.38	1.38	Beil. EIII1, 2643
δ	25	7.8	15.9	653
soly	in aq,25	5%w	5%w	4026
		0.75M	0.75M	7219
aq az	82.6	85.5%w	85.5%w	3497;3498
fl pt	CC	36	309	5712;152

Beil	78,603
uv	6506,1282
ir	5534,4250,2564,1097
Raman	5536,1069,4632,5188
ms	2573,4453,4768,154,7408
nmr	3003,2564,6567,5933

182. Acetaldehyde

CH_3CHO 75–07–0 C_2H_4O

		cgs	SI	
mw		44.053	44.053	3468
bp	1 atm	20.4	293.6	7259;1645,2580,6850,7058,7472
dt/dp	1 atm	0.0351	0.263	7259;6850
dp/dt	1 atm	28.4	3.80	7259
p	25	910	121.3	3779;7058
eq 2.5	A B C	7.0565 1070.6 236.0	6.1814 1070.6 236.0	7258;6850,1914
fp		-123	150	7259;3048,7069,7472
d	15 20	0.7846 0.7780	784.6 778.0	6850;1585,1645,2580 7259;6850
dd/dt	0 - 30	0.001325	1.325	6850
α	0 - 30	0.00169	0.00169	4027
n_D	20	1.3311	1.3311	7259;6850
dn/dt	0 - 20	0.0005635	0.0005635	6850;1645
η	0.1 15 20.3	0.2711 0.2456 0.2237	0.0002711 0.0002456 0.0002237	2580 4027 Beil. EIII1, 2621
	Table 2.2			
γ	20	21.2	0.0212	4027
ΔH_v	25 bp	6.24 6.500	26.11 27.196	7826 7269;1585,7850,7755
ΔH_m		0.775	3.243	4027
ΔH_f°	25(l) 25(g)	-45.96 -39.72	-192.30 -166.19	7826 7826,2036;5877,5469
ΔH_c°	25(l) 25(g)	-278.77 -285.01	-1166.37 -1192.48	2036;7446,3975,5469 2036
C_p°	25	13.216	55.296	4624;6863
C_p	24.9	14.8	61.9	1585;5877
t_c		188	461	4299;3470
P_c		63.2	6.40	4027
pK_b	in CF_3CO_2H $-H_2SO_4$,26	-10.2	-10.2	4558
κ	0	1.20×10^{-6}	0.000120	7842;7843
ε	21	21.1	21.1	4937;7850,2100,7458
μ	20 in 38	2.51	2.51	1645;3595
δ		10.3	21.1	652;6723
soly	in aq	inf	inf	6850
aq az		none	none	3497;6850

182.   Acetaldehyde

(Continued)

		cgs	SI	
fl pt	TCC	−38	235	5712
Beil	77,594			
uv	155,4378,1634,8159			
ir	153,4015,2713,5267,7431,2324,8159,5932			
Raman	3692,5267,1783,5797,4104,2324			
ms	154,2050,7408,1221			
nmr	1396,7026,710,5933			

183.   Propionaldehyde

Propanal

CH_3CH_2CHO 123–38–6 C_3H_6O

		cgs	SI	
mw		58.080	58.080	3468
bp	1 atm	48.0	321.2	7259;1645,2052,2090,2580 4778,6850
dt/dp	1 atm	0.0380	0.285	7259;2090,6850
dp/dt	1 atm	26.3	3.51	7259
p	25	317.3	42.3	131;6850,2091,1685
eq 2.5	A	7.1087	6.2336	131;7258,2090,6850
	B	1180	1180	
	C	231	231	
fp		−80	193	7259;2090,4778
d	20	0.7970	797.0	7259;1645,2090,2580,6850
	25	0.7912	791.2	
dd/dt	0 – 30	0.001165	1.165	6850
α	55	0.00201	0.00201	7619
n_D	20	1.3619	1.3619	7259;1645,2090,4778,6850
	25	1.3593	1.3593	

183. Propionaldehyde

(Continued)

		cgs	SI	
dn/dt	0 - 20	0.0005180	0.0005180	6850
η	15.4	0.3568	0.0003568	2580
	26.7	0.3167	0.0003167	
ΔH_v	25	7.0813	29.628	131;1627,1231,1685
	bp	6.7631	28.297	131,1685
ΔH_f°	25(1)	-51.55	-215.7	1627;8316,1231,2036
	25(g)	-44.19	-184.9	1231,1627
ΔH_c°	25(1)	-434.1	-1816	2036;3975,2446,7506,8316
C_p°	25	18.52	77.50	1231,1685
C_p	25	32.79	137.2	2593
κ	25	9.5×10^{-5}	0.0095	7843
ε	17	18.5	18.5	4937;1141,2100,7458
μ	20 in 35	2.54	2.54	1645;3595
δ	25	9.4	19.3	653
soly	in aq,25	30.6%w	30.6%w	6850
	aq in,25	13.0%w	13.0%w	
aq az	47.79	98.1%w	98.1%w	6850;3498
fl pt	TOC	<-6	<267	7619
Beil	82,629			
uv	155,4808,903,1634,1343,8159,1282			
ir	153,2250,7224,7225,4015,8159,2524,5932			
Raman	1783,4104,2524			
ms	154,7408,1221			
nmr	5554,7026,5933			

184. Butyraldehyde

Butanal

| CH₃CH₂CH₂CHO | | | 123–72–8 | C₄H₈O |

$$CH_3CH_2CH_2CHO \qquad 123\text{--}72\text{--}8 \qquad C_4H_8O$$

		cgs	SI	
mw		72.107	72.107	3468
bp	1 atm	74.8	348.0	7259;2056,2588,4298,6850 7475,7476
dt/dp	1 atm	0.0411	0.308	7259;6850
dp/dt	1 atm	24.3	3.24	7259
p	25	118	15.7	7619;6850
eq 2.5	A	7.0212	6.1461	7258;6850,4298
	B	1233.0	1233.0	
	C	223.0	223.0	
fp		-96.4	176.8	7259,5683;7476,7475
d	20	0.8016	801.6	7259,6850;1645,2580,7507
	25	0.7964	796.4	
dd/dt	0 – 30	0.001040	1.040	6850
α	55	0.00119	0.00119	7619
n_D	20	1.3791	1.3791	7259;6850
	25	1.3766	1.3766	
dn/dt	0 – 20	0.0005045	0.0005045	6850;617,1645
η	18.35	0.4575	0.0004575	2580
	38.7	0.3673	0.0003673	
γ	24	29.9	0.0299	4027
ΔH_v	25	8.05	33.68	1231
	bp	7.528	31.497	4027
ΔH_m		2.65	11.09	5683
ΔH°_f	25(1)	-56.90	-238.07	1231;7507,8316,2036
	25(g)	-48.85	-204.39	1231
ΔH°_c	25(1)	-592.58	-2479.34	1231;7507,8316,2036
C°_p	25	23.94	100.16	1231
C_p	26.85	39.2	164.0	5683
ε	26	13.4	13.4	4937
μ	40(1)	2.450	2.450	4739;1645,3595
δ		9.0	18.4	652,4593
soly	in aq,25	7.1%w	7.1%w	6850
	aq in,25	3.0%w	3.0%w	
aq az	67.8	93.3%w	93.3%w	3498;6850
fl pt	TCC	-6.7	266.5	152,5712

184.   Butyraldehyde

(Continued)

Beil	87,662
uv	4808,1634,8159
ir	153,466,2250,7225,953,4015,8159,5932
Raman	1069,4104
ms	154,4949,7408,1221,3188
nmr	1396,3794,6748,5933

185.   Isobutyraldehyde

2-Methylpropanal

$(CH_3)_2CHCHO$ 78-84-2 C_4H_8O

		cgs	SI	
mw		72.107	72.107	3468
bp	1 atm	64.1	337.3	7259;3251,1645,5673,3064,6579
dt/dp	1 atm	0.0396	0.297	7259
dp/dt	1 atm	25.2	3.37	
eq 2.15	A	6.01614	5.14104	6579
	B	1946.999	1946.999	
	C	0.005297	0.005297	
fp		-65	208	7259;3064
d	20	0.7891	789.1	7259;3251,1645,1189,3064
	25	0.7836	783.6	6579,7507
dd/dt	25	0.0011	1.1	7259
α	55	0.00146	0.00146	7619
n_D	20	1.3727	1.3727	7259;3251,1645,1189,3064,6579
	25	1.3698	1.3698	
dn/dt	25	0.00058	0.00058	7259
η	28.2	0.5382	0.0005382	
ΔH_f°	25(1)	-34.45	-144.14	7507

185. Isobutyraldehyde

(Continued)

		cgs	SI	
ΔH°_c	25(1)	-546.71	-2287.43	7507;7446,7506,3975
μ	20 in 35	2.58	2.58	1645
δ	25	8.8	17.9	653
soly	in aq,20	9.1%w	9.1%w	6999;3064
aq az	64.3	93.3%w	93.3%w	3498;3251,3064
fl pt	OC	<-6	<267	7619
Beil	87,671			
uv	903,1634,1343			
ir	953,5932			
Raman	4104			
nmr	4527,5933,6748			

186. Benzaldehyde

C_6H_5CHO 100-52-7 C_7H_6O

		cgs	SI	
mw		106.124	106.124	3468
bp	1 atm	178.75	451.90	117;4665,7058,2776
dt/dp	1 atm	0.05228	0.3922	117;4665
dp/dt	1 atm	19.13	2.550	117
p	25	1.27	0.169	117;7058,1645,3248,1943
eq 2.5	A	7.07761	6.20251	117;1943
	B	1611.217	1611.217	
	C	205.166	205.166	
fp		-55.6	217.6	7482;7058,7473
melts		-26	247	7651

186. Benzaldehyde

(Continued)

		cgs	SI	
ṭp		−57.13	216.02	117
d	20	1.04463	1044.63	117;1645
	25	1.04013	1040.13	117;5429,2807
dd/dt	25	0.00090	0.90	<u>117</u>
n_D	20	1.5455	1.5455	1645;3248
	25	1.5437	1.5437	5850;2807
η	25	1.321	0.001321	2123;1432
γ	15.4	39.19	0.03919	4636
	Table 2.7			
ΔH_v	25	12.0	50.3	117;7850,4683,1943
	bp	10.2	42.5	117,7755;4665
ΔH_m		2.228	9.320	117
ΔH_f°	25(1)	−20.8	−87.1	117;4450
	25(g)	−8.80	−36.8	117;4450
ΔH_c°	25(1)	−842.50	−3525.0	117;3975,4450
C_p°	25	26.70	111.7	117
C_p	0	39.67	166.0	117
	25	41.11	172.0	117;4153,4665,4683,2593
t_c		422	695	117;2776,7848
P_c		45.89	4.650	117;2776
d_c		0.330215	330.215	117
V_c		0.321378	0.000321378	<u>117</u>
pK_{BH}^+	in aq H_2SO_4, 25	−7.10	−7.10	8210;1744
ε	20	17.8	17.8	4937;7843,7850,2100,7458
μ	20(1)	2.77	2.77	4739;3691,8210,2807,1645,6478
	25 in <u>38</u>	3.02	3.02	3765;5850,4292,3908
S		9.4	19.2	652
soly	in aq,20	0.3%w	0.3%w	6999;987
fl pt	TCC	64	337	152

Beil	<u>622</u>,174
uv	2054,5276,5954,7090,7891,1251,1830,6686,6749
ir	2250,2558,7724,3791,3888,6345,2675,2937,5932
Raman	5797,5103,1782,1790,640,2937
ms	16,4949
nmr	7227,1429,7026,3238,6748

187. Acrolein

2-Propenal

$CH_2{=}CHCHO$			107–02–8	C_3H_4O
		cgs	SI	
mw		56.064	56.064	3468
bp	1 atm	52.69	325.84	6635;1645,7058,7475,5294
dt/dp	1 atm	0.00355	0.0266	6635
dp/dt	1 atm	282	37.6	6635
p	25	265	35.3	7619; 5294
eq 2.4	A	7.750	6.875	7058;1645,3248,5294
	B	1584	1584	
fp		−86.95	186.20	6635;7058,7475,5294,7525
d	20	0.8389	838.9	5294;1645
$d_4^t{=}0.86205/(1{+}\alpha t)$				5294
$\alpha(0{-}50){=}0.001318{+}3.3 \times 10^{-6}t$				5294
n_D	15	1.4048	1.4048	6635;1645
	20	1.4017	1.4017	
dn/dt		0.00057	0.00057	7619
ΔH_v	bp	6.77	28.33	5294;6635
ΔH_c°	25(1)	−390.7	−1634.7	3975
C_p°	25	17.1	71.5	1617
C_p	17 − 44	28.65	119.87	5294
κ		1.55×10^{-7}	1.55×10^{-5}	5294
μ	25 in 38	2.90	2.90	782;1645
δ	25	9.8	20.0	653
soly	in aq,20	20.8%w	20.8%w	3498;1020
	aq in,20	6.8%w	6.8%w	
aq az	52.4	97.4%w	97.4%w	3498;4038
fl pt	TOC	<−17	<256	7619;5712
Beil	90,725			
uv	155,1275,4795,4808,903,7229,1594			
ir	4386,5849,826,1036,5932			
Raman	1017			
ms	154,1035			
nmr	7026,2055			

188. *trans*-Crotonaldehyde

(*E*)-2-Butenal

CH$_3$CH
‖
HCCHO   123–73–9   C$_4$H$_6$O

		cgs	SI	
mw		70.091	70.091	3468
bp	1 atm	104.1	377.3	7475;2035,2296
p	25	38	5.1	7619
fp		−76.5	196.7	7475;2035
d	15.2	0.8575	857.5	589;2296
	20	0.8516	851.6	7619
α		0.00123	0.00123	7619
n$_D$	10	1.44361	1.44361	589;2035,6180,2296,905
	20	1.4369	1.4369	7619
dn/dt		0.000556	0.000556	6180
ΔH$_v$	av	8.62	36.1	4027
ΔH$^\circ_c$	25(1)	−547.0	−2289	3975;7134
C$^\circ_p$	25	22.8	95.4	1617
C$_p$	20	49.4	207	7619
μ	25 in 38	3.49	3.49	4292;3595,782,2296
δ	25	10.0	20.5	653
soly	in aq,20	15.6%w	15.6%w	7619
	aq in,20	9.6%w	9.6%w	7619
aq az	84.0	75.2%w	75.2%w	7619
fl pt	TCC	13	286	152;5712

Beil	90,728
uv	3233,4808,903,1413,905,2598
ir	4015,5932
Raman	2920,6393,3971,742,3176
ms	1035
nmr	7227,5913,7026,8051,2055,5933

189. Acetone

2-Propanone

CH_3COCH_3 67–41–1 C_3H_6O

		cgs	SI	
mw		58.080	58.080	3468
bp	1 atm	56.067	329.217	121;7262,2090,2287,7471,7495 8237,8304
dt/dp	1 atm	0.03852	0.2889	121;7262,2090,7470,7495,8304
dp/dt	1 atm	25.96	3.461	121
p	20	181.72	24.227	6019;2091,6461,7058,8304
	25	231.06	30.806	121;1009
eq 2.5	A	7.12987	6.25478	121;1009,7264,2090,5752
	B	1216.689	1216.689	
	C	230.275	230.275	
fp		-94.7	178.5	7262;2090,2827,4797,5104 5682,7471
d	20	0.78998	789.98	7262;2090,2287,2677,7397,4650
	25	0.78440	784.40	7470,7495,1150,948
	30	0.78033	780.33	6416
dd/dt	25	0.00112	1.12	7262
α	20	0.00143	0.00143	2348;1965
κ_T	20	1.692×10^{-7}	1.269×10^{-6}	6150;1880,1965
	25	1.765×10^{-7}	1.324×10^{-6}	
κ_S	35	1.367×10^{-7}	1.025×10^{-6}	5945;1965
n_D	20	1.35868	1.35868	7262;1690,2090,2287,4969,5647
	25	1.35596	1.35596	7495,7549,8081,6104,5289 4650
dn/dt		0.00050	0.00050	7495
η	15	0.3371	0.0003371	7495;4570,6165,2677,3947,4367
	25	0.3029	0.0003029	3062;6396,3724
	30	0.2954	0.0002954	7495
	Table 2.2			
γ	20	23.32	0.02332	3283;28,1690,2287,5632,7522
	25	22.68	0.02268	6958
	30	22.01	0.02201	3283
	Table 2.7			
ΔH_v	25	7.48	31.3	121;5682,7850,5752,1595 4969,5153
	bp	6.952	29.09	5752;121
ΔH_m		1.360	5.690	5682;3934
ΔH_f°	25(1)	-58.99	-246.81	8316;5752,1232,1703
	25(g)	-51.90	-217.15	1420;1703
ΔH_c°	25(1)	-427.77	-1789.79	5132;7446,3975,8316
	25(g)	-435.32	-1821.38	1703
C_p°	25	17.81	74.52	1420
	65.1	19.48	81.50	5752
	Table 2.14			

189. Acetone

(Continued)

		cgs	SI	
C_p	25	29.85	124.9	2593
	30	30.87	129.16	8081;7549,6977,768,1595 5682,5817
λ	20	0.000428	0.1791	2348
t_c		234.95	508.10	116,121;1328,3345,7263,4299 5153,4086
P_c		46.35	4.696	121;7842,7850,4086,3345,116 1328,4299,5153
d_c		0.278	278	7263;4299,1328,4086,3345
v_c		0.209	0.000209	7263;4299,4086
K_f	eq 2.67	2.40	2.40	3621;5606
K_B	eq 2.79	1.71	1.71	726,3406;5060,2798,1370,7590
pK_a		24.2	24.2	994
pK_{BH}^+	in aq H_2SO_4,25	−2.85	−2.85	5755;1335
pK_s	25	32.5	32.5	1300;4259
κ	25	4.9×10^{-9}	4.9×10^{-7}	6396;3586,7843,3975,4703,7842
ϵ	20	20.90	20.90	4650
	25	20.56	20.56	2309;5236,8184,5117,2953,4251 4703,4937
	30	20.20	20.20	6416
μ	20(1)	2.69	2.69	4739;3461,1229,4603,5860,6191 4970,6411,6428,8141
δ		10.0	20.5	1271;6631,6905,3565,7159,6240 1399,7635
ER	BuOAc	5.59	5.59	6631;1399,6905,6628,2190
	90%	82	82	6628,6631;1399
soly		inf	inf	4027
aq az		none	none	3497
fl pt	TOC	−9	264	2348
	TCC	−17	256	2348;152
Beil	83,635			
uv	155,5883,4388,1879,4378,7016,4898,6749			
ir	7225,153,2250,6368,2560,5849,7077,6675,5932			
Raman	1069,4927,3921,742,1458,1537,6776			
ms	154,6199,6608,99,2050,4949,7408			
nmr	7026,4694,1454,2464,5933			

190. 2-Butanone

$CH_3CH_2COCH_3$ 78-93-3 C_4H_8O

		cgs	SI	
mw		72.107	72.107	3468
bp	1 atm	79.583	79.583	121;7471,7495,1590,7262,2090 2783,6625
dt/dp	1 atm	0.04121	0.3091	121;1590,4665,7262,2090 7453,7495
dp/dt	1 atm	24.26	3.235	121
p	25	90.60	12.079	121;6636,2091,6019
eq 2.5	A	7.05954	6.18444	121;7264,7293,1590
	B	1259.223	1259.223	
	C	221.758	221.758	
fp		−86.69	186.46	7262,1590;7495,6767,351 3048,7471
d	20	0.8049	804.9	7262;4288,2090,2677,2765,1690
	25	0.7997	799.7	4969,6636,7470,7495 1720,948
	30	0.7946	794.6	1414;6416
dd/dt		0.00084	0.84	6905
α		0.00119	0.00119	2348;6631,7635,6905,4902
κ_T	20	1.488×10^{-7}	1.116×10^{-6}	6150;1880
	25	1.584×10^{-7}	1.188×10^{-6}	
n_D	20	1.3788	1.3788	7262;4969,7495,5289,1069,1405 1690,2090
	25	1.37685	1.37685	1414
dn/dt		0.00048	0.00048	7495
η	20	0.399	0.000399	4288;485
	25	0.378	0.000378	4288
	30	0.366	0.000366	6946;7495,2129,2677,1720
	Table 2.2			
γ	0	26.9	0.0269	4902;1690
	20	24.6	0.0246	2348
	24.8	23.97	0.02397	5632
	Table 2.7			
ΔH_v	25	8.248	34.51	7086;121,5484
	bp	7.60	31.8	121;1590
ΔH_m		2.017	8.439	6767;7478,5683,351
ΔH_f°	25(1)	−66.68	−278.99	5686;5484,6767,1232
	25(g)	−57.02	−238.57	1420
ΔH_c°	25(1)	−582.80	−2438.44	5686;6767,3975
C_p°	25	24.68	103.26	1420;6767,5484
C_p	25	37.98	158.91	6767;768,4153,351,4665
	Table 2.11			5683,2593
λ	20	0.000358	0.150	

190.   2-Butanone

(Continued)

		cgs	SI	
t_c		263.63	536.78	116;121,4299,7263,4086,2302
p_c		41.52	4.207	116,121;7263,4299,2302,4086
d_c		0.270	270	7263;4086,4299
v_c		0.267	0.000267	7263;4086,4299
K_B	eq 2.79	2.28	2.28	3406
pK_a	25	14.7	14.7	8158
$pK_{BH}{}^+$	in aq H_2SO_4	−7.2	−7.2	1335
pK_s	25	25.94	25.94	4259
κ		3.6×10^{-9}	3.6×10^{-7}	3567;3621,5325
ε	20	18.51	18.51	4937;2100,7458,5236,7842 7850,5117
	30	17.71	17.71	6416
μ	25 in 38	2.76	2.76	2297;8141,2302,5484
	30 in 38	2.86	2.86	5860
δ		9.3	19.0	7159;1271,4593,6631,1399
ER	BuOAc	3.8	3.8	6628,6631;1399,2190
	90%	121	121	6628;1399
soly	in aq,20	24.00%w	24.00%w	3745;7219,2348,794,2765,6636
	aq in,20	10.00%w	10.00%w	3745;2348
aq az	73.41	88.73%w	88.73%w	6636;3497,3498
fl pt	TCC	−2	271	2348;152
	TOC	1	274	2348

Beil	87,666
uv	155,858,883,5883,6384,1879,5289
ir	153,1897,2560,4475,7077,7438,4419,5932
Raman	883,1069,7228,6773,7542,4104,4419
ms	154,6608,7572,99,1364,7408
nmr	7227,7026,5790,5933,6748

191. Cyclopentanone

CH₂CH₂CH₂CH₂CO   120–92–3   C₅H₈O

		cgs	SI	
mw		84.118	84.118	3468
bp	1 atm	130.7	403.9	7491,8136;7774,5506
p	25	11.6	1.54	8136
eq 2.5	A	7.2097	6.3346	8136
	B	1505.80	1505.80	
	C	220	220	
fp		−51.3	221.9	7491
d	20	0.94865	948.65	4650;7491,7774,506,5506
	30	0.93902	939.02	7491
dd/dt	0 – 30	0.000966	0.966	7491
n_D	20	1.43746	1.43746	4650;5506,7491,506
	25	1.4354	1.4354	8136
η	15	2.253	0.002253	7491;3553
	30	0.995	0.000995	7491;6946
γ	20	33.87	0.03387	7491;7774
	30	32.58	0.03258	
ΔH_v	25	10.19	42.63	8136
ΔH_f°	25(1)	−56.74	−237.40	8136
	25(g)	−46.55	−194.77	
ΔH_c°	25(1)	−686.788	−2873.52	8136;6563
C_p	25	36.9	154.5	2593
t_c		353	626	8136
P_c		51.4	5.21	8136
pK_{BH^+}	in aq H₂SO₄,25	−7.5	−7.5	1335
ε	20	13.60	13.60	4650;7497,2994
μ	in 38	2.86	2.86	3026;2040
δ		10.4	21.3	1271,4593
aq az	94.6	57.6%w	57.6%w	3498
fl pt	CC	26	299	5415

Beil	612,5
uv	2040,883,6428,6749
ir	3093,2159,6140,3522,3863,5932
Raman	4120,2159,3522,3863,883
ms	840,6608
nmr	368,4478,5933,6748

192.   2-Pentanone

$CH_3CH_2CH_2COCH_3$ 107-87-9 $C_5H_{10}O$

		cgs	SI	
mw		86.133	86.133	3468
bp	1 atm	102.262	375.412	121;1590,7260,7473,5910,1405
dt/dp	1 atm	0.04380	0.3285	121;1590,7260
dp/dt	1 atm	22.83	3.044	121
p	25	35.40	4.720	121
eq 2.5	A	7.01435	6.13925	121;1590,7261,5484,5007
	B	1309.592	1309.592	
	C	214.561	214.561	
fp		-76.86	196.29	1590,7260;7473,2043,5560
tp		-76.84	196.31	351
d	20	0.8064	806.4	7260;5910,1690,1582,978
	25	0.8015	801.5	1405,5632
dd/dt	25	0.00098	0.98	7260
κ_T	25	1.456×10^{-7}	1.092×10^{-6}	6150
n_D	20	1.39080	1.39080	4086;7260,5910,978,2043
	25	1.38849	1.38849	1690,1405
dn/dt	25	0.000469	0.000469	4086
$\dot{\eta}$	20	0.489	0.000489	8169;6946,7404
	30	0.438	0.000438	
γ	20	25.09	0.02509	978;1690,5632,4637
	30	23.98	0.02398	
ΔH_v	25	9.18	38.4	121
	bp	8.03	33.6	121;1590
ΔH_m		2.541	10.63	351;5560
ΔH_t	-163	0.03291	0.1377	351
ΔH_f°	25(1)	-71.05	-297.29	3195;2036,1703
	25(g)	-61.91	-259.05	3195;1703
ΔH_c°	25(1)	-740.78	-3099.41	3195;2036,1703
C_p°	25	28.91	120.96	5484
C_p	25	44.06	184.34	5560;2593
t_c		287.93	561.08	121,116;7288,4086
P_c		36.46	3.694	121,116;7288,4086
d_c		0.286	286	7288,4086
v_c		0.301	0.000301	7288;4086
pK_s	25	25.62	25.62	4259
ϵ	20	15.38	15.38	2994;1582
μ	22 in 38	2.70	2.70	8141;7810
δ		8.9	18.2	2190,6905,1399,4593
ER	BuOAc	2.50	2.50	2190,6905,1399;663‡
	90%	204	204	1399;6631

192. 2-Pentanone

(Continued)

		cgs	SI	
soly	in aq,20	5.95%w	5.95%w	2765;2985
	aq in,20	3.30%w	3.30%w	2765;794
aq az	79	13%w	13%w	3497
fl pt	TOC	14	287	1399,7635
Beil	87,676			
uv	4898,883,5883			
ir	1459,4898,4471,7438,5932			
Raman	883,4104			
ms	6608			
nmr	5933			

193. 3-Pentanone

$CH_3CH_2COCH_2CH_3$			96–22–0		$C_5H_{10}O$

		cgs	SI	
mw		86.133	86.133	3468
bp	1 atm	101.96	375.110	121;6186,7476,1590,7262,2090
				1690,5632
dt/dp	1 atm	0.04361	0.3271	121;7262,2090,4665,1590
dp/dt	1 atm	22.93	3.057	121
p	25	35.43	4.723	121;4902,2091,6186,7058
eq 2.5	A	7.02080	6.14570	121;7264,2090,6186,1590
	B	1307.927	1307.927	
	C	213.966	213.966	
fp		−38.97	234.18	7262,5982;7476,351,2090,7058
d	20	0.81430	814.30	7262;7489,1690,2090,2677,5632
	25	0.80945	809.45	
dd/dt		0.00082	0.82	6905
α		0.00124	0.00124	6905

193. 3-Pentanone

(Continued)

		cgs	SI	
n_D	20	1.39227	1.39227	7262;1069,1690,2090,5289
	25	1.39002	1.39002	
dn/dt	25	0.00045	0.00045	7262
η	15	0.493	0.000493	7482;2677
	25	0.442	0.000442	7404
	30	0.424	0.000424	6946;7404
	Table 2.2			
γ	20	25.26	0.02526	7482;1690,5632
	30	24.37	0.02437	
	Table 2.7			
ΔH_v	25	9.216	38.56	3195;121,3083
	bp	8.05	33.7	121;1590,7191
	Table 2.9			
ΔH_m		2.771	11.594	351
ΔH_f°	25(1)	−70.868	−296.51	3195
	25(g)	−61.652	−257.95	3195;1232
ΔH_c°	25(1)	−740.96	−3100.19	3195;3975,7134
C_p°	25	30.516	127.68	3083
	Table 2.14			
C_p	25	45.43	190.1	2593;4665
t_c		288.31	561.46	116,121;4299,4086,7263
P_c		36.80	3.729	116,121;4299,7263
d_c		0.256	256	7263,4086;4299
v_c		0.336	0.000336	7263,4086;4299
pK_a	in 462	27.1	27.1	4998
ε	20	17.00	17.00	4937;5251,5117,2100
μ	25 in 317	2.82	2.82	450;2900,5251
δ		8.8	18.0	1271,4593,6905
ER	BuOAc	2.23	2.23	6631;6628,6905
	90%	205	205	6631;6628
soly	in aq,20	3.4%w	3.4%w	4902;2985,987,7219
	aq in,20	2.6%w	2.6%w	
aq az	82.9	86%w	86%w	3497
fl pt	TOC	13	286	6905,6631

193. 3-Pentanone

(Continued)

Beil	87,679
uv	858,883,5883,1879,5289
ir	153,2560,3201,4475,7077,7438,3772,5932
Raman	883,1069,1458,6776,6773,4104,3772
ms	154,6608,2745,2050
nmr	3805,7026,3003,5933

194. Cyclohexanone

$$\underset{\llcorner\underline{\hspace{4.5cm}}\lrcorner}{CH_2CH_2CH_2CH_2CH_2CO} \qquad\qquad 108\text{-}94\text{-}1 \qquad\qquad\qquad C_6H_{10}O$$

		cgs	SI	
mw		98.144	98.144	3468
bp	1 atm	155.65	428.80	7491;506,1729,7058,2776,8136
dt/dp	1 atm	0.0048	0.036	7491
dp/dt	1 atm	210	28	7491
p	25	4.8	0.64	95;6019,7058,5078
eq 2.5	A	6.978401	6.103304	5078;8136
	B	1495.511	1495.511	
	C	209.5517	209.5517	
fp		−32.1	241.1	7491;1729,7479
d	15	0.95099	950.99	7491;506,7816
	20	0.9452	945.2	4195
	30	0.93761	937.61	7491
dd/dt		0.00089	0.89	7491
α		0.000955	0.000955	4195;6905,6631,7635
κ_S	35	7.187×10^{-8}	5.391×10^{-7}	5945
n_D	15	1.45203	1.45203	7491;506,7758,5289
	20	1.45097	1.45097	1729
	25	1.4500	1.4500	8136

194. Cyclohexanone

(Continued)

		cgs	SI	
η	15	2.453	0.002453	7491;7404
	30	1.810	0.001810	6946;6029,7491
γ	20	35.05	0.03505	4195;7491,5556
ΔH_v	25	10.73	44.89	8136;4902
	bp	9.62	40.25	4902
ΔH_f°	25(1)	−65.163	−272.642	8136;8316,6013
	25(g)	−54.43	−227.74	8136
ΔH_c°	25(1)	−840.735	−3517.635	8136;8316,6563,6013
C_p	25	42.85	179.3	2593;5817
t_c		381	654	8136;2776,4299
p_c		46.0	4.66	8136;2776,4299
$pK_{BH}{}^+$	in aq H_2SO_4	−6.8	−6.8	1335;749
κ	25	5×10^{-18}	5×10^{-16}	4902
ε	20	16.10	16.10	2994;4937,1729,5251,8075
μ	25 in 38	3.08	3.08	817;8075,6159,782,3026,2040 3105,5251
δ		9.9	20.3	1271,4593,6632;1399,6240
ER	BuOAc	0.29	0.29	6628,6631;1399
	90%	1566	1566	6628,6631;1399
soly	in aq,20	2.3%w	2.3%w	2044
	aq in,20	8.0%w	8.0%w	
aq az	96.3	45%w	45%w	3498
fl pt	TCC	44	317	6631;152,5712
Beil	612,8			
uv	883,2040,2487,6428,1413,7229,5289,6749			
ir	2560,6311,6384,1747,633,3093,870,5932			
Raman	883,1458,6355			
ms	154,6608,840,99			
nmr	7026,4623,4478,5933,6748			

195. 2-Hexanone

$CH_3(CH_2)_3COCH_3$ 591-78-6 $C_6H_{12}O$

		cgs	SI	
mw		100.160	100.160	3468
bp	1 atm	127.583	400.733	7151;7260,7058,7473,1405
dt/dp	1 atm	0.04639	0.3479	7151;7260
dp/dt	1 atm	21.56	2.874	7151
p	25	11.62	1.549	7151
eq 2.5	A	7.03740	6.16230	7151;7261,5007
	B	1401.738	1401.738	
	C	209.646	209.646	
fp		−55.8	217.4	2043,7260;7473,7058
tp		−55.46	217.69	352
d	20	0.8113	811.3	7260;1690,5632,1405,6186
	25	0.8067	806.7	
dd/dt	25	0.00092	0.92	7260
κ_T	25	1.349×10^{-7}	1.012×10^{-6}	6150
n_D	20	1.4007	1.4007	7260;1405,1690,2043
	25	1.3987	1.3987	
dn/dt	25	0.00040	0.00040	7260
η	25	0.584	0.000584	7404
γ	24.80	25.50	0.02550	5632;1690
	34.85	24.32	0.02432	
ΔH_v	25	10.3	42.9	7151
	bp	8.60	36.0	
ΔH_m		3.56	14.90	352
ΔH_f°	25(1)	−76.96	−322.01	3195;1703
	25(g)	−66.87	−279.79	
ΔH_c°	25(1)	−897.23	−3754.02	3195;1703
C_p	25	51.0	213.4	352;2593
t_c		313.8	587.0	116;7151
p_c		32.8	3.32	116,7151
pK_a	in aq H_2SO_4	−8.3	−8.3	546
pK_s	25	25.30	25.30	4259
ε	20	14.56	14.56	2994
μ	22 in 38	2.66	2.66	8141
δ		8.6	17.6	6905,1399,2190
ER	BuOAc	1.00	1.00	6631;1399,2190
	90%	482	482	6631;1399
soly	in aq,20	1.75%w	1.75%w	2765;2985
	aq in,20	2.12%w	2.12%w	
aq az	90.5	74%v	74%v	3497

195. 2-Hexanone

(Continued)

		cgs	SI	
fl pt	TOC	28	301	2190

Beil	<u>87</u>,689
uv	4898,883
ir	4898,4471,5932
Raman	883,4104
ms	6608
nmr	5933,6748

196. 4-Methyl-2-pentanone

$(CH_3)_2CHCH_2COCH_3$ 108–10–1 $C_6H_{12}O$

		cgs	SI	
mw		100.160	100.160	3468
bp	1 atm	117.4	117.4	121;7472,3861,2599,7262,1690 2765,6625
dt/dp	1 atm	0.0453	0.340	121;7262,6636
dp/dt	1 atm	22.1	2.94	121
p	25	18.8	2.51	121;6636,6186,7058,2599
eq 2.5	A	6.9727	6.0976	7264;6186,6636,2599
	B	1190.69	1190.69	
	C	195.45	195.45	
fp		−84	189	7262;7058,7472
d	20	0.8010	801.0	6176,6175;7262,6636,1690,2765
	25	0.7963	796.3	6625,6186
dd/dt		0.00078	0.78	6905
α		0.000116	0.000116	4902;2348,6905,6631
n_D	20	1.39576	1.39576	6176,6175;7262,6636,1069
	25	1.39361	1.39361	
dn/dt	25	0.00043	0.00043	<u>6176</u>

196. 4-Methyl-2-pentanone

(Continued)

		cgs	SI	
η	20	0.5848	0.0005848	3062;6175,563
	25	0.5463	0.0005463	3062;3861,6175
	30	0.49751	0.00049751	1798
γ	20	23.64	0.02364	1690;2599
	23.7	23.29	0.02329	
	62.1	19.62	0.01962	
ΔH_v	25	9.80	41.0	121;6625,6636
	bp	8.37	35.0	121
ΔH_f°	25(g)	−69.60	−291.21	2104
ΔH_c°	25(1)	−735.6	−3077.8	3975;7190
C_p	25	51.58	215.8	2593;6625
t_c		298.3	571.5	7263,4086;4299
P_c		32.3	3.27	7263,4086;4299
κ	35	$<5.2 \times 10^{-8}$	$<5.2 \times 10^{-6}$	4633
ϵ	20	13.11	13.11	4937,2994;5251,5117,4633
δ		8.4	17.2	7159,1271,4593,6905
ER	BuOAc	1.62	1.62	6631;6628,1399
	90%	282	282	6628,6631;1399,6704
soly	in aq,25	1.7%w	1.7%w	6636,2765;2985
	aq in,25	1.9%w	1.9%w	
aq az	87.9	75.7%w	75.7%w	3497
fl pt	TCC	16	289	2348;152
	TOC	23	296	2348

Beil	87,691
uv	155,2560,4388,6749
ir	4471,4419,5932
Raman	1069,1458,4104,4419
ms	2745
nmr	7227,7026,8051,2464,5933

197. 2-Heptanone

$CH_3(CH_2)_4COCH_3$ 110–43–0 $C_7H_{14}O$

		cgs	SI	
mw		114.187	114.187	3468
bp	1 atm	151.058	424.208	121;7260,8169,6650,6733,1405
dt/dp	1 atm	0.04869	0.3652	121;7260
dp/dt	1 atm	20.54	2.738	121
p	25	3.86	0.514	121;6298
eq 2.5	A	7.02544	6.15034	121;7261,7057
	B	1462.981	1462.981	
	C	201.929	201.929	
fp		−35.0	238.2	2043,7260;6733
d	20	0.81537	815.37	7260;1690,1582,1405,6733,6650
	25	0.81123	811.23	6298;7260,5632
dd/dt	25	0.000860	0.860	7260
κ_T	25	1.28×10^{-7}	9.57×10^{-7}	6150
n_D	20	1.40869	1.40869	7260;2043,8169,1690,1405,6733
	25	1.40655	1.40655	
dn/dt	25	0.000428	0.000428	7260
η	20	0.815	0.000815	8169;1313,7404,6733
	30	0.709	0.000709	
γ	20	26.60	0.02660	1690
	24.80	26.17	0.02617	5632
ΔH_v	25	11.29	47.24	7086;121
	bp	9.15	38.3	121;4971
C_p		58.22	243.6	2593
t_c		338.3	611.5	116,121
P_c		29.5	2.99	121;116
ε	20	11.98	11.98	1313;1582,2291
μ	20(1)	2.97	2.97	1313
	22 in 38	2.59	2.59	2291
δ		8.5	17.4	4593,1271,1399,6631
ER	BuOAc	0.33	0.33	6631,6628;1399,2190
	90%	1376	1376	6631,6628;1399
soly	in aq,25	0.43%w	0.43%w	2785;7219,6404
	aq in,25	1.41%w	1.41%w	2785
aq az	95	52%w	52%w	3497
fl pt	TOC	48	321	6402

Beil	87,699
uv	1879,883,5883
ir	4471,1459,7438,5932
Raman	883,4104
ms	4949,6608
nmr	5933

198. 3-Heptanone

$CH_3(CH_2)_3COCH_2CH_3$ 106–35–4 $C_7H_{14}O$

		cgs	SI	
mw		114.187	114.187	3468
bp	1 atm	147.4	420.6	7476,7265;6650,5632
dt/dp	1 atm	0.043	0.32	7265
dp/dt	1 atm	23	3.1	7265
p	25	5.60	0.746	6298
eq 2.5	A	8.4634	7.5883	7266
	B	2347	2347	
	C	273	273	
fp		−39.0	234.2	7265,7476
d	20	0.8181	818.1	7762;7265,6650
	25	0.81344	813.44	6298;7265,5632
dd/dt	25	0.0008	0.8	<u>7265</u>
n_D	20	1.4088	1.4088	7265;7762
	25	1.4066	1.4066	
dn/dt	25	0.00044	0.00044	<u>7265</u>
γ	20	26.30	0.02630	7762
	24.80	25.72	0.02572	5632
ε	22	12.88	12.88	2291
μ	22 in <u>38</u>	2.78	2.78	2291
δ		8.52	17.43	3185;2190
ER	BuOAc	0.43	0.43	6628,6631
	90%	1075	1075	6628,6631
soly	in aq,20	1.43%w	1.43%w	4902
	aq in,20	0.78%w	0.78%w	
aq az	94.6	57.8%w	57.8%w	3498
fl pt	TOC	46	319	6402,5415

Beil	<u>87</u>,699
uv	1879,883
ir	4471,5932
Raman	883,5156
ms	6608,4238
nmr	5933

199.   2,4-Dimethyl-3-pentanone

$(CH_3)_2CHCOCH(CH_3)_2$ 565–80–0 $C_7H_{14}O$

		cgs	SI	
mw		114.187	114.187	3468
bp	1 atm	125.25	398.40	2090,2091;7265,8169
dt/dp	1 atm	0.04610	0.3458	2090;7265
dp/dt	1 atm	21.69	2.892	2090
p	19.4	10.0	1.33	7266
eq 2.5	A	7.14587	6.27077	7266;2090,5007
	B	1463.53	1463.53	
	C	218.75	218.75	
fp		−69.03	204.12	2090;7265
tp		−68.34	204.81	352
d	20	0.80302	803.02	2090;7265,6562
	25	0.79863	798.63	
dd/dt	25	0.00088	0.88	2090
n_D	20	1.39995	1.39995	2090;7265,8169,6562
	25	1.39759	1.39759	
dn/dt	25	0.00047	0.00047	2090
η	20	0.632	0.000632	8169
	30	0.559	0.000559	
ΔH_v	25	9.93	41.55	6562,7821
ΔH_m		2.67	11.18	352
ΔH_f°	25(1)	−84.34	−352.88	6562
	25(g)	−74.41	−311.33	
ΔH_c°	25(1)	−1052.22	−4402.49	
C_p	25	55.9	233.7	352
pK_a		21.6	21.6	8308
μ	25 in 38	2.740	2.740	3384;450
δ		8.0	16.4	4593,1271
soly	in aq,20	0.59%w	0.59%w	2765;6404
	aq in,20	0.76%w	0.76%w	2765

Beil	87,703
uv	5025,4898
ir	5025,4898,1459,4471,5932
Raman	1069,4104
ms	6608
nmr	3656,5933

200. 2-Octanone

$CH_3(CH_2)_5COCH_3$ 111–13–7 $C_8H_{16}O$

		cgs	SI	
mw		128.214	128.214	3468
bp	1 atm	173.0	446.2	121;7260,1896,1405,7058
dt/dp	1 atm	0.0509	0.382	121;7260
dp/dt	1 atm	19.7	2.62	121
p	25	1.35	0.18	121
eq 2.5	A	7.8953	7.0202	7261;2699
	B	1965.2	1965.2	
	C	219.3	219.3	
fp		−20.29	252.86	5560;7260,1896,2043
d	20	0.8185	818.5	7260;1405,1896,1690,1582
	25	0.8143	814.3	
dd/dt	25	0.00084	0.84	7260
κ_T	25	1.20×10^{-7}	8.99×10^{-7}	6150
n_D	20	1.4153	1.4153	7260;2043,1690,1405,7727
	25	1.4133	1.4133	
dn/dt	25	0.00040	0.00040	7260
η	15	1.118	0.001118	1896;485
	30	0.869	0.000869	
γ	20	25.73	0.02573	1690;3283
ΔH_v	25	12.4	51.8	121;2699
	bp	9.6	40.3	
ΔH_m		5.8363	24.419	5560
ΔH_f°	25(1)	−91.98	−384.84	2699
	25(g)	−82.47	−345.05	
ΔH_c°	25(1)	−1207.01	−5050.13	2699
	25(g)	−1216.52	−5089.92	
C_p	25	66.2	276.9	2593;5560
ε	20	10.39	10.39	1582;8140
μ	in 38	2.70	2.70	8140
soly	in aq,25	0.113%w	0.113%w	7219
Beil	87,704			
uv	8140			
ir	7438,1459,4471,5932			
Raman	4104			
nmr	5933,6748			

201. 2,6-Dimethyl-4-heptanone

$(CH_3)_2CHCH_2COCH_2CH(CH_3)_2$ 103–83–8 $C_9H_{18}O$

		cgs	SI	
mw		142.241	142.241	3468
bp	1 atm	168.24	441.39	7052,121
dt/dp	1 atm	0.0505	0.379	121
dp/dt	1 atm	19.8	2.64	121;7052
p	25	1.65	0.22	121
eq 2.5	A	6.94539	6.07029	7052
	B	1476.40	1476.40	
	C	195.0	195.0	
fp		−46.04	227.11	7052
d	20	0.8069	806.9	6562;7052,978,1690
	25	0.8022	802.2	
dd/dt	25	0.00094	0.94	6562
n_D	20	1.41224	1.41224	978;7052,1690
	25	1.4106	1.4106	6562
η	20	0.903	0.000903	7052
	30	0.765	0.000765	
γ	20	24.54	0.02454	978;1690
	30	23.59	0.02359	
ΔH_v	25	12.17	50.92	6562;121
	bp	9.54	39.92	7052;121
ΔH_f°	25(l)	−97.65	−408.57	6562
	25(g)	−85.48	−357.65	
ΔH_c°	25(l)	−1363.65	−5705.50	6562
t_c		340	613	7052
p_c		30	3.0	7052
pK_a		21.0	21.0	8308
μ	25 in 317	2.66	2.66	450
δ		7.8	16.0	652,7159,4593
ER	BuOAc	0.19	0.19	6631,6628;6905,2190
	90%	2437	2437	6631,6628;1399
soly	in aq,25	0.043%w	0.043%w	7052
	aq in,23	0.4%w	0.4%w	
aq az	97.0	48.1%w	48.1%w	3498
fl pt	CC	48	321	6402

Beil	87,710
uv	4898,6749
ir	4471,4898,5932
Raman	1537,4104
ms	6608
nmr	5933

202.   *d*-Camphor

1,7,7-Trimethyl-(1*R*)-bicyclo[2.2.1]heptan-2-one

464–49–3 $C_{10}H_{16}O$

		cgs	SI	
mw		152.236	152.236	3468
bp	1 atm	207.42	480.57	7482;3012,3013,4026,4391,8057
p	24.2(s)	0.39	0.052	7483;1828,3012,7058,8057
eq 2.4 A	A	11.352	10.477	8296
0 – 25 B	B	3439.2	3439.2	
fp		178.75	451.90	4525;2152,3012,3013,4026
tp	385.8 Torr	180.1	453.3	7482
d	0	1.0000	1000.0	3013;4026,5331
	20	0.9920	992.0	
ΔH_v	bp	14.22	59.50	4026;8057
ΔH_f°	25(s)	−78.4	−328	1926
ΔH_c°	25(s)	−1391	−5822	1926
	25(1)	−1413.2	−5912.8	3975
ΔH_m		1.635	6.841	2514;4026,7167
$[\alpha]_D^{20}$	in 79	+44.22	+44.22	4027;3013,4026,6755,1431
K_f	eq 2.67	37.7	37.7	2514;2152,4026,5050,8057
K_B	eq 2.79	5.611	5.611	726
ϵ	20	11.35	11.35	1525
μ	25 in 38	3.10	3.10	4496;3690
soly	in aq	ca 0.01%w	ca 0.01%w	4027,4026;3013
Beil	618,101			
uv	2040,1641			
ir	6368,848			
Raman	333,966,7455,3693			
ms	7144,7954			
nmr	5280,7026,3647,3409,5933			

203. Acetophenone

1-Phenylethanone

$C_6H_5COCH_3$ 98–86–2 C_8H_8O

		cgs	SI	
mw		120.151	120.151	3468
bp	1 atm	202.0	475.2	2969;7058,7490,7762
dt/dp	1 atm	0.055	0.41	7490
dp/dt	1 atm	18	2.4	7490
p	25	0.37	0.049	2087;7058
eq 2.5	A	7.15738	6.28228	2087
	B	1723.46	1723.46	
	C	201	201	
fp		19.62	292.77	7540;4639,7490,7859
d	20	1.02810	1028.10	2087;2807,7072,7490,7540,7762
	25	1.02382	1023.82	7848,4639
dd/dt		0.00066	0.66	6905
α		0.00087	0.00087	6905
n_D	15	1.53631	1.53631	7482;2807,7490,7762
	20	1.53423	1.53423	
dn/dt		0.00045	0.00045	7482
η	15	2.015	0.002015	7490
	25.50	1.642	0.001642	2772
	30	1.511	0.001511	7490
γ	15	40.09	0.04009	7490;7540,7762,7848,7850
	30	38.21	0.03821	
	Table 2.7			
	Table 2.8			
ΔH_v	25	12.76	53.39	2087,4682
	bp	9.275	38.807	4685;7848
ΔH_f°	25(1)	−34.06	−142.51	1598
ΔH_c°	25(1)	−992.9	−4154.3	3975;1598
C_p	25	48.90	204.6	2593
	30	54.42	227.69	5817;4685,4682
t_c		456	729	7848
K_B	eq 2.79	5.65	5.65	2662;2354
pK_a	25	21.55	21.55	2182;994,5525
pK_{BH}^+	in aq H_2SO_4	6.15	6.15	7014
κ	25	3.1×10^{-9}	3.1×10^{-7}	3567;7072,5236,3947,4202,5325
ε	25	17.39	17.39	7072;4937,5236,7850,828,2100
μ	25 in 38	2.95	2.95	3765;782,5860,2807,4489
				8210,1322
δ		10.58	21.65	6240;652

203. Acetophenone

(Continued)

		cgs	SI	
ER	BuOAc	0.03	0.03	6905
soly		sl sol	sl sol	7033
aq az	99.1	18.5%w	18.5%w	
fl pt	TCC	105	378	5712

Beil	<u>639</u>,271
uv	1337,2054,4930,7659,6932,1830,6380,6749
ir	1460,5250,7438,3791,3888,633,2675,6548,2645,2939
Raman	4101,5250,3972,1783,742,5103,1782,2939
ms	6608,1276
nmr	7227,4886,6358,920,5933,6748

204. 4-Methyl-3-penten-2-one

$(CH_3)_2C{=}CHCOCH_3$ 141-79-7 $C_6H_{10}O$

		cgs	SI	
mw		98.144	98.144	3468
bp	1 atm	129.76	402.91	7053;121,7474,2599
dt/dp	1 atm	0.04682	0.3512	7053
dp/dt	1 atm	21.36	2.848	7053;121
p	25	10.43	1.39	121;2599
eq 2.5	A	7.01266	6.13756	7053;2599
	B	1399.09	1399.09	
	C	208.85	208.85	
fp		−52.85	220.30	7053;7474
d	20	0.85482	854.82	7053;2599
	40	0.83663	836.63	
dd/dt	20 - 40	0.00091	0.91	<u>7053</u>
n_D	20	1.44575	1.44575	7053
	25	1.4414	1.4414	2599
η	20	0.639	0.000639	7053
γ	20	22.9	0.0229	7053;2599

204. 4-Methyl-3-penten-2-one

(Continued)

		cgs	SI	
ΔH_v	25 bp	10.4 8.63	43.4 36.11	121;2599 7053;121,2599
t_c		330	603	7053
P_c		35	3.5	7053
pK_a	in aq H_2SO_4,20	−5.36	−5.36	2952;3713
ε	0	15.6	15.6	5251
μ	25 in 38	2.79	2.79	2297;782
δ	25	9.0	18.4	653,652,4593
ER	BuOAc 90%	0.85 537	0.85 537	6628,6631;7635,2190 6628,6631
soly	in aq,20 aq in,20	2.89%w 2.85%w	2.89%w 2.85%w	2042
aq az	91.8	65.2%w	65.2%w	3497
fl pt	TOC	37	310	2190
Beil	90,736			
uv	927,2897,2912,5274,6427,6749			
ir	2912,7438,5932			
Raman	6895,4111			
ms	2745			
nmr	8220,566,4925,5933,6748			

205.   4-Methyl-4-penten-2-one

$CH_2=C(CH_3)CH_2COCH_3$ 3744–02–3 $C_6H_{10}O$

		cgs	SI	
mw		98.144	98.144	3468
bp	1 atm	121.49	394.64	7053;121
dt/dp	1 atm	0.04556	0.3417	7053
dp/dt	1 atm	21.95	2.926	7053;121
p	25	15.08	2.01	121
eq 2.5	A	7.01301	6.13791	7053
	B	1361.50	1361.50	
	C	208.0	208.0	
fp		−72.60	200.55	7053
d	20	0.8411	841.1	7053
n_D	20	1.42130	1.42130	7053
η	20	0.634	0.000634	7053
γ	20	23.0	0.0230	7053
ΔH_v	25	10.0	41.9	121
	bp	8.62	36.07	7053;121
t_c		280	553	7053
p_c		35	3.5	7053
Beil	90,EI 382			
uv	2912			
ir	2912			

206.   2,4-Pentanedione

$CH_3COCH_2COCH_3$ 123-54-6 $C_5H_8O_2$

		cgs	SI	
mw		100.117	100.117	3468
bp	1 atm	138.3	411.5	5389;5052,507,3664
p	22.76	8.62	1.15	5389
eq 2.5	A	6.86495	5.98985	5389;5052
	B	1377.34	1377.34	
	C	207.35	207.35	
fp		−23.2	250.0	7473;3664
d	20	0.976	976	507
	25	0.9721	972.1	3664
n_D	20	1.4512	1.4512	507
η	65	0.4831	0.0004831	7597
	75	0.4347	0.0004347	
γ	11	32.13	0.03213	7071;3664
	39.5	28.50	0.02850	
ΔH_v	25	10.0	41.8	3636
	bp	8.42	35.23	5389
ΔH_f°	25(1)	−101.7	−425.5	3058;5052,7743,5482
	25(g)	−92.5	−387.1	5052
	25(keto,1)	−99.5	−416.3	3058;5052
	25(enol,1)	−102.2	−427.6	
	25(keto,g)	−89.5	−374.4	
	25(enol,g)	−91.9	−384.4	
ΔH_c°	25(1)	−641.8	−2685.4	3058;5482
pK_a	25	8.933	8.933	2227;493,6513,7872
pK_s	25	19.3	19.3	6350;7871
ε	20	25.7	25.7	4937
μ	22 in 38	2.78	2.78	8139
soly	in aq,20	16.6%w	16.6%w	7623;987,5213
	aq in,20	4.5%w	4.5%w	7623
aq az	94.4	59%w	59%w	3498
fl pt	TOC	40	313	7623

Beil	95,777
uv	919,6071,8206,4677
ir	5022,6071,2289,5932
Raman	4109,2289
ms	1032,6424
nmr	6103,5933

207. Formic acid

HCOOH   64-18-6 CH_2O_2

		cgs	SI	
mw		46.026	46.026	3468
bp	1 atm	100.56	373.71	7267;7488,1743,2090,3784,7058
dt/dp	1 atm	0.0442	0.332	7267,2090;1897,7488
dp/dt	1 atm	22.6	3.02	7267,2090
p	25	43.1	5.75	7030;889,7058,8166,1644
eq 2.5	A	7.37790	6.50280	7268;2090
	B	1563.28	1563.28	
	C	247.06	247.06	
fp		8.27	281.42	7950;7488,8166,7030,666 2090,3793
d	15	1.22647	1226.47	7482;2090,2677,7488,7555,7590 7594,373,7848
	25	1.21405	1214.05	
dd/dt		0.00124	1.24	7488
κ_T	25	8.630×10^{-8}	6.473×10^{-7}	4198
κ_S	25	6.67×10^{-8}	5.00×10^{-7}	2873
n_D	20	1.37140	1.37140	2090;7583,373
	25	1.36938	1.36938	
dn/dt	20 - 40	0.00042	0.00042	7629;7488
η	25	1.966	0.001966	7488;2129,2677,7453,7594
	30	1.443	0.001443	
	Table 2.2			
γ	20	37.58	0.03758	7488;7848,7850
	30	36.48	0.03648	
	Table 2.7 Table 2.8			
ΔH_v	25	4.804	20.10	4178;7030,1644,7842,7848 7908,2355
	bp	5.540	23.18	1154
ΔH_m		3.031	12.682	7030
ΔH_f°	25(1)	-101.51	-424.72	7826;6765,2928,8316,4446
ΔH_c°	25(1)	-60.86	-254.64	6765;7446,3975,8316,4446
C_p°	25	10.918	45.679	1421;2928,7908
C_p	25	23.67	99.04	7826;7030,6440
	Table 2.11			
λ	25	0.000643	0.269	3032
K_f	eq 2.67	2.77	2.77	7590;6061
K_B	eq 2.79	2.4	2.4	722;724,1201,1743
K_a	25	0.0001772	0.0001772	3172;3174,1840
K_s	25	1.7×10^{-6}	1.7×10^{-6}	5033
pK_a	25	3.7515	3.7515	3172

207.   Formic acid

(Continued)

		cgs	SI	
pK_s	25	5.77	5.77	5033;1181
κ		6.08×10^{-5}	0.00608	7950;7842,7594
ε	16	58.5	58.5	4937;7594,7842,7850,2100
μ	30 in 38	1.82	1.82	5899;8139,2283,8252,4002
δ		12.1	24.8	652
soly	in aq	inf	inf	4027
aq az	107.65	74.5%w	74.5%w	3498;3497,7595
fl pt	TOC	59	332	7629

Beil	155,8
uv	858,1544,3147,7069,1390,7148
ir	153,1643,3328,4021,6448,7428,5153,5932
Raman	7228,680,681,973,5360,1785,1781,922
ms	154,1081,3148,99,6251,7408,5592
nmr	1345,7026,2464,5933,6748

208.   Acetic acid

CH_3COOH 64-19-7 $C_2H_4O_2$

		cgs	SI	
mw		60.052	60.052	3468
bp	1 atm	117.885	391.035	123;7267,2090,4390,7037 7488,7768
dt/dp	1 atm	0.04298	0.3224	123;7267,2090,7488
dp/dt	1 atm	23.27	3.102	123
p	25	15.59	2.079	123;7058,8167,114,3514,2091 4761,6461

208. Acetic acid

(Continued)

		cgs	SI	
eq 2.5	A	7.54196	6.66686	123;7268,2090,4337,5926,4760
	B	1633.288	1633.288	
	C	232.885	232.885	
fp		16.66	289.81	7267;8167,4772,3096,3355
				3746,5682
tp		16.54	289.69	4919;123
d	20	1.04955	1049.55	3081;2090,2677,3774,7590,7768
	25	1.04392	1043.92	7488,5371,4357,7267
dd/dt	10 - 40	0.00114	1.14	7617
α	20	0.001078	0.001078	3081;1043
	55	0.00111	0.00111	7610
κ_T	25	1.223×10^{-7}	9.175×10^{-7}	4198
κ_S	25	1.02×10^{-7}	7.68×10^{-7}	2873
n_D	20	1.3719	1.3719	7267;2090,5995,6873,7768
	25	1.3698	1.3698	8199,3476
dn/dt		0.00038	0.00038	7488;7617
η	15	1.314	0.001314	7488;2129,2677,3774,3947
	25	1.1302	0.0011302	3062;5037
	30	1.024	0.001024	6029;7488,4357
	Table 2.2			
	Table 2.6			
γ	20	27.42	0.02742	3283;5168,7488,7768,6762
	30	26.34	0.02634	
ΔH_v	25	5.583	23.36	4178;2355,4682,7754,4761,5153
				7848,7784
	bp	5.825	24.37	1154
ΔH_m		2.801	11.72	4919;5682,7167
ΔH_f°	25(1)	-115.80	-484.5	4446,7826;7960,8316
	25(g)	-103.31	-432.25	7826,1421
ΔH_c°	25(1)	-208.94	-874.20	4446;7960,7446,3975,8316
C_p°	25	15.162	63.438	1421;7826
C_p	21.5	29.22	122.3	4919;6439,6440,6461,8231
				4682,5682
	25	29.42	123.1	4919;7826
	Table 2.12			
t_c		319.56	592.71	123;6082,1671,4018,4299,7269
				8231,8233
P_c		57.10	5.786	7269,4299,123,2089;3216,5153
d_c		0.3506	350.6	2089,7269;4299
v_c		0.1713	0.0001713	2089,7269;4018,1671,8233,4299
K_f	eq 2.67	3.90	3.90	7590;6061
K_B	eq 2.79	2.530	2.530	7590;722,731,3353,718,5088
				4772,7843

208. Acetic acid

(Continued)

		cgs	SI	
K_a	25	1.754×10^{-5}	1.754×10^{-5}	3169,3170;3174,2013,1840
pK_a	25	4.7560	4.7560	3169,3170
pK_s		14.45	14.45	1181
κ	25	6×10^{-9}	6×10^{-7}	5371;6014,7843,2775,4772 3353,3947
ε	20	6.170	6.170	1043;7518,5846,2100,2416,4937 6873,7842,7497
μ	30 in <u>38</u>	1.68	1.68	5899;3382,7850,8139,3854
δ		10.1	20.7	652;4628,6240
soly	in aq	inf	inf	4027
aq az		none	none	3499
fl pt	TCC TOC	42 44	315 317	1043;5712 7617

Beil	<u>158</u>,96
uv	859,858,1544,3147,3773,5883,1830,1390,7148
ir	1283,5849,6368,153,3472,1663,4021,5932
Raman	679,680,681,810,4218,7715,1785
ms	154,3422,3148,99,7408,5592
nmr	3609,3562,1396,1345,6670,7026,4852,2464,5933,6748

209. Propionic acid

Propanoic acid

CH_3CH_2COOH			79-09-4	$C_3H_6O_2$
		cgs	SI	
mw		74.079	74.079	3468
bp	1 atm	141.163	414.313	124;7267,877,2090,3049,4684 7476,7768
dt/dp	1 atm	0.04236	0.3178	124;7264,7488,1899,2090,4684
dp/dt	1 atm	23.60	3.147	124
p	25	3.38	0.451	124
eq 2.5	A	7.51844	6.64334	124;7268,2090
	B	1594.273	1594.273	
	C	202.605	202.605	
fp		-20.7	252.5	7267;745,7484,7488,2090 7058,7476
tp		-20.50	252.65	4919
d	20	0.99349	993.49	3081;875,2090,2677,3049,5956
	25	0.98808	988.08	4198,6452,7488,755,7768 4357,7267
dd/dt		0.00106	1.06	7488
α	20	0.001090	0.001090	3081;7610
κ_T	25	1.239×10^{-7}	9.295×10^{-7}	4198
κ_S	20	1.00×10^{-7}	7.50×10^{-7}	2873
n_D	20	1.3865	1.3865	7267;2090,3049,7555,7768,7921
	25	1.3843	1.3843	8003,373
dn/dt	20 - 40	0.00044	0.00044	7634;7488
η	15	1.175	0.001175	7488;6165,2677,3049
	25	1.024	0.001024	5037
	30	0.974	0.000974	6029;7488,4357
	Table 2.6			
γ	20	26.70	0.02670	3283,7488;7768
	25	26.2	0.0262	7634
	30	25.71	0.02571	3283,7488
	Table 2.8			
ΔH_v	25	7.443	31.14	4178
	bp	7.716	32.28	4969;1154,4684
ΔH_m		2.548	10.66	4919;7478,4952
ΔH_f°	25(1)	-122.12	-510.95	8316;4446
ΔH_c°	25(1)	-364.98	-1527.08	8316;7134,3975,4446,7446
C_p	20	36.16	151.3	4919;6440,4684,6439,4952
	25	36.52	152.8	4919;4179
	Table 2.12			
t_c		339.5	612.7	2089;4299,3049,7269,1671,7900

209. Propionic acid

(Continued)

		cgs	SI	
p_c		53.0	5.37	2089;4299,7269,758
d_c		0.315	315	2089;7269,4299
v_c		0.235	0.000235	2089;7269,1671,4299
K_B	eq 2.79	3.51	3.51	722
K_a	25	1.336×10^{-5}	1.336×10^{-5}	3171;5622,745,2090,7724 1040,2013
pK_a	25	4.8742	4.8742	3171
κ	25	$<1 \times 10^{-9}$	$<1 \times 10^{-7}$	3947
ϵ	40	3.435	3.435	5846;4937,1124,2100,7458
μ	30 in 38	1.68	1.68	5899;8139,4502,4222
δ	25 in 317	10.0	20.5	4628
soly	in aq	inf	inf	7610
aq az	99.9	17.7%w	17.7%w	7610;3497,1797
fl pt	TOC	57	330	7634;7610
Beil	162,234			
uv	859,858,3328,5784,1390			
ir	2462,153,3201,6448,7877,3472,1663,5932			
Raman	680,3821,5360,7715,1789,1781,1790,3671			
ms	154,3148,99,7408,5592			
nmr	7227,1396,1345,7026,376,2464,6748,5933			

210. Butyric acid

Butanoic acid

$CH_3CH_2CH_2COOH$ 107-92-6 $C_4H_8O_2$

		cgs	SI	
mw		88.106	88.106	3468
bp	1 atm	163.718	436.868	124;5329,6873,7489,7175,2013 2090,4390
dt/dp	1 atm	0.04315	0.3236	124;7175,1897,2090,7489
dp/dt	1 atm	23.18	3.090	124

210. Butyric acid

(Continued)

		cgs	SI	
p	25	0.765	0.102	124;2089,2091,7058
eq 2.5	A	7.43153	6.55643	124;7268,2090
	B	1563.444	1563.444	
	C	179.843	179.843	
fp		−5.2	268.0	7175;745,2090,7476,7489
tp		−5.12	268.03	4919
d	20	0.9582	958.2	7175;875,2677,3774,5329,5956
	25	0.9532	953.2	6873,7489,7584,2090
				3604,4357
dd/dt	10 − 40	0.00100	1.00	7622,7633
α	20	0.00103	0.00103	7610;7622
n_D	20	1.3980	1.3980	7175;2090,5329,7584,7768
	25	1.3958	1.3958	7921,8003
dn/dt		0.000382	0.000382	7482
η	15	1.814	0.001814	7489;2129,2677,3774,5329
	25	1.529	0.001529	5037
	30	1.328	0.001328	6029;4357,7489
	Table 2.2			
	Table 2.3			
	Table 2.5			
	Table 2.6			
γ	20	26.74	0.02674	7489;5329,7768
	25	26.2	0.0262	7622
	30	25.57	0.02557	7489
	Table 2.8			
ΔH_v	25	9.668	40.45	4178;5675
	bp	10.040	42.01	1154
ΔH_m		2.770	11.59	4919;7167
ΔH_t	−51	0.249	1.040	4919
ΔH_f°	25(1)	−127.9	−535.1	8316;4446
ΔH_c°	25(1)	−521.6	−2182.4	8316;4446,3975,7134
C_p	20	42.04	175.9	4919;6440,5675,6439
	25	42.47	177.7	4919;4179
	Table 2.12			
t_c		355.0	628.2	2089,7269;1671,4299,7900
				1156,6854
P_c		52	5.3	7269;2089,4299
d_c		0.304	304	4299;2089,7269
v_c		0.290	0.000290	4299;2089,7269,1671
K_B	eq 2.79	3.94	3.94	722
K_a	25	1.508×10^{-5}	1.508×10^{-5}	745;5196,8116,2013,2012
				4390,1040

210. Butyric acid

(Continued)

		cgs	SI	
pK$_a$	25	4.8216	4.8216	745
ε	20	2.97	2.97	4937;7458,6873,7518,5846,2100
μ	30 in 38	1.65	1.65	5899;8139,4502,4222
δ		10.5	21.5	652;4628
soly	in aq	inf	inf	7610
aq az	94.4	18.5%w	18.5%w	3497
fl pt	TOC	76	349	7622;5712

Beil	162,264
uv	859,858,3147,5883,5784,1390
ir	7225,2552,6384,153,3472,6697,1663,5932
Raman	680,3280,7665,7715,1790
ms	3148,99,7408,5592
nmr	1396,6670,7026,5933

211. Isobutyric acid

2-Methylpropanoic acid

$(CH_3)_2CHCOOH$ 79-31-2 $C_4H_8O_2$

		cgs	SI	
mw		88.106	88.106	3468
bp	1 atm	154.70	427.85	7482,7485;4891,7770
dt/dp	1 atm	0.032	0.24	7470
dp/dt	1 atm	31	4.2	7470
p	25	1.39	0.185	2089
fp		−45.97	227.18	6012;7482,7485,4891
d	20	0.96815	968.15	2089;7485,4891,7768,7770
	25	0.94288	942.88	
dd/dt	25	0.00505	5.05	2089
κ$_T$	25	1.3736×10^{-7}	1.0303×10^{-6}	4198

211. Isobutyric acid

(Continued)

		cgs	SI	
n_D	15	1.39525	1.39525	949;4891,7485
	20	1.39300	1.39300	7768
η	25	1.213	0.001213	4901;7404
	30	1.126	0.001126	7482
γ	20	25.55	0.02555	2089
	26.49	24.45	0.02445	5377
	30.10	24.13	0.02413	5377;2089
	Table 2.8			
ΔH_v	25	8.437	35.30	4178;1922
	bp	10.620	44.434	7754
ΔH_m		1.200	5.021	7478
ΔH_c°	25(1)	-521.0	-2179.9	3975
C_p	25	41.3	173	4179;6012
t_c		336	609	7269;4299
P_c		40	4.1	7269;4299
d_c		0.304	304	7269;4299
v_c		0.2898	0.0002898	7269;4299
K_a	25	1.38×10^{-5}	1.38×10^{-5}	2013;2518,8054,5622,2098,8006
pK_a	25	4.860	4.860	2013
ε	40	2.73	2.73	4937,5846;7518,2100
μ	25(1)	1.08	1.08	5812;4216
δ	25	10.3	21.1	653
soly	in aq,20	22.8%w	22.8%w	3745;1841
	aq in,20	44.6%w	44.6%w	
aq az	98.8	28.2%w	28.2%w	3498
fl pt		77	350	5713

Beil	162,288
uv	1390
ir	3472,6697,2239,754
Raman	4106,558,754
ms	3148,99,5592
nmr	1396,1345,6670,4852,5933

212. Valeric acid

Pentanoic acid

$CH_3CH_2CH_2CH_2COOH$ 109-52-4 $C_5H_{10}O_2$

		cgs	SI	
mw		102.133	102.133	3468
bp	1 atm	185.5	458.7	7267;7768,3130,5379,7472,7489
dt/dp	1 atm	0.045	0.34	7267;1897,7489
dp/dt	1 atm	22	3.0	7267
p	25	0.14	0.019	4364;875,2013,7058
eq 2.5	A	7.6569	6.7818	7268
	B	1777.2	1777.2	
	C	186.6	186.6	
fp		-33.67	239.48	4762;38,7058,6929,7489
d	20	0.9390	939.0	7267;7489,7768,4357,875,7677
	25	0.9345	934.5	3774,5329
dd/dt	10 - 40	0.00090	0.90	7636
α	20	0.00098	0.00098	7636
	55	0.00101	0.00101	7610;7636
n_D	20	1.4080	1.4080	7267;5329,7768,7921,3130
	25	1.4060	1.4060	
dn/dt	20 - 40	0.00040	0.00040	7636
η	15	2.359	0.002359	7489;5329,2129,2677,3774
	25	1.975	0.001975	5037
	30	1.774	0.001774	7489;4357
	Table 2.5			
	Table 2.6			
γ	15	27.83	0.02783	7489;5329,7768
	25	26.1	0.0261	7636
	30	26.35	0.02635	7489
ΔH_v	25	16.56	69.29	2089;1922
	bp	10.53	44.06	1154
ΔH_m		3.385	14.163	4762;7478,38
ΔH_f°	25(1)	-133.88	-560.15	8316;3130,4446
ΔH_c°	25(1)	-677.95	-2836.54	8316;39,3975,7134,3130,4446
C_p°	25	50.28	210.37	4762;6439,6440
C_p	25	47.1	197	4179
t_c		378	651	7269;1156,4299,6854,7900,2089
p_c'		46.1	4.67	2089
d_c		0.352	352	2089
v_c		0.290	0.000290	2089
A	eq 2.72	0.02497	0.02497	2089
K_a	25	1.38×10^{-5}	1.38×10^{-5}	2012;2098,5622,2518,2013

212.   Valeric acid

(Continued)

		cgs	SI	
pK$_a$	25	4.860	4.860	2012
ε	20	2.66	2.66	4937;2100,7458
μ	20 in 172	1.61	1.61	4222
δ		9.5	19.4	4628
soly	in aq,20	2.4%w	2.4%w	7610;6999,418
	aq in,20	13.0%w	13.0%w	7610;6999
aq az	99.8	11%w	11%w	7610
fl pt	COC	96	369	7636;7610

Beil	162,299
uv	5992,1390
ir	1643,3328,154,3472,6697,1663,5932
Raman	3280
ms	3148
nmr	7026,5933

213.   Isovaleric acid

3-Methylbutanoic acid

(CH$_3$)$_2$CHCH$_2$COOH   503–74–2   C$_5$H$_{10}$O$_2$

		cgs	SI	
mw		102.133	102.133	3468
bp	1 atm	176.50	449.65	7489;7768,3130,2013,7058,7471
dt/dp	1 atm	0.046	0.35	7489
dp/dt	1 atm	22	2.9	7489
p	20	0.19	0.025	7633
	34.5	1	0.1	7058;2013
fp		-29.3	243.9	7482,7479;7489,7471,7058,7472
d	15	0.93080	930.80	7489;5956,7555,7768
	30	0.91708	917.08	
dd/dt	15 - 30	0.000915	0.915	7489
n$_D$	15	1.4064	1.4064	7555;7768
	20	1.4063	1.4063	
	25	1.4022	1.4022	3130

213. Isovaleric acid

(Continued)

		cgs	SI	
η	15	2.731	0.002731	7489
	30	1.967	0.001967	
γ	15	25.78	0.02578	7489;7768
	30	24.45	0.02445	
	Table 2.8			
ΔH_v	25	11.21	46.91	4178;1922
	bp	10.32	43.18	1154
ΔH_m		1.750	7.322	7478
ΔH_f°	25(1)	−134	−561	3130
ΔH_c°	25(1)	−678	−2837	3130
C_p	25(s)	42.5	178	4179
t_c		361	634	7269;1156,4299
K_a	25	1.67×10^{-5}	1.67×10^{-5}	2013;2518,1706,2098,8054
pK_a	25	4.777	4.777	2013
κ	0 − 80	$<4 \times 10^{-13}$	$<4 \times 10^{-11}$	6028
ϵ	20	2.64	2.64	4937;2100
μ	25 in 38	0.63	0.63	8139;5812
δ	25 in 317	9.6	19.6	4628
soly	in aq,21	4.1%w	4.1%w	4891
aq az	99.5	18.4%w	18.4%w	3497

Beil	162,309
uv	1390
ir	609
Raman	680,681,3280
nmr	6670,7026

214. Hexanoic acid

$CH_3(CH_2)_3CH_2COOH$ 142-62-1 $C_6H_{12}O_2$

		cgs	SI	
mw		116.160	116.160	3468
bp	1 atm	205.02	478.17	124;6733,7130,7267,5326
				5912,7058
dt/dp	1 atm	0.04538	0.3404	123;7267
dp/dt	1 atm	22.04	2.938	124
p	25	0.04	0.005	124
	61.7	1	0.1	7130
eq 2.5	A	7.63833	6.76323	124;7268,951
	B	1789.425	1789.425	
	C	171.22	171.22	
fp		-3.44	269.71	38;1897,7130
d	20	0.9272	927.2	7267;7555,6733,7130,2677
				3774,5329
	25	0.9230	923.0	5956;2048
	30	0.91832	918.32	6733;4357
dd/dt	0 - 30	0.000865	0.865	6733
α	15 - 30	0.000974	0.000974	6733
n_D	20	1.4168	1.4168	7267;1127,5329,7555,6733,6256
	25	1.4148	1.4148	951,7130
dn/dt		0.000388	0.000388	6733,2048
η	15	3.525	0.003525	6733;2129,2677,5329,3774
	25	2.826	0.002826	5037
	30	2.511	0.002511	6733;4357
	Table 2.5			
	Table 2.6			
γ	20	28.05	0.02805	5329;5956,7130
	25	27.55	0.02755	
ΔH_v	-1.7	18.2	76.1	1922
	94	15.45	64.64	1674
	190	13.1	54.8	4891
ΔH_m		3.68	15.4	38;2669
ΔH_f°	25(1)	-139.04	-581.75	2405;4446
ΔH_c°	25(1)	-835.16	-3494.29	2405;39,3975,7134,4446
C_p	0 - 23	59.30	248.11	7130;2669
K_a	25	1.32×10^{-5}	1.32×10^{-5}	2013;5622,2518
pK_a	25	4.879	4.879	2013
κ	mp	7×10^{-13}	7×10^{-11}	4482
	bp	2.1×10^{-9}	2.1×10^{-7}	4482
$\ln \kappa(ohm^{-1} \cdot cm^{-1}) = -13.007 + 0.0274t - 0.000042t^2$				4482
ε	71	2.63	2.63	4937
μ	25(1)	1.13	1.13	4739;5812,4222

214.   Hexanoic acid

(Continued)

		cgs	SI	
δ	25 in 317	9.5	19.4	4628
soly	in aq,20	0.958%w	0.958%w	6028;418
aq az	99.8	7.9%w	7.9%w	
fl pt		102	375	5713

Beil	162,321
uv	6030,1390
ir	5549,1663,3648,5932
Raman	3280
nmr	7026,5933

215.   Octanoic acid

$CH_3(CH_2)_5CH_2COOH$ 124–07–2 $C_8H_{16}O_2$

		cgs	SI	
mw		144.213	144.213	3468
bp	1 atm	239.9	513.1	7267;1896,5329,5912,7058
dt/dp	1 atm	0.49	3.7	7267
dp/dt	1 atm	2.0	0.27	7267
p	92	1	0.1	5713
eq 2.5	A	7.3486	6.4735	7268;6255
	B	1723.8	1723.8	
	C	145.9	145.9	
fp		16.55	289.70	38;3444,7130
tp		16.51	289.660	6413
d	20	0.9106	910.6	7267;7130
	25	0.9066	906.6	
	80	0.8615	861.5	2048

215. Octanoic acid

(Continued)

		cgs	SI	
n_D	20	1.4280	1.4280	7267;6256,6255,7130
	25	1.4261	1.4261	
dn/dt		0.00037	0.00037	7482;2048
η	20	5.828	0.005828	1896;3774,2201,2129,2677
	30	4.690	0.004690	5329,7130
	Table 2.5			
γ	20	29.2	0.0292	5329;5956,7130
	25	28.7	0.0287	
ΔH_v	20.6	19.9	83.3	1922
	134	16.73	70.00	1709
	bp	13.97	58.45	7130
ΔH_m		5.10	21.35	6413;2670,7167
ΔH_f°	25(l)	-152.22	-636.89	4446
ΔH_c°	25(l)	-1146.73	-4797.92	4446;39
C_p	25	71.20	297.92	6413;2669,7130
K_a	25	1.27×10^{-5}	1.27×10^{-5}	2013;2518,8054
pK_a	25	4.896	4.896	2013
κ	mp to 80	$<3.7 \times 10^{-13}$	$<3.7 \times 10^{-11}$	6028
$\dot{\epsilon}$	20	2.45	2.45	4937
μ	25(l)	1.15	1.15	4739;5812,4222
δ	25 in 317	9.4	19.2	4628
soly	in aq,25	0.0798%w	0.0798%w	2219;1636,6028,4891,418
Beil	162,347			
uv	5883,6315			
ir	5549,3472,1663,3648,4805,5932			
Raman	810			
nmr	7227,5933			

216.   Nonanoic acid

$CH_3(CH_2)_6CH_2COOH$ 112–05–0 $C_9H_{18}O_2$

		cgs	SI	
mw		158.240	158.240	3468
bp	1 atm	255.6	528.8	7267;5912,5329,1896
dt/dp	1 atm	0.051	0.38	7267
dp/dt	1 atm	19.6	2.6	7267
p	25	0.00140	0.000187	529
eq 2.5	A	7.2484	6.3733	7268;529
	B	1702.0	1702.0	
	C	134.1	134.1	
fp		12.54	285.69	38;7267,5329,2219,5912,1896
tp		12.384	285.534	6412
d	20	0.9052	905.2	5329;7267,70,1896,2124,2671
	25	0.9013	901.3	
dd/dt	25	0.00078	0.78	5329
n_D	20	1.4322	1.4322	5329,2048,7267;7921
	25	1.4302	1.4302	
dn/dt	25	0.00040	0.00040	7267
η	20	8.30	0.00830	5329;70,1896,2124
	25	7.15	0.00715	
γ	27.4	28.94	0.02894	2056;5329,69
	40.0	27.87	0.02787	
ΔH_v	25	19.7	82.4	529;1703
	30.83	20.4	85.3	1923
ΔH_m		4.738	19.823	6412;38
ΔH_t	-9.3	1.2	4.9	38
ΔH_f°	25(l)	-157.67	-659.69	2036;1703,4446
	25(g)	-137.98	-577.31	1703
ΔH_c°	25(l)	-1303.62	-5454.35	2036;1703,4446,39
C_p	25	78.00	326.37	6412
K_f	eq 2.67	8.0	8.0	7756
K_a	25	1.12×10^{-5}	1.12×10^{-5}	2013
pK_a	25	4.951	4.951	2013
μ	20	0.792	0.792	2697
δ	25	9.3	19.0	4628
soly	in aq,20	0.0284%w	0.0284%w	692;2219

Beil	162,352
ir	3768,1663,6174,5932
Raman	4106
nmr	5933

217. Acrylic acid

2-Propenoic acid

$CH_2{=}CHCOOH$ 79–10–7 $C_3H_4O_2$

		cgs	SI	
mw		72.063	72.063	3468
bp	1 atm	141.2	414.2	7610;5295,5296,7603
dt/dp	1 atm	0.042	0.32	7610
dp/dt	1 atm	24	3.2	7610
p	20	7.76	1.03	7607;5296,5295,6077
eq 2.4	A	7.6021	6.7270	1873
	B	1954.6	1954.6	
fp		13.5	286.7	6238;5296,5295
d	12(1)	1.0600	1060.0	5295;5296,6077
	20	1.0511	1051.1	
dd/dt	10 – 40	0.00110	1.10	7633
α	20	0.00079	0.00079	7610
n_D	20	1.4224	1.4224	5295;1,6077
	25	1.4185	1.4185	6236
η	25	1.10cSt	1.10×10^{-6}	6236
γ	25	27.6	0.0276	7633
	30	28.1	0.0281	7610
ΔH_v	25	13.70	57.32	7744
	136	8.9	37.2	7607
ΔH_m		2.66	11.13	7607
ΔH_f°	25(1)	−91.77	−383.97	8316;7744
	25(g)	−80.56	−337.06	8316
ΔH_c°	25(1)	−327.01	−1368.21	8316;3975
	25(g)	−338.22	−1415.11	8316
ΔH_p		18.49	77.36	2305;5212
K_a	25	5.56×10^{-5}	5.56×10^{-5}	2012;5295,5622,1077
pK_a	25	4.255	4.255	2012;597
δ	25	12.0	24.5	653
soly	in aq	inf	inf	7610;4027
aq az		none	none	3498
fl pt	COC	68	341	6238
	TOC	50	323	7633

Beil	163,397
uv	3298
ir	1661,2462,2387,5932
Raman	1015,4122,2387
ms	3148
nmr	6104,1387.2174,401,1187,7026,5933

218.   Crotonic acid

(*E*)-2-Butenoic acid

$CH_3CH{=}CHCOOH$ 107–93–7 $C_4H_6O_2$

		cgs	SI	
mw		86.090	86.090	3468
bp	1 atm	185.0	458.2	7033;7603
p	30	0.18	0.024	7603;7033
eq 2.4	A	9.08047	8.20537	7033
	B	2845.6842	2845.6842	
fp		71.4	344.6	6403;7607
d	72	0.973	973	7607
	77	0.9604	960.4	Beil. EIII2, 1255
n_D	77	1.4249	1.4249	Beil. EIII2, 1255
	80	1.4228	1.4228	7033,7609
ΔH_m		3.102	12.98	7609
ΔH_c°	25(1)	−481.9	−2016.3	3975
C_p	80 – 95	42.6	178.2	4027
K_a	25	1.975×10^{-5}	1.975×10^{-5}	6403;2012,5622,1077,8116
pK_a	25	4.7044	4.7044	6403;4747
μ	30 in 38	2.13	2.13	4936
soly	in aq,25	$94g{\cdot}L^{-1}$	$94kg{\cdot}m^{-3}$	7033
aq az	99.9	2.2%w	2.2%w	3498

Beil	163,408
uv	3233,6315,5206,2598
ir	2462,2386,5932
Raman	3792
ms	3473
nmr	1907,7026,4916,1594,4223,6310,5933

219. Methacrylic acid

2-Methyl-2-propenoic acid

$CH_2{=}C(CH_3)COOH$ 79–41–4 $C_4H_6O_2$

		cgs	SI	
mw		86.090	86.090	3468
bp	1 atm	160.5	433.7	7603;2629
p	25.5	1	0.1	6238
eq 2.4	A	9.13473	8.25963	7603
	B	2712.325	2712.325	
fp		15.5	288.7	6365;6236,6012
d	20	1.0153	1015.3	7603;2629
n_D	20	1.4314	1.4314	7603;8014
	25	1.4288	1.4288	6236
η	25	1.30cSt	1.30×10^{-6}	6236
γ	25	26.5	0.0265	6238
ΔH_p		15.84	66.27	2305;6012
C_p	27	38.5	161.1	6012
A	eq 2.72	0.0120	0.0120	6365
K_a	25	3.72×10^{-5}	3.72×10^{-5}	6238;4418
pK_a	25	4.429	4.429	6238;597
μ	0 in 38	1.65	1.65	4739;4216
δ	25	11.2	22.9	653
soly	in aq,20	8.9%w	8.9%w	7603
	aq in,20	28.5%w	28.5%w	
aq az	99.3	23.1%w	23.1%w	2585
fl pt	COC	77	350	6236
Beil	163,421			
ir	2386,755,5932			
ms	3473			
nmr	6891,5933			

220.   Oleic acid

(Z)-9-Octadecenoic acid

$CH_3(CH_2)_7CH=CH(CH_2)_6CH_2COOH$ 112-80-1 $C_{18}H_{34}O_2$

		cgs	SI	
mw		282.465	282.465	3468
bp	1 atm	360.0(d)	633.2	7058
p	176.5	1	0.1	7058;1153
	223.0	10	1.3	
	264	50	6.7	4237
	286.0	100	13.3	7058
	309.8	200	26.7	
	334.7	400	53.3	
fp		13.38	286.53	3442
mp	α-form	13.38	286.53	3442;829,7058,4893,6832,1153
	β-form	16.30	289.45	
d	15	0.8939	893.9	3924;5670,7555,3641
	25	0.8870	887.0	
	90	0.8429	842.9	
dd/dt	30 - 90	0.00068	0.68	3934
n_D	20	1.45947	1.45947	6584;1153,3442,5670,4893,6832
	35	1.4544	1.4544	1707
dn/dt		0.000354	0.000354	1707
η	20	38.80	0.03880	3924;2201
	25	23.000	0.02300	3641;3924
	45	14.45	0.01445	2207
	80	4.85	0.00485	3924
γ	20	32.8	0.0328	69;7131,3161
	90	27.94	0.02794	6568
	180	21.6	0.0216	69
ΔH_v	bp	16.10	67.36	4480;4481

$\Delta H_v = (26660+1.75 \times 1.987T-0.0304T^2)(1-p/p_c)$ cal·mol^{-1}

		cgs	SI	
ΔH_f°	25(g)	169.7	710.2	5588
ΔE_c	25(l)	-2663.9	-11146	3924;2257,3975,3925,3923
C_p	50	138	577	7131;5040,4481
	100	155	649	
	150	180	753	
P_c		30	3.0	4891
pK_a	25,in aq79	5.02	5.02	8012
κ	mp	3×10^{-13}	3×10^{-11}	4482,7131
	bp	2.8×10^{-9}	2.8×10^{-7}	

$\ln \kappa(\text{ohm}^{-1} \cdot \text{cm}^{-1}) = -12.843+0.0275t-0.000042t^2$

		cgs	SI	
ε	20	2.46	2.46	4937;5670
μ	20 in 317	1.18	1.18	5977;7779,6994,6995,7778,4222
δ	25	7.6	15.6	653
soly	in aq	ins	ins	3013

220.   Oleic acid

(Continued)

Beil	163,463
uv	635,6315,3314,5992
ir	3771,6754,6694,5548,49,3472,7225
Raman	8246,7666
nmr	4477,5986,511,2588,6748

221.   Acetic anhydride

Acetic acid anhydride

$(CH_3CO)_2O$ 108-24-7 $C_4H_6O_3$

		cgs	SI	
mw		102.090	102.090	3468
bp	1 atm	140.0	413.2	7488;7473,4760,8169,4026 4562,7058
dt/dp	1 atm	0.044	0.33	7488
dp/dt	1 atm	23	3.0	7488
p	25	5.1	0.68	3779;4026,7058
eq 2.5	A	7.12165	6.24655	4760
	B	1427.77	1427.77	
	C	198.037	198.037	
fp		-73.1	200.1	7488,4026;3746,7058,7473
d	15	1.08712	1087.12	7488;4026,4562,5371
	30	1.06911	1069.11	
dd/dt	10 - 40	0.00121	1.21	4605
α	20	0.00112	0.00112	7610
n_D	15	1.39299	1.39299	7488;8169,5371
	20	1.3904	1.3904	4026
dn/dt		0.00041	0.00041	7488

221. Acetic anhydride

(Continued)

		cgs	SI	
η	15	0.971	0.000971	7488;4026,4562,8169
	30	0.783	0.000783	
	Table 2.2			
γ	20	32.56	0.03256	4562;3283,4026,7488
	25	31.90	0.03190	
ΔH_v	bp	9.13	38.20	<u>7610</u>;4330,3621,4026,7850,7755
ΔH_f°	25(l)	−149.16	−624.09	8316
	25(g)	−137.62	−575.80	8316
ΔH_c°	25(l)	−431.99	−1807.45	8316;3975,7446
	25(g)	−443.53	−1855.73	8316
C_p	30	45.77	191.50	5817;4026,4330
t_c		296	569	4026;4299
p_c		46.2	4.70	4299;4026,758
K_B	eq 2.79	3.53	3.53	731
pK_s	25	9.85	9.85	2336
κ	25	5×10^{-9}	5×10^{-7}	5371;3096,4026,7843,7205
ϵ	19	20.7	20.7	4937;4026,7842,7850,7205
	32	20.55	20.55	1330
μ	25 in <u>451</u>	2.82	2.82	5846;8255
δ		10.3	21.1	4593,652
fl pt	TOC	58	331	7617
	TCC	53	326	5712
Beil	<u>159</u>,166			
uv	155,3773			
ir	633,1640,5166,5932			
Raman	7222,4116,5166			
ms	6663			
nmr	7227,5933,6748			

222. Propionic anhydride

Propanoic acid anhydride

$(CH_3CH_2CO)_2O$ 123–62–6 $C_6H_{10}O_3$

		cgs	SI	
mw		130.143	130.143	3468
bp	1 atm	169.0	442.2	7610;4562,8003,8169,7058
dt/dp	1 atm	0.049	0.37	7634;7610
dp/dt	1 atm	20	2.7	7634
p	20	0.87	0.12	7610
eq 2.4	A	8.30	7.43	7610
	B	2390	2390	
fp		−43.0	230.2	7610
d	20	1.0110	1011.0	4562;7555
	25	1.0057	1005.7	
dd/dt	10 – 40	0.00109	1.09	7634
α		0.00108	0.00108	7610;7634
n_D	20	1.4045	1.4045	7610;8003,7555,8169
dn/dt	20 – 40	0.00042	0.00042	7634
η	20	1.144	0.001144	4562;8169
	25	1.061	0.001061	
	Table 2.2			
	Table 2.3			
γ	20	30.30	0.03030	4562
	25	29.70	0.02970	
ΔH_v	bp	9.98	41.7	7634
C_p	25	56.1	235	7610
t_c		349	622	7634
P_c		33	3.3	7634
d_c		0.325	325	7634
v_c		0.400	0.000400	7634
ε	16	18.3	18.3	4937
δ		10.0	20.5	4593
fl pt	TOC	69	342	7634;7610
Beil		162,242		
ir		637,5932		
Raman		4116		
nmr		6748		

223. Butyric anhydride

Butanoic acid anhydride

$(CH_3CH_2CH_2CO)_2O$ 106–31–0 $C_8H_{14}O_3$

		cgs	SI	
mw		158.197	158.197	3468
bp	1 atm	199.5	472.7	4027;4562,5803,7476,8169
dt/dp	1 atm	0.051	0.38	7610
dp/dt	1 atm	19.6	2.6	7610
p	20	0.3	0.04	7610
fp		−65.7	207.5	4027
d	20	0.96677	966.77	4562;5803,7555
	25	0.96199	961.99	
dd/dt	25	0.00096	0.96	4562
α	20	0.00100	0.00100	7610
n_D	9	1.4148	1.4148	5803;8003,7555,8169
	20	1.4125	1.4125	6180;4027
dn/dt	10 − 60	0.000416	0.000416	6180
η	20	1.615	0.001615	4562;8169
	25	1.486	0.001486	
γ	20	28.93	0.02893	4562
	25	28.44	0.02844	
ΔH_v	bp	11.95	50.00	7610
C_p	20	67.8	283.7	7610
ε	20	12.9	12.9	4937;7842
δ	25	9.2	18.8	653
fl pt	OC	88	361	7610
Beil	162,274			
ir	3648,5932			
Raman	4116			
ms	5862			
nmr	6748			

224. Methyl formate

HCOOCH$_3$ 107-31-3 C$_2$H$_4$O$_2$

		cgs	SI	
mw		60.052	60.052	3468
bp	1 atm	31.75	304.90	7270;7488,4969,7470,7471,8236
dt/dp	1 atm	0.0349	0.262	7270;7488,7470,8236
dp/dt	1 atm	28.7	3.82	7270
p	25	585.5	78.06	7273;3781
eq 2.5	A	7.17039	6.29529	7273;5446
	B	1125.2	1125.2	
	C	230.56	230.56	
fp		-99.0	174.2	7488,7270;3793,7058,7471
d	0	1.00317	1003.17	7488,7271,2677,4674,4969,7470
	15	0.98149	981.49	7555,7848,8231,8236
	20	0.97421	974.21	
	25	0.9664	966.4	7271
dd/dt		0.00156	1.56	7271
n$_D$	15	1.34648	1.34648	7488;7555,4969
	20	1.34332	1.34332	7488;7270,1488
	25	1.3415	1.3415	7270
dn/dt		0.00044	0.00044	7488
η	15	0.360	0.000360	7488;2677
	25	0.328	0.000328	
	Table 2.2			
γ	20	24.62	0.02462	7848
	30	23.09	0.02309	
	Table 2.7			
ΔH$_v$	25	7.31	30.59	565
	bp	7.103	29.72	1488;4969,1154,5153,7842
ΔH$_m$		1.780	7.448	7842;7754
ΔH$_f^\circ$	25(1)	-92.28	-386.10	565;8316,7826
	25(g)	-84.97	-355.51	565
ΔH$_c^\circ$	25(1)	-232.46	-972.61	565;3975,7446,8316
C$_p^\circ$	25	15.37	64.31	7697
C$_p$	25	28.61	119.7	2594;7826,5315
t$_c$		214.0	487.2	4088,7272;7054,923,8233
				8236,1671
p$_c$		59.2	6.00	4088;5163,7842,8231,8236,7272
d$_c$		0.349	349	4088,7272
v$_c$		0.172	0.000172	4088,7272;1671,8233
K$_B$	eq 2.79	1.649	1.649	726
κ	17	1.92 x 10^{-6}	0.000192	651
ε	20	8.5	8.5	4937;7842,7850,2100

224. Methyl formate

(Continued)

		cgs	SI	
μ	(g)	1.77	1.77	1764;6248,5452
δ		10.2	20.9	4593
soly	in aq,25	23%w	23%w	4026
aq az		none	none	3497
fl pt	TCC	−19	254	5712

Beil	<u>155</u>,18
uv	859,3138
ir	153,3201,7427,7438,7526,5937,7225,5932
Raman	7405,1785,7228,3186
ms	154,839,1103,6609,2745
nmr	7227,5933

225. Ethyl formate

$HCOOC_2H_5$ 109-94-4 $C_3H_6O_2$

		cgs	SI	
mw		74.079	74.079	3468
bp	1 atm	54.31	327.46	7270;7489,4969,5329,5446,6877 7058,7438
dt/dp	1 atm	0.0379	0.284	7270;7489,8236
dp/dt	1 atm	26.4	3.52	7270
p	25	242.8	32.37	<u>7273</u>
eq 2.5	A	6.95409	6.07899	7273;5446
	B	1101.0	1101.0	
	C	215.98	215.98	
fp		−79.6	193.6	7270;6972,3048,7058,7471,7479
d	15	0.92892	928.92	7489;2677,4674,4969,5329,6877 7555,7594,8231
	20	0.9220	922.0	7271
	25	0.9153	915.3	7271

225. Ethyl formate

(Continued)

		cgs	SI	
dd/dt	25	0.00135	1.35	<u>7271</u>
α		0.00141	0.00141	7489
n_D	20	1.35994	1.35994,	4969;7270
	25	1.3575	1.3575	7270
dn/dt	25	0.00046	0.00046	<u>7270</u>;7489
η	15	0.419	0.000419	7489;2677,5329,7594
	30	0.358	0.000358	
	Table 2.2			
γ	15	24.37	0.02437	7489;5329
	30	22.38	0.02238	
ΔH_v	30.85	7.561	31.64	1488;4969,1154,5153,7842,7754
	bp	7.156	29.94	7109
ΔH_m		2.200	9.205	7478
ΔH_c°	25(1)	-391.7	-1638.9	3975;7446
C_p	25	34.48	144.3	2594;4126,768
t_c		235.3	508.5	4088,8231,8233,8236,7272; 923,1671
p_c		46.8	4.74	4088,7842,8236;5153,7850 8231,7272
d_c		0.323	323	4088,7272
v_c		0.229	0.000229	4088,7272;1671,8233
K_B	eq 2.79	2.080	2.080	726
κ	20	1.45×10^{-9}	1.45×10^{-7}	4902;651,7594
ε	25	7.16	7.16	4937;7594,7842,7850,2100
μ	25 in <u>38</u>	1.94	1.94	6877;4491
δ		9.4	19.2	4593,652
soly	in aq,25	11.8%w	11.8%w	4902;4026
	aq in,20	17.0%w	17.0%w	4902
aq az	52.6	95%w	95%w	3498;3497
fl pt	TCC	-20	253	5712

Beil	<u>156</u>,19
uv	859,7229,1390
ir	153,3201,7438,5937,7225,5932
Raman	1789,1781,1790,7228
ms	154,839,1235,6609,2745,99
nmr	2464,5933

226. Propyl formate

$HCOOCH_2CH_2CH_3$ 110–74–7 $C_4H_8O_2$

		cgs	SI	
mw		88.106	88.106	3468
bp	1 atm	80.82	353.97	7270;7483,7058,4969,5446,7453 7475,7764
dt/dp	1 atm	0.0412	0.309	7270;8236
dp/dt	1 atm	24.3	3.24	7270
p	25	82.77	11.03	7273;3779
eq 2.5	A	6.84518	5.97008	7273;5446
	B	1132.3	1132.3	
	C	204.80	204.80	
fp		-92.9	180.3	7475,7058;7484,7270
d	15	0.91109	911.09	7483;887,888,2677,3138,4674 4969,7764,8231,8236,4649
	20	0.9055	905.5	7271
	25	0.8996	899.6	7271
	30	0.89406	894.06	7483;7271
dd/dt	25	0.00117	1.17	7271
n_D	15	1.37898	1.37898	7483;3138,4969,2540
	20	1.37693	1.37693	7764;7270
	25	1.3750	1.3750	
dn/dt		0.00048	0.00048	7483
η	15	0.544	0.000544	7483;2677,7453
	30	0.460	0.000460	
	Table 2.2			
γ	20	29.49	0.02949	7483;7764
	30	23.28	0.02328	
ΔH_v	25	8.960	37.49	
	bp	8.031	33.60	7109;4969,1154,5153,6669,7754
ΔH_f°	25(1)	-116	-484	1926
ΔH_c°	25(1)	-530	-2217	1926;7446
C_p	9 - 57	42.7	178.7	6439
t_c		264.9	538.1	4088,8231,8233,8236,7272; 1671,5153
P_c		40.1	4.06	4088;5153,8231,8236,7272
d_c		0.309	309	4088,7272
v_c		0.285	0.000285	4088,7272;1671,8233
κ	17	5.5×10^{-5}	0.0055	651
ε	19	7.72	7.72	4937;2100
μ	22 in 38	1.89	1.89	8141
δ		9.2	18.8	4593
soly	in aq,22	2.05%w	2.05%w	6999
aq az	71.6	97.7%w	97.7%w	3497;3432

226. Propyl formate

(Continued)

		cgs	SI	
fl pt	TCC	−3	270	5712
Beil	156,21			
uv	8298			
ir	3201,4524,7438,4470,5937			
Raman	4109			
ms	154,839,1103,6609,2745			

227. Butyl formate

$HCOOCH_2CH_2CH_2CH_3$ 592-84-7 $C_5H_{10}O_2$

		cgs	SI	
mw		102.133	102.133	3468
bp	1 atm	106.1	379.3	7270;7476,3138,4969,5446,7058 7475,7764
dt/dp	1 atm	0.0429	0.322	7270
dp/dt	1 atm	23.3	3.11	7270
p	25	28.8	3.84	7273;3779
eq 2.5	A	7.4028	6.5277	7273;5446
	B	1533.4	1533.4	
	C	233.0	233.0	
fp		−91.9	181.3	7270;7475,7476
d	20	0.8919	891.9	7271;7764,2677,3138,4969,5330
	25	0.8869	886.9	7271
dd/dt	25	0.00101	1.01	7271
n_D	20	1.38903	1.38903	7764;3138,4969,5330,7270
	25	1.3874	1.3874	4902;7270,1488
dn/dt	25	0.00040	0.00040	7270

227.   Butyl formate

(Continued)

		cgs	SI	
η	16.3	0.7244	0.0007244	Beil. EIII2, 39
	20	0.704	0.000704	2677
	27	0.625	0.000625	3030
γ	21.3	24.89	0.02489	7764
	41.3	22.87	0.02287	
ΔH_v	25	9.859	41.25	7109
	bp	8.380	35.06	7109;4969
κ		small	small	3659
ϵ	ca 80	2.43	2.43	Beil. EII2, 31
μ	25 in 38	2.03	2.03	930
δ		8.7	17.8	4593
aq az	83.8	83.5%w	83.5%w	3497;3138
fl pt	TOC	76	349	5712;4193
Beil	156,21			
ir	3201,4465,4524,7438			
ms	839,1103			

228.   Isobutyl formate

2-Methylpropyl formate

$HCOOCH_2CH(CH_3)_2$ 542–55–2 $C_5H_{10}O_2$

		cgs	SI	
mw		102.133	102.133	3468
bp	1 atm	98.07	371.22	7274;3138,7476,4969,5446,7058
				7475,7764
dt/dp	1 atm	0.0432	0.324	7274
dp/dt	1 atm	23.1	3.09	7274
p	25	40.08	5.343	7277;3779,7058,4970
eq 2.5	A	6.8596	5.9845	7727;5446
	B	1195.9	1195.9	
	C	202.50	202.50	

228. Isobutyl formate

(Continued)

		cgs	SI	
fp		−95.8	177.4	7274;7484,7058,7475,7476
d	20	0.8776	877.6	7275;887,2677,3138,4969,4970
				5330,7764
	25	0.8732	873.2	7275
dd/dt	25	0.00090	0.9	7275
n_D	20	1.38546	1.38546	7764;7274,1069,3138,4969,4970
	25	1.3835	1.3835	7274
dn/dt	25	0.00038	0.00038	7274
η	16.1	0.688	0.000688	Beil. EIII2, 41
	20	0.680	0.000680	2677;4970
	88.0	0.311	0.000311	Beil. EIII2, 41
γ	15.4	24.47	0.02447	Beil. EIII2, 41
	22.3	23.66	0.02366	7764
	59.6	19.48	0.01948	Beil. EIII2, 41
ΔH_v	bp	8.018	33.55	4969;1154
ΔH_c		−719.90	−3012.06	7446
C_p	58.2	50.3	210.5	6439
t_c		277.8	551.0	1671;6851,7276,4299,4970
P_c		38.29	3.880	4970;7276,4299,758
d_c		0.2881	288.1	1671;7276,4299
v_c		0.3545	0.0003545	1671;7276,4299
ε	19	6.41	6.41	4937;2100
μ	22 in 38	1.88	1.88	8141;6248
δ	25	8.2	16.8	653
soly	in aq,22	1.0%w	1.0%w	6999
aq az	79.5	81.1%w	81.1%w	3497

Beil	156,21
ir	3201,4465,7438,5937,5932
Raman	1069
nmr	5933

229. Methyl acetate

CH$_3$COOCH$_3$			79-20-9	C$_3$H$_6$O$_2$
		cgs	SI	
mw		74.079	74.079	3468
bp	1 atm	56.868	330.018	124;5077,7278,8133,8091,1406 3138,4969,5329,7471
dt/dp	1 atm	0.03767	0.2826	124;7278,8133,8091,8236
dp/dt	1 atm	26.54	3.539	124
p	25	216.23	28.828	124;7058,1057,6019,6461,8236
eq 2.5	A	7.11920	6.24410	124;7281,5077,5903
	B	1183.700	1183.700	
	C	222.414	222.414	
fp		-98.05	175.10	7471;7278,3048,7058,7484
d	20	0.9342	934.2	5329;1406,2677,3138,4674,4969
	25	0.9279	927.9	7764,8133,8134,7279
dd/dt	25	0.00138	1.38	7279
α	20	0.00139	0.00139	4902
κ$_S$	30	1.122 x 10^{-7}	8.416 x 10^{-7}	4371
n$_D$	20	1.3614	1.3614	5329;7278,3138,4969,7764
	25	1.3589	1.3589	7278
dn/dt	25	0.00050	0.00050	7278
η	20	0.385	0.000385	5329;1406,2677,2750,7505
	25	0.364	0.000364	
	Table 2.2			
γ	20	24.8	0.0248	5329;7764
	25	24.1	0.0241	
ΔH$_v$	25	7.720	32.30	7109;7086,1629,7784,6669
	bp	7.249	30.33	7109;4969,1154,5153,6493
ΔH$_f^°$	25(1)	-106.57	-445.89	565
	25(g)	-98.00	-410.03	565
ΔH$_c^°$	25(1)	-380.54	-1592.18	565;3975,7446
C$_p^°$	25	20.39	85.31	7697;1629
	Table 2.14			
C$_p$	25	34.39	143.9	2594;7968,768,6461
λ	37.8	0.000385	0.161	6352
t$_c$		233.40	506.55	124;4088,8231,8233,8236,7280 5153,4299
P$_c$		46.88	4.750	124;4088,8236,5153,8231 7280,4299
d$_c$		0.325	325	4088,7280;4299
v$_c$		0.228	0.000228	4088,7280;8233,4299
K$_B$	eq 2.79	2.061	2.061	6493;726,5060
pK$_s$	25	22.50	22.50	4259
κ	20	3.4 x 10^{-6}	0.00034	4902;651

229. Methyl acetate

(Continued)

		cgs	SI	
ε	25	6.68	6.68	4937;4251,5251,7842,7497,2100
μ	25 in 38	1.72	1.72	4098;6950,4275,3381,5251,5318 8141,5452
δ		9.6	19.6	4593,1271,6905,3174;3182
ER	BuOAc 90%	11.8 93(90% purity)	11.8 93	2190;6628,6905,3455,778 6628
soly	in aq,20 aq in,20	24.5%w 8.2%w	24.5%w 8.2%w	6905;2601,4902
aq az	56.4	96.3-96.8%w	96.3-96.8%w	3497;3498
fl pt	TCC	-10	263	5712;4193
Beil	159,124			
uv	858,1390			
ir	2250,3201,6384,6811,7438,633,6368,5932			
Raman	1789,1453,1781,1790,7228			
ms	154,839,1103,6609,2745,99,7408,78			
nmr	7227,6252,2464,5933,6748			

230. Vinyl acetate

Ethenyl acetate

$CH_3COOCH{=}CH_2$ 108-05-4 $C_4H_6O_2$

		cgs	SI	
mw		86.090	86.090	3468
bp	1 atm	72.5	345.7	7116,4903,7058;2035,7519,4966
p	25	106	14.1	3779;7116,7058
eq 2.4	A B	8.091 1798.4	7.216 1798.4	4903
fp		-92.8	180.4	7637;2035
d	20	0.9312	931.2	2922;4903,4495
dd/dt		0.001302	1.302	2922

230. Vinyl acetate

(Continued)

		cgs	SI	
n_D	20	1.3959	1.3959	5270;4495,2035,7116,6974,4966
	25	1.3934	1.3934	5531
η	20	0.4213	0.0004213	4026;5531
γ	20	23.95	0.02395	2922
	30	22.54	0.02254	
ΔH_v	25	8.88	37.2	7744
	bp	8.211	34.35	4903
ΔH_f°	25(1)	-82.67	-345.89	7744
	25(g)	-74.5	-311.7	7744
ΔH_p		21.3	89.1	4026
C_p	20(g)	22.5	94.1	4026
	1 atm			
t_c		228.3	501.5	4903
P_c		22.4	2.27	4026
μ	25 in 38	1.79	1.79	4495;5750
δ	25	9.0	18.4	653
soly	in aq,20	2.0%w	2.0%w	7637
	aq in,20	1.0%w	1.0%w	
fl pt	TCC	-8	265	5712

Beil	159,136
uv	3799
ir	633,3201,1866,153,7225,6292,5932
Raman	4031,3973,742,5103,6292
ms	78
nmr	1387,7227,1187,5933

231. Ethylene glycol diacetate

1,2-Ethanediol diacetate

$CH_3COOCH_2CH_2OOCCH_3$ 111–55–7 $C_6H_{10}O_4$

		cgs	SI	
mw		146.143	146.143	3468
bp	1 atm	190.9	464.1	7614;7058,7467,7192,1766
dt/dp	1 atm	0.049	0.37	7614
dp/dt	1 atm	20	2.7	7614
p	25	0.2	0.03	7192;7058
eq 2.4	A	8.5945	7.7194	7058
	B	2649.43	2649.43	
fp		−41.5	231.7	7614;7058,7192
d	20	1.1043	1104.3	7614;7192,7467
	25	1.09911	1099.11	4339
dd/dt	10 – 40	0.00113	1.13	7614
α	55	0.00106	0.00106	7614
n_D	15	1.4183	1.4183	7192,7467
	20	1.4159	1.4159	1766
	25	1.4139	1.4139	4339
η	20	2.9	0.0029	7614
H_v	25	14.59	61.04	4342
	bp	10.88	45.52	7614
ε	−54	∿10	∿10	4165
μ	30 in 38	2.34	2.34	4739
δ		10.0	20.5	2190,1399
ER	BuOAc	0.026	0.026	1399;2190
	90%	18000	18000	1399
soly	in aq,20	21.3%w	21.3%w	2042
	aq in,20	21.2%w	21.2%w	
aq az	99.7	15.4%w	15.4%w	3498
fl pt	COC	96	369	7614
	TOC	104	377	1399
Beil	159,142			

232. Ethyl acetate

CH$_3$COOCH$_2$CH$_3$ 141-78-6   C$_4$H$_8$O$_2$

		cgs	SI	
mw		88.106	88.106	3468
bp	1 atm	77.111	350.261	124;8134,8133,2754,3948,4969
				5329,7488,7278
dt/dp	1 atm	0.04001	0.3001	124;8134,8133,2090,7470
dp/dt	1 atm	24.99	3.332	124
p	25	94.508	12.600	124;2091,2759,6090,6461
				7058,8236
eq 2.5	A	7.06309	6.18799	124;5903,2090,7281
	B	1224.673	1224.673	
	C	215.712	215.712	
fp		-83.55	189.60	7278;3946,5104,6783,7058,7480
				7499,2090
d	20	0.90063	900.63	2090;13,5077,1406,2754,4674
	25	0.89455	894.55	4969,5329,6877,7470,7764
				8133,8134,7279
dd/dt		0.00120	1.20	7488
α	20	0.00139	0.00139	4902
κ_S	30	1.198 x 10^{-7}	8.987 x 10^{-7}	4371
n_D	20	1.37239	1.37239	2090;1069,4969,5329,6877,7764
	25	1.36978	1.36978	7278,4650
dn/dt		0.00049	0.00049	7488
η	15	0.473	0.000473	7488;2677,2750,3947,3948,5329
	20	0.4508	0.0004508	326
	25	0.426	0.000426	1406
	30	0.400	0.000400	7488
	Table 2.2			
γ	20	23.75	0.02375	3283,7488;2266,5329,7764,7848
	30	22.55	0.02255	
	Table 2.7			
ΔH_v	25	8.513	35.62	7109;7821,7784,7754
	bp	7.641	31.97	7109;1629,4969
ΔH_m		2.51	10.50	1610
ΔH_f°	25(1)	-114.44	-478.82	2405;7805,1294,8316
	25(g)	-106.0	-443.7	2405;8316
ΔH_c°	25(1)	-535.02	-2238.54	2405;3975,7446,7134,8316
	25(g)	-541.5	-2265.6	8316
C_p°		86.85	30.07	125.82	1629
	Table 2.14			
C_p	20.4	40.4	169.0	5681;6439,7805
	25	40.1	167.7	2594
λ	37.8	0.000341	0.143	6352

232. Ethyl acetate

(Continued)

		cgs	SI	
t_c		250.15	523.30	124;4088,8231,8233,8236,7280 923,7848
p_c		38.31	3.882	124;4088,5153,7842,7850,8231 8236,7280
d_c		0.308	308	4088,7280;4299
v_c		0.286	0.000286	4088,7280;1671,8233,4299
K_B	eq 2.79	2.583	2.583	726;1370,5060,718
K_i	25 in H_2O	3×10^{-25}	3×10^{-25}	5735
pK_s	25	22.83	22.83	4259
κ		$<1 \times 10^{-9}$	$<1 \times 10^{-7}$	4902;651,3947,3948,3965
ϵ	20	6.053	6.053	4650
	25	6.02	6.02	4937;4251,5251,6657,7850,7497
	30	5.984	5.984	6983
μ	25 in 38	1.82	1.82	4098;6877,3854,5251,5318,6657 8128,6950
δ		9.1	18.6	4593,1271,6631,1399,6905
ER	BuOAc	3.90	3.90	6628;6631,1399,2190
	90%	110	110	6704;6628,1399,6631
soly	in aq,25	8.08%w	8.08%w	6556;3745,2601,7219
	aq in,25	2.94%w	2.94%w	
aq az	70.38	91.53%w	91.53%w	3497
fl pt	TCC	-4	269	5712
	TOC	-1	272	7625
Beil	159,125			
uv	858,3147,1113,1390			
ir	2250,3201,4524,6384,7438,1097,633,5932			
Raman	1069,7228,5357,5359,1453,3973,1781			
ms	154,6199,839,1103,6609,7262,2745,78			
nmr	6252,7227,2464,6017,5933			

233. Propargyl acetate

2-Propyn-1-ol acetate

$CH_3COOCH_2C \equiv CH$ 627–09–8 $C_5H_6O_2$

		cgs	SI	
mw		98.101	98.101	3468
bp	1 atm	121.5	394.7	3008
d	20	0.9982	998.2	3008;1188
	40	0.9761	976.1	
dd/dt		0.0011	1.1	3008
n_D	20	1.41866	1.41866	3008;1188
γ	20	32.81	0.03281	3008
	40	30.20	0.03020	
Beil	159,140			
uv	7181			
ms	1028			

234. Allyl acetate

2-Propenyl acetate

$CH_3COOCH_2CH = CH_2$ 591–87–7 $C_5H_8O_2$

		cgs	SI	
mw		100.117	100.117	3468
bp	1 atm	103.5	376.7	5583;650,3699
p	30	46.0	6.13	650;648
eq 2.4	A	7.908	7.033	650
	B	1893.5	1893.5	
d	18	0.9279	927.9	5583
	20.9	0.9267	926.7	3699
	28.0	0.9190	919.0	
dd/dt	25	0.00108	1.08	3699
n_D	18	1.4048	1.4048	5583
	20	1.4040	1.4040	3699;558
	24	1.3985	1.3985	648
η	30	0.2068	0.0002068	650

234. Allyl acetate

(Continued)

		cgs	SI	
γ	20.9	26.25	0.02625	3699
	28.0	25.40	0.02540	
ΔH$_v$		8.66	36.3	650
ΔH$_c$		-654.9	-2740.1	7174
δ	25	9.2	18.8	653
aq az	83	83.3%w	83.3%w	3498
Beil	159,136			
uv	7174			
ir	633,5932			
ms	1028			
nmr	5933			

235. Propyl acetate

$CH_3COOCH_2CH_2CH_3$ 109-60-4 $C_5H_{10}O_2$

		cgs	SI	
mw		102.133	102.133	3468
bp	1 atm	101.536	374.686	124;7278,8133,8134,504,3138 4969,5329
dt/dp	1 atm	0.04286	0.3214	124;8133,8134,8236
dp/dt	1 atm	23.33	3.111	124
p	25	33.73	4.497	124;7058,6019,8236
eq 2.5	A	7.01872	6.14362	124;5077,5903,7281
	B	1284.080	1284.080	
	C	208.786	208.786	
fp		-95.0	178.2	7278;7614,3048,7058,7484

235.   Propyl acetate

(Continued)

		cgs	SI	
d	15	0.89377	893.77	7483;504,887,2677,3138 4674,4969
	25	0.88303	883.03	8134;4969,5329,5330,7764,8133 8231,5077
	30	0.87716	877.16	7483;7279
dd/dt	25	0.00114	1.14	7279
α	20	0.00131	0.00131	7614
κ_S	30	1.157×10^{-7}	8.675×10^{-7}	4371
n_D	15	1.38656	1.38656	7483;504,3138,4969,4970,5329
	20	1.38442	1.38442	7764;7278
	25	1.3828	1.3828	7278
dn/dt		0.00048	0.00048	7483
η	20	0.585	0.000585	5329;1911,4257,4970,7453
	25	0.551	0.000551	
	Table 2.2			
γ	20	24.28	0.02428	2266;5329,7764
	30	23.10	0.02310	7483
	40	22.39	0.02239	1239
ΔH_v	25	9.520	39.83	7109;7821
	bp	8.093	33.86	7109;1629,4969,1154,5153 7754,6669
C_p°	126.85	38.798	162.33	1629
	Table 2.14			
C_p	20	46.9	196.2	7614;6439,7540
λ	37.8	0.000329	0.138	6352
t_c		276.58	549.73	124;4088,8231,8233,8236,7280 1671,5153
P_c		33.2	3.36	7280;4088,5153,7850,8231,8236
d_c		0.296	296	4088,7280;4299
v_c		0.345	0.000345	4088,7280;1671,8233,4299
κ	17	2.2×10^{-7}	2.2×10^{-5}	4902;651
ε	20	6.002	6.002	5818;4671,4937,7850,2100
μ	22 in 38	1.78	1.78	8141;5318
δ		8.8	18.0	7159,4593,1271,6905,2190
ER	BuOAc	2.1	2.1	6628;6631,1399
	90%	220	220	6628;6631,1399
soly	in aq,20	2.3%w	2.3%w	7614;4026,2601,7219
	aq in,20	2.9%w	2.9%w	4902
aq az	82.2	86%w	86%w	3497;3138
fl pt	TCC	14	287	5712
	TOC	18	291	7625

235. Propyl acetate

(Continued)

Beil	159,129
uv	859,858,860,1390
ir	3201,4524,6384,6811,7438,5932
Raman	3459,1453
ms	154,839,1103,6609,2745,78,5108
nmr	6252,7227

236. Isopropyl acetate

1-Methylethyl acetate

$CH_3COOCH(CH_3)_2$ 108-21-4 $C_5H_{10}O_2$

		cgs	SI	
mw		102.133	102.133	3468
bp	1 atm	88.601	361.751	124;3067,5330,7058,7475 7764,7278
dt/dp	1 atm	0.04159	0.3119	124;7278
dp/dt	1 atm	24.05	3.206	124
p	25	59.19	7.892	124;3067,7058
eq 2.5	A	7.00443	6.12933	124;7281
	B	1237.232	1237.232	
	C	211.436	211.436	
fp		−73.4	199.8	7475;7484,4026,7278
d	20	0.8773	877.3	7279;7764
	25	0.8702	870.2	7279
dd/dt	25	0.00139	1.39	7279
α	20	0.00131	0.00131	6631
	55	0.00140	0.00140	2190,7614
κ_S	30	1.279×10^{-7}	9.592×10^{-7}	4371
n_D	20	1.37730	1.37730	7764;1069,4026,5330,7278
	25	1.3750	1.3750	7278

236. Isopropyl acetate

(Continued)

		cgs	SI	
dn/dt	25	0.00040	0.00040	<u>7278</u>
η	20	0.569	0.000569	7121;2677
	25	0.52	0.00052	6631
γ	20	21.2	0.0212	6631
	22	22.10	0.02210	7764
	41.4	20.08	0.02008	
ΔH$_v$	25	8.89	37.20	7821
	bp	7.900	33.05	3067
ΔH$_f^\circ$	25(1)	-125.94	-526.93	1294
C$_p$	20	53.2	222.6	<u>4902</u>
	25	46.99	196.6	2594
t$_c$		257.85	531.00	124
pK$_s$	25	22.25	22.25	4259
δ		8.4	17.2	4593,1271,6905,1399
ER	BuOAc	3.40	3.40	6628;6905,1399,3455,778
	90%	134	134	6628;1399
soly	in aq,20	2.9%w	2.9%w	7637;4026,2601
	aq in,20	1.8%w	1.8%w	
aq az	17.6	89.4%w	89.4%w	3497
fl pt	TCC	2	275	152;6051
	TOC	16	289	2270;7614
Beil	<u>159</u>,130			
uv	1390			
ir	3201,7438,5932			
Raman	1069,3459,1453			
ms	154,839,1103,2745			
nmr	6252,7227,2464,5933			

237. Triacetin

1,2,3-Propanetriol triacetate

$(CH_3COO)_3C_3H_5$			102–76–1	$C_9H_{14}O_6$
		cgs	SI	
mw		218.206	218.206	3468
bp	1 atm	258.0	531.2	7625
dt/dp	1 atm	0.051	0.38	7625
dp/dt	1 atm	20	2.7	7625
p	140	10.0	1.33	7625
fp		−37	236	7625
d	20	1.1583	1158.3	7625
dd/dt	10 − 40	0.00109	1.09	7625
α	55	0.00097	0.00097	7625
n_D	20	1.4312	1.4312	7625
	25	1.429	1.429	425
η	20	22	0.022	7625
	25	16	0.016	423
ΔH_v	bp	13.8	57.8	7625
ΔH_f°	25(1)	−318.3	−1331.8	2036
ΔH_c°	25(1)	−1006.4	−4210.8	2036
C_p	25(1)	96.1	402	2594
soly	in aq,25	5.8%w	5.8%w	7625
	aq in,25	3.6%w	3.6%w	
fl pt	COC	143	416	425
Beil	159,147			

238. Butyl acetate

$CH_3COOCH_2CH_2CH_2CH_3$ 123-86-4 $C_6H_{12}O_2$

		cgs	SI	
mw		116.160	116.160	3468
bp	1 atm	126.061	399.211	5077;8133,3138,5329,7473,7764 7911,6450,7278
dt/dp	1 atm	0.0452	0.339	7278;8133,8134,6450
dp/dt	1 atm	22.1	2.95	7278
p	25	12.48	1.664	7281
eq 2.5	A	7.026542	6.151445	5077;6620,7281,4064,6450
	B	1368.051	1368.051	
	C	203.9298	203.9298	
fp		-73.5	199.7	7121;7473,7278
d	15	0.88652	886.52	7483;3138,5329,5330,7764 7911,8122
	25	0.87636	876.36	8134;5007
	30	0.87129	871.29	7483;3887,4371,7279
dd/dt	10 - 40	0.00102	1.02	7614
α	20	0.00117	0.00117	1737
	55	0.00121	0.00121	7614;4026,6450
κ_S	30	1.119×10^{-7}	8.390×10^{-7}	4371
n_D	15	1.39636	1.39636	7483
	20	1.3942	1.3942	7278;563
	25	1.3918	1.3918	7278
dn/dt		0.00047	0.00047	7483
η	15	0.770	0.000770	7483;2750,5329,6450
	20	0.7375	0.0007375	563
	30	0.6444	0.0006444	563;7483
γ	20	25.09	0.02509	7483;2266,5329,7764,7911 1239,6450
	30	23.98	0.02398	
ΔH_v	25	10.43	43.64	2596;7821
	bp	8.559	35.81	7109;1154,4026
	Table 2.9			
ΔH_f°	20(1)	-128.97	-539.61	6630
ΔH_c°	20(1)	-840.65	-3517.3	6630;6450
C_p	20	53.3	223.1	1737;4153
	25	54.59	228.4	
λ	37.8	0.000328	0.137	6352
t_c		306	579	7900;4299,7280
pK_s	25	23.28	23.28	4259
κ	25(85%)	1.6×10^{-8}	1.6×10^{-6}	3965
ε	20	5.01	5.01	4937;4671,2100
μ	25 in 38	1.87	1.87	4098;8141,5318
δ		8.5	17.4	4593,1271,7159,6905

238. Butyl acetate

(Continued)

		cgs	SI	
ER	BuOAc	0.98	0.98	6628;1399,7635,6905
	90%	468	468	6704;6628,1399
soly	in aq,20	0.68%w	0.68%w	6630,7625,2042,7219
	aq in,20	1.2%w	1.2%w	
aq az	90.2	71.3%w	71.3%w	3497,3138
fl pt	TCC	23	296	1737;5712
	TOC	37	310	7625,6630;2270,4193

Beil	159,130
uv	859,858,4388,1113,1390
ir	2250,3201,4524,6384,7438,5932
Raman	3459,1453,4172
ms	154,839,1103,6609,1812,99,78
nmr	7227,6252,5933

239. Isobutyl acetate

2-Methylpropyl acetate

$CH_3COOCH_2CH(CH_3)_2$ 110–19–0 $C_6H_{12}O_2$

		cgs	SI	
mw		116.160	116.160	3468
bp	1 atm	116.6	389.8	7278;7058,504,887,4969,5329
				7472,7764
dt/dp	1 atm	0.0442	0.332	7278;4902
dp/dt	1 atm	22.6	3.02	7278
p	25	17.92	2.389	7281
eq 2.5	A	7.2297	6.3546	7281
	B	1462.4	1462.4	
	C	219.7	219.7	
fp		-98.85	174.30	7472;7058,7278

239. Isobutyl acetate

(Continued)

		cgs	SI	
d	20	0.8728	872.8	7279;5329,504,887,2677,3138
	25	0.8677	867.7	4969,5330,7764
dd/dt	10 - 40	0.00105	1.05	7614
α	20	0.0012	0.0012	1738
	55	0.00126	0.00126	7614
κ_S	30	1.206×10^{-7}	9.042×10^{-7}	4371
n_D	20	1.39018	1.39018	7764;504,1069,3138,5329,5330
	25	1.3880	1.3880	4902;7278
dn/dt	25	0.00040	0.00040	7278
η	20	0.697	0.000697	5329;2677,3367
	25	0.651	0.000651	
γ	20	23.7	0.0237	5329;7764
	25	23.15	0.02315	
ΔH_v	25	9.38	39.2	7058
	bp	8.568	35.85	4969;1154
ΔH_f°	20(1)	-130.89	-547.64	6629
ΔH_c°	20(1)	-838.7	-3509.0	6629
C_p	20	53.3	223.1	6629,1738
	70.2	58.5	244.8	6439;768
t_c		287.8	561.0	1671;7697,4299
p_c		31	3.1	1738
d_c		0.2809	280.9	1671
v_c		0.4135	0.0004135	1671
pK_S	25	22.00	22.00	4259
κ	19	0.000255	0.0255	651
ϵ	20	5.29	5.29	4937;4671,1249
μ	25 in 38	1.86	1.86	4098;8141,5318
δ		8.3	17.0	7159,4593,1271,1399,2190
ER	BuOAc	1.50	1.50	6628;1399,7635,6905
	90%	305	305	6628;1399
soly	in aq,20	0.63%w	0.63%w	7625,6629;2042,2601
	aq in,20	1.02%w	1.02%w	
aq az	87.44	80.5%w	80.5%w	6629;3497,3138
fl pt	CC	18	291	152
	TOC	31	304	2270,6629;7625,4193

Beil	159,131
uv	1390
ir	3201,7438,7979,5932
Raman	1069,1453
ms	7572
nmr	6252,7227

240.   *sec*-Butyl acetate

1-Methylpropyl acetate

CH$_3$COOCH(CH$_3$)CH$_2$CH$_3$ 105–46–4   C$_6$H$_{12}$O$_2$

		cgs	SI	
mw		116.160	116.160	3468
bp	1 atm	112.34	385.49	6636;7674,5330,7764,5836
p	25	24	3.2	4902
fp		-98.9	174.3	6631
d	20 25	0.8748 0.8694	874.8 869.4	7279;7764,5836,2042,5330
dd/dt	25	0.00109	1.09	7279
α	10 – 30	0.00118	0.00118	2042;4902
n$_D$	20 25	1.38941 1.3875	1.38941 1.3875	7764;1069,3510,5836 7279;5330
dn/dt	25	0.00038	0.00038	7279
η	25	0.65	0.00065	6631
γ	21.1 41.9	23.33 21.24	0.02333 0.02124	7764
[α]$_D$	20	(+)25.43 (-)20.19	(+)25.43 (-)20.19	5836 3951
μ	25 in 38	1.87	1.87	4098
δ·		8.2	16.8	4593,1271,6631,1399,6905
ER	BuOAc 90%	1.80 257	1.80 257	6628,6905;778,1399 6628;1399
soly	in aq,20 aq in,20	0.62%w 1.65%w	0.62%w 1.65%w	2042;6631
aq az	87	77.5%w	77.5%w	3497
fl pt	TOC	32	305	2270;152

Beil	159,131
uv	1390
ir	3201,7438,5932
Raman	1069,3459
ms	839,7572
nmr	839,6252,7227,5933

241. Pentyl acetate

$CH_3COOCH_2(CH_2)_3CH_3$ 　　　　　628–63–7　　　　　　　　　$C_7H_{14}O_2$

		cgs	SI	
mw		130.186	130.186	3468
bp	1 atm	149.2	422.4	5329;3138,7764,7278
dt/dp	1 atm	0.048	0.36	7278
dp/dt	1 atm	21	2.8	7278
p	25	9.70	1.29	7281
eq 2.5	A	6.3066	5.4315	7281;1856
	B	1197	1197	
	C	200	200	
fp	1 atm	−70.8	202.4	7278
	(gl)	<−100	<173	4902
d	20	0.8766	876.6	7279;5077,5329,3138,6461
	25	0.8719	871.9	7764,1239
	70	0.8282	828.2	7279;1239
dd/dt	40 − 100	0.000997	0.997	1239
	10 − 40	0.000947	0.947	7279
α		0.00115	0.00115	4902
κ_S	30	1.071×10^{-7}	8.036×10^{-7}	4371
n_D	20	1.4028	1.4028	5329;7278,3138,7764
η	20	0.924	0.000924	5329;2750,3714
	25	0.862	0.000862	
γ	20	25.68	0.02568	7764;5329,1239,3031
	25.9	25.13	0.02513	
ΔH_v	bp	9.8	41.0	4902
$\Delta H_c°$	25(1)	−1058.5	−4428.8	3975
C_p	30.1	66.0	276.1	5817;6439,6461
λ	37.8	0.000323	0.135	6352
t_c		332	605	6352
K_B	eq 2.79	4.83	4.83	723
pK_S	25	24.25	24.25	4259
κ	25	1.6×10^{-9}	1.6×10^{-7}	3965
ε	20	4.75	4.75	4937;3714,2100
μ	25 in 38	1.87	1.87	4098;3714,6950
δ		8.5	17.4	4593,1271,6632
ER	BuOAc	0.67	0.67	6631;6905
	90%	689	689	6631
soly	in aq,20	0.17%w	0.17%w	2042
	aq in,20	1.15%w	1.15%w	
aq az	95.2	59%w	59%w	3497;3138
fl pt	TCC	25	298	5712
	TOC	41	314	2270;4193

241. Pentyl acetate

(Continued)

Beil	159,131
uv	3147,1113,1390
ir	6811,7225,7215,5932
Raman	1453,1781,1790
ms	839,2745,7215
nmr	6252,7227,7215,5933

242. Isopentyl acetate

3-Methyl-1-butyl acetate

$CH_3COOCH_2CH_2CH(CH_3)_2$ 123–92–2 $C_7H_{14}O_2$

		cgs	SI	
mw		130.186	130.186	3468
bp	1 atm	142.1	415.3	7278;7058,504,3138,5330 7476,7764
dt/dp	1 atm	0.047	0.35	4902
dp/dt	1 atm	21	2.8	4902
p	20	4.5	0.60	4902
eq 2.4	A	8.4163	7.5412	3779
	B	2289.2	2289.2	
fp		−78.5	194.7	4026
d	20	0.8709	870.9	7279;7764,504,3138,4026,5330 6161,2365
	25	0.8664	866.4	3138,7279
	40	0.8529	852.9	7279;7764
dd/dt	20 – 40	0.00090	0.90	7279
α		0.00119	0.00119	4902
κ_S	30	1.109×10^{-7}	8.321×10^{-7}	4371
n_D	20	1.4005	1.4005	7278;2546,6452,504,7142 5330,2365
	25	1.3981	1.3981	372;7278

242. Isopentyl acetate

(Continued)

		cgs	SI	
dn/dt	25	0.00048	0.00048	7278
η	19.91	0.872	0.000872	5136;3031
	25	0.7895	0.0007895	Beil. EII2, 60
γ	21.1	24.62	0.02462	7764
	30	23.6	0.0236	Beil. EIII2, 252
ΔH_v	bp	8.97	37.53	1154;7848
C_p	20	59.73	249.91	6438
t_c		326	599	7280,4299
K_B	eq 2.79	4.83	4.83	726
pK_s	25	18.80	18.80	4259
ϵ	30	4.63	4.63	4937;4671,6657
μ	25 in 38	1.86	1.86	4098;8141,6657
δ		8.1	16.6	7159,114
ER	BuOAc	0.83	0.83	1736
soly	in aq,20	0.2%v	0.2%v	1736;4026
	aq in,20	1.0%v	1.0%v	1736
aq az	93.6	63.7%w	63.7%w	3497;3138
fl pt	CC	33	306	152
Beil	159,132			
uv	3147,1390			
ir	3201,7215,5932			
Raman	1453			
ms	839,7215			
nmr	7227,7215,5933			

243. Hexyl acetate

$CH_3COOCH_2(CH_2)_4CH_3$ 142–92–7 $C_8H_{16}O_2$

		cgs	SI	
mw		144.213	144.213	3468
bp	1 atm	170.5	443.7	5329;871,2678,373
p	45.6	5.0	0.67	6661
eq 2.5	A	8.331027	7.455930	6661
30 - 100	B	2289.093	2289.093	
	C	254.17	254.17	
fp		-80.9	192.3	871
d	20	0.8726	872.6	5329;4209,871,6661
	25	0.8681	868.1	
dd/dt	25	0.00090	0.90	5329
α		0.00104	0.00104	2270
κ_S	30	1.036×10^{-7}	7.774×10^{-7}	4371
n_D	20	1.4096	1.4096	5329;6661,4209
	40	1.4002	1.4002	6661
dn/dt	20 - 40	0.00045	0.00045	6661
η	20	1.17	0.00117	5329;871
	25	1.075	0.001075	
γ	20	26.55	0.02655	5329;871
	25	26.0	0.0260	
δ	25	8.65	17.7	653
ER	BuOAc	0.18	0.18	6628,6631;2270
	90%	2567	2567	6628,6631
soly	in aq,20	0.02%w	0.02%w	2270
	aq in,20	0.66%w	0.66%w	
aq az	97.4	61%w	61%w	3498
fl pt	TOC	59	332	2270
Beil	159,132			

244.   4-Methyl-2-pentyl acetate

$CH_3COOCH(CH_3)CH_2CH(CH_3)_2$ 108–84–9 $C_8H_{16}O_2$

		cgs	SI	
mw		144.213	144.213	3468
bp	1 atm	146.2	419.4	7625
dt/dp	1 atm	0.048	0.36	7625
dp/dt	1 atm	21	2.8	7625
p	20	4.0	0.53	7625
fp		−63.8	209.4	7625
d	20	0.8580	858.0	7625
	25	0.8523	852.3	7625
dd/dt	25	0.00094	0.94	7625
α	55	0.00114	0.00114	7625;2270
n_D	20	1.4012	1.4012	7625
η	20	0.93	0.00093	7625
ΔH_v	bp	8.97	37.54	7625
μ		1.9	1.9	2190,1399
δ	25	8.0	16.4	653
ER	BuOAc	0.46	0.46	6631,6628;2270,7625
	90%	1004	1004	6631,6628;1399
soly	in aq,20	0.13%w	0.13%w	2270,7625
	aq in,20	0.57%w	0.57%w	
aq az	94.8	63.3%w	63.3%w	3497
fl pt	COC	43	316	6402
Beil	159,133			

245.   2-Ethylhexyl acetate

$$\underset{\displaystyle CH_3COOCH_2CH(CH_2)_3CH_3}{\overset{\displaystyle CH_2CH_3}{\big|}}$$

103–09–3

$C_{10}H_{20}O_2$

		cgs	SI	
mw		172.267	172.267	3468
bp	1 atm	198.6	471.8	4902;2042,3220
dt/dp	1 atm	0.053	0.40	7614
dp/dt	1 atm	19	2.5	7614
p	20	0.40	0.053	4902
eq 2.4	A	8.4184	7.5433	3779
	B	2521.8	2521.8	
fp		-93	180	5713;6905
d	20	0.8718	871.8	7614
dd/dt	10 - 40	0.00087	0.87	7625;7614
α	10 - 30	0.00099	0.00099	2042
	55	0.00103	0.00103	7625,7635
n_D	20	1.4204	1.4204	7674,2042
	25	1.4173	1.4173	3220
dn/dt		0.00062	0.00062	3220
η	20	1.5	0.0015	7614
ΔH_v	25	11.5	48.1	3779
	bp	10.4	43.5	7614
C_p	20	85.8	359	7614
μ		1.8	1.8	2190
δ		7.9	16.1	2190,1399;6631,7635
ER	BuOAc	0.03	0.03	6905,2190,6631,7635
	90%	13750	13750	6631;1399
soly	in aq,20	<0.3%w	<0.3%w	4902;6905,7635,7625
	aq in,20	0.55%w	0.55%w	
aq az	99.0	26.5%w	26.5%w	7614;3498
fl pt	COC	88	361	4902;2042,7625
Beil		159,135		
nmr		6748		

246. Benzyl acetate

Phenylmethyl acetate

$CH_3COOCH_2C_6H_5$ 140-11-4 $C_9H_{10}O_2$

		cgs	SI	
mw		150.177	150.177	3468
bp	1 atm	215.5	488.7	3779;2659,4026,7058
dt/dp	1 atm	0.052	0.39	3779
dp/dt	1 atm	19	2.6	3779
p	60	1.42	0.189	3779;2659,7058
eq 2.4	A	8.1616	7.2865	3779
150 - 220	B	2583.3	2583.3	
46 - 150	A	9.1700	8.2949	
	B	3000.00	3000.00	
10 - 55	A	9.670	8.795	6581
	B	3157	3157	
fp		-51.5	221.7	7058,7473;4026
d	18	1.0563	1056.3	5583
	25	1.0515	1051.5	3896,2659,5583,1632;2027,4026
	30	1.0482	1048.2	3896;7555
dd/dt		0.000673	0.673	3896,2659,5583,1632
n_D	20	1.5232	1.5232	4027;2659,7555,6313
η	45	1.399	0.001399	7412,3896
ΔH$_v$	10 - 55	14.4	60.2	6581
	bp	11.8	49.4	3779
C$_p$	25	35.49	148.5	2594
	32.8	36.79	153.9	5817
ε	21	5.1	5.1	4937;2027,5428
μ	25 in 9	1.22	1.22	4963;5318
soly	in aq	sl sol	sl sol	4026
aq az	99.60	12.5%w	12.5%w	3497
fl pt	CC	102	375	5712;4026
Beil	528,435			
uv	6749			
ir	4524,6069			
Raman	5357,5359,4112			
ms	2258			
nmr	7227			

247. Ethyl propionate

Ethyl propanoate

$CH_3CH_2COOCH_2CH_3$ 105-37-3 $C_5H_{10}O_2$

		cgs	SI	
mw		102.133	102.133	3468
bp	1 atm	99.10	372.25	7488,7058;4969,4970,6877,7453
dt/dp	1 atm	0.0426	0.320	7283;7488,7470,8236
dp/dt	1 atm	23.5	3.13	7283
p	25	37.25	4.966	5903
eq 2.5	A	7.009966	6.134869	5903;7284
	B	1268.942	1268.942	
	C	208.301	208.301	
fp		-73.85	199.30	7484;3048,7058,7471,7488
d	15	0.89574	895.74	7488;504,887,2027,2677,4969
				4970,5330,6877,7470,7764
				8235,6158
	20	0.8898	889.8	7283
	25	0.8840	884.0	7283
	30	0.87903	879.03	7488
dd/dt		0.00112	1.12	7488
α	10 - 30	0.00125	0.00125	2042
n_D	15	1.38643	1.38643	7488;504,4969,4970,5330
				6877,6896
	20	1.38394	1.38394	7764;7283
	25	1.3814	1.3814	7283
dn/dt		0.00046	0.00046	7488
η	15	0.564	0.000564	7488;2677,2750,4970
	30	0.473	0.000473	
	Table 2.2			
γ	20	24.27	0.02427	3283,7488;2266,7764,6158
	30	23.16	0.02316	
ΔH_v	25	9.381	39.25	7109;4876
	bp	8.076	33.79	7109;4969,1154,5153,7754
ΔH_f°	25(l)	-120.15	-502.70	4876
	25(g)	-110.80	-463.59	4876
ΔH_c°	25(l)	-691.68	-2894.00	4876;3975
C_p	25	46.9	196.1	4876
λ	37.8	0.000331	0.138	6352
t_c		272.9	546.1	7282,8233;923,8231,8236
				1671,6352
P_c		33.18	3.362	7282;5153,8231,8235,8236
d_c		0.297	297	7282,8233;4299
v_c		0.344	0.000344	7282;1671,4299
κ	17	0.000833	0.0833	651

247.   Ethyl propionate

(Continued)

		cgs	SI	
ε	19	5.65	5.65	4937;2027,2026,4671,2100
μ	22 in 38	1.74	1.74	8141;6877,6861
δ	25	8.4	17.2	653
soly	in aq,20	1.92%w	1.92%w	2042;4026,6896,7219
	aq in,20	1.22%w	1.22%w	
aq az	81.2	90%w	90%w	3497;3498
fl pt	TCC	12	285	5712

Beil	162,240
uv	859,860,1113,1390
ir	2250,3201,6811,7438,153,7656,5932
Raman	5357,1781,1790,7656
ms	154,839,2745,6609,5335,7408,5108
nmr	5933

248.   Ethyl butyrate

Ethyl butanoate

$CH_3CH_2CH_2COOCH_2CH_3$ 105-54-4 $C_6H_{12}O_2$

		cgs	SI	
mw		116.160	116.160	3468
bp	1 atm	121.55	394.70	7489;3012,3779,2042,4970,4969
dt/dp	1 atm	0.045	0.34	3012
dp/dt	1 atm	22	3.0	3012
p	25	17	2.3	3779;3012,2042,4970
eq 2.4	A	8.903	8.028	3779
	B	2053.58	2053.58	
fp		−98.0	175.2	7483;7489
d	15	0.88440	884.40	7489;7764,2042,4970,6896
				6157,4969
	25	0.87394	873.94	7489
dd/dt		0.001046	1.046	7489

248. Ethyl butyrate

(Continued)

		cgs	SI	
α	10 - 30	0.00116	0.00116	2042
n_D	20	1.3928	1.3928	7489;7764,4970,6896,4969
dn/dt		0.00047	0.00047	7489
η	15	0.771	0.000771	7489;4970,2131
	25	0.6127	0.0006127	2750
	30	0.595	0.000595	7489
γ	20	24.58	0.02458	7489;7764,6157
	30	23.26	0.02326	
ΔH_v	25	10.04	42.01	7821
	bp	8.679	36.31	4969
ΔH_c°	25(1)	−850.1	−3556.8	3975
C_p	24.08	52.62	220.16	4154
	25	54.49	228.0	2594
λ	37.8	0.000323	0.135	6352
t_c		292.8	566.0	4970;6352,7285,4299
P_c		30.24	3.064	4970
ϵ	18	5.10	5.10	4937;2100
μ	22 in 38	1.74	1.74	8141;6861
δ		8.5	17.4	4593
soly	in aq,20	0.49%w	0.49%w	2042;2601,6844
	aq in,20	0.75%w	0.75%w	
aq az	87.9	78.5%w	78.5%w	3497
fl pt	TCC	26	299	5712

Beil	162,270
uv	857,858,1390
ir	153,7438,320,5932
Raman	4107,1781,1790
ms	6609,2745
nmr	5933

249. Isobutyl isobutyrate

2-Methylpropyl-2-methylpropanoate

$(CH_3)_2CHCOOCH_2CH(CH_3)_2$ 97–85–8 $C_8H_{16}O_2$

		cgs	SI	
mw		144.213	144.213	3468
bp	1 atm	147.51	420.66	3779;574
dt/dp	1 atm	0.061	0.46	3779
dp/dt	1 atm	16	2.1	3779
p	25	4.8	0.64	3779
eq 2.4	A	7.9086	7.0335	3779
60–150	B	2111	2111	
10–60	A	8.8522	7.9771	3779
	B	2433	2433	
fp		−80.7	192.5	7603
d	0	0.87496	874.96	7603
	20	0.8542	854.2	5713
	148.5	0.7249	724.9	6438
n_D	18.2	1.3986	1.3986	574
	20	1.3999	1.3999	6631
η	25	0.83	0.00083	6631
γ	−76.5	33.8	0.0338	3663
	134.5	13.1	0.0131	
ΔH_v	25	11.1	46.4	3779
	bp	9.14	38.2	1154,7755
t_c		328.74	601.89	1156;7286,4299
μ		1.9	1.9	1399
δ		8.0	16.4	6905,2190,1399;6631
ER	BuOAc	0.47	0.47	3182,6628;6905,2190,1399
	90%	965	965	3182,6628;1399
soly	in aq	0.5%w	0.5%w	5713;6905
	aq in,20	<0.2%w	<0.2%w	6905
aq az	95.5	60.6%w	60.6%w	3497
fl pt	TOC	44	317	1399,6905;6631
Beil	162,291			
uv	4953			
ir	7438			
Raman	574			

250. Ethyl isovalerate

Ethyl-3-methylbutanoate

$(CH_3)_2CHCH_2COOCH_2CH_3$ 108–64–5 $C_7H_{14}O_2$

		cgs	SI	
mw		130.186	130.186	3468
bp	1 atm	134.7	407.9	7472;887,5330,7058,7764,3130
dt/dp	1 atm	0.047	0.35	3779
dp/dt	1 atm	21	2.8	
p	25	7.9	1.1	3779
eq 2.4	A	8.3403	7.4652	3779
	B	2216.8	2216.8	
fp		−99.3	173.9	7472;7058
d	20	0.8652	865.2	7764;5330,887,2027,504,3163
	40.9	0.8456	845.6	
n_D	20	1.39621	1.39621	7764;504
	25	1.3944	1.3944	3130
γ	21.1	23.82	0.02382	7764
	41.4	21.46	0.02146	
ΔH_v	25	11.3	47.3	3779
	bp	8.832	36.95	1154
ΔH_f°	25(1)	−129	−540	3130
ΔH_c°	25(1)	−845	−3535	3130
C_p	40	64.82	271.2	4486
	Table 2.11			
t_c		314.87	588.02	1156;7287,4299
ε	18	4.71	4.71	7850
δ		8.65	17.70	50
soly	in aq,20	0.2%w	0.2%w	6999
	aq in,20	0.3%w	0.3%w	6999
aq az	92.2	69.8%w	69.8%w	3497
Beil	162,312			
uv	1390			
ir	3201,4470,5932			
Raman	4107,7438			
nmr	5933			

251. Isopentyl isovalerate

3-Methylbutyl-3-methylbutanoate

$(CH_3)_2CHCH_2COOCH_2CH_2CH(CH_3)_2$ 659–70–1 $C_{10}H_{20}O_2$

		cgs	SI	
mw		172.267	172.267	3468
bp	1 atm	194.0	467.2	7058;504,5330
dt/dp	1 atm	0.0521	0.391	7058
dp/dt	1 atm	19.2	2.56	7058
p	27.0	1	0.1	7058
eq 2.4	A	8.13721	7.26211	7058
25–100	B	2440.11	2440.11	
	A	8.01057	7.13547	
100–200	B	2394.85	2394.85	
d	18.7	0.8583	858.3	504;2026
	25	0.8541	854.1	5330
n_D	18.7	1.41300	1.41300	504
	25	1.4100	1.4100	5330
ΔH_v	25	11.17	46.74	7058
	bp	10.96	45.86	
ε	19	3.62	3.62	2026
aq az	98.8	25.9%w	25.9%w	3497
Beil		162,312		
ir		7979,4470		

252. Butyl stearate

Butyl octadecanoate

$CH_3(CH_2)_{16}COOCH_2CH_2CH_2CH_3$ 123–95–5 $C_{22}H_{44}O_2$

		cgs	SI	
mw		340.588	340.588	3468
bp	25 Torr	220–225	493–498	4026
	1 atm	~350	~623	1606
p	20	1.27×10^{-6}	1.69×10^{-7}	1049,1050;5776
eq 2.4	A	14.88	14.00	5776
	B	5220	5220	

252. Butyl stearate

(Continued)

		cgs	SI	
fp		26.3	299.5	6831;4026,7525,8003
d	25	0.8540	854.0	7525
	30	0.8501	850.1	
dd/dt	25	0.00078	0.78	7525
α	10 - 40	0.00080	0.00080	6905;1606
n_D	20	1.4441	1.4441	6165
	25	1.4422	1.4422	
	50	1.4328	1.4328	8003
dn/dt	20 - 50	0.00038	0.00038	6165
η	25	8.26	0.00826	7525;2797
	50	4.9	0.0049	2797
γ	25	33.0	0.0330	7525
	30	32.7	0.0327	
ΔH_v	75 - 125	23.9	100.0	5776;6246
ΔH_m		13.6	56.9	1606
κ	30	2.1×10^{-13}	2.1×10^{-11}	5960
ε	30	3.111	3.111	4937;5251
μ	24 in 38	1.88	1.88	4274;6248
δ	25	7.5	15.3	653
soly	in aq,25	∿0.17%w	∿0.17%w	1606
	aq in,25	∿0.35%w	∿0.35%w	
fl pt	COC	191	464	7525;4026,5712
Beil	162,380			
ir	5932			

253. Methyl acrylate

Methyl 2-propenoate

$CH_2{=}CHCOOCH_3$ 96–33–3 $C_4H_6O_2$

		cgs	SI	
mw		86.090	86.090	3468
bp	1 atm	80.2	353.4	7058;6169
p	0	23.4	3.12	6236;7058
	10	40.5	5.40	
	20	68.2	9.09	
	25	84	11	3779
	30	109	14.5	6236
	50	225	30.0	
	70	530	70.7	
fp		<−75	<198	2068;6169
d	15	0.9564	956.4	5296;3813
	20	0.9535	953.5	6107
n_D	18	1.4117	1.4117	5296
	25	1.4003	1.4003	6236
η	20	1.3984	0.0013984	3814
ΔH_v	25	6.98	29.20	565
	bp	7.9	33.1	6236
ΔH_f°	25(1)	−86.57	−362.21	565
	25(g)	−79.59	−333.00	
ΔH_c°	25(1)	−494.59	−2069.36	565
ΔH_p		20.19	84.47	2305
C_p	25	38.60	161.5	2594
μ	25 in 38	1.77	1.77	2905
δ		8.9	18.21	50
soly	in aq,25	4.94%w	4.94%w	6236;6107
	aq in,30	2.65%w	2.65%w	
aq az	71	92.8%w	92.8%w	3497
fl pt	TCC	−3.0	270.2	6236
Beil	163,399			
uv	587,1542			
ir	6236,1097,3379,633,7526,7225,1356,5932			
Raman	3379,4115,7526,4122,7228			
nmr	1387,1187,5933			

254. Ethyl acrylate

Ethyl 2-propenoate

$CH_2{=}CHCOOCH_2CH_3$ 140–88–5 $C_5H_8O_2$

		cgs	SI	
mw		100.117	100.117	3468
bp	1 atm	99.5	372.7	7058;6169
p	25	38	5.1	3470;6236
eq 2.4	A	8.0854	7.2103	7058
	B	1939.49	1939.49	
fp		−71.2	202.0	7058;3470
d	20	0.9234	923.4	6107;5296
n_D	20	1.4068	1.4068	6107;5296
	25	1.4034	1.4034	6236
ΔH_v	bp	8.3	34.7	6236
ΔH_f°	25(l)	−88.58	−370.62	7744
	25(g)	−81.4	−340.6	
ΔH_p		18.6	77.8	6236
C_p		47	197	6236
μ	25 in 38	1.96	1.96	2905
δ		8.40	17.19	50
soly	in aq,25	1.50%w	1.50%w	6236;6169
	aq in,25	1.50%w	1.50%w	
aq az	81.1	15%w	15%w	3498;6169
fl pt	TCC	9	282	6236;3470

Beil	163,399
uv	1542,5932
ir	3379,633,7225
Raman	3379,4122,7228
nmr	7227,401,1187,3319

255. Methyl methacrylate

Methyl 2-methyl-2-propenoate

$CH_2{=}C(CH_3)COOCH_3$ 80-62-6 $C_5H_8O_2$

		cgs	SI	
mw		100.17	100.17	3468
bp	1 atm	100.3	273.5	2146
p	25	38	5.1	3779;4510,6236
eq 2.4	A	8.71369	7.83859	6236
0 - 30	B	2126.21	2126.21	
fp		-48.2	225.0	5012;4987,2146
d	20	0.94331	943.31	2280;2146,6106,4510
	30	0.93174	931.74	
dd/dt	25	0.00116	1.16	2280
n_D	20	1.4146	1.4146	2280;2146,5012,6106,4510
	25	1.4120	1.4120	6236
η	20	0.6322	0.0006322	2280
γ	20	28-29	0.028-0.029	Beil. EIII2, 1280;5069
ΔH_v	25	9.73	40.7	6236
	bp	8.6	36.0	6236;1303
ΔH_s		14.5	60.7	6236;1303
ΔH_f°	25(1)	-92.93	-388.82	7744
	25(g)	-82.4	-344.8	
ΔH_c		-645.9	-2702.4	5069
ΔH_p		13.8	57.7	6236;5636
C_p	25	45.7	191.2	5051;2280
ε		2.9	2.9	7089
μ	25 in 38	1.675	1.675	4510
δ	25	8.8	18.0	653
soly	in aq,20	1.56%w	1.56%w	6236;6441
	aq in,20	1.14%w	1.14%w	6236;6604
aq az	83	86%w	86%w	3498
fl pt	TCC	10	283	6236

Beil	163,422
uv	2862,8218,1542
ir	6236,633,6106,153,8218,4873,5932
Raman	6067,3369,4873
nmr	7227,8218,5933

256. Methyl oleate

Methyl (Z)-9-octadecenoate

$$CH_3(CH_2)_7CH{=}CH(CH_2)_7COOCH_3 \quad 112\text{--}62\text{--}9 \qquad\qquad C_{19}H_{36}O_2$$

		cgs	SI	
mw		296.492	296.492	3468
bp	16 Torr	217(d)	490(d)	102
p	128.28	0.2031	0.02708	6536;6442
	185.31	4.709	0.6278	
eq 2.5	A	6.8805	6.0054	6536;5514
128 – 185	B	1963.5	1963.5	
	C	131	131	
fp		19.9	293.1	7993
d	25	0.8702	870.2	3924;7993,4892,2894
	90	0.8234	823.4	
dd/dt	30 – 90	0.00072	0.72	3924
n_D	20	1.45214	1.45214	2894;7993,1707,4892
	40	1.44450	1.44450	
dn/dt		0.000377	0.000377	1707
η	30	4.88	0.00488	3924
	60	2.62	0.00262	
	90	1.64	0.00164	
γ	25	31.3	0.0313	69
	100	25.4	0.0254	
	180	19.1	0.0191	
ΔH_v	1 Torr	20.170	84.391	6536;5514
	25	25.53	106.82	2596
ΔH_f°	25(1)	−173.91	−727.64	6227
ΔE_c	25(1)	−2837.3	−11871	3924;3923,3925
K_f	eq 2.67	3.4	3.4	6536
ε	20	3.211	3.211	2894
	40	3.117	3.117	
δ	25	7.6	15.5	653

Beil	163,467
uv	5172,6805,1940
ir	1906,633,5932
Raman	1844
ms	3100
nmr	7024,3204

257.   Butyl oleate

Butyl (Z)-9-octadecenoate

$CH_3(CH_2)_7CH{=}CH(CH_2)_7COOC_4H_9$ 142-77-8 $C_{22}H_{42}O_2$

		cgs	SI	
mw		338.573	338.573	3468
bp	4 Torr	204	477	424
fp		-15	258	424,423
d	20	0.86645	866.45	70;7473
	25	0.86125	861.25	
dd/dt	25	0.00104	1.04	
n_D	25	1.450	1.450	424,423
η	20	9.751	0.009751	70;7473
	25	7.910	0.007910	
ΔH_f°	25(1)	-195.2	-816.7	1703
ΔH_c°	25(1)	-3308.5	-13843	1703
ε	25	4.0	4.0	5251
soly	in aq,25	ins	ins	424
fl pt	OC	180	453	5415,6402
Beil	163,467			

258.   Methyl benzoate

$C_6H_5COOCH_3$ 93-58-3 $C_8H_8O_2$

		cgs	SI	
mw		136.150	136.150	3468
bp	1 atm	199.50	472.65	7490;2090,7058,7472,7766
dt/dp	1 atm	0.05328	0.3996	2090;7490
dp/dt	1 atm	18.77	2.502	2090
p	25	0.3944	0.05258	2087;2091,7058
eq 2.5	A	7.48253	6.60743	2090
	B	1974.6	1974.6	
	C	230	230	

258. Methyl benzoate

(Continued)

		cgs	SI	
fp	(st)	-12.10	261.05	1897;2090,3946,7472,7479
	(ms)	-13.9	259.3	7490
d	15	1.09334	1093.34	7490;2090,2677,4695,7103,7766
	30	1.07901	1079.01	
dd/dt	25	0.000955	0.955	7490
α		0.000876	0.000876	4902
n_D	20	1.51679	1.51679	2087;2090,4695,7766
	25	1.51457	1.51457	
	50	1.50298	1.50298	
dn/dt		0.00046	0.00046	7482
η	15	2.298	0.002298	7490;2677
	30	1.673	0.001673	
γ	20	38.14	0.03814	7490;7766
	30	36.91	0.03691	
ΔH_v	25	13.281	55.568	4344;2087,7850,565,6081
	bp	10.32	43.18	2087
ΔH_m		2.327	9.736	2087
ΔH_f°	25(1)	-82.11	-343.55	565
	25(g)	-71.66	-299.83	
ΔH_c°	25(1)	-943.57	-3947.90	565;3975
C_p	20	51.92	217.2	6440;6439
	25	52.89	221.3	
t_c		438	711	2087
P_c		39.5	4.00	2087
d_c		0.37	370	2087
v_c		0.369	0.000369	2087
A	eq 2.72	0.01722	0.01722	2087
κ	22	1.37×10^{-5}	0.00137	651
ε	20	6.59	6.59	4937;3106,4671,7850,2100
μ	25 in 38	1.94	1.94	3765;3311,2295,7103,3106
δ	25	10.5	21.5	653
soly	in aq,20	0.21%w	0.21%w	4902;987
	aq in,20	0.74%w	0.74%w	4902
aq az	99.08	20.8%w	20.8%w	3497
fl pt	CC	250	523	152

258. Methyl benzoate

(Continued)

Beil	900,109
uv	8144,3501,6380
ir	6069,6811,1097,3888,6548,6486,2939,5932
Raman	5356,3158,4112,5982,1784,6486,2939
ms	16,2258,99,78
nmr	7227

259. Ethyl benzoate

$C_6H_5COOCH_2CH_3$ 93–89–0 $C_9H_{10}O_2$

		cgs	SI	
mw		150.177	150.177	3468
bp	1 atm	212.40	485.55	7490;3948,4329,5329,7058 7471,7766
dt/dp	1 atm	0.057	0.43	7490;4329
dp/dt	1 atm	18	2.3	7490
p	20	0.18	0.024	4902
eq 2.4 90 - 140	A B	8.6330 2750.0	7.7579 2750.0	3779
140 - 220	A B	8.0350 2500.0	7.1599 2500.0	
fp		−34.7	238.5	7490,7475;3946,7058,7471,7479
d	15 30	1.05112 1.03718	1051.12 1037.18	7490;2125,2669,3948,4695,5329 7103,7766,5583
dd/dt	25	0.000929	0.929	7490
α	20	0.00089	0.00089	4902
n_D	15 20 25	1.50748 1.50570 1.50348	1.50748 1.50570 1.50348	7490;4695,5329,7766,5583 7482 7490
dn/dt		0.00040	0.00040	7482

259.   Ethyl benzoate

(Continued)

		cgs	SI	
η	15 30	2.407 1.751	0.002407 0.001751	7490;2125,2669,3947,3948,5329
γ	20 25	35.4 34.75	0.0354 0.03475	7483;5329,7490,7766
ΔH_v	av	9.67	40.5	4329
ΔH_c°	25(1)	-1107.1	-4632.1	3975
C_p	25	58.80	246.0	2594;4153,6439,6440
κ	25	$<1 \times 10^{-9}$	$<1 \times 10^{-7}$	3947;3948
ε	20	6.02	6.02	4937;3106,4671,6657,7850 7497,2100
μ	in 38	1.99	1.99	1651;452,4275,3602,3106 6657,7103
δ		9.75	19.95	50
soly	in aq,20 aq in,20	0.05%w 0.5%w	0.05%w 0.5%w	4902;987 4902
aq az	99.40	16.0%w	16.0%w	3497
fl pt	CC	96	369	152
Beil	900,110			
uv	155,583,5207,1113			
ir	2250,3527,7438,3791,4604,3888,4283,2939,5932			
Raman	5356,5257,5359,3158,5103,5982,1784,2939			
ms	2258			
nmr	6017			

260. Propyl benzoate

$C_6H_5COOCH_2CH_2CH_3$ 2315–68–6 $C_{10}H_{12}O_2$

		cgs	SI	
mw		164.204	164.204	3468
bp	1 atm	231.2	504.4	7472;7058,7766
dt/dp	1 atm	0.056	0.42	7058
dp/dt	1 atm	18	2.4	
p	54.6	1	0.1	7058
eq 2.4	A	8.30275	7.42765	7058
80 - 160	B	2712.71	2712.71	
	A	8.03341	7.15831	
160 - 230	B	2597.91	2597.91	
fp		-51.6	221.6	7472,7058
d	15	1.0274	1027.4	4580
	20	1.0232	1023.2	7766
	41.5	1.0046	1004.6	
dd/dt	25	0.00087	0.87	7766
n_D	15	1.50139	1.50139	4695
	20	1.50031	1.50031	7766
γ	23.5	38.87	0.03887	7766
	41.5	32.04	0.03204	
ΔH_v	25	12.41	51.92	7058
	bp	11.89	49.75	
ΔH_c°	25(1)	-1263.4	-5286.1	3975
C_p	20	65.35	273.42	6440;6439
aq az	99.70	9.1%w	9.1%w	3497
Beil	900,112			
uv	6749			
nmr	6748			

261. Benzyl benzoate

Phenylmethyl benzoate

$C_6H_5COOCH_2C_6H_5$ 120–51–4 $C_{14}H_{12}O_2$

		cgs	SI	
mw		212.248	212.248	3468
bp	1 atm	323.24	596.39	3475;3779,1499,3948,4130
dt/dp	1 atm	0.065	0.49	3779
dp/dt	1 atm	15	2.1	3779
p	150	3.46	0.461	3779
eq 2.5	A	6.42726	5.55216	3475
	B	1594.49	1594.49	
	C	126.36	126.36	
fp		19.4	292.6	3948
d	19.5	1.1200	1120.0	2027;1499
	25	1.1121	1112.1	3948
n_D	21	1.5686	1.5686	373
η	25	8.292	0.008292	880;3947,3948
	40	5.243	0.005243	
γ	210.5	26.6	0.0266	Beil. EI9, 68
ΔH_v	12 – 60	18.6	77.8	6582
	bp	12.8	53.6	3475
κ	25	$<1 \times 10^{-9}$	$<1 \times 10^{-7}$	3947;3948
ε	20	4.9	4.9	4937;2027,5428
μ	30 in 38	2.06	2.06	7641;1085
fl pt	CC	148	421	5712

Beil	900,121
uv	5207,6749
ir	4699,5932
Raman	5103
ms	16
nmr	7227,6748

262. Ethyl cinnamate

Ethyl 3-phenyl-2-propenoate

$C_6H_5CH{=}CHCOOCH_2CH_3$ 103–36–6 $C_{11}H_{12}O_2$

		cgs	SI	
mw		176.215	176.215	3468
bp,trans	1 atm	271.0	544.2	7058;7864,8086,788
p,trans	87.6	1	0.1	7058;3699
cis	131 – 134	17	2.3	788
eq 2.4	A	8.3708	7.4957	7058
	B	2989.4	2989.4	
fp,trans		12	285	7058;7864
d	20	1.0494	1049.4	3699
	25	1.0469	1046.9	7864
	91.1	0.98815	988.15	5763
n_D	20	1.55983	1.55983	3699;7864
	91.1	1.5250	1.5250	5763
dn/dt		0.000492	0.000492	5763
η	20	8.7	0.0087	7791
γ	19	37.08	0.03708	7864;3699
	24.7	36.57	0.03657	
	Table 2.8			
ΔH_v	10 – 100 Torr	14.1	59.0	7058
	400 – 760 Torr	14.0	58.6	7058
κ	curve given			3621;651
ε	18	6.1	6.1	4937;4671,2100
μ,trans	20.2 in 38	1.84	1.84	788
cis	15.4 in 38	1.77	1.77	788
δ	25	10.1	20.6	658
aq az	99.93	3%w	3%w	3497
Beil	948,581			
uv	583,2413			
ir	4464			
Raman	5355,5357,5359,5982			
ms	2258			

263. γ-Butyrolactone

2(3*H*)-Dihydrofuranone

$$\begin{array}{cc} H_2C\!-\!\!-\!\!-\!CH_2 \\ | \qquad\quad | \\ H_2C \qquad C\!=\!O \\ \diagdown O \diagup \end{array}$$

96–48–0

$C_4H_6O_2$

		cgs	SI	
mw		86.090	86.090	3468
bp	1 atm	204	477	4780,3306
dt/dp	1 atm	0.0477	0.358	4780
dp/dt	1 atm	21.0	2.80	
p	25	3.2	0.43	2626
	79	10	1.3	2707
eq 2.4	A	8.60195	7.72685	4780
	B	2728.28	2728.28	
fp		−43.37	229.78	4444;4780,3306
d	25	1.1254	1125.4	4780
dd/dt		0.000969	0.969	Beil. H17, 234
n_D	25	1.4348	1.4348	4780
η	25	1.7	0.0017	2707
ΔH_v	bp	12.48	52.22	4780
ΔH_m		2.29	9.57	4444
C_p	25	33.80	141.4	2594,4444
t_c		436	709	2707
P_c		34	3.4	2707
pK_b		6.12	6.12	6672
ε	20	39	39	2707
μ	25 in 38	4.12	4.12	4905;4657
δ		12.6	25.8	652;4593,2626
ER	BuOAc	<0.1	<0.1	2626
soly	in aq	inf	inf	2707
aq az		none	none	3498
fl pt	OC	98	371	2707

Beil	2457,234
ir	957,3093,2159
Raman	2159
ms	4767,2577
nmr	2707,4477,5933

264. Ethylene carbonate

1,3-Dioxolan-2-one

96–49–1

$C_3H_4O_3$

		cgs	SI	
mw		88.063	88.063	3468
bp	1 atm	248.2	521.4	3476;3940,3695
dt/dp	1 atm	0.043	0.32	7789
dp/dt	1 atm	23	3.1	7789
p	95.21	3.371	0.4494	3476;5806
	129.32	17.92	2.389	
	175.86	103.51	13.800	
eq 2.5	A	7.11916	6.24407	3476
	B	1821.590	1821.590	
	C	181.076	181.076	
fp		36.37	309.52	7442;6586,3940,5454,7920,976
				2971,1763,3695
tp		36.34	309.49	7687
d	20	1.3417	1341.7	7920
	25	1.3383	1338.3	1763;1817
	40	1.3215	1321.5	2971;3476,1817,3940,3042
dd/dt	25	0.00111	1.11	7920
n_D	40	1.4195	1.4195	6544;3940,2971,3042
	50	1.4158	1.4158	4902,3695
η	40	1.930	0.001930	6544
	50	1.69	0.00169	2971
ΔH_v	150	13.5	56.5	3476;3695
	bp	12.2	51.0	3695
ΔH_s	25	18.77	78.53	1470
ΔH_m		3.153	13.19	7442;7687,7686,976
ΔH_f°	25(s)	−140.13	−586.30	1318;7688,1470
ΔH_c°	25(s)	−278.65	−1165.87	1318;7688,1470,6724
C_p	25(s)	28.07	117.44	7687
	100(1)	40.60	169.9	3695
	150(1)	42.18	176.5	3695
K_f	eq 2.67	5.32	5.32	7442;976,3941
κ		$10^{-8} - 10^{-7}$	$10^{-6} - 10^{-5}$	5728;6586
ε	40	89.78	89.78	5728;3042,1817,3695,6544
				3940,6586
μ	25 in 38	4.87	4.87	3940;6586,4657,409,3042
δ		14.7	30.1	4593,1271,652,3565
soly	in aq,40	inf	inf	4902
aq az		none	none	3498
fl pt	OC	160	433	3695,4902

264. Ethylene carbonate

(Continued)

Beil	2376,100
ir	371,3093,2491,2159,5791,1302,5932
Raman	2491,2159
nmr	6366,408,5791,5933

265. Propylene carbonate

4-Methyl-1,3-dioxolan-2-one

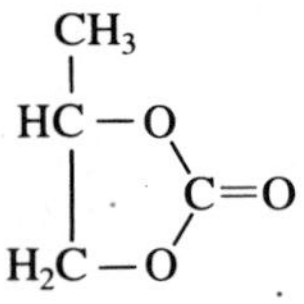

108-32-7

$C_4H_6O_3$

		cgs	SI	
mw		102.090	102.090	
bp	1 atm	241.7	514.9	779,3679;2971,3940
p	55.0	1.20	0.160	1470
	164.56	89.83	11.98	3476;5806
eq 2.5	A	7.08000	6.20491	3476;5806
90 - 100	B	1793.655	1793.655	
	C	185.312	185.312	
fp		-54.53	218.62	3744;3679,2971,779
d	20	1.2006	1200.6	5352;2971,6363,513,779
	25	1.1951	1195.1	4282,3187
dd/dt	20	0.000963	0.963	5352;6729
α	25	0.00084	0.00084	779
κ_T	25	7.9×10^{-8}	5.9×10^{-7}	1777
n_D	20	1.4215	1.4215	5352;3940,2971,779
	25	1.4199	1.4199	
dn/dt	20	0.000375	0.000375	6729
η	20	2.76	0.00276	2971
	25	2.530	0.002530	3187;3679

265. Propylene carbonate

(Continued)

		cgs	SI	
γ	20	41.93	0.04193	513
	25	41.39	0.04139	
ΔH_v	150	13.2	55.2	3476;779
	bp	11.9	49.8	779
ΔH_m		2.299	9.617	7686
ΔH_f°	25(l)	-151.01	-631.83	1470;7688
	25(g)	-143.68	-601.16	
ΔH_c°	25(l)	-430.14	-1799.71	1470;7688
K_f	eq 2.67	4.38	4.38	3744
κ	25	2×10^{-8}	2×10^{-6}	8176
ε	25	64.92	64.92	5728;6729,6586,4282,7920,4574
μ	18	4.94	4.94	3040,4278
δ	25	13.3	27.3	653
soly	in aq,25	17.5%w	17.5%w	8087,779;3476,2971
	aq in,25	8.3%w	8.3%w	779;8087
fl pt	TOC	132	405	779;2971
Beil	2738,104			
ir	5791,3082,6729			
Raman	3680			
nmr	5791,2433,6729			

266. Ethyl carbonate

Diethyl carbonate

$(CH_3CH_2O)_2CO$ 105-58-8 $C_5H_{10}O_3$

		cgs	SI	
mw		118.132	118.132	3468
bp	1 atm	126.8	400.0	7488;1024,4679,7771,4186,4665
dt/dp	1 atm	0.042	0.32	7488,4665
dp/dt	1 atm	24	3.2	7488,4665
p	23.8	10	1.3	7058
	35.1	21.97	2.929	1470
eq 2.4	A	8.2645	7.3894	7058
	B	2148.10	2148.10	
fp		-43.0	230.2	7488,7484;7058
d	15	0.98043	980.43	7488;7771
	25	0.96926	969.26	1024,7448
dd/dt	25	0.00110	1.10	7488
n_D	15	1.38654	1.38654	7488;1069,7771,4186
	25	1.38287	1.38287	1024;7448
dn/dt		0.00039	0.00039	7488
η	15	0.868	0.000868	7488;2129,4186
	25	0.748	0.000748	1024
γ	20	26.44	0.02644	3283,7488;809,7771,3161
	30	25.47	0.02547	
ΔH_v	25	10.42	43.60	4876;1470,1916
	bp	8.64	36.15	4665
ΔH_f°	25(l)	-162.89	-681.53	4876;1470
	25(g)	-152.47	-637.93	4876;1470
ΔH_c°	25(l)	-648.94	-2715.17	4876;1470,3975,7446,7134
C_p	25	50.76	212.4	4876;7675
κ	25	9.1×10^{-10}	9.1×10^{-8}	3965;651,6367
ε	20	2.820	2.820	5818,7497
μ	25 in 38	0.90	0.90	7448
δ		8.8	18.0	652
ER	BuOAc	1.1	1.1	3455
soly	in aq,25	ins	ins	5713
	aq in,25	1.4%w	1.4%w	4902
aq az	91	70%w	70%w	3497
fl pt	CC	25	298	152

Beil	199,5
uv	4794
ir	6811,7438,633,6250,6730,3794,1302,5932
Raman	1069,5188,6730,1785
ms	1161
nmr	3794,5933

267. Ethyl oxalate

Diethyl ethanedioate

$(COOCH_2CH_3)_2$		95–92–1		$C_6H_{10}O_4$
		cgs	SI	
mw		146.143	146.143	3468
bp	1 atm	185.4	458.6	7488;7058,7471,7764,6339,4665
dt/dp	1 atm	0.052	0.39	7482;4665
dp/dt	1 atm	19	2.6	7482
p	25	0.2	0.03	7058
eq 2.4	A	9.3626	8.4875	7058
	B	2974.49	2974.49	
fp		−40.6	232.6	7488,7058,7479;5658,7471,3946
d	15	1.08426	1084.26	7488;1024,2125,5658,6877,7764
	30	1.06687	1066.87	6339,6788
	95.65	0.98829	988.29	7482,7483
dd/dt		0.00116	1.16	7488
n_D	15	1.41239	1.41239	7488;5658,6877,6339,6788
	20	1.41023	1.41023	7764
dn/dt		0.00042	0.00042	7488
η	15	2.311	0.002311	7488;2129,2125,1093
	30	1.618	0.001618	
γ	20	32.22	0.03222	3283,7488;1024,7764
	30	31.03	0.03103	
	Table 2.8			
ΔH_v	bp	10.04	42.01	4665;4683,7755
ΔH_c	25(1)	−723.3	−3026.3	3975
C_p	20	63.29	264.81	6440;4683,4486,4665
	Table 2.11			
κ	25	7.12×10^{-7}	7.12×10^{-5}	4607
ε	21	1.8	1.8	4937;7850,2100
μ	25 in 38	2.49	2.49	6877;4491
δ	25	8.6	17.6	653
soly	in aq,25	<3.6%w	<3.6%w	6165
	aq in,25	<1.2%w	<1.2%w	
aq az		hydrolyzes	hydrolyzes	6165
fl pt	OC	76	349	152
Beil	170,535			
uv	3147,3332,1113			
ir	4465,4466,1097,2226,5932			
Raman	6354,6356,1780,3792,5669,1784			

268. Ethyl malonate

Diethyl propanedioate

$CH_2(COOCH_2CH_3)_2$ 105–53–3 $C_7H_{12}O_4$

		cgs	SI	
mw		160.169	160.169	3468
bp	1 atm	199.30	472.45	7485;5329,7058,7471,7757
dt/dp	1 atm	0.058	0.44	7482
dp/dt	1 atm	17	2.3	7482
p	40	1	0.1	7058;2545
eq 2.4	A	8.6110	7.7359	7058
	B	2705.20	2705.20	
fp		−48.9	224.3	5658;7058,7471,7485
d	15	1.06040	1060.40	7485;2027,2125,2545,5329,5658
	30	1.04460	1044.60	6107,6877,7757,6896
dd/dt		0.001053	1.053	7485
n_D	20	1.41363	1.41363	2545;5329,5658,6877,6896
dn/dt		0.00039	0.00039	6877
η	20	2.15	0.00215	5329;1093,2129,2125
	25	1.94	0.00194	
γ	20	31.71	0.03171	3283;5329,7757
	30	30.49	0.03049	
	Table 2.8			
ΔH_v	bp	13.10	54.81	7058
ΔH_c°	25(1)	−866.6	−3625.9	3975
C_p	21.44	68.06	284.8	1917;6440
K_a	25	5×10^{-14}	5×10^{-14}	5736
pK_a	25	13.3	13.3	5736
ε	25	7.87	7.87	6657;2027,2100
μ	25 in 38	2.54	2.54	6877;6657
soly	in aq,20	2.7%w	2.7%w	6896
	aq in,25	<1.9%w	<1.9%w	6165
aq az	99.5	14%w	14%w	6165
fl pt		93	366	5713
Beil	171,573			
uv	6032			
ir	2250,4465,4466,7438,7979,153,7916,5932			
Raman	6354,6356,3792,1784			
ms	1033			
nmr	14,6017,5933			

269. Methyl maleate

(Z)-Dimethyl 2-butenedioate

HCCOOCH₃
‖
HCCOOCH₃   624–48–6 $C_6H_8O_4$

		cgs	SI	
mw		144.127	144.127	3468
bp	1 atm	200.4	473.6	7525,3699,7058
dt/dp	1 atm	0.046	0.35	7525
dp/dt	1 atm	22	2.9	7525
p	25	0.30	0.040	7525
eq 2.4	A	8.6530	7.7779	7525
	B	2733.74	2733.74	
fp		−17.5	255.7	7525
d	20	1.1513	1151.3	7525;3699
	25	1.1462	1146.2	
dd/dt	25	0.00102	1.02	7525
α	20 – 30	0.00089	0.00089	7525
n_D	20	1.4422	1.4422	7525;3699
	25	1.4405	1.4405	
	30	1.4385	1.4385	
dn/dt	25	0.00037	0.00037	7525
η	20	3.54	0.00354	7525
	25	3.21	0.00321	
γ	20	42.3	0.0423	7525;3699,7074
	25	41.2	0.0412	
ΔH_v	av	12.2	51.0	7525
ΔH_s	54	10.6	44.4	8145
ΔH_m		3.52	14.73	7914
ΔH_c°	25(1)	−674.8	−2823.4	3975
C_p	0.41 – 24	61.98	259.32	7917
μ	25 in 317	2.48	2.48	4491
δ	25	10.9	22.3	653
soly	in aq,25	8.0%w	8.0%w	7525
	aq in,25	4.4%w	4.4%w	
aq az	99.3	12.3%w	12.3%w	6165
fl pt	COC	113	386	7525
Beil	179,751			
uv	6313			
ir	633			
nmr	7227			

270. Ethyl maleate

(Z)-Diethyl 2-butenedioate

HĊCOOCH$_2$CH$_3$
‖
HĊCOOCH$_2$CH$_3$ 141–05–9   C$_8$H$_{12}$O$_4$

		cgs	SI	
mw		172.180	172.180	3468
bp	1 atm	225.3	498.5	7525;3699,7058,7864,7519
dt/dp	1 atm	0.053	0.40	7525
dp/dt	1 atm	19	2.5	7525
p	30	2.2	0.29	650
eq 2.4	A	8.4193	7.5442	7525
	B	2760.9	2760.9	
fp		-8.8	264.4	7519;7525
d	20	1.0687	1068.7	7525;1101,3699,6877,7864
	25	1.0637	1063.7	
dd/dt	25	0.0010	1.0	7525
α	20 - 30	0.00094	0.00094	7525
n$_D$	20	1.4400	1.4400	7525;1101,3699,6877,7519
	25	1.4383	1.4383	
dn/dt	25	0.00034	0.00034	7525
η	20	3.57	0.00357	7525;1093,650
	25	3.14	0.00314	
γ	20	37.0	0.0370	7525;3699,7864
	25	36.7	0.0367	
ΔH$_v$	av	12.5	52.3	7525
ε	23	8.58	8.58	4937;1101
μ	25 in 38	2.54	2.54	6877;4491
δ	25	9.9	20.3	653
soly	in aq,30	ca 1.4%w	ca 1.4%w	7525
	aq in,30	ca 1.9%w	ca 1.9%w	
aq az	99.65	11.8%w	11.8%w	3497
fl pt		121	394	5713
Beil	179,751			
uv	7575			
ir	633,5932			
Raman	1099,1100,6354,6356			
nmr	5933			

271. Butyl maleate

(Z)-Dibutyl 2-butenedioate

HCCOOCH$_2$CH$_2$CH$_2$CH$_3$
‖
HCCOOCH$_2$CH$_2$CH$_2$CH$_3$ 105–76–0   C$_{12}$H$_{20}$O$_4$

		cgs	SI	
mw		228.288	228.288	3468
bp	1 atm	ca 280(d)	ca 553(d)	7525;3699
p	25	0.016	0.0021	5947;7525
eq 2.4	A	8.5064	7.6313	7525
140 - 225	B	3094.71	3094.71	
fp	(gl)	<-80	<193	7525;5713
d	20	0.9950	995.0	7525;3699
	25	0.9907	990.7	
	86.9	0.9376	937.6	3699
dd/dt	25	0.00086	0.86	7525
α	20 - 30	0.00088	0.00088	7525
n$_D$	20	1.4454	1.4454	7525;3699
	25	1.4435	1.4435	
dn/dt	25	0.00038	0.00038	7525
η	20	5.63	0.00563	7525
	25	4.76	0.00476	
γ	20	29.5	0.0295	7525;3699
	25	28.7	0.0287	
ΔH$_v$	140 - 225	14.2	59.4	7525
δ	25	9.0	18.4	653
soly	in aq,25	<0.05%w	<0.05%w	7525
	aq in,25	<0.05%w	<0.05%w	
aq az	99.9	1.6%w	1.6%w	3498
fl pt	COC	141	414	7525
Beil	179,152			
ir	633,5932			

272. Butyl phthalate

Dibutyl 1,2-benzenedicarboxylate

$1,2\text{-}C_6H_4(COOC_4H_9)_2$ 84–74–2 $C_{16}H_{22}O_4$

		cgs	SI	
mw		278.347	278.347	3468
bp	1 atm	340.0	613.2	7058;1608,2659,5086
p	89	0.1	0.01	2659;1049,1275,3374,3853
				7058,6804
	198.2	10	1.3	1608
eq 2.11	A	7.065	6.190	3779;6804
134 – 340	B	1666	1666	
	C	-5.477×10^5	-5.477×10^5	
fp		−35	238	1608;1093,5086
d	20	1.0465	1046.5	1348;2659,1093,6123,5086
	25	1.0426	1042.6	6165;1608
dd/dt		0.00082	0.82	1348
α		0.00086	0.00086	1608
n_D	20	1.4926	1.4926	1348;2659,6123,5086
	25	1.4901	1.4901	6165
η	−195.7(gl)	8.7×10^{66}	8.7×10^{63}	4614;1348,1093,5086,6561
	25	15.4	0.0154	2184;1608
	100	2.2	0.0022	2184
γ	20	33.40	0.03340	411
ΔH_v	bp	18.93	79.20	1348;3374
ΔH_f	(1)	245	1025	112
ΔH_c°	(1)	−2058	−8611	112
C_p	21	119	498	6561;5960
κ	30	4.2×10^{-8}	4.2×10^{-6}	1608;5960
ε	30	6.436	6.436	4937
μ	20 in 38	2.82	2.82	8311
δ		9.3	19.0	4593,652;1271
soly	in aq,20	<0.01%w	<0.01%w	1608;5099
	aq in,20	0.46%w	0.46%w	
aq az		none	none	6165
fl pt	COC	171	444	1608
Beil	972,798			
uv	155,7229			
ir	633			
Raman	686			
ms	2022			

273. Bis(2-ethylhexyl) phthalate

Bis(2-ethylhexyl) 1,2-benzenedicarboxylate

1,2-C_6H_4[COOCH$_2$CH(C$_2$H$_5$)C$_4$H$_9$]$_2$ 117–81–7 $C_{24}H_{38}O_4$

		cgs	SI	
mw		390.562	390.562	3468
bp	1 atm	384	657	2186
p	183.0	0.5	0.07	3375;6804
	231	5	0.7	7628;5086
eq 2.4	A	12.47	11.59	3779;6804
190 - 277	B	5757	5757	
fp		-50	223	5086
d	20	0.9843	984.3	6123;5086
	25	0.979	979	5869
dd/dt	10 - 40	0.00075	0.75	7628
α	20	0.00076	0.00076	7628
n$_D$	20	1.4859	1.4859	7628;6123,5086
	25	1.4836	1.4836	2186;3375,7548
dn/dt	20 - 40	0.00037	0.00037	7628
η	0	381	0.381	7628;6123,5086
	20	81.4	0.0814	7628
	25	56.5	0.0565	2186
ε	20	5.3	5.3	6248
μ	20	2.84	2.84	6248
δ		7.9	16.2	652
soly	in aq,20	<0.01%w	<0.01%w	7628;5086
	aq in,20	0.20%w	0.20%w	
fl pt	COC	218	491	7628
Beil	972,798			

274. Butyl sebacate

Dibutyl decanedioate

$C_4H_9OOC(CH_2)_8COOC_4H_9$ 109–43–3 $C_{18}H_{34}O_4$

		cgs	SI	
mw		314.464	314.464	3468
bp	1 atm	349	622	2185;2479,4026,5086
p	71	0.001	0.0001	5776;6804
eq 2.4	A	14.10	13.22	5776;6804
	B	4850	4850	
fp		−11	262	2185;5086,1093,4026
d	28.1	0.9382	938.2	5819
	57.9	0.9172	917.2	
n_D	20	1.4415	1.4415	6165;4026,6123,5086,2247,2479
	25	1.4397	1.4397	
dn/dt	25	0.00036	0.00036	6165
η	20	9.03	0.00903	2479;5086
	25	8.6	0.0086	2185
ΔH_v	bp	22.2	92.9	5776
C_p	39	148	619	5820
	64	152	636	
κ	30	1.7×10^{-11}	1.7×10^{-9}	5960
ε	30	4.540	4.540	4937;2247
μ	20	2.48	2.48	6248
δ	25	9.2	18.8	653
soly	in aq,20	0.004%w	0.004%w	5086;4026,2479
fl pt	COC	202	475	2185;4026,6123
Beil	178,719			
ir	5932			

275. Butyl borate

Tributyl borate

$(CH_3CH_2CH_2CH_2O)_3B$ 688–74–4 $C_{12}H_{27}O_3B$

		cgs	SI	
mw		230.154	230.154	3468
bp	1 atm	233.5	506.7	1476;6410,1472,7525,3070
dt/dp	1 atm	0.051	0.38	1476
dp/dt	1 atm	20	2.6	1476
p	103.8	10	1.3	7525;412,136
eq 2.4	A B	8.60626 2902.14	7.73116 2902.14	1476;136
fp		<−70	<203	7525
d	20 25 30	0.8580 0.85366 0.84938	858.0 853.66 849.38	7525;412,5628,6410,3070 1477;7525 1477
dd/dt	25 − 40	0.000851	0.851	1477
n_D	20 25	1.40910 1.40714	1.40910 1.40714	1473;7525,412,5628,6410,3070 1473
dn/dt	10 − 30	0.0004074	0.0004074	1473
η	20 25 30	1.776 1.64437 1.51033	0.001776 0.00164437 0.00151033	7525;3070 1477;7525 1477
γ	20 25 30	26.2 23.00 22.63	0.0262 0.02300 0.02263	7525;412 1474;7525 1474
ΔH_v	25 bp	12.968 13.418	54.258 56.141	1476;1472,7525 1476
	Table 2.9			
ΔH_f°	25(1) 25(g)	−284.7 −274.2	−1191.2 −1147.3	2632;1427,1703 1703
ΔH_c°	25(1)	−1925.2	−8055.0	1703,2632
t_c		470.0	743.2	1472
p_c		19.63	1.989	1472
d_c		0.2666	266.6	1472
v_c		0.8633	0.0008633	1472
μ	25 in 38	0.77	0.77	453;5628,2909
soly		hydrolyzes	hydrolyzes	6165
aq az		hydrolyzes	hydrolyzes	6165
fl pt	TOC	77	350	7525
Beil	24,369			
ir	7915,7985			
Raman	3722,3819,3820			
nmr	5828			

276. Methyl phosphate

Trimethyl phosphate

$(CH_3O)_3PO$ 512–56–1 $C_3H_9O_4P$

		cgs	SI	
mw		140.076	140.076	3468
bp	1 atm	197.2	470.4	6732;2101,2310,7948,1395
p	79	12.0	1.60	5458
eq 2.5	A	8.1440	7.2689	2310
	B	2468.4	2468.4	
	C	273	273	
fp	(st)	−46.1	227.1	5646
	(ms)	−62.5	210.7	
d	20	1.2144	1214.4	7772,5646,7777,1395,7948
	25	1.2052	1205.2	2310
dd/dt	20 − 120	0.001371	1.371	7772
α	25 − 40	0.00104	0.00104	2310;7948
n_D	20	1.39630	1.39630	7772;5646,7777
	25	1.3950	1.3950	7777;6203,2310
η	20	2.258	0.002258	2311
	25	2.030	0.002030	
γ	17.5	37.76	0.03776	7772
	60.9	32.88	0.03288	
ΔH_v		11.3	47.3	2310
ΔH_f°	25(1)	−268.2	−1122.1	5499
ε	25	16.39	16.39	7824;1478
μ	25 in 38	3.18	3.18	445;5792,403,7815
δ	25	12.4	25.3	653
fl pt	COC	107	380	7079

Beil	19,286
uv	6583,775
ir	5271,790,4774,1776,2674,7825,5570
Raman	686,2674,6732,7825
ms	6381
nmr	2154,521,3122,4420,2464,6701

277. Ethyl phosphate

Triethyl phosphate

$(CH_3CH_2O)_3PO$ 78–40–0 $C_6H_{15}O_4P$

		cgs	SI	
mw		182.156	182.156	3468
bp	1 atm	216	489	2543;2310,2101,4601,1395
p	90	10.0	1.33	2310
eq 2.5	A	8.3283	7.4532	2310
	B	2658.4	2658.4	
	C	273	273	
d	20	1.0695	1069.5	7772;7777,1395,2101,7073
	25	1.06826	1068.26	2543;2310,2298
dd/dt	25 – 55	0.000995	0.995	2543
α	25 – 40	0.00096	0.00096	2310
n_D	20	1.40527	1.40527	2543;7772
	25	1.4034	1.4034	2298;2310,3447
η	25	2.147	0.002147	2543;2311
	40	1.684	0.001684	
γ	18.3	30.22	0.03022	7772
	38.5	28.30	0.02830	7073
ΔH_v	25	13.7	57.3	1703;2310
ΔH_f°	25(1)	−298.2	−1247.7	1703;1456,2036
	25(g)	−284.5	−1190.3	1703
ΔH_c°	25(1)	−983.4	−4114.5	1703;2036
κ	25	1.19×10^{-8}	1.19×10^{-6}	2543
ϵ	25	10.79	10.79	7824;2543,1478
μ	25 in 38	3.12	3.12	445;5792,403,2298,7104
				3959,5002
δ	25	10.9	22.3	653
fl pt		115	388	5415,6402
Beil	21,332			
ir	751,7413,5570,790,5271,4774			
Raman	686			
ms	4783			
nmr	5320,6701,7679,2154,521			

278. Butyl phosphate

Tributyl phosphate

$(CH_3CH_2CH_2CH_2O)_3PO$		126–73–8		$C_{12}H_{27}O_4P$
		cgs	SI	
mw		266.317	266.317	3468
bp	1 atm	289	562	2310;1260,7772,2298
p	114	0.8	0.1	2506
	160 – 162	15	2.0	5507
eq 2.5	A	8.5861	7.7110	2310
	B	3206.5	3206.5	
	C	273	273	
fp		<−80	<193	1616
d	25	0.9760	976.0	3153;2506,1260,2310,5215
	40	0.9640	964.0	7772,2298
	55	0.9540	954.0	
dd/dt		0.00081	0.81	396
α		0.00093	0.00093	2310
n_D	20	1.42496	1.42496	7772;2506,2310,2298
	25	1.42256	1.42256	1260;396
dn/dt		0.00040	0.00040	396
η	25	3.39	0.00339	5215;1260,7576,7224
	35	2.77	0.00277	
	65	1.61	0.00161	
γ	20	27.55	0.02755	2506
	40	25.96	0.02596	
	60	24.44	0.02444	
ΔH_v	bp	14.680	61.421	2310
ΔH_f°	25(1)	−348.6	−1458.5	5498;6963,3954
ΔH_c°	25(1)	−1905.7	−7973.4	5498;6963
ε	25	8.91	8.91	7824
	30	7.959	7.959	4937;2298
μ	20	3.32	3.32	5792
	25 in 38	3.07	3.07	2298;5101,5002,3596
δ	25	8.8	18.1	653
soly	in aq,25	0.039%w	0.039%w	1260;3385,3153,5120
	aq in,25	4.67%w	4.67%w	1260;6220
aq az	100	0.6%w	0.6%w	3498
fl pt	COC	146	419	1616
Beil	24,369			
ir	3500,2693,5570,5932			
nmr	7227,6701,7679			

279.   Tricresyl phosphate

Tris(methylphenyl) phosphate
(Mixed isomers)

$(CH_3C_6H_4O)_3PO$ 1330-78-5 $C_{21}H_{21}O_4P$

		cgs	SI	
mw		368.368	368.368	3468
bp	1 atm	313.0	586.2	7058;1274
p	154.6	1.0	0.13	7058
eq 2.4	A	12.69	11.81	6804
	B	5925	5925	
fp		-33	240	2190
d	20	1.164	1164	2613
	25	1.159	1159	
α	10 - 40	0.00067	0.00067	6905
n_D	20	1.5571	1.5571	2613;8061
	25	1.5548	1.5548	2247
dn/dt		0.000388	0.000388	
η	20	94.8	0.0948	8088;5048,2832,2246
	30	48.0	0.0480	
ΔH_v		27.1	113.4	6804
λ	28	0.000317	0.133	1398,1397
κ	40	5×10^{-12}	5×10^{-10}	2246
ε	25	6.7	6.7	2247;2246,2832
μ	20 in 317	2.92	2.92	5792;4739,5886
δ	25	11.3	23.1	653
soly	in aq,20	<0.01%w	<0.01%w	7628,2190
	aq in,20	0.4%w	0.4%w	
fl pt	CC	238	511	6402
ir	2818,3944,5521			
nmr	7679,3994,6701			

280.   Tri-*o*-cresyl phosphate

Tris(2-methylphenyl) phosphate

(*o*-CH₃C₆H₄O)₃PO   78–30–8 $C_{21}H_{21}O_4P$

		cgs	SI	
mw		368.368	368.368	3468
bp	1 atm	206	479	1756
eq 2.4	A	9.44	8.56	2028;4239,3373
	B	4535	4535	
d	25	1.17176	1171.76	2311;8217,878
	30	1.16760	1167.60	
dd/dt	25	0.000756	0.756	2311
η	25	86.64	0.08664	2311;8217
	30	61.00	0.06100	
ΔH$_v$		20.7	86.6	2028
ε	25	7.36	7.36	7824;8217
μ	25 in 38	2.87	2.87	1756;7824
Beil	525,358			
ir	751,2652,5521,3142			
nmr	5320,7679			

281.   Tri-*m*-cresyl phosphate

Tris(3-methylphenyl) phosphate

(*m*-CH₃C₆H₄O)₃PO   563–04–2 $C_{21}H_{21}O_4P$

		cgs	SI	
mw		368.368	368.368	3468
bp	0.5 Torr	216	489	1756
eq 2.4	A	11.55	10.67	2028;6804,7726,5947
	B	5787	5787	
ΔH$_v$		26.5	110.9	2028;6804
ε	25	7.50	7.50	7824;8217
μ	25 in 38	3.05	3.05	1756;7824

281. Tri-*m*-cresyl phosphate

(Continued)

Beil	526,EIII 1312
ir	751,2652,5521,3142
nmr	5320,7679

282. Tri-*p*-cresyl phosphate

Tris(4-methylphenyl) phosphate

(p-CH$_3$C$_6$H$_4$O)$_3$PO   78–32–0   C$_{21}$H$_{21}$O$_4$P

		cgs	SI	
mw		368.368	368.368	3468
eq 2.4	A	14.12	13.24	5776;6804,7726
	B	5480	5480	
fp		77	350	1756
ΔH$_v$		25.1	105.0	5776;6804
μ	25 in 38	3.18	3.18	1756;7824
Beil	527,401			
ir	751,2652,5521,3142			
nmr	5320,7679			

283. 1,1-Difluoroethane

CH_3CHF_2 75–37–6 $C_2H_4F_2$

		cgs	SI	
mw		66.050	66.050	3468
bp	1 atm	−24.7	248.5	3291,7305,2142;5019
dt/dp	1 atm	0.0302	0.227	2089
dp/dt	1 atm	33.1	4.41	2089
p	25	4437.1	591.56	2089
eq 2.5	A	5.7654	4.8903	7313
	B	474.2	474.2	
	C	190.2	190.2	
fp		−117	156	7305,2142
d	20	0.909	909	7305;5019,2704
	25	0.896	896	
n_D	−72	1.3011	1.3011	2990
	20	1.2444	1.2444	2510
	25	1.2413	1.2413	
η	20	0.251	0.000251	5019
	30	0.227	0.000227	
γ	20	11.25	0.01125	2089
	30	11.02	0.01102	
ΔH_v	25	4.561	19.08	2089
	bp	5.154	21.56	
ΔH_f°	25(g)	−119.70	−500.82	7314,1442;4366,4143
ΔH_c°	25(g)	−290.7	−1216.3	1442;4143
C_p°	25	16.24	67.94	1442,7307;5162,6818
t_c		113.5	386.7	7306,2142
p_c		44.37	4.496	7306,2142
d_c		0.365	365	7306,2142
v_c		0.181	0.000181	7306
μ	(g)	2.24	2.24	2733

Beil	7,EIII 130
ir	6818,2375
Raman	6818
ms	4599
nmr	2466,2243

284. 1,1,1-Trifluoroethane

CH_3CF_3 420–46–2 $C_2H_3F_3$

		cgs	SI	
mw		84.091	84.091	3468
bp	1 atm	−47.6	225.6	5019,7309;6316
dt/dp	1 atm	0.031	0.23	7309
dp/dt	1 atm	32	4.3	7309
p	−65.17	300	40.0	7313
eq 2.5	A	6.98466	6.10956	7313
	B	814.36	814.36	
	C	245.84	245.84	
fp		−111.3	161.9	7309;6316
d	25.3	0.962	962	5019;7309
	30	0.942	942	
n_D	20	1.22	1.22	7309
ΔH_v	−48.8	4.583	19.175	6316
	20	3.54	14.81	5019
ΔH_m		1.48	6.19	6316
ΔH_t	−116.81	0.0709	0.297	6316
ΔH_f°	25(g)	−178.20	−745.59	7314,1442;7826,4142
ΔH_c°	25(g)	−241.0	−1008.3	1442,4142
C_p°	25	18.76	78.49	7311,1442;7826,5162,6815
C_p	−53	26.21	109.66	6316
t_c		73.1	346.3	5019,7310;6316
p_c		37.09	3.758	7310;5019,6316
d_c		0.434	434	5019,7310
v_c		0.194	0.000194	7310
μ	(g)	2.321	2.321	6702;2614
Beil		8,EIII 131		
ir		1240,6815,5490,1692,7434		
Raman		1240,5490,3225,6993		
ms		2817,6983		
nmr		1240,3045,2243		

285.   1,1,1,2,2-Pentafluoropropane

$CH_3CF_2CF_3$ 1814–88–66 $C_3H_3F_5$

		cgs	SI	
mw		134.049	134.049	3468
bp	745 Torr	−17.9	255.3	6603;7312
dt/dp	1 atm	0.032	0.24	7312
dp/dt	1 atm	31	4.2	7312
p	−24.9	550.0	73.33	6603
d	−40	1.3709	1370.9	6603
	0	1.2578	1257.8	
C_p°	25	28.71	120.12	370
t_c		106.96	380.11	6603
P_c		30.96	3.137	6603
d_c		0.491	491	6603
v_c		0.273	0.000273	6603
ir	370			
Raman	370			

286.   Octafluorocyclobutane

$\overline{CF_2CF_2CF_2CF_2}$ 115–25–3 C_4F_8

		cgs	SI	
mw		200.031	200.031	3468
bp	1 atm	−5.99	267.16	2063,2616;1733,586,484
p	−26.17	303.3	40.44	2616
	30	2741	365.5	2063
eq 2.12	A	14.75709	13.88200	2063;2616
	B	2025.845	2025.845	
	C	0.021857	0.021857	
	D	2.1655×10^{-5}	2.1655×10^{-5}	
fp		−38.7	234.5	484;1733

286. Octafluorocyclobutane

(Continued)

		cgs	SI	
tp		-40.20	232.95	2616,586
d	25	1.500	1500	2142
	30	1.4506	1450.6	2063
n_D	25	1.217	1.217	2142
η	-11.14	1.1	0.0011	3955
	25	0.42	0.00042	2142
γ	0	10.57	0.01057	586
	20	8.41	0.00841	
ΔH_v	-11.9	5.669	23.721	2616
	bp	5.555	23.241	1733
ΔH_m		0.6616	2.7682	2616
ΔH_t	-131.9	0.9743	4.0764	2616
	-98.6	1.1967	5.0071	
	-58.32	1.0993	4.5997	
	-56.17	0.3296	1.3792	
ΔH_f°	25(g)	-355.7	-1488.2	4145;2169
ΔH_c°	25(g)	-271.8	-1137.4	2169
C_p°	10	36.19	151.42	4946
	25	38.0	159.0	2142;4946,2200
C_p	-4.64	50.14	209.77	2616
	25	53.21	222.62	2142
λ	25	0.00010	0.042	2142
t_c		115.22	388.37	7304,2063;586,4999
P_c		27.412	2.7775	7304,2063;586,4999
d_c		0.6159	615.9	7304,2063;4999
v_c		0.3248	0.0003248	7304,2063
μ		0	0	2006
δ	25	5.7	11.7	6535;6371

Beil	452,EIII 8
ir	1495,688,5138,2200
Raman	1495,688,5138,2200
ms	5203
nmr	688

287. Decafluorobutane

CF$_3$CF$_2$CF$_2$CF$_3$ 355-25-9   C$_4$F$_{10}$

		cgs	SI	
mw		238.028	238.028	3468
bp	1 atm	-2.19	270.96	6747;1152,6740,2503
p	-39.92	121	16.1	6747
eq 2.13	A	26.1077	25.2326	6747;1152
	B	2039.6	2039.6	
	C	6.4528	6.4528	
fp		-128.19	144.96	6747
d	0	1.592	1592	1152
	20	1.517	1517	
γ	-21.8	11.96	0.01196	4791
	-7.8	10.60	0.01060	
ΔH_v	0	5.46	22.84	1152;2503,6740
	20	5.03	21.05	
ΔH_f°	25(g)	-505.5	-2115.0	1218
t_c		113.2	386.4	1152,7303;2503
p_c		22.93	2.323	1152,7303;2503
d_c		0.629	629	7303;2503,1152
v_c		0.378	0.000378	7303;2503
δ	-13.2	6.16	12.60	6747;6535,6723
Beil	10,EIII 274			
ms	5203			
nmr	5321			

288. Fluorobenzene

C_6H_5F 462-06-6 C_6H_5F

		cgs	SI	
mw		96.104	96.104	3468
bp	1 atm	84.734	357.884	2062,6527;2537,5235,7059 7765,1671
dt/dp	1 atm	0.040	0.30	7482
dp/dt	1 atm	25	3.3	7482
p	25	78.63	10.48	6527
eq 2.5	A	6.95208	6.07698	6527;8298
	B	1248.083	1248.083	
	C	221.827	221.827	
fp		-42.21	230.94	6527;7059,7479
d	15	1.03091	1030.91	7482;487,1969,2537,7765,8231
	30	1.01314	1013.14	
dd/dt	25	0.00118	1.18	7482
α		0.00192	0.00192	4195
n_D	15	1.46837	1.46837	7482;487,2537,7059,7765,2664
dn/dt		0.00050	0.00050	7482
η	15	0.620	0.000620	7482
	30	0.517	0.000517	
γ	20	27.78	0.02778	7482;1969,7765,4195
	30	26.47	0.02647	
ΔH_v	45.25	7.996	33.46	6527;5153,7754,8298
	bp	7.457	31.20	
	Table 2.9			
ΔH_m		2.702	11.31	6527;7059,7478
ΔH_f°	25(1)	-34.75	-145.39	2856,6527;8316
ΔH_c°	25(1)	-741.86	-3103.94	2856;3975,8316
C_p°	25	22.54	94.31	1289;6527
	Table 2.14			
C_p	25	34.98	146.36	6527;7059
	Table 2.12			
t_c		286.94	560.09	4299;2062,5153,8231,8233,1671 119.5971
P_c		44.910	4.5505	2062;4299,5971,4088,2314
d_c		0.2688	268.8	2062;4299,5153,8233,1671
v_c		0.3541	0.0003541	5971;4299,4088
A	eq 2.72	0.02549	0.02549	6527
ϵ	25	5.42	5.42	4937;5905,7497,5023
μ	25 in 9	1.48	1.48	447;4488,2537,489,3593,4526 4533,5235
δ		9.11	9.11	6240;8111

288.   Fluorobenzene

(Continued)

		cgs	SI	
soly	in aq,30	0.153%w	0.153%w	2986
	aq in,25	0.0316%w	0.0316%w	8111
Beil	464,198			
uv	155,461,6825,6936,8152,6207,1978,6686,6749			
ir	4472,7077,3791,153,6777,5893,2961,140,106			
Raman	3825,4126,8122,4112,6777,5304			
ms	4785			
nmr	6295,1978,2318,5202,5645,5933,6748			

289.   *o*-Fluorotoluene

1-Fluoro-2-methylbenzene

o-CH$_3$C$_6$H$_4$F   95–52–3   C$_7$H$_7$F

		cgs	SI	
mw		110.131	110.131	3468
bp	1 atm	114.40	387.55	5927;5235,7058
p	21.4	20	2.7	7058
eq 2.5	A	6.9732	6.0981	5927
	B	1356.8	1356.8	
	C	217.15	217.15	
fp		−62.0	211.2	5927;7058,4056
d	13.2	1.0041	1004.1	Beil. H5, 290
	17.3	1.00124	1001.24	7482
n$_D$	21	1.4738	1.4738	7126
η	20	0.680	0.000680	7127
	30	0.601	0.000601	
ΔH$_v$	bp	8.456	35.38	5927
ΔH$^\circ_c$	25(1)	−903.2	−3779.0	2746
ε	30	4.22	4.22	4937;5353
μ	30	1.26	1.26	4739;4900,5235

289. *o*-Fluorotoluene

(Continued)

Beil	466,290
uv	582,6825,7502,576,6570,7172,3787,6749
ir	7435,2932,3786,5932
Raman	3336,4101,2932
ms	4785
nmr	6154,5319,5933,6748

290. *m*-Fluorotoluene

1-Fluoro-3-methylbenzene

m-CH₃C₆H₄F   352-70-5   C₇H₇F

		cgs	SI	
mw		110.131	110.131	3468
bp	1 atm	116.5	389.7	5927;5235,6435,7058
p	23.4	20	2.7	7058
eq 2.5	A	7.0095	6.1344	5927
	B	1382.7	1382.7	
	C	218.34	218.34	
fp		-87.7	185.5	5927;7058,7473,4056
d	20	0.9974	997.4	1969
n_D	27	1.46524	1.46524	6435
η	20	0.608	0.000608	7127
	30	0.534	0.000534	
γ	20	27.97	0.02797	1969
ε	30	5.42	5.42	4937;5353
μ	30	1.66	1.66	4739;5235

290. *m*-Fluorotoluene

(Continued)

Beil	466,290
uv	582,576,6570,7172,6749
ir	7435,4473,2931,5932
Raman	2125,2152,2931
ms	154,4785
nmr	6295,6154,6417,5933,6748

291. *p*-Fluorotoluene

1-Fluoro-4-methylbenzene

p-CH₃C₆H₄F   352–32–9   C₇H₇F

		cgs	SI	
mw		110.131	110.131	3468
bp	1 atm	116.6	389.8	5927;5235,6435,7058,7765
p	24.0	20	2.7	7058
eq 2.5	A	7.0574	6.1823	5927
	B	1410.8	1410.8	
	C	221.19	221.19	
fp		−56.8	216.4	5927;4056
d	20	0.9975	997.5	7765;2188
	26.9	0.9911	991.1	
dd/dt		0.00107	1.07	87
α	30 – 40	0.00108	0.00108	2188
n_D	20	1.46884	1.46884	7765;6435
η	20	0.622	0.000622	7127
	30	0.522	0.000522	
γ	26.9	27.34	0.02734	7765
	31.5	27.2	0.0272	87
	59.5	23.7	0.0237	
ΔH_f°	25(1)	−44.07	−184.39	8316

291.   *p*-Fluorotoluene

(Continued)

		cgs	SI	
ΔH_c°	25(1)	−884.03	−3698.78	8316;2746
ε	30	5.86	5.86	4937;5353
μ	30	1.76	1.76	4739;4060,5235,6304
Beil	466,290			
uv	582,576,6570,7172,6749			
ir	2407,7435,5305,2930,2961,7722,5911,5932			
Raman	3341,5720,2930			
ms	154,4785			
nmr	6418,6154,5319,3614,7918,5933,6748			

292.   α,α,α-Trifluorotoluene

Trifluoromethylbenzene

$C_6H_5CF_3$ 98-08-8 $C_7H_5F_3$

		cgs	SI	
mw		146.112	146.112	3468
bp	1 atm	102.050	375.200	6521;5927,2422,2537,986,7123
dt/dp	1 atm	0.04468	0.3351	2087
dp/dt	1 atm	22.38	2.984	2087
p	25	38.55	5.140	2087;986
eq 2.5	A	6.96911	6.09401	6521;5927,2087,2422,986
	B	1305.509	1305.509	
	C	217.280	217.280	
fp		−28.16	244.99	3103;5927,2087,986,6541
tp		−29.01	244.14	6521
d	20	1.18838	1188.38	2087;1969,6317,1163,2537
	25	1.18129	1181.29	1010,7123
	30	1.17351	1173.51	
dd/dt	25	0.00149	1.49	2087
α	30 − 40	0.00121	0.00121	6317

292. α,α,α-Trifluorotoluene

(Continued)

		cgs	SI	
n_D	20	1.41458	1.41458	2087;5749,1163,1010,1461
	25	1.41225	1.41225	2537,7125
dn/dt	25	0.000466	0.000466	2087
η	19.8	0.5740	0.0005740	7126
	60	0.3873	0.0003873	
γ	20	23.41	0.02341	2087;1969
	30	22.34	0.02234	
ΔH_v	25	8.960	37.489	6521;2087,2422,986
	bp	7.800	32.635	
ΔH_s		12.99	54.35	6541
ΔH_m		3.218	13.46	2087;6541
ΔH_f°	25(1)	−150.06	−627.85	8316;2856
	25(g)	−141.08	−590.28	8316;6521
ΔH_c°	25(1)	−771.00	−3225.86	8316;2856,4139
	25(g)	−779.99	−3263.48	8316
C_p°	25	31.17	130.42	6521
C_p	25	45.04	188.45	6521
t_c		289.5	562.7	2087
P_c		35.13	3.559	2087
d_c		0.427	427	2087
v_c		0.342	0.000342	2087
A	eq 2.72	0.02781	0.02781	652
K_f	eq 2.67	4.90	4.90	3103
ε	25	9.035	9.035	2087;1881,1461
μ	25 in 38	2.56	2.56	2537,1010,3311;1163,1312
	(g)	2.86	2.86	1635
soly	in aq	dec	dec	2087
fl pt	CC	12	285	6402

Beil	466,290
uv	5137,6273,4063,6749,7229
ir	7435,2940,4440,5397,3791,5932
Raman	2940,5397,5749
ms	7699
nmr	3045,4210,2607,5933,6748

293. Hexafluorobenzene

392–56–3 C_6F_6

		cgs	SI	
mw		186.056	186.056	3468
bp	1 atm	80.255	353.405	2069;1682,115,2314,5980,5694 8264,5707
dt/dp	1 atm	0.04072	0.3054	1682
dp/dt	1 atm	24.56	3.274	1682
p	24.005	80.504	10.733	115
eq 2.5	A	7.01741	6.14231	1682;2314,5707
	B	1219.410	1219.410	
	C	214.525	214.525	
fp		5.10	278.25	1682;5980,2120
tp		5.15	278.30	5066
d	20	1.61866	1618.66	3085;1863,1682,5980,8264
	25	1.60732	1607.32	5707,3079
	30	1.59588	1595.88	
dd/dt	25	0.002278	2.278	3085
α		0.001412	0.001412	1682
n_D	20	1.37761	1.37761	1863;5980,5694,8264
	25	1.37482	1.37482	
dn/dt	25	0.000558	0.000558	1863
η	20	0.931	0.000931	1863
	25	0.860	0.000860	
γ	20	22.23	0.02223	1863;5980
	25	21.64	0.02164	1863;5369,4788
ΔH_v	25	8.530	35.690	1682;4252,1702,5707
	bp	7.569	31.670	3504;1682
	Table 2.9			
ΔH_m		2.769	11.585	5066;1682,5714
ΔH_f°	25(1)	−237.02	−991.69	1703
	25(g)	−228.49	−956.00	1703;4252,1702,2935
ΔH_c°	25(1)	−584.03	−2443.58	1703;1702
C_p°	25	37.40	156.48	1289;1682,3504,2935
	Table 2.14			
C_p	25	52.966	221.610	5066;1682
t_c		243.63	516.78	3085;4299,1450,5707,2314,2285 2060,5300,1682
p_c		32.304	3.273	2060;5300,2285,4299,2314,1450

293. Hexafluorobenzene

(Continued)

		cgs	SI	
d_c		0.5504	550.4	3085;2285,5300
v_c		0.3351	0.0003351	2060
μ	25 in 38	0.33	0.33	2119
	25 in 9	0	0	2119,447
δ	25	8.1	16.6	6535

Beil	464,199
uv	5694,576
ir	1928,1863,5779,1929,1937,6982,5932
Raman	1928,1929,6982
ms	1995,1863,5942
nmr	3474,7021,3488,3614

294. 1,1-Difluoroethylene

1,1-Difluoroethene

CH$_2$=CF$_2$ 75-38-7   C$_2$H$_2$F$_2$

		cgs	SI	
mw		64.035	64.035	3468
bp	1 atm	-85.7	187.5	5019,2142;3292
eq 2.4	A	7.1766	6.3015	5019
	B	805	805	
d	3.76	0.795	795	5019
	25	0.585	585	2142
ΔH_v	25	0.980	4.100	5019
ΔH_f°	25(g)	-78.6	-328.9	7826;5460,4141,1703
ΔH_c°	25(g)	-262.15	-1096.84	5460;4141,1703
C_p°	25	14.36	60.08	7826
t_c		30.1	303.3	5019,2142,7308
P_c		43.75	4.433	5019,2142,7308
d_c		0.417	417	5019,2142,7308
v_c		0.1535	0.0001535	7308
μ	25 in 38	1.96	1.96	7690

294.   1,1-Difluoroethylene

(Continued)

Beil	11,186
uv	744
ir	6816,7530
Raman	6816,2201
ms	4598
nmr	2467,4746

295.   Chloromethane

CH_3Cl 74-87-3 CH_3Cl

		cgs	SI	
mw		50.488	50.488	3468
bp	1 atm	−24.20	248.95	7315;4771,5069,2731
dt/dp	1 atm	0.03500	0.2625	7315;2088
dp/dt	1 atm	28.57	3.810	7315
p	25	4309.7	574.6	2088
eq 2.5	A	7.04043	6.16533	7320;4577,7184,2651,5069
	B	920.86	920.86	
	C	245.58	245.58	
fp		−97.70	175.45	7315;5069,2778,4771,3966
tp		−97.72	175.43	5069
d	20	0.9214	921.4	7315;4771,5250
	25	0.9111	911.1	
dd/dt	25	0.00206	2.06	7315
α	−30 − 30	0.00209	0.00209	4771
n_D	20	1.3384	1.3384	7315;2510,2988,4771
	25	1.3344	1.3344	
dn/dt	25	0.00080	0.00080	7315
η	20	0.244	0.000244	4771
	30	0.197	0.000197	518
ΔH_v	25	4.521	18.92	2088
	bp	5.114	21.40	2088;4771,5069
ΔH_m		1.537	6.431	5069

295. Chloromethane

(Continued)

		cgs	SI	
ΔH_f°	25(g)	-19.590	-81.965	6221,7318,2459;4087,4362
ΔH_c°	25(g)	-182.60	-764.01	2459
C_p°	25	9.737	40.740	6221,7334;2706,3949
C_p	-27.9	18.06	75.56	5069;2303
	20	19.2	80.3	4771
t_c		143.1	416.3	7316,5838,1102,7184,4771
P_c		65.9	6.68	7316,5838,1102,7184,4771;758
d_c		0.353	353	7316,1102;5838
v_c		0.143	0.000143	7316,1102
A	eq 2.72	0.02512	0.02512	2779
ε	-25	12.93	12.93	4771;5250,7602
μ	(g)	1.869	1.869	6702;3111,4771
δ	25	9.7	19.8	653;6723
soly	in aq,30	0.648%w	0.648%w	7100
fl pt	OC	<0	<273	6402

Beil	5,59
uv	3300,3351,6038
ir	3469,5488,5845
Raman	5494,1785,7988,3278
ms	4785,3035,7199
nmr	4740,98,3256

296. Chloroethane

CH_3CH_2Cl 75-00-3 C_2H_5Cl

		cgs	SI	
mw		64.515	64.515	3468
bp	1 atm	12.27	285.42	7315;2870,7472,7491,4771,3525
dt/dp	1 atm	0.03500	0.2625	7315;2870,7491
dp/dt	1 atm	28.57	3.810	7315
p	25	1199.22	159.88	7320;2088,7058,7319,3525

296. Chloroethane

(Continued)

		cgs	SI	
eq 2.5	A	6.96598	6.09088	7320;4577,2870
	B	1020.63	1020.63	
	C	237.57	237.57	
fp		−136.4	136.8	7315;2870,7058,7472,7491,7479
d	0	0.92390	923.90	7491;4771
	20	0.89600	896.00	7315
	25	0.88891	888.91	
α	20	0.0021	0.0021	4902
n_D	0	1.3790	1.3790	2088;4771
	10	1.3738	1.3738	
	20	1.3680	1.3680	7315;2510
	25	1.3652	1.3652	
dn/dt	25	0.00056	0.00056	7315
η	5	0.292	0.000292	7491;4771
	10	0.279	0.000279	
γ	5	21.20	0.02120	7491;4771
	10	20.64	0.02064	
ΔH_v	30 Torr	6.64	27.8	2088;4771
	bp	5.892	24.652	2870;3706,1417,3871
ΔH_m		1.064	4.452	2870;516
ΔH_f°	25(1)	−32.63	−136.52	7826;362
	25(g)	−26.83	−112.26	7318,2459,362;1417,3525,4394
ΔH_c°	25(1)	−322.8	−1350	3975;7446,8316
	25(g)	−337.73	−1413.05	2459
C_p°	25	14.97	62.63	7334,1417;2946,7826,3525,3949
C_p	16.90	24.69	103.30	2870;4327,5817,4771
	25	24.94	104.35	7826
	Table 2.11			
λ	0(g)	2.220×10^{-5}	0.009288	5373
t_c		187.2	460.4	7322,822;4299,4088,4771
P_c		52	5.3	7322;4299,4771,288
K_B	eq 2.79	1.95	1.95	729
κ	0	$<3 \times 10^{-9}$	$<3 \times 10^{-7}$	7987
ε	20	9.45	9.45	5483;4937,4771
μ	20(1)	1.96	1.96	5483;489,2592,3701,4915,6862
δ	25	8.5	17.4	3395,4593
soly	in aq,0	0.447%w	0.447%w	6999;4771,2601
fl pt	TCC	−50	223	5712

Beil	8,82
uv	3573,6038
ir	2251,5849,4757,153,1947,89,5140,5932,2374
Raman	6643,7781,7832,4757,1783,89,5140
ms	4785
nmr	4740,3256,4852

297. 1-Chloropropane

$CH_3CH_2CH_2Cl$ 540-54-5 C_3H_7Cl

		cgs	SI	
mw		78.541	78.541	3468
bp	1 atm	46.52	319.67	7315;7488,4971,5329,7058 7472,7759
dt/dp	1 atm	0.03900	0.2925	7315;7488
dp/dt	1 atm	25.64	3.419	7315
p	25	344.4	45.92	7315;7058
eq 2.5	A	6.95165	6.07655	7315;4577,3938
	B	1125.09	1125.09	
	C	229.86	229.86	
fp		-122.8	150.4	7472,7488;7315,1256,7058
d	20	0.8899	889.9	7315;2027,5329,7488,7759
	25	0.8830	883.0	
dd/dt		0.00123	1.23	7488
n_D	20	1.3878	1.3878	7315;5329,7759
	25	1.3851	1.3851	
dn/dt		0.00054	0.00054	7480
η	15	0.372	0.000372	7480;5329,7404
	30	0.318	0.000318	
	Table 2.2			
γ	20	21.78	0.02178	7480,3283;5329,7759
	30	20.48	0.02048	
ΔH_v	25	6.812	28.50	4878;4971
	bp	6.51	27.24	2089
ΔH_m		1.325	5.544	Beil. EIII1, 219
ΔH_f°	25(1)	-38.2	-159.8	8316
	25(g)	-31.83	-133.18	7318;2459
ΔH_c°	25(1)	-483.0	-2020.9	8316;6840,7446,3975
	25(g)	-495.25	-2072.11	2459
C_p°	25	20.38	85.29	7317;2946,3949
C_p	25	31.6	132.2	4327
	Table 2.11			
t_c		230.0	503.2	822,7322;4299
p_c		45.18	4.578	822,7322;4299
ε	20	7.7	7.7	4937;2027,3532
μ	(1)	1.97	1.97	4915;6862,4499
δ		8.5	17.4	4593,652
soly	in aq,20	0.271%w	0.271%w	5712;2601
aq az	44	97.8%w	97.8%w	3498;3497
fl pt	TCC	<-18	<255	5712

297.   1-Chloropropane

(Continued)

Beil	10,104
uv	3573,3574,6038,7229
ir	7077,4757,5711
Raman	5385,6018,4757,1783
ms	154,4785
nmr	1396,4886,6748

298.   2-Chloropropane

CH$_3$CHClCH$_3$ 75-29-6   C$_3$H$_7$Cl

		cgs	SI	
mw		78.541	78.541	3468
bp	1 atm	35.74	308.89	7321;2999,3564,7759,7495,3525
dt/dp	1 atm	0.039	0.29	7321;7495
dp/dt	1 atm	26	3.4	7321
p	25	515.3	68.70	2089;7058
eq 2.5	A	6.2394	5.3643	7323
	B	779.7	779.7	
	C	196.5	196.5	
fp		−117.18	155.97	7321;7472,7495
d	20	0.8617	861.7	7321
	25	0.8563	856.3	
dd/dt	25	0.00108	1.08	7321
n$_D$	20	1.3777	1.3777	7321;3525
	25	1.3749	1.3749	
dn/dt		0.00056	0.00056	7495

298. 2-Chloropropane

(Continued)

		cgs	SI	
η	15	0.335	0.000335	7495
	30	0.286	0.000286	
	Table 2.2			
γ	20	18.09	0.01809	2089;7759
	30	17.13	0.01713	
ΔH_v	25	6.42	26.9	2089
	bp	6.28	26.3	
ΔH_f°	25(1)	-39.2	-164.0	8316;3525
	25(g)	-34.66	-145.02	2459;5510
ΔH_c°	25(1)	-482.0	-2016.7	8316;6840
	25(g)	-492.26	-2059.62	2459
C_p°	25	20.93	87.57	3525
t_c		212	485	2088
p_c		46.6	4.72	2088
d_c		0.341	341	2088
v_c		0.230	0.000230	2088
A	eq 2.72	0.03661	0.03661	2088
ε	20	9.82	9.82	Beil. EIII1, 222
μ	(1)	2.02	2.02	4915;2999,6862
δ	25	8.1	16.6	653
soly	in aq,12.5	0.342%w	0.342%w	2601
aq az	35.0	99%w	99%w	3498;3497
fl pt	CC	-32	241	152

Beil	10,105
uv	3573,3574,6038,7229
ir	4757,153,4048,1765
Raman	6018,4757,4048
ms	154,4785
nmr	8008,1396,3256,4852

299. 1-Chlorobutane

$CH_3CH_2CH_2CH_2Cl$ 109-69-3 C_4H_9Cl

		cgs	SI	
mw		92.568	92.568	3468
bp	1 atm	78.43	351.58	7315;4971,5329,6869,7581 7759,8119
dt/dp	1 atm	0.04200	0.3150	7315;2090,6968
dp/dt	1 atm	23.81	3.175	7315
p	24.47	100	13.3	7319;1348,2091,6864,7058
eq 2.5	A	6.92664	6.05154	7320;4577,2090,3938
	B	1216.82	1216.82	
	C	222.33	222.33	
fp		-123.1	150.1	7315,1256,7473,6968,7581
d	20	0.8857	885.7	7315;489,2090,2216,2999,5329 7495,7759,3130
	25	0.88095	880.95	1687,3318;7315
dd/dt	20 - 30	0.001111	1.111	7616
α		0.00080	0.00080	4902
n_D	20	1.4023	1.4023	7315,2090,489,5329,6864,6869 6874,7759,8119
	25	1.39996	1.39996	1687,3318
dn/dt		0.00051	0.00051	<u>7482</u>;6968
η	15	0.469	0.000469	7495,6968;5329
	25	0.426	0.000426	1687
γ	15	24.40	0.02440	7495,6968;3283,5329,7759
	30	22.88	0.02288	
ΔH_v	25	8.011	33.52	7822;4971
	bp	7.17	30.0	<u>2088</u>
ΔH_f°	25(1)	-44.97	-188.15	7047
	25(g)	-36.83	-154.10	7318;7047
ΔH_c°	25(1)	-644.32	-2695.83	7047;6840
C_p°	25	25.80	107.93	7317;2946
C_p	20	41.75	174.68	4902
t_c		269.0	542.2	2088
p_c		36.37	3.685	<u>2088</u>
d_c		0.207	207	2088
v_c		0.3119	0.0003119	<u>2088</u>
κ	30	10^{-10}	10^{-8}	5485;3130
ε	20	7.39	7.39	4937;5251,6874,3130
μ	-90 - 70 in 16	1.90	1.90	6874;487,2999,5697,6862 6869,8119
δ	25	8.4	17.1	653
soly	in aq,20	0.11%w	0.11%w	1348;2601,7219
	aq in,20	0.08%w	0.08%w	

299. 1-Chlorobutane

(Continued)

		cgs	SI	
aq az	68.1	93.4%w	93.4%w	3497
fl pt	CC	−28	245	5712

Beil	10,118
uv	3573,3574,6038,7229
ir	7077,5932
Raman	7781
ms	4785
nmr	64,5933,6748

300. 2-Chlorobutane

$CH_3CH_2CHClCH_3$ 78-86-4 C_4H_9Cl

		cgs	SI	
mw		92.568	92.568	3468
bp	1 atm (active)	68.25	341.40	7495,7476;2216,6877,8119
dt/dp	1 atm	0.041	0.31	7495;7321
dp/dt	1 atm	24	3.3	7495
p	25	151.6	20.21	2089
eq 2.5	A	6.9973	6.1222	7323
	B	1245.2	1245.2	
	C	234.4	234.4	
fp	racemic	−113.3	159.9	7472,1256,7495
	active	−140.5	132.7	
d	0,racemic	0.87556	875.56	7495;2216,7759
	15,racemic	0.87880	878.80	
	20,racemic	0.87323	873.23	
	30,racemic	0.86210	862.10	
	0,active	0.89497	894.97	7495
n_D	20	1.3971	1.3971	7321,7495,7759,8119
	25	1.3942	1.3942	

300. 2-Chlorobutane

(Continued)

		cgs	SI	
dn/dt		0.00048	0.00048	7495
η	15	0.439	0.000439	7495;4349
	30	0.363	0.000363	
γ	19.7	21.84	0.02184	7759
	26.0	21.32	0.02132	
ΔH_v	25	7.543	31.56	7822
	bp	6.98	29.2	2089
ΔH_c	18(1)	−642.10	−2686.54	6840
t_c		247.5	520.7	7322
$[\alpha]_D$	20	−8.48	−8.48	7495
κ	30	$<10^{-7}$	$<10^{-5}$	5485
ε	30	7.090	7.090	6874
μ	25 in 38	2.07	2.07	6234;5697,6862,6874,8119,653
δ	25	8.1	16.6	
soly	in aq,25	0.1%w	0.1%w	Beil. EIII1, 278
	aq in,25	<0.1%w	<0.1%w	6165
aq az	61.1	94.3%	94.3%	6165
fl pt	TCC	−29	244	3125
Beil	10,119			
uv	3574,7229			
ir	4757,5384,5932			
Raman	575,4757,1793,5384			
ms	4785			
nmr	3805,6499,5933,6748			

301. 1-Chloro-2-methylpropane

$(CH_3)_2CHCH_2Cl$ 513-36-0 C_4H_9Cl

		cgs	SI	
mw		92.568	92.568	3468
bp	1 atm	68.85	342.0	7321,7494,7472;2216,5329 7866,7759
dt/dp	1 atm	0.042	0.32	7321,7494
dp/dt	1 atm	24	3.2	
eq 2.4	A	7.6672	6.7921	7058
	B	1641.4	1641.4	
fp		−130.3	142.9	7321;956,7432,7494,7581
d	20	0.8773	877.3	7321;2216,5329,7759,7494,4349
	25	0.8717	871.7	
dd/dt	25	0.00112	1.12	7321
n_D	20	1.3980	1.3980	7321;5329,7494,7759,8119
	25	1.3951	1.3951	
dn/dt		0.00052	0.00052	7494
η	15	0.471	0.000471	7494;5329,7453,4349
	30	0.395	0.000395	
	Table 2.2			
γ	20	21.99	0.02199	3283;5329,7759,3163
	30	20.87	0.02087	
ΔH_v	25	7.569	31.67	7822
	bp	7.91	33.1	7058
ΔH_c°	25(1)	−635.5	−2658.9	3975;6840,7446
C_p	20	41.75	174.68	7490;4327
	Table 2.11			
κ	30	$<10^{-7}$	$<10^{-5}$	5485
ε	14	6.49	6.49	4937;4671
μ	10 - 50 in 38	2.09	2.09	5697,6862;8119
δ		8.1	16.6	4593
soly	in aq,12.5	0.092%w	0.092%w	2601
aq az	61.9	92.1%w	92.1%w	6165;3497

Beil	10,124
uv	4953,7229
ir	2251,4757,5932
Raman	575,4757,4183
ms	4785
nmr	4221,5933,6748

302. 2-Chloro-2-methylpropane

$(CH_3)_3CCl$ 507–20–0 C_4H_9Cl

		cgs	SI	
mw		92.568	92.568	3468
bp	1 atm	50.7	323.9	7321,7485;4346
dt/dp	1 atm	0.040	0.30	7321
dp/dt	1 atm	25	3.3	7321
p	25	294.8	39.30	2089
eq 2.4	A	7.4644	6.5893	7058
	B	1486.61	1486.61	
fp		−25.4	247.8	7321;4346,7479,2018,7485
d	20	0.8420	842.0	7321;4349,7485
	25	0.8361	836.1	
dd/dt	25	0.00118	1.18	7321
n_D	20	1.3857	1.3857	7321;4346
	25	1.3828	1.3828	
dn/dt		0.00052	0.00052	7482;3524
η	15	0.543	0.000543	7485;4349
	30	0.439	0.000439	
γ	15	20.06	0.02006	7485;7759,3163
	30	18.35	0.01835	
ΔH_v	25	6.926	28.98	7822
	bp	6.55	27.4	2089
ΔH_m		0.500	2.09	7478;4346,2018
ΔH_f°	25(g)	−42.99	−179.87	3525;3524,8316
ΔH_c°	25(g)	−640.6	−2680.3	8316;6840
C_p°	25	27.30	114.22	3525
κ	30	$<10^{-7}$	$<10^{-5}$	5485
ε	20	9.961	9.961	487;4937,5251,7497
μ	20 in 38	2.15	2.15	5697;6914,489,525,6862,7581
				8119,4492
aq az	47 - 48	hydrolyzes	hydrolyzes	6165
fl pt	TCC	−5	268	6168

Beil	10,125
ir	4757,2325,4530,4475
Raman	4757,1783
ms	4785
nmr	3256,4852,5933,6748

303.   1-Chloropentane

$CH_3(CH_2)_3CH_2Cl$ 543-59-9 $C_5H_{11}Cl$

		cgs	SI	
mw		106.595	106.595	
bp	1 atm	108.39	381.54	7315,6186;2999,4971,5329 6733,7759
dt/dp	1 atm	0.04700	0.3525	7315
dp/dt	1 atm	21.28	2.837	7315
p	25	31.07	4.142	2089;7319,6186
eq 2.5	A B C	6.81151 1271.16 215.00	5.93641 1271.16 215.00	7320;4577,6186
fp		-99.0	174.2	7315;6733
d	20 25	0.8820 0.87700	882.0 877.00	7315;2999,5329,6186,6733,7759 1687,3318;7315
n_D	20 25	1.4126 1.41000	1.4126 1.41000	7315;5329,6733,7759 1687,3318;7315
η	20 25	0.580 0.5461	0.000580 0.0005461	5329;6733 1687,3318;5329
γ	20 25	25.15 24.55	0.02515 0.02455	5329;3283,7759
ΔH_v	25 bp	9.140 7.82	38.24 32.7	7822;4971 <u>2089</u>
ΔH_f°	25(1) 25(g)	-51.01 -41.83	-213.44 -175.02	7047 7318;7047
ΔH_c°		-800.64	-3349.88	7047;6840
C_p°	25	31.21	130.57	7317;2946
C_p	20	46.90	196.23	<u>6352</u>
λ	37.8	0.000279	0.117	<u>6352</u>
ε	11	6.6	6.6	4937
μ	20 in <u>38</u>	1.94	1.94	4739;6564,2999,6862
δ	25	8.3	17.0	653
soly	in aq,25	0.02%w	0.02%w	Beil. EIII1, 339
aq az	82	67.9%w	67.9%w	3497
fl pt	CC	13	286	5712

Beil	<u>10</u>,130
uv	7229
ir	4993
Raman	1793,4993
ms	4785
nmr	6748

304. 1-Chloro-3-methylbutane

$(CH_3)_2CHCH_2CH_2Cl$ 107–84–6 $C_5H_{11}Cl$

		cgs	SI	
mw		106.595	106.595	3468
bp	1 atm	98.2	371.4	7321;7582,6264,1621
dt/dp	1 atm	0.045	0.34	7321
dp/dt	1 atm	22	3.0	7321
fp		−104.4	168.8	7321,7582
d	20	0.8750	875.0	7321;7759,3163,6264,7540
	25	0.8700	870.0	
dd/dt	25	0.0010	1.0	7321
n_D	20	1.4087	1.4087	7321;7759,7582
	25	1.4063	1.4063	
dn/dt	25	0.00048	0.00048	7321
η	25	5.00	0.00500	1626
γ	18.3	23.57	0.02357	7759;3163
	25	22.90	0.02290	7540
ΔH_c°	18	−798.6	−3341.3	6840
C_p	25	41.86	175.15	4327
	Table 2.11			
ε	23.6	5.97	5.97	7582
μ	20 in 38	1.92	1.92	5699
Beil	10,135			
ir	4467			
Raman	1793,3164			
ms	4785			

305. 1-Chlorooctane

$CH_3(CH_2)_6CH_2Cl$ 111–85–3 $C_8H_{17}Cl$

		cgs	SI	
mw		148.675	148.675	3468
bp	1 atm	183.47	456.62	7315;5329,1621,1503,8301
dt/dp	1 atm	0.05300	0.3975	7315;2089
dp/dt	1 atm	18.87	2.516	7315
p	25	0.95	0.13	2089
eq 2.5	A	6.96168	6.08658	7320;3938,4577
	B	1541.08	1541.08	
	C	194.16	194.16	
fp		−57.8	215.4	7315,2089
d	20	0.8734	873.4	7315;5329,7759,7047,3318
	25	0.8692	869.2	
dd/dt	25	0.00084	0.84	<u>7315</u>
n_D	20	1.4309	1.4309	7315;7047,7759,3318,5911
	25	1.4287	1.4287	
dn/dt	25	0.00044	0.00044	<u>7315</u>
η	20	1.23	0.00123	5329;3318,5911,3361
	25	1.135	0.001135	
γ	20	27.65	0.02765	5329;7759
	25	27.15	0.02715	
ΔH_v	25	12.73	53.25	2089;7822
	bp	9.37	39.20	
ΔH_f°	25(1)	−69.62	−291.30	7047
	25(g)	−57.09	−238.88	7047;7318
ΔH_c°	25(1)	−1269.13	−5310.04	7047
C_p°	25	47.44	198.49	7317
κ	30	1×10^{-7}	1×10^{-5}	5485
ϵ	25	5.05	5.05	3361
μ	20(1)	1.999	1.999	2697,3361
Beil	<u>10</u>,159			
ir	5932			
Raman	4102			
ms	4785,7226			
nmr	5933			

306.   3-Chloromethylheptane

$CH_3(CH_2)_3CH(CH_2Cl)CH_2CH_3$ 123–04–6 $C_8H_{17}Cl$

		cgs	SI	
mw		148.675	148.675	3468
bp	1 atm	173.0	446.2	7616;5932
dt/dp	1 atm	0.052	0.39	7616
dp/dt	1 atm	19	2.6	7616
p	20	1.2	0.16	7616
fp		−135	138	7616
d	20	0.8817	881.7	7616;2839
dd/dt	20 – 30	0.00086	0.86	7616
n_D	20	1.4324	1.4324	7616;5932
η	20	1.0	0.0010	7616
ΔH_v	bp	9.42	39.41	7616
soly	in aq,20	0.01%w	0.01%w	7623
	aq in,20	0.04%w	0.04%w	
aq az	97.3	45%w	45%w	3498
fl pt	TOC	58	331	7623
Beil	10,EIII 475			
ir	5932			
ms	7226,4785			

307. Chlorobenzene

C_6H_5Cl 108–90–7 C_6H_5Cl

		cgs	SI	
mw		112.559	112.559	3468
bp	1 atm	131.687	404.837	8304;2986,5235,7059,7478,7494 2314,4535
dt/dp	1 atm	0.0488	0.366	8304;2090,7494
dp/dt	1 atm	20.5	2.73	8304
p	25	11.75	1.567	2087;2026,4912,6090,7058,8304
eq 2.5	A	7.18473	6.30963	2090;8298,4535,4011
	B	1556.6	1556.6	
	C	230	230	
fp		−45.58	227.57	2090;1090,6783,7473,7480 7484,7499
d	15	1.11172	1111.72	7494;887,2090,2537,2739,3333
	20	1.10630	1106.30	4001,4969,5329,7499,7707 7766,6791
	25	1.1009	1100.9	1965,5429;3183
	30	1.09550	1095.50	7494
dd/dt	15 – 30	0.001081	1.081	7494
α	25	0.000990	0.000990	1965;4195,4902
κ_T	20	9.75×10^{-8}	7.31×10^{-7}	1880
	25	1.028×10^{-7}	7.71×10^{-7}	1965
κ_S	25	7.58×10^{-8}	5.68×10^{-7}	1966,1965
	30	7.753×10^{-8}	5.815×10^{-7}	7157
n_D	15	1.52748	1.52748	6639;1057,1850,2090,3361 4969,5329
	20	1.52481	1.52481	7766;6584,2664
	25	1.52185	1.52185	4535;1414
dn/dt		0.00054	0.00054	7494
η	20	0.799	0.000799	7505;2125,3361,4257,5329 7494,2488
	30	0.7184	0.0007184	6759;7505
γ	20	32.96	0.03296	4195;5495,3283,5329,7766 1985,7864
	30	31.98	0.03198	46;3283
	Table 2.7			
ΔH_v	25	9.792	40.97	7822;3759,5120,7848,7754
	bp	8.735	36.55	4969
ΔH_m		2.284	9.556	7059;7478
ΔH_f°	25(1)	2.55	10.67	4147,8316
ΔH_c°	25(1)	−743.04	−3108.88	6840,4147,3807,7446,8316
C_p°	25	23.42	97.98	1289;2806
C_p	25	35.57	148.83	4011
	26.84	35.90	150.21	7059;5817,6440,8081

307. Chlorobenzene

(Continued)

		cgs	SI	
t_c		359.2	632.4	4088;4299,113,923,7848,8231 8233,1671
p_c		44.6	4.52	4088;4299,4026,5153,8231 5971,2314
d_c		0.365	365	4088;4299,5971
v_c		0.308	0.000308	4088;4299,1671,8233
K_B	eq 2.79	4.15	4.15	3406;7973
κ	25	7×10^{-11}	7×10^{-9}	6027;7986,6791,3043
ϵ	25	5.621	5.621	4937;1518,3360,3361,5251,8084 7497,3624
	30	5.552	5.552	6483;6760
μ	25 in 38	1.62	1.62	3765;4488,1518,2442,3462,5251 5318,6497,5858,5354,4490 1716,1123,3538
δ		9.68	19.81	4024,6240,3759;8111,7159,898 1271,3395,4593,652
soly	in aq,30 aq in,25	0.0488%w 0.0327%w	0.0488%w 0.0327%w	2986;987,400,7219 8111;3759,4024
aq az	90.2	71.6%w	71.6%w	3497
fl pt	CC	29	302	5712

Beil	464,199
uv	155,460,2054,5276,6936,6207,1830,6686,6749
ir	153,627,1044,3942,4472,5890,7077,5171,5241,3158,633,228,8188,6777,2961
Raman	1967,2390.3825,4126,6938,7754,8122,6599,4401,1793,5103,6777,1782,1790
ms	4785
nmr	1429,710,4477,2318,5933,6748

308.   α-Chlorotoluene

(Chloromethyl)benzene

$C_6H_5CH_2Cl$ 100–44–7 C_7H_7Cl

		cgs	SI	
mw		126.585	126.585	3468
bp	1 atm	179.4	452.6	7058;7765,6847
p	47.8	4.81	0.641	465;7058
eq 2.5	A	6.801	5.926	465;3087
	B	1477	1477	
	C	193.7	193.7	
fp		−39.2	234.0	7490;6847
d	20	1.1004	1100.4	7765;7490
	40.7	1.0806	1080.6	
dd/dt	20 – 40	0.00096	0.96	7765
α		0.000972	0.000972	7490
n_D	15	1.54124	1.54124	7490
	20	1.53887	1.53887	7765
η	15	1.501	0.001501	7490
γ	20.6	37.46	0.03746	7765;7490
	26.3	36.76	0.03676	
ΔH_v	25	12.3	51.5	1703;465
ΔH_f°	25(1)	−7.8	−32.6	4028,1703
	25(g)	4.5	18.8	1703
ΔH_c°	25(1)	−38.2	−159.8	1703
C_p	25	43.6	182.4	6847
ε	13	7.0	7.0	4937
μ	25 in 38	1.82	1.82	7093;5700
δ	25	9.9	20.3	653
soly	in aq,20	0.0493%w	0.0493%w	5569
fl pt	CC	60	333	6402

Beil	466,292
uv	7229,7114,1337,6686,6749
ir	4440,3790,752,5932
Raman	6130,1783
ms	8213
nmr	7227,1410,8282,6748,5933

309. *o*-Chlorotoluene

1-Chloro-2-methylbenzene

o-CH₃C₆H₄Cl   95–49–8   C₇H₇Cl

		cgs	SI	
mw		126.585	126.585	3468
bp	1 atm	159.15	432.30	7488;7058,7057
dt/dp	1 atm	0.0517	0.388	2087
dp/dt	1 atm	19.3	2.58	2087
p	25	3.619	0.4825	2087
eq 2.5	A	7.1545	6.2794	2396;7057,2087
	B	1637.6	1637.6	
	C	223.3	223.3	
fp		−36.5	236.7	7477
d	20	1.08245	1082.45	2087;4373,7488
		1.07762	1077.62	
dd/dt	25	0.000972	0.972	2087
n$_D$	20	1.52680	1.52680	2087;3789
	25	1.52221	1.52221	
dn/dt	20 – 30	0.00047	0.00047	3789
η	20	1.022	0.001022	879;7488
	30	0.8928	0.0008928	
γ	20	33.44	0.03344	3283
	30	32.33	0.03233	
ΔH$_v$	25	10.905	45.628	2087
	bp	8.971	37.535	2087;4969
ΔH$_m$		2.000	8.368	7477
ΔH$_c^\circ$	18	−895.98	−3748.79	6840
C$_p$	0	39.9	166.8	2087
t$_c$		385.9	659.1	2087
P$_c$		37.98	3.848	2087
d$_c$		0.348	348	2087
v$_c$		0.364	0.000364	2087
ε	25	4.73	4.73	2087
μ	20 in 38	1.43	1.43	7461

Beil	466,290
uv	7229,7115,2476,8142,6749
ir	5305,2932,2532,5932
Raman	4112,2932
ms	4785,4243
nmr	7227,8200,5383,6986,6748,5933

310. *m*-Chlorotoluene

1-Chloro-3-methylbenzene

m-CH$_3$C$_6$H$_4$Cl   108–41–8   C$_7$H$_7$Cl

		cgs	SI	
mw		126.585	126.585	3468
bp	1 atm	162.3	435.5	7058;7057
p	30.3	5.0	0.67	7058
eq 2.5	A	7.62515	6.75005	7057
	B	1887.31	1887.31	
	C	239.75	239.75	
fp		−26.25	246.90	7477
d	20	1.0728	1072.8	4373
	30	1.0628	1062.8	
dd/dt	25	0.0010	1.0	4373
n$_D$	20	1.5224	1.5224	3789
	30	1.5179	1.5179	
dn/dt	20 – 30	0.00045	0.00045	3789
η	20	0.8771	0.0008771	879
	30	0.7641	0.0007641	
ΔH$_m$		2.500	10.460	7477
ΔH$^{\circ}_c$	18	−896.49	−3750.90	6840
ε	20	5.55	5.55	4937
μ	25 in 38	1.82	1.82	3765;7461,1311
Beil	466,291			
uv	7229,7115,8142,2476,6749			
ir	2931,5305,2532,5932			
Raman	3335,2931,4112,5394			
ms	4785,4243			
nmr	7227,6986,5383,8200,6748,5933			

311. *p*-Chlorotoluene

1-Chloro-4-methylbenzene

p-CH$_3$C$_6$H$_4$Cl   106-43-4   C$_7$H$_7$Cl

		cgs	SI	
mw		126.585	126.585	3468
bp	1 atm	161.99	435.14	7057;7765,7058
p	31.0	5.0	0.67	7058
eq 2.5	A	7.3210	6.4459	2396;7057
	B	1776.5	1776.5	
	C	237.2	237.2	
fp		7.30	280.45	7057;748
d	20	1.0697	1069.7	4373;7765
	30	1.0596	1059.6	
dd/dt		0.00100	1.00	2396
n$_D$	20	1.5209	1.5209	3789;6180,7765
	30	1.5165	1.5165	
dn/dt	20	0.000504	0.000504	6180;3789
η	20	0.8925	0.0008925	879
	30	0.7843	0.0007843	
	Table 2.6			
γ	18.3	32.90	0.03290	7765
	26.0	32.28	0.03228	
ΔH$_v$	bp	9.258	38.737	4969
ΔH$_c^\circ$	18	-897.59	-3755.51	6840
K$_f$	eq 2.67	5.53	5.53	748
ε	20	6.08	6.08	4937
μ	20 in 38	1.94	1.94	7461
δ	25	8.8	18.0	653
aq az	95	na	na	3497

Beil	466,292
uv	7229,2476,8142,7115,6749
ir	2961,5305,2930,7722,2532,5932
Raman	5394,4112,2930,5720
ms	4243,4785
nmr	7227,8200,5383,6986,6748,5933

312.   1-Chloronaphthalene

HC=C−C=C
HC=C−C=CH (structure diagram)

90–13–1

$C_{10}H_7Cl$

		cgs	SI	
mw		162.618	162.618	3468
bp	1 atm	259.3	532.5	3815;6722,505,7058
dt/dp	1 atm	0.062	0.47	3815
dp/dt	1 atm	16	2.2	3815
p	60	0.39	0.052	3779;3373,7058,3815,4056,458
fp		−2.3	270.9	4026;5698
d	15	1.1976	1197.6	6722;505,3361
	20	1.19382	1193.82	3815
	80	1.1426	1142.6	4195
	40	1.1778	1177.8	458
dd/dt		0.00077	0.77	458
α		0.000699	0.000699	4195
n_D	20	1.63321	1.63321	3815;458
η	25	2.940	0.002940	3361;3295
γ	20	41.04	0.04104	4195;458
	40	39.97	0.03997	458
ΔE_v	105–140	13.8	57.7	3815
	bp	12.44	52.05	3815
ΔH_c°	18	−1197.3	−5009.5	6840
ε	25	5.04	5.04	3361;4937
μ	25(1)	1.33	1.33	3361;5390,5698
	25 in 38	1.57	1.57	2605
aq az	99.9	3.4%w	3.4%w	6165
fl pt	TCC	79	352	6168

Beil	477,541
uv	5990,2408,2379,6749
ir	1515,1887,2409,6757,5098,5932
Raman	4704,1515,1887,6547,7172, 5098,6748
ms	3717
nmr	2261

313. Chloroethylene

Chloroethene

$CH_2 = CHCl$ 75-01-4 C_2H_3Cl

		cgs	SI	
mw		62.499	62.499	3468
bp	1 atm	-13.80	259.35	4760;7324,2091,1803,1815
dt/dp	1 atm	0.03423	0.2567	2088;7324
dp/dt	1 atm	29.21	3.895	2088
p	25	2660	354.6	2088
eq 2.5	A	6.86108	5.98598	4760,2088,1803,1815,3871
	B	892.757	892.757	
	C	238.099	238.099	
fp		-153.79	119.36	2088,7324;1803
d	20	0.9106	910.6	7324;1803
	25	0.9013	901.3	
dd/dt	25	0.00191	1.91	1803
n_D	20	1.3682	1.3682	2510;7324
	25	1.3642	1.3642	
η	-20	0.2734	0.0002734	2088;6378
γ	-20	22.27	0.02227	2088
	-10	20.88	0.02088	
ΔH_v	25	4.454	18.636	2088;1803,3871,1815
	bp	4.971	20.799	
ΔH_m		1.134	4.745	2088
ΔH_f°	25(1)	3.5	14.6	7826
	25(g)	8.072	33.77	4363;4087,4537,7826,2822,85
ΔH_c°	25(g)	-298.7	-1249.8	8316
ΔH_p	25	-31.5	-131.8	6768
C_p°	25	12.84	53.72	7826,3020;6153
C_p	27	21.47	89.83	4443
t_c		156.5	429.7	2088
P_c		55.3	5.60	2088
d_c		0.370	370	2088
v_c		0.169	0.000169	2088
A	eq 2.72	0.04002	0.04002	2088
ε	17.2	6.26	6.26	2088
μ	in 172	1.3	1.3	4508
δ	25	7.8	16.0	653
soly	in aq,25 (g)	0.27%w	0.27%w	3244
fl pt	COC	-78	195	6402

313. Chloroethylene

(Continued)

Beil	11,186
uv	4364
ir	3020,8173
Raman	3020,2323,4124,7988
ms	7199
nmr	8002

314. 3-Chloropropene

$CH_2{=}CHCH_2Cl$ 107–05–1 C_3H_5Cl

		cgs	SI	
mw		76.526	76.526	3468
bp	1 atm	45.10	318.25	7489;3699,5329,6625,7471,8169
dt/dp	1 atm	0.042	0.32	7489
dp/dt	1 atm	24	3.2	7489
p	25	366.8	48.90	2089;3626,6625,7058
eq 2.4	A	7.848	6.973	3779
	B	1577	1577	
fp		−134.5	138.7	7484,7489;6625,7058,7471,7479
d	15	0.94419	944.19	7489;3699,5329,6625
	30	0.92454	924.54	
dd/dt	15 – 30	0.00131	1.31	7489
n_D	20	1.41566	1.41566	3699;8169,5329,6625
	25	1.4130	1.4130	4026
dn/dt		0.00069	0.00069	7482
η	15	0.347	0.000347	7489;5329
	25	0.313	0.000313	8169
	30	0.300	0.000300	7489
	Table 2.2			

314. 3-Chloropropene

(Continued)

		cgs	SI	
γ	15 30	28.85 21.83	0.02885 0.02183	7489;3699,5329
ΔH_v	25 bp	6.75 6.940	28.24 29.036	2089 4026;3626
ΔH_f°	25(g)	−7	−29	4948
ΔH_c°	25(g)	−440.8	−1844.3	3975;7446
C_p°	0	24.1	100.8	6625
t_c		ca 241	ca 514	4026
ε	20	8.2	8.2	4937;2027,7850
μ	104(g)	1.98	1.98	3136
δ	25	8.8	18.0	653
soly	in aq,20 aq in,20	0.36%w 0.08%w	0.36%w 0.08%w	4026;6625
aq az	43.0	97.8%w	97.8%w	6625
fl pt	CC	−32	241	5712

Beil	11,198
uv	757
ir	7441,5711,153,2982,5932
Raman	5797
ms	7750
nmr	3883,5933,6748

315. Dichloromethane

CH$_2$Cl$_2$		75–09–2		CH$_2$Cl$_2$
		cgs	SI	
mw		84.933	84.933	3468
bp	1 atm	39.64	312.79	7332;6758,1009,4935,4969,5329 5777,7058,7771
dt/dp	1 atm	0.0369	0.277	7332;7489
dp/dt	1 atm	27.1	3.61	7332
p	25	435.8	58.10	2089;7058,5777,7333
eq 2.5	A	6.95132	6.07622	7333;1009,2651
	B	1070.07	1070.07	
	C	223.24	223.24	
fp		−94.92	178.23	7332;5277,2089,7325,3534,1090 7471,7489,7479
d	15	1.33479	1334.79	7489;612,2125,2216,2677 5250,5329
	20	1.3256	1325.6	7332;4195
	25	1.31678	1316.78	7489;7771,2958,4771,7332
	30	1.30777	1307.77	7489;6483
dd/dt		0.00180	1.80	7489
α		0.001391	0.001391	4195;7489,4771
κ$_T$	25	1.368 x 10^{-7}	1.026 x 10^{-6}	1262
n$_D$	20	1.42416	1.42416	2088;4935,5250,5251,5329
	25	1.42115	1.42115	7771,7332
dn/dt		0.00055	0.00055	7482
η	15	0.449	0.000449	7489;2125,2677,5329,7453 2958,4771
	27.61	0.4043	0.0004043	5827
	30	0.393	0.000393	7489
	Table 2.2			
γ	20	27.89	0.02789	4195;7489,3283,5329,7771 4771,3163
	30	26.54	0.02654	7489
ΔH$_v$	25	6.825	28.56	2089;7648
	bp	6.688	27.98	4969;5777,4771
ΔH$_m$		1.472	6.160	5277;7478,7671
ΔH$_f^\circ$	25(1)	−29.03	−121.46	8316;3536,4948
	25(g)	−22.80	−95.40	6221,7336;4361
ΔH$_c^\circ$	25(1)	−133.34	−557.89	8316;3536,3975,6840
C$_p^\circ$	25	12.16	50.88	7334;2706,808,2792,2790
	Table 2.14			
C$_p$	19.3	24.11	100.88	5771;4327,4771
	Table 2.11			
t$_c$		237	510	7328,2179;4771,5777

315. Dichloromethane

(Continued)

		cgs	SI	
p_c		60.9	6.17	7328;4026,4771
K_B	eq 2.79	2.60	2.60	7839
κ	25	4.3×10^{-11}	4.3×10^{-9}	6014
ε	25	8.93	8.93	612;4827,4937,5250,5251,4771
	30	8.649	8.649	6483
μ	(1)	1.14	1.14	612;2997,4824,4826,5250
				5317,5318
δ		9.88	20.21	6240;6905,4593,1271
ER	BuOAc	14.5	14.5	6905
	90%	24	24	6905
soly	in aq,25	1.30%w	1.30%w	4040;5496,7100,4771,987
	aq in,25	0.198%w	0.198%w	4040;6979
aq az	38.1	98.5%w	98.5%w	4026;3497
fl pt		none	none	5713
Beil	5,60			
uv	4954,7229,7571,8306			
ir	8188,3917,2480,603,1662,2259,5894,5932			
Raman	4125,5494,7782,7833,5653,7228,3770			
ms	3035,2050			
nmr	4740,4921,4922,4852,527,2464,5933,6748			

316. Chloroform

Trichloromethane

CHCl$_3$ 67–66–3   CHCl$_3$

		cgs	SI	
mw		119.378	119.378	3468
bp	1 atm	61.178	334.328	7332;8304,4935,4969,5251,6160 7059,7771
dt/dp	1 atm	0.0404	0.303	7332;8304,7470,7494
dp/dt	1 atm	24.8	3.30	7332
p	25	194.8	25.97	2089;6019,6407,6461,7058,8304
eq 2.5	A	6.83798	5.96288	7333;1009,4398
	B	1106.94	1106.94	
	C	218.552	218.552	
fp		−63.52	209.63	7332;7059,725,1090,5104,6783 7471,7480
d	15	1.49845	1498.45	7494;2677,4969,5250,6980 7470,7499
	20	1.48911	1489.11	7332
	25	1.47970	1479.70	7332;7771,4771,464
	30	1.47060	1470.60	7494
dd/dt		0.001857	1.857	7494
α		0.00126	0.00126	4902;4771,1179
κ_T	20	1.331×10^{-7}	9.980×10^{-7}	1262;1880
κ_S	35	1.009×10^{-7}	7.571×10^{-7}	5945
n_D	15	1.44858	1.44858	7494;4935,4969,5251,5329 6160,7771
	20	1.4459	1.4459	7332
	25	1.44293	1.44293	2088;7332,8080,8081,4771
dn/dt		0.00059	0.00059	7494
η	15	0.596	0.000596	7494;2677,4570,5136,5329 7505,4771
	16.36	0.5939	0.0005939	
	25	0.5357	0.0005357	464
	30	0.514	0.000514	7494;5410
	Table 2.2			
γ	20	27.16	0.02716	2366;28,3135,3283,5329 7771,4771
	25	26.53	0.02653	
	30	25.25	0.02525	6762;46
ΔH_v	20	7.97	33.35	7136;4340,5872,4771
	bp	7.020	29.37	4969;5153,7592,7842,7754
ΔH_m		2.280	9.540	7478;7167
ΔH_f°	25(1)	−32.14	−134.47	8316;3536,4948,7826
	25(g)	−24.60	−102.93	7336,6221;7826
ΔH_c°	25(1)	−96.07	−401.96	8316;6840,3536,3807,7446,3975
C_p°	25	15.63	65.40	7334;6221,2706,808,2792,2790
	Table 2.14			

316. Chloroform

(Continued)

		cgs	SI	
C_p	30 Table 2.11	27.94	116.90	8080;5817,6160,8081,6977 5872,4771
t_c		263.4	536.6	7329,4088;1328,5153,7848,1671 923,4771
P_c		52.59	5.329	1328;7329,7842,3345,4771,5153
d_c		0.458	458	1328;7329,4088,3345
v_c		0.261	0.000261	1328;7329,4088,1671
K_f	eq 2.67	4.90	4.90	725;735
K_B	eq 2.79	3.62	3.62	726,734;730,7588,4437,5595,865
κ	25	$<1 \times 10^{-10}$	$<1 \times 10^{-8}$	7853
ε	20	4.806	4.806	4937;4251,4703,5250,5251 5455,7497
μ	25 in 38	1.15	1.15	4581;612,3854,4935,5318,6867 6872,6118
δ	25	9.49	19.42	4024;7159,6371,652,4187,3395 1271,4593,6240
ER	BuOAc	10.45	10.45	3455
soly	in aq,20 aq in,25	0.815%w 0.093%w	0.815%w 0.093%w	6556;2986,4771,7100,5725,591 6979
aq az	56.12	97.8%w	97.8%w	3497;3498
fl pt	nonflam			152
Beil	5,61			
uv	4062,4954,7229,7571,8306			
ir	767,4927,7440,7537,8188,3917,7225,5932			
Raman	2390,5494,6046,6058,8293,6067,4171,1783,3770,4172			
ms	2050			
nmr	4740,4921,4922,3794,4852,2464,5933,6748			

317. Carbon tetrachloride

Tetrachloromethane

CCl_4 · · · · · · · · · · · · · 56–23–5 · · · · · · · · · · · · · CCl_4

		cgs	SI	
mw		153.823	153.823	3468
bp	1 atm	76.638	349.788	7332;2090,5329,7494,6160,7470 8154,8305
dt/dp	1 atm	0.04320	0.3240	2090;7332,1910,7470,7494,8304
dp/dt	1 atm	23.15	3.086	2090
p	25	115.2	15.36	2088;2091,6625,6019,6864 7058,8154
eq 2.5	A	6.97955	6.10445	7333;1009,3396,2090
	B	1265.632	1265.632	
	C	232.148	232.148	
fp		−22.82	250.33	7332;1897,7482,7480,5104,6783 6976,7471
d	15	1.60370	1603.70	7494,8305,6121;887,1696,2090 2677,4969
	20	1.59402	1594.02	326;7332,4650,6121
	25	1.58436	1584.36	4381,6095,6096;1149,8063,7332 5250,5710,7470,3317,7494 7771,5329
	30	1.57480	1574.80	7494,6121;6074,1150,4771,8156
dd/dt	20	0.001931	1.931	6121
α	25	0.001229	0.001229	8156,6096;4195,1965,1179,4902
κ_T	25	1.4397×10^{-7}	1.0799×10^{-6}	6096;6151,1262,8156,1965 6487,3897
κ_S	25	9.96×10^{-8}	7.47×10^{-7}	6096;1965,1966,5411
n_D	15	1.46305	1.46305	7494;1850,1892,2090,4969 5329,5647
	20	1.46018	1.46018	4381;4650,7332
	25	1.45739	1.45739	1149;5758,7332,6160,6864,6980 7771,8081,8305,3317
dn/dt		0.00055	0.00055	7482
η	0	1.329	0.001329	7482;1024,2677,4570,4955,5136 5329,7494,7505,4771
	20	0.9785	0.0009785	326
	25	0.9004	0.0009004	3317;6095
	30	0.845	0.000845	7482;5412,5410
	Table 2.2			
γ	20	26.92	0.02692	4195;5329,3283,28,7771,7848 7850,7864
	25	26.13	0.02613	4381;6762,46
	Table 2.7			
	Table 2.8			
ΔH_v	25	7.746	32.41	3396;5153,7592,7842,7848 4186,6669

317. Carbon tetrachloride

(Continued)

		cgs	SI	
ΔH_v	bp	7.161	29.96	4969;4340,1647,7754,4771
ΔH_m		0.581	2.43	7059;3376,7039
ΔH_t	-22.87	0.170	0.170	417
ΔH_f°	25(l)	-32.37	-135.44	7826,8316;548,3536
	25(g)	-22.90	-95.81	7336,6221;5057,7826
ΔH_c°	25(l)	-61.68	-258.07	8316;6840,3536,7446,3975
C_p°	25	19.94	83.43	7334,6221;7826,2706,808
	Table 2.14			2871,2790
C_p	25	31.395	131.36	2490,8063;7178,7738,7826,7780
				3376,3692,4153,5817
	Table 2.11			8284,6977
λ	23	0.00028	0.12	1855
t_c		283.4	556.6	7511;1329,7331,8231,8233,923
				5153,7848,1671
P_c		44.57	4.516	7511;1329,7331,5153,4088,7739
				4771,8231
d_c		0.558	558	7331,4088;1329
v_c		0.275	0.000275	7331;1671,8233
K_f	eq 2.67	29.8	29.8	735
K_B	eq 2.79	4.88	4.88	723;1370,5060,7845,734,2798
κ	18	4×10^{-18}	4×10^{-16}	2685
ε	20	2.23790	2.23790	7019;4650,4937,1518,2027,2953
				4251,5250,7497,3199,5023
				5062,4771,7672
	25	2.2288	2.2288	5758;5240,2438,5411,3624
μ	(1)	0	0	612;6374,4771
δ	25	8.551	17.49	9112;6240,898,4024,4187,4593
				1271,652,6371
ER	BuOAc	6.0	6.0	6905
	90%	60	60	6905
soly	in aq,25	0.077%w	0.077%w	6556;7100,591,2986,5513,4771
	aq in,30	0.0135%w	0.0135%w	6737;6556,6979,4024,2782,3734
aq az	66	95.9%w	95.9%w	3497
fl pt	nonflam			152

Beil	5,64
uv	3773,4062,4954,7229,7571,8306
ir	2686,5695,5894,153,1162,7225,5932
Raman	1471,2170,2390,5259,5494,6046,7719,8293,6599,3921,4655,3770,4172
ms	6199,5217
nmr	4694,4852,710

318. 1,1-Dichloroethane

CH_3CHCl_2 75–34–3 $C_2H_4Cl_2$

		cgs	SI	
mw		98.960	98.960	3468
bp	1 atm	57.30	330.45	7332,4935;2983,5329,5710 7471,7771
dt/dp	1 atm	0.0396	0.297	7332;7494
dp/dt	1 atm	25.3	3.37	7332
p	25	227.7	30.36	2089;6090,7058
eq 2.5	A	7.04290	6.16780	7333;4576
	B	1201.05	1201.05	
	C	231.27	231.27	
fp		−96.96	176.19	7332;4576,3289,7058,7494,7471
d	15	1.18350	1183.50	7494;2027,2216,3289,5329 5710,7771
	20	1.1755	1175.5	7332
	25	1.1679	1167.9	7332;5329,6791
	30	1.16010	1160.10	7494
dd/dt		0.00156	1.56	7494
n_D	15	1.41975	1.41975	7494;3289,5329,7771,658
	20	1.4164	1.4164	7332
	25	1.4138	1.4138	
dn/dt		0.00052	0.00052	7494
η	25	0.505	0.000505	7494;5329,7453
	30	0.430	0.000430	
	Table 2.2			
γ	20	24.75	0.02475	3283,5329;7771
	30	23.62	0.02362	
ΔH_v	25	7.318	30.62	4435;1417,4576,4186,1688
	bp	6.83	28.6	2089
ΔH_m		1.881	7.870	4576
ΔH_f°	25(1)	−38.45	−160.87	362;7826
	25(g)	−31.10	−130.12	7336,1417;7826,4361,362
ΔH_c°	25(1)	−261.94	−1095.95	4877;3975,7446,6840
C_p°	25	18.24	76.32	7334;7826,1417,1774,4576
C_p	25	30.18	126.27	7826
	Table 2.11			
t_c		250.0	523.2	888;758,7328
p_c		50.0	5.07	4088,7058,7328;758
d_c		0.42	420	758
v_c		0.24	0.00024	758
K_B	eq 2.79	3.13	3.13	726
κ		2.0×10^{-9}	2.0×10^{-7}	6791
ε	18	10.0	10.0	4937;2027,6791

318. 1,1-Dichloroethane

(Continued)

		cgs	SI	
μ	20 in 38	1.82	1.82	6680;2984,7661,4826,4935
δ	25	8.9	18.3	653
soly	in aq,20	5.03%w	5.03%w	2983;7662
	aq in,25	0.096%w	0.096%w	6979;6165
aq az	53.0	95.3%w	95.3%w	6165
Beil	8,83			
uv	4954,7229			
ir	2259,4927,7440,7441,5849,1774,89,5932			
Raman	894,1783,4103,89,2162			
nmr	4740,7711,5461,4623,5933,6748			

319. 1,2-Dichloroethane

CH_2ClCH_2Cl 107–06–2 $C_2H_4Cl_2$

		cgs	SI	
mw		98.960	98.960	3468
bp	1 atm	83.483	356.633	6823;7332,657,2983,6025,7471
				7495,866
dt/dp	1 atm	0.0419	0.314	7332;2090,7470,7495,6823
dp/dt	1 atm	23.9	3.18	7332
p	20	83.35	11.11	2089;2425,6019,6823,1853
eq 2.5	A	7.15866	6.28356	7333;2090,2639
	B	1341.37	1341.37	
	C	230.05	230.05	
fp		−35.66	237.49	7327;657,2425,5870,7471
				2090,7479

319. 1,2-Dichloroethane

(Continued)

		cgs	SI	
d	0	1.28164	1281.64	7495;553,657,2027,6791 2216,2425
	20	1.25209	1252.09	7332;8064,2431
	25	1.24637	1246.37	7332;8063,8315,5710,7470,7732 7765,866
	30	1.23831	1238.31	7495;5703
dd/dt		0.00144	1.44	7495;4771,5329
α	20	0.001141	0.001141	8064;2431
	55	0.00121	0.00121	7616
κ_T	30	1.13×10^{-7}	8.46×10^{-7}	2431
n_D	15	1.44759	1.44759	7495;657,2090,3361,4969 5251,5329
	20	1.4448	1.4448	7332
	25	1.4421	1.4421	7332;5647,7765,866,2637 4771,7482
dn/dt		0.00051	0.00051	7495
η	15	0.887	0.000887	7495;553,2125,2425,3361,3464
	30	0.730	0.000730	5329,8315,2488,4771
	Table 2.2			
γ	20	32.23	0.03223	3283;2425,2638,3464,5329
	30	30.84	0.03084	7765,4771
	Table 2.8			
ΔH_v	25	8.401	35.15	7822;4771
	bp	7.654	32.02	4969;2425,2639,7848
ΔH_m		2.112	8.837	5870;6260,7478
ΔH_f°	25(1)	−40.55	−169.66	3536,362;6768,8316,7826
	25(g)	−30.30	−126.78	57;4361,7826
ΔH_c°	25(1)	−265.65	−1111.48	6768;7446,3975,8316,6840,3536
C_p°	25	18.48	77.32	7334;3053
C_p	25	30.83	128.99	8063;8064,5870,2425,3692,4327
	Table 2.11			6025,6977,4771
t_c		288	561	7328;923,2639,3465,3464 4771,561
p_c		53	5.4	7328,4088;4771,7850
d_c		0.44	440	7328,4088
v_c		0.22	0.00022	7328;1671
K_B	eq 2.79	3.44	3.44	7869;726,734
κ	25	4×10^{-11}	4×10^{-9}	6027;4248,7852,8315,6791
ϵ	25	10.37	10.37	3624;4937,1518,3360,3361,6791 7850,7497
μ	20 in 38	1.83	1.83	4045;2,2984,7661,1518,3381 5251,6980,8079

319. 1,2-Dichloroethane

(Continued)

		cgs	SI	
δ	25	9.78	20.00	4024;652,4187,7159,1271 6905,3395
ER	BuOAc 90%	4.46 78	4.46 78	6905;3455 6905
soly	in aq,20	0.81%w	0.81%w	1348;591,7100,1853,2983 2986,4771
	aq in,25	0.187%w	0.187%w	6979;3734
aq az	72	91.8%w	91.8%w	6165;1853,3121,3497,3498
fl pt	CC	13	286	5712;2425

Beil	8,84
uv	4954,3574,6038
ir	4927,5890,6666,6728,6811,153,5849,104,5932
Raman	1783,456,5187,5260,5457,5014,8175,5191,6566,104
ms	154
nmr	4740,6122,2464,5933,6748

320. 1,2-Dichloropropane

$CH_2ClCHClCH_3$ 78-87-5 $C_3H_6Cl_2$

		cgs	SI	
mw		112.986	112.986	3468
bp	1 atm	96.37	369.52	2088,7327;2090,4771,5448
dt/dp	1 atm	0.04442	0.3332	2088
dp/dt	1 atm	22.51	3.001	2088
p	25	49.67	6.622	2088;5448
eq 2.5	A B C	6.96395 1295.9 221	6.08885 1295.9 221	2088;5448
fp		−100.44	172.71	2088,7327;2090

320.   1,2-Dichloropropane

(Continued)

		cgs	SI	
d	20	1.15597	1155.97	2088;7327,4373,5448
	25	1.14936	1149.36	
dd/dt	20 - 30	0.00131	1.31	7616
α	10 - 30	0.000813	0.000813	4771
n_D	20	1.43937	1.43937	2088;7327,2090,7765
	25	1.43679	1.43679	
dn/dt	25	0.000516	0.000516	2088
η	20	0.8572	0.0008572	2088;4771
γ	20	28.65	0.02865	2088;7765
	30	27.37	0.02737	
ΔH_v	25	8.700	36.400	2088;5448
	bp	7.649	32.004	
ΔH_m		1.529	6.397	2088
ΔH_f°	25(1)	−47.3	−197.9	1703
	25(g)	−38.71	−161.96	1703;4361
ΔH_c°	25(1)	−450.1	−1883.2	1703;6840
C_p	25	36.80	153.98	4327
	Table 2.11			
t_c		304.3	577.5	2088
P_c		43.8	4.44	2088
d_c		0.410	410	2088
v_c		0.276	0.000276	2088
A	eq 2.72	0.02592	0.02592	2088
ε	26.1	8.925	8.925	5999;4771
μ	25 in 38	1.85	1.85	2984
δ	25	9.0	18.4	653;4593,1271
ER	BuOAc	3.22	3.22	6905
	90%	108	108	
soly	in aq,25	0.274%w	0.274%w	2088;2983,4771
	aq in,25	0.132%w	0.132%w	2088;4771
aq az	78	88%w	88%w	3497
fl pt	CC	21	294	6402
	OC	24	297	7616
Beil	10,105			
ir	104,7440			
Raman	104,3827,4129			
nmr	5807,2432,8008,6748			

321. 1,1,1-Trichloroethane

CH_3CCl_3 71-55-6 $C_2H_3Cl_3$

		cgs	SI	
mw		133.405	133.405	3468
bp	1 atm	74.083	347.233	132,7332;7482,659,3289,4935
				7581,7771,8119
dt/dp	1 atm	0.0426	0.320	7332;7482
dp/dt	1 atm	23.5	3.13	7332
p	25	123.7	16.49	132;2089,6297,7058
eq 2.5	A	6.86265	5.98755	132;7333,5703
	B	1182.527	1182.527	
	C	222.894	222.894	
fp		-30.4	242.80	3289;659,6297,7581,7671
				1728,7479
tp		-30.02	243.13	353
d	15	1.34587	1345.87	7482;659,3286,4935,7771,866
	20	1.3381	1338.1	7332;4877,7771
	25	1.3299	1329.9	7332;4877,1059
	30	1.32096	1320.96	7482;5703
dd/dt	0 - 30	0.001657	1.657	7482
n_D	20	1.4380	1.4380	7332;7771,7482,1728,866,7581
	25	1.4359	1.4359	2089;7332,4877
dn/dt		0.00052	0.00052	7482
η	15	0.903	0.000903	7482
	25	0.795	0.000795	1626
	30	0.725	0.000725	7482
γ	15	26.17	0.02617	7482;7771
	20	25.56	0.02556	
	30	24.25	0.02425	
ΔH_v	25	7.741	32.390	353;4435,1417,4877,7648
	bp	7.100	29.708	353;6297
ΔH_m		0.5617	2.350	353;7671,6297,1728
ΔH_t	-48.35	1.790	7.490	353
ΔH_f°	25(1)	-42.74	-178.82	3537;3536,4877
	25(g)	-34.01	-142.30	7336,4877,1417;3537
ΔH_c°	25(1)	-264.83	-1108.05	3537;4877,3536
C_p°	25	22.07	92.34	7334,1417
C_p	25	34.51	144.4	353;7826
t_c		272	545	132;4435
P_c		42.4	4.30	132;4435
κ		7.3×10^{-9}	7.3×10^{-7}	2067
ε	0	7.953	7.953	2438;7672
	19.8	7.252	7.252	7672;4937,1626,7581,7497,5905
μ	in 38	1.70	1.70	2438;7672,7093,4935,8119
δ		8.5	17.4	6905

321.   1,1,1-Trichloroethane

(Continued)

		cgs	SI	
ER	BuOAc 90%	6.0 60	6.0 60	6905 6905
soly	in aq,20 aq in,25	0.132%w 0.034%w	0.132%w 0.034%w	7662;591 6979
aq az	65.2	91.7%w	91.7%w	6165
fl pt		none	none	2067
Beil	8,85			
uv	3574			
ir	153,3794,89,5932			
Raman	3578,6050,4103,6993			
nmr	4740,3794,2464,6748,5933			

322.   1,1,2-Trichloroethane

$CH_2ClCHCl_2$ 79-00-5 $C_2H_3Cl_3$

		cgs	SI	
mw		133.405	133.405	3468
bp	1 atm	113.85	387.00	7332;2088,2090,6953,4771
dt/dp	1 atm	0.04610	0.3458	2090;7332
dp/dt	1 atm	21.69	2.892	2090
p	25	22.49	2.998	2088
eq 2.5	A	6.97811	6.10301	7333
	B	1332.6	1332.6	
	C	211.38	211.38	
fp		-36.53	236.62	7332;2088,7492,3289,1728,4771
d	20	1.43931	1439.31	7332;2088,5329,2090,7492,3289
	25	1.43213	1432.13	
dd/dt	20 - 30	0.00156	1.56	7616

322. 1,1,2-Trichloroethane

(Continued)

		cgs	SI	
α	0 - 25	0.00100	0.00100	
n_D	20	1.47124	1.47124	2088;7332,2090,3289
	25	1.46868	1.46868	
dn/dt	25	0.000524	0.000524	2088
η	20	1.19	0.00119	5329;7492,5328,2088
	25	1.10	0.00110	
γ	20	33.75	0.03375	5329;2088,7492,1969
	25	33.00	0.03300	
ΔH_v	25	9.63	40.28	4435;2088,8089
	bp	8.18	34.23	2088;4771
ΔH_m		2.759	11.543	2088;1728
ΔH_f°	25(1)	-43.5	-182.0	7826
	25(g)	-33.94	-142.01	7826;7335,7336,2822
C_p°	25	20.29	84.89	7334;3193,7826
C_p	20	35.5	148.5	4771;1728
t_c		339	612	2088
P_c		47.71	4.834	2088
d_c		0.497	497	2088
v_c		0.268	0.000268	2088
A	eq 2.72	0.02484	0.02484	2088
ϵ	20	7.29	7.29	7492;2088,1728
μ	in 38	1.55	1.55	7661
	(g)	1.41	1.41	2007
δ	25	9.6	19.6	653,4593
ER	BuOAc	1.9	1.9	6905
	90%	180	180	
soly	in aq,20	0.44%w	0.44%w	4771,7662
	aq in,25	0.118%w	0.118%w	6979
aq az	86.0	83.6%w	83.6%w	3498
fl pt	nonflam			7616

Beil	8,85
ir	4850,4324,5932
Raman	4103,3578,2712
nmr	4740,6647,348,6748

323. 1,2,3-Trichloropropane

$CH_2ClCHClCH_2Cl$ 96–18–4 $C_3H_5Cl_3$

		cgs	SI	
mw		147.432	147.432	3468
bp	1 atm	156.85	430.00	7473,7330;5329
dt/dp	1 atm	0.0509	0.382	2089;7330
dp/dt	1 atm	19.6	2.62	2089
p	25	3.69	0.492	2089
eq 2.5	A	6.98716	6.11206	2089;7644
	B	1502.3	1502.3	
	C	209	209	
fp		−14.7	258.5	7473,7330
d	20	1.3888	1388.8	7330;5329,911
	25	1.3832	1383.2	
dd/dt	20 – 30	0.00112	1.12	7616
n_D	20	1.4832	1.4832	7330;5329
	25	1.4812	1.4812	
dn/dt	25	0.00040	0.00040	<u>7330</u>
η	20	2.505	0.002505	5329
	25	2.23	0.00223	
γ	20	37.80	0.03780	5329;2089
	25	37.05	0.03705	
ΔH_v	25	11.22	46.94	2089
	bp	8.872	37.122	
ΔH_f°	25(l)	−55.17	−230.83	1703
	25(g)	−43.8	−183.3	
ΔH_c°	25(l)	−414.77	−1735.40	1703;901,6840
C_p	25	41.25	172.59	4327;5442
	Table 2.11			
ε	21	7.45	7.45	2027
δ	25	10.1	20.6	653
soly	in aq,20	0.19%w	0.19%w	7616
	aq in,20	0.05%w	0.05%w	
aq az	96	58%w	58%w	7616
fl pt	OC	82	355	7616

Beil	<u>10</u>,106
ir	3033,7440,5932
Raman	3033,3157,2712,4129
ms	7658
nmr	8088,6748

324. 1,1,2,2-Tetrachloroethane

$CHCl_2CHCl_2$ 79–34–5 $C_2H_2Cl_4$

		cgs	SI	
mw		167.850	167.850	3468
bp	1 atm	145.1	418.3	7332;7494,4969,5329,7472,7771 4771,2216
dt/dp	1 atm	0.0485	0.364	7332;7494
dp/dt	1 atm	20.6	2.75	7332
p	25	5.95	0.793	2089;380,3347,4997,5447,7058
eq 2.5	A B C	7.0046 1444.3 205.1	6.1295 1444.3 205.1	7333;5703
fp		−43.8	229.4	7332,7494;380,3347,7058 7472,4771
d	15 20 25 30	1.60255 1.59449 1.58666 1.57860	1602.55 1594.49 1586.66 1578.60	7494 7332 7332;380,4969,5329,5710 7771,7863 7494;5703,4771
dd/dt		0.00159	1.59	7494
α		0.000998	0.000998	3349
n_D	15 20 25	1.49678 1.4940 1.4914	1.49678 1.4940 1.4914	7494 7332 7332;2089,4969,5329,7771,4771
dn/dt		0.00051	0.00051	7494
η	15 30	1.844 1.456	0.001844 0.001456	7494;3342,5329,7863,2486,4771
γ	20 40	36.04 33.30	0.03604 0.03330	2216;5329,7771,4771
ΔH_v	25 bp	10.94 9.236	45.78 38.64	4435;2089,380,3346,4997,4771 4969
ΔH_f°	25(1) 25(g)	−47.0 −35.6	−196.6 −149.0	7826 7336;7826
ΔH_c°	18(1)	−232.12	−971.19	6840
C_p°	25	23.66	98.99	7334;7826
C_p	20 25	45.0 39.6	188.3 165.7	3347;380,4327,4771 7820
	Table 2.11			
κ	not measurable			7863
ε	20	8.20	8.20	4937;7867
μ	20 in 38	1.71	1.71	4045,5189;7661,971,7411
δ	25	9.80	20.05	4024;652,4593
soly	in aq,20 aq in,25	0.287%w 0.110%w	0.287%w 0.110%w	7662;4771,591 6979;3734
aq az	94.3	66.0%w	66.0%w	6165

324. 1,1,2,2-Tetrachloroethane

(Continued)

		cgs	SI	
fl pt	nonflam			380

Beil	8,86
uv	4956,3574
ir	153,5890,6811,7077,7440,4374,3794,1062
Raman	337,5112,6050,1783,890,640
nmr	3044,4740,3794,7155,2464,5933,6748

325. Pentachloroethane

$CHCl_2CCl_3$ 76-01-7 C_2HCl_5

		cgs	SI	
mw		202.295	202.295	3468
bp	1 atm	159.88	433.03	7332;7494,5329,7472,4771
dt/dp	1 atm	0.0537	0.403	7332;7494,1897
dp/dt	1 atm	18.6	2.48	7332
p	25	4.4	0.59	5447;380,7058
eq 2.5	A	6.5950	5.7199	7333
	B	1295.67	1295.67	
	C	188.96	188.96	
fp		−29.0	244.2	7494,7472,7332;380,4771
d	0	1.71100	1711.00	7494,380,3346,5329,4771
	15	1.68813	1688.13	7494
	20	1.6808	1680.8	7332
	25	1.6732	1.6732	7332
dd/dt	0 - 30	0.00152	1.52	7494
n_D	15	1.50542	1.50542	7494;5329,4771
	20	1.5030	1.5030	7483;7332
	25	1.5005	1.5005	7332
dn/dt		0.00046	0.00046	7494

325. Pentachloroethane

(Continued)

		cgs	SI	
η	15	2.751	0.002751	7494;3342,5329,4771
	30	2.070	0.002070	
γ	20	34.72	0.03472	3283;5329
	30	33.58	0.03358	
ΔH_v	av	9.75	40.79	3779;4771
	bp	8.829	36.941	3346;380
ΔH_m		2.710	11.339	7671
ΔH_f°	25(1)	−44.9	−187.9	7826
	25(g)	−34.80	−145.60	7336;1417
ΔH_c°	18(1)	−205.88	−861.38	6840
C_p°	25	28.219	118.07	1417,7334;7826
C_p	18 − 53	44.31	185.4	7968;380,3347,4327,3973
	25	46.7	195.4	7826
	Table 2.11			
t_c		373.0	646.2	923
κ	not measurable			7853
ε	20	3.73	3.73	4937;7853,7867,3973
μ	25 − 85(1)	0.94	0.94	3689;7661,7411,3973
δ		9.4	19.2	4593
soly	in aq,25	0.05%w	0.05%w	4935;3973
	aq in,25	0.035%w	0.035%w	6979
aq az	95.8	64.9%w	64.9%w	4935;3497
fl pt	nonflam			380

Beil	8,87
uv	5314
ir	5863,7077,7440,6957,89,5932
Raman	2714,5112,5261,89
nmr	4740,4921,4922,5933,6748

326.   o-Dichlorobenzene

1,2-Dichlorobenzene

$o\text{-}C_6H_4Cl_2$			95–50–1		$C_6H_4Cl_2$
		cgs	SI		
mw		147.004	147.004	3468	
bp	1 atm	180.48	453.63	2090;1375,3593,5235,4760	
dt/dp	1 atm	0.05270	0.3953	2090	
dp/dt	1 atm	18.98	2.530	2090	
p	25	1.28	0.171	2087;2091,7058,4760,8315	
eq 2.5	A	7.07028	6.19518	4760;2090	
	B	1649.55	1649.55		
	C	213.314	213.314		
fp		−17.01	256.14	8121;2090,1375,3593,4760	
d	20	1.30589	1305.89	2090;1375,8315,2958	
	25	1.30033	1300.33		
dd/dt	25	0.001112	1.112	2090	
α		0.00085	0.00085	4902	
n_D	20	1.55145	1.55145	2090;1375,3593,3789	
	25	1.54911	1.54911		
dn/dt		0.00049	0.00049	7482	
η	25	1.324	0.001324	8315	
	31.9	1.2018	0.0012018	2958	
γ	20	26.84	0.02684	2087	
	30	35.55	0.03555		
ΔH_v	25	12.00	50.21	3759;1703	
	bp	9.48	39.66	4026	
ΔH_m		3.01	12.59	4026	
ΔH_f°	25(1)	−4.32	−18.07	1703;6768,1852	
	25(g)	7.1	29.7	1703	
ΔH_c°	25(1)	−707.13	−2958.63	1703;6840,6768,3975,1852	
C_p°	25	27.14	113.55	1289	
C_p	25	52.98	221.67	4026	
t_c		424.1	697.3	2087	
P_c		40.5	4.10	2087	
d_c		0.408	408	2087	
v_c		0.360	0.000360	2087	
A	eq 2.72	0.02215	0.02215	2087	
κ	25	3×10^{-11}	3×10^{-9}	6027	
ε	25	9.93	9.93	4937;5251,8315	
μ	20 in 38	2.14	2.14	6680;4739,7461,2442,3593,5235	
				5251,6871	
δ	25	10.05	20.56	4024;4593,8111,6905,652,3759	
ER	BuOAc	0.15	0.15	6905	
	90%	2280	2280	6905	

326.   *o*-Dichlorobenzene

(Continued)

		cgs	SI	
soly	in aq,25	0.0156%w	0.0156%w	<u>591</u>
	aq in,25	0.309%w	0.309%w	<u>8111</u>;4024,4026,3759
aq az	98.2	33.1%w	33.1%w	6165
fl pt	CC	66	339	5712

Beil	<u>464</u>,201
uv	582,5954,6706,6936,7172,379,6749
ir	5890,633,153,3887,6433,6957,2932,5932
Raman	3336,6938,1793,6433,1782,2932
nmr	7227,83,7021,5914,4918,2901,5933,6748

327.   *m*-Dichlorobenzene

1,3-Dichlorobenzene

m-C$_6$H$_4$Cl$_2$ 541–73–1   C$_6$H$_4$Cl$_2$

		cgs	SI	
mw		147.004	147.004	3468
bp	1 atm	173.00	446.15	2090;2091,5258,7058,7765
dt/dp	1 atm	0.05206	0.3905	2090
dp/dt	1 atm	19.21	2.561	2090
p	25	1.89	0.252	2087;2091,7058
eq 2.5	A	7.30364	6.42854	2090
	B	1782.4	1782.4	
	C	230	230	
fp		−24.76	248.39	2090;7058
d	20	1.28844	1288.44	2090;7765,2958
	25	1.28280	1282.80	
dd/dt	25	0.001128	1.128	<u>2090</u>
n$_D$	20	1.54586	1.54586	2090;7765,3789
	25	1.54337	1.54337	

327. *m*-Dichlorobenzene

(Continued)

		cgs	SI	
dn/dt	25	0.000498	0.000498	2090
η	23.3	1.0450	0.0010450	2958
	32.8	0.9551	0.0009551	
	40.4	0.8807	0.0008807	
γ	20	36.16	0.03616	7765
	41.8	33.53	0.03353	
ΔH_v	25	11.61	48.58	2087;1703
	bp	9.23	38.62	2087
ΔH_f°	25(1)	−5.00	−20.92	1703;1852
	25(g)	6.1	25.5	1703
ΔH_c°	25(1)	−706.44	−2955.75	1703,1852;6840
C_p°	25	27.21	113.85	1289
t_c		410.8	684.0	2087
P_c		38.3	3.88	2087
d_c		0.410	410	2087
v_c		0.359	0.000359	2087
ϵ	25	5.04	5.04	4937;5251
μ	25 in 38	1.54	1.54	3765;1311,4739,5251,6871
δ	25	9.8	20.0	653
soly	in aq,20	0.0111%w	0.0111%w	Beil. EIII5, 542;591
fl pt		72	345	2087

Beil	464,202
uv	582,5276,6936,7172,6749
ir	7077,633,3890,6433,5893,2931,5932
Raman	3335,3339,6938,1793,6433,2931
nmr	2455,5933,6748

328. *p*-Dichlorobenzene

1,4-Dichlorobenzene

p-$C_6H_4Cl_2$ 106–46–7 $C_6H_4Cl_2$

		cgs	SI	
mw		147.004	147.004	3468
bp	1 atm	174.12	447.27	2090;7058
dt/dp	1 atm	0.05217	0.3913	2090
dp/dt	1 atm	19.17	2.556	2090
p	25	1.76	0.235	2087;2091,7058,8298
eq 2.5	A	6.99800	6.12290	4760;8298,7896
	B	1575.11	1575.11	
	C	208.513	208.513	
fp		53.13	326.28	7896,4760;1394,2986,4913
				5250,7167
tp		53.005	326.155	2171
d	55	1.24750	1247.50	2090;5250,6293,2958,4987
	60	1.24166	1241.66	
n_D	60	1.52849	1.52849	2090
η	55.4	0.8394	0.0008394	2958;4987
	69.3	0.7202	0.0007202	
	79.4	0.6678	0.0006678	
γ	68	30.69	0.03069	7482
	96	27.58	0.02758	
ΔH_v	25	11.7	49.0	2087;8298
	bp	9.27	38.79	
ΔH_s	α-form	15.48	64.77	7896
	β-form	15.06	63.01	
ΔH_m		4.347	18.187	2171;3533,7167,5880
ΔH_t	−1.38	0.300	1.256	2171
	31.20	0.05127	0.2145	
ΔH_f°	25(s)	−10.24	−42.84	1703;1852
ΔH_c°	25(s)	−701.27	−2934.11	1703,1852;6840,6841
C_p°	25	27.22	113.89	1289
C_p	25(s)	ca 40	ca 167	4026;3532
	57(1)	44.02	184.20	2171
ε	50	2.41	2.41	4937;1394,5250
μ	24 in 38	0	0	4739,3443,4913,8078;7461
				4489,6014
δ	25	9.7	19.8	653
soly	in aq,25	0.00872%w	0.00872%w	400;591
α-form	in aq,35	0.008%w	0.008%w	4055;987,2986,7933
β-form	in aq,55	0.016%w	0.016%w	
(1)	in aq,59.2	0.021%w	0.021%w	7933
fl pt	OC	67	340	5713

328.   *p*-Dichlorobenzene

(Continued)

Beil	464,203
uv	582,6936,7172,379,6749
ir	6155,7077,633,3889,6433,1938,5893,2930,5911,2961,5932
Raman	4108,5720,6938,7802,1793,6433,7510,2930,2961
ms	4783,7010
nmr	2455,6748

329.   α,α-Dichlorotoluene

(Dichloromethyl)benzene

$C_6H_5CHCl_2$ 98-87-3 $C_7H_6Cl_2$

		cgs	SI	
mw		161.030	161.030	3468
bp	1 atm	214.0	487.2	7058;4457
p	64.0	5.0	0.67	7058
fp		-16.1	257.1	7058;7093
d	20	1.2536	1253.6	6432
	25	1.2472	1247.2	
dd/dt	25	0.00119	1.19	6432
n_D	20	1.5503	1.5503	3976
ε	20	6.9	6.9	4937
μ	25 in 317	2.07	2.07	4497;7093,5700
soly	in aq,30	0.025%w	0.025%w	5569

Beil	466,297
uv	7229,6749
ir	752,3791,4440
nmr	6986,6748

330.   2,4-Dichlorotoluene

2,4-Dichloro-1-methylbenzene

2,4-Cl$_2$C$_6$H$_3$CH$_3$ 95–73–8   C$_7$H$_6$Cl$_2$

		cgs	SI	
mw		161.030	161.030	3468
bp	1 atm	199.6	472.8	2396
p	25	0.416	0.0555	2396
eq 2.5	A	6.4950	5.6199	2396
	B	1330.4	1330.4	
	C	168.5	168.5	
fp		−13.5	259.7	7834
d	20	1.2476	1247.6	7834
dd/dt		0.00106	1.06	2396
n$_D$	22	1.5480	1.5480	7834
η	25	1.532	0.001532	2396
	Table 2.6			
μ	25 in 317	1.70	1.70	1311;5401

Beil	295,466
uv	5065,6749
ir	2945,5043,1888,5932
Raman	2945,4123,1888
ms	4243
nmr	1919,7227,6748,5933

331.   3,4-Dichlorotoluene

1,2-Dichloro-4-methylbenzene

3,4-Cl$_2$C$_6$H$_3$CH$_3$ 95–75–0   C$_7$H$_6$Cl$_2$

		cgs	SI	
mw		161.030	161.030	3468
bp	1 atm	208.92	482.07	2087
dt/dp	1 atm	0.0563	0.422	2087
dp/dt	1 atm	17.8	2.37	2087
p	25	0.315	0.0420	2087

331.   3,4-Dichlorotoluene

(Continued)

		cgs	SI	
eq 2.5	A B C	6.97925 1655.44 195.0	6.10415 1655.44 195.0	2087
fp		-15.25	257.90	2087;7834
d	20 25	1.25256 1.24751	1252.56 1247.51	2087;7834
dd/dt	25	0.001011	1.011	2087
n_D	20 25	1.54712 1.54494	1.54712 1.54494	2087;7834
dn/dt	25	0.000436	0.000436	2087
η	20	1.571	0.001571	2087
γ	20 30	36.50 35.61	0.03650 0.03561	2087
ΔH_v	25 bp	13.33 10.13	55.77 42.39	2087
ΔH_m		2.552	10.68	2087
t_c		451.2	724.4	2087
p_c		36.82	3.731	2087
d_c		0.407	407	2087
v_c		0.396	0.000396	2087
A	eq 2.72	0.01932	0.01932	2087
ε	25	8.970	8.970	2087
μ	(1)	2.95	2.95	5401
soly	aq in,30	0.0026%w	0.0026%w	2087
Beil	466,296			
ir	2945,1888,5932			
Raman	2945,1888			
nmr	7227,5933			

332. α,α,α-Trichlorotoluene

(Trichloromethyl)benzene

$C_6H_5CCl_3$ 98–07–7 $C_7H_5Cl_3$

		cgs	SI	
mw		195.476	195.476	3468
bp	1 atm	213.5	486.7	7058;7120
p	73.7	5.0	0.67	7058;2537
fp		−4.4	268.8	3290;7120
d	20	1.3741	1374.1	3290;502,1969
n_D	20	1.5580	1.5580	6180;3290
	30	1.5540	1.5540	2537
dn/dt	20	0.000482	0.000482	6180
γ	20	23.39	0.02339	1969
ε	21	6.9	6.9	4937
μ	25 in 317	2.03	2.03	4497;2537,5700,7093
soly	in aq,5	0.0053%w	0.0053%w	5569

Beil	466,300
uv	6686,7229,6749
ir	7002,4440,3791,752,5932
Raman	7002,1785
nmr	3238,6748,5933

333. 1,1-Dichloroethylene

1,1-Dichloroethene

$CH_2{=}CCl_2$ 75–35–4 $C_2H_2Cl_2$

		cgs	SI	
mw		96.944	96.944	3468
bp	1 atm	31.56	304.71	3397;6117
dt/dp	1 atm	0.037	0.28	3397
dp/dt	1 atm	27	3.6	3397
p	25	599.0	79.86	2300

333. 1,1-Dichloroethylene

(Continued)

		cgs	SI	
eq 2.5	A	6.98200	6.10690	3397
	B	1104.29	1104.29	
	C	237.697	237.697	
fp		−122.56	150.59	3397;6117
d	0	1.2517	1251.7	2073;2300,6229,6117
	20	1.2132	1213.2	
	25	1.1747	1174.7	
dd/dt		0.00192	1.92	2073
n_D	10	1.43062	1.43062	2073;2300,6117,6229,4365
	15	1.42777	1.42777	
	20	1.42468	1.42468	
η	0	0.422	0.000422	2300
	20	0.358	0.000358	
ΔH_v	25	6.328	26.48	3397
	bp	6.257	26.18	2073
ΔH_m		1.557	6.514	3397
ΔH_f°	25(1)	−5.81	−24.31	4877;7826,6768,5091
	25(g)	0.58	2.43	7826
ΔH_c°	25(1)	−261.93	−1095.95	2073,6768,4877
ΔH_p	25	−18.0	−75.3	2073
C_p°	25	16.03	67.07	7826;3397,2073
C_p	25.15	26.745	111.90	2073;3397,7826
t_c		222	495	3397
p_c		51.3	5.20	3397
d_c		0.443	443	2073
v_c		0.219	0.000219	2073
μ	20 in 38	1.28	1.28	929;6229,2290,1055
δ		9.1	18.6	652
soly	in aq,25	0.021%w	0.021%w	2073
	aq in,25	0.035%w	0.035%w	
aq az	30.4	93.0%w	93.0%w	6165
fl pt	OC	−10	263	2300

Beil	11,186
uv	4364,5314,7895
ir	2320,5932
Raman	4655,2320
nmr	8002,5933,6748

334. *cis*-1,2-Dichloroethylene

(Z)-1,2-Dichloroethene

$$\begin{matrix} \text{HCCl} \\ \| \\ \text{HCCl} \end{matrix}$$
156-59-2
$C_2H_2Cl_2$

		cgs	SI	
mw		96.944	96.944	3468
bp	1 atm	60.63	333.78	3960;3347,4935,5529,7472 4771,3535
dt/dp	1 atm	0.038	0.29	7058
dp/dt	1 atm	26	3.5	7058
p	25	200	26.7	3960;3347,7058,3535
eq 2.4	A	7.8522	6.9771	7053
	B	1651.52	1651.52	
fp		-80.0	193.2	1116;7058,7472,7484,4771
d	15	1.2917	1291.7	3958;2216,3347,5329,7863,4771
	20	1.2837	1283.7	3958
dd/dt		0.0016	1.6	3958
α	15-45	0.00127	0.00127	4771
n_D	15	1.45189	1.45189	3958;3342
	20	1.4490	1.4490	7482
dn/dt		0.00059	0.00059	3958
η	20	0.467	0.000467	3958;3342,3347,5329,7863,4771
	25	0.444	0.000444	
γ	ca 20	28	0.028	3958;5329
ΔH_v	0	7.545	31.57	3960;3347
	bp	7.225	30.23	3960;4771
ΔH_m		1.722	7.205	1116
ΔH_f°	25(1)	-6.6	-27.6	7826
	25(g)	0.90	3.77	
ΔH_c°	18(1)	-261.63	-1094.67	6840
C_p°	25	15.55	65.06	5876,7826;3785
C_p	15	27.24	113.97	5041;4771
	Table 2.11			
t_c		271.0	544.2	4771
p_c		57.9	5.87	758
κ	25	8.5×10^{-9}	8.5×10^{-7}	7863
ϵ	25	9.20	9.20	4937;3960,7867,4771
μ	20 in 38	1.80	1.80	6680;4739,971,4935,5318,4771 2290,1055
δ		9.1	18.6	4593
soly	in aq,25	0.35%w	0.35%w	4771
	aq in,25	0.55%w	0.55%w	4771

334. *cis*-1,2-Dichloroethylene

(Continued)

		cgs	SI	
aq az	55.3	96.65%w	96.65%w	3497
fl pt	CC	4	277	152;4771

Beil	11,187
uv	4364,3574,7893
ir	809,2259,5863,7077,5876,6673,7173,5932
ms	154,2047
nmr	4436,3531,8002

335. *trans*-1,2-Dichloroethylene

(*E*)-1,2-Dichloroethene

ClCH
‖
HCCl

156-60-5

$C_2H_2Cl_2$

		cgs	SI	
mw		96.944	96.944	3468
bp	1 atm	47.67	320.82	3960;3347,5329,7472,4771,3535
dt/dp	1 atm	0.031	0.23	3960
dp/dt	1 atm	32	4.3	3960
p	25	340	45.3	3960;3347,7058,3535
eq 2.4	A	7.55657	6.68147	3960
	B	1498.42	1498.42	
fp		−49.8	223.4	1116;7058,7472,7484,4771
d	15	1.2631	1263.1	3958;2216,3347,5329,4771
	20	1.2547	1254.7	3958
dd/dt		0.00168	1.68	3958
α	15-45	0.00136	0.00136	4771
n_D	15	1.45189	1.45189	3342
	20	1.4462	1.4462	2216
η	15	0.423	0.000423	3958;3342,5329,4771
	20	0.404	0.000404	
γ	ca 20	25	0.025	3958;5329
ΔH_v	0	7.180	30.04	3960;3347
	bp	6.905	28.89	3960;4771

335. *trans*-1,2-Dichloroethylene

(Continued)

		cgs	SI	
ΔH_m		2.864	11.98	1116
ΔH_f°	25(1) 25(g)	-5.53 1.47	-23.14 6.15	7826
ΔH_c°	18(1)	-261.06	-1092.28	6840
C_p°	25	15.93	66.65	5876,7826,4290;3785,736
C_p	15	26.95	112.76	5041;4771
t_c		243.3	516.5	4771
P_c		54.4	5.51	4771
ε	25	2.14	2.14	4937;3960,4771
μ	25 in 38	0.70	0.70	4739;971,5318,4771,2290,1055
δ		9.0	18.4	3395,4593
soly	in aq,25 aq in,25	0.63%w 0.55%w	0.63%w 0.55%w	6169 4771
aq az	45.3	98.1%w	98.1%w	3497
fl pt	OC	4	277	4902,4771
Beil	11,187			
uv	4364,7892,3574			
ir	809,2259,5863,7077,5876,6673,7173,5932			
ms	154,2047			
nmr	4436,3531,8002,5933,6748			

336.   Trichloroethylene

Trichloroethene

$CHCl{=}CCl_2$ 79-01-6 C_2HCl_3

		cgs	SI	
mw		131.389	131.389	3468
bp	1 atm	87.19	360.34	4759;380 , 1352,2216,4969 5329,7472
dt/dp	1 atm	0.041	0.31	2088
dp/dt	1 atm	24	3.3	2088
p	25	47.31	6.307	2088;380,7058
eq 2.5	A	7.02808	6.15298	4398;4759
	B	1315.04	1315.04	
	C	230.0	230.0	
fp		-86.4	186.8	7472;380,1352,7058,4771
d	15	1.4762	1476.2	1352;380,2216,3347,4759,4969 5329,7549,4771
	30	1.4514	1451.4	
dd/dt		0.001649	1.649	4307
α	0-40	0.00117	0.00117	6975;3346,4771
n_D	21.4	1.4767	1.4767	4759;4771,1352,4969,5329 7549,4307
	24.6	1.4750	1.4750	
dn/dt		0.0005675	0.0005675	4759
η	20	0.566	0.000566	5329;1352,3342,7549,4771
	25	0.532	0.000532	
	30	0.517	0.000517	5410
γ	20	29.5	0.0295	5329;4771
	25	28.8	0.0288	
ΔH_v	25	8.190	34.27	2088
	bp	7.521	31.47	4969;380,1352,3346,4759,4771
ΔH_f°	25(1)	-10.1	-42.3	7826
	25(g)	-1.86	-7.78	7826;4948
ΔH_c°	18(1)	-230.01	-962.36	
C_p°	25	19.18	80.25	7826;736,808,4307,88
C_p	20	29.3	122.6	3347;380,1352,4329,7549 4771,4307
	25	28.8	120.5	7826
t_c		298	571	2088;4771
p_c		48.5	4.91	2088;4771
d_c		0.513	513	2088
v_c		0.256	0.000256	2088
K_B	eq 2.79	4.43	4.43	7869
κ		8×10^{-12}	8×10^{-10}	4307;7853
ε	ca 16	3.42	3.42	4937;1352,7853,7867,4771
μ	in 317	0.8	0.8	7661;971,1215,4771,1055

336.   Trichloroethylene

(Continued)

		cgs	SI	
δ		9.3	19.0	4593,1271,6905;652
ER	BuOAc	4.46	4.46	6905
	90%	78	78	6905
soly	in aq,25	0.137%w	0.137%w	7219;591,4026,1352,4771
	aq in,25	0.32%w	0.32%w	4026
aq az	73.6	94.6%w	94.6%w	3497
fl pt	nonflam			6975

Beil	11,187
uv	4388,4364,7894
ir	807,5863,5889,6673,5932
Raman	6050,6382,6811,88
nmr	714,4922,8002,5933,6748

337.   Tetrachloroethylene

Tetrachloroethene

$CCl_2\!=\!CCl_2$ 127–18–4 C_2Cl_4

		cgs	SI	
mw		165.834	165.834	3468
bp	1 atm	121.07	394.22	1009;7494,7488,4969,5329
				7472,4771
dt/dp	1 atm	0.050	0.38	7494,7488
dp/dt	1 atm	20	2.7	7494,7488
p	25	18.47	2.462	2088;380,2091,6090,7058,7059
eq 2.5	A	6.97680	6.10170	1009;4398
	B	1386.90	1386.90	
	C	217.52	217.52	
fp		−22.35	250.80	7484,7472;7488,380,4771
d	15	1.63109	1631.09	7494;380,3346,4969,5329,7488
	20	1.62283	1622.83	4650
	25	1.61432	1614.32	4381
	30	1.60640	1606.40	7494

337.   Tetrachloroethylene

(Continued)

		cgs	SI	
dd/dt		0.001646	1.646	7494
α	0-25	0.00102	0.00102	6975;4771
κ_S	20	7.39×10^{-8}	5.55×10^{-7}	5411
n_D	15	1.50759	1.50759	7488;5329,4771
	20	1.50576	1.50576	4381;4969;4650
	25	1.5032	1.5032	5411
dn/dt		0.00053	0.00053	7488
η	15	0.932	0.000932	7494,7488;3342,5329,4771
	30	0.798	0.000798	7494,7488,5412
γ	15	32.86	0.03286	7494;7488,5329,4771
	25	31.30	0.03130	4381
ΔH_v	25	9.467	39.61	2088;4771
	bp	8.299	34.72	4969;380,3346
ΔH_m		2.525	10.56	7671
ΔH_f°	25(l)	-12.5	-52.3	7826
	25(g)	-2.9	-12.1	7826;8316,2106
ΔH_c°	25(l)	-162.5	-679.9	3975;7446,8316
C_p°	25	22.69	94.93	7826;808
C_p	20	35.8	149.8	3347;380
	25	35.01	146.49	2977;7826
	Table 2.11			
t_c		347.1	620.3	4026;4771
p_c		44.325	4.491	2088
d_c		0.573	573	2088;4771
v_c		0.2909	0.0002909	2088
K_B	eq 2.79	5.50	5.50	697
κ	20	0.000555	0.0555	4902
ε	25	2.280	2.280	5411;4650,4937,7867,4771
μ	25 in 317	0	0	4937;7867,4771,1055,971
δ		9.3	19.0	652,4593,6905
ER	BuOAc	2.10	2.10	6905
	90%	168	168	6905
soly	in aq,25	0.015%w	0.015%w	4026;591,4771
	aq in,25	0.0105%w	0.0105%w	4026;6737
aq az	87.7	84.2%w	84.2%w	3497
fl pt	nonflam			4902;380

337. Tetrachloroethylene

(Continued)

Beil	11,187
uv	4956,4364,2884,3582
ir	2107,5889,7529,5932
Raman	6050,6382,8123,1783,7711,3770,640

338. Bromomethane

CH_3Br 74-83-9 CH_3Br

		cgs	SI	
mw		94.939	94.939	3468
bp	1 atm	3.55	276.70	4301,7351;2218,2088
dt/dp	1 atm	0.03372	0.2529	4301;7346,2088,2218
dp/dt	1 atm	29.66	3.954	4301
p	25	1633.0	217.7	2088
eq 2.5	A	7.08823	6.21313	4301,7351;4577
	B	1044.42	1044.42	
	C	244.684	244.684	
fp		-94.07	179.08	2088;7346,2218
d	20	1.6758	1675.8	7350,4300;5250
	25	1.6622	1662.2	
n_D	20	1.4164	1.4164	2510;7346,2988
	25	1.4130	1.4130	
η	-10	0.409	0.000409	2088
ΔH_v	25	5.56	23.26	4301;2088
	bp	5.760	24.098	
ΔH_f°	25(l)	-14.6	-61.0	7352;8316
	25(g)	-9.020	-37.740	4304;7352,2499,6789,8316
ΔH_c°	25(l)	-182.5	-763.6	8316
	25(g)	-188.1	-787.0	

338. Bromomethane

(Continued)

		cgs	SI	
$C_p^°$	25	10.147	42.456	4304;7352,7972,2706,7007
C_p	25	27.98	117.07	4327
	Table 2.11			
t_c		191	464	4299;2088
A	eq 2.72	0.02120	0.02120	2088
ε	-20	10.91	10.91	5250
μ	25 in 317	1.73	1.73	4514
	(g)	1.797	1.797	6702;3111
δ	25	9.6	19.6	653

Beil	5,66
uv	3300,3351,4006
ir	1168,3750,7972
Raman	1785,5494
ms	3035,4785
nmr	98,3256

339. Bromoethane

CH_3CH_2Br 74-96-4 C_2H_5Br

		cgs	SI	
mw		108.966	108.966	3468
bp	1 atm	38.35	311.50	7346;7494,7759,5329,6784 6864,7470
dt/dp	1 atm	0.0381	0.286	8304;7470,7494
dp/dt	1 atm	26.2	3.50	8304
p	25	468.6	62.47	2089;7058,6090,8304
eq 2.5	A	6.91995	6.04485	3856
	B	1090.810	1090.810	
	C	231.71	231.71	
fp		-118.6	154.6	7484,7479;7058,7494,1090 6383,6784

339. Bromoethane

(Continued)

		cgs	SI	
d	0	1.50136	1501.36	7494;5710,6784,2999,5250 5251,5329
	15	1.47080	1470.80	7494;6872,7470,7487,7759,6864
	25	1.4505	1450.5	7346
	30	1.44030	1440.30	
n_D	15	1.42756	1.42756	7494;4694,3361,5329,6864,7759
	20	1.4239	1.4239	7346
	25	1.4212	1.4212	
dn/dt		0.00056	0.00056	7494;2992
η	15	0.418	0.000418	7494;3361,3367
	25	0.379	0.000379	5329
	30	0.348	0.000348	7494
	Table 2.2			
γ	15	24.83	0.02483	3283;5329,7759,3163
	20	24.15	0.02415	
	30	22.83	0.02283	
ΔH_v	31.40	6.585	27.55	7108
	bp	6.328	26.48	2089;7850,7754,7593
ΔH_m		1.400	5.858	7478
ΔH_f°	25(1)	-21.99	-92.01	7826;8316,4394,7352
	25(g)	-15.42	-64.52	7826;2499,7352
ΔH_c°	25(1)	-335.0	-1401.6	3975;7446,8316
C_p°	25	15.42	64.52	7826,2946;3949,7352
C_p	25	24.1	100.8	7826
	Table 2.11			
t_c		230.7	503.9	7344;923,1671,3345,4299
p_c		61.5	6.23	7344,3345;7850,4299
d_c		0.507	507	7344,3345;4299
v_c		0.215	0.000215	7344;4088,1671,4299
K_B	eq 2.79	2.53	2.53	7850;726
κ	25	$<2 \times 10^{-8}$	$<2 \times 10^{-6}$	5708
ϵ		9.39	9.39	4937;8184,7497,3360,3361 5251,6873
μ	25(1)	1.90	1.90	3361;4514,489,5251,6478,6873
δ		8.91	18.23	6240;4593,3395,652
soly	in aq,20	0.91%w	0.91%w	5713;2601
aq az	37	99.1%w	99.1%w	3497

339. Bromoethane

(Continued)

Beil	<u>8</u>,88
uv	472,3633
ir	5272,5888,7077,4757,1947,1158,2687,5932
Raman	4400,6643,7832,4757,1783,2687
ms	1592,4785
nmr	1396,4694,3256,4852,1795,5933,6748

340. 1-Bromopropane

$CH_3CH_2CH_2Br$ 106–94–5 C_3H_7Br

		cgs	SI	
mw		122.992	122.992	3468
bp	1 atm	70.97	344.12	6784;7494,7759,2986,2999 4971,7472
dt/dp	1 atm	0.042	0.32	7346;7494
dp/dt	1 atm	24	3.2	7346
p	25	138.3	18.44	2089;7058
eq 2.5	A	6.91470	6.03960	7108;4577
	B	1194.33	1194.33	
	C	225.223	225.223	
fp	(st)	−108.1	165.1	Beil EIII 1,240;7472,7494
	(ms)	−109.8	163.4	6784,7058
d	20	1.3537	1353.7	7346;7759,7494,1697,2027,2677 2999,6748
	25	1.3452	1345.2	
dd/dt	25	0.0017	1.7	<u>7346</u>
n_D	20	1.4343	1.4343	7346;1697,7759,3361
	25	1.4317	1.4317	
dn/dt		0.00052	0.00052	7494
η	15	0.539	0.000539	7494;2677,3361,7404
	30	0.459	0.000459	7494;2178

340. 1-Bromopropane

(Continued)

		cgs	SI	
γ	20	25.85	0.02585	3283;7759
	30	24.59	0.02459	
ΔH_v	25	7.620	31.88	7822;2089,7821
	68.8	7.137	29.86	4971;7108
ΔH_m	(st)	1.560	6.527	7478;7477
	(ms)	2.160	9.037	Beil III 1,240
ΔH_f°	25(l)	−27.5	−115.1	8316
	25(g)	−21.13	−88.41	6789
ΔH_c°	25(l)	−491.54	−2056.60	7483,899;3975,7446,8316
C_p°	25	20.66	86.44	2946;3949
C_p	25	33.46	140.00	4327
	Table 2.11			
λ	37.8	0.000236	0.0987	6352
ε	25	8.09	8.09	4937;2027,3360,3361
μ	20 in 38	1.93	1.93	1697;2999,6862,4514
δ		8.9	18.2	4593
soly	in aq,30	0.230%w	0.230%w	2986;7057
aq az	62.8	91.3%w	91.3%w	6165
fl pt	TCC	<79	<352	6168

Beil	10,108
uv	3582,3574,7229
ir	5272,7077,4505,4757,8224,6957,1158,3240,5932
Raman	5385,6165,4757,1783,8224,640
ms	4785
nmr	1396,6122,5933,6748

341. 2-Bromopropane

$CH_3CHBrCH_3$ 75–26–3 C_3H_7Br

		cgs	SI	
mw		122.992	122.992	3468
bp	1 atm	59.41	332.56	6784;7494,7759,1697,4971 5329,7473
dt/dp	1 atm	0.0404	0.303	7346;2089
dp/dt	1 atm	24.8	3.30	7346
p	25	236.3	31.50	2089
eq 2.5	A	6.80251	5.92741	7108
	B	1106.82	1106.82	
	C	222.851	222.851	
fp		−89.0	184.2	7346;7473,7495,6784
d	20	1.3140	1314.0	7346;7759,7494,7495,1697 2999,6784
	25	1.3060	1306.0	
dd/dt	25	0.0016	1.6	7346
n_D	15	1.42847	1.42847	7494,7495;7759,1697,3361,5329
	20	1.4251	1.4251	7346
	25	1.4221	1.4221	2089,7346
dn/dt		0.00057	0.00057	7494,7495
η	15	0.536	0.000536	7494,7495;3361,5329
	30	0.437	0.000437	
	Table 2.2			
γ	20	22.90	0.02290	3283;5329,7759
	30	21.74	0.02174	
ΔH_v	25	7.21	30.16	7822;7821
	58.6	6.788	28.40	4971;7108
ΔH_f°	25(1)	−29.9	−125.1	8316;3471
ΔH_c°	25(1)	−490.42	−2051.92	899;8316
ε	25	9.46	9.46	4937;3360,3361
μ	20 in 38	2.04	2.04	1697,6862;2999
soly	in aq,18	0.286%w	0.286%w	2601
aq az	54.0	95.3%w	95.3%w	6165
Beil	10,108			
uv	3573,3574,7229			
ir	5272,7077,4757,4048,5932			
Raman	4757,6165,4048			
ms	4785			
nmr	1396,3256,4852,6748			

342.   1-Bromobutane

$CH_3CH_2CH_2CH_2Br$ 109-65-9 C_4H_9Br

		cgs	SI	
mw		137.019	137.019	3468
bp	1 atm	101.60	374.75	7494,7346;4971,6784
dt/dp	1 atm	0.0454	0.341	7346,2089
dp/dt	1 atm	22.0	2.94	7346,2089
p	25	41.27	5.502	
eq 2.5	A	6.92254	6.04744	7345;4577,7108
	B	1298.608	1298.608	
	C	219.70	219.70	
fp		-112.4	160.8	7494,7346;6784,1895
d	20	1.2758	1275.8	5329,7346;7494,4373,6874
	25	1.2687	1268.7	
dd/dt		0.00144	1.44	7494
κ_T	25	1.368×10^{-7}	1.026×10^{-6}	4376,1987
n_D	20	1.4401	1.4401	7346;900,7494,5329
	25	1.4378	1.4378	
dn/dt		0.00050	0.00050	7494
η	20	0.633	0.000633	5329;2178,7494,3361
	25	0.597	0.000597	
γ	20	25.37	0.02537	2089;3283,5329,7759
	30	24.25	0.02425	
ΔH_v	25	8.75	36.60	7822;2089,2946,4152
	bp	7.61	31.85	2089;4971,7108
ΔH_m		1.600	6.694	7472;1895
ΔH_f°	25(1)	-34.47	-144.22	900,2946;3471
	25(g)	-25.65	-107.32	2946
ΔH_C°	25(1)	-649.17	-2716.13	900
C_p°	25	26.13	109.33	2946
C_p	25	40.19	168.15	4327;1895
	Table 2.11			
λ	37.8	0.000240	0.100	6352
ε	20	7.099	7.099	489;3361,6874,2027
μ	25 in 38	1.956	1.956	3245;489,1698,3117,4514
δ	25	8.7	17.8	653,4593
soly	in aq,30	0.0608%w	0.0608%w	2986;2601,7219
fl pt	CC	18	291	5415

342.   1-Bromobutane

(Continued)

Beil	<u>10</u>,119
ir	1158,6957,5563,5932
Raman	3824,1718,1784,5563
ms	4785
nmr	7227,6748

343.   2-Bromobutane

$CH_3CH_2CHBrCH_3$ 78–76–2 C_4H_9Br

		cgs	SI	
mw		137.019	137.019	3468
bp	1 atm	91.22	364.37	7346;2088,6784,7476
dt/dp	1 atm	0.04510	0.3383	2088;7346
dp/dt	1 atm	22.17	2.956	2088
p	25	64.65	8.619	2088
eq 2.5	A	6.82724	5.95214	2088
	B	1229.08	1229.08	
	C	220	220	
fp		-112.65	160.50	2088;7476,6784,7346
d	20	1.26085	1260.85	2088;7346,7759,7485,3317
	25	1.25354	1253.54	
dd/dt	25	0.00146	1.46	<u>2088</u>
n_D	20	1.43705	1.43705	2088;7346,3317,7759
	25	1.43453	1.43453	
dn/dt	25	0.000504	0.000504	<u>2088</u>
η	25	0.563	0.000563	3317;7485,1626,3361
γ	20	25.01	0.02501	2088;3283,7759,7485
	30	23.90	0.02390	
ΔH_v	25	8.239	34.470	7822;7821,2088
	bp	7.354	30.768	2088

343. 2-Bromobutane

(Continued)

		cgs	SI	
ΔH_m		1.646	6.887	2088
ΔH_f°	25(1)	−37.06	−155.06	1703;3471
	25(g)	−28.85	−120.71	1703
ΔH_c°	25(1)	−646.56	−2705.21	1703;899
t_c		286	559	2088
P_c		35.21	3.568	2088
d_c		0.4124	412.4	2088
v_c		0.3322	0.0003322	2088
A	eq 2.72	0.03217	0.03217	2088
ε	25	8.64	8.64	3361
μ	in 38	2.12	2.12	5697
δ	25	8.4	17.2	653
fl pt		21	294	6402

Beil	10,119
uv	4006,7229
ir	5384,4757,5932
Raman	5384,1793,4757,1069
ms	4785
nmr	6499,6748

344. 2-Bromo-2-methylpropane

$(CH_3)_3CBr$ 507–19–7 C_4H_9Br

		cgs	SI	
mw		137.019	137.019	3468
bp	1 atm	73.25	346.40	7346,7485;1219,4346
dt/dp	1 atm	0.0416	0.312	7346
dp/dt	1 atm	24.0	3.20	7346
p	25	141.2	18.83	2089
eq 2.5	A	6.66850	5.79340	2089
	B	1129.7	1129.7	
	C	225	225	

344.   2-Bromo-2-methylpropane

(Continued)

		cgs	SI	
fp		−16.20	256.95	7479,7346;1219,4346
d	20	1.2209	1220.9	7346;7485,489,1219
	25	1.2132	1213.2	
dd/dt	25	0.00154	1.54	7346
n_D	20	1.4278	1.4278	7346;4346,3361,489,1219
	25	1.4252	1.4252	
dn/dt	25	0.00052	0.00052	7346
η	25	0.750	0.000750	3361
	30	0.702	0.000702	7485
γ	20	21.23	0.02123	2089;7485
	30	20.15	0.02015	
ΔH_v	25	7.603	31.81	7822;2089
	bp	6.577	27.52	2089
ΔH_m		0.47	1.97	4346
ΔH_t	−64.5	1.35	5.65	4346
	−41.6	0.25	1.05	
ΔH_f°	25(1)	−39.3	−164.4	1703;3471
	25(g)	−31.9	−133.5	1703
C_p	−8.0	36.1	151.0	4346
ε	25	10.30	10.30	3361;489,4346
μ	20 in 317	2.17	2.17	489;3014,6914,5697
aq az	75.3	92.7%w	92.7%w	3497

Beil	10,127
uv	4006
ir	823,6644,4757,5932
Raman	823,1792,1069,4757
ms	4785
nmr	3256,6748

345. 1-Bromopentane

$CH_3(CH_2)_3CH_2Br$ 110–53–2 $C_5H_{11}Br$

		cgs	SI	
mw		151.046	151.046	3468
bp	1 atm	129.58	402.73	7346;6733,6784,5329,4346
dt/dp	1 atm	0.0482	0.362	7346,2089
dp/dt	1 atm	20.7	2.77	7346,2089
p	25	12.60	1.680	2089
eq 2.5	A	6.95580	6.08070	4577,7345
	B	1401.634	1401.634	
	C	214.38	214.38	
fp		−87.9	185.3	6784,7346;4346,1895,6733
d	20	1.2182	1218.2	7346;5329,900,6733,4591
	25	1.2119	1211.9	1895,7759
dd/dt	25	0.00126	1.26	7346
n_D	20	1.4443	1.4443	7346;900,6733,4346,5329
	25	1.4420	1.4420	
dn/dt	25	0.00046	0.00046	7346
η	20	0.803	0.000803	5329;2178,3361,1626,6733
	25	0.753	0.000753	
γ̇	20	27.35	0.02735	5329;3283,7759,2089
	25	26.80	0.02680	
ΔH_v	25	9.902	41.43	7822;2946,2089,7943
	bp	8.243	34.49	2089
ΔH_m		2.74	11.46	4346;1895
ΔH_f°	25(1)	−40.72	−170.37	2946,900
	25(g)	−30.70	−128.45	2946
ΔH_c°	25(1)	−805.30	−3369.38	900
C_p°	25	31.60	132.21	2946
C_p	17.5	41.04	171.71	1895;4346
λ	37.8	0.000248	0.104	6352
ε	25	6.31	6.31	3361;4346
μ	25 in 317	1.96	1.96	4514;1698,6654,3361
soly	in aq,25	0.0127%w	0.0127%w	7219
fl pt	CC	32	305	6402
Beil	10,131			
uv	7229			
ir	1158,4993,780,5932			
Raman	1793,4993			
ms	2046,4785			

346. 1-Bromodecane

$CH_3(CH_2)_8CH_2Br$ 112–29–8 $C_{10}H_{21}Br$

		cgs	SI	
mw		221.180	221.180	3468
bp	1 atm	240.6	513.8	7343
dt/dp	1 atm	0.0570	0.428	7343,2089
dp/dt		17.5	2.34	7343,2089
p	25	0.04	0.005	2089
eq 2.5	A	7.2336	6.3585	4577,7345
	B	1888.67	1888.67	
	C	193.3	193.3	
fp		−29.2	244.0	7343;5083
d	20	1.0702	1070.2	7343;7759,4169
	25	1.0656	1065.6	
dd/dt	25	0.00092	0.92	7343
n_D	20	1.4557	1.4557	7343;8033,3361,4169
	25	1.4538	1.4538	
dn/dt	25	0.00038	0.00038	7343
η	20	2.37	0.00237	2178;3361
	30	2.02	0.00202	
γ	26.5	29.09	0.02909	7759
ε	25	4.44	4.44	3361
μ	25 in 317	1.93	1.93	4514;3301,3361
Beil	10,EII 130			
ir	1158,780,5932			
ms	4785			

347. Bromobenzene

C_6H_5Br 108-86-1 C_6H_5Br

		cgs	SI	
mw		157.010	157.010	3468
bp	1 atm	155.908	429.058	8304;7494,2314,2090,2986 4969,7059
dt/dp	1 atm	0.05165	0.3874	8304;2090,7494
dp/dt	1 atm	19.36	2.581	8304
p	25	4.182	0.5576	8304;8298
eq 2.5	A	7.25422	6.37912	2090;8298
	B	1688.4	1688.4	
	C	230	230	
fp		-30.82	242.33	2090;7059,7471,7494
tp		-30.749	242.401	4947
d	15	1.50170	1501.70	7549;2739,4969,7766,8231,487 879,887,888,2027,2090
	20	1.4959	1495.9	4195;5429
	25	1.48820	1488.20	7549;5429
	30	1.48150	1481.50	7549;7157
dd/dt	15-30	0.00135	1.35	<u>7549</u>
α		0.000933	0.000933	4195
κ_T	20	8.37×10^{-8}	6.28×10^{-7}	1880
κ_S	30	6.96×10^{-8}	5.219×10^{-7}	7157
n_D	15	1.56252	1.56252	7494;4069,7766,487,879 2090,3361
	20	1.55680	1.55680	6584
	25	1.55709	1.55709	2087
dn/dt		0.00049	0.00049	7494
η	15	1.196	0.001196	7494;3361,5136
	30	1.010	0.001010	6029;7494
γ	20	35.80	0.03580	4195;3283,7766,7070
	30	35.09	0.03509	3283;46
ΔH_v	20	10.507	43.963	4947
	25	10.65	44.54	7822;3759,8298,5153,7848,7850
	bp	9.049	37.861	4969;7755,7754
ΔH_m		2.558	10.702	4947;7059,7478
ΔH_f°	25(1)	16.5	69.0	8316;3471
ΔH_c°	25(1)	-751.6	-3144.7	8316;6839,3807
C_p°		23.819	99.658	4947;1289
C_p	25	36.88	154.29	4947
t_c		397	670	8233;923,5971,4088,5153 8231,1671
P_c		44.6	4.52	5153;8231,5971,4088,2314
d_c		0.458	458	4088,8231;5971

347. Bromobenzene

(Continued)

		cgs	SI	
v_c		0.343	0.000343	4088;1671,8233
A	eq 2.72	0.02039	0.02039	8304
K_B	eq 2.79	6.26	6.26	3406
κ	25	1.2×10^{-11}	1.2×10^{-9}	6014;1998,3043
ε	25	5.40	5.40	487;2027,5905,7497,3360 3361,7850
μ	25 in 38	1.55	1.55	4488;3854,5318,7461,1170,489 1338,3002,3593,4490,1758 5354,1123,3538
δ		9.87	20.19	6240;652,8111,4187,3759
soly	in aq,30 aq in,25	0.0446%w 0.0424%w	0.0446%w 0.0424%w	2986 8111;3759
aq az	95.5	51.3%w	51.3%w	6165
fl pt	CC	65	338	5712

Beil	464,206
uv	155,460,2054,5954,6207,1830,5944,6686,6749
ir	627,1044,4472,5272,5888,5810,7077,3888,3791,153,8188,6777,5893,2961
Raman	3825,4126,8122,4112,5103,6777,1782
ms	4785
nmr	1429,4922,2318,5933,6748

348.   1-Bromonaphthalene

```
    H      Br
    C      C
 HC    C      CH
    C      C
 HC    C      CH
    C      C
    H      H
```

90-11-9 $C_{10}H_7Br$

		cgs	SI	
mw		207.069	207.069	3468
bp	1 atm	280.66	553.81	3475;7766,458,7058
dt/dp	1 atm	0.064	0.48	3779
dp/dt	1 atm	16	2.1	3779
p	25	0.00535	0.000713	7646
eq 2.5	A	5.38175	4.50665	3475
	B	929.64	929.64	
	C	91.06	91.06	
fp	(st)	6.20	279.35	3763;7058
	(ms)	0.2-0.7	273.4-273.9	3262
d	20	1.4834	1483.4	3763;7491,7766,646,458
	25	1.4785	1478.5	
	55	1.4480	1448.0	3361
dd/dt		0.000934	0.934	458
n_D	20	1.6580	1.6580	7096;3763,7491,7766,458
	40	1.6490	1.6490	
	60	1.6401	1.6401	
dn/dt	20-60	0.000448	0.000448	7096
η	25	4.52	0.00452	3361
	40	3.200	0.003200	
	55	2.330	0.002330	
γ	27	44.19	0.04419	7766;646,458
	41	43.03	0.04303	
	60.4	40.88	0.04088	
ΔH_v	56	13	56	7646
	bp	9.4	39.3	3475
ΔE_c		-1200	-5020	6841;6839
κ	25	3.66×10^{-11}	3.66×10^{-9}	6014
ε	25	4.83	4.83	3361;2100
μ	25(1)	1.29	1.29	3361;890
	25 in 38	1.55	1.55	2605
δ		10.6	21.7	4593;652

Beil	477,547
uv	5990,6749
ir	1515,2409,6611,5098,5932
Raman	1515,6547,6611,5098
ms	4785
nmr	4623,2261,6748,7227

349. Bromoethylene

Bromoethene

$CH_2{=}CHBr$ 593-60-2 C_2H_3Br

		cgs	SI	
mw		106.950	106.950	3468
bp	1 atm	15.80	288.95	7347;5042,3052
dt/dp	1 atm	0.038	0.29	7347
dp/dt	1 atm	26	3.5	7347
p	25	1033	137.7	2088;3052
eq 2.5	A	6.66715	5.79205	2088
	B	953.4	953.4	
	C	236.0	236.0	
fp		-139.54	133.61	2088;7347
d	20	1.4933	1493.3	7347;3052
	25	1.4738	1473.8	
dd/dt	25	0.0039	3.9	7347
n_D	20	1.4288	1.4288	2510;7347
	25	1.4260	1.4260	
dn/dt	25	0.00056	0.00056	2510
η	0	0.2393 cSt	2.393×10^{-7}	2088
γ	20	20.04	0.02004	2088
	25	18.97	0.01897	
ΔH_v	25	5.40	22.60	2088;3052
	bp	5.60	23.43	2088;5042
ΔH_m		1.224	5.121	2088
ΔH_f°	25(l)	13.3	55.6	7826,8316
	25(g)	18.7	78.2	8316
ΔH_c°	25(l)	-303.9	-1271.5	8316
	25(g)	-309.3	-1294.1	
C_p°	25	13.27	55.52	7826;3020,6153
C_p	15	25.8	107.9	5042
A	eq 2.72	0.03453	0.03453	2088
ε	5	5.628	5.628	2088
μ	20 in 38	1.36	1.36	929
fl pt	nonflam			150

Beil	11,188
ir	4917,7439,3020,5932
Raman	3020,3278,4124
nmr	4917

350. 2-Bromopropene

$CH_2{=}CBrCH_3$ 557-93-7 C_3H_5Br

		cgs	SI	
mw		120.977	120.977	3468
bp	1 atm	48.4	321.6	7347,1434;3222,6229
dt/dp	1 atm	0.041	0.31	7347
dp/dt	1 atm	24	3.3	
fp		-124.75	148.40	7473;1434
d	20	1.3890	1389.0	7347;1434,2372,6229
	25	1.3803	1380.3	
dd/dt	25	0.00174	1.74	7347
n_D	20	1.4426	1.4426	7347;1434,6229,2372
	25	1.4374	1.4373	
dn/dt	25	0.00104	0.00104	7347
μ	25 in 38	1.51	1.51	6229
Beil	11,200			
ir	4917,5087,5932			
Raman	3823,4032			
nmr	4917,8001			

351. Bromoform

Tribromomethane

$CHBr_3$ 75-25-2 $CHBr_3$

		cgs	SI	
mw		252.731	252.731	3468
bp	1 atm	149.21	422.36	4301;5251,7058,7771,7351,7494
				7495,1244
dt/dp	1 atm	0.05012	0.3759	4301;7494
dp/dt	1 atm	19.95	2.660	4301
p	25	5.9	0.79	3779

351. Bromoform

(Continued)

		cgs	SI	
eq 2.5	A	7.03141	6.15631	7351,4301;1009
	B	1511.50	1511.50	
	C	214.959	214.959	
fp		8.05	281.20	7494,7495;1244,2986,7058
d	20	2.8909	2890.9	7350,4300;7495,7771,1244,5251
				5710,7494,1969
	25	2.8779	2877.9	
dd/dt	25	0.00261	2.61	7350
n_D	15	1.60053	1.60053	7494;7495,7771
	20	1.59763	1.59763	7771
	25	1.5956	1.5956	7348
dn/dt		0.00055	0.00055	7494
η	15	2.152	0.002152	7494,7495
	30	1.741	0.001741	
γ	24.8	45.10	0.04510	7771;1969
	41.7	42.30	0.04230	
ΔH_v	25	10.68	44.67	4301;4435
	bp	9.30	38.91	4301;7167
ΔH_f°	25(l)	-6.8	-28.5	7826
	25(g)	4.000	16.736	4304;7352,7826,4948
ΔE_c		-90.3	-377.8	3807
C_p°	25	16.964	70.977	4304,7352;2792,2706,2790,7007
	Table 2.14			
C_p	17-21	31.75	132.84	1569;4327
	Table 2.11			
K_f	eq 2.67	14.4	14.4	330;7840,734,716
κ	25	$<2 \times 10^{-8}$	$<2 \times 10^{-6}$	5708
ϵ	20	4.39	4.39	4937;7458,1394,2027,5251,2100
μ	10-70 in 38	0.99	0.99	6874;5251,5318
δ		10.7	21.9	4187
soly	in aq,30	0.318%w	0.318%w	2986;987
aq az	94.2	66.1%w	66.1%w	6165
fl pt	TCC	<79	<352	6168,5713

351. Bromoform

(Continued)

Beil	5,68
uv	4955,3573,3574
ir	2259,5695,5894,7077,7537,3917,2653,5932
Raman	4125,7782,4200,1783,4655,4623
nmr	4740,4921,4852,2464,5933,6748

352. 1,2-Dibromoethane

CH_2BrCH_2Br 106-93-4 $C_2H_4Br_2$

		cgs	SI	
mw		187.862	187.862	3468
bp	1 atm	131.36	404.51	7348;6865,7765,2090,4969 5251,5329
dt/dp	1 atm	0.058	0.44	7348;1910,7494,2090
dp/dt	1 atm	17	2.3	7348
p	25	7.79	1.04	2089;1321
eq 2.5	A	7.19807	6.32297	2090
	B	1560.3	1560.3	
	C	230	230	
fp		9.79	282.94	7348;1321,5104,2986,5870 6025,7494
tp		9.7	282.9	1321
d	20	2.1791	2179.1	7348;5710,6865,7494,7765,2090 2125,4969,5329,3009,1321 2089,6791,8064,2431
	25	2.1687	2168.7	
dd/dt	25	0.00208	2.08	7348
α	20	0.000943	0.000943	8064;2431
κ_T	27	8.67×10^{-8}	6.50×10^{-7}	2431
n_D	15	1.54160	1.54160	7494;7765,2090,3361,5329,6865 3009,1321 7348
	20	1.53874	1.53874	
	25	1.5360	1.5360	

352. 1,2-Dibromoethane

(Continued)

		cgs	SI	
dn/dt		0.00058	0.00058	7494
η	15	1.880	0.001880	7494;2125,3361,5329,7453
	30	1.490	0.001490	
	Table 2.2			
γ	20	38.91	0.03891	3283;5329,7765,3009,3163
	30	37.61	0.03761	
ΔH_v	25	9.974	41.73	7822
	bp	8.688	36.35	4969;7848
ΔH_m		2.616	10.945	5870;6260,7167
ΔH_s		14.075	58.890	5551
ΔH_f°	25(1)	−19.4	−81.2	7826
	25(g)	−9.16	−38.33	
ΔH_c°		−296.5	−1240.6	6839;5916,6841
C_p°	25	20.7	86.6	7826
	110	23.14	96.82	3053
C_p	21.22	32.48	135.90	5870;8064,4327,6025,7500
	25	32.51	136.02	7826
	Table 2.11			
t_c		309.8	583.0	7349;7739,2088
P_c		70.6	7.15	7349;7739,2088
d_c		0.776	776	2088
v_c		0.242	0.000242	<u>2088</u>
K_f	eq 2.67	12.5	12.5	5216;734
K_B	eq 2.79	6.608	6.608	726;716
κ	19	$<2 \times 10^{-10}$	$<2 \times 10^{-8}$	2685;6791
ε	30	4.7503	4.7503	1983;3480,8004,6791,4937,2027 3361,5251
μ	20 in <u>38</u>	1.19	1.19	4045;103,4974,5251,8079,6865
δ		9.7	19.8	652
soly	in aq,30	0.429%w	0.429%w	<u>2986</u>;987
	aq in,25	0.071%w	0.071%w	2088
aq az	91.0	73.7%w	73.7%w	6165
fl pt	nonflam			5713

352. 1,2-Dibromoethane

(Continued)

Beil	8,90
ir	2259,5272,5890,6728,8175,104,5932
Raman	894,895,3277,5187,5190,8175,5551,104,3416
ms	154
nmr	4740,6122,2464,6748

353. 1,2-Dibromopropane

$CH_3CHBrCH_2Br$ 78–75–1 $C_3H_6Br_2$

		cgs	SI	
mw		201.888	201.888	3468
bp	1 atm	141.99	415.14	2088;7348,5865,7765
dt/dp	1 atm	0.05044	0.3783	2088;7348
dp/dt	1 atm	19.83	2.643	2088
p	25	8.037	1.072	2088;3815
eq 2.5	A	6.89105	6.01595	2088;5865
	B	1419.6	1419.6	
	C	212	212	
fp		−55.50	217.65	2088;7348
d	20	1.93268	1932.68	2088;7348,4373,7765
	25	1.92344	1923.44	
dd/dt	25	0.001849	1.849	2088
n_D	20	1.52004	1.52004	2088;7348,3978,3982
	25	1.51737	1.51737	
dn/dt	25	0.000534	0.000534	2088
η	20	1.623	0.001623	4817;2088
	30	1.399	0.001399	
γ	20	34.14	0.03414	2088;7765
	30	32.95	0.03295	
ΔH_v	25	10.14	42.43	2088;5865
	bp	8.49	35.52	

353. 1,2-Dibromopropane

(Continued)

		cgs	SI	
ΔH_m		2.136	8.937	2088
ΔH_f°	25(l)	−29.27	−122.47	1703
	25(g)	−17.00	−71.13	
C_p	25	38.23	159.95	4327
	Table 2.11			
t_c		371	644	2088
p_c		40.42	4.096	2088
d_c		0.605	605	2088
v_c		0.337	0.000337	2088
A	eq 2.72	0.02272	0.02272	2088
ε	25	4.369	4.369	2088
μ	25 in 38	1.43	1.43	103
	(g)	1.13	1.13	5182
soly	in aq,25	0.143%w	0.143%w	2088
	aq in,25	0.052%w	0.052%w	

Beil	10,109
uv	494,7229
ir	6906,104,5932
Raman	6906,104,2801,3827,3239
nmr	2432,4740,6748

354. 1,1,2,2-Tetrabromoethane

$CHBr_2CHBr_2$ 79-27-6 $C_2H_2Br_4$

		cgs	SI	
mw		345.654	345.654	3468
bp	1 atm	243.5	516.7	7058;7771
dt/dp	1 atm	0.0555	0.416	2089
dp/dt	1 atm	18.0	2.40	2089

354.   1,1,2,2-Tetrabromoethane

(Continued)

		cgs	SI	
p	25	0.02	0.003	2089;7472
eq 2.4	A	8.35595	7.48085	7058
	B	2821.95	2821.95	
fp		0.0	273.2	7472,4532
d	19.3	2.9656	2965.6	7482;646,4373,7771,5710
	25	2.9529	2952.9	
	29.95	2.9462	2946.2	1575
dd/dt	25	0.00225	2.25	7482
n_D	20	1.63533	1.63533	7771
	25	1.6323	1.6323	2089
η	10.97	13.950	0.013950	5136
	19.91	9.797	0.009797	
	24.93	9.687	0.009687	1575
γ	20	49.44	0.04944	646;3163
	25	48.65	0.04865	
ΔH_v	25	16.73	70.00	2089
	bp	11.63	48.65	
C_p°	25	25.64	107.28	1288
	Table 2.14			
C_p	25	39.6	165.7	7826
	Table 2.11			
K_f	eq 2.67	21.7	21.7	4532
ε	22	7.0	7.0	4937;7850
μ	20 in 9	1.38	1.38	4045;4230
δ	25	12.1	24.7	653
soly	in aq,30	0.0651%w	0.0651%w	2987;987,2845
aq az	99.8	11.4%w	11.4%w	6165

Beil	8,94
uv	3574
ir	6666,3794
Raman	337,1783
ms	154
nmr	4740,3794,6748

355. Iodomethane

CH_3I			74–88–4		CH_3I
		cgs	SI		
mw		141.939	141.939	3468	
bp	1 atm	42.43	315.58	7359,4301;7452,7485,7759	
				1009,1697,2999,5251	
dt/dp	1 atm	0.03885	0.2914	4301;7359,7485	
dp/dt	1 atm	25.74	3.432	4301	
p	25	405.9	54.12	2089;7058,7432	
eq 2.5	A	6.97241	6.09731	4301;1009,4577	
	B	1138.29	1138.29		
	C	235.774	235.774		
fp		−66.45	206.7	7359;7058,7504,7485,7479	
d	20	2.2792	2279.2	7362,5250,4300,5710,7452,7759	
				487,1697,2677	
	25	2.2650	2265.0		
dd/dt	25	0.00285	2.85	7362	
n_D	20	1.5308	1.5308	7359;487,1697,5251,7759,3190	
	25	1.5270	1.5270		
dn/dt	25	0.00076	0.00076	7359	
η	15	0.518	0.000518	7485;2677,7452	
	30	0.460	0.000460		
	Table 2.2				
γ	20	30.14	0.03014	7485,3283;7759	
	30	28.61	0.02861		
ΔH_v	25	6.685	27.970	4301	
	bp	6.534	27.338	4301;7848	
ΔH_f°	25(1)	−3.4	−14.3	7363;8316,7826,1373,4948,2895	
	25(g)	3.290	13.77	4304;7363,7826	
ΔH_c°	25(1)	−194.7	−814.6	3975;7446,8316	
C_p°	25	10.536	44.083	4304,7826,7363;2706,3949,2202	
	Table 2.14				
C_p	25	19.8	82.8	7363	
	30.1	19.95	83.47	3190;4327	
	Table 2.11				
t_c		255	528	4088;4561,4299	
K_B	eq 2.79	4.19	4.19	7845;726	
ε		7.00	7.00	4937;489,2999,5250,5251	
μ	25 in 317	1.48	1.48	4490;489,8079,1697,5251	
				7845,6897	
δ	25	10.2	20.9	653	
soly	in aq,20	1.4%w	1.4%w	4026;7662,2601	
aq az	40.0	99.3%w	99.3%w	6165	

355. Iodomethane

(Continued)

Beil	<u>5</u>,69
uv	3633,5327,6426,3214,472,2260,3574
ir	1688,4374,5919,7077,7472,5955,1168,5932
Raman	7831,4623
ms	4785,5984,2947,6781,1994
nmr	1396,3256,4852,2464,7257

356. Iodoethane

CH₃CH₂I   75-03-6   C₂H₅I

		cgs	SI	
mw		155.966	155.966	3468
bp	1 atm	72.30	345.45	7359;7759,7932,2986,2999 6864,7495
dt/dp	1 atm	0.043	0.32	7359;7495
dp/dt	1 atm	23	3.1	7359
p	25	136.2	18.16	2089;6090,6864,7058
eq 2.5	A B C	6.83198 1175.709 225.26	5.95688 1175.709 225.26	4566,7358
fp		-111.1	162.1	7359,7495;7058,7471
d	20	1.9357	1935.7	7359;7495,487,7759,2677,7848 2999,5251,1697,7932,5329
	25	1.9244	1924.4	
dd/dt	25	0.00226	2.26	<u>7359</u>
n_D	20	1.5133	1.5133	7359;7495,7759,7932,487,1697 4969,5329
	25	1.5101	1.5101	
dn/dt		0.00063	0.00063	7495
η	15 30	0.617 0.540	0.000617 0.000540	7495;2129,2677,5329
	Table 2.2			

356. Iodoethane

(Continued)

		cgs	SI	
γ	20	28.83	0.02883	3283;5329,7759,7848
	30	27.57	0.02757	
	Table 2.7			
ΔH_v	25	7.631	31.93	7822;7848,7850
	bp	7.116	29.77	4969
ΔH_f°	25(1)	-9.61	-40.2	7363;7826
	25(g)	-1.84	-7.70	7826;7363
ΔH_c°	25(1)	-356.0	-1489.5	3975;7446
C_p°	25	15.61	65.33	7363;3949,7826
C_p	25	27.5	115.1	7363;3808
	Table 2.11			
t_c		281.0	554.2	923;7848
K_B	eq 2.79	5.16	5.16	726;2798
ε	20	7.82	7.82	4937;2100,487,5251,6876,7850
μ	20 in 38	1.78	1.78	1697;489,4498,4826,5251,6862
δ	25	9.4	19.2	3395,4593
soly	in aq,30	3.88%w	3.88%w	2986;2601
aq az	66	96-97%v	96-97%v	3497

Beil	8,96
uv	3633,6426,3214,472,2260,3574
ir	5888,6911,4757,1947,5932
Raman	6643,7832,4757,1785,2163
ms	1592,4785,5984,2947
nmr	1396,4694,3256,4852,1795,5933,6748

357.   1-Iodopropane

$CH_3CH_2CH_2I$ 107-08-4 C_3H_7I

		cgs	SI	
mw		169.993	169.993	3468
bp	1 atm	102.45	375.60	7359;7485,7759,1697,2986
				2999,5329
dt/dp	1 atm	0.047	0.35	7359
dp/dt	1 atm	21	2.8	7359
p	25	43.09	5.745	2089,7058
eq 2.5	A	6.81603	5.94093	4577,7358
	B	1267.062	1267.062	
	C	219.53	219.53	
fp		-101.3	171.9	7359,7485;7479,3048,7058,7471
d	20	1.7489	1748.9	7359;7485,1697,7759,2677
				2999,5329
	25	1.7394	1739.4	
dd/dt	25	0.00190	1.90	7359
n_D	20	1.5058	1.5058	7359;1697,5329,7759
	25	1.5028	1.5028	
dn/dt	25	0.00060	0.00060	7359
η	15	0.837	0.000837	7485;2129,2677,5329,6456
	30	0.670	0.000670	
	Table 2.2			
γ	15	30.11	0.03011	936;3283,5329,7759
	30	28.47	0.02847	
ΔH_v	25	8.664	36.25	7822
ΔH_f°	26.85(1)	-16.00	-66.94	3808
	25(g)	-7.4	-31.0	2619
ΔH_c°	25(1)	-514.3	-2151.8	3975
C_p°	25	20.49	85.73	3949
C_p		34.68	145.10	6352;3808
λ	37.8	0.000208	0.0870	6352
t_c		323	596	6352
ε	20	7.00	7.00	6967
μ	20 in 38	1.84	1.84	1697;2806,6862,4498
soly	in aq,30	0.104%w	0.104%w	2986;2601,2094
Beil	10,113			
uv	3214,3216,2260,3574			
ir	7077,5932			
Raman	5385,6018			
ms	4785,5984			
nmr	1396,6122,5933,6748			

358. 2-Iodopropane

CH₃CHICH₃   75-30-9   C₃H₇I

		cgs	SI	
mw		169.993	169.993	3468
bp	1 atm	89.50	362.65	7359;7485,1697,7759,3564 7058,7472
dt/dp	1 atm	0.045	0.34	7359
dp/dt	1 atm	22	3.0	7359
p	25	43	5.7	3779
eq 2.4	A	7.75002	6.87492	3779
	B	1765.15	1765.15	
fp		-90.0	183.2	7359;7058,3048,7472,7775,7485
d	20	1.7042	1704.2	7359;1697,7485,7759
	25	1.6946	1694.6	
dd/dt	25	0.00192	1.92	7359
n_D	20	1.4991	1.4991	7767;1697,7759
	25	1.4961	1.4961	
dn/dt	25	0.00060	0.00060	7762
η	15	0.732	0.000732	7485;2129
	30	0.620	0.000620	
	Table 2.2			
γ	20	27.42	0.02742	936;7759
	30	26.13	0.02613	
ΔH_v	25	8.141	34.06	7822
ΔH°_f	26.85(1)	-18.40	-76.99	3808
	25(g)	-9.4	-39.3	2619
ΔH°_c	25(1)	-509.1	-2130.1	3975
C_p	26.85	21.77	91.09	3808
ε	20	8.19	8.19	4937
μ	20 in 38	1.95	1.95	1697
soly	in aq,20	0.140%w	0.140%w	6999

Beil	10,114
uv	3214,3574,2260
ir	7077,4757,4048,5932
Raman	6018,4757,4048
ms	7199,4785,5984
nmr	1396,3256,4852,5933,6748

359. 1-Iodobutane

$CH_3CH_2CH_2CH_2I$ 542-69-8 C_4H_9I

		cgs	SI	
mw		184.020	184.020	3468
bp	1 atm	130.53	403.68	7359;7485,5329
dt/dp	1 atm	0.0500	0.375	7359,2089
dp/dt	1 atm	20.0	2.67	7359,2089
p	25	13.86	1.848	2089
eq 2.5	A	6.82262	5.94752	7358,4577
	B	1358.860	1358.860	
	C	214.20	214.20	
fp		-103.0	170.2	7359,7485
d	20	1.6154	1615.4	7359;5329,4373,6874,7485
	25	1.6072	1607.2	
dd/dt	25	0.00164	1.64	7359
n_D	20	1.5001	1.5001	7359;5329,488,7485
	25	1.4973	1.4973	
dn/dt	25	0.00056	0.00056	7359
η	20	0.877	0.000877	5329;1626,7485
	25	0.826	0.000826	
γ	20	29.25	0.02925	5329;3283,7759,7485
	25	28.70	0.02870	
ΔH_v	25	9.711	40.63	7822;2089
	bp	7.983	33.40	2089;4969
ε	20	6.288	6.288	488;6874
μ	20 in 317	1.93	1.93	489;5677,1698,3111
δ	25	8.6	17.6	653
soly	aq in,17.5	0.012%w	0.012%w	2601

Beil	10,123
uv	4006,2260,3214
ir	5563,5932
Raman	3824,5563,3826,1788
ms	4785,7199,5580
nmr	6748

360.   1-Iodo-2-methylpropane

$(CH_3)_2CHCH_2I$ 513–38–2 C_4H_9I

		cgs	SI	
mw		184.020	184.020	3468
bp	1 atm	120.4	393.6	7359,7058;402,7485
dt/dp	1 atm	0.048	0.36	7359
dp/dt	1 atm	21	2.8	7359
p	29.8	20.0	2.67	7058
fp		-90.7	182.5	7359,3048;7485
d	20 25	1.6035 1.5952	1603.5 1595.2	7359;488,7759,7485
dd/dt	25	0.00166	1.66	7359
n_D	20 25	1.4960 1.4935	1.4960 1.4935	7359;488,7759
dn/dt	25	0.00050	0.00050	7359
η	30	0.777	0.000777	7485
γ	20 30	27.97 27.04	0.02797 0.02704	3283;7759
ΔH_v	25	9.28	38.83	7822
ΔH_f°	25(1) 25(g)	-25.9 -17.4	-108.4 -72.8	1703 1703;8316
ΔH_c°	25(g)	-666.4	-2788.2	8316
ε	20	6.466	6.466	488
μ	in 38	1.87	1.87	5697
aq az	95-96	79%v	79%v	3497

Beil	10,128
uv	4006,2260,3214
ir	1727,4757
Raman	3826,1727,5648,4183,4757
ms	4785
nmr	6748

361. Diiodomethane

CH_2I_2 75-11-6 CH_2I_2

		cgs	SI	
mw		267.836	267.836	3468
bp	1 atm	182 d	455	7360
dt/dp	1 atm	0.05359	0.4019	4301;7360
dp/dt	1 atm	18.66	2.488	4301
p	25	1.2	0.16	7361;7771
eq 2.5	A	6.961	6.086	7361;4301
	B	1575	1575	
	C	204	204	
fp	(st)	6.1	279.3	7360,1897
	(ms)	5.60	278.75	7482,7489,1897
d	20	3.3212	3321.2	7362;7489,2958,7771,4195
	25	3.3079	3307.9	
dd/dt	25	0.00266	2.66	7362
α		0.000833	0.000833	4195
n_D	20	1.7411	1.7411	7360
	25	1.7380	1.7380	
dn/dt	25	0.00062	0.00062	7360
η	15	3.043	0.003043	7482;2958
	30	2.392	0.002392	
γ	20	50.88	0.05088	4195;3160
ΔH_v	25	11.80	49.38	4301
	bp	10.16	42.49	
ΔH_m	(st)	10.7	44.8	721
	(ms)	11.2	46.9	
ΔH_f°	25(1)	15.9	66.5	7363;7826,8316,4948
	25(g)	28.100	117.57	4304,7363;7826,2618
ΔH_c°	25(1)	-178.4	-746.4	3975,8316;3621
C_p°	25	13.798	57.731	4304,7363;2790,7826
	Table 2.14			
C_p	25	32.0	133.9	7827;4327
K_f	(st)	14.4	14.4	721
eq 2.67	(ms)	13.7	13.7	
ε	25	5.316	5.316	6874
μ	25-50 in 38	1.08	1.08	6874;4273
δ		10.2	20.9	652
soly	in aq,30	0.124%w	0.124%w	2986

361. Diiodomethane

(Continued)

Beil	5,71
uv	3214,2260,3573,3574
ir	2480,611,2259,5932
Raman	1786,530
ms	7226
nmr	4852,527,6748

362. Iodobenzene

C_6H_5I 591-50-4 C_6H_5I

		cgs	SI	
mw		204.010	204.010	3468
bp	1 atm	188.33	461.48	2087;7058,7059
dt/dp	1 atm	0.0554	0.416	2087
dp/dt	1 atm	18.1	2.41	2087
p	25	1.009	0.1345	2087;7058
eq 2.5	A	6.89506	6.01996	2087;8296
	B	1562.87	1562.87	
	C	201.0	201.0	
fp		−31.35	241.80	7489;7059,2087
d	20	1.8308	1830.8	2087;4373,7489,7766,5082
	25	1.8229	1822.9	
dd/dt	25	0.00158	1.58	2087.
α		0.000856	0.000856	4195
κ	20	7.49×10^{-8}	5.62×10^{-7}	1880
n_D	20	1.6200	1.6200	2087;7766,489,5082,6584
	25	1.6172	1.6172	
dn/dt	25	0.00056	0.00056	2087
η	15	1.740	0.001740	7489;5082
	30	1.417	0.001417	

362. Iodobenzene

(Continued)

		cgs	SI	
γ	18.4 41.3	39.38 36.97	0.03938 0.03697	7766;2087,4195
ΔH_v	25 bp	11.85 9.44	49.58 39.50	2087;8296
ΔH_m		2.330	9.748	2087,7059
ΔH_f°	25(1) 25(g)	29.0 40.5	121.3 169.5	8316;1372 8316
ΔH_c°	25(1) 25(g)	−764.1 −775.6	−3197.0 −3245.1	8316;6838 8316
C_p°	25	24.09	100.79	1289,7999
C_p	26.85	37.96	158.82	7059
t_c		448	721	4299,2087
P_c		44.6	4.52	4299,2087
d_c		0.581	581	4299,2087
v_c		0.351	0.000351	4299
κ		2.40×10^{-10}	2.40×10^{-8}	3043
ε	20	4.486	4.486	7497;489,3538
μ	25 in 38	1.40	1.40	4634;7461,3261,4051,489,3538
δ	25	10.1	20.7	653;6240,8111,3759
soly	in aq,30 aq in,25	0.034%w 0.0276%w	0.034%w 0.0276%w	2987,365;7219 8111;3759

Beil	464,215
uv	7229,6686,7606,4068,6749
ir	3791,5890,610,5888,4440,2961,5932
Raman	1788,3825
ms	4785
nmr	6931,1429,8169,5933,6748

363. Chlorotrifluoromethane

$CClF_3$ 75-72-9 $CClF_3$

		cgs	SI	
mw		104.459	104.549	3468
bp	1 atm	-81.44	191.71	4301;7338,2142,5147,7450,2451
dt/dp	1 atm	0.02464	0.1848	4301
dp/dt	1 atm	40.58	5.410	4301
p	-111.9	104	13.9	7450
	25	27100	3613	2437
eq 2.5	A	6.80316	5.92806	4301,7338;72,7450
	B	663.370	663.370	
	C	250.537	250.537	
fp		-189.0	84.2	5147;2142,2451
d	-30	1.296	1296	5505;2142,72
	20	0.9224	922.4	7337
n_D	-73	1.199	1.199	2142
η	25	0.016	1.6×10^{-5}	2142
γ	0.50	4.416	0.004416	6078
	20	0.574	0.000574	5340
ΔH_v	bp	3.764	15.749	4301;7450,2142
ΔH_f°	25(g)	-169.2	-707.9	7342,1444;1703,4148,548,4029
				3906,7910,4659
ΔH_c°	25(g)	-345.9	-1447.2	1703
C_p°	25	15.98	66.86	7340,1444;2706,71
C_p	-30	25.8	108.0	2142
λ	25	8.3×10^{-5}	0.035	2142
t_c		28.73	301.88	5566;5340,6078,2142,7337,2451
				72,5105
P_c		39.55	4.007	5566;6078,2142,2451,72
d_c		0.5824	582.4	5566;6078,72,7337,2142
v_c		0.1794	0.0001794	5566
ϵ		2.32	2.32	8092
μ	(g)	0.50	0.50	3356;2006,2614
soly	in aq,25	0.009%w	0.009%w	2142
Beil	5,EIII 42			
ir	5894,4463,7434,6667			
Raman	7198,3828,1496,6667			
ms	1996,2817,709,7226			
nmr	5564			

364. Dichlorodifluoromethane

CCl_2F_2 75-71-8 CCl_2F_2

		cgs	SI	
mw		120.914	120.914	3468
bp	1 atm	−29.77	243.38	4301,7338;2142,1242,2891
dt/dp	1 atm	0.03090	0.2318	4301
dp/dt	1 atm	32.36	4.314	4301
p	20	4250	566.6	2749;8276
eq 2.5	A	6.82100	5.94591	4301,7338;2749
	B	839.622	839.622	
	C	242.861	242.861	
fp		−158.2	115.0	5147;2142
d	20	1.3292	1329.2	7337;2142,851
	25	1.3113	1311.3	
n_D	20	1.2950	1.2950	2510;2142
	25	1.2909	1.2909	
η	−15	0.32	0.00032	518
	30	0.22	0.00022	
γ	25	9	0.009	2142
ΔH_v	bp	4.807	20.112	4301;2142,5439,1242
ΔH_m		0.99	4.14	516
ΔH_f°	25(g)	−117.9	−493.3	1444,7342;1703,4148,4029,7910
ΔH_c°	25(g)	−354.1	−1481.6	1703
C_p°	25	17.31	72.43	1444,7340;7826,2142,2706
				71,4945
C_p	25	28.0	117.2	2142;1242
λ	25	0.00017	0.071	2142
t_c		111.8	385.0	7337,5105;2142,851,2891
P_c		40.71	4.125	5105;2142,851,2891
d_c		0.558	558	7337,2142,2891;851
v_c		0.217	0.000217	2142
ε	29	2.13	2.13	2142
μ	29(g)	0.55	0.55	2614;6867
δ	25	5.5	11.3	653;6723,6641,6371
soly	in aq,25	0.028%w	0.028%w	2142
	aq in,25	0.009%w	0.009%w	

Beil	5,61
uv	2873,4364
ir	5894,4463,7434,6667
Raman	1047,6667,1496
ms	709
nmr	12

365. Trichlorofluoromethane

CCl_3F 75-69-4 CCl_3F

		cgs	SI	
mw		137.368	137.368	3468
bp	1 atm	23.63	296.780	4301,7338;2142,5612,770
dt/dp	1 atm	0.03715	0.2786	4301;771
dp/dt	1 atm	26.92	3.589	4301
p	24.02	766.8	102.2	771;8276
eq 2.5	A	6.89396	6.01886	4301,7338;5612,771
	B	1047.04	1047.04	
	C	237.276	237.276	
fp		−110.48	162.67	5612;2142,770,5147
d	20	1.4879	1487.9	7337;7004,772,2142,1969
	25	1.4760	1476.0	
dd/dt	20	0.00237	2.37	772
n_D	20	1.3824	1.3824	2510;7004
	25	1.3794	1.3794	
dn/dt	25	0.00060	0.00060	2510
η	0	0.548	0.000548	5148
	20	0.443	0.000443	
γ	15	19.09	0.01909	1969
	25	18	0.018	2142
ΔH_v	25	5.98	25.02	4301;5612,2142
	bp	5.99	25.06	
ΔH_m		1.648	6.895	5612
ΔH_f°	25(1)	−72.02	−301.33	8316,7826;5500
	25(g)	−68.1	−284.9	1444,7342;548,4029,7910
C_p°	25	18.66	78.07	1444,7340,7826;71,2706,806,2142
C_p	25	29.05	121.55	7826;2142,774,5612
	Table 2.11			
λ	25	0.00021	0.088	2142
t_c		198.0	471.2	7337,770,2142
p_c		43.5	4.41	2142;770
d_c		0.554	554	7337,770,2142
v_c		0.248	0.000248	2142
ε	20	2.303	2.303	5148
μ	(g)	0.46	0.46	6118;5148,2614,6867
δ	25	7.6	15.5	653;6723,6641,6371
soly	in aq,25	0.11%w	0.11%w	2142
	aq in,25	0.011%w	0.011%w	
Beil	5,64			
uv	2873,4364			
ir	806,5894,4463,7434,6667			
Raman	8293,1496,6667,2793			
ms	1769			

366. Chlorodifluoromethane

$CHClF_2$ 75-45-6 $CHClF_2$

		cgs	SI	
mw		86.469	86.469	3468
bp	1 atm	-40.83	232.32	4301;7338,770,5438,2142
dt/dp	1 atm	0.02807	0.2106	4301;771
dp/dt	1 atm	35.62	4.749	4301
p	-61.26	263.4	35.12	771
	4.5	4279	570.5	8276
eq 2.5	A	6.93892	6.06382	4301,7338;771
	B	808.919	808.919	
	C	240.161	240.161	
fp		-157.42	115.73	5438;770,2142
d	20	1.2136	1213.6	7337;772,2142,5827
	25	1.1942	1194.2	
dd/dt	20	0.00374	3.74	772
n_D	20	1.267	1.267	2510;2142
	25	1.264	1.264	
η	12.39	0.2058	0.0002058	5827
	25.97	0.1834	0.0001834	
γ	25	8	0.008	2142
ΔH_v	bp	4.837	20.238	4301;2142,5439,5438
ΔH_m		0.98547	4.1232	5438
ΔH_t	-214	0.016	0.067	5438
ΔH_f°	25(g)	-115.6	-483.7	1444,7342;3906
ΔH_c°	25(g)	-29.1	-121.8	1703
C_p°	25	13.35	55.86	7340,7826,1444,2706;2286,2791
C_p	-40.7	22.234	93.027	5438;774
	Table 2.11			
λ	25	0.00021	0.090	2142
t_c		96.0	369.2	7337,770,2142;989
P_c		49.12	4.977	2142;989,770
d_c		0.525	525	7337,770,2142
v_c		0.165	0.000165	2142
ϵ	24	6.11	6.11	2142
μ	25(g)	1.41	1.41	2614;6867,4934
δ	25	7.3	14.9	653;6371,6723,6641
soly	in aq,25	0.30%w	0.30%w	2142
	aq in,25	0.13%w	0.13%w	

366.   Chlorodifluoromethane

(Continued)

Beil	5,EIII 41
ir	4463,5894
Raman	2794,2789,2791
ms	7226
nmr	5322,4552,7227

367.   Dichlorofluoromethane

$CHCl_2F$ 75–43–4 $CHCl_2F$

		cgs	SI	
mw		102.923	102.923	3468
bp	1 atm	8.90	282.05	4301,7338;771,770,2142,2891
dt/dp	1 atm	0.03363	0.2523	4301;771
dp/dt	1 atm	29.73	3.964	4301
p	32.2	1688	225.0	8276
eq 2.5	A	7.39128	6.51618	7338;771
	B	1195.68	1195.68	
	C	256.19	256.19	
fp		−135	138	770,2142
d	20	1.3783	1378.3	7337;5827,7004,2142,772
	25	1.3662	1366.2	
dd/dt	20	0.00242	2.42	772
n_D	20	1.3602	1.3602	2510;2142,7004
	25	1.3569	1.3569	
η	18.03	0.3367	0.0003367	5827;4367
	32.88	0.2978	0.0002978	
γ	25	18	0.018	2142
ΔH_v	25	5.79	24.23	4301
	bp	6.01	25.15	
ΔH_f°	25(g)	−68.1	−284.9	1444,7342;3906

367. Dichlorofluoromethane

(Continued)

		cgs	SI	
C_p°	25	14.58	61.00	1444,7340;7826,2706,2791
C_p	25	25.9	108.4	774;2142
	Table 2.11			
λ	25	0.00026	0.11	2142
t_c		178.4	451.6	7337,7800;770,2142,2891
P_c		51.29	5.197	7800;770,2142,2891
d_c		0.522	522	7337,770,2142;7800
v_c		0.197	0.000197	7337,2142
ε	28	5.34	5.34	2142
μ	30(g)	1.34	1.34	2614
δ	25	8.3	17.0	653;6723,6371
soly	in aq,25	0.95%w	0.95%w	2142
	aq in,25	0.13%w	0.13%w	
Beil	5,61			
uv	2873			
ir	7434,4463,5894			
Raman	1047,2791,2794			
nmr	4552,5322			

368. Dibromofluoromethane

$CHBr_2F$ 1868-53-7 $CHBr_2F$

		cgs	SI	
mw		191.825	191.825	3468
bp	739 Torr	63	336	3410;7118
d	20	2.421	2421	7004;7118
	27.7	2.2533	2253.3	3410
n_D	20	1.4685	1.4685	7004
	27.7	1.4628	1.4628	3410

368.   Dibromofluoromethane

(Continued)

		cgs	SI	
ΔH_f°	25(g)	−45.80	−191.63	7353,4305
C_p°	25	15.569	65.141	7353,4305;2706,7826
Beil	5,EI 16			
Raman	2794			

369.   Bromochloromethane

CH_2ClBr 74–97–5 CH_2ClBr

		cgs	SI	
mw		129.384	129.384	3468
bp	1 atm	68.06	341.21	7354;4760,2088,1976
dt/dp	1 atm	0.04074	0.3056	2088
dp/dt	1 atm	24.55	3.273	2088
p	25	147.2	19.6	2088
eq 2.5	A	6.41307	5.53797	7354;4760,2088
	B	903.38	903.38	
	C	187.69	187.69	
fp		−87.95	185.20	4760;1976
d	20	1.93439	1934.39	2088;7004,4300
	25	1.92292	1922.92	
dd/dt	25	0.002296	2.296	2088
n_D	20	1.48376	1.48376	2088;7004
	25	1.48076	1.48076	
dn/dt	25	0.000615	0.000615	2088
η	20	0.6743	0.0006743	2088
γ	20	33.32	0.03332	2088
	30	31.87	0.03187	

369. Bromochloromethane

(Continued)

		cgs	SI	
ΔH_v	25	7.851	32.849	2088
	bp	7.172	30.007	
ΔH_f°	25(g)	-10.71	-44.81	7355,4305
C_p°	25	12.60	52.71	7944,7355,4305;2706
t_c		297	570	2088
P_c		60.0	6.08	2088
d_c		0.625	625	2088
v_c		0.207	0.000207	2088
μ	25 in 38	1.66	1.66	931
δ	25	9.1	18.6	653
soly	in aq,25	1.7%w	1.7%w	7219
Beil	5,67			
ir	7944,5897,2259,5932			
Raman	7944,7833,1941			
nmr	4740,2787,5933			

370. Bromochlorofluoromethane

CHBrClF   593-98-6   CHBrClF

		cgs	SI	
mw		147.374	147.374	3468
bp	756 Torr	36.11	309.26	815
fp		-115	158	815
d	20	1.921	1921	7356
	25	1.907	1907	
dd/dt	25	0.0029	2.9	7356
n_D	25	1.4143	1.4143	3411
ΔH_f°	25(g)	-54.93	-229.83	4305,7357;3906

370.   Bromochlorofluoromethane

(Continued)

		cgs	SI	
C_p°	25	15.02	62.85	4305,7357;2706,7826
Beil	5,67			
ir	5896,2004			
Raman	2794,2004			
nmr	2308			

371.   Chloropentafluoroethane

$CClF_2CF_3$ 76-15-3 C_2ClF_5

		cgs	SI	
mw		154.467	154.467	3468
bp	1 atm	−39.11	234.04	481;2142,3230,4643
p	−29.77	1120	149	5018
fp		−106	167	1315,2142
tp		−99.44	173.71	481
d	−8.65	1.4341	1434.1	5018;3230
	25	1.291	1291	2142
n_D	20	1.2950	1.2950	2510
	25	1.2909	1.2909	
η	25	0.26	0.00026	2142
γ	25	5	0.005	2142
ΔH_v	bp	4.639	19.410	481;2142
ΔH_m		0.4489	1.878	481
ΔH_t	−192.9	0.6276	2.626	481
ΔH_f°	25	−267.4	−1118.8	1233;7341,3906
C_p°	25	25.70	107.53	2635;2286,7339
C_p	−44	34.44	144.10	481
	25	44.0	184.2	2142

371.   Chloropentafluoroethane

(Continued)

		cgs	SI	
λ	25	0.00011	0.046	2142
t_c		79.9	353.1 ·	5018,481,2142;4647
p_c		31.16	3.157	5018;481,2142,4647
d_c		0.6131	613.1	5018;2142
v_c		0.2519	0.0002519	5018;2142,4647
ϵ	27.4(g)	1.0018	1.0018	2142
μ	(g)	0.52	0.52	2006;2614
soly	in aq,25	0.006%w	0.006%w	2142
Beil	8,EIII 139			
ir	6187,5492,605,3230			
Raman	6187,5492			
ms	7226			

372.   1,2-Dichlorotetrafluoroethane

1,2-Dichloro-1,1,2,2-tetrafluoroethane

$CClF_2CClF_2$ 76-14-2 $C_2Cl_2F_4$

		cgs	SI	
mw		170.922	170.922	3468
bp	1 atm	3.77	276.92	2142;4920,4643,7450
p	32.2	2014	269	8276
eq 2.4	A	7.568	6.693	7450;4920
	B	1300	1300	
fp		-92.53	180.62	4140;2142
d	25	1.455	1455	4643;2142,4920
	35	1.422	1422	
n_D	20	1.2905	1.2905	2510;2142,4643
	25	1.2865	1.2865	

372.   1,2-Dichlorotetrafluoroethane

(Continued)

		cgs	SI	
η	25	0.38	0.00038	2142
	30	0.354	0.000354	7623
γ	25	12	0.012	2142
ΔH_v	bp	5.557	23.250	2142;5772,7450
ΔH_m		1.510	6.318	4140
ΔH_t	−164	1.212	5.071	4140
	−139	2.628	10.996	
ΔH_f°	25(1)	−220.8	−923.8	1703
	25(g)	−212.8	−890.4	7826;7341,4029,4146,3906
ΔH_c°	25(g)	−57.3	−239.7	1703
C_p°	25	28.02	117.24	7339;2635
C_p	25	39.2	164.2	4140
λ	25	0.00014	0.059	
t_c		145.7	418.9	2142,4920;2891
P_c		32.2	3.26	2142,4920;2891
d_c		0.582	582	2142,4920,2891
v_c		0.293	0.000293	2142
ε	25	2.26	2.26	2142
μ	(g)	0.533	0.533	4821;941
δ	25	6.3	12.8	653;6723
soly	in aq,25	0.013%w	0.013%w	2142
	aq in,25	0.009%w	0.009%w	
Beil	8,EIII 152			
ir	4920,5492,6752			
Raman	2796,4920,5492			
nmr	3184			

373. 1,1,2-Trichloro-1,2,2-Trifluoroethane

$CFCl_2CF_2Cl$ 76-13-1 $C_2F_3Cl_3$

		cgs	SI	
mw		187.376	187.376	3468
bp	1 atm	47.633	320.783	7685;6171,343,2360,2142,4643 3516,770
dt/dp	1 atm	0.0403	0.302	6171
dp/dt	1 atm	24.8	3.31	6171
p	25	363.6	48.48	3420,3421;3516
eq 2.5	A	7.5534	6.6783	6171;5163,3516
	B	1499.2	1499.2	
fp		-36.4	236.8	4643;770,6171,2360
d	0	1.6200	1620.0	4643;3516,3171,2360,1059
	25	1.56354	1563.54	
	35	1.53982	1539.82	
dd/dt		0.00229	2.29	4643
κ_T	25	2.365×10^{-7}	1.774×10^{-6}	6151
n_D	25	1.35572	1.35572	4643;2360
	35	1.35124	1.35124	
dn/dt	25	0.000448	0.000448	4643
η	0	0.925	0.000925	3516;2144
	20	0.711	0.000711	
	30	0.627	0.000627	
γ	0	19.85	0.01985	3516
	20	17.75	0.01775	
	30	16.56	0.01656	
ΔH_v	25	6.86	28.70	3421,3420;774,6171
	bp	6.4157	26.843	7685;3516,770
ΔH_f°	25(1)	-188.37	-788.14	7826
	25(g)	-181.5	-759.4	7826;4146,7341,3906
C_p°	25(g)	30.37	127.1	7339;3386,2634
	60(g)	31.26	130.8	2286
C_p	25	41.5	173.6	7826;4643,770,6171
	Table 2.11			
λ	25	0.000157	0.0657	2142
t_c		214.3	487.5	4691;4299,2142,770
P_c		33.66	3.411	4961;4299,2142,770
d_c		0.5699	569.9	4961;4299,2142,770
v_c		0.3288	0.0003288	4961;2142
K_B	eq 2.79	5.75	5.75	6139
ϵ	25	2.41	2.41	2142
δ	25	7.25	14.8	3421;6371,652
	bp	6.91	14.1	6723

373.   1,1,2-Trichloro-1,2,2-Trifluoroethane

(Continued)

		cgs	SI	
soly	in aq,25 aq in,25	0.017%w 0.011%w	0.017%w 0.011%w	2142
aq az	44.5	99.0%w	99.0%w	2142

Beil	8,86
ir	4050,3794,3833,5932
Raman	4050,2796,4276
ms	7226
nmr	7070,3794,3614

374.   1,1,2,2-Tetrachlorodifluoroethane

1,1,2,2-Tetrachloro-1,2-difluoroethane

CFCl₂CFCl₂   76–12–0   C₂F₂Cl₄

		cgs	SI	
mw		203.831	203.831	3468
bp	1 atm	92.8	366.0	4643;3454,343
p	28.1	65.8	8.77	3454
eq 2.4	A B	8.1723 1929.27	7.2972 1929.27	3454
fp		26.55	299.70	4204;2141,4643
tp		24.76	297.91	4039
d	25 35	1.64470 1.62516	1644.70 1625.16	4643;3454,1059
dd/dt	25	0.001954	1.954	4643
n_D	25 35	1.41297 1.40831	1.41297 1.40831	4643;3454
η	25 30 50	1.21 1.208 0.908	0.00121 0.001208 0.000908	2142 3454
γ	30 40	22.73 21.56	0.02273 0.02156	3454

374. 1,1,2,2-Tetrachlorodifluoroethane

(Continued)

		cgs	SI	
ΔH_v	bp	8.36	35.0	2141;3454
ΔH_m		0.8762	3.666	4039;4204
C_p^o		28.70	120.1	2634
C_p	25(s)	41.95	175.52	4204
	25(l)	42.68	178.59	4039
λ	25	0.000165	0.0692	2141
t_c		278	551	2141;3454,4299
p_c		34	3.4	2141
d_c		0.55	550	2141
v_c		0.370	0.000370	2141
K_f	eq 2.67	37.7	37.7	813;3992
ε	25	2.52	2.52	2142
δ		7.9	16.2	2141
soly	in aq,25	0.012%w	0.012%w	2142
	aq in,27.8	0.0099%w	0.0099%w	2141
Beil	8,86			
ir	3811,3833			
Raman	3811			
nmr	364,7460,5473,5474,12			

375. 1-Chloro-1,1-difluoroethane

CH_3CClF_2 75-68-3 $C_2H_3ClF_2$

		cgs	SI	
mw		100.496	100.496	3468
bp	1 atm	-9.8	263.4	5019,2142;3291
eq 2.4	A	7.3776	6.5025	5019
	B	1184.5	1184.5	

375. 1-Chloro-1,1-difluoroethane

(Continued)

		cgs	SI	
fp		−130.8	142.4	5772;2142
d	0	1.170	1170	5019
	30	1.096	1096	
n_D	20	1.2912	1.2912	2510
	25	1.2882	1.2882	
η	20	0.334	0.000334	5019
	25	0.317	0.000317	2142
ΔH_v	20	4.94	20.67	5019;5772
ΔH_m		0.642	2.687	5772
ΔH_f°	25(g)	−126.6	−529.7	5667
C_p°	25	19.74	82.59	7340;7826,6815
C_p	25	31.4	131.4	7826;5772
t_c		137.1	410.3	5019,4299
p_c		40.7	4.12	5019,4299
d_c		0.435	435	5019,4299
v_c		0.231	0.000231	5019,4299
μ	(g)	2.14	2.14	2007;2614
δ		6.8	13.9	6371
Beil	8,EIII 138			
ir	6815			
Raman	6815			
ms	7226			
nmr	2243			

376.   1,2-Dibromotetrafluoroethane

1,2-Dibromo-1,1,2,2-tetrafluoroethane

CF_2BrCF_2Br 124–73–2 $C_2F_4Br_2$

		cgs	SI	
mw		259.824	259.824	3468
bp	1 atm	47.26	320.41	2143;2796,4643,6307,343
dt/dp	1 atm	0.031	0.23	6307
dp/dt	1 atm	32	4.3	6307
$\bar{p}$	25	434	57.9	2143
eq 2.4	A	7.768	6.893	6307
	B	1567.0	1567.0	
fp		−110.5	162.7	2143;4643,6307
d	10	2.209	2209	2143
	25	2.164	2164	1059;2142,4643,2705
dd/dt		0.00304	3.04	6307
n_D	25	1.367	1.367	2142
η	25	0.72	0.00072	2142
	38	0.65	0.00065	2143
γ	20	18.9	0.0189	2143
	25	18.1	0.0181	
ΔH_v	bp	7.166	29.98	6307
ΔH_f°	25(g)	−186.5	−780.3	7826
C_p	25	43.1	180.3	2142
λ	25	0.000112	0.0467	2142
t_c		214.65	487.80	7800;2142
P_c		33.49	3.393	7800;2142,1682
d_c		0.7616	761.6	7800;2142
v_c		0.3411	0.0003411	7800;2142
ϵ	25	2.34	2.34	2142
Beil	8,92			
ir	3833,6705			
Raman	2796,6705			
nmr	3184,2306,5474,5473			

377. 1,2-Dibromo-1,1-difluoroethane

$CH_2BrCBrF_2$ 75–82–1 $C_2H_2Br_2F_2$

		cgs	SI	
mw		223.843	223.843	3468
bp	1 atm	93.2	366.4	3292
fp		-61.3	211.9	3292
d	20	2.2238	2223.8	3292
	25	2.209	2209	1574
n_D	20	1.4456	1.4456	3292
	25	1.4452	1.4452	1574
Beil	8,92			
ir	4047,3810,1226			
Raman	3327,3810,1226			
nmr	6479			

378.   1-Bromo-2-chloroethane

CH_2BrCH_2Cl 107–04–0 C_2H_4BrCl

		cgs	SI	
mw		143.411	143.411	3468
bp	1 atm	106.7	379.9	7058
p	29.7	40.0	5.33	7058
fp		-16.7	256.5	6025;7058
d	20	1.7392	1739.2	6863
	30	1.7211	1721.1	
dd/dt	10–30	0.00184	1.84	6863
n_D	20	1.49174	1.49174	6863;3980
ΔH_v	25	9.11	38.13	7822
ΔH_m		2.30	9.62	6025
ΔH_t	-93	0.74	3.10	6025
ΔH_f°	(g)	-20.4	-85.4	1279

378.   1-Bromo-2-chloroethane

(Continued)

		cgs	SI	
C_p	25	31.1	130.1	7826;6025
ε	30	6.92	6.92	6863
μ	30 in 16	1.19	1.19	6863
	66(g)	1.09	1.09	8253
soly	in aq,30	0.683%w	0.683%w	2987
Beil	8,89			
ir	604,909			
Raman	8098,1452,4127			
nmr	5914,6748			

379.   1,1-Dichloro-2,2-difluoroethylene

1,1-Dichloro-2,2-difluoroethene

$CCl_2{=}CF_2$ 79–35–6 $C_2Cl_2F_2$

		cgs	SI	
mw		132.925	132.925	3468
bp	1 atm	19.0	292.2	3294;5150,4643
fp		−115	158	3294;5150
d	−20	1.555	1555	5150
n_D	−20	1.388	1.388	5150
ΔH_f°	25(g)	−76.3	−319.2	7826;3906
μ	(g)	0.50	0.50	2690
Beil	11,EIII 657			
uv	4364			
ir	7530,5491			
Raman	3225,5491			
nmr	2307			

380. Nitromethane

CH_3NO_2 75-52-5 CH_3O_2N

		cgs	SI	
mw		61.040	61.040	3468
bp	1 atm	101.20	374.35	7524,1075;4969,7489,8119,2090 3206,3436
dt/dp	1 atm	0.0427	0.320	7524;1897,2090,7489
dp/dt	1 atm	23.4	3.12	7524
p	25	36.66	4.888	3777;2091,3467,7058,800
eq 2.5	A	7.274170	6.399073	7524;7433,1148,1344,4754,4604
	B	1441.610	1441.610	3436,2090,6090
	C	226.939	226.939	
fp		-28.55	244.60	7524;7484,7489,2090,6881 7473,3777
d	20	1.13816	1138.16	7524;3589,4649,4969,6881,1038
	25	1.13128	1131.28	1191,2090,2999,3205,7489
	30	1.12439	1124.39	7771,7848,8168,1148,1149 7638,7423,800,3025
dd/dt		0.001377	1.377	7524
α		0.00124	0.00124	6102
κ_T	20	7.9×10^{-8}	5.9×10^{-7}	1115
n_D	20	1.38188	1.38188	7524;3508,3589,4649,7725,1038
	25	1.37964	1.37964	1191,2090,3467,6881,7771
	30	1.37738	1.37738	1148,1149,7638,7423
dn/dt		0.000450	0.000450	7524
η	20	0.647	0.000647	6168;8168,7423,3467,7489,7838
	25	0.614	0.000614	917;6168,5089
	30	0.5737	0.0005737	7990;6168
γ	20	37.19	0.03719	7087;7771,1075,6886,1038,3283
	30	35.50	0.03550	7087;6886,3024
	Table 2.8			
ΔH_v	25	9.147	38.27	3777;7848,7755,3467,4969,5816
	bp	8.225	34.41	5875
	Table 2.9			
ΔH_m		2.319	9.703	3777
ΔH_f°	25(1)	-27.03	-113.09	8316,7826;4451,3467,1383
	25(g)	-17.86	-74.73	7826;4072
ΔH_c°	25(1)	-169.3	-708.4	8316;7095,1383,4451,4072,3975 7446,3467
C_p°	25	13.70	57.32	4754
	Table 2.14			
C_p	23.97	25.28	105.77	3777;3508,7846
	35	25.58	107.03	797
λ	37.8	0.000484	0.203	6352

380. Nitromethane

(Continued)

		cgs	SI	
t_c		315	588	2957;1038,4299,1075
p_c		62.3	6.31	2957;4299,758,1075
d_c		0.352	352	2957
v_c		0.173	0.000173	4299
K_B	eq 2.79	1.86	1.86	5816,7845
K_i	acid, 25	0.00055	0.00055	7584;7996,3802,6892,3801
K_a	NO_2, 25	6.15×10^{-11}	6.15×10^{-11}	7584
pK_a	NO_2, 25	10.211	10.211	7584;3758
	in 462	15.1	15.1	994;3758
κ	25	5×10^{-9}	5×10^{-7}	7638;7843,8168,4633
ε	30	35.87	35.87	4937;7497,4633,6881,7842,7850
μ	20 in 10	3.56	3.56	5452;7964,4490,2999,3589 6862,6866
δ		12.7	26.0	652,4593;6240,6631,2190
ER	BuOAc 90%	1.3 361	1.3 361	6628;6631,2190 6628,6631
soly	in aq, 25 aq in, 25	11.1 %w 2.09%w	11.1 %w 2.09%w	6168;1613,766,987,1075
aq az	83.59	76.4%w	76.4%w	7525;3498,3497
fl pt	TOC	44	317	1613
Beil	6,75			
uv	155,7433,8274,694,2470,2684,8074,4658,6749			
ir	5493,7977,1942,6817,397,6368,153,5932			
Raman	4977,5750,7977,8124,6817,742,4849			
ms	99,5094			
nmr	1396,2522,3435,3454,8120,2972,2464,5933,6748			

381. Nitroethane

$CH_3CH_2NO_2$		79-24-3		$C_2H_5O_2N$
		cgs	SI	
mw		75.067	75.067	3468
bp	1 atm	114.07	387.22	7524;3205,3206,3436,7771
dt/dp	1 atm	0.0445	0.334	7524
dp/dt	1 atm	22.5	3.00	7524
p	25	20.93	2.790	7524;3467
eq 2.5	A	7.175154	6.300057	7524;1344,3436,3467,2860
	B	1435.402	1435.402	
	C	220.184	220.184	
fp		-89.52	183.63	7524;2621,7054,4635
d	20	1.05057	1050.57	7524;7771,842,1516,3469,1191
	25	1.04464	1044.64	3205,4649
	30	1.03870	1038.70	7524;3025
dd/dt		0.001187	1.187	7524
n_D	20	1.39193	1.39193	7524;1516,4649,7771,842,1191
	25	1.38973	1.38973	3467,3594
	30	1.38754	1.38754	
dn/dt		0.000439	0.000439	7524
η	20	0.677	0.000677	6168;3467
	25	0.638	0.000638	
	30	0.6062	0.0006062	7990;6168
γ	20	32.66	0.03266	6886;1516,7771
	30	31.50	0.03150	6886;3024
	Table 2.8			
ΔH_v	25	9.94	41.59	3467
	bp	9.075	37.970	2860
ΔH_m		2.355	9.853	4635
ΔH_f°	25(1)	-33.9	-141.8	8316;7826,4451,3467,1383
	25(g)	-23.56	-98.58	7826
ΔH_c°	25(1)	-325.6	-1362.3	3975;8316,4451,7446,3467,1383
C_p	25	33.1	138.5	5109;7826,4635
λ	37.8	0.000398	0.167	6352
t_c		284	557	6352;1516
K_B	eq 2.79	2.60	2.60	7845;726
K_i	acid, 25	3.93×10^{-5}	3.93×10^{-5}	7584;3800,7996
K_a	NO_2, 25	3.50×10^{-9}	3.50×10^{-9}	7584
pK_a	NO_2, 25	8.456	8.456	7584;3758
	in 462	16.7	16.7	4998
κ	30	5×10^{-7}	5×10^{-5}	5960
ε	30	28.06	28.06	4937;7842,7850,7458
μ	20 in 10	3.60	3.60	5452;6862,3589,410,2999,3594

381. Nitroethane

(Continued)

		cgs	SI	
δ		11.1	22.7	652,4593;6631,2190
ER	BuOAc	1.0	1.0	6628;6631,2190
	90%	446	446	6628;6631
soly	in aq, 25	4.68%w	4.68%w	6168;1613,987
	aq in, 25	1.05%w	1.05%w	
aq az	87.22	71%w	71%w	7525;3498,6165
fl pt	TOC	41	314	1613
Beil	9,99			
uv	8274,2861,3215,7647,2470,2684,8074,6749			
ir	153,5493,3215,6817,397,1235,7077,5932			
Raman	4977,8124,3215,6817,2696			
ms	154,2528			
nmr	1396,3454,8120,2972,5933,6748			

382. 1-Nitropropane

$CH_3CH_2CH_2NO_2$ 108-03-2 $C_3H_7O_2N$

		cgs	SI	
mw		89.094	89.094	3468
bp	1 atm	131.18	404.33	7524;2090,2999,3436,7771
dt/dp	1 atm	0.0467	0.350	7524;2090
dp/dt	1 atm	21.4	2.85	7524
p	25	10.23	1.364	7524;2091,3467,7058
eq 2.5	A	7.127539	6.252442	7524;2090,3436
	B	1474.299	1474.299	
	C	215.986	215.986	
fp		-103.99	169.16	7524;2090
d	20	1.00144	1001.44	7524;1038,1191,2090,2999,7771
	25	0.99609	996.09	
	30	0.99073	990.73	

382. 1-Nitropropane

(Continued)

		cgs	SI	
dd/dt		0.001071	1.071	7524
α	20	0.00101	0.00101	6905
n_D	20	1.40160	1.40160	7524;1038,1191,2090,3467,7771
	25	1.39956	1.39956	
	30	1.39755	1.39755	
dn/dt		0.000405	0.000405	7524
η	20	0.844	0.000844	6168;3467
	25	0.790	0.000790	
	30	0.740	0.000740	
γ	20	30.64	0.03064	6886;1038,7771
	30	29.61	0.02961	
	Table 2.8			
ΔH_v	25	10.37	43.39	3467
	bp	9.194	38.47	7524
ΔH_f°	25(1)	−40.15	−167.99	8316;3467,1383,4451
ΔH_c°	25(1)	−481.9	−2016.3	3975;4451,3467,7095,1383,8316
C_p	25	41.9	175.3	5109
λ	37.8	0.000361	0.151	6352
t_c		402.0	675.2	1038
pK_a	NO_2, 25	8.98	8.98	7996,4990,37858
κ	35	3.3×10^{-7}	3.3×10^{-5}	5960
ε	30	23.24	23.24	4937
μ	20 in 10	3.59	3.59	5452;2806,6862,8119
δ	25	10.5	21.5	7159;4593,6632,2190
ER	BuOAc	0.71	0.71	6628,6631;2190
	90%	644	644	6628,−631
soly	in aq, 25	1.50%w	1.50%w	6168;1613
	aq in, 25	0.62%w	0.62%w	
aq az	91.63	63.5%w	63.5%w	7058;3498,6165
fl pt	TOC	49	322	1613
Beil	10,115			
uv	8274,3215,7647,2470,8074,6749			
ir	5849,3215,6817,397,2696,1235,5493,5932			
Raman	4977,8124,3215,6817,2696			
nmr	1396,3454,8120,2972,6748			

383.   2-Nitropropane

$CH_3CH(NO_2)CH_3$ 79-46-9 $C_3H_7O_2N$

		cgs	SI	
mw		89.094	89.094	3468
bp	1 atm	120.25	393.40	7524;3205,3436,7771,8119
dt/dp	1 atm	0.0460	0.345	7524
dp/dt	1 atm	21.7	2.90	7524
p	25	18.0	2.40	7524;2621,3467,7058
eq 2.5	A	7.083240	6.208143	7524;3436
	B	1422.898	1422.898	
	C	218.341	218.341	
fp		-91.32	181.83	7524;860,7058
d	20	0.98839	988.39	7524;1038,3205,3467,7771
	25	0.98290	982.90	
	30	0.97740	977.40	
dd/dt		0.001099	1.099	7524
α	20	0.00104	0.00104	6631
n_D	20	1.39439	1.39439	7524;1038,3467,7771,8119
	25	1.39235	1.39235	
	30	1.39028	1.39028	
dn/dt		0.000411	0.000411	7524
η	20	0.770	0.000770	6168;3467
	25	0.721	0.000721	
	30	0.677	0.000677	
γ	20	29.87	0.02987	6886;7771,1038
	30	28.68	0.02868	
	Table 2.8			
ΔH_v	25	9.88	41.34	3467
	bp	8.793	36.790	7524
ΔH_f°	25(1)	-43.2	-180.7	1383;3467,8316
ΔH_c°	25(1)	-478.0	-2000.0	8316;3467,1383,3823
t_c		344.7	617.9	1038
K_i	acid, 25	7.73×10^{-6}	7.73×10^{-6}	7584;7996,3802
K_a	NO_2,25	2.12×10^{-8}	2.12×10^{-8}	7584
pK_a	NO_2,25	7.674	7.674	7584;4990
κ	30	5×10^{-7}	5×10^{-5}	5960
ε	30	25.52	25.52	4937
μ	120-180(g)	3.73	3.73	8119
	20 in 317	3.73	3.73	410
δ		9.9	20.3	652,6632;2190,1399
ER	BuOAc	1.10	1.10	6628,6631;2190,1399
	90%	415	415	6628,6631;1399
soly	in aq,25	1.71%w	1.71%w	6168;1613
	aq in,25	0.53%w	0.53%w	

383.   2-Nitropropane

(Continued)

		cgs	SI	
aq az	88.55	70.6%w	70.6%w	7058;3498,6165
fl pt	TOC	39	312	1613

Beil	<u>10</u>,116
uv	8274,3215,7647,2470,2684,8074,6749
ir	153,5493,5849,633,3215,6817,397,5932
Raman	8124,3215,6817,640
nmr	1396,3454,8120,5933,6748

384.   Nitrobenzene

$C_6H_5NO_2$ 98–95–3 $C_6H_5O_2N$

		cgs	SI	
mw		123.111	123.111	3468
bp	1 atm	210.80	483.95	7490;7470,7771,4325,4608 4684,7058
dt/dp	1 atm	0.05573	0.4180	2087;4684,7470,7490
dp/dt	1 atm	17.94	2.392	2087
p	25	0.284	0.0379	2087;4721,4846,1185,5975 6090,7058
eq 2.5	A	7.545	6.670	4721
	B	2064	2064	
	C	230	230	
eq 2.4	A	8.680	7.805	6797
10–30	B	2741	2741	
fp		5.76	278.91	1567;8121,6196,2362,2986 6645,5015
d	15	1.20824	1208.24	7490;1057,4913,4958,1567 4608,4649
	20	1.20331	1203.31	3702;6314,7158,4195
	25	1.19833	1198.33	3702;6833,7848,7858,7470 7555,7771
	30	1.19341	1193.41	7490;6314,6416,7158

384. Nitrobenzene

(Continued)

		cgs	SI	
dd/dt		0.000989	0.989	<u>7490</u>
α	25	0.0008333	0.0008333	3702;4195
κ_S	20	5.117×10^{-8}	3.838×10^{-7}	7158
n_D	15	1.55457	1.55457	7490;4649,4913,1057,1176 3180,3815
	25	1.54997	1.54997	<u>7490</u>;6833,7771,4731
dn/dt		0.00046	0.00046	<u>7482</u>
η	15	2.165	0.002165	7490;1024,4958,7838,7858
	30	1.619	0.001619	6029;7490
γ	20	42.70	0.04270	4195;3283,7490,7771,7848
	30	42.17	0.04217	3283;46
	Table 2.7 Table 2.8			
ΔH_v	25	13.148	55.013	4344
	bp	9.744	40.769	4684;7848
ΔH_m		2.78	11.63	7167;5689
ΔH_f°	25(1)	-2.32	-9.71	4449;7128
ΔH_c°	25(1)	-738.068	-3088.08	4449;3975,2668,7128
C_p	30	42.37	177.28	8080;4684,5689,7846
t_c		459	732	7848
K_f	eq 2.67	6.852	6.852	7590;733,6196,6061
K_B	eq 2.79	5.04	5.04	873;874,724,872,531
κ	25	2.05×10^{-10}	2.05×10^{-8}	638;6014,7843,7852,7860,7858
ε	25	34.78	34.78	3624;8184,3199,3677,4937,6416 2953,3180,8084
μ	25 in <u>38</u>	4.00	4.00	4138;3106,4913,5317,6478,4488 1518,2362,2882,2998,5858 2562,1170,1322,2907
δ	25	10.0	20.5	652,3395,4593
soly	in aq,20	0.19%w	0.19%w	6556;7662,987,591,7219,5725
	aq in,20	0.24%w	0.24%w	6556
aq az	98.6	12%w	12%w	3497
fl pt	CC	88	361	5712

Beil	<u>465</u>,233
uv	155,827,1176,2054,1830,5497,2379,6686
ir	1044,3942,4524,6311,7077,3888,633,4353,5932
Raman	3971,1782,926,8171,4353
ms	99,5922
nmr	1429,1981,1954,4477,6358,5933

385. Acetonitrile

CH$_3$CN   75–05–8   C$_2$H$_3$N

		cgs	SI	
mw		41.052	41.052	3468
bp	1 atm	81.60	354.75	7488;7470,7845,2090,3049 4564,5061
dt/dp	1 atm	0.04192	0.3144	7373;7470,7488,2090
dp/dt	1 atm	23.85	3.180	7373
p	25	88.81	11.84	5997;4345,7058,1726,2091,2339
eq 2.5	A	7.12257	6.24747	2090;4345,5299,5997
	B	1315.2	1315.2	
	C	230	230	
fp		−43.835	229.315	5997;2090,7471,7488,4978
d	20	0.7822	782.2	7373;1695,4564,7470,1819 2090,2199
	25	0.77649	776.49	4681;1777,7856,7373,7488,7848
	30	0.77125	771.25	5061
dd/dt		0.001078	1.078	5061
α	20	0.001368	0.001368	4195
κ_T		1.42 x 10^{-7}	1.07 x 10^{-6}	1777
n_D	20	1.34411	1.34411	2090;4564,5061,1695,1819,7488 3049,2199
	25	1.34163	1.34163	
dn/dt		0.00045	0.00045	7488
η	15	0.374	0.000374	57;1555,2580,7488,3049 7838,7856
	25	0.341	0.000341	57
	30	0.324	0.000324	57;7488
γ	20	29.29	0.02929	4195;7488,3283,3700,7848
	25	28.25	0.02825	3797
	30	27.80	0.02780	4195
	Table 2.7			
ΔH_v	25	7.873	32.94	3521;4682,4686,7754,5997
	80.5	7.127	29.819	3816
ΔH_m		1.952	8.167	5997;7478
ΔH_f°	25(1)	9.694	40.56	332;565,8316,5200,7826
	25(g)	17.70	74.04	332;565,7826
ΔH_c°	25(1)	−300.27	−1256.33	332;3975,7135,565,7446 7137,8316
C_p°	25	12.48	52.22	2337,7826
	Table 2.14			
C_p	25	21.86	91.46	7826;4685,7846,1726,5997,4686 2337,3816
t_c		272.3	545.5	6225;7848,4088,4299,3051,3049
P_c		47.7	4.83	6225;4088,3049,4299,3051

385.   Acetonitrile

(Continued)

		cgs	SI	
d_c		0.237	237	4088;4299
v_c		0.173	0.000173	4088;4299
K_B	eq 2.79	1.30	1.30	7845;3817,4600,7590,5313,3816
pK_a	in 462	29.1	29.1	994
pK_{BH^+}	in H_2SO_4	-10.12	-10.12	1953;1956,4519
pK_s	25	32.2	32.2	4156;1300,1557
κ	25	6×10^{-10}	6×10^{-8}	3681;7856,5333,4090,7842,7843
ε	25	35.94	35.94	4681,6947;7850,828,7497,1555 3677,6947,8177,4937
μ	25 in 38	3.53	3.53	1748;8099,4490,7667,1695 4092,4121,6862
δ		12.11	24.77	6240;652,1271,4593
soly	in aq	inf	inf	7033
aq az	76.7	84.2%w	84.2%w	3787;3497,3498
fl pt	CC	6	279	152
Beil	159,183			
uv	2886,6230,3573,6507,7229			
ir	536,7077,633,5435,7667,2877,747,3107,3927			
Raman	893,6129,7718,1077,5435,2322,3770,3107			
ms	154,4785,339,4196,3257			
nmr	7227,1396,3764,7572,1953,4852,2464,1795,5933,6748			

386.   Propionitrile

Propanenitrile

CH_3CH_2CN 107-12-0 C_3H_5N

		cgs	SI	
mw		55.079	55.079	3468
bp	1 atm	97.35	370.50	7373;5061,7058,1695,2090 2806,3049
dt/dp	1 atm	0.0445	0.334	7373;2090
dp/dt	1 atm	22.5	3.00	7373

386. Propionitrile

(Continued)

		cgs	SI	
p	25	44.63	5.950	2089
eq 2.5	A	7.15217	6.27707	2090
	B	1398.2	1398.2	
	C	230	230	
eq 2.4	A	8.0269	7.15180	7943
9-25	B	1894.1	1894.1	
fp		-92.78	180.37	7943;7473,7485,2090,3793,7058
d	20	0.78182	781.82	2090;3049,3700,5061,7848,1191 1695,1696,1819,2999
	25	0.77682	776.82	
	30	0.77196	771.96	7485
dd/dt		0.00103	1.03	1814
n_D	20	1.3658	1.3658	7373;3049,3700,8169,459,1191 1695,1696,1819,2090
	25	1.3636	1.3636	
dn/dt		0.00045	0.00045	5061
η	15	0.454	0.000454	7485;2580,3049,7838,8169
	30	0.389	0.000389	
γ	20	27.25	0.02725	3283;3700
	30	26.19	0.02619	
	Table 2.7			
ΔH_v	22.85	8.611	36.03	3521;3266,4682,7943
	bp	7.40	30.96	7848
ΔH_m		1.206	5.046	7943;7478
ΔH_f°	25(1)	3.70	15.48	565
	25(g)	12.30	51.46	
ΔH_c°	25(1)	-456.65	-1910.62	565;3975,7137,7446,7135
C_p	26.85	28.61	119.70	7945;7846,4682
t_c		291.2	564.4	4088;3051,7739,4299,3049,7848
p_c		41.3	4.18	4088;3049,3051,4299,7739
d_c		0.240	240	4088;4299
v_c		0.230	0.000230	4088;4299
K_B	eq 2.79	1.87	1.87	7845;726
pK_s	25	33.54	33.54	4263
κ	25	8.51×10^{-8}	8.51×10^{-6}	7843;7731,7851,5236
ε	20	28.86	28.86	3676;4937,7842,7850,1814 1696,5236
μ	25 in 38	3.50	3.50	1748;3595,6862,7558,8079 4716,1695,6027,2999
δ		10.8	22.1	652,4593
soly	in aq, 25	10.3%w	10.3%w	987

386.   Propionitrile

(Continued)

		cgs	SI	
aq az	81.5-83.0	76%w	76%w	3497
fl pt	OC	17	290	3580
Beil	162,245			
uv	1770,3573			
ir	7077,153,2118,2877,4049,8180,3271,5932			
Raman	893,6129,2118,1786,4049,8180,3271			
ms	154,4785,99,4196,3257			
nmr	1396,4694,1953,1795,5933			

387.   Succinonitrile

Butanedinitrile

$NCCH_2CH_2CN$ 110-61-2 $C_4H_4N_2$

		cgs	SI	
mw		80.089	80.089	3468
bp	1 atm	267	540	4026
dt/dp	1 atm	0.06	0.45	6166
dp/dt	1 atm	17	2.2	6166
p(s)	25.14	0.007774	0.001036	8157;4565
eq 2.4	A	8.987	8.112	4026,7033
	B	3259	3259	
fp		57.88	331.03	8179;2455,8157
tp		58.0805	331.2305	4870;8179
d	60	0.98669	986.69	7482
	70	0.97860	978.60	
	80	0.97059	970.59	
dd/dt	70	0.000805	0.805	7482
n_D	60	1.41734	1.41734	7482
dn/dt		0.00037	0.00037	7482
η	60	2.591	0.002591	7482
	75	2.008	0.002008	

387. Succinonitrile

(Continued)

		cgs	SI	
γ	60	46.78	0.04678	7482
	70	45.72	0.04572	
	80	44.62	0.04462	
ΔH_v	25	15.29	63.97	8157
	bp	11.6	48.5	eq 2.4
ΔH_s	25	16.730	69.998	8179;8157
ΔH_m		0.8851	3.703	8179;8157
ΔH_f°	25(s)	33.38	139.66	6066
	25(g)	50.11	209.66	6066;5200
ΔH_c°	25(s)	−546.22	−2285.38	6066
C_p	61.85	38.37	160.54	8179
	Table 2.12			
K_f	eq 2.67	18.26	18.26	3621
κ		0.000564	0.0564	7842
ε	57.4	56.5	56.5	4937;8004
μ	30 in 39	3.68	3.68	4565;7558,8079,6916
soly	in aq,25	11.5%w	11.5%w	3621

Beil	172,615
ir	2455,747,5932,3682
Raman	1779,4562
nmr	5933,6748

388. Butyronitrile

Butanenitrile

CH$_3$CH$_2$CH$_2$CN   109–74–0   C$_4$H$_7$N

		cgs	SI	
mw		69.106	69.106	3468
bp	1 atm	117.62	390.770	5078;3700,7476,7485,7482 2999,3049,4684
dt/dp	1 atm	0.0463	0.347	7373;4684
dp/dt	1 atm	21.6	2.88	7373
p	25	19.10	2.546	2089
eq 2.5	A	7.129067	6.253970	5078
	B	1452.076	1452.076	
	C	224.1855	224.1855	
fp		–111.9	161.3	7485,7476;3793,7472
d	15	0.79544	795.44	7485;1695,1819,2999,3049 3700,5061
	20	0.7911	791.1	7373
	25	0.7865	786.5	7373
	30	0.78183	781.83	7485
n$_D$	15	1.38600	1.38600	5061;1695,1819,2049,3049,3700
	20	1.3838	1.3838	7373
	25	1.3820	1.3820	7373
	30	1.37954	1.37954	5061
dn/dt		0.00043	0.00043	5061
η	15	0.624	0.000624	7485;3049,2580
	30	0.515	0.000515	
γ	20	27.33	0.02733	3283;3700
	30	26.24	0.02624	
	Table 2.8			
ΔH$_v$	25	9.40	39.33	3521;2316
	bp	8.23	34.43	2089
ΔH$_m$		1.200	5.021	7478
ΔH$_f^\circ$	25(l)	–1.39	–5.82	2316;1703
	25(g)	8.12	33.97	1703
ΔH$_c^\circ$	25(l)	–613.93	–2568.68	2316;7137,3975,1703
t$_c$		309.1	582.3	3049;3051,7900,4299
p$_c$		37.4	3.79	3049;3051,4299
ε	20	24.83	24.83	3676;6456,7850,4937
μ	25 in 38	3.50	3.50	1748;4716;4503,1695,6862,6916
δ		10.5	21.5	4593,1271
soly	in aq	3.3%w	3.3%w	2189
	aq in	trace	trace	
aq az	88.7	67.5%w	67.5%w	3498;3497
fl pt	TOC	29	302	2189

388. Butyronitrile

(Continued)

Beil	162,275
uv	4062
ir	3498,5932
Raman	4590,8171
ms	154,4785,99,3204,3257,4196
nmr	1396,5933,6748

389. Isobutyronitrile

2-Methylpropanenitrile

$CH_3CH(CH_3)CN$ 78–82–0 C_4H_7N

		cgs	SI	
mw		69.106	69.106	3468
bp	1 atm	103.85	377.00	7485;459
dt/dp	1 atm	0.0460	0.345	7374,7485
dp/dt	1 atm	21.7	2.90	7374,7485
fp		−71.5	201.7	7485
d	20	0.7704	770.4	7374;3451
	25	0.7656	765.6	
n_D	15	1.37563	1.37563	7485;3451,459
	20	1.3734	1.3734	7374
	25	1.3712	1.3712	2216
dn/dt		0.00043	0.00043	7485
η	15	0.551	0.000551	7485
	30	0.456	0.000456	
γ	20	24.93	0.02493	7485
	30	23.84	0.02384	
ΔH_v	25	8.76	36.65	565;2316,3521
ΔH_f°	25(1)	−3.31	−13.85	565;2316
	25(g)	5.44	22.76	
ΔH_c°	25(1)	−612.01	−2560.65	565;2316
ε	24	20.4	20.4	6456
μ	25 in 38	3.61	3.61	6228;6916
δ	25	9.8	20.1	653
aq aq	82.7	77%w	77%w	3497

Beil	162,294
ir	4048,5932
Raman	6129,4048
ms	4785,2458,3258 nmr: 1396,5933,6748

390.   Valeronitrile

Pentanenitrile

$CH_3CH_2CH_2CH_2CN$ 110–59–8 C_5H_9N

		cgs	SI	
mw		83.133	83.133	3468
bp	1 atm	141.3	414.5	7373;7476,7485,2090,3700 5061,5329
dt/dp	1 atm	0.0489	0.367	7373;2090
dp/dt	1 atm	20.4	2.73	7373
p	25	7.07	0.943	2089;2091,3266,7058
eq 2.5	A	7.26060	6.38550	2090
	B	1626.0	1626.0	
	C	230	230	
fp		−96.2	177.0	7373;2090,3793,7476,7485
d	20	0.7993	799.3	7373;1695,1819,2090,3700 5329,7485
	25	0.7950	795.0	
	50	0.7730	773.0	592
n_D	20	1.3971	1.3971	7373;1695,1819,2090,2049 3700,5061,5329
	25	1.3951	1.3951	
dn/dt		0.00040	0.00040	5061
η	15	0.779	0.000779	7485;2580,5329
	25	0.6928	0.0006928	592
	50	0.5084	0.0005084	
γ	20	27.44	0.02744	3283;3700,5329,7485
	30	26.33	0.02633	
	Table 2.8			
ΔH_c	25	10.58	44.27	4177;3521
	bp	7.98	33.39	3816
ΔH_m		1.130	4.728	7478
ΔH_f°	25(1)	−7.93	−33.18	4177
	25(g)	2.65	11.09	
ΔH_c°	25(1)	−769.76	−3220.68	4177
C_p		43.22	180.83	3816
κ		1.2×10^{-8}	1.2×10^{-6}	592
ε	25	19.709	19.709	1961;6456,7497,492,3676
μ	25 in 38	3.57	3.57	1695;6916,6862
δ	25	9.6	19.6	653

Beil	162,301
ir	7077,2877,747,5932
Raman	6129
ms	4785,99,3257
nmr	5933,6748

391. Hexanenitrile

$CH_3(CH_2)_4CN$ 628–73–9 $C_6H_{11}N$

		cgs	SI	
mw		97.160	97.160	3468
bp	1 atm	163.463	436.613	5078;5061,6733,8169,7373,2090 3049,3699
dt/dp	1 atm	0.0508	0.381	7373;2090
dp/dt	1 atm	19.7	2.62	7373
p	25	2.66	0.355	2089;2091,3266,7058
eq 2.5	A	7.071220	6.196123	5078;2090
	B	1553.811	1553.811	
	C	207.3386	207.3386	
fp		−80.3	192.9	7373;2090,6733,7484
d	20	0.8052	805.2	7373;1819,3049,3700,5061 6733,1191
	25	0.8012	801.2	
dd/dt	25	0.00080	0.80	7373
n_D	20	1.4069	1.4069	7373,3699;1191,1819,2049 3049,5061
	25	1.4048	1.4048	7373,2090;6733,8169
dn/dt	25	0.00042	0.00042	7373
η	15	1.041	0.001041	6733;2580,3049,8169
	30	0.830	0.000830	
γ	20	27.85	0.02785	3283;3700
	30	26.99	0.02699	
ΔH_v	25	11.45	47.91	3521;3266,4686
	bp	9.089	38.028	2089
C_p	18–155	52.63	220.20	4686
t_c		348.8	622.0	3049;3051,6854,7900
P_c		32.1	3.25	3046;3051
ε	25	17.263	17.263	1961;6456
δ		9.4	19.2	4593

Beil	162,324
ir	5932
Raman	6127
ms	4785,4783,3257
nmr	5933,6748

392. 4-Methylvaleronitrile

4-Methylpentanenitrile

$(CH_3)_2CHCH_2CH_2CN$ 542-54-1 $C_6H_{11}N$

		cgs	SI	
mw		97.160	97.160	3468
bp	1 atm	154	427	7374,3700;8169,7473
dt/dp	1 atm	0.050	0.38	7374
dp/dt	1 atm	20	2.7	7374
p	65.2	35	4.7	6228
fp		-51.1	222.1	7374,7484;7473
d	20	0.8035	803.5	7374;3700,7589
	25	0.7993	799.3	
dd/dt	25	0.00084	0.84	7374
n_D	20	1.4061	1.4061	7374;8169
	25	1.4040	1.4040	
dn/dt	25	0.00042	0.00042	7374
η	20	0.980	0.000980	8169
	30	0.843	0.000843	
γ	20	26.53	0.02653	7589;3700
	30	25.61	0.02561	
ε	22	15.5	15.5	4937
μ	25 in 38	3.53	3.53	6228;3137
Beil	162,329			
uv	6108			
ir	747			
Raman	1779,6127			
ms	4785			

393. Octanenitrile

$CH_3(CH_2)_6CN$ 124–12–9 $C_8H_{15}N$

		cgs	SI	
mw		125.213	125.213	3468
bp	·1 atm	205.2	478.4	7373;1896,7058
dt/dp	1 atm	0.0547	0.410	7373
dp/dt	1 atm	18.3	2.44	7373
p	25	0.39	0.052	2089;3266,7058
fp		−45.6	227.6	7373;1896,7484,7479
d	20	0.8135	813.5	7373;1819,1896,5061
	25	0.8097	809.7	
dd/dt	25	0.00076	0.76	7373
n_D	20	1.4202	1.4202	7373;2049,5061,1819
	25	1.4182	1.4182	
dn/dt	25	0.00040	0.00040	7373
η	15	1.811	0.001811	1896
	30	1.356	0.001356	
γ	20	28.10	0.02810	2089
	30	37.96	0.03796	
ΔH_v	25	13.58	56.80	7048
	bp	9.869	41.29	2089
ΔH_f°	25(1)	−25.7	−107.4	7048
	25(g)	−12.1	−50.6	
ΔH_c°	25(1)	−1239.1	−5184.5	
t_c		348.8	622.0	4299
P_c		82.1	8.32	4299
ε	25	13.897	13.897	1961

Beil	162,349
uv	5932
ir	7077,8163
ms	4785
nmr	5933,6748

394. α-Tolunitrile

Benzeneacetonitrile

$C_6H_5CH_2CN$ 140-29-4 C_8H_7N

		cgs	SI	
mw		117.150	117.150	3468
bp	1 atm	233.5	506.7	7058;7474,2969,8169
dt/dp	1 atm	0.053	0.40	7058
dp/dt	1 atm	19	2.5	7058
p	60	1	0.1	7058
eq 2.4	A	8.2285	7.3534	7058
	B	2710.0	2710.0	
fp		-23.8	249.4	7058;7474
d	20	1.0155	1015.5	3700;2025,6878,7070
	25	1.0125	1012.5	7848
n_D	20	1.52327	1.52327	6878;3700,8169
	25	1.52086	1.52086	
dn/dt	25	0.000482	0.000482	6878
η	20	2.161	0.002161	8169;7791,2025,7838,4093
	30	1.779	0.001779	
γ	20	41.36	0.04136	7589;7070
	30	40.27	0.04027	
ΔH_v	bp	12.62	52.80	7058
ΔH_c°	25(1)	-1029.4	-4307.0	3975
κ	25	$<0.5 \times 10^{-7}$	$<0.5 \times 10^{-5}$	7852;7843
ε	27	18.7	18.7	4937;6456,2969,7842,7850
μ	25 in 38	3.47	3.47	6878;1467

Beil	941,441
uv	155,2609,5207,4281
ir	5932
Raman	3973,3971,1780,5796

395. Benzonitrile

C_6H_5CN 100–47–0 C_7H_5N

		cgs	SI	
mw		103.123	103.123	3468
bp	1 atm	191.10	464.25	7490;7471,8169,2719,3049 7058,7470
dt/dp	1 atm	0.047	0.35	7490;7470
dp/dt	1 atm	21	2.8	
p	28.2	1	0.1	7058;3815,4912,6090
eq 2.5	A B C	6.74631 1436.72 181.0	5.87121 1436.72 181.0	4398
fp		−12.75	260.40	8121;7471,7490,7058
d	15	1.00948	1009.48	7490;1191,1696,2125,2719 3049,4487
	25	1.0006	1000.6	
	30	0.99628	996.28	7490;4911,4912,7470,7848
dd/dt		0.00088	0.88	7490;3345
κ_S	35	7.08×10^{-8}	5.31×10^{-7}	3868
n_D	20	1.52823	1.52823	3700;1696,3049,7490,8169
	25.5	1.52570	1.52570	1191
dn/dt		0.00048	0.00048	7490
η	15	1.444	0.001444	57;7490,2125,3049,4911,7791 7838,8169,1555
	25	1.237	0.001237	57
	30	1.161	0.001161	57;7490
γ	15	38.65	0.03865	7490;7848
	27	38.43	0.03843	3700
	Table 2.7 Table 2.8			
ΔH_v	25	13.26	55.48	2316;3816,7848
	bp	10.98	45.94	5816
ΔH_m		2.60	10.88	8121;7708
ΔH_f°	25(1)	39.00	163.18	2316
ΔH_c°	25(1)	−868.15	−3632.34	2316,3975,7135
C_p		45.48	190.29	4685;3816
t_c		426.2	699.4	2049;3051,4299,7848
p_c''		41.6	4.22	3049;3051,4299
K_f	eq 2.67	5.34	5.34	8121;7708
K_B	eq 2.72	3.87	3.87	5816,3816;7845,7982,3817,4600
κ	25	5×10^{-8}	5×10^{-6}	4911;7843,7852,5236
ε	25	25.20	25.20	4911;2100,1696,5236,7850,8184 7491,3677
μ	25 in 38	4.01	4.01	1748;1758,1696,2330,4501,5905

395.   Benzonitrile

(Continued)

		cgs	SI	
δ		8.4	17.2	4593,652
soly	in aq,25	∿0.2%w	∿0.2%w	6165;7708
	aq in,28	∿1%w	∿1%w	6165
aq az	98.9	16.7%w	16.7%w	6165
fl pt	COC	none seen	none seen	7708
Beil	926,275			
uv	155,583,2054,3429,5276,5954,1830,6749			
ir	4472,7537,153,3791,3888,633,4283,2938,5932			
Raman	893,3972,1783,5797,4112,3971,742,2938			
ms	154,99,5049,4196			
nmr	1953,4627,4623,6748			

396.   Acrylonitrile

2-Propenenitrile

$CH_2{=}CHCN$ 107–13–1 C_3H_3N

		cgs	SI	
mw		53.063	53.063	3468
bp	1 atm	77.3	350.5	1857;3595,6230,7773
dt/dp	1 atm	0.043	0.32	1857
dp/dt	1 atm	23	3.1	1857
p	20	83	11	7611
eq 2.4	A	7.518	6.643	143
	B	1644.7	1644.7	
fp		−83.55	189.60	143;1857
tp		−83.52	189.63	2437
d	20	0.8060	806.0	1857;3595,6230,7773
	25	0.8004	800.4	
dd/dt	0–30	0.001106	1.106	1857
α	55	0.00146	0.00146	7611

396. Acrylonitrile

(Continued)

		cgs	SI	
n_D	25	1.3888	1.3888	143;1857,3595,6230,7773
dn/dt		0.000539	0.000539	143
η	0	0.43	0.00043	7611
	20	0.362	0.000362	57;7611
	25	0.342	0.000342	57;143
γ	17.8	27.53	0.02753	7773
	24	27.3	0.0273	143
ΔH_v	0–80	7.800	32.635	1857
	bp	7.78	32.55	7611
ΔH_m		1.4890	6.2300	2437;143
ΔH_t	-110.65	0.2840	1.1883	2437
ΔH_f°	25(1)	35.16	147.11	565;143,5200,8316
	25(g)	42.95	179.70	565;2437
ΔH_c°	25(1)	-419.79	-1756.40	565;143,1857,8316
ΔH_p		17.3	72.4	143
C_p°	25(g)	15.34	64.18	2437;3104
C_p	25(1)	26.003	108.80	2437;1857
	Table 2.12			
t_c		246	519	143
p_c		34.9	3.54	143
A	eq 2.72	0.02083	0.02083	2437
K_f	eq 2.67	2.7%m/°C	2.7%m/°C	143
ε	20	33.0	33.0	1814;1857,6457
μ	25 in 38	3.67	3.67	1748;6230,6916,3595,3836,929
δ		10.5	21.5	652,4593
soly	in aq,20	7.35%w	7.35%w	143;1857
	aq in,20	3.1%w	3.1%w	143
aq az	69.6	86.7%w	86.7%w	7611;2542,1857
fl pt	TOC	0	273	143;1857
Beil	163,400			
uv	4046,143,1857,3265			
ir	143,3104,7437,633,153,5932			
Raman	143,3104,6128,7469,4046			
ms	143,4196			
nmr	2530,1387,5341,5933,6748			

397. Methacrylonitrile

2-Methyl-2-propenenitrile

$CH_2=C(CH_3)CN$ 126–98–7 C_4H_5N

		cgs	SI	
mw		67.090	67.090	3468
bp	1 atm	90.3	363.5	5788;821,7753
dt/dp	1 atm	0.0433	0.325	5788
dp/dt	1 atm	23.1	3.08	
p	25	70	9.3	3779;5788,7058
eq 2.5	A	6.98021	6.10511	5788
	B	1274.959	1274.959	
	C	220.734	220.734	
fp		-35.8	237.4	5788;7058,7753
d	20	0.8001	800.1	5788,7753
	30	0.7896	789.6	
dd/dt	25	0.00105	1.05	5788
α	20–25	0.00133	0.00133	7753
n_D	20	1.4007	1.4007	5788;7753
	30	1.3954	1.3954	
dn/dt	25	0.00053	0.00053	5788
η	20	0.392	0.000392	5788
γ	20	24.4	0.0244	5788
ΔH_v	bp	7.6	31.8	5788
ΔH_p		15.3	64.0	7753
μ	(g)	3.69	3.69	6916
soly	in aq,20	2.57%w	2.57%w	5788
	aq in,20	1.62%w	1.62%w	
aq az	77	84%w	84%w	5788
fl pt	TOC	13	286	5788,7753
Beil	163,423			
uv	3265,1214			
ir	633,1052,5932			
Raman	1052			
nmr	6094,6748			

398. Methylamine

Methanamine

CH_3NH_2 74-89-5 CH_5N

		cgs	SI	
mw		31.057	31.057	3468
bp	1 atm	-6.33	266.82	7364;1607,2399,2731
dt/dp	1 atm	0.0292	0.219	7364
dp/dt	1 atm	34.2	4.57	7364
p	25	2680	357.3	2089
eq 2.5	A	7.4969	6.6218	7365;2399,2519
	B	1079.15	1079.15	
	C	240.23	240.23	
fp		-93.46	179.69	7365;1607,6237
d	20	0.6624	662.4	7365;1607,2399,3663
	25	0.6562	656.2	
α	25-35	0.000136	0.000136	1607
n_D	20	1.3527	1.3527	7365
	25	1.3491	1.3491	
η	0	0.236	0.000236	1607;3637
γ	25	19.19	0.01919	1607;6958,3663
	35	17.65	0.01765	
ΔH_v	25	5.792	24.234	2089;4815,2399
	bp	6.210	25.982	2089;1607,2399
ΔH_m		1.466	6.134	1607
ΔH_f°	25(l)	-11.3	-47.3	6221;7826,8316
	25(g)	-5.49	-22.97	7826,8316;6789,6177
ΔH_c°	25(l)	-253.5	-1060.6	6221,8316
	25(g)	-259.35	-1085.12	8316
C_p°	25	11.77	49.25	6554;2398,7826,1607
t_c		156.9	430.1	7367,822
P_c		73.6	7.46	7367,822
K_b	25	1.75×10^{-5}	1.75×10^{-5}	1222,5774
pK_b	25	4.757	4.757	1222,5774
pK_{BH^+}	25	10.66	10.66	4194;1935,3174
κ	33.5	1.0×10^{-6}	0.00010	1607
ϵ	25	9.4	9.4	1607;7602,6457,6948
μ	25 in 38	1.47	1.47	4506;6987
	25(l)	1.09	1.09	4506;6987
δ	25	11.2	22.9	653
aq az		none	none	3499
fl pt	CC	0	273	5712

398. Methylamine

(Continued)

Beil	335,32
ir	4965,4660,8148,8191
Raman	8147,8149
ms	1061,4211,1593
nmr	4965,3071,4964,7076

399. Propylamine

1-Propanamine

$CH_3CH_2CH_2NH_2$ · 107–10–8 · C_3H_9N

		cgs	SI	
mw		59.111	59.111	3468
bp	1 atm	47.229	320.379	2059;7770,2776,1672,4842,7366 1191,7058
dt/dp	1 atm	0.0376	0.282	7366;7621
dp/dt	1 atm	26.6	3.55	
p	25	307.9	41.05	7096;1653,7058
eq 2.5	A	6.92646	6.05136	5608;4842,7365
	B	1044.028	1044.028	
	C	210.833	210.833	
fp		−83.0	190.2	7366;7058,7621
tp		−84.79	188.36	2437
d	20	0.7173	717.3	7366;1191,7770,1693,1672,6845
	25	0.7121	712.1	
dd/dt	10–40	0.00104	1.04	7621
n_D	20	1.3882	1.3882	7366;1191,7770,1693
	25	1.3851	1.3851	
dn/dt	25	0.00062	0.00062	7366
η	20	0.405	0.000405	6598
	25	0.353	0.000353	5365
	30	0.350	0.000350	6598

399. Propylamine

(Continued)

		cgs	SI	
α	55	0.00156	0.00156	7621
γ	19.2 26.4	22.21 21.64	0.02221 0.02164	7770
ΔH_v	25 bp	7.471 7.060	31.26 29.54	4842;6845,972 4842
ΔH_m		2.6229	10.974	2437
ΔH_f°	25(1) 25(g)	−24.26 −16.7	−101.50 −69.9	6845;4522,8316 6517;6177
ΔH_c°	25(1)	−565.31	−2365.26	6845;4522,7446,7137,3975 7135,8316
C_p°	25	21.8	91.2	6517;6177
C_p	25	38.84	162.51	2437;4179,6845,972
	Table 2.12			
t_c		223.8	497.0	822;2776,1672,4299
P_c		46.76	4.738	822;2776,4299
A	eq 2.72	0.03719	0.03719	2437
K_i	25	0.00046	0.00046	7621
pK_{BH^+}	25	10.568	10.568	670;3090,1078,4194,2331 5829,3092
ε	23.0	5.08	5.08	1723;1693,2663
μ	20 in 38	1.33	1.33	1693;6449
soly		inf	inf	5713
aq az		none	none	3498
fl pt	CC	−12	261	152

Beil	337,136
uv	155
ir	2557,6384,5605,862,608,8191,153,8148,6517,5932
Raman	2206,4100,6018,8150,8149,8147
ms	2812
nmr	1396,5345,5933,6748

400. Isopropylamine

2-Propanamine

$CH_3CHNH_2CH_3$ 75-31-0 C_3H_9N

		cgs	SI	
mw		59.111	59.111	3468
bp	1 atm	31.766	304.916	2059;7366,1672
dt/dp	1 atm	0.033	0.25	7366
dp/dt	1 atm	30	4.0	
p	25	574.9	76.65	2089
eq 2.5	A	6.89017	6.01507	5608;4842
	B	985.650	985.650	
	C	214.071	214.071	
fp		-95.2	178.0	7366;7621
tp		-95.16	177.99	2437
d	20	0.6875	687.5	7366;1672,6845
	25	0.6821	682.1	
dd/dt	10-40	0.00108	1.08	7621
α	55	0.00171	0.00171	7621
n_D	20	1.37363	1.37363	4089;7366
	25	1.37058	1.37058	
dn/dt	25	0.00061	0.00061	
η	20	0.344	0.000344	6598
	30	0.302	0.000302	
γ	20	19.53	0.01953	2089;3663
	30	18.34	0.01834	
ΔH_v	25	6.778	28.36	4842;6845,972
	bp	6.654	27.84	4842;2089
ΔH_m		1.7506	7.3245	2437
ΔH_f°	25(1)	-26.83	-112.26	6845
	25(g)	-20.0	-83.7	6517
ΔH_c°	25(1)	-562.74	-2354.50	6845
C_p°	25	22.6	94.6	6517
C_p	25	39.16	163.85	2437;6845,972
	Table 2.12			
t_c		198.7	471.9	4089;7749,1672
p_c		44.8	4.54	4089
d_c		0.268	268	4089
v_c		0.221	0.000221	4089
A	eq 2.72	0.02794	0.02794	2437
K_i	25	0.00043	0.00043	7621
pK_{BH+}	25	10.71	10.71	5075;4194,3090,3092
ϵ	20	5.45	5.45	6457
	23	5.11	5.11	1723
μ	25 in 38	1.45	1.45	4271

400. Isopropylamine

(Continued)

		cgs	SI	
soly	in aq	inf	inf	5713
fl pt	OC	-26	247	152

Beil	337,153
uv	155
ir	3417,5605,153,6517,5932
nmr	1396,5933,6748

401. Butylamine

1-Butanamine

$CH_3CH_2CH_2CH_2NH_2$ 109-73-9 $C_4H_{11}N$

		cgs	SI	
mw		73.138	73.138	3468
bp	1 atm	77.07	350.22	4842;2776,2415,2315,4535,7366 6228,7770,1672
dt/dp	1 atm	0.0419	0.314	7366
dp/dt	1 atm	23.9	3.18	7366
p	25	91.75	12.23	2089;1653
eq 2.5	A	6.94519	6.07009	4842;4535,7365
	B	1157.810	1157.810	
	C	207.800	207.800	
fp		-49.1	224.1	7366;2663,720
d	20	0.7392	739.2	7366;1693,6288,7770,1672,2415
	25	0.73683	736.83	6945;7366,4535
dd/dt	10-40	0.00092	0.92	7621
α	55	0.00135	0.00135	7621
κ_S	30	1.18×10^{-7}	8.85×10^{-7}	6945
n_D	20	1.40087	1.40087	6716;7366,1693,6288,7770,2315
	25	1.3987	1.3987	7366,4535

401. Butylamine

(Continued)

		cgs	SI	
η	25	0.578	0.000578	6598;5365
	30	0.501	0.000501	6946
γ	19.2	24.03	0.02403	7770
	27.9	23.17	0.02317	
ΔH_v	25	8.542	35.74	4842;2823,2089,972,2315
	bp	7.600	31.80	4842;2089
ΔH_f°	25(1)	-30.53	-127.74	2315;8316,4522,6845
	25(g)	-22.53	-94.27	6177
ΔH_c°	25(1)	-721.43	-3018.46	2315;4522,7137,3975,7135,8316
C_p°	25	27.97	117.03	6177
C_p	25	44.9	188	4179;972
t_c		287.9	561.1	1672;2776
p_c		41	4.2	2776
K_i	25	0.00041	0.00041	7621
pK_{BH+}	25	10.77	10.77	5075;1950,4194,2331,670,3443
				5333,3095,5737,3090
ε	20	4.88	4.88	1693;6457,4937,2663
μ	20 in 38	1.37	1.37	1693;615,6288,6449,6612
				1530,4272
δ		8.7	17.8	652
soly	in aq	inf	inf	5713
aq az	575 Torr, 69	98.7%w	98.7%w	3498
fl pt	TOC	-1	272	7637;6114,5313

Beil	337,156
uv	7709
ir	2557,633,2203,5605,608,8191,5932
Raman	4102,4103,8150
ms	2812
nmr	5933,6748

402. Isobutylamine

2-Methyl-1-propanamine

(CH$_3$)$_2$CHCH$_2$NH$_2$ 78-81-9   C$_4$H$_{11}$N

		cgs	SI	
mw		73.138	73.138	3468
bp	1 atm	67.568	340.718	5610;7770,1672,7474,4842,7366 1191,7058
dt/dp	1 atm	0.0402	0.302	7366
dp/dt	1 atm	24.9	3.32	7366
p	25	140.7	18.76	2089;2090,7058,5610
eq 2.5	A	6.95456	6.07946	4842
	B	1142.039	1142.039	
	C	212.800	212.800	
fp		−84.6	188.6	7366;6731,7058,7474
d	20	0.7346	734.6	7366;1191,7770,1672
	25	0.7297	729.7	
dd/dt	25	0.00098	0.98	7366
n$_D$	20	1.3972	1.3972	7366;1191,7770
	25	1.3945	1.3945	
dn/dt	25	0.00054	0.00054	7366
η	20	0.609	0.000609	6598
	25	0.569	0.000569	6598;5365
γ	19.7	22.25	0.02225	7770
	25.5	21.71	0.02171	
ΔH$_v$	25	8.090	33.85	4842;7823,2089,972
	bp	7.314	30.60	4842
ΔH$_f^\circ$	25(1)	−31.68	−132.55	2855;4522
	25(g)	−23.57	−98.62	
ΔH$_c^\circ$	25(1)	−720.25	−3013.53	2855;3975,4522,7446,7137,7135
C$_p$	25(1)	46.4	194	4179;972
t$_c$		266.7	539.9	1672
pK$_{BH^+}$	25	10.42	10.42	3095;1078,3443,1077
ε	21	4.43	4.43	6457;4937,7850
μ	25 in 38	1.27	1.27	4739;615
soly	in aq	inf	inf	5713
aq az		none	none	6165
fl pt	CC	<−6	<267	5713

Beil	337,163
ir	2557,4965,5932
Raman	575,4183
ms	2812
nmr	4965,5933,6748

403. *sec*-Butylamine

2-Butanamine

$CH_3CH_2CHNH_2CH_3$ 13952–84–6 $C_4H_{11}N$

		cgs	SI	
mw		73.138	73.138	3468
bp	1 atm	62.73	335.88	4842;2315,7366,7422,1189 6228,7770
dt/dp	1 atm	0.043	0.32	7366
dp/dt	1 atm	23	3.1	7366
p	25	172.6	23.01	2089
eq 2.5	A B C	6.89445 1106.675 213.000	6.01935 1106.675 213.000	4842
d	20 25	0.7246 0.7201	724.6 720.1	7366;568,7422,1191,7770
dd/dt	25	0.00090	0.90	7366
n_D	20 25	1.3934 1.3907	1.3934 1.3907	7366;1191,7770,6228,2315
dn/dt	25	0.00054	0.00054	7366
γ	21.0 27.4	21.49 20.91	0.02149 0.02091	7770
$ΔH_v$	25 bp	7.815 7.153	32.70 29.93	4842;2315,7823 4842
$ΔH_f^\circ$	25(1)	−32.88	−137.57	2315;8316,4522,6845
$ΔH_c^\circ$	25(1)	−719.08	−3008.63	2315;4522,3975,7135,8316
$[α]_D^{20}$	(+) (S) (−) (R)	+7.44 −7.40	+7.44 −7.40	7422;4542 7422;568
pK_{BH^+}	25	10.56	10.56	3095;3092,3090,9,1078
μ	25 in 38	1.28	1.28	6228
aq az		none	none	6165
fl pt		−29	244	4027

Beil	337,160
ir	5932
ms	1907
nmr	5933,6748

404.   *tert*-Butylamine

2-Methyl-2-propanamine

(CH$_3$)$_3$CNH$_2$ 74-64-9   C$_4$H$_{11}$N

		cgs	SI	
mw		73.138	73.138	3468
bp	1 atm	44.040	317.190	7366;2415,3856,6228,2315
dt/dp	1 atm	0.041	0.31	7366
dp/dt	1 atm	24	3.3	7366
p	25	362.1	48.28	7366
eq 2.5	A	6.78204	5.90694	5608
	B	992.719	992.719	
	C	210.423	210.423	
fp		-72.65	200.50	3856
tp		-66.96	206.19	2437
d	20	0.6958	695.8	7366;2415,3856,6845
	25	0.6908	690.8	
dd/dt	25	0.00100	1.00	7366
n$_D$	20	1.37764	1.37764	4089;7366,2415,3856,2315
	25	1.37462	1.37462	
dn/dt	25	0.00058	0.00058	4089
ΔH$_v$	25	7.084	29.64	7823;6845,2315
	bp	6.784	28.38	2089
ΔH$_m$		0.2108	0.8820	2437
ΔH$_t$	-181.85	0.02713	0.1135	2437
	-70.88	1.4466	6.0526	
ΔH$_f^\circ$	25(1)	-36.00	-150.62	6845;2315,4522,8316
	25(g)	-28.8	-120.5	6517
ΔH$_c^\circ$	25(1)	-715.94	-2995.49	6845;4522,3975,2315,7135,8316
C$_p^\circ$	25	28.8	120.5	6517
C$_p$	25	45.82	191.71	2437;4179
	Table 2.12			
t$_c$		210.7	483.9	4089
P$_c$		37.9	3.84	4089
d$_c$		0.250	250	4089
v$_c$		0.293	0.000293	4089
A	eq 2.72	0.00249	0.00249	2437
pK$_{BH^+}$	25	10.45	10.45	3095;3092,3090,1077
ε	25	58.5	58.5	6228
μ	25 in 38	1.29	1.29	6228;2415
soly	in aq	inf	inf	4027
aq az		none	none	6165
fl pt	TCC	-9	264	6237,4027

404. *tert*-Butylamine

(Continued)

Beil	337,173
uv	155
ir	3417,153,6517,5932
ms	154,6717
nmr	5933

405. Pentylamine

1-Pentanamine

$CH_3(CH_2)_3CH_2NH_2$ 110-58-7 $C_5H_{13}N$

		cgs	SI	
mw		87.164	87.164	3468
bp	1 atm	104.5	377.7	7364
dt/dp	1 atm	0.0445	0.334	7364
dp/dt	1 atm	22.5	3.00	7364
p	25	30.53	4.07	2089
eq 2.5	A	7.198	6.323	7365
	B	1396.9	1396.9	
	C	219.2	219.2	
fp		−43.2	230.0	7364
d	20	0.7544	754.4	7364
	25	0.7502	750.2	
dd/dt	25	0.00084	0.84	7364
α	20–60	0.00116	0.00116	4902
n_D	20	1.4118	1.4118	7364
	25	1.4093	1.4093	
dn/dt	25	0.00050	0.00050	7364
η	25	0.718	0.000718	6598
	35	0.622	0.000622	
γ	20	48.87	0.04887	2089
	30	46.71	0.04671	

405. Pentylamine

(Continued)

		cgs	SI	
ΔH_v	25	9.579	40.08	7823;2089
	bp	8.129	34.01	2089
ΔH_f°	25(g)	−26.95	−112.76	6177
C_p°	25	33.44	139.91	6177
C_p	25	52.1	218	4179
$pK_{BH}+$	25	10.64	10.64	2769;4194
ε	22	4.5	4.5	4937
δ	25	8.7	17.8	653
soly	in aq	inf	inf	4902
fl pt	OC	7	280	5415
Beil	337,175			
ir	5932			
nmr	6748,5933			

406. Cyclohexylamine

Cyclohexanamine

$CH_2CH_2CH_2CH_2CH_2CHNH_2$ 108-91-8 $C_6H_{13}N$

		cgs	SI	
mw		99.175	99.175	3498
bp	1 atm	133.96	407.11	4842;6197,7410,5524,1376,4564
dt/dp	1 atm	0.050	0.38	7410
dp/dt	1 atm	20	2.7	7410
p	25	8.8	1.2	5524,1376
eq 2.5	A	6.68954	5.81444	4842;5524
	B	1229.418	1229.418	
	C	188.802	188.802	
fp		−17.7	255.5	1376

406. Cyclohexylamine

(Continued)

		cgs	SI	
d	15	0.87128	871.28	7483;1693,1376,4564,7770
	30	0.85782	857.82	1980;7483
α		0.001164	0.001164	1693
κ_S	30	8.09×10^{-8}	6.07×10^{-7}	1980
n_D	20	1.45926	1.45926	7770;1693
	25	1.4565	1.4565	1376
dn/dt		0.00058	0.00058	7483
η	15	2.517	0.002517	7483
	20	1.662	0.001662	
γ	20	31.51	0.03151	7483;7770
	30	30.29	0.03029	
ΔH_v	25	10.926	45.714	5524
	bp	8.576	35.88	4842;7749
ΔH_f°	25(1)	-35.3	-147.7	6984
	25(g)	-25.1	-104.9	
ΔH_c°	25(1)	-973.1	-4071.3	6984
t_c		341.7	614.9	7749
K_b		0.00044	0.00044	6
pK_b		3.36	3.36	6
pK_{BH^+}	25	10.68	10.68	4194;3092,3090,7835,3095
ε	20	4.73	4.73	1693;4937
μ	20 in 38	1.26	1.26	1693;4564,6449,2951
δ	25	9.2	18.9	653
soly	in aq	inf	inf	1376
aq az	96.4	44.2%w	44.2%w	1396
fl pt	OC	27	300	6
Beil	1594,5			
ir	2245,8148,5932			
nmr	5437,7227,983,1795,5933,6748			

407. 2-Ethylhexylamine

2-Ethyl-1-hexanamine

$CH_3(CH_2)_3CH(C_2H_5)CH_2NH_2$ 104-75-6 $C_8H_{19}N$

		cgs	SI	
mw		129.245	129.245	3468
bp	1 atm	169.2	442.4	7620;2187
dt/dp	1 atm	0.050	0.38	7620
dp/dt	1 atm	20	2.7	7620
p	54	10.0	1.33	7620
d	20	0.7880	788.0	7620;2187
dd/dt	10-40	0.00081	0.81	7620
α	20	0.00106	0.00106	7620
n_D	20	1.4308	1.4308	7620;2187
η	20	1.1	0.0011	7620
ΔH_v	bp	9.55	39.96	7620
K_b	25	0.000212	0.000212	7620
pK_b	25	3.674	3.674	7620
δ	25	8.5	17.3	653
soly	in aq,20 aq in,20	0.25%w 25.3%w	0.25%w 25.3%w	7620
aq az	98.2	36%w	36%w	3499
fl pt	TOC	60	333	7620;2187
Beil	337,EIII 388			

408. Aniline

$C_6H_5NH_2$ | 62-53-3 | C_6H_7N

		cgs	SI	
mw		93.128	93.128	3468
bp	1 atm	184.40	457.55	7490;7472,4760,4333,2090,4078 4684,7470
dt/dp	1 atm	0.05042	0.3782	2090;4328,4684,7470,7490
dp/dt	1 atm	19.83	2.644	2090
p	25	0.671	0.0895	2087;4828
eq 2.5	A	7.57170	6.69660	2090;4760
	B	1941.7	1941.7	
	C	230	230	
fp		-5.98	267.17	398;7479,4396,5104,7472 8121,4760
d	15	1.02613	1026.13	7490;398,1057,2414,2739 3049,4078
	20	1.02173	1021.73	2090;4649,4912,6833,6990,7470
	25	1.01750	1017.50	7770,7848,6902
	30	1.01306	1013.06	6416;3900
dd/dt		0.000864	0.864	7490
κ_S	25	4.91×10^{-8}	3.68×10^{-7}	1966
n_D	20	1.58628	1.58628	2087;398,1057,3049,3508,4078
	25	1.58364	1.58364	4649,6833,7490,7770,2090 7081,6902
dn/dt	25	0.000528	0.000528	2087
η	20	4.400	0.004400	6990;398,2125,3049,7403,7490
	25	3.770	0.003770	5365,6902,3900
	Table 2.4			
γ	19	43.38	0.04338	2395;7490,7770,7848
	25	42.79	0.04279	369
	32.0	42.01	0.04201	7482
	Table 2.7			
ΔH_v	25	13.345	55.834	4344;3228,517,4328,5681 7848,1801
	bp	10.643	44.530	3228;7754
ΔH_f°	25(1)	7.47	31.25	3228;4522,7801,8316
ΔH_c°	25(1)	-810.5	-3391.1	3975;4522,7137,7117,7446 2069,7135
C_p°	25	25.91	108.41	3228
C_p	25	45.65	191.01	5478;3228,904,2406,4328 5681,6440
λ	15	0.00044	0.18	1855
t_c		426	699	2087;3051,3049,7848
P_c		52.39	5.308	2087;3051,3049
d_c		0.324	324	4375;2087

408. Aniline

(Continued)

		cgs	SI	
v_c		0.287	0.000287	4375;2087
K_f	eq 2.67	5.87	5.87	331;2179
K_B	eq 2.79	3.22	3.22	727;7845,720,5088,732
pK_{BH^+}	25	4.60	4.60	4194;4308
	25,in aq HCl.	4.606	4.606	867,5773;869,4766,7049 1121,1951
κ	25	2.4×10^{-8}	2.4×10^{-6}	5730;4703
ε	30	6.71	6.71	6416;2416,4937,6457,1518,2027 3106,4703,7497
μ	25 in 38	1.51	1.51	4504,4488;2254,2295,6612 4320,6411
δ		10.3	21.1	652;4593
soly	in aq,25	3.38%w	3.38%w	4026;987,5413
	aq in,25	4.76%w	4.76%w	
aq az	742 Torr, 98.6	19.2%w	19.2%w	3498
fl pt	CC	76	349	5413;5712,152
Beil	1598,59			
uv	155,827,1177,2054,4067,4202,5276,5989,4388,1251,5497,6686,4308			
ir	8208,3791,633,7997,4283,3408,5605,5413,1660,2463,4021,4472,6311,8207			
Raman	1791,3825,4126,7713,8122,742,1787			
ms	5221,99			
nmr	3805,1429,4050,7227,4720,2318,8195,6358,5933,6748			

409. *o*-Toluidine

2-Methylbenzenamine

o-CH$_3$C$_6$H$_4$NH$_2$ 99–94–5   C$_7$H$_9$N

		cgs	SI	
mw		107.155	107.155	3468
bp	1 atm	200.40	473.55	7490;2776,4333,796,2090
				4684,7472
dt/dp	1 atm	0.054	0.41	7490;2090,4684
dp/dt	1 atm	19	2.5	7490
p	25	0.32	0.043	2087;2091,3402,3815,6090,7058
eq 2.5	A	7.60681	6.73171	2090;796
	B	2033.6	2033.6	
	C	230	230	
fp	(st)	-16.10	257.05	1897;2090,7058,7472,7473
	(ms)	-24.5	248.7	7490,7479
d	20	0.99843	998.43	2090;1191,4649,7403,7490
	25	0.99430	994.30	
dd/dt	25	0.000826	0.826	2090
n$_D$	20	1.57246	1.57246	2090;1191,4649,7490
	25	1.56987	1.56987	
dn/dt	25	0.000518	0.000518	2090
η	15	5.195	0.005195	7490;7403,5365
	25	3.390	0.003390	369
	30	3.183	0.003183	7490
γ	20	40.03	0.04003	7482;936,4518
	30	38.93	0.03893	
	Table 2.8			
ΔH$_v$	25	13.561	56.739	2087;4684
	bp	10.659	44.597	
	Table 2.9			
ΔH$_c^{\circ}$	25(1)	-963.3	-4030.4	3975;7117,7135
C$_p$	20	51.92	217.23	6440;4133,4684
t$_c$		421	694	2776;4299
P$_c$		37	3.7	2776;4299
pK$_{BH^+}$	25,in aq HCl	4.45	4.45	869;7835,3095,1121,1951
				429,4013
κ	25	3.792 x 10^{-7}	3.792 x 10^{-5}	5325
ε	18	6.34	6.34	4937;2027,6457,6024
μ	25 in 38	1.60	1.60	615;6024,1753
soly	in aq,20	1.66%w	1.66%w	6556
	aq in,20	2.44%w	2.44%w	
aq az	99.6	19%w	19%w	6165
fl pt	CC	85	358	5712

409. *o*-Toluidine

(Continued)

Beil	1672,772
uv	582,4067,5989,4308,6749,1755
ir	2557,633,3887,8191,4965,4369,4285,5932
Raman	3336,4112,4101
ms	6341
nmr	4050,7227,4965,8195,8219,6144,8200,5933,6748

410. *m*-Toluidine

3-Methylbenzenamine

m-CH$_3$C$_6$H$_4$NH$_2$ 108-44-1   C$_7$H$_9$N

		cgs	SI	
mw		107.155	107.155	3468
bp	1 atm	203.40	476.55	7490;796,2090,7058,2776
dt/dp	1 atm	0.055	0.41	7490;2090
dp/dt	1 atm	18	2.4	7490
p	25	0.27	0.036	2087;2091,3402,3815,5949,7058
eq 2.13	A	18.5043	17.6292	796;2090
	B	3200.9	3200.9	
	C	3.323	3.323	
fp		-30.40	242.75	2090;7058,7490,7479
d	15	0.99302	993.02	7490;1191,2090,3815,7403
	25	0.98444	984.44	6902
	30	0.98096	980.96	7490
dd/dt	15-30	0.000804	0.804	7490
n$_D$	20	1.56811	1.56811	2090;1191,3815
	25	1.56570	1.56570	2090;9027
dn/dt	25	0.000482	0.000482	2090
η	15	4.418	0.004418	7490;7403,881
	30	2.741	0.002741	
	130	0.506	0.000506	5365

410. *m*-Toluidine

(Continued)

		cgs	SI	
γ	20	38.02	0.03802	936
	30	37.16	0.03716	
ΔH_v	25	13.691	57.283	2087
	bp	10.719	44.848	
	Table 2.9			
ΔH_m		0.930	3.891	7478
ΔH_c°	25(1)	−963.3	−4030.4	3975;7117
C_p	15-30	51.84	226.90	4153
t_c		436	709	2776;4299
P_c		41	4.2	2776;4299
pK_{BH+}	25	4.70	4.70	4194
	25,in aq	4.73	4.73	869;3095,3189,5195,1156
	HCl			429,6226
κ	25	5.5×10^{-10}	5.5×10^{-8}	3262
ε	18	5.95	5.95	4937;2027,6457,6024
μ	25 in 38	1.45	1.45	615;6024,1753
soly	in aq	sl sol	sl sol	4026
aq az	99.6	17%w	17%w	6165

Beil	1682,853
uv	582,5989,6749,1755
ir	2557,633,7997,3890,8191,4285,5932
Raman	4110,4101
ms	6341
nmr	4050,7227,8195,8219,6144,8200,1167,5933,6748

411. *p*-Toluidine

4-Methylbenzenamine

p-CH$_3$C$_6$H$_4$NH$_2$ 106-49-0   C$_7$H$_9$N

		cgs	SI	
mw		107.155	107.155	3468
bp	1 atm	200.55	473.70	7491;796,4090,7058,2776
dt/dp	1 atm	0.054	0.41	7491
dp/dt	1 atm	19	2.5	7491
p	25	0.34	0.045	2087;3402,3815,6090,7058
eq 2.13	A	20.1569	19.2818	796;4090
	B	3269.3	3269.3	
	C	3.877	3.877	
fp		43.75	316.90	7491;796,1394,2783,7058
				7590,6075
d	45	0.96589	965.89	7491;1191,7403,7590
	50	0.96155	961.55	
dd/dt		0.000803	0.803	7491
n_D	45	1.55397	1.55397	2087;1191
	50	1.55348	1.55348	
dn/dt		9.8 x 10^{-5}	9.8 x 10^{-5}	2087
η	45	1.945	0.001945	7491
	55	1.557	0.001557	7403
	60	1.425	0.001425	7491
	130	0.522	0.000522	5365
γ	45	36.06	0.03606	936;7491
	60	34.10	0.03410	
	Table 2.8			
ΔH_v	25	13.431	56.195	2087;796,4090,4330
	bp	10.581	44.271	
	Table 2.9			
ΔH_m		4.520	18.91	6075;7117,7167,2356
ΔH_c°	25(1)	-963.3	-4030.4	3975;7117,7135
C_p	43	64.1	268.2	3621;3532
t_c		394	667	4299;7848,2776
P_c		23.5	2.38	2776;4299
K_f	eq 2.67	5.372	5.372	7590;2356
K_B	eq 2.79	4.14	4.14	727
pK_{BH}^+	25	5.08	5.08	4194;4308
	25 in aq	5.08	5.08	869;4766,3095,1077,1121
	HCl			4672,3092
κ	100	6.2 x 10^{-8}	6.2 x 10^{-6}	661
ε	54	4.98	4.98	4937;1394,5795,6024
μ	25 in 38	1.52	1.52	615;4690,4818,6024,762
soly	in aq,20.8	7.35%w	7.35%w	4672;2212
fl pt	CC	87	360	5712

411. *p*-Toluidine

(Continued)

Beil	1683,880
uv	155,582,4067,7503,4308,6749,1755
ir	2557,633,8191,4369,4965,4285,7932
Raman	3341,5720,4112,4016
ms	6341
nmr	2743,7227,4965,6167,8200,6144,8219,8195,5933,6748

412. 2,4-Xylidene

2,4-Dimethylbenzenamine

2,4-(CH₃)₂C₆H₃NH₂   95-68-1   C₈H₁₁N

$2,4\text{-}(CH_3)_2C_6H_3NH_2 \qquad 95\text{-}68\text{-}1 \qquad C_8H_{11}N$

		cgs	SI	
mw		121.182	121.182	3468
bp	1 atm	211.5	484.7	7058;2087
dt/dp	1 atm	0.05330	0.3998	2087
dp/dt	1 atm	18.76	2.501	2087
p	25	0.1539	0.0205	2087
eq 2.5	A	7.63281	6.75771	2087
	B	2035.4	2035.4	
	C	216	216	
fp		-14.3	258.9	884;2087
d	20	0.9763	976.3	884
	25	0.9723	972.3	2087
n_D	20	1.5576	1.5576	884
	25	1.55689	1.55689	2087
γ	25	36.75	0.03675	2087
	50	34.46	0.03446	
ΔH_v	25	14.26	59.66	2087
	bp	11.01	46.07	
pK_{BH^+}	25	4.89	4.89	977,701;7981
ε	20	4.9	4.9	4937
μ	25 in 38	1.40	1.40	2444;4714

412. 2,4-Xylidene

(Continued)

Beil	<u>1704</u>,1111
uv	1633,8144
ir	633
Raman	4101
nmr	8195,8219,6144,1167,6898

413. 2,6-Xylidene

2,6-Dimethylbenzenamine

2,6-(CH₃)₂C₆H₃NH₂ $2,6\text{-}(CH_3)_2C_6H_3NH_2$ 87-62-7 $C_8H_{11}N$

		cgs	SI	
mw		121.182	121.182	3468
bp	1 atm	217.9	491.1	341,7058
p	72.6	5.0	0.67	7058
fp		11.2	284.4	884;341
d	20	0.9842	984.2	884;341
n_D	20	1.5610	1.5610	884;341
pK_{BH^+}	25	3.95	3.95	701,977;7981
μ	25 in <u>38</u>	1.63	1.63	444;2471
fl pt	CC	96	369	5415

Beil	<u>1704</u>,1107
uv	1633,5642,7981
ir	4369,2955
Raman	4101
nmr	2527,8195,8219

414. Allylamine

2-Propen-1-amine

$CH_2{=}CHCH_2NH_2$ 107–11–9 C_3H_7N

		cgs	SI	
mw		57.095	57.095	3468
bp	1 atm	53.3	326.5	5762;1191,5783,7770,3979,4027
fp		-88.2	185.0	4027
d	20	0.7629	762.9	5762,7770;1191,4027,2774
	25	0.7575	757.5	
dd/dt		0.00107	1.07	5762,7770
n_D	20	1.42051	1.42051	7770;1191,3979,2774
	22	1.41943	1.41943	3262
η	25	0.3745	0.0003745	5365
γ	24.5	24.27	0.02427	7770;7849
	31.1	23.57	0.02357	
ΔH_f°		-2.4	-10.0	4522
ΔE_c		-524.8	-2195.8	7137,7446;4522
pK_a	25	9.49	9.49	3090,4194;3092,1077,4027
κ	25	5.7×10^{-5}	0.0057	3262
μ	25 in 38	1.31	1.31	6234
soly	in aq	inf	inf	3262
aq az		none	none	3497
fl pt	CC	-12	261	152,4027

Beil	338,205
ir	2557,2982,8191,6725,5932
Raman	3823,6725
nmr	84,2743,5933,6748

415. Ethylenediamine

1,2-Ethanediamine

$H_2NCH_2CH_2NH_2$ 107–15–3 $C_2H_8N_2$

		cgs	SI	
mw		60.099	60.099	3468
bp	1 atm	116.918	390.068	5067;1182,6414,7493,1191,8100 1512,7700
dt/dp	1 atm	0.045	0.34	7493
dp/dt	1 atm	22	3.0	7493
p	26.51	13.1	1.75	7483;7058,8100
eq 2.4	A	8.100	7.225	3779
	B	2038.2	2038.2	
fp		11.3	284.5	5789;8100,6414,5950
tp		11.14	284.29	5067
d	15	0.90001	900.01	7490;1191,3508,7770,8100 6414,5950
	25	0.8931	893.1	3651
	30	0.88595	885.95	7490;6054
dd/dt	10–40	0.00088	0.88	7627;8100
α	25	0.001024	0.001024	3651
κ_T	25	6.77×10^{-8}	5.08×10^{-7}	1280
κ_S	30	5.55×10^{-8}	4.16×10^{-7}	6054
n_D	15	1.45887	1.45887	7493;1191,3508,8100,5950
	20	1.45677	1.45677	7770
	30	1.4513	1.4513	1512
dn/dt		0.00050	0.00050	7493
η	15	1.722	0.001722	7493;924,6414,8100
	25	1.54	0.00154	2125
	30	1.226	0.001226	7493
γ	20	40.77	0.04077	7493;7770
	30	39.49	0.03949	
ΔH_v	25	10.921	45.693	5067;7823,2855,7483
	bp	10.0	41.8	8100
ΔH_m		5.3975	22.583	5067;3380,8100
ΔH_t	-84.2	0.1165	0.4874	5067
ΔH_f°	25(1)	-15.06	-63.01	2855
	25(g)	-4.07	-17.03	2855;5067
ΔH_c°	25(1)	-446.30	-1867.32	2855;8100
C_p	25	41.25	172.59	5067
	30	42.31	177.03	3508
	Table 2.12			
t_c		319.8	593.0	5404
p_c		62.1	6.29	5404

415. Ethylenediamine

(Continued)

		cgs	SI	
d_c		0.292	292	5404
v_c		0.206	0.000206	5404
A	eq 2.72	0.03361	0.03361	5067
K_f	eq 2.67	2.43	2.43	5808;5307
pK_{BH+}	25,1st	6.838	6.838	3363;5829,8100,2332,944
				3090,670
	25,2nd	10.960	10.960	
κ	25	9×10^{-8}	9×10^{-6}	1118;1182,6414,5950,1715
ε	25	12.9	12.9	5950;6414,8004,8100
μ	25 in 38	1.90	1.90	7556;4007
δ		12.3	25.2	652
soly	in aq	inf	inf	5307
aq az	119	81.6%w	81.6%w	3498;3497
fl pt	TOC	43	316	7627;152,8100
Beil	343,230			
ir	6550,1000,2011,6551,5932			
Raman	4103,2723,1787,7228,2011,2768			
nmr	4964,4965,5933,6748			

416. Diethylenetriamine

N-(2-Aminoethyl)-1,2-ethanediamine

$(NH_2CH_2CH_2)_2NH$ 111–40–0 $C_4H_{13}N_3$

		cgs	SI	
mw		103.167	103.167	3468
bp	1 atm	206.9	480.1	7627
dt/dp	1 atm	0.052	0.39	7627
dp/dt	1 atm	19	2.6	7627
p	89	10.0	1.33	7627
eq 2.5	A	8.15242	7.27732	6780
	B	2257.3	2257.3	
	C	230	230	

416.   Diethylenetriamine

(Continued)

		cgs	SI	
fp		-39	234	7627
d	20	0.9525	952.5	7627
	25	0.9492	949.2	6287
dd/dt	10-40	0.00083	0.83	7627
α	55	0.00091	0.00091	7627
n_D	20	1.4859	1.4859	7627
	25	1.4815	1.4815	6287
η	20	7	0.007	7627
ΔH_v	bp	11.3	47.3	7627
pK_b	20	3.55	3.55	5743
pK_{BH^+}	20	9.94	9.94	5972,4773;6249,556
μ	25 in 38	1.89	1.89	4007
δ	25	12.6	25.8	653
soly	in aq	inf	inf	7627
fl pt	TOC	102	375	4879
Beil	343,255			
ir	6551			
nmr	7064			

417.   Ethylenimine

Aziridine

CH_2CH_2NH 151-56-4 C_2H_5N

		cgs	SI	
mw		43.068	43.068	3468
bp	1 atm	57	330	2071;1245,3448,2622
p	-21.16	15.050	2.0065	5610
	20	160	21.3	5713
fp		-78.0	195.2	2071;7479

417. Ethylenimine

(Continued)

		cgs	SI	
d	10	0.846	846	2071;2622
	25	0.832	832	
dd/dt	10–25	0.00093	0.93	2071
n_D	25	1.4123	1.4123	2071
η	25	0.418	0.000418	2071
γ	25	32.8	0.0328	2071
ΔH_v	20	7.9	33.1	2071
ΔH_f°	25(1)	21.97	91.92	7826;5449
	25(g)	30.2	126.4	7826;3788
ΔH_c°	25(1)	−380.86	−1593.52	5449
C_p°	25	12.23	51.17	3788
pK_a	25	8.04	8.04	6539;1245,5596,6640
κ	25	8×10^{-6}	0.0008	3911
ε	25	18.3	18.3	3911
μ	25 in 38	1.77	1.77	3911;3737
soly	in aq	inf	inf	5713;2622
fl pt		−11	262	2071

Beil	3035,1
uv	1502
ir	153,3448,7429,2358,6622,5174,4456
Raman	4119,6622,4456
ms	5925,5924
nmr	3046

418.   Dimethylamine

N-Methylmethanamine

$(CH_3)_2NH$ 124–40–3 C_2H_7N

		cgs	SI	
mw		45.084	45.084	3468
bp	1 atm	6.88	280.03	7368;7058,1607
dt/dp	1 atm	0.0335	0.251	2089;7368
dp/dt	1 atm	29.9	3.98	2089
p	25	1476	196.8	2089
eq 2.5	A	7.06396	6.18886	2089
	B	1024.4	1024.4	
	C	238	238	
fp		−92.19	180.96	7368;7058
d	20	0.6556	655.6	7368;1607,3663
	25	0.6496	649.6	
α	25–35	0.00193	0.00193	1607
n_D	20	1.3597	1.3597	7368
	25	1.3566	1.3566	
η	15	0.207	0.000207	1607
	25	0.186	0.000186	
γ	15	17.6	0.0176	1607;2089
	25	16.33	0.01633	
ΔH_v	25	5.652	23.65	2089
	bp	5.882	24.61	2089;1607
ΔH_m		1.420	5.941	2089
ΔH_f°	25(1)	−10.50	−43.93	8316;2036,7826
	25(g)	−4.41	−18.45	7826;2036
ΔH_c°	25(1)	−416.70	−1743.47	8316;2036
	25(g)	−422.70	−1768.58	8316
C_p°	25	16.92	70.79	6554;7826,1607,2398
C_p	25	32.9	137.7	7826
	0(g)	2.53×10^{-5}	0.0106	5373
t_c		164.5	437.7	7369;1607
P_c		52.4	5.31	7369;1607
pK_{BH^+}	25	10.732	10.732	5774;3110,2333
ε	25	5.26	5.26	1607;6948
μ	25 in $\underline{38}$	1.18	1.18	4506;6987
	25(1)	1.14	1.14	4506
aq az		none	none	3499
fl pt	CC	20	293	6237

418.   Dimethylamine

(Continued)

Beil	335,39
ir	804,636,1291,5754,2646,2427
Raman	2646
ms	4211,1061
nmr	3795,7076,3071,5933

419.   Diethylamine

N-Ethylethanamine

$(CH_3CH_2)_2NH$ 109-89-7 $C_4H_{11}N$

		cgs	SI	
mw		73.138	73.138	3468
bp	1 atm	55.55	328.70	4842;7770,7932,1672,7368,4535 897,7058,7473
dt/dp	1 atm	0.0385	0.289	7368
dp/dt	1 atm	26.0	3.46	7368
p	25	233.5	31.13	2089;1653,5900,7058
eq 2.5	A	5.80159	4.92649	4842;4535,1654
	B	583.297	583.297	
	C	144.145	144.145	
fp		-49.8	223.4	7368;5900,7058,7473,2663
d	20	0.7070	707.0	7368;1191,2580,4649,7139,7776 7932,1672,4399
	25	0.7016	701.6	
dd/dt	10-40	0.00108	1.08	7621
α	55	0.00156	0.00156	7621
n_D	20	1.38455	1.38455	6716;7368,1191,4649,7770,7932
	25	1.3825	1.3825	7368;4535
η	10.2	0.3878	0.0003878	2586;5365,7404
	20	0.3060	0.0003060	4399;6598
	30	0.2729	0.0002729	4399;6598

419. Diethylamine

(Continued)

		cgs	SI	
γ	15	20.63	0.02063	7140;7770
	25	19.39	0.01939	
ΔH_v	25	7.486	31.32	4842;7823
	bp	6.948	29.07	4842;897,7754
ΔH_f°	25(1)	−24.79	−103.72	8316;5422,7826
	25(g)	−17.07	−71.42	7826
ΔH_c°	25(1)	−727.15	−3042.40	8316;4522,7446,7137,7135,3975
C_p	20	42.244	176.75	3902
	Table 2.11			
t_c		223.8	497.0	3345;822,1672,758
P_c		40.0	4.05	3345;822,758
d_c		0.243	243	3345
v_c		0.301	0.000301	3345
pK_{BH^+}	25	10.933	10.933	670,2305;5333,3090,5773,1077 5737,4194,1950
ϵ	20	3.894	3.894	4399;6457,2663
μ	25(1)	1.11	1.11	614;3384
	25 in 38	1.19	1.19	4716
δ		8.0	16.4	652
soly		inf	inf	7033
aq az		none	none	3498
fl pt	TOC	<−17	<256	7621,5712

Beil	336,95
uv	862,3414,4062
ir	1660,2557,608,636,2646,5932
Raman	1471,1791,3825,1787,2646
ms	1593,4211
nmr	6943,5345,5933

420. Dipropylamine

N-Propyl-1-propanamine

$(CH_3CH_2CH_2)_2NH$ 142–84–7 $C_6H_{15}N$

		cgs	SI	
mw		101.191	101.191	3468
bp	1 atm	109.2	382.4	7368;1191,7473,7770
dt/dp	1 atm	0.045	0.34	7368
dp/dt	1 atm	22	3.0	7368
p	25	24.09	3.212	2089
fp		−63	210	7368;7473,2663
d	20	0.7375	737.5	7368;1693,1191,4649,7770
	25	0.7329	732.9	
dd/dt	10–40	0.00096	0.96	7621
α	55	0.00129	0.00129	7621
n_D	20	1.4043	1.4043	7368;1693,1191,4649,7770
	25	1.4018	1.4018	
dn/dt	25	0.00050	0.00050	7368
η	20.1	0.5335	0.0005335	2580
	25	0.525	0.000525	6598
	37.2	0.4267	0.0004267	2580
γ	16.9	23.13	0.02313	7770
	24.5	22.28	0.02228	
ΔH_v	25	9.570	40.04	7823
	bp	8.335	34.87	7403;3816
ΔH_f°	25(1)	−37.33	−156.19	4448
	25(g)	−27.43	−114.77	
ΔH_c°	25(1)	−1039.36	−4348.68	4448
C_p		60.43	252.84	3816
t_c		277	550	4299
p_c		31	3.1	4299
K_i	25	0.000804	0.000804	7621
pK_a	25	11.00	11.00	3090;4194,3092,1077,4937
ε	20	3.068	3.068	1693;6457,1078,2663
μ	20 in 38	1.03	1.03	1693;4272
δ	25	8.0	16.3	653
soly	in aq,20	2.5%w	2.5%w	7621;1865
	aq in,20	28.7%w	28.7%w	
aq az	86.7	na	na	3453
fl pt	TOC	17	290	7621

420. Dipropylamine

(Continued)

Beil	337,138
uv	862
ir	2557,633,608,5932
Raman	7454
ms	2812
nmr	2812,5933,6748

421. Diisopropylamine

N-(1-Methylethyl)-2-propanamine

[(CH₃)₂CH]₂NH   108–18–9   C₆H₁₅N

$[(CH_3)_2CH]_2NH$ 108–18–9 $C_6H_{15}N$

		cgs	SI	
mw		101.191	101.191	3468
bp	1 atm	83.57	356.72	5799;7637,1672,7770
dt/dp	1 atm	0.042	0.32	7749
dp/dt	1 atm	24	3.2	7749
p	25	81.157	10.820	5799
eq 2.5	A	6.573222	5.698125	5799
	B	1038.183	1038.183	
	C	197.600	197.600	
fp		-96.3	176.9	7637
d	20	0.7153	715.3	1672;7770
	25	0.7100	710.0	1672
dd/dt	10–40	0.00080	0.80	7621
α	55	0.00140	0.00140	7621
n_D	20	1.39236	1.39236	7770
η	20	0.415	0.000415	6598
	25	0.40	0.00040	7749
γ	16.0	20.04	0.02004	7770
	26.3	19.11	0.01911	
ΔH_v	25	8.265	34.58	5799;7823
	bp	7.263	30.39	5799;7749

421. Diisopropylamine

(Continued)

		cgs	SI	
ΔH_f°	25(1) 25(g)	−42.64 −34.44	−178.41 −144.10	4448
t_c		249.0	522.2	7749
pK_a	25	11.07	11.07	4194;3090,3095,3092
μ	25 in 38	1.15	1.15	4716;4271
δ	25	7.4	15.2	653
soly	in aq,30 aq in,30	11%w 40%w	11%w 40%w	1865
aq az	74.1	90.8%w	90.8%w	3497
fl pt	TOC	−6	267	7637

Beil	337,154
ir	153,5932
Raman	7454
ms	2812,4212
nmr	5345,5933,6748

422. Dibutylamine

N-Butyl-1-butanamine

$(CH_3CH_2CH_2CH_2)_2NH$ 111−92−2 $C_8H_{19}N$

		cgs	SI	
mw		129.245	129.245	3468
bp	1 atm	159.6	432.8	7368;1672,7770
dt/dp	1 atm	0.051	0.38	7368
dp/dt	1 atm	20	2.6	7368
p	25	2.28	0.304	2089
eq 2.5	A B C	7.26603 1616.4 209	6.39093 1616.4 209	2089
fp		−62	211	7368

422. Dibutylamine

(Continued)

		cgs	SI	
d	20	0.7619	761.9	7368;1693,1672,7770,2580
	25	0.7577	757.7	
dd/dt	10--40	0.00084	0.84	7621
α	55	0.00112	0.00112	7621
n_D	20	1.4177	1.4177	7368;1693,7770
	25	1.4152	1.4152	
dn/dt	25	0.00050	0.00050	7368
η	20	0.95	0.00095	7749;2580
	25	0.946	0.000946	6598
	30	0.83	0.00083	7749;2580
γ	20	24.57	0.02457	7368
	40.9	22.63	0.02263	7770
ΔH_v	25	11.82	49.44	7823;2089
	bp	9.523	39.84	2089
ΔH_f°	25(l)	-49.27	-206.15	4448
	25(g)	-37.77	-158.03	
ΔH_c°	25(l)	-1352.16	-5657.44	4448
t_c		222.6	495.8	1672
pK_a	25	11.25	11.25	4194;3095,3092,3090,5737
ϵ	20	2.978	2.978	1693;2663
μ	25 in 38	0.98	0.98	4716,1693,1530
δ	25	8.1	16.6	653
soly	in aq,20	0.47%w	0.47%w	7637
	aq in,20	6.2%w	6.2%w	
aq az	97	49.5%w	49.5%w	3498
fl pt	TOC	51	324	7621;152
Beil	337,157			
ir	4079,633,608,5932			
ms	2812			
nmr	5345,5933			

423. Pyrrole

1 *H*-Pyrrole

HC———CH
‖   ‖
HC   CH
  \   /
   N
   |
   H

109-97-7 C_4H_5N

		cgs	SI	
mw		67.090	67.090	3468
bp	1 atm	129.765	402.915	2059;6519,3275,7564,5128
dt/dp	1 atm	0.0429	0.322	6519
dp/dt	1 atm	23.3	3.11	6519
p	25	8.2	1.1	6519;5128
eq 2.5	A	7.30275	6.42765	5608;6519
	B	1506.877	1506.877	
	C	210.995	210.995	
fp		−23.4146	249.7354	6519;5128
tp		−23.41	249.74	3275
d	20	0.96985	969.85	3275;7564,4889,1694
	25	0.96565	965.65	
	30	0.96133	961.33	
dd/dt	25	0.000852	0.852	3275
n_D	20	1.51015	1.51015	3275;7564,4889,1694
	25	1.50779	1.50779	
	30	1.50543	1.50543	
dn/dt	25	0.000472	0.000472	3275
η	20	1.352	0.001352	3275
	25	1.233	0.001233	
	30	1.129	0.001129	
γ	20	37.61	0.03761	3275;4866
	25	37.06	0.03706	
	30	36.51	0.03651	
ΔH_v	25	10.79	45.15	6519;5128,2276
	bp	9.261	38.75	6519
	Table 2.9			
ΔH_m		1.890	7.908	6519
ΔH_c°	25(1)	−562.07	−2351.70	6519;3975,2908
C_p°	25	17.05	71.34	6519
C_p	25	30.64	128.20	6519
	Table 2.12			
t_c		366.5	639.7	1451;3497,4299
p_c		56	5.7	3497
A	eq 2.72	0.01525	0.01525	6519;3275
	B	0.00208	0.00208	
pK_a	25,in aq H_2SO_4	−3.80	−3.80	1462;5396,4162,7168,3790

423. Pyrrole

(Continued)

		cgs	SI	
ε	25	8.13	8.13	5026;7497,2563
μ	25 in 38	1.80	1.80	1228;4312,6212,1694,2842,4493 4889,1166,825,4175
soly	in aq,25	ca 4.5%w	ca 4.5%w	913
fl pt	TCC	39	312	4026
Beil	3047,159			
uv	155,4199,5396,5056,1002,1462,3503,6749			
ir	153,3275,4462,6098,4661,4269,5244,4428,6035			
Raman	968,970,6125,7003,4882,4108,961,4428			
ms	3275,3430,1238,3709			
nmr	2567,2559,1462,3778,2980,6092,2016,5933,6748			

424. Pyrrolidine

$$H_2C-CH_2$$
$$H_2C \quad CH_2$$
$$\underset{H}{N}$$

123-75-1 C_4H_9N

		cgs	SI	
mw		71.122	71.122	3468
bp	1 atm	86.558	359.708	2059,3275
p	23.98	60.00	7.999	2059
eq 2.5	A	6.92527	6.05017	5608
	B	1180.409	1180.409	
	C	205.299	205.299	
tp		-57.85	215.30	3275
d	20	0.85859	858.59	3275
	25	0.85380	853.80	
dd/dt	25	0.000961	0.961	3275
n_D	20	1.44283	1.44283	3275
	25	1.44020	1.44020	
dn/dt	25	0.000530	0.000530	3275
η	20	0.756	0.000756	3275
	25	0.702	0.000702	
γ	20	29.65	0.02965	3275
	25	29.23	0.02923	
ΔH_v	25	8.98	37.57	1703

424. Pyrrolidine

(Continued)

		cgs	SI	
ΔH_f°	25(1)	-9.87	-41.30	2036;8316
	25(g)	-0.89	-3.72	
ΔH_c°	25(1)	-673.75	-2818.97	2036;8316
	25(g)	-682.73	-2856.54	
t_c		295.4	568.6	4299
P_c		55.4	5.61	4299
d_c		0.286	286	4299
v_c		0.249	0.000249	4299
A	eq 2.72	0.0221	0.0221	3275
pK_{BH^+}	25	11.27	11.27	6539;4194,5840,1708,3363
μ	20 in 38	1.57	1.57	6212
fl pt	TCC	3	276	6402

Beil	3037,4
uv	5840
ir	965,3275
ms	2112,6392,3275
nmr	4964,4965,6748

425. Piperidine

$HNCH_2CH_2CH_2CH_2CH_2$ 110-89-4 $C_5H_{11}N$

		cgs	SI	
mw		85.149	85.149	3468
bp	1 atm	106.219	379.369	2059;7491,3049
dt/dp	1 atm	0.044	0.33	7491
dp/dt	1 atm	23	3.0	7491
p	23.93	29.1	3.88	2339;6181

425. Piperidine

(Continued)

		cgs	SI	
eq 2.5	A	6.85699	5.98189	5608
	B	1239.577	1239.577	
	C	205.528	205.528	
fp		-10.5	262.7	2339
d	15	0.86591	865.91	7491;2339,3049,4636,2580
	25	0.85664	856.64	2847
	30	0.85215	852.15	7491
dd/dt		0.001157	1.157	7491;2339
n_D	20	1.4525	1.4525	5;1185,3049
dn/dt		0.00048	0.00048	7491
η	15	1.679	0.001679	7491;5365,7404,3049,2580
	25	1.362	0.001362	4331
	30	1.224	0.001224	7491
γ	20	30.05	0.03005	7491
	30	28.81	0.02881	
	Table 2.7			
ΔH_v	25	9.39	39.29	2847;1703,7755,1932
	bp	7.572	31.68	4686
ΔH_f°	25(1)	-21.15	-88.49	1703;717
	25(g)	-11.23	-46.99	6516
ΔH_c°	25(1)	-825.33	-3453.18	2847;3975,1932,7446,717
C_p°	25	25.17	105.31	6516
C_p	20-100	44.56	186.44	4686;1932
t_c		320.95	594.10	113;1458,4299
P_c		45.9	4.65	3049;758
K_B	eq 2.79	2.84	2.84	Beil. H20, 7
pK_{BH^+}	25	11.123	11.123	3362,4036,670,5773;3092,3090 3091,3094,3069,1077,6539 7982,4606,4194
ε	20	5.8	5.8	6457
μ	25 in 38	1.19	1.19	451;6612,7577
δ	25	8.7	17.8	653
soly	in aq	inf	inf	3262;359
aq az	92.8	65%w	65%w	3497
fl pt		3	276	5

425. Piperidine

(Continued)

Beil	3037,6
ir	608,2551,2203,7901,153,3408,4551,4965,7703,5932
Raman	4551,7703
ms	1775,2112,1029,6392,3986
nmr	5345,4965,4964,4623,5933,6748

426. *N*-Methylaniline

N-Methylbenzenamine

$C_6H_5NHCH_3$ 100-61-8 C_7H_9N

		cgs	SI	
mw		107.155	107.155	3468
bp	1 atm	196.25	469.40	7482;7058
dt/dp	1 atm	0.053	0.40	7482
dp/dt	1 atm	19	2.5	7482
p	62.8	5.0	0.67	7058
fp		−57	216	7058
d	15	0.99018	990.18	7482
	25	0.98221	982.21	
dd/dt	0–30	0.000798	0.798	7482
n_D	15	1.57367	1.57367	7472
	25	1.5684	1.5684	
η	15	2.568	0.002568	7482
	30	1.766	0.001766	
γ	15	39.97	0.03997	7482
	30	39.54	0.03954	
ΔH_v	25	12.7	53.1	7801
ΔH_f°	25(l)	7.7	32.2	7801
	25(g)	20.4	85.4	7801
ΔH_c°	25(l)	−973.5	−4073.1	7801

426. *N*-Methylaniline

(Continued)

		cgs	SI	
t_c		428	701	4299
p_c		51.3	5.20	4299
pK_{BH^+}	25	4.848	4.848	528;3034,4194,4308
ε	25	6.06	6.06	2416;6457
Beil	1599,135			
uv	2571,4308,6749			
ir	5932			
nmr	6144,6358,6748,5933			

427. Trimethylamine

N,N-Dimethylmethanamine

$(CH_3)_3N$ 75-50-3 C_3H_9N

		cgs	SI	
mw		59.111	59.111	3468
bp	1 atm	2.87	276.02	7371;1607
dt/dp	1 atm	0.0331	0.248	2089;7371
dp/dt	1 atm	30.2	4.03	2089
p	25	1699	226.5	2089
eq 2.5	A	6.97038	6.09528	2089
	B	968.7	968.7	
	C	234	234	
fp		-117.3	155.9	7371;1607,6237
d	20	0.6331	633.1	7371;3663,1607
	25	0.6270	627.0	
α	25-35	0.00202	0.00202	1607
n_D	20	1.3476	1.3476	7371
	25	1.3443	1.3443	
η	15	0.194	0.000194	1607
	25	0.177	0.000177	

427. Trimethylamine

(Continued)

		cgs	SI	
γ	15	14.5	0.0145	1607;3663,2089
	25	13.47	0.01347	
ΔH_v	25	5.461	22.850	2089
	bp	6.618	27.690	2089;1607
ΔH_m		1.564	6.544	2089
ΔH_f°	25(1)	-10.94	-45.77	8316;7826,2036
	25(g)	-5.70	-23.85	8316;7826
ΔH_c°	25(1)	-578.63	-2420.99	8316;2036
	25(g)	-583.87	-2442.91	8316
C_p°	25	21.73	90.92	6554;1607
C_p	25	32.31	135.19	7826
t_c		160.1	433.3	7372
P_c		40.2	4.07	7372
d_c		0.233	233	7372
v_c		0.254	0.000254	7372
pK_b	20	4.24	4.24	3139
$pK_{BH}{}^+$	25	9.752	9.752	5774;4906,6907
κ	33.5	2.2×10^{-10}	2.2×10^{-8}	1607
ϵ	25	2.44	2.44	1607;6457
μ	25 in 38	0.87	0.87	4506;6987
	25(1)	0.72	0.72	
aq az		none	none	3499
fl pt	CC	>20	>293	6237
Beil	335,43			
ir	2820			
ms	1061,4211,3606			
nmr	3071,7445			

428. Triethylamine

N,N-Diethylethanamine

$(CH_3CH_2)_3N$ 121–44–8 $C_6H_{15}N$

		cgs	SI	
mw		101.191	101.191	3468
bp	1 atm	88.87	362.02	7795,7932,7472,1655,4535,4842 897,7370,1680,7479
dt/dp	1 atm	0.043	0.32	7370
dp/dt	1 atm	23	3.1	7370
p	25	57.07	7.609	2089;1655,6717
eq 2.5	A	6.09424	5.21914	4842;4535,1655,5900,1654,2167
	B	834.442	834.442	
	C	170.800	170.800	
fp		−114.7	158.5	7370,7479;7472,7489,3793
d	20	0.7276	727.6	7370;1191,4649,7139,7489,7770 7032,2580,1655,1672
	25	0.72305	723.05	4535;7370
dd/dt		0.00092	0.92	<u>1217</u>
α		0.00126	0.00126	7489
κ_S	25	1.465×10^{-7}	1.099×10^{-6}	4315
n_D	20	1.4010	1.4010	7370;1191,4649,7770,7932 1217,1655
	25	1.3980	1.3980	
dn/dt	25	0.00060	0.00060	<u>7370</u>
η	15	0.394	0.000394	7489;2580
	20	0.3658	0.0003658	4399;6598
	25	0.363	0.000363	5365
	30	0.3295	0.0003295	4399;7489,6598
γ	20	20.66	0.02066	7489;5064,7140,7770
	30	19.62	0.01962	
ΔH_v	25	8.320	34.81	4842;897,7823
	bp	7.412	31.01	4842;7754
ΔH_f°	25(1)	−30.54	−127.78	4447;4502,4447,8316
ΔH_c°	25(1)	−1046.15	−4377.09	4447;3975,4502,7446,7137 7135,8316
C_p	20	53.396	223.41	3902
	Table 2.11			
t_c		262.2	535.4	3345;7848,1602,4299
P_c		30.0	3.04	3345;4299
d_c		0.257	257	3345;4299
v_c		0.394	0.000394	<u>3345</u>
K_B	eq 2.79	3.45	3.45	<u>7370,7754</u>
pK_a	25	10.778	10.778	1950;4194,670,2331,5773,3090 1077,3094,5333

428. Triethylamine

(Continued)

		cgs	SI	
ε	20	2.423	2.423	4399;4937,7850,7497,1330
μ	25 in 38	0.87	0.87	451;4716,614,3384,6612
δ		7.42	15.18	6240
soly	in aq,20	5.5%w	5.5%w	7542
	aq in,20	4.6%w	4.6%w	
aq az	75	90%w	90%w	3497;1185
fl pt	CC	−6	267	5712
Beil	336,99			
uv	862,4062,7571			
ir	1660,7077,153,6368,5932			
Raman	1471,1791,4110,1787,7228			
ms	2812,4211			
nmr	2744,1795,5933,6748			

429. Tributylamine

N,*N*-Dibutyl-1-butanamine

$(CH_3CH_2CH_2CH_2)N$ 102-82-9 $C_{12}H_{27}N$

		cgs	SI	
mw		185.352	185.352	3468
bp	1 atm	214.0	487.2	7371;1672,7770,7749
dt/dp	1 atm	0.054	0.41	7371
dp/dt	1 atm	18.5	2.5	7371
p	119.0	40.0	5.33	7749
fp		−70.0	203.2	7749
d	20	0.7781	778.1	7371;1672,7770,5199
	25	0.7743	774.3	
n_D	20	1.4291	1.4291	7371;7770
	25	1.4286	1.4286	
η	15	1.600	0.001600	6598;881
	25	1.313	0.001313	
γ	20	24.76	0.02476	7749;406,7770
	42.9	22.91	0.02291	7770
ΔH_v	bp	11.2	46.9	7749

429. **Tributylamine**

(Continued)

		cgs	SI	
ΔH_f°	25(1)	-67.35	-281.79	4447
ΔH_c°	25(1)	-1983.55	-8299.17	4447
t_c		365.2	638.4	1672
pK_b	25	3.11	3.11	5739;1802
pK_{BH^+}	25	9.93	9.93	5774;3095
μ	25 in 38	0.78	0.78	451;5199,4716
δ	25	7.7	15.8	653
soly	in aq,18	0.004%w	0.004%w	1802
aq az	99.65	20.7%w	20.7%w	3499
fl pt	OC	85	358	6402
Beil	337,157			
uv	5199			
ms	5108,2812			
nmr	5399,6748			

430. *N,N*-Dimethylaniline

N,N-Dimethylbenzenamine

$C_6H_5N(CH_3)_2$ 121-69-7 $C_8H_{11}N$

		cgs	SI	
mw		121.182	121.182	3468
bp	1 atm	194.05	467.20	7482;7058
dt/dp	1 atm	0.052	0.39	7482
dp/dt	1 atm	19	2.6	7482
p	56.3	5.0	0.67	7058
fp		2.45	275.60	7482;7058
d	20	0.9559	955.9	4195
	25	0.95232	952.32	3900,6902;7482
α		0.000865	0.000865	4195

430.   *N, N*-Dimethylaniline

(Continued)

		cgs	SI	
n_D	25	1.55661	1.55661	3900,6902
	30	1.55389	1.55389	3900
dn/dt	30	0.000504	0.000504	3900
η	25	1.288	0.001288	3900,6902
	30	1.173	0.001173	3900
γ	20	36.04	0.03604	4195;937
	60	32.33	0.03233	937
ΔH_v	25	11.9	49.8	7801
ΔH_f°	25(1)	8.2	34.3	7801
	25(g)	20.1	84.1	
ΔH_c°	25(1)	−1136.3	−4754.3	7801
t_c		414	687	4299
p_c		35.8	3.63	4299;758
pK_{BH^+}	25	5.150	5.150	528;2418,4194,3034,4308
ϵ	20	4.91	4.91	4937;6457
fl pt	CC	62	335	5712

Beil	1601,141
uv	2571,4308,6749
ir	5932
nmr	6144,6748

431. Pyridine

110–86–1

C_5H_5N

		cgs	SI	
mw		79.101	79.101	3468
bp	1 atm	115.254	388.404	6915; 2059,854,4578,3325,3275 6181,7491,7470
dt/dp	1 atm	0.04535	0.3402	854;7470,7491
dp/dt	1 atm	22.05	2.940	854
p	24.8	20	2.7	4902;4091,3380,6090,6181,7058
eq 2.5	A	7.06105	6.18595	6915;5608,3325,4578
	B	1386.683	1386.683	
	C	216.469	216.469	
fp		–41.55	231.60	854;8121,430,5689,7471 7473,7491
tp		–41.67	231.48	3275
d	20	0.98319	983.19	3275;854,430,4517,4969,553
	25	0.97824	978.24	5111,4937,6254,7470,7491
	30	0.97319	973.19	8104,6902
dd/dt		0.00099	0.99	<u>1217</u>
α		0.001070	0.001070	4195
n_D	20	1.51016	1.51016	3275;4091,854,1217,2790,4518
				4969,5111,4937,6254,7491
	25	1.50745	1.50745	8104,6902
	30	1.50466	1.50466	
dn/dt	25	0.000550	0.000550	<u>3275</u>
η	20	0.952	0.000952	3275;553,5996,7491,6962
	25	0.884	0.000884	
	30	0.815	0.000815	
γ	20	36.88	0.03688	3275;7491,4195
	25	36.33	0.03633	
	30	35.78	0.03578	
	Table 2.8			
ΔH_v	25	9.659	40.41	4578;7755,4686,1932,3816 4969,7848
	bp	8.697	36.39	854
ΔH_m		1.772	7.414	3275;4793,5689
ΔH_f°	25(1)	23.95	100.21	4578,1701;8316
ΔH_c°	25(1)	–665.00	–2782.36	4578,1701;3540,3975,7135,7446
C_p°	25	19.06	79.75	4578
	Table 2.14			
C_p	17	32.4	135.6	4949;1054,1932,3816,5689,4686

431. Pyridine

(Continued)

		cgs	SI	
t_c		344.0	617.2	3345;4091,125,119,1451,6854
p_c		60.0	6.08	3345;4091,125,4299
d_c		0.312	312	3345;4091,4299
v_c		0.253	0.000253	3345;4299
A	eq 2.72	0.01858	0.01858	3275
K_f	eq 2.67	4.75	4.75	854;735
K_B	eq 2.79	2.710	2.710	7590,7862;5595,7982,7845,3816
pK_{BH^+}	25	5.17	5.17	1136;5736,4616,2829,3094,5773 3069,3439,641,5195,31 3092,2718
κ	25	4.0×10^{-8}	4.0×10^{-6}	7855;5730
ε	25	12.91	12.91	2416;6456,828,7497,553 1436,7850,8184
μ	25 in 317	2.37	2.37	6612,4493;4517,5111,1227 4716,4175,6191
δ		10.62	21.7	6240;652,4593
soly	in aq	inf	inf	4026
aq az	93.6	58.7%w	58.7%w	3498;3497
fl pt	CC	23	296	5712

Beil	3051,181
uv	155,3299,6929,6939,4388,4724,3275
ir	4468,7583,854,6368,153,8188,3275,5932
Raman	742,691,4881,640
ms	6199,3275
nmr	7232,5351,2746,2561,1386,2834,4427,1795,2464,5933,6748

432. 2-Picoline

2-Methylpyridine

$$\begin{array}{c} \text{H} \\ \text{C} \\ HC \diagup \quad \diagdown CH \\ \| \qquad | \\ HC \diagdown \quad \diagup CCH_3 \\ N \end{array}$$

109–06–8 C_6H_7N

		cgs	SI	
mw		93.128	93.128	3468
bp	1 atm	129.408	402.558	854,3325;2059,2535,3275,8192
dt/dp	1 atm	0.04695	0.3522	854,3325
dp/dt	1 atm	21.30	2.840	854,3325
p	22.82	10.00	1.333	2059
eq 2.5	A	7.03450	6.15940	3325;5608,6525
	B	1417.578	1417.578	
	C	211.874	211.874	
fp		−66.74	206.41	854;2535
tp		−66.704	206.446	6525;3275
d	20	0.94440	944.40	3275;854,2535,4359,4227,6232
	25	0.93981	939.81	2130,3776
dd/dt	25	0.000926	0.926	3275
α	25	0.000989	0.000989	854;2535
n_D	20	1.50102	1.50102	3275;4089,2535,854,1136,4359
	25	1.49839	1.49839	6232,2222,1063
dn/dt	25	0.000520	0.000520	3275
η	20	0.805	0.000805	3275;2535,2130
	25	0.753	0.000753	
γ	20	33.18	0.03318	3275;3249,4356,3663
	25	32.78	0.03278	
ΔH_v	25	10.258	42.919	361;6525
	bp	8.669	36.271	361;6525,854,3325
ΔH_m		2.324	9.724	6525;854,4520
ΔH_f°	25(1)	13.83	57.86	2036;1701,6525,8316
	25(g)	24.05	100.63	
ΔH_c°	25(1)	−827.46	−3462.09	2036;1701,6525,8316
	25(g)	−817.24	−3419.33	
C_p°	25	23.90	100.0	6525
C_p	28	38.051	159.205	6525
t_c		347.9	621.1	4089;4299,119
P_c		45.4	4.60	4089
d_c		0.278	278	4089
v_c		0.335	0.000335	4089
A	eq 2.72	0.02744	0.02744	6525

432. 2-Picoline

(Continued)

		cgs	SI	
K_f	eq 2.67	3.36	3.36	854;4520
pK_{BH^+}	25	5.997	5.997	5815;5760,5343,1186,1136,1513 4616,2718,3141
ε	20	9.8	9.8	4937
μ	25 in 38	1.97	1.97	1760;4716,3141,6232,4175,6079
soly	in aq,25	inf	inf	2222,359
aq az	93.5	52%w	52%w	3499
fl pt	OC	29	302	6402

Beil	3052,234
uv	2571,3319
ir	1639,7682,5663,4065
Raman	783,6338,4065
ms	1583
nmr	5351,4758

433. 3-Picoline

3-Methylpyridine

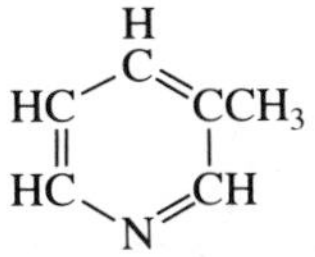

108–99–6

C_6H_7N

		cgs	SI	
mw		93.128	93.128	3468
bp	1 atm	144.143	417.293	854;3325,2059,8192,1679
dt/dp	1 atm	0.04873	0.3655	854,3325
dp/dt	1 atm	20.52	2.736	854,3325
p	33.38	10.00	1.333	2059
eq 2.5	A	7.03247	6.15737	3325;5608,6524
	B	1469.894	1469.894	
	C	209.907	209.907	
fp		−18.09	255.06	6524;854,1679
tp		−18.14	255.01	6524
d	20	0.95658	956.58	854;2222,1679,4227,2130,3776
	25	0.95197	951.97	

433. 3-Picoline

(Continued)

		cgs	SI	
dd/dt	25	0.000922	0.922	<u>854</u>
α	25	0.000969	0.000969	854
n_D	20	1.50685	1.50685	4359;1136,1679
	25	1.50682	1.50682	854
η	25	0.8723	0.0008723	2130
γ	20	35.06	0.03506	4354;3249
	40	32.71	0.03271	
ΔH_v	25	10.811	45.233	361;6524,3325
	bp	9.016	37.723	361;854,3325
ΔH_m		3.389	14.180	6524;854,4520
ΔH_f°	25(1)	15.57	65.14	3141;1701,6524,8316
	25(g)	26.30	110.04	
ΔH_c°	25(1)	-818.98	-3426.61	3141;1701,8316
	25(g)	-829.71	-3471.51	
C_p	26.4	38.013	159.046	6524
t_c		371.7	644.9	119;4299
K_f	eq 2.67	4.88	4.88	854;4520
pK_{BH^+}	25	5.750	5.750	5813;1513,1136,3140,360,3666 5343,5760,2292,3141
μ	25 in <u>38</u>	2.40	2.40	1760;4716,4175,6079,3141
soly	in aq,25	inf	inf	351
aq az	97	40%w	40%w	3499
Beil	<u>3052</u>,239			
uv	2571,3319,2292			
ir	5663,7682,1639,1677			
Raman	6338,783			
ms	1583			
nmr	4233			

434. 4-Picoline

4-Methylpyridine

108-89-4 C_6H_7N

		cgs	SI	
mw		93.128	93.128	3468
bp	1 atm	145.356	418.506	854;3325,2059,1063,8192
dt/dp	1 atm	0.04887	0.3666	854,3325
dp/dt	1 atm	20.46	2.728	854,3325
p	25	5.68	0.757	3325
eq 2.5	A	7.04250	6.16740	3325;5608
	B	1481.204	1481.204	
	C	210.558	210.558	
fp		3.65	276.80	854
d	20	0.95478	954.78	854;4227,4359
	25	0.95020	950.20	
dd/dt	25	0.000917	0.917	854
α	25	0.000965	0.000965	854
n_D	20	1.50566	1.50566	4059;854,1136,4359,1063,4517
	25	1.50322	1.50322	
dn/dt	25	0.000503	0.000503	4089
γ	20	35.45	0.03545	4359
ΔH_v	25	10.71	44.81	3504;2847,361
	bp	8.968	37.573	3504;854,361,3325
ΔH_m		2.766	11.573	854;4520
ΔH_f°	25(1)	13.58	56.82	3141,1701;2847
	25(g)	24.44	102.26	
ΔH_c°	25(1)	-816.99	-3418.29	3141,1701
	25(g)	-827.85	-3463.72	
C_p°	128	31.89	133.45	3504
t_c		373.1	646.3	4059;4299,119
P_c		46.0	4.66	4059
d_c		0.286	286	4059
v_c		0.326	0.000326	4059
K_f	eq 2.67	5.13	5.13	854;4520
pK_{BH^+}	25	6.062	6.062	5815;5760,1136,1513,5343,3666
				360,3140,3141
μ	25 in 38	2.60	2.60	1760,1759;4716,4175,4517
soly	in aq,25	inf	inf	359
aq az	97.35	37.2%w	37.2%w	3499
fl pt	OC	57	330	5415

434. 4-Picoline

(Continued)

Beil	3052,240
uv	2571,3319
ir	1639,7682,5663,1677
Raman	6338
ms	1583,5436
nmr	5351,4758,8174

435. 2,4-Lutidine

2,4-Dimethylpyridine

108–47–4 C_7H_9N

		cgs	SI	
mw		107.155	107.155	3468
bp	1 atm	158.403	431.553	341;8192,2222,1063
dt/dp	1 atm	0.04984	0.3738	341
dp/dt	1 atm	20.06	2.675	341
fp		−63.96	209.19	341
d	20	0.9310	931.0	341;6039,2222
	25	0.9271	927.1	
dd/dt	25	0.00078	0.78	341
n_D	25	1.4984	1.4984	1063;2222
η	20	0.887	0.000887	6039
γ	20	33.17	0.03317	5526
ΔH_v	25	11.42	47.78	1703
ΔH_f°	25(l)	3.85	16.11	2036
	25(g)	15.27	63.89	1703
ΔH_c°	25(l)	−969.63	−4056.93	2036
t_c		374	647	4299;119
pK_{BH^+}	25	6.63	6.63	360;5760,1106,2718,1513,3141
ε	20	9.60	9.60	6039
μ	25 in 38	2.30	2.30	1759,1760

435. 2,4-Lutidine

(Continued)

		cgs	SI	
soly	in aq,23	inf	inf	3776
aq az	96.5	34%w	34%w	3776

Beil	3053,244
uv	2571,545,3338
ir	1639,1677,5031,2939,3338
Raman	5031,2936,3338
nmr	4758

436. 2,6-Lutidine

2,6-Dimethylpyridine

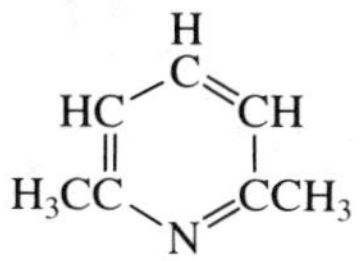

108–48–5

C_7H_9N

		cgs	SI	
mw		107.155	107.155	3468
bp	1 atm	144.045	417.195	854,3325;8192,1063
dt/dp	1 atm	0.04818	0.3614	854,3325
dp/dt	1 atm	20.76	2.767	854,3325
p	25	5.58	0.744	3325
eq 2.5	A	7.05246	6.17736	3325
	B	1467.362	1467.362	
	C	207.701	207.701	
fp		−6.10	267.05	854;1134
d	20	0.92257	922.57	854;6039,4354,1134,2130
	25	0.91806	918.06	
dd/dt	25	0.000902	0.902	854
α	25	0.000983	0.000983	854
n_D	20	1.4971	1.4971	1134;854,4354,2222,1063
	25	1.4953	1.4953	
η	20	0.869	0.000869	6039;2130
	40	0.668	0.000668	

436.   2,6-Lutidine

(Continued)

		cgs	SI	
γ	20	31.65	0.03165	4354
	40	29.18	0.02918	
ΔH_v	25	11.009	46.062	361
	bp	9.029	37.777	361;3325,854
ΔH_m		2.40	10.04	854
ΔH_f°	25(1)	2.41	10.08	1701;3141
	25(g)	13.43	56.19	1701
ΔH_c°	25(1)	−968.20	−4050.95	1701;3141
	25(g)	−979.22	−4097.06	1701
t_c		350.6	623.8	4299,119
K_f	eq 2.67	6.32	6.32	854
pK_{BH^+}	25	6.72	6.72	360;1136,1134,1513,5760,2718
ϵ	20	7.33	7.33	6039
μ	25 in 38	1.66	1.66	1759,1760;4716,6232,3141
soly	in aq,25	inf	inf	359
aq az	96.02	48.2%w	48.2%w	3499
Beil	3053,244			
uv	545,2571			
ir	5031,1639,2936			
Raman	5031,2936			
nmr	4758,5351			

437.   2,4,6-Trimethylpyridine

108–75–8

$C_8H_{11}N$

		cgs	SI	
mw		121.182	121.182	3468
bp	1 atm	171.0	444.2	8192;1063,1134,2264
p	75.0	38.8	5.17	1134
fp		−44.192	228.958	4959;1134,2264

437.   2,4,6-Trimethylpyridine

(Continued)

		cgs	SI	
d	25	0.9103	910.3	2264;1134
n_D	20	1.4981	1.4981	1134
	25	1.4959	1.4959	1134,1063;2264
dn/dt	25	0.00044	0.00044	1134
ΔH_m		2.279	9.535	4959
A	eq 2.72	0.02188	0.02188	4959
pK_{BH^+}	25	7.43	7.43	5958,2718;3141,5591,1134
μ	20 in 38	2.05	2.05	2810;6079,3141
Beil	3054,250			
uv	545,2571			
ir	1677,1639			
nmr	5351,4758,4623			

438.   Quinoline

91-22-5

C_9H_7N

		cgs	SI	
mw		129.161	129.161	3468
bp	1 atm	237.10	510.25	7491;7471,8226,2969,2776,4847
dt/dp	1 atm	18	2.3	7491;4847
p	25.16	0.0840	0.0112	7669
	59.7	1	0.1	5713;8226
eq 2.5	A	6.80189	5.92679	4847;3779
	B	1656.3	1656.3	
	C	184.78	184.78	
fp		-14.85	258.30	8121;4404,7471,7479,5689
d	15	1.09771	1097.71	7491;2027,1191,7405,5249,1053
	30	1.08579	1085.79	
dd/dt		0.000794	0.794	7491

438. Quinoline

(Continued)

		cgs	SI	
n_D	15	1.62928	1.62928	7491;1191
	20	1.6273	1.6273	4847
	25	1.62475	1.62475	7847
η	15	4.354	0.004354	7491;7405,1053
	25	3.145	0.003145	2273;5365
	30	2.997	0.002997	7491;3642
γ	20	45.65	0.04565	7491;5249
	30	44.82	0.04482	
ΔH_v	bp	11.88	49.71	3779;2969
ΔH_m		2.58	10.79	5689
ΔH_c°	25(1)	-1123.5	-4700.72	7506;1931
C_p	26.8	47.80	200.00	5689;1054
t_c		521.3	794.5	5249;113,2776
P_c		57	5.8	2776
K_f	eq 2.67	1.95	1.95	8121
K_B	eq 2.79	5.84	5.84	6477;728,727
pK_{BH^+}	25, in aq EtOH	4.94	4.94	5738;429,66,7787,2829,3094
κ	25	2.2×10^{-8}	2.2×10^{-6}	3621
ε	25	8.95	8.95	2416;4937,6456,2027,2969,1227
μ	25 in 38	2.184	2.184	4507;4493,789,1227,6612 4716,7943
δ		10.8	22.1	652
soly	in aq,20	0.609%w	0.609%w	65

Beil	3077,339
uv	155,4724,2377
ir	153,633,1886,329,5932
Raman	1886,4623,329
ms	1583
nmr	907,2561,1795,4623,5933,6748

439. Isoquinoline

119–65–3

C_9H_7N

		cgs	SI	
mw		129.161	129.161	3468
bp	1 atm	243.24	516.39	4847;2536,8192,3181
dt/dp	1 atm	0.060	0.45	4847;2536
dp/dt	1 atm	16.7	2.2	4847
p	92.7	5.0	0.67	7058
eq 2.5	A	6.90714	6.03204	4847
	B	1719.5	1719.5	
	C	183.83	183.83	
fp		26.48	299.63	2536;3181
d	30	1.09101	1091.01	2536
	40	1.08309	1083.09	
dd/dt	40	0.000791	0.791	2536
n_D	30	1.62077	1.62077	4847;2536
η	30	3.253	0.003253	2536
	100	1.023	0.001023	
γ	26.8	46.28	0.04628	341
ΔH_v	bp	11.70	48.96	341
ΔH_m		1.78	7.45	341
ΔH_f°	25(s)	37.9	158.6	2036
	31(l)	34.68	145.10	2847
ΔH_c°	25(s)	−1123.5	−4700.7	2036
	31(l)	−1120.09	−4686.5	2847
t_c		530	803	113,4299
pK_{BH}^+	20	5.38	5.38	5611;3862,67
ε	25	10.7	10.7	4937
μ	25 in 38	2.61	2.61	1227;4716,789,4507,1166

Beil	3078,380
uv	5611,2571,2376,6749
ir	329
Raman	4707,329
nmr	1795,908,6748

440. Formamide

$HCONH_2$ 75-12-7 CH_3ON

		cgs	SI	
mw		45.041	45.041	3468
bp	1 atm	210.5 (d)	483.7 (d)	7058,685;3663,4718
p	129.4	29.7	3.96	2140
eq 2.4	A	10.1366	9.2615	<u>7058</u>
	B	3509	3509	
fp		2.55	275.70	6828,7490,685;4438,7194
				3663,5523
d	20	1.13339	1133.39	6828;1191,2269,3663,7194
				7490,1870
	25	1.12915	1129.15	2183;6828,1777,948
α		0.000775	0.000775	4195
κ_T		5.32×10^{-8}	3.99×10^{-7}	2183
n_D	20	1.44754	1.44754	6829;1191,1212,7490,3625
	25	1.44682	1.44682	2721
η	20	3.764	0.003764	6828;2269,3947,7490,7791
				1870,2126
	25	3.302	0.003302	
γ	20	58.35	0.05835	6828;4887
	25	58.15	0.05815	513;6828
	Table 2.8			
ΔH_v	25	15.530	64.978	685
ΔH_m		1.600	6.694	685;7167,2721,6910
ΔH_f°	25(1)	−60.7	−254.0	8316,7826
ΔH_c°	25(1)	−135.8	−568.2	8316;3975
C_p°	25	10.84	45.35	7826;6554,685,2319
C_p	25	25.72	107.62	6798;7846
K_f	eq 2.67	3.85	3.85	1203;2097,2721
pK_a	20	−0.48	−0.48	3569;3390
κ		$<2 \times 10^{-7}$	$<2 \times 10^{-5}$	5502;3947,4438,6828,7725
				7842,7843
ε	20	111.0	111.0	663;1253,4438,4937,7842
				1870,4718
μ	30 in <u>38</u>	3.37	3.37	676;1253,8254,4718,455,6073
δ		19.2	39.3	652
soly	in aq	inf	inf	5713
fl pt	COC	154	427	4027

440. Formamide

(Continued)

Beil	156,26
uv	3350,155,3587
ir	2557,4523,7427,5932
Raman	4523,6368,2319,5983,1783,5466,1787,1791,3829,4115,6052,659,5983,6814
ms	2760
nmr	8037,3837,3838

441. *N*-Methylformamide

$HCONH(CH_3)$ 123–39–7 C_2H_5ON

		cgs	SI	
mw		59.068	59.068	3468
bp	1 atm	180–185	453–458	Beil. H4, 58
p	44	0.4	0.05	3292,2541;3625,1860
	47	0.6	0.08	
	55	1.5	0.20	
fp		−3.8	269.4	4439;1800,3292
d	15	1.0075	1007.5	2541;1800,3625,4439,2867
	25	0.9988	998.8	2541
dd/dt	0–50	0.0008674	0.8674	6771
α	25	0.000869	0.000869	6771
n_D	20	1.4319	1.4319	3625;4439
	25	1.4300	1.4300	1800
η	15	1.99	0.00199	2541;2867
	25	1.65	0.00165	
γ	25	39.46	0.03946	513
	30	37.96	0.03796	2867
	30	36.50	0.03650	
C_p	25	29.59	123.8	6798
pK_a	20	−0.04	−0.04	3569
pK_s	25	10.74	10.74	5589

441.   *N*-Methylformamide

(Continued)

		cgs	SI	
κ	25	8×10^{-7}	8×10^{-5}	2541;4439,3292
ε	25	182.4	182.4	4439
	−40	308	308	663
μ	25 in 38	3.86	3.86	2834;455
δ		9.93	20.3	6240;4593
soly	in aq	inf	inf	Beil. H4, 58
Beil	335,58			
uv	5489			
ir	5185,3254,1907,3766,5984,4407,5932			
Raman	1907,5984			
ms	2760			
nmr	4225,4226,4224,4407,5933			

442.   *N,N*-Dimethylformamide

$HCON(CH_3)_2$ 68–12–2 C_3H_7ON

		cgs	SI	
mw		73.094	73.094	3468
bp	1 atm	153.0	426.2	6309;2089,2149,3625,2660,1074
dt/dp	1 atm	0.0476	0.357	4902
dp/dt	1 atm	21.0	2.80	4902
p	25	3.7	0.49	2149;1192,5947
eq 2.5	A	7.1085	6.2334	5102;2868
	B	1537.78	1537.78	
	C	210.39	210.39	
fp		−60.43	212.72	2089;6309,1074
d	20	0.94873	948.73	2089;4439,6309,2541,1192
				3625,2703
	25	0.94387	943.87	1777;2089
	30	0.9412	941.2	2867
	40	0.9310	931.0	
dd/dt		0.00072	0.72	6905

442. *N, N*-Dimethylformamide

(Continued)

		cgs	SI	
α		0.00100	0.00100	6905
κ_T		8.6×10^{-8}	6.5×10^{-7}	1777
n_D	20	1.43047	1.43047	2089;4439,6309,1868,1192
	25	1.42817	1.42817	3625,2703
dn/dt	25	0.00046	0.00046	2089
η	20	0.9243	0.0009243	2089;2541,2867
	25	0.802	0.000802	2149
	40	0.7386	0.0007386	2089
γ	20	36.76	0.03676	2089;2867,6958
	25	36.42	0.03642	513;2149
	40	34.40	0.03440	2089
ΔH_v	25	11.356	47.514	2089;2139,2703
	bp	9.164	38.342	
ΔH_m		3.861	16.154	2089
ΔH_f°	25(1)	-57.21	-239.37	7689
ΔH_c°	25(1)	-464.06	-1941.63	7689;2139
C_p°	100	26	109	2139
C_p	25	35.46	148.36	8269;7792,2703,2134
t_c		323.4	596.6	2149;2867
P_c		51.5	5.22	2149;2703
d_c		0.294	294	2703
v_c		0.249	0.000249	2703
pK_a	20	-0.01	-0.01	3569;7223
pK_s	25	23.10	23.10	4259;1300,7220
κ	25	6×10^{-8}	6×10^{-6}	2541;1868,5554,6545
ϵ	25	36.71	36.71	4439;6985,1074
	-40	52.1	52.1	663
μ	25 in 38	3.24	3.24	6073;3603,2711,7099,455
δ		12.1	24.8	652,4593,6905,1399;6240
ER	BuOAc	0.20	0.20	6628,6631;1399,6905
	90%	2280	2280	6628,6631;1399
soly	in aq	inf	inf	2139
aq az		none	none	3498
fl pt	TOC	67	340	2139

442. *N*, *N*-**Dimethylformamide**

(Continued)

Beil	335,58
uv	3587,7878,5489
ir	3201,3254,3766,5984,2156,3902,5932
Raman	5984,3902
ms	2760,4783
nmr	7227,4981,4225,2507,3315,5462,2464,5933

443. **Acetamide**

CH₃CONH₂ $\qquad$ 60–35–5 $\qquad$ C₂H₅ON

		cgs	SI	
mw		59.068	59.068	3468
bp	1 atm	221.15	494.30	7482;2582
dt/dp	1 atm	0.054	0.41	7038
dp/dt	1 atm	19	2.5	7038
p	65.0	1	0.1	7038;7033
eq 2.4	A	8.80919	7.93409	7038
150–bp	B	2936.07	2936.07	
65–150	A	9.12026	8.24516	
	B	3082.80	3082.80	
mp*		69.5	342.7	5789;5313,676,4321,1846
fp		80.00	353.15	5789
d	91.1	0.9892	989.2	2582;7875,7589
	111.8	0.9711	971.1	
	131.7	0.9538	953.8	
dd/dt	90–130	0.00087	0.87	2582
n_D	80	1.4270	1.4270	5433;4027
	110	1.4158	1.4158	7482
	130	1.4079	1.4079	
η	91.1	2.182	0.002182	2582;2126
	111.8	1.461	0.001461	
	131.7	1.056	0.001056	

443. Acetamide

(Continued)

		cgs	SI	
γ	85 95 105	38.96 37.95 36.96	0.03896 0.03795 0.03696	7589
ΔH_v	bp	13.4	56.1	<u>7038</u>
ΔH_s	25	18.80	78.66	622;1846
ΔH_m	80	3.754	15.707	5789;3533,4027
ΔH_f°	25(s) 25(g)	-75.76 -56.96	-316.99 -238.33	622;1323,7826 622
ΔH_c°	25(s) 25(1)	-283.13 -286.5	-1184.60 -1198.7	622 3975
C_p	25(s) 80-150	21.6 39.67	90.3 165.98	6798 5789
K_f	eq 2.67	4.04	4.04	6414
pK_{BH}^+	25, in aq H_2SO_4	0.59	0.59	2210;3094,3439,3390,5735,1956 4519,2819,7787,3569,31 7033,3969
κ	83.2 100	8.8×10^{-7} 4.3×10^{-5}	8.8×10^{-5} 0.0043	3418;7841,750 7875
ε	83	59	59	4937;4439,1973
μ	30 in <u>38</u>	3.44	3.44	664;1974,4321,3433,4318 455,6073
soly	in aq,20	40.8%w	40.8%w	6999;7033
aq az		none	none	4937

* see Ch V, amides, general

Beil	<u>159</u>,175
uv	6507,3332,4572,6344,862,5489
ir	3489,4503,4474,4527,1281,1284,2157,5932
Raman	1783,4115,7223,5984
ms	2760
nmr	2507,3837,6748,5933

444. *N*-Methylacetamide

CH$_3$CONH(CH$_3$)   79–16–3   C$_3$H$_7$ON

		cgs	SI	
mw		73.094	73.094	3468
bp	1 atm	206	479	4069;5374,739
dt/dp	90 Torr	140.5	413.7	1800
p	56	1.5	0.20	4602
eq 2.4	A	11.063	10.188	2868
40–90	B	3606	3606	
fp		30.55	303.70	974;1800,4439,4069,5374,2542
d	30	0.9500	950.0	471;3512,1871,4439,4069
				4602,2867
	35	0.94604	946.04	2542
	45	0.93764	937.64	
dd/dt		0.00084	0.84	2542
n$_D$	28	1.4286	1.4286	1800;4602,4439,739,3512
	35	1.4253	1.4253	
dn/dt		0.00047	0.00047	1800
η	30	3.65	0.00365	471
	35	3.23	0.00323	2542;1871,4069,2867
	40	3.012	0.003012	471
γ	30	33.67	0.03367	2867
	40	32.17	0.03217	
	50	30.62	0.03062	
ΔH$_v$	115–205	14.2	59.4	6145
ΔH$_m$		2.323	9.719	4256;974,471
C$_p$	25(l)	36.2	151.4	6798
t$_c$		417	690	2867
K$_f$	eq 2.67	5.77	5.77	4256;974
pK$_{BH}$+	25,in aq H$_2$SO$_4$	1.25	1.25	2210;2819,5194,3569,4207
κ	40	2 x 10^{-7}	2 x 10^{-5}	4069;1871,4439,2542
ε	32	191.3	191.3	974;1800,4069,4602,2542
				4439,3512
	40	164.5	164.5	471
μ	30 in 172	4.27	4.27	6076;5194,5382,5376
δ		14.6	29.9	4593
soly	in aq	soluble	soluble	5374

444.   *N*-Methylacetamide

(Continued)

Beil	335,58
uv	5194,6343,5489
ir	5194,3884,3254,5185,739,3766,6036,5932
Raman	5194,6036,4114
ms	2760
nmr	4981,4982,2507,4448,3837,5933

445.   *N,N*-Dimethylacetamide

$CH_3CON(CH_3)_2$ 127-19-5 C_4H_9ON

		cgs	SI	
mw		87.121	87.121	3468
bp	1 atm	166.1	439.3	2138;4321,2582,4718,7772,3625
dt/dp	1 atm	0.043	0.32	5228
dp/dt	1 atm	23	3.1	5228
p	25	1.3	0.17	2138
eq 2.5	A	7.76228	6.88718	5228;2868
	B	1889.10	1889.10	
	C	221.0	221.0	
fp		-20	253	2149;7772
d	15.5	0.9448	944.8	5228;2582,31,4439,7772,3625
	20	0.9415	941.5	6985
	25	0.936337	936.337	948;5228,471
	30	0.93169	931.69	5859;2867
	40	0.9232	923.2	2867
n_D	20	1.4384	1.4384	3625
	25	1.4356	1.4356	2138
η	20.4	2.141	0.002141	2582
	25	0.927	0.000927	471
	30	0.871	0.000871	5859;2867
	40	0.766	0.000766	2867

445. *N*, *N*-Dimethylacetamide

(Continued)

		cgs	SI	
γ	30	32.43	0.03243	2867
	40	30.92	0.03092	
	50	29.50	0.02950	
ΔH_v	25	11.75	49.15	5805
	bp	10.360	43.346	2138;5228
ΔH_m		2.490	10.418	5228
ΔH_f°	25(1)	-66.52	-278.32	7689
ΔH_c°	25(1)	-617.11	-2581.99	7689
C_p°	27	26	109	2138
C_p	20	42	176	2138
λ	22.2	0.000416	0.174	2138
t_c		364	637	2867
P_c		38.7	3.92	2138
K_f	eq 2.67	4.46	4.46	5228
K_B	eq 2.79	3.22	3.22	5228
pK_a	<25, in aq H_2SO_4	-0.19	-0.19	2819;3390,31,4207
pK_s	25	23.95	23.95	4259
ε	25	37.78	37.78	4439;4718,6985
μ	30 in 38	3.71	3.71	6076;7424,4321,5064,4693,6073
δ		10.8	22.1	4593,652;2626
soly	in aq	inf	inf	5228
aq az		none	none	5228
fl pt	TOC	77	350	5228

Beil	335,59
uv	5489
ir	3884,4523,3093,3766,3501,4319,2157,2676,5932
Raman	4114,2157,2676
ms	2760
nmr	4225,2507,5462,2464,5933

446.   *N*-Methylpropionamide

N-Methylpropanamide

CH$_3$CH$_2$CONHCH$_3$ 1187–58–2   C$_4$H$_9$ON

		cgs	SI	
mw		87.121	87.121	3468
bp	100 Torr	148	421	1800;739,6260,4439
dt/dp	100 Torr	0.30	2.3	6166
dp/dt	100 Torr	3.3	0.44	6166
p	10	94	13	6166
eq 2.4 A		8.7813	7.9062	6166;2868
10-100 Torr B		2855.5	2855.5	
fp		-30.9	242.3	3482;1800
d	20	0.93452	934.52	5151;4439,1869,1800,2867,3482
	25	0.93050	930.50	
	30	0.92650	926.50	
	40	0.91850	918.50	
dd/dt	20-40	0.00060	0.600	5151
n$_D$	25	1.4345	1.4345	1800;739,6260,4439,4087
η	20	6.016	0.006016	5151;1869,3952,3485
	25	5.215	0.005215	
	30	4.554	0.004554	
	40	3.533	0.003533	5151,3952
γ	30	31.20	0.03120	2867
	40	30.11	0.03011	
	50	29.12	0.02912	
ΔH$_v$	10-100 Torr	13.0	54.4	7842
C$_p$	25	79.8	334	4179
t$_c$		412	685	2867
κ	25	8 x 10^{-8}	8 x 10^{-6}	3483;6272,1869,3483
ε	25	175.7	175.7	996;4439,1869,663,3484
	-40	384	384	663
μ	110 (g)	3.59	3.59	5044
soly	in aq	soluble	soluble	
aq az		none	none	
Beil	335,59			
ir	739,3254,4407,6037,3884,3767			
Raman	6037			
nmr	1918			

447. 1,1,3,3-Tetramethylurea

Tetramethylurea

$(CH_3)_2NCON(CH_3)_2$ 632–22–4 $C_5H_{12}ON_2$

		cgs	SI	
mw		116.163	116.163	3468
bp	1 atm	175.2	448.4	4718;5107,8263,4114
dt/dp	1 atm	0.03	0.2	4718
dp/dt	1 atm	33	4.4	4718
p	71	13.9	1.85	55;740
eq 2.4	A	8.973	8.098	55
	B	2687.40	2687.40	
fp		−1.2	272.0	4718
d	15	0.972	972	2508
	20	0.9687	968.7	4718
	25	0.9619	961.9	5577
n_D	25	1.4493	1.4493	8263;4718
η	25	139.50	0.13950	5577
ΔH_v	60–175	12.22	51.12	55
ΔH_m		3.37	14.10	8309
ΔH_f°		−60.8	−254.4	4718
ΔH_c°		−819.3	−3428.0	2015
pK_{BH}^+	in 380	0.40	0.40	31;5024,4718
κ		<6 x 10^{-8}	<6 x 10^{-6}	4718
ε	25	23.60	23.60	5577;2688,4718
μ	25 in 38	3.50	3.50	4691;740,4719,741
δ	25	10.6	21.7	653
soly	in aq	inf	inf	4718
fl pt		ca 77	ca 350	4718

Beil	335,74
uv	4718
ir	741,5024,6421,6924,5932
Raman	4114,6924
ms	567
nmr	8059,6748,5933

448. 2-Pyrrolidinone

$$\begin{array}{ccc} H_2C & \!\!\!\!-\!\!\!\! & CH_2 \\ | & & | \\ H_2C & & C\!\!=\!\!O \\ & \diagdown N \diagup & \\ & H & \end{array}$$

616–45–5

C_4H_7ON

		cgs	SI	
mw		85.105	85.105	3468
bp	1 atm	245	518	390;7154
p	122	10	1.3	390
fp		25	298	390
d	25	1.107	1107	390,2441
	30	1.1020	1102.0	5368
	50	1.087	1087	390,2441
	100	1.046	1046	
n_D	25	1.486	1.486	390,2441
	30	1.4840	1.4840	390,2441;5368
η	25	13.3	0.0133	390
	30	10.282	0.010282	5368
ΔH_f°	25(1)	−68.4	−286.2	1703
ΔH_c°	25(1)	−546.9	−2288.2	1703;7044
$pK_{BH}{}^+$	33	−0.65	−0.65	1704
μ	25 in 38	3.55	3.55	3571;1973,850,1974,4483,2441
δ		14.7	30.1	652;4593
soly	in aq	inf	inf	390
fl pt	OC	129	402	390
Beil	3177,236			
uv	5489			
ir	3093,6596,4057,5932			
Raman	6687			
ms	2111			
nmr	5933			

449.   1-Methyl-2-pyrrolidinone

$$\begin{array}{c} \text{H}_2\text{C}\!\!-\!\!\text{CH}_2 \\ \text{H}_2\text{C}\diagdown_{\!\!\text{N}}\diagup\text{C}{=}\text{O} \\ \text{CH}_3 \end{array}$$

872-50-4 C_5H_9ON

		cgs	SI	
mw		99.132	99.132	3468
bp	1 atm	202	475	2625;2626,2708,7154
dt/dp	1 atm	0.49	3.7	2625
dp/dt	1 atm	2.0	0.27	2625
p	25	0.334	0.0445	8209
	100	24	3.2	2626,2625
eq 2.4	A	8.291	7.416	921
60-200	B	2572.7	2572.7	
fp		-24.4	248.8	2626,2625;2708
d	20	1.0304	1030.4	5352;1073
	25	1.0259	1025.9	5352;6542,2441,2708
dd/dt	20-40	0.00092	0.92	5352
n_D	20	1.4700	1.4700	5352;1073
	25	1.4675	1.4675	5352;6542,2441,2708,1073
dn/dt	25	0.00050	0.00050	
η	25	1.666	0.001666	6542;471,2708
γ	25	40.7	0.0407	2625
	100	31.2	0.0312	
ΔH_v		12.62	52.80	2625
ΔH_f°	25(1)	-62.7	-262.3	1703
ΔH_c°	25(1)	-715.0	-2991.6	1703;4231,2708
ΔH_p		0.8	3.3	4231
C_p	20	39.7	166.1	2625
λ	38	0.000455	0.190	2625
pK_{BH}^+	33	-0.75	-0.75	1704;1076,7752
pK_s	25	25.6	25.6	4261;4259
κ	25	$1\text{-}2 \times 10^{-8}$	$1\text{-}2 \times 10^{-6}$	2172
ϵ	25	32.2	32.2	6142;6542,1073,2625
μ	30 in 38	4.09	4.09	2441;1073,2625
δ		11.3	23.1	6521;2625
ER	BuOAc	<0.1	<0.1	2626
soly	in aq	inf	inf	7154
fl pt	OC	95	368	2708,2645,2626

449. 1-Methyl-2-pyrrolidinone

(Continued)

Beil	<u>3177</u>,237
ir	3884,2111,2707,4701,2625,2626,4313,5932
nmr	5933

450. ε-Caprolactam

Hexahydro-2 *H*-azepin-2-one

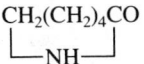

105–60–2 $C_6H_{11}ON$

		cgs	SI	
mw		113.159	113.159	3468
bp	50 Torr	180	453	<u>94</u>
p	100	2.9	0.39	94
eq 2.4 80–140	A B	6.78 2344	5.90 2344	5241
fp		69.207	342.357	4959;94,4843,6532
d	77	1.02	1020	94
n_D	31 40	1.4965 1.4935	1.4965 1.4935	94
η	70 78	19.7 9	0.0197 0.009	3125 94
ΔH_v		13.1	54.8	94
ΔH_m		3.856	16.131	4959
ΔH_c°		−861.6	−3604.9	7044
ΔH_p		20	84	94
C_p	25	37.33	156.19	94
A	eq 2.72 B	0.01655 0.0007	0.01655 0.0007	4959
K_f	eq 2.67	7.30	7.30	3125

450. ε-Caprolactam

(Continued)

		cgs	SI	
pK$_a$	25	-0.46	-0.46	1704
	25 in 208	0.36	0.36	3570
μ	25 in 38	3.88	3.88	3571
δ	25	12.7	26.0	653
soly	in aq,25	84.0%w	84.0%w	94
fl pt	COC	125	398	94

Beil	3179,240
uv	7229,3570
ir	3884,3093,3570,94,6596,5932
Raman	6687
ms	5176
nmr	5254,5933

451. Carbon disulfide

CS$_2$ 75–15–0   CS$_2$

		cgs	SI	
mw		76.131	76.131	3468
bp	1 atm	46.225	319.375	7818;7488,1009,2754,4959,7470 7471,8304
dt/dp	1 atm	0.0407	0.305	2089;1970,8304
dp/dt	1 atm	24.6	3.28	2089
p	25	361.6	48.21	2089;6090,6461,7058,8304,1009
eq 2.5	A	6.94194	6.06684	7818;1009
	B	1168.623	1168.623	
	C	241.534	241.534	
fp		-111.57	161.58	7059;1090,1159,5104,7471 7495,7498
d	0	1.29270	1292.70	7495;1696,2754,4969,6980,7470 7499,787,1243
	15	1.27005	1270.05	
	20	1.26311	1263.11	4650
	25	1.2555	1255.5	5240

451. Carbon disulfide

(Continued)

		cgs	SI	
α	20	0.001218	0.001218	5941
κ_T	25	1.27×10^{-7}	9.50×10^{-7}	6151
n_D	15	1.63189	1.63189	7495;3617,6980,3089,3961 787,5991
	20	1.62746	1.62746	5827;1243
	25	1.62409	1.62409	7495
dn/dt		0.00078	0.00078	7495
η	15	0.380	0.000380	7495;4570,5136,7453
	20	0.363	0.000363	
	Table 2.2			
γ	20	32.25	0.03225	3283;7848
	30	30.79	0.03079	
ΔH_v	25	6.578	27.522	7818;1159,4969,5153
	bp	6.390	26.736	7818
	Table 2.9			
ΔH_m		1.049	4.389	1159;7203,7478
ΔH_f°	25(1)	21.44	89.70	7826;2853,3014,4810
	25(g)	28.05	117.36	7826;2853
ΔH_c°	25(1)	-402.69	-1684.86	2853;3975,7446,4810
C_p°	25	10.914	45.664	1430;7818,7826
	Table 2.14			
C_p	24.27	18.17	76.02	1159;2099,5817,6461,6977
t_c		279	552	4088;5153,7848
P_c		78	7.90	4088;5153
d_c		0.44	440	4088
v_c		0.170	0.000170	4088
K_B	eq 2.79	2.35	2.35	726;5060,718,5595
κ	25	0.0037	0.37	4902
ϵ	20	2.643	2.643	4650;4937,5240,1330,1696,2027 2953,7850,8084,7497
μ	25 in 38	0.06	0.06	8084,4487;3854,5317,1094
δ		9.92	20.29	6240;4593,3395,652,6535,6610 6723,3421
soly	in aq,20	0.210%w	0.210%w	7265;5496,6556
	aq in,25	0.0142%w	0.0142%w	7265;5496
aq az	42.6	97.2%w	97.2%w	3497;3498
fl pt	CC	-30	243	5712,5941

451.   Carbon disulfide

(Continued)

Beil	218,197
uv	327,827,3229,4599,5952,6338,4054
ir	549,618,4021,5895,6368,153,7229,5932
Raman	1314,5000,6046,4172
ms	154,6199,2050
nmr	3794

452.   Dimethyl disulfide

CH$_3$SSCH$_3$ 624-92-0   C$_2$H$_6$S$_2$

		cgs	SI	
mw		94.189	94.189	3468
bp	1 atm	109.745	382.895	8317;8013,1731,5826
p	25.89	30.00	4.000	8317
eq 2.5	A	6.97792	6.10282	8317;8013
	B	1346.342	1346.342	
	C	218.863	218.863	
fp		-84.7	188.5	1731,5826
d	20	1.0625	1062.5	1731,5826;1976
n_D	20	1.52592	1.52592	1731;5826,1970
	25	1.52298	1.52298	
dn/dt	25	0.000588	0.000588	1731
η	20	0.619	0.000619	5826,1731
	25	0.585	0.000585	
γ	20	33.6	0.0336	5826;1731
	25	32.8	0.0328	
ΔH_v	25	9.16	38.33	5826
	bp	8.089	33.844	8013
ΔH_f°	25(l)	-14.91	-62.38	8316
	25(g)	-5.74	-24.02	
ΔH_c°	25(l)	-630.60	-2638.43	8316
	25(g)	-639.77	-2676.80	

452. Dimethyl disulfide

(Continued)

		cgs	SI	
C_p°	15.6	23.44	98.08	5826
C_p	15.6	34.78	145.52	5826
t_c		333	606	8013,5826
P_c		47.8	4.84	8013,5826
d_c		0.341	341	8013
v_c		0.276	0.000276	8013
ϵ	25	9.6	9.6	1731
μ	20 in 38	1.85	1.85	448;446
fl pt	TOC	24	297	1731
Beil	12,291			

453.   1-Butanethiol

$CH_3CH_2CH_2CH_2SH$ 109-75-5 $C_4H_{10}S$

		cgs	SI	
mw		90.183	90.183	3468
bp	1 atm	98.455	371.605	2059;4979,342,303,4975 1958,7769
dt/dp	1 atm	0.04470	0.3353	303
dp/dt	1 atm	22.37	2.983	303
p	25	45.5	6.07	6523
eq 2.5	A	6.92754	6.05244	6523;304,5607,8013
	B	1281.018	1281.018	
	C	218.100	218.100	
fp		-115.67	157.48	303;2253,1958

453. 1-Butanethiol

(Continued)

		cgs	SI	
tp		-115.69	157.46	6523
d	20	0.84159	841.59	303;2253,4975,1958,7769,4973
	25	0.83674	836.74	3541,3588
dd/dt	25	0.000970	0.970	<u>303</u>
α	25	1.38×10^{-6}	1.38×10^{-6}	3541
n_D	20	1.44295	1.44295	303;2253,4973,1958,3155
	25	1.44033	1.44033	7769,3588
dn/dt	25	0.000524	0.000524	<u>303</u>
η	0	0.648	0.000648	878
	20	0.535	0.000535	47
	30	0.479	0.000479	
γ	21.9	25.57	0.02557	7769
	41.2	23.41	0.02341	
ΔH_v	25	8.73	36.53	308,305;4669,8013
	bp	7.702	32.225	308,305
	Table 2.9			
ΔH_m		2.500	10.460	6523,308,305
ΔH_f°	25(1)	-29.61	-123.89	6523,3541;307,4750,4810,8316
ΔH_c°	25(1)	-831.77	-3480.13	3541;4870,8316
C_p°	25	28.24	118.16	6523,306
	Table 2.14			
C_p	25	41.18	172.30	6523,3541
A	eq 2.72	0.05073	0.05073	308
K_a	20,in aq NaOH	1.65×10^{-11}	1.65×10^{-11}	8189;2461
ε	20	5.204	5.204	47;5041,4937,999
μ	25 in <u>38</u>	1.53	1.53	5041;3588,7890
soly	in aq,20	0.0597%w	0.0597%w	<u>8189</u>
fl pt		3	276	5824
Beil	<u>24</u>,370			
uv	2609,3573,3073,1501			
ir	2875,7557,8072,153,3073,5935			
Raman	4973			
ms	154,3073,4551			
nmr	5933			

454. Benzenethiol

C_6H_5SH 108–98–5 C_6H_6S

		cgs	SI	
mw		110.173	110.173	3468
bp	1 atm	169.138	442.288	320;5268,3588,6528,2059,7058 7769,4976
dt/dp	1 atm	0.05176	0.3882	320;4976
dp/dt	1 atm	19.32	2.576	320
p	25	2.98	0.397	2087;7058
eq 2.5	A	6.99019	6.11509	321,6528,5607
	B	1529.46	1529.46	
	C	203.05	203.05	
fp		-14.94	258.21	320;5689,8121,5268
tp		-14.88	258.27	6528
d	20	1.07755	1077.55	320;4976,1057,7769,7864
	25	1.07273	1072.73	5268,3588
dd/dt	25	0.000964	0.964	320
n_D	20	1.59008	1.59008	320;4976,1057,3155,7769
	25	1.58722	1.58722	5268,3588
dn/dt	25	0.000572	0.000572	320
η	20	1.239	0.001239	5268
	25	1.144	0.001144	
γ	20	39.5	0.0395	5268;7769,7864
	25	38.7	0.0387	
	Table 2.8			
ΔH_v	25	10.84	45.35	2087
	bp	8.836	36.970	
	Table 2.9			
ΔH_m		2.736	11.447	6528;5689
ΔH_f°	25(1)	15.02	62.84	4810,6528;8316
ΔH_c°	25(1)	-910.78	-3810.70	8316;4810,6528
C_p°	25	25.07	104.89	6528
	Table 2.14			
C_p	25	41.40	173.22	6528;5689
	Table 2.12			
A	eq 2.72	0.02064	0.02064	6528;5268
	B	0.00255	0.00255	
pK_a	25	6.615	6.615	1944;4253,1838,6892,2461,3060
ϵ	25	4.382	4.382	4976;999
μ	25 in 38	1.23	1.23	4976;569,999,3588

454. Benzenethiol

(Continued)

Beil	524,294
uv	945,4094,6205,5268,4041,5497,780,3757
ir	2875,153,3791,5268,6528,61
Raman	3825,4126,5030,8150,6528,61
ms	5268,2644,4431
nmr	4981,4982,2318,6053

455. Methyl sulfide

Thiobismethane

$(CH_3)_2S$ 75–18–3 C_2H_6S

		cgs	SI	
mw		62.129	62.129	3468
bp	1 atm	37.333	310.483	314;5544,4729,342,312,1749 7772,3588
dt/dp	1 atm	0.0376	0.282	312
dp/dt	1 atm	26.6	3.55	312
p	25	484.9	64.65	2089
eq 2.5	A	6.94879	6.07369	5607;314
	B	1090.755	1090.755	
	C	230.799	230.799	
fp		−98.27	174.88	312;5544
d	20	0.84823	848.23	312;1749,7772,3588,5544
	25	0.84228	842.28	
α		0.0013	0.0013	1732
n_D	20	1.43542	1.43542	312;1749,7772,3588,5544
	25	1.43228	1.43228	
dn/dt	25	0.000628	0.000628	312
η	20	0.289	0.000289	2089
	25	0.279	0.000279	
γ	11.1	26.50	0.02650	1732
	20	24.48	0.02448	2089;7772
	30	23.06	0.02306	

455. Methyl sulfide

(Continued)

		cgs	SI	
ΔH_v	25	6.609	27.65	315;7755,4751
	bp	6.453	27.00	315
	Table 2.9			
ΔH_m		1.9084	7.9847	315;5544
ΔH_f°	25(1)	−15.55	−65.06	7826;4751,4810,309,4750,8316
ΔH_c°	25(1)	−521.09	−2180.24	4751;4810,3975,7446,8316
C_p°	25	17.37	72.68	4751;311,644
	Table 2.14			
C_p	25	28.23	118.11	7826
t_c		229.9	503.1	313;4299,3345
P_c		54.6	5.53	313;4299,3345
d_c		0.309	309	313;4299,3345
v_c		0.201	0.000201	313;4299
A	eq 2.72	0.03140	0.03140	315;5544,4729
K_B	eq 2.79	1.85	1.85	7982
$pK_{BH}{}^+$	25,in aq H_2SO_4	−6.99	−6.99	5755;980
ε	20	6.2	6.2	4937
μ	25 in 38	1.45	1.45	1749;3588,5847
δ	25	9.4	19.2	653
soly	in aq,25	2%w	2%w	1732;1734
	aq in,25	0.4%w	0.4%w	
fl pt		−34	239	1734

Beil	19,288
uv	3073,1060,3233,1750,7444,1501,639
ir	153,3073,5754,4569,6681,8240,97,7049
Raman	6681,8240,97
ms	3073,4557,1745
nmr	1396,892,6366,6288,5933

456. Ethyl sulfide

1,1'-Thiobisethane

| $(CH_3CH_2)_2S$ | | 352-93-2 | | $C_4H_{10}S$ |

		cgs	SI	
mw		90.183	90.183	3468
bp	1 atm	92.102	365.252	314,312;3072,7490,5544,2059 1749,6522,1970
dt/dp	1 atm	0.0439	0.329	312;5083
dp/dt	1 atm	22.8	3.04	312
p	25	58.37	7.782	2089;7058
eq 2.5	A	6.92836	6.05326	314,6522,5607
	B	1257.833	1257.833	
	C	218.662	218.662	
fp		-103.93	169.22	312,3072;7490,7472,7479,5544
tp		-103.95	169.20	6522
d	20	0.83621	836.21	312;1749,4729,1970,522,5208
	25	0.83118	831.18	5329,7490,7772,5544
α	25	1.45×10^{-6}	1.45×10^{-6}	3541
n_D	20	1.44294	1.44294	312;1749,4729,1970,3072,3155
	25	1.44015	1.44015	5329,7490,7678,7770,5544
dn/dt	25	0.000558	0.000558	312
η	20	0.440	0.000440	3072;2129,5329,7490
	25	0.417	0.000417	
	Table 2.2			
γ	20	25.2	0.0252	3072;5208,5329,7490,840,7772
	25	24.5	0.0245	
ΔH_v	25	8.55	35.77	315;1675,4809,3541
	bp	7.591	31.761	315,6522
	Table 2.9			
ΔH_m		2.845	11.903	315,6522;5544
ΔH_f°	25(1)	-28.34	-118.57	3541;310,4750,6522,3542 4810,7085
ΔH_c°	25(1)	-833.04	-3485.44	3541;7446,4810,3542,3975
C_p°	25	27.97	117.03	311;6522
	Table 2.14			
C_p	25	40.97	171.42	3541;6522
t_c		284	557	313;4299,2089,7848
p_c		39.1	3.96	313;4299,758,2089
d_c		0.284	284	313;4299,2089
v_c		0.318	0.000318	313;4299,2089
A	eq 2.72	0.05001	0.05001	315;6596,4729,3072,5344
K_B	eq 2.79	3.23	3.23	7982
pK_a	27 in 38	relative	relative	8115

456. Ethyl sulfide

(Continued)

		cgs	SI	
ε	25	5.72	5.72	4937,7497
μ	25 in 38	1.61	1.61	1749;3588,7094,7890
δ	25	8.5	17.4	653

Beil	23,344
uv	155,5205,5209,1060,4571,2393,1750,639
ir	7557,3072,153,633,4569,5932
Raman	6775,7775,7228
ms	3072,4557,2644
nmr	1396,892,6288,5933

457. Butyl sulfide

1,1'-Thiobisbutane

$(CH_3CH_2CH_2CH_2)_2S$ 544–40–1 $C_8H_{18}S$

		cgs	SI	
mw		146.290	146.290	3468
bp	1 atm	188.91	462.06	8013;5826,5824,7058
p	51.8	5.0	0.67	7058
eq 2.5	A	6.98807	6.11297	8013
	B	1574.633	1574.633	
	C	194.473	194.473	
fp		−75.03	198.12	5826;5545,7058
d	20	0.83849	838.49	5826;5545
n_D	20	1.45293	1.45293	5826;5545
	25	1.45062	1.45062	
dn/dt	25	0.000462	0.000462	5826
η	20	1.054	0.001054	5826
	25	0.978	0.000978	
γ	20	27.35	0.02735	5329;5826
	25	26.8	0.0268	
ΔH_v	25	12.8	53.6	5826
	bp	9.867	41.284	8013

457. Butyl sulfide

(Continued)

		cgs	SI	
ΔH_f°	25(1)	−52.67	−220.37	1703
	25(g)	−40.02	−167.44	8316;1703
ΔH_c°	25(1)	−1458.0	−6100.3	1703
	25(g)	−1453.46	−6081.28	8316
C_p°	15.6	48.34	202.25	5826
C_p	15.6	67.2	281.2	5826
t_c		444	717	8013
P_c		267	27.1	8013
d_c		0.258	258	8013
v_c		0.568	0.000568	8013
μ	25 in 38	1.61	1.61	1010
fl pt	COC	76	349	5824
Beil	24,370			

458. Thiophene

110–0–1 C_4H_4S

		cgs	SI	
mw		84.136	84.136	3468
bp	1 atm	84.16	357.31	316;7817,3072,5544,2384 3957,7058
dt/dp	1 atm	0.0428	0.321	316;5544,2384
dp/dt	1 atm	23.4	3.12	316
p	25	79.68	10.62	316;4091,7058,7817

458. Thiophene

(Continued)

		cgs	SI	
eq 2.5	A	6.95926	6.08416	318;8023,7817
	B	1246.02	1246.02	
	C	221.35	221.35	
fp		−38.24	234.91	316,3072;5544,2384,7058,7817
d	20	1.06482	1064.82	316;3072,1970,3177,5544,2384
	25	1.05884	1058.84	3957,5238
dd/dt	25	0.001196	1.196	316
n_D	20	1.52890	1.52890	316,3072;2664,4091,1970,5544
	25	1.52572	1.52572	2384,3155,3957,5238
dn/dt	25	0.000636	0.000636	316
η	20	0.654	0.000654	3072;5544,2384
	25	0.613	0.000613	
	Table 2.2			
γ	20	32.8	0.0328	3072
	25	32.0	0.0320	
ΔH_v	25	8.27	34.60	319;8013,7817,2276
	bp	7.522	31.472	319
ΔH_s		11.199	46.857	5130
ΔH_m		1.216	5.088	319;5544,5676,3661
ΔH_f°	25(1)	19.52	81.67	4810,7817;7085,3543,8316
ΔH_c°	25(1)	−675.55	−2826.50	3543;7446,3975,4810,7817,8316
C_p°	25	17.42	72.89	3543
C_p	26.84	32.29	135.10	5689;3661,7817
t_c		306.2	579.4	317;4091,5727,1451,8013,7848
P_c		56.2	5.69	4091;317,5727,8013
d_c		0.385	385	4091;317
v_c		0.219	0.000219	4091;317
A	eq 2.72	0.01109	0.01109	319;5544,3072
ε	25	2.705	2.705	3177;4937
μ	25 in 38	0.52	0.52	3177;4687,4493,3208,3952
				4689,6212
δ	25	9.8	20.1	653
soly	in aq	miscible	miscible	4026
aq az	minimum bp			3497

458. Thiophene

(Continued)

Beil	2364,29
uv	155,2802,5129,3072,5056,1051,947,3503,6749
ir	153,3072,633,5841,4462,4883,5326
Raman	967,964,4121,6126,61,6125
ms	154,6199,3072,3430
nmr	2901,9,6419,5933

459. Tetrahydrothiophene

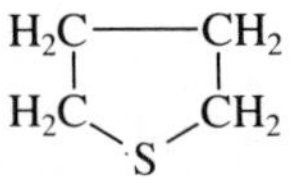

$$H_2C-CH_2$$
$$H_2C\diagdown_{S}\diagup CH_2$$

110-01-0

C_4H_8S

		cgs	SI	
mw		88.167	88.167	3468
bp	1 atm	120.9	394.1	3072;3554
p	25	18.4	2.45	3554
eq 2.5	A	7.39514	6.52004	3554
	B	1628.8	1628.8	
	C	240.7	240.7	
fp		-96.16	176.99	3072
d	20	0.99869	998.69	3072;3542
	25	0.99379	993.79	
	30	0.98928	989.28	
dd/dt	25	0.000941	0.941	3072
n_D	20	1.52890	1.52890	3072;3554
	25	1.52572	1.52572	
	30	1.52257	1.52257	
dn/dt	25	0.000633	0.000633	3072
η	20	1.042	0.001042	3072
	25	0.971	0.000971	
	30	0.914	0.000914	
γ	20	35.8	0.0358	3072
	25	35.0	0.0350	
	30	34.6	0.0346	

459. Tetrahydrothiophene

(Continued)

		cgs	SI	
ΔH_v	25	9.23	38.62	3542;2276
ΔH_f°	25(1)	−17.31	−72.42	3542;4750,8316
ΔH_c°	25(1)	−758.22	−3172.39	8316;3542
t_c		358.8	632.0	1451;4299
A	eq 2.72	0.028	0.028	3072
pK_a	in aq H_2SO_4	−4.52	−4.52	436
μ	25 in 38	1.90	1.90	1757;6212
δ	25	10.0	20.5	653
Beil	2362,10			
uv	3072,1501			
ir	3072,153,7561,4569,2915			
Raman	7561,61,6033			
ms	3072,2641,2113			
nmr	4806,5933			

460. 2-Methylthiophene

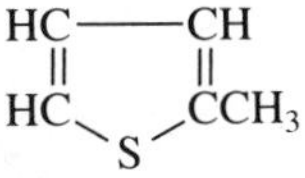

HC——CH
‖   ‖
HC   CCH₃
    S

554–14–3 C_5H_6S

		cgs	SI	
mw		98.162	98.162	3468
bp	1 atm	112.56	385.71	8013,5751;3073,2383
dt/dp	1 atm	0.0460	0.345	2383
dp/dt	1 atm	21.7	2.90	2383
p	25	24.89	3.318	2087
eq 2.5	A	6.93897	6.06387	8013
	B	1326.474	1326.474	
	C	214.309	214.309	
fp		−63.386	209.764	5751;3073,2383
tp		−63.37	209.78	5751

460. 2-Methylthiophene

(Continued)

		cgs	SI	
d	20	1.01965	1019.65	3073;3698,8245,1970,2383
	25	1.01422	1014.22	
dd/dt	25	0.001074	1.074	3073
n_D	20	1.52035	1.52035	3073;2383,3698,8245,1970
	25	1.51744	1.51744	
dn/dt	25	0.000584	0.000584	3073
η	20	0.703	0.000703	3073;2383
	25	0.660	0.000660	
γ	20	30.95	0.03095	3698;3073
	40	28.17	0.02817	
ΔH_v	25	9.407	39.358	2087
	bp	8.147	34.087	2087;8013,5751
ΔH_m		2.200	9.205	2383
ΔH_f°	25(1)	10.75	44.98	2036;5751
	25(g)	4.74	19.83	5751
ΔH_c°	25(1)	-829.83	-3472.01	2036
C_p°	25	22.80	95.40	5751
C_p	25	35.18	147.19	5751
t_c		333.0	606.2	2087;8013
P_c		46.12	4.673	2087;8013
d_c		0.351	351	2087;8013
v_c		0.280	0.000280	2087
A	eq 2.72	0.0259	0.0259	5751;3073
δ	25	9.5	19.5	653
Beil	2364,37			
uv	2571			
ir	2571			

461. 3-Methylthiophene

HC——CCH$_3$
‖   ‖
HC   CH
  ＼S／

616-44-4

C$_5$H$_6$S

		cgs	SI	
mw		98.162	98.162	3468
bp	1 atm	115.44	388.59	8013;3073,2383
dt/dp	1 atm	0.0462	0.347	2087;2383
dp/dt	1 atm	21.6	2.89	2087
p	25	22.15	2.953	2087
eq 2.5	A	6.98611	6.11101	8013
	B	1363.862	1363.862	
	C	216.784	216.784	
fp		-68.94	204.21	3073;2383
tp		-68.97	204.18	2383
d	20	1.02183	1021.83	3073;3698,8245,2383,1970
	25	1.01647	1016.47	
dd/dt	25	0.001047	1.047	<u>3073</u>
n$_D$	20	1.52042	1.52042	3073;2383,3698,8245,1970
	25	1.51758	1.51758	
dn/dt	25	0.000575	0.000575	<u>3073</u>
η	20	0.674	0.000674	3073;2383
	25	0.635	0.000635	
γ	20	32.37	0.03237	3698;3073
	40	29.88	0.02988	
ΔH$_v$	25	9.465	39.601	2087
	bp	8.187	34.253	2087,4755;8013
ΔH$_m$		2.518	10.535	2087,4755;2383
ΔH$_f^\circ$	25(1)	10.38	43.43	2036;4755,1703
	25(g)	19.82	82.93	1703
ΔH$_c^\circ$	25(1)	-829.46	-3470.46	2036;1703,4755
C$_p^\circ$	25	22.67	94.85	4755
C$_p$	24.6	35.98	150.54	
t$_c$		337.6	610.8	2086;8013
p$_c$		47.2	4.79	2086;8013
d$_c$		0.357	357	2086;8013
v$_c$		0.275	0.000275	2086
A	eq 2.72	0.0304	0.0304	4755;3073
δ	25	9.5	19.5	653

461.   3-Methylthiophene

(Continued)

Beil	2364,38
ir	3391

462.   Dimethyl sulfoxide

Sulfinylbismethane

$(CH_3)_2SO$ 67–68–5 C_2H_6OS

		cgs	SI	
mw		78.129	78.129	3468
bp	1 atm	189.0	462.2	6453;1749
p	25	0.600	0.0800	2057
eq 2.5	A	7.59677	6.72167	3669;3783,3269
	B	1962.05	1962.05	
	C	225.892	225.892	
fp		18.54	291.69	2672;2057,6453,6859,6544,3269
tp		18.52	291.67	1540
d	20	1.10041	1100.41	852;1327,4195
	25	1.09537	1095.37	1777;3513,1733,6543,4452,6542
				513,1327,1385
	40	1.08043	1080.43	852;1538,1385,6544
	60	1.0616	1061.6	1538
dd/dt		0.00099887	0.99887	852
α		0.000928	0.000928	4195;4902
κ_T	25	7.0×10^{-8}	5.2×10^{-7}	1777
n_D	20	1.47933	1.47933	1539;1733,4452,6542,852
	25	1.47754	1.47754	1539;1733,1385
dn/dt	25	0.000358	0.000358	1539

462. Dimethyl sulfoxide

(Continued)

		cgs	SI	
η	20	2.2159	0.0022159	852
	25	1.991	0.001991	1385;6542,6543,3513,2488,4452
	35	1.654	0.001654	6543
γ	20	43.72	0.04372	4195;1538,6453
	25	42.98	0.04298	6958;513,1538
	50	40.05	0.04005	46
ΔH_v	25	12.64	52.88	2057
	bp	10.31	43.14	1733;3269
ΔH_m		3.434	14.368	1540;6787,4614
ΔH_s		18.4	77.0	1733
ΔH_f°	25(1)	-48.73	-203.89	1703;7826,4810,4811
	25(g)	-36.09	-151.01	1703;7826
ΔH_c°	25(1)	-472.7	-1977.7	2449,1733
C_p°	25	21.39	89.50	1540;7826,4811
	Table 2.14			
C_p	25	36.61	153.18	1540;1733,6787,7826
d_c		0.366	366	1327
v_c		0.2377	0.0002377	1327
K_f	eq 2.67	4.07	4.07	2672;4609,6787
pK_a	in 221	1.4	1.4	6892;6989,4160,339
	25, in aq	-1.54	-1.54	5755
	H_2SO_4			
pK_s	25	31.8	31.8	1300;4160
κ	25	2×10^{-9}	2×10^{-7}	1214;5729,6453
ϵ	25	46.45	46.45	1385;998,2672,449,6542
				3513,6859
μ	25 in 317	4.06	4.06	5745;5793,1673,6453,449,1749
δ		12.0	24.5	652,1733
soly	in aq	25.3%w	25.3%w	1733,4914
aq az		none	none	1733
fl pt	OC	95	368	1733

Beil	19,289
uv	1750,1501
ir	153,6364,6453,5932
Raman	6364,4623
ms	1034
nmr	4436,892,6366,2464,4623,852,5933

463. Sulfolane

Tetrahydrothiophene 1,1-dioxide

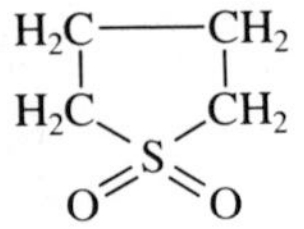

H₂C——CH₂ structure

126–33–0 $C_4H_8O_2S$

		cgs	SI	
mw		120.171	120.171	3468
bp	1 atm	287.3 d	560.5 d	6633;7751,5264,7943
dt/dp	1 atm	0.065	0.49	6633
dp/dt	1 atm	15	2.1	6633
p	118	5.0	0.67	6633
eq 2.13	A	28.6824	27.8073	6633
118–285	B	4350.7	4350.7	
	C	6.5633	6.5633	
fp		28.45	301.60	2672;885,431,1277,4151
				7751,5264
d	30	1.2604	1260.4	5352;1385,7694,7751,5264
	35	1.2568	1256.8	4694
	50	1.2447	1244.7	4694;1385
	150	1.158	1158	6633
n_D	25	1.4833	1.4833	5352
	30	1.4816	1.4816	5352;1385,4694,7751
	35	1.4798	1.4798	4694
	50	1.4742	1.4742	1385
dn/dt	30–50	0.00034	0.00034	1385
η	30	10.286	0.010286	7694;2236,5264,1277,1385,3678
	35	9.033	0.009033	
	50	6.312	0.006312	
	150	1.46	0.00146	6633
	Table 2.6			
γ	30	35.5	0.0355	3125
ΔH_v	100	15.0	62.8	6633
	200	14.7	61.5	
ΔH_m		0.341	1.43	2672;1277,6633
C_p	30	43	180	6633
K_f	eq 2.67	64.1	64.1	6633;431,1277
pK_a	25	15.3	15.3	8158
pK_s	25	25.45	25.45	4259
κ	30	$<2 \times 10^{-8}$	$<2 \times 10^{-6}$	1277;5225
ε	30	43.26	43.26	1385;998,4694,449,2236,3675
				1757,5264,1277,3676
μ	25 in 38	4.81	4.81	1757;4657,2236,449,998
soly	in aq,30	miscible	miscible	7033
aq az		none	none	6165
fl pt	OC	177	450	6633;7033

463. Sulfolane

(Continued)

Beil	2362,9
ir	4622,2029,42,5932
ms	7955
nmr	2389,5933

464. 2-Methoxyethanol

CH$_3$OCH$_2$CH$_2$OH   109–86–4   C$_3$H$_8$O$_2$

		cgs	SI	
mw		76.095	76.095	3468
bp	1 atm	124.6	397.8	958;7768,8035,5832,1716 2639,3456
dt/dp	1 atm	0.042	0.32	5832
dp/dt	1 atm	24	3.2	5832
p	25	9.7	1.3	2069;1839,7058
eq 2.5	A	7.7085	6.8334	5832;2639
	B	1711.2	1711.2	
	C	230	230	
fp		−85.1	188.1	4902
d	20	0.96459	964.59	1480;3461,1839,2638,3456,5654
	25	0.96024	960.24	7768,8035,1298,2637,5832
dd/dt		0.000780	0.780	6905
α	20	0.00095	0.00095	4902,6631;6905,7635
n$_D$	20	1.4021	1.4021	1480;3461,1839,3456,7768
	25	1.4002	1.4002	5832,8035
dn/dt	25	0.00038	0.00038	1480
η	20	1.72	0.00172	4902
	25	1.60	0.00160	
γ	14.9	31.82	0.03182	7768;6538
	41.0	29.28	0.02928	

464.   2-Methoxyethanol

(Continued)

		cgs	SI	
γ	Table 2.8			
ΔH_v	25	10.80	45.17	4343,958,2639
	bp	9.43	39.46	2069
ΔH_c		-440.6	-1843.5	6735
C_p	25	41.80	174.9	4341;2069
pK_a	25	15.0	15.0	8158;578
pK_s	25	20.7	20.7	4265
κ	20	1.09×10^{-6}	1.09×10^{-4}	4902;578
ε	25	16.93	16.93	3461
μ	25 in 38	2.04	2.04	3461;2813,1298
δ		10.8	22.1	4593,6631,1399,2190,2346
ER	BuOAc	0.52	0.52	6628,6631;6905,1399,3455,2190
	90%	884	884	6628,6631;1399
soly	in aq	inf	inf	2069;3456
aq az	99.9	15.3%w	15.3%w	3498;3497
fl pt	CC	42	315	4902;3456,2069
	OC	46	319	2346

Beil	30,467
ir	8312,633,1097,4041,4287,3654,5932
Raman	3526
ms	7935
nmr	5933

465.   2-Ethoxyethanol

$CH_3CH_2OCH_2CH_2OH$ 110-80-5 $C_4H_{10}O_2$

		cgs	SI	
mw		90.122	90.122	3468
bp	1 atm	135.6	408.8	7637;1716,3456,5654,7768,8035
dt/dp	1 atm	0.042	0.32	5832
dp/dt	1 atm	24	3.2	5832
p	25	5.3	0.71	2069;1839
eq 2.5	A	7.8191	6.9440	5832
	B	1801.9	1801.9	
	C	230	230	
fp	gl	<-90	<183	7637
d	20	0.92945	929.45	1480;1298,1839,3456,5654,7768
	25	0.92520	925.20	8035,760
dd/dt		0.00070	0.70	6905
α	20	0.00097	0.00097	4902,6631;6905
n_D	20	1.4077	1.4077	1480;1839,3456,7768,8035
	25	1.4057	1.4057	
dn/dt	25	0.00040	0.00040	1480
η	20	2.05	0.00205	4902
	25	1.85	0.00185	2069
γ	25	28.2	0.0282	2069;6538
	75	23.6	0.0236	
ΔH_v	25	11.52	48.21	4343;958
	bp	9.64	40.35	2346
C_p	25	50.38	210.8	4341
pK_a		14.8	14.8	8158
κ		9.3×10^{-8}	9.3×10^{-6}	700;3965
ε	24	29.6	29.6	700
μ	25 in 38	2.08	2.08	1298;2807,5209
δ		9.9	20.3	2346,6905,6631,4593,1399,2190
ER	BuOAc	0.38	0.38	6628,6631;6905,7635,3455,1399
	ER	1213	1213	6628,6631;1399
soly	in aq	inf	inf	3456
aq az	99.4	28.8%w	28.8%w	3497;4027
fl pt	CC	44	317	3456
	TOC	49	322	7637

465.   2-Ethoxyethanol

(Continued)

Beil	30,467
uv	4164
ir	642,3654,5932
Raman	3526
nmr	7227,7543

466.   2-Butoxyethanol

$CH_3CH_2CH_2CH_2OCH_2CH_2OH$ 111–76–2 $C_6H_{14}O_2$

		cgs	SI	
mw		118.175	118.175	3468
bp	1 atm	170.2	443.4	6470;6408,1716,7164,2042,5471
p	25	0.852	0.114	6408;6470,5122,4523,2042
eq 2.4	A	8.6405	7.7654	6470
	B	2548.0	2548.0	
fp		−75	198	2346,6631
d	20	0.90075	900.75	6470;4789,2042,5471
	25	0.89625	896.25	4789
dd/dt		0.00066	0.66	6905
α	20	0.00092	0.00092	4902
n_D	20	1.41980	1.41980	5659;7768,6470,5122,3461,7164
	25	1.4176	1.4176	4902,2042
η	25	3.15	0.00315	2069
	60	1.51	0.00151	
γ	25	27.4	0.0274	2069
	75	23.3	0.0233	
ΔH_v	25	13.53	56.59	4343;6470
C_p	25	65.27	273.1	4343
t_c		370	643	6470

466.   2-Butoxyethanol

(Continued)

		cgs	SI	
P_c		38.5	3.90	6470
κ	20	4.32×10^{-7}	4.32×10^{-5}	4902
ε	25	9.30	9.30	3461
μ	25 in 38	2.08	2.08	3461;1298
δ		8.9	18.2	4593,1271,6631,1399,2190
ER	BuOAc 90%	0.07 6750	0.07 6750	6628,6631;1399,3455,2190 6628,6631;1399
soly	in aq	inf	inf	1699;2042
aq az	98.8	20.8%w	20.8%w	3497;6470
fl pt	CC OC	61 74	334 347	152;2042
Beil	30,468			
ir	5932			
nmr	5933			

467.   Furfuryl alcohol

2-Furanmethanol

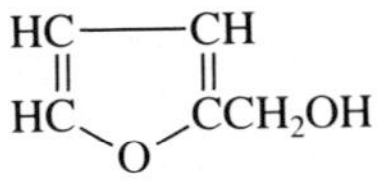

HC———CH
‖   ‖
HC   _O_   CCH$_2$OH

98-00-0   C$_5$H$_6$O$_2$

		cgs	SI	
mw		98.101	98.101	3468
bp	1 atm	170.0	443.2	7058;7550,8052
dt/dp	1 atm	0.040	0.30	7058
dp/dt	1 atm	25	3.3	7058
p	25	0.6	0.08	7058;3248,8052
eq 2.4	A B	9.2086 2804.5	8.3335 2804.5	7058

467. Furfuryl alcohol

(Continued)

		cgs	SI	
fp	(ms)	−29	244	4026
	(st)	−14.63	258.52	6002
d	20	1.1285	1128.5	3563;6002,3508,7550,8052,5398
	30	1.1238	1123.8	7971
n_D	20	1.4868	1.4868	3563;3248,3508,7550,8052,5398
	30	1.4801	1.4801	7971
dn/dt		0.00041	0.00041	3563
η	25	4.62	0.00462	6002
γ	20	ca 38	ca 0.038	4902
ΔH_v	25	15.40	64.43	1703
	bp	12.82	53.64	7058
ΔH_m		3.138	13.13	5683;4026
ΔH_f°	25(l)	−66.05	−276.35	5686;1703
	25(g)	−50.62	−211.79	1703
ΔH_c°	25(l)	−609.16	−2548.73	5686,6002;1703
C_p	25	49.2	205.9	6002;5683,3508
ε		No values in abstract.		5398
μ	25 in 38	1.92	1.92	5432;334,5398
δ		12.5	25.6	4593
soly	in aq	inf	inf	6002
aq az	98.5	20%w	20%w	3497
fl pt	CC	65	338	6002

Beil	2382,112
uv	5056,4808,6022,3563,7229,6749
ir	1169,3990,8172,6575,5932
Raman	6575
ms	154,1591,3368,4664
nmr	9,5987,4627,4623,5933

468. Tetrahydrofurfuryl alcohol

Tetrahydro-2-furanmethanol

$$H_2C \text{------} CH_2$$
$$H_2C \diagdown \quad CHCH_2OH$$
$$O$$

97-99-4

$C_5H_{10}O_2$

		cgs	SI	
mw		102.133	102.133	3468
bp	1 atm	178	451	6003;8052,7550,3905
dt/dp	1 atm	0.049	0.37	6003
dp/dt	1 atm	20	2.7	6003
p	25	0.8	0.1	6003;8052
eq 2.4	A	8.1085	7.2334	6003;1913,921
140-178	B	2358.7	2358.7	
eq 2.4	A	8.9228	8.0477	
below 120	B	2692.9	2692.9	
fp		<80	<353	6003
d	20	1.0524	1052.4	6003;8052,7550,3905,5398
	24	1.0483	1048.3	
	31	1.0402	1040.2	
dd/dt		0.00110	1.10	6003
α	20-37.8	0.00052	0.00052	6003
n_D	20	1.4520	1.4520	6003;8052,7530,3905,5398
	25	1.4499	1.4499	
dn/dt	25	0.00042	0.00042	6003
η	20	6.24	0.00624	6003
γ	25	37	0.037	6003
ΔH_v	25	12.32	51.55	6003
	bp	10.80	45.19	
ΔH_f°	25(1)	-104.1	-435.6	1703
	25(g)	-88.2	-369.0	
ΔH_c°	25(1)	-707.7	-2961.0	1703;6003
C_p	20-27	43.3	181.2	6003
ϵ	23	13.61	13.61	6003;5398
μ	35(1)	2.12	2.12	4739;2079,5398
ER	BuOAc	0.07	0.07	6003
soly	in aq,25	inf	inf	6003
fl pt	TOC	84	357	6003
Beil	2380,107			
ir	5932			
ms	1591			

469.   Diethylene glycol

2,2'-Oxybisethanol

HOCH$_2$CH$_2$OCH$_2$CH$_2$OH   111–46–6   C$_4$H$_{10}$O$_3$

		cgs	SI	
mw		106.121	106.121	3468
bp	1 atm	245.69	518.84	126;7058,2637,7630,1839,6183 2639,5623
dt/dp	1 atm	0.0499	0.375	126;7630
dp/dt	1 atm	20.0	2.67	126
p	25	0.0045	0.00060	126;7630,6183,7058,4133 5209,4983
eq 2.5	A	7.54621	6.67111	126
	B	1897.637	1897.637	
	C	161.067	161.067	
fp		−7.8	265.4	7630;4902,6183,6184,4133,6184
d	20	1.1164	1116.4	7630,1839,6183,4983
dd/dt	10–40	0.00072	0.72	7630;6905
α	20	0.000635	0.000635	4902;6905
n$_D$	15	1.4490	1.4490	6183;4133,6184,4983,7630
	20	1.4475	1.4475	
	25	1.4461	1.4461	
η	20	36	0.036	7630
	25	30	0.030	6183
γ	20	48.5	0.0485	4902;6501
	bp	26.28	0.02628	2638
	Table 2.8			
ΔH$_v$	bp	12.491	52.262	2639;6183,958,7710
ΔH$_c$	(1)	−568.5	−2378.6	5297;8123,100
C$_p$	20	58.0	242.7	7000
κ	20	5.86 x 10^{-7}	5.86 x 10^{-5}	4902,4135
ε	20	31.69	31.69	4135
μ	in 38	2.31	2.31	5209;4135
δ		14.24	29.13	6240;652,6631,4593,1271
ER	BuOAc	<0.001	<0.001	6631
soly	in aq	inf	inf	6983
aq az		none	none	3497;4027
fl pt	COC	143	416	7630
Beil	30,468			
uv	5209			
ir	3654			

470.   Triethylene glycol

2,2'-[1,2-Ethanediylbis(oxy)]bisethanol

$HOCH_2CH_2OCH_2CH_2OCH_2CH_2OH$ 112-27-6 $C_6H_{14}O_4$

		cgs	SI	
mw		150.174	150.174	3468
bp	1 atm	288.0	561.2	7630;2639,1839,7058,4983
dt/dp	1 atm	0.055	0.41	7630
dp/dt	1 atm	18	2.4	7630
p	25	0.00134	0.000179	5975;1839,6184,7058,4983,4133
eq 2.4	A	9.6396	8.7645	2639
	B	3726.2	3726.2	
fp		−4.3	268.9	7630;4133,6184
d	20	1.1235	1123.5	7630
	25	1.1195	1119.5	1463
dd/dt	10–40	0.00078	0.78	7630
α	55	0.00071	0.00071	7630
n_D	15	1.4578	1.4578	4983;4133,6184
	20	1.4558	1.4558	1463;7630
	25	1.4541	1.4541	1463
η	20	49.0	0.0490	7630
	60	8.5	0.0085	5048
γ	20	45.2	0.0452	4902;6511
	bp	22.45	0.02245	2638
	Table 2.8			
ΔH_v	bp	17.066	71.404	2639
ΔH_c	25(1)	−851.1	−3561.0	5297
C_p	20	77.6	324.8	7000
pK_a	25	14.5	14.5	8158
κ	20	8.4×10^{-8}	8.4×10^{-6}	4135
ε	20	23.69	23.69	4135
μ	20(1)	5.58	5.58	4135
δ		10.7	21.9	6905,652;6631
ER	BuOAc	0.001	0.001	6631
soly	in aq	inf	inf	7630
aq az		none	none	3497
fl pt	COC	166	439	7630
Beil	30,468			
ir	628,5932			
Raman	628			
nmr	5933			

471. 2-(2-Methoxyethoxy)ethanol

$CH_3OCH_2CH_2OCH_2CH_2OH$ 111-77-3 $C_5H_{12}O_3$

		cgs	SI	
mw		120.148	120.148	3468
bp	1 atm	194.1	467.3	2069;8055
dt/dp	1 atm	0.045	0.34	Beil EIII 1, 2095
dp/dt	1 atm	22	3.0	Beil EIII 1, 2095
p	25	0.18	0.024	2069;8055,2872
eq 2.4	A	8.69464	7.81954	
	B	2717.39	2717.39	Beil EIII 1, 2095
fp	supercooled	−76	197	055
	(st)	−85	188	6631
d	20	1.0210	1021.0	7568;1464
	25	1.0167	1016.7	
dd/dt		0.00068	0.68	6905
α	20	0.00088	0.00088	6631;4902,6905
n_D	20	1.4264	1.4264	7568;1464,2872
	25	1.4245	1.4245	
	40	1.4188	1.4188	
dn/dt	20-40	0.00038	0.00038	7568
η	25	3.48	0.00348	2069
	60	1.61	0.00161	
γ	25	28.49	0.02849	3898;2069
	30	28.23	0.02823	3898
ΔH_v	bp	11.13	46.57	2069;958
ΔH_c		−719.3	−3009.6	6735
C_p	25	64.8	271.1	2069
μ		1.6	1.6	1399,2190
δ		8.5	17.4	652;6965,2190,1399,7635
ER	BuOAc	0.02	0.02	6631,1399,2190,6905,7635
	90%	26260	26260	6631;1399
soly	in aq	inf	inf	2069
aq az		none	none	3494
fl pt	OC	93	366	2069

Beil	30,468
ir	633,5932
nmr	5933

472.   2-(2-Ethoxyethoxy)ethanol

$CH_3CH_2OCH_2CH_2OCH_2CH_2OH$ 111–90–0 $C_6H_{14}O_3$

		cgs	SI	
mw		134.175	134.175	3468
bp	1 atm	202.0	475.2	2069;7058,8035,1839,3456 5623,6559
dt/dp	1 atm	0.045	0.34	7058
dp/dt	1 atm	22	3.0	7058
p	25	0.13	0.017	2069;1839,2659,7058
eq 2.4	A	8.7300	7.8549	7058
	B	2779.0	2779.0	
fp		-76	197	6631
d	20	0.9885	988.5	7568;1464,1839,2659,3456 6559,8035
	25	0.9841	984.1	
dd/dt		0.00064	0.64	6905
α	20	0.00090	0.00090	6631;6905
n_D	20	1.4273	1.4273	7568;1464,1839,2659,6559,8035
	25	1.4254	1.4254	
	40	1.4194	1.4194	
dn/dt	20–40	0.00039	0.00039	7568
η	25	3.85	0.00385	4902
	60	1.72	0.00172	
γ	25	29.53	0.02953	3898;2069
	75	27.2	0.0272	2069
ΔH_v	bp	11.34	47.45	2069
C_p	25	72	301	2069
κ	25	2.5×10^{-8}	2.5×10^{-6}	4902
μ		1.6	1.6	2190
δ		9.6	19.6	1271,4593,1399,2190,6905
ER	BuOAc	0.02	0.02	6631
	90%	27800	27800	6631
soly	in aq	inf	inf	3456
aq az		none	none	3498
fl pt	OC	96	369	2069

Beil	30,468
ir	633,3654,5932
nmr	5933

473. Salicylaldehyde

2-Hydroxybenzaldehyde

OH
|
HC—C=CCHO
||
HC—C=CH
|
H

90–02–8 $C_7H_6O_2$

		cgs	SI	
mw		122.123	122.123	3468
bp	1 atm	196.7	469.9	4666;958,497,4667,1767 5765,4026
dt/dp	1 atm	0.055	0.41	4666
dp/dt	1 atm	18	2.4	4666
p	33	1	0.1	7058;3012,7414
eq 2.4	A	7.9407	7.0656	3012
170–200	B	2360.6	2360.6	
fp		−7	266	4026;1377,3663,7844
d	25	1.1525	1152.5	497;7844,4026,1377,5765,7847 3663,7401,2357
	50	1.1282	1128.2	
dd/dt		0.000979	0.979	6166
n_D	20	1.5718	1.5718	1767;4026,2357,3663,497,4765
	25	1.57017	1.57017	7847
η	20	2.90	0.00290	5311,7844
	45	1.669	0.001669	7401
ΔH_v	bp	9.14	38.24	958,4666,4667;7755,
ΔH_c°	25(1)	−794.5	−3324.2	3975;4026
C_p	20–195	55.4	231.8	4666;4667
K_B	eq 2.79	4.96	4.96	3012
pK_a	25	8.38	8.38	6216;4823,6211,891,48,443
κ	25	1.64×10^{-7}	1.64×10^{-5}	7842
ε	20	13.9	13.9	7842;2100,7458,7847
μ	20 in 38	2.86	2.86	4687;8097,1767,5020,4693,6693
soly	in aq,85.8	1.68%w	1.68%w	6709;4026

Beil	744,31
uv	6056,3268,15,1883,340,2897,6380,1804
ir	1565,2165,633,1466,6345,3887,5932
Raman	1466,5103,5981,969,6242
nmr	5921,3115,4972,1920

474. 4-Hydroxy-4-methyl-2-pentanone

$(CH_3)_2C(OH)CH_2COCH_3$ 123-42-0 $C_6H_{12}O_2$

		cgs	SI	
mw		116.160	116.160	3468
bp	1 atm	168.1 d	441.3 d	3056;3456,7058
dt/dp	1 atm	0.044	0.33	2599
dp/dt	1 atm	23	3.0	2599
p	25	1.7	0.23	2599;6636,7058,2042
eq 2.4 to 115	A B	8.5552 2482.93	7.6801 2482.93	2599;6636,3056
fp		-42.8	230.4	6905;4403,7058
d	20 25	0.9387 0.9342	938.7 934.2	6905;3456,4403 2599
α	20	0.00094	0.00094	6631;4902
n_D	20 25	1.4235 1.4213	1.4235 1.4213	4403;3456,3056,2599
dn/dt	25	0.00044	0.00044	4403
η	-1.8 25	7.5 2.9	0.0075 0.0029	6636 6631
γ	20	31.0	0.0310	2599
ΔH_v	30-110	11.4	47.7	2599;958
ΔH_f°	25(1)	-141.69	-592.83	7743
ΔH_c		-1000	-4184	4902
C_p	20	52.2	218.4	4902
ε	25	18.2	18.2	4937
μ	20 in 38	3.24	3.24	4273;4937
δ		9.2	18.8	4593,1271,6905,6631,1399
ER	BuOAc 90%	0.12 3840	0.12 3840	6628,6631;1399,7635 6628,6631;1399
soly	in aq	inf	inf	7015;3456
aq az	99.50	15.7%w	15.7%w	3056;3497,6636
fl pt*	TOC	58	331	1399;6095

* Commercial products may flash lower due to presence of acetone.

Beil	113,836
uv	7229
ir	6071,537,5932
ms	154
nmr	6606,8194

475. 2,2,2-Trifluoroethanol

CF_3CH_2OH 75-89-8 $C_2H_3F_3O$

		cgs	SI	
mw		100.040	100.040	3468
bp	1 atm	74.05	347.20	7122,7482
p	25	75.68	10.09	6215
eq 2.5	A	6.8407	5.9656	6215;5035
	B	952.466	952.466	
	C	166.587	166.587	
fp		−43.5	229.7	7122,7482
d	0	1.4106	1410.6	7122,7482
	22	1.3736	1373.6	
n_D	22	1.2907	1.2907	7122,7482
η	20	1.995	0.001995	7122,7482
	30	1.543	0.001543	
ΔH_v	25	10.51	43.97	6215
	107	10.12	42.34	
ΔH_f°	25(1)	−207.4	−867.8	1703;7826
ΔH_c°	25(1)	−211.6	−885.3	1703
t_c		249.3	522.5	6215
P_c		552.8	56.01	6215
d_c		0.489	489	6215
v_c		0.205	0.000205	6215
K_a	25	4.3×10^{-13}	4.3×10^{-13}	577
pK_a	25	12.37	12.37	577
Beil	22, EIII 1342			
ir	3213			

476. 2-Chloroethanol

HOCH$_2$CH$_2$Cl   107–07–3   C$_2$H$_5$OCl

		cgs	SI	
mw		80.514	80.514	3468
bp	1 atm	128.6	401.8	7495;3456,4969,6879,7058,3843
dt/dp	1 atm	0.040	0.30	7495
dp/dt	1 atm	25	3.3	7495
eq 2.15	A	7.83951	6.96441	4025
	B	2727.40	2727.40	
	C	0.0067805	0.0067805	
fp		−67.5	205.7	7495,7058
d	15	1.20720	1207.20	7495;3456,4969,6879,3843
	20	1.20190	1201.90	
	30	1.19118	1191.18	
dd/dt	15–30	0.001068	1.068	7495
n$_D$	15	1.44380	1.44380	7494;3456,4969,6879,4615,3843
dn/dt		0.00039	0.00039	7494
η	15	3.913	0.003913	7495;4615
	30	2.688	0.002688	
γ	20	38.9	0.0389	4025
ΔH$_v$	bp	9.901	41.43	4969
ΔH$_f^\circ$	25(1)	−70.6	−295.4	7826
ΔE$_c$		−290.2	−1214.2	6841;5916
pK$_a$	25	14.31	14.31	578
ε	25	25.8	25.8	4937
μ	25	1.88	1.88	6879
δ	25	13.0	26.5	653
soly	in aq	inf	inf	3456
aq az	97.75 748 Torr	42%w	42%w	3497;8254,6055
fl pt	OC	41	314	152;3456

Beil	22,337
uv	2609,4423
ir	535,8300,8312,642,1169,5781,5932
Raman	4859,5385,3159,6566,5781
ms	2528
nmr	4923,4748,4978,5108,5933

477. 2-Cyanoethanol

3-Hydroxypropanenitrile

NCCH$_2$CH$_2$OH   109–78–4   C$_3$H$_5$ON

		cgs	SI	
mw		71.079	71.079	3468
bp		220 d	493 d	4128;6258
dt/dp	1 atm	0.05	0.4	4128
dp/dt	1 atm	20	2.7	4128
p	25	0.08	0.01	6258
eq 2.4	A	8.822	7.947	4128
	B	2929	2929	
fp		−46	227	6258
d	0	1.059	1059	3262
	25	1.0404	1040.4	6258
ΔH$_v$	av	13.4	56.1	4128
δ	25	15.2	31.0	653
soly	in aq	inf	inf	6258
Beil	222,298			
uv	7426,4988			
ir	4128			
nmr	7543			

478. 2-Aminoethanol

HOCH$_2$CH$_2$NH$_2$ 141–43–5   C$_2$H$_7$ON

		cgs	SI	
mw		61.083	61.083	3468
bp	1 atm	170.95	444.10	7493;6124,4760,2064
dt/dp	1 atm	0.046	0.35	7493
dp/dt	1 atm	22	2.9	7493
p	20	0.36	0.048	7612,4515,4997,831,4595

478. 2-Aminoethanol

(Continued)

		cgs	SI	
eq 2.5	A	7.73800	6.86290	4760;4997,2070,1816
	B	1732.11	1732.11	
	C	186.215	186.215	
fp		10.53	283.68	1104;2064,4760,6124,4997
d	15	1.01949	1019.49	7493;6124,4515,4997,7569
				4595,4076
	20	1.0147	1014.7	2064
	25	1.0127	1012.7	5352
dd/dt	20-30	0.00078	0.78	7626;4997
α	55	0.00079	0.00079	7612
n_D	20	1.4545	1.4545	5352;7569,1104,6124,4515,4997
				4595,4076,7626
	25	1.4525	1.4525	5352,2064;7569
dn/dt	20-40	0.00034	0.00034	7626;4997
η	15	30.855	0.030855	7493
	25	19.346	0.019346	1104;2064
	30	14.417	0.014417	7493
γ	15.80	49.39	0.04939	7493;6124
	20	48.89	0.04889	
	30	47.74	0.04774	
ΔH_v	0	22.010	92.090	4997
	bp	11.910	49.831	4997
ΔH_m		4.900	20.502	2070
ΔH_c	25(1)	-220.8	-923.8	7483
C_p	30	30.4	127.2	2070
t_c		341.3	614.5	2070
P_c		44.1	4.47	2070
$pK_{BH}+$	25	9.50	9.50	4194
	25, in aq HCl	9.498	9.498	670;3092,5773,3090,674
				3095,2785
κ	25	1.10×10^{-5}	0.00110	1087
ε	25	37.72	37.72	1104;4370
μ	25 in 172	2.27	2.27	5731
δ	25	15.5	31.8	653
soly	in aq	inf	inf	7612
aq az		none	none	3498
fl pt	CC	85	358	152
	COC	93	366	7626

478. 2-Aminoethanol

(Continued)

Beil	353,274
ir	7426,4286,7225,5932
Raman	2205,2723
ms	4783,7797
nmr	2743,1878

479. 2-Amino-2-methyl-1-propanol

$$\overset{NH_2}{\underset{|}{CH_3C(CH_3)CH_2OH}}$$
                        124–68–5 $C_4H_{11}NO$

		cgs	SI	
mw		89.137	89.137	3468
bp	1 atm	165	438	2589,1615
p	67.4	10.0	1.33	2589
fp		30–31	303–304	2589,1615
d	40	0.921	921	2589
α		0.00095	0.00095	2589
n_D	20	1.449	1.449	2589
η	30	102.4	0.1024	2589
	50	24.6	0.0246	
ΔH_v	110	13.2	55.2	2589
	bp	12.1	50.6	
pK_a	20	9.82	9.82	2589
soly	in aq,20	inf	inf	2589,1615
aq az		none	none	2589
fl pt	TOC	68	341	2589
Beil	354,EIII 783			

480.   Diethanolamine

2,2'-Iminobisethanol

$(HOCH_2CH_2)_2NH$ 111-42-2 $C_4H_{11}O_2N$

		cgs	SI	
mw		105.136	105.136	3468
bp	1 atm	268.39 d	541.54 d	2070;7612,4074,1741,5731 4595,1008
p	20	0.0002	3×10^{-5}	1008;4074,4595
eq 2.5	A	8.13968	7.26458	4760;1816,2064
	B	2328.56	2328.56	
	C	174.399	174.399	
fp		27.95	301.10	4760;1741,4074,2064
d	25	1.0936	1093.6	5352
	30	1.0909	1090.9	5352;4074,1741,1008,4076
	40	1.0838	1083.8	5352;2070
dd/dt	30-40	0.00065	0.65	7626;7612,4595
α	55	0.00060	0.00060	7626
n_D	25	1.4735	1.4735	5352
	30	1.4721	1.4721	5352;7612,4074,4076
dn/dt	20-40	0.00025	0.00025	7626
η	30	351.9	0.3519	6905;7612,1741
	40	196.4	0.1964	2070
ΔH_v	160.5	16.8	70.3	7612
	bp	15.590	65.229	2070
ΔH_m		6.000	25.10	2070
C_p	30	55.8	233.5	2070
t_c		442.1	715.3	2070
P_c		32.3	3.27	2070
$pK_{BH}{}^+$	25	8.88	8.88	4194
	25, in aq	8.883	8.883	670;1027,3095
	HCl			
μ	25 in 172	2.81	2.81	5731
soly	in aq,20	95.4%w	95.4%w	7626;7612
aq az		none	none	3498
fl pt	COC	138	411	152,7626
Beil	353,283			
ir	7225,5932			
ms	4783,7297			

481. Triethanolamine

2,2′,2″-Nitrilotrisethanol

$(HOCH_2CH_2)_3N$ 102-71-6 $C_6H_{15}O_3N$

		cgs	SI	
mw		149.189	149.189	3468
bp	1 atm	335.39	608.54	4760;1741,7626,2064
dt/dp	1 atm	0.073	0.55	7612
dp/dt	1 atm	14	1.8	7612
p	20	<0.01	<0.0013	7612;4760,5731,4075,850
eq 2.5	A	8.55499	7.67989	4760;2064
	B	2962.73	2962.73	
	C	186.750	186.750	
fp		21.57	294.72	4760;1741,4595,2064
d	25	1.1196	1119.6	2070;4075,2715,4515,8182
				1008,4076
	40	1.1116	1111.6	
dd/dt	30	0.00055	0.55	7626;7612
α	55	0.00053	0.00053	7626
n_D	25	1.4835	1.4835	2070;4075,4076
	40	1.4798	1.4798	
dn/dt	20-40	0.00020	0.00020	
η	25	613.6	0.6136	2070;4515,1741
	40	208.1	0.2081	
ΔH_v	bp	16.127	67.475	2070
ΔH_m		6.500	27.20	2070
C_p	30	74.1	310.0	2070;5040
t_c		514.3	787.5	2070
p_c		24.2	2.45	2070
pK_{BH}^+	25	7.76	7.76	4194
	25, in aq	7.762	7.762	670,671;3090,3095,7951,675
	HCl			
ε	25	29.36	29.36	850
μ	25 in 172	3.57	3.57	5731
soly	in aq	inf	inf	7612;4075,1008,1741
fl pt	COC	179	452	7612;152
Beil	353,285			
ir	5932			
ms	4783,7797			

482. 2,2'-Thiodiethanol

2,2'-Thiobisethanol

$(HOCH_2CH_2)_2S$ 111–48–8 $C_4H_{10}O_2S$

		cgs	SI	
mw		122.182	122.182	3468
bp	1 atm	282	555	4027;2301,1520,8022,5208
dt/dp	25 Torr	∿0.9	∿7	Beil EIII 1,2122*
p	115.5	9	1.2	3610
eq 2.4	A	10.242	9.367	Beil EIII 1,2122
1–50 Torr	B	3963	3963	
fp		−10.2	263.0	3610;1520,2301,4027
d	0	1.1973	1197.3	1520;2301,4027,4266
	20	1.18723	1187.23	3610;5208
	25	1.1793	1179.3	1520
dd/dt		0.00072	0.72	1520
n_D	20	1.52108	1.52108	3610;2301,4027
	30	1.51770	1.51770	
dn/dt	30	0.000322	0.000322	3610
γ	20	53.8	0.0538	5208
ΔH_v	av	∿18	∿75	Beil EIII 1,2122
soly	in aq,20	inf	inf	1520;4027
fl pt	TOC	160	433	4027

* Selected values.

Beil	30,470
uv	5209,4426,6085,5205
ir	6490,5932

483. 2-Furaldehyde

2-Furancarboxaldehyde

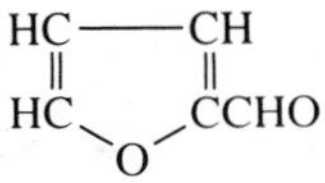

HC————CH
‖   ‖
HC   CCHO
   O

98-01-1

$C_5H_4O_2$

		cgs	SI	
mw		96.085	96.085	3468
bp	1 atm	161.8	435.0	7058;4969,5218
p	25	2.5	0.33	6001;4997,7475,7058
eq 2.4	A	7.959	7.084	4575;4997
	B	2209	2209	
fp		-36.5	236.7	7058;7475
d	20	1.1598	1159.8	6001;7848,5218,7552,3563,5398
	25	1.1545	1154.5	
dd/dt		0.001057	1.057	4575
n_D	20	1.52608	1.52608	4575;7552,5218,3563,5398
	25	1.52345	1.52345	
dn/dt	25	0.000526	0.000526	4575
η	0	2.48	0.00248	6001;7838,4575
	25	1.49	0.00149	
	38	1.35	0.00135	
γ	30	41.1	0.0411	6001;7850
	Table 2.8			
ΔH_v	bp	10.330	43.221	4969;4997,7850,7755
ΔH_m		3.43	14.35	4575
ΔH_f°	25(1)	-47.85	-200.20	515;1703
	25(g)	-36.1	-151.0	1703
ΔH_c°	25(1)	-559.04	-2339.02	515;1703,4575,3975
C_p	25	37.7	157.7	5587
	Table 2.9			
λ	37.8	0.000630	0.264	6001
t_c		397	670	4575
p_c		54.4	5.51	4575
ϵ	25	38	38	4575;4054,2100
μ	25 in 38	3.54	3.54	824,5850;2960,6693,4054
δ		11.2	22.9	652
soly	in aq,20	8.2%w	8.2%w	3745;6556,987,7219
	aq in,20	6.3%w	6.3%w	
aq az	97.85	35%w	35%w	3497
fl pt	CC	60	333	152

483. 2-Furaldehyde

(Continued)

Beil	2461,272
uv	827,2025,7891,5056,3234,4808
ir	1773,633,6115,7066,2941,6575,5786,17,5932
Raman	7066,2941,6575
nmr	7227,2981,9,5987,2464

484. Bis(2-chloroethyl) ether

1,1'-Oxybis[2-chloroethane]

(ClCH$_2$CH$_2$)$_2$O　　　　　　111-44-4　　　　　　C$_4$H$_8$OCl$_2$

		cgs	SI	
mw		143.013	143.013	3468
bp	1 atm	178.75	451.90	2639;6879,7763,656,3456
				5209,5329
dt/dp	1 atm	0.0575	0.431	2639
dp/dt	1 atm	17.4	2.32	2639
p	25	1.55	0.207	2639;2425,7058,2740
eq 2.4	A	8.1040	7.2289	2639
	B	2359.6	2359.6	
fp		−46.8	226.4	7560;2425,2735,5644,7562
d	20	1.2192	1219.2	5329;656,2638,3456,6879
				7763,3843
	25	1.2130	1213.0	
dd/dt	20–30	0.00118	1.18	7615,7616
α	10–30	0.00097	0.00097	2042
n$_D$	20	1.45750	1.45750	6879;656,2425,3456,5329,7763
	25	1.45534	1.45534	7560
dn/dt	25	0.00043	0.00043	6879
η	20	2.41	0.00241	5329;2425,7562,5644,2637,3843
	25	2.14	0.00214	
	Table 2.6			

484.   Bis(2-chloroethyl) ether

(Continued)

		cgs	SI	
γ	20	37.6	0.0376	5329;2425,2638,7763
	25	37.0	0.0370	
	Table 2.8			
ΔH_v	bp	10.81	45.23	2639;2425,5644
ΔH_m		2.070	8.661	7562;7560
C_p	30	52.8	220.9	2425;5644
pK_{BH}^+	25	−2.18	−2.18	4556
ε	20	21.2	21.2	4937;7562
μ	25 in 38	2.58	2.58	6879;5209,7562
δ		9.8	20.0	4593,6905
ER	BuOAc	0.1	0.1	6905
	90%	3500	3500	6905
soly	in aq,20	1.02%w	1.02%w	2042;5644,2425,3456
	aq in,20	0.1%w	0.1%w	
aq az	98	34.5%w	34.5%w	3498
fl pt	CC	55	328	152
	OC	85	358	7615,7616
Beil	22, EII 337			
uv	3612,5209			
ir	5932			
Raman	3879			
nmr	5532,5933			

485. Epichlorohydrin

(Chloromethyl)oxirane

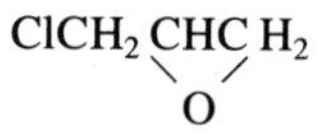

$ClCH_2 CHC H_2$
$\diagdown \diagup$
O

106–89–8

C_3H_5OCl

		cgs	SI	
mw		92.525	92.525	3468
bp	1 atm	116.11	389.26	6626
dt/dp	1 atm	0.044	0.33	6626
dp/dt	1 atm	23	3.0	6626
p	25	18	2.4	7613
eq 2.5	A	7.4709	6.5958	7628
	B	1587.9	1587.9	
	C	230	230	
fp		−57.2	216.0	6626;3663
d	15	1.18683	1186.83	6626;7075
	20	1.18066	1180.66	
	25	1.17455	1174.55	
dd/dt	15–25	0.00123	1.23	6626
α	55	0.00104	0.00104	7613
n_D	20	1.43805	1.43805	6626
	25	1.4358	1.4358	
η	0	1.56	0.00156	6626;7839
	25	1.03	0.00103	
γ	20	37.00	0.03700	6626;7075
	31	35.48	0.03548	
ΔH_v	bp	9.060	37.907	6626;7850
t_c		351	624	7848
pK_b		8.9	8.9	2880
κ	25	3.4×10^{-8}	3.4×10^{-6}	6626;7842
ε	22	22.6	22.6	4937;2027,6626,7850
μ	in 317	1.8	1.8	7661
δ		11.0	22.5	652
soly	in aq,20	6.58%w	6.58%w	6626
	aq in,20	1.47%w	1.47%w	
aq az	88	75%w	75%w	3497
fl pt	OC	41	314	152

Beil	2362,6
ir	633,5711,3835,5932
Raman	575,4533,3835
nmr	7227,5977,6113,4803

486. *o*-Nitroanisole

1-Methoxy-2-nitrobenzene

o-CH$_3$OC$_6$H$_4$NO$_2$ 91-23-6   C$_7$H$_7$O$_3$N

		cgs	SI	
mw		153.137	153.137	3468
bp	737 Torr	265	538	1057;3001,6710
fp		10.45	283.60	2960;3143,1057,1694,1249,6710
d	20	1.2527	1252.7	1057;3001
	25	1.2408	1240.8	1694
n$_D$	20	1.56188	1.56188	1057;6464
	25	1.5597	1.5597	1694
γ	26	45.7	0.0457	6710
	55	41.93	0.04193	
ε	30	5.230	5.230	3694
μ	25 in 38	4.98	4.98	2343;4138,1694,3466,2039,3001
soly	in aq,30	0.169%w	0.169%w	2987
Beil	523,217			
uv	5993,581,6799,4712,1657,6749			
ir	3114,5932			
Raman	8125			
ms	3166			
nmr	6362,6607,6851,5933			

487. Morpholine

OCH$_2$CH$_2$NHCH$_2$CH$_2$ 110-91-8   C$_4$H$_9$ON

		cgs	SI	
mw		87.121	87.121	3468
bp	1 atm	128.94	402.09	7493;4091
dt/dp	1 atm	0.045	0.34	7632;7612,7493
dp/dt	1 atm	22	3.0	7632
p	25	10.08	1.344	7612;952

487. Morpholine

(Continued)

		cgs	SI	
eq 2.5	A	7.71813	6.84303	4398
0–44	B	1745.8	1745.8	
	C	235.0	235.0	
44–170	A	7.16030	6.28520	
	B	1447.70	1447.70	
	C	210.0	210.0	
fp		−4.8	268.4	7632;7612
d	15	1.00495	1004.95	7493;952
	25	0.99547	995.47	7493
dd/dt	10–40	0.00095	0.95	7632
α	55	0.00097	0.00097	7632
n_D	15	1.45733	1.45733	7493;4091
	20	1.4542	1.4542	7612
dn/dt	20–40	0.00044	0.00044	7632
η	15	2.534	0.002534	7493
	30	1.792	0.001792	
γ	15	38.27	0.03827	7493;952
	20	37.63	0.03763	
	30	36.24	0.03624	
ΔH_v	25	10.508	43.965	2087
	bp	8.855	37.049	2087;952
$pK_{BH}{}^+$	25	8.492	8.492	3362;7984,3092,3090,4194
ε	25	7.42	7.42	5555;7483
μ	25 in 38	1.56	1.56	4716;5555
δ	25	10.8	22.1	653
soly	in aq	inf	inf	7612
aq az		none	none	3498
fl pt	TOC	39	312	7632;152

Beil	4190,5
ir	7901,2570,5814,7703,5932
Raman	5814,7703
nmr	3185,2464,5933

488. Ethyl lactate

Ethyl 2-hydroxypropionate

$CH_3CH(OH)COOCH_2CH_3$ 97–64–3 $C_5H_{10}O_3$

		cgs	SI	
mw		118.132	118.132	3468
bp	1 atm	154.5	427.7	6492;6843,3456,8182
dt/dp	1 atm	0.04	0.3	4902
dp/dt	1 atm	25	3.3	4902
p	30	5	0.7	2042;1868,8182
eq 2.4	A	8.7020	7.8269	Beil H 3,280
	B	2489.7	2489.7	
fp		−26	247	4902
d	20	1.0328	1032.8	8153;6843,3456,7864
	25	1.0272	1027.2	
dd/dt		0.00112	1.12	8153
α	10–30	0.00098	0.00098	2042
n_D	20	1.4124	1.4124	5219;3456,6843,7864,2042,1264
η	25	2.44	0.00244	4902
γ	17.3	28.90	0.02890	7864;5219,2266
	38.4	26.80	0.02680	
ΔH_v	25	11.8	49.4	Beil H 3,280
	bp	11.1	46.4	
ΔH_c°	25(1)	−654.7	−2739.3	3975
C_p	20–30	60.6	253.6	4902
$[\alpha]_D$	18.9	+11.26	+11.26	8153;7575
κ	25	1.0×10^{-6}	1.0×10^{-4}	3965
ε	20	13.1	13.1	5219
μ	20 in 38	2.4	2.4	5219
δ		10.0	20.5	4593,1271,6631,2190,1399
ER	BuOAc	0.18	0.18	6628,6631;1399,3455
	90%	2568	2568	6628,6631;1399
soly	in aq	inf	inf	2042;3456
fl pt	CC	47	320	3456

Beil	221,280
uv	6811,7575
ir	6811,949,5252,5253
Raman	1264

489. Methyl salicylate

Methyl 2-hydroxybenzoate

o-HOC$_6$H$_4$COOCH$_3$ 119–36–8   C$_8$H$_8$O$_3$

		cgs	SI	
mw		152.149	152.149	3468
bp	1 atm	233.3	506.5	7472;3012,5765,2959
dt/dp	1 atm	0.057	0.43	2959
dp/dt	1 atm	18	2.3	2959
p	25	0.11	0.015	4997;497,3012
eq 2.12	A	36.5929	35.7178	4997
	B	4410.9	4410.9	
	C	9.1999	9.1999	
fp		−8.6	264.6	7472
d	20	1.1831	1183.1	5765;497,2027
	25	1.1782	1178.2	
dd/dt		0.00100	1.00	5765
n$_D$	18.1	1.53773	1.53773	497;4997
	20	1.5365	1.5365	2773
γ	−19.8	44.2	0.0442	3663
	212.2	19.8	0.0198	
ΔH$_v$	bp	11.155	46.673	4997;958
ΔH$^\circ_c$	25(1)	−902.2	−3774.8	3975
C$_p$	15–30	59.46	248.78	4153
ε	30	9.41	9.41	4937;2027
μ	25 in 38	2.47	2.47	4164;1651
δ		10.6	21.7	652
soly	in aq,30	0.74%w	0.74%w	6999
fl pt	TCC	99	372	2773
Beil	1061,70			
uv	3583,7229,6380			
ir	2565,3528,7225,5252,5932			
ms	4909			
nmr	4924,2531,2567			

490.   2-Methoxyethyl acetate

2-Methoxyethanol acetate

$CH_3COOCH_2CH_2OCH_3$ 110–49–6 $C_5H_{10}O_3$

		cgs	SI	
mw		118.132	118.132	3468
bp	1 atm	143.2	416.4	4339;4902,3456,2042
dt/dp	1 atm	0.044	0.33	3779
dp/dt	1 atm	23	3.0	3779
p	25	5	0.7	3779;2042
eq 2.4	A	8.3312	7.4561	3779
	B	2277.68	2277.68	
fp		−65.1	208.1	7637
d	20	1.0049	1004.9	7637;3456,2042
	25	1.00033	1000.33	4339
α	10–30	0.00110	0.00110	2042
n_D	20	1.4022	1.4022	5219
	25	1.3999	1.3999	4339;3456
ΔH_v	25	12.01	50.27	4342
	bp	10.5	43.9	3779
ε	20	8.25	8.25	5219
μ	30 in 38	2.13	2.13	2813;5219
δ		9.2	18.8	4593,1399,2190,6905
ER	BuOAc	0.20	0.20	2190,6905;1399,7635
	90%	1395	1395	1399
soly	in aq	inf	inf	7637
aq az	97.0	48.5%w	48.5%w	3497
fl pt	TOC	52	325	2190
	TCC	57	330	7635;152
Beil	159,141			
uv	4388			
ir	5932			
nmr	5933			

491. 2-Ethoxyethyl acetate

2-Ethoxyethanol acetate

$CH_3COOCH_2CH_2OCH_2CH_3$ 111–15–9 $C_6H_{12}O_3$

		cgs	SI	
mw		132.159	132.159	3468
bp	1 atm	156.1	429.3	4339;835
p	20	1.09	0.145	835
fp		−61.7	211.5	7637
d	20	0.9730	973.0	$\frac{4908}{4339}$
	25	0.96761	967.61	
dd/dt		0.00086	0.86	6905
α	20	0.00112	0.00112	6631;6905
n_D	25	1.4032	1.4032	4339
	30	1.4003	1.4003	5813
dn/dt		0.00042	0.00042	5813
η	25	1.025	0.001025	4902
γ	25	31.8	0.0318	4902
ΔH_V	25	12.59	52.69	4342
ε	30	7.567	7.567	5813;5219
μ	30 in 38	2.25	2.25	2813;5813,5219
δ		8.7	17.8	4593,1271,6905,6631,2190
ER	BuOAc	0.20	0.20	1399,2190,6905,6631,7635
	90%	2370	2370	1399;6631
soly	in aq,20	22.9%w	22.9%w	7637
	aq in,20	6.5%w	6.5%w	
aq az	97.5	44.4%w	44.4%w	3498;3497
fl pt	TOC	57	330	7637
Beil	159,141			
ir	5932			
nmr	5933			

492.   2-(2-Ethoxyethoxy)ethyl acetate

2-(2-Ethoxyethoxy)ethanol acetate

$CH_3COOCH_2CH_2OCH_2CH_2OCH_2CH_3$ $C_8H_{16}O_4$

		cgs	SI	
mw		176.214	176.214	3468
bp	1 atm	217.4	490.6	7637;3456
dt/dp	1 atm	0.05	0.4	4902
dp/dt	1 atm	20	3	4902
p	20	0.1	0.01	8056,2042
eq 2.4	A	8.589	7.714	4902
	B	2800	2800	
fp		−25	248	7637
d	20	1.0096	1009.6	7637;3456
dd/dt		0.00078	0.78	6905;4902
α	20	0.00101	0.00101	4902;6905
n_D	20	1.4213	1.4213	4902;3456
η	20	2.8	0.0028	4902
ΔH_v		21.8	91.2	4902
C_p	20	94.8	396.6	4902
μ	(1)	1.8	1.8	2190
δ		8.5	17.4	4593,1399,2190,6905
ER	BuOAc	0.0063	0.0063	1399
	ER	74280	74280	1399
soly	in aq	inf	inf	7637
aq az		none	none	3498;3497
fl pt	COC	110	383	7637
Beil	159,141			

493. Methyl acetoacetate

Methyl 3-oxobutanoate

$CH_3COCH_2COOCH_3$ 105–45–3 $C_5H_8O_3$

		cgs	SI	
mw		116.116	116.116	3468
bp	1 atm	171.7	444.9	Beil EIII 3,1182;2677,3016
p	25	6	0.8	Beil EIII 3,1182
eq 2.4	A	7.1066	6.2315	Beil EIII 3,1182
	B	1880.4	1880.4	
fp		−80	193	7033
d	15	1.0800	1080.0	5764;500,2677
	20	1.0747	1074.7	
	25	1.0724	1072.4	
n_D	20	1.4186	1.4186	500
η	20	1.704	0.001704	2677
ΔH_v	av	8.60	36.0	Beil EIII 3,1182
ΔH_c		−594.0	−2485.3	3017
δ	25	10.7	21.8	653
fl pt	CC	77	350	152

Beil	280,632
uv	861
ir	5932
ms	1030,1029,6100
nmr	1252,6748

494. Ethyl acetoacetate

Ethyl 3-oxobutanoate

$CH_3COCH_2COOCH_2CH_3$ 141–97–9 $C_6H_{10}O_3$

		cgs	SI	
mw		130.143	130.143	3468
bp	1 atm	180.8	454.0	7058
p	keto,40–41	2	0.3	4077;1193,2127
eq 2.4	A	8.40179	7.52669	7058
	B	2506.57	2506.57	

494. Ethyl acetoacetate

(Continued)

		cgs	SI	
fp	keto	−39	234	4077
	enol	−44	229	4027
d	keto,10	1.0368	1036.8	5085;1193,7541,2127
	enol,10	1.0119	1011.9	
	25	1.02126	1021.26	6514
dd/dt		0.00097	0.97	6514
n_D	keto,10	1.4225	1.4225	4077;5085,1193
	enol,10	1.4480	1.4480	
	20	1.4192	1.4192	5289
η	25	1.5081	0.0015081	2127
γ	14.8	32.47	0.03247	6040
	46.4	29.09	0.02909	
	Table 2.8			
ΔH_f°	25(1)	−146.33	−612.24	7743
ΔH_c°	25(1)	−755.4	−3160.6	3975
C_p	24.5	59.8	250.2	4027
t_c		400	673	1310
pK_a	25	10.68	10.68	68;4027,2227,2828
κ	25	4×10^{-8}	4×10^{-6}	3818
ϵ	22	15.7	15.7	2100
μ	enol,−80 in 451	2.04	2.04	835
	keto,18.2 in 38	3.22	3.22	
δ	25	9.8	20.1	653
soly	in aq,25	12%w	12%w	7637
	aq in,25	4.9%w	4.9%w	
fl pt	OC	84	357	152

Beil	280,632
uv	861,5289,7872
ir	633,5932
Raman	6662
ms	1030,1029,1031,6100
nmr	1252,3839

495. Methyl cyanoacetate

NCCH$_2$COOCH$_3$ 105–34–0   C$_4$H$_5$O$_2$N

		cgs	SI	
mw		99.089	99.089	3468
bp	1 atm	205.09	478.24	2089;7842,3869,3016
dt/dp	1 atm	0.0490	0.368	2089
dp/dt	1 atm	20.4	2.72	2089
p	25	0.14	0.019	2089
eq 2.5	A	7.60624	6.73114	2089
	B	1914.22	1914.22	
	C	200	200	
fp		−13.07	260.08	2089
d	0	1.1492	1149.2	7844;3869,3016
	25	1.1225	1122.5	
n$_D$	20	1.41791	1.41791	2089;3869
	25	1.41662	1.41662	
dn/dt	25	0.000258	0.000258	2089
η	20	2.793	0.002793	2089;7844
	40	1.764	0.001764	
γ	20	42.32	0.04232	2089
	30	41.16	0.04116	
ΔH$_v$	25	14.740	61.672	2089
	bp	11.509	48.154	
ΔH$_c$	25(1)	−475.8	−1990.7	3975;3017
A	eq 2.72	0.02195	0.02195	2089
κ	25	4.49 x 10^{-7}	4.49 x 10^{-5}	7842;7851
ε	20	29.30	29.30	7868;7842
soly		sl sol	sl sol	3918
Beil	171,584			
Raman	1784,1780			

496. Ethyl cyanoacetate

$NCCH_2COOCH_2CH_3$ 105–56–6 $C_5H_7O_2N$

		cgs	SI	
mw		113.116	113.116	3468
bp	1 atm	206.0	479.2	7058;7842,2580,3869,5468
p	67.8	1	0.1	7058;1691,7493
eq 2.4	A	9.94022	9.06512	7058
100–200	B	3382.89	3382.89	
fp		−22.5	250.7	3918
d	15	1.06652	1066.52	7493;2580,3869,1191,1691,7864
	25	1.0564	1056.4	7844,5468
	30	1.05106	1051.06	7493
dd/dt		0.00103	1.03	7493
n_D	20	1.41555	1.41555	3869;5468,1191,1691,7493
η	0	5.06	0.00506	7844,7838;2580
	15	3.256	0.003256	7493
	20	2.63	0.00263	7791
	25	2.50	0.00250	4093
	30	2.148	0.002148	7493
γ	20	36.48	0.03648	7493,1691,7864
	30	35.42	0.03542	
ΔH_v	100–200	15.5	64.9	7058
ΔH_c	25(1)	−635.1	−2657.3	3975;3017
K_a	25	$<10^{-9}$	$<10^{-9}$	5735
κ	25	6.9×10^{-7}	6.9×10^{-5}	7842;7843
ε	18	26.7	26.7	2100;4093,7842,7850,4937,4370
μ		2.17	2.17	6861
δ	25	11.0	22.5	653
soly	in aq	25.9%w	25.9%w	4093
fl pt		110	383	3918
Beil	171,585			
uv	6507			
ir	6069			
nmr	6017			

497. Trifluoroacetic acid

CF₃COOH → CF_3COOH

76–05–1

$C_2HO_2F_3$

		cgs	SI	
mw		114.024	114.024	3468
bp	1 atm	71.78	344.93	4254;2733,3901,7119,3287
p	25	108	14.4	7197
eq 2.4	A	8.389	7.514	7197
	B	1895	1895	
fp		−15.216	257.934	3194;6746,7119,4254,3287,3179
d	20	1.4890	1489.0	4026;3901,7119
	25	1.4785	1478.5	4254;3101
	30	1.4671	1467.1	3688
dd/dt		0.002346	2.346	3688;3101
n_D	20	1.2850	1.2850	4026
η	20	0.926	0.000926	3901
	25	0.855	0.000855	
γ	24	13.63	0.01363	3688
	Table 2.8			
ΔH_v	av	8.672	36.28	7197
	bp	7.949	33.26	7197;4163
ΔH_f°	25(l)	−253.4	−1060.2	4144;1703
	25(g)	−244.2	−1021.7	4144;1703
ΔH_c°	25(l)	−94.93	−397.19	4144
t_c		218.1	491.3	4299;93
P_c		32.15	3.258	4299;93
d_c		0.559	559	4299
v_c		0.204	0.000204	4299
K_f	eq 2.67	6.25	6.25	3194
K_a	25	0.65	0.65	5154
pK_a	25	0.19	0.19	5154;68,3287,3478
ε	20	8.55	8.55	3179;4937,6746
μ	100(g)	2.28	2.28	2733
soly	in aq	inf	inf	3901
aq az	105.46	79.4%w	79.4%w	7119;299,3293,3498

Beil	160,194
uv	7148
ir	760,7225,5932
Raman	2473
nmr	6295,1345,3478,5933

498.   Heptacosafluorotributylamine

$(C_4F_9)_3N$ 311–89–7 $C_{12}F_{27}N$

		cgs	SI	
mw		671.096	671.096	3468
bp	1 atm	177	450	6386
p	25.05	0.552	0.0736	6281
d	25	1.8839	1883.9	6281;2175,6386
n_D	25	1.2910	1.2910	6281
η	25	5.3	0.0053	6386
γ	25	16	0.016	6386
ΔH_v	25	13.250	55.438	6281
	bp	11.1	46.4	6386
C_p	25	100	418.4	6386
t_c		355	628	6386
p_c		12.2	1.24	6386
ε		1.9	1.9	6281
δ		5.9	12.1	6281,5334,2175
nmr	92			

499.   o-Chloroaniline

2-Chlorobenzenamine

$o\text{-ClC}_6H_4NH_2$ 95–51–2 C_6H_6NCl

		cgs	SI	
mw		127.573	127.573	3468
bp	1 atm	208.84	481.99	2090;395,3815,7058,1754
dt/dp	1 atm	0.05277	0.3958	2090
dp/dt	1 atm	18.95	2.526	2090
p	25	0.2533	0.03377	2087
eq 2.5	A	7.63311	6.75801	2090
	B	2085.5	2085.5	
	C	230	230	

499. *o*-Chloroaniline

(Continued)

		cgs	SI	
fp		-1.94	271.21	2090;395
	α-form	-11.92	261.23	395;4073
	β-form	-1.78	271.37	
d	20	1.21251	1212.51	2090;395,3815,7403
	25	1.20775	1207.75	
dd/dt	25	0.000952	0.952	2090
n_D	20	1.58807	1.58807	2090;395,3815,1754
	25	1.58586	1.58586	
dn/dt	25	0.000442	0.000442	2090
η	20	2.9157	0.0029157	2087,7403
	40	1.8458	0.0018458	
γ	20	43.66	0.04366	2087
	30	42.54	0.04254	
ΔH_v	25	13.565	56.756	2087
	bp	10.60	44.35	
ΔH_m		2.84	11.88	6712
A	eq 2.72	0.01785	0.01785	2087
$pK_{BH}{}^+$	25	2.64	2.64	867;5773,7049,7725,6226
ε		13.4	13.4	7850
μ	25 in 38	1.77	1.77	1754;3209,7462,6837
soly	in aq,25	0.876%w	0.876%w	2087;6712
	aq in,19	0.56%w	0.56%w	6999

Beil	1670,597
uv	582,5993,8142,6749
ir	2557,6763,4285,4369,5932
Raman	3336
nmr	2743,4720,6607,7919,5933

500.   Hexamethylphosphoric triamide

$[(CH_3)_2N]_3PO$ 680–31–9 $C_6H_{18}ON_3P$

		cgs	SI	
mw		179.201	179.201	3468
bp	1 atm	233	506	6192;5511,7515
dt/dp	25 Torr	0.903	6.77	6192
dp/dt	25 Torr	1.11	0.148	6192
p	30	0.07	0.009	5512;5511,1392,5947
eq 2.4	A	8.7227	7.8476	6192
	B	2956	2956	
fp		7.20	280.35	5511;2109,6192
d	20	1.027	1027	6192;7515,7698
	25	1.02022	1020.22	1777;844,3956
dd/dt	10-75	0.00085	0.85	844
κ_T		1.05×10^{-7}	7.90×10^{-7}	1777
n_D	20	1.4588	1.4588	991;7515
	25	1.4570	1.4570	5512,2630
η	20	3.47	0.00347	6192
	25	3.10	0.00310	844;3956
γ	20	33.8	0.0338	6192
ΔH_v	25	14.6	61.1	6331
	av	13.5	56.5	1206
ΔH_m		3.414	14.28	5719
ΔH_f		No values in abstract		5512
C_p	25	76.79	321.3	7792
K_f	eq 2.67	8.4	8.4	5719;5511,2109
κ	25	$2\text{-}3 \times 10^{-7}$	$2\text{-}3 \times 10^{-5}$	844;5719
ε	20	29.30	29.30	2630;5511,6192,7698,3956
μ	25(1)	4.31	4.31	6420;2105,7698,2630
	25 in 38	5.54	5.54	7058
δ		10.5	21.5	652
soly	in aq	inf	inf	6192

ir	3142,2041,3312,6021,5665,2481,5932
Raman	6021,5665,2481
nmr	7679,7936,4420,5933

501. Tetramethylsilane

$(CH_3)_4Si$ 75-76-3 $C_4H_{12}Si$

		cgs	SI	
mw		88.225	88.225	3468
bp	1 atm	26.64	299.79	476;7183,8257,109,8225
dt/dp	1 atm	0.0380	0.285	476
dp/dt	1 atm	26.3	3.51	476
p	25	717.3	95.63	476
fp	(st)	−99.04	174.11	476;7183,6678
	(ms)	−102.12	171.03	
d	20	0.6462	646.2	8257;4149,6389
n_D	20	1.3582	1.3582	7183;109,4149,8257,6389
ΔH_v	bp	5.785	24.204	476;7183,6529
ΔH_m	(st)	1.648	6.895	476;6678,5509
	(ms)	1.4268	5.970	
ΔH_f°	25(1)	−63.0	−263.6	7826;5931,7183,7182,7664,7209
	25(g)	−57.149	−239.11	7826;7183,7182,7209
ΔH_c°	25(1)	−938.9	−3928.4	4429;7183
C_p°	25	34.39	143.89	6529,7826;6671,5279
C_p	25	48.78	204.10	7826;6678,476
ε	20	1.921	1.921	109
μ		0	0	109;5380,4739
soly	in aq,25	0.00196%w	0.00196%w	1936
Beil	420,625			
uv	3150,5867			
ir	6048,4262,8225,7463,4220,6671			
Raman	7928,2221,4149,6655,4220,6047,6671			
ms	6213,492,7668,7166,1962,1714			
nmr	6930,5794,2078,5620,1160			

502. Trichloromethylsilane

CH_3SiCl_3 75-79-6 CH_3Cl_3Si

		cgs	SI	
mw		149.479	149.479	3468
bp	1 atm	66.12	339.27	3074,3075;3707,988,2583
				2624,6901
p	25	167.0	22.26	3707
eq 2.5	A	6.87213	5.99703	3707;5917,988
	B	1167	1167	
	C	226	226	
fp		-77.8	195.4	988
tp		-75.78	197.37	6369
d	20	1.2761	1276.1	3074,3075,2624;6901
	25	1.2672	1267.2	6901
dd/dt	25	0.00173	1.73	570
n_D	20	1.4123	1.4123	3074,3075,2624
	25	1.4085	1.4085	4026
γ	20	19.82	0.01982	6901;4379
	25	19.30	0.01930	
ΔH_v	bp	7.240	30.292	988
ΔH_m		2.138	8.945	6369
ΔH_f°	25(1)	-150.5	-629.7	2624;7664
	25(g)	-139.8	-584.9	3075,45
ΔH_c°	25(1)	-261.7	-1095.0	2624
C_p°	25	21.72	90.88	5114
C_p	25	38.98	163.09	6369
t_c		244.6	517.8	6996;6901
P_c		34.8	3.53	6996;6901
d_c		0.440	440	6996;6901
v_c		0.340	0.000340	6996;6901
μ	25 in 38	1.87	1.87	3864,4739;6415,4277
soly	in aq	ins	ins	7938
fl pt		8.3	281.5	4026

Beil	423,EIII 1896
uv	5867
ir	1483,1266,6810,7463
Raman	4150,6668,2887,6655,2221
ms	3428,6899,8278
nmr	1172,1160

503. Trichloroethylsilane

$C_2H_5SiCl_3$ 115-21-9 $C_2H_5Cl_3Si$

		cgs	SI	
mw		163.506	163.506	3468
bp	1 atm	99.1	372.3	6900;2623,984,3707,5336
p	25	47.18	6.290	6900
eq 2.5	A	6.92739	6.05229	6900;3707,984
	B	1305.166	1305.166	
	C	223.43	223.43	
fp		-105.6	167.6	984
d	20	1.2349	1234.9	2623
	24	1.2342	1234.2	1768
n_D	20	1.4270	1.4270	2623
ΔH_v	bp	7.70	32.2	984
ΔH_m		1.663	6.958	5378
ΔH_f°	25(g)	-132.8	-555.6	7663;2623
C_p	25	46.03	192.59	5378
t_c		286.8	560.0	6997
p_c		32.9	3.33	6997
d_c		0.406	406	6997
v_c		0.403	0.000403	6997
μ	25 in 38	2.04	2.04	1768;4739
fl pt	OC	22	295	6402

Beil	423,630
ir	6302
Raman	2888,5336,6397,6302
ms	6899
nmr	4775,6472

504. Tetraethylsilane

$(C_2H_5)_4Si$ 631–36–7 $C_8H_{20}Si$

		cgs	SI	
mw		144.332	144.332	3468
bp	1 atm	153.0	426.2	8029,7058;109,7183
p	23.9	5.0	0.67	7058
fp		−83.79	189.36	6978;7183
d	20	0.7662	766.2	7183,8029;7210
	25	0.7635	763.5	347
n_D	20	1.4268	1.4268	8029;7183,109,7210
	25	1.4249	1.4249	347
η	20	0.649	0.000649	8029
ΔH_v	bp	9.500	39.75	8029;3638
ΔH_m		3.110	13.01	6978
ΔH_f°	25(1)	−49	−205	7210,7209,7208;7183,7182
	25(g)	−38	−159	7210,7209;7183,7182,7663
ΔH_c°	25(1)	−1595	−6673	7210,7208;7183
C_p	−63.2	63.3	264.8	6978
ε	20	2.090	2.090	109
μ	20 in 38	0	0	4739;7050
soly	in aq,25	3.25 x 10^{-5}%w	3.25 x 10^{-5}%w	1936
Beil	420,625			
ir	7989,3916,4262,1483			
Raman	4855,2221,347,5336			
ms	1457			
nmr	4769,5620,4775			

Boiling Point Index

Boiling Point 1 atm	Compound No.	Compound
−85.7	294	1,1-Difluoroethylene
−81.44	363	Chlorotrifluoromethane
−47.6	284	1,1,1-Trifluoroethane
−42.070	1	Propane
−40.83	366	Chlorodifluoromethane
−39.11	371	Chloropentafluoroethane
−29.77	364	Dichlorodifluoromethane
−24.84	149	Methyl ether
−24.7	283	1,1-Difluoroethane
−13.80	313	Chloroethylene
−11.730	3	2-Methylpropane
−9.8	375	1-Chloro-1,1-difluoroethane
−6.33	398	Methylamine
−5.99	286	Octafluorocyclobutane
−2.19	287	Decafluorobutane
−0.50	2	Butane
2.87	427	Trimethylamine
3.55	338	Bromomethane
3.77	372	1,2-Dichlorotetrafluoroethane
6.88	418	Dimethylamine
8.90	367	Dichlorofluoromethane
9.500	7	2,2-Dimethylpropane
12.27	296	Chloroethane
15.80	349	Bromoethylene
19.0	379	1,1-Dichloro-2,2-difluoro-ethylene
20.4	182	Acetaldehyde
23.63	365	Trichlorofluoromethane
27.875	6	2-Methylbutane
29.962	57	1-Pentene
31.36	169	Furan
31.56	333	1,1-Dichloroethylene
31.75	224	Methyl formate
31.766	400	Isopropylamine
33.9	165	Propylene oxide
34.431	151	Ethyl ether
35.72	150	Ethylvinyl ether
35.74	298	2-Chloropropane
36.065	5	Pentane
36.344	60	trans-2-Pentene
36.7	58	2-Pentene
36.922	59	cis-2-Pentene
37.333	455	Methyl sulfide
38.35	339	Bromoethane
39.64	315	Dichloromethane
42.30	180	Dimethoxymethane
42.43	355	Iodomethane
44.040	404	tert-Butylamine
45.10	314	3-Chloropropene
46.225	451	Carbon disulfide
46.52	297	1-Chloropropane

Boiling Point Index (Continued)

Boiling Point 1 atm	Compound No.	Compound
47.26	376	1,2-Dibromotetrafluoroethane
47.633	373	1,1,2-Trichloro-1,2,2-trifluoro ethane
47.67	335	trans-1,2-Dichloroethylene
47.229	399	Propylamine
48.0	183	Propionaldehyde
48.4	350	2-Bromopropene
49.262	4	Cyclopentane
49.741	13	2,2-Dimethylbutane
50.7	302	2-Chloro-2-methylpropane
52.69	187	Acrolein
53.3	414	Allylamine
54.31	225	Ethyl formate
55.55	419	Diethylamine
56.067	189	Acetone
56.868	229	Methyl acetate
57	417	Ethylenimine
57.30	318	1,1-Dichloroethane
57.988	14	2,3-Dimethylbutane
59.41	341	2-Bromopropane
60.271	11	2-Methylpentane
60.63	334	cis-1,2-Dichloroethylene
61.178	316	Chloroform
62.73	403	sec-Butylamine
63.282	12	3-Methylpentane
63.424	167	1,2-Epoxybutane
63.478	61	1-Hexene
64.1	185	Isobutyraldehyde
64.546	78	Methanol
65.965	170	Tetrahydrofuran
67.568	402	Isobutylamine
68.06	369	Bromochloromethane
68.25	300	2-Chlorobutane
68.51	153	Isopropyl ether
68.736	10	Hexane
68.85	301	1-Chloro-2-methylpropane
70.97	340	1-Bromopropane
71.78	497	Trifluoroacetic acid
71.812	8	Methylcyclopentane
72.30	356	Iodoethane
72.5	230	Vinyl acetate
73.25	344	2-Bromo-2-methylpropane
74.05	475	2,2,2-Trifluoroethanol
74.083	321	1,1,1-Trichloroethane
74.8	184	Butyraldehyde
75.6	166	1,3-Dioxolane
76.638	317	Carbon tetrachloride
77.07	401	Butylamine
77.111	232	Ethyl acetate
77.3	396	Acrylonitrile
78.293	79	Ethanol

Boiling Point Index (Continued)

Boiling Point 1 atm	Compound No.	Compound
78.43	299	1-Chlorobutane
79.583	190	2-Butanone
79.9	171	2-Methyltetrahydrofuran
80.094	38	Benzene
80.2	253	Methyl acrylate
80.255	293	Hexafluorobenzene
80.500	20	2,4-Dimethylpentane
80.730	9	Cyclohexane
80.82	226	Propyl formate
81.60	385	Acetonitrile
82.242	81	2-Propanol
82.347	85	2-Methyl-2-propanol
82.979	71	Cyclohexene
83.483	319	1,2-Dichloroethane
83.57	421	Diisopropylamine
84.16	458	Thiophene
84.50	159	1,2-Dimethoxyethane
84.734	288	Fluorobenzene
86.558	424	Pyrrolidine
87.19	336	Trichloroethylene
88	173	Tetrahydropyran
88.601	236	Isopropyl acetate
88.87	428	Triethylamine
89.50	358	2-Iodopropane
89.784	19	2,3-Dimethylpentane
90.052	17	2-Methylhexane
90.08	152	Propyl ether
90.3	397	Methacrylonitrile
91.22	343	2-Bromobutane
91.850	18	3-Methylhexane
92.102	456	Ethyl sulfide
92.24	155	Butylethyl ether
92.8	374	1,1,2,2-Tetrachlorodifluoroethane
93.2	377	1,2-Dibromo-1,1-difluoroethane
93.639	62	1-Heptene
93.82	154	Butylvinyl ether
96.37	320	1,2-Dichloropropane
97.08	125	2-Propen-1-ol
97.151	80	1-Propanol
97.35	386	Propionitrile
98.07	228	Isobutyl formate
98.2	304	1-Chloro-3-methylbutane
98.424	16	Heptane
98.455	453	1-Butanethiol
99.10	247	Ethyl propionate
99.238	27	2,2,4-Trimethylpentane
99.5	254	Ethyl acrylate
99.512	83	2-Butanol
100.000	0	Water
100.3	255	Methyl methacrylate

Boiling Point Index   (Continued)

Boiling Point 1 atm	Compound No.	Compound
100.56	207	Formic acid
100.934	15	Methylcyclohexane
101.20	380	Nitromethane
101.320	172	p-Dioxane
101.536	235	Propyl acetate
101.60	342	1-Bromobutane
101.96	193	3-Pentanone
102.0	91	2-Methyl-2-butanol
102.050	292	α,α,α-Trifluorotoluene
102.262	192	2-Pentanone
102.45	357	1-Iodopropane
102.090	265	Propylene carbonate
103.5	234	Allyl acetate
103.6	181	Acetal
103.85	389	Isobutyronitrile
104.1	188	Crotonaldehyde
104.5	405	Pentylamine
106.1	227	Butyl formate
106.219	425	Piperidine
106.29	28	2,2,3,3-Tetramethylbutane
107.886	84	2-Methyl-1-propanol
108.39	303	1-Chloropentane
109.2	420	Dipropylamine
109.745	452	Dimethyl disulfide
109.844	26	2,2,3-Trimethylpentane
110.630	39	Toluene
111.5	92	3-Methyl-2-butanol
112.34	240	sec-Butyl acetate
112.56	460	2-Methylthiophene
113.1	93	2,2-Dimethyl-1-propanol
113.6	128	2-Propyn-1-ol
113.85	322	1,1,2-Trichloroethane
114.07	381	Nitroethane
114.40	289	o-Fluorotoluene
115.254	431	Pyridine
115.3	88	3-Pentanol
115.44	461	3-Methylthiophene
116.11	485	Epichlorohydrin
116.5	290	m-Fluorotoluene
116.6	291	p-Fluorotoluene
116.6	239	Isobutyl acetate
116.918	415	Ethylenediamine
117.4	196	4-Methyl-2-pentanone
117.62	388	Butyronitrile
117.725	82	1-Butanol
117.885	208	Acetic acid
119.0	87	2-Pentanol
120.25	383	2-Nitropropane
120.4	360	1-Iodo-2-methylpropane
120.9	459	Tetrahydrothiophene

Boiling Point Index (Continued)

Boiling Point 1 atm	Compound No.	Compound
121.07	337	Tetrachloroethylene
121.2	127	trans-2-Buten-1-ol
121.2	160	1,2-Diethoxyethane
121.286	63	1-Octene
121.4	97	2-Methyl-2-pentanol
121.49	205	4-Methyl-4-penten-2-one
121.5	233	Propargyl acetate
121.55	248	Ethyl butyrate
123.419	23	trans-1,2-Dimethylcyclohexane
123.6	126	cis-2-Buten-1-ol
124.093	30	2,2,5-Trimethylhexane
124.25	64	2-Octene (mixed isomers)
125.0	66	trans-2-Octene
125.25	199	2,4-Dimethyl-3-pentanone
125.64	65	cis-2-Octene
125.673	25	Octane
126.061	238	Butyl acetate
126.8	266	Ethyl carbonate
127	21	1,2-Dimethylcyclohexane (mixed isomers)
127.583	195	2-Hexanone
128.6	476	2-Chloroethanol
128.7	89	2-Methyl-1-butanol
128.94	487	Morpholine
129.408	432	2-Picoline
129.58	345	1-Bromopentane
129.728	22	cis-1,2-Dimethylcyclohexane
129.76	204	4-Methyl-3-penten-2-one
129.765	423	Pyrrole
130.5	90	3-Methyl-1-butanol
130.53	359	1-Iodobutane
130.7	191	Cyclopentanone
131.18	382	1-Nitropropane
131.36	352	1,2-Dibromoethane
131.687	307	Chlorobenzene
131.7	98	4-Methyl-2-pentanol
131.795	24	Ethylcyclohexane
132	133	trans-2-Butene-1,4-diol
133.96	406	Cyclohexylamine
134.7	250	Ethyl isovalerate
135.6	465	2-Ethoxyethanol
136.193	43	Ethylbenzene
137.983	86	1-Pentanol
138.3	206	2,4-Pentanedione
138.359	42	p-Xylene
139.120	41	m-Xylene
140.0	221	Acetic anhydride
140.29	156	Butyl ether
140.163	209	Propionic acid
141.2	217	Acrylic acid
141.3	390	Valeronitrile

Boiling Point Index   (Continued)

Boiling Point 1 atm	Compound No.	Compound
141.99	353	1,2-Dibromopropane
142.1	242	Isopentyl acetate
143.2	490	2-Methoxyethyl acetate
144.045	436	2,6-Lutidine
144.143	433	3-Picoline
144.429	40	o-Xylene
145.1	324	1,1,2,2-Tetrachloroethane
145.14	75	Styrene
145.356	434	4-Picoline
146.2	244	4-Methyl-2-pentyl acetate
146.5	99	2-Ethyl-1-butanol
146.883	67	1-Nonene
147.4	198	3-Heptanone
147.51	249	Isobutyl isobutyrate
147.93	96	2-Methyl-1-pentanol
148	446	N-Methylpropionamide
149.2	241	Pentyl acetate
149.21	351	Bromoform
150.818	29	Nonane
151.058	197	2-Heptanone
152.411	44	Isopropylbenzene
153.0	442	N,N-Dimethylformamide
153.60	175	Anisole
154	392	4-Methylvaleronitrile
154.5	488	Ethyl lactate
154.70	211	Isobutyric acid
155.65	194	Cyclohexanone
155.9	76	2-Pinene
155.908	347	Bromobenzene
156.1	491	2-Ethoxyethyl acetate
156.7	111	3-Heptanol
156.85	323	1,2,3-Trichloropropane
ca 157	100	1-Methylcyclohexanol
157.0	95	1-Hexanol
158.403	435	2,4-Lutidine
159.15	309	o-Chlorotoluene
159.6	422	Dibutylamine
159.7	110	2-Heptanol
159.76 d	162	Bis(2-methoxyethyl) ether
159.88	325	Pentachloroethane
160.5	219	Methacrylic acid
161.10	94	Cyclohexanol
161.8	483	2-Furaldehyde
161.99	311	p-Chlorotoluene
162.3	310	m-Chlorotoluene
163.463	391	Hexanenitrile
163.718	210	Butyric acid
164.743	45	Mesitylene
165	102	cis-2-Methylcyclohexanol
165	479	2-Amino-2-methyl-1-propanol

Boiling Point Index (Continued)

Boiling Point 1 atm	Compound No.	Compound
166.0	77	2(10)-Pinene
166.1	445	N,N-Dimethylacetamide
166.5	103	trans-2-Methylcyclohexanol
ca 167	106	trans-3-Methylcyclohexanol
167.6	101	2-Methylcyclohexanol (mixed isomers)
ca 168	105	cis-3-Methylcyclohexanol
168.1 d	474	4-Hydroxy-4-methyl-2-pentanone
168.24	201	2,6-Dimethyl-4-heptanone
169.0	222	Propionic anhydride
169.119	54	tert-Butylbenzene
169.138	454	Benzenethiol
169.2	407	2-Ethylhexylamine
169.84	176	Phenetole
170.0	467	Furfuryl alcohol
170.2	466	2-Butoxyethanol
170.5	243	Hexyl acetate
170.597	68	1-Decene
170.95	478	2-Aminoethanol
171	108	cis-4-Methylcyclohexanol
171	109	trans-4-Methylcyclohexanol
171.0	437	2,4,6-Trimethylpyridine
171.7	493	Methyl acetoacetate
172	104	3-Methylcyclohexanol (mixed isomers)
172	107	4-Methylcyclohexanol (mixed isomers)
172.759	53	Isobutylbenzene
173.0	200	2-Octanone
173.0	306	3-Chloromethylheptane
173.00	327	m-Dichlorobenzene
173.305	52	sec-Butylbenzene
173.4	158	Isopentyl ether
174.12	328	p-Dichlorobenzene
174.155	34	Decane
174.6	72	Dipentene
175.0	73	d-Limonene
175.2	447	1,1,3,3-Tetramethylurea
176.0	168	Cineole
176.50	213	Isovaleric acid
176.7	142	dl-2,3-Butanediol
177	498	Heptacosafluorotributylamine
177.10	55	p-Cymene
177.6-177.8	74	l-Limonene
178	468	Tetrahydrofurfural alcohol
178.75	484	Bis(2-chloroethyl) ether
178.75	186	Benzaldehyde
179-182	141	d-2,3-Butanediol
179.0	140	l-2,3-Butanediol
179.4	308	α-Chlorotoluene
179.8	113	2-Octanol
180-185	441	N-Methylformamide

Boiling Point Index　(Continued)

Boiling Point 1 atm	Compound No.	Compound
180.48	326	o-Dichlorobenzene
180.8	494	Ethyl acetoacetate
181.839	116	Phenol
182 d	361	Diiodomethane
182.0	139	2,3-Butanediol (mixed isomers)
182.3	143	meso-2,3-Butanediol
183.270	51	Butylbenzene
183.47	305	1-Chlorooctane
184.34	114	2-Ethyl-1-hexanol
184.40	408	Aniline
185.0	174	Benzylethyl ether
185.0	218	Crotonic acid
185.4	267	Ethyl oxalate
185.5	212	Valeric acid
186.8	157	Pentyl ether
187.273	33	trans-Decahydronaphthalene
187.6	135	1,2-Propanediol
188.33	362	Iodobenzene
188.9	163	Bis(2-ethoxyethyl) ether
188.91	457	Butyl sulfide
189.0	462	Dimethyl sulfoxide
190.9	231	Ethylene glycol diacetate
191.004	117	o-Cresol
191.10	395	Benzonitrile
191.7	31	Decahydronaphthalene (mixed isomers)
194.0	251	Isopentyl isovalerate
194.05	430	N,N-Dimethylaniline
194.1	471	2-(2-Methoxyethoxy) ethanol
195.156	112	1-Octanol
195.774	32	cis-Decahydronaphthalene
196.25	426	N-Methylaniline
196.7	473	Salicylaldehyde
197.2	276	Methyl phosphate
197.5	145	Hexylene glycol
197.54	134	1,2-Ethanediol
198.6	245	2-Ethylhexyl acetate
199.30	268	Ethyl malonate
199.5	223	Butyric anhydride
199.50	258	Methyl benzoate
199.6	330	2,4-Dichlorotoluene
200.4	269	Methyl maleate
200.40	409	o-Toluidine
200.55	411	p-Toluidine
201.030	122	2,6-Xylenol
201.940	119	p-Cresol
202	449	1-Methyl-2-pyrrolidinone
202.0	203	Acetophenone
202.0	472	2-(2-Ethoxyethoxy) ethanol
202.232	118	m-Cresol
203.3	161	1,2-Dibutoxyethane

Boiling Point Index   (Continued)

Boiling Point 1 atm	Compound No.	Compound
203.40	410	m-Toluidine
204	263	γ-Butyrolactone
205.09	495	Methyl cyanoacetate
205.2	393	Octanenitrile
205.45	115	Benzyl alcohol
205.02	214	Hexanoic acid
206	444	N-Methylacetamide
206.0	496	Ethyl cyanoacetate
206.25	179	Veratrole
206.9	416	Diethylenetriamine
207.42	202	Camphor
207.5	137	1,3-Butanediol
207.65	50	1,2,3,4-Tetrahydronaphthalene
208.84	499	o-Chloroaniline
208.92	331	3,4-Dichlorotoluene
210.5 d	440	Formamide
210.80	384	Nitrobenzene
210.931	120	2,4-Xylenol
211.132	121	2,5-Xylenol
211.5	412	2,4-Xylidene
212.40	259	Ethyl benzoate
213.401	69	1-Dodecene
213.5	332	α,α,α-Trichlorotoluene
214.0	329	α,α-Dichlorotoluene
214.0	429	Tributylamine
214.4	136	1,3-Propanediol
215.5	246	Benzyl acetate
216	277	Ethyl phosphate
216.323	36	Dodecane
217.4	492	2-(2-Ethoxyethoxy)ethyl acetate
217.5	129	dl-α-Terpineol
217.9	413	2,6-Xylidene
217.942	46	Naphthalene
220 d	477	2-Cyanoethanol
221.15	443	Acetamide
221.692	124	3,5-Xylenol
225.3	270	Ethyl maleate
226.947	123	3,4-Xylenol
228	138	1,4-Butanediol
231	273	Bis(2-ethylhexyl) phthalate
231.2	260	Propyl benzoate
232.837	70	1-Tridecene
233	500	Hexamethylphosphoric triamide
233.3	489	Methyl salicylate
233.5	275	Butyl borate
233.5	394	α-Tolunitrile
235	132	cis-2-Butene-1,4-diol
235.466	37	Tridecane
237.10	438	Quinoline
239.04	35	1,1'-Bicyclohexyl

Boiling Point Index (Continued)

Boiling Point 1 atm	Compound No.	Compound
239.9	215	Octanoic acid
240.12	56	Cyclohexylbenzene
240.6	346	1-Bromodecane
242.4	144	1,5-Pentanediol
243.24	439	Isoquinoline
243.5	354	1,1,2,2-Tetrabromoethane
244.2	146	2-Ethyl-1,3-hexanediol
244.685	47	1-Methylnaphthalene
245	448	2-Pyrrolidinone
248.2	264	Ethylene carbonate
245.69	469	Diethylene glycol
254.6	164	Bis(2-butoxyethyl) ether
255.6	216	Nonanoic acid
258.0	237	Triacetin
258.06	178	Phenyl ether
259.3	312	1-Chloronaphthalene
265(737 Torr)	486	o-Nitroanisole
265.6	49	1,6-Dimethylnaphthalene
266.3	48	1,2-Dimethylnaphthalene
267	387	Succinonitrile
268.39 d	480	Diethanolamine
271.0	262	Ethyl cinnamate
ca 280 d	271	Butyl maleate
280.66	348	1-Bromonaphthalene
282	482	2,2'-Thiodiethanol
287.3 d	463	Sulfolane
288.0	470	Triethylene glycol
288.30	177	Benzyl ether
289	278	Butyl phosphate
290.0	147	Glycerol
313.0	279	Tricresyl phosphate
323.24	261	Benzyl benzoate
335.39	481	Triethanolamine
340.0	272	Butyl phthalate
349	274	Butyl sebacate
ca 350	252	Butyl stearate
360.0 d	220	Oleic acid
384	273	Bis(2-ethylhexyl) phthalate

Freezing Point Index

Freezing Point	Compound No.	Compound
−189.0	363	Chlorotrifluoromethane
−165.219	57	1-Pentene
−159.900	6	2-Methylbutane
−159.600	3	2-Methylpropane
−158.2	364	Dichlorodifluoromethane
−157.42	366	Chlorodifluoromethane
−153.79	313	Chloroethylene
−153.670	11	2-Methylpentane
−151.390	59	cis-2-Pentene
−150	167	1,2-Epoxybutane
−142.434	8	Methylcyclopentane
−141.49	149	Methyl ether
−140.5 (active)	300	2-Chlorobutane
−140.244	60	trans-2-Pentene
−139.813	61	1-Hexene
−139.54	349	Bromoethylene
−138.350	2	Butane
−138	58	2-Pentene (mixed isomers)
−137.2	171	2-Methyltetrahydrofuran
−136.4	296	Chloroethane
−135	367	Dichlorofluoromethane
−135	306	3-Chloromethylheptane
−134.5	314	3-Chloropropene
−130.8	375	1-Chloro-1,1-difluoroethane
−130.3	301	1-Chloro-2-methylpropane
−129.730	5	Pentane
−129	125	2-Propen-1-ol
−128.538	14	2,3-Dimethylbutane
−128.19	287	Decafluorobutane
−126.593	15	Methylcyclohexane
−126.2	80	1-Propanol
−124.75	350	2-Bromopropene
−124 ± 2	31	Decahydronaphthalene (mixed isomers)
−123.3 (ms)	151	Ethyl ether
−123.2	152	Propyl ether
−123.1	299	1-Chlorobutane
−123	182	Acetaldehyde
−122.8	297	1-Chloropropane
−122.56	333	1,1-Dichloroethylene
−119.4	18	3-Methylhexane
−119.242	20	2,4-Dimethylpentane
−118.856	62	1-Heptene
−118.6	339	Bromoethane
−118.276	17	2-Methylhexane
−117.3	427	Trimethylamine
−117.2	90	3-Methyl-1-butanol
−117.18	298	2-Chloropropane
−117	283	1,1-Difluoroethane
−116.3 (st)	151	Ethyl ether
−115.8	150	Ethylvinyl ether

Freezing Point Index   (Continued)

Freezing Point	Compound No.	Compound
−115.67	453	1-Butanethiol
−115	379	1,1-Dichloro-2,2-difluoro-ethylene
−115	370	Bromochlorofluoromethane
−114.7	83	2-Butanol
−114.7	428	Triethylamine
−114.49	79	Ethanol
−114.4	99	2-Ethyl-1-butanol
−113.3 (racemic)	300	2-Chlorobutane
−112.65	343	2-Bromobutane
−112.4	342	1-Bromobutane
−112.27	26	2,2,3-Trimethylpentane
−111.9	388	Butyronitrile
−111.57	451	Carbon disulfide
−111.323	24	Ethylcyclohexane
−111.3	284	1,1,1-Trifluoroethane
−111.1	356	Iodoethane
−110.5	376	1,2-Dibromotetrafluoroethane
−110.48	365	Trichlorofluoromethane
−109.8 (ms)	340	1-Bromopropane
−108.39	170	Tetrahydrofuran
−108.1 (st)	340	1-Bromopropane
−108	84	2-Methyl-1-propanol
−107.388	27	2,2,4-Trimethylpentane
−106.8	64	2-Octene (mixed isomers)
−106	371	Chloropentafluoroethane
−105.780	30	2,2,5-Trimethylhexane
−105.15	180	Dimethoxymethane
−104.4	165	Propylene oxide
−104.4	304	1-Chloro-3-methylbutane
−103.99	382	1-Nitropropane
−103.93	456	Ethyl sulfide
−103.512	71	Cyclohexene
−103.0	359	1-Iodobutane
−102	97	2-Methyl-2-pentanol
−101.690	63	1-Octene
−101.3	357	1-Iodopropane
−100.44	320	1,2-Dichloropropane
−100.2	65	cis-2-Octene
−99.865	13	2,2-Dimethylbutane
−99.3	250	Ethyl isovalerate
−99.0	224	Methyl formate
−99.0	303	1-Chloropentane
−98.9	240	sec-Butyl acetate
−98.85	239	Isobutyl acetate
−98.44	126	cis-2-Buten-1-ol
−98.27	455	Methyl sulfide
−98.05	229	Methyl acetate
−98.0	248	Ethyl butyrate
−97.68	78	Methanol
−97.22	166	1,3-Dioxolane

Freezing Point Index (Continued)

Freezing Point	Compound No.	Compound
-96.96	318	1,1-Dichloroethane
-96.9	73	d-Limonene
-96.4	184	Butyraldehyde
-96.3	421	Diisopropylamine
-96.2	390	Valeronitrile
-96.16	459	Tetrahydrothiophene
-96.033	44	Isopropylbenzene
-95.8	228	Isobutyl formate
-95.322	10	Hexane
-95.2	156	Butyl ether
-95.2	400	Isopropylamine
-95.0	235	Propyl acetate
-94.991	39	Toluene
-94.975	43	Ethylbenzene
-94.92	315	Dichloromethane
-94.7	189	Acetone
-94.07	338	Bromomethane
-93.879	4	Cyclopentane
-93.46	398	Methylamine
-93	245	2-Ethylhexyl acetate
-92.9	226	Propyl formate
-92.8	230	Vinyl acetate
-92.78	386	Propionitrile
-92.53	372	1,2-Dichlorotetrafluoroethane
-92.19	418	Dimethylamine
-91.9	227	Butyl formate
-91.32	383	2-Nitropropane
-90.7	360	1-Iodo-2-methylpropane
-90.582	16	Heptane
<-90	465	2-Ethoxyethanol
-90	98	4-Methyl-2-pentanol
-90.0	358	2-Iodopropane
-89.52	381	Nitroethane
-89.44	126	cis-2-Buten-1-ol
-89.0	341	2-Bromopropane
-88.62	82	1-Butanol
-88.2	414	Allylamine
-88.194	23	trans-1,2-Dimethylcyclohexane
-88.01 (ms)	51	Butylbenzene
-88.0	81	2-Propanol
-87.95	369	Bromochloromethane
-87.9	345	1-Bromopentane
-87.85 (st)	51	Butylbenzene
-87.7	290	m-Fluorotoluene
-87.7	66	trans-2-Octene
-86.95	187	Acrolein
-86.69	190	2-Butanone
-86.4	336	Trichloroethylene
-85.61	169	Furan
-85.5	153	Isopropyl ether

Freezing Point Index (Continued)

Freezing Point	Compound No.	Compound
-85.1	464	2-Methoxyethanol
-85 (st)	471	2-(2-Methoxyethoxy) ethanol
-84.7	398	Methylamine
-84.6	402	Isopropylamine
-84.0	196	4-Methyl-2-pentanone
-83.55	232	Ethyl acetate
-83.55	396	Acrylonitrile
-83.0	399	Propylamine
-81.338	67	1-Nonene
-80.9	243	Hexyl acetate
-80.7	249	Isobutyl isobutyrate
-80.3	391	Hexanenitrile
<-80	278	Butyl phosphate
-80	183	Propionaldehyde
-80	493	Methyl acetoacetate
<-80.0	271	Butyl maleate
-80.0	334	cis-1,2-Dichloroethylene
-79.6	225	Ethyl formate
-78.5	242	Isopentyl acetate
-78.2	86	1-Pentanol
-78.0	417	Ethylenimine
-76.86	192	2-Pentanone
-76.5	188	Crotonaldehyde
-76	472	2-(2-Ethoxyethoxy) ethanol
-76	114	2-Ethyl-1-hexanol
-76 (supercooled)	471	2-(2-Methoxyethoxy) ethanol
-75.03	457	Butyl sulfide
-75	466	2-Butoxyethanol
-74.0	160	1,2-Diethoxyethane
-73.85	247	Ethyl propionate
-73.5	238	Butyl acetate
-73.4	236	Isopropyl acetate
-73.1	221	Acetic anhydride
-72.65	404	tert-Butylamine
-72.60	205	4-Methyl-4-penten-2-one
-71.5	389	Isobutyronitrile
-71.2	254	Ethyl acrylate
-70.8	241	Pentyl acetate
<-70	275	Butyl borate
<-70	89	2-Methyl-1-butanol
-70	111	3-Heptanol
-70.0	429	Tributylamine
-69.4	157	Pentyl ether
-69.1	161	1,2-Butoxyethane
-69.03	199	2,4-Dimethyl-3-pentanone
-69	159	1,2-Dimethoxyethane
-68.94	461	3-Methylthiophene
-67.935	55	p-Cymene
-67.5	476	2-Chloroethanol
-66.74	432	2-Picoline

Freezing Point Index (Continued)

Freezing Point	Compound No.	Compound
-66.45	355	Iodomethane
-66.307	68	1-Decene
-65.7	223	Butyric anhydride
-65.1	490	2-Methoxyethyl acetate
-65	185	Isobutyraldehyde
-64	162	Bis(2-methoxyethyl) ether
-64.0	76	2-Pinene
-63.96	435	2,4-Lutidine
-63.8	244	4-Methyl-2-pentyl acetate
-63.52	316	Chloroform
-63.386	460	2-Methylthiophene
-63	420	Dipropylamine
-62.5 (ms)	276	Methyl phosphate
-62	422	Dibutylamine
-62.0	289	o-Fluorotoluene
-61.7	491	2-Ethoxyethyl acetate
-61.54	77	2(10)-Pinene
-61.3	377	1,2-Dibromo-1,1-difluoroethane
-60.43	442	N,N-Dimethylformamide
-60.2	164	Bis(2-butoxyethyl) ether
-60	135	Propanediol
-58	159	1,2-Dimethoxyethane
-57.85	424	Pyrrolidine
-57.850	54	tert-Butylbenzene
-57.8	305	1-Chlorooctane
-57.2	485	Epichlorohydrin
-57	426	N-Methylaniline
-56.8	291	p-Fluorotoluene
-56.765	25	Octane
-55.85	195	2-Hexanone
-55.6	186	Benzaldehyde
-55.50	353	1,2-Dibromopropane
-54.53	265	Propylene carbonate
-53.489	29	Nonane
-52.85	204	4-Methyl-3-penten-2-one
-51.80	128	2-Propyn-1-ol
-51.6	260	Propyl benzoate
-51.5	246	Benzyl acetate
-51.48	53	Isobutylbenzene
-51.3	191	Cyclopentanone
-51.1	392	4-Methylvaleronitrile
-50.023	22	cis-1,2-Dimethylcyclohexane
<-50	137	1,3-Butanediol
-50	273	Bis(2-ethylhexyl) phthalate
-50 (gl)	145	Hexylene glycol
-49.8	335	trans-1,2-Dichloroethylene
-49.8	419	Diethylamine
-49.1	401	Butylamine
-48.9	268	Ethyl malonate
-48.2	255	Methyl methacrylate

Freezing Point Index (Continued)

Freezing Point	Compound No.	Compound
-47.872	41	m-Xylene
-46.8	484	Bis(2-chloroethyl) ether
-46.1 (st)	276	Methyl phosphate
-46.04	201	2,6-Dimethyl-4-heptanone
-46	477	2-Cyanoethanol
-45.97	211	Isobutyric acid
-45.6	393	Octanenitrile
-45.58	307	Chlorobenzene
ca -45	173	Tetrahydropyran
-44.720	45	Mesitylene
-44.6	95	1-Hexanol
-44.3	163	Bis(2-ethoxyethyl) ether
-44.192	437	2,4,6-Trimethylpyridine
-44 (enol)	494	Ethyl acetoacetate
-43.835	385	Acetonitrile
-43.8	324	1,1,2,2-Tetrachloroethane
-43.5	475	2,2,2-Trifluoroethanol
-43.37	263	γ-Butyrolactone
-43.2	405	Pentylamine
-43.01	32	cis-Decahydronaphthalene
-42.8	474	4-Hydroxy-4-methyl-2-pentanone
-43.0	222	Propionic anhydride
-43.0	266	Ethyl carbonate
-42.21	288	Fluorobenzene
-41.55	431	Pyridine
-41.5	231	Ethylene glycol diacetate
-40.6	267	Ethyl oxalate
<-40	72	Dipentene
-40 (gl)	146	2-Ethyl-1,3-hexanediol
-39.2	308	α-Chlorotoluene
-39 (keto)	494	Ethyl acetoacetate
-39	416	Diethylenetriamine
-39.0	198	3-Heptanone
-38.97	193	3-Pentanone
-38.7	286	Octafluorocyclobutane
-38.24	458	Thiophene
-37.5	175	Anisole
-37	237	Triacetin
-36.53	322	1,1,2-Trichloroethane
-36.5	483	2-Furaldehyde
-36.5	309	o-Chlorotoluene
-36.4	373	1,1,2-Trichloro-1,2,2-tri-fluoroethane
-35.8	397	Methacrylonitrile
-35.749	50	1,2,3,4-Tetrahydronaphthalene
-35.66	319	1,2-Dichloroethane
-35.209	69	1-Dodecene
-35	272	Butyl phthalate
-35.0	197	2-Heptanone
-34.7	259	Ethyl benzoate
-33.67	212	Valeric acid

Freezing Point Index (Continued)

Freezing Point	Compound No.	Compound
-33	279	Tricresyl phosphate (mixed isomers)
-32.1	194	Cyclohexanone
-32	113	2-Octanol
-31.35	362	Iodobenzene
-30.9	446	N-Methylpropionamide
-30.82	347	Bromobenzene
-30.628	75	Styrene
-30.45	47	1-Methylnaphthalene
-30.400	33	trans-Decahydronaphthalene
-30.40	410	m-Toluidine
-30.4	321	1,1,1-Trichloroethane
-29.644	34	Decane
-29.52	176	Phenetole
-29.3	213	Isovaleric acid
-29.2	346	1-Bromodecane
-29.0	325	Pentachloroethane
-29 (ms)	467	Furfuryl alcohol
-28.55	380	Nitromethane
-28.16	292	α,α,α-Trifluorotoluene
-26.7	136	1,3-Propanediol
-26.25	310	m-Chlorotoluene
-26 (melts)	186	Benzaldehyde
-26	488	Ethyl lactate
-25.4	302	2-Chloro-2-methylpropane
-25.182	40	o-Xylene
-25	492	2-(2-Ethoxyethoxy)ethyl acetate
-24.76	327	m-Dichlorobenzene
-24.5 (ms)	409	o-Toluidine
-24.4	449	1-Methyl-2-pyrrolidinone
-23.8	394	α-Tolunitrile
-23.4146	423	Pyrrole
-23.2	206	2,4-Pentanedione
-23.056	70	1-Tridecene
-22.82	317	Carbon tetrachloride
-22.5	496	Ethyl cyanoacetate
-22.35	337	Tetrachloroethylene
-20.7	209	Propionic acid
-20.29	200	2-Octanone
-20	445	N,N-Dimethylacetamide
-18.09	433	3-Picoline
-20 (gl)	148	1,2,6-Hexanetriol
-17.7	406	Cyclohexylamine
-17.5	269	Methyl maleate
-17.01	326	o-Dichlorobenzene
-16.9	49	1,6-Dimethylnaphthalene
-16.7	378	1-Bromo-2-chloroethane
-16.58	7	2,2-Dimethylpropane
-16.20	344	2-Bromo-2-methylpropane
-16.1	329	α,α-Dichlorotoluene
-16.10 (st)	409	o-Toluidine

Freezing Point Index   (Continued)

Freezing Point	Compound No.	Compound
-15.6	144	1,5-Pentanediol
-15.3	115	Benzyl alcohol
-15.25	331	3,4-Dichlorotoluene
-15.216	497	Trifluoroacetic acid
-14.97	112	1-Octanol
-14.94	454	Benzenethiol
-14.85	438	Quinoline
-14.7	323	1,2,3-Trichloropropane
-14.63 (st)	467	Furfuryl alcohol
-14.3	412	2,4-Xylidene
-13.9 (ms)	258	Methyl benzoate
-13.5	330	2,4-Dichlorotoluene
-13.07	495	Methyl cyanoacetate
-12.75	395	Benzonitrile
-12.6	134	1,2-Ethanediol
-12.10 (st)	258	Methyl benzoate
-11.92 (α)	499	o-Chloroaniline
-11	274	Butyl sebacate
-10.5	425	Piperidine
-10.2	482	2,2'-Thiodiethanol
-9.582	36	Dodecane
-8.8	91	2-Methyl-2-butanol
-8.8	270	Ethyl maleate
-8.6	489	Methyl salicylate
-7.8	469	Diethylene glycol
-7	473	Salicylaldehyde
-6.10	436	2,6-Lutidine
-5.98	408	Aniline
ca -5.5	105	cis-3-Methylcyclohexanol
-5.390	37	Tridecane
-5.2	210	Butyric acid
-4.8	487	Morpholine
-4.4	332	α,α,α-Trichlorotoluene
-4.3	470	Triethylene glycol
ca -4.0	103	trans-2-Methylcyclohexanol
-3.8	441	N-Methylformamide
-3.44	214	Hexanoic acid
-2.3	312	1-Chloronaphthalene
-1.94	499	o-Chloroaniline
-1.78 (β)	499	o-Chloroaniline
-1.2	447	1,1,3,3-Tetramethylurea
-1.0	48	1,2-Dimethylnaphthalene
-0.5	106	trans-3-Methylcyclohexanol
0.00	0	Water
0.0	354	1,1,2,2-Tetrabromoethane
ca 0.4 (ms)	348	1-Bromonaphthalene
1.3	168	Cineole
2.45	430	N,N-Dimethylaniline
2.55	440	Formamide
3.60	177	Benzyl ether

Freezing Point Index (Continued)

Freezing Point	Compound No.	Compound
3.63	35	1,1'-Bicyclohexyl
3.65	434	4-Picoline
5.10	293	Hexafluorobenzene
5.533	38	Benzene
5.60 (ms)	361	Diiodomethane
5.76	384	Nitrobenzene
6.1 (st)	361	Diiodomethane
6.20	348	1-Bromonaphthalene
6.72	9	Cyclohexane
ca 7.0	102	cis-2-Methylcyclohexanol
7.20	500	Hexamethylphosphoric triamide
7.30	311	p-Chlorotoluene
7.35	56	Cyclohexylbenzene
7.6	142	dl-2,3-Butanediol
8.05	351	Bromoform
8.27	207	Formic acid
9.79	352	1,2-Dibromoethane
10.45	486	o-Nitroanisole
10.53	478	2-Aminoethanol
11.0	132	cis-2-Butene-1,4-diol
11.2	413	2,6-Xylidene
11.3	415	Ethylenediamine
11.80	172	p-Dioxane
12 (trans)	262	Ethyl cinnamate
12.22	118	m-Cresol
13.263	42	p-Xylene
12.54	216	Nonanoic acid
13.38 (α)	220	Oleic acid
13.5	217	Acrylic acid
15.5	219	Methacrylic acid
16.30 (β)	220	Oleic acid
16.55	215	Octanoic acid
16.66	208	Acetic acid
18.18	147	Glycerol
18.54	462	Dimethyl sulfoxide
19	140	1-2,3-Butanediol
19.4	261	Benzyl benzoate
19.62	203	Acetophenone
19.65	138	1,4-Butanediol
19.9	256	Methyl oleate
20.9 (ms)	179	Veratrole
21.57	481	Triethanolamine
22.5	139	2,3-Butanediol (mixed isomers)
22.5 (st)	179	Veratrole
24.54	120	2,4-Xylenol
25	448	2-Pyrrolidinone
25.15	94	Cyclohexanol
25.62	85	2-Methyl-2-propanol
26	100	1-Methylcyclohexanol
26.3	252	Butyl stearate

Freezing Point Index (Continued)

Freezing Point	Compound No.	Compound
26.48	439	Isoquinoline
26.55	374	1,1,2,2-Tetrachlorodifluoro-ethane
26.87	178	Phenyl ether
27.3	133	trans-2-Butene-1,4-diol
27.95	480	Diethanolamine
28.45	463	Sulfolane
30-31	479	2-Amino-2-methyl-1-propanol
30.55	444	N-Methylacetamide
30.944	117	o-Cresol
34.4	143	meso-2,3-Butanediol
34.739	119	p-Cresol
35	129	dl-α-Terpineol
36.37	264	Ethylene carbonate
36.9	130	d-α-Terpineol
37.0	131	l-α-Terpineol
40.90	116	Phenol
43.75	411	p-Toluidine
45.62	122	2,6-Xylenol
52	93	2,2-Dimethyl-1-propanol
53.13	328	p-Dichlorobenzene
58.0805	387	Succinonitrile
63.24	124	3,5-Xylenol
65.11	123	3,4-Xylenol
69.207	450	ε-Caprolactam
71.4	218	Crotonic acid
74.85	121	2,5-Xylenol
77	282	Tri-p-cresyl phosphate
<80	468	Tetrahydrofurfuryl alcohol
80.00	443	Acetamide
80.290	46	Naphthalene
100.69	28	2,2,3,3-Tetramethylbutane
178.75	202	Camphor

Density Index at 25°

d	Compound No.	Compound
0.4927	1	Propane
0.55092	3	2-Methylpropane
0.57287	2	Butane
0.585	294	1,1-Difluoroethylene
0.5852	7	2,2-Dimethylpropane
0.6142	6	2-Methylbutane
0.62139	5	Pentane
0.6270	427	Trimethylamine
0.63533	57	1-Pentene
0.6431	60	trans-2-Pentene
0.64446	13	2,2-Dimethylbutane
0.64852	11	2-Methylpentane
0.6496	418	Dimethylamine
0.6504	59	cis-2-Pentene
0.6545 (20°)	58	2-Pentene (mixed isomers)
0.65484	10	Hexane
0.6562	398	Methylamine
0.65702	14	2,3-Dimethylbutane
0.65976	12	3-Methylpentane
0.6612	149	Methyl ether
0.66832	20	2,4-Dimethylpentane
0.66848	61	1-Hexene
0.67439	17	2-Methylhexane
0.67946	16	Heptane
0.6821	400	Isopropylamine
0.68295	18	3-Methylhexane
0.68781	27	2,2,4-Trimethylpentane
0.6908	404	tert-Butylamine
0.69091	19	2,3-Dimethylpentane
0.69267	62	1-Heptene
0.69862	25	Octane
0.7016	419	Diethylamine
0.70322	30	2,2,5-Trimethylhexane
0.70782	151	Ethyl ether
0.7100	421	Diisopropylamine
0.71085	63	1-Octene
0.71207	26	2,2,3-Trimethylpentane
0.7121	399	Propylamine
0.71375	29	Nonane
0.7154	64	2-Octene (mixed isomers)
0.7157	66	trans-2-Octene
0.71854	153	Isopropyl ether
0.7201	403	sec-Butylamine
0.7201	65	cis-2-Octene
0.72305	428	Triethylamine
0.72531	67	1-Nonene
0.72635	34	Decane
0.7297	402	Isobutylamine
0.7329	420	Dipropylamine
0.73683	401	Butylamine

Density Index at 25° (Continued)

d	Compound No.	Compound
0.73693	68	1-Decene
0.74045	4	Cyclopentane
0.7419	152	Propyl ether
0.74394	8	Methylcyclopentane
0.7448	155	Butylethyl ether
0.74518	36	Dodecane
0.7502	405	Pentylamine
0.75271	37	Tridecane
0.7531	150	Ethylvinyl ether
0.75474	69	1-Dodecene
0.7575	414	Allylamine
0.7577	422	Dibutylamine
0.76168	70	1-Tridecene
0.7641	156	Butyl ether
0.76506	15	Methylcyclohexane
0.7656	389	Isobutyronitrile
0.7713	158	Isopentyl ether
0.77204	23	trans-1,2-Dimethylcyclohexane
0.7727	154	Butylvinyl ether
0.77389	9	Cyclohexane
0.7743	429	Tributylamine
0.77649	386	Propionitrile
0.7766	385	Acetonitrile
0.7780 (20°)	182	Acetaldehyde
0.7790	157	Pentyl ether
0.7812	85	2-Methyl-2-propanol
0.78126	81	2-Propanol
0.7836	185	Isobutyraldehyde
0.78390	24	Ethylcyclohexane
0.78440	189	Acetone
0.78493	79	Ethanol
0.78637	78	Methanol
0.7880 (20°)	407	2-Ethylhexylamine
0.7893 (20°)	21	1,2-Dimethylcyclohexane (mixed isomers)
0.7912	183	Propionaldehyde
0.79222	22	cis-1,2-Dimethylcyclohexane
0.7948	397	Methacrylonitrile
0.7950	390	Valeronitrile
0.7963	196	4-Methyl-2-pentanone
0.7964	184	Butyraldehyde
0.7978	84	2-Methyl-1-propanol
0.79863	199	2,4-Dimethyl-3-pentanone
0.7993	392	4-Methylvaleronitrile
0.7997	190	2-Butanone
0.79960	80	1-Propanol
0.8004	396	Acrylonitrile
0.8012	391	Hexanenitrile
0.8015	192	2-Pentanone
0.8022	201	2,6-Dimethyl-4-heptanone
0.80241	83	2-Butanol

Density Index at 25° (Continued)

d	Compound No.	Compound
0.8036	98	4-Methyl-2-pentanol
0.8050	91	2-Methyl-2-butanol
0.8054	87	2-Pentanol
0.80575	82	1-Butanol
0.80609	71	Cyclohexene
0.8067	195	2-Hexanone
0.8071	90	3-Methyl-1-butanol
0.80945	193	3-Pentanone
0.8095	97	2-Methyl-2-pentanol
0.8097	393	Octanenitrile
0.81080	86	1-Pentanol
0.81123	197	2-Heptanone
0.8134	110	2-Heptanol
0.81344	198	3-Heptanone
0.8138	92	3-Methyl-2-butanol
0.8143	200	2-Octanone
0.8150	89	2-Methyl-1-butanol
0.81534	95	1-Hexanol
0.8160	88	3-Pentanol
0.8170	111	3-Heptanol
0.8171	113	2-Octanol
0.8206	96	2-Methyl-1-pentanol
0.82157	112	1-Octanol
0.8213	181	Acetal
0.8215 (s)	28	2,2,3,3-Tetramethylbutane
0.823	165	Propylene oxide
0.824	167	1,2-Epoxybutane
0.8290	114	2-Ethyl-1-hexanol
0.8295	99	2-Ethyl-1-butanol
0.83118	456	Ethyl sulfide
0.83189	161	1,2-Dibutoxyethane
0.832	417	Ethylenimine
0.83510	160	1,2-Diethoxyethane
0.8361	302	2-Chloro-2-methylpropane
0.83674	453	1-Butanethiol
0.8370	72	Dipentene
0.8383	73	d-Limonene
0.8384	74	l-Limonene
0.83849 (20°)	457	Butyl sulfide
0.8389 (20°)	187	Acrolein
0.8411 (20°)	205	4-Methyl-4-penten-2-one
0.84209 (30°)	125	2-Propen-1-ol
0.84228	455	Methyl sulfide
0.8454	127	trans-2-Buten-1-ol
0.84745 (30°)	180	Dimethoxymethane
0.84907	53	Isobutylbenzene
0.8516 (20°)	188	Crotonaldehyde
0.8523	244	4-Methyl-2-pentyl acetate
0.8533	55	p-Cymene
0.85366	275	Butyl borate

Density Index at 25° (Continued)

d	Compound No.	Compound
0.8539	76	2-Pinene
0.85380	424	Pyrrolidine
0.854	126	cis-2-Buten-1-ol
0.8540	252	Butyl stearate
0.8540 (20°)	171	2-Methyltetrahydrofuran
0.8541	251	Isopentyl isovalerate
0.8542 (20°)	249	Isobutyl isobutyrate
0.85482	204	4-Methyl-3-penten-2-one
0.85607	51	Butylbenzene
0.8563	298	2-Chloropropane
0.85661	42	p-Xylene
0.85664	425	Piperidine
0.85743	44	Isopropylbenzene
0.85782	406	Cyclohexylamine
0.85797	52	sec-Butylbenzene
0.86009	41	m-Xylene
0.86111	45	Mesitylene
0.86125	257	Butyl oleate
0.8621	159	1,2-Dimetoxyethane
0.86219	39	Toluene
0.86240	54	tert-Butylbenzene
0.86253	43	Ethylbenzene
0.86370	159	1,2-Dimethoxyethane
0.8652 (20°)	250	Ethyl isovalerate
0.86596	33	trans-Decahydronaphthalene
0.8664	242	Isopentyl acetate
0.8667	77	2(10)-Pinene
0.867	300	2-Chlorobutane, active
0.86766	300	2-Chlorobutane, racemic
0.8681	243	Hexyl acetate
0.8692	305	1-Chlorooctane
0.8694	240	sec-Butyl acetate
0.8700	304	1-Chloro-3-methylbutane
0.8702	236	Isopropyl acetate
0.8702	239	Isobutyl acetate
0.8702	256	Methyl oleate
0.8717	301	1-Chloro-2-methylpropane
0.8718	245	2-Ethylhexyl acetate
0.8719	241	Pentyl acetate
0.8732	228	Isobutyl formate
0.87360	38	Benzene
0.87394	248	Ethyl butyrate
0.87594	40	o-Xylene
0.87636	238	Butyl acetate
0.87700	303	1-Chloropentane
0.8772	173	Tetrahydropyran
0.8789	31	Decahydronaphthalene (mixed isomers)
0.88095	299	1-Chlorobutane
0.8817 (20°)	306	3-Chloromethylheptane
0.88249	35	1,1'-Bicyclohexyl

Density Index at 25° (Continued)

d	Compound No.	Compound
0.8830	297	1-Chloropropane
0.88303	235	Propyl acetate
0.8837	164	Bis(2-butoxyethyl) ether
0.8840	247	Ethyl propionate
0.8869	227	Butyl formate
0.8870	220	Oleic acid
0.88891	296	Chloroethane
0.8892 (20°)	170	Tetrahydrofuran
0.89288	32	cis-Decahydronaphthalene
0.8931	415	Ethylenediamine
0.89455	232	Ethyl acetate
0.896	283	1,1-Difluoroethane
0.89625	466	2-Butoxyethanol
0.8996	226	Propyl formate
0.90122	75	Styrene
0.9013	216	Nonanoic acid
0.9013	313	Chloroethylene
0.9063 (20°)	163	Bis(2-ethoxyethyl) ether
0.9066	215	Octanoic acid
0.9080	109	trans-4-Methylcyclohexanol
0.9103	437	2,4,6-Trimethylpyridine
0.9129(21.5°)	108	cis-4-Methylcyclohexanol
0.9153	225	Ethyl formate
0.9155 (20°)	105	cis-3-Methylcyclohexanol
0.9163 (20°)	104	3-Methylcyclohexanol (mixed isomers)
0.91708 (30°)	213	Isovaleric acid
0.91806	436	2,6-Lutidine
0.9181	145	Hexylene glycol
0.9183(22.5°)	107	4-Methylcyclohexanol (mixed isomers)
0.9192	168	Cineole
0.92058	103	trans-2-Methylcyclohexanol
0.921 (40°)	479	2-Amino-2-methyl-1-propanol
0.9214 (20°)	106	trans-3-Methylcyclohexanol
0.9215(24.7°)	101	2-Methylcyclohexanol (mixed isomers)
0.9224 (20°)	363	Chlorotrifluoromethane
0.9230	214	Hexanoic acid
0.9234	254	Ethyl acrylate
0.92454 (30°)	314	3-Chloropropene
0.9251(24.65)	100	1-Methylcyclohexanol
0.92520	465	2-Ethoxyethanol
0.9271	435	2,4-Lutidine
0.9279	229	Methyl acetate
0.93050	446	N-Methylpropionamide
0.9312	230	Vinyl acetate
0.93181	102	cis-2-Methylcyclohexanol
0.9337 (20°)	129	dl-α-Terpineol
0.9342	474	4-Hydroxy-4-methyl-2-pentanone
0.9345	212	Valeric acid
0.936337	445	N,N-Dimethylacetamide
0.9364	131	1-α-Terpineol

Density Index at 25° (Continued)

d	Compound No.	Compound
0.93781 (20°)	169	Furan
0.9382(28.1°)	274	Butyl sebacate
0.9384	162	Bis(2-methoxyethyl) ether
0.93874	56	Cyclohexylbenzene
0.93971 (20°)	146	2-Ethyl-1,3-hexanediol
0.93981	432	2-Picoline
0.94288	211	Isobutyric acid
0.94331 (20°)	255	Methyl methacrylate
0.94387	442	N,N-Dimethylformamide
0.9446	174	Benzylethyl ether
0.9450	128	2-Propyn-1-ol
0.9452 (20°)	194	Cyclohexanone
0.9475(11.5°)	130	d-α-Terpineol
0.94865 (20°)	191	Cyclopentanone
0.9492	416	Diethylenetriamine
0.9500 (30°)	444	N-Methylacetamide
0.95020	434	4-Picoline
0.95197	433	3-Picoline
0.95232	430	N,N-Dimethylaniline
0.9532	210	Butyric acid
0.9535	253	Methyl acrylate
0.96024	464	2-Methoxyethanol
0.96049	176	Phenetole
0.9619	447	1,1,3,3-Tetramethylurea
0.96199	223	Butyric anhydride
0.962(25.3°)	284	1,1,1-Trifluoroethane
0.96565	423	Pyrrole
0.96589 (45°)	411	p-Toluidine
0.9660	50	1,2,3,4-Tetrahydronaphthalene
0.9664	224	Methyl formate
0.96761	491	2-Ethoxyethyl acetate
0.9684	94	Cyclohexanol
0.96926	266	Ethyl carbonate
0.9721	206	2,4-Pentanedione
0.9723	412	2,4-Xylidene
0.973 (72°)	218	Crotonic acid
0.9760	278	Butyl phosphate
0.9782 (80°)	46	Naphthalene
0.97824	431	Pyridine
0.979	273	Bis(2-ethylhexyl) phthalate
0.98221	426	N-Methylaniline
0.98290	383	2-Nitropropane
0.9841	472	2-(2-Ethoxyethoxy) ethanol
0.9842 (20°)	413	2,6-Xylidene
0.98444	410	m-Toluidine
0.9858	144	1,5-Pentanediol
0.98669 (60°)	387	Succinonitrile
0.9869	140	1-2,3-Butanediol
0.9872	141	d-2,3-Butanediol
0.98808	209	Propionic acid

Density Index at 25° (Continued)

d	Compound No.	Compound
0.9892(91.1°)	443	Acetamide
0.98932	175	Anisole
0.9907	271	Butyl maleate
0.9920	202	Camphor
0.993 (20°)	142	dl-2,3-Butanediol
0.99379	459	Tetrahydrothiophene
0.9939	143	meso-2,3-Butanediol
0.99430	409	o-Toluidine
0.99547	487	Morpholine
0.99609	382	1-Nitropropane
0.99705	0	Water
0.9974 (20°)	290	m-Fluorotoluene
0.9975 (20°)	291	p-Fluorotoluene
0.9982 (20°)	233	Propargyl acetate
0.9984	49	1,6-Dimethylnaphthalene
0.9988	441	N-Methylformamide
1.0000	137	1,3-Butanediol
1.00033	490	2-Methoxyethyl acetate
1.0006	395	Benzonitrile
1.0033 (20°)	139	2,3-Butanediol (mixed isomers)
1.0041	289	o-Fluorotoluene
1.0046	492	2-(2-Ethoxyethoxy)ethyl acetate
1.0057	222	Propionic anhydride
1.0125	394	α-Tolunitrile
1.0127	478	2-Aminoethanol
1.01289	138	1,4-Butanediol
1.01314 (30°)	288	Fluorobenzene
1.01422	460	2-Methylthiophene
1.0143	48	1,2-Dimethylnaphthalene
1.0153 (20°)	219	Methacrylic acid
1.01601	120	2,4-Xylenol
1.01647	461	3-Methylthiophene
1.0167	471	2-(2-Methoxyethoxy) ethanol
1.01676	47	1-Methylnaphthalene
1.01750	408	Aniline
1.02 (77°)	450	ε-Caprolactam
1.02022	500	Hexamethylphosphoric triamide
1.02126	494	Ethyl acetoacetate
1.0232 (20°)	260	Propyl benzoate
1.02382	203	Acetophenone
1.0259	449	1-Methyl-2-pyrrolidinone
1.0272	488	Ethyl lactate
1.02797	172	p-Dioxane
1.03019	118	m-Cresol
1.0328	135	1,2-Propanediol
1.03718 (30°)	259	Ethyl benzoate
1.04013	186	Benzaldehyde
1.0404	477	2-Cyanoethanol
1.04127	115	Benzyl alcohol
1.0426	272	Butyl phthalate

Density Index at 25° (Continued)

d	Compound No.	Compound
1.0428 (20°)	177	Benzyl ether
1.04392	208	Acetic acid
1.04460 (30°)	268	Ethyl malonate
1.04464	381	Nitroethane
1.0460 (20°)	117	o-Cresol
1.0483 (24°)	468	Tetrahydrofurfuryl alcohol
1.0494	262	Ethyl cinnamate
1.050	136	1,3-Propanediol
1.0511 (20°)	217	Acrylic acid
1.0515	246	Benzyl acetate
1.0564	496	Ethyl cyanoacetate
1.05884	458	Thiophene
1.0625 (20°)	452	Dimethyl disulfide
1.0637	270	Ethyl maleate
1.0647 (20°)	166	1,3-Dioxolane
1.0656	346	1-Bromodecane
1.06611 (30°)	178	Phenyl ether
1.06687 (30°)	267	Ethyl oxalate
1.06826	277	Ethyl phosphate
1.0685 (20°)	133	trans-2-Butene-1,4-diol
1.0697 (20°)	311	p-Chlorotoluene
1.0724	493	Methyl acetoacetate
1.07273	454	Benzenethiol
1.0728	310	m-Chlorotoluene
1.0740 (20°)	132	cis-2-Butene-1,4-diol
1.07762	309	o-Chlorotoluene
1.07901 (30°)	258	Methyl benzoate
1.0819	179	Veratrole
1.08579 (30°)	438	Quinoline
1.08712	221	Acetic anhydride
1.09101 (30°)	439	Isoquinoline
1.0936	480	Diethanolamine
1.09537	462	Dimethyl sulfoxide
1.096 (30°)	375	1-Chloro-1,1-difluoroethane
1.09911	231	Ethylene glycol diacetate
1.1004 (20°)	308	α-Chlorotoluene
1.1009	307	Chlorobenzene
1.1030 (20°)	148	1,2,6-Hexanetriol
1.107	448	2-Pyrrolidinone
1.1100	134	1,2-Ethanediol
1.1121	261	Benzyl benzoate
1.115 (s)	124	3,5-Xylenol
1.1164 (20°)	469	Diethylene glycol
1.1195	470	Triethylene glycol
1.1196	481	Triethanolamine
1.1225	495	Methyl cyanoacetate
1.1254	263	γ-Butyrolactone
1.1285 (20°)	467	Furfuryl alcohol
1.12915	440	Formamide
1.13128	380	Nitromethane

Density Index at 25° (Continued)

d	Compound No.	Compound
1.132 (s)	116	Phenol
1.132 (s)	122	2,6-Xylenol
1.138 (s)	123	3,4-Xylenol
1.1462	269	Methyl maleate
1.14936	320	1,2-Dichloropropane
1.1525	473	Salicylaldehyde
1.154 (s)	119	p-Cresol
1.1545	483	2-Furaldehyde
1.1583 (20°)	237	Triacetin
1.159	279	Tricresyl phosphate (mixed isomers)
1.1679	318	1,1-Dichloroethane
1.17176	280	Tri-o-cresyl phosphate
1.17455	485	Epichlorohydrin
1.1747	333	1,1-Dichloroethylene
1.1782	489	Methyl salicylate
1.1793	482	2,2'-Thiodiethanol
1.18129	292	α,α,α-Trifluorotoluene
1.189 (s)	121	2,5-Xylenol
1.19382 (20°)	312	1-Chloronaphthalene
1.1942	366	Chlorodifluoromethane
1.1951	265	Propylene carbonate
1.19835	384	Nitrobenzene
1.2052	276	Methyl phosphate
1.20775	499	o-Chloroaniline
1.2119	345	1-Bromopentane
1.2130	484	Bis(2-chloroethyl) ether
1.2132	344	2-Bromo-2-methylpropane
1.21405	207	Formic acid
1.2408	486	o-Nitroanisole
1.24637	319	1,2-Dichloroethane
1.2472	329	α,α-Dichlorotoluene
1.24750(55°)	328	p-Dichlorobenzene
1.24751	331	3,4-Dichlorotoluene
1.2476	330	2,4-Dichlorotoluene
1.25354	343	2-Bromobutane
1.2547 (20°)	335	trans-1,2-Dichloroethylene
1.2555	451	Carbon disulfide
1.2559	147	Glycerol
1.2578 (0°)	285	1,1,1,2,2-Pentafluoropropane
1.2604 (30°)	463	Sulfolane
1.2687	342	1-Bromobutane
1.28280	327	m-Dichlorobenzene
1.2837 (20°)	334	cis-1,2-Dichloroethylene
1.291	371	Chloropentafluoroethane
1.30033	326	o-Dichlorobenzene
1.3060	341	2-Bromopropane
1.3113	364	Dichlorodifluoromethane
1.31678	315	Dichloromethane
1.3299	321	1,1,1-Trichloroethane
1.3383	264	Ethylene carbonate

Density Index at 25° (Continued)

d	Compound No.	Compound
1.3452	340	1-Bromopropane
1.3662	367	Dichlorofluoromethane
1.3736 (22°)	475	2,2,2-Trifluoroethanol
1.3741	332	α,α,α-Trichlorotoluene
1.3803	350	2-Bromopropene
1.3832	323	1,2,3-Trichloropropane
1.43213	322	1,1,2-Trichloroethane
1.4505	339	Bromoethane
1.4514 (30°)	336	Trichloroethylene
1.455	372	1,2-Dichlorotetrafluoroethane
1.4738	349	Bromoethylene
1.476	365	Trichlorofluoromethane
1.4785	348	1-Bromonaphthalene
1.4785	497	Trifluoroacetic acid
1.47970	316	Chloroform
1.48820	347	Bromobenzene
1.500	286	Octafluorocyclobutane
1.517 (20°)	287	Decafluorobutane
1.555(-20°)	379	1,1-Dichloro-2,2-difluoro-ethylene
1.56354	373	1,1,2-Trichloro-1,2,2-tri-fluoroethane
1.58436	317	Carbon tetrachloride
1.58666	324	1,1,2,2-Tetrachloroethane
1.5952	360	1-Iodo-2-methylpropane
1.6072	359	1-Iodobutane
1.60732	293	Hexafluorobenzene
1.61432	337	Tetrachloroethylene
1.64470	374	1,1,2,2-Tetrachlorodifluoro-ethane
1.6622	338	Bromomethane
1.6732	325	Pentachloroethane
1.6946	358	2-Iodopropane
1.7392 (20°)	378	1-Bromo-2-chloroethane
1.7394	357	1-Iodopropane
1.8229	362	Iodobenzene
1.8839	498	Heptacosafluorotributylamine
1.907	370	Bromochlorofluoromethane
1.92292	369	Bromochloromethane
1.92344	353	1,2-Dibromopropane
1.9244	356	Iodoethane
2.164	376	1,2-Dibromotetrafluoroethane
2.1687	352	1,2-Dibromoethane
2.209	377	1,2-Dibromo-1,1-difluoroethane
2.2650	355	Iodomethane
2.421 (20°)	368	Dibromofluoromethane
2.8779	351	Bromoform
2.9529	354	1,1,2,2-Tetrabromoethane
3.3079	361	Diiodomethane

Ebullioscopic Constant Index

K_B	Compound No.	Compound
0.515	0	Water
0.785	78	Methanol
1.160	79	Ethanol
1.30	385	Acetonitrile
1.59	80	1-Propanol
1.649	224	Methyl formate
1.71	189	Acetone
1.745	85	2-Methyl-2-propanol
1.824	151	Ethyl ether
1.85	455	Methyl sulfide
1.86	380	Nitromethane
1.87	386	Propionitrile
1.95	296	Chloroethane
2.061	229	Methyl acetate
2.080	225	Ethyl formate
2.125	180	Dimethoxymethane
2.166	84	2-Methyl-1-propanol
2.255	91	2-Methyl-2-butanol
2.28	190	2-Butanone
2.35	451	Carbon disulfide
2.4	207	Formic acid
2.53	38	Benzene
2.530	208	Acetic acid
2.53	339	Bromoethane
2.583	232	Ethyl acetate
2.65	90	3-Methyl-1-butanol
2.60	315	Dichloromethane
2.60	381	Nitroethane
2.710	431	Pyridine
2.75	9	Cyclohexane
2.84	425	Piperidine
2.86	132	cis-2-Butene-1,4-diol
3.13	318	1,1-Dichloroethane
3.22	408	Aniline
3.22	445	N,N-Dimethylacetamide
3.23	456	Ethyl sulfide
3.270	172	p-Dioxane
3.29	39	Toluene
3.43	16	Heptane
3.44	319	1,2-Dichloroethane
3.45	428	Triethylamine
3.51	209	Propionic acid
3.53	221	Acetic anhydride
3.60	116	Phenol
3.62	316	Chloroform
3.87	395	Benzonitrile
3.94	210	Butyric acid
4.14	411	p-Toluidine
4.15	307	Chlorobenzene
4.19	355	Iodomethane

Ebullioscopic Constant Index (Continued)

K_B	Compound No.	Compound
4.43	336	Trichloroethylene
4.502	175	Anisole
4.83	241	Pentyl acetate
4.83	242	Isopentyl acetate
4.88	317	Carbon tetrachloride
4.96	473	Salicylaldehyde
5.0	176	Phenetole
5.04	384	Nitrobenzene
5.16	356	Iodoethane
5.50	337	Tetrachloroethylene
5.52	55	p-Cymene
5.582	50	1,2,3,4-Tetrahydronaphthalene
5.611	202	Camphor
5.65	203	Acetophenone
5.75	373	1,1,2-Trichloro-1,2,2-trifluoro-ethane
5.762	31	Decahydronaphthalene (mixed isomers)
5.80	46	Naphthalene
5.84	438	Quinoline
6.26	347	Bromobenzene
6.52	147	Glycerol
6.608	352	1,2-Dibromoethane
7.01	264	Ethylene carbonate

Cryoscopic Constant Index, K_f

K_f	Compound No.	Compound
1.79	151	Ethyl ether
1.853	0	Water
1.95	438	Quinoline
2.40	189	Acetone
2.43	415	Ethylenediamine
2.77	207	Formic acid
3.36	432	2-Picoline
3.4	256	Methyl oleate
3.27-3.69	147	Glycerol
3.85	440	Formamide
3.90	208	Acetic acid
4.04	443	Acetamide
4.07	462	Dimethyl sulfoxide
4.3	42	p-Xylene
4.38	265	Propylene carbonate
4.46	445	N,N-Dimethylacetamide
4.63	172	p-Dioxane
4.75	431	Pyridine
4.88	433	3-Picoline
4.90	316	Chloroform
4.90	292	α,α,α-Trifluorotoluene
5.12	38	Benzene
5.13	434	4-Picoline
5.32	264	Ethylene carbonate
5.34	395	Benzonitrile
5.372	411	p-Toluidine
5.53	311	p-Chlorotoluene
5.60	117	o-Cresol
5.77	444	N-Methylacetamide
5.87	408	Aniline
6.25	497	Trifluoroacetic acid
6.27	177	Benzyl ether
6.32	436	2,6-Lutidine
6.38	179	Veratrole
6.7	168	Cineole
6.852	384	Nitrobenzene
6.94	46	Naphthalene
6.96	119	p-Cresol
7.15	176	Phenetole
7.30	450	ε-Caprolactam
7.40	116	Phenol
7.88	178	Phenyl ether
8.0	216	Nonanoic acid
8.28	85	2-Methyl-2-propanol
8.4	500	Hexamethylphosphoric triamide
10.4	91	2-Methyl-2-butanol
11.0	93	2,2-Dimethyl-1-propanol
12.5	352	1,2-Dibromoethane
13.7 (ms)	361	Diiodomethane
14.13	15	Methylcyclohexane

Cryoscopic Constant Index, K_f (Continued)

K_f	Compound No.	Compound
14.4	351	Bromoform
14.4 (st)	361	Diiodomethane
14.52	35	1,1'-Bicyclohexyl
16.50	76	2-Pinene
18.26	387	Succinonitrile
19.47	32	cis-Decahydronaphthalene
20.0	9	Cyclohexane
20.81	33	trans-Decahydronaphthalene
21.7	354	1,1,2,2-Tetrabromoethane
29.8	317	Carbon tetrachloride
37.7	202	Camphor
37.7	374	1,1,2,2-Tetrachlorodifluoro-ethane
39.3	94	Cyclohexanol
64.1	463	Sulfolane

Cryoscopic Constant Index, A

A	Compound No.	Compound
0.00228	4	Cyclopentane
0.002319	13	2,2-Dimethylbutane
0.0024	79	Ethanol
0.00249	404	tert-Butylamine
0.004196	9	Cyclohexane
0.00460	14	2,3-Dimethylbutane
0.00575	7	2,2-Dimethylpropane
0.00613	28	2,2,3,3-Tetramethylbutane
0.0077	91	2-Methyl-2-butanol
0.0090	85	2-Methyl-2-propanol
0.01097	119	p-Cresol
0.0110	35	1,1'-Bicyclohexyl
0.01109	458	Thiophene
0.0120	219	Methacrylic acid
0.0125	78	Methanol
0.0130	169	Furan
0.01340	118	m-Cresol
0.01376	71	Cyclohexene
0.01404	116	Phenol
0.01418	47	1-Methylnaphthalene
0.01525	423	Pyrrole
0.01528	38	Benzene
0.01655	450	ε-Caprolactam
0.01684	117	o-Cresol
0.01722	258	Methyl benzoate
0.01785	499	o-Chloroaniline
0.01827	46	Naphthalene
0.01850	124	3,5-Xylenol
0.01858	431	Pyridine
0.0190	81	2-Propanol
0.01932	331	3,4-Dichlorotoluene
0.01940	438	Quinoline
0.02039	347	Bromobenzene
0.02064	454	Benzenethiol
0.02083	396	Acrylonitrile
0.02120	338	Bromomethane
0.02154	32	cis-Decahydronaphthalene
0.02173	54	tert-Butylbenzene
0.02177	178	Phenyl ether
0.02188	437	2,4,6-Trimethylpyridine
0.02193	45	Mesitylene
0.02195	495	Methyl cyanoacetate
0.0221	424	Pyrrolidine
0.02215	326	o-Dichlorobenzene
0.02272.	353	1,2-Dibromopropane
0.02319	13	2,2-Dimethylbutane
0.02484	322	1,1,2-Trichloroethane
0.02497	212	Valeric acid
0.02509	42	p-Xylene
0.02515	39	Toluene

Cryoscopic Constant Index, A (Continued)

A	Compound No.	Compound
0.02549	288	Fluorobenzene
0.0259	460	2-Methylthiophene
0.02592	320	1,2-Dichloropropane
0.02657	50	1,2,3,4-Tetrahydronaphthalene
0.0266	30	2,2,5-Trimethylhexane
0.02660	40	o-Xylene
0.02742	41	m-Xylene
0.02744	432	2-Picoline
0.02759	55	p-Cymene
0.02794	400	Isopropylamine
0.028	459	Tetrahydrothiophene
0.0289	80	1-Propanol
0.02941	33	trans-Decahydronaphthalene
0.0295	56	Cyclohexylbenzene
0.02985	44	Isopropylbenzene
0.0303	52	sec-Butylbenzene
0.0304	461	3-Methylthiophene
0.0306	53	Isobutylbenzene
0.03084	2	Butane
0.0311	86	1-Pentanol
0.03140	455	Methyl sulfide
0.03217	343	2-Bromobutane
0.0331	82	1-Butanol
0.03361	415	Ethylenediamine
0.03425	149	Methyl ether
0.03453	349	Bromoethylene
0.03476	20	2,4-Dimethylpentane
0.03479	43	Ethylbenzene
0.03569	151	Ethyl ether
0.0361	95	1-Hexanol
0.03661	298	2-Chloropropane
0.03664	23	trans-1,2-Dimethylcyclohexane
0.0370	69	1-Dodecene
0.03719	399	Propylamine
0.0377	153	Isopropyl ether
0.03780	15	Methylcyclohexane
0.03780	170	Tetrahydrofuran
0.03828	24	Ethylcyclohexane
0.03857	29	Nonane
0.03931 (st)	51	Butylbenzene
0.03951 (ms)	51	Butylbenzene
0.04002	313	Chloroethylene
0.0401	26	2,2,3-Trimethylpentane
0.04026	27	2,2,4-Trimethylpentane
0.04234	3	2-Methylpropane
0.0439	70	1-Tridecene
0.04605	17	2-Methylhexane
0.04780	37	Tridecane
0.04830	6	2-Methylbutane
0.04877	8	Methylcyclopentane

Cryoscopic Constant Index, A (Continued)

A	Compound No.	Compound
0.04910	5	Pentane
0.04974	10	Hexane
0.05001	456	Ethyl sulfide
0.05065	16	Heptane
0.05073	453	1-Butanethiol
0.0509	61	1-Hexene
0.05247	12	3-Methylpentane
0.05287	11	2-Methylpentane
0.05328	25	Octane
0.05687	60	trans-2-Pentene
0.05770	59	cis-2-Pentene
0.05802	1	Propane
0.05826	34	Decane
0.0588	67	1-Nonene
0.0594	68	1-Decene
0.05996	57	1-Pentene
0.06208	36	Dodecane
0.06264	62	1-Heptene
0.06268	63	1-Octene
0.077	112	1-Octanol

Dielectric Constant Index

Temp.	ε	Compound No.	Compound
27.4	1.0018 (g)	371	Chloropentafluoroethane
25	1.2898	1	Propane
25	1.734	3	2-Methylpropane
25	1.776	2	Butane
21	1.8	267	Ethyl oxalate
20	1.801	7	2,2-Dimethylpropane
20	1.8275	6	2-Methylbutane
20	1.841	5	Pentane
20	1.87	240	sec-Butyl acetate
	1.873	13	2,2-Dimethylbutane
25	1.8799	10	Hexane
	1.881	11	2-Methylpentane
	1.890	14	2,3-Dimethylbutane
	1.895	12	3-Methylpentane
	1.9	498	Heptacosafluorotributylamine
20	1.914	20	2,4-Dimethylpentane
20	1.919	17	2-Methylhexane
20	1.9246	16	Heptane
20	1.927	18	3-Methylhexane
20	1.939	19	2,3-Dimethylpentane
20	1.940	27	2,2,4-Trimethylpentane
20	1.948	25	Octane
20	1.962	26	2,2,3-Trimethylpentane
20	1.96875	4	Cyclopentane
20	1.970	29	Nonane
20	1.989	34	Decane
30	2.002	36	Dodecane
25	2.002	66	trans-2-Octene
20	2.017	57	1-Pentene
20	2.020	15	Methylcyclohexane
20	2.02431	9	Cyclohexane
25	2.03	37	Tridecane
25	2.036	23	trans-1,2-Dimethylcyclohexane
20	2.051	61	1-Hexene
	2.054	24	Ethylcyclohexane
25	2.062	22	cis-1,2-Dimethylcyclohexane
25	2.062	65	cis-2-Octene
20	2.071	62	1-Heptene
20	2.084	63	1-Octene
29	2.13	364	Dichlorodifluoromethane
25	2.14	335	trans-1,2-Dichloroethylene
25	2.1542	31	Decahydronaphthalene (mixed isomers)
20	2.172	33	trans-Decahydronaphthalene
20	2.197	32	cis-Decahydronaphthalene
25	2.209	172	p-Dioxane
20	2.220	71	Cyclohexene
20	2.23790	317	Carbon tetrachloride
20	2.253	55	p-Cymene
30	2.2588	76	2-Pinene
25	2.26	372	1,2-Dichlorotetrafluoroethane

Dielectric Constant Index (Continued)

Temp.	ε	Compound No.	Compound
20	2.2699	42	p-Xylene
25	2.273	45	Mesitylene
25	2.27401	38	Benzene
25	2.280	337	Tetrachloroethylene
20	2.303	365	Trichlorofluoromethane
20	2.319	53	Isobutylbenzene
	2.32	363	Chlorotrifluoromethane
25	2.34	376	1,2-Dibromotetrafluoroethane
20	2.359	51	Butylbenzene
20	2.364	52	sec-Butylbenzene
20	2.366	54	tert-Butylbenzene
25	2.3738	74	1-Limonene
20	2.3742	41	m-Xylene
25	2.3746	73	d-Limonene
25	2.3807	39	Toluene
25	2.3810	72	Dipentene
20	2.3833	44	Isopropylbenzene
20	2.4042	43	Ethylbenzene
50	2.41	328	p-Dichlorobenzene
25	2.41	373	1,1,2-Trichloro-1,2,2-tri-fluoroethane
25	2.423	428	Triethylamine
20	2.4257	75	Styrene
80	2.43	227	Butyl formate
25	2.44	427	Trimethylamine
20	2.45	215	Octanoic acid
20	2.46	220	Oleic acid
25	2.4970	77	2(10)-Pinene
25	2.52	374	1,1,2,2-Tetrachlorodifluoro-ethane
80	2.54	46	Naphthalene
20	2.568	40	o-Xylene
25	2.583	49	1,6-Dimethylnaphthalene
25	2.613	48	1,2-Dimethylnaphthalene
71	2.63	214	Hexanoic acid
20	2.64	213	Isovaleric acid
20	2.643	451	Carbon disulfide
20	2.645	180	Dimethoxymethane
20	2.66	212	Valeric acid
20	2.68 (s)	178	Phenyl ether
25	2.705	458	Thiophene
40	2.73	211	Isobutyric acid
25	2.77	157	Pentyl ether
20	2.773	50	1,2,3,4-Tetrahydronaphthalene
20	2.82	158	Isopentyl ether
20	2.820	266	Ethyl carbonate
	2.9	255	Methyl methacrylate
20	2.915	47	1-Methylnaphthalene
25	2.942	169	Furan
20	2.97	210	Butyric acid
20	2.978	422	Dibutylamine
20	3.068	420	Dipropylamine

Dielectric Constant Index (Continued)

Temp.	ε	Compound No.	Compound
20	3.083	156	Butyl ether
30	3.111	252	Butyl stearate
20	3.211	256	Methyl oleate
26	3.39	152	Propyl ether
16	3.42	336	Trichloroethylene
40	3.435	209	Propionic acid
19	3.62	251	Isopentyl isovalerate
20	3.686 (1)	178	Phenyl ether
20	3.73	325	Pentachloroethane
25	3.80	181	Acetal
25	3.88	153	Isopropyl ether
20	3.894	419	Diethylamine
20	3.9	174	Benzylethyl ether
25	4.0	257	Butyl oleate
25	4.09	179	Veratrole
30	4.22	289	o-Fluorotoluene
20	4.22	176	Phenetole
25	4.33	175	Anisole
20	4.335	151	Ethyl ether
25	4.369	353	1,2-Dibromopropane
25	4.382	454	Benzenethiol
20	4.39	351	Bromoform
90	4.41	114	2-Ethyl-1-hexanol
21	4.43	402	Isobutylamine
25	4.44	346	1-Bromodecane
20	4.486	362	Iodobenzene
22	4.5	405	Pentylamine
30	4.540	274	Butyl sebacate
23.5	4.57	168	Cineole
30	4.63	242	Isopentyl acetate
18	4.71	250	Ethyl isovalerate
20	4.73	406	Cyclohexylamine
25	4.73	309	o-Chlorotoluene
20	4.75	241	Pentyl acetate
30	4.7503	352	1,2-Dibromoethane
20	4.806	316	Chloroform
25	4.83	348	1-Bromonaphthalene
20	4.88	401	Butylamine
20	4.9	412	2,4-Xylidene
20	4.9	261	Benzyl benzoate
20	4.91	430	N,N-Dimethylaniline
40	4.90	122	2,6-Xylenol
54	4.98	411	p-Toluidine
20	5.01	238	Butyl acetate
25	5.02	149	Methyl ether
25	5.04	312	1-Chloronaphthalene
25	5.04	327	m-Dichlorobenzene
25	5.05	305	1-Chlorooctane
23	5.08	399	Propylamine
21	5.1	246	Benzyl acetate

Dielectric Constant Index (Continued)

Temp.	ε	Compound No.	Compound
	5.10	160	1,2-Diethoxyethane
18	5.10	248	Ethyl butyrate
20	5.204	453	1-Butanethiol
30	5.23	486	o-Nitroanisole
20	5.29	239	Isobutyl acetate
20	5.3	273	Bis(2-ethylhexyl) phthalate
25	5.316	361	Diiodomethane
28	5.34	367	Dichlorofluoromethane
25	5.40	347	Bromobenzene
25	5.42	288	Fluorobenzene
30	5.42	290	m-Fluorotoluene
20	5.45	400	Isopropylamine
20	5.55	310	m-Chlorotoluene
25	5.61	173	Tetrahydropyran
25	5.621	307	Chlorobenzene
5	5.628	349	Bromoethylene
19	5.65	247	Ethyl propionate
	5.70	163	Bis(2-ethoxyethyl) ether
25	5.72	456	Ethyl sulfide
20	5.8	425	Piperidine
25	5.78	91	2-Methyl-2-butanol
30	5.86	291	p-Fluorotoluene
18	5.95	410	m-Toluidine
23.6	5.97	304	1-Chloro-3-methylbutane
20	6.002	235	Propyl acetate
25	6.02	232	Ethyl acetate
20	6.02	259	Ethyl benzoate
25	6.06	426	N-Methylaniline
20	6.08	311	p-Chlorotoluene
18	6.1	262	trans-Ethyl cinnamate
24	6.11	366	Chlorodifluoromethane
30	6.16	120	2,4-Xylenol
20	6.170	208	Acetic acid
90	6.19	99	2-Ethyl-1-butanol
20	6.2	455	Methyl sulfide
17.2	6.26	313	Chloroethylene
20	6.288	359	1-Iodobutane
40	6.31	117	o-Cresol
25	6.31	345	1-Bromopentane
18	6.34	409	o-Toluidine
65	6.36	121	2,5-Xylenol
19	6.41	228	Isobutyl formate
30	6.436	272	Butyl phthalate
20	6.466	360	1-Iodo-2-methylpropane
14	6.49	301	1-Chloro-2-methylpropane
20	6.59	258	Methyl benzoate
11	6.6	303	1-Chloropentane
25	6.68	229	Methyl acetate
25	6.7	279	Tricresyl phosphate (mixed isomers)
30	6.71	408	Aniline

Dielectric Constant Index (Continued)

Temp.	ε	Compound No.	Compounds
22	6.86	111	3-Heptanol
20	6.9	329	α,α-Dichlorotoluene
21	6.9	332	α,α,α-Trichlorotoluene
30	6.92	378	1-Bromo-2-chloroethane
25	6.97	171	2-Methyltetrahydrofuran
13	7.0	308	α-Chlorotoluene
22	7.0	354	1,1,2,2-Tetrabromoethane
	7.00	355	Iodomethane
20	7.00	357	1-Iodopropane
30	7.090	300	2-Chlorobutane
20	7.099	342	1-Bromobutane
25	7.16	225	Ethyl formate
19.8	7.252	321	1,1,1-Trichloroethane
25	7.20	159	1,2-Dimethoxyethane
20	7.29	322	1,1,2-Trichloroethane
20	7.33	436	2,6-Lutidine
25	7.36	280	Tri-o-cresyl phosphate
20	7.39	299	1-Chlorobutane
25	7.42	487	Morpholine
21	7.45	323	1,2,3-Trichloropropane
25	7.50	281	Tri-m-cresyl phosphate
30	7.567	491	2-Ethoxyethyl acetate
25	7.58	170	Tetrahydrofuran
20	7.7	297	1-Chloropropane
19	7.72	226	Propyl formate
20	7.82	356	Iodoethane
25	7.87	268	Ethyl malonate
20	8.05	106	trans-3-Methylcyclohexanol
25	8.09	340	1-Bromopropane
25	8.13	423	Pyrrole
20	8.173	113	2-Octanol
20	8.19	358	2-Iodopropane
20	8.2	314	3-Chloropropene
20	8.20	324	1,1,2,2-Tetrachloroethane
20	8.25	490	2-Methoxyethyl acetate
60	8.35	93	2,3-Dimethyl-1-propanol
20	8.5	224	Methyl formate
20	8.55	497	Trifluoroacetic acid
23	8.58	270	Ethyl maleate
25	8.64	343	2-Bromobutane
25	8.91	278	Butyl phosphate
26.1	8.925	320	1,2-Dichloropropane
25	8.93	315	Dichloromethane
25	8.95	438	Quinoline
25	8.970	331	3,4-Dichlorotoluene
60	9.02	123	3,4-Xylenol
25	9.03	292	α,α,α-Trifluorotoluene
50	9.06	124	3,5-Xylenol
25	9.20	334	cis-1,2-Dichloroethylene
22	9.21	110	2-Heptanol

Dielectric Constant Index   (Continued)

Temp.	ε	Compound No.	Compound
25	9.30	466	2-Butoxyethanol
	9.39	339	Bromoethane
25	9.4	398	Methylamine
30	9.41	489	Methyl salicylate
20	9.45	296	Chloroethane
25	9.46	341	2-Bromopropane
25	9.6	452	Dimethyl disulfide
20	9.60	435	2,4-Lutidine
60	9.78	116	Phenol
20	9.8	432	2-Picoline
20	9.82	298	2-Chloropropane
25	9.93	326	o-Dichlorobenzene
20	9.961	302	2-Chloro-2-methylpropane
-54	ca 10	231	Ethylene glycol diacetate
18	10.0	318	1,1-Dichloroethane
25	10.30	344	2-Bromo-2-methylpropane
20	10.34	112	1-Octanol
25	10.37	319	1,2-Dichloroethane
	10.39	200	2-Octanone
25	10.7	439	Isoquinoline
25	10.79	277	Ethyl phosphate
-20	10.91	338	Bromomethane
40	11.07	119	p-Cresol
20	11.35	202	Camphor
40	11.6	116	Phenol
30	11.916	115	Benzyl alcohol
20	11.98	197	2-Heptanone
20	12.3	104	3-Methylcyclohexanol (mixed isomers)
25	12.44	117	m-Cresol
25	12.47	85	2-Methyl-2-propanol
22	12.88	198	3-Heptanone
20	12.9	223	Butyric anhydride
25	12.9	415	Ethylenediamine
25	12.91	431	Pyridine
20	13.1	488	Ethyl lactate
20	13.11	196	4-Methyl-2-pentanone
25	13.3	95	1-Hexanol
20	13.3	101	2-Methylcyclohexanol (mixed isomers)
20	13.3	107	4-Methylcyclohexanol (mixed isomers)
25	13.35	88	3-Pentanol
26	13.4	184	Butyraldehyde
	13.4	499	o-Chloroaniline
20	13.60	191	Cyclopentanone
23	13.61	468	Tetrahydrofurfuryl alcohol
22	13.71	87	2-Pentanol
25	13.897	393	Octanenitrile
25	13.9	86	1-Pentanol
20	13.9	473	Salicylaldehyde
20	14.56	195	2-Hexanone

Dielectric Constant Index (Continued)

Temp.	ε	Compound No.	Compound
25	15.0	94	Cyclohexanol
25	15.19	90	3-Methyl-1-butanol
20	15.38	192	2-Pentanone
22	15.5	392	4-Methylvaleronitrile
0	15.6	204	4-Methyl-3-penten-2-one
22	15.7	494	Ethyl acetoacetate
20	16.10	194	Cyclohexanone
25	16.39	276	Methyl phosphate
20	16.47	105	cis-3-Methylcyclohexanol
25	16.56	83	2-Butanol
25	16.93	464	2-Methoxyethanol
20	17.00	193	3-Pentanone
25	17.263	391	Hexanenitrile
25	17.39	203	Acetophenone
25	17.51	82	1-Butanol
20	17.8	186	Benzaldehyde
25	17.93	84	2-Methyl-1-propanol
25	18.2	474	4-Hydroxy-4-methyl-2-pentanone
16	18.3	222	Propionic anhydride
25	18.3	417	Ethylenimine
17	18.5	183	Propionaldehyde
20	18.51	190	2-Butanone
27	18.7	394	α-Tolunitrile
20	18.73	146	2-Ethyl-1,3-hexanediol
25	19.709	390	Valeronitrile
25	19.92	81	2-Propanol
24	20.4	389	Isobutyronitrile
25	20.45	80	1-Propanol
25	20.56	189	Acetone
19	20.7	221	Acetic anhydride
21	21.1	182	Acetaldehyde
20	21.2	484	Bis(2-chloroethyl) ether
15	21.6	125	2-Propen-1-ol
22	22.6	485	Epichlorohydrin
30	23.24	382	1-Nitropropane
	23.60	447	1,1,3,3-Tetramethylurea
20	23.69	470	Triethylene glycol
25	24.5	128	2-Propyn-1-ol
25	24.55	79	Ethanol
20	24.83	388	Butyronitrile
25	25.20	395	Benzonitrile
30	25.52	383	2-Nitropropane
20	25.7	206	2,4-Pentanedione
25	25.8	476	2-Chloroethanol
20	25.86	145	Hexylene glycol
18	26.7	496	Ethyl cyanoacetate
30	28.06	381	Nitroethane
20	28.86	386	Propionitrile
20	29.30	495	Methyl cyanoacetate
20	29.30	500	Hexamethylphosphoric triamide
25	29.36	481	Triethanolamine

Dielectric Constant Index (Continued)

Temp.	ε	Compound No.	Compound
30	30.2	138	1,4-Butanediol
20	31.69	469	Diethylene glycol
20	32.0	135	1,2-Propanediol
25	32.2	449	1-Methyl-2-pyrrolidinone
25	32.66	78	Methanol
20	33.0	396	Acrylonitrile
25	34.78	384	Nitrobenzene
20	35.0	136	1,3-Propanediol
30	35.87	380	Nitromethane
25	35.94	385	Acetonitrile
25	36.71	442	N,N-Dimethylformamide
25	37.7	134	1,2-Ethanediol
25	37.72	478	2-Aminoethanol
25	37.78	445	N,N-Dimethylacetamide
25	38	483	2-Furaldehyde
20	39	263	γ-Butyrolactone
25	42.5	147	Glycerol
30	43.26	463	Sulfolane
25	46.45	462	Dimethyl sulfoxide
-40	52.1	442	N,N-Dimethylformamide
57.4	56.5	387	Succinonitrile
30	58.5	404	tert-Butylamine
16	58.5	207	Formic acid
83	59	443	Acetamide
25	64.92	265	Propylene carbonate
25	78.39	0	Water
40	89.78	264	Ethylene carbonate
20	111.0	440	Formamide
25	175.7	446	N-Methylpropionamide
25	182.4	441	N-Methylformamide
32	191.3	444	N-Methylacetamide
-40	308	441	N-Methylformamide
-40	384	446	N-Methylpropionamide

Dipole Moment Index

"Conditions" refers to the state of the compound for which the dipole moment was determined; e.g., (g) or (l)—purified compound in gaseous or liquid state, respectively; "in 38" signifies a solution in solvent corresponding to the number given; in the case cited the solvent is benzene.

μ	Temp.	Conditions	Compound No.	Compound
0.00	25		4	Cyclopentane
0.00	25		5	Pentane
0.0	25		16	Heptane
0.0	25	in 9	293	Hexafluorobenzene
0		(g)	2	Butane
0	25		15	Methylcyclohexane
0	25	(1)	17	2-Methylhexane
0	25	(1)	18	3-Methylhexane
0	25	(1)	19	2,3-Dimethylpentane
0	25	(1)	20	2,4-Dimethylpentane
0			21	1,2-Dimethlcyclohexane (mixed isomers)
0	100	(g)	24	Ethylcyclohexane
0	25	(1)	25	Octane
0	25	(1)	27	2,2,4-Trimethylpentane
0	25	(1)	31	Decahydronaphthalene (mixed isomers)
0			286	Octafluorocyclobutane
0	20–60	(1)	38	Benzene
0	20–30	(1)	45	Mesitylene
0	25	in 38	46	Naphthalene
0	20–30	(1)	55	p-Cymene
0	25	(1)	317	Carbon tetrachloride
0	24	in 38	328	p-Dichlorobenzene
0	25	in 317	337	Tetrachloroethylene
ca0	20–100	in 38	32	cis-Decahydronaphthalene
ca0	20–100	in 38	33	trans-Decahydronaphthalene
ca0	–30–170	(1)	34	Decane
ca0	–10–210	(1)	36	Dodecane
0.02	20–60	(1)	42	p-Xylene
0.06	25	in 38	451	Carbon disulfide
0.083		(g)	1	Propane
0.085	25	(1)	10	Hexane
0.13	20–60	(1)	75	Styrene
0.132		(g)	3	2-Methylpropane
0.28	20	(1)	71	Cyclohexene
0.30	20–60	(1)	41	m-Xylene
0.31	20–60	(1)	39	Toluene
0.32			49	1,6-Dimethylnaphthalene
0.33	25	in 38	293	Hexafluorobenzene
0.34	20	(1)	57	1-Pentene
0.34	20	(1)	61	1-Hexene
0.34	20	(1)	62	1-Heptene
0.34	20	(1)	63	1-Octene
0.36	20–30	(1)	51	Butylbenzene
0.36	20–30	(1)	54	tert-Butylbenzene
0.36	30	in 38	76	2-Pinene
0.37	25	in 38	47	1-Methylnaphthalene
0.37	20–60	(1)	43	Ethylbenzene
0.37	20–30	(1)	52	sec-Butylbenzene
0.39	20–60	(1)	44	Isopropylbenzene
<0.4	25		35	1,1'-Bicyclohexyl

Dipole Moment Index (Continued)

μ	Temp.	Conditions	Compound No.	Compound
0.41	25	(1)	69	1-Dodecene
0.42	20	in 38	68	1-Decene
0.43	25	in 38	75	Styrene
0.44			53	Isobutylbenzene
0.45	20-60	(1)	40	o-Xylene
0.45			172	p-Dioxane
0.46		(g)	365	Trichlorofluoromethane
0.50		(g)	379	1,1-Dichloro-2,2-difluoro-ethylene
0.50		(g)	363	Chlorotrifluoromethane
0.52		(g)	371	Chloropentafluoroethane
0.52	25	in 38	458	Thiophene
0.533		(g)	372	1,2-Dichlorotetrafluoroethane
0.55	29	(g)	364	Dichlorodifluoromethane
0.59	20	in 38	67	1-Nonene
0.60	20-30	(1)	50	1,2,3,4-Tetrahydronaphthalene
0.61	25	in 38	73	d-Limonene
0.63	25	in 38	213	Isovaleric acid
0.68			48	1,2-Dimethylnaphthalene
0.70	25	in 38	335	trans-1,2-Dichloroethylene
0.71	25	in 38	169	Furan
0.72	25	(1)	427	Trimethylamine
0.74	34	(g)	180	Dimethoxymethane
0.77	25	in 317	275	Butyl borate
0.78	25	in 38	429	Tributylamine
0.792	20		216	Nonanoic acid
0.8		in 317	336	Trichloroethylene
0.87	25	in 38	427	Trimethylamine
0.87	25	in 38	428	Triethylamine
0.90	25	in 38	266	Ethyl carbonate
0.94	25-85	(1)	325	Pentachloroethane
0.98	25	in 38	422	Dibutylamine
0.99	10-70	in 38	351	Bromoform
1.03	20	in 38	420	Dipropylamine
1.08	25	(1)	211	Isobutyric acid
1.08	25-50	in 38	361	Diiodomethane
1.09	25	(1)	398	Methylamine
1.09	66	(g)	378	1-Bromo-2-chloroethane
1.11	25	(1)	419	Diethylamine
1.13	25	(1)	214	Hexanoic acid
1.13		(g)	353	1,2-Dibromopropane
1.14		(1)	315	Dichloromethane
1.14	25	(1)	418	Dimethylamine
1.15	20		151	Ethyl ether
1.15	25	(1)	215	Octanoic acid
1.15	25	in 38	421	Diisopropylamine
1.15	25	in 38	316	Chloroform
1.16	25	in 38	178	Phenyl ether
1.18	20		156	Butyl ether
1.18	25	in 38	418	Dimethylamine
1.18	20	in 317	220	Oleic acid

Dipole Moment Index   (Continued)

μ	Temp.	Conditions	Compound No.	Compound
1.19	30	in 16	378	1-Bromo-2-chloroethane
1.19	25	in 38	425	Piperidine
1.19	25	in 38	419	Diethylamine
1.20	25	in 317	157	Pentyl ether
1.19	20	in 38	352	1,2-Dibromoethane
1.22	25	in 9	246	Benzyl acetate
1.22	20		153	Propyl ether
1.23			158	Isopentyl ether
1.23	25	in 38	454	Benzenethiol
1.24			155	Butylethyl ether
1.245	25	in 317	175	Anisole
1.25	25	in 38	154	Butylvinyl ether
1.26	20	in 38	150	Ethylvinyl ether
1.26	30	(1)	289	o-Fluorotoluene
1.26	20	in 38	406	Cyclohexylamine
1.27	25	in 38	402	Isobutylamine
1.28		(g)	149	Methyl ether
1.28	25	in 38	403	sec-Butylamine
1.29	25	(1)	348	1-Bromonaphthalene
1.28	20	in 38	333	1,1-Dichloroethylene
1.29	25	in 38	404	tert-Butylamine
1.29	25	in 38	179	Veratrole
1.3		in 172	313	Chloroethylene
1.31	25	in 38	414	Allylamine
1.32	25	in 317	152	Propyl ether
1.33	25	(1)	312	1-Chloronaphthalene
1.33	20	in 38	399	Propylamine
1.34	30	(g)	367	Dichlorofluoromethane
1.35	25	in 38	149	Methyl ether
1.36	20	in 38	349	Bromoethylene
1.36	25	in 317	176	Phenetole
1.37	20	in 38	401	Butylamine
1.38	20	in 9	354	1,1,2,2-Tetrabromoethane
1.38		in 38	181	Acetal
1.38	25	in 38	153	Isopropyl ether
1.40	20	in 9	120	2,4-Xylenol
1.40	20	in 9	122	2,6-Xylenol
1.40	25	in 38	362	Iodobenzene
1.40	25	in 38	412	2,4-Xylidene
1.41		(g)	322	1,1,2-Trichloroethane
1.41	25	(g)	366	Chlorodifluoromethane
1.43	20	in 38	309	2-Chlorotoluene
1.45	20	in 38	121	2,5-Xylenol
1.45	40	(1)	122	2,6-Xylenol
1.45	30	in 38	116	Phenol
1.45	25	in 38	400	Isopropylamine
1.45	25	in 38	117	o-Cresol
1.45	25	in 38	410	m-Toluidine
1.45	25	in 38	455	Methyl sulfide
1.47	25	in 38	398	Methyamine

Dipole Moment Index (Continued)

μ	Temp.	Conditions	Compound No.	Compound
1.47		in 38	166	1,3-Dioxolane
1.48	25	in 38	288	Fluorobenzene
1.48	20	in 38	119	p-Cresol
1.48	25	in 317	355	Iodomethane
1.48	20	in 38	118	m-Cresol
1.51	25	in 38	408	Aniline
1.51	25	in 38	350	2-Bromopropene
1.52	25	in 38	411	p-Toluidine
1.53	25	in 38	453	1-Butanethiol
1.54	20	in 38	327	m-Dichlorobenzene
1.55	20	in 38	95	1-Hexanol
1.55	25	in 38	347	Bromobenzene
1.55		in 38	322	1,1,2-Trichloroethane
1.55	20	in 38	124	3,5-Xylenol
1.56	20	in 38	123	3,4-Xylenol
1.57	20	in 38	424	Pyrrolidine
1.59	30	in 38	116	Phenol
1.60	25	in 38	409	o-Toluidine
1.60			471	2-(2-Methoxyethoxy) ethanol
1.60			472	2-(2-Ethoxyethoxy) ethanol
1.61	25	in 38	457	Butyl sulfide
1.61	20	in 172	212	Valeric acid
1.61	25	in 38	456	Ethyl sulfide
1.62	25	in 38	307	Chlorobenzene
1.63	25	in 38	173	Tetrahydropyran
1.63			125	2-Propen-1-ol
1.63	25	in 38	413	2,6-Xylidene
1.64	30		117	o-Cresol
1.64	22	in 38	88	3-Pentanol
1.65	30	in 38	210	Butyric acid
1.65	0	in 38	219	Methacrylic acid
1.66	20	(1)	79	Ethanol
1.66	30	in 38	81	2-Propanol
1.66	30	in 38	85	2-Methyl-2-propanol
1.66	25	in 38	436	2,6-Lutidine
1.66	25	in 38	369	Bromochloromethane
1.66	70	(1)	121	2,5-Xylenol
1.66	22	in 38	87	2-Pentanol
1.66			115	Benzyl alcohol
1.66	30	(1)	290	m-Fluorotoluene
1.675	25	in 38	255	Methyl methacrylate
1.68	30	in 38	208	Acetic acid
1.68	30	in 38	209	Propionic acid
1.7	25	in 317	86	1-Pentanol
1.7			98	4-Methyl-2-pentanol
1.7			91	2-Methyl-2-butanol
1.70	20	(1)	120	2,4-Xylenol
1.70	25	in 317	330	2,4-Dichlorotoluene
1.70		in 38	321	1,1,1-Trichloroethane
1.707	20		113	2-Octanol

Dipole Moment Index   (Continued)

μ	Temp.	Conditions	Compound No.	Compound
1.71	25	in 38	159	1,2-Dimethoxyethane
1.71	22	in 38	110	2-Heptanol
1.71	22	in 38	111	3-Heptanol
1.71	25	in 38	324	1,1,2,2-Tetrachloroethane
1.72	28	in 38	229	Methyl acetate
1.73	25	in 317	338	Bromomethane
1.74	25	in 38	114	2-Ethyl-1-hexanol
1.74	22	in 38	247	Ethyl propionate
1.74	22	in 38	248	Ethyl butyrate
ca1.75			106	trans-3-Methylcyclohexanol
1.75	25	in 317	82	1-Butanol
1.75	25	in 38	170	Tetrahydrofuran
1.76	25	in 317	112	1-Octanol
1.76	30	(1)	291	p-Fluorotoluene
1.77		(g)	224	Methyl formate
1.77	15.4	in 38	262	cis-Ethyl cinnamate
1.77	25	in 38	417	Ethylenimine
1.77	25	in 38	499	o-Chloroaniline
1.77		in 38	125	2-Propen-1-ol
1.77	25	in 38	253	Methyl acrylate
1.78	25	in 38	128	2-Propyn-1-ol
1.78	22	in 38	235	Propyl acetate
1.78	20	in 38	356	Iodoethane
1.79			84	2-Methyl-1-propanol
1.79	25	in 38	230	Vinyl acetate
1.797		(g)	338	Bromomethane
1.8			245	2-Ethylhexyl acetate
1.8		in 317	485	Epichlorohydrin
1.8		(1)	492	2-(2-Ethoxyethoxy)ethyl acetate
1.80	25	in 38	246	Benzyl acetate
1.80	20	in 38	334	cis-1,2-Dichloroethylene
1.80	25	in 38	423	Pyrrole
1.82			90	3-Methyl-1-butanol
1.82	30	in 38	207	Formic acid
1.82	25	in 38	232	Ethyl acetate
1.82	20	in 38	318	1,1-Dichloroethane
1.82	25	in 38	308	α-Chlorotoluene
1.82	25	in 38	310	m-Chlorotoluene
1.83	20	in 38	319	Dichloroethane
1.84	20.2	in 38	262	trans-Ethyl cinnamate
1.84	20	in 38	357	1-Iodopropane
1.85	25	in 38	320	1,2-Dichloropropane
1.85	20	in 38	452	Dimethyl disulfide
1.86	25	in 38	242	Isopentyl acetate
1.86	25	in 317	94	Cyclohexanol
1.86	22	in 38	239	Isobutyl acetate
1.87	22	in 38	238	Butyl acetate
1.87	25	in 38	241	Pentyl acetate
1.87		in 38	360	1-Iodo-2-methylpropane
1.88	22	in 38	228	Isobutyl formate

Dipole Moment Index (Continued)

μ	Temp.	Conditions	Compound No.	Compound
1.88	24	in 38	252	Butyl stearate
1.88	20	in 38	89	2-Methyl-1-butanol
1.88	25		476	2-Chloroethanol
1.89	22	in 38	226	Propyl formate
1.89	25	in 38	415	Diethylenetriamine
1.89–1.90			261	Benzyl benzoate
1.9			104	3-Methylcyclohexanol (mixed isomers)
1.9			107	4-Methylcyclohexanol (mixed isomers)
1.9			244	2-Methyl-2-pentyl acetate
1.90	-90–70	in 16	299	1-Chlorobutane
1.90	25	(1)	339	Bromoethane
1.90	25	in 38	415	Ethylenediamine
1.90	25	in 38	459	Tetrahydrothiophene
1.91			105	cis-3-Methylcyclohexanol
1.92	20	in 38	304	1-Chloro-3-methylbutane
1.92	25	in 38	467	Furfuryl alcohol
1.93	20	in 38	340	1-Bromopropane
1.93	25	in 317	346	1-Bromodecane
1.93	20	in 317	359	1-Iodobutane
1.94	25	in 38	225	Ethyl formate
1.94	25	in 38	258	Methyl benzoate
1.94	20	in 38	311	p-Chlorotoluene
1.95			101	2-Methylcyclohexanol (mixed isomers)
1.95	20	in 38	358	2-Iodopropane
1.956	25	in 38	342	1-Bromobutane
1.96	25	in 38	254	Ethyl acrylate
1.96	25	in 38	294	1,1-Difluoroethylene
1.96	20	(1)	296	Chloroethane
1.96	25	in 317	345	1-Bromopentane
1.97	25	in 38	162	Bis(2-methoxyethyl) ether
1.97		(1)	297	1-Chloropropane
1.97	25	in 38	432	2-Picoline
1.98	104	(g)	314	3-Chloropropene
1.99		in 38	259	Ethyl benzoate
1.999	20	(1)	305	1-Chlorooctane
2.00	25	in 38	165	Propylene oxide
2.01	20	in 38	167	1,2-Epoxybutane
2.02		(1)	298	2-Chloropropane
2.03	25	in 38	227	Butyl formate
2.03	25	in 317	332	α,α,α-Trichlorotoluene
2.04	20	in 38	341	2-Bromopropane
2.04	25	in 38	464	2-Methoxyethanol
2.04	-80	in 451	494	Ethyl acetoacetate, enol
2.05	20	in 38	437	2,4,6-Trimethylpyridine
2.06	30	in 38	261	Benzyl benzoate
2.07	25	in 38	300	2-Chlorobutane
2.07	25	in 317	329	α,α-Dichlorotoluene
2.08	25	in 38	465	2-Ethoxyethanol
2.08	25	in 38	466	2-Butoxyethanol
2.09	10–50	in 38	301	1-Chloro-2-methylpropane

Dipole Moment Index (Continued)

μ	Temp.	Conditions	Compound No.	Compound
2.12	35	(1)	468	Tetrahydrofurfuryl alcohol
2.12		in 38	343	2-Bromobutane
2.13	30	in 38	218	Crotonic acid
2.13	30	in 38	490	2-Methoxyethyl acetate
2.14	20	in 38	326	o-Dichlorobenzene
2.14		(g)	375	1-Chloro-1,1-difluoroethane
2.15	20	in 38	302	2-Chloro-2-methylpropane
2.17			496	Ethyl cyanoacetate
2.17	20	in 317	344	2-Bromo-2-methylpropane
2.184	25	in 38	438	Quinoline
2.24		(g)	283	1,1-Difluoroethane
2.25			135	1,2-Propanediol
2.25	30	in 38	491	2-Ethoxyethyl acetate
2.27	25	in 172	478	2-Aminoethanol
2.28	100	(g)	497	Trifluoroacetic acid
2.29	50	(1)	124	3,5-Xylenol
2.30	25	in 38	435	2,4-Lutidine
2.31		in 38	469	Diethylene glycol
2.31	25	in 172	134	1,2-Ethanediol
2.32	60	(1)	123	3,4-Xylenol
2.34	30	in 38	231	Ethylene glycol diacetate
2.321		(g)	284	1,1,1-Trifluoroethane
2.37	25	in 317	431	Pyridine
2.4	20	in 38	488	Ethyl lactate
2.40	25	in 38	433	3-Picoline
2.45			133	trans-2-Butene-1,4-diol
2.450	40	(1)	184	Butyraldehyde
2.47	25	in 38	489	Methyl salicylate
2.48			132	cis-2-Butene-1,4-diol
2.48	25	in 317	269	Methyl maleate
2.48	20		274	Butyl sebacate
2.49	25	in 38	267	Ethyl oxalate
2.51	25	in 172	144	1,5-Pentanediol
2.51	20	in 38	182	Acetaldehyde
2.54	20	in 35	183	Propionaldehyde
2.54	25	in 38	268	Ethyl malonate
2.54	25	in 38	270	Ethyl maleate
2.55	25	in 172	136	1,3-Propanediol
2.56		in 172	147	Glycerol
2.56	25	in 38	292	α,α,α-Trifluorotoluene
2.58	20	in 35	185	Isobutyraldehyde
2.58	25	in 38	484	Bis(2-chloroethyl) ether
2.58	25	in 172	138	1,4-Butanediol
2.59	22	in 38	197	2-Heptanone
2.60	25	in 38	434	4-Picoline
2.61	25	in 38	439	Isoquinoline
2.66	22	in 38	195	2-Hexanone
2.66	25	in 317	201	2,6-Dimethyl-4-heptanone
2.69	20	(1)	189	Acetone

Dipole Moment Index (Continued)

μ	Temp.	Conditions	Compound No.	Compound
2.70	22	in 38	192	2-Pentanone
2.70		in 38	200	2-Octanone
2.740	25	in 38	199	2,4-Dimethyl-3-pentanone
2.76	25	in 38	190	2-Butanone
2.77	20	(1)	186	Benzaldehyde
2.78	22	in 38	198	3-Heptanone
2.78	22	in 38	206	2,4-Pentanedione
2.79	25	in 38	204	4-Methyl-3-penten-2-one
2.81	25	in 172	480	Diethanolamine
2.82	25	in 317	193	3-Pentanone
2.82	20	in 38	272	Butyl phthalate
2.82	25	in 451	221	Acetic anhydride
2.84	20		273	Bis(2-ethylhexyl) phthalate
2.86		(g)	292	α,α,α-Trifluorotoluene
2.86	20	in 38	473	Salicylaldehyde
2.86		in 38	191	Cyclopentanone
2.87	25	in 38	280	Tri-o-cresyl phosphate
2.87	20	(1)	78	Methanol
2.9		in 172	145	Hexylene glycol
2.90	25	in 38	187	Acrolein
2.92	20	in 317	279	Tricresyl phosphate
2.95		(1)	331	3,4-Dichlorotoluene
2.95	25	in 38	203	Acetophenone
2.97	20	(1)	197	2-Heptanone
3.02	25	in 38	186	Benzaldehyde
3.05	25	in 38	281	Tri-m-cresyl phosphate
3.07	25	in 38	278	Butyl phosphate
3.08	25	in 38	194	Cyclohexanone
3.09	20	(1)	80	1-Propanol
3.10	25	in 38	202	Camphor
3.12	25	in 38	277	Ethyl phosphate
3.18	25	in 38	282	Tri-p-cresyl phosphate
3.18	25	in 38	276	Methyl phosphate
3.22	18.2	in 38	494	Ethyl acetoacetate, keto
3.24	20	in 38	474	4-Hydroxy-4-methyl-2-pentanone
3.37	30	in 38	440	Formamide
3.24	25	in 38	442	N,N-Dimethylformamide
3.44	30	in 38	443	Acetamide
3.47	25	in 38	394	α-Tolunitrile
3.49	25	in 38	188	Crotonaldehyde
3.50	25	in 38	447	1,1,2,2-Tetramethylurea
3.50	25	in 38	386	Propionitrile
3.50	25	in 38	388	Butyronitrile
3.53	25	in 38	385	Acetonitrile
3.53	25	in 38	392	4-Methylvaleronitrile
3.54	25	in 38	483	2-Furaldehyde
3.55	25	in 38	448	2-Pyrrolidinone
3.56	20	in 10	380	Nitromethane
3.57	25	in 38	390	Valeronitrile
3.57	25	in 172	481	Triethanolamine

Dipole Moment Index   (Continued)

μ	Temp.	Conditions	Compound No.	Compound
3.59	110	(g)	446	N-Methylpropionamide
3.59	20	in 10	382	1-Nitropropane
3.60	20	in 10	381	Nitroethane
3.61	25	in 38	389	Isobutyronitrile
3.67	25	in 38	396	Acrylonitrile
3.68	30	in 39	387	Succinonitrile
3.69		(g)	397	Methacrylonitrile
3.71	30	in 38	445	N,N-Dimethylacetamide
3.73	120–180		383	2-Nitropropane
3.86	25	in 38	441	N-Methylformamide
3.88	25	in 38	450	ε-Caprolactam
4.00	25	in 38	384	Nitrobenzene
4.01	25	in 38	395	Benzonitrile
4.06	25	in 317	462	Dimethyl sulfoxide
4.09	30	in 38	449	1-Methyl-2-pyrrolidinone
4.12	25	in 38	263	γ-Butyrolactone
4.27	30	in 172	444	N-Methylacetamide
4.31	25	(1)	500	Hexamethylphosphoric triamide
4.81	25	in 38	463	Sulfolane
4.87	25	in 38	264	Ethylene carbonate
4.94	18		265	Propylene carbonate
4.98	25	in 38	486	o-Nitroanisole
5.54	25	in 38	500	Hexamethylphosphoric triamide
5.58	20	(1)	470	Triethylene glycol

Chapter IV

CRITERIA OF PURITY.
DRYING AND DETERMINATION OF WATER

The purity of a substance can be defined either thermodynamically or kinetically. Purity in a thermodynamic sense implies that the substance behaves in a system of more than one phase as one and only one independent component, so that it possesses a sharply defined chemical potential that is the same in all operations. According to the kinetic definition, the substance is pure when it consists of molecules of one type only. These definitions are not rigorous and must be modified in certain cases, as in the thermodynamic definition for racemates and the kinetic definition for isotopes. The concept of purity and the difficulties that arise in its definition are fully discussed by Timmermans [7496] and in the Symposium on Purity and Identity of Organic Compounds [629].

These definitions set an ideal that can only be experimentally approached more or less closely. There are, in fact, no criteria of ideal purity. Ostwald considered a pure substance to be the point of convergence of various methods of purification. In practice, the substance is obtained from as many different sources as possible, subjected to a variety of purification processes and the constancy of the physical properties determined. Timmermans [7496, p. 64] gives the following general rules: (1) The starting material should be the best specimen of a substance commercially available (for precision work this must always be subjected to some process of purification); (2) impurities may often be removed more simply by transforming them chemically than by separating them in the form in which they are initially present; and (3) if a product of a high degree of purity is to be obtained synthetically by several distinct operations, each intermediate product should be purified as far as possible.

The term *pure* has a relative connotation and, for practical use, must be defined in explicit terms. The explicit definition must be adequate to define the purity to the extent desired. This definition must be based upon the specific use for which the substance is intended. For instance, water for the calibration of thermometers at the steam point must boil at 100° at 760 Torr. This water may contain impurities that make it entirely unsuited for conductivity measurements, and water completely satisfactory for conductivity measurements may contain so much particulate matter that it is not suitable as a solvent for light-scattering photometry. The definition of purity for practical purposes may be stated as: *A material is sufficiently pure if it does not contain impurities of such nature and in such quantity as to interfere with the use for which it is intended.*

CRITERIA OF PURITY

The criteria of purity are standards by which the purity of a substance may be judged. The standards are based on results of tests of known reliability. Some tests have proven thermodynamic and stoichiometric accuracy, and some are empirical. The kind and amount of information that may be deduced from chemical and physical tests vary. These tests must be sensitive enough to characterize the purity to the desired accuracy.

Recent advances in electronic instrumentation have resulted in a great improvement in the sensitivity of test methods and in the ability to separate impurities bearing close structural resemblance to the main compound. Gas chromatography is the greatest, and certainly the most spectacular test method to be developed for the characterization of organic solvents. It has been followed in importance by significant advances in mass spectrometry and in spectrophotometry.

The extension of the sensitivity of tests has made it possible to prepare purer substances; this has given rise to a new grade of purity called *ultrapure*. Although the degree of impurity implied by the new term has been the subject of some confusion, Wilcox et al. [8056], and Janke et al. [6613] have presented lucid discussions that serve as excellent guides. The former group proposed 500 ppm for the upper limit for impurities in organic compounds classed as ultrapure.

It is sometimes desirable to express the composition as the percentage of the major component present instead of as concentration values of the impurities. Since the percentage figure for a substance containing, for example, 50 parts per billion (ppb) of impurities would be somewhat cumbersome, 99.99995%, the "nine" or "N" method of expression has been proposed and apparently has been well received. This method gives the number of successive nines in the percentage as an Arabic figure followed by an N, which in turn is followed by the Arabic figure(s) following the nines. Thus, 50 ppb becomes 99.99995% or 6N5, 99.992% becomes 4N2, and 99.9% becomes 3N.

Caution should be exercised when composition is expressed on the basis of results from impurity tests. A purity of 99.992%, for example, implies that the purity of the principal component is 99.992% within the accuracy of the test method. This situation rarely exists when the figure is based on impurity analyses, particularly for organic solvents. The expression "effective purity" is compatible with the stated definition of purity. It implies that the objectionable impurities are present to no greater extent than 0.008%.

Chemical Tests

Chemical tests are usually based on the reaction of a functional group and have certain limitations as criteria of purity. Generally, it is not feasible to analyze a substance of high purity for the principal component. Functional group analysis usually has an accuracy of about ±0.2%, and it is therefore desirable to determine at least one physical property, and preferably two, to use in conjunction with the chemical results.

Empirical composition obtained from elemental or functional group analyses is no guarantee of purity. For instance, isomers that may be present will give the same elemental, and possibly functional, analysis; in high-molecular-weight compounds, adjacent homologues will have to be present in fairly large amounts to be detected by the usual chemical methods.

The application of chemical methods of analysis in the testing of high-purity organic compounds presupposes a knowledge of the types of impurities present. This is a broad field; specific information may be obtained from Rosin [6263], ACS *Reagent Chemicals* [137], and Chapter V.

Physical Tests

The physical tests may be used as a quantitative or qualitative indication of the purity, and most of them are not specific for the chemical nature of the impurity present.

Test methods furnish four types of standards for judging purity:

1. Total amount of impurity present, but no differentiation as to kind.
2. The amount of a general class of impurities present.
3. The amount of a specific impurity present.
4. Empirical tests based on some artificial standard.

The selection of the type of test to be used depends on the material being tested, the accuracy desired to define the required purity, knowledge of impurities present, and available equipment. The more common standards and tests are discussed in the following pages.

Cryoscopy

Cryoscopy, according to the *Random House Dictionary of the English Language*, is ". . . the determination of the freezing points of liquids or solutions or the lowering of the freezing point by dissolved substances." Glasgow and Ross [4159, Chap. 88] ". . . have included melting points because the freezing and the melting point designate the same temperature according to thermodynamic definition." The cryoscopy concept of Glasgow and Ross is logical and brings together a number of methods derived from the same basic considerations. Cryoscopy is based on the assumption that the solid phase is in equilibrium with the liquid phase and that the temperature measured depends on the change in composition of the liquid phase.

The several cryoscopic procedures are alike in that the temperatures for solid–liquid equilibria are measured. They differ in the method used to determine the relative amounts of the liquid and solid phases present and in the precision obtained. The procedures now widely used are:

1. Thermometric—relates temperature and time.
2. Calorimetric—relates temperature and heat input.
3. Dilatometric—relates temperature and volume change.

Recently Ross and Frolen [6267] developed a method based on the determination of the dielectric constant during melting.

Glasgow and Ross [4159, Chap. 88] point out that, "notwithstanding the wide variation in apparatus and procedures for cryoscopic measurements as to cost, complexity, accuracy, sample size, and time of determination, each cryoscopic procedure finds important application as an analytical tool." The procedure and apparatus that is used must characterize the solvent to the desired precision.

METHODS BASED ON THE FREEZING POINT

The *thermometric method*, or time–temperature method, is used to designate the open method of White [8015], as improved and simplified by Rossini and coworkers [2781, 4838, 7202], and the variations that are now in common use. The less precise method in which a mercury or alcohol-filled thermometer is generally used will be referred to as the *freezing curve method*.

The thermometric method is one of the most effective and accurate means of evaluating the purity of materials, and particularly of high-purity substances. It is a thermodynamic method based on the phase rule interpretation of Raoult's law. For an ideal or sufficiently dilute solution, it measures the equilibrium that exists between a single crystalline phase consisting of the major component and a liquid phase consisting of one or more components, one of which is the major component. It is applicable to all substances that form ideal solutions, or whose solutions are so dilute that the departure from the ideal state is not significant. It is concerned only with the amount of impurity and not the kind. Rossini [6274] gives a relationship for those cases in which a significant departure from the ideal state occurs. The method is applicable when the purity is above 95%m, and its reliability increases as the purity increases. It has an accuracy in some cases of 0.002%m; the sensitivity for some systems has been reported to be in the 0.0002–0.0005%m range.

Hüttner and Tammann [3604] in 1905 observed that it was possible, from the behavior during crystallization, to draw conclusions about the amount of impurities. White [8015], 15 years later, proposed a method for estimating the amount of impurity by means of the melting curve. Rossini and coworkers [4838] in 1941 developed the method to its present high sensitivity and also expanded it to enable the true freezing point to be calculated for the major component, even though it contained an appreciable amount of impurity.

When only information about the degree of purity is desired, irrespective of the nature of the contaminants, this method has the advantage that no other knowledge of the system is necessary except that the substance to be tested should have high purity. The temperature-indicating device need not be calibrated but must measure precisely throughout the temperature range. The values are uncertain if a solid solution is formed or if there is a eutectic point near 100%m. It also has the disadvantage that the equipment required for the most precise measurements is expensive. Where the high precision of resistance thermometry is not required, the apparatus of White [8015] should find wide use.

The *closed thermometric method* was developed somewhat later than the open method. Glasgow and Ross [4159, Chap. 88] give the following advantages of the closed cell: There is less chance of sample contamination during transfer and subsequent measurements. Noxious or highly flammable substances may be measured without discomfort or danger. The error due to dissolved gases (air or other blanketing gases) may be eliminated. The error due to dissolved gases becomes appreciable in the ultrapure range, particularly >99.99%m.

The cell described by Ross and Dixon [6266] permits cleaning the system of adsorbed gases and water, effective outgassing of the sample, and transferring of the liquid from the sample vial to the freezing chamber. Substances with a purity range of 99.0 to 99.999%m were used for the study.

A *simplified freezing point method* was introduced recently by Ross and Glasgow [6268]. It utilizes a semiclosed system apparatus, which makes it possible to follow a purification or other operation, depending on change of composition with a precision of about 0.01°. This precision in a typical purity analysis corresponds to an estimated standard deviation of 0.01%m. The apparatus is inexpensive, rugged, and easy to operate. It is easily cleaned and has a cell capacity of 8 mL.

The *freezing curve method* includes various procedures common in the laboratory for determining the freezing point with a precision of 0.05° or less, usually less. The sample is generally placed in a double-walled test tube. A thermometer may be used as a stirrer or held in place with a stopper that also serves as a guide for a stirrer. The bath may be any convenient medium that may be held 5–10° below the temperature of the sample. The temperature–time data are plotted and the section where Δ-temp/Δ-time $= 0$ is taken as the freezing point. This is a convenient, rapid, and inexpensive test that is used for characterizing many commercially pure substances. It is also used with specific tests for impurities to characterize a high functional purity.

The *freezing point* was early recognized as a specific property of a pure substance. There are many methods that have been reported, and many of these have been used extensively but have not been mentioned.

A differential thermometer may be used to follow the purification of a substance. If a reference sample is available, the purity may be calculated from the difference in freezing points. The use of the freezing point constant has been extensively studied; see references for K_f [Eq. (2.67)] and particularly Brancker et al. [1056], Meldrum et al. [5050], Kraus and Vingee [4249], Rall and Smith [6026], Menzies and Wright [5060], Adams [20], and Washburn and Read [7912].

METHODS BASED ON THE MELTING POINT

The *adiabatic calorimetric method* is the most accurate and the most sensitive of the cryoscopic methods. The basic technique was proposed by Giauque and Wiebe [2727] and was further improved by Aston and Messerly [479]. Aston and Fink [473] describe the apparatus and give typical data. The method is based on the measurement of the temperatures of the liquid–solid equilibrium at known liquid-to-solid ratios, precisely evaluated from the total heat input, heat of fusion, and heat capacity. The equipment is costly and the experiment is time-consuming, but there is no other method of characterization that gives as much information about the system being studied. The method has an accuracy of at least 0.001 %m and a sensitivity of at least 0.0001 %m.

The *simple melting curve method* has been developed by Gunn [3023] to a sensitivity of the order of 10^{-5} mol fraction impurity and an accuracy of 10–20 % using about 0.3 gram sample. The apparatus is simple and a determination requires about 2 hr. The data may be used to calculate not only purity but also heat of fusion and heat capacity.

The *capillary tube* and *melting block techniques* are approximate methods for the determination of melting points and are not suitable for characterizing a very pure substance. These techniques serve a purpose when their possibilities and limitations are considered. They are dependent on such factors as the state of transition taken as the melting point, the rate of heat transfer from the bath medium to the substance, the rate of response of the thermometer to change of temperature, and the rate of heating. When the freezing point of a solvent is high enough to permit the use of either of these techniques, they are useful devices to approximate the degree of purity.

There is a voluminous literature on capillary tube and melting block melting points. One of the more significant studies is that of Francis and Collins [2513]. They observed that a single observer can determine a capillary melting point to within 0.1–0.2°. Using the same apparatus and the same materials, different observers may differ by 0.2–0.3°. Freezing points can be determined by different observers to within 0.02°. The melting and freezing points of the same material should be the same, but, when determined by this technique, the former are invariably higher. These authors further observed that capillary melting points determined with different apparatus may differ by 2°.

Ebulliometry

Ebulliometry is defined by Leslie and Kuehner [4159, Chap. 89] as the science of measuring and using boiling points. They continue, "properly measured, the boiling point has definite thermodynamic significance and can be used to identify materials, determine or indicate their purity, determine molecular weights, study azeotropes and the association of molecules. . . ."

The boiling point of a liquid is that temperature at which the vapor pressure of the liquid equals the pressure on the system at the vapor–liquid equilibrium.

The *differential ebulliometric method* proposed by Swietoslawski is a sensitive test that is easy to make with inexpensive apparatus. A comprehensive discussion of this method is presented by Swietoslawski in [7131], and an excellent summary suitable for general use was prepared by Swietoslawski and Anderson [7958]. Swietoslawski has set up an arbitrary scale of purity based on the difference between the boiling temperatures (t_b) and the condensing temperatures (t_c; see Table 4.1).

It is possible, when this scale is used, for two samples of liquid to have the same amount of different impurities but to exhibit different degrees of purity. The effect of an impurity is dependent on the vapor pressure curve of the mixture. This does not detract from the usefulness of the method. If an impurity forms an azeotrope, it is the effect of the azeotrope that is manifest and not that of the impurity itself. If the impurity is known, the amount present may be estimated from the vapor pressures, provided that the concentration is low enough for Raoult's law to be valid.

This is the only ebullioscopic method by which the purity may be estimated with any degree of certainty. It is an indication of the amount, and is fairly independent of the kind, of impurity. It is considered possible to detect 0.002% of impurity by the differential ebulliometer. Zepalova-Mikhailova [7131, p. 94] found almost a linear relationship between the amount of impurity present and Δt for several systems. This method has the further advantage that only the temperature-measuring instrument need be precise. The actual temperature is of no importance, only the difference between the boiling and condensing temperatures.

The *equilibrium boiling point* has long been regarded as a criterion of purity, although two possible circumstances may negate its usefulness for this purpose. An impurity whose

Table 4.1 Swietoslawski Degree of Purity Scale

Degree of Purity	$\Delta t = t_b - t_c$
I	1.00–0.10°
II	0.10–0.05°
III	0.050–0.020°
IV	0.020–0.005°
V	0.005–0.000°

boiling point lies very close to that of the principal component or which forms an azeotrope having a proximate boiling point will probably go undetected. The equilibrium boiling point must therefore be used in conjunction with one, and preferably two, other tests such as refractive index and density. The accuracy of this method is dependent upon how closely the pressure may be controlled or measured. The precision of the measurements of the auxiliary properties must be comparable.

The boiling point may be used in one of two ways as a test of purity. If the boiling point of the substance has been established to an accuracy suitable for the particular characterization, the substance may be purified until it has this boiling point. If the boiling point of the substance is unknown or has not been determined with sufficient accuracy, the substance may be purified until there is no further change in its boiling point.

It should be noted that the boiling points of relatively few substances have been accurately determined. This may be attributed to several factors, namely, impure material, inaccurate temperature-measuring devices, inaccurate pressure measurements, and improperly designed and operated ebulliometers.

A brief description and diagrams of the several ebulliometers are presented by Swietoslawski and Anderson [7958]. The Quiggle, Tongberg, and Fenske [6009] modification of the Cottrell [1674] apparatus offers a convenient and rapid means of determining the boiling point, and to a limited extent the distillation range.

The determination of the *distillation range* (boiling range) is an arbitrary test that is suitable for characterizing commercial grade materials. The results are qualitative but will show the presence of low and high boiling impurities. The sensitivity depends on the amount of impurities present and the difference between their boiling points and those of the principal constituents. It requires as much as, or more material than, the differential ebulliometric method, about the same amount of time, and is less informative. ASTM D 86-67 and D 1078-67 are the most precise modifications of this technique and are recommended when the distillation range is desired for characterization.

The distillation range is a useful and convenient test for many solvents for general laboratory use. It is a criterion used by Rosin [6263] and in the ACS *Reagent Chemicals* [137]. It is one of the standard industrial tests. In conjunction with other data, such as density and refractive index, it may be used to characterize a liquid but not to define its purity.

Fractional distillation is a convenient means of detecting and estimating impurities. For instance, 10 μg of benzene per liter of ethanol may be determined readily with an accuracy of ± 1 μg by removing the lower boiling azeotrope and analyzing for benzene.

Gas Chromatography

Gas chromatography was invented by A. J. P. Martin in 1952. The growth and development of this versatile analytical tool may be characterized as phenomenal. It is the most useful method of analysis for solvents that exists and may be used to analyze both the major component and the volatile impurities in a solvent system. Allen [91] and his associates have analyzed most of the solvents in this book by gas chromatography, either as a major component or as an impurity using a thermal conductivity detector.

Gas chromatography is the simplest and usually the fastest means of analyzing a solvent and is a convenient and rapid means of following purification. When the purification has reached the 3N to 4N range, or better, the impurities are best analyzed with a high sensitivity detector. Some impurities may be determined in the low nano-range.

The sensitivity obtained from a gas chromatograph for a given solvent system depends on the characteristics of the instrument and on the operating conditions. Such factors as the sensitivity of the recorder, the length of column, the kind of column packing and the method used in packing the column, sample port and column temperatures, and the carrier gas and carrier gas pressure play important roles.

It has been reported by Allen [91] that a Teflon solid phase with either Carbowax 20-M, Carbowax 1540, or diisodecyl phthalate liquid phase is the most versatile packing used at the present time. Good separation is obtained for polar compounds, low boiling hydrocarbons, and mixtures of polar and nonpolar compounds. A cross-linked polystyrene with no liquid phase makes a good packing for intermediate boiling compounds. These packings may or may not be the best for a particular solvent, but they are excellent for general use.

The detection or sensitivy characteristics of several of the detectors are given.

Thermal conductivity is the most commonly used detection system for routine and research analyses of solvents when a solvent analysis of 100% is desired, that is, major and minor components. The lowest response that has been reported is about 3×10^{-4} mV mg^{-1} mL^{-1} with a noise level of <2 μV, or about 50 ppb. With a lower noise level, the limit of detection can be reduced to about 20 ppb, all other factors being optimal. A sensitivity of at least 0.01% may be expected for routine handling of most solvent systems.

Gas density is reported to have a sensitivity probably in the 1–10 ppm range.

Hydrogen flame ionization has a detection limit of about 8×10^{-14} gram-atoms of carbon sec^{-1} (nearly 10^{-14} mol heptane sec^{-1}) with a clean system.

Electron capture has about the same or better sensitivity for favorable samples as the hydrogen flame ionization detector.

Spectrometry

Spectrometric methods are widely used in the detection and determination of impurities in solvents; criteria for certain grades of solvents are defined in terms of absorption in the infrared and ultraviolet regions (See Chapter V, General Comments). Some manufacturers supply individual absorption curves and gas chromatograms showing mole per cent purity for their best grades of the more common solvents, such as acetone, benzene, carbon tetrachloride, and methanol. Mass spectrometry and nuclear magnetic resonance spectrometry are also useful for establishing the purity of organic solvents.

Other Methods

The determination of *density* (specific gravity) is one of the standard means of characterizing a substance. Bauer [7958] defines and discusses density and several methods for its determination. Density is easy to determine, but the simplicity of the determination is deceptive—considerable care and experience are required to secure accurate data. Bauer [7958] gives the accuracy of the thermometer-type pycnometer, without thermostating, as ± 0.001. This is the maximum accuracy that can be expected; usually the error is nearer ± 0.005. Several pycnometers are capable of a sensitivity of $\pm 5 \times 10^{-6}$ if properly handled. The specific gravity or density balance is capable of an accuracy of

±0.001 if the thermometer plummet is used. The density balance modified similarly to that described by Forziati, Mair, and Rossini [2495] has a sensitivity of ±2 to 3 × 10^{-5} and an overall uncertainty of ±5 × 10^{-5} g/mL.

That the density does not change on successive purifications is no indication that a pure material has been achieved. If the successive purifications are made by different methods, such as fractional distillation followed by fractional crystallization, and there is no change in density, it is probable that the substance is as pure as may be detected by the sensitivity of the density determination. This property has the same limitations, when used alone, for characterizing purity as the boiling point and refractive index.

The density–composition relationship for dilute solutions may be considered additive. The most convenient method of determining the relationship is to determine the density of the solution and plot the two points on suitable graph paper. The calculation may be made algebraically:

$$d_{\text{soln}} = (C_1 \times d_1) + (C_2 \times d_2)$$

where d_{soln} is the density of the solution, d_1 and d_2 are the densities of pure solvent and solute, respectively, and C_1 and C_2 are weight fractions of each present.

The sensitivity of density as a quantitative measure depends upon the sensitivity of the measurement and the difference in the densities of the solvent and the solute. For example, the densities of cyclohexane and benzene differ by 0.09917 g/mL at 30°. If the method used to determine the density is sensitive to 5 × 10^{-5} g/mL, benzene can be determined in cyclohexane with an accuracy of slightly less than 1 part per 1000. The density of hexane and that of 3-methylpentane differ by 0.00495 g/mL at 25°. A density method sensitive to 5 × 10^{-5} g/mL would detect only 1 part of 3-methylpentane in 50 parts of n-hexane, or 2%. The density of water and that of methanol differ by 0.21033 g/mL at 25°. Density will determine as little as 0.05%w if the method is sensitive to 5 × 10^{-5} g/mL.

The *refractive index* affords one of the easiest and most rapid tests that can be made. Bauer and Fajans [7959] discuss the theory and practice in a concise and practical manner. If the refractive index of the substance has been previously established with sufficient accuracy, a material may be purified until it has the proper refractive index. If the refractive index has not been determined, or has not been determined with sufficient reliability, the substance may be purified to a constant refractive index. The refractive index alone is not a sufficient criterion of purity but, together with the density and equilibrium boiling point, it characterizes the purity of a liquid quite closely.

The refractive index is a convenient test to control the progress of a purification. For instance, the refractive index of the distillate, together with the reflux temperature, is very useful in fractional distillations but its value depends upon the sensitivity of the refractometer and the purity desired.

Refractive indexes may be considered additive for dilute solutions. An ethanol ($n_D 25°$ 1.35941) and benzene ($n_D 25°$ 1.49790) mixture can be analyzed accurately with an Abbé or dipping refractometer; a 1% benzene solution of alcohol could be analyzed to ±0.1%. Ethanol, as an impurity in iodoethane, can be determined with considerably greater accuracy. The refractive index of iodoethane differs from that of ethanol by 0.56512. One part ethanol in 2500 parts of iodoethane can be detected with an Abbé refractometer. On the other hand, refractometry is almost useless for measuring ethanol in ethyl ether.

Systems that do not adhere fairly closely to Raoult's law may offer some anomalies that make the refractive index useless when the concentration increases beyond a few percent. The data of Carr and Riddick [1365] show that little information can be obtained from the refractive index of the methanol–water system.

The small Abbé refractometer, precise to $\pm 2 \times 10^{-4}$ units, is satisfactory for characterizing solvents for the majority of laboratory uses. The Precision Abbé Refractometer has a precision of $\pm 2 \times 10^{-5}$ and the dipping refractometer of about $\pm 3.5 \times 10^{-5}$ units. The differential refractometer has a precision of $\pm 3 \times 10^{-6}$ units, but has the disadvantage that the refractive index of the sample is compared to the refractive index of a standard substance whose refractive index must be within 0.01 unit of the sample.

The *specific conductance* has been used widely as a criterion of purity. If the impurity and the solvent are both nonconductive, the specific conductance is of little, if any, use. For example, benzene cannot be detected in toluene by this method, but ethanol can be detected because its specific conductance is much greater than that of toluene. Water, salts, inorganic and organic acids, and basic impurities may be readily detected by the conductance of the solution.

The specific conductance is a most useful criterion for the presence or absence of ionizable impurities. It is a sensitive test for water in hydrocarbons and in most halogenated hydrocarbons, and for acids and alkaline substances in alcohols, ketones, aldehydes, and some esters. Its reliability as a means of detecting decomposition products during the purification of formamide [7725] illustrates its usefulness. The increased interest in electrochemical studies in nonaqueous systems has brought about an increase in the use of specific conductance as a rapid and sensitive method for the detection and determination of water and other ionic substances; for example, ethylenediamine [1715], nitromethane [7638], acetonitrile [4155], and acetamide [3418].

The *critical solution temperature* is again being recognized as a convenient criterion that has wide application. The amount of impurities that will be detected by this method depends on the system used and the exactness of the determination. A simple application of the critical solution temperature is the "cloud point" test for water in many solvents. This is considered in more detail later in the chapter.

A. W. Francis [2509] has compiled a table of more than 6000 critical solution temperatures. The book includes an introduction to the subject, description of methods of determination, and other pertinent information. It is well documented.

Specifications are criteria set up, generally between the seller and the buyer, based on specific test methods to assure a uniform product. Certain organizations and individuals have published specifications for certain types of materials. The United States Pharmacopoeia (USP) lists purity and limits of impurities with prescribed tests of determined reliability that are considered safe for substances to be used as pharmaceuticals. It is not the intention of the USP Committee to set standards and devise tests to characterize a material as suitable for other than pharmaceutical use, although many solvents meeting USP specifications are suitable for general laboratory use. Each group of specifications and tests is generally designed to fit the substance for a specific use. The *ACS Reagent Chemicals* [137] lists purity, limits of impurities and tests to qualify the substance as a general analytical reagent. Rosin [6263] includes many reagents in *Reagent Chemical and Standards* that are in the American Chemical Society (ACS) list, and many that are not, together with specifications and tests.

The American Society for Testing and Materials (ASTM) lists specifications and tests for a variety of solvents for industrial use in the *1968 Book of ASTM Standards*, Part 20.

Manufacturers have developed specifications and test methods to characterize their products. Some specifications are as stringent as those of the ACS, while others characterize a "technical" or "commercial" product. Unless the solvent is produced by a number of manufacturers, for example, methanol and acetone, the specifications and tests generally characterize the product of the particular manufacturer. The impurities present often vary from manufacturer to manufacturer, particularly if the product is made by different processes.

Standard or organizational specifications are excellent criteria of purity for a solvent for a specific type of use. The manufacturer or seller does not guarantee that the solvent will be suitable for all uses. For instance, "absolute alcohol" is guaranteed to meet certain minimum requirements. Some manufacturers remove the water from 95% ethanol by the benzene–water–ethanol azeotrope. A very small amount of benzene remains in the absolute alcohol. This alcohol is not suitable for a solvent in ultraviolet spectroscopy. The manufacturers or vendors have not stated, nor implied, that the dry ethanol meets ultraviolet clarity as one of the criteria of purity.

The level of impurities sometimes lies in a range above the maximum permitted by the criteria and below the minimum determinable by direct application of available analytical methods. In such cases it is convenient to carry out a *concentration of impurities* prior to the analysis of the sample. An example is the concentration of benzene in ethanol, which is mentioned under fractional distillation and discussed under **79**. Ethanol in Chapter V.

Concentration by fractional distillation has many applications. It is most useful when the impurity and the principal component (C) form a positive azeotrope boiling several degrees below C; or when another component that can be added to the system forms a positive ternary azeotrope with the impurity and C or forms a positive binary azeotrope with the impurity and not with C. The distilled system permits a greater accuracy if the azeotrope is homogeneous. Generally, impurities that are higher boiling than C and do not form an azeotrope are difficult to concentrate unless the distillation alpha is large; microamounts usually distill with C.

Concentration by fractional solidification, see Zief and Wilcox [8290], offers a general, more positive means than fractional distillation. Fractional freezing and zone melting are two techniques that have been frequently used.

The following references are recommended for additional information: *Cryoscopy:* Glasgow and Ross [4159, Chap. 88], Skau, Arthur, and Wakeham [7958, Chap. 7], Smit [6809]; *Ebulliometry:* Swietoslawski [7131], Swietoslawski and Anderson [7958, Chap. 8], Lestie and Kuehner [4159, Chap. 89]; *Temperature measurements*: Sturtevant [7958, Chap. 6], and Corruccini [4159, Chap. 87].

DRYING OF SOLVENTS AND DETERMINATION OF WATER

Water is the universal impurity in all organic solvents because it usually remains in the solvent during its preparation and because practically all organic solvents are hygroscopic to some extent. Small amounts of water are often difficult to detect because of the limited number of reliable and delicate chemical tests and because the presence of water does not cause an appreciable change in many physical properties. The literature concerning the drying of organic liquids and the detection of water and its effect on the physical properties is voluminous.

Drying of organic liquids may be accomplished by either chemical or physical means or both. The method selected depends on the solvent, on the material or equipment available, and the degree of dryness desired.

The chemical method of drying depends on the use of substances generally classed as *drying agents*. These agents remove water by either chemical or physical combination. The efficiency of drying agents depends upon the vapor pressure of the hydrated compound, the physical state of the agent, and the technique of use. Other factors being equal, the lower the vapor pressure, the more efficient is the drying agent. However, it is seldom possible to fulfill the basic requirement of "other factors being equal." Drying agents usually are rated by the amount of residual water left in air, under specified test conditions, when moist air is passed over the agent. A number of comparative investigations are listed in Table 4.2 (cf. the recent work of Trusell and Diehl [7559] and Bower [1026]).

The *residual water* criterion provides an excellent method to rate drying agents for removing moisture from gases. There is assumed to be a correlation of a drying agent's efficiency for drying gases and its ability to dry organic solvents. Pearson and Ollerenshaw [5733] used a rapid, near-infrared method for the rating of the efficiency of drying agents for organic liquids. The rate and the extent of drying for five agents with three solvents have been reported. Small samples were withdrawn at intervals and analyzed to determine the rate. The data are given in Table 4.3.

Table 4.2 Comparative Efficiency of Some Drying Agents (micrograms residual water per liter of dried air)

Alumina, Al_2O_3	2.9
Barium perchlorate	599
Barium oxide	0.6
Calcium bromide	200
Calcium chloride, anhydrous	360
Calcium chloride, anhydrous, technical	1250
Calcium chloride, granular	1500
Calcium oxide	3
Calcium sulfate, Drierite	5
Magnesium perchlorate, anhydrous	2
Magnesium perchlorate, Anhydrone	1.5
Magnesium perchlorate, $Mg(ClO_4)_2 \cdot 3H_2O$	4.4
Molecular sieves, 5A	3.9
Sodium hydroxide, sticks	513
Phosphorus pentoxide	$<0.2^a$
Silica gel	30
Sulfuric acid, 100%	8
Sulfuric acid, 95.1%	300

[a] The reported values vary from 0.02 to 2.9.

Table 4.3 Comparative Efficiency of Five Drying Agents for Three Solvents[a-d]

	$CaCl_2$	Drierite	Molecular Sieves	$MgSO_4$	Na_2SO_4
Benzene					
C_1 (%w)	0.054	0.028	0.048	0.023	0.030
C_2 (%w)	0.000	0.009	0.000	0.019	0.021
T_{50} (min)	2	3	30	38	30
T_{95} (min)	40	20	9 hr	75	4 days
Ethyl ether					
C_1 (%w)	1.468	1.580	1.468	1.570	2.070
C_2 (%w)	0.011	0.320	0.009	0.220	1.830
T_{50} (min)	0.5	1	3.5	1	0.5
T_{95} (min)	25	20	60	30 hr	2
Ethyl acetate					
C_1 (%w)	0.370	0.515	0.370	0.495	1.490
C_2 (%w)	0.042	0.195	0.039	0.346	1.280
T_{50} (min)	5	1.5	26	9 hr	6 days
T_{95} (min)	2 hr	25	4 hr	7 days	12 days

[a] C_1 is the initial water concentration.
[b] C_2 is the concentration of water after the solution had been in contact with the drying agent for 3-4 weeks.
[c] T_{50} is the time required to reduce the concentration, C_1, by 50% based on the final concentration, C_2.
[d] T_{95} is the time required to reduce C_1 by 95% based on C_2, i.e., to reduce C_1 to $[0.05(C_1 - C_2) + C_2]$.

The data of Pearson and Ollerenshaw point out that the several drying agents have different drying efficiencies for the same solvent and for different solvents. These differences presumably are due to such factors as hydrogen bonding, the rate of contact of the water molecule with the drying agent, and the rate of diffusion of the water from the surface to the interior of the drying agent particle.

Bower [1026] classified drying agents in five groups according to the type of reaction with water:

1. $A(H_2O)_{y(s)} + xH_2O \rightarrow A(H_2O)_{x+y(s)}$ (e.g., calcium chloride, calcium sulfate, and copper sulfate.)
2. $A_{(s)} + H_2O \rightarrow$ saturated solution (e.g., sodium chloride). This type of drying is used primarily for removing large amounts of water from lower-molecular-weight compounds such as alcohols.
3. $A_{(1)} + H_2O \rightarrow$ solution (e.g., sulfuric acid).
4. Adsorption (e.g., silica gel, alumina).
5. $M + (H_2O)_x \rightarrow M(OH)_x + xH_2$ (e.g., sodium).

A chemical classification of types is also possible:

1. Direct union (e.g., phosphorus pentoxide and calcium oxide).
2. Double decomposition (e.g., lithium aluminum hydride, calcium carbide, and alkali alkoxides).
3. Displacement (e.g., sodium).

4. Hydrate formation (e.g., calcium sulfate, magnesium perchlorate, and sulfuric acid).
5. Adsorption (e.g., silica gel and alumina).

This system does not account for the removal of water by concentrated solutions of calcium chloride and potassium carbonate.

Phosphorus pentoxide has long been the standard for drying agents. It is among the best of the agents but it has numerous disadvantages. It is hard to handle; it reacts with many solvents such as amines; it dehydrates many compounds or catalyzes reactions in others. Very dry substances may be obtained using phosphorus pentoxide. The most successful method is to reflux the material with the oxide and distill. It is particularly recommended for hydrocarbons, acids, anhydrides, ethers, and halohydrocarbons.

Metal hydrides such as lithium hydride, calcium hydride, and lithium aluminum hydride are excellent drying agents for solvents with which there is no deleterious reaction. Riddick [6165] has found calcium hydride to be particularly effective. The reaction products are not a problem in many solvents; hydrogen escapes as a gas and calcium hydroxide is insoluble. Methanol, ethanol, and several hydrocarbons have been dried to the available lower limit of detection, about 0.0005% water.

Magnesium perchlorate is one of the most efficient drying agents. However, it is soluble in many solvents and may cause explosions if improperly used.

Molecular sieves, particularly 4A and 3A, are the most universally useful and efficient of the drying agents. Riddick [6165] has reduced the water in alcohols to 0.5% or less and then slowly passed them through long columns of molecular sieves. The water content was found to be equal to, or less than, 0.001%.

The *alkali hydroxides* are good agents for drying organic bases to 0.1% or less water. They are inexpensive and easy to use.

Calcium chloride, long a favorite drying agent, has several excellent characteristics. It is inexpensive, easy to use, and insoluble in most substances. It is an excellent agent for removing water to 0.1%, or in some instances to 0.05% or lower.

Potassium carbonate finds its primary use in drying alcohols. The carbonate is added until it remains solid on prolonged shaking.

Magnesium sulfate as a drying agent does not have any outstanding characteristics. It is included because it is used in several methods of purification given in Chapter V. An anhydrous grade is available commercially. Riddick [6165] studied the effectiveness of magnesium sulfate, together with several other drying agents for the drying of solvents. The heptahydrate must be dehydrated at reduced pressure to prevent fusion of the particles at the dehydration temperature of 300°. The particles are free-flowing, porous, and somewhat fragile. If the temperature of dehydration is below 300°, the water is not removed completely; if it is heated much above 300°, its effectiveness as a drying agent is decreased. It is soluble in many organic solvents.

Silica gel and *alumina* are effective drying agents. Finely divided silica gel was used to dry some API hydrocarbons to a very low water content (see Chapter V). The effectiveness depends upon the amount of water to be removed, the fineness of the agent, and the length of time and effectiveness of contact.

Wymore [8185] found that the *sulfonic-type cation-exhange resins* are excellent desiccants for drying organic liquids. The resins used were production grade and, after conversion to the desired form, were dried at 110° *in vacuo*. The potassium form of Dowex 50W-X8 resin, 20–50 mesh, was the best tested. The rate of flow depends on the polarity of the solvent and the amount of water in the liquid. The rate for nonpolar substances may be quite high, and the resin swelling is small. Nonpolar solvents are easily dried to less than 5 ppm of water. Polar solvents may be dried to 10 ppm or less of water but the flow rate per unit of resin is less.

Drierite, a form of anhydrous *calcium sulfate*, was the most practical laboratory drying agent until the advent of molecular sieves. It still is a convenient and efficient agent.

Fractional distillation is a convenient and generally applicable method for the drying of solvents. If the solvent and water do not form a positive azeotrope, the greater the difference in boiling points between the water and the organic liquid and the more efficient the fractionating column, the drier will be the distilled compound, other factors remaining constant. The xylenes, nitrobenzene, aniline, and *N*-methylacetamide may be dried to at least 1–2 ppm of water by removing the water and distilling a small amount of the solvent to clear out the distillation apparatus. Methanol containing 2–5 ppm of water can be prepared by fractional distillation. Generally, about one-quarter of the charge must be kept in the still pot to retain the water if a particular dry product is desired. Acetone could not be dried to less than about 0.02 % water in a 100-plate still at a reflux ratio of 50 : 1 and better.

Many organic compounds form azeotropes with water. A large number of these boil several degrees below the boiling point of the solvent and the latter may be dried by distilling the azeotrope. If the boiling points of the compound and the azeotrope are too close together to permit separation by fractional distillation, a third substance sometimes may be found that forms a lower boiling ternary azeotrope with the solvent and water. For instance, the boiling point of the ethanol–water azeotrope (78.174°) is too near that of ethanol (78.325°) for efficient removal of water by fractional distillation. If benzene is added, it forms a ternary azeotrope (bp64.86°) that is easily removed. The alcohol may be dried to less than 0.01 % water.

Barabanov [596] describes an electrolytic method for the removal of water and other ionic impurities from organic solvents. The apparatus is a modification of the Hittorf tube, and electrical conductance is the criterion of ionic impurities present.

THE DETERMINATION OF WATER

A critical review of the methods for the determination of water has been prepared by Tranchant [7539]. He presents the critique in three main sections: (1) review of the principal methods for the determination of water; (2) precautions to take for the determination of water; (3) and the elements of choice of the method of determination.

Cloud Point

Generally, the solubility of water in solvents partially miscible with water decreases with decrease in temperature. In those solvents in which water is only slightly soluble, the amount of water may be determined quite accurately by lowering the temperature of the liquid until a faint cloudiness is observed. This method may be used to within a few degrees of the freezing point of the liquid. Knowledge of the solubility curve for water in the solvent is necessary. Many of these temperature-composition systems are cited by Francis [2509].

Babko and Schevchenko [526] found that cobalt(II) iodide was the most sensitive of the cobalt halides for the determination of water in alcohols, ketones, nitriles, and some esters.

Knight and Weiss [4071] developed a *calcium carbide–gas chromatography* method for the determination of traces of water in hydrocarbons. The precision at 3 to 20 ppm levels is about 15 and 7%, respectively. Meda and Bertino [5028] describe the apparatus and method for the determination of water in glycerol, phenol, ethanol, essential oils, syrups, and so forth, based on measurements of the amount of acetylene liberated from calcium carbide. The inadequacies of other methods are pointed out.

Pearson and Ollerenshaw [5733] point out that water determination in many organic solvents by means of the overtone band of water near 5300 cm^{-1} has become a simple matter. The method is rapid and only a small volume of sample is required. An accuracy of $\pm0.02\%$ has been suggested for the range of 0.02 and 1.00% water, but probably can be improved. Good agreement has been reported between this and other methods that have been proven reliable; a comparison with the Karl Fischer Method has been made by Meeker and coworkers [5034] for several functional group types. Streim and coworkers [7043] have compared results with those obtained by gas chromatography for amines and alcohols. Cordes and Tait [1658] report the results of a study of the determination of water in hydrazines, and Keyworth [3968] reports the application for alcohols.

Forbes [2475] devised an infrared method for the determination of water in liquids in the few parts per million range; examples are given from 2.37 to 16.8 ppm. The liquid was treated with calcium carbide, and the acetylene was collected in a dry gas carrier, transferred to a cell, dissolved in carbon tetrachloride, and then measured.

The *Karl Fischer Method* is the most universal and accurate method for the determination of water. Mitchell and Smith incorporated a study of accuracy, precision, and application of the Karl Fischer method in [5171].

Swensen and Keyworth [7129] found that iodine can be coulometrically generated from potassium iodide dissolved in a solvent containing pyridine, formamide, and sulfur dioxide, and it can be used in a Karl Fischer-type reaction for the determination of water below the 10-ppm range in benzene and related-type solvents.

Water cannot be determined in vinyl ethers by the regular Karl Fischer method. The reagent reacts rapidly with the ethers. It is believed by Barnes and Parvlak [625] that the methanol present in the reagent as the sample solvent takes part in the iodoacetal reaction:

$$ROCH{=}CH_2 + I_2 + CH_3OH \rightarrow ROCH(OCH_3)CH_2I + HI$$

When methanol is eliminated, the water in vinyl ethers can be determined. A commercially available, stabilized Karl Fischer reagent was used containing 2-methoxyethanol instead of methanol as the solvent.

The *acetyl pyridinium chloride method* developed by Smith and Bryant [6819, 6820] is a convenient method for determining water in liquids that do not react with the reagent. It may be preferred to the modified Karl Fischer method, which uses hydrocyanic acid for aldehydes and ketones.

Electrical conductivity offers a rapid and convenient means of determining water in many organic liquids. The electrical conductivity is very sensitive to small amounts of water in most solvents. Schmidt and Jones [6463], Lund and Bjerrum [4698], and de Brouckère and Prigogine [1891] have used electrical conductivity as a criterion of purity. Some factors that increase the precision of the method for measuring the conductivity of dilute solutions have been reported by de Brouckère and Prigogine [1890].

Other methods, such as the measurement of density, refractive index, viscosity, boiling point, freezing point, infrared spectroscopy, and mass spectroscopy, are convenient for the determination of the amount of water in some organic liquids. The physical property difference caused by the water must be of sufficient magnitude to give the accuracy desired.

Chapter V

PURIFICATION METHODS

General Comments

GENERAL COMMENTS

Introduction

The compounds are discussed in this chapter in the order established in Chapter I.

General information concerning an entire class of compounds is given, when available, following the title of the class or subgroup class. Purification and preparative methods applicable to several members of a functional group or subgroup are given in the General Comments section. Sections pertaining to specific compounds follow the General Comments section for each group and contain information on preparation, purification, criteria of purity for specific applications, and safety. When the number of compounds prepared or purified by a method is small, or if compounds from several functional groups or subgroups are involved, the method is given under one compound and reference is made thereto under the other compounds; for example, purification of ethyleneimine, Searles and coworkers [6539] and piperidine. The primary entry is under ethyleneimine.

Reference should be made to related compounds when using the directions, since one method of purification is applicable to several compounds.

Since the publication of the second edition of *Organic Solvents*, interest in the information, presence, and determination of *peroxides* has greatly increased. Formerly associated principally with ethers, peroxide formation is now known to occur among such functional-group types as hydrocarbons (particularly unsaturated ones), alcohols, ethers, esters, and nitriles.

The general discussion of peroxides is in the *Ether* section. When a method or procedure is confined, either chemically, or by the author, to one compound or one functional-group type, the information appears under the compound or group.

The sections on *safety* have been included as a convenient resumé of available information on the safe handling of the several solvents. Patty's *Industrial Hygiene and Toxicology* [5712, 5713], the *Handbook of Organic Industrial Solvents* [152], Sax's *Dangerous Properties of Industrial Materials* [6402], and the *Threshold Limit Values* published by the American Conference of Governmental Industrial Hygienists [139–142, 7022] have been drawn upon for most of the information presented. More extensive information on specific solvents is available from publications of government agencies such as the National Institute for Occupational Safety and Health.

The modes of exposure given primary consideration are the inhalation of vapors and absorption of liquid through the skin. Eye damage from some liquids and vapors is an important consideration.

The *threshold limit value* (TLV), often referred to as maximum permissible concentration (MPC) or maximum allowable concentration (MAC), is generally accepted to mean that the average person can be exposed to the stated concentration of the vapors for 8 hr per day, 5 days per week without harm. These values should be considered as guides to good practice.

The concentration of vapors in the air that is fatal to 100% and 50% of the test animals in the stated time is represented by LC_{100} and LC_{50}, respectively. The maximum concentration in the air that was not fatal to any of the test animals is represented by LC_0. LD_{100}, LD_{50}, and LD_0 represent lethal dosages, the subscript having the same meaning as for lethal concentrations.

The *flammable limits* represent the concentration range in air in which explosions occur.

The *minimum ignition temperature* is the minimum temperature at which rapid combustion becomes independent of externally supplied heat.

Although distillation remains the most important general method of purification of organic solvents, newer techniques involving solid–liquid equilibria deserve wider consideration. Specific applications include the purification of solvents for use as reference standards and as high purity materials for electrical, physical, and thermal studies. The method of *zone melting* has several ramifications of which *zone refining*, the purification technique developed by Pfann in 1952, is but one. Zone refining is the phase of zone melting of particular interest for solvent purification. The general practice of using the less specific term, zone melting, will be followed in most instances. Zief and Wilcox [8290] introduced the term *fractional solidification* as a general term for a group of processes including zone melting.

There are excellent books and reviews available, for example, *Fractional Solidification* [8290], *Zone Melting of Organic Compounds* [3321], *Zone Melting* [5809], *Zonenschmelzen* [6445], and "Zone Melting of Organic Compounds" [8056].

The principal limitation of zone melting and related techniques is that mixtures of eutectic composition present difficulties in separation. The equipment becomes expensive to obtain and operate when the temperature required to maintain the substance in the solid state is lower than that obtainable from commercially available refrigeration units.

The technique has shown itself capable of removing impurities to the parts per billion range. The need of ultrapure germanium for transistors brought zone refining into being. It is feasible to extend zone refining to achieve a purity in many solvents comparable to that which has been reached with metals.

Dickinson and Eaborn [2000] found that *purification by progressive freezing* is superior to the zone melting technique involving a single zone.

Dugacheva and Anikin [2114] describe a zone melting method for the purification of solvents crystallizing between $0°$ and $-100°$ in a hermetically sealed chamber without contact with air. Fivefold melting results in eight- to nine-fold purification of the sample. Examples are: acetone, mp $-56.798°$, and hexane, mp $-95.347°$.

Solvents of Known Quality Available

There are a number of solvents available in laboratory quantities that are of good purity and have been precisely and/or accurately characterized. Some of these materials are designed for specific end use. The small amount of impurities generally present permit them to be used for any purpose except: (1) where a very pure material is required (research grade excepted), (2) when a certain impurity, or type of impurity, is present that is deleterious to the intended use. These solvents may be further purified to the desired degree of purity relatively easily.

Some of the solvent grades available follow.

American Chemical Society reagent grade is often designated as ACS reagent grade and sometimes as ACS when the meaning is obvious. The requirements that serve as the set of minimum standards for this grade of solvents may be found in *Reagent Chemicals* [137]. Some solvents, for example, hexane, meet the reagent grade requirements, and a second grade has, in addition, an optical absorbance requirement. Rosin's *Reagent Chemicals and Standards* [6263] is similar to *Reagent Chemicals* but includes more compounds. In general, if a solvent will pass the requirements of one, it will pass those of the other.

Research grade is believed to have been first used by Phillips Petroleum Company to designate its high-purity research and reference standards characterized by the thermodynamic freezing curve [5822]. It is used herein to include solvents of similar purity from other sources. A research-grade solvent usually has a purity of 99.9 + % m. Often the impurities, or at least some of them, will be given.

Spectrophotometric grade was originally conceived as suitable for the ultraviolet region of the spectrum. It has been extended by several suppliers to include the visible, near-infrared and infrared regions. Some suppliers will furnish copies of the spectrum of the batch shipped. Generally, these solvents are also of reagent-grade quality. Eastman's Distillation Products Industries' brochure, *Spectrophotometric Solvents*, and Matheson Coleman & Bell's brochure, *Spectroquality Solvents*, list 30+ solvents each and give typical spectra and related information.

Chromatographic-quality solvents are 99+%m determined by gas chromatography and are available from a limited number of sources. A chromatogram may be obtained with each unit that will also give information for the conditions of analysis.

Pesticide grade is also known by the registered trademark name *Nanograde*. These solvents are suitable for pesticide residue analysis. They meet ACS reagent-grade requirements and, in addition, do not contain more than 10 ng/L of a chlorinated pesticide or an equivalent amount of an impurity that would produce a comparable chromatographic peak at similar operating conditions.

Nonaqueous titration solvent-grade solvents are low in water and in substances that are acidic or basic in the solvent. They are suitable for normal use as received. They often will have to be dried for high-precision titrations. The Eastman Kodak Company lists 54 solvents suitable for normal titrimetric use in their 1968 brochure *Titrants, Indicators, Solvents for Non-Aqueous Titrimetry*.

Electronic-grade solvents are used for precleaning and drying metal parts of electric circuitry. They have low boron and metallic impurities and low specific conductance. The specifications for three electronic-grade solvents are given by the Fisher Scientific Co. in *Technical Data Sheet*, TD-143, 10/63. The specifications are given in Table 5.1.

There are other specialized grades of solvents, but those cited are the most common and easily obtained.

Table 5.2 lists the solvents available in 1968 in several categories. This list has been compiled from the following sources: Matheson Coleman & Bell, *Laboratory Chemical Catalogue*, 1967 and MC/B *Spectroquality Solvents*, 1965; *Phillips 66 Hydrocarbons and Petro-Sulfur Compounds*, Bulletin 522, 6th ed.; *Fisher Chemical Index 67-C*; E. H. Sargent & Co., *Chemical Reagents*; Eastman's Distillation Products Industries, *Spectrophotometric Solvents*; Eastman Organic Chemicals, *Titration, Indicators, Solvents for Non-Aqueous Titrimetry*; Mallinckrodt, *Product List*, 1965. Matheson Coleman & Bell was the primary reference source because of the large number of solvents tabulated in the appendix of their Chemical Catalogue according to the several grades. Phillips 66 was the primary reference for hydrocarbons and petrosulfur compounds, and Eastman's Distillation Products for nonaqueous titration solvents.

HYDROCARBONS (1–77)

General Comments

Aromatic hydrocarbons were available in England about the middle of the nineteenth century. The discovery of the first synthetic dye *mauve* by Perkin in 1856 ushered in the golden age of the aromatics. Stillman's analysis of the surface petroleum at Titusville was made in 1859. About three-quarters of a century passed before scientific technology became sufficiently sophisticated to utilize the potentials of petroleum hydrocarbons. W.

Table 5.1 Specification for Electronic-Grade Solvents

	Acetone	Methanol	Trichloroethylene
Assay, min (%)	95.5	99.5	99.5
Color, APHA	5	5	5
Boiling point (°C, 760 Torr)	56.1	64.6	87.0
Boiling range (°C, 760 Torr)	±0.5	±0.5	±0.5
Specific gravity (25°C)	0.7871	0.7886	1.460 to 1.464
Specific conductance, max, 25°C (ohm^{-1} cm^{-1})	1.0	4.0	0.08
Residue on evaporation (%)	0.001	0.001	0.001
Water, Karl Fischer (%, max)	0.5	0.10	0.01
Solubility in water (ACS)	PT^{1a}	PT^1	—
Acidity as acetic acid (%)	0.002	0.002	0.0 as HCl
Alkalinity as NH_3 (%)	0.001	0.0003	PT^{2b}
Methanol (%)	0.05	—	—
Acetone (%)	—	0.001	—
Aldehydes (%)	PT^1	—	—
Substances reducing $KMnO_4$ (ACS)	PT^1	PT^1	—
Subtances darkened by H_2SO_4	—	PT^1	—
Heavy metals, as lead (%)	—	—	0.001
Free halogen	—	—	none
Al, Fe, Cu, As, Sb, Pb, Ni, Ag, Au, and B, each (max, ppm)	0.1	0.1	0.1

[a] PT^1, to pass test.
[b] PT^2, to pass test as NaOH.

Nelson Axe believes that, "As a chemical class, hydrocarbons must rank close to air, water, carbohydrates, and protein in basic importance to the human race."

Improved technology in the petroleum industry has provided additional aromatic and many paraffinic and naphthenic hydrocarbons in several grades of purity at a reasonable cost. The impurities present in the different grades may be characterized and evaluated by gas chromatography and mass spectroscopy. The amount of impurities present, but not the kind, is most accurately evaluated in high purity materials by the thermometric freezing or the adiabatic calorimetric method (see Chapter IV).

The American Petroleum Institute Research Project 6 was begun in 1927 as a comprehensive investigation of hydrocarbons in petroleum [6274]. This project was the real start of the intensive and systematic separation and study of the several components of petroleum. The success of this study was made possible by the development of needed methods and equipment to solve the many problems associated with such a complex system. The methods used for separating the hydrocarbons have been mainly the physical processes of distillation, crystallization, extraction, adsorption, and the group of processes characterized as fractional solidification.

Several variations of distillation such as regular, reduced pressure, and azeotropic, have been used. Regular distillation has normally been the first step in the purification.

Table 5.2 Solvents Available for Special Purposes

Code		Code	
A	American Chemical Society reagent grade	C	Chromatographic grade
R	Research grade	P	Pesticide grade
S	Spectrophotometric grade	N	Nonaqueous titration solvent grade
		E	Electronic grade

Number	Compound	A	R	S	C	P	N	E
4	Cyclopentane	·	×	×	·	·	·	·
5	Pentane	·	×	×	×	·	·	·
6	2-Methylbutane	·	×	×	×	·	·	·
7	2,2-Dimethylpropane	·	×	·	×	·	·	·
8	Methylcyclopentane	·	×	·	×	·	·	·
9	Cyclohexane	·	×	×	×	×	×	·
10	Hexane	×	×	×	×	×	·	·
11	2-Methylpentane	·	×	·	×	×	·	·
12	3-Methylpentane	·	×	·	×	·	·	·
13	2,2-Dimethylbutane	·	×	·	×	·	·	·
14	2,3-Dimethylbutane	·	×	·	×	·	·	·
15	Methylcyclohexane	·	×	×	×	·	·	·
16	Heptane	·	×	×	×	×	·	·
18	3-Methylhexane	·	×	·	×	·	·	·
20	2,4-Dimethylpentane	·	·	×	·	·	·	·
25	Octane	·	×	·	×	·	·	·
27	2,2,4-Trimethylpentane	×	×	×	×	×	·	·
29	Nonane	·	×	·	×	·	·	·
30	2,2,5-Trimethylhexane	·	×	·	×	·	·	·
31	Decahydronaphthalene	·	·	×	·	·	·	·
32	cis-Decahydronaphthalene	·	×	·	·	·	·	·
33	trans-Decahydronaphthalene	·	×	·	·	·	·	·
34	Decane	·	×	·	·	·	·	×

Table 5.2 (Continued)

Code

Code	Description
A	American Chemical Society reagent grade
R	Research grade
S	Spectrophotometric grade

Code

Code	Description
C	Chromatographic grade
P	Pesticide grade
N	Nonaqueous titration solvent grade
E	Electronic grade

Number	Compound	A	R	S	C	P	N	E
36	Dodecane	·	X	·	·	·	·	·
38	Benzene	X	X	X	X	X	X	·
39	Toluene	X	X	X	X	X	X	·
40	o-Xylene	·	X	·	X	·	·	·
41	m-Xylene	·	X	X	X	·	·	·
42	p-Xylene	·	X	·	X	·	·	·
43	Ethylbenzene	·	X	·	·	·	·	·
44	Isopropylbenzene	·	X	·	·	·	·	·
46	Naphthalene	·	X	·	·	·	·	·
51	Butylbenzene	·	X	·	·	·	·	·
52	sec-Butylbenzene	·	X	·	·	·	·	·
54	tert-Butylbenzene	·	X	·	·	·	·	·
57	1-Pentene	·	X	·	·	·	·	·
58	2-Pentene (mixed isomers)	·	X	·	·	·	·	·
59	cis-2-Pentene	·	X	·	·	·	·	·
61	1-Hexene	·	X	·	·	·	·	·
63	1-Octene	·	X	·	·	·	·	·
71	Cyclohexene	·	X	·	·	·	·	·
78	Methanol	X	·	X	X	X	X	X
79	Ethanol	X	·	·	·	X	·	·
80	1-Propanol	·	·	·	·	·	X	·
81	2-Propanol	X	·	X	X	X	X	·
82	1-Butanol	X	·	X	X	X	X	·
83	2-Butanol	X	·	·	X	·	·	·

No.	Compound
84	2-Methyl-1-propanol
85	2-Methyl-2-propanol
86	1-Pentanol
90	3-Methyl-1-butanol
94	Cyclohexanol
116	Phenol
125	2-Propen-1-ol
128	2-Propyn-1-ol
134	1,2-Ethanediol
135	1,2-Propanediol
137	1,3-Butanediol
147	Glycerol
151	Ethyl ether
153	Isopropyl ether
155	Butyl ethyl ether
162	Bis(2-methoxyethyl) ether
169	Furan
170	Tetrahydrofuran
172	p-Dioxane
189	Acetone
190	2-Butanone
194	Cyclohexanone
196	4-Methyl-2-pentanone
203	Acetophenone
207	Formic acid
208	Acetic acid
209	Propionic acid
221	Acetic anhydride
222	Propionic anhydride
224	Methyl formate
225	Ethyl formate
229	Methyl acetate

Table 5.2 (Continued)

Legend:

Code
- A — American Chemical Society reagent grade
- R — Research grade
- S — Spectrophotometric grade

Code
- C — Chromatographic grade
- P — Pesticide grade
- N — Nonaqueous titration solvent grade
- E — Electronic grade

Code		A	R	S	C	P	N	E
232	Ethyl acetate	×	·	·	×	×	·	×
238	Butyl acetate	·	·	·	×	×	·	·
241	Pentyl acetate	·	·	·	·	·	·	·
247	Ethyl propionate	·	·	·	×	×	·	·
254	Ethyl acrylate	·	·	·	·	·	·	·
255	Methyl methacrylate	·	·	·	×	·	·	·
264	Ethylene carbonate	·	·	·	×	×	·	·
298	2-Chloropropane	·	·	·	×	·	·	·
307	Chlorobenzene	·	·	·	·	·	·	×
315	Dichloromethane	·	·	·	×	×	×	×
316	Chloroform	×	·	·	×	×	×	×
317	Carbon tetrachloride	×	·	·	×	×	×	×
319	1,2-Dichloroethane	·	·	·	×	×	·	·
324	1,1,2,2-Tetrachloroethane	·	·	·	×	·	·	·
328	p-Dichlorobenzene	·	×	·	·	·	·	·
336	Trichloroethylene	·	·	·	×	×	·	×
337	Tetrachloroethylene	·	·	·	×	×	·	·
341	2-Bromopropane	·	·	·	·	·	·	·
351	Bromoform	·	·	·	×	×	·	·
355	Iodomethane	·	·	·	×	·	×	×
380	Nitromethane	·	·	·	×	×	×	×

No.	Compound						
384	Nitrobenzene	·	×	·	·	·	·
385	Acetonitrile	·	×	×	×	×	·
395	Benzonitrile	·	×	·	·	×	·
396	Acrylonitrile	·	·	·	·	·	·
401	Butylamine	·	×	·	·	·	·
408	Aniline	·	×	·	×	·	×
419	Diethylamine	·	×	·	·	·	·
428	Triethylamine	·	×	·	·	·	·
431	Pyridine	·	×	×	·	×	×
441	N-Methylformamide	·	·	·	·	·	·
442	N,N-Dimethylformamide	·	×	×	×	×	·
445	N,N-Dimethylacetamide	·	·	×	×	×	·
449	1-Methyl-2-pyrrolidinone	·	·	×	×	×	·
451	Carbon disulfide	×	·	·	×	·	×
462	Dimethyl sulfoxide	·	×	×	×	×	·
463	Sulfolane	·	×	·	×	×	·
464	2-Methoxyethanol	·	×	×	×	×	×
465	2-Ethoxyethanol	·	×	×	×	·	·
466	2-Butoxyethanol	·	·	×	×	·	·
467	Furfuryl alcohol	·	·	·	·	·	·
468	Tetrahydrofurfuryl alcohol	·	×	×	×	·	×
469	Diethylene glycol	·	·	×	·	·	·
472	2-(2-Ethoxyethoxy)ethanol	·	·	×	×	×	×
476	2-Chloroethanol	·	×	×	·	×	·
478	2-Aminoethanol	·	·	·	·	·	·
483	2-Furaldehyde	·	×	×	·	×	·
487	Morpholine	·	×	·	·	×	·

In the absence of azeotropic mixtures, it removes all except the very close boiling impurities.

The primary purposes of azeotropic distillation are: (1) to remove impurities of a type different from the main component; (2) to remove isomeric or closely related impurities; (3) to reduce the column hold-up by selecting a suitable low-boiling, azeotrope-forming compound; and (4) to permit complete distillation of the main component by using an excess of the azeotropic-forming compound. The distillation equipment and operational techniques used by the API-NBS group have been discussed by Willingham and Rossini [8090].

Adsorption has been used to remove water and nonhydrocarbon impurities. It is also applicable for removing aromatic hydrocarbons from paraffinic and cycloparaffinic hydrocarbons and will, in some cases, remove isomeric impurities from hydrocarbon compounds. The chromatographic method for separating the aromatic hydrocarbons from paraffinic and cycloparaffinic compounds has been described by Mair and Forziati [4837]. This method has been extended by Mair [4835] to allow the separation of aromatic, paraffinic, and cycloparaffinic hydrocarbons from a mixture of the three. It may be used analytically to determine the amount of each type of hydrocarbon in a mixture [4836].

The results in Table 5.3 summarize the purification of paraffins and monoolefins as carried out by the API-NBS group [7036]. In connection with the purification, the authors state: "(1) A logical, simple purification following the original synthesis or other preparation of a given hydrocarbon concentrate will usually remove all impurities except more or less close-boiling isomers, (2) an impurity of several percent of close-boiling isomers will have relatively little effect on the boiling point, refractive index, or density, but will normally affect the freezing point appreciably, (3) in a series of fractions obtained from a distillation at a high efficiency of a hydrocarbon containing close-boiling isomers as impurity, the fraction of highest purity will be beyond or ahead of the middle portion of the distillate as frequently as in the middle portion, and (4) for producing material of highest purity, the blending of the fractions of a distillate can be done safely only on the basis of the freezing points of the selected fractions."

Additional purification data from API Project 6 [2494] are summarized in Table 5.4.

Table 5.3 Purification of Paraffins and Mono-Olefins by Fractional Distillation

Number	Compound	Initial Purity (% m)	Kind of Distillation	Azeotrope-Forming Substance	Final Purity (% m)
17	2-Methylhexane	—	Azeotropic	Methanol	99.82 ± 0.07
18	3-Methylhexane	—	Azeotropic	Ethanol	99.80 ± 0.15
19	2,3-Dimethylpentane	—	Azeotropic	Ethanol	99.80 ± 0.15
20	2,4-Dimethylpentane	99.35	Azeotropic	Ethanol	99.88 ± 0.05
25	Octane	98.78	Regular	—	99.95 ± 0.04
26	2,2,3-Trimethylpentane	91.4	Regular	—	99.68 ± 0.20
57	1-Pentene	95.0	Regular	—	99.49 ± 0.40
59	*cis*-2-Pentene	—	Azeotropic	Methanol	99.80 ± 0.15
60	*trans*-2-Pentene	99.90	Regular	—	99.93 ± 0.05

Table 5.4 Purification of Hydrocarbons by Fractional Distillation

Number	Compound	Kind of Distillation	Azeotrope- Forming Substance	Final Purity (% m)
5	Pentane	Azeotropic	Methanol	99.86
6	2-Methylbutane	Regular	—	99.46
8	Methylcyclopentane	Azeotropic	Methanol	99.87
9	Cyclohexane	Regular	—	99.997
10	Hexane	Azeotropic	Methanol	99.91
11	2-Methylpentane	Azeotropic	Methanol	99.89
12	3-Methylpentane	Azeotropic	Methanol	—
13	2,2-Dimethylbutane	Azeotropic	Methanol	99.94
14	2,3-Dimethylbutane	Azeotropic	Methanol	99.90
15	Methylcyclohexane	Azeotropic	Methanol	99.90
16	Heptane	Azeotropic	Methanol	99.93
25	Octane	Azeotropic	Ethanol	99.96
26	2,2,3-Trimethylpentane	Azeotropic	2-Methoxyethanol	99.3
27	2,2,4-Trimethylpentane	Azeotropic	Ethanol	99.88
29	Nonane	Azeotropic	2-Ethoxyethanol	>99.7
34	Decane	Azeotropic	2-Butoxyethanol	>99.7
38	Benzene	Regular	—	99.96
39	Toluene	Regular	—	99.90
40	o-Xylene	Azeotropic	2-Ethoxyethanol	99.90
41	m-Xylene	Azeotropic	2-Ethoxyethanol	99.71
42	p-Xylene	Azeotropic	2-Ethoxyethanol	99.92
43	Ethylbenzene	Regular	—	99.54
44	Isopropylbenzene	Azeotropic	2-Ethoxyethanol	99.96

For further information, see Glasgow, Murphy, and coworkers [2780] and Streiff, Murphy, and coworkers [7037].

Mair and coworkers [4841] have described a method of purification and sealing in vacuum of standard samples of hydrocarbons. The technique and apparatus described generally are useful for storing pure solvents.

The freezing point of a pure substance can be calculated from the freezing point of the impure substance by the thermometric freezing point method developed by Streiff and Rossini [7039], Taylor and Rossini [7202], and Glasgow, Streiff, and Rossini [2781]. The purity must be above 95%m for the best accuracy, and the impurity must be a similar substance or substances. The purity of the material is determined at the same time. This method has been used as the criterion of purity for all hydrocarbons isolated and purified by the API-NBS group.

Fenske, Quiggle, and Tongberg [2403] isolated from Pennsylvania straight-run gasoline the normal paraffins as high as tridecane in a pure state. The gasoline was distilled in a 27-ft column, 3 in. in diameter [2402]. The cuts containing the normal paraffins were treated with chlorosulfonic acid according to the method of Shepard and Henne [6638]. The remaining hydrocarbon was washed, dried, and fractionally distilled. Tongberg and Fenske [7520] isolated cyclohexane from the hexane fraction of gasoline, after removing benzene, by fractional distillation in a 52-ft, $\frac{3}{4}$-in.-diameter column [2402].

Vondráček and Dostál [7786] found that the high solubility of aromatic hydrocarbons in phenol was useful in the purification of paraffins; pure hexane could be fractionally distilled from a petroleum fraction after it had been extracted with phenol. 2-Methyl-pentane, methylcyclohexane, heptane, 3-methylpentane, and cyclohexane were also obtained in a high state of purity.

Bruun, Hicks-Bruun, and Faulconer [1210] isolated 2-methylpentane, 3-methylpentane, and 2,3-dimethylbutane from the 55–65° commercial cut from natural gas of the Clendener Gas Field of West Virginia by fractional distillation in a Bruun-type column [1208] constructed by Bruun and Faulconer [1209].

Nonoxygenated extractive solvents have been prepared *carbonyl-free* by a simple process that is adaptable to laboratory or production quantities by Hornstein and Crowe [3494]. A uniform blend was made of 60 mL of concentrated sulfuric acid and 100 grams of Celite 545 and packed into a chromatographic tube as follows: some of the solvent was poured into the tube and approximately 10 cm of granular reagent-grade sodium sulfate added. Then 160 g of the Celite mixture was added in portions, tamping each portion when added and the column topped with 7–8 cm of crystalline sodium sulfate. A flow rate of 3–5 mL per min was found satisfactory. The results for one pass of six solvents, hexane, Skelly F, benzene, dichloromethane, chloroform, and carbon tetra-chloride were reported. The carbonyl content of the unpurified solvent varied from very small to 280 μM/L; that of the purified solvent from 0.000 to approximately 0.45 μM/L. Celite impregnated with 2,4-dinitrophenylhydrazine will also remove carbonyl compounds [6512].

Saturated Aliphatic Hydrocarbons (1–37)

General Comments

Methods of Preparation

Howard and coworkers [2412] prepared 65 paraffinic hydrocarbons in pure form, many by chemical means. The procedures included several techniques for chemical preparation in large laboratory quantities of intermediates and final compounds. Several paraffins were prepared by hydrogenating alkenes. The alkenes were prepared by dehydrating the alcohols made by the Grignard synthesis.

Aschan [463] observed that chlorosulfonic acid reacted rather rapidly with branched-chain hydrocarbons. Young [8229] reported that the acid reacted much more readily with branched-chain than with straight-chain hydrocarbons. The acid has long been known to react readily with unsaturated hydrocarbons. Shepard, Henne, and Midgley [6639] isolated the normal saturated aliphatic hydrocarbons, pentane through decane, by removing the unsaturated and branched-chain components by treating an American gasoline fraction with chlorosulfonic acid until there was no change in density. The normal paraffins were separated from the naphthenes by fractional distillation. The distillation and freezing curves were used as the criteria of purity.

Isomeric hexanes were prepared by Cramer and Mulligan [1711] by the general alcohol–olefin–paraffin method. For example, hexane was prepared by passing 3-hexanol over an aluminum oxide catalyst at 350° and hydrogenating the olefin in the presence of platinum black, according to the method of Adams and Shriner [23].

A general method for the preparation of branched-chain paraffinic hydrocarbons having one carbon atom less than the parent compound is given under 2,2-dimethylpentane [3061]. Another general method for preparing branched-chain compounds is described in [1225].

Pines [5859] has reported a general method for alkylation, isomerization, autodestructive alkylation, and dealkylation. Some examples cited are: methylcyclohexane to dimethylcyclopentane or ethylcyclopentane; methylcyclopentane to cyclohexane; pentane to methylpropane, 2-methylbutane and some unsaturates; benzene and 2-chloro-2-methylbutane to tert-butylbenzene, p-tert-dibutylbenzene, and other tert-butylbenzenes. Van Pelt and Wibaut [7674] prepared esters by Spassow's method [6920]: The esters were decomposed by passing them through a tube filled with glass wool at 500–520°. Alkenes were formed that could be hydrogenated to paraffins.

Fischer and Klemm [2445] prepared aliphatic hydrocarbons by the Grignard synthesis and by decomposing nitriles with sodium.

The Wurtz synthesis has been used to prepare several compounds; see, for example, hexane, octane, and decane.

Cyclopentane, cyclohexane, and methylcyclohexane appear to be gaining status as standards in Raman spectroscopy. Aleksanyan and Sterin [80] determined the spectra of these compounds at 25–35° under precise conditions.

Hesse and Schildknecht [3358] prepared hydrocarbons for ultraviolet spectrographic study by treating them with sulfuric acid and filtering through a suitable grade of alumina. Some technical-grade substances, for example, cyclohexane, can be purified by successive passage through alumina or silica gel.

Drying and Criteria of Purity

All of the drying agents listed in Table 4.1 have been used to dry hydrocarbons. Phosphorus pentoxide, sodium, molecular sieves, and metallic hydrides are all very efficient. Silica gel has been used by Glasgow and coworkers [2780] to dry aliphatic hydrocarbons. The freezing point of the dried materials indicates that the amount of water remaining was less than 0.005%m. For the usual laboratory work, calcium chloride, sodium hydroxide, and most of the other common drying agents can be used.

The most accurate criterion of purity of hydrocarbons is the adiabatic calorimetric method. The thermometric freezing point method is somewhat more rapid and has an accuracy for some systems to ±0.002%m. Boiling point, refractive index, and density are inadequate when isomeric impurities are present, but these impurities usually have an appreciable effect on the freezing point.

The boiling point, refractive index, and density, in combination, are satisfactory for detecting impurities in the 0.01%w and higher range, particularly for nonhydrocarbons. The ebullioscopic technique of Swietoslawski [7131] is a rapid and convenient method with good sensitivity.

Safety

"Pharmacologically, the hydrocarbons above ethane can be grouped with the general anesthetics.... The vapors of these hydrocarbons are mildly irritating to the mucous membranes, the irritation increasing in intensity from pentane to octane. The liquid paraffin hydrocarbons are fat solvents and primary skin irritants.'' The *threshold limit value* is available only for the first eight members of the normal, or straight-chain, homologous series. The value for pentane is 1000 ppm; for hexane, heptane, and octane it is 500 ppm [5713].

Specific Solvents

1. Propane

Timmermans [7481] prepared nearly pure propane by heating propylcyanide with sodium. Other reduction methods for preparing propane are given in Tafel [7151], Burrell and Robertson [1268], Sabatier and Senderens [6327], and Mailhe [4833].

Hornyi [3495] prepared propane by the electrochemical hydrogenation of propionaldehyde in perchloric acid solution.

Propane was purified for *thermodynamic studies* by Kemp and Egan [3939] by bubbling it through 12 N sodium hydroxide and 36 N sulfuric acid, then through a tube containing phosphorus pentoxide, and then condensing in a previously evacuated bulb. It was fractionally distilled twice in a vacuum-jacketed column, with the middle 30% saved each time. The amount of liquid-soluble, solid-insoluble impurity was estimated to be less than 0.001%m from the heat capacity curve just below the melting point.

A propane of greater than 99% purity was further purified by Grosse and Linn [2989] for *physical property studies* by fractional distillation through a Podbielniak low-temperature distillation apparatus. Five cuts were taken on the plateau corresponding to the boiling point of the pure substance. Cuts whose refractive index did not vary more than 0.0002 from the adjacent cut were used.

Hicks-Bruun and Bruun [3377] shook propane for 10 hr under pressure at 70° with chlorosulfonic acid, which reacts with tertiary carbon atoms. The mixture was treated with sodium hydroxide, followed by distillation in a 20-plate, all-glass still.

Casado and coworkers [1380] purified propane for *molecular weight measurements* by catalytic hydrogenation followed by oxidation in a potassium permanganate tower, drying, and fractional distillation.

Murzin and coworkers [5361] removed ethyl mercaptan and hydrogen sulfide from propane with NaX zeolite. Kel'tsev and coworkers [3937] dried liquid propane by passing it through columns of synthetic zeolites and reported a drying effect twice as great as for silica gel.

A potential hazard was encountered by Kimura and Freeman [4010] during the purification of propane to a degree that would allow electron mobilities to be measured in the liquid. The gas, at about 2 atm pressure, was circulated over the surface of sodium-potassium alloy at room temperature. The alloy was stirred overnight to provide a fresh surface. The next morning the speed of the stirrer was increased. A red glow immediately formed around one end of the stirring bar, which was beneath the surface of the alloy and touching the glass wall of the trap. The stirring motor was stopped but the flow increased. Within a few seconds the glass softened and the pressure blew a hole in the container. It is believed that the reaction might have been initiated by triboelectric discharge between the alloy and the glass through the gas bubble around the end of the stirrer where it touched the wall. Kimura and Freeman recommended that the gas be circulated over a large film of alkali metal rather than over the alloy.

SAFETY

Aviado and Smith [514] reported that propane increased the pulmonary resistance in rhesus monkeys at 10–20%v.

Concentrations of 1%v in air exert only anesthetic effects, but concentrations of about 10%v may produce dizziness and much higher concentrations usually produce complete anesthesia. There are no chronic effects from continued exposure to concentrations in the range of 0.05–0.2%v [5825].

The *flammable limits* in air are 2.2 and 9.5%v and the *minimum ignition temperature* in air is 504° [5825].

2. Butane

Frankland [2525] prepared butane in 1849 by heating ethyl iodide with zinc at 150°, and Lowig [4673] reported an 1860 preparation by the action of sodium amalgam on ethyl iodide.

Calingaert and Hitchcock [1319] reduced *n*-butyl iodide with sodium in liquid ammonia. The crude butane was purified by successive distillations at 0°. Other reduction methods for preparing butane are given in Seibert and Burrell [6555], Lebeau [4441], and Faillebin [2359].

Tafel and Jurgens [7153] and Bermejo and Blas [801] prepared butane by electrochemical methods. Tanaka and Ichikawa [7176] reported a 90% conversion to butane when 1-butene was hydrogenated over a graphite–potassium–alumina complex catalyst.

Coffin and Maass [1561] purified butane, obtained by hydrogenation of 1-butene over a nickel catalyst, by slowly passing the gas successively through bromine, potassium hydroxide, sulfuric acid, and phosphorus pentoxide. After condensing the gas, the final product was fractionally distilled.

Carefully purified ethyl iodide and sodium wire were condensed by Aston and Messerly [480] to butane to assure a product that was free from substances boiling near zero. It was fractionally distilled twice in a low-temperature distillation apparatus. The portion taken for *calorimetric measurements* was dried with phosphorus pentoxide and freed from traces of air.

Peters and Lohmar [5787] adsorbed butane on activated carbon at −185° followed by stepwise desorption *in vacuo* from −80° to 100°; bp −0.7°. Vyakhirev and co-workers [7809] removed isobutane, butylene, and propane from butane by adsorption of the impurities on high-activity silica gel at 18–22°; the final product contained less than 0.2% total hydrocarbon impurities. For drying with zeolites, see Kel'tsev [3937], propane.

SAFETY

For exposure effects, see propane [5825].

The *threshold limit value* is 600 ppm, 1430 mg/m^3 [142].

The *flammable limits* in air are 1.9 and 8.5%v and the *minimum ignition temperature* in air is 430° [5825].

3. 2-Methylpropane

Butlerov [1292] prepared 2-methylpropane in 1867 by the reduction of isobutyl iodide with zinc. Other reduction methods of alkyl halides for preparing 2-methylpropane are given in Burrell and Robertson [1269], Lebeau [4441], and Noyes [5529].

Aston and coworkers [475] prepared 2-methylpropane by the hydrogenation of iso-butylene over Raney nickel at 67 atm and 16°. For *thermodynamic studies* they purified their crude product by successively passing it through concentrated sulfuric acid solution, sulfuric acid scrubbing towers, a potassium hydroxide tube, and a phosphorus pentoxide tube. Then the condensed gas was fractionally distilled over a low-temperature column. For a similar purification technique, see Coffin and Maass [1561], butane.

Landa and Mostecky [4389] hydrogenated *tert*-butyl alcohol over tungsten disulfide catalyst at 76 atm and 320–30° for 30 min; the yield of 2-methylpropane was 63%.

For purification with activated carbon, see Peters and Lohmar [5787], butane.

SAFETY

Aviado and Smith [514] reported that 2-methylpropane influenced the respiratory or circulatory system or both in rhesus monkeys at 10–20%v.

For exposure effects, see propane [5825].

The *flammable limits* in air are 1.8 and 8.4%v and the *minimum ignition temperature* in air is 476° [5825].

4. Cyclopentane

Vogel [7758] reduced pure cyclopentene in ethanol with hydrogen in the presence of Adams' catalyst [2757, p. 430] and purified the product; bp 48.4–48.6° at 763 Torr. For preparation from dicyclopentadiene, see Jacobs and Parks [3661].

Barker and Al-Madfai [620] used a continuous circular chromatography machine to produce 99.999% pure cyclopentane.

SAFETY

The toxicity resembles that of pentane [5713]. The *lower flammable limit* in air is about 1.4%v [322]. The *minimum ignition temperature* in air is 380° [1411].

5. Pentane

Directions for the preparation of pentane from 2-bromopentane are given in Marvel [4931, p. 84] and Nollar [912].

Pitzer [5872] purified a commercial product for *thermodynamic study* by fractionally freezing about 40% on a copper coil through which cold air was circulating. It was then washed with concentrated sulfuric acid and fractionally distilled. The middle 60% was used.

Murray and Keller [5346] purified pentane and other hydrocarbons by passing the solvents through columns filled with silver nitrate on alumina to remove olefins and aromatics. Spectroscopically pure hydrocarbons were obtained.

Shinkarenko and Aleskovskiĭ [6676] purified organic solvents for *ultraviolet spectroscopy* with calcined CaX zeolite.

Hesse and coworkers [3357] purified pentane for *optical studies* by column chromatography. Adsorbents used were silica gel and basic alumina.

SAFETY

Fairhall [7022] concluded that narcosis and irritation are effects of inhaling pentane; see also Stoughton and Lamson [7028].

The *threshold limit value* is 1000 ppm, 2950 mg/m^3 [5713].

The *flammable limits* in air are 1.42 and 7.80%v. The vapor is 2.49 times as heavy as air [5713]. The *minimum ignition temperature* in air is 287° [5712].

6. 2-Methybutane

See Table 5.4.

Purified 2-methylbutane was found by Potts [5929] to have remarkable transmission in the *far ultraviolet*. He studied the transmission of four promising solvents as received and treated by the following purification methods:

1. Vigorous stirring with concentrated sulfuric acid for four hours; washing twice with water and drying with anhydrous calcium sulfate. Longer or repeated treatment with sulfuric acid was ineffective.
2. Passing through a water-jacketed, 1-in.-diameter column of Davidson 200-mesh silica gel 18-in. long. The method was most effective when the hydrocarbon and the adsorbent were absolutely dry. The silica gel was dried in place at 350° for 12 hr before use. The hydrocarbon was refluxed with sodium for 1 hr and fractionally distilled from sodium through a 40 theoretical plate Podbielniak column directly into the silica gel column.
3. Same as No. 2 except the 1-hr reflux period with sodium was omitted.

Methods 2 and 3 were about equally effective. 2-Methylbutane has a useful transmission to 1790 Å in a 1-cm cell and to 1720 Å in a 0.13-cm cell.

Nilsson [5503] purified 2-methylbutane and other hydrocarbons by treating them initially with fuming sulfuric acid and alkali followed by distillation *in vacuo* at 200 mL/min. Dissolved oxygen was removed by degassing *in vacuo* at −196°.

SAFETY

The *flammable limits* in air are 1.2 and 8.2%v [7012]. The *minimum ignition temperature* in air is 420° [5712].

7. 2,2-Dimethylpropane

Whitmore and Fleming [8023] prepared 2,2-dimethylpropane by reacting methylmagnesium chloride and 2-chloro-2-methylpropane in toluene at 45–50°; yield 42–50%. The crude product was freed from olefins, fractionally distilled, scrubbed with 85% sulfuric acid and then 25% potassium hydroxide, dried with phosphorus pentoxide, vaporized and condensed, and finally fractionally distilled. The freezing point indicated a high order of purity.

The compound has also been prepared in larger yields from zinc dimethyl and 2-chloro-2-methylpropane in toluene at 5° [3520].

Haensel and Ipatieff [3061] developed a method for the preparation of hydrocarbons having one less carbon atom than the parent compound. Branched-chain hydrocarbons were reacted with hydrogen in the presence of a nickel or cobalt catalyst [3629] to yield hydrocarbons of one less carbon atom. The carbon attached to the secondary carbon is removed more rapidly than those attached to a tertiary or quaternary carbon. Conditions are given for the production of 2,2-dimethylpropane, 2,3-dimethylbutane, and 2,3-dimethylpentane.

Buck and coworkers [1225] have described a method for preparing any branched-chain hydrocarbon, including those containing a quaternary carbon atom, by starting with the appropriate ketone or iso-ester.

Heat capacity measurements were made on a high-purity 2,2-dimethylpropane purified by Aston and Messerly [479]. A partially purified sample from Whitmore and Fleming [8023] was fractionally sublimed, twice distilled in the absence of air in an all-glass system, and the middle fraction collected in a glass bulb. The material was distilled from the bulb for use, cooled to liquid air temperature, and the air pumped out to less than 10^{-5} Torr, melted, cooled, and repumped.

SAFETY

The *flammable limits* in air are 1.4 and 7.5%v and the *minimum ignition temperature* is 450° [5415].

8. Methylcyclopentane

See Table 5.4.

Tooke [7523] separated a mixture of hexane and methylcyclopentane. The procedure is described under hexane.

9. Cyclohexane

Cyclohexane is produced in large quantities by the hydrogenation of synthetic benzene.

Palfray [5651] hydrogenated commercial, thiophene-free benzene, raising the temperature every 15 min and obtained 97.5% yield of a very pure cyclohexane. The hydrogen adsorption is rapid at 155°. The pressure for benzene was not given but toluene was hydrogenated at 115 kg.

Seyer and coworkers [6594] also prepared cyclohexane by hydrogenating benzene. The hydrogenated product was fractionally distilled in a Penn State-type column 4-meters long filled with No. 18 jack chain. The distillate of 3 L was taken off in 30 fractions at a rate of 5 mL/hr at a reflux ratio of 35 : 1. The density, refractive index, and freezing point were in good agreement with API values. See also [559, 7758, 3983].

Cyclohexane as a *solvent for spectroscopy in the ultraviolet* was prepared by Ashmore [467] by passing the liquid through a 1-meter-long, 10-mm id column packed with silica. This appeared to be an application of the extensive study of Mair and Forziati [4836] on the analytical determination of hydrocarbons by adsorption on silica gel.

de Pauw and Limido [1959] prepared cyclohexane containing less than 0.01% benzene for *spectrographic use* by chlorinating in the dark at room temperature catalyzed by anhydrous iron (III) chloride. It was washed with sodium carbonate and distilled from the benzene–iron salt complex.

Samsanova and coworkers [6370] purified cyclohexane as a solvent for *ultraviolet and infrared spectroscopy* by passing it through three 11-mm-diameter, 50-mL burets filled with 40 grams of silica gel. This removed benzene, paraffinic hydrocarbons, and carbonyl compounds. The treated solvent was fractionally distilled in a 25-plate column and the first 15% of the distillate discarded.

Zaugg and Schaefer [8265] used four purified solvents to study the *effect on the ultraviolet spectra* of salts of phenols and enols. Spectrographic-grade cyclohexane was stored over lithium aluminum hydride and twice distilled from the hydride in an atmosphere of nitrogen just before use.

Timmermans and Martin [7494] treated cyclohexane containing benzene with nitrating acid. Cyclohexane is practically insoluble, whereas the nitrobenzene produced is appreciably soluble. After this treatment, any remaining benzene can be removed by the method of Mair and Forziati [4836].

Crowe and Smyth [1729] purified cyclohexane for *dielectric constant measurements.* It was washed several times in the cold with a mixture of concentrated nitric and sulfuric acids to nitrate the benzene that may have been present. After repeated washings with distilled water, it was fractionally distilled over sodium; bp 80.6°, mp 6.5°, n_D 20° 1.42650.

The use of *ultraviolet spectrophotometry for the comparison of the purity of* different samples of several solvents (cyclohexane among them), for the *detection of impurities* originally present or formed by deterioration and for the testing of purification procedures has been described by Maclean, Jencks, and Acree [4812].

Forster [2486] purified cyclohexane in a similar manner as he did benzene [2484] for *electrical conductance studies.*

Cyclohexane was purified by Brown [1170] as a solvent for *dipole moment studies* by passing it through a 60-cm column of acid-washed alumina followed by careful fractional distillation from phosphorus pentoxide. It was stored over sodium wire; d 24° 0.77410.

Aston and Mastrangelo [478] designed the apparatus and purified several compounds by *fractional melting.* Cyclohexane of 85% was purified to 99.22%m with a 37.8% yield. Another pass increased the purity of 99.75%m. The results are of doubtful validity because about 1%m methylcyclopentane, which forms solid solutions with cyclohexane, was present.

Staudhammer and Saver [6972] purified cyclohexane as a solvent for *dielectric constant measurements* by crystallizing 13 times; fp 6.51°. Before use, the cyclohexane was allowed to stand over silica gel with the hope that any polar compounds would be removed.

A spectrographic grade cyclohexane was purified by Cheng [1449] for *critical constant study* by fractional crystallization until the refractive indices of the melted crystals and the liquor were the same. It was dried by storing over sodium wire; bp 80.7–80.8°, n_D 20° 1.4261, t_c 280.2°.

For removal of dissolved oxygen, see Nilsson [5503], 2-methylbutane. For purification with silver nitrate on alumina columns, see Murray and Keller [5346], pentane.

Barker and Al-Madfai [620] used a continuous chromatography machine to produce 99.99% pure cyclohexane.

Hesse and coworkers [3357] purified cyclohexane for *optical studies* by column chromatography. Adsorbents used were silica gel and basic alumina.

Mascarelli and coworkers did considerable work with cyclohexane as a *cryoscopic and ebullioscopic solvent.* Mascarelli and Musatty's [4938] value is given for K_B. They concluded that it is not reliable as a solvent for the determination of molecular weight because it caused association with solutes containing the hydroxyl, carboxyl, nitro, and carbonyl groups [4940]. Some molecular weight values are given in [4939]; see also [4941].

The *nature of the hydrogen bonded ion-pair* was studied by Yerger [8214] by using cyclohexane that was stirred with sulfuric acid, washed, dried, and distilled.

Brown and Ives [1128] prepared cyclohexane for *dielectric constant measurements* by recrystallizing three times followed by fractional distillation; bp 80.8°, d 25° 0.77379.

A rapid test for cyclohexane is given in [1724].

Determination in air; range 0.01–100 mg/L with $\pm 3\%$ error [4085, p. 63].

SAFETY

Cyclohexane is absorbed by inhalation [5713]. The *threshold limit value* is 300 ppm, 1050 mg/m^3 [7022].

The *flammable limits* in air are 1.33 and 8.35%v [5713]. The *minimum ignition temperature* in air is 260° [5712].

10. Hexane

Petroleum ether, bp 60–70°, is principally hexane. This product may be used for hexane for many solvent purposes. The aromatics that absorb in the ultraviolet may be removed by one of the procedures described below. Fuchs [2591] describes a procedure for purifying "benzin" or hexane contaminated with benzene.

Kelso and Felsing [3936] prepared hexane by treating iodopropane successively with quantities of sodium. The hexane was washed, dried, and repeatedly treated with fresh sodium until an excess of the metal remained unchanged. The solvent was allowed to stand over sodium for 24 hr and then distilled. The distillate was heated repeatedly with concentrated sulfuric acid, washed with aqueous sodium carbonate, and then dried and distilled.

Timmermans and Martin [7495] prepared hexane by the Wurtz reaction from 1-bromopropane and sodium. They removed unsaturated compounds from the crude product by treating with nitrating acid. After being boiled with potassium carbonate until the halide was completely decomposed, the solvent was fractionated to a common density and a constant critical solution temperature in nitrobenzene. See also Vogel [7760].

Shepard, Henne, and Midgley [6639] treated crude hexane with chlorosulfonic acid for 20 days, fractionally distilled the mixture, and obtained a product distilling with a constant density to ± 0.00001 g/cm^3.

Tongberg and Johnston [7521] prepared pure hexane from petroleum ether, bp 60–70°, by fractional distillation in a 6-ft column. Benzene was removed by treating with an equal volume of nitrating mixture (58%w concentrated sulfuric acid, 25%w concentrated nitric acid, and 17%w water) and shaking for 8 hr. The hydrocarbon layer was washed with concentrated sulfuric acid, then with water, and dried. It was distilled over sodium and the distillate dried over sodium. See also Castille and Henri [1389].

Tooke [7523] separated hexane and methylcyclopentane by azeotropic distillation with methanol in a 35-plate batch still at a pressure of 10–50 psi and a reflux ratio of 20:1.

The *heat capacity* of hexane was measured by Waddington and Douslin [7816] on hexane purified as follows: The aromatics were first removed by passing through silica gel. It was fractionally distilled through a 100-plate column at a reflux ratio of 50:1. The product had a purity of 99.77%m calculated from the freezing curve.

Castille and Henri [1389] prepared hexane, suitable for *optical measurements*, from petroleum ether. See also Weigert [7952]. Ley and Hünecke [4573] purified hexane for *optical purposes* by shaking hexane repeatedly with fuming sulfuric acid until the acid was only slightly colored.

Nikuradse [5502] purified hexane to obtain a product of low *specific conductance*. After being dried with phosphorus pentoxide and filtered, the suspended and electrolytic impurities still present were removed by the application of a high potential. See also Jaffé [3665].

Forster [2486] purified hexane in the same manner as he did benzene [2484] for *specific conductance studies*.

For purification by zone melting, see Dugacheva and Anikin [2114].

Hartmann and coworkers [3199] purified hexane for *dielectric constant studies*; see benzyl alcohol.

For use as a solvent for *dipole measurements*, Morgan and Lowry [5250] treated hexane by shaking it several times with concentrated sulfuric acid, then with a 0.1 N solution of potassium permanganate in 10% sulfuric acid and finally with a 0.1 N solution of permanganate in 10% sodium hydroxide. The hexane was washed with water, dried over sodium wire, and distilled. The fraction boiling between 65–70° was used.

For *optical measurements*, hexane must have a transparency down to 1900 Å, according to Castille and Henri [1389]. The copper 1944 Å line must appear without sensible reduction in intensity through a 10- to 15-mm layer and must be detectable through a 40-mm layer. As a solvent for ordinary ultraviolet absorption measurements, hexane need only be optically transparent to about 2100 Å in a 10-cm cell. The procedure described by Maclean, Jencks, and Acree [4812] provides a means of determining the suitability of the hexane and of evaluating the effectiveness of the purification procedure.

Phillips 99%m hexane was purified by Clever and Taylor [1539] for *refractive index studies by the method of maximum deviation*. It was washed with concentrated sulfuric acid until the acid layer was no longer colored and then with distilled water to remove all traces of acidity. It was dried with Drierite and distilled from calcium hydride through a 2-ft column packed with stainless steel helices; boiling point and density were used as the criteria of purity.

For purification with silver nitrate on alumina columns, see Murray and Keller [5346], pentane. For removal of dissolved oxygen, see Nilsson [5503], 2-methylbutane.

Barker and Al-Madfai [620] used a continuous circular chromatography machine to produce 99.98% pure hexane.

For purification with zeolite, see Shinkarenko and Aleskovskiĭ [6676], pentane. Hesse and coworkers [3357] purified hexane for *optical studies* by column chromatography, using silica gel and basic alumina as adsorbents.

The initial temperature of thermal decomposition of hexane is 230–235° [7591].

SAFETY

Hexane is three times as toxic to mice as pentane. The *threshold limit value* is 500 ppm, 1800 mg/m^3 [5713, 5444].

The *flammable limits* in air are 1.18 and 7.43%v [5713]; see also [41]. The *minimum ignition temperature* in air is 225° [1411].

11. 2-Methylpentane

Cramer and Mulligan [1711] prepared 2-methylpentane by heating 2-methyl-1-pentanol with iodine at 122°. The product was hydrogenated and purified.

SAFETY

The *flammable limits* in air are 1.18 and 6.95%v [5712]; see also [8248]. The *minimum ignition temperature* in air is 306° [5712].

12. 3-Methylpentane

Howard and coworkers [3520] prepared 3-methyl-3-pentanol from ethylmagnesium chloride and ethyl acetate. The alcohol was dehydrated by refluxing with 0.2% β-naphthalene sulfonic acid. The alkenes were hydrogenated to alkanes. The mixture was fractionally distilled, filtered through silica gel, and refractionated. The criteria of purity were the boiling point and the density. See also Cramer and Mulligan [1711].

Finke and Messerly [2463] apparently induced crystallization in 3-methylpentane for the first time by cooling the liquid through the liquid-to-glass transition temperature of about 77 K followed by slow warming to temperatures between the transition and triple point, 110.25 K.

For purification with silver nitrate in alumina columns, see Murray and Keller [5346], pentane. For removal of dissolved oxygen, see Nilsson [5503], 2-methylbutane.

13. 2,2-Dimethylbutane

Brooks and coworkers [1125] dehydrated pinacolyl alcohol with alumina to 3,3-dimethyl-1-butene, 2,3-dimethyl-1-butene, and 2,3-dimethyl-2-butene. After separation by fractional distillation, the 3,3-dimethyl-1-butene was hydrogenated in the presence of Raney nickel to 2,2-dimethylbutane. See also Cramer and Mulligan [1711].

Waddington and Douslin [7816] measured the *heat capacity* of 2,2-dimethylbutane purified by fractional distillation through a 77-plate column at a reflux ratio of 60:1; the purity was found to be 99.70% from the freezing curve.

SAFETY

The *flammable limits* in air are 1.2 and 7.0%v [5415]. The *minimum ignition temperature* in air is 425° [5712].

14. 2,3-Dimethylbutane

Kay [3912] prepared 2,3-dimethylbutane by catalytic alkylation of 2-methylpropane and ethylene. The product was purified by fractional distillation through a 15-plate column at a high reflux ratio. It was used to determine the *critical temperature* and *critical pressure*.

Kotlyarevskiĭ and coworkers [4219] found the optimum conditions for alkylation to be 1.5% aluminum chloride catalyst and a temperature of 50–55°; yield 42–47%. Knight and Kelly [4070] studied various catalysts and found iron (III) pyrophosphate tetrahydrate $(Fe_4(P_2O_7)_3 \cdot 4H_2O)$ and boron trifluoride in a 1:1 mol ratio to be the most efficient. See also Haensel and Ipatieff [3061] and Brooks and coworkers [1125].

SAFETY

The *flammable limits* in air are 1.2 and 7.0%v [5415]. The *minimum ignition temperature* in air is 420° [5712].

15. Methylcyclohexane

Palfray [5651] hydrogenated toluene at 150° and 115 kg pressure for 5 hr and obtained practically pure methylcyclohexane. Vogel [7758] prepared methylcyclohexane from the 2-, 3-, and 4-methylcyclohexanols.

The ultraviolet absorption spectrum was used by Maclean, Jencks, and Acree [4812] to characterize the effectiveness of purification.

Forster [2486] purified methylcyclohexane in the same manner as he did benzene [2484] for *specific conductance studies.*

Methylcyclohexane was one of the four solvents studied by Potts for their transmission in the far ultraviolet; see 2-methylbutane [5929].

For purification with silver nitrate in alumina columns, see Murray and Keller [5346], pentane.

SAFETY

Methylcyclohexane is absorbed by inhalation. Rabbits exposed to 15,000 ppm (1.5%) of methylcyclohexane in air died in 70 min. The symptoms were conjunctival congestion, salivation, labored breathing, narcosis, and convulsions [7545]. All rabbits died after exposure to 10,000 ppm (1.0%) for 2 weeks for 6 hr/day and 5 days a week. Four weeks of exposure to about 0.5% caused no deaths [5713].

The *threshold limit value* is 500 ppm, 2000 mg/m^3 [7022].

The *lower flammable limit* in air is 1.15%v. The *minimum ignition temperature* in air is 285° [5712].

16. Heptane

Heptane has been gaining recognition as a standard reference substance. It is relatively easy to obtain or prepare and to maintain in a high state of purity. It has many attributes to recommend it as a standard substance. Many of its physical and thermodynamic properties have been accurately determined.

Vogel [7760] prepared heptane by reducing 4-heptanone with zinc and hydrochloric acid. The hydrocarbon was steam-distilled, and washed successively with water, 10% sodium carbonate solution, and water. After a preliminary drying with anhydrous magnesium sulfate, it was distilled from sodium. Toussaint [7536] obtained heptane by hydrogenating 2-ethyl-2-hexenal.

The dipole moment was measured by Smyth and Rogers [6874] on heptane obtained from the Ethyl Gasoline Corporation. No improvement was observed when the material was subjected to purification.

Heptane for *optical purposes* was purified by Herold and Wolf [3333]. Petroleum heptane was shaken twice with concentrated sulfuric acid (12 hr each time); for 12 hr more with sulfuric monohydrate; and then for 1 hr each with water, dilute potassium hydroxide, and again with water. After standing over potassium hydroxide for 24 hr, it was fractionally distilled.

Forster [2486] purified heptane in the same manner as he did benzene [2484] for *specific conductance studies.*

Heptane was one of the four solvents that Potts [5929] studied for the transmission in the far ultraviolet; see 2-methylbutane.

Aston and Mastrangelo [478] purified heptane by *fractional melting.* Four successive meltings raised the purity from 99.5 to 99.996%m in 70% yield.

For purification with zeolite, see Shinkarenko and Aleskovskiĭ [6676], pentane. Hesse and coworkers [3357] purified heptane for *optical studies* by column chromatography using silica gel and basic alumina as adsorbents.

The initial temperature of thermal decomposition is 210–215° [7591].

SAFETY

Slight dizziness develops in humans after breathing air containing 1000 ppm (0.1%) of heptane for 6 min and for 4 min at 2000 ppm [5713].

The *threshold limit value* is 400 ppm, 1600 mg/m^3 [142].

The *flammable limits* in air are 1.10 and 6.70%v [5712]. The *maximum ignition temperature* in air is 215° [1411].

17.   2-Methylhexane

Streiff and coworkers [7036] purified 2-methylhexane by azeotropic distillation with methanol at a reflux ratio of 290:1. Azeotropic distillation with ethanol in a 135-plate column at a 145:1 reflux ratio was also used. The purity was 99.82%m.

18.   3-Methylhexane

Optically active 3-methylhexane was prepared by Gordon and Burwell [2869] from optically active 4-methyl-2-hexanol, which had been prepared from 2-methyl-2-bromobutane and acetaldehyde by the Grignard reaction; $[\alpha]_D$ 25° +9.12°, d 25° 0.6824, and n_D 25° 1.3889. Heller [3274] converted $d(-)$-2-methyl-1-butanol to the $d(+)$-2-methyl-1-bromobutane. The yield of the optically active hydrocarbon by the Grignard reagent was low. He also prepared the $d(+)$-hydrocarbon by the malonic ester synthesis; bp 91.9°, n_D 20° 1.3886, d 23° 0.6855, and $[\alpha]_D$ 23° +8.02°.

Easton and Hargreaves [2191] prepared *optically active* 3-methylhexane by three methods starting with optically active 2-methyl-1-butanol. One method was a slight modification of that of Heller [3274]. Another method involved the use of ethyl lithium and 1-bromo-2-methylbutane. The third method involved the preparation of pentyl lithium and its reaction with dimethyl sulfate to yield 44.4% purified product with a rotation of $[\alpha]_D$ 19.9° +9.43°, n_D 19.9° 1.3892, d 19.9° 0.6898.

Lardicci and coworkers [4412] prepared a number of *optically active* paraffins and alcohols by the addition of ethylene to (D)(+)-2-methyl-1-butyl lithium. The lithium reagent was prepared from (L)(+)-1-chloro-2-methylbutane in a petroleum ether solution in 80–90% yield. The reaction product of the lithium alkyl and acetylene was hydrogenated and the optically active paraffin purified; bp 91–92°, at 760 Torr, n_D 25° 1.3861, d 25° 0.6826, and $[\alpha]_D$ 25° +9.43°.

19.   2,3-Dimethylpentane

Streiff and coworkers [7036] purified 2,3-dimethylpentane by azeotropic distillation with ethanol in a 130-plate column at a reflux ratio of 145 : 1; 99.80%m from the freezing curve.

For preparation, see Haensel and Ipatieff [3061].

Lardicci and coworkers [4411] reacted (−)-(s)-3-methylpentanal with formaldehyde and dimethylammonium chloride at 80° and obtained 60% (+)-(s)-2-methylene-3-methylpentanal. This was reduced with lithium aluminum hydride to (+)-(s)-2-methylene-3-methylpentanol, which was transformed into (−)-(s)-2,3-dimethylpentane; $[\alpha]_D$ 25° −10.39°.

SAFETY

The *explosive limits* in air are 1.12 and 6.75%v. The *minimum ignition temperature* in air is 337° [5712].

20.   2,4-Dimethylpentane

Fawcett [2382] prepared 2,4-dimethylpentane by alkylating 2-methylbutane with propylene in the presence of aluminum chloride. The reaction product was distilled, filtered through silica gel to remove halogen-containing compounds, and then fractionally distilled; bp 80.7°, d 20° 0.6738, n_D 20° 1.3821.

21.   1,2-Dimethylcyclohexane (Mixed Isomers)
22.   *cis*-1,2-Dimethylcyclohexane
23.   *trans*-1,2-Dimethylcyclohexane

Sabatier and Senderens [6326] prepared the mixed isomer by reacting o-xylene and hydrogen over nickel at 180°; Chavanne and Simon [1435] carried out the reaction over platinum black.

Auwers [499] prepared the mixed isomer by the reduction of 1-chloro-1,2-dimethylcyclohexane with sodium in moist ether.

Siegel and Dunkel [6719] hydrogenated 1,2-dimethylcyclohexene over Adams' platinum(IV) oxide in acetic acid to give 77% *cis*-isomer at 2 atm and 86% *cis*-isomer at 150 atm.

Lapporte and Schuett [4410] used a complex nickel catalyst in the hydrogenation of o-xylene at 150° to give 64.5% *cis*- and 35.5% *trans*-1,2-dimethylcyclohexane.

Forziati and coworkers [2494] purified the two isomers for *physical property studies* by azeotropic distillation with isopropanol.

SAFETY

The *lower flammable limit* in air is 0.95%v for each isomer [5826].

24.   Ethylcyclohexane

Glasgow and coworkers [2780] purified ethylcyclohexane by azeotropic distillation with 2-ethoxyethanol. The purity determined from the freezing curve was 99.90%m. See also Forziati et al. [2494].

SAFETY

The *flammable limits* in air at elevated temperatures are 0.95 and 6.60%v. The *minimum ignition temperature* in air is 262° [5712].

25. Octane

Doolittle and Peterson [2045] prepared octane for the determination of *physical properties*. Constant density octanol, obtained by fractional distillation, was converted to the bromo compound by saturation with anhydrous hydrogen bromide at 100° and the reaction mixture distilled. Constant density fractions were converted to octane by the Grignard reaction. The product was fractionally distilled twice, and the heart cuts of identical densities were used for the measurememts.

Vogel [7760] prepared octane for *parachor studies* by two methods: (1) the Wurtz synthesis from sodium and 1-bromobutane, and (2) the Clemmensen reduction of 2-octanone. The boiling point, refractive index, and density were used as the criteria of purity. Several references to other methods of preparation are given.

According to Lewis and coworkers [4568], large quantities of octane can be obtained from 1-bromobutane by the Wurtz synthesis; 450 g yield.

Pomerantz and coworkers [5910] prepared a high purity 1-octene by fractional distillation of a commercial product and reduced it to octane by hydrogenation. The octane was passed through silica gel and fractionally distilled through a 2.5 × 127-cm glass Heligrid packed column. The hydrocarbon was 99.7%m and was used for *physical property studies*.

Purification by zone melting; see Dugacheva and Anikin [2114].

For purification with zeolite, see Shinkarenko and Aleskovskiĭ [6676], pentane.

The initial temperature of thermal decomposition is 195–200° [7591].

SAFETY

The *threshold limit value* has been established at 300 ppm, 1450 mg/m^3 [142].

The *flammable limits* in air are 0.96 and 4.66%v [5713]; see also [41, 8248]. The *minimum ignition temperature* in air is 220° [5712].

26. 2,2,3-Trimethylpentane

Howard and coworkers [3520] prepared this compound in a 31% yield from isopropylmagnesium chloride and 2-chloro-2-methylbutane.

27. 2,2,4-Trimethylpentane

Samsanova and coworkers [6370] freed 2,2,4-trimethylpentane from isomeric compounds and a small amount of toluene by passing it through an 11-mm diameter, 50-mL buret packed with 40 g of silica gel and found it *spectrographically pure* as a solvent for *ultraviolet* spectroscopy. It was found suitable for *infrared* use when passed through a second column.

A 99.6%m 2,2,4-trimethylpentane was purified by one pass of fractional melting by Aston and Mastrangelo [478] to a purity of 99.99%m in a 78.0% yield. An improved apparatus [4960] permitted a purification from 94.4 to 99.998%m in six passes in 80% yield.

For purification with silver nitrate in alumina columns, see Murray and Keller [5346], pentane. For purification with zeolite, see Shinkarenko and Aleskovskiĭ [6676], pentane.

Hesse and coworkers [3357] purified 2,2,4-trimethylpentane by column chromatography. Adsorbents used were silica gel, basic alumina, and acidic alumina.

The initial temperature of decomposition is 295–300° [7591].

SAFETY

The *flammable limits* in air at elevated temperature are 1.00 and 6.03%v [5712]; see also [8248]. The *minimum ignition temperature* in air is 434° [5712].

28. 2,2,3,3-Tetramethylbutane

Henry [3305] prepared 2,2,3,3-tetramethylbutane by reacting pentamethylethyl bromide and methylmagnesium bromide.

Whitmore and coworkers [8031] mixed pentamethylethyl bromide and dimethylzinc in xylene at 32°, heated the mixture at 43° for 15 min and decomposed the excess dimethylzinc with water and concentrated HCl solution. The xylene layer was separated, dried over anhydrous potassium carbonate, and fractionally distilled to give a 50% yield of 2,2,3,3-tetramethylbutane.

Flood and Calingaert [2465] mixed *tert*-butyl chloride, a few flakes of iodine, and magnesium turnings in ether. The ether layer was separated and dried over calcium chloride. After removing the ether by rapid distillation, the residue was brominated to remove unsaturated hydrocarbons, and 2,2,3,3-tetramethylbutane was obtained by distillation up to 120°; yield 10%, mp 100.7–101.4°.

Marker and Oakwood [4890] reacted *tert*-butylmagnesium chloride in ether, *tert*-butyl chloride, and *tert*-butyl iodide with copper(I) iodide to give 16% 2,2,3,3-tetramethylbutane.

SAFETY

The *lower flammable limit* in air is 1.0%v [5826].

29. Nonane

Nonane was first synthesized by Krafft [4235] in 1882 by heating pelargonic (nonanoic) acid with hydroiodic acid and phosphorus in a sealed tube at 240°.

Bazhulin and coworkers [698] prepared nonane from 5-nonanol, ethyl formate, and butylmagnesium bromide by dehydration and hydrogenation over a platinum catalyst.

For *physical property measurements*, Doolittle and Peterson [2045] isolated 1-decanol, fp > 6.5°, by fractionally distilling du Pont "Antifoam" LF and hydrogenating to nonane; yield 85%. It was fractionally distilled twice and the final center cuts of identical density were used.

Mears [5016] and coworkers prepared 2.38 kg of high-purity nonane in three steps: 5-nonanol was prepared by a modification of the method of Coleman and Craig [912, p. 179] by reacting methyl formate with butylmagnesium chloride. The 5-nonanol was dehydrated with aluminum oxide catalyst at 285–325° to nonenes, which were hydrogenated with nickel-on-kieselguhr at 140–160° at an initial pressure of 1350 psi. The nonane was purified by fractional distillation in a Podbielniak Hypercal Heligrid packed column 2.5 × 250-cm. The freezing point was the criterion of purity.

Kieffer [3996] obtained 99.8% pure nonane with a zone melting technique. For purification with zeolite, see Shinkarenko and Aleskovskiĭ [6676], pentane.

The initial temperature of thermal decomposition is 185–190° [7591].

SAFETY

The *flammable limits* in air are 0.8 and 2.9%v [5415]. The *minimum ignition temperature* in air is 206° [5712].

The *threshold limit value* is 200 ppm, 1050 mg/m^3 [142].

31. Decahydronaphthalene (Mixed Isomers)
32. *cis*-Decahydronaphthalene
33. *trans*-Decahydronaphthalene

Commercial decahydronaphthalene is a mixture of the *cis*- and *trans*-isomers produced by the hydrogenation of naphthalene. The purity of the hydrogenated product depends on the purity of the naphthalene and the purification process of the manufacturer.

Several processes for the purification of mixed isomers differ slightly from one another [3348, 6497, 8093, 8291]. The following method is based on those processes cited and the experience of the authors in purifying a similar material. Shake with several portions of about 7% sulfuric acid until no appreciable darkening is observed. Wash with water, dilute sodium hydroxide, and three portions of water, and then dry with Drierite. Fractionally distill under reduced pressure. The boiling points of the *cis*- and *trans*-isomers at a particular pressure may be calculated from the vapor pressure–temperature equation. The purity of the product will depend on the efficiency of the column used, the reflux ratio, and the closeness of the cut to the calculated boiling point of the isomers.

The washed material may be dried with calcium chloride and then allowed to stand over sodium wire for several days with occasional shaking, or the last traces of water may be removed from the distillate by passing it through a column of silica gel.

Decahydronaphthalene from different sources will show considerable variation in composition of the *cis*- and *trans*-isomers. Seyer and Walker [6593] found that the material decomposed when distilled at atmospheric pressure. The solvents from various sources were combined and fractionally distilled into five fractions. Fractions one and two were mixed and refractionated as were fractions four and five. The respective distillates were then fractionally crystallized to a constant melting point. Nine solutions of *cis*- and *trans*-isomers were prepared by mixing weighed amounts. It was found that density and refractive index were linear functions of composition. Other *physical property measurements* have been reported [6590, 6591].

For preparation by hydrogenating naphthalene, see Baker and Schuetz [559] and Palfray [5651].

Musser and Adkins [5366] studied the selective hydrogenation of naphthalene and biphenyl with Raney nickel and copper–chromium oxide catalysts, respectively, to 1,2,3,4-tetrahydronaphthalene, decahydronaphthalene, cyclohexylbenzene, and bicyclohexyl. Raney nickel is the more active catalyst, requiring the lowest temperature and the shortest time of hydrogenation.

Commercial decalin was fractionally distilled by Allinger and Coke [96] through an all-glass Penn State-type still with a 4-ft column packed with glass helices. The fractions distilling at 192.0–193.0° at 740 Torr were refractionated in the same equipment and gave *cis*- and *trans*-isomers that did not show impurities by gas chromatography.

Hydrogenation of *cis*- or *trans*-decahydronaphthalene with Raney nickel catalyst at 250–300° and 200 atm was reported by Boelhouwer and coworkers [938] to give an equilibrium mixture containing 10% of the *cis*-isomer. At 275° and 15% catalyst, the *cis*-isomer was transformed into the equilibrium mixture in 4 hr and the *trans*-isomers in 1 hr.

Zelinsky and Turowa-Pollak [8275] found that aluminum chloride at room temperature converts the *cis*- into the more stable *trans*-form.

Fenske, Myers, and Quiggle [2401] separated the *cis*- and *trans*-isomers from decalin by fractional distillation through a 75-plate column at a reflux ratio of 40:1. *Physical properties* and *vapor-liquid equilibrium data* were determined on the center cut of each isomer. The boiling point and refractive index were used as the criteria of purity.

The isometric decahydronaphthalenes were purified by Staudhammer and Sayer [6972] for *dielectric constant study*. The materials were fractionally distilled, followed by fractional crystallization. The *freezing point* of the *cis*-isomer after 12 successive crystallizations was −43.15°, that of the *trans*-isomer after 11 successive crystallizations was −30.56°.

Streiff and coworkers [7041] purified *cis*-decahydronaphthalene for *physical and thermodynamic property studies* by fractionally distilling in a 130 theoretical plate column at an approximate reflux ratio of 145:1; purity 99.93%m calculated from the freezing curve. The *trans*-isomer was distilled in a 200 theoretical plate column at a reflux ratio of 160:1; purity 99.97%m.

Seyer's [6587] study of *heat capacity* of *cis*-decahydronaphthalene definitely suggests a *transition point* at 50.1–50.5°. He did not report a similar irregularity for the *trans*-isomer. No anomalous behavior of the *trans*-isomer was detected from the temperature-viscosity relationship [7642] by Urazovskii and Chernyavskii.

SAFETY

Decahydronaphthalene vapor is irritating to the eyes, nose, and throat. Colored urine has been reported by workers exposed to a mixture of decalin and tetralin. Decalin and each isomer is absorbed through the skin. Dermatitis, but no systemic poisoning, has been reported by painters using decalin. Prolonged application to the skin caused the death of guinea pigs [5992].

The *threshold limit value* of 25 ppm, 140 mg/m^3, has been suggested by Stokinger [7022].

The *flammable limits* in air at 100° are 0.7 and 4.9%v for the mixed isomers [5415]. The *minimum ignition temperature* in air is 262° [5712].

34.   Decane

Vogel [7760] prepared decane by the Wurtz synthesis from 1-bromopentane and sodium. The material was used for *parachor studies*.

Chu [1479] purified "olefin-free"-grade decane and bis(2-chloroethyl) ether for *critical opalescence studies* by essentially the same method. The decane was dried over sodium. Both distilled liquids were then purified by preparative gas chromatography on an Aerograph Model A-700 Autoprep Vapor Fractometer. A 20-ft × $\frac{3}{8}$-in. od column packed with 30% SE-30 (General Electric methylsilicone rubber) on 42/60 chromosorb P, operated at 150° for decane (165° for the ether) and 40 psig was used. The helium carrier gas rate was about 200 mL per min.

Decane containing 1.40% impurity was given two passes of zone refining and attained a purity of 99.73% in an automatic zone melting apparatus capable of operating between 0 and −170°, designed by Kieffer and coworkers [3997].

Mears and coworkers [5016] prepared decane by hydrogenating 1-decene at 170° with nickel-on-kieselguhr catalyst at a hydrogen pressure of 500 psi. The decane was fractionally distilled in a 2.5 × 250-cm Podbielniak Hypercal Heligrid column. The freezing point was the criterion of purity.

Karabinos [3855] prepared decane by the Kolbe synthesis by the electrolysis of hexanoic acid. The reduction product was shown to be essentially free from branched-chain hydrocarbons.

For purification with zeolite, see Shinkarenko and Aleskovskiĭ [6676], pentane.

SAFETY

The *flammable limits* in air are 0.72 and 5.0%v [41]; see also [8248]. The *minimum ignition temperature* in air is 208° [5712].

35. 1,1′-Bicyclohexyl

Ipatiew [3632] reported the preparation of bicyclohexyl in 1907 from diphenyl at 191 atm and 260° with a reduced nickel catalyst. In 1912, Sabatier and Murat [6325] reduced diphenyl in two stages at 180° and 160° with nickel and hydrogen.

Bicyclohexyl was prepared in 40% yield by Gardner and Borgstrom [2660] from cyclohexylmagnesium bromide and an equivalent amount of dry silver bromide in dry ethyl ether.

A commercially available bicyclohexyl was purified by Vogel [7767] for *physical property and chemical constitution studies* by shaking with one-half of its volume of concentrated sulfuric acid. The acid was separated and the hydrocarbon was washed repeatedly with water, dried with calcium chloride, heated 5 hr with sodium at 110°, filtered, and distilled.

Mears and coworkers [5017] prepared bicyclohexyl for *physical property study* to a purity of 99.96%m by catalytic hydrogenation of selected fractions of cyclohexylbenzene with nickel-on-kieselguhr catalyst at 180° at an initial hydrogen pressure of about 1850 psig. The hydrogenated material was filtered and percolated through silica gel. The portion of material boiling at 148.5–150° at 58 Torr with a refractive index at 20° of 1.4796–1.4797 was collected and again percolated through the silica gel. See also Musser and Adkins [5366], decahydronaphthalene.

Schrauth and Gorig [6489] present data to support the existence of three stereoisomers of bicyclohexyl.

The initial thermal decomposition temperature is >240° [2028].

36. Dodecane

Garach [2655] prepared saturated and unsaturated hydrocarbons to study their properties, and especially the *infrared spectra*, for optical analysis. The unsaturated compounds were prepared by reacting heptanal with the proper Grignard reagent, hydrolyzing, and dehydrating the alcohol over kaolin just below 300°. The saturated compounds were made by hydrogenation over nickel at 170–180° and purifying with sulfuric acid.

Deanesly and Carleton [1882] prepared and purified dodecane for precise *physical property study*. Lauryl alcohol (dodecanol) was purified by fractional distillation. It was converted to iododecane [5443, p. 29], which was filtered, washed, and reduced to dodecane [5443, p. 27]. The alkane was extracted with ethyl ether and fractionally distilled into 100-mL cuts. Those whose n_D 20° was 1.42156 were combined and crystallized five times without solvent. The freezing point rose only 0.05° after the fifth crystallization. It was washed with 98% sulfuric acid containing 3% silver sulfate. It was crystallized twice, and the temperature rose less than 0.002°. The freezing point was the principal criterion of purity.

Mears and coworkers [5017] prepared dodecane for *physical property studies* by dehydrating 1-dodecanol over alumina at 350–375°; 80% yield per pass. The mixed dodecenes were washed, dried, and fractionally distilled through a short helices-packed column at 210–215°. They were hydrogenated with nickel-on-kieselguhr catalyst at 140° with an initial pressure of 1800 psig. The dodecane was freed from nickel, percolated through silica gel and fractionally distilled; purity 99.92%m calculated from the freezing curve.

Vogel [7760] prepared dodecane by Wurtz synthesis for *physical property studies*. See also Timmermans and Martin [7495, under hexane].

SAFETY

The *flammable limits* in air at 25° are 0.61 and 4.7%v [41]; see also [8248]. The *minimum ignition temperature* in air is 204° [5712].

37. Tridecane

Krafft [4235] prepared tridecane in 1882 by heating tridecanoic acid with HI and phosphorus at 210–240°; bp 234°, fp −6.2°. In 1889 Mai [4829] prepared tridecane from the vacuum distillation of the barium salt of myristic acid (tetradecanoic acid) with sodium methoxide.

Wojcik and Adkins [8135] prepared tridecane by hydrogenolysis of 1-tetradecanol at 100–200 atm and 250° for 5 hr over a nickel catalyst; bp 84–85° at 3 Torr.

Garach [2655] hydrogenated 1-tridecene over nickel at 170–180° to prepare tridecane, which was then purified with sulfuric acid. Petrov and coworkers [5802] prepared tridecane by reacting allyl bromide with decylmagnesium bromide in ether and then hydrogenating the reaction product over Raney nickel at 100 atm and 150°.

Tridecane is available in at least three grades. Phillips Petroleum [5823] lists the following grades: research, 99.9%w; pure, 99.8%w, and technical, 99.2%w. The impurities are isoparaffins.

Aromatic Hydrocarbons (38–56)

General Comments

Benzene and several of its simple alkyl and alkene derivatives are now prepared synthetically from petroleum. The impurities in the synthetic compounds differ from those present in products from the destructive distillation of coal. This is an important consideration when the nature of the impurity is a factor.

Alkyl derivatives of benzene can be prepared from 1 mol each of the corresponding alcohol and benzene and 20–65 g of boron trifluoride, depending on the alcohol used. Normal, iso-, secondary, and tertiary alcohols react.

Simons and Archer [6739] prepared the corresponding alkyl benzene from the olefin and benzene with hydrogen fluoride as the condensing agent. Some of the compounds prepared were isopropylbenzene, *tert*-butylbenzene, and cyclohexylbenzene.

A method for preparing *alkyl derivatives of benzene* by catalyzing a mixture of benzene and an alkyl bromide with aluminum was reported by Turova-Pollack and Maslova [7591]. The mixture of benzene and the alkyl bromide is heated with aluminum shavings 1.5–2 hours until the reaction starts, followed by cooling to 5–10° until the reaction ceases. The reaction mixture was heated to 70–80° and let stand at room temperature for 2 hr. The reddish complex was broken with dilute acid, and the alkyl benzene was separated and purified. Examples are: 96% yield of ethylbenzene from 8 mol of benzene, 1 mol bromoethane, and 0.01 gram-atom of aluminum; isopropyl benzene was obtained from the same ratio.

Forster [2484, 2485] purified saturated and unsaturated hydrocarbons and methyl-substituted benzenes for *specific conductance studies* from 99% commercially available material by fractional distillation in a nitrogen atmosphere. The fraction that distilled within ±0.1° of the reported boiling point was fractionally crystallized three times and degassed twice. No impurities were detected using a gas chromatograph at the highest sensitivity.

Specific Solvents

38.   Benzene

Benzene, because of its ease of purification, is one of the secondary standards for physical measurements and is a common solvent for *dipole moment* determinations. It has been used extensively as a solvent in acid-base titrations in nonaqueous solvents [6164].

Benzene, sufficiently pure for most purposes, may be purchased in quantity. ACS *Reagent Chemicals* [137] and Rosin [6263] list specifications and tests for benzene for use as an analytical solvent or test medium; boiling range 79.5–81.0°, minimum fp 5.2°. Some sulfur and traces of other hydrocarbons remain. It is not sufficiently pure for a physical property standard or for the preparation of solutions for most accurate physical measurements.

A Research grade, 99.94%m, is available from Phillips Petroleum Co. Other high-purity grades are available.

An anomalous behavior was deduced from a viscosity study; see *p*-dioxane [7642].

Anderson and Engelder [345] isolated and identified the nonbenzenoid hydrocarbons and toluene from a sample believed to be typical of the type called "nitration grade" from the destructive distillation of coal. The benzene met ACS *Reagent* grade specifications for boiling range and freezing point. The saturated nonbenzenoid hydrocarbons were obtained by fractional crystallization, fractional adsorption, and fractional distillation. It was found that the impurities were a complex mixture that is predominantly naphthenic.

The estimated impurities were: saturated nonbenzenoid hydrocarbon, 0.6%v; toluene, 0.004%v; unidentified, 0.1%v. The results indicated a correlation between the freezing point depression of 0.67° and a paraffin content of 1%. Another investigation showed that European nitration grade benzene from coke plant production contained mainly naphthenes and methylcyclohexane.

Forziati and coworkers [2494] distilled benzene, after preliminary treatment, to a purity of 100.00%m. National Bureau of Standards Standard Samples of benzene were prepared by Mair and coworkers [4841] by a combination of crystallization, filtration through silica gel, and distillation. The crystallization was performed as follows: "A volume of about 200 mL of benzene was mixed with 50 mL of ethanol in a cylindrical brass container (5 cm in diameter and 20 cm long), which was placed in a cooling bath. of ice and salt (temperature about −10°). With vigorous hand-stirring and scraping, a thick mush, or slurry, of hydrocarbon and alcohol was produced. This slurry was then transferred to the basket of a centrifuge having a jacket cooled to nearly −10° with a mixture of ice and salt. The centrifuge was operated for about 5 min, leaving in the basket about half of the benzene in crystalline form. The crystals were removed from the basket of the centrifuge and allowed to melt. The purified liquid benzene was then washed three times with distilled water and then filtered through silica gel to remove any alcohol and water remaining." The purity was determined from the freezing curve [2781].

Swietoslawski [7133] stated that benzene cannot be purified, after removal of thiophene and unsaturated hydrocarbons, by fractional distillation and crystallization alone. Crystallization from methanol or methanol-water produces 99.998% purity. The purity test was carried out in a dilatometric cryometer. The benzene was frozen in the apparatus and the decrease in volume due to partial solidification determined. The decrease in volume was then plotted against the equilibrium temperature. Once the curve had been established, the purity of a sample of benzene could be determined by obtaining only one point on the curve.

Stull [7059] purified thiophene-free benzene for the determination of the *thermodynamic properties*. The material was crystallized six times; one-fourth was discarded each time as unfrozen liquid. It was then dried over phosphorus pentoxide for 2 weeks and finally fractionally distilled through a 5-ft column packed with glass helices. The purity was above 99.95% determined from the freezing curve.

Washburn and Read [7912] purified thiophene-free benzene by allowing it to stand in contact with pure sulfuric acid, with frequent shaking, for 4 weeks. The treated benzene was then distilled from sulfuric acid into a distilling flask containing metallic calcium, and from there into the apparatus in which it was to be used. The freezing point was 5.43°.

The determination of sulfur compounds in crude and refined benzene from coke has been thoroughly studied by Claxton and Hoffert [1517]. It was shown that crude benzene contained about half of its sulfur content as carbon disulfide. While the remainder of the sulfur content is usually considered to be thiophene and its derivatives, other sulfur compounds such as mercaptans, sulfides, and disulfides are also present. The reader is referred to the original article for the recommended analytical procedure. See also French and Claxton [2548].

Wojciechowski [8131] purified ACS reagent-grade benzene by three different methods.

1. Two liters were fractionally distilled through a Swietoslawski-type, vacuum-jacketed column. Five middle fractions of 200 mL each were collected.
2. A second portion was recrystallized and fractionally distilled.
3. A third portion was subjected to azeotropic distillation with ethanol and water.

The alcohol was removed from the distillate by washing with water and the benzene redistilled. The maximum deviation in the boiling point between samples purified by the three methods was 0.006°.

Bender and coworkers [761] purified benzene by two methods.

1. Research-grade benzene was treated with concentrated sulfuric acid until thiophene-free. After repeated water washes, it was dried first with calcium chloride, then with sodium wire. It was recrystallized by the second method of Schwab and Wiehers [6509] after distillation from sodium wire.
2. Thiophene-free, reagent-grade benzene was dried with sodium wire and distilled at a reflux ratio of 20:1 through a 30-plate Oldershaw column. The constant boiling portion was retained.

Leonard and Sutton [4526] purified an analytical grade of benzene for use as a solvent for determining *dielectric constants*. The benzene was frozen three times, boiled under reflux over phosphorus pentoxide, and finally distilled in a stream of dry air. The purified material was stored under dry air.

Thiophene-free benzene was dried with sodium sulfate by Brown and Ives [1128] for *dielectric constant studies*. It was fractionally crystallized three times, dried with sodium, and fractionally distilled in a 4-ft, vacuum-jacketed column packed with helices; bp 80.1°, d 25° 0.87365.

Meighan and Cole [5044] used reagent-grade benzene in *dielectric studies* after refluxing with calcium hydride for 48 hr and distilling. Subsequent handling was in a closed system.

Pearce and Berhenke [5731] made benzene thiophene-free and purified it in the usual way for *dielectric moment measurements*. It was finally dried over sodium wire and fractionally crystallized; bp 760 Torr 80.1°, fp 5.4°.

A spectrographic-quality, reagent-grade benzene was used as such or further purified by Forster [2484] for study of *specific conductance* in liquid hydrocarbons. It was further purified by fractional distillation in an atmosphere of nitrogen. The fraction distilling at 80.00–80.20° was collected and fractionally crystallized followed by two degassing cycles. The highest sensitivity available for gas chromatography did not show any impurities. Thallium shavings removed traces of oxygen without impairing the purity of the benzene. The material was stored under vacuum in the freezing compartment of a refrigerator to prevent decomposition.

For *specific conductance measurements*, Walden [7853] dried benzene with sodium and distilled in an all-glass apparatus closed with a calcium chloride–soda lime tube. The middle fraction was used for measurements.

Bruckenstein and Saito [1183] purified ACS reagent-grade benzene for *acid–base equilibria studies* by storing the solvent over 4A molecular sieves ($\frac{1}{16}$ in. pellets), filtering and finally fractionally distilling. The water by Karl Fischer method was 5×10^{-3} M.

Maruyama [4929] describes a method for purifying benzene for use in *polarography*.

Röck [6217] refined benzene containing about 0.2% impurity by zone melting and obtained a fraction containing $<0.008\%$m impurity.

Thiophene in benzene was reduced from 0.1% to 0.01% in 15 *zone refining* passes [1939].

Dickinson and Eaborn [2000] studied *progressive freezing* purification of benzene and concluded that it was as effective as the more complex zone melting technique.

Molinari [5217] converted benzene feed stock, mp 4.7°, containing 3.8% paraffins and 0.3% total sulfur, into 20% benzene, mp 5.53° and 55% nitration grade, mp 5.2°, by using the Probad process.

Graul and Karabinos [2910] developed a rapid method for removing *thiophene from benzene*. They refluxed 100 mL of benzene containing 1% thiophene for 15 min with 10 g of Raney nickel prepared according to the method of Pavlic and Adkins [5721]. The isatin-sulfuric acid test was negative, indicating the removal of thiophene. When 5 g of Raney nickel was used, not all of the thiophene was removed.

Cherkasova and Gorin [1455] reported thiophene can be removed from benzene by condensing with formaldehyde in the presence of sulfuric acid.

Mathieu [4979] prepared benzene by the thermal decomposition of calcium benzoate and purified it by fusion and crystallization for direct measurement of *pressure coefficient of temperature (dt/dp) and change of volume on melting* in a Swietoslawski-type dilatometric cryometer; purity 99.997%m, tp 5.690°.

Hartmann and coworkers [3199] purified benzene; see benzyl alcohol.

Total sulfur in benzene; combustion, oxidation to sulfate and determination by conductometric titration is given in [2547].

Benzene sufficiently pure for all except measurements of the highest accuracy is most conveniently prepared by shaking a *good commercial grade* successively with concentrated sulfuric acid until free from thiophene, then with water, dilute sodium hydroxide, and finally with two portions of water. The benzene may be dried by first shaking with anhydrous calcium chloride and removing the last traces of water by distilling over phosphorus pentoxide, shaking with molecular sieves or passing through a column of silica gel. Kraus and Vingee [4249] used a final washing of only one portion of water, distilling from phosphorus pentoxide when preparing benzene for use as a *cryoscopic solvent*. The purified benzene was stored over sodium–lead alloy and distilled just prior to use.

CRITERIA OF PURITY

The *freezing point* is, perhaps, the most easily determined property and the most sensitive to impurities. The amount of impurity may be calculated from the freezing curve [2781].

The *dilatometric cryometric method* of Swietoslawski [7133] is sensitive enough for use as a criterion of high purity; see Mathieu [4979] who characterized benzene to 99.997%m.

The *boiling point, density,* and *refractive index* offer a means of estimating the purity.

The following specifications and tests, which are easy to make, characterize benzene sufficiently well for general solvent and laboratory use:

> *Freezing point:* not less than 5.2°. Fill a double-walled test tube half full of benzene, immerse in an ice-water slurry, and determine the freezing point. See Skau and coauthors [7958, p. 287 ff.] or [137, p. 98].
>
> *Nonvolatile material:* transfer 100 mL of benzene to a tared evaporating dish, evaporate on a steam bath, dry at 110° for 30 min, and weigh. Maximum increase in weight must not exceed 0.001 g.

Substances darkened by sulfuric acid: shake 25 mL of benzene with 15 mL of ACS reagent-grade sulfuric acid about 20 sec and allow the phases to separate. There should be no darkening in either phase [137, 6263].

Thiophene: add about 3 mg of isatin mixture from the preceding test and shake well. A blue-green color should not develop in the acid phase within 1 hr.

Water: the water content is most conveniently determined by the Karl Fischer method.

Gas chromatography may be used to characterize any grade of benzene.

SAFETY

Benzene may be regarded as a cumulative poison due to the slow build-up in the body tissues and fluids because of its low solubility in the circulating blood. After exposure the depletion is equally slow. Benzene poisoning results almost exclusively from breathing the vapors. Small quantities of benzene may be absorbed through the skin, but Gerarde [5713] does not believe that systemic poisoning can arise from immersing the hands in benzene. However, skin defatting, erythema, scaling, and even secondary infection may occur.

In 1977, clinical epidemiological data established that employee exposure to benzene presented a leukemia hazard. OSHA set an employee exposure to 1 ppm in an 8-hr time-weighted average concentration with a ceiling level of 5 ppm for any 15-min period [382]. A detailed analysis of benzene is given in the 1981 Second Annual Report on Carcinogens [5424].

The oral LD_{50} for rats has been reported as 4080 mg/kg [3708]; see also [7195].

The danger inherent in inhaling benzene has been controversial [1122, 2095, 2242]. The *threshold limit value* has been successively reduced to 10 ppm, 30 mg/m^3 [142].

Many methods have been proposed for the determination of benzene in air; see Patty [5713].

The *flammable limits* in air are given by Huff [3556] as 1.4 and 7.1%v; see also [959]. The *minimum ignition temperature* in air is 508° [3576].

39. Toluene

Toluene has a limited use as a secondary standard for physical and thermal measurements. It may be purchased in quantities sufficiently pure for most purposes and in limited quantity with a purity of 99.97%m. The specifications for Analytical grade are given in *Reagent Chemicals* [137] and by Rosin [6263].

Toluene, like benzene, is available from two different basic manufacturing sources, the coking of coal and synthesis from petroleum. The toluene from each source has its own characteristic impurities. There is little published information concerning the impurities in the synthetic material. The impurities in coke toluene are of the same type as in benzene [345], and they are more difficult to remove completely.

Toluene may be purified by a combination of chemical treatments, fractional distillation, and fractional crystallization, but the freezing point of toluene, −95°, is too low for convenient crystallization; see Ramsay and Steele [6041]. The sulfur impurities may be removed by methods similar to those described for benzene.

Vogel [7758] prepared toluene for *physical property measurements* as follows: 100 g of purified and redistilled benzaldehyde, bp 179° at 757 Torr, was reduced with 200 g of amalgamated zinc and hydrochloric acid; yield 43 g toluene boiling at 109–110° at

748 Torr. It was washed twice with 10% of its volume of concentrated sulfuric acid, then with sodium carbonate solution, and finally with water. After being dried over anhydrous magnesium chloride, it was twice distilled from sodium and passed over constantly at 110° at 763 Torr. The middle fraction was used.

Ramsay and Steele [6041] (see also Mitsukuri and Nakatsuchi [5178]) prepared pure toluene from p-toluidine by diazotization; Perkin [5765] (see also Orton and Jones [5602]) purified the potassium salt of p-toluenesulfonic acid by recrystallization and obtained pure toluene by hydrolysis with steam in a sulfuric acid solution.

Hardy [3154] studied the synthesis of toluene from benzene and chloromethane by the Friedel–Crafts reaction. The best yields were obtained at moderate pressures of 4–10 atmospheres and temperatures of 4–15°. The best benzene conversion to toluene under these conditions was 63%, corresponding to a chloromethane conversion of 43%.

Schwable [6510] recommended the following process for the removal of thiophene and its homologues: Nitrous oxide and nitric oxide are passed into the toluene until the concentration is 0.2–0.5%; after standing for 2–3 hr, the mixture is shaken with 2% of its volume of concentrated sulfuric acid for 5 min, allowed to stand for 15 min, and separated. The operation is repeated two or three times until the acid is only slightly yellow in color; the toluene is then shaken with water and dilute sodium hydroxide until fresh portions are no longer colored.

According to Timmermans and Martin [7494], to obtain toluene pure it is necessary to convert it into a crystalline substance from which the hydrocarbon can be recovered after purification, or to prepare it in the first place from a substance purified by crystallization.

Mathews [4969] purified commercial toluene for the determination of *heat of vaporization* by successive shaking with sulfuric acid, sodium hydroxide, and mercury, followed by drying over phosphorus pentoxide and then fractionally distilling. A similar process was used by Williams and Krchma [8083] for purifying toluene for the determination of the *dielectric constant*. See also Richards and Wallace [6160].

Forster [2485] purified toluene in the same manner as he did benzene for *specific conductance studies*. For *conductance measurements*, Walden [7853] purified a good grade of Kahlbaum's toluene by the same method that he used for benzene.

For *thermoregulators*, Beal and Souther [699] washed toluene first with sulfuric acid, then with water, and then boiled it under reflux with 1% sodium amalgam. After decanting, the solvent was washed with water and distilled.

For an additional method of purification, see Duncan and coworkers [2119, under mesitylene].

Kuss and coworkers [4350] prepared "very pure" toluene by reducing benzyl alcohol at 200–300° with a chromium–copper catalyst.

CRITERIA OF PURITY

The freezing curve method has been successfully used [2781, 2494, 4841].

Timmermans and Martin [7494] found that the boiling point should not be used as a criterion of purity. Fractions having the same boiling point were found to have different densities.

Lumsden [4696] recommended the following criteria: (1) density; (2) nitration followed by the determination of the amount of nitration in a nitrometer; (3) miscibility with acetic acid and the determination of the critical solution temperature. See also Orton and Jones [5602].

Rosin [6263] lists the following specifications for reagent-grade toluene:

Distillation range: 100-mL sample to distill between 110° and 111°.

Nonvolatiles: 0.001% maximum.

Acidity: none.

Alkalinity: none.

Substances darkened by sulfuric acid: shake 15 mL of toluene with 15 mL of sulfuric acid for 15 min. There should be no darkening of the toluene phase, the acid phase should be no darker than a mixture of 2 parts of water and 1 part of a solution containing 5 g of cobalt chloride, 40 g iron [III] chloride and 20 mL of hydrochloric acid per liter of solution.

Sulfur compounds (as S): 0.003% maximum.

Water: no cloudiness should be produced at 0°.

There are no simple tests for homologues, naphthenic, and paraffinic hydrocarbons except gas chromatography.

The water is most conveniently determined by the Karl Fischer method [5171].

SAFETY

Gerarde [5713] states that toluene is a more powerful narcotic and is more acutely toxic than benzene. Controlled exposure of humans to concentrations of 50 to 800 ppm indicates that 200 ppm for a period of 8 hr produces mild fatigue, weakness, confusion, and paresthesia of the skin. Toluene, like benzene, is most dangerous by inhalation. It is irritating to the skin, and contact should be avoided when possible.

Kawasaki and coworkers [3907] reported an inhalation toxicity LC_{50} of 289 mg/L.

The *threshold limit value* on skin has been set at 100 ppm, 375 mg/m^3 [142].

The *flammable limits* in air at elevated temperature are 1.17 and 7.10%v [5712]. The *minimum ignition temperature* in air is 473° [3576].

40. *o*-Xylene
41. *m*-Xylene
42. *p*-Xylene

Less than 2% of the xylene produced today comes from coal tars. The remainder is obtained by refining petroleum naphtha. The aromatics are extracted and the benzene, toluene, and C_8-aromatics separated by fractional distillation. *o*-Xylene and ethylbenzene are separated by distillation, and *p*-xylene is fractionally crystallized from the residue. *m*-Xylene is separated from part of the residue by distillation and/or clathration. Mixed xylenes usually contain about 40% *meta*- and 20% each of the *ortho*- and *para*-isomers and ethylbenzene.

The composition of coal tar xylene varies in about the same range as the petroleum product. The principal impurities are ethylbenzene and paraffins [7131, p. 141].

Analytical reagent-grade xylene, boiling range of 137–140° [137], is available from most supply houses. Research-grade *o*-, *m*-, and *p*-xylenes are available in 99.9+%m purity. Other grades of the isomers are available in 99+%m purity. A 95%w minimum *p*-xylene is available in quantity.

The xylenes were purified by Forster [2485] for *specific conductance studies* in the same manner as was benzene [2484].

A method for separating the xylenes has been developed by Arnold [441]. The mixture of *m*- and *p*-xylenes is diluted with a relatively volatile, inert solvent having a freezing point below −58.5°, the eutectic point of the xylenes, and cooled until a large part of the *p*-xylene has crystallized. The diluent is removed by distillation and the *m*-xylene crystallized from the remaining liquid. Suitable solvents are methanol, ethanol, 2-propanol, acetone, butanol, toluene, the pentanes, and the pentenes.

Aromatic isomers were separated by Radzitzky and Hanotier [6020] by clathration with Werner complexes of the form $Ni(SCN)_2(RC_6H_4CHR'NH_2)_4$.

Timmermans and Hennaut-Roland [7488] sulfonated mixed xylenes and separated the alkali salts of the sulfonic acids by fractional crystallization. The purification was attended with considerable loss.

An extraction method was developed by McCaulay and coworkers [4738] for separating *m*-xylene from commercial xylene; they used a mixture of hydrogen fluoride and boron trifluoride. The reactions involved are rapid and sufficiently selective, so that 95% *m*-xylene can be separated from mixed xylenes. An inert hydrocarbon diluent further improves the selectivity of the extraction. The authors state that, when their method is used in conjunction with fractional distillation or crystallization, each of the xylene isomers can be separated in a high state of purity.

Clarke and Taylor [1510] sulfonated 4400 grams of technical *o*-xylene, 95% of which boiled between 143 and 144°, by stirring for 4 hr with 2.5 L of concentrated sulfuric acid at 95°; after cooling the mixture the product was separated from unsulfonated material, diluted with 3 L of water, and neutralized with 40% sodium hydroxide. The precipitate that formed on cooling was separated and crystallized from half its weight of water. A further quantity was obtained by concentrating the mother liquor to a third of its volume and cooling; this was recrystallized from half its weight of water and purified with the first fraction. A further crystallization of the whole material produced a homogeneous product, and the mother liquor was worked up as before. The purified sodium salt was dissolved in the necessary quantity of cold water, mixed with the same volume of concentrated sulfuric acid and steam distilled after heating to 110°. The yield of *o*-xylene was 1980 g or 43%. The boiling point was 144–145°.

Pure *o*-xylene was prepared by Skita and Schneck [6796] from *o*-toluidine (purified by means of the acetyl compound) by conversion into *o*-bromotoluene, followed by the Grignard reaction with dimethyl sulfate.

Moldavskiĭ and Turetskaya [5214] prepared *o*-xylene by hydrogenating phthalic anhydride or phthalide at 300–330° and 120 atm initial pressure with 10% molybdenum disulfide as catalyst. The phthalic anhydride was dissolved in 1,2,3,4-tetrahydronaphthalene and gave a 32% yield. The phthalide was dissolved in two volumes of 1,2,3,4-tetrahydronaphthalene and hydrogenated for 5 hr; the yield was 85%.

Clarke and Taylor [1510] started with a technical product boiling within 1° and boiled it with dilute nitric acid (one part acid to three parts water). The *o*- and *p*-xylenes were oxidized, but the *m*-xylene is unattacked. After washing with water and alkali, the product was distilled, first with steam and then through a column, and finally sulfonated by stirring at 95° with half of its volume of concentrated sulfuric acid. The resulting solution was separated from the unsulfonated material, and an equal volume of concentrated sulfuric acid and twice its volume of water was added. Pure *m*-xylene passes over from 110 to 120° during steam distillation.

Cole [1580] selectively sulfonated *m*-xylene in a mixture of xylenes. The mixture was refluxed with the theoretical amount of 50–70% sulfuric acid at 85–95° *in vacuo*. The water–xylene vapors were condensed, the xylene returned to the reaction vessel, and the water withdrawn until the amount present in the sulfuric acid and the theoretical amount formed from the reaction were collected. Water was then added to the reaction mixture and the unreacted xylenes removed at reduced pressure; *m*-xylene monosulfonic acid was hydrolyzed by steam distillation up to 140°. The unreacted xylenes, chiefly *ortho* and *para*, may be separated into material of fair purity by fractional distillation.

Mathews [4969] purified *m*- and *p*-xylenes for the determination of the *heat of vaporization* by successive shaking with sulfuric acid, sodium hydroxide, and mercury, followed by drying over phosphorus pentoxide and fractional distillation. The *p*-xylene was fractionally crystallized six times.

The *meta*- and *para*-xylenes form a binary eutectic of 87% *meta*. It is not possible to get a pure *m*-xylene by *fractional solidification* of mixtures containing less than 87% *meta*. Laconte [8290] states that Egan and Luthy took advantage of the solid compound formed between carbon tetrachloride and *p*-xylene. When the mixture was cooled, the solid complex separated and the mother liquor contained 98% *meta*.

Skita and Schneck [6796] prepared pure *m*-xylene from asymmetric *m*-xylidine (purified by means of the acetate) by reduction of its diazo compound with tin in an alkaline solution. By the same method, they prepared pure *p*-xylene from *p*-xylidine which had been purified by means of the benzylidene compound.

According to Timmermans and Martin [7494] *p*-xylene can be obtained pure quite easily by fractional crystallization. *p*-Xylene was purified in this way until the freezing point was constant (13.35 ± 0.03°).

Greenburg [2948] isolated *p*-xylene from mixtures containing at least 16% *para* from close-boiling constituents such as *o*- and *m*-xylenes or ethylbenzene by fractional melting. About equal volumes of the hydrocarbon mixture were mixed with methanol or ethanol and partially solidified. The solid was fractionally melted into steps not exceeding 40°. The liquid was withdrawn at each step. The last fraction, melting above 10°, consisted substantially of *p*-xylene.

Cole and Burtt [1581] developed a method for separating *m*- and *p*-xylenes. A mixture of coke oven xylenes was fractionally distilled to remove the *o*-isomer. The mixture (100 parts) of the *m*- and *p*-isomers was treated with 120 parts of 26% fuming sulfuric acid. The mixture was partially hydrolyzed by steam distillation. The first distillate contained paraffinic hydrocarbons, and ethylbenzene was discarded. It was followed by very pure *m*-xylene. The residue was cooled to 10°, the crystals of *p*-xylenesulfonic acid were collected by filtration, washed with aqueous sulfuric acid, and then steam distilled. Pure *p*-xylene was obtained. The separated *m*- and *p*-isomers may be washed with dilute sodium carbonate and further purified.

McArdle and Mason [4732] developed a crystallization procedure similar to that of Arnold [441] for separating *m*- and *p*-xylenes. It was stated that 56.6% of the *m*-xylene was crystallized from a mixture with C_8-hydrocarbons.

Sue and coworkers [7065] effectively removed *o*-xylene from *p*-xylene by *zone melting*.

Notaro and coworkers [5522] separated p-xylene from m-xylene by chlorinating the latter and distilling the unreacted p-xylene.

CRITERIA OF PURITY

ACS Reagent Chemicals [137] lists the following specifications for reagent-grade xylene (mixed isomers):

Boiling range: from 137 to 140°.
Residue after evaporation: not more than 0.002%.
Sulfur compounds (as S): not more than 0.003%.
Water (H_2O): not more than 0.05%.

Mair and coworkers [4841] used the freezing curve method to determine the purity of the NBS Standard Samples of the xylenes.

According to Timmermans and Martin [7494], density and melting point are the most satisfactory criteria of purity. Swietoslawski [7131] found that differential ebulliometric measurements gave a lower accuracy for the xylenes than for other compounds.

All of the known hydrocarbon impurities that have been found in the xylenes have been resolved by gas chromatography.

SAFETY

The acute toxicity of the xylenes probably is greater than that of benzene or toluene. The isomers differ in their acute toxicity, but their relationship has not been definitely established. The chronic toxicity appears to be less than that of benzene. The absorption of xylene takes place chiefly by breathing the vapors. Skin irritation is more serious than for benzene or toluene [5713].

There is definite irritation to the eyes, nose, and throat at 200 ppm and olfactory fatigue is rapid. The *threshold limit value* has been established at 100 ppm, 435 mg/m^3 [7022]. The *flammable limits* in air for the *ortho-*, *meta-*, and *para-*isomers, respectively, are 1.0 and 6.0, 1.1 and 7.0, and 1.1 and 7.0%v [152]; at elevated temperatures the corresponding values are 1.09 and 6.40, 1.09 and 6.4, and 1.08 and 6.60%v. The respective *minimum ignition temperatures* in air are 464°, 528°, and 529° [5712].

43. Ethylbenzene

Most of the ethylbenzene is obtained from petroleum (see the introduction to the xylenes). It is used principally to make styrene. It is available in research grade, 99.90%m, and in several special grades.

Ethylbenzene was prepared by O'Kelly and coworkers [5574] by passing benzene and ethylene over a silica–alumina catalyst at temperatures of 448–496° and pressures of 57–75 psig; conversion was as high as 81.1% based on ethylene.

Hanai [3129] established the best condition for the formation of ethylbenzene from benzene and ethylene. The yield was 30–40%. The conditions were: 100 mol of benzene, 40–60 mol of ethylene, 8–10% aluminum chloride (benzene = 100%), saturated with hydrogen chloride. The mixture was kept under a slight pressure.

Hammick and Roberts [3119] catalyzed benzene and ethylene at atmospheric pressure to give predominantly ethylbenzene. Although the yield is low per pass of gas, the method is of interest because pressure equipment is not necessary.

Buu-Hoï and Janicaud [1296] synthesized aromatic ethyl compounds by the Friedel-Crafts reaction using ethyl chloroformate. The alkylating agent is used in the usual manner in the presence of an aluminum or iron(III) chloride catalyst.

Passino [5702] reacted benzene (two parts), ethylene (one part) and hydrogen fluoride, equivalent to 12% of the benzene, in a shaking autoclave for 16 hr at 500 psi and 70–80°F. Ethylbenzene equivalent to 10% of the benzene was obtained.

Vogel [7761] reduced 100 g of acetophenone with 200 g of zinc amalgam and hydrochloric acid; yield 53 g, bp 134.5–135° at 758 Torr. The ethylbenzene was purified by shaking with 6-mL portions of concentrated sulfuric acid until the acid layer was colorless, then with sodium carbonate solution, then with water, and finally drying twice with anhydrous magnesium sulfate. After two distillations over sodium, the middle fraction of the second distillate was used for *physical property determinations*.

For preparation from bromoethane and benzene, see Aromatic Hydrocarbons, General Comments [7591].

Kuss and coworkers [4350] prepared "very pure" ethylbenzene by reducing acetophenone with a chromium–copper catalyst at 180–200°.

Scott and Brickwedde [6533] further purified a sample of ethylbenzene obtained from M. R. Fenske as follows: The material was partially frozen three times, one-fourth of the liquid being discarded each time. It was passed over finely divided silica gel to remove water. No cloudiness was observed near the freezing point. It was used to determine *thermodynamic properties*.

Neuzil and Rosback [5464] purified ethylbenzene by adsorbing xylene isomers on Sr-K-X-type zeolites.

SAFETY

The acute toxic effects in mice are comparable to toluene and *m*-xylene. Men exposed to 1000 ppm experienced eye iritation, which gradually decreased on continued exposure. Animals exposed to 400–2200 ppm 7–8 hr/day for 5 days a week for as long as 6 months were not affected, judged by the usual criteria of injury. Absorption is chiefly by inhalation [5713].

The *threshold limit value* in air has been set at 100 ppm, 435 mg/m^3, [7022].

The *explosive limits* in air at elevated temperature are 0.99 and 6.70%v. The *minimum ignition temperature* in air is 432° [5712].

44. Isopropylbenzene

Isopropylbenzene is a significant commercial petroleum solvent in the boiling range of 150–160°. It is readily synthesized from benzene and propylene or benzene and 2-propanol. See, for example, O'Kelly and coworkers [5574], Huston and Kaye [3601], Simons and Hart [6743], Brun [1195], and Vermillion and Hill [7730].

Schmerling [6458] developed a method for the alkylation of aromatic hydrocarbons by olefins in a solution of aluminum chloride with a nitroparaffin. The temperature was from −10 to 100° and pressure from 1 to 100 atm. Suggestions are given for batch or continuous operations.

Rueggeberg and coworkers [6305] used chlorosulfonic acid to catalyze the reaction between an aryl hydrocarbon and an aliphatic alcohol. The method was not thoroughly investigated, but it is simple and the yields are fair.

Two moles of benzene and 1 mol of 2-chloropropane that were reacted 4 hr at 25° with 75% iron(III) chloride saturated with hydrogen chloride gave 66% isopropylbenzene [3710]. For a preparation from benzene and 1-bromopropane, see Aromatic Hydrocarbons, General Comments [7591].

Eastman Kodak isopropylbenzene was fractionally distilled by Hentz [3307] in a 100-theoretical plate column operating at 85% efficiency. Approximately one-fifth of the middlecut fraction was used for *γ-radiation studies*, n_D 20° 1.4913. Not more than 2 days prior to each experiment, 20 mL was passed through 10 mL of Alcoa F-20 alumina in a 50-mL buret with a Teflon stopcock to remove the peroxides.

Thizy [7398] reports that olefins can be removed from isopropylbenzene by the addition of chlorine for 5–7 min. The liquid was washed with water and then steam-distilled in a glass column in a 95% yield.

For purification, see mesitylene, Duncan and coworkers [2119].

Vogel [7761] purified a commercial product by washing it five times with concentrated sulfuric acid in a ratio of 10:1, then with water, sodium carbonate solution, and water again. This was followed by drying with anhydrous magnesium sulfate. The isopropylbenzene was then fractionally distilled in a three-section Young and Thomas column and about 90% collected. It was refractionated over sodium and the middle fraction collected for *physical property measurements*.

Gruden and Zief [3004] purified isopropylbenzene by preparative gas chromatography on 20% Carbowax M on a Chromosorb P support. Purity was 99.97%.

Isopropylbenzene of 99.9% purity was obtained by removing alkenes through adsorption on silica–alumina clay having a specific surface of 280–300 m^2/g [3019]. Isopropylbenzene was treated with active charcoal to prevent the increase of peroxides [3485].

SAFETY

Isopropylbenzene is a depressant to the central nervous system; its narcotic action is slowly induced and of long duration. Considered a primary eye and skin irritant, it is absorbed through the skin more rapidly than is benzene, toluene, the xylenes, or ethylbenzene.

A *threshold limit value* of 50 ppm, 250 mg/m^3 has been recommended [7022]. Gerarde [5713] states that, if this concentration elicits no mucous membrane effects, it should be a safe working atmosphere.

The *flammable limits* in air are 0.9 and 6.5%v [152]. An upper limit of 8.8%v has been reported [1290]. The *minimum ignition temperature* in air is 424° [5712].

A method for the determination of isopropylbenzene in air has been reported [83].

45. Mesitylene

Adams and Hufferd gave a method for preparing mesitylene from acetone and concentrated sulfuric acid in 13–15% yield in Conant [1618]. Ipatiew and coworkers [3631] studied the acetone-to-mesitylene reaction, using hydrogen chloride as the catalyst, and found the optimum conditions for maximum yield of 43%. Sucharda and Kuczyński [7063] found that lower pressures gave a 47% conversion in 48 hr.

Mesitylene was prepared by Norris and Ingraham [5516] from toluene and aluminum chloride cooled to 10–15°, to which a mixture of toluene and methanol had been added.

The reaction mixture was heated to 110° for 3 hr and poured over chipped ice and purified by chemical and physical means; purity 99 + %. Norris and Sturgis [5518] heated aluminum chloride, toluene, and methanol in a 2, 2.5, and 1 mol ratio, respectively, at 100° for 3.5 hr and obtained 53% mesitylene.

Duncan and coworkers [2119] purified reagent or commercial grade mesitylene, cyclohexene, toluene, p-xylene, isopropylbenzene, and cyclohexane for *thermodynamic studies*. The substances were twice fractionally distilled through a 40-theoretical plate column retaining the middle one-third of each distillate. The purity was found to be better than 99% by gas chromatography for all except mesitylene. The latter compound was "extemely difficult to purify," and was further fractionally crystallized and distilled. The final analysis was: mesitylene 98.52%, m- and/or p-ethyltoluene 0.65%, ethylbenzene 0.03%, and nonaromatic compounds 0.80%.

Mesitylene was purified by Streiff and coworkers [7036] for *physical and thermodynamic property studies* by distilling about 8 L through a 125-plate column at a 125:1 reflux ratio. The criterion of purity was the freezing curve, 99.8%m.

Spectral studies in the 1700–2300 Å range were made by Platt and Klevens [5881] using mesitylene, butylbenzene, sec-butylbenzene, and tert-butylbenzene by crystallizing several times, then distilling twice through a 50-plate Podbielniak column taking the center cut each time. The boiling point and refractive index were used as the criteria of purity.

Forster [2485] purified mesitylene in the same manner as he did benzene [8223] for *specific conductance studies*.

Amir [328] treated an aromatic hydrocarbon fraction (bp 80–180°) with 1%w aluminum chloride and 25%m 2-chlorobutane to give a product of 99.44%m mesitylene and trace amounts of other compounds. This is a purification process using an alkylating agent in the presence of a Friedel–Crafts catalyst.

46. Naphthalene

The commercial source of naphthalene until 1961 was the high-temperature coking of coal. Since then, naphthalene has been recovered from petroleum. About 43% of the U.S. production in 1965 was from the new source.

The impurities in coke oven naphthalene have been extensively studied. Thianaphthene, which forms mixed crystals with naphthalene, is the most universally objectionable and is difficult to remove. The methyl- and dimethylnaphthalenes are usually present, as in anthracene. Koptyug and coworkers [4184] studied the impurities in coke oven naphthalene by *infrared spectroscopy* and *gas chromatography*; they found thianaphthene, α- and β-methylnaphthalene, indene, indane, indole, o-, m-, and p-xylene, phenol, 2,4-, 2,5-, 3,5-, and 2,3-xylenol, and unidentified higher boiling components. Weissgerber and Kruber [7970] found thianaphthene in a pure commercial product.

Coke oven naphthalene was purified by Garcia [2658] to 99.9% and in 95% yield with 10% sodium hydroxide to remove phenols, with 50% sodium hydroxide to remove nitriles, with 10% sulfuric acid to remove organic bases, and with 0.8 grams of aluminum chloride per 100 grams of naphthalene to remove thianaphthene and alkyl derivatives of cumarin and indene. All extractions were done at 85°. The material was then treated with 20% sulfuric acid, then with 15% sodium carbonate, and finally distilled.

Armstrong and coworkers [428] found thianaphthene to be the main impurity in commercially pure naphthalene from coke. Their purification consisted in heating at 145° for 18 hr with 5% sodamid. The residual liquid was fractionally distilled through a 15-in. glass helices-packed column jacketed with naphthalene vapors at the boiling point. The sulfur content was reduced to 0.015%. A second similar treatment reduced the sulfur to less than 0.005%. The product was recrystallized twice from redistilled methanol and fractionally distilled again; mp 80.32°. An *infrared study* suggested the presence of small amounts of other impurities believed to be methylnaphthalenes.

The sulfur content of naphthalene was reduced from 0.15 to 0.031% by one crystallization from alcohol [8094].

Herington and coworkers [3322] report that recrystallization and ''simple'' distillation do not remove thianaphthene very effectively from naphthalene. They purified commercial crystals for *freezing point determination* by keeping the naphthalene overnight under a stream of nitrogen to remove water. It was stirred 24 hr at 140° with 5% sodamid, decanted, stirred 24 hr at 180° with 1% sodamid, and filtered through glass wool. It was fractionally distilled in a 10-plate adiabatic column. The middle fraction was twice crystallized from methanol and fractionally distilled again. The sulfur content was less than 0.002%; fp 80.278 ± 0.002°, purity 99.978 ± 0.10%m.

Mair and Streiff [4839] isolated naphthalene from a kerosene and purified it by systematic crystallization from ethanol. The residues rich in naphthalene were dissolved in a minimum of hot alcohol, the solution cooled to about 10°, and the crystals separated by filtration with suction. Second crops of crystals were obtained by removing some of the alcohol in a fractional distillation unit. Before determining the freezing point, the naphthalene was fractionally distilled at 56 Torr; mp 80.24°, n_D 85° 1.5898.

Washburn and Read [7912] purified naphthalene by recrystallization from the melt, discarding 75 mL of the liquid from 500 grams of the original melt. Distillation at reduced pressure gave a middle fraction melting at 80.09°.

The *thermodynamic properties* of naphthalene were determined by Barrow and McClellan [643] on resublimed material that had been repeatedly recrystallized from ether and dried by vacuum.

Rosanoff and Dunphy [6253] purified naphthalene by twice subliming, twice recrystallizing from ethanol, washing with water, and drying over phosphorus pentoxide.

Cheng and coworkers [1451] purified naphthalene for *critical tempreature studies* by chromatography on alumina using benzene as the eluent. It was then crystallized from benzene, which was removed by vacuum sublimation; mp 80.3°.

For the determination of the *absorption spectrum*, Baly and Tuck [584] purified naphthalene by heating with concentrated sulfuric acid and manganese dioxide (see also Lunge [4702]), distilling with steam, and, after a repetition of the process, several recrystallizations of the picrate. The regenerated naphthalene was then distilled with steam and recrystallized from dilute ethanol.

Mastrangelo [4959] purified refined liquid naphthalene for *cryoscopic studies* by slow cooling in an insulated Dewar flask and discarding the center core. The procedure was repeated four times in successively smaller flasks.

Naphthalene was purified as the picrate for *hydrogen bonding studies* [8223].

More work has been done on the purification of naphthalene by *zone melting* and related techniques than on any other organic compound. Anthracene is often used as the criterion of efficacy for purification because of its undesirableness as an impurity in naphthalene, for some uses even in the low parts per billion range, and its ability to fluoresce at parts per billion concentrations. Wolf and Deutsch [8138] reduced the concentration of anthracene from 0.1% to 1 ppm in two zone passes. It was suggested that further purification be done *in vacuo* or a nitrogen atmosphere. Herington and coworkers [3323] reduced the anthracene concentration from 0.2% to <200 ppb in seven zone passes. Some of the other references that have made significant contributions to the refining of naphthalene by zone melting are [3132, 838, 8186, 6444, 6446, 3743, 2590].

Grazuliene and coworkers [2916] purified naphthalene for luminescence studies by a combination of recrystallization, extraction, column chromatography, and zone melting.

Paûlopoulos and El-Sayed [5722] purified naphthalene by crystallization followed by 40 zone melting passes.

A commercial naphthalene containing 1% thianaphthene was recrystallized from ethanol by Miller [5141] and purified by zone melting for *vapor pressure measurements*. A freezing point of 80.1° from the cooling curve corresponded to 99.7%m.

Mattox [5000] prepared naphthalene by passing the vapors of ethylbenzene over a catalyst of silica and alumina, thoria, or zirconia at 525–700°.

Naphthalene decomposes 1%m per hour at 480° [3727].

SAFETY

Naphthalene has a characteristic well-known odor. A concentration of 25 ppm in air is easily detectable by the odor. The inhalation of the vapors may cause headache, confusion, nausea, and profuse perspiration. Severe exposures to vapors of naphthalene may cause vomiting, optic neuritis, and hematuria. It is irritating to the skin, and hypersensitivity of certain individuals has been reported [5713].

It is reported that concentrations in excess of about 15 ppm result in noticeable irritation of the eyes. The *threshold limit value* was, therefore, set at 10 ppm, 50 mg/m^3 [7022].

The *flammable limits* in air at elevated temperature are 0.88 and 5.90%v [5712]; see also [3755]. The *minimum ignition temperature* in air is 515° [5712].

47. 1-Methylnaphthalene

Fittig and Remsen [2454] prepared 1-methylnaphthalene in 1870 by reacting 1-bromonaphthalene, methyl iodide, and sodium. Darzens [1821] reported a rapid preparation of 1-methylnaphthalene by heating 1-methyl-3-naphthoic acid with calcium carbonate.

Hata [3218] prepared 1-methylnaphthalene by the catalytic hydrogenation of 1-cyanonaphthalene over nickel at 270–290°.

Grummitt and Buck [3005] hydrolyzed 1-naphthylcarbinylmagnesium chloride in ether with ammonium chloride solution under a nitrogen atmosphere. After separating the ether layer, 1-methylnaphthalene was distilled at 235–245°. Petroleum ether was added to the product and the mixture was dried over potassium carbonate. The dried solution was distilled over barium oxide at 238–240°.

A mixture of 600 grams of 1-methylnaphthalene and 18 grams of sodium was stirred at 100–110° for 30 min and then the temperature was raised to 180° for 4–6 hr. The slurry was distilled *in vacuo* to give 515 g of 1-methylnaphthalene; hydrogen index 90.2 and rubber content 0.06% [5231].

Streiff and coworkers [7035] purified 95%m 1-methylnaphthalene for *physical and thermodynamic property study* by fractional distillation through a 125 theoretical plate column at the rate of 8 mL/hr. A sample containing 96.8%m was fractionally crystallized. The impurities calculated from the freezing curve varied from 0.02 to 0.17%m.

Cumper and coworkers [1751] purified commercial 1-methylnaphthalene by recrystallization of its picrate at constant melting point 141–142°, followed by regeneration of the compound by steam distillation from sodium hydroxide.

48.   1,2-Dimethylnaphthalene

Mayer and Sieglitz [5008] prepared 1,2-dimethylnaphthalene by treating 1-bromo-2-methylnaphthalene and magnesium in ether with dimethyl sulfate. A similar preparation using lithium rather than magnesium is given in Plattner and Ronco [5884].

Darzens and Levy [1822] prepared 1,2-dimethylnaphthalene by the reaction of aluminum on 1-chloromethyl-2-methylnaphthalene in the presence of alcoholic hydrochloric acid. A similar preparation using pulverized zinc rather than aluminum is given in Buu-Hoi and Cagniant [1295].

Buchta and Meyer [1223] heated 1,2-dimethyl-3,4-dihydronaphthalene and powdered sulfur and distilled 1,2-dimethylnaphthalene.

49.   1,6-Dimethylnaphthalene

Weissgerber and Kruber [7969] treated a coal-tar fraction boiling at 260–265° with sulfuric acid at 40–45°. The resulting crystals of sodium 1,6-dimethylnaphthalene-4-sulfonate were heated with 60% sulfuric acid solution to recover 1,6-dimethylnaphthalene.

Clemo and Dickenson [1526] dehydrogenated 1,1,6-trimethyltetralin with selenium at 280–300° to obtain 60% yield of 1,6-dimethylnaphthalene. A similar dehydrogenation of vitamin A is given in Heilbron [3264].

Kipping and Wild [4022] prepared 1,6-dimethylnaphthalene by heating 2,5-dimethyltetralin with selenium in a sealed tube for 40 hr at 320–350°.

50.   1,2,3,4-Tetrahydronaphthalene

Mair and Streiff [4840] purified 1,2,3,4-tetrahydronaphthalene for *physical property studies* by five fractional crystallizations from a mixture of 50% dichlorodifluoromethane. The freezing curve was the criterion of purity; 99.985%m.

1,2,3,4-Tetrahydronaphthalene was purified by Morton and de Gouveia [5275] for *spectral studies* by preparing the barium sulfate salt by the Willstätter and Seitz [8095] method and recrystallizing several times. It was then converted to the sodium salt and recrystallized. The solvent was regenerated by adding 66% sulfuric acid and fractionally distilling with superheated steam at 160–180°. The principal impurities found were dihydronaphthalene, hexahydronaphthalene, decahydronaphthalene, and naphthalene. They are the impurities that are normal for the tetrahydro compound prepared from a high-purity naphthalene.

Bass [662] purified 1,2,3,4-tetrahydronaphthalene by extracting 2.0 mol of commercially available material with 3–30 mL of concentrated sulfuric acid, washing with 250 mL of water, and drying with sodium sulfate. It was then distilled from sodium through a short, lagged Vigreux column; yield 230 mL. The partially purified material was vigorously shaken with 150 mL of concentrated sulfuric acid and then heated on a water bath for about 2 hr with stirring. The warm mixture was poured into a solution of 120 g ammonium chloride in 400 mL of water. The crystals of ammonium 1,2,3,4-tetrahydronaphthalene-6-sulfonate were separated, recrystallized, washed with 50% aqueous ethanol, and dried at 100°. Traces of naphthalene were removed by steam distillation. The salt was recrystallized from water until it was pure. One mole of salt was mixed with 5 mol of concentrated sulfuric acid and steam distilled from an oil bath at 165–170°. The distillate was extracted with ether, and was washed with sodium carbonate solution and then with water. The oil layer was then dried and the ether removed by distillation. The liquid was finally refluxed over sodium and distilled through a Vigreux column.

Tanaka [7177] hydrogenated a mixture of 96 grams of phenol and 120 grams of naphthalene at 96 atm in the presence of nickel. Normally, the hydrogenation of naphthalene gives decahydronaphthalene, but phenol limits the reaction to the addition of four hydrogen atoms. Naphthalene can also be reduced to tetrahydronaphthalene in the presence of sodium hydroxide with activated aluminum–nickel catalyst. Some other metal catalysts are also effective [544].

Palfray [5651] hydrogenated naphthalene in the presence of Raney nickel at 125° and 115 kg pressure for 30 min. The tetrahydronaphthalene was separated from the decahydronaphthalene by fractional distillation. (See also under decahydronaphthalene [5366].)

SAFETY

The odor of 1,2,3,4-tetrahydronaphthalene is similar to that of naphthalene, and 110 ppm in the air is irritating and offensive. Dermatitis has been reported by painters working with tetralin. They also complained of headache, malaise, and irritation of the eyes, throat, and mucous membranes of the nasal passage. Severe exposure has resulted in the elimination of green urine [5713]. Smyth and coworkers [6882] report LD_{50} for rats.

The *threshold limit value* of 25 ppm, 135 mg/m^3 has been recommended [5713].

51. Butylbenzene

Read and coworkers [6085] prepared butylbenzene as follows: A mixture of 411 grams of 1-bromobutane and 471 grams of bromobenzene was added to 300 mL of dry ether containing 161 grams of sodium over a period of about 2.5 hr. The temperature was maintained at about 20°. The mixture was allowed to stand for 2 days, the liquid decanted, 300 mL of methanol added, and the mixture refluxed for 4 hr. Eight hundred milliliters of water was added and the aqueous layer extracted with ether. The decanted liquid, the hydrocarbon layer, and the ether extract were combined, dried, and fractionally distilled; yield 65–70%.

Vogel [7761] prepared butylbenzene by reducing two ketones to the hydrocarbon.

1. Seventy-five grams of butyrophenone was reduced with 150 grams of zinc amalgam and concentrated hydrochloric acid; yield, 50 grams of crude product upon steam distillation. The crude material was distilled from sodium and washed with 7-mL portions of concentrated sulfuric acid until the acid was no longer colored. After additional

washings with water and sodium carbonate, it was dried with anhydrous magnesium sulfate. After it was fractionally distilled twice from sodium, the middle fraction from the second distillation was used for *physical property measurements.*

2. Seventy-five grams of benzylethyl ketone was used for the second preparation. The same procedure was used as described for (1).

Platt and Klevens [5881] purified butylbenzene for *ultraviolet studies*; see mesitylene.

Farroha and coworkers [2373] reported the separation of butylbenzene isomers on dinonylphthalate, diethylhexyl sebacate, squalane, 1,2,3-tris(2-cyanoethoxy)propane, and Apiezon L.

SAFETY

The *flammable limits* in air are 0.8 and 5.8%v [5415]. The *minimum ignition temperature* in air is 412° [5712].

52. *sec*-Butylbenzene

Huston and Kaye [3601] prepared 20 secondary aliphatic derivatives of benzene by condensing benzene with secondary alcohols. The crude *sec*-butylbenzene was dried over anhydrous sodium sulfate and the remaining benzene removed by vacuum. The residue was distilled through a 10-in Vigreux column; yield 71%. A much purer product may be prepared by drying with Drierite or molecular sieves and fractionally distilling at a high reflux ratio in an efficient fractionation column. A second fractional distillation should give a product approaching 100% purity if the starting materials were of good purity.

Simons and Hart [6743] prepared *sec*-butylbenzene by condensing 1-chlorobutane and benzene with hydrogen chloride at 100 psi and 195°; yield 30%.

Platt and Klevens 5881 purified *sec*-butylbenzene for *ultraviolet studies*; see mesitylene.

For separation of isomers, see Farroha [2373], butylbenzene.

SAFETY

The *flammable limits* in air are 0.8 and 6.9%v [5415]. The *minimum ignition temperature* in air is 418° [5712].

53. Isobutylbenzene

Köhler and Aronheim [4099] prepared isobutylbenzene in 1875 from isopropyl iodide, benzyl chloride, and sodium in ether. Schramm [6488] reacted bromobenzene, isobutyl iodide, and sodium in benzene and reported a 27% yield of isobutylbenzene.

Vavon and Mottez [7696] prepared isobutylbenzene by the reaction of phenylmagnesium bromide and isobutyl bromide in ether with an iron(III) chloride catalyst; see also Späth [6921].

Brown and coworkers [1175] prepared isobutylbenzene by hydrogenation of 2-methylpropenylbenzene over a platinum catalyst at 2 atm. The final product was fractionally distilled; bp 170–173°.

Hennion and Auspos [3296] purified isobutylbenzene by heating with phosphorus pentoxide for 5 hr and then distilling. A second distillation was done over sodium.

Glasgow and coworkers [2780] purified isobutylbenzene for *freezing point studies* by azeotropic distillation with methyl carbitol with 130 theoretical plates at total reflux and 145 : 1 reflux ratio; purity 99.86%m.

Romadane and Pelchers [6245] carried out the alkylation of benzene with 1-butanol and boron trifluoride in an autoclave at 230° and 100–120 atm to give 70% isobutyl-benzene.

Several grades are available. Phillips [5823] lists the following: research, 99.97%w, 0.01%w toluene, 0.02%w butylbenzene; pure, 99.6%w, 0.2% toluene, 0.02%w butyl-benzene, < 100 ppm water; technical, 99.1%w, 0.5%w toluene, 0.4%w butylbenzene, < 100 ppm water.

SAFETY

The *lower flammable limit* in air is 0.8%v [5826]. The *minimum ignition temperature* is 428° [5712].

54. *tert*-Butylbenzene

Huston and coworkers [3600] prepared *tert*-butylbenzene by condensing benzene and 2-methyl-2-propanol. The crude product was purified by fractional distillation through a 30-cm column; bp 740 Torr was 168–170°, yield 65–70%.

Simons and Hart [6743] prepared *tert*-butylbenzene by condensing 2-chloro-2-meth-ylpropane and benzene in the presence of hydrogen chloride at 100 psi and 150°; yield 45.5%, bp 168.0–168.8°.

Norris and Sturgis [5518] prepared *tert*-butylbenzene in 84% yield by allowing 0.6 mol of aluminum chloride, 0.5 mol of 2-methyl-2-propanol, and 3 mol of benzene to react at room temperature for 24 hr.

Bowman and coworkers prepared *tert*-butylbenzene [1037] by direct alkylation of benzene with 2-methylpropene in the presence of boron trifluoride hydrate ($BF_3 \cdot H_2O$).

Platte and Klevens [5881] purified *tert*-butylbenzene for *ultraviolet studies*; see mesitylene.

For separation of isomers, see Farroha [2373], butylbenzene.

SAFETY

The *flammable limits* in air at 100° are 0.7 and 5.7%v [5415]. The *minimum ignition temperature* in air is 450° [5712].

55. *p*-Cymene

Le Fèvre and coworkers [4492] purified *p*-cymene as follows: Several liters of the technical product was refluxed over powdered sulfur for 2 days. It was then shaken with successive quantities of concentrated sulfuric acid until the latter was no longer colored, then twice with small amounts of chlorosulfonic acid and, after a water wash, with potassium permanganate solution, followed by dilute aqueous sodium hydroxide. The material was dried over sodium sulfate and fractionally distilled.

Lombard [4652] obtained *p*-cymene from the fraction of pine-tar oil of the cluster pine boiling at 200°. For additional natural sources, see Katasura [3877] and Hasebe [3202].

Simons and Hart [6743, 6744] prepared isopropyltoluenes by condensing 2-chloropropane and toluene in the presence of hydrogen chloride at 300 psi and 235°; yield 67% mixed isomers.

Plate and Tarasova [5879] converted turpentine sulfate into *p*-cymene at 380–520° in contact with three catalysts: chromium oxide on aluminum oxide, 1:3; molybdenum oxide on aluminum oxide, 1:3; and synthetic Houdry aluminosilicate catalyst. The first two catalysts gave as much as 66% *p*-cymene at 400° calculated from turpentine, and 77% calculated on the basis of the pinene-carene content. The aluminosilicate promoted the formation of lower aromatic compounds.

Korous and Neuzil [4197] separated *p*-cymene from the other cymene isomers with type-X zeolite which had been exchanged with potassium and barium ions. After desorption with toluene, an extract stream containing 99.7% *p*-cymene was obtained. Nixon [5504] used a similar method with molecular sieve 13X.

SAFETY

The *lower flammable limit* in air at 100° is 0.7%v [5415]. The *minimum ignition temperature* in air is 436° [5712].

56. Cyclohexylbenzene

The alkylation of benzene was studied by Carson and Ipatieff [1374]. Fifty-eight grams of cyclohexylbenzene was prepared by stirring an ice-cooled mixture of 176 grams of benzene and 60 grams of anhydrous aluminum chloride, adding 246 grams of cyclohexane over a 2.8-hr period and stirring an additional hour. The catalyst was decomposed with dilute hydrochloric acid. The hydrocarbon was purified by fractional distillation and fractional crystallization. They also prepared cyclohexylbenzene by using sulfuric acid as the catalyst. Ipatieff and Grosse [3630] used aluminum chloride and hydrogen chloride. Bodroux [934] used only aluminum chloride. Pajeau [5650] used aluminum chloride, cyclohexanol, and benzene.

The reaction of chlorocyclohexane and benzene to give cyclohexylbenzene was studied by Neunhoeffer [5463]. It was found that the larger the ratio of benzene to chlorocyclohexane, the greater the yield. The yield was 0.8 mol at a ratio of 12:1.

Metallic chlorides were found to be alkylating catalysts for aromatic compounds. One mole of chlorocyclohexane, 2 mol of benzene, and 100 mL of 83.5% zinc chloride saturated with hydrogen chloride at 75° for 6 hr gave 50% cyclohexylbenzene; the substitution of cyclohexane gave 43% and cyclohexanol 70%.

For compound from bicyclohexyl, see [695] under decahydronaphthalene; from benzene and cyclohexene, see Aromatic Hydrocarbons, General Comments [6739].

Mears and coworkers [5017] purified commercially available cyclohexylbenzene for *physical property studies* by fractionally distilling at 60 Torr in a Heligrid column. Fractions of about 50 mL were collected and classified according to refractive index. The purest material was 99.83%m. The less pure fractions were hydrogenated to bicyclohexyl.

Unsaturated Hydrocarbons (57–77)

General Comments

Methods of Preparation

A general method of preparing alkenes by the decomposition of xanthates was developed by Tschugaeff [7565] and applied to the preparation of several compounds [7566]. Whitmore and Simpson [8028] have studied this method and have made several improvements.

Kirrmann [4033] prepared 1-alkenes from Grignard compounds of suitable alkyls and bromopropene in ethyl ether. Propyl ether was used for 1-pentene. The yields are reported to be excellent.

Wilkinson [8067] treated 3-bromopropene with the appropriate Grignard reagent and obtained high purity 1-alkenes. Moisture and carbon dioxide were excluded from the system. The five straight-chain 1-alkenes of C_5, C_6, C_7, C_8, and C_9 were prepared.

Normant [5513] found that aliphatic, cyclic, or vinyl magnesium halides are obtained in good yields by reacting magnesium with the corresponding bromo compounds (chloro compounds will work) in tetrahydrofuran, 2-methyltetrahydrofuran, or tetrahydropyran. The reaction of bromo compounds with magnesium is promoted by iodine. The vinyl magnesium halides apparently form oxonium salts with the solvent. Hydrolysis of the Grignard compound gives the corresponding alkene.

Swallen and Boord [7112] found that β-bromoalkylethyl ethers react with magnesium to form alkenes. The method was further developed by Dykstra and coworkers [2173]. The latter quote an earlier author concerning the preparation of olefins, "High temperature and many chemical reagents, particularly acids, cause such α-olefins to rearrange or the double bond to shift position. Only reactions employing low temperatures and absence of isomerization agents can be expected to produce α-olefins in any degree of purity."

Bourguel [1014] reported that the reduction of alkynes to the corresponding alkenes with hydrogen and colloidal platinum at room temperature gave purer compounds than previously described methods.

Van Pelt and Wibaut [7674] reported a method for the preparation of alkenes from the corresponding alkyl acetate by passing through a 50-cm tube 2.5 cm in diameter packed with glass wool in a tube furnace heated to the temperature most suited to the particular decomposition. A preheater and a source of nitrogen were provided to prevent undesirable decomposition and polymerization. Information for the preparation of pertinent alkenes is given in Table 5.5. See also [6920].

Olefins can be prepared by catalytically hydrogenating acetylenes. Campbell and Eby [1336] studied this procedure and the sodium–liquid ammonia method. During catalytic hydrogenation, it is difficult to stop precisely at the olefin stage; therefore the product usually contains traces of saturated hydrocarbons that are difficult, if not impossible, to separate by fractional distillation. When the sodium–ammonia method is used, there is no danger of forming saturated hydrocarbons and a much purer product will be obtained. 1-Hexene, 1-heptene, and 1-octene were prepared by both methods and *physical properties* were determined.

A simple general method for preparing alkenes was reported by Brandenberg and

Table 5.5 Conditions for van Pelt and Wibaut Preparation of Alkenes

Alkene	Acetate Ester	Temp. (°C)	Feed Rate Ester (g/hr)	Feed Rate N₂, L/hr	Yield (%)
1-Hexene	Pentyl	500°	22	4	66
1-Heptene	Heptyl	515°	17	—	72
1-Octene	Octyl	500°	14.7	—	77
2-Pentene	3-Pentyl	430°	20.4	—	71

Galat [1058]. Boric acid or anhydride will dehydrate primary, secondary, or tertiary alcohols. The boric ester, $B(OCH_2CH_2R)_3$, which is formed is decomposed by heat, regenerating the catalyst, and producing an alkene. A typical example is: 1 mol of 1-octanol and 1 mol of boric acid were gradually heated to 350°. Water was continually separated from the azeotrope and the alcohol returned to the reaction vessel. When 1 mol of water was removed, the water–octene azeotrope started to distill. After the alkene azeotrope had all distilled, a second mole of the alcohol was added, then the third, and so forth; 90% yield.

No proof has been obtained that 1-alcohols give only 1-alkenes.

Purification

Several alkenes are now produced commercially. Riddick [6165] obtained samples from the pilot plants of the C_6–C_{10} 1-alkene cut of the Gulf Oil Corp., Petrochemical Department; and the California Chemical Company. The individual 1-alkenes were separated and purified to 99.7–99.9 + % for *azeotrope studies* as follows: The primary separation was made in an 8-ft, Penn State-type column packed with 0.16 × 0.16-in. protruded stainless steel packing and fitted with an automatic take-off head. The pressure was maintained at 760 Torr under nitrogen. The desired product was taken at a reflux ratio of 25:1 at the beginning and end of the fraction, decreasing to 10:1 during the take-off of the main part. The partially purified (98 + %) product was fractionally distilled in the same or a similar still. The reflux ratio at the beginning and end of the distillation was high. The main cut was taken at either a 10:1 or a 15:1 reflux ratio, depending on the difficulty of separating the impurities. The reflux ratios and the cut points were determined by gas chromatographic analysis of the product stream. The final analysis of the individual 1-alkenes was verified by mass spectroscopy.

Safety

Only a few unsaturated aliphatic hydrocarbons have been available in commercial quantities until recently. Consequently, there is little safety information available other than a modicum for the C_2, C_3, and C_4's The lower members are simple asphixiants or weak anesthetics. Anesthetic potency increases with increasing chain length in the ethylene series. Pentene has been used for surgical anesthesia [5992]. Caution is recommended when extending conclusions based on C_1–C_4 and C_5 to higher homologs.

Specific Solvents

57.   1-Pentene

Whitmore and Simpson [8028] made 1-pentene from 1-pentanol by the Tschugaeff method [7565].

Preparing 1-alkenes for standards for the Soviet oil industry, Kazanskiĭ and coworkers [3919] obtained 1-pentene from the Grignard reaction in 40–45% yield; boiling point, density, and refractive index indicated a product of high purity.

Streiff and coworkers [7042] purified 1-pentene from Phillips Petroleum Co. by fractionally distilling in a 200-theoretical-plate column at a reflux ratio of 160:1 to a purity of 99.82%m.

Research, pure, and technical grades are available.

SAFETY

The *threshold limit value* is 1000 ppm, 2950 mg/m^3 [6402].

The *flammable limits* in air are 1.5 and 7.7%v [5415, 6402]. The *minimum ignition temperature* in air is 273° [5415].

58.   2-Pentene (Mixed Isomers)
59.   *cis*-2-Pentene
60.   *trans*-2-Pentene

Directions for the preparation of 2-pentene by the dehydration of 2-pentanol in 65–80% yield are given in Gilman and Blatt [2757, p. 430].

Jeffery and Vogel [3699] prepared 2-pentene for *physical property measurements* from the low boiling product obtained during the preparation of 3-bromopentane.

Norris [2757] prepared 2-pentene by dehydrating 2-pentanol with sulfuric acid and water mixed 1:1.

Alkenes were prepared by thermal decomposition of alkyl acetates by Wibaut and van Pelt [8041]. 2-Pentene was obtained from 2-pentyl acetate by heating at 440°; heptyl acetate gave 1-heptene in 75% yield. See also [7674] under Unsaturated Hydrocarbons, General Comments.

Pomerantz and coworkers [5910] prepared 2-pentene from 3-pentanol for *physical property studies* by passing the alcohol over alumina at 300–320° with an 86% yield. The pentene layer was washed with water, dried over anhydrous sodium carbonate, and fractionally distilled in a 2.5 × 250-cm glass-Heligrid packed column. The center cut had a purity of 99.4%m by mass spectrometry analysis.

Aston and Mastrangelo [477] purified 95.7% *trans*-2-pentene by two passes of fractional melting to a purity of 99.99%m in 60% yield.

61.   1-Hexene

Van Risseghem [7677] prepared 1- and 2-hexene in several different ways to study the possibility of *stereoisomer* existence.

Mears and coworkers [5016] prepared 1-hexene by dehydrating 1-hexanol over alumina at 400°. Thirty-two liters of commercial hexanol was fractionally distilled in a 2.5

× 183-cm total reflux, variable take-off column packed with $\frac{3}{16}$-in. glass helices to remove 2-ethyl-1-butanol, which would dehydrate to form hexenes boiling close to 1-hexene. The crude olefin was distilled in a column similar to the one described and 4.2 L of 1-hexene boiling at 63.47° was obtained.

Waterman and de Kok [7922] prepared pure 1-hexene by reacting paraldehyde, ethanol, and hydrogen chloride to 1-chloroethylethyl ether, which was brominated to 1,2-dibromoethylethyl ether. The ether was reacted with butylmagnesium bromide to 1-butyl-2-bromoethylethyl ether. The latter was reduced with zinc in ethanol to 1-hexene "in a high state of purity." They also prepared 1-hexene from 3-bromopropane and butylmagnesium bromide. It was stored over sodium for several months and fractionally distilled from sodium.

Komarewsky and coworkers [4166] prepared the 1-alkene from 1-hexanol without other unsaturated compounds being present; see 1-octanol.

See also [1336, 7674] under Unsaturated Hydrocarbons, General Comments.

Streiff and coworkers [7042] purified 1-hexene for API Research Project 45 for *physical property* and *thermodynamic studies* by fractional distillation in a 135-theoretical-plate column at a reflux ratio of 165:1 to a purity of 99.87%m.

62. 1-Heptene

A pure 1-heptene was prepared by Waterman and de Kok [7923] from 3-bromo-1-propene and butylmagnesium bromide. The crude material was kept over sodium for several months until the white deposit no longer formed. It was then fractionally distilled from sodium.

For preparations by heating heptyl acetate at 525°, see 2-pentene [5516] and Unsaturated Hydrocarbons, General Comments [7674], also by hydrogenation and sodium–ammonia reduction of acetylenes in the same section [1336].

Streiff and coworkers [7042] purified 1-heptene from the same source, and by the same method they used for 1-hexene, to a purity of 99.84%m.

Sherrill and coworkers [6651] purified 1-heptene by azeotropic distillation with ethanol.

63. 1-Octene

Whitmore and Herndon [8024] showed that octene prepared by the dehydration of either 1- or 2-octanol (capryl alcohol) was a mixture of 1- and 2-octene. Komarewsky and coworkers [4166] reported that they passed 1-octanol over aluminum oxide catalyst in a 12-mm-diameter, Pyrex-brand glass tube at 350° at 35 mL/hr and obtained 1-octene and about 2% of an impurity they surmised to be 2-, 3-, or 4-*trans*-octene. The criteria of purity were: bp 121–121.4° at 750 Torr , n_D 20° 1.4095, d 20° 0.7152, and the Raman spectra. Shchekin [6616] reported that 1- and 2-octanols may be dehydrated to the corresponding octenes over aluminum silicate or platinized aluminum silicate without accompanying polymerization or hydrogen redistribution.

Waterman and de Kok [7924] prepared a high purity 1-octene from 3-bromopropene and butylmagnesium bromide and studied the *molecular refraction*.

Pomerantz and coworkers [5910] prepared this alkene to a purity of 99.2% for *phys-*

ical property studies by fractionally distilling 1-octene in the same still used to purify 2-pentene.

Streiff and coworkers [7038] purified 99.70%m 1-octene by fractional distillation in a 130-theoretical-plate column at a reflux ratio of 155:1 to a purity of 99.77%m determined from the freezing curve.

See Unsaturated Hydrocarbons, General Comments, for preparation from octyl acetate [7674], hydrogenation and sodium-ammonia reduction of acetylenes [1336], and Grignard preparation [8067].

64. 2-Octene (Mixed Isomers)
65. *cis*-2-Octene
66. *trans*-2-Octene

Detoeuf [1971] prepared 2-octene by reacting alcoholic potassium hydroxide and 2-iodooctane. Bourguel [1014] prepared 2-octene by reducing 2-octyne with hydrogen on colloidal platinum at room temperature.

Seifert [6558] prepared 2-octene by distilling 3-bromo-2-ethoxyoctane with zinc dust in methanol.

Campbell and Eby [1336] prepared *cis*-2-octene by partial hydrogenation of 2-octyne on Raney nickel in ethanol and *trans*-2-octene by reduction of 2-octyne with sodium in liquid ammonia.

Gredy [2918] prepared *cis*-2-octene by catalytic hydrogenation of 2-octyne over colloidal palladium.

Hoff and coworkers [3446] used an inversion method to convert *trans*-2-octene to *cis*-2-octene.

SAFETY

The *lower flammable limit* in air for each isomer is 0.9%v [5826].

67. 1-Nonene

Anomalous behavior was found in the melting region during *low-temperature thermal studies* of 1-nonene [4748].

Adamson and Kenner [27] prepared 1-nonene from nonyl-ammonium nitrate in 83.4% yield.

Streiff and coworkers [7041] purified 1-nonene for *physical and thermodynamic property studies* by fractional distillation in a 200-theoretical-plate column at a reflux ratio of 160:1 to a purity of 99.76%m calculated from the freezing curve.

68. 1-Decene

1-Decene was isolated from cracked petroleum in 1949 [7465].

The 1-alkene was purified in the same column that was used for 1-hexanol; see 1-hexene, by Mears and coworkers [5016] for *physical property studies* and the preparation of *n*-decane by hydrogenation.

Thermodynamic and physical property studies were made on 1-decene purified by Streiff and coworkers [7041] in a 125-theoretical-plate column at a reflux ratio of 125:1 to a purity of 99.91%m calculated from the freezing curve.

69. 1-Dodecene

Krafft [4236] reported a preparation of 1-dodecene in 1883 by the distillation of dodecyl palmitate at 600 Torr.

Hunig and Kiessel [3585] reacted 1-bromododecane with ethyldicyclohexylamine for 20 hr and, by extraction with petroleum ether, produced 98% yield of a product containing 98.6% 1-dodecene; bp 94.5% at 12 Torr. For other preparations using 1-bromododecane, see Dehmlow and Lissel [1908], Zafiriadis and Mastagli [8250], and Asinger [469].

Bertsch and coworkers [830] dehydrated 1-dodecanol by passage through neutral alumina at 280°; 93.3% yield. For other preparations by dehydration of 1-dodecanol, see Freidlin and coworkers [2534], Dupont and coworkers [2136], and Walker [7873].

Welch and Walters [7976] prepared 1-dodecene by the reduction of the enol phosphate ester (from 2-dodecanone) with activated titanium in tetrahydrofuran.

Clement and coworkers [1524] stirred 1-dodecene with tin(IV) chloride and anhydrous HCl for 1 hr at 25°. The organic layer was washed twice with sodium bicarbonate solution and three times with water, dried, and distilled at 52–75° and 1.2 Torr. They reported a similar purification using iron(III) chloride; purity 96.1% [1523].

Stridh [7046] purified 1-dodecene for *thermodynamic studies* by fractional distillation at 15 Torr in a 0.7-m all-glass column filled with helices.

Streiff and coworkers [7035] purified 1-dodecene for *cryoscopic studies* by regular distillation with 200 plates; purity 99.47%.

70. 1-Tridecene

Mabery [4728] distilled Gilsonite, a bitumen, and obtained 1-tridecene from the fraction collected at 20 Torr and 100–102°.

Cainelli and coworkers [1306] prepared 1-tridecene by reacting dodecanal with methylene bromide or iodide and magnesium turnings in ether. Kozacik and Reid [4229] prepared the compound from the conversion of decylmagnesium bromide with allyl bromide in ether.

Welch and Walters [7976] prepared 1-tridecene by the reduction of the enol phosphate ester (from 2-tridecanone) with activated titanium in tetrahydrofuran.

Streiff and coworkers [7035] purified 1-tridecene for *cryoscopic studies* with an azeotropic methyl carbitol and 30% tridecene mixture; purity 99.60%.

71. Cyclohexene

Most of the commercially available cyclohexene has been separated from petroleum. It contains small amounts of other hydrocarbons that are difficult to remove. If a purity of better than 99.9% is desired, it is advisable to prepare the cyclohexene from a high-purity cyclohexanol. Directions are given in Marvel [4930] for the preparation of large quantities of cyclohexene by dehydrating cyclohexanol with concentrated sulfuric acid. See also Senderens [6573]. Dehn and Jackson [1910] prepared cyclohexene from cyclohexanol and phosphorus pentoxide with a 96% yield.

Mousseron and coworkers [5301] reported that cyclohexene may be prepared by heating cyclohexylmethyl and butyl ethers, gradually adding sulfuric acid.

Kharasch and Urry [3983] prepared a mixture of cyclohexane and cyclohexene (35%) by a modification of the Grignard reaction.

Waterman and Van Westen [7926] purified cyclohexene prepared from cyclohexanol by fractional distillation through a 60-cm Vigreux tube, refluxing the cyclohexene cut over sodium for 7 hr and redistilling. Bromine and hydrogen bromide absorption was used as a test for purity.

Bodroux [934] dehydrated cyclohexanol with phosphoric acid and obtained cyclohexene.

Balandin and coworkers [562] obtained cyclohexene in 100% yield by passing cyclohexanol over anhydrous magnesium sulfate with a space velocity of 0.2–0.4.

Mathews [4969] dried commercial cyclohexene over phosphorus pentoxide and fractionally distilled for the determination of *heat of vaporization*.

Cyclohexene was washed several times with iron(II) sulfate by Chen and coworkers [1440], dried over magnesium perchlorate, and distilled for the investigation of *molecular polarizability* and *dipole moment*. The portion boiling at 83–83.4° at 760 Torr was collected and stored under nitrogen until ready to be used. Air was excluded to prevent oxidation. The *critical temperature* was determined with material that was washed with aqueous iron(II) sulfate and fractionally distilled at 83.0° by Cheng and coworkers [1451].

Streiff and coworkers [7042] of API Research Project 44 purified commercially available cyclohexene by fractional distillation in a 135-theoretical-plate column at a reflux ratio of 165:1 to a purity of 99.97%m for *physical property thermodynamic studies*.

Forster purified cyclohexene in the same manner as he did benzene [2486]. See also Duncan and coworkers [2119] under mesitylene.

SAFETY

The *threshold limit value* is 300 ppm, 1015 mg/m^3 [142].

72. Dipentene
73. *d*-Limonene
74. *l*-Limonene

Dipentene is obtained from the distillation of wood turpentine and other sources. There are several grades available. The impurities present depend on the source and the manufacturer and are mainly other terpenes or essential oils occurring in the plant source. Purification is difficult. The available data are often difficult to evaluate. The criteria of purity are not definitive for indicating the purity required for precision physical and thermodynamic values. Detailed descriptions of dipentene are given in Guenther [3012, 3013] and Simonsen [6751].

dl-Limonene occurs in Ajowan oil to about 40% of the distillate. It is found in petit grain lemon oil and in citrus peel of most kinds. It is widely distributed in many plants. The odor is not unpleasant and the hazards may be considered about the same as turpentine.

The name *dipentene* is used in the solvent field almost exclusively and will be used to designate the racemate.

d- and *l*-Limonene and the *dl*-racemic form (dipentene) are possibly the most widely distributed terpenes in nature with the possible exception of α-pinene.

Brühl [1190] studied the *dl*-, *d*-, and *l*-forms and concluded that within experimental errors the physical properties he studied for the three forms were nearly the same except for optical activity.

Ginsberg [2767] prepared dipentene by heating α-terpineol for 10 hr with acetic anhydride, saturating the resulting mixture with potassium hydroxide solution and steam distilling. Perkin [5767] and Wallach [7882] heated α-terpineol with potassium hydrogen sulfate at 180–190° and distilled dipentene.

Wallach [7888] heated dipentene dihydrochloride with aniline, then added glacial acetic acid to the mixture, and steam-distilled dipentene. The product was dried over potassium carbonate and distilled several times over sodium.

Vig and coworkers [7741] prepared dipentene by reacting 1-methyl-4-bromo-1-cyclohexene with lithium diisopropenyl copper below 0°; 80% yield.

Kremers [4258] prepared *d*-limonene by heating *d*-α-terpineol with potassium hydrogen sulfate at 180°. Colonge and Crabalona [1602] and Svirbely and coworkers [7103] extracted *d*-limonene from oil of orange.

Braun and Lemke [1066] prepared the pure limonene tetrabromides, treated them with magnesium in ether, and obtained the high-purity isomers.

A similar preparation of *d*-limonene using zinc dust rather than magnesium is given in Godlewski and Roshanowitsch [2805].

Padmanabhan and Jatkar [5643] purified *d*-limonene by fractionating over sodium. Widmark [8046] purified *d*-limonene by a methanol extraction.

Goldblatt and Palkin [2815] prepared *l*-limonene by heating α-pinene in a nitrogen stream at 375°; 42% yield.

SAFETY

For *d*-limonene, Tsuji and coworkers [7574] reported an oral LD_{50} in mice as 5.6–6.6 g/kg.

For dipentene, the *flammable limits* in air at elevated temperatures are 0.7 and 6.1%v and the *minimum ignition temperature* in air is 236° [5415].

75. Styrene

Styrene is difficult to purify and to keep pure. Purified styrene must be used immediately, unless it contains an inhibitor.

Pitzer and coworkers [5874] added a few milligrams of hydroquinone to about 80 g of styrene to prevent polymerization for *thermodynamic studies*.

Adams [22, p. 84] gives a method for the preparation from cinnamic acid; yield 40–42 grams. Paul and Tchelitcheff [5718] prepared styrene in 81% yield by heating benzaldehyde, malonic acid, and quinoline in a water bath. Styrene may be prepared in about 25% yield by reacting acetophenone with formamide. The reaction product, 1-phenyl-1-ethylamine, is converted to the hydrochloride and the salt heated to give styrene [3640].

Cheney and McAllister [1448] dehydrogenated isopropylbenzene by passing it through an unpacked column at 750° with a residence time of 0.1 sec at atmospheric pressure. A dehydrogenation of 68% per pass of feed was reported with a yield of 33% styrene. The styrene was separated by fractional distillation. When activated alumina was used as the catalyst in the dehydrogenation column, operating temperatures could be reduced to 450–650°, but dehydrogenation per pass was lower.

A number of methods were used by Hippel and Wesson [3413] to prepare highly purified styrene for *dielectric loss measurements*. The best results were obtained by drying Dow Chemical Company's N-100 with Drierite for 2 days, followed by vacuum distillation at 25° in the presence of an inhibitor (*p-tert*-butylcatechol, 0.005%). The dielectric loss increases if the inhibitor is removed before drying and distilling, or if the liquid is overheated, or if air is admitted. The removal of all traces of water was found to be difficult as determined by the Karl Fischer method.

Kienitz [3998] studied the purification of organic substances by fractional crystallization. The method essentially consisted of partially crystallizing the liquid, withdrawing the liquid, melting the solid, and then repeating the process. A mathematical treatment shows the determination of absolute purity from the change in the equilibrium temperature with time and the cryoscopic constant. Examples of purification of 99.9_6–99.9_9%m styrene and 98.4_0–99.9_0%m pyridine are given with their time–temperature curves.

For the determination of less than 0.01% of *o*-xylene in styrene, see Marquardt and Luce [4899].

SAFETY

Guinea pigs exposed to 650 ppm of styrene repeatedly for a long period of time appeared as well in all respects as the controls. The vapors in concentrations of 200–400 ppm have a transient irritating effect on the eyes and mucous membranes of the nose. The initial odor is easily noticeable and diminishes to faint or very faint [5713, 7022].

The *threshold limit value* has been set at 100 ppm, 420 mg/m^3 [7022].

The *flammable limits* in air at elevated temperature are 1.10 and 6.10%v. The *minimum ignition temperature* in air is 490° [5712].

Ivanov [3650] describes a rather fast method for the determination of styrene in air. Another method [83] gives the sensitivity, specificity, and likely interferences.

76. 2-Pinene
77. 2(10)-Pinene

The nomenclature of the pinenes is somewhat confused in the literature. The most commonly used names for these compounds appear to be *α-pinene* and *β-pinene*, respectively. See also synonyms in Chapter I.

d-2-Pinene is the principal constituent of Greek turpentine oil, Russian turpentine oil, and hinoki wood oil. American turpentine oil contains a small amount. *l*-2-Pinene is the principal constituent of Spanish, Austrian, French, and American turpentine oils. It is also found in oil of lemon, American peppermint oil, coriander seed oil, and other essential oils.

The pinenes are isomerized by heat, acids, and certain catalysts. It is advisable to distill pinenes in an atmosphere of nitrogen and store them in the dark. The material should be used shortly after preparation if the presence of isomerization or oxidation products is objectionable.

2-Pinene and 2(10)-pinene can be separated by fractional distillation. 2(10)-Pinene has a constant optical rotation, regardless of the source and time of collection.

Svirbely and coworkers [7103] obtained chemically pure 2-pinene from the de Haen Company and used it without further purification for the determination of *dipole moment*.

Guenther [3013] describes several methods for the isolation of 2-pinene and its separation into optically active isomers. 2-Pinene can be isolated from essential oils by fractional distillation and subsequent purification.

Mirov [5167] has reported that turpentine from *Pinus muricata* contains 98–99% *d*-2-pinene. The physical properties of the turpentine fractions obtained by distillation are given.

2-Pinene and 2(10)-pinene were purified from commercial products by fractional distillation in a Lecky and Ewell column [4460] at a reflux ratio of about 100:1. The refractive index was used as the criterion of purity; 2-pinene 1.4634 and 2(10)-pinene 1.4768.

Fuguitt and coworkers [2600] purified commercial 2- and 2(10)-pinenes for *physical property studies* by fractional distillation in a Lecky and Ewell column [4460] at 20 Torr pressure and a reflux ratio of 40:1, 75 mL fractions were collected. All fractions of 2-pinene having a refractive index of 1.4631–1.4633 at 25° were redistilled under the same conditions and 50-mL fractions collected. Those fractions that have constant refractive indices and optical rotation were considered pure. The purified 2(10)-pinene was prepared in the same manner.

2-Pinene was purified from American turpentine oil by fractional distillation, followed by purification through nitrosochloride by Waterman and coworkers [7925]

Widmark [8045] found the most efficient way to purify 2-pinene is by fractional distillation at reduced pressure.

Hawkins and Eriksen [3236] purified 2- and 2(10)-pinene for *heat of combustion studies* in a 120-plate column. The purity was calculated from the freezing point.

Schorger [6480] isolated *d*-2-pinene from the Port Oxford cedar, *Chamaecyparis lawsoniana* (Murr.) *Parlatore*, by rectifying, followed by fractionation over sodium. He found the highest specific rotation $[\alpha]_D$ 51.21° to that date; Grecian turpentine oil gives a *d*-2-pinene with $[\alpha]_D$ of 48.7°.

Lynn [4722] *purified the optically active form* of 2-pinene by mixing equal volumes of pinene, ethyl nitrate, and absolute ethanol, and then cooling in a freezing mixture. An amount of hydrogen chloride in dry ethanol equivalent to the pinene was added at such a rate that the temperature of the mixture was maintained below −5°. It was allowed to stand for about $\frac{1}{2}$ hr and the crystals separated by filtration. The filtrate was diluted with one or two volumes of ethanol and the temperature reduced to below −10°. Crystals separated from the filtrate; they were washed with cool ethanol.

The crystals were dissolved in ethanol and aniline added, and the resulting red solution was distilled with steam. The isolated pinenes had an optical rotation of ±53.75°. This was slightly higher than $[\alpha]_D$ 51.21° for *d*-2-pinene reported by Schorger [6480].

Gasopoulos [2680] reports that 2-pinene immediately precipitated mercury(I) acetate from an ethanolic solution, whereas the mixture containing 2(10)-pinene remains clear for 2 or 3 days.

Thomas and Hawkins [7409] stored purified 2(10)-pinene for *physical and thermodynamic property studies* under nitrogen at −20° until ready to use.

2-Pinene is available commercially. The Hercules Powder Company specifications are:

Specific gravity, 15.6/15.6°: 0.863.
Refractive index, 20°: 1.466.
Unpolymerized residue: 0.8%.
Distillation range, 5%: 156°; 95%: 158°.

SAFETY

It is suggested that the safety practices observed for oil of turpentine be followed for the pinenes until specific information becomes available.

The *minimum ignition temperature* of pinene in oxygen is 275° [5991].

COMPOUNDS WITH ONE TYPE OF CHARACTERISTIC ATOM OR GROUP (78–463)

HYDROXY COMPOUNDS (78–148)

Aliphatic Monohydric Alcohols (78–114)

General Comments

Preparation

Brown and Subba Rao [1138, 1140] and Brown and Zweifel [1144] prepared primary alcohols from the appropriate olefins using sodium borohydride and aluminum chloride, first obtaining the trialkylborane. The latter was hydrolyzed in ethanol with sodium hydroxide and hydrogen peroxide. The boron is preferentially attached to the terminal carbon in the olefin.

Pohoryles and coworkers [5901] prepared ethanol, 2-butene-1-ol and 1,3- and 1,4-butanediol by lithium aluminum hydride reduction.

Sladkov and coworkers [6802] and Bolotiv and Dolgov [954] prepared various alcohols from ethylene and other unsaturated hydrocarbons by reaction with alkyl aluminums, followed by oxidation and acid hydrolysis.

Olefins [2293] were condensed with perfluorocarboxylic acids to produce esters. These were hydrolyzed to alcohols.

Purification

Biddiscombe and coworkers [853] purified and characterized 1- and 2-propanol, 1- and 2-butanol, 2-methyl-1-propanol, and 2-methyl-2-propanol for *physical and thermodynamic property studies*. Commercially available materials were fractionally distilled twice in a column of 50–75 theoretical plates and the best fractions were selected by gas chromatographic analysis. Purity was evaluated by comparison with samples containing known amounts of impurities. Further purification of 2-methyl-2-propanol was made by fractional crystallization. Water was removed by treatment with calcium hydride. The purity of 2-propanol, 1-butanol, and 2-methyl-2-propanol were checked by the freezing curve. The other alcohols did not freeze but turned to glasses. The purities found are shown in Table 5.6.

Ballinger and Long [578] purified a number of alcohols for *ionization studies*, starting with the best material available. They were treated with anhydrous sodium carbonate to remove acidic materials, then with a suitable drying agent, and fractionally distilled.

Table 5.6 Purities of Alcohols for Property Studies

	Gas Chromatography	fp
1-Propanol	99.94%	—
2-Propanol	99.96	99.96%
1-Butanol	99.94	99.92
2-Butanol	99.95	—
2-Methyl-1-propanol	99.93	—
2-Methyl-2-propanol	—	99.96

Gas chromatography was used whenever possible during the purification procedure to establish the purity. Traces of any materials stronger than the alcohol being studied would lead to spurious values for the pK, more particularly those greater than 14. The *criterion of acceptability* was specific conductance.

McCurdy and Laidler [4756] purified 14 certified reagent grade C_1–C_5 alcohols by refluxing with freshly ignited calcium oxide for 4 hr and then fractionally distilling through a 40-cm column packed with 4-mm Raschig rings. The middle fraction was used to determine the *heat of vaporization.*

Purification of alcohols, particularly the optically active compounds, often can be most satisfactorily accomplished by the methods of Ingersoll [23]. A pure acid anhydride, usually phthalic anhydride, that reacts rather readily with the alcohol is used to form the ester or acid ester in an appropriate solvent such as pyridine. The acid ester is converted to the sodium salt with sodium carbonate. The salt is filtered and washed with immiscible solvents to remove the nonreactive impurities, then reprecipitated by acidification with a mineral acid.

Resolution of optically active alcohols may be accomplished by forming the alkaloid salts of the acid ester, crystallizing, washing, and so forth. After reaching the desired state of purity, the salt is decomposed with hydrochloric acid and the ester saponified with aqueous alkali. The alcohol is recovered and purified by customary methods.

Campbell [1331] has prepared the acid ester by heating the alcohol and anhydride gently to fusion. In some cases, this simplifies the preparation.

McEwen [4766] determined the *ionization constants* of alcohols that were carefully fractionally distilled from their sodium salts and kept in a sealed bottle in a desiccator over phosphorus pentoxide. Different samples of the same compound were used in the various runs to guard against any one of the samples being contaminated.

Halpern and Westley [3102] found that many optically active mixtures of secondary alcohols could be resolved by esterification with L-amino acids in the presence of *p*-toluenesulfonic acid. The crystal sulfonates of the ester are produced in 60–80% yield.

Pfenninger [5811] found that all aliphatic alcohols to four carbon atoms can be separated by gas chromatography using dioctyl sebacate on Celite.

Nagai [5375] reports that the R_f values of alcohols used for paper chromatography generally increase with the length of the carbon chain. Gasparič [2682] presented the R_f values of a large number and variety of alcohols.

Peltonen [5748] reported a method for the determination of sulfur in alcohols.

Safety

Zagradnik and coworkers [8251] determined the toxicity of saturated aliphatic alcohols to C_8 and tabulated the LD_{50} values.

Specific Solvents

78. Methanol

Practically all methanol available today is synthetic. The commercial grade is sufficiently pure to meet all except the most exacting requirements. The impurities that have been identified in a representative manufacturer's product are carbon dioxide, methyl ether, methylal, methylol, methyl formate, methyl acetate, formaldehyde, acetaldehyde, acetone, ethanol, 2-propanol, 1-butanol, 2-butanol, 2-methyl-2-propanol, and water. The concentration of all impurities is small. Methanol as it comes from the still usually contains from 0.01 to 0.04 % water, the impurity normally present in the largest amount. Each time the methanol is handled, unless extraordinary care is exercised, the water content increases. The acetone content of synthetic methanol is low. There are numerous articles in the literature concerning the removal of acetone from methanol. This was a major problem when methanol was obtained by the destructive distillation of wood. Water is the most objectionable substance present for most users. Recently, it has been found that ethanol, in normal impurity concentrations, poisons certain catalysts. A grade of methanol is available that contains 10 ppm or less ethanol.

The following manufacturer's specifications are typical of United States production:

Purity, min: 99.85%.
Specific gravity at 20/20°, max 25/25°: 0.7926–0.7889.
Acidity as acetic acid, max: 0.003%.
Acetone, max: 0.003%.
Alkalinity as ammonia, max: 0.0003%.
Distillation range including 64.5°, max: 1.0°.
Nonvolatile matter, g/100 mL, max: 0.0010.
Water, max: 0.1%.
Color, max: 5 APHA.
Permanganate time at 15°, max: 50 min.
Carbonizable substances (H_2SO_4 test) max: 50 APHA.
Chlorides: to pass test.
Sulfur: to pass test.
Appearance: clear and free from suspended matter.
Odor: characteristic, free from foreign odor, nonresidual.
Water solubility: no turbidity after 1 hr at 25° when 1 volume of methanol is mixed with 3 volumes of distilled water.

The purity is calculated from the specific gravity.

Gas chromatography could be used to determine the impurities present in methanol, but it would be slow because a high-precision method, such as flame ionization, would be required for those components in the low ppm range.

The acidity of commercial methanol is mainly due to carbon dioxide and methyl formate. The alkalinity is due to ammonia or perhaps methylamines.

Purified methanol must be kept in an atmosphere free of oxygen to prevent the formation of formaldehyde.

The removal of impurities from methanol was studied by Morton and Mark [5273]. The alcohol was refluxed with 2-furaldehyde and aqueous sodium hydroxide and fractionally distilled. A yield of 95% of the methanol treated gave no test with Nessler's reagent. The reagent is sensitive to 6 ppm (0.0006%) acetone. Tests with aniline acetate indicated that 2-furaldehyde has been removed to at least 0.0001%. Kretschmer and Wiebe [4268] used the Morton–Mark treatment, fractionally distilled through a 50-plate column, and dried the distillate with magnesium; d 25° 0.78653.

Evers and Knox [2334] purified methanol for *specific conductance studies.* Approximately 4.5 L of synthetic methanol was refluxed for 24 hr over 50 grams of magnesium. Four liters were removed by distillation and refluxed for 24 hr over silver nitrate, excluding atmospheric moisture and carbon dioxide. After distilling, the solvent was shaken for 24 hr with activated alumina and filtered through a sintered glass funnel into a pot containing 1 atm of pure nitrogen. The methanol was carefully distilled in an all-glass distilling column having a conductivity cell sealed between the column condenser and the receiver. Methanol with a low specific conductance may be stored for several weeks without noticeable effect.

Lund and Bjerrum [4698] dried methanol for *specific conductance studies* by means of magnesium activated with iodine. To a flask provided with a reflux condenser are added 0.5 grams of iodine, 5 grams of magnesium, and 50–75 mL of methanol. The flask is warmed until the iodine disappears. If vigorous evolution of hydrogen does not occur, add 0.5 grams more of iodine and heat until all of the magnesium has been converted into methylate. An additional 900 mL of methanol is then added and the mixture boiled for $\frac{1}{2}$ hr under reflux; the product is distilled with the exclusion of moisture and redistilled over tribromobenzoic acid to remove basic impurities.

Maryott [4933] purified methanol for measurement of the *dielectric constant* by drying over magnesium ribbon and then fractionally distilling through a 180-cm Dufton column; bp 64.51°, critical solution temperature in carbon disulfide 35.2°.

Thermal properties of carefully purified methanol were determined by Fiock and coworkers [2439].

Bruun [1207] collected a 1000-mL center fraction from 2500 mL of chloroform-free methanol by distilling through a 30-plate, bubble-cap column. The center fraction was refractionated in an atmosphere of dry carbon dioxide.

Pesce [5782] stated that the best way to purify methanol is to start with a product containing no acetone and fractionally distill through an efficient column, dehydrate with calcium hydride, and repeat the distillation and drying three times. This process will yield about 50% of the starting material.

Gillo [2753] found that one distillation of methanol over sodium reduced the water content to 0.003%; after the second distillation, it was 0.00005%. Acetone cannot be removed satisfactorily by distillation; chemical means must be used. Bates and coworkers [667] removed 0.2% acetone from methanol by treating with sodium hypoiodite. Twenty-five grams of iodine was dissolved in 1 L of methanol and the solution slowly poured, with constant stirring, into 500 mL of 1 N sodium hydroxide. The iodoform was precipitated upon the addition of 150 mL of water. After standing overnight, the solution was filtered and the filtrate boiled under reflux until the smell of iodoform disappeared. A single fractional distillation produced 800 mL of acetone-free methanol.

Bredig and Bayers [1079] obtained methanol *free from aldehydes and ketones* by treatment with sodium hypoiodite and silver oxide. See also Pearce and Mortimer [5732] and the methods described under ethanol.

Weigert [7952] recommended that a sodium hydroxide solution containing iodine be added to the alcohol, and that after standing for 1 day the mixture be poured slowly into about one-quarter of its volume of a 10% silver nitrate solution. After being shaken for several hours, it is distilled.

For *electrochemical measurements*, Hartley and Raikes [3197] used methanol freed from acetone with sodium hypoiodite. The alcohol was dried with aluminum amalgam, and ammonia and other volatile impurities were removed by boiling for 6 hr with freshly dehydrated copper sulfate (2 g/L) while passing through a current of dry air. Sulfanilic acid was found to be less efficient for removing basic impurities. Distillation over silver nitrate to remove reducing impurities did not seem to be necessary. See also Walden and coworkers [7865].

For *optical purposes*, Herold and Wolf [3333] dried methanol with magnesium or calcium and distilled over sulfanilic acid.

Methanol free from impurities may be prepared from carefully purified methyl esters. The formation of formaldehyde may be reduced materially by working in an atmosphere of nitrogen. Haller [3098] used potassium methyl phthalate to prepare pure methanol. See also Wohler [8127], Carius [1350], and Kraemer and Grodzki [4234].

Coetzee and Simon [1588] purified methanol for *voltammetry studies* by using a mixed strong acid–strong base ion-exchange resin followed by distillation.

For *high-pressure liquid chromatography studies*, Wolfhagen and Soltes [8151] refluxed reagent-grade methanol, 5% sodium hydroxide solution, and sodium borohydride for 30 min and distilled the mixture through a 47-cm Vigreux column.

REMOVAL OF WATER

Appreciable amounts of water are most easily removed by fractional distillation. Efficient distillation can reduce the water to 0.01%. Methanol absorbs water rather readily from the air. Hampton and Riddick [3125] found that the convenient way to prepare a methanol containing not more than 0.0007% water, the sensitivity limit of the test method, was by a combination of techniques. If a purity of components other than water, greater than that of a good commercial methanol, is desired, a suitable purification is made and the water content reduced to at least 0.04%. The purified material, or a commercial product containing not more than 0.04% water, is slowly percolated through a column of type-3A or -4A molecular sieves. The effluent methanol is stored over calcium hydride. The hydride is effective as long as it remains in a lump. The solubility of calcium hydride or calcium oxide in methanol is very small. *Drierite is not suitable for drying methanol.*

Hartley and Raikes [3197] concluded that the use of calcium oxide for drying methanol is tedious and wasteful. Timmermans and Hennaut-Roland [7488] found that calcium and barium oxides do not remove all of the water; calcium also leads to contamination with ammonia. Sodium is not efficient because the hydrolysis of sodium methylate is reversible. Goldschmidt and Thuesen [2831] dried methanol repeatedly with calcium and then removed ammonia with sulfanilic acid. Bjerrum and Zechmeister [902] recommended magnesium for the preparation of absolute methanol. Calcium carbide was recommended as a drying agent for alcohols by Cook and Haines [1638].

CRITERIA OF PURITY

ACS Reagent Chemicals [137] lists the following requirements for reagent-grade methanol:

Appearance: clear and colorless.

Water: not more than 0.20%.

Boiling range: 1 mL–95 mL, not more than 1°; 95 mL to dryness, not more than 1°.

Residue after evaporation: not more than 0.001%.

Solubility in water: to pass test.

Acetone, aldehydes: to pass test, limit about 0.001% acetone.

Acidity (as HCOOH): not more than 0.002%.

Alkalinity (as NH$_3$): not more than 0.0003%.

Substances darkened by sulfuric acid: to pass test.

Substances reducing permanganate: to pass test.

The test methods are given.

A comparison of the reagent-grade specifications with the typical industrial specifications reveals that the latter are "more demanding."

Gas chromatography is a convenient method for determining the purity of methanol for impurities present in the range of 0.01–0.02% or greater. High sensitivity detectors may be used for impurities in the ppm or ppb range.

SAFETY

Methanol does not have suitable warning or irritating properties except at high concentrations. The literature on the toxicology of methanol is voluminous, indicating a wide variation in tolerance to methanol among human beings. Apparently there is not much danger in breathing moderate concentrations of methanol vapor for a short time.

Treon [5713] concludes that industrial exposures to 200 ppm are not considered hazardous, even if the variation in individual response is kept in mind. In contrast to the many cases of blindness and fatalities from ingestion of methanol, little difficulty has been encountered in industry from inhalation or skin contact. A comprehensive study of the toxicological effects of methanol is reported by Cooper and coworkers [1649, 4016, 4017] and Chang [1415].

Alekseeva [82] developed a method for the determination of methanol and formaldehyde in air that is sensitive to 0.008 μg/mL.

The *threshold limit value* has been established as 200 ppm, 260 mg/m^3 [7022].

The *flammable limits* in air are 6.72 and 36.50%v [5712]; the upper limit is at elevated temperature. The *minimum ignition temperature* in air is 470° [5712].

79. Ethanol

Commercial ethanol is manufactured either by chemical synthesis or by fermentation. The major portion manufactured in the United States is synthetic. It may be obtained as

Table 5.7 SDA Numbers and Formulas

SDA Number	Formulation
2-B	100 gal ethanol + 0.5 gal benzene
3-A	100 gal ethanol + 5 gal methanol
13-A	100 gal ethanol + 10 gal ethyl ether
23-A	100 gal ethanol + 10 gal USP acetone

the water azeotrope containing about 5% water or as absolute alcohol which contains 0.1% or less water. Internal Revenue Department restrictions on the use of alcohol, make it more convenient for private laboratories to use Specially Denatured Alcohols (SDA). The formulas listed in Table 5.7 have been found [6165] to meet practically all laboratory solvent requirements of ethanol. The major producers of ethanol will supply, upon request, booklets on industrial alcohol containing a variety of useful information, the SDA formulas, and a synopsis of the United States Internal Revenue Department regulations.

NATURE AND DETECTION OF IMPURITIES

The usual impurities in fermentation alcohol are fusel oil, aldehydes, ketones, esters, and water. The exact composition and amounts of each component depend on the composition of the mash and the rectifying process. Synthetic alcohol contains acids, aldehydes, ketones, and other oxygenated compounds and possibly organic sulfur and some cyclic compounds, olefinic hydrocarbons, and "unknown" polymerics.

The most satisfactory method of analysis for the impurities is gas chromatography with a sensitivity adequate for the end use of the solvent.

Deckenbrock [1893] states that "grain spirits" can be identified by the methanol content. Mohr [5210] found that 0.05% methanol could be detected by dimethyldihydroresorcinol. See also [6263, 137]. Zapletálek [8261] developed a polarographic method for the determination of methanol (1–10%) in ethanol.

Fusel oil is the trade name for a mixture that includes alcohols boiling higher than ethanol, principally the pentyl alcohols, which cause the optical activity of fusel oil. If present in more than trace amounts, fusel oil can be detected by its distinctive odor by this procedure: Wet a piece of filter paper with the alcohol and allow the alcohol to evaporate. If fusel oil is present, it can be detected as a distinctive odor on the filter paper. See also [137, 6263]. The Komarowsky color reaction [4167] has been studied by Coles and Tournay [1589]. This method is sensitive enough to detect 1 gram of higher alcohols in 1000 L of ethanol. The higher alcohols may be removed by filtering through a bed of Fuller's earth [1588]. Aldehydes may be removed by the method of Stout and Schuette [7029].

Megaloikonomos [5038] developed a method using Tollens and Deniges' reagent for the determination of acetone in ethanol in the presence of aldehydes.

Orchin [5594] has reported a practical test for detecting ethanol, 1-propanol, and acetone in the presence of each other.

Raposo and coworkers [6064] studied the minor components of fermentation alcohol and found formaldehyde, acetaldehyde, methanol, ethyl acetate, 1-propanol, and 3-methyl-1-butanol.

PURIFICATION

Ethanol is available primarily in two grades based on water content. The 190 proof, or about 95%v, is the composition of the ethanol–water azeotrope. The 200 proof is commonly referred to as *absolute alcohol*. It is termed "anhydrous" to the trade and contains not more than 0.01%v water. The water from 190 proof usually is removed by azeotropic distillation.

If it is not expedient to purchase anhydrous ethanol, or if a dryer material is desired than is available, there are a number of ways whereby a dry alcohol may be prepared in the laboratory.

The removal of the major portion of the water from 190 proof is easily accomplished by distilling the water as one of the components of a ternary heteroazeotrope. Many azeotroping agents have been used. The use of benzene to form the ternary azeotrope, benzene–ethanol–water, bp 64.48°, is the basis of a widely used commercial process for the production of anhydrous ethanol [8230]. It is also a convenient means for drying ethanol in the laboratory. A 10-plate, Penn State column has been found adequate for laboratory use. About 5% benzene is added to the alcohol and the column adjusted so that the take-off of the lower aqueous phase is less than the amount of water condensing.

After most of the water has been removed, the rate of the aqueous phase distillation will decrease. When this stage has been reached, the lower aqueous phase is withdrawn intermittently as it collects in the head. After the water has been removed, the homogeneous benzene–ethanol azeotrope (bp 68.24°, 32.4%w ethanol) is distilled until benzene no longer separates upon dilution with distilled water. If the ethanol dried by the benzene–azeotrope process is to be used as a solvent for *ultraviolet measurements*, the column should be operated at intermediate take-off until the distillate shows no absorption in ultraviolet. About 10 mL of distillate is taken off every $\frac{1}{2}$ hr per liter of pot charge.

Hexane, cyclohexane, and 2,2,4-trimethylpentane are also convenient to use as ternary azeotroping agents.

Ethanol may be dried to a water content approaching 0.005% if adequate precautions are used to exclude water during the distillation. It has been found more convenient to take only normal precautions and get an ethanol containing not more than 0.05% water. The final drying may be accomplished with type-3A or -4A molecular sieves of Drierite; see methanol.

The classical method for the preparation of dry ethanol is to reflux over calcium oxide, then distill or fractionally distill. The literature for this method is especially voluminous. See, for example, Noyes [5530] and Danner and Hildebrand [1810], and others cited later in the text. Sometimes cloudiness is observed in absolute ethanol prepared from calcium oxide. This is probably due to the hydrolysis of calcium ethylate, which can be prevented by careful distillation to remove all calcium salts.

Smyth and Stoops [6876] determined the *dielectric constant* of ethanol that had been refluxed over freshly ignited calcium oxide. The calcium oxide was changed once during the 24 hr of refluxing. The anhydrous ethanol was then fractionally distilled.

For *electrical measurements*, Danner and Hildebrand [1810] distilled 95% ethanol with 5 mL of concentrated sulfuric acid and 20 mL of water per liter. Acetaldehyde was removed by boiling the distillate for several hours under reflux with 10 grams of silver nitrate and 1 gram of potassium hydroxide per liter. After distillation, the ethanol was boiled for 8 hr with 600–700 grams/L of the best commercial lime, and the mixture was shaken vigorously at room temperature for 24–36 hr. It was boiled for a further 4–6 hr with 100–150 grams/L of special lime (see below) and distilled into a special all-glass vacuum distillation apparatus connected to the conductivity vessel where it was subjected to its final purification. Preparation of calcium oxide: A mixture of calcium hydroxide and carbonate obtained from the hydroxide by drying in air was heated until completely converted to the oxide without sintering.

Thermal properties of ethanol were determined on carefully purified material by Fiock and coworkers [2439]. Low aldehyde alcohol was refluxed over freshly burned lime and distilled. The remainder of the water was removed as the ternary benzene azeotrope in a modified Bruun still [1207].

Locquin [4644] states that commercial ethanol may be dried with aluminum foil slightly amalgamated with mercury. The alcohol thus dried was used to *dehydrate biological specimens*. It was found that amalgamated aluminum foil in the bottom of the jar kept the ethanol anhydrous.

Maryott [4933] purified absolute ethanol for *dielectric constant* determinations by drying over magnesium ribbon and distilling in a 180-cm Dufton column, bp 78.34°. The *critical solution temperature* in carbon disulfide was −23.5°, indicating less than 0.1% water.

According to Winkler [8113] 98–99% ethanol is best dried with metallic calcium. Technical calcium is cleaned with a wire brush under 70% ethanol and rasped; the raspings are added to the alcohol (20 grams/L), which is then warmed on a water bath for several hours until the evolution of hydrogen is almost over. The alcohol obtained by distillation is 99.9%; further distillation over a few grams of calcium filings, the first part of the distillate being discarded, gives absolute alcohol.

Konek [4176] obtained absolute ethanol by boiling commercial absolute ethanol for 1–2 hr with 2–10% of magnesium amalgam; see also Evans and Fetsch [2329] and Andrews [366]. Lund and Bjerrum [4698] dried ethanol with magnesium activated with iodine by the method described for methanol.

Walden and coworkers [7865] used amalgamated aluminum chippings for the preparation of dry ethanol for *specific conductance measurements*. For this purpose, the aluminum chippings were degreased, treated with alkali until there was a vigorous evolution of hydrogen, washed with a little water until the washings were weakly alkaline; then they were treated with 1% mercury(II) chloride solution and after 2 min washed quickly with water, with alcohol, with ether, and then dried with filter paper. If the product is satisfactory, it will become warm during this treatment. The amalgam was then added to the alcohol to be dried, gently warmed until the evolution of hydrogen ceased (requiring a few hours), then distilled and pure air passed through the distillates for a long time. See also Thomas and Maxum [7412], Kraus and Callis [4247], Wislicenus and Kaufmann [8118], Osborne and coworkers [5614], Brunel and coworkers [1198], and Müller and coworkers [5324].

According to Leighton and coworkers [4516], the best method for the preparation of ethanol for *optical measurements* is the following: 95% ethanol is distilled over a period of several hours with 25 mL of 12 N sulfuric acid per liter, and the distillate heated under reflux with 20 grams of potassium hydroxide and 10 grams of silver nitrate per liter and distilled. After standing for 1 week over activated aluminum amalgam, it is filtered and distilled again. The material so obtained showed a smaller absorption than ethanol dried over lime. See also Lüthy [4710].

Goldschmidt [2825] prepared absolute ethanol from ethanol dried with lime by boiling several hours with a quantity of nitride-free calcium, in pieces the size of a pea, corresponding to 10 times the water content. For removing traces of ammonia, a stream of dried air, dried by passing over sodium wire, was led into the vapor during the boiling.

According to Riiber [6178], the last traces of water can be removed from 99.9% ethanol, obtained by drying with lime, by means of calcium hydride. This, he claims, is more suitable for the rapid, convenient, and certain preparation of absolute ethanol than calcium oxide or metallic calcium. The process is carried out as follows: Three and one-half grams of calcium hydride is crushed to a powder and dissolved in 200 mL of absolute alcohol by gently boiling. Two-thirds of the alcohol is then distilled off to remove ammonia. The calcium ethylate solution so obtained is poured into 2 L of 99.9% alcohol in a distillation apparatus fitted with a double condenser. The mixture is boiled for 20 hr so slowly that only about 100 mL, which contains all aldehydes, distills; at the same time a slow stream of pure and carefully dried hydrogen is bled in. Slow distillation, in a current of hydrogen, the first and last 100 mL of the distillate being discarded, gives pure ethanol; d 20° 0.789334 ± 0.000003.

Smith and Bonner [6849] purified absolute alcohol for *refractive index studies*. It was first tested for methanol, 2-propanol, 2-methyl-2-butanol, pentanol, carbonyls, acids, benzene, water, and nonvolatile matter. The only appreciable impurity found was water. The alcohol was dried for 3 days with calcium sulfate and distilled in a dry nitrogen atmosphere; bp 78.32°.

Othmer and Wentworth [5625] prepared anhydrous ethanol by using ethyl ether as the entrainment agent in the distilling column. The column was operated under pressure, and under these conditions the azeotrope contained only ether and water. No ternary azeotrope was formed. See also Wentworth [7980].

The use of dichloromethane for drying ethanol was proposed in [525]. It forms an azeotrope with water boiling at 38.1° and containing 1.8% water.

Arnold [439] found C_8–C_{12} hydrocarbons more effective for dehydrating ethanol than benzene; 2,2,4-trimethylpentane is preferred. When it is used, 3–5% of the aqueous phase is entrained in the hydrocarbon phase, up to 25% in benzene. The stratification is faster when aliphatic hydrocarbons are used. The data indicate a saving in heat requirement of 9–19% for the hydrocarbons over benzene.

Several chloroparaffins have been proposed to remove water as the ternary azeotrope. The ternary azeotropes of ethanol, water, and other compounds are given by Horsley [3497, 3498].

Maruyama [4929] describes a method for purifying alcohol for use in *polarography*. Ethanol is discussed as a *polarographic solvent* [7532]. Maruta and Matubora [4928] determined as little as 0.0007% acetaldehyde and 0.0005% diacetal furfural in ethanol *polarographically*.

Hassion and Cole [3210] prepared alcohol for *electrical measurements* in the same manner as they did 2-propanol, collecting the fraction distilling at 78.3°, n_D 25° 1.3594, and specific conductance 8.2×10^{-8} ohm^{-1} cm^{-1} at room temperature.

Teramoto and coworkers [7213] purified alcohol from sweet potatoes, barley, molasses, and so forth by passing over an acid and then an alkaline ion-exchange resin. Up to 95% of the aldehyde and 40% of the fusel oil were removed.

For purification for *dielectric constant measurements*, see [5905], also [5117, 3210].

Izergin [3655] purified ethanol by zone melting.

CRITERIA OF PURITY

The criteria of purity for an analytical reagent grade are given by Rosin [6263] and *ACS Reagent Chemicals* [137]. Water is easily determined by the Karl Fischer method. If water is the principal impurity, as it is in any good grade of ethanol, it may be determined from density measurements and density tables [5414].

Gas chromatography of the required precision to characterize ethanol for the desired end use is to be recommended.

Gillo [2754] of the Bureau des Etalons Physico-Chemiques proposed a method for the purification and criteria of purity for ethanol as an organic standard. Five hundred milliliters of ethanol and 75 mL of carbon disulfide are placed in an all-glass distilling column and distilled in an atmosphere of nitrogen. Three hundred milliliters of ethanol can be obtained, which will meet the following specifications: bp 78.30°, density 0° 0.80624, ethyl acetate 0.003%, acetaldehyde 0.004%, water 0.001%, 2-propanone 0.001%, methanol 0.0005%, and carbon disulfide 0.0001–0.0002%. Careful fractional distillation will remove the ethyl acetate; the acetaldehyde and methanol are difficult to remove.

Dreisbach and Martin [2090] determined the purity from the freezing curve. Swietoslawski [7131] used the differential ebulliometric method for determining both volatile and nonvolatile impurities.

SAFETY

It is practically impossible to produce any toxic effects by inhalation of ethanol vapors under usual industrial conditions. The minimum identifiable odor is about 350 ppm. Concentrations of 6000–9000 ppm have an intense odor that may be practically intolerable at first, but one becomes acclimated soon. Concentrations of about 1000 ppm cause slight irritation of the mucous membranes and other symptoms [5713, 7022].

Alstott and coworkers [101] reported a 24-hr LD_{50} in mice as 6.66 mg/kg.

The *threshold limit value* has been set at 1000 ppm, 1900 mg/m^3 [7022]. The *flammable limits* in air are 3.28 and 18.95%v; the upper limit is at elevated temperature. The *minimum ignition temperature* in air is 439° [5712].

80. 1-Propanol

1-Propanol is produced commercially by the hydration of propylene. It is available as either the anhydrous alcohol or as the water azeotrope that contains 72% 1-propanol. Some manufacturers produce a product comparable in purity to the best grade of ethanol. One of the principal impurities is 2-propen-1-ol.

1-Propanol is also produced by the partial oxidation of propane [8053].

Water may be removed from the 72% alcohol by azeotropic distillation similar to the procedures described for ethanol. Ternary azeotropes suitable for this purpose are given by Horsley [3497, 3498].

De Brouckère [1891] purified technical 1-propanol by refluxing over lime for 5 hr and then distilling through a 1-meter column. The fraction boiling at 97–97.3° was redistilled in a stream of hydrogen at 40–50-mm pressure. The best fraction had a specific conductance of 4.4×10^{-9} ohm^{-1} cm^{-1}.

Brunel [1197] purified the alcohol by means of the phthalate ester.

Brown and Ives [1128] prepared 1-propanol for *dielectric constant measurements*. Some trouble was experienced in obtaining material that did not give a slight haze with water. This was overcome by bromination to remove 2-propen-1-ol, refluxing with sodium hydroxide treatment with, and distillation from, Hewett's reagent, and a final fractional distillation through a 4-ft column; bp 97.1°, d 25° 0.79995.

Kretschmer [4267] found that 1-propanol contained as its principal impurity about 1.5% 2-propen-1-ol. He carefully purified the alcohol for the determination of the *thermal expansion* as follows: About 1.5 mL of bromine was added to a liter of alcohol. It was fractionally distilled from a small amount of potassium carbonate through a 75-plate column. The middle fraction of 600 mL was dried with 1 gram of magnesium ribbon, freshly cleaned with steel wool, in a storage flask attached to a vacuum system. The dried sample was then withdrawn as needed by vacuum distillation at room temperature. Before the flask was sealed, 1 gram of 2,4-dinitrophenylhydrazine was added to react with propionaldehyde that had not been removed by distillation.

Mathews and McKetta [4967] studied the *thermodynamic properties* of 1-propanol purified by fractional distillation twice through a 1.2-meter column packed with glass helices. The product has a boiling range of less than 0.02°.

Purification to 99.9%m for *physical and thermodynamic property studies*, see Biddiscombe [853] in Alcohols, General Comments.

For *specific conductance measurements*, Keyes and Winninghoff [3967] dried with metallic sodium and fractionally distilled. See also Kraus and Bishop [4245]. Goldschmidt and Thomas [2830] dried 1-propanol with aluminum amalgam and, to remove the basic impurities, distilled over sulfanilic or tartaric acid.

For *refractive index measurements*, Berner [805] boiled 1-propanol with lime for 6 hr and, after distilling, warmed the middle fraction of about 80% with calcium hydride in a stream of hydrogen for 3 hr. He finally fractionated in a stream of hydrogen. The treatment with calcium hydride and the fractional distillation were repeated until the density and refractive index were constant.

For additional purifications, see Brunel and coworkers [1198] and Lund and Bjerrum [4698].

SAFETY

1-Propanol is one of the less toxic and one of the least irritating solvents as indicated by the almost complete absence of ill effects reported from industrial usage. Two of six rats died when exposed for 6 hr to a concentration of the alcohol of 4000 ppm. Some of the symptoms that may become apparent if the concentration in the air is sufficient are intoxication, ataxia, and irritation of the mucous membranes [5713]. The LD$_{50}$ for rats, single dose, is 1.87 grams/kg [6883]; see also [7195].

The *threshold limit value* of 200 ppm, 490 mg/m^3 has been recommended [7022, undated supplement].

The *flammable limits* in air are 2.15 and 13.50%v at elevated temperature. The *minimum ignition temperature* in air is 439° [5712].

81. 2-Propanol

2-Propanol is manufactured by absorption of propylene in sulfuric acid and subsequent hydrolysis of the mono- and diisopropyl sulfate to propanol and sulfuric acid. The crude mixture is distilled as the 91% alcohol–water azeotrope. The process is basically the same as reported by Bertholet in 1855. The ternary heteroazeotrope is used to produce anhydrous 2-propanol commercially. Isopropyl ether is often used as the heteroazeotroping agent, but ethyl ether, ethyl-*tert*-butyl ether, and several other agents are also used [2271].

An especially high-purity 2-propanol for such uses as perfumes is made by the hydrogenation of acetone.

The availability of 2-propanol of a high purity and at a reasonable price has resulted in the use of large volumes as a solvent to replace ethanol for many of its uses. The anhydrous alcohol from several sources, including several grades such as spectrographic, reagent, and essence, were analyzed by gas chromatography and by the Karl Fischer method for water. The gas chromatograph had a sensitivity for most substances in the range of 0.01–0.02%. Only four impurities, including water, were detected. Water was present in the greatest amount, 0.05–0.06% [6165].

Specifications are available for several grades of 2-propanol. Some of these are included in Table 5.8.

For purification to 99.9 + %m for *physical and thermodynamic properties*, see Biddiscombe [853], Alcohols, General Comments.

Itakura [3644] prepared 2-propanol by the catalytic hydration of propylene with sulfuric acid. The maximum yield was 76% obtained by using 99% sulfuric acid at 15°. At 100° with 85% acid, hardly any propylene was absorbed.

Reynolds and Grudgings [6143] obtained an increased yield of 2-propanol and less polymer formation by reducing propylene with tungsten oxide catalyst. When the catalyst was pelletized at 62 tons psi pressure, the conversion at 250° was 16%. If the pellets were formed at only 32 tons pressure, the conversion at the same temperature was 10%.

For additional preparations, see de Melle [1945], Hunter [3590], and Brunel [1197].

Ozol and Masterson [5637] purified 2-propanol, produced from olefins such as propylene, by first treating with aqueous sodium hydroxide solution, followed by preliminary distillation. Before final distillation, the alcohol was stabilized and the odor improved by adding a small amount of copper(I) chloride (0.5% or less).

Frejacques [2539] prepared anhydrous 2-propanol by saturating an aqueous solution with ammonia and carbon dioxide. Two layers were formed. The water-rich layer was distilled, yielding ammonia, carbon dioxide, and the azeotrope. The water-poor layer was fractionally distilled to give the azeotrope and anhydrous alcohol.

Maryott [4933] purified a good commercial grade by drying over magnesium ribbon and fractionally distilling in a 180-cm Dufton column. The best fraction, boiling at 82.33 to 82.39°, was used for *dipole moment studies*.

Mathews [4969] purified 2-propanol for the determination of the *heat of vaporization* by drying first with calcium chloride and then with barium oxide, and then three careful fractional distillations.

Lebo [4454] prepared "absolute" 2-propanol by distillation over freshly burned lime through an efficient column. The fraction boiling at 82–82.4° was collected and shaken

Table 5.8 Specifications for Several Grades of 2-Propanol

	ACS Reagent Grade	ASTM D 770-59	Commercial 99%	Commercial 91%	Essence[a] Grade	British Standard
Density (25°C)	0.781–0.783	—	—	—	—	—
Sp. Gr. (20°/20°C)	—	0.785–0.790	0.7862–0.7867	0.8175–0.8185	0.785–0.789	0.785–0.789
Color, not greater than	Co–Pt 10	Colorless	Co–Pt 10	Co–Pt 10	Water-white[b]	Colorless
Distillation range (°C)	See [137]	1.5°C[c]	0.5°C[c]	1.0°C[d]	95% 80–82	95% 81–83%
Nonvolatile matter (max)	0.001%	0.005 g/100 mL	0.001 g/100 mL	0.002 g/100 mL	0.001%	0.01%
Odor, characteristic	—	Nonresidual	Nonresidual	Nonresidual	—	—
Water (max %)	0.50[e]	0.5[f]	0.1	—	1.0	0.5
Miscibility with water	PT[g]	PT[g]	Clear	Clear	Complete	Complete
Acidity (as H_2SO_4 max %)	PT[g]	0.002[h]	0.002[h]	0.0024[h]	None	0.002
Alkalinity[i]	PT[g]	—	—	—	None	None
Carbonyls (max)[j]	—	—	—	—	—	0.2
Permanganate time (15°C)	—	—	20 min	30 min	—	—
2-Propanol (min %)	—	99%	99.8% w	87.50% w	98	—
Bibliography (No.)	137	See title above	7631	7631	4902	4902

[a] Essence or Cosmetic grade.
[b] This is an indefinite term. One of the definitions of (Water-white) is a Hazen or Co–Pt value of 25–30.
[c] To include 82.3°.
[d] To include 80.4 ± 0.1°C.
[e] Karl Fischer method.
[f] Based on cloud test.
[g] Pass test.
[h] Acetic acid.
[i] To phenolphthalein.
[j] As acetone.

for 2 days with anhydrous copper sulfate. It was then fractionally distilled several times until the boiling point remained constant. The alcohol so obtained contained less than 0.010% water.

For drying with calcium oxide or aluminum amalgam, see Brunel and coworkers [1198].

Aerov and Motina [40] report that isopropyl ether is economically advantageous for removing the water from 2-propanol. They rectified alcohol containing 45.6%m water in a 44-sieve plate column at a reflux ratio of 8.6:1.

Aluminum isopropylate, $Al(OCH(CH_3)_2)_3$, was found by Bezborodko [843] to be effective for drying 2-propanol and removing traces of catalyst.

Catterall [1393] dried 91%v 2-propanol to 99.9% by continuous azeotropic distillation with ethyl ether in a column of about 50 plates at 150–170 psig.

Berman and coworkers [798] purified commercial 2-propanol by storing over magnesium ribbon, followed by distillation through a 1.2-meter, glass-ring packed column. The middle cut boiling over a range of 0.04° was stored under vacuum.

The *solubility* of normally gaseous paraffinic hydrocarbons was studied by Kretschmer and Wiebe [4268] in a 2-propanol that was purified by adding a small quantity of 2,4-dinitrophenylhydrazine and fractional distillation through a 50-plate column; d 25° 0.7881.

For *optical measurements*, Herold and Wolf [3333] dried with magnesium and distilled from sulfanilic acid.

A laboratory method for purifying 91% 2-propanol may be found in [2762].

Métra and coworkers [5073] developed a method for the determination of 2-propanol in the presence of other alcohols. An oxidimetric method for the determination of 2-propanol in a mixture of methanol, ethanol, 1-propanol, and acetone is described by Wehle [7949].

For use as a polarographic solvent, see [7532].

CRITERIA OF PURITY

Test methods are available for all of the specifications given in Table 5.8.

Swietoslawski [7131] characterized 2-propanol by differential ebulliometric measurements.

Gas chromatography procedure of suitable sensitivity for the end use of the solvent is common practice.

SAFETY

The effect of exposure to 800 ppm of 2-propanol in air is not severe but most people find it objectionable. Mild irritation to the eyes, nose, and throat is experienced by most human subjects to 400 ppm for 3–5 min. The minimum identifiable concentration is about 200 ppm [5713].

The *threshold limit value* has been established at 400 ppm. This is low enough to prevent narcosis but slight irritation may occur [7022].

The *flammable limits* in air are 2.02 and 11.80%v at elevated temperature [5712]. The *minimum ignition temperature* in air is 456° [5713].

Butyl Alcohols

82. 1-Butanol

Bremner [1080] prepared 1-butanol in 70% yield by the hydrogenation of furan at 100 atm and 220° in the presence of copper–barium oxalate catalyst. The production of tetrahydrofuran or its derivatives is suppressed if a small amount of an acid, or acid salt, is added.

For *dielectric constant measurements*, Smyth and Stoops [6876] refluxed 1-butanol, prepared by fermentation, over freshly ignited lime and fractionally distilled the product.

Jones and Christian [3751] refluxed 1000 grams of a good commercial grade of 1-butanol over 50 grams of freshly ignited calcium oxide for 4 hr. The outlet tube was protected from atmospheric moisture. The alcohol was decanted from the lime, refluxed with magnesium turnings, and then distilled through a 75-cm column. A small, low-boiling fraction was discarded. The bulk of the material distilled at 117.70° ± 0.01°, d 0° 0.82460, d 25° 0.80572.

Williams and Daniels [8080] digested 1-butanol over unslaked lime and distilled. The center portion was further digested over barium oxide, then treated with sodium and fractionally distilled. The purified alcohol was used for *specific heat measurments*.

Clarke and coworkers [1511] washed 1-butanol with dilute sulfuric acid and with sodium bisulfite solution to remove bases, aldehydes, and ketones. Esters were removed by boiling for 1.5 hr with 20% sodium hydroxide. The alcohol was dried with potassium carbonate, followed by barium oxide, and finally distilled through an efficient column.

Harkins and Wampler [3165] treated 1-butanol with sodium bisulfite solution, then boiled for 4 hr with 10% sodium hydroxide solution, washed the separated alcohol with water, and neutralized with hydrochloric acid. The alcohol was dried overnight with lime and boiled three times with fresh lime for 3 hr. It was finally fractionally distilled.

The usual purification did not appear satisfactory to Orton and Jones [5602]. They purified by means of the sodium salt of the salicylic ester, which separates out immediately when the ester is dropped, with vigorous stirring, into a slight excess of sodium hydroxide. It is pressed out and converted back to the alcohol by treatment with very dilute hydrochloric acid. The purification process is then repeated. The 1-butanol is dried with potassium carbonate and distilled.

Reagent grade 1-butanol was refluxed for several hours with lime by Venkatasetty and Brown [7712] and distilled three times. The fraction boiling between 117.3 and 117.5° was collected from the final distillation.

For purification to >99.9%m for *physical and thermodynamic properties*, see Biddiscombe [853], under Alcohols, General Comments.

1-Butanol containing 9% water can be dried by fractional distillation. The water passes over in the first fraction as the binary azeotrope containing about 27% water [6165].

For *optical measurements* Herold and Wolf [3333] dried 1-butanol with magnesium ribbon and distilled over sulfanilic acid. See also Berner [805].

For *specific conductance measurements*, Goldschmidt and Mathiesen [2827] used 1-butanol, boiling between 116 and 117°, dried by Walden's method with aluminum amalgam. The dried alcohol was distilled from tartaric acid.

For drying with barium oxide or with calcium and aluminum amalgam, see Brunel and coworkers [1198].

For 1-butanol as a polarographic solvent, see [7532].

CRITERIA OF PURITY

The specifications for reagent-grade 1-butanol are given by Rosin [6263] including distillation range, nonvolatiles, and free acidity.

Orton and Jones [5602] used the critical solution temperature in hydrochloric acid to characterize the purity.

Dreisbach and Martin [2090] determined the purity by the Swietoslawski differential ebulliometric method [7131].

SAFETY

1-Butanol is potentially more toxic than any of the lower alcohols, but the actual hazard is reduced by its low vapor pressure. Humans briefly exposed to concentrations of 25 ppm showed mild irritation of the nose, throat, and eyes. A concentration of 50 ppm is objectionable because it produces pronounced irritation of the throat of all persons and mild headaches in some. The minimum concentration with identifiable odor is 15 ppm [5713].

Rumyantsev and coworkers [6312] reported an acute oral LD_{50} of 2.68 g/kg for mice.

The *threshold limit value* on skin has been set at 50 ppm, 150 mg/m^3 [142].

The *flammable limits* in air are 1.45 and 11.24%v; the upper limit is at elevated temperature. The *minimum ignition temperature* in air is 345° [5713].

83.   2-Butanol

Whitman [8018] prepared 2-butanol by hydrogenation of 3-hydroxy-1-butene in the presence of titanium hydride at 175° and pressures of 1000–2000 psi. The reaction mixture contained 26% 2-butanol.

Houtman and coworkers [3510] prepared 2-butanol by hydrogenating butanone at 200° in the presence of a nickel catalyst. See also Brunel [1197].

Dannhauser and Cole [1813] prepared 2-butanol for *electrical measurements* by drying over calcium hydride and distilling; $\kappa = 1.0 \times 10^{-7}$ ohm^{-1} cm^{-1}.

For purification to 99.9+%m for *physical and thermodynamic properties*, see Biddiscombe [853], Alcohols, General Comments.

Berman and McKetta [799] purified 2-butanol for *thermodynamic study* by repeated fractional distillation through a 1-in., 90-plate Oldershaw column to a purity of 99.92% by gas chromatography analysis.

Pickard and Kenyon [5833] have described a method for *resolving racemic 2-butanol* into its optically active isomers and have recorded some of their *physical properties*.

The method is further refined and described [5835] wherein ethylmagnesium bromide and acetaldehyde are employed in large-scale preparations. The resolution employs the 2-butyl hydrogen phthalate and brucine. The salt was recrystallized from acetone, then methanol, followed by hydrolysis.

2-Butanol and 2-pentanol were resolved into their *optical isomers* by Adembri [32] by slowly reacting the racemate with phosgene at −60°, allowing the phosgene and hydrogen chloride to escape, and then purging with air. The chlorocarbonates were reacted with 1-asparagine and the isomers separated by crystallization. The salts were hydrolyzed with 25% potassium hydroxide and the alcohols isolated and purified. The optical rotation was taken in ethanol and was found to be; L(+)-2-butanol 20° $[\alpha]_D$ + 13.8°, D(−)-2-butanol 20° $[\alpha]_D$ − 13.5°, (−)-2-pentanol 20° $[\alpha]_D$ − 16.8°, and (+)-2-pentanol 20° $[\alpha]_D$ + 16.6°.

Hargreaves [3156] has purified 2-butanol by conversion to the alkyl hydrogen phthalate and recrystallization to constant values.

Levene and coworkers [4543] describe a method for preparing the optical isomers of 2-butanol.

Optically active 2-butanol was prepared from material obtained from a chemical supply house by Kantor and Hauser [3852]. The alcohol was converted to the acid phthalate ester and 2.01 mol thoroughly mixed with 2.0 mol of brucine. The mix was charged into a 6-L Erlenmeyer flask with 2-L of acetone and refluxed for 24 hr. The solid changed in appearance during the reflux period and was kept broken up so that the acetone could come in contact with all of the solid. The mixture was filtered hot and recrystallized two or three times from methanol. The brucine salt was hydrolyzed with sodium hydroxide to give a high purity (+)2-butanol; bp 98–99.5°, d 30° 0.799, n_D 23° 1.3955, $[\alpha]_D$ 27° + 13.28° corresponding to an optical purity of at least 98.2% based on the highest reported value, +13.52, found by Timmermans and Martin [7495].

Ambrose and Sprake [130] demonstrated that the racemate of 2-butanol and the enantiomer components behave ideally. Enantiomer components of racemates have the same physical properties except optical activity, which is equal but opposite in direction. Some racemates of liquids are fairly strongly hydrogen bonded and the enantiomer components differ in many of their physical properties; for instance, (±)-2-octanol and the *d*- and *l*-forms have boiling points about 2° apart. The boiling points of the *d*-form and racemic mixture of 2-butanol were found to be 372.660 and 372.662 K, respectively. This difference is not significant.

See Andon and Connett [350] for preparation and purification of the optical isomers.

Cody and Jones [1549] reported that (±)-2-butanol solidified as a glass at −114.7°.

Synthetic 2-butanol may have traces to 1–2% of one or more of the following substances depending on the method of manufacture and purification: 2-propanol, C_8 (mostly) hydrocarbons, butanone, *sec*-butyl ether, 2-methyl-2-propanol, and other alcohols.

SAFETY

Limited data for the acute toxicity of 2-butanol indicate that it is less toxic than the normal homologue. The physiological response of animals to the vapors of the alcohol in air is about the same. There does not appear to be any difference in the toxicity of the optical isomers of 2-butanol [5713; 7022, undated supplement].

The *threshold limit value* has been set at 150 ppm, 450 mg/m³. This limit has been set on meager data mainly to prevent narcotic and irritative effects [7022, undated supplement].

The *flammable limits* in air are 1.4 and 11.2%v [2351]. The *minimum ignition temperature* in air is 406° [5712].

84. 2-Methyl-1-propanol

Itakura [3644] prepared 2-methyl-1-propanol by the catalytic hydration of isobutylene with sulfuric acid. A maximum yield of 25% was obtained by using 50% sulfuric acid at 105°.

The preparation of 2-methyl-1-propanol and 3-methyl-1-butanol is easily accomplished by fractional distillation in the presence of sodium chloride, according to Sukhodol and Chatskii [7078]. Further purification is obtained by extraction with benzene and carbon tetrachloride.

Sarancha and Dubovikova [6385] studied the products of synthesis of 2-methyl-1-propanol from carbon monoxide and hydrogen on a 2-meter chromatograph column of diatomite saturated with dinonyl phthalate. The by-products were identified as 1-butanol, 3-methyl-1-butanol, 1-propanol, and methanol.

Brunel [1196] boiled commercial 2-methyl-1-propanol with lime for 5 hr and then with calcium shavings for 1.5 hr. The dried alcohol was fractionally distilled with care several times through an efficient column. See also Brunel and coworkers [1198].

Michael and coworkers [5096] fractionally distilled a good commercial grade through a long Hempel column until the boiling point was constant to within $0.15°$ and obtained a 40% yield. After being shaken with dilute sulfuric acid to remove basic substances, the alcohol was washed with water and dried, first with ignited potassium carbonate and then with lime. The alcohol was further purified by conversion to the borate ester by heating for 6 hr in an autoclave at $160–175°$ with a quarter of its weight of boric acid. The ester was fractionally distilled several times in a vacuum, then hydrolyzed by heating for a short time with aqueous alkali. After drying with lime, it was distilled.

For purification to $99.9 + \%$m for *physical and thermodynamic properties*, see Biddiscombe [853], under Alcohols, General Comments.

Hückel and Ackermann [3544] converted 2-methyl-1-propanol into the acid phthalate by heating with phthalic anhydride. The ester was recrystallized from petroleum ether until the melting point was constant at $65°$ and then hydrolyzed with 15% potassium hydroxide. The alcohol was distilled as the water azeotrope, and dried with potassium carbonate, anhydrous copper sulfate, and finally with magnesium turnings. It was then fractionally distilled.

Mathews [4969] dried commercial 2-methyl-1-propanol repeatedly with barium oxide and carefully distilled. The dried material was used for the determination of the *heat of vaporization*.

For *specific conductance measurements*, Goldschmidt [2826] dried and purified the alcohol by repeated treatment with calcium or by a single treatment with aluminum amalgam. See Walden [7865]. The dried material was then fractionally distilled from sulfanilic or tartaric acid.

SAFETY

2-Methyl-1-propanol, generally, may be considered of the same order of toxicity as 1- and 2-butanol. Its primary action is narcotic. No evidence of eye irritation was noted with repeated 8-hr exposures. It may be a skin irritant [5713].

Kushneva and coworkers [4347] reported the following median lethal concentrations in mg/m^3: mice, 15,500; rats, 19,200; guinea pigs, 19,900; and rabbits, 26,250.

The *threshold limit value* is 50 ppm, 150 mg/m^3 [142].

The *lower flammable limit* in air is 1.68%v. The *minimum ignition temperature* in air is $434°$ [5713].

85. 2-Methyl-2-propanol

Remiz and Frost [6131] prepared 2-methyl-2-propanol by hydrating isobutylene with solutions containing 10–30%w of sulfuric acid and 2–3% silver sulfate at $85–95°$. When the gas was passed over the catalytic solution rapidly, the yield of the alcohol was 2–3 kg per kilogram of the sulfuric acid–silver sulfate mixture. An increase in the temperature or acid concentration caused an increase in the polymerization processes. See also Katsuno [3892].

Getman [2720] determined the *cryoscopic constant* of 2-methyl-2-propanol that was dried over lime and fractionally distilled.

Bigelow [864] recommended 2-methyl-2-propanol as a solvent for *cryoscopic molecular weight determination.* It has a convenient melting point at, or just below, room temperature and is available in adequate purity at a reasonable cost. The use of this alcohol, in general, is limited to substances that do not associate extensively with it such as hydrocarbons, carbonyls, tertiary amines, ethers; and nitro- and halogen compounds. A simple, rapid method is described that utilizes common laboratory equipment.

Fritz and Marple [2584] found that 2-methyl-2-propanol was a solvent that was well suited for the differentiating of titration-weak acids and weakly acidic compounds.

For purification to 99.9 + %m for *physical and thermodynamic properties*, see Biddiscombe [853], under Alcohols, General Comments.

Parks and Anderson [5675] purified a good grade of a commercial alcohol by distilling from lime and fractionally crystallizing the distillate eight times; mp 25.4°.

Simonsen and Washburn [6750] reported evidence of a second form of 2-methyl-2-propanol melting at 25.0°.

Natradze and Novikova [5425] extracted 2-methyl-2-propanol from its aqueous solution with chloroform. The extracted mixture was fractionated several times to free it from water. The chloroform was distilled and 2-methyl-2-propanol containing 0.1–0.15% water was obtained.

Maryott [4933] distilled 2-methyl-2-propanol from sodium and fractionally crystallized from its own melt until no change in melting point occurred; mp 24.85–25.0°.

Parks and coworkers [5691] purified 2-methyl-2-propanol melting at about 20° for *cryoscopic studies* by two fractional distillations, followed by two fractional crystallizations; mp 25.4°.

DeVries and Soffer [1975] proposed the use of 2-methyl-2-propanol as a *thermometric standard.* A commercial grade of the alcohol, melting at 24.08°, was first crystallized in an open beaker. This resulted in the lowering of the freezing point due to the absorption of moisture from the air. The alcohol was then dried over lime, decanted, and distilled into a cell, to the inlet tube of which a short length of rubber tubing fitted with two pinch clamps was then attached. The alcohol in the cell was frozen until only 5–10 mL of liquid remained. The cell was inverted and the liquid was withdrawn by alternately opening and closing the pinch clamps. This crystallization procedure was repeated 13 times. Thirty-seven cooling curves were made with an NBS-calibrated platinum resistance thermometer, or with one calibrated at the ice and steam points. After the addition of a small amount of calcium hydride, the freezing point was increased by 0.06° to 25.66°.

Smyth and McNeight [6870] purified 2-methyl-2-propanol for *electrical measurement studies* by washing with a saturated solution of sodium bicarbonate, refluxing with 10% sodium hydroxide, distilling, and drying with sodium sulfate or potassium carbonate. It was then refluxed over lime and distilled. It was again distilled after standing over calcium turnings for 2 days. A final distillation was made from tartaric acid; bp 8.23°.

A Certified reagent-grade 2-methyl-2-propanol was purified by Cocivera [1546] for *kinetic measurements* by refluxing for 24 hr with 1 gram of calcium hydride per liter of alcohol and then distilling. The center portion was redistilled with about 25 mg of benzoic acid per liter. The center fraction contained less than 10^{-5} M acid and less than 0.002 M water.

Jannelli and coworkers [3674] did not consider 2-methyl-2-propanol as a reliable solvent for cryoscopic studies since there may be three crystalline forms, each producing a different cryoscopic constant.

CRITERIA OF PURITY

Dreisbach and Martin [2090] determined the purity from the freezing curve.

SAFETY

Animals exposed to vapors of 2-methyl-2-propanol show symptoms of intoxication similar to other butyl alcohols. It has a stronger narcotic action on mice than the normal or iso-homologues. Skin contact by humans resulted in only slight erythemia and hyperemia [7022].

The *threshold limit value* has been set at 100 ppm, 300 mg/m^3 to prevent narcosis [7022].

The *lower flammable limit* in air at 25° in 2.35%v; the *upper flammable limit* in air at 55° is 8%v. The *minimum ignition temperature* in air is 478° [5713].

Pentyl Alcohols

All of the eight pentyl alcohols, $C_5H_{11}OH$, have been included in the Fourth Edition.

There are three commercial sources of the pentyl alcohols. The earliest of these sources was fusel oil; see 3-methyl-1-butanol. All of the pentyl alcohols are obtained, except 2,2-dimethyl-1-propanol, when a mixture of pentane and 2-methylbutane is chlorinated and the chloropentanes hydrolyzed to the alcohols [523]. These alcohols are available in varying degrees of purity; their composition may be determined by gas chromatography. The pentyl alcohols are also produced by the Oxo process. The impurities are other monomeric alcohols, dimeric alcohols, acetals, and several other miscellaneous substances. These impurities vary from producer to producer.

Several references in this section discuss the impurities in the pentanols from different sources.

SAFETY

The physiological action of the several pentyl alcohols does not differ to an appreciable extent. 3-Methyl-1-butanol (isoamyl) appears to have been studied the most. The most important effect to be expected from vapor inhalation is narcosis. The narcotic action for rabbits decreases from 2-pentanol to 2-methyl-2-butanol to 3-methyl-1-butanol; the toxicity of the same isomers decreases in the following order: tertiary, secondary, primary. The toxicity for rats decreases from 2-methyl-2-butanol to 2-pentanol to 1-pentanol. The pentanols may be irritating to the skin [5713, 7022].

The *threshold limit value* and *flammability* for four pentyl alcohols are given in Table 5.9.

Table 5.9 Safe Air Concentration and Flammability of the Pentyl Alcohols

Isomer	Threshold Limit Value (ppm)[a]	Minimum Ignition Temperature (°C)	Lower Flammability Limit (% v)
86. 1-Pentanol	—	391	1.19
87. 2-Pentanol	—	343	—
89. 3-Methyl-1-butanol	100	343	1.2
91. 2-Methyl-2-butanol	—	437	6.0

[a]360 mg/m^3.   [5713, 7022]

86. 1-Pentanol

1-Pentanol can be prepared in 56–61% yields from ethyl valerate [25].

Brown and Subba Rao [1140] prepared primary alcohols from internally unsaturated hydrocarbons through hydroboration and oxidation.

Medoks and Ozerskaya [5032] prepared 1-pentanol by the Grignard reaction between butylmagnesium bromide and trioxymethylene.

Olivier [5579] purified 1-pentanol by esterifying with *p*-hydroxybenzoic acid, crystallizing the ester from carbon disulfide, and saponifying with alcoholic potassium hydroxide. Further purification by drying with Drierite and fractionally distilling will give a high purity product.

1-Propanol, 2-methyl-2-propanol, and 1-pentanol were separated from purified fermented sulfite liquor [1246].

87. 2-Pentanol

Brauns [1071] prepared 4.5 kg of the inactive 2-pentanol from bromopropane and acetaldehyde by the Grignard method. The directions are given by Pickard and Kenyon [5833]. The alcohol was kept for a few months over dry potassium carbonate and fractionally distilled; bp 119.4°, yield 40%. The racemic mixture was then separated into the *optically active* forms. Physical properties are given.

Brunel [1197] prepared 2-pentanol from the corresponding ketone and purified it by fractional distillation.

Brown and Subba Rao [1138, 1140] reported a method for the preparation of primary alcohols from olefins by aluminum borohydride. See Aliphatic Monohydric Alcohols, General Comments.

Timmermans [7482] recommended the preparation of 2-pentanol from acetaldehyde and propylmagnesium bromide. Paraldehyde, bp 124°, forms an azeotrope boiling at 118.5°, which is difficult to separate. The alcohol is reported to be unstable at its boiling point. Brown and Nakagawa [1137] obtained 82% yield from methylmagnesium iodide and butyraldehyde.

Norton and Hass [5520] prepared 2-pentanol by reacting 1,2-epoxypropane with (1) diethylmagnesium, yield 23%, or (2) ethylmagnesium bromide, yield 11.7%.

88. 3-Pentanol

Sabatier and Senderens [6329] prepared 3-pentanol by the reduction of 3-pentanone with hydrogen on nickel catalyst at 130–140°.

Grignard [2965] prepared 3-pentanol from ethylmagnesium bromide and ethyl formate; see also Brown and Nakagawa [1137].

Hamelin [3113] found that ethylmagnesium bromide and propionaldehyde gave 3-pentanol, propanol, and *tert*-heptanol; the distribution depended on the reaction conditions.

For separation of the isomers, see 2-butanol [32].

89. 2-Methyl-1-butanol

Houtman and coworkers [3510] prepared 2-methyl-1-butanol from formaldehyde and 2-butylmagnesium bromide. Brown and Zweifel [1142] obtained the alcohol from internally unsaturated hydrocarbons by hydrobromination.

Birun [889] found that the inactive 2-methyl-1-butanol reacted 2.5 times as fast with hydrogen chloride as the active form. A mixture was enriched with the active form in this manner.

Kortuem and Faltusz [4201] found that 2,3-dichloro-1-propanol was the most effective entrainer tested for the separation of 2-methyl- and 3-methyl-1-butanols by extractive distillation. The average *alpha* for the mixture could be increased from 1.078 to 1.2 with 92 % of the entrainer.

2-Methyl-1-butanol may be prepared in a high state of purity by fractional crystallization of a suitable ester such as the trinitrophthalate [7482]. Terry and coworkers [7216] prepared the 3-nitrophthalic ester of the alcohol by the method of Marckwald and McKenzie [4885]. The acid ester was converted to the cinchoidine salt in acetone and then recrystallized from chloroform by adding pentane. The salt was saponified, extracted with ether, and fractionally distilled: bp 128.5°, n_D 25° 1.4082, d 25° 0.8161, and $[\alpha]_D$ 25° − 9.50°. *d*-2-Methyl-1-butanol was prepared by Ehrlich [2224] from *d*-isoleucine.

Easton and coworkers [2192] found that fusel oil contained up to 20 % dextrorotatory 2-methyl-1-butanol. They isolated the compound to measure the *optical rotation* by first drying a commercial product with calcium oxide and then fractionally distilling in a 73–117 theoretical plate Stedman column operating at 70 : 1 to 120 : 1 reflux ratio, depending on the separation required. The distillate was converted to the (1) 3-nitrophthalate acid ester and to (2) the acid phthalate ester. It was then crystallized, hydrolyzed, and the active alcohol separated and dried; $[\alpha]_D$ 20.7° − 5.900°.

Brauns [1070] and Whitmore and Olewine [8026] prepared the *optically active isomer* from fusel oil by fractional distillation. Brauns used a column packed with glass Raschig rings. Whitmore and Olewine used a Penn State packing. Apparently, both prepared a high-purity optical product.

90. 3-Methyl-1-butanol

3-Methyl-1-butanol is the principal component of *fusel oil*, a by-product of alcoholic fermentation of starches and sugars. The composition of fusel oil varies considerably

Table 5.10   Average Composition of Fusel Oil

3-Methyl-1-butanol	60–65%
l-2-Methyl-1-butanol	8–10%
Ethanol and 1-propanol	4%
Isobutyl alcohol	20%
Water	0.1–1%
1-Butanol, pentanols, and higher	Traces
Aldehydes, acids, pyridines, pyrazines, 2-furaldehyde	Traces

with the fermentation organism, the fermentation medium, and the process and equipment used for separating it from crude spirits. The composition shown in Table 5.10 is an average.

2-Furaldehyde may be detected by shaking the alcohol with an equal quantity of concentrated sulfuric acid. No more than a pale yellow or reddish color should be produced. The alcohol should not become colored on shaking with an equal quantity of potassium hydroxide solution. For a sensitive test for 2-furaldehyde with β-naphthol and sulfuric acid, see Udránsky [7598].

Refined fusel oil is produced by chemical treatment and rectification of the crude material.

Refined amyl alcohol is produced by further chemical treatment and more careful rectification. It contains principally about 85% 3-methyl-1-butanol and 15% 2-methyl-1-butanol. The 2-methyl-1-butanol may be removed from the higher boiling isomer by fractional distillation in an efficient column of at least 70–100 plates at a high reflux ratio, fractional crystallization, or preparative gas chromatography.

For preparation from fusel oil, see Sukhodol and Chatskii [7078] under 2-methyl-1-propanol.

Hargreaves [3156] prepared 3-methyl-1-butanol through the alkyl hydrogen phthalate.

Brown and Zweifel [1142] obtained the alcohol by hydroboration of an internally unsaturated hydrocarbon.

For separation of 3-methyl-1-butanol from 2-methyl-1-butanol, see 2-methyl-1-butanol [4201].

Timmermans and Hennaut-Roland [7487] prepared 3-methyl-1-butanol by allowing the magnesium derivative of 1-bromo-2-methylpropane to react with pure, dry trioxymethylene in dry air. After distilling in a Crismer column, the alcohol had a boiling point of 132.00°. See also Veibel et al. [7704] and Timmermans [7482].

Udránsky [7599] purified 3-methyl-1-butanol by recrystallizing potassium amyl sulfate several times. Alcohol practically free from 2-furaldehyde can be obtained by heating fermentation 3-methyl-1-butanol with half its volume of concentrated sulfuric acid for 8 hr on a steam bath, separating the acid, shaking with calcium carbonate, and steam distilling. The process is repeated several times.

Kablukov and Malischeva [3806] dried 3-methyl-1-butanol over anhydrous copper(II) sulfate and distilled twice from calcium for *solubility studies*.

For *electrical measurements*, Müller et al. [5324] boiled 3-methyl-1-butanol with concentrated potassium hydroxide solution, washed it with a dilute solution of phosphoric acid, dried first over potassium carbonate and then over anhydrous copper sulfate, and finally fractionated.

Williams and Daniels [8080] purified 3-methyl-1-butanol for *specific heat measurements* by digesting over unslaked lime and fractionally distilling. The center portion of the distillate was further digested over barium oxide and redistilled. See also Meyer [5081] and Kraus and Bishop [4246].

91. 2-Methyl-2-butanol

Adams and coworkers [24] prepared 2-methyl-2-butanol in high yields from amylene.

A relatively pure material can be prepared from the mixed synthetic pentanols by fractional distillation. Fractional crystallization, zone melting, or preparative gas chromatography may be used for further purification.

Ginnings and Baum [2764] prepared 2-methyl-2-butanol for *solubility studies* from ethylmagnesium bromide and acetone.

92. 3-Methyl-2-butanol

The preparation of 3-methyl-2-butanol from 600 grams of 2-bromopropane, 146 grams of magnesium turnings, and 200 grams of acetaldehyde has been described in Whitmore [8021, p. 48]. The yield is 210–215 grams. It may also be prepared from methylmagnesium halides and isobutyraldehyde, by the reduction of 3-methyl-2-butanone [8020, p. 130], or from 2-chloropropane and acetaldehyde [5834].

3-Methyl-2-butanol can be separated from synthetic pentanols by careful fractional distillation [3858].

93. 2,2-Dimethyl-1-propanol

This alcohol has been known for about a century. Tissier [7504] reported its preparation from 2,2-dimethyl propionyl chloride and sodium amalgam. Despite the fact that this compound has some unique characteristics, there has been only casual interest shown in its physical properties. However, the literature for its preparation is quite extensive. This alcohol is a solid at ambient temperatures; the crystalline material sublimes and has an odor of peppermint.

Purification by fractional distillation is difficult unless the entire still is in an atmosphere whose temperature is greater than 55°. Fractional crystallization, zone melting, and related methods are well suited for the purification of this alcohol.

Whitmore and Rothrock [8027] reported that 2,2-dimethyl-1-propanol is stable to heat and that 5% water liquefies the solid at room temperature.

Trimethylacetaldehyde reacts with *tert*-butylmagnesium chloride in an ether solution to give 2,2-dimethyl-1-propanol [1623]. Greenwood et al. [2950] obtained an almost theoretical yield from these reagents. Beattie [8027], Ginnings and Baum [2764], and Conant and coworkers [1623] used formaldehyde with the magnesium alkyl reagent.

Hoffman and Boord [3449] reported the preparation of 2,2-dimethyl-1-propanol from 30% hydrogen peroxide and diisobutylene in cold 95% sulfuric acid solution. The alcohol and acetone peroxide are formed; the latter is removed by suction filtration. The acid filtrate is poured over ice and the alcohol separated and fractionally distilled. The heart cut was redistilled and sublimed twice; mp 54.5–55.5°.

D'Adams and Kienle [1778] reduced 2,2-dimethylpropionic acid with lithium aluminum hydride to 2,2-dimethyl-1-propanol in 85% yield; mp 51–52°; Kornblum and Iffland [4190] obtained 85–90% yield.

Ethyl 2,2-dimethylpropionate was reduced with hydrogen at 350° and 220 atm in the presence of copper chromate catalyst in 88% yields [5540, 35].

Sommer and coworkers [6908] reacted *tert*-butylmagnesium chloride and methyl formate for a 72% yield of 2,2-dimethyl-1-propanol.

Other Aliphatic Alcohols

94. Cyclohexanol

Cyclohexanol is produced principally by the catalytic air oxidation of cyclohexane. An appreciable amount of cyclohexanone is produced at the same time. A small amount of the alcohol is produced by the catalytic hydrogenation of phenol. Richards and Shipley [6159] report that cyclohexanol is very hygroscopic and state "...exposure to air for a few minutes lowers the freezing point several tenths of a degree." Kelley [3935] reports two crystalline forms of cyclohexanol.

Brown [1132] gave a general method for preparing alcohols from carbonyls with diborane.

Foresti [2482] hydrogenated cyclohexanone in an acid medium in the presence of platinized pumice and obtained cyclohexanol along with small quantities of cyclohexene. In an alkaline medium, only the alcohol was obtained. Ferrier [2412] hydrogenated either cyclohexylaldehyde or cyclohexanone with aluminum *sec*-butoxide in 75% yield.

Zal'kind and Markov [8259] mixed 1 mol of cyclohexene and 1–1.5 mol of 70–75% sulfuric acid at 55° for 1 hr, diluted the mixture with five volumes of distilled water, and steam-distilled. The cyclohexanol was salted out with sodium chloride, dried with potassium carbonate, and fractionally distilled; yield 55%.

The literature for preparing cyclohexanol from phenol is voluminous; a few methods are cited. Sasa [6387] reports that the catalyst obtained by boiling nickel(II) formate in biphenyl, phenyl ether, or a mixture of the two, was most active. Phenol was completely hydrogenated at 160° and 15 atm in 6.3 hr to cyclohexanol. Yeh and coworkers [8212] quantitatively hydrogenated 940 grams of phenol to cyclohexanol over 60 grams of nickel–kaolin catalyst at 140° and 80 atm. Bag et al. [544] reported the reduction of phenol at 63° in 95% yield in the presence of fragments of sodium hydroxide, activated nickel–aluminum and other metal catalysts. A more detailed description was given the following year [543].

Vogel [7758] purified cyclohexanone through the bisulfite complex and reduced the ketone to the alcohol with sodium for the *determination of physical properties*.

Cyclohexanol was purified for *dielectric constant measurements* by fractional crystallization in the absence of moist air [6159].

Wilson and Heron [8103] state that cyclohexanol of sufficient purity for a *cryoscopic solvent* may be prepared by fractional distillation. The melting point may be used as the *criterion of purity*.

Mikulak and Runquist [5124] found cyclohexanol to be an ideal *cryoscopic solvent* for molecular weight determinations because of its large freezing point constant, fairly general solvent properties, and a melting point slightly above room temperature. They have devised a simple and rapid method that overcomes the difficulty encountered due to hygroscopicity. The method is accurate to $\pm 2\%$ using a thermometer graduated by 0.1° and is suitable as an experiment in general chemistry, or for use in an analytical or physical chemistry laboratory. The commercially available cyclohexanol was distilled at reduced pressure at the rate of 10 mL/min. The material that freezes at 24° or higher was collected and placed in vials that were stoppered and sealed with paraffin. A found K_f of 39.3 $\pm$ 0.5° compared favorably with the value calculated from the heat of fusion, 39.6°.

Crowe and Smyth [1729] purified commercial cyclohexanol by refluxing 24 hr over freshly ignited lime, followed by fractional distillation. The distillate was heated with small pieces of sodium and fractionally distilled a second time. Since cyclohexanol is extremely hygroscopic, the purified material was kept in a desiccator until needed. The *dielectric constant* was determined.

For purification by vacuum distillation and fractional crystallization, see Schreiner and Frivold [6491], Sidgwick and Sutton [6714], Herz and Bloch [3343], Kelley [3935], and Lange [4397].

CRITERIA OF PURITY

The freezing point is the most convenient means of determining the purity of cyclohexanol. Aldehydes, ketones, and other common impurities may be determined by tests in Rosin [6263] or *ACS Chemical Reagents* [137]. For a cryoscopic test of purity, see Lange [4397]. Gas chromatography is a convenient means for determining the number and amount of the impurities.

SAFETY

Cyclohexanol has a low vapor pressure at room temperature. It has a faint odor that has been described as resembling menthol or camphor. Nelson et al. [5444] found the highest concentration tolerable in air for an 8-hr day to be less than 100 ppm. Treon and coworkers [7546] showed that 1000 ppm caused narcosis but 100 ppm has no effect on rabbits and monkeys.

The *threshold limit value* has been chosen as 50 ppm, 200 mg/m^3 to reduce objectionable irritation [7022].

The *minimum ignition temperature* in air is 300° [5712].

95. 1-Hexanol

1-Hexanol is available commercially. Generally it has been separated from a mixture of alcohols; it contains other alcohols that are quite difficult to separate. If a material of high purity is required, synthesis that does not produce homologues is preferred.

Hovorka et al. [3518] prepared 1-hexanol by the Grignard synthesis. Preliminary purification was carried out in a 40-cm column packed with glass beads. A final fractional distillation was performed in a similar column. Aluminum amalgam was used as the drying agent. Some decomposition to the corresponding olefin occurred with the drying agent. The final distillation was done in the absence of the amalgam. The purified alcohol had a boiling range not greater than 0.04°.

The preparation of 1-hexanol by the Grignard synthesis is given in [2756, p. 54].

Brown and Subba Rao [1140] prepared 1-hexanol; see 1-pentanol.

Decker and Halz [1894] prepared 2-ethylbutanol and 1-hexanol from butyraldehyde and acetaldehyde. Detailed preparation and recovery are given.

Brown and Subba Rao [1141] added diborane to hexane and then oxidized the trihexylboron produced with alkaline hydrogen peroxide to produce hexanol.

Olivier [5579] purified 1-hexanol by esterification with hydroxybenzoic acid, recrystallization of the ester, followed by saponification.

Hill and White [3405] dried 1-hexanol over a molecular sieve and further purified the solvent by vacuum distillation in a spinning band column. The middle fraction boiling at 62° and 20 Torr was collected; purity was 99.5%.

Adkins and Billica [33] reduced hexaldehyde with a specially prepared Raney nickel catalyst. For additional information, see 2-propanol.

SAFETY

Korbakova [4185] reported a *threshold limit value* in air as 100.0 mg/m^3.

The *minimum ignition temperature* in air is 300° [5712].

97. 2-Methyl-2-pentanol

Hovorka and Lankelma [3517] prepared 2-methyl-2-pentanol by the Deschamps method [1963] for *physical properties study*. 1-Bromopropane was slowly added to a mixture of dry ether and magnesium turnings to prepare 14.7 g of propylmagnesium bromide. It was cooled to 0° and a mixture of 5.8 grams of acetone and an equal volume of ethyl ether was added dropwise. The reaction was violent and the product was partially soluble. The mixture was allowed to stand 24 hr and was poured into a mixture of 9 g of ice and 9 g of glacial acetic acid. It was extracted with ether, washed, and fractionally distilled.

Osokin and Fel'dblyum [5619] prepared 2-methyl-2-pentanol by treating either 2-methyl-2-pentene or 2-methyl-1-pentene with 60–70% sulfuric acid solution at 0–10° and diluting the product with water to a residual acid concentration of 40%. Then the product was hydrolyzed at 15–25° to yield nearly 100% alcohol.

van Risseghem [7676] prepared 2-methyl-2-pentanol by a Grignard reaction with methyl propyl ketone, and Huston and Bailey [3598] used a similar preparation method with butanoic acid and methylmagnesium bromide.

98. 4-Methyl-2-pentanol

Brown and Nakagawa [1137] prepared 4-methyl-2-pentanol in 84% yield by sodium borohydride reduction in methanol of 4-methyl-2-pentanone.

Brown and Zweifel [1143] prepared 4-methyl-2-pentanol in 57% yield from *trans*-isopropylmethylethylene and bis-3-methyl-2-butylborane.

Levene and Walti [4549] prepared *optically active* 4-methyl-2-pentanol from 0.6 mol of *dextro*-propylene oxide dissolved in 100 mL of dry ethyl ether which had been added dropwise to a cooled solution of 0.6 mol of isopropylmagnesium bromide in 250 mL of dry ethyl ether with stirring. The mixture was allowed to stand 7 days, and the ether was removed by distillation. To the residue was added 70 mL of dry pyridine and 40 g of phthalic anhydride. The purified phthalate was steam-distilled in the presence of sodium hydroxide. The distillate was extracted with ether and the extract dried and fractionated; bp 65–66° at 60 Torr.

Hovorka and coworkers [3518] prepared 4-methyl-2-pentanol for *thermodynamic and physical property studies* from 4-methyl-2-pentanone and sodium in ethanol.

SAFETY

The *threshold limit value* was reported as 25 ppm [6905].
The *flammable limits* in air are 1.0 and 5.5%v [2349].

99.   2-Ethyl-1-butanol

Matsui [4989] fractionally distilled the hydrogenated polymer of acetaldehyde. From 82 grams of volatile distillate, he obtained 15.6 grams of ethanol, 16.2 grams of 1-butanol, 7.5 grams of 2-ethyl-1-butanol, and 1.2 grams of 1-octanol. The polymerization of the acetaldehyde was carried out with cooling in the presence of barium hydroxide, which was then neutralized by passing carbon dioxide through the polymerization product. The reduction was carried out in the presence of a nickel catalyst at pressures below 80 atm and temperatures up to 200°.

Hovorka et al. [3519] allowed 2-ethyl-1-butanol to stand over Drierite for at least 2 weeks, filtered, and then fractionally distilled. A fraction boiling over a range of 0.02° was used for *physical property measurements*.

Methylcyclohexanols

There are four methylcyclohexanols, three of which exist in *cis-* and *trans*-configurations; in the case of 2-methylcyclohexanol and 3-methylcyclohexanol, each possesses an asymmetric carbon atom and exhibits optical activity. A disagreement in the literature concerning the correct identification of the several isomers apparently has been resolved by Noyce and Denney [5528] who identified *cis*-3-methylcyclohexanol as the isomer of lower refractive index and lower density in contradiction to the prediction of the von Auwers–Skita rule. Discussions of *cis-* and *trans*-methylcyclohexanols have been presented by Skita and Faust [6793], Goering and Serres [2809], and Jackman and coworkers [3657]. No attempt has been made in this chapter to reevaluate configurational assignments.

Gough et al. [2892] have studied extensively the various forms of the methylcyclohexanols and present considerable data. They express the belief that their α-form is *trans* and the β-form is *cis*.

Practically all references to the stereoisomers give methods for their preparation and separation. Only a few of these methods are presented here or under the appropriate compound because of the complexity. Some unique reduction methods with lithium aluminum hydride, sodium borohydride, and aluminum isopropoxide are reported by Noyce and coworkers [5527, 1831] and Siegel [6718].

Skita [6792] and Skita et al. [6795] concluded from their work that *cis-* and *trans*-alcohols could be produced in predominating amounts depending on the method of preparation. Peppiatt and Wicker [5753] and Wicker [8042, 8043] have shown that the Skita rule is not too reliable for the methylcyclohexanols. The three articles [5753, 8042, 8043] constitute a good reference source up to 1957.

Vogel [7758] prepared 2- and 3-methylcyclohexanol from the corresponding hexanones. Palfray [5651] hydrogenated the corresponding cresol in the presence of Raney nickel at 95° and 120 kg pressure.

The cresols were reduced by Brode and Van Dolah [1114] under hydrogen using copper chromate catalyst at 500–2600 psi at 250–300°. Shigeru and coworkers [6660] employed 1 : 1 nickel–aluminum catalyst to reduce the same compounds.

Calas and coworkers [1307] reduced 2-, 3-, and 4-methylcyclohexanone with trichlorosilane to the respective *trans*-methylcyclohexanols in 60, 30, and 75% yields.

Eliel and Haber [2237] separated the isomers from commercially available 2-, 3-, and 4-methylcyclohexanols by fractional distillation in a 4-ft Podbielniak column at a reflux ratio of 100 : 1 to 120 : 1. A charge of 300–500 grams gave about one-third to one-half quite pure epimer by infrared examination. It was pointed out that the alcohols must be free from the corresponding cyclohexanone, since it is not satisfactorily separated from the lower boiling alcohol.

Komers and coworkers [4168] have achieved good separation of the *cis*- and *trans*-2-methyl, 3-methyl, and 4-methylcyclohexanols using gas–liquid partition where the stationary phase is capable of hydrogen bond formation.

Brownstein and Miller [1178] have used nuclear magnetic resonance to determine the *cis*- and *trans*-methylcyclohexanols.

Elliott and coworkers [2248] gave methods of preparing optically inactive and active isomeric methylcyclohexanols for *biological and enzymic studies.*

CRITERIA OF PURITY

The freezing curve may be used with its usual precision and accuracy. The *cis*- and *trans*-isomer distributions may be determined by gas chromatography and by nuclear magnetic resonance.

The British Standards Institute [1108] lists the standards for methylcyclohexanols.

SAFETY

Most of the toxicological and safety information available is for a commercial product usually listed as "methylcyclohexanol," made by hydrogenating *m*- and *p*-cresols. The isomers may occur as the *cis*- and *trans*-forms. The vapors of methylcyclohexanol are recognizable at 500 ppm. They are capable of causing upper respiratory irritation at this concentration. It has been concluded that the 2-compound is more toxic than the other two isomers [5713, 7022].

The *threshold limit value* has been set at 100 ppm, 470 mg/m^3 [7022].

The *minimum ignition temperatures* in air of the 2- and 4-isomers are, respectively, 296° and 295° [5712].

100. 1-Methylcyclohexanol

Markownikaw and Tscherdynzew [4896] prepared 1-methylcyclohexanol from 1-amino-1-methylcyclohexanol and sodium nitrite.

Zelinsky [8273], Sabatier and Mailhe [6322], and Wallach [7885] prepared 1-methylcyclohexanol from cyclohexanone and methylmagnesium iodide.

Grignard and Vegnon [2968] treated 1,5-bis-pentamethylenemagnesium bromide with ethyl acetate and hydrolyzed the product to 1-methylcyclohexanol.

Jerkunica and Traylor [3715] added 1-methylcyclohexene to mercury(II) acetate in a water–ether solvent. Then 6 *M* sodium hydroxide was added, followed by a sodium borohydride solution. The yield was 70–75% 1-methylcyclohexanol.

896   PURIFICATION METHODS

101.   2-Methylcyclohexanol (Mixed Isomers)
102.   *cis*-2-Methylcyclohexanol
103.   *trans*-2-Methylcyclohexanol

Anziani and coworkers [394] found that reduction of 2-substituted cyclohexanones with 2 mol of isobutylmagnesium bromide at 35° gave the corresponding alcohols, mostly the *cis*-isomer. Similar reduction with *tert*-butylmagnesium chloride, *tert*-pentylmagnesium chloride, or *tert*-butylmagnesium bromide gave mainly the *trans*-isomer. Methylcyclohexanone reduced with isobutylmagnesium bromide gave a 37% yield of 87% of *cis*-isomer.

Cornubert and coworkers [1665] produced chiefly *trans*-isomers with mild reducing agents.

McQuillin and Ord [4798] reduced 1-methylcyclohexene oxide in ethyl acetate containing perchloric acid and obtained 58% *trans*-isomer of 2-methylcyclohexanol.

Brown and Zweifel [1145] report that hydroboration of 1-methylcyclohexene, followed by oxidation with alkaline hydrogen peroxide produces pure *trans*-2-methylcyclohexanol.

Yamamoto and coworkers [8203] reduced 2-methylcyclohexene with a borabicyclononane complex to give *cis*-2-methylcyclohexanol (35% yield, 92% isomeric purity) and, in the presence of lithium methoxide, *trans*-2-methylcyclohexanol (59% yield, 89% isomeric purity).

104.   3-Methylcyclohexanol (Mixed Isomers)
105.   *cis*-3-Methylcyclohexanol
106.   *trans*-3-Methylcyclohexanol

Ebersole [2196] prepared 3-methylcyclohexanol by hydrogenating *m*-cresol in the presence of nickel-on-kieselguhr catalyst. A temperature of 160° was used with maximum pressure of 1800 psi. The theoretical yield was obtained in 12 hr.

Peppiatt and Wicker [5753] prepared the *trans*-isomers as 77% of the mixture by hydrogenating 3-methylcyclohexanone using a nickel catalyst. The mixed isomers were converted to the 3,5-dinitrobenzyl ester. Repeated crystallizations produced a pure *trans*-ester, which on hydrolysis gave the *trans*-alcohol; bp 170–177° at 772 Torr, n_D 20° 1.4581. They also reduced *m*-cresol and obtained a product containing 55% of the *cis*-isomer, which was converted to the hydrogen phthalate ester, then to the potassium salt. The salt was refluxed with excess potassium hydroxide until completely hydrolyzed. The *cis*-alcohol was steam distilled: bp 175–177° at 777 Torr, n_D 20° 1.4577.

Goering and Seres [2809] and Noyce and Denney [5528] worked through *p*-toluenesulfonic acid to isolate the *cis*- and *trans*-compounds.

Macbeth and Mills [4799] prepared *dl-cis*-3-methylcyclohexanol, d 30° 0.917, n_D 20° 1.4583, and *dl-trans*-3-methylcyclohexanol, d 30° 0.9022, n_D 20° 1.4573. Included are the derivatives such as hydrogen phthalate, *p*-nitrobenzoate, phenylurethane, and α-naphthylurethane.

Hückel and Kurz [3551] isolated (−)*cis* and (+)*trans*-3-methylcyclohexanol but gave no data for the physical constants.

Yamamoto and coworkers [8203] reduced 3-methylcyclohexanone with a borabicyclononane complex in the presence of methanol to give *trans*-3-methylcyclohexanol (100% yield, 86% isomeric purity).

107. 4-Methylcyclohexanol (Mixed Isomers)
108. *cis*-4-Methylcyclohexanol
109. *trans*-4-Methylcyclohexanol

Schmidt and Seydel [6466] report 58% yield of 4-methylcyclohexanol by heating with stirring for 5 hr at 260–280° and 200–250 atm 1 part of chromium (2%) activated catalyst, 15 parts 4-aminotoluene, and 10 parts of water.

Wheeler and Mateos [7994] reduced 4-methylcyclohexanone with lithium tri-*tert*-butoxyaluminum hydride (Li(tert-$C_4H_9O)_3$AlH); LiAlH$_4$ in excess *tert*-butyl alcohol and obtained 84% of the *trans*-alcohol.

Kuivila and Beumel [4309] reduced 4-methylcyclohexanone with organotin hydrides and obtained an 84% yield of 4-methylcyclohexanol, containing 70 ± 2% *trans*- and 30 ± 2% *cis*-isomer.

Stork and White [7025] report for *cis*-4-methylcyclohexanol a bp of 183–184° and 174° for the *trans*-isomer. There is uncertainty which compound is *cis*- and which is *trans*-.

Yamamoto and coworkers [8203] reduced 4-methylcyclohexanone with a borabicyclononane complex to prepare the isomers of 4-methylcyclohexanol. In the presence of methanol, the *cis*-isomer was produced (91% yield, 84% isomeric purity) and, in the presence of lithium methoxide, the *trans*-isomer was produced (94% yield, 90% isomeric purity).

Other Aliphatic Alcohols

110. 2-Heptanol

Directions for the preparation of 2-heptanol from 228 grams of 2-heptanone, 600 mL of 95% ethanol, and 130 grams of sodium are given in [1509, p. 60]. The yield is 145–150 grams of distilled alcohol. Farkas and Stribley [2371] prepared 2-heptanol by controlled oxidation of methylcyclohexane with air. Fractionation of the oxidation product gave nearly equal parts of methylcyclohexanol and 2-heptanone. The ketone was separated by extraction with sodium bisulfite and hydrogenated to the alcohol.

Malinovskiĭ and coworkers [4853] prepared 2-heptanol in 30% yield by the Grignard reaction.

Thomas and Meatyard [7416] used bromopentane and acetaldehyde to prepare the alcohol for *molecular association studies*.

Prout and Spikner [5970] reacted pentylmagnesium bromide with acetaldehyde and obtained 85% yield of *dl*-2-heptanol. Proceeding through the acid phthalate brucine salt, they obtained (+)-2-heptanol; bp 157–159°, n_D 25° 1.4178, d 25° 0.835, and [α]$_D$ 25° 10.11°.

Yamamoto and coworkers [8203] reduced a mixture of 2-heptanone and 4-heptanone with a borabicyclononane complex to yield 91% 2-heptanol and 9% 4-heptanol.

111. 3-Heptanol

Dillon and Lucas [2008] prepared 3-heptanol from butylmagnesium bromide and propionaldehyde in ether and purified it by fractional distillation.

Thomas and Meatyard [7416] purified 3-heptanol for *physical property studies* by fractional distillation in a column packed with Dixon gauze rings and having total reflux efficiency of 12 theoretical plates.

SAFETY

The single dose LD_{50} in rats was reported as 1.87 g/kg [6882].

112. 1-Octanol

Deffet [1896] decomposed 1-heptylmagnesium iodide with formaldehyde to obtain 1-octanol.

Smyth and Stoops [6876] purified a commercial product by fractional distillation for determination of the *dielectric constant*.

Ueno and Komori [7601] fractionally distilled commerical 1-octanol and treated the distillate with boric anhydride. The treated alcohol was fractionally distilled at 5 Torr pressure and the portion distilling at 195–205° was collected. The alcohol was neutralized with sodium hydroxide and again fractionally distilled. The purified alcohol boiled at 98° at 19 Torr pressure.

Brown and Subba Rao [1140] prepared the alcohol, see 1-pentanol.

Kondo [4173] converted butanol to octanol and octanoic acid by heating with metallic sodium at 320° and 1.00 atm.

Takao and Kumamoto [7161] prepared 1-octanol from octyl caprylate in a 97.1% yield by catalytic reduction with a copper chromate catalyst at 220°.

Rose and coworkers [6257] purified 1-octanol for *vapor–liquid equilibria studies*. It was precision vacuum-distilled through a 35-in. long, 2-in.-diameter column packed with 0.16 × 0.16-in. protruded packing at 20–40 Torr pressure and a reflux ratio of 40:1. Criterion of purity: distillation at several pressures to give product of constant refractive index at 25°, 1.42727.

CRITERIA OF PURITY

Dreisbach and Martin [2090] determined the purity from the freezing curve.

O'Connor and Norris [5547] report a colorimetric method for the analysis of acetals and aldehydes in C_8–C_{10} alcohols below the 100-ppm range.

SAFETY

The *threshold limit value* of 100.0 mg/m^3 has been suggested for Russia in 1964 [4185].

113. 2-Octanol

Gilman and Blatt [2757] prepared 2-octanol by treating castor oil with concentrated sodium hydroxide solution and fractionally distilling the alcohol from the mixture at 175–185°. The alcohol was then shaken with 15% sodium bisulfite solution, separated, and steam-distilled.

Svetlov and Vul'fson [7101] purified crude 2-octanol, also prepared by treating castor oil with sodium hydroxide, by heating it with phthalic anhydride at 110–115° for 12 hr. After recrystallizing the ester, the solid was decomposed with fused potassium carbonate and distilled under vacuum. Finally, 2-octanol was redistilled at atmospheric pressure; bp 179.2°.

Ellis and Reid [2253] prepared 2-octanol by the Grignard reaction using methyl iodide and *n*-heptaldehyde.

McGinn [4770] prepared high-purity *d*- and *l*-2-octanol and the racemate. It was found that the optical isomers had identical physical properties (within experimental error), but the properties of the racemate were different.

SAFETY

Marsden and Mann [4902] stated that 2-octanol is not generally regarded as offering an industrial hazard under usual conditions of use. Care should be taken where mists or sprays are formed or the alcohol is heated to elevated temperatures.

114. 2-Ethyl-1-hexanol

Kobayashi and coworkers [4083] prepared 2-ethyl-1-hexanol by the condensation of 1-butanol, using sodium or potassium hydroxide as the principal catalyst and magnesium as the cocatalyst. Two moles of butanol were condensed by Guerbet's reaction [2550] with 1 mol of tetrabutoxysilane, $(C_4H_9O)_4Si$, with 1–2 mol of alkaline catalyst and various cocatalysts at 200–290° in 80–93% yield [4084]; see also [5149].

Some other condensations involving 1-butanol are obtained by: refluxing butanol in toluene with sodium hydroxide, heating butanol at high temperature and pressure with sodium or potassium butyrate [6111] or with copper and potassium carbonate [6531], and heating butanol in the presence of alkaline earth alcoholates [2549]. Condensations are obtained in 50% yield from butanol with sodium and zinc chloride at 195°. Yields of 71–72% were obtained from butyric acid, sodium butyrate, and butanol with a nickel oxide catalyst [8241].

Pearce and Berhenke [5731] purified commercial 2-ethyl-1-hexanol for *dipole moment studies* by treating with sodium to remove water and lower alcohols. It was fractionally distilled, and the middle fraction boiling at 185.5° at 760 Torr was retained.

Levene and Taylor [4546] reduced 2-ethylhexylate with sodium and absolute alcohol by the method of Levene and Cretcher [4538] to prepare 2-ethyl-1-hexanol in 78% yield; bp 743 Torr 181–183°, d 20° 0.8328, n_D 20° 1.4328.

SAFETY

Large quantities of 2-ethyl-1-hexanol have been used in industry without reports of injury [720]. Hodge [3437] found the median lethal dose for rats to be about 0.8 mL/kg interperitoneally or about 3.9 mL/kg by stomach tube.

Aromatic Monohydric Alcohols (115–124)

115. Benzyl Alcohol

Palfray [5651] hydrogenated benzaldehyde to benzyl alcohol at 105° and 115 kg pressure for 20 min in the presence of Raney nickel. Adkins and Billica [33] reduced benzaldehyde with a specially prepared Raney nickel catalyst. For additional information, see 2-propanol.

Hazlet and Callison [3248] prepared benzyl alcohol and furfuryl alcohols as follows:

53 grams of benzaldehyde and 48 grams of 2-furaldehyde were treated, with cooling, with 22.5 grams of sodium hydroxide and 45 mL of water, shaken for 2 hr, and then allowed to stand at room temperature for 24–48 hr. Sufficient water was added to dissolve the solid and the solution extracted with ether for 8–12 hr. After acidification of the extracted alkaline solution, the solution was again extracted with ether for an additional 8–10 hr. The average mole ratio of benzyl to furfuryl alcohols was 0.56:1. The ether extract was dried over anhydrous magnesium sulfate, the ether removed by distillation, and the alcohols distilled under reduced pressure in an atmosphere of nitrogen.

Nystrom and Brown [5541] reduced benzoic acid with lithium aluminum hydride in a dry ether solution to benzyl alcohol. The yield was 81%. Chaikin and Brown [1407] reduced benzoyl chloride to benzyl alcohol with sodium borohydride.

For benzyl alcohol from benzaldehyde and diborane, see cyclohexanol, Brown [1132] for a general method.

The general method for purification is careful fractional distillation of the commercial alcohol, which, according to Mathews [4969] is best carried out at reduced pressure with the exclusion of air. Martin and George [4913] purified the alcohol by shaking with aqueous potassium hydroxide and extracting with ether that had been freed from peroxides with silver nitrate and sodium hydroxide. After being washed, the extract was treated with saturated sodium hydrogen sulfide solution, filtered, washed, and then dried over potassium carbonate. After removal of the ether, the alcohol was distilled under reduced pressure and the middle fraction dried over lime that has been burned in an atmosphere of nitrogen. Schiff's reagent indicated a barely detectable quantity of benzaldehyde.

Benzaldehyde was tested for purity by gas chromatography by Hartmann and coworkers [3199] and purified for *dielectric constant studies* by fractional distillation in a 25-theoretical plate column at a reflux ratio of 30:1 to 20:1. The last traces of water were removed with molecular sieves. Gas chromatography was used as the criterion of purity.

CRITERIA OF PURITY

Rosin lists the following specifications: boiling range, 202–206°; nonvolatile matter, not more than 0.005%; chlorine, about 0.02%; 2 mL of alcohol shaken with 20 mL of water should give a clear solution.

Dreisbach and Martin [2090] determined the purity from the freezing curve.

Rees and Anderson [6101] determined benzaldehyde in benzyl alcohol by ultraviolet measurements at 283 nm.

Hartmann and coworkers [3199] determined the purity by gas chromatography.

SAFETY

Animal experiments indicate that benzyl alcohol is 2.5–3 times as toxic as 2-propanol or 1-butanol. Inhalation of the vapors at a concentration of 200–300 ppm, about 880–1325 mg/m^3, by rats gave a LC_0 for 2 hr and a LC_{33} for 4 hr with a LC_{100} for 8 hr. It is believed that benzene and benzyl alcohol that were present in a poorly ventilated area were the cause of violent headaches, vertigo, nausea, and other symptoms [5713].

The *minimum ignition temperature* in air is 426° [5713].

Phenols

General Comments

Keading and coworkers [3920] report an air oxidation of benzoic acid to phenol in the presence of copper(II) benzoate. Of particular interest is the production of pure *meta*-cresol by the air oxidation of either *p*- or *o*-toluic acids. *m*-Toluic acid gives about 40% *ortho*-cresol.

Andon and coworkers [349] prepared high-purity phenols for the determination of *high-precision physical and thermodynamic properties*. All specimens were similarly purified. Steam was passed into a boiling solution containing 1 mol of the phenol and 1.5 to 2.0 mol of sodium hydroxide in 5 L of water until no nonacidic material distilled. The cooled residue was made acidic with 20% sulfuric acid. The phenol was separated and dried over calcium sulfate. The aqueous layer was extracted with ether, and the recovered phenol was again treated with steam. The dried phenol was distilled at approximately 10 Torr pressure. It was then fractionally distilled at a pressure of 100–212 Torr through a 45-theoretical-plate column at a reflux ratio of 50:1. Batches of 1200 grams were distilled; when the column had reached a steady state of operation, fractions of 150 mL were collected. A series of fractions with an overall boiling range of 0.5° was examined by infrared spectroscopy, and the best fractions were combined. The fractional distillation of the best material was repeated if it seemed desirable. The final purification was done by fractional crystallization or zone melting, or both, several times. In the latter stages of the purification, all material was protected from light and manipulations made under dry nitrogen. The final colorless material was stored in the dark in a vacuum.

Chen and Laidler [1441] purified phenols for the determination of *ionization constants* by starting with the highest-purity product obtainable and then vacuum distilling or crystallizing from water. The water was prepared in an all-glass still in a carbon dioxide-free atmosphere.

Ashworth and Lazik [468] purified phenol and the cresols by distillation under ordinary pressure and 2,4- and 2,6-xylenols at 20 Torr. They also used sublimation at 20 Torr for the compounds of melting points exceeding 60°. The criteria of purity were melting point and gas chromatography.

Ambrose [113] reported that phenol, the cresols, the xylenols, and quinoline decompose fairly rapidly at the critical temperature.

Winters and Shelmerdine [8114] report that sulfur can be removed from phenolic materials by passing the vapors and hydrogen over a cobalt–molybdate catalyst. For example, *m*-cresol containing 0.08% sulfur was treated with hydrogen at atmospheric pressure at 238° over the catalyst at a liquid space velocity of 0.5 vol/vol of catalyst per hour.

Phenols and alkylated phenols were identified by Gasparic [2683] through their R_f values as 3,5-dinitrobenzoates.

Specific Solvents

116. Phenol

The majority of phenol is synthetically produced. Natural sources, that is, degeneration of natural products such as carbonization of coal, accounts for only a small portion of the United States' total production. Four sources account for practically all phenol in the United States: carbonization of coal, alkali fusion of benzene sulfonic acid, hydrolysis of chlorobenzene, and the cumene process, which involves the cleavage of cumene hydroperoxide.

Phenol is a hygroscopic solid at normal room temperature. Synthetic material usually is at least 99.5% phenol, and the majority of the 0.5% is water. In storage, phenol tends to acquire a yellow, pink to red, or brown discoloration. The formation of the color is promoted by trace amounts of substances acting as catalyst, such as iron or copper. Commercial phenol may contain a preservative.

Boorman and coworkers [982] found that phenol from different sources contained characteristic impurities: (1) that prepared from alkali fusion of benzene sulfonic acid contained o- and p-hydroxydiphenyls; (2) that from high temperature hydrolysis of chlorobenzene contained diphenyl ether; and (3) phenol obtained from coal tar contained naphthalene. The following methods are given for testing the purity and identifying the source: (1) "Distill the sample at 1–1.5 Torr in a bath at 58–60° until the distillation ceases. Take up the residue from 1 kg in ether, filter, evaporate, and heat the residue to 100° in a current of air to remove traces of phenol." Class 1) yields 0.3–10 grams of residue, and class 2) or 3) yields a trace of residue. The weighed residue from class 1) was identified by infrared analysis. (2) "Dissolve 500 grams in 800 mL of 30% wt/vol sodium hydroxide and distill in a current of steam until 200 mL has been collected. Add 80 mL of the alkali to this, and repeat the steam distillation. Collect 100 mL of the distillate allowing the condenser to become warm toward the end of the operation. Extract the final distillate with 20 mL of carbon tetrachloride and examine the infrared spectrum of the solution."

Filar [2426] states that phenol prepared from cumene contains generic impurities not separable by distillation. He recommended that the phenol be treated with nonpolar adsorbents such as activated carbon, talc, or precipitated calcium carbonate, the solids separated, and the final purification made by fractional distillation. Impurities in phenol from cumene are reported to be mesityl oxide, methylstyrene, cumene, and benzofuran. One method of purifying phenol consists of treating with hydrogen peroxide, neutralizing with sodium hydroxide, and steam distilling [3479]. Another method recommends neutralizing with sodium hydroxide to a pH 7, heating to 108° for 90 hr and fractionally distilling [2017].

Yamamoto and Ohara [8199] prepared phenol by sulfonating commercial chlorobenzene (bp 127–133°) with sulfuric acid. The product was then treated with sodium hydroxide, acidified with sulfuric acid, and heated to 230°. On cooling, the crude phenol appeared at the surface. It was extracted with ether and distilled. The yield was 54–58%.

Cresols [561] were demethylated when passed over nickel-on-aluminum catalyst in steam at 410–470°; see also [6660].

Timmermans [7482] states that phenol is purified by fractional distillation *in vacuo* and by fractional fusion. The criteria of purity are the consistency of the melting point and the critical solution point in water for successive fractions.

Commercial-grade phenol, mp 40.3°, was purified by Wille and Rappen [8069] by mixing 7380 grams with 2620 mL of water. The mixture was cooled to 11 ° and seeded with $C_6H_5OH \cdot \frac{1}{2}H_2O$. After crystallization was complete, the crystals were centrifuged and washed at 5° with 500 mL of water saturated with phenol at 0–2; yield 6590 grams with a mp of 40.9°.

Davison [1864] purified phenol, o- and p-cresols, and nitrobenzene by distilling through a small column and then fractionally crystallizing until a constant melting point was obtained. The material was stored in a desiccator and distilled, just prior to use, to remove water and any oxidation products that might have accumulated.

Draper and Pollard [2081] purified phenol by adding 12% water, 0.1% aluminum, and 0.05% sodium bicarbonate. The mixture was distilled at atmospheric pressure until the azeotrope was removed. The pure phenol was then distilled at 25 Torr until 20 mL of the black residue remained in the distilling flask.

Williams and Ogg [8084] determined the *dielectric constant* and *dipole moment* on phenol that was purified by crystallization and fractional distillation; bp 179.5–180.0°.

Judson and Kilpatrick [3798] purified a good commerical product by fractional distillation for *dissociation constant measurements*; bp 182.0–182.5°, mp 39.8–39.9°.

Phenol was purified of catalytic poisons with mixed sodium and potassium salts of ethylenediamine tetracetic acid [2116].

Phenols were purified by distillation with steam at 6 and 60 Torr. The recovery was better than that obtained at atmospheric distillation [4059].

The purification of phenol by *zone melting* has been studied. Deluzarche and co-workers [1939] used phenol containing methyl red to test the operation and efficiency of an automatic zone melting apparatus. Schildknecht and Vetter [6447] also used phenol to test the efficiency of a zone melting apparatus. Sorensen [6912] studied the speed of zone travel and number of passes to remove impurities from phenol.

CRITERIA OF PURITY

ACS reagent-grade phenol meets the following specifications [137]:

> *Water:* to pass test (limit about 0.5%).
> *Insoluble matter:* to pass test.
> *Residue on evaporation:* not more than 0.05%.

[Note: Phenol that conforms to this specification may contain preservatives. If a preservative is present, it should be stated on the label.]

Water in phenol may be determined by the Karl Fischer method [5171].

Dolique [2034] proposed the ciritcal solution temperature in water as a criterion of purity; see also Timmermans [7482].

The freezing point is the most exact criterion of purity. Dreisbach and Martin [2090] used the freezing curve to determine the purity of a high purity (99.96%m) material; see also Timmermans [7482].

SAFETY

The minimum detection by odor is 0.022 mg/m^3; the minimum concentration affecting light sensitivity of the eye is 0.0155 mg/ m^3; the minimum concentration affecting electrocortical activity is 0.0156 mg/m^3; the maximum concentration not affecting cholinesterase activity, porphyrin metabolism, and motor chronoxy of rats during continuous exposure for 61 days is 0.01 mg/m^3. On the basis of these findings, the recommended threshold limit value for single or continued exposure was 1 ppm or 3.84 mg/ m^3 [5308].

Phenol vapors or liquid are readily absorbed through the skin or any mucous membrane. A concentrated solution caused bad acid burns. Acute and chronic poisoning may result from skin exposures. Repeated skin contamination by low concentrations of phenol, either vapor or solution, will cause chronic poisoning. Many individuals are allergic to phenol. Phenol is not sufficiently volatile to constitute a respiratory hazard under normal conditions [7022, 5713].

Nomiyama and coworkers [5508] reported a subcutaneous LD_{50} in mice of 0.34 g/kg. Horikawa and Okada [3490] reported an LD_{50} in mice of 282.0 mg/kg.

The *threshold limit value* of 5 ppm, 19 mg/m^3 contains a sufficiently large safety factor to prevent systemic poisoning, provided there is no skin contact [7022].

The *minimum ignition temperature* in air is 715° [5712].

Cresols (117–119)

The cresols, which are by-products of the coking and petroleum industries, are available as crude or refined materials containing varying amounts of the three isomers. *o*-Cresol is easily separated and purified. There is a large literature on the separation and purification of the isomeric cresols.

Gluzman and coworkers [2799] obtained 96–99% *m*-cresol from a mixture of the *meta*- and *para*-isomers by forming a complex between sodium acetate and the *m*-cresol.

Cislak and Otto [1491] found that the *m*- and *p*-cresols selectively formed crystalline addition compounds with cyclic nitrogen compounds. *m*-Cresol forms addition compounds with 2,4-xylenol and 2,6-lutidine; *p*-cresol crystallizes with 4-picoline or 2,3,6-collidine. Both addition compounds, when regenerated, give cresols of about 80% purity. Additional directions are given for obtaining products in high states of purity.

o-Cresol may be isolated from cresol mixtures by fractional distillation. *m*-Cresol and *p*-cresol must be separated by chemical, or a combination of chemical and physical means. The data available indicate that the cresol mixtures follow Raoult's law fairly well. Usually, three distillations are required with efficient equipment and a high reflux ratio to produce material approaching 99.9 + % *o*-cresol.

Several techniques have been developed for separating *m*- and *p*-cresols by differences of their sulfonates. Brückner [1184] states that a convenient way to carry out the sulfonation is as follows: Mix 100 grams of the cresol with an equal volume of sulfuric acid and stir with a glass rod until no more streaking is noticeable. Heat for 3 hr at 103–105°. Dilute carefully with 200–300 mL of water, heat to the boiling point, and steam-distill until all of the unsulfonated cresol has been removed. Cool and extract the resins with ether. Evaporate the solution until the boiling point reaches the decomposition temperature of the sulfonate (133–135°), and then introduce steam. Shorygin and coworkers [6688] tried several methods for separating *m*- and *p*-cresols and found that Brückner's method gave the purest *meta* compound.

Stevens [7005] isolated 250 g of mixed cresols with a boiling range of 1° as a middle cut from 450 g of a commercial mixture of cresols by fractional distillation. To this fraction 12.5 g of concentrated sulfuric acid was added and isobutylene was passed into the mixture, with rapid agitation, as fast as it could be absorbed. The excess isobutylene was washed from the mixture with natural gas. The mixture was neutralized with sodium carbonate and washed. The washed material was fractionally distilled through a 15-plate column at 20 Torr pressure and a reflux ratio of 20:1. The tertiary butylated cresols boil 20° apart at this pressure. Debutylation was accomplished by heating with 0.2% of concentrated sulfuric acid, neutralizing with sodium carbonate, and again fractionally distilling.

Kotake [4215] treated a mixture of *m*- and *p*-cresols with concentrated phosphoric acid or with a phosphate and strong acid, and treated the resulting tolyl phosphate with petroleum ether; the *para*-compound is insoluble, the *meta*-compound soluble.

Golumbic [2840] developed a countercurrent distribution method suitable for partial separation of the *m*- and *p*-cresols.

Widiger [8044] found that impurities in alkyl phenols prepared by alkaline hydrolysis of aryl chlorides may be removed by treating with aqueous alkali and air at 60–120°.

CRITERIA OF PURITY

The freezing point is the most convenient and sensitive measurement for the determination of the purity of cresol isomers. Dreisbach and Martin [2090] determined the purity of *o*- and *p*-cresols from the freezing curve. The infrared spectrum was used by Whiffen and Thompson [8000] for the analysis of cresols.

SAFETY

The hazards for the cresols are about the same as described for phenol. Deichmann and Witherup [1911] found *m*-cresol the least toxic of the isomers. *p*-Cresol did not seem to be any more toxic than phenol.

The *threshold limit value* on skin has been reported as 5 ppm, 22 mg/m^3 for all isomers [142].

117. *o*-Cresol

Carney and Sanford [1357] purified *o*-cresol for use in developing an *ultraviolet method* for the isomeric cresols. It was recrystallized several times from a petroleum solvent with a distillation range of 80–120°, followed by a distillation. The fp was 30.937° corresponding to 99.99%m and a 100% mp of 30.994°.

For purification by distillation, see Ashworth and Lazik [468], Phenols, General Comments.

SAFETY

The oral LD$_{50}$ for mice was found to be 344 mg/kg [7652].
The *minimum ignition temperature* in air is 480.0° [5309].

118. *m*-Cresol

A high-purity *m*-cresol was prepared by Carney and Sanford [1357] for *ultraviolet analytical method studies* of the isomeric cresols from commercial 12° mp material. It was fractionally distilled through a Podbielniak high-temperature distillation analyzer and a heart cut collected; mp 11.67° corresponding to 99.93%m purity and a 100% mp of 12.20°.

m-Cresol can be prepared nitrogen free [8070] by forming the *m*-cresol–phenol complex, washing, distilling, and regenerating; purity 99.6%.

Millikan [5152] reports a method for the separation of *m*-cresol from *p*-cresol by azeotropic distillation with benzyl alcohol. A mixture of 29 parts of *p*-cresol, 29 parts of *m*-cresol, and 42 parts of benzyl alcohol by volume was fractionally distilled at 30 Torr and a reflux of 6:1 in a 10-theoretical-plate still; 70–75% *m*-cresol was obtained. A more efficient column and a higher reflux ratio will produce up to 98% *m*-cresol.

The best grades available of *m*-cresol and benzyl acetate were dried over anhydrous sodium sulfate for long periods and fractionally distilled for *physical property studies* of mixtures by Katti and Chaudhri [3895]; the density was used as the criterion of purity.

Chivate and Shah [1469] reported the separation of *m*- and *p*-cresols by extractive crystallization of the eutectic with acetic acid.

Keicher and Klopfer [3926] removed *m*-cresol from other phenols by conversion to the urea additives, washing with solvents, then decomposing the additive compounds by treatment with chlorinated hydrocarbons at elevated temperatures. The liberated urea is insoluble.

Rigamonti and Schiavina [6173] developed a process for separating *m*- and *p*-cresols based on the ternary system *m*-cresol–ligroin–methanol (aq) and *p*-cresol–ligroin–methanol (aq). A complicated column operation was employed.

m-Cresol was treated [7600] with *o*-toluidine, 1:4, at 90–100°, cooled to room temperature, and allowed to stand. The crystals of the addition compound were filtered, washed with benzene, and distilled. The purity was 95–97%.

Engel [2262] added phenol to a mixture containing at least 35% *m*-cresol and *p*-cresol. A paraffinic petroleum solvent, boiling at 90–150°, was added, and the resulting mixture cooled. The phenol–*m*-cresol addition compound crystallized out and was separated.

For purification by distillation, see Ashworth and Lazik [468], Phenols, General Comments.

SAFETY

The oral LD_{50} for mice was found to be 828 mg/kg [7652].

The *minimum ignition temperature* in air is 550.0° [5309].

119.   *p*-Cresol

The phenol *p*-cresol was purified by Carney and Sanford [1357] for *ultraviolet studies* in the same manner described for *o*-cresol except that it was not distilled. The bp was 34.734°, corresponding to 99.99%m and a 100% bp of 34.739°.

Erić and coworkers [2282] purified *p*-cresol by distillation and recrystallization from light petroleum cut; bp 40–60°.

A mixture of *m*- and *p*-cresols can be separated by crystallizing from a paraffinic hydrocarbon containing not more than 5%w aromatics and not more than 10%w of olefins. Engel [2263] obtained *p*-cresol in 98.6% purity from a mixture containing 85% *p*-cresol. The remainder was principally the *meta*-isomer with a small amount of mixed xylenols. See also Fox and Barker [2504].

Golumbic [2840] developed a counter distribution method suitable for the partial separation of *m*- and *p*-cresol.

Kendall and Beaver [3945] prepared *p*-cresol by diazotizing pure *p*-toluidine and boiling the diazo solution. The product was steam-distilled, dried, and fractionally distilled. However, Hartman and coworkers [1508, p. 37] state that the quality of *p*-cresol prepared by diazotizing *p*-toluidine is almost always poor.

For *optical measurements*, Savard [6394] converted *p*-cresol to sodium *p*-cresoxyacetate, which was crystallized from water. The purified salt was decomposed by heating with hydrochloric acid in an autoclave. The yield was 60%.

Gibbs [2732] describes a method for purifying technical *p*-cresol, which depends on the fact that *o*- and *m*-cresols react with 2,6-dichloroquinone-chlorimide to form indophenols, while the *p*-isomer does not.

For purification by distillation, see Ashworth and Lazik [468], Phenols, General Comments.

SAFETY

The oral LD_{50} for mice was found to be 344 mg/kg [7652].
The *minimum ignition temperature* in air is 555.0° [5309].

Xylenols (120–124)

120. 2,4-Xylenol

Cocker and Lipman [1547] refluxed *m*-xylene-4-sulfonyl chloride with potassium hydroxide in methanol for 6 hr. After removing the methanol, more potassium hydroxide was added and the resultant mixture was stirred for 1 hr at 280°. 2,4-Xylenol was then obtained by acidifying the mixture and steam-distilling.

Shuikin and coworkers [6700] treated *p*-cresol with aqueous formaldehyde, diethylamine, and methanol for 3 hr at 60–70° to produce 2-(*N*,*N*-diethylaminomethyl) methylphenol. The latter was hydrogenated over Raney nickel to give 2,4-xylenol.

Andon and coworkers [349] purified 2,4-xylenol for *thermodynamic studies*. Steam was passed through a boiling mixture of the xylenol and sodium hydroxide until no nonacidic material distilled. After acidifying the residue with 20% sulfuric acid, the phenolic compound was dried over calcium sulfate and distilled at low pressure to remove the last trace of water and high-boiling materials. Then a fractional distillation was done in the range 100–212 Torr. The final step involved crystallization of the melt or zone refining or both methods; purity 99.972%m.

SAFETY

The oral LD_{50} for mice found to be 809 mg/kg [7652].

121. 2,5-Xylenol

Rericha and Protiva [6137] treated *m*-cresol with formaldehyde and diethylamine and the resultant product was hydrogenated to give 2,5-xylenol.

Oblasova and coworkers [5543] treated *p*-xylene with sulfuric acid to prepare *p*-xylenesulfonic acid which was converted to 2,5-xylenol by alkali fusion.

For purification for *thermodynamic studies*, see Andon [349], 2,4-xylenol; purity 99.896%m.

SAFETY

The oral LD_{50} for mice was found to be 1140 mg/kg [7652].

122. 2,6-Xylenol

de Feu and coworkers [2110] prepared 2,6-xylenol by dehydrogenation of 2,6-dimethylcyclohexanone over palladium and charcoal at 330–335°.

Carlin and Landerl [1351] prepared 2,6-xylenol by hydrogenation of 2-methyl-6-dimethylaminomethylphenol at 200° and 18,000 psi. After shaking a benzene solution of the crude product with 15% hydrochloric acid solution and then with 15% sodium hydroxide solution, the alkaline mixture was acidified with 20% hydrochloric acid solution and the xylenol was separated as an oil.

For purification for *thermodynamic studies*, see Andon [349], 2,4-xylenol; purity 99.886%m.

2,6-Xylenol was purified by crystallization from an ethylene glycol–water mixture. After three subsequent crystallizations, the xylenol contained less than 0.002% impurities [5590].

SAFETY

The oral LD_{50} for mice was found to be 980 mg/kg [7652]. Larionov [4413] reported that "inhalation of 270 mg/m^3 for 2 hr by mice and 4 hr by rats disturbed breathing and caused spasmodic shaking and agitation."

The *minimum ignition temperature* in air is 470.0° [5309].

123. 3,4-Xylenol

Kharlampovich and Oblasova [3984] treated *o*-xylene with sulfuric acid to prepare *o*-xylenesulfonic acid, which was converted to 3,4-xylenol by alkali fusion.

For purification for *thermodynamic studies*, see Andon [349], 2,4-xylenol; 99.971%m.

SAFETY

The oral LD_{50} for mice was found to be 948 mg/kg [7652].

124. 3,4-Xylenol

Horning and coworkers [3493] refluxed a mixture of 5% palladium–carbon catalyst and 3,5-dimethyl-2-cyclohexen-1-one in trimethylbenzene for 2 hr. The solution was extracted with 10% sodium hydroxide solution. Subsequent acidification with dilute hydrochloric acid solution produced crystalline 3,5-xylenol, which was washed with water and air-dried.

Ikarashi and coworkers [3611] oxidized 3,5-dimethylcumene to the hydroperoxide, which was partially decomposed with an acid and then hydrogenated to 3,5-xylenol; purity 99.5%.

Nair and coworkers [5381] treated *m*-xylene with sulfuric acid to give *m*-xylenesulfonic acid, which was converted to 3,5-xylenol by alkali fusion.

For purification for *thermodynamic studies*, see Andon [349], 2,4-xylenol; purity 99.960%m.

Mastrangelo [4959] purified commercial-grade 3,5-xylenol by selective sulfonation and recrystallization from a petroleum solvent.

SAFETY

The oral LD_{50} for mice was found to be 836 mg/kg [7652].

Unsaturated Monohydric Alcohols (125–131)

125. 2-Propen-1-ol

Directions are given in Adams [21, p. 15] for the preparation from glycerol and formic acid. The yield is 845 grams, 45–47% of theory, of 2-propen-1-ol. It may be dried to 98–99% by refluxing with successive portions of fused potassium carbonate until no further action is observed. An azeotropic method for producing anhydrous material is described. See also Coffey and Ward [1560].

Biedermann and Raichle [546] heated 1,3-propanediol with ethyl carbonate and potassium carbonate to 200–210° until evolution ceased; yield 80.2%.

Bharuchka and Weedon [847] reduced 1-butyn-1-ol with lithium aluminum hydride to *trans*-2-propen-1-ol.

2-Propen-1-ol may be prepared by the high-temperature chlorination of propylene [2361]. 2-Propen-1-ol is produced from the chloride by alkaline hydrolysis. The process is used commercially.

Ballard and coworkers [573] prepared 2-propen-1-ol as follows: Vapors of ethanol and propenealdehyde were passed over uncalcined magnesium oxide and zinc oxide in a mole ratio of 2.3:1 and at a rate of 0.099 mol/100 mL of catalyst per minute. The yield of 2-propen-1-ol at 391–401° was 77.5%.

Shokal and Evans [6685] heated an aqueous solution of acetic acid, propylene, and selenium dioxide almost to the boiling point. An excess of aqueous sodium hydroxide was added and the ester saponified. The 2-propen-1-ol was distilled as the azeotrope.

Rapean and coworkers [6062] found that pure commercial-grade 2-propen-1-ol is relatively stable on heating. Less than 1% decomposes per hour at 225° in stainless steel. Iron, iron salts, and copper have no significant effect on the decomposition rate. It is stable when heated in the presence of sodium carbonate. Extensive decomposition takes place in the presence of sodium hydroxide, starting as low as 100°. At about 135°, the reaction becomes exothermic and becomes explosive at about 190°.

A commercial 2-propen-1-ol was fractionally distilled by Kepner and Andrews [3953] through a 1.5 × 120-cm, vacuum-jacketed column packed with 3-mm glass helices to study its *complex with copper*: bp 97.1° at 761 Torr.

Azeotropic distillation may be used to dry 2-propen-1-ol [3133].

SAFETY

2-Propen-1-ol has a sharp pungent odor and is a powerful lachrymator. The threshold odor concentration is reported to be less than 0.78 ppm. Nasal irritation starts in the same range as odor perception. It is absorbed through the skin [5713]. Dunlap and coworkers [2122] found that the LD_{50} for rabbits is 45–105 mg/kg. Chronic exposure of 50 ppm was fatal to four animals.

Torkelson et al. [7528] suggest that the average exposure concentration of 2-propen-1-ol not exceed 2 ppm and that concentrations always be kept below 5 ppm.

The *threshold limit value* of 1 ppm, 3 mg/m^3, has been suggested [7022].

The *flammable limits* in air are 2.52 and 18.00%v; the upper value at elevated temperature. The *minimum ignition temperature* in air is 378° [5712].

126.   *cis*-2-Buten-1-ol
127.   *trans*-2-Buten-1-ol

Lieben and Zeisel [4592] reduced crotonaldehyde with iron and acetic acid to 2-buten-1-ol in 1881.

Yada and coworkers [8190] selectively reduced crotonaldehyde with hydrogen over Raney copper–cadmium catalyst at 225–245°.

Clarke and Crombie [1505] reduced 2-butyn-1-ol to *cis*-2-buten-1-ol (3,5-dinitroben-zoate, mp 52–53°) using a copper–zinc couple in boiling ethanol. Apparently this is a specific *cis*-reduction. Ingold and Ingold [3620] used the same method of preparation. They purified and dried the alcohol to a constant refractive index. See also Charon [1428].

Petrov [5800] saturated 1,3-butadiene with hydrogen bromide in acetic acid at −10° and obtained 60% of the bromobutane. The latter was digested 1.5–2 hr with 10% sodium hydroxide. A mixture of 2-buten-1-ol and 3-buten-1-ol was obtained.

Hatch and Nesbitt [3224] prepared *cis*-2-buten-1-ol from 25 grams of 2-butyn-1-ol stirred with 8 grams of Schmidt [6460] catalyst (palladium on barium sulfate) in 290 mL of methanol and treated with 0.357 mol of hydrogen for 3 hr in a 1-L suction flask. The methanol was removed by fractional distillation, and the *cis*-2-buten-1-ol was fractionally distilled at reduced pressure; yield 76%. It was also prepared from isocrotonic acid. *trans*-2-Buten-1-ol was prepared from *trans*-crotonaldehyde by reduction with lithium aluminum hydride.

Young and Andrews [8238] prepared *trans*-2-buten-1-ol from the corresponding chloro-compound by hydrolysis with silver oxide.

Smets [6807] prepared *cis*-ethylenic alcohols from acetylenic magnesium halide and formaldehyde. The triple-bonded alcohol was reduced to the ethylenic alcohol by mild reduction with Bourgel's catalyst. The *trans*-alcohols are formed from Grignard reagent and an unsaturated aldehyde.

Meerwein and coworkers [5036] reduced 2-butenal with $Mg(OC_2H_5)_2 \cdot Mg(OC_2H_5)Cl$ in 60% yield to 2-buten-1-ol. No butanol was found.

Young and coworkers [8239] studied methods to 1936 for the preparation of 2-buten-1-ol and found that all gave low yields. They report a method that gave 60% theoretical of 93% unsaturated alcohol. Aluminum isopropoxide is first prepared in the flask and then reacted with crotonaldehyde.

128.   2-Propyn-1-ol

2-Propyn-1-ol has a mild geranium-like odor. Commercial material is of good purity and contains a stabilizer. It is miscible with water and a large number of organic compounds such as benzene, chloroform, 1,2-dichloroethane, ethyl ether, acetone, *p*-diox-ane, tetrahydrofuran, and pyridine. It is partially miscible with carbon tetrachloride and immiscible with aliphatic hydrocarbons.

2-Propyn-1-ol is considered to be a stable compound. Manufacturers' experience indicates a decrease in assay of about 1% per 6 months' storage. Alkalies appear to cause polymerization and development of color on heating or storage over a period of time. Acids stabilize against polymerization during the heating of aqueous solutions [389].

2-Propyn-1-ol has been prepared from acetylene and formaldehyde. Yamamoto and coworkers [8198] used methanol and acetone as the reaction medium in the presence of copper chromate; Suzuki [7098] used tetrahydrofuran, and Hanford and Fuller [3134] used a copper–bismuth catalyst.

Hatch and Moore [3223] studied the preparation of 2-propyn-1-ol from *cis*- and *trans*-3-chloro-2-propen-1-ol. The lower boiling chloroalcohol was refluxed with 10% excess 12.5% sodium hydroxide for 3 hr and a 69.3% yield of the alcohol was obtained. The water–alcohol azeotrope was obtained. It was found that repeated extraction with ether was the most satisfactory way of separating the alcohol.

Henry [3302] in 1872 described the preparation of 2-propyn-1-ol by the action of potassium hydroxide on monobromoalkyl alcohols.

Johnson [3729] prepared 2-propyn-1-ol by heating 2-butyne-1,4-diol with potassium carbonate at 140–150° and Kreimeier [4255] reacted sodium acetylide with paraformaldehyde. See also Smets [6807] under 2-buten-1-ol.

Shiga and Tsuruta [6659] concentrated 2-propyn-1-ol by feeding the aqueous solution to the top of an extraction tower and two volumes of water saturated ethyl acetate into the bottom in countercurrent flow. Distillation of the acetate phase collected at the top gave 99% 2-propyn-1-ol.

Lespieau [4531] purified 2-propyn-1-ol by freezing the aqueous solution, forming the monohydrate, mp 16.52°. The triiodide was used to characterize the alcohol.

An aqueous solution of 2-propyn-1-ol can be concentrated by azeotropic distillation with either butanol or butyl acetate [5571].

McKinley [4779] found that 0.25–0.5% of succinic acid or aqueous hydrogen chloride prevents decomposition of 2-propyn-1-ol during distillation.

SAFETY

2-Propyn-1-ol is primarily a skin irritant, but not a skin sensitizer. The LD_{50} for white rats and guinea pigs, respectively, is 0.07 and 0.06 mL/kg [389].

The *minimum ignition temperature* in air is 374° [5712].

129. *dl*-α-Terpineol
130. *d*-α-Terpineol
131. *l*-α-Terpineol

α-Terpineol occurs in nature as *d*-, *l*-, and *dl*-forms. Available data indicate that the optically active forms are enantiomers and are identical in all respects except rotation of polarized light. Detailed descriptions of the compound are given in Simonsen [6751].

Teeple [7206] reported in 1908 that steam-distilled oil from *Pinus palustris* (long leaf pine) has a high *dl*-α-terpineol content.

Alder and Vogt [76] prepared the *dl*-form by reacting 1-methyl-4-acetylcyclohexene with methylmagnesium bromide in ether. Matsuura and Aratani [4995] heated a mixture of 1,8-cineole and sodium peroxide for 20 hr at 160° to produce *dl*-α-terpineol.

Arbuzov and coworkers [404] treated terpinolene in petroleum ether with monoperphthalic acid for 2 days. The yield was 55.5% 4,8-epoxy-*p*-menth-1-ene which was reduced with lithium aluminum hydride to give 74.3% *dl*-α-terpineol.

Fuller and Kenyon [2610] resolved *dl*-α-terpineol into its optical isomers by fractional crystallization of the brucine and morphine salts of α-terpineol hydrogen phthalate.

Wallach [7884] shook *d*-limonenehydrochloride with 2% aqueous potassium hydroxide solution at 50° to prepare the *d*-isomer. He also prepared the *l*-isomer by shaking a solution of *l*-pinene hydride in ether with 5% sulfuric acid solution for 1 hr [7886].

Matsuura [4994] heated linalool with 30% sulfuric acid solution at 100° and recovered the *d*-isomer in the alcoholic fraction.

Perkin and Fisher [5768] synthesized the *d*- and *l*-isomers from the reaction of *d*- and ethyl-1-methyl-Δ^1-cyclohexene-4-carboxylate, respectively, with methylmagnesium iodide.

Polyhydric Alcohols (132–148)

132.   *cis*-2-Butene-1,4-diol
133.   *trans*-2-Butene-1,4-diol

Reppe [6135] partially reduced *cis*-2-butyne-1,4-diol with hydrogen and palladium on calcium carbonate to *cis*-2-butene-1,4-diol. He also prepared the alcohol from acetylene and aqueous formaldehyde at 100° under pressure.

Clarke and Crombie [1505] reduced 2-butyne-1,4-diol to *cis*-2-butene-1,4-diol, mp 8–10°, using a copper–zinc couple in boiling ethanol. Apparently this is a specific *cis*-reduction.

Valette [7657] prepared *trans*-2-butene-1,4-diol by lengthy heating of *trans*-1,4-dibromo-2-butene with water. He also prepared the *cis*-alcohol by partial reduction of 2-butyne-1,4-diol in ethanol with hydrogen and Raney nickel catalyst.

Kato and Sakuma [3880] used polarographic reduction of 2-butyne-1,4-diol to 2-butene-1,4-diol on platinum, palladium, silver, and copper cathodes. No 1,4-butanediol was found.

Kato et al. [3881] reduced 2-butyne-1,4-diol on spongy silver and obtained almost pure *cis*-2-butene-1,4-diol.

134.   1,2-Ethanediol

Most of the 1,2-ethanediol is produced by the hydration of ethylene oxide by one of two processes. Diethylene glycol and triethylene glycol are by-products. The principal impurities found in 1,2-ethanediol are 1,2-propanediol, butanediol, and water. It is very hygroscopic.

Pukirev [5978] prepared 1,2-ethanediol in a 34% yield by saponifying 1,2-dibromoethane with potassium carbonate.

Ghosh and coworkers [2726] prepared the diol and ethylene chlorohydrin by the electrolytic reduction of ethylene on porous carbon anodes. Ethylene was fed to a porous carbon tube anode at the rate of 57mL/hr per cm^2 of anode surface, with a current density of 0.023 amp/cm^2 in 10% sodium chloride solution: 1,2-ethanediol and ethylene chlorohydrin were formed. The current efficiency of the chlorohydrin was 91% at 1° and 1.1% at 91°. The glycol efficiency was 5.4% at 1° and 16.5% at 91°.

Fogg and coworkers [2469] purified technical-grade 1,2-ethanediol, 99.5%, by fractional distillation at 7–10 Torr at a reflux ratio of 10:1 in a nitrogen-blanketed, adiabatically operated, $\frac{1}{16}$-in. glass helices packed 1 × 36-in. column. Only the middle third distilling at constant temperature was collected. Eleven lots were purified for *density and refractive indices studies of aqueous solutions;* d 25° varied from 1.1098 to 1.1099, n 25° varied from 1.43179 to 1.43182.

Koizumi and Hanai [4135] purified commercial 1,2-ethanediol for *electrical property studies* by drying over dehydrated sodium sulfate and then running three successive fractional distillations at 8 Torr pressure, retaining only the middle fractions. It was then fractionally crystallized twice.

Smyth and Walls [6877] fractionally distilled 1,2-ethanediol in a vacuum, dried the middle fraction over sodium sulfate, and, after decanting, fractionated repeatedly. The water was removed by fractional distillation, the first fraction containing all of the water.

Taylor and Rinkenbach [7192] concluded, after studying the results of careful investigators and comparing their own experience, that the true melting point of 1,2-ethanediol is intermediate between -17.4 and $-11.5°$ due to its tendency to supercool.

Kanbayashi and Nukada [3846] *studied the infrared* and *Raman spectra* of 1,2-ethanediol as liquids and in solutions and their temperature response. A *rotational isomer* was found that exists only in the gauche form in the liquid state.

1,2-Ethanediol and 1,2-propanediol were simultaneously determined by Warshowsky and Elving [7909] by oxidation with periodic acid and then analyzing polarographically for the formaldehyde in the presence of acetaldehyde by the technique of Whitnack and Moshier [8036]. See also Francis [2512], Reinke and Luce [6120], and Hoepe and Treadwell [3441].

Gardner and Hussain [2661] purified 1,2-ethanediol for *standard enthalpy of formation study* by the same method described for 1,2-propanediol.

CRITERIA OF PURITY

The boiling point, density, and refractive index may be used as a qualitative measure of purity for 99–99.8% material. Water can be determined by the Karl Fischer method [5171]. Gas chromatography may be used for the analysis of 1,2-ethanediol.

SAFETY

1,2-Ethanediol is a colorless, odorless liquid with a bittersweet taste. It has a low vapor pressure at normal temperature. It presents negligible hazards to health except, possibly, when being used at elevated temperature. It has a low acute oral toxicity. Bornmann [997] found the LD_{50} for mice to be 13.79 mg/kg. It was fed to monkeys at a dietary level of 0.2 and 0.5% for 3 years without any toxic effects [918]. See also [5266]. There does not appear to be any significant irritation from skin contact. Toxic amounts can be absorbed through the skin [5713]. Bornmann [997] reports that all glycols have a narcotic effect and, in toxic amounts, cause a central paralysis of respiration and the heart.

The *threshold limit value* of 100 ppm, 274 mg/m^3, appears to be generally acceptable [152, 5713].

The *minimum ignition temperature* in air is 413° [5712].

Several methods have been published for the determination in air. Mandric [4868] oxidized the glycol to formaldehyde and determined with chromatropic acid. He also reports the determination in air to a sensitivity of 0.001 mg per 7 mL of solution. See also [4902].

135.   1,2-Propanediol

Pukirev [5978] converted 2-propanol (free of tertiary alcohols) to propylene with sulfuric acid and aluminum sulfate. The propylene was passed into cold bromine to yield dibromopropane, which, upon heating with dry silver acetate in 100% acetic acid for 75 hr, gave the diacetate. The diacetate was redistilled and saponified with barium hydroxide in water to give 1,2-propanediol.

The conditions for the production of 1,2-propanediol and glycerol from propylene, air, and steam were studied by Newitt and Mene [5469].

Curme and Johnson [1766] hydrated propylene oxide and obtained 1,2-propanediol.

Levene and Walti [4547] prepared *levo*-1,2-propanediol from *levo*-1-amino-2-hydroxypropane.

MacBeth and Thompson [4737] purified a 99 + % commercial product for *refractive index and density studies of aqueous solutions*. It was fractionally distilled in a glass system vented to the atmosphere through a tube of anhydrous calcium sulfate. The column was 18 in. long and 0.5 in. in diameter packed with $\frac{3}{16}$-in. glass helices. The middle fraction was used; 0.04%w water was determined by a Karl Fischer method.

Several aliphatic diols were purified by Gardner and Hussain [2661] for *standard enthalpy of formation study*. They were fractionally distilled at atmospheric pressure through a 35-cm column packed with $\frac{1}{8}$-in. stainless steel helices at a reflux ratio of 10:1. The center cut was used. Water was determined by a Karl Fischer method accurate to 0.002%w. The criterion of purity was gas chromatography with a flame ionization detector.

Nakanishi and coworkers [5388] purified several glycols for *physical property study*. They were dried with salts such as anhydrous sodium sulfate and fractionally distilled until the gas chromatography analysis showed a minimum of impurities estimated to be 99.8%.

SAFETY

Rowe [5713] states, ''The hazards to health in industrial handling and use of propylene glycol would seem to be negligible. Its systemic toxicity is especially low and, since 1942, it has been considered a proper ingredient for pharmaceutical products. The Food and Drug Administration does not object to its use in food products or cosmetics. The inhalation of atmospheres containing propylene glycol vapor presents no health hazard.'' Rowe further finds it hard to visualize how a 1,2-propanediol fog in the air would be a health hazard.

Walther [7898] found that the liquid diol was not irritating to the skin of animals or of human beings. When taken orally, vomiting results; but even single large doses cause no ill effects. It has been firmly established, however, that this glycol is no more toxic than glycerol.

Guchok and Zborovskaya [3010] reported a subcutaneous LD_{50} of 17.37 g/kg in mice and an intravenous LD_{50} of 13.1 g/kg in rats.

A *threshold limit value* is not considered necessary [5713].

The *flammable limits* in air are 2.62 and 12.55%v at elevated temperature [5712]. The *minimum ignition temperature* in air is 216° [3555].

136.   1,3-Propanediol

Hatch and Evans [3221] hydrated acrolein in dilute sulfuric acid, followed with reduction over Raney nickel at 32–40°. 1,3-Propanediol constituted about 25% of the product and may be separated by fractional distillation.

Senkus [6578] recovered 1,3-propanediol by treating with formaldehyde to form the 1,3-dioxane.

Gardner and Hussain [2661] found that 1,3-propanediol showed considerable decomposition, which was indicated by development of color when fractionally distilled at atmospheric pressure. Fractional distillation at reduced pressure gave a clear distillate that analyzed 99.9%w by gas chromatography with the injection port at 373 ± 3 K and the column at 381 K. They reported that 1,3-propanediol slowly decomposed at room temperature becoming pale yellow. A pure sample will remain colorless indefinitely when kept at 273 K. The diol was used for *standard enthalpy of formation study.*

For purification for *physical property study,* see Nakanishi and coworkers [5388], 1,2-propanediol.

SAFETY

Van Winkle [7680] found 1,3-propanediol to be about twice as toxic as the 1,2-isomer.

A *threshold limit value* does not seem necessary because of the low vapor pressure at room temperature and the relatively low toxicity [5713].

137.   1,3-Butanediol

1,3-Butanediol is a relatively new commercial product. It is produced by essentially the same type of hydration process of the corresponding oxide as the commercial synthetic glycols.

Kling and Roy [4066] prepared 1,3-butanediol and other glycols similarly by the action of magnesium amalgam on acetaldehyde in benzene solution. Hibbert [3371] condensed acetaldehyde in the presence of potassium carbonate and reduced with aluminum amalgam. See also [1106].

Carothers and Adams [1359] reduced aldol to the butanediol with hydrogen and platinum oxide in 86% yield. Palfray [5651] used nickel as the catalyst. Levene and coworkers [4550] reduced freshly prepared aldol by biochemical means by the method of Neuberg and Kerb [5459] to *dextro*-1,3-butanediol. The electrolytic reduction of aldol has been reported [1015]. Acetaldol has also been used as the starting material for reduction [1107].

Gardner and Hussain [2661] purified 1,3-butanediol for *standard enthalpy of formation study* by the same method described for 1,2-propanediol.

For purification for *physical property study,* see Nakanishi and coworkers [5388], 1,2-propanediol.

SAFETY

Ingestion of 1 mL/100 g body weight twice weekly for 45–185 days: no damage to vital organs, teeth, and bones. It did not cause globinuria or proteinuria in man or experimental animals [4182]. Loeser found butanediol to be no more toxic than glycerol on oral administration. It was not harmful to the skin or mucous membranes [4648].

There does not seem to be a necessity for a *threshold limit value* due to the low vapor pressure at room temperature and low toxicity.

138.   1,4-Butanediol

Adkins and Folkers [35] prepared 1,4-butanediol by hydrogenation of diethyl succinate over copper chromite at 250° and 220 atm; 80.5% yield. Other preparations using diethyl succinate are given in Hamonet [3123], Müller [5312], and Hill and Hibbert [3407].

Reppe and coworkers [6136] reported an industrial preparation of 1,4-butanediol. 2-butyne-1,4-diol was hydrogenated at 200–400 atm over nickel or cobalt catalyst.

1,4-Butanediol was purified by Gardner and Hussain [2661] for *enthalpy of formation study*. It was dried over Drierite, fractionally distilled at 95° and 0.5 Torr, and fractionally frozen at 20°; purity 99.8%w.

Evans and coworkers [2326] purified 1,4-butanediol by crystallizing the compound from an anhydrous ether–acetone mixture.

Water was removed from 1,4-butanediol by multistage distillation at 130° and 110 Torr, followed by steam distillation. Purity was 99.66% with traces of low-boiling and high-boiling organic materials, according to du Pont [2147].

Clendenning and coworkers [1529] purified 1,4-butanediol for *physical property studies* by fractional distillation in an 18-in. Stedman column and then distillation *in vacuo* over sodium hydroxide pellets.

SAFETY

Stasenkova [6964] reported an oral LD_{100} of 5 g/kg for rats and 1 g/kg for mice. A recommended maximum permissible air concentration is 0.5 mg/L. Knyshova [4082] reported the following LD_{50} values in mg/kg: mice, 2062; rats, 1525; rabbits, 2531.

139. 2,3-Butanediol (Mixed Isomers)
140. *l*-2,3-Butanediol
141. *d*-2,3-Butanediol
142. *dl*-2,3-Butanediol
143. *meso*-2,3-Butanediol

2,3-Butanediol has four isomers: *levo (l)*, *dextro (d)*, racemic *(dl)*, and *meso*. The physical property table in Chapter III of 2,3-butanediol (mixed isomers) contains properties found in the literature under "2,3-butanediol," rather than a specific isomer.

Most commercially available 2,3-butanediol consists of a mixture of *meso* and racemic isomers. One commercial sample was found to be 64% racemic and 36% *meso* [4871].

The chemical methods of preparing 2,3-butanediol (mixed isomers) will be given first, followed by biochemical methods of preparing specific isomers.

Sabatier and Mailhe [6324] prepared 2,3-butanediol by reducing biacetyl with hydrogen over a nickel catalyst at 140–150°. Böeseken and Cohen [939] reacted silver acetate and 2,3-dibromobutane to produce a diacetate, which was hydrolyzed to give the diol.

Schierholtz and Staples [6436] reported a detailed preparation by hydrolysis of 2,3-dibromobutane over lead(II) oxide.

Shida and coworkers [6656] reported that 2,3-butanediol formed a transparent glassy solid at 77 K suitable for trapping electrons. They studied the glass by infrared spectroscopy.

Manius and coworkers [4871] studied the gas chromatographic *separation and determination* of the isomers of 2,3-butanediol. They could not separate the enantiomers but were able to separate the *meso* and racemic forms.

Grivsky [2973] studied the preparation of the isomers by chemical and biological methods. For details of methods and techniques, see the reference article. Thaker and Dave [7221] described in detail the preparation of the *d*- and *l*-isomers by the reaction of 2,3-butanediol with phthalic anhydride and the subsequent separations.

The biological process for the production of the four isomers has been extensively studied. Many of the citations in Chapter III are from the glycol produced by fermentation. The following references are considered pertinent:

Connor and Bulgrin [1625]; *Aerobacillus polymyxa* and *Aerobacter aerogenes* with corn mash to produce the *l*- and *meso*-isomers, respectively.

Kooi and coworkers [4181]; *Aerobacillus polymyxa* with a corn starch mash.

Morell and Auernheimer [5246]; *Aerobacter aerogenes* with a glucose mash for the *d*-isomer; *Bacillus polymyxa* for the *l*-isomer.

Ward and coworkers [7905]; *Bacillus polymyxa* on grain mash substrates for the *l*-isomer.

Neish [5440]; *Aerobacillus polymyxa* with whole wheat mash for the *l*-isomer.

Rubin and coworkers [6296] synthesized the D(−)- and L(+)-forms of 2,3-butanediol from the corresponding D- and L-mannitol isomers "by a series of unequivocal reactions." The reaction conditions were mild and the configuration was guarded. Danger of racemization was almost eliminated.

Wilson and Lucas [8101] studied the preparation and conversion of the racemic and *meso*-forms through the Walden inversion. They found that the diacetates offered a good method for purification.

Taufen and coworkers [7189] prepared a high-purity racemic isomer for *Raman study* by saponification of *dl*-2,3-diacetoxybutane with alcoholic sodium hydroxide solution. Dry ether was added to precipitate sodium acetate and the diol was collected by fractional distillation.

Senkus [6577] proposed a method for recovering racemic and *meso*-2,3-butanediol from fermentation beer.

Stepanova and coworkers [6998] hydrated a 64:36 *cis–trans* mixture of 2-butene oxide at 20° to produce the racemic and *meso*-isomers.

Racemic and *meso*-2,3-butanediol were purified by Nasybullina and coworkers [5406] using preparative gas chromatography.

meso-2,3-Butanediol from Lucidol Corp. was separated from the racemic isomer by four crystallizations from diethyl ether [7189].

Bosnich and Watts [1005] found *l*-2,3-butanediol to be a "remarkably efficient" *enantiomerizing medium*. They concluded that the solubility measurements could provide precise thermodynamic data on diastereoptic interactions.

SAFETY

The *minimum ignition temperature* in air for 2,3-butanediol is 376° [6402].

144. 1,5-Pentanediol

Hamonet [3124] prepared 1,5-pentanediol by saponification of the diacetate which was synthesized from 1,5-dibromopentane and silver acetate.

Kaufmann and Adams [3905] reduced furfural with hydrogen using a platinum black catalyst and iron(II) chloride in ethanol. 1,5-Pentanediol was the high-boiling fraction obtained upon distillation of the mixture; bp 237–239° at 751 Torr.

Kaufman and Reeve [3903] reported a detailed synthesis of 1,5-pentanediol by the hydrogenation of tetrahydrofurfuryl alcohol over copper chromite catalyst at 255° and

230–240 atm. Another hydrogenation process to prepare 1,5-pentanediol is given in Schniepp and Geller [6473].

For purification of 1,5-pentanediol for *physical property studies*, see Clendenning [1529], 1,4-butanediol.

1,5-Pentanediol was purified by Gardner and Hussain [2661] for *enthalpy of formation study*. It was fractionally distilled at atmospheric pressure through a 35-cm column packed with $\frac{1}{8}$-in. stainless steel helices at reflux ratio of 10:1. The center cut was used. Water was determined by a Karl Fischer method. The criterion of purity was gas chromatography with a flame ionization detector.

SAFETY

The LD_{50} single dose for rats is reported to be 5.89 g/kg. No preliminary skin irritation and only a trace of eye injury was found for rabbits [7630].

145. Hexylene Glycol

Francke [2516] prepared hexylene glycol in 1901 by reducing diacetone alcohol in water with aluminum amalgam. The compound was also prepared by hydrogenation of diacetone alcohol over finely divided nickel at 120–130° and 15–20 atm [2368].

Chiao and Thompson [1463] purified hexylene glycol for *density and refractive index studies* by fractionation at 10–15 Torr with a reflux ratio of 25:1 in a packed column of glass helices. It was highly hygroscopic and the purified material contained 0.04% water (Karl Fischer).

Yamamoto and Kambara [8197] purified hexylene glycol for *infrared spectra* by successively drying with sodium sulfate and calcium hydride. They collected the fraction boiling at 107.5–108.5° and 25 Torr.

SAFETY

According to Shell Chemical Co. [6627], hexylene glycol is slightly toxic by oral administration and mildly irritating on prolonged skin contact. It is readily removed by washing with water.

The single dose LD_{50} in rats was reported as 4.76 g/kg [6882].

The *threshold limit value* has been set at 100 ppm, 410 mg/m^3 [142].

146. 2-Ethyl-1,3-hexanediol

Grignard and Fluchaire [2967] prepared 2-ethyl-1,3-hexanediol by boiling 2-ethyl-1,3-hexanediol butyrate with 25% alcoholic potassium hydroxide over a water bath.

Wear [7937] prepared 2-ethyl-1,3-hexanediol by reducing 3-hydroxymethyl-4-heptanone with lithium aluminum hydride. The diol was distilled from the crude product.

Svirbely and coworkers [7105] purified 2-ethyl-1,3-hexanediol by vacuum distillation in a Vigreux column.

SAFETY

The LD_{50} oral single dose for rats is 6.50 mL/kg. Its skin penetration is less than ethylene glycol for rabbits and it showed a slight skin irritation and moderate eye injury [7630].

147. Glycerol

All glycerol was obtained from glycerides of fats and oils until 1949, when the synthetic product was first manufactured. The synthetic material now accounts for more than 60% of the total production. It is made by several processes.

Milas [5126] prepared glycerol in 60% yield by hydroxylating 2-propen-1-ol with substantially anhydrous solutions of hydrogen peroxide. The preferred catalysts were oxides of osmium, rubidium, vanadium, molybdenum, or chromium. The temperature was an important variable. The formation of glycols in high yields was favored by a temperature of 0–21°.

2-Propen-1-ol is converted to glycerol in 70% yield when treated with hydrogen peroxide in the presence of phosphotungstic acid at 80–82° for 2 hr [5013].

Hass and Patterson [3207] purified glycerol, prepared by hydrogenolysis, by dissolving it in an equal volume of 1-butanol. The solution was placed in a watertight bottle, cooled, seeded, and slowly revolved in an ice-water slurry until crystallization had taken place. The impurities and most of the solvent were removed by centrifuging. The crystals were washed with cold acetone or isopropyl ether. Over 60% of the fraction boiling at 290° was isolated as crystalline glycerol. The crystalline material passes the USP specifications. Other solvents that may be used are 1-propanol, the pentanols, or liquid ammonia.

Eeckelaers [2213] purified glycerol by passing it through a cation-exchange resin followed by an ion-exchange resin and then concentrating in vacuum.

Glycerol was purified for *viscosity studies* by Segur and Oberstar [6552] by distilling C.P. material from a 5-L flask through a 25-cm Vigreux tube. The middle portion, weighing 4330 grams and distilling at 140° at 1 Torr, was used. The specific gravity at 25/25° was 1.26192, which corresponded to a purity of 99.97%.

Hoyt [3530] purified by digesting with hot concentrated sulfuric acid, saponifying with lime paste, acidifying with sulfuric acid, filtering, treating with an anion absorbent resin, and finally fractionally distilling. The recovery was 91%.

Wallerstein and coworkers [7889] purified glycerol, produced by fermentation of carbohydrates, by distilling. Before distillation, the impure alcohol was treated with a small quantity of formaldehyde and kept for approximately 1 hr in an alkaline condition.

Evans [2328] removed the color from distilled and substantially dry glycerol by extracting with 2,2,4-trimethylpentane.

Cammenga and coworkers [1326] vacuum-distilled analytical grade glycerol at 140° and 4 Torr for *vapor pressure studies*. The vapor was passed over 3 Å molecular sieve and then collected and stored over activated molecular sieve.

The 1.93 μm band is characteristic of water and has been used to determine water in glycerol [1422].

Glycerol containing ionic and/or nonionic materials may be purified by ion exchange or ion exclusion or a combination of both [5053, p. 149].

Glycerol is very hygroscopic and can absorb water up to 50% of its weight from air [5053, p. 150].

Glycerol decomposes at its boiling point.

CRITERIA OF PURITY

Requirements and test methods for ACS reagent grade are given in [137]. A dynamite grade is available. It has a higher density than the ACS reagent grade, that is, less water.

USP and BP grades are available in the United States and the British Commonwealth, respectively.

Glycerol is difficult to keep dry. After other impurities have been removed, the density and the critical solution point in 3-methyl-1-butanol are good criteria of the amount of water present [7482].

SAFETY

Glycerol is the standard of comparison for substances that are considered to be safe to use in medicinals, cosmetics, and similar products; for example, see Safety for 1,2-propanediol and 1,3-butanediol.

Kudo and Ito [4306] determined the LD_{50} in mice as 19,300 mg/kg.

148.   1,2,6-Hexanetriol

Zelinski and Eichel [8271] heated 2-hydroxymethyl-2,3-dihydro-4*H*-pyran with hydrochloric acid solution and made the mixture alkaline with potassium carbonate. Sodium boron hydride was stirred into the mixture, which was then neutralized with hydrochloric acid solution. Subsequent filtration produced 1,2,6-hexanetriol.

Whetstone and Seaver [7998] dimerized acrolein and then hydrolyzed the dimer to produce 2-hydroxy-1,6-hexanedial. The aldehyde was hydrogenated over Raney nickel to give 1,2,6-hexanetriol.

SAFETY

Kuramoto and coworkers [4323] reported the LD_{50} for mice and rats, respectively, to be 5.8 and 10.9 mL/kg subcutaneously. Daily skin applications of 25% for a month showed no serious behavior in food intake and blood or serum enzyme activity. Small lesions were found in the lungs, spleen, and liver. Smyth and coworkers [6885] reported the oral LD_{50} in rats to be 16 mL/kg.

ETHERS (149–179)

General Comments

Preparation and Purification

Reppe [6134, p. 28 ff.] gives methods, or references thereto, for a number of method for the preparation of vinyl ethers.

Kobe et al. [4091] purified 20 compounds by the same general procedure for use in the determination of *critical properties and vapor pressures*. The compounds were dried with Drierite and those suspected of forming peroxides were treated with iron(II) sulfate. The compounds were fractionally distilled three times and 80% heart cuts were taken. Reflux ratios of 60 to 80:1 were used. The criterion of purity was the boiling range measured with a double-junction thermopile sensitive to about 1/80°.

Peroxides

Ethers form peroxides under normal storage conditions. Unless a test shows peroxide to be absent, ethers should not be evaporated or distilled to near dryness. The rate of formation of peroxides in a given ether depends upon the storage conditions. Peroxide formation is accelerated by heat, light, and air. A small amount of moisture will accel erate peroxide formation in most ethers. Ethers kept in the laboratory should be in a dark bottle in an atmosphere of nitrogen in contact with iron filings or copper powder. Storage in a refrigerator is further recommended.

Removal or Destruction of Peroxides.   Bodea and Silberg [932] prepared peroxide-free ethers by heating 1 L of the ether with about 50 grams of phenothiazine, refluxing for 5 min, and distilling.

Peroxides may be removed from water-insoluble ethers by shaking the ether with a solution of iron(II) sulfate acidified with sulfuric acid until the ether phase no longer gives a test for peroxides.

Fierz-David [2424] freed ethyl ether from peroxides by shaking 1 L with 10 grams of copper-zinc couple. The couple was prepared by suspending 10 grams of zinc dust in 50 mL of hot water, adding 5 mL of 2 N hydrochloric acid and decanting after 20 s. The dust was then washed twice with water and covered with 50 mL of water and 5 mL of 5% copper(I) sulfate with swirling. The liquid was decanted and the residue washed three times with a small amount of water, three times with 20 mL of ethanol, and twice with 20 mL of ethyl ether. The treated ether was stabilized by adding 2 mg of pure diphenylamine. It is not necessary to remove the zinc dust before stabilizing.

A summary [4880] of methods of destroying peroxides has been prepared. The peroxide content is lowered markedly by passing the ether over an alumina or Dowex 1 column.

Dasler and Bauer [1827] removed peroxides from ethyl ether, butyl ether, and p-dioxane by passing through a column containing activated aluminum oxide. This treatment also reduced the aldehyde content. The ratio of aluminum oxide to the volume of ether varied with the amount of peroxides present; 82 grams of aluminum oxide in a column 1.9 $\times$ 23 cm was sufficient for the complete purification of 700 mL of ethyl ether containing 127 μmoles of peroxides per liter.

Peroxides may be destroyed by sodium bisulfite, or more slowly by an alkali.

Fisher and Baxter [2447] report that peroxides in p-dioxane can be decreased by shaking with tin(II) chloride or by refluxing with lead(IV) oxide and filtering through a slow filter paper. Isopropyl, ethyl, and probably isopentyl ethers can be freed from peroxides by lead(IV) oxide. No lead was found in the ethers after treatment.

Ramsey and Aldridge [6044] efficiently removed peroxides from ethers with cerium(III) hydroxide. The reagent has the advantage that it is a solid and is insoluble in ethers. It apparently is quite safe. Some reagents, such as lithium aluminum hydride, are reported to have caused fire [5201]. Cerium(III) hydroxide is recommended and is prepared in the following manner: Sodium hydroxide is added to a cerium(III) salt solution until the supernatant liquid is slightly alkaline. The white precipitate is separated and washed by centrifugation. It is used undried. The amount of the hydroxide to use is not given. Water is required in the ratio of 1 part to 15 parts of ether. The cerium(III) hydroxide presumably is shaken or stirred vigorously with the ether. The peroxide removal is complete in about 15 min. The hydroxide changes color from white to reddish brown, depending upon the amount of peroxide present. The solid is removed from the ether by decantation and centrifugation. The clear ether can be tested by the benzidine test [2395].

Nineteen ethers were tested and only two, allylethyl and benzyl-n-butyl, gave positive tests for cerium. Starch-iodide paper for testing for the presence of peroxides is not as reliable as a slightly acidified solution of potassium iodide and starch.

Detection and Determination of Peroxides.   Rinse a small glass-stoppered cylinder with the ether; then add 10 mL of the ether to the tube and 1 mL of a 10% solution of freshly prepared potassium iodide. No yellow color should be observed in either layer at the end of 1 min [137, 6263]. This test may be made more sensitive by adding a drop of starch indicator and shaking.

Prepare an approximately 1% solution of iron(II) ammonium sulfate. Transfer 5 mL to each of two test tubes and add 0.5 mL of 1 N sulfuric acid and 0.5 mL of 0.1 N thiocyanate solution to each tube. Add 10 mL of ether to one tube and shake well. The aqueous phase in the ether tube should not have a redder color than the blank. When testing water-soluble ethers, such as p-dioxane, add 10 mL more of water to the blank.

A rapid method for the detection of microgram quantities of peroxides in ethers was reported by Dugan [2115]. N,N-Dimethyl-p-phenylenediamine was dissolved in 10 mL of water and brought to 100.0 mL with methanol. Two milliliters of the reagent was mixed with 2–5 mL of the ether to be tested and a like amount of ether that had been distilled from sodium and allowed to stand for 5 min in subdued light. A red-blue reaction product develops when peroxides are present. The reagent is an amine that will form Schiff bases with aldehydes. If the ether contains aldehydes in concentrations of 1000 mg/L, or higher, interference will develop for the test.

Five different methods for determining the amount of peroxides in ethers were studied by Rothenbach [6285]. He found that the dichromate method was the least sensitive, that the potassium iodide test was somewhat better, and that the method using reduced alkaline phenolphthalein (sensitive to 0.1 mg/L) was the most sensitive. The basic titanium sulfate–sulfuric acid test was found suitable for both general qualitative and quantitative determinations, although the sensitivity was only one-hundredth that of reduced phenolphthalein.

A spectrographic method for the determination of hydroperoxides in ethers using acidic titanium tetrachloride was reported by Wolfe [8146]. The agreement with the iodometric method is good.

Wagner and coworkers [7829] report a modification of the iodometric method for peroxides that has a wide application. The iodine does not react with olefinic compounds. Add 40 mL of dry 2-propanol, 2 mL of acetic acid, and the sample (5 to 10 mL, depending on the peroxide content) to a 250-mL Erlenmeyer flask. Heat until reflux starts, add 10 mL of saturated 2-propanol–potassium iodide solution, reflux 5 min, add 5 mL of water, and titrate with 0.1 or 0.01 N sodium thiosulfate. Prepare the potassium iodide solution by refluxing 25 grams of the salt with 100 mL of 2-propanol. Blank determination on all reagents will be nil unless oxidizing impurities are present in the alcohol. A blank determination on each new lot of 2-propanol is sufficient.

Banerjee and Budke [588] describe a sensitive spectrophotometric method for the determination of traces of organic peroxides in organic solvents. Iodine is liberated from potassium iodide and measured at 470 nm. Active oxygen in the range of 5–80 ppm can be detected in a 1.0-cm cell; this range can be lowered to 1–5 ppm in a 1.5-cm cell at 410 nm. They extended their method, so that it is applicable to the parent method of Wagner and coworkers [7829] and, in addition to hydrocarbons, ethers, esters, alcohols, and so forth, include α, β-conjugated unsaturates such as acrolein, isoprene, and sorbic acid.

Inhibition of Peroxide Formation. Bailey and Roy [552] found that peroxide formation can best be inhibited by storing ethers over sodium amalgam in brown bottles.

In France, du Pont has patented a stabilizer for cyclic ethers, 4,4'-thiobis(6-*tert*-butyl-*m*-cresol) effective in concentrations of 0.01–1% [2145].

Hamstead and Van Delinder [3128] present an excellent discussion of autoxidation and inhibition of peroxides in ethers. They studied the inhibition of isopropyl ether and present information that should be valuable for the storage of all ethers.

Robertson and Jones [6200] state that aldehydes, ethers, and esters of unsaturated acids are stabilized from peroxide formation by the addition of 0.0001–1% of 2,5-bis(dimethylaminomethyl)hydroquinone.

Jones [3747] found 3,3',5,5'-tetralkylstilbenequinone to be a stabilizer for aldehydes, ethers, and esters of unsaturated acids.

N-Benzyl-*p*-aminophenol, 16 ppm; diethylenetriamine, 50 ppm; triethylenetetramine, 50 ppm; tetraethylenepentamine, 50 ppm; morpholine, 300 ppm; and ethylenediamine, 50 ppm were found effective for peroxide inhibition in isopropyl ether [1518]. The last two amines were the least effective.

Aliphatic Open-Chain Ethers (149–164)

149. Methyl Ether

Erlenmeyer and Kriechbaumer [2284] prepared methyl ether in 1874 by heating 1.3 parts of methanol with 2 parts of concentrated sulfuric acid solution at 140°. Similar preparations with methanol and concentrated acid solutions are given in Newth [5476] and Senderens [6573].

Kennedy and coworkers [3950] prepared methyl ether for *calorimetric studies* by adding methyl iodide to a solution of sodium in aldehyde-free methanol. The gas was dried over activated alumina and then the ether was distilled in a low-temperature column of 12 theoretical plates. Impurities were determined from a freezing point determination as 1 ppm.

SAFETY

Anesthesia is caused when the ether concentration is 65%v in air; profound anesthesia occurs at 85%v concentration [1174]. The LC_{50} in mice was reported as 494.4 ppm for a 15-min inhalation [1347].

The *flammable limits* in air are 3.3 and 27.3%v [2456]; see also [6402, 1609]. The *minimum ignition temperature* in air is 350° [5415, 6402].

150. Ethyl Vinyl Ether

Dolliver and coworkers [2035] passed acetal over 5% platinized asbestos at 280–290°. Repeated fractional distillation, followed by drying with calcium chloride, gave a pure ethyl vinyl ether.

Böhme and Bentler [943] prepared ethyl vinyl ether in 49% yield by refluxing 21.7 grams of 2-ethoxy-2-chloroethane with 36 grams of *N*,*N*-dimethylaniline on a water bath until the reaction starts and fractionally distilling the upper layer.

Wacker-Chemie [7812] reported the preparation of ethyl vinyl ether by stirring for 30 min at −15 to −8° 340 grams of vinyl acetate, 92 grams of ethanol, 3 grams of mercury acetate, and 3.05 grams of concentrated sulfuric acid. The mixture was cooled to −15°, and 128 grams of ethyl vinyl ether was distilled at 8 Torr over a 3-hr period; 96% yield, 90% conversion.

Twenty-four grams of 1-chloroacetyl ether was added to 40 mL of pyridine at room temperature. Fourteen grams distilled below 100° and was redistilled through a 12-cm Vigreux column to yield 6.8 grams of ethyl vinyl ether [3591].

Yuki and coworkers [8243] prepared ethyl vinyl ether in 70% yield by reacting equal moles of ethanol and methyl vinyl ether for 1 hr at 30° in the presence of molecular sieve type MS-4A and mercury(II) acetate catalyst.

Pilcher and coworkers [5857] purified ethyl vinyl ether for the determination of *heat of combustion* by fractional distillation from sodium in a still with a 50-cm column packed with glass helices. The center fraction was redistilled using a 1 in. × 5 ft column packed with stainless steel Knitmesh Multifil. The center fraction of 100 mL from a charge of 700 mL had a purity of 99.99%m determined in a melting point calorimeter. Portions were distilled from calcium hydride before use.

Water may be determined in ethyl vinyl ether by a modified Karl Fischer method.

SAFETY

Ethyl vinyl ether has a narcotic action similar to, but slightly weaker than, ethyl ether. Its action disappears faster and apparently does not produce any after effects [6689].

The *flammable limits* in air are 1.7 and 28%v [5415].

151. Ethyl Ether

The principal impurities found in commercial ethyl ether are ethanol, water, aldehydes, peroxides, and sometimes acetone. The following method of purification has been found to produce an ether of high purity satisfactory for exacting *analytical techniques*. Shake the ether with one-tenth its volume of 10% sodium bisulfite solution intermittently for 1 hr, withdraw the aqueous phase, and wash the ether with a saturated sodium chloride solution containing 0.5% sodium hydroxide. Then wash with a saturated sodium chloride solution containing a small amount of sulfuric acid, then with two portions of saturated sodium chloride solution, and finally distill in an atmosphere of nitrogen.

A substantially dry and alcohol-free ether may be prepared by fractional distillation. The distillate generally contains some aldehydes.

Rabinowitsch [6014] prepared ether for *specific conductance measurements* by distilling the ether over calcium chloride and then several times over sodium. Fresh distillate was always used for the measurements. Vogel [7763] determined some *physical properties* on ethyl ether purified in the same manner, but distilled only once from sodium.

Bruce [1180] prepared absolute ethyl ether by the action of flaked, technical-grade sodium hydroxide on USP ether for a period of 2 weeks. One pound of sodium hydroxide was used for each 3 L of ether. Comparison with commercial ether by the method of Forbes and Coolidge [2474] gave the following reaction times with strongly alkaline potassium permanganate: USP, 3 s; absolute ether, 12 s; and the product prepared with sodium hydroxide, 20 s.

Lamond [4387] states that ethyl ether may be rendered completely free of alcohol by washing with 10% sodium chloride solution. One liter of ether may be conveniently treated in a 2-L separatory funnel with three successive 300-mL portions of salt solution. The ether is then dried and distilled. Lassar-Cohn [4421] recommends drying the ether and refluxing for 24 hr with sodium potassium amalgam to remove alcohol. Aldehydes were removed by shaking with an alkaline solution of a mercury(II) salt for several days.

Ziebell [8289] freed ethyl ether from aldehydes by adding hydrazine hydrogen sulfate, phenylhydrazine, or $NH_2CSNHNH_2$ and distilling the ether from the nonvolatile hydrazone compounds.

It is reported that ethanol may be removed by repeated washing with water. Wade and Finnemore [7820] washed 700 mL of commercial ether 16 times with 50-mL portions of water before the iodoform test became negative.

For *optical measurements*, Castille and Henri [1389] treated ether for 12 hr with a 10% solution of sodium carbonate, dried over calcium chloride, and distilled slowly over sodium through a long column in an apparatus with ground joints; the first and last portions of the distillate were discarded. Ether purified in this manner was transparent in a 10-mm layer to 2050 Å.

Strecker and Spitaler [7034] treated ether with potassium permanganate and potassium hydroxide, with water, and then with 50% sulfuric acid, and dried over calcium chloride. The dried ether was shaken with a few grams of mercury, further dried over sodium, and fractionally distilled.

Scheibe and coworkers [6429] recommended shaking several times with sulfuric acid, sodium hydroxide solution, and water. After drying with sodium, the ether is distilled through an efficient column.

Clague and Danti [1497] purified ether by vacuum distillation over sodium for *infrared and Raman studies*.

Shinkarenko and Aleskovskiĭ [6676] purified ether; see pentane.

Werner [7983] added 4 grams of silver nitrate dissolved in 30 mL of water to 500 mL of ether in a liter bottle. Fifty mL of a 4% solution of sodium hydroxide was then introduced and the mixture shaken for 6 min. The ether was removed and found to be free of peroxides, aldehydes, and unsaturated substances.

A *reference standard* of ethyl ether for *infrared absorption* studied was prepared by Colon and Frediani [1600]. Residual ethanol and water were removed from absolute ether by drying with anhydrous calcium chloride and then adding sodium ribbon. The ether was repeatedly decanted and fresh sodium added until evolution of hydrogen ceased. It was filtered and distilled in an essentially dry system.

Gillo [2754] proposed a method for the purification and the criteria of purity for ethyl ether as a *standard*. Two methods of purification were recommended.

1. Ether is shaken with 50% sulfuric acid, distilled, dried 8–15 days over calcium chloride and for 1 month over sodium, and then distilled.
2. This method is the same as the first except that the ether is washed with a saturated potassium permanganate solution containing 5% sodium hydroxide. The final drying period over sodium is 2 months in the absence of light and air.

The products of these two methods differ in boiling point by 0.006–0.014°. The water and acetaldehyde content varied from 0.0008 to 0.0010%. The peroxide content of ether prepared by method 1 was about 1 mg/100 mL, that of the ether prepared by method 2 about 0.01–0.1 mg/100 mL. Ether is best preserved by purifying by the second method and storing in a sealed container in the absence of both air and light. Phosphorus pentoxide is a good preservative.

The preparation of ethyl ether as a polarographic solvent is described [7748].

Kablukov and Malischeva [3806] state that the presence of a third substance, even in insignificant amounts, may have a decided influence on the *mutual solubility* of ethyl ether and water. They washed ether five times with water, dried over calcium chloride, and distilled from sodium. A portion of 100–200 mL was redistilled from sodium for each determination; the first fraction being rejected.

CRITERIA OF PURITY

ACS Reagent Chemicals [137] and Rosin [6263] list specifications and test methods for both anhydrous ether and that containing 2–3% ethanol and 0.5% water.

Lamond [4387] modified the method of Duke and Smith [2117] for the determination of ethanol in ether. The results are accurate to 0.01 grams of ethanol in 100 mL of ether.

An infrared absorption method for determining ethanol in ethyl ether has been reported by Colon and Frediani [1600]. The water is first determined by the Karl Fischer method, and then ethanol and water are determined at 2.83 μm. Once the calibration curve has been established, a determination can be made in a short time with a relative accuracy of $\pm 2\%$.

Pesez [5783] has described a method for detecting acetaldehyde in ethyl ether that is sensitive to 10 ppm in 1 hr and 2 ppm in 12 hr.

Gas chromatography is a convenient method for the determination of impurities in ethyl ether.

SAFETY

The vapor pressure of ethyl ether at normal temperatures is high, the flammable limits in air are wide, and the minimum ignition temperature is low. Therefore the principal hazard with ethyl ether is considered its explosibility with air. Concentrations of ethyl ether in air of 3.6–6.5%v produce anesthesia in humans. Concentrations of over 10%v are usually fatal. Repeated-dose inhalation is often the result of intentional exposure to an ''ether jag,'' which results in loss of appetite, dizziness, headache, exhaustion, psychic disturbances, and damage or malfunction to some body organ [5713].

A *threshold limit value* of 400 ppm, 1200 mg/m^3, is suggested [7022].

Ethyl ether has wide *flammable limits* of 1.85 and 36.50%v in air. The *minimum ignition temperature* in air is low, 160° [1411].

152. Propyl Ether

Clague and Danti [1497] purified propyl ether by vacuum distillation over sodium for *infrared and Raman studies*.

SAFETY

Marsden and Mann [4902] state that detailed safety data are not available. Propyl ether is more toxic than ethyl ether. Exposure to vapors of 1%v does not seem to produce any ill effect other than an occasional intoxication.

Propyl ether is a highly flammable substance.

153. Isopropyl Ether

Nekrasov and Krentsel [5441] prepared isopropyl ether by passing 2-propanol vapor over an ascarite clay catalyst. Under optimum conditions, the yield was 52%w.

Tabuteau and Gautier [7149] absorbed propylene in concentrated sulfuric acid in the presence of a copper and copper sulfate catalyst. The ether was obtained by adding either 2-propanol or water.

Calderazzo [1309] dried tetrahydrofuran and isopropyl ether over sodium and distilled over lithium aluminum hydride and stored under nitrogen for *carbonylation of amines studies.*

Andon and coworkers [354] purified the ether by fractional freezing. The impurity content was 0.14%m.

Clague and Danti [1497] purified isopropyl ether by vacuum distillation over sodium for *infrared and Raman studies.*

For purification, see Ethers, General Comments [4091].

Peroxide formation in isopropyl ether is rapid. It is advisable to test for peroxides, even though it has been freed of them only a short time before. Katsuno [3893] found that the addition of 0.001%w of such dihydric phenols as pyrocatechol, resorcinol, and hydroquinone prevented the formation of peroxides in isopropyl ether for 6 months.

Hamstead [3127] conducted an extensive investigation of destroying peroxides in isopropyl ether. Five procedures were given.

1. In cold washing operations, high peroxide concentrations were completely removed by treatment with 25%w of triethylenetetramine and reduced to very low concentrations by treatment with the same amount of diethylenetriamine or propylenediamine.
2. Complete removal of peroxides was effected by refluxing the ether with 25%w of undiluted triethylenetetramine or diethylenetriamine.
3. Aqueous 20% solutions of hydroxylamine hydrochloride destroyed peroxides when used as a cold wash.
4. A 5% aqueous sodium metabisulfite solution, mixed one volume to five volumes of ether, was the most effective inorganic solution. It reacted rapidly with the peroxide when shaken for 15 min.
5. Tin(II) chloride, ~4%w/v, was the only solid reagent that effectively destroyed the peroxide.

SAFETY

The vapors of isopropyl ether in air are an explosive hazard. The physiological properties of isopropyl ether are similar to those of ethyl ether but the former is more toxic. It is not absorbed through the skin in harmful amounts. A concentration of 500 ppm in air produced no complaints after 15-min exposure to human volunteers, but 35% of the subjects objected to the odor at 300 ppm [5713, 7022].

Pavlova and coworkers [5723] reported a maximum permissible concentration of isopropyl ether as 100 mg/m^3.

A *threshold limit value* has been set at 250 ppm, 1050 mg/m^3 [142].

The *flammable limits* in air are 1.38 and 7.90%v. The *minimum ignition temperature* in air is 189° [5712].

154. Butyl Vinyl Ether

Amagasa and coworkers [111] prepared butyl vinyl ether by mixing 5 grams of 1-butanol and 1 gram of potassium hydroxide, and adding 40 grams of a solution of acetylene in liquid ammonia. The mixture was heated for 8 hr at 100° and extracted with chloroform.

Alkyl vinyl ethers were prepared by a continuous vapor-phase reaction by Kozlov and Chumakov [4233]. The reactor contained a zinc acetate or copper acetalide catalyst and was fitted with an acetylene meter and a device so that the alcohol could be dropped on the catalyst. The alkyl vinyl ethers were prepared from 1-butanol, 2-methyl-1-propanol, and 3-methyl-1-butanol. The yield of butyl vinyl ether at 200–220° was 47.6%.

Reppe [6133] reported on the reaction of acetylene and alkyl alcohols over soda lime in the vapor phase.

Schildknecht and coworkers [6442] developed an improved method of purification and determined more precisely additional *physical properties* of butyl vinyl ether. The ether was washed with water at room temperature to remove alcohols and the stabilizing alkali until no further volume decrease was noted. The pH was kept below 8. It was dried with a neutral or alkaline drying agent and, finally, by standing over sodium wire. The finished product was obtained by fractional distillation over sodium wire through a 10–20-plate column packed with glass helices. The distillate should be stored over potassium hydroxide or with triethanolamine to prevent hydrolysis or polymerization.

Water can be determined by a modified Karl Fischer method.

SAFETY

The oral LD_{50} for rats is 3.10 g/kg; the vapor exposure is 4 hr for no deaths, when the animal is exposed to vapors obtained by passing air through the ether at room temperature. A LD_{50} of 3.00 mL/kg was found for absorption through the skin of rabbits [5713]. Butyl vinyl ether tested on frogs showed a similar reaction to ethyl ether [6689].

155. Butyl Ethyl Ether

Vogel [7763] prepared butyl ethyl ether for *physical property measurements* from 1-butanol, sodium, and iodoethane. The ether was purified by repeated refluxing over sodium and then distilling.

Norris and Rigby [5517] prepared butyl ethyl ether from sodium butylate and bromoethane; Kranzfelder and coworkers [4240] used ethyl sulfate instead of bromoethane.

SAFETY

Six rats survived inhaling air containing 1000 ppm for 4 hr [5713].

Butyl ethyl ether may be determined in air with a sensitivity of 5 mg in 2.7 mL of solution [4354].

156. Butyl Ether

Vogel [7763] prepared butyl and higher homologous ethers for *physical property measurements* by removing the theoretical amount of water as the alcohol–water azeotrope from the alcohol–sulfuric acid mixture, and then steam-distilling the ether.

Smith and coworkers [6846] prepared butyl ether by the Williamson synthesis with

1-butanol, 1-chlorobutane, and sodium hydroxide in dimethyl sulfoxide. The yield was 95% alcohol converted to ether.

Technical-grade butyl ether was purified by Kusama and Koike [4337] by treating with carbon disulfide and sodium hydroxide. The excess sulfide was expelled by heating. The ether layer was washed with water, dried with sodium hydroxide, and vacuum-distilled.

Dasler and Bauer [1827] freed the ether of peroxides and aldehydes by adsorption of the impurities on alumina.

For purification; see Ethers, General Comments [4091].

SAFETY

Butyl ether seems to have a higher toxicity by inhalation than the lower members of the series. It is more irritating to the skin, eyes, and nose. Rats survived air saturated with the ether for 30 min. It has been estimated that 100 ppm, 533 mg/m^3 should be the highest concentration to which workmen should be exposed for an 8-hr day [5713].

Kosyan [4214] reported an LD$_{50}$ of 567 mg/kg for rats.

157.   Pentyl Ether

Vogel [7763] prepared pentyl ether in the same manner as he did butyl ether.

Kranzfelder and coworkers [4240] prepared pentyl ether from 1-pentanol and pentyl sulfate.

SAFETY

No specific data are available. The low vapor pressure of pentyl ether makes it a relatively safe solvent at ambient temperatures with normal handling care. One should avoid breathing the vapors at elevated temperature over a period of time [4902].

158.   Isopentyl Ether

Vogel [7763] prepared isopentyl ether in the same manner as he did butyl ether.

Schorigin and Makaroff-Semljanski [6482] freed isopentyl ether of the alcohol with boric acid.

Underwood and Toone [7605] decomposed the peroxides in isopentyl ether by shaking with potassium iodide solution.

Späth [6922] prepared alcohol-free isopentyl ether by boiling the commercial product for 5 hr under reflux with sodium followed by distillation. See also Zappi and Degiorgi [8262]. For the removal of 3-methyl-1-butanol from isopentyl ether with barium, see [6482].

For *electrochemical measurements*, Bruus [1211] dried isopentyl ether over calcium chloride, purified by careful fractional distillation, and then dried over sodium or phosphorus pentoxide.

159.   1,2-Dimethoxyethane

Capinjola [1346] prepared 1,2-dimethoxyethane by refluxing 1830 grams of 2-methoxyethanol and 138 grams of sodium until the latter dissolved. Chloromethane was passed into the liquid only as fast as it would react. The weight of the flask was used to determine when sufficient chloromethane was added. The portion that distilled below 123° was collected and fractionally distilled.

1,2-Dimethoxyethane was prepared in 81% yield by Kranzfelder and Vogt [4241] by treating the monomethyl ether in the presence of sodium with methyl sulfate.

Koyano and coworkers [4228] prepared 1,2-dimethoxyethane by reacting methanol and ethylene dichloride at 200° with magnesium hydroxide or zinc hydroxide as HCl scavengers; 85% of the dichloride was converted to the ether.

Commercial 1,2-methoxyethane was treated with lithium aluminum hydride by Wallace and Mathews [7881] and fractionally distilled immediately before making aqueous solutions to study some of their *physical properties*.

For the *study of solvent effect* of alkali salts of phenols and enols, 1,2-dimethoxyethane was purified by Zaugg and Schaefer [8265] in the same way they purified hexane.

Agami [44] discusses 1,2-dimethoxyethane as a solvent for inorganic, organometallic, and organic substances, and as a medium for several types of reactions.

The glycol diethers may contain inhibitors such as N-methylpyrrolidine, N-butylpyrrole, and similar compounds [8277].

SAFETY

1,2-Dimethoxyethane has an acute oral LD_{50} of approximately 7 g/kg for adult albino rats [7703].

160. 1,2-Diethoxyethane

Liston and Dehn [4630] reacted ethylene glycol ethyl ether with sodium and then converted the sodium compound to 1,2-diethoxyethane by adding ethyl chloride.

Gist and Mason [2771] prepared 1,2-diethoxyethane by adding ethyl chloromethyl ether slowly onto sodium over sodium-dried ether. After overnight stirring, the mixture was filtered, the solvent was removed, and the ether was distilled. A similar preparation using powdered copper is given in Teramura and Oda [7214].

Kusano [4339] purified 1,2-diethoxyethane by fractionation over a sodium wire.

SAFETY

The 4-hr inhalation of 8000 ppm was reported fatal to rats [1361].
The *minimum ignition temperature* in air is 208° [6402].

161. 1,2-Dibutoxyethane

Field and Hardy [2419] dissolved potassium hydroxide pellets in 2-butoxyethanol and added 1-bromobutane to this mixture. The fraction boiling in the range 170–195° was collected and washed first with nitric acid solution and then with water. The 1,2-dibutoxyethane was dried over calcium chloride.

For preparation of 1,2-dibutoxyethane from butyl chloromethyl ether, see Gist and Mason [2771] and Teramura and Oda [7214].

Liston and Dehn [4630] reacted ethylene glycol butyl ether with sodium and then converted the sodium compound to 1,2-dibutoxyethane by adding butyl chloride.

Kusano [4339] purified 1,2-dibutoxyethane by fractionation over a sodium wire.

SAFETY

The single oral LD_{50} dose was reported as 3.25 g/kg [6883].

162. Bis(2-methoxyethyl) Ether

Hamamoto and Tone [3109] prepared bis(2-methoxyethyl) ether by a general method for the preparation of bis-ethers of mono-, di-, or triethylene glycols. Ethylene oxide reacts readily with methanol or ethanol in the presence of sodium methoxide or ethoxide with the generation of heat (to about 90°). The reaction is complete in about an hour.

Gallaugher and Hibbert [2637] prepared bis(2-methoxyethyl) ether by the action of bis(2-chloroethyl) ether on sodium methylate in methanol at 25°. Thomas and Weisse [7421] used sodium hydroxide to bring about the same reaction.

Bis(2-methoxyethyl) ether starts to decompose at 60° [2639].

SAFETY

Bis(2-methoxyethyl) ether has an acute oral LD_{50} of approximately 7 g/kg for adult albino rats [388].

163. Bis(2-ethoxyethyl) Ether

Cretcher and Pittenger [1717] prepared bis(2-ethoxyethyl) ether by heating dichlorodiethyl ether with sodium ethylate in ethanol. After filtering out the sodium chloride, the ether was purified by distillation; 41% yield.

Turner and coworkers [7587] used bis(2-ethoxyethyl) ether as a solvent in *calorimetric studies* of aromatic compounds. Three distillation methods for purification were used. The first was done at 10 Torr and 85–86°. The second method was initiated by treating the ether with lithium aluminum hydride at 50° for 4 days, followed by distillation at 2–3 Torr and 65–66°. The third method was distillation through a helix-packed column under nitrogen at 37 Torr and 100°.

SAFETY

Union Carbide [7615] reported the single oral LD_{50} dose as 4.97 g/kg for rats.

164. Bis(2-butoxyethyl) Ether

Cretcher and Pittenger [1717] prepared bis(2-butoxyethyl) ether by heating dichlorodiethyl ether with sodium butylate in ethanol.

Milyukova and coworkers [5157] reported that bis(2-butoxyethyl) ether is used for the *extraction of plutonium* from aqueous solutions. Besides the advantages of high flash point and low solubility in water, this ether is stable to nitric acid, which is used as a salting-out agent.

Morris and Ali Khan [5265] described a purification process for bis(2-butoxyethyl) ether used for the *extraction of gold* from aqueous solutions. The ether was treated twice with 2% potassium permanganate in aqueous sulfuric acid solution. After washing the organic layer with water, the ether was treated with 30% iron(II) sulfate in sulfuric acid solution. Another washing with water was followed by treatment of the ether with 5% sodium carbonate solution and finally with 4 *M* hydrochloric acid solution.

Rimmer [6182] described a production plant for *extraction of gold* from hydrochloric acid solution with bis(2-butoxyethyl) ether.

Union Carbide [7615] reported the single oral LD_{50} dose as 3.90 g/kg for rats.

Aliphatic Cyclic Ethers (165–173)

165. Propylene Oxide

Propylene oxide was purified by Oetting [5561] for *heat capacity measurements* by drying with Drierite and fractionally distilling in a 100-plate still. Melting curves showed the purity to be 99.94%m. The sample was transferred to a high vacuum line, degassed, and loaded into the sample container by a one-plate distillation. Helium was admitted to the container before sealing.

Price and Osgan [5951] purified commerical propylene oxide by refluxing for 2 hr over potassium hydroxide, then fractionally distilling, and discarding the first and last 15%.

Barlett [647] described a method for purifying epoxides containing 2–12 carbon atoms by treating with ammonia, amines, hydrazine, a hydroxylamine, or a semicarbazide in such amounts that not more than 10% of the epoxide will be consumed.

Levine and Walti [4547] prepared *dextro*-propylene oxide: 40 mL of water and 19 grams of sodium hydroxide were put in a double-necked distilling flask immersed in a 40° water bath. The flask was provided with a condenser through which 0° water circulated. 3-Bromo-1,2-propanediol was added and the temperature was raised to 60°. The flask was shaken vigorously, while the propylene oxide was distilled.

Rogers [6230] fractionally distilled a commercial sample of propylene oxide through an efficient column for determination of the *dipole moment*. The boiling point, refractive index, and density were used as the criteria of purity.

Abderhalden and Eichwald [7] prepared *dextro*-propylene oxide from *d*-2-chloro-1-propanol.

A crystalline hydrate, $C_3H_6O \cdot 16H_2O$, mp $-3°$, has been reported by Zimakov and Sokolova [8299].

Proplylene oxide is a highly reactive substance. The liquid is relatively stable. It reacts exothermally with any substance that has a labile hydrogen such as water, alcohols, amines, and organic acids. Certain substances, such as acids, alkalies, and certain salts, act as catalysts. It also polymerizes exothermally. No acetylide-forming metals, such as copper or copper alloys, should be in touch with propylene oxide.

Physiologically, it is primarily an irritant, a mild protoplasmic poison, and a mild depressant of the central nervous system. Contact with the skin, even in dilute solutions, may cause irritation and narcosis. Excessive exposure to the vapor may cause irritation to the eyes, respiratory tract, and lungs [5713].

All animals tolerate 100 ppm without adverse effects.

A survey [925] of the several methods for the determination of propylene oxide in air indicates that the method based on oxidation to formaldehyde is the most rapid and sensitive.

The oral LD_{50} for rats has been reported as 550 mg/kg [1247].

The *threshold limit value* of 100 ppm, 240 mg/m^3 has been suggested [7022].

The *flammable limits* in air are 2.1 and 21.5%v [5415, 6402]. The *minimum ignition temperature* in air is 421° [5712].

166.   1,3-Dioxolane

Verley [7729] prepared 1,3-dioxolane in 1899 by heating 100 grams of ethylene glycol with 50 grams of 40% formaldehyde solution and 50 grams of syrupy phosphoric acid over a water bath and steam-distilling the ether. Delepine [1933] and Clarke [1507] prepared the ether by reacting ethylene glycol and polyoxymethylene in acidic media.

Brikenstein and coworkers [1098] obtained 1,3-dioxolane in 90% yield by mixing ethylene glycol, paraformaldehyde, and sulfuric acid solution.

Pihlaja and Heikkila [5852] synthesized 1,3-dioxolane for *calorimetric studies* by boiling ethylene glycol and paraformaldehyde with *p*-toluenesulfonic acid in toluene. After the mixture was neutralized with potassium carbonate solution, the ether was separated by distillation. Any aldehydes were removed by allowing the crude product to stand over a saturated solution of sodium bisulfite. The final product was obtained by distillation over sodium; purity 99.98%m.

Barker and coworkers [621] purified 1,3-dioxolane for *vibrational spectral studies*. A commercial sample was refluxed for 2 hr with lead(IV) oxide, cooled, and filtered. Xylene and more lead(IV) oxide were added and the mixture was fractionally distilled. The main fraction, collected at 70–71°, was treated with xylene and sodium wire and then redistilled. More sodium was added to the product and finally a sample boiling at 74–75° was collected.

SAFETY

Stasenkova and coworkers [6970] reported that 1,3-dioxolane has low toxicity and recommended a maximum permissible concentration in air as 50 mg/m^3.

The *minimum ignition temperature* in air is 274° [2903].

167.   1,2-Epoxybutane

Houtman and coworkers [3511] describe a procedure for straight-chain epoxybutane substantially free of isobutylene oxide where the oxygen is attached to the adjacent carbon. Agitate to create an intimate depression of the butylene oxide and add 30–50% aqueous alkali hydroxide, allow the phases to separate, draw off the aqueous phase, and fractionally distill the oxide. The small amount of water and isobutylene oxide remaining will distill as the azeotrope.

De Montmollin and Matile [1948] prepared 1,2-epoxybutane for *infrared studies* by treating 2-chloro- or 2-bromo-1-butanol with sodium hydroxide; bp 58.5–59°.

SAFETY

The *explosive limits* in air are 3.1 and 25.1%v. The *minimum ignition temperature* in air is 515° [2072].

168.   Cineole

Cineole is a syrupy oil with a camphor-like odor and a pungent taste. It occurs in many essential oils. The oils from eucalyptus species contain from 30–92% cineole; hence the synonym, *oil of eucalyptol*. It is also found in oils of ginger, cajuput, niouli, cardamom, spike lavender, laurel leaf, rosemary, and certain of the *Ocimum artemesia* and *O. alpina* [3013]. It has been reported in oil from leaves of *Chamaelyparis obtusa formosana* [3877].

If the cineole content of the oil is high, it can be separated by fractional distillation. The 170–180° fraction is cooled and the cineole separated by fractional crystallization. The first crystallization is reported to be of high purity. The authors experienced some difficulty in fractionally distilling California bay tree oil at atmospheric pressure, but cineole and other components were separated without difficulty at reduced pressure. Johnson [3731] has described a method for separating cineole from hydrocarbons by fractional azeotropic distillation with phenol.

Cineole forms addition compounds with hydrogen halides. The hydrochloride compound is quite soluble in the other constituents of the essential oil, but the hydrobromide is less soluble and therefore suitable for isolation. The procedure is as follows: Dilute refined cineole, bp 175–180°, with an equal volume of petroleum ether, and saturate the solution with dry hydrogen bromide. Filter the white precipitate, wash with several small portions of petroleum ether, and regenerate the cineole by stirring the crystals with water.

If the cineole content of the oil is high, it may be isolated by forming the addition compound with resorcinol. Low cineole oils first must be fractionally distilled. The details of this and other methods are given by Guenther [3013].

CRITERIA OF PURITY

See Guenther [3013].

SAFETY

Cineole is used widely in pharmaceuticals for both internal and topical application.

169.   Furan

Furan can be prepared by treating 2-furaldehyde with steam at elevated temperatures in the presence of a catalyst containing a dehydrogenating oxide, a mixture of dehydrogenating oxides, or a chromite of a dehydrogenating oxide [2137]. One volume of 2-furaldehyde and five volumes of steam were passed over manganese chromite pellets containing potassium chromate at about 400°. After stripping, drying, and redistilling, one-half of the theoretical amount of pure furan was obtained.

Katsuno [3894] prepared furan by converting 2-furaldehyde to 2-furoic acid and furfuryl alcohol by the Cannizzaro reaction. The furoic acid was decarboxylated to furan in 83–88% yield. See also Whitmore [8021, p. 40].

A method has been described by Wilson [2757] for the preparation of furan by the decarboxylation of 2-furancarboxylic acid.

Guthrie and coworkers [3037] distilled technical furan through a 70-plate column packed with steel helices under an atmosphere of nitrogen. Hydroquinone was added to each cut as an inhibitor. The heart cuts were 99.98%m. Mass spectra indicated 0.01%m isopentane and a trace of an unidentified impurity.

For purification, see Ethers, General Comments [4091].

Cheng and coworkers [1451] purified furan for *critical temperature studies* by distilling commercially available material, treating with 5% potassium hydroxide, drying over anhydrous sodium sulfate, and distilling *in vacuo* over sodium; bp 31.33° at 765 Torr.

Cass [1382] reported that 0.01 to 1.0% hydroquinone will stabilize furan against polymerization and color formation.

SAFETY

The Quaker Oats Company [6005] states that the concentration of furan vapors in the air always should be kept to a minimum both as a health and as a fire precaution. Chronic exposure of dogs to concentrations in air of 10 ppm produces noticeable circulatory disturbances. Furan is readily absorbed through the skin. Continual exposure of humans to low concentrations may cause fatigue, headache, and gastrointestinal disturbances.

Stasenkova and coworkers [6969] reported a *threshold limit value* as 0.5 mg/m^3.

The *explosive limits* in air are 2.3 and 14.3%v [6005].

170. Tetrahydrofuran

Ivanskii and Dolgov [3652] developed two catalysts: (1) 60% Cu, 15% Al_2O_3, 25% SiO_2, and (2) 60% NiO, 15% Al_2O_3, 25% SiO_2 for converting 1,4-butanediol to tetrahydrofuran and γ-butyrolactone. The catalysts were made by adding sodium hydroxide and sodium silicate to a hot solution of the metal nitrates, drying the precipitate, heating to 250–450°, and reducing in a stream of hydrogen. At 190°, catalyst 1 converts 1,4-butanediol quantitatively to γ-butyrolactone. At 190–260°, catalyst 2 converts the diol quantitatively to tetrahydrofuran.

Cyanuric chloride and 1,4-butanediol will form tetrahydrofuran, according to Matuszko and Chang [5001]; see also tetrahydropyran.

Nalesnik and Holy [5391] oxidized 1,4-butanediol in the presence of palladium chloride catalyst to produce tetrahydrofuran in 97% yield; see also tetrahydropyran.

Starr and Hixon [3733, p. 77] prepared tetrahydrofuran by catalytic hydrogenation of furan using palladium oxide.

Mikhailov [5118] prepared furan by catalytic decarboxylation of furaldehyde in hydrogen or steam. Hydrogenation of the furan over nickel, palladium, or osmium gave tetrahydrofuran.

Kirrman and Hamaide [4035] prepared tetrahydrofuran in high yields by heating a *gamma-, delta-* or *epsilon*-methoxylated alkyl bromide with a few milligrams of iron(III) chloride.

Furan containing 0.01% dibutylamine was hydrogenated in the vapor phase to tetrahydrofuran at 85° in almost quantitative yield by Manly [4874].

Arnett and Wu [438] report that, despite the widely different nature of acidic systems, the same order of basicity is always found for the cyclic ethers. When 2 pKs are plotted against any other function of basicity, linear correlations are nearly always obtained.

The cyclic ethers are generally five to ten times more basic than the simple acylic ones. Tetrahydrofuran is very basic in all systems, which probably explains, at least in part, its superior solvent ability among the ethers for organometallic systems.

Commercial tetrahydrofuran was distilled from sodium for *vapor–liquid equilibrium studies* [2689].

For purification, see Ethers, General Comments [4091]; for *carbonylation of amines studies*, see Calderazzo [1309] under isopropyl ether.

Bhattacharyya and coworkers [849] purified tetrahydrofuran for *conductance studies* by refluxing overnight with sodium–potassium alloy and fractionally distilling. In some cases, the ether was further purified by stirring with the alloy, to which benzophenone had been added, and then vacuum-distilling.

Commercial tetrahydrofuran was prepared for electrochemical studies by Perichon and Buvet [5759] by allowing it to stand for 15 days over potassium hydroxide and then over naphthalene sodium for several hours. It was distilled in an inert atmosphere, re-fluxed over lithium aluminum hydride, and distilled directly into the electrolytic cell in an argon atmosphere. They discuss the *possibilities and limitations of electrochemical reactions* in this ether.

Coates [1545] discussed a reported explosion when tetrahydrofuran was refluxed with calcium hydride and then described the removal of peroxides with sodium hydroxide. For samples containing more than 0.5%w peroxide, the caustic should be added in small increments.

Tetrahydrofuran was purified for the study of the *solvent effect on ultraviolet spectra* of alkali phenols and enols by Zaugg and Schaefer [8265] in exactly the same way that they purified cyclohexane.

Pickett and coworkers [5841] purified tetrahydrofuran from three sources for *vacuum ultraviolet absorption studies* by drying over calcium sulfate and metallic sodium, and then fractionally distilling through a 30-cm Penn State glass helices packed column. Because of the unique appearance of the spectrum and the possibility that impurities might be present, two samples were prepared by the catalytic hydrogenation of furan.

Cheng and coworkers [1451] purified tetrahydrofuran for *critical temperature studies* by distilling commercial material, allowing it to stand 48 hr over freshly fused sodium hydroxide and 24 hr over sodium wire, over which it was refluxed. It was fractionally distilled in an atmosphere of dry nitrogen, and finally vacuum-distilled from lithium aluminum hydride; bp 66.2–66.4° at 766 Torr.

Paul and coworkers [5715] purified tetrahydrofuran as a *solvent for sodium metal resection study* with aromatic hydrocarbons as follows: It was distilled from a mixture of activated alumina and sodium hydroxide pellets into a receiver containing sodium chips and anthracene. It was allowed to stand at least 1 hr and distilled into a receiver containing the reactants in active form.

An anonymous note [381] suggested that the method of purifying tetrahydrofuran with potassium hydroxide be avoided due to danger of explosions, particularly from perox-ides. An alternate method was distillation under nitrogen from lithium aluminum hy-dride, provided that the sample is peroxide-free and dry. The purified compound should be stored under nitrogen with 0.025%w of 2,6-di-*tert*-butyl-4-methylphenol if storage is more than a few days.

Hernandez [3329] purified tetrahydrofuran for *ultraviolet studies* by repeated vacuum distillation, retaining the middle cut each time. Samples were stored in vacuum until ready for use.

Morikawa and Yoshida [5255] studied the purification and aging of tetrahydrofuran, and identified eight aging products: namely, tetrahydrofuran α-hydroperoxide, butyraldehyde, 4-hydroxybutyric acid, 4-hydroxyperbutyric acid, butyric acid, 4-hydroxybutyraldehyde, α-oxotetrahydrofuran, and water.

A sample of du Pont tetrahydrofuran was analyzed by gas chromatography for *azeotrope studies* [6165]. It was found to be 99.9% with less than 0.1% water and traces of two impurities that eluted prior to the ether.

Some tetrahydrofurans contain impurities such as 2- or 3-*tert*-butyl-4- hydrofuran, or 0.025% butylated hydroxytoluene, introduced to inhibit peroxidation.

Bordner [992] found 0.01 to 1.0% *p*-cresol will inhibit peroxide formation in tetrahydrofuran. Hinegardner [3412] found 0.05 to 0.10% hydroquinone effective.

A photometric method [8310] is given for the determination of peroxides in tetrahydrofuran that compares favorably with the iodometric method; range 0.002 to 0.8%.

SAFETY

The Quaker Oats Company [6004] states that tetrahydrofuran is only moderately toxic; normal precautions recommended for volatile solvents are satisfactory.

Tetrahydrofuran has a hypnotic effect at 6%. The onset is slower then ethyl ether and the effect is longer lasting. Jockmann reported 20–50% aqueous solutions irritated the skin of rabbits. However, it has been concluded that it is not irritating to the skin and is not a skin sensitizer [7022].

The *threshold limit value* of 200 ppm, 590 mg/m^3, has been recommended [7022]; see also [140].

The *explosive limits* in air are 1.8 and 11.8%v at 25°. The *minimum ignition temperature* in air is 321° [6004].

A TLV of 100.0 mg/m^3 was suggested in 1964 for Russia [4185]. Pozdnyakova [5939] reported the LD$_{50}$ as 2.3 g/kg for mice and guinea pigs and 3.0 g/kg for rats. Katahira [3876] reported that the median lethal concentration upon a 3-hr inhalation by rats is 21,000 ppm.

Deyanova [1979] reported a method for the determination of tetrahydrofuran in the air; sensitivity, 1 μg/mL solution; on a sample collected by adsorption on silica gel.

171. Methyltetrahydrofuran

Lipp [4621] prepared 2-methyltetrahydrofuran in 1889 by heating one part 1,4-pentanediol with three parts 60% sulfuric acid solution for 1 hr in a pressure tube at 100°. An equal volume of water was added to the mixture which was then distilled. The final product was distilled over potash.

Whitehead and coworkers [8016] prepared 2-methyltetrahydrofuran by hydrogenation of 2-methylfuran at 175–185° over Raney nickel and fractionation of the crude product in a 50-plate column. Bolestova and coworkers [950] carried out a similar preparation with an ionic hydrogenation of 2-methylfuran.

Paul and Normant [5717] distilled 4-penten-1-ol with concentrated sulfuric acid solution to give 88% yield of 2-methyltetrahydrofuran.

172. *p*-Dioxane

Urazovskii and Chernyavskii [7642] determined the viscosity of benzene, *cis*- and *trans*-decahydronaphthalene, and *p*-dioxane over a wide temperature range and plotted the data as log η versus $1/T$. Only the data for *trans*-decahydronaphthalene gave a straight line. Data for the other compounds gave two intersecting straight lines. Explanations are offered. Urazovskii and Ezhik [7643] continued the study of the behavior of *p*-dioxane and found that the temperature dependence curves of several properties exhibit an anomaly at about 26°. The dipole moment is negligible at 20° but is about 0.30 at 30°. They conclude that the *p*-dioxane molecule started to assume the boat form above this temperature.

Hess and Frahm [3354] found that the chief impurities in commercial *p*-dioxane are acetic acid, water, and glycol acetal ($CH_3CH\!-\!O\!-\!CH_2CH_2O$). Depending upon the composition of the dioxane, heating with hydrochloric acid may be sufficient to remove the acetal. When the acetal is present in only small amounts, prolonged heating with sodium is adequate. Additional purification is effected by careful fractional distillation and crystallization. Once purified, *p*-dioxane is stable indefinitely when protected from atmospheric oxygen.

According to Hepworth [3310], *p*-dioxane prepared from 1,2-ethanediol by heating with concentrated sulfuric acid contains aldehydes. These can be removed by repeated boiling with silver oxide. After several distillations over freshly fused potassium hydroxide, it is fractionally distilled over sodium.

Rieche [6170] and Milas [5125] state that *p*-dioxane is auto-oxidized, but according to experiments of Eigenberger [2228] it is not auto-oxidized when it is pure and free from acetal. As the best method for the preparation of acetal-free *p*-dioxane, they recommended that ethylene acetal be decomposed by acid, whereby a better yield is obtained than by fractional distillation. Crude *p*-dioxane is boiled under reflux for 7 hr with 10% of its volume of 1 *N* hydrochloric acid, a slow stream of air being passed in through the condenser to remove the acetaldehyde formed. To remove water, the distilled *p*-dioxane is treated with potassium hydroxide, and the aqueous layer removed. It is then allowed to stand over pellets of potassium hydroxide for 1 day and dried over sodium for several hours. It is then distilled from the sodium.

p-Dioxane was purified for *dielectric constant studies* by the method of Hess and Frahm [3354], except that recrystallization was omitted. It had a freezing point of 11.73° and a purity of not less than 99.95%m from the freezing curve [1721].

Vinson and Martin [7748] allowed commercial-grade *p*-dioxane to stand over calcium chloride for several weeks for *heat of vaporization* and *vapor pressure studies*. The boiling temperature showed no rising or declining trend over several consecutive determinations; mass spectral analysis showed 0.04% water.

Smith and Wojciechowski [6824] purified a good commercial grade of *p*-dioxane by distilling through a 40-bulb Swietoslawski-type column of about 30-theoretical plates. McKinney and coworkers [4781] purified by refluxing over sodium and distilling through a 24-in. column packed with Raschig rings.

Kraus and Vingee [4249] purified *p*-dioxane as a solvent for a *study of nonpolar solvent solutions of electrolytes* as follows: The *p*-dioxane was refluxed over sodium hydroxide, distilled, and refluxed for an extended period over fresh sodium, fresh sodium being added from time to time.

Kraus and Fuoss [4248] boiled technical p-dioxane with sodium hydroxide, dried over barium oxide, and fractionally distilled from a sodium-lead alloy. The middle fraction was dissolved in liquid ammonia and the ammonia boiled off. The p-dioxane was redistilled, stored over sodium-lead alloy, and distilled only when needed.

Oglukian [5565] discovered, when *following kinetic reactions by ultraviolet spectroscopy* in p-dioxane as the solvent, that decomposition of the solvent was causing an increase in absorbency in the spectral region of interest. A study of the purification was made, including more than 50 published methods. The following method was evolved.

Step 1.   p-Dioxane, any good grade, was let stand over iron(II) sulfate for at least 2 days.

Nitrogen was used continually in all subsequent operations at a light bubbling rate unless otherwise specified. It was water-pumped grade and passed through a concentrated sulfuric acid drying tower, then through a 1 × 24-in. tube of sodium hydroxide pellets.

Step 2.

(a) 200 mL of water and 28 mL of concentrated hydrochloric acid were added to 2 L of p-dioxane. A pale yellow color developed.

(b) It was refluxed not less than 8 nor more than 12 hr with heavy nitrogen bubbling.

(c) Potassium hydroxide pellets were added to the solution, while still warm, until two layers formed. The yellow color disappeared. The contents of the flask were cooled, often rapidly in an ice bath.

(d) It was magnetically stirred and potassium hydroxide pellets added until no more dissolved. The solution was cooled when the flask became hot to the touch.

(e) The flask was allowed to sit for 4–12 hr until the two phases were sharply separated. If the lower phase was black, the lot was discarded and the purification started again with step 1.

(f) The upper phase was decanted rapidly into a clean flask and then into a still pot containing sodium.

Step 3.   The p-dioxane was refluxed in a still with an 18-plate Oldershaw column until the sodium remained very shiny for some time. After the system had been under total reflux for an hour, 30–50 mL were slowly distilled until the head temperature reached 101° with the nitrogen bubbling shut off during the reading. The spectrum was checked every 4–6 hr after the appearance of the shiny sodium to assure that a minimum absorbency below 2500 Å had been attained. The material was then fractionally distilled into a round bottom flask and stoppered.

Step 4.   The distillate was recrystallized three times, discarding 15–20% each time. The flask was placed in an 8° refrigerator and shaken frequently or magnetically stirred. Nitrogen was not bubbled through the p-dioxane during the freezing. It was bubbled through during the melting. If the spectral cut-off was higher than 2170 Å, a crystallization would not help and the process was started over. By the third crystallization, there was little or no spectral difference between the mother liquor and the melted solid therefrom.

The material was stored in a refrigerator until ready to use. It was thawed, refluxed over sodium for 48 hr, and distilled as needed directly into the container in which it was to be used. The first 5–10 mL were discarded from each lot distilled.

All joints were Teflon-tape-clad and *p*-dioxane was its own lubricant for the stopcocks.

The *p*-dioxane has zero absorbency at 2200 Å.

The density at 25° was 1.02803. It was very hygroscopic, so that any operation requiring exposure to air was difficult.

Katti and Chaudhri [3895] purified AR grade *p*-dioxane for *physical property of mixtures studies* by refluxing with hydrochloric acid, treating with potassium hydroxide, allowing to stand over sodium and fractionally distilling; d 30° 1.0232.

For purification, see Ethers, General Comments [4091].

Palyi and Peter [5660] prepared *polarographically pure p*-dioxane from a commercial product. It was mixed with concentrated hydrochloric acid in a ratio of 5:2 ether to acid and refluxed at 120–130° for 10 hr under nitrogen The resinous material was separated by filtration when cool and the filtrate neutralized with solid sodium carbonate and 25 grams of sodium hydroxide per liter. The aqueous phase was allowed to separate; 30 grams of iron(II) sulfate or chloride was added per liter of ether. The mixture was shaken mechanically for 3–5 hr. The solids were filtered and the two phases of the filtrate allowed to separate. The *p*-dioxane was separated from the aqueous phase. It was distilled twice from solid hydroxide and once from sodium under an oxygen and water free atmosphere. The purified material was stored in a glass bottle.

p-Dioxane was purified by Brown [1170] as a solvent for *dipole moment studies* by drying with anhydrous magnesium sulfate, refluxing with sodium, and fractionally distilling. It was stored over sodium wire; d 24° 1.0292.

Sideri and Osol [6707] reported that *p*-dioxane that had been shaken with 20 grams per liter of Powminco brand asbestos (used for filtering mats) did not develop a color when used as a *solvent for standard perchloric acid titrant*. Riddick [6165] checked this purification and found one lot of Powminco brand asbestos out of four was not effective. No other brand of asbestos was found to be effective.

Levi and coworkers [4553] found that Amberlite IRC 50 was effective for removing the objectionable impurities from *p*-dioxane that caused discoloration when used as a solvent to prepare *standard perchloric acid solutions*.

Fisher and Baxter [2447] found that the peroxide content of *p*-dioxane could be decreased either by shaking with tin(II) chloride and distilling, or by refluxing with lead(IV) oxide and filtering through a slow filter paper.

SAFETY

The liquid is painful to the eyes and irritating to the skin upon prolonged contact. It can be absorbed through the skin in toxic amounts. *p*-Dioxane is insidious. Its vapors have poor warning properties; they are faint and inoffensive. Concentrations in air of 300 ppm cause irritation of the eyes, nose, and throat of humans. The vapors can be inhaled in amounts that cause serious systemic injury, particularly to the liver and kidneys [5713]. Johnstone [3741] reports a death that was supposedly due to *p*-dioxane after exposure during working hours for 1 week to concentrations of 208–650 ppm in an unventilated room. It had been used to dissolve glue from the victim's hands.

The *threshold limit value* has been set at 100 ppm, 360 mg/m^3, and is believed to be low enough to prevent systemic poisoning with a good margin of safety [7022].

The *flammable limits* in air are 1.97 and 22.25%v [5713]. The *minimum ignition temperature* in air is 180° [7624].

173. Tetrahydropyran

Dihydropyran containing 0.01% aniline was hydrogenated to tetrahydropyran in the vapor phase at 85° over a highly active nickel catalyst in quantitative yield by Manly [4874]. Andrus and Johnson [6482] present a laboratory method for hydrogenation in the liquid phase that is almost quantitative.

Allen and Hibbert [90] prepared tetrahydropyran in 90% yield for *electrical moment measurements* by heating at 150° for 4 hr in a sealed tube 25 grams of purified 1,5-dibromopentane, 25 grams of distilled water, and 8 grams of pure zinc oxide. The reaction mixture was extracted three times with ether and dried over sodium sulfate. The fractions from three preparations boiling at 85–90° were further purified by fractional distillation.

Matuszko and Chang [5001] studied the reaction of cyanuric chloride and diols. Tetrahydropyran was prepared in 77% yield from 1,5-pentanediol; see also tetrahydrofuran.

Nalesnik and Holy [5391] oxidized 1,5-pentanediol in the presence of palladium chloride catalyst to produce tetrahydropyran in 100% yield; see also tetrahydrofuran.

Kirrmann and Hamaide [4035] prepared tetrahydropyran in high yields by heating *gamma-*, *delta-*, or *epsilon-* methoxylated alkyl bromide with a few milligrams of iron(III) chloride.

Pickett and coworkers [5841] purified two samples of tetrahydropyran for *vacuum ultraviolet absorption studies* by drying over calcium sulfate; and refluxing over, and distilling from, sodium. Both samples were found to contain benzene (0.0005% estimated) which was removed by passing through a graphon column.

Hernandez [3329] purified tetrahydropyran for *ultraviolet studies* by repeated vacuum distillation retaining the middle cut each time. Samples were stored in vacuum until ready for use.

Aromatic Ethers (174–179)

174. Benzyl Ethyl Ether

Braun [1065] prepared benzyl ethyl ether by the action of ethanol on benzyl bromide and removed the unchanged bromide by adding a small quantity of a base, for example, piperidine, warming for a short time, and shaking with dilute hydrochloric acid. After drying over calcium chloride, the ether was fractionally distilled.

White and coworkers [8007] prepared the ether by the action of potassium ethylate on benzyl chloride. The ether solution was dried with sodium hydroxide and fractionally distilled.

The unchanged ester in benzyl ethyl ether, prepared by electrolytic reduction of ethyl benzoate, can be removed by boiling with alcoholic potassium hydroxide [7152].

For *optical measurements*, Baly and Collie [580] purified the ether by fractional distillation until the boiling point was constant.

According to Schorigin [6481] benzyl ethyl ether is decomposed by sodium at 140°.

175. Anisole

King and Wright [4014] prepared anisole in 75% yield by adding 3.91 grams of potassium to 30 mL of ice-cold methanol and then adding 9.4 grams of phenol and 19.4 grams of methyl phthalate. The excess methanol was removed by warming at reduced pressure and the syrupy liquid was refluxed with an air condenser in an oil bath (a heating mantle may be used) at 190–200° for 3 hr and allowed to cool. The solid mass was dissolved in water and ethyl ether. The ether soluble portion was washed with 2 N sodium hydroxide and fractionally distilled.

Hiers and Hager [2757] prepared anisole from phenol, sodium hydroxide, and methyl sulfate.

Anisole was prepared in the same manner as phenetole in 83% yield [2080]; see phenetole.

Cullinane and Davies [1746] treated phenol with a methylating agent, such as methyl ether or methanol, in the presence of activated alumina at 175–225°. By recycling the phenol, the yield of anisole was increased to 43%. See also Conant [1619] and Vogel [7763].

Vaughn and coworkers [7695] purified anisole for *dielectric relaxation studies* by distilling from sodium and storing over sodium sulfate. Roberti and Smyth [6193] distilled from barium oxide.

SAFETY

Data are not available to suggest a *threshold limit value*. The absence of reported adverse effects indicates that there probably is little need for this information [5713].

176. Phenetole

Phenetole was prepared by Drahowzal and Klamann [2080] in 79% yield by heating phenol and ethyl *p*-toluenesulfonate in the presence of sodium hydroxide. The organic product was extracted with ether, washed with dilute sodium hydroxide, dried, and distilled.

Phenetole may be prepared in 66% yield by the same method used for anisole [4014] by substituting an equivalent amount of ethyl phthalate for methyl ether.

For preparation from sodium phenolate and bromoethane in liquid ammonia, see [8007].

SAFETY

The limited information that is available indicates that phenetole is quite low in acute toxicity. It is not believed that this ether would present any unusual hazards [5713].

177. Benzyl Ether

Bennett and Willis [765] refluxed 30 grams of benzyl alcohol and 25 grams of α-chlorotoluene with 25 grams of potassium hydroxide for 5 hr and fractionally distilled to prepare benzyl ether for *freezing point studies*. Repeated distillation at reduced pressure did not reduce the odor of benzaldehyde. It was concluded that the ether decomposed even at temperatures below 200°. It was further purified by fractional freezing. The purified substance had a faint odor quite distinct from that of the aldehyde.

It was found necessary to seed benzyl ether when using it as a cryoscopic solvent.

178. Phenyl Ether

Yamamoto [8202] prepared phenyl ether by treating phenol with iodobenzene over a copper catalyst at 220–230° for 12 hr and obtained a yield of 94%.

Phenyl ether has been suggested by Furukawa and coworkers [2615] of the National Bureau of Standards as a *calorimetric standard*. They accurately determined the *thermal properties* of a high-purity material. It is a relatively stable substance and can be obtained in a high state of purity by fractional distillation, followed by fractional crystallization. It melts close to room temperature, 26.87°.

The phenyl ether used in the thermal studies was fractionally distilled and then fractionally crystallized 25 times. Dissolved air and water were removed by slowly freezing in a glass bulb during evacuation. The material was remelted and the freezing-evacuation process repeated. It was then distilled into another bulb and the freezing–evacuation process repeated. The purity was determined from melting point studies.

Phenyl ether decomposes 1%m per hr at 538° [3727].

Merz and Weith [5063] prepared phenyl ether by heating phenol with zinc chloride.

SAFETY

Hake and Rowe [5713, pp. 1698 ff.] state that information regarding the physiological effects of phenyl ether is not extensive. It appeared to be relatively nonirritating to the skin and of a low order of toxicity. Dowtherm A is an eutectic mixture of phenyl ether and biphenyl. It is used as a heat-transfer medium below 750°F (399°). Industrial experience with this eutectic is a good criterion of the hazards of phenyl ether. Its odor is disagreeable and it sometimes causes nausea. Pecchiai and Saffotti [5740] investigated the oral toxicity.

It is not an explosion or fire hazard due to its low vapor pressure.

The *threshold limit value* probably should be 1.0 ppm, 6.79 mg/m^3, or perhaps less than 0.1 ppm for some persons. The exact level will depend on the individual concerned [5713].

The *spontaneous ignition temperature* is 646° [5713].

179. Veratrole

Veratrole was prepared in 95% yield by Perkin and Weizmann [5769] by dissolving 100 grams of 1,2-dihydroxybenzene in 200 mL of methanol containing 75 grams of dimethyl sulfate at −50°. A solution of 150 grams of potassium hydroxide in 350 mL of water was added rapidly. The reaction was fast. The mixture was diluted and extracted with ether.

Curran [1767] prepared veratrole for *dipole moment studies* by treating catechol with sodium hydroxide and methyl sulfate. The organic phase was washed with successive portions of sodium hydroxide solution to remove guaiacol before drying and distilling.

Roberti and Smyth [6193] purified veratrole for *dielectric relaxation studies* by fractional distillation from barium oxide. Vaughn and coworkers [7695] distilled from sodium and stored over anhydrous sodium sulfate for the same purpose.

Cauwood and Turner [1394] purified veratrole for *dielectric constant studies* by repeated crystallization from a low boiling petroleum fraction to a constant melting point.

Veratrole was distilled over calcium hydride and the fraction boiling at 206–207° was used for the study of the *variation of the dipole moment* with temperature by Di Bell and coworkers [1997].

SAFETY

The oral LD_{50} for rats is reported to be 1360 mg/kg [3708].

ACETALS (180–181)

General Comments

Croxall and Neher [1735] have described a general method for preparation of acetals Acetals are formed when a vinyl ester and a primary or secondary alcohol are heated ϵ 20–50° with a mercury compound and a strongly acidic catalyst. Mercury(II) sulfat will serve as a combined catalyst. An excellent combination is mercury(II) oxide an boron trifluoride. A ratio of mercury compound to acid catalyst of 1 gram to 10 per mol of ester and an excess of alcohol are preferred. See also Shostakovskiĭ and Gershteĭ [6691].

Another general method for the preparation of acetals has been described by Heine mann [3267]. It is applicable to alcohols from methanol to the pentanols and also t higher homologues that are liquid at the reaction temperature; suitable aldehydes are th lower aliphatics and some aromatics. A catalyst bed, prepared from a clay absorbent such as bauxite, Fuller's earth, or bentonite, is activated at 600–1200° for a sufficien time to reduce the water to 5% or less. The alcohol–aldehyde mixture in a mole ratio o 2:1 to 4:1 is passed through the bed, which has been heated to a suitable temperatur between 0 and 50°, at a velocity of 0.01 to 0.1 volume per volume of catalyst per hour The yield per pass is reported to be about 50% of the theoretical.

Renault [6132] described a general method for preparing acetals from alcohols, al dehydes, and hydrochloric acid in the presence of calcium chloride.

Helferich and Hausen [3273] reported a general method for the preparation by treatin aldehydes or ketones with the appropriate tetralkyl silicate, using dry hydrogen chlorid as the catalyst.

Specific Solvents

180.   Dimethoxymethane

According to Bourgom [1013], the chief impurity in dimethoxymethane is methanol which can be removed by treatment with sodium. Fractional distillation over sodiun gives a 50% yield of a product boiling at 42.3°. See also the methods of purificatio given for acetal.

Rambraud and Besserre [6034] describe a laboratory method for dimethoxymethan in 84% yield from trioxymethylene and methanol in the presence of calcium chlorid and hydrogen chloride. Emde and Hornemann [2255] prepared dimethoxymethane b distilling a formaldehyde solution of methanol containing ammonium chloride.

Buehler and coworkers [3492] boiled formaldehyde and methanol with calcium chlo ride and a trace of hydrogen chloride to produce dimethoxymethane.

Vinokurov [7747] prepared a technical formaldehyde in 95% yield and gave thre methods for further purifying.

1. The technical dimethoxymethane, containing about 8% methanol, was allowed to stand with paraformaldehyde and a few drops of sulfuric acid 24 hr at room temperature and then distilled.
2. The technical material was shaken with an equal volume of 20–30% sodium hydroxide at 20°. The phases were separated at the end of 30 min and the acetal distilled.
3. Methanol-free formaldehyde was added to the middle of the distilling column during the preparation of the technical product.

Haarer and Rühl [3055] produced methanol-free dimethoxymethane by introducing water into the lower third of a fractionating column. The product contained 0.5% water.

McEachern and Kilpatrick [4764] purified commercial material for *thermodynamic study* by fractional distillation. The distilled liquid was dried over sodium; purity 99.9% by gas chromatography.

Pilcher and Fletcher [5855] fractionally distilled dimethoxymethane twice from calcium hydride for *calorimetric studies*. Their column was 1.5-m long with a 0.025-m diameter and was packed with stainless steel wire.

A commercial grade should meet the following specifications: ASTM distillation range 42.0–43.5°, sp. gr. 20/20° 0.863–0.865, n_D 20° 1.353–1.355, purity 97%w min, color APHA 15 max, aldehydes 0.1% max, water 1.5%w max, acidity as acetic acid 0.1%w max [4902].

SAFETY

The acetals only recently are becoming industrially important. There is only a meager amount of toxicological and other safety data available.

Dimethoxymethane has a chloroform-like odor and a pungent taste. The lethal concentration in air for rats is approximately 15,000 ppm. It may cause liver and kidney damage in humans; in higher concentration, narcosis [7033].

Fassett [5713, pp. 1984 ff.] has summarized the data that has been reported. Dimethoxymethane has been studied as a possible anesthetic. The LD_{50} for mice was found to be about 18,000 ppm. The investigators concluded that the threshold toxic concentration for mice and guinea pigs is of the order of 11,000 ppm.

The *threshold limit value* has been set at 1000 ppm, 3100 mg/m^3 [7022].

The *flammable limits* in air are 2.95 and 17.40%v. The *minimum ignition temperature* in air is 237° [5712].

181. Acetal

Reichert and coworkers [6109] prepared acetal by passing acetylene into a mixture of ethanol, sulfuric acid, and mercury(II) sulfate.

Bozel-Maletra [1042] prepared acetal with a purity of 98.5% by stirring acetaldehyde, ethanol, and ammonium chloride for 6 hr at room temperature.

For preparation from alkyl vinyl ethers, see [6690, 5431, 2757].

For *physical property measurements*, Walden [7848] dried acetal over sodium and purified by distillation. Adams and Adkins [18], who studied catalysis in acetal formation, purified acetal by shaking with hydrogen peroxide at 65° to remove aldehydes. The acetal was then washed with water, dried first with anhydrous potassium carbonate and then with sodium, and distilled. Fouque and Cabanac [2498] allowed acetal to stand over sodium for several weeks to remove alcohols and water and to polymerize aldehyde

Commercial acetal was purified by Vogel [7763] for *physical property measurements* by treating with alkaline hydrogen peroxide solution at 40–45°, saturating the solution with sodium chloride, drying with anhydrous potassium carbonate, and distilling over sodium.

SAFETY

Acetal produces only minor symptoms of weakness on rats; even at high dose levels no typical anesthesia was noted [5713].

The *explosive limits* in air are 1.6 and 10.4%v. The *minimum ignition temperature* in air is 230° [5712].

CARBONYLS (182–206)

Aldehydes (182–188)

General Comments

Aldehydes are quite reactive and their use as solvents is, therefore, limited. They are easily oxidized by oxygen in the air.

Zakharkin and Khorlina [8256] obtained aldehydes by the reduction of nitriles with diisobutylammonium hydroxide.

Brown and Subba Rao [1139] prepared aldehydes of widely varying structure by reduction of acid chlorides with lithium tri-*tert*-butoxyaluminum hydroxide in bis(2-methoxyethyl) ether solution at −78°. Aromatic aldehydes were obtained in better yields than aliphatic aldehydes but the experimental simplicity of the method makes it preferable when there is no advantage in yield. Stabilization of aldehydes from peroxide formation; see Ethers, General Comments, Robertson and Jones [6200].

Specific Solvents

Saturated Aliphatic Aldehydes

182. Acetaldehyde

Mita [5169] prepared acetaldehyde from acetylene and a special copper catalyst. Three grams of 1-chloroethylene (vinyl chloride) and 40 grams of acetaldehyde were obtained by passing 300 L of acetylene through a solution at 78° containing 100 grams of copper(I) chloride, 50 grams of ammonium chloride, 32 grams of leucine hydrochloride, 150 mL of water, and 30 mL of concentrated hydrochloric acid.

Smith and Bonner [6850] purified acetaldehyde by fractional distillation through a 100-cm column packed with glass helices. During the distillation, the column was kept under a positive pressure of dry nitrogen (1 in. of water pressure or 2.5 cm). The heads were removed at a reflux ratio of 20:1, the main fraction at 8:1.

Coleman and De Vries [1585] purified by distillation through a 4-ft Fenske-type column. The distilled fraction was sealed in Pyrex-brand glass containers and stored in the dark until used.

Muroi and coworkers [5344] developed a modification of the Karl Fischer method for the determination of water in acetaldehyde and other aldehydes. Pyridine and 1,2

ropanediol in 3:1 volume ratio were used as the sample solvent. The titrant was Karl
Fischer Reagent SS from Mitsubishi.

AFETY

The vapor pressure of acetaldehyde is above 1 atm at normal laboratory temperature;
herefore the material must be handled with caution because of the explosion hazard.
The odor of the vapor can be detected well below 50 ppm; some persons can detect it
elow 25 ppm. Volunteers were exposed to vapors of 50 ppm for 15 min and some
howed signs of eye irritation. At 200 ppm, all subjects had red eyes and conjunctivitis.
The LD_{50} for rats is about 20,000 ppm for a 30-min exposure. At this level, the animals
evelop pronounced excitement, followed by an anesthetic-like state after about 15 min;
urvivors recover rapidly. The principal finding at autopsy is pulmonary edema [5713].

Appelman and coworkers [399] determined the 4-hr median lethal concentration of
cetaldehyde in rats to be 13,300 ppm (24 g/m^3 air).

Although acetaldehyde is irritating in higher concentrations, it is rapidly metabolized
7027].

The *threshold limit value* is 100 ppm, 180 mg/m^3 [142].

The *explosive limits* in air are 3.97 and 57.00%v [5712]; 4.1 and 55%v have also
een reported [152]. The *minimum ignition temperature* in air is 176° [5712].

183. Propionaldehyde

Hurd and Meinert [912] oxidized 1-propanol with acid dichromate to obtain propion-
ldehyde; see also [6937, p. 64].

Hoffmann [3453] passed propylene oxide and steam over silica gel at 300° and ob-
ained a yield of 70% propionaldehyde.

McKenna and coworkers [4778] purified propionaldehyde by distillation under nitro-
en. The distillate was collected in glass containers that had been boiled with water for
0–30 min and dried.

Smith and Bonner [6850] purified both propionaldehyde and butyraldehyde by care-
ully drying three times with Drierite and then distilling under nitrogen in a Podbielniak
olumn. The heads were removed at a 50:1 reflux ratio, the main fraction at 20:1.

AFETY

Propionaldehyde has about the same skin and eye irritation as acetaldehyde [7619].

184. Butyraldehyde

Purification; see Smith and Bonner [6850] under propionaldehyde.

AFETY

Butyraldehyde appears to have no primary skin irritation for rabbits but it is a severe
ye irritant [7619].

The *threshold limit value* of 1.0 mg/m^3 has been suggested for Russia in 1964 [4185].

The *lower explosive limit* in air is 2.5%v [152].

185. Isobutyraldehyde

Pfeiffer [5810] oxidized 2-methyl-1-propanol with acid dichromate, distilled, and purified through the acid bisulfite derivative; see also Lipp [4620].

SAFETY

Zolotov and Svintukhovskii [8307] reported that the maximum permissible concentration of isobutyraldehyde in industrial areas is 5 mg/m^3.

It did not cause skin irritation to rabbits but did cause moderate eye injury [7619].

Aromatic Aldehydes

186. Benzaldehyde

Highet and Wildman [3389] prepared benzaldehyde by oxidation of benzyl alcohol in chloroform, ether, or hexane with manganese dioxide; yields of 61–89% were obtained depending upon the solvent used and the reaction time.

Yamashita and Matsumura [8204] oxidized primary alcohols with p-benzoquinone to obtain aldehydes. Two moles of p-benzoquinone and $\frac{1}{3}$ mole of aluminum phenolate were reacted with 1 mole of alcohol in benzene; benzyl alcohol gave benzaldehyde and furfuryl alcohol gave 2-furaldehyde.

Sheikh and Eadon [6621] prepared benzaldehyde by oxidation of benzyl alcohol mixed with helium, in a column packed with copper(II) oxide at 250–300°; yield was 92%.

Santaniello and coworkers [6379] oxidized benzyl alcohol with potassium dichromate solution for 1–3 hr at 90–100° and obtained a yield of 80% benzaldehyde; see also acetophenone.

Hesse and Schrödel [3359] prepared benzaldehyde in 92% yield by the reduction of benzonitrile with sodium hydrogen aluminum ethoxide ($NaHAl-(C_2H_5O)_3$), in tetrahydrofuran at room temperature.

Meyers [5093] reported that potassium hypochlorite oxidized benzyl alcohol to the corresponding aldehyde without any side reactions in 77% yield.

For additional preparations see [8283, 8244].

Benzaldehyde may be purified through its bisulfite addition compound or by fractional distillation at reduced pressure [3248].

For *thermodynamic studies*, Ambrose and coworkers [117] fractionally distilled benzaldehyde through a 1.5-m Podbielniak column with a Heligrid packing under a nitrogen pressure of 8 kPa. Purity was over 99.9%.

CRITERIA OF PURITY

Rosin [6263] describes the reagent grade as containing at least 98% benzaldehyde when analyzed by the hydroxylammonium chloride method and not more than 0.01% chlorine compounds. It should give no reaction for nitrobenzene. He cautions to keep the aldehyde protected from the light. The *National Formulary* [5417] also specifies 98% and no hydrocyanic acid.

SAFETY

The oral LD_{50} for rats was determined to be 3265 mg/kg [6940]. Peresdov [5757] recommended the maximum permissible concentration of benzaldehyde to be 5 mg/m^3, although the human eye and respiratory tract tolerated 20 mg/m^3.

The *minimum ignition temperature* in air is 192° [5712].

Unsaturated Aldehydes

187. Acrolein

Acrolein polymerizes easily. The dimer, 2-formyl-3,4-dihydro-2 *H*-pyran, is present in all except freshly prepared and stabilized monomers.

Gilman [2756, p. 1] gives directions for the preparation of acrolein from 1 kg of potassium hydrogen sulfate, 200 grams potassium sulfate, and 600 grams dry glycerol. Hydroquinone is used as a stabilizer; yield 240–350 grams.

Malinovski et al. [4854] prepared acrolein by the gas-phase aldol condensation of formaldehyde and acetaldehyde. A yield of 56.5% was obtained using a silica gel catalyst saturated with an aqueous solution of sodium and potassium silicates, tungstates, and titinates in the temperautre range of 250–320°.

Bremner and Jones [1081] and Wilson [7960] prepared acrolein from tetrahydrofuruyl alcohol over a catalyst in two stages.

Parry [5693] stabilized the monomer by the synergistic action exerted by copper and phenolic antioxidants in the presence of water. The rate of polymer formation in monomeric acrolein containing 0.1%w in contact with copper metal and in the presence of 0.001–0.002% hydroquinone decreased by a factor of 100 or more with respect to the same system without copper. Hydroquinone has been reported to be effective as a monomer stabilizer [4027] and Union Carbide includes the hydroquinone in their specifications, 0.10–0.25% [7619].

Blacet and coworkers [905] purified acrolein by distilling in a nitrogen-filled, 90-cm column packed with glass rings. To avoid the formation of diacryl, the vapor was passed through an ice-cooled condenser into a receiver containing 0.5 grams catechol in an ice-salt bath. The acrolein was twice distilled from anhydrous copper sulfate at low pressure. During the distillation, catechol was placed in both the distilling flask and the receiver to avoid polymerization.

SAFETY

The committee on Threshold Limit Values [7022] has summarized the pertinent information concerning the toxicity of acrolein: Exposure of cats to 10 ppm for 3.5 hr caused respiratory irritation, salivation, lacrimation, and mild narcosis; 1 ppm is immediately detectable and produces marked irritation to the eyes and nose; and 5.5 ppm is intensely irritating and 10.0 ppm and over is lethal in a short time. Di Macco [2009] reviewed the general toxic effects and states that acrolein inhibits the action of the enzymatic oxidation process and causes disturbances in the endocrine and neuroregenerative mechanisms.

Sinkuviene [6769] recommended a maximum permissible concentration of acrolein as 0.03 mg/m^3.

The *threshold limit value* is 0.1 ppm, 0.25 mg/m^3 [7022].

The *flammable limits* in air are 2.85 and 30.5%v. The *minimum ignition temperature* in air is 234° [5712].

188. Crotonaldehyde

Commercial crotonaldehyde is more than 95% *trans*-form [4027]. It polymerizes rather readily and becomes colored. Water and hydroquinone are two of the substances that have been found to inhibit polymerization, which is really an oxidation and resinification, according to some reports.

Horsley [3496] excluded oxygen and carried out the aldol condensation in a faintly alkaline solution buffered with sodium acetate and acetic acid at about 30°. A British patent [6897] describes a method for making crotonaldehyde by mixing equal quantities of acetaldehyde and water, cooling, gradually adding sodium hydroxide solution, and then acidifying. The aldehyde is distilled from the mixture. There have been other reported modifications of the two methods cited but apparently their principal novelty was mostly to evade existing patents and not to change significantly the yield or cost.

Raphael and Sondheimer [6063] hydrogenated 3,3-diethoxypropyne using a palladium-calcium carbonate catalyst to obtain *cis*-crotonaldehyde diethyl acetal. Hydrolysis of the latter with hot aqueous oxalic acid gave *trans*-crotonaldehyde.

Blacet and coworkers [905] converted the *cis*-portion of a commercial crotonaldehyde to the *trans*-isomer by exposing to bright sunlight a sample containing hydrochloric acid. This was distilled and the portion distilling between 102.6 and 102.8° was collected.

Hünig [3584] prepared acid-free crotonaldehyde to study its *condensation with secondary amines* by distilling a technical product over sodium pyrogallate and collecting the fraction boiling at 95–105°. It was redistilled in a Widmer column in a nitrogen atmosphere.

The *hydrogen-bonding properties* of crotonaldehyde were studied by Chandra and Sannigrahi [1413] by the ultraviolet spectra. The aldehyde was purified by drying with anhydrous calcium chloride and then distilling.

Dolliver and coworkers [2035] carefully fractionated commercial crotonaldehyde, saving the fraction boiling at 102.29–102.30° to study the *heat of hydrogenation*.

SAFETY

The Committee on Threshold Limit Values [7022] summarized pertinent information concerning the toxicity of crotonaldehyde. It produces the same symptoms as acrolein. The LD_{50} for rats for a 30-min exposure was 1500 ppm, 135 ppm for acrolein, and 825 ppm for formaldehyde. A case of apparent sensitization of a person handling small amounts of the aldehyde has been reported. Union Carbide [7619] states that crotonaldehyde is not a primary skin irritant to rabbits but causes severe eye injuries.

The *threshold limit value* has been set as 2 ppm, 5.6 mg/m^3 [7022].

The *flammable limits* in air are 2.12 and 15.5%v, the latter value at elevated temperature. The *minimum ignition temperature* in air is 232° [5712].

Ketones (189–206)

General Comments

Harrison and Eisenbraun [3192] pointed out that steam-volatile ketones may be purified through their bisulfite addition product, semicarbazone, hydrazones of *p*-hydra

Table 5.11   Amount of Enol Form Normally Present in Some Ketones

No.	Ketone	Enol Present (%)	Reference
189	Acetone	1.5×10^{-4}	2716
190	2-Butanone	1.2×10^{-1}	2716, 2717
193	3-Pentanone	7×10^{-2}	2716, 2717
194	Cyclohexanone	1.18	2716, 2717
196	4-Methyl-2-pentanone	7.4×10^{-2}	2716
202	Camphor	0.123	2716
203	Acetophenone	3.5×10^{-2}	2716
268	Ethyl malonate	7.7×10^{-3}	2716
493	Methyl acetoacetate	5.0	2716
494	Ethyl acetoacetate	13.2	2716, 2001
496	Ethyl cyanoacetate	2.5×10^{-1}	2716

zinobenzenesulfonic acid [7544], or 2,4-dinitrophenylhydrazones. The bisulfite or semicarbazone products are preferred if they form a good yield, since they are readily cleaved to the ketone. The bisulfite addition product is limited to aldehydes and methyl ketones but they are readily split with dilute acid or alkali. The semicarbazones are hydrolyzed by steam distillation in the presence of oxalic acid [2023] or by exchange with pyruvic acid [3334]. 2,4-Dinitrophenylhydrazones are impractical because the carbonyl is not readily recovered. Several procedures have been reported in which pyruvic acid or levulinic acid is used as an acceptor for 2,4-dinitrophenylhydrazine, thereby freeing the ketone [1960, 2021].

A method [3192] has been specifically designed for steam-volatile ketones using the steam-nonvolatile α-ketoglutaric acid as the acceptor for 2,4-dinitrophenylhydrazine. The procedure consists of steam-distilling of 2,4-dinitrophenylhydrazone in the presence of an equivalent amount of α-ketoglutaric acid and an excess of 30–50% sulfuric acid.

Ketones exist in the tautomeric *keto* and *enol* forms, both forms being present in the normal material. The amount of the enol form present in several ketones has been reported and is tabulated in Table 5.11.

A general route of ketones from esters was discovered by Corey and Chaykovsky [1659]. The esters are treated with the sodium salt of dimethyl sulfoxide to form a β-keto-sulfoxide which was reacted with aluminum amalgam in 90% tetrahydrofuran–10% water mixture to give ketones. Acetophenone, for example, was formed from ethyl benzoate in 77% yield.

Specific Solvents

Aliphatic Ketones

189.   Acetone

More than 95% of the acetone produced in the United States is from four synthetic processes. The remainder is from fermentation, and a small amount is still produced from pyroligneous acid. A good grade of commercial acetone should contain less than 0.1% organic impurities. The water content is generally below 0.4% and often below

0.2%. Acetone is much more reactive than is generally supposed. Such mildly basic material as alumina gel induces the aldol condensation to 4-hydroxy-4-methyl-2-pentanone, and an appreciable quantity is formed in a short time if the acetone is warm. Small amounts of acidic material, even as mild as anhydrous magnesium sulfate, cause acetone to condense [6165].

Acetone is available in a high state of purity at such a reasonable cost that it is impractical to prepare it in the laboratory. References to methods of preparation may be found in [664, 2103, 2265, 3577].

Livingston [4640] purified acetone as follows: The acetone is saturated with dry sodium iodide at 25–30°, the solution decanted from the excess solid, cooled to about −10° and the mother liquor removed from the crystals by filtration. The cold salt is transferred to a flask and warmed to above 26° (about 30° is optimum). The resulting liquid is transferred to a distilling flask and distilled, the last 10% being rejected. This is a modification of the Shipsey–Werner method of purification [6679].

Timmermans and Gillo [7486] found that it was difficult to produce a dry acetone; they questioned the possibility of preparing a very pure material and of storing it in a high state of purity if it were produced. They were working with acetone from pyroligneous acid, which contained relatively large amounts of various impurities, some of which are also present in synthetic and fermentation acetone. Since it was impossible to free acetone completely from certain hydrocarbons, homologues, methanol, acetals, and water by simple distillation, other methods of purification were necessary if a very pure product was to be produced. They suggested precipitation as the water-insoluble complex with sodium bisulfite or sodium iodide.

Methanol may be eliminated as its bromomethane azeotrope boils at 35°. The last traces of methanol may be eliminated by treating with acetyl chloride [2553] or by heating with acidified permanganate. The last traces of aldehydes may be removed by treatment with ammoniacal silver nitrate solution. Timmermans and Gillo [7486] found that water was the most difficult impurity to remove. They could not separate water by distillation, and the dehydrating agents such as sodium, phosphorus pentoxide, and calcium chloride caused condensation of the acetone. Dehydration with such salts as calcium nitrate and potassium carbonate does not dry to better than 0.1% water. The following procedure was finally adopted. The acetone was treated with calcium chloride, distilled, treated with phosphorus pentoxide, and fractionally distilled. The distillate contained 0.01–0.02% water. A second distillation from phosphorus pentoxide reduced the water content to 0.001% or lower.

An analytical-grade acetone was dried for *heat of mixing studies* by fractional distillation from anhydrous copper sulfate. Molecular sieves 4A were also used. The same refractive index was obtained with both drying agents [5245].

A number of methods involve a treatment with potassium permanganate, distilling, drying, and then usually fractional distillation. Some of these methods may be found in [1053, 6334, 6346, 7946].

Werner [7983] has recommended the following method: To 700 mL of acetone in a liter bottle is added 3 grams of silver nitrate dissolved in 20 mL of water. Twenty milliliters of 1 N sodium hydroxide is introduced and the mixture shaken for about 10 min. The acetone is then dried with calcium chloride and distilled.

Duclaux and Lanzenberg [2108] purified acetone by azeotropic distillation with carbon disulfide.

Lannung [4402] purified acetone for *solubility determinations* by drying for several days over potassium carbonate and distilling in a vacuum over fresh potassium carbonate.

Frankforter and Cohen [2520] dried over calcium chloride, treated with sodium amalgam, and fractionally distilled. For *electrical measurements*, acetone was distilled over anhydrous copper sulfate [886]. It was dried by treating for a long time with anhydrous potassium carbonate and distilling several times [7857].

For *optical measurements*, Scheibe and coworkers [6429] allowed acetone from the bisulfite compound to stand for 3 days over potassium permanganate, then boiled for 2 hr; after distillation it was dried over potassium carbonate and then fractionally distilled through an efficient column.

According to Lannung [4402], the most suitable drying agent is anhydrous potassium carbonate. Calcium chloride forms a compound with acetone and limits its efficiency, resulting in an appreciable loss of acetone. Phosphorus pentoxide results in contamination by condensation products.

Maruyama [4929] describes a method for purifying acetone for use as a *polarographic solvent*; see also [7532].

Smith and coworkers [6848] allowed reagent-grade acetone to percolate slowly through a 2-ft column packed with $\frac{1}{16}$-inch pellets of Linde type 4A molecular sieves. The dried acetone was distilled from 4A sieves through an 18-in. Vigreux column and was used for *solvolysis rate studies*. Savedoff [6396] found that acetone prepared for *conductivity studies* by this method had a specific conductance of 1.66×10^{-8} ohm^{-1} cm^{-1} and contained less than 0.005% water.

For purification for *dielectric constant studies*, see [3199], benzyl alcohol.

An extensive study has been made by Riddick [6165] on the preparation of high-purity acetone with a water content of 0.05% or less. The acetone used was a special middle cut, containing less than 0.25% water, from a 10,000-gallon distillation of fermentation acetone. The acetone contained 0.01% or less methanol, no aldehydes as tested by a modified Tollen's reagent, and no detectable hydrocarbons. The acidity and alkalinity was 0.002% or less, and the permanganate time was 30 min or more.

The acetone was fractionally distilled through all-glass-type, Penn State-type columns packed with $\frac{1}{8}$-in. single-turn glass helices (see Table 5.12).

Silica gel and alumina increased the water content of the acetone, presumably through aldol condensation and subsequent dehydration. The water content of acetone was increased from 0.24 to 0.46% by one pass over alumina. All other drying agents tried,

Table 5.12 Purification of Acetone by Fractional Distillation

Column	Reflux Ratio	Distilled (%)	Water Content of Distillate (%)
100 Plates	25:1	6–16	0.002
	25:1	16–25	0.012
35 Plates	10:1	47	0.04
	2:1	94	0.05
25 Plates	2:1	94	0.16

including copper sulfate, potassium carbonate, calcium chloride, sodium sulfate, and phosphorus pentoxide, caused some condensation. Drierite and anhydrous magnesium sulfate were the most satisfactory. Magnesium sulfate prepared by heating in a furnace to 400° was unsatisfactory. The most satisfactory method of preparation was to heat Epsom salt crystals gradually to 300° at approximately 10 Torr pressure. The dried salt contained about 0.2% water. Weight for weight, Drierite was found to be 2–4 times as efficient as magnesium sulfate in drying acetone to about 0.05% water. A large excess of the magnesium salt will not dry acetone below 0.04% water.

The water content of acetone can be reduced to 0.001% or less with Drierite. It is not practical to reduce the water below the 0.01% range, except in relatively small amounts, because the drying agent required is large. The acetone should first be distilled to reduce the water content to 0.25% or less water. Any chemical treatment to remove impurities should be done before the drying operation.

Acetone dried with Drierite is suitable for most purposes without distillation. If a product 99.95 + % is required, the acetone must be distilled. There is a trace of 4-hydroxy-4-methyl-2-pentanone in all dried acetone. Acetone distilled from a vessel whose wall temperature is greater than 20° above the boiling point of acetone will result in a distilled product that has a greater color and a shorter permanganate time than if the wall temperature were kept within 20° of the boiling point. The most satisfactory way to control the wall temperature is to immerse the distillation flask in a water bath. The following procedure has been found to produce a distillate from dry 99.95% acetone that had a color no greater than water distilled from alkaline permanganate and a permanganate time of 12 + hr. The previously dried acetone is distilling from an all-glass still fitted with a 15-in. spray trap containing $\frac{1}{4} \times \frac{1}{4}$-in. Raschig rings. The still pot is immersed in a water bath and a slow stream of dry nitrogen is passed into the receiver. The temperature of the water bath is brought to 70–75°, the still flushed with acetone vapors, and the condenser water started. The receiver is washed with several portions of distilled acetone before the main fraction is collected. The distillation is continued until about 100 mL remains in the pot. The temperature of the water bath must be reduced as the volume of acetone in the still pot decreases to keep the volume of distillate within the capacity of the condenser.

A more recent investigation indicates that 3A or 4A molecular sieves may be used instead of Drierite without causing condensation of the acetone. Weight for weight, the sieves are more efficient than Drierite.

Connett [1628] reported the formation of highly purified acetone when standing over molecular sieves for several months. The same impurities were detected by gas chromatography in a sample that had been boiled with molecular sieve 4A for 2 hr. He reported that Union Carbide advises that acetone could be dried using special sieve AW 500.

CRITERIA OF PURITY

The specifications for a good commercial-grade [1603] and ACS reagent-grade [137] acetone are given in Table 5.13.

Swietoslawski [7131] used the differential ebulliometer to detect impurities.

Timmermans and Gillo [7486] studied the determination of the several impurities present in acetone and recommended the following procedures.

Table 5.13 Specifications for a Commercial and ACS Reagent-Grade Acetone

	Commercial	ACS Reagent
Purity (min)	99.5%	—
Density (25°)	0.784–0.786	0.7857, max
Specific gravity (25/25°)	0.786–0.788	—
Distillation range	1° Including 56.1°	1 mL to 95 mL, not more than 0.5°; 95 mL to dryness, not more than 0.5°
Acidity (max)	0.002% as CO_2	0.002% as acetic acid
Alkalinity (max)	None to *p*-nitrophenol	0.001% as ammonia
Aldehydes (max)	—	0.002%
Permanganate time at 25°	2 hr	15 min
Methanol, not more than	—	40 mg methanol/100 mL
Nonvolatile matter (max)	0.001 g/100 mL	0.001% max
Water solubility	Miscible with no turbidity	To pass test
Color (max)	5 APHA[a]	—

[a] ASTM now includes this test as D 2108-64.

Aldehydes: Schiff–Villavecchia reaction, sensitive to 0.001%.

Acidity: titration with barium hydroxide to phenolphthalein end point, sensitivity to better than 0.001%.

Action of permanganate: sensitive to less than 0.003% methanol.

Total impurities: add known amounts of water, methanol, and acetic acid to pure acetone and observe the differential boiling–condensing temperature, sensitivity to about 0.001%.

Water: differential boiling–condensing temperature, sensitive to 0.001%; infrared absorption at 3 μm range, sensitive to better than 0.001%.

Dreisbach and Martin [2090] determined the purity from the freezing curve.

SAFETY

Acetone presents a relatively low degree of health hazard. Acetone has a high vapor pressure at normal room temperatures [5713].

The method of Alekseeva [83] for the determination in air gives the sensitivity, specificity, and likely interferences.

The *threshold limit value* has been set at 1000 ppm, 2400 mg/m^3 [7022].

The *flammable limits* in air are 2.55 and 12.80%v [5712]. The *minimum ignition temperature* in air is 465° [1411].

190. 2-Butanone

Bewley [834] prepared 2-butanone by the continuous dehydration of 2,3-butanediol. See also Robertson [6201].

2-Butanone generally can be purified by the methods described for acetone.

Newman and coworkers [5471] dried 2-butanone by extractive distillation and solvent extraction. 2-Butoxyethanol was found suitable for extractive distillation. Suitable solvents for solvent extraction were 1,1,2-trichloroethane, trichloroethylene, and chloro-benzene.

Bresler and Drasky [1084] give directions for purifying 2-butanone, which is contaminated and discolored by diketones. The color-forming diketones can be removed by agitation with an aqueous solution of ammonia or a primary amine.

Lochte [4642] purified 2-butanone, like acetone and other ketones, by means of the addition compound with sodium iodide. The impure product was saturated with sodium iodide by boiling under reflux, and filtering while hot. It was then cooled and the white needles of the addition compound separated and were removed by filtration. The crystals melt at 73–74° and yield the ketone quantitatively upon distillation.

For *electrical measurements*, Müller and coworkers [5325] treated 2-butanone several times with potassium carbonate solution to remove acidic impurities, separated the two layers, and distilled to remove most of the water. The ketone was then dried for several days over sodium sulfate and anhydrous potassium carbonate and, after decanting, fractionally distilled several times. Walden and Birr [7857] purified the ketone by means of the bisulfite compound. Commercial 2-butanone was fractionally distilled and the bisulfite compound precipitated from the middle fraction with concentrated sodium bisulfite solution. After filtering, the crystals were dissolved in water and precipitated by the addition of sodium bisulfite. The precipitate was filtered, pressed out, and decomposed with concentrated potassium carbonate solution. The nonaqueous layer was separated (ketones remaining in the solution were separated by distillation), dried over potassium carbonate, distilled, and air passed through the liquid for 24 hr. After drying twice more with potassium carbonate, it was poured off and distilled very carefully, great care being taken to insure that evaporation took place only from the surface, so that splashing through bubble formation was avoided.

Collerson and coworkers [1590] purified 2-butanone to determine *thermodynamic properties* as follows: Commercial material was fractionally distilled at 700 Torr in a 100-theoretical-plate still at a reflux ratio of 100:1. The fractions of highest purity by gas chromatographic analysis were combined. Boiling range and gas chromatography indicated adequate purity; 99.95%m from the freezing curve.

For *optical measurements*, Rice [6148] treated 2-butanone with a saturated solution of sodium bisulfite until no more reacted. The mixture was cooled to 0°, filtered, washed twice with ether, and dried on a porous plate. The dried addition compound was added to a sodium bicarbonate solution and steam-distilled. The ketone was salted out of the distillate with potassium carbonate, separated, and treated with potassium hydroxide to remove sulfur dioxide or carbon dioxide. After standing for 4 hr over calcium chloride, it was distilled. See also Mathews [4969].

Sinke and Oetting [6767] purified a commercial 2-butanone for *low-temperature studies* by drying with calcium sulfate and fractionally distilling in a 100-plate still; 99.78%m shown by variation in melting point with fraction melted.

CRITERIA OF PURITY

Tests for acetone may be applied to 2-butanone. Dreisbach and Martin [2090] determined the purity from the freezing curve.

SAFETY

2-Butanone presents a low degree of hazard with normal handling. It has a low order of toxicity. The liquid may produce moderate skin irritation when exposures are frequent and prolonged. It may be classed as a mild eye irritant. The threshold odor detection concentration is less than 25 ppm. There have been no recorded instances of human illness due to the use of 2-butanone [5713].

The *threshold limit value* has been set at 200 ppm, 590 mg/m^3, to prevent objectionable irritation [7022].

The *flammable limits* in air are 2.05 and 11.00%v. The *minimum ignition temperature* in air is 505° [5712].

191. Cyclopentanone

Wallach [7887] treated cyclohexanone with bromine in cold acetic acid and eventually produced 1-hydroxycyclopentanecarboxylic acid by reaction with potassium hydroxide. The organic acid was then heated with lead(IV) oxide and sulfuric acid solution to produce cyclopentanone.

Thorpe and Kon [7451] obtained a yield of 75–80% cyclopentanone by heating adipic acid with crystalline barium hydroxide at 285–295°. Frey [2554] reported an industrial preparation of cyclopentanone produced by passing cyclopentanol vapor over a zinc–copper alloy at 300°.

Dem'yanov and Shuikina [1949] prepared cyclopentanone by oxidizing cyclopentylamine with oxygen in the presence of pulverized copper or osmium tetroxide.

Noller and Adams [5506] purified cyclopentanone by distillation over calcium oxide. Turk [7580] purified the ketone by heating it with either phosphoric acid, chromic acid, or hydrochloric acid and distilling the mixture at reduced pressure.

192. 2-Pentanone

Perkin [5761] prepared 2-pentanone in 1886 by heating sodium acetate with butyric anhydride at 200°. Senderens [6572] prepared the ketone by passing an equimolar mixture of acetic acid and butyric acid over thorium oxide at 400–430°.

Brown [1130] patented a preparation of 2-pentanone by passing acetone and ethylene over catalysts such as nickel(II) oxide and zinc oxide.

Yoke and Louder [8216] reported a detailed preparation of 2-pentanone from 2-pentanol.

Collerson and coworkers [1590] purified 2-pentanone for *vapor pressure studies* by distilling a commercial sample at 700 Torr through a 100-theoretical plate column at a reflux ratio of 100:1. Purity was 99.93%m from the freezing curve.

Nickerson and coworkers [5484] purified reagent-grade 2-pentanone by fractional distillation of a sample that had stood over anhydrous calcium sulfate. The distillate had a boiling range of less than 0.1°. The material was vacuum-distilled into a receiver suitable for introduction into the apparatus used for the determination of *thermodynamic properties*. No evidence of polymerization or decomposition was observed during the experimental work.

Cowan and coworkers [1690] purified 2-pentanone for *physical property studies* by converting the ketone, initially fractionally distilled, to the semicarbazone. The latter was recrystallized, dried over calcium chloride and paraffin wax, and then refluxed with oxalic acid solution. The ketone was separated by steam distillation, dried with anhydrous sodium sulfate, and distilled twice.

SAFETY

The *threshold limit value* of 200 ppm, 700 mg/m^3, has been suggested [142].

The *flammable limits* in air are 1.5 and 8%v. The *minimum ignition temperature* in air is 505° [5415].

193. 3-Pentanone

Nwaukwa and Keehn [5538] oxidized 3-pentanol with calcium hypochlorite at 0° to produce 3-pentanone in 97% yield; see also cyclohexanone and 2-octanone.

Collerson and coworkers [1590] distilled and tested 3-pentanone as described for 2-pentanone to determine *thermodynamic properties*. Further purification was necessary. Fractional freezing gave a satisfactory product; 99.95%m from the freezing curve.

SAFETY

The LD$_{50}$ for rats, single dose, was found to be 2.14 g/kg [6883].

The *minimum ignition temperature* in air is 452° [5712].

194. Cyclohexanone

Joris and Vitrone [3782] heated phenol containing 0.1% palladium on charcoal at 140–145° under hydrogen at atmospheric pressure for 30 hr to obtain cyclohexanone in 79% yield.

For the preparation of cyclohexanone from phenol through cyclohexanol see cyclohexanol, Yeh and coworkers [8212].

Nwaukwa and Keehn [5538] oxidized cyclohexanol with calcium hypochlorite at 0° to produce cyclohexanone in 91% yield; see also 3-pentanone and 2-octanone.

According to Auwers and coworkers [506], commercial cyclohexanone can be freed from cyclohexanol with chromic acid. It can be purified by means of the semicarbazone. Herz and Bloch [3343] dried a good grade of commercial material over anhydrous sodium sulfate for 2 days and fractionally distilled. Garland and Reid [2666] purified by means of the bisulfite addition complex, which they decomposed with sodium carbonate and then steam-distilled.

Crowe and Smyth [1729] purified cyclohexanone for *dielectric constant measurements;* it was dried for 24 hr over anhydrous sodium sulfate and fractionally distilled twice.

Cyclohexanone was purified for use in *ultraviolet studies of hydrogen bonding* by Chandra and Sannigrahi [1413] by preparation of the bisulfite derivative and subsequent release with 80% sodium hydroxide solution. The ketone layer was dried with anhydrous sodium sulfate and distilled.

Takamitsu and Sumida [7160] identified 1-butanol, 1-pentanol, cyclohexanol, caproaldehyde, 3-pentanone, 2-hexanone, 2-heptanone, 3-heptanone, 4-heptanone, cyclopentanone, cyclohexane, and benzene as impurities in cyclohexanone made by the autoxidation of cyclohexane in the liquid phase.

SAFETY

Cyclohexanone is considered to be of a low degree of hazard under normal use conditions. It has strong warning properties at low concentrations. It defats the skin, and frequent contact of the skin and the liquid should be avoided. Contact of the liquid with the eyes causes marked irritation and some transient corneal injury. The principal health hazard is the inhalation of the vapors [5713]. Cyclohexanone was found [803] to cause vascular and degenerative lesions of the lungs and liver. See also [1675].

Gupta and coworkers [3028] reported an LD_{50} range of 0.93 to 1.54 g/kg for mice, rats, and guinea pigs.

The *threshold limit value* is 50 ppm, 200 mg/m^3 [7022].

The *lower flammable limit* in air at elevated temperature is 1.11%v. The *minimum ignition temperature* in air is 453° [5712].

195. 2-Hexanone

Michael and Hartman [5095] prepared 2-hexanone by oxidizing 2-hexanol with a chromic acid solution. Grignard and Chambret [2966] prepared the ketone by decomposing methyldibutyl carbinol or methylbutylcrotyl carbinol on glasswool at 600°.

Newman and Smith [5472] reacted acetic anhydride with butylmagnesium bromide in ether at −82° to obtain an 80% yield of 2-hexanone.

Heath and Skomoroski [3252] reported a patent to prepare 2-hexanone by dehydrogenation of 2-hexanol over a catalyst of activated carbon containing platinum. An Esso Research patent [2294] stated that "dimethyltetrahydrofuran was isomerized to 2-hexanone over an activated carbon–copper catalyst at 399° and 400 psi hydrogen."

2-Hexanone prepared by the method of Yoke, Louder, et al. [8216] was purified by Ginnings and Plonk [2765] for *solubility studies* by fractionally distilling through a 1-m column from barium oxide and finally purifying through a sodium bisulfite complex.

Andon and coworkers [352] purified 2-hexanone for *thermodynamic studies* by distillation and fractional freezing.

For purification for *physical property studies*, see Cowan [1690], 2-pentanone.

SAFETY

The single dose oral LD_{50} was reported as 2.59 g/kg in rats [6683].

The *threshold limit value* of 25 ppm, 100 mg/m^3, has been suggested [142].

The *flammable limits* in air are 1.2 and 8%v. The *minimum ignition temperature* in air is 553° [5415].

196. 4-Methyl-2-pentanone

Riggio and coworkers [6175] purified 4-methyl-2-pentanone by distillation twice over dried potassium carbonate under a nitrogen atmosphere with a Vigreux column. The middle cut of each distillation was used for *density, refractive index, and viscosity measurements*.

Karr and coworkers [3861] purified a special commercial material (99.5%) for *phase-equilibrium studies* by taking a 70% middle cut from a fractional distillation at a reflux ratio of 25:1 from a 33-theoretical-plate still.

SAFETY

4-Methyl-2-pentanone is not considered to be hazardous to health under normal conditions of handling. Inhalation does not normally constitute a health problem because of its warning properties. Workers exposed to about 100-pm complained of headache and nausea [5713].

Batyrova [684] reported that the oral lethal doses of 4-methyl-2-pentanone were 2.85 and 4.6 g/kg for mice and rats, respectively, and recommended that the maximum permissible concentration be 5 mg/m^3.

The *threshold limit value* is 100 ppm, 410 mg/m^3 [152].

The *flammable limits* in air are 1.35 and 7.60%v, the latter value at elevated temperature [5712]; see also [4902]. The *minimum ignition temperature* in air is 465° [5712].

197. 2-Heptanone

Dakin [1796] prepared 2-heptanone in 1908 by oxidizing the ammonium salt of caprylic acid with a 3% hydrogen peroxide solution.

Johnson and Hager [3735] reported a detailed synthesis of 2-heptanone beginning with ethyl-*n*-butyl acetoacetate and sodium hydroxide solution. After stirring these reagents for 4 hr at room temperature, the aqueous layer was separated and then acidified with 50% sulfuric acid solution. The mixture was distilled and the distillate was made alkaline with solid sodium hydroxide and then redistilled. The yield after subsequent distillations was 52–61% ketone.

Sherrill [6650] prepared 2-heptanone by oxidizing 2-heptanol with potassium dichromate in concentrated sulfuric acid solution. The ketone was purified by preparing the sodium bisulfite addition product and subsequent recovery by reaction with sodium carbonate. The final product was collected by distillation.

For purification for *solubility studies*, see Ginnings and Plonk [2765], 2-hexanone, and Saylor [6404], 2,4-dimethyl-3-pentanone. For purification for *physical property studies*, see Cowan [1690], 2-pentanone.

198. 3-Heptanone

Vogel [7762] prepared 3-heptanone for *physical property studies* by passing pentanoic and propanoic acids over thoria catalyst for 9 hr at 430–450°. The organic layer was separated, washed with 10% sodium hydroxide solution and water, and finally dried over calcium sulfate. The ketone was separated by fractional distillation and subsequently purified by distillation. See also Pickard and Kenyon [5835].

Newman and Smith [5472] reacted propanoic anhydride with butylmagnesium bromide in ether at −70° to obtain a 70–80% yield of 3-heptanone.

For preparation by oxidation of 3-heptanol and purification, see Sherrill [6650], 2-heptanone.

Paul [5716] dehydrogenated 3-heptanol over Raney nickel at 150° to give a yield of 89% 3-heptanone.

SAFETY

A reported LD$_{50}$ in rats is 2.76 g/kg [6884].

199. 2,4-Dimethyl-3-pentanone

Senderens [6571] prepared 2,4-dimethyl-3-pentanone by heating isobutyric acid over water-free thorium oxide at 400°. Ivanov [3649] prepared 2,4-dimethyl-3-pentanone and 2,6-dimethyl-4-heptanone by pyrogenic decomposition of mixed magnesium carbonates.

Hauser and Renfrow [3232] added isobutyronitrile to isopropylmagnesium bromide in ether and then mixed the white precipitate formed with 10 N sulfuric acid solution and crushed ice. The mixture was steam-distilled and the distillate was extracted with ether. 2,4-Dimethyl-3-pentanone was collected by distillation; bp 123–125°.

Bolotov and coworkers [955] obtained a 50–56% yield of 2,4-dimethyl-3-pentanone by passing isobutyl alcohol and hydrogen over a copper catalyst activated with thorium oxide at 425–450°.

For purification for *thermodynamic studies*, see Andon [352], 2-hexanone. For purification for *solubility studies*, see Ginnings and Plonk [2765], 2-hexanone.

Sellers [6562] purified 2,4-dimethyl-3-pentanone for *thermodynamic studies* by preparative gas chromatography (Autoprep 700, 6-m by 1-cm, 25% Carbowax 20M on Chromosorb W).

Saylor and coworkers [6404] purified 2,4-dimethyl-3-pentanone for *solubility studies* by shaking a commercial sample with ammoniacal silver nitrate until no discoloration appeared. Then the ketone was washed, dried over calcium sulfate, and fractionally distilled in a 50-cm Widmer still.

200.   2-Octanone

Verhulst and Glorieux [7727] prepared 2-octanone by oxidizing 2-octanol with dichromate ion in sulfuric acid solution. Moureau and Mignonac [5291] prepared the ketone by passing a mixture of 2-octanol vapor and oxygen over a silver catalyst at 25–30 Torr and 330–340°.

Nwaukwa and Keehn [5538] oxidized 2-octanol with calcium hypochlorite at 0° in a solvent containing acetic acid and reported a yield of 80% 2-octanone.

Paul [5716] dehydrogenated 2-octanol over Raney nickel at 120° to give a yield of 95% 2-octanone.

Sabatier and Senderens [6328] prepared 2-octanone by passing methylhexylcarbinol over finely divided copper at 250–300°.

For purification for *physical property studies*, see Cowan [1690], 2-pentanone.

SAFETY

Abbasov and coworkers [4] reported the oral LD_{50} for mice and rats as 3.1 and 9.2 g/kg, respectively. A nonhazardous level for 2-octanone was set at 200 mg/m^3.

201.   2,6-Dimethyl-4-heptanone

Delepine and Horeau [1934] prepared 2,6-dimethyl-4-heptanone by the hydrogenation of phorone in the presence of Raney nickel and platinum in ethanol. For the preparation from pyrogenic decomposition of magnesium carbonate, see Ivanov [3649], 2,4-dimethyl-3-pentanone.

For the preparation from isoamyl alcohol, see Bolotov [955], 2,4-dimethyl-3-pentanone.

The ketone was purified by Stross and coworkers [7052] for *physical property studies*. After distillation in an 80-plate Oldershaw column at 200 Torr, samples were passed through a silica gel column to remove alcohols, unsaturated ketones, and water. Peroxides and diketones were removed by flash distillation over sodium hydroxide. The purified ketone was stored over freshly calcined calcium sulfate.

For purification for *thermodynamic studies*, see Sellers [6562], 2,4-dimethyl-3-pentanone.

SAFETY

A reported LD_{50} in rats is 5.75 g/kg [6884].

The *threshold limit value* is 25 ppm, 150 mg/m^3 [142].

The *flammable limits* in air at 100° are 0.8 and 6.2%v [5415]. The *minimum ignition temperature* in air is 388° [5712].

202. Camphor

Camphor occurs naturally in its optically active forms and as the racemate. Synthetic camphor is optically inactive. The commercial nomenclature is confusing. The *d*-form is called Japan camphor, Formosa camphor, laurel camphor, or gum camphor and is obtained principally from the camphor tree, *Cinnamomum camphora* Nees and Ebermaier (fam. Lauracea) [3013]. *l*-Camphor has been reported in the oil from *Salvia triloba, Blumea balsamfera, Artemisia austrachanica, Lippa adoensis,* and *Lavedula pedunculata*. The *l*-camphor is generally called "matricaria camphor." *dl*-Camphor occurs in the oil of *Chrysanthemum sinense* var. *japonicum* and in oil of sage.

Although *d*- and *l*-camphor are optical antipodes, the melting points are the same within 0.2°, the boiling points of the *d*- and *l*-forms as given in the literature differ by 3°. The density of the *l*-form has been reported to be about 0.001 g/cm^3 lower than that of the *d*-form. Synthetic camphor is just as suitable for the Rast melting point method for molecular weights as *d*-camphor. The *l*-form is not used as a solvent.

Camphor shows most of the characteristic reactions of the ketones. The *dl*-form, which is less reactive in some instances than the separate optically active forms, does not give an addition compound with sodium bisulfite.

Commercial camphor is a material of high purity. It may be further purified by sublimation, fractional crystallization, or crystallization from 50% ethanol–water solution or from a low-boiling petroleum solvent. A French patent [5230] states that the impurities can be removed from camphor by dissolving it in concentrated acetic acid and then adding water until the camphor precipitates.

CRITERIA OF PURITY

The optical rotation and freezing point have been used most frequently as criteria of purity.

SAFETY

The *threshold limit value* for synthetic camphor is 2 ppm, 12 mg/m^3 [142].

The *minimum ignition temperature* in air is 466° [5415, 6402].

Aromatic Ketones

203. Acetophenone

Synthetic acetophenone is prepared by the Friedel–Crafts reaction from benzene and acetic anhydride [2975, 2976, 6361]. Smeets and Verhulst [6806] report that continuous elimination of hydrogen chloride with a stream of dry air in the Friedel–Crafts reaction using acyl chlorides gives a yield of 97%, or better, ketone with mild temperature conditions in about 0.5 hr with no side reactions.

Acetophenone occurs naturally as the main constituent of oil of *Stirlinga latifolia*.

Webb and Webb [7941] prepared acetophenone by the catalytic decomposition of methyl or ethyl benzoate and acetic acid over thorium oxide at an elevated temperature. See also Porter and Cosby[5923].

Petrov and Lagucheva [5804] describe a method for preparing aromatic and alkaryl ketones using silicon tetrachloride and aluminum chloride. An acid is reacted with silicon tetrachloride to form $(RCO_2)_4Si$, which is reacted with the aromatic hydrocarbon in the presence of aluminum chloride to form the ketone.

Hauser and coworkers [3231] prepared acetophenone in 33% yield by the Grignard reaction with acetonitrile.

Daniher [1809] mixed ethylbenzene and $(NH_4)_2S_2O_8$ in water at 50–55° and then treated the mixture with silver nitrate solution at 65–70° for 2 hr. The yield was 73% acetophenone.

Santaniello and coworkers [6379] oxidized 1-phenylethanol with potassium dichromate solution for 1–3 hr at 90–100° and obtained a yield of 80% acetophenone; see also benzaldehyde.

Kornblum and coworkers [4189] treated α-phenyl-α-nitroethane with potassium permanganate and obtained a 90% yield of acetophenone.

The criteria of purity and methods of purification of acetophenone have been studied by Livingston and coworkers [4639]. An extensive bibliography is given to 1924. They recommend subjecting commercial acetophenone to filtration, followed by careful, slow, and repeated crystallization with exclusion of light and moisture. Since fairly large quantities of water can be removed more efficiently by distillation than by crystallization, the mother liquors were frequently distilled at atmospheric pressure through a Vigreux column, and further purified by crystallization.

Sugden [7072] froze acetophenone until the ternary freezing point in contact with water was constant to 0.2°. He then dried over calcium chloride and repeatedly distilled at 2 Torr.

CRITERIA OF PURITY

Chemical purity as the ketone can be determined by the hydroxylammonium chloride method. The freezing point is the most sensitive and accurate physical criterion.

SAFETY

Acetophenone has been used as a hypnotic [7033]. A method for the determination in air has been reported [83]. The sensitivity, specificity, and likely interferences are given.

The oral LD_{50} in rats was reported as 3 g/kg [6981].

The *minimum ignition temperature* in air is 571° [5712].

Unsaturated Ketones

204.   4-Methyl-3-penten-2-one
205.   4-Methyl-4-penten-2-one

Mesityl oxide was independently reported as existing in two forms by Harris [3718] and Hibbert [3370] in 1899. Hibbert called 4-methyl-4-penten-2-one the *alpha*-form and 4-methyl-3-penten-2-one the *beta*-form. The two isomers are also called isomesityl oxide and conjugated mesityl oxide, respectively. Doeuvre [2030] reported that mesityl oxide was a mixture of these two isomers in variable proportions. He studied the reaction products and concluded that it was at least 95% *beta*-form.

Methods of preparation are reported in the literature for the mixture rather than the pure isomers. Freer and Lachman [2533] prepared mesityl oxide in 1897 by saturating acetone with HCl gas and washing the resultant oil with water and alkali.

Gasopoulos [2681] prepared mesityl oxide by treating dry acetone with phosphorus oxychloride. Conant and Tuttle [1622] gave a detailed synthesis using diacetone alcohol and iodine. Kohn [4131] reported a preparation by the dehydration of diacetone alcohol with concentrated sulfuric acid.

Stross and coworkers [7053] studied the *preparation of high-purity isomers*, their *physical properties*, and the *equilibrium* of the *alpha* and *beta* forms. They noted that the rate of isomerization increased by the addition of hydrogen ions or alkali, and this finding was used to obtain a good yield of 4-methyl-4-penten-2-one. The separation and purification of the isomers are given in detail.

Oxidation products, acids, and other impurities were removed from a commercial mesityl oxide by washing several times with 0.5 N sodium hydroxide solution at a temperature below 10° to avoid condensation and oxidation. It was washed with water, dried with anhydrous calcium sulfate, and fractionally distilled at atmospheric pressure in an 85-actual-plate column in the presence of 0.2% *p*-toluenesulfonic acid monohydrate used to maintain isomerization equilibrium. The lower-boiling isomesityl oxide was distilled and cuts with refractive index between 1.4210 and 1.4230 were redistilled at 50 Torr and reflux ratio of 40:1. The purest samples contained less than 0.05% 4-methyl-3-penten-2-one.

Another commercial sample, washed as before, was distilled at atmospheric pressure in a 20-plate column to remove higher boiling contaminants. A second distillation at 200 Torr in a 60-plate still produced pure 4-methyl-3-penten-2-one.

Estok and Sikes [2297] purified Eastman Kodak white label mesityl oxide by careful fractional distillation through a metal helix-packed column at reduced pressure for *dipole moment determination*.

Bryant [1216] reduced the peroxide content from 272 to 54 mmol/L by percolating mesityl oxide through a column of activated alumina.

4-Methyl-3-penten-2-one has been prepared by self-condensation of acetone catalyzed by aluminum oxide. There are extensive studies of this method; for example, see Muzart [5367].

Hagemeyer [3036] reacted ketene with acetylacetone in the presence of boron trifluoride in ether at 0° and distilled out 4-methyl-4-penten-2-one.

SAFETY

Mesityl oxide is highly irritating to all tissues on contact and its vapors are also irritating. The *threshold limit value* has been set at 15 ppm, 60 mg/m^3 [142].

The *minimum ignition temperature* in air is 344° [6402].

Diketones

206. 2,4-Pentanedione

At room temperature, 2,4-pentanedione has a composition of 18.6% keto-form and 81.4% enol-form in the liquid phase. The vapor composition is 93.3% enol and only 6.7% keto [3636]. Physical property studies of the two tautomers are given in Irving and Wadso [3636], Hacking and Pilcher [3058], Melia and Merrifield [5052], and Reeves [6103].

Claisen [1498] prepared 2,4-pentanedione in 1894 by reacting acetone and ethyl acetate in the presence of sodium ethoxide. See Adkins and Rainey [37] for details of the complete synthesis.

Denoon [1957] reported a detailed synthesis of 2,4-pentanedione involving the passage of boron trifluoride through a mixture of acetone and acetic anhydride.

Sieglitz and Horn [6720] prepared 2,4-pentanedione by reacting acetyl chloride and vinyl acetate with aluminum chloride in carbon disulfide.

Eidenhoff [2227] purified 2,4-pentanedione for *dissociation constant studies* by fractional distillation at 138.8–138.9° and 750 Torr. Melia and Merrifield [5052] dried reagent-grade 2,4-pentanedione over sodium sulfate and collected the fraction boiling at 135–137° for *thermodynamic studies*.

Nakanishi and Toyama [5389] purified 2,4-pentanedione for *vapor pressure measurements*. The solvent was first distilled under reduced nitrogen pressure and then in a column with glass packing.

SAFETY

The *minimum ignition temperature* in air is 340° [5712].

ACIDS (207–220)

Saturated Acids (207–216)

207. Formic Acid

Formic acid usually is produced from sodium formate, which is made by reacting carbon monoxide and sodium hydroxide under pressure. It is also obtained as a by-product of the pentaerythritol process and the acetic acid process from butane. In the acetic acid process, the formic acid is most efficiently separated from the acetic acid by azeotropic distillation with an aliphatic chlorohydrocarbon. A dark color develops in the formic acid in the subsequent process. Höfermann [3445] found that the coloration was due to an ill-defined polymerization product of isopropenyl ketone, which boils only 3° from formic acid.

It is available in several concentrations and grades of purity. ACS reagent grade is not less than 88% formic acid; high-purity commercial grades are available in 85 and 90%. An anhydrous grade is available in limited quantities. Some chemical supply houses list a 97% grade.

Formic acid slowly decomposes into carbon monoxide and water at room temperature. The decomposition can be prevented by keeping the acid frozen [1643]. Hammett and Dietz [3116] found it a stronger "super acid" than acetic acid. Its solvent use is limited because of its rapid decomposition. Formic acid generally is referred to as "very" or "extremely" hygroscopic. Hammett and Dietz [3116], however, exposed a sample freezing at 8.05° to air for 13 min and could not detect a change in the freezing point.

Coolidge [1643] purified the best grade of acid available for *vapor density measurements*. The acid was distilled in vacuum at room temperature to avoid decomposition. The vapors were condensed in a bulb packed with ice. Five fractional distillations yielded a product having a vapor pressure of 11.15 Torr at 0°. It was then fractionally crystallized and sublimed three times. The vapor pressure at 0° (11.16 Torr) and the melting point of 8.26° were used as the criteria of purity. It was found that dehydrating agents were unsatisfactory for preparing an anhydrous product.

Hammett and Dietz [3116] distilled in vacuum the best commercial grade of formic acid and then followed the method of Coolidge [1643]. The purified acid was used as a solvent for *potentiometric studies*. Precautions were taken to exclude moist air. See also Ewins [2341], Auerbach and Zeglin [491], Beckmann [722], and Timmermans and Hennaut-Roland [7488].

Wehman [7950] purified formic acid by bulk fractional freezing with temperatures ranging from 0° to 20° for freezing and melting, respectively. The limiting purity was checked by means of conductance, 6.08×10^{-5} ohm^{-1} cm^{-1}. The autoprotolysis constant was determined to be 2.19×10^{-7}.

Stout and Fisher [7030] prepared a high-purity formic acid for *thermodynamic studies*. An 85% solution of phosphoric acid was concentrated by heating to 250°. It was then cooled and added to ammonium formate and the formic acid distilled; the temperature of the flask was kept below 50°. The formic acid–water mixture was distilled into a bulb containing Drierite. The dried formic acid was fractionally distilled through a small column and the middle fraction used. The apparatus was an all-glass system from which air had been eliminated. The sample was kept frozen except during transfer and during the measurement of melting. The criteria of purity were the heat capacity and the change in melting point with fraction melted.

Boric anhydride and anhydrous copper sulfate have been used as drying agents but phosphorus pentoxide and calcium chloride are unsuitable, since they react with formic acid. Schlesinger and Martin [6455] dried formic acid in an all-glass apparatus for several days with powdered boric anhydride and distilled under reduced pressure at 22–25°. For the preparation of boric anhydride, boric acid was melted in an oven at a high temperature, poured on an iron plate, cooled in a desiccator, and powdered. Garner and coworkers [2667] distilled under reduced pressure over anhydrous copper sulfate.

Formic acid of greater than 99% purity can be produced from a dilute aqueous solution by a two-step distillation process involving butyric acid [1105]. The first distillation removes the bulk of the water and leaves a residue containing about 77% of the acid. The residue is then distilled with three to six times its amount of butyric acid as an azeotroping agent.

Sinke [6765] purified formic acid for *heat of combustion* determination by fractional crystallization to a purity of better than 99.9%m determined from the freezing curve.

Beckmann [722] froze the available formic acid, centrifuged, dried over boric anhydride, and fractionally distilled. The distillation system was protected against moisture; mp 5.6°.

Harned and Embree [3172] twice fractionally distilled the best formic acid available for the determination of the *ionization constant*.

CRITERIA OF PURITY

Specifications and test methods are given for the minimum 88% reagent-grade formic acid by Rosin [6263] and ACS reagent grade [137].

Coolidge [1643] used the vapor pressure at 0° and the freezing point as the criteria of purity. Hammett and Dietz [3116] and Dreisbach and Martin [2090] used the freezing point.

Stout and Fisher [7030] used the heat capacity and the change in melting point with fraction melted.

SAFETY

Formic acid is very corrosive to the skin, eyes, and mucous membranes. The vapors are irritating to the eyes, respiratory tract, and the skin [5713,7022].

The LD_{50} for rats, single dose, was reported as 1.21 g/kg [7633].

The *threshold limit* has been set at 5 ppm, 9 mg/m^3 [7022].

Care should be exercised when opening formic acid containers. When the acid has been stored for a long period of time or subjected to elevated temperatures, considerable pressure can develop as a result of decomposition.

208. Acetic Acid

Acetic acid is available at a reasonable cost and in good purity in quantities as desired. The ACS reagent grade is 99.7% based on the freezing point. The adjective *glacial* usually used to characterize the better grades of the acid is an indefinite description. Monnier [4027] states, "*glacial* is a term applied to acetic acid of high purity (above 99%) which congeals to ice-like crystals at 58–60°F."; (14.4–15.6°). The Random House *Dictionary of the English Language* defines it as at least 99.5% and solidifying at 16.7°. *USP* XVII lists the assay based on total acid titration as 99.4% and the freezing point is not a requirement. Marsden and Mann [4902] list a 99% minimum assay and a minimum crystallizing point of 15.5°.

Acetic acid is the most extensively used of the nonaqueous solvents for the titration of bases [6164].

Othmer and White [5626] developed a method for drying acetic acid by removing the water as the butyl acetate–water azeotrope. The method is applicable to other low-molecular-weight aliphatic acids.

Acetic acid was dried over phosphorus pentoxide for 24 hr, then refluxed over phosphorus pentoxide for 5 hr, and finally distilled. About 450 mL were distilled and then 2–10 mL portions collected for study of the *conductance behavior* of the acid in *N*-methylacetamide. It was shown that small quantities of acetic acid and/or the acetate remained with persistent tenacity in the amide even after it had been subjected to a rather extensive purification procedure [1872].

For *kinetic investigation*, Dimroth [2010] used glacial acetic acid distilled in a quartz apparatus. Commercial glacial acetic acid always contains traces of copper, which may act as a catalyst.

Kahane [3812] determined the heat of mixing of perchloric acid ($HClO_4 \cdot 2.5H_2O$) and acetic anhydride, and found that the mixture was perchloric acid and acetic acid. He cautioned that care should be exercised when using mixtures of perchloric acid and acetic acid.

MacInnes and Shedlovsky [4772] purified acetic acid for *ionization constant study* by fractionally distilling from 2% potassium permanganate, then from "a little chromic anhydride," followed by a final fractional distillation. The first quarter of the final distillation was discarded. The remaining fractions were kept or discarded on the basis of their freezing points.

Hutchinson and Chandlee [3602] determined the *activity coefficient* of sulfuric acid in anhydrous acetic acid purified as follows: USP acetic acid was refluxed for 10 hr with an excess of chromium(III) oxide to remove any easily oxidizable material. The acid was distilled from the solid residue and then fractionally distilled at a reflux ratio of 10:1 in a Marshall column [4907] packed with small Raschig rings. The distillate was protected from atmospheric moisture. McDougall [4761] purified acetic acid by the above method. A fraction of the distillate was further purified by fractional crystallization; mp $16.60 \pm 0.01°$.

According to Orton and coworkers [5601], impurities that react with chlorine or bromine, if present in an acid not melting below 16°, can be removed by a single distillation over phosphorus pentoxide. The distillate will contain a little acetic anhydride due to the action of phosphorus pentoxide. According to Orton and Bradfield [5599], technical acetic acid is best purified by distillation with chromium(III) oxide, preferably with the addition of a quantity of acetic anhydride corresponding to the water content of the acetic acid.

Harned and Ehlers [3169] purified acetic acid for *dissociation constant studies* by fractionally distilling three times in an all-glass still from 2% chromium(III) oxide. The first and last quarters were discarded. A fourth like distillation was made without the chromium(III) oxide.

Vogel [7768] found that acetic acid that had been distilled from potassium permanganate was not suitable for *physical property measurements*. The acid used was prepared by partially freezing and rejecting about one-half as unfrozen liquid, then fractionally distilling from permanganate. The portion distilling between 116.5 and 117.5° at 765 Torr was collected, partially frozen, and about one-half of the acid discarded as liquid. The solid was melted and again fractionally distilled from permanganate.

Turner and Pollard [7590] purified commercial glacial acetic acid for use as a solvent for *molecular weight determinations*. The acid was distilled from potassium permanganate and the middle portion collected and dried over phosphorus pentoxide and redistilled. The middle portion from the second distillation was subjected to fractional freezing. See also Hess [3353].

For *specific conductance measurements*, Rabinowitsch [6014] dried acetic acid before distilling by shaking for 20 min with phosphorus pentoxide. The acid was filtered into the distillation flask through a layer of glass wool and phosphorus pentoxide. Schall and Thieme-Wiedtmarckter [6422] treated the purest commercial acetic acid, free from known impurities, with a considerable quantity of boron triacetate (Pictet and Geleznoff [5844]), allowed it to stand for 3 days in a sealed flask at 30–40°, distilled under reduced pressure, and purified by freezing once. The acid was kept in a desiccator over sulfuric acid, particular care being taken when samples were withdrawn.

Myers [5371] purified acetic acid for *electrical property studies*. Reagent-grade acid was analyzed for water by a colorimetric method similar to that of Greathause and coworkers [2917]. A solution of 6 mL of perchloric acid in 39 mL of acetic acid was used as the catalyst to avoid rise in temperature. See [5371] for details of the colorimetric method. The amount of acetic anhydride plus 2% was added to the acetic acid to react with the water present. The reaction was catalyzed by 1 gram of anhydrous 5-sulfosalicylic acid per liter. The solution was kept overnight at near 100° and distilled through a 1-m column packed with glass helices. The distillate had a *specific conductance* of 6×10^{-9} ohm^{-1} cm^{-1}.

Lundin [4700] purified acetic acid by mixing a commercial acid with its own volume of water and adding 1% of activated carbon or 2% of charcoal. The mixture was boiled 10–15 min, filtered, and distilled.

Korpela [4198] purified acetic acid for *compressibility studies*. Acetic anhydride equivalent to the water present was added to the acid. After boiling the mixture for 24 hr, acetic acid was recrystallized twice from the melt (mp 16.6° and water content, 0.005%w).

Suitable drying agents for acetic acid are phosphorus pentoxide, boron triacetate, magnesium perchlorate [3353], anhydrous copper sulfate, and chromium(III) acetate. Acetic acid may be dried with acetic anhydride and used without further treatment if the small amount of excess anhydride is not objectionable. For the determination of acetic anhydride in acetic acid, see Benson and Kitchen [777].

CRITERIA OF PURITY

The specifications and test methods for Reagent grade acetic acid are given by Rosin [6263] and for ACS reagent grade in *Reagent Chemicals* [137]. The reagent-grade "glacial" acetic acid from most suppliers meets more stringent specifications than those of ACS reagent grade. The melting point is the most critical criterion of purity. Jones and Betts [3746] suggest the critical solution temperature with benzene; Bousfield and Lowry [1019] suggest the specific conductance. The freezing curve was used by Dreisbach and Martin [2090].

SAFETY

The odor of acetic acid has been described as sharp and penetrating. It can be detected by odor below 10 ppm. Normally, the chief hazard from acetic acid is the liquid on the skin or in the eyes. Concentrations greater than 80% produced severe burns on the skin of guinea pigs. Concentrations below 50% produce relatively mild injuries [5713].

Takhirov [7163] reported a maximum permissible concentration of 0.06 mg/m^3.

The LD_{50} for rats, single dose, was reported as 3.53 g/kg [7633].

The *threshold limit value* is 10 ppm, 25 mg/m^3 [7022].

The *lower explosion limit* in air is 4.05%v [1043]. The *minimum ignition temperature* in air is 550° [5712].

209. Propionic Acid

Child [1465] states that propionic acid is thermally unstable at the normal boiling point.

Weizmann [7974] prepared propionic acid by the air oxidation of propionaldehyde in the presence of a catalyst containing lead, manganese, and cobalt oleates. The air was passed through a stirred mixture containing 288 parts of the aldehyde and 5.76 parts of the catalyst for 8 hr. No external heat was used, but the temperature increased to 45° over the first 2 hr and remained there for the remaining reaction time. The propionic acid was isolated by distillation; yield 62.5%.

Vogel [7768] purified commercial propionic acid for *physical property determinations*. About 1 L was dried over anhydrous sodium sulfate and fractionally distilled through a Young and Thomas column. The fraction distilling at 139–141° was collected and refractionated over potassium permanganate. The middle fraction distilling at 140.7° at 760 Torr was collected.

Korpela [4198] purified propionic acid for *compressibility measurements* by boiling the acid over potassium permanganate, distilling, and crystallizing twice from the melt (mp $-22.2°C$ and water content, 0.06%w).

Martin and Andon [4919] purified propionic acid for *heat capacity measurements* by fractional freezing with a dry ice–ethanol bath and drying over calcium sulfate. Purity was 99.93%m.

CRITERIA OF PURITY

Rosin [6263] gives the specifications and test methods for a reagent-grade propionic acid.

Dreisbach and Martin [2090] determined the purity from the freezing curve.

SAFETY

The chief hazard from propionic acid is the local damage of the liquid on the skin or in the eyes. No accumulative effects are known from industrial exposures [5713].

The LD_{50} for rats, single dose, was reported as 4.29 g/kg [7633]; see also [6617].

210. Butyric Acid

Vogel [7768] purified a commercial grade of butyric acid to determine some *physical properties*. About 250 mL of redistilled acid was mixed with 5 grams of potassium permanganate and refractionated. The first one-third was discarded and the remainder distilled constantly at 162.5° at 767 Torr.

Martin and Andon [4919] purified butyric acid for *heat capacity measurements* by fractional distillation with a spinning-band distillation apparatus at 500 Pa. Purity was 99.93%m.

CRITERIA OF PURITY

Dreisbach and Martin determined the purity from the freezing curve [2090].

SAFETY

Stasenkova and Kochetkova [6966] concluded, from their test with animals, that butyric acid is not very toxic. Lethal vapor concentrations could not be obtained but the vapors are irritating to the skin and eyes. The chronic threshold concentration was found to be 0.1 mg/L and a permissible trial concentration of 0.01 mg/L was recommended. A *threshold limit value* of 10.0 mg/m^3 was recommended for Russia in 1964 [4185]. For oral toxicity, see [6883].

The *minimum ignition temperature* is 452° [5415].

211. Isobutyric Acid

Isobutyric acid is prepared by the oxidation of 2-methyl-1-propanol or from 2-propanol through the chloride or nitrile. It has the same characteristic odor of rancid butter as butyric acid.

Lippincott and Hass [4626] prepared isobutyric acid by the hydrolysis of 1-nitro-2-methylpropane. Groll and Tamele [2979] prepared the acid from isobutyraldehyde by air oxidation at less than 45° in the presence of iron isobutyrate.

Konicek and Wadso [4178] purified isobutyric acid by drying the acid over molecular sieve 4A and fractionally distilling; see also isovaleric acid.

212. Valeric Acid

Valeric acid has a putrid odor. It is made by the action of carbon dioxide and water with olefins, by the electrolytic oxidation of 1-pentanol, and by the reaction of formic acid and butene.

Valeric acid can be prepared by treating the nitrile with sodium hydroxide, removing the excess alcohol and freeing the acid with sulfuric acid [25]. About 300 grams with an 81% yield was prepared.

Vogel [7768] prepared valeric acid from 1-bromobutane. A large sample was distilled from the middle fraction, bp 184° at 768 Torr, and collected for *physical property measurements*.

Two samples of valeric acid were purified by McDougall and Kilpatrick [4762] for *entropy and related thermodynamic properties study* using a Nester and Faust spinning band column at reduced pressure. The refinements in collection and storage of the samples reduced the impurity from 1.79%m in the first sample to 0.41%m ± 0.01 in the second, based on the sharpness of the melting under 1 atm of helium. The samples analyzed 99.6 and 99.8% by gas chromatography.

CRITERIA OF PURITY

Adriaanse and coworkers [38] determined the purity from the freezing curve.

SAFETY

The single dose LD_{50} for rats was reported as 1.12 g/kg [7633].

213. Isovaleric Acid

Isovaleric acid has a disagreeable rancid cheese odor. A limited amount is obtained from pyroligneous acid, which is one of the compounds formed during the high-pressure reaction in the manufacture of alcohols. It is made by the electrolytic oxidation of 3-methyl-1-butanol and the catalytic oxidation of the alcohol by carbon dioxide.

Vogel [7768] dried about 500 grams of a pure commercial acid over anhydrous sodium sulfate and fractionally distilled; bp 176.5° at 762 Torr.

Konicek and Wadso [4178] purified isovaleric acid by drying the acid over molecular sieve 4A and fractionally distilling; see also isobutyric acid.

214. Hexanoic Acid

Hexanoic (caproic) acid has the characteristic odor of goats, *Capra* L. It is manufactured by the oxidation of 1-hexanol and by the hydrolysis of hexanenitrile.

Butyl malonic ester can be saponified to hexanoic acid in a 74% yield [25].

Vogel [7768] dried a pure grade of hexanoic acid and fractionally distilled. The fraction collected at 203° at 756 Torr was used for *physical property measurements*.

Simon [6733] saponified hexanenitrile that was prepared for *physical property studies* (see hexanenitrile) with an alcoholic solution of a large excess of sodium carbonate by refluxing in a water bath for 8 hr. The alcohol was separated by distillation. The remaining mixture was heated almost to dryness in a large evaporating dish with a flame and then dissolved in water. Hexanoic acid was liberated with a moderate excess of 50% sulfuric acid and then washed, dried, and distilled. The portion that boiled within a 2° range was distilled three times.

Rose and coworkers [6256] purified hexanoic and octanoic acids for *liquid–vapor equilibrium studies*. Commercial-grade materials with a purity of 85–95% were fractionally distilled at a high reflux ratio in a 30-plate packed column at a pressure of about 50 Torr. The heart cuts were used. The criteria of purity were the acid number, the iodine number, carbon and hydrogen analysis, water by Karl Fischer, and partial crystallization, followed by comparison of fractions.

Garner and coworkers [2669] studied some *heat values* of hexanoic acid purified by fractional distillation or crystallization at a constant setting point of −3.6°.

CRITERIA OF PURITY

Adriaanse and coworkers [38] determined the purity from the freezing curve.

SAFETY

Skin and eye irritation are relatively severe [5713]. The LD_{50} single dose for rats is 5.19 g/kg [6683].

215.   Octanoic Acid

Armour [427] lists the approximate chemical compositions of two grades of octanoic acid purified by crystallization and fractional distillation and analyzed by gas chromatography. Commercially pure octanoic acid contains 92.0% octanoic acid, 5.0% hexanoic acid, and 3.0% decanoic acid whereas 98% minimum commerically pure octanoic acid contains 99.2% octanoic acid, 0.4% hexanoic acid, and 0.4% decanoic acid.

Vogel [7768] carefully fractionally distilled commercial octanoic acid for *physical property studies;* bp 236° at 769 Torr.

See Rose and coworkers [6256] under hexanoic acid for purification.

SAFETY

Octanoic (caprylic) acid is a rather mild irritant. It can be detected by odor at about 0.008 ppm [5713].

The LD_{50} for rats is 10,080 mg/kg [3708].

216.   Nonanoic Acid

Nonanoic acid is a colorless fatty liquid with a heavy rancid odor. It is prepared from unsaturated hydrocarbons by the oxo process, from tall oil unsaturated fatty acids, and from rice bran fatty acids. It probably was recognized as an entity during the study of oleic acid and is one of the oxidation products of oleic acid.

Kraaft [4235] prepared nonanoic acid by the reaction of undecylenic acid with potassium hydroxide, followed by treatment with hydrochloric acid solution and distillation.

Port and Riser [5920] esterified commercial nonanoic acid and recovered the pure ester by vacuum distillation. The pure acid was recovered by treating the ester with 33% sulfuric acid solution, followed by distillation.

Adriaanse and coworkers [38] purified nonanoic acid for *thermodynamic studies* by a recrystallization process; purity 99.75%m.

Reid and Ruhoff [6112] reported a detailed synthesis of nonanoic acid by reacting sodium butoxide, ethyl malonate, and heptyl bromide.

Unsaturated Acids (217–220)

General Comments

Ratchford and coworkers [6077] found that unsaturated acids could be prepared in high yields by passing a suitable corresponding ester through a Pyrex-brand glass tube heated to 500–580°. The ethyl ester of acrylic and methacrylic acids pyrolyzed to 92 and 87%, respectively. The secondary esters gave the highest yields. Butyl crotonate at 537° and a residence time of 7.2 s gave 84% crotonic acid. The purification of the acids is discussed.

Specific Solvents

217. Acrylic Acid

Rheberg [3492] described a method for the preparation of acrylic acid by acidolysis of ethyl acrylate with formic acid.

Moureau [5292] prepared acrylic acid for refluxing 1 mol of 3-bromopropionic acid with 2 mol of aqueous potassium hydroxide. The acrylic acid was liberated with sulfuric acid and distilled.

Rohm & Haas [6238] lists 0.020% p-methoxyphenol or 0.10% N,N'-diphenyl-p-phenylenediamine as polymerization inhibitors. Ababi and Mihailia [1] used p-methoxyphenol as a stabilizer and removed it by distilling at 15 Torr.

SAFETY

Acrylic acid has an acrid odor. It is a severe skin, eye, and respiratory irritant as a liquid or in concentrated solutions. It should be handled with the usual precautions used for acetic or propionic acids [5713].

The LD_{50} for rats, single dose, was reported as 2.50 g/kg [7633]. A permissible concentration in air has been suggested as 3600 mg/m^3 [4844].

218. Crotonic Acid

Dutt [2166] states that malonic acid easily condenses with aldehydes in the presence of piperidine in pyridine solution to alkylidine- and arylidenemalonic acids, which when heated lose carbon dioxide to give the α-unsaturated carboxylic acid. Acetaldehyde gave 75% crotonic acid; see also [6430].

Crotonic acid may be purified by crystallization either by partial freezing or from a solvent such as hexane. It may be sublimed.

219. Methacrylic Acid

Crawford and McGrath [1713] heated acetone cyanohydrin and concentrated sulfuric acid at 130°. The methacrylamide was hydrolyzed in boiling aqueous solution to the acid.

Rohm & Haas [6238] lists 0.025% p-methoxyphenol, 0.10% hydroquinone, and 0.05% N,N'-diphenyl-p-phenylene diamine as polymerization inhibitors.

SAFETY

Lobanov and coworkers [4641] determined the acute oral LD_{50} as 1250 mg/kg in mice and 1600 mg/kg in rats. They recommended a maximum permissible vapor concentration of 0.44 mg/m^3.

220. Oleic Acid

Oleic acid is considered the most prevalent acid in nature. It comprises 50% or more of the total acids of many fats and oils. Few fats are known to contain less than 10% of this acid [4891]. The fatty acids of olive oil are 80% oleic. Markley [4893] states that all natural fats contain unsaponifiable matter such as hydrocarbons, long-chain aliphatic alcohols, and sterols. The concentration may vary from 0.2% in lard and tallow to 20% in shark liver oil. These impurities tend to separate in the alcoholic fractions during the lead salt purification.

Markley [4893, Chap. 20] has an excellent 140-page discussion, including 569 references, concerning the techniques of separating naturally occurring fatty acids. The discussion includes consideration of hydrolysis and esterification for preparation from natural fats, oils, and other material. The separation and purification of the acids are discussed during the presentation of the three principal separation techniques, namely, distillation, salt solubility, and low-temperature crystallization.

Armak, formerly known as Armour, lists the approximate chemical compositions of four grades of oleic acid purified by crystallization and fractional distillation and analyzed by gas chromatography. The grades are low polyunsaturated oleic, 5°C maximum titer crystallized white oleic, 5°C maximum titer crystallized red oil, and 8–11°C titer crystallized red oil [423].

The iodine number was calculated as 89.87 [1707].

See methyl oleate for preparation and purification of the ester which is used as part of the purification of oleic acid.

Moreno [5248] separated oleic acid by extracting olive oil with 96% ethanol in a tower packed with Raschig rings or pieces of solid glass. The latter was more effective. On passing through the tower, the mixture separated into an oil phase and an alcohol phase, the latter containing 75–90% oleic acid.

Hartsuch [3200] separated oleic acid from linoleic and saturated acids by successive crystallization from acetone at −20, −40, and −60°. Part of the remaining saturated acids were precipitated in ethanol as lead soaps. The product was then distilled at 1 Torr in a 4-ft column packed with glass helices. The oleic acid has a purity of 97.8% as determined by the iodine and thiocyanogen numbers.

Ware and Dunell [7907] purified a commercial "purified-grade" oleic acid by crystallization from acetone at −11°. The crystals were pulverized and dried *in vacuo* at −11° for 1 week for *proton magnetic resonance studies*.

Keffler and McLean [3924] made an extensive study of the *purification* and *thermochemical properties* of oleic acid. The "solid" acids were separated from the "liquid" acids by the lead salt method repeated two or three times. The "liquid" acids were converted into the lithium salts and recrystallized from 80% ethanol as many as 12 times, until the iodine value became and remained identical within 0.2% for the successive precipitates and corresponding mother liquors. The oleic acid was free of linoleic (9,10-octadecadienoic) and linolenic (9,12,15-octadecatrienoic) acids but contained about 4% saturated acids. The acid was fractionally distilled at 0.1 Torr several times. It was found that a less laborious method was to convert the oleic acid containing the 4% saturates to the methyl or ethyl ester and fractionally distill at a pressure obtained with two mercury vapor pumps in series.

Two lots of oleic acid were prepared, one from the methyl ester and the other from the ethyl ester. The iodine value and the density were the criteria of purity.

Wheeler and Riemeneschneider [7993] prepared 99.6% methyl ester. The acid was obtained from the ester. The methyl esters from the acids from olive oil were distilled. The unsaturated esters were crystallized from acetone at different concentrations and temperatures at −60, −35, and −60°. The material was again fractionally distilled and crystallized twice from a petroleum naphtha at about −65°.

Brown and Shinowara [1153] concluded from previous work that high-purity oleic acid could be obtained by direct crystallization from acetone. Their method involved the removal of the saturated acids by a −20° acetone precipitation, crystallization of the unsaturated acids at −60° four or more times, followed by a partial crystallization from acetone at −35° to remove the small amount of palmitic acid not removed in the initial −20° crystallization. Three specimens melted at 13.0°, mean molecular weight 282.4, mean iodine number 89.90 (calculated value 89.93), mean n_D 20° 1.4585.

Smith [6832] carefully prepared oleic acid from methyl oleate that was fractionally distilled. The acid was crystallized by the Brown and Shinowara [1153] method to a constant melting point. Smith is also of the opinion that pure oleic acid may be obtained by crystallizing from acetone only but he recommends a preliminary esterification and fractional distillation. He prepared and measured the *freezing points* of the α- and β-forms.

The methyl and ethyl esters are generally used in the separation of naturally occurring fatty acids by fractional distillation. Metcalf and Schmitz [5072] converted the acid to the ester by boiling 2 min with boron trifluoride–methanol reagent and obtained practically quantitative conversion. The ester was hydrolyzed and converted to the lead salt and crystallized from ethanol or ethyl ether.

Lithium salts are crystallized from acetone and barium salts from benzene [4893].

Hoerr and Harwood [3442] prepared a high-purity oleic acid from olive oil for *solubility studies* by acetone crystallization and fractional distillation. They concur with, and explain, Hartsuch's [3200] conclusion that oleic acid exceeding 98% purity had not been prepared because stearic and oleic acids form an eutectic at 98% oleic acid. Hartsuch's statement, therefore, applies only to oleic acid prepared by crystallization. The solubility of the acid was determined in 23 solvents of various functional group types at seven temperatures from −40 to 20°. This article is a good reference source for pertinent work done on oleic acid prior to 1952. Hoerr and Harwood are of the opinion that the freezing point (not the melting point) is the best criterion of purity.

SAFETY

There are no known hazards in the use of oleic acid in industry [5713].

ACID ANHYDRIDES (221–223)

General Comments

The acid anhydrides are hygroscopic and slowly react with water to form acids.

Allen and coworkers [3492] found that an acid chloride in a dry solution of pyridine in benzene reacts with an acid to form the anhydride. Mixed anhydrides may be prepared by this method.

Kostyuk and coworkers [4213] report a method for preparing the anhydrides of carboxylic acids by dehydrating the acids with acetic anhydride. An equimolar mixture of the anhydride desired and acetic anhydride (an excess may be used) was refluxed in a 20-theoretical-plate column until the head temperature came to the boiling point of acetic acid. The mixture was distilled at a reflux ratio, so that the distillation took 2.4 hr. Propionic, isovaleric, hexanoic, undecanoic, and benzoic anhydrides were prepared.

Lewis [4562] purified acetic, propionic, and hexanoic anhydrides and determined their *boiling points, surface tensions, densities,* and *viscosities.* The anhydrides were fractionated through a glass bead column until they gave a fairly constant boiling point and were relatively free from the lower boiling acid. They were refractionated twice over the fused sodium salt of the corresponding fatty acid and twice under reduced pressure. The higher anhydrides decompose slowly when distilled at atmospheric pressure.

Cockerille [1548] described a method for purifying aliphatic acid anhydrides. About 0.2–2% of sulfuric acid was added to the impure anhydride in two stages. In the first stage, the reaction mixture was heated to between 60° and the boiling point; in the second stage, the mixture was heated to 50–55° for some time. An alkali or alkaline earth metal salt of the corresponding acid was added to the anhydride to neutralize all traces of the sulfuric acid; the mixture was then distilled.

Specific Solvents

221. Acetic Anhydride

Acetic anhydride may be purified by fractional distillation through an efficient column; see Orton and Jones [5604] and Jones and Betts [3746]. To remove acid, Walton and Withrow [7899] allowed 97% anhydride to stand several days over thin slices of sodium, boiled it in a vacuum under reflux for several hours, and finally distilled it over a mixture of sodium and sodium acetate. It was further purified by fractional distillation. Since sodium reacts vigorously with acetic anhydride at 65–70°, Calcott et al. [1308] treated 95% anhydride (1450 g) with coarse magnesium filings (200 g) under reflux at 80–90° for 5 days. After boiling moderately for 17 hr, it was fractionally distilled through a 30-cm Hempel column.

Myers [5371] purified acetic anhydride for *electrical properties study* by adding toluene to the anhydride and removing the toluene–acetic acid azeotrope in an efficient distillation column; bp 100.6°. The toluene was removed and the anhydride distilled at a reflux ratio of 20:1. The distillate reached the lowest conductivity when about one-half of the charge had distilled; 5×10^{-9} ohm^{-1} cm^{-1}.

Rollett and Leimüller [6243] freed acetic anhydride from most of the acetic acid present by refluxing with calcium carbide and distilling the anhydride. The calcium acetate showed no tendency to decompose. Aluminum chloride also could be used; however, it decomposes slightly during distillation.

McClure and coworkers [4744] report two methods for the direct determination of acetic acid in acetic anhydride. One is a colorimetric and the other a thermometric titration method. Neither method is applicable if amines are present in the anhydride.

A method for the determination of acetic acid in acetic anhydride was reported by Mitra and coworkers [5177].

CRITERIA OF PURITY

ACS reagent-grade acetic anhydride is not less than 97% anhydride by alkali titration of the acidity in aqueous solution [137]. Rosin's specifications [6263] are essentially the same as those for the ACS reagent grade. A 99+% grade is available.

Timmermans and Hennaut-Roland [7488] used the consistency of the density of fractions as a criterion of purity.

Acetic anhydride is a severe eye and skin irritant. It has a pungent odor and is a lachrymator. Bronchial and lung injuries are likely to occur from inhalation of the vapor [7022,5713].

Takhirov [7163] reported a maximum permissible concentration of 0.03 mg/m^3.

The *threshold limit value* is 5 ppm, 20 mg/m^3 [7022].

The *flammable limits* in air at elevated temperature are 2.90 and 10.30%v [5712]; see also [152]. The *minimum ignition temperature* in air is 392° [5712].

222. Propionic Anhydride

Smith and Hunter [6827] reduced crotonic anhydride in the presence of nickel, cobalt, or Raney nickel at 100 psi pressure at 80°.

The effects are similar to those of acetic anhydride [5713].

223. Butyric Anhydride

Butyric anhydride is a skin and eye irritant [5713]. A *threshold limit value* of 0.001 mL/L has been recommended [6967].

ESTERS (224–282)

General Comments

The same methods of preparation and purification are usually applicable to a large number of esters, particularly to members of a homologous series.

Spassow [6920] developed a method for preparation of esters from aliphatic primary, secondary, and tertiary alcohols, and aromatic alcohols including phenol (except triphenyl methanol). The method is simple and rapid and produces esters in good yield. The general directions are: The alcohol and magnesium are mixed with ether in the ratio of 0.1 mol of alcohol to 0.1 atom weight of magnesium to 10–15 g ether; some tertiary alcohols require a ratio of ether to alcohol of 15–20:0.1 mol. Acid chloride in ether (0.1 mol: 5–10 g) is slowly dropped, with moderate cooling if necessary, into the alcohol mixed in slight excess (1.25–1.5 equivalents). The mixture is allowed to stand for 1 hr, heated 1–2 hr in a water bath, or poured into water or dilute sodium carbonate solution, extracted several times with ether, dried, and fractionally distilled to the desired purity.

Lippincott [4625] devised a method of preparing esters by heating acetals with an aliphatic acid of low molecular weight with, or without, 10% sulfuric acid. For example, 56 parts of dimethoxymethane, 92 parts of formic acid and 30 parts of 10% sulfuric acid were placed in a reactor equipped with an efficient fractionating column. Upon heating, 120 parts of methyl formate were obtained by distillation.

Gonzalez [2844] prepared a series of esters of saturated and unsaturated mono- and dicarboxylic and iso- and heterocyclic acids. In general, 1 mol of the acid, 3.5 mol of the alcohol, and 0.02 mol of chlorosulfinic acid were heated for 2 hr at the boiling point of the mixture. The esters were neutralized, washed with sodium carbonate solution and then with water, and dried over anhydrous sodium sulfate. Solid esters were purified by crystallization. Typical yields were: ethyl acetate 82.5%; isopentyl acetate, 78.8%; ethyl propionate, 77.7%; ethyl benzoate, 81.8%, and ethyl oxalate, 14.9%.

Extensive work on purifying low-molecular-weight fatty acid esters has been done by Young and Thomas [8236]. The esters were obtained from Kahlbaum and also prepared from the acid or acid anhydride and the alcohol. Materials from both sources were carefully purified and the physical properties compared. It was assumed that the impurities present were alcohols, acids, water, and homologous esters. The acids were removed by repeated agitation with strong aqueous solutions of potassium carbonate; it was observed that solid potassium carbonate was useless for this purpose. The esters were then washed several times with water, except those compounds with an appreciable water solubility. The lower alcohols were largely removed during the water wash. Drying was necessary before fractional distillation. Potassium carbonate was useful only for clearing the esters when turbid with suspended water. The final drying was made with phosphorus pentoxide, which combined with the water and any residual alcohols. If appreciable amounts of water and alcohols were present, the oxide would hiss and liquefy rapidly. The ester was repeatedly decanted and more phosphorus pentoxide added until the oxide merely became pasty; the ester was then distilled. Phosphorus pentoxide was added to the distillate; by the next day a jelly had formed with the ester. This is a common phenomenon when phosphorus pentoxide is allowed to stand with a dry, alcohol-free ester. The ester was again distilled in a dry atmosphere.

Esters, alcohols, and water are noted for azeotrope formation. For example, ethanol and water, ethanol and ethyl acetate, and ethyl acetate and water form binary azeotropes; the three compounds form a ternary azeotrope. Esters, as a rule, do not form azeotropes with esters, particularly with those of the same homologous series. Esters boiling below 100° usually can be distilled at atmospheric pressure. Some esters decompose above 100°, but many can be distilled as high as 150° without decomposition.

Safety

The simple aliphatic esters produce irritation and a slight to marked anesthetic action when inhaled in sufficiently high concentration. Some irritate the upper respiratory tract. They produce a considerable fatigue of the olfactory sense, so that the sense of smell is effective only in detecting initial exposure.

The formates are the least toxic of the simple esters, the toxicity increasing slightly from formates to acetates to propionates, and so forth. Within each series, the toxicity increases with molecular weight and boiling point, with the exception of the methyl derivatives. The formates, especially the methyl and ethyl esters, are quite irritating to the eyes and respiratory tract, probably because they hydrolyze readily in contact with water to formic acid. The benzyl fatty acid esters are more toxic than the aliphatic compounds.

The chronic effects of the simpler aliphatic esters are relatively slight.

The *threshold limit value*, the *flammable limits* in air, and the *minimum ignition temperature* in air for aliphatic esters of some saturated and unsaturated monocarboxylic acids are given in Table 5.14.

Table 5.14 Safety Data for Esters of Aliphatic Monocarboxylic Acids

Ester	TLV ppm	TLV mg/m^3	FL in Air (% v) Lower	FL in Air (% v) Upper	MIT in Air (°)
224. Methyl formate	100	250	5.90	21.70	456
225. Ethyl formate	100	300	3.15	16.40	455
226. Propyl formate	100	360	—	—	455
227. Butyl formate	—	—	1.73	8.15	334
228. Isobutyl formate	—	—	2.00	8.90	—
229. Methyl acetate	200	610	3.85	15.20	502
230. Vinyl acetate	—	—	—	—	427
232. Ethyl acetate	400	1400	2.18	11.40	484
235. Propyl acetate	200	840	1.77	8.00^a	450
236. Isopropyl acetate	250	950	1.78	7.80^a	476
238. Butyl acetate	150	700	1.40	8.00^a	421
239. Isobutyl acetate	—	—	—	—	423
241. Pentyl acetate	100	525	1.03	7.13^a	358
247. Ethyl propionate	—	—	1.85	11.05	440
248. Ethyl butyrate	—	—	—	—	463
252. Butyl stearate	—	—	—	—	355
253. Methyl acrylate	10	35	—	—	—
254. Ethyl acrylate	25	100	—	—	—
255. Methyl methacrylate	100	410	—	—	—

TLV is threshold limit value; FL is flammable limits in air; and MIT is minimum ignition temperature. TLV reference [7022, 152]; FL and MIT reference [5712].

^{a}At elevated temperature.

Esters of Saturated Aliphatic Monocarboxylic Acids [224–252]

224. Methyl Formate

For the determination of *physical constants*, Timmermans and Hennaut-Roland [7488] treated methyl formate with sodium carbonate and distilled several times in a water bath from phosphorus pentoxide.

225. Ethyl Formate

Bodroux [935] prepared ethyl formate by distilling a mixture of ethanol, formic acid, and water. The yield depends on the proportion of the reactants.

Schiff [6439] found that ethyl formate, like methyl acetate, could not be dried with calcium chloride because it reacts rapidly with the ester to form a crystalline compound.

226. Propyl Formate

Vogel purified propyl formate for *physical property measurements*. The distillate boiling below 85° was washed sucessively with saturated sodium chloride solution and saturated sodium bicarbonate solution in the presence of solid sodium chloride, dried with anhydrous magnesium sulfate, and distilled [7764].

227. Butyl Formate

Vogel [7764] prepared and purified butyl formate for *physical property measurements* by refluxing 46 grams of formic acid and 37 grams of 1-butanol for 26 hr. The reaction mixture was successively washed with a solution saturated with sodium bicarbonate and sodium chloride until effervescence ceased, and then dried and distilled. The yield was 40 grams of ester boiling at 105–107° at 765 Torr.

228. Isobutyl Formate

Isobutyl formate was prepared and purified by Vogel [7764] in a similar manner to butyl formate.

229. Methyl Acetate

Hurd and Strong [3592] purified methyl and ethyl acetates by the following procedure: One liter of the ester was refluxed for 6 hr with 85 mL of acetic anhydride and then distilled through a Vigreux column. The distillate was shaken with 20 grams of anhydrous potassium carbonate and redistilled.

Vogel [7764] purified methyl acetate for *physical property measurements*. The ester was washed with saturated sodium chloride solution, dried with anhydrous magnesium sulfate, and distilled.

To remove methanol, Schiff [6439] treated methyl acetate with acetyl chloride, washed with concentrated sodium chloride solution, and dried over calcium oxide. It must not be dried with calcium chloride, since it reacts rapidly to form a crystalline compound.

230. Vinyl Acetate

Mitsutani and Kominami [5179] studied the vinyl acetate reaction product and identified several compounds from the mass spectra. The following compounds may be present in the commercial product as impurities: acetaldehyde, acetone, 2-butenal, ethyl acetate, isopropyl acetate, 1-butene, butenyne ($CH \equiv CCH = CH_2$), 1,3-butadiene, 1-hexene, 1-octene, benzene, toluene, and 1,7-octadiene.

Ushakov and Feinstein [7649] prepared vinyl acetate by reacting acetylene and acetic acid in the vapor state in the presence of zinc or cadmium acetate on activated carbon at 180–250°.

Acetylene and glacial acetic acid react to form the ester in the presence of mercury(II) sulfate [1439].

Nozaki and Bartlett [5531] fractionally distilled vinyl acetate through a 100+-plate column with stainless steel Lecky-Ewell packing and a total reflux, intermittent take-off head designed to avoid contact with mercury, grease, air, or metal other than the packing. The fraction distilling at 72.3° at 756 Torr was used to study the *rate constant in* addition polymerization.

Vinyl acetate was found to be stabilized by 0.05% anthracene [793].

SAFETY

The *minimum ignition temperature* in air is 427° [1411].

231. Ethylene Glycol Diacetate

Taylor and Rinkenbach [7192] dried a commercial product boiling at 188–189.6° at 739 Torr with calcium chloride, filtered with the exclusion of moist air, and fractionally distilled at reduced pressure.

Kusano [4339] dried the ester over molecular sieve type 4A and fractionally distilled at reduced pressure.

SAFETY

Rowe [5713] believes that ethylene glycol diacetate should be considered like ethylene glycol, based on available limited information.

232. Ethyl Acetate

This ester is available commercially with a composition corresponding to the water azeotrope of about 85–92% ester. It is also available in an anhydrous, or absolute grade, usually guaranteed to be at least 99.5% ester by saponification.

Vogel [7764] prepared ethyl acetate for *physical property measurements*. The preparation and purification were similar to those described for butyl formate.

Gillo [2754] proposed a method of purification and criteria of purity for ethyl acetate as an *organic standard*. The ethyl acetate was dried over potassium carbonate for 2 months, filtered, distilled, and the first and last portions of the distillate discarded. The entire center fraction was distilled over phosphorus pentoxide (10–20 g/kg) from an all-glass apparatus. The first 10 mL was discarded before the main fraction was collected. The instability and decomposition of the ester stored in the absence of air for 4, 6, and 13 years are discussed.

Hurd and Strong [3592] purified ethyl acetate in the same manner as they did methyl acetate.

CRITERIA OF PURITY

The ACS reagent-grade [137] ethyl acetate is characterized by density and distillation range, plus tests for impurities. Rosin [6263] bases the characterization on a saponification assay and a distillation range. Dreisbach and Martin [2090] determined the purity from the freezing curve. Maclean and Jencks [4812] characterized the effectiveness of their purification procedure by the ultraviolet absorption spectra.

233. Propargyl Acetate

Yields of propargyl acetate up to 50% were reported by Golendeev and Okrokova [2835] by esterification of acetic acid and 2-propyl-1-ol in the presence of concentrated sulfuric acid and with benzene as the solvent.

234. Allyl Acetate

Wislicenus [8117] prepared allyl acetate from 3-iodopropene and sodium acetate. Palomaa and Juvala [5657] reacted allyl alcohol and acetyl chloride to obtain the ester.

Bartlett and Altschul [648] prepared allyl acetate for *polymerization studies* by heating for 2 hr under reflux 1.5 mol of allyl alcohol and 0.75 mol of acetic anhydride to which

5 mL of 96% sulfuric acid had been added. The mixture was poured slowly into ice water with stirring. It was then extracted with ice-cold sodium carbonate and cold saturated calcium chloride solutions. The ester was dried and then distilled in an all-glass apparatus in a nitrogen atmosphere. The middle fraction was collected.

235.   Propyl Acetate

Vogel [7764] prepared and purified propyl acetate for *physical property measurements* by refluxing 40 grams of 1-propanol, 120 grams of glacial acetic acid, and 2 grams of sulfuric acid for 12 hr. An equal volume of water was then added to the reaction mixture and the ester separated and purifed as described for butyl formate.

236.   Isopropyl Acetate

Haggerty and Weiler [3067] purified isopropyl acetate for *vapor pressure determinations*. It was shaken with potassium carbonate (50 g/100 mL of ester) to remove acids, and then with a strong solution of sodium chloride to remove alcohols. It was allowed to stand over calcium chloride overnight and then carefully distilled. A fraction boiling within 0.1° was collected for measurements. See also Vogel [7764].

237.   Triacetin

SAFETY

The *minimum ignition temperature* in air is 433° [5415].

238.   Butyl Acetate

Butyl acetate can be purchased in a relatively pure state. For laboratory preparation Vogel [7764] passed dry hydrogen chloride into 37 grams of 1-butanol until 1 gram was absorbed; 120 grams of acetic acid was then added and the mixture refluxed 8 hr. The reaction mixture was diluted with an excess of water saturated with sodium chloride to separate the ester. It was washed with aqueous saturated sodium bicarbonate solution until neutral. It was dried with anhydrous magnesium sulfate and fractionally distilled.

241.   Pentyl Acetate

Vogel [7764] prepared and purified pentyl acetate for *physical property measurements* by the procedure described for propyl acetate.

242.   Isopentyl Acetate

Vogel [7764] prepared and purified isopentyl acetate for *physical property measurements* by the procedure described for propyl acetate.

SAFETY

The *flammable limits* in air at elevated temperature are 1.16 and 6.96%v [5712]. The *minimum ignition temperature* in air is 560–600° [1736].

243. Hexyl Acetate

SAFETY

The chief hazard of hexyl acetate is inhalation and its most important effect is narcosis. Prolonged skin contact will defat and dry the skin, according to Enjay [2270].

244. 4-Methyl-2-pentyl Acetate

SAFETY

For toxicity, see Enjay [2270], hexyl acetate.
The *exposive limits* in air are 0.9 and 5.7%v [6905].

245. 2-Ethylhexyl Acetate

Van Pelt and Wibaut [7674] prepared 2-ethylhexyl acetate from 2-ethyl-1-hexanol and acetyl chloride; Hatch and Adkins [3220] prepared it with acetic anhydride.

246. Benzyl Acetate

Three methods of preparation of benzyl acetate were studied by Su and P'an [7061]. Potassium acetate in an excess of α-chlorotoluene in acetic acid gave a 71.9% yield. Potassium or sodium acetate with α-chlorotoluene and trimethylamine as a catalyst gave 80% ester. Benzyl alcohol, acetic acid, and phosphoric acid gave 84% ester.

Benzyl acetate was prepared in large batches by Bogdanov and Antonov [940] in 98.2 to 99.9% yield of high-purity material. They reacted benzyl alcohol, acetic acid, and acetic anhydride catalyzed by sulfuric acid.

Joshi and Merchant [3785] heated 256 grams of benzyl alcohol, 1200 mL of water, and 640 grams of sodium acetate for 8 hr in an oil bath at 115° with continuous stirring. The top oil layer was washed with twice its volume of water, dried over calcium chloride, and distilled to give 68% yield of benzyl acetate; bp 208–215°.

Katti and Chaudhri [3895] purified benzyl acetate for the study of *physical properties of solutions*; see *m*-cresol.

SAFETY

The oral LD_{50} for rats was found to be 2490 mg/kg [3708].

247. Ethyl Propionate

Vogel [7764] purified ethyl propionate for *physical property measurements* in the same manner that he used for ethyl formate.

249. Isobutyl Isobutyrate

Pierre and Puchot [5848] prepared isobutyl isobutyrate by partial oxidation of 2-methyl-1-propanol with potassium dichromate and strong sulfuric acid.

Grünzweig [3007] prepared the ester from the alcohol and the acid with sulfuric acid as the catalyst.

Isobutyl isobutyrate is considered to be practically nontoxic orally and only slightly toxic intraperitoneally. Only slight skin irritation results from prolonged contact with the liquid [5713].

250. Ethyl Isovalerate

Vogel [7764] prepared and purified ethyl isovalerate for *physical property measurements*. A mixture of 51 grams of isovaleric acid, 11 grams of absolute ethanol, and 1 gram of sulfuric acid was refluxed for 20 hr and then fractionally distilled to a temperature of 155°. After being washed with saturated sodium bicarbonate and sodium chloride solutions, the ester was dried and distilled.

252. Butyl Stearate

Whitby [8003] prepared butyl stearate from silver stearate and 1-iodobutane by heating for 1.5 hr at 100° and extracting with hot ethanol.

Butyl stearate has been purified by Riddick [6165] by fractional freezing and fractional crystallization from solvents whose boiling points are below 100°. Acidic impurities were removed by shaking with 0.05 N sodium hydroxide or a 2% sodium bicarbonate solution, followed by several water washes. The acidity may be determined by titration of an aqueous wash with sodium hydroxide to the phenolphthalein end point. A more sensitive and accurate method is to titrate with a quarternary ammonium hydroxide in methanol-benzene solution. Butyl stearate may be distilled at 1–2 Torr.

SAFETY

Two years of feeding rats at 1.25 and 6.25% produced no effects on growth, mortality, or fertility. There are no ill effects known from industrial handling.

Esters of Unsaturated Aliphatic Monocarboxylic Acids (253–257)

General Comments

Stabilization of esters of unsaturated acids from peroxide formation; see Ethers, General Comments, Robertson and Jones [6200].

Specific Solvents

254. Ethyl Acrylate

Ethyl acrylate has a persistent acrid odor and usually contains polymerization inhibitors.

Moureu and coworkers [5296] warmed acrylic acid, dry ethanol, and 10% sulfuric acid to prepare the ester.

SAFETY

Ethyl acrylate is moderately toxic and highly irritating by several routes. It is absorbed through the skin [3608].

255. Methyl Methacrylate

Methacrylate esters were prepared by Ruzicka [6318] by heating 50 grams of the corresponding γ-bromoisobutyric ester with 50 grams of quinoline to boiling under reflux until the heat of reaction was sufficient to maintain the reflux. Heat was again applied at the end to complete the reaction. The methacrylate ester was distilled in 1 atm of hydrogen.

Methyl methacrylate was prepared by Bruson and Washburne [1205] by leading methyl α-hydroxyisobutyrate over phosphoric acid-wetted alumina.

Methyl methacrylate of 0.2% impurities was purified to 0.01% impurities by Shelepin and coworkers [6623] by washing with 25% alkali solution, the alkali washed out, and the material distilled *in vacuo* in a stream of nitrogen. The carbonyl compounds were removed by boiling 12–14 hr with powdered potassium hydroxide, filtering, and distilling at 90 Torr in oxygen-free nitrogen. The middle fraction was collected and repeatedly distilled until carbonyl-free to 2,4-dinitrophenylhydrazine. The peroxide content was reduced to less than 1×10^{-4} mol/L by iodometric determination. The freezing point was the criterion of purity.

Methyl methacrylate was purified by zone melting and the purity determined cryoscopically by Anikin and Dugacheva [375]. The monomer was heated above its melting point in a hermetically sealed vessel and then subjected to the action of a cooling agent. The purity was determined from a cooling curve; see also [376, 377, 378].

Makarov and coworkers [4845] stabilized methyl methacrylate by passing it through a sorbent of silica gel or KIL clay. The addition of either sorbent to the stabilized ester increases its storage time.

Yates and Ihrig [8211] have studied the retardation of methyl methacrylate by aromatic amines. They conclude that the amines (like phenols) act as antioxidants by conserving the supply of oxygen inhibitor present.

SAFETY

The odor of methyl methacrylate is readily detectable at 50 ppm. A person can become tolerant to lower concentrations.

A method for the determination of methyl methacrylate in air is described and gives the sensitivity, specificity, and likely interferences.

256. Methyl Oleate

Methyl oleate is an unusual compound in several respects [4891]. The use of the methyl ester is an important part of the preparation of high-purity oleic acid. A combination of fractional distillation and crystallization at different temperatures appears to be the most satisfactory method of preparing a high-purity methyl oleate as well as oleic acid. Much of the information concerning the purification of this ester is given under oleic acid.

It appears that Wheeler and Riemeneschneider [7993] developed a method for preparing a high-purity methyl oleate; see oleic acid.

Humko [3579] lists typical acid carbon chain composition of methyl oleate to be: myrisate, 4; palmitate, 12; stearate, 4; oleate, 73; and linolate, 7%.

Markley [4892, pp. 778 ff.] discusses ''Esterification Practices'' on both industrial and laboratory scales. The more recent methods for laboratory esterification of fatty acids commonly used are modifications of the Fischer-Speier method [2443] of 1895. Their method consists of heating the acid to be esterified with an excess of alcohol containing 3% dry hydrogen chloride.

There is no generally accepted method for esterifying a specific acid with a specific alcohol. However, Markley [4892, pp. 778 ff.] states that, for methanolic esterification of fatty acids, the concentration of the hydrogen chloride catalyst is usually 3–5%w of the alcohol. Apparently, it makes little difference if the hydrogen chloride is added to the methanol or to the mixture of alcohol and oleic acid.

Senderens and Aboulenc [6574] demonstrated that sulfuric acid could be used as a catalyst.

Hilditch [3398] gives three variations for esterification:

1. About twice the weight of oleic acid to methanol was added in the presence of 2% concentrated sulfuric acid. Then add ethyl ether and wash with dilute potassium carbonate solution to remove unesterified acids. Conversion usually is 97–98%.

2. The acid is converted to the methyl ester by boiling with four times its weight of methanol in the presence of 1% concentrated sulfuric acid. About 70–80% of the methanol is distilled from the reaction mixture, ethyl ether is added, and the free acid removed by shaking with dilute potassium carbonate [3399].

3. To esterify acids separated by the lead salt method (see oleic acid), add about four times the weight of methanol containing 2% concentrated sulfuric acid and reflux 2 hr. Remove as much of the alcohol as possible by distillation from a water bath. Dissolve the residue in ethyl ether, wash with water, and then cautiously with dilute sodium carbonate to remove unesterified acids [3400].

Other workers use hydrogen chloride with equal success.

The conditions for preparing methyl oleate may be varied widely without sacrificing yield.

One precaution must be observed. The unesterified acids must be removed to a quite low concentration. If they are present in appreciable amount, they will emulsify with the alkali carbonates and hydroxides during the washing out of the acids. Buxton and Kapp [1297] report a more satisfactory procedure for removing substantially all of the unreacted fatty acid. Distill the unreacted alcohol from the reaction mixture, dissolve the residue in 2–5 parts of such a solvent as 1,2-dichloroethane to one part of ether. Determine the free acid in the mixture by titration. Add the calculated equivalent amount of 38% potassium hydroxide slowly with constant stirring. Allow the mixture to stand: the potassium soap will rise to the top as a floc and the mineral acid salt will settle to the bottom. Filter without suction and wash the precipitate with a small amount of 1,2-dichloroethane. Distill the filtrate.

The ester may be purified by a combination of fractional distillation at reduced pressure and crystallization by, or similar to, that of Wheeler and Riemeneschneider [7993].

Craig [1707] prepared methyl oleate for *refractive index studies* from olive oil in the usual manner and distilled the ester. The C$_{18}$ fraction was recrystallized repeatedly from acetone until the precipitate and filtrate had the same refractive index.

The purification of methyl oleate usually is done by a combination of fractional dis-

Table 15.5 Solubility of Methyl Oleate in Acetone

t	Solubility[a]	t	Solubility[a]
$-27.5°$	208.0	$-35.3°$	12.0
-29.4	98.0	-40.0	5.42
-31.0	44.9	-55.8	0.60
-33.0	27.1		

[a] Grams ester per 100 grams of acetone.

tillation and crystallization from acetone. The solubilities of the ester in acetone are given in Table 5.15 [5961].

Althouse and Triebold [102] determined that methyl oleate decomposes at 217° at 16 Torr.

CRITERIA OF PURITY

The iodine number, refractive index, and melting or freezing point are usually used to characterize the purity. The capillary tube melting point is reported to be -19.9 to $-19.6°$ [7993].

The iodine number has been calculated as 85.62 [1707, 6536, 3924].

Keffler and McLean [3924] point out that the density of methyl oleate, and other unsaturated fatty acid esters, is not very sensitive to the presence of saturated fatty acid ester impurities; for example, for the 97.4% ester the density is 0.8773, and for the 99.8% ester the density is 0.8774.

Esters of Aromatic Monocarboxylic Acids (258–262)

General Comments

The toxicological properties of benzoates have not been studied extensively. The information available indicates a low order of toxicity. Their low vapor pressure at ambient temperature greatly reduces any danger from inhalation of the vapors. They may be slight or moderate skin irritants.

Specific Solvents

258. Methyl Benzoate

Vogel [7766] prepared methyl benzoate for *physical property measurements* from 1 mol of benzoic acid and 10 mol of absolute methanol and concentrated sulfuric acid (5%w of the methanol). The boiling point was 199° at 775 Torr.

Kusano and Wadso [4344] purified methyl benzoate for *calorimetric studies*. It was dried over molecular sieve 4A and fractionally distilled at reduced pressure at 100°. Purity was 99.9%m.

CRITERIA OF PURITY

Dreisbach and Martin [2090] determined the purity from the freezing curve.

259. Ethyl Benzoate

Vogel [7766] prepared ethyl benzoate for *physical property measurements* in the same manner as described for methyl benzoate.

Cohen and Mier [1571] reacted benzoic acid and triethyl orthoformate for 45 min at 165–192° and obtained an 80–94% yield of ethyl benzoate.

260. Propyl Benzoate

Vogel [7766] prepared propyl benzoate by refluxing for 35 hr 30.5 grams of benzoic acid, 30 grams 1-propanol, and 50 mL of dry benzene and concentrated sulfuric acid. The yield was 37 grams; bp 229.5° at 766 Torr. Some *physical constants* were determined.

261. Benzyl Benzoate

Benzyl benzoate is reported to decompose slightly when distilled at atmospheric pressure [880].

Kamm and Mathews [3844] recommend that benzyl benzoate be prepared if a high-purity material is desired. They prepared it by dissolving 3 grams of sodium in 70 grams of benzyl alcohol and, after cooling, gradually added 454 grams of benzaldehyde, the temperature being kept at 50–60° by cooling. After warming on a water bath for 1 hr, the product was washed with water and distilled in vacuum. The benzyl alcohol and benzaldehyde passed over in the first portion of the distillate; yield 410–420 grams ester boiling at 183–185° at 15 Torr. The ester had a purity of 99%.

Guerin [3015] heated and stirred 80 grams of sodium benzoate with 125 mL of 1,2-ethanediol for a few minutes, then added 63 grams of α-chlorotoluene, refluxed for 15 min, and then cooled. A little water was added to dissolve the salt and the benzyl benzoate upper phase separated from the 1,2-ethanediol. The ester was washed with dilute ammonium hydroxide and distilled.

Hon and coworkers [3475] purified commercial benzyl benzoate by fractional distillation in a 20-plate sieve column of 2.54-cm id at a 5:1 reflux ratio. A purity of 99.6% was determined by density measurement of 20°.

262. Ethyl Cinnamate

Jeffery and Vogel [3699] washed commercial ethyl cinnamate with 10% sodium carbonate, then with water; the product was dried with anhydrous magnesium sulfate and distilled. The purified ester was saponified with aqueous potassium hydroxide and the acid isolated, washed, and dried. A mixture of 25 grams of the purified cinnamic acid, 23 grams of absolute ethanol, 4 grams of concentrated sulfuric acid, and 100 mL of dry benzene was refluxed for 15 hr. The ester was isolated and purified. The yield of pure ester was 23.5 grams; bp 127° at 6 Torr. See [6952] for a slight modification of this method.

Cinnamic acid when reacted with triethyl orthoformate at 180–222° for 1 hr yielded 94% ethyl cinnamate.

Lactones (263)

263. γ-Butyrolactone

This lactone is a powerful solvent. Its viscosity is low for its volatility and it has good heat stability. γ-Butyrolactone is stable at pH 7; it hydrolyzes rapidly in alkaline solution and less rapidly in acid solution. The hydrolysis is reversible and the lactone is reformed when the pH is restored to 7 [2707].

For the quantitative preparation of γ-butyrolactone from 1,4-butanediol; see tetra-hydrofuran, Ivanskii and Dolgov [3652].

Fittig and Chanlarow [2452] studied the formation of γ-butyrolactone from γ-hy-droxybutyric acid. The lactone forms slowly at room temperature and goes to completion rapidly when distilled.

SAFETY

γ-Butyrolactone is stated to be neither a primary irritant nor skin-sensitizer [2707]. It probably has a low toxicity. It appears to be absorbed readily through the skin. Inhalation of the vapors and absorption of the liquid through the skin should be avoided because of possible delayed effects [5713].

Esters of Dicarboxylic Acids (264–274)

264. Ethylene Carbonate

Ethylene carbonate exhibits some unusual solvent properties for inorganic compounds; for instance, the alkali metal chlorides are almost insoluble and the chlorides of iron(II), mercury(II), and other "heavy metals" are soluble. The carbonate ester dissolves electrolytes and forms electrolytic solutions as expected from its large dielectric constant.

Vorländer [7789] and Nemirowsky [5454] prepared ethylene carbonate by reacting equimolecular amounts of 1,2-ethanediol and phosgene in a closed tube at room temperature.

Bonner and Kim [975] determined that ethylene carbonate has an average of 8.3 monomer units per cluster in pure liquid.

Ethylene carbonate is commercially produced by the reaction of ethylene oxide and carbon dioxide at 190–200° and 80 atm using Et_4NBr as a catalyst. The catalyst is soluble in ethylene oxide making a homogeneous liquid-phase reaction possible. A conversion of 96–98% has been obtained [6944].

According to Jefferson Chemical [3695], ethylene carbonate is a solvent of a number of polymeric materials and plasticizers.

Ethylene carbonate is stable under normal storage conditions. It shows little change in freezing point, nor is the "water white" color deepened when stored at 40° for 2 months, even in the presence of traces of acid or alkali. The dry ester is slowly decomposed at 200–250°. It hydrolyzes rapidly above 125° in the presence of alkalies and more slowly in the presence of strong acids. Pure ethylene carbonate is stable in the presence of water at 100°. Salts such as sodium chloride accelerate hydrolysis. Alkalies and strong acids cause rapid hydrolysis [3695].

Ethylene carbonate may be rapidly distilled at atmospheric pressure with only slight decomposition. Distillation at reduced pressure is recommended.

Kempa and Lee [3940] purified ethylene carbonate for *dielectric constant* and *dipole moment studies* by fractional distillation at reduced pressure and fractional crystallization from dry ethyl ether; mp 36.2°, n_D 40° 1.4199.

Cislan and Cornilescu [1493] studied a process for the continuous purification of cyclic ethylene carbonate by continuous crystallization in a column with thermal gradient; the product was 99.93%.

Choi and Joncich [1470] purified a commercially available material by the zone refining method described by Pfann [5809] for *heat of combustion and vapor pressure study*.

Sears and coworkers [6544] purified ethylene carbonate, N-methylformamide and dimethyl sulfoxide by a sequence of fractional freezings under a nitrogen atmosphere until a constant maximum freezing point was obtained. The solvents were used for *physical property studies of binary aqueous solutions*.

The *dielectric properties* of ethylene carbonate were studied by Payne and Theodorou [5728]. A commerical sample was dried with 4A molecular sieves and fractionally distilled twice at reduced pressure through a 4-cm heated glass column packed with glass helices. Chromatographic examination indicated less than 10 ppm water and no other low boiling constituents; mp 36.6°.

Seward and Vierira [6586] purified commerical ethylene carbonate for *dielectric constant study* by distilling several times at 5 Torr; fp 36.4°.

Bonner and coworkers [976] fractionally distilled ethylene carbonate three times in a column of single-turn glass helices under 3 Torr. Analysis by nmr, vpc, and ir showed no detectable impurities.

SAFETY

Inhalation by rats of the concentrated vapor for 8 hr resulted in no deaths. It is only a slight irritant to the skin of rabbits and produces moderate irritation to their eyes [5713].

It has a low acute toxicity: LD_{50} in male albino rats was found to be 11.2–15.6 g/kg. Inhalation tests showed no significant toxicity. Patch tests with 40% aqueous solution showed no irritation or skin sensitization. Repeated feeding of ethylene carbonate over a period of time resulted in liver damage to the test animals [3695].

Weisburger and coworkers [7956] have determined that ethylene carbonate is not carcinogenic.

265. Propylene Carbonate

Propylene carbonate is an odorless, colorless, aprotic solvent with a wide liquid range and a relatively high dielectric constant. Simeral and Amey [6729] found it to behave as a normal polar liquid with strong dipole–dipole interactions but with little or no specific association. It is not very hygroscopic. At 25°, 8.3 grams of water can be dissolved in 100 grams of propylene carbonate and the solubility increases with rising temperature. The upper critical solution temperature is 61.1° [8087]. Many organic compounds and some inorganic salts are soluble in propylene carbonate. It is stable under ordinary conditions of storage. Appreciable decomposition occurs above 200°. Traces of acids or bases will promote degradation to propylene oxide, propionaldehyde, allyl alcohol, and carbon dioxide even at ambient temperatures. It is fairly stable toward oxidation and reduction.

Propylene carbonate has received considerable attention as an electrochemical solvent, particularly in the development of aprotic battery electrolytes. A solvent of high purity is required for electrochemical use and the presence of impurities to the extent of less than 100 ppm can have a marked effect on its performance. Jasinski and Kirkland [3683] studied the purification and characterization of propylene carbonate suitable as an electrochemical solvent. A practical grade, bp 120–122° at 17 Torr, was dried with 5A molecular sieve pellets. It was fractionally distilled in a Podbielniak Semi-Cal (Series 3650) adiabatic distillation apparatus under a pressure of 0.5–1 Torr. This gave a pot temperature of 110–111° at which the carbonate was stable. A reflux ratio of 10:1 to 20:1 was used and performance was relatively independent thereof. A boil-up rate of 200 mL/hr was used. The head of the distillation apparatus was modified so that the reflux and product would not come in contact with any material that would contaminate it. Analysis was by gas chromatography. The principal impurities found were propylene oxide, allyl alcohol, propylene glycol, and water.

The *dielectric properties* of propylene carbonate were studied by Payne and Theodorou [5728]. A commerical sample was fractionally distilled in an apparatus similar to that used by Jasinski and Kirkland resulting in a product containing 10 ppm water and having a specific conductivity of 10^{-7} to 10^{-8} ohm^{-1} cm^{-1}.

A technical grade of propylene carbonate was purified by Mukherjee and Boden [5306] by distilling twice from anhydrous calcium carbonate and keeping the middle fraction. Gas chromatography showed no impurities and the Karl Fischer method showed no water. The carbon dioxide content was estimated to be less than 0.02%m.

Simeral and Amey [6729] purified a commercial grade of propylene carbonate by first passing through a 30-cm column of type 4A molecular sieve under a nitrogen atmosphere. The effluent was fractionally distilled through a 91-cm spinning band column at –7 Torr. The middle 70% was collected under dry nitrogen. The water content was found to be less than 20 ppm.

L'Her and Courtot-Coupez [4574] state that propylene carbonate can be chosen as a reference to evaluate the solvation of electrolytes because it is a nonstructural aprotic solvent with low basic and acidic properties.

Fujinaga and Izutsu [2603] reported detailed methods of purification and tests for purity.

SAFETY

Kuramoto and coworkers [4323] reported the LD_{50} for mice and rats to be 15.8 and 1.1 cm^3/kg subcutaneously. Daily skin application of 17.5% solution for a month showed no serious changes to behavior, food intake, or blood or serum enzyme activity. Small lesions were found in lung, spleen, and liver and there was some hyperkerotosis.

266. Ethyl Carbonate

SAFETY

Brown and coworkers [1131] gave mice drinking water containing ethyl carbonate in concentrations up to 1000 ppm for a period of 83 weeks. No carcinogenic effect was detected. The daily no-untoward-effect level is approximately 140 mg/kg.

267. Ethyl Oxalate

Dutt [2168] obtained good yields of ethyl oxalate by passing alcohol vapor through oxalic acid until water ceased to distill.

Vogel [7764] described a general method for preparing dialkyl oxalates. Finely ground AR-grade oxalic acid dihydrate was spread thinly on a large watch glass and heated to 105° for 6 hr to produce the anhydrous acid. One mole of the anhydrous acid, 2.5–3.5 mol of alcohol, dry AR-grade benzene (twice the volume of the alcohol used), and concentrated sulfuric acid (60%w of the oxalic acid) were refluxed for 6–12 hr and poured into a large excess of water. For the actual preparation cited, the reflux period was 20–34 hr. The benzene phase was separated, the aqueous phase extracted with ether and the extracts added to the benzene phase. The benzene–ether extract was washed first with aqueous saturated sodium bicarbonate until free from acid and then with water. It was dried with anhydrous magnesium sulfate and distilled.

Ethyl oxalate was prepared by Clarke and Davis [2757] from oxalic acid and 95% ethanol by removing the water as the carbon tetrachloride azeotrope.

Palomaa and Mikkila [5658] determined the *melting point* of ethyl oxalate that was purified by distillation under vacuum.

269. Methyl Maleate

Toops [7525] fractionally distilled 1500 mL of technical ester (99.26% ester by saponification) in a Podbielniak Hyper-Cal column at 6 Torr. The column was preflooded and operated under total reflux for 1 hr. A head fraction of 200 mL was removed at a reflux ratio of 45:1. One liter of distillate was removed at a reflux ratio of 30:1 for *physical property measurements*.

270. Ethyl Maleate

Toops [7525] fractionally distilled 3 L of diethyl maleate (99.25% ester by saponification) in a 20-plate, Penn-State-type column packed with $\frac{1}{8}$-in. single-turn glass helices. The column was operated at total reflux for 1 hr. The first one-third was removed at a 10:1 reflux ratio and discarded. The center fraction was distilled at a reflux ratio of 5:1 and used for the determination of *physical constants*. The distilled ester was 99.53% by saponification.

271. Butyl Maleate

Toops [7525] determined some *physical constants* on a technical grade of this ester. The purity was 99.98% by saponification; 0.01% acidity as maleic acid; 0.08% water.

272. Butyl Phthalate

Butyl phthalate is easily purified, due largely to its relative chemical inertness, its insolubility in water, and its high boiling point. It may be freed from alcohol by water washing and from acids and any butyl hydrogen phthalate with dilute caustic. A careful fractional distillation at 10 Torr or less will produce a product approaching 100%, as nearly as can be determined by several analytical approaches [6165].

Kuskov and Zhukova [4348] refluxed 30 grams of phthalic anhydride and 25 grams of 1-butanol for 2 hr at 200°, cooled it to 100°, added 2.5 grams of boric acid, heated it to 200° and added 27 grams of 1-butanol during 12 hr; yield, 80% butyl phthalate.

SAFETY

Butyl phthalate is quite toxic when breathed. Spasovski [6919] states that it is not absorbed through the skin; Elizarova [2240] reports that it is. The Committee on Threshold Limit Values [7022] considers it a substance of low acute and chronic oral toxicity under conditions of normal use.

The *threshold limit value* of 5 mg/m^3 has been recommended more to control airborne mist rather than as a health measure [7022].

The *lower explosive limit* in air at 236° is 0.47%v and the *minimum ignition temperature* in air is 404° [2184].

273. Bis(2-ethylhexyl) Phthalate

Hickman and Trevoy [3375] purified bis(2-ethylhexyl) phthalate by high vacuum evaporation. They describe the apparatus in the article cited and discuss the technique and apparatus in articles immediately following. Trevoy and Torpey [7548] purified the ester by high vacuum fractional distillation in a falling-stream still.

SAFETY

Fassett [5713] concluded, after studying all available toxicological information, that he sees no reason for predicting that the use of this ester in industry would be associated with any health hazard. See Carpenter and coworkers [1362] for a study of animal feeding of the ester.

The *lower explosive limit* in air at 246° is 0.28%v and the *minimum ignition temperature* in air is 391° [2186].

274. Butyl Sebacate

SAFETY

The *lower explosive limit* in air at 243° is 0.44%v and the *minimum ignition temperature* in air is 366° [2185].

Esters of Polybasic Acids (275–282)

275. Butyl Borate

The chief impurities in butyl borate are 1-butanol and boric acid from hydrolysis. The ester hydrolyzes readily on contact with atmospheric moisture. Purification must be carried out in a closed system and transfers of the dry ester made in a dry box [6165].

Butyl borate undergoes some decomposition at its boiling point [7915].

Carothers [1358, p. 16] gives directions for the preparation from 124 grams of boric acid and 666 grams of 1-butanol. The yield is 400–425 grams of vacuum-distilled product.

Toops [7525] fractionally distilled 1500 mL of commercial butyl borate in a Podielniak Hyper-Cal column at 25 Torr. A center fraction of the distillate was removed at a reflux ratio of 20:1 and used for *physical property measurements*.

Lappert [4409] gives a good review of the methods of preparation. Washburn and coworkers [7915] present an extensive discussion of the methods of preparation. Kuskov and Zhukova [4348] prepared butyl borate in 95.6% yield. O'Brien [5546] prepared the trialkyl orthoborates by removing the water formed during esterification as the benzene azeotrope. Butyl borate was prepared in 80% yield. See also [3070].

SAFETY

Little information has been published on the toxicity of butyl borate. The toxicity of boric acid is considered to be the gauge [7915].

276. Methyl Phosphate

For preparation from sodium methoxide and phosphorus oxychloride, see Evans [2310], butyl phosphate.

Petkovic and coworkers [5792] purified methyl phosphate by fractional distillation at reduced pressure for *dipole moment studies*.

SAFETY

Deichmann and Witherup [1912] reported lethal doses to rabbits and rats as 1.05 and 1.65 mL/kg, respectively.

277. Ethyl Phosphate

For preparation from sodium ethoxide and phosphorus oxychloride, see Evans [2310], butyl phosphate.

French and coworkers [2543] purified ethyl phosphate for *conductance studies* by drying the solvent over sodium sulfate and distilling it under reduced pressure. The middle fraction was stored for several weeks over sodium sulfate, fractionated, and repeatedly distilled under reduced pressure to a constant specific conductance.

For purification for *dipole moment studies*, see Petkovic [5792], methyl phosphate.

SAFETY

Pyatlin [5998] reported an average oral dose LD_{50} in mice as 1370 mg/kg.

278. Butyl Phosphate

Commerical butyl phosphate is generally made by reacting 1-butanol with phosphorus oxychloride. The ester is washed with water and then with sodium hydroxide or sodium carbonate, and finally with water. The ester may be dried at a reduced pressure of about 600 Torr and a temperature of not more than 125° with good agitation. The likely impurities are phosphates of the alcohol impurities present in the 1-butanol, sodium dibutyl phosphate, disodium butyl phosphate, and traces of butanol and water. Butyl phosphate may be distilled at less than 1 Torr if an inert gas, such as nitrogen, is bled into the flask. Decomposition will be quite small under these conditions [6165]. Dyrssen [2177] reported that this ester formed a complex with dibutyl phosphate. A more recent study has been by Liem [4596].

Commercial butyl phosphate has been purified by stirring with 5% sodium carbonate or a 1% sodium hydroxide solution for several hours, washing well with water, and drying at reduced pressure. The ester is then vacuum-distilled, discarding the first and last 10% [6165].

Evans and coworkers [2310] prepared butyl phosphate from sodium butoxide and phosphorus oxychloride. Forty-six grams of sodium were added to 600 mL of 1-butanol and the excess alcohol distilled off under reduced pressure. The butoxide was mixed with 200 mL of dry ether, and 60 mL of phosphorus oxychloride in 100 mL of ethyl ether was added with cooling. The mixture was boiled gently for 1 hr, water was added and the ether layer was dried over anhydrous sodium sulfate. The ether was removed and the ester distilled at 50 Torr.

Butyl phosphate was purified for *solubility studies* [75] by boiling 100 mL of the ester with 500 mL of 0.4% sodium hydroxide solution, followed by distillation of 200 mL to remove steam volatile impurities. The remaining ester was washed repeatedly with water and dried under vacuum. See also Hardy and coworkers [3153].

Butyl phosphate was purified by Liem [4596] for *distribution studies* by washing with an equal volume of 0.1 M sodium hydroxide, then with 0.1 M nitric acid, and finally with water. It was dried with the heat of an infrared lamp at about 120° for $\frac{1}{2}$ hour at reduced pressure in 1 atm of nitrogen.

The 7170 and 6960 cm^{-1} absorption bands were used to determine the water in butyl phosphate in the range of 0.5 to 5.0% with an error of ≥ 3 to 4%. This is as accurate as the Karl Fischer method and much faster [3870].

SAFETY

Vapors from the steam distillation and from the hot ester are respiratory irritants. Available information indicates that butyl phosphate should be handled with caution [5713].

279. Tricresyl Phosphate (Mixed Isomers)
280. Tri-*o*-cresyl Phosphate
281. Tri-*m*-cresyl Phosphate
282. Tri-*p*-cresyl Phosphate

Tricresyl phosphate is used as a plasticizer in lacquers and coating materials. It is an excellent solvent for nitrocellulose and most resins [6703].

Ohashi [5568] prepared tricresyl phosphate by treating cresol with phosphorus pentachloride. Similar preparations are given in Faith [2364], Celluloid Corp. [1401], and Bonstedt [979].

For purification of tri-*p*-cresyl phosphate, see Krupicka [4289], quinoline.

SAFETY

Staub [6971] reported that the *meta* and *para* compounds are harmless for humans but 2% of the *ortho* used in a plasticizer has a toxic effect on the skin. Due to its solubility in fat, tri-*o*-cresyl phosphate can be accumulated in the lipoid substances of the human organism and less than 0.15 grams may become harmful.

Bock and coworkers [928] reported that absorption of tricresyl phosphate through the skin can cause damage to peripheral nerves. The oral LD$_{50}$ in mice is 4.64 g/kg [5337].

The *minimum ignition temperature* in air is 385° [6402].

HALOGENATED HYDROCARBONS (283–379)

General Comments

In their study of the kinetics of dehydrochlorination of substituted chlorohydrocarbons, Barton and Howlett [657] developed a general procedure for purifying chlorohydrocarbons. The compound is repeatedly shaken with concentrated sulfuric acid until no further color develops in the acid. The chloro compound is then washed with a solution of sodium bicarbonate and then with water, dried with calcium chloride, and fractionally distilled through an efficient column. The final purification is made by fractional crystallization until the freezing range is constant to 0.1°. In each crystallization only half of the liquid is frozen, the remainder being discarded.

Barton [655] has shown that trace impurities have a marked influence on the thermal dehydrochlorination of chlorohydrocarbons. This was further demonstrated by Barton and Howlett [657] and Barton and coworkers [656].

For the preparation of *carbonyl-free* nonoxygenated extractive solvents, see Hornstein and Crowe [3494], Hydrocarbons, General Comments.

Wadso [7822] determined the *heat of vaporization* of a number of alkyl and aromatic halides. Commercially available materials were purified by fractional distillation, when possible, to a purity of over 99.9% as determined by gas chromatography. Iodides were distilled and stored in the dark over silver wool. 1-Chlorobutane, 2-chloro-2-methylpropane, 1-chloropentane, and 1-chloroheptane were purified by preparative gas chromatography followed by drying and simple distillation.

Safety

The toxicity of halogenated hydrocarbons varies from dichlorodifluoromethane, which is among the most physiologically inert organic substances, to 3-chloropropene (allyl chloride), which is quite toxic both acutely and chronically. It seems necessary to consider the halogenateds individually, instead of grouping them together or even dividing them into subgroups [5713].

Fluorinated Hydrocarbons (283–294)

Aliphatic Fluorinated Hydrocarbons

General Comments

Sargent and La Zerte [6386] state that the fluorocarbons are nonflammable and nonexplosive and have low toxicity. They are stable in the presence of metals and other materials of construction during prolonged exposures to temperatures in excess of 300°. They do not form gums or sludges, even under extreme rigorous conditions, and are not attacked by strong acids of bases, oxidizing agents, or reducing agents.

Simons [6738] states that one of the most striking characteristics of fluorocarbons is their high degree of thermal stability. CF_4 has been found to react only slightly even at the temperature of the carbon arc. At room temperature they are singularly nonreactive even to alkali metals. They are not attacked by strong oxidizing agents.

Scott observed that completely fluorine substituted hydrocarbons, usually referred to as "fluorocarbons," show lower solubility and solvent power than normally associated with nonpolar organic liquids [6535].

Specific Solvents

283. 1,1-Difluoroethane

Henne and Renoll [3291] reacted 1,1-dibromoethane with mercury(II) fluoride in an ice-cold metal container with an ice reflux condenser and collected 1,1-difluoroethane in a receiver cooled with dry ice.

Industrial preparations have included the reaction of acetylene and HF in the presence of catalysts such as boron trifluoride [1261] and a mixture of mercury(II) oxide and activated carbon [2369] and the reaction of 1,1-dichloroethane with HF at high temperatures and pressures [2367].

Gibbs and Smyth [2733] purified 1,1-difluoroethane for *dipole moment studies* by passing a sample through ascarite and calcium oxide to remove water and HF. Other impurities were pumped out while 1,1-difluoroethane was kept in a chilled trap below its normal boiling point.

284. 1,1,1-Trifluoroethane

Henne and Renoll [3291] prepared 1,1,1-trifluoroethane by reacting 1,1,1-trichloroethane with a mixture of antimony trifluoride and antimony dichloridetrifluoride. One industrial preparation was the reaction of 1,1,1-trichloroethane and HF at 210–365° [2367].

Stevens [7006] cooled acetonitrile and anhydrous HF in a liquid nitrogen bath and then condensed boron trifluoride into the mixture. Gaseous 1,1,1-trifluoroethane which was produced in the reaction was passed through soda lime and Drierite and was condensed at −196°.

Russell and Golding [6316] used a material that was furnished by du Pont and was a part of a redistilled batch for *thermodynamic studies*. It was dried with phosphorus pentoxide, pumped to remove small amounts of dissolved air, and fractionally distilled at a reflux ratio of 40:1 through a 150-cm vacuum jacketed column packed with glass helices.

SAFETY

The *flammable limits* in air are 9.2 and 18.4%v [600].

285. 1,1,1,2,2-Pentafluoropropane

Yale [8193] prepared 1,1,1,2,2-pentafluoropropane by treating tetrafluoroethylene with a fluoride salt such as cesium fluoride.

Shank [6603] purified 1,1,1,2,2-pentafluoropropane for *physical property studies* by first scrubbing it with water and then sodium hydroxide solution. The vapor was passed through a calcium chloride tower and then condensed and distilled through a glass column of 30 theoretical plates; purity 99.9%.

286. Octafluorocyclobutane

Harmon [3168] prepared octafluorocyclobutane by heating tetrafluoroethylene under pressure at 125–500°.

Weeks [7947] passed octafluorocyclobutane over an alumina catalyst which had been pretreated with sodium hydroxide and reduced olefinic impurities to 2 ppm. Furukawa

and coworkers [2616] purified octafluorocyclobutane for *thermodynamic studies* by frac
tional crystallization.

Masi [4946] evaporated octafluorocyclobutane near the normal boiling point and the
condensed it into a liquid nitrogen trap to remove any decafluorobutane impurity.

SAFETY

Aviado and Smith [514] reported that octafluorocyclobutane influenced the respiratory
or circulatory systems or both in rhesus monkeys at 10–20%v.

287. Decafluorobutane

Decafluorobutane was prepared electrochemically by Simons and Mausteller [6747]
using the method reported by Simons and Dresdner [6742]. It was fractionally distilled
in a column of about 100-theoretical plates. The freezing curve showed a purity o
99.8%m.

Fowler and coworkers [2502] prepared decafluorobutane by reacting butane and co
balt(III) fluoride at 150–300°.

Brown and Mears [1152] removed low-boiling impurities from decafluorobutane by
vacuum distillation. Noncondensable gases were pumped off after freezing the com
pound with liquid nitrogen.

Aromatic Fluorinated Hydrocarons

288. Fluorobenzene

Balz and Schiemann [585] prepared fluoroaryl compounds from diazonium fluoro
borates. The method involves the diazotization of the appropriate aryl amine in as con
centrated hydrochloric acid as possible, precipitation of the fluoroborate with 40% fluo
roboric acid (HBF_4), washing and drying, and decomposition of the solid to the corre
sponding fluoro aryl compound. The yields are high, some being almost theoretical.
is stated that there are no side reactions. The preparation of fluorobenzene and *p*-fluo
rotoluene was reported.

Carothers [1358, p. 46] gives directions for preparing 780–870 grams of fluoroben
zene from 16 mol of aniline.

Vogel [7765] prepared fluorobenzene for *physical property measurements* according
to the method described by Flood [1358]; bp 84.5° at 760 Torr. The sodium borofluorid
can replace the fluoroboric acid.

Stull [7059] fractionally distilled a commercial product to 99.9+%m. Moore and
Hobbs [5235] fractionally distilled a commercial grade through a 60-cm Widmer colum
and used the middle fraction for *dipole moment measurements;* bp 84.74–84.76° at 76(
Torr.

CRITERIA OF PURITY

Stull [7059] determined the purity from the freezing curve.

SAFETY

Lapik [4406] reported the median lethal concentration for mice for a 2-hr exposure
to be 45 mg/L.

289. *o*-Fluorotoluene
290. *m*-Fluorotoluene
291. *p*-Fluorotoluene

For preparation, see Balz and Schiemann [585] under fluorobenzene.

Vogel [7766] prepared the *para*-isomer for *physical property measurements* from pure *p*-toluidine; bp 111.5° at 756 Torr.

Moore and Hobbs [5235] purified the three isomers for *dipole moment measurements* by the method described for fluorobenzene.

The *vapor pressure* and *freezing point* of the three isomers were determined by Potter and Saylor [5927] on the best commercial grades, which had been dried over Drierite and fractionally distilled at a high reflux ratio. The material distilling within 0.01° was used.

CRITERIA OF PURITY

Potter and Saylor [5927] determined the purity from the freezing curve.

SAFETY

Lapik [4406] reports the median lethal concentration for mice for a 2-hr exposure to be 55 mg/L.

292. α,α,α-Trifluorotoluene

Booth and coworkers [986] prepared α,α,α-trifluorotoluene by reacting benzotri-chloride with excess sublimed antimony trifluoride. The compound was washed with 6 *N* hydrochloric acid solution, then sodium bicarbonate solution, and finally 20% sodium hydroxide solution and then dried with barium oxide. The final product was obtained after distillation at a high reflux ratio. Other preparations with benzotrichloride and HF are given in Simons and Lewis [6745] and Kinetic Chemicals [4012].

Conradi and Li [1653] purified α,α,α-trifluorotoluene for *dielectric study* by repeat-edly treating with boiling sodium carbonate solution until no tests for chlorides were obtained in the solution, drying over anhydrous potassium carbonate and then over phosphorus pentoxide, and fractionally distilling.

SAFETY

Khalepo [3970] reported the LD_{50} for rats as 15,000 mg/kg.

293. Hexafluorobenzene

An easy, though low-yield, method for preparing hexafluorobenzene was reported by Vorozhtsov and coworkers [7799]. Dry hexachlorobenzene and potassium fluoride were heated in an autoclave at 450–500°; yield 21%.

Evans and Tiley [2314] purified a relatively high-purity material by deaerating by a three-stage distillation at 10^{-5} Torr; purity 99.8% by gas chromatography on a do-decylphthalate column. The material was used to determine *critical* and *other physical properties*.

Duncan and Swinton [2120] obtained a commercial product that was about 98% hex-afluorobenzene. The main impurities were incompletely fluorinated benzenes, primarily pentafluorobenzene. The material was purified for *thermodynamic property studies* of systems containing hexafluorobenzene. The commercial material was contacted with oleum for 4 hr at room temperature. The treatment was repeated until the oleum was no discolored. It was washed several times with distilled water and dried with phosphorus pentoxide. The hexafluorobenzene was fractionally crystallized 15 times. The criterion of purity was the melting point, $5.082 \pm 0.005°$, which indicated a purity of $99.95 \pm 0.05\%$m.

Davis and Morris [1863] purified hexafluorobenzene for *spectra studies* by preparative gas chromatography. Some physical properties were determined. The purified material was examined by its mass spectra, infrared spectra, nuclear magnetic resonance, and gas chromatography. Freezing point data indicated a purity of 99.9%m.

Cox and Gundry [1702] purified a sample of hexafluorobenzene for *calorimetric studies* by 12 fractional freezings with rejection of $\sim 1\%$ of charge after each freezing. Gas chromatographic examination of the purified specimen using polyethylene glycol 200 at 85° as the stationary phase and a flame ionization detector showed traces of two impurities assumed to be partially fluorinated benzene. Calorimetric evaluation indicated 99.97%m purity.

Evans and coworkers [2313] purified commercially available hexafluorobenzene by directional freezing. After the sixth freezing at 3°, purity was 99.993%m.

Hexafluorobenzene decomposes at a temperature greater than 650° in the vapor phase [3727].

SAFETY

Lapik [4406] reports the median lethal concentration for mice for a 2-hr exposure to be 95 mg/L compared to 37 mg/L for benzene.

Unsaturated Fluorinated Hydrocarbons
294. 1,1-Difluoroethylene

Gilman and Jones [2758] prepared 1,1-difluoroethylene by reacting 2,2,2-trifluoro-1 iodoethane with magnesium in ether. Calfee and Miller [1316] passed a mixture of 1,2 dichloro-1,1-difluoroethane and hydrogen over nickel wire at 490–500° and achieved 43% conversion to 1,1-difluoroethylene.

Edgell and Byrd [2201] prepared 1,1-difluoroethylene for *Raman spectral studies* by pyrolyzing 1-chloro-1,1-difluoroethane in a platinum tube at 600° and purified it by distillation in a Podbelniak column.

SAFETY

The *flammable limits* in air are 5.8 and 20.3%v [600].

Monochlorinated Hydrocarbons (295–314)
Aliphatic Chlorinated Hydrocarbons
295. Chloromethane

Groves [2996] prepared chloromethane in 1874 by passing HCl into a solution of methanol and zinc chloride at the boiling point. He was the first to observe methyl ether in chloromethane prepared from methanol and HCl [2995].

Norris and Sturgis [5518] reacted methanol and aluminum chloride at 140° to produce chloromethane, which was redistilled through a potassium hydroxide tube to remove HCl; 100% yield.

Lacy [4368] mixed chlorine gas with excess methane at 400° using bone-black or other carbon forms as a catalyst. The resultant gaseous mixture was passed through water to absorb HCl, dried, and then subjected to cold and pressure to liquefy chloromethane.

Boulin and Simon [1011] prepared chloromethane by adding dimethyl sulfate to warm 27% hydrochloric acid solution; 90% purity. Shamshurin [6602] reported a similar preparation using anhydrous aluminum chloride rather than hydrochloric acid solution.

Chloromethane was purified from a 95% cylinder product by Morgan and Lowry [5250] for *dielectric polarization study*. It was passed successively through bottles containing concentrated sulfuric acid solution, 10% potassium hydroxide solution, and phosphorus pentoxide and then fractionally distilled several times.

Messerly and Aston [5069] purified chloromethane for *thermodynamic studies* by distillation in a vacuum-jacketed, low-temperature fractionating column. The middle portion was redistilled and dried over phosphorus pentoxide.

Thronson and Mendolia [7456] purified chloromethane by successive scrubbings with water, sodium hydroxide solution, and concentrated sulfuric acid solution.

Shcherbak and coworkers [6618] passed chloromethane vapor at 150° over palladium(II) chloride and carbon or activated alumina and neutralized the sample with granular calcium hydroxide. The compound was dried over calcium oxide and then a zeolite.

SAFETY

Toxicity studies are given in White and Somers [8011] and Smith and von Oettingen [6855]. The *threshold limit value* has been set at 100 ppm, 210 mg/m^3 [142].

The *flammable limits* in air are 8.1 and 17.2%v [2538]. The *minimum ignition temperature* in air is 632° [6402].

296. Chloroethane

Commercial chloroethane is generally prepared by one of three methods: (1) the addition of hydrogen chloride to ethylene, (2) chlorination of ethane, or (3) the reaction of ethanol and hydrogen chloride. The first method is the most widely used. Technical chloroethane has a purity of at least 99%.

Chloroethane was prepared in 97.6% yield by Norris and Sturgis [5518] by mixing equal molecular amounts of ethanol and aluminum chloride and heating to 110°, then to 120° at which temperature the reaction was rapid, and finally to 150°. The chloro compound was purified by distillation.

Groves [2996] passed chloroethane vapors through sulfuric acid to obtain a dry and alcohol-free product.

Gordon and Giauque [2870] purified a commercial material by twice fractionally crystallizing, pumping free from air, fractionally distilling, and retaining the middle fraction. From the amount of premelting in *heat capacity measurements*, the purity was estimated to be 99.98%m.

SAFETY

The vapors of chloroethane produce narcosis and slight symptoms of irritation. Less toxic than chloroform, it has been used extensively as an anesthetic [5713]. Troshina

[7556] reported LC_{100}, LC_{50}, and LC_{min} of 180, 152, and 145 mg/L, respectively, for 2-hr exposure for rats.

The *threshold limit value* has been set at 1000 ppm, 2600 mg/m^3 [7022, 5713].

The *flammable limits* in air are 3.85 and 15.40%v. The *minimum ignition temperature* in air is 494° [5712].

297. 1-Chloropropane

SAFETY

1-Chloropropane may be considered relatively safe. Liver and kidney injury are likely if the time of administration is long or repeated [5713].

The *flammable limits* in air are 2.60 and 11.10%v, the latter value at elevated temperature [5712].

298. 2-Chloropropane

Norris and Sturgis [5518] prepared 2-chloropropane in 70% yield in a similar way to chloroethane.

Hughes and Shapiro [3564] dried 2-chloropropane over phosphorus pentoxide and fractionated from a small amount of anhydrous sodium carbonate for *kinetic studies;* b 36.5° at 760 Torr.

SAFETY

2-Chloropropane has been used as an anesthetic. Some histological changes were observed in livers and kidneys of animals exposed 127 times 7 hr a day and 5 days a week over a period of 181 days [5713].

The *threshold limit value* should be well below 1000 ppm, 3210 mg/m^3 [5713].

2-Chloropropane is highly flammable [5713]. The *explosive limits* in air are 2.2 and 10.7%v [6402].

299. 1-Chlorobutane

1-Chlorobutane is manufactured from 1-butanol and hydrogen chloride. It is available at reasonable cost with a 1° distillation range.

Deb [1885, p. 27] gives directions for the preparation of about 60 g of 1-chlorobutane from 1-butanol; see also [2757].

Anderson and coworkers [338] prepared 1-chlorobutane by reacting 1-butanol with a mixture of dimethylformamide and phosphorus trichloride. The yield was 63–86%, depending on the alcohol/trichloride ratio.

Fuchs and Cole [2595] passed hydrogen chloride through hexamethylphosphoric triamide and reacted this mixture with 1-butanol to give a yield of 74% 1-chlorobutane.

Smyth and McAlpine [6869] purified a commercial product for the determination of *dipole moment* by gently refluxing with concentrated sulfuric acid, washing several times with water, drying over two portions of calcium chloride, and distilling.

Stridh and Sunner [7047] purified 1-chlorobutane for *heat of combustion studies;* see 1-chloropentane.

For purification for *heat of vaporization study*, see Wadso [7822], Halogenated Hydrocarbons, General Comments.

See Barton and coworkers [656] and Barton and Howlett [657] under Halogenated Hydrocarbons, General Comments, for additional methods of purification.

Smyth and Rogers [6874] used the boiling point and the refractive index as the criteria of purity.

Tomashevskaya and Zholdakova [7517] reported the LD_{50} in mice as 5600 mg/kg.

The *flammable limits* in air are 1.85 and 10.10%v, the latter value at elevated temperature. The *minimum ignition temperature* in air is 460° [5712].

300. 2-Chlorobutane

Anderson and coworkers [338] reacted 2-butanol with a mixture of dimethylformamide and phosphorus trichloride and obtained a yield of 63% 2-chlorobutane.

Hooz and Gilani [3486] prepared 2-chlorobutane by reacting 2-butanol for 5 min with tri-*n*-octylphosphine in carbon tetrachloride; 60% yield.

For purification for *heat of vaporization study*, see Wadso [7822], Halogenated Hydrocarbons, General Comments.

301. 1-Chloro-2-methylpropane

Anderson and coworkers [338] reacted 2-methyl-1-propanol with a mixture of dimethylformamide and phosphorus trichloride and obtained 27% 1-chloro-2-methylpropane.

For purification for *heat of vaporization study*, see Wadso [7822], Halogenated Hydrocarbons, General Comments.

The *flammable limits* in air are 2.05 and 8.75%v, the latter value at elevated temperature [5713].

302. 2-Chloro-2-methylpropane

Adams [22, p. 50] describes a simple method of preparation by shaking 2-methyl-2-propanol with concentrated hydrochloric acid. The 2-chloro-2-methylpropane is separated, shaken with 5% sodium bicarbonate and then with water until it is neutral to litmus, dried with calcium chloride, and fractionally distilled.

Turkevich and Smyth [7581] washed a commercial material with ice water, dried over fused calcium chloride, and fractionally distilled; bp 50.5°, fp −25.4°, n_D 20° 1.38786.

Howlett [3524] determined *equilibrium constants* on a commercial product that was distilled. The main fraction, bp 50.3–50.5°, was fractionally crystallized six times. The freezing point was constant at −25.4° for the last three crystallizations until almost the entire bulk was frozen. The material was finally dried over calcium chloride and a small amount of calcium oxide; bp 50.4°, fp −25.4°, n_D 20° 1.3852.

For purification for *heat of vaporization study*, see Wadso [7822], Halogenated Hydrocarbons, General Comments.

Turkevich and Smyth [7581] and Howlett [3524] used the boiling point, freezing point, and refractive index as the criteria of purity.

303. 1-Chloropentane

Hooz and Gilani [3486] prepared 1-chloropentane by reacting 1-pentanol and tri-*n*-octylphosphine in carbon tetrachloride; 94% yield. Hepburn and Hudson [3308] reacted 1-pentanol with the Vilsmeier-Haack reagent chloromethylene dimethyl ammonium chloride in dioxane for 1 hr and 100° to give 77% 1-chloropentane.

Simon [6733] prepared 1-chloropentane for *physical property study* from a portion of 1-pentanol prepared from purified 1-butanol which distilled at a constant temperature. The alcohol was saturated with dry hydrogen chloride, sealed in a tube, and heated for 12–15 hr; 85% of the chloropentane distilled within 1°. The unreacted alcohol was again treated with hydrogen chloride and the total conversion was 94%. It was finally purified by distillation and treated with phosphorus pentoxide to remove traces of alcohol; the portion distilling at a constant temperature was used.

Stridh and Sunner [7047] purified commercially available 1-chlorobutane and 1-chloropentane for *heat of combustion studies* by fractional distillation in a Neaster/Faust Annular Teflon spinning band apparatus. Gas chromatography was used as the criterion of purity. Two columns with different stationary phases at two temperatures were used.

For purification for *heat of vaporization study*, see Wadso [7822], Halogenated Hydrocarbons, General Comments.

304. 1-Chloro-3-methylbutane

Norris and Taylor [5519] prepared 1-chloro-3-methylbutane by refluxing isoamyl alcohol, concentrated hydrochloric acid solution, and zinc chloride for 1 hr. The oily mixture was then refluxed with an equal volume of concentrated sulfuric acid solution 60% yield. Other preparation methods by the chlorination of isoamyl alcohol are given in Darzens [1820], Mouneyrat [5286], Anderson et al. [338], and Dehn and Davis [1909]

Transue and coworkers [7540] purified 1-chloro-3-methylbutane by refractionation of a commercial sample through a 21-stage Snyder column.

Turkevich and Smyth [7582] purified 1-chloro-3-methylbutane for *dielectric constant studies*. It was treated with concentrated sulfuric acid solution and then washed with water until the oil formed no precipitate with alcoholic silver nitrate solution. After drying over potassium carbonate and calcium chloride, it was repeatedly fractionated in a 30-in. column.

305. 1-Chlorooctane

Zincke [8301] prepared 1-chlorooctane in 1869 by saturating 1-octanol with HCl and heating at 120°. Clark and Streight [1503] reported that best results occurred when 1-octanol was boiled with excess thionyl chloride; they also used phosphorus trichloride and phosphorus pentachloride as chlorinating agents. Other preparation methods by chlorination of 1-octanol are given in Hodge and Richardson [3438], Bouveault and Blanc [1021], and Fuchs and Cole [2595].

1-Chlorooctane was purified by Stridh and Sunner [7074] for *thermodynamic studies*. It was rapidly fractionally distilled at atmospheric pressure in a vacuum-jacketed, all-glass column packed with helices. One of the original impurities was identified as 1-dodecane. The distillate was refractionated at a high reflux ratio. After about 10% of the charge was distilled (about 30 hr), all impurities except 1-dodecane had been removed. Further distillation did not change the composition of the distillate. It was analyzed by gas chromatography using two columns with different stationary phases at two temperatures. The 1-dodecane impurity was about 0.03%m. Further removal of the dodecane was not tried because of the small difference in energy of combustion of the two substances. The water content was less than 0.005%m. Type 4A molecular sieves could not be used because they caused slight decomposition.

For purification for *heat of vaporization studies*, see Wadso [7822], Halogenated Hydrocarbons, General Comments.

306. 3-Chloromethylheptane

Weizmann and coworkers [7975] and Kyrides [4358] prepared 3-chloromethylheptane by heating 3-hydroxymethylheptane with thionyl chloride.

Downie and coworkers [2075] reacted 3-hydroxymethylheptane with triphenyl phosphine in excess carbon tetrachloride to obtain a yield of 82% 3-chloromethylheptane.

SAFETY

The single dose LD_{50} in rats was reported as 7.34 g/kg [6882].

Aromatic Chlorinated Hydrocarbons

307. Chlorobenzene

The aromatic chloro compounds are, in general, relatively stable compounds and are much less reactive than the aliphatic ones.

Dubchenko [2120] prepared chlorobenzene by direct substitution of chlorine in a gas-phase reaction of nitrobenzene with carbon tetrachloride at 310–450° in the presence of activated carbon.

Brackman and Smit [1045] prepared chlorobenzene by reacting aniline with copper(II) chloride and nitrogen(II) oxide. Voronkov and coworkers [7798] reported a 96% yield of chlorobenzene after treating benzene with nitrosyl chloride.

Grossert and Chip [2993] treated a solution of benzene and titanium tetrachloride in dry dichloromethane with an equivalent amount of anhydrous trifluoroperacetic acid at ° for 4 hr. A yield of 60% chlorobenzene was obtained with no detectable dichlorobenzene.

Ralph and Gilkerson [6027] purified chlorobenzene in the same manner as they did -dichlorobenzene; $\kappa = 7 \times 10^{-11}$ ohm^{-1} cm^{-1}.

Chlorobenzene was purified by Bayles and Chetwyn [693] for use as a solvent for the determination of *acid–base function in nonaqueous media*. It was distilled and then fractionally distilled, and the middle fraction was dried over phosphorus pentoxide for 2–9 days and again fractionally distilled. The portion distilling at 132 ± 0.5° was collected for use. The solvent was used only if the aqueous extract showed a pH of 7 with a British Drug House Universal indicator.

For purification, see Hartmann and coworkers [3199] under benzyl alcohol.

Vogel [7766] prepared chlorobenzene from AR-grade aniline. The crude chloroben-zene was washed with sodium hydroxide, cold concentrated sulfuric acid, and then with water; dried and distilled; bp 131.5° at 765 Torr.

McAlpine and Smyth [4731] determined the *dipole moment* of commercial chloro-benzene that was purified by shaking repeatedly with portions of sulfuric acid until the acid was no longer colored. It was washed with water and with dilute potassium bicar-bonate solution, dried with calcium chloride, and fractionally distilled; bp 131.1°, n_D 20° 1.52459. After redrying over phosphorus pentoxide and again distilling, the boiling point dropped to 130.7°, while the refractive index did not change.

Stull [7059] fractionally distilled a commercial product through a 5-ft column and obtained a purity of 99.9 + %, calculated from the freezing curve.

Moore and Hobbs [5235] purified the best commercial grade, as described under fluorobenzene, for *dipole moment measurements;* bp 131.24–131.25° at 745.9 Torr.

Variations of the method of McAlpine and Smyth have been described. Often, chem-ical treatment is omitted and the commercial material, usually a good grade, is dried with either calcium chloride or phosphorus pentoxide and fractionally distilled; see also Timmermans and Martin [7494], Bramley [1053], Sugden [7072], and Williams and Ogg [8084].

For purification for *heat of vaporization study*, see Wadso [7822], Halogenated Hy-drocarbons, General Comments.

CRITERIA OF PURITY

Stull [7059] and Dreisbach and Martin [2090] determined the purity from the freezing curve.

Timmermans and Martin [7494] purified by fractional distillation until the physical properties of the fractions were constant. McAlpine and Smyth [4731] used the boiling point and the refractive index as their criteria of purity.

Swietoslawski [7131] recommended the differential ebulliometric method.

Maclean and coworkers [4812] characterized the effectiveness of their purification by ultraviolet absorption spectra.

SAFETY

Voluntary acute exposure by inhalation to chlorobenzene is unlikely because eye and nasal irritation begins at about 200 ppm, at which concentration the odor is unpleasant. There are dangers of chronic exposure. Repeated skin contact should be avoided.

Vecerek and coworkers [7701] reported an oral LD_{50} of 3.4 g/kg for rats.

The *threshold limit value* has been established at 75 ppm, 345 mg/m^3 [145].

The *flammable limits* in air are 1.35 and 7.05%v at elevated temperature. The *mini-mum ignition temperature* in air is 638° [5712].

308.   α-Chlorotoluene

Cannizzaro [1339] prepared α-chlorotoluene in 1853 by reacting benzyl alcohol with HCl. Hedayatullah and coworkers [3366] reacted sulfuryl chloride with benzyl alcohol in ether at −60° and in the presence of pyridine to produce α-chlorotoluene.

Methods to prepare the compound by photochlorination of toluene are given in Dorr [2053], Ellis [2249], and Asolkar and Guha [470].

SAFETY

The *threshold limit value* is 1 ppm, 5 mg/m^3 [142].

The *lower flammable limit* in air is 1.1%v. The *minimum ignition temperature* in air is 585° [5415].

309. o-Chlorotoluene

Erdmann [2279] prepared o-chlorotoluene by mixing o-toluidine and water with concentrated hydrochloric acid solution and cooling the mixture with ice. Sodium nitrite solution was added for the diazotization process and the resulting diazonium solution was poured into cold copper(I) chloride solution. Finally, the aqueous layer was separated from the o-chlorotoluene layer. A similar preparation is given in Oda [5550].

Autenrieth and Geyer [496] heated tri-o-cresyl phosphite dichloride, prepared from o-cresol and phosphorus pentoxide, at 180° to produce o-chlorotoluene.

BASF [540] patented a preparation carried out by passing chlorine through p-toluenesulfonic acid in sulfuric acid solution in the presence of iron(III) chloride. o-Chlorotoluene was obtained by hydrolysis with superheated steam.

SAFETY

The values of LD$_{50}$ in mice and rats are 4400 and 5700 mg/kg, respectively [5866]. The *threshold limit value* for skin absorption is 50 ppm, 250 mg/m^3 [142].

310. m-Chlorotoluene

Wynne [8187] prepared m-chlorotoluene in 1892 by reacting 3-chloro-4-aminotoluene with hydrazine. The hydrochloric acid salt was removed with copper(II) sulfate solution.

For preparation with m-cresol at 210°, see Autenrieth and Geyer, [496], o-chlorotoluene.

Oda and coworkers [5550] prepared m-chlorotoluene by refluxing its diazonium salt in ether for 5 hr at 35°.

311. p-Chlorotoluene

Miller [5135] reported a photochemical preparation of p-chlorotoluene. p-Toluenesulfonyl chloride in chloroform was photochlorinated to 4-chlorobenzyl chloride which was reduced with iron. The resultant p-chlorotoluene was extracted with methylene chloride, washed with water and dilute sodium bicarbonate solution, and dried over magnesium sulfate.

For preparation using p-toluidine, see Erdman [2279], o-chlorotoluene. For preparation with p-cresol at 140°, see Autenrieth and Geyer [496], o-chlorotoluene.

Oda and coworkers [5550] prepared p-chlorotoluene by refluxing its diazonium salt in benzene for 5 hr.

SAFETY

The values of LD$_{50}$ in mice and rats are 4000 and 5500 mg/kg, respectively [5866].

312. 1-Chloronaphthalene

1-Chloronaphthalene is prepared commercially by the chlorination of naphthalene in the presence of a catalyst. Some isomeric monochloronaphthalenes are formed together with small amounts of dichloronaphthalenes.

Brackman and Smit [1045] prepared 1-chloronaphthalene by reacting 1-naphthyl-amine with copper(II) chloride and nitrogen(II) oxide.

Satisfactory purification can be attained by fractional distillation of commercial 1-chloronaphthalene. A higher purity may be achieved by fractionally crystallizing the middle fraction from the distillation.

SAFETY

Some information is available for chloronaphthalenes as a group but very little is available specifically for 1-chloronaphthalene. The following information is from the *Hygienic Guide Series* [149]: Nothing is known regarding short exposure tolerance for humans. There is no recommended threshold limit value but data for tri- and penta-chloronaphthalenes would indicate the value for the monochloro compounds should be 1–2 ppm or greater. The chloronaphthalenes cause dermatitis. The areas most fre-quently affected are the face and neck. Probably liver damage would result from repeated inhalation of 1-chloronaphthalene. Fortunately, its vapor pressure is quite low.

Unsaturated Chlorinated Hydrocarbons

313. Chloroethylene

Regnault [6105] prepared chloroethylene in 1835 by heating a mixture of 1,2-di-chloroethane and alcoholic potassium. Wang and Shih [7902] passed 1,2-dichloroethane through a quartz tube packed with glass rings and active carbon at 530° to give a yield of 85% chloroethylene.

Ostromislensky [5621] prepared chloroethylene by passing acetylene through concen-trated hydrochloric acid solution containing mercury(II) chloride.

Berg and Mader [784] passed chloroethylene vapor through an aqueous solution of sodium hydroxide or potassium hydroxide to remove impurities which may promote polymer formation.

SAFETY

Chloroethylene has been recognized to have carcinogenic or carcinogenic potential with an assigned *threshold limit value* of 5 ppm, 10 mg/m^3 [142].

The *flammable limits* in oxygen are 3.6 and 72.0%v while the *flammable limits* in air are 3.6 and 26.4%v [5537]. The *minimum ignition temperatures* in air is 415° [7645].

314. 3-Chloropropene

This solvent is manufactured by the chlorination of propylene. The commercial prod-uct is likely to contain 2-chloropropene, 2-chloropropane, 1-chloropropane, 3,3-di-chloropropene, and 1,3-dichloropropene. It usually has a purity of about 97%.

3-Chloropropene is more reactive than the saturated chloroparaffins and less active than propene. Alcoholic potassium hydroxide reacts to form small amounts of allene (propadiene) and large amounts of allylethyl ether. Aqueous alkalies are reported to cause some hydrolysis.

SAFETY

Allyl chloride is a highly toxic and irritating material. It is very irritating to the eyes and upper respiratory tract. Lung irritation and kidney injury are the primary responses to acute exposure. Similar injuries are observed for chronic exposures but liver injuries become more important. It is readily absorbed through the skin. The threshold odor concentration for all persons is considered to be 25 ppm; for one-half of the people from 3 to 6 ppm.

Shamilov and Abasov [6600] reported an LC_{50} of 6.6 mg/L for mice after inhalation of 3-chloropropene for 2 hr and 8.2 mg/L for rats after inhalation for 4 hr.

The *threshold limit value* is 1 ppm, 3 mg/m^3 [7022].

The *flammable limits* in air are 2.90 and 11.30%v. The *minimum ignition temperature* in air is 487° [5712].

Polychlorinated Hydrocarbons (315–337)
General Comments

Carbonyl-free halogenated compounds were prepared by Hornstein and Crowe [3494] for *extractive studies;* see Hydrocarbons, General Comments.

STABILITY AND STABILIZATION

Chlorinated hydrocarbons, particularly chlorinated aliphatic hydrocarbons, are unstable under normal storage conditions. They generally react with oxygen in the presence of moisture to give hydrogen chloride, phosgene, and other reaction products, depending on the chlorinated compound. Light, heat, and common metals catalyze the reaction. Skeeters [6786] believes that the products of oxidation catalyze the oxidation reaction, and the reaction once initiated is self-catalyzed and self-sustaining. Some of the impurities in the commercial material often are more reactive than the principal component and hence initiate the reaction.

High-purity chlorinated compounds are more stable than the commercial grades that have not been stabilized. The most convenient means of purifying most chlorinated hydrocarbons is to neutralize the acid impurities, wash well with water, and fractionally distill. Wash the chlorinated compounds with one-fourth their volume of 1% aqueous sodium carbonate solution and then three times with the same amounts of distilled water. Thorough agitation is necessary for efficient washing. Select a still and operating conditions that will give a product of the purity desired. Practically all chlorinated hydrocarbons form positive azeotropes with water, as do many of their stabilizers; therefore these impurities are generally removed as heads. Flush the still and continually bleed nitrogen into the receiver to keep out oxygen and moisture. Remove at least 10% heads and collect the 80% middle fraction. Store the product in a nitrogen atmosphere in the dark, preferably in a refrigerator or freezer. Most chlorinated hydrocarbons can be obtained with a purity of 99.8–99.9+% [6165].

The manufacturer's problem is a two-phase one: (1) to retard or prevent the oxygen-moisture reaction and (2) to retard or prevent the reaction of the commercial polychlorinateds with metals, particularly aluminum.

Rapp [6065] stated that various metals are attacked by polychlorinated hydrocarbons and that aluminum is the most reactive of all of the common metals.

A large amount of work has been done to develop stabilizers for the large-volume industrial chlorinated solvents. The use and purification of these solvents usually is facilitated if some knowledge of the stabilizers is available.

Stabilizers usually consist of two and sometimes three substances. It has been found that certain substances exhibit synergistic properties. Most stabilizers are synergistic combinations. Some of the stabilizers are included in the following list.

1. For chlorinated hydrocarbons: an amide and an epoxide [1652].
2. Polychlorinated aliphatic hydrocarbons: nitromethane and *p*-dioxane [6065].
3. 1,1,1-Trichloroethane; nitromethane and amines or acetylenic alcohols [6753]; nitromethane and 1,3-dioxolane and epichlorohydrin [2898]; tetrahydrofuran and *N*-methylpyrrole [1334].
4. Trichloroethylene; tetrahydrothiophene and an amine or phenol or a 1,2-epoxide [1986]
5. Tetrachloroethylene; lower-molecular-weight C_1–C_3, nitroalkanes, and chloro derivatives thereof (purification greatly enhances the effect).

Sims [6753] found ''. . . that the discoveries relative to the stabilization of particular chlorinated hydrocarbon solvents are neither applicable nor translatable to the problem of stabilizing other chlorinated hydrocarbon solvents.'' Rapp [6065], however, studied the inhibition of dichloromethane and translated the findings to other hydrocarbon solvents; he states, ''Hydrocarbon solvents which are inhibited in the same manner are, in general, any polychlorinated aliphatic hydrocarbon, such as, for example, carbon tetrachloride, 1,2-dichloroethane, 1,1,1-trichloroethane, tetrachloroethylene, chloroform, and trichloroethylene.'' The works and conclusions of Copelin [1652] and Dial [1986] seem to support the findings of Rapp.

Bromo- and Iodoalkanes. It was reported [4422] that the iodoalkanes are less stable than the bromoalkanes. Copper and zinc are effective for stabilizing the iodo-compounds and copper silicate for the bromo-compounds.

Aliphatic Chlorinated Hydrocarbons

315. Dichloromethane

Dichloromethane is manufactured commercially by the chlorination of methane or chloromethane, generally the former. All possible chlorine substitution products are formed. The purity of the methane determines the amount of C_2 and higher chlorine compounds present in the crude product.

Dichloromethane appears to be stable to the common laboratory reagents. It reacts slowly with oxygen and must be stored in a colored container out of direct light. There is no appreciable hydrolysis due to dissolved water in the compound, but excess water causes hydrolysis, the rate being rather rapid at elevated temperatures. At low temperature it forms a hydrate that is stable to 2°.

Treszczanowicz and Bakowski [7547] describe a method and apparatus for the continuous preparation of dichloromethane from methane and chlorine and from chloromethane and chlorine.

Mathews [4969] purified dichloromethane by washing with water and sodium carbonate solution, drying over calcium chloride, and fractionally distilling. For *dielectric constant measurements*, Morgan and Lowry [5250] repeatedly fractionally distilled the commercial product until the conductivity of the middle fraction was constant. Maryott and coworkers [4935] washed a commercial product with concentrated sulfuric acid, then with dilute sodium hydroxide, and finally with water. The washed material was left standing overnight over sodium hydroxide and calcium chloride and fractionally distilled in a 60-cm Widmer column; bp 39.93–40.12°, n_D 20° 1.4249.

Vogel [7771] fractionated a commercial product and collected the fraction boiling at 40–41°. This was washed with 5% sodium bicarbonate solution, then with water, dried, and fractionally distilled in a Widmer column.

Vershinin and coworkers [7733] purified dichloromethane by treatment with calcium hypochlorite and passage through silica gel. Koshelev and coworkers [4206] reduced the water content to less than 0.006% by adsorption with zeolites KA-3Ts-112.

CRITERIA OF PURITY

Maryott and coworkers [4935] characterized the purity by the boiling point and refractive index. Morgan and Lowry [5250] fractionally distilled to a minimum conductivity. Massol and Faucon [4954] used the optical clarity as a criterion.

SAFETY

Irish states [5713] "Methylene chloride is by far the least toxic of the four chlorinated methanes. The toxic effect is predominantly narcosis. . . . It does not cause significant organic injury." It is mildly irritating to the skin on repeated contact if free to evaporate. It is painful to the eyes but no permanent damage may be expected [148].

Adams and Erickson [19] reported that methylene chloride stressed the heart and cardiovascular system of dogs. They recommended that the 2-hr acute exposure limit should be less than 500 ppm.

Aviado and Smith [514] reported that methylene chloride caused tachycardia and hypotension, and it anesthetized rhesus monkeys when inhaled at 2.5–5.0%v.

The *threshold limit value* of 200 ppm, 700 mg/m^3, has been suggested [142].

NIOSH has a booklet on the criteria for a recommended standard of occupational exposure to methylene chloride [5422].

The *flammable limits* in oxygen are 15.5 and 66%v [5415]. The *minimum ignition temperature* in air is 642° [5712].

Dichloromethane has no flash point but does give an effect that has been called a "halo," particularly in the TCC Flash Point Tester [6165].

316.   Chloroform

Chloroform reacts slowly with oxygen, or oxidizing agents, when exposed to the air and light. The principal products of this decomposition are carbonyl chloride (phosgene), chlorine, and hydrogen chloride.

Serious accidents have been reported from phosgene poisoning from chloroform that had stood in a warm place for a long time. Potassium dichromate and sulfuric acid oxidizes chloroform to phosgene and chlorine. Concentrated nitric acid and chloroform produce a small amount of chloropicrin. Alkalies react with chloroform in a number of ways. It is not advisable to heat chloroform with an alkali when purifying.

Chloroform formerly was made almost entirely from acetone and ethanol. It is made now principally by the chlorination of methane. A limited amount is made by the reduction of carbon tetrachloride [4027].

The many procedures for the purification of chloroform in the literature can be divided, in general, into those that use concentrated sulfuric acid for the removal of alcohols and those that do not. There are slight variations in the individual methods. Several modifications for each group are given.

REMOVAL OF CARBONYL CHLORIDE

This impurity reacts slowly with water, alcohols, and alkalies and rapidly with sodium phenolate, even in small concentrations. Gillo [2753] found chloroform hard to purify because of its instability and sensitivity to light and oxygen. He found that the concentration of carbonyl chloride could not be reduced below 0.0003%.

REMOVAL OF ETHANOL

For *preparative and analytical use*, ethanol may be removed by washing five or six times with about one-half its volume of water and then drying.

Buddle [1236] recommends shaking with concentrated acid as the best method for removing alcohol. Morgan and Lowry [5250] washed several times with concentrated sulfuric acid, then with dilute sodium hydroxide, and then with ice water; then dried over potassium carbonate, stored in a completely filled brown flask, and distilled shortly before use. Williams and Daniels [8080] washed with dilute sodium hydroxide and then with water, dried over fused calcium chloride and then over phosphorus pentoxide, and fractionally distilled for *specific heat measurements*. Vogel [7771] washed with water until neutral to litmus, dried over calcium sulfate, and distilled in an all-glass Widmer column for *physical property measurements*. See also Maryott and coworkers [4935] under dichloromethane, Ball [572], Gibby and Hall [2734], Hantzsch and Hofmann [3146], Mathews [4969], Peddle and Turner [5744], and Cohen [1572].

Yerger [8214] washed commercial chloroform with sulfuric acid, dilute sodium carbonate solution, and water. It was dried with calcium chloride and distilled in an atmosphere of nitrogen in a 30-cm column. Freshly distilled material was used to study the *nature of hydrogen-bonded ion pairs*.

Gross and Saylor [2986] washed chloroform for 1 hr with sodium hydroxide solution and twice with distilled water. It was then treated three times with concentrated sulfuric acid, twice with distilled water, once with mercury, and finally with distilled water. It was dried with calcium chloride and distilled through a 1-meter column for use in *solubility measurements*. According to Richards and Wallace [6160], washing with sulfuric acid, sodium hydroxide, and water leads to partial oxidation to carbonyl chloride. They fractionally distilled over calcium chloride until the boiling point and refractive index were constant for *specific heat measurements*.

For *electrical measurements*, Walden and coworkers [7866] allowed chloroform to stand several weeks over calcium chloride, then distilled and dried a number of times over either sodium sulfate or potassium carbonate, distilling after each drying. They recommended passing carbon dioxide-free air through the liquid, preferably during a slow distillation. Walden [7853] made *specific conductance measurements* on chloroform that was washed with water, dried with calcined sodium carbonate or phosphorus pentoxide, and fractionally distilled over calcined sodium carbonate.

Scheibe and coworkers [6429] purified chloroform for *optical measurements* by the method reported by Anschütz [384].

Chloroform was refluxed by Brown [1170], simply distilled, and stored over sodium for *dipole moment studies;* d 25° 1.4754.

Coomber and Rose [1646] found that chloroform prepared from acetone and bleaching powder does not contain the impurities present in commercially available material. It should be used in the assay of strychnine and other organic bases.

Goodspeed and Millson [2864] reported that reagent grade chloroform kept under anhydrous calcium chloride for several hours has little infrared absorption at 1600 cm^{-1}.

Hesse and coworkers [3357] purified chloroform for *optical studies* by column chromatography. Adsorbents used were silica gel and basic alumina.

Technical-grade chloroform may be stabilized by absolute ethanol, methylated spirits, *tert*-butylphenol, *n*-octylphenol, or thymol [4027]. Reagent-grade chloroform contains about 0.75% ethanol [137]. Anesthetic grade contains 0.5–1.0% alcohol (USP XVI).

The impurities in a typical technical grade chloroform are: 300 ppm of water, 2 ppm of acid as HCl, 200 ppm of dichloromethane, 555 ppm of bromochloromethane, 1500 ppm carbon tetrachloride, 20 ppm of residue on evaporation, and no detectable free halogen [4027].

CRITERIA OF PURITY

ACS *Reagent Chemicals* [137] lists two grades: the regular reagent grade and one suitable for spectrophotometry. Requirements and test methods are given.

Timmermans and Martin [7494] consider consistency of the density of distilled fractions a sufficient criterion of purity.

Swietoslawski [7131] recommends the differential ebulliometric method.

Maryott and coworkers [4935] used the boiling point and the refractive index as the criteria.

SAFETY

The USP XVI cautions "Care should be taken not to vaporize chloroform in the presence of a naked flame, because of the production of noxious gases."

Rosin [6263] states "Keep in tightly closed containers, protected from light, in a cool place." *ACS Reagent Chemicals* [137] adds "Chloroform should be supplied and stored in amber containers · · ·"

Kawasaki and coworkers [3907] reported an inhalation toxicity LC_{50} in rats of 74.6 mg/L.

Chloroform can produce kidney and liver injuries and heart irregularities; its industrial use has not caused many difficulties.

NIOSH has a booklet on potential occupational hazards of chloroform [5421].

The *threshold limit value* has been set at 10 ppm, 50 mg/m^3 [142].

317. Carbon Tetrachloride

Most carbon tetrachloride produced in the United States is made by direct chlorination of methane. The impurities present are similar to those found in chloroform. The chlorination of carbon disulfide is still used extensively in other countries and, to a limited extent, in the United States.

A good portion of the literature concerning the purification of carbon tetrachloride deals with the removal of carbon disulfide from the carbon disulfide process, since this is the most objectionable impurity. Carbon disulfide is removed by boiling with dilute alkali. Schmitz-Dumont [6467] recommended that carbon tetrachloride be shaken with

$1\frac{1}{2}$ times the amount of potassium hydroxide (dissolved in an equal volume of water and 100 mL ethanol) necessary to react with the carbon disulfide. The mixture is shaken thoroughly for one half hr at 50–60°. Klein [4053] refluxed for one half hr with 5% sodium hydroxide. Günther and coworkers [3027] treated carbon tetrachloride with three successive portions of 100 mL each of alcoholic sodium hydroxide; the carbon disulfide-free solvent was then washed several times with water, dried, and distilled.

For *thermal measurements*, Williams and Daniels [8080] purified carbon tetrachloride by refluxing over mercury to remove sulfides, followed by washing with dilute sodium hydroxide and water. It was dried over fused calcium chloride and distilled. Hicks and coworkers [3376] purified two lots by the same method. They were fractionally distilled twice in a vacuum-jacketed column, the first and last quarters being discarded each time. The second sample was protected from light. See also Williams and Krchma [8083].

Zimm [8300] purified a reagent grade for *critical point studies*. It was fractionally distilled through a 20-plate column. The first one-fifth was discarded and the remainder collected; bp 76.84°.

Stull [7059] fractionally distilled a good grade twice through a 5-ft column and obtained a purity of 99.98%, presumably calculated from the freezing curve.

Purification for *hydrogen-bonded ion-pair studies;* see Yerger [8214] under chloroform. See also Hartman and coworkers [3199] under benzyl alcohol.

Hesse and coworkers [3357] purified carbon tetrachloride for *optical studies* by column chromatography. Adsorbents used were silica gel and acidic alumina.

Bonner and coworkers [974] purified carbon tetrachloride for *freezing point depression measurements* by passing it over alumina and then storing it for several days over alumina that had been heated to 1000°. It was distilled prior to use.

Taylor [7201] reports that carbon tetrachloride reacts with water on silica gel at 110° to produce phosgene and hydrogen chloride; see also Rao and Rao [6057].

CRITERIA OF PURITY

Carbon tetrachloride is available in a number of different grades. It may be purchased in a purity that is adequate for most solvent uses.

Dreisbach and Martin [2090] and Stull [7059] determined the purity from the freezing curve. Maclean and coworkers [4812] characterized the effectiveness of their purification by the ultraviolet absorption spectra. Swietoslawski [7131] recommended the differential ebulliometric method for estimating the purity.

Ingold and Powell [3619] distilled commercial carbon tetrachloride and tested for chloroform in the constant-boiling middle fraction by warming to 60° with Fehling's solution. The latter is reduced by chloroform but not by the tetrachloride.

SAFETY

Carbon tetrachloride is a toxic substance, both acutely and chronically. It is an excellent solvent for many purposes and it is easy to obtain in the purity desired. Therefore, since it is widely used, one tends to become careless, which increases the toxic hazard. Carbon tetrachloride may enter the body by inhalation, ingestion, and absorption through the skin. The threshold concentration of detection by odor is about 80 ppm, and the odor is strong at about 175 ppm. There have been several reports of illness from breathing air containing 25 ppm. The concentration, therefore, is too high when the odor is detectable [5713, 7022].

Kutepov [4351] reported the LD_{50} as 5650–6200 mg/kg in rats, 900–9200 in mice, and 5760 in guinea pigs.

The *threshold limit value* is 10 ppm, 64 mg/m^3 [7022].

Carbon tetrachloride is not flammable; it has no flash or fire point; it will neither explode nor support combustion.

318.   1,1-Dichloroethane

Barton [461] prepared and purified 1,1-dichloroethane for *kinetic studies*. Paraformaldehyde was treated with phosphorus pentachloride, and the reaction product was shaken with concentrated sulfuric acid or aqueous permanganate solution and then fractionally distilled; bp 58.0-58.7°.

Maryott and coworkers [4935] dried a good commercial grade with anhydrous potassium carbonate and distilled over Drierite through a Dúfton still.

Vogel [7771] washed a commercial product with a saturated sodium bicarbonate solution until effervescence ceased, then with water; dried and fractionally distilled.

Timmermans and Martin [7494] purified 1,1-dichloroethane by fractional distillation until the density and boiling point were constant. About 200 grams were obtained from a 500-gram commercial product.

Li and Pitzer [4576] purified a commercially available 1,1-dichloroethane for *thermodynamic property study* by fractional distillation and crystallization three times. The criterion of purity was the change in melting point as a function of the fraction melted.

Mansson and coworkers [4877] purified 1,1-dichloroethane for *thermodynamic studies* by preparative gas chromatography; purity was 99.99%m from melting behavior.

SAFETY

Heppel and coworkers [3309] concluded that 1,1-dichloroethane is one of the most toxic of the commonly used chlorinated hydrocarbons. It has a chronic toxicity somewhat less than carbon tetrachloride, and its most important effect seems to be on the liver.

The *threshold limit value* of 100 ppm, 400 mg/m^3, may be low enough to prevent injury, considering the distinctive odor and irritating properties [7022].

The *explosive limits* in air are 5.6 and 11.4%v [6402]. The *minimum ignition temperature* in air is 458° [5713].

319.   1,2-Dichloroethane

1,2-Dichloroethane is a very useful solvent with a favorable boiling point for crystallization and extraction. It is more reactive than many of the commonly used laboratory chloro solvents. It is generally prepared commercially by chlorinating ethylene; besides the principal addition product, some substitution takes place. The other impurities present depend on the impurities in the ethylene. Generally, there are several other chlorinated hydrocarbons present in varying amounts. It is available in a variety of purities.

Vogel [7765] prepared 1,2-dichloroethane for *physical property measurements* from dry 1,2-ethanediol and thionyl chloride. The crude material was purified by washing with 5% sodium hydroxide solution and water; after drying it was distilled; bp 83° at 756 Torr. A commercial product was purified in a similar manner; bp 83° at 762 Torr.

Pitzer [5870] washed 1,2-dichloroethane with water and repeatedly fractionally distilled for *heat capacity* and *related physical property studies*. The impurities were estimated from the rise in the heat capacity below the melting point; 0.058% impurities; see also Railing [6025].

The spectra of *trans*-1,2-dichloroethane prepared by fractional distillation, followed by several crystallizations just below the melting point was studied by Tokuhiro [7514].

This compound has been purified by Götz [2885], Smyth and coworkers [6863], and Sugden [7072] by washing with dilute potassium hydroxide and water, drying over calcium chloride or phosphorus pentoxide, and fractionally distilling through an efficient column.

For *electrical measurements*, Walden and coworkers [7866] alternately dried and distilled over sodium sulfate for several days, finally dried for one day over phosphorus pentoxide, and collected the fractions distilling within 0.5° of their measurement. Walden and Busch [7861] fractionally distilled commercially pure material, dried the middle fraction for about 4 weeks with anhydrous potassium carbonate, and distilled in a distillation apparatus painted black.

Barton [654] purified by fractional distillation to a boiling point of 83.5–83.7° for *kinetic studies*.

For purification for *heat of vaporization study*, see Wadso [7822], Halogenated Hydrocarbons, General Comments.

A good commercial grade of 1,2-dichloroethane was fractionally distilled by Zwolenik and Fuoss [8315] in a Todd still at a reflux ratio of 5:1. The first 10% was discarded and the main fraction distilled sharply at 83.7°. It was used for *specific conductance measurements*.

Hammond [3121] purified the crude compound by shaking with sodium hydroxide solution to remove the acid, heating to vaporize and remove material boiling below 71.5°, adding water, and distilling the azeotrope at 71.5°. The lower phase was dried by distilling over potassium or sodium hydroxide.

Railing [6025] purified by the procedure described for 1,2-dibromoethane; premelting indicated a purity of 99.6%.

Davis and Schumann [1860] found that 1,2-dichloroethane stored in a colorless glass container in direct sunlight acquired an acid reaction within a few weeks. Deterioration was very slow when the solvent was protected from light.

1,2-Dichloroethane was purified by Coetzee and Lok [1556] as a *nonhydrogen-bonding solvent* to study *self-association of acids and bases* as follows: Reagent material was shaken with Brockman Activity 1 basic alumina, 1 gram/L, to remove traces of hydrochloric acid. It was refluxed over a like amount of alumina and fractionally distilled at a reflux ratio of 8:1; bp 83.5°.

Ralph and Gilkerson [6027] purified 1,2-dichloroethane in the same manner that they used for *o*-dichlorobenzene; $\kappa = 4 \times 10^{-11}$ ohm^{-1} cm^{-1}.

Beckers [711] found that purified 1,2-dichloroethane could be stabilized more than 207 days by the addition of 0.05–0.10% diisopropyl amine.

SAFETY

The odor of 1,2-dichloroethane is definitely detectable at 100 ppm and pronounced at 200 ppm. One can become adapted to the odor at low concentrations; therefore it cannot be considered as a reliable warning [147]. The acute and chronic effects of 1,2-dichloroethane can be significant [5713].

The *threshold limit value* has been set at 50 ppm, 200 mg/m^3 [7022].

The *flammable limits* in air are 6.20 and 15.90%v, the value at elevated temperature. The *minimum ignition temperature* in air is 413° [5712].

320. 1,2-Dichloropropane

Reboul [6088] prepared 1,2-dichloropropane in 1873 by reacting allyl chloride with concentrated hydrochloric acid solution at 100°. Mouneyrat [5287] and Lebedev [4445] chlorinated 1-chloropropane to prepare 1,2-dichloropropane.

Goudet and Schenker [2890] and Curme [1765] prepared 1,2-dichloropropane by chlorination of propylene.

Nelson and Young [5448] purified 1,2-dichloropropane for *vapor pressure measurements* by distillation in a 32-plate rectifying still at 165 Torr.

SAFETY

The value of LD$_{50}$ in rats was reported as 2200 mg/kg [2235].

The *threshold limit value* is 75 ppm, 350 mg/m^3 [142].

The *flammable limits* in air are 3.4 and 14.5%v. The *minimum ignition temperature* in air is 1035° [3574].

321. 1,1,1-Trichloroethane

1,1,1-Trichloroethane reacts with calcium hydroxide and sodium ethoxide at ordinary temperatures.

Barton and Onyon [659] purified a commercial material for *pyrolysis studies*. It was washed with concentrated hydrochloric acid, 10% potassium carbonate solution, and 10% sodium chloride solution, dried over calcium chloride, and distilled through an 18-in., Penn State-type column. The main fraction boiling at 73.9° was fractionally crystallized three times, the lower freezing fraction being rejected each time.

Hardies [3151] stabilized 1,1,1-trichloroethane with 0.1–1.0%w dimethoxyethane, 1.0–5%w 2-methyl-2-propanol, 0.5–5%w nitromethane, and 0.1–1%w butylene oxide. Hardies and Pray [3152] found 0.5%w tetrahydrofuran, 1.5%w 2-methyl-2-butanol, and 1.5%w 3-hydroxy-3-methylbutyne to be effective. *p*-Dioxane, nitromethane, and 2-butanol and a vicinyl C$_4$–C$_6$ monoepoxide were found effective [2066].

Air oxidation of tri- and tetrachloroethane can be prevented, or delayed for 22 hr or longer, during the purification by fractional distillation by the addition of 100 mg of phenol or 25 mg of pyrocatechol. When both were present the chloro compound was stabilized for 10 days [6904].

Connett [1628] reports the formation of an apparently unsaturated impurity in 1,1,1-trichloroethane when left in contact with 4A molecular sieves for a week. The impurity was formed rather rapidly when the molecular sieves were added to about three times their weight of the trichloroethane, reaching about 2%m in 24 hr.

Mansson and coworkers [4877] purified 1,1,1-trichloroethane for *thermodynamic studies* by fractional distillation followed by preparative gas chromatography; purity was 99.99%m from melting behavior.

Stuchkov and coworkers [7056] adsorbed water from 1,1,1-trichloroethane with such zeolites as NaA powder, NaA granular, and NaA-2KT granules.

Archer and coworkers [414] reported that "the reaction of 1,1,1-trichloroethane with aluminum was completely inhibited during 24 hr of refluxing by adding 2-, 3-, or 4-cyanopyridine in concentrations of 0.05, 0.04, or 0.04 M, respectively."

SAFETY

The American Industrial Hygiene Association introduced a new toxicological factor in 1964 [146], the *Emergency Exposure Limit* (EEL). This is the concentration of a substance in air that can be tolerated without adversely affecting health, but not necessarily without acute discomfort or other evidence of irritation or intoxication. These values are intended to give guidance in the management of single brief exposures. Only a few substances have had their EEL's established. The EEL's for 1,1,1-trichloroethane are: 5 min at 2500 ppm (13,500 mg/m^3); 15 min at 2000 ppm (10,800 mg/m^3); 60 min at 1000 ppm (5400 mg/m^3). Exposure to 2–3%v for more than a few minutes will cause complete incoordination, unconsciousness, and possible death.

Aviado and Smith [514] reported that 1,1,1-trichloroethane caused tachycardia and hypotension and anesthetized rhesus monkeys when inhaled at 2.5–5%v.

The *threshold limit value* in air is 350 ppm, 1900 mg/m^3 [4509].

The *flammable limits* in air are 10–15.5%v by hot wire ignition. It will not sustain combustion [5713].

322. 1,1,2-Trichloroethane

Delacre [1925] prepared 1,1,2-trichloroethane by the reaction of dichloroethanol with phosphorus pentachloride. Meyer and Muller [5092] prepared the compound by heating 1,1-dichloroethane with antimony pentachloride, while Hamai [3108] reacted 1,1-dichloroethane with chlorine at room temperature.

Prins [5957] added HCl to dichloroethylene in the presence of powdered aluminum chloride at 30–40° to produce 1,1,2-trichloroethane.

Crowe and Smyth [1728] purified 1,1,2-trichloroethane for *physical property studies*. A sample was washed twice with 6 N sodium carbonate solution and several portions of water. After drying over anhydrous calcium chloride for 24 hr, it was fractionally distilled at atmospheric pressure.

SAFETY

The *threshold limit value* for skin absorption is 10 ppm, 45 mg/m^3 [142].

323. 1,2,3-Trichloropropane

Fittig and Pfeffer [2453] prepared 1,2,3-trichloropropane in 1865 by saturating a mixture of glycerol and glacial acetic acid with HCl, distilling at 130°, and washing the residue with water and soda. The crude dichlorohydrin produced was dried over calcium chloride and then treated further with phosphorus pentachloride to give 1,2,3-trichloropropane.

Kharasch and Brown [3977] chlorinated allyl chloride with sulfuryl chloride in the presence of dibenzyl peroxide to give a yield of 80% 1,2,3-trichloropropane.

Preparation methods by the reaction of propylene chloride and chlorine are given in Saunders [6391] and Levine [4555].

Mikhailov and coworkers [5121] treated commercial 1,2,3-trichloropropane with fuming 17–20% sulfuric acid at 115–120° and removed the purified sample by filtration.

SAFETY

The values of LD_{50} in mice and rats are 369 and 505 mg/kg, respectively [6619].

The *threshold limit value* has been set at 50 ppm, 300 mg/m^3 [6402].

The *flammable limits* in air are 3.2 and 12.6%v. The *minimum ignition temperature* in air is 303° [6402].

324. 1,1,2,2-Tetrachloroethane

Field and Kovacic [2423] reacted 1,1-dichloroethylene with trichloroamine in methylene chloride at 25° to give 89% 1,1,2,2-tetrachloroethane.

Thomas and Gwinn [7411] repeatedly washed a technical grade with concentrated sulfuric acid, steam-distilled, and then fractionally distilled in a 30-plate column. The center portion boiling within 0.20° was stored over anhydrous calcium sulfate until it was ready for use in determining *rotational configuration* and *dipole moment*.

Physical properties were determined by Vogel [7771] on technical material that was purified by stirring 135 mL with 17 mL of concentrated sulfuric acid for 10 min at 80–90°. The discolored acid was removed and the acid washing repeated two more times. The 1,1,2,2-tetrachloroethane was washed with water, steam-distilled, again washed with water, dried with potassium carbonate, and fractionally distilled throught a Widmer column.

For *kinetic studies*, Barton [654] purified by shaking with concentrated sulfuric acid and fractionally distilling; bp 146–147°.

Walden and Werner [7867] determined the *dielectric constant* on material that was shaken with potassium carbonate solution, dried over anhydrous potassium carbonate, fractionated under reduced pressure at 60°, and the middle fraction again fractionated at atmospheric pressure. The distillate was further dried over potassium carbonate for 1 day before use.

Walden and Gloy [7863] washed commercial material with potassium carbonate solution, dried, and distilled; bp 145.6°. The *specific conductance* was found to be immeasurably low.

Stabilization during fractional distillation; see 1,1,1-trichloroethane [6904].

SAFETY

It is reported that 1,1,2,2-tetrachloroethane has a noticeable odor at 3 ppm; however, the odor is not particularly striking and the acceptable level is 5 ppm. Neither level is of any significance as a warning property. This compound is considered to be highly toxic [5713, 7022].

The *threshold limit value* is 5 ppm, 35 mg/m^3 [7022].

325. Pentachloroethane

Pentachloroethane is manufactured by the chlorination of trichloroethylene. The crude product contains about 5% unreacted trichloroethylene.

Field and Kovacic [2423] reacted trichloroethylene with trichloroamine at 23° to give 41% pentachloroethane.

This compound cannot be distilled at atmospheric pressure without some decomposition. Water in excess of that dissolved by the compound causes appreciable hydrolysis, even at room temperature. Sodium ethylate, alcoholic potassium hydroxide, and aqueous calcium hydroxide slurry split out hydrogen chloride. Sodium hydroxide in an aqueous-alcoholic solution is reported to form an aldehyde resin. Ammonia and primary and secondary amines also react [4026].

Walden and Werner [7867] treated pentachloroethane with potassium carbonate solution, dried over potassium carbonate, fractionally distilled at reduced pressure, and finally fractionated the middle portion at atmospheric pressure. The fraction boiling at 157.5–158.5° was used for *dielectric constant measurements* after being dried over potassium carbonate.

Thomas and Gwinn [7411] repeatedly washed a commercial product with concentrated sulfuric acid, steam-distilled, and then fractionally distilled through a 30-plate column. The center portion distilling within 0.2° was stored over anhydrous calcium sulfate and used for *rotational configuration* and *dipole moment measurements*.

Barton [654] purified pentachloroethane for *kinetic studies* by shaking with concentrated sulfuric acid and fractionally distilling.

SAFETY

Pentachloroethane has a strong narcotic effect, perhaps even greater than chloroform. Exposure to this compound may result in injury to the liver, lungs, and kidneys. It is irritating to the eyes and upper respiratory tract. Cats exposed to 121 ppm, 8–9 hr daily for 23 days showed significant pathological changes in the liver, lungs, and kidneys [5713].

Aromatic Chlorinated Hydrocarbons

326. *o*-Dichlorobenzene

This hydrocarbon is a by-product in the manufacture of chlorobenzene and *p*-dichlorobenzene. The technical grade approximated the composition of the *ortho–para* eutectic, about 84% *ortho* and 15–16% *para*, plus as much as 1% trichlorobenzene. The purified product may be obtained with a purity of 98–99+%. A 99.9%m material is available but it is expensive.

o-Dichlorobenzene appears to be stable to many of the common reagents and has some excellent characteristics as a solvent.

Dubchenko [2102] prepared *o*-dichlorobenzene by direct substitution of chlorine in a gas-phase reaction of *o*-chloronitrobenzene with carbon tetrachloride.

Moore and Hobbs [5235] purified Eastman's best grade, as described under fluorobenzene, for *dipole moment measurements*; bp 179.7° at 746.2 Torr.

A technical *o*-dichlorobenzene was passed through Alcoa grade F-20 alumina in a 30 × 2-cm column by Ralph and Gilkerson [6027] and then fractionally distilled through a 35-cm column packed with glass helices. The middle fraction, bp 48° at 5 Torr, was stored over alumina which had been heated to about 800° for 1 hr. Just before use, the solvent was passed through a 30 × 2-cm column, the bottom of which was heated alumina and the top one-half of which was 5A molecular sieves. The specific conductance was 3–8 × 10^{-11} ohm^{-1} cm^{-1}. See also [974].

The isomeric dichlorobenzenes were separated by Radzitzky and Hanotier [6020] by clatheration with Werner complexes of the form

$$Ni(SCN)_2(RC_6H_4CHR'NH_2)_4.$$

SAFETY

The physiological effect of o-dichlorobenzene is primarily injury to the liver, secondary to the kidneys. Short exposures to high concentrations may result in depression of the central nervous system. The odor becomes strong and irritating at concentrations of 100 ppm.

The *threshold limit value* is 50 ppm, 300 mg/m^3 [7022].

The *minimum ignition temperature* in air is 648° [5712].

327.   m-Dichlorobenzene

Dubchenko [2102] prepared m-dichlorobenzene by direct substitution of chlorine in a gas-phase reaction of m-chlorobenzene with carbon tetrachloride. Brackman and Smit [1045] prepared m-dichlorobenzene by reacting m-phenylenediamine with copper(II) chloride and nitrogen(II) oxide.

Vogel [7765] further purified a pure commercial product for *physical property measurements*. It was washed successively with 10% sodium hydroxide solution, then with water until the washings were neutral, dried, and distilled; bp 172.5° at 760 Torr.

The compound attacks rubber stoppers.

Kiba and coworkers [3993] heated a mixture of m- and p-dichlorobenzene with 95% sulfuric acid for 5 hr at 100°. The precipitated p-dichlorobenzene was removed and the remaining acid layer was washed with hexane. After adding water, the mixture was autoclaved 6 hr at 180° to give 99.2% purity m-dichlorobenzene.

328.   p-Dichlorobenzene

Brackman and Smit [1045] prepared p-dichlorobenzene by reacting either p-chloroaniline or p-phenylenediamine with copper(II) chloride and nitrogen(II) oxide.

The principal impurity is o-dichlorobenzene. The presence of 0.5%w of the *ortho*-compound depresses the freezing point 0.25°.

p-Dichlorobenzene is stable to the common laboratory reagents. It reacts with concentrated sulfuric acid and other reagents that react with benzene.

For separation of the dichlorobenzene isomers, see o-dichlorobenzene, Radzitzky and Hanotier [6020].

SAFETY

p-Dichlorobenzene has a very distinctive aromatic odor detectable at 15–30 ppm in air. It is painful to the eyes and nose at concentrations of 80–160 ppm. A person may become accustomed to and tolerate high concentrations. The physiological symptoms are similar to those of the o-isomer but this isomer is somewhat less toxic [5713, 7022].

The *threshold limit value* is 75 ppm, 450 mg/m^3 [7022].

329.   α,α-Dichlorotoluene

Cahours [1304] prepared α,α-dichlorotoluene in 1849 by reacting benzaldehyde with phosphorus pentachloride. Kharasch and Brown [3976] treated toluene with sulfuryl chloride in the presence of benzoyl peroxide catalyst to produce 90% α,α-dichloro-toluene and 10% α-chlorotoluene.

Scheraga and Hobbs [6432] purified α,α-dichlorotoluene for *kinetics studies* by fractional crystallization in vacuum.

SAFETY

The reported LD_{50} in mice is 467 mg/kg with a maximum permissible dose of 0.003 mg/L [6960].

The *threshold limit value* is 0.5 mg/m^3 [4185].

330. 2,4-Dichlorotoluene

Erdman [2278] treated 2,4-diaminotoluene in hydrochloric acid solution with sodium nitrite solution and added copper(I) chloride solution to the mixture to produce 2,4-dichlorotoluene.

Cohen and Dakin [1573] diazotized 3-chloro-*p*-toluidine with sodium nitrite and reduced the resulting product with tin(II) chloride to produce 2,4-dichlorotoluene. Gelfand [2702] chlorinated *p*-cholorotoluene with antimony trichloride at room temperature and obtained a 68% yield of 2,4-dichlorotoluene.

331. 3,4-Dichlorotoluene

Cohen and Dakin [1573] diazotized 2-chloro-*p*-toluidine with sodium nitrite and reduced the resulting product with tin(II) chloride to produce 3,4-dichlorotoluene.

332. α,α,α-Trichlorotoluene

Schischkoff and Rosing [6451] prepared α,α,α-trichlorotoluene in 1858 by the reaction of benzoyl chloride and phosphorus pentachloride. Henne and Newman [3290] reacted benzotrifluoride and acetyl chloride with anhydrous aluminum chloride and isolated α,α,α-trichlorotoluene by fractional distillation.

Stephenson and Coburn [7002] purified a commercial sample for *spectral studies*. They bubbled dry nitrogen through the sample which was then extracted three times with a saturated sodium bicarbonate solution and dried with magnesium sulfate. The final product was obtained by fractional distillation at atmospheric pressure.

Unsaturated Chlorinated Hydrocarbons
333. 1,1-Dichloroethylene

Rogers [6229] prepared 1,1-dichloroethylene by dehydrohalogenation of 1,1,2-trichloroethane with aqueous calcium hydroxide. It was purified by fractional distillation for *dipole moment studies*.

Hildenbrand and coworkers [3397] purified, 1,1-dichloroethylene for *thermodynamic and spectrographic studies* by fractional distillation; purity 99.97%m by calorimetric melting studies. Precautions were taken to avoid elevated temperatures and contact with air.

1,1-Dichloroethylene must be stored away from light and air. When stored in the presence of oxygen, a peroxide is formed [2300].

The Dow Chemical Company [2073] states, ''The safe handling of vinylidine chloride monomer is dependent on knowledge of the characteristics of the material." The observance of this principle contributes to the safe handling of all solvents.

The Ethyl Corporation [2299] reports that 1 ppm of *p*-methoxyphenol is an effective stabilizer for the monomer; it is about 50 times as effective as phenol under similar experimental conditions.

Dow Chemical Company [2073] furnishes the monomer inhibited either with 0.6–0.8%w phenol or with 200 ppm of monomethoxyhydroquinone. The latter inhibitor usually is preferred for most purposes because it is present in smaller amount and does not interfere with many uses.

SAFETY

The odor is the sweet smell characteristic of chloroform. Most people can detect it at 1000 ppm but some can detect the odor at 500 ppm [5713]. Ethyl Corporation [2299] states that 1,1-dichloroethylene is irritating to the skin and eyes and may cause burns. Overexposure to the vapors may cause anesthesia.

Irish [5713] suggests that the *threshold limit value* be less than 25 ppm, 99 mg/m^3.

The *flammable limits* in air are 7 and 16%v [2300]. The *minimum ignition temperature* in air is 570° [5713].

334.   *cis*-1,2-Dichloroethylene
335.   *trans*-1,2-Dichloroethylene

Commercial 1,2-dichloroethylene is a mixture of the *cis*-and *trans*-isomers. The individual isomers are available at a reasonable cost. There is some confusion in the literature as to which is the *cis*- and which is the *trans*-form. Some of the properties listed in the tables may be interchanged because it was not possible to identify positively the isomer in all publications.

Knox and Riddick [4081] separated a commercial product into the isomers by fractional distillation. The *cis*-fraction contained about 0.01% *trans*-isomer and 0.01% of an unidentified substance. The *trans*-fraction contained less than 0.0001% *cis*-isomer. The purified material was stored in dark bottles with a small amount of hydroquinone to inhibit oxidation and polymerization.

Maryott and coworkers [4935] purified Eastman's *cis*-product for *dielectric constant* and *dipole moment measurements* by shaking with mercury, drying over anhydrous potassium carbonate, and distilling from Drierite in a Dufton still; bp 60.33–60.38°.

Walden and Glory [7863] separated the *cis*-isomer from technical 1,2-dichloroethylene by fractional distillation from fused potassium carbonate; bp 59–61°; the specific conductance at 25° was 8.5 × 10^{-9} ohm^{-1} cm^{-1}.

SAFETY

Apparently 1,2-dichloroethylene does not have a high toxicity. The major response to both isomers is one of central nervous system depression. Liver and kidney injury does not appear to be a major factor [5713].

The *threshold limit value* is 200 ppm, 790 mg/m^3 [7022].

The *flammable limits* in air are 9.7 and 12.8%v [6402].

336.   Trichloroethylene

Like chloroform, trichloroethylene undergoes autoxidation, forming, among other components, hydrogen chloride, carbon monoxide, and carbonyl chloride.

It has been reported that trichloroethylene and potassium or sodium hydroxide may explode. Trichloroethylene reacts with 90% sulfuric acid, and dropwise addition of nitric acid produces, among other substances, trichloronitromethane (chloropicrin).

Bruyne and coworkers [1215] treated commercial trichloroethylene containing carbonyl chloride with potassium carbonate solution and then with water. After drying over potassium carbonate and calcium chloride, it was fractionally distilled through a Widmer column in an atmosphere of dry carbon dioxide.

McDonald [4759] distilled a commercial material in an Ewell column of 20 theoretical plates. The 40% middle fraction was used for *vapor pressure measurements*.

Commercial trichloroethylene was purified by Carlisle and Levine [1352] for *physical property measurements* and *stability studies*. It was steam-distilled from a 10%w calcium hydroxide slurry. The oil phase was cooled to −30 to −50° and the ice removed by filtration. The filtrate was fractionally distilled at 252 Torr at a reflux ratio of 3:1. The middle fraction was collected in a receiver covered with black paper; bp 86.7°, fp −88°. Neither purified nor commercial material showed any evidence of decomposition in 1 year when stored in glass in the dark, or in steel. It behaved like chloroform toward light and air.

For *conductivity measurements*, Walden [7853] distilled technical material several times, dried over sodium sulfate, distilled again, and dried the fraction distilling at exactly 87° for a long time over potassium carbonate.

Trichloroethylene may be stabilized against oxidation by 0.01–0.02% 1-ethoxy 2-iminoethane [5963], and against heat and oxidation by mixtures of pyrrole or its derivatives and an epoxy compound [6903]; 0.05–1% methacrylonitrile is claimed to be an effective stabilizer [4910], and 0.001–1.0% acetylenic alcohol such as 3-methyl-3-hydroxypentyne and a light-stabilizing substituted phenol such as isoeugenol is claimed to be effective [6785].

SAFETY

The predominant physiological response to trichloroethylene is that of depression of the central nervous system, particularly from acute exposures. Visual disturbances and mental confusion accompanied by incoordination may lead to accidents.

The National Cancer Institute found that trichloroethylene induced tumors in mice, predominately liver cancer with some transfer to the lungs [6565].

An OSHA review recommended that the 8-hr time weighted average exposure limit be 100 ppm and the 15-min ceiling limit be 150 ppm [2388].

NIOSH has a booklet on potential occupational hazards of trichloroethylene [5420].

The *threshold limit value* is 100 ppm, 520 mg/m^3 [7022].

The *minimum ignition temperature* in air is 463° [5712].

337. Tetrachloroethylene

Tetrachloroethylene is an excellent solvent and is less reactive than many of the other halogenated hydrocarbons. It reacts with concentrated sulfuric and nitric acids. It is oxidized to phosgene and trichloroacetic acid [551] under the same conditions of storage given for chloroform. Bailey [550] found that ethanol, ethyl ether, or thymol could be used to retard the oxidation, the effectiveness increasing in the order given. Little oxidation was observed in material containing 2–5 ppm thymol when stored in a dark glass container. The pure material is more stable than the commercial material.

Timmermans and Hennaut-Roland [7488] treated a commercial product with sodium carbonate and fractionally distilled until the density was constant. Walden and Werner [7867] dried a commercial product over sodium sulfate or calcium chloride and fractionally distilled. See also Mathews [4969].

Wimer and coworkers [8108] purified tetrachloroethylene by washing with 10% aqueous sodium hydroxide solution, neutralizing, and then mixing and storing 16–24 hr with ammonia. The resultant solution was then washed with water and dried to less than 40 ppm water.

Bretscher [1085] distilled the material in a vacuum in order to avoid phosgene formation and stored it in the dark out of contact with air. Greenwald [2949] removed 1,1,2-trichloroethane and 1,1,1,2-tetrachloroethane by countercurrent extraction with an ethanol–water mixture. The purified product is more stable to light, heat, moisture, and oxidation.

Tetrachloroethylene is stabilized against heat and oxidation by mixtures of pyrrole and its derivatives and an epoxy compound [6903]. It has also been stabilized by the addition of 0.005–1% of diallylamine or tripropyleneamine. Baldridge [564] found 0.01–1.0% acetylenic halides effective as a stabilizer such as 3-chloro- or 3-iodo-1-propyne, 4-bromo-1-butyne, or 1,4-dichloro-1-butyne.

SAFETY

Tetrachloroethylene is a central nervous system depressant at high concentrations. Although tetrachloroethylene is considered a relatively safe solvent, the available information indicates that the word *relatively* should be remembered [5713].

Dmitrieva [2024] reported that a maximum concentration of 0.03 mg/L in air is recommended.

The *threshold limit value* is 100 ppm, 670 mg/m^3 [7022].

It is nonflammable, nonexplosive, and will not support combustion.

Brominated Hydrocarbons (338–354)

Aliphatic Monobrominated Hydrocarbons

338. Bromomethane

Steinkopf and Schwen [6992] obtained a yield of 77% bromomethane by heating a mixture of methanol, bromine, and red phosphorus. Boulin and Simon [1011] prepared bromomethane by adding dimethyl sulfate in an acidified sodium bromide solution.

Bygden [1299] added methanol to a concentrated sulfuric acid solution followed by addition of powdered potassium bromide. The mixture was heated until no gases were evolved. The gaseous bromomethane produced was passed through sodium hydroxide solution to remove HBr and then through concentrated sulfuric acid solution.

Egan and Kemp [2218] purified bromomethane by bubbling the condensed phase through 36 N sulfuric acid solution and then through phosphorus pentoxide.

Eastman Kodak practical-grade bromoethane was purified by Morgan and Lowry [5250] for *dielectric polarization studies* by distillation. The distilling column was cooled with ice water and refluxed to allow the hydrogen bromide to boil off. The product was distilled into a closed receiver immersed in an ice–salt-water mixture. The process was repeated three times.

The *threshold limit value* for skin absorption is 15 ppm, 60 mg/m^3 [142].

The *flammable limits* in air are 13.5 and 14.5%v. The *minimum ignition temperature* in air is 537° [6402].

339. Bromoethane

Adams [21, p. 6] gives directions for preparing about 1 kg from hydrobromic acid, ethanol, and concentrated sulfuric acid. Kamm and Marvel [3842] prepared about 1 kg in about 90–95% yield from ethanol and bromine. Rodionov [6224] claims good yields from ethyl-*p*-toluene sulfonate, and potassium bromide.

Jung and Hatfield [3803] prepared bromoethane by reacting ethanol with trimethyl-silyl bromide in chloroform at 25°.

Smyth and Engel [6864] purified a commercial product by washing several times with cold concentrated sulfuric acid and then with water until all the sulfate was removed. The washed material was dried with calcium chloride and fractionally distilled; bp38.3–38.4°.

The principal impurities in bromoethane are ethanol and water, both of which form azeotropes with bromoethane; the three form a ternary azeotrope. Material prepared according to Adams [21] can be further purified by rapidly agitating the interphase of cold concentrated sulfuric acid and bromoethane, 1:5 by volume, for 2 hr [6165]. The alcohol-free material is washed with water and then agitated as described with 5% sodium bicarbonate solution, washed several times with water, shaken with calcium chloride, and dried overnight with Drierite. It is then fractionally distilled in a 25-plate, all-glass, Penn State-type column. The column is operated under total reflux for 2 hr, 5%v removed at a reflux ratio of 25:1 and 90%v removed at a reflux ratio of 10-15:1. The middle 90% distills within 0.1° at a constant density.

Kosenko [4205] reported an LD$_{50}$ for mice as 2.8 g/kg.

The *threshold limit value* has been set at 200 ppm, 890 mg/m^3. It is believed that this is sufficiently low to prevent systemic effects and narcosis [7022].

The *flammable limits* in air are 6.75 and 11.25%v. The *minimum ignition temperature* in air is 511° [5712].

340. 1-Bromopropane
341. 2-Bromopropane

Goshorn and coworkers [2757] give directions for the preparation of the isomeric bromopropanes from the corresponding alcohols, red or yellow phosphorus, and bromine. The yield for 1-bromopropane is 143–152 grams or 69–73%; for the 2-isomer the yield is 170–186 grams or 82–90%.

Jung and Hatfield [3803] prepared 2-bromopropane by reacting 2-propanol with tri-methylsilyl bromide in chloroform at 50°.

Bjellerup [899] purified a commercial sample of 1-bromopropane for *heat of combustion study*. It was treated with phosphorus pentoxide and successively distilled at 755, 360, and 755 Torr. The purified material distilled over a 0.002° range and had a corresponding density range of 0.0003 g/mL.

For purification of 1-bromopropane for *heat of vaporization study*, see Wadso [7822], Halogenated Hydrocarbons, General Comments.

A commercial sample of 2-bromopropane was purified by Bjellerup [899] for *heat of combustion study*. It was washed in order with concentrated hydrochloric acid and 10% sodium bicarbonate. It was dried with calcium chloride and then phosphorus pentoxide. It was successively fractionally distilled at 740, 752, and 760 Torr. The final distillation had a boiling range of 0.01° and a corresponding density range of 0.0004 g/mL.

Maccoll and Thomas [4801] purified 2-bromopropane for *pyrolysis studies* by fractional distillation through a 1-meter glass helices-packed column; bp 59.32° at 760 Torr, $n_D15°$ 1.4285.

SAFETY

For 1-bromopropane, Kosenko [4205] reported an LD_{50} for mice as 2.5 g/kg.

342.   1-Bromobutane

Kamm and Marvel [3842] prepared 1-bromobutane by refluxing 1-butanol, HBr, and concentrated sulfuric acid solution for 2 hr. The product was removed by direct distillation and then washed successively with water, concentrated sulfuric acid solution, and sodium carbonate solution. Finally it was dried over calcium chloride and distilled. Other preparation techniques for bromination of 1-butanol are given in Tseng [7567], Wiley [8058], and Frazer [2529].

Vaughan and coworkers [7693] obtained a 99% yield of 1-bromobutane by photochemically reacting 1-butene with HBr in the liquid phase at −78°. The ultraviolet radiation source was a mercury arc.

Brown and Lane [1135] reacted tributylborane with bromine in tetrahydrofuran in methanol. The yield of 1-bromobutane was 91% at 0°.

Svoboda and coworkers [7108] purified 1-bromobutane for *vapor pressure measurements*. A sample was shaken with concentrated sulfuric acid solution, washed with water, neutralized by sodium carbonate or sodium bicarbonate solution, and dried over calcium chloride.

Bjellerup [900] purified a commercial 1-bromobutane for *heat of combustion studies*. It was treated with phosphorus pentoxide and successively fractionally distilled at 750 and 340 Torr. Eighty per cent of the final distillate was collected over a density range of 0.0003 g/mL. The second half of this fraction was collected at a constant density. It was treated with calcium sulfate and distilled at atmospheric pressure with the middle portion used.

For purification for *heat of vaporization study*, see Wadso [7822], Halogenated Hydrocarbons, General Comments.

Skau and McCullough [6784] purified 1-bromobutane for *physical property measurements*. A sample was washed with constant boiling HBr solution, dried over anhydrous calcium sulfate, and successively fractionally and vacuum distilled.

SAFETY

Kosenko [4205] reported an LD_{50} for mice as 6.7 g/kg.

The *flammable limits* in air at 100° are 2.6 and 6.6%v. The *minimum ignition temperature* in air is 265° [5415].

343. 2-Bromobutane

Meyer and Muller [5092] prepared 2-bromobutane in 1892 by heating 2-butanol and HBr. Other preparation techniques for bromination of 2-butanol are given in Goshorn and Degering [2881] and Frazer [2529].

Brown and Lane [1135] reacted the organoborane from 2-butene with bromine and sodium methoxide simultaneously at 25°; the yield was 74% 2-bromobutane.

For purification for *physical property measurements*, see Skau and McCullough [6784], 1-bromobutane.

A commercial 2-bromobutane was purified by Bjellerup [899] for *heat of combustion study* by successive fractional distillation at 750 and 170 Torr. The purified sample has a density range of 0.0001 g/mL. It was dried with calcium sulfate and distilled at 290 Torr.

For purification for *heat of vaporization study*, see Wadso [7822], Halogenated Hydrocarbons, General Comments.

2-Bromobutane has two optical isomers. D(+)-2-Bromobutane has been prepared by various methods of bromination of L(−)-2-butanol; details are given in Franke and Dworzak [2517], Helmkamp et al. [3276], Kenyon et al. [3951], Levene and Marker [4541], and Goodwin and Hudson [2865]. Jensen and coworkers [3712] treated 2-butylmercury(II) bromide with bromine in pyridine to give D(+)-2-bromobutane.

L(−)-2-Bromobutane has been prepared by Levene and Marker [4541] and Goodwin and Hudson [2865] using the methods of bromination of L(−)-2-butanol. Kenyon et al. [3951] reacted D(+)-*p*-toluenesulfonate and ethylmagnesium bromide in ether to prepare L(−)-2-bromobutane.

Reported values of the specific rotation at 25° vary significantly for D(+)-2-bromobutane; the angles range from 10.81° [4541] to 28.45° [3276]. For L(−)-2-bromobutane the angles range from −12.24° [3951] to −13.79° [4541].

344. 2-Bromo-2-methylpropane

Reboul [6089] prepared 2-bromo-2-methylpropane by reacting 2-methyl-2-propanol with phosphorus pentabromide. Other preparation techniques for bromination of 2-methyl-2-propanol are given in Goshorn and Degering [2881], Tseng and Hou [7567], and Wiley et al. [8058].

Vaughan and coworkers [7693] reacted isobutene and HBr in the gas phase in the dark for 18 hr to produce 2-bromo-2-methylpropane.

Michael and Zeidler [5097] observed an isomerization reaction between 2-bromo-2-methylpropane and 1-bromo-2-methylpropane. When 1-bromo-2-methylpropane was irradiated with ultraviolet light at ordinary pressures and temperatures, 13.5% 2-bromo-2-methylpropane was recovered [6013]. Kharasch and Hinckley [3981] reported that 1-bromo-2-methylpropane is produced by the reaction of isobutene and HBr in the presence of peroxides.

Bertie and Sunder [823] purified 2-bromo-2-methylpropane for *spectral studies*. Isobutene and HBr impurities were removed by washing with a 10% sodium bicarbonate solution and water. After drying over anhydrous calcium carbonate, the sample was fractionated on a Teflon spinning band column.

For purification for *heat of vaporization study*, see Wadso [7822], Halogenated Hydrocarbons, General Comments. For purification for *thermodynamic studies*, see Kushner [4346], 1-bromopentane.

Kosenko [4205] reported an LD_{50} for mice as 4.4 g/kg.

345. 1-Bromopentane

Lieben and Rossi [4591] prepared 1-bromopentane in 1871 by treating 1-pentanol with HBr. Other preparation techniques for bromination of 1-butanol are given in Olivier [5579] and Hooz and Gilani [3486].

Braun and Sobecki [1067] distilled pentylbenzamide with phosphorus pentabromide to produce 1-bromopentane.

Kushner and coworkers [4346] purified 1-bromopentane for *thermodynamic studies*. A commercial sample was washed with 6 N sodium carbonate solution and then with water. After drying over anhydrous calcium chloride for 24 hr, the solvent was fractionally distilled at atmospheric pressure in a 5-ft column packed with glass helices.

For purification for *physical property studies*, see Skau and McCullough [6784], 1-bromobutane.

A commercial sample of 1-bromopentane was purified by Bjellerup [900] for *heat of combustion studies* by treating it with phosphorus pentoxide and successively fractionally distilling at 760 and 41 Torr. It was dried over calcium sulfate and then redistilled; the middle portion was used.

346. 1-Bromodecane

Whitmore and coworkers [8033] prepared 1-bromodecane by heating a mixture of 1-decanol, HBr, and sulfuric acid at 110°. The organic layer was separated from the acid by steam distillation and the crude product was washed with cold 95% sulfuric acid solution, then with ammoniacal methanol–water solution, and finally with water. After it was dried over anhydrous potassium carbonate, pure 1-bromodecane was distilled through an all-glass column of 25–30 plates.

Aromatic Monobrominated Hydrocarbons

347. Bromobenzene

Bromobenzene was purified as a *solute for freezing-point depression measurements* by passing through alumina and then storing over alumina that had been heated to 1000°. It was distilled prior to use [974].

Masi and Scott [4947] purified bromobenzene for *thermodynamic property studies* by redistillation at atmospheric pressure in a 75-cm column packed with stainless steel helices. The middle third was further purified by four successive freezing and melting *in vacuo*; purity was 99.998%.

Vogel [7766] prepared bromobenzene for *physical property measurements* from AR-grade aniline, according to the method of Marvel [4930] for bromotoluene. The crude material was purified in the manner described for chlorobenzene; for specific directions, see Timmermans and Martin [7494], Mathews [4969], Rabinowitsch [6014], Williams and Krchma [8083], and Hurdis and Smyth [3593].

Brackman and Smit [1045] prepared bromobenzene by reacting aniline with copper(II) bromide and nitrogen(II) oxide.

Stull [7059] fractionally distilled a commercial product through a 5-ft column and obtained a purity of 99.9 + %m, as calculated from the freezing curve.

For purification for *heat of vaporization study*, see Wadso [7822], Halogenated Hydrocarbons, General Comments.

CRITERIA OF PURITY

Stull [7059] and Dreisbach and Martin [2090] determined the purity from the freezing curve.

Swietoslawski [7131] reports that the differential ebulliometric method may be used as the criterion of purity.

348.   1-Bromonaphthalene

Wright and Mura [8170] describe a classroom experiment for preparing 1-bromonaphthalene; 6.4 grams (0.05 mol) of naphthalene is mixed with 100 mL of ethyl acetate and 4–5 grams of anhydrous iron(III) chloride. A solution of 8 grams (0.05 mol) of bromine (apparently in ethyl acetate) is added to the mixture with efficient stirring. The reaction mass is refluxed 3–4 hr during the release of hydrogen bromide. The residue is poured into 200 mL of ice water after removing 125 mL of ethyl acetate by distillation. The product is steam distilled. No 2-bromonaphthalene was detected by the authors by gas chromatography.

Unsaturated Monobrominated Hydrocarbons

349.   Bromoethylene

Swarts [7124] prepared bromoethylene by heating a mixture of 1,2-dibromoethane and 20% alcoholic potassium hydroxide solution and refluxing it at 60%. Kozlov [4232] prepared the compound by passing acetylene through a mixture of copper(I) bromide and 15% HBr solution at 60°.

Kryszewski [4293] catalytically decomposed 1,2-dibromoethane with antimony trichloride at 330° to produce bromoethylene.

SAFETY

The *threshold limit value* is 250 ppm, 1100 mg/m^3 [142].

350.   2-Bromopropene

Reboul [6087] prepared 2-bromopropene by heating 2,2-dibromopropane with sodium acetate for 6 hr at 100°. Chavanne [1434] and Hatch and coworkers [3222] treated 1,2-dibromopropane with sodium phenolate in ethanol and separated 2-bromopropene by fractional distillation.

Aliphatic Polybrominated Hydrocarbons

351.   Bromoform

The stability on storage and reactions of bromoform are similar to those of chloroform.

For *dielectric constant measurements*, Smyth and Rogers [6874] washed bromoform with concentrated sulfuric acid, dilute sodium hydroxide, and water. After drying over anhydrous potassium carbonate, it was distilled at reduced pressure. The density to the third decimal place is given as the criterion of purity. Cauwood and Turner [1394] fractionally crystallized eight times; fp 7.5°.

Vogel [7771] washed bromoform containing 4% ethanol with saturated calcium chloride solution, dried with calcium chloride, and fractionally distilled. The colorless product, after standing several days, developed a yellow cloudiness at the surface.

SAFETY

Relatively little is known of the toxicity of bromoform. It has been shown to cause liver damage [5712, 7022].

The *threshold limit value* has been set at 0.5 ppm, 5.0 mg/m^3 [7022].

352. 1,2-Dibromoethane

Vogel [7765] prepared 1,2-dibromoethane from 48% hydrobromic acid, concentrated sulfuric acid, and 1,2-ethanediol. The crude material was washed twice with concentrated hydrochloric acid, water, sodium bicarbonate solution, and finally with water. It was dried and distilled; bp 130.5° at 758 Torr.

Smyth and Kamerling [6865] washed commercial 1,2-dibromoethane with concentrated sulfuric acid, sodium carbonate solution and water, dried over calcium chloride, and fractionally distilled.

Meisenheimer and Dorner [5045] dried 1,2-dibromoethane over phosphorus pentoxide and fractionated repeatedly under reduced pressure until the melting point was 9.98°. Repeated fractional distillation, followed by fractional crystallization several times, and a final distillation produced a material melting at 10.06°, but the melting point began to fall after a short time.

For *cryoscopic measurements* Moles [5216] dried 1,2-dibromoethane over calcium chloride, fractionally distilled, and dried over sulfuric acid in a desiccator for 6 days, and finally fractionally crystallized.

For purification for *heat of vaporization study*, see Wadso [7822], Halogenated Hydrocarbons, General Comments.

Railing [6025] purified a commercial product for *specific heat* and *heat of fusion measurements* by fractionally distilling twice through a 5-ft column, retaining a portion boiling over a 0.02° range. This narrow boiling range material was fractionally crystallized twice and then fractionally distilled. The amount of premelting indicated a purity of 99.92%m.

Timmermans and Martin [7494] purified the material by repeated distillation and fractional crystallization; see also Gross and Saylor [2986].

Biron [888] stated that fractional crystallization must be carried out with the exclusion of light, because even in diffused light a small quantity of a deep red substance is formed.

CRITERIA OF PURITY

Railing [6025] used the amount of premelting, and Meisenheimer and Dorner [5045] the melting point as their criteria of purity.

Injury to the lungs, liver, and kidneys is generally observed following acute exposure. The difference between the concentration that is tolerable, 25 ppm, and that which causes severe injury and death, 50 ppm, is not great. 1,2-Dibromoethane is injurious and irritating to the skin. Any clothing upon which the liquid has been spilled should be removed immediately. It will penetrate through several kinds of protective clothing, particular neoprene-type rubber and several types of plastic [5713, 7022].

NIOSH has a booklet on potential occupational hazards of 1,2-dibromoethane [5418]. The *threshold limit value* of 20 ppm, 155 mg/m^3, has been suggested [142].

353. 1,2-Dibromopropane

Mouneyrat [5288] prepared 1,2-dibromopropane in 1892 by slowly adding a solution of aluminum bromide in bromine to a mixture of 1-bromopropane and bromine at 45–50°. A similar preparation is given in Lebedev [4445].

Dewael [1977] added concentrated HBr solution to 2-chloro-1-propanol to obtain 1,2-dibromopropane. Kharasch and coworkers added HBr to 1-bromopropene [3978] or methylacetylene [3982], each in the presence of the peroxide, ascaridole, to produce essentially pure 1,2-dibromopropane.

Purification methods include washing the solvent with either sodium carbonate or sodium bicarbonate solution, drying it over calcium chloride, and fractionally distilling [3982, 7765].

354. 1,1,2,2-Tetrabromoethane

Vogel [7771] washed a commercial product three times with 12% of its volume of concentrated sulfuric acid and three times with water, dried with potassium carbonate and calcium sulfate, and distilled.

SAFETY

The vapor pressure is low at room temperature, and the material may be handled with reasonable precautions and ordinary ventilation [5713].

The *threshold limit value* is 1 ppm, 14 mg/m^3 [7022].

Iodinated Hydrocarbons (355–362)
Aliphatic Iodinated Hydrocarbons

355. Iodomethane

Carothers [1358, p. 60] gives directions for the preparation from iodine, red and yellow phosphorus, and methanol. The method is suitable for the preparation of iodoethane, 1-iodopropane, 2-iodopropane, and the iodobutyl and iodopentyl compounds, excluding the tertiaries.

Adams and Voorhees [26] give a general method for preparing alkyl iodides. They prepared 3.15 kg of iodomethane in 94% yields, 3.4 kg of iodoethane in 92% yield, and 0.48 kg of 1-iodopropane in 96% yield.

Rahman and Gilman [6023] reacted CH$_3$Cu and solid iodine in ether at −78° for 2 hr to produce 65–70% yield of iodomethane.

Jung and Ornstein [3804] prepared iodomethane by reacting methanol with trimethylsilyl iodide in a chlorinated hydrocarbon solvent at 25° under a nitrogen atmosphere.

Cowley and Partington [1697] purified the lower members of the aliphatic bromides and iodides by shaking with dilute sodium carbonate solution and washing repeatedly with water. After a preliminary drying with calcium chloride, the material was allowed to stand for a day over phosphorus pentoxide and fractionally distilled twice. Iodomethane, iodoethane, iodopropane, and bromopropane so purified were used for *dipole moment measurements*.

Iodomethane is slightly unstable. A drop of mercury, copper turnings, or silver powder generally will keep the compound clean. Discolored material may be decolorized by shaking with mercury, sodium sulfite, or other reducing agents. It may be kept for an extended period of time if it is fractionally distilled and collected in 1 atm of oxygen-free nitrogen in a dark bottle, preferably a Corning Low Actinic or similar type. The bottle must be stoppered and sealed without admitting oxygen and stored in a dark cool place.

A method for purifying alkyl iodides is described by Gand [2647, 2648, 2649, 2650]; see iodoethane.

CRITERIA OF PURITY

Rosin [6263] gives specifications and test methods for reagent grade.

SAFETY

Iodomethane should be considered as a toxic substance, more toxic than carbon tetrachloride and chronically dangerous [7022, 5713].

The *threshold limit value* is 5 ppm, 28 mg/m^3 [7022].

356. Iodoethane

Taylor and Grant [7196] discovered that anhydrous hydrogen iodide reacts immediately with ethyl ether at room temperature to produce ethanol and iodoethane as a two-phase system that can be readily separated.

Jung and Ornstein [3804] prepared iodoethane in 99% yield by reacting ethanol with trimethylsilyl iodide.

For purification for *heat of vaporization study*, see Wadso [7822], Halogenated Hydrocarbons, General Comments.

Smyth and Stoops [6876] purified iodoethane for *dielectric constant determinations* by repeatedly washing CP material with water, drying over calcium chloride, and fractionally distilling. The boiling point was used as the criterion of purity. See also Cowley and Partington [1697] under iodomethane.

For *vapor pressure measurements*, Smyth and Engel [6864] purified a commercial product by washing with very dilute potassium hydroxide solution until decolorized. Several drops of the caustic solution were added in excess and the iodoethane separated, thoroughly washed with water, dried over fused calcium chloride, and fractionally distilled. The fraction distilling without any appreciable change in temperature at 73.2° was collected.

Gand [2647] developed a method for removing free iodine from alkyl iodides. The iodides are distilled through a LeBel-Henninger-type column containing copper turnings.

Purification by shaking with alkaline solutions and storing over silver powder are adversely criticized. Storing over silver is unsatisfactory for preserving iodoethane because surface films of silver iodide are formed. The best preservative is metallic calcium [2650].

Further experiments [2648] indicated that distillation through copper turnings removes all traces of hydrogen iodide. Investigations of drying agents showed that calcium chloride does not remove all of the water. Calcium carbide removes all of the water but makes the iodoethane impure, and phosphorus pentoxide is unsatisfactory because of several physical difficulties. The iodoethane also becomes contaminated with ethyl phosphate. Sodium wire was found to be effective but some sodium iodide was formed. Iodoethane, distilled after being dried with sodium, was reported to be very pure. The advantage of calcium for drying was stressed.

Gand concluded [2649] that, after iodoethane had been purified by distilling through copper turnings, photolysis can be prevented by suspending metallic calcium in the liquid. No photolysis was detected even after a period of several months.

SAFETY

Iodoethane causes central nervous sytem depression. It may affect the kidneys, thyroid, lungs, and liver. A concentration of 150 ppm, 960 mg/m^3, for 24 hr was fatal to mice [5713].

357. 1-Iodopropane
358. 2-Iodopropane

Jung and Ornstein [3804] prepared 2-iodopropane by reacting 2-propanol with trimethylsilyl iodide.

Gross and Saylor [2986] decolorized 1-iodopropane by shaking with sodium thiosulfate solution. After being washed twice with distilled water, it was allowed to stand for 10 days over aluminum oxide and then fractionally distilled. A fraction whose distillation range was 102.28–102.58° was collected for *solubility measurements*. See also Adams and Voorhees [26] and Cowley and Partington [1697] under iodoethane.

For *kinetic studies*, Hughes and Shapiro [3564] treated 2-iodopropane with mercury to remove iodine, dried over phosphorus pentoxide, and fractionated twice; bp 89.5° at 760 Torr. No color was observed on storing in a brown bottle in the dark.

See Gand's studies on purification of iodoalkanes under iodoethane [2647, 2648, 2649, 2650].

For purification of 1-iodopropane for *heat of vaporization study*, see Wadso [7822], Halogenated Hydrocarbons, General Comments.

359. 1-Iodobutane

Linnemann [4617] prepared 1-iodobutane in 1872 by treating 1-butanol with HI. Dangyan [1807] treated 1-butanol and iodine with either magnesium or aluminum and distilled 1-iodobutane from the mixture.

Rahman and Gilman [6023] reacted dibutylcopper and iodine in tetrahydrofuran to obtain 64% 1-iodobutane. Stone and Shechter [7023] reported a synthesis of 1-iodobutane with butyl ether and potassium iodide in the presence of phosphoric acid.

Mathews [4969] purified 1-iodobutane for *heat of vaporization studies* by washing with sodium hydroxide solution, drying over calcium chloride, and fractionally distilling three times.

Smyth and Rogers [6874] purified 1-iodobutane for *dielectric constant measurements*. A commercial sample was shaken with ice-cold concentrated sulfuric acid solution and then with dilute sodium hydroxide solution. After washing with water and drying over calcium chloride, it was distilled at reduced pressure.

360.   1-Iodo-2-methylpropane

Wurtz [8181] prepared 1-iodo-2-methylpropane in 1855 by reacting isobutyl alcohol with phosphorus and iodine. A similar preparation is given in Hirao [3419].

Stone and Shechter [7023] reported a synthesis of 1-iodo-2-methylpropane by reacting isobutyl alcohol with potassium iodide and phosphoric acid.

For purification for *heat of vaporization study*, see Wadso [7822], Halogenated Hydrocarbons, General Comments.

361.   Diiodomethane

Adams and Marvel [2757] describe a method for converting iodoform to diiodomethane with alkaline arsenic(III) oxide.

Lifshits and coworkers [4597] heated acetone and anhydrous sodium iodide in an autoclave at 30–40°, followed by the addition of dichloromethane. The mixture was heated 4 hr at 5 atm and 4 hr at 10 atm. After the acetone was distilled off, the residue was poured into water and diiodomethane was separated. The product was rinsed with water, dried with calcium chloride, and distilled *in vacuo;* yield was 47%.

Aromatic Iodinated Hydrocarbons
362.   Iodobenzene

Datta and Chatterjee [1829] prepared iodobenzene by heating a mixture of benzene and iodine. Nitric acid was added until most of the iodine was consumed. The crude product was washed with dilute alkali, dried over calcium chloride, and distilled; bp 186–188°. The iodobenzene was then washed with dilute caustic soda and water to remove nitrophenol. A similar preparation is given in Varma [7681].

Barker and Waters [619] stirred benzene into an emulsion of silver sulfate in concentrated sulfuric acid solution and then treated the mixture with iodine. The residue was stirred with sodium thiosulfate solution and iodobenzene was extracted with ether.

Gozzelino and coworkers [2896] iodinated benzene in a single stage with iodine and mercury(II) acetate. When perchloric acid was used as a catalyst, the yield of iodobenzene was 80%.

Audsley and Goss [489] purified iodobenzene for *dielectric constant studies*. A sample was washed successively with sodium thiosulfate solution, dilute sulfuric acid solution, sodium hydroxide solution, and water. It was dried over calcium chloride and fractionally distilled twice at 16 Torr.

Mixed Halogenated Compounds (363–379)
General Comments

The mixed fluorohalogenated aliphatic compounds in this section, in general, are stable to a degree not ordinarily found in organic compounds. Each compound has, of course, its own degree of stability. The presence of fluorine atoms in the molecule is responsible for the stability; as a general rule, the more fluorine the greater the stability [2142].

Specific Solvents

363.   Chlorotrifluoromethane

Simons and coworkers [6741] prepared chlorotrifluoromethane by reacting dichlorodifluoromethane and fluorine over a mercury catalyst. Whalley [7991] produced a 95% yield by heating dichlorodifluoromethane with antimony pentachloride and HF at 160° and 1000 psi.

Murray [5349] passed anhydrous dichlorodifluoromethane or trichlorofluoromethane through aluminum bromide or aluminum chloride at 100–175°. The disproportionation reaction followed by fractional distillation gave a yield of 96% chlorotrifluoromethane.

Haszeldine [3212] reacted silver trifluoroacetate and chlorine in sealed tubes to produce a yield of 90% chlorotrifluoromethane.

364.   Dichlorodifluoromethane

Dichlorodifluoromethane has several properties that are advantageous for industrial and commercial usage. It is nonflammable, nonexplosive, and nontoxic, and does not support combustion. When used as a refrigerant, it is not absorbed by food [7443].

Daudt and Youker [1833] prepared dichlorodifluoromethane by passing carbon tetrachloride and hydrogen fluoride into a mixture of antimony trichloride and antimony pentachloride. Henne [3285] used a similar preparation technique by reacting carbon tetrachloride with antimony trifluoride.

Gleave [2786] prepared dichlorodifluoromethane by passing a gaseous mixture of HF, carbon disulfide, and chlorine over copper (I) chloride at 200–400°.

Gilkey and coworkers [2749] bubbled dichlorodifluoromethane through sulfuric acid solution and dried it over phosphorus pentoxide for *vapor pressure measurements*.

Chalupa and coworkers [1409] removed carbon disulfide and acids containing sulfur by treating dichlorodifluoromethane with diethylenetriamine or triethylenetetramine. The complex dithiocarbamates that are produced are insoluble in the solvent.

SAFETY

Shugaev [6698, 6699] conducted toxicity tests with mice and reported that a 3-hr exposure of dichlorodifluoromethane produced an LD_{50} of 3.348 g/L. Aviado and Smith [514] reported that dichlorodifluoromethane increased the pulmonary resistance in rhesus monkeys at 10–20%v.

The *threshold limit value* has been established as 1000 ppm, 4950 mg/m³ [142].

365.   Trichlorofluoromethane

Daudt and Mattison [1832] prepared trichlorofluoromethane by treating carbon tetrachloride with a heavy-metal fluoride in the presence of antimony trichloride and antimony pentachloride. Brown and Whalley [1157] heated carbon tetrachloride and anhydrous HF at 230° and 1000 psi to obtain a 63% yield of trichlorofluoromethane.

Simons and coworkers [6741] reacted carbon tetrachloride and arsenic with a small amount of bromine to produce pure trichlorofluoromethane.

Gleave [2786] prepared trichlorofluoromethane by passing a gaseous mixture of HF, carbon disulfide, and chlorine over chromium(III) fluoride at 450-500°.

Osborne and coworkers [5612] purified trichlorofluoromethane for *heat capacity measurements*. A commercial sample was fractionated twice at atmospheric pressure in a 70-cm column packed with glass helices. The middle portion was distilled through phosphorus pentoxide and then frozen and melted under high vacuum.

For removal of carbon disulfide, see Chalupa [1409], dichlorodifluoromethane.

SAFETY

Aviado and Smith [514] reported that trichlorofluoromethane caused tachycardia and hypotension, and it anesthesized rhesus monkeys when inhaled at 2.5-5%v.

The *threshold limit value* has been established as 1000 ppm, 4200 mg/m^3 [142].

366. Chlorodifluoromethane

Kwasnik [4356] prepared chlorodifluoromethane by heating chloroform with antimony pentachloride and HF at 60 atm and 80°. A similar preparation is given in Booth and Bixby [985].

Konao and coworkers [4170] removed acid impurities from chlorodifluoromethane with 17% sodium carbonate solution at 30-40°.

SAFETY

Aviado and Smith [514] reported that chlorodifluoromethane increased the pulmonary resistance in rhesus monkeys at 10-20%v.

The *threshold limit value* has been established as 1000 ppm, 3500 mg/m^3 [142].

367. Dichlorofluoromethane

Booth and Bixby [985] prepared dichlorofluoromethane by fluorination of chloroform with a mixture of antimony trifluoride and antimony pentafluoride. Daudt and Youker [1835] reported a similar preparation by treatment of chloroform with HF.

Benning and coworkers [773] removed sulfur dioxide and other acidic compounds from dichlorofluoromethane by scrubbing with 15% sodium sulfite solution.

Gordus and Bernstein [2873] degassed dichlorofluoromethane in vacuum and bubbled it through concentrated sulfuric acid solution prior to *spectral studies*.

SAFETY

Aviado and Smith [514] reported that dichlorofluoromethane caused tachycardia and hypotension, and it anesthesized rhesus monkeys when inhaled at 2.5-5%v.

The *threshold limit value* has been established as 1000 ppm, 4200 mg/m^3 [142].

368. Dibromofluoromethane

Swarts [7118] prepared dibromofluoromethane by heating bromoform, bromine, and antimony trifluoride at 110-120°. Henne [3284] reported a similar preparation under 4 atm pressure.

Haszeldine [3217] heated a silver salt of bromofluoroacetic acid with bromine at 100° to give a 64% yield of dibromofluoromethane.

The *threshold limit value* has been set at 100 ppm, 860 mg/m^3 [6402].

369. Bromochloromethane

Henry [3304] prepared bromochloromethane in 1885 by reacting chloroiodomethane and bromine; bp 68–69°.

Bromochloromethane was prepared by the bromination of methylene chloride in the presence of aluminum [6434] or aluminum bromide [4377].

De Vries and Vanderkooi [1976] purified bromochloromethane by distillation in a multiple-plate fractionating column.

SAFETY

The LD$_{50}$ in mice was reported as 2994 ppm [7102]. Torkelson and coworkers [7527] recommended that repeated human exposure to bromochloromethane should be below 400 ppm and gave a *threshold limit value* of 200 ppm.

370. Bromochlorofluoromethane

Berry and Sturtevant [815] prepared bromochlorofluoromethane by treating dibromochloromethane with antimony trifluoride and bromine. After the product was washed with dilute alkali solution and water and dried over calcium chloride, it was fractionated from phosphorus pentoxide through a 48-cm vacuum-jacketed column. The fraction boiling at 36.3–36.8° was collected and stored in the dark.

Hine and coworkers [3411] mixed dibromochloromethane with mercury(II) fluoride at 40° and fractionally distilled bromochlorofluoromethane at 36.0–36.7°.

Haszeldine [3217] heated the silver salt of bromofluoroacetic acid with chlorine at 40° to give a 67% yield of bromochlorofluoromethane.

371. Chloropentafluoroethane

Daudt and Youker [1834] prepared chloropentafluoroethane by reacting 1,2,2-trifluoro-1,1,2-trichloroethane and HF over an activated carbon catalyst at 500–600°. Clark and Rectenwald [1500] used a similar preparation by reacting 1,1,1-trichloro-2,2,2-trifluoroethane and HF over anhydrous chromium(III) oxide at 350°.

Benning [769] bubbled 1-chloro-1,2,2-trifluoroethene through antimony pentafluoride at 95° and 500–700 Torr for 4 hr to produce pure chloropentafluoroethane.

Chloropentafluoroethane was purified by Mears and Rosenthal [5018] for *vapor pressure and critical constant study*. It was fractionally distilled through a vacuum-jacketed Helipak column. The heads were removed and the distillate was periodically analyzed by gas chromatography until the impurities began to increase. The sample was purged for noncondensables by freezing in liquid nitrogen, pumping off the residual gas, and allowing the system to come to ambient temperature. The process was repeated until no pressure was observed on the manometer when the system was cooled in liquid nitrogen; impurities by gas chromatography <0.05%, purity >99.9%m.

372.   1,2-Dichlorotetrafluoroethane

Roberts [6198] prepared 1,2-dichlorotetrafluoroethane by irradiating carbon tetra-chloride, chlorine, and tetrafluoroethylene with a mercury-vapor lamp; yield 95%.

SAFETY

Aviado and Smith [514] reported that 1,2-dichlorotetrafluoroethane influenced either respiration or circulation or both in rhesus monkeys at 10–20%v.

The *threshold limit value* has been set at 1000 ppm, 7000 mg/m^3 [142].

373.   1,1,2-Trichloro-1,2,2-trifluoroethane

Locke and coworkers [4643] purified laboratory-prepared material to study its *physical properties* by washing with water and then with a weak caustic solution. The compound was dried with calcium chloride or sulfuric acid, and finally fractionally distilled.

1,1,2-Trichloro-1,2,2-trifluoroethane shows the first trace of decomposition in quartz at about 300° [2142].

Hiraoka and Hildebrand [3420] purified a commercial product for *solubility studies* by degassing by repeated freezing and melting *in vacuo*.

SAFETY

Michelson and Huntsman [5100] found the oral LD$_{50}$ for rats to be 43.0 ± 4.8 g/kg or about 2.5 times less toxic than ethanol.

Aviado and Smith [514] reported that the compound caused tachycardia and hypoten-sion and anesthesized rhesus monkeys when inhaled at 2.5–5%v.

The *threshold limit value* is 1000 ppm, 7600 mg/m^3 [142].

The compound is not flammable or explosive in air [2142].

374.   1,1,2,2-Tetrachlorodifluoroethane

Kishimoto and coworkers [4039] purified 1,1,2,2-tetrachlorodifluoroethane by tak-ing the middle fraction of an atmospheric distillation and fractionally melting it, followed by molecular sieve Linde 3A purification. In the final step, the material was fractionally distilled *in vacuo*. Purity was 99.93%m.

SAFETY

Toxicity studies by Clayton and coworkers [1519] indicate a low order of acute, dermal, and inhalation toxicity. The approximate lethal dose for rats by inhalation for single 4-hr exposure is approximately 15,000 ppm. Ten repeated daily exposures of 4-hr each to 3000 ppm and 31 4-hr exposures of rats, guinea pigs, and mice to 1000 ppm were not lethal.

A *threshold limit value* of 500 ppm has been recommended [1519].

1,1,2,2-Tetrachlorodifluoroethane is nonflammable and nonexplosive [2141].

375.   1-Chloro-1,1-difluoroethane

Henne and Renoll [3291] prepared 1-chloro-1,1-difluoroethane by fluorination of ,1,1-trichloroethane with a mixture of antimony trifluoride and antimony dichloridetrifluoride. A similar fluorination method using antimony pentachloride and HF is given in Tatsuya [7188].

Calfee and Smith [1317] prepared 1-chloro-1,1-difluoroethane by photochlorination of 1,1-difluoroethane.

SAFETY

Aviado and Smith [514] reported that 1-chloro-1,1-difluoroethane influenced either respiration or circulation or both in rhesus monkeys at 10–20%v.

The *flammable limits* in air are 9.8 and 16.1%v [600].

376.   1,2-Dibromotetrafluoroethane

Madai [4816] removed water from 1,2-dibromotetrafluoroethane by distillation with methanol which forms a ternary azeotrope.

For purification for *physical properties*, see Locke and coworkers [4643], 1,1,2-trichlorotrifluoroethane.

SAFETY

The LD_{50} for rats exposed for 15 min is approximately 126,000 ppm [2143].

1,2-Dibromotetrafluoroethane is nonflammable [2143].

377.   1,2-Dibromo-1,1-difluoroethane

Cohn and Bergmann [1574] prepared 1,2-dibromo-1,1-difluoroethane by passing 1,1-difluoroethene through bromine. The mixture was illuminated and cooled with ice water; purity 99.8%.

378.   1-Bromo-2-chloroethane

James [3672] prepared 1-bromo-2-chloroethane in 1882 by passing bromine into a mixture of equal parts concentrated hydrochloric acid solution and water. Then the resultant mixture was saturated first with chlorine and later with ethylene at 0°.

Kharasch and Hannum [3980] added HBr to vinyl chloride to produce 1-bromo-2-chloroethane rapidly. Lorenz and Auer [4662] heated chloroethyl phosphite and (bromomethyl)benzene at 120° and distilled 1-bromo-2-chloroethane from the mixture.

Smyth and coworkers [6863] purified commercial 1-bromo-2-chloroethane for *dipole moment studies*. The sample was washed with dilute sodium hydroxide solution and water, dried over calcium chloride, and fractionally distilled.

379.   1,1-Dichloro-2,2-difluoroethylene

Padbury and Tarrant [5640] prepared 1,1-dichloro-2,2-difluoroethylene by dehydrochlorination of 1,2,2-trichloro-1,1-difluoroethane at 582–590°. Henne and Wiest [3294] prepared the compound by treating 1,1,1,2-tetrachloro-2,2-difluoroethane with zinc in alcohol.

SAFETY

Sakharova and Tolgskaya [6351] reported that a concentration of 13.4 mg/m^3 and long-term exposure affected the nervous system, liver, and kidney of animals. A maximum permissible concentration of 1 mg/m^3 was suggested.

NITROGEN COMPOUNDS (380–450)

Nitro Compounds (380–384)

> 380. Nitromethane
> 381. Nitroethane
> 382. 1-Nitropropane
> 383. 2-Nitropropane

The four nitroalkanes (nitroparaffins) nitromethane, nitroethane, 1-nitropropane, and 2-nitropropane are available commercially. They are manufactured by the vapor-phase nitration of propane, purification of the reaction product, and separation of the nitrocompounds. The principal impurities in the nitroalkanes are isomers or homologues. Water is present to 0.1% or less, usually less. The nitroalkanes may be analyzed by mass spectrometry or gas chromatography. The latter is more rapid and offers a wide range of sensitivities. Allen [91] found that a thermistor or hot wire detector is sensitive to 0.005–0.05% impurity. All known possible impurities are detectable in the 0.005–0.01% range, except the carboxylic acids. Allen has summarized the impurity analyses of the nitroalkanes on a water-free basis as follows.

> **380.** Nitromethane: Nitroethane, 4–5%; 2-nitropropane and acrylonitrile, less than 0.1%; acetonitrile and propionitrile, traces.
> **381.** Nitroethane: 2-Nitropropane, 3–4%; nitromethane, about 1%; 1-nitropropane and 1- and 2-propanol, traces.
> **382.** 1-Nitropropane: 2-Nitropropane, 3–4%; nitroethane, trace to 0.5%; 2-methyl-2-nitropropane, 1- and 2-nitrobutane, 1- and 2-propanol, traces.
> **383.** 2-Nitropropane: 1-Nitropropane, 4–5%; nitroethane, 0.5–1.0%; acetone, trace to 0.1%; 2-methyl-2-nitropropane. trace to 0.5%.

"Trace" is used to express an amount near the limit of detectability.

These solvents are quite reactive. The primary and secondary nitroalkanes exist in equilibrium with their tautomeric aci-form, a pseudo acid. They are sparingly soluble in water; the aci-form is more soluble and reacts with alkalies to form a water-soluble salt that may be regenerated with a weak acid, particularly carbonic acid. Riddick [6165] has quantitatively titrated the nitroalkanes, except nitromethane, with alkali in nonaqueous systems. The best titration that was obtained for nitromethane was 99%.

MUTUAL SOLUBILITY WITH WATER

The mutual solubility of nitroethane, 1-nitropropane, and 2-nitropropane with water was determined by Hampton and Riddick [3125] from 0–80° and that of nitromethane and water from 0° to the critical solution temperature.

The nitroalkanes used for this study were prepared by Toops [7524] and were at least 99.9% the principal component. The solubilities were determined by a precision cloud point technique. Selected points were checked by: (1) water in the nitroalkane by the Karl Fischer method, and (2) nitroalkanes in water in concentrations less than 10% by the appropriate colorimetric method of Jones and Riddick [3762]. The data were plotted and a smooth curve was drawn through the points. The solubilities read from the curve are given in Table 5.16.

Table 5.16 Mutual Solubilities of Nitroalkanes and Water (% w)

t(°C)	Nitromethane		Nitroethane		1-Nitropropane		2-Nitropropane	
	NM in aq	aq in NM	NE in aq	aq in NE	1-NP in aq	aq in 1-NP	2-NP in aq	aq in 2-NP
0	8.9	1.0	4.6	0.5	1.4	0.2	1.7	0.2
10	9.6	1.3	4.6	0.6	1.4	0.4	1.7	0.3
20	10.5	1.8	4.6	0.9	1.5	0.6	1.7	0.5
25	11.1	2.1	4.7	1.0	1.5	0.6	1.7	0.5
30	11.6	2.5	4.8	1.2	1.5	0.7	1.7	0.6
40	13.1	3.6	5.0	1.7	1.6	0.9	1.8	0.8
50	14.8	4.7	5.4	2.0	1.8	1.1	1.9	1.1
60	16.9	6.0	5.9	2.4	1.9	1.4	2.1	1.3
70	19.3	7.6	6.6	3.0	2.2	1.7	2.3	1.6
80	22.1	9.4	7.3	3.5	2.4	2.0	2.6	1.9
90	29.	14.	—	—	—	—	—	—
100	40.	25.	—	—	—	—	—	—
105	58.	42.	—	—	—	—	—	—

SYNTHESES

Nitromethane is best prepared from chloroacetic acid and sodium nitrite, see Clarke [1508, p. 83]. It may be prepared by the Walden synthesis [7836] from sodium nitrite and dimethyl sulfate. Toops [7525] prepared it from chloroacetic acid and sodium nitrite, distilled in a 25-plate, Penn State-type column at 100 Torr, and obtained a 25% yield with a purity of 99.98%m from the freezing curve.

The other nitroalkanes may be prepared by the Victor Meyer [5091] synthesis. An alkyl halide is reacted with silver nitrite; the bromo compound has been found to be the most satisfactory. Average yields for replicate preparations are: nitroethane, 51%; 1-nitropropane, 57%; and 2-nitropropane, 13%. McCombie and coworkers [4745] prepared nitroethane in 46% yield by the Walden synthesis.

Kornblum and coworkers [4191] found that primary and secondary alkyl nitro compounds could be prepared in a better yield, 55–62%, in dimethylformamide than in water as the reaction medium. A bromo- or iodoalkyl or cycloalkyl was reacted with sodium nitrite in dimethylformamide. Chloro compounds react very slowly. Hardie, an associate of Kornblum, found that the reaction took place in dimethyl sulfoxide and that the latter is a better solvent for the sodium salts.

PURIFICATION

The purification of commercial nitroalkanes has been studied by Riddick [6165]. Nitromethane, 1-nitropropane, and 2-nitropropane may be prepared in 99.9%m purity by fractional distillation in a 35–50-theoretical-plate column at a pressure sufficient to give a pot temperature not greater than 90°; 100 Torr is a convenient pressure if the pressure drop in the column is not too great. There is some decomposition of nitroalkanes above 90°. The purity depends on the reflux ratio used during the distillation.

Wright and coworkers [8168] made conductivity measurements on commercial nitromethane that was dried over calcium chloride and the remaining water removed by azeotropic distillation. Volatile impurities were removed by boiling under reflux in a stream of dry air for several hours.

Unni and coworkers [7638] purified 5 L of commercial nitromethane by distillation through a 3.6×300-cm column packed with glass helices. The distillation was made at 4–6 Torr and a 10:1 reflux ratio. The first and last 1.5 L were discarded. The middle 2 L were fractionally crystallized. After repeated distillations and crystallizations, the specific conductance was $5. \times 10^{-9}$ ohm^{-1} cm^{-1}.

Coetzee and Cunningham [1552] studied the *conductivity of single ions* in nitromethane purified by the following modification of Unni and coworkers [7638]. A spectrographic-grade material was fractionally distilled four times at reduced pressure in the presence of Drierite. It was fractionally crystallized six times. It was their opinion that fractional crystallization was more effective than fractional distillation as a means of purification.

Mathews [4969] purified nitromethane by drying over phosphorus pentoxide and fractionally distilling. Smyth and McAlpine [6866] purified nitromethane for the determination of the *dipole moment* from a commercial material that was dried over calcium chloride and fractionally distilled. The fraction boiling at 100.8–101.1° was further dried over phosphorus pentoxide and fractionally distilled. The fraction distilling at 100.5° at 748 Torr, n_D 1.38195 was used. Smyth and Walls [6881] determined that the *dipole moment* on commercial material that was distilled from phosphorus pentoxide twice and fractionally distilled; mp $-28.6°$, bp 101.5° at 762 Torr, n_D 25° 1.37970, d 25° 1.1314. For the effects of phosphorus pentoxide, see Wright and coworkers [8168].

Thompson and coworkers [7423] purified nitromethane essentially according to Riddick and determined the *physical properties;* 99.9+%m from the freezing curve. McCullough and coworkers [4754] dried the Thompson material by passing the vapors through magnesium perchlorate. The purity was redetermined by the freezing curve and found to be unchanged. Impurities of different volatility were not present in significant amount because the difference in the boiling and condensing temperatures was only 0.006° at 760 Torr. The material was used to study *thermodynamic properties.*

Aplin and coworkers [397] purified the nitroalkanes for *mass spectral studies* by preparative gas chromatography in a Wilkins Aerograph instrument with a polybutylene column operating at 30 psi helium pressure.

Nitromethane for *voltammetric reaction studies* was purified [4415] by passing 1 L of a "certified reagent" grade, boiling range 100–102°, through a 2.5-cm id × 35-cm column of Woelm chromatographic neutral alumina, activity grade 1, protected from air by a calcium chloride tube. Three passes over fresh alumina reduced the water to 3 × 10^{-3} M. Nitromethane treated in the same manner passed through a fine-porosity, fritted glass was used for *dielectric constant measurements* [4151].

Nitromethane is a differentiating solvent for the titration of weakly basic substances. It was purified by Clarke and Sandler [1506] using a strongly acidic cation exchange resin in the hydrogen form, washing several times, first with anhydrous methanol and then with two 300-mL portions of nitromethane, and discarding the eluate. A total of 15–25 L of nitromethane can be treated with 75 grams of the resin.

Nitroethane was purified by Pearson [5734] for *kinetics of neutralization studies* by drying with magnesium sulfate, refluxing with urea, and drying with phosphorus pentoxide. It was finally distilled through a 10-plate column.

The purification of nitromethane by progressive freezing was studied by Dickinson and Eaborn [2000]. The 4% impurities, other than water, were removed by eight passes.

Wright and coworkers [8168] found phosphorus pentoxide unsuited for drying nitromethane. When it is dried with the pentoxide and distilled, a white residue collects in the condenser. (This is characteristic of nitroalkanes containing an appreciable amount of formaldehyde or when a reaction takes place forming formaldehyde.)

Turnbull and Maron [7584] purified nitromethane, nitroethane, and 2-nitropropane before determining the *ionization constants* of the aci- and nitro-forms. The individual nitroparaffins were carefully purified by fractional distillation. Water and possible oxides of nitrogen were eliminated by treating first with urea and then with anhydrous sodium sulfate, and finally separating from the drying agent by low-temperature distillation.

Hesse and coworkers [3357] purified nitromethane for *optical studies* by column chromatography with alumina.

SAFETY

The nitroalkanes act chiefly as moderate irritants when inhaled. The nitropropanes are somewhat more toxic than nitromethane and nitroethane. There is no evidence of sufficient absorption through the skin to produce systemic injuries [5713]. Prolonged exposure to concentrations of vapors greater than the threshold limit values may produce

Table 5.17   The Threshold Limit Values and Lower Explosive Limits of the Nitroalkanes

	Threshold Limit Values		Lower Explosive Limits (% v)
	ppm	*mg/m³*	
Nitromethane	100	250	7.3
Nitroethane	100	310	3.4
1-Nitropropane	25	90	2.2
2-Nitropropane	25	90	2.5

warning symptoms such as irritation of the respiratory tract, headache, and nausea, which clear promptly when the concentration is reduced by proper ventilation.

Avoid evaporation of alkaline mixtures containing nitroalkanes because of the formation of nitronate salts, some of which are explosive. Observe the special precautions recommended by the manufacturer [1614] for handling nitromethane.

The *threshold limit values* [7022] and the *lower explosive limits* in air [8247] are given in Table 5.17.

Subbotin [7062] reported the LD_{50} in mice for nitromethane, nitroethane, and 1-nitropropane as 950, 860, and 800 mg/kg, respectively, and recommended the maximum permissible concentration of each compound should be 0.005 mg/L.

384. Nitrobenzene

Nitrobenzene is made by the direct nitration of benzene. The impurities likely to be present are the nitro derivatives of the cyclic impurities that were present in the benzene. The most common are nitrotoluene and dinitrothiophene. Free acidity may be present in some improperly purified material.

For *electrical measurements*, McAlpine and Smyth [4731] purified a good commercial material crystallizing five times from itself, drying with phosphorus pentoxide, and distilling at reduced pressure. The criteria of purity were the freezing point and the refractive index; see also Roberts and Bury [6196], Murray-Rust and coworkers [5350], and Walden [7852].

The aim of a considerable amount of investigation has been the production of nitrobenzene free from nitrotoluene, dinitrobenzene, and dinitrothiophene. Cohen and Boekhorst [1567] made an exhaustive study of purification. A "pure" commercial product was repeatedly fractionally crystallized, then fractionally distilled at 2 Torr several times, dried with phosphorus pentoxide, and finally fractionally crystallized in a dry atmosphere. The melting point in an atmosphere of dry hydrogen was 5.76 ± 0.01°. The material was kept in the dark in a sealed vessel for 5 months. Its melting point changed less than 0.01°.

Masson [4957] fractionally crystallized thiophene-free benzene several times, nitrated and purified. The melting point was the same as that reported by Cohen and Boekhorst. No allotropy was observed.

Sidgwick and Ewbank [6711] and Davy and Sidgwick [1867] nitrated thiophene-free benzene with the theoretical amount of nitric acid in the presence of a large excess of sulfuric acid. The temperature was kept below 40° to avoid the formation of dinitro compounds. It was distilled at reduced pressure and fractionally crystallized.

Hantzsch [3143] observed that nitrobenzene prepared from pure benzene and purified by repeated fractional distillation and crystallization is colorless. Impure nitrobenzene rapidly becomes colored when standing over phosphorus pentoxide, but a pure product shows hardly any color, even after standing in contact with the pentoxide for several months; see also Roberts and Bury [6196] and Sidgwick and Ewbank [6711].

For nitrobenzene as a *polarographic solvent*, see [7532].

Catalyst poisons were removed from nitrobenzene with mixed sodium and potassium salts of EDTA.

Fractional distillation followed by fractional crystallization and a final distillation was used by Brown and Ives [1128] to purify nitrobenzene for *dielectric constant measurements;* bp 210.8°, d 25° 1.19817.

Nitrobenzene was purified by Barreira and Hills [638] to study the *kinetics of ion migration.* It was considered necessary to remove two kinds of impurities, water and dissolved ionic impurities. The main impurities, other than water, found in nitrobenzene are believed to be o- and p-nitrophenols. These cannot be removed by fractional distillation because of the closeness of the boiling points to the boiling point of nitrobenzene. They state, "White claims to have purified nitrobenzene to 10^{-12} ohm^{-1} cm^{-1} by a succession of distillations and fractional freezings." They purified AR-grade nitrobenzene by drying over calcium chloride for 5 days and distilling slowly at 2 Torr to avoid local heating and consequent decomposition. The middle half of the distillate was collected and kept in 1 atm of dry nitrogen. It was passed through a column of dried chromatographic alumina 3 cm × 20 cm long. After about 50 mL of the nitrobenzene had passed through the column, a yellow ring was clearly discernible at the top of the column. The final product was a straw color but it darkened on exposure to light. The solvent was stored in the dark to avoid darkening. It had a specific conductance of 2.05 × 10^{-10} ohm^{-1} cm^{-1}.

For *cryoscopic measurements*, Meisenheimer and Dorner [5045] dried a commercial nitrobenzene with calcium chloride and fractionated at reduced pressure. The nitrobenzene was distilled fresh for each series of measurements. Roberts and Bury [6196] purified a commercial product by fractional distillation and fractional crystallization five times. The middle fraction of the last crystallization was preserved with phosphorus pentoxide. Just before it was to be used, it was distilled into a carefully cleaned freezing apparatus in a vacuum, only the middle fraction being used.

For *solubility determinations*, Smith and coworkers [6833] purified nitrobenzene by vacuum distillation and drying over calcium chloride. The refractive index was used as the criterion of purity.

Williams and Ogg [8084] fractionally crystallized nitrobenzene repeatedly and then fractionally distilled; bp 210.6–210.8°; see also Davison [1864] under phenols.

For *optical measurements*, Brand and Kranz [1057] nitrated carefully purified benzene, steam-distilled the nitrobenzene, dried with calcium chloride, and distilled several times under reduced pressure. Scheibe and coworkers [6429] prepared nitrobenzene from thiophene-free benzene and distilled and recrystallized from ethanol to remove the dinitro compounds. See also Möller [5220] and Hehlgans [3259].

CRITERIA OF PURITY

Rosin [6263] lists the accompanying specifications for a reagent chemical grade. The acidity is determined by shaking 16 mL of nitrobenzene with 50 mL of water for 1 min,

Table 5.18   Specifications for Reagent Grade
Nitrobenzene

Distillation range	210–212°
Freezing point (min)	5°
Acid (as nitric) (max)	0.003%
Water (max)	0.05%

separating the water phase and adding 2 drops of bromophenol blue indicator solution. The yellow color of the solution is changed to a bluish violet by not more than 0.5 mL of 0.02 *N* sodium hydroxide; see Table 5.18.

The freezing point established by Cohen and Boekhorst [1567] has been used as the criterion of purity by many investigators.

SAFETY

Nitrobenzene is a powerful methemoglobin former [5713]. The effects of overexposure are grave, resulting in methemoglobinemia, central nervous system injury, and effects on the spleen and liver. It is readily absorbed through the skin [7022].

The *threshold limit value* is 1 ppm, 5 mg/m^3 [7022].

The *minimum ignition temperature* in air is 482° [5712].

Nitriles (385–397)

General Comments

Nitriles are relatively inert to many of the common reagents. They can be hydrolyzed under certain conditions; the ease of hydrolysis depends upon the nature of the R-group attached to the CN. Nitriles should not be used with strong alkalies or boiled with dilute acids. Strong dilute acids, and weaker acids, such as carboxylic acid, slowly hydrolyze nitriles. This can be demonstrated by periodically analyzing a 0.1 *N* solution of perchloric acid in acetonitrile.

A general method for the preparation of the lower-molecular-weight aliphatic nitriles was given by Pyryalova and Zil'berman [6000]. Adiponitrile NC(CH$_2$)$_4$CN, was heated with the acid corresponding to the nitrile desired in the presence of a catalyst; for example, 54 grams of adiponitrile and 64 grams of valeric acid was refluxed for 3 hr with 1.7 grams *p*-toluenesulfonic acid and then distilled during a 4-hr period through a 35-cm packed column. The distillate was washed with sodium carbonate solution, dried with phosphorus pentoxide and redistilled at 93° at 20 Torr; yield, 91% valeronitrile.

The refluxing of amides with sulfamic acid at 180–220° is a generally applicable method for the preparation of nitriles reported by Kirsanov and Zolotov [4037].

Friedman and Stechter [2576] found that primary and secondary chloro compounds react with sodium cyanide in dimethyl sulfoxide rapidly and efficiently. Advantages are also realized for the reaction of the corresponding bromo compounds to nitriles. Rearrangement does not occur. Typical procedures illustrating the various experimental methods are given.

For the preparation of *carbonyl-free*, nonoxygenated, extractive solvents, see Hornstein and Crowe [3494], Hydrocarbons, General Comments.

Aliphatic Nitriles

385   Acetonitrile

Acetonitrile was first produced commercially in 1952. The unusual solvent potential and high selectivity make it outstanding for liquid–liquid extraction, selective recrystallization, and extractive distillation. Its high solubility parameter and low hydrogen bonding ability make it an excellent polymer solvent. Acetonitrile is the preferred solvent for many reactions [7700].

The favorable dielectric, solvent, and optical properties of acetonitrile have permitted it to be used widely in spectrophotometric and electrochemical experiments and in peptide chemistry. The impurities generally present in the commercial-grade material are water, unsaturated nitriles, toluene, and some aldehydes and amines. If the nitrile were subjected to hydrolytic conditions, it would contain acetic acid and ammonia. O'Donnell and coworkers [5553] found that the reagent-grade material has a higher chemical purity than the commercial grade but may actually have poorer optical properties and is not suitable for electrochemical use. The spectrographic grade does not show optimum optical or electrochemical properties.

Coetzee [1551] prepared a report for the Analytical Chemistry Division, Commission of Electroanalytical Chemistry, International Union of Pure and Applied Chemistry. This report included the effect of impurities, purification, handling and storage of purified material, and the detection and determination of impurities and purification of other nitriles.

Acetonitrile, a relatively inert material, is readily available in several purity grades that are suitable for much of its solvent use. It has a favorable boiling point for general use and is relatively easy to purify to a higher state of purity. It is a weaker base and a much weaker acid than water. Coetzee [1551] points out that many possible impurities are sufficiently reactive to modify its properties significantly, even when present in extremely low concentrations.

The most widely used methods of purification prior to the late 1950s were to treat the commercial material with a saturated aqueous solution of potassium hydroxide or with anhydrous potassium or sodium carbonate and subsequent distillation from phosphorus pentoxide. Water removal by azeotropic distillation slightly improved the purity. The final purification was by fractional distillation.

Coetzee [1551] states, "There can be little doubt that the pretreatment with potassium hydroxide does more harm than good, since undesirable base hydrolysis may occur . . . . Repeated distillation of acetonitrile from phosphorus pentoxide has the disadvantage that it usually causes extensive polymerization of the solvent."

O'Donnell and coworkers [5553] in their comments on the sodium carbonate–phosphorus pentoxide treatment state, "MeCN prepared by us with this method . . . contains less than 0.05% water, and is a satisfactory voltammetric solvent. GLC analysis does reveal, however, a substantial amount ($\sim$ 0.5%) of an impurity which apparently forms an azeotrope with MeCN and which is not removed until the greater part of the charge has been distilled over."

Walden and Birr [7856] purified a commercial product for *conductivity measurements* by distilling from phosphorus pentoxide in an all-glass apparatus protected from air by a phosphorus pentoxide tube. The same distillation procedure was repeated until the phosphorus pentoxide in the still pot no longer became colored. The acetonitrile was then distilled over anhydrous potassium carbonate to remove traces of the phosphorus pentoxide and finally distilled without a drying agent.

Wawzonek and Runner [7934] shook a good grade of acetonitrile with a saturated solution of potassium hydroxide. It was decanted and dried over anhydrous sodium carbonate. It was then fractionally distilled from phosphorus pentoxide in a nitrogen atmosphere through a 3-ft, Penn State-type column. The first 150–200 mL of distillate from a 2-L batch was discarded. The nitrile was used as a *polarographic solvent* in the study of alkali and alkaline earth and heavy metal ions. Kolthoff and Coetzee [4157] used essentially the same method of purification for further *polarographic studies* of metal ions. The water content was always less than 2 mM. Coetzee and Kolthoff [1554] used the same purification method for further polarographic studies: *amperometric titration* of amines with perchloric acid.

Putnam and coworkers [5997] purified a commercial product for *thermodynamic studies* by treatment with anhydrous potassium carbonate, followed by several distillations from phosphorus pentoxide. It was then fractionally distilled in a 90-cm glass helices-packed column. The center cut was redistilled twice and stored over anhydrous calcium sulfate. Apparently the purified acetonitrile attacked slightly the inside of the copper sample container of the calorimeter.

The *ionization processes* of hydrogen halides were studied by Janz and Danyluk [3681] in acetonitrile that had been purified to a water content of less than 0.01% and free from such contaminants as acetamide, acetic acid, and methyl isonitrile. A reagent-grade material was allowed to stand over anhydrous calcium chloride for several days. It was distilled from a Hyper-Cal distillation unit, and the middle fraction was stored over phosphorus pentoxide. A bright orange coloration was noted. This had been reported previously by French and Roe [2544]. Coetzee [1551] stated that the orange-colored material is a polymer. The nitrile was decanted and redistilled. It was necessary to repeat the phosphorus pentoxide treatment-distillation process at least six times before coloration no longer developed. The nitrile was distilled into a vessel containing fused potassium carbonate and then distilled from the fused salt. It was then fractionally distilled; $\kappa = 0.6$–2.5×10^{-9} ohm^{-1} cm^{-1}.

Lewis and Smyth [4565] dried acetonitrile over calcium chloride, refluxed over phosphorus pentoxide repeatedly until no color appeared on the oxide, distilled onto freshly fused potassium carbonate, distilled, and finally fractionally distilled. The middle portion was used for *dipole moment measurements*.

Pratt [5946] purified acetonitrile by azeotropic distillation. The azeotrope can be dried by using a third component such as benzene or trichloroethylene. At 776.5 Torr, the ternary azeotrope, benzene–acetonitrile–water, boils at about 66°; trichloroethylene–acetonitrile–water boils at 67.5°.

Levi and coworkers [4553] found that Amberlite IRC-50 would remove the interfering basic substances from acetonitrile when it was used as a solvent in *nonaqueous titrimetry*.

Kolthoff and coworkers [4155] purified acetonitrile, which contained traces of water, acetamide, ammonium acetate, and ammonia, by treating with activated alumina, then distilling through a 1-meter column packed with stainless steel wool. The picric acid test indicated that the ammonia had been reduced to 10^{-6} M. The specific conductance was 0.7–1.5 $\times 10^{-7}$ ohm^{-1} cm^{-1}.

Coetzee and coworkers [1553] discussed the purification of acetonitrile as a solvent for exact measurements. A number of methods of purification are given.

Coetzee and McGuire [1555] purified acetonitrile by preliminary drying with silica gel, then with calcium hydride, followed by distillation from phosphorus pentoxide and again from calcium hydride; see [1553].

Muney and Coetzee [5333] studied the *properties of bases by electrical measurements* in acetonitrile that was purified as follows: It was shaken successively with two lots of silica gel for 30–45 min each to remove most of the water, then successively with two lots of Fisher 80–200-mesh alumina to remove acetic acid. It was finally shaken with phosphorus pentoxide to remove most of the remaining water and then fractionally distilled from 5–10 grams of phosphorus pentoxide per liter at a high reflux ratio through a 30-in. column packed with glass helices. It was noted that large amounts of phosphorus pentoxide caused polymerization of the nitrile during distillation. Usually the solvent was fractionally distilled twice with no appreciable improvement from the second distillation. The middle portion of the distillate was stored in a black-painted container from which it was siphoned as needed. The liquid was protected from atmospheric contamination. The main reason for the extreme weakness of ionization in acetonitrile as a solvent is discussed; see also [1557].

Coetzee and Padmanabhan [1557] purified acetonitrile by three methods to study the *properties of bases* therein and to determine the *protolysis constant*.

> *Method A*. The acetonitrile was dried with silica gel and then phosphorus pentoxide, and fractionally distilled twice from phosphorus pentoxide.
>
> *Method B*. The nitrile was dried with silica gel and then with calcium hydride. It was fractionally distilled from phosphorus pentoxide and then from calcium hydride.
>
> *Method C*. This method removed traces of unsaturated nitriles from the solvent. It was refluxed with 1 mL of 1% aqueous potassium hydroxide per liter, fractionally distilled, dried with calcium hydride, then fractionally distilled from phosphorus pentoxide and finally from calcium hydride.

The water content of acetonitrile prepared by Method A was generally between 1 and 2 mmol; from Methods B and C it was below 1 mmol.

O'Donnell and coworkers [5553] describe a relatively rapid procedure that produces an acetonitrile that is more generally satisfactory for *electrochemical experimentation*, as a *spectrometric solvent* and in *peptide chemistry*. Reflux 800 mL of acetonitrile containing less than 0.2% water with benzoyl chloride for 1 hr. Distill into a receiver containing 10 mL of water at a rate of 5–10 mL/min. Add 20 grams of sodium carbonate, reflux 2 hr, and distill into a receiver with a drying tube. Add 10 grams of anhydrous sodium carbonate and 15 grams of potassium permanganate and distill at 5–10 mL/min into a receiver protected from moisture. Make the distillate slightly acidic with concentrated sulfuric acid. Decant from the crystals and distill from a 30 plate column at a reflux ratio of 20:1 at a rate of 10 mL/hr, discarding the first 40 mL of the distillate.

The purified solvent has a transmittancy of at least 85% at 200 nm and a uv cutoff at 189 nm. Only one impurity was discernible in the gas chromatogram using a flame ionization detector. Material prepared using only the latter part of the purification procedure starting with the permanganate treatment is recommended for *voltammetric* and *coulometric* purposes.

Coetzee [1551] recommends two methods of purification of acetonitrile based on extensive investigations with purifications and use. The method used depends on the use of the nitrile. For the majority of uses, Procedure I is recommended. When the presence of traces of unsaturated nitriles would be objectionable, Procedure II is recommended. The procedures are given in the proper sequence of operations.*

"*Procedure I.*   (1) If a considerable amount of water is present (more than approx. 0.1 M), preliminary drying by shaking with silica gel or molecular sieves (e.g. Linde Type 3A) is advisable. This step can often be omitted.

(2) Shake or stir (a magnetic stirrer is particularly effective) with sufficient calcium hydride to remove most of the remaining water, until hydrogen evolution ceases (relieve pressure). Decant.

(3) Distill in an all-glass apparatus under a high reflux ratio from phosphorus pentoxide (not more than 5 g per liter, otherwise gel formation may be excessive). Protect the distillate from atmospheric moisture by means of a drying tube packed with anhydrous magnesium perchlorate. If extensive gel formation occurs, take care not to overheat the distillation flask. Discard the first 5 percent of the distillate, collect the next 80 to 85 percent (if possible).

(4) Reflux the distillate over calcium hydride (5 g per liter) for at least an hour, and then distill slowly under maximum reflux ratio (b.p. 81.6°C). Again collect the 5-90 percent fraction. (*Note:* Careful fractionation is required to reduce the acrylonitrile content.)"

"*Procedure II.*   (1) Reflux the acetonitrile with a small amount of potassium hydroxide (1 mL of 1 percent solution per liter) for at least an hour, longer if it proves necessary.

(2) Distill fractionally. Collect the 10-90 percent fraction.

(3) Follow Procedure I, steps 2 through 4."

"*Alternate Procedure II.*   (1) Reflux over sodium hydride (1 g per liter) for 10 min.

(2) Distill rapidly.

(3) Reflux over potassium bisulfite (2 g per liter) for 20 min. in order to remove amines produced in step 1.

(4) Distill rapidly.

(5) Follow Procedure I, step 4."

Sherman and Olson [6649] report that acetonitrile is a difficult solvent to purify for *electrochemical applications*. The most bothersome impurities are unsaturated nitriles, particularly acrylonitrile, which requires drastic treatment for removal. They found that the method of Coetzee and coworkers [1553] is long and involved and may introduce ammonia into the nitrile. The ammonia can be removed by the second step of Forcier and Olver [2477]. They have developed a method that is relatively short and simple for purifying large batches to approximately the same unsaturated concentration, 10^{-3} M, as the longer and more involved method of Coetzee [1551]. A weak acid remains at about 3×10^{-5} M, which will have to be considered for certain uses.

*Reprinted by permission of the International Union of Pure and Applied Chemistry and Butterworths Scientific Publications.

They charged 3.8 L of a purified grade of acetonitrile (such as Mallinckrodt's Nanograde or Matheson, Coleman, and Bell's Pesticide quality) into a 5-L round-bottom flask fitted with a calcium chloride tube. Six milliliters of nitrogen tetroxide was added. The solution was heated for about 2 hr at 50°. Thirty-eight grams of calcium hydride (equal parts of Metal Hydride's -40 and $-4 + 40$ mesh) was added and the mixture was purged with nitrogen at the boiling point to remove the nitrogen tetroxide. The nitrile was rapidly distilled through a short column in an atmosphere of nitrogen and a 250-mL forecut and a 200-mL pot residue were discarded. The distillate was refluxed with calcium hydride and the acetonitrile fractionally distilled from calcium hydride through a Nester/Faust (Model NF-136) 36-in. spinning band column. The 1-in. delivery tube for the distillation apparatus was packed for 20 in. with F-20 chromatographic alumina, which was freshly activated at 225° for 12 hr. The first 400 mL distilled at a reflux ratio of 15:1 was discarded. The remainder was distilled at a reflux ratio of 5:1, leaving a 200-mL pot residue.

Walter and Ramaley [7897] remarked that "the use of acetonitrile as a solvent for acid–base, electrochemical, and spectroscopic studies has markedly increased over the last decade." They observed that "it is difficult to purify, particularly with respect to acrylonitrile and benzene." They have developed two methods of purification that gave "a product of higher purity than those previously reported." It was found that the obtainable purity was dependent on the method of manufacture. They used the anodic current density, cathodic current, uv cutoff and absorbance at 200 nm, and gas chromatography as the criteria of purity.

Method A.   (1) Reflux 1 L of acetonitrile over 15 grams of anhydrous aluminum chloride for 1 hr and rapidly distill.
  (a) Reflux over lithium carbonate (10 grams/L of acetonitrile) for 1 hr and rapidly distill.
  (3) Reflux over calcium hydride (2 grams/L of acetonitrile) for 1 hr and fractionally distill through a helices-packed column at a high reflux ratio, retaining the middle 80% of the distillate. Recovery was 70%.

Method B.   (1) Same as Step (1), Method A.
  (2) Reflux over alkaline permanganate (10 grams of potassium permanganate and 10 grams of lithium carbonate per liter of acetonitrile) for 15 min and rapidly distill.
  (3) Reflux over potassium bisulfite (15 grams/L of acetonitrile) for 1 hr and rapidly distill.
  (4) Same as Step 3, Method A. Recovery was 60%.

The specific conductance was less than 10^{-7} ohm^{-1} cm^{-1} by both methods and the water content was less than 10 ppm.

Method A produced acetonitrile suitable for most polarographic studies. However, the unfavorable uv cutoff and high anodic current showed it to be unsuitable for spectroscopic and anodic electrochemical use.

Kiesele [4000] described a four-step process of purification consisting of an azeotropic distillation of acetonitrile and methanol, distillation from sodium hydride, filtration through acidic aluminum oxide, and distillation over calcium hydride. Water content was below 5×10^{-4} M.

Carlsen and coworkers [1353] described a three-step process consisting of distillation of acetonitrile with sodium *n*-octanolate in *n*-octanol, followed by distillation with sulfuric acid and adsorption over activated alumina. Purity was suitable for electrochemical and spectroscopic studies.

Moe [5198] added 300 mL of 96% ethanol to 3.5 L of commercial acetonitrile and carried out a fractional distillation. Polarography residual currents were in the 0.01–0.02 μamp range.

Hofmanova and Angelis [3457] used a three-step procedure to purify acetonitrile for *electrochemical measurements*. First the solvent was boiled for 5 hr with 10% aqueous KOH solution. Then the potassium salts of any carboxylic acids were removed by adsorption with alumina and then the acetonitrile was dried over phosphorus pentoxide or molecular sieve. The last step was a fractional distillation over phosphorus pentoxide. The polarographic determination showed a concentration of impurities as less than 2 × 10^{-5} M and the uv spectrum showed no bands in the range of 200–300 nm.

Addison and coworkers [30] purified acetonitrile as a solvent for the *determination of infrared and Raman spectra of metal salts* by shaking with 5 grams/L of phosphorus pentoxide, followed by distillation from calcium hydride. The distilled solvent was free from fluorescent impurities.

The impurities in acetonitrile were reduced from·0.43 to 0.03% by 10 zone melting passes in a closed tube [374].

Acetonitrile is a strongly differentiating solvent for acids and bases [5333, 4157, 1544]. It is an excellent solvent for many inorganic substances [3681].

CRITERIA OF PURITY

Lewis and Smyth [4565] used the physical properties such as the boiling point, density, and refractive index as the criteria of purity; Timmermans and Hennaut-Roland distilled until the density was constant [7488].

Dreisbach and Martin [2090] determined the purity from the freezing curve.

Rosin [6263] and *ACS Reagent Chemicals* give specifications and test methods for reagent-grade acetonitrile. The specifications are listed in Table 5.19.

O'Donnell and coworkers used the transmittancy at 200 nm and the uv cutoff and the gas chromatographic examination with a flame ionization detector for spectrographic solvent use [5553]. An electrolytic test was used for voltammetric and coulometric purposes. The specific conductance is a criterion for all acetonitrile used for conductometric investigations.

Coetzee [1551] gives the methods for the detection and determination of important impurities in acetonitrile. These are water, acetic acid, ammonia, and unsaturated nitriles.

Nemes and Orville-Thomas [5453] determined the purity of acetonitrile by examining the infrared spectrum for impurities.

SAFETY

Animal studies indicate that acetonitrile is only moderately toxic, necessitating the use of ordinary precautionary measures. The odor is easily detectable at 40 ppm [82]. Human volunteers were studied at levels of 40, 80, and 160 ppm of acetonitrile vapors for periods of 4 hr. No specific subjective responses were noted. Several accidents have

Table 5.19 Specifications for Reagent Grade Acetonitrile

Specification	ACS[a]	Rosin
Distillation range	1–95 mL 1°	81–82°
95 mL to dryness	95 mL to dryness	1°
Boiling point	81.6°	—
Density (25°)	0.775–0.780	—
Residue on evaporation (max)	0.005%	0.01%
Neutrality to red and blue litmus	—	Neutral
Acidity as CH_3COOH (max)	0.05%	—
Alkalinity as NH_3 (max)	0.001%	—
Water (max)	0.3%	0.2%
Permanganate time (max)	—	30 min
Hydrogen cyanide (max)	—	0.005%
Miscibility with water	—	To pass test[b]
Color (APHA)	10	—
Appearance and odor	To pass test	—

[a] From *Fisher Chemical Index*, 67-C.
[b] 5 mL acetonitrile and 15 mL of water; remains clear for 30 min.

been reported from extended exposure to undetermined concentrations. Marked delay of the onset of symptoms appears to be characteristic [5713].

Willhite [8071] reported a lethal concentration to male mice as 2693 ppm.

The *threshold limit value* is 40 ppm, 70 mg/m^3 [7022].

The *flammable limits* in air are 4.35 and 14.80%v [1946].

386.   Propionitrile

Teter and Merwin [7218] found that acetronitrile and propionitrile, when prepared by high-temperature ammoniation of olefins, contained hydrocarbon impurities that were difficult to remove because of the formation of azeotropes with the nitriles. The hydrocarbons may be removed from propionitrile by adding acetonitrile, which forms lower-boiling azeotropes with the hydrocarbons than does propionitrile. After the removal of the excess acetonitrile, the higher nitrile may be obtained in substantially pure condition.

Jeffrey and Vogel [3700] prepared propionitrile for *physical property measurements* by an elaboration of the Walden method [7836]. The upper phase was distilled from the reaction mixture, treated with saturated calcium chloride solution, and cooled in ice to remove cyanide impurities. The isocyanides were destroyed by shaking twice with 5-mL portions of concentrated hydrochloric acid for 5 min. It was shaken with saturated potassium carbonate solution and then with calcium chloride solution, dried with anhydrous magnesium sulfate, and fractionally distilled three times; bp 97.5° at 765 Torr.

Gresham [2956] prepared propionitrile in a 91% yield by dehydrating propionamide at 320–450° over a catalyst of iron, nickel, or manganese chromite.

Propionitrile was purified by Duncan and Janz [2118] for *molecular structure studies* by fractional distillation of a commercial product after drying over phosphorus pentoxide; $n_D25°$ 1.3633, d 25° 0.755.

Dreisbach and Martin [2090] determined the purity from the freezing curve.

SAFETY

Propionitrile is quite toxic. It is believed that the toxicity is produced through the metabolism of the cyanide ion. The LD_{50} oral dose for rats is reported to be about 39 mg/kg; 500 ppm in air was fatal to one-third of the rats in 4 hr [5726, see also 5713].

Willhite [8071] reported a lethal concentration to male mice as 163 ppm.

387. Succinonitrile

Succinonitrile was purified by Wulff and Westrum [8179] for *thermodynamic property studies* by two vacuum sublimations. The purity was determined by fractional fusion.

Lewis and Smyth [4565] determined the *dipole moment* of succinonitrile that was twice recrystallized from acetone and distilled four times at about 1 Torr; bp about 108°.

Succinonitrile may be heated for 72 hr at 200° without decomposition [7033].

SAFETY

The acute toxicity appears to be somewhat lower than that of propio- or butyronitriles. Adequate precautions should be taken against skin and eye contact. The inhalation hazard is uncertain [5713].

388. Butyronitrile

Jeffery and Vogel [3700] prepared butyronitrile for *physical property measurements* from sodium cyanide and 1-bromopropane. The product was purified in a manner similar to propionitrile.

Teter and Merwin [7218] used the method of purification described for propionitrile; see also acetonitrile.

SAFETY

Eastman Chemical Products [2189] states that butyronitrile is an extremely toxic compound producing symptoms similar to these caused by other nitriles such as isobutyronitrile. Results of screening studies in their Laboratory of Industrial Medicine show this compound to be extremely hazardous on oral ingestion, skin contact, or inhalation.

Willhite [8071] reported a lethal concentration to male mice as 249 ppm.

The *lower explosive limit* in air is 1.65%v. The *autoignition temperature* is 502° [2189].

389. Isobutyronitrile

SAFETY

Eastman Chemical Products [2188] states that isobutyronitrile is capable of being absorbed by the intact skin and in full strength can cause systemic toxicity and even death. Fassett [5713] cautions about the inhalation of the vapors.

390. Valeronitrile

Jeffery and Vogel [3700] prepared valeronitrile for *physical property measurements* by a modification of the method given by Adams and Marvel [25]. The crude nitrile was washed twice with half its volume of concentrated hydrochloric acid and then with saturated sodium bicarbonate solution, dried with anhydrous magnesium sulfate, and distilled; bp 141° at 764 Torr.

Valeronitrile was prepared in 54–75% yield by Ferguson [2411] by heating 51 grams of 96% sodium cyanide, 2 grams of potassium iodide, and 65 mL of water; then adding 130 mL of 2-ethoxyethanol and 118 grams of 1-bromobutane and refluxing vigorously for 2 hr. Then 200 mL of water was added and distilled until a single phase was reached. The upper phase was washed with 40% calcium chloride solution, water, cold 50% sulfuric acid, and saturated sodium bicarbonate solution, and then was dried with anhydrous calcium chloride and distilled. The yield dropped to 30–40% when 1-chlorobutane was used.

Banewicz and coworkers [592] purified valeronitrile that was prepared by the method of Ferguson [2411] by drying with anhydrous calcium chloride and distilling. The distillate stood over phosphorus pentoxide and was fractionally distilled. The fraction distilling at 140–141° was used as a solvent for *specific conductance studies* of tetraethylammonium perchlorate. Some *physical properties* were also determined.

CRITERIA OF PURITY

Dreisbach and Martin [2090] determined the purity from the freezing curve.

391. Hexanenitrile

Jeffery and Vogel [3700] prepared hexanenitrile for *physical property measurements* from sodium cyanide and 1-iodopentane. The material was purified in the manner described for valeronitrile.

Vvedenskiĭ and coworkers [7808] prepared hexanenitrile and octanenitrile in 97–100% yield by treatment of the respective acids with ammonia at 350–370° over activated alumina.

CRITERIA OF PURITY

Dreisbach and Martin [2090] determined the purity from the freezing curve.

392. 4-Methylvaleronitrile

Jeffery and Vogel [3700] prepared 4-methylvaleronitrile for *physical property measurements* by refluxing 30 grams of sodium cyanide, 40 mL of water, and 77 grams of 1-bromo-3-methylbutane in 150 mL of methanol for 27 hr. It is assumed that the material was purified in the same manner as described for propionitrile; bp 153.5° at 756 Torr.

393. Octanenitrile

Vvedenskiĭ and coworkers prepared octanenitrile; see hexanenitrile [7808].

Stridh and coworkers [7048] purified octanenitrile for *thermochemistry studies* "by fractional distillation at reduced pressure in a 0.7-m all-glass column filled with helices."

394. α-Tolunitrile

This aliphatic nitrile may be purchased at a reasonable cost, or it may be easily prepared [1618, p. 9]. Further purification is possible by fractional distillation.

Gomberg and Buchler [2841] prepared the nitrile from benzyl chloride and aqueous sodium cyanide in a 50–60% yield as compared to 80–90% yield when the compound was prepared according to Conant [1618].

Jeffery and Vogel [3700] prepared α-tolunitrile by a method similar to the one used by Conant [1618] for *physical property measurements;* bp 108.5° at 15 Torr.

Benzyl isocyanide, which gives the product a disagreeable odor, may be removed after the first distillation by shaking vigorously with an equal volume of 50%w sulfuric acid at 60°, washing with saturated sodium bicarbonate solution, and then with half-saturated sodium chloride solution. The material is then dried and fractionally distilled under reduced pressure. The product is colorless and will remain so for some time [3733, p. 89].

SAFETY

The LD_{50} for rats was found to be less than 350 mg/kg body weight [2003]. Galibin and coworkers [2633] recommended the maximum concentration of the vapor as 0.0008 mg/L.

Aromatic Nitriles

395. Benzonitrile

Benzonitrile is a strongly polar aromatic solvent that dissolves a wide variety of organic substances. It also dissolves many anhydrous inorganic salts and organometallic compounds. Its low vapor pressure, low mutual solubility with water, and wide solvent ability make it a uniquely versatile solvent.

Dangyan [1806] prepared benzonitrile in 85% yield by refluxing dicyandiamide and benzoic acid in a mole ratio of 1:3 for 15 hr.

Jeffery and Vogel [3700] prepared benzonitrile for *physical property measurements* by heating 45 grams of benzamide and 75 grams of phosphorus pentoxide. The product was washed with potassium carbonate solution, dried, and distilled; bp 189° at 758 Torr.

Everard and coworkers [2330] found that fractional distillation of two commercial samples of benzonitrile gave unsatisfactory material for *dipole moment determinations*. They prepared the nitrile from benzamide (mp 127–128°) by heating with phosphorus pentoxide for 15 min and distilling. Before use, four-fifths was slowly frozen and the rest discarded.

Martin [4911] purified commercial material by steam distillation to remove small quantities of carbylamine, washed the distillate with sodium carbonate solution to remove benzoic acid, dissolved in ether, washed with water, and dried with calcium chloride. The solution was filtered and the ether removed by distillation. The residue was fractionally distilled under reduced pressure. The middle fraction that distilled at 81° and 22 Torr was shaken with a mixture of potassium carbonate and calcium chloride and then with calcium chloride, and was allowed to stand several days over fresh calcium chloride. After each of the drying operations, the material was distilled. It was used for *electrical measurements*.

Benzonitrile was fractionally distilled three times through a 2-ft column by Brown and Ives [1128] for *dielectric constant measurements;* bp 191.0°, d 25° 1.00061.

The *solvent effects* on the *polarographic reduction* of metal ions in benzonitrile-acetonitrile system was studied by Larson and Iwamoto [4416]. A practical-grade benzonitrile was dried over calcium sulfate for several days. It was transferred to a distilling flask and distilled from fresh calcium sulfate. The distillate was repeatedly redistilled from phosphorus pentoxide in an all-glass apparatus, the center portions only being collected, until almost no black residue remained in the distilling flask; bp 188 ± 0.5° at 735 Torr. The distillate was kept in the receiving flask under a nitrogen atmosphere.

Walden and Werner [7868] fractionated a commercial product in vacuum several times and dried for several days with phosphorus pentoxide between distillations. Müller and coworkers [5324] dried with lime and fractionated several times. Sugden [7072] fractionated repeatedly at atmospheric pressure in an all-glass apparatus.

Coetzee and McGuire [1555] purified benzonitrile to study the *relative basicity of solvents* by first shaking with silica gel, then stirring with calcium hydride for 2 hr, carefully decanting, and vacuum-distilling from a small amount of phosphorus pentoxide.

CRITERIA OF PURITY

Velsicol Chemical Corp. [7708] uses the freezing point and calculates the mole per cent impurity by the method of Witschonke [8121]. The assay method is: Add about 25 mL of the benzonitrile sample to a large test tube. Place the test tube in a dry ice bath and agitate continuously with a thermometer (range −36 to +54°, graduated in 0.2°). Take point just before supercooling begins, where the temperature remains constant the longest, as the freezing point. Calibrate the thermometer by determining the freezing point of the water.

The assay is based on the calculated freezing point of 100%m benzonitrile of −12.75° and a freezing point constant, $K_f' = 1.93\%m/1°$ [8121]. This method is discussed in Chapter II; see eqs. (2.77) and (2.78) and Table 5.16. Plot the data in Table 5.20 and read the purity from the chart.

SAFETY

The Velsicol Chemical Corp. in its *Technical Data Bulletin* [7708] states that the physiological properties of benzonitrile have not been completely established. It may be toxic to some persons. The symptoms of exposure are headache, nausea, weakness, increased heart beat, and dizziness. There have been no indications of chronic toxic effects in handling and use where the normal precautions have been observed.

Table 5.20 Purity Assay of Benzonitrile

Purity (%m)	fp
100.0	−12.75°
99.6	−13.0
97.9	−14.0
96.1	−15

Agaev [43] reported the LC_{50} for mice and rats as 6.0 and 38.6 mg/L, respectively, and the oral LD_{50} for mice and rats as 1.4 and 1.5 g/kg, respectively. The recommended concentration in air is 1.0 mg/m^3.

The *explosive limits* in air are 3.1 and 17%v [6402].

Unsaturated Nitriles

396. Acrylonitrile

Acrylonitrile was first prepared by Moureau [5293] in 1893 from acrylamide and twice its weight of phosphorus pentoxide at 150–250° in 30% yield.

Present industrial production starts with either ethylene cyanohydrin or acetylene [143]. Ethylene cyanohydrin decomposes spontaneously and rapidly in the presence of a suitable catalyst and operating conditions. The wet nitrile may be dried by binary azeotropic distillation of the two components or by the use of a ternary azeotrope using dichloromethane or chloroform. Acrylonitrile is formed when acetylene and hydrogen cyanide are passed over a catalyst at 400–500°. The reaction also takes place rapidly in an aqueous solution of ammonium chloride and copper(I) chloride.

Cyanamid [143] states that highly purified acrylonitrile may polymerize spontaneously, particularly in the absence of oxygen or on exposure to visible light. Traces of ammonia, ammonium carbonate, or one of the products resulting from the reaction of ammonia and acrylonitrile will stabilize it. Other materials that appear in the literature as stabilizing agents are: alkylamines, hydrazines, mercaptans and sulfides, anthracene, aromatic nitro compounds, benzoquinone, chlorine, bromine, iodine, ethylene cyanohydrin, metals (such as copper, zinc, lead, mercury), pyridine, and unsaturated amines. It was found to be stabilized effectively by 0.01% of methoxyhydroquinone, 2-aminoethanol, diethanolamine, triethanolamine, and monoethanoldiethylamine.

Acrylonitrile polymerizes violently in the presence of concentrated alkali. Caution is urged if it is necessary to treat acrylonitrile with strongly alkaline materials.

It is miscible with most organic liquids, including many from the following functional group classes: aliphatic and aromatic hydrocarbons, alcohols, ethers, and esters.

Butskus [1293] prepared acrylonitrile in 60% yield by treating 22 grams of β-bromopropionitrile with 9.4 grams of potassium hydroxide in 30 mL of water.

Baum and Hermann [687] prepared acrylonitrile by reacting acetylene and hydrogen cyanide in the vapor phase in the presence of a catalyst. Heuser [3365] reacted acetylene and hydrogen cyanide in an aqueous solution of ammonium chloride and copper(I) chloride.

Sevryugova and coworkers [6585] purified crude acrylonitrile by a three-stage process. The low boiling components, such as hydrogen cyanide, were removed by fractional distillation in a 50-plate column at 150 Torr. The acrylonitrile was distilled in a 100-plate column at 50–100 Torr. Four volumes of water and 0.001% hydroquinone were added to the distilled acrylonitrile; the azeotrope, which contained water, acrylonitrile, acrolein, and hydrogen cyanide was fractionally distilled from a 50-plate column at 50–100 Torr. The acrylonitrile residue contained 5×10^{-4}% hydrogen cyanide.

Hurdis and Smyth [3595] purified 200 grams of commercial acrylonitrile for *structural studies* by *dipole moment* by drying over calcium chloride and fractionally distilling; bp 77.30° at 750 Torr.

Thompson and Torkington [7437] used a purified sample of acrylonitrile for *infrared studies* that was furnished by the manufacturer and was stabilized by copper powder. It was distilled before use.

CRITERIA OF PURITY

Union Carbide's [7611] test method for acrylonitrile is mass spectrometry using the 41 mass-to-charge ratio as a measure of the impurity.

Cyanamid [143] describes an assay method based on the addition of 1-dodecanethiol to the double bond, followed by the titration of the excess mercaptan with bromate-iodide solution.

SAFETY

Acrylonitrile is toxic by inhalation, oral ingestion, and skin penetration. The toxicity of acrylonitrile results from the release of the cyanide radical [7611]. There is no evidence to indicate that repeated exposures produce a cumulative effect [143]. Acrylonitrile has a characteristic unpleasant odor resembling that of pyridine. Fatigue develops rapidly, and the odor cannot be used as a reliable exposure index.

Stasenkova and coworkers [6965] reported an oral LD_{50} of 69.6 mg/kg for rats and recommended a maximum permissible concentration of 0.5 mg/m^3.

The *threshold limit value* is 20 ppm, 45 mg/m^3 [7022].

The *explosive limits* in air are 2.42 and 17.34%v [1946]. The *minimum ignition temperature* in air is 481° [5713].

397.   Methacrylonitrile

Peters and coworkers [5788] describe a process for the manufacture of methacrylonitrile by passing methallylamine, air, and steam over a silver oxide catalyst at 450–600°. The reaction product was slightly acidified, and the nitrile and other volatile products were removed. The methacrylonitrile distilled in this manner was about 96%; it contained about 2% of other nitriles. Further purification may be accomplished by fractional distillation. The acidification of the reaction product is important because, in the absence of acid, by-product ammonia and unreacted amines react rather rapidly with unsaturated nitriles to form high-boiling nitrogen bases.

SAFETY

The toxicity of methacrylonitrile is similar to that of acrylonitrile. There is no appreciable or easily recognizable odor, and it is only mildly irritating to the eyes, nose, and skin. Odor and irritation, therefore, cannot be relied upon as warning devices [5713, 7753].

There has been no threshold limit established. Osborn [7344] believes it should be less than 20 ppm, 34.8 mg/m^3. Pozzani and coworkers [5940] in a later and more extensive study, including human volunteers, believe the threshold limit value should not be greater than 3 ppm. They found that "it is very likely that most people will not be able to detect the odor of dangerous concentration of methacrylonitrile vapors after 2 or 3 min."

Amines (398–439)
General Comments

The primary aliphatic amines, C_1–C_5, are soluble in alcohol, ether, and water. As the molecular weight increases from *n*-hexylamine, the solubility in water decreases.

The secondary amines, R_2NH, have a limited solubility in water, the solubility decreasing rapidly above C_3, but are generally soluble in ether and alcohol. Trimethylamine is the only tertiary amine that is completely soluble in water. The lower-molecular-weight amines form stable hydrates with water, which make some of them difficult to purify. Aromatic amines are only slightly soluble in water [4027].

Amines are bases and form coordinated covalent compounds with acids. The complexes with the strong and moderately strong acids are quite stable. Most amines, C_6 or less, can be titrated in water with a strong acid, provided that the proper indicator is used or the titration made potentiometrically; the latter gives more accurate results. Amines are strong bases in acetic acid and may be titrated in this medium with a precision of $\pm 0.05\%$ [4897, 6164].

When an alcohol and ammonia are passed over a dehydration catalyst, such as nickel on keiselguhr, at 200°, the corresponding nitrile is formed. The mixture, when passed over a hydrogenation catalyst, such as copper at 180° or nickel at 100°, gives an equilibrium mixture of primary, secondary, and tertiary amines, nitriles, alcohol, and water. The desired amine can be recovered by rectification and the other components recycled. The yield is nearly 100% [7650].

Primary amines can be prepared uncontaminated with secondary or tertiary amines by the method of Brown and Jones [1164].

$$2RMgX + CH_3ONH_2 \rightarrow RNH_2MgX + RH + CH_3OMgX$$

$$X = Cl, Br, \text{ but not } I$$

The reaction mixture is acidified with hydrochloric acid and the amines separated by adding potassium hydroxide. (Sodium hydroxide is equally effective.) The purification depends on the amine prepared and the purpose for which it will be used.

Delépine [1930] prepared primary amines by converting the halide of the desired alkyl amine to a quaternary salt with hexamethylenetetramine. The salt was hydrolyzed and the amine isolated.

SAFETY

The physiological action of, and response to, aliphatic and alicyclic amines varies only little among the individual amines. The toxicological information available for amines is not great. Acute oral, cutaneous, and eye toxicities of the amines do not show marked change with increasing chain length because these are influenced to a major degree by the acute local irritating properties. Amine vapors are irritating to the eyes, producing tears, inflammation, and a halo around lights. The vapors irritate the mucous membranes of the respiratory tract. Polyamines may cause asthmatic symptoms and perhaps respiratory tract sensitization. The vapors may cause skin irritation and dermatitis. Liquid amines splashed in the eye will cause severe pain and may result in permanent eye damage. The liquid will produce burns on the skin if allowed to remain long enough. Amines may be absorbed through the skin amd mucous membranes either as vapor or as liquid [5713].

The *threshold limit values* that have been established or suggested, the *flammable limits*, and the *minimum ignition temperatures* are tabulated in Tables 5.21 and 5.22.

Table 5.21   Threshold Limit Value for Aliphatic Amines

Compound Number	Compound Name	TLV ppm	mg/m³	Reference Number
398.	Methylamine	10	12	142
400.	2-Propylamine	5	12	7022
401.	Butylamine	5	15	7022
406.	Cyclohexylamine	20	81	5713
415.	Ethylenediamine	10	25	142
418.	Dimethylamine	10	18	142
419.	Diethylamine	25	75	7022
421.	Diisopropylamine	10	41	5713
428.	Triethylamine	25	100	7022

Table 5.22   Flammable Limits and Minimum Ignition Temperatures of Aliphatic Amines in Air

Compound Number	Compound Name	Flammable Limits (% v) Lower	Upper	Minimum Ignition Temp. (°C)
399.	Propylamine	2.01	10.35	318
400.	2-Propylamine	—	—	402
401.	Butylamine	1.70	9.75[a]	312
402.	Isobutylamine	—	—	378
417.	Ethylenimine	3.60	45.80	322
419.	Diethylamine	1.77	10.10	312
428.	Triethylamine	1.25	7.90[a]	232

[a] Determination made at elevated temperatures.

Primary Aliphatic Amines

398.   Methylamine

Felsing and Thomas [2399] prepared purified methylamine for *physical property studies* by crystallizing the best available methylammonium chloride from absolute ethanol, from butanol, and from triple-distilled water. The recrystallized salt was introduced into a closed system and the amine regenerated with concentrated sodium hydroxide solution. The gas was passed over solid sodium hydroxide and soda lime into a reservoir filled with sodium wire surrounded by liquid air. The flask was warmed with liquid ammonia after a considerable quantity had collected and finally with crushed ice. The sodium reacted with the water present at this temperature. The contents of the flask were again frozen with liquid air, the system was slightly evacuated, and the methylamine distilled into another reservoir containing sodium metal. The same general procedure as described was followed and the amine was finally collected in a reservoir surrounded by liquid air.

Stairs and coworkers [6958] purified commercial methylamine for *surface tension studies* by distilling it from the cylinder onto lithium and letting it stand at −78° for 24 hr. It was then degassed and subjected to several bulb-to-bulb fractional distillations. Gibbs [2731] used sodium instead of lithium.

SAFETY

The *flammable limits* in air are 4.95 and 20.75%v. The *minimum ignition temperature* in air is 430° [1607].

399. Propylamine

Majer and coworkers [4842] purified propylamine for *vapor pressure studies* by distillation over sodium on a 1.5-m packed column.

400. Isopropylamine

Majer and coworkers [4842] purified isopropylamine for *vapor pressure studies* by first converting the amine to the sulfate for separation from any isopropanol and acetone and then reconverting to the amine which was distilled over sodium on a 1.5-meter packed column.

401. Butylamine

Butylamine was dried over potassium carbonate for 1 day and fractionally distilled. The portion distilling at 77.0° was collected by Bayles and Chetwyn [693] to study *acid–base function in nonaqueous systems*.

Cluett [1543] purified a good commercial butylamine by distilling over potassium hydroxide, collecting the middle 80%, and storing under dry carbon dioxide-free nitrogen. The material was used to study the *differentiating titration* of phenylureas as acids.

Muney and Coetzee [5333] purified butyl-, diethyl-, and triethylamines obtained from supply houses for *conductivity studies* in acetonitrile as follows: The amines were dried over sodium hydroxide, then over sodium, followed by a 2-hr refluxing over sodium, and finally fractional distillation. The criterion of purity was the boiling point.

Letcher and Bayles [4535] purified butylamine, diethylamine, and triethylamine for *physical property study*. The solvents were allowed to stand over sodium hydroxide pellets for several days, twice fractionally distilled through a 3-m × 30-mm glass column packed with borosilicate "glass pieces" and the middle fraction retained. The authors state that there is a confusion in the literature concerning the physical constants for diethyl and butylamines. The criteria of purity selected was concurrent consisting of boiling point, density, and refractive index.

Wadso [7823] purified amines for *calorimetric studies* by drying the solvents over calcium hydride or molecular sieve 4A and then fractionally distilling until gas chromatographs showed purities greater than 99.9%.

For purification for *vapor pressure studies*, see Majer [4842], propylamine.

SAFETY

Cheever and coworkers [1438] reported a 14-day oral single dose LD_{50} for male and female rats as 365.4 and 382.7 mg/kg, respectively.

402. Isobutylamine

For purification for *vapor pressure studies*, see Majer [4842], propylamine. For purification for *calorimetric studies*, see Wadso [7823], butylamine.

SAFETY

Cheevers and coworkers [1438] reported a 14-day oral single dose LD_{50} for male and female rats as 224.4 and 231.8 mg/kg, respectively.

403. *sec*-Butylamine

For purification for *vapor pressure studies*, see Majer [4842], propylamine. For purification for *calorimetric studies*, see Wadso [7823], butylamine.

SAFETY

Cheevers and coworkers [1438] reported a 14-day oral single dose LD_{50} for male and female rats as 157.5 and 146.8 mg/kg, respectively.

404. *tert*-Butylamine

Karabinos and Serijan [3856] reported a convenient method for the preparation of a relatively large quantity of *tert*-butylamine by catalytic hydrogenation of 2,2-dimethyl-eneimine; see also Horning [3492, p. 148].

Pearson et al. [3492] prepared *tert*-butylamine by the hydrolysis of *tert*-butylurea in 71–78% yields.

For purification for *calorimetric studies*, see Wadso [7823], butylamine.

SAFETY

Cheever and coworkers [1438] reported a 14-day single oral dose LD_{50} for male and female rats as 82.3 and 78.1 mg/kg, respectively.

The *flammable limits* in air are 1.73 and 8.9%v [6239].

405. Pentylamine

For purification for *calorimetric studies*, see Wadso [7823], butylamine.

406. Cyclohexylamine

Auwers [501] prepared and purified cyclohexylamine for *optical measurements* by reducing cyclohexanone oxime with sodium in boiling alcohol. The reaction product was steam-distilled, acidified with hydrochloric acid, evaporated to dryness, made alkaline, and extracted with ether. The ether extract was dried over potassium hydroxide and rectified in a stream of hydrogen; a Linnemann column was used.

Sabatier and Senderens [6330] separated this amine from aniline by distillation at 30 Torr. Baker and Schuetz [559] prepared cyclohexylamine from aniline in the same manner as cyclohexane.

Palfray [5651] hydrogenated aniline at 150° and 60 kg pressure in the presence of Raney nickel. Fractional distillation of 199 grams of product gave 96 grams of cyclohexylamine for a 61% yield.

Adkins and Billica [33] reduced cyclohexanone oxime with a specially prepared Raney nickel catalyst; see 2-propanol.

Carswell and Morrill [1376] determined the *physical properties* of cyclohexylamine purified by crystallizing the hydrochloride several times from water, liberating the amine with alkali, and fractionally distilling under nitrogen.

Lewis and Smyth [4564] distilled a commercial product, dried it with calcium chloride and then with sodium in an atmosphere of nitrogen, and redistilled; bp 133.8°, d 25° 0.86253.

Steele [6984] dried cyclohexylamine over molecular sieve and fractionally distilled over sodium, followed by two fractional distillations under dry nitrogen. For purification for *vapor pressure studies*, see Majer [4842], propylamine.

SAFETY

According to Abbott Chemicals [6], "cyclohexylamine is a toxic chemical, showing irritation and caustic action when brought into contact with the lungs, skin, or eyes." The acute inhalation LD_{50} for rabbits is 360 mg/kg.

Primary Aromatic Amines

408. Aniline

Aniline, the simplest of the aromatic amines, exhibits the chemical properties of both the amine and the aromatic nucleus. It has some amine reactions not common to aliphatic amines. It forms salts readily, and the amine group is easily oxidized, as shown by darkening on exposure to air. It is a weak base in water and a strong base in some solvents such as acetic acid, benzene, and carbon tetrachloride.

Aniline is manufactured by: (1) vapor-phase reduction of nitrobenzene with hydrogen at temperatures up to 350° in the presence of catalyst, (2) reduction of nitrobenzene with hydrochloric acid and iron filings or powder, and (3) heating under pressure chlorobenzene, aqueous ammonia, and copper(I) oxide. Distillation of the "aniline oil" obtained from the three reactions yields a product of high purity because of its boiling point and scarcity of side reactions.

Bag and coworkers [544] reported reducing nitrobenzene almost stoichiometrically to aniline in the presence of fragments of sodium hydroxide activated with nickel and aluminum plus other metals as a catalyst.

Palfray [5651] rapidly hydrogenated quantitatively 1 mol of nitrobenzene at 75° and 95-kg pressure (the temperature rose to 115°) in the presence of Raney nickel.

In 1920, Knowles made a critical study [4078] of the more common properties of aniline reported in the literature. He concluded that it was not possible to evaluate the data because of the wide variation in the reported values. He found it a "matter of considerable difficulty" to purify aniline. He concluded that the following method would produce a high-purity product. Distill 2 L of a colorless aniline four times under reduced pressure, rejecting the first and last portions each time. Allow the product to drop slowly into a solution of recrystallized oxalic acid. Filter the oxalate, wash several times with water, and recrystallize three times from 95% ethanol. Regenerate the aniline from the salt with a saturated solution of recrystallized sodium carbonate and distill the aniline from the solution. Distill the aniline three times under reduced pressure. When possible, use the aniline immediately; otherwise keep it in a dark place in a vacuum desiccator; bp 184.32–184.39°, fp −6.24°, d 15°/15° 1.0268, n_D 20° 1.5850. Aniline is extremely hygroscopic, and precautions must be taken to keep moisture from an anhydrous product.

Few and Smith [2414] purified aniline by dissolving in 40% sulfuric acid. Nonbasic materials were removed by passing steam through the solution for 1 hr. The base was liberated by the addition of sodium hydroxide pellets, steam-distilled, and dried over potassium hydroxide. It was distilled twice from zinc dust at 20 Torr, dried over freshly prepared barium oxide, and finally distilled from barium oxide in an all-glass apparatus.

Smith and coworkers [6833] distilled aniline under vacuum and dried over sodium hydroxide. The purified aniline was used for *solubility studies*.

Keyes and Hildebrand [3964] dissolved a pure commercial material in hydrochloric acid, steam-distilled to remove traces of nitro compounds, made alkaline to free the base, and steam-distilled. The aniline was separated from the water, dried with potassium hydroxide, and fractionally distilled at reduced pressure.

Jones and Kenner [3748] reported that aromatic amine hydrochlorides form double salts with copper(I) chloride that may be used for purification.

Oddo and Tognacchini [5552] crystallized aniline several times from ethyl ether and distilled at reduced pressure.

Hantzsch and Freese [3145] offered an explanation for the rapid coloration of aniline in air. They found that *sulfur-free* aniline remained colorless on standing exposed to air for weeks, whereas technical aniline became brown. The sulfur-free aniline was prepared by refluxing aniline with 10% acetone for 10 hr, the mixture was acidified with hydrochloric acid to Congo Red, and extracted with ethyl ether until it no longer showed any color. The hydrochloride was purified by repeating crystallizations; the aniline was freed, dried, and distilled.

Weissberger and Strasser [7966] found that the tendency of aromatic amines to become colored by oxidation was greatly reduced by treatment with tin(II) chloride followed by precipitating the tin as the sulfide.

Aniline has been stabilized by the addition of mercaptobenzimidazoline [5636].

LaPaglia [4405] purified aniline for *fluorescent studies* by distilling twice in an argon atmosphere and then freezing and degassing in successive cycles. No impurities were found by gas chromatography.

Pearce [5730] allowed a commercial product to stand several weeks over potassium hydroxide, distilled and treated the portion distilling at 181–182° by the Hantzsch–Freese procedure (see above). The middle fraction was distilled twice over zinc. It was used for *electrical measurements*. See also Pound [5934] and Müller and coworkers [5324] for other purifications for *electrical measurements*.

Aniline was purified by Armanet and Pouyet [426] by fractional distillation, crystallization, and zone melting. The improvement in purity was followed by the melting point, specific conductance, and near-ultraviolet and visible spectroscopy.

CRITERIA OF PURITY

At least two grades are available from most supply houses but most list three: a 99+%m, a reagent grade to meet, at least, ACS reagent-grade specifications, and a purified grade. Fisher Scientific Company's specifications for their reagent grade are given in Table 5.23.

Table 5.23 Specifications for Reagent-Grade Aniline[a]

Boiling point	184.4°
Distillation range (max)	3°
Residue after ignition (max)	0.005%
Chlorobenzene (max)	0.010%
Hydrocarbons	Pass test
Nitrobenzene (max)	0.0010%
Appearance	Pass test

[a] From *Fisher Chemical Index* 67-C.

SAFETY

The toxicology of aniline in animals and its effect on humans have been extensively studied. It is considered a hazardous substance; it is an acute and a chronic poison; and it is absorbed through the respiratory tract and through the skin [5713]. Tkachev [7508] found the threshold odor concentration to be 0.37 mg/m^3.

Kondrashov [4174] reported an oral LD_{50} for mice as 464 mg/kg. The *threshold limit value* is 5 ppm, 19 mg/m^3 [7022].

409. *o*-Toluidine
410. *m*-Toluidine
411. *p*-Toluidine

The toluidines undergo about the same reactions as aniline. They form salts with most acids, and are weak bases in water and strong bases in acetic acid and other organic solvents. The methods of purification for aniline are generally applicable to the toluidines.

Adkins and Billica [33] reduced *p*-nitrotoluene with a specially prepared Raney nickel catalyst; see under 2-propanol.

Berliner and May [796] purified the three isomeric toluidines as follows for *vapor pressure studies:*

1. *o-Toluidine.*   The material was twice distilled and then dissolved in four times its volume of ethyl ether. An equivalent amount of oxalic acid to form the dioxalate was added to the ether solution and the crystals of *p*-toluidine dioxalate removed by filtration. The ether solution was evaporated and the crystals of *o*-toluidine dioxalate formed were separated by filtration. They were recrystallized five times from water containing a small amount of oxalic acid to prevent hydrolysis. The purified crystals were treated with a dilute solution of sodium carbonate and the amine separated, dried with calcium chloride, and distilled three times at reduced pressure; bp 199.84°.

2. *m-Toluidine.*   The amine was twice distilled and heated with a slight excess of hydrochloric acid. The hydrochloride was fractionally crystallized five times from 95% ethanol and twice from distilled water. In each case, the material that crystallized first was discarded. The amine was regenerated and distilled as described above; bp 202.86°.

3. *p-Toluidine.*   The amine was distilled three times and sublimed twice at 30°. It was then dissolved in five volumes of ethyl ether containing an equivalent amount of oxalic acid. The filtered and washed crystals were recrystallized three times from hot distilled water and the base regenerated with sodium carbonate solution. The base was

recrystallized three times from distilled water; mp 43.5°. Seven recrystallizations from ethanol did not produce any change in the melting point.

Rastogi and coworkers [6075] purified *p*-toluidine for *thermodynamic studies* by fractional crystallization.

o-Toluidine has been stabilized by the addition of mercaptobenzimidazoline [5636].

SAFETY

Hamblin [5713] states that *o*- and *p*-toluidines have a toxicity similar to that of aniline, but in addition cause a transient hematuria which, in observed experience, cleared completely upon removal from further exposure.

The *threshold limit value* has been set at 5 ppm, 22 mg/m^3 [7022].

The LD_{50} for rats, intraperitoneal injection for the toluidines has been reported as: *ortho* 0.083, *meta* 0.1166, and *para* 0.150 g/kg [1494].

412.   2,4-Xylidene

Markov and coworkers [4894] prepared 2,4-xylidene by adding 0.1 mol of 2,4-dimethylnitrobenzene in 50 mL of absolute ether and 0.4 mol of absolute alcohol to 250 g of dry ammonia to which 3.2 mol of magnesium had been added. The mixture was stirred until all of the magnesium reacted. The ammonia was evaporated and 2000 mL of ether was added. The mixture was acidified with hydrochloric acid and the amine was isolated by conventional procedure.

For purification for *ionization studies*, see Beale [701], 2,6-xylidene. For purification for *photochemical work*, see Pouyet [5935, 5936], 2,6-xylidene.

SAFETY

For toxicity studies, see Magnusson [4822], 2,6-xylidene.

413.   2,6-Xylidene

Commercial 2,6-xylidene was purified by Pouyet [5935,5936] for *photochemical work* by fractional distillation to remove ionic impurities followed by column crystallization. Zone melting and column chromatography were also used for purification.

Beale [701] purified 2,6-xylidene for *ionization studies* by pouring a sample into 18% hydrochloric acid solution. The precipitate was dissolved by heating and the solution was allowed to cool for 1–2 hr. The crystals were collected and washed with a small amount of hydrochloric acid solution. The salt was dissolved in water and the solution was made alkaline with 40% sodium hydroxide solution, cooled, and extracted with ether. After removing the ether, the residue was fractionally distilled. Small middle cuts were used.

SAFETY

Magnusson and coworkers [4822] reported that 2–50 mg of 2,6-xylidene given orally for 4 weeks induced fatty degeneration of the livers of dogs.

Primary Unsaturated Amines

414.   Allylamine

Tollens [7516] reported the preparation of allylamine in 1872.

Clavier [1516] prepared allylamine in 90% yield from 3-bromopropene and hexamethylenetetramine and in 25% yield from 3-chloropropene. A modification raised the yields for the chloro compound to 55–60%.

Fuson [2620, p. 5] gives directions for preparation from allyl isothiocyanate, hydrochloric acid, and potassium hydroxide.

SAFETY

The *flammable limits* in air are 2.20 and 22.05%v and the *minimum ignition temperature* in air is 374° [5712].

Primary Polyamines

415.   Ethylenediamine

Ethylenediamine is a water-soluble, colorless liquid with an ammoniacal odor. It is very hygroscopic and forms a carbamate with carbon dioxide. It is commonly available in two grades: a "practical" grade of 91–93% and a 98–100% grade often referred to as "anhydrous."

The amine has found a variety of solvent uses, such as an ionizing solvent for electrochemical investigation, a reaction medium particularly for ionic reactions, and an enhancing solvent for acids in nonaqueous titrations.

Ethylenediamine always should be protected from atmospheric moisture and carbon dioxide during purification. It should be protected during use if the moisture and carbon dioxide absorbed from the air would cause a greater error in the results than should be tolerated.

The "anhydrous"-grade ethylenediamine is suitable for some solvent uses. It is recommended that this grade be used as the starting material for purification. The water and carbon dioxide may be reduced by shaking the amine with sodium or potassium hydroxide pellets for several hours, transferring the liquid to a still, and distilling the amine through a spray trap. The amount of water may be reduced by treating with a suitable drying agent, preferably 3A or 4A molecular sieves, or with alumina. The amine is then distilled from the sieves or alumina.

The water may be removed by azeotropic distillation. Ethylenediamine and water form a negative azeotrope that has a boiling point less than 2° above that of the amine. Therefore, it is not feasible to try to separate the amine from the azeotrope by fractional distillation. A third substance that will form an azeotrope with water but not with ethylenediamine, and that boils sufficiently below the amine to permit the two to be easily separated by fractional distillation, may be used. There has not been much information published concerning azeotropes of ethylenediamine; however, Creamer and Chambers [1715] have used both benzene and isopropyl ether as azeotropic separating agents. Benzene is preferable because it removes about twice the amount of water on a weight basis.

The preparation of high-purity ethylenediamine requires an additional step, treatment with a substance that will react quantitatively with water. The alkali metals, sodium hydride or boron hydride, and subsequent efficient fractional distillation have been successfully used. The following procedure has given ethylenediamine that analyzed 99.9+%w by mass spectrometry and gas chromatography: Stir commercially "anhydrous" amine for 2–4 hr with 10%w freshly opened sodium hydroxide pellets. Transfer the amine to a still pot and add about 2%w of sodium wire, ribbon, or fine slices, connect to an efficient distillation column, and reflux for about 3 hr. Fractionally distill at a reflux ratio sufficient to give the purity desired. A 7-ft (2.1 m) × 25 mm-id, Penn State-type column packed with 0.16 × 0.16-in. protruded packing and fitted with an automatic vacuum-jacketed head has been used to obtain a 99.9 + % amine when operated as follows. It was refluxed for 3 hr just under flooding, and 50 mL was distilled at a flux ratio of 50:1; about 50 mL more was removed at a reflux ratio of 35:1 and the remainder of a 10%v heads fraction volume of a 3500-mL pot charge was taken at a reflux ratio of 25:1. The middle 80% was distilled at a reflux ratio of 10:1. Samples were analyzed from time-to-time to be sure the desired purity was distilling.

Several methods of purification are given that have some unique features, or that have been used for specific purposes requiring special criteria.

Heumann and coworkers [3364] purified "anhydrous" grade for *ion-exchange studies* in the amine. The amine was dried for several days with porous barium oxide (Barium and Chemicals, Inc.). It was decanted into a vacuum distillation apparatus fitted with Teflon-sleeved ground-glass joints and Teflon stopcocks. Liquid potassium (75%)–sodium alloy was added until a persistent dark blue color developed. The amine was distilled at reduced pressure at a temperature not greater than 40°.

Windwer and Sundheim [8110] purified the amine by drying over calcium oxide, then filtering in a cellophane bag enclosure in an atmosphere of nitrogen. It was refluxed overnight with sodium and fractionally distilled through a 10-ft column, and the middle fraction was collected. Sodium–potassium alloy was introduced under vacuum and then repeatedly cycled through shaking, freezing, evacuating, and remelting until a permanent blue color developed. This color is evidence of a high purity.

For removal of acidic impurities as a solvent for *titration of acids*, see Harlow et al. [3167] under piperidine.

Bruckenstein and Mukherjee [1182] extensively studied the purification of ethylenediamine to obtain the best purity as a solvent for *equilibrium studies*. The 114–117° fraction was shaken with either Alcoa F-20 or Kaiser KA 101 activated alumina, and the liquid was decanted and fractionally distilled in a dry carbon dioxide-free atmosphere at a reflux ratio of 20:1. This procedure was as satisfactory as sodium treatment. The fraction boiling at 117.2° was used.

Creamer and Chambers [1715] prepared anhydrous ethylenediamine for *electrochemical investigation* by removing the water as the benzene azeotrope in a fractionating still with a phase-separating stillhead; the specific conductance was 1×10^{-5} ohm^{-1} cm^{-1}.

Schaap and coworkers [6414] found that they obtained the best product for *polarographic studies* by twice refluxing over sodium and vacuum-distilling.

Mukherjee and Bruckenstein [5307] continued their study of the purification of ethylenediamine and developed a method for producing "pure" amine suitable for *electroanalytical chemistry*. In addition to the purification method, it is an excellent review and discussion article. Their recommended procedure is: Shake 98% "anhydrous" commercial ethylenediamine with 70 grams/L of activated Linde Type 5A molecular sieves for about 12 hr. Decant the amine from the molecular sieves and shake approximately 12 hr with 50 grams/L of calcium oxide and 15 grams/L of potassium hydroxide. Keep in an atmosphere free of carbon dioxide and water during the remainder of the purification. Fractionally distill the supernatant liquid at a reflux ratio of 20:1 from freshly activated molecular sieves and collect the material that distills at 117.2° at 760 Torr. Fractionally distill the 117.2° cut from sodium.

For purification for *calorimetric studies*, see Wadso [7823], butylamine.

The vapors of ethylenediamine are irritating to the eyes, mucous membranes, and respiratory tract. This liquid causes severe skin and eye injuries. Dermatitis is a common result of repeated exposure to the vapors [5713]. It is both irritating and allergenic, and some individuals are hypersensitive [7022].

The single oral dose LD_{50} for rats was reported as 1.85 g/kg [7627].

The *threshold limit value* is 10 ppm, 25 mg/m^3 [142].

416. Diethylenetriamine

Pecsok and Garber [5743] purified a "technical"-grade diethylenetriamine for *base strengths study* by double distilling from metallic sodium at 1 Torr pressure and 64–69°. About 300 grams of the distillate was dissolved in 750 mL of 95% ethanol; 150 mL of 8.5 N hydrochloric acid was added and the mixture was cooled to 0°. The white crystals were collected by filtration, washed with ether, dried by suction, and stored in a vacuum dessicator. The crystals were 16.70% monochloride and 83.30% dichloride.

Diethylenetriamine was purified by Rouleau and Thompson [6287] for *dissociation constant studies*. It was fractionally distilled through a 36 × 1-in. column packed with $\frac{3}{16}$-in. glass helices at a reflux ratio of 25:1. The system was protected from atmospheric moisture. The middle 20–90% was used.

SAFETY

Fujino [2604] observed that histological changes in rats due to diethylenetriamine were mainly in the kidneys and liver and to a lesser extent in the spleen and adrenals.

Diethylenetriamine has a relatively high order of toxicity. It causes severe skin and eye burns. It has sensitizing properties of the skin and respiratory tract, and a sensitized person retains the sensitivity for several years [7627].

The *threshold limit value* for skin absorption is 1 ppm, 4 mg/m^3 [142].

Secondary Monoamines

417. Ethylenimine

Ethylenimine is the simplest of the heterocyclic nitrogen compounds. It is a colorless mobile liquid with an amine-like odor. Traces of acid cause exothermic polymerization to a water soluble polyimine [2071]

$$n\text{H}_2\text{C}\overset{\displaystyle\diagdown\diagup}{\underset{\underset{\text{H}}{\overset{|}{\text{N}}}}{}}\text{CH}_2 \xrightarrow{\text{H}^+} -(-\text{CH}_2-\text{CH}_2-\text{NH}-)_n-$$

A few pellets of sodium hydroxide will react with any acid absorbed by ethylenimine and prevent ring opening and polymerization.

Gray and Jones [2914] purified ethylenimine by distilling twice in an atmosphere of nitrogen and retaining the constant boiling fraction each time. A gas chromatogram showed the distillate to be pure.

Ethylenimine and piperidine were purified by Searles and coworkers [6539] for *hydrogen bonding and basicity studies* by drying over barium oxide and sodium and distilling through a 5-plate column in a nitrogen atmosphere. See also Kay and coworkers [3911].

Clark and Pickett [1502] distilled commercial ethylenimine over sodium hydroxide in a nitrogen atmosphere using a 40-cm fractionating column packed with glass helices. The middle fraction boiling at 55.8° was collected and then redistilled *in vacuo*.

Allen and coworkers [6015] prepared ethylenimine from β-aminoethyl-sulfuric acid.

SAFETY

The Dow Chemical Company [2071] recommends special precautions when handling and using ethylenimine. It can polymerize violently in the presence of acids or acid-forming substances. The liquid is corrosive to the skin and is readily absorbed in toxic amounts. Short contact will cause marked irritation and longer contact is likely to result in severe burns that heal slowly. Repeated exposure to vapors not concentrated enough to cause immediate discomfort may result in delayed lung injury.

The *threshold limit value* is 0.5 ppm, 1 mg/m^3 [7022].

The *flammable limits* in air are 3.60 and 45.80%v. The *minimum ignition temperature* in air is 322° [5712]

418. Dimethylamine

SAFETY

The *flammable limits* in air are 2.80 and 14.40%v. The *minimum ignition temperature* in air is 402° [1607].

419. Diethylamine

Swift [7139] converted a commerical diethylammonium chloride into *p*-toluenesulfonamide and recrystallized it three times from pure dry ligroin (bp 90–120°) until a sharp melting point was obtained. The amide was hydrolyzed with hydrochloric acid, excess sodium hydroxide added, and the amine distilled through a tower of activated alumina. The diethylamine was then distilled through a 5-ft adiabatic column at a reflux ratio of 20:1. A fraction distilling over a range of about 0.01° was collected under nitrogen and sealed under vacuum. It was dried with activated alumina before use.

For purification for *conductivity studies*, see Muney and Coetzee [5333], butylamine; for *calorimetric studies*, see Wadso [7823], butylamine; and for *vapor pressure studies*, see Majer [4842], propylamine.

SAFETY

Tkachev [7509] reported that short-term inhalation of diethylamine disrupts the human nervous system and recommended a maximum permissible atmospheric concentration of 0.47 mg/m^3.

420. Dipropylamine

For purification for *calorimeteric studies*, see Wadso [7823], butylamine.

SAFETY

Teplyakova and coworkers [7212] reported the LD_{50} for rats and mice as 280 and 320 mg/kg, respectively.

421. Diisopropylamine

Petros and coworkers [5799] dried diisopropylamine by boiling with calcium hydride for 2–3 hr. For purification for *calorimetric studies*, see Wadso [7823], butylamine.

SAFETY

The single dose oral LD_{50} for rats was reported as 0.77 g/kg [7621].

422. Dibutylamine

Dibutylamine was dried over sodium hydroxide and fractionally distilled. The portion distilling from 158–159° was collected by Bayles and Chetwyn [693] to study *acid–base functions* in nonaqueous systems.

For purification for *calorimetric studies*, see Wadso [7823], butylamine.

SAFETY

The single-dose oral LD_{50} for rats was reported as 0.36 g/kg [7621].

423. Pyrrole

Pyrrole has unusual solvent properties that have been investigated only slightly because of its ease of reacting with atmospheric oxygen. It should be handled, used, and stored in an inert atmosphere. Its existence as an entity was first surmised by Runge in 1834. Anderson isolated it from coal tar in 1857. It is a good solvent for donor molecules of high molecular weight. This is illustrated by the strong tendency for the formation of $=N-H\leftarrow O$ bonds [4932]. It is an extremely weak base, but strong proton donor acids form

protonated coordinate covalent compounds. The protonated ion of pyrrole is highly unstable and undergoes polymerization readily [5396].

Ciamician [1485] presented a historical summary of pyrrole in 1905 and gave several methods of preparation. The original method of Schwanert [1485] published in 1860 of heating dry ammonium mucate is still practical. Goldschmidt in 1867 found the addition of glycerol helpful; Khotinsky [3991] added ammonia; Blicke et al. [913, 914] further modified the method for the preparation of large or small amounts in yields of 40–52%. McElvain and Bolligen [2757] have further refined the basic method of Blicke and Powers.

Physical properties were determined by Helm and coworkers [3275] on a commercially available pyrrole that was fractionally distilled in a 1-in., 80-plate (65 theoretical plates) Oldershaw all-glass column. The product was fractionally redistilled in a 200-theoretical-plate Helipac still. The highest purity material was 99.994%m determined from the freezing curve.

They found that pyrrole was stable for at least 1 year in sealed glass ampules, whether stored exposed on the roof or in the dark at room temperature. A sample of the pyrrole in a loosely capped vial was exposed to air and light in the laboratory. The sample darkened with exposure, and a decrease in purity was found from the freezing curve. The change was followed by infrared spectroscopy. There was an increased absorbance at 5.87 μm, the carbonyl region.

Scott and coworkers [6519] used a portion of the pyrrole prepared by Helm et al. [3275] for their *thermodynamic studies*. Scrupulous care was taken to avoid exposing the material to atmospheric oxygen. It was stored and handled *in vacuo* or blanketed with dry, oxygen-free helium.

SAFETY

Scant evidence indicates that pyrrole can be considered of low toxicity, both systemically and locally [5713].

424. Pyrrolidine

Helm and coworkers [3275] purified a commercial pyrrolidine for *physical property studies* by distillation in an 80-plate Oldershaw column. The distillate near the end of the charge was refractionated; criterion of purity was low-temperature calorimetric measurements. The compound was stable in sealed Pyrex glass ampules stored in the dark at ambient laboratory temperature and in the open on the roof for a period of approximately 1 year.

Craig and Hixon [1708] studied the ionization constant of pyrrolidine and found that the amine dried with solid potassium hydroxide contained as much as 2% water. They dried the substance by distilling from sodium and protecting the distillate from moisture.

The best commercial grade of pyrrolidine was purified by Searles and Tamres [6539] for *base strength study* by drying over barium oxide and distilling through a 5-plate column.

425. Piperidine

Pyridine is readily reduced to piperidine by catalytic reduction in acid solution with a platinum catalyst, in neutral solution with Raney nickel catalyst, electrolytically, or by sodium and ethanol. All the reduction products contain some impurities. Davies and McGee [1854] studied the impurities in commercial piperidine. Their initial study consisted of fractionally distilling a "pure" 98–99% material in a 46-theoretical-plate still at a reflux ratio varying from 30:1 to 15:1. A 6% fraction that distilled at 92° was supposed to have been "tetrahydropyridine." However, it was shown to be the piperidine–water azeotrope. The fraction distilling at 106° and comprising 70% of the charge was piperidine. A 20% fraction distilling at 107° had been assumed to be 1,2,3,6-tetrahydropyridine but it was shown to be 1,2,5,6-tetrahydropyridine. They found that not all commercial compounds contained the 1,2,5,6- partial reduction compound. None was found in a laboratory-prepared piperidine with nickel catalyst. Available information indicated that the electrolytically reduced product contains the 1,2,5,6-impurity.

Schildknecht [6443] reduced the pyridine in piperidine form 0.1 to 0.001 mol fraction by zone melting.

Werner and coworkers [7982] purified piperidine as a solvent for the *study of constitution of inorganic compounds* by removing the water and carbon dioxide with potassium hydroxide. It was warmed over sodium, allowed to stand, and distilled; bp 104.5–105.0°. The *ebullioscopic constant* was found to be 2.842.

Bates and Bower [672] purified "CP" and "purified" grades of piperidine in the following ways to determine the *dissociation constant* of the piperidinium ion and *related thermodynamic quantities:* (1) piperidine was distilled from sodium hydroxide pellets and redistilled, rejecting the first and the last tenths of the distillate; (2) a portion of the redistilled base was frozen and the last 5% of the liquid rejected, and the main portion was redistilled; and (3) a portion of the finished (2) sample was twice redistilled, and the first and last fractions rejected.

Searles and coworkers [6539] purified piperidine and ethylenediamine for *hydrogen bonding studies;* see ethylenediamine.

Harlow and coworkers [3167] purified piperidine, pyridine, and ethylenediamine as solvents for the *potentiometric titration study* of very weak acids by passing them through a column of activated Alcoa F-20 alumina.

Hudson and coworkers [3554] allowed "purified" piperidine to stand over freshly fused barium oxide for 3 days and then fractionally distilling, as did Davies and McGee [1854] for *vapor diffusion coefficient and collision parameter studies.*

SAFETY

The oral toxicity of the salts of piperidine is very low; for instance, the LD_{50} of piperidinium citrate for mice is 11 g/kg. Piperidine is a strong base and has the inherent hazards of such compounds [5713]. First-degree burns have been reported by Linch [4606] when it was in contact with the skin for 3 min. The vapors are irritating to the respiratory tract [5].

426. *N*-Methylaniline

SAFETY

The *threshold limit value* is 2 ppm, 9 mg/m^3 [6402].

Tertiary Monoamines

427. Trimethylamine

Hannson [3139] purified trimethylamine by recrystallization of the oxalate.

A mixture of ammonia and trimethylamine can be separated by adding 1–4% of water to the mixture and cooling to about −70° [3054].

SAFETY

Rotenberg and Mashbits [6282] studied the inhalation toxicity with white mice and rats. A *threshold limit value* of 0.005 mg/L was suggested.

The *flammable limits* in air are 2.00 and 11.60%v. The *minimum ignition temperature* in air is 190° [1607].

428. Triethylamine

Swift [7139] distilled commercial anhydrous triethylamine from acetic anhydride to remove traces of primary and secondary amines, dried with activated alumina, and distilled three times under reduced pressure.

Counsell and coworkers [1680] purified triethylamine for *phase diagram studies* by refluxing a high-purity commercial product containing a small amount of lower-molecular-weight amines with *p*-toluenesulfonyl chloride and potassium hydroxide to remove primary and secondary amines. It was then fractionally distilled from metallic sodium at a reflux ratio of about 15:1; bp 89.50° at 760 Torr.

Muney and Coetzee [5333] purified triethylamine for *conductivity studies*; see butylamine.

Commercial triethylamine was purified by Bryant and Wardrop [1217] by converting to the triethylammonium chloride and recrystallizing from ethanol to a sharp and constant melting point, 254°. The amine was regenerated with aqueous sodium hydroxide, dried over potassium hydroxide, and twice fractionally distilled over sodium in an atmosphere of nitrogen. It was used for *conductometric studies* in a nonaqueous system.

Copp and Findlay [1655] purified triethylamine for *thermodynamic studies of binary systems* by fractionally distilling at a reflux ratio of 8:1 in a 30-plate column that was functioning adiabatically. The amine was first distilled from potassium hydroxide pellets. A middle fraction was dried with sodium and again distilled.

For purification for *calorimetric studies*, see Wadso [7823], butylamine. For purification for *vapor pressure studies*, see Majer [4842], propylamine.

SAFETY

See also Amines, General Comments. Kulagina and Kochetkova [4311] found the lethal concentrations of triethylamine in air to be: LC_{100} 7.5, LC_{50} 6, and LC_{min} 5 mg/L. Concentrations as low as 0.03 g/L caused chronic effects. Tkachev [7509] reported that short-term inhalation of triethylamine disrupts the human nervous system and recommended a maximum permissible atmospheric concentration of 0.16 mg/m^3.

429. Tributylamine

Vogel [7770] purified tributylamine for *physical property studies*. A commercial sample was shaken with about one-half its volume of potassium hydroxide pellets, filtered, and then distilled from sodium into a flask with a distillation side arm.

SAFETY

Loit and coworkers [4651] found that inhalation of tributylamine vapors by rats was not toxic. Inhalation of 0.3 mg/L by humans was not irritating.

430. *N, N*-Dimethylaniline

SAFETY

The *threshold limit value* is 5 ppm, 25 mg/m^3 [142].
The *minimum ignition temperature* in air is 371° [6402].

431. Pyridine

Pyridine is obtained from the destructive distillation of coal and from petroleum. A limited amount is found in "bone oil." It is available commercially in three regular grades and several grades for special purposes. Pyridine is a weak base in water and a strong base in acetic acid. It forms pyridinium salts with a wide variety of acids. It is very hygroscopic.

Coulson and Ditcham [1676] studied the impurities occurring in pyridine. *o*-Xylene and tetrahydrothiophene were both isolated and identified. No other sulfur compounds were positively identified, although others were present. Methods of purification were given, particularly for reducing the sulfur-bearing impurities.

Whitford [8017] separated pyridine from some of its homologues by utilizing the difference in the solubilities of the oxalates. The impure pyridine was slowly added to a stirred solution of oxalic acid in acetone. The pyridinium oxalate was precipitated, filtered, washed with cold acetone, regenerated, and isolated. The oxalate has a low density and consequently a large bulk; the method is convenient only for small quantities.

Arndt and Nachtwey [430] found that pyridinium perchlorate could easily be crystallized to a high purity. Unlike the perchlorates of pyridine homologues, it is sparingly soluble in water. Technical pyridine was dried with potassium hydroxide and fractionally distilled. The distillate was neutralized with 6 N hydrochloric acid until the odor of pyridine could not be detected. A slight excess of 6 N perchloric acid, based on the hydrochloric acid used, was added and the mixture allowed to stand for several hours in the cold. The crystals were removed by filtration, washed with cold water or 20% perchloric acid solution, and dried; mp 288°. The base was liberated with 50% sodium hydroxide or ammonia; the latter is preferable because no water is involved. A slow stream of ammonia was passed into the cooled flask containing the dry pyridinium perchlorate until only a fraction of the salt remained. Dry air was passed through the warmed flask to remove the ammonia. The product was distilled at reduced pressure. Zecherl [8268] stated that this method yielded a very dry pyridine *but is very dangerous*.

Wilson and Hughes [8104] purified 800 mL of technical pyridine by stirring for 24 hr with a mixture prepared by gently grinding 20 grams of technical cerium(IV) sulfate with 15 grams of anhydrous potassium carbonate. Heating was not necessary. The mixutre was filtered through a Buchner funnel and fractionally distilled. A column of at least 12–14 theoretical plates was recommended.

Heap and coworkers [3249] dried crude pyridine over sodium hydroxide, fractionally distilled several times, and purified by means of the coordinate compound with zinc chloride. The double salt was prepared as follows: One liter of pyridine was added to a mixture of 845 grams of zinc chloride dissolved in a mixture of 600 mL of water, 345 mL of concentrated hydrochloric acid, and 690 mL of absolute alcohol. The crystalline mass was filtered with suction, recrystallized twice from dry ethanol, dried, and decomposed with 26.7 grams of sodium hydroxide per 100 grams of dry precipitate. The pyridine was dried over sodium hydroxide and distilled. Bryant and Wardrop [1217] used the same method of purification except that the freed pyridine was dried with potassium hydroxide and distilled from barium oxide. They commented that the method was "tedious and wasteful." The pyridine was used for *specific conductance studies*.

Ralph and Gilkerson [6027] distilled a reagent-grade pyridine from potassium hydroxide and then from barium oxide for *ion–solvent interaction studies*. Stewart and O'Donnell [7013] distilled a like-grade material from barium oxide for use as a solvent for the *determination of the H-function* of strongly basic systems.

Leis and Currans [4517] purified AR-grade pyridine for the *determination of dipole moments* by repeated fractional freezing, refluxing over freshly heated barium oxide, and distilling through a column at least 1 meter long packed with glass helices. The middle fraction, with a constant refractive index, was used for *physical measurements;* n_D 20° 1.5094, d 20° 0.9832. Another portion, from which the lower boiling material had been removed by fractional distillation, was purified as described; bp 114.5° at 760 Torr, n_D 25° 1.5067, d 25° 0.9786.

Müller and coworkers [5324] purified pyridine for *specific conductance studies* by allowing it to stand over freshly fused potassium hydroxide for a long time and by repeated fractional distillation.

The principal acidic impurity present in most basic solvents used in nonaqueous titrimetry is carbon dioxide. The solvent is adequate for precision titrations, in most cases, if this impurity is removed [6165]. A column of alumina such as that used by Harlow and coworkers [3167] is convenient (see piperidine). If the column is connected to a reservoir of the solvent, an adequate supply of low acid solvent is available at all times.

Pyridine and pyridine mixed with another solvent are widely used as the sample solvent in acid–base titrimetry because certain of the mixtures are such excellent differentiating solvents for acids. The technical 2° material is usually used with or without a blank correction, depending on the accuracy desired. Often a purified solvent with a low, or negligible, blank is desired or is necessary. Cundiff and Markunas [1762] purified technical-grade pyridine by letting it stand overnight over sodium hydroxide pellets and then flash-distilling for their study of the titration of strong acids. They purified the solvent by flash distillation from barium oxide for their study of the determination of the alkoxyl group.

Banick [593] developed a simple and rapid procedure for purifying large volumes of pyridine to a blank of less than 0.05 mL of 0.1 N titrant. Shake intermittently for 5 min in a 1000-mL, iodine-type flask 500 mL of pyridine and 20 grams of alumina and filter by suction through a 9.0-cm, glass-fiber filter paper containing a layer of 5.0 grams of diatomaceous earth (Celite grade is satisfactory).

Brown and Barbaras [1133] purified a reagent-grade pyridine by fractional distillation through a 15-theoretical plate column. The middle portion boiling at 114.9° at 750 Torr was collected and used for the study of the *steric nature of the ortho effect.*

Vapor diffusion coefficients and *collision parameters* were studied [3554] with pyridine purified as follows: It stood over freshly fused potassium hydroxide for 2 months and was distilled. It was fractionally frozen three times, each time rejecting one-third of the mother liquor and remelting the crystals. The final material was dried over freshly fused potassium hydroxide and distilled.

For purification for *critical properties*, see Ethers, General Comments [4091].

Kienitz [3998] partially froze 98.4% pyridine and obtained 99.9%m material after three cycles.

For purification by progressive freezing, see General Comments, beginning of this chapter, Dickinson and Eaborn [2000].

Hesse and coworkers [3357] purified pyridine for *optical studies* by column chromatography. Adsorbents used were silica gel, basic alumina, acidic alumina, and 1-1 silica gel–alumina.

Trachuk [7538] reported the pyridine could be effectively dried to less than 0.05%

Table 5.24 ACS Specifications for Reagent-Grade Pyridine

Distillation range (max)	$2°^{a}$
Solubility in water	To pass test
Residue on evaporation (max)	0.002%
Water (max)	0.10%
Chloride (Cl) (max)	0.001%
Sulfate (SO_4) (max)	0.001%
Ammonia (NH_3) (max)	0.002%
Copper (Cu) (max about)	0.0005%
Reducing substances	To pass test

^{a}Some supply houses now list a reagent grade with a 1° distillation range.

water by distilling as the ternary pyridine–benzene–water azeotrope; bp 69.3°. Berg and coworkers [785] found that toluene could be used successfully as a ternary azeotroping agent for drying pyridine.

The water may be reduced to less than 0.01% with Linde 3A or 4A molecular sieves [6165].

For special techniques of drying, see Zerewitinoff [8280], Müller and Brenneis [5323], and Timmermans [7471].

Eberius and Kowalski [2195] described a method for the dehydration of a methanol–pyridine mixture for the preparation of Karl Fischer reagent. Coulson and coworkers [1678] developed an infrared procedure for the determination of water in pyridine down to 0.002%v/v that is rapid and requires only a 0.1-mL sample. The 3450 cm^{-1} band depends on water concentration. The disadvantages of the other methods for the determination of water in pyridine are discussed.

Waldron [7870] reported a good color stability for pyridine that was refluxed for 0.5 hr with 20% of its volume of 28% sodium hydroxide and then fractionated.

Rosin [6263] and *ACS Reagent Chemicals* [137] give the specifications and test methods for reagent-grade pyridine. The ACS requirements are given in Table 5.24.

Cox and coworkers [1701] characterized pyridine for *heat of combustion studies* by the freezing curve.

The titratable purity may be determined potentiometrically to +0.05% in an acetic acid solvent system by the method of Markunas and Riddick [4897].

SAFETY

The vapor is irritating to the mucous surfaces and causes eye and nasal irritation. The inhalation of low concentrations that produce chronic poisoning is considered the greatest hazard. It affects the kidneys, liver, and bone marrow. Pyridine is absorbed through the skin and can cause acute or chronic conditions [7022].

The *threshold limit value* is 5 ppm, 15 mg/m^3 [7022].

The *explosive limits* in air are 1.81 and 12.40%v, the latter value at elevated temperature. The *minimum ignition temperature* in air is 482° [5712].

432. 2-Picoline

Wilkie and Shaw [8066] were the first to demonstrate that a pyridine base from selected cuts could be fractionally distilled to high purity in an efficient column. A lengthy purification process by preparing zinc chloride and mercury(II) complexes with 2-picoline had been previously reported as the best process; see Heap [3249], pyridine.

Kowalski and coworkers [4227] dried 2-picoline over potassium hydroxide and fractionally distilled it for *excess volume studies with alcohols.*

A commercially "pure" 2-picoline sample was purified by Biddiscombe and coworkers [854] for *thermodynamic and physical property studies.* Steam was passed through a boiling solution of the base and 22.6% sulfuric acid solution until 10% of the picoline had distilled with contaminating impurities. The pot residue was made strongly alkaline with sodium hydroxide and the picoline was then separated from the supernatant liquid and dried over sodium hydroxide. The final step was fractional distillation through a 50-theoretical-plate, gauze-packed column at a reflux ratio of 50:1. Purity was 99.87%m as determined from the freezing curve.

Kyte and coworkers [4359] purified 2-picoline for *physical property study.* A commercial sample was dried over powdered barium oxide and fractionally distilled. A portion of the distillate was added to an anhydrous solution of zinc chloride and HCl in absolute ethanol. The resultant white crystals of the picoline complex were filtered off and recrystallized twice from ethanol. The base was liberated by mixing with sodium hydroxide solution and steam distilling. Subsequent fractional and steam distillation produced pure 2-picoline; bp 129.5° at 770 Torr.

For determination of water, see Coulson [1678], pyridine.

SAFETY

Veselov [7736,7737] reported the LD_{50} of 2-picoline for mice, rats, and guinea pigs as 674, 790, and 900 mg/kg, respectively, and concluded that the compound is relatively nontoxic.

The *minimum ignition temperature* in air is 538° [6402].

433. 3-Picoline

Coulson and Jones [1679] studied the purification of commercial 3-picoline and found the 143–145° fraction contained about equal amounts of 3- and 4-picoline and 2,6-dimethylpyridine. These three components could not be effectively separated by straight or aqueous azeotropic fractional distillation. The acetic or propionic acid azeotropes boiled far enough apart to permit separation to 95–98% purity by fractional distillation. Further purification by fractional freezing gave much purer products.

Coulson and Ditcham [1676] developed the method for removing hydrocarbons and sulfur compounds from pyridine by passing steam through a boiling solution of the base and 22.6% sulfuric acid solution until 10% of the pyridine had distilled with the contaminants. They combined this method with azeotropic distillation and fractional freezing to purify 3-picoline; purity was 99.97%m.

Several fractional distillation methods have been reported for the separation of 3-picoline from other bases. A French patent [5907] stated that a 5% aqueous solution of ammonia or ammonium sulfate will effect separation. 3-Picoline may be separated from a mixture of pyridine bases by azeotropic rectification with aqueous formaldehyde solution [1484].

Cislak and Karnatz [1490] separated mixtures of bases by distillation with water to obtain a 90% purity. Each base was recovered from water by addition of solid potassium hydroxide. A similar technique is given in Privalov [5959].

Cislak and Wheeler [1492] heated 3-picoline containing 4-picoline and/or 2,6-lutidine with an oxide, oxyhalide, or halide of phosphorus or sulfur. This treatment converted the contaminants into complex, nonvolatile reaction products. The reaction mixture was made alkaline with sodium carbonate and 3-picoline was separated by distillation. Riethof and coworkers [6172] removed contaminants with phthalic anhydride rather than the phosphorus or sulfur compounds.

For purification by crystallization from the melt, see Krupicka [4289], quinoline. For purification for *excess volume studies*, see Kowalski [4227], 2-picoline.

434. 4-Picoline

4-Picoline from a primary producer was fractionally distilled by Leis and Currans [4517]. A middle fraction was converted to the 4-picolinium chloride salt which was recrystallized from hydrochloric acid solution. An aqueous solution of the salt was neutralized with excess potassium hydroxide, and the picoline was extracted with ethyl ether. The solvent was dried with anhydrous potassium carbonate, refluxed over barium oxide, and distilled; bp 143–145°.

Kyte and coworkers [4359] purified 4-picoline for *physical property study*. A commercial sample was added to an aqueous solution of zinc chloride and the mixture was then steam-distilled. The picoline complex residue was treated with sodium hydroxide and the base was isolated by steam distillation followed by the addition of sodium hydroxide, ether extraction, drying, and distillation. The base was further purified by five fractional freezing operations.

For removal of hydrocarbons and sulfur compounds, see Coulson and Ditcham [1676], 3-picoline. For separation by fractional distillation, see the French patent [5907], Cislak and Karnatz [1490], and Privalov [5959], 3-picoline.

For purification by crystallization from the melt, see Krupicka [4289], quinoline. For purification for *excess volume studies*, see Kowalski [4227], 2-picoline.

435. 2,4-Lutidine

Cislak and Cunningham [1489] separated 2,4- and 2,5-lutidine by fractional distillation as the hydrochlorides. For another separation by fractional distillation, see Cislak and Karnatz [1490], 3-picoline.

436. 2,6-Lutidine

Coulson and Ditcham [1676] purified 2,6-lutidine for *thermodynamic and physical property studies*. Equal amounts of a commercial sample, urea, and water were mixed and then heated under reflux until all solid dissolved and two liquid phases formed. The urea complex crystallized as the liquid was cooled and stirred. After another crystallization, the purified crystals were refluxed with a concentrated sodium hydroxide solution for 6 hr and then cooled. The lutidine phase was removed, dried with sodium hydroxide, and distilled; yield 78% with a purity of about 99%m. The final purification was made by slow fractional freezing followed by distillation to remove water; purity 99.89%m.

A similar purification with urea is given in Kyte [4359].

Brown and coworkers [1134] purified a commercial coal tar 2,6-lutidine for *study of its base strength*. It was fractionally distilled in a 70-plate column and a narrow boiling center fraction was collected. To 529 grams of the center cut was added 21 grams of boron trifluoride. The mixture was redistilled through the 70-plate column until little material remained in the flask except the addition compounds. The product distilled at a constant temperature and had a constant refractive index; yield 80% with a 99.8% purity by freezing curve.

Brown and Mihm [1136] purified 2,6-lutidine for *base strength determination* by fractional crystallization; purity 99.6% from the freezing curve. For another purification by crystallization from the melt, see Krupicka [4289], quinoline.

Vymetal and Kulhanek [7811] separated 2,6-lutidine from 3-picoline. A sample was mixed with 80% urea solution and heated until all solid dissolved. After cooling to 20°, crystallization occurred. The crystals were filtered, washed with benzene, and decomposed with 30% sodium hydroxide solution. 2,6-Lutidine was then extracted with benzene and fractionally distilled.

For separation by fractional distillation, see the French patent [5907], Cislak and Karnatz [1490], and Privalov [5959], 3-picoline.

437. 2,4,6-Trimethylpyridine

Wille and coworkers [8068] isolated 2,4,6-trimethylpyridine from a mixture of pyridine bases. The mixture was mixed with o-cresol, cooled to −10°, and centrifuged. The crystals were centrifuged at 3° with 5% of a wash liquid prepared from 2 mol of o-cresol, 1 mol of 2,4,6-trimethylpyridine, and 2% water. The final product was obtained by treating the crystals with sodium hydroxide and steam-distilling.

Engel [2264] separated 2,4,6-trimethylpyridine from all other bases as a hydrochloride salt in an acid medium.

2,4,6-Trimethylpyridine was purified by Mastrangelo [4959] for *calorimetric study*. A commercial-grade sample was fractionally distilled through a 25-mm × 3-ft Podbielniak column. The heart cut of the distillate was distilled three times in a nitrogen atmosphere.

A detailed purification using boron trifluoride is given in Brown [1134]; see also 2,6-lutidine.

438. Quinoline

Quinoline is produced from coal tar. It is made by several synthetic methods; one of the best known is Skraup's synthesis from aniline, glycerol, nitrobenzene, and sulfuric acid [2757].

Mastrangello [4959] purified refined quinoline by repeated precipitation of the phosphate, followed by washing the crystalline mass three times with methanol after filtering and drying overnight at 55°. The quinoline was liberated with sodium hydroxide. The wet base was fractionally distilled and the first 10% and the last 20% rejected. The material used for *calorimetric studies* was distilled *in vacuo* to remove the last traces of water.

Malanowski [4847] purified several quinoline bases for precise *vapor pressure studies*. They were distilled through a 30-plate adiabatic column with a Swietoslawski ebulliometric head. One-half of each fraction distilling within ±0.1° was converted to the chloride and twice recrystallized from the remaining base. The quinoline base was regenerated with sodium hydroxide and again fractionally distilled. The distillate condensing within 0.05° was collected. The ebulliometric purity by a Swietoslawski differential ebulliometer was 0.002°.

Krupicka [4289] purified quinoline by crystallization from the melt followed by fractional melting. Although the method yielded a sample of lower purity than those obtained by crystallization from a solvent, it was considered suitable for industrial use. A lab-scale purification of quinoline gave one cycle of greater than 99% purity.

Quinoline decomposes 1% per hr at 510–535° [3727].

SAFETY

Limited animal tests indicate that quinoline has a moderate degree of toxicity by ingestion, skin penetration, or inhalation [5713].

439. Isoquinoline

Malanowski [4847] purified commercial isoquinoline by fractional distillation through a 30-plate adiabatically jacketed column fitted with a Swietoslawski ebulliometric head. Half of each fraction collected within ±0.1° was converted into the chloride crystallized from the remaining solution of the free base. The isoquinolium chloride was recrystallized and then isoquinoline was recovered with sodium hydroxide, dried over solid sodium hydroxide, and redistilled as previously described. The fractions condensing within 0.05° were collected and used for *vapor pressure study*.

For purification by crystallization from the melt, see Krupicka [4289], quinoline.

Amides (440–450)
General Comments

The amides are excellent solvents both for organic and inorganic compounds with properties that make them of particular interest to both research and production. The second edition of *Organic Solvents* contained two amides; this edition contains 11, distributed among several structural types.

Peterson [5789] found that aliphatic amides are characterized by two melting points (different melting and freezing points). If the melt is allowed to cool slowly and is seeded just below the lower melting point, crystallization takes place at a sharply defined temperature. The melting point on remelting is higher. The curve for the melting point for acetamide was extrapolated to 69.5°. The slopes of the freezing and melting curves are different.

Gopal and Rizvi [2867] purified six amides as follows: Freshly ignited calcium oxide was added to the liquid and it was distilled at reduced pressure. The middle fraction was collected. The process was repeated until the specific conductance of the distillate was reduced to 10^{-5} to 10^{-6} ohm^{-1} cm^{-1}. They were stored in an amber bottle in a dry nitrogen box and used, when possible, immediately after distillation. *Physical properties* at different temperatures were determined on the amides. Five of the six amides that were used are included in this section: N-methylformamide, N,N-dimethylformamide, N-methylacetamide, N,N-dimethylacetamide, and N-methylpropionamide.

Campbell [1332] developed a method for the vapor-phase reaction at atmospheric pressure of carboxylic acids with a boiling point such that it can be vaporized without difficulty, not over 250°, and primary or secondary amines to give the corresponding amide with a high purity. This reaction is

$$RCOOH + HNR_1R_2 \xrightarrow{190-330°} RCONR_1R_2 + H_2O$$

R and R_1 may be H or an aliphatic group. R_2 is an aliphatic group.

There are little if any side reactions. The reaction takes place in a heated Pyrex-brand glass tube, 100 cm long by 22 mm id with an 8-mm od glass tube thermowell running the length of the reactor. The reactor is packed with Alcoa E-10 alumina or Davis Chemical Co. catalyst SMR-55-5138. The preferred amine–carboxylic acid ratio is 1.05. The optimum temperature and feed rate depends upon the reactants. If the reaction effluent is fed into, or near, the middle of a fractionating column, the lower section of which is maintained between the boiling point of water and that of the amide, conditions can be maintained to give an amide with a purity of 99 + % in the still pot.

For a method of preparing N-methylamides of C_1–C_9 carboxylic acids, see D'Alelio and Reid [1800] under N-methylpropionamide.

Loder [4646] describes a process for the preparation of N-alkylformamides in which an alcohol reacts with ammonia and carbon monoxide at 900-atm pressure and 210–260°.

Hobbs and Bates [3433] conclude from dielectric and dipole moment studies that N-unsubstituted amides exist principally in the *cis*-configuration and dimerize similarly to carboxylic acids:

$$
\begin{array}{c}
\qquad\qquad H \\
\qquad\qquad | \\
\quad O\cdots H-N \\
R-C \diagup \qquad\qquad \diagdown C-R \\
\quad\diagdown N-H\cdots O \diagup \\
\qquad | \\
\qquad H
\end{array}
$$

Carpenter and Donohue [1363] studied the crystal structure of peptides and concluded that N-monosubstituted amides, CONH, existed principally in the keto form. Shortly thereafter, Mizushima and Simanouti [5194] surmised from infrared and dipole moment studies that CONH existed mainly in the *trans* configuration. Worsham and Hobbs [8162] inferred from dipole moment that a chain association of the N-monosubstituted amides exists in the following form:

$$
\begin{array}{c}
\qquad R_1 \\
\qquad | \\
\quad C=O\cdots H-N \diagup R \\
H-N \diagup \qquad\qquad\qquad \diagdown C=O \\
| \qquad\qquad\qquad\qquad | \\
R \qquad\qquad\qquad\qquad R_1
\end{array}
$$

Kotera and Shibata [4217] confirmed the *trans* configuration from ultraviolet and dipole moment studies. Davies and Thomas [1848] studied energies and entropies of association of amides in benzene solution and stated that the chain association of CONH amides is a consequence of the *trans* configuration of the monomer. The *cis* configuration of monomeric amides, for steric reasons, would be essentially confined to cyclic dimerization. Suzuki [7097] questioned whether amides, such as *N*-methylformamide, contained any of the *cis* form. LaPlanche and Rogers [4407] continued the configuration study with nuclear magnetic resonance; they believe that the CONH amides exist substantially as the *trans* configuration.

Machida and coworkers [4805] studied the infrared spectra of *n*-fatty acid amides and their *N*-deuterated compounds. The characteristic absorption frequencies for the $-CONH_2$ and $-COND_2$ groups are given and discussed.

Reid and Vincent [6110] analyzed the criteria for purity of amides for usage as electrochemical solvents.

Specific Solvents

440. Formamide

Röhler [6235] in 1910 pointed out that formamide resembles water in that both have a high dielectric constant. From his study of the amide as a *solvent for inorganic salts* and as an *electrolysis solvent*, he concluded that the salts were solvated in a manner similar to water. Walden [7837] studied formamide as an ionizing solvent and concluded that it imitates, in a remarkable manner, the physical characteristics and constants of water. He found that, when binary salts were used as solutes, the salts may exhibit a degree of ionization greater than in aqueous solution. Strong organic acids do not become markedly ionized in this solvent.

Formamide is an excellent ionizing solvent soluble in water, lower alcohols, and glycols and insoluble in hydrocarbons, chlorinated hydrocarbons, and nitrobenzene. It dissolves casein, gelatin, zein, animal glue, and related water-soluble gums and resins. The chlorides and some sulfates and nitrates of copper, lead, zinc, tin, nickel, cobalt, iron, aluminum, and magnesium are soluble. Toops [7525] found that Drierite could not be used as a drying agent. The salt dissolved and the solution set to a gel on standing overnight.

Formamide is a very weak base, even weaker than water. It is easily hydrolyzed to the parent acid and ammonia by acids, bases, and enzymes. It reacts with peroxides, and the amide hydrogen may be replaced by acid halides, acid anhydrides, and esters. On heating with alcohols, it gives formic esters. Strong dehydrating agents remove a molecule of water to form the corresponding nitrile. Maxim and Mavrodineanu [5004] reported that it reacts with Grignard compounds. Berthelot and Gaudechon [818] found that formamide was decomposed by exposure to an ordinary incandescent lamp.

Smith [6828] found that neither fractional distillation nor fractional crystallization would purify formamide to a high state of purity. He recommended that it be purified by fractional distillation to a freezing point of 2.25° and then by fractional crystallization to a freezing point of 2.55°. Formamide is very hygroscopic, and precautions must be taken to exclude water so that the purified product will have a freezing point of 2.55° and a specific conductance of 1.98×10^{-6} ohm^{-1} cm^{-1}.

Practically all of the simpler amides possess the above described properties to a greater or lesser degree.

Berger and Dawson [786] studied the *determination of chlorides* in formamide solutions using a commercial material that had been purified by the same general method of Smith [6828] above, except that the order of operations was reversed. The material was fractionally frozen several times in a specially constructed flask. The impure mother liquor was drained and the solid melted. The liquid was dried overnight with calcium oxide and then slowly distilled at about 1 Torr. The middle 80% was collected and stored in "low actinic" glass-stoppered bottles that had been thoroughly purged with nitrogen.

Verhoek [7725] found that commercial formamide often contains acids and ammonium formate. The latter is so largely solvated that the ammonia distills, leaving the formic acid, which cannot be separated satisfactorily from the amide by distillation. Water present during the distillation hydrolyzes the formamide. He found the following method of purification to be the most satisfactory for preparing a material for a solvent for the *determination of the strength of acids*.

A few pieces of bromothymol blue were added to the formamide and the acidity exactly neutralized with sodium hydroxide. The neutral liquid was heated to 80–90° under reduced pressure, the ammonia and water pumped off, and the amide again neutralized. The procedure was repeated four or five times until the liquid remained neutral when distillation began. Sodium formate was added to the still pot and the formamide distilled between 80 and 90°. The distillate was neutralized and distilled as before and the last four-fifths collected; mp ~2.2°, $\kappa = 5 \times 10^{-5}$ ohm^{-1} cm^{-1}. The distillate was then fractionally crystallized in a water- and carbon dioxide-free atmosphere $\kappa = 1-2 \times 10^{-6}$ ohm^{-1} cm^{-1}. The liquid from the crystallization was again neutralized, distilled, and fractionally crystallized.

The pure material is not stable. If a stream of air, purified by passing through sodalime, calcium chloride, phosphorus pentoxide, and cotton, is bubbled rapidly through the formamide, the stability is greatly improved. The lowest *specific conductance* found was 6.2×10^{-7} ohm^{-1} cm^{-1}. The solvent was considered sufficiently pure when the specific conductance was 3.5×10^{-6} ohm^{-1} cm^{-1}.

A measure of the hygroscopicity was made by weighing 5.6 grams into a small beaker with a free surface of about 9 cm^2. The beaker was exposed to the air in the laboratory (humidity not given) and the following gains in weight noted: 1 hr 0.2%, 5.5 hr 1%, 141 hr 10%. It was found that water up to 1% had no appreciable effect on conductometric or potentiometric experiments.

"Chemically pure"-grade formamide was found by Letaw and Gropp [4534] to be *polarographically pure* in the range they studied, after being held at a pressure of 1 Torr, or less, for 2 hr.

Conductometric studies of solutions of hydrogen chloride and potassium chloride in formamide were made by Dawson and coworkers [1870] purified from a commercial material by distilling several times from calcium oxide at a pressure of less than 0.1 Torr between 51 and 58°. Decomposition of the vapors was held to a minimum by using a fractionating column in the region where the vapors were carried over a short path and condensed immediately at room temperature. The specific conductance ranged from 3×10^{-5} to 1×10^{-5} ohm^{-1} cm^{-1} at 20°.

Bates and Hobbs [676] determined the *dipole moment* of a good commercial grade purified as follows: The amide was dried over anhydrous sodium sulfate and vacuum-distilled. The middle portion of the distillate was fractionally crystallized three times, about one-half of the material being discarded in each crystallization; fp $2.50 \pm 0.06°$.

Leader [4438] purified a commercial material for *dielectric constant measurements*. The material was treated with 5 grams of calcium oxide per liter and distilled at a pressure of 1 Torr in a still, affording essentially no fractionation. Two additional distillations of the middle fractions were made with lime treatment each time. The final distilled product was statically fractionally crystallized three times in a container that was protected from carbon dioxide and water; fp $2.3-2.4°$; $\kappa = 4 \times 10^{-6}$ ohm^{-1} cm^{-1}. The conductivity increased to 1×10^{-5} in 5–6 hr.

Notley and Spiro [5523] developed a method to produce formamide of low water and ion content in large laboratory amounts for *electrochemical investigation*. The chief impurities in high-grade commercial formamide were found to be about 0.04 M water and an electrolyte concentration of about 0.01 M.

Skold and coworkers [6798] further purified analytical-grade formamide for *heat capacity studies* by fractional freezing three times, followed by drying over 4A molecular sieve and fractionally distilling under nitrogen at reduced pressure.

Removal of Water. A 5-cm column was packed with 1 kg of 3A molecular sieves in $\frac{1}{16}$-in. pellets. The column was heated electrically to about 60° to increase the rate of water removal. Formamide was passed through at the rate of 0.2–0.25 L/hr. The water content was found to be about 0.008 to 0.01 M. The column will dry about 20 L of the liquid. The sieves were washed with water, dried at 100°, and heated at 360° in a column through which a stream of nitrogen was flowing.

Removal of Ions. Ions were removed by a mixed bed resin that had been especially treated. Two hundred-fifty grams of Amberlite IR-120 cation exchange resin was well washed with water and taken through several regenerations with 2 M hydrochloric acid and sodium hydroxide. It was washed with ethanol until the effluent was colorless, taken through a further cycle and let soak in water. Amberlite IRA-400 anion exchange resin was similarly treated. The IR-120 was drained of water, put in a column, and washed with formamide (~ 7 bed-volumes) until the effluent contained less than 0.05 M water. One bed-volume of 2 M sulfuric acid in formamide was allowed to stand 1 day on the resin. Two liters of 2 M sulfuric acid in the solvent was passed through the resin. The anion resin was similarly treated with 2 M sodium formamide in formamide. The two resins were washed with formamide until the respective effluents were only slightly acidic or basic. The resins were mixed and about 6 L of formamide were passed through: $\kappa = 2 \times 10^{-7}$ ohm^{-1} cm^{-1}, fp 2.3°. (It is recommended that one read the article if the use of the procedure is considered. Only the essential steps are in the above description.)

Cole and coworkers [1578] purified formamide for use as a solvent for the *paper chromatography of steroids* by shaking 4 hr with a 20% charcoal suspension and filtering. Taylor and Davis [7194] used formamide as a solvent for the study of the *velocity of esterification*. A commercial material was dried over anhydrous sodium sulfate for several days and distilled 15 times at 30–40 Torr. It was slightly alkaline and odorless; fp 2.20°.

CRITERIA OF PURITY

Rosin [6263] gives specifications and test methods for reagent-grade formamide. The requirements are listed in Table 5.25.

Table 5.25   Rosin's Specifications for Reagent-Grade
Formamide

Specific gravity (25°/25°)	1.132–1.134
Freezing temperature (not less than)	2°
Miscibility with water	To pass test
Ammonia, amines, etc.	To pass test
Evaporation residue (max)	0.1%
Water (max)	0.3%

Smith [6828] found a freezing point of 2.55° to characterize a very pure substance with a specific conductance of 1.98×10^{-6} ohm^{-1} cm^{-1}. Notley and Spiro [5523] found a specific conductance of less than 2×10^{-7} ohm^{-1} cm^{-1} for a material especially prepared for electrochemical investigation; however, its freezing point was 2.3°.

SAFETY

No human injuries have been reported for formamide. It was absorbed through the skin of guinea pigs. The LD_{50} was found to be <5 mL/kg with only slight irritation [5713].

Zaeva and coworkers [8249] reported the oral LD_{50} for mice and rats as 3.15 and 6.10 g/kg.

The *threshold limit value* is 20 ppm, 30 mg/m^3 [142].

The *minimum ignition temperature* is 601° [6402].

441.   *N*-Methylformamide

Leader and Gormley [4439] led gaseous methylamine into 98–99% formic acid until its proper weight had been absorbed. The salt solution was heated to remove water. The resulting amide was vacuum-fractionated through a 1×35-cm glass-helices-packed column and successively crystallized to a constant melting point. Further purification using dried Amberlite IR-120 resin in the hydrogen form gave a satisfactory product for *conductivity measurements*; $\kappa = \sim 5 \times 10^{-5}$ ohm^{-1} cm^{-1}.

Held and Criss [3272] studied the *heat of solution* of selected alkali metal halides in anhydrous *N*-methylformamide purified as follows: It was treated with sodium hydroxide pellets and barium oxide for at least 4 hr. It was decanted into a distilling flask containing barium oxide and distilled at 1 Torr through a 35-cm Vigreux tube. The specific conductance was $3–5 \times 10^{-6}$ ohm^{-1} cm^{-1}. Water by the Karl Fischer method was found to be 0.003%.

LaPlanche and Rogers [4407] purified a commercial material for *nuclear magnetic resonance studies* by drying with anhydrous sodium sulfate and fractionally distilling *in vacuo*.

French and Glover [2541] purified *N*-methylformamide for the study of *specific conductance* in solvents of high dielectric constant. It was shaken with phosphorus pentoxide, filtered through glass wool, and distilled at about 1 Torr; bp 51°. This treatment was repeated three times, followed by two distillations without treatment. The specific conductance normally was in the range of $1.0–2.0 \times 10^{-6}$ ohm^{-1} cm^{-1}. This method of purification reduced the change of conductance with time to a very small amount.

442.   *N,N*-Dimethylformamide

N,N-Dimethylformamide is a colorless mobile liquid miscible with water and organic solvents [4902, 6309]. The slight amine odor results from hydrolysis by absorbed water. It is a uniquely versatile and powerful solvent that has a wide liquid range, good chemical and thermal stability, a high polarity, and a wide solubility range for both organic and inorganic compounds. It is only slightly basic. It is considered to be thermally stable at its boiling point. Above 350°, decomposition may occur to dimethylamine and carbon monoxide [2139]. Thomas and Rochow [7406] found that it decomposes slightly at the normal boiling point. Acidic and basic materials catalyze the decomposition at lower temperatures. It will decompose at room temperature when in contact with basic substances such as solid sodium or potassium hydroxide. It is somewhat sensitive to ultraviolet radiation; dimethylamine and formaldehyde are the main decomposition products.

N,N-Dimethylformamide was prepared by Campbell [1332] using an alumina catalyst; see Amides, General Comments.

Dawson and coworkers [1868] used *N,N*-dimethylformamide as a solvent for studying *limiting equivalent conductance*. They purified by distilling under reduced pressure; κ = 1.83×10^{-6} ohm^{-1} cm^{-1}, n_D 25° 1.4294.

Chan and Valleau [1412] observed that it was the usual practice when preparing water-free *N,N*-dimethylformamide to treat it with a drying agent and distill under reduced pressure. They shook a reagent grade (water below 0.05%) for 12 hr with Linde-type 4A molecular sieves. The liquid was decanted and distilled in a nitrogen atmosphere in a 50-plate bubble cap column with a head pressure of 20 Torr; κ = 3.5×10^{-7} ohm^{-1} cm^{-1} at 25°.

Specific conductance studies of quaternary ammonium bromides were made by Sears and coworkers [6545] in *N,N*-dimethylformamide that had been purified by allowing it to stand in contact with potassium hydroxide for several days, taking a large center cut from an atmospheric distillation, and finally retaining a cut from a 10 Torr fractional distillation. Criterion of purity; κ = $0.6–2.0 \times 10^{-7}$ ohm^{-1} cm^{-1}.

Brummer [1194] purified commercial *N,N*-dimethylformamide as a solvent for *temperature and pressure coefficient of ionic conductance*. It was dried for 3 days over type 4A molecular sieves. During this time, several changes of sieves were made; finally the *N,N*-dimethylformamide was run slowly down a column of fresh sieves into a still that had been flushed with dry nitrogen. It was fractionally distilled through a 1-meter, Penn State-type column packed with "Fenske helices" and collected as soon as the specific conductance had reached 0.8×10^{-7} ohm^{-1} cm^{-1}, usually after 200 mL had distilled from a 2000-mL charge. The conductance steadily decreased during the course of the distillation, and the final liter had a specific conductance of $0.2–0.5 \times 10^{-7}$ ohm^{-1} cm^{-1}. Dry nitrogen was passed through the apparatus during the distillation.

Thomas and Rochow [7406] studied the *conductance of ionophoric compounds* in the solvent and state "... it is very difficult to purify." Trace amounts of proton-releasing impurities have a profound effect on the observed conductometric behavior of any solute that yields a hydrogen halide as a solvolysis product. Every effort was made to eliminate such impurities or to estimate their concentrations, so that their effects could be taken into account in the interpretation of the results. Alcohols, water, and primary and secondary amines are deleterious impurities in this respect.

It is doubtful that distillation alone can remove water from *N,N*-dimethylformamide.

When 20% water is added to the amide, a chemical reaction takes place above 20°. It is recommended that a chemical method be used to remove protonic impurities, and necessarily one that does not cause decomposition.

Technical-grade N,N-dimethylformamide was first dried by adding about 10%v of benzene, which had been dried over calcium hydride. The benzene–water azeotrope was removed by distillation at atmospheric pressure. (*Author's note*: slow heating recommended to prevent superheating.) N,N-Dimethylformamide dried in this manner is referred to as "benzene-dried."

ACS reagent-grade magnesium sulfate was heated overnight at 300–400° and 25 grams was shaken with each liter of benzene-dried N,N-dimethylformamide for 1 day. A comparable amount of the heated magnesium sulfate was added and the amide was distilled at 15–20 Torr in a 3-ft, vacuum-jacketed column packed with "steel" helices. The middle fraction of the distillate was used; $\kappa = 0.9\text{--}1.5 \times 10^{-7}$ ohm^{-1} cm^{-1}. The specific conductance was reduced to $0.3\text{--}0.9 \times 10^{-7}$ when a middle fraction was treated with 50 grams/L of Merck chromatographic-grade alumina powder that had been heated overnight at 500–600°. The solvent was distilled from an additional portion of the heated alumina at 5–10 Torr and the middle fraction retained.

Libbey and Stock [4583] purified N,N-dimethylformamide for use as a solvent in *acid–base titrimetry* by twice azeotropically distilling a reagent-grade sample with dry benzene, drying over barium oxide, and then vacuum distilling over nitrogen at 10–12 Torr. The purified amide was collected in a blackened receiver and dispensed under dry nitrogen. Batch yields were about 50%.

It is pointed out that specific conductance is no criterion of purity with respect to freedom from covalent impurities. Methods are given for estimating these impurities.

Prue and Sherrington [5973] removed the major part of the water from reagent-grade N,N-dimethylformamide as the benzene azeotrope [7406]. The product was stored in glass bottles with glass stoppers. About 800-mL portions were shaken with phosphorus pentoxide for 3 days; about 10 grams of fresh oxide was added each morning. The amide was decanted and shaken with potassium hydroxide pellets to neutralize the formic acid. It was fractionally distilled at 50–64° at 15–30 Torr with a slow stream of dry nitrogen passing into the liquid. The middle fraction was collected for use. The water was estimated to be 4×10^{-3} M by reaction with triphenylsilyl chloride. The purified material was used for *conductance studies to test the Fuoss–Onsager equation* and to determine the *size of ions* in N,N-dimethylformamide.

The high solvent capacity of N,N-dimethylformamide combined with its transparency in the visible and near ultraviolet regions of the spectrum has accounted for its serviceability in spectrographic work. Zaugg and Schaefer [8264] purified N,N-dimethylformamide by the modification of a combination of the methods of Thomas and Rochow [7406] and Prue and Sherrington [5973] for *spectrophotometric studies*. Two liters of reagent grade N,N-dimethylformamide was mixed with 400 mL of reagent-grade benzene and distilled at atmospheric pressure until the vapor temperature reached 130°. The undistilled liquid was transferred to two 1-L bottles that had been flushed with dry nitrogen. About 25 grams of phosphorus pentoxide was added to each bottle and the bottles closed with Teflon-covered rubber stoppers, then shaken for about 3.5 hr. After the solid had settled, the liquid phase was transferred to two other dry nitrogen-filled bottles and 50 grams of potassium hydroxide pellets was added to each. They were shaken for about 1.5 hr. The liquid was decanted into a 3-L, round-bottom flask with side arm, filled with nitrogen, and distilled at 20–25 Torr in 1 atm of dry nitrogen. The dry nitrogen was bled into the flask through a fine capillary tube during the distillation. The distillation was repeated and the center cut taken for use.

The *vacuum ultraviolet spectrum* (1500–2340 Å) was determined by Hunt and Simpson [3587] on a "quite pure" N,N-dimethylformamide that was dried over magnesium sulfate for 24 hr and then over potassium hydroxide to remove water and formic acid. It was distilled at atmospheric pressure and the middle fraction used; bp 152.6° at 760 Torr, $n_D25°$ 1.4281.

Crompton and Buckley [1722] used N,N-dimethylformamide as a *solvent for polarographic studies*. The high-purity material obtained from the chemical supply houses is sufficiently pure. A blank polarogram will show whether or not a purification is necessary. The method of Sears and coworkers [6545] will produce satisfactory material.

A polarographically active impurity, present even in high-purity grades of N,N-dimethylformamide, was removed by Oelschelaeger and coworkers [5557]. The amide was passed at a rate of 100 mL hr^{-1} through a glass column 60-cm × 22-mm packed with three sections of 16 grams of crystalline silica gel (0.05–0.2 mm) separated by two sections of 25 grams each of basic alumina (activity 10). The first 50 mL of effluent was discarded. The purified dimethylformamide was stable for several months. No interference was noted on polarographic waves given by diazepam in Britton-Robinson buffer solution at pH 1.8.

N,N-Dimethylformamide has been purified as a *solvent for nonaqueous titrations* by treating with air-dried Dowex 1-X10(OH form) resin [5281].

Bender and coworkers [759] purified N,N-dimethylformamide, bp 148–150°, for *solvent use in thin-layer chromatography* by distilling over sodium bicarbonate.

The du Pont de Nemours Co. [2139] found that molecular sieves can be used to lower the water content from 0.5–1.0% to a few parts per million. Ion-exchange resins have been used successfully to remove ionic impurities. Calcium hydride has been used successfully as a drying agent.

Water in N,N-dimethylformamide may be determined by the Karl Fischer method [2139]. Infrared absorption, generally at 9.2 μm, provides an accurate and fast method of analysis.

SAFETY

N,N-Dimethylformamide was found by Martelli [4908] to affect the kidneys, liver, and digestive tract of guinea pigs when given orally or subcutaneously. Liver disturbances were found in workers subjected to vapors. There were numerous complaints of stomachache, headache, loss of appetite, and nausea from workers exposed to less than 20 ppm [7022]. The Russians apparently have done considerable toxicological investigation of N,N-dimethylformamide. In 1962, 0.001 mg/L was recommended as the MPC (TLV) and 10.0 mg/m^3 was recommended in 1964 [3848,4185, respectively]. An extensive study of the toxicological properties has been made by du Pont [2139]. N,N-Dimethylformamide is capable of producing both acute and chronic effects. It is absorbed through the skin and by breathing the vapors.

The *threshold limit value* is 10 ppm, 30 mg/m^3 [7022].

The *flammable limits* in air are 2.2 and 15.2%v. The *minimum ignition temperature* in air is 445° [2139].

443. Acetamide

Acetamide is a colorless deliquescent crystalline substance. It is odorless when pure but otherwise has a mousy odor. It is an excellent solvent for both organic and inorganic compounds.

Walker and Johnson [7875] in 1905 apparently were the first to recognize the unusual solvent and ionizing ability of acetamide. They *studied the compound as a solvent and as an ionizing solvent*. It was recrystallized twice to remove the chief impurities, ammonium acetate and water. The specific conductance at $100°$ was 4.3×10^{-5} ohm^{-1} cm^{-1}. They were of the opinion that the conductivity could be lowered comparable to that of other liquids by additional crystallizations. See also Walden [7841]. Menschutkin [5058] in 1907 followed with additional information of acetamide as a solvent for inorganic compounds and showed that acetamidates are similar in nature to hydrates. Stafford [6955] determined the approximate solubility of about 400 organic and 200 inorganic compounds and reported his findings in 1933. He concluded that acetamide has a wider solvent ability than any other substance that had been reported.

An extensive study of liquid acetamide as a solvent was made by Jander and Winkler [3673] in 1959. There is not much new that could not have been predicted by considering earlier work and the general characterization of amides. The study does present a useful panorama of acetamide as an ionizing solvent, showing its similarity to water in many respects. The amide ionizes and solvates, therefore, giving rise to a system of acids and bases that permits conductometric, potentiometric, and similar types of measurements.

Müller [5313] studied the *stable and unstable forms* of acetamide and determined the *heats of fusion and heats of crystallization*. Many of the melting points that had been reported were low because acetamide absorbs water from the air. See also Peterson [5789], Amides, General Comments.

A refinement of the classical method for the preparation of acetamide is presented by Coleman and Alvarado [1587] who reacted acetic acid with the proper amount of ammonium carbonate. The product was fractionally distilled and the desired fraction recrystallized from benzene–ethyl acetate solvent.

Kirsanov and Zolotov [4036] heated acetic acid and sulfamide in pyridine 3 hr at $100°$. It was treated with sodium carbonate and extracted with acetone; yield 84%.

Wagner [7830] is of the opinion that complete purification of acetamide by distillation at atmospheric pressure probably is impossible. Hitch and Gilbert [3431] observed that acetamide distilled at atmospheric pressure had a lower melting point than that purified by crystallization. The high temperature causes some dehydration to acetonitrile and water, which lowers the melting point of the distillate. Wagner agrees and presents a purification procedure by crystallization that yields a product of crisp, well-formed, needle crystals that are dry without odor. He discusses the disadvantages of the several solvents that are used. Ethyl ether gives the best material but the solubility is so low at the solvent's boiling point that the preparation of only small amounts of the amide is practical. The solubility in chloroform, benzene, and a mixture of benzene and ethyl acetate is so great that their saturated solution "sets" and the crystals and mother liquor cannot be separated satisfactorily. He combined the solvents of excessive and sparse solubilities to produce the following simple purification procedure.

Dissolve the acetamide in hot methanol in the ratio of 1 gram of amide to each 0.5 mL of solvent. If the solution is not clear, filter rapidly with suction or through cotton into a vessel that will permit a 10-fold dilution. Wash the flask and filter with an amount of warm methanol to bring the ratio to 1 gram of amide per 0.8 mL of alcohol. Measure a volume of ethyl ether of 8–10 mL for each gram of acetamide. Add the ether slowly at first and with stirring, but avoid splashing. Crystallization usually begins when about one-fourth, or slightly more, of the ether has been added. Seed at this point, if necessary, to start crystallization. When crystallization starts, stop stirring and stop ether addition; allow the liquid to rest until the crystallization appears to be about complete. Add the remainder of the ether rapidly with vigorous stirring and cool in an ice bath. Filter the crystals rapidly with suction. Wash once with cold ether. Continue the suction only long enough to remove most of the ether. Finish the solvent removal and drying *in vacuo* or over sulfuric acid.

Reagent-grade acetamide was purified by Bates and Hobbs [676] for *dipole moment studies* by crystallization from methyl acetate, an equal volume of ethyl acetate and benzene, and finally from benzene. It was then sublimed and stored over phosphorus pentoxide for 2 weeks. The very dry material was difficult to handle during weighing because a high static electrical charge had accumulated.

Acetamide was purified by Kumler and Porter [4321] for *electrical measurements* by distillation, once crystallizing from ethyl acetate, subliming in vacuum, crystallizing once from ethyl acetate, and three times from chloroform. It was dried at 70° and kept in a desiccator over phosphorus pentoxide. The product was odorless and melted at 80.0–80.1°. Belladen [750] measured the *conductance of some electrolytes* in liquid acetamide. The amide was prepared from ammonium acetate and acetic acid. It was distilled repeatedly at 220 Torr, shaking the middle fraction with anhydrous ether and vacuum-drying; specific conductance at 90° was 1.5×10^{-5} ohm^{-1} cm^{-1}.

Hirano [3418] measured the *specific conductance* of acetamide that had been fractionally distilled, recrystallized three times from benzene by the Lewin and Vance [4560] method, and dried in an Abderhalden drier for 5 hr; 83.2° $\kappa = 8.8 \times 10^{-7}$, 87.5° $\kappa = 1.0 \times 10^{-6}$ ohm^{-1} cm^{-1}.

Friend and Hargreaves [2582] purified acetamide for *structural studies by viscosity function* from material that was crystallized twice from a mixture of benzene and dry ethyl acetate.

Davis [1858] recommends acetamide as the solvent for laboratory experiments in general and physical chemistry for freezing point depression for *molecular weight determination* and for *solubility experiments*. Reasonably accurate results have been obtained by high school and college freshmen chemistry students with simple equipment and mass measurements to 1 mg and temperature to 0.1°. The curve for acetamide characteristically exhibits a short supercooling period and a well-defined freezing plateau. He points out that acetamide can serve as a solvent for research-oriented experiments.

Acetamide was successfully purified by Schildknecht and Vetter [6447] by zone-refining.

A commercial sample of acetamide was purified twice by crystallizing from ethanol followed by zone refining for *thermodynamic study*. No impurities were found by gas chromatography analysis. The purified amide was stored over phosphorus pentoxide [622].

444. N-Methylacetamide

N-Methylacetamide like all N-alkyl amides of the lower fatty acids has wide solvent application. It is easy to prepare and to purify. It is characterized by a high dielectric constant (165.5 at 40°) and extensive solubility and dissociating ability [1874]. The characteristics of solutions of a number of salts in this solvent have been studied; most soluble salts are completely ionized.

Bonner and Kim [975] determined that pure liquid N-methylacetamide has an average chain length of 4.8 units.

Dawson and coworkers [1871] prepared N-methylacetamide by reacting methylamine and acetic acid and subsequently heating to split out and distill water. The product was purified by fractional distillation, followed by five, or more, fractional freezing cycles using the procedure of Berger and Dawson [786]; see formamide.

Knecht and Kolthoff [4069] found that commercially available N-methylacetamide was not sufficiently pure for *polarographic studies*. At best, it contains methylamine and acetic acid. They studied methods of preparation and purification that would yield large volumes of a solvent sufficiently pure as a polarographic solvent. It was prepared by allowing an excess of acetic acid to react with 40% methylamine under reflux for 12–15 hr. A preliminary purification gave a "crude product." Two purification procedures were developed.

Purification Procedure 1. The crude product contained only acetic acid and methylamine. It was fractionally distilled under an atmosphere of nitrogen through an 18-in. heated column packed with porcelain saddles. The major fraction distilling at 204–206° was found to contain three impurities: methylamine, acetic acid, and impurity "X." These impurities were present in small amounts and were removed by extracting "X" with petroleum ether. Water was added to separate the phases. The water and dissolved petroleum ether were removed by simple distillation to 130°. The pot temperature was then allowed to cool to room temperature. About 10 mL of concentrated sulfuric acid was added and the mixture distilled at about 80° at 3–5 Torr. The acetic acid and most of the remaining water were removed by allowing the mixture to stand several hours over calcium oxide, after which it was filtered and vacuum-distilled.

Purification Procedure 2. The crude product was distilled slowly to 130° to remove as much of the methylamine, acetic acid, and water as possible. The methylamine was removed by adding sulfuric acid and distilling as in Procedure 1. There was too much acetic acid to remove with calcium oxide. Potassium oxide was added to a known excess and the mixture allowed to stand several hours, filtered, and vacuum-distilled. When potassium acetate started to precipitate, the mixture was cooled and filtered. The distillation of the filtrate was continued.

The criteria of purity were: (1) the polarographic range, determined with 0.1 $M(C_2H_5)_4NClO_4$ supporting electrolyte, between +0.35 and −2.75 volts; (2) water content by the Karl Fischer method; and (3) specific conductance less than 3×10^{-7} ohm^{-1} cm^{-1}.

It is recommended that the original article be consulted if this method is to be used.

N-Methylacetamide was vacuum-distilled from calcium oxide and fractionally crystallized through several freezing cycles by Dawson and coworkers [1872] as a *solvent for the determination of the dielectric constant of acetic acid;* $\kappa = 5 \times 10^{-8}$ ohm^{-1} cm^{-1} and n_D 32° 1.4282.

Lin and Dannhauser [4602] made *dielectric constant studies* with *N*-methylacetamide that was purified by distilling at reduced pressure through a 1.5×80-cm glass-helices-packed column at a high reflux rate.

French and Glover [2542] purified *N*-methylacetamide as a *solvent for the conductance studies of salts* as follows: The liquid was shaken with phosphorus pentoxide, filtered through glass wool, and vacuum-distilled. This treatment was repeated three times, followed by two distillations without the phosphorus pentoxide treatment. The normal specific conductance was 6×10^{-7}; the lowest found was 4.2×10^{-7} ohm^{-1} cm^{-1}. Their criterion of purity, other than low electrical conductance, and the one that determined the selection of the method of purification, was the constancy of the conductance with time. The electrical conductance of the purified material rose with time by all other methods tried.

Commercial *N*-methylacetamide was purified by Bonner and coworkers [974] by vacuum distillation, followed by zone refining in an apparatus designed especially for this research. Atmospheric contaminations were avoided during transfer by the use of a 50-mL hypodermic syringe. *N*-Methylacetamide is unstable in the presence of air. The purified material was used to determine the *freezing point, freezing point constant, heat of fusion, and dielectric constant.* Appreciable changes are believed to occur in the hydrogen bonding of the solvent when small quantities of impurities are present.

For purification by fractional freezing to constant maximum melting point, see Sears and coworkers [6544], ethylene carbonate.

Pure *N*-methylacetamide has no detectable odor.

Rey-Lafon [6141] determined the *infrared spectra* of *N*-methylacetamide in 33 inert solvents, including proton donors and acceptors. The bands for the CO and NH groups were similar to those in cyclohexanone and pyrrole in dilute solutions of the same solvents.

445.   *N,N*-Dimethylacetamide

High-molecular-weight resins and polymers are soluble in *N,N*-dimethylacetamide. The lower paraffinic hydrocarbons have only a limited solubility at ambient temperatures. It is thermally stable at its atmospheric boiling point and in the presence of acidic and alkaline substances. It decomposes above 350° to dimethylamine and acetic acid. It is not very hygroscopic. It acts as a dehydrohalogenating agent with halogen compounds. The heat of reaction with highly halogenated substances is large and the reaction may become violent, particularly in the presence of iron [2138].

Campbell prepared *N,N*-dimethylacetamide (see Amides, General Comments) over alumina catalyst at 230° [1332]. The crude amide was fractionally distilled to remove the impurities; yield 99.5% containing very small amounts of acetic acid and water.

Friend and Hargreaves [2582] prepared *N,N*-dimethylacetamide for *structural studies through viscosity* by heating dimethylammonium chloride with an excess of alkali, drying with solid potassium hydroxide. The dimethylamine was passed into a solution of acetyl chloride in benzene. The precipitated amide was filtered and fractionally distilled.

Schmulbach and Drago [6468] found that Eastman white label *N,N*-dimethylacetamide contained two impurities after shaking with barium oxide and fractionally distilling through a 2.0 × 40-cm, vacuum-jacketed Podbielniak column at reduced pressure. The center fraction boiled at 58.0–58.5° at 11.4 Torr and was retained. Chromatography indicated that the impurities were below 0.01%.

Drago and coworkers [2077] purified *N,N*-dimethylacetamide for *solvolysis studies of iron(III) complexes* by shaking with barium oxide and distilling at reduced pressure from barium oxide. The middle fraction was distilled at reduced pressure with calcium hydride.

Gutmann and coworkers [3040] purified *N,N*-dimethylacetamide as a *polarographic solvent* by fractionally distilling twice from calcium hydride at 10 Torr under nitrogen in an 80-cm Vigreux column, and finally without the hydride.

SAFETY

Toxicity by dermal contact and by chronic inhalation was investigated by Horn [3491]. It was found that the amide was readily absorbed into the bloodstream by both routes and that repeated exposure to relatively low levels resulted in liver damage. In [2138] du Pont cautioned that *N,N*-dimethylacetamide should be handled with care and that contact of the liquid with the skin and the eyes should be avoided. Spills on the skin should be immediately and thoroughly flushed with water; if splashed in the eye, irrigate for 30 min, and get medical care. Breathing *N,N*-dimethylacetamide vapors should be avoided.

Morenkova [5247] recommended a maximum permissible concentration of 0.2 mg/m^3. Kafyan [3809] reported the acute experimental LD_{50} of 3560 and 2580 mg/kg for rats and mice, respectively.

The *threshold limit value* of 10 ppm, 35 mg/m^3, has been recommended by du Pont [2138].

The *flammable limits* in air are: at 100°, 1.70-18.5%v; at 200°, 1.45–15.2%v; and the upper limit at 160° and 740 Torr, 11.5%v. The *minimum ignition temperature* in air is 420° [2138].

446. *N*-Methylpropionamide

Only the *trans* isomer of *N*-methylpropionamide was found by *configurational studies by infrared* made by La Planche and Rogers [4407] in a sample of the amide that had been purified by drying over anhydrous sodium sulfate and fractionally distilling *in vacuo*.

Leader and Gromley [4439] passed gaseous methylamine into propionic acid until an equivalent weight had been absorbed. It was heated at 160–170° to split out water. The pressure was gradually reduced and it was fractionally distilled *in vacuo* through a 1 × 35-cm glass-helices-packed column. It was further purified by successive crystallizations for *dielectric constant studies*. Hoover [3482] used the same preparation and purification for *conductance studies*, except that the final purification was by repeated fractional distillation at 5 Torr in a 4 × 1000-cm column packed with 3-mm helices.

D'Alelio and Reid [1800] prepared the *N*-methyl amides of the first nine members of the homologous series of *n*-RCOOH acids by slowly dropping the corresponding acid chloride into 3 mol of concentrated aqueous solution of methylamine at − 10 to − 20°

during the addition. Solid potassium hydroxide was added and the amide separated. It was then dried with solid potassium hydroxide and distilled. Ross and Labes [6272] prepared N-methylpropionamide by the D'Alelio and Reid method and distilled from calcium oxide before use for *solvolyses studies;* $\kappa = 8.3 \times 10^{-5}$ ohm^{-1} cm^{-1}, 0.002–0.008% water by Karl Fischer.

N-Methylpropionamide was prepared for Beer and coworkers [739] for *infrared spectra studies* by the reaction of propionic anhydride with aqueous methylamine. The amide was salted out, dried, and distilled; bp 104° at 16 Torr, n_D 24.1° 1.4313.

Dawson and coworkers [1869] prepared N-methylpropionamide as a solvent for the *study of solutions of potassium halides* by preparing the methylammonium propionate. Xylene was added and the mixture heated to split out the water. The water and unreacted propionic acid were removed as their respective xylene azeotropes. The amide was fractionally distilled several times at 5 Torr through an efficient column; specific conductance at 30° was 3×10^{-6} ohm^{-1} cm^{-1}.

The amide was prepared by Meighan and Cole [5044] for *dielectric property studies* from 40% aqueous methylamine and propionic acid. Water was removed by distillation. It was dried over calcium oxide and vacuum-distilled three times. Bass and coworkers [663] prepared the amide in the same manner but distilled five times.

447.   1,1,3,3-Tetramethylurea

1,1,3,3-Tetramethylurea is unique among the solvents. It is an amide with a relatively low dielectric constant. It is an excellent solvent for organic substances, particularly aromatics. It is a suitable reaction medium for the base-catalyzed isomerization, alkylation, acylations, and other condensation reactions. It is inert toward Grignard compounds and makes a useful solvent for these reactions and for compounds sparingly soluble in ethers [4718]. It is a solvent for acetylene and for polyacrylonitrile. It dissolves inorganic compounds to about the same extent as does acetone. 1,1,3,3-Tetramethylurea has the dissolving power and miscibility of pyridine. It has a more convenient boiling point and can be made practically anhydrous by simple distillation. It has a mild and pleasant odor.

Michler and Escherich [5107] first reported the preparation of 1,1,3,3-tetramethylurea in 1897 from dimethylamine and phosgene. Lüttringhaus and Dirksen [4718] reported that the reaction proceeded via N,N-dimethylcarbamyl chloride (bp 164–165°), which is easy to isolate. This procedure is not recommended [4718] for large-scale preparative purposes because of the amount of heat evolved and the formation of a troublesome by-product salt. There have been several modifications of the original method that would be of interest to anyone using the phosgene method. Of greatest interest is circumventing the formation of dimethylammonium chloride [6303,6988].

Dirksen [2014] has developed a method that overcomes the troublesome features of the phosgene method. He charged 1.17 mol (135 g) of 40% aqueous dimethylamine into a 500-mL four-neck flask, equipped with two dropping funnels and a stirrer, and immersed into a cooling bath. One mole (107 g) of dimethylcarbamyl chloride was added drop by drop, so that the temperature did not get above − 10°. After the first 20 mL of the chloride was added, 100 mL of 40% sodium hydroxide was added at such a rate that the two dropping funnels were emptied at about the same rate. Sixty grams of sodium hydroxide was added and two layers formed. The upper layer was separated and freed from salts by reduced pressure distillation. The lower layer was extracted three times with benzene and combined with the distillate. The benzene, water, and dimethylamine were removed by distillation. The residue was fractionally distilled at 10–20 Torr through a short column; yield 78–85% based on the acid chloride.

Lüttringhaus and Dirksen [4718] describe a method of Farbenfabriken Bayer: Introduce gaseous dimethylamine into 10 mol (2140 g) of diphenyl carbonate externally cooled with water until the increase in weight is 25 mol (1150 g). Heat the mixture 4 hr at 200° in an autoclave at about 20 atm. Add 7 L of 5 N sodium hydroxide after cooling and extract the mixture with ether for at least 24 hr. Fractionally distill through a short column and collect the fraction boiling at 60–66° at 13 Torr; yield 850 grams. Further distillation of the residue yields 345 grams of the urethane, which can be reused.

1,1,3,3-Tetramethylurea was purified by Foerster and coworkers [2468] to study its *solvent properties* of inorganic substances by drying over porous barium oxide and then distilling in an atmosphere of nitrogen.

SAFETY

Dixon and coworkers [2019] made a study of the toxicity of tetramethylurea to mice and monkeys. Monkeys tolerated 150 mg/kg/day for 5 days without showing signs of toxicity. The LD_{50} was found by Hotovy [4718, ref. 56] to be 1.1 g/kg for rats.

448.   2-Pyrrolidinone

Sakai and Tsunakawa [6348] prepared 2-pyrrolidinone by mixing 53.2 grams of γ-butyrolactone, 53.5 grams of ammonium chloride, and 40 grams of sodium hydroxide in an autoclave, heating to 210°, filtering, and distilling *in vacuo;* yield 69.4 grams, bp 139° at 22 Torr.

Lynn [4723] prepared 2-pyrrolidinone in 91% yield by hydrogenating 254 grams of ethyl 3-cyanopropionate prepared by a combination of the Kurtz method [*Ann. Chem.* **572,** 52(1951)], 750 mL methanol, 50 grams of wet Raney nickel, and 117 grams of ammonia at 90–95° and 100 psi for 1 hr. It was purified by distillation.

The *dipole moments* were determined by Fischer [2441] on commercial 2-pyrrolidinone and 1-methyl-2-pyrrolidinone by vacuum distillation immediately before use.

SAFETY

Antara Chemicals [390] describes 2-pyrrolidinone as a mild primary skin irritant with definite sensitizing properties. Prompt removal from the skin with soap and water is recommended.

The oral toxicities for white rats and guinea pigs are LD_0 4.0 mL/kg, LD_{50} 6.5 mL/kg, and LD_{100} 9.5 mL/kg [390].

449.   1-Methyl-2-pyrrolidinone

1-Methyl-2-pyrrolidinone is a stable, hygroscopic liquid completely miscible with water. It can be hydrolyzed by 4% aqueous sodium hydroxide to 50–70% in 8 hr; the lactam ring can be split by concentrated hydrochloric acid when exposed for an extended period of time.

It is a lactam with good stability that is a versatile solvent with a wide utility. The Antara Division of General Aniline [2708] further characterizes it as a maximum safety solvent with a high flash point, low vapor hazard, high boiling and low freezing points, and chemical and thermal stability.

Tafel and Wassmuth [7154] prepared 1-methyl-2-pyrrolidinone in benzene solution from the sodium salt of pyrrolidinone and iodomethane by shaking for 2 hr at 60–70°. The sodium iodide was removed by filtration. The filtrate was distilled and the portion distilling at 197–202° and 736 Torr was saved.

Eilingsfeld and coworkers [2230] prepared 1-methyl-2-pyrrolidinone by dropwise addition at 20° of 200 grams of carbonyl chloride in 250 mL of benzene to 198 grams of methylpyrrolidone in 500 mL of benzene, A paste was formed and it was liquefied by refluxing 3 hr with hydrogen sulfide. It was distilled and 190 grams of product was obtained at 145° and 15 Torr.

Gal'pern and coworkers [2642] found 1-methyl-2-pyrrolidinone to be a good differentiating solvent for the titration of acids and phenols. Kreshkov and coworkers [4260] reported it is increasingly used as a solvent for acid–base titration of electrolytes.

Sears and coworkers [6542] purified 1-methyl-2-pyrrolidinone as one of the solvents for studying the *dipole moment and solution viscosity* of sulfamic acid. The lactam was dried by removing the water as the benzene azeotrope. It was fractionally distilled at 10 Torr through a 100-cm column packed with glass helices and the middle 60% retained.

Fischer [2441] purified for *dipole moment;* see 2-pyrrolidinone.

SAFETY

1-Methyl-2-pyrrolidinone is of low order of toxicity [2625]. It must not be taken internally. The acute toxicity expressed as the LD_{50} is 7 mL/kg for white rats. Tests show that the solvent is not a primary irritant when in contact with the skin for less than 24 hr. Repeated and prolonged contact with the skin produces a mild transient irritation. It is not a sensitizing agent. It is definitely irritating to the eyes. Inhalation tests indicate no special hazard.

Stasenkova and Kochetkova [6968] reported the LD_{100} and LD_{min} as 5 and 3 g/kg, respectively. General Aniline [2708] gives LD_{100} and LD_0 as 10 and 3 mL/kg, respectively, for albino rats for acute oral toxicity. They also reported the inhalation of saturated vapor to be nontoxic and suggested a threshold limit value of 0.05 mg/L. A TLV of 100.0 mg/m^3 was suggested for Russia in 1964 [4185].

The *flammable limits* in air are 2.18 and 12.24%v [2625]. The *minimum ignition temperature* in air is 346° [6954].

450. ε-Caprolactam

ε-Caprolactam, hereinafter called caprolactam, is a high-melting (69.2°) solvent that slowly polymerizes at its melting point. It is very hygroscopic. It is an excellent solvent, particularly for high-molecular-weight polymeric materials that are difficult to make soluble, or that are insoluble, in other solvents.

Scott and coworkers [6532] prepared caprolactam by rearranging cyclohexanone oxime. The caprolactam–acid phase was neutralized with ammonia and the oily upper layer separated, and then dried and purified by distillation at 2.5 Torr. The distillate was crystallized from a mixture of 185 mL of petroleum ether, bp 70°, and 30 mL of 2-methyl-2-propanol; yield 88%, mp 70.5–71.5°.

Either nitromethane, nitroethane, or 1-nitropropane was treated by Kimura and coworkers [4004] in sulfuric acid in a mole ratio of 1:5 at 125–130° and hydroxylammonium hydrogen sulfate was formed in about 90% yield. Cyclohexanone was added at 120–125°. The oxime was formed and the rearrangement proceeded smoothly.

Caprolactam deteriorates on long standing; in the liquid state it turns yellow in contact with air. The yellow color formation may be suppressed by the addition of basic substances such as sodium hydroxide, carbonate, or tetraborate; or lithium acetate in concentrations of about 0.2 grams/L [3616]. Indest and coworkers [3615] found that caprolactam may be stabilized with 0.001–1% of sodium trichloroacetate. Vereinigte Glanzstoff-Fabriken [7723] found 0.07% hexamethylenetetramine to be effective.

A purification procedure was reported [981] that gives a product that does not discolor on exposure to the air and light: Dissolve 250 grams of discolored caprolactam in 750 grams of nitromethane and 6 mL of 30% hydrogen peroxide, heat 1 hr and evaporate. Distill the lactam *in vacuo* in the presence of 0.5% sodium hydroxide; yield 288 grams.

Kornev and coworkers [4192] purified caprolactam by mixing 600 grams of the amide with 21 mL of 40% sodium hydroxide and 6 grams of paraformaldehyde. Light ends and water were distilled at 65–80 Torr until foaming ceased. About 10% of the caprolactam was distilled at 5–10 Torr at 125–135° and added to the next purification batch. The distillation was continued rapidly to produce pure caprolactam in good yield.

Commercial Solvents Corporation [1604] substantially freed caprolactam from volatile bases by refluxing 1 hr in a stream of nitrogen at 30 Torr or for a shorter time with 0.1% sodium hydroxide present, and then distilling at 10 Torr.

Snider and coworkers [6890] reported that oxidizable impurities may be removed from caprolactam by heating at 50° with about 15% water and 25% sodium hydroxide solution. The aqueous phase was separated and the lactam distilled *in vacuo* at 145° and crystallized.

Grekov [2954] purified caprolactam by zone melting.

Hampton and Riddick [3125] found caprolactam to be an excellent solvent for the determination of molecular weights by the freezing point method for substances that are not sufficiently soluble in other cryoscopic solvents and for many substances that are difficult to make soluble, where the degree of polymerization is not large. It was found suitable for routine laboratory use for molecular weight determination for polymers in the range of $(M)_2$–$(M)_{20}$. It was not tried for polymers above $(M)_{20}$. Caprolactam was the only solvent that would dissolve some polymers.

The freezing point constant was determined using naphthalene and biphenyl. The freezing point apparatus was essentially the freezing tube and stirrer described by Glasgow and coworkers [2781] actuated by a windshield wiper motor. The freezing tube was unsilvered and the inner tube of uniform diameter. A differential thermometer was used.

The freezing point constant was determined in the following manner: The melting point tube was heated to 80–100°, the heat turned off, and the tube stoppered with a cork. It was allowed to cool to about 60° and 25–30 grams of caprolactam, accurately weighed, was added. The heat was turned on, so that the rate of heating was about 1° per min to about 70°. The heat was then reduced and the lactam melted. When all of the solid was melted, the heat was shut off, the stirrer started, 0.5 grams of 20–40 mesh 3A or 4A molecular sieves added. The apparatus was evacuated to about 5 Torr and the sample added. There was only about 0.1° supercooling. The temperature–time data were taken, from which the freezing point was determined. The freezing point of the solvent was determined in the same manner without the addition of the sample.

The range and standard deviation of the melting points of 10 replicates were 0.332° and 0.110°, respectively, for caprolactam without molecular sieves. The same values for eight replicas with molecular sieves added were 0.084 and 0.029°.

The average freezing point constant for 28 replicates, using naphthalene and biphenyl as the solute, was 7.3° with a standard deviation of 0.2°. Some values were determined by adding only one increment of solute to the solvent, and others by adding several increments. The precision of replicates is not nearly as good with caprolactam as it is with such solvents as benzene and camphor, for instance. The standard deviation may be reduced to below 0.1° if the system is kept under 1 atm of nitrogen.

The K_f calculated from the heat of fusion is 6.8 [4959].

SAFETY

Caprolactam has a low order of toxicity and presents no appreciable health hazard if handled properly [94].

SULFUR COMPOUNDS (451–463)

Sulfides (451–452)

451. Carbon Disulfide

Prior to about 1950, carbon disulfide was manufactured by the high-temperature reaction of carbon and sulfur vapors. The natural gas-sulfur process is used almost exclusively in the United States at the present time.

Carbon disulfide is an extremely volatile and flammable liquid. The pure substance has a weak odor and keeps well in the dark. Gillo [2754] proposed a method in 1939 for the purification of carbon disulfide and criteria of purity for its use as an *organic standard*. Waddington and coworkers [7818] reported in 1962 that carbon disulfide met the requirements as a *reference standard* for vapor flow calorimetry. They purified CP carbon disulfide in an efficient fractional distillation column. The purity was determined from the freezing curve to be 99.98%m. The Swietoslawski ebulliometric $\triangle t$ was 0.001°.

Brown and Manov [1159] purified carbon disulfide for *heat capacity measurements*. It was dried with calcium chloride and fractionally distilled several times; then 2 L was fractionally distilled in an atmosphere of helium through a 60-cm column packed with brass shoe eyelets. About half of the distillate was collected as the middle fraction. The fractionation procedure was repeated twice. The carbon disulfide contained less than 0.001% impurities as determined by fractional melting.

Obach [5542] purified carbon disulfide by first distilling over lime and then treating with potassium permanganate, 5 grams/L. The carbon disulfide was allowed to stand until the hydrogen sulfide was completely removed. After separating, the liquid was shaken for some time with mercury to remove sulfur, poured off and shaken with 25 grams/L of mercury(II) sulfate until the unpleasant smell of carbon disulfide had disappeared. The liquid was poured off and distilled from calcium chloride, bright daylight being excluded. The purified material was stored in the dark. See also Hammick and Howard [3117].

Chenevier [1447] allowed carbon disulfide to stand with 0.5 mL of bromine per liter for 3–4 hr, removed the bromine by shaking with potassium hydroxide solution or copper turnings, and dried over calcium chloride. McKelvy and Simpson [4777] purified 12 L of a technical product by distilling from 900 grams of ceresin and distilling the middle fraction from fused calcium chloride.

For the *measurement of vapor pressure*, Stock and Seelig [7017] agitated carbon disulfide with mercury, distilled, dried with phosphorus pentoxide, and fractionally distilled in vacuum, avoiding all greased joints. The first fraction always had a rather high vapor pressure, probably because of the presence of carbon dioxide or carbonyl sulfide.

For *complete removal of hydrocarbons*, Ruff and Golla [6308] mechanically agitated a solution of 130 grams of sodium sulfide and 150 grams of water with 45–50 grams of technical carbon disulfide for 24 hr at 35–40°. The sodium thiocarbonate solution was separated from the excess carbon disulfide, precipitated with 140 grams of copper sulfate in 350 grams of water with cooling; after the copper thiocarbonate was separated, it was decomposed with steam. The distillate was separated from water and distilled from phosphorus pentoxide.

According to Staudinger [6973] shock can cause an explosive reaction between alkali metals and carbon disulfide.

CRITERIA OF PURITY

McKelvy and Simpson [4777] recommended the critical solution point in ethanol as a criterion of purity. Maclean and coworkers [4812] characterized the effectiveness of their purification by the ultraviolet spectrum. Gillo [2754] reported the boiling point and refractive index as criteria for the material they studied as an organic standard. Waddington and coworkers [7818] characterized the material they studied as a reference standard for vapor flow calorimetry by the freezing curve and the Swietoslawski Δt.

ACS Reagent Chemicals [137] and Rosin [6263] give the requirements and the test method for reagent-grade carbon disulfide. They are very nearly the same. The combined requirements are given in Table 5.26.

SAFETY

The high vapor pressure of carbon disulfide, together with its low minimum ignition temperature, wide flammable limits, and toxicity, makes it a hazardous substance. Reagent-grade carbon disulfide has a somewhat unpleasant, slightly ethereal odor that does not offer adequate warning for unsafe concentrations. Carbon disulfide is readily ab-

Table 5.26 Specifications for Reagent-Grade Carbon Disulfide

Distillation range	46–47°
1 mL to 95 mL, not more than 0.5°; 95 mL	
to dryness not more than 0.5°.	
Residue on evaporation (max)	0.002%
Foreign sulfides and dissolved sulfur	TPT[a,b]
Sulfites and sulfates	0.002%
Water (max)	0.05%

[a]By Karl Fischer method or cool 10 mL in a test tube to 0°; no turbidity or drops of water develop.

[b]To pass test.

sorbed through the lungs. It has a limited absorption through the skin. Liquid or vapors may cause dermatitis or blistering of the skin or mucous membranes [5713].

The *threshold limit value* is 20 ppm, 60 mg/m^3 [7022]. Kashin [3873] studied workers under actual conditions and concluded that the then threshold limit value (Russian) of 10 mg/m^3 was too high.

The *flammable limits* in air are 1.25 and 50.00%v. The *minimum ignition temperature* in air is 120° [5712].

Thiols (453–454)

453.   1-Butanethiol

Denyer and coworkers [1958] isolated 1-butanethiol from mixed thiols obtained from petroleum naphtha purification. The mixture was fractionally distilled through a 25-plate column into fractions of definite boiling ranges. The 70–110° fraction was refractionated in a 100-plate still. The 1-butanethiol was distilled as a series of azeotropes with various hydrocarbons. The thiol was separated from the hydrocarbons by extractive distillation with aniline. The purity was 99.1%m calculated from the freezing curve.

1-Butanethiol was prepared by Mathias [4973] for *refractivity studies* by the method of Backer and Dijkstra [532]. It was purified by converting to the lead salt, regenerating, and thrice fractionally distilling in a 22-theoretical plate still.

Ellis and Reid [2253] prepared aliphatic thiols (ethyl through heptyl) by adding the appropriate alkyl bromide to alcoholic potassium hydroxide solution saturated with hydrogen sulfide. The solution was gently stirred by a stream of hydrogen sulfide. The reaction mixture was maintained at 50–60° for 2 hr or at reflux for 1 hr. The alkyl thiol separated when the solution was diluted with a large volume of water. The impurities present were: bromoalkane, ethanol, and alkyl sulfide. The crude thiol was separated, dissolved in 20% sodium hydroxide, extracted with a small portion of benzene, and steam-distilled until the alkaline solution became clear. The solution was slightly acidified with 15% sulfuric acid when cooled and the thiol distilled.

The material used by Scott and coworkers [6523] in their *thermodynamic studies* between 12 and 500 K was prepared by the Laramie Station of the Bureau of Mines. The purity was 99.990 ± 0.005%m determined by the calorimetric melting point.

SAFETY

Stanley [6961] reported that 1-butanethiol may be highly corrosive under certain conditions.

It has been reported that butanethiol caused the death of some animals in 2 hr at a concentration of 60 mg/L [915].

454.   Benzenethiol

Adams [21, p.71] gives directions for preparation from 7200 grams of cracked ice, 2400 grams of concentrated sulfuric acid, 600 grams of crude benzene sulfonyl chloride, and 1200 grams of zinc dust. The yield is 359 grams of crude benzenethiol or 340 grams of purified material. The crude product is purified by drying over calcium chloride and distilling at 15 Torr.

Morris and coworkers [5268] purified Eastman Kodak Co. benzenethiol for *physical property study* by twice fractionally distilling at 90 Torr. The best fractions were combined and had a purity of 99.98%m determined from the freezing curve.

Scott and coworkers [6528] used a portion of the standard sample prepared at the Laramie Station of the Bureau of Mines for *thermodynamic property study*. The purity was found to be 99.98 ± 0.01%m by the calorimetric melting point as a function of the fraction melted. The sample was dried by passing the vapors through magnesium perchlorate.

A commercial product was purified by Mathias and coworkers [4976] for *physical and electrical property study* by distillation at reduced pressure in a Claisen apparatus, dried with calcium chloride, and successively fractionally distilled.

SAFETY

Stanley [6961] reported that benzenethiol may be highly corrosive under certain conditions.

Thioethers (455–461)

455. Methyl Sulfide

Methyl sulfide is a solvent for most classes of organic compounds, for many organic resins, and for certain inorganic compounds. It also is a good solvent for a number of reactions. It is mutually only slightly soluble in water [1734].

Cumper and coworkers [1749] prepared methyl sulfide for *dipole moment studies* by slowly adding an aqueous solution of sodium sulfide to a boiling solution of iodomethane and 95% ethanol, refluxing the mixture for 2 hr, and distilling. Methyl sulfide separated as a yellow oil when water was added to the mixture. The oil was allowed to stand over sodium hydroxide pellets for 24 hr, washed until neutral, and dried. It was fractionally distilled and the middle fraction repeatedly washed with 15% sodium hydroxide, then with water, and dried. It was fractionally distilled from sodium.

The *preparation, purification, and physical properties* of 14 dialkyl sulfides were studied by McAllan and coworkers [4729]. Two general methods for the preparation are given.

Method I.   Dissolve 1.5 mol of sodium sulfide, $Na_2S \cdot 9H_2O$, in 250 mL water in a three-necked flask fitted with a condenser, a nonreactive sealed stirrer, and a dropping funnel. Add slowly 2 mol of the desired alkyl halide, or 1 mol each of the corresponding alkyl halides for a mixed thioether, to the vigorously stirred sodium sulfide solution. Regulate the addition to maintain a reasonable reflux. Apply external heat to continue the reflux for 3 hr after the alkyl halide has been added. Steam-distill the thioether, wash well with water, then with 10% sodium hydroxide, and finally with water; then dry with calcium chloride.

Method II.   Dissolve 1.025 mol of sodium hydroxide in 250 mL of water in the same apparatus described in Method I. Slowly add 1 mol of mercaptan with vigorous stirring. When the reaction has ceased, add 1 mol of alkyl halide as rapidly as the reflux will permit. Continue as described for Method I.

Purification.   Fractionally distill in an efficient still. Purify the desired alkyl sulfide fraction through the mercury(II) chloride complex. Dissolve 1 mol of mercury(II) chloride in 1250 mL of ethanol and slowly add the boiling alcoholic solution of the proper ratio of sulfide to give the complex: $2(CH_3)_2S \cdot 3HgCl_2$ or $(C_2H_5)_2S \cdot 2HgCl_2$. Recrystallize the complex to a constant melting point. Heat 500 grams of the purified complex with 250 mL of concentrated hydrochloric acid in 750 mL water. Separate the sulfide, wash with water, and dry with calcium chloride; purity, methyl sulfide 99.995%m, ethyl sulfide 99.8%m from the melting curve.

The equilibrium $R_1R_2S(1) \rightleftharpoons R_1R_2S(s)$ apparently is difficult to establish during freezing. The melting curves were used to calculate the purity.

SAFETY

Crown Zellerbach [1732] reports preliminary toxicity studies for methyl sulfide. It has a moderate acute oral, dermal, and inhalation toxicity.

The *flammable limits* in air are 2.2 and 19.7%v [5415]. The *minimum ignition temperature* in air is 206° [1732].

456. Ethyl Sulfide

Ethyl sulfide for *physical property and chemical constitution studies* was prepared by Vogel and Cowan [7772] by distilling an aqueous solution of sodium ethyl sulfate with an excess of sodium sulfide until the temperature of the reaction mixture reached 120°. The sulfide layer of the distillate was separated, shaken several times with 20% sodium hydroxide solution, and kept over sodium hydroxide pellets for 24 hr. It was washed with water until neutral, dried over calcium chloride, and distilled over sodium.

Drefahl and Schick [2082] prepared ethyl sulfide from thiourea and bromoethane in 85% yield. Eighty-six grams of thiourea was refluxed 6 hr with 50 mL of ethanol and 109 grams bromoethane. The solution was concentrated, 100 mL water added drop by drop, made alkaline with 4 mol of sodium hydroxide in water, and refluxed 1 hr. It was cooled, 164 grams bromoethane was added drop by drop, and it was refluxed 5 hr.

For preparation and purification, see McAllan and coworkers, methyl sulfide [4729].

Cumper and coworkers [1749] prepared ethyl sulfide by the reaction

$$Na_2S + 2C_2H_5I \rightarrow 2NaI + (C_2H_5)_2S$$

but could not free it from iodoethane; see Cumper et al., under methyl sulfide. It was prepared by the method of Gray and Gutekunst [2913] for butyl sulfide from fuming sulfuric acid, anhydrous ethanol, and sodium sulfide.

Approximately 3 L of commercially available ethyl sulfide was fractionally distilled by Haines and coworkers [3072]. Fractions were selected on the basis of refractive index and composited. The freezing curve of the 1.68 L showed a purity of 99.94%m. Samples of purified ethyl sulfide and thiophene were sealed in vacuum in clear glass ampules and exposed to sunlight for 1 year. Other samples were refluxed in an all-glass apparatus at atmospheric pressure for 240 hr. The apparatus was vented through traps to collect any decomposition products. Decomposition or decrease in purity was not noted for either compound in either test.

458. Thiophene

Whitmore [8022, p. 72] gives directions for preparation from sodium succinate and phosphorus trisulfide; yield 20–30% after fractional distillation; see also [912].

Keswani and Freiser [3957] washed 99%m thiophene with hydrochloric acid and then with water, dried, and carefully distilled in a 3-ft column for *dipole moment and structure studies*.

Fawcett and Rasmussen [2384] purified a commercial-grade material for *physical property studies* by both physical and chemical means. The thiophene was washed successively with dilute hydrochloric acid, sodium hydroxide, and distilled water, then dried over calcium chloride. About 2 L were fractionally distilled at atmospheric pressure at a 50:1 reflux ratio through a 235-cm column packed with 2.4-mm stainless steel helices. The first and last quarters of the distillate were discarded. Mass spectra analysis indicated 0.37%m benzene. The distillate was then fractionally crystallized six times. The purified thiophene was degassed and sealed in Pyrex-brand glass flasks. A portion of the material rejected from the fractional crystallization was treated with mercury(I) chloride in an ethanol-sodium acetate solution. The solid obtained was heated under reflux with dilute hydrochloric acid and the thiophene extracted from the cooled liquid with pentane. The pentane solution was dried over calcium chloride and fractionally distilled through a 28-plate column; fp $-38.5°$.

Thiophene was purified of its hydrogen sulfide impurity by Brady [1051] to use in the development of a *spectrophotometric (uv) method* for its determination in producer gas by adding piperidine to the sulfur compound in the ratio of 1:50. The mixture was fractionally distilled and the heads and residue discarded.

For purification for *critical properties*, see Ethers, General Comments [4091].

Haines and coworkers [3072] purified commercial thiophene by fractional distillation through a 1-in. $\times$ 9-ft column packed with stainless steel helices. The product was 99.99%m and shown to be benzene-free by its mass spectrum. The freezing curve failed to reveal benzene because benzene and thiophene form a solid solution [2384]. Stability tests are given under Haines and coworkers; see ethyl sulfide.

SAFETY

Mikhailets [5116] obtained a LC_{50} of 9.5 mg/L for mice by inhalation for 2 hr for thiophene.

459. Tetrahydrothiophene

Whitehead and coworkers [8016] prepared 1,4-dibromobutane from tetrahydrofuran in 71% yield in an estimated purity of 99.85%m. A solution of 5 mol (1203 g) of sodium sulfide, $Na_2S \cdot 9H_2O$, was made with 1100 mL water and 1350 mL ethanol. An 1100-mL portion of this solution was refluxed in a 5-L, three-necked flask with a reflux condenser and two dropping funnels. The remainder of the solution was placed in one of the funnels and 3.5 mol of dibromide in the other. The liquids were added to the flask at such a rate that both funnels were emptied at the same time, about 30 min. The crude material was purified by crystallization of the mercury(II) chloride complex to a constant melting point; see methyl sulfide, McAllan and coworkers [4729]. The regenerated tetrahydrothiophene was washed, dried, and fractionally distilled in a 20-plate, glass-packed still at reduced pressure. There was decomposition when the distillation was done in a 100-plate still at atmospheric pressure.

Pipparelli et al. [5861] prepared tetrahydrothiophene from tetrahydrofuran and hydrogen sulfide using an aluminum oxide gel catalyst at 400°. Yields of 88–90% were obtained under optimum conditions.

Lawson and coworkers [4528, p. 89] describe a method of preparation from 1,4-dichlorobutane and sodium sulfide; yield 160–170 grams or 73–78%.

Commercially available material from the Oronite Chemical Co. was fractionally distilled by Haines and coworkers [3072]. Several 30-mL cuts were made. The freezing points and mass spectra were used to select succeeding cuts that were composited; purity 99.95%m.

460. 2-Methylthiophene
461. 3-Methylthiophene

Fawcett [2383] purified pilot plant products of about 95% of the respective methylthiophene isomer for *physical property study*. Each was washed successively with dilute hydrochloric acid solution, sodium hydroxide solution, and distilled water and then dried by distilling part of the material. The dried isomer was fractionally distilled at atmospheric pressure through a column of 95 theoretical plates at a reflux ratio of 50:1. Subsequently the sample was evacuated to remove gases, distilled, and collected in ampules which were sealed.

Haines and coworkers [3073] purified each isomer for *physical property study* by fractional distillation. 3-Methylthiophene discolored after standing several weeks and later deposited a dark brown film on the storage bottle. It was redistilled and stored under vacuum.

Oxo–Sulfur Compounds (462–463)

462. Dimethyl Sulfoxide

Dimethyl sulfoxide has been known for more than a century, but only recently has it been available commercially. It is a versatile solvent. Schläfer and Schaffernicht [6453] discuss it as a solvent for inorganic compounds; see also [3]. Martin and others [4914] discuss its use as a solvent for organic reactions and for organic compounds. It is a useful solvent for a number of gases [3], and a variety of polymers [2741, 3068, 4872]. Kononenko and Herstein [4180] found dimethyl sulfoxide probably to be the best nonprotogenic solvent for sucrose. Gutman and Schober [3041] used dimethyl sulfoxide as a *solvent in polarographic studies* for inorganic ions. It has favorable attributes as a cryoscopic solvent for molecular weight determination and has been used as the stationary phase in paper chromatographic separation.

Dimethyl sulfoxide is a colorless, odorless, hygroscopic liquid with a slightly bitter taste. It has an equilibrium moisture content of 10% with air at 20°. The partial decomposition observed during boiling at atmospheric pressure is dependent on the kind and amount of impurities present. Thermal decomposition is catalyzed by acids. Many weak bases and basic and neutral salts inhibit decomposition.

Lebel and Goring [4452] purified commercial material for *physical property studies* of the dimethyl sulfoxide–water system. The solvent was kept in contact with Drierite or barium oxide for several days and shaken from time to time. It was distilled near 90° under reduced pressure.

Menashi and coworkers [5055] purified dimethyl sulfoxide for *iron(II) and iron(III) isotope exchange studies*. Commercial material from the manufacturer, or a 99.9% analyzed material from a chemical supply house, gave identical products when purified. It was fractionally distilled at 2–3 Torr and about 50°. The center two-thirds was collected and shaken overnight with alumina, followed by fractionation under the previously given conditions. The center one-half was saved for use. It was stored at 5° in a flask fitted with a magnesium perchlorate guard tube. The purified dimethyl sulfoxide contained 0.015%m water by Karl Fischer analysis.

Johnson and coworkers [3730] purified dimethyl sulfoxide as a solvent for the study of the *polarographic reduction of oxygen* by holding a commercially purified material over sodium hydroxide for 3 hr at 90°, distilling at reduced pressure, and using the center fraction.

Kolthoff and Reddy [4161] studied the *polarography and voltammetry* in dimethyl sulfoxide as a solvent. The sulfoxide was shaken overnight with freshly heated and cooled Woelm chromatographic-grade alumina in a sealed container. The liquid was siphoned into a distilling flask with suction. It was fractionally distilled at a few Torr from alumina through a 20-in. column packed with porcelain saddles. The middle 60% of the distillate was saved. The purified fraction was subjected to the same procedure twice more or until the specific conductance was 2×10^{-8} ohm^{-1} cm^{-1}. The water was 0.01%, or less, determined by the Karl Fischer method.

Kolthoff and Reddy [4160] also *determined acid–base strengths* in dimethyl sulfoxide purified as described above. They found the autoprotolysis constant to be of the order of 5×10^{-18}, that dimethyl sulfoxide is a base of strength comparable to water but it is a much weaker acid, and that water is an extremely weak base in dimethyl sulfoxide.

Stewart and O'Donnell [7013] purified reagent-grade dimethyl sulfoxide as a *solvent for determining the H-function* of strongly basic systems by distilling from sodium hydroxide pellets before use.

Franzen and Driesen [2526] purified dimethyl sulfoxide as a solvent to study the *reaction of sulfonium ylides with polar double bonds* by shaking for 24 hr with calcium sulfate, cooling to 5° until about 75% had crystallized, and filtering. The molten crystals were fractionally distilled at 12 Torr. It was finally shaken 24 hr with 4A molecular sieves.

Commercial dimethyl sulfoxide was purified as a solvent for the study of the *chemical shift of the hydroxyl proton of phenols* by distilling from calcium hydride under reduced pressure and storing over molecular sieves by Ouellette [5630].

A spectrographic-grade dimethyl sulfoxide was purified by Smyrl and Tobias [6859] for lithium chloride *solution thermodynamics* by distilling under reduced pressure at 80°; mp 18.58° and water < 50 ppm.

Dimethyl sulfoxide and sulfolane were both purified in the same manner by Garnsey and Prue [2672] for *cryoscopic determination of osmotic coefficients*. The solvent was fractionally distilled through a column packed with "small glass rings" at 20 Torr with dry nitrogen bled into the pot. The middle 70% of the distillate was collected over granular calcium hydride. The collected material was fractionally distilled two or three times from calcium hydride at 0.05 Torr. The middle 80% was collected each time, the equilibrium temperature with 50% of the solid phase present was only 0.01° lower than when it finally disappeared for dimethyl sulfoxide. This corresponds to a 0.012%m impurity or 0.002%w if the impurity is water.

Dimethyl sulfoxide was purified by Krueger and Johnson [4284] as a *solvent for hydrolysis studies* by slowly heating a liter of commercial material containing 1 gram of potassium hydroxide for 1 hr at 125°. It was distilled at about 72° at 10 Torr through a 16-in. column packed with glass helices. The 70% center cut was used.

Johnson [3736] purified dimethyl sulfoxide by treating with BDH 5A molecular sieves, followed by drying with barium oxide, and fractionally distilling at 12 Torr as solvent for the study of the reaction of the *bromide and p-nitrobenzenediazonium ions.*

Willson and coworkers [8093] purified commerical dimethyl sulfoxide for a *toxicological study* by filtering it through solvent-resistant Millipore filters, type OH, 1.5 μm pore size.

A reagent-grade dimethyl sulfoxide was purified by Clever and Taylor [1539] to determine the *refractive index by the method of minimum deviation*. It was distilled at reduced pressure; d 20°C, 1.09996 g/mL. Three fractional crystallizations were then made for the study.

For purification by fractional freezing to constant maximum melting point, see Sears and coworkers [6544], ethylene carbonate.

Koch and Purdy [4095] found that analytical reagent-grade dimethyl sulfoxide has to be purified further for use as a solvent in the *polarographic analysis of amino acids*. It was distilled at about 1 Torr pressure from 8–14 mesh activated alumina through a 36-in. column packed with porcelain saddles. The first 20–40 and the last 300 mL of a 2-L charge were discarded. The purified material was collected in a sealed evacuated flask from which it was dispensed by pressurizing the flask with "ultra"-pure nitrogen. It gave no polarographic waves between -0.2 and -2.74 V and a small anodic wave between -0.2 and $+0.5$ V. Water by Karl Fischer was 3.15 ppm.

For *dielectric constant studies*, Casteel and Sears [1385] dried dimethyl sulfoxide over Linde 3A molecular sieves and fractionally distilled the liquid at pressures below 1 Torr.

Clever and Westrum [1540] purified commercial dimethyl sulfoxide for *thermodynamic studies* by multistage contact with activated charcoal and subsequent fractional distillation through a 20-plate glass-packed column at 8 Torr and a 4:1 reflux ratio. Freezing point, gas chromatography, and cooling curve were used as criteria of purity.

Reddy [6097] gives a thorough review of the methods of purification and the tests for purity.

Hesse and coworkers [3357] purified dimethyl sulfoxide for *optical studies* by column chromatography with basic alumina.

SAFETY

Crown Zellerbach [1733] states that preliminary toxicity studies show a low degree of toxicity. All reports confirm their statement; for example, [3870,8093]. Buckley [1230] points out that dimethyl sulfoxide dehydrates and defats the skin, but seems to be relatively free from toxic effects. Schläfer and Schaffernicht [6453] point out that dimethyl sulfoxide penetrates the skin and that toxic solutes are carried with it into the body fluid.

The *flammable limits* in air at 100° are 3–3.5 and 42–63%v and the *minimum ignition temperature* in air is 300–302° [1733].

Several substances have been reported to have produced an explosion when mixed with dimethyl sulfoxide. A violent pressure explosion occurred at Mount Zion Hospital, Cancer Chemotherapy Department, after about 1 hr when 4.5 mol of sodium hydride was added to 18.4 mol of dimethyl sulfoxide [*C & EN* **1966**, *44*, 18]. Rowe and co-workers [6289] describe an explosion that occurred 5–10 min after about 70 mL of dimethyl sulfoxide was added to 34.0 grams of periodic acid. An explosion of dimethyl sulfoxide with magnesium perchlorate occurred at Washington State University following a procedure in *J. Chem. Soc.* **1964**, 2991, and another with 70% perchloric acid following a procedure reported by N. Karash in *Organic Sulfur Compounds*. Patricia Able [8] reported a mild explosion during the addition of 70% perchloric acid to dimethyl sulfoxide to prepare a 0.05 *N* solution.

463. Sulfolane

Sulfolane is an aprotic solvent with a moderately high dielectric constant. It possesses very weak acidic and basic characteristics and Morman and Harlow [5264] state that this makes it suitable for both very weak acid and very weak base titrations, while at the same time it is nonlevelling for strong acids and bases. They found that the commercial material contains an acidic impurity and had to be purified before it could be used as a solvent for *acid–base titrimetric studies*. It could have been purified by fractional distillation at reduced pressure. It is more easily purified for acid–base titrimetry by passing through a column of freshly activated alumina (Alcoa F-20 chromatographic).

Sulfolane is a good solvent for most classes of organic compounds and many common polymers. It is also soluble in most solvents except paraffins. It is stable to 220° and decomposes slightly at 285° [6633]; see also [7943].

It has a very high cryoscopic constant [2672].

A commercial-grade sulfolane was purified by Stewart and O'Donnell [7013] as a *solvent for determining H-function of strongly basic systems* by distilling from sodium hydroxide.

Arnett and Douty [431] vacuum-distilled sulfolane from sodium hydroxide pellets repeatedly until 1 mL did not develop a visible color within 5 min after the addition of an equal part of 100% sulfuric acid. The purified material was used as a *weakly basic aprotic solvent for high dielectric constant study*.

Monica and coworkers [5225] purified sulfolane for *transport number and ion conductivity studies* by distilling from solid potassium hydroxide at 10^{-4} Torr until the specific conductance reached 2.0×10^{-8} ohm^{-1} cm^{-1}.

Monica and coworkers [5224] purified commercial sulfolane for *cryoscopic studies* by repeated fractional distillation from phosphorus pentoxide at 10^{-4} Torr through a 1.5-m Podbielniak column packed with glass helices.

See Garnsey and Prue [2672] under dimethyl sulfoxide for purification of sulfolane for *cryoscopic determination of osmotic coefficients*. They doubt if the fraction melted *versus* temperature is a reliable criterion of purity for sulfolane because the solid and the liquid phases are probably so similar that any impurity not removed by distillation is likely to be soluble in the solid phase. Batch samples of about 500 mL were redistilled until the melting point was not raised further by distillation. The Karl Fischer method is of limited sensitivity because of a solvent–iodine complex, but it indicated that the water content was below 0.007%w.

Casteel and Sears [1385] purified sulfolane for *dielectric constant studies* by first drying the solvent over Linde 3A molecular sieves and then "two fractional distillations of solid potassium hydroxide below 1 Torr followed by a distillation in the absence of potassium hydroxide.''

Weatherall and Bathurst [7940] crystallized sulfolane from a saturated ether solution at $-70°$ with dry ice. The cold sulfolane was filtered at $-10°$ with hourly crystallizations.

Coetzee and coworkers [1559] purified sulfolane as a solvent for *voltammetry studies*. ''After removing 3-sulfolene by thermal decomposition, commercial sulfolane was heated in 1-liter batches with 10–15 g of solid sodium hydroxide at 170–180° for 24 hr while bubbling nitrogen through the solvent. Acidic and basic impurities were removed with Rohm and Haas Amberlyst-15 and Amberlyst A-21 ion exchange resins. The final product was obtained after vacuum distillation from calcium hydride (pressure 0.01–0.02 Torr and oil bath temperature 90–100°) and had a conductivity of 2×10^{-9} ohm^{-1} cm^{-1}, a dielectric constant of 43.3, and a density of 1.2625 g/mL. Water content was 0.001 M.''

SAFETY

No vapor toxicity has been reported because of the low vapor pressure of sulfolane. Shell Chemical Corp. [6633] classifies it as a slightly toxic compound. It is nonirritating to the skin.

Brown and coworkers [1173] reported an acute oral LD_{50} in rats and mice as 2.1 and 1.9–2.5 g/kg, respectively.

COMPOUNDS WITH MORE THAN ONE TYPE OF CHARACTERISTIC ATOM OR GROUP (464–504)

Ether Alcohols (464–472)

464.   2-Methoxyethanol

Ruch and Critchfield [6300] employed 2-methoxyethanol as the solvent of choice in the *titration of small amounts of tertiary amine* in primary and secondary amines.

Chu and Thompson [1480] purified 2-methoxyethanol and 2-ethoxyethanol by fractional distillation in a closed system in an adiabatically operated, 36-in. column packed with $\frac{3}{16}$-in. glass helices at a reflux ratio of 25:1. The 20–90% constant boiling fraction was used for *density and refractive index study* of aqueous solutions.

Ballinger and Long [578] determined the *ionization constant* of 2-methoxyethanol that had been treated with anhydrous sodium carbonate to remove mineral or carboxylic acids. It was dried and fractionally distilled. Gas chromatography and specific conductance were criteria of purity.

Kusano and Wadso [4343] dried 2-methoxy-, 2-ethoxy-, 2-butoxy-, and 2-isopropoxy-ethanol over 4A molecular sieves and fractionally distilled three or more times through a 15-plate column at approximately 100°. The last fraction collected was dried over molecular sieves and finally simply distilled. They were distilled into dry glass ampules and sealed under dry nitrogen; 99.85%m by gas chromatography, water <0.02%w. They observed that all are hygroscopic and 2-methoxy- and 2-ethoxyethanol absorb water at a very fast rate. They were used for *calorimetric study*.

2-Methoxyethanol in concentrations estimated to be as low as 25 ppm caused neurological and hematological changes in 19 exposed workers. Postmortem findings of hemorrhagic gastritis and kidney and liver changes were reported following a death by ingestion. It does not cause skin irritation but toxic amounts are readily absorbed through the skin. It causes immediate pain when introduced into the eye [5713, 7022].

The *threshold limit value* is 25 ppm, 80 mg/m^3 [7022].

The *flammable limits* in air are 2.50 and 19.80%v at elevated temperature. The *minimum ignition temperature* in air is 383° [5712].

465.   2-Ethoxyethanol

Beal and Mann [700] purified 2-ethoxyethanol as a solvent for the study of the *electrolysis of metal perchlorates* by drying with Drierite, filtering, and distilling *in vacuo* from fresh calcium turnings. It was fractionally distilled and the fraction distilling at 133.8° at 738 Torr was used; $\kappa = 9.3 \times 10^{-8}$ ohm^{-1} cm^{-1}.

Chu and Thompson [1480] purified 2-ethoxyethanol in the same manner described for 2-methoxyethanol. The distillate contained 0.06%w water by the Karl Fischer method. For purification for *calorimetric study*, see Kusano and Wadso [4343], 2-methoxy-ethanol.

Miller [5143] purified CP 2-ethoxyethanol for *boiling point–composition studies* by fractionally distilling from a 3-ft, all-glass column; the center cut was retained.

2-Ethoxyethanol is less toxic than its methyl homologue, 2-methoxyethanol.

The *threshold limit value* is 100 ppm, 370 mg/m^3 [142]. Rowe [5713, p. 1550] believes this to be a reasonable figure and the margin of safety to be small.

The *flammable limits* in air are 2.6 and 15.7%v [2346]. The *minimum ignition temperature* in air is 238° [5712].

466.   2-Butoxyethanol

This ether alcohol is soluble in water in all proportions at 25° but exhibits partial miscibility between 48.6 and 128° [1699].

Commercial 2-butoxyethanol was purified by Scatchard and Wilson [6408] by repeated distillation at 45 Torr in a 4-ft glass-helices-packed column with nitrogen bubbling into the pot. The midfraction of the fourth distillate, d 27°/4° 0.89473, was used for *vapor pressure studies* of the binary water system.

Mikhant'ev and Pyrakhina [5122] added 70 grams of sodium to 500 grams of 1,2-ethanediol at 50° and then added 300 grams of 1-chlorobutane. The product was heated on the steam bath for 3–4 hr to obtain a 61% yield of 2-butoxyethanol.

Chu and Thompson [1480] purified 2-butoxyethanol in the same manner as described for 2-methoxyethanol. The distillate contained 0.04%w water by the Karl Fischer method. For purification for *calorimetric study*, see Kusano and Wadso [4343], 2-methoxyethanol.

2-Butoxyethanol is moderately toxic orally, appreciably irritating and injurious to the eyes, and not significantly irritating to the skin. It is readily absorbed through the skin in toxic amounts. It is moderately toxic when inhaled; Rowe [5713, pp. 1552–53]. He cautions [5713, p. 1558] concerning the possible absorption of toxic quantities through the skin.

The *threshold limit value* is 50 ppm, 240 mg/m^3 [7022].

The *flammable limits* in air are 1.1 and 10.6%v [152]. The *minimum ignition temperature* is 245° [5712].

467. Furfuryl Alcohol

Furfuryl alcohol is stable in the presence of alkaline substances. It is extremely sensitive to strong acids and resinifies in their presence [6002].

Brown and Hixon [1146] prepared furfuryl alcohol by the continuous vapor-phase hydrogenation of 2-furaldehyde, using a calcium-stabilized copper chromite catalyst prepared according to the directions of Burnette and coworkers [1267]. Yields of furfuryl alcohol up to 95% were recovered after one pass. In a laboratory model, at least 68 grams of the aldehyde could be hydrogenated for each gram of catalyst in the reactor.

Hazlet and Callison [3248] prepared benzyl and furfuryl alcohols by the crossed Cannizzaro reaction of benzaldehyde and 2-furaldehyde. For details see benzyl alcohol. The purification was carried out in the same manner as for benzyl alcohol.

Nystrom and Brown [5541] reduced furoic acid with lithium aluminum hydride in dry ether solution to furfuryl alcohol with an 85% yield.

Issoire [3643] prepared furfuryl alcohol in 80% yield by the drop by drop addition of an aqueous solution of 2.5 M sodium hydroxide to a stirred mixture of 1 mol of 2-furaldehyde and 2 mol of formaldehyde at 25–30°.

Furfuryl alcohol apparently is only moderately toxic. Rats and mice exposed to 19 ppm were restless the first 5–10 min and drowsy the remainder of the 6 hr period. Changes were observed only in the respiratory tract of rats repeatedly exposed to 19 ppm. Treon [5713] states that a threshold limit value of 5 ppm was established in 1958.

The Committee on Threshold Limits of American Conference of Governmental Industrial Hygienists established the *threshold limit value* at 50 ppm in 1961, Treon [5713] and [7022].

The *flammable limits* in air are 1.8 and 16.3%v in the temperature range of 72.5–122°. The *minimum ignition temperature* in air is 391° [6002].

468. Tetrahydrofurfuryl Alcohol

The Quaker Oats Company [6003] states that there has been no explosion reported associated with the heating or distilling of tetrahydrofurfuryl alcohol due to peroxide formation. This does not preclude the possibility that there is an inherent danger from peroxide formation.

The Quaker Oats Company also reports that this alcohol has been in use in substantial quantities for over 20 years without evidence of harmful effect to the health. They caution to provide good ventilation and avoid contact with the skin; good safety practices should be observed.

The *flammable limits* in air are 1.5 and 9.7%v. The *minimum ignition temperature* in air is 282° [6003].

469.   Diethylene Glycol

Rinkenbach [6183] purified diethylene glycol for *physical property determinations* by distilling 1650 mL at reduced pressure. The first fraction of 480 mL was discarded. The next fraction of 1000 mL was fractionally crystallized and gave a final volume of 700 mL.

Jen [3705] reacted 1,2-ethanediol, 2-chloroethanol, and sodium hydroxide for 1 hr at 120° and 3 hr at 130°, obtaining a 74.8% yield of diethylene glycol. He [3704] prepared it in a 71.6% yield by condensing 1,2-ethanediol with ethylene oxide in the presence of a small amount of sulfuric acid. The most favorable conditions were found to be a mole ratio of 1,2-ethanediol to ethylene oxide of 16, a temperature of 120–130°, and a reaction time of 3 hr.

Koizumi and Hanai [4135] purified diethylene glycol in the same manner as they did 1,2-ethanediol for *dielectric property study*.

SAFETY

Rowe [5713, pp. 1502] states that diethylene glycol presents negligible hazards to health in industrial handling except possibly when it is being used at elevated temperature. There does not appear to be a need to establish a threshold limit value because of the low vapor pressure.

The *minimum ignition temperature* in air is 229° [5712].

470.   Triethylene Glycol

Koizumi and Hanai [4135] purified triethylene glycol in the same manner as they did 1,2-ethanediol for *dielectric property studies*.

Chiao and Thompson [1463] fractionally distilled a pure grade of triethylene glycol at 10–15 Torr at a reflux ratio of 25:1 in an adiabatically operated system through a column 36 × 1-in. packed with $\frac{3}{16}$-in. glass helices. The system was vented through a tube of anhydrous activated alumina. The purified glycol was used to determine the *densities and refractive indices of aqueous solutions* and contained 0.02%w water as determined by Karl Fischer titration.

SAFETY

Rowe [5713, pp. 1507 ff.] believes that the industrial handling of triethylene glycol should present no significant problem from ingestion, skin contact, or vapor inhalation. He does not consider a hygienic standard necessary for triethylene glycol.

The *flammable limits* in air at elevated temperature are 0.89 and 9.20%v. The *minimum ignition temperature* in air is 371° [5712].

471.   2-(2-Methoxyethoxy)ethanol

2-(2-Methoxyethoxy)ethanol was purified by Chiao and Thompson [1464] for *study of aqueous solutions* by fractional distillation through a 36 × 1-in. column packed with $\frac{3}{16}$-in. glass helices. The system was protected from atmospheric moisture. The middle 20–90% distilled was used and contained 0.023%w water by Karl Fischer.

SAFETY

2-(2-Methoxyethoxy)ethanol is low in single-dose oral toxicity, moderate in repeated dose oral toxicity, and appreciably irritating to the skin. It is fairly irritating to the mucous membranes. It is toxic when inhaled and is readily absorbed through the skin in toxic amounts [5713].

The *threshold limit value* is 25 ppm [5713].

472.   2-(2-Ethoxyethoxy)ethanol

Seikel [6559] found 1,2-ethanediol difficult to remove from 2-(2-ethoxyethoxy)ethanol by fractional distillation alone. It could best be removed by extracting 250 grams of alcohol in 750 mL benzene with 5-mL portions of water, allowing 10 min for separation of the phases. The volumes of the aqueous extracts were accurately measured, the increase in volume calculated, and the extraction continued until the increase in volume became constant.

Chiao and Thompson [1464] purified 2-(2-ethoxyethoxy)ethanol by the same procedure described for 2-(2-methoxyethoxy)ethanol.

SAFETY

Rowe [5713, pp. 1561 ff.] states that it is generally agreed that the relatively pure ether does not present any serious industrial hazard. Reasonable precautions are adequate to insure safe handling. Impurities, such as 1,2-ethanediol, in appreciable amounts increase the toxicity.

A hygienic standard is not considered necessary because of the low toxicity and low vapor pressure.

Carbonyl Alcohols (473–474)

473.   Salicylaldehyde

Casnati and coworkers [1381] found that phenoxymagnesium halide heated with an excess of triethyl orthoformate, $HC(OC_2H_5)_3$, gave salicylaldehyde in 42% yield.

Salicylaldehyde was prepared in 92% yield by Davies and Hodgson [1851] by heating under pressure salicylic acid, titanium dioxide, and formic acid at a temperature of 250–256°. They [1852] also reduced salicylic acid with sodium amalgam in 64% yield.

Carswell and Pfeifer [1377] fractionally distilled a commercial product, fp −7°, in a 4-ft column packed with Raschig rings. The midcut has a freezing point of 1°. Three hundred grams of the 1° cut was slowly added, with agitation, to an ethanolic solution of sodium bisulfite such that a slight excess of the bisulfite was present. When the yellow bisulfite addition product settled, it was filtered, washed with ethanol, and recrystallized from 10% ethanol. The bisulfite complex was mixed with sodium carbonate. The yellowish-white precipitate was filtered, stirred with water, and treated with hydrochloric acid to the Congo Red end point. The yellow oil was fractionally distilled *in vacuo* and the middle cut refractionated at atmospheric pressure. The distillate was used to determine *physical properties*.

Bratus, Voronin, et al. [1064] precipitated the sodium salt of salicylaldehyde from the benzene, toluene, or carbon tetrachloride solution of a technical grade material by adding sodium carbonate and mixing. The insoluble sodium salt was filtered, washed with solvent, and added with cooling to 50% sulfuric acid. It was steam-distilled and gave an 86% yield of 92–95% purity.

Perkins [5770] purified a crude salicylaldehyde that contained phenol by dissolving in water and adding the hydroxide of an alkaline earth metal, separating the metal salt and regenerating the aldehyde by acidification.

474.   4-Hydroxy-4-methyl-2-pentanone

The preparation of 4-hydroxy-4-methyl-2-pentanone is a long procedure. It can be purchased commercially at reasonable cost.

Adams [21, p. 45] gives directions for preparation from 1190 grams of acetone and barium hydroxide; yield 850 grams. See also Locquin [4645], Edmonds [2204], and Conant and Tuttle [2757, p. 199].

4-Hydroxy-4-methyl-2-pentanone is very reactive and loses water when heated. It is doubted that a high-purity material has been prepared and kept in such a state long enough to determine highly accurate physical properties. It is best purified by fractional distillation at reduced pressure. It can be dried with Drierite.

Commercial-grade material was fractionally distilled three times at 20 Torr and the middle 80% retained each time by Hack and Van Winkle [3056] for *vapor–liquid equilibria studies*.

Fuge and coworkers [2599] three times fractionally distilled the purest obtainable 4-hydroxy-4-methyl-2-pentanone at reduced pressure for *physical property studies;* bp 61.7° at 13 Torr. They reported that the vapor pressure indicated that decomposition starts to become evident at about 130°.

SAFETY

4-Hydroxy-4-methyl-2-pentanone is only slightly toxic by skin absorption; however, it defats the skin and may cause dermatitis on prolonged or frequent contact. The liquid causes irritation of the eyes. It is readily absorbed through the lungs, but a relatively high concentration is required to be injurious to health. A much lower concentration is uncomfortable to the eyes, nose, and throat. It is a low health hazard; Rowe and Wolf [5713, pp. 1750 ff.].

The *threshold limit value* of 50 ppm, 238 mg/m^3, has been suggested [5713, p. 1752].

Halogenated Alcohols (475–476)
475.   2,2,2-Trifluoroethanol

Rochester and Symonds [6215] stated that "the electronegative inductive effects of the fluorine atom in fluorocarbon alcohols makes the hydrolytic hydrogen atom considerably more acidic than the corresponding hydrogen atom in the hydrocarbon alcohol." They concluded that the extent of self-association of fluoroalcohols in the pure liquid or in solution in an aprotic solvent should be higher than the corresponding hydrocarbon alcohol.

Sherry and Purcell [6654] purified 2,2,2-trifluoroethanol by fractional distillation in a 52-theoretical-plate spinning band column. The solvent was used for determining *enthalpies of reactions with Lewis acids*.

Ballinger and Long [577] fractionally distilled a commercial 99.9 + % 2,2,2-trifluoroethanol sample from anhydrous calcium sulfate and a small amount of sodium bicarbonate. The dry acid-free alcohol was used to determine the *ionization constant*.

2,2,2-Trifluoroethanol is one of the solvents used for gel permeation chromatography [7927].

476.   2-Chloroethanol

Ghosh et al. [2726] prepared 2-chloroethanol and 1,2-ethanediol by the electrolytic reduction of ethylene on porous carbon anodes; see 1,2-ethanediol.

Heard [3250] treated a mixture of ethylene and 1-chloro-2-propanol (propylene chlorohydrin) at 25° with an amount of calcium hydroxide equivalent to the 1-chloro-2-propanol. After 5 min, the mixture was stripped to remove the propylene oxide. Almost pure 2-chloroethanol was obtained.

Ballinger and Long [578] used the best 2-chloroethanol available, treated it with anhydrous sodium carbonate, and fractionally distilled for *ionization constant determination*. The specific resistance was used as a criterion for freedom from electrolytes. A unique method for determining the specific resistance is given.

Mathews [4969], de Laszlo [4423], and Smyth and Walls [6879] used fractional distillation for the purification of 2-chloroethanol.

SAFETY

2-Chloroethanol is considered quite toxic. The toxicity by skin absorption has been found to be greater than by inhalation. Inhalation by animals has resulted in nasal irritation, incoordination, convulsions, prostration, and respiratory failure.

Semenova and coworkers [6567] reported the oral LD_{50} for rats and mice as 71 and 91 mg/kg, respectively.

The *threshold limit value* for skin absorption has been established at 1 ppm, 3 mg/m^3 [142].

The *minimum ignition temperature* in air is 425° [5712].

Cyano Alcohols (477)

477.   2-Cyanoethanol

Sekino prepared 2-cyanoethanol [6560] of 80–95% concentration in 95–96% yield by the reaction of liquefied ethylene oxide and hydrogen cyanide using alkali or alkaline earth oxides as catalysts.

Kendall and Mackenzie [2757] described the preparation of 2-cyanoethanol by the reaction of sodium cyanide and 2-chloroethanol.

SAFETY

Fassett [5713, pp. 2018–19] believes 2-cyanoethanol to be of a low order of toxicity compared with some nitriles. There has been little evidence of skin irritation and no significant skin absorption. A hygienic standard does not seem warranted.

Amino Alcohols (478–481)

478. 2-Aminoethanol

Reitmeier and coworkers [6124] purified 2-aminoethanol for *physical property measurements* as follows: The alcohol was given a preliminary distillation, followed by repeated washings with ether and then recrystallization from ethanol. This was followed by fractional distillation, care being taken to prevent the absorption of carbon dioxide. Two successive crystallizations and distillations gave a product with a constant melting point of 10.51°. The purity was determined by potentiometric titration with standard hydrochloric acid using a quinhydrone electrode, and by weight titration using methyl orange as the indicator. The results were 99.97 and 99.96%, respectively.

2-Aminoethanol was purified by vacuum distillation by Tseng and Thompson [7569] for *physical property measurements of aqueous solutions*. Water was determined by the Karl Fischer method.

Arnold and Churms [442] distilled 2-aminoethanol at 82° and 20 Torr and stored it under dry carbon dioxide-free nitrogen. The water content could be reduced only to 0.05%. The aminoalcohol was used to study the *swelling and exchange equilibria of cation exchange resins*.

Brewster and coworkers [1086] found that there was a slight decomposition of 2-aminoethanol when distilled at atmospheric pressure, resulting in the formation of conducting substances. They did not find a suitable drying agent. The drying and purification was done by vacuum distillation at 5 Torr.

2-Aminoethanol was purified by Bates and Pinching [674] for the study of the *acid dissociation constant and related thermodynamic quantities* by twice fractionally distilling and collecting the middle third.

SAFETY

2-Aminoethanol can cause severe injuries to the eyes and skin and the vapors are irritating to the eyes, skin, and mucous membranes. Its low vapor pressure reduces the hazard by inhalation [5713].

The *threshold limit value* is 3 ppm, 6 mg/m^3 [7022].

479. 2-Amino-2-methyl-1-propanol

Tindall [7501] found the impurities present in 2-amino-2-methyl-1-propanol (AMP) are from several sources and may vary in amount and kind. Homologues, 2-aminoethanol, 2-aminopropanol, and 2-aminobutanol are generally present in small amounts to "traces" resulting from other nitroalkanes present as impurities in the 2-nitropropane; see *Organic Solvents*, 3rd ed., p. 791. Other nitrogen compounds are from side reactions during each step of the manufacture and refining. 2-Nitro-2-methyl-1-propanol reacts with AMP during the reduction in the presence of a sluggish catalyst to form a nitro-aminoalcohol which is reduced to the diamine and in turn loses a molecule of water to cyclize to 1-isopropyl-4,4'-dimethylimadazoline. This compound boils within 2° of AMP. AMP dehydrates and cyclizes during the distillation in the tetramethylpyrazine which boils about 10° higher than AMP.

SAFETY

The vapor pressure of 2-amino-2-methyl-1-propanol is low; therefore, it is not considered hazardous at ambient temperature. It is irritating to the eyes and is considered mildly irritating to the skin [1615].

481. Triethanolamine

Pearce and Berhenke [5731] purified commercially available triethanolamine for *dipole moment studies* by fractional distillation; bp 175° at 2 Torr.

Bates and Schwarzenbach [675] gave the thermodynamic acidity constants, pK, of triethanolamine at 20°, 25°, and 30° as 7.87, 7.77, and 7.68, respectively. They *recommend it as a buffer substance* in the pH range of 7–8.5.

SAFETY

Triethanolamine is considered to have a low acute and chronic toxicity. Any toxic effects probably would be due to the alkalinity. The greatest hazard is from spills and splashes on the skin and in the eyes [5713].

Thioether Alcohols (482)

482. 2,2-Thiodiethanol

Victor Meyer [5090] prepared 2,2′-thiodiethanol from 2-chloroethanol and concentrated aqueous potassium sulfide.

Faber and Miller [8022] prepared 2,2′-thiodiethanol by adding sodium sulfide to aqueous 2-cyanoethanol. Water was removed by distillation at 30–40 Torr after the reaction was completed and the sulfide extracted from the residue with hot ethanol.

Clayton and Reid [1520] prepared 2,2′-thiodiethanol from cyanohydrin. The crude material was freed from dithiane, $S(CH_2CH_2)_2S$, and its polymers by distillation at 8 Torr. The distillate was diluted with sufficient water to bring the boiling point to 165°, and superheated steam was passed through the mixture. After evaporation of the water, the 2,2′-thiodiethanol was distilled at 147.5° at 6 Torr.

Ross [6270] distilled a commercial product and used the middle cut boiling at 137° at 5 Torr to study the *role of oxidation*.

Aldehyde Ethers (483)

483. 2-Furaldehyde

This aldehyde ether, available commercially at a reasonable cost, is unstable and must be purified by fractional distillation under reduced pressure before use. 2-Furaldehyde turns dark red in the presence of air, light, and acids. It undergoes oxidation by absorbing oxygen that stops when the aldehyde decomposition proceeds to 7–8%. Formic acid, β-formylacrylic acid, and furan-α-carboxylic acid have been found in the oxidized material. Decomposition and apparent polymerization increase with increasing temperature. Polymerization takes place at 230° regardless of time.

Several reviews on its preparation and properties are available; see, for example, Monroe [5227], Miner and coauthors [5160], and van Os [7673].

Adams [21, p. 49] gives directions for the preparation from 1500 grams of corn cobs, 5 L of 10% sulfuric acid, and 2000 grams of sodium chloride; yield 180–220 grams.

For preparation from furfuryl alcohol, see Zetzsche and Zala [8283] and Yamashita and Matsumura [8204].

2-Furaldehyde may be separated from impurities other than carbonyl compounds by the bisulfite addition compound [3013].

Mayland [5011] purified 2-furaldehyde that was contaminated with close boiling hydrocarbons by water extraction in the presence of a C_3–C_8 hydrocarbon. The amount of light hydrocarbon added corresponded to 50–100 times the weight of the contaminating hydrocarbon present in the 2-furaldehyde. The amount of water for the extraction was regulated to give a solution of 10–50% of the aldehyde in water. The extraction temperature was from 32 to 65°. The aqueous solution was distilled and the 2-furaldehyde separated from the azeotrope.

Miller [5142] recovered 2-furaldehyde from a mixture with its polymer by treatment with a large volume of water containing a surface active agent that caused coagulation and settling of the polymer. The aldehyde was separated from the supernatant aqueous solution by distillation.

2-Furaldehyde was purified by Harris and Zoch [3182] for *analytical spectrophotometric studies* by saving the middle one-third from successive fractional distillations. It was stored in sealed glass ampules in a freezer for 4 years. It was originally colorless but developed a pale yellow color during storage.

Chueh and Briggs [1482] purified 2-furaldehyde by distilling twice *in vacuo* and storing *in vacuo* for no more than 4 days.

Impurities that develop during storage of 2-furaldehyde may be removed by passing through a chromatographic-grade alumina [3988]; the resulting product is reported to be nearly equivalent to that produced by vacuum distillation.

Furamide, 0.08%, was used to stabilize 2-furaldehyde [6903].

Khol'kin and Chernyaeva [3989] reduced resin formation by storing technical 2-furaldehyde under an inert atmosphere. The amount of resin after 40 days under oxygen, air, carbon dioxide, and nitrogen was 1.7, 0.3, 0.1, and 0.05%, respectively. The acidity was lower in the inert atmosphere. *p*-Hydroxydiphenylamine, diphenylamine, cadmium iodide, hydroquinone, pyrogallol and β-naphthol were found to be the most effective oxidation inhibitors at 0.1% level of 160 compounds tested.

Oosterhout and Roddy [5593] stabilized 2-furaldehyde by the addition of 0.001–0.1% of *N*-phenyl-substituted guanidines, thioureas, or naphthalamines. Polymer formation was prevented at 60–170°. Stabilized 2-furaldehyde is *useful for the extraction of unsaturated materials from hydrocarbon mixtures*.

The presence of formaldehyde in 2-furaldehyde may be detected spectrophotometrically [7706]. For the determination of unsaturated aldehydes, see Wearn and coworkers [7938].

SAFETY

The odor threshold of 2-furaldehyde has been reported to be 0.25–0.38 ppm. Some hygienists believe that the odor becomes readily noticeable near 5 ppm. Nasal irritation would indicate that the concentration in the air is too high. Human eye injury has been reported from 2-furaldehyde vapors. Full strength liquid resulted in immediate injury to the cornea of a rabbit's eye. Contact with the skin should be avoided [3607].

Kuznetsov [4355] reported the LD_{50} in rats and mice as 126.7 and 425 mg/kg, respectively.

The *threshold limit value* is 5 ppm, 20 mg/m^3 [7022].

The *lower flammable limit* in air at 125° and 740 Torr is 2.1%v. The *minimum ignition temperature* in air is 315° [3607].

Chloro Ethers (484–485)

484.   Bis(2-chloroethyl)ether

Barton and coworkers [656] purified commercial bis-(2-chloroethyl) ether for *kinetic studies* by washing several times with concentrated hydrochloric acid and then fractionally distilling at reduced pressure.

Kamm and Waldo [3843] heated 2-chloroethanol and concentrated sulfuric acid to produce bis(2-chloroethyl)ether.

For purification for *critical opalescence studies*, see Chu [1479] under decane.

SAFETY

Most people can detect the odor of bis(2-chloroethyl)ether at its *threshold limit value* of 15 ppm, 90 mg/m^3. Hake and Rowe [5713, pp. 1673 ff.] caution that the odor should not be relied upon for monitoring purposes. Brief exposures of human volunteers to concentrations above 550 ppm resulted in severe irritation to the eyes and nasal passages and was considered intolerable; coughing, retching, and nausea were also reported. Concentrations from 100–260 ppm were not considered intolerable but the irritating effects were still present to some extent. Although it is not injurious to the intact skin, it is rapidly absorbed in lethal amounts through the skin. Its vapors can cause delayed response resulting in lung lesions.

The *minimun ignition temperature* in air is 369° [5712].

485.   Epichlorohydrin

Epichlorohydrin was prepared by Clarke and Hartman [2757] from glycerol-1,3-dichlorohydrin by treatment with solid, powdered sodium hydroxide in ether; see also [1508, p. 47; 3733, p. 30].

Epichlorohydrin may be purified by fractional distillation.

SAFETY

Systemically, epichlorohydrin is moderately toxic by whatever way it enters the body. It is very irritating to the eyes, nose, throat, mucous membranes, and skin [5713].

Fedyanina [2391] reported the LD$_{50}$ in rats and mice as 141 and 195 mg/kg, respectively.

The *threshold limit value* is 5 ppm, 19 mg/m^3 [7022]. Two recommended threshold limit values from Russia are 1.0 mg/m^3 [4185] in 1964 and 0.26 ppm [3848] in 1962.

Nitro Ethers (486)

486.   *o*-Nitroanisole

According to Hantzsch [3143], the purest nitroanisole is colorless. It can be purified by repeated vacuum distillation of a freshly prepared sample as free from air as possible, followed by distillation in a high vacuum in a sealed apparatus.

Amino Ethers (487)

487.   Morpholine

Morpholine has a penetrating odor and has been described as a mild base. Its double function as both an ether and an amine makes it an unusually interesting solvent.

Kobe and coworkers [4091] purified morpholine for *critical property and vapor pressure studies*. It was dried with Drierite and fractionally distilled three times at a reflux ratio of 60 to 80:1 and the 80% heart cut taken. The first two distillations were in a 12-mm × 36-in.-column packed with 0.25-in. glass helices. The third distillation was in a 20-mm × 48-in.-column packed with the same size helices.

Friedel and McKinney [2570] purified morpholine for *infrared studies* by refluxing with sodium for 1 hr and distilling onto fresh sodium in an atmosphere of pure dry nitrogen. The middle fraction was redistilled from sodium and collected in a nitrogen atmosphere.

SAFETY

Morpholine has the usual amine hazards to the eyes, skin, and mucous membranes. Concentrated morpholine penetrates the skin. The hazards diminish as the base is diluted to less than 25% with water [5713].

The *threshold limit value* is 20 ppm, 70 mg/m^3 [7022].

Esters of Hydroxy Acids (488–489)

488.   Ethyl Lactate

Ethyl lactate was prepared by D'Ianni and Adkins [1992] by refluxing 250 grams of 85% lactic acid, 500 mL dry ethanol, and 250 mL carbon tetrachloride for 24 hr using an automatic separator that returned the heavier liquid to the reaction flask. Additional ethanol was added and the reaction continued as long as water was formed. The ester was purified by distillation at 50 Torr.

Wood and coworkers [8153] resolved commercial lactic acid into the *l*-form by the means found most suitable from the studies of Patterson and Forsyth [5709] and converted it to the zinc ammonium *l*-lactate. A mixture of 36 grams of the zinc ammonium salt, 195 grams of anhydrous ethanol, and 27.8 grams of concentrated sulfuric acid were heated under reflux for 4 hr. The excess sulfuric acid was neutralized with potassium carbonate and the salts filtered off. The filtrate was extracted with dry ethyl ether. The ether was evaporated and the ester fractionally distilled four times at 15 Torr; yield 72%.

Smith and Claborn [6843] studied methods of preparing 18 lactate esters.

SAFETY

Fassett [5713, p. 1865] states that little toxicity data are available and there does not seem to be any problem in industrial handling. No cumulative effects are expected.

The *lower flammable limit* at 100° in air is 1.5%v. The *minimum ignition temperature* in air is 400° [5415, 6402].

489.   Methyl Salicylate

Copeland and Rigg [1651] determined the *dipole moment* of methyl salicylate that had been fractionally distilled under reduced pressure as needed. A middle, constant boiling fraction, free from water and acid, was used.

SAFETY

Methyl salicylate long has been used as a flavoring agent as "oil of wintergreen." It is a counterirritant used in some ointments and linaments and formerly was considered helpful for "rheumatism." It is absorbed through the skin. The oral lethal dose is about 0.5 g/kg [5713].

The *minimum ignition temperature* in air is 454° [5712].

Ether Esters (490–492)

490.   2-Methoxyethyl Acetate

Kusano [4339] prepared 2-methoxyethyl acetate by reacting 2-methoxyethanol with glacial acetic acid. The product was dried with Drierite and then fractionally distilled under reduced pressure.

SAFETY

This ether ester is not significantly irritating to the eyes or skin. It is poorly absorbed through the skin and is moderately toxic when inhaled [5713].

The *threshold limit value* is 25 ppm, 120 mg/m^3 [7022].

The *flammable limits* in air are 1.75 and 8.20%v at elevated temperature [5712].

491.   2-Ethoxyethyl Acetate

SAFETY

High concentrations of the vapors are irritating to the eyes and nose. It is not appreciably irritating to the skin and is poorly absorbed through it. 2-Ethoxyethyl acetate is not especially toxic when inhaled in concentrations likely to be encountered. It is capable of causing central nervous system depression and lung and kidney damage [5713].

The *flammable limits* in air are 1.7 and 6.7%v [2344].

492.   2-(2-Ethoxyethoxy)ethyl Acetate

SAFETY

No adverse human experience has been reported, nor would any be expected under ordinary conditions with usual care [5713].

The *flammable limits* in air are 1.0 and 6.9%v [6905].

Esters of Keto Acids (493–494)

493.   Methyl Acetoacetate

Several high boiling impurities in methyl acetoacetate, which was made from diketene and methanol, were reported by Motodo and Yoshie [5284] from gas chromatographic studies. The following were found: methylacetoxy-3-butenoate, methyl diacetoacetate, *trans*-enolacetate and *cis*-enolacetate of methyl acetoacetate, and the methyl ether of the subject compound.

SAFETY

No deaths were reported from rats exposed to the concentrated vapors of methyl acetoacetate for 8 hr [5713].

494.   Ethyl Acetoacetate

Ethyl acetoacetate from a chemical supply house was purified for *dissociation constant determinations* by fractional distillation at 18 Torr by Eidenhoff [2227] and the portion distilling at 79.5–80.0° was collected. The narrow range fraction was refractionated under the same conditions.

SAFETY

No deaths were reported from rats exposed to the concentrated vapors of ethyl acetoacetate for 8 hr [5713].

Esters of Cyano Acids (495–496)

SAFETY

Fassett [5713, p. 2029] gives the oral LD_{50} to be 400 mg/kg or greater for these two esters. They are toxic from skin contact and inhalation of the vapors should be avoided.

495. Methyl Cyanoacetate

Cowan and Vogel [1691] purified a commercially available pure product for *physical and chemical constitution studies* by shaking with 10% sodium carbonate solution, washing well with water, drying with anhydrous sodium sulfate and distilling from a fractionating Claissen flask at 6 Torr.

Methyl cyanoacetate may be prepared by heating methyl monochloroacetate with sodium cyanide in the presence of acetic acid at 80–85°.

496. Ethyl Cyanoacetate

Inglis [2757] reported the preparation of cyanoacetic acid from chloroacetic acid, followed by esterification with ethanol.

Clarke [1508, p. 53] gives the directions for preparation from 625 grams chloroacetic acid and 940 grams of cracked ice. The mixture is neutralized to litmus with about 825 mL of sodium hydroxide, 333 grams/L. A solution of 387.5 grams of sodium cyanide in 780 mL water is gently heated and boiled. The sodium chloroacetate solution is added to the hot cyanide solution in portions of 300 mL at such a rate that, without the application of heat, a gentle reaction is maintained. After purification, the yield is about 490–550 grams.

Stephens [7001] prepared the ester from ethyl chloroacetate and sodium cyanide and by esterifying cyanoacetic acid with ethanol.

Newitt and coworkers [5468] purified ethyl cyanoacetate for *hydrolysis studies* by distilling at 46 Torr.

Fluoro Acids (497)

497. Trifluoroacetic Acid

Trifluoroacetic acid is a strong acid that is not easily oxidized or reduced. It has a high thermal stability and can be heated in borosilicate glass to 400° without significant decomposition [3901]. It is as highly ionized in water as hydrochloric acid and forms salts and esters readily. The CF_3 group is not hydrolyzed by acids or bases [2434]. It is very hygroscopic [93].

Yerger [8214] distilled trifluoroacetic acid in a nitrogen atmosphere and a constant boiling center cut was used for studying the *nature of the hydrogen-bond ion-pair.*

SAFETY

Trifluoroacetic acid fumes strongly in air and has a sharp biting odor [7119]. When spilled or splashed on the skin, it can cause painful burns that are slow to heal. The vapors are irritating to the eyes, nose, and mucous membranes. Vapors should not be breathed [5713]. Kheilo and Kremneva [3985] studied the toxicity of trifluoroacetic acid with mice, rats, and humans. They considered that 0.025–0.05 mg/L would be a safe threshold concentration range.

Halogenated Amines (498–499)

498.   Heptacosafluorotributylamine

Hiraoka and Hildebrand [3420] purified the amine for *solubility studies* by distillation and degassing by repeated freezing and melting under vacuum.

Solubility studies were made by Dymond and Hildebrand [2175] on the amines after purification of a commercial sample that was refluxed with alkaline potassium permanganate solution. The fraction distilling at 177° was used.

Heptacosafluorotributylamine was fractionally distilled through an 8-mm × 91-cm Helipak 90-plate column at a reflux ratio of 20:1. The fraction distilling at 176–177° was used for *solubility studies* by McLaughlin and Scott [4787].

Rotariu and Hanrahan [6281] reported that the amine is very inert. It has no basic properties, forms no salts with acids or bases, and does not react with oxidizing or reducing agents.

499.   *o*-Chloroaniline

Sidgwick and Rubie [6712] freed *o*-chloroaniline from small amounts of the *p*-isomer by the method of Beilstein and Kurbatow [743]; see also Jones and Orton [3775]. The amine was dissolved in an equivalent quantity of sulfuric acid and steam-distilled; the *para* compound remained behind as the sulfate. For the separation of *o*-chloroaniline from 2,4-dichloroaniline, see Jones and Orton [3775].

For *o*-chloroaniline that contains only a small amount of the *p*-isomer, King and Orton [4013] recommended dissolving in warm 10% hydrochloric acid (11 mL/gram of the amine). On cooling, the hydrochloride of *o*-chloroaniline separates out and is further purified by recrystallization until the melting point of the acetyl compound is constant. The yields of pure product obtained in this way are better than those obtained by recrystallization of the picrate from ethanol or of the acetyl compound from petroleum ether.

Hale and Cheney [3077] prepared *o*-chloroaniline by heating *o*-dichlorobenzene with alcoholic ammonia in the presence of copper and copper(I) chloride at 150–250°.

SAFETY

o-Chloroaniline is readily absorbed through the skin and may cause kidney, and to a lesser extent, liver damage. Avoid breathing the vapors [5713].

Kondrashov [4174] reported an oral LD_{50} of 256 mg/kg in mice.

Phosphoramides (500)

500.   Hexamethylphosphoric Triamide

Normant [5511] characterized hexamethylphosphoric triamide as the most remarkable of the polar aprotic solvents. It is a colorless mobile liquid, miscible in all proportions with water. It is miscible with many polar and nonpolar solvents, but not with saturated hydrocarbons. It can be extracted from aqueous solution with chlorinated solvents with which it forms a complex. It has been studied as a solvent for gases, organic and inorganic salts, and for polymers. It has some unique properties as a reaction medium and shows promise for electrochemical processes. The symmetrical distribution of the positive charge over the N_3P grouping and the high electron density on the oxygen give rise to a large dipole moment and a high basicity. Normant [5511, 5512] and Robert [6192] present reviews of the physical and chemical properties and uses of hexamethylphosphoric triamide.

Catterall and coworkers [1392] purified hexamethylphosphoric triamide as a solvent for *spectral study of solutions of alkali metals*. The amide was refluxed over calcium oxide for 24 hr and then distilled from sodium at 10^{-3} Torr.

Brooks and Dewald [1126] purified hexamethylphosphoric triamide supplied by Dow Chemical Company as a *solvent* for studying *absorption of alkali metals*. An all-Pyrex-brand-glass purification assembly with only break-seals and seal-offs was used. The system could be evacuated to 5×10^{-6} Torr after flaming. The triamide was stored over sodium for 1 week, transferred to the stillpot, and refluxed at about 0.5 Torr in an atmosphere of nitrogen for 24 hr. The charge was fractionally distilled through a 3-ft column packed with helices. The middle 450 mL was collected on a potassium mirror. It immediately formed a blue solution. The excess metal was allowed to react (about two days). The distillation and collection on a potassium mirror was repeated, allowing the excess metal to react. The solvent was distilled into the reservoir and transferred to proper size break-seal ampules.

They found it difficult to purify hexamethylphosphoric triamide in large quantities. The solvent was transferred without exposure to the atmosphere for recovery.

Wayland and Drago [7936] obtained hexamethylphosphoric triamide commercially and distilled it at reduced pressure from barium oxide. The middle fraction that distilled at 127° at 20 Torr was used for *spectra studies*.

Hanson and Bouck [3142] purified a number of phosphoryl compounds by fractional distillation. The hexamethylphosphoric triamide contained less than 5% $[(C_2H_5)_2N]_2POCl$ and decomposed in attempts at further purification.

Robert [6192] states that hexamethylphosphoric triamide is generally prepared by the reaction of an excess dimethylamine with phosphorus oxychloride in a solvent such as isopropyl ether. The dimethylammonium chloride is removed by filtration and the phosphorus amide is obtained by distilling the solvent from the filtrate.

Hexamethylphosphoric triamide was purified by Drago and coworkers [2079] for *chloride-36 exchange studies* by distilling through a micropath column just prior to use and storing in a desiccator containing activated alumina or Drierite in the absence of light.

Basic impurities were found to be present in hexamethylphosphoric triamide from three suppliers by Dietrich and coworkers [2005]. Reduced pressure distillation gave a neutral material of low specific conductance. This displayed some basicity toward phenol by widening the nuclear magnetic resonance peaks. The purified amide was stored over 4A molecular sieves before use.

Ducom and Normant [2109] purified hexamethylphosphoric triamide by two vacuum distillations over lithium aluminum hydride with argon bleeding into the flask by capillary. The distillate was kept in an inert atmosphere and was used to determine the *freezing point* and *cryoscopic constant*.

Hexamethylphosphoric triamide was purified by Gremmo and Randles [2955] to study the *conductivity of solutions of alkali metals*. A sample from one source contained naphthalene and one from another source contained triphenylmethane. Both impurities were reported difficult to remove. "The liquid was fractionally distilled under reduced pressure, refluxed 12 hr over calcium hydride, distilled, and collected in an argon atmosphere. Quantities of about 300 mL at a time were finally purified by transferring without contact with air to a distillation flask carrying a side arm containing sodium. The liquid was circulated over the sodium until it acquired a pale but permanent blue color. The final distillation was carried out at about 0.2 Torr pressure between 55 and 60°."

Brusset and coworkers [1206] purified hexamethylphosphoric triamide by passing the solvent through a column of Linde molecular sieve 13X or 4A and then distilling under a reduced pressure of argon.

Paul and coworkers [5719] purified hexamethylphosphoric triamide for *cryoscopic constant determination* by distillation under reduced pressure and fractional crystallization. "The solvent was kept over calcium hydride for 48 hr, decanted, and degassed for 2–3 hr under vacuum. It was distilled over 2–3 g of anthracene and 1–2 g of sodium three times."

Itaya and coworkers [3645] reported an extensive purification procedure for *chemiluminescence studies*. "The solvent was predried with calcium hydride for more than one week and was fractionally distilled over calcium hydride at 5–7 Torr nitrogen pressure. The middle fraction was passed through a column of activated alumina and then poured into a vacuum-tight flask containing pyrene and an equivalent amount of sodium metal. After standing for one day to remove moisture and reducible impurities, the solution was degassed on a high-vacuum line. The solvent was transferred under vacuum by cooling the flask with liquid nitrogen and then stored in glass tubes."

SAFETY

Kimbrough and Gaines [4003] found the LD_{50} for male and female rats by oral administration to be 2650 and 3360 mg/kg, respectively. Dermal administration for both sexes was about 4000 mg/kg. Testicular atrophy in male rats was noted at a single dose of 1000 mg/kg. Prolonged dietary intake of 40–80 mg/kg produced partial or complete atrophy.

NIOSH has published a booklet on potential occupational hazards [5419].

Silicon Compounds (501–504)

501. Tetramethylsilane

Aston and coworkers [476] prepared tetramethylsilane for *thermodynamic studies* from silicon tetrachloride and methyl iodide by the Grignard reagent in butyl ether. It was separated from the reaction mixture and purified by dissolving and reclaiming from a dilute Grignard reagent. It was passed through two concentrated sulfuric acid traps, a potassium hydroxide tube, and a phosphorus pentoxide tube. It was condensed in a low-temperature fractional distillation apparatus. A middle cut of about 44 mL, boiling range 0.1°, was used in the calorimeter. The melting range and the heat capacity showed an impurity of 20.1%m methyl iodide present. The azeotrope, bp 26.1°, is very close to the eutectic composition. The methyl iodide was removed as the pyridine complex. The purified sample contained no iodide.

Whitmore and Sommer [8030] prepared and purified a large laboratory amount of tetramethylsilane in 63% yield. Dry ice–acetone traps and vapor-tight stirrers were used to prevent loss of the low boiling product. Methylmagnesium bromide (19.1 equivalents) was prepared in a 12-L flask fitted with stirrer, a dropping funnel, and an efficient condenser. The flask was cooled in an ice bath and a solution of 748 grams (4.4 mol) of silicon tetrachloride in 1 L of dry ether was added over a period of 8 hr. The reactants were stirred at room temperature for 14 hr. The tetramethylsilane–ether mixture was slowly distilled into 1800 mL of concentrated hydrochloric acid in a 3-neck flask in an ice bath with stirring and a reflux condenser. The clear tetramethylsilane above the ether–sulfuric acid layer was slowly distilled into a receiver in dry ice–acetone and fractionally distilled.

A commercially available material was purified by Shinoda and coworkers [6678] for *heat capacity study*. Unsuccessful attempts were made to remove the impurities by fractional distillation and zone refining. It was purified by a preparation gas chromatography technique to a purity of 99.995%m determined by melting point measurement.

Tetramethylsilane and tetraethylsilane were purified by Altschuller and Rosenblum [109] for *electrical property studies* by distilling through a 3-ft (91-cm) vacuum-jacketed distillation column packed with $\frac{1}{4}$-in. (0.64-cm) stainless steel saddles. It was then percolated through a 2-ft (61-cm) column of alumina in an inert atmosphere.

Davis and Morris [1863] purified tetramethylsilane for *spectra studies* by fractional distillation in an Oldershaw column to a purity of 99.8% from mass spectra, gas chromatography, and freezing point. Trouble was encountered as all joint lubricants were attacked by the tetramethylsilane resulting in losses from leakage.

SAFETY

Tetramethylsilane is nonexplosive at 450° below 9%v in dry air [4277].
The *spontaneous ignition temperature* in air is 450° [2962].

502. Trichloromethylsilane

Trichloromethylsilane was prepared by Zemany and Price [8278] by adding trimethylchlorosilane to methylmagnesium bromide in butyl ether. Material boiling below 142° was stripped out and fractionally distilled in a 10-theoretical plate column. The fraction collected at 26.0–26.5° was used for *mass spectra studies*.

Sauer and Hadsell [6390] report that the chloromethylsilanes possess reasonable thermal stability.

Booth and Martin [988] prepared and purified trichloromethylsilane for *physical property study*. They believe that trichloromethylsilane is best prepared by adding 2 mol of methyl Grignard reagent in 1 L of ether to 1 L of silicon tetrachloride in 500 mL of ether. The ether and silicon tetrachloride were removed in a liquid-type column and the trichlorosilane was purified further by distilling at reduced pressure as if it were a gas. It is difficult to separate dichlorodimethylsilane from the main product since both have similar boiling points.

Yellow-colored trichloromethylsilane containing boron trichloride and aluminum chloride was purified by Mlavsky [5197] by passing it through a column of a tertiary-amine anion-exchange resin in the base form. He also purified it by passing the vapors through a quaternary ammonium anion exchange resin in the chloride form and then condensing the vapors.

The purification of trichloromethylsilane for manufacturing *semiconductor-grade silicon carbide* was described by Mikhailov and coworkers [5119]. It was fractionally distilled through a 30-theoretical plate column at a reflux ratio of 30:1 while removing the low-boiling impurities at a 1.4:1 ratio.

Mixtures of methylchlorosilanes may be separated or purification of a methylchlorosilane containing others may be accomplished by fractional distillation through a Penn State-type column at least 100 cm long and packed with stainless steel helices [2608].

Hajiev [3074] designed a glass laboratory fractional distillation column for moisture sensitive and chemically active liquids. It has an efficiency of 112 theoretical plates. Methylchlorosilanes were prepared in purities of 99.5–99.8%.

Golosova and Korobov [2837] fractionally distilled laboratory prepared silanes in a 50-theoretical-plate column at a high reflux ratio to a purity of 99.95–99.99% characterized by gas chromatography and density.

SAFETY

The introduction of chlorine into silicon-containing compounds enhances the toxicity [4310].

Korlyakova [4188] reported that trichloromethylsilane causes acute mucosal irritation and disturbed respiration and central nervous system functions. A dosage of 0.03 mL/L, 2 hr/day for 60 days causes irreparable damage. A maximum tolerance dose of 0.08–0.1 mg/L was recommended.

The *lower flammable limit* in air is 6.5%v [599]; a value 7.6%v had been reported earlier [571]. The *minimum igintion temperature* in air is greater than 400° [6138].

503. Trichloroethylsilane

Booth and Carnell [984] prepared and purified trichloroethylsilane for *physical property study*. Grignard reagent from 1 mol of bromoethane was added to 5 mol of silicon tetrachloride. When the reaction was completed, all of the liquid in the generator was distilled from the solid residue. The distillate was fractionally distilled in a column packed with nichrome wire. A 200-gram lot was fractionally distilled several times using the middle cut for the distillation; yield, 47% based on Grignard reagent. They found the silane was hydrolyzed slowly by sodium hydroxide.

Curran and coworkers [1768] purified trichloroethylsilane obtained from Dow for *dipole moment studies* by fractionally distilling through a 53-in. (1.35-m) glass helices packed column.

For preparation in high purity, see Golosova and Korobov [2837], trichloromethylsilane.

Trichloroethylsilane was purified by Gadshiev and Agarunov [2623] for *combustion studies* by fractional distillation.

SAFETY

A reported LD_{50} in rats is 1.33 grams/kg [6884].

Trichloroethylsilane in the eyes of rabbits produces chemical burns followed by inflammation. Immediate washing with water markedly reduces the effect [756].

504. Tetraethylsilane

For purification for *electrical properties*, see Altschuller and Rosenblum [109], tetramethylsilane.

SAFETY

The *spontaneous ignition temperature* in air is 333° [2962].

Chapter VI

BIBLIOGRAPHY

A

1. Ababi, V., and G. Mihaila, *Studia Univ. Babes-Bolyai, Ser. Chem.* **1963**, *8*, 429; *C.A.* **1964**, *61*, 12694.
2. Abadie, A., C. Michel, et al., *C.R. Acad. Sci.* (Paris), *Ser. C.* **1967**, *264*, 1433.
3. Abadie-Maumert, F. A., *Papeterie* **1959**, *81*, 187; from [4725].
4. Abbasov, D. M., K. G. Zenalova, et al., *Tr. Azerb. Nauchno-Issled. Inst. Gig. Tr. Prof. Zabol.* **1977**, *11*, 145; *C.A.* **1979**, *91*, 152209.
5. Abbott Laboratories, *Technical Information Bull. No. 325*, "Piperidine," North Chicago.
6. Abbott Laboratories, Chemical Marketing Division, *Bull. 68-56*, "Cyclohexylamine," North Chicago, **1968.**
7. Abderhalden, E., and E. Eichwald, *Ber. Deut. Chem. Gesell.* **1918**, *51*, 1312.
8. Abel, P., St. Mary-of-the-Woods College, private communication.
9. Abraham, R. J., and H. J. Bernstein, *Can. J. Chem.* **1959**, *37*, 1056, 2095.
10. Abraham, R. J., and H. J. Bernstein, *Can. J. Chem.* **1961**, *39*, 216.
11. Abraham, R. J., and K. G. R. Pachler, *Mol. Phys.* **1963–64**, *7*, 165; *C.A.* **1964**, *60*, 14034.
12. Abraham, R. J., and D. F. Wileman, *J. Chem. Soc., Perkin Trans. 2,* **1973**, 1521.
13. Abraham, T., V. Bery, et al., *J. Chem. Eng. Data* **1971**, *16*, 355.
14. Abramovitch, R. A., J. B. Rajan, et al., *J. Chem. Eng. Data* **1967**, *12*, 594.
15. Achyuta, I., and V. R. Rao, *Indian J. Phys.* **1960**, *34*, 196; *C.A.* **1960**, *54*, 21996.
16. Aczel, T., and H. E. Lumpkin, *Anal. Chem.* **1960**, *32*, 1819; **1962**, *34*, 32.
17. Adamek, P., K. Volka, et al., *J. Mol. Spectrosc.* **1973**, *47*, 252.
18. Adams, E. W., and H. Adkins, *J. Amer. Chem. Soc.* **1925**, *47*, 1358.
19. Adams, J. C., and H. H. Erickson, *Prepr. Annu. Sci. Meet.–Aerosp. Med. Assoc.* **1976,** 61; *C.A.* **1976**, *85*, 187359.
20. Adams, L. H., *J. Amer. Chem. Soc.* **1915**, *37*, 481.
21. Adams, R., *Organic Syntheses, Vol. 1*, Wiley, New York, 1921.
22. Adams, R., *Organic Syntheses, Vol. 8*, Wiley, New York, 1928.
23. Adams, R., and co-editors, *Organic Syntheses Collection, Vol. 2*, Wiley, New York, 1943.
24. Adams, R., O. Kamm, et al., *J. Amer. Chem. Soc.* **1918**, *40*, 1950.
25. Adams, R., and C. S. Marvel, *J. Amer. Chem. Soc.* **1920**, *42*, 310.
26. Adams, R., and V. Voorhees, *J. Amer. Chem. Soc.* **1919**, *41*, 789.
27. Adamson, D. W., and J. Kenner, *J. Chem. Soc.* **1934,** 838.
28. Addison, C. C., *J. Chem. Soc.* **1943,** 535.
29. Addison, C. C., *J. Chem. Soc.* **1945,** 98.
30. Addison, C. C., D. W. Amon, et al., *J. Chem. Soc. A* **1968,** 2285.
31. Adelman, R. L., *J. Org. Chem.* **1964,** *29*, 1837.
32. Adembri, G., *Ann. Chim.* (Rome) **1956**, *46*, 62; *C.A.* **1956**, *50*, 13714.
33. Adkins, H., and H. R. Billica, *J. Amer. Chem. Soc.* **1948**, *70*, 695.
34. Adkins, H., and H. R. Billica, *J. Amer. Chem. Soc.* **1948**, *70*, 3121.
35. Adkins, H., and K. Folkers, *J. Amer. Chem. Soc.* **1931**, *53*, 1095.
36. Adkins, H., and B. J. Nissen, *J. Amer. Chem. Soc.* **1922**, *44*, 2749.
37. Adkins, H., and J. L. Rainey, *Organic Syntheses, Vol. 20*, Wiley, New York, 1940.
38. Adriaanse, J., H. Dekker, et al., *Rec. Trav. Chim. Pays-Bas* **1964**, *83*, 557.
39. Adriaanse, J., H. Dekker, et al., *Rec. Trav. Chim. Pays-Bas* **1965**, *84*, 393.
40. Aerov, M. E., and G. L. Motina, *Khim. Prom.* **1962,** 368; *C.A.* **1963**, *58*, 2356.

41. Affens, W. A., *J. Chem. Eng. Data* **1966**, *11*, 197.

42. AF Materials Laboratory, Research, and Technology Division, RTD-TDR-63-4016, "Infrared Spectra of Organosulfur Compounds Between 2000 and 250 cm^{-1};" Wright-Patterson Air Force Base, 1963.

43. Agaev, F. B., *Azerb. Med. Zh.* **1975**, *52*, 60; *C.A.* **1976**, *85*, 41685.

44. Agami, C., *Bull. Soc. Chim. Fr.* **1968**, 1205.

45. Agarunov, J. Ya., and S. N. Gadzhiev, *Proc. Acad. Sci. USSR* (Eng. Trans.) **1969**, *185*, 221.

46. Agarwal, D. K., R. Gopal, et al., *J. Chem. Eng. Data* **1979**, *24*, 181.

47. Agarwal, V. K., A. K. Sharma, et al., *J. Chem. Eng. Data* **1977**, *22*, 127.

48. Agren, A., *Acta Chem. Scand.* **1955**, *9*, 39; *C.A.* **1955**, *49*, 13008.

49. Ahlers, N. H. E., R. A. Brett, et al., *J. Appl. Chem.* (London) **1953**, *3*, 433.

50. Ahmad, H., and M. Yaseen, *J. Oil Colour Chem. Assoc.* **1977**, *60*, 99; *C.A.* **1977**, *87*, 38283.

51. Ahn, M. K., private communication.

52. Aicart, E., G. Tardajos, et al., *J. Chem. Eng. Data* **1980**, *25*, 140.

53. Aicart, E., G. Tardajos, et al., *J. Chem. Eng. Data* **1981**, *26*, 283.

54. Aicart, E., G. Tardajos, et al., *J. Chem. Thermodyn.* **1980**, *12*, 283.

55. Airoldi, C., A. Chagas, et al., *J. Chem. Soc.* **1980**, 1823.

56. Akerlöf, G., *J. Amer. Chem. Soc.* **1932**, *54*, 4125.

57. Akhmetkarimov, K. A., I. I. Mai, et al., *Zh. Obshch. Khim.* **1973**, *43*, 458; *J. Gen. Chem.* (USSR) **1973**, *43*, 460.

58. Akhundov, T. S., *Izv. Vyssh. Ucheb. Zaved., Neft Gaz* **1973**, *13*, No. 9, 26, 42; *C.A.* **1974**, *80*, 7305.

59. Akhundov, T. S., and F. G. Abdullaev, *Zh. Fiz. Khim.* **1971**, *45*, 1862; *Russ. J. Phys. Chem.* **1971**, *45*, 1063.

60. Akhundov, T. S., and Ch. I. Sultanov, *Izv. Vyssh. Uchebn. Zaved., Neft Gaz* **1974**, *17*, No. 11, 72; *C.A.* **1975**, *82*, 145973.

61. Akishin, P. A., N. G. Rambidi, et al., *Sbornik Trudov Mezhvuz. Soveshchaniya po Khim. Nefti*, (Moscow) **1956**, 146; *C.A.* **1961**, *55*, 17218.

62. Akishin, P. A., N. G. Rambidi, et al., *Vestnik Moskov. Univ.* **1955**, *10*, No. 12, *Ser. Fiz. Mat. i Estestven. Nauk* **1955**, No. 8, 103; *C.A.* **1956**, *50*, 8329.

63. Aksnes, D. W., and P. Albriktsen, *Acta Chem. Scand.* **1970**, *24*, 3754.

64. Aksnes, D. W., and J. Støgård, *Acta Chem. Scand.* **1972**, *26*, 2552.

65. Albersmeyer, W., *Gas u. Wasserfach* **1958**, *99*, 269; *C.A.* **1958**, *52*, 10553.

66. Albert, A., and R. Goldacre, *Nature* **1944**, *153*, 467.

67. Albert, A., and J. N. Phillips, *J. Chem. Soc.* **1956**, 1294.

68. Albert, A., and E. P. Serjeant, *Ionization Constants of Acids and Bases*, Wiley, New York, 1962.

69. Albert, O., *Osterr. Chem.-Ztg.* **1938**, *41*, 287; *Beil.* **E III 2**, 1388.

70. Albert, O., *Z. Phys. Chem.* (Leipzig) **1938**, *182A*, 421.

71. Albright, L. F., W. C. Galegar, et al., *J. Amer. Chem. Soc.* **1954**, *76*, 6017.

72. Albright, L. F., and J. J. Martin, *Ind. Eng. Chem.* **1952**, *44*, 188.

73. Albright, P. S., *J. Amer. Chem. Soc.* **1937**, *59*, 2098.

74. Albright, P. S., and L. J. Gosting, *J. Amer. Chem. Soc.* **1946**, *68*, 1061.

75. Alcock, K., S. S. Grimley, et al., *Trans. Faraday Soc.* **1956**, *52*, 39.

76. Alder, K., and W. Vogt, *Ann. Chem.* **1949**, *564*, 109.

77. Aleksandrov, V. V., B. N. Bezpalyi, et al., *Zh. Fiz. Khim.* **1976**, *50*, 524; *Russ. J. Phys. Chem.* **1976**, *50*, 305.

78. Aleksankin, M. M., P. S. Dibrova, et al., *Teor. Eksp. Khim.* **1972**, *8*, 490; *Theor. Expt. Chem.* **1972**, *8*, 404.

79. Aleksanyan, V. T., Kh. E. Sterin, et al., *Izv. Akad. Nauk SSSR* **1955**, *19*, 225; *C.A.* **1956**, *50*, 3892.

80. Aleksanyan, V. T., and Kh. E. Sterin, *Opt. Spektrosk.* **1957**, *2*, 562; *C.A.* **1957**, *51*, 16097.

81. Aleksanyan, V. T., Kh. E. Sterin, et al., *Izv. Akad. Nauk SSSR, Otd. Khim. Nauk* **1961**, 1437; *C.A.* **1963**, *58*, 1066.

82. Alekseeva, M. V., *Predel'no Dopustimye Kontsentratsii Atm. Zagryagnenii* **1960**, No. 4, 150.

83. Alekseeva, M. V., *Predel'no Dopustimye Kontsentratsii Atm. Zagryagnenii* **1962**, No. 6, 187; *C.A.* **1963**, *58*, 8344.

84. Alexander, S., *J. Chem. Phys.* **1960**, *32*, 1700.

85. Alfassi, Z. B., D. M. Golden, et al., *J. Chem. Thermodyn.* **1973**, *5*, 411.

86. Allen, B. B., S. P. Lingo, et al., *J. Phys. Chem.* **1939**, *43*, 425.

87. Allen, E. C., and S. Sugden, *J. Chem. Soc.* **1932**, 76.
88. Allen, G., and H. J. Bernstein, *Can. J. Chem.* **1954**, *32*, 1044.
89. Allen, G., and H. J. Bernstein, *Can. J. Chem.* **1954**, *32*, 1124.
90. Allen, J. S., and H. Hibbert, *J. Amer. Chem. Soc.* **1934**, *56*, 1398.
91. Allen, J. T., private communication.
92. Alley, S. K., and R. L. Scott, *J. Chem. Eng. Data* **1963**, *8*, 117.
93. Allied Chemical, General Chemical Division, "Trifluoroacetic acid (3FA*)," Product Data sheet, New York, 1962.
94. Allied Chemical Corp., National Aniline Division, "ε-Caprolactam," Technical Bull. I-14R; "Molten Caprolactam."
95. Allied Chemical Corp., National Aniline Division, "Tech. Bull. I-19."
96. Allinger, N. L., and J. L. Coke, *J. Amer. Chem. Soc.* **1959**, *81*, 4080.
97. Allkins, J. R., and P. J. Hendra, *Spectrochim. Acta* **1966**, *22*, 2075.
98. Allred, A. L., and E. G. Rochow, *J. Amer. Chem. Soc.* **1957**, *79*, 5361.
99. Alpin, R. T., H. Budzikiewicz, et al., *J. Amer. Chem. Soc.* **1965**, *87*, 3180.
100. Al'shits, I. M., and I. E. Flis, *Zh. Prikl. Khim.* **1961**, *34*, 644; *C.A.* **1961**, *55*, 16115.
101. Alstott, R. L., M. E. Tarrant, et al., *Toxicol. Appl. Pharmacol.* **1973**, *24*, 393.
102. Althouse, P. M., and H. O. Triebold, *Ind. Eng. Chem., Anal. Ed.* **1944**, *16*, 605.
103. Altona, C., and H. J. Hageman, *Rec. Trav. Chim. Pays-Bas* **1968**, *87*, 279.
104. Altona, C., and H. J. Hageman, *Rec. Trav. Chim. Pays-Bas* **1969**, *88*, 33.
105. Altschul, M., *Z. Phys. Chem.* (Leipzig) **1893**, *11*, 571.
106. Altschuller, A. P., *J. Phys. Chem.* **1953**, *57*, 538.
107. Altschuller, A. P., *J. Phys. Chem.* **1954**, *58*, 392.
108. Altschuller, A. P., *J. Phys. Chem.* **1955**, *59*, 32.
109. Altschuller, A. P., and L. Rosenblum, *J. Amer. Chem. Soc.* **1955**, *77*, 272.
110. Altsybeeva, A. I., V. P. Belousov, et al., *Zh. Fiz. Khim.* **1964**, *38*, 1242; *C.A.* **1964**, *61*, 7757.
111. Amagasa, M., I. Yamaguchi, et al., Japanese Patent 954, April 26, 1962; *C.A.* **1963**, *58*, 3399.
112. Ambler, R. H., *J. Soc. Chem. Ind.* **1936**, *44*, 291T.
113. Ambrose, D., *Trans. Faraday Soc.* **1963**, *59*, 1988.
114. Ambrose, D., *J. Chem. Thermodyn.* **1979**, *11*, 183.
115. Ambrose, D., *J. Chem. Thermodyn.* **1981**, *13*, 1161.
116. Ambrose, D., B. Broderick, et al., *J. Appl. Chem. Biotechnol.* **1974**, *24*, 359.
117. Ambrose, D., J. E. Connett, et al., *J. Chem. Thermodyn.* **1975**, *7*, 1143.
118. Ambrose, D., J. F. Counsell, et al., *J. Chem. Thermodyn.* **1978**, *10*, 1033.
119. Ambrose, D., J. D. Cox, et al., *Trans. Faraday Soc.* **1960**, *56*, 1452.
120. Ambrose, D., J. H. Ellender, et al., *J. Chem. Thermodyn.* **1974**, *6*, 909.
121. Ambrose, D., J. H. Ellender, et al., *J. Chem. Thermodyn.* **1975**, *7*, 453.
122. Ambrose, D., J. H. Ellender, et al., *J. Chem. Thermodyn.* **1976**, *8*, 165.
123. Ambrose, D., J. H. Ellender, et al., *J. Chem. Thermodyn.* **1977**, *9*, 735.
124. Ambrose, D., J. H. Ellender, et al., *J. Chem. Thermodyn.* **1981**, *13*, 795.
125. Ambrose, D., and D. G. Grant, *Trans. Faraday Soc.* **1957**, *53*, 771.
126. Ambrose, D., and D. J. Hall, *J. Chem. Thermodyn.* **1981**, *13*, 61.
127. Ambrose, D., and I. J. Lawrenson, *J. Chem. Thermodyn.* **1972**, *4*, 755.
128. Ambrose, D., I. J. Lawrenson, et al., *J. Chem. Thermodyn.* **1975**, *7*, 1173.
129. Ambrose, D., and C. H. S. Sprake, *J. Chem. Thermodyn.* **1970**, *2*, 631.
130. Ambrose, D., and C. H. S. Sprake, *J. Chem. Thermodyn.* **1971**, *3*, 1261.
131. Ambrose, D., and C. H. S. Sprake, *J. Chem. Thermodyn.* **1974**, *6*, 453.
132. Ambrose, D., C. H. S. Sprake, et al., *J. Chem. Soc., Faraday Trans. 1* **1973**, *69*, 839.
133. Ambrose, D., C. H. S. Sprake, et al., *J. Chem. Thermodyn.* **1972**, *4*, 247.
134. Ambrose, D., and R. Townsend, *J. Chem. Soc.* **1963**, 3614.
135. Ambrose, D., and R. Townsend, *Trans. Faraday Soc.* **1968**, *64*, 2622.
136. American Chemical Society, *Metal Organic Compounds* (Advances in Chemistry Series), American Chemical Society, Washington, 1959.
137. American Chemical Society, *Reagent Chemicals*, American Chemical Society, Washington, 1960.
138. American Chemical Society, *Reagent Chemicals*, 4th ed., American Chemical Society, Washington, 1968.

139. American Conference of Governmental Industrial Hygienists, *Arch. Ind. Hyg. and Occupational Med.* **1953,** *8,* 296; or **1954,** *9,* 530.

140. American Conference of Governmental Industrial Hygienists, *Threshold Limit Values for 1963,* Cincinnati, Ohio, 1963.

141. American Conference of Governmental Industrial Hygienists, *Threshold Limit Values,* Cincinnati, Ohio, 1976.

142. American Conference of Governmental Industrial Hygienists, *Threshold Limit Values,* Cincinnati, Ohio, 1979.

143. American Cyanamid Co., *The Chemistry of Acrylonitrile,* 2nd ed., New York, 1959.

144. American Industrial Hygiene Association, *Am. Ind. Hyg. Assoc. J.* **1964,** *25,* 94.

145. American Industrial Hygiene Association, *Am. Ind. Hyg. Assoc. J.* **1964,** *25,* 97.

146. American Industrial Hygiene Association, *Am. Ind. Hyg. Assoc. J.* **1964,** *25,* 578.

147. American Industrial Hygiene Association, *Am. Ind. Hyg. Assoc. J.* **1965,** *26,* 435.

148. American Industrial Hygiene Association, *Am. Ind. Hyg. Assoc. J.* **1965,** *26,* 633, 636, 640.

149. American Industrial Hygiene Association, *Am. Ind. Hyg. Assoc. J.* **1966,** *27,* 89.

150. American Institute of Chemical Engineers, ''Dow's Process Safety Guide, 1967,'' a reprint CEP manual.

151. American Institute of Physics, *Temperature, Its Measurement and Control in Science and Industry,* Reinhold, New York, 1941.

152. American Mutual Insurance Alliance (formerly National Association of Mutual Casualty Companies), *Handbook of Organic Industrial Solvents,* 3rd ed., Chicago, 1966.

153–321. American Petroleum Institute Research Project 44, Thermodynamics Research Center, College Station, Texas A & M University. Specific references:

153. *Catalog of Infrared Spectral Data.*

154. *Catalog of Mass Spectral Data.*

155. *Catalog of Ultraviolet Spectral Data.*

156. Table Ov-G, June 30, 1949.

157. Table 2-1-d, Apr. 30, 1967.

158. Table 2-1-(1.02)-k, Oct. 31, 1964.

159. Table 23-2-(1.101)-a, Apr. 30, 1956.

160. Table 23-2-(1.101)-a, Oct. 31, 1976.

161. Table 23-2-(1.101)-a, Oct. 31, 1977.

162. Table 23-2-(1.101)-c, Oct. 31, 1952.

163. Table 23-2-(1.101)-c, Apr. 30, 1955.

164. Table 23-2-(1.101)-d, June 30, 1973.

165. Table 23-2-(1.101)-e, Oct. 31, 1955.

166. Table 23-2-(1.101)-i, Apr. 30, 1969.

167. Table 23-2-(1.101)-k, Oct. 31, 1971.

168. Table 23-2-(1.101)-k, June 30, 1974.

169. Table 23-2-(1.101)-m, Oct. 31, 1968.

170. Table 23-2-(1.101)-p, Oct. 31, 1954.

171. Table 23-2-(1.200)-a, Oct. 31, 1952.

172. Table 23-2-(1.200)-c, Oct. 31, 1952.

173. Table 23-2-(1.200)-d, June 30, 1973.

174. Table 23-2-(1.200)-e, Oct. 31, 1952.

175. Table 23-2-(1.200)-i, Oct. 31, 1968.

176. Table 23-2-(1.200)-k, Oct. 31, 1974.

177. Table 23-2-(1.200)-m, Oct. 31, 1965.

178. Table 23-2-(1.200)-n, Dec. 31, 1952.

179. Table 23-2-(1.200)-n, Apr. 30, 1955.

180. Table 23-2-(1.200)-p, Dec. 31, 1952.

181. Table 23-2-(1.200)-p, Apr. 30, 1955.

182. Table 23-2-(1.200)-v, Apr. 30, 1975.

183. Table 23-2-(1.201)-a, Oct. 31, 1952.

184. Table 23-2-(1.201)-e, Oct. 31, 1952.

185. Table 23-2-(1.201)-i, Oct. 31, 1968.

186. Table 23-2-(1.201)-k, Apr. 30, 1975.
187. Table 23-2-(1.201)-m, Oct. 31, 1965.
188. Table 23-2-(1.201)-n, Apr. 30, 1975.
189. Table 23-2-(1.201)-p, Oct. 31, 1975.
190. Table 23-2-(1.201)-v, Apr. 30, 1974.
191. Table 23-2-(1.201)-vc, Oct. 31, 1975.
192. Table 23-2-(1.202)-a, Oct. 31, 1952.
193. Table 23-2-(1.202)-e, Oct. 31, 1952.
194. Table 23-2-(1.202)-i, Oct. 31, 1968.
195. Table 23-2-(1.202)-k, Apr. 30, 1975.
196. Table 23-2-(1.202)-m, Oct. 31, 1964.
197. Table 23-2-(1.202)-n, Apr. 30, 1975.
198. Table 23-2-(1.202)-p, Oct. 31, 1975.
199. Table 23-2-(1.202)-v, Dec. 31, 1973.
200. Table 23-2-(1.202)-vc, Oct. 31, 1975.
201. Table 23-2-(1.203)-a, Apr. 30, 1956.
202. Table 23-2-(1.203)-e, Oct. 31, 1952.
203. Table 23-2-(1.203)-i, Oct. 31, 1968.
204. Table 23-2-(1.203)-k, Apr. 30, 1975.
205. Table 23-2-(1.203)-m, Oct. 31, 1965.
206. Table 23-2-(1.203)-n, Apr. 30, 1975.
207. Table 23-2-(1.203)-p, Oct. 31, 1975.
208. Table 23-2-(1.203)-vc, Oct. 31, 1976.
209. Table 23-2-(1.204)-a, Apr. 30, 1958.
210. Table 23-2-(1.204)-e, Oct. 31, 1953.
211. Table 23-2-(1.204)-i, Apr. 30, 1969.
212. Table 23-2-(1.204)-m, Oct. 31, 1965.
213. Table 23-2-(1.204)-n, Apr. 30, 1975.
214. Table 23-2-(1.204)-v, Dec. 31, 1973.
215. Table 23-2-(1.204)-vc, Oct. 31, 1975.
216. Table 23-2-(1.205)-n, Apr. 30, 1975.
217. Table 23-2-(1.1101)-v, June 30, 1974.
218. Table 23-2-(1.1200)-v, June 30, 1974.
219. Table 23-2-(1.1201)-v, June 30, 1974.
220. Table 23-2-(1.1202)-v, Apr. 30, 1974.
221. Table 23-2-(3.100)-a, Oct. 31, 1968.
222. Table 23-2-(3.100)-e, Oct. 31, 1975.
223. Table 23-2-(3.100)-i, Apr. 30, 1975.
224. Table 23-2-(3.100)-k, Apr. 30, 1975.
225. Table 23-2-(3.100)-m, Oct. 31, 1968.
226. Table 23-2-(3.100)-n, Oct. 31, 1968.
227. Table 23-2-(3.100)-p, Oct. 31, 1968.
228. Table 23-2-(3.1030)-a, Apr. 30, 1956.
229. Table 23-2-(3.1030)-c, Dec. 31, 1948.
230. Table 23-2-(3.1030)-k, Oct. 31, 1967.
231. Table 23-2-(3.1030)-m, Oct. 31, 1965.
232. Table 23-2-(3.1030)-n, Apr. 30, 1962.
233. Table 23-2-(3.1032)-e, Oct. 31, 1975.
234. Table 23-2-(3.1032)-i, Apr. 30, 1967.
235. Table 23-2-(3.1032)-v, Dec. 31, 1952.
236. Table 23-2-(3.1110)-a, Apr. 30, 1956.
237. Table 23-2-(3.1110)-c, Dec. 31, 1968.
238. Table 23-2-(3.1110)-m, Oct. 31, 1965.
239. Table 23-2-(3.1110)-n, Apr. 30, 1962.
240. Table 23-2-(3.1110)-p, Dec. 31, 1952.

241. Table 23-2-(3.1112)-a, Oct. 31, 1952.
242. Table 23-2-(3.1112)-e, Oct. 31, 1975.
243. Table 23-2-(3.1112)-i, Apr. 30, 1967.
244. Table 23-2-(3.1112)-k, Apr. 30, 1970.
245. Table 23-2-(3.1112)-m, Oct. 31, 1965.
246. Table 23-2-(3.1112)-n, Oct. 31, 1966.
247. Table 23-2-(3.1112)-p, Oct. 31, 1948.
248. Table 23-2-(3.1112)-v, Apr. 30, 1947.
249. Table 23-2-(5.1101)-a, Apr. 30, 1956.
250. Table 23-2-(5.1101)-c, Aug. 31, 1949.
251. Table 23-2-(5.1101)-c, Apr. 30, 1955.
252. Table 23-2-(5.1101)-e, Oct. 31, 1955.
253. Table 23-2-(5.1101)-i, Oct. 31, 1976.
254. Table 23-2-(5.1101)-k, June 30, 1972.
255. Table 23-2-(5.1101)-m, Oct. 31, 1965.
256. Table 23-2-(5.1200)-a, Apr. 30, 1954.
257. Table 23-2-(5.1200)-e, Oct. 31, 1955.
258. Table 23-2-(5.1200)-i, Apr. 30, 1977.
259. Table 23-2-(5.1200)-k, Apr. 30, 1972.
260. Table 23-2-(5.1200)-m, Oct. 31, 1965.
261. Table 23-2-(5.1200)-n, Oct. 31, 1972.
262. Table 23-2-(5.1200)-p, Dec. 31, 1952.
263. Table 23-2-(5.1200)-v, Dec. 31, 1952.
264. Table 23-2-(5.1203)-a, Oct. 31, 1952.
265. Table 23-2-(5.5210)-e, Oct. 31, 1976.
266. Table 23-2-(9.11100)-n, Apr. 30, 1960.
267. Table 23-2-(9.1112)-a, Oct. 31, 1961.
268. Table 23-2-(9.1112)-v, Dec. 31, 1952.
269. Table 23-2-(9.1112)-w, Dec. 31, 1952.
270. Table 23-2-(33.1100)-c, Oct. 31, 1948.
271. Table 23-2-(33.1100)-n, Apr. 30, 1962.
272. Table 23-2-(33.1100)-p, Dec. 31, 1952.
273. Table 23-2-(33.1100)-v, Dec. 31, 1952.
274. Table 23-2-(33.1110)-a, Apr. 30, 1956.
275. Table 23-2-(33.1110)-c, Nov. 30, 1949.
276. Table 23-2-(33.1110)-d, Apr. 30, 1977.
277. Table 23-2-(33.1110)-e, Oct. 31, 1952.
278. Table 23-2-(33.1110)-e, Oct. 31, 1976.
279. Table 23-2-(33.1110)-k, Oct. 31, 1977.
280. Table 23-2-(33.1110)-m, Apr. 30, 1966.
281. Table 23-2-(33.1110)-n, Apr. 30, 1960.
282. Table 23-2-(33.1110)-p, Nov. 30, 1945.
283. Table 23-2-(33.1110)-v, Nov. 30, 1945.
284. Table 23-2-(33.1111)-a, Oct. 31, 1957.
285. Table 23-2-(33.1111)-e, Oct. 31, 1976.
286. Table 23-2-(33.1111)-i, Oct. 31, 1968.
287. Table 23-2-(33.1111)-k, Oct. 31, 1969.
288. Table 23-2-(33.1111)-m, Apr. 30, 1966.
289. Table 23-2-(33.5200)-a, Apr. 30, 1969.
290. Table 23-2-(33.5200)-e, Oct. 31, 1976.
291. Table 23-2-(33.5210)-a, Apr. 30, 1969.
292. Table 23-2-(33.5210)-a, Apr. 30, 1970.
293. Table 23-2-(33.5210)-i, Oct. 31, 1968.
294. Table 23-2-(33.5210)-k, Apr. 30, 1969.
295. Table 23-2-(33.5210)-n, Apr. 30, 1977.

296. Table 23-2-(33.5210)-p, Apr. 30, 1977.
297. Table 23-2-(33.5210)-v, Apr. 30, 1977.
298. Table 23-2-(35.5212)-a, Oct. 31, 1967.
299. Table 23-2-(37.1110)-a, Oct. 31, 1955.
300. Table 23-2-(37.1110)-n, Apr. 30, 1960.
301. Table 23-2-(37.1110)-p, Feb. 28, 1949.
302. Table 23-2-(37.1110)-v, Sept. 30, 1947.
303. Table 23-14-2-(1.1000)-a, Oct. 31, 1964.
304. Table 23-14-2-(1.1000)-k, Apr. 30, 1961.
305. Table 23-14-2-(1.1000)-m, Oct. 31, 1966.
306. Table 23-14-2-(1.1000)-v, Oct. 31, 1963.
307. Table 23-14-2-(1.1000)-w, Oct. 31, 1963.
308. Table 23-14-2-(1.1020)-m, Dec. 31, 1965.
309. Table 23-14-2-(1.2100)-w, Oct. 31, 1963.
310. Table 23-14-2-(1.2101)-w, Oct. 31, 1963.
311. Table 23-14-2-(1.2111)-v, Oct. 31, 1963.
312. Table 23-14-2-(1.2120)-a, Oct. 31, 1964.
313. Table 23-14-2-(1.2120)-i, Oct. 31, 1968.
314. Table 23-14-2-(1.2120)-k, Apr. 30, 1969.
315. Table 23-14-2-(1.2120)-m, Oct. 31, 1965.
316. Table 23-14-2-(9.2020)-a, Oct. 31, 1964.
317. Table 23-14-2-(9.2020)-i, Oct. 31, 1968.
318. Table 23-14-2-(9.2020)-k, Oct. 31, 1953.
319. Table 23-14-2-(9.2020)-m, Oct. 31, 1965.
320. Table 23-14-2-(33.1020)-a, Oct. 31, 1964.
321. Table 23-14-2-(33.1020)-k, Oct. 31, 1954.
322. American Society for Testing and Materials, ASTM Special Technical Publication No. 109A, "Physical Constants of Hydrocarbons C_1 to C_{10}," Philadelphia, 1963.
323. American Society for Testing and Materials, ASTM Data Series DS 4A, "Physical Constants of Hydrocarbons C_1 to C_{10}," Philadelphia, 1971.
324. American Society for Testing and Materials, "1984 Annual Book of ASTM Standards," Sect. 5, Vol. 05.01, 667.
325. American Society for Testing and Materials, Metric Practice Guide E 380-72.
326. Aminabhavi, T. M., R. C. Patel, et al., *J. Chem. Eng. Data* **1982**, *27*, 125.
327. Amiot, O., and H. Marsac, *J. Phys. Radium* **1937**, *8*, 498; *C.A.* **1938**, *32*, 2834.
328. Amir, E. M., U.S. Patent 3,418,389, Dec. 24, 1968; *C.A.* **1969**, *70*, 67838.
329. Ammini Amma, R., S. N. Thakur, et al., *Appl. Spectros.* **1970**, *24*, 344.
330. Ampola, G., and C. Manuelli, *Gazzetta* **1895**, *25b*, 91; *J. Chem. Soc.* **1896Aii**, 238.
331. Ampola, G., and C. Rimatori, *Gazz. Chim. Ital.* **1897**, *27I*, 35; *J. Chem. Soc.* **1897Aii,** 306.
332. An, S. W., and M. Maansson, *J. Chem. Thermodyn.* **1983**, *15*, 287.
333. Anand, B. M., and S. Narain, *Indian J. Phys.* **1939**, *13*, 159; *C.A.* **1939**, *33*, 9140.
334. Anantakrishnan, S. V., and S. Soundararajan, *Proc. Indian Acad. Sci.* **1964,** *59A*, 365; *C.A.* **1965,** *63*, 114.
335. Ananthakrishnan, R., *Proc. Indian Acad. Sci.* **1936,** *3A*, 527; *C.A.* **1936,** *30*, 7040.
336. Ananthakrishnan, R., *Proc. Indian Acad. Sci.* **1937,** *5A*, 200; *C.A.* **1937,** *31*, 4596.
337. Ananthakrishnan, R., *Proc. Indian Acad. Sci.* **1937,** *5A*, 285; *C.A.* **1937,** *31*, 4596.
338. Anderson, A. G., N. E. T. Owen, et al., *Synthesis* **1976,** 398.
339. Andersen, K. K., W. H. Edmonds, et al., *J. Org. Chem.* **1966,** *31*, 2859.
340. Andersen, L., *Suomen Kemistilehti* **1956,** *29B*, No. 3, 94; *C.A.* **1957,** *51*, 5551.
341. Anderson, H. C., and W. R. K. Wu, *Properties of Compounds in Coal-Carbonization Products*, Bur. Mines, Washington, 1963.
342. Anderson, H. H., *J. Chem. Eng. Data* **1964,** *9*, 272.
343. Anderson, H. H., *J. Chem. Eng. Data* **1966,** *11*, 117.
344. Anderson, J. A., *Anal. Chem.* **1948,** *20*, 80.
345. Anderson, J. R., and C. J. Engelder, *Ind. Eng. Chem.* **1945,** *37*, 541.

346. Anderson, M. M., and P. M. Henry, *Chem. Ind.* (London) **1961**, 2053.
347. Anderson, T. F., *J. Chem. Phys.* **1936**, *4*, 161.
348. Anderson, W. A., *Phys. Rev.* **1956**, *102*, 151.
349. Andon, R. J. L., D. P. Biddiscombe, et al., *J. Chem. Soc.* **1960**, 5246.
350. Andon, R. J. L., J. E. Connett, et al., *J. Chem. Soc. A* **1971**, 661.
351. Andon, R. J. L., J. F. Counsell, et al., *J. Chem. Soc. A* **1968**, 1894.
352. Andon, R. J. L., J. F. Counsell, et al., *J. Chem. Soc. A* **1970**, 833.
353. Andon, R. J. L., J. F. Counsell, et al., *J. Chem. Soc., Faraday Trans. 1* **1973**, *69*, 1721.
354. Andon, R. J. L., J. F. Counsell, et al., *J. Chem. Soc., Faraday Trans. 1* **1974**, *70*, 1914.
355. Andon, R. J. L., J. F. Counsell, et al., *J. Chem. Thermodyn.* **1975**, *7*, 587.
356. Andon, R. J. L., J. F. Counsell, et al., *Trans. Faraday Soc.* **1963**, *59*, 830.
357. Andon, R. J. L., J. F. Counsell, et al., *Trans. Faraday Soc.* **1963**, *59*, 1555.
358. Andon, R. J. L., J. F. Counsell, et al., *Trans. Faraday Soc.* **1967**, *63*, 1115.
359. Andon, R. J. L., and J. D. Cox, *J. Chem. Soc.* **1952**, 4601.
360. Andon, R. J. L., J. D. Cox, et al., *Trans. Faraday Soc.* **1954**, *50*, 918.
361. Andon, R. J. L., J. D. Cox, et al., *Trans. Faraday Soc.* **1957**, *53*, 1074.
362. Andreevskii, D. N., *Zh. Fiz. Khim.* **1972**, *46*, 1420; *Russ. J. Phys. Chem.* **1972**, *46*, 819.
363. Andreevskii, D. N., and M. M. Brazhnikov, *Vestn. Beloruss. Univ.* **1970**, *2*, 14; *C.A.* **1972**, *76*, 158651.
364. Andrew, E. R., and D. P. Tunstall, *Proc. Phys. Soc.* (London) **1963**, *81*, 986; *C.A.* **1963**, *59*, 2299.
365. Andrews, L. J., and R. M. Keefer, *J. Amer. Chem. Soc.* **1950**, *72*, 3113.
366. Andrews, L. W., *J. Amer. Chem. Soc.* **1908**, *30*, 353.
367. Anet, F. A. L., *Can. J. Chem.* **1961**, *39*, 2262.
368. Anet, F. A. L., *Can. J. Chem.* **1961**, *39*, 2316.
369. Angelescu, E., and C. Eustatiu, *Z. Phys. Chem.* (Leipzig) **1936**, *177A*, 263.
370. Angell, C. L., *J. Chem. Eng. Data* **1964**, *9*, 341.
371. Angell, C. L., *Trans. Faraday Soc.* **1956**, *52*, 1178; *C.A.* **1957**, *51*, 5555.
372. Angus, W. R., and W. K. Hill, *Trans. Faraday Soc.* **1943**, *39*, 190.
373. Angus, W. R., G. I. W. Llewelyn, et al., *Trans. Faraday Soc.* **1954**, *50*, 1311.
374. Anikin, A. G., and G. M. Dugacheva, *Vestn. Mosk. Univ. Ser. II, Khim.* **1965**, *20*, 16; *C.A.* **1965**, *63*, 5528.
375. Anikin, A. G., and G. M. Dugacheva, *Zh. Fiz. Khim.* **1964**, *38*, 1372; *C.A.* **1964**, *61*, 7101.
376. Anikin, A. G., G. M. Dugacheva, et al., *Russ. J. Phys. Chem.* **1962**, *36*, 1115; *Chem. Zentr.* **1964**, *XIX*, 0768; see also [6244], [6247].
377. Anikin, A. G., Ya. I. Gerasimov, et al., *Plasticheskie Massy* **1962**, No. 12, 13; *Chem. Zentr.* **1964**, *VIII*, 2231; see also [6247].
378. Anikin, A. G., and H. V. Ormanetz, *Dokl. Akad. Nauk. SSSR* **1962**, *142*, 817; *Chem. Zentr.* **1965**, *IV*, 0895; see also [6247].
379. Anno, T., and I. Matsubara, *J. Chem. Phys.* **1955**, *23*, 796.
380. Anonymous, *Chem. Age* (London) **1928**, *18*, 606.
381. Anonymous, *Org. Syn.* **1966**, *46*, 105; *C.A.* **1967**, *66*, 75855.
382. Anonymous, *Fed. Regist.* **May 3, 1977**, *42*, No. 85, 22516.
383. Anonymous, *Plastics* (London) **1946**, *10*, 579; *C.A.* **1947**, *41*, 2269.
384. Anschütz, R., *Ber. Deut. Chem. Gesell.* **1892**, *25*, 3512.
385. Ansul Company, Data Sheet E-242, Marinette, Wisconsin, 1979.
386. Ansul Company, Data Sheet, 1,2-Dimethoxyethane, Form No. C-76237, Marinette, Wisconsin, 1977.
387. Ansul Company, Data Sheet E-444, Marinette, Wisconsin, 1978.
388. Ansul Company, Data Sheet, Ansul E-141, Marinette, Wisconsin, 1977.
389. Antara Chemicals Division, General Aniline and Film Corp., "Propargyl Alcohol," New York, 1961.
390. Antara Chemicals Division, General Aniline and Film Corp., "2-Pyrrolidone," New York, 1957.
391. Antoine, C., *Compt. Rend.* **1888**, *107*, 681.
392. Antoine, C., *Compt. Rend.* **1888**, *107*, 836.
393. Antonoff, G., *J. Phys. Chem.* **1944**, *48*, 80.
394. Anziani, P., and A. Aubry, *Bull. Soc. Chim. Fr.* **1955**, 408; *C.A.* **1956**, *50*, 2449.
395. Aoyama, S., and I. Morita, *J. Pharm. Soc. Jpn.* **1933**, *53*, 1089; *C.A.* **1935**, *29*, 7303.

396. Apelblat, A., *J. Chem. Soc. B* **1969**, *2*, 175.
397. Aplin, R. T., M. Fischer, et al., *J. Amer. Chem. Soc.* **1965**, *87*, 4888.
398. Applebey, M. P., and P. G. Davies, *J. Chem. Soc.* **1925**, 1836.
399. Appelman, L. M., R. A. Woutersen, et al., *Toxicology* **1982**, *23*, 293; *C.A.* **1982**, *97*, 86590.
400. Aquan-Yuen, M., D. Mackay, et al., *J. Chem. Eng. Data* **1979**, *24*, 30.
401. Arata, Y., *J. Phys. Soc. Jpn.* **1960**, *15*, 2119; *C.A.* **1961**, *55*, 10076.
402. Arbusov, A. E., *Z. Phys. Chem.* (Leipzig) **1927**, *131*, 49; *C.A.* **1928**, *22*, 896.
403. Arbuzov, A. E., and P. I. Rakov, *Izv. Akad. Nauk SSSR, Ser. Khim.* **1950**, *3*, 237; *C.A.* **1950**, *44*, 8713.
404. Arbuzov, B. A., I. S. Andreeva, et al., *Izv. Akad. Nauk SSSR, Ser. Khim.* **1970**, 2058; *C.A.* **1971**, *74*, 42498.
405. Arbuzov, B. A., L. A. Grozina, et al., *Teor. Eksp. Khim.* **1968**, *4*, 367; *C.A.* **1968**, *69*, 100605.
406. Arbuzov, B. A., L. M. Guzhivina, et al., *Dokl. Akad. Nauk SSSR* **1948**, *61*, 63; *C.A.* **1948**, *42*, 8559.
407. Arbuzov, B. A., Z. G. Isaeva, et al., *Dokl. Akad. Nauk SSSR* **1961**, *137*, 589; *C.A.* **1961**, *55*, 18311.
408. Arbuzov, B. A., Y. Y. Samitov, et al., *Izv. Nauk SSSR, Ser. Fiz.* **1963**, *27*, No. 1, 89; *C.A.* **1963**, *59*, 147.
409. Arbuzov, B. A., and T. G. Sharisha, *Dokl. Akad. Nauk SSSR* **1949**, *68*, 1045; *C.A.* **1950**, *44*, 886.
410. Arbuzov, G. A., A. P. Timosheva, et al., *Izv. Akad. Nauk SSSR, Ser. Khim.* **1976**, No. 11, 2457; *Bull. Acad. Sci. USSR, J. Div. Chem. Sci.* **1976**, No. 11, 2290.
411. Arbuzov, B. A., and Z. Z. Valeeva, *Zh. Fiz. Khim.* **1953**, *27*, 713; *C.A.* **1954**, *48*, 13.
412. Arbuzov, B. A.. and V. S. Vinogradov, *Compt. Rend. Acad. Sci. URSS* **1947**, *55*, 411; *C.A.* **1947**, *41*, 7183.
413. Archer, G., and J. H. Hildebrand, *J. Phys. Chem.* **1963**, *67*, 1830.
414. Archer, W. L., and E. L. Simpson, U.S. Patent 3,452,108, June 24, 1969; *C.A.* **1969**, *71*, 60668.
415. Archibald, R. C., *J. Amer. Chem. Soc.* **1931**, *53*, 4452.
416. ARCO Chemical Company, Division of Atlantic Richfield Co., "Petrochemicals from ARCO," Philadelphia.
417. Arentsen, J. G., and J. C. Van Miltenburg, *J. Chem. Thermodyn.* **1972**, *4*, 789.
418. Aritan, A., and A. R. Berkem, *Chim. Acta. Turc.* **1973**, *1*, 91; *C.A.* **1974**, *81*, 177333.
419. Arizona Chemical Company, "ACINTENE, A Alpha Pinene," New York, 1967.
420. Arizona Chemical Company, "ACINTENE, B Beta Pinene," New York, 1967.
421. Arizona Chemical Company, "ACINTENE, DP Dipentene; ACINTENE DP High Purity Dipentene," New York, 1967.
422. Arizona Chemical Company, "Terpenes," New York, 1968.
423. Armak, Industrial Chemicals Div., "Chemicals for Industry," Chicago, 1975.
424. Armak, Chemical Division, Tech. Data Sheet, "Kessco Butyl Oleate Distilled," Chicago.
425. Armak, Chemical Division, Data Sheet, Bull. 76-3, "Kessco Triacetin," Chicago.
426. Armanet, G., and B. Pouyet, *Bull. Soc. Chim. Fr.* **1966**, 1931; *C.A.* **1966**, *65*, 13581.
427. Armour Industrial Chemical Company, "*Neo-Fat Fatty Acids Manual*," Chicago, 1971.
428. Armstrong, W. E., A. B. Denshaw, et al., *J. Chem. Soc.* **1950**, 3359.
429. Arnall, F., *J. Chem. Soc.* **1920**, 835.
430. Arndt, F., and P. Nachtwey, *Ber. Deut. Chem. Gesell.* **1926**, *59B*, 448.
431. Arnett, E. M., and C. F. Douty, *J. Amer. Chem. Soc.* **1964**, *86*, 409.
432. Arnett, E. M., E. J. Mitchell, et al., *J. Amer. Chem. Soc.* **1974**, *96*, 3875.
433. Arnett, E. M., and C. Y. Wu, *Chem. Ind.* (London) **1959**, 1488.
434. Arnett, E. M., and C. Y. Wu, *J. Amer. Chem. Soc.* **1960**, *82*, 4999.
435. Arnett, E. M., and C. Y. Wu, *J. Amer. Chem. Soc.* **1960**, *82*, 5660.
436. Arnett, E. M., and C. Y. Wu, *J. Amer. Chem. Soc.* **1962**, *84*, 1674.
437. Arnett, E. M., and C. Y. Wu, *J. Amer. Chem. Soc.* **1962**, *84*, 1680.
438. Arnett, E. M., and C. Y. Wu, *J. Amer. Chem. Soc.* **1962**, *84*, 1684.
439. Arnold, C., British Patent 590,311, July 14, 1947; *C.A.* **1948**, *42*, 200.
440. Arnold, D. S., C. A. Plank, et al., *Ind. Eng. Chem., Chem. Eng. Data Series* **1958**, *3*, 253.
441. Arnold, J. C., British Patent 585,076, Jan. 29, 1947; *C.A.* **1947**, *41*, 5897.
442. Arnold, R., and S. C. Churms, *J. Chem. Soc.* **1965**, 325.
443. Arnold, R. T., and J. Sprung, *J. Amer. Chem. Soc.* **1939**, *61*, 2475.

444. Arntz, H., and G. M. Schneider, *Faraday Discuss. Chem. Soc.* **1980,** *69*, 139.
445. Aroney, M. J., L. H. L. Chia, et al., *J. Chem. Soc.* **1964,** 2948.
446. Aroney, M. J., H. Chio, et al., *Aust. J. Chem.* **1970,** *23*, 199.
447. Aroney, M. J., G. Cleaver, et al., *J. Chem. Soc., Perkin Trans. 2* **1974,** 3.
448. Aroney, M. J., S. W. Filipuzuk, et al., *J. Chem. Soc., Perkin Trans. 2* **1975,** 695.
449. Aroney, M. J., L. R. Fisher, et al., *J. Chem. Soc.* **1963,** 4450.
450. Aroney, M. J., D. Izsak, et al., *J. Chem. Soc.* **1961,** 4148.
451. Aroney, M. J., and R. J. W. Le Fèvre, *J. Chem. Soc.* **1958,** 3002.
452. Aroney, M. J., R. J. W. Le Fèvre, et al., *J. Chem. Soc.* **1960,** 3173.
453. Aroney, M. J., R. J. W. Le Fèvre, et al., *J. Chem. Soc.* **1961,** 4140.
454. Aroney, M. J., R. J. W. Le Fèvre, et al., *J. Chem. Soc.* **1962,** 2886.
455. Aroney, M. J., R. J. W. Le Fèvre, et al., *J. Chem. Soc.* **1965,** 3179.
456. Aronov, O. L., V. M. Tatevskii, et al., *Dokl. Akad. Nauk SSSR* **1948,** *60*, 387, 1177, 1343; *C.A.* **1948,** *42*, 7235.
457. Arrhenius, S., *Medd. Vetensk. Nobel.* **1916,** *3*, No. 20.
458. Arrowsmith, G. B., G. H. Jeffery, et al., *J. Chem. Soc.* **1965,** 2072.
459. Arthur, P., D. C. England, et al., *J. Amer. Chem. Soc.* **1954,** *76*, 5364.
460. Asagoe, K., and Y. Ikemato, *Proc. Phys.-Math. Soc. Jpn.* **1940,** *22*, 677, 685; *C.A.* **1941,** *35*, 1317.
461. Asagoe, K., and Y. Ikemato, *Proc. Phys.-Math. Soc. Jpn.* **1941,** *23*, 829; *C.A.* **1947,** *41*, 6150.
462. Asagoe, K., Y. Simokawa, et al., *Jpn. J. Phys.* **1940–41,** *14*, 11; *C.A.* **1941,** *35*, 7827.
463. Aschan, O., *Ber. Deut. Chem. Gesell.* **1898,** *31*, 1801.
464. Asfour, A. F. A., and F. A. L. Dullien, *J. Chem. Eng. Data* **1981,** *26*, 312.
465. Ashcroft, S. J., *J. Chem. Eng. Data* **1976,** *21*, 397.
466. Ashdown, A., and T. A. Kletz, *J. Chem. Soc.* **1948,** 1454.
467. Ashmore, S. A., *Analyst* **1947,** *72*, 206.
468. Ashworth, M. R. F., and G. Lazik, *Anal. Chim. Acta* **1972,** *60*, 464.
469. Asinger, F., *Ber. Deut. Chem. Gesell.* **1942,** *75B*, 660; *C.A.* **1942,** *36*, 6136.
470. Asolkar, G. V., and P. C. Guha, *J. Indian Chem. Soc.* **1946,** *23*, 47; *C.A.* **1946,** *40*, 5406.
471. Assarsson, P., and F. R. Eirech, *J. Phys. Chem.* **1968,** *72*, 2710.
472. Astoin, N., and J. Granier, *J. Phys. Radium* **1958,** *19*, 507; *C.A.* **1959,** *53*, 8805.
473. Aston, J. G., and H. L. Fink, *Anal. Chem.* **1947,** *19*, 218.
474. Aston, J. G., H. L. Fink, et al., *J. Amer. Chem. Soc.* **1943,** *65*, 341.
475. Aston, J. G., R. M. Kennedy, et al., *J. Amer. Chem. Soc.* **1940,** *62*, 2059.
476. Aston, J. G., R. M. Kennedy, et al., *J. Amer. Chem. Soc.* **1941,** *63*, 2343.
477. Aston, J. G., and S. V. R. Mastrangelo, *Anal. Chem.* **1950,** *22*, 197.
478. Aston, J. G., and S. V. R. Mastrangelo, *Anal. Chem.* **1950,** *22*, 636.
479. Aston, J. G., and G. H. Messerly, *J. Amer. Chem. Soc.* **1936,** *58*, 2354.
480. Aston, J. G., and G. H. Messerly, *J. Amer. Chem. Soc.* **1940,** *62*, 1917.
481. Aston, J. G., P. E. Wills, et al., *J. Amer. Chem. Soc.* **1955,** *77*, 3939.
482. Atkins, W. R. G., *J. Chem. Soc.* **1911,** 10.
483. Atkins, W. R. G., and T. A. Wallace, *J. Chem. Soc.* **1913,** 1461.
484. Atkinson, B., and A. B. Trenwith, *J. Chem. Soc.* **1953,** 2082.
485. Atoyan, V. A., and I. A. Mamedov, *Zh. Fiz. Khim.* **1980,** *54*, 856; *Russ. J. Phys. Chem.* **1980,** *54*, 496.
486. Auclair, M., and N. Hameau, *Compt. Rend. Soc. Biol.* **1964,** *158*, 245; *C.A.* **1964,** *61*, 12535.
487. Audsley, A., and F. R. Goss, *J. Chem. Soc.* **1941,** 864.
488. Audsley, A., and F. R. Goss, *J. Chem. Soc.* **1942,** 358.
489. Audsley, A., and F. R. Goss, *J. Chem. Soc.* **1942,** 497.
490. Audsley, A., and F. R. Goss, *J. Chem. Soc.* **1951,** 120.
491. Auerbach, F., and H. Zeglin, *Z. Phys. Chem.* (Leipzig) **1923,** *103*, 178.
492. Aulinger, F., and W. Reerink, *Colloq. Spectrosc. Inter. 9th Lyon* **1961,** *3*, 556; *C.A.* **1963,** *56*, 7085.
493. Ault, J. L., H. J. Harries, et al., *Inorg. Chim. Acta* **1977,** *25*, 65.
494. Aumüller, W., H. Fromherz, et al., *Z. Phys. Chem.* (Leipzig) **1937,** *37B*, 30.
495. Austerweil, G., *Chim. Ind.* Sept. 1926, Special No. 603; *C.A.* **1927,** *21*, 818.
496. Autenrieth, W., and A. Geyer, *Ber. Deut. Chem. Gesell.* **1908,** *41*, 146; *C.A.* **1908,** *2*, 1435.

497. Auwers, D. V., *Ann. Chem.* **1915**, *408*, 212; *Chem. Zentr.* **1915**, *I*, 936.
498. Auwers, K. V., *Ann. Chem.* **1915**, *408*, 270; *Beil.* **EI17**, 16.
499. Auwers, K. V., *Ann. Chem.* **1920**, *420*, 84; *C.A.* **1920**, *14*, 2161.
500. Auwers, K. V., *Ber. Deut. Chem. Gesell.* **1913**, *46*, 494.
501. Auwers, K. V., *Ges. Wiss.* (Marburg) **1927**, *62*, No. 4, 113.
502. Auwers, K. V., *J. Prakt. Chem.* **1938**, *150*, 166; *C.A.* **1938**, *32*, 4582.
503. Auwers, K. V., *Z. Phys. Chem.* (Leipzig) **1895**, *18*, 595.
504. Auwers, K. V., and F. Eisenlohr, *Z. Phys. Chem.* (Leipzig) **1913**, *83*, 429.
505. Auwers, K. V., and A. Frühling, *Ann. Chem.* **1921**, *422*, 192; *C.A.* **1921**, *15*, 2085.
506. Auwers, K. V., R. Hinterseber, et al., *Ann. Chem.* **1915**, *410*, 257; *C.A.* **1916**, *10*, 458; *Chem. Zentr.* **1916**, *I*, 99.
507. Auwers, K. V., and H. Jacobsen, *Ann. Chem.* **1922**, *426*, 161; *C.A.* **1922**, *16*, 1742.
508. Auwers, K. V., W. A. Roth, et al., *Ann. Chem.* **1910**, *373*, 267; *C.A.* **1910**, *4*, 2143.
509. Auwers, K. V., W. A. Roth, et al., *Ann. Chem.* **1911**, *385*, 102; *C.A.* **1912**, *6*, 604.
510. Averin, V. A., and V. Y. Schol, *Gazov. Khromatogr.* **1969**, No. 9, 142; *C.A.* **1970**, *72*, 2733b.
511. Aversa, M. C., M. Crisafulli, et al., *Gazz. Chim. Ital.* **1967**, *97*, 200.
512. Aveyard, R., *Trans. Faraday Soc.* **1967**, *63*, 2778.
513. Aveyard, R., and Y. Thompson, *Can. J. Chem.* **1979**, *57*, 856.
514. Aviado, D. M., and D. G. Smith, *Toxicology* **1975**, *3*, No. 2, 241; *C.A.* **1975**, *82*, 133688.
515. Avramescu, F., and D. A. Isacescu, *Rev. Roum. Chim.* **1978**, *23*, 655.
516. Awbery, J. H., *Phil. Mag.* **1941**, *31*, 247; *C.A.* **1941**, *35*, 3887.
517. Awbery, J. H., and E. Griffiths, *Proc. Phys. Soc.* (London) **1924**, *36*, 303; *C.A.* **1924**, *18*, 2979.
518. Awbery, J. H., and E. Griffiths, *Proc. Phys. Soc.* (London) **1936**, *48*, 372; *C.A.* **1936**, *30*, 4728.
519. Awwad, A. M., and R. A. Pethrick, *J. Chem. Thermodyn.* **1984**, *16*, 131.
520. Axford, D. W. E., and D. H. Rank, *J. Chem. Phys.* **1949**, *17*, 430.
521. Axtmann, R. C., W. E. Shuler, et al., *J. Chem. Phys.* **1959**, *31*, 850.
522. Ayers, G. W., and M. S. Agruss, *J. Amer. Chem. Soc.* **1939**, *61*, 83.
523. Ayres, E. E., *Ind. Eng. Chem.* **1929**, *21*, 899.
524. Azarnoosh, A., and J. J. McKetta, *Petrol. Refiner* **1958**, *37*, 275; *C.A.* **1959**, *53*, 1896.

B

525. B., S., *J. Soc. Chem. Ind.* (London) **1942**, 61; *Chem. Ind.* (London) **1942**, *20*, 120.
526. Babko, A. K., and L. L. Shevchenko, *Dopovidi Akad. Nauk Ukr. RSR* **1958**, 1212; *C.A.* **1959**, *53*, 9890.
527. Babushkina, T. A., and G. K. Semin, *Zh. Strukt. Khim.* **1966**, *7*, 114; *C.A.* **1966**, *64*, 18730.
528. Bacarella, A. L., E. Grunwald, et al., *J. Org. Chem.* **1955**, *20*, 747.
529. Baccanari, D. P., J. A. Novinski, et al., *Trans. Faraday Soc.* **1968**, *64*, 1201.
530. Bacher, W., and J. Wagner, *Z. Phys. Chem.* (Leipzig) **1939**, *B43*, 191; *C.A.* **1939**, *33*, 6719.
531. Bachmann, P., and K. Dziewònski, *Ber. Deut. Chem. Gesell.* **1903**, *36*, 971.
532. Backer, H. J., and N. D. Dijkstra, *Rec. Trav. Chim. Pays-Bas* **1932**, *51*, 290; Backer, H. J., *ibid* **1935**, *54*, 215; from [5122].
533. Badachhape, R. B., M. K. Gharpurey, et al., *J. Chem. Eng. Data* **1965**, *10*, 143.
534. Baddeley, G., *J. Chem. Soc.* **1944**, 330.
535. Badger, R. M., *J. Chem. Phys.* **1940**, *8*, 288.
536. Badger, R. M., and S. H. Bauer, *J. Amer. Chem. Soc.* **1937**, *59*, 303.
537. Badger, R. M., and S. H. Bauer, *J. Chem. Phys.* **1937**, *5*, 839.
538. Badin, E. J., and E. Pacsu, *J. Amer. Chem. Soc.* **1944**, *66*, 1963.
539. Badin, E. J., and E. Pascu, *J. Amer. Chem. Soc.* **1945**, *67*, 1352.
540. Badische (BASF), German Patent 294,638, July 5, 1914; *C.A.* **1917**, *11*, 2582.
541. Badische Anilin and Soda-Fabrik Akt.-Ges. (E. Rotter and E. Haaren), German Patent 957,212, Jan. 31, 1957; *C.A.* **1959**, *53*, 18864.
542. Badoche, M., *Bull. Soc. Chim. Fr.* **1941**, *8*, 212; *C.A.* **1942**, *36*, 1840.
543. Bag, A., T. Egupov, et al., *Masloboino Zhirovae Delo* **1937**, *13*, No. 2, 27; *C.A.* **1937**, *31*, 7410.
544. Bag, A., T. Egupov et al., *Org. Chem. Ind.* (USSR) **1936**, *2*, 141; *C.A.* **1937**, *31*, 1006.
545. Bag, S. C., *Indian J. Phys.* **1971**, *45*, 155; *C.A.* **1972**, *76*, 133667.

546. Bagal, L. I., G. F. Tereshenko, et al., *Zh. Org. Khim.* **1969**, *5*, 2106.
547. Baggett, N., S. A. Barker, et al., *J. Chem. Soc.* **1960**, 4565.
548. Baibuz, V. F., *Dokl. Akad. Nauk SSSR* **1961**, *140*, 1358; *C.A.* **1962**, *56*, 10999.
549. Bailey, C. R., *Nature* **1937**, *140*, 851.
550. Bailey, K. C., *J. Chem. Soc.* **1939**, 767.
551. Bailey, K. C., and W. E. S. Hickson, *J. Chem. Soc.* **1941**, 145.
552. Bailey, W. G., and A. Roy, *J. Soc. Chem. Ind.* (London) **1946**, *65*, 421.
553. Bair, E. J., and C. A. Kraus, *J. Amer. Chem. Soc.* **1951**, *73*, 2459.
554. Bak, B., *Kgl. Danske Videnskab. Selskab, Mat.-fys. Medd.* **1948**, *24*, 15; *C.A.* **1948**, *42*, 4821.
555. Bak, B., S. Brodersen, et al., *Acta Chem. Scand.* **1955**, *9*, 749; *C.A.* **1956**, *50*, 2296.
556. Bakaroluga, O., and S. Fallab, *Helv. Chim. Acta* **1963**, *46*, 2120.
557. Baker, F., *J. Chem. Soc.* **1912**, 1409.
558. Baker, R. H., *J. Amer. Chem. Soc.* **1938**, *60*, 2673.
559. Baker, R. H., and R. D. Schuetz, *J. Amer. Chem. Soc.* **1947**, *69*, 1250.
560. Baker, W., and F. B. Field, *J. Chem. Soc.* **1932**, 86.
561. Balandin, A. A., T. A. Slovokhotova, et al., *Dokl. Akad. Nauk SSSR* **1961**, *141*, 839; *C.A.* **1962**, *56*, 14138.
562. Balandin, A. A., M. B. Turova-Pollak, et al., *Dokl. Akad. Nauk SSSR* **1957**, *114*, 773; *C.A.* **1958**, *52*, 3701.
563. Baldauf, W., and H. Knapp, *Ber. Bunsen-Ges. Phys. Chem.* **1983**, *87*, 304.
564. Baldridge, J. R., British Patent 795,891, June 4, 1958; *C.A.* **1959**, *53*, 290.
565. Baldt, J. H., and H. K. K. Hall, *J. Amer. Chem. Soc.* **1971**, *93*, 140.
566. Baldwin, J. E., *J. Org. Chem.* **1965**, *30*, 2423.
567. Baldwin, M. A., A. M. Kirkien-Konasiewicz, et al., *J. Chem. Soc. B* **1968**, 34.
568. Baldwin, W. C. G., *Proc. Roy. Soc.* (London) **1937**, *162A*, 215; *Chem. Zentr.* **1938**, *II*, 1567.
569. Baliah, V., and M. Uma, *Indian J. Chem.* **1972**, *10*, 395.
570. Balis, E. W., W. F. Gilliam, et al., *J. Amer. Chem. Soc.* **1948**, *70*, 1654.
571. Balis, E. W., and H. A. Liebhafsky, *Ind. Eng. Chem.* **1946**, *38*, 538.
572. Ball, A. O., *J. Chem. Soc.* **1930**, 570.
573. Ballard, S. A., H. de V. Finch, et al., British Patent 619,014, March 2, 1949; *C.A.* **1949**, *43*, 7497.
574. Ballaus, O., *Monatsh. Chem.* **1943**, *74*, 88; *C.A.* **1943**, *37*, 6191.
575. Ballaus, O., and J. Wagner, *Z. Phys. Chem.* (Leipzig) **1939**, *45B*, 272; *C.A.* **1940**, *34*, 4991.
576. Ballester, M., J. Riera, et al., *J. Amer. Chem. Soc.* **1964**, *86*, 4276.
577. Ballinger, P., and F. A. Long, *J. Amer. Chem. Soc.* **1959**, *81*, 1050.
578. Ballinger, P., and F. A. Long, *J. Amer. Chem. Soc.* **1960**, *82*, 795.
579. Balson, E. W., *Trans. Faraday Soc.* **1947**, *43*, 48; *C.A.* **1947**, *41*, 5346.
580. Baly, E. C. C., and J. N. Collie, *J. Chem. Soc.* **1905**, 1332.
581. Baly, E. C. C., W. H. Edwards, et al., *J. Chem. Soc.* **1906**, 514.
582. Baly, E. C. C., and E. K. Ewbank, *J. Chem. Soc.* **1905**, 1347, 1355.
583. Baly, E. C. C., and F. G. Tryhorn, *J. Chem. Soc.* **1915**, 1058.
584. Baly, E. C. C., and W. B. Tuck, *J. Chem. Soc.* **1908**, 1902.
585. Balz, G., and G. Schiemann, *Ber. Deut. Chem. Gesell.* **1927**, *60B*, 1186.
586. Bambach, G., *Kaeltetechnik* **1956**, *8*, 334; *C.A.* **1957**, *51*, 4601; *Beil.* **EIV5**, 8.
587. Bandow, F., *Biochem. Z.* **1938**, *298*, 81; *Chem Zentr.* **1939**, *I*, 2389.
588. Banerjee, D. K., and C. C. Budke, *Anal. Chem.* **1964**, *36*, 792.
589. Banerjee, D. K., and C. C. Budke, *Anal. Chem.* **1964**, *36*, 2367.
590. Banerjee, S. B., *Indian J. Phys.* **1956**, *30*, 106; *C.A.* **1957**, *51*, 9313.
591. Banerjee, S., S. H. Yalkowsky, et. al., *Environ. Sci. Technol.* **1980**, *14*, 1227.
592. Banewicz, J. J., J. A. Maguire, et. al., *J. Phys. Chem.* **1968**, *72*, 1960.
593. Banick, W. M., *Anal. Chem.* **1962**, *34*, 296.
594. Bapat, R. N., *Indian J. Phys.* **1970**, *44*, No. 5, 273.
595. Bapat, R. N., *Proc. Indian Acad. Sci.* **1958**, *48A*, 119; *C.A.* **1959**, *53*, 10954.
596. Barabanov, V. P., *Zh. Fiz. Khim.* **1963**, *37*, 710; *C.A.* **1963**, *59*, 3551.
597. Barbanov, V. P., Tsentovskii, V. M., et al., *Zh. Fiz. Khim.* **1973**, *47*, 2316; *Russ. J. Phys. Chem.* **1973**, *47*, 1304.

598. Barak, M., and H. Hartley, *Z. Phys. Chem.* (Leipzig) **1933**, *165A*, 272.
599. Baratov, A. N., and V. V. Bulanova, *Khim. Prom.* (Moscow) **1970**, *46*, 735; *C.A.* **1971**, *74*, 33203.
600. Baratov, A. N., and V. M. Kucher, *Zh. Prikl. Khim.* **1965**, *38*, 1068; *C.A.* **1965**, *63*, 4088.
601. Barbaudy, J., *Bull. Soc. Chim. Fr.* **1926**, *39*, 371.
602. Bârcă-Gălăteanu, D., *Bull. Soc. Roumaine Phys.* **1938**, *38*, No. 69, 109; *C.A.* **1938**, *32*, 4433.
603. Bârcă-Gălăteanu, D., *Bull. Soc. Roumaine Phys.* **1945**, *46*, No. 83, 9; *C.A.* **1947**, *41*, 6812.
604. Barceló Matutano, J. R., *Anales. Real Soc. Espan. Fis. y Quim* **1949**, *45A*, 315; *C.A.* **1950**, *44*, 6723.
605. Barceló Matutano, J. R., *J. Res. Nat. Bur. Stand.* **1950**, *44*, 521.
606. Barchewitz, P., *Ann. Combustibles Liquides* **1938**, *13*, 501; *C.A.* **1939**, *33*, 2412.
607. Barchewitz, P., *Ann. Office Nat. Combustibles Liquides* (May/June 1938), *30*, 501, Paris, Sorbonne; *Chem. Zentr.* **1939**, *I*, 1743.
608. Barchewitz, P., *Compt. Rend.* **1937**, *204*, 246; *Chem. Zentr.* **1937**, *II*, 3591.
609. Barchewitz, P., *Compt. Rend.* **1939**, *208*, 807; *C.A.* **1939**, *33*, 3694.
610. Barchewitz, P., and M. Parodi, *Compt. Rend.* **1938**, *207*, 903; *C.A.* **1939**, *33*, 481.
611. Barchewitz, P., and M. Parodi, *J. Phys. Radium* **1939**, *10*, 143; *Chem. Zentr.* **1939**, *II*, 827.
612. Barclay, G. A., and R. J. W. Le Fèvre, *J. Chem. Soc.* **1950**, 556.
613. Barclay, G. A., and R. J. W. Le Fèvre, *J. Chem. Soc.* **1952**, 1643.
614. Barclay, G. A., R. J. W. Le Fèvre, et al., *Trans. Faraday Soc.* **1950**, *46*, 812.
615. Barclay, G. A., R. J. W. Le Fèvre, et al., *Trans. Faraday Soc.* **1951**, *47*, 357; *C.A.* **1951**, *45*, 8828.
616. Bardyshev, I. I., *Zh. Priklad. Khim.* **1948**, *21*, 1019; *C.A.* **1949**, *43*, 5249.
617, Bardyshev, I. I., A. L. Piryatinskiĭ, et al., *Zh. Priklad. Khim.* **1947**, *20*, 1308; *C.A.* **1948**, *42*, 9201.
618. Barker, E. F., *Rev. Mod. Phys.* **1942**, *14*, 198; *C.A.* **1943**, *37*, 1331.
619. Barker, I. R. L., and W. A. Waters, *J. Chem. Soc.* **1952**, 150.
620. Barker, P. E., and S. Al-Madfai, *J. Chromatogr. Sci.* **1969**, *7*, 425.
621. Barker, S. A., E. J. Bourne, et al., *J. Chem. Soc.* **1959**, 802.
622. Barnes, D. S., and G. Pilcher, *J. Chem. Thermodyn.* **1975**, *7*, 377.
623. Barnes, J., and W. H. Fulweiler, *J. Amer. Chem. Soc.* **1927**, *49*, 2034.
624. Barnes, J., and W. H. Fulweiler, *J. Amer. Chem. Soc.* **1929**, *51*, 1750.
625. Barnes, L., and M. S. Parvlak, *Anal. Chem.* **1959**, *31*, 1875.
626. Barnes, R. B., *Nature* **1929**, *124*, 300.
627. Barnes, R. B., *Phys. Rev.* **1930**, *35*, 1524.
628. Barnes, R. B., L. G. Bonner, et al., *J. Chem. Phys.* **1936**, *4*, 772.
629. Barnes, R. B., H. Eyring, et al., *Anal. Chem.* **1948**, *20*, 96.
630. Barnes, R. B., R. C. Gore, et al., *Anal. Chem.* **1948**, *20*, 402.
631. Barnes, R. B., R. C. Gore, et al., *Infrared Spectroscopy*, Reinhold, New York, 1944.
632. Barnes, R. B., U. Liddel, et al., *Ind. Eng. Chem.*, *Anal. Ed.* **1943**, *15*, 83.
633. Barnes, R. B., U. Liddel, et al., *Ind. Eng. Chem.*, *Anal. Ed.* **1943**, *15*, 659.
634. Barnes, R. B., R. S. McDonald, et al., *J. Appl. Phys.* **1945**, *16*, 77.
635. Barnes, R. H., I. I. Rusoff, et al., *Ind. Eng. Chem.*, *Anal. Ed.* **1944**, *16*, 385.
636. Barr, D. A., and R. N. Haszeldine, *J. Chem. Soc.* **1955**, 4169.
637. Barr, E. S., and E. K. Plyler, *J. Chem. Phys.* **1936**, *4*, 90.
638. Barreira, F., and G. J. Hills, *Trans. Faraday Soc.* **1968**, *64*, 1359.
639. Barrett, J., and M. J. Hitch, *Spectrochim. Acta, Part A* **1968**, *24*, 265.
640. Barrett, J. J., and M. C. Tobin, *J. Opt. Soc. Amer.* **1966**, *56*, 129.
641. Barron, E. S. G., *J. Biol. Chem.* **1937**, *121*, 313.
642. Barrow, G. M., *J. Phys. Chem.* **1955**, *59*, 1129.
643. Barrow, G. M., and A. L. McClellan, *J. Amer. Chem. Soc.* **1951**, *73*, 573.
644. Barrow, G. M., and K. S. Pitzer, *Ind. Eng. Chem.* **1949**, *41*, 2737.
645. Barrow, G. M., and S. Searles, *J. Amer. Chem. Soc.* **1953**, *75*, 1175.
646. Bartell, F. E., and A. D. Wooley, *J. Amer. Chem. Soc.* **1933**, *55*, 3518.
647. Bartlett, P. D., U.S. Patent 3,149,131, Sept. 15, 1964; *C.A.* **1964**, *61*, 14639.
648. Bartlett, P. D., and R. Altschul, *J. Amer. Chem. Soc.* **1945**, *67*, 812.
649. Bartlett, P. D., and J. D. McCollum, *J. Amer. Chem. Soc.* **1956**, *78*, 1441.
650. Bartlett, P. D., and K. Nozaki, *J. Amer. Chem. Soc.* **1946**, *68*, 1495.
651. Bartoli, A., *Gazzetta* **1894**, *24*, *II*, 156.
652. Barton, A. F. M., *Chem. Reviews* **1975**, *75*, 731.

653. Barton, A. F. M., *CRC Handbook of Solubility Paramters*, CRC Press, Boca Raton, 1983.
654. Barton, D. H. R., *J. Chem. Soc.* **1949**, 148.
655. Barton, D. H. R., *Nature* **1946**, *157*, 626.
656. Barton, D. H. R., A. J. Head, et al., *J. Chem. Soc.* **1951**, 2039.
657. Barton, D. H. R., and K. E. Howlett, *J. Chem. Soc.* **1949**, 155.
658. Barton, D. H. R., and K. E. Howlett, *J. Chem. Soc.* **1949**, 165.
659. Barton, D. H. R., and P. F. Onyon, *J. Chem. Soc.* **1950**, *72*, 988.
660. Bartoszewicz, E., *Rocz. Chem.* **1931**, *11*, 90.
661. Baskov, A., *J. Russ. Phys. Chem. Soc.* **1913**, *45*, 1604.
662. Bass, K. C., *J. Chem. Soc.* **1964**, 3498.
663. Bass, S. J., W. I. Nathan, et al., *J. Phys. Chem.* **1964**, *68*, 509.
664. Bataafsche, de N. V., Petroleum Maatschappij, Dutch Patent 63,335, June 15, 1949; *C.A.* **1949**, *43*, 7502.
665. Batalin, V. S., and P. G. Ugryumov, *Sintet. Kauchuk* **1936**, No. 6, 8; *C.A.* **1936**, *30*, 6701.
666. Bateman, L. C., and E. D. Hughes, *J. Chem. Soc.* **1937**, 1187.
667. Bates, H. H., J. M. Mullaly, et al., *J. Chem. Soc.* **1923**, 401.
668. Bates, O. K., *Ind. Eng. Chem.* **1936**, *28*, 494.
669. Bates, R. G., *Ann. N.Y. Acad. Sci.* **1961**, *92*, Art. 2, 341.
670. Bates, R. G., *Determination of pH, Theory and Practice*, Wiley, New York, 1964.
671. Bates, R. G., and G. F. Allen, *J. Res. Nat. Bur. Stand.* **1960**, *64A*, 343.
672. Bates, R. G., and V. E. Bower, *J. Res. Nat. Bur. Stand.* **1956**, *57*, 153.
673. Bates, R. G., and H. B. Hetzer, *J. Phys. Chem.* **1961**, *65*, 667.
674. Bates, R. G., and G. D. Pinching, *J. Res. Nat. Bur. Stand.* **1951**, *46*, 349.
675. Bates, R. G., and G. Schwarzenbach, *Helv. Chim. Acta* **1954**, *37*, 1437.
676. Bates, W. W., and M. E. Hobbs, *J. Amer. Chem. Soc.* **1951**, *73*, 2151.
677. Batisheheva, M. G., *Inzhener-Fiz. Zh.*, *Akad. Nauk Belorus. SSR* **1959**, *2*, No. 5, 107; *C.A.* **1960**, *54*, 5245.
678. Batuev, M. I., *Bull. Acad. Sci. URSS, Classe Sci. Chim.* **1947**, No. 1, 3; *C.A.* **1948**, *42*, 4463.
679. Batuev, M. I., *Compt. Rend. Acad. Sci. URSS* **1946**, *52*, 401; *C.A.* **1947**, *41*, 1936.
680. Batuev, M. I., *Compt. Rend. Acad. Sci. URSS* **1946**, *53*, 507; *C.A.* **1947**, *41*, 2995.
681. Batuev, M. I., *Dokl. Akad. Nauk SSSR* **1948**, *59*, 1117; *C.A.* **1948**, *42*, 6664.
682. Batuev, M. I., A. A. Akhrem, et al., *Izvest. Akad. Nauk SSSR, Otdel. Khim. Nauk* **1960**, 2201; *C.A.* **1961**, *55*, 13056.
683. Batuev, M. I., E. N. Prilezhaeva, et al., *Bull. Acad. Sci. URSS, Classe Sci. Chim.* **1947**, 123; *C.A.* **1948**, *42*, 4464.
684. Batyrova, T. F., *Gig. Tr. Prof. Zabol.* **1973**, 52; *C.A.* **1974**, *80*, 78925.
685. Bauder, A., and H. H. Günthard, *Helv. Chim. Acta* **1958**, *41*, 670.
686. Baudler, M., *Z. Elektrochem.* **1955**, *59*, 173; *C.A.* **1955**, *49*, 11418.
687. Baum, E., and W. O. Hermann, German Patent 559,734, May 20, 1930; *C.A.* **1933**, *27*, 735.
688. Bauman, R. P., and B. J. Bulkin, *J. Chem. Phys.* **1966**, *45*, 496.
689. Baume, G., *J. Chim. Phys.* **1908**, *6*, 1.
690. Bauveault, L., *Compt. Rend.* **1904**, *138*, 1108; *Beil.* **H1**, 406.
691. Bayard, P., *Bull. Soc. Roy. Sci. Liege* **1943**, *12*, 179; *Chem. Zentr.* **1943**, *II*, 409; *C.A.* **1944**, *38*, 5147.
692. Baykut, S., *Rev. Fac. Sci. Univ. Istanbul* **1956**, *21C*, 36; *Beil.* **EIV2**, 1019.
693. Bayles, J. W., and A. Chetwyn, *J. Chem. Soc.* **1958**, 2328.
694. Bayliss, N. S., and C. J. Breckenridge, *J. Amer. Chem. Soc.* **1955**, *77*, 3959.
695. Bazhulin, P. A., A. V. Koperina, et al., *Izvest. Akad. Nauk SSSR, Otdel. Khim. Nauk* **1954**, 709; *C.A.* **1955**, *49*, 10866.
696. Bazhulin, P. A., A. F. Plate, et al., *Bull. Acad. Sci. URSS, Classe, Sci. Chim.* **1941**, 13; *C.A.* **1943**, *37*, 5315.
697. Bazhulin, P. A., K. E. Sterin, et al., *Izvest. Akad. Nauk SSSR, Otdel. Khim. Nauk* **1946**, 7; *C.A.* **1948**, *42*, 6238.
698. Bazhulin, P. A., S. A. Ukholin, et al., *Izvest. Akad. Nauk SSSR, Otdel. Khim. Nauk* **1949**, 481; *C.A.* **1950**, *44*, 1331.
699. Beal, G. D., and B. L. Souther, *J. Amer. Chem. Soc.* **1927**, *49*, 1994.

700. Beal, J. L., and C. A. Mann, *J. Phys. Chem.* **1938,** *42*, 283.
701. Beale, R. N., *J. Chem. Soc.* **1954,** 4494.
702. Beattie, J. A., D. R. Douslin, et al., *J. Chem. Phys.* **1951,** *19*, 948.
703. Beattie, J. A., and D. G. Edwards, *J. Amer. Chem. Soc.* **1948,** *70*, 3382.
704. Beattie, J. A., D. G. Edwards, et al., *J. Chem. Phys.* **1949,** *17*, 576.
705. Beattie, J. A., and W. C. Kay, *J. Amer. Chem. Soc.* **1937,** *59*, 1586.
706. Beattie, J. A., and S. W. Levine, *J. Amer. Chem. Soc.* **1951,** *73*, 4431.
707. Beattie, J. A., N. Poffenberger, et al., *J. Chem. Phys.* **1935,** *3*, 96.
708. Beattie, J. A., G. L. Simard, et al., *J. Amer. Chem. Soc.* **1939,** *61*, 24.
709. Beattie, W. H., *Appl. Spectrosc.* **1975,** *29*, 334.
710. Becconsall, J. K., and P. Hampson, *Mol. Phys.* **1965,** *10*, 21; *C.A.* **1965,** *64*, 13578.
711. Beckers, N. L., U.S. Patent 2,903,488, Sept. 8, 1959; *C.A.* **1960,** *54*, 2165.
712. Beckett, C. W., and K. S. Pitzer, *J. Amer. Chem. Soc.* **1946,** *68*, 2213.
713. Beckett, C. W., K. S. Pitzer, et al., *J. Amer. Chem. Soc.* **1947,** *69*, 2488.
714. Beckey, H. D., and P. Schulze, *Z. Naturforsch.* **1965,** *20A*, 1329, 1335; *C.A.* **1966,** *64*, 4411.
715. Beckey, H. D., and P. Schulze, *Z. Naturforsch.* **1966,** *21A*, 214; *C.A.* **1966,** *65*, 4816.
716. Beckmann, E., *Arch. Pharm.* (Weinheim) *245*, 211; *Chem. Zentr.* **1907,** *II*, 191.
717. Beckmann, E., *Z. Phys. Chem.* (Leipzig) **1888,** *2*, 638, 715.
718. Beckmann, E., *Z. Phys. Chem.* (Leipzig) **1890,** *6*, 437.
719. Beckmann, E., *Z. Phys. Chem.* (Leipzig) **1891,** *7*, 323.
720. Beckmann, E., *Z. Phys. Chem.* (Leipzig) **1891,** *8*, 223.
721. Beckmann, E., *Z. Phys. Chem.* (Leipzig) **1903,** *46*, 853.
722. Beckmann, E., *Z. Phys. Chem.* (Leipzig) **1907,** *57*, 129.
723. Beckmann, E., *Z. Phys. Chem.* (Leipzig) **1907,** *58*, 543.
724. Beckmann, E., *Z. Phys. Chem.* (Leipzig) **1908,** *63*, 177.
725. Beckmann, E., and O. Faust, *Z. Phys. Chem.* (Leipzig) **1915,** *89*, 247.
726. Beckmann, E., G. Fuchs, et al. *Z. Phys. Chem.* (Leipzig) **1895,** *18*, 473.
727. Beckmann, E., and W. Gabel, *Ber. Deut. Chem. Gesell.* **1906,** *39*, 2611
728. Beckmann, E., and W. Gabel, *Z. Anorg. Allg. Chem.* **1906,** *51*, 236; *Chem. Zentr.* **1906,** *II*, 1804.
729. Beckmann, E., and F. Junker, *Z. Anorg. Allg. Chem.* **1907,** *55*, 371; *J. Chem. Soc.* **1907Aii,** 927.
730. Beckmann, E., and O. Liesche, *Z. Phys. Chem.* (Leipzig) **1914,** *88*, 23.
731. Beckmann, E., and O. Liesche, *Z. Phys. Chem.* (Leipzig) **1914,** *88*, 419.
732. Beckmann, E., and O. Liesche, *Z. Phys. Chem.* (Leipzig) **1915,** *89*, 111.
733. Beckmann, E., and G. Lockemann, *Z. Phys. Chem.* (Leipzig) **1907,** *60*, 385.
734. Beckmann, E., and A. Stock, *Z. Phys. Chem.* (Leipzig) **1895,** *17*, 107.
735. Beckmann, E., and P. Waentig, *Z. Anorg. Allg. Chem.* **1910,** *67*, 17; *Chem. Zentr.* **1910,** *II*, 277.
736. Beckwith, W. F., and R. W. Fakien, *Chem. Eng. Prog. Symposium, Ser. 44, Thermodyn.* **1963,** 75.
737. Bedford, A. F., A. E. Beezer, et al., *J. Chem. Soc.* **1963,** 2039.
738. Bednenko, P. F., *Vrachebnoe Delo* **1947,** *27*, 85; *C.A.* **1948,** *42*, 4675.
739. Beer, M., H. B. Kessler, et al., *J. Chem. Phys.* **1958,** *29*, 1097.
740. Béguin, C., and T. Gaümann, *Helv. Chim. Acta* **1958,** *41*, 1971.
741. Béguin, C., and H. H. Günthard, *Helv. Chim. Acta* **1959,** *42*, 2262.
742. Behringer, J., *Z. Elektrochem.* **1958,** *62*, 544; *C.A.* **1958,** *52*, 16872.
743. Beilstein, F., and A. Kurbatow, *Ann. Chem.* **1875,** *176*, 27.
744. Belanger, G., and C. Sandorfy, *Chem. Phys. Lett.* **1969,** *3*, 661.
745. Belcher, D., *J. Amer. Chem. Soc.* **1938,** *60*, 2744.
746. Belinskii, B. A., and E. V. Ergopulo, *Zh. Fiz. Khim.* **1968,** *42*, 1520.
747. Bell, F. K., *J. Amer. Chem. Soc.* **1935,** *57*, 1023.
748. Bell, R. P., E. C. Baughan, et al., *J. Chem. Soc.* **1934,** 1969.
749. Bell, R. P., and P. W. Smith, *J. Chem. Soc., Phys. Org. B* **1966,** 241.
750. Belladen, L., *Gazz. Chim. Ital.* **1927,** *57*, 407; *C.A.* **1927,** *21*, 3298.
751. Bellamy, I. L. J., and L. Beecher, *J. Chem. Soc.* **1952,** 475.
752. Bellanato, J., and E. D. Schmid, *Anales Real Soc. Espan. Fis. y Quim.* (Madrid) **1960,** *56B*, 949; *C.A.* **1961,** *55*, 16144.
753. Bellanato, J., and E. D. Schmid, *Anales Real Soc. Espan. Fis. y Quim.* (Madrid) **1961,** *57A*, No. 11/12, 319; *C.A.* **1962,** *57*, 5481.

754. Bellocq, A. M., R. Martegoutes, et al., *J. Chim. Phys. Physicochim. Biol.* **1969**, *66*, 449.
755. Belopol'skaya, T. V., *Vestn. Leningr. Univ.*, *Fiz.*, *Khim.* **1976**, *22*, 44.
756. Belova, S. F., and E. A. Korlyakova, *Gig. i Sanit.* **1958**, *23*, No. 9, 72; *C.A.* **1959**, *53*, 10560.
757. Bencze, K., *Chem. Zvesti* **1961**, *15*, 571; *C.A.* **1962**, *57*, 53.
758. Benko, J., *Acta Chim. Acad. Sci. Hungr.* **1962**, *34*, 217.
759. Bender, D. F., E. Sawicki, et al., *Anal. Chem.* **1964**, *36*, 1011.
760. Bender, M. L., *J. Amer. Chem. Soc.* **1953**, *75*, 5986.
761. Bender, P., G. T. Furukawa, et al., *Ind. Eng. Chem.* **1952**, *44*, 387.
762. Bennett, G. M., and S. Glasstone, *Proc. Roy. Soc.* (London) **1934**, *145A*, 71.
763. Bennett, G. M., and W. G. Philip, *J. Chem. Soc.* **1928**, 1930.
764. Bennett, G. M., and W. G. Phillip, *J. Chem Soc.* **1928**, 1937.
765. Bennett, G. M., and G. H. Willis, *J. Chem Soc.* **1928**, 2305.
766. Bennett, O. F., *Anal. Chem.* **1964**, *36*, 684.
767. Bennett, W. H., and F. Daniels, *J. Amer. Chem. Soc.* **1927**, *49*, 50.
768. Bennewitz, K., and W. Rossner, *Z. Phys. Chem.* (Leipzig) **1938**, *39B*, 126; *C.A.* **1938**, *32*, 4868.
769. Benning, A. F., U.S. Patent 2,426,172, Aug. 26, 1947; *C.A.* **1948**, *42*, 198.
770. Benning, A. F., and R. C. McHarness, *Ing. Eng. Chem.* **1939**, *31*, 912.
771. Benning, A. F., and R. C. McHarness, *Ing. Eng. Chem.* **1940**, *32*, 497.
772. Benning, A. F., and R. C. McHarness, *Ing. Eng. Chem.* **1940**, *32*, 814.
773. Benning, A. F., R. C. McHarness, et al., U.S. Patent 2,470,088, May 17, 1949; *C.A.* **1949**, *43*, 5789.
774. Benning, A. F., R. C. McHarness, et al., *Ind. Eng. Chem.* **1940**, *32*, 976.
775. Bennoson, M., and D. J. Williams, *J. Phys. Chem.* **1972**, *76*, 3673.
776. Benoit, J., and G. Ney, *Compt. Rend.* **1939**, *208*, 1888; *C.A.* **1939**, *33*, 6103.
777. Benson, G., and R. M. Kitchen, *Can. J. Res.* **1949**, *27F*, 266; *C.A.* **1949**, *43*, 6947.
778. Bent, F. A., and S. N. Wik, *Ind. Eng. Chem.* **1936**, *28*, 312.
779. Bentley, F. E., private communication.
780. Bentley, F. F., N. T. McDevitt, et al., *Spectrochim Acta* **1964**, *20*, 105.
781. Bentley, F. F., and E. F. Wolfarth, *Spectrochim. Acta* **1959**, 165.
782. Bentley, J. B., K. B. Everard, et al., *J. Chem. Soc.* **1949**, 2957.
783. Berezina, R. I., and N. K. Sidorov, *Opt. Spektrosk.* **1972**, *32*, 297; *C.A.* **1972**, *76*, 119421.
784. Berg, H., and H. Mader, U.S. Patent 2,356,562, Aug. 22, 1944; *C.A.* **1945**, *39*, 221.
785. Berg, L., J. M. Harrison, et al., *Ind. Eng. Chem.* **1945**, *37*, 585.
786. Berger, C., and L. R. Dawson, *Anal. Chem.* **1952**, *24*, 994.
787. Berghoff, V., *Z. Phys. Chem.* (Leipzig) **1894**, *15*, 422.
788. Bergmann, E., *J. Chem Soc.* **1936**, 402.
789. Bergmann, E., L. Engel, et al., *Ber. Deut. Chem. Gesell.* **1932**, *65*, 446.
790. Bergmann, E. D., U. Z. Littauer, et al., *J. Chem Soc.* **1952**, 847.
791. Bergmann, E. D., and S. Pinchas, *Rec. Trav. Chim. Pays-Bas* **1952**, *71*, 161; *C.A.* **1953**, *47*, 6397.
792. Bergmann, M., A. Miekeley, et al., *Ber. Deut. Chem. Gesell.* **1929**, *62*, 1467.
793. Bergmeister, E., and G. Hübner, German Patent 1,005,953, April 11, 1957; *C.A.* **1959**, *53*, 16597.
794. Berkengeïm, T. I., *Zavodskaya Lab.* **1941**, *10*, 592; *C.A.* **1946**, *40*, 6961.
795. Berkman, Y. P., and L. A. Polonskaya, *Nauch. Zapiski L'vov. Politekh. Inst.* **1956**, *22*, 49; *Referat. Zh. Khim.* **1957**, Abstr. No. 756; *C.A.* **1959**, *53*, 4867.
796. Berliner, J. F. T., and O. E. May, *J. Amer. Chem Soc.* **1927**, *49*, 1007.
797. Berman, H. A., and E. D. West, *J. Chem. Eng. Data.* **1969**, *14*, 107.
798. Berman, N. S., C. W. Larkam, et al., *J. Chem. Eng. Data.* **1964**, *9*, 218.
799. Berman, N. S., and J. J. McKetta, *J. Phys. Chem.* **1962**, *66*, 1444.
800. Berman, H. A., and E. D. West, *J. Chem. Eng. Data.* **1967**, *12*, 197.
801. Bermejo, L., and L. Blas, *Anales Soc. Espan. Fis. Quim.* **1930**, *28*, 706; *C.A.* **1930**, *24*, 4226.
802. Bernal, P., and W. A. Van Hook, *J. Chem. Thermodyn.* **1984**, *16*, 197.
803. Bernard, B., S. Pallade, et al., *Prom. Toksikol. i Klinika Prof. Zabol. Khim. Etiol. Sb* **1962**, 133; *C.A.* **1964**, *61*, 3600.
804. Bernard-Houlplain, M. C., and C. Sandorfy, *J. Chem. Phys.* **1972**, *56*, 3412.
805. Berner, E., *Z. Phys. Chem.* (Leipzig) **1929**, *141A*, 91.
806. Bernstein, R. B., J. P. Zietlow, et al., *J. Chem. Phys.* **1953**, *21*, 1778.
807. Bernstein, H. J., *Can. J. Res.* **1950**, *28B*, 132.

808. Bernstein, H. J., *J. Chem. Phys.* **1956,** *24,* 911.

809. Bernstein, H. J., and D. A. Ramsay, *J. Chem. Phys.* **1949,** *17,* 556.

810. Bernstein, H. J., R. G. Romans, et al., *Trans. Roy. Soc. Can.* **1936,** *III,* 33, 49, 57; *C.A.* **1937,** *31,* 1701.

811. Bernstein, H. J., and W. C. Schneider, *Proc. Roy. Soc.* (London) **1956,** *236A,* 515; *C.A.* **1957,** *51,* 76.

812. Bernstein, H. J., W. G. Schneider, et al., *J. Chem. Phys.* **1957,** *26,* 957.

813. Bernstein, J., and W. T. Miller, *J. Amer. Chem. Soc.* **1940,** *62,* 948.

814. Berry, C. E., *J. Chem. Phys.* **1949,** *17,* 1164.

815. Berry, K. L., and J. M. Sturtevant, *J. Amer. Chem. Soc.* **1942,** *64,* 1599.

816. Berry, R. S., S. A. Rice, et al., *Physical Chemistry,* Wiley, New York, 1980.

817. Bertelli, D. J., and T. G. Andrews, *Tetrahedron Lett.* **1967,** 4467.

818. Berthelat, D., and H. Goudechon, *Compt. Rend.* **1910,** *151,* 478; *C.A.* **1911,** *5,* 2629.

819. Berthelon, G., M. Giray, et al., *Bull. Soc. Chim. Fr.* **1971,** 3180.

820. Berthelot, D., and J. Ogier, *Ann. Chim. Phys.* **1881,** *23,* 201.

821. Berthet, M., *Bull. Classe Sci. Acad. Roy. Belges* **1941,** *27,* 212; *Chem. Zentr.* **1942,** *I,* 2115; *C.A.* **1943,** *37,* 3400.

822. Berthoud, A., *J. Chim. Phys.* **1917,** *15,* 3; *C.A.* **1917,** *11,* 1922.

823. Bertie, J. E., and S. Sunder, *Can. J. Chem.* **1973,** *51,* 3344.

824. Bertin, D. M., C. Chatain-Cathaud, et al., *C. R. Acad. Sci.,* Ser C. **1972,** *274,* 1112.

825. Bertin, D. M., M. Farnier, et al., *Bull. Soc. Chim. Fr.* **1974,** No. 12, Part 1, 2677.

826. Bertoluzza, A., G. Battista, et al., *J. Chim. Phys.* **1966,** *63,* 395; *C.A.* **1966,** *65,* 1595.

827. Berton, A., *Ann. Chim.* **1944,** *19,* 394; *C.A.* **1946,** *40,* 1733.

828. Bertram, R., and K. Cruse, *Ber. Bunsenges. Phys. Chem.* **1963,** *67,* 98; *C.A.* **1963,** *58,* 12038.

829. Bertram, S. H., *Rec. Trav. Chim. Pays-Bas* **1927,** *46,* 397.

830. Bertsch, H., A. Greiner, et al., *J. Prakt. Chem.* **1964,** *25,* 184; *C.A.* **1964,** *61,* 15967.

831. Beskov, S. D., L. I. Kochetkova, et al., *Uchen. Zapiski Moskov. Gosudarst. Pedagog. Inst. im. V. I. Lenina* **1957,** *99,* 147; *C.A.* **1959,** *53,* 20993.

832. Besley, L., and G. A. Bottomley, *J. Chem. Thermodyn.* **1973,** *5,* 397.

833. Besley, L., and G. A. Bottomley, *J. Chem. Thermodyn.* **1974,** *6,* 577.

834. Bewley, T., British Patent 584,788, Jan. 23, 1947; *C.A.* **1947,** *41,* 5145.

835. Beyaert, M., *Natuurwetensch. Tijdschr.* **1937,** *19,* 197; *Chem. Zentr.* **1938,** *I,* 1768.

836. Beynon, E. T., and J. J. McKetta, *J. Phys. Chem.* **1963,** *67,* 2761.

837. Beynon, J. H., G. R. Lester, et al., *J. Phys. Chem.* **1959,** *63,* 1861.

838. Beynon, J. H., and R. A. Saunders, *Brit. J. Appl. Phys.* **1960,** *11,* 128; *C.A.* **1960,** *54,* 14805; Ref [3321].

839. Beynon, J. H., R. A. Saunders, et al., *Anal. Chem.* **1961,** *33,* 221

840. Beynon, J. H., R. A. Saunders, et al., *Appl. Spectrosc.* **1960,** *14,* 95; *C.A.* **1960,** *54,* 20464.

841. Beynon, W. J. G., *Phil. Mag.* **1938,** *25,* 443; *C.A.* **1938,** *32,* 4077.

842. Beynon, W. J. G., and E. J. Evans, *Phil. Mag.* **1938,** *25,* 476; *C.A.* **1938,** *32,* 4076.

843. Bezborodko, G. L., *Plast. Massy* **1961,** No. 4, 63; *C.A.* **1963,** *58,* 5496.

844. Bhadani, S. N., *Indian J. Chem.* **1972,** 88.

845. Bhagvantam, S., *Indian J. Phys.* **1932,** *6,* 595.

846. Bhagvantam, S., and S. Venkataswaran, *Nature* **1930,** *125,* 237; *C.A.* **1930,** 24, 2052.

847. Bharucha, K. R., and B. C. L. Weedon, *J. Chem Soc.* **1953,** 1584.

848. Bhatnagar, V. M., *Chem Ind.* (London) **1966,** 731.

849. Bhattacharyya, D. N., C. L. Lee, et al., *J. Phys. Chem.* **1965,** *69,* 608.

850. Bhattacharyya, S. K., and S. H. Nakhati, *J. Indian Chem. Soc.* **1947,** *24,* 1; *Beil.* **EIII4,** 697.

851. Bichowsky, F. R., and W. K. Gilkey, *Ind. Eng. Chem.* **1931,** *23,* 366.

852. Bicknell, R. T. M., D. B. Davies, et al., *J. Chem. Soc., Faraday Trans. 1* **1982,** *78,* 1595.

853. Biddiscombe, D. P., R. R. Collerson, et al., *J. Chem Soc.* **1963,** 1954.

854. Biddiscombe, D. P., E. A. Coulson, et al., *J. Chem Soc.* **1954,** 1957.

855. Biddiscombe, D. P., and J. F. Martin, *Trans. Faraday Soc.* **1958,** *54,* 1316.

856. Biedermann, W., and K. Raichle, German Patent 964,044, May 16, 1957; *C.A.* **1959,** *53,* 11223.

857. Bielecki, J., and V. Henry, *Ber. Deut. Chem. Gesell.* **1912,** *45,* 2819.

858. Bielecki, J., and V. Henry, *Compt. Rend.* **1912**, *155*, 456; *Chem. Zentr.* **1912**, *II*, 1330.
859. Bielecki, J., and V. Henry, *Compt. Rend.* **1912**, *155*, 1617; *Chem. Zentr.* **1913**, *I*, 587.
860. Bielecki, J., and V. Henry, *Compt. Rend.* **1913**, *156*, 550.
861. Bielecki, J., and V. Henry, *Compt. Rend.* **1913**, *156*, 1322; *Chem. Zentr.* **1913**, *II*, 238.
862. Bielecki, J., and V. Henry, *Compt. Rend.* **1913**, *156*, 1861; *Chem. Zentr.* **1913**, *II*, 649.
863. Bien, G. S., C.A. Kraus, et al., *J. Amer. Chem. Soc.* **1934**, *56*, 1860.
864. Bigelow, M. J., *J. Chem. Ed.* **1968**, *45*, 108.
865. Bigelow, S. L., *Amer. Chem. J.* **1899**, *22*, 280.
866. Bigg, D. C., S. C. Banerjee, et al., *J. Chem. Eng. Data* **1964**, *9*, 17.
867. Biggs, A. I., *J. Chem Soc.* **1961**, 2572.
868. Biggs, A. I., *Trans. Faraday Soc.* **1956**, *52*, 35.
869. Biggs, A. I., and R. A. Robinson, *J. Chem. Soc.* **1961**, 388.
870. Billeter, E., T. Bürer, et al., *Helv. Chim. Acta* **1957**, *40*, 2046.
871. Bilterys, R., and J. Gisseleire, *Bull. Soc. Chim. Belges* **1935**, *44*, 574.
872. Biltz, H., *Ber. Deut. Chem. Gesell.* **1903**, *36*, 1110.
873. Biltz, H., *Monatsh. Chem.* **1901**, *22*, 627; *J. Chem. Soc. A* **1903**, 411.
874. Biltz, H., *Z. Phys. Chem.* (Leipzig) **1896**, *19*, 385.
875. Biltz, W., W. Fischer, et al., *Z. Phys. Chem.* (Leipzig) **1930**, *151*, 13.
876. Bingham, E. C., *Fluidity and Plasticity*, 1st ed., McGraw-Hill, New York, 1922.
877. Bingham, E. C., *J. Rheology* **1931**, *2*, 403.
878. Bingham, E. C., and H. J. Fornwalt, *J. Rheology* **1929–30**, *1*, 372.
879. Bingham, E. C., and J. A. Geddes, *Physics* **1934**, *5*, 42.
880. Bingham, E. C., and L. A. Sarver, *J. Amer. Chem. Soc.* **1920**, *42*, 2011.
881. Bingham, E. C., and L. W. Spooner, *J. Rheology* **1932**, *3*, 221.
882. Binns, E. H., *Trans. Faraday Soc.* **1959**, *55*, 1900.
883. Biquard, D., *Bull. Soc. Chim. Fr.* **1940**, *7*, 894; *C.A.* **1942**, *36*, 2207.
884. Birch, S. F., R. A. Dean, et al., *J. Amer. Chem. Soc.* **1949**, *71*, 1362.
885. Birch, S. F., and D. T. McAllan, *J. Chem. Soc.* **1951**, 2556.
886. Birkenstock, W., *Z. Phys. Chem.* (Leipzig) **1928**, *138A*, 432.
887. Biron, E., *J. Russ. Phys. Chem. Soc.* *42*, 135, 167; *C.A.* **1911**, *5*, 609; *Chem. Zentr.* **1910**, *I*, 1914.
888. Biron, E., *Z. Phys. Chem.* (Leipzig) **1913**, *81*, 590.
889. Birun, A. M., *Nauch. Zapiski Inst. Naradnogo Khoz. im. Plekhanova* **1938**, No. 1, 120; *C.A.* **1940**, *34*, 1969.
890. Bisanz, T., *Rocz. Chem.* **1963**, *37*, 133; *C.A.* **1963**, *59*, 7348.
891. Bisanz, T., and B. Dybowska, *Rocz. Chem.* **1959**, *33*, 975; *C.A.* **1960**, *54*, 6656.
892. Biscarini, P., F. Taddei, et al., *Boll. Sci. Fac. Chim Ind. Bologna* **1963**, *21*, 169; *C.A.* **1964**, *60*, 2464.
893. Bishui, B. M., *Indian J. Phys.* **1948**, *22*, 167; *C.A.* **1948**, *42*, 8651.
894. Bishui, B. M., *Indian J. Phys.* **1948**, *22*, 253; *C.A.* **1949**, *43*, 2514.
895. Bishui, B. M., and S. B. Senyal, *Indian J. Phys.* **1947**, *21*, 233; *C.A.* **1948**, *42*, 6238.
896. Bittrich, H., A. Fahl, et al., *Z. Phys. Chem.* (Leipzig) **1983**, *264*, 891.
897. Bittrich, H. J., E. Kauer, et al., *J. Prakt. Chem.* **1962**, *17*, 250.
898. Bittrich, H. J., and R. Kind, *Wiss. Z. Tech. Hochsch. Chem. Leuna-Merseburg* **1969**, *11*, 222; *C.A.* **1970**, *72*, 36586.
899. Bjellerup, L., *Acta Chem. Scand.* **1961**, *15*, 121; *C.A.* **1961**, *55*, 19446.
900. Bjellerup, L., *Acta Chem. Scand.* **1961**, *15*, 231.
901. Bjellerup, L., and L. Smith, *Kgl. Fysiograf. Sallskap Lund. Forh.* **1954**, *24*, 21; *C.A.* **1955**, *49*, 14464.
902. Bjerrum, N., and L. Zechmeister, *Ber. Deut. Chem. Gesell.* **1923**, *56*, 894.
903. Blacet, F. E., *J. Phys. Colloid Chem.* **1948**, *52*, 534.
904. Blacet, F. E., P. A. Leighton, et al., *J. Phys. Chem.* **1931**, *35*, 1935.
905. Blacet, F. E., W. E. Young, et al., *J. Amer. Chem. Soc.* **1937**, *59*, 608.
906. Black, C., G. G. Joris, et al., *J. Chem. Phys.* **1948**, *16*, 537.
907. Black, P. J., and M. L. Heffernan, *Aust. J. Chem.* **1964**, *17*, 558; *C.A.* **1964**, *61*, 1735.
908. Black, P. J., and M. L. Heffernan, *Aust. J. Chem.* **1966**, *19*, 1287.
909. Blaine, L. R., *J. Res. Nat. Bur. Stand.* **1963**, *67A*, 207.

910. Blanc, P. A., F. O. Guelacar, et al., *Org. Mass Spectrom.* **1978,** *13,* 135.
911. Blanchard, L., *Bull. Soc. Chim. Fr.* **1928,** *45,* 1194; *C.A.* **1929,** *23,* 2152.
912. Blatt, A. H., Editor, *Organic Syntheses, Collective Volume II,* 1st ed., Wiley, New York, 1943.
913. Blicke, F. F., and E. S. Blake, *J. Amer. Chem. Soc.* **1930,** *52,* 235.
914. Blicke, F. F., and J. L. Powers, *Ind. Eng. Chem.* **1927,** *19,* 1334; addn. notes, *J. Amer. Chem. Soc.* **1930,** *52,* 235.
915. Blinova, E. A., *Toksikol. Seraorgan. Soldinenii, Ufa, Sb.* **1964,** 43; *C.A.* **1965,** *63,* 4854.
916. Block, H., *Z. Phys. Chem.* (Leipzig) **1912,** *78,* 385; **1913,** *82,* 403.
917. Blokhra, R. L., and M. L. Parmar, *Z. Physik. Chem.* (Frankfurt am Main) **1975,** *95,* 310.
918. Blood, F. R., G. A. Elliott, et al., *Toxicol. Appl. Pharmacol.* **1962,** *4,* 489.
919. Blout, E. R., V. W. Eager, et al., *J. Amer. Chem. Soc.* **1946,** *68,* 566.
920. Bloxsidge, J., J. R. Jones, et al., *Org. Magn. Resonance* **1970,** *2,* 337.
921. Bludilina, V. I., A. K. Baev, et al., *Zh. Fiz. Khim.* **1979,** *53,* 1052; *Russ. J. Phys. Chem.* **1979,** *53,* 600.
922. Blumenfeld, S. M., and M. Fast, *Spectrochim. Acta, Part A,* **1968,** *24,* 1449.
923. Boas, A., *Ind. Eng. Chem.* **1948,** *40,* 2202.
924. Boas-Traube, S. G., E. M. Dresel, et al., *Nature* **1948,** *162,* 960.
925. Bobbio, G., L. Pozzoli, et al., *Lavoro Umano* **1963,** *15,* No. 12, 600; *C.A.* **1964,** *61,* 6255.
926. Bobovitch, Ya. S., *Opt. Spektrosk.* **1966,** *20,* 252; *C.A.* **1966,** *65,* 2105.
927. Boccara, N., and P. Maitte, *Tetrahedron Lett.* **1977,** 4031.
928. Bock, E., V. Bocková, et al., *Lekarske Listy* **1952,** *7,* 490; *C.A.* **1954,** *48,* 13990.
929. Bock, E., and E. F. Dojack, *Can. J. Chem.* **1966,** *45,* 1097.
930. Bock, E., and D. Iwacha, *Can. J. Chem.* **1967,** *45,* 3177.
931. Bock, E., and D. Iwacha, *Can. J. Chem.* **1968,** *46,* 523.
932. Bodea, C., and I. Silberg, Belgian Patent 625,714, Mar. 29, 1963; *C.A.* **1963,** *59,* 10066.
933. Bodor, E., G. Bor, et al., *Veszprémi Vegyipari Egyetem Kozleményei* **1957,** *1,* 89; *C.A.* **1961,** *55,* 3176.
934. Bodroux, D., *Ann. Chim.* (Paris) **1929,** *11,* No. 10, 511; *C.A.* **1929,** *23,* 4936.
935. Bodroux, F., *Compt. Rend.* **1915,** *160,* 204; *C.A.* **1915,** *9,* 1309.
936. Bodson, H., *Annual Tables of Physical Constants* **1941,** Sec. 700C.
937. Bodson, H., Thesis, Bruxelles, 1937, and J. Timmermans, *Physico-Chemical Constants of Pure Organic Compounds,* Elsevier, New York, 1950.
938. Boelhouwer, C., G. A. M. Diepen, et al., *Brennst.-Chem.* **1958,** *39,* 173; *C.A.* **1958,** *52,* 17193.
939. Böeseken, J., and R. Cohen, *Rec. Trav. Chim. Pays-Bas* **1928,** *47,* 837.
940. Bogdanov, K. A., and A. P. Antonov, *Masloboĭon-Zh. Prom.* **1953,** *18,* No. 11, 19; *C.A.* **1955,** *49,* 932.
941. Boggs, J. E., and A. P. Deam, *J. Chem. Phys.* **1960,** *32,* 315.
942. Bohlmann, F., R. Zeisberg, et al., *Org. Magn. Reson.* **1975,** *7,* 426.
943. Böhme, H., and H. Bentler, *Chem. Ber.* **1956,** *89,* 1468.
944. Böhme, H., and W. Schürhoff, *Chem. Ber.* **1951,** *84,* 28.
945. Böhme, H., and J. Wagner, *Ber. Deut. Chem. Gesell.* **1942,** *75,* 606.
946. Bohon, R. L., and W. F. Claussen, *J. Amer. Chem. Soc.* **1951,** *73,* 1571.
947. Boig, F. S., G. W. Costa, et al., *J. Org. Chem.* **1953,** *18,* 775.
948. Bøje, L., and A. Hvidt, *J. Chem. Thermodyn.* **1971,** *3,* 663.
949. Bolard, J., *J. Chim. Phys.* **1965,** *62,* 887; *C.A.* **1966,** *64,* 160.
950. Bolestova, G. I., Z. N. Parnes, et al., *Zh. Org. Khim.* **1979,** *15,* 1264; *C.A.* **1979,** *91,* 140651.
951. Bolle, J., and L. Bourgeois, *Compt. Rend.* **1951,** *233,* 1466; *C.A.* **1952,** *46,* 7514.
952. Bolling, J. M., A. R. Collett, et al., *Proc. W. Virginia Acad. Sci.* **1947,** *19,* 61; *C.A.* **1951,** *45,* 8314.
953. Bologna, L., and F. Albonico, *Chim. Ind.* (Milan) **1951,** *33,* 141; *C.A.* **1951,** *45,* 6475.
954. Bolotiv, B. A., and B. N. Dolgov, *Kataliz v Vysshei Shkole, Min. Vyssh. i Srednego Spets. Obrazov. SSSR, Tr 1-go (Pervogo) Mezhuuz. Soveshch. po Katalizu* **1958,** *1,* Part 2, 248; *C.A.* **1963,** *59,* 8579.
955. Bolotov, B. A., K. P. Katkova, et al., *Zh. Priklad. Khim.* **1957,** *30,* 131.
956. Bond, D. L., and G. Thodos, *J. Chem. Eng. Data* **1960,** *5,* 289.
957. Bond, R. P. M., T. Cairns, et al., *J. Chem. Soc.* **1965,** 3958.
958. Bondi, A., and D. J. Simkin, *AIChEJ* **1957,** *3,* 473.

959. Bondroit, C., *Rev. Universelle Mines* **1939**, *15*, 197; *C.A.* **1940**, *34*, 6076.
960. Bonichon, P., *Bull. Inst. Pin.* **1933**, 249; *C.A.* **1934**, *28*, 4403.
961. Bonino, G. B., *Gazz. Chim. Ital.* **1936**, *66*, 316; *C.A.* **1937**, *31*, 2099.
962. Bonino, G. B., and L. Brull, *Gazz. Chim. Ital.* **1929**, *59*, 728; *C.A.* **1930**, *24*, 839.
963. Bonino, G. B., and P. Cella, *Mem. Accad. Ital.*, *Chim.* **1931**, *2*, 5; *C.A.* **1932**, *26*, 2656.
964. Bonino, G. B., and R. Manzoni-Ansidei, *Atti Accad. Lincei, Classe Sci. Fis. Mat. Nat.* **1936**, *24*, 207; *C.A.* **1937**, *31*, 420.
965. Bonino, G. B., and R. Manzoni-Ansidei, *Proc. Indian Acad. Sci.* **1938**, *8A*, 405; *C.A.* **1939**, *33*, 4874.
966. Bonino, G. B., and R. Manzoni-Ansidei, *Ric. Sci.* **1935**, *6*, *II*, No. 5-6, 1; *C.A.* **1937**, *31*, 3389.
967. Bonino, G. B., and R. Manzoni-Ansidei, *Ric. Sci.* **1936**, *7*, *I*, No. 7-8, 2; *C.A.* **1937**, *31*, 3388.
968. Bonino, G. B., and R. Manzoni-Ansidei, *Ric. Sci.* **1936**, *7*, *I*, No. 11-12, 2; *C.A.* **1937**, *31*, 3388.
969. Bonino, G. B., R. Manzoni-Ansidei, et al., *Ric. Sci.* (reprint) **1937**, *8*, *II*, No. 5-6; *C.A.* **1938**, *32*, 4846.
970. Bonino, G. B., R. Manzoni-Ansidei, et al., *Z. Phys. Chem.* (Leipzig) **1933**, *22B*, 21; *Chem. Zentr.* **1933**, *II*, 1149.
971. Bonino, G. B., and P. Mascherpa, *Arch. Ital. Sci. Farmacol. Vol. Giubilare A. Benedicenti* **1937**, *15*, 22; *C.A.* **1938**, *32*, 4662.
972. Bonnefoi, J., *Ann. Chim. Phys.*, *23*, 317; *Chem. Zentr.* **1901**, *II*, 396.
973. Bonner, L. G., and J. S. Kirby-Smith, *Phys. Rev.* **1939**, *57*, 1078; *C.A.* **1941**, *35*, 4688.
974. Bonner, O. D., C. F. Jordan, et al., *J. Phys. Chem.* **1964**, *68*, 2450.
975. Bonner, O. D., and S. J. Kim, *J. Chem. Thermodyn.* **1970**, *2*, 63.
976. Bonner, O. D., S. J. Kim, et al., *J. Phys. Chem.* **1969**, *73*, 1968.
977. Bonner, T. G., and M. Barnard, *J. Chem. Soc.* **1958**, 4181.
978. Bonnet, J. C., and F. P. Pike, *J. Chem. Eng. Data* **1972**, *17*, 145.
979. Bonstedt, K. P. (to Koppers Co.), U.S. Patent 2,870,192, Jan. 20, 1959; *C.A.* **1959**, *53*, 10128.
980. Bonvicini, P., A. Levi, et al., *J. Chem. Soc., Perkin Trans. 2* **1972**, 2267.
981. Boon, J. W. P., and J. W. F. Kampschmidt, U.S. Patent 2,752,336, June 26, 1956.
982. Boorman, E. J., C. G. Daubney, et al., *Analyst* **1947**, *72*, 246.
983. Booth, H., N. C. Franklin, et al., *Tetrahedron* **1965**, *21*, 1077.
984. Booth, H. H., and P. H. Carnell, *J. Amer. Chem. Soc.* **1946**, *68*, 2650.
985. Booth, H. S., and E. M. Bixby, *Ind. Eng. Chem.* **1932**, *24*, 637.
986. Booth, H. S., H. M. Elsey, et al., *J. Amer. Chem. Soc.* **1935**, *57*, 2066.
987. Booth, H. S., and H. E. Everson, *Ind. Eng. Chem.* **1948**, *40*, 1491.
988. Booth, H. S., and W. F. Martin, *J. Amer. Chem. Soc.* **1946**, *68*, 2655.
989. Booth, H. S., and C. F. Swinehart, *J. Amer. Chem. Soc.* **1935**, *57*, 1337.
990. Borden, A., and E. F. Barker, *J. Chem. Phys.* **1938**, *6*, 553.
991. Borden Chemical Company, "Preliminary Data Sheet No. 5482, Hexamethylphosphoramide," Philadelphia, 1964.
992. Bordner, C. A., U.S. Patent 2,489,260, Nov. 29, 1949; *C.A.* **1950**, *44*, 1543.
993. Bordwell, F. G., and G. D. Cooper, *J. Amer. Chem. Soc.* **1952**, *74*, 1058.
994. Bordwell, F. G. and W. S. Matthews, *J. Amer. Chem. Soc.* **1974**, *96*, 1216.
995. Borisov, G. K., and S. G. Chugunova, *Zh. Fiz. Khim.* **1976**, *50*, 3004; *Russ. J. Phys. Chem.* **1976**, *50*, 1791.
996. Borka, Y., and R. G. Bates, *Anal. Chem.* **1975**, *47*, 1110.
997. Bornmann, G., *Arzneim. Forsch.* **1954**, *4*, 643, 710; **1955**, *5*, 38; *C.A.* **1955**, *49*, 7131.
998. Borovikov, Y. Y., Y. P. Egorov, et al., *Zh. Obshch. Khim.* **1973**, *43*, 2476; *J. Gen. Chem.* (USSR) **1973**, *43*, 2462.
999. Borovikov, Y. Y., and S. I. Vdovenko, *Zh. Obshch. Khim.* **1978**, *48*, 1876; *J. Gen. Chem.* (USSR) **1978**, *48*, 1703.
1000. Borring, A. L., and K. Rasmussen, *Spectrochim. Acta, Part A* **1975**, *31A*, 889.
1001. Bortfeld, D. P., and M. Gelder, *J. Chem. Phys.* **1964**, *40*, 1770.
1002. Borynice, A., and L. Marchewski, *Bull. Inter. Acad. Polonaise* **1931A**, 392; *C.A.* **1933**, *27*, 229.
1003. Bosart, L. W., and A. O. Snoddy, *Ind. Eng. Chem.* **1927**, *19*, 506.
1004. Bose, E., and A. Müller, *Z. Phys. Chem.* (Leipzig) **1907**, *58*, 586.

1005. Bosnich, B., and D. W. Watts, *Inorg. Chem.* **1975,** *14,* 47.
1006. Bothner-By, A. A., and C. Naar-Colin, *J. Amer. Chem. Soc.* **1961,** *83,* 231.
1007. Bott, T. R., and H. N. Sadler, *J. Chem. Eng. Data* **1966,** *11,* 25.
1008. Bottoms, R. R., *Ind. Eng. Chem.* **1931,** *23,* 501.
1009. Boublik, T., and K. Aim, *Collect. Czech. Chem. Commun.* **1972,** *37,* 3513.
1010. Boud, A. H., and J. W. Smith, *J. Chem. Soc.* **1956,** 4507.
1011. Boulin, C., and L. J. Simon, *Compt. Rend.* **1920,** *170,* 595; *C.A.* **1920,** *14,* 2623.
1012. Bourchardat, G., *Compt. Rend.* **1885,** *100,* 452; *Ber. Deut. Chem. Gesell.* **1885,** *18R,* 178.
1013. Bourgom, A., *Bull. Soc. Chim. Belges* **1924,** *33,* 101; *C.A.* **1924,** *18,* 1814.
1014. Bourguel, M., *Bull. Soc. Chim. Fr.* **1927,** *41,* 1475; *C.A.* **1928,** *22,* 1324.
1015. Bourguel, M., *Compt. Rend.* **1932,** *194,* 129; *C.A.* **1932,** *26,* 4252.
1016. Bourguel, M., B. Gredy, et al., *Compt. Rend.* **1932,** *195,* 129; *C.A.* **1932,** *26,* 5544.
1017. Bourguel, M., and L. Piaux, *Bull. Soc. Chim. Fr.* **1935,** *2,* No. 5, 1458; *C.A.* **1936,** *30,* 962.
1018. Bourguignon, A., *Bull. Soc. Chim. Belges* **1908,** *22,* 87; *J. Chem. Soc.* **1908Ai,** 280.
1019. Bousfield, W. R., and T. M. Lowry, *J. Chem. Soc.* **1911,** 1432.
1020. Boutaric, A., *J. Chim. Phys.* **1920,** *18,* 126; *C.A.* **1921,** *15,* 790.
1021. Bouveault, L., and G. L. Blare, *Bull. Soc. Chim. Fr.* **1904,** *31,* 673.
1022. Bovey, F. A., F. P. Hood, et al., *J. Amer. Chem. Soc.* **1965,** *87,* 2060.
1023. Bowden, K., *Can. J. Chem.* **1965,** *43,* 2624.
1024. Bowden, S. T., and E. T. Butler, *J. Chem. Soc.* **1939,** *75,* 79.
1025. Bowden, S. T., and W. J. Jones, *J. Soc. Chem. Ind.* (London) **1947,** *66,* 342.
1026. Bower, J. H., *J. Res. Nat. Bur. Stand.* **1934,** *12,* 241.
1027. Bower, V. E., R. A. Robinson, et al., from R. G. Bates and H. B. Hetzer, *J. Phys. Chem.* **1961,** *65,* 667.
1028. Bowie, J. H., R. G. Cooks, et al., *Aust. J. Chem.* **1967,** *20,* 689.
1029. Bowie, J. H., R. Grigg, et al., *Chem. Commun.* **1965,** 403; *C.A.* **1965,** *63,* 16183.
1030. Bowie, J. H., R. Grigg, et al., *J. Amer. Chem. Soc.* **1966,** *88,* 1699.
1031. Bowie, J. H., S. O. Lawesson, et al., *J. Amer. Chem. Soc.* **1965,** *87,* 5742.
1032. Bowie, J. H., D. H. Williams, et al., *J. Org. Chem.* **1966,** *31,* 1384.
1033. Bowie, J. H., D. H. Williams, et al., *J. Org. Chem.* **1966,** *31,* 1792.
1034. Bowie, J. H., D. H. Williams, et al., *Tetrahedron* **1966,** *22,* 3515.
1035. Bowles, A. J., E. F. H. Brittain, et al., *Org. Mass Spectrom.* **1969,** *2,* 809.
1036. Bowles, A. J., W. O. George, et al., *J. Chem. Soc. B* **1969,** 810.
1037. Bowman, R. S., D. R. Stevens, et al., *J. Amer. Chem. Soc.* **1957,** *79,* 87.
1038. Boyd, G. E., and G. E. Copeland, *J. Amer. Chem. Soc.* **1942,** *64,* 2540.
1039. Boyd, R. H., S. N. Sanwal, et al., *J. Phys. Chem.* **1971,** *75,* 1264.
1040. Boyd, R. H., and C. H. Wang, *J. Amer. Chem. Soc.* **1965,** *87,* 430.
1041. Boyer, M., M. M. Claudon, et al., *Bull. Soc. Chim. Fr.* **1966,** 1139; *C.A.* **1966,** *65,* 3180.
1042. Bozel-Maletra, British Patent 607,130, Aug. 26, 1948; *C.A.* **1949,** *43,* 2222.
1043. BP Chemicals (U.K.) Limited, BISOL Technigram, Dovenshire House, London.
1044. Brachewitz, P., *Compt. Rend.* **1936,** *203,* 930; *C.A.* **1937,** *31,* 319.
1045. Brackman, W., and P. J. Smit, *Rec. Trav. Chim. Pays-Bas* **1966,** *85,* 857.
1046. Bradford, M. L., and G. Thodos, *J. Chem. Eng. Data* **1967,** *12,* 373.
1047. Bradley, C. A., *Phys. Rev.* **1932,** *40,* 908.
1048. Bradley, R. S., and T. G. Cleasby, *J. Chem. Soc.* **1953,** 1690.
1049. Bradley, R. S., and M. G. Evans, *Proc. Roy. Soc.* (London) **1946,** *186A,* 368.
1050. Bradley, R. S., M. G. Evans, et al., *Proc. Roy. Soc.* (London) **1946,** *186A,* 368; *Beil.* **EIII2,** 1016.
1051. Brady, L. J., *Anal. Chem.* **1948,** *20,* 512.
1052. Bragin, J., K. L. Kizer, et al., *J. Mol. Spectrosc.* **1971,** *38,* 289.
1053. Bramley, A., *J. Chem. Soc.* **1916,** 434.
1054. Bramley, A., *J. Chem. Soc.* **1916,** 496.
1055. Bramley, R., C. G. Le Fèvre, et al., *J. Chem. Soc.* **1959,** 1183.
1056. Brancker, A. V., S. J. Leach, et al., *Nature* **1944,** *153,* 407.
1057. Brand, K., and K. W. Kranz, *J. Prakt. Chem.* **1927,** *115,* 143; *Chem. Zentr.* **1927,** *I,* 2726.
1058. Brandenburg, W., and A. Galat, *J. Amer. Chem. Soc.* **1950,** *72,* 3275.

1059. Brandreth, D. A., and R. E. Johnson, *J. Chem. Eng. Data* **1971**, *16*, 325.

1060. Brandt, G. R. A., H. J. Emeléus, et al., *J. Chem. Soc.* **1952**, 2549.

1061. Branson, H., *Struct. Chem. Mol. Biol.* **1968**, 671; *C.A.* **1968**, *69*, 95763.

1062. Brasch, J. W., *J. Chem. Phys.* **1965**, *43*, 3473.

1063. Bratton, A. C., and J. R. Baily, *J. Amer. Chem. Soc.* **1937**, *59*, 175.

1064. Bratus, I. N., V. G. Voronin, et al., *Tr. Vses. Nauch.-Issled. Inst. Sintetich, i Natural'n. Dushistykh Veschchestv* **1961**, No. 5, 111; *C.A.* **1962**, *57*, 8398.

1065. Braun, J. v., *Ber. Deut. Chem. Gesell.* **1910**, *43*, 1350, 2594.

1066. Braun, J. v., and G. Lemke, *Ber. Deut. Chem. Gesell.* **1932**, *56B*, 1562; *C.A.* **1924**, *18*, 65.

1067. Braun, J. v., and W. Sobecki, *Ber. Deut. Chem. Gesell.* **1910**, *43*, 3596; *C.A.* **1911**, *5*, 1285.

1068. Braun, W. G., and M. R. Fenske, *Anal. Chem.* **1949**, *21*, 12.

1069. Braun, W. G., D. F. Spooner, et al., *Anal. Chem.* **1950**, *22*, 1074.

1070. Brauns, D. H., *J. Res. Nat. Bur. Stand.* **1937**, *18*, 315.

1071. Brauns, D. H., *J. Res. Nat. Bur. Stand.* **1943**, *31*, 83.

1072. Bravo, R., et al., *J. Chem. Thermodyn.* **1984**, *16*, 73.

1073. Bréant, M., *Bull. Soc. Chim. Fr.* **1971**, 723.

1074. Bréant, M., and G. Demange-Guérin, *Bull. Soc. Chim. Fr.* **1969**, 2935.

1075. Bréant, M., and G. Demange-Guérin, *Bull. Soc. Chim. Fr.* **1975**, 163.

1076. Bréant, M., and M. Dupin, *C.R. Acad. Sci., Ser. C* **1969**, *269*, 306.

1077. Bredig, G., *Z. Phys. Chem.* (Leipzig) **1894**, *13*, 191.

1078. Bredig, G., *Z. Phys. Chem.* (Leipzig) **1894**, *13*, 289.

1079. Bredig. G., and R. Bayers, *Z. Phys. Chem.* (Leipzig) **1927**, *130A*, 15.

1080. Bremner, J. G. M., British Patent 586,222, March 11, 1947; *C.A.* **1947**, *41*, 6892.

1081. Bremner, J. G. M., and D. G. Jones, U.S. Patent 2,451,712, Oct. 19, 1948; British Patent 573,507, Nov. 23, 1945; *C.A.* **1949**, *43*, 1795.

1082. Bremner, J. G. M., D. G. Jones, et al., British Patent 619,976, Aug. 14, 1946; *C.A.* **1949**, *43*, 5650.

1083. Brescia, F., *J. Chem. Ed.* **1947**, *24*, 123.

1084. Bresler, F., and A. Drasky, U.S. Patent 2,204,956, June 18, 1940; *C.A.* **1940**, *34*, 6944.

1085. Bretscher, E., *Phys. Z.* **1931**, *32*, 765.

1086. Brewster, P. W., F. C. Schmidt, et al., *J. Amer. Chem. Soc.* **1959**, *81*, 5532.

1087. Brewster, P. W., F. C. Schmidt, et al., *J. Phys. Chem.* **1961**, *65*, 990.

1088. Brickwedde, F. G., M. Moskow, et al., *J. Res. Nat. Bur. Stand.* **1946**, *37*, 263.

1089. Bridgman, O. C., and E. W. Aldrich, *J. Heat Transfer* **1965**, Paper No. 64-WA/HT-2.

1090. Bridgman, P. W., *J. Chem. Phys.* **1941**, *9*, 794.

1091. Bridgman, P. W., *Proc. Amer. Acad. Arts Sci.* **1932**, 7, 1; *C.A.* **1932**, *26*, 2096.

1092. Bridoux, M., A. Deffontaine, et al., *C.R. Hebd. Seances Acad. Sci., Ser. C* **1976**, *282*, No. 17, 771.

1093. Bried, E. M., H. F. Kidder, et al., *Ind. Eng. Chem.* **1947**, *39*, 484.

1094. Briegleb, G., *Z. Phys. Chem.* (Leipzig) **1932**, *16B*, 276; *Chem. Zentr.* **1932**, *II*, 1272.

1095. Briegleb, G., and W. Lauppe, *Z. Phys. Chem.* (Leipzig) **1937**, *37B*, 260; *C.A.* **1938**, *32*, 48.

1096. Briggs, D. K. H., *Ind. Eng. Chem.* **1957**, *49*, 418.

1097. Briggs, L. H., L. D. Colebrook, et al., *Anal. Chem.* **1957**, *29*, 904.

1098. Brikenshtein, A. A., V. P. Volkov, et al., *Khim. Prom.* **1966**, *42*, 500; *C.A.* **1966**, *65*, 15359.

1099. Briner, E., E. Perrottet, et al., *Helv. Chim. Acta* **1936**, *19*, 558.

1100. Briner, E., E. Perrottet, et al., *Helv. Chim. Acta* **1936**, *19*, 1163.

1101. Briner, E., E. Perrottet, et al., *Helv. Chim. Acta* **1936**, *19*, 1354.

1102. Brinkman, C. H., Thesis, Amsterdam (1904), and J. Timmermans, *Physico-Chemical Constants of Pure Organic Compounds*, Elsevier, New York, 1954.

1103. Brion, C. E., and W. J. Dunning, *Trans. Faraday Soc.* **1963**, *59*, 647.

1104. Briscol, H. T., and T. P. Dirkse, *J. Phys. Chem.* **1904**, *44*, 388.

1105. British Celanese Limited, British Patent 597,078, Jan. 16, 1948; *C.A.* **1948**, *42*, 4603.

1106. British Patent 17,259, July 28, 1911; *C.A.* **1913**, 7, 417.

1107. British Patent 853,266, Feb. 11, 1960; *Chem. Zentr.* **1962**, 13941.

1108. British Standards Institute, *British Standards* **1956**, 2173.

1109. Broad, D. W., and A. G. Foster, *J. Chem. Soc.* **1946**, 446.

1110. Broadhurst, M. G., *J. Res. Nat. Bur. Stand.* **1962**, *66A*, 241.

1111. Broadhurst, M. G., *J. Res. Nat. Bur. Stand.* **1966,** *70A,* 481.
1112. Brode, W. R., *Chemical Spectroscopy,* 2nd ed., Wiley, New York, 1943.
1113. Brode, W. R., *J. Phys. Chem.* **1926,** *30,* 56.
1114. Brode, W. R., and R. W. Van Dolah, *Ind. Eng. Chem.* **1947,** *39,* 1157.
1115. Brodskii, I. A., V. S. Libov, et al., *Zh. Fiz. Khim.* **1980,** *54,* 1187; *Russ. J. Phys. Chem.* **1980,** *54,* 678.
1116. Broeres, G. H. J., J. A. A. Ketelaar, et al., *Rec. Trav. Chim. Pays-Bas* **1950,** *69,* 1122; *C.A.* **1951,** *45,* 2761.
1117. Bromiley, E. C., and D. Quiggle, *Ind. Eng. Chem.* **1931,** *25,* 1136.
1118. Bromley, W. H., and W. F. Luder, *J. Amer. Chem. Soc.* **1944,** *66,* 107.
1119. Brønsted, J. N., *Chem. Rev.* **1928,** *5,* 231.
1120. Brønsted, J. N., *Rec. Trav. Chim. Pays-Bas* **1923,** *42,* 718.
1121. Brønsted, J. N., and H. C. Duus, *Z. Phys. Chem.* (Leipzig) **1925,** *117,* 299.
1122. Brooks, A. L., *Ind. Med.* **1937,** *6,* 299; *C.A.* **1937,** *31,* 8064.
1123. Brooks, C. S., and M. E. Hobbs, *J. Amer. Chem. Soc.* **1940,** *62,* 2851.
1124. Brooks, D. B., *J. Res. Nat. Bur. Stand.* **1938,** *21,* 847.
1125. Brooks, D. B., F. L. Howard, et al., *J. Res. Nat. Bur. Stand.* **1940,** *24,* 33.
1126. Brooks, J. M., and R. R. Dewald, *J. Chem. Phys.* **1968,** *72,* 2655.
1127. Broughton, G., *Trans. Faraday Soc.* **1934,** *30,* 367; *Chem. Zentr.* **1934,** *II,* 2382.
1128. Brown, A. C., and D. J. G. Ives, *J. Chem. Soc.* **1962,** 1608.
1129. Brown, A. S., P. M. Levin, et al., *J. Chem. Phys.* **1951,** *19,* 1226.
1130. Brown, B. K. (to Commercial Solvents Corp.), U.S. Patent 1,757,830, May 6, 1930; *C.A.* **1930,** *24,* 3247.
1131. Brown, D., I. F. Gault, et al., *Toxicology* **1978,** *10,* 291; *C.A.* **1978,** *89,* 161667.
1132. Brown, H. C., U.S. Patent 2,709,704, May 31, 1955; *C.A.* **1956,** *50,* 7156.
1133. Brown, H. C., and G. K. Barbaras, *J. Amer. Chem. Soc.* **1947,** *69,* 1137.
1134. Brown, H. C., S. Johnson, et al., *J. Amer. Chem. Soc.* **1954,** *76,* 5556.
1135. Brown, H. C., and C. F. Lane, *J. Amer. Chem. Soc.* **1970,** *92,* 6660.
1136. Brown, H. C., and X. R. Mihm, *J. Amer. Chem. Soc.* **1955,** *77,* 1723.
1137. Brown, H. C., and M. Nakagawa, *J. Amer. Chem. Soc.* **1955,** *77,* 3614.
1138. Brown, H. C., and B. C. Subba Rao, *J. Amer. Chem. Soc.* **1956,** *78,* 5694.
1139. Brown, H. C., and B. C. Subba Rao, *J. Amer. Chem. Soc.* **1958,** *80,* 5377.
1140. Brown, H. C., and B. C. Subba Rao, *J. Amer. Chem. Soc.* **1959,** *81,* 6423, 6428, 6434.
1141. Brown, H. C., and B. C. Subba Rao, *J. Org. Chem.* **1957,** *22,* 1136.
1142. Brown, H. C., and G. Zweifel, *J. Amer. Chem. Soc.* **1960,** *82,* 1504.
1143. Brown, H. C., and G. Zweifel, *J. Amer. Chem. Soc.* **1960,** *82,* 3222.
1144. Brown, H. C., and G. Zweifel, *J. Amer. Chem. Soc.* **1961,** *83,* 1241.
1145. Brown, H. C., and G. Zweifel, *J. Amer. Chem. Soc.* **1961,** *83,* 2544.
1146. Brown, H. D., and R. M. Hixon, *Ind. Eng. Chem.* **1949,** *41,* 1382.
1147. Brown, I., W. Fock, et al., *J. Chem. Thermodyn.* **1969,** *1,* 273.
1148. Brown, I., and F. Smith, *Aust. J. Chem.* **1955,** *8,* 62.
1149. Brown, I., and F. Smith, *Aust. J. Chem.* **1955,** *8,* 501.
1150. Brown, I., and F. Smith, *Aust. J. Chem.* **1957,** *10,* 423.
1151. Brown, I., and F. Smith, *Aust. J. Chem.* **1959,** *12,* 407.
1152. Brown, J. A., and W. H. Mears, *J. Phys. Chem.* **1958,** *62,* 960.
1153. Brown, J. B., and G. Y. Shinowara, *J. Amer. Chem. Soc.* **1937,** *59,* 6.
1154. Brown, J. C., *J. Chem. Soc.* **1903,** 987.
1155. Brown, J. C., *J. Chem. Soc.* **1905,** 265.
1156. Brown, J. C., *J. Chem. Soc.* **1906,** 311.
1157. Brown, J. H., and W. B. Whalley, *J. Soc. Chem. Ind.* (London) **1948,** *67,* 331; *C.A.* **1949,** *43,* 2924.
1158. Brown, J. K., and N. Sheppard, *Trans. Faraday Soc.* **1954,** *50,* 535.
1159. Brown, O. L. I., and G. G. Manov, *J. Amer. Chem. Soc.* **1937,** *59,* 500.
1160. Brown, M. P., and D. E. Webster, *J. Phys. Chem.* **1960,** *64,* 698.
1161. Brown, P., and C. Djerassi, *J. Amer. Chem. Soc.* **1966,** *88,* 2469.
1162. Brown, P., and C. Djerassi, *Tetrahedron* **1968,** *24,* 2949.
1163. Brown, P. E., and T. DeVries, *J. Amer. Chem. Soc.* **1951,** *73,* 1811.

1164. Brown, R., and W. E. Jones, *J. Chem. Soc.* **1946**, 781.

1165. Brown, R. C., *Phil. Mag.* **1932**, *13*, 578.

1166. Brown, R. D., and B. A. W. Coller, *Theor. Chim.-Acta* **1967**, *7*, 259.

1167. Brown, R. F. C., and L. Radom, *Can. J. Chem.* **1968**, *46*, 2578.

1168. Brown, R. G., and T. H. Edwards, *J. Chem. Phys.* **1962**, *37*, 1029, 1035.

1169. Brown, T. L., *J. Amer. Chem. Soc.* **1958**, *80*, 6489.

1170. Brown, T. L., *J. Amer. Chem. Soc.* **1959**, *81*, 3232.

1171. Brown, T. L., and M. T. Rogers, *J. Amer. Chem. Soc.* **1957**, *79*, 577.

1172. Brown, T. L., and K. Stark, *J. Phys. Chem.* **1965**, *69*, 2679.

1173. Brown, V. K. H., L. W. Ferrigan, et al., *Brit. J. Ind. Med.* **1966**, *23*, 302; *C.A.* **1967**, *66*, 9577.

1174. Brown, W. E., *J. Pharmacol.* **1924**, *23*, 485; *C.A.* **1924**, *18*, 2921.

1175. Brown W. G., C. J. Mighton, et al., *J. Org. Chem.* **1938**, *3*, 62.

1176. Brown, W. G., and H. Reagen, *J. Amer. Chem. Soc.* **1947**, *69*, 1032.

1177. Brownlie, I. A., *J. Roy Tech. Coll.* (Glasgow) **1950**, *5*, 161.

1178. Brownstein, S., and R. Miller, *J. Org. Chem.* **1959**, *24*, 1886.

1179. Bruce, G. R., and G. N. Malcolm, *J. Chem. Thermodyn.* **1969**, *1*, 183.

1180. Bruce, W. F., *Science* **1938**, *87*, 171.

1181. Bruckenstein, S., and I. M. Kolthoff, *Treatise on Analytical Chemistry, Part I, Vol. 1*, 475ff, Interscience, New York, 1959.

1182. Bruckenstein, S., and L. M. Mukherjee, *J. Phys. Chem.* **1960**, *64*, 1601.

1183. Bruckenstein, S., and A. Saito, *J. Amer. Chem. Soc.* **1965**, *87*, 698.

1184. Brückner, H., *Z. Anal. Chem.* **1928**, *75*, 289; *C.A.* **1929**, *23*, 1738.

1185. Brückner, H., *Z. Anorg. Allg. Chem.* **1931**, *199*, 91.

1186. Bruehlman, R. J., and F. H. Verhoek, *J. Amer. Chem. Soc.* **1948**, *70*, 1401.

1187. Brügel, W., T. Ankel, et al., *Z. Elektrochem.* **1960**, *64*, 1121; *C.A.* **1961**, *55*, 10075.

1188. Brühl, J. W., *Ann. Chem.* **1880**, *200*, 139; *J. Chem. Soc. A* **1880**, 295.

1189. Brühl, J. W., *Ann. Chem.* **1880**, *203*, 1; *J. Chem. Soc. A* **1880**, 781.

1190. Brühl, J. W., *J. Chem. Soc.* **1907**, 121.

1191. Brühl, J. W., *Z. Phys. Chem.* (Leipzig) **1895**, *16*, 193.

1192. Brühl, J. W., *Z. Phys. Chem.* (Leipzig) **1897**, *22*, 373.

1193. Brühl, J. W., and H. Schröder, *Z. Phys. Chem.* (Leipzig) **1905**, *51*, 1.

1194. Brummer, S. B., *J. Phys. Chem.* **1965**, *42*, 1636.

1195. Brun, *Bull. Soc. Chim. Fr.* **1945**, *12*, 452; *C.A.* **1946**, *40*, 847.

1196. Brunel, R. F., *Ber. Deut. Chem. Gesell.* **1911**, *44*, 1000.

1197. Brunel, R. F., *J. Amer. Chem. Soc.* **1923**, *45*, 1334.

1198. Brunel, R. F., J. L. Crenshaw, et al., *J. Amer. Chem. Soc.* **1921**, *43*, 561.

1199. Brunel, L., *Compt. Rend.* **1895**, *120*, 912.

1200. Bruni, G., and M. Amadori, *Gazz. Chim. Ital.* **1910**, *40*, *II*, 1; *Chem. Zentr.* **1910**, *II*, 1437.

1201. Bruni, G., and P. Berti, *Atti Real Accd. Lincei* **1900**, *9*, *V*, 393; *J. Chem. Soc. A* **1900**, 592.

1202. Bruni, G., and B. Sala, *Gazz. Chim. Ital.* **1905**, *34*, *II*, 479; *Chem. Zentr.* **1905**, *I*, 673; *J. Chem. Soc. A* **1905**, 146.

1203. Bruni, G., and A. Trovanelli, *Atti Real Accad. Lincei Roma* **1904**, *13*, *II*, 176; *Chem. Zentr.* **1904**, *II*, 944; from [6114].

1204. Brunjes, A. S., and C. C. Furnas, *Ind. Eng. Chem.* **1935**, *27*, 396.

1205. Bruson, H. A., and R. N. Washburne, U. S. Patent 2,184,934, Dec. 26, 1939; *Official Gaz. U.S. Patent Office*, 509.

1206. Brusset, H., P. Delvalle, et al., *Bull. Soc. Chim. Fr.* **1969**, 3800.

1207. Bruun, J. H., *Ind. Eng. Chem. Anal. Ed.* **1929**, *1*, 212.

1208. Bruun, J. H., *Ind. Eng. Chem. Anal. Ed.* **1936**, *8*, 224.

1209. Bruun, J. H., and W. B. M. Faulconer, *Ind. Eng. Chem. Anal. Ed.* **1937**, *9*, 192.

1210. Bruun, J. H., M. M. Hicks-Bruun, et al., *J. Amer. Chem. Soc.* **1937**, *59*, 2355.

1211. Bruus, B. P., *Z. Anorg. Allg. Chem.* **1927**, *163*, 126.

1212. Bruylants, P., *Annual Tables of Physical Constants* **1941**, Sec. 301c, 514c, 921c.

1213. Bruylants, A., and J. Timmermans, *Physico-Chemical Constants of Pure Organic Compounds*, Elsevier, New York, 1950.

1214. Bruylants, P., and A. Castille, *Bull. Soc. Chim. Belges* **1925**, *34*, 261; *Chem. Zentr.* **1926**, *I*, 1962.

1215. Bruyne, I. M. A., R. M. Davis, et al., *Phys. Z.* **1932**, *33*, 719; *C.A.* **1933**, *27*, 1248.
1216. Bryant, F., *Aust. J. Sci.* **1956**, *19*, 116; *C.A.* **1957**, *51*, 4268.
1217. Bryant, P. J. R., and A. W. H. Wardrop, *J. Chem. Soc.* **1957**, 895.
1218. Bryant, W. M. D., *J. Polymer Sci.* **1962**, *56*, 277.
1219. Bryce-Smyth, D., and K. E. Howlett, *J. Chem. Soc.* **1951**, 1141.
1220. Bryostowski, W., and S. Warycha, *Bull. Acad. Polon. Sci., Ser. Sci. Chim.* **1963**, *11*, 539; *C.A.* **1964**, *60*, 4868.
1221. Buback, M., and H. Lendle, *Z. Naturforsch., A.* **1981**, *36A*, 1371.
1222. Buchanan, J., and S. D. Hamann, *Trans. Faraday Soc.* **1953**, *49*, 1425.
1223. Buchta, E., and K. Meyer, *Chem. Ber.* **1962**, *95*, 213.
1224. Buck, F. R., K. F. Coles, et al., *J. Chem. Soc.* **1949**, 2377.
1225. Buck, F. R., B. B. Elsner, et al., *J. Inst. Petroleum* **1948**, *34*, 339; *C.A.* **1948**, *42*, 7011.
1226. Bucker, H. P., and J. R. Nielsen, *J. Mol. Spectrosc.* **1963**, *11*, 47.
1227. Buckingham, A. D., J. Y. H. Chau, et al., *J. Chem. Soc.* **1956**, 1405.
1228. Buckingham, A. D., B. Harris, et al., *J. Chem. Soc.* **1953**, 1626.
1229. Buckingham, A. D., and R. J. W. Le Fèvre, *J. Chem. Soc.* **1953**, 4169.
1230. Buckley, A., *J. Chem. Ed.* **1965**, *42*, 674.
1231. Buckley, E., and J. D. Cox, *Trans. Faraday Soc.* **1967**, *63*, 895.
1232. Buckley, E., and E. F. G. Herrington, *Trans. Faraday Soc.* **1965**, *61*, 1618.
1233. Buckley, G. S., W. G. F. Ford, et al., *Thermochim. Acta* **1981**, *49*, 199.
1234. Buckley, P., and P. A. Giguère, *Can. J. Chem.* **1967**, *45*, 397.
1235. Buczkowski, Z., and T. Urbanski, *Spectrochim. Acta* **1962**, *18*, 1187; *C.A.* **1963**, *58*, 5166.
1236. Budde, T., *Pharm. Zent.* **1913**, *54*, 1054; *Apoth. Ztg.* **1913**, *28*, 709.
1237. Budke, C. C., and D. K. Banerjee, *Encyclopedia of Industrial Chemical Analysis, Vol. 13*, 533ff, Wiley, New York, 1971.
1238. Budzikiewicz, H., C. Djerassi, et al., *J. Chem. Soc.* **1964**, 1949.
1239. Buehler, C. A., T. S. Gardner, et al., *J. Org. Chem.* **1937–38**, *2*, 167.
1240. Buerger, H., H. Niepel, et al., *Spectrochim. Acta, Part A* **1980**, *36A*, 7.
1241. Buess, C. M., J. V. Karabinos, et al., *Natl. Advisory Comm. Aeronaut.* **1946**, Tech. Note No. 1021; *C.A.* **1947**, *41*, 4113.
1242. Buffington, R. M., and J. Fleischer, *J. Ind. Eng. Chem.* **1931**, *23*, 1290.
1243. Bugarszky, S., *Z. Phys. Chem.* (Leipzig) **1910**, *71*, 705.
1244. Buhmann, H., *Arch. Pharm.* **1928**, *266*, 123; *C.A.* **1928**, *22*, 4717.
1245. Buist, G. J., and H. J. Lucas, *J. Amer. Chem. Soc.* **1957**, *79*, 6157.
1246. Bukala, M., B. Burczyk, et al., *Chem. Stos.* **1959**, *3*, 497; *C.A.* **1960**, *54*, 14568; ibid., *4*, 129; *C.A.* **1960**, *54*, 22318.
1247. Bukhalovskii, A. A., *Toksikol. Gig. Prod. Neftekhim. Neftekhim. Proizvod.* **1972**, 70; *C.A.* **1974**, *80*, 128925.
1248. Bulkley, R., *J. Res. Nat. Bur. Stand.* **1931**, *6*, 89.
1249. Bunton, C. A., G. J. Minkoff, et al., *J. Chem. Soc.* **1947**, 1416.
1250. Burawoy, A., and I. Markowitsch-Burawoy, *J. Chem. Soc.* **1936**, 36.
1251. Burawoy, A., and A. R. Thompson, *J. Chem. Soc.* **1956**, 4314.
1252. Burdett, J. L., and M. T. Rogers, *J. Amer. Chem. Soc.* **1964**, *86*, 2105.
1253. Burdun, G. D., and P. B. Kantor, *Dokl. Akad. Nauk SSSR* **1949**, *67*, 985; *C.A.* **1949**, *43*, 8679.
1254. Bureau International des Etalons, *Annual Tables of Physical Constants* **1941**, Sec. 301c.
1255. Bureau International des Etalons, *Annual Tables of Physical Constants* **1941**, Sec. 381c.
1256. Bureau International des Etalons, *Annual Tables of Physical Constants* **1941**, Sec. 505c.
1257. Bureau International des Etalons, *Annual Tables of Physical Constants* **1941**, Sec. 514c.
1258. Bureau International des Etalons, *Annual Tables of Physical Constants* **1941**, Sec. 921c.
1259. Burger, L. L., *J. Chem. Eng. Data* **1964**, *9*, 112.
1260. Burger, L. L., and R. M. Wagner, *Ind. Eng. Chem., Chem. Eng. Data Series* **1958**, *3*, 310.
1261. Burk, R. E., D. D. Cokkman, et al., U.S. Patent 2,425,991, Aug. 19, 1947; *C.A.* **1948**, *42*, 198.
1262. Burkat, R. K., and A. J. Richard, *J. Chem. Thermodyn.* **1975**, *7*, 271.
1263. Burke, J. J., and P. C. Lauterbur, *J. Amer. Chem. Soc.* **1964**, *86*, 1870.
1264. Burkhard, O., and L. Kahovec, *Monatsh. Chem. Wien* **1938**, *71*, 333; *Chem. Zentr.* **1939**, *I*, 79; *Beil.* **EIII3**, 473.

1265. Burlew, J. S., *J. Amer. Chem. Soc.* **1940,** *62,* 681, 690.

1266. Burnelle, L., and J. Duchesne, *J. Chem. Phys.* **1952,** *20,* 1324.

1267. Burnett, L. W., J. B. Johns, et al., *Ind. Eng. Chem.* **1948,** *40,* 1324.

1268. Burrell, G. A., and I. W. Robertson, *J. Amer. Chem. Soc.* **1915,** *37,* 2188.

1269. Burrell, G. A., and I. W. Robertson, *J. Amer. Chem. Soc.* **1915,** *37,* 2482.

1270. Burrell, H., *Interchem. Rev.* **1955,** *14,* 31.

1271. Burrell, H., *Interchem. Rev.* **1955,** *14,* 3.

1272. Burrell, H., *Official Digest* **1955,** *27,* 726.

1273. Burriel-Marti, F., *Bull. Soc. Chim. Belges* **1930,** *39,* 590; *Chem. Zentr.* **1931,** *I,* 3442.

1274. Burrows, G., *J. Soc. Chem. Ind.* **1946,** 360; *C.A.* **1947,** *41,* 1902.

1275. Burrows, G., *J. Soc. Chem. Ind.* (London) **1946,** *65,* 360.

1276. Bursey, M. M., and F. W. McLafferty, *J. Amer. Chem. Soc.* **1966,** *88,* 529.

1277. Burwell, R. L., and C. H. Langford, *J. Amer. Chem. Soc.* **1959,** *81,* 3799.

1278. Buss, G., *Z. Phys.* **1933,** *82,* 445; *C.A.* **1933,** *27,* 4228.

1279. Bushneva, L. I., S. V. Levanova, et al., *Izv. Vyssh. Uchebn. Zaved., Khim. Khim. Tekhnol.* **1979,** *22,* 533; *C.A.* **1979,** *91,* 56137.

1280. Buslaeva, M. P., V. P. Kartsev, et al., *Zh. Fiz. Khim.* **1982,** *56,* 1254; *Russ. J. Phys. Chem.* **1982,** *56,* 761.

1281. Buswell, A. M., J. R. Downing, et al., *J. Amer. Chem. Soc.* **1940,** *62,* 2759.

1282. Buswell, A. M., E. C. Dunlop, et al., *J. Amer. Chem. Soc.* **1940,** *62,* 325.

1283. Buswell, A. M., W. H. Rodebush, et al., *J. Amer. Chem. Soc.* **1938,** *60,* 2239.

1284. Buswell, A. M., W. H. Rodebush, et al., *J. Amer. Chem. Soc.* **1938,** *60,* 2244.

1285. Buswell, A. M., W. H. Rodebush, et al., *J. Amer. Chem. Soc.* **1947,** *69,* 770.

1286. Butler, J. A. V., C. N. Ramchandani, et al., *J. Chem. Soc.* **1935,** 280.

1287. Butler, J. A. V., D. W. Thomson, et al., *J. Chem. Soc.* **1933,** 674.

1288. Butler, J. B., and J. Lielmezs, *J. Chem. Eng. Data* **1968,** *13,* 122.

1289. Bulter, J. B., and J. Lielmezs, *J. Chem. Eng. Data* **1969,** *14,* 335.

1290. Butler, J. C., and W. P. Webb, *Ind. Eng. Chem., Chem. Eng. Data Series* **1957,** *2,* 42.

1291. Butler, M. J., and V. C. McKean, *Spectrochim. Acta* **1965,** *21,* 465.

1292. Butlerov, A. M., *Annalen* **1867,** *144,* 10.

1293. Butskus, P. F., *Zh. Obshch. Khim.* **1960,** *30,* 1816; *C.A.* **1961,** *55,* 7408.

1294. Butwill, M. E., and J. D. Rockenfeller, *Thermochim. Acta* **1970,** *1,* 289.

1295. Buu-Hoï, and P. Cagniant, *Rev. Sci.* **1942,** *80,* 271; *C.A.* **1945,** *39,* 3275.

1296. Buu-Hoï, and J. Janicaud, *Bull. Soc. Chim. Fr.* **1945,** *12,* 640; *C.A.* **1946,** *40,* 3733.

1297. Buxton, L. O., and R. Kapp, *J. Amer. Chem. Soc.* **1940,** *62,* 986.

1298. Byers, W. J., *J. Chem. Phys.* **1939,** *7,* 175.

1299. Bygden, A., *J. Prakt. Chem.* **1922,** *104,* 285; *C.A.* **1923,** *17,* 723.

1300. Bykova, L. N., and S. I. Petrov, *J. Anal. Chem. USSR* (Eng. Trans.) **1972,** *27,* 965.

1301. Bylewski, T., *Rocz. Chem.* **1932,** *12,* 311.

1302. Byrne, J. S., P. E. Jackson, et al., *J. Chem. Soc., Perkin Trans. 2* **1976,** 1800.

1303. Bywater, S., *J. Poly. Sci.* **1953,** *9,* 417; *C.A.* **1953,** *47,* 2004.

C

1304. Cahours, A., *Justus Liebigs Ann. Chem.* **1849,** *70,* 39.

1305. Cailletet, L., and E. Mathias, *Compt. Rend.* **1886,** *102,* 1202; *J. Chem. Soc. A* **1886,** 758.

1306. Cainelli, G., F. Bertini, et al., *Tetrahedron Lett.* **1967,** 5153.

1307. Calas, R., M. L. Josien, et al., *Compt. Rend.* **1958,** *247,* 2008; *C.A.* **1959,** *53,* 14020.

1308. Calcott, W. S., F. L. English, et al., *Ind. Eng. Chem.* **1935,** *17,* 942.

1309. Calderazzo, B., *Inorg. Chem.* **1965,** *4,* 293.

1310. Calderbank, K. E., and R. J. W. Le Fevre, *J. Chem. Soc.* **1949,** 1462.

1311. Calderbank, K. E., R. J. W. Le Fevre, et al., *J. Chem. Soc. B* **1969,** 968.

1312. Calderbank, K. E., and R. K. Pierens, *J. Chem. Soc., Perkin Trans. 2* **1972,** 293.

1313. Calderwood, J. H., and C. P. Smyth, *J. Amer. Chem. Soc.* **1956,** *78,* 1295.

1314. Caldirola, P., and L. Giulotto, *Nuovo Cimento* **1941,** *18,* 45; *C.A.* **1942,** *36,* 5092.

1315. Calfee, J. D., N. Fukuhara, et al., *J. Amer. Chem. Soc.* **1939,** *62,* 267.

1316. Calfee, J. D., and C. B. Miller, U.S. Patent 2,734,090, Feb. 7, 1956; *C.A.* **1956,** *50,* 9441.

1317. Calfee, J. D., and L. B. Smith, U.S. Patent 2,499,629, May 7, 1950; *C.A.* **1950,** *44,* 44881.
1318. Calhoun, W. L., *J. Chem. Eng. Data* **1983,** *29,* 146.
1319. Calingaert, G., and L. B. Hitchcock, *J. Amer. Chem. Soc.* **1927,** *49,* 750.
1320. Calingaert, G., H. Soroos, et al., *J. Amer. Chem. Soc.* **1944,** *66,* 1389.
1321. Call, F., *J. Sci. Food Agr.* **1957,** *8,* Sect. 10, 81.
1322. Calus, H., and B. Zyczynska, *Zeszyty Nauk. Politech. Warszaw-Chem.* **1957,** No. 2, 61; *C.A.* **1959,** *53,* 20984.
1323. Calvert, E., *J. Chim. Phys.* **1933,** *30,* 140.
1324. Camin, D. L., and F. D. Rossini, *J. Phys. Chem.* **1955,** *59,* 1173.
1325. Camin, D. L., and F. D. Rossini, *J. Phys. Chem.* **1956,** *60,* 1446.
1326. Cammenga, H. K., F. W. Schulze, et al., *J. Chem. Eng. Data* **1977,** *22,* 131.
1327. Campbell, A. N., *Can. J. Chem.* **1979,** *57,* 705.
1328. Campbell, A. N., and R. M. Chatterjee, *Can. J. Chem.* **1968,** *46,* 575.
1329. Campbell, A. N., and R. M. Chatterjee, *Can. J. Chem.* **1969,** *47,* 3893.
1330. Campbell, A. N., and E. M. Katzmark, *Can. J. Chem.* **1969,** *47,* 619.
1331. Campbell, A. W., previously unpublished observations.
1332. Campbell, A. W., U.S. Patent 3,015,674, Jan. 2, 1962.
1333. Campbell, A. W., and P. F. Tryon, *Ind. Eng. Chem.* **1953,** *45,* 125.
1334. Campbell, D. H., U.S. Patent 2,998,461, Aug. 29, 1961.
1335. Campbell, H. J., and J. T. Edwards, *Can. J. Chem.* **1960,** *38,* 2109.
1336. Campbell, K. N., and L. T. Eby, *J. Amer. Chem. Soc.* **1941,** *63,* 2683.
1337. Campbell, T. W., S. Linden, et al., *J. Amer. Chem. Soc.* **1947,** *69,* 880.
1338. Canals, E., M. Mousseron, et al., *Bull. Soc. Chim. Fr.* **1937,** *4,* 2048; *C.A.* **1938,** *32,* 8269.
1339. Cannizzaro, S., *Justus Liebigs Ann. Chem.* **1853,** *88,* 130.
1340. Cannon, M. R., and M. R. Fenske, *Ind. Eng. Chem. Anal. Ed.* **1938,** *10,* 297.
1341. Cannon, M. R., and M. R. Fenske, *Oil Gas J.* **1935,** *33,* 52.
1342. Cannon, M. R., and M. R. Fenske, *Oil Gas J.* **1936,** *34,* 45.
1343. Cantacuzene, J., *Bull. Soc. Chim. Fr.* **1962,** 741; *C.A.* **1962,** *57,* 5787.
1344. Cantoni, A., and J. Feldman, *Ind. Eng. Chem.* **1953,** *45,* 2580.
1345. Capalbi, A., C. Franconi, et al., *SIPS, Sci. Tec.* **1961,** *5,* No. 3-4, 135; *C.A.* **1964,** *61,* 5108.
1346. Capinjola, J. V., *J. Amer. Chem. Soc.* **1945,** *67,* 1615.
1347. Caprino, L., and G. Togna, *Eur. J. Toxicol. Environ. Hyg.* **1975,** *8,* 287; *C.A.* **1976,** *84,* 85154.
1348. Carbide and Carbon Chemicals Corp., *Synthetic Organic Chemicals,* 12th ed., New York, 1945.
1349. Cardoso, E., and A. Bruno, *J. Chim. Phys.* **1923,** *20,* 347.
1350. Carius, L., *Ann. Chem.* **1859,** *110,* 210.
1351. Carlin, R. B., and H. P. Landerl, *J. Amer. Chem. Soc.* **1950,** *72,* 2762.
1352. Carlisle, P. J., and A. A. Levine, *Ind. Eng. Chem.* **1932,** *24,* 1164.
1353. Carlsen, L., H. Egsgaard, et al., *Anal. Chem.* **1979,** *51,* 1593.
1354. Carlson, H. G., and E. F. Westrum, Jr., *J. Chem. Phys.* **1971,** *54,* 1464.
1355. Carmichael, L. T., V. M. Berry, et al., *J. Chem. Eng. Data* **1969,** *14,* 27.
1356. Carmona, P., and J. Moreno, *J. Mol. Struc.* **1982,** *82,* 1977.
1357. Carney, G. E., and J. K. Sanford, *Anal. Chem.* **1953,** *25,* 1417.
1358. Carothers, W. H., *Organic Syntheses, Vol. XIII,* Wiley, New York, 1933.
1359. Carothers, W. H., and R. Adams, *J. Amer. Chem. Soc.* **1924,** *46,* 1675.
1360. Carothers, W. H., C. F. Bickford, et al., *J. Amer. Chem. Soc.* **1927,** *49,* 2908.
1361. Carpenter, C. P., H. F. Smyth, et al., *J. Ind. Hyg. Toxicol.* **1949,** *31,* 343.
1362. Carpenter, C. P., C. S. Weil, et al., *Arch. Ind. Hyg. Occupational Med.* **1953,** *8,* 219; *C.A.* **1954,** *48,* 272.
1363. Carpenter, G. B., and J. Donohue, *J. Amer. Chem. Soc.* **1950,** *72,* 2315.
1364. Carpenter, W., A. M. Duffield, et al., *J. Amer. Chem. Soc.* **1967,** *89,* 6167.
1365. Carr, C., and J. A. Riddick, *Ind. Eng. Chem.* **1951,** *43,* 692.
1366. Carr, E. P., and H. Stucklen, *J. Chem. Phys.* **1936,** *4,* 760.
1367. Carr, E. P., and H. Stucklen, *Z. Phys. Chem.* (Leipzig) **1934,** *23B,* 57.
1368. Carr, E. P., and G. F. Walker, *J. Chem. Phys.* **1936,** *4,* 756.
1369. Carrington, R. A. G., *Anal. Chem.* **1959,** *31,* 1117.

1370. Carroll, B. H., G. K. Rollefson, et al., *J. Amer. Chem. Soc.* **1925,** *47*, 1791.

1371. Carruth, G. F., and R. Kobayashi, *J. Chem. Eng. Data* **1973,** *18*, 115.

1372. Carson, A. S., E. M. Carson, et al., *Nature* **1952,** *170*, 320.

1373. Carson, A. S., W. Carter, et al., *Proc. Roy. Soc.* (London) **1961,** *260A*, 550; *C.A.* **1961,** *55*, 15101.

1374. Carson, B. B., and V. N. Ipatieff, *J. Amer. Chem. Soc.* **1937,** *59*, 645.

1375. Carswell, T. S., *Ind. Eng. Chem.* **1928,** *20*, 728.

1376. Carswell, T. S., and H. L. Morrill, *Ind. Eng. Chem.* **1937,** *29*, 1247.

1377. Carswell, T. S., and C. E. Pfeifer, *J. Amer. Chem. Soc.* **1928,** *50*, 1765.

1378. Cartwright, C. H., *Phys. Rev.* **1930,** *35*, 415.

1379. Carvajal, C., K. J. Tolle, et al., *J. Amer. Chem. Soc.* **1965,** *87*, 5548.

1380. Casado, F. L., D. S. Massie, et al., *J. Chem. Soc.* **1949,** 1746.

1381. Casnati, G., M. Crisafulli, et al., *Tetrahedron Lett.* **1965,** 243; *C.A.* **1965,** *62*, 9049.

1382. Cass, O. W., U.S. Patent 2,489,265, Nov. 29, 1949; *C.A.* **1950,** *44*, 1543.

1383. Cass, R. C., S. E. Fletcher, et al., *J. Chem. Soc.* **1958,** 958.

1384. Cass, R. C., S. E. Fletcher, et al., *J. Chem. Soc.* **1958,** 1406.

1385. Casteel, J. F., and P. G. Sears, *J. Chem. Eng. Data* **1974,** *19*, 196.

1386. Catellano, S., and R. Kostelnik, *J. Chem. Phys.* **1967,** *46*, 327.

1387. Castellano, S., and J. S. Waugh, *J. Chem. Phys.* **1964,** *34*, 295.

1388. Castelli, V. J., and E. M. Stanley, *J. Chem. Eng. Data* **1974,** *19*, 8.

1389. Castille, A., and V. Henri, *Bull. Soc. Chim. Biol.* **1924,** *6*, 229; *C.A.* **1924,** *18*, 3165.

1390. Caswell, L. R., M. F. Howard, et al., *J. Org. Chem.* **1976,** *41*, 3312.

1391. Catherall, N. F., and A. G. Williamson, *J. Chem. Eng. Data* **1971,** *16*, 335.

1392. Catterall, R., L. P. Stodulski, et al., *J. Chem. Soc. A* **1968,** 437.

1393. Catterall, W. E., U.S. Patent 2,787,586, Apr. 2, 1957.

1394. Cauwood, J. D., and W. E. S. Turner, *J. Chem. Soc.* **1915,** 276.

1395. Cavalier, J., *Ann. Chim. et Phys.* **1899,** *18*, 449.

1396. Cavanaugh, J. R., and B. P. Dailey, *J. Chem. Phys.* **1961,** *34*, 1094, 1099.

1397. Cecil, O. B., W. E. Keomer, et al., *Chem. Eng. Data Ser. 2* **1957,** 54.

1398. Cecil, O. B., and R. H. Munch, *Ind. Eng. Chem.* **1956,** *48*, 437.

1399. Celanese Chemical Company, "Celanese Solvent Chart," New York.

1400. Celiano, A. V., P. S. Gentile, et al., *J. Chem. Eng. Data* **1962,** *7*, 391.

1401. Celluloid Corporation, British Patent 455,014, Oct. 12, 1936; *C.A.* **1937,** *31*, 1427.

1402. Cencelj, L., and D. Hadzi, *Spectrochim. Acta* **1955,** *7*, 274.

1403. Centnerszwer, M., *Z. Phys. Chem.* (Leipzig) **1904,** *49*, 199.

1404. Cepelak, J., *Plasta u. Kautschuk* **1955,** *2*, 159; *C.A.* **1957,** *51*, 7052.

1405. Ceurerick, P., *Bull. Soc. Chim. Belges* **1936,** *45*, 545.

1406. Chadwell, H. M., *J. Amer. Chem. Soc.* **1926,** *48*, 1912.

1407. Chaikin, S. W., and W. G. Brown, *J. Amer. Chem. Soc.* **1949,** *71*, 122.

1408. Chalmers, W., *Can. J. Res.* **1932,** *7*, 464; *C.A.* **1933,** *27*, 701.

1409. Chalupa, J., V. Matecha, et al., Czechoslovakian Patent 135,265, Feb. 15, 1970; *C.A.* **1971,** *74*, 124756.

1410. Chamberlain, N. F., *Anal. Chem.* **1959,** *31*, 56.

1411. Chamlee, R. D., and S. G. Woinsky, *IEEE Trans. Ind. Appl.* **1974,** *10*, 288.

1412. Chan, S. C., and J. P. Valleau, *Can. J. Chem.* **1968,** *46*, 853.

1413. Chandra, A. K. and A. B. Sannigrahi, *J. Phys. Chem.* **1965,** *69*, 2494.

1414. Chandrashekara, M. N., and D. N. Seshadri, *J. Chem. Eng. Data* **1979,** *24*, 6.

1415. Chang, C.-C., *Predel'no Dopustimye Kontsentratsii Atm. Zagraznenii Sb.* **1961,** No. 5, 94; *C.A.* **1962,** *57*, 17011.

1416. Chao, J., K. R. Hall, et al., *Thermochim. Acta* **1983,** *64*, 285.

1417. Chao, J., A. S. Rodgers, et al., *J. Phys. Chem. Ref. Data* **1974,** *3*, 141.

1418. Chao, J., and F. D. Rossini, *J. Chem. Eng. Data* **1965,** *10*, 374.

1419. Chao, J., R. C. Wilhoit, et al., *J. Phys. Chem. Ref. Data* **1973,** *2*, 427.

1420. Chao, J., and B. J. Zwolinski, *J. Phys. Chem. Ref. Data* **1976,** *5*, 319.

1421. Chao, J., and B. J. Zwolinski, *J. Phys. Chem. Ref. Data* **1978,** *7*, 363.

1422. Chapman, D., and J. F. Nacey, *Analyst* **1958,** *83*, 377.

1423. Chapman, O. L., and R. W. King, *J. Amer. Chem. Soc.* **1964**, *86*, 1258; see also [7543].
1424. Chappelow, C. C., P. S. Snyder, et al., *J. Chem. Eng. Data* **1971**, *16*, 440.
1425. Charlampowicz, B., and K. I. Marchlews, *L. Bull Intern. Acad. Polonaise* **1930A**, 376; *C.A.* **1931**, *25*, 5096.
1426. Charlot, R., and B. Tremillion (translated by P. J. J. Harvey), *Chemical Reactions in Solvents and Melts*, Pergamon Press, New York, 1969.
1427. Charnley, T., H. A. Skinner, et al., *J. Chem. Soc.* **1952**, 2288.
1428. Charon, E., *Bull. Soc. Chim. Fr.* **1896**, *15*, 390; *Chem. Zentr.* **1896**, *I*, 992.
1429. Charton-Koechlin, M., and M. A. Leroy, *J. Chim. Phys.* **1959**, *56*, 850; *C.A.* **1961**, *55*, 21808.
1430. Chase, M. W., J. L. Curnutt, et al., *J. Phys. Chem. Ref. Data* **1982**, *11*, 695.
1431. Chatterjie, N., J. G. Umans, et al., *J. Org. Chem.* **1978**, *43*, 1003.
1432. Chandhuri, P. M., R. A. Stager, et al., *J. Chem. Eng. Data* **1968**, *13*, 9.
1433. Chavanne, G., *Annual Tables of Physical Constants* **1941**, Sec 514c.
1434. Chavanne, G., *Compt. Rend.* **1914**, *158*, 1698; *C.A.* **1914**, *8*, 3010.
1435. Chavanne, G., and L. Simon, *Compt. Rend.* **1919**, *168*, 1324; *C.A.* **1919**, *13*, 3182.
1436. Chavanne, G., and G. Tock, *Bull. Soc. Chim. Belges* **1932**, *41*, 630.
1437. Chavez, M., J. M. Palacios, et al., *J. Chem. Eng. Data* **1982**, *27*, 350.
1438. Cheever, K. L., D. E. Richards, et al., *Toxicol. Appl. Pharmacol.* **1982**, *63*, 150; *C.A.* **1982**, *96*, 175601.
1439. Chemische Fabrik Griesheim-Elektron, German Patent 271,381, March 13, 1914; *Chem. Zentr.* **1914**, *I*, 1316.
1440. Chen, C.-Y., R. J. W. Le Fevre, et al., *J. Chem. Soc.* **1965**, 553.
1441. Chen, D. T. Y., and L. J. Laidler, *Trans. Faraday Soc.* **1962**, *58*, 480.
1442. Chen, S. S., A. S. Rodgers, et al., *J. Phys. Chem. Ref. Data* **1975**, *4*, 441.
1443. Chen, S. S., R. C. Wilhoit, et al., *J. Phys. Chem. Ref. Data* **1975**, *4*, 859.
1444. Chen, S. S., R. C. Wilhoit, et al., *J. Phys. Chem. Ref. Data* **1976**, *5*, 571.
1445. Chen, S. S., R. C. Wilhoit, et al., *J. Phys. Chem. Ref. Data* **1977**, *6*, 105.
1446. Chen, T.-L., *J. Chin. Chem. Soc.* (Taipei) **1969**, *16*, 115; *C.A.* **1970**, *72*, 116305.
1447. Chenevier, A., *Z. Anal. Chem.* **1892**, *31*, 68.
1448. Cheney, H. A., and S. H. McAllister, U.S. Patent 2,441,095, May 4, 1948; *C.A.* **1948**, *42*, 6849.
1449. Cheng, D. C.-H., *Chem. Eng. Sci.* **1963**, *18*, 715.
1450. Cheng, D. C.-H., and J. C. McCoubrey, *J. Chem. Soc.* **1963**, 4993.
1451. Cheng, D. C.-H., J. C. McCoubrey, et al., *Trans. Faraday Soc.* **1962**, *58*, 224.
1452. Cheng, H.-C., *J. Chim. Phys.* **1935**, *32*, 715; *C.A.* **1936**, *30*, 2549.
1453. Cheng, H.-C., *Z. Phys. Chem.* (Leipzig) **1934**, *24B*, 293; *Chem. Zentr.* **1934**, *I*, 3440.
1454. Chenon, M. T., and N. Lumbroso-Bader, *J. Chim. Phys.* **1965**, *62*, 1075; *C.A.* **1966**, *64*, 4908.
1455. Cherkasova, L. M., and L. K. Gorin, *Koks i Khim.* **1963**, No. 1, 44; *C.A.* **1963**, *58*, 10020.
1456. Chernick, C. L., and H. A. Skinner, *J. Chem. Soc.* **1956**, 1401.
1457. Chernyak, N. Y., R. A. Khmel'nitskii, et al., *J. Gen. Chem.* (Eng. trans.) **1966**, *36*, 93.
1458. Cherrier, C., *Compt. Rend.* **1947**, *225*, 930; *C.A.* **1948**, *42*, 2519.
1459. Cherrier, C., *Compt. Rend.* **1947**, *225*, 997.
1460. Cherrier, C., *Compt. Rend.* **1949**, *228*, 379; *C.A.* **1949**, *43*, 4955.
1461. Cherry, L. V., M. E. Hobbs, et al., *J. Phys. Chem.* **1957**, *61*, 465.
1462. Chiang, Y., and E. B. Whipple, *J. Amer. Chem. Soc.* **1963**, *85*, 2763.
1463. Chiao, T.-T., and A. R. Thompson, *Anal. Chem.* **1957**, *29*, 1678.
1464. Chiao, T.-T., and A. R. Thompson, *J. Chem. Eng. Data* **1961**, *6*, 192.
1465. Child, W., Thesis, Univ. of Wisconsin, Madison, 1955.
1466. Chiorboli, P., and P. Mirone, *Ann. Chim.* (Rome) **1958**, *48*, 363; *C.A.* **1958**, *52*, 15247.
1467. Chiu, K. K., H. H. Huang, et al., *J. Chem. Soc. B* **1970**, 304.
1468. Chiurodoglu, G., and W. Masschelein, *Bull. Soc. Chim. Belges* **1959**, *68*, 484; *C.A.* **1960**, *54*, 8284.
1469. Chivate, M. R., and S. M. Shah, *Trans. Indian Inst. Chem. Engrs.* **1953–4**, *6*, 170; *C.A.* **1956**, *50*, 1382.
1470. Choi, J. K., and M. J. Joncich, *J. Chem. Eng. Data* **1971**, *16*, 87.
1471. Choudhuri, B. K., *Indian J. Phys.* **1937**, *11*, 203; *C.A.* **1937**, *31*, 8376.
1472. Christopher, P. M., *J. Chem. Eng. Data* **1960**, *5*, 568.

1473. Christopher, P. M., P. B. Deshpande, et al., *J. Chem. Eng. Data* **1979,** *24,* 67.

1474. Christopher, P. M., and G. V. Guerra, *J. Chem. Eng. Data* **1971,** *16,* 468.

1475. Christopher, P. M., W. L. S. Laukhuf, et al., *J. Chem. Eng. Data* **1976,** *21,* 443.

1476. Christopher, P. M., and A. Schilman, *J. Chem. Eng. Data* **1967,** *12,* 333.

1477. Christopher, P. M., and H. W. Washington, Jr., *J. Chem. Eng. Data* **1969,** *14,* 437.

1478. Chromy, I., and A. Budniok, *Pr. Nauk. Univ. Slask. Katowiczch* **1973,** *41,* 71; *C.A.* **1975,** *82,* 50359.

1479. Chu, B., *J. Chem. Phys.* **1964,** *41,* 226.

1480. Chu, K.-Y., and A. R. Thompson, *J. Chem. Eng. Data* **1960,** *5,* 147.

1481. Chuchani, G., and A. Frolich, *J. Chem. Soc. B* **1971,** 1417.

1482. Chueh, P. L., and S. W. Briggs, *J. Chem. Eng. Data* **1964,** *9,* 207.

1483. Chumaeskii, M. A., *Opt. Spektrosk.* (Eng. trans.) **1961,** *10,* 33.

1484. Chumakov, Y. I., and Y. P. Shapavalova, Russian Patent 196,884, May 14, 1967; *C.A.* **1968,** *68,* 105004.

1485. Ciamician, G., *Ber. Deut. Chem. Gesell.* **1904,** *37,* 4200.

1486. Ciccone, A., *Nuovo Cimento* **1948,** *5,* 489; *C.A.* **1949,** *43,* 4573.

1487. Cidlinsky, J., and J. Polak, *Collect. Czech. Chem. Commun.* **1969,** *34,* 1317.

1488. Cihlar, J., et al., *Collect Czech. Chem. Commun.* **1976,** *41,* 1.

1489. Cislak, F. E., and O. D. Cunningham, U.S. Patent 2,320,322, May 25, 1943; *C.A.* **1943,** *37,* 6280.

1490. Cislak, F. E., and F. A. Karnatz, British Patent 580,048, Aug. 26, 1946; *C.A.* **1947,** *41,* 2447.

1491. Cislak, F. E., and M. M. Otto, U.S. Patent 2,432,062, Dec. 2, 1947; *C.A.* **1948,** *42,* 1967; 2,456,581, Dec. 14, 1948; *C.A.* **1949,** *43,* 3459.

1492. Cislak, F. E., and W. R. Wheeler, U.S. Patent 2,338,571, Jan. 4, 1944.

1493. Cislan, I., and A. Cornilescu, *Rev. Chim.* (Bucharest) **1965,** *16,* 98; *C.A.* **1965,** *63,* 6842.

1494. Citroni, M., *Arch. Ital. Sci. Farmacol.* **1951,** *1,* 284; *C.A.* **1956,** *50,* 1213.

1495. Claassen, H. H., *J. Chem. Phys.* **1950,** *18,* 543.

1496. Claassen, H. H., *J. Chem. Phys.* **1954,** *22,* 50.

1497. Clague, A. D. H., and A. Danti, *Spectrochim. Acta, Part A* **1968,** *24,* 439.

1498. Claisen, L., *Ann. Chem.* **1894,** *277,* 168.

1499. Claisen, L., *Ber. Deut. Chem. Gesell.* **1887,** *20,* 646.

1500. Clark, J. W., and C. E. Rectenwald, British Patent 896,068, May 9, 1962; *C.A.* **1962,** *57,* 11018.

1501. Clark, L. B., and W. T. Simpson, *J. Chem. Phys.* **1965,** *43,* 3666.

1502. Clark, P. A., and L. W. Pickett, *J. Chem. Phys.* **1976,** *64,* 2062.

1503. Clark, R. H., and H. R. L. Streight, *Trans. Roy. Soc. Can.* **1929,** *23,* Sec. 3, 77; *C.A.* **1930,** *24,* 586.

1504. Clark, T., T. Knox, et al., *J. Chem. Soc., Faraday Trans 1* **1975,** *71,* 2107.

1505. Clarke, A. J., and L. Crombie, *Chem. Ind.* (London) **1957,** 143.

1506. Clarke, G. A., and S. Sandler, *Chemist-Analyst* **1961,** *50,* 76.

1507. Clarke, H. T., *J. Chem. Soc.* **1912,** 1788.

1508. Clarke, H. T., *Organic Syntheses, Vol. III,* Wiley, New York, 1923.

1509. Clarke, H. T., *Organic Syntheses, Vol. X,* Wiley, New York, 1930.

1510. Clarke, H. T., and E. R. Taylor, *J. Amer. Chem. Soc.* **1923,** *45,* 830.

1511. Clarke, J., R. Robinson, et al., *J. Chem. Soc.* **1927,** 2647.

1512. Clarke, J. T., and E. R. Blout, *J. Polymer Sci.* **1946,** *1,* 419.

1513. Clarke, K., and K. Rothwell, *J. Chem. Soc.* **1960,** 1885.

1514. Claussen, W. F., and M. F. Polglase, *J. Amer. Chem. Soc.* **1952,** *74,* 4817.

1515. Claverie, N., and C. Garrigou-Lagrange, *J. Chim. Phys.* **1964,** *61,* 889; *C.A.* **1964,** *61,* 11480.

1516. Clavier, A., *Bull. Soc. Chim. Fr.* **1954,** 646; *C.A.* **1954,** *48,* 10540.

1517. Claxton, G., and W. H. Hoffert, *J. Soc. Chem. Ind.* **1946,** *65,* 333, 341.

1518. Clay, J., A. J. Dekker, et al., *Physica* **1943,** *10,* 768; *C.A.* **1944,** *38,* 5120.

1519. Clayton, J. W., H. Sherman, et al., *Amer. Ind. Hygiene Assn. J.* **1966,** *27,* 332.

1520. Clayton, W. R., and E. E. Reid, *J. Amer. Chem. Soc.* **1942,** *64,* 908.

1521. Cleaves, A. P., and M. E. Sherrick, *Natl. Advisory Comm. Aeronaut., ARR* **1945,** No. E5F27; *C.A.* **1948,** *42,* 7960.

1522. Clegg, G. A., and T. P. Melia, *Polymer* **1969,** *10,* 912; *C.A.* **1970,** *72,* 90955.

1523. Clement, W. H., H. M. Peters, et al., U.S. Patent 3,317,624, May 2, 1967; *C.A.* **1964,** *67,* 63694.

1524. Clement, W. H., H. M. Peters, et al., U.S. Patent 3,318,969, May 9, 1967; *C.A.* **1967,** *67,* 21433.

1525. Clemett, C., and M. Davies, *Trans. Faraday Soc.* **1962,** *58,* 1718.
1526. Clemo, G. R., and H. G. Dickenson, *J. Chem. Soc.* **1935,** 735.
1527. Clendenning, K. A., *Can. J. Res.* **1946,** *24B,* 269.
1528. Clendenning, K. A., *Can. J. Res.* **1946,** *24F,* 249.
1529. Clendenning, K. A., F. J. MacDonald, et al., *Can. J. Res.* **1950,** *28B,* 608; *C.A.* **1951,** *45,* 7007.
1530. Clerbaux, T., and P. Huyskens, *Ann. Soc. Sci. Bruxelles, Ser. I* **1966,** *80,* 130; *C.A.* **1966,** *65,* 13517.
1531. Cleveland, F. F., *J. Chem. Phys.* **1943,** *11,* 301.
1532. Cleveland, F. F., *J. Chem. Phys.* **1943,** *11,* 1; *C.A.* **1943,** *37,* 831.
1533. Cleveland, F. F., and D. E. Lee, *Phys. Rev.* **1944,** *65,* 350.
1534. Cleveland, F. F., J. E. Lamport, et al., *J. Chem. Phys.* **1950,** *18,* 1320.
1535. Cleveland, F. F., and M. J. Murray, *J. Chem. Phys.* **1937,** *5,* 752.
1536. Cleveland, F. F., M. J. Murray, et al., *J. Chem. Phys.* **1940,** *8,* 153.
1537. Cleveland, F. F., M. J. Murray, et al., *J. Chem. Phys.* **1942,** *10,* 18.
1538. Clever, H. L., and C. C. Snead, *J. Phys. Chem.* **1963,** *67,* 918.
1539. Clever, H. L., and M. L. Taylor, *J. Chem. Eng. Data* **1971,** *16,* 91.
1540. Clever, H. L., and E. F. Westrum, *J. Phys. Chem.* **1970,** *74,* 1309.
1541. Cline, J. K., and D. H. Andrews, *J. Amer. Chem. Soc.* **1931,** *53,* 3668.
1542. Closson, W. D., S. F. Brady, et al., *J. Org. Chem.* **1965,** *30,* 4026.
1543. Cluett, M. L., *Anal. Chem.* **1962,** *34,* 1491.
1544. Cluzet, J., and T. Kofman, *Compt. Rend. Soc. Biol.* **1930,** *103,* 783; *Chem. Zentr.* **1930,** *II,* 12.
1545. Coates, J. S., *Chem. Eng. News* **1978,** *56,* 3.
1546. Cocivera, M., *J. Amer. Chem. Soc.* **1966,** *88,* 672.
1547. Cocker, W., and C. Lipman, *J. Chem. Soc.* **1947,** 533.
1548. Cockerille, F. O., U.S. Patent 2,352,253, June 27, 1944; *C.A.* **1944,** *38,* 5508.
1549. Cody, H. P., and E. A. Jones, *J. Phys. Chem.* **1933,** *37,* 310.
1550. Coe, J. R., and T. B. Godfrey, *J. Appl. Phys.* **1944,** *15,* 625.
1551. Coetzee, J. F., *Pure Appl. Chem.* **1966,** *13,* 429.
1552. Coetzee, J. F., and G. P. Cunningham, *J. Amer. Chem. Soc.* **1965,** *87,* 2529.
1553. Coetzee, J. F., G. P. Cunningham, et al., *Anal. Chem.* **1963,** *34,* 1139.
1554. Coetzee, J. F., and I. M. Kolthoff , *J. Amer. Chem. Soc.* **1957,** *79,* 6110.
1555. Coetzee, J. F., and D. K. McGuire, *J. Phys. Chem.* **1963,** *67,* 1810.
1556. Coetzee, J. F., and R. Mei-Shun Lok, *J. Phys. Chem.* **1965,** *69,* 2690.
1557. Coetzee, J. F., and G. R. Padmanabham, *J. Phys. Chem.* **1962,** *66,* 1708.
1558. Coetzee, J. F., and J. M. Simon, *Anal. Chem.* **1972,** *44,* 1129.
1559. Coetzee, J. F., J. M. Simon, et al., *Anal. Chem.* **1969,** *41,* 766.
1560. Coffey, S., and C. F. Ward, *J. Chem. Soc.* **1921,** 1301.
1561. Coffin, C. C., and O. Maass, *J. Amer. Chem. Soc.* **1928,** *50,* 1427.
1562. Coggeshall, N. D., *J. Chem. Phys.* **1960,** *33,* 1247.
1563. Coggeshall, N. D., *J. Phys. Chem.* **1963,** *67,* 183.
1564. Coggeshall, N. D., and A. S. Glessner, *J. Amer. Chem. Soc.* **1949,** *71,* 3150.
1565. Cogrossi, C., *Ann. Chim.* (Rome) **1966,** *56,* 270; *C.A.* **1966,** *64,* 19368.
1566. Cohen, C. A., U.S. Patent 3,196,174, July 20, 1965; *C.A.* **1965,** *63,* 11434.
1567. Cohen, E., and L. C. J. te Boekhorst, *Z. Phys. Chem.* (Leipzig) **1934,** *24B,* 241.
1568. Cohen, E., and J. S. Buij, *Z. Phys. Chem.* (Leipzig) **1937,** *35B,* 270; *C.A.* **1937,** *31,* 4174.
1569. Cohen, E., and A. L. T. Moesveld, *Z. Phys. Chem.* (Leipzig) **1922,** *100,* 151.
1570. Cohen, E. R., and B. N. Taylor, *Dimensions/NBS,* Jan. 1974, pp. 4, 5.
1571. Cohen, H., and J. D. Mier, *Chem. Ind.* (London) **1965,** 349.
1572. Cohen, I., *J. Amer. Chem. Soc.* **1930,** *52,* 2827.
1573. Cohen, J. B., and H. D. Dakin, *J. Chem. Soc.* **1901,** 1111.
1574. Cohn, H., and E. D. Bergmann, *Israel J. Chem.* **1965,** *2,* 355.
1575. Cokelet, G. R., F. J. Hollander, et al., *J. Chem. Eng. Data* **1969,** *14,* 470.
1576. Colarusso, V. G., and M. A. Semon, *Anal. Chem.* **1968,** *40,* 1521.
1577. Cole, A. R. H., P. R. Jefferies, et al., *J. Chem. Soc.* **1959,** 1222.
1578. Cole, J. R., J. O. Know, et al., *Chemist-Analyst* **1959,** *48,* 38.
1579. Cole, P. J., *J. Chem. Eng. Data* **1960,** *5,* 367.

1580. Cole, P. J., U.S. Patent 2,393,888, Jan. 29, 1946; *C.A.* **1946,** *40,* 2468.
1581. Cole, P. J., and G. W. Burtt, U.S. Patent 2,348,329, May 9, 1944; *C.A.* **1945,** *39,* 532.
1582. Cole, R. H., *J. Chem. Phys.* **1941,** *9,* 251.
1583. Cole, W. G., D. H. Williams, et al., *J. Chem. Soc. B* **1968,** 1284.
1584. Coleman, C. F., *J. Phys. Chem.* **1968,** *72,* 365.
1585. Coleman, C. F., and T. De Vries, *J. Amer. Chem. Soc.* **1949,** *71,* 2839.
1586. Coleman, D. J., and G. Pilcher, *Trans. Faraday Soc.* **1966,** *62,* 821.
1587. Coleman, G. H., and A. M. Alvarado, *Organic Syntheses, Vol. III,* Wiley, New York, 1923.
1588. Coles, H. W., and W. E. Tournay, *Anal. Chem.* **1947,** *19,* 936.
1589. Coles, H. W., and W. E. Tournay, *Ind. Eng. Chem., Anal. Ed.* **1942,** *14,* 20.
1590. Collerson, R. R., J. F. Counsell, et al., *J. Chem. Soc.* **1965,** 3697.
1591. Collin, J., *Bull. Soc. Chim. Belges* **1960,** *69,* 575; *C.A.* **1961,** *55,* 14048.
1592. Collin, J., *Bull. Soc. Roy. Sci. Liege* **1956,** *25,* 520; *C.A.* **1957,** *51,* 17418.
1593. Collin, J. E., and M. J. Franklin, *Bull. Roy. Soc. Leige* **1966,** *35,* 285; *C.A.* **1966,** *65,* 12983.
1594. Collin, P. J., and S. Sternhell, *Aust. J. Chem.* **1966,** *19,* 317; *C.A.* **1966,** *64,* 16867.
1595. Collins, B. T., C. F. Coleman, et al., *J. Amer. Chem. Soc.* **1949,** *71,* 2929.
1596. Collins, R. L., and J. R. Nielsen, *J. Chem. Phys.* **1955,** *23,* 531.
1597. Colomina, M., P. Jimenez, et al., *J. Chem. Thermodyn.* **1982,** *14,* 779.
1598. Colomina, M., C. Latorre, et al., *Pure Appl. Chem.* **1961,** *2,* 133.
1599. Colomina, M., A. S. Pell, et al., *Trans. Faraday Soc.* **1965,** *61,* 2641.
1600. Colon, A. A., and H. A. Frediani, *J. Amer. Pharm. Assoc., Sci. Ed.* **1951,** *40,* 607.
1601. Colonge, J., and J. Crabalona, *Bull. Soc. Chim. Fr.* **1959,** 1505.
1602. Colonge, J., and J. Crabalona, *Bull. Soc. Chim. Fr.* **1960,** 102.
1603. Commerical Solvents Corp., "Acetone," Tech. Data Sheet No. 13, IC Ser., New York, 1962.
1604. Commercial Solvents Corp., British Patent 898,388, June 6, 1962.
1605. Commercial Solvents Corp., "Butyl Lactate," Tech. Data Sheet No. 1, New York.
1606. Commercial Solvents Corp., "Butyl Stearate," Tech. Data Sheet No. 27, IC Ser., New York, 1963.
1607. Commercial Solvents Corp., "CSC Methylamines," New York, 1965.
1608. Commercial Solvents Corp., "Dibutyl Phthalate," Tech. Data Sheet No. 14, IC Ser., New York, 1963.
1609. Commercial Solvents Corp., "Dimethyl Ether," Tech. Data Sheet No. 32, New York, 1961.
1610. Commercial Solvents Corp., "Ethyl Acetate," Tech. Data Sheet No. 20, IC Ser., New York, 1962.
1611. Commercial Solvents Corp., "Industrial Ethyl Alcohol," New York, 1967.
1612. Commercial Solvents Corp., "Methanol," New York, 1960.
1613. Commercial Solvents Corp., "The Nitroparaffins," New York.
1614. Commercial Solvents Corp., "Storage and Handling of Nitromethane," NP Tech. Data Sheet, Tech. Data Sheet No. 2, New York, 1961.
1615. Commercial Solvents Corp., "The Aminohydroxy Compounds and Oral and Skin Toxicity of Aminohydroxy Compounds," NP Tech. Data Sheets No. 10 and 13, New York, 1970 and 1967 (respt.).
1616. Commercial Solvents Corp., "Tributyl Phosphate," Tech. Data Sheet No. 7, IC Ser., New York, 1965.
1617. Compton, D. A. C., *J. Chem. Soc., Perkin Trans. 2* **1977,** 1307.
1618. Conant, J. B., *Organic Syntheses, Vol. II,* Wiley, New York, 1922.
1619. Conant, J. B., *Organic Syntheses, Vol. IX,* Wiley, New York, 1929.
1620. Conant, J. B., and G. H. Carlson, *J. Amer. Chem. Soc.* **1929,** *51,* 3464.
1621. Conant, J. B., and R. E. Hussey, *J. Amer. Chem. Soc.* **1925,** *47,* 476.
1622. Conant, J. B., and N. Tuttle, *Organic Syntheses, Vol. I,* Wiley, New York, 1921.
1623. Conant, J. B., C. N. Webb, et al., *J. Amer. Chem. Soc.* **1929,** *51,* 1249.
1624. Condirston, D. A., and J. D. Laposa, *J. Mol. Spectrosc.* **1976,** *63,* 466.
1625. Conner, J. M., and V. C. Bulgrin, *J. Inorg. Nucl. Chem.* **1967,** *29,* 1953.
1626. Conner, W. P., and C. P. Smyth, *J. Amer. Chem. Soc.* **1943,** *65,* 382.
1627. Connett, J. E., *J. Chem. Thermodyn.* **1972,** *4,* 233.
1628. Connett, J. E., *Lab. Practice* **1972,** *545,* 553.
1629. Connett, J. E., J. F. Counsell, et al., *J. Chem. Thermodyn.* **1976,** *8,* 1199.
1630. Connolly, J. F., *J. Chem. Eng. Data* **1966,** *11,* 13.

1631. Connolly, J. F., and G. A. Kandalic, *J. Chem. Eng. Data* **1962,** *7,* 137.
1632. Conrad, M., and W. R. Hodgkinson, *Ann. Chem.* **1885,** *193,* 320; *Beil.* **EI6,** 435.
1633. Conrad-Billroth, H., *Z. Phys. Chem.* (Leipzig) **1933,** *20B,* 222.
1634. Conrad-Billroth, H., *Z. Phys. Chem.* (Leipzig) **1933,** *23B,* 315; *C.A.* **1934,** *28,* 948.
1635. Conradi, J. J., and N. C. Li, *J. Amer. Chem. Soc.* **1953,** *75,* 1785.
1636. Constable, F. H., and S. Tegul, *Nature* **1946,** *157,* 735.
1637. Cook, A. N., *J. Amer. Chem. Soc.* **1904,** *26,* 303.
1638. Cook, A. N., and A. L. Haines, *Proc. Iowa Acad. Sci.* **1901,** *9,* 86; *J. Amer. Chem. Soc.* **1903,** *25R,* 559.
1639. Cook, G. L., and F. M. Church, *J. Phys. Chem.* **1957,** *61,* 458; *Anal. Chem.* **1956,** *28,* 993.
1640. Cooke, R. G., *Chem. Ind.* **1955,** 142.
1641. Cookson, R. C., *J. Chem. Soc.* **1954,** 282.
1642. Cookson, R. C., and S. H. Dandegaonker, *J. Chem. Soc.* **1955,** 1651.
1643. Collidge, A. S., *J. Amer. Chem. Soc.* **1928,** *50,* 2166.
1644. Coolidge, A. S., *J. Amer. Chem. Soc.* **1930,** *52,* 1874.
1645. Coomber, D. I., and J. R. Partington, *J. Chem. Soc.* **1938,** 1444.
1646. Coomber, D. I., and B. A. Rose, *J. Pharm. Pharmacol.* **1959,** *11,* 670; *C.A.* **1960,** *54,* 9204.
1647. Coon, E. D., and F. Daniels, *J. Phys. Chem.* **1933,** *37,* 1.
1648. Coop, I. E., and L. E. Sutton, *J. Chem. Soc.* **1938,** 1869.
1649. Cooper, J. R., M. Kini, et al., *J. Neurochem.* **1962,** *9,* 119; *C.A.* **1962,** *57,* 8844.
1650. Coops, J., D. Mulder, et al., *Rev. Trav. Chim. Pays-Bas,* **1946,** *65,* 129; *C.A.* **1946,** *40,* 3972.
1651. Copeland, C. A., and M. W. Rigg, *J. Amer. Chem. Soc.* **1951,** *73,* 3584.
1652. Copelin, H. B., U.S. Patent 2,904,600, Sept. 15, 1959; *C.A.* **1961,** *55,* 851.
1653. Copley, M. J., E. Ginsberg, et al., *J. Amer. Chem. Soc.* **1941,** *63,* 254.
1654. Copp, J. L., and D. H. Everett, *Discussion Faraday Soc.* **1953,** No. 15, 174; from [7483].
1655. Copp, J. L., and T. J. V. Findlay, *Trans. Faraday Soc.* **1960,** *56,* 13.
1656. Coppock, J. B. M., and F. R. Goss, *J. Chem. Soc.* **1939,** 1789.
1657. Corbett, J. F., *Spectrochim. Acta, Part A* **1967,** *23,* 2315.
1658. Cordes, H. F., and C. W. Tait, *Anal. Chem.* **1957,** *29,* 485.
1659. Corey, C. J., and M. Chaykovsky, *J. Amer. Chem. Soc.* **1964,** *86,* 1639.
1660. Corin, C., *Bull. Soc. Roy. Sci. Liege* **1941,** *10,* 99; *C.A.* **1945,** *39,* 4548.
1661. Corin, C., *J. Chim. Phys.* **1936,** *33,* 448; *C.A.* **1936,** *30,* 7038.
1662. Corin, C., and G. B. B. M. Sutherland, *Proc. Roy. Soc.* (London) **1938,** *165A,* 43.
1663. Corish, P. J., and D. Chapman, *J. Chem. Soc.* **1957,** 1746.
1664. Cornu, A., *Bull. Soc. Chim. Fr.* **1959,** 721; *C.A.* **1959,** *53,* 15762.
1665. Cornubert, R., G. Barraud, et al., *Bull. Soc. Chim. Fr.* **1955,** 400; *C.A.* **1956,** *50,* 2449.
1666. Cornubert, R., V. Kondrachov, et al., *Bull. Soc. Chim. Fr.* **1959,** 385; *C.A.* **1959,** *53,* 16016; *Beil.* **EIII6,** 69, 73.
1667. Cornubert, R., and M. Lafont-Lemoine, *Compt. Rend.* **1953,** *237,* 469; *C.A.* **1954,** *48,* 13643.
1668. Corruccini, R. J., and D. C. Ginnings, *J. Amer. Chem. Soc.* **1947,** *69,* 2291.
1669. Corval, M., and P. Masclet, *Org. Mass Spectrom.* **1972,** *6,* 511.
1670. Cosner, J. L., J. E. Gagliardo, et al., *J. Chem. Eng. Data* **1961,** *6,* 360.
1671. Costello, J. M., and S. T. Bowden, *Chem. Ind.* **1956,** 1041.
1672. Costello, J. M., and S. T. Bowden, *Rec. Trav. Chim. Pays-Bas* **1959,** *78,* 391; from [7483].
1673. Cotton, F. A., and R. Francis, *J. Amer. Chem. Soc.* **1960,** *82,* 2986.
1674. Cottrell, F. G., *J. Amer. Chem. Soc.* **1919,** *41,* 721.
1675. Coujolle, F., and G. Roux, *Compt. Rend.* **1954,** *239,* 680; *C.A.* **1955,** *49,* 4181.
1676. Coulson, E. A., and J. B. Ditcham, *J. Appl. Chem.* (London) **1952,** *2,* 236.
1677. Coulson, E. A., J. L. Hales, et al., *J. Appl. Chem.* **1952,** 2, 71.
1678. Coulson, E. A., J. L. Hales, et al., *J. Chem. Soc.* **1951,** 2125.
1679. Coulson, E. A., and J. I. Jones, *J. Soc. Chem. Ind.* (London) **1946,** *65,* 169.
1680. Counsell, J. F., D. H. Everett, et al., *Pure Appl. Chem.* **1961,** *2,* 335.
1681. Counsell, J. F., J. O. Fenwick, et al., *J. Chem. Thermodyn.* **1970,** *2,* 367.
1682. Counsell, J. F., J. H. S. Green, et al., *Trans. Faraday Soc.* **1965,** *61,* 212.
1683. Counsell, J. F., J. L. Hales, et al., *Trans. Faraday Soc.* **1965,** *61,* 1869.

1684. Counsell, J. F., D. A. Lee, et al., *J. Chem. Soc. A* **1971**, 313.
1685. Counsell, J. F., and D. A. Lee, *J. Chem. Thermodyn.* **1972**, *4*, 915.
1686. Counsell, J. F., E. B. Lees, et al., *J. Chem. Soc. A* **1968**, 1819.
1687. Coursey, B. M., and E. L. Heric, *J. Chem. Eng. Data* **1969**, *14*, 426.
1688. Courtoy, C., *Ann. Soc. Sci. Bruxelles, Ser. I, Sci. Math. Phys. Astron.* **1946**, *60*, 122; *C.A.* **1947**, *41*, 4717.
1689. Courtoy, C., *Ann. Soc. Sci. Bruxelles, Ser. I* **1947**, *61*, 139; *C.A.* **1949**, *43*, 4140.
1690. Cowan, D. M., G. H. Jeffery, et al., *J. Chem. Soc.* **1940**, 171.
1691. Cowan, D. M., and A. I. Vogel, *J. Chem. Soc.* **1940**, 1528.
1692. Cowan, R. D., G. Herzberg, et al., *J. Chem. Phys.* **1950**, *18*, 538.
1693. Cowley, E. G., *J. Chem. Soc.* **1952**, 3557.
1694. Cowley, E. G., and J. R. Partington, *J. Chem. Soc.* **1933**, 1257.
1695. Cowley, E. G., and J. R. Partington, *J. Chem. Soc.* **1935**, 604.
1696. Cowley, E. G., and J. R. Partington, *J. Chem. Soc.* **1936**, 1184.
1697. Cowley, E. G., and J. R. Partington, *J. Chem. Soc.* **1938**, 977.
1698. Cowley, E. G., and J. R. Partington, *Nature* **1937**, *140*, 1100.
1699. Cox, H. L., and L. H. Cretcher, *J. Amer. Chem. Soc.* **1926**, *48*, 451.
1700. Cox, J. D., *Pure Appl. Chem.* **1961**, *2*, 125.
1701. Cox, J. D., A. R. Challoner, et al., *J. Chem. Soc.* **1954**, 265.
1702. Cox, J. D., H. A. Gundry, et al., *Trans. Faraday Soc.* **1964**, *60*, 653.
1703. Cox, J. D., and G. Pilcher, *Thermochemistry of Organic and Organometallic Compounds*, Academic Press, London, 1970.
1704. Cox, R. A., L. M. Druet, et al., *Can. J. Chem.* **1981**, *59*, 1568.
1705. Crable, G. F., G. L. Kearns, et al., *Anal. Chem.* **1960**, *32*, 13.
1706. Crafts, J. M., *Ber. Deut. Chem. Gesell.* **1887**, *20*, 709.
1707. Craig, B. M., *Can. J. Chem.* **1953**, *31*, 499.
1708. Craig, L. C., and R. M. Hixon, *J. Amer. Chem. Soc.* **1931**, *53*, 4307; from [8242].
1709. Cramer, J. S. N., *Rec. Trav. Chim. Pays-Bas* **1943**, *62*, 606; *C.A.* **1944**, *38*, 6148.
1710. Cramer, P. L., and V. A. Miller, *J. Amer. Chem. Soc.* **1940**, *62*, 1452.
1711. Cramer, P. L., and M. J. Mulligan, *J. Amer. Chem. Soc.* **1936**, *58*, 373.
1712. Crawford, B. L., and L. Joyce, *J. Chem. Phys.* **1939**, *7*, 307.
1713. Crawford, J. W. C., and J. McGrath, U.S. Patent 2,140,469, Dec. 13, 1938.
1714. Crawford, R. W., P. C. Crawford, et al., *U.S. At. Energy Comm.* **1969**, *UCRL-50808*, 26; *C.A.* **1971**, *74*, 58435.
1715. Creamer, R. M., and D. H. Chambers, *J. Electrochem. Soc.* **1954**, *101*, 162.
1716. Cretcher, L. H., and W. H. Pittenger, *J. Amer. Chem. Soc.* **1924**, *46*, 1503.
1717. Cretcher, L. H., and W. H. Pittenger, *J. Amer. Chem. Soc.* **1925**, *47*, 163.
1718. Crigler, E. A., *J. Amer. Chem. Soc.* **1932**, *54*, 4199.
1719. Crisler, R. O., and A. M. Burrill, *Anal. Chem.* **1959**, *31*, 2055.
1720. Crisp, S., S. R. C. Hughes, et al., *J. Chem. Soc. A* **1968**, 603.
1721. Critchfield, F. E., J. A. Gibson, et al., *J. Amer. Chem. Soc.* **1953**, *75*, 1991.
1722. Crompton, T. R., and D. Buckley, *Analyst* **1965**, *90*, 76.
1723. Cronenwett, W. T., and L. W. Hoogendoorn, *J. Chem. Eng. Data* **1972**, *17*, 298.
1724. Cropper, F. R., and S. Kaminsky, *Anal. Chem.* **1963**, *35*, 735.
1725. Crossley, J., *J. Chem. Phys.* **1973**, *58*, 5315.
1726. Crowder, G. A., and B. R. Cook, *J. Chem.* **1967**, *71*, 914.
1727. Crowder, G. A., and M. R. Jalilian, *J. Mol. Struct.* **1978**, *49*, 287.
1728. Crowe, R. W., and C. P. Smyth, *J. Amer. Chem. Soc.* **1950**, *72*, 4009.
1729. Crowe, R. W., and C. P. Smyth, *J. Amer. Chem. Soc.* **1951**, *73*, 5406.
1730. Crowley, J. D., G. S. Teague, et al., *J. Paint Technology* **1966**, *38*, 269.
1731. Crown Zellerbach Corp., *Dimethyl Disulfide*, Camas, Washington.
1732. Crown Zellerbach Corp., Chemical Products Div., ''Dimethyl Sulfide (DMS),'' Tech. Bulletin, Camas, Washington.
1733. Crown Zellerbach Corp., Chemical Products Div., ''Dimethyl Sulfoxide,'' Tech. Bulletin, Camas, Washington, 1963 and 1966.

1734. Crown Zellerbach Corp., Chemical Products Div., "Properties and Suggested Uses of Dimethyl Sulfide (DMS)," Camas, Washington, 1957; Looseleaf Revision, 1965.

1735. Croxall, W. J., and H. T. Neher, U.S. Patent 2,448,660, Sept. 7, 1948; *C.A.* **1949**, *43*, 1433.

1736. C. R. S. Chemicals, Ltd., "Amyl Acetate," Data sheet No. 207, Sydney, 1971.

1737. C. R. S. Chemicals, Ltd., "*n*-Butyl Acetate," Data sheet No. 205, Sydney, 1969.

1738. C. R. S. Chemicals, Ltd., "Isobutyl Acetate," Data sheet No. 213, Sydney, 1969.

1739. Cruickshank, A. J. B., T. Ackermann, et al., *Exp. Thermodyn.* **1968**, *1*, 421.

1740. Cruickshank, A. J. B., and A. J. B. Cutler, *J. Chem. Eng. Data* **1967**, *12*, 326.

1741. Cryder, D. S., and J. O. Maloney, *Trans. Amer. Inst. Chem. Eng.* **1941**, *37*, 827.

1742. Crymble, C. R., A. W. Stewart, et al., *J. Chem. Soc.* **1911**, 1262.

1743. Cuisa, *Gazz. Chim.* **1911**, *41a*, 688; from [6114], p. 607.

1744. Culbertson, G., and R. Pettit, *J. Amer. Chem. Soc.* **1963**, *85*, 741.

1745. Cullen, W. R., D. C. Frost, et al., *Inorg. Chem.* **1970**, *9*, 1976.

1746. Cullinane, N. M., and W. C. Davies, British Patent 600,837, Apr. 20, 1948; *C.A.* **1948**, *42*, 7334.

1747. Cummins, E. G., and J. E. Page, *J. Chem. Soc.* **1957**, 3847.

1748. Cumper, C. W. N., S. K. Dev, et al., *J. Chem. Soc., Perkin Trans. 2* **1973**, 537.

1749. Cumper, C. W. N., J. F. Read, et al., *J. Chem. Soc.* **1965**, 5323.

1750. Cumper, C. W. N., J. F. Read, et al., *J. Chem. Soc. A* **1966**, 239.

1751. Cumper, C. W. N., D. G. Redford, et al., *J. Chem. Soc.* **1963**, 1176.

1752. Cumper, C. W. N., and A. Singleton, *J. Chem. Soc. B* **1967**, 1096.

1753. Cumper, C. W. N., and A. Singleton, *J. Chem. Soc. B* **1967**, 1100.

1754. Cumper, C. W. N., and A. Singleton, *J. Chem. Soc. B* **1968**, 645.

1755. Cumper, C. W. N., and A. Singleton, *J. Chem. Soc. B* **1968**, 649.

1756. Cumper, C. W. N., and A. P. Thurston, *J. Chem. Soc. B* **1971**, 422.

1757. Cumper, C. W. N., and A. I. Vogel, *J. Chem. Soc.* **1959**, 3531.

1758. Cumper, C. W. N., and A. I. Vogel, *J. Chem. Soc.* **1960**, 4723.

1759. Cumper, C. W. N., A. I. Vogel, et al., *J. Chem. Soc.* **1956**, 3221.

1760. Cumper, C. W. N., A. I. Vogel, et al., *J. Chem. Soc.* **1956**, 3621.

1761. Cumper, C. W. N., A. I. Vogel, et al., *J. Chem. Soc.* **1957**, 3640.

1762. Cundiff, R. H., and P. C. Markunas, *Anal. Chem.* **1958**, *30*, 1450; **1960**, *33*, 1028.

1763. Cunningham, G. P., G. A. Vidulich, et al., *J. Chem. Eng. Data* **1967**, *12*, 336.

1764. Curl, R. F., *J. Chem. Phys.* **1959**, *30*, 1529.

1765. Curme, G. O., U.S. Patent 1,315,547, Sept. 9, 1919; *C.A.* **1919**, *13*, 2883.

1766. Curme, G. O., and F. Johnston, eds., *Glycols*, Reinhold, New York, 1952.

1767. Curran, C., *J. Amer. Chem. Soc.* **1945**, *67*, 1835.

1768. Curran, C., R. M. Witucki, et al., *J. Amer. Chem. Soc.* **1950**, *72*, 4471.

1769. Curran, R. K., *J. Chem. Phys.* **1961**, *34*, 2007.

1770. Cutler, J. A., *J. Chem. Phys.* **1948**, *16*, 136.

1771. Cuypers, R., *Annual Tables of Physical Constants*, Sec. 301c, 1941.

1772. Czapska, W., *Compt. Rend.* **1929**, *189*, 32; *C.A.* **1929**, *23*, 5108.

1773. Czerny, M., and P. Mollet, *Z. Phys.* **1937**, *108*, 85; *C.A.* **1938**, *32*, 4432.

D

1774. Daasch, L. W., *J. Chem. Phys.* **1954**, *22*, 1293.

1775. Daasch, L. W., *J. Phys. Chem.* **1965**, *69*, 3196.

1776. Daasch, L. W., and D. C. Smith, *Anal. Chem.* **1951**, *23*, 853.

1777. Dack, M. J. R., K. J. Bird, et al., *Aust. J. Chem.* **1975**, *28*, 955.

1778. D'Adams, A. F., and R. H. Kienle, *J. Amer. Chem. Soc.* **1955**, *77*, 4408.

1779. Dadieu, A., *Monatsh. Chem.* **1931**, *57*, 437; *C.A.* **1931**, *25*, 4794.

1780. Dadieu, A., and K. W. F. Kohlrausch, *Ber. Deut. Chem. Gesell.* **1930**, *63*, 251.

1781. Dadieu, A., and K. W. F. Kohlrausch, *Monatsh. Chem.* **1929**, *52*, 220; *Chem. Zentr.* **1929**, *II*, 697, 1776.

1782. Dadieu, A., and K. W. F. Kohlrausch, *Monatsh. Chem.* **1929**, *52*, 379; *Chem. Zentr.* **1930**, *I*, 2218.

1783. Dadieu, A., and K. W. F. Kohlrausch, *Monatsh. Chem.* **1930**, *55*, 58; *C.A.* **1930**, *24*, 2105; ibid. **1929**, *53/54*, 282; *C.A.* **1930**, *24*, 552.

1784. Dadieu, A., and K. W. F. Kohlrausch, *Monatsh. Chem.* **1930**, *55*, 201; *C.A.* **1930**, *24*, 3437.

1785. Dadieu, A., and K. W. F. Kohlrausch, *Monatsh. Chem.* **1930**, *55*, 379; *Chem. Zentr.* **1930**, *II*, 1340.

1786. Dadieu, A., and K. W. F. Kohlrausch, *Monatsh. Chem.* **1930**, *56*, 461; *Chem. Zentr.* **1931**, *I*, 2170.

1787. Dadieu, A., and K. W. F. Kohlrausch, *Monatsh. Chem.* **1930**, *57*, 225; *Chem. Zentr.* **1931**, *I*, 2170.

1788. Dadieu, A., and K. W. F. Kohlrausch, *Monatsh. Chem.* **1931**, *57*, 488; *C.A.* **1931**, *25*, 4795.

1789. Dadieu, A., and K. W. F. Kohlrausch, *Naturwissenschaften* **1929**, *17*, 336; *C.A.* **1929**, *23*, 5108.

1790. Dadieu, A., and K. W. F. Kohlrausch, *Phys. Z., 30*, 384; *Chem. Zentr.* **1929**, *II*, 970.

1791. Dadieu, A., and K. W. F. Kohlrausch, *Sitzb. Akad. Wiss. Wien* **1930**, *Abt. IIa, 139*, 459; *Monatsh. Chem.* **1931**, *57*, 225; *C.A.* **1931**, *25*, 2365.

1792. Dadieu, A., A. Pongratz, et al., *Sitzb. Akad. Wiss. Wien Math.-Naturw. Klasse* **1932**, *141*, 267; *Monatsh. Chem.* **1932**, *61*, 383; *C.A.* **1933**, *27*, 902.

1793. Dadieu, A., A. Pongratz, et al., *Monatsh. Chem.* **1932**, *61*, 409, 426; *Sitzb. Akad. Wiss. Wien Math.-Naturw. Klasse* **1932**, *Abt. II, 141*, 477; *C.A.* **1933**, *27*, 4172.

1794. Dahlgard, M., and R. Q. Brewster, *J. Amer. Chem. Soc.* **1958**, *80*, 5861.

1795. Dailey, B. P., A. Gawer, et al., *Discussions Faraday Soc.* **1962**, *34*, 18.

1796. Dakin, H. D., *J. Biol. Chem.* **1908**, *4*, 221; *C.A.* **1908**, *2*, 2564.

1797. Dakshinamurty, P., G. J. Rao, et al., *J. Appl. Chem.* (London) **1961**, *11*, 226.

1798. Dakshinamurty, P., K. V. Rao, et al., *J. Chem. Eng. Data* **1973**, *18*, 39.

1799. Dale, J., *Acta Chem. Scand.* **1957**, *11*, 971; *C.A.* **1958**, *52*, 14325.

1800. D'Alelio, G. F., and E. E. Reid, *J. Amer. Chem. Soc.* **1937**, *59*, 109.

1801. Dalton, F., G. Meakins, et al., *J. Chem. Soc.* **1962**, 1566.

1802. Damsgaard-Sørensen, P., *Z. Phys. Chem.* (Leipzig) **1935**, *172*, 389.

1803. Dana, L. I., J. N. Burdick, et al., *J. Amer. Chem. Soc.* **1927**, *49*, 2801.

1804. Dandegaonker, S. H., *J. Indian Chem. Soc.* **1969**, *46*, 148.

1805. Danforth, W. E., *J. Phys. Rev.* **1931**, *38*, 1224; *Chem. Zentr.* **1931**, *II*, 3582.

1806. Dangyan, M. T., *Bull. Armenian Branch. Acad. Sci. USSR* **1942**, No. 9/10, 53; *C.A.* **1946**, *40*, 3399.

1807. Dangyan, M. T., *J. Gen. Chem.* (USSR) **1941**, *11*, 616; *C.A.* **1941**, *35*, 6925.

1808. Daniels, F., J. H. Mathews, et al., *Experimental Physical Chemistry*, 3rd ed., McGraw-Hill, New York, 1941.

1809. Daniher, F. A., *Org. Prep. Proced.* **1970**, *2*, 207.

1810. Danner, P. S., and J. H. Hildebrand, *J. Amer. Chem. Soc.* **1922**, *44*, 2824.

1811. Dannhauser, W., and L. W. Bahe, *J. Chem. Phys.* **1964**, *40*, 3058.

1812. Dannhauser, W., L. W. Bahe, et al., *J. Chem. Phys.* **1965**, *43*, 257.

1813. Dannhauser, W., and R. H. Cole, *J. Chem. Phys.* **1955**, *23*, 1762.

1814. Dannhauser, W., and A. F. Fleuckinger, *J. Phys. Chem.* **1964**, *68*, 1814.

1815. Danov, S. M., and Yu. D. Golubev, *Tr. Khim. Khim. Tekhnol.* **1967**, 52; *C.A.* **1968**, *68*, 95253.

1816. Danov, S. M., N. B. Matin, et al., *Zh. Fiz. Khim.* **1969**, *43*, 733; *Russ. J. Phys. Chem.* **1969**, *43*, 401.

1817. D'Aprano, A., *Gazz. Chim. Ital.* **1974**, *104*, 91.

1818. D'Aprano, A., D. I. Donato, et al., *J. Solution Chem.* **1981**, *10*, 673.

1819. Daragan, B., *Bull. Soc. Chim. Belges* **1935**, *44*, 597.

1820. Darzens, G., *Compt. Rend.* **1911**, *152*, 1314; *C.A.* **1911**, *5*, 3410.

1821. Darzens, G., *Compt. Rend.* **1926**, *183*, 748; *C.A.* **1927**, *21*, 581.

1822. Darzens, G., and A. Levy, *Compt. Rend.* **1936**, *202*, 73; *C.A.* **1936**, *30*, 2561.

1823. Das, T. R., C. O. Reed, Jr., et al., *J. Chem. Eng. Data* **1977**, *22*, 3.

1824. Das, T. R., C. O. Reed, Jr., et al., *J. Chem. Eng. Data* **1977**, *22*, 9.

1825. Das, T. R., C. O. Reed, Jr., et al., *J. Chem. Eng. Data* **1977**, *22*, 16.

1826. DaSilva, D. J., *Rec. Trav. Chim. Pays-Bas* **1934**, *53*, 1097.

1827. Dasler, W., and C. D. Bauer, *Ind. Eng. Chem., Anal. Ed.* **1946**, *18*, 52.

1828. Datin, P., *Ann. Phys.* **1916**, *5*, 218.

1829. Datta, R. L., and N. R. Chatterjee, *J. Amer. Chem. Soc.* **1917**, *39*, 435.

1830. Daub, L., and J. M. Vandenbelt, *J. Amer. Chem. Soc.* **1947**, *69*, 2714.

1831. Dauben, W. G., G. J. Fonken, et al., *J. Amer. Chem. Soc.* **1956**, *78*, 2579.

1832. Daudt, H. W., and E. L. Mattison, U.S. Patent 2,004,931, June 18, 1935; *C.A.* **1935**, *29*, 5123.

1833. Daudt, H. W., and M. A. Youker, U.S. Patent 2,005,705, June 18, 1935; *C.A.* **1935**, *29*, 5123.

1834. Daudt, H. W., and M. A. Youker, U.S. Patent 2,005,706, June 18, 1935; *C.A.* **1935,** *29,* 5123.
1835. Daudt, H. W., and M. A. Youker (to Kinetic Chemical), U.S. Patent 2,005,713, June 18, 1935; *C.A.* **1935,** *29,* 5123.
1836. Daure, P., *Ann. Phys.* **1929,** *12,* 435.
1837. Davanathan, T., *Proc. Indian Acad. Sci.* **1963,** Sect. A, *57,* No. 2, 109; *C.A.* **1963,** *59,* 2250.
1838. David, J. G., and H. E. Hallam, *Trans. Faraday Soc.* **1964,** *60,* 2013.
1839. Davidson, J. G., *Ind. Eng. Chem.* **1926,** *18,* 669.
1840. Davies, C. W., *Phil. Mag.* **1927,** *4,* 244.
1841. Davies, D. G., *J. Chem. Soc.* **1935,** 1166.
1842. Davies, G. F., and E. C. Gilbert, *J. Amer. Chem. Soc.* **1941,** *63,* 1585.
1843. Davies, G. F., and E. C. Gilbert, *J. Amer. Chem. Soc.* **1941,** *63,* 2730.
1844. Davies, J. E. D., P. Hodge, et al., *J. Chem. Soc., Perkin Trans. 2* **1972,** 1557.
1845. Davies, M., *J. Chem. Phys.* **1948,** *16,* 267.
1846. Davies, M., A. H. Jones, et al., *Trans. Faraday Soc.* **1959,** *55,* 1100.
1847. Davies, M., and B. Kybett, *Trans. Faraday Soc.* **1965,** *61,* 1608.
1848. Davies, M., and D. K. Thomas, *J. Phys. Chem.* **1956,** *60,* 767.
1849. Davies, M. M., *J. Chem. Phys.* **1940,** *8,* 577.
1850. Davies, R. M., *Phil. Mag.* **1936,** *21,* 1, 1008.
1851. Davies, R. R., and H. H. Hodgson, *J. Chem. Soc.* **1943,** 84.
1852. Davies, R. R., and H. H. Hodgson, *J. Soc. Chem. Ind. Trans.* **1943,** *62,* 128.
1853. Davies, W., J. B. Jagger, et al., *J. Soc. Chem. Ind.* (London) **1949,** *68,* 26.
1854. Davies, W. H., and L. L. McGee, *J. Chem. Soc.* **1950,** 678.
1855. Davis, A. H., *Phil. Mag.* **1924,** *47,* 972; *C.A.* **1924,** *18,* 2457.
1856. Davis, D. S., *Ind. Eng. Chem.* **1941,** *33,* 401.
1857. Davis, H. S., and O. F. Wiedeman, *Ind. Eng. Chem.* **1945,** *37,* 482.
1858. Davis, J. C., *J. Chem. Ed.* **1966,** *43,* 611.
1859. Davis, M. M., *Acid–Base Behavior in Aprotic Organic Solvents*, U.S. Government Printing Office, Washington, 1968.
1860. Davis, M. M., and P. J. Schumann, *J. Res. Nat. Bur. Stand.* **1947,** *39,* 221.
1861. Davis, R., H. S. Bridge, et al., *J. Amer. Chem. Soc.* **1943,** *65,* 857.
1862. Davis, R. T., and R. W. Schiessler, *Anal. Chem.* **1955,** *27,* 1824.
1863. Davis, T. C., and J. C. Norris, *Bur. Mines Rept. Investigation* **1965,** 6633.
1864. Davison, J. A., *J. Amer. Chem. Soc.* **1945,** *67,* 228.
1865. Davison, R. R., W. H. Smith, et al., *J. Chem. Eng. Data* **1960,** *5,* 420.
1866. Davison, W. H. T., and G. R. Bates, *J. Chem. Soc.* **1953,** 2607.
1867. Davy, L. G., and N. V. Sidgwick, *J. Chem. Soc.* **1933,** 281.
1868. Dawson, L. R., M. Golben, et al., *J. Electrochem. Soc.* **1952,** *99,* 28.
1869. Dawson, L. R., R. H. Graves, et al., *J. Amer. Chem. Soc.* **1957,** *79,* 298.
1870. Dawson, L. R., T. M. Newell, et al., *J. Amer. Chem. Soc.* **1954,** *76,* 6024.
1871. Dawson, L. R., P. G. Sears, et al., *J. Amer. Chem. Soc.* **1955,** *77,* 1986.
1872. Dawson, L. R., J. W. Vaughn, et al., *J. Phys. Chem.* **1962,** *66,* 2684.
1873. Dawson, L. R., E. D. Wilhoit, et al., *J. Amer. Chem. Soc.* **1956,** *78,* 1569.
1874. Dawson, L. R., E. D. Wilhoit, et al., *J. Amer. Chem. Soc.* **1957,** *79,* 3004.
1875. Dawson, P. P., Jr., I. H. Silberberg, et al., *J. Chem. Eng. Data* **1973,** *18,* 7.
1876. Day, H. O., and W. A. Felsing, *J. Amer. Chem. Soc.* **1952,** *74,* 1951.
1877. Day, H. O., D. E. Nicholson, et al., *J. Amer. Chem. Soc.* **1948,** *70,* 1784.
1878. Day, J. H., and A. Joachim, *J. Org. Chem.* **1965,** *30,* 4107.
1879. Day, R. A., A. E. Robinson, et al., *J. Amer. Chem. Soc.* **1950,** *72,* 1379.
1880. Dayantis, J., *C.R. Acad. Sci.* (Paris) Ser. C. **1968,** *267,* 223.
1881. Deal, C. H., Dissertation, Duke Univ., Durham, North Carolina; A. A. Maryott, et al., *Table of Dielectric Constants of Pure Liquids*, Nat. Bur. Stand. Circular 514, Washington, 1951.
1882. Deanesly, R. M., and L. T. Carleton, *J. Phys. Chem.* **1941,** *45,* 1104.
1883. Dearden, J. C., and W. F. Forbes, *Can. J. Chem.* **1959,** *37,* 1294.
1884. Dearden, J. C., and W. F. Forbes, *Can. J. Chem.* **1959,** *37,* 1305.
1885. Deb, A. R., *Indian J. Phys.* **1953,** *27,* 457; *Sci. Abstr.* **1954,** *57A,* 698; *C.A.* **1957,** *51,* 7856.

1886. Deb, K. K., *Indian J. Phys.* **1961**, *35*, 535; *C.A.* **1962**, *57*, 300.

1887. Deb, K. K., *Indian J. Phys.* **1962**, *36*, 557; *C.A.* **1963**, *58*, 12084.

1888. Deb, K. K., and S. B. Banerjee, *Indian J. Phys.* **1960**, *34*, 554; *C.A.* **1961**, *55*, 12038.

1889. de Beule, P., *Bull. Soc. Chim. Belges* **1931**, *40*, 195.

1890. de Brouckère, L., and A. Prigogine, *Bull. Soc. Chim. Belges* **1938**, *47*, 382; *C.A.* **1938**, *32*, 8888.

1891. de Brouckère, L., and A. Prigogine, *Bull. Soc. Chim. Belges* **1938**, *47*, 399; *C.A.* **1938**, *32*, 8888.

1892. de Brayne, J. M. A., R. M. Davis, et al., *J. Amer. Chem. Soc.* **1933**, *55*, 3936.

1893. Deckenbrock, W., *Deut. Lebensm.-Rundschau* **1953**, *49*, 30; *C.A.* **1955**, *49*, 7804.

1894. Decker, R., and H. Halz, German Patent 837,995, May 5, 1952; *C.A.* **1957**, *51*, 15548.

1895. Deese, R. F., *J. Amer. Chem. Soc.* **1931**, *53*, 3673.

1896. Deffet, L., *Bull. Soc. Chim. Belges* **1931**, *40*, 385; *C.A.* **1932**, *26*, 352.

1897. Deffet, L., *Bull. Soc. Chim. Belges* **1935**, *44*, 41, 97; *Chem. Zentr.* **1935**, *II*, 33.

1898. Deffet, L., *Bull. Soc. Chim. Belges* **1938**, *47*, 446; *Chem. Zentr.* **1939**, *I*, 3149.

1899. Deffet, L., *Bull. Soc. Chim. Belges* **1940**, *49*, 223; *C.A.* **1942**, *36*, 322.

1900. de Forcrand, R., *Compt. Rend.* **1892**, *114*, 1062; *J. Chem. Soc. A* **1892**, 1066.

1901. de Forcrand, R., *Compt. Rend.* **1900**, *130*, 1622; *Chem. Zentr.* **1900**, *II*, 160.

1902. de Forcrand, R., *Compt. Rend.* **1901**, *132*, 569; *Chem. Zentr.* **1901**, *I*, 818.

1903. de Forcrand, R., *Compt. Rend.* **1903**, *136*, 1037; *J. Chem. Soc.* **1903Ai**, 455.

1904. de Forcrand, R., *Compt. Rend.* **1912**, *154*, 1327; *J. Chem. Soc.* **1912Ai**, 548.

1905. de Forcrand, R., *Compt. Rend.* **1912**, *154*, 1767; *J. Chem. Soc.* **1912Aii**, 735.

1906. De Fracesco, F., *Rev. Ital. Sostanze Grasse* **1964**, *41*, 20; *C.A.* **1964**, *61*, 2070.

1907. De Graff, D. E., and G. B. B. M. Sutherland, *J. Chem. Phys.* **1957**, *26*, 716.

1908. Dehmlow, E. W., and M. Lissel, *Synthesis* **1979**, 372.

1909. Dehn, W. M., and G. T. Davis, *J. Amer. Chem. Soc.* **1907**, *29*, 1328.

1910. Dehn, W. M., and K. E. Jackson, *J. Amer. Chem. Soc.* **1933**, *55*, 4284.

1911. Deichmann, W. B., and S. Witherup, *J. Pharmacol.* **1944**, *80*, 223; *C.A.* **1944**, *38*, 2389.

1912. Deichmann, W. B., and S. Witherup, *J. Pharnacol. Exp. Therap.* **1946**, *88*, 338.

1913. Deizenrot, I. V., V. B. Kogan, et al., *Zh. Prikl. Khim.* **1966**, *39*, 1880; *C.A.* **1966**, *65*, 16119.

1914. Deizenrot, I. V., and V. B. Kogan, *Zh. Prikl. Khim.* **1972**, *45*, 1137.

1915. Dekker, H., and C. Mosselman, *Rec. Trav. Chim. Pays-Bas* **1970**, *89*, 1276.

1916. de Kolossowski, N., and A. Alemov, *Bull. Soc. Chim. Fr.* **1934**, *1*, 877; *Beil.* **EIII3**, 5.

1917. de Kolossowski, N., and W. W. Udowensko, *Compt. Rend.* **1933**, *197*, 519; *Chem. Zentr.* **1933**, *II*, 2375; *Beil.* **EIII2**, 1610.

1918. de Kowalewski, D. G., *J. Phys. Radium* **1962**, *23*, 255; *C.A.* **1962**, *57*, 11096.

1919. de Kowalewski, D. G., R. Buirago, et al., *J. Mol. Struct.* **1973**, *16*, 395.

1920. de Kowalewski, D. G., and V. J. Kowalewski, *Anales. Asoc. Quim. Arg.* **1960**, *48*, 157; *C.A.* **1961**, *55*, 18311.

1921. De Kruif, C. G., T. Kuipers, et al., *J. Chem. Thermodyn.* **1981**, *13*, 1081.

1922. De Kruif, C. G., and H. A. Oonk, *J. Chem. Thermodyn.* **1979**, *11*, 287.

1923. De Kruif, C. G., R. C. F. Schaake, et al., *J. Chem. Thermodyn.* **1982**, *14*, 791.

1924. Delaby, R., and J. Lecomte, *Bull. Soc. Chim. Fr.* **1937**, *4*, 738.

1925. Delacre, M., *Compt. Rend.* **1887**, *104*, 1184.

1926. Delafontaine, J., R. Sabbah, et al., *Z. Phys. Chem.* (Frankfurt) **1973**, *84*, 157.

1927. Delaplace, R., *Compt. Rend.* **1937**, *204*, 493.

1928. Delbouille, L., *Bull. Classe Sci. Acad. Roy. Belges* **1958**, *44*, 791; *C.A.* **1959**, *53*, 19571.

1929. Delbouille, L., *J. Chem. Phys.* **1956**, *25*, 182.

1930. Delépine, M., *Compt. Rend.* **1897**, *124*, 292; *J. Chem. Soc.* **1897Ai**, 394.

1931. Delépine, M., *Compt. Rend.* **1898**, *126*, 964; *Chem. Zentr.* **1898**, *I*, 989.

1932. Delépine, M., *Compt. Rend.* **1898**, *126*, 1794; *J. Chem. Soc.* **1898Aii**, 559.

1933. Delépine, M., *Compt. Rend.* **1900**, *131*, 745.

1934. Delépine, M., and A. Horeau, *Bull. Soc. Chim. Fr.* **1937**, *4*, 31; *C.A.* **1937**, *31*, 3003.

1935. De Ligny, C. L., *Rec. Trav. Chim. Pays-Bas* **1960**, *79*, 731.

1936. De Ligny, C. L., and N. G. Van der Veen, *Rec. Trav. Chim. Pays-Bas* **1971**, *90*, 984.

1937. Delorme, P., F. Denisselle, et al., *J. Chim. Phys.* **1967**, *64*, 591.

1938. Delorme, P., V. Lorenzelli, et al., *Compt. Rend.* **1965**, *261* (Group 6), 3331; *C.A.* **1966**, *64*, 6450.

1939. Deluzarche, A., A. Maillard, et al., *Bull. Soc. Chim. Fr.* **1963**, 89; *C.A.* **1963**, *58*, 13463; Ref. [8056].
1940. Delvaux, E., *J. Pharm. Belges* **1936**, *18*, 101, 131, 153; *Chem. Zentr.* **1936**, *I*, 3769.
1941. Delwaulle, M. L., *Compt. Rend.* **1944**, *217*, 172; *C.A.* **1944**, *38*, 3908.
1942. de Maine, P. A. D., L. H. Daly, et al., *Can. J. Chem.* **1960**, *38*, 1921.
1943. De Mare, G. R., T. Lehman, et al., *J. Chem. Thermodyn.* **1973**, *5*, 829.
1944. De Maria, P., and A. Fini, *J. Chem. Soc., Perkin Trans. 2* **1975**, 1540.
1945. de Melle, U., French Patent 871,565, Apr. 30, 1942; *C.A.* **1949**, *43*, 5032; British Patent 614,165, Dec. 10, 1948; *C.A.* **1949**, *43*, 4682.
1946. De Micheli, S., and V. Tartari, *J. Chem. Eng. Data* **1982**, *27*, 273.
1947. Demidenkova, I. V., *Tr. Gos. Inst. Prikl. Khim* **1964**, No. 52, 157; *C.A.* **1965**, *63*, 14230.
1948. de Montmollin, M., and P. Matile, *Helv. Chim. Acta* **1924**, *7*, 106.
1949. Dem'yanov, N. Ya., and Z. I. Shuikina, *J. Gen. Chem.* (USSR) **1936**, *6*, 350; *C.A.* **1936**, *30*, 6337.
1950. Denis, G., P. Adomenas, et al., *Zh. Fiz. Khim.* **1970**, *44*, 2070; *Russ. J. Phys. Chem.* **1970**, *44*, 1174.
1951. Denison, R. B., and B. D. Steele, *J. Chem. Soc.* **1906**, 999, 1386.
1952. Deno, N., T. Edwards, et al., *J. Amer. Chem. Soc.* **1957**, *79*, 2108.
1953. Deno, N. C., R. W. Gaugler, et al., *J. Org. Chem.* **1966**, *31*, 1967.
1954. Deno, N. C., R. W. Gaugler, et al., *J. Org. Chem.* **1966**, *31*, 1968.
1955. Deno, N. C., and J. O. Turner, *J. Org. Chem.* **1966**, *31*, 1969.
1956. Deno, N. C., and M. J. Wisotsky, *J. Amer. Chem. Soc.* **1963**, *85*, 1735.
1957. Denoon, C. E., *Organic Syntheses, Vol. 20*, Wiley, New York, 1940.
1958. Denyer, R. L., F. A. Fidler, et al., *Ind. Eng. Chem.* **1949**, *41*, 2727.
1959. de Pauw, F., and G. E. Limido, *Compt. Rend. 27 Congr. Intern. Chim. Ind. Brussels* **1954**, 2; *Ind. Chim. Belges* **1955**, *20*, Spec No. 178; *C.A.* **1956**, *50*, 14573.
1960. De Puy, C. H., and B. W. Ponder, *J. Amer. Chem. Soc.* **1959**, *81*, 4629.
1961. Dereppe, J. M., and M. van Meerssche, *Bull. Soc. Chim. Belges* **1960**, *69*, 466; from [7483].
1962. de Ridder, J. J., and G. Dijkstra, *Rec. Trav. Chim. Pays-Bas* **1967**, *86*, 737; *C.A.* **1967**, *67*, 103569.
1963. Deschamps, A., *J. Amer. Chem. Soc.* **1920**, *42*, 2670.
1964. Deschner, W. W., and G. G. Brown, *Ind. Eng. Chem.* **1940**, *32*, 836.
1965. Deshpande, D. D., and L. G. Bhatgadde, *Aust. J. Chem.* **1971**, *24*, 1817.
1966. Deshpande, D. D., and L. G. Bhatgadde, *J. Phys. Chem.* **1968**, *72*, 261.
1967. Deslandres, H., *Compt. Rend.* **1941**, *212*, 832; *C.A.* **1944**, *38*, 5455.
1968. Deslandres, H., *Compt. Rend.* **1941**, *213*, 749; *C.A.* **1943**, *37*, 4013.
1969. Desreux, V., *Bull. Soc. Chim. Belges* **1935**, *44*, 11, 249; *C.A.* **1936**, *30*, 344.
1970. Desty, D. G., and F. A. Fidler, *Ind. Eng. Chem.* **1951**, *43*, 905.
1971. Detoeuf, A., *Bull. Soc. Chim. Belges* **1922**, *31*, 169; *C.A.* **1922**, *16*, 2133.
1972. Dever, D. F., A. Finch, et al., *J. Phys. Chem.* **1955**, *59*, 668.
1973. Devoto, G., *Gazz. Chim. Ital.* **1933**, *63*, 495; *Chem. Zentr.* **1933**, *II*, 3394.
1974. Devoto, G., and R. DiNola, *Gazz. Chim. Ital.* **1933**, *63*, 495; *C.A.* **1934**, *28*, 744.
1975. De Vries, T., and H. Soffer, *J. Phys. Colloid Chem.* **1951**, *55*, 406.
1976. De Vries, T., and W. N. Vanderkooi, *J. Amer. Chem. Soc.* **1953**, *75*, 2253.
1977. Dewael, A., *Bull. Soc. Chim. Belges* **1930**, *39*, 87; *C.A.* **1930**, *24*, 3984.
1978. Dewar, M. J. S., R. C. Fahey, et al., *Tetrahedron Lett.* **1963**, 343; *C.A.* **1963**, *59*, 2604.
1979. Deyanova, E. V., *Novoe v Oblasti Sanit.-Khim. Analiza (Raboty po Prom.-Sanit. Khim)* **1962**, 120; *C.A.* **1963**, *59*, 2094.
1980. Dharmarju, G., G. N. Swamy, et al., *Indian J. Chem., Sect. A* **1981**, *20A*, 1109.
1981. Dharmatti, S. S., G. Govil, et al., *Proc. Indian Acad. Sci.* **1962**, *56A*, 86; *C.A.* **1963**, *58*, 4069.
1982. Dhillon, M. S., and H. S. Chugh, *J. Chem. Eng. Data* **1977**, *22*, 262.
1983. Dhillon, M. S., and H. S. Chugh, *J. Chem. Eng. Data* **1978**, *23*, 263.
1984. Dhillon, M. S., and J. Kaur, *Z. Phys. Chem.* (Leipzig) **1983**, *264*, 525.
1985. Dhillon, M. S., and B. S. Mahl, *Z. Phys. Chem.* (Leipzig) **1978**, *259*, 249.
1986. Dial, W., U.S. Patent 3,025,331, March 13, 1962; *C.A.* **1962**, *57*, 1182.
1987. Diaz Pena, M., A. Lainez, et al., *J. Chem. Thermodyn.* **1983**, *15*, 807.
1988. Diaz Pena, M. D., and G. Tardajos, *J. Chem. Thermodyn.* **1978**, *10*, 19.
1989. Diaz Pena, M. D., and G. Tardajos, *J. Chem. Thermodyn.* **1979**, *11*, 441.
1990. Diaz Pena, M. D., G. Tardajos, et al., *J. Chem. Thermodyn.* **1979**, *11*, 67.

1991. Diaz Pena, M. D., G. Tardajos, et al., *J. Chem. Thermodyn.* **1979,** *11,* 951.

1992. D'Ianni, J., and H. Adkins, *J. Amer. Chem. Soc.* **1939,** *61,* 1675.

1993. Dibeler, V. H., *J. Res. Nat. Bur. Stand.* **1952,** *49,* 235.

1994. Dibeler, V. H., and R. M. Reese, *J. Res. Nat. Bur. Stand.* **1955,** *54,* 127.

1995. Dibeler, V. H., R. M. Reese, et al., *J. Chem. Phys.* **1957,** *26,* 304.

1996. Dibeler, V. H., R. M. Reese, et al., *J. Res. Nat. Bur. Stand.* **1956,** *57,* 113.

1997. Di Bello, L. M., H. M. McDevitt, et al., *J. Phys. Chem.* **1968,** *72,* 1405.

1998. Di Ciommox, *Nuovo Cim.* **1903,** *3,* 97.

1999. Dickinson, E., *J. Phys. Chem.* **1977,** *81,* 2108.

2000. Dickinson, J. D., and C. Eaborn, *Chem. Ind.* (London) **1956,** 959.

2001. Dieckmann, W., *Ber. Deut. Chem. Gesell.* **1922,** *55,* 2470.

2002. Diehl, P., and G. Svegliado, *Helv. Chim. Acta* **1963,** *46,* 461.

2003. Dieke, S. H., G. S. Allen, et al., *J. Pharmacol. Exp. Therap.* **1949,** *90,* 260.

2004. Diem, M., and D. F. Burow, *J. Chem. Phys.* **1976,** *64,* 5179.

2005. Dietrich, M. W., J. S. Nash, et al., *Anal. Chem.* **1966,** *38,* 1479.

2006. Di Giacomo, A., and C. P. Smyth, *J. Amer. Chem. Soc.* **1955,** *77,* 774.

2007. Di Giacomo, A., and C. P. Smyth, *J. Amer. Chem. Soc.* **1955,** *77,* 1361.

2008. Dillon, R. T., and H. J. Lucas, *J. Amer. Chem. Soc.* **1928,** *50,* 1711.

2009. Di Macco, G., *Scientia Med. Ital.* **1955,** *4,* 100; *C.A.* **1956,** *50,* 14111.

2010. Dimroth, O., *Angew. Chem.* **1933,** *46,* 571.

2011. Diot, A., and T. Theophanides, *Can. J. Spectrosc.* **1972,** *17,* 67.

2012. Dippy, J. F. J., *Chem. Rev.* **1939,** *25,* 151.

2013. Dippy, J. F. J., *J. Chem. Soc.* **1938,** 1222.

2014. Dirksen, H.-W., Diploma Thesis, Universitat Freiberg, 1959; from [5253].

2015. Dirksen, H.-W., Ph.D. Thesis, Universitat Freiberg, 1961; from [5253].

2016. Discher, B., *Z. Naturforsch.* **1965,** *20a,* 888; *C.A.* **1965,** *63,* 17343.

2017. Distillers Co. Ltd., British Patent 883,746, Dec. 6, 1961; *C.A.* **1962,** *57,* 13684.

2018. Divorkin, A., and M. Guillamin, *J. Chim. Phys.* **1966,** *63,* 53, 57; *C.A.* **1966,** *64,* 18517.

2019. Dixon, R. L., R. H. Adamson, et al., *Arch. Intern. Pharmacodyn.* **1966,** *160,* 333.

2020. Dizechi, M., and E. Marschall, *J. Chem. Eng. Data* **1982,** *27,* 358.

2021. Djerassi, C., *J. Amer. Chem. Soc.* **1949,** *71,* 1003.

2022. Djerassi, C., and C. Fenselau, *J. Amer. Chem. Soc.* **1965,** *87,* 5756.

2023. Djerassi, C., J. Osiecki, et al., *J. Amer. Chem. Soc.* **1961,** *83,* 4433.

2024. Dmitrieva, N. V., *Gig. Sanit.* **1966,** 31; *C.A.* **1967,** *66,* 13873.

2025. Dobrinskaya, A. A., M. B. Neiman, et al., *Dokl. Akad. Nauk SSSR* **1948,** *63,* 549; *C.A.* **1949,** *43,* 2865.

2026. Dobrosserdow, D., *J. Russ. Phys. Chem. Soc.* **1911,** *43,* 73.

2027. Dobrosserdow, D., *Sapiske Kasaner Univ.* **1909,** 1; *Chem. Zentr.* **1911,** *I,* 953.

2028. Dobry, A., and R. Keller, *J. Phys. Chem.* **1957,** *61,* 1448.

2029. Doerffel, K., and K. Dathe, *Weis. Z. Tech. Hochsch. Chem. Leuna-Merseburg* **1965,** 7, 306; *C.A.* **1966,** *64,* 17385.

2030. Doeuvre, J., *Bull. Soc. Chim.* **1926,** *39,* 1594; *C.A.* **1927,** *21,* 565.

2031. Doldi, S., *Ann. Chim. Applicata* (Rome) **1938,** *28,* 454; *C.A.* **1939,** *33,* 5250.

2032. Dolique, R., *Ann. Chim.* (Paris) **1931,** *15,* 425; *C.A.* **1932,** *26,* 2791.

2033. Dolique, R., *Bull Sci. Pharmacol.* **1932,** *39,* 129.

2034. Dolique, R., *Compt. Rend.* **1932,** *194,* 289.

2035. Dolliver, M. A., T. L. Gresham, et al., *J. Amer. Chem. Soc.* **1938,** *60,* 440.

2036. Domalski, E. S., *J. Phys. Chem. Ref. Data* **1972,** *1,* 221.

2037. Donaldson, N., W. H. Powell, et al., *J. Chem. Doc.* **1974,** *14,* 3.

2038. Donaldson, R. E., and O. R. Quayle, *J. Amer. Chem. Soc.* **1950,** 72, 35.

2039. Donle, H. L., and K. A. Gehrckens, *Z. Phys. Chem.* (Leipzig) **1932,** *18B,* 316.

2040. Donle, H. L., and G. Volkert, *Z. Phys. Chem.* (Leipzig) **1930,** *8B,* 60.

2041. Donoghue, J. T., and R. S. Drago, *Inorg. Chem.* **1962,** *1,* 866; **1963,** *2,* 572.

2042. Doolittle, A. K., *Ind. Eng. Chem.* **1935,** *27,* 1169.

2043. Doolittle, A. K., *Ind. Eng. Chem.* **1944,** *36,* 239.

2044. Doolittle, A. K., *The Technology of Solvents and Plasticizers*, Wiley, New York, 1954.
2045. Doolittle, A. K., and R. H. Peterson, *J. Amer. Chem. Soc.* **1951**, *73*, 2145.
2046. D'Or, L., H. Heyns, et al., *Bull. Soc. Chim. Belges* **1957**, *66*, 512; *C.A.* **1957**, *51*, 16085.
2047. D'Or, L., J. Momigny, et al., *Advan. Mass Spectrometry, Proc. Conf.*, 2nd, Oxford, 1961, *2*, 370 (Pub. 1963); *C.A.* **1963**, *59*, 6225.
2048. Dorinson, A., M. R. McCorkle, et al., *J. Amer. Chem. Soc.* **1942**, *64*, 2739.
2049. Dorinson, A., and A. W. Ralston, *J. Amer. Chem. Soc.* **1944**, *66*, 361.
2050. Dorman, F. H., *J. Chem. Phys.* **1966**, *44*, 3856.
2051. Dornte, R. W., and C. P. Smyth, *J. Amer. Chem. Soc.* **1930**, *52*, 3546.
2052. Dorough, G. L., H. B. Glass, et al., *J. Amer. Chem. Soc.* **1941**, *63*, 3100.
2053. Dory, I., *Magyar Kemikusok Lapja* **1954**, *9*, 276; *C.A.* **1957**, *51*, 14575.
2054. Doub, L., and J. M. Vandenbelt, *J. Amer. Chem. Soc.* **1947**, *69*, 2714.
2055. Douglas, A. W., and J. H. Golstein, *J. Mol. Spectrosc.* **1965**, *16*, 1; *C.A.* **1965**, *63*, 1374.
2056. Douglas, D. G., and C. A. MacKay, *Can. J. Res.* **1946**, *24A*, 8.
2057. Douglas, T. B., *J. Amer. Chem. Soc.* **1948**, *70*, 2001.
2058. Douglas, T. B., G. T. Furukawa, et al., *J. Res. Nat. Bur. Stand.* **1954**, *53*, 139.
2059. Douslin, D. R., Bureau of Mines, Bartlesville Energy Research Center, "Special Boiling Point Tables."
2060. Douslin, D. R., R. H. Harrison, et al., *J. Chem. Thermodyn.* **1969**, *1*, 305.
2061. Douslin, D. R., and H. M. Huffman, *J. Amer. Chem. Soc.* **1946**, *68*, 1704.
2062. Douslin, D. R., R. T. Moore, et al., *J. Amer. Chem. Soc.* **1958**, *80*, 2031.
2063. Douslin, D. R., R. T. Moore, et al., *J. Phys. Chem.* **1959**, *63*, 1959.
2064. Dow Chemical Co., *Alkanolamines Handbook*, 3rd ed., Midland, Michigan, 1964.
2065. Dow Chemical Co., "Alkylene Oxides from Dow," Midland, Michigan, 1969.
2066. Dow Chemical Co., British Patent 916,129, Jan. 23, 1963; *C.A.* **1963**, *58*, 13792.
2067. Dow Chemical Co., "Chlorothene NU," Midland, Michigan, 1962.
2068. Dow Chemical Co., "Dow Products 1964–65," Midland, Michigan, 1964.
2069. Dow Chemical Co., "Dowanol. Glycol-ether Solvents," Midland, Michigan, 1958.
2070. Dow Chemical Co., "Ethanolamines," Midland, Michigan, 1954.
2071. Dow Chemical Co., "Ethylenimine," Midland, Michigan.
2072. Dow Chemical Co., "Physical Properties of the Dow Alkene Oxides," data sheet, Midland, Michigan, 1969.
2073. Dow Chemical Co., "Vinylidene Chloride Monomer," Midland, Michigan, 1966.
2074. Down, J. L., J. Lewis, et al., *J. Chem. Soc.* **1959**, 3767.
2075. Downie, I. M., J. B. Holmes, et al., *Chem. Ind.* (London) **1966**, 900; *C.A.* **1966**, *65*, 8741.
2076. Draeger, J. A., and D. W. Scott, *J. Chem. Phys.* **1981**, *74*, 4748.
2077. Drago, R. S., R. L. Carlson, et al., *Inorg. Chem.* **1965**, *4*, 15.
2078. Drago, R. S., and N. A. Matwiyoff, *J. Organometal. Chem.* **1965**, *3*, 62; *C.A.* **1965**, *62*, 7612.
2079. Drago, R. S., V. A. Mode, et al., *J. Amer. Chem. Soc.* **1965**, *87*, 5010.
2080. Drahowzal, F., and D. Klamann, *Monatsh. Chem.* **1951**, *82*, 588; *C.A.* **1952**, *46*, 8036.
2081. Draper, O. J., and A. L. Pollard, *Science* **1949**, *109*, 448.
2082. Drefahl, G., and H. Schick, *Z. Chem.* **1964**, *4*, 347; *C.A.* **1964**, *61*, 13178.
2083. Dreisbach, D., *Liquids and Solutions*, Houghton Mifflin, Boston, 1966.
2084. Dreisbach, R. R., *Ind. Eng. Chem., Anal. Ed.* **1940**, *12*, 160.
2085. Dreisbach, R. R., *Ind. Eng. Chem.* **1948**, *40*, 2269.
2086. Dreisbach, R. R., *J. Amer. Chem. Soc.* **1951**, *73*, 3147.
2087. Dreisbach, R. R., *Physical Properties of Chemical Compounds*, American Chemical Society, Washington, 1955.
2088. Dreisbach, R. R., *Physical Properties of Chemical Compounds II*, American Chemical Society, Washington, 1959.
2089. Dreisbach, R. R., *Physical Properties of Chemical Compounds III*, American Chemical Society, Washington, 1961.
2090. Dreisbach, R. R., and R. A. Martin, *Ind. Eng. Chem.* **1949**, *41*, 2875.
2091. Dreisbach, R. R., and S. A. Shrader, *Ind. Eng. Chem.* **1949**, *41*, 2879.
2092. Dreisbach, R. R., and R. S. Spencer, *Ind. Eng. Chem.* **1949**, *41*, 176.

2093. Dreisbach, R. R., and R. S. Spencer, *Ind. Eng. Chem.* **1949,** *41,* 1363.

2094. Drenth, W., G. L. Hekkert, et al., *Rec. Trav. Chim. Pays-Bas* **1960,** *79,* 1056; *C.A.* **1961,** *55,* 6361

2095. Drinker, P., and W. A. Cook, *J. Ind. Hyg. Toxicol.* **1949,** *31,* 51; *C.A.* **1949,** *43,* 4400.

2096. Drozdov, V. A., A. P. Kreshkov, et al., *J. Gen. Chem.* (USSR) (Eng. trans.) **1970,** *40,* 94.

2097. Drucker, C., *Ostwald-Luther Handund Hilfobuch zur Auofuhruna physikochemischer Messungen,* Dover, New York, 1943.

2098. Drucker, C., *Z. Phys. Chem.* (Leipzig) **1905,** *52,* 641.

2099. Drucker, C., and H. Weissbach, *Z. Phys. Chem.* (Leipzig) **1925,** *117,* 209.

2100. Drude, P., *Z. Phys. Chem.* (Leipzig) **1897,** *23,* 267.

2101. Drushel, W. A., *Amer. J. Sci* **1915,** *40,* 643.

2102. Dubchenko, V. N., *Zh. Prikl. Khim.* (Leningrad) **1973,** *46,* 397; *C.A.* **1973,** *78,* 135784.

2103. Dubois, J. E., *Compt. Rend.* **1947,** *224,* 1234; *C.A.* **1947,** *41,* 6525.

2104. Dubois, J. E., and H. Herzog, *J. Amer. Chem. Soc., Chem. Commun.* **1972,** 932.

2105. Dubois, J. E., and H. Viellard, *J. Chim. Phys.* **1965,** *62,* 699; *C.A.* **1965,** *63,* 14178.

2106. Duchense, J., *Nature* **1946,** *157,* 733.

2107. Duchesne, J., *Physica* **1941,** *8,* 144; *C.A.* **1941,** *35,* 7832.

2108. Duclaux, J., and A. Lanzenberg, *Bull. Soc. Chim.* **1920,** *27,* 779; *Chem. Zentr.* **1921,** *I,* 277.

2109. Ducom, J., and M. H. Normant, *Compt. Rend.* **1967,** *264,* 722.

2110. du Feu, E. C., F. J. Mc Quillin, et al., *J. Chem. Soc.* **1937,** 53.

2111. Duffield, A. M., H. Budzikiewicz, et al., *J. Amer. Chem. Soc.* **1964,** *86,* 5536.

2112. Duffield, A. M., H. Budzikiewicz, et al., *J. Amer. Chem. Soc.* **1965,** *87,* 810.

2113. Duffield, A. M., H. Budzikiewicz, et al., *J. Amer. Chem. Soc.* **1965,** *87,* 2920.

2114. Dugacheva, G. M., and A. G. Anikin, *Zh. Fiz. Khim.* **1964,** *38,* 208; *C.A.* **1964,** *60,* 10523.

2115. Dugan, P. R., *Anal. Chem.* **1961,** *33,* 1630.

2116. Duggan, R. J., and L. O. Winstrom, French Patent 1,316,057, Jan. 25, 1963; *C.A.* **1963,** *59,* 2719.

2117. Duke, F. R., and G. F. Smith, *Ind. Eng. Chem., Anal. Ed.* **1940,** *12,* 201.

2118. Duncan, N. E., and G. J. Janz, *J. Chem. Phys.* **1955,** *23,* 434.

2119. Duncan, W. A., J. P. Sheridan, et al., *Trans. Faraday Soc.* **1966,** *62,* 1090.

2120. Duncan, W. A., and F. L. Swinton, *Trans. Faraday Soc.* **1966,** *62,* 1082.

2121. Dunken, H., H. Klapproth, et al., *Kolloid-Z.* **1940,** *92,* 232; *Beil.* **EIII3,** 2173.

2122. Dunlap, M. K., J. K. Dodama, et al., *A.M.A. Arch. Ind. Health* **1958,** *18,* 303; *C.A.* **1959,** *53,* 1591.

2123. Dunstan, A. E., *J. Chem. Soc.* **1904,** 817.

2124. Dunstan, A. E., *J. Chem. Soc.* **1915,** 667.

2125. Dunstan, A. E., T. P. Hilditch, et al., *J. Chem. Soc.* **1913,** 133.

2126. Dunstan, A. E., and A. G. Mussell, *J. Chem. Soc.* **1910,** 1935.

2127. Dunstan, A. E., and J. A. Stubbs, *J. Chem. Soc.* **1908,** 1919.

2128. Dunstan, A. E., and F. B. Thole, *J. Chem. Soc.* **1908,** 1815.

2129. Dunstan, A. E., and F. B. Thole, *J. Chem. Soc.* **1913,** 127.

2130. Dunstan, A. E., F. B. Thole, et al., *J. Chem. Soc.* **1907,** 1728.

2131. Dunstan, A. E., F. B. Thole, et al., *J. Chem. Soc.* **1914,** 782.

2132. Dupont, G., and M. Barraud, *Bull. Soc. Chim. Fr.* **1924,** *35,* 625; *C.A.* **1924,** *18,* 2425.

2133. Dupont, G., and M. Darmon, *Bull. Soc. Chim. Fr.* **1939,** *6,* 1208; *C.A.* **1939,** *33,* 8174.

2134. Dupont, G., P. Daure, et al., *Bull. Soc. Chim. Fr.* **1932,** *51,* 921; *C.A.* **1933,** *27,* 26.

2135. Dupont, G., R. Dulou, et al., *Bull. Soc. Chim. Fr.* **1936,** *3,* 1639; *C.A.* **1936,** *30,* 8026.

2136. Dupont, G., R. Dulou, et al., *Compt. Rend.* **1958,** *246,* 128; *C.A.* **1958,** *52,* 13607.

2137. du Pont de Nemours & Co., E.I., British Patent 575,362, Feb. 14, 1946; *C.A.* **1947,** *41,* 5551.

2138. du Pont de Nemours & Co., E.I., "du Pont DMAC," Wilmington, Delaware.

2139. du Pont de Nemours & Co., E.I., "du Pont DMF," Wilmington, Delaware, 1967.

2140. du Pont de Nemours & Co., E.I., "du Pont Formamide," Wilmington, Delaware, 1967.

2141. du Pont de Nemours & Co., E.I., "Freon," Technical Bulletin B-1, Wilmington, Delaware, 1965.

2142. du Pont de Nemours & Co., E.I., "Freon," Technical Bulletin B-2, Wilmington, Delaware, 1966.

2143. du Pont de Nemours & Co., E.I., "Freon," Technical Bulletin B-4B, Wilmington, Delaware, 1964.

2144. du Pont de Nemours & Co., E.I., "Freon-113," Wilmington, Delaware, 1938.

2145. du Pont de Nemours & Co., E.I., French Patent 1,311,076, Dec. 7, 1962; *C.A.* **1963,** *58,* 13916.

2146. du Pont de Nemours & Co., E.I., *Ind. Eng. Chem.* **1936,** *28,* 1160.

2147. du Pont de Nemours & Co., E.I., Japanese Patent 76 56,405, May 18, 1976; *C.A.* **1977,** *86,* 4931.

2148. du Pont de Nemours & Co., E.I., "Properties and Uses of THF du Pont Tetrahydrofuran," Wilmington, Delaware, 1961.

2149. du Pont de Nemours & Co., E.I., "A Review of Catalytic and Synthetic Applications for DMF and DMAC," Wilmington, Delaware.

2150. du Pont de Nemours & Co., E.I., "Tetrahydrofuran as a Reaction Solvent," Wilmington, Delaware.

2151. du Pont de Nemours & Co., E.I., "Tetralin and Decalin Solvents," Wilmington, Delaware.

2152. Durand, J. E., *Bull. Soc. Chim. Fr.* **1937,** *4,* 67; *C.A.* **1937,** *31,* 4185.

2153. Durand, F. J., and E. Rouge, *Bull. Soc. Chim. Fr.* **1925,** *37,* 697; *C.A.* **1925,** *19,* 2903.

2154. Durand, M., and J.-P. Laurent, *Bull. Soc. Chim. Fr.* **1969,** *1,* 48.

2155. Durand, P., J. Parello, et al., *Bull. Soc. Chim. Fr.* **1963,** 2438.

2156. Durgaprasad, G., D. N. Sathyanarayana, et al., *Bull. Soc. Chem. Jpn.* **1971,** *44,* 316.

2157. Durgaprasad, G., D. N. Sathyanarayana, et al., *Spectrochim. Acta, Part A* **1972,** *28,* 2311.

2158. Durig, J. R., and D. A. C. Compton, *J. Phys. Chim.* **1979,** *83,* 265.

2159. Durig, J. R., G. L. Coulter, et al., *J. Mol. Spectrosc.* **1968,** *27,* 285.

2160. Durig, J. R., and D. J. Gerson, *J. Mol. Struct.* **1981,** *71,* 131.

2161. Durig, J. R., K. L. Kizer, et al., *J. Raman Spectrosc.* **1973,** *1,* 17.

2162. Durig, J. R., A. E. Sloan, et al., *J. Phys. Chem.* **1972,** *76,* 3591.

2163. Durig, J. R., J. W. Thompson, et al., *J. Mol. Struct.* **1975,** *24,* 41.

2164. Durmis, J., M. Karvas, et al., *Collect. Czech. Chem. Commun.* **1973,** *38,* 215.

2165. Durocher, G., and C. Sandorfy, *J. Mol. Spectrosc.* **1965,** *15,* 22; *C.A.* **1965,** *62,* 4787.

2166. Dutt, I. S., *Quart. J. Chem. Soc.* **1925,** *1,* 297; *C.A.* **1925,** *19,* 2475.

2167. Dutt, N. V. K., A. P. Kahol, et al., *J. Chem. Eng. Data* **1982,** *27,* 369.

2168. Dutt, P. K., *J. Chem. Soc.* **1923,** 2714.

2169. Duus, H. C., *Ind. Eng. Chem.* **1955,** *47,* 1445.

2170. Duyckaerts, G., and G. Michel, *Anal. Chim. Acta* **1948,** *2,* 750; *C.A.* **1949,** *43,* 7858.

2171. Dworkin, A., P. Figiuere, et al., *J. Chem. Thermodyn.* **1976,** *8,* 835.

2172. Dyke, M. D., P. G. Sears, et al., *J. Phys. Chem.* **1967,** *71,* 4140.

2173. Dykstra, H. B., J. F. Lewis, et al., *J. Amer. Chem. Soc.* **1930,** *52,* 3396.

2174. Dykyj, J., M. Seprakova, et al., *Chem. Zvesti* **1961,** *15,* 465; *C.A.* **1962,** *56,* 10930.

2175. Dymond, J. H., and J. H. Hildebrand, *J. Phys. Chem.* **1967,** *71,* 1145.

2176. Dymond, J. H., K. L. Young, et al., *J. Chem. Thermodyn.* **1979,** *11,* 887.

2177. Dyrssen, D., *Acta Chem. Scand.* **1957,** *11,* 1771: from [4596].

2178. Dzhamalov, R. M., A. A. Namedov, et al., *Zh. Fiz. Khim.* **1973,** *47,* 1000; *Russ J. Phys. Chem.* **1973,** *47,* 565.

2179. Dzung, L. S., *Brown Boveri Rev.* **1946,** *33,* 158; *C.A.* **1947,** *41,* 4372.

E

2180. Earle, J. C., *J. Soc. Chem. Ind.* (London) **1918,** *37,* 2749.

2181. Earle, R. B., and H. L. Jackson, *J. Amer. Chem. Soc.* **1906,** *28,* 104.

2182. Earls, D. W., J. R. Jones, et al., *J. Chem. Soc., Perkin Trans. 2* **1975,** 878.

2183. Easteal, A. J., and L. A. Woolf, *J. Chem. Thermodyn.* **1983,** *15,* 195.

2184. Eastman Chemical Products, Inc., Publication No. L-139, Plasticizers, "Kodaflex DBP, Dibutyl Phthalate," Kingsport, Tennessee, 1970.

2185. Eastman Chemical Products Inc., Publication No. L-147, Plasticizers, "Kodaflex DBS, Dibutyl Sebacate," Kingsport, Tennessee, 1970.

2186. Eastman Chemical Products Inc., Publication No. L-142, Plasticizers, "Kodaflex DOP, Dioctyl Phthalate," Kingsport, Tennessee, 1970.

2187. Eastman Chemical Products Inc., "3-Ethylhexylamine," Technical Data sheet, Kingsport, Tennessee, 1962.

2188. Eastman Chemical Products Inc., "Isobutyronitrile, Properties, Applications," Technical Data Report No. N-102, Kingsport, Tennessee, 1957.

2189. Eastman Chemical Products Inc., Technical Data, "*n*-Butyronitrile," Kingsport, Tennessee.

2190. Eastman Chemical Products Inc., "Solvents, Diluents and Plasticizers," Kingsport, Tennessee, 1971.

2191. Easton, B. C., and M. K. Hargreaves, *J. Chem. Soc.* **1959,** 1413.

2192. Easton, B. C., M. K. Hargreaves, et al., *J. Appl. Chem.* (London) **1957,** *7,* 198.

2193. Ebbighausen, W., E. Breitmaier, et al., *Z. Naturforsch. B* **1970,** *25,* 1239; *C.A.* **1971,** *74,* 42494.

2194. Eberhardt, H., and O. Osberghaus, *Z. Naturforsch.* **1958,** *13a,* 16; *C.A.* **1958,** *52,* 8750.

2195. Eberius, E., and W. Kowalski, *Chem.-Ztg.* **1957,** *81,* 75; *C.A.* **1957,** *51,* 9417.

2196. Ebersole, E. R., *Natl. Advisory Comm. Aeronaut. Tech. Notes* **1946,** *1164,* 6; *C.A.* **1947,** *41,* 2010.

2197. Edgar, G., *Ind. Eng. Chem.* **1927,** *19,* 145.

2198. Edgar, G., and G. Calingaert, *J. Amer. Chem. Soc.* **1929,** *51,* 1483.

2199. Edgar, G., and G. Calingaert, *J. Amer. Chem. Soc.* **1929,** *51,* 1540.

2200. Edgell, W. F., *J. Amer. Chem. Soc.* **1947,** *69,* 660.

2201. Edgell, W. F., and W. E. Byrd, *J. Chem. Phys.* **1950,** *18,* 892.

2202. Edgell, W. F., and G. Glockler, *J. Chem. Phys.* **1941,** *9,* 484.

2203. Edgell, W. F., M. T. Yang, et al., *J. Amer. Chem. Soc.* **1965,** *87,* 3080.

2204. Edmonds, W. J., U.S. Patent 1,550,792, Aug. 25, 1925.

2205. Edsall, J. T., *J. Amer. Chem. Soc.* **1943,** *65,* 1767.

2206. Edsall, J. T., *J. Chem. Phys.* **1937,** *5,* 225.

2207. Eduljee, G. H., and A. P. Boyes, *J. Chem. Eng. Data* **1980,** *25,* 249.

2208. Edward, J. T., *Chem. Ind.* (London) **1963,** 489.

2209. Edward, J. T., J. B. Leane, et al., *Can. J. Chem.* **1962,** *40,* 1521.

2210. Edward, J. T., and I. C. Wang, *Can. J. Chem.* **1962,** *40,* 966.

2211. Edwards, D. A., and C. F. Bonilla, *Ind. Eng. Chem.* **1944,** *36,* 1038.

2212. Edwards, P. R., *J. Chem. Soc.* **1925,** 744.

2213. Eeckelaers, R., Belgian Patent 505,125, **1952;** *C.A.* **1955,** *49,* 15950.

2214. Efremov, Y. V., *Zh. Fiz. Khim.* **1966,** *40,* 1240; *Russ. J. Phys. Chem.* **1966,** *40,* 667.

2215. Efremov, Y. Y., *Dokl. Akad. Nauk SSSR* **1966,** *170,* No. 3, 597; *C.A.* **1967,** *66,* 76162.

2216. Eftring, E., *Annual Tables of Physical Constants,* Frick Chemical Laboratory, Princeton, New Jersey, 1941.

2217. Eftring, E., Thesis, Lund, 1938; and J. Timmermans, *Physico-Chemical Constants of Pure Organic Compounds,* Elsevier, New York, 1950.

2218. Egan, C. J., and J. D. Kemp, *J. Amer. Chem. Soc.* **1938,** *60,* 2097.

2219. Eggenberger, D. N., F. K. Broome, et al., *J. Org. Chem.* **1949,** *14,* 1108.

2220. Egloff, G., *Physical Constants of Hydrocarbons,* Reinhold, New York, 1946.

2221. Egorov, Y. P., *Bull. Acad. Sci. USSR, Div. Chem. Sci.* (Eng. trans.) **1960,** No. 9, 1447.

2222. Eguchi, T., *Bull. Chem. Soc. Jpn.* **1927,** *2,* 176.

2223. Ehrhardt, E., and O. Osberghaus, *Z. Naturforsch.* **1960,** *15a,* 575; *C.A.* **1960,** *54,* 23719.

2224. Ehrlich, F., *Ber. Deut. Chem. Gesell.* **1907,** *40,* 2538.

2225. Eicher, L. D., and B. J. Zwolinski, *J. Phys. Chem.* **1972,** *76,* 3295.

2226. Eichmann, O., *Z. Phys.* **1933,** *82,* 470; *C.A.* **1933,** *27,* 4228.

2227. Eidenhoff, M. L., *J. Amer. Chem. Soc.* **1945,** *67,* 2072.

2228. Eigenberger, E., *J. Prakt. Chem.* **1931,** *130,* 75.

2229. Eilingsfeld, H., M. Seefelder, et al., *Angew. Chem.* **1960,** *72,* 836; from [4718].

2230. Eilingsfeld, H., M. Seefelder, et al., *Chem. Ber.* **1963,** *96,* 2671.

2231. Eisen, O., A. Elvelt, et al., *Eesti NSV Tead. Akad. Tiom., Keem., Geol.* **1972,** *21,* No. 3, 219; *C.A.* **1972,** *77,* 156643.

2232. Eisenbach, M., S. R. Caplan, et al., *Biochim. Biophys. Acta* **1979,** *554,* No. 2, 269.

2233. Eisenbrand, J., and K. Baumann, *Z. Lebensm.-Unters.-Forsch.* **1969,** *140,* No. 4, 210; *C.A.* **1969,** *71,* 74003.

2234. Eisenlohr, F., *Fortschritte der Chem., Phys. Chem.* **1925,** *18,* 521; *C.A.* **1926,** *20,* 171.

2235. Ekshtat, B. Y., N. G. Kurysheva, et al., *Uch. Zap.-Mosk. Nauchn.-Issled. Inst. Gig.* **1975,** *22,* 46; *C.A.* **1977,** *86,* 134512.

2236. Eliassaf, J., R. M. Fuoss, et al., *J. Phys. Chem.* **1963,** *67,* 1724.

2237. Eliel, E. L., and R. G. Haber, *J. Org. Chem.* **1958,** *23,* 2041.

2238. Eliel, E. L., and C. A. Lukach, *J. Amer. Chem. Soc.* **1957,** *79,* 5986.

2239. Eliel, E. L., C. C. Price, et al., *Spectrochim. Acta* **1958,** *10,* 423; *C.A.* **1961,** *55,* 25466.

2240. Elizarova, O. N., *Uch. Zap., Mosk. Nauchn.-Issled. Inst. Gig.* **1961,** No. 9, 105; *C.A.* **1964,** *61,* 3605.

2241. Elkarim, A. A., *Rev. Sci. Instruments* **1948**, *19*, 833.
2242. Elkins, H. B., *Ind. Med.* **1939**, *8*, 426; *C.A.* **1940**, *34*, 5196.
2243. Elleman, D. D., L. C. Brown, et al., *J. Mol. Spectrosc.* **1961**, *7*, 307.
2244. Ellestad, O. H., P. Klaboe, et al., *Spectrochim. Acta, Part A* **1971**, *27*, 1025.
2245. Elliot, J. J., and S. F. Mason, *J. Chem. Soc.* **1959**, 1275.
2246. Elliott, M. A., and R. M. Fuoss, *J. Amer. Chem. Soc.* **1939**, *61*, 294.
2247. Elliott, M. A., A. R. Jones, et al., *Anal. Chem.* **1947**, *19*, 10.
2248. Elliott, T. H., E. Jacob, et al., *J. Pharm. Pharmacol.* **1969**, *21*, 561.
2249. Ellis, C., U.S. Patent 1,202,040, Oct. 24, 1917; *C.A.* **1917**, *11*, 187.
2250. Ellis, J. W., *J. Amer. Chem. Soc.* **1929**, *51*, 1384.
2251. Ellis, J. W., *Phys. Rev.* **1923**, *22*, 200.
2252. Ellis, J. W., *Phys. Rev.* **1924**, *23*, 48; *C.A.* **1924**, *18*, 1086.
2253. Ellis, L. M., and E. E. Reid, *J. Amer. Chem. Soc.* **1932**, *54*, 1674.
2254. Emblem, H. G., and C. A. McDowell, *J. Chem. Soc.* **1946**, 641.
2255. Emde, H., and T. Hornemann, *Helv. Chem. Acta* **1931**, *14*, 892.
2256. Emeleus, H. J., and C. J. Wilkins, *J. Chem. Soc.* **1944**, 454.
2257. Emery, A. G., and F. G. Benedict, *Amer. J. Physiol.*, *28*, 301; *C.A.* **1912**, *6*, 235.
2258. Emery, E. M., *Anal. Chem.* **1960**, *32*, 1495.
2259. Emschwiller, G., and J. Lecomte, *J. Phys. Radium* **1937**, *8*, 130; *C.A.* **1937**, *31*, 5680.
2260. Emschwiller, M. G., *Compt. Rend.* **1930**, *191*, 208.
2261. Emsley, J. W., S. R. Salman, et al., *J. Chem. Soc. B* **1970**, 1513.
2262. Engel, K. H., U.S. Patent 2,339,388, Jan. 18, 1944; *C.A.* **1944**, *38*, 3672.
2263. Engel, K. H., U.S. Patent 2,382,142, Aug. 14, 1945; *C.A.* **1945**, *39*, 4893.
2264. Engel, K. H., U.S. Patent 2,426,442, Aug. 26, 1947; *C.A.* **1948**, *42*, 226.
2265. Engel, W. F., Dutch Patent 59,179, April 15, 1947; *C.A.* **1947**, *41*, 5893.
2266. Engelhard, H., H. Schilfarth, et al., *Ann. Chim.* **1949**, *563*, 239.
2267. Engelhardt, G., and K. Licht, *Z. Chem.* **1970**, *10*, No. 7, 266; *C.A.* **1970**, *73*, 71885.
2268. Englin, B. A., A. F. Plate, et al., *Khim. i Tekhnol. Topliv i Masel* **1965**, 10, *42*; *C.A.* **1965**, *63*, 14608.
2269. English, S., and W. E. S. Turner, *J. Chem. Soc.* **1914**, 1656.
2270. Enjay Chemical Co., "Enjay Acetates," Humble Oil and Refining Co., New York.
2271. Enjay Chemical Co., "Isopropyl Alcohol," New York, 1961.
2272. Enjay Chemical Co., "Exxon Isopropyl Alcohol," Houston, 1972.
2273. Ennan, A. A., V. G. Samoilenko, et al., *Zh. Prikl. Khim.* (Leningrad) **1971**, *44*, No. 6, 1427; *C.A.* **1971**, *75*, 133287.
2274. Enokido, H., T. Shinoda, et al., *Bull. Chem. Soc. Jpn.* **1969**, *42*, 84.
2275. Ens, A., and F. E. Murray, *Can. J. Chem.* **1957**, *35*, 170.
2276. Eon, C., C. Pommier, et al., *J. Chem. Eng. Data* **1971**, *16*, 408.
2277. Eötvös, R., *Ann. Phys.* (Leipzig) **1886**, *27*, 448.
2278. Erdmann, H., *Ber. Deut. Chem. Gesell.* **1891**, *24*, 2769.
2279. Erdmann, H., *Justus Liebigs Ann. Chem.* **1893**, *272*, 145.
2280. Erdös, E., L. Jager, et al., *Chem. Listy* **1952**, *46*, 770; *C.A.* **1953**, *47*, 4183.
2281. Eremenko, L. T., and A. M. Korolev, *Bull. Acad. Sci. USSR, Div. Chem. Sci.* (Eng. trans.) **1972**, *1*, No. 1, 172.
2282. Erić, B., E. U. Goode, et al., *J. Chem. Soc.* **1960**, 55.
2283. Erlandsson, G., and H. Selen, *Arkiv. Fysik.* **1958**, *14*, 61; *C.A.* **1959**, *53*, 5785.
2284. Erlenmeyer, E., and A. Kriechbaumer, *Ber. Deut. Chem. Gesell.* **1874**, *7*, 699.
2285. Ermakov, G. V., and V. P. Skripov, *Zh. Fiz. Khim.* **1969**, *43*, 1308; *Russ. J. Phys. Chem.* **1969**, *43*, 726.
2286. Ernst, G., and J. Busser, *J. Chem. Thermodyn.* **1970**, *2*, 787.
2287. Ernst, R. C., E. E. Litkenhous, et al., *J. Phys. Chem.* **1932**, *36*, 842.
2288. Ernst, R. C., C. H. Watkins, et al., *J. Phys. Chem.* **1936**, *40*, 627.
2289. Ernstbrunner, E. E., *J. Chem. Soc. A* **1970**, 1558.
2290. Errera, J., *Phys. Z.* **1926**, *27*, 764; *Chem. Zentr.* **1927**, *1*, 1929.
2291. Errera, J., and M. L. Sherrill, *J. Amer. Chem. Soc.* **1930**, *52*, 1933.

2292. Essery, J. M., and K. Schofield, *J. Chem. Soc.* **1961,** 3939.

2293. Esso Research and Engineering Co., British Patent 797,242, **1958.**

2294. Esso Research and Engineering Co., British Patent 875,856, Aug. 4, 1959; *C.A.* **1962,** *56,* 7138.

2295. Estermann, J., *Z. Phys. Chem.* (Leipzig) **1928,** *Abt. B1,* 134.

2296. Estok, G. K., and J. S. Dehn, *J. Amer. Chem. Soc.* **1955,** *77,* 4769.

2297. Estok, G. K., and J. H. Sikes, *J. Amer. Chem. Soc.* **1953,** *75,* 2745.

2298. Estok, G. K., and W. W. Wendlandt, *J. Amer. Chem. Soc.* **1955,** *77,* 4767.

2299. Ethyl Corp., British Patent 935,315, Aug. 28, 1963; *C.A.* **1963,** *59,* 14171.

2300. Ethyl Corp., "Vinylidine Chloride," Technical Bulletin, New York, 1965.

2301. Ettel, V., and A. Kohlik, *Collect. Trav. Chim. Tchecoslovaquie* **1931,** *3,* 585; *Chem. Zentr.* **1932,** *I,* 1513.

2302. Eubank, P. T., and J. M. Smith, *J. Chem. Eng. Data* **1962,** *7,* 75.

2303. Eucken, A., and F. Hauck, *Z. Phys. Chem.* (Leipzig) **1928,** *134,* 161; *C.A.* **1928,** *22,* 4288.

2304. Evans, A. G., and S. D. Hamann, *Trans. Faraday Soc.* **1951,** *47,* 34; R. E. Bates, and H. B. Hetzer, *J. Phys. Chem.* **1961,** *65,* 667.

2305. Evans, A. G., and E. Tyrrall, *J. Polym. Sci.* **1947,** *2,* 387.

2306. Evans, D. F., *J. Chem. Soc.* **1959,** 877.

2307. Evans, D. F., *J. Chem. Soc.* **1960,** 877.

2308. Evans, D. F., *J. Chem. Soc.* **1963,** 5575.

2309. Evans, D. F., J. Thomas, et al., *J. Phys. Chem.* **1971,** *75,* 1714.

2310. Evans, D. P., W. C. Davies, et al., *J. Chem. Soc.* **1930,** 1310.

2311. Evans, D. P., and W. J. Jones, *J. Chem. Soc.* **1932,** 985.

2312. Evans, E. B., *J. Inst. Petrol. Technol.* **1938,** *24,* 537; *Beil.* **EIII5,** 1257.

2313. Evans, F. D., M. Bogan, et al., *Anal. Chem.* **1968,** *40,* 224.

2314. Evans, F. D., and P. F. Tiley, *J. Chem. Soc. B* **1966,** 134.

2315. Evans, F. W., D. M. Fairbrother, et al., *Trans. Faraday Soc.* **1959,** *55,* 399.

2316. Evans, F. W., and H. A. Skinner, *Trans. Faraday Soc.* **1959,** *55,* 255.

2317. Evans, F. W., and H. A. Skinner, *Trans. Faraday Soc.* **1959,** *55,* 260.

2318. Evans, H. B., A. R. Tarpley, et al., *J. Phys. Chem.* **1968,** *72,* 2552.

2319. Evans, J. C., *J. Chem. Phys.* **1954,** *22,* 1228.

2320. Evans, J. C., *J. Chem. Phys.* **1959,** *30,* 934.

2321. Evans, J. C., *Spectrochim. Acta.* **1960,** *16,* 1382; *C.A.* **1961,** *55,* 12039.

2322. Evans, J. C., and H. J. Bernstein, *Can. J. Chem.* **1955,** *33,* 1746.

2323. Evans, J. C., and H. J. Bernstein, *Can. J. Chem.* **1955,** *33,* 1792.

2324. Evans, J. C., and H. J. Bernstein, *Can. J. Chem.* **1956,** *34,* 1083.

2325. Evans, J. C., and G. Y.-S. Lo, *J. Amer. Chem. Soc.* **1966,** *88,* 2118.

2326. Evans, R. D., H. R. Mighton, et al., *J. Amer. Chem. Soc.* **1950,** *72,* 2018.

2327. Evans, R. F., and J. C. Smith, *J. Inst. Petroleum* **1953,** *39,* 716; *C.A.* **1954,** *48,* 3319.

2328. Evans, T., U. S. Patent 2,154,930, April 18, 1939; *C.A.* **1939,** *33,* 5411.

2329. Evans, T., and W. C. Fetsch, *J. Amer. Chem. Soc.* **1904,** *26,* 1158.

2330. Everard, L. B., L. Kumar, et al., *J. Chem. Soc.* **1951,** 2807.

2331. Everett, D. H., unpublished results; R. G. Bates, et al., *J. Phys. Chem.* **1961,** *65,* 667.

2332. Everett, D. H., and B. R. W. Pinsent, *Proc. Roy. Soc.* (London) **1952,** *215A,* 416; Bates, R. G., and H. B. Hetzer, *J. Phys. Chem.* **1961,** *65,* 667.

2333. Everett, D. H., and W. G. K. Wynn-Jones, *Proc. Roy. Acad.* (London) **1941,** *177A,* 499.

2334. Evers, E. C., and A. G. Knox, *J. Amer. Chem. Soc.* **1951,** *73,* 1739.

2335. Evseeva, L. A., and L. M. Sverdlov, *Zh. Fiz. Khim.* **1969,** *43,* 519; *Russ. J. Phys. Chem.* **1969,** *43,* 468.

2336. Evstratova, K. I., and N. A. Goncharova, *Zh. Fiz. Khim.* **1969,** *43,* 519; *Russ. J. Phys. Chem.* **1969,** *43,* 284.

2337. Ewell, R. H., and J. F. Bourland, *J. Chem. Phys.* **1940,** *8,* 635.

2338. Ewert, M., *Bull. Soc. Chim. Belges* **1936,** *45,* 509.

2339. Ewert, M., *Bull. Soc. Chim. Belges* **1937,** *46,* 90.

2340. Ewing, M. B., and K. N. Marsh, *J. Chem. Thermodyn.* **1977,** *9,* 371.

2341. Ewins, A. J., *J. Chem. Soc.* **1917,** 350.

2342. Exner, O., and V. Jehlicka, *Collect. Czech. Chem. Commun.* **1983,** *48,* 1030.
2343. Exner, O., J. Koudelka, et al., *Collect. Czech. Chem. Commun.* **1983,** *48,* 735.
2344. Exxon Chemical Co., "Ethylene Glycol Monoethyl Ether Acetate," Houston.
2345. Exxon Chemical Co., "Exxon Petroleum Solvents," Houston, 1973.
2346. Exxon Chemical Co., "Glycol-Ethers," New York, 1964.
2347. Exxon Chemical Co., "Handling Petroleum Solvents," Houston, 1973.
2348. Exxon Chemical Co., "Ketones," Houston, 1970.
2349. Exxon Chemical Co., "Methyl Amyl Alcohol (MIBC)," Houston.
2350. Exxon Chemical Co., "Oxo Alcohols," Houston.
2351. Exxon Chemical Co., "Sec-Butyl Alcohol," Houston.
2352. Exxon Chemical Co., Industrial Chemical Div., *Ethyl Alcohol Handbook*, New York, 1969.
2353. Eykman, J. F., *Rec. Trav. Chim. Pays-Bas* **1895,** *14,* 185; *J. Chem. Soc.* **1896Aii,** 133.
2354. Eykman, J. F., *Z. Phys. Chem.* (Leipzig) **1889,** *3,* 113.
2355. Eykman, J. F., *Z. Phys. Chem.* (Leipzig) **1889,** *3,* 203.
2356. Eykman, J. F., *Z. Phys. Chem.* (Leipzig) **1889,** *4,* 49.
2357. Eykman, J. F., F. Bergema, et al., *Chem. Weekblad,* *2,* 59; *Chem. Zentr.* **1905,** *I,* 814.
2358. Eyster, E. H., *J. Chem. Phys.* **1938,** *6,* 576.

F

2359. Faillebin, M., *Compt. Rend.* **1922,** *175,* 1077.
2360. Fainberg, A. H., and W. T. Miller, *J. Org. Chem.* **1965,** *30,* 864.
2361. Fairbairn, A. W., H. A. Cheney, et al., *Chem. Eng. Progr., 43,* No. 6; *Trans. A.I.C.H.E.* **1947,** 280; *C.A.* **1947,** *41,* 5090.
2362. Fairbrother, F., *J. Chem. Soc.* **1934,** 1846.
2363. Fairbrother, F., *Proc. Roy Soc.* **1933,** *142A,* 173.
2364. Faith, W. L., D. B. Keyes, et al., *Industrial Chemicals*, Wiley, 1957.
2365. Falk, K. G., *J. Amer. Chem. Soc.* **1909,** *31,* 806.
2366. Fanelli, R., *J. Amer. Chem. Soc.* **1948,** *70,* 1796.
2367. I. G. Farbenind. A.-G., French Patent 798,421, May 16, 1936; *C.A.* **1936,** *30,* 7121.
2368. I. G. Farbenind. A.-G., German Patent 486,767, April 1, 1925; *C.A.* **1930,** *24,* 1870.
2369. I. G. Farbenind. A.-G., German Patent 641,878, Feb. 16, 1937; *C.A.* **1937,** *31,* 5809.
2370. Farbwerke Hoechst AG, *HOECHST Solvents*, 4th ed., Frankfurt, 1969.
2371. Farkas, A., and A. F. Stribley, U. S. Patent 2,410,642, Nov. 5, 1946; *C.A.* **1947,** *41,* 1833.
2372. Farrell, J. K., and G. B. Bachman, *J. Amer. Chem. Soc.* **1935,** *57,* 1281.
2373. Farroha, S. M., A. E. Habboush, et al., *Talanta* **1972,** *19,* 1711.
2374. Fateley, W. G., F. E. Kiviat, et al., *Spectrochim. Acta, Part A* **1970,** *26,* 315.
2375. Fateley, W. G., and F. A. Miller, *Spectrochim. Acta* **1961,** *17,* 857.
2376. Favini, G., and R. Bellabono, *Rend. 1st Lombardo Sci. Lettre* **1965,** *A99,* 381; *C.A.* **1966,** *64,* 18711.
2377. Favini, G., S. Carra, et al., *Nouvo Cim.* **1958,** *8,* 60; *C.A.* **1958,** *52,* 16031.
2378. Favini, G., A. Gamba, et al., *Spectrochim. Acta* **1967,** *23A,* 89; *C.A.* **1967,** *66,* 50516.
2379. Favini, G., and M. Simonetta, *Gazz. Chim. Ital.* **1959,** *89,* 2111; *C.A.* **1961,** *55,* 8036.
2380. Favret, A. G., and R. Meister, *J. Chem. Phys.* **1964,** *41,* 1011.
2381. Favrot, J., and J. M. Lebas, *J. Chim. Phys. Physicochim. Biol.* **1970,** *67,* 1705; *C.A.* **1971,** *74,* 81338.
2382. Fawcett, F. S., *Ind. Eng. Chem.* **1946,** *38,* 338.
2383. Fawcett. F. S., *J. Amer. Chem. Soc.* **1946,** *68,* 1420.
2384. Fawcett. F. S., and H. E. Rasmussen, *J. Amer. Chem. Soc.* **1945,** *67,* 1705.
2385. Fawsitt, C. E., *J. Chem. Soc.* **1919,** 790.
2386. Feairheller, W. R., and J. E. Katon, *Appl. Spectrosc.* **1968,** *22,* 488.
2387. Feairheller, W. R., and J. E. Katon, *Spectrochim. Acta* **1967,** *23A,* 2225; *Chem. Zentr.* **1968,** 0834.
2388. Fed. Regist., *Oct. 20, 1975,* 40, 49032; *C.A.* **1976,** *84,* 54922.
2389. Fedin, E. I., P. V. Petrovskii, et al., *Neftekhimiya* **1962,** *2,* 270; *C.A.* **1963,** *58,* 2042.
2390. Fedosow, J., *Soc. Sci. Fennica Commentationes Phys.-Math.* **1948,** *14,* No. 1, 1; *C.A.* **1948,** *42,* 6692.
2391. Fedyanina, V. N., *Gig. Sanit.* **1968,** *33,* 46; *C.A.* **1968,** *68,* 62536.
2392. Feeny, J., A. Ledwith, et al., *J. Chem. Soc.* **1962,** 2021.

2393. Fehnel, E. A., and M. Carmack, *J. Amer. Chem. Soc.* **1949**, *71*, 84.

2394. Fehnel, E. A., and M. Carmack, *J. Amer. Chem. Soc.* **1949**, *71*, 2932.

2395. Feigel, F., *Spot Tests in Inorganic Analysis*, 5th ed., Elsevier, New York, 1958.

2396. Fel'dman, I. N., V. V. Savko, et al., *Zh. Fiz. Khim.* **1973**, *47*, 2715; *Russ. J. Phys. Chem.* **1973**, *47*, 1531.

2397. Felsing, W. A., A. M. Cuellar, et al., *J. Amer. Chem. Soc.* **1947**, *69*, 1972.

2398. Felsing, W. A., and F. W. Jessen, *J. Amer. Chem. Soc.* **1933**, *55*, 4418.

2399. Felsing, W. A., and A. R. Thomas, *Ind. Eng. Chem.* **1929**, *21*, 1269.

2400. Fenske, M. R., W. G. Braun, et al., *Ind. Eng. Chem., Anal. Ed.* **1947**, *19*, 700.

2401. Fenske, M. R., and H. S. Myers, *Ind. Eng. Chem.* **1950**, *42*, 649.

2402. Fenske, M. R., and D. Quiggle, et al., *Ind. Eng. Chem.* **1932**, *24*, 408.

2403. Fenske, M. R., D. Quiggle, et al., *Ind. Eng. Chem.* **1932**, *24*, 542.

2404. Fenwick, J. O., D. Harrop, et al., *J. Chem. Thermodyn.* **1975**, *7*, 943.

2405. Fenwick, J. O., D. Harrop, et al., *J. Chem. Thermodyn.* **1978**, *10*, 687.

2406. Ferguson, A., and J. T. Miller, *Proc. Phys. Soc.* (London) **1933**, *45*, 194; *C.A.* **1933**, *27*, 2372.

2407. Ferguson, E. E., R. L. Hudson, et al., *J. Chem. Phys.* **1953**, *21*, 1936.

2408. Ferguson, J., *J. Chem. Soc.* **1954**, 304.

2409. Ferguson, J., and R. L. Werner, *J. Chem. Soc.* **1954**, 3645.

2410. Ferguson, J. B., M. Freed, et al., *J. Phys. Chem.* **1933**, *37*, 87.

2411. Ferguson, J. W., *Proc. Indiana Acad. Sci.* **1953**, *63*, 131; *C.A.* **1955**, *49*, 8094.

2412. Ferrier, R., *Compt. Rend.* **1945**, *220*, 460; *C.A.* **1946**, *40*, 1475.

2413. Fetisova, M. M., and V. G. Tischenko, *Teor. Eksp. Khim.* **1977**, *13*, 244; *Theor. Expt. Chem.* **1977**, *13*, 185.

2414. Few, A. V., and J. W. Smith, *J. Chem. Soc.* **1949**, 753.

2415. Few, A. V., and J. W. Smith, *J. Chem. Soc.* **1949**, 2663.

2416. Fialkov, Y. Y., and Y. Y. Borovikov, *Zh. Obshch. Khim.* **1966**, *36*, 1554; *J. Gen. Chem.* (USSR) **1966**, *36*, 1557.

2417. Fialkovskaya, O. V., *Izv. Akad. Nauk SSSR, Ser. Fiz.* **1958**, *22*, 1093; *C.A.* **1959**, *53*, 862.

2418. Fickling, M. M., A. Fischer, et al., *J. Amer. Chem. Soc.* **1959**, *81*, 4226.

2419. Field, B. O., and C. J. Hardy, *Chem. Ind.* (London) **1962**, 212; *C.A.* **1962**, *57*, 7091.

2420. Field, F. H., and M. S. B. Munson, *J. Amer. Chem. Soc.* **1967**, *89*, 4272.

2421. Field, F. H., M. S. B. Munson, et al., *Advan. Chem. Ser.* **1966**, *58*, 167; *C.A.* **1966**, *65*, 17868.

2422. Field, F. H., and J. H. Saylor, *J. Amer. Chem. Soc.* **1946**, *68*, 2649.

2423. Field, K. W., and P. Kovacic, *J. Org. Chem.* **1971**, *36*, 3566.

2424. Fierz-David, H. E., *Chimia* (Switz.) **1947**, *1*, 246; *C.A.* **1948**, *42*, 2228.

2425. Fife, H. R., and E. W. Reid, *Ind. Eng. Chem.* **1930**, *22*, 513.

2426. Filar, L. J., U.S. Patent 2,948,758, Aug. 9, 1960; *C.A.* **1961**, *55*, 3521.

2427. Finch, A., I. J., Hyams, et al., *J. Mol. Spectrosc.* **1965**, *16*, 103.

2428. Finch, J. N., and E. R. Lippincott, *J. Phys. Chem.* **1957**, *61*, 894.

2429. Findenegg, G. H., *Monatsh. Chem.* **1970**, *101*, 1081.

2430. Findenegg, G. H., *Monatsh. Chem.* **1973**, *104*, 998.

2431. Findenegg, G. H., and F. Kohler, *Trans. Faraday Soc.* **1967**, *63*, 870.

2432. Finegold, H., *J. Chem. Phys.* **1964**, *41*, 1808.

2433. Finegold, H., *J. Phys. Chem.* **1968**, *72*, 3244.

2434. Finger, G. C., and F. H. Reed, *Trans. Illinois State Acad. Sci.* **1936**, *29*, 89; *C.A.* **1937**, *31*, 2999.

2435. Finke, H. L., M. E. Gross, et al., *J. Amer. Chem. Soc.* **1954**, *76*. 333.

2436. Finke, H. L., and J. F. Messerly, *J. Chem. Thermodyn.* **1973**, *5*, 247.

2437. Finke, H. L., and J. F. Messerly, et al., *J. Chem. Thermodyn.* **1972**, *4*, 359.

2438. Finsy, R., and R. Van Loon, *J. Phys. Chem.* **1976**, *80*, 2783.

2439. Fiock, E. F., D. C. Ginnings, et al., *J. Res. Nat. Bur. Stand.* **1931**, *6*, 881.

2440. Fischer, A., G. J. Leary, et al., *J. Chem. Soc., Phys. Org. B* **1966**, 782.

2441. Fischer, E., *J. Chem. Soc.* **1955**, 1382.

2442. Fischer, E., and F. Royowski, *Phys. Z.* **1939**, *40*, 331; *C.A.* **1939**, *33*, 7633.

2443. Fischer, E., and A. Speier, *Ber. Deut. Chem. Gesell,* **1895**, *28*, 3252.

2444. Fischer, I., *Acta Chem. Scand.* **1950**, *4*, 1197.

2445. Fischer, W., and W. Klemm, *Z. Phys. Chem.* (Leipzig) **1930**, *147A*, 275.

2446. Fishbein, L., and J. A. Gallaghan, *J. Amer. Chem. Soc.* **1956**, *78*, 1218.

2447. Fisher, F. R., and R. A. Baxter, *Mines. Mag.* (Colo. School of Mines) **1940**, *30*, 447; *C.A.* **1940**, *34*, 8111.

2448. Fisher, K., and W. H. Perkin, *J. Chem. Soc.* **1908**, 1871.

2449. Fisher Scientific Co., "Dimethyl Sulfoxide," TD-176, Pittsburgh, 1963.

2450. Fishman, E., and T. L. Chen. *Spectrochim. Acta, Part A* **1969**, *25*, 1231.

2451. Fiske, D. L., *Refrig. Eng.* **1949**, *57*, 336; *C.A.* **1950**, *44*, 1288.

2452. Fittig, R., and W. G. Chanlarow, *Ann. Chem., Justus Liebig* **1884**, *226*, 331.

2453. Fittig, R., and W. Pfeffer, *Liebigs Ann. Chem.* **1865**, *135*, 357.

2454. Fittig, R., and J. Remsen, *Ann. Chem.* **1870**, *155*, 114.

2455. Fitzgerald, W. E., and G. J. Janz, *Mol. Spectrosc.* **1957**, *51*, 11856.

2456. Fiurmara, A., *Riv. Combust.* **1971**, *25*, 327; *C.A.* **1971**, *75*, 153481.

2457. Flanagan, P. W., and H. F. Smith, *Anal. Chem.* **1965**, *37*, 1699.

2458. Flannery, J. B., and G. J. Janz, *J. Chem. Eng. Data* **1965**, *10*, 387.

2459. Fletcher, R. A., and G. Pilcher, *Trans. Faraday Soc.* **1971**, *67*, 3191.

2460. Fletcher, S. E., C. T. Mortimer, et al., *J. Chem. Soc.* **1959**, 580.

2461. Fletcher, W. H., *J. Amer. Chem. Soc.* **1946**, *68*, 2726.

2462. Flett, M. St. C., *J. Chem. Soc.* **1951**, 962.

2463. Flett, M. St. C., *Trans. Faraday Soc.* **1948**, *44*, 767; *C.A.* **1949**, *43*, 3715.

2464. Fletton, R. A., and J. E. Page, *Analyst* (London) **1971**, *96*, 370.

2465. Flood, D. T., and G. Calingaert, *J. Amer. Chem. Soc.* **1934**, *56*, 1211.

2466. Flynn, G. W., and J. D. Baldeschwieler, *J. Chem. Phys.* **1962**, *37*, 2907.

2467. Flynn, G. W., and J. D. Baldeschwieler, *J. Chem. Phys.* **1963**, *38*, 226.

2468. Foerster, D. R., R. Miller, et al., "10th Annual Report on Research, Petroleum Research Fund," p. 15, American Chemical Society, Washington, 1966.

2469. Fogg, E. T., A. N. Hixson, et al., *Anal Chem.* **1955**, *27*, 1609.

2470. Foley, R. A., W.-M. Lee, et al., *Anal. Chem.* **1964**, *36*, 1100.

2471. Fong, F. K., and C. P. Smyth, *J. Chem. Phys.* **1964**, *40*, 2404.

2472. Fontaine, H., Longueville W., et al., *C.R. Acad. Sci., Ser. B* **1972**, *274*, 641.

2473. Fonteyne, R., *Nutuurwetensch. Fijdschr.* **1942**, *24*, 161; *Chem. Zentr.* **1943**, *I*, 265.

2474. Forbes, G. S., and A. S. Coolidge, *J. Amer. Chem. Soc.* **1919**, *41*, 150.

2475. Forbes, J. W., *Anal. Chem.* **1962**, *34*, 1125.

2476. Forbes, W. F., *Can. J. Chem.* **1960**, *38*, 1104.

2477. Forcier, G. A., and J. W. Olver, *Anal. Chem.* **1965**, *37*, 1447.

2478. Fordham, S., *Research* (London) **1948**, *1*, Res. Suppl. 336; *C.A.* **1948**, *42*, 5292.

2479. Fordyce, C. R., and L. W. A. Meyer, *Ind. Eng. Chem.* **1940**, *32*, 1053.

2480. Forel, M. T., A. Lafaix, et al., *J. Chim. Phys.* **1963**, *60*, 875; *C.A.* **1963**, *59*, 10894.

2481. Forel, M. T., S. Volf, et al., *Ann. Chim.* (Paris) **1972**, *7*, 295.

2482. Foresti, B., *Ann. Chim. Appl.* **1937**, *27*, 359; *C.A.* **1938**, *32*, 2089.

2483. Forman, E. J., and D. N. Hume, *J. Phys. Chem.* **1959**, *63*, 1949.

2484. Forster, E. O., *J. Chem. Phys.* **1962**, *37*, 1021.

2485. Forster, E. O., *J. Chem. Phys.* **1964**, *40*, 86.

2486. Forster, E. O., *J. Chem. Phys.* **1964**, *40*, 91.

2487. Forster, G., R. Skrabal, et al., *Z. Electrochem.* **1937**, *43*, 290; *C.A.* **1937**, *31*, 5237.

2488. Fort, R. F., and W. R. Moore, *Trans. Faraday Soc.* **1966**, *62*, 1112.

2489. Fortier, J. L., and G. C. Benson, *J. Chem. Eng. Data* **1979**, *24*, 34.

2490. Fortier, J. L., G. C. Benson, et al., *J. Chem. Thermodyn.* **1976**, *8*, 289.

2491. Fortunato, B., P. Mirone, et al., *Spectrochim. Acta* **1971**, *27A*, 1917.

2492. Forziati, A. F., *J. Res. Nat. Bur. Stand.* **1950**, *44*, 373.

2493. Forziati, A. F., D. L. Camin, et al., *J. Res. Nat. Bur. Stand.* **1950**, *45*, 406.

2494. Forziati, A. F., A. R. Glasgow, et al., *J. Res. Nat. Bur. Stand.* **1946**, *36*, 129.

2495. Forziati, A. F., B. J. Mair, et al., *J. Res. Nat. Bur. Stand.* **1945**, *36*, 513.

2496. Forziati, A. F., W. R. Norris, et al., *J. Res. Nat. Bur. Stand.* **1949**, *43*, 555.

2497. Forziati, A. F., and F. D. Rossini, *J. Res. Nat. Bur. Stand.* **1949**, *43*, 473; Timmermans, J., *Physicochemical Constants of Pure Organic Compounds*, Vol. 2, Elsevier, New York, 1965.

2498. Fouque, G., and M. Cabanac, *Bull. Soc. Chim. Fr.* **1926**, *39*, 1184; *C.A.* **1926**, *20*, 3687.

2499. Fowell, P., J. R. Lacher, et al., *Trans. Faraday Soc.* **1965**, *61*, 1324.

2500. Fowler, F. W., and A. R. Katritzky, *J. Chem. Soc.* **1971**, 460.

2501. Fowler, L., W. N. Trump, et al., *J. Chem. Eng. Data* **1968**, *13*, 209.

2502. Fowler, R. D., W. B. Burford, et al., *Ind. Eng. Chem.* **1947**, *39*, 292.

2503. Fowler, R. D., J. M. Hamilton, et al., *Ind. Eng. Chem.* **1947**, *39*, 375.

2504. Fox, J. J., and M. F. Barker, *J. Soc. Chem. Ind.* **1918**, *37*, 268T.

2505. Fox, J. J., and A. E. Martin, *Proc. Roy. Soc.* (London) **1937**, *162A*, 419; *C.A.* **1937**, *31*, 8374.

2506. Foxton, A. A., G. H. Jeffery, et al., *J. Chem. Soc. A* **1966**, 249.

2507. Fraenkel, G., *Proc. Natl. Acad. Sci. U.S.* **1958**, *44*, 688; *C.A.* **1959**, *53*, 52.

2508. Franchimont, A. P. N., *Rec. Trav. Chim. Pays-Bas* **1884**, *3*, 226; from [4718].

2509. Francis, A. W., *Critical Solution Temperatures*, American Chemical Society, Washington, 1961.

2510. Francis, A. W., *J. Chem Eng. Data* **1960**, *5*, 534.

2511. Francis, A. W., and G. W. Robbins, *J. Amer. Chem. Soc.* **1933**, *55*, 4339.

2512. Francis, C. V., *Anal. Chem.* **1949**, *21*, 1238.

2513. Francis, F., and F. J. E. Collins, *J. Chem. Soc.* **1936**, 137.

2514. Frandsen, B. S., *J. Res. Nat. Bur. Stand.* **1931**, 7, 477.

2515. Franeau, J., and G. Birnbaum, *J. Appl. Phys.* **1949**, *20*, 817; *C.A.* **1949**, *43*, 8769.

2516. Franke, A., *Monatsh. Chem.* **1901**, *22*, 1070.

2517. Franke, A., and R. Dworzak, *Monatsh. Chem.* **1923**, *43*, 661; *C.A.* **1923**, *17*, 2103.

2518. Franke, E., *Z. Phys. Chem.* (Leipzig) **1895**, *16*, 463.

2519. Frankel, S., *Arzenimittelsynthese*, Julius Springer, Berlin, 1927.

2520. Frankforter, G. B., and L. Cohen, *J. Amer. Chem. Soc.* **1914**, *36*, 1103.

2521. Frankiss, S. G., *J. Chem. Soc., Faraday Soc. 2* **1974**, *70*, 1516.

2522. Frankiss, S. G., *J. Phys. Chem.* **1963**, *67*, 752.

2523. Frankiss, S. G., *J. Phys. Chem.* **1967**, *71*, 3418.

2524. Frankiss, S. G., and W. Kynaston, *Spectrochim. Acta* **1972**, *28A*, 2149.

2525. Frankland, E., *Ann. Chem.* **1849**, *71*, 173.

2526. Franzen, V., and H.-E. Driesen, *Chem. Ber.* **1963**, *96*, 1881.

2527. Fraser, R. R., *Can. J. Chem.* **1960**, *38*, 2226.

2528. Fraser, R. T. M., and N. C. Paul, *J. Chem. Soc. B* **1968**, 1407.

2529. Frazer, M. J., W. Gerrard, et al., *Chem. Ind.* **1954**, 931; *C.A.* **1955**, *49*, 11586.

2530. Freeman, R., *J. Chem. Phys.* **1964**, *40*, 3571.

2531. Freeman, R., N. S. Bhacca, et al., *J. Chem. Phys.* **1963**, *38*, 293.

2532. Freeman, S. K., *Anal. Chem.* **1957**, *29*, 63.

2533. Freer, P. C., and A. Lachman, *Amer. Chem. J.* **1897**, *19*, 878.

2534. Freidlin, L. Kh., V. Z. Sharf, et al., *Neftekhimiya* **1963**, *3*, 10; *C.A.* **1963**, *59*, 3753.

2535. Freiser, H., and W. L. Glowacki, *J. Amer. Chem. Soc.* **1948**, *70*, 2575.

2536. Freiser, H., and W. L. Glowacki, *J. Amer. Chem. Soc.* **1949**, *71*, 514.

2537. Freiser, H., M. E. Hobbs, et al., *J. Amer. Chem. Soc.* **1949**, *71*, 111.

2538. Freitag, *Oberflachentech* **1940**, *17*, 7; *C.A.* **1940**, *34*, 2174.

2539. Frejacques, J. L. M., U.S. Patent 2,461,048, Feb. 8, 1949; *C.A.* **1949**, *43*, 7275.

2540. French, C. M., *Trans. Faraday Soc.* **1947**, *43*, 357.

2541. French, C. M., and K. H. Glover, *Trans. Faraday Soc.* **1955**, *51*, 1418.

2542. French, C. M., and K. H. Glover, *Trans. Faraday Soc.* **1955**, *51*, 1427.

2543. French, C. M., P. B. Hart, et al., *J. Chem. Soc.* **1959**, 3582.

2544. French, C. M., and I. G. Roe, *Trans. Faraday Soc.* **1953**, *49*, 314, 791.

2545. French, C. M., and U. C. Trew, *Trans. Faraday Soc.* **1951**, *47*, 365; *C.A.* **1951**, *45*, 8827.

2546. French, H. E., and G. G. Wrightsman, *J. Amer. Chem. Soc.* **1938**, *60*, 50.

2547. French, K. H. V., chairman, *Anal. Chem.* **1964**, *36*, 339.

2548. French, K. H. V., and G. Claxton, *J. Soc. Chem. Ind.* **1946**, *65*, 344.

2549. French Patent 784,656, July 22, 1935; *C.A.* **1936**, *30*, 108.

2550. French Patent 1,045,726, Dec. 1, 1953; *C.A.* **1958**, *52*, 11888.

2551. Frenzel, C. A., D. W. Scott, et al., *U.S. Bur. Mines, Rept. Invest.* **1960**, *5658*, 17 pp; *C.A.* **1961**, *55*, 2277.

2552. Freri, M., *Gazz. Chim. Ital.* **1948**, *78*, 286; *C.A.* **1948**, *42*, 8647.

2553. Freudenberg, K. J., German Patent 281,473, Jan. 8, 1915; *Chem. Zentr.* **1915,** *1,* 230.
2554. Frey, F. E. (to Phillips Petroleum), U.S. Patent 2,377,417, June 5, 1945; *C.A.* **1946,** *40,* 1541.
2555. Frey, P. R., and E. C. Gilbert, *J. Amer. Chem. Soc.* **1937,** *59,* 1344.
2556. Freyer, F. A., *Mfg. Chemist* **1962,** *33,* 100; *C.A.* **1962,** *57,* 1296; Ref. [3321].
2557. Freymann, M., *Ann. Chim.* **1939,** *11,* 11; *C.A.* **1939,** *33,* 3694.
2558. Freymann, M., and R. Freymann, *Compt. Rend.* **1944,** *219,* 515; *C.A.* **1946,** *40,* 1393.
2559. Freymann, M., and R. Freymann, *Compt. Rend.* **1959,** *248,* 677; *C.A.* **1959,** *53,* 14694.
2560. Freymann, M., and R. Freymann, *Groupement Fr. Develop. Res. Aeronaut.* **1943,** Note Tech. No. 11; *C.A.* **1949,** *43* 3715.
2561. Freymann, M., R. Freymann, et al., *Arch. Sci.* **1960,** *13,* 506; *C.A.* **1963,** *58,* 3289.
2562. Freymann, R., *Compt. Rend.* **1932,** *194,* 1471; *C.A.* **1932,** *26,* 3991.
2563. Freymann, R., *Compt. Rend.* **1936,** *202,* 952; *C.A.* **1936,** *30,* 3691.
2564. Freymann, R., *Compt. Rend.* **1965,** *261,* 2637; *C.A.* **1966,** *64,* 5943.
2565. Freymann, R., *J. Phys. Radium* **1939,** *10,* 1; *C.A.* **1939,** *33,* 3694.
2566. Freymann, R., and P. Duhamel, *Bull. Soc. Chim. Fr.* **1961,** 1238.
2567. Freymann, R., M. Freymann, et al., *Arch. Sci.* (Geneva) **1959,** *12,* Fasc. Spec. 204; *C.A.* **1960,** *54,* 11705.
2568. Fried, V., J. Pick, et al., *Chem. Listy* **1954,** *48,* 774; *C.A.* **1954,** *48,* 10398.
2569. Friedel, R. A., *J. Amer. Chem. Soc.* **1951,** *73,* 2881.
2570. Friedel, R. A., and D. S. McKinney, *J. Amer. Chem. Soc.* **1947,** *69,* 604.
2571. Friedel, R. A., and M. Orchin, *Ultraviolet Spectra of Aromatic Compounds*, Wiley, New York, 1951.
2572. Friedel, R. A., J. L. Schultz, et al., *Anal. Chem.* **1956,** *28,* 926.
2573. Freidel, R. A., and A. G. Sharkey, *Anal. Chem.* **1956,** *28,* 940.
2574. Friedman, H. M., *Appl. Spectrosc.* **1970,** *24,* 44.
2575. Friedman, H. M., *Spectrochim. Acta* **1966,** *22,* 1465.
2576. Friedman, I., and H. Stechter, *J. Org. Chem.* **1960,** *25,* 877.
2577. Friedman, L., and F. A. Long, *J. Amer. Chem. Soc.* **1953,** *75,* 2832.
2578. Friedman, L., F. A. Long, et al., *J. Chem. Phys.* **1957,** *27,* 613.
2579. Friend, J. N., and W. D. Hargreaves, *Phil. Mag.* **1943,** *34,* 643; *Beil.* **EIII2,** 39.
2580. Friend, J. N., and W. D. Hargreaves, *Phil. Mag.* **1944,** *35,* 619; *C.A.* **1945,** *39,* 1098.
2581. Friend, J. N., and W. D. Hargreaves, *Phil. Mag.* **1946,** *37,* 120.
2582. Friend, J. N., and W. D. Hargreaves, *Phil. Mag.* **1946,** *37,* 201.
2583. Fritz, G., *Z. Naturforsch.* **1952,** *7b,* 379.
2584. Fritz, J. F., and L. W. Marple, *Anal. Chem.* **1962,** *34,* 921.
2585. Frolov, A. F., M. A. Loginova, et al., *Zh. Fiz. Khim.* **1962,** *36,* 2282; *C.A.* **1963,** *58,* 3957.
2586. Frolov, Y. L., A. V. Kalabina, et al., *Zh. Strukt. Khim.* **1965,** *6,* 397; *C.A.* **1965,** *63,* 9780.
2587. Frost, A. A., and D. R. Kalkwarf, *J. Chem. Phys.* **1953,** *21,* 264.
2588. Frost, D. J., and J. Barzilay, *Anal. Chem.* **1971,** *43,* 1316.
2589. Frump, J. A., Commercial Solvents Corp., private communication.
2590. Fryer, F. A., *Mfg. Chemist* **1962,** *33,* 100; *C.A.* **1962,** *57,* 1296; Ref. [3321].
2591. Fuchs, L., *Spectrochim. Acta* **1942,** *2,* 243; *C.A.* **1943,** *37,* 5313.
2592. Fuchs, O., *Z. Phys.* **1930,** *63,* 824.
2593. Fuchs, R., *Can. J. Chem.* **1980,** *58,* 2305.
2594. Fuchs, R., *J. Chem. Thermodyn.* **1979,** *11,* 959.
2595. Fuchs, R., and L. L. Cole, *Can. J. Chem.* **1975,** *53,* 3620.
2596. Fuchs, R., and L. A. Peacock, *Can. J. Chem.* **1980,** *58,* 2796.
2597. Fueki, K., D. A. Feng, et al., *J. Chem. Phys.* **1972,** *56,* 5351.
2598. Fueno, T., and K. Yamaguchi, *Bull. Chem. Soc. Jpn.* **1972,** *45,* 3290.
2599. Fuge, E. T. J., S. T. Bowden, et al., *J. Phys. Chem.* **1952,** *56,* 1013.
2600. Fuguitt, R. E., W. D. Stallcup, et al., *J. Amer. Chem Soc.* **1942,** *64,* 2978.
2601. Fuhner, H., *Ber. Deut. Chem. Gesell.* **1924,** *57,* 510.
2602. Fujimura, T., and K. Kamiyoshi, *Sci. Rept. Res. Inst. Tohoku Univ. Ser. A* **1961,** *13,* 320; *C.A.* **1962,** *56,* 12411.
2603. Fujinaga, T., and K. Izutsu, *Pure Appl. Chem.* **1971,** *27,* 273.
2604. Fujino, M., *Igaku Kenkyu* **1970,** *40,* 139; *C.A.* **1971,** *75,* 33279.

2605. Fujita, T., K. Koshimizu, et al., *Tetrahedron* **1967**, *23*, 2633.

2606. Fujiwara, S., and T. Wainai, *Anal. Chem.* **1961**, *33*, 1085.

2607. Fukuii, K., S. Hattori, et al., *Nippon Kagaku Zasshi* **1959**, *80*, 541; *C.A.* **1961**, *55*, 4445.

2608. Fukukawa, S., *Kagaku to Kogoyo* (Osaka) **1957**, *31*, 77; *C.A.* **1957**, *51*, 12818.

2609. Fukumoto, Y., *Sci. Repts. Tohoku Imp. Univ., First Ser.* **1937**, *25*, 1162; *C.A.* **1937**, *31*, 5273.

2610. Fuller, A. T., and J. Kenyon, *J. Chem. Soc.* **1924**, 2304.

2611. Fuller, W. R., *Federation Series on Coating Technology*, Unit Six, Federation of Societies for Paint Technology, Philadelphia, 1967.

2612. Funk, E. W., F. C. Chai, et al., *J. Chem. Eng. Data* **1972**, *17*, 24.

2613. Fuoss, R. M., *J. Amer. Chem. Soc.* **1938**, *60*, 451.

2614. Fuoss, R. M., *J. Amer. Chem. Soc.* **1938**, *60*, 1633.

2615. Furukawa, G. T., D. C. Ginnings, et al., *J. Res. Nat. Bur. Stand.* **1951**, *46*, 195.

2616. Furukawa, G. T., R. E. McCoskey, et al., *J. Res. Nat. Bur. Stand.* **1954**, *52*, 11.

2617. Furutsuka, T., T. Imura, et al., *Technol. Rep. Osaka Univ.* **1974**, *24*, 367; *C.A.* **1974**, *81*, 160915.

2618. Furuyama, S., D. M. Golden, et al., *J. Phys. Chem.* **1968**, *72*, 4713.

2619. Furuyama, S., D. M. Golden, et al., *J. Chem. Thermodyn.* **1969**, *1*, 363.

2620. Fuson, R. C., *Organic Syntheses, Vol. XVIII*, Wiley, New York, 1938.

G

2621. Gabriel, C. L., *Ind. Eng. Chem.* **1940**, *32*, 887.

2622. Gabriel, S., and R. Stelzner, *Ber. Deut. Chem. Gesell.* **1895**, *28*, 2929.

2623. Gadshiev, S. N., and M. Ya. Agarunov, *Russ. J. Phys. Chem.* (Eng. trans.) **1965**, *39*, 130.

2624. Gadshiev, S. N., M. Ya. Agarunov, et al., *Izv., Akad. Nauk Azerb. SSR Fiz.-Tekn. Nat. Nauk* **1966**, 57; *C.A.* **1967**, *66*, 94620.

2625. GAF Corp., "M-Pyrol (*N*-methyl-2-pyrrolidone)," New York, 1972.

2626. GAF Corp., "M-Pyrol (*N*-methyl-2-pyrrolidone)," Tech. Bull. 7543-038, 4573-026, New York, 1965.

2627. GAF Corp., "1,4-Butanediol," New York, 1980.

2628. Gage, J. C., *Brit. J. Ind. Med.* **1970**, *27*, 1; *C.A.* **1970**, *73*, 12650.

2629. Gakhokidse, A., *J. Gen. Chem.* (USSR) **1947**, *17*, 1329; *Beil.* **EIII2**, 1278.

2630. Gal, J. Y., and C. Moliton-Bouchetout, *Bull. Soc. Chim. Fr.* **1973**, 464.

2631. Gălăteanu, D. B., *Bull. Soc. Roumain Phys.* **1942**, *43*, 5; *Chem. Zentr.* **1943**, *I*, 2189; *C.A.* **1944**, *38*, 4512.

2632. Gal'chenko, G. L., M. M. Ammar, et al., *Vestn. Mosk. Univ., Ser. II, Khim.* **1965**, *20*, 3; *C.A.* **1965**, *63*, 7703.

2633. Galibin, G. P., V. I. Fedorova, et al., *Gig. Sanit.* **1967**, *32*, 20; *C.A.* **1967**, *67*, 98674.

2634. Galkin, N. P., V. T. Orekhov, et al., *Russ. J. Phys. Chem.* (Eng. trans.) **1973**, *47*, 276.

2635. Galkin, N. P., V. T. Orekhov, et al., *Zh. Fiz. Khim.* **1972**, *46*, 2681; *Russ. J. Phys. Chem.* **1972**, *46*, 1537.

2636. W. A. Gallant, *Hydrocarbon Process.* **1969**, *48*(1), 153.

2637. Gallaugher, A. F., and H. Hibbert, *J. Amer. Chem. Soc.* **1936**, *58*, 813.

2638. Gallaugher, A. F., and H. Hibbert, *J. Amer. Chem. Soc.* **1937**, *59*, 2514.

2639. Gallaugher, A. F., and H. Hibbert, *J. Amer. Chem. Soc.* **1937**, *59*, 2521.

2640. Gallegos, E. J., and R. W. Kiser, *J. Amer. Chem. Soc.* **1961**, *83*, 773.

2641. Gallegos, E. J., and R. W. Kiser, *J. Phys. Chem.* **1962**, *66*, 136.

2642. Gal'pern, G. D., Ya. A. Gurrich, et al., *Zh. Anal. Khim.* **1970**, *25*, 1819.

2643. Gal'pern, G. D., M. M. Kusakov, et al., *Fiz. Sbornik L'vov. Univ.* **1957**, No. 3, 334; *C.A.* **1962**, *56*, 3378.

2644. Gal'perin, Ya. V., G. M. Bogolyubov, et al., *Zh. Obshch. Khim.* **1969**, *39*, 1599; *J. Gen. Chem. USSR* **1969**, *39*, 1567.

2645. Gambi, A., S. Giorgianni, et al., *Spectrochim. Acta, Part A* **1980**, *36A*, 871.

2646. Gamer, G., and H. Wolff, *Spectrochim. Acta, Part A* **1973**, *29*, 129.

2647. Gand, E., *Ann. Faculte Sci. Marseille* **1939**(2), *12*, 134; *C.A.* **1941**, *35*, 681.

2648. Gand, E., *Ann. Faculte Sci. Marseille* **1941**, *15*, 29; *C.A.* **1946**, *40*, 3967.

2649. Gand, E., *Bull. Soc. Chim. Fr.* **1943**, *10*, 465; *C.A.* **1944**, *38*, 3951.

2650. Gand, E., *Compt. Rend. Trav. Faculte Sci. Marseille* **1943**, *1*, 129; *C.A.* **1946**, *40*, 3967.

2651. Ganeff, J. M., and J. C. Jungers, *Bull. Soc. Chim. Belges* **1948**, *57*, 82.
2652. Ganz, E., *Helv. Chim. Acta* **1945**, *28*, 1580.
2653. Gaponova, N. E., M. P. Lisitso, et al., *Opt. Spektrosk.* **1960**, *8*, 465; *C.A.* **1962**, *57*, 298.
2654. Garach, J., *Compt. Rend.* **1953**, *236*, 1414; *C.A.* **1953**, *47*, 9152.
2656. Garach, J., *Groupement. Franc. Develop. Recherches. Aeronaut.*, Note Tech. No. 61, **1946**; *C.A.* **1949**, *43*, 2175.
2656. Garach, J., *Groupement. Franc. Develop. Recherches. Aeronaut.*, Note Tech. No. 61, **1946**; *C.A.* 1949, *43*, 2175.
2657. Garach, J., and J. Lecomte, *Compt. Rend.* **1946**, *222*, 74; *C.A.* **1947**, *41*, 1933.
2658. Garcia, M. O. G., *Rec. Trav. Cienc. Univ. Oviedo* **1964**, *5*, 143; *C.A.* **1965**, *62*, 7547.
2659. Gardner, G. S., and J. E. Brewer, *Ind. Eng. Chem.* **1937**, *29*, 179.
2660. Gardner, J. H., and P. Borgstrom, *J. Amer. Chem. Soc.* **1929**, *51*, 3375.
2661. Gardner, P. J., and K. S. Hussain, *J. Chem. Thermodyn.* **1972**, *4*, 819.
2662. Garelli, F., and C. Montanari, *Gazz. Chim. Ital.* **1894**, *24B*, 229.
2663. Garg, S. K., and P. K. Kadaba, *J. Phys. Chem.* **1964**, *68*, 737.
2664. Garg, S. K., and C. P. Smyth, *J. Chem. Phys.* **1965**, *42*, 1397.
2665. Garg, S. K., and C. P. Smyth, *J. Phys. Chem.* **1965**, *69*, 1294.
2666. Garland, C. E., and E. E. Reid, *J. Amer. Chem. Soc.* **1925**, *47*, 2333.
2667. Garner, J. B., B. Saxton, et al., *Amer. Chem. J.* **1911**, *46*, 236.
2668. Garner, W. E., and C. L. Abernethy, *Proc. Roy. Soc.* (London) **1921**, *99A*, 213; *Chem. Zentr.* **1921**, *III*, 866; *Beil.* **EII5,** 172.
2669. Garner, W. E., F. C. Madden, et al., *J. Chem. Soc.* **1926**, 2491.
2670. Garner, W. E., and F. C. Randall, *J. Chem. Soc.* **1924**, 881.
2671. Garner, W. E., and E. A. Ryder, *J. Chem. Soc.* **1925**, 720.
2672. Garnsey, R., and J. E. Prue, *Trans. Faraday Soc.* **1968**, *64*, 1206.
2673. Garrick, F. J., *Trans. Faraday Soc.* **1927**, *23*, 560.
2674. Garrigou-Lagrange, C., and O. Boloussa, *C.R. Hebd. Seances Acad. Sci., Ser. C* **1976**, *282*, 15.
2675. Garrigou-Lagrange, C., N. Claverie, et al., *J. Chim. Phys.* **1961**, *58*, 559; *C.A.* **1961**, *55*, 23047.
2676. Garrigou-Lagrange, C., C. DeLoze, et al., *J. Chim. Phys. Physicochim. Biol.* **1970**, *67*, 1936.
2677. Gartenmeister, R., *Z. Phys. Chem.* (Leipzig) **1890**, *6*, 524.
2678. Gartenmeister, R., *Justus Liebigs Ann. Chem.* **1886**, *233*, 262; *Beil.* **H2,** 132.
2679. Gary, J. T., and L. W. Pickett, *J. Chem. Phys.* **1954**, *22*, 1266.
2680. Gasopoulos, I., *Ber. Deut. Chem. Gesell.* **1926**, *59*, 2184.
2681. Gasopoulos, I., *Ber. Deut. Chem. Gesell.* **1926**, *59*, 2188; *C.A.* **1927**, *21*, 387.
2682. Gasparic, J., and J. Borecky, *J. Chromatog.* **1961**, *5*, 466; *C.A.* **1962**, *56*, 9393.
2683. Gasparic, J., J. Petranek, et al., *J. Chromatog.* **1961**, *5*, 408; *C.A.* **1962**, *56*, 9393.
2684. Gast, J. H., and F. L. Estes, *Anal. Chem.* **1960**, *32*, 1712.
2685. Gates, C. B., *J. Phys. Chem.* **1911**, *15*, 97.
2686. Gates, D. M., *J. Chem. Phys.* **1949**, *17*, 393.
2687. Gaufres, R., and M. Bejaud-Bianchi, *Spectrochim. Acta, Part A* **1971**, *27*, 2249.
2688. Gaumann, T., *Helv. Chim. Acta* **1958**, *41*, 1956.
2689. Gause, E. M., and F. M. Ernsberger, *Ind. Eng. Chem., Chem. Eng. Data Series* **1957**, *2*, 28.
2690. Gauss, E. J., and T. S. Gilman, *J. Phys. Chem.* **1969**, *73*, 3969.
2691. Gavat, I., I. Irimescu, et al., *Bull. Soc. Roumaine Phys.* **1941**, *42*, 63; *C.A.* **1943**, *37*, 3670.
2692. Gayles, J. N., Jr., and W. T. King, *Spectrochim. Acta* **1965**, *21*, 543.
2693. Geddes, A. L., *J. Phys. Chem.* **1954**, *58*, 1062.
2694. Geddes, J. A., and E. C. Bingham, *J. Amer. Chem. Soc.* **1934**, *56*, 2625.
2695. Geiseler, G., J. Fruwert, et al., *Chem. Ber.* **1966**, *99*, 1594.
2696. Geiseler, G., and H. Kessler, *Ber. Bunsenges. Phys. Chem.* **1964**, *68*, 571; *C.A.* **1964**, *61*, 15540.
2697. Geiseler, G., and E. Manz, *Monatsh. Chem.* **1969**, *100*, 1133.
2698. Geiseler, G., and E. Pilz, *Chem. Ber.* **1962**, *95*, 96; *C.A.* **1962**, *56*, 12717.
2699. Geiseler, G., and M. Ratzsch, *Ber. Bunsenges. Phys. Chem.* **1965**, *69*, 485.
2700. Geist, J. M., and M. R. Cannon, *Ind. Eng. Chem., Anal. Ed.* **1946**, *18*, 611.
2701. Geldof, H., and J. P. Wibaut, *Rec. Trav. Chim. Pays-Bas* **1948**, *67*, 105; *C.A.* **1948**, *42*, 5409.
2702. Gelfand, S., German Patent 2,258,747, June 7, 1973; *C.A.* **1973**, *79*, 42126.

2703. Geller, B. E., *Zh. Fiz. Khim.* **1961,** *35,* 2210; *C.A.* **1962,** *56,* 14940.

2704. Geller, V. Z., E. G. Porichanskii, et al., *Kholod. Tekh. Tekhnol.* **1979,** *29,* 43; *C.A.* **1980,** *93,* 192150.

2705. Geller, Z. I., R. K. Nikul'shin, et al., *Zh. Prikl. Khim.* (Leningrad) **1969,** *42,* 1121; *C.A.* **1969,** *71,* 74212.

2706. Gelles, E., and K. S. Pitzer, *J. Amer. Chem. Soc.* **1953,** *75,* 5259.
    General Aniline and Film Corp.; see also Antara Chemicals.

2707. General Aniline and Film Corp., "Butyrolactone," New York, 1964.

2708. General Aniline and Film Corp., "Methylpyrrolidone," New York, 1961.

2709. Gensler, W. J., and C. M. Samoier, *J. Org. Chem.* **1953,** *18,* 9.

2710. Gent, W. L. G., *J. Chem. Soc.* **1957,** 58.

2711. George, M. V., and G. F. Wright, *Can. J. Chem.* **1958,** *36,* 189.

2712. Gerding, H., and H. G. Haring, *Rec. Trav. Chim. Pays-Bas* **1955,** *74,* 841; *C.A.* **1955,** *49,* 15485.

2713. Gerding, H., and J. Lecomte, *Rec. Trav. Chim. Pays-Bas* **1939,** *58,* 614; *C.A.* **1939,** *33,* 9141.

2714. Gerding, H., and G. W. A. Pijnders, *Rec. Trav. Chim. Pays-Bas* **1946,** *65,* 143; *C.A.* **1946,** *40,* 5644.

2715. Germann, F. E. E., and O. S. Knight, *J. Amer. Chem. Soc.* **1933,** *55,* 4150.

2716. Gero, A., *J. Org. Chem.* **1954,** *19,* 469, 1960.

2717. Gero, A., *J. Org. Chem.* **1961,** *26,* 3156.

2718. Gero, A., and J. J. Markham, *J. Org. Chem.* **1951,** *16,* 1835.

2719. Gershtein, N. A., and M. F. Shostakovskii, *J. Gen. Chem.* (USSR) **1948,** *18,* 451; *C.A.* **1948,** *42,* 7243.

2720. Getman, F. H., *J. Amer. Chem. Soc.* **1940,** *62,* 2179.

2721. Getman, F. H., *Rec. Trav. Chim. Pays-Bas* **1936,** *55,* 231; *C.A.* **1936,** *30,* 4743.

2722. Ghanem, N. A., M. Marek, et al., *Int. J. Appl. Radiat. Istop.* **1970,** *21,* 239; *C.A.* **1970,** *73,* 7862.

2723. Ghazanfar, S. A. S., J. T. Edsall, et al., *J. Amer. Chem. Soc.* **1964,** *86,* 559.

2724. Gheorghiu, A., and T. Gheorghiu, *Ann. Combustibles Liquides* **1933,** 8, 451; *C.A.* **1935,** *29,* 2448.

2725. Ghiringhelli, L., *Med. Lavora* **1956,** *47,* 192; *C.A.* **1957,** *51,* 1461.

2726. Ghosh, J. C., S. K. Bhattacharyya, et al., *Current Sci.* **1947,** *16,* 88; *C.A.* **1947,** *41,* 4725.

2727. Giauque, W. F., and R. Wiebe, *J. Amer. Chem. Soc.* **1928,** *50,* 101.

2728. Gibbard, H. F., and J. L. Creek, *J. Chem. Eng. Data* **1974,** *19,* 308.

2729. Gibbons, L. C., J. F. Thompson, et al., *J. Amer. Chem. Soc.* **1946,** *68,* 1130.

2730. Gibbons, W. A., and V. M. S. Gil, *Mol. Phys.* **1965,** *9,* 163; *C.A.* **1965,** *63,* 9783.

2731. Gibbs, H. D., *J. Amer. Chem. Soc.* **1905,** *27,* 851.

2732. Gibbs, H. D., *J. Amer. Chem. Soc.* **1927,** *49,* 839.

2733. Gibbs, J. H., and C. P. Smyth, *J. Amer. Chem. Soc.* **1951,** *73,* 5115.

2734. Gibby, C. W., and J. Hall, *J. Chem. Soc.* **1931,** 691.

2735. Gibson, C. S., and J. B. A. Johnson, *J. Chem. Soc.* **1931,** 266.

2736. Gibson, G. E., and W. F. Giauque, *J. Amer. Chem. Soc.* **1923,** *45,* 93.

2737. Gibson, R. E., and J. F. Kincaid, *J. Amer. Chem. Soc.* **1937,** *59,* 579.

2738. Gibson, R. E., and J. F. Kincaid, *J. Amer. Chem. Soc.* **1938,** *60,* 511.

2739. Gibson, R. E., and O. H. Loeffler, *J. Amer. Chem. Soc.* **1939,** *61,* 2515.

2740. Gibson, S. G., and J. D. A. Johnson, *J. Chem. Soc.* **1930,** 2525.

2741. Giertz, H. W., and J. MacPhersson, *Norsk Skogsin* **1956,** *10,* 348.

2742. Gierut, J. A., F. J. Sowa, et al., *J. Amer. Chem. Soc.* **1936,** *58,* 897.

2743. Giessner-Prettre, C., *Compt. Rend.* **1961,** *252,* 3238; *C.A.* **1961,** 55, 21808.

2744. Giessner-Prettre, C., *Compt. Rend.* **1962,** *254,* 4165; *C.A.* **1963,** *59,* 149.

2745. Gifford, A. P., S. M. Rock, et al., *Anal. Chem.* **1949,** *21,* 1026.

2746. Gil, V. M. S., and J. N. Murrell, *Trans. Faraday Soc.* **1964,** *60,* 248.

2747. Gil'denblat, I. A., A. S. Furmanov, et al., *Zh. Prikl. Khim.* **1960,** *33,* 246; *C.A.* **1960,** *54,* 9410.

2748. Giles, G. D., and C. F. Wells, *Nature* **1964,** *201,* 606.

2749. Gilkey, W. K., F. W. Gerard, et al., *Ind. Eng. Chem.* **1931,** *23,* 364.

2750. Gill, A. H., and F. P. Dexter, *Ind. Eng. Chem.* **1934,** *26,* 881.

2751. Giller, E. B., and H. G. Driekamer, *Ind. Eng. Chem.* **1949,** *41,* 2067.

2752. Gillespie, D. T. C., A. K. Macbeth, et al., *J. Chem. Soc.* **1940,** 280.

2753. Gillo, L., *Ann. Chim.* **1939,** *12,* 281; *C.A.* **1940,** *34,* 1530.

2754. Gillo, L., *Bull. Soc. Chim. Belges* **1939,** *48,* 341; *C.A.* **1940,** *34,* 2221.

2755. Gillois, M., and P. Rumpf, *Bull. Soc. Chim. Fr.* **1954,** 112.
2756. Gilman, H., *Organic Syntheses, Vol. VI*, Wiley, New York, 1926.
2757. Gilman, H., and A. M. Blatt, *Organic Syntheses, Coll. Vol. 1*, 2nd ed., Wiley, New York, 1941.
2758. Gilman, H., and R. G. Jones, *J. Amer. Chem. Soc.* **1943,** *65*, 2037.
2759. Gilman, H. H., and P. Gross, *J. Amer. Chem. Soc.* **1938,** *60*, 1525.
2760. Gilpin, J. A., *Anal. Chem.* **1959,** *31*, 935.
2761. Gilson, D. F. R., and C. A. McDowell, *Mol. Phys.* **1961,** *4*, 125.
2762. Gilson, L. E., *J. Amer. Chem. Soc.* **1932,** *54*, 1445.
2763. Ginnings, D. C., and G. T. Furukawa, *J. Amer. Chem. Soc.* **1953,** *75*, 522.
2764. Ginnings, P. M., and R. Baum, *J. Amer. Chem. Soc.* **1937,** *59*, 1111.
2765. Ginnings, P. M., D. Plonk, et al., *J. Amer. Chem. Soc.* **1940,** *62*, 1923.
2766. Ginnings, P. M., and R. Webb, *J. Amer. Chem. Soc.* **1938,** *60*, 1388.
2767. Ginsberg, A. S., *Chem. Zentr.* **1897,** *II*, 417.
2768. Giorgini, M. G., M. R. Pelletti, et al., *J. Raman Spectrosc.* **1983,** *14*, 16.
2769. Girault-Vexlearschi, G., *Bull. Soc. Chim. Fr.* **1956,** 589.
2770. Gisseleire, J., and R. Bilterys, *Bull. Soc. Chim. Belges* **1935,** *44*, 567; *Chem. Zentr.* **1936,** *I*, 3999.
2771. Gist, L. A., and C. T. Mason, *J. Amer. Chem. Soc.* **1954,** *76*, 3728.
2772. Giua, M., and G. Guastalla, *Chem. Ind.* **1933,** *29*, 268.
2773. Givaudan-Delarvanna, Inc., *The Givaudan Index*, 2nd ed., New York, 1961.
2774. Gladstone, J. H., *J. Chem. Soc.* **1884,** 241.
2775. Glagoleva, A. A., *Zh. Obshch. Khim.* **1948,** *18*, 1005; *C.A.* **1949,** *43*, 6900.
2776. Glaser, R., and H. Rulard, *Chem. Ing.-Tech.* **1957,** *29*, 772.
2777. Glasgow, A. R., *J. Res. Nat. Bur. Stand.* **1940,** *24*, 509.
2778. Glasgow, A. R., N. Krouskop, et al., *Anal. Chem.* **1948,** *20*, 410.
2779. Glasgow, A. R., N. Krouskop, et al., *Anal. Chem.* **1949,** *21*, 688.
2780. Glasgow, A. R., E. T. Murphy, et al., *J. Res. Nat. Bur. Stand.* **1946,** *37*, 141.
2781. Glasgow, A. R., A. J. Streiff, et al., *J. Res. Nat. Bur. Stand.* **1945,** *35*, 355.
2782. Glasoe, P. K., and S. D. Schultz, *J. Chem. Eng. Data* **1972,** *17*, 66.
2783. Glass, H. M., and W. M. Madgin, *J. Chem. Soc.* **1934,** 1292.
2784. Glasstone, S., *Textbook of Physical Chemistry*, Van Nostrand, New York, 1940.
2785. Glasstone, S., and A. F. Schram, *J. Amer. Chem. Soc.* **1947,** *69*, 1213.
2786. Gleave, W. W., U.S. Patent 2,104,695, Jan. 4, 1938; *C.A.* **1938,** *32*, 1712.
2787. Glick, R. E., and D. F. Kates, *J. Phys. Chem.* **1958,** *62*, 1469.
2788. Glidewell, C., D. W. H. Rankin, et al., *Trans. Faraday Soc.* **1969,** *65*, 2801.
2789. Glockler, G., and J. H. Bachman, *Phys Rev.* **1939,** *55*, 669.
2790. Glockler, G., and W. F. Edgell, *Ind. Eng. Chem.* **1942,** *34*, 532.
2791. Glockler, G., and W. F. Edgell, *J. Chem. Phys.* **1941,** *9*, 224.
2792. Glockler, G., and W. F. Edgell, *J. Chem. Phys.* **1941,** *9*, 527.
2793. Glockler, G., and G. R. Leader, *J. Chem. Phys.* **1939,** *7*, 278.
2794. Glockler, G., and G. R. Leader, *J. Chem. Phys.* **1940,** *8*, 697.
2795. Glockler, G., and R. E. Peck, *J. Chem. Phys.* **1936,** *4*, 624.
2796. Glockler, G., and C. Sage, *J. Chem. Phys.* **1941,** *9*, 387.
2797. Gloor, W. E., *Ind. Eng. Chem.* **1947,** *39*, 1125.
2798. Glover, C. A., and C. P. Hill, *Anal. Chem.* **1953,** *25*, 1379.
2799. Gluzman, L. D., V. A. Ivanushkina, et al., *Org. Chem. Ind.* (USSR) **1938,** *5*, 682; *C.A.* **1939,** *33*, 5158.
2800. *Gmelins Handbuch der Anorganischen Chemie*, 8th ed., Pt. 5, Weinheim/Bergstr., Verlag Chemie, 1963.
2801. Gockel, H., *Z. Phys. Chem.* (Leipzig) **1935,** *29B*, 79; *C.A.* **1935,** *29*, 6143.
2802. Godart, J., *J. Chim. Phys.* **1937,** *34*, 70; *C.A.* **1937,** *31*, 4595.
2803. Godchot, M., and G. Cauquil, *Compt. Rend.* **1934,** *198*, 663; *C.A.* **1934,** *28*, 2686.
2804. Godfrey, N. B., *Chem. Tech.* **1972.**
2805. Godlewski, I. O., and K. Roshanowitsch, *Chem. Zentr.* **1889,** *I*, 1241.
2806. Godnev, I. N., A. S. Sverdlin, et al., *Zh. Fiz. Khim.* **1950,** *24*, 807; *C.A.* **1951,** *45*, 4128.
2807. Goebel, H. L., and H. H. Wenzke, *J. Amer. Chem. Soc.* **1937,** *59*, 2301.

2808. Goering, H. L., and F. H. McCarron, *J. Amer. Chem. Soc.* **1958**, *80*, 2287.

2809. Goering, H. L., and C. Serres, *J. Amer. Chem. Soc.* **1952**, *74*, 5908.

2810. Goethals, C. A., *Rec. Trav. Chim.* **1953**, *54*, 299.

2811. Goetz, H., F. Nerdel, et al., *Ann. Chem,* **1965**, *681*, 1; *C.A.* **1965**, *62*, 11666.

2812. Gohlke, R. S., and F. W. McLafferty, *Anal. Chem.* **1962**, *34*, 1281.

2813. Gokhale, S. D., N. L. Phalnikar, et al., *J. Univ. Bombay* **1943**, *11A*, Pt. 5, 56; *C.A.* **1943**, *37*, 5293.

2814. Gold, P. I., and R. L. Perrine, *J. Chem. Eng. Data* **1967**, *12*, 4.

2815. Goldblatt, L. A., and S. Palkin, *J. Amer. Chem. Soc.* **1941**, *63*, 3517.

2816. Goldblum, K. B., R. W. Martin, et al., *Ind. Eng. Chem.* **1947**, *39*, 1474.

2817. Gol'denfel'd, I. V., I. Z. Korostyshevskii, et al., *Teor. Eksp. Khim.* **1970**, *6*, 830; *C.A.* **1971**, *75*, 26815.

2818. Goldenson, J., *Appl. Spectrosc.* **1964**, *18*, 155.

2819. Goldfarb, A. R., A. Mele, et al., *J. Amer. Chem. Soc.* **1955**, *77*, 6194.

2820. Goldfarb, T. D., and B. N. Khare, *J. Chem. Phys.* **1967**, *46*, 3379.

2821. Gol'dfarb, Ya. L., and F. M. Stoyanovich, *Bull. Acad. Sci. USSR, Div. Chem. Sci.* **1970**, No. 8, 1774.

2822. Goldfinger, P., and G. Martens, *Trans. Faraday Soc.* **1961**, *57*, 2220.

2823. Goldman, S., *Can. J. Chem.* **1974**, *52*, 1668.

2824. Goldschmid, O., *J. Amer. Chem. Soc.* **1953**, *75*, 3780.

2825. Goldschmidt, H., *Z. Phys. Chem.* (Leipzig) **1915**, *89*, 130.

2826. Goldschmidt, H., *Z. Phys. Chem.* (Leipzig) **1926**, *124*, 23.

2827. Goldschmidt, H., and E. Mathiesen, *Z. Phys. Chem.* (Leipzig) **1926**, *121*, 153.

2828. Goldschmidt, H., and L. Oslan, *Ber. Deut. Chem. Gesell.* **1900**, *33*, 1140.

2829. Goldschmidt, H., and R. Salcher, *Z. Phys. Chem.* (Leipzig) **1899**, *29*, 89.

2830. Goldschmidt, H., and L. Thomas, *Z. Phys. Chem.* (Leipzig) **1927**, *126*, 24.

2831. Goldschmidt, H., and A. Thuesen, *Z. Phys Chem.* (Leipzig) **1913**, *81*, 30.

2832. Goldschmidt, R., *Schweiz. Arch. angew. Wiss. u. Tech.* **1947**, *13*, 21; *C.A.* **1947**, *41*, 7088.

2833. Gol'dshtein, I. P., E. N. Gur'yanova, et al., *Zh. Strukt. Khim.* **1966**, *7*, 222; *C.A.* **1966**, *65*, 16840.

2834. Gol'dshtein, I. P., Yu. M. Kessler, et al., *Zh. Strukt. Khim.* **1963**, *4*, 445; *C.A.* **1963**, *59*, 5894.

2835. Golendeev, V. P., and I. S. Okrokova, *Tr. Khim. Khim. Tekhnol.* **1964**, 531; *C.A.* **1966**, *64*, 19401.

2836. Golik, A. Z., and S. D. Ravikovich, *Zh. Fiz. Khim.* **1949**, *23*, 86; *C.A.* **1949**, *43*, 4064.

2837. Golosova, R. M., and V. V. Korobov, *Russ. J. Phys. Chem.* (Eng. trans.), **1971**, *45*, 460.

2838. Golosova, R. M., A. M. Mosin, et al., *Russ. J. Phys. Chem.* (Eng. trans.) **1967**, *41*, 781.

2839. Golubkov, Yu. V., A. P. Tsurkin, et al., *Zh. Prikl. Khim.* **1980**, *53*, 1894; *C.A.* **1980**, *93*, 225761.

2840. Golumbic, C., *J. Amer. Chem. Soc.* **1949**, *71*, 2627.

2841. Gomberg, M., and C. C. Buchler, *J. Amer. Chem. Soc.* **1920**, *42*, 2059.

2842. Gomel, M., and H. Lumbroso, *Bull. Soc. Chim. Fr.* **1962**, 2200; *C.A.* **1963**, *58*, 13762.

2843. Gomel, M., H. Lumbroso, et al., *Bull. Soc. Chim. Fr.* **1959**; *C.A.* **1960**, *54*, 14171.

2844. Gonzalez R., E., *Ciencia* (Mex.) **1947**, *8*, 175; *C.A.* **1949**, *43*, 127.

2845. Gooch, J. P., E. K. Landis, and J. S. Browning, *U.S. Nat. Tech. Inform. Serv., PB Report* **1972**, No. 211354; *C.A.* **1973**, *78*, 34523.

2846. Good, W. D., *J. Chem. Eng. Data* **1969**, *14*, 231.

2847. Good, W. D., *J. Chem. Eng. Data* **1972**, *17*, 28.

2848. Good, W. D., *J. Chem. Thermodyn.* **1970**, *2*, 237.

2849. Good, W. D., *J. Chem. Thermodyn.* **1972**, *4*, 709.

2850. Good, W. D., *J. Chem. Thermodyn.* **1973**, *5*, 707.

2851. Good, W. D., *J. Chem. Thermodyn.* **1975**, *7*, 49.

2852. Good, W. D., *J. Chem. Thermodyn.* **1975**, *8*, 67.

2853. Good, W. D., J. L. Lacina, et al., *J. Phys. Chem.* **1961**, *65*, 2229.

2854. Good, W. D., and S. H. Lee, *J. Chem. Thermodyn.* **1976**, *8*, 643.

2855. Good, W. D., and R. T. Moore, *J. Chem. Eng. Data* **1970**, *15*, 150.

2856. Good, W. D., D. W. Scott, et al., *J. Phys. Chem.* **1956**, *60*, 1080.

2857. Good, W. D., and N. K. Smith, *J. Chem. Eng. Data* **1969**, *14*, 102.

2858. Good, W. D., and N. K. Smith, *J. Chem. Thermodyn.* **1979**, *11*, 111.

2859. Goode, E. V., and D. A. Ibbitson, *J. Chem. Soc.* **1960**, 4265.

2860. Goodeve, J. W., *Trans. Faraday Soc.* **1934**, *30*, 501; *Chem. Zentr.* **1934**, *II*, 3232.

2861. Goodeve, J. W., *Trans. Faraday Soc.* **1934**, *30*, 504; *Chem. Zentr.* **1934**, *II*, 3228.

2862. Goodeve, J. W., *Trans. Faraday Soc.* **1938**, *34*, 1239; *Brit. Chem. Abstr.* **1938AI**, 598.

2863. Goodman, M. A., and S. L. Whittenburg, *J. Chem. Eng. Data* **1983**, *28*, 350.

2864. Goodspeed, F. E., and M. F. Millson, *Chem. Ind.* (London) **1967**, 1594; *C.A.* **1968**, *68*, 7985.

2865. Goodwin, D. G., and H. R. Hudson, *J. Chem. Soc. B* **1968**, 1333.

2866. Goodwin, R. D., *J. Res. Nat. Bur. Stand.* **1978**, *83*, 449.

2867. Gopal, R., and S. A. Rizvi, *J. Indian Chem. Soc.* **1966**, *43*, 179; *C.A.* **1966**, *65*, 3739.

2868. Gopal, R., and S. A. Rizvi, *J. Indian Chem. Soc.* **1968**, *45*, 13.

2869. Gordon, G. S., and R. L. Burwell, *J. Amer. Chem. Soc.* **1949**, *71*, 2355; **1948**, *70*, 3128.

2870. Gordon, J., and W. F. Giauque, *J. Amer. Chem. Soc.* **1948**, *70*, 1506.

2871. Gordon, J. S., *J. Chem. Phys.* **1958**, *29*, 889.

2872. Gordon, M., J. G. Miller, et al., *J. Amer. Chem. Soc.* **1949**, *71*, 1245.

2873. Gordus, A. A., and R. B. Bernstein, *J. Chem. Phys.* **1954**, *22*, 790.

2874. Gordy, W., *J. Chem. Phys.* **1939**, *7*, 93.

2875. Gordy, W., and S. C. Stanford, *J. Amer. Chem. Soc.* **1940**, *62*, 497.

2876. Gordy, W., and S. C. Stanford, *J. Chem. Phys.* **1941**, *9*, 204.

2877. Gordy, W., and D. Williams, *J. Chem. Phys.* **1936**, *4*, 85.

2878. Gornowski, E. J., E. H. Amick, et al., *Ind. Eng. Chem.* **1947**, *39*, 1348.

2879. Gorodetskii, I. Y. A., and V. M. Dlevskii, *Vestn. Leningrad Univ. 15*, No. 16, *Ser. Fiz. i. Khim.* **1960**, No. 3, 102; Leslie, R. T. and E. C. Kuehner, *Anal. Chem.* **1962**, *34*, 5OR.

2880. Gorshkova, G. N., Z. B. Barinova, et al., *Izv. Akad. Nauk. SSSR, Ser. Khim.* **1968**, No. 2, 312; *Acad. Sci. USSR, Div. Chem. Sci.* **1968**, No. 2, 303.

2881. Goshorn, R. H., and E. F. Degering, *Proc. Indiana Acad. Sci.* **1935**, *45*, 139; *C.A.* **1937**, *31*, 1004.

2882. Goss, F. R., *J. Chem. Soc.* **1940**, 888.

2883. Gosse, J. P., and B. Rose, *C. R. Acad. Sc. Paris, Ser. C* **1968**, *267*, 927.

2884. Goto, K., *Sci. Light* (Tokyo) **1962**, *11*, 119; *C.A.* **1964**, *61*, 183.

2885. Götz, I. D., *Z. Phys. Chem.* (Leipzig) **1920**, *94*, 181.

2886. Goubeau, J., *Reichsamt Wirtschaftsausbau, Pruf-Nr.* **1940**, *43*, (PB 52003), 129; *C.A.* **1947**, *41*, 6491.

2887. Goubeau, J., H. Siebert, et al., *Z. Anorg. Allg. Chem.* **1949**, *259*, 240; *C.A.* **1950**, *44*, 1807.

2888. Goubeau, J., and H. Siebert, *Z. Anorg. Allg. Chem.* **1950**, *261*, 62.

2889. Goubeau, J., and V. von Schneider, *Angew. Chem.* **1940**, *53*, 531; *C.A.* **1941**, *35*, 1355.

2890. Goudet, H., and F. Schenker, *Helv. Chim. Acta.* **1927**, *10*, 132; *C.A.* **1927**, *21*, 2657.

2891. Goue, T. H., and R. E. Jentoft, *J. Chromatog.* **1972**, *68*, 303.

2892. Gough, G. A. C., H. Hunter, et al., *J. Chem. Soc.* **1926**, 2052.

2893. Gouravelev, D. J., *J. Phys. Chem.* (USSR) **1937**, *9*, 877.

2894. Gouw, T. H., and J. C. Vlugter, *J. Amer. Oil Chem. Soc.* **1964**, *41*, 142, 426, 675.

2895. Goy, C. A., and H. O. Pritchard, *J. Phys. Chem.* **1965**, *69*, 3040.

2896. Gozzelino, G., F. Ferrero, and M. Panetti, *Chim. Ind.* (Milan) **1980**, *62*, 13; *C.A.* **1980**, *93*, 71154.

2897. Grammaticakis, I. P., *Bull. Soc. Chim. Fr.* **1953**, 821; *C.A.* **1954**, *48*, 1148.

2898. Grammer, G. N., and P. W. Trotter, U.S. Patent 3,238,137, Mar. 1, 1966.

2899. Granett, P., and H. L. Haynes, *J. Econ. Entomol.* **1945**, *38*, 671; *C.A.* **1946**, *40*, 2918.

2900. Granier, J., *Compt. Rend.* **1946**, *223*, 893; *C.A.* **1947**, *41*, 1511.

2901. Grant, D. M., and R. C. Hirst, *J. Chem. Phys.* **1963**, *38*, 470.

2902. Grant, D. M., and E. G. Paul, *J. Amer. Chem. Soc.* **1964**, *86*, 2984.

2903. Grant Chemical Division, "1,3-Dioxolane," Baton Rouge, Louisiana, 1982.

2904. Grant Chemical Division, "Glymes," Baton Rouge, Louisiana, 1982.

2905. Granzhan, V. A., *Zh. Vses. Khim. Obshchest.* **1969**, *14*, 223; *C.A.* **1969**, *71*, 21591.

2906. Granzhan, V. A., and S. K. Laktionova, *Zh. Fiz. Khim.* **1973**, *47*, 515; *Russ. J. Phys. Chem.* **1973**, *47*, 294.

2907. Granzhan, V. A., L. M. Savenko, et al., *Zh. Fiz. Khim.* **1970**, *44*, 2445; *Russ. J. Phys. Chem.* **1970**, *44*, 1388.

2908. Grasshof, H., *Chem. Ber.* **1951**, *84*, 916.

2909. Grau, A., and H. Lumbroso, *Bull. Soc. Chim. Fr.* **1961**, 1860; *C.A.* **1962**, *57*, 5404.

2910. Graul, R. J., and J. V. Karabinos, *Science* **1946**, *104*, 557.

2911. Gray, G. A., and T. A. Albright, *J. Amer. Chem. Soc.* **1977,** *99*, 3243.

2912. Gray, H. F., R. S. Rasmussen, et al., *J. Amer. Chem. Soc.* **1947,** *69*, 1630.

2913. Gray, H. LeB., and G. O. Gutekunst, *J. Amer. Chem. Soc.* **1920,** *42*, 856.

2914. Gray, P., and A. Jones, *Trans. Faraday Soc.* **1965,** *61*, 2161.

2915. Grazia Giorgini, M., G. Paliani, et al., *Spectrochim. Acta, Part A* **1977,** *33A*, 1083.

2916. Grazuliene, S., G. F. Telegin, et al., *Zh. Prikl. Khim.* **1977,** *50*, 885; *C.A.* **1977,** *87*, 39156.

2917. Greathause, L. H., H. J. Janssen, et al., *Anal. Chem.* **1956,** *28*, 357.

2918. Grédy, B., *Bull. Soc. Chim.* **1935,** [5], *2*, 1029; *C.A.* **1935,** *29*, 5807.

2919. Grédy, B., *Bull. Soc. Chim.* **1936,** [5], *3*, 1101; *C.A.* **1936,** *30*, 7038.

2920. Grédy, B., and L. Piaux, *Bull. Soc. Chim. Pays-Bas* **1934,** *1*, 1481; *C.A.* **1935,** *29*, 2449.

2921. Grédy, B., and L. Piaux, *Compt. Rend.* **1934,** *198*, 1235; *C.A.* **1934,** *28*, 3379.

2922. Green, C., J. Marsden, et al., *Can. J. Res.* **1933,** *9*, 396; *Beil.* **EIII2,** 269, 1042, 1509a.

2923. Green, J. H. S., *Chem. Ind.* (London) **1960,** 1215.

2924. Green, J. H. S., *Chem. Ind.* (London) **1961,** 369, 1218.

2925. Green, J. H. S., *Chem. Ind.* (London) **1962,** 1575.

2926. Green, J. H. S., *J. Appl. Chem.* (London) **1961,** *11*, 397.

2927. Green, J. H. S., *J. Chem. Soc.* **1961,** 2236.

2928. Green, J. H. S., *J. Chem. Soc.* **1961,** 2241.

2929. Green, J. H. S., *Spectrochim. Acta* **1962,** *18*, 39; *C.A.* **1962,** *57*, 6765.

2930. Green, J. H. S., *Spectrochim. Acta, Part A* **1970,** *26*, 1503.

2931. Green, J. H. S., *Spectrochim. Acta, Part A* **1970,** *26*, 1523.

2932. Green, J. H. S., *Spectrochim. Acta, Part A* **1970,** *26*, 1913.

2933. Green, J. H. S., *Trans. Faraday Soc.* **1961,** *57*, 2132.

2934. Green, J. H. S., *Trans. Faraday Soc.* **1963,** *59*, 1559.

2935. Green, J. H. S., and D. J. Harrison, *J. Chem. Thermodyn.* **1976,** *8*, 529.

2936. Green, J. H. S., and D. J. Harrison, *Spectrochim. Acta, Part A* **1970,** *26*, 2113; *C.A.* **1971,** *74*, 59078.

2937. Green, J. H. S., and D. J. Harrison, *Spectrochim. Acta, Part A* **1976,** *32A*, 1265.

2938. Green, J. H. S., and D. J. Harrison, *Spectrochim. Acta, Part A* **1976,** *32A*, 1279.

2939. Green, J. H. S., and D. J. Harrison, *Spectrochim. Acta, Part A* **1977,** *33A*, 583.

2940. Green, J. H. S., and D. J. Harrison, *Spectrochim. Acta, Part A* **1977,** *33A*, 837.

2941. Green, J. H. S., and D. J. Harrison, *Spectrochim. Acta, Part A* **1977,** *33A*, 843.

2942. Green, J. H. S., D. J. Harrison, et al., *Spectrochim. Acta, Part A* **1971,** *27*, 2199.

2943. Green, J. H. S., D. J. Harrison, et al., *Spectrochim. Acta, Part A* **1972,** *28*, 33.

2944. Green, J. H. S., D. J. Harrison, et al., *Spectrochim. Acta, Part A* **1971,** *27*, 793.

2945. Green, J. H. S., D. J. Harrison, et al., *Spectrochim. Acta, Part A* **1971,** *27*, 807.

2946. Green, J. H. S., and D. J. Holden, *J. Chem. Soc.* **1962,** 1794.

2947. Green, J. H., and K. R. Ryan, *Proc. Roy. Soc.* (London) **1965,** Ser. *A286*, No. 1405, 178; *C.A.* **1965,** *63*, 2511.

2948. Greenburg, R. B., U.S. Patent 2,398,526, April 16, 1946; *C.A.* **1946,** *40*, 3774.

2949. Greenwald, W. C., U.S. Patent 2,456,184, Dec. 17, 1948; *C.A.* **1949,** *43*, 3437.

2950. Greenwood, F. L., F. C. Whitmore, et at., *J. Amer. Chem. Soc.* **1938,** *60*, 2028.

2951. Gregory, M. D., H. E. Affsprung, et al., *J. Phys. Chem.* **1968,** *72*, 1748.

2952. Greig, C. C., and C. D. Johnson, *J. Amer. Chem. Soc.* **1968,** *90*, 6453.

2953. Greinacher, H., *Helv. Phys. Acta* **1948,** *21*, 261; *C.A.* **1948,** *42*, 8554.

2954. Grekov, A. P., USSR Patent 168,296, Feb. 18, 1965; *C.A.* **1965,** *62*, 16068.

2955. Gremmo, N., and J. E. B. Randles, *J. Chem. Soc., Faraday Trans. 1* **1974,** *70*, 1480.

2956. Gresham, W. F., U.S. Patent 2,439,426, April 13, 1948; *C.A.* **1948,** *42*, 6841.

2957. Griffin, D. M., *J. Amer. Chem. Soc.* **1949,** *71*, 1423.

2958. Griffing, V., M. A. Cargyle, et al., *J. Phys. Chem.* **1954,** *58*, 1054.

2959. Griffiths, E. H., *Phil Trans.* **1891,** *182A*, 43; from [7482].

2960. Griffiths, P. H., W. A. Walkey, et al., *J. Chem. Soc.* **1934,** 631.

2961. Griffiths, P. R., and H. W. Thompson, *Proc. Roy. Soc., Ser. A* **1967,** *298*, 51.

2962. Griffiths, S. T., and R. R. Wilson, *Combustion & Flame* **1958,** *2*, 224.

2963. Griffiths, V. S., *J. Chem. Soc.* **1954,** 860.

2964. Grignard, V., *Compt. Rend.* **1900,** *130*, 1322; *Chem. Zentr.* **1901,** *II*, 622.

2965. Grignard, V., *Compt. Rend.* **1901,** *132,* 336; *Chem. Zentr.* **1901,** *I,* 612.
2966. Grignard, V., and F. Chambret, *Compt. Rend.* **1926,** *182,* 299; *C.A.* **1926,** *20,* 1602.
2967. Grignard, V., and M. Fluchaire, *Ann. Chim.* **1928,** *9,* 5; *C.A.* **1928,** *22,* 1951.
2968. Grignard, V., and G. Vegnon, *Compt. Rend.* **1907,** *144,* 1358; *Chem. Zentr.* **1907,** *II,* 681; *C.A.* **1907,** *1,* 2553.
2969. Grimm, F. V., and W. A. Patrick, *J. Amer. Chem. Soc.* **1923,** *45,* 2794.
2970. Grimm, H. G., *Z. Phys. Chem.* (Leipzig) **1929,** *140A,* 321.
2971. Grishchenko, N. F, and V. N. Pokorskii, *Neftepererab. Neftekhim.* **1966,** No. 1, 33; *C.A.* **1967,** *66,* 54960.
2972. Griswold, A. A., and P. S. Starcher, *J. Org. Chem.* **1965,** *30,* 1687.
2973. Grivsky, E., *Bull. Soc. Chim. Belges* **1942,** *51,* 63.
2974. Grodde, K.-H., *Phys. Z.* **1938,** *39,* 772, 777; *Beil.* **EIII5,** 915.
2975. Groggins, P. H., and R. H. Nagel, *Ind. Eng. Chem.* **1934,** *26,* 1313.
2976. Groggins, P. H., R. H. Nagel, et al., *Ind. Eng. Chem.* **1934,** *26,* 1317.
2977. Grolier, J. P. E., A. Inglese, et al., *J. Chem. Thermodyn.* **1982,** *14,* 523.
2978. Grolier, J. P. E., A. Inglese, et al., *J. Chem. Thermodyn.* **1984,** *16,* 67.
2979. Groll, H. P. A., and M. W. Tamele, U.S. Patent 2,010,358, Aug. 6, 1935; *Official Gaz. U.S. Patent Office,* No. 457.
2980. Gronowitz, S., A. B. Hornfeldt, et al., *Arkiv Kemi* **1961,** *18,* 133; *C.A.* **1962,** *56,* 10075.
2981. Gronowitz, S., G. Sorlin, et al., *Arkiv Kemi* **1962,** *19,* 483; *C.A.* **1963,** *58,* 2041.
2982. Gross, B., and M.-T. Forel, *J. Chim. Phys.* **1965,** *62,* 1163; *C.A.* **1966,** *64,* 9089.
2983. Gross, P., *J. Amer. Chem. Soc.* **1939,** *51,* 2362.
2984. Gross, P., *Z. Phys.* **1931,** *32,* 587; *Chem. Zentr.* **1931,** *102,* 2700.
2985. Gross, P., J. C. Rintelen, et al., *J. Phys. Chem.* **1939,** *43,* 197.
2986. Gross, P. M., and J. H. Saylor, *J. Amer. Chem. Soc.* **1931,** *53,* 1744.
2987. Gross, P. M., J. H. Saylor, et al., *J. Amer. Chem. Soc.* **1933,** *55,* 650.
2988. Grosse, A. V., *J. Amer. Chem. Soc.* **1937,** *59,* 2739.
2989. Grosse, A. V., and C. B. Linn, *J. Amer. Chem. Soc.* **1939,** *61,* 751.
2990. Grosse, A. V., and C. B. Linn, *J. Amer. Chem. Soc.* **1943,** *64,* 2289.
2991. Grosse, A. V., E. J. Rosenbaum, et al., *Ind. Eng. Chem., Anal. Ed.* **1940,** *12,* 191.
2992. Grosse, A. V., R. C. Wackher, et al., *J. Chem. Phys.* **1940,** *44,* 275.
2993. Grossert, J. S., and G. K. Chip, *Tetrahedron Lett.* **1970,** 2611.
2994. Grove, E. L., and G. E. Walden, *J. Chem. Eng. Data* **1965,** *10,* 98.
2995. Groves, C. E., *Justus Liebigs Ann. Chem.* **1874,** *174,* 372.
2996. Groves, C. E., *Justus Liebigs Ann. Chem.* **1874,** *174,* 373.
2997. Groves, L. G., *J. Chem. Soc.* **1938,** 1195.
2998. Groves, L. G., and S. Sugden, *J. Chem. Soc.* **1934,** 1094.
2999. Groves, L. G., and S. Sugden, *J. Chem. Soc.* **1937,** 158.
3000. Groves, L. G., and S. Sugden, *J. Chem. Soc.* **1937,** 1779.
3001. Groves, L. G., and S. Sugden, *J. Chem. Soc.* **1937,** 1782.
3002. Groves, L. G., and S. Sugden, *J. Chem. Soc.* **1937,** 1992.
3003. Grude, K., J. Haupt, et al., *Z. Naturforsch.* **1966,** *21A,* 1231; *C.A.* **1966,** *65,* 17936.
3004. Gruden, J. R., and M. Zief, *Ultrapurity* **1972,** 131; *C.A.* **1974,** *80,* 108078.
3005. Grummitt, O., and A. C. Buck, *J. Amer. Chem. Soc.* **1943,** *65,* 295.
3006. Grunwald, E., and S. Winstein, *J. Amer. Chem. Soc.* **1948,** *70,* 846.
3007. Grünzweig, C., *Justus Liebigs Ann. Chem.* **1872,** *162,* 213.
3008. Grzeskowiak, R., G. H. Jeffery, et al., *J. Chem. Soc.* **1960,** 4719.
3009. Grzeskowiak, R., G. H. Jeffery, et al., *J. Chem. Soc.* **1960,** 4728.
3010. Guchok, V. M., and E. A. Zborovskaya, *Kriobiol. Kriomed.* **1981,** *8,* 46; *C.A.* **1981,** *95,* 197592.
3011. Gudzinowicz, B. J., R. H. Campbell, et al., *J. Chem. Eng. Data* **1963,** *8,* 201.
3012. Guenther, E., *The Essential Oils, Vol. I,* Van Nostrand, New York, 1948.
3013. Guenther, E., *The Essential Oils, Vol. II,* Van Nostrand, New York, 1949.
3014. Guérin, H., and J. Adam-Gironne, *Bull. Soc. Chim. Fr.* **1949,** 607; *C.A.* **1950,** *44,* 1788.
3015. Guerin, J., *Bull. Soc. Pharm. Lille* **1956,** No. 2, 71; *C.A.* **1957,** *51,* 3504.
3016. Guinchant, J., *Ann. Chim.* **1918,** *9,* 49; *Chem. Zentr.* **1918,** *II,* 609.

3017. Guinchant, J., *Ann. Chim.* **1918**, *10*, 30; *Chem. Zentr.* **1919**, *I*, 532.
3018. Guldberg, C. M., *Z. Phys. Chem.* (Leipzig) **1890**, *5*, 374.
3019. Gulf Research and Development Co., Dutch Patent 74 06,008, Nov. 6, 1974; *C.A.* **1975**, *83*, 27846.
3020. Gullikson, C. W., and J. R. Nielson, *J. Mol. Spectrosc.* **1957**, *1*, 158.
3021. Gundry, H. A., D. Harrop, et al., *J. Chem. Thermodyn.* **1969**, *1*, 321.
3022. Gundry, H. A., A. J. Head, et al., *Trans. Faraday Soc.* **1962**, *59*, 1309.
3023. Gunn, S. R., *Anal. Chem.* **1962**, *34*, 1292.
3024. Gunter, C. R., R. D. Madding, et al., *Trans. Ill. State Acad. Sci.* **1971**, *64*, 55.
3025. Gunter, C. R., J. F. Wettaw, et al., *J. Chem. Eng. Data* **1967**, *12*, 472.
3026. Gunthard, M. M., and T. Gaumann, *Helv. Chim. Acta* **1951**, *34*, 39.
3027. Günther, P., H. D. von der Horst, et al., *Z. Elektrochem.* **1928**, *34*, 616.
3028. Gupta, P. K., W. H. Lawrence, et al., *Toxicol. Appl. Pharmacol* **1979**, *49*, 525; *C.A.* **1979**, *91*, 187466.
3029. Gurevich, B. S., and V. M. Bednov, *Zh. Fiz. Khim.* **1972**, *46*, 2673; *Russ. J. Phys. Chem.* **1972**, *46*, 1532.
3030. Guseinov, K. D., and G. A. Aslanov, *Zh. Fiz. Khim.* **1977**, *51*, 2091; *Russ. J. Phys. Chem.* **1977**, *51*, 1221.
3031. Guseinov, K. D., and T. F. Klimova, *Zh. Fiz. Khim.* **1983**, *57*, 1951; *Russ. J. Phys. Chem.* **1983**, *57*, 1183.
3032. Guseinov, K. D., and S. G. Magevramov, *Russ. J. Phys. Chem.* (Eng. trans.) **1974**, *48*, 332.
3033. Gustavsen, J. E., P. Klaeboe, et al., *J. Mol. Struct.* **1978**, *50*, 285; *C.A.* **1979**, *90*, 46197.
3034. Gutbezahl, B., and E. Grunwald, *J. Amer. Chem. Soc.* **1953**, *75*, 559.
3035. Gutbier, H., *Z. Naturforsch.* **1954**, *9A*, 348; *C.A.* **1954**, *48*, 9806.
3036. Guthrie, G. B., and H. M. Huffman, *J. Amer. Chem. Soc.* **1943**, *65*, 1139.
3037. Guthrie, G. B., D. W. Scott, et al., *J. Amer. Chem. Soc.* **1952**, *74*, 4662.
3038. Gutmann, F., and L. M. Simmons, *J. Appl. Phys.* **1952**, *23*, 977.
3039. Gutmann, F., and L. M. Simmons, *J. Chem. Phys.* **1950**, *18*, 696.
3040. Gutmann, V., M. Michlmayr, et al., *Anal. Chem.* **1968**, *40*, 619.
3041. Gutmann, V., and G. Schober, *Z. Anal. Chem.* **1959**, *171*, 339; S. Wawzonek, *Anal. Chem.* **1962**, *34*, 182R.
3042. Gutmann, V., and K. H. Wegleitner, *Z. Phys. Chem.* (Frankfurt) **1972**, *77*, 77.
3043. Gutorova, L. D., *Elektron, Obrab. Mater.* **1974**, 59; *C.A.* **1974**, *91*, 96906.
3044. Gutowsky, H. S., *Pure Appl. Chem.* **1963**, *7*, 93.
3045. Gutowsky, H. S., and C. J. Hoffman, *Phys. Rev.* **1950**, *80*, 110.
3046. Gutowsky, H. S., R. L. Rutledge, et al., *J. Amer. Chem. Soc.* **1954**, *76*, 4242.
3047. Gutt, J., *Ber. Deut. Chem. Gesell.* **1907**, *40*, 2061.
3048. Guttmann, L. F., *J. Amer. Chem Soc.* **1907**, *29*, 345.
3049. Guye, P. A., and E. Mallet, *Arch. Sci. Phys. Nat. Geneve* **1902**, *13*, 274; *Chem. Zentr.* **1902**, *I*, 1314.
3050. Guye, P. A., and E. Mallet, *Compt. Rend.* **1901**, *133*, 1287; *J. Chem. Soc.* **1902Aii**, 195.
3051. Guye, P. A., and E. Mallet, *Compt. Rend.* **1902**, *134*, 168; *J. Chem. Soc.* **1902Aii**, 243.
3052. Guyer, A., M. Schutze, et al., *Helv. Chim. Acta* **1937**, *20*, 936; *C.A.* **1937**, *31*, 8337.
3053. Gwinn, W. D., and K. S. Pitzer, *J. Chem. Phys.* **1948**, *16*, 303.

H

3054. Haarer, E., and R. Plass, German Patent 1,190,950, April 15, 1965; *C.A.* **1965**, *63*, 3929.
3055. Haarer, E., and K. Rühl, German Patent 1,002,305, Feb. 14, 1957; *C.A.* **1960**, *54*, 1305.
3056. Hack, C. W., and M. Van Winkle, *Ind. Eng. Chem.* **1954**, *46*, 2392.
3057. Hackel, W., *Phys. Z.* **1937**, *38*, 195.
3058. Hacking, J. M., and G. Pilcher, *J. Chem. Thermodyn.* **1979**, *11*, 1015.
3059. Hadni, A., *Compt. Rend.* **1954**, *239*, 349.
3060. Haefele, J. W., and R. W. Broge, *Kosmetik-Parfum-Drogen Rundschau* **1961**, *8*, 1; *C.A.* **1962**, *56*, 4591.
3061. Haensel, V., and V. N. Ipatieff, *Ind. Eng. Chem.* **1947**, *39*, 853.
3062. Hafez, M., and S. Hartland, *J. Chem. Eng. Data* **1976**, *21*, 179.
3063. Hagemeyer, H. J., *Ind. Eng. Chem.* **1949**, *41*, 765.

3064. Hagemeyer, H. J., and G. C. DeCroes, "The Chemistry of Isobutyraldehyde and Its Derivatives," Eastman, Kingsport, Tennessee, 1953.

3065. Haggenmacher, J. E., *Ind. Eng. Chem.* **1948,** *40,* 436.

3066. Haggenmacher, J. E., *J. Amer. Chem. Soc.* **1946,** *68,* 1633.

3067. Haggerty, C. J., and J. F. Weiler, *J. Amer. Chem. Soc.* **1929,** *51,* 1623.

3068. Hagglund, E., B. Lindberg, et al., *Acta Chem. Scand.* **1956,** *10,* 1160.

3069. Hahn, F. L., and R. Klockmann, *Z. Phys. Chem.* (Leipzig) **1930,** *146,* 373.

3070. Haider, S. Z., M. H. Khundkar, et al., *J. Appl. Chem.* (London) **1954,** *4,* 93.

3071. Haigh, P. J., P. C. Canapa, et al., *J. Chem. Phys.* **1968,** *48,* 4234.

3072. Haines, W. E., R. V. Helm, et al., *J. Phys. Chem.* **1954,** *58,* 270.

3073. Haines, W. E., R. V. Helm, et al., *J. Phys. Chem.* **1956,** *60,* 549.

3074. Hajiev, S. N., *Ind. Eng. Chem.*, *Process Des. Develp.* **1970,** *9,* 229.

3075. Hajiev, S. N., and M. J. Agarunov, *J. Organometal. Chem.* **1970,** *22,* 305.

3076. Hakala, R. W., *Chem. Eng. News*, March 16, 1959, 43.

3077. Hale, W. J., and G. H. Cheney, U.S. Patent 1,729,775, Oct. 1, 1929.

3078. Hales, J., J. Cox, et al., *Trans. Faraday Soc.* **1963,** *59,* 1544.

3079. Hales, J. L., *J. Phys. E.* **1970,** *3,* 855.

3080. Hales, J. L., and J. H. Ellender, *J. Chem. Thermodyn.* **1976,** *8,* 1187.

3081. Hales, J. L., H. A. Gundry, et al., *J. Chem. Thermodyn.* **1983,** *15,* 211.

3082. Hales, J. L., J. J. Jones, et al., *J. Chem. Soc.* **1957,** 618.

3083. Hales, J. L., E. B. Lees, et al., *Trans. Faraday Soc.* **1967,** *63,* 1876.

3084. Hales, J. L., and R. Townsend, *J. Chem. Thermodyn.* **1972,** *4,* 763.

3085. Hales, J. L., and R. Townsend, *J. Chem. Thermodyn.* **1974,** *6,* 111.

3086. Halford, J. O., L. C. Anderson, et al., *J. Chem. Phys.* **1937,** *5,* 927.

3087. Halford, J. O., and E. B. Reid, *J. Amer. Chem. Soc.* **1941,** *63,* 1873.

3088. Halford, R. S., and O. A. Schaeffer, *J. Chem. Phys.* **1946,** *14,* 141.

3089. Hall, E. E, and A. R. Payne, *Phys. Rev.* **1922,** *20,* 249; *Chem. Zentr.* **1923,** *III,* 1380.

3090. Hall, H. K., *J. Amer. Chem. Soc.* **1957,** *79,* 5441.

3091. Hall, H. K., *J. Amer. Chem. Soc.* **1957,** *79,* 5444.

3092. Hall, H. K., *J. Phys. Chem.* **1956,** *60,* 63.

3093. Hall, H. K., and R. Zbinden, *J. Amer. Chem. Soc.* **1958,** *80,* 6428.

3094. Hall, N. F., *J. Amer. Chem. Soc.* **1930,** *52,* 5115.

3095. Hall, N. F., and M. R. Sprinkle, *J. Amer. Chem. Soc.* **1932,** *54,* 3469.

3096. Hall, N. F., and H. H. Voge, *J. Amer. Chem. Soc.* **1933,** *55,* 239.

3097. Hall, W. P., and E. E. Reid, *J. Amer. Chem. Soc.* **1943,** *65,* 1466.

3098. Haller, A., *Compt. Rend.* **1892,** *114,* 1326.

3099. Haller, W., and H. C. Duecker, *J. Res. Nat. Bur. Stand.* **1960,** *64A,* 527.

3100. Hallgren, B., R. Ryhage, et al., *Acta Chem. Scand.* **1959,** *13,* 845; *C.A.* **1961,** *55,* 12024.

3101. Halocarbon Products Corp., "Trifluoroacetic Acid," Hackensack, New Jersey, 1963.

3102. Halpern, B., and J. W. Westley, *Aust. J. Chem.* **1966,** *19,* 1533; *C.A.* **1966,** *65,* 19996.

3103. Hals, L. J., and H. G. Bryce, *Anal. Chem.* **1951,** *23,* 1694.

3104. Halverson, F., R. F. Stamm, et al., *J. Chem. Phys.* **1948,** *16,* 808.

3105. Halverstadt, I. F., and W. D. Kulmer, *J. Amer. Chem. Soc.* **1942,** *64,* 1982.

3106. Halverstadt, I. F., and W. D. Kulmer, *J. Amer. Chem. Soc.* **1942,** *64,* 2988.

3107. Hamada, K., and H. Morishita, *Spect. Lett.* **1980,** *13,* 15.

3108. Hamai, S., *Bull. Chem. Soc. Jpn.* **1934,** *9,* 542; *C.A.* **1935,** *29,* 1771.

3109. Hamamoto, K., and H. Tone, *Ann. Repts. Shionogi Res. Lab.* **1952,** *1,* 107; *C.A.* **1957,** *51,* 8004.

3110. Hamann, S. D., and W. Strauss, *Trans. Faraday Soc.* **1955,** *51,* 1684.

3111. Hamano, H., *Bull. Chem. Soc. Jpn.* **1958,** *31,* 832; *C.A.* **1959,** *53,* 15674.

3112. Hambly, A. N., and B. V. O'Grady, *Aust. J. Chem.* **1964,** *17,* 860.

3113. Hamelin, R., *Bull. Soc. Chem. Fr.* **1961,** 926; *C.A.* **1961,** *55,* 27027.

3114. Hamer, J., L. Placek, et al., *Nitro Compounds, Proc. Intern. Symp., Warsaw* **1963,** 395 (Pub. 1964); *C.A.* **1966,** *64,* 162.

3115. Hammaker, R. M., *Can. J. Chem.* **1965,** *43,* 2916.

3116. Hammett, L. P., and N. Dietz, *J. Amer. Chem. Soc.* **1930,** *52,* 4795.

3117. Hammick, D. L., and J. Howard, *J. Chem. Soc.* **1932,** 2915.
3118. Hammick, D. L., A. Norris, et al., *J. Chem. Soc.* **1938,** 1755.
3119. Hammick, D. L., and M. Roberts, *J. Chem. Soc.* **1948,** 73.
3120. Hammond, B. R., and R. H. Stokes, *Trans. Faraday Soc.* **1955,** *51,* 1641.
3121. Hammond, J. A. S., U.S. Patent 2,356,758, Aug. 15, 1944; *C.A.* **1945,** *39,* 90.
3122. Hammond, P. R., *J. Chem. Soc.* **1962,** 1370.
3123. Hamonet, J. L., *Compt. Rend.* **1901,** *132,* 632.
3124. Hamonet, J. L., *Compt. Rend.* **1904,** *139,* 59.
3125. Hampton, J., and J. A. Riddick, unpublished data from the Research and Development Division, Commercial Solvents Corp.
3126. Hampton, R. R., *Anal. Chem.* **1949,** *21,* 923.
3127. Hamstead, A. C., *Ind. Eng. Chem.* **1964,** *56,* 37.
3128. Hamstead, A. C., and L. S. Van Delinder, *J. Chem. Eng. Data* **1960,** *5,* 383.
3129. Hanai, S., *J. Chem. Soc. Jpn.* **1941,** *62,* 1208; *C.A.* **1947,** *41,* 3436.
3130. Hancock, C. K., G. M. Watson, et al., *J. Phys. Chem.* **1954,** *58,* 127.
3131. Handa, Y. P., C. J. Halpin, et al., *J. Chem. Thermodyn.* **1981,** *13,* 875.
3132. Handley, R., *Ind. Chem. Mfg.* **1955,** *31,* 535.
3133. Hands, C. H. G., and W. S. Norman, *Ind. Chemist* **1945,** *21,* 307; *C.A.* **1945,** *39,* 4273.
3134. Hanford, W. E., and D. L. Fuller, *Ind. Eng. Chem.* **1948,** *40,* 1171.
3135. Hannan, M. C., and M. C. Markham, *J. Amer. Chem. Soc.* **1947,** *71,* 1120.
3136. Hannay, N. B., and C. P. Smyth, *J. Amer. Chem. Soc.* **1946,** *68,* 1005.
3137. Hannay, N. B., and C. P. Smyth, *J. Amer. Chem. Soc.* **1946,** *68,* 1357.
3138. Hannotte, T., *Bull. Soc. Chim. Belges* **1926,** *35,* 86; *Chem. Zentr.* **1926,** *II,* 741; *C.A.* **1926,** *20,* 2657.
3139. Hannson, J., *Svensk. Kem. Tid.* **1955,** *67,* 256.
3140. Hanson, C., and M. W. T. Pratt, *Trans. Inst. Chem. Eng.* **1971,** *49,* 95; *C.A.* **1971,** *75,* 50919.
3141. Hanson, J., *Svensk. Kem. Tid.* **1955,** *67,* 246.
3142. Hanson, M. W., and J. B. Bouck, *J. Amer. Chem. Soc.* **1957,** *79,* 5631.
3143. Hantzsch, A., *Ber. Deut. Chem. Gesell.* **1906,** *39,* 1084.
3144. Hantzsch, A., *Ber. Deut. Chem. Gesell.* **1912,** *45,* 553; *C.A.* **1912,** *6,* 1435.
3145. Hantzsch, A., and H. Freese, *Ber. Deut. Chem. Gesell.* **1894,** *27,* 2529, 2966.
3146. Hantzsch, A., and O. K. Hofmann, *Ber. Deut. Chem. Gesell.* **1911,** *44,* 1776.
3147. Hantzsch, A., and E. Scharf, *Ber. Deut. Chem. Gesell.* **1913,** *46,* 3570.
3148. Happ, G. P., and D. W. Stewart, *J. Amer. Chem. Soc.* **1952,** *74,* 4404.
3149. Harada, I., H. Takeuchi, et al., *Bull. Chem. Soc. Jpn.* **1977,** *50,* 102.
3150. Harada, Y., J. N. Murrell, et al., *Chem. Phys. Lett.* **1968,** *1,* 595; *C.A.* **1968,** *69,* 14328.
3151. Hardies, D. E., Belgian Patent 620,486, Aug. 14, 1962.
3152. Hardies, D. E., and B. O. Pray, U.S. Patent 3,070,634, Dec. 25, 1962; *C.A.* **1963,** *58,* 10075.
3153. Hardy, C. J., D. Fairhurst, et al., *Trans. Faraday Soc.* **1964,** *60,* 1626.
3154. Hardy, D. V. N., *J. Soc. Chem. Ind.* (London) **1948,** *67,* 81.
3155. Haresnap, D., F. A. Fidler, et al., *Ind. Eng. Chem.* **1949,** *41,* 2691.
3156. Hargreaves, M. K., *J. Chem. Soc.* **1956,** 3679.
3157. Hariharan, T. A., *Indian J. Phys.* **1953,** *27,* 323; *C.A.* **1957,** *51,* 7862.
3158. Hariharan, T. A., *J. Indian Inst. Sci.* **1954,** *36,* 189; *C.A.* **1955,** *49,* 724.
3159. Hariharan, T. A., *J. Indian Inst. Sci.* **1954,** *36A,* 215; *C.A.* **1955,** *49,* 2871.
3160. Harkins, W. D., *Z. Phys. Chem.* (Leipzig) **1928,** *139A,* 647.
3161. Harkins, W. D., F. E. Brown, et al., *J. Amer. Chem. Soc.* **1917,** *39,* 354.
3162. Harkins, W. D., and Y. C. Cheng, *J. Amer. Chem. Soc.* **1921,** *43,* 35.
3163. Harkins, W. D., G. L. Clark, et al., *J. Amer. Chem. Soc.* **1920,** *42,* 700.
3164. Harkins, W. D., and R. R. Haun, *J. Amer. Chem. Soc.* **1932,** *54,* 3920.
3165. Harkins, W. D., and R. W. Wampler, *J. Amer. Chem. Soc.* **1931,** *53,* 850.
3166. Harley-Mason, J., et al., *J. Chem. Soc. B* **1966,** 396.
3167. Harlow, G. A., G. M. Noble, et al., *Anal. Chem.* **1956,** *28,* 787.
3168. Harmon, J., U.S. Patent 2,404,374, July 23, 1946; *C.A.* **1946,** *40,* 7234.
3169. Harned, H. S., and R. W. Ehlers, *J. Amer. Chem. Soc.* **1932,** *54,* 1350.

3170. Harned, H. S., and R. W. Ehlers, *J. Amer. Chem. Soc.* **1933,** *55,* 652.

3171. Harned, H. S., and R. W. Ehlers, *J. Amer. Chem. Soc.* **1933,** *55,* 2379.

3172. Harned, H. S., and N. D. Embree, *J. Amer. Chem. Soc.* **1934,** *56,* 1042.

3173. Harned, H. S., and J. O. Morrison, *J. Amer. Chem. Soc.* **1936,** *58,* 1908.

3174. Harned, H. S., and B. B. Owen, *J. Amer. Chem. Soc.* **1930,** *52,* 5079.

3175. Harned, H. S., and R. A. Robinson, *Trans. Faraday Soc.* **1940,** *36,* 973; from Covington, A. K., R. A. Robinson, et al., *J. Phys. Chem.* **1966,** *70,* 3820.

3176. Harrand, M., and H. Martin, *Bull. Soc. Chim. Fr.* **1956,** 1383; *C.A.* **1957,** *51,* 4143.

3177. Harris, B., R. J. W. Le Févre, et al., *J. Chem. Soc.* **1953,** 1622.

3178. Harris, C., *Ber. Deut. Chem. Gesell.* **1899,** *32,* 1326; *Justus Liebigs Ann. Chem.*, **1904,** *330,* 185.

3179. Harris, F. E., and C. T. O'Konski, *J. Amer. Chem. Soc.* **1954,** *76,* 4317.

3180. Harris, H., *J. Chem. Soc.* **1925,** 1049.

3181. Harris, J. E. G., and W. J. Pope, *J. Chem. Soc.* **1922,** 1029.

3182. Harris, J. F., and L. L. Zoch, *Anal. Chem.* **1962,** *34,* 201.

3183. Harris, K. R., and P. J. Dunlop, *J. Chem. Thermodyn.* **1970,** *2,* 813.

3184. Harris, R. K., and N. Sheppard, *Trans. Faraday Soc.* **1963,** *59,* 606.

3185. Harris, R. K., and R. A. Spragg, *Chem. Commun.* **1966,** 314.

3186. Harris, W. C., D. A. Coe, et al., *Spectrochim. Acta, Part A* **1976,** *32A,* 1.

3187. Harris, W. S., Ph.D. Thesis, Univ. Calif., Berkeley, 1958.

3188. Harrison, A. G., *Org. Mass. Spectrom.* **1970,** *3,* 549.

3189. Harrison, A. J., B. J. Cederholm, et al., *J. Chem. Phys.* **1959,** *30,* 355.

3190. Harrison, D., and E. A. Moelwyn-Hughes, *Proc. Roy. Soc.* **1957,** *239A,* 230; from [7483].

3191. Harrison, G. R., R. C. Lord, et al., *Practical Spectroscopy*, Prentice-Hall, New York, 1948.

3192. Harrison, H. R., and E. J. Eisenbraun, *J. Org. Chem.* **1966,** *31,* 1294.

3193. Harrison, R. H., and K. A. Kobe, *J. Chem. Phys.* **1957,** *26,* 1411.

3194. Harriss, M. G., and J. B. Milne, *Can. J. Chem.* **1976,** *54,* 3031.

3195. Harrop, D., A. J. Head, et al., *J. Chem. Thermodyn.* **1970,** *2,* 203.

3196. Harteck, P., and R. Edse, *Z. Phys. Chem.* (Leipzig) **1938,** *182A,* 220.

3197. Hartley, H., and H. R. Raikes, *J. Chem. Soc.* **1925,** 524.

3198. Hartley, W. N., and J. J. Dobbie, *J. Chem. Soc.* **1900,** 846.

3199. Hartmann, H., A. Neumann, et al., *Z. Phys. Chem.* (Leipzig) **1965,** *44,* 204.

3200. Hartsuch, P. J., *J. Amer. Chem. Soc.* **1939,** *61,* 1142.

3201. Hartwell, E. J., R. E. Richards, et al., *J. Chem. Soc.* **1948,** 1436.

3202. Hasebe, N., *J. Chem. Soc. Jpn.* **1943,** *64,* 1041; *C.A.* **1947,** *41,* 3586.

3203. Hasegawa, K., M. Murata, et al., *Yukagaku* **1972,** *21,* 383; *C.A.* **1972,** *77,* 96632.

3204. Hashimoto, T., H. Shina, et al., *Kogyo Kagaku Zasshi* **1965,** *68,* 1434; *C.A.* **1965,** *63,* 17826.

3205. Hass, H. B., E. B. Hodge, et al., *Ind. Eng. Chem.* **1936,** *28,* 339.

3206. Hass, H. B., and J. A. Patterson, *Ind. Eng. Chem.* **1938,** *30,* 67.

3207. Hass, H. B., and J. A. Patterson, *Ind. Eng. Chem.* **1941,** *33,* 615.

3208. Hassel, O., and E. Naeshagen, *Tids. Kjemoig Bergvesen* **1930,** *10,* 81; *C.A.* **1931,** *25,* 2698.

3209. Hassel, O., and E. Naeshagen, *Z. Phys. Chem.* (Leipzig) **1931,** *12B,* 79.

3210. Hassion, F. X., and R. H. Cole, *J. Chem. Phys.* **1955,** *23,* 1756.

3211. Hastings, S. H., and D. E. Nicholson, *J. Phys. Chem.* **1957,** *61,* 730.

3212. Haszeldine, R. N., *J. Chem. Soc.* **1951,** 584.

3213. Haszeldine, R. N., *J. Chem. Soc.* **1953,** 1757.

3214. Haszeldine, R. N., *J. Chem. Soc.* **1953,** 1764.

3215. Haszeldine, R. N., *J. Chem. Soc.* **1953,** 2525.

3216. Haszeldine, R. N., *J. Chem. Soc.* **1953,** 2622.

3217. Haszeldine, R. N., U.S. Patent 2,716,668, Aug. 30, 1955; *C.A.* **1956,** *50,* 7845.

3218. Hata, K., *Bull. Inst. Phys. Chem. Res.* (Tokyo), *Chem. Ed.* **1944,** *23,* 296; *C.A.* **1948,** *42,* 6783.

3219. Hatada, K., M. Takeshita, et al., *Tetrahedron Lett.* **1968,** *44,* 4621.

3220. Hatch, G. B., and H. Adkins, *J. Amer. Chem. Soc.* **1937,** *59,* 1694.

3221. Hatch, L. F., and T. W. Evans, U.S. Patent 2,434,110, Jan. 6, 1948; *C.A.* **1948,** *42,* 2267.

3222. Hatch, L. F., and K. E. Harwell, *J. Amer. Chem. Soc.* **1953,** *75,* 6002.

3223. Hatch, L. F., and A. C. Moore, *J. Amer. Chem. Soc.* **1944,** *66,* 285.

3224. Hatch, L. F., and S. S. Nesbitt, *J. Amer. Chem. Soc.* **1950,** *72,* 727.

3225. Hatcher, J. B., and D. M. Yost, *J. Chem. Phys.* **1937,** *5,* 992.

3226. Hatschek, E., *The Viscosity of Liquids,* Van Nostrand, New York, 1928.

3227. Hatton, J. V., and R. E. Richards, *Trans. Faraday Soc.* **1960,** *56,* 315.

3228. Hatton, W. E., D. L. Hildenbrand, et al., *J. Chem. Eng. Data* **1962,** *7,* 229.

3229. Hauptman, E., *Acta Phys. Polon.* **1938,** *7,* 86; *C.A.* **1939,** *33,* 1594.

3230. Hauptschein, M., E. A. Nodiff, et al., *J. Amer. Chem. Soc.* **1952,** *74,* 1347.

3231. Hauser, C. R., W. J. Humphlett, et al., *J. Amer. Chem. Soc.* **1948,** *70,* 426.

3232. Hauser, C. R., and W. B. Renfrow, *J. Amer. Chem. Soc.* **1937,** *59,* 1823.

3233. Hausser, K. W., R. Kuhn, et al., *Z. Phys. Chem.* (Leipzig) **1935,** *29B,* 371; Lewis, G. N., and M. Calvin, *Chem. Rev.* **1939,** *25,* 273; *Chem. Zentr.* **1936,** *1,* 2326.

3234. Hausser, K. W., R. Kuhn, et al., *Z. Phys. Chem.* (Leipzig) **1935,** *29B,* 378; *Chem. Zentr.* **1936,** *1,* 2327.

3235. Hawkins, J. E., and G. T. Armstrong, *J. Amer. Chem. Soc.* **1954,** *76,* 3756.

3236. Hawkins, J. E., and W. T. Eriksen, *J. Amer. Chem. Soc.* **1954,** *76,* 2669.

3237. Haworth, W. N., *J. Chem. Soc.* **1913,** 1242.

3238. Hayamizu, K., and O. Yamamoto, *J. Mol. Spectrosc.* **1968,** *28,* 89.

3239. Hayashi, M., *Nippon Kagaku Zasshi* **1957,** *78,* 222; *C.A.* **1957,** *51,* 6339.

3240. Hayashi, M., K. Ohno, et al., *Bull. Chem. Soc. Jpn.* **1973,** *41,* 2332.

3241. Hayashi, S., and K. Mizoguchi, *Nippon Kagaku Zasshi* **1959,** *80,* 308; *C.A.* **1961,** *55,* 1295.

3242. Hayashida, H., K. Teruya, et al., *Kagaku Kogaku* **1972,** *36,* 221; *C.A.* **1972,** *76,* 104496.

3243. Hayden, A. L., O. R. Sammul, et al., *J. Assoc. Offic. Agr. Chemist* **1962,** *45,* 797.

3244. Hayduk, W., and H. Laudie, *J. Chem. Eng. Data* **1974,** *19,* 253.

3245. Hayman, H. J. G., and I. Eliezer, *J. Chem. Phys.* **1961,** *35,* 644.

3246. Haynes, W. M., *J. Chem. Thermodyn.* **1983,** *15,* 801.

3247. Haynes, W. M., and M. J. Itiza, *J. Chem. Thermodyn.* **1977,** *9,* 179.

3248. Hazlet, S. E., and R. B. Callison, *J. Amer. Chem. Soc.* **1944,** *66,* 1248.

3249. Heap, J. G., W. J. Jones, et al., *J. Amer. Chem. Soc.* **1921,** *43,* 1936.

3250. Heard, J. R., U.S. Patent 2,417,685, Mar. 18, 1947; *C.A.* **1947,** *41,* 3481.

3251. Hearne, G., M. Tamele, et al., *Ind. Eng. Chem.* **1941,** *33,* 805.

3252. Heath, C. E., and R. M. Skomoroski (to Esso Research), U.S. Patent 3,053,898, Sept. 11, 1962; *C.A.* **1963,** *58,* 1352.

3253. Hebeisen, J., *Ann. Phys.* **1925,** *77,* No. 4, 216; *J. Chem Soc.* **1925Aii,** 763.

3254. Hecht, L. T., and D. L. Wood, *Proc. Roy. Soc.* **1956,** *235A,* 174.

3255. Hedlund, A. I., Swedish Patent 119,077, June 25, 1947; *C.A.* **1948,** *42,* 1752.

3256. Heel, H., and W. Zeil, *Z. Elektrochem.* **1960,** *64,* 962; *C.A.* **1961,** *55,* 115.

3257. Heerma, W., J. J. de Ridder, et al., *Org. Mass Spectrom.* **1969,** *2,* 1103.

3258. Heerma, W., and J. J. de Ridder, *Org. Mass Spectrom.* **1970,** *3,* 1439.

3259. Hehlgans, F., *Phys. Z.* **1929,** *30,* 942.

3260. Heigl, J. J., J. F. Black, et al., *Anal. Chem.* **1949,** *21,* 556.

3261. Heil, L. M., *Phys. Rev.* **1932,** *39,* 666.

3262. Heilbron, I., ed., *Dictionary of Organic Compounds,* 4th ed., Oxford Univ. Press, New York, 1965.

3263. Heilbron, I., ed., *Dictionary of Organic Compounds,* 4th rev., Eyre and Spottiswoode, London, 1965 ff. and 5th and following supplements.

3264. Heilbron, I. M., R. A. Morton, et al., *Biochem. J.* **1932,** *26,* 1194.

3265. Heilmann, R., J. M. Bonnier, et al., *Compt. Rend.* **1957,** *244,* 1787; *C.A.* **1957,** *51,* 14538.

3266. Heim, G., *Bull. Soc. Chim. Belges* **1933,** *42,* 461.

3267. Heinemann, H., U.S. Patent 2,451,949, Oct. 19, 1948; *C.A.* **1949,** *43,* 2222.

3268. Heinert, D., and A. E. Martell, *J. Amer. Chem. Soc.* **1963,** *85,* 183.

3269. Heinrich, J., and J. Surovy, *Sb. Pr. Chem. Fak. Svst.* **1966,** 207.

3270. Heinze, A. J., *Hydrocarbon Process. Petrol. Refiner.* **1962,** *41,* No. 3, 187; *C.A.* **1963,** *58,* 1281.

3271. Heise, H. M., F. Winther, et al., *J. Mol. Spectrosc.* **1981,** *90,* 531.

3272. Held, R. P., and C. M. Criss, *J. Phys. Chem.* **1965,** *69,* 2611.

3273. Helferich, B., and J. Hausen, *Ber. Deut. Chem. Gesell.* **1924,** *57,* 795.

3274. Heller, H. E., *J. Amer. Chem. Soc.* **1952,** *74,* 4858.

3275. Helm, R. V., W. J. Lanum, et al., *J. Phys. Chem.* **1958,** *62,* 858.

3276. Helmkamp, G. K., C. D. Joel, et al., *J. Org. Chem.* **1956,** *21,* 844.

3277. Hemptinne, M. de, *Contrib. Étude Structure Mol, Vol. Commem. Victor Henri* 1947/48, 151; *C.A.* **1949,** *43,* 2865.

3278. Hemptinne, M. de, *Trans. Faraday Soc.* **1946,** *42,* 5.

3279. Henchman, M. J., H. T. Otwinowska, et al., *Advan. Mass Spectrom.* **1966,** *3,* 359.

3280. Hende, A. van den, and R. Fonteyne, *Natuurw. Tijdschr.* (Ghent) **1943,** *25,* 24; *Chem. Zentr.* **1943,** *I,* 1463.

3281. Hendra, P. J., and D. B. Powell, *J. Chem. Soc.* **1960,** 5105.

3282. Hendricks, S. B., and M. E. Jefferson, *J. Opt. Soc. Amer.* **1933,** *23,* 299; *C.A.* **1933,** *27,* 5594.

3283. Hennaut-Roland, M., and M. Lek, *Bull. Soc. Chim. Belges* **1931,** *40,* 177; *C.A.* **1931,** *25,* 5322.

3284. Henne, A. L., *J. Amer. Chem. Soc.* **1937,** *59,* 1200.

3285. Henne, A. L., U.S. Patent 1,900,692, Feb. 12, 1935; *C.A.* **1935,** *29,* 2174.

3286. Henne, A. L., and T. Alderson, *J. Amer. Chem. Soc.* **1945,** *67,* 918.

3287. Henne, A. L., and V. J. Fox, *J. Amer. Chem. Soc.* **1951,** *73,* 2323.

3288. Henne, A. L., and K. W. Greenlee, *J. Amer. Chem. Soc.* **1945,** *67,* 484.

3289. Henne, A. L., and D. M. Hubbard, *J. Amer. Chem. Soc.* **1936,** *58,* 404.

3290. Henne, A. L., and M. S. Newman, *J. Amer. Chem. Soc.* **1938,** *60,* 1697.

3291. Henne, A. L., and M. W. Renoll, *J. Amer. Chem. Soc.* **1936,** *58,* 889.

3292. Henne, A. L., and R. P. Ruh, *J. Amer. Chem. Soc.* **1948,** *70,* 1025.

3293. Henne, A. L., and P. Trott, *J. Amer. Chem. Soc.* **1947,** *69,* 1820.

3294. Henne, A. L., and E. G. Wiest, *J. Amer. Chem. Soc.* **1940,** *62,* 2051.

3295. Hennelly, E. J., W. M. Heston, et al., *J. Amer. Chem. Soc.* **1948,** *70,* 4102.

3296. Hennion, G. F., and L. A. Auspos, *J. Amer. Chem. Soc.* **1943,** *65,* 1603.

3297. Hennion, G. F., and W. S. Murray, *J. Amer. Chem. Soc.* **1942,** *64,* 1220.

3298. Henri, M. W., *Compt. Rend.* **1924,** *178,* 846.

3299. Henri, V., and P. Angenot, *J. Chim. Phys.* **1936,** *33,* 641; *C.A.* **1937,** *31,* 3786.

3300. Henrici, A., *Z. Phys.* **1932,** *77,* 35; *C.A.* **1932,** *26,* 5847.

3301. Henrion, J., *Bull. Soc. Roy. Sci. Liege* **1941,** *10,* 414; *C.A.* **1943,** *37,* 2235.

3302. Henry, L., *Ber. Deut. Chem. Gesell.* **1872,** *5,* 569.

3303. Henry, L., *Ber. Deut. Chem. Gesell.* **1873,** *6,* 728.

3304. Henry, L., *J. Prakt. Chemie* **1885,** *32,* No. 2, 431.

3305. Henry, L., *Rec. Trav. Chim.* **1906,** *26,* 106; *C.A.* **1907,** *1,* 2077.

3306. Henry, P., *Z. Phys. Chem.* (Leipzig) **1892,** *10,* 96.

3307. Hentz, R. R., *J. Phys. Chem.* **1962,** *66,* 1622.

3308. Hepburn, D. R., and H. R. Hudson, *J. Chem. Soc., Perkin Trans. 1* **1976,** 754.

3309. Heppel, L. A., P. A. Neal, et al., *J. Ind. Hyg. Toxicol.* **1946,** *28,* 113; *C.A.* **1946,** *40,* 6170.

3310. Hepworth, H., *J. Chem. Soc.* **1921,** 1249.

3311. Hepworth, J. D., J. A. Hudson, et al., *J. Chem. Soc., Perkin Trans. 2* **1972,** 1905.

3312. Herail, F., *Compt. Rend.*, Ser. **1966,** *262C,* 22; *C.A.* **1966,** *64,* 18706.

3313. Herald, A. E., and H. A. Price, *Oil Gas J.* **1947,** *46,* 94; *C.A.* **1947,** *41,* 5818.

3314. Herb, S. F., *J. Amer. Oil Chem. Soc.* **1955,** *32,* 153.

3315. Hervison-Evans, D., and R. E. Richards, *Trans. Faraday Soc.* **1962,** *58,* 845.

3316. Hercules Inc., Pine and Paper Chemicals Dept., "Terpenes and Related Pine Chemicals," Wilmington, Delaware, 1967.

3317. Heric, E. L., and J. G. Brewer, *J. Chem. Eng. Data* **1967,** *12,* 574.

3318. Heric, E. L., and B. M. Coursey, *J. Chem. Eng. Data* **1971,** *16,* 185.

3319. Herington, E. F. G., *Discussions Faraday Soc.* **1950,** *9,* 26.

3320. Herington, E. F. G., *J. Amer. Chem. Soc.* **1951,** *73,* 5883.

3321. Herington, E. F. G., *Zone Melting of Organic Compounds*, Wiley, New York, 1963.

3322. Herington, E. F. G., A. B. Densham, et al., *J. Chem. Soc.* **1954,** 2643.

3323. Herington, E. F. G., R. Handley, et al., *Chem. Ind.* (London) **1956,** 292.

3324. Herington, E. F. G., and W. Kynaston, *Trans. Faraday Soc.* **1957,** *53,* 138.

3325. Herington, E. F. G., and J. F. Martin, *Trans. Faraday Soc.* **1953,** *49,* 154; *C.A.* **1953,** *47,* 8442.

3326. Herlan, A., *Brennst-Chem.* **1965,** *46,* 264; *C.A.* **1966,** *64,* 10541.

3327. Herman, M., *Ind. Chim. Belges* **1951,** *16,* 86; *C.A.* **1951,** *45,* 6488.

3328. Herman, R. C., *J. Chem. Phys.* **1940**, *8*, 252.
3329. Hernandez, G. J., *J. Chem. Phys.* **1963**, *38*, 2233.
3330. Hernandez, G. J., *J. Chem. Phys.* **1963**, *39*, 1355.
3331. Hernandez, G. J., and A. B. F. Duncan, *J. Chem. Phys.* **1962**, *36*, 1504.
3332. Herold, W., *Z. Phys. Chem.* (Leipzig) **1932**, *18B*, 265; *Chem. Zentr.* **1932**, *II*, 2807; *C.A.* **1932**, *26*, 5257.
3333. Herold, W., and L. L. Wolf, *Z. Phys. Chem.* (Leipzig) **1931**, *12B*, 194; *C.A.* **1931**, *25*, 2921.
3334. Hershberg, E. B., *J. Org. Chem.* **1948**, *13*, 542.
3335. Herz, E., *Monatsh. Chem.* **1943**, *74*, 160; *C.A.* **1944**, *38*, 1428.
3336. Herz, E., *Monatsh. Chem.* **1946**, *76*, 1; *C.A.* **1947**, *41*, 1558.
3337. Herz, E., L. Kahovec, et al., *Monatsh. Chem.* **1946**, *76*, 100; *C.A.* **1947**, *41*, 3370.
3338. Herz, E., L. Kahovec, et al., *Z. Phys. Chem.* (Leipzig) **1943**, *538*, 124.
3339. Herz, E., and K. W. F. Kohlrausch, *Monatsh. Chem.* **1943**, *74*, 175; *C.A.* **1944**, *38*, 1428.
3340. Herz, E., K. W. F. Kohlrausch, et al., *Monatsh. Chem.* **1946**, *76*, 112; *C.A.* **1947**, *41*, 3370.
3341. Herz, E., K. W. F. Kohlrausch, et al., *Monatsh. Chem.* **1947**, *76*, 200; *C.A.* **1947**, *41*, 6153.
3342. Herz, W., *Z. Elektrochem.* **1917**, *23*, 24; *C.A.* **1917**, *11*, 2631.
3343. Herz, W., and W. Bloch, *Z. Phys. Chem.* (Leipzig) **1924**, *110*, 23.
3344. Herz, W., and E. Lorentz, *Z. Phys. Chem.* (Leipzig) **1929**, *140*, 406.
3345. Herz, W., and E. Neukirch, *Z. Phys. Chem.* (Leipzig) **1923**, *104*, 433.
3346. Herz, W., and W. Rathmann, *Chem.-Zeit.* **1912**, *36*, 1417; *J. Chem. Soc.* **1913Aii**, 26.
3347. Herz, W., and W. Rathmann, *Chem.-Zeit.* **1913**, *37*, 621; *C.A.* **1913**, *7*, 3866.
3348. Herz, W., and P. Schuftan, *Z. Phys. Chem.* (Leipzig) **1922**, *101*, 269.
3349. Herzberg, G., *Infrared and Raman Spectra of Polyatomic Molecules*, Van Nostrand, New York, 1945.
3350. Herzberg, G., and R. Kölsch, *Z. Elektrochem.* **1933**, *39*, 572; *C.A.* **1933**, *27*, 4734.
3351. Herzberg, G., and G. Scheibe, *Z. Phys. Chem.* (Leipzig) **1930**, *Abt. B*, *7*, 390; *C.A.* **1930**, *24*, 3710.
3352. Herzog, R., *Ind. Eng. Chem.* **1944**, *36*, 997.
3353. Hess, K., *Ber. Deut. Chem. Gesell.* **1930**, *63*, 518.
3354. Hess, K., and H. Frahm, *Ber. Deut. Chem. Gesell.* **1938**, *71B*, 2627.
3355. Hess, K., and H. Haber, *Ber. Deut. Chem. Gesell.* **1937**, *70*, 2205.
3356. Hess, W. W., R. L. Mather, et al., *J. Chem. Eng. Data* **1962**, *7*, 317.
3357. Hesse, G., B. P. Engelbrecht, et al., *Fresenius' Z. Anal. Chem.* **1968**, *241*, 91.
3358. Hesse, G., and H. Schildknecht, *Angew. Chem.* **1955**, *67*, 737; *C.A.* **1956**, *50*, 3675.
3359. Hesse, G., and R. Schrödel, *Angew. Chem.* **1956**, *68*, 438; *C.A.* **1957**, *51*, 11341.
3360. Heston, W. M., E. J. Hennelly, et al., *J. Amer. Chem. Soc.* **1948**, *70*, 4093.
3361. Heston, W. M., E. J. Hennelly, et al., *J. Amer. Chem. Soc.* **1950**, *72*, 2071.
3362. Hetzer, H. B., R. G. Bates, et al., *J. Phys. Chem.* **1966**, *70*, 2869.
3363. Hetzer, H. B., R. A. Robinson, et al., *J. Phys. Chem.* **1968**, *72*, 2081.
3364. Heumann, W. R., O. Dionne, et al., *Can. J. Chem.* **1965**, *43*, 1096.
3365. Heuser, R. V., U.S. Patent 2,409,124, Oct. 8, 1946; *C.A.* **1947**, *41*, 1235.
3366. Hedayatullah, M., J. C. Leveque, et al., *C. R. Acad. Sci.*, *Ser. C* **1972**, *274*, 1937.
3367. Heydweiller, A., *Ann. Phys.* **1896**, *59*, 193.
3368. Heyns, K., R. Stute, et al., *Tetrahedron* **1966**, *22*, 2223.
3369. Hibben, J. H., *J. Chem. Phys.* **1937**, *5*, 706.
3370. Hibbert, H., *Ber. Deut. Chem. Gesell.* **1899**, *32*, 2533; from Doeuvre, J., *Bull. Soc. Chim.* **1926**, *39*, 1594.
3371. Hibbert, H., U.S. Patent 1,008,333, Nov. 14, 1911; *Official Gaz. U.S. Patent Office* **1911**, *172*, 280.
3372. Hibbert, H., and J. S. Allen, *J. Amer. Chem. Soc.* **1932**, *54*, 4115.
3373. Hickman, K. C. D., *J. Phys. Chem.* **1930**, *34*, 627.
3374. Hickman, K. C. D., J. C. Hecker, et al., *Ind. Eng. Chem. Anal. Ed.* **1937**, *9*, 264.
3375. Hickman, K. C. D., and D. J. Trevoy, *Ind. Eng. Chem.* **1952**, *44*, 1882.
3376. Hicks, J. F. G., J. G. Hooley, et al., *J. Amer. Chem. Soc.* **1944**, *66*, 1064.
3377. Hicks-Bruun, M. M., and J. H. Bruun, *J. Amer. Soc.* **1936**, *58*, 810.
3378. Hicks-Bruun, M. M., and J. H. Bruun, *J. Res. Nat. Bur. Stand.* **1932**, *8*, 525.
3379. Hidalgo, A., and M. T. Sardina, *An. Rea. Soc. Espan. Fis y Quim* (Madrid) **1956**, *52B*, 627; *C.A.* **1960**, *54*, 4153.
3380. Hieber, W., and A. Woerner, *Z. Elektrochem.* **1934**, *40*, 252, 256; *Chem. Zentr.* **1934**, *II*, 290.

3381. Higashi, K., *Bull. Inst. Phys.-Chem. Res.* (Tokyo) **1933,** *12,* 2; *C.A.* **1933,** *27,* 2358.

3382. Higashi, K., *Bull. Inst. Phys.-Chem. Res.* (Tokyo) **1933,** *12,* 771; *C.A.* **1934,** *28,* 386.

3383. Higashi, K., *Bull. Inst. Phys.-Chem. Res.* (Tokyo) **1936,** *15,* 766; *C.A.* **1937,** *31,* 4863.

3384. Higashi, K., *Sci. Papers Inst. Phys.-Chem. Res.* (Tokyo) **1937,** *31,* 317; *C.A.* **1937,** *31,* 6559.

3385. Higgins, C. E., W. H. Baldwin, et al., *J. Phys. Chem.* **1959,** *63,* 112.

3386. Higgins, E. R., and J. Lielmezo, *J. Chem. Eng. Data* **1965,** *10,* 178.

3387. High, M. E., *Phys. Rev.* **1931,** *38,* 1837; *C.A.* **1932,** *26,* 920.

3388. Highet, R. J., and P. F. Highet, *J. Org. Chem.* **1965,** *30,* 902.

3389. Highet, R. J., and W. C. Wildman, *J. Amer. Chem. Soc.* **1955,** *77,* 4399.

3390. Higuchi, T., C. H. Barnstein, et al., *Anal. Chem.* **1962,** *34,* 400.

3391. Hildago, A., *Compt. Rend.* **1954,** *237,* 253; *C.A.* **1955,** *49,* 12130.

3392. Hildebrand, J. H., *Chem. Rev.* **1949,** *44,* 37.

3393. Hildebrand, J. H., *J. Chem. Phys.* **1949,** *17,* 1346.

3394. Hildebrand, J. H., J. M. Prausnitz, et al., *Regular and Related Solutions,* Van Nostrand, New York, 1970.

3395. Hildebrand, J. H., and R. L. Scott, *The Solubility of Nonelectrolytes,* Reinhold, New York, 1950.

3396. Hildenbrand, D. L., and R. A. McDonald, *J. Phys. Chem.* **1959,** *63,* 1521.

3397. Hildenbrand, D. L., R. A. McDonald, et al., *J. Chem. Phys.* **1959,** *30,* 930.

3398. Hilditch, T. P., *The Chemical Constitution of Natural Fats,* 2nd ed., Wiley, New York, 1940; from Markley [4892].

3399. Hilditch, T. P., *The Chemical Constitution of Natural Fats,* 3rd ed., Chapman & Hall, London, 1956, from Markley [4892].

3400. Hilditch, T. P., *The Industrial Chemistry of Fats and Waxes,* 2nd ed., Balliere, Tindall & Cox, London, 1941, from Markley [4892].

3401. Hill, A. E., *J. Amer. Chem. Soc.* **1923,** *45,* 1143.

3402. Hill, A. E., and T. B. Fitzgerald, *J. Amer. Chem. Soc.* **1935,** *57,* 250.

3403. Hill, A. E., and W. N. Malisoff, *J. Amer. Chem. Soc.* **1926,** *48,* 918.

3404. Hill, A. V., *Proc. Roy. Soc.* **1930,** *127A,* 9.

3405. Hill, D. J. T., and L. R. White, *Aust. J. Chem.* **1974,** *27,* 1905.

3406. Hill, F. N., and A. Brown, *Anal. Chem.* **1950,** *22,* 562.

3407. Hill, H. S., and H. Hibbert, *J. Amer. Chem. Soc.* **1923,** *45,* 3124.

3408. Hill, R. D., and G. D. Meakins, *J. Chem. Soc.* **1958,** 760.

3409. Hinckley, C. C., *J. Org. Chem.* **1970,** *35,* 2834.

3410. Hine, J., R. Butterworth, et al., *J. Amer. Chem. Soc.* **1958,** *80,* 819.

3411. Hine, J., A. M. Dowell, et al., *J. Amer. Chem. Soc.* **1956,** *78,* 479.

3412. Hinegardner, W. S., U.S. Patent 2,525,410, Oct. 10, 1950; British Patent 642,969, Dec. 19, 1946; *C.A.* **1951,** *45,* 2025.

3413. Hippel, A. V., and L. G. Wesson, *Ind. Eng. Chem.* **1946,** *38,* 1121.

3414. Hipple, J. A., and M. Shepherd, *Anal. Chem.* **1949,** *21,* 32.

3415. Hipsher, H. F., and P. H. Wise, *J. Amer. Chem. Soc.* **1954,** *76,* 1747.

3416. Hiraishi, J., and T. Shinoda, *Bull. Chem. Soc. Jpn.* **1975,** *48,* 2385, 2732.

3417. Hirakawa, A. Y., and M. Tsuboi, *Proc. Intern. Symp. Mol. Struct. Spectry. Tokyo* **1962,** *A107*; *C.A.* **1964,** *61,* 1396.

3418. Hirano, K., *Bull. Chem. Soc. Jpn.* **1965,** *38,* 842.

3419. Hirao, N., *J. Chem. Soc. Jpn.* **1931,** *52,* 269; *C.A.* **1932,** *26,* 5062.

3420. Hiraoka, H., and J. H. Hildebrand, *J. Phys. Chem.* **1963,** *67,* 916.

3421. Hiraoka, H., and J. H. Hildebrand, *J. Phys. Chem.* **1963,** *67,* 1919.

3422. Hirota, K., K. Nagoshi, et al., *Bull. Soc. Chem. Jpn.* **1961,** *34,* 226; *C.A.* **1961,** *55,* 15105.

3423. Hirota, K., and J. Takezaki, *Bull. Chem. Soc. Jpn.* **1968,** *41,* 76.

3424. Hirota, K., J. Takezaki, et al., *Nippon Hoshasan Kobunshi Kenkyu Kyokai Nenpo* **1964–65,** *6,* 247; *C.A.* **1966,** *64,* 18636.

3425. Hiroto, Z., Y. Niwa, et al., *Z. Phys. Chem.* (Frankfurt) **1969,** *67,* 244; *C.A.* **1970,** *72,* 84024.

3426. Hirschler, A. E., and W. B. M. Faulconer, *J. Amer. Chem. Soc.* **1946,** *68,* 210.

3427. Hirsjarvi, P., *Suomen Kemistilehti* **1956,** *29B,* 200; *C.A.* **1957,** *51,* 4820.

3428. Hirt, C. A., *Anal. Chem.* **1961,** *33,* 1786.

3429. Hirt, R. C., and J. P. Howe, *J. Chem. Phys.* **1948**, *16*, 480.
3430. Hissel, J., *Bull. Soc. Roy. Sci. Liege* **1952**, *21*, 457; *C.A.* **1953**, *47*, 11961.
3431. Hitch, E. F., and H. N. Gilbert, *J. Amer. Chem. Soc.* **1913**, *35*, 1780.
3432. Hite, H. B., *Amer. Chem. J.* **1895**, *17*, 507.
3433. Hobbs, M. E., and W. W. Bates, *J. Amer. Chem. Soc.* **1952**, *74*, 746.
3434. Hobson, R. W., R. J. Hartman, et al., *J. Amer. Chem. Soc.* **1941**, *63*, 2094.
3435. Hochstrasser, G., *Helv. Phys. Acta* **1961**, *34*, 189; *C.A.* **1961**, *55*, 23059.
3436. Hodge, E. B., *Ind. Eng. Chem.* **1940**, *32*, 748.
3437. Hodge, H. C., *Proc. Soc. Exp. Biol. Med.* **1943**, *53*, 20; *C.A.* **1943**, *37*, 4472.
3438. Hodge, P., and G. Richardson, *J. Chem. Soc., Chem. Commun.* **1975**, 622.
3439. Hodgman, C. D., ed., *Handbook of Chemistry and Physics*, 44th ed., Chemical Rubber Publishing, Cleveland, Ohio, 1962.
3440. Hodgson, H. H., *J. Chem. Soc.* **1943**, 380.
3441. Hoepe, G., and W. D. Treadwell, *Helv. Chim. Acta* **1942**, *25*, 353.
3442. Hoerr, C. W., and H. J. Harwood, *J. Phys. Chem.* **1952**, *56*, 1068.
3443. Hoerr, C. W., M. R. McCorkle, et al., *J. Amer. Chem. Soc.* **1943**, *65*, 328.
3444. Hoerr, C. W., and R. S. Sedgwick, *J. Org. Chem.* **1946**, *11*, 603.
3445. Höfermann, H., *Chem. Ing. Tech.* **1964**, *36*, 422.
3446. Hoff, M. C., K. W. Greenlee, et al., *J. Amer. Chem. Soc.* **1951**, *73*, 3329.
3447. Hoffman, F. W., T. R. Moore, et al., *J. Amer. Chem. Soc.* **1956**, *78*, 6413.
3448. Hoffman, H. T., G. E. Evans, et al., *J. Amer. Chem. Soc.* **1951**, *73*, 3028.
3449. Hoffman, J., and C. E. Boord, *J. Amer. Chem. Soc.* **1955**, *77*, 3139.
3450. Hoffman, R. A., B. Gestblom, et al., *J. Chem. Phys.* **1963**, *39*, 486.
3451. Hoffmann, C. de, and E. Barbier, *Bull. Soc. Chim. Belges* **1936**, *45*, 565; *Chem. Zentr.* **1937**, *I*, 2590; *Beil.* **EIII2**, 655.
3452. Hoffmann, E. C., *Z. Phys. Chem.* (Leipzig) **1943**, *B53*, 179; *C.A.* **1944**, *38*, 300.
3453. Hoffmann, U., German Patent 618,972, Sept. 19, 1935; *C.A.* **1936**, *30*, 1066.
3454. Hofman, W., L. Stenaniak, et al., *J. Amer. Chem. Soc.* **1964**, *86*, 554.
3455. Hofmann, H. E., *Ind. Eng. Chem.* **1932**, *24*, 135.
3456. Hofmann, H. E., and E. W. Reid, *Ind. Eng. Chem.* **1929**, *21*, 955.
3457. Hofmanova, A., and K. Angelis, *Chem. Listy* **1978**, *72*, 306; *C.A.* **1978**, *88*, 179367.
3458. Hogfeldt, E., and F. Fredlund, *Acta Chem. Scand.* **1970**, *24*, 1858.
3459. Hogler, R., and L. Kahovec, *Monatsh. Chem.* **1946**, *76*, 27; *C.A.* **1947**, *41*, 1558.
3460. Hogness, T. R., A. E. Sidwell, et al., *J. Biol. Chem.* **1937**, *120*, 239.
3461. Hoigne, J., and T. Gaumann, *Helv. Chim. Acta* **1958**, *41*, 1933.
3462. Højendahl, K., *Beret. 18th Skand. Naturforskendmode Copenhagen* **1929**, 344; *C.A.* **1930**, *24*, 2649.
3463. Højendahl, K., Dissertation, Copenhagen, 1928; from R. N. Kerr, *J. Chem. Soc.* **1926**, 2798.
3464. Højendahl, K., *Kgl. Danske Videnskab. Selskab. Mat.-Fys. Medd.* **1946**, *24*, No. 2, 11 pp; *C.A.* **1947**, *41*, 4985.
3465. Højendahl, K., *Nord. Kemikermode, Forh.* **1939**, *5*, 209; *C.A.* **1944**, *38*, 2249.
3466. Højendahl, K., *Phys. Z.*, *30*, 394; *Chem. Zentr.* **1929**, *II*, 1898; *Beil.* **EII6**, 209.
3467. Holcomb, D. E., and C. L. Dorsey, *Ind. Eng. Chem.* **1949**, *41*, 2788.
3468. Holden, N. E., *Pure Appl. Chem.* **1980**, *52*, 3249.
3469. Holladay, T. M., and A. M. Neilsen, *J. Mol. Spectry.* **1964**, *14*, 371.
3470. Hollmann, R., *Z. Phys. Chem.* (Leipzig) **1963**, *43*, 129.
3471. Holm, T., *J. Org. Chem.* **1973**, *56*, 89.
3472. Holman, R. T., and P. R. Edmondson, *Anal. Chem.* **1956**, *28*, 1533.
3473. Holmes, J. L., J. K. Terlouw, et al., *Org. Mass. Spectrom.* **1979**, *14*, 204.
3474. Homer, J., and L. F. Thomas, *J. Chem. Soc. B* **1966**, 141.
3475. Hon, M. C., R. P. Jingh, et al., *J. Chem. Eng. Data* **1976**, *21*, 430.
3476. Hong, C. S., R. Wakslak, et al., *J. Chem. Eng. Data* **1982**, *27*, 146.
3477. Honing, R. E., *J. Chem. Phys.* **1948**, *16*, 105.
3478. Hood, G. C., O. Redlich, et al., *J. Chem. Phys.* **1955**, *23*, 2229.
3479. Hood, H. E., U.S. Patent 2,971,893, Feb. 14, 1961; *Official Gaz. U.S. Patent Office* **1961**, *763*, 454.
3480. Hooper, G. S., and C. A. Kraus, *J. Amer. Chem. Soc.* **1934**, *56*, 2265.

3481. Hoosier Solvents and Chemical Corp., "Physical Properties of Modern Organic Solvents," Indianapolis.
3482. Hoover, T. B., *J. Phys. Chem.* **1964,** *68,* 876.
3483. Hoover, T. B., *J. Phys. Chem.* **1964,** *68,* 3003.
3484. Hoover, T. B., *J. Phys. Chem.* **1969,** *73,* 57.
3485. Hoover, T. S., German Patent 2,029,042, Jan. 7, 1971; *C.A.* **1971,** *74,* 64060.
3486. Hooz, J., and S. S. H. Gilani, *Can. J. Chem.* **1968,** *46,* 86.
3487. Hopke, E. R., and G. W. Sears, *J. Amer. Chem. Soc.* **1948,** *70,* 3801.
3488. Hopton, F. J., A. J. Rest, et al., *J. Chem. Soc. A* **1966,** 1326.
3489. Horak, M., and O. Exner, *Chem. Listy* **1958,** *52,* 1451; *C.A.* **1958,** *52,* 19894.
3490. Horikawa, E., and T. Okada, *Shika Gahuko* **1975,** *75,* 934; *C.A.* **1976,** *84,* 38667.
3491. Horn, H. J., *Toxicol. Appl. Pharmacol.* **1961,** *3,* 12.
3492. Horning, E. C., ed., *Organic Syntheses, Collective Vol. III,* Wiley, New York, 1955.
3493. Horning, E. C., M. G. Horning, et al., *J. Amer. Chem. Soc.* **1949,** *71,* 169.
3494. Hornstein, I., and P. F. Crowe, *Anal. Chem.* **1962,** *34,* 1037.
3495. Hornyi, G., and M. Novak, *Acta Chim.* (Budapest) **1973,** *75,* No. 4, 369; *C.A.* **1973,** *78,* 167750.
3496. Horsley, G. F., and Imperial Chemical Industries, British Patent 313,466, Feb. 6, 1928; *C.A.* **1930,** *24,* 1125.
3497. Horsley, L. H., *Azeotropic Data,* American Chemical Society, Washington, 1952.
3498. Horsley, L. H., *Azeotropic Data II,* American Chemical Society, Washington, 1962.
3499. Horsley, L. H., *Azeotropic Data III,* American Chemical Society, Washington, 1973.
3500. Horton, C. A., and J. C. White, *Talanta* **1961,** *7,* 215; *C.A.* **1961,** *55,* 23045.
3501. Hosoya, H., J. Tanaka, et al., *J. Mol. Spectry.* **1962,** *8,* 257; *C.A.* **1962,** *57,* 2990.
3502. Horvath, A. L., *J. Chem. Doc.* **1972,** *12,* 163.
3503. Horvath, G., and A. I. Kiss, *Spectrochim. Acta, Part A* **1967,** *23,* 921.
3504. Hossenlopp, I. A., and D. W. Scott, *J. Chem. Thermodyn.* **1981,** *13,* 405.
3505. Hossenlopp, I. A., and D. W. Scott, *J. Chem. Thermodyn.* **1981,** *13,* 415.
3506. Hossenlopp, I. A., and D. W. Scott, *J. Chem. Thermodyn.* **1981,** *13,* 423.
3507. Hougen, O. A., and K. M. Watson, *Chemical Process Principles,* Wiley, New York, 1943.
3508. Hough, E. W., D. M. Mason, et al., *J. Amer. Chem. Soc.* **1950,** *72,* 5775.
3509. Houston, D. F., *J. Amer. Chem. Soc.* **1933,** *55,* 4131.
3510. Houtman, J. P. W., J. van Steenis, et al., *Rec. Trav. Chim. Pays-Bas* **1946,** *65,* 781; *C.A.* **1947,** *41,* 3049.
3511. Houtman, T., T. E. Zurawic, et al., U.S. Patent 2,779,721, Jan. 29, 1957; *C.A.* **1957,** *51,* 7073; *Official Gaz. U.S. Patent Office* **1957,** *714,* 1018.
3512. Hovermale, R. A., P. G. Sears, et al., *J. Chem. Eng. Data* **1963,** *8,* 490.
3513. Hovermale, R. A., and P. G. Sears, *J. Phys. Chem.* **1956,** *60,* 1579.
3514. Hovorka, F., and D. Dreisbach, *J. Amer. Chem. Soc.* **1934,** *56,* 1664.
3515. Hovorka, F., D. Dreisbach, et al., *J. Amer. Chem. Soc.* **1936,** *58,* 2264.
3516. Hovorka, F., and F. E. Geiger, *J. Amer. Chem. Soc.* **1933,** *55,* 4759.
3517. Hovorka, F., H. P. Lankelma, et al., *J. Amer. Chem. Soc.* **1933,** *55,* 4820.
3518. Hovorka, F., H. P. Lankelma, et al., *J. Amer. Chem. Soc.* **1938,** *60,* 820.
3519. Hovorka, F., H. P. Lankelma, et al., *J. Amer. Soc.* **1940,** *62,* 2372.
3520. Howard, F. L., T. W. Mears, et al., *J. Res. Nat. Bur. Stand.* **1947,** *38,* 365.
3521. Howard, P. B., and I. Wadso, *Acta Chem. Scand.* **1970,** *24,* 145.
3522. Howard-Lock, H. E., and G. W. King, *J. Mol. Spectrosc.* **1970,** *35,* 393.
3523. Howell, O. W., *Proc. Roy. Soc.* **1932,** *137A,* 418.
3524. Howlett, K. E., *J. Chem. Soc.* **1951,** 1409.
3525. Howlett, K. E., *J. Chem. Soc.* **1955,** 1784.
3526. Howlett, L. E., *Can. J. Res.* **1931,** *4,* 79; *Chem. Zentr.* **1931,** *I,* 3437.
3527. Hoyer, H., *Z. Phys. Chem.* (Leipzig) **1940,** *45B,* 389; *C.A.* **1940,** *34,* 4990.
3528. Hoyer, H., *Z. Elektrochem.* **1941,** *47,* 451; *C.A.* **1942,** *36,* 4415.
3529. Hoyt, C. A., and C. K. Fink, *J. Phys. Chem.* **1937,** *41,* 453.
3530. Hoyt, H. E., U.S. Patent 2,381,055, Aug. 7, 1945; *C.A.* **1945,** *39,* 5106.
3531. Hruska, F., E. Bock, et al., *Can. J. Chem.* **1936,** *41,* 3034.

3532. Hrynakowski, K., and A. Smoczkiewiczowa, *Rocz. Chem.* **1937**, *17*, 140; *C.A.* **1937**, *31*, 4883.
3533. Hrynakowski, K., and A. Smoczkiewiczowa, *Rocz. Chem.* **1937**, *17*, 165; *C.A.* **1937**, *31*, 6096.
3534. Hrynakowski, K., and M. Szmyt, *Z. Phys. Chem.* (Leipzig) **1938**, *182A*, 405.
3535. Hsia, A. W., *Z. Tech. Phys.* **1931**, *12*, 550.
3536. Hu, A. T., and G. C. Sinke, *J. Chem. Thermodyn.* **1969**, *1*, 507.
3537. Hu, A. T., G. C. Sinke, et al., *J. Chem. Thermodyn.* **1972**, *4*, 239.
3538. Huang, H. H., and E. P. A. Sullivan, *Aust. J. Chem.* **1968**, *21*, 1721.
3539. Huang, T. C., and K. P. Sung, *Science Repts. Nat. Tsinghua Univ.* **1934**, *2*, 303; *Beil.* **EIII6**, 1454.
3540. Hubbard, W. N., F. R. Frow, et al., *J. Phys. Chem.* **1961**, *65*, 1326.
3541. Hubbard, W. N., W. O. Good, et al., *J. Phys. Chem.* **1958**, *62*, 614.
3542. Hubbard, W. N., C. Katz, et al., *J. Phys. Chem.* **1954**, *58*, 142.
3543. Hubbard, W. N., D. W. Scott, et al., *J. Amer. Chem. Soc.* **1955**, *77*, 5857.
3544. Hückel, W., and P. Ackermann, *J. Pr. Chem.* **1933**, *136*, 15; *Brit. Chem. Abs.* **1933**, 372.
3545. Hückel, W., J. Datow, et al., *Z. Phys. Chem.* (Leipzig) **1940**, *186A*, 129; *C.A.* **1941**, *35*, 1688.
3546. Hückel, W., and K. Hagenguth, *Ber. Deut. Chem. Gesell.* **1931**, *64B*, 2892.
3547. Hückel, W., and H. Harder, *Chem. Ber.* **1947**, *80*, 357.
3548. Hückel, W., and A. Hubele, *Ann. Chem.* **1958**, *613*, 27; *Chem. Zentr.* **1959**, 6411.
3549. Hückel, W., C. Jennewein, et al., *Ann. Chem.* **1965**, *686*, 51; *C.A.* **1965**, *63*, 16181.
3550. Hückel, W., K. Kumetat, et al., *Ann. Chem.* **1935**, *517*, 184; *C.A.* **1935**, *29*, 6126.
3551. Hückel, W., and J. Kurz, *Chem. Ber.* **1958**, *91*, 1290.
3552. Hückel, W., and W. Rothkegel, *Chem. Ber.* **1948**, *81*, 71.
3553. Hückel, W., and H. W. Wunsch, *J. Prakt. Chem.* **1935**, *142*, 225; *Beil.* **III7**, 6.
3554. Hudson, G. H., J. C. McCoubrey, et al., *Trans. Faraday Soc.* **1960**, *56*, 1144.
3555. Huff, W. J., *U.S. Bur. Mines, Rept. Invest.* **1945**, 3794; *C.A.* **1945**, *39*, 1291.
3556. Huff, W. J., *U.S. Bur. Mines, Rept. Invest.* **1946**, 4031; *C.A.* **1947**, *41*, 6719.
3557. Huffman, H. M., M. E. Gross, et al., *J. Phys. Chem.* **1961**, *65*, 495.
3558. Huffman, H. M., G. S. Parks, et al., *J. Amer. Chem. Soc.* **1930**, *52*, 1547.
3559. Huffman, H. M., G. S. Parks, et al., *J. Amer. Chem. Soc.* **1930**, *52*, 3241.
3560. Huffman, H. M., G. S. Parks, et al., *J. Amer. Chem. Soc.* **1931**, *53*, 3876.
3561. Huffman, H. M., S. S. Todd, et al., *J. Amer. Chem. Soc.* **1949**, *71*, 584.
3562. Huggins, C. M., G. C. Pimentel, et al., *J. Phys. Chem.* **1956**, *60*, 1311.
3563. Hughes, E. C., and J. R. Johnson, *J. Amer. Chem. Soc.* **1931**, *53*, 737.
3564. Hughes, E. D., and U. G. Shapiro, *J. Chem. Soc.* **1937**, 1177.
3565. Hughes, L. J., and G. E. Britt, *J. Appl. Polymer Sci.* **1961**, *5*, 337.
3566. Hughes, O. L., and H. Hartley, *Phil. Mag.* **1933**, *15*, 610.
3567. Hughes, S. R. C., *J. Chem. Soc.* **1957**, 634.
3568. Hugill, J. A., M. L. McGlashan, et al., *J. Chem. Thermodyn.* **1978**, *10*, 95.
3569. Huisgen, R., and H. Brade, *Chem. Ber.* **1957**, *90*, 1432.
3570. Huisgen, R., H. Brade, et al., *Chem. Ber.* **1957**, *90*, 1437.
3571. Huisgen, R., and H. Walz, *Chem. Ber.* **1956**, *89*, 2616.
3572. Huisman, J., and B. H. Sage, *J. Chem. Eng. Data* **1964**, *9*, 223.
3573. Hukumoto, Y., *J. Chem. Phys.* **1935**, *3*, 165.
3574. Hukumoto, Y., *Sci. Rep. Tohoku Imp. Univ.* **1933**, *22*, 23; *Chem. Zentr.* **1933**, *II*, 1968.
3575. Hukumoto, Y., *Sci. Rep. Tohoku Imp. Univ. Ser. I* **1937**, *25*, 1162; *Chem. Zentr.* **1937**, *II*, 1548.
3576. Hulanicki, S., and M. Klincewicz, *Ochr. Przec. Przem. Chem.* **1972**, No. 4, 1; *C.A.* **1973**, *79*, 7561.
3577. Hull, D. C., and A. H. Agett, U.S. Patent 2,422,016, June 10, 1947; *C.A.* **1947**, *41*, 6275.
3578. Hull, G. F., *J. Chem. Phys.* **1935**, *3*, 534.
3579. Humko Products, Chemical Division, "Fatty Acids," Memphis, Tennessee.
3580. Humko Products, Chemical Division, Tech. Bulletin, "Kemamine N-200," Memphis, Tennessee.
3581. Humphery Chemical Co., The, "Chemical Specialities, Physical Properties," North Haven, Connecticut.
3582. Humphries, C. M., A. D. Walsh, et al., *Trans. Faraday Soc.* **1967**, *63*, 513.
3583. Hunecke, H., *Ber. Deut. Chem. Gesell.* **1927**, *60*, 1451.
3584. Hünig, S., *Ann. Chem.* **1950**, *569*, 198; *C.A.* **1951**, *45*, 7951.
3585. Hünig, S., and M. Kiessel, *Chem. Ber.* **1958**, *91*, 380; *C.A.* **1958**, *52*, 18187.

3586. Hunt, H., and H. T. Briscoe, *J. Phys. Chem.* **1929**, *33*, 1495.

3587. Hunt, H. D., and W. T. Simpson, *J. Amer. Chem. Soc.* **1953**, *75*, 4540.

3588. Hunter, E. C. E., and J. R. Partington, *J. Chem. Soc.* **1932**, 2812.

3589. Hunter, E. C. E., and J. R. Partington, *J. Chem. Soc.* **1933**, 309.

3590. Hunter, W., British Patent 606,608, Aug. 17, 1948; *C.A.* **1949**, *43*, 4692.

3591. Hurd, C. D., and D. G. Botteron, *J. Amer. Chem. Soc.* **1946**, *68*, 1200.

3592. Hurd, C. D., and J. S. Strong, *Anal. Chem.* **1951**, *23*, 542.

3593. Hurdis, E. C., and C. P. Smyth, *J. Amer. Chem. Soc.* **1942**, *64*, 2212.

3594. Hurdis, E. C., and C. P. Smyth, *J. Amer. Chem. Soc.* **1942**, *64*, 2829.

3595. Hurdis, E. C., and C. P. Smyth, *J. Amer. Chem. Soc.* **1943**, *65*, 89.

3596. Hurwic, J., and J. Michakczyk, *Nukleonika* **1965**, *10*, 221; *C.A.* **1968**, *69*, 55331.

3597. Hust, J. C., and R. E. Schramm, *J. Chem. Eng. Data*, **1976**, *21*, 7.

3598. Huston, R. C., and D. L. Bailey, *J. Amer. Chem. Soc.* **1946**, *68*, 1382.

3599. Huston, R. C., and C. O. Bastwick, *J. Org. Chem.* **1948**, *13*, 331.

3600. Huston, R. C., W. B. Fox, et al., *J. Org. Chem.* **1938–1939**, *3*, 251.

3601. Huston, R. C., and I. A. Kaye, *J. Amer. Chem. Soc.* **1942**, *64*, 1576.

3602. Hutchinson, A. W., and G. C. Chandler, *J. Amer. Soc.* **1931**, *53*, 2881.

3603. Hutchinson, M. H., and L. E. Sutton, *J. Chem. Soc.* **1958**, 4382.

3604. Hüttner, K., and G. Tammann, *Z. Anorg. Chem.* **1905**, *43*, 218.

3605. Hutzler, J. A., R. J. Colton, et al., *J. Chem. Eng. Data* **1972**, *17*, 324.

3606. Hvistendahl, G., and K. Undheim, *Org. Mass Spectrom.* **1970**, *3*, 821.

3607. Hygienic Guide Series, *Amer. Ind. Hyg. Assoc. J.* **1965**, *26*, 196.

3608. Hygienic Guide Series, *Amer. Ind. Hyg. Assoc. J.* **1966**, *27*, 571.

I

3609. Ievskaya, N. M., and R. M. Umarkhodzhaev, *Paramagnitn. Rezonans, Kazansk. Univ., Sbornik* **1960**, 137; *C.A.* **1962**, *56*, 8197.

3610. Ikada, E., *J. Phys. Chem.* **1971**, *75*, 1240.

3611. Ikarashi, T., M. Goto, et al., German Patent 2,650,416, May 12, 1977; *C.A.* **1977**, *87*, 134536.

3612. Imanishi, S., and Y. Kanda, *J. Sci. Research Inst.* (Tokyo) **1949**, *43*, 1; *C. A.* **1949**, *43*, 8885.

3613. Imanov, L. M., *Dokl. Akad. Nauk Azerbaidzhan SSR* **1956**, *12*, 531; *C.A.* **1957**, *51*, 2345.

3614. Imbaud, J. P., and G. Berthet, *Compt. Rend.* **1965**, *261*, 953; *C.A.* **1966**, *64*, 190.

3615. Indest, H., H. Massat, et al., German Patent 1,111,193, Dec. 20, 1955; *C.A.* **1962**, *56*, 2338.

3616. Indest, H., H. Massat, et al., German Patent 1,139,122, Nov. 8, 1962; *C.A.* **1963**, *58*, 5840.

3617. Ingersoll, L. R., *J. Opt. Soc. Amer.* **1922**, *6*, 663.

3618. Ingle, J. D., and H. P. Cady, *J. Phys. Chem.* **1938**, *42*, 397.

3619. Ingold, C. K., and W. J. Powell, *J. Chem. Soc.* **1921**, 1222.

3620. Ingold, E. H., and C. K. Ingold, *J. Chem. Soc.* **1932**, 756.

3621. *International Critical Tables*, McGraw-Hill, New York, 1930.

3622. IUPAC Information Bulletin No. 32, **Aug. 1968**, 14.

3623. IUPAC Physical Chemistry Division, *Pure Appl. Chem.* **1974**, *40*, 463.

3624. IUPAC, *Pure Appl. Chem.* **1981**, *53*, 1847.

3625. Ioffe, B. V., *Zh. Obshch. Khim.* **1955**, *25*, 902; *C.A.* **1955**, *49*, 13717.

3626. Ioffe, I. I., and E. S. Yampol'skaya, *J. Appl. Chem.* (USSR) **1944**, *17*, 527; *C.A.* **1945**, *39*, 3986.

3627. Iogansen, A. V., *Doklady Akad. Nauk SSSR* **1951**, *81*, 1077; *C.A.* **1952**, *46*, 7983.

3628. Iogansen, A. V., *Zavodskaya Lab.* **1959**, *25*, 302; *C.A.* **1960**, *54*, 17050.

3629. Ipatieff, V. N., and B. B. Corson, *Ind. Eng. Chem.* **1938**, *30*, 1039.

3630. Ipatieff, V., and A. V. Grosse, U.S. Patent 2,104,424, Jan. 4, 1938.

3631. Ipatiew, V., B. Dolgow, et al., *Ber. Deut. Chem. Gesell.* **1930**, *63*, 3072.

3632. Ipatiew, W., *Ber. Deut. Chem. Gesell.* **1907**, *40*, 1281.

3633. Iredale, T., and A. G. Mills, *Nature* **1930**, *126*, 604; *Chem. Zentr.* **1931**, *I*, 23.

3634. Irving, H. M. N. H., and R. B. Simpson, *J. Inorg. Nucl. Chem.* **1972**, *34*, 2241.

3635. Irving, R. J., *J. Chem. Thermodyn.* **1972**, *4*, 793.

3636. Irving, R. J., and I. Wadso, *Acta Chem. Scand.* **1970**, *24*, 589.

3637. Isakova, N. P., and L. A. Oshueva, *Zh. Prikl. Khim.* (Leningrad) **1968**, *14*, 2805; *C.A.* **1969**, *70*, 61265.

3638. Iseard, B. S., J. B. Pedley, et al., *J. Chem. Soc. A* **1971,** 3095.
3639. Ishida, K., *Bunko Kenkyu* **1960,** *9,* 27; *C.A.* **1962,** *57,* 15808.
3640. Ishihara, M., and R. Oda, *J. Soc. Chem. Ind. Jpn.* **1943,** *46,* 1268; *C.A.* **1948,** *42,* 6333.
3641. Islam, N., and I. Ahmad, *Indian J. Chem., Sect. A* **1983,** *22A,* 229.
3642. Islam, N., B. Jamil, et al., *Indian J. Chem.* **1973,** *11,* 266.
3643. Issoire, J., *Mem. Poudres* **1960,** *42,* 333; *C.A.* **1961,** *55,* 14421.
3644. Itakura, T., *J. Chem. Soc. Jpn.* **1942,** *63,* 1400; *C.A.* **1947,** *41,* 3041.
3645. Itaya, K., M. Kawai, et al., *J. Amer. Chem. Soc.* **1978,** *100,* 5996.
3646. Ito, M., *J. Mol. Spectrosc.* **1960,** *4,* 125; *C.A.* **1961,** *55,* 13048.
3647. Ito, S., and I. Miura, *Bull. Chem. Soc. Jpn.* **1965,** *38,* 2197.
3648. Ivanchev, S. S., A. I. Yurzhenko, et al., *Zh. Fiz. Khim.* **1965,** *39,* 1900; *C.A.* **1966,** *64,* 166.
3649. Ivanov, D., *Bull. Soc. Chim.* **1928,** *43,* 441; *C.A.* **1928,** *22,* 4464.
3650. Ivanov, V. A., *Tr. Voronezhsk. Gos. Med. Inst.* **1957,** *29,* 23.
3651. Ivanova, E. F., A. I. Kruglyak, et al., *Zh. Fiz. Khim.* **1976,** *50,* 817; *Russ. J. Phys. Chem.* **1976,** *50,* 489.
3652. Ivanskii, V. I., and B. N. Dolgov, *Kinetika i Kataliz* **1963,** *4,* 165; *C.A.* **1963,** *59,* 538.
3653. Iwai, I., H. Shindo, et al., *Yakugaku Zasshi* **1960,** *80,* 1588; cf. *Chem. Pharm. Bull.* **1960,** *8,* 815; *C.A.* **1961,** *55,* 10063.
3654. Iwamoto, R., *Spectrochim. Acta, Part A* **1971,** *27,* 2385.
3655. Izergin, A. P., *Izv. Vyssh. Ucheb. Zaved. Fiz.* **1958,** No. 5, 115; *C.A.* **1959,** *53,* 6719; Ref. [8056].

J

3656. Jackman, L. M., and D. P. Kelly, *J. Chem. Soc. B* **1970,** 102.
3657. Jackman, L. M., A. K. Macbeth, et al., *J. Chem. Soc.* **1949,** 1717.
3658. Jackowski, A. W., *J. Chem. Thermodyn.* **1974,** *6,* 49.
3659. Jackson, L. C., *Phil. Mag.* **1922,** *43,* 481.
3660. Jackson, M. D., and P. G. Sears, *J. Chem. Eng. Data* **1979,** *24,* 85.
3661. Jacobs, C. J., and G. S. Parks, *J. Amer. Chem. Soc.* **1934,** *56,* 1513.
3662. Jacobs, T. L., J. D. Roberts, et al., *J. Amer. Chem. Soc.* **1944,** *66,* 656.
3663. Jaeger, F. M., *Z. Anorg. Allg. Chem.* **1917,** *101,* 1.
3664. Jaeger, F. M., and J. Kahn, *J. Chem. Soc.* **1915,** *108,* 747.
3665. Jaffé, G., *Ann. Physik* **1909,** *28,* 326.
3666. Jaffe, H. H., and G. O. Doak, *J. Amer. Chem. Soc.* **1955,** *77,* 4441.
3667. Jain, D. V. S., and R. Sidhu, *J. Chem. Thermodyn.* **1984,** *16,* 111.
3668. Jain, D. V. S., S. Singh, et al., *J. Chem. Soc., Faraday Trans. 1* **1974,** *70,* 961.
3669. Jakli, G., and W. A. Van Hook, *J. Chem. Thermodyn.* **1972,** *4,* 857.
3670. Jakobsen, R. J., *U.S. Dept. Commerce, Office Tech. Serv., PB Rept. 171,342* **1960;** *C.A.* **1963,** *58,* 4055.
3571. Jakobsen, R. J., Y. Mikawa, et al., *J. Mol. Struct.* **1971,** *10,* 300.
3672. James, J. W., *J. Prakt. Chem.* **1882,** *26,* 378.
3673. Jander, G., and G. Winkler, *J. Inorg. Nucl. Chem.* **1959,** *9,* 24, 32, 39; *C.A.* **1959,** *53,* 8775.
3674. Jannelli, L., A. Lopez, et al., *J. Chem. Eng. Data* **1979,** *24,* 174.
3675. Jannelli, L., A. Lopez, et al., *J. Chem. Eng. Data* **1980,** *25,* 259.
3676. Jannelli, L., A. Lopez, et al., *J. Chem. Eng. Data* **1983,** *28,* 166.
3677. Jannelli, L., A. Lopez, et al., *J. Chem. Eng. Data* **1983,** *28,* 169.
3678. Jannelli, L., A. K. Rakshit, et al., *Z. Naturforsch., Teil A* **1974,** *29,* 355.
3679. Jansen, M. L., and H. L. Yeager, *J. Phys. Chem.* **1973,** *77,* 3089.
3680. Janz, G. J., J. Ambrose, et al., *Spectrochim. Acta* **1979,** *35A,* 175.
3681. Janz, G. J., and S. S. Danyluk, *J. Amer. Chem. Soc.* **1959,** *81,* 3846.
3682. Janz, G. J., and W. E. Fitzgerald, *J. Chem. Phys.* **1955,** *23,* 1973.
3683. Jasinski, R. J., and S. Kirkland, *Anal. Chem.* **1967,** *39,* 1663.
3684. Jasper, J. J., *J. Phys. Chem. Ref. Data* **1972,** *1,* 841.
3685. Jasper, J. J., and E. R. Kerr, *J. Amer. Chem. Soc.* **1954,** *76,* 2659.
3686. Jasper, J. J., E. R. Kerr, et al., *J. Amer. Chem. Soc.* **1953,** *75,* 5252.
3687. Jasper, J. J., and E. V. Kring, *J. Phys. Chem.* **1955,** *59,* 1019.
3688. Jasper, J. J., and H. L. Wedlick, *J. Chem. Eng. Data* **1964,** *9,* 446.

3689. Jatkar, S. K. K., and C. M. Deshpande, *J. Indian Chem. Soc.* **1960**, *37*, 1; *C.A.* **1960**, *54*, 18000.
3690. Jatkar, S. K. K., and C. M. Deshpande, *J. Indian Chem. Soc.* **1960**, *37*, 15; *C.A.* **1960**, *54*, 18001.
3691. Jatkar, S. K. K., and C. M. Deshpande, *J. Univ. Poona Sci. and Technol.* **1954**, 48; *C.A.* **1955**, *49*, 15316.
3692. Jatkar, S. K. K., and D. Lakshminarayanan, *J. Indian Inst. Sci.* **1946**, *28A*, 1; *C.A.* **1947**, *41*, 1901.
3693. Jatkar, S. K. K., and R. Padmanabhan, *Indian J. Phys.* **1936**, *10*, 55; *C.A.* **1936**, *30*, 6284.
3694. Jatkar, S. K. K., V. K. Phansalkar, et al., *Indian J. Chem.* **1969**, *7*, 93.
3695. Jefferson Chemical Co. Inc., "Ethylene Carbonate," Houston.
3696. Jefferson Chemical Co. Inc., "Polyethylene Glycols," Houston, 1967.
3697. Jefferson Chemical Co. Inc., "Propylene Oxide," Houston, 1960.
3698. Jeffery, G. H., R. Parker, et al., *J. Chem. Soc.* **1961**, 570.
3699. Jeffery, G. H., and A. I. Vogel, *J. Chem. Soc.* **1948**, 658.
3700. Jeffery, G. H., and A. I. Vogel, *J. Chem. Soc.* **1948**, 674.
3701. Jelatis, J. G., *J. Appl. Phys.* **1948**, *19*, 419.
3702. Jelinek, R. M., and H. Leopold, *Monatsh. Chem.* **1978**, *109*, 387.
3703. Jen, M., and D. R. Linde, Jr., *J. Chem. Phys.* **1962**, *36*, 2525.
3704. Jen, T.-Ko., *Union Ind. Res. Inst. Rept. No. 17* (Taiwan) **1956**; *C.A.* **1961**, *55*, 22116.
3705. Jen, T.-Ko., *Union Ind. Res. Inst. Rept. No. 35* (Taiwan) **1958**.
3706. Jenkin, C. F., *Trans. Faraday Soc.* **1922**, *18*, 197.
3707. Jenkins, A. C., and G. F. Chembers, *Ind. Eng. Chem.* **1954**, *46*, 2367.
3708. Jenner, P. M., E. C. Hagan, et al., *Food Cosmet. Toxicol.* **1964**, *2*, 327; *C.A.* **1965**, *62*, 4516.
3709. Jennings, A. L., and J. E. Boggs, *J. Org. Chem.* **1964**, *29*, 2065.
3710. Jenny, R., *Compt. Rend.* **1958**, *246*, 3477; *C.A.* **1959**, *53*, 2122.
3711. Jensen, F. R., D. S. Noyce, et al., *J. Amer. Chem. Soc.* **1962**, *84*, 386.
3712. Jensen, F. R., L. D. Whipple, et al., *J. Amer. Chem. Soc.* **1959**, *81*, 1262.
3713. Jensen, J. L., and A. T. Thibeault, *J. Org. Chem.* **1977**, *42*, 2168.
3714. Jensen, O., and R. A. Gortner, *J. Phys. Chem.* **1932**, *36*, 3138.
3715. Jerkunica, J. M., and T. G. Traylor, *Org. Syn.* **1973**, *53*, 94.
3716. Jesse, R. H., *J. Amer. Chem. Soc.* **1912**, *34*, 1337.
3717. Jessup, R. S., *J. Res. Nat. Bur. Stand.* **1937**, *18*, 115.
3718. Jezewski, M., *J. Phys. E* **1970**, *3*, 663.
3719. Jhon, M. S., E. R. Van Artsdalen, et al., *J. Chem. Phys.* **1967**, *47*, 2231.
3720. Jocelyn, P. C., and N. Polgar, *J. Chem. Soc.* **1953**, 132.
3721. Jockmann, W., *Arch. Toxicol.* **1961**, *18*, 698; *C.A.* **1962**, *56*, 10497.
3722. Joglekar, M. S., and V. N. Thatte, *Z. Physik.* **1936**, *98*, 692; *Chem. Zentr.* **1937**, *I*, 568.
3723. Johannsen, R. B., F. E. Brinckman, et al., *J. Phys. Chem.* **1968**, *72*, 660.
3724. Johari, G. P., *J. Chem. Eng. Data* **1968**, *13*, 541.
3725. Johari, G. P., and W. Dannhauser, *J. Chem. Phys.* **1968**, *48*, 5114.
3726. Johari, G. P., and W. Dannhauser, *J. Chem. Phys.* **1969**, *51*, 1626.
3727. Johns, I. B., E. A. McElhill, et al., *J. Chem. Eng. Data* **1962**, *7*, 277.
3728. Johnsen, S. E. J., *Anal. Chem.* **1947**, *19*, 305.
3729. Johnson, A. W., *J. Chem. Soc.* **1946**, 1014.
3730. Johnson, E. L., K. H. Pool, et al., *Anal. Chem.* **1966**, *38*, 183.
3731. Johnson, H. E., U.S. Patent 2,459,432, Jan. 18, 1949; *C.A.* **1949**, *43*, 2476.
3732. Johnson, J. F., and R. L. LeTourneau, *J. Amer. Chem. Soc.* **1953**, *75*, 1743.
3733. Johnson, J. R., *Organic Syntheses, Vol. 16*, Wiley, New York, 1936.
3734. Johnson, J. R., S. D. Christian, et al., *J. Chem. Soc. A* **1966**, 77.
3735. Johnson, J. R., and F. D. Hager, *Org. Syntheses* **1927**, *7*, 60.
3736. Johnson, M. D., *J. Chem. Soc.* **1965**, 805.
3737. Johnson, R. D., R. J. Meyers, et al., *J. Chem. Phys.* **1953**, *21*, 1425.
3738. Johnson, W. H., E. J. Prosen, et al., *J. Res. Nat. Bur. Stand.* **1946**, *37*, 51.
3739. Johnson, W. H., E. J. Prosen, et al., *J. Res. Nat. Bur. Stand.* **1947**, *38*, 419.
3740. Johnson, W. H., E. J. Prosen, et al., *J. Res. Nat. Bur. Stand.* **1947**, *39*, 49.
3741. Johnstone, R. T., *A. M. A. Arch. Ind. Health* **1959**, *20*, 445; *C.A.* **1960**, *54*, 9130.
3742. Jonathan, N., S. Gordon, et al., *J. Chem. Phys.* **1962**, *36*, 2443.
3743. Joncich, M. J., and D. R. Bailey, *Anal. Chem.* **1960**, *32*, 1578.

3744. Jones, A. R., and D. A. Alkens, *J. Chem. Eng. Data* **1982**, *27*, 25.

3745. Jones, D. C., *J. Chem. Soc.* **1929**, 799.

3746. Jones, D. C., and H. F. Betts, *J. Chem. Soc.* **1928**, 1177.

3747. Jones, D. G., British Patent 699,079, Oct. 28, 1953; *C.A.* **1955**, *49*, 3262.

3748. Jones, E. C. S., and J. Kenner, *J. Chem. Soc.* **1932**, 711.

3749. Jones, E. R., *J. Phys. Chem.* **1927**, *31*, 1316.

3750. Jones, E. W., R. J. L. Popplewell, et al., *Spectrochim. Acta* **1966**, *22*, 647.

3751. Jones, G., and S. M. Christian, *J. Amer. Chem. Soc.* **1939**, *61*, 82.

3752. Jones, G., and H. J. Fornwalt, *J. Amer. Chem. Soc.* **1938**, *60*, 1683.

3753. Jones, G. T., M. Randic, et al., *Croat. Chem. Acta* **1964**, *36*, 111; *C.A.* **1965**, *62*, 8541.

3754. Jones, G. W., W. E. Miller, et al., *Ind. Eng. Chem.* **1933**, *25*, 771.

3755. Jones, G. W., and G. S. Scott, *U.S. Bur. Mines, Rept. Invest. 3881*, **1946**; *C.A.* **1946**, *40*, 5919.

3756. Jones, H. C., *Z. Phys. Chem.* (Leipzig) **1899**, *31*, 114.

3757. Jones, I. W., and J. C. Tebby, *J. Chem. Soc., Perkin Trans.* 2 **1973**, 1125.

3758. Jones, J. R., *Quarterly Rev.* **1971**, *3*, 365.

3759. Jones, J. R., and C. B. Monk, *J. Chem. Soc.* **1963**, 2633.

3760. Jones, L. C., and L. W. Taylor, *Anal. Chem.* **1955**, *27*, 228.

3761. Jones, L. C., and R. A. Friedel, *Ind. Eng. Chem., Anal. Ed.* **1945**, *17*, 349.

3762. Jones, L. R., and J. A. Riddick, *Anal. Chem.* **1951**, *23*, 349; **1952**, *24*, 1533; **1956**, *28*, 1137, 1493.

3763. Jones, M., and A. Lapworth, *J. Chem. Soc.* **1914**, 1804.

3764. Jones, R. A. Y., A. R. Katritzky, et al., *J. Chem. Soc.* **1962**, 2576.

3765. Jones, R. A. Y., A. R. Katritzky, et al., *J. Chem. Soc. B* **1971**, 1795.

3766. Jones, R. L., *J. Mol. Spectrosc.* **1963**, *11*, 411; *C.A.* **1964**, *60*, 3620.

3767. Jones, R. L., *Spectrochim. Acta* **1966**, *22*, 1555; *C.A.* **1966**, *65*, 16827.

3768. Jones, R. N., *Can. J. Chem.* **1962**, *40*, 321.

3769. Jones, R. N., *Nat. Res. Council Bull.* (Canada) **1957**, No. 5; *C.A.* **1958**, *52*, 3516.

3770. Jones, R. N., J. B. DiGiorgio, et al., *J. Org. Chem.* **1965**, *30*, 1822.

3771. Jones, R. N., A. F. McKay, et al., *J. Amer. Chem. Soc.* **1952**, *74*, 2575.

3772. Jones, R. N., and K. Noack, *Can. J. Chem.* **1961**, *39*, 2214.

3773. Jones, R. N., and G. D. Thorn, *Can. J. Res.* **1949**, *27B*, 580; *C.A.* **1949**, *43*, 8273.

3774. Jones, W. J., and S. T. Bowden, et al., *J. Phys. Colloid Chem.* **1948**, *52*, 753.

3775. Jones, W. J., and K. J. P. Orton, *J. Chem. Soc.* **1909**, 1056.

3776. Jones, W. J., and J. B. Speakman, *J. Amer. Chem. Soc.* **1921**, *43*, 1867.

3777. Jones, W. M., and W. J. Giauque, *J. Amer. Chem. Soc.* **1947**, *69*, 983.

3778. Joop, N., and H. Zimmerman, *Z. Elektrochem.* **1962**, *66*, 440; *C.A.* **1962**, *57*, 10681.

3779. Jordan, T. E., *Vapor Pressure of Organic Compounds*, Interscience, New York, 1950.

3780. Jordan, T. E., *Vapor Pressure of Organic Compounds*, Interscience, New York, 1954.

3781. Joris, G. G., and H. S. Taylor, *J. Chem. Phys.* **1948**, *16*, 45.

3782. Joris, G. G., and J. Vitrone, U.S. Patent 2,829,166, April 23, 1956.

3783. Jose, J., R. Philippe, et al., *Bull. Soc. Chim. Fr.* **1971**, 2860.

3784. Josefowicz, E., *Rocz. Chem.* **1938**, *18*, 577.

3785. Joshi, D. V., and J. R. Merchant, *J. Sci. Ind. Res.* (India) **1955**, *14B*, 482; *C.A.* **1956**, *50*, 11272.

3786. Joshi, G., and N. L. Singh, *Spectrochim. Acta, Part A* **1967**, *23*, 1341.

3787. Joshi, G. J., *Indian J. Pure Appl. Phys.* **1966**, *4*, 40; *C.A.* **1966**, *64*, 16843.

3788. Joshi, R. M., *J. Polym. Sci., Part A-2* **1970**, *8*, 679.

3789. Joshi, S. S., and G. D. Tuli, *J. Indian Chem. Soc.* **1951**, *28*, 450.

3790. Josien, M. L., *Compt. Rend.* **1953**, *237*, 175; *C.A.* **1954**, *48*, 4977.

3791. Josien, M. L., and J. M. Lebas, *Bull. Soc. Chim. Fr.* **1956**, 53; *C.A.* **1956**, *50*, 6920.

3792. Jost, F., and M. Harrand, *J. Phys. Radium* **1962**, *23*, 308; *C.A.* **1962**, *57*, 8085.

3793. Joukovsky, I., *Bull. Soc. Chim. Belges* **1934**, *43*, 397; *Chem. Zentr.* **1935**, *I*, 3126.

3794. Jouve, P., *Ann. Phys.* (Paris) **1966**, *1*, 127; *C.A.* **1966**, *65*, 1633.

3795. Jouve, P., G. Widenlocher, et al., *Compt. Rend.* **1963**, *257*, 1079; *C.A.* **1963**, *59*, 14769.

3796. Ju, T. Y., C. E. Wood, et al., *J. Inst. Petrol.* **1942**, *28*, 159; *C.A.* **1943**, *37*, 4886.

3797. Juarez Araoz, A. J., B. O'Donell, et al., *An. Assoc. Quim. Argent.* **1983**, *71*, 287; *C.A.* **1983**, *99*, 219129.

3798. Judson, C. M., and M. Kilpatrick, *J. Amer. Chem. Soc.* **1949**, *71*, 3110.

3799. Jullander, I., and K. Brune, *Acta Chem. Scand.* **1948**, *2*, 204; *C.A.* **1949**, *43*, 4184.
3800. Junell, R., *Ark. Kem., Mineral. Geol., Ser.* **1934**, *B11*, No. 30, 6/3; *Chem. Zentr.* **1934**, *I*, 3727.
3801. Junell, R., *Svensk Kem. Tidskr.* **1934**, *46*, 125; Dissertation, Univ. of Uppsala, 1935.
3802. Junell, R., *Z. Phys. Chem.* (Leipzig) **1929**, *141A*, 71.
3803. Jung, M. E., and G. L. Hatfield, *Tetrahedron Lett.* **1978**, 4483.
3804. Jung, M. E., and P. L. Ornstein, *Tetrahedron Lett.* **1977**, 2659.
3805. Jungnickel, J. L., and J. W. Forbes, *Anal. Chem.* **1963**, *35*, 938.

K

3806. Kablukov, I. A., and V. T. Malischeva, *J. Amer. Chem. Soc.* **1925**, *47*, 1553.
3807. Kablukow, I. A., and F. M. Perelman, *Dokl. Akad. Nauk SSSR* **1930**, 519; *Chem. Zentr.* **1931**, *II*, 23.
3808. Kabo, G. Y., and D. N. Andreevskii, *Izv. Vysshikh Uchebn. Zaved., Khim. i Khim. Tekhnol.* **1965**, *8*, 574; *C.A.* **1966**, *64*, 4343.
3809. Kafyan, V. B., *Zh. Ekzp. Klin. Med.* **1971**, *11*, 39; *C.A.* **1971**, *75*, 128127.
3810. Kagarise, R. E., *J. Chem. Phys.* **1956**, *24*, 1264.
3811. Kagarise, R. E., and L. W. Daasch, *J. Chem. Phys.* **1955**, *23*, 113.
3812. Kahane, E., *Compt. Rend.* **1948**, *227*, 841; *C.A.* **1949**, *43*, 1982.
3813. Kahlbaum, G. W. A., *Ber. Deut. Chem. Gesell,* **1880**, *13*, 2348.
3814. Kahlbaum, G. W. A., *Ber. Deut. Chem. Gesell.* **1885**, *18*, 2108.
3815. Kahlbaum, G. W. A., and K. Arndt, *Z. Phys. Chem.* (Leipzig) **1898**, *26*, 577.
3816. Kahlenberg, L., *J. Phys. Chem.* **1901**, *5*, 215; *J. Chem. Soc.* **1901Aii**, 492.
3817. Kahlenberg, L., *J. Phys. Chem.* **1902**, *6*, 45; *J. Chem. Soc.* **1902Aii**, 310.
3818. Kahlenberg, L., and A. T. Lincoln, *J. Phys. Chem.* **1899**, *3*, 12; *I.C.T.*, *6*, 144.
3819. Kahovec, L., *Z. Phys. Chem.* (Leipzig) **1938**, *40B*, 135; *C.A.* **1938**, *32*, 6549.
3820. Kahovec, L., *Z. Phys. Chem.* (Leipzig) **1939**, *43B*, 109; *C.A.* **1939**, *33*, 6718.
3821. Kahovec, L., and K. W. F. Kohlrausch, *Monatsh. Chem.* **1936**, *68*, 359; *C.A.* **1937**, *31*, 36.
3822. Kahovec, L., and K. W. F. Kohlrausch, *Z. Phys. Chem.* (Leipzig) **1937**, *35B*, 29; *C.A.* **1937**, *31*, 293.
3823. Kahovec, L., and K. W. F. Kohlrausch, *Z. Phys. Chem.* (Leipzig) **1940**, *46B*, 165; *C.A.* **1941**, *35*, 977.
3824. Kahovec, L., and K. W. F. Kohlrausch, *Z. Phys. Chem.* (Leipzig) **1940**, *48B*, 7; *C.A.* **1941**, *35*, 3172.
3825. Kahovec, L., and A. W. Reitz, *Monatsh. Chem.* **1936**, *69*, 363; *C.A.* **1937**, *31*, 2515.
3826. Kahovec, L., and J. Wagner, *Z. Phys. Chem.* (Leipzig) **1939**, *42B*, 123; *C.A.* **1939**, *33*, 4129.
3827. Kahovec, L., and J. Wagner, *Z. Phys. Chem.* (Leipzig) **1940**, *47B*, 48; *C.A.* **1941**, *35*, 977.
3828. Kahovec, L., and J. Wagner, *Z. Phys. Chem.* (Leipzig) **1941**, *48B*, 188; *C.A.* **1942**, *36*, 2477.
3829. Kahovec, L., and H. Wassmuth, *Z. Phys. Chem.* (Leipzig) **1940**, *48B*, 70; *C.A.* **1941**, *35*, 3172.
3830. Kahre, L. C., *J. Chem. Eng. Data* **1973**, *18*, 267.
3831. Kahre, L. C., and R. J. Livingston, *Hydrocarbon Process. Petrol. Refin.* **1964**, *43*, 119; *C.A.* **1964**, *61*, 499.
3832. Kailan, A., and K. Melkus, *Monatsh, Chem.* **1927**, *48*, 9.
3833. Kakac, B., and M. Hudlicky, *Collect. Czech. Chem. Commun.* **1965**, *30*, 745; *C.A.* **1965**, *63*, 1343.
3834. Kalafati, D. D., D. S. Rasskazov, et al., *Zh. Fiz. Khim.* **1967**, *41*, 1357; *Russ. J. Phys. Chem.* **1967**, *41*, 720.
3835. Kalasinsky, V. F., and C. J. Wurrey, *J. Raman Spectrosc.* **1980**, *9*, 315.
3836. Kambara, S., *J. Soc. Chem. Ind. Jpn.* **1944**, *47*, 518; *C.A.* **1949**, *43*, 1594.
3837. Kamei, H., *Bull. Chem. Soc. Jpn.* **1965**, *38*, 1212; *C.A.* **1965**, *63*, 12539.
3838. Kamei, H., *Jpn. J. Appl. Phys.* **1965**, *4*, 212; *C.A.* **1965**, *63*, 2543.
3839. Kameo, T., T. Hirashima, et al., *Kagaku To Kogyo* (Osaka) **1966**, *40*, 135; *C.A.* **1966**, *65*, 10466.
3840. Kamiyoshi, K., and T. Fujimura, *J. Phys. Radium* **1962**, *23*, 311; *C.A.* **1962**, *57*, 9322.
3841. Kamm, O., *Qualitative Organic Analysis*, 2nd ed., Wiley, New York, 1932.
3842. Kamm, O., and C. S. Marvel, *J. Amer. Chem. Soc.* **1920**, *42*, 299.
3843. Kamm, O., and J. H. Waldo, *J. Amer. Chem. Soc.* **1921**, *43*, 2225.
3844. Kamm, W. F., and A. O. Mathews, *J. Amer. Pharm. Assoc.* **1922**, *11*, 599.
3845. Kanazawa, Y., and K. Nukada, *Bull. Chem. Soc. Jpn.* **1962**, *35*, 612.

3846. Kanbayashi, U., and K. Nukada, *Nippon Kagaku Zasshi* **1963**, *84*, 297.

3847. Kandiyoti, R., E. McLaughlin, et al., *J. Chem. Soc., Faraday Trans. 1* **1975**, *69*, 1953.

3848. Kanerevskaya, A. A., *Prom. Toksikol., Moscow, Sb.* **1960**, 15; *C.A.* **1962**, *57*, 6270.

3849. Kankaanpera, A., *Acta Chem. Scand.* **1969**, *23*, 1723.

3850. Kankaanpera, A., *Acta Chem. Scand.* **1969**, *23*, 2211.

3851. Kanomata, I., *Bull. Chem. Soc. Jpn.* **1961**, *34*, 1596; *C.A.* **1962**, *57*, 239.

3852. Kantor, S. W., and C. R. Hauser, *J. Amer. Chem. Soc.* **1953**, *75*, 1744.

3853. Kapff, S. F., and R. B. Jacobs, *Rev. Sci. Instrum.* **1947**, *18*, 581.

3854. Kapustin, A. P., *J. Exptl. Theoret. Phys.* (USSR) **1947**, *17*, 30; *C.A.* **1948**, *42*, 1092.

3855. Karabinos, J. V., *Science Counselor* **1954**, *17*, 3; *C.A.* **1955**, *49*, 4508.

3856. Karabinos, J. V., and K. T. Serijan, *J. Amer. Chem. Soc.* **1945**, *67*, 1856.

3857. Karapet'yants, M. Kh., *Khim. i Tekhnol. Topliva* **1956**, No. 9, 22; *C.A.* **1957**, *51*, 836.

3858. Karnatz, F. A., and F. C. Whitmore, *J. Amer. Chem. Soc.* **1938**, *60*, 3082.

3859. Karpov, B. D., *Farmakol. i Toksikol.* **1954**, *17*, No. 1, 49; *C.A.* **1954**, *48*, 13984.

3860. Karpov, B. D., *Gig. Sanit.* **1955**, No. 8, 19; *C.A.* **1956**, *50*, 2199.

3861. Karr, A. E., W. M. Bowes, et al., *Anal. Chem.* **1951**, *23*, 459.

3862. Karrer, P., and H. Schmid, *Helv. Chim. Acta* **1946**, *29*, 1853.

3863. Kartha, V. B., H. H. Mantsch, et al., *Can. J. Chem.* **1973**, *51*, 1749.

3864. Kartsev, G. N., Ya. K. Syrkin, et al., *Proc. Acad. Sci., USSR Phys. Chem. Sect.* (Eng. trans.) **1958**, *122*, 631.

3865. Kartsev, V. N., *Zh. Fiz. Khim.* **1976**, *50*, 764; *Russ. J. Phys. Chem.* **1976**, *50*, 449.

3866. Kartsev, V. N., O. Ya. Samoilov, et al., *Zh. Fiz. Khim.* **1979**, *53*, 757; *Russ. J. Phys. Chem.* **1979**, *53*, 429.

3867. Kartsev, V. N., and V. A. Zabelin, *Zh. Fiz. Khim.* **1978**, *52*, 1221; *Russ. J. Phys. Chem.* **1978**, *52*, 1221.

3868. Karunakar, J., K. D. Reddy, et al., *J. Chem. Eng. Data* **1982**, *27*, 348.

3869. Karvonen, A., *Ann. Acad. Sci. Fennicae* **1924**, *20*, No. 9; *C.A.* **1924**, *18*, 1981.

3870. Karyakin, A. V., and A. V. Petrov, *Zh. Anal. Khim.* **1964**, *19*, 1234; *C.A.* **1965**, *62*, 3405.

3871. Karyakin, N. V., and I. B. Rabinovich, *Zh. Fiz. Khim.* **1969**, *43*, 2141; *Russ. J. Phys. Chem.* **1969**, *43*, 1201.

3872. Karyakin, N. V., I. B. Rabinovich, et al., *Zh. Fiz. Khim.* **1968**, *42*, 1814; *Russ. J. Phys. Chem.* **1968**, *42*, 954.

3873. Kashin, L. M., *Gig. Sanit.* **1965**, *30*, 23; *C.A.* **1965**, *63*, 7558.

3874. Kassel, L. S., *J. Amer. Chem. Soc.* **1936**, *58*, 670.

3875. Katagiri, S., *Sci. Repts. Tohoku Univ. First Ser.* **1960**, *44*, 165; *C.A.* **1961**, *55*, 25406.

3876. Katahira, T., *Osaka-shiritsu Daigaku Igaku Zasshi* **1982**, *31*, 221; *C.A.* **1983**, *99*, 65455.

3877. Katasura, S., *J. Chem. Soc. Jpn.* **1942**, *63*, 1483; *C.A.* **1947**, *41*, 3449.

3878. Katayama, M., *Sci. Repts. Tohoku Imp. Univ.* **1916**, *4*, 373; *C.A.* **1916**, *10*, 994.

3879. Katayama, M., and Y. Morino, *Repts. Radiation Chem. Res. Inst., Tokyo Univ.* **1949**, *1*, No. 1, 254.

3880. Kato, J., and M. Sakuma, *Denki Kagaku* **1957**, *25*, 126; *C.A.* **1958**, *52*, 4468.

3881. Kato, J., M. Sakuma, et al., *Denki Kagaku* **1957**, *25*, 331; *C.A.* **1958**, *52*, 4469.

3882. Kato, S., and F. Someno, *Sci. Papers Inst. Phys. Chem. Res.* (Tokyo) **1938**, *34*, 950; *C.A.* **1938**, *32*, 8267.

3883. Kato, Y., *J. Chem. Phys.* **1966**, *44*, 2824.

3884. Katon, J. E., W. R. Feairheller, et al., *Anal. Chem.* **1964**, *36*, 2126.

3885. Katon, J. E., W. R. Feairheller, et al., *J. Mol. Spectrosc.* **1964**, *13*, 72; *C.A.* **1964**, *60*, 15316.

3886. Katritzky, A. R., and N. A. Coats, *J. Chem. Soc.* **1959**, 2062.

3887. Katritzky, A. R., and R. A. Jones, *J. Chem. Soc.* **1959**, 3670.

3888. Katritzky, A. R., and J. M. Lagowski, *J. Chem. Soc.* **1958**, 4155.

3889. Katritzky, A. R., and P. Simmons, *J. Chem. Soc.* **1959**, 2051.

3890. Katritzky, A. R., and P. Simmons, *J. Chem. Soc.* **1959**, 2058.

3891. Katsuno, M., *J. Soc. Chem. Ind. Jpn.* **1938**, *41*, 75; *C.A.* **1938**, *32*, 5372.

3892. Katsuno, M., *J. Soc. Chem. Ind. Jpn.* **1941**, *44*, 898; *C.A.* **1948**, *42*, 2575.

3893. Katsuno, M., *J. Soc. Chem. Ind. Jpn.* **1941**, *44*, 903; *C.A.* **1948**, *42*, 2575.

3894. Katsuno, M., *J. Soc. Chem. Ind. Jpn.* **1943**, *46*, 1028; *C.A.* **1948**, *42*, 7285.

3895. Katti, P. K., and M. M. Chaudhri, *J. Chem. Eng. Data* **1964**, *9*, 128.
3896. Katti, P. K., and M. M. Chaudhri, *J. Chem. Eng. Data* **1964**, *9*, 442.
3897. Katti, P. K., and S. K. Shil, *J. Chem. Eng. Data* **1966**, *11*, 601.
3898. Katti, S. S., and S. Pathak, *J. Chem. Eng. Data* **1969**, *14*, 73.
3899. Katz, D. L., and W. Saltman, *Ind. Eng. Chem.* **1939**, *31*, 91.
3900. Katz, M., P. W. Lobo, et al., *Can. J. Chem.* **1971**, *49*, 2605.
3901. Kauck, E. A., and A. R. Diesslin, *Ind. Eng. Chem.* **1951**, *43*, 2332.
3902. Kauer, E., L. Grote, et al., *Z. Phys. Chem.* (Leipzig) **1966**, *232*, 356; *Wies. Z. Tech. Hochsh. Chem.*, *Leuna-Merseberg* **1965**, *7*, No. 2, 73; *C.A.* **1965**, *63*, 17222.
3903. Kaufman, D., and W. Reeve, *Org. Syntheses* **1946**, *26*, 83.
3904. Kaufmann, G., and M. J. F. Leroy, *Bull. Soc. Chim. Fr.* **1967**, *2*, 402.
3905. Kaufmann, W. E., and R. Adams, *J. Amer. Chem. Soc.* **1923**, *45*, 3029.
3906. Kauschka, G., and L. Kolditz, *Z. Chem.* **1976**, *16*, 377.
3907. Kawasaki, Y., S. Suzuki, et al., *Eisei Shikenjo Hokoku* **1975**, *93*, 72; *C.A.* **1976**, *85*, 14287.
3908. Kawashima, Y., and K. Kozima, *Bull. Chem. Soc. Jpn.* **1974**, *47*, 2879.
3909. Kawazoe, Y., and M. Ohnishi, *Chem. Pharm. Bull.* (Tokyo) **1963**, *11*, 243; *C.A.* **1963**, *59*, 14774.
3910. Kay, F. W., and W. H. Perkins, Jr., *J. Chem. Soc.* **1980**, 839.
3911. Kay, R. L., G. A. Vidulich, et al., *J. Chem. Phys.* **1967**, *47*, 866.
3912. Kay, W. B., *J. Amer. Chem. Soc.* **1946**, *68*, 1336.
3913. Kay, W. B., *J. Amer. Chem. Soc.* **1947**, *69*, 1273.
3914. Kay, W. B., and S. K. Pak, *J. Chem. Thermodyn.* **1980**, *12*, 673.
3915. Kay, W. B., and F. M. Warzel, *Ind. Eng. Chem.* **1951**, *43*, 1150.
3916. Kaye, S., and S. Tannenbaum, *J. Org. Chem.* **1958**, *18*, 1750.
3917. Kaye, W., *Spectrochim. Acta* **1954**, *6*, 257; *C.A.* **1955**, *49*, 769.
3918. Kay-Fries Chemicals, Inc., "Technical Data," New York.
3919. Kazanskiĭ, B. A., A. L. Liberman, et al., *J. Gen. Chem.* (USSR) **1947**, *17*, 1503; *C.A.* **1948**, *42*, 2225.
3920. Keading, W. W., R. O. Lindblom, et al., *Ind. Eng. Chem.* **1961**, *53*, 805.
3921. Kecki, Z., *Spectrochim. Acta* **1962**, *18*, 1155; *C.A.* **1963**, *58*, 5162.
3922. Kecki, Z., and H. Wincel, *Nukleonika* **1963**, *8*, No. 2, 117; *C.A.* **1964**, *60*, 3595.
3923. Keffler, L., *Bull. Soc. Chim. Belges* **1935**, *44*, 429; *C.A.* **1936**, *30*, 719.
3924. Keffler, L., and J. H. McLean, *J. Chem. Soc. Ind.* (London) **1935**, *54*, 178T.
3925. Keffler, L. J. P., *J. Phys. Chem.* **1930**, *34*, 1319.
3926. Keicher, G., and O. Klopfer, German Patent 943,886, Jan. 1, 1956; *Chem. Zentr.* **1956**, 13257.
3927. Kelen, G. P. van der, *Bull. Soc. Chim. Belges* **1962**, *71*, 421.
3928. Kell, G. S., *J. Chem. Eng. Data* **1967**, *12*, 66.
3929. Kell, G. S., *J. Chem. Eng. Data* **1975**, *20*, 97.
3930. Keller, G. H., L. Bauer, et al., *J. Heterocycl. Chem.* **1968**, *5*, 647.
3931. Keller, H., and H. V. Halban, *Helv. Chim. Acta* **1944**, *27*, 1439.
3932. Kelley, K. K., *J. Amer. Chem. Soc.* **1929**, *51*, 180.
3933. Kelley, K. K., *J. Amer. Chem. Soc.* **1929**, *51*, 779.
3934. Kelley, K. K., *J. Amer. Chem. Soc.* **1929**, *51*, 1145.
3935. Kelley, K. K., *J. Amer. Chem. Soc.* **1929**, *51*, 1400.
3936. Kelso, E. A., and W. A. Felsing, *J. Amer. Chem. Soc.* **1940**, *62*, 3132.
3937. Kel'tsev, N. V., Yu. K. Titova, et al., *Zh. Vses. Khim. Obshch. im. D.I. Mendeleeva* **1964**, *9*, 588; *C.A.* **1965**, *62*, 3676.
3938. Kemme, H. R., and S. I. Kreps, *J. Chem. Eng. Data* **1969**, *14*, 98.
3939. Kemp, J. D., and C. J. Egan, *J. Amer. Chem. Soc.* **1938**, *60*, 1521.
3940. Kempa, R., and W. H. Lee, *J. Chem. Soc.* **1958**, 1936.
3941. Kempa, R. F., and W. H. Lee, *Talanta* **1964**, *9*, 325; *C.A.* **1962**, *56*, 13630.
3942. Kempton, H., *Z. Phys.* **1940**, *116*, 1; *C.A.* **1941**, *35*, 3170.
3943. Kemula, W., and S. Mrazek, *Compt. Rend.* **1932**, *195*, 1004.
3944. Kendall, D. N., R. R. Hampton, et al., *Appl. Spectrosc.* **1953**, *7*, 179.
3945. Kendall, J., and J. J. Beaver, *J. Amer. Chem. Soc.* **1921**, *43*, 1853.
3946. Kendall, J., and J. E. Booge, *J. Amer. Chem. Soc.* **1916**, *38*, 1712.
3947. Kendall, J., and P. M. Gross, *J. Amer. Chem. Soc.* **1921**, *43*, 1426.

3948. Kendall, J., and A. H. Wright, *J. Amer. Chem. Soc.* **1920,** *42,* 1776.

3949. Kennedy, H., and J. Lielmezs, *Ind. Eng. Chem., Fundamentals* **1967,** *6,* 310.

3950. Kennedy, R. M., M. Sagenkahn, et al., *J. Amer. Chem. Soc.* **1941,** *63,* 2267.

3951. Kenyon, J., H. Phillips, et al., *J. Chem. Soc.* **1935,** 1072.

3952. Kenyon, J., and H. E. Strauss, *J. Chem. Soc.* **1949,** 2153.

3953. Kepner, R. E., and L. J. Andrews, *J. Org. Chem.* **1948,** *13,* 208.

3954. Kertes, A. S., *J. Inorg. Nucl. Chem.* **1972,** *34,* 796; *C.A.* **1972,** *76,* 145735.

3955. Kessel'man, P. M., E. G. Porichanskii, et al., *Kholod. Tekh. Tekhnol.* **1976,** *22,* 48; *C.A.* **1977,** *86,* 21902.

3956. Kessler, Yu. M., M. G. Fomicheva, et al., *Zh. Strukt. Khim.* **1972,** *13,* 517; *C.A.* **1972,** *77,* 105718.

3957. Keswani, R., and H. Freiser, *J. Amer. Chem. Soc.* **1949,** *71,* 218.

3958. Ketelaar, J. A. A., L. deVries, et al., *Rec. Trav. Chim. Pays-Bas* **1947,** *66,* 733; *C.A.* **1948,** *42,* 3632; *Beil.* **EIII1,** 651.

3959. Ketelaar, J. A. A., H. R. Gersman, et al., *Rec. Trav. Chim. Pays-Bas* **1958,** *77,* 982; *C.A.* **1959,** *53,* 10920.

3960. Ketelaar, J. A. A., P. F. Van Velden, et al., *Rec. Trav. Chim. Pays-Bas* **1947,** *66,* 721; *C.A.* **1948,** *42,* 3631; *Beil.* **EIII1,** 651.

3961. Ketteler, E., *Ann. Phys. Chem.* **1888,** *2,* 33, 353, 506; *J. Chem. Soc.* **1888A,** 541.

3962. Kettering, C. F., and W. W. Sleator, *Physics* **1933,** *4,* 39; *C.A.* **1933,** *27,* 2095.

3963. Kevan, L., *J. Chem. Phys.* **1972,** *56,* 838.

3964. Keyes, D. B., and J. H. Hildebrand, *J. Amer. Chem. Soc.* **1917,** *39,* 2126.

3965. Keyes, D. B., S. Swann, et al., *Trans. Amer. Electrochem. Soc.* **1928,** *54,* (pre-print), 7 pp; *C.A.* **1928,** *22,* 3843.

3966. Keyes, F. G., B. Townshend, et al., *J. Math. Phys. Mass. Inst. Tech.* **1922,** *1,* 243; *C.A.* **1923,** *17,* 668.

3967. Keyes, F. G., and W. J. Winninghoff, *J. Amer. Chem. Soc.* **1916,** *38,* 1178.

3968. Keyworth, D. A., *Talanta* **1961,** *8,* 461.

3969. Kezdy, F., and A. Bruylants, *Bull. Soc. Chim. Fr.* **1959,** 947; *C.A.* **1960,** *54,* 4370.

3970. Khalepo, A. I., *Toksikol, Nov. Prom. Khim.* **1968,** No. 10, 131; *C.A.* **1969,** *71,* 47808.

3971. Khalilov, A. K., *Izv. Akad. Nauk SSSR, Ser. Fiz.* **1953,** *17,* 586; *C.A.* **1954,** *48,* 5652.

3972. Khalilov, A. K., and S. Z. Rzaeva, *Dokl. Akad. Nauk Azerbaidzhan SSR* **1956,** *12,* 441; *C.A.* **1957,** *51,* 885.

3973. Khalilov, A. K., and P. P. Shorygin, *Dokl. Akad. Nauk SSSR* **1951,** *78,* 1177; *C.A.* **1952,** *46,* 4368.

3974. Khalilov, K., *J. Exptl. Theoret. Phys.* (USSR) **1939,** *9,* 335; *C.A.* **1939,** *33,* 8069.

3975. Kharasch, M. S., *J. Res. Nat. Bur. Stand.* **1929,** *2,* 359.

3976. Kharasch, M. S., and H. C. Brown, *J. Amer. Chem. Soc.* **1939,** *61,* 2142.

3977. Kharasch, M. S., and H. C. Brown, *J. Amer. Chem. Soc.* **1939,** *61,* 3432.

3978. Kharasch, M. S., H. Engelman, et al., *J. Org. Chem.* **1937,** *2,* 288.

3979. Kharasch, M. S., and C. F. Fuchs, *J. Org. Chem.* **1945,** *10,* 159.

3980. Kharasch, M. S., and C. W. Hannum, *J. Amer. Chem. Soc.* **1934,** *56,* 712.

3981. Kharasch, M. S., and J. A. Hinckley, *J. Amer. Chem. Soc.* **1934,** *56,* 1243.

3982. Kharasch, M. S., J. G. McNab, et al., *J. Amer. Chem. Soc.* **1935,** *57,* 2463.

3983. Kharasch, M. S., and W. H. Urry, *J. Org. Chem.* **1948,** *13,* 101.

3984. Kharlampovich, G. D., and L. Z. Oblasova, USSR Patent 577,202, Oct. 25, 1977; *C.A.* **1978,** *88,* 50486.

3985. Kheilo, G. I., and S. N. Kremneva, *Gigiena Truda i Prof. Zabolevaniya* **1966,** *10,* 13; *C.A.* **1966,** *65,* 1281.

3986. Khmel'nitskii, R. A., N. A. Klyuev, et al., *Zh. Org. Khim.* **1971,** *7,* 391.

3987. Khokhlovkin, M. A., and A. V. Kalacheva, *Sintet. Kauchuk* **1936,** No. 1, 25; *C.A.* **1936,** *30,* 4080.

3988. Khol'kin, Y. I., *Izvest. Vysshikh Ucheb. Zavedenii Lesnoi Zhur.* **1960,** *3,* No. 4, 131; *C.A.* **1961,** *55,* 14421.

3989. Khol'kin, Y. I., and G. N. Chernyaeva, *Gidrolizn. i Lesokhim. Prom.* **1963,** *16,* 6; *C.A.* **1964,** *60,* 1674.

3990. Khol'kin, Y. I., L. S. Solov'er, et al., *Izv. Sibirsk. Odt. Akad. Nauk SSSR Ser. Khim. Nauk.* **1964,** 105; *C.A.* **1965,** *63,* 2873.

3991. Khotinsky, E., *Ber. Deut. Chem. Gesell.* **1909,** *42,* 2506.

3992. Khramchenkov, V. A., and E. N. Kokoreva, *Zavod. Lab.* **1962**, *28*, 1355; *C.A.* **1963**, *59*, 4990.

3993. Kiba, T., S. Arimoto, et al., Japanese Patent 77 44,528, April 21, 1978; *C.A.* **1978**, *89*, 75309.

3994. Kichkin, G. I., A. I. Kupreev, et al., *Neft. Gaz.* **1973**, *16*, 49; *C.A.* **1973**, *79*, 105.

3995. Kieffer, F., and P. Rumpf, *Compt. Rend.* **1950**, *230*, 2302.

3996. Kieffer, R., *Bull. Soc. Chim. Fr.* **1967**, 3024.

3997. Kieffer, R., J. C. Marie, et al., *Bull. Soc. Chim. Fr.* **1965**, 3271; *C.A.* **1966**, *64*, 6473.

3998. Kienitz, H., *Z. Elektrochem.* **1955**, *59*, 168; *C.A.* **1955**, *49*, 15341.

3999. Kierstead, H. A., and J. Turkevich, *J. Chem. Phys.* **1944**, *12*, 24.

4000. Kiesele, M., *Anal. Chem.* **1980**, *52*, 2230.

4001. Kilpatrick, J. E., E. J. Prosen, et al., *J. Res. Nat. Bur. Stand.* **1946**, *36*, 559.

4002. Kim, H., R. Keller, et al., *J. Chem. Phys.* **1962**, *37*, 2748.

4003. Kimbrough, R., and T. B. Gaines, *Nature* **1966**, *211*, 146.

4004. Kimura, C., K. Kashiwaya, et al., *Yuki Gosei Kagaku Kyokoi Shi.* **1965**, *33*, 254; *C.A.* **1965**, *62*, 13041.

4005. Kimura, K., and R. Fijishiro, *Bull. Chim. Soc. Jpn.* **1966**, *39*, 608.

4006. Kimura, K., and S. Nagakura, *Spectrochim. Acta* **1961**, *17*, 166.

4007. Kimura, K., Y. Toshiyasu, et al., *Bull. Chim. Soc. Jpn.* **1966**, *39*, 1681.

4008. Kimura, O., *J. Chem. Soc. Jpn.* **1942**, *63*, 98; *C.A.* **1947**, *41*, 2951.

4009. Kimura, O., *J. Chem. Soc. Jpn.* **1942**, *63*, 423; *C.A.* **1947**, *41*, 2951.

4010. Kimura, T., and G. R. Freeman, *J. Chem. Ed.* **1973**, *50*, 85A.

4011. Kind, R., G. Kahnt, et al., *Z. Phys. Chem.* (Leipzig) **1968**, *238*, 277; *C.A.* **1968**, *69*, 110578.

4012. Kinetic Chemicals, Inc., British Patent 391,168, April 13, 1933; *C.A.* **1933**, *27*, 4813.

4013. King, H., and K. J. P. Orton, *J. Chem. Soc.* **1911**, 1377.

4014. King, H., and E. V. Wright, *J. Chem. Soc.* **1939**, 1168.

4015. King, J. F., and B. Vig, *Can. J. Chem.* **1962**, *40*, 1023.

4016. Kini, M. M., and J. R. Cooper, *Biochem. et Biophys. Acta* **1960**, *44*, 599; *C.A.* **1961**, *55*, 13670.

4017. Kini, M. M., and J. R. Cooper, *Biochem. J.* **1962**, *82*, 164.

4018. Kinney, C. R., *Ind. Eng. Chem.* **1940**, *32*, 559.

4019. Kinney, C. R., *Ind. Eng. Chem.* **1941**, *33*, 791.

4020. Kinney, C. R., *J. Amer. Chem. Soc.* **1938**, *60*, 3032.

4021. Kinsey, E. L., and J. W. Ellis, *J. Chem. Phys.* **1937**, *5*, 399.

4022. Kipping, F. B., and F. Wild, *J. Chem. Soc.* **1940**, 1239.

4023. Kirchmer, H. H., *Z. Phys. Chem.* (Frankfurt) **1963**, *39*, 273; *C.A.* **1964**, *60*, 14013.

4024. Kirchnerova, J., G. G. B. Cave, et al., *Can. J. Chem.* **1976**, *54*, 3909.

4025. Kirejew, W. A., and W. A. Nikiforowa, *Zh. Obshch. Khim.* **1936**, *6*, 75; *Beil.* **EIII1**, 1344.

4026. Kirk, R. E., and D. F. Othmer, *Encyclopedia of Chemical Technology*, Interscience, New York, 1947–51.

4027. Kirk-Othmer, *Encyclopedia of Chemical Technology*, 2nd ed., Interscience-Wiley, New York, 1963.

4028. Kirkbride, F. W., *J. Appl. Chem.* **1956**, *6*, 11.

4029. Kirkbride, F. W., and F. G. Davidson, *Nature* **1954**, *174*, 79.

4030. Kirkland, J. J., ed., *Modern Practice of Liquid Chromatography*, Wiley-Interscience, New York, 1971.

4031. Kirrmann, A., *Bull. Soc. Chim. Fr.* **1938**, *5*, 915; *C.A.* **1938**, *32*, 6619.

4032. Kirrmann, A., *Bull. Soc. Chim. Fr.* **1939**, *6*, 841; *C.A.* **1939**, *33*, 9139.

4033. Kirrmann, A., *Bull. Soc. Chim. Fr.* **1926**, *39*, 988; *C.A.* **1926**, *20*, 3443.

4034. Kirrmann, A., *Compt. Rend.* **1939**, *308*, 353; *C.A.* **1939**, *33*, 2413.

4035. Kirrmann, A., and N. Hamaide, *Bull. Soc. Chim. Fr.* **1957**, 789; *C.A.* **1957**, *51*, 16496.

4036. Kirsanov, A. V., and Y. M. Zolotov, *Zh. Obshch. Khim.* **1949**, *19*, 2201; *C.A.* **1950**, *44*, 4446.

4037. Kirsanov, A. V., and Y. M. Zolotov, *Zh. Obshch. Khim.* **1950**, *44*, 6384.

4038. Kirsanov, R. P., and S. Byk, *Zh. Prikl. Khim.* **1961**, *34*, 1373; *Anal. Chem.* **1964**, *36*, 56.

4039. Kishimoto, K., M. Suga, et al., *Bull. Chem. Soc. Jpn.* **1978**, *51*, 1691.

4040. Kiss, A., J. Molnar, et al., *Compt. Rend.* **1948**, *277*, 724; *C.A.* **1949**, *43*, 2519.

4041. Kiss, A. J., and B. R. Muth, *Magyar Tudomanyos Akad. Kozponti Fiz. Kutato Intezetenek Kozlemenyei* **1959**, *1*, 147; *C.A.* **1960**, *54*, 14932.

4042. Kistiakowsky, G. B., and W. W. Rice, *J. Chem. Phys.* **1940**, *8*, 618.

4043. Kistiakowsky, G. B., J. R. Ruhoff, et al., *J. Amer. Chem. Soc.* **1936**, *58*, 137.

4044. Kiyama, R. K., K. Suzuki, et al., *Rev. Phys. Chem. Jpn.* **1951**, *21*, 50; *C.A.* **1952**, *46*, 3818.
4045. Kiyohara, O., and K. Higasi, *Bull. Chem. Soc. Jpn.* **1969**, *42*, 1158.
4046. Klaboe, P., *Acta Chem. Scand.* **1963**, *17*, 1179; *C.A.* **1964**, *60*, 10062.
4047. Klaboe, P., *Acta Chem. Scand.* **1969**, *23*, 3473.
4048. Klaboe, P., *Spectrochim. Acta, Part A* **1970**, *26*, 87.
4049. Klaboe, P., and J. Grundnes, *Spectrochim. Acta, Part A* **1968**, *24*, 1905.
4050. Klaboe, P., and J. R. Nielsen, *J. Mol. Spectrosc.* **1961**, *6*, 379; *C.A.* **1961**, *55*, 19479.
4051. Klages, G., and E. Klopping, *Z. Electrochem.* **1953**, *57*, 369.
4052. Kland-English, M. J., R. K. Summerbell, et al., *J. Amer. Chem. Soc.* **1953**, *75*, 3709.
4053. Klein, A. K., *J. Assoc. Offic. Agr. Chemists* **1949**, *32*, 349.
4054. Kleman, B., *Can. J. Phys.* **1963**, *41*, 2034; *C.A.* **1964**, *60*, 3626.
4055. Klemenc, A., and M. Low, *Rec. Trav. Chim. Pays-Bas* **1930**, *49*, 626; *Beil.* **EII5**, 545.
4056. Klemm, L., W. Kleman, et al., *Z. Phys. Chem.* (Leipzig) **1933**, *165A*, 379; *Beil.* **EII5**, 223, 1570.
4057. Klemperer, W., M. W. Cronyn, et al., *J. Amer. Chem. Soc.* **1954**, *76*, 5846.
4058. Klepper, J., *Chimie et Industrie* **1929**, 261; *C.A.* **1929**, *23*, 3897.
4059. Klesment, I. R., and K. M. Soo, *Khim i Tekhnol. Goryuch. Slantsev i Produktov Pererabotki* **1962**, *11*, 307; *C.A.* **1963**, *58*, 13681.
4060. Kletz, T. A., and W. C. Price, *J. Chem Soc.* **1947**, 644.
4061. Klevens, H. B., *J. Phys. Colloid Chem.* **1950**, *54*, 283.
4062. Klevens, H. B., and J. R. Platt, *J. Amer. Chem. Soc.* **1947**, *69*, 3055.
4063. Klevens, H. B., and L. J. Zemring, *J. Chim. Phys.* **1952**, *49*, 377; *C.A.* **1953**, *47*, 11979.
4064. Kliment, V., V. Fried, et al., *Collect. Czech. Chem. Commun.* **1964**, *29*, 2008; *C.A.* **1964**, *61*, 12689.
4065. Kline, C. H., and J. Turkevich, *J. Chem. Phys.* **1944**, *12*, 300.
4066. Kling, A., and P. Roy, *Compt. Rend.* 144, 111; *C.A.* **1907**, *1*, 2233.
4067. Klingstedt, F. W., *Compt. Rend.* **1923**, *176*, 248.
4068. Klingstedt, F. W., *Z. Phys. Chem.* (Leipzig) **1933**, *20B*, 125; *C.A.* **1933**, 27, 2095.
4069. Knecht, L. A., and I. M. Kolthoff, *Inorg. Chem.* **1962**, *1*, 195.
4070. Knight, H. M., and J. T. Kelly, *Ind. Eng. Chem.* **1959**, *51*, 1355.
4071. Knight, H. S., and F. T. Weiss, *Anal. Chem.* **1962**, *34*, 749.
4072. Knobel, Y. K., E. A. Miroshmichenko, et al., *Izv. Akad. Nauk. SSSR, Ser. Khim.* **1971**, *3*, 485; *Acad. Sci. USSR Div. Chem. Sci.* **1971**, *20*, 425.
4073. Knoevenagel, E., *Ber. Deut. Chem. Gesell.* **1907**, *30*, 508.
4074. Knorr, L., *Ber. Deut. Chem. Gesell.* **1897**, *30*, 915.
4075. Knorr, L., *Ber. Deut. Chem. Gesell.* **1897**, *30*, 918.
4076. Knorr, L., *Ber. Deut. Chem. Gesell.* **1897**, *30*, 1492.
4077. Knorr, L., O. Rothe, et al., *Ber. Deut. Chem. Gesell.* **1911**, *44*, 1138.
4078. Knowles, C. L., *Ind. Eng. Chem.* **1920**, *12*, 881.
4079. Knowlton, J. W., and F. D. Rossini, *J. Res. Nat. Bur. Stand.* **1939**, *22*, 415.
4080. Knowlton, J. W., N. C. Schieltz, et al., *J. Amer. Chem. Soc.* **1946**, *68*, 208.
4081. Knox, J. H., and J. Riddick, *Trans. Faraday Soc.* **1966**, *62*, 1190.
4082. Knyshova, S. P., *Gig. Sanit.* **1968**, *33*, 37; *C.A.* **1968**, *68*, 62516.
4083. Kobayashi, R., Y. Fukuda, et al., *J. Chem. Soc. Jpn.*, *Ind. Chem. Sect.* **1953**, *56*, 102; *C.A.* **1954**, *48*, 8581.
4084. Kobayashi, R., T. Saito, et al., *J. Chem. Soc. Jpn.*, *Ind. Chem. Sect.* **1953**, *56*, 169; *C.A.* **1954**, *48*, 9099.
4085. Kobayashi, Y., *J. Soc. Org. Synthet. Chem. Jpn.* **1954**, *12*, 63; *C.A.* **1957**, *51*, 952.
4086. Kobe, K. A., M. R. Crawford, et al., *Ind. Eng. Chem.* **1955**, *47*, 1767.
4087. Kobe, K. A., and M. R. Crawford, *Pet. Refiner* **1958**, *37*, 125.
4088. Kobe, K. A., and R. E. Lynn, *Chem. Rev.* **1953**, *52*, 117.
4089. Kobe, K. A., and J. F. Mathews, *J. Chem. Eng. Data* **1970**, *15*, 182.
4090. Kobe, K. A., T. S. Okabe, et al., *J. Amer. Chem. Soc.* **1941**, *63*, 3251.
4091. Kobe, K. A., A. E. Ravicz, et al., *Chem. Eng. Data Series* **1956**, *1*, 50.
4092. Koch, F. K. V., *J. Chem. Soc.* **1927**, 647.
4093. Koch, F. K. V., *J. Chem. Soc.* **1928**, 269.
4094. Koch, H. P., *J. Chem. Soc.* **1949**, 387.

4095. Koch, T. R., and W. C. Purdy, *Anal. Chim. Acta* **1971,** *54,* 271.
4096. Koefoid, J., and J. V. Villadsen, *Acta Chem. Scand.* **1958,** *12,* 1124.
4097. Koffel, M. B., and R. A. Lad, *J. Chem. Phys.* **1948,** *16,* 420.
4098. Koga, Y., H. Takahashi, et. al., *Bull. Chem. Soc. Jpn.* **1974,** *47,* 84.
4099. Kohler, F., and F. Aronheom, *Ber. Deut. Chem. Gesell.* **1875,** *8,* 509.
4100. Kohlrausch, K. W. F., *Monatsh. Chem.* **1936,** *68,* 349; *C.A.* **1937,** *31,* 36.
4101. Kohlrausch, K. W. F., *Monatsh. Chem.* **1947,** *76,* 231; *C.A.* **1947,** *41,* 6153.
4102. Kohlrausch, K. W. F., and F. Koppl, *Monatsh. Chem.* **1933,** *63,* 255; *Chem Zentr.* **1934,** *I,* 3440.
4103. Kohlrausch, K. W. F., and F. Koppl, *Monatsh. Chem.* **1935,** *65,* 185; *C.A.* **1935,** *29,* 2850.
4104. Kohlrausch, K. W. F., and F. Koppl, *Z. Phys. Chem.* (Leipzig) **1934,** *24B,* 370; *Chem. Zentr.* **1934,** *I,* 3440.
4105. Kohlrausch, K. W. F., and F. Koppl, *Z. Phys. Chem.* (Leipzig) **1934,** 209; *C.A.* **1934,** *28,* 6065.
4106. Kohlrausch, K. W. F., and F. Koppl, *Z. Phys. Chem.* (Leipzig) **1933,** *21B,* 242; *Chem. Zentr.* **1933,** *II,* 336.
4107. Kohlrausch, K. W. F., F. Koppl, et al., *Z. Phys. Chem.* (Leipzig) **1933,** *22B,* 359; *Chem. Zentr.* **1933,** *II,* 3666.
4108. Kohlrausch, K. W. F., and O. Paulsen, *Monatsh. Chem.* **1939,** *72,* 268; *C. A.* **1939,** *33,* 8117.
4109. Kohlrausch, K. W. F., and A. Pongrantz, *Ber. Deut. Chem. Gesell.* **1933,** *66,* 1357.
4110. Kohlrausch, K. W. F., and A. Pongrantz, *Ber. Deut. Chem. Gesell.* **1934,** *67,* 976.
4111. Kohlrausch, K. W. F., and A. Pongrantz, *Ber. Deut. Chem. Gesell.* **1934,** *67B,* 1465.
4112. Kohlrausch, K. W. F., and A. Pongrantz, *Monatsh. Chem.* **1934,** *63,* 427; *Chem. Zentr.* **1934,** *I,* 3441.
4113. Kohlrausch, K. W. F., and A. Pongrantz, *Monatsh. Chem.* **1934,** *65,* 6; *C.A.* **1935,** *29,* 1325.
4114. Kohlrausch, K. W. F., and A. Pongrantz, *Monatsh. Chem.* **1937,** *70,* 226; *C.A.* **1937,** *31,* 4902; *Beil.* **EIII4,** 145.
4115. Kohlrausch, K. W. F., and A. Pongrantz, *Z. Phys. Chem.* (Leipzig) **1934,** *27B,* 176.
4116. Kohlrausch, K. W. F., and A. Pongrantz, et al., *Ber. Deut. Chem. Gesell.* **1933,** *66,* 1.
4117. Kohlrausch, K. W. F., and A. Pongrantz, et al., *Monatsh. Chem.* **1937,** *70,* 213; *C.A.* **1937,** *31,* 4902.
4118. Kohlrausch, K. W. F., and A. Reitz, *Z. Phys. Chem.* (Leipzig) **1940,** *45B,* 269.
4119. Kohlrausch, K. W. F., and H. W. Reitz, *Z. Phys. Chem.* (Leipzig) **1939,** *45B,* 249; *C.A.* **1940,** *34,* 4990.
4120. Kohlrausch, K. W. F., A. W. Reitz, et al., *Z. Phys. Chem.* (Leipzig) **1936,** *32B,* 229; *C.A.* **1936,** *30,* 4761.
4121. Kohlrausch, K. W. F., and R. Seka, *Ber. Deut. Chem. Gesell.* **1938,** *71,* 985.
4122. Kohlrausch, K. W. F., and R. Skrabl, *Monatsh. Chem.* **1937,** *70,* 377; *C.A.* **1947,** *31,* 7758.
4123. Kohlrausch, K. W. F., W. Stockmair, et al., *Monatsh. Chem.* **1935,** *67,* 80; *C.A.* **1936,** *30,* 2848.
4124. Kohlrausch, K. W. F., and W. Stockmair, *Z. Phys. Chem.* (Leipzig) **1935,** *29B,* 274; *C.A.* **1935,** *29,* 7190.
4125. Kohlrausch, K. W. F., and J. Wagner, *Z. Phys. Chem.* (Leipzig) **1939,** *45B,* 93; *C.A.* **1940,** *34,* 4990.
4126. Kohlrausch, K. W. F., and H. Wittek, *Monatsh. Chem.* **1941,** *74,* 1; *C.A.* **1942,** *36,* 6083.
4127. Kohlrausch, K. W. F., and H. Wittek, *Z. Phys. Chem.* (Leipzig) **1940,** *47B,* 55; *C.A.* **1940,** *35,* 977.
4128. Kohlrausch, K. W. F., and G. P. Ypsilanti, *Z. Phys. Chem.* (Leipzig) **1935,** *29B,* 274; *Chem. Zentr.* **1936,** *I,* 2065.
4129. Kohlrausch, K. W. F., and G. P. Ypsilanti, *Z. Phys. Chem.* (Leipzig) **1936,** *32B,* 407; *C.A.* **1936,** *30,* 5127.
4130. Kohn, C. A., and W. Tranton, *J. Chem. Soc.* **1899,** 1155.
4131. Kohn, M., *Monatsh. Chem.* **1913,** *34,* 781; *C.A.* **1913,** *7,* 2542.
4132. Koizumi, N., *J. Chem. Phys.* **1953,** *21,* 1898.
4133. Koizumi, N., *J. Chem. Phys.* **1957,** *27,* 625.
4134. Koizumi, N., and T. Hanai, *Bull. Inst. Chem. Research, Kyoto Univ.* **1955,** *33,* 14; *C.A.* **1956,** *50,* 634.
4135. Koizumi, N., and T. Hanai, *J. Phys. Chem.* **1956,** *60,* 1496.
4136. Koizumi, E., and S. Ouchi, *Nippon Kagaku Zasshi* **1970,** *91,* 501; *C.A.* **1970,** *73,* 91620.
4137. Kokoshko, Z., V. G. Yui, et al., *Zh. Obshch. Khim.* **1967,** *37,* 58; *C.A.* **1967,** *66,* 89338.
4138. Kolesnik, M. I., *Zh. Fiz. Khim.* **1974,** *48,* 114; *Russ. J. Phys. Chem.* **1974,** *48,* 648.

4139. Kolesov, V. P., L. S. Ivanov, et al., *Zh. Fiz. Khim.* **1970,** *44,* 2965; *Russ. J. Phys. Chem.* **1970,** *44,* 1688.

4140. Kolesov, V. P., E. A. Kosarukina, et al., *J. Chem. Thermodyn.* **1981,** 115.

4141. Kolesov, V. P., A. M. Martynov, et al., *Zh. Fiz. Khim.* **1962,** *36,* 2078; *C.A.* **1963,** *58,* 2909.

4142. Kolesov, V. P., A. M. Martynov, et al., *Zh. Fiz. Khim.* **1965,** *39,* 435; *C.A.* **1963,** *58,* 12515.

4143. Kolesov, V. P., S. M. Shtekher, et al., *Zh. Fiz. Khim.* **1968,** *42,* 1847; *Russ. J. Phys. Chem.* **1968,** *42,* 975.

4144. Kolesov, V. P., G. M. Slavutskaya, et al., *Zh. Fiz. Khim.* **1972,** *46,* 815; *Russ. J. Phys. Chem.* **1972,** *46,* 474.

4145. Kolesov, V. P., O. G. Talakin, et al., *Zh. Fiz. Khim.* **1968,** *42,* 2307; *Russ. J. Phys. Chem.* **1968,** *42,* 1218.

4146. Kolesov, V. P., O. G. Talakin, et al., *Zh. Fiz. Khim.* **1968,** *42,* 3033; *Russ. J. Phys. Chem.* **1968,** *42,* 1617.

4147. Kolesov, V. P., E. M. Tomareva, et al., *Zh. Fiz. Khim.* **1967,** *41,* 1528; *Russ. J. Phys. Chem.* **1967,** *41,* 817.

4148. Kolesov, V. P., I. D. Zenkov, et al., *Zh. Fiz. Khim.* **1963,** *37,* 720; *C.A.* **1963,** *59,* 1149.

4149. Kolesova, V. A., E. V. Kukharskaya, et al., *Izvest. Akad. Nauk. SSSR, Odtel. Khim. Nauk* **1953,** 294.

4150. Kolesova, V. A., and M. G. Voronkov, *Chem. Listy* **1957,** *51,* 686; *C.A.* **1957,** *51,* 11857.

4151. Kolling, O. W., and K. K. O'Hare, et al., *Trans. Kansas Acad. Sci.* **1963,** *66,* No. 3, 435.

4152. Kolosovskii, N. A., and A. Alimov, *Bull Soc. Chim. Fr.* **1935,** [5], *2,* 686; *C.A.* **1935,** *29,* 534.

4153. Kolosovskii, N. A., and V. V. Udovenko, *J. Gen. Chem.* (USSR) **1934,** *4,* 1029; *C.A.* **1935,** *29,* 3588.

4154. Kolossowskii, N., and W. W. Udowenko, *Compt. Rend.* **1933,** *197,* 519; *Beil.* **EII2,** 593; *Chem. Zentr.* **1933,** *II,* 2375.

4155. Kolthoff, I. M., S. Bruckenstein, et al., *J. Amer. Chem. Soc.* **1961,** *83,* 3927.

4156. Kolthoff, I. M., and M. K. Chantooni, *J. Phys. Chem.* **1968,** *72,* 2270.

4157. Kolthoff, I. M., and J. F. Coetzee, *J. Amer. Chem. Soc.* **1957,** *79,* 870.

4158. Kolthoff, I. M., and P. J. Elving, eds., *Treatise on Analytical Chemistry, Part 1, Vol. 1,* Interscience-Wiley, New York, 1959.

4159. Kolthoff, I. M., and P. J. Elving, eds., *Treatise on Analytical Chemistry, Part 1, Vol. 8,* Interscience-Wiley, New York, 1968.

4160. Kolthoff, I. M., and T. B. Reddy, *Inorg. Chem.* **1962,** *1,* 189.

4161. Kolthoff, I. M., and T. B. Reddy, *J. Electrochem. Soc.* **1961,** *108,* 980.

4162. Kolthoff, J. M., *Biochem. Z.* **1925,** *162,* 289.

4163. Kolysko, L. E., and E. Y. Gorodinskaya, *Zh. Fiz. Khim.* **1977,** *51,* 2215; *Russ. J. Phys. Chem.* **1977,** *51,* 1300.

4164. Komandin, A. V., and A. K. Bonetskaya, *Zh. Fiz. Khim.* **1954,** *28,* 1113; *C.A.* **1955,** *49,* 7905.

4165. Komandin, A. V., and B. D. Shimit, *Zh. Fiz. Khim.* **1963,** *37,* 510; *C.A.* **1963,** *59,* 4623.

4166. Komarewsky, V. I., S. C. Uhlick, et al., *J. Amer. Chem. Soc.* **1945,** *67,* 557.

4167. Komarowsky, A., *Chem.-Ztg.* **1903,** *27,* 807; *Chem. Zentr.* **1903,** *II,* 742, 1396.

4168. Komers, R., K. Kochloefl, et al., *Chem. Ind.* (London) **1958,** 1405.

4169. Komppa, G., and Y. Talvitie, *J. Prakt. Chem.* **1932,** *135,* 193; *C.A.* **1933,** *27,* 951.

4170. Konao, T., K. Watanabe, et al., Japanese Patent 72 11,963, April 14, 1972; *C.A.* **1972,** *72,* 33.

4171. Kondilenko, I. I., P. A. Korotokov, et al., *Opt. Spektrosk.* **1961,** *11,* 169; *C.A.* **1962,** *56,* 3038.

4172. Kondilenko, I. I., and V. E. Poporelov, *Opt. Spektrosk.* **1965,** *19,* 41; *C.A.* **1966,** *64,* 188.

4173. Kondo, S., Japanese Patent 5312, Oct. 16, 1953; *C.A.* **1955,** *49,* 6987.

4174. Kondrashov, V. A., *Gig. Tr. Prof. Zabol.* **1969,** *13,* 29; *C.A.* **1969,** *71,* 68958.

4175. Kondratov, V. K., *Russ. J. Phys. Chem.* (Eng. trans.) **1968,** *42,* 838.

4176. Konek, v. Fr., *Ber. Deut. Chem. Gesell.* **1906,** *39,* 2263.

4177. Konicek, J., M. Prochazka, et al., *Collect. Czech. Commun.* **1969,** *34,* 2249.

4178. Konicek, J., and I. Wadso, *Acta Chem. Scand.* **1970,** *24,* 2612.

4179. Konicek, J., and I. Wadso, *Acta Chem. Scand.* **1971,** *25,* 1541.

4180. Kononenko, O. K., and K. M. Herstein, *Ind. Eng. Chem., Chem Eng. Series* **1956,** *1,* 87.

4181. Kooi, E. R., E. I. Fulmer, et al., *Ind. Eng. Chem.* **1948,** *40,* 1440.

4182. Kopf, R., A. Loeser, et al., *Arch. Exptl. Path. Pharmakol.* **1950,** *210,* 346; *C.A.* **1951,** *45,* 5308.

4183. Kopper, H., R. Seka, et al., *Monatsh. Chem.* **1932,** *61,* 397; *Sitzber. Akad. Wiss. Wien, Math.-Naturw. Klasse* **1932,** *Abt. IIa, 141,* 465; *C.A.* **1933,** *27,* 4172.
4184. Koptyug, V. A., A. G. Khmel'nitskii, et al., *Izv. Sib. Otd. Akad. Nauk SSSR, Ser. Khim. Nauk* **1964,** 116; *C.A.* **1964,** *61,* 11813.
4185. Korbakova, A. I., *Vestn. Akad. Nauk SSSR* **1964,** *19,* 17; *C.A.* **1964,** *61,* 16694.
4186. Kordes, E., *Z. Elektrochem.* **1954,** *58,* 424; *C.A.* **1955,** *49,* 2140.
4187. Korenman, Y. I., *Zh. Fiz. Khim.* **1972,** *46,* 578; *Russ. J. Phys. Chem.* **1972,** *46,* 334.
4188. Korlyakova, E. A., *Toksikol. Novykh. Prom. Khim. Veshchestv.* **1961,** *3,* 23; *C.A.* **1962,** *57,* 12828.
4189. Kornblum, N., A. S. Erickson, et al., *J. Org. Chem.* **1982,** *47,* 453.
4190. Kornblum, H., and D. C. Iffland, *J. Amer. Chem. Soc.* **1955,** *77,* 6653.
4191. Kornblum, N., H. O. Larson, et al., *J. Amer. Chem. Soc.* **1956,** *78,* 1497.
4192. Kornev, K. A., A. P. Grekov, et al., *Khim. Prom. Nauk. Tekhn. Zb.* **1963,** *1,* 63; *C.A.* **1963,** *59,* 6252.
4193. Korlchenko, A. Y., G. N. Kravchuck, et al., *Russ. J. Phys. Chem.* **1980,** *54,* 896.
4194. Korolev, B. A., M. A. Maltseva, et al., *Zh. Obsch. Khim.* **1974,** *44,* 864; *J. Gen. Chem.* (USSR) **1974,** *44,* 833.
4195. Korosi, G., and E. Kovats, *J. Chem. Eng. Data* **1981,** *26,* 232.
4196. Korostyshevskii, I. Z., M. M. Aleksankin, et. al., *Teor. Eksp. Khim.* **1972,** *8,* 777; *Theor. Expt. Chem.* **1972,** *8,* 640.
4197. Korous, D. J., and R. W. Neuzil, U.S. Patent 4,044,062, Aug. 23, 1977; *C.A.* **1978,** *88,* 6514.
4198. Korpela, J., *Acta Chem. Scand.* **1971,** *25,* 2852.
4199. Korschun, T. W., and K. W. Roll, *J. Russ. Phys. Chem. Gesell.* **1917,** *49,* 153; *Chem. Zentr.* **1923,** *III,* 775.
4200. Korshunov, A. V., and P. S. Sarapkin, *Trudy Sibir. Tekhnol. Inst.* **1959,** *24,* 13; *C.A.* **1962,** *56,* 3038.
4201. Kortuem, G., and E. Faltusz, *Chem.-Ing.-Tech.* **1961,** *33,* 599; *C.A.* **1962,** *56,* 283.
4202. Kortum, G., *Z. Phys. Chem.* (Leipzig) **1939,** *42B,* 39; *C.A.* **1939,** *33,* 3261.
4203. Kosakewitsch, P., *Z. Phys. Chem.* (Leipzig) **1928,** *133A,* 1.
4204. Kosarukina, E. A., V. P. Kolesov, et al., *Zh. Fiz. Khim.* **1978,** *52,* 509; *Russ. J. Phys. Chem.* **1978,** *52,* 294.
4205. Kosenko, A. M., *Mater, Povolzh. Konf. Fiziol. Uchasteim. Biokhim., Farmakol., Morfol., 6th* **1973,** *2,* 36; *C.A.* **1975,** *82,* 81353.
4206. Koshelev, V. S., S. Sh. Byk, et al., *Khim. Prom.* (Moscow) **1975,** 636; *C.A.* **1976,** *84,* 123908.
4207. Koskikallio, J., and S. Syrjaepalo, *Soumen Kemistilehti* **1964,** *37B,* 120; *C.A.* **1965,** *63,* 14119.
4208. Kosower, E. M., *J. Amer. Chem. Soc.* **1958,** *80,* 3253.
4209. Kost, A. N., and A. M. Yurkevich, *Zh. Obshch. Khim.* **1953,** *23,* 1738; *C.A.* **1954,** *48,* 13622.
4210. Kostelnik, R., M. P. Williamson, et al., *Can. J. Chem.* **1969,** *47,* 3313.
4211. Kostyanovsky, R. G., and V. G. Plekhanov, *Org. Mass. Spectrom.* **1972,** *6,* 1183.
4212. Kostyanovsky, R. G., V. G. Plekhanov, et al., *Org. Mass. Spectrom.* **1972,** *6,* 1199.
4213. Kostyuk, N. G., S. V. L'vov, et al., *Zh. Prikl. Khim.* **1962,** *35,* 698; *C.A.* **1962,** *57,* 2068.
4214. Kosyan, Sh. A., *Tr. Erevan. Gos. Inst. Usoversh. Vrachei* **1967,** *3,* 617; *C.A.* **1969,** *71,* 37213.
4215. Kotake, M., Japanese Patent 133,250, Nov. 14, 1939; *C.A.* **1941,** *35,* 4042.
4216. Kotera, A., S. Nishmura, et al., *J. Chem. Soc. Jpn.* **1944,** *65,* 527; *C.A.* **1947,** *41,* 3335.
4217. Kotera, A., S. Shibata, et al., *J. Amer. Chem. Soc.* **1955,** *77,* 6183.
4218. Koteswaram, P., *Z. Physik.* **1938,** *110,* 118; *C.A.* **1938,** *32,* 8938.
4219. Kotlyarevskiĭ, I. L., A. N. Volkov, et al., *Izvest. Sibir. Odtel. Akad. Nauk. SSSR* **1959,** *3,* 62; *C.A.* **1959,** *53,* 21605.
4220. Kovalev, I. F., *Optika i Spectroskopiya* **1959,** *6,* 594; *C.A.* **1959,** *53,* 16695.
4221. Kovats, E., *Helv. Chem. Acta* **1958,** *411,* 1915.
4222. Kovrigina, L. P., and L. I. Bogdanov, *Zh. Fiz. Khim.* **1970,** *44,* 1571; *Russ. J. Phys. Chem.* **1970,** *44,* 881.
4223. Kowalewski, V. J., *J. Mol. Spectrosc.* **1969,** *31,* 256.
4224. Kowalewski, V. J., and D. G. de Kowalewski, *Arkiv. Kemi.* **1960,** *16,* 373.
4225. Kowalewski, V. J., and D. G. de Kowalewski, *J. Chem. Phys.* **1960,** *32,* 1272.
4226. Kowalewski, V. J., and D. G. de Kowalewski, *J. Chem. Phys.* **1960,** *33,* 1794.
4227. Kowalski, B., A. Orszagh, et al., *J. Chem. Thermodyn.* **1976,** *8,* 425.

4228. Koyano, T., K. Kaneko, et al., *Kogyo Kagaku Zasshi* **1971**, *74*, 203.

4229. Kozacik, A. P., and E. E. Reid, *J. Amer. Chem. Soc.* **1938**, *60*, 2436.

4230. Kozima, K., and S. Mizushima, *Sci. Papers Inst. Phys. Chem. Research* (Tokyo) **1937**, *31*, 296; *C.A.* **1937**, *31*, 6558.

4231. Kozima, M. P., and S. M. Skuratov, *Doklady Akad. Nauk. SSSR* **1959**, *127*, 561; *C.A.* **1960**, *54*, 284.

4232. Kozlov, N. S., *J. Appl. Chem.* (USSR) **1937**, *10*, 116; *C.A.* **1937**, *31*, 4263.

4233. Kozlov, N. S., and S. Ya. Chumakov, *Zh. Prikl. Khim.* **1957**, *30*, 318; *C.A.* **1957**, *51*, 12816.

4234. Kraemer, G., and M. Grodzki, *Ber. Deut. Chem. Gesell.* **1928**, *9*, 1876.

4235. Krafft, F., *Ber. Deut. Chem. Gesell.* **1882**, *15*, 1687.

4236. Krafft, F., *Ber. Deut. Chem. Gesell.* **1883**, *16*, 3020.

4237. Krafft, F., and H. Noerdlinger, *Ber. Deut. Chem. Gesell.* **1889**, *22*, 816.

4238. Kraft, M., and G. Spiteller, *Justus Liebigs Ann. Chem.* **1968**, *712*, 28.

4239. Krajewski, G. A., and G. Krajewski, *Pharmazie* **1950**, *5*, 324; *C.A.* **1950**, *44*, 10.

4240. Kranzfelder, A. L., and F. J. Sowa, *J. Amer. Chem. Soc.* **1937**, *59*, 1490.

4241. Kranzfelder, A. L., and R. R. Vogt, *J. Amer. Chem. Soc.* **1938**, *60*, 1714.

4242. Krase, N. W., and J. B. Goodman, *Ind. Eng. Chem.* **1930**, *22*, 13.

4243. Krasnoshchek, A. P., *Zh. Org. Khim.* **1972**, *8*, 2100; *C.A.* **1973**, *78*, 28664.

4244. Kratzke, H., E. Spillner, et al., *J. Chem. Thermodyn.* **1982**, *14*, 1175.

4245. Kraus, C. A., and J. E. Bishop, *J. Amer. Chem. Soc.* **1921**, *43*, 1568.

4246. Kraus, C. A., and J. E. Bishop, *J. Amer. Chem. Soc.* **1922**, *44*, 2206.

4247. Kraus, C. A., and C. C. Callis, *J. Amer. Chem. Soc.* **1923**, *45*, 2624.

4248. Kraus, C. A., and R. M. Fuoss, *J. Amer. Chem. Soc.* **1933**, *55*, 21.

4249. Kraus, C. A., and R. A. Vingee, *J. Amer. Chem. Soc.* **1934**, *56*, 511.

4250. Krausz, F., *Ann. Chim.* **1949**, *4*, 811; *C.A.* **1950**, *44*, 5811.

4251. Krchma, I. J., and J. W. Williams, *J. Amer. Chem. Soc.* **1927**, *49*, 2408.

4252. Krech, M., J. W. Price, et al., *Can. J. Chem.* **1972**, *50*, 2935.

4253. Kreevoy, M. M., B. E. Eichinger, et al., *J. Org. Chem.* **1964**, *29*, 1641.

4254. Kreglewski, A., *Bull. Acad. Pol. Sci. Ser. Sci. Chim.* **1962**, *10*, 629; *C.A.* **1963**, *59*, 2180.

4255. Kreimeier, O. R., U.S. Patent 2,106,181, Jan. 25, 1938; *Official Gaz. U.S. Pat Office* **1938**, *486*, 726.

4256. Kreis, R. W., and R. W. Wood, *J. Chem. Thermodyn.* **1969**, *1*, 523.

4257. Kremann, R., F. Gugl, et al., *Monatsh. Chem.* **1914**, *35*, 1365.

4258. Kremers, E., *Chem. Zentr.* **1909**, *I*, 21.

4259. Kreshkov, A. P., N. Sh. Aldarova, et al., *Zh. Fiz. Khim.* **1970**, *44*, 2089; *Russ. J. Phys. Chem.* **1970**, *44*, 1186.

4260. Kreshkov, A. P., Ya. A. Gurvich, et al., *J. Anal. Chem.* USSR (Eng. trans.) **1972**, *27*, 1043.

4261. Kreshkov, A. P., Ya. A. Gurvich, et al., *Zh. Obshch. Khim.* **1972**, *42*, 2513; *J. Gen. Chem.* (USSR) **1972**, *42*, 2503.

4262. Kreshkov, A. P., Yu. Ya. Makhailenko, et al., *Zh. Fiz. Khim.* **1954**, *28*, 537.

4263. Kreshkov, A. P., T. Niyazkhonov, et al., *Zh. Fiz. Khim.* **1980**, *54*, 756; *Russ. J. Phys. Chem.* **1980**, *54*, 433.

4264. Kreshkov, A. P., N. T. Smolova, et al., *Russ. J. Phys. Chem.* (Eng. trans.) **1972**, *46*, 382.

4265. Kreshkov, A. P., N. T. Smolova, et al., *Zh. Fiz. Khim.* **1977**, *51*, 1827; *Russ. J. Phys. Chem.* **1977**, *51*, 1073.

4266. Kretow, A. E., *J. Russ. Phys. Chem. Soc.* **1929**, *61*, 2350; *Chem. Zentr.* **1930**, *II*, 370.

4267. Kretschmer, C. B., *J. Phys. Colloid Chem.* **1951**, *55*, 1351.

4268. Kretschmer, C. B., and R. Wiebe, *J. Amer. Chem. Soc.* **1952**, *74*, 1276.

4269. Kretzberger, A., and P. A. Kalter, *J. Phys. Chem.* **1961**, *65*, 624.

4270. Kreuzer, J., and R. Mecke, *Z. Phys. Chem.* (Leipzig) **1941**, *49B*, 309; *C.A.* **1942**, *36*, 6898.

4271. Krishna, B., and B. Prakash, *Aust. J. Chem.* **1969**, *22*, 2679.

4272. Krishna, B., and A. H. Srivastava, *Aust. J. Chem.* **1966**, *19*, 1847; *C.A.* **1966**, *65*, 18475.

4273. Krishna, B., and K. K. Srivastava, *J. Chem. Phys.* **1960**, *32*, 663.

4274. Krishna, K. V. G., *Indian J. Phys.* **1957**, *31*, 283; *C.A.* **1957**, *51*, 17293.

4275. Krishna, K. V. G., *Trans. Faraday Soc.* **1957**, *53*, 767.

4276. Krishnaji, and R. K. Laloraya, *Indian J. Pure Appl. Phys.* **1974**, *12*, 585; *C.A.* **1975**, *82*, 42509.

4277. Krishnamurty, V. N., and S. Soundararjan, *J. Inorg. Nucl. Chem.* **1965**, *27*, 2341.
4278. Krishnan, C. V., and H. L. Friedman, *J. Phys. Chem.* **1971**, *75*, 3598.
4279. Krishnan, K., *Proc. Indian Acad. Sci.* **1961**, *53A*, 151; *C.A.* **1961**, *55*, 19478.
4280. Kronberger, H., and J. Weiss, *J. Chem. Soc.* **1944**, 464.
4281. Krongauz, V. A., *Dokl. Akad. Nauk. SSSR* **1965**, *162*, 1300; *C.A.* **1965**, *63*, 9780.
4282. Kronick, P. L., and R. M. Fuoss, *J. Amer. Chem. Soc.* **1955**, *77*, 6114.
4283. Kross, R. D., V. A. Fassel, et al., *J. Amer. Chem. Soc.* **1956**, *78*, 1332.
4284. Krueger, J. H., and W. A. Johnson, *Inorg. Chem.* **1968**, *7*, 679.
4285. Krueger, P. J., *Can. J. Chem.* **1962**, *40*, 2300; **1963**, *41*, 363.
4286. Krueger, P. J., and H. D. Mettee, *Can. J. Chem.* **1965**, *43*, 2970.
4287. Krueger, P. J., and H. D. Mettee, *J. Mol. Spectrosc.* **1965**, *18*, 131; *C.A.* **1966**, *64*, 4906.
4288. Krumgal'z, B. S., Yu. I. Gerzhberg, et al., *Zh. Obshch. Khim.* **1971**, *41*, 39; *J. Gen. Chem.* (USSR) **1971**, *41*, 35.
4289. Krupicka, J., *Prom. Krystalizace, Sb. Prednasek Semin.* **1971**, 90; *C.A.* **1972**, 76, 129351.
4290. Krupp, R. H., E. A. Piotrowski, et al., *Spectrosc. Mol.* **1974**, *23*, 20; *C.A.* **1974**, *81*, 17623.
4291. Krygowski, T. M., and W. R. Fawcett, *J. Amer. Chem. Soc.* **1975**, *97*, 2143.
4292. Kryshtal, G. V., L. A. Yanovskaya, et al., *Izv. Akad. Nauk SSSR, Ser. Khim.* **1972**, *21*, 2362; *Bull. Acad. Sci. USSR, Div. Chem. Sci.* **1972**, *21*, 2307.
4293. Kryszewski, M., *Rocz. Chem.* **1952**, *26*, 350; *C.A.* **1954**, *48*, 8634.
4294. Kubicek, A. J., and P. T. Eubank, *J. Chem. Eng. Data* **1972**, *17*, 232.
4295. Kubo, M., *Sci. Papers Inst. Phys. Chem. Res.* (Tokyo) **1935**, *27*, 65; *C.A.* **1935**, *29*, 648.
4296. Kubo, M., *Sci. Papers Inst. Phys. Chem. Res.* (Tokyo) **1936**, *30*, 238; *C.A.* **1937**, *31*, 1669.
4297. Kubo, V. M., *Sci. Papers Inst. Phys. Chem. Res.* (Tokyo) **1936**, *29*, 181; *C.A.* **1936**, *30*, 7002.
4298. Kuchinskaya, K., *Sbornik Trudov Optynogo Zavoda im. Akad. S.V. Lebedeva* **1938**, 27; *Khim. Referat. Zh.* **1939**, *2*, 23; *C.A.* **1940**, *34*, 3147.
4299. Kudchadker, A. P., G. H. Alani, et al., *Chem. Rev.* **1968**, *68*, 659.
4300. Kudchadker, A. P., S. A. Kudchadker, et al., *J. Phys. Chem. Ref. Data* **1978**, *7*, 425.
4301. Kudchadker, A. P., S. A. Kudchadker, et al., *J. Phys. Chem. Ref. Data* **1979**, *8*, 499.
4302. Kudchadker, A. P., and J. J. McKetta, *AIChEJ* **1961**, *7*, 707.
4303. Kudchadker, S. A., A. P. Kudchadker, et al., *J. Phys. Chem. Ref. Data* **1978**, *7*, 417.
4304. Kudchadker, S. A., and A. P. Kudchadker, *J. Phys. Chem. Ref. Data* **1975**, *4*, 457.
4305. Kudchadker, S. A., and A. P. Kudchadker, *J. Phys. Chem. Ref. Data* **1978**, *7*, 1285.
4306. Kudo, K., and R. Ito, *Toho Igakkai Zasshi* **1972**, *19*, 415; *C.A.* **1973**, *78*, 80486.
4307. Kudryashov, I. V., and R. I. Savachenko, *Izv. Vysshikh Uchebn. Zavedenii, Khim. i. Khim. Tekhnol.* **1965**, *8*, 602; *C.A.* **1966**, *64*, 18438.
4308. Kudryashava, N. I., and N. V. Khromov-Borisov, *Zh. Org. Khim.* **1966**, *2*, 578; *C.A.* **1966**, *65*, 8716.
4309. Kuivila, H. G., and O. F. Beumel, *J. Amer. Chem. Soc.* **1961**, *83*, 1246.
4310. Kulagina, N. K., and A. I. Korbakova, *Toksikol. Novykh Prom. Khim. Vestchestv* **1961**, *3*, 81; *C.A.* **1962**, *57*, 12828.
4311. Kulagina, N. K., and T. A. Kochetkova, *Toksikol. Novykh Prom. Khim. Vestchestv* **1965**, *7*, 56; *C.A.* **1965**, *63*, 7548.
4312. Kofod, H., L. E. Sutton, et al., *J. Chem. Soc.* **1952**, 1467.
4313. Kulevsky, N. I., *J. Chem. Eng. Data* **1967**, *11*, 492.
4314. Kulkarni, P. S., M. M. Khan, et al., *J. Heterocycl. Chem.* **1980**, *17*, 929.
4315. Kumar, A., O. Prakash, et al., *J. Chem. Eng. Data* **1981**, *26*, 64.
4316. Kumar, S., V. P. Gupta, et al., *Acta Cienc. Indica, (Ser.) Phys.* **1979**, *5*, 8.
4317. Kumaran, M. K., G. C. Benson, et al., *J. Chem. Eng. Data* **1983**, *28*, 66.
4318. Kumler, W. D., *J. Amer. Chem. Soc.* **1952**, *74*, 261.
4319. Kumler, W. D., *J. Amer. Chem. Soc.* **1961**, *83*, 4983.
4320. Kumler, W. D., and T. F. Halverstadt, *J. Amer. Chem. Soc.* **1941**, *63*, 2182.
4321. Kumler, W. D., and C. W. Porter, *J. Amer. Chem. Soc.* **1934**, *56*, 2549.
4322. Kundu, K. K., P. K. Chattopadhyay, et al., *J. Phys. Chem.* **1970**, *74*, 2633.
4323. Kuramoto, M., A. Kimura, et al., *Shikoku Igaku Zasshi* **1972**, *28*, 276.
4324. Kuratani, K., and S. Mizushima, *J. Chem. Phys.* **1954**, *22*, 1403.
4325. Kurbatow, V. F. J., *Comm. Trav. Scient. Techn. Rep. Leningrad* **1925**, *17*, 24.

4326. Kurbatow, V. Y., *J. Gen. Chem.* (USSR) **1947**, *17*, 999; *C.A.* **1948**, *42*, 4829.

4327. Kurbatow, V. Y., *Zh. Obshch. Khim.* **1948**, *18*, 372; *C.A.* **1949**, *43*, 30.

4328. Kurbatow, W. A., *J. Russ. Phys. Chem. Soc.* **1902**, *34*, 776; *J. Chem. Soc.* **1903Aii**, 246.

4329. Kurbatow, W. A., *J. Russ. Phys. Chem. Soc.* **1903**, *35*, 319; *Chem. Zentr.* **1903**, *II*, 323; *J. Chem. Soc.* **1903Aii**, 710.

4330. Kurbatow, W. A., *J. Russ. Phys. Chem. Soc.* *40*, 1471; *Chem. Zentr.* **1909**, 635; *J. Chem. Soc.* **1909Aii**, 119.

4331. Kurnakow, N., and S. Zemcuzny, *Z. Phys. Chem.* (Leipzig) **1913**, *83*, 481.

4332. Kuroda, Y., and M. Kubo, *J. Polymer Sci.* **1957**, *26*, 323; *C.A.* **1958**, *52*, 10715.

4333. Kurtyka, Z. M., and E. A. Kurtyka, *J. Chem. Eng. Data* **1979**, *24*, 15.

4334. Kurtz, S. S., S. Amon, et al., *Ind. Eng. Chem.* **1950**, *42*, 174.

4335. Kurtz, S. S., and M. R. Lipkin, *J. Amer. Chem. Soc.* **1941**, *63*, 2158.

4336. Kurz, J. L., and M. A. Stein, *J. Phys. Chem.* **1976**, *80*, 154.

4337. Kusama, T., and D. Koike, *J. Chem. Soc. Jpn., Pure Chem. Sect.* **1951**, *72*, 779; *C.A.* **1953**, *47*, 2155.

4338. Kusano, T., and D. Koike, *J. Chem. Soc. Jpn.* **1952**, *73*, 270; *C.A.* **1953**, *47*, 3118.

4339. Kusano, K., *J. Chem. Eng. Data* **1978**, *23*, 141.

4340. Kusano, K., *Nippon Kagaku Zasshi* **1958**, *78*, 614; *C.A.* **1959**, *53*, 3786.

4341. Kusano, K., J. Suurkuusk, et al., *J. Chem. Thermodyn.* **1973**, *5*, 757.

4342. Kusano, K., and I. Wadso, *Acta Chem. Scand.* **1970**, *24*, 2037.

4343. Kusano, K., and I. Wadso, *Acta Chem. Scand.* **1971**, *25*, 219.

4344. Kusano, K., and I. Wadso, *Bull. Soc. Chem. Jpn.* **1971**, *44*, 1705.

4345. Kushchenko, V. V., and K. P. Mishchenko, *Zh. Prikl. Khim.* **1968**, *41*, 646; *J. Appl. Chem.* (USSR) **1968**, *41*, 620.

4346. Kushner, L. M., R. W. Crowe, et al., *J. Amer. Chem. Soc.* **1950**, *72*, 1091.

4347. Kushneva, V. S., G. A. Koloskova, et al., *Gig. Tr. Prof. Zabol.* **1983**, *1*, 46; *C.A.* **1983**, *98*, 120807.

4348. Kuskov, C. K., and V. A. Zhukova, *Izv. Akad. Nauk SSSR Otdel. Khim. Nauk* **1956**, 733; *C.A.* **1957**, *51*, 1877.

4349. Kuss, E., *Z. Angew. Phys.* **1955**, *7*, 372; *C.A.* **1956**, *50*, 3028.

4350. Kuss, E., J. Moos, et al., *Naturwissenschaften* **1961**, *48*, 73; *C.A.* **1961**, *55*, 19748.

4351. Kutepov, E. N., *Gig. Sanit.* **1968**, *33*, 32; *C.A.* **1968**, *68*, 62489.

4352. Kutsenko, A., and V. Lyubomilov, *Zh. Prikl. Khim.* **1958**, *31*, 1419; *C.A.* **1959**, *53*, 2070.

4353. Kuwae, A., and K. Machida, *Spectrochim. Acta, Part A* **1979**, *35*, 27.

4354. Kuz'micheva, M. N., *Gig. Sanit.* **1962**, *27*, 41; *C.A.* **1963**, *58*, 9546.

4355. Kuznetsov, P. I., *Gig. Sanit.* **1967**, *32*, 7; *C.A.* **1967**, *67*, 36298.

4356. Kwasnik, W., U.S. Patent 2,658,927, Nov. 10, 1953; *C.A.* **1954**, *48*, 12788.

4357. Kynaston, W., and J. F. Martin, *J. Appl. Chem. Biotechnol.* **1977**, *27*, 296.

4358. Kyrides, L. P., U.S. Patent 2,161,826, June 13, 1939; *C.A.* **1939**, *33*, 7315.

4359. Kyte, C. T., G. H. Jeffery, et al., *J. Chem. Soc.* **1960**, 4454.

L

4360. Labbauf, A., J. R. Greenshields, et al., *J. Chem. Eng. Data* **1961**, *6*, 261.

4361. Lacher, J. R., A. Amador, et al., *Trans. Faraday Soc.* **1967**, *63*, 1608.

4362. Lacher, J. R., E. Emery, et al., *J. Phys. Chem.* **1956**, *60*, 429; *Trans. Faraday Soc.* **1956**, *52*, 1500.

4363. Lacher, J. R., H. B. Gottlieb, et al., *Trans. Faraday Soc.* **1962**, *58*, 2348.

4364. Lacher, J. R., L. E. Hummel, et al., *J. Amer. Chem. Soc.* **1950**, *72*, 5486.

4365. Lacher, J. R., R. E. Scruby, et al., *J. Amer. Chem. Soc.* **1949**, *71*, 1797.

4366. Lacher, J. R., and M. A. Skinner, *J. Chem. Soc. A* **1968**, 1034.

4367. Lacher, J. R., C. H. Walden, et al., *J. Amer. Chem. Soc.* **1949**, *71*, 3026.

4368. Lacy, B. S., British Patent 16,194, July 7, 1914; *C.A.* **1916**, *10*, 92.

4369. Lady, J. H., and K. B. Whetsel, *Spectrochim. Acta* **1965**, *21*, 1669.

4370. Lafontaine, I., *Bull. Soc. Chim. Belges* **1958**, *67*, 153; *C.A.* **1959**, *53*, 791; see also [7483].

4371. Lagemann, R. T., E. G. McLeroy, et al., *J. Amer. Chem. Soc.* **1951**, *73*, 5891.

4372. Lagemann, R. T., D. R. McMillan, et al., *J. Chem. Phys.* **1948**, *16*, 247.

4373. Lagemann, R. T., D. R. McMillan, et al., *J. Chem. Phys.* **1949**, *17*, 369.

4374. Lagemann, R. T., and H. H. Nielsen, *J. Chem. Phys.* **1942,** *10,* 668.
4375. Lagutkin, O. D., and E. I. Kuropatkin, *Zh. Fiz. Khim.* **1981,** *55,* 1329; *Russ. J. Phys. Chem.* **1981,** *55,* 746.
4376. Lainez, A., I. Escudero, et al., *J. Chem. Thermodyn.* **1984,** *16,* 7.
4377. Lake, D. E., and A. A. Asadorian, U.S. Patent 2,553,518, May 15, 1951; *C.A.* **1952,** *46,* 2561.
4378. Lake, J. S., and A. J. Harrison, *J. Chem. Phys.* **1959,** *30,* 361.
4379. Lakomy, J., and L. Lehar, *Chem. Listy* **1965,** *59,* 985; *C.A.* **1965,** *63,* 106721.
4380. Lalande, A., *J. Chim. Phys.* **1934,** *31,* 583; *C.A.* **1935,** *29,* 5339.
4381. Lam, V. T., and G. C. Benson, *Can. J. Chem.* **1970,** *48,* 3773.
4382. Lamb, A. B., and E. E. Roper, *J. Amer. Chem. Soc.* **1940,** *62,* 806.
4383. Lambert, P., and J. Lecomte, *Ann. Phys.* **1932,** *18,* 329; *C.A.* **1933,** *27,* 1825.
4384. Lambert, P., and J. Lecomte, *Ann. Phys.* **1938,** *10,* 503; *C.A.* **1939,** *33,* 4127.
4385. Lambert, P., and J. Lecomte, *Compt. Rend.* **1932,** *77,* 960.
4386. Lambert, P., and J. Lecomte, *Compt. Rend.* **1939,** *208,* 740; *C.A.* **1939,** *33,* 3693.
4387. Lamond, J., *Analyst* **1949,** *74,* 560.
4388. Lamprecht, W., *Fette, Seifen, Anstrichmittle* **1959,** *61,* 96; *Chem. Zentr.* **1960,** 5263.
4389. Landa, S., and J. Mostecky, *Collect. Czech. Chem. Commun.* **1955,** *20,* 430; *Chem. Listy* **1955,** *49,* 67; *C.A.* **1956,** *50,* 771.
4390. Landee, F. A., and I. B. Johns, *J. Amer. Chem. Soc.* **1941,** *63,* 2891.
4391. Landolt, H., *Ann. Chem.* **1877,** *189,* 333.
4392. Landolt, H., and H. John, *Z. Phys. Chem.* (Leipzig) **1892,** *10,* 293.
4393. Landsberg, G. S., and S. A. Ukholin, *Compt. Rend. Acad. Sci. URSS* **1937,** *16,* 391.
4394. Lane, M. R., J. W. Linnett, et al., *Proc. Roy. Soc.* (London) **1953,** *216A,* 316.
4395. Lane, W. H., *Ind. Eng. Chem., Anal. Ed.* **1946,** *18,* 295.
4396. Lang, H. R., *Proc. Roy. Soc.* (London) **1928,** *118A,* 138.
4397. Lange, J., *Z. Phys. Chem.* (Leipzig) **1932,** *161A,* 77; *Chem. Zentr.* **1932,** *II,* 2848.
4398. Lange, N. A., ed., *Handbook of Chemistry,* 11th ed., McGraw-Hill, New York, 1967.
4399. Langguth, U., and H.-J. Bittrich, *Z. Phys. Chem.* (Leipzig) **1970,** *244,* 327.
4400. Langseth, A., and B. Bak, *Kgl. Danske Videnskab. Selskab, Mat.-Fys. Medd.* **1947,** *24,* No. 3, 16; *C.A.* **1948,** *42,* 459.
4401. Langseth, L. H., *Nature* **1929,** *124,* 92; *C.A.* **1929,** *23,* 5108.
4402. Lannung, A., *Z. Phys. Chem.* (Leipzig) **1932,** *161A,* 255.
4403. Lantz, V., *J. Amer. Chem. Soc.* **1940,** *62,* 3260.
4404. Lanum, W. J., and J. C. Morris, *J. Chem. Eng. Data* **1969,** *14,* 93.
4405. LaPaglia, S. R., *Trans. Faraday Soc.* **1964,** *60,* 1210.
4406. Lapik, A. S., *Izv. Sibirsk. Odt. Akad. Nauk SSSR, Ser. Biol.-Med. Nauk* **1965,** 91; *C.A.* **1966,** *64,* 20497.
4407. LaPlanche, L. A., and M. T. Rogers, *J. Amer. Chem. Soc.* **1964,** *86,* 337.
4408. LaPlanche, L. A., H. B. Thompson, et al., *J. Phys. Chem.* **1965,** *69,* 1482.
4409. Lappert, M. F., *Chem. Rev.* **1956,** *56,* 959.
4410. Lapporte, S. J., and W. R. Schuett, *J. Org. Chem.* **1963,** *28,* 1947.
4411. Lardicci, L., R. Rossi, et al., *Chim. Ind.* (Milan) **1962,** *44,* 1002; *C.A.* **1964,** *60,* 13138.
4412. Lardicci, L., P. Salvadori, et al., *Ann. Chim.* (Rome) **1962,** *52,* 652; *C.A.* **1963,** *58,* 5494.
4413. Larionov, A. G., *Uch. Zap.-Mosk. Nauch.-Issled. Inst. Gig.* **1975,** *22,* 64; *C.A.* **1977,** *86,* 66420.
4414. LaRochelle, J. H., and A. A. Vernon, *J. Amer. Chem. Soc.* **1950,** *72,* 3293.
4415. Larson, R. C., and R. T. Iwamoto, *Inorg. Chem.* **1962,** *1,* 316.
4416. Larson, R. C., and R. T. Iwamoto, *J. Amer. Chem. Soc.* **1960,** *82,* 3239.
4417. Larson, R. G., and H. Hunt, *J. Phys. Chem.* **1939,** *43,* 417.
4418. Larsson, E., *Z. Phys. Chem.* (Leipzig) **1932,** *159A,* 315; *Chem. Zentr.* **1932,** *II,* 678.
4419. Lascombe, J., and M. L. Josien, *Bull. Soc. Chim. Fr.* **1955,** 1227; *C.A.* **1956,** *50,* 2296.
4420. Laskorin, B. N., V. V. Yakshin, et al., *Dokl. Akad. Nauk SSSR* **1974,** *215,* 595; *C.A.* **1974,** *80,* 145159.
4421. Lassar-Cohn, *Justus Liebigs Ann. Chem.* **1895,** *284,* 226.
4422. Lastovskii, R. P., and V. Y. Temkina, *Khim. Prom.* **1958,** 219; *C.A.* **1959,** *53,* 1101.
4423. Laszlo, H. de, *J. Amer. Chem. Soc.* **1927,** *49,* 2106.

4424. Laszlo, H. G. de, *Z. Phys. Chem.* (Leipzig) **1925**, *118*, 369; *C.A.* **1926**, *20*, 1178.
4425. Lauer, K., *Ber. Deut. Chem. Gesell.* **1937**, *70*, 1127, 1288.
4426. Laughlin, R. C., and F. C. Whitmore, *J. Amer. Chem. Soc.* **1933**, *55*, 2607.
4427. Lauterbur, P. C., *J. Chem. Phys.* **1965**, *43*, 360.
4428. Lautie, A., and A. Novak, *J. Chim. Phys. Physicochim. Biol.* **1972**, *69*, 1332.
4429. Lautsch, W. F., *Chem. Tech.* (Berlin) **1958**, *10*, 419.
4430. Lavalley, J. C., and N. Sheppard, *Spectrochim. Acta, Part A* **1972**, *28*, 2091.
4431. Lawesson, S. O., J. O. Madsen, et al., *Acta Chem. Scand.* **1966**, *20*, 2325.
4432. Lawrenson, I. J., and D. A. Lee, *J. Chem. Thermodyn.* **1978**, *10*, 1111.
4433. Lawrie, J. W., "Glycerol and the Glycols," Chemical Catalog Co., New York, 1928.
4434. Lawson, A. W., R. Lowell, et al., *J. Chem. Phys.* **1959**, *30*, 643.
4435. Laynez, J., and I. Wadso, *Acta Chem. Scand.* **1972**, *26*, 3148.
4436. Lazlo, P., *Bull. Soc. Chim. Fr.* **1964**, 2658; *C.A.* **1965**, *62*, 8540.
4437. Lazniewski, M., *Rocz. Chem.* **1934**, *14*, 560; *Chem. Zentr.* **1936**, *I*, 986.
4438. Leader, G. R., *J. Amer. Chem. Soc.* **1951**, *73*, 856.
4439. Leader, G. R., and G. F. Gormley, *J. Amer. Chem. Soc.* **1951**, *73*, 5731.
4440. Lebas, J. M., *J. Chim. Phys.* **1962**, *59*, 1072; *C.A.* **1963**, *58*, 5168.
4441. Lebeau, P., *Bull. Acad. Roy. Belges* **1908**, 300; *C.A.* **1909**, *3*, 1013.
4442. Lebedev, B. V., I. B. Rabinovich, et al., *J. Chem. Thermodyn.* **1978**, *10*, 321.
4443. Lebedev, B. V., I. B. Rabinovich, et al., *Vysokomol. Soedin., Ser. A* **1967**, *9*, 488; *C.A.* **1967**, *67*, 11845.
4444. Lebedev, B. V., and A. A. Yevstropov, *J. Chem. Thermodyn.* **1983**, *15*, 115.
4445. Lebedev, N. N., *Zh. Obshch. Khim.* **1954**, *24*, 1959; *C.A.* **1955**, *49*, 13876.
4446. Lebedeva, N. D., *Zh. Fiz. Khim.* **1964**, *38*, 2648; *Russ. J. Phys. Chem.* **1964**, *38*, 1435.
4447. Lebedeva, N. D., *Zh. Fiz. Khim.* **1966**, *40*, 2725; *Russ. J. Phys. Chem.* **1966**, *40*, 1465.
4448. Lebedeva, N. D., Yu. A. Katin, et al., *Zh. Fiz. Khim.* **1971**, *45*, 1357; *Russ. J. Phys. Chem.* **1971**, *45*, 771.
4449. Lebedeva, N. D., Yu. A. Katin, et al., *Zh. Fiz. Khim.* **1971**, *45*, 2103; *Russ. J. Phys. Chem.* **1971**, *45*, 1192.
4450. Lebedeva, N. D., and Yu. A. Katin, *Russ. J. Phys. Chem.* **1972**, *46*, 1088.
4451. Lebedeva, N. D., and V. L. Ryadnenko, *Zh. Fiz. Khim.* **1973**, *47*, 2442; *Russ. J. Phys. Chem.* **1973**, *47*, 1382.
4452. Lebel, R. G., and D. A. I. Goring, *J. Chem. Eng. Data* **1962**, *7*, 100.
4453. LeBlanc, R. B., *Anal. Chem.* **1958**, *30*, 1797.
4454. Lebo, R. B., *J. Amer. Chem. Soc.* **1921**, *43*, 10056.
4455. LeBreton, J. G., J. J. McKetta, et al., *Hydrocarbon Process. Petrol. Refiner* **1964**, *43*, 136.
4456. LeBrumant, J., *C.R. Acad. Sci., Paris. Ser. A, B* **1969**, *268B*, 1424.
4457. Lecat, M., *Ann. Soc. Sci. Bruxelles* **1927**, *47BI*, 63; *C.A.* **1928**, *22*, 4296.
4458. Lecat, M., *Mem. Acad. Belges* **1949**, *23*, 3.
4459. Lecat, M., *Annales de Chemie* **1947**, *2*, 169; *Beil.* **EIII2**, 255.
4460. Lecky, H. S., and R. H. Ewell, *Ind. Eng. Chem., Anal. Ed.* **1940**, *12*, 544.
4461. Leclercq, M., *Mem. Poudres* **1961**, *43*, 369; *C.A.* **1962**, *57*, 6766.
4462. Lecomte, J., *Bull. Soc. Chim. Fr.* **1946**, 415; *C.A.* **1947**, *41*, 1933.
4463. Lecomte, J., *Cahiers Phys.* **1957**, *11*, 77; *C.A.* **1960**, *54*, 14946.
4464. Lecomte, J., *Compt. Rend.* **1924**, *178*, 1530.
4465. Lecomte, J., *Compt. Rend.* **1924**, *178*, 1698.
4466. Lecomte, J., *Compt. Rend.* **1924**, *178*, 2073.
4467. Lecomte, J., *Compt. Rend.* **1934**, *198*, 65; *C.A.* **1934**, *28*, 1926.
4468. Lecomte, J., *Compt. Rend.* **1938**, *207*, 395; *C.A.* **1938**, *32*, 7821.
4469. Lecomte, J., *Groupement Fr. Develop. Recherches Aerinaut. Note Tech.* **1947**, 70; *C.A.* **1949**, *43*, 3715.
4470. Lecomte, J., *J. Phys. Radium* **1942**, *3*, 193; *C.A.* **1944**, *38*, 3549.
4471. Lecomte, J., *J. Phys. Radium* **1945**, *6*, 127; *C.A.* **1946**, *40*, 1394.
4472. Lecomte, J., *J. Phys. Radium* **1937**, *8*, 489; *C.A.* **1938**, *32*, 2833.
4473. Lecomte, J., *J. Phys. Radium* **1938**, *9*, 13; *C.A.* **1938**, *32*, 2833.

4474. Lecomte, J., and R. Freymann, *Bull. Soc. Chim. Fr.* **1941,** *8,* 601; *C.A.* **1942,** *36,* 2477.
4475. Lecomte, J., E. Gray, et al., *Bull. Soc. Chim. Fr.* **1947,** 774; *C.A.* **1948,** *42,* 1129.
4476. Lecomte, J., and L. Piaux, *Bull. Soc. Chim. Belges* **1934,** *43,* 239; *C.A.* **1934,** *28,* 4311.
4477. Ledaal, T., *Tetrahedron Lett.* **1966,** 1653; *C.A.* **1966,** *64,* 18736.
4478. Ledaal, T., *Tetrahedron Lett.* **1968,** 651.
4479. Lederer, E. L., *Allgem. Ol.-u Fittzig* **1930,** *27,* 237; *C.A.* **1931,** *25,* 2013.
4480. Lederer, E. L., *Seifensieder-Ztg.* **1930,** *57,* 67; *C.A.* **1930,** *24,* 2625.
4481. Lederer, E. L., *Seifensieder-Ztg.* **1930,** *57,* 68; *Beil.* **EIII2,** 1388; *Chem. Zentr.* **1932,** *II,* 1095.
4482. Lederer, E. L., and D. Hartleb, *Seifensie der-Ztg.* **1929,** *56,* 345; *C.A.* **1930,** *24,* 741.
4483. Lee, C. M., and W. D. Kumber, *J. Amer. Chem. Soc.* **1961,** *83,* 4593.
4484. Lee, D. G., and R. Cameron, *Can. J. Chem.* **1972,** *50,* 445.
4485. Lee, S., *J. Soc. Chem. Ind. Jpn.* **1940,** *43,* Suppl. Binding 190; *C.A.* **1940,** *34,* 7677.
4486. Leech, J. W., *Proc. Phys. Soc.* (London) **1949,** *62B,* 390; *Chem. Zentr.* **1950,** *I,* 2339.
4487. Le Fèvre, C. G., and R. J. W. Le Fèvre, *J. Chem. Soc.* **1936,** 487.
4488. Le Fèvre, C. G., and R. J. W. Le Fèvre, *J. Chem. Soc.* **1936,** 1130.
4489. Le Fèvre, C. G., and R. J. W. Le Fèvre, *J. Chem. Soc.* **1950,** 1829.
4490. Le Fèvre, C. G., and R. J. W. Le Fèvre, *J. Chem. Soc.* **1954,** 1577.
4491. Le Fèvre, C. G., R. J. W. Le Fèvre, et al., *Aust. J. Chem.* **1957,** *10,* 218; *C.A.* **1958,** *52,* 819.
4492. Le Fèvre, C. G., R. J. W. Le Fèvre, et al., *J. Chem. Soc.* **1935,** 480.
4493. Le Fèvre, C. G., R. J. W. Le Fèvre, et al., *J. Chem. Soc.* **1959,** 1188.
4494. Le Fèvre, C. G., R. J. W. Le Fèvre, et al., *J. Chem. Soc.* **1960,** 123.
4495. Le Fèvre, C. G., R. J. W. Le Fèvre, et al., *J. Chem. Soc.* **1960,** 1814.
4496. Le Fèvre, R. J. W., and F. Maramba, *J. Chem. Soc.* **1952,** 235.
4497. Le Fèvre, R. J. W., B. J. Orr, et al., *Aust. J. Chem.* **1966,** *19,* 2175.
4498. Le Fèvre, R. J. W., and B. J. Orr, *J. Chem. Soc.* **1965,** 5349.
4499. Le Fèvre, R. J. W., and B. J. Orr, *J. Chem. Soc. B* **1966,** 37.
4500. Le Fèvre, R. J. W., B. J. Orr, et al., *J. Chem. Soc.* **1965,** 2499.
4501. Le Fèvre, R. J. W., B. J. Orr, et al., *J. Chem. Soc.* **1965,** 3619.
4502. Le Fèvre, R. J. W., R. K. Pierens, et al., *Aust. J. Chem.* **1966,** *19,* 1325; *C.A.* **1966,** *65,* 14572.
4503. Le Fèvre, R. J. W., G. L. D. Ritchie, et al., *J. Chem. Soc. B* **1967,** 819.
4504. Le Fèvre, R. J. W., W. P. H. Roberts, et al., *J. Chem. Soc.* **1949,** 902.
4505. Le Fèvre, R. J. W., R. Roper, et al., *Aust. J. Chem.* **1959,** *12,* 743; *C.A.* **1960,** *54,* 5246.
4506. Le Fèvre, R. J. W., and P. Russell, *Trans. Faraday Soc.* **1947,** *43,* 374.
4507. Le Fèvre, R. J. W., and J. W. Smith, *J. Chem. Soc.* **1932,** 2810.
4508. Le Fèvre, R. J. W., and A. Sundaram, *J. Chem. Soc.* **1962,** 1494.
4509. Le Fèvre, R. J. W., and A. Sundaram, *J. Chem. Soc.* **1962,** 4756.
4510. Le Fèvre, R. J. W., and K. M. S. Sundaram, *J. Chem. Soc.* **1963,** 1880.
4511. Le Fèvre, R. J. W., and A. J. Williams, *J. Chem. Soc.* **1960,** 108.
4512. Le Fèvre, R. J. W., and A. J. Williams, *J. Chem. Soc.* **1960,** 115.
4513. Le Fèvre, R. J. W., and A. J. Williams, *J. Chem. Soc.* **1960,** 1825.
4514. Le Fèvre, R. J. W., and A. J. Williams, *J. Chem. Soc.* **1965,** 4185.
4515. Leibush, A. G., and E. D. Shorina, *J. Appl. Chem.* (USSR) **1947,** *20,* 69; *C.A.* **1947,** *41,* 5446; *Beil.* **EIII4,** 696.
4516. Leighton, P. A., R. W. Crary, et al., *J. Amer. Chem. Soc.* **1931,** *53,* 3017.
4517. Leis, D. G., and B. C. Currans, *J. Amer. Chem. Soc.* **1945,** *67,* 79.
4518. Lek, L., *Annual Tables of Physical Constants*, Sec. 700C, 1941.
4519. Lemaire, H., and H. J. Liecas, *J. Amer. Chem. Soc.* **1951,** *73,* 5198.
4520. Lemmerman, L. V., A. W. Davidson, et al., *J. Amer. Chem. Soc.* **1946,** *68,* 1361.
4521. Lemons, J. F., and W. A. Felsing, *J. Amer. Chem. Soc.* **1943,** *65,* 46.
4522. Lemoult, P., *Compt. Rend.* **1906,** *143,* 746.
4523. Lenormant, H., *Bull. Soc. Chim. Fr.* **1948,** 33; *C.A.* **1948,** *42,* 4060.
4524. Lenormant, H., and P. L. Clement, *Bull. Soc. Chim. Fr.* **1946,** 559; *C.A.* **1947,** *41,* 2642.
4525. Lenz, W., *Arch. Pharm.* **1911,** *249,* 289.
4526. Leonard, N. J., and L. E. Sutton, *J. Amer. Chem. Soc.* **1948,** *70,* 1564.
4527. Lenormant, H., *Compt. Rend.* **1949,** *228,* 1861; *C.A.* **1949,** *43,* 8274.

4528. Leonard, L. J., *Organic Syntheses, Vol. 36*, Wiley, New York, 1956.

4529. Leroux, P. J., and H. J. Lucas, *J. Amer. Chem. Soc.* **1951**, *73*, 41.

4530. Leroy, Y., and E. Constant, *Compt. Rend., Ser. A, B* **1966**, *262B*, 1391; *C.A.* **1966**, *65*, 8198.

4531. Lespieau, R., *Bull. Soc. Chim. Fr.* **1908**, *3*, 638; *Chem. Zentr.* **1909**, *II*, 151.

4532. Lespieau, R., *Compt. Rend.* **1919**, *169*, 31.

4533. Lespieau, R., and B. Gredy, *Compt. Rend.* **1933**, *196*, 399.

4534. Letaw, H., and A. H. Gropp, *J. Phys. Chem.* **1953**, *57*, 964.

4535. Letcher, T. M., and J. W. Bayles, *J. Chem. Eng. Data* **1971**, *16*, 266.

4536. Letcher, T. M., and F. Marsicano, *J. Chem. Thermodyn.* **1974**, *6*, 509.

4537. Levanova, S. V., Yu. A. Treger, et al., *Zh. Fiz. Khim.* **1976**, *50*, 1901; *Russ. J. Phys. Chem.* **1976**, *50*, 1148.

4538. Levene, P. A., and L. H. Cretcher, *J. Biol. Chem.* **1918**, *33*, 505.

4539. Levene, P. A., and H. L. Haller, *J. Biol. Chem.* **1926**, *69*, 569.

4540. Levene, P. A., and H. L. Haller, *J. Biol. Chem.* **1929**, *83*, 579.

4541. Levene, P. A., and R. E. Marker, *J. Biol. Chem.* **1931**, *91*, 405.

4542. Levene, P. A., and A. Rothen, *J. Biol. Chem.* **1936**, *115*, 415.

4543. Levene, P. A., and A. Rothen, *J. Biol. Chem.* **1936**, *116*, 209.

4544. Levene, P. A., A. Rothen, et al., *J. Biol. Chem.* **1936**, *115*, 401.

4545. Levene, P. A., A. Rothen, et al., *J. Biol. Chem.* **1937**, *120*, 759.

4546. Levene, P. A., and F. A. Taylor, *J. Biol. Chem.* **1922**, *54*, 351.

4547. Levene, P. A., and A. Walti, *J. Biol. Chem.* **1926**, *68*, 415.

4548. Levene, P. A., and A. Walti, *J. Biol. Chem.* **1931**, *94*, 361.

4549. Levene, P. A., and A. Walti, *J. Biol. Chem.* **1931**, *94*, 367.

4550. Levene, P. A., A. Walti, et al., *J. Biol. Chem.* **1926**, *71*, 465.

4551. Levi, G., *Compt. Rend.* **1965**, *261*, 4007; *C.A.* **1966**, *64*, 4905.

4552. Levi, G., and M. Chalaye, *C.R. Acad. Sci. Paris, Ser. A,B* **1967**, *264B*, 857.

4553. Levi, L., L. G. Chatten, et al., *J. Amer. Pharm. Assoc. Sci. Ed.* **1955**, *44*, 61.

4554. Levin, B. Ya., *Zh. Fiz. Khim.* **1954**, *28*, 1399; *C.A.* **1955**, *49*, 7980.

4555. Levine, A. A., and O. W. Cass, British Patent 471,188, Aug. 30, 1937; *C.A.* **1938**, *32*, 957.

4556. Levitt, L. S., and B. W. Levitt, *Z. Naturforsch., B: Anorg. Chem., Org. Chem.* **1979**, *34B*, 614.

4557. Levy, E. J., and W. A. Stahl, *Anal. Chem.* **1961**, *33*, 707.

4558. Levy, G. C., and J. D. Cargioli, *Tetrahedron Lett.* **1970**, *12*, 919.

4559. Lewin, A. H., and S. Winstein, *J. Amer. Chem. Soc.* **1962**, *84*, 2464.

4560. Lewin, S. Z., and J. E. Vance, *J. Amer. Chem. Soc.* **1952**, *74*, 1433.

4561. Lewis, D. T., *J. Chem. Soc.* **1938**, 1056.

4562. Lewis, D. T., *J. Chem. Soc.* **1940**, 32.

4563. Lewis, E., *J. Chem. Ind.* **1922**, *41*, 97T.

4564. Lewis, G. L., and C. P. Smyth, *J. Amer. Chem. Soc.* **1939**, *61*, 3067.

4565. Lewis, G. L., and C. P. Smyth, *J. Chem. Phys.* **1939**, *7*, 1085.

4566. Lewis, G. N., "Valence and Structure of Atoms and Molecules," Chemical Catalog Co., New York, 1923.

4567. Lewis, G. N., *J. Franklin Inst.* **1938**, *266*, 293.

4568. Lewis, H. F., R. Hendricks, et al., *J. Amer. Chem. Soc.* **1928**, *50*, 1993.

4569. Lewis, J., J. R. Miller, et al., *J. Chem. Soc.* **1965**, 5850.

4570. Lewis, J. R., *J. Amer. Chem. Soc.* **1925**, *47*, 626.

4571. Ley, H., and B. Arendo, *Z. Phys. Chem.* (Leipzig) **1932**, *15B*, 311; *Chem. Zentr.* **1932**, *I*, 2549.

4572. Ley, H., and B. Arendo, *Z. Phys. Chem.* (Leipzig) **1932**, *17B*, 177; *Chem. Zentr.* **1932**, *II*, 835.

4573. Ley, H., and H. Hünecke, *Ber. Deut. Chem. Gesell.* **1926**, *59*, 510.

4574. L'Her, M., and J. Courtot-Coupez, *Bull. Soc. Chim. Fr.* **1972**, 3645.

4575. Li, C.-C., *Hua Hsueh Tung Pao* **1960**, No. 3, 128; *C.A.* **1961**, *55*, 24167.

4576. Li, J. C. M., and K. S. Pitzer, *J. Amer. Chem. Soc.* **1956**, *78*, 1077.

4577. Li, J. C. M., and F. D. Rossini, *J. Chem. Eng. Data* **1961**, *6*, 268.

4578. Li, K., *J. Phys. Chem.* **1957**, *61*, 782.

4579. Li, N. C. C., *J. Chem. Phys.* **1939**, *7*, 1068.

4580. Li, N. C. C., and P. C. Hsu, *J. Chinese Chem. Soc.* **1946**, *13*, 11; *C.A.* **1947**, *41*, 3672.

4581. Li, N. C. C., and T. D. Terry, *J. Amer. Chem. Soc.* **1948**, *70*, 344.
4582. Liardon, R., and T. Gaeumann, *Helv. Chim. Acta* **1969**, *52*, 1042.
4583. Libbey, A. J., and J. T. Stock, *Anal. Chem.* **1970**, *42*, 526.
4584. Liberman, A. L., Z. N. Parnes, et al., *Bull. Acad. Sci. URSS, Classe Sci. Chim.* **1948**, 101; *C.A.* **1948**, *42*, 5291.
4585. Licht, K., *Z. Chem.* **1967**, *7*, 321.
4586. Liddel, U., *Ann. N.Y. Acad. Sci.* **1957**, *69*, 70.
4587. Liddel, U., and C. Kasper, *J. Res. Nat. Bur. Stand.* **1934**, *11*, 599; *C.A.* **1934**, *28*, 1601.
4588. Lide, D. R., *J. Chem. Phys.* **1960**, *33*, 1514.
4589. Lide, D. R., and D. E. Mann, *J. Chem. Phys.* **1958**, *29*, 914.
4590. Liebaert, R., A. Lebrun, et al., *Compt. Rend.* **1961**, *253*, 2496; *C.A.* **1962**, *56*, 8070.
4591. Lieben, A., and A. Rossi, *Justus Liebigs Ann. Chem.* **1871**, *159*, 73.
4592. Lieben, A., and S. Zeisel, *Ber. Deut. Chem. Gesell.* **1881**, *14*, 515.
4593. Lieberman, E. P., *Official Digest* **1962**, *34*, 30.
4594. Liebermann, L. N., *Phys. Rev.* **1941**, *60*, 496.
4595. Liebush, A. G., and E. D. Shorina, *J. Appl. Chem.* (USSR) **1947**, *20*, 69; *C.A.* **1947**, *41*, 5447.
4596. Liem, H. D., *Acta Chem. Scand.* **1968**, *22*, 753.
4597. Lifshits, A. L., E. P. Ageev, et al., *Metody Poluch. Khim. Reaktivov Prep.* **1967**, No. 17, 80; *C.A.* **1969**, *70*, 114494.
4598. Lifshitz, C., and F. A. Long, *J. Phys. Chem.* **1963**, *67*, 2463.
4599. Lifshitz, C., and F. A. Long, *J. Phys. Chem.* **1965**, *69*, 3731.
4600. Liler, M., and D. Kosanovic, *J. Chem. Soc.* **1958**, 1084.
4601. Limpricht, H., *Ann. Chem.* **1985**, *134*, 347.
4602. Lin, R.-Y., and W. Dannhauser, *J. Phys. Chem.* **1963**, *67*, 1805.
4603. Lin. T. F., S. D. Christian, et al., *J. Phys. Chem.* **1967**, *71*, 1133.
4604. Lin. W.-C., and F.-T. Tuan, *J. Chinese Chem. Soc.* **1958**, *5*, 33; *C.A.* **1959**, *53*, 10872.
4605. Linard, J., *Bull. Soc. Chim. Belges* **1925**, *34*, 363.
4606. Linch, A. L., *Amer. Ind. Hyg. Assoc. J.* **1965**, *26*, 95.
4607. Lincoln, A. T., *J. Phys. Chem.* **1899**, *3*, 457.
4608. Lind, E. L., and T. F. Young, *J. Chem. Phys.* **1933**, *1*, 266.
4609. Lindberg, J. J., J. Kenttama, et al., *Soumen Kemistilehti* **1961**, *34B*, 98 (Eng. trans.); *C.A.* **1963**, *58*, 1946.
4610. Lindenberg, A. B., *Compt. Rend., Ser. C* **1966**, *262*, 1504; *C.A.* **1966**, *65*, 6450.
4611. Linder, E. G., *J. Phys. Chem.* **1931**, *35*, 531.
4612. Ling, A. C., and J. E. Martin, *J. Phys. Chem.* **1968**, *72*, 3349.
4613. Ling, A. C., and J. E. Willard, *J. Phys. Chem.* **1968**, *72*, 1918.
4614. Ling. A. C., and J. E. Willard, *J. Phys. Chem.* **1968**, *72*, 3349.
4615. Ling, T. D., and M. Van Winkle, *Ind. Eng. Chem., Chem. Eng. Data Series* **1958**, *3*, 88.
4616. Linnell, R. H., *J. Org. Chem.* **1960**, *25*, 290.
4617. Linnemann, E., *Justus Liebigs Ann. Chem.* **1872**, *161*, 196.
4618. Lipkin, M. R., J. A. Davidson, et al., *Ind. Eng. Chem.* **1942**, *34*, 976.
4619. Lipnik, R. L., *J. Amer. Chem. Soc.* **1974**, *96*, 2941.
4620. Lipp, A., *Ann. Chem.* **1880**, *205*, 1; *J. Chem. Soc.* **1881A**, 84.
4621. Lipp, A., *Ber. Deut. Chem. Gesell.* **1889**, *22*, 2567.
4622. Lippert, E., and H. Prigge, *Ann. Chem.* **1962**, *659*, 81; *C.A.* **1963**, *58*, 8878.
4623. Lippincott, E. R., C. E. Meyers, et al., *Spectrochim. Acta* **1966**, *22*, 1493.
4624. Lippincott, E. R., G. Nagarajan, et al., *Bull. Soc. Chim. Belges* **1966**, *75*, 655.
4625. Lippincott, S. B., U.S. Patent 2,423,783, July 8, 1947; *C.A.* **1947**, *41*, 6275.
4626. Lippincott, S. B., and H. B. Hass, *Ind. Eng. Chem.* **1939**, *31*, 118.
4627. Lippmaa, E., A. Olivson, et al., *Eesti NSV Tead. Akad. Toim., Fuusik.-Mat. ja Tehnik. Seer.* **1965**, *14*, 473; *C.A.* **1966**, *64*, 7552.
4628. Liron, Z., and S. Cohen, *J. Pharm. Sci.* **1983**, *72*, 499.
4629. Lister, M. W., *J. Amer. Chem. Soc.* **1941**, *63*, 143.
4630. Liston, L., and W. M. Dehn, *J. Amer. Chem. Soc.* **1938**, *60*, 1264.
4631. Litovitz, T. A., R. Higgs, et al., *J. Chem. Phys.* **1954**, *22*, 1281.

4632. Little, M. H., and A. E. Martell, *J. Phys. Colloid Chem.* **1949**, *53*, 472.

4633. Little, R. C., and C. R. Sengleterry, *J. Phys. Chem.* **1964**, *68*, 2709.

4634. Littlejohn, A. C., and J. W. Smith, *J. Chem. Soc.* **1953**, 2456.

4635. Liu, K. F., and W. T. Ziegler, *J. Chem. Eng. Data* **1966**, *11*, 187.

4636. Livingston, J., R. Morgan, et al., *J. Amer. Chem. Soc.* **1911**, *33*, 1713.

4637. Livingston, J., R. Morgan, et al., *J. Amer. Chem. Soc.* **1913**, *35*, 1505.

4638. Livingston, J., R. Morgan, et al., *J. Amer. Chem. Soc.* **1913**, *35*, 1821.

4639. Livingston, J., R. Morgan, et al., *J. Amer. Chem. Soc.* **1924**, *46*, 881.

4640. Livingston, R., *J. Amer. Chem. Soc.* **1947**, *69*, 1220.

4641. Lobanov, E. Ya., S. A. Astapova, et al., *Khim. Prom.-st., Ser.: Toksikol. Sanit. Khim. Plastmass* **1979**, No. 3, 21; *C.A.* **1980**, *93*, 38848.

4642. Lochte, H. L., *Ind. Eng. Chem.* **1924**, *16*, 956.

4643. Locke, E. G., W. R. Brode, et al., *J. Amer. Chem. Soc.* **1934**, *56*, 1726.

4644. Locquin, M., *Bull. Mens. Soc. Linneenne Lyon* **1942**, *11*, 95; *C.A.* **1946**, *40*, 427.

4645. Locquin, R., *Ann. Chim.* **1923**, *19*, 32; *C.A.* **1923**, *17*, 2559.

4646. Loder, D. J., U.S. Patent 2,204,371, June 11, 1940; *Official Gaz. U.S. Patent Office* **1940**, *515*, 494.

4647. Loeffler, H. J., and H. Matthias, *Kaeltetechnik* **1966**, *18*, 408; *C.A.* **1967**, *67*, 36886.

4648. Loeser, A., *Pharmazie* **1949**, *4*, 263; *C.A.* **1949**, *43*, 8558.

4649. Loewenherz, R., *Z. Phys. Chem.* (Leipzig) **1890**, *6*, 552.

4650. Loiseleur, H., J. C. Merlin, et al., *J. Chim. Phys.* **1967**, *64*, 634.

4651. Loit, A. O., and V. A. Filov, *Gig. Truda i Prof. Zabol.* **1964**, *8*, 23; *C.A.* **1965**, *62*, 12353.

4652. Lombard, R., *Bull. Soc. Chim. Fr.* **1946**, 598; *C.A.* **1947**, *41*, 28754.

4653. Lomonova, G. V., *Prom. Toksikol. i Klinika Prof. Zabol. Khim. Etiol. Sb.* **1962**, 160; *C.A.* **1964**, *61*, 3600.

4654. Long, F. A., and P. Ballinger, *Electrolytes, Proc. Intern. Symp., Trieste, Yugoslavia* **1959**, 152 (Pub. 1962); *C.A.* **1964**, *61*, 10098.

4655. Long, T. V., and R. A. Plane, *J. Chem. Phys.* **1965**, *43*, 457.

4656. Longinow, W., and A. Prjaschnikow, *Trans. Inst. Pure Chem. Reagents* (Moscow) **1931**, 48.

4657. Longster, G. F., and E. E. Walker, *Trans. Faraday Soc.* **1953**, *49*, 228; *C.A.* **1953**, *47*, 9697.

4658. Loos, K. R., and U. P. Wild, *Spectrochim. Acta, Part A* **1969**, *25*, 275.

4659. Lord, A., C. A. Goy, et al., *J. Phys. Chem.* **1967**, *71*, 2705.

4660. Lord, R. C., *U.S. Dept. Com., Office Tech., Serv., P.B. Rept.* **1960**, *161*, 738; *C.A.* **1962**, *56*, 9589.

4661. Lord, R. C., and F. A. Miller, *J. Chem. Phys.* **1942**, *10*, 328.

4662. Lorenz, J., and J. Auer, *Angew. Chem.* **1965**, *77*, 218.

4663. Louder, E. A., T. R. Briggs, et al., *Ind. Eng. Chem.* **1924**, *16*, 932.

4664. Loughran, E. D., E. M. Wewerka, et al., *J. Heterocycl. Chem.* **1972**, *9*, 57.

4665. Louguinine, W., *Ann. Chim. Phys.* **1898**, *13*, 289; *Chem. Zentr.* **1898**, *I*, 824.

4666. Louguinine, W., *J. Chim. Phys.* **1904**, *2*, 1; *Chem. Zentr.* **1904**, *II*, 436.

4667. Louguinine, W., *J. Chim. Phys.* **1904**, *2*, 1; *Chem. Zentr.* **1904**, *II*, 436.

4668. Louguinine, W., and G. Dupont, *Bull. Soc. Chim.* **1911**, *9*, 219; *C.A.* **1911**, *5*, 1857.

4669. Lovering, E. G., and K. J. Laidler, *Can. J. Chem.* **1960**, *38*, 2367.

4670. Lovering, E. G., and O. B. M. Nor, *Can. J. Chem.* **1962**, *40*, 199.

4671. Lowe, K. F., *Ann. Phys.* **1898**, *66*, 390.

4672. Lowenherz, R., *Z. Phys. Chem.* (Leipzig) **1898**, *25*, 385.

4673. Lowig, C., *Jahr. Fort. Chem.* **1860**, 397.

4674. Lowry, T. M., *J. Chem. Soc.* **1914**, 81.

4675. Lowry, T. M., *Trans. Faraday Soc.* **1924**, *20*, 13.

4676. Lowry, T. M., and C. B. Allsopp, *Proc. Roy. Soc.* (London) **1931**, *113A*, 26.

4677. Lowry, T. M., and R. E. Lishmund, *J. Chem. Soc.* **1935**, 1313.

4678. Lozac'h, N., *Bull. Soc. Chim. Fr.* **1949**, 286; *C.A.* **1949**, *43*, 6162.

4679. Lucatu, E., *Compt. Rend.* **1938**, *207*, 1403; *C.A.* **1939**, *33*, 1563.

4680. Luder, W. F., and S. Zuffanti, *The Electronic Theory of Acids and Bases*, Wiley, New York, 1946.

4681. Leuhrs, C., and G. Schwitzgebel, *Ber. Bunsenges. Phys. Chem.* **1979**, *83*, 623.

4682. Luginin, W., *Arch. Sci. Phys. Nat. Geneve 9*, 5; *J. Chem. Soc.* **1900Aii**, 334.

4683. Luginin, W., *Ann. Chim. Phys.* (Paris) **1902**, *26*, 288; *J. Chem. Soc.* **1902Aii**, 547.

4684. Luginin, W. F., *Ann. Chim. Phys.* (Paris) **1902**, *27*, 105; *J. Chem. Soc.* **1903Aii,** 7.
4685. Luginin, W. F., *Arch. Sci. Phys. Nat.* **1900,** *9*, 5.
4686. Luginin, W. F., *Compt. Rend.* **1899,** *128*, 366; *J. Chem. Soc.* **1899Aii,** 354.
4687. Lumbroso, H., *Ann. Fac. Sci. Univ. Toulouse Sci.-Math. et Sci. Phys.* **1950,** *14*, 21, 35, 56, 108; *C.A.* **1954,** *48*, 5781.
4688. Lumbroso, H., *Compt. Rend.* **1947,** *225*, 1003; *C.A.* **1948,** *42*, 2827.
4689. Lumbroso, H., *Compt. Rend.* **1949,** *228*, 77; *C.A.* **1949,** *43*, 4061.
4690. Lumbroso, H., and J. Barassin, *J. Soc. Chim. Fr.* **1965,** 3143.
4691. Lumbroso, H., and D. M. Bertin, et al., *Bull. Soc. Chim. Fr.* **1970,** 1728.
4692. Lumbroso, H., D. M. Bertin, et al., *Bull. Soc. Chim. Fr.* **1966,** 540; *C.A.* **1966,** *64*, 16752.
4693. Lumbroso, H., and P. Rumpf, *Bull. Soc. Chim. Fr.* **1950,** 371; *C.A.* **1950,** *44*, 7106.
4694. Lumbroso, N., T. K. Wu, et al., *J. Phys. Chem.* **1963,** *67*, 2469.
4695. Lumsden, J. S., *J. Chem. Soc.* **1905,** 90.
4696. Lumsden, J. S., *J. Chem. Soc.* **1919,** 1366.
4697. Lunazzi, L., and F. Taddei, *Bull. Soc. Chim. Fac. Chim. Ind. Bologna* **1964,** *22*, 91; *C.A.* **1965,** *63*, 12538.
4698. Lund, H., and J. Bjerrum, *Ber. Deut. Chem. Gesell.* **1931,** *64*, 210.
4699. Lundin, H. G., and M. L. Crowder, *Proc. Sci. Sec. Toilet Goods Assoc.* **1952,** *18*, 1; *C.A.* **1953,** *47*, 3698.
4700. Lundin, N. P., *Farmatsiya* **1946,** *9*, No. 2, 39; *C.A.* **1947,** *41*, 4445.
4701. Lunelli, B., and C. Pecile, *Gazz. Chim. Ital.* **1966,** 96; *C.A.* **1966,** *65*, 9946.
4702. Lunge, G., *Ber. Deut. Chem. Gesell.* **1881,** *14*, 1755.
4703. Lunt, R. W., and M. A. G. Rau, *Proc. Roy. Soc.* (London) **1930,** *126A*, 213.
4704. Luther, H., *Z. Electrochem.* **1948,** *52*, 210; *C.A.* **1949,** *43*, 8888.
4705. Luther, H., and E. Lohrengel, *Brennstoff-Chem.* **1954,** *35*, 338; *C.A.* **1955,** *49*, 3736.
4706. Luther, H., and J. Operskalski, *Naturwissenschaften* **1950,** *37*, 376.
4707. Luther, H., and C. Reichel, *Z. Phys. Chem.* (Leipzig) **1950,** *195*, 103; *C.A.* **1950,** *44*, 9257.
4708. Luther, H., and G. Wachter, *Chem. Ber.* **1949,** *82*, 161.
4709. Luther, R., and F. Weigert, *Z. Phys. Chem.* (Leipzig) **1905,** *51*, 297.
4710. Lüthy, A., *Z. Phys. Chem.* (Leipzig) **1923,** *107*, 285.
4711. Lutskiĭ, A. E., *Zh. Obshch. Khim.* **1954,** *24*, 440; *C.A.* **1954,** *48*, 8609.
4712. Lutskiĭ, A. E., V. V. Bocharova, et al., *Zh. Obshch. Khim.* **1975,** *45*, 2276; *J. Gen. Chem* (USSR) **1975,** *45*, 2236.
4713. Lutskiĭ, A. E., and S. A. Mikhailenko, *Zh. Strukt. Khim.* **1962,** *3*, 523; *C.A.* **1963,** *59*, 4622.
4714. Lutskiĭ, A. E., and E. M. Obukhova, *Russ. J. Phys. Chem.* **1969,** *43*, 736.
4715. Lutskiĭ, A. E., and A. N. Panova, *Zh. Fiz. Khim.* **1959,** *33*, 970; *C.A.* **1960,** *54*, 8261.
4716. Lutskiĭ, A. E., and V. V. Prezhdo, *Zh. Fiz. Khim.* **1971,** *45*, 1273; *Russ. J. Phys. Chem.* **1971,** *45*, 723.
4717. Luttke, W., *Chem. Ber.* **1950,** *83*, 571.
4718. Lüttringhaus, A., and H. W. Dirksen, *Angew. Chem. Intern. Ed. Engl.* **1964,** *3*, 260.
4719. Lüttringhaus, A., and J. Grohmann, *Z. Naturforsch.* **1955,** *10B*, 367; from [4718].
4720. Lynch, B. M., B. C. Macdonald, et al., *Tetrahedron* **1968,** *24*, 3595.
4721. Lynch, E. J., and C. R. Wilke, *J. Chem. Eng. Data* **1960,** *5*, 300.
4722. Lynn, E. V., *J. Amer. Chem. Soc.* **1919,** *41*, 361.
4723. Lynn, J. W., *J. Org. Chem.* **1956,** *21*, 578.
4724. Lyons, L. E., *J. Proc. Roy. Soc. N.S. Wales* **1949,** *83*, 75; *C.A.* **1951,** *45*, 7430.
4725. Lyubomilov, V. I., and N. M. Merkula, *Zh. Obshch. Khim.* **1963,** *33*, 22; *C.A.* **1963,** *59*, 425.

M

4726. Maass, O., and E. H. Boomer, *J. Amer. Chem. Soc.* **1922,** *44*, 1708.
4727. Maass, O., and C. H. Wright, *J. Amer. Chem. Soc.* **1921,** *43*, 1098.
4728. Mabery, C. F., *J. Amer. Chem. Soc.* **1917,** *39*, 2015.
4729. McAllan, D. T., T. V. Cullum, et al., *J. Amer. Chem. Soc.* **1951,** *73*, 3626.
4730. McAlpine, K. B., and C. P. Smyth, *J. Amer. Chem. Soc.* **1933,** *55*, 453.
4731. McAlpine, K. B., and C. P. Smyth, *J. Chem. Phys.* **1935,** *3*, 55.

4732. McArdle, E. H., and D. M. Mason, U.S. Patent 2,435,792, Feb. 10, 1948; *C.A.* **1948,** *42,* 2988.
4733. McAuliffe, C., *J. Phys. Chem.* **1966,** *70,* 1267.
4734. McAuliffe, C., *Nature* **1963,** *200,* 1092.
4735. McAuliffe, C., *Science* **1969,** *163,* 478.
4736. McBee, E. T., V. V. Lindgren, et al., *Ind. Eng. Chem.* **1947,** *39,* 378.
4737. MacBeth, G., A. R. Thompson, et al., *Anal. Chem.* **1951,** *23,* 618.
4738. McCaulay, D. A., B. H. Shoemaker, et al., *Ind. Eng. Chem.* **1950,** *42,* 2103.
4739. McClellan, A. L., *Tables of Experimental Dipole Moments,* W. H. Freeman, San Francisco, 1963.
4740. McClellan, A. L., and S. W. Nicksic, *J. Phys. Chem.* **1965,** *69,* 446.
4741. McCleod, D. B., *Trans. Faraday Soc.* **1923,** *19,* 38; *C.A.* **1923,** *17,* 2072.
4742. McClune, C. R., *Cryogenics* **1976,** *16,* 289.
4743. McClure, H. B., *Chem. Eng. News* **1944,** *22,* 416.
4744. McClure, J. H., T. M. Roder, et al., *Anal. Chem.* **1955,** *27,* 1599.
4745. McCombie, H., B. C. Saunders, et al., *J. Chem. Soc.* **1944,** 24.
4746. McConnell, H. M., A. D. McLean, et al., *J. Chem. Phys.* **1955,** *23,* 1152.
4747. McCoy, E. A., and L. L. McCoy, *J. Org. Chem.* **1968,** *33,* 2354.
4748. McCullough, J. P., H. L. Finke, et al., *J. Phys. Chem.* **1957,** *61,* 289.
4749. McCullough, J. P., H. L. Finke, et al., *J. Phys. Chem.* **1957,** *61,* 1105.
4750. McCullough, J. P., and W. D. Good, *J. Phys. Chem.* **1961,** *65,* 1430.
4751. McCullough, J. P., W. N. Hubbard, et al., *J. Amer. Chem. Soc.* **1957,** *79,* 561.
4752. McCullough, J. P., R. E. Pennington, et al., *J. Amer. Chem. Soc.* **1959,** *81,* 5880.
4753. McCullough, J. P., W. B. Person, et al., *J. Amer. Chem. Soc.* **1951,** *73,* 4069.
4754. McCullough, J. P., D. W. Scott, et al., *J. Amer. Chem. Soc.* **1954,** *76,* 4791.
4755. McCullough, J. P., S. Sunner, et al., *J. Amer. Chem. Soc.* **1953,** *75,* 5075.
4756. McCurdy, K. G., and K. J. Laidler, *Can. J. Chem.* **1963,** *41,* 1867.
4757. McDevitt, N. T., A. L. Rozek, et al., *J. Chem. Phys.* **1965,** *42,* 1173.
4758. McDonald, F. M., A. W. Decora, et al., *Appl. Spectrosc.* **1968,** *22,* 325, 329.
4759. McDonald, H. V., *J. Chem. Phys.* **1944,** *48,* 47.
4760. McDonald, R. A., S. A. Schrader, et al., *J. Chem. Eng. Data* **1959,** *4,* 311.
4761. McDougall, F. H., *J. Amer. Chem. Soc.* **1936,** *58,* 2585.
4762. McDougall, L. A., and J. E. Kilpatrick, *J. Chem. Phys.* **1965,** *42,* 2307.
4763. McDuffie, G. E., J. W. Forbes, et al., *J. Chem. Eng. Data* **1969,** *14,* 176.
4764. McEachern, D. M., and J. E. Kilpatrick, *J. Chem. Phys.* **1964,** *41,* 3127.
4765. McEwen, B., *J. Chem. Soc.* **1923,** 2284.
4766. McEwen, W. K., *J. Amer. Chem. Soc.* **1936,** *58,* 1124.
4767. McFadden, W. H., E. A. Day, et al., *Anal. Chem.* **1965,** *37,* 89.
4768. McFadden, W. H., J. Wasserman, et al., *Anal. Chem.* **1964,** 1031.
4769. McGarvey, B. R., and G. Slomp, *J. Chem. Phys.* **1959,** *30,* 1586.
4770. McGinn, C. J., *J. Phys. Chem.* **1961,** *65,* 1896.
4771. McGovern, E. W., *Ind. Eng. Chem.* **1943,** *35,* 1230.
4772. MacInnes, D. A., and T. Shedlovsky, *J. Amer. Chem. Soc.* **1932,** *54,* 1429.
4773. McIntire, G. H., B. P. Block, et al., *J. Amer. Chem. Soc.* **1959,** *81,* 529.
4774. McIvor, R. A., C. E. Hubley, et al., *Can. J. Chem.* **1958,** *36,* 820.
4775. MacKay, K. M., A. E. Watt, et al., *J. Organometal. Chem.* **1968,** *12,* 49.
4776. McKay, R. A., and B. N. Sage, *J. Chem. Eng. Data* **1960,** *5,* 21.
4777. McKelvy, E. C., and D. H. Simpson, *J. Amer. Chem. Soc.* **1922,** *44,* 105.
4778. McKenna, F. E., H. V. Tartar, et al., *J. Amer. Chem. Soc.* **1949,** *71,* 729.
4779. McKinley, C., British Patent 705,670, Mar. 17, 1954; *C.A.* **1955,** *49,* 9026.
4780. McKinley, C., and J. P. McKinley, *J. Amer. Chem. Soc.* **1950,** *72,* 5331.
4781. McKinney, D. S., C. E. Leberknight, et al., *J. Amer. Chem. Soc.* **1937,** *59,* 481.
4782. McKinnon, G. P., ed., *Fire Protection Handbook,* 15th ed., National Fire Protection Association, Quincy, Mass., 1981.
4783. McLafferty, F. W., *Anal. Chem.* **1956,** *28,* 306.
4784. McLafferty, F. W., *Anal. Chem.* **1957,** *29,* 1782.
4785. McLafferty, F. W., *Anal. Chem.* **1962,** *34,* 2, 16, 26.

4786. McLafferty, F. W., *Atlas of Mass Spectra*, Wiley-Interscience, New York, 1968.

4787. McLaughlin, E. P., and R. T. Scott, *J. Amer. Chem. Soc.* **1954**, *76*, 5276.

4788. McLure, I. A., B. Edmonds, et al., *J. Colloid Interface Sci.* **1983**, *91*, 361.

4789. McLure, I. A., F. Guzman-Figueroa, et al., *J. Chem. Eng. Data* **1982**, *27*, 398.

4790. McLure, I. A., J. T. Sipowska, et al., *J. Chem. Thermodyn.* **1982**, *14*, 733.

4791. McLure, I. A., V. A. Soares, et al., *J. Chem. Soc., Faraday Trans. 1* **1982**, *78*, 2251.

4792. McMillan, D. R., *Phys. Rev.* **1939**, *57*, 941.

4793. McMullan, R. K., and J. D. Corbett, *J. Chem. Ed.* **1956**, *33*, 313.

4794. McMurry, H. L., *J. Chem. Phys.* **1941**, *9*, 231.

4795. McMurry, H. L., *J. Chem. Phys.* **1941**, *9*, 241.

4796. McMurray, H. S., V. Thornton, et al., *J. Chem. Phys.* **1949**, *17*, 918.

4797. McNeight, S. A., and C. P. Smyth, *J. Amer. Chem. Soc.* **1936**, *58*, 1718.

4798. McQuillin, F. J., and W. O. Ord, *J. Chem. Soc.* **1959**, 3169.

4799. Macbeth, A. K., and J. A. Mills, *J. Chem. Soc.* **1945**, 709.

4800. Macbeth, A. K., and J. A. Mills, *J. Chem. Soc.* **1947**, 205.

4801. Maccoll, A., and P. J. Thomas, *J. Chem. Soc.* **1955**, 979.

4802. Macdonald, C. J., and W. F. Reynolds, *Can. J. Chem.* **1970**, *48*, 1046.

4803. Macdonald, C. J., and T. Schaefer, *Can. J. Chem.* **1970**, *48*, 1033.

4804. Mace, G. E., *Pipeline Eng.* **1969**, *41*, 45, 48; *C.A.* **1969**, *71*, 95044.

4805. Machida, K. S., S. Kojima, et al., *Spectrochim. Acta* **1972**, *28A*, 235.

4806. Maciel, G. E., and G. B. Savitsky, *J. Chem. Phys.* **1965**, *69*, 3925.

4807. Mackay, D., and W. J. Shiu, *J. Chem. Eng. Data* **1977**, *22*, 399.

4808. Mackinney, G., and O. Temmer, *J. Amer. Chem. Soc.* **1948**, *70*, 3586.

4809. Mackle, H., and R. T. B. McClean, *Trans. Faraday Soc.* **1964**, *60*, 817.

4810. Mackle, H., and P. A. G. O'Hare, *Tetrahedron* **1963**, *19*, 961.

4811. Mackle, H., and P. A. G. O'Hare, *Trans. Faraday Soc.* **1962**, *58*, 1912.

4812. Maclean, M. E., P. J. Jencks, et al., *J. Res. Nat. Bur. Stand.* **1945**, *34*, 271.

4813. Macknick, A. B., and J. M. Prausnitz, *J. Chem. Eng. Data* **1979**, *24*, 175.

4814. Mackor, E. L., A. Hofstra, et al., *Trans. Faraday Soc.* **1958**, *54*, 186.

4815. Macrise, R. A., *J. Chem. Eng. Data* **1961**, *12*, 28.

4816. Madai, H., East German Patent 83,984, Aug. 20, 1971; *C.A.* **1973**, *78*, 42843.

4817. Madge, F. W., *J. Phys. Chem.* **1930**, *34*, 1599.

4818. Madic, C., *Chim. Anal.* **1972**, *54*, 102.

4819. Maesen, F. van der, *Physica* **1944**, *15*, 481; *C.A.* **1950**, *44*, 3319.

4820. Magill, P. L., *Ind. Eng. Chem.* **1934**, *26*, 611.

4821. Magnuson, D. W., *J. Chem. Phys.* **1956**, *24*, 344.

4822. Magnusson, G., N. Bodin, et al., *Acta. Pathol. Microbiol. Scand. Sect. A* **1971**, *79*, 639; *C.A.* **1972**, *76*, 21667.

4823. Magnusson, L. B., C. Postmus, et al., *J. Amer. Chem. Soc.* **1963**, *85*, 1711.

4824. Mahanti, P. C., *J. Indian Chem. Soc.* **1928**, *5*, 673.

4825. Mahanti, P. C., *J. Indian Chem. Soc.* **1929**, *6*, 743; *C.A.* **1930**, *24*, 1001.

4826. Mahanti, P. C., and R. N. Das-Gupta, *J. Indian Chem. Soc.* **1929**, *6*, 411; *C.A.* **1930**, *24*, 1001.

4827. Mahanti, P. C., and D. N. Sen-Gupta, *J. Indian Chem. Soc.* **1928**, *5*, 673; *C.A.* **1929**, *23*, 2615.

4828. Maher, P. J., and B. D. Smith, *J. Chem. Eng. Data* **1979**, *24*, 16.

4829. Mai, J., *Ber. Deut. Chem. Gesell.* **1889**, *22*, 2134.

4830. Maibaum, B. K., *J. Exptl. Theoret. Phys.* (URSS) **1939**, *9*, 1383; *C.A.* **1941**, *35*, 5006.

4831. Maier, W., and K. R. Frohner, *Spectrochim. Acta* **1959**, 977; *C.A.* **1960**, *54*, 7341.

4832. Maill, L. M., and D. W. A. Sharp, *A New Dictionary of Chemistry*, Wiley, New York, 1968.

4833. Mailhe, A., *Chemiker-Zeitung* **1909**, *33*, 253.

4834. Mair, B. J., *J. Res. Nat. Bur. Stand.* **1932**, *9*, 457.

4835. Mair, B. J., *J. Res. Nat. Bur. Stand.* **1945**, *34*, 435.

4836. Mair, B. J., and A. F. Forziati, *J. Res. Nat. Bur. Stand.* **1944**, *32*, 151.

4837. Mair, B. J., and A. F. Forziati, *J. Res. Nat. Bur. Stand.* **1944**, *32*, 165.

4838. Mair, B. J., A. R. Glasgow, et al., *J. Res. Nat. Bur. Stand.* **1941**, *26*, 591.

4839. Mair, B. J., and A. J. Streiff, *J. Res. Nat. Bur. Stand.* **1940**, *24*, 395.

4840. Mair, B. J., and A. J. Streiff, *J. Res. Nat. Bur. Stand.* **1941**, *27*, 343.

4841. Mair, B. J., D. J. Termimi, et al., *J. Res. Nat. Bur. Stand.* **1946**, *37*, 229.

4842. Majer, V., V. Svoboda, et al., *Collect. Czech. Chem. Commun.* **1979**, *44*, 3521.

4843. Majewska, J., and S. Urbanowicz, *Polimery* **1964**, *9(1)*, 18; *C.A.* **1964**, *61*, 8409.

4844. Majka, J., K. Knoblock, et al., *Med. Pr.* **1974**, *25*, 427.

4845. Makarov, Yu. A., L. N. Voronina, et al., USSR Patent 162,835, May 27, 1964; *C.A.* **1964**, *61*, 8485.

4846. Maksimov, Yu. Ya., *Zh. Fiz. Khim.* **1968**, *42*, 2921; *Russ. J. Phys. Chem.* **1968**, *42*, 1550.

4847. Malanowski, S., *Bull. Acad. Pol. Sci. Ser. Sci. Chim.* **1961**, *9*, No. 2, 71.

4848. Malecki, J., *Acta Phys. Pol.* **1962**, *21*, 13; *C.A.* **1963**, *58*, 3985.

4849. Malewski, G., M. Peiffer, et al., *J. Mol. Struct.* **1969**, *3*, 419.

4850. Malherbe, F. E., and H. J. Bernstein, *J. Amer. Chem. Soc.* **1952**, *74*, 1859.

4851. Malherbe, F. E., and H. J. Bernstein, *J. Amer. Chem. Soc.* **1952**, *74*, 4408.

4852. Malinowski, E. R., T. Vladimiroff, et al., *J. Phys. Chem.* **1966**, *70*, 2046.

4853. Malinovskiĭ, M. S., E. E. Volkova, et al., *Zh. Obshch. Khim.* **1949**, *19*, 114; *C.A.* **1949**, *43*, 6155.

4854. Malinowski, S., H. Jedrzejewska, et al., *Rocz. Chem.* **1957**, *31*, 71; *C.A.* **1957**, *51*, 14557.

4855. Malkovich, R. S., and V. A. Kolesova, *Zh. Fiz. Khim.* **1954**, *28*, 926; *C.A.* **1955**, *49*, 6730.

4856. de Mallemann, R., F. Suhner, et al., *Compt. Rend.* **1951**, *232*, 1385.

4857. de Mallemann, R., F. Suhner, et al., *Compt. Rend.* **1952**, *234*, 2247.

4858. Malmberg, C. G., and A. A. Maryott, *J. Res. Nat. Bur. Stand.* **1956**, *56*, 1.

4859. Malyshev, V. I., *Bull. Acad. Sci. URSS, Ser. Phys.* **1941**, *5*, No. 1, 13; *Khim. Referat. Zhur.* **1941**, *4*, No. 9, 6; *C.A.* **1944**, *38*, 679.

4860. Malyshev, V. I., and V. N. Murzin, *Izv. Akad. Nauk SSSR, Ser. Fiz.* **1958**, *22*, 1107; *C.A.* **1959**, *53*, 789.

4861. Malyshev, V. I., and M. V. Shishkina, *Dokl. Akad. Nauk SSSR* **1949**, *66*, 833; *C.A.* **1949**, *43*, 7345.

4862. Maman, A., *Compt. Rend.* **1938**, *207*, 1401; *C.A.* **1939**, *33*, 1657.

4863. Maman, A., *Pub. Sci. Tech. Ministere Air* (France) **1935**, No. 66; *C.A.* **1936**, *30*, 7095.

4864. Mamedov, A. M., T. S. Akhundov, et al., *Teploenergetika* **1967**, *14(5)*, 81.

4865. Mamedov, A. M., T. S. Akhundov, et al., *Zh. Fiz. Khim.* **1970**, *44*, 1565; *Russ. J. Phys. Chem.* **1970**, *44*, 877.

4866. Mandravel, C., and M. Butel, *Rev. Roum. Chim.* **1972**, *17*, 1841.

4867. Mandric, Gh., *Igiena* (Bucharest) **1961**, *9*, 271; *C.A.* **1962**, *57*, 8847.

4868. Mandric, Gh., *Rev. Chim.* (Bucharest) **1961**, *12*, 503; *C.A.* **1962**, *56*, 6317.

4869. Mangini, A., *Gazz. Chim. Ital.* **1958**, *88*, 1063; *C.A.* **1959**, *53*, 18624.

4870. Mangum, B. W., *Clin. Chem.* (Winston-Salem, N.C.) **1983**, *29*, 1380; *C.A.* **1983**, *99*, 84726.

4871. Manius, G., F. P. Mahn, et al., *J. Chromatographic Sci.* **1971**, *9*, 367.

4872. Manley, R. St. J., *Svensk Papperstidn.* **1958**, *61*, 96.

4873. Manley, T. R., and C. G. Martin, *Spectrochim. Acta, Part A* **1976**, *32A*, 357.

4874. Manly, D. G., U.S. Patent 3,021,342, Feb. 13, 1962; *C.A.* **1962**, *56*, 15489; *Official Gaz. U.S. Patent Office* **1962**, *775*, 558.

4875. Manly, D. G., and E. D. Amstutz, *J. Org. Chem.* **1957**, *22*, 323.

4876. Mansson, M., *J. Chem. Thermodyn.* **1972**, *4*, 865.

4877. Mansson, M.. B. Ringner, et al., *J. Chem. Thermodyn.* **1971**, *3*, 547.

4878. Mansson, M., P. Sellers, et al., *J. Chem. Thermodyn.* **1977**, *9*, 91.

4879. Manufacturing Chemists Association Chemical Safety Sheets, Washington.

4880. Manufacturing Chemists Association Data Sheet SD-29, *J. Chem. Ed.* **1964**, *41*, 578A.

4881. Manzoni-Ansidei, R., *Boll. Sci. Fac. Chim. Ind. Bologna* **1940**, 137; *C.A.* **1942**, *36*, 6904.

4882. Manzoni-Ansidei, R., *Ric. Sci.* **1939**, *10*, 328; *Chem. Zentr.* **1940**, *I*, 35.

4883. Manzoni-Ansidei, R., and M. Rolla, *Atti. Accad. Naz. Lincei, Rend Cl. Sci. Fis. Mat. Nat.* **1938**, *27*, 410; *C.A.* **1938**, *32*, 8933.

4884. Marchlewski, L., and A. Moroz, *Bull. Soc. Chim.* **1924**, *35*, 473; *C.A.* **1924**, *18*, 1995.

4885. Marckwald, W., and A. McKenzie, *Ber. Deut. Chem. Gesell.* **1901**, *34*, 485.

4886. Marcus, S. H., W. F. Reynolds, et al., *J. Org. Chem.* **1966**, *31*, 1872.

4887. Mardles, E. W. J., *Trans. Faraday Soc.* **1933**, *29*, 476; *C.A.* **1933**, *27*, 4150.

4888. Marenzi, A. D., and F. Vilallonga, *Rev. Soc. Argentina Biol.* **1941**, *17*, 232; *C.A.* **1942**, *36*, 500.

4889. Marinangeli, A., *Ann. Chim.* (Rome) **1954**, *44*, 211, 219; *C.A.* **1955**, *49*, 2158.

4890. Marker, R. E., and T. S. Oakwood, *J. Amer. Chem. Soc.* **1938,** *60,* 2598.

4891. Markley, K. S., ed., *Fatty Acids,* 2nd ed., pt. 1, Interscience, New York, 1960.

4892. Markley, K. S., ed., *Fatty Acids,* 2nd ed., pt. 2, Interscience, New York, 1961.

4893. Markley, K. S., ed., *Fatty Acids,* 2nd ed., pt. 3, Interscience, New York, 1964.

4894. Markov, P., V. Dryanska, et al., *C.R. Acad. Bulg. Sci.* **1966,** *19,* 815.

4895. Markova, S. V., and P. A. Bazhulin, *Opt. Spectrosk.* **1956,** *1,* No. 1, 41; *C.A.* **1956,** *50,* 13413.

4896. Markownikow, W., and W. Tscherdynzew, *J. Russ. Phys. Chem. Ges. 32,* 302; *Chem. Zentr.* **1900,** *II,* 630.

4897. Markunas, P. C., and J. A. Riddick, *Anal. Chem.* **1951,** *23,* 337.

4898. Maroni, P., *Ann. Chim.* (Paris) **1957,** *2,* 757; *C.A.* **1958,** *52,* 8728.

4899. Marquardt, R. P., and E. N. Luce, *Ind. Eng. Chem., Anal. Ed.* **1944,** *16,* 751.

4900. Marrison, L. W., *J. Chem. Soc.* **1951,** 1614.

4901. Marschalks, B., and J. Barna, *Acta Tech. Acad. Sci. Hung.* **1957,** *19,* 85; *C.A.* **1958,** *52,* 15994.

4902. Marsden, C., and S. Mann, *Solvents Guide,* 2nd ed., Interscience, New York, 1963.

4903. Marsden, J., and A. C. Cuthbertson, *Can. J. Res.* **1933,** *9,* 419; *C.A.* **1934,** *28,* 696; and ref. [7116].

4904. Marsden, J., and O. Maass, *Can. J. Res.* **1935,** *13B,* 296.

4905. Marsden, R. J. B., and L. E. Sutton, *J. Chem. Soc.* **1936,** 1383.

4906. Marshall, H. P., and E. Grunwald, *J. Amer. Chem. Soc.* **1954,** *76,* 2000.

4907. Marshall, M. J., *Ind. Eng. Chem.* **1928,** *20,* 1379.

4908. Martelli, D., *Med. Lavoro* **1960,** *51,* 123; *C.A.* **1960,** *54,* 23008.

4909. Martens, J., K. Praefcke, et al., *Z. Naturforsch., B: Anorg. Chem., Org. Chem.* **1975,** *30B,* 259.

4910. Martens, T. F., Belgian Patent 613,121, July 25, 1962; *C.A.* **1963,** *58,* 8902.

4911. Martin, A. R., *J. Chem. Soc.* **1928,** 3270.

4912. Martin, A. R., and B. Collie, *J. Chem. Soc.* **1932,** 2658.

4913. Martin, A. R., and C. M. George, *J. Chem. Soc.* **1933,** 1413.

4914. Martin, D., A. Weise, et al., *Angew. Chem. Intern. Ed.* **1967,** *6,* 318.

4915. Martin, D. R., *J. Phys. Colloid Chem.* **1947,** *51,* 1400.

4916. Martin, G. J., and M. L. Martin, *Bull. Soc. Chim. Fr.* **1966,** 1636; *C.A.* **1966,** *65,* 13511.

4917. Martin, G. J., and M. L. Martin, *J. Chim. Phys.* **1964,** *61,* 1222; *C.A.* **1965,** *62,* 6033.

4918. Martin, J., and B. P. Dailey, *J. Chem. Phys.* **1962,** *37,* 2594.

4919. Martin, J. F., and R. J. L. Andon, *J. Chem. Thermodyn.* **1982,** *14,* 679.

4920. Martin, J. J., *J. Chem. Eng. Data* **1960,** *5,* 334.

4921. Martin, M., *Ann. Phys.* (Paris) **1962,** *7,* 35; *C.A.* **1963,** *59,* 148.

4922. Martin, M., *Bull. Soc. Roy. Sci. Liege* **1962,** *31,* 434; *C.A.* **1962,** *57,* 6772.

4923. Martin, M., *J. Chim. Phys.* **1962,** *59,* 736; *C.A.* **1963,** *58,* 158.

4924. Martin, M., and F. Herail, *Compt. Rend.* **1959,** *248,* 1994; *C.A.* **1962,** *56,* 10983.

4925. Martin, M., and G. Martin, *Compt. Rend.* **1959,** *249,* 884.

4926. Martin, M., and M. Quilbeuf, *Compt. Rend.* **1961,** *252,* 4151; *C.A.* **1962,** *56,* 3046.

4927. Martin, R., *Rev. Inst. Franc. Petrole* **1947,** *2,* 323; *C.A.* **1948,** *42,* 4497.

4928. Maruta, Y., and I. Matubora, *Nippon Nogei-Kagaku Kaishi* **1954,** *28,* 125; *C.A.* **1957,** *51,* 9416.

4929. Maruyama, M., *Polarography* **1955,** *3,* 22; *C.A.* **1956,** *50,* 7348.

4930. Marvel, C. S., *Organic Syntheses, Vol. 5,* Wiley, New York, 1925.

4931. Marvel, C. S., *Organic Syntheses, Vol. 11,* Wiley, New York, 1931.

4932. Marvel, C. S., J. Harkema, et al., *J. Amer. Chem. Soc.* **1941,** *63,* 16096.

4933. Maryott, A. A., *J. Amer. Chem. Soc.* **1941,** *63,* 3079.

4934. Maryott, A. A., and F. Buckley, *Table of Dielectric Constants and Electric Dipole Moments of Substances in the Gaseous State,* NBS Circular 537, U.S. Govt. Printing Office, Washington, 1953.

4935. Maryott, A. A., M. E. Hobbs, et al., *J. Amer. Chem. Soc.* **1941,** *63,* 659.

4936. Maryott, A. A., M. E. Hobbs, et al., *J. Amer. Chem. Soc.* **1949,** *71,* 1671.

4937. Maryott, A. A., and E. R. Smith, *Table of Dielectric Constants of Pure Liquids,* NBS Circular 514, U.S. Govt. Printing Office, Washington, 1951.

4938. Mascarelli, C., and I. Musatty, *Gazz. Chim. Ital.* **1911,** *41a,* 80; from [6114], p. 607.

4939. Mascarelli, L., *Atti R. Accad. Lincei* **1907,** *16(V),* 924; *J. Chem. Soc.* **1907Aii,** 602.

4940. Mascarelli, L., *Atti R. Accad. Lincei* **1908,** *17(V),* 494; *Chem. Zentr.* **1909,** *I,* 169.

4941. Mascarelli, L., and F. Benati, *Gazz. Chem. Ital.* **1909,** *39II,* 642; *Chem. Zentr.* **1910,** *I,* 1024.

4942. Mascarelli, L., and L. Vecchiotti, *Atti R. Accad. Lincei* **1910**, *19II*, 410; *Chem. Zentr.* **1911**, *I*, 75.
4943. Mashiko, Y., *Nippon Kagaku Zasshi* **1958**, *79*, 470; *C.A.* **1958**, *52*, 10721.
4944. Mashiko, Y., *Nippon Kagaku Zasshi* **1959**, *80*, 593; *C.A.* **1959**, *53*, 13778.
4945. Masi, J. F., *J. Amer. Chem. Soc.* **1952**, *74*, 4738.
4946. Masi, J. F., *J. Amer. Chem. Soc.* **1953**, *75*, 5082.
4947. Masi, J. F., and R. B. Scott, *J. Res. Nat. Bur. Stand., Sect. A* **1975**, *79A*, 619.
4948. Maslov, P. G., and Y. P. Maslov, *Khim. Tekhnol. Topl. Masel* **1958**, *3*, No. 10, 50; *C.A.* **1959**, *53*, 1910.
4949. Mason, M. E., B. Johnson, et al., *Anal. Chem.* **1965**, *37*, 760.
4950. Massaldi, H. A., and C. J. King, *J. Chem. Eng. Data* **1973**, *18*, 393.
4951. Massart, L., *Bull. Soc. Chim. Belges* **1936**, *45*, 76; *C.A.* **1936**, *30*, 4062.
4952. Massol, G., and A. Faucon, *Compt. Rend.* **1909**, *149*, 345; *J. Chem. Soc.* **1909Aii**, 791.
4953. Massol, G., and A. Faucon, *Compt. Rend.* **1913**, *157*, 386.
4954. Massol, G., and A. Faucon, *Compt. Rend.* **1914**, *159*, 314; *C.A.* **1915**, *9*, 10.
4955. Massol, G., and A. Faucon, *Compt. Rend.* **1916**, *163*, 92; *Chem. Zentr.* **1916**, *II*, 725.
4956. Massol, G., and A. Faucon, *Compt. Rend.* **1917**, *164*, 308; *C.A.* **1917**, *11*, 2560.
4957. Masson, I., *Nature* **1931**, *128*, 726.
4958. Massy, N. B., F. L. Warren, et al., *J. Chem. Soc.* **1932**, 91.
4959. Mastrangelo, S. V. R., *Anal. Chem.* **1957**, *29*, 841.
4960. Mastrangelo, S. V. R., and J. G. Aston, *Anal. Chem.* **1954**, *26*, 764.
4961. Mastrolanni, M. J., R. F. Stahl, et al., *J. Chem. Eng. Data* **1978**, *23*, 113.
4962. Masuno, M., T. Asahara, et al., *J. Soc. Chem. Ind. Jpn.* **1946**, *49*, 192; *C.A.* **1948**, *42*, 6736.
4963. Mata, F., and F. Garcia, *Z. Phys. Chem.* (Leipzig) **1981**, *262*, 49.
4964. Mateos, J. L., R. Cetain, et al., *Chem. Commun.* **1965**, *21*, 519.
4965. Mateos, J. L., A. Rodrigues, et al., *Bol. Inst. Quim. Univ. Nacl. Auton. Mex.* **1964**, *16*, 30; *C.A.* **1965**, *63*, 5142.
4966. Matheson, M., E. Auer, et al., *J. Amer. Chem. Soc.* **1949**, *71*, 2610.
4967. Mathews, J. F., and J. J. McKetta, *J. Phys. Chem.* **1961**, *65*, 758.
4968. Mathews, J. H., *J. Amer. Chem. Soc.* **1917**, *39*, 1125.
4969. Mathews, J. H., *J. Amer. Chem. Soc.* **1926**, *48*, 562.
4970. Mathews, J. H., and K. E. Faville, *J. Phys. Chem.* **1918**, *22*, 1.
4971. Mathews, J. H., and P. R. Fehlandt, *J. Amer. Chem. Soc.* **1931**, *53*, 3212.
4972. Mathias, A., *Tetrahedron* **1966**, *22*, 217.
4973. Mathias, S., *J. Amer. Chem. Soc.* **1950**, *72*, 1897.
4974. Mathias, S., and R. G. Cecchini, *Anales Real Soc. Espan. Fis. Quim.* (Madrid) **1964**, *60*, 241; *C.A.* **1965**, *62*, 1149.
4975. Mathias, S., and E. de C. Filho, *J. Phys. Chem.* **1958**, *62*, 1427.
4976. Mathias, S., E. de C. Filho, et al., *J. Phys. Chem.* **1961**, *65*, 425.
4977. Mathieu, V. P., and D. Massignon, *Ann. Phys.* **1941**, *16*, 5; *Chem. Zentr.* **1942**, *II*, 24; *C.A.* **1943**, *37*, 4304.
4978. Mathieu, M. P., *Acad. Roy. Belg., Classe Sci. Mem. Collect. in-8* **1953**, *28*, No. 2, 224 pp.; *C.A.* **1953**, *47*, 10288.
4979. Mathieu, M. P., *Bull. Soc. Chim. Belges* **1952**, *61*, 683; *C.A.* **1954**, *48*, 7365.
4980. Mathis, A., *Anal. Chim. Acta* **1964**, *31*, 598.
4981. Mathur, R., E. D. Becker, et al., *J. Phys. Chem.* **1963**, *67*, 2190.
4982. Mathur, R., S. M. Wang, et al., *J. Phys. Chem.* **1964**, *68*, 2140.
4983. Matignon, C., H. Moureu, et al., *Bull. Soc. Chim. Fr.* **1934**, *1*, 1308; *Chem. Zentr.* **1935**, *I*, 1855.
4984. Matossi, F., *Phys. Z.* **1944**, *45*, 304; *C.A.* **1946**, *40*, 6993.
4985. Matsen, F. A., N. Ginsburg, et al., *J. Chem. Phys.* **1945**, *13*, 309.
4986. Matsen, F. A., W. W. Robertson, et al., *Chem. Rev.* **1947**, *41*, 273.
4987. Matsen, J. M., and E. F. Johnson, *J. Chem. Eng. Data* **1960**, *5*, 531.
4988. Matsui, H., and S. Ishimoto, *Tetrahedron Lett.* **1966**, 1827; *C.A.* **1966**, *65*, 593.
4989. Matsui, K., *J. Chem. Soc. Jpn.* **1943**, *64*, 1417; *C.A.* **1947**, *41*, 3753.
4990. Matsui, T., and L. G. Hepler, *Can. J. Chem.* **1973**, *51*, 1941.
4991. Matsuno, K., and K. Han, *Bull. Chem. Soc. Jpn.* **1936**, *11*, 321; *C.A.* **1936**, *30*, 5887.

4992. Matsuoka, S., S. Hattori, et al., *J. Phys. Soc. Jpn.* **1965**, *20*, 1212; *C.A.* **1965**, *63*, 7799.
4993. Matsuura, H., S. Imazeki, et al., *Bull. Chem. Soc. Jpn.* **1979**, *52*, 2512.
4994. Matsuura, T., *J. Sci. Hiroshima Univ.* **1938**, *8A*, 303; *C.A.* **1939**, *33*, 2104.
4995. Matsuura, T., and T. Aratani, *J. Sci. Hiroshima Univ., Ser. A* **1954**, *18*, 253; *C.A.* **1956**, *50*, 4048.
4996. Matterson, A. H. S., and L. A. Woodward, *Proc. Roy. Soc.* (London) **1955**, *231A*, 514; *C.A.* **1956**, *50*, 676.
4997. Matthews, J. B., J. F. Sumner, et al., *Trans. Faraday Soc.* **1950**, *46*, 797.
4998. Matthews, W. S., J. E. Bares, et al., *J. Amer. Chem. Soc.* **1975**, *97*, 7006.
4999. Matthias, M., and M. J. Loeffler, *Kaeltetechnik* **1966**, *18*, 240; *C.A.* **1967**, *66*, 10471.
5000. Mattox, W. J., U.S. Patent 2,436,932, Mar. 2, 1948; *Official Gaz. U.S. Patent Office* **1948**, *608*, 119.
5001. Matuszko, A. J., and M. S. Chang, *Chem. Ind.* (London) **1963**, 822.
5002. Mauret, P., J. P. Fayet, et al., *J. Chim. Phys. Physicochim. Biol.* **1968**, *65*, 549.
5003. Mavel, G., *Compt. Rend.* **1960**, *250*, 1477.
5004. Maxim, N., and R. Mavrodineanu, *Bull. Soc. Chim. Fr.* **1935**, *2*, 591; *C.A.* **1935**, *29*, 4328.
5005. May and Black, Ltd., *Diethyl Ether, Its Properties and Uses,* Dagenham, Essex.
5006. May, W. E., S. P. Wasik, et al., *Anal. Chem.* **1978**, *50*, 997.
5007. Mayberry, M. G., and J. G. Aston, *J. Amer. Chem. Soc.* **1934**, *56*, 2682.
5008. Mayer, F., and A. Sieglitz, *Ber. Deut. Chem. Gesell.* **1922**, *55B*, 1835; *C.A.* **1923**, *17*, 101.
5009. Mayer, U., V. Gutmann, et al., *Monatsh. Chem.* **1975**, *106*, 1235.
5010. Mayer-Pitsch, E., H. Duftschmid-Hinrichs, et al., *Z. Elektrochem.* **1943**, *49*, 368; *C.A.* **1944**, *38*, 21.
5011. Mayland, B. J., U.S. Patent 2,412,823, Dec. 17, 1946; *C.A.* **1947**, *41*, 1711.
5012. Mayo, F. R., and F. M. Lewis, *J. Amer. Chem. Soc.* **1944**, *66*, 1596.
5013. Mazonski, T., D. Gasztych, et al., *Przemysl Chem.* **1962**, *41*, 251; *C.A.* **1963**, *58*, 11202.
5014. Mazumder, M., *Indian J. Phys.* **1953**, *27*, 406; *C.A.* **1957**, *51*, 7862; *C.A.* **1956**, *50*, 384; *C.A.* **1957**, *51*, 9319.
5015. Mazur, I. J., *Acta Phys. Polon.* **1938**, *7*, 285; *C.A.* **1939**, *33*, 8068.
5016. Mears, T. W., A. Fookson, et al., *J. Res. Nat. Bur. Stand.* **1950**, *44*, 299.
5017. Mears, T. W., C. L. Stanley, et al., *J. Res. Nat. Bur. Stand.* **1963**, *67A*, 475.
5018. Mears, W. H., E. Rosenthal, et al., *J. Chem. Eng. Data* **1966**, *11*, 338.
5019. Mears, W. H., R. F. Stahl, et al., *Ind. Eng. Chem.* **1955**, *47*, 1449.
5020. Mecka, R., and A. Reuter, *Z. Naturforsch.* **1949**, *4A*, 368; *C.A.* **1950**, *44*, 8182.
5021. Mecke, R., *Z. Phys. Chem.* (Leipzig) **1937**, *36B*, 347; *C.A.* **1937**, *31*, 8376.
5022. Mecke, R., and E. Funck, *Z. Elektrochem.* **1956**, *60*, 1124.
5023. Mecke, R., and R. Joeckle, *Z. Elektrochem.* **1962**, *66*, 255; *C.A.* **1962**, *57*, 9322.
5024. Mecke, R., R. Mecke, et al., *Z. Naturforsch.* **1955**, *10b*, 367; from [5070].
5025. Mecke, R., and K. Noack, *Chem. Ber.* **1960**, *93*, 210.
5026. Mecke, R., and G. Rossmy, *Z. Elektrochem.* **1955**, *59*, 866; *Chem. Zentr.* **1956**, 5506.
5027. Mecke, R., and H. Speckt, *Z. Elektrochem.* **1958**, *62*, 500; *C.A.* **1958**, *52*, 16819.
5028. Meda, E., and A. Bertino, *Minerva Farm.* **1962**, *11*, 45; *C.A.* **1962**, *57*, 11313.
5029. Medard, L., *J. Chim. Phys.* **1936**, *33*, 626; *Chem. Zentr.* **1936**, *II*, 3531.
5030. Medard, L., and F. Deguillon, *Compt. Rend.* **1936**, *203*, 1518.
5031. Medhi, K. C., and D. K. Mukherjee, *Spectrochim. Acta* **1965**, *21*, 895.
5032. Medoks, G. V., and L. E. Ozerskaya, *Zh. Obshch. Khim.* **1960**, *30*, 1643; *C.A.* **1961**, *55*, 1410.
5033. Medvetskaya, I. M., L. P. Andreeva, et al., *Zh. Fiz. Khim.* **1980**, *54*, 2304; *Russ. J. Phys. Chem.* **1980**, *54*, 1313.
5034. Meeker, R. L., F. E. Critchfield, et al., *Anal. Chem.* **1962**, *34*, 1510.
5035. Meeks, A. C., and I. J. Goldfarb, *J. Chem. Eng. Data* **1967**, *12*, 196.
5036. Meerwein, H., B. J. Bock, et al., *J. Prakt. Chem.* **1946**, *147*, 211; *C.A.* **1937**, *31*, 656.
5037. Meeussen, E., C. Debeuf, et al., *Bull. Soc. Chim. Belges* **1967**, *76*, 145.
5038. Megaloikonomos, J. G., *Praktika Akad. Athenon* **1937**, *12*, 226; *C.A.* **1939**, *33*, 5321.
5039. Megson, N. J. L., *Trans. Faraday Soc.* **1938**, *34*, 525.
5040. Mehl, W., *Chem. Fabrik.* **1934**, *7*, 240; *Chem. Zentr.* **1934**, *II*, 1958.
5041. Mehl, W., *Z. Ges. Kalte-Ind.* **1934**, *41*, 152; *Chem. Zentr.* **1934**, *II*, 3736.
5042. Mehl, W., *Z. Phys. Chem.* (Leipzig) **1934**, *169A*, 312; *Beil.* **EIII1**, 669; *C.A.* **1934**, *28*, 6606.
5043. Mehrotra, V. K., *Indian J. Pure Appl. Phys.* **1968**, *6*, 691; *C.A.* **1969**, *70*, 72335.

5044. Meighan, R. M., and R. H. Cole, *J. Phys. Chem.* **1964,** *68,* 503.
5045. Meisenheimer, J., and O. Dorner, *Justus Liebigs Ann. Chem.* **1930,** *482,* 130.
5046. Meissner, H. P., and A. S. Michaels, *Ind. Eng. Chem.* **1949,** *41,* 2782.
5047. Meissner, H. P., and E. M. Redding, *Ind. Eng. Chem.* **1942,** *34,* 521.
5048. Meissner, W., *Z. Angew. Phys.* **1948,** *1,* 75.
5049. Mekhiev, S. D., R. G. Rizaev, et al., *Azerb. Khim. Zh.* **1965,** 70; *C.A.* **1966,** *64,* 10539.
5050. Meldrum, W. B., L. P. Saxer, et al., *J. Amer. Chem. Soc.* **1943,** *65,* 2023.
5051. Melia, T. P., *Polymer* **1962,** *3,* 317, Part 8.
5052. Melia, T. P., and R. Merrifield, *J. Appl. Chem.* (London) **1969,** *19,* 79.
5053. Mellan, I., *Polyhydric Alcohols,* Spartan Books, Washington, 1962.
5054. Melton, C. E., and P. S. Rudolph, *J. Chem. Phys.* **1959,** *31,* 1485.
5055. Menashi, J., W. L. Reynolds, et al., *Inorg. Chem.* **1965,** *4,* 299.
5056. Menczel, S., *Z. Phys. Chem.* (Leipzig) **1927,** *125,* 161.
5057. Mendenhall, G. D., D. M. Golden, et al., *J. Phys. Chem.* **1973,** *77,* 2707.
5058. Menschutkin, B. N., *J. Russ. Phys. Chem. Soc.* **1907,** *39,* 121; *C.A.* **1907,** *1,* 1547; ibid., **1909,** *40,* 1415; *C.A.* **1910,** *4,* 2262.
5059. Menschutkin, N., *J. Chem. Soc.* **1906,** *89,* 1532.
5060. Menzies, A. W. C., and S. L. Wright, *J. Amer. Chem. Soc.* **1921,** *43,* 2314.
5061. Merckx, R., J. Verhulst, et al., *Bull. Soc. Chim. Belges* **1933,** *42,* 177; *Chem. Zentr.* **1933,** *II,* 1173.
5062. Meredith, C. C., and G. F. Wright, *Can. J. Chem.* **1960,** *38,* 1177.
5063. Merz, V., and W. Weith, *Ber. Deut. Chem. Gesell.* **1881,** *14,* 187.
5064. Merzlin, R. W.. *J. Russ. Phys. Chem. Soc.* **1935,** *67,* 161.
5065. Mesorana, F. P., and F. Marquet, *Anales Real Soc. Espan. Fis. Quim.* **1953,** *49B,* 5; *C.A.* **1953,** *47,* 11000.
5066. Messerly, J. F., and H. L. Finke, *J. Chem. Thermodyn.* **1970,** *2,* 867.
5067. Messerly, J. F., H. L. Finke, et al., *J. Chem. Thermodyn.* **1975,** *7,* 1029.
5068. Messerly, J. F., G. B. Guthrie, et al., *J. Chem. Eng. Data* **1967,** *12,* 338.
5069. Messerly, G. H., and J. G. Aston, *J. Amer. Chem. Soc.* **1940,** *62,* 886.
5070. Messerly, G. H., and R. M. Kennedy, *J. Amer. Chem. Soc.* **1940,** *62,* 2988.
5071. Messerly, J. F., S. S. Todd, et al., *J. Phys. Chem.* **1965,** *69,* 4303.
5072. Metcalf, L. D., and A. A. Schmitz, *Anal. Chem.* **1961,** *33,* 363.
5073. Métra, M., L. Lesage, et al., *Compt. Rend.* **1938,** *206,* 1026; *C.A.* **1938,** *32,* 4908.
5074. Metz, D. J., and A. Glines, *J. Phys. Chem.* **1967,** *71,* 1158.
5075. Metzler, C. M., A. Cahill, et al., *J. Amer. Chem. Soc.* **1980,** *102,* 6075.
5076. Meyer, C. F., D. W. Bronk, et al., *J. Opt. Soc. Am.* **1927,** *15,* 257.
5077. Meyer, E. F., and M. J. Awe, *J. Chem. Eng. Data* **1980,** *25,* 371.
5078. Meyer, E. F., and R. D. Hotz, *J. Chem. Eng. Data* **1973,** *18,* 359.
5079. Meyer, F., and A. G. Harrison, *Can. J. Chem.* **1964,** *42,* 2008.
5080. Meyer, H., *Analyse und Konstitutionsermittlung Organischer Verbindungen,* Edwards Brothers, Inc., Ann Arbor, Michigan, reproduced 1943 (1st German ed. 1938).
5081. Meyer, H., *Lehrb. der Organ.-Chem. Methodik,* 4th ed., *Vol. 1,* Berlin, 1922.
5082. Meyer, J., and B. Mylius, *Z. Phys. Chem.* (Leipzig) **1920,** *95,* 349; *C.A.* **1920,** *14,* 3176.
5083. Meyer, J. D., and E. E. Reid, *J. Amer. Chem. Soc.* **1933,** *55,* 1574.
5084. Meyer, K. H., and R. Luhdemann, *Helv. Chim. Acta* **1935,** *18,* 307.
5085. Meyer, K. H., and F. G. Willson, *Ber. Deut. Chem. Gesell.* **1914,** *47,* 837.
5086. Meyer, L. W. A., and W. M. Gearhart, *Ind. Eng. Chem.* **1948,** *40,* 1478.
5087. Meyer, R., and H. H. Guenthard, *Spectrochim. Acta, Part A* **1967,** *23,* 2341.
5088. Meyer, R., and J. Jaeger, *Ber. Deut. Chem. Gesell.* **1903,** *36,* 1555.
5089. Meyer, R., M. Meyer, et al., *Bull. Soc. Chim. Fr.* **1968,** 2692.
5090. Meyer, V., *Ber. Deut. Chem. Gesell.* **1886,** *19,* 3259.
5091. Meyer, V., and coworkers, *Ber. Deut. Chem. Gesell.* **1872,** *5,* 203, 399, 514, 1029, 1034.
5092. Meyer, V., and F. Muller, *J. Prakt. Chemie* **1892,** [2], *46,* 161.
5093. Meyers, C. Y., *J. Org. Chem.* **1961,** *26,* 1046.
5094. Meyerson, S., and E. K. Fields, *Org. Mass. Spectrom.* **1974,** *9,* 485.
5095. Michael, A., and R. N. Hartman, *Ber. Deut. Chem. Gesell.* **1907,** *40,* 140; *C.A.* **1907,** *1,* 982.

5096. Michael, A., E. Scharf, et al., *J. Amer. Chem. Soc.* **1916**, *38*, 653.
5097. Michael, A., and F. Zeidler, *Justus Liebigs Ann. Chem.* **1912**, *393*, 81; *C.A.* **1913**, *7*, 784.
5098. Michaelian, K. H., and S. M. Ziegler, *Appl. Spectrosc.* **1973**, *27*, 13.
5099. Michaelis, L., *Ber. Deut. Chem. Gesell.* **1913**, *46*, 3683.
5100. Michaelson, J. B., and D. J. Huntsman, *J. Med. Chem.* **1960**, *7*, 378.
5101. Michalczyk, J., *Rocz. Chem.* **1964**, *38*, 697; *C.A.* **1965**, *62*, 1152.
5102. Michalski, H., S. Michalowski, et al., *Zeszyty Nauk. Politech. Lodz. Chem.* **1962**, *12*, 73; *C.A.* **1964**, *61*, 6445.
5103. Michel, G., *Bull. Soc. Chim. Belges* **1959**, *68*, 643; *C.A.* **1960**, *54*, 16178.
5104. Michel, J., *Bull. Soc. Chim. Belges* **1939**, *48*, 105; *C.A.* **1939**, *33*, 7650.
5105. Michels, A., T. Wassenaar, et al., *J. Chem. Eng. Data* **1966**, *11*, 449.
5106. Michio, N., and T. Atsuchi, Japanese Patent 71 30 176 02, Sept. 1971; *C.A.* **1972**, *76*, 13951.
5107. Michler, W., and C. Escherich, *Ber. Deut. Chem. Gesell.* **1879**, *12*, 1162.
5108. Michnowicz, J., and B. Munson, *Org. Mass. Spectrom.* **1970**, *4*, 481.
5109. *Microfilm Abstract* **1947**, *7*, No. 2, 42.
5110. Middelhock, J., and C. J. F. Bottcher, *Chem. Soc.* (London), *Spec. Publ.* **1966**, *20*, 69; *C.A.* **1966**, *65*, 8130.
5111. Middleton, B. A., and J. R. Partington, *Nature* **1938**, *141*, 516; *C.A.* **1938**, *32*, 4399.
5112. Midzushima, S., Y. Morino, et al., *J. Chem. Soc. Jpn.* **1944**, *65*, 127.
5113. Mikawa, Y., *Bull. Chem. Soc. Jpn.* **1956**, *29*, 110; *C.A.* **1957**, *51*, 7143.
5114. Mikawa, Y., *Nippon Kagaku Zasshi* **1960**, *61*, 1512.
5115. Mikaya, A. I., V. G. Zaikin, et al., *Izv. Akad. Nauk. SSSR, Ser. Khim.* **1979**, (10), 2221; *Bull. Acad. Sci. USSR, Div. Chem. Sci.* **1979**, (10), 2043.
5116. Mikhailets, G. A., *Toksikol. Serasgan. Soedin. Ufa., Sb.* **1964**, 4; *C.A.* **1965**, *63*, 7550.
5117. Mikhailov, G. P., and B. I. Sazhin, *Zh. Tekh. Fiz.* **1955**, *25*, 1696; *C.A.* **1956**, *50*, 634.
5118. Mikhailov, L., *Khim. Ind.* (Sofia) **1961**, *33*, No. 4, 117; *C.A.* **1962**, *57*, 8525.
5119. Mikhailov, V. A., S. K. Kharchenko, et al., *Int. Symp.* "Reinstoffe Wiss. Tech.," Tagungsber., 2nd. **1965**, *1*, 21; *C.A.* **1969**, *70*, 68464.
5120. Mikhailov, V. A., S. K. Kharchenko, et al., *Izv. Sibirsk. Otd. Akad. Nauk SSSR* **1962**, No. 7, 50; *C.A.* **1963**, *58*, 976.
5121. Mikhailov, V. S., M. S. Dyaminov, et al., USSR Patent 196,762, May 31, 1967; *C.A.* **1968**, *68*, 39078.
5122. Mikhant'ev, B. J., and E. A. Pyrakhina, *Zh. Obshch. Khim.* **1960**, *30*, 958; *C.A.* **1961**, *55*, 369.
5123. Mikiewicz, M., *J. Inst. Petroleum* **1952**, *38*, 425; *C.A.* **1952**, *46*, 8960.
5124. Mikulak, R., and O. Runquist, *J. Chem. Ed.* **1961**, *38*, 557.
5125. Milas, N. A., *J. Amer. Chem. Soc.* **1931**, *53*, 221.
5126. Milas, N. A., U.S. Patent 2,414,385, Jan. 14, 1947; *C.A.* **1947**, *41*, 5894.
5127. Milas, N. A., and L. H. Perry, *J. Amer. Chem. Soc.* **1946**, *68*, 1938.
5128. Milazzo, G., *Boll. Sci. Fac. Chim. Ind. Bologna* **1941**, 94; *Chem. Zentr.* **1942**, *I*, 1995.
5129. Milazzo, G., *Rend. 1st Super. Sanita* **1948**, *11*, 372; *C.A.* **1949**, *43*, 2094.
5130. Milazzo, G., *Rend. 1st Super. Sanita* **1956**, *19*, 322, 342; *C.A.* **1957**, *51*, 7791.
5131. Milburn, A. H., and E. V. Truter, *J. Chem. Soc.* **1954**, 3344.
5132. Miles, C. B., and H. Hunt, *J. Phys. Chem.* **1941**, *45*, 1346.
5133. Millard, B. J., and D. F. Shaw, *J. Chem. Soc. B* **1966**, 664.
5134. Miller, A., and D. W. Scott, *J. Chem. Phys.* **1978**, *68*, 1317.
5135. Miller, B., *J. Org. Chem.* **1973**, *38*, 1243.
5136. Miller, C. C., *Proc. Roy. Soc.* (London) **1924**, *106A*, 724.
5137. Miller, C. M., and H. W. Thompson, *J. Chem. Phys.* **1949**, *17*, 845.
5138. Miller, F. A., and R. J. Capwell, *Spectrochim. Acta, Part A* **1971**, *27*, 1113.
5139. Miller, F. A., and R. G. Inskeep, *J. Chem. Phys.* **1950**, *18*, 1519.
5140. Miller, F. A., and F. E. Kiviat, *Spectrochim. Acta, Part A* **1969**, *25*, 1363.
5141. Miller, G. A., *J. Chem. Eng. Data* **1963**, *8*, 69.
5142. Miller, G. H., U.S. Patent 2,428,120, Sept. 30, 1947; *C.A.* **1948**, *42*, 610.
5143. Miller, K. J., *Ind. Eng. Chem., Chem. Eng. Data Series* **1958**, *3*, 239.
5144. Miller, O., *Bull. Soc. Chim. Belges* **1932**, *41*, 217.

5145. Miller, O., *Bull. Soc. Chim. Belges* **1933**, *42*, 238; *C.A.* **1932**, *26*, 5080.

5146. Miller, O., and L. Piaux, *Compt. Rend.* **1933**, *197*, 412; *C.A.* **1933**, *27*, 5248.

5147. Miller, R. C., and C. P. Smyth, *J. Amer. Chem. Soc.* **1957**, *79*, 20.

5148. Miller, R. C., and C. P. Smyth, *J. Chem. Phys.* **1956**, *24*, 814.

5149. Miller, R. E., and G. E. Bennett, *Ind. Eng. Chem.* **1961**, *53*, 33.

5150. Miller, W. T., *Nat. Nuclear Energy Ser., Div. 7* **1951**, *1*, 567; *C.A.* **1952**, *46*, 7988.

5151. Millero, F. J., *J. Phys. Chem.* **1968**, *72*, 3209.

5152. Millikan, A. F., U.S. Patent 3,031,383, Apr. 24, 1962; *C.A.* **1962**, *57*, 3722.

5153. Mills, J. E., *J. Amer. Chem. Soc.* **1909**, *31*, 1099.

5154. Milne, J. B., and T. J. Parker, *J. Solution Chem.* **1981**, *10*, 479.

5155. Milone, M., *IX Congr. Intern. Quim. Pura Aplicada* **1934**, *2*, 191; *C.A.* **1935**, *29*, 7139.

5156. Milone, M., *Gazz. Chim. Ital.* **1934**, *64*, 876; *C.A.* **1935**, *29*, 3310.

5157. Milyukova, M. A., N. J. Gusev, et al., *Analytical Chemistry of Plutonium*, Ann Arbor-Humphrey Science Publishers, Ann Arbor, Michigan, 1969.

5158. Minami, T., and T. Ando, *Osaka Kogyo Gijyutsu Shikenjyo Kiho* **1957**, *8*, 51; *C.A.* **1959**, *53*, 13649.

5159. Miner, C. S., and N. N. Dalton, *Glycerol*, Reinhold, New York, 1953.

5160. Miner, C. S., J. P. Trickey, et al., *Chem. Met. Eng.* **1922**, *299*, 362.

5161. Mingaleva, K. S., K. G. Golodova, et al., *Zh. Org. Khim.* **1965**, *1*, 2078; *C.A.* **1966**, *64*, 9750.

5162. Minkin, D. M., and A. A. Vvedenskii, *Zh. Fiz. Khim.* **1967**, *41*, 1571; *Russ. J. Phys. Chem.* **1967**, *41*, 740.

5163. Minkov, P., *Mashinostroene* **1962**, *11*, 19; *C.A.* **1962**, *57*, 9295.

5164. Mirone, P., *Atti Accad. Naz. Lincei, Rend., Cl. Sci. Fis., Mat. Nat.* **1954**, *16*, 483; *C.A.* **1955**, *49*, 5967.

5165. Mirone, P., and G. F. Fabbri, *Gazz. Chim. Ital.* **1954**, *84*, 187; *C.A.* **1955**, *49*, 10741.

5166. Mirone, P., B. Fortunato, et al., *J. Mol. Struct.* **1970**, *5*, 283.

5167. Mirov, N. T., *J. Forestry* **1947**, *45*, 659; *C.A.* **1948**, *42*, 386.

5168. Miskidzh'yan, S. P., and N. A. Trifonov, *J. Gen. Chem.* (USSR) **1947**, *17*, 1231; *C.A.* **1949**, *43*, 922.

5169. Mita, I., *J. Chem. Soc. Jpn.* **1942**, *63*, 760; *C.A.* **1947**, *41*, 3038.

5170. Mitchell, A. D., and C. Smith, *J. Chem. Soc.* **1913**, 489.

5171. Mitchell, J., and D. M. Smith, *Aquametry*, Interscience Publishers, New York, 1948.

5172. Mitchell, J. H., and H. R. Kraybill, *J. Amer. Chem. Soc.* **1942**, *64*, 988.

5173. Mitchell, J. J., and F. F. Coleman, *J. Chem. Phys.* **1949**, *17*, 44.

5174. Mitchell, R. W., J. C. Burr, et al., *Spectrochim. Acta* **1967**, *23A*, 195.

5175. Mitchell, S., *J. Chem. Soc.* **1926**, 1333.

5176. Mitera, J., and V. Kubelka, *Org. Mass. Spectrom.* **1971**, *5*, 651.

5177. Mitra, B. C., P. Ghosh, et al., *Anal. Chem.* **1964**, *36*, 673.

5178. Mitsukuri, S., and A. Nakatsuchi, *Sci. Rep. Tohoku Imp. Univ.* **1926**, *15*, 45; *Chem. Zentr.* **1926**, *II*, 545; *C.A.* **1926**, *20*, 1020.

5179. Mitsutani, A., and T. Kominami, *Nippon Kagaku Zasshi* **1959**, *80*, 895; *C.A.* **1961**, *55*, 4349.

5180. Mitzner, B. M., and S. Lemberg, *Am. Perfumer Cosmet.* **1966**, *81*, 25; *C.A.* **1966**, *64*, 18701.

5181. Mitzner, B. M., E. T. Theimer, et al., *Appl. Spectrosc.* **1965**, *19*, 169; *C.A.* **1966**, *64*, 4443.

5182. Miyagawa, I., T. Chiba, et al., *Bull. Chem. Soc. Jpn.* **1957**, *30*, 218.

5183. Miyake, A., *J. Amer. Chem. Soc.* **1960**, *82*, 3040.

5184. Miyazawa, T., and K. S. Pitzer, *J. Amer. Chem. Soc.* **1958**, *80*, 60.

5185. Miyazawa, T., T. Shimanouchi, et al., *J. Chem. Phys.* **1955**, *24*, 408.

5186. Mizuhara, S., and W. F. Seyer, *J. Amer. Chem. Soc.* **1953**, *75*, 3274.

5187. Mizushima, S., and Y. Morino, *Proc. Indian Acad. Sci.* **1938**, *8A*, 315; *C.A.* **1939**, *33*, 6158.

5188. Mizushima, S., Y. Morino, et al., *Phys. Z.* **1937**, *38*, 459; *C.A.* **1937**, *31*, 6111.

5189. Mizushima, S., Y. Morino, et al., *Sci. Paper Inst. Phys. Chem. Res.* (Tokyo), **June 1936**, *29*, 111; *Chem. Zentr.* **1937**, *I*, 4489; *Beil.* EIII1, 161.

5190. Mizushima, S., Y. Morino, et al., *Sci. Papers Inst. Phys. Chem. Res.* (Tokyo) **1939**, *36*, 281; *C.A.* **1940**, *34*, 2707.

5191. Mizushima, S., T. Shimanouchi, et al., *Can. J. Phys.* **1975**, *53*, 2085.

5192. Mizushima, S., Y. Uehara, et al., *Bull. Chem. Soc. Jpn.* **1937**, *12*, 132.

5193. Mizushima, S.-I., and T. Simanouti, *J. Amer. Chem. Soc.* **1949**, *71*, 1320.
5194. Mizushima, S.-I., T. Simanouti, et al., *J. Amer. Chem. Soc.* **1950, 72**, 3490.
5195. Mizutani, M., *Z. Phys. Chem.* (Leipzig) **1925**, *116*, 350.
5196. Mizutani, M., *Z. Phys. Chem.* (Leipzig) **1925**, *118*, 327.
5197. Mlavsky, A. I., U.S. Patent 3,414,603, Dec. 3, 1968.
5198. Moe, N. S., *Acta Chem. Scand.* **1967**, *21*, 1389.
5199. Moede, J. A., and C. Curran, *J. Amer. Chem. Soc.* **1949**, *71*, 852.
5200. Moffat, J. B., *J. Chem. Eng. Data* **1968**, *13*, 36.
5201. Moffett, R. B., and B. D. Aspergen, *Chem. Eng. News* **1954**, *32*, 4328.
5202. Mohanty, S., and P. Venkateswarlu, *Mol. Phys.* **1966**, *11*, 329.
5203. Mohler, F. L., V. H. Dibeler, et al., *J. Res. Nat. Bur. Stand.* **1952**, *49*, 343.
5204. Mohler, F. L., L. Williamson, et al., *J. Res. Nat. Bur. Stand.* **1950**, *44*, 291.
5205. Mohler, H., *Helv. Chim. Acta* **1937**, *20*, 1188.
5206. Mohler, H., and H. Lohr, *Helv. Chim. Acta* **1938**, *21*, 485.
5207. Mohler, H., and J. Polya, *Helv. Chim. Acta* **1937**, *20*, 96.
5208. Mohler, H., and J. Sorge, *Helv. Chim. Acta* **1940**, *23*, 119.
5209. Mohler, H., and J. Sorge, *Helv. Chim. Acta* **1940**, *23*, 1200.
5210. Mohr, O., *Mikrochemie* **1930**, *2*, 154; *C.A.* **1930**, *24*, 4242.
5211. Moiseev, V. D., and N. D. Antonova, *Zh. Fiz. Khim.* **1970**, *44*, 2912; *Russ. J. Phys. Chem.* **1970**, *44*, 1659.
5212. Moiseev, V. D., A. N. Girshov, et al., *Zh. Fiz. Khim.* **1978**, *52*, 477.
5213. Moiseeva, L. M., and N. M. Kuznetsova, *Zh. Anal. Khim.* **1971**, *26*, 2094; *J. Anal. Chem. USSR* **1971**, *26*, 1872.
5214. Moldavskiĭ, B. L., and L. E. Turetskaya, *J. Gen. Chem.* (USSR) **1946**, *16*, 445; *C.A.* **1947**, *41*, 951.
5215. Mole, J. F., W. S. Holmes, et al., *J. Chem. Soc.* **1964**, 5144.
5216. Moles, E., *Z. Phys. Chem.* (Leipzig) **1912**, *80*, 531.
5217. Molinari, J. G. D., *Ind. Chemist* **1961**, *37*, 323; Ref. [8290].
5218. Molinski, S., F. Nowotny, et al., *Przem. Chem.* **1939**, *23*, 30; *C.A.* **1939**, *33*, 4593.
5219. Moll, W. L. H., *Kolloid-Beihefte* **1939**, *49*, 1; *C.A.* **1939**, *33*, 2018; *Beil.* **EIII3**, 474; *Beil.* **EIII2**, 305.
5220. Möller, R., *Phys. Z.* **1931**, *32*, 697.
5221. Momigny, J., *Bull. Soc. Roy. Sci. Liege* **1953**, *22*, 541; *C.A.* **1954**, *48*, 9806.
5222. Momigny, J., *Bull. Soc. Roy. Sci. Liege* **1955**, *24*, 111; *C.A.* **1956**, *50*, 1455.
5223. Monahan, J. E., and H. E. Stanton, *J. Chem. Phys.* **1962**, *37*, 2654.
5224. Monica, M. D., L. Jannelli, et al., *J. Phys. Chem.* **1968**, *72*, 1068.
5225. Monica, M. D., U. Lamanna, et al., *J. Phys. Chem.* **1968**, *72*, 2124.
5226. Moniz, W. B., and J. A. Dixon, *J. Amer. Chem. Soc.* **1961**, *83*, 1671.
5227. Monroe, K. P., *Ind. Eng. Chem.* **1921**, *13*, 133.
5228. Monsanto Chemical Co., "Dimethylacetamide (DMAC)," Technical Data Sheet, St. Louis, Missouri, 1960.
5229. Monsanto Chemical Co., U.S. Patent 2,426,691, 1944; *Beil.* **EIII1**, 1205.
5230. "Montecatini" Soc. Gen. per l'Industria Mineraria et Agricola, French Patent 808,057, Jan. 28, 1937; *C.A.* **1937**, *31*, 6262.
5231. "Montecatini" Soc. Gen. per l'Industria Mineraria e Chimica, Italian Patent 605,541, Mar. 3, 1959; *C.A.* **1962**, *56*, 1410.
5232. Montgomery, R. L., F. D. Rossini, et al., *J. Chem. Eng. Data* **1978**, *23*, 125.
5233. Moor, V. G., E. K. Kanep, et al., *Trans. Exptl. Research Lab. Khemgas, Materials on Cracking and Chemical Treatment of Cracking Products* USSR **1937**, *3*, 320; *C.A.* **1937**, *31*, 6072.
5234. Moore, D. W., L. A. Burkardt, et al., *J. Chem. Phys.* **1956**, *25*, 1235.
5235. Moore, E. M., and M. E. Hobbs, *J. Amer. Chem. Soc.* **1949**, *71*, 411.
5236. Moore, F. J., and I. B. Johns, *J. Amer. Chem. Soc.* **1941**, *63*, 3336.
5237. Moore, G. E., and G. S. Parks, *J. Amer. Chem. Soc.* **1939**, *61*, 2561.
5238. Moore, G. E., M. L. Renquist, et al., *J. Amer. Chem. Soc.* **1940**, *62*, 1505.
5239. Mopsik, F. I., *J. Res. Nat. Bur. Stand.* **1967**, *71A*, 287.
5240. Mopsik, F. I., *J. Chem. Phys.* **1969**, *50*, 2559.

5241. Moravek, J., *Chem. Prum.* **1957,** *7,* 49; *C.A.* **1957,** *51,* 14351.

5242. Morawetz, E., *J. Chem. Thermodyn.* **1972,** *4,* 139.

5243. Morawetz, E., *J. Chem. Thermodyn.* **1972,** *4,* 455.

5244. Morcillo, J., and J. M. Orza, *Anal. Real. Soc. Espan. Fis. Quim.* (Madrid) **1960,** *56B,* 231, 253; *C.A.* **1960,** *54,* 19126.

5245. Morcom, K. W., and D. N. Travers, *Trans. Faraday Soc.* **1965,** *61,* 230.

5246. Morell, S. A., and A. H. Auernheimer, *J. Amer. Chem. Soc.* **1944,** *66,* 792.

5247. Morenkova, N. V., *Gig. Sanit.* **1981,** 7; *C.A.* **1982,** *96,* 15715.

5248. Moreno, J. M. M., *Anal. Fis. Quim.* (Madrid) **1947,** *43,* 261; *C.A.* **1947,** *41,* 6420.

5249. Morgan, J. L. R., and A. McD. McAfee, *J. Amer. Chem. Soc.* **1911,** *33,* 1275.

5250. Morgan, S. O., and H. H. Lowry, *J. Phys. Chem.* **1930,** *34,* 2385.

5251. Morgan, S. O., and W. A. Yager, *Ind. Eng. Chem.* **1940,** *32,* 1519.

5252. Mori, N., S. Omura, et al., *Bull. Chem. Soc. Jpn.* **1965,** *38,* 2149.

5253. Mori, N., S. Omura, et al., *Bull. Chem. Soc. Jpn.* **1965,** *38,* 2199.

5254. Moriarty, R. M., and J. M. Kliegman, *J. Org. Chem.* **1966,** *31,* 3007.

5255. Morikawa, T., and K. Yoshida, *Kagaku to Kogyo* (Osaka) **1963,** *37,* 107; *C.A.* **1964,** *60,* 3161.

5256. Morin, M. G., C. C. de Perez, et al., *Rev. Col. Quim. Puerto Rico* **1959,** *16,* 19; *C.A.* **1960,** *54,* 1996.

5257. Morin, M. G., C. C. de Perez, et al., *Rev. Col. Quim. Puerto Rico* **1959,** *16,* 21; *C.A.* **1962,** *57,* 1612.

5258. Morino, Y., *Inst. Phys. Chem. Res.* (Tokyo) *Sci. Pap.* **1933,** *23,* 49.

5259. Morino, Y., and I. Watanabe, *J. Chem. Soc. Jpn.* **1942,** *63,* 377; *C.A.* **1947,** *41,* 2995.

5260. Morino, Y., I. Watanabe, et al., *Sci. Papers Inst. Phys. Chem. Res.* (Tokyo) **1942,** *39,* 396; *C.A.* **1947,** *41,* 6154.

5261. Morino, Y., S. Yamaguchi, et al., *Sci. Papers Inst. Phys. Chem. Res.* (Tokyo) **1944,** *42,* Chemistry, 1; *C.A.* **1947,** *41,* 6089.

5262. Moriyoshi, T., S. Kaneshina, et al., *J. Chem. Thermodyn.* **1975,** *7,* 537.

5263. Morlet, J., *Rev. Inst. Franc. Petrole Ann. Combust. Liquides* **1963,** *18,* 127; *C.A.* **1963,** *59,* 2181.

5264. Morman, D. H., and G. A. Harlow, *Anal. Chem.* **1967,** *39,* 1869.

5265. Morris, D. F. C., and M. Ali Khan, *Talanta* **1968,** *15,* 1301.

5266. Morris, H. J., A. A. Nelson, et al., *J. Pharmacol.* **1942,** *74,* 266.

5267. Morris, J. C., *J. Chem. Phys.* **1943,** *11,* 230.

5268. Morris, J. C., W. J. Lanum, et al., *J. Chem. Eng. Data* **1960,** *5,* 112.

5269. Morris, W. W., *J. Ass. Off. Anal. Chem.* **1973,** *56,* 1037.

5270. Morrison, G. O., and T. P. G. Shaw, *Trans. Electrochem. Soc.* **1933,** *63,* 443; *Chem. Zentr.* **1933,** *II,* 131; *Beil.* **EIII2,** 269.

5271. Mortimer, F. S., *Spectrochim. Acta* **1957,** *9,* 270.

5272. Mortimer, F. S., R. B. Blodgett, et al., *J. Amer. Chem. Soc.* **1947,** *69,* 822.

5273. Morton, A. A., and J. G. Mark, *Ind. Eng. Chem., Anal. Ed.* **1934,** *6,* 151.

5274. Morton, R. A., *J. Chem. Soc.* **1926,** 719.

5275. Morton, R. A., and A. J. A. de Gouveia, *J. Chem. Soc.* **1934,** 916.

5276. Morton, R. A., and A. L. Stubbs, *J. Chem. Soc.* **1940,** 1347.

5277. Moseeva, E. M., I. B. Rabinovich, et al., *Thermodin. Org. Soedin.* **1978,** *7,* 8; *C.A.* **1980,** *92,* 83508.

5278. Mosher, W. A., *J. Amer. Chem. Soc.* **1940,** *62,* 552.

5279. Mosin, A. M., *Russ. J. Phys. Chem.* (Eng. trans.) **1972,** *46,* 318.

5280. Moskalev, V. V., and F. I. Skripov, *Fiz. Probl. Spektroakopii, Akad. Nauk SSSR Materialy 13-go* [*Trinadtsatogo*] *Soveshoh* Leningrad, **1960,** *2,* 148; *C.A.* **1963,** *59,* 13504.

5281. Moskalyk, R. E., L. G. Chatten, et al., *J. Pharm. Sci.* **1961,** *50,* 179.

5282. Mosselman, C., and H. Dekker, *Rec. Trav. Chim. Pays-Bas* **1969,** *88,* 257.

5283. Mosselman, C., and H. Dekker, *J. Chem. Soc., Faraday Trans. 1* **1975,** *71,* 417.

5284. Motodo, T., and Y. Yoshie, *Kogyo Kagaku Zasshi* **1965,** *68,* 1669; *C.A.* **1966,** *64,* 195.

5285. Motoyama, I., and C. H. Jarboe, *J. Phys. Chem.* **1966,** *70,* 3226.

5286. Mouneyrat, A., *Ann. Chim. Phys.* **1900,** [7], *20,* 538.

5287. Mouneyrat, A., *Bull. Soc. Chim. Fr.* **1899,** [3], *21,* 618.

5288. Mouneyrat, A., *Compt. Rend.* **1898,** *127,* 274.

5289. Mountcastle, W. R., D. F. Smith, et al., *J. Phys. Chem.* **1960,** *64,* 1342.

5290. Mouradoff, L., and E. Darmois, *Bull. Soc. Chim. Fr.* **1944,** 446D; *C.A.* **1950,** *44,* 2301.

5291. Moureau, C., and G. Mignonac, *Compt. Rend.* **1920,** *171,* 652; *C.A.* **1921,** *15,* 235.

5292. Moureu, C., *Bull. Soc. Chim. Fr.* *9,* 386; *J. Chem. Soc.* **1893Ai,** 548.

5293. Moureu, C., *Bull. Soc. Chim. Fr.* *9,* 424; *J. Chem. Soc.* **1893Ai,** 682.

5294. Moureu, A., A. Boutaric, et al., *J. Chim. Phys.* **1921,** *18,* 333; *C.A.* **1921,** *15,* 2417; *Beil.* **EII1,** 783.

5295. Moureu, C., and A. Boutaric, *J. Chim. Phys.* **1920,** *18,* 348; *C.A.* **1921,** *15,* 2220.

5296. Moureu, C., M. Murat, et al., *Compt. Rend.* **1921,** *172,* 1267; *C.A.* **1922,** *16,* 55; *Beil.* **EII2,** 386.

5297. Moureu, H., and M. Dode, *Bull. Soc. Chim. Fr.* **1937,** [5], *4,* 637.

5298. Mousa, A. El H. N., *J. Chem. Thermodyn.* **1977,** *9,* 1063.

5299. Mousa, A. El H. N., *J. Chem. Thermodyn.* **1981,** *13,* 201.

5300. Mousa, A. El H. N., W. B. Kay, et al., *J. Chem. Thermodyn.* **1972,** *4,* 301.

5301. Mousseron, M., R. Granger, et al., *Bull. Soc. Chim. Fr.* **1947,** 459; *C.A.* **1948,** *42,* 1895.

5302. Mueler, R., and C. Dathe, Belgian Patent 631,145, Aug. 16, 1963; *C.A.* **1964,** *61,* 5691.

5303. Mukerji, S. K., *Phil. Mag.* **1935,** *19,* 1079; *C.A.* **1935,** *29,* 6505.

5304. Mukherjee, D. K., *Indian J. Phys.* **1960,** *34,* 402; *C.A.* **1961,** *55,* 7041.

5305. Mukherjee, D. K., P. K. Bishui, et al., *Indian J. Phys.* **1965,** *39,* 537; *C.A.* **1966,** *64,* 16838.

5306. Mukherjee, L. M., and D. P. Boden, *J. Phys. Chem.* **1969,** *73,* 3964.

5307. Mukherjee, L. M., and S. Bruckenstein, *Pure Appl. Chem.* **1966,** *13,* 421.

5308. Mukhitov, B. M., *Gig. Sanit.* **1962,** *27,* 16; *C.A.* **1962,** *57,* 12830.

5309. Mullayanov, F. I., I. S. Salikov, et al., *Tr., Nauch.-Issled. Inst. Neftekhim. Proizvod.* **1970,** No. 2, 97; *C.A.* **1971,** *74,* 128300.

5310. Müller, A., *Fette Seifen* **1942,** *49,* 572; *C.A.* **1943,** *37,* 6510.

5311. Müller, A., *Fette Seifen* **1951,** *53,* 462; *C.A.* **1952,** *46,* 2757.

5312. Müller, A., *Monatsh. Chem.* **1928,** *49,* 27; *C.A.* **1928,** *22,* 3391.

5313. Müller, A. H., *Z. Phys. Chem.* (Leipzig) **1914,** *86,* 177.

5314. Müller, E., and K. Ehrmann, *Ber. Deut. Chem. Gesell.* **1936,** *69,* 2207.

5315. Müller, F., and L. Bornstein, *Physikalisch Chemische Tabellen,* 5 ed., Suppl. 2, pt. 2, Julius Springer, Berlin, 1931.

5316. Müller, F. H., *Phys. Z.* **1937,** *38,* 283; *C.A.* **1937,** *31,* 4551.

5317. Müller, H., *Phys. Z.* **1934,** *35,* 346.

5318. Müller, H., and H. Sack, *Phys. Z.* **1930,** *31,* 815; *Chem. Zentr.* **1930,** *II,* 3374.

5319. Müller, N., *J. Chem. Phys.* **1965,** *42,* 4309.

5320. Müller, N., P. C. Lauterbur, et al., *J. Amer. Chem. Soc.* **1956,** *78,* 3557.

5321. Müller, N., P. C. Lauterbur, et al., *J. Amer. Chem. Soc.* **1957,** *79,* 1807.

5322. Müller, N., and P. I. Rose, *J. Amer. Chem. Soc.* **1962,** *84,* 3973.

5323. Müller, R., and H. Brenneis, *Z. Elektrochem.* **1932,** *38,* 450; *Chem. Zentr.* **1932,** *II,* 1420.

5324. Müller, R., F. Griengl, et al., *Monatsh. Chem.* **1926,** *47,* 88.

5325. Müller, R., V. Raschka, et al., *Monatsh. Chem.* **1927,** *48,* 659.

5326. Mulliken, R. S., *J. Chem. Phys.* **1939,** *7,* 339.

5327. Mulliken, R. S., and E. Teller, *Phys. Rev.* **1942,** *61,* 283.

5328. Mumford, S. A., and J. W. C. Phillips, *J. Chem. Soc.* **1928,** 155.

5329. Mumford, S. A., and J. W. C. Phillips, *J. Chem. Soc.* **1950,** 75.

5330. Munch, J. C., *J. Amer. Chem. Soc.* **1926,** *48,* 994.

5331. Muncke, *Berz. Jb.* 27, 453.

5332. Munday, E. B., J. C. Mullins, et al., *J. Chem. Eng. Data* **1980,** *25,* 191.

5333. Muney, W. S., and J. F. Coetzee, *J. Phys. Chem.* **1962,** *66,* 89.

5334. Munson, M. S. B., *J. Phys. Chem.* **1964,** *68,* 796.

5335. Munson, M. S. B., and F. H. Field, *J. Amer. Chem. Soc.* **1966,** *88,* 4337.

5336. Murata, H., *J. Chem. Phys.* **1950,** *18,* 1308.

5337. Murata, S., *Igaku Kenkyu* **1960,** *30,* 3388; *C.A.* **1961,** *55,* 27644.

5338. Murata, S., M. Sakiyama, et al., *J. Chem. Thermodyn.* **1982,** *14,* 707.

5339. Muratov, G. N., *Russ. J. Phys. Chem.* **1980,** *54,* 1185; *Zh. Fiz. Khim.* **1980,** *54,* 2088.

5340. Muratov, G. N., and V. P. Skripov, *Zh. Fiz. Khim.* **1975,** *49,* 2148; *Russ. J. Phys. Chem.* **1975,** *49,* 1262.

5341. Murayama, K., and K. Nukada, *Bull. Chem. Soc. Jpn.* **1963,** *36,* 1223; *C.A.* **1964,** *60,* 1249.

5342. Murkerjee, D. K., P. K. Bishui, et al., *Indian J. Phys.* **1965,** *39,* 537; *C.A.* **1966,** *64,* 16838.

5343. Murmann, R. K., and F. Basolo, *J. Amer. Chem. Soc.* **1955,** *77,* 3484.

5344. Muroi, K., K. Ogawa, et al., *Bull. Soc. Chem. Jpn.* **1965,** *38,* 1176.

5345. Murphy, R. A., and J. C. Davis, *J. Phys. Chem.* **1968,** *72,* 3111.

5346. Murray, E. C., and R. N. Keller, *J. Org. Chem.* **1969,** *34,* 2234.

5347. Murray, J. W., and D. H. Andrews, *J. Chem. Phys.* **1933,** *1,* 406.

5348. Murray, M. J., and F. F. Cleveland, *J. Chem. Phys.* **1941,** *9,* 129.

5349. Murray, W. S., U.S. Patent 2,426,637, Sept. 2, 1947; *C.A.* **1948,** *42,* 199.

5350. Murray-Rust, D. M., H. J. Hadow, et al., *J. Chem. Soc.* **1931,** 215.

5351. Murrell, J. N., and V. M. S. Gil, *Trans. Faraday Soc.* **1965,** *6,* 402.

5352. Murrieta-Guevara, F., and A. T. Rodriguez, *J. Chem. Eng. Data* **1984,** *29,* 204.

5353. Murty, C. R. K., *J. Sci. Ind. Res* (India) **1959,** *18B,* 268.

5354. Murty, C. R., and D. V. G. L. N. Rao, *J. Sci. Ind. Res.* (India) **1956,** *15B,* 350; *C.A.* **1957,** *51,* 5484.

5355. Murty, G. V. L. N., and T. R. Seshadri, *Proc. Indian Acad. Soc.* **1938,** *8A,* 519; *C.A.* **1939,** *33,* 4874.

5356. Murty, G. V. L. N., and T. R. Seshadri, *Proc. Indian Acad. Sci.* **1939,** *10A,* 307; *C.A.* **1940,** *34,* 2707.

5357. Murty, G. V. L. N., and T. R. Seshadri, *Proc. Indian Acad. Sci.* **1940,** *11A,* 32; *C.A.* **1940,** *34,* 6881.

5358. Murty, G. V. L. N., and T. R. Seshadri, *Proc. Indian Acad. Sci.* **1940,** *11A,* 424; *C.A.* **1940,** *34,* 7744.

5359. Murty, G. V. L. N., and T. R. Seshadri, *Proc. Indian Acad. Sci.* **1941,** *14A,* 593; *C.A.* **1942,** *36,* 4026.

5360. Murty, G. V. L. N., and T. R. Seshadri, *Proc. Indian Acad. Sci.* **1942,** *16A,* 264; *C.A.* **1943,** *37,* 3669.

5361. Murzin, V. I., A. A. Anisonyan, et al., *Gazov. Prom-st* **1975,** (1), 50; *C.A.* **1975,** *82,* 142430.

5362. Musher, J. I., *Spectrochim. Acta* **1960,** *16,* 835.

5363. Musher, J. I., and R. G. Gordon, *J. Chem. Phys.* **1962,** *36,* 3097.

5364. Musher, J. I., and R. E. Richards, *Proc. Chem. Soc.* **1958,** 230.

5365. Mussell, A. G., F. B. Thole, et al., *J. Chem. Soc.* **1912,** *101,* 1008.

5366. Musser, D. M., and H. Adkins, *J. Amer. Chem. Soc.* **1938,** *60,* 664.

5367. Muzart, J., *Synthesis* **1982,** 60.

5368. Myasnikova, L. F., V. I. Bushinskii, et al., *Zh. Prikl. Khim.* (Leningrad) **1971,** *44,* 1849; *J. Appl. Chem.* (USSR) **1971,** *44,* 1867.

5369. Myers, R. S., B. A. Berenbach, et al., *J. Chem. Eng. Data* **1969,** *13,* 91.

5370. Myers, R. S., and H. L. Clever, *J. Chem. Thermodyn.* **1974,** *6,* 949.

5371. Myers, R. T., *J. Phys. Chem.* **1965,** *69,* 700.

N

5372. Nadeschdin, *J. Russ. Phys. Chem. Soc.,* *14,* No. 2, 538; *Beil.* **EIII1,** 1873.

5373. Nadi, M. El., and E. Salam, *Z. Phys. Chem.* (Leipzig) **1960,** *215,* 121.

5374. Naegeli, C., L. Gruntuch, et al., *Helv. Chim. Acta* **1929,** *12,* 22.

5375. Nagai, H., *Kumamoto J. Sci.* **1954,** *Ser. A, 2,* 100; *C.A.* **1956,** *50,* 13649.

5376. Nagakura, S., and A. Kuboyama, *Rept. Inst. Sci. Technol., Univ. Tokyo* **1951,** *5,* 27; *C.A.* **1952,** *46,* 1315.

5377. Nagarajan, N., W. W. Webb, et al., *J. Chem. Phys.* **1982,** *77,* 5771.

5378. Nagier, M. F., O. I. Dzhafarov, et al., *Dokl. Akad. Nauk SSSR* **1969,** *184,* No. 3, 648; *C.A.* **1969,** *70,* 91354.

5379. Nagornow, N., and L. Rotinjanz, *Ann. Inst. Anal. Physio-Chim.*, Leningrad **1926,** *3,* 162; *Chem. Zentr.* **1927,** *I,* 2648.

5380. Nagy, J., and J. Reffy, *J. Organometal. Chem.* **1970,** *22,* 565.

5381. Nair, C. S. B., K. K. Tiwari, et al., Indian Patent 143,793, Feb. 4, 1978; *C.A.* **1980,** *92,* 110657.

5382. Nakagawa, I., *Nippon Kagaku Zasshi* **1958,** *79,* 1353; *C.A.* **1959,** *53,* 7699.

5383. Nakagawa, N., and S. Fujiwara, *Bull. Chem. Soc. Jpn.* **1961,** *34,* 143.
5384. Nakamura, K., *Nippon Kagaku Zasshi* **1957,** *78,* 1164; *C.A.* **1958,** *52,* 1048.
5385. Nakamura, S., *J. Chem. Soc. Jpn.* **1939,** *60,* 1010; *C.A.* **1940,** *34,* 1563.
5386. Nakamura, S., and E. Kanda, *J. Chem. Soc. Jpn.* **1939,** *60,* 1275.
5387. Nakanishi, K., N. Kato, et al., *J. Chem. Phys.* **1967,** *71,* 814.
5388. Nakanishi, K., T. Matsumoto, et al., *J. Chem. Eng. Data* **1971,** *16,* 44.
5389. Nakanishi, K., and O. Toyama, *Bull. Chem. Soc. Jpn.* **1972,** *45,* 3210.
5390. Nakata, N., *Ber. Deut. Chem. Gesell.* **1931,** *64,* 2059.
5391. Nalesnik, T. E., and N. L. Holy, *J. Org. Chem.* **1977,** *42,* 372.
5392. Nametkin, S. S., *J. Russ. Phys. Chem. Soc.* **1923–4,** *55,* 75.
5393. Nametkin, S. S., and E. S. Pokrovskaya, *Zh. Obshch. Chim.* **1937,** *7,* 962; *Beil.* **EIII5,** 1256.
5394. Nandy, S. K., *Indian J. Phys.* **1966,** *40,* 415.
5395. Naphthachimie, *Naphthachimie*, Foubourg Saint-Honore, Paris.
5396. Naqvi, N., and Q. Fernando, *J. Chem. Phys.* **1960,** *25,* 551.
5397. Narasimham, N. A., J. R. Nielsen, et al., *J. Chem. Phys.* **1957,** *27,* 740.
5398. Narasimhan, P. T., *J. Indian Inst. Sci.* **1951,** *37A,* 30; *C.A.* **1955,** *49,* 7906.
5399. Narasimhan, P. T., L. Laine, et al., *J. Chem. Phys.* **1958,** *28,* 1257.
5400. Narasimhan, P. T., N. Laine, et al. *J. Chem. Phys.* **1958,** *29,* 1184.
5401. Narasimhan, Rao, D. V. G. L., *Indian J. Phys.* **1956,** *30,* 582; *C.A.* **1957,** *51,* 9241.
5402. Narayanaswamy, P. K., *Proc. Indian Acad. Sci.* **1947,** *26A,* 121; *C.A.* **1948,** *42,* 2519.
5403. Narayanaswamy, P. K., *Proc. Indian Acad. Sci.* **1948,** *27A,* 336; *C.A.* **1948,** *42,* 6665.
5404. Narsimham, G., *Chem. Process Eng.* **1965,** *46,* 498; *C.A.* **1965,** *63,* 16196.
5405. Nasir, P., S. C. Hwang, et al., *J. Chem. Eng. Data* **1980,** *25,* 298.
5406. Nasybullina, R. K., R. Kh. Mar'yakbin, et al., *Bull. Acad. Sci. USSR, Div. Chem. Sci.* **1973,** *22,* No. 4, 778 (Eng. trans.).
5407. Natalis, P., *Bull. Soc. Chim. Belges* **1960,** *69,* 519.
5408. Natarajan, G. S., and K. A. Venkatachlam, *J. Chem. Eng. Data* **1972,** *17,* 328.
5409. Natelson, S., and J. E. Bonas, *Microchem. J.* **1965,** *9,* 68; *C.A.* **1965,** *63,* 4920.
5410. Nath, J., and S. N. Dubey, *J. Phys. Chem.* **1981,** *85,* 886.
5411. Nath, J., and B. Narain, *J. Chem. Eng. Data* **1982,** *27,* 308.
5412. Nath, J., and B. Narain, *J. Chem. Eng. Data* **1983,** *28,* 296.
5413. National Aniline Division, "Aniline," Allied Chemical Corp., New York, 1964.
5414. National Bureau of Standards, *Standard Density and Volumetric Tables*, Circular No. 19, 6th ed., Government Printing Office, Washington, 1924.
5415. *National Fire Codes, Vol. 1*, National Fire Protection Assoc., Boston, 1964.
5416. *National Fire Codes, Vol. 2*, National Fire Protection Assoc., Boston, 1980.
5417. *National Formulary, The*, 12th ed., American Pharmaceutical Association, Washington, 1965.
5418. National Institute for Occupational Safety and Health, "Background Information in Ethylene Dibromide," Rockville, Maryland, 1975.
5419. National Institute for Occupational Safety and Health, "Background Information on Hexamethylenephosphoric Triamide," Rockville, Maryland, 1975.
5420. National Institute for Occupational Safety and Health, "Background Information on Trichloroethylene," Rockville, Maryland, 1975.
5421. National Institute for Occupational Safety and Health, "Current Intelligence Bulletin: Chloroform," Rockville, Maryland, 1976.
5422. National Institute for Occupational Safety and Health, "Criteria for a Recommended Standard . . . Occupational Exposure to Methylene Chloride," Rockville, Maryland, 1976.
5423. National Paint, Varnish and Lacquer Association, *Paint Industry Smog Chamber*, Final Tech. Report, Washington, 1970.
5424. National Toxicology Program, *Second Annual Report on Carcinogens*, U.S. Dept. of Health and Human Services, Dec. 1981.
5425. Natradze, A. G., and K. E. Novikova, *Med. Prom SSSR* **1958,** *12,* No. 4, 33; *C.A.* **1959,** *53,* 11714.
5426. Natural Gas Association of America, *Ind. Eng. Chem.* **1942,** *34,* 1240.
5427. Nauruzov, M. Kh., *Russ. J. Phys. Chem.* (Eng. trans.) **1973,** *47,* 1269.
5428. Naves, Y. R., *Mfg. Chemist* **1946,** *17,* 187; *C.A.* **1948,** *42,* 6056.

5429. Nayar, S., and A. P. Kudchadker, *J. Chem. Eng. Data* **1973**, *18*, 356.

5430. Nazarov, I. N., L. D. Bergel'son, et al., *Izvest. Akad. Nauk SSSR, Otdel. Khim. Nauk* **1953**, 889; *C.A.* **1955**, *49*, 1082.

5431. Nazarov, I. N., S. M. Makin, et al., *Zh. Obshch. Khim.* **1959**, *29*, 111; *C.A.* **1960**, *54*, 255.

5432. Nazarova, L. M., and Y. K. Syrkin, *Izvest. Akad. Nauk SSSR, Otdel. Khim. Nauk* **1949**, 35; *C.A.* **1949**, *43*, 4913.

5433. Nechai, F., *Zh. Tekhnol. Fiz.* **1956**, *26*, 436; *C.A.* **1957**, *51*, 29.

5434. Neelakantan, P., *Proc. Indian Acad. Sci., Sect. A* **1963**, *57*, 94; *C.A.* **1963**, *58*, 13317.

5435. Neelakantan, P., *Proc. Indian Acad. Sci., Sect. A* **1964**, *60*, 422; *C.A.* **1965**, *62*, 15608.

5436. Neeter, R., N. M. M. Nibberling, et al., *Org. Mass Spectrom.* **1970**, *3*, 597.

5437. Neikam, W. C., and B. P. Dailey, *J. Chem. Phys.* **1963**, *38*, 445.

5438. Neilson, E. F., and D. White, *J. Amer. Chem. Soc.* **1957**, *79*, 5618.

5439. Neilson, E. F., and D. White, *J. Phys. Chem.* **1959**, *63*, 1363.

5440. Neish, A. C., *Can. J. Res.* **1945**, *23B*, 10.

5441. Nekrasov, A. S., and B. A. Krentsel, *Zh. Obshch. Khim.* **1949**, *19*, 948; *C.A.* **1950**, *44*, 1006.

5442. Nelson, E. W., and R. F. Newton, *J. Amer. Chem. Soc.* **1941**, *63*, 2178.

5443. Noller, C. R., ed., *Organic Syntheses, Vol. 15*, Wiley, New York, 1935.

5444. Nelson, K. W., J. F. Ege, et al., *J. Ind. Hyg. Toxicol.* **1943**, *25*, 282; *C.A.* **1945**, *39*, 5001.

5445. Nel'son, K. V., and Z. D. Stepanova, *Kolebatel'nye Spektry i Molekul. Protsessy v Kauchukakh, Vses. Nauchn.-Issled. Inst. Sintekich. Kauchuka* **1965**, 127; *C.A.* **1966**, *64*, 9476.

5446. Nelson, O. A., *Ind. Eng. Chem.* **1928**, *20*, 1382.

5447. Nelson, O. A., *Ind. Eng. Chem.* **1930**, *22*, 971.

5448. Nelson, O. A., and M. D. Young, *J. Amer. Chem. Soc.* **1933**, *55*, 2429.

5449. Nelson, R. A., and R. S. Jessup, *J. Res. Nat. Bur. Stand.* **1952**, *48*, 206.

5450. Nelson, R. C., E. Q. Hemwall, et al., *J. Paint Technol.* **1970**, *42*, 636.

5451. Nelson, R. C., V. F. Rigurelli, et al., *J. Paint Technol.* **1970**, *42*, 644.

5452. Nelson, R. D., C. A. Billings, et al., *J. Phys. Chem.* **1967**, *71*, 2742.

5453. Nemes, L., and W. J. Orville-Thomas, *Trans. Faraday Soc.* **1965**, *61*, 1839.

5454. Nemirowsky, J., *J. Prakt. Chem.*, *28*, No. 2, 439; *J. Chem. Soc.* **1884A**, 419.

5455. Nernst, W., *Z. Phys. Chem.* (Leipzig) **1894**, *14*, 622.

5456. Nesmeyanov, A. N., R. Kh. Freidlina, et al., *Izvest. Akad. Nauk SSSR, Otdel. Khim. Nauk* **1949**, 623; *C.A.* **1950**, *44*, 3920.

5457. Neu, J. T., A. Ottenberg, et al., *J. Chem. Phys.* **1948**, *16*, 1004.

5458. Neuberg, C., and K. P. Jacobsohn, *Biochem. Z.* **1928**, *199*, 498.

5459. Neuberg, C., and E. Kerb, *Biochem. Z.* **1918**, *92*, 96.

5460. Neugebauer, C., and J. L. Margrave, *J. Phys. Chem.* **1956**, *60*, 1318.

5461. Neuman, R. C., and M. L. Rahn, *J. Org. Chem.* **1966**, *31*, 1857.

5462. Neuman, R. C., W. Snider, et al., *J. Phys. Chem.* **1968**, *72*, 2469.

5463. Neunhoeffer, O., *J. Prakt. Chem.* **1932**, *133*, 95; *C.A.* **1932**, *26*, 2435.

5464. Neuzil, R. W., and D. H. Rosback, U.S. Patent 3,998,901, Dec. 21, 1976; *C.A.* **1977**, *86*, 89359.

5465. Nevgi, G. V., and S. K. K. Jatkar, *Indian J. Phys.* **1934**, *8*, 397; *C.A.* **1934**, *28*, 6635.

5466. Nevgi, G. V., and S. K. K. Jatkar, *J. Indian Inst. Sci.* **1934**, *17A*, 175; *C.A.* **1935**, *29*, 6505.

5467. Nevgi, G. V., and S. K. K. Jatkar, *J. Indian Inst. Sci.* **1934**, *17A*, 189; *C.A.* **1935**, *29*, 6505.

5468. Newitt, D. M., R. P. Linstead, et al., *J. Chem. Phys.* **1937**, 876.

5469. Newitt, D. M., and P. S. Mene, *J. Chem. Phys.* **1946**, 97.

5470. Newkirk, A. E., *J. Amer. Chem. Soc.* **1946**, *68*, 2736.

5471. Newman, M., C. B. Hayworth, et al., *Ind. Eng. Chem.* **1949**, *41*, 2039.

5472. Newman, M. S., and A. S. Smith, *J. Org. Chem.* **1948**, *13*, 592.

5473. Newmark, R., AEC Accession No. *5934*, Rept. No. *UCRL-11649*, Avail. OTS, 1964, 194 pp.; *C.A.* **1965**, *63*, 163.

5474. Newmark, R. A., and C. H. Sederholm, *J. Chem. Phys.* **1965**, *43*, 602.

5475. Newsham, D. M. T., and E. J. Mendez-Leccanda, *J. Chem. Thermodyn.* **1982**, *14*, 291.

5476. Newth, G., *J. Chem. Soc.* **1901**, 917.

5477. Nichols, D., C. Sutphen, et al., *J. Phys. Chem.* **1968**, *72*, 1021.

5478. Nichols, N., and I. Wadso, *J. Chem. Thermodyn.* **1975**, *7*, 329.

5479. Nicholson, D. E., *Anal. Chem.* **1960,** *32,* 1372.
5480. Nicholson, D. E., *Anal. Chem.* **1960,** *32,* 1634.
5481. Nicholson, D. E., *Anal. Chem.* **1962,** *34,* 370.
5482. Nicholson, G. R., *J. Chem. Soc.* **1957,** 2431.
5483. Nickerson, J. D., and R. McIntosh, *Can. J. Chem.* **1957,** *35,* 1325.
5484. Nickerson, J. K., K. A. Kobe, et al., *J. Phys. Chem.* **1961,** *65,* 1037.
5485. Nicolescu, I. V., and O. Serban, *Analele Univ. Bucuresti, Ser. Stiint. Nat.* **1963,** *12,* 165; *C.A.* **1966,** *64,* 8012.
5486. Nicolini, E., *Ann. Chim.* **1951,** *6,* 582; *C.A.* **1952,** *46,* 796.
5487. Nicolini, E., and P. Laffitte, *Compt. Rend.* **1949,** *229,* 757; *C.A.* **1950,** *44,* 4409.
5488. Nielsen, A. H., and E. F. Barker, *Phys. Rev.* **1934,** *46,* 970.
5489. Nielsen, E. B., and J. A. Schellman, *J. Phys. Chem.* **1967,** *71,* 2297.
5490. Nielsen, J. R., and H. H. Claasen, *J. Chem. Phys.* **1950,** *18,* 1471.
5491. Nielsen, J. R., H. H. Claasen, et al., *J. Chem. Phys.* **1950,** *18,* 485.
5492. Nielsen, J. R., C. Y. Liang, et al., *J. Chem. Phys.* **1953,** *21,* 383.
5493. Nielsen, J. R., and D. C. Smith, *Ind. Eng. Chem., Anal. Ed.* **1943,** *15,* 609.
5494. Nielsen, J. R., and N. E. Ward, *J. Chem. Phys.* **1942,** *10,* 81.
5495. Nigham, R. K., and M. S. Dhillon, *J. Chem. Thermodyn.* **1971,** *3,* 819.
5496. Niini, A., *Suomen Kemistilehti* **1938,** *11A,* 19; *C.A.* **1938,** *32,* 4861.
5497. Nikitina, A. N., V. A. Petukhov, et al., *Optika i Spetrosk.* **1964,** *16,* 976; *C.A.* **1964,** *61,* 12803.
5498. Nikolaev, A. V., Yu. A. Afanas'ev, et al., *Dokl. Akad. Nauk SSSR* **1966,** *168,* 351; *C.A.* **1966,** *65,* 9826.
5499. Nikolaev, A. V., Yu. A. Afanas'ev, et al., *Izv. Sib. Atd. Akad. Nauk SSSR, Ser. Khim. Nauk* **1968,** *6,* 3.
5500. Nikolaev, N. I., *Zh. Fiz. Khim.* **1977,** *51,* 1235; *Russ. J. Phys. Chem.* **1977,** *51,* 1253.
5501. Nikolaev, P. N., and I. B. Rabinovich, *Zh. Fiz. Khim.* **1967,** *41,* 2191; *Russ. J. Phys. Chem.* **1967,** *41,* 1179.
5502. Nikuradse, A., *Z. Phys. Chem.* (Leipzig) **1931,** *155A,* 59.
5503. Nilsson, O., *Acta Chem. Scand.* **1967,** *21,* 1501.
5504. Nixon, J. R., Jr., German Patent 2,239,423, Feb. 15, 1973; *C.A.* **1973,** *78,* 124235.
5505. Nodiff, E. A., A. V. Grosse, et al., *J. Org. Chem.* **1953,** *18,* 235.
5506. Noller, C. R., and R. Adams, *J. Amer. Chem. Soc.* **1926,** *48,* 1080.
5507. Noller, C. R., and G. R. Dutton, *J. Amer. Chem. Soc.* **1933,** *55,* 424.
5508. Nomiyama, K., M. Minai, et al., *Ind. Health* (Kawasaki, Jpn.) **1967,** *5,* 143; *C.A.* **1968,** *69,* 9440.
5509. Nordwal, H. J. De, and L. A. K. Staveley, *Trans. Faraday Soc.* **1956,** *52,* 1207.
5510. Noren, I., and S. Sunner, *J. Chem. Thermodyn.* **1907,** *2,* 597.
5511. Normant, H., *Angew. Chem. Intern. Ed. Engl.* **1967,** *6,* 1046.
5512. Normant, H., *Bull. Soc. Chim. Fr.* **1968,** 791.
5513. Normant, H., *Compt. Rend.* **1954,** *239,* 1510; *C.A.* **1956,** *50,* 228.
5514. Norris, F. A., and D. E. Terry, *Oil and Soap* **1945,** *22,* 41.
5515. Norris, J. F., and F. Cartese, *J. Amer. Chem. Soc.* **1927,** *49,* 2640.
5516. Norris, J. F., and J. N. Ingraham, *J. Amer. Chem. Soc.* **1938,** *60,* 1421.
5517. Norris, J. F., and G. W. Rigby, *J. Amer. Chem. Soc.* **1932,** *54,* 2088.
5518. Norris, J. F., and B. M. Sturgis, *J. Amer. Chem. Soc.* **1939,** *61,* 1413.
5519. Norris, J. F., and H. B. Taylor, *J. Amer. Chem. Soc.* **1924,** *46,* 753.
5520. Norton, F. H., and H. B. Hass, *J. Amer. Chem. Soc.* **1936,** *58,* 2147.
5521. Noskov, A. M., and V. V. Moskovskikh, *Zh. Prikl. Spektrosk.* **1968,** *9,* No. 2, 235; *C.A.* **1969,** *70,* 15549.
5522. Notaro, V. A., C. M. Selwitz, et al., U.S. Patent 3,644,552, Feb. 22, 1972; *C.A.* **1972,** *76,* 99311.
5523. Notley, J. M., and M. Spiro, *J. Chem. Soc. B* **1966,** 362.
5524. Novak, J., J. Matous, et al., *Collect. Czech. Chem. Commun.* **1960,** *25,* 583; *C.A.* **1960,** *54,* 16067; and [7483].
5525. Novak, M., and G. M. Loudon, *J. Org. Chem.* **1977,** *42,* 2494.
5526. Novikov, E. G., *Zh. Fiz. Khim.* **1973,** *47,* 2257; *Russ. J. Phys. Chem.* **1973,** *47,* 1274.
5527. Noyce, D. S., and D. B. Denney, *J. Amer. Chem. Soc.* **1950,** *72,* 5743.

5528. Noyce, D. S., and D. B. Denney, *J. Amer. Chem. Soc.* **1952**, *74*, 5912.
5529. Noyes, W. A., *J. Amer. Chem. Soc.* **1908**, *30*, 142.
5530. Noyes, W. A., *J. Amer. Chem. Soc.* **1923**, *45*, 857.
5531. Nozaki, K., and P. D. Bartlett, *J. Amer. Chem. Soc.* **1946**, *68*, 2377.
5532. Nukada, K., *Bull. Chem. Soc. Jpn.* **1960**, *33*, 1606; *C.A.* **1961**, *55*, 13064.
5533. Nukada, K., *Nippon Kagaku Zasshi* **1959**, *80*, 976; *C.A.* **1959**, *53*, 21156.
5534. Nukada, K., *Nippon Kagaku Zasshi* **1959**, *80*, 1112; *C.A.* **1960**, *54*, 1071.
5535. Nukada, K., *Spectrochim. Acta* **1962**, *18*, 745; *C.A.* **1962**, *57*, 8085.
5536. Nukada, K., and U. Maeda, *Bull. Chem. Soc. Jpn.* **1959**, *32*, 655; *C.A.* **1960**, *54*, 6316.
5537. Numano, T., and T. Kitagawa, *Kogyo Kagaku Zasshi* **1962**, *65*, 182; *C.A.* **1962**, *57*, 5753.
5538. Nwaukwa, S. O., and P. M. Keehn, *Tetrahedron Lett.* **1982**, *23*, 35.
5539. Nyquist, R. A., *Spectrochim. Acta, Part A* **1971**, *27*, 2513.
5540. Nystrom, R. F., and W. G. Brown, *J. Amer. Chem. Soc.* **1947**, *69*, 1197.
5541. Nystrom, R. F., and W. G. Brown, *J. Amer. Chem. Soc.* **1947**, *69*, 2548.

O

5542. Obach, E., *J. Prakt. Chem.* **1882**, *26*, 299.
5543. Oblasova, L. Z., V. V. Moskaovskikh, et al., *Izv. Vyssh. Ucheb. Zaved., Khim. Khim. Tekhnol.* **1973**, *16*, 1047; *C.A.* **1973**, *79*, 104845.
5544. Oblentsev, R. D., N. G. Marina, et al., *Khim. Ser. Soedin., Soderzhashch. v Neft. i Nefteprod., Akad. Nauk SSSR, Bashkirsk. Filial* **1964**, *6*, 220; *C.A.* **1964**, *61*, 9369.
5545. Oblentsev, R. D., S. V. Netupskava, et al., *Izv. Vostoch. Filial. Akad. Nauk SSSR* **1957**, No. 10, 60.
5546. O'Brien, K. G., *Aust. J. Chem.* **1957**, *10*, 91; *C.A.* **1957**, *51*, 10367.
5547. O'Connor, J. G., and M. S. Norris, *Anal. Chem.* **1964**, *36*, 1391.
5548. O'Connor, R. T., *J. Amer. Oil Chemists' Soc.* **1955**, *32*, 624.
5549. O'Connor, R. T., E. T. Field, et al., *J. Amer. Oil Chemists' Soc.* **1951**, *28*, 154; *C.A.* **1951**, *45*, 4558.
5550. Oda, R., K. Nakano, et al., *Bull. Inst. Chem. Res., Kyoto Univ.* **1950**, *20*, 62; *C.A.* **1951**, *45*, 7037.
5551. Odan, M., S. Midzushima, et al., *Sci. Papers Inst. Phys. Chem. Res.* (Tokyo) **1944**, *42*, 27; *C.A.* **1947**, *41*, 6089.
5552. Oddo, B., and F. Tognacchini, *Gazz. Chim. Ital.* **1922**, *52II*, 347; *Chem. Zentr.* **1923**, *III*, 923.
5553. O'Donnell, J. F., J. T. Ayres, et al., *Anal. Chem.* **1965**, *37*, 1161.
5554. Oehme, F., *Chem-Ztg.* **1958**, *82*, 33; *C.A.* **1958**, *52*, 8419.
5555. Oehme, F., *Farbe u. Lack* **1958**, *64*, 183; *C.A.* **1959**, *53*, 790.
5556. Oehme, F., *Fette, Seifen, Anstrichmittel* **1960**, *62*, 910; *C.A.* **1961**, *55*, 20460.
5557. Oelschelaeger, H., H.-P. Oehr, et al., *Pharm. Acta Helv.* **1973**, *48*, No. 11–12, 662; *Anal. Abst.* **1974**, *27*, 11.
5558. Oestling, G. J., *J. Chem. Soc.* **1912**, 457.
5559. Oetjen, R. A., H. M. Randall, et al., *Rev. Mod. Phys.* **1944**, *16*, 260.
5560. Oetting, F. L., *J. Chem. Eng. Data* **1965**, *10*, 122.
5561. Oetting, F. L., *J. Chem. Phys.* **1964**, *41*, 149.
5562. Oetting, F. L., *J. Phys. Chem.* **1963**, *67*, 2757.
5563. Ogawa, Y., S. Imazeki, et al., *Bull. Chem. Soc. Jpn.* **1978**, *51*, 748.
5564. Ogden, J. S., and J. J. Turner, *Chem. Ind.* (London) **1966**, 1295; *C.A.* **1966**, *65*, 11576.
5565. Oglukian, R. L., Ph.D. Dissertation, Tulane Univ., 1968.
5566. Oguchi, K., I. Tanishita, et al., *Nippon Kikai Gakkai Rombunshu* **1975**, *41*, 1234; *C.A.* **1978**, *88*, 94956.
5567. Oguri, S., S. Hinonishi, et al., *Mem. Faculty. Sci. and Eng., Waseda Univ.*, Tokyo **1937**, 99; *C.A.* **1938**, *32*, 6520.
5568. Ohashi, K., to East Asia Synthetic Chem. Ind. Co., Japanese Patent 7660, Dec. 14, 1951; *C.A.* **1954**, *48*, 1415.
5569. Ohnishi, R., and K. Tanabe, *Bull. Chem. Soc. Jpn.* **1971**, *44*, 2647.
5570. Ohwada, K., *Appl. Spectrosc.* **1968**, *22*, 209; *C.A.* **1968**, *69*, 47785.
5571. Oka, E., Y. Numata, et al., Japanese Patent 5769, Aug. 18, 1955; *C.A.* **1957**, *51*, 15549.
5572. Okazaki, H., *J. Chem. Soc. Jpn.* **1942**, *63*, 1314, 1500; *C.A.* **1947**, *41*, 3370.

5573. Okazaki, H., *J. Chem. Soc. Jpn.* **1943,** *64,* 1225; *C.A.* **1947,** *41,* 3740.
5574. O'Kelly, A. A., J. Kellett, et al., *Ind. Eng. Chem.* **1947,** *39,* 154.
5575. Oki, M., and H. Iwamura, *Bull. Chem. Soc. Jpn.* **1959,** *32,* 567; *C.A.* **1960,** *54,* 6307.
5576. Oki, M., and H. Iwamura, *Bull. Chem. Soc. Jpn.* **1960,** *33,* 1632; *C.A.* **1961,** *55,* 23055.
5577. Okpala, C., A. Guiseppi-Elie, et al., *J. Chem. Eng. Data* **1980,** *25,* 384.
5578. Oliver, G. D., M. Eaton, et al., *J. Amer. Chem. Soc.* **1948,** *70,* 1502.
5579. Olivier, S. C. J., *Rec. Trav. Chim. Pays-Bas* **1936,** *55,* 1027.
5580. Olmstead, J., K. Street, et al., *J. Chem. Phys.* **1964,** *40,* 2114.
5581. Olsen, A. L., and E. R. Washburn, *J. Amer. Chem. Soc.* **1935,** *57,* 303.
5582. Olsen, W. T., H. F. Hipsher, et al., *J. Amer. Chem. Soc.* **1947,** *69,* 2451.
5583. Olsson, H., Z. *Phys. Chem.* (Leipzig) **1927,** *125,* 243.
5584. O'Malley, J. A., C. Owens, et al., *J. Phys. Chem.* **1968,** *72,* 3584.
5585. Omarov, T. T., and O. V. Agashkin, et al., *Izv. Akad. Nauk Kaz. SSR, Ser. Khim.* **1968,** *18,* No. 4, 38; *C.A.* **1969,** *70,* 15548.
5586. Omel'chenko, F. S., *Izv. Vyssh. Uchebn. Zaved., Pishch. Tekhnol.* **1962,** 97; *C.A.* **1962,** *57,* 11928.
5587. Omel'chenko, F. S., *Izv. Vyssh. Uchebn. Zaved., Pishch. Tekhnol.* **1962,** 151; *C.A.* **1962,** *57,* 10591.
5588. Omel'chenko, F. S., *Izv. Vyssh. Uchebn. Zaved., Pishch. Tekhnol.* **1967,** 37; *C.A.* **1967,** *67,* 4006.
5589. Oncescu, T., A. M. Oancea, et al., *J. Phys. Chem.* **1980,** *84,* 3090.
5590. N. V. Onderzoekingsinstitut Res., French Patent 1,396,197, April 16, 1965; *C.A.* **1965,** *63,* 8261.
5591. Ong, K. V., B. Douglas, et al., *J. Chem. Eng. Data* **1966,** *11,* 574.
5592. Ono, Y., T. Makita, et al., *Bull. Chem. Soc. Jpn.* **1968,** *41,* 1793.
5593. Oosterhout, J. C. D., and T. C. Roddy, U.S. Patent 2,419,499, April 22, 1947; *C.A.* **1947,** *41,* 5151.
5594. Orchin, M., *J. Assoc. Official Agr. Chem.* **1942,** *25,* 839.
5595. Orndorff, W. R., and F. K. Cameron, *Amer. Chem. J.* **1895,** *17,* 517.
5596. O'Rourke, C. E., L. B. Clapp, et al., *J. Amer. Chem. Soc.* **1956,** *78,* 2159.
5597. O'Rourke, D. F., and S. C. Mraw, *J. Chem. Thermodyn.* **1983,** *15,* 489.
5598. Ortega, J., *J. Chem. Eng. Data* **1982,** *27,* 312.
5599. Orton, K. J. P., and A. E. Bradfield, *J. Chem. Soc.* **1924,** 960.
5600. Orton, K. J. P., and A. E. Bradfield, *J. Chem. Soc.* **1927,** 983.
5601. Orton, K. J. P., M. G. Edwards, et al., *J. Chem. Soc.* **1911,** 1178.
5602. Orton, K. J. P., and D. C. Jones, *J. Chem. Soc.* **1919,** 1055, 1194.
5603. Orton, K. J. P., and M. Jones, *J. Chem. Soc.* **1912,** 1708.
5604. Orton, K. J. P., and M. Jones, *J. Chem. Soc.* **1912,** 1720.
5605. Orville-Thomas, W. J., A. E. Parsons, et al., *J. Chem. Soc.* **1958,** 1047.
5606. Osaka, H., *Bull. Inst. Phys. Chem. Res.* (Tokyo) **1928,** *7,* 873; *English Ed.* **1928,** *I,* 80; *C.A.* **1929,** *23,* 20.
5607. Osborn, A. G., and D. R. Douslin, *J. Chem. Eng. Data* **1966,** *11,* 50.
5608. Osborn, A. G., and D. R. Douslin, *J. Chem. Eng. Data* **1968,** *13,* 534.
5609. Osborn, A. G., and D. R. Douslin, *J. Chem. Eng. Data* **1974,** *19,* 114.
5610. Osborn, A. G., and D. W. Scott, *J. Chem. Thermodyn.* **1980,** *12,* 429.
5611. Osborn, A. R., K. Schofield, et al., *J. Chem. Soc.* **1956,** 4191.
5612. Osborne, D. W., C. S. Garner, et al., *J. Amer. Chem. Soc.* **1941,** *63,* 3496.
5613. Osborne, N. S., and D. C. Ginnings, *J. Res. Nat. Bur. Stand.* **1947,** *39,* 453.
5614. Osborne, N. S., E. C. McKelvey, et al., *Bull. Bur. Stand.* **1913,** *9,* 327; *J. Wash. Acad. Sci.,* 2, 95; *C.A.* **1912,** *6,* 1085; **1913,** *7,* 2889.
5615. Osborne, N. S., and C. H. Meyers, *J. Res. Nat. Bur. Stand.* **1934,** *13,* 1.
5616. Osborne, N. S., H. F. Stimson, et al., *J. Res. Nat. Bur. Stand.* **1939,** *23,* 197.
5617. Osina, T. M., *Trudy Nauch. Sessii Leningrad. Nauch.-Issled. Inst. Gig. Truda i Profzabolev.* **1958,** 256; *C.A.* **1962,** *56,* 2681.
5618. Osipova, L. I., and S. G. Krasnova, *Zh. Fiz. Khim.* **1979,** *53,* 2863; *Russ. J. Phys. Chem.* **1979,** *53,* 1636.
5619. Osokin, Yu. G., and V. Sh. Fel'dblyum, *Khim. Prom.* **1965,** *41,* 806; *C.A.* **1966,** *64,* 17406.
5620. Ostdick, T., and P. A. McCusker, *Inorg. Chem.* **1967,** 98.
5621. Ostromislensky, I., U.S. Patent 1,541,174, June 9, 1925; *C.A.* **1925,** *19,* 2210.
5622. Ostwald, W., Z. *Phys. Chem.* (Leipzig) **1889,** *3,* 369.

5623. Othmer, D. F., S. A. Savitt, et al., *Ind. Eng. Chem.* **1949**, *41*, 572.

5624. Othmer, D. F., N. Shlecter, et al., *Ind. Eng. Chem.* **1945**, *37*, 895.

5625. Othmer, D. F., and T. O. Wentworth, *Ind. Eng. Chem.* **1940**, *32*, 1588.

5626. Othmer, D. F., and R. E. White, U.S. Patent 2,275,802, Mar. 10, 1942; *Offic. Gaz. U.S. Pat. Office* **1942**, *536*, 359.

5627. Ott, J. B., J. R. Goates, et al., *J. Chem. Thermodyn.* **1979**, *11*, 1167.

5628. Otto, M. M., *J. Amer. Chem. Soc.* **1935**, *57*, 1476.

5629. Oualline, C. M., and M. Van Winkle, *Ind. Eng. Chem.* **1952**, *44*, 1668.

5630. Ouellette, R. J., *Can. J. Chem.* **1965**, *43*, 707.

5631. Owen, K., O. R. Quayle, et al., *J. Amer. Chem. Soc.* **1939**, *61*, 900.

5632. Owen, K., O. R. Quayle, et al., *J. Amer. Chem. Soc.* **1942**, *64*, 1294.

5633. Owen, N. L., and R. E. Hester, *Spectrochim. Acta, Part A* **1969**, *25*, 343.

5634. Owen, N. L., and G. O. Sørensen, *J. Phys. Chem.* **1979**, *83*, 1483.

5635. Oxford, A. E., *Biochem. J.* **1934**, *28*, 1325.

5636. Ozeki, T., and T. Nishida, Japanese Patent 7060, June 14, 1960; *C.A.* **1961**, *55*, 7354.

5637. Ozol, R. J., and C. R. Masterson, U.S. Patent 2,356,689, Aug. 22, 1944; *C.A.* **1945**, *39*, 86.

P

5638. Paal, C., *Ber. Deut. Chem. Gesell.* **1892**, *25*, 1202.

5639. Packendorff, K., *Ber. Deut. Chem. Gesell.* **1934**, *67*, 905.

5640. Padbury, J. J., and P. Tarrant, U.S. Patent 2,566,807, Sept. 4, 1951; *C.A.* **1952**, *46*, 2561.

5641. Padhye, M. R., and V. V. Bhujle, *Indian J. Pure Appl. Phys.* **1970**, *8*, 479; *C.A.* **1971**, *74*, 47601.

5642. Padhye, M. R., and P. B. Desai, *Proc. Nat. Inst. Sci. India, Part A* **1964**, *30*, 460.

5643. Padmanabhan, R., and S. K. K. Jatkar, *J. Amer. Chem. Soc.* **1935**, *57*, 334.

5644. Page, J. M., C. C. Buchler, et al., *Ind. Eng. Chem.* **1933**, *25*, 418.

5645. Page, T. F., *Mol. Phys.* **1967**, *13*, 523.

5646. Pagel, H. A., and W. A. Schroeder, *J. Amer. Chem. Soc.* **1940**, *62*, 1837.

5647. Pahlavouni, E., *Bull. Soc. Chim. Belges* **1927**, *36*, 533; *Chem. Zentr.* **1928**, *1*, 477.

5648. Pai, N. G., *Indian J. Phys.* **1933**, *7*, 519; *C.A.* **1933**, *27*, 4736.

5649. Pajeau, R., *Bull. Soc. Chim. Fr.* **1946**, 544; *C.A.* **1947**, *41*, 2643.

5650. Pajeau, R., *Bull. Soc. Chim. Fr.* **1948**, 59; *C.A.* **1948**, *42*, 4958.

5651. Palfray, L., *Bull. Soc. Chim. Fr.* **1940**, *7*, 401; *C.A.* **1942**, *36*, 2837.

5652. Pall, D. B., and O. Maass, *Can. J. Res.* **1936**, *14B*, 96.

5653. Palma, F. E., E. A. Piotrowski, et al., *J. Mol. Spectrosc.* **1964**, *13*, 119; *C.A.* **1964**, *61*, 196.

5654. Palomaa, M. H., *Ber. Deut. Chem. Gesell.* **1909**, *42*, 3873.

5655. Palomaa, M. H., and V. Aalto, *Ber. Deut. Chem. Gesell.* **1933**, *66*, 471.

5656. Palomaa, M. H., and I. Honkanen, *Ber. Deut. Chem. Gesell.* **1937**, *70*, 2199.

5657. Palomaa, M. H., and A. Juvala, *Ber. Deut. Chem. Gesell.* **1928**, *61*, 1770.

5658. Palomaa, M. H., and I. Mikkila, *Ber. Deut. Chem. Gesell.* **1942**, *75*, 1659.

5659. Palomaa, M. H., and T. A. Siitonen, *Ber. Deut. Chem. Gesell.* **1930**, *63*, 3117.

5660. Palyi, G., and F. Peter, *Magy. Kem. Lapja* **1962**, *17*, 354; *C.A.* **1963**, *58*, 6820.

5661. Pan, W. P., M. H. Mady, et al., *AIChEJ* **1975**, *21*, 283.

5662. Panchenkov, G. M., and V. F. Oreshko, *J. Phys. Chem.* (USSR) **1937**, *9*, 704; *C.A.* **1937**, *31*, 8282.

5663. Pandrey, B. R., and R. S. Tripathi, *Indian J. Pure Appl. Phys.* **1971**, *9*, 346.

5664. Panov, N., and I. Dudnikov, *Khim. Referat. Zh.* **1939**, *2*, No. 4, 24; *C.A.* **1940**, *34*, 1514.

5665. Pantzer, R., W. D. Burkhardt, et al., *Z. Anorg. Allg. Chem.* **1975**, *416*, 297.

5666. Paolini, V., *Gazz.* **1925**, *55*, 804.

5667. Papina, T. S., and V. P. Kolesov, *Vestn. Mosk. Univ., Ser. 2: Khim* **1978**, *19*, 500; *C.A.* **1978**, *89*, 214763.

5668. Papini, G., and S. Cuomo, *Antincendio* **1956**, *8*, 338; *C.A.* **1957**, *51*, 4118.

5669. Papousek, D., *Collect. Czech. Chem. Commun.* **1959**, *24*, 2666; *C.A.* **1960**, *54*, 1005.

5670. Paranjpe, G. R., and D. J. Kavar, *Indian J. Phys.* **1938**, *12*, 283; *C.A.* **1939**, *33*, 3222.

5671. Pardee, W. A., and W. Weinrich, *Ind. Eng. Chem.* **1944**, *36*, 595.

5672. Parker, A. J., *Chem. Rev.* **1969**, *69*, 1.

5673. Parkinson, A. E., and E. C. Wagner, *Ind. Eng. Chem., Anal. Ed.* **1934**, *6*, 434.

5674. Parks, G. S., *J. Amer. Chem. Soc.* **1925,** *47,* 338.

5675. Parks, G. S., and C. T. Anderson, *J. Amer. Chem. Soc.* **1926,** *48,* 1506.

5676. Parks, G. S., and B. Barton, *J. Amer. Chem. Soc.* **1928,** *50,* 24.

5677. Parks, G. S., and J. A. Hatton, *J. Amer. Chem. Soc.* **1949,** *71,* 2773.

5678. Parks, G. S., and H. M. Huffman, *J. Amer. Chem. Soc.* **1926,** *48,* 2788.

5679. Parks, G. S., and H. M. Huffman, *J. Amer. Chem. Soc.* **1930,** *52,* 4381.

5680. Parks, G. S., H. M. Huffman, et al., *J. Amer. Chem. Soc.* **1930,** *52,* 1032.

5681. Parks, G. S., H. M. Huffman, et al., *J. Amer. Chem. Soc.* **1933,** *55,* 2733.

5682. Parks, G. S., and K. K. Kelly, *J. Amer. Chem. Soc.* **1925,** *47,* 2089.

5683. Parks, G. S., W. D. Kennedy, et al., *J. Amer. Chem. Soc.* **1956,** *78,* 56.

5684. Parks, G. S., K. E. Manchester, et al., *J. Chem. Phys.* **1954,** *22,* 2089.

5685. Parks, G. S., and G. E. Moore, *J. Chem. Phys.* **1939,** *7,* 1006.

5686. Parks, G. S., J. R. Mosley, et al., *J. Chem. Phys.* **1950,** *18,* 152.

5687. Parks, G. S., and W. K. Nelson, *J. Phys. Chem.* **1928,** *32,* 61.

5688. Parks, G. S., and S. S. Todd, *Ind. Eng. Chem.* **1929,** *21,* 1235.

5689. Parks, G. S., S. S. Todd, et al., *J. Amer. Chem. Soc.* **1936,** *58,* 398.

5690. Parks, G. S., S. S. Todd, et al., *J. Amer. Chem. Soc.* **1936,** *58,* 2505.

5691. Parks, G. S., G. E. Warren, et al., *J. Amer. Chem. Soc.* **1935,** *57,* 616.

5692. Parks, G. S., T. J. West, et al., *J. Amer. Chem. Soc.* **1946,** *68,* 2524.

5693. Parry, H. L., U.S. Patent 2,653,172, Sept. 22, 1953; *C.A.* **1954,** *48,* 12169.

5694. Parshall, G. W., *J. Org. Chem.* **1962,** *27,* 4649.

5695. Parti, Y. P., and R. Samuel, *Proc. Roy. Soc.* (London) **1937,** *49,* 568; *C.A.* **1937,** *31,* 8371.

5696. Partington, J. R., *Trans. Faraday Soc.* **1922,** *17,* 734.

5697. Parts, A., *Z. Phys. Chem.* (Leipzig) **1930,** *7B,* 327.

5698. Parts, A., *Z. Phys. Chem.* (Leipzig) **1930,** *10B,* 264.

5699. Parts, A., *Z. Phys. Chem.* (Leipzig) **1931,** *12B,* 312; *C.A.* **1931,** *25,* 3885.

5700. Parts, A., *Z. Phys. Chem.* (Leipzig) **1931,** *12B,* 323; *C.A.* **1931,** *25,* 3885.

5701. Pascual, O. S., and E. Almeda, *Phillippine At. Energy Comm. (Rept.) PAEC(D)CH-634,* 1963, 9 pp; *C.A.* **1964,** *60,* 10521.

5702. Passino, H. J., U.S. Patent 2,396,966, March 19, 1946; *C.A.* **1946,** *40,* 3595.

5703. Patel, H. R., S. Sundaram, et al., *J. Chem. Eng. Data* **1979,** *24,* 40.

5704. Paterno, E., *Gazz. Chem. Ital.* **1899,** *19,* 640, from [6114].

5705. Paterno, E., and C. Montemartini, *Gazz. Chem. Ital.* **1894,** *24b,* 197; *J. Chem. Soc.* **1895 Aii,** 207.

5706. Paterson, W. G., and N. R. Tipman, *Can. J. Chem.* **1962,** *40,* 2122.

5707. Patrick, C. R., and G. S. Prosser, *Trans. Faraday Soc.* **1964,** *60,* 700.

5708. Patten, H. E., *J. Phys. Chem.* **1902,** *6,* 554.

5709. Patterson, T. S., and W. C. Forsyth, *J. Chem. Soc.* **1913,** 2263.

5710. Patterson, T. S., and D. Thompson, *J. Chem. Soc.* **1908,** 355.

5711. Patterson, W. A., *Anal. Chem.* **1954,** *26,* 823.

5712. Patty, F. A., *Industrial Hygiene and Toxicology,* 2nd rev. ed., *Vol. 1,* Interscience, New York, 1958.

5713. Patty, F. A., *Industrial Hygiene and Toxicology,* 2nd rev. ed., *Vol. 2,* Interscience, New York, 1963.

5714. Paukov, I. E., and L. K. Glukhikh, *Zh. Vses. Khim. Obshch.* **1967,** *12,* 236; *C.A.* **1967,** *67,* 53530.

5715. Paul, D. E., D. Lipkin, et al., *J. Amer. Chem. Soc.* **1956,** *78,* 116.

5716. Paul, R., *Bull. Soc. Chim. Fr.* **1941,** *8,* 507; *C.A.* **1942,** *36,* 1588.

5717. Paul, R., and H. Normant, *Compt. Rend.* **1943,** *216,* 689.

5718. Paul, R., and S. Tchelitcheff, *Bull. Soc. Chim. Fr.* **1947,** 453; *C.A.* **1948,** *42,* 1258.

5719. Paul, R. C., J. S. Banait, et al., *Indian J. Chem.* **1975,** *13,* 954.

5720. Paulsen, O., *Monatsh. Chem.* **1939,** *72,* 244; *C.A.* **1939,** *33,* 9138.

5721. Pavlic, A. A., and H. Adkins, *J. Amer. Chem. Soc.* **1946,** *68,* 1471.

5722. Pavlopoulos, T., and M. A. El-Sayed, *J. Chem. Phys.* **1964,** *41,* 1082.

5723. Pavlova, L. P., V. V. Lagunova, et al., *Gig. Tr. Prof. Zabol.* **1975,** *10,* 55; *C.A.* **1976,** *84,* 110930.

5724. Pavlova, S. P., S. Yu. Pavlov, et al., *Prom. Sin. Kauch.* **1966,** No. 3, 18; *C.A.* **1968,** *69,* 61997.

5725. Pavlovskaya, E. M., A. K. Charykov, et al., *Zh. Obshch. Khim.* **1977,** *47,* 2439; *J. Gen. Chem.* (USSR) **1977,** *47,* 2230.

5726. Pawlewski, B., *Ber. Deut. Chem. Gesell.* **1882,** *15,* 2460, 3034.

5727. Pawlewski, B., *Ber. Deut. Chem. Gesell.* **1888**, *21*, 2141.

5728. Payne, R., and I. E. Theodorou, *J. Phys. Chem.* **1972**, *76*, 2892.

5729. Peach, M. E., and T. C. Waddington, *J. Chem. Soc.* **1962**, 2680.

5730. Pearce, J. N., *J. Phys. Chem.* **1915**, *19*, 14.

5731. Pearce, J. N., and L. F. Berhenke, *J. Phys. Chem.* **1935**, *39*, 1005.

5732. Pearce, J. N., and J. S. Mortimer, *J. Amer. Chem. Soc.* **1918**, *40*, 509.

5733. Pearson, B. D., and J. E. Ollerenshaw, *Chem. Ind.* **1966**, 370.

5734. Pearson, R. G., *J. Amer. Chem. Soc.* **1948**, *70*, 204.

5735. Pearson, R. G., and R. L. Dillon, *J. Amer. Chem. Soc.* **1953**, *75*, 2439.

5736. Pearson, R. G., and J. M. Mills, *J. Amer. Chem. Soc.* **1950**, *72*, 1692.

5737. Pearson, R. G., and D. C. Vogelsong, *J. Amer. Chem. Soc.* **1958**, 1038.

5738. Pearson, R. G., and F. V. Williams, *J. Amer. Chem. Soc.* **1953**, *75*, 3073.

5739. Pearson, R. G., and F. V. Williams, *J. Amer. Chem. Soc.* **1954**, *76*, 258.

5740. Pecchiai, L., and U. Saffotti, *Med. Lavoro* **1957**, *48*, 247; *C.A.* **1957**, *51*, 15040.

5741. Pechalin, L. I., and G. M. Panchenko, *Vestn. Mosk. Univ., Ser. II: Khim.* **1966**, *21*, 30; *C.A.* **1966**, *65*, 11374.

5742. Pechalin, L. I., and G. M. Panchenkov, *Zh. Fiz. Khim.* **1966**, *40*, 1140; *C.A.* **1966**, *65*, 6328.

5743. Pecsok, R. L., R. A. Garber, et al., *Inorg. Chem.* **1965**, *4*, 447.

5744. Peddle, C. J., and W. E. S. Turner, *J. Chem. Soc.* **1913**, 1202.

5745. Pekary, A. E., *J. Phys. Chem.* **1974**, *78*, 1744.

5746. Pell, A. S., and G. Pilcher, *Trans. Faraday Soc.* **1965**, *61*, 71.

5747. Pelliccia-Galand, M.-F., and J. Hurwic, *C.R. Acad. Sci., Ser. C.* **1973**, *277*, 137.

5748. Peltonen, R. J., P. Neuenschwander, et al., *Z. Lebensm-Untersuch. u.-Forsch.* **1955**, *102*, 245; *C.A.* **1956**, *50*, 1259.

5749. Pendl, E., and G. Radinger, *Monatsh. Chem.* **1939**, *72*, 378; *C.A.* **1940**, *34*, 2256.

5750. Pendl, E., A. W. Reitz, et al., *Proc. Indian Acad. Sci.* **1938**, *8A*, 508.

5751. Pennington, R. E., H. L. Finke, et al., *J. Amer. Chem. Soc.* **1956**, *78*, 2055.

5752. Pennington, R. E., and K. A. Kobe, *J. Amer. Chem. Soc.* **1957**, *79*, 300.

5753. Peppiatt, E. G., and R. J. Wicker, *J. Chem. Soc.* **1955**, 3122.

5754. Perchard, J. P., M. T. Forel, et al., *J. Chim. Phys.* **1964**, *61*, 645; *C.A.* **1964**, *61*, 10188.

5755. Perdoncin, G., and G. Scorrano, *J. Amer. Chem. Soc.* **1977**, *99*, 6983.

5756. Peregudov, G. V., S. V. Markova, et al., *Izv. Akad. Nauk SSSR, Otdel. Khim. Nauk* **1957**, 37; *C.A.* **1957**, *51*, 12657.

5757. Peresedov, V. P., *Gig. Tr. Prof. Zabol.* **1974**, No. 11, 40; *C.A.* **1975**, *82*, 150057.

5758. Perez, P., T. E. Block, et al., *J. Chem. Eng. Data* **1971**, *16*, 333.

5759. Perichon, J., and R. Buvet, *Bull. Soc. Chim. Fr.* **1968**, 1279, 1282.

5760. Perkampus, H.-H., and G. Prescher, *Ber. Bunsenges. Phys. Chem.* **1968**, *72*, 429.

5761. Perkin, W. H., *J. Chem. Soc.* **1886**, 326.

5762. Perkin, W. H., *J. Chem. Soc.* **1889**, 680.

5763. Perkin, W. H., *J. Chem. Soc.* **1892**, 287.

5764. Perkin, W. H., *J. Chem. Soc.* **1894**, 815.

5765. Perkin, W. H., *J. Chem. Soc.* **1896**, 1025.

5766. Perkin, W. H., *J. Chem. Soc.* **1902**, 315.

5767. Perkin, W. H., *J. Chem. Soc.* **1904**, 654.

5768. Perkin, W. H., and K. Fisher, *J. Chem. Soc.* **1908**, 1871; *C.A.* **1909**, *3*, 886.

5769. Perkin, W. H., and C. Weizmann, *J. Chem. Soc.* **1906**, 1649.

5770. Perkins, R. P., U.S. Patent 2,190,607, Feb. 13, 1940; *C.A.* **1940**, *34*, 4081.

5771. Perlick, A., *Bull. Inst. Intern. Froid* **1937**, *1*, 4.

5772. Perlick, A., *Bull. Intern. Inst. Refrig.* **1937**, *18*, Al; *C.A.* **1938**, *32*, 4063; *Beil.* **EIII138**, 152.

5773. Perrin, D. D., *Aust. J. Chem.* **1964**, *17*, 484.

5774. Perrin, D. D., *Dissociation Constants of Organic Bases in Aqueous Solution*, Butterworth, London, 1965.

5775. Perrin, R., and P. Issartel, *Bull. Soc. Chim. Fr.* **1967**, 1083.

5776. Perry, E. S., and W. H. Weber, *J. Amer. Chem. Soc.* **1949**, *71*, 3726.

5777. Perry, J. H., *J. Phys. Chem.* **1927**, *31*, 1737.

5778. Perry, R. E., and G. Thodos, *Ind. Eng. Chem.* **1952**, *44*, 1649.
5779. Person, W. B., D. A. Olsen, et al., *Spectrochim. Acta* **1966**, *22*, 1733.
5780. Person, W. B., and G. C. Pimentel, *J. Amer. Chem. Soc.* **1953**, *75*, 532.
5781. Perttila, M., J. Murto, et al., *Spectrochim. Acta, Part A* **1978**, *34A*, 469.
5782. Pesce, B., *Gazz. Chim. Ital.* **1940**, *70*, 710; *C.A.* **1941**, *35*, 5091.
5783. Pesez, M., *Ann. Phar. Fr.* **1948**, *6*, 279; *C.A.* **1949**, *43*, 5341.
5784. Pestemer, M., and A. Alslev-Klinker, *Z. Electrochem. Agnew. Phys. Chem.* **1949**, *53*, 387; *Chem. Zentr.* **1950**, *II*, 515.
5785. Pestemer, M., and H. Flaschka, *Monatsh. Chem.* **1938**, *71*, 325; *C.A.* **1938**, *32*, 8266.
5786. Pete, B., A. I. Kiss, et al., *Spectrochim. Acta, Part A* **1980**, *36A*, 633.
5787. Peters, K., and W. Lohmar, *Beiheft Z. Ver. Deut. Chem., No. 2, Agnew Chem.* **1937**, *50*, 40; *C.A.* **1937**, *31*, 2579.
5788. Peters, L. M., K. E. Marple, et al., *Ind. Eng. Chem.* **1948**, *40*, 2046.
5789. Peterson, B. H., *Proc. Iowa Acad. Sci.* **1943**, *50*, 253.
5790. Peterson, P. E., *J. Org. Chem.* **1966**, *31*, 439.
5791. Pethrick, R. A., E. Wyn-Jones, et al., *J. Chem. Soc. A* **1969**, 1852.
5792. Petkovic, D. M., B. A. Kezele, et al., *J. Phys. Chem.* **1973**, *77*, 922.
5793. Petkovic, D. M., and M. S. Pavlovic, *Z. Phys. Chem.* (Frankfurt) **1974**, *88*, 54.
5794. Petrakis, L., and C. H. Sederholm, *J. Chem. Phys.* **1961**, *35*, 1174.
5795. Petrescu, V., *Ann. Sci. Univ. Jassy, Sect. I* **1940**, *26*, 233; *C.A.* **1940**, *34*, 4626.
5796. Petrikaln, A., *Z. Phys. Chem.* (Leipzig) **1929**, *B3*, 362.
5797. Petrikaln, A., and J. Hochberg, *Z. Phys. Chem.* (Leipzig) **1929**, *Abt. 4B*, 299; *C.A.* **1929**, *23*, 4620.
5798. Petro, A. J., and C. P. Smyth, *J. Amer. Chem. Soc.* **1958**, *80*, 73.
5799. Petros, L., V. Majer, et al., *Collect. Czech. Chem. Commun.* **1979**, *44*, 3533.
5800. Petrov, A. A., *J. Gen. Chem.* (USSR) **1941**, *11*, 713; *C.A.* **1942**, *36*, 404.
5801. Petrov, A. A., and G. I. Semenov, *Zh. Obshch. Khim.* **1957**, *27*, 2941; *C.A.* **1958**, *52*, 8102.
5802. Petrov, A. A., S. R. Sergienko, et al., *Izvest. Akad. Nauk SSSR, Otdel. Khim. Nauk* **1958**, 575; *C.A.* **1958**, *52*, 19898.
5803. Petrov, A. A., and M. I. Itkina, *J. Gen. Chem.* (USSR) **1947**, *17*, 220; *C.A.* **1948**, *42*, 516.
5804. Petrov, K. D., and E. S. Lagucheva, *Zh. Obshch. Khim.* **1948**, *18*, 1150; *C.A.* **1949**, *43*, 1742.
5805. Petrov, V. M., *Tr. Khim. Khim. Tekhnol.* **1974**, No. 1, 100; *C.A.* **1975**, *82*, 72425.
5806. Petrov, V. M., and L. E. Sandler, *Zh. Fiz. Khim.* **1975**, *49*, 2797; *Russ. J. Phys. Chem.* **1975**, *49*, 1649.
5807. Petrova, R. G., I. I. Kandor, et al., *Org. Magn. Reson.* **1978**, *11*, 406; *C.A.* **1979**, *90*, 167422.
5808. Pettit, L. D., and S. Bruckenstein, *J. Inorg. Nucl. Chem.* **1962**, *24*, 1478.
5809. Pfann, W., *Zone Melting*, 2nd ed., Wiley, New York, 1966.
5810. Pfeiffer, A., *Ber. Deut. Chem. Gesell.* **1872**, *5*, 699.
5811. Pfenninger, H., *Helv. Chim. Acta* **1962**, *45*, 460.
5812. Phadke, R. P., *J. Indian Inst. Sci.* **1953**, *34*, 189; *C.A.* **1953**, *47*, 3060.
5813. Phadke, S. R., S. D. Gokhale, et al., *J. Indian. Chem. Soc.* **1945**, *22*, 235; *C.A.* **1946**, *40*, 5310; *Beil.* **EIII2**, 306.
5814. Pham, T. M., M. H. Herzog-Cance, et al., *Can. J. Chem.* **1982**, *60*, 2777.
5815. Phan-Tau-Luu, R., J.-M. Surzur, et al., *Bull. Soc. Chim. Fr.* **1967**, 3274.
5816. Philip, J. C., and S. C. Waterton, *J. Chem. Soc.* **1930**, 2783.
5817. Philip, N. M., *Proc. Indian Acad. Sci.* **1939**, *9A*, 109; *C.A.* **1939**, *33*, 4484.
5818. Philippe, R., and A. M. Piette, *Bull. Soc. Chim. Belges* **1955**, *64*, 600; *C.A.* **1956**, *50*, 15151; from [7483].
5819. Phillips, J. C., and G. J. Mattamal, *J. Chem. Eng. Data* **1978**, *23*, 1.
5820. Phillips, J. C., and M. M. Mattamal, *J. Chem. Eng. Data* **1976**, *21*, 228.
5821. Phillips Petroleum Co., Data Sheets, Bartlesville, Oklahoma, 1973.
5822. Phillips Petroleum Co., Chem. Dept., "Phillips Hydrocarbons," Bull. 522, 6th ed., Bartlesville, Oklahoma, 1964.
5823. Phillips Petroleum Co., Special Products Div., Chem. Dept., "Hydrocarbons Catalog," Bartlesville, Oklahoma.
5824. Phillips Petroleum Co., Special Products Div., Chem. Dept., "Petro-Sulfur Compound Catalog," Bartlesville, Oklahoma.

5825. Phillips Petroleum Co., Special Products Div., Chem. Dept., "Phillips Hydrocarbon Aerosol Propellants," Bartlesville, Oklahoma, 1969.

5826. Phillips Petroleum Co., Special Products Div., Chem. Dept., "Reference Data for Hydrocarbons and Petro-Sulfur Compounds," Bartlesville, Oklahoma, 1962.

5827. Phillips, T. W., and K. P. Murphy, *J. Chem. Eng. Data* **1970**, *15*, 304.

5828. Phillips, W. D., H. C. Miller, et al., *J. Amer. Chem. Soc.* **1959**, *81*, 4496.

5829. Philpot, J. P., E. C. Rhodes, et al., *J. Chem. Soc.* **1940**, 84.

5830. Philpotts, A. R., W. Thain, et al., *Anal. Chem.* **1951**, *23*, 268.

5831. Piatti, L., *Z. Phys. Chem.* (Leipzig) **1931**, *152A*, 36.

5832. Pick, J., V. Fried, et al., *Chem. Listy* **1955**, *49*, 1720; *C.A.* **1956**, *50*, 636.

5833. Pickard, R. H., and J. Kenyon, *J. Chem. Soc.* **1911**, 45.

5834. Pickard, R. H., and J. Kenyon, *J. Chem. Soc.* **1912**, 620.

5835. Pickard, R. H., and J. Kenyon, *J. Chem. Soc.* **1913**, 1923.

5836. Pickard, R. H., and J. Kenyon, *J. Chem. Soc.* **1914**, 830.

5837. Pickering, R. A., and C. C. Price, *J. Amer. Chem. Soc.* **1958**, *80*, 4931.

5838. Pickering, S. F., *J. Phys. Chem.* **1924**, *28*, 97.

5839. Pickett, L. W., *J. Chem. Phys.* **1942**, *10*, 660.

5840. Pickett, L. W., M. E. Corning, et al., *J. Amer. Chem. Soc.* **1953**, *75*, 1618.

5841. Pickett, L. W., N. J. Hoeflich, et al., *J. Amer. Chem. Soc.* **1951**, *73*, 4865.

5842. Pickett, L. W., M. Muntz, et al., *J. Amer. Chem. Soc.* **1951**, *73*, 4862.

5843. Pickett, O. A., and J. M. Peterson, *Ind. Eng. Chem.* **1929**, *21*, 325.

5844. Pictet, A., and A. Geleznoff, *Ber. Deut. Chem Gesell.* **1903**, *36*, 2219.

5845. Pieck, R., and C. Courtoy, *Bull. Soc. Chim. Belges* **1947**, *56*, 63.

5846. Piekara, A., and B. Piekara, *Compt. Rend.* **1934**, *198*, 1018; *C.A.* **1934**, *28*, 2961.

5847. Pierce, L., and M. Hayashi, *J. Chem. Phys.* **1961**, *35*, 479.

5848. Pierre, I., and E. Puchot, *Ann. Chem.* **1872**, *163*, 283.

5849. Pierson, R. H., A. N. Fletcher, et al., *Anal Chem.* **1956**, *28*, 1218.

5850. Pigenet, C., J. P. Morizur, et al., *Bull. Soc. Chim. Fr.* **1969**, 361.

5851. Pihlaja, K., and J. Heikkila, *Acta Chem. Scand.* **1968**, *22*, 2731.

5852. Pihlaja, K., and J. Heikkila, *Acta Chem. Scand.* **1969**, *23*, 1053.

5853. Pikkarainen, L., *J. Chem. Eng. Data* **1983**, *28*, 344.

5854. Pilcher, G., and J. D. M. Chadwick, *Trans. Faraday Soc.* **1967**, *63*, 2357.

5855. Pilcher, G., and R. A. Fletcher, *Trans. Faraday Soc.* **1969**, *65*, 2326.

5856. Pilcher, G., A. S. Pell, et al., *Trans. Faraday Soc.* **1964**, *60*, 499.

5857. Pilcher, G., H. A. Skinner, et al., *Trans. Faraday Soc.* **1963**, *59*, 316.

5858. Pilpel, N., *J. Amer. Chem. Soc.* **1955**, *77*, 2949.

5859. Pines, H., U.S. Patent 2,422,798, June 24, 1947; *C.A.* **1947**, *41*, 6270.

5860. Pinkus, A. G., and H. C. Custard, *J. Phys. Chem.* **1970**, *74*, 1042.

5861. Pipparelli, E., A. Balducci, et al., *Ann. Chim.* (Rome) **1956**, *46*, 112; *C.A.* **1956**, *50*, 13862.

5862. Piretti, M., P. Capella, et al., *Riv. Ital. Sostanze Grasse* **1969**, *46*, 235; *C.A.* **1969**, *71*, 65320.

5863. Pirlot, G., *Bull. Soc. Chim. Belges* **1949**, *58*, 28; *C.A.* **1949**, *43*, 7366.

5864. Pirlot, G., *Bull. Soc. Roy. Sci. Liege* **1949**, *18*, 115; *C.A.* **1949**, *43*, 7817.

5865. Pisarev, V. V., A. M. Rozhnov, et al., *Zh. Fiz. Khim.* **1975**, *49*, 2464; *Russ. J. Phys. Chem.* **1975**, *49*, 1450.

5866. Pis'ko, G. T., G. V. Tolstopyatova, et al., *Gig. Sanit.* **1981**, 67; *C.A.* **1981**, *94*, 186355.

5867. Pitt, C. G., and H. M. Habercom, *J. Organomet. Chem.* **1968**, *15*, 359.

5868. Pittam, D. H., and G. Pilcher, *J. Chem. Soc., Faraday Trans. 1* **1972**, *68*, 2224.

5869. Pittsburgh Industrial Chemicals, "PX-138 Dioctyl Phthalate," Pittsburgh.

5870. Pitzer, K. S., *J. Amer. Chem. Soc.* **1940**, *62*, 331.

5871. Pitzer, K. S., *J. Amer. Chem. Soc.* **1940**, *62*, 1224.

5872. Pitzer, K. S., *J. Amer. Chem. Soc.* **1941**, *63*, 2413.

5873. Pitzer, K. S., *J. Chem. Phys.* **1937**, *5*, 473.

5874. Pitzer, K. S., L. Guttman, et al., *J. Amer. Chem. Soc.* **1946**, *68*, 2209.

5875. Pitzer, K. S., and W. D. Gwinn, *J. Amer. Chem. Soc.* **1941**, *63*, 3313.

5876. Pitzer, K. S., and J. L. Hollenberg, *J. Amer. Chem. Soc.* **1954**, *76*, 1493.

5877. Pitzer, K. S., and W. Weltner, *J. Amer. Chem. Soc.* **1949**, *71*, 2842.

5878. Plamondon, J. E., R. J. Buenker, et al., *Proc. Iowa Acad. Sci.* **1963**, *70*, 163; *C.A.* **1964**, *61*, 11424.
5879. Plate, A. F., and G. A. Tarasova, *J. Appl. Chem.* (USSR) **1944**, *17*, 576; *C.A.* **1946**, *40*, 2318.
5880. Plato, C., and A. R. Glasgow, *Anal. Chem.* **1969**, *41*, 330.
5881. Platt, J. R., and H. B. Klevens, *J. Chem. Phys.* **1948**, *16*, 832.
5882. Platt, J. R., H. B. Klevens, et al., *J. Chem. Phys.* **1949**, *17*, 466.
5883. Platt, J. R., I. Rusoff, et al., *J. Chem. Phys.* **1943**, *11*, 535.
5884. Plattner, P. A., and A. Ronco, *Helv. Chim. Acta* **1944**, *27*, 400.
5885. Plebanski, T., *Bull. Acad. Polon. Sci., Ser. Sci. Chim.* **1960**, *8*, 239; *C.A.* **1961**, *55*, 20553.
5886. Plessner, K. W., and R. B. Richards, *Trans. Faraday Soc.* **1946**, *42A*, 206; *C.A.* **1949**, *43*, 6020.
5887. Pliva, J., and V. Herout, *Collect. Czech. Chem. Commun.* **1950**, *15*, 160; *C.A.* **1951**, *45*, 8481.
5888. Plum, W. B., *J. Chem. Phys.* **1937**, *5*, 172.
5889. Plyler, E. K., *J. Chem. Phys.* **1948**, *16*, 1008.
5890. Plyler, E. K., *J. Chem. Phys.* **1949**, *17*, 218.
5891. Plyler, E. K., *J. Opt. Soc. Amer.* **1947**, *37*, 746.
5892. Plyler, E. K., and N. Acquista, *J. Res. Nat. Bur. Stand.* **1949**, *43*, 37.
5893. Plyler, E. K., H. C. Allen, et al., *J. Res. Nat. Bur. Stand.* **1957**, *58*, 255.
5894. Plyler, E. K., and W. S. Benedict, *J. Res. Nat. Bur. Stand.* **1951**, *47*, 202.
5895. Plyler, E. K., and C. J. Humphreys, *J. Res. Nat. Bur. Stand.* **1947**, *39*, 59.
5896. Plyler, E. K., and M. A. Lamb, *J. Res. Nat. Bur. Stand.* **1951**, *46*, 382.
5897. Plyler, E. K., W. H. Smith, et al., *J. Res. Nat. Bur. Stand.* **1950**, *44*, 503.
5898. Plyler, E. K., R. Stair, et al., *J. Res. Nat. Bur. Stand.* **1947**, *38*, 211.
5899. Pohl, H. A., M. E. Hobbs, et al., *J. Chem. Phys.* **1941**, *9*, 408.
5900. Pohland, E., and W. Mehl, *Z. Phys. Chem.* (Leipzig) **1933**, *164A*, 48.
5901. Pohoryles, L. A., S. Sarel, et al., *J. Org. Chem.* **1959**, *24*, 1878.
5902. Polak, J., and G. C. Benson, *J. Chem. Thermodyn.* **1971**, *3*, 235.
5903. Polak, J., and I. Mertl, *Collect. Czech. Chem. Commun.* **1965**, *30*, 3526; *C.A.* **1965**, *63*, 14072.
5904. Polak, L. S., R. A. Khmel'nitskii, et al., *Petrol. Chem. USSR* **1963**, *2*, No. 1, 1; *C.A.* **1964**, *61*, 13166.
5905. Poley, J. P., *Appl. Sci. Res.* **1955**, *4B*, 337; *C.A.* **1955**, *49*, 15316.
5906. Polonskii, T. M., *Khim. Referat. Zh.* **1941**, *1*, No. 3, 6; *C.A.* **1943**, *37*, 3668.
5907. Polska Akademia Nauk Inst. Chemii Fiz., French Patent 1,484,583, June 9, 1967.
5908. Polyakova, A. A., I. M. Lukashenko, et al., *Zh. Org. Khim.* **1969**, *5*, 1720; *C.A.* **1970**, *72*, 16638.
5909. Polyakova, A. A., T. I. Popova, et al., *Neftekhimiya* **1966**, *6*, 150; *C.A.* **1966**, *64*, 18635.
5910. Pomerantz, P., A. Fookson, et al., *J. Res. Nat. Bur. Stand.* **1954**, *52*, 59.
5911. Pommez, P., M. Lafaix, et al., *J. Chim. Phys.* **1967**, *64*, 1450.
5912. Pool, W. O., and A. W. Ralston, *Ind. Eng. Chem.* **1942**, *34*, 1104.
5913. Pople, J. A., and T. Schaefer, *Mol. Phys.* **1960**, *3*, 547; *C.A.* **1961**, *55*, 8045.
5914. Pople, J. A., W. G. Schneider, et al., *Can. J. Chem.* **1957**, *35*, 1060.
5915. Poponova, R. V., I. M. Lukashenko, et al., *Zh. Org. Khim.* **1971**, *7*, 2032.
5916. Popov, M. M., and P. K. Shirokich, *Z. Phys. Chem.* (Leipzig) **1933**, *167A*, 183; *C.A.* **1934**, *28*, 2573.
5917. Popov, V. V., *Plast. Massy* **1960**, No. 8, 26; *C.A.* **1961**, *55*, 11006.
5918. Porcaro, P. J., and V. D. Johnston, *Anal. Chem.* **1961**, *33*, 361.
5919. Porret, D., and C. F. Goodeve, *Trans. Faraday Soc.* **1937**, *33*, 690; *C.A.* **1937**, *31*, 7337.
5920. Port, W. S., and G. R. Riser, U.S. Patent 2,890,230, June 9, 1959; *C.A.* **1959**, *53*, 19890.
5921. Porte, A. L., and H. S. Gutowsky, *J. Org. Chem.* **1963**, *28*, 3216.
5922. Porter, C. J., J. H. Beynon, et al., *Org. Mass. Spectrom.* **1981**, *16*, 101.
5923. Porter, F., and J. N. Cosby, U.S. Patent 2,507,048, May 9, 1950; *C.A.* **1950**, *44*, 7879.
5924. Porter, Q. N., and R. J. Spear, *Org. Mass. Spectrom.* **1970**, *3*, 1259.
5925. Potapov, V. K., O. A. Yuzhakova, et al., *Dokl. Akad. Nauk SSSR* **1972**, *203*, 379; *Dokl. Phys. Chem.*, *Proc. Acad. Sci. USSR* **1972**, *203*, 210.
5926. Potter, A. E., and H. L. Ritter, *J. Phys. Chim.* **1954**, *58*, 1040.
5927. Potter, J. C., and J. H. Saylor, *J. Amer. Chem. Soc.* **1951**, *73*, 90.
5928. Potts, W. A., *Anal. Chem.* **1955**, *27*, 1027.
5929. Potts, W. J., *J. Chem. Phys.* **1952**, *20*, 809.
5930. Potts, W. J., *J. Chem. Phys.* **1955**, *23*, 65.

5931. Potzinger, P., and F. W. Lampe, *J. Phys. Chem.* **1970**, *74*, 719.
5932. Pouchert, C. J., *The Aldrich Library of Infrared Spectra*, Aldrich Chemical Co., Milwaukee, Wisconsin, 1970.
5933. Pouchert, C. J., and J. R. Campbell, *The Aldrich Library of NMR Spectra*, Aldrich Chemical Co., Milwaukee, Wisconsin, 1974.
5934. Pound, J. R., *J. Phys. Chem.* **1927**, *31*, 547.
5935. Pouyet, B., *Anal. Chim. Acta* **1967**, *38*, 291.
5936. Pouyet, B., *Purif. Inorg. Org. Mater.* **1969**, 121; *C.A.* **1969**, *71*, 60893.
5937. Powers, R. M., M. T. Tetenbaum, et al., *Anal. Chem.* **1962**, *34*, 1132.
5938. Powles, J. G., and A. Hartland, *Arch. Sci.* (Geneva) **1960**, *13*, Spec. No. 475.
5939. Pozdnyakova, A. G., *Gig. Sanit.* **1969**, *34*, 114; *C.A.* **1969**, *71*, 116316.
5940. Pozzani, U. C., E. R. Kinkead, et al., *Amer. Ind. Hyg. Assoc. J.* **1968**, *29*, 202; *C.A.* **1968**, *69*, 69530.
5941. PPG Industries Inc., "Carbon Disulfide," Pittsburgh, 1971.
5942. Praet, M. T., M. J. Hubin-Franskin, et al., *Org. Mass. Spectrom.* **1977**, *12*, 297.
5943. Prakash, S., *J. Sci. Res. Banaras Hindu Univ.* **1962–3**, *13*, No. 1, 144; *C.A.* **1963**, *59*, 2296.
5944. Prakash, S., and N. L. Singh, *J. Sci. Ind. Res.* (India) **1962**, *21B*, 512; *C.A.* **1963**, *58*, 5155.
5945. Prakash, S., and S. B. Srivastav, *J. Chem. Thermodyn.* **1975**, *7*, 997.
5946. Pratt, H. R. C., *Ind. Chemist* **1947**, *23*, 658, 727; *C.A.* **1948**, *42*, 3233.
5947. Preston, S. T., Jr., *J. Gas. Chromatog.* **1963**, *1*, 8; *C.A.* **1963**, *59*, 5811.
5948. Prevost, C., *Compt. Rend.* **1926**, *183*, 1292.
5949. Prevost, C., and A. Valelte, *Compt. Rend.* **1946**, *222*, 326; *C.A.* **1946**, *40*, 3719.
5950. Prey, V., and W. Unger, *Ann. Chem.* **1962**, *651*, 154; *C.A.* **1962**, *57*, 3540.
5951. Price, C. C., and M. Osgan, *J. Amer. Chem. Soc.* **1957**, *78*, 4787.
5952. Price, W. C., and D. M. Simpson, *Proc. Roy. Soc.* (London) **1938**, *165A*, 272.
5953. Price, W. C., and A. D. Walsh, *Proc. Roy. Soc.* (London) **1941**, *179A*, 201; *C.A.* **1942**, *36*, 336.
5954. Price, W. C., and A. D. Walsh, *Proc. Roy. Soc.* (London) **1947**, *191A*, 22; *C.A.* **1948**, *42*, 1504.
5955. Prichard, W. H., and W. J. Orville-Thomas, *Trans. Faraday Soc.* **1965**, *61*, 1549.
5956. Prideaux, E. B. R., and R. N. Coleman, *J. Chem. Soc.* **1936**, 1346.
5957. Prins, H. J., *Rec. Trav. Chim. Pays-Bas* **1926**, *45*, 80; *C.A.* **1926**, *20*, 1977.
5958. Pritchard, J. G., and F. A. Long, *J. Amer. Chem. Soc.* **1957**, *79*, 2365.
5959. Privalov, V. E., L. D. Glazman, et al., *Koks Khim.* **1970**, No. 5, 38; *C.A.* **1970**, *73*, 25252.
5960. Private communication to the authors.
5961. Privett, O. S., E. Breault, et al., *J. Amer. Oil Chem. Soc.* **1958**, *35*, 366.
5962. Prochazka, M., and M. Palacek, *Collect. Czech. Chem. Commun.* **1967**, *32*, 3149.
5963. Produits Chimiques Pechiney-Saint-Gobain, *French Addn.* 83,361, July 31, 1964; *C.A.* **1964**, *61*, 15974.
5964. Prosen, E. J., W. H. Johnson, et al., *J. Res. Nat. Bur. Stand.* **1946**, *36*, 455.
5965. Prosen, E. J., W. H. Johnson, et al., *J. Res. Nat. Bur. Stand.* **1947**, *39*, 173.
5966. Prosen, E. J., F. W. Maron, et al., *J. Res. Nat. Bur. Stand.* **1951**, *46*, 106.
5967. Prosen, E. J., and F. D. Rossini, *J. Res. Nat. Bur. Stand.* **1944**, *33*, 255.
5968. Prosen, E. J., and F. D. Rossini, *J. Res. Nat. Bur. Stand.* **1945**, *34*, 59.
5969. Prosen, E. J., and F. D. Rossini, *J. Res. Nat. Bur. Stand.* **1945**, *34*, 263.
5970. Prout, F. S., and J. E. Spikner, *J. Org. Chem.* **1962**, *27*, 1488.
5971. Prud'hommer, M., *Bull. Soc. Chim. Fr.* **1922**, *31*, 295; *Chem. Zentr.* **1922**, *III*, 1160.
5972. Prue, J. E., and G. Schwarzenbach, *Helv. Chim. Acta* **1950**, *33*, 985.
5973. Prue, J. E., and P. J. Sherrington, *Trans. Faraday Soc.* **1961**, *57*, 1795.
5974. Pryor, W. A., and R. E. Jentoft, *J. Chem. Eng. Data* **1961**, *6*, 36.
5975. Puck, T. T., and H. Wise, *J. Phys. Chem.* **1946**, *50*, 329.
5976. Pudowik, A. N., *Zh. Obshch. Khim.* **1949**, *19*, 1187.
5977. Pugacheva, A. I., *Uchen. Zap. Mosv. Oblast. Pedagog. Inst.* **1960**, *92*, 43; *C.A.* **1962**, *56*, 2073.
5978. Pukirev, A. G., *Trans. Inst. Pure Chem. Reagents* (Moscow) **1937**, *15*, 45; *C.A.* **1938**, *32*, 5378.
5979. Pulfrich, C., *Neues Jahar. Beilageband* **1887**, *V*, 167; from [7482].
5980. Pummer, W. J., *J. Chem. Eng. Data* **1961**, *6*, 76.
5981. Puranik, P. G., *J. Chem. Phys.* **1955**, *23*, 761.

5982. Puranik, P. G., *Proc. Indian Acad. Sci.* **1955**, *42A*, 326; *C.A.* **1956**, *50*, 7595.
5983. Puranik, P. G., and K. V. Ramiah, *J. Mol. Spectrosc.* **1959**, *3*, 486; *C.A.* **1961**, *55*, 10064.
5984. Puranik, P. G., and K. V. Ramiah, *Proc. Indian Acad. Sci.* **1961**, *54*, 69; *C.A.* **1962**, *56*, 9592.
5985. Puranik, P. G., and D. V. Ramiah, *Proc. Indian Acad. Sci.* **1961**, *54A*, 121; *C.A.* **1962**, *56*, 8183.
5986. Purcell, J. M., S. G. Morris, et al., *Anal. Chem.* **1966**, *38*, 588.
5987. Purcell, J. M., and H. Susi, *Appl. Spectrosc.* **1965**, *19*, 105; *C.A.* **1965**, *63*, 13032.
5988. Purcell, R. F., Commercial Solvents Corp., "The Nitroparaffins and Their Uses in Coatings," Tech. Bull. No. 1, New York, 1958.
5989. Purvis, J. E., *J. Chem. Soc.* **1910**, 1546.
5990. Purvis, J. E., *J. Chem. Soc.* **1912**, 1315.
5991. Purvis, J. E., *J. Chem. Soc.* **1924**, 406.
5992. Purvis, J. E., *Proc. Cambridge Phil Soc.* **1925**, *23*, 588; *Chem. Zentr.* **1927**, *II*, 379.
5993. Purvis, J. E., and N. P. McCleland, *J. Chem. Soc.* **1913**, 1088.
5994. Puschin, N. A., and A. A. Glakgoleva, *J. Chem. Soc.* **1922**, 2813.
5995. Puschin, N. A., and P. G. Malavulj, *Z. Phys. Chem.* (Leipzig) **1932**, *158A*, 161, 162, 290, 341, 415.
5996. Puschin, N. A., and T. Pinter, *Z. Phys. Chem.* (Leipzig) **1929**, *142A*, 211.
5997. Putnam, W. E., D. M. McEachern, et al., *J. Chem. Phys.* **1965**, *42*, 749.
5998. Pyatlin, V. N., *Sb. Nauch. Tr. Kuibyshev. Nauch.-Issled. Inst. Epidemial. Gig.* **1968**, No. 5, 117.
5999. Pyle, W. R., *Phys. Rev.* **1931**, *38*, 1057.
6000. Pyryalova, P. S., and E. N. Zil'berman, *Tr. po Khim. i Khim. Tekhnol.* **1963**, No. 2, 353; *C.A.* **1964**, *61*, 4211.

Q

6001. Quaker Oats Company, The, "QO Furfural," Bulletin 203-A, Chicago.
6002. Quaker Oats Company, Chemicals Division, "Furfuryl Alcohol," Bulletin 205-A, Chicago, 1963.
6003. Quaker Oats Company, The, "QO Tetrahydrofurfuryl Alcohol," Bulletin 206-A, Chicago, 1960.
6004. Quaker Oats Company, Tech. Bulletin No. 148, "QO Tetrahydrofuran," Chicago, 1964.
6005. Quaker Oats Company, Tech. Bulletin No. 149, "QO Furan," Chicago, 1965.
6006. Quaker Oats Company, The, "QO Tetrahydrofuran (THF), Recovery Techniques," Tech. Bulletin No. 159, Chicago.
6007. Quayle, O. R., *Chem. Rev.* **1953**, *53*, 439.
6008. Quayle, O. R., and R. A. Day, *J. Amer. Chem. Soc.* **1944**, *66*, 938.
6009. Quiggle, D., C. O. Tongberg, et al., *Ind. Eng. Chem., Anal. Ed.* **1934**, *6*, 466.
6010. Quilico, A., P. Grunnanger, et al., *Tetrahedron* **1957**, *1*, 186.
6011. Quinn, H. W., J. S. McIntyre, et al., *Can. J. Chem.* **1965**, *43*, 2896.

R

6012. Rabinovich, I. B., B. V. Lebedev, et al., *Vysokomol. Soedin., Ser. A* **1967**, *9*, 1699; *C.A.* **1967**, *67*, 91172.
6013. Rabinovitch, I. B., V. D. Tel'noi, et al., *Dokl. Akad. Nauk SSSR* **1962**, *143*, 133; *C.A.* **1962**, *57*, 4557.
6014. Rabinowitsch, N., *Z. Phys. Chem.* (Leipzig) **1926**, *119*, 59, 70.
6015. Rabjohn, N., *Organic Syntheses, Collective Vol. IV*, 1st ed., Wiley, New York, 1963.
6016. Radchenko, L. G., and A. I. Kitaigorodski, *Zh. Fiz. Khim.* **1974**, *48*, 2702; *Russ. J. Phys. Chem.* **1974**, *48*, 1595.
6017. Radeglia, R., S. Scheithauer, et al., *Z. Naturforsch. B* **1969**, *24*, 283.
6018. Radinger, B., and H. Wittek, *Z. Phys. Chem.* (Leipzig) **1939**, *45B*, 329; *C.A.* **1940**, *34*, 4991.
6019. Radulescu, D., and M. Alexa, *Bull. Soc. Chim. Romania* **1938**, *20A*, 89; *C.A.* **1940**, *34*, 934.
6020. Radzitzky, P. de, and J. Hanotier, *Ind. Chim. Belges* **1962**, *27*, 125; *C.A.* **1963**, *58*, 4464.
6021. Raeuchle, F., W. Pohl, et al., *Ber. Bunsenges. Phys. Chem.* **1971**, *75*, 66.
6022. Raffauf, R. F., *J. Amer. Chem. Soc.* **1950**, *72*, 753.
6023. Rahman, H. T., and H. Gilman, *J. Indian Chem. Soc.* **1974**, *51*, 1018.
6024. Rahman, S. M. F., A. M. Roy, et al., *J. Nat. Sci. Math.* **1965**, *5*, 203; *C.A.* **1966**, *65*, 129.
6025. Railing, W. E., *J. Amer. Chem. Soc.* **1939**, *61*, 3349.
6026. Rall, H. T., and H. M. Smith, *Ind. Eng. Chem., Anal. Ed.* **1936**, *8*, 324.

6027. Ralph, E. D., and W. A. Gilkerson, *J. Amer. Chem. Soc.* **1964,** *86,* 4783.

6028. Ralston, A. W., and C. W. Hoerr, *J. Org. Chem.* **1942,** 7, 546.

6029. Raman, G. K., and P. R. Naidu, *Proc. Indian Acad Sci., Sect. A* **1972,** *75,* 68; *C.A.* **1972,** *77,* 39427.

6030. Ramart-Lucas, Mme., Mlle. Beguard, et al., *Compt. Rend.* **1930,** *190,* 1196.

6031. Ramart-Lucas, Mme., and F. Salmon-Legagneur, *Compt. Rend.* **1928,** *186,* 39.

6032. Ramart-Lucas, de Mme., and M. F. Salmon-Legagneur, *Compt. Rend.* **1930,** *190,* 492.

6033. Rambidi, N. G., *Dokl. Akad. Nauk SSSR* **1955,** *102,* 747; *C.A.* **1955,** *49,* 15485.

6034. Rambraud, R., and D. Besserre, *Bull. Soc. Chim. Fr.* **1955,** 45; *C.A.* **1956,** *50,* 11940.

6035. Ramiah, K. V., and V. V. Chalapathi, *Current Sci.* (India) **1966,** *35,* 301; *C.A.* **1966,** *65,* 8732.

6036. Ramiah, K. V., and V. V. Chalapathi, *Proc. Indian Acad. Sci., Sect. A* **1964,** *60,* 242; *C.A.* **1965,** *63,* 155.

6037. Ramiah, K. V., V. V. Chalapathi, et al., *Current Sci.* (India) **1966,** *35,* 350; *C.A.* **1966,** *65,* 13038.

6038. Ramonda, J. W., L. O. Edwards, et al., *J. Amer. Chem. Soc.* **1974,** *96,* 1708.

6039. Rampolla, R. W., and C. P. Smyth, *J. Amer. Chem. Soc.* **1958,** *80,* 1057.

6040. Ramsay, W., and J. Shields, *Z. Phys. Chem.* (Leipzig) **1893,** *12,* 433.

6041. Ramsay, W., and B. D. Steele, *Z. Phys. Chem.* (Leipzig) **1903,** *44,* 348.

6042. Ramsey, B. G., *J. Amer. Chem. Soc.* **1966,** *88,* 5358.

6043. Ramsey, D. A., *Proc. Roy. Soc.* (London) **1947,** *190A,* 562; *C.A.* **1948,** *42,* 4059.

6044. Ramsey, J. B., and F. T. Aldridge, *J. Amer. Chem. Soc.* **1955,** *77,* 2561.

6045. Randall, H. M., R. G. Fowler, et al., *Infrared Determination of Organic Structures,* Van Nostrand, New York, 1949.

6046. Rank, D. H., *J. Optical Soc. Amer.* **1947,** *37,* 798.

6047. Rank, D. H., and E. R. Bordner, *J. Chem. Phys.* **1935,** *3,* 248.

6048. Rank, D. H., B. D. Saksena, et al., *Disc. Faraday Soc.* **1950,** *9,* 187.

6049. Rank, D. H., R. W. Scott, et al., *Ind. Eng. Chem., Anal. Ed.* **1942,** *14,* 816.

6050. Rank, D. H., N. Sheppard, et al., *J. Chem. Phys.* **1948,** *16,* 698.

6051. Rao, A. L. S., *Indian J. Phys.* **1941,** *14,* Pt. 5, 365; *C.A.* **1941,** *35,* 6187.

6052. Rao, A. L. S., *J. Indian Chem. Soc.* **1941,** *18,* 337; *C.A.* **1942,** *36,* 967.

6053. Rao, B. D. N., P. Venkateswarlu, et al., *Can. J. Chem.* **1962,** *40,* 965.

6054. Rao, D. N., A. Krishnaiah, et al., *J. Chem. Thermodyn.* **1981,** *13,* 677.

6055. Rao, D. V. G. L. N., *Current Sci.* (India) **1956,** *25,* 217; *C.A.* **1957,** *51,* 2348.

6056. Rao, I. A., *J. Sci. Ind. Res.* (India) **1961,** *20B,* 145; *C.A.* **1961,** *55,* 24228.

6057. Rao, M. R. A., and B. S. Rao, *J. Indian Chem. Soc.* **1935,** *12,* 322; *Chem. Zentr.* **1936,** *I,* 1389; *C.A.* **1935,** *29,* 7148.

6058. Rao, M. V., *Proc. Indian Acad. Sci.* **1946,** *24A,* 510; *C.A.* **1947,** *41,* 3697.

6059. Rao, P. R., *Proc. Indian Acad. Sci.* **1962,** *55A,* 232; *C.A.* **1962,** *57,* 6755.

6060. Rao, P. R., and N. L. Singh, *Proc. Indian Acad. Sci.* **1962,** *55A,* 178; *C.A.* **1962,** *57,* 5475.

6061. Raoult, F. M., *Ann. Chim. Phys.* **1884,** *2,* 66; *J. Chem. Soc.* **1884,** 952.

6062. Rapean, J. C., D. L. Pearson, et al., *Ind. Eng. Chem.* **1959,** *51,* 77.

6063. Raphael, R. A., and F. Sondheimer, *J. Chem. Soc.* **1951,** 2693.

6064. Raposo, M. R., C. Estevens, et al., *Rev. Port. Farm.* **1961,** *11,* 79; *C.A.* **1962,** *56,* 5217.

6065. Rapp, D. E., U.S. Patent 2,923,747, Feb. 2, 1960.

6066. Rapport, N. J., E. F. Westrum, et al., *J. Amer. Chem. Soc.* **1971,** *93,* 4363.

6067. Raskin, S. S., and F. I. Skripov, *Zh. Eksptl. i Teoret. Fiz.* **1954,** *26,* 479; *C.A.* **1955,** *49,* 15484.

6068. Rasmussen, R. S., *J. Chem. Phys.* **1948,** *16,* 712.

6069. Rasmussen, R. S., and R. R. Brattain, *J. Amer. Chem. Soc.* **1949,** *71,* 1073.

6070. Rasmussen, R. S., R. R. Brattain, et al., *J. Chem. Phys.* **1947,** *15,* 135.

6071. Rasmussen, R. S., D. O. Tunnieliff, et al., *J. Amer. Chem. Soc.* **1949,** *71,* 1068.

6072. Rast, K., *Ber. Deut. Chem. Gesell.* **1922,** *55,* 1051.

6073. Rastogi, R. P., P. C. Tewari, et al., *J. Indian Chem. Soc.* **1977,** *54,* 1063.

6074. Rastogi, R. P., J. Nath, et al., *J. Phys. Chem.* **1967,** *71,* 2524.

6075. Rastogi, R. P., R. K. Nigam, et al., *J. Phys. Chem.* **1963,** *39,* 3042.

6076. Ratan, R., and M. Singh, *Indian J. Chem., Sect. A* **1979,** *18A,* 67.

6077. Ratchford, W., C. Rehberg, et al., *J. Amer. Chem. Soc.* **1944,** *66,* 1864.

6078. Rathjen, W., and J. Straub, *Heat Transfer Boiling* **1977,** 425; *C.A.* **1978,** *89,* 186217.

6079. Rau, M. A. G., and B. N. Narayanaswamy, *Z. Phys. Chem.* (Leipzig) **1934**, *26B*, 23.
6080. Rauf, M. A., G. H. Stewart, et al., *J. Chem. Eng. Data* **1983**, *28*, 324.
6081. Ravdel, A. A., and V. V. Danilov, *Izv. Vyssh. Ucheb Zaved., Khim. Khim. Tekhnol.* **1968**, *11*, 642; *C.A.* **1968**, *69*, 90232.
6082. Ravikovich, S. D., and V. P. Solomko, *Ukrain. Khim. Zh.* **1958**, *24*, 7; *C.A.* **1958**, *52*, 15173.
6083. Raymonda, J. W., and W. T. Simpson, *J. Chem. Phys.* **1967**, *47*, 430.
6084. Rea, D. G., *Anal. Chem.* **1960**, *32*, 1638.
6085. Read, R. R., L. S. Foster, et al., *Org. Syntheses* **1945**, *25*, 11; *C.A.* **1946**, *40*, 320.
6086. Reay, W. H., and G. T. Merrall, British Patent 891,823, Mar. 21, 1962; *C.A.* **1962**, *57*, 4515.
6087. Reboul, E., *Ann. Chim. Phys.* **1878**, *5*, 14, 475.
6088. Reboul, E., *Jahr. Fort. Chem.* **1873**, 321.
6089. Reboul, E., *Jahr. Fort. Chem.* **1881**, 409.
6090. Rechenberg, C. v., *Einfache und Fraktionierte Destillation in Theorie und Praxis*, Selbstverlag von Schimmel and Co., Leipzig, 1923.
6091. Reddy, G. S., *Z. Naturforsch* **1966**, *21A*, 609; *C.A.* **1966**, *65*, 13505.
6092. Reddy, G. S., and J. H. Goldstein, *J. Amer. Chem. Soc.* **1961**, *83*, 5020.
6093. Reddy, G. S., and J. H. Goldstein, *J. Amer. Chem. Soc.* **1962**, *84*, 583.
6094. Reddy, G. S., J. H. Goldstein, et al., *J. Amer. Chem. Soc.* **1961**, *83*, 1300.
6095. Reddy, K. S., and P. R. Naidu, *Indian J. Chem., Sect. A* **1981**, *20A*, 503.
6096. Reddy, K. S., M. Sreenivasulu, et al., *Z. Phys. Chem.* (Wiesbaden) **1981**, *124*, 149.
6097. Reddy, T. B., *Pure Appl. Chem.* **1971**, *25*, 459.
6098. Redlich, O., and W. Stricks, *Monatsh. Chem.* **1936**, *68*, 47; *Chem. Zentr.* **1936**, *II*, 2521.
6099. Reed, C. D., and J. J. McKetta, *Petrol. Refiner* **1959**, *38*, 159.
6100. Reed, R. I., and V. V. Takhistov, *Tetrahedron* **1967**, *23*, 2807.
6101. Rees, H. L., and D. H. Anderson, *Anal. Chem.* **1949**, *21*, 989.
6102. Reese, B. O., L. B. Seely, et al., *J. Chem. Eng. Data* **1970**, *15*, 140.
6103. Reeves, L. W., *Can. J. Chem.* **1957**, *35*, 1351.
6104. Reeves, L. W., and W. G. Schneider, *Trans. Faraday Soc.* **1958**, *54*, 314.
6105. Regnault, H. V., *Justus Liebigs Ann. Chem.* **1835**, *14*, 30.
6106. Rehberg, C. E., and C. H. Fisher, *Ind. Eng. Chem.* **1948**, 40, 1429.
6107. Rehberg, C. E., and C. H. Fisher, *J. Amer. Chem. Soc.* **1944**, *66*, 1203.
6108. Reichel, L., and O. Strasser, *Ber. Deut. Chem. Gesell.* **1931**, *64*, 1997.
6109. Reichert, J. S., J. H. Bailey, et al., *J. Amer. Chem. Soc.* **1923**, *45*, 1552.
6110. Reid, D. S., and C. A. Vincent, *J. Electroanal. Chem.* **1968**, *18*, 427.
6111. Reid, E. E., U.S. Patent 1,821,667, Sept. 1, 1931; *Official Gaz. U.S. Patent Office* **1931**, *410*, 221.
6112. Reid, E. E., and J. R. Ruhoff, *Org. Syntheses* **1936**, *16*, 60.
6113. Reilley, C. A., and J. D. Swalen, *J. Chem. Phys.* **1961**, *35*, 1522.
6114. Reilly, J., and W. N. Rae, *Physico-chemical Methods, Vol. II,* 3rd ed., Van Nostrand, Princeton, 1939.
6115. Reilly, J., R. Wolter, et al., *Sci. Proc. Roy. Dublin. Soc.* **1930**, *19*, 467; *Chem. Zentr.* **1931**, *I*, 1899.
6116. Rein, R., N. Fukuda, et al., *J. Chem. Phys.* **1966**, *45*, 4743.
6117. Reinhardt, R. C., *Ind. Eng. Chem.* **1943**, *35*, 422.
6118. Reinhardt, P. B., Q. Williams, et al., *J. Chem. Phys.* **1970**, *53*, 1418.
6119. Reinisch, L., *J. Chim. Phys.* **1954**, *51*, 113; *C.A.* **1955**, *49*, 2801.
6120. Reinke, R. C., and E. N. Luce, *Ind. Eng. Chem., Anal. Ed.* **1946**, *18*, 244.
6121. Reiser, E., H. Eisenberg, et al., *J. Chem. Soc., Faraday Trans. 2* **1972**, *68*, 1001.
6122. Reisse, J., and R. Ottinger, *Spectrochim. Acta* **1966**, *22*, 969; *C.A.* **1966**, *65*, 1935.
6123. Reith, H., and H. Eckardt, *Freiberger Forschungsh. A* **1962**, *250*, 39; *C.A.* **1963**, *59*, 11158.
6124. Reitmeier, R. E., V. Sivertz, et al., *J. Amer. Chem. Soc.* **1940**, *62*, 1943.
6125. Reitz, A. W., *Z. Phys. Chem.* (Leipzig) **1937**, *38B*, 179; *Chem. Zentr.* **1937**, *II*, 367.
6126. Reitz, A. W., *Z. Phys. Chem.* (Leipzig) **1937**, *38B*, 275; *C.A.* **1938**, *32*, 2837.
6127. Reitz, A. W., and R. Sabathy, *Monatsh. Chem.* **1937**, *71*, 100; *C.A.* **1938**, *32*, 2027.
6128. Reitz, A. W., and R. Sabathy, *Monatsh. Chem.* **1938**, *71*, 131; *C.A.* **1938**, *32*, 4076.
6129. Reitz, A. W., and R. Skrabal, *Monatsh. Chem.* **1937**, *70*, 398; *C.A.* **1937**, *31*, 7758.
6130. Reitz, A. W., and W. Stockmair, *Monatsh. Chem.* **1935**, *67*, 92; *C.A.* **1936**, *30*, 2848.
6131. Remiz, E. K., and A. V. Frost, *J. Appl. Chem.* (USSR) **1940**, *13*, 210; *C.A.* **1941**, *35*, 433.

6132. Renault, L., French Patent 868,182, Dec. 23, 1941; *C.A.* **1949,** *43,* 5034.

6133. Reppe, B. W., *Chemie und Technik der Acelylene-Druckvaktionen.*, Weinheim, 1951.

6134. Reppe, J. W., *Acetylene Chemistry*, Chas. A. Meyer, New York, 1949.

6135. Reppe, J. W., *Justus Liebigs Ann. Chem.* **1955,** *596,* 25; *C.A.* **1956,** *50,* 16774.

6136. Reppe, W., W. Schmidt, et al. (to General Aniline and Film), U.S. Patent 2,319,707, May 18, 1943; *C.A.* **1943,** *37,* 6274.

6137. Rericha, V., and M. Protiva, *Chem. Listy* **1951,** *45,* 157; *C.A.* **1952,** *46,* 1497.

6138. Reuther, H., *Chem. Tech.* (Berlin) **1953,** *5,* 330; *C.A.* **1954,** *48,* 11304.

6139. Reuther, H., and G. Reichel, *Plaste Kautschuk* **1963,** *10,* 720; *C.A.* **1964,** *60,* 11389.

6140. Rey-Lafon, M., and M. T. Forel, *Bull. Soc. Chim. Fr.* **1967,** 1145.

6141. Rey-Lafon, M., J. Lascombe, et al., *Ann. Chim.* (Paris) **1963,** *8,* 493; *C.A.* **1964,** *60,* 10065.

6142. Reynaud, R., *C.R. Acad. Sci., Ser. C* **1968,** *266,* 489.

6143. Reynolds, P. W., and D. M. Grudgings, British Patent 622,937, May 10, 1949; *C.A.* **1949,** *43,* 8397.

6144. Reynolds, W. F., and T. Schaefer, *Can. J. Chem.* **1963,** *41,* 2339.

6145. Reynolds, W. L., and R. H. Weiss, *J. Amer. Chem. Soc.* **1959,** *81,* 1790.

6146. Rhodes, R. E., M. Barber, et al., *Anal. Chem.* **1966,** *38,* 48.

6147. Riccardi, R., and P. Franzosini, *Ann. Chim.* (Rome) **1957,** *47,* 977; *C.A.* **1958,** *52,* 842.

6148. Rice, F. O., *Proc. Roy. Soc.* (London) **1914,** *91A,* 76.

6149. Rice, P. A., R. P. Gale, et al., *J. Chem. Eng. Data* **1976,** *21,* 204.

6150. Richard, A. J., *J. Phys. Chem.* **1978,** *82,* 1265.

6151. Richard, A. J., and P. B. Fleming, *J. Chem. Thermodyn.* **1981,** *13,* 863.

6152. Richard, A. J., K. T. McCrickard, et al., *J. Chem. Thermodyn.* **1979,** *11,* 93.

6153. Richards, R. E., *J. Chem. Soc.* **1948,** 1931.

6154. Richards, R. E., and T. Schaefer, *Trans. Faraday Soc.* **1958,** *54,* 1447.

6155. Richards, R. E., and H. W. Thompson, *Proc. Roy. Soc.* (London) **1948,** *195A,* 1; *C.A.* **1949,** *43,* 4140.

6156. Richards, T. W., and F. Barry, *J. Amer. Chem.* **1915,** *37,* 993.

6157. Richards, T. W., and L. B. Coombs, *J. Amer. Chem. Soc.* **1915,** *37,* 1656.

6158. Richards, T. W., and J. H. Mathews, *J. Amer. Chem. Soc.* **1908,** *30,* 8.

6159. Richards, T. W., and J. W. Shipley, *J. Amer. Chem. Soc.* **1919,** *41,* 2002.

6160. Richards, W. T., and J. H. Wallace, *J. Amer. Chem. Soc.* **1932,** *54,* 2705.

6161. Richardson, G. M., and P. W. Robertson, *J. Chem. Soc.* **1928,** 1775.

6162. Richter, F., and W. Wolff, *Ber. Deut. Chem. Gesell* **1930,** *63B,* 1721; *C.A.* **1930,** *24,* 5032.

6163. Rico, M., M. Barrachina, et al., *J. Mol. Spectrosc.* **1967,** *24,* 133.

6164. Riddick, J. A., *Anal. Chem.* **1952,** *24,* 41; **1954,** *26,* 77; **1956,** *28,* 679; **1958,** *30,* 793; **1960,** *32,* 172.

6165. Riddick, J. A., unpublished data from Research and Development Dept., Commercial Solvents Corp.

6166. Riddick, J. A., and W. B. Bunger, calculated value.

6167. Riddick, J. A., and W. B. Bunger, *Organic Solvents*, 3rd ed. Wiley-Interscience, New York, 1970.

6168. Riddick, J. A., and J. Hampton, unpublished data from Research and Development Div., Commercial Solvents Corp.

6169. Riddle, E., "Monomeric Acrylic Esters," Reinhold, New York, 1954.

6170. Rieche, A., *Z. Angew. Chem.* **1931,** *44,* 896.

6171. Riedel, L., *Z. Geo. Kälti-Ind.* **1938,** *45,* 221; *C.A.* **1939,** *33,* 5735; *Chem. Zentr.* **1939,** *I,* 3349.

6172. Riethof, G., S. G. Richards, et al., *Ind. Eng. Chem., Anal. Ed.* **1946,** *18,* 458.

6173. Rigamonti, R., and G. Schiavina, *Chimica e Industria* (Milan) **1954,** *36,* 611; *C.A.* **1955,** *49,* 3510.

6174. Rigaux, R., J. F. Ramos, et al., *Can. J. Chem.* **1981,** *59,* 3305.

6176. Riggio, R., M. H. Ubeda, et al., *J. Chem. Eng. Data* **1980,** *25,* 318.

6177. Rihani, D. N., *Hydrocarbon Process.* **1968,** *47,* 111.

6178. Riiber, C. N., *Z. Electrochem.* **1923,** *29,* 334; *C.A.* **1923,** *17,* 3434.

6179. Riiber, C. N., T. Sorensen, et al., *Ber. Deut. Chem. Gesell.* **1925,** *58,* 964.

6180. Rikovski, I., and V. Carichi, *Glasnik Khem Drushtva, Beograd* **1957,** *22,* 87; *C.A.* **1960,** *54,* 2859.

6181. Riley, F. T., and K. C. Bailey, *Proc. Roy. Irish Acad.* **1929,** *38B,* 450; *Brit. Abstr.* **1929A,** 1137.

6182. Rimmer, B. F., *Chem. Ind.* (London) **1974,** *2,* 63.

6183. Rinkenbach, W. H., *Ind. Eng. Chem.* **1927,** *19,* 474.

6184. Rinkenbach, W. H., and H. A. Aaronson, *Ind. Eng. Chem.* **1931,** *23,* 160.

6185. Rinse, J. (to J. W. Ayers and Co.), U.S. Patent 2,886,586, May 12, 1959; *C.A.* **1959,** *53,* 17903.
6186. Rintelen, J. C., J. H. Saylor, et al., *J. Amer. Chem. Soc.* **1937,** *59,* 1129.
6187. Risgin, O., and R. C. Taylor, *Spectrochim. Acta* **1959,** *15,* 1036.
6188. Risseghem, H. Van, *Bull. Soc. Chim. Belges* **1938,** *47,* 194, 221.
6189. Risseghem, H. Van, *Bull. Soc. Chim. Fr.* **1959,** 1661; *C.A.* **1961,** *55,* 27248.
6190. Rizk, H. A., and I. M. Elanwar, *Can. J. Chem.* **1968,** *46,* 507.
6191. Rizk, H. A., N. Youssef, et al., *Can. J. Chem.,* **1969,** *47,* 3767.
6192. Robert, L., *Chim. Ind. Génie Chim.* **1967,** *97,* 337.
6193. Roberti, D. M., and C. P. Smyth, *J. Amer. Chem. Soc.* **1960,** *82,* 2106.
6194. Roberts, D. E., W. W. Walton, et al., *J. Res. Nat. Bur. Stand.* **1947,** *38,* 627.
6195. Roberts, G. N., *J. Amer. Chem. Soc.* **1953,** *75,* 2264.
6196. Roberts, H. M., and C. R. Bury, *J. Chem. Soc.* **1923,** 2037.
6197. Roberts, J. D., and V. C. Chambers, *J. Amer. Chem. Soc.* **1951,** *73,* 5030.
6198. Roberts, R., British Patent 698,127, Oct. 7, 1953; *C.A.* **1954,** *48,* 6459.
6199. Roberts, R. H., and S. E. J. Johnsen, *Anal. Chem.* **1948,** *20,* 690, 1225.
6200. Robertson, A., and D. G. Jones, British Patent 698,677, Oct. 21, 1953; *C.A.* **1954,** *48,* 5878.
6201. Robertson, N. C., U.S. Patent 2,477,087, July 25, 1949; *C.A.* **1949,** *43,* 9079.
6202. Robertson, P. W., N. T. Clare, et al., *J. Chem. Soc.* **1937,** 335.
6203. Robertson, R. E., and P. M. Laughton, *Can. J. Chem.* **1957,** *35,* 1319.
6204. Robertson, W. W., N. Ginsburg, et al., *Ind. Eng. Chem., Anal. Ed.* **1946,** *18,* 746.
6205. Robertson, W. W., and F. A. Matsen, *J. Amer. Chem. Soc.* **1950,** *72,* 5248.
6206. Robertson, W. W., and F. A. Matsen, *J. Amer. Chem. Soc.* **1950,** *72,* 5250.
6207. Robertson, W. W., and F. A. Matsen, *J. Amer. Chem. Soc.* **1950,** *72,* 5252.
6208. Robertson, W. W., F. A. Matsen, et al., *Phys. Rev.* **1947,** *73,* 652; *C.A.* **1949,** *43,* 4094.
6209. Robertson, W. W., J. F. Music, et al., *J. Amer. Chem. Soc.* **1950,** *72,* 5260.
6210. Robertson, W. W., A. J. Seriff, et al., *J. Amer. Chem. Soc.* **1950,** *72,* 1539.
6211. Robinson, R. A., and A. K. Kiang, *Trans. Faraday Soc.* **1956,** *52,* 327.
6212. Robles, H. de V., *Rec. Trav. Chim. Pays-Bas* **1939,** *58,* 111; *C.A.* **1939,** *33,* 2785.
6213. Robrock, B. G., and R. W. Kiser, *J. Phys. Chem.* **1961,** *65,* 2186.
6214. Rochester, C. H., *Trans. Faraday Soc.* **1966,** *62,* 355.
6215. Rochester, C. H., and J. R. Symonds, *J. Chem. Soc., Faraday Trans.* **1973,** *69,* 1267.
6216. Rochester, C. H., and D. N. Wilson, *J. Chem. Soc., Faraday Trans. 1* **1976,** *72,* 2930.
6217. Röck, H., *Naturwissenschaften* **1956,** *43,* 81; Ref. [3321], [8056].
6218. Rockenfeller, J. D., and F. D. Rossini, *J. Phys. Chem.* **1961,** *65,* 267.
6219. Roddy, J. W., and C. F. Coleman, *Talanta* **1968,** *15,* 1281.
6220. Roddy, J. W., and J. Mrochek, *J. Inorg. Nucl. Chem.* **1966,** *28,* 3019.
6221. Rodgers, A. S., J. Chao, et al., *J. Phys. Chem. Ref. Data* **1974,** *3,* 117.
6222. Rodgers, R. C., and G. E. Hill, *Br. J. Anaesth.* **1978,** *50,* 415.
6223. Rodionov, P. P., and G. I. Borodkin, *Izv. Vyssh. Ucheb. Zaved., Khim. Khim. Tekhnol.* **1969,** *12,* 1214; *C.A.* **1970,** *72,* 25799.
6224. Rodionov, V., *Bull. Soc. Chim. Fr.* **1926,** *30,* 305; *C.A.* **1926,** *20,* 1795.
6225. Rodriguez, A. T., and I. A. McLure, *J. Chem. Thermodyn.* **1979,** *11,* 1113.
6226. Rodziewicz, W., *Rocz. Chem.* **1965,** *39,* 539; *C.A.* **1965,** *63,* 8169.
6227. Rogers, D. W., and N. A. Siddiqui, *J. Phys. Chem.* **1975,** *79,* 574.
6228. Rogers, M. T., *J. Amer. Chem. Soc.* **1947,** *69,* 457.
6229. Rogers, M. T., *J. Amer. Chem. Soc.* **1947,** *69,* 1243.
6230. Rogers, M. T., *J. Amer. Chem. Soc.* **1947,** *69,* 2544.
6231. Rogers, M. T., *J. Amer. Chem. Soc.* **1955,** *77,* 3681.
6232. Rogers, M. T., and T. W. Campbell, *J. Amer. Chem. Soc.* **1953,** *75,* 1209.
6233. Rogers, M. T., and M. B. Panish, *J. Amer. Chem. Soc.* **1955,** *77,* 3684.
6234. Rogers, M. T., and M. B. Panish, *J. Amer. Chem. Soc.* **1955,** *77,* 4230.
6235. Röhler, H., *Z. Elektrochem., 16,* 419; *C.A.* **1910,** *4,* 2597.
6236. Rohm & Haas Company, "Acrylic and Methacrylic Acids and Esters," Philadelphia, 1963.
6237. Rohm & Haas Company, Industrial Chemicals Dept., "Methylamines," Philadelphia, 1965.
6238. Rohm & Haas Company, Special Products Dept., "Glacial Methacrylic Acid–Glacial Acrylic Acid," Philadelphia, 1960.

6239. Rohm & Haas Company, Special Products Dept., "*t*-Alkyl Primary Amines," Philadelphia, 1963.

6240. Rohrschneider, L., *Anal. Chem.* **1973,** *45,* 1241.

6241. Roland, M., *Bull. Soc. Chim. Belges* **1928,** *37,* 117.

6242. Rolla, M., *Atti Accad. Italia, Rend. Cl. Sci. Fis., Mat. Nat.* **1940,** *1,* No. 7, 756; *C.A.* **1943,** *37,* 832.

6243. Rollett, A., and A. Leimuller, *Monatsh. Chem.* **1940,** *73,* 197; *C.A.* **1941,** *35,* 4347.

6244. Rollier, M. A., *Gazz. Chim. Ital.* **1947,** *77,* 366; *C.A.* **1948,** *42,* 2826.

6245. Romadane, I., and J. Pelchers, *Zh. Obshch. Khim.* **1959,** *29,* 103; *C.A.* **1959,** *53,* 18890.

6246. Román, U. D., and A. M. Górriz, *An. Real. Soc. Espan. Fis. Quim.* (Madrid) **1958,** *54B,* 559; *C.A.* **1959,** *53,* 16628.

6247. Romanova, A. D., and V. A. Drozdov, *Tr. Mosk. Khim-Tekhnol. Inst.* **1969,** No. 62, 242; *C.A.* **1970,** *73,* 109119.

6248. Romans, J. B., and C. R. Singleterry, *J. Chem. Eng. Data* **1961,** *6,* 56.

6249. Rometsch, R., A. Marxer, et al., *Helv. Chim. Acta* **1951,** *34,* 1611.

6250. Roper, E. E., *J. Amer. Chem. Soc.* **1938,** *60,* 1693.

6251. Ropp, G. A., and C. E. Melton, *J. Amer. Chem. Soc.* **1958,** *80,* 3509.

6252. Rosado-Lojo, O., C. K. Hancock, et al., *J. Org. Chem.* **1966,** *31,* 1899.

6253. Rosanoff, M. A., and R. A. Dunphy, *J. Amer. Chem. Soc.* **1914,** *36,* 1411.

6254. Roschdestwensky, M. S., A. Pukirew, et al., *Trans. Inst. Pure Chem. Reagents* (USSR) **1935,** *14,* 58; *C.A.* **1936,** *30,* 597.

6255. Rose, A., J. A. Acciarri, et al., *Ind. Eng. Chem.* **1957,** *49,* 104.

6256. Rose, A., J. A. Acciarri, et al., *Ind. Eng. Chem., Chem. Eng. Data Series* **1958,** *3,* 210.

6257. Rose, A., B. T. Papahornis, et al., *Ind. Eng. Chem., Chem. Eng. Data Series* **1958,** *3,* 216.

6258. Rose, A., and E. Rose, *The Condensed Chemical Dictionary*, 6th ed., Reinhold, New York, 1961.

6259. Rose, A., and W. R. Suoina, *J. Chem. Eng. Data* **1961,** *6,* 173.

6260. Rose, F. W., *J. Res. Nat. Bur. Stand.* **1937,** *19,* 143.

6261. Rosenbaum, E. J., A. V. Grosse, et al., *J. Amer. Chem. Soc.* **1939,** *61,* 689.

6262. Rosenbaum, J., and M. C. R. Symons, *J. Chem. Soc.* **1961,** 1.

6263. Rosin, J., *Reagent Chemicals and Standards*, 5th ed., Van Nostrand, Princeton, 1967.

6264. Ross, A. A., and F. E. Bibbins, *Ind. Eng. Chem.* **1939,** *31,* 255.

6265. Ross, G. R., and W. J. Heideger, *J. Chem. Eng. Data* **1962,** *7,* 505.

6266. Ross, G. S., and H. D. Dixon, *J. Res. Nat. Bur. Stand.* **1963,** *67A,* 247.

6267. Ross, G. S., and L. J. Frolen, *J. Res. Nat. Bur. Stand.* **1963,** *67A,* 607.

6268. Ross, G. S., and A. R. Glasgow, *Anal. Chem.* **1964,** *36,* 700.

6269. Ross, J. D. M., and I. C. Somerville, *J. Chem. Soc.* **1926,** 2770.

6270. Ross, S., *J. Amer. Chem. Soc.* **1946,** *68,* 1484.

6271. Ross, S., and R. E. Patterson, *J. Chem. Eng. Data* **1979,** *24,* 111.

6272. Ross, S. D., and M. M. Labes, *J. Amer. Chem. Soc.* **1957,** *79,* 4155.

6273. Ross, S. D., M. Markarian, et al., *J. Amer. Chem. Soc.* **1953,** *75,* 4967.

6274. Rossini, F. D., *Anal. Chem.* **1948,** *20,* 110.

6275. Rossini, F. D., *J. Res. Nat. Bur. Stand.* **1933,** *11,* 553; *C.A.* **1934,** *28,* 961.

6276. Rossini, F. D., *J. Res. Nat. Bur. Stand.* **1934,** *12,* 735.

6277. Rossini, F. D., *J. Res. Nat. Bur. Stand.* **1934,** *13,* 189.

6278. Rossini, F. D., *Pure Appl. Chem.* **1964,** *9,* 453.

6279. Rossini, F. D., B. J. Mair, et al., *Hydrocarbons from Petroleum*, Reinhold, New York, 1953.

6280. Rossini, F. J., *J. Res. Nat. Bur. Stand.* **1935,** *15,* 357.

6281. Rotariu, G. J., R. J. Hanrahan, et al., *J. Amer. Chem. Soc.* **1954,** *76,* 3752.

6282. Rotenberg, Y. S., and F. D. Mashbits, *Gig. Tr. Prof. Zabol.* **1967,** *11,* 26.

6283. Roth, R., *Diplomarbeiten* **1956,** from [6408].

6284. Roth, W. A., and I. Meyer, *Z. Elektrochem.* **1933,** *39,* 35; *Chem. Zentr.* **1933,** I, 1418.

6285. Rothenbach, E., *Wochschr. Brau.* **1939,** *56,* 273; *C.A.* **1940,** *34,* 6076.

6286. Rotinjantz, L., and N. Nagornow, *Z. Phys. Chem.* (Leipzig) **1934,** *169A,* 20; *C.A.* **1934,** *28,* 5743.

6287. Rouleau, D. J., and A. R. Thompson, *J. Chem. Eng. Data* **1962,** *7,* 356.

6288. Rousselot, M. M., and M. Martin, *Compt. Rend., Ser. C* **1966,** *262,* 1445; *C.A.* **1966,** *65,* 6545.

6289. Rowe, J. J. M., K. B. Gibney, et al., *J. Amer. Chem. Soc.* **1968,** *90,* 1924.

6290. Rowe, V. K., H. C. Spencer, et al., *J. Ind. Hyg. Toxicol.* **1948,** *30,* 332.

6291. Rowley, H. H., and W. R. Reed, *J. Amer. Chem. Soc.* **1951,** *73,* 2960.

6292. Roy, N. K., *Indian J. Phys.* **1977,** *51B*, 65.
6293. Rozental, D., *Bull. Soc. Chim. Belges* **1936,** *45*, 585; *C.A.* **1937,** *31*, 1686.
6294. Rozett, R. W., and E. M. Petersen, *Anal. Chem.* **1975,** *47*, 2377.
6295. Rubin, H., and R. E. Swarbrick, *Anal. Chem.* **1961,** *33*, 217.
6296. Rubin, L. J., J. A. Lardy, et al., *J. Amer. Chem. Soc.* **1952,** *74*, 425.
6297. Rubin, T. R., B. H. Le Vedahl, et al., *J. Amer. Chem. Soc.* **1944,** *66*, 279.
6298. Rubio, R. G., J. A. R. Renuncio, et al., *J. Chem. Thermodyn.* **1983,** *15*, 779.
6299. Ruby, W. R., and R. P. Loveland, *J. Phys. Chem.* **1946,** *50*, 345.
6300. Ruch, J. E., and F. E. Critchfield, *Anal. Chem.* **1961,** *33*, 1569.
6301. Rudakov, G. A., and S. Ya. Korotov, *J. Appl. Chem.* (USSR) **1937,** *10*, 312; *C.A.* **1937,** *31*, 4554.
6302. Rudakova, S. E., and Yu. A. Pentin, *Optika i Spectroskopiya* **1965,** *18*, 592; *C.A.* **1965,** *63*, 6472.
6303. Rudenko, V. A., A. Ya. Yakubovich, et al., *J. Gen. Chem.* (USSR) **1947,** *17*, 2256; *C.A.* **1948,** *42*, 4918.
6304. Rudolph, H. D., and H. Seiler, *Z. Naturforsch.* **1965,** *20A*, 1682; *C.A.* **1966,** *64*, 13573.
6305. Rueggeberg, W. H. C., M. L. Cushing, et al., *J. Amer. Chem. Soc.* **1946,** *68*, 191.
6306. Ruehrwein, R. A., and H. M. Huffmann, *J. Amer. Chem. Soc.* **1943,** *65*, 1620.
6307. Ruff, O., and O. Bretschneider, *Z. Anorg. Allg. Chem.* **1933,** *210*, 173; *Chem. Zentr.* **1933,** *I*, 1925.
6308. Ruff, O., and H. Golla, *Z. Anorg. Allg. Chem.* **1924,** *138*, 17.
6309. Ruhoff, J. R., and E. E. Reid, *J. Amer. Chem. Soc.* **1937,** *59*, 401.
6310. Rummens, F. H. A., and J. W. DeHaan, *Org. Magn. Resonance* **1970,** *2*, 351.
6311. Rumpf, K., and R. Mecke, *Z. Phys. Chem.* (Leipzig) **1939,** *44B*, 299; *C.A.* **1940,** *34*, 676.
6312. Rumyantsev, A. P., N. A. Ostroumova, et al., *Khim. Prom-st., Ser.: Toksikol. Sanit. Khim. Plastmass* **1979,** 24; *C.A.* **1980,** *92*, 175337.
6313. Ruppol, E., *Bull. Acad. Roy. Belges Cl. Sci.* **1935,** *21*, 236; *Chem. Zentr.* **1935,** *II*, 2659; *Bull. Soc. Chim. Belges* **1935,** *44*, 551; *Chem. Zentr.* **1936,** *I*, 3316.
6314. Rusanov, A. I., S. A. Levichev, et al., *Zh. Fiz. Khim.* **1969,** *43*, 2636; *Russ. J. Phys. Chem.* **1969,** *43*, 1481.
6315. Rusoff, I. I., J. R. Platt, et al., *J. Amer. Chem. Soc.* **1945,** *67*, 673.
6316. Russell, H., D. R. V. Golding, et al., *J. Amer. Chem. Soc.* **1944,** *66*, 16.
6317. Rutledge, G. P., and W. T. Smith, *J. Amer. Chem. Soc.* **1953,** *75*, 5762.
6318. Ruzicka, L., *Helv. Chim. Acta* **1919,** *2*, 144; *C.A.* **1919,** *13*, 1320.
6319. Ryland, G., *Amer. Chem. J.* **1899,** *22*, 384.
6320. Ryskin, Y. I., and M. G. Voronkov, *Khim. i Primenenie Kremneorg. Soedinenii, Trudy Konf.* (Leningrad) **1958,** No. 3, *42*; *C.A.* **1959,** *53*, 12832.

S

6321. Sabatier, P., and A. Mailhe, *Ann. Chim. Phys.* **1907,** [8], *10*, 527; *Chem. Zentr.* **1907,** *I*, 1695; *C.A.* **1908,** *2*, 793.
6322. Sabatier, P., and A. Mailhe, *Compt. Rend.* **1904,** *138*, 1321; *J. Chem. Soc.* **1904Ai,** 666.
6323. Sabatier, P., and A. Mailhe, *Compt. Rend.* **1905,** *141*, 21.
6324. Sabatier, P., and A. Mailhe, *Compt. Rend.* **1907,** *144*, 1068; *C.A.* **1907,** *1*, 2234.
6325. Sabatier, P., and M. Murat, *Compt. Rend.* **1912,** *154*, 1390; *C.A.* **1912,** *6*, 2082.
6326. Sabatier, P., and J. B. Senderens, *Compt. Rend.* **1901,** *132*, 568.
6327. Sabatier, P., and J. B. Senderens, *Compt. Rend.* **1902,** *134*, 1127.
6328. Sabatier, P., and J. B. Senderens, *Compt. Rend.* **1903,** *136*, 985.
6329. Sabatier, P., and J. B. Senderens, *Compt. Rend.* **1903,** *137*, 302; *Chem. Zentr.* **1903,** *II*, 708.
6330. Sabatier, P., and J. B. Senderens, *Compt. Rend.* **1904,** *138*, 457.
6331. Sabbah, R., R. Chastel, et al., *Thermochim. Acta* **1972,** *5*, 117.
6332. Sachek, A. I., A. D. Peshchenko, et al., *Vestsi Akad. Navuk BSSR, Ser. Khim. Navuk* **1978,** 124; *C.A.* **1978,** *89*, 5775.
6333. Sachek, A. I., A. D. Peshchenko, et al., *Zh. Fiz. Khim.* **1974,** *48*, 1057; *Russ. J. Phys. Chem.* **1974,** *48*, 617.
6334. Sachs, F., *Ber. Deut. Chem. Gesell.* **1901,** *34*, 494.
6335. Sackman, H., and F. Sauerwald, *Z. Phys. Chem.* (Leipzig) **1950,** *195*, 295; *C.A.* **1951,** *45*, 8314.
6336. Safir, L. I., *Izv. Vyssh. Uchebn. Zaved., Neft Gaz* **1978,** *21*, 14, 82; *C.A.* **1979,** *91*, 45225.
6337. Sage, B. H., and W. N. Lacey, *Ind. Eng. Chem.* **1938,** *30*, 673.

6338. Sagitova, E. V., F. K. Tukhvatullin, et al., *Opt. Spectrosk.* **1970**, 465; *C.A.* **1970**, *73*, 65578.
6339. Sah, P. P. T., and S.-L. Chien, *J. Amer. Chem. Soc.* **1931**, *53*, 3901.
6340. Sah, P. P. T., and H. H. Lei, *Sci. Repts. Nat. Tsing Hua Univ. Ser. A* **1932**, *1*, 193; *Beil.* **EIII6**, 1454.
6341. Sahini, V. E., V. Constantin, et al., *Rev. Roum. Chim.* **1967**, *12*, 1405.
6342. Sahney, R. C., R. M. Beri, et al., *J. Indian Chem. Soc.* **1949**, *26*, 329.
6343. Saidel, L. J., *Arch. Biochem. Biophys.* **1955**, *54*, 184.
6344. Saidel, L. J., *Nature* **1953**, *172*, 955.
6345. Saier, E. L., L. R. Cousins, et al., *Anal. Chem.* **1962**, *34*, 824.
6346. Saint-Gilles, L. P. de, *Ann. Chim.* **1859**, *55*, 396.
6347. Sakaguchi, M., A. Hirakata, et al., *Koryo* **1972**, No. 102, 41; *C.A.* **1973**, *78*, 84556.
6348. Sakai, S., and N. Tsunakawa, Japanese Patent 6757, May 22, 1963; *C.A.* **1963**, *59*, 11435.
6349. Sakakibara, M., F. Inagaki, et al., *Bull. Chem. Soc. Jpn.* **1976**, *49*, 46.
6350. Sakamoto, I., K. Masuda, et al., *Bunseki Kagaku* **1982**, *31*, E49; *C.A.* **1982**, *96*, 111123.
6351. Sakharova, L. N., and M. S. Tolgskaya, *Gig. Tr. Prof. Zabol.* **1977**, 36; *C.A.* **1977**, *87*, 63600.
6352. Sakiadis, B. C., and J. Coates, *AIChEJ* **1955**, *1*, 275.
6353. Saksena, B. D., *Proc. Indian Acad. Sci.* **1939**, *10A*, 333; *C.A.* **1940**, *34*, 3173.
6354. Saksena, B. D., *Proc. Indian Acad. Sci.* **1940**, *12A*, 312; *C.A.* **1941**, *35*, 3172.
6355. Saksena, B. D., *Proc. Indian Acad. Sci.* **1940**, *12A*, 321; *C.A.* **1941**, *35*, 3173.
6356. Saksena, B. D., *Proc. Indian Acad. Sci.* **1940**, *12A*, 416; *C.A.* **1941**, *35*, 3173.
6357. Sakurai, M., and T. Nakagawa, *J. Chem. Thermodyn.* **1984**, *16*, 171.
6358. Salier, B., and J. Paviot, *J. Chim. Phys. Physico-Chim. Biol.* **1968**, *65*, 1676.
6359. Salimov, M. A., L. A. Erivanskaya, et al., *Azerb. Khim. Zhur.* **1961**, No. 3, 123; No. 4, 93; *C.A.* **1962**, *56*, 8184.
6360. Salimov, M. A., E. A. Viktorova, et al., *Azerb. Khim. Zhur.* **1960**, 99; *C.A.* **1961**, *55*, 19481.
6361. Salmi, E. J., and E. Vaihkonen, *Suomen Kemistilehti* **1946**, *19B*, 132; *C.A.* **1947**, *41*, 5481.
6362. Salmon, M., E. Cortes, et al., *Bol. Inst. Quim. Univ. Nacl. Auton. Mex.* **1965**, *17*, 151; *C.A.* **1966**, *65*, 13506.
6363. Salomon, M., *J. Phys. Chem.* **1969**, *73*, 3299.
6364. Salonen, A. K., *Ann. Acad. Sci. Fennicae Ser.* **1961**, *A6*, No. 67; *C.A.* **1961**, *55*, 20624.
6365. Samarina, A. V., V. I. Arulin, et al., *Tr. Khim. Khim. Tekhnol.* **1967**, No. 2, 148; *C.A.* **1969**, *70*, 77270.
6366. Samitov, Y. Y., and R. M. Aminova, *Zh. Strukt. Khim.* **1964**, *5*, 538; *C.A.* **1964**, *61*, 15555.
6367. Sammis, J. L., *J. Phys. Chem.* **1906**, *10*, 593.
6368. Sammul, O. R., W. L. Brannon, et al., *J. Assoc. Offic. Agr. Chemist* **1964**, *47*, 918.
6369. Samorukov, O. P., and V. N. Kostryukov, *Russ. J. Phys. Chem.* (Eng. trans.) **1971**, *45*, 747.
6370. Samsanova, N. I., N. S. Kalyazina, et al., *Tr. Vses. Nauchn.-Issled. Inst. Khim. Reaktivov* **1963**, No. 25, 434; *C.A.* **1964**, *61*, 2984.
6371. Sanders, P. A., *Principles of Aerosol Technology*, Van Nostrand, New York.
6372. Sanemasa, I., M. Araki, et al., *Bull. Chem. Soc. Jpn.* **1982**, *55*, 1054.
6373. Sanger, R., *Helv. Phys. Acta* **1931**, 3, 162; *C.A.* **1931**, *25*, 4452.
6374. Sanger, R., *Phys. Z.* **1926**, *27*, 556.
6375. Sangers, J. M. H. L., W. L. Greer, et al., *J. Phys. Chem. Ref. Data* **1976**, *5*, 1.
6376. Sangewald, R., and A. Weissberger, *Phys. Z.* **1929**, *30*, 268.
6377. San Jose, J. L., G. Mellinger, et al., *J. Chem. Eng. Data* **1976**, *21*, 414.
6378. Sano, T., Y. Ogo, et al., *Mem. Fac. Eng., Osaka City Univ.* **1971**, *12*, 133; *C.A.* **1972**, *77*, 168908.
6379. Santaniello, E., P. Ferraboschi, et al., *Synthesis* **1980**, 646.
6380. Santavy, F., D. Walterova, et al., *Collect. Czech. Chem. Commun.* **1972**, *37*, 1825.
6381. Santoro, E., *Org. Mass Spectrom.* **1973**, *7*, 589.
6382. Sanyal, S. B., *Indian J. Phys.* **1950**, *24*, 151; *C.A.* **1951**, *45*, 450.
6383. Sapgir, S., *Bull. Soc. Chim. Belges* **1929**, *38*, 392.
6384. Sappenfield, J. W., *Phys. Rev.* **1929**, *33*, 37; *Chem. Zentr.* **1929**, *I*, 1419.
6385. Sarancha, E. T., and A. P. Dubovikova, *Zavodsk. Lab.* **1961**, *27*, 398; *C.A.* **1962**, *57*, 6588.
6386. Sargent, J. W., and J. D. La Zerte, *ASTM Special Tech. Publ.* **1964**, No. 346, 51.
6387. Sasa, T., *J. Soc. Org. Synthesis Chem.* (Japan) **1953**, *11*, 463; *C.A.* **1955**, *49*, 919.
6388. Sastry, C. R., and K. R. Rao, *Indian J. Phys.* **1947**, *21*, 313; *C.A.* **1948**, *42*, 6237.

6389. Sauer, R. O., *J. Amer. Chem. Soc.* **1946,** *68,* 954.
6390. Sauer, R. O., and E. M. Hadsell, *J. Amer. Chem. Soc.* **1948,** *70,* 3590.
6391. Saunders, H. F., and L. T. Sutherland, U.S. Patent 1,362,355, Dec. 14, 1921; *C.A.* **1921,** *15,* 537.
6392. Saunders, R. A., and A. E. Williams, *Advn. Mass Spectrom., Proc. Conf.* **1966,** *3,* 681; *C.A.* **1968,** *69,* 100938.
6393. Saunders, R. H., M. J. Murray, et al., *J. Amer. Chem. Soc.* **1943,** *65,* 1309.
6394. Savard, J., *Ann. Chim.* (Paris) **1929,** *11,* 287.
6395. Savard, J., *Compt. Rend.* **1928,** *187,* 540; *C.A.* **1929,** *23,* 128.
6396. Savedoff, L. G., *J. Amer. Chem. Soc.* **1966,** *88,* 664.
6397. Savidan, L., *Bull. Soc. Chim. Fr.* **1953,** *411,* 533.
6398. Savkovic-Stevanovic, J., and D. Simonovic, *J. Chem. Eng. Data* **1976,** *21,* 456.
6399. Sawodny, W., K. Niedenzu, et al., *Spectrochim. Acta, Part A* **1967,** *23,* 799.
6400. Sawyer, D. T., and J. R. Brannan, *Anal. Chem.* **1966,** *38,* 192.
6401. Sawyer, R. A., *Experimental Spectroscopy*, Prentice-Hall, New York, 1944.
6402. Sax, N. I., *Dangerous Properties of Industrial Materials*, Reinhold, New York, 1963.
6403. Saxton, B., and G. Waters, *J. Amer. Chem. Soc.* **1937,** *59,* 1048.
6404. Saylor, J. H., V. J. Baxt, et al., *J. Amer. Chem. Soc.* **1942,** *64,* 2742.
6405. Scaife, W. G., *J. Phys. A* **1972,** *5,* 897.
6406. Scatchard, G., *Chem. Rev.* **1931,** *8,* 321.
6407. Scatchard, G., and C. L. Raymond, *J. Amer. Chem. Soc.* **1938,** *60,* 1278.
6408. Scatchard, G., and G. M. Wilson, *J. Amer. Chem. Soc.* **1964,** *86,* 133.
6409. Scatchard, G., and S. E. Wood, *J. Phys. Chem.* **1939,** *43,* 119.
6410. Scattergood, A., W. H. Miller, et al., *J. Amer. Chem. Soc.* **1945,** *67,* 2150.
6411. Schaaffs, W., *Z. Phys. Chem.* (Leipzig) **1944,** *194,* 170; *C.A.* **1948,** *42,* 8550.
6412. Schaake, R. C. F., J. C. Van Miltenburg, et al., *J. Chem. Thermodyn.* **1982,** *14,* 763.
6413. Schaake, R. C. F., J. C. Van Miltenburg, et al., *J. Chem. Thermodyn.* **1982,** *14,* 771.
6414. Schaap, W. B., R. E. Bayer, et al., *Record Chem. Progr.* **1961,** *22,* 197.
6415. Schaarschmidt, K., *Z. Anorg. Allg. Chem.* **1961,** *310,* 78; *C.A.* **1962,** *56,* 2899.
6416. Schadow, E., and R. Steiner, *Z. Phys. Chem.* (Frankfurt) **1969,** *66,* 105.
6417. Schaefer, T., W. Danchura, et al., *Can. J. Chem.* **1978,** *56,* 2233.
6418. Schaefer, T., and W. G. Schneider, *Can. J. Chem.* **1959,** *37,* 2078; *J. Chem. Phys.* **1960,** *32,* 1218.
6419. Schaefer, T., and W. G. Schneider, *J. Chem. Phys.* **1960,** *32,* 1224.
6420. Schafer, Sr. M., and C. Curran, *Inorg. Chem.* **1965,** *4,* 623.
6421. Schafer, Sr. M., and C. Curran, *Inorg. Chem.* **1966,** *5,* 265.
6422. Schall, C., and C. Thieme-Wiedtmarckter, *Z. Electrochem.* **1929,** *35,* 337.
6423. Schalla, R. L., and G. E. McDonald, *Nat. Advisory Comm. Aeronaut. Tech. Note* **1955,** No. 3405; *C.A.* **1955,** *49,* 7248.
6424. Schamp, N., and M. Vandewalle, *Bull. Soc. Chim. Belges* **1966,** *75,* 539.
6425. Schatzberg, P., *J. Phys. Chem.* **1963,** *67,* 776.
6426. Schauffle, R. F., and L. Goodman, *Trans. Faraday Soc.* **1965,** *61,* 597.
6427. Scheibe, G., *Ber. Deut. Chem. Gesell.* **1925,** *58,* 586; *C.A.* **1925,** *19,* 1662.
6428. Scheibe, G., and H. Grieneisen, *Z. Phys. Chem.* (Leipzig) **1934,** *25B,* 52.
6429. Scheibe, G., F. May, et al., *Ber. Deut. Chem. Gesell.* **1924,** *57,* 1330.
6430. Scheibler, H., and J. Magasanik, *Ber. Deut. Chem. Gesell.* **1915,** *48,* 1810.
6431. Schep, R. A., S. Norval, et al., *Inorg. Chem.* **1973,** *12,* 2711.
6432. Scheraga, H. A., and M. E. Hobbs, *J. Amer. Chem. Soc.* **1948,** *70,* 3015.
6433. Scherer, J. R., and J. C. Evans, *Spectrochim. Acta* **1963,** *19,* 1739; *C.A.* **1963,** *59,* 13482.
6434. Scherer, O., F. Dostal, et al., German Patent 727,690, Oct. 8, 1942; *C.A.* **1943,** *37,* 6676.
6435. Schiemann, G., *Z. Phys. Chem.* (Leipzig) **1931,** *156A,* 397.
6436. Schierholtz, O. J., and M. L. Staples, *J. Amer. Chem. Soc.* **1935,** *57,* 2709.
6437. Schiessler, R. W., C. H. Herr, et al., *Proc. Amer. Petroleum Inst.* **1946,** *26III,* 254.
6438. Schiff, R., *Ann. Chim.* (Paris) **1884,** *220,* 278; *J. Chem. Soc.* **1884A,** 386.
6439. Schiff, R., *Ann. Chim.* (Paris) **1886,** *234,* 300; *J. Chem. Soc.* **1887A,** 6.
6440. Schiff, R., *Z. Phys. Chem.* (Leipzig) **1887,** *1,* 376.
6441. Schildknecht, C. E., *Vinyl and Related Polymers*, Wiley, New York, 1952.

6442. Schildknecht, C. E., A. O. Zoss, et al., *Ind. Eng. Chem.* **1947**, *39*, 180.
6443. Schildknecht, H., *Chimia* (Aarua) **1963**, *17*, 145; *Chem. Zentr.* **1965**, 1738; Ref. [8056].
6444. Schildknecht, H., Z. *Anal. Chem.* **1961**, *181*, 254; Ref. [3321].
6445. Schildknecht, H., *Zonenschmelzen*, Verlag Chemie, Weiheim/Bergstr., 1964.
6446. Schildknecht, H., and V. Hopf, Z. *Anal. Chem.* **1963**, *193*, 401; *C.A.* **1963**, *59*, 6324.
6447. Schildknecht, H., and H. Vetter, *Angew. Chem.* **1961**, *73*, 240; *C.A.* **1961**, *55*, 22934.
6448. Schiller, A., Z. *Phys.* **1937**, *105*, 175; *C.A.* **1937**, *31*, 5271.
6449. Schindlbauer, H., and G. Hajek, *Chem. Ber.* **1963**, *96*, 2601.
6450. Schioper, I. M., O. Bot, et al., *Bul. Mst. Politeh. Iasi* **1961**, *7*, 115; *C.A.* **1963**, *58*, 10793.
6451. Schischkoff, L., and A. Rosing, *Jahr. Fort. Chemie* **1858**, 279.
6452. Schjanberg, E., Z. *Phys. Chem.* (Leipzig) **1935**, *172A*, 197.
6453. Schläfer, H. L., and W. Schaffernicht, *Angew. Chem.* **1960**, *72*, 618; *Chem. Zentr.* **1961**, 2543; *C.A.* **1961**, *55*, 14036.
6454. Schlegel, H., *J. Chim. Phys.* **1934**, *31*, 517.
6455. Schlesinger, H. I., and A. W. Martin, *J. Amer. Chem. Soc.* **1914**, *36*, 1589.
6456. Schlundt, H., *J. Phys. Chem.* **1901**, *5*, 157; *Chem. Zentr.* **1901**, *I*, 1135.
6457. Schlundt, H., *J. Phys. Chem.* **1901**, *5*, 503; *Chem. Zentr.* **1902**, *I*, 3.
6458. Schmerling, L., U.S. Patent 2,385,303, Sept. 18, 1945; *C.A.* **1946**, *40*, 601.
6459. Schmidt, A. W., V. Schoeller, et al., *Ber. Deut. Chem. Gesell.* **1941**, *74B*, 1313; *C.A.* **1942**, *36*, 5131.
6460. Schmidt, E., *Ber. Deut. Chem. Gesell.* **1919**, *52*, 400.
6461. Schmidt, G. C., Z. *Phys. Chem.* (Leipzig) **1926**, *121*, 221.
6462. Schmidt, M., thesis (1934), Paris; *Annual Table of Physical Constants*, Sec. 514(c), 1941; 513(c), 1941.
6463. Schmidt, M. R., and H. C. Jones, *Amer. Chem. J.* **1909**, *42*, 37.
6464. Schmidt, O., Z. *Phys. Chem.* (Leipzig) **1907**, *58*, 513.
6465. Schmidt, R. L., J. C. Randall, et al., *J. Phys. Chem.* **1966**, *70*, 3912.
6466. Schmidt, W., and K. Seydel, German Patent 872,342, March 30, 1953.
6467. Schmitz-Dumont, W., *Chem.-Ztg.* **1897**, *21*, 511.
6468. Schmulbach, C. D., and R. S. Drago, *J. Amer. Chem. Soc.* **1960**, *82*, 4484.
6469. Schnaible, H. W., and J. M. Smith, *Chem. Eng. Progr. Symp. Ser.* **1953**, *49*, No. 7, 159.
6470. Schneider, G., and G. Wilhelm, Z. *Phys. Chem.* (Frankfurt) **1959**, *20*, 219; *Chem. Zentr.* **1960**, 1453.
6471. Schneider, R. L., *Eastman Organic Chemical Bulletin* **1975**, *47*, No. 1, Eastman Kodak Co., Rochester, New York.
6472. Schnell, E., and E. G. Rochow, *J. Inorg. Nucl. Chem.* **1958**, *6*, 303.
6473. Schniepp, L. E., and H. H. Geller, *J. Amer. Chem. Soc.* **1946**, *68*, 1646.
6474. Schnurmann, R., and W. J. Morris, *Research* (London) **1949**, *2*, Suppl. 394; *C.A.* **1950**, *44*, 3759.
6475. Schnurmann, R., and S. Whincup, *Petroleum* (London) **1945**, *8*, 122, 142; *C.A.* **1946**, *40*, 4000.
6476. Schober, G., and V. Gutmann, *Monatsh. Chem.* **1958**, *89*, 649; *C.A.* **1959**, *53*, 7826; Ref. [5307].
6477. Scholl, R., and H. Berblinger, *Ber. Deut. Chem. Gesell.* **1903**, *36*, 3427.
6478. Scholte, T. G., *Physica* **1949**, *15*, 450; *C.A.* **1950**, *44*, 3319.
6479. Schoolery, J. N., and B. Crawford, *J. Mol. Spectrosc.* **1957**, *1*, 270.
6480. Schorger, A. W., *J. Ind. Eng. Chem.* **1914**, *6*, 631.
6481. Schorigin, P., *Ber. Deut. Chem. Gesell.* **1923**, *56*, 186.
6482. Schorigin, P., and Y. Makaroff-Semljanski, *Ber. Deut. Chem. Gesell.* **1932**, *65*, 1293.
6483. Schornack, L. G., and C. A. Eckert, *J. Phys. Chem.* **1970**, *74*, 3014.
6484. Schott, H., *J. Chem. Eng. Data* **1961**, *6*, 19.
6485. Schouteden, F., and C. J. Deveux, *Natuurwetensch. Tijds.* **1936**, *18*, 242.
6486. Schrader, B., and W. Meier, *Fresenius Z. Anal. Chem.* **1972**, *260*, 248.
6487. Schraiber, L. S., *Isr. J. Chem.* **1975**, *13*, 185.
6488. Schramm, T., *Monatsh. Chem.* **1888**, *9*, 616.
6489. Schrauth, W., and K. Gorig, *Ber. Deut. Chem. Gesell.* **1923**, *56B*, 1900.
6490. Schreiber, K. C., *Anal. Chem.* **1949**, *21*, 1168.
6491. Schreiner, E., and O. E. Frivold, Z. *Phys. Chem.* (Leipzig) **1926**, *124*, 1.
6492. Schreiner, L., *Ber. Deut. Chem. Gesell.* **1879**, *12*, 179.

6493. Schroeder, J., and H. Steiner, *J. Prakt. Chem.* **1909,** *79,* 49; *C.A.* **1909,** *3,* 1374.
6494. Schroeder, M. R., B. E. Poling, et al., *J. Chem. Eng. Data* **1982,** *27,* 256.
6495. Schroer, E., *Z. Phys. Chem.* (Leipzig) **1929,** *140,* 381.
6496. Schroer, E., *Z. Phys. Chem.* (Leipzig) **1941,** *49B,* 271; *C.A.* **1942,** *36,* 6389.
6497. Schroeter, G., *Ann. Chem.* **1922,** *426,* 1.
6498. Schroeter, G., L. Lichtenstadt, et al., *Ber. Deut. Chem. Gesell.* **1918,** *51,* 1587; *C.A.* **1919,** *13,* 991.
6499. Schrumpf, G., *Chem. Ber.* **1973,** *106,* 246.
6500. Schulek, E., E. Pungor, et al., *Mikrochim. Acta* **1958,** 52; *C.A.* **1959,** *53,* 977.
6501. Schulman, F., and W. A. Zisman, *J. Colloid Sci.* **1952,** *7,* 465.
6502. Schulz, H., and H. Wagner, *Angew. Chem.* **1950,** *62,* 105; *C.A.* **1950,** *44,* 5802.
6503. Schumann, S. C., Univ. Microfilms (Ann Arbor, Michigan) 1941, Pub. No. 241; *C.A.* **1941,** *35,* 4666.
6504. Schumann, S. C., J. G. Aston, et al., *J. Amer. Chem. Soc.* **1942,** *64,* 1039.
6505. Schupp, R. L., and R. Mecke, *Z. Elektrochem.* **1948,** *52,* 54; *C.A.* **1949,** *43,* 6480.
6506. Schurz, J., and E. Kienzl, *Monatsh. Chem.* **1957,** *88,* 78; *C.A.* **1957,** *51,* 11852.
6507. Schurz, J., H. Zah, et al., *Z. Phys. Chem.* (Frankfurt) **1959,** *21,* 185.
6508. Schutz, G., *Z. Phys. Chem.* (Leipzig) **1938,** *40B,* 156.
6509. Schwab, F. W., and E. Wiehers, *J. Res. Nat. Bur. Stand.* **1940,** *25,* 747.
6510. Schwable, C., *Z. Farben Textilechemie* **1904,** *3,* 461; *Chem. Zentr.* **1905,** 360.
6511. Schwarcz, A., and R. S. Farinato, *J. Polym. Sci.*, Part A-2 **1972,** *10,* 2025.
6512. Schwartz, D. P., and O. W. Parks, *Anal. Chem.* **1961,** *33,* 1396.
6513. Schwarzenbach, H. S., and K. Lutz, *Helv. Chim. Acta* **1940,** *23,* 1191.
6514. Schwers, F., *J. Phys. Chem.* **1911,** *9,* 15; *I.C.T.*, *3,* 29, 33.
6515. Scott, D. W., *J. Chem. Thermodyn.* **1970,** *2,* 833.
6516. Scott, D. W., *J. Chem. Thermodyn.* **1971,** *3,* 649.
6517. Scott, D. W., *J. Chem. Thermodyn.* **1971,** *3,* 843.
6518. Scott, D. W., W. T. Berg, et al., *J. Phys. Chem.* **1960,** *64,* 906.
6519. Scott, D. W., W. T. Berg, et al., *J. Phys. Chem.* **1967,** *71,* 2263.
6520. Scott, D. W., D. R. Douslin, et al., *J. Amer. Chem. Soc.* **1952,** *74,* 883.
6521. Scott, D. W., D. R. Douslin, et al., *J. Amer. Chem. Soc.* **1959,** *81,* 1015.
6522. Scott, D. W., H. L. Finke, et al., *J. Amer. Chem. Soc.* **1952,** *74,* 4656.
6523. Scott, D. W., H. L. Finke, et al., *J. Amer. Chem. Soc.* **1957,** *79,* 1062.
6524. Scott, D. W., W. D. Good, et al., *J. Phys. Chem.* **1963,** *67,* 685.
6525. Scott, D. W., W. H. Hubbard, et al., *J. Phys. Chem.* **1963,** *67,* 680.
6526. Scott, D. W., J. P. McCullough, et al., *J. Amer. Chem. Soc.* **1951,** *73,* 1707.
6527. Scott, D. W., J. P. McCullough, et al., *J. Amer. Chem. Soc.* **1956,** *78,* 5457.
6528. Scott, D. W., J. P. McCullough, et al., *J. Amer. Chem. Soc.* **1956,** *78,* 5463.
6529. Scott, D. W., J. F. Messerly, et al., *J. Phys. Chem.* **1961,** *65,* 1320.
6530. Scott, D. W., and A. G. Osborn, *J. Phys. Chem.* **1979,** *83,* 2714.
6531. Scott, N. D., U.S. Patent 2,004,350, June 11, 1935; *Official Gaz. U.S. Pat. Office* **1935,** *455,* 363.
6532. Scott, P. T., D. E. Pearson, et al., *Org. Chem.* **1954,** *19,* 1815.
6533. Scott, R. B., and F. G. Brickwedde, *J. Res. Nat. Bur. Stand.* **1945,** *35,* 501.
6534. Scott, R. B., and J. W. Mellors, *J. Res. Nat. Bur. Stand.* **1945,** *34,* 243.
6535. Scott, R. L., *J. Amer. Chem. Soc.* **1948,** *70,* 4090.
6536. Scott, T. A., D. Macmillan, et al., *Ind. Eng. Chem.* **1952,** *44,* 172.
6537. Scudder, H., from Miner and Dalton [5159], ref. 145, p. 333.
6538. Searles, S., and C. F. Butler, *J. Amer. Chem. Soc.* **1954,** *76,* 56.
6539. Searles, S., M. Tamres, et al., *J. Amer. Chem. Soc.* **1956,** *78,* 4917.
6540. Sears, G. W., and E. R. Hopke, *J. Amer. Chem. Soc.* **1949,** *71,* 1632.
6541. Sears, G. W., and E. R. Hopke, *J. Phys. Colloid Chem.* **1948,** *52,* 1137.
6542. Sears, P. G., W. H. Fortune, et al., *J. Chem. Eng. Data* **1966,** *11,* 406.
6543. Sears, P. G., W. Siegfried, et al., *J. Chem. Eng. Data* **1964,** *9,* 261.
6544. Sears, P. G., T. M. Stoeckinger, et al., *J. Chem. Eng. Data* **1971,** *16,* 220.
6545. Sears, P. G., E. D. Wilhoit, et al., *J. Phys. Chem.* **1955,** *59,* 373.
6546. Sechkarev, A. V., and E. G. Brutan, *Spektrosk. Metody Primen.* **1973,** 98; *C.A.* **1974,** *80,* 21097.

6547. Sechkarev, A. V., and N. I. Dvorovenko, *Izv. Vyssh. Uchebn. Zaved. Fiz.* **1966**, *9*, 111; *C.A.* **1966**, *65*, 6539.

6548. Sedlacek, P., J. Stokr, et al., *Collect. Czech. Chem. Commun.* **1981**, *46*, 1646.

6549. Seeman, F. W., and M. Urban, *Erdoel-Kohle-Erdgas-Petrochem.* **1963**, *16*, 117; *C.A.* **1965**, *62*, 3430.

6550. Segal, L., and F. V. Eggerton, *Appl. Spectrosc.* **1961**, *15*, 116; *C.A.* **1962**, *56*, 4264.

6551. Segal, L., and F. B. Eggerton, *Appl. Spectrosc.* **1961**, *15*, 148.

6552. Segur, J. B., and H. E. Oberstar, *Ind. Eng. Chem.* **1951**, *43*, 2117.

6553. Seha, Z., *Chem. Listy* **1955**, *49*, 1569.

6554. Seha, Z., *Collect. Czech. Chem. Commun.* **1961**, *26*, 2435.

6555. Seibert, F. M., and G. A. Burrell, *J. Amer. Chem. Soc.* **1915**, *37*, 2683.

6556. Seidell, A., *Solubilities of Organic Compounds*, Vol. *II*, 3rd ed., Van Nostrand, New York, 1941.

6557. Seidman, J., *Anal. Chem.* **1951**, *23*, 559.

6558. Seifert, H., *Monatsh. Chem.* **1948**, *79*, 198; *C.A.* **1950**, *44*, 2910.

6559. Seikel, M. K., *Ind. Eng. Chem., Anal. Ed.* **1941**, *13*, 388.

6560. Sekino, M., *Repts. Res. Lab., Asahi Glass Co.* **1950**, *1*, 96; *C.A.* **1956**, *50*, 1584.

6561. Seliverstov, V. M., and M. P. Sergievskaya, *Tr. Leningr. Inst. Vodn. Transp.* **1964**, No. 75, 47; *C.A.* **1966**, *64*, 5828.

6562. Sellers, P., *J. Chem. Thermodyn.* **1970**, *2*, 211.

6563. Sellers, P., and S. Sunner, *Acta Chem. Scand.* **1962**, *16*, 46; *C.A.* **1962**, *56*, 15000.

6564. Sellier, G., and B. Wojtkowiak, *C.R. Acad. Sci., Paris, Ser. C* **1966**, *263*, 1273.

6565. Seltzer, R. J., *Chem. Eng. News* **1975**, *May 19*, 41.

6566. Selvarajan, A., and K. Krishnan, *J. Indian Inst. Sci.* **1971**, *53*, 13; *C.A.* **1972**, *76*, 19720.

6567. Semenova, V. N., S. S. Kazanina, et al., *Gig. Sanit.* **1971**, *36*, 37; *C.A.* **1971**, *75*, 74267.

6568. Semeria, G. B., and G. Ribotti-Lissone, *Gazz. Chim. Ital.* **1930**, *60*, 862; *C.A.* **1931**, *25*, 1800.

6569. Sen, S. K., *Indian J. Phys.* **1956**, *30*, 321; *C.A.* **1957**, *51*, 9312.

6570. Sen, S. K., *Indian J. Phys.* **1960**, *34*, 237; *C.A.* **1960**, *54*, 23728.

6571. Senderens, J. B., *Compt. Rend.* **1909**, *148*, 927; *C.A.* **1909**, *3*, 2145.

6572. Senderens, J. B., *Compt. Rend.* **1909**, *149*, 995; *C.A.* **1910**, *4*, 1478.

6573. Senderens, J. B., *Compt. Rend.* **1923**, *176*, 813; **1923**, *177*, 15.

6574. Senderens, J. B., and J. Aboulenc, *Compt. Rend.* **1911**, *152*, 1671; **1913**, *155*, 1012, 1254; **1913**, *156*, 1620; from [4892].

6575. Senechal, M., and P. Saumagne, *J. Chim. Phys. Physicochim. Biol.* **1972**, *69*, 1246.

6576. Sengers, J. M. H., B. Kamgar-Parsi, et al., *J. Chem. Eng. Data* **1983**, *28*, 354.

6577. Senkus, M., *Ind. Eng. Chem.* **1946**, *38*, 913.

6578. Senkus, M., U.S. Patent 2,406,713, Aug. 27, 1946; *C.A.* **1947**, *41*, 150.

6579. Seprakova, M., J. Paulech, et al., *Chem. Z.* **1959**, *13*, 313; *Chem. Zentr.* **1962**, *133*, 12605.

6580. Serijan, K. T., I. A. Goodman, et al., *Nat. Advisory Committee Aeronaut.* Tech. Note. No. 2557, **1951**; *C.A.* **1952**, *46*, 3406.

6581. Serpinskii, V. V., S. A. Voitkevich, et al., *Zh. Fiz. Khim.* **1954**, *28*, 810; *C.A.* **1955**, *49*, 6677.

6582. Serpinskii, V. V., S. A. Voitkevich, et al., *Zh. Fiz. Khim.* **1957**, *31*, 1278; *C.A.* **1958**, *52*, 3270.

6583. Serra, M., and P. Malatesta, *Ann. Chim.* **1955**, *45*, 911.

6584. Seucan, S., E. Moraru, et al., *Metrol. Apl.* (Bucharest) **1967**, *14*, 313; *C.A.* **1968**, *69*, 6387.

6585. Sevryugova, N. N., V. A. Sokol'skii, et al., *Khim. Prom.* **1963**, 572; *C.A.* **1964**, *60*, 1580.

6586. Seward, R. P., and E. C. Vierira, *J. Phys. Chem.* **1958**, *62*, 127.

6587. Seyer, W. F., *J. Amer. Chem. Soc.* **1953**, *75*, 616.

6588. Seyer, W. F., and G. M. Barrow, *J. Amer. Chem. Soc.* **1948**, *70*, 802.

6589. Seyer, W. F., R. B. Bennett, et al., *J. Amer. Chem. Soc.* **1949**, *71*, 3447.

6590. Seyer, W. F., and C. H. Davenport, *J. Amer. Chem. Soc.* **1941**, *63*, 242.

6591. Seyer, W. F., and J. D. Leslie, *J. Amer. Chem. Soc.* **1942**, *64*, 1912.

6592. Seyer, W. F., and C. W. Mann, *J. Amer. Chem. Soc.* **1945**, *67*, 328.

6593. Seyer, W. F., and R. D. Walker, *J. Amer. Chem. Soc.* **1938**, *60*, 2125.

6594. Seyer, W. F., M. M. Wright, et al., *Ind. Eng. Chem.* **1939**, *31*, 759.

6595. Seyewetz, A., and P. Trawitz, *Compt. Rend.* **1903**, *136*, 241.

6596. Shablygin, M. V., D. N. Shigorin, et al., *Zh. Prikl. Spektrosk. Akad. Nauk Belorussk. SSR* **1965**, *3*, 56; *C.A.* **1966**, *64*, 5939.

6597. Shafer, P. R., D. R. Davis, et al., *Proc. Nat. Acad. Sci. U.S.* **1961**, *47*, 49; *C.A.* **1961**, *55*, 12044.

6598. Shah, J. K., K. J. DeWitt, et al., *J. Chem. Eng. Data* **1969**, *14*, 333.
6599. Shakhparonov, M. I., and N. G. Shlenkina, *Dokl. Akad. Nauk SSSR* **1954**, *96*, 55; *C.A.* **1956**, *50*, 62.
6600. Shamilov, T. A., and D. M. Abasov, *Tr. Azerb. Nauch.-Issled. Inst. Gig. Tr. Prof. Zabol.* **1973**, *8*, 12; *C.A.* **1975**, *82*, 93797.
6601. Shamonin, Yu. Yu., and K. A. Gold'gammer, *Dokl. Akad. Nauk SSSR* **1961**, *140*, 1136; *C.A.* **1962**, *56*, 12449.
6602. Shamshurin, A. A., *J. Gen. Chem.* (USSR) **1939**, *9*, 2207; *C.A.* **1940**, *34*, 4052.
6603. Shank, R. L., *J. Chem. Eng. Data* **1967**, *12*, 474.
6604. Shanley, E. S., and F. P. Greenspan, *Ind. Eng. Chem.* **1947**, *39*, 1536.
6605. Shannon, T. W., T. E. Mead, et al., *Anal. Chem.* **1967**, *39*, 1748.
6606. Shapet'ko, N. N., D. N. Shigorin, et al., *Opt. Spectrosk.* **1964**, *17*, 459; *C.A.* **1965**, *62*, 145.
6607. Shapiro, B. L., and L. E. Mohrmann, *J. Phys. Chem. Ref. Data* **1977**, *6*, 919.
6608. Sharkey, A. G., J. L. Schultz, et al., *Anal. Chem.* **1956**, *28*, 934.
6609. Sharkey, A. G., J. L. Schultz, et al., *Anal. Chem.* **1959**, *31*, 87.
6610. Sharma, B. K., *Indian J. Pure Appl. Phys.* **1977**, *15*, 633.
6611. Sharma, O. P., and R. D. Singh, *Indian J. Pure Appl. Phys.* **1972**, *10*, 885; *C.A.* **1973**, *78*, 166558.
6612. Sharpe, A. N., and S. Walker, *J. Chem Soc.* **1961**, 2974.
6613. Sharpless, N. E., and M. Flavin, *Biochem.* **1966**, *5*, 2963.
6614. Shaw, C. F., and A. L. Allred, *J. Organomet. Chem.* **1971**, *28*, 53.
6615. Shay, J. F., S. Skilling, et al., *Anal. Chem.* **1954**, *26*, 652.
6616. Shchekin, V. V., *Tr. Inst. Neft. Akad. Nauk SSSR* **1952**, *1*, No. 2, 33; *C.A.* **1955**, *49*, 832.
6617. Shchepetova, G. A., *Gig. Sanit.* **1970**, *35*, 96; *C.A.* **1970**, *72*, 136133.
6618. Shcherbak, S. K., G. A. Skorokod, et al., French Patent 1,402,565, June 11, 1965; *C.A.* **1965**, *63*, 9807.
6619. Shcherban, N. G., and N. N. Piten'ko, *Tr. Khar'k. Med. Inst.* **1975**, *124*, 27; *C.A.* **1978**, *89*, 191916.
6620. Sheehan, R., and S. H. Langer, *J. Chem. Eng. Data* **1969**, *14*, 248.
6621. Sheikh, M. Y., and G. Eadon, *Tetrahedron Lett.* **1972**, 257.
6622. Sheinker, Y. N., E. M. Peresleni, et al., *Zh. Fiz. Khim.* **1955**, *29*, 518.
6623. Shelepin, I. V., G. M. Dugacheva, et al., *Plast. Massy.* **1963**, *9*, 50; *C.A.* **1963**, *59*, 15176.
6624. Shell Chemical Co., "Allyl Alcohol," New York, 1946.
6625. Shell Chemical Co., "Organic Chemicals," 2nd. ed., San Francisco, 1942.
6626. Shell Chemical Co., "Tech. Booklet," SC-49-30, 1949.
6627. Shell Chemical Co., Ind. Chem. Div., "Ethylene Glycol to Hexylene Glycol," New York, 1967.
6628. Shell Chemical Co., Ind. Chem. Div., "Evaporation Rate of Solvents as Determined Using the Shell Automatic Thin Film Evaporometer," Tech. Bull., Houston, 1969.
6629. Shell Chemical Co., Ind. Chem. Div., "Isobutyl Acetate," New York, 1964.
6630. Shell Chemical Co., Ind. Chem. Div., "Normal Butyl Acetate," New York, 1963.
6631. Shell Chemical Co., Ind. Chem. Div., "Shell Chart of Solvent Properties," Houston, 1972.
6632. Shell Chemical Co., Ind. Chem. Div., "Solvent Notes," New York, 1967.
6633. Shell Chemical Co., Ind. Chem. Div., "Sulfolane," Tech. Bull. No. IC: 63-13R, New York, 1964.
6634. Shell Chemical Co., Petrochem. Div., "Hydrocarbons Unlimited, Solvents," Houston, 1969.
6635. Shell Development Co. "Acrolein," Emeryville, California, 1947.
6636. Shell Development Co., unpublished data, private communications.
6637. Shell Research Ltd., British Patent 990,486, April 28, 1965; *C.A.* **1965**, *63*, 1730.
6638. Shepard, A. F., and A. L. Henne, *Ind. Eng. Chem.* **1930**, *22*, 356.
6639. Shepard, A. F., A. L. Henne, et al., *J. Amer. Chem. Soc.* **1931**, *53*, 1948.
6640. Shepherd, E. J., and J. A. Kitchener, *J. Chem Soc.* **1956**, 2448.
6641. Shepherd, H. R., ed., *Aerosols, Science and Technology*, Interscience, New York, 1961.
6642. Sheppard, N., *J. Chem. Phys.* **1948**, *16*, 690.
6643. Sheppard, N., *J. Chem. Phys.* **1949**, *17*, 79.
6644. Sheppard, N., *Trans. Faraday Soc.* **1950**, *46*, 527.
6645. Sheppard, N., and G. B. B. M. Southerland, *Proc. Roy. Soc.* (London) **1949**, *196A*, 195; *C.A.* **1950**, *44*, 1806.
6646. Sheppard, N., and G. J. Szasz, *J. Chem. Phys.* **1949**, *17*, 86.
6647. Sheppard, N, and J. J. Turner, *Proc. Roy. Soc.* (London) **1959**, *252A*, 506.
6648. Shergina, N. I., V. P., Kuznetsova, et al., *Hua Hsueh Hsueh Pao* **1959**, *25*, 236; *C.A.* **1960**, *54*, 14937.
6649. Sherman, E. O., and D. C. Olson, *Anal. Chem.* **1968**, *40*, 1174.

6650. Sherrill, M. L., *J. Amer. Chem. Soc.* **1930,** *52,* 1982.

6651. Sherrill, M. L., K. E. Mayer, et al., *J. Amer. Chem. Soc.* **1934,** *56,* 926.

6652. Sherrill, M. L., and P. Mollet, *J. Chim. Phys.* **1936,** *33,* 701; *C.A.* **1937,** *31,* 3784.

6653. Sherrill, M. L., and G. F. Walters, *J. Amer. Chem. Soc.* **1936,** *58,* 742.

6654. Sherry, A. D., and K. F. Purcell, *J. Amer. Chem. Soc.* **1972,** *94,* 1853.

6655. Shevchenko, I., V. Kovalev, et al., *Dokl. Akad. Nauk SSSR* **1969,** *184,* 824.

6656. Shida, T., S. Iwata, et al., *J. Phys. Chem.* **1972,** *76,* 3683.

6657. Shidlovskaya, A. N., and Ya. K. Syrkin, *Zh. Fiz. Khim.* **1948,** *22,* 913; *C.A.* **1949,** *43,* 454.

6658. Shieh, N., and C. C. Price, *J. Org. Chem.* **1959,** *24,* 1169.

6659. Shiga, K., and T. Tsuruta, Japanese Patent 223, Jan. 21, 1953; *C.A.* **1954,** *48,* 8812.

6660. Shigeru, T., A. Hisashi, et al., *Technol. Repts. Osaka Univ.* **1956,** *6,* 163.

6661. Shigley, J. W., C. W. Bonhorst, et al., *J. Amer. Oil Chem. Soc.* **1955,** *32,* 213; *Beil.* **EIV2,** 159.

6662. Shigorin, D. N., *Zh. Fiz. Khim.* **1953,** *27,* 689; *C.A.* **1954,** *48,* 13633.

6663. Shigorin, D. N., A. D. Filyugina, et al., *Zh. Fiz. Khim.* **1967,** *41,* 2336.

6664. Shikhaliev, Ya. A., A. A. Mamedov, et al., *Zh. Fiz. Khim.* **1974,** *48,* 1696; *Russ. J. Phys. Chem.* **1974,** *48,* 1002.

6665. Shimanko, N. A., M. V. Shishkina, et al., *Izv. Anal. Nauk SSSR, Ser. Fiz.* **1962,** *26,* 1252; *C.A.* **1963,** *58,* 7511.

6666. Shimanouchi, T., *J. Chem. Soc. Jpn.* **1941,** *62,* 1264; *C.A.* **1947,** *41,* 2994.

6667. Shimanouchi, T., *J. Phys. Chem. Ref. Data* **1974,** *3,* 269.

6668. Shimanouchi, T., I. Tsuchiya, et al., *J. Chem. Phys.* **1950,** *18,* 1306.

6669. Shimanskii, Yu. I., *Nauk Povidomlennya Kiiv. Univ.* **1956,** No. 1, 44; *C.A.* **1960,** *54,* 9441.

6670. Shimizu, H., and S. Fujiwara, *Chem. Pharm. Bull.* (Tokyo) **1960,** *8,* 272; *C.A.* **1961,** *55,* 9048.

6671. Shimizu, K., and H. Murata, *J. Mol. Spectrosc.* **1960,** *5,* 44.

6672. Shimizu, T., T. Tsuda, et al., *Kogyo Kagaku Zasshi* **1965,** *68,* 2473; *C.A.* **1966,** *65,* 5343.

6673. Shimozawa, T., and M. K. Wilson, *Spectrochim. Acta* **1966,** *22,* 1599; *C.A.* **1966,** *65,* 16261, 16262.

6674. Shindo, Y., and K. Kusano, *J. Chem. Eng. Data* **1979,** *24,* 106.

6675. Shiner, V. J., *J. Amer. Chem. Soc.* **1952,** *74,* 5285.

6676. Shinkarenko, N. V., and V. B. Aleskovskii, *Zh. Prikl. Khim.* (Leningrad) **1974,** *47,* 912; *C.A.* **1974,** *81,* 3271.

6677. Shinoda, K., T. Yamanaka, et al., *J. Phys. Chem.* **1959,** *63,* 648.

6678. Shinoda, T., H. Enokido, et. al., *Bull. Chem. Soc. Jpn.* **1973,** *46,* 48.

6679. Shipsey, K., and E. A. Werner, *J. Chem Soc.* **1913,** 1255.

6680. Shiratori, N., H. Takahashi, et al., *Bull. Chem. Soc. Jpn.* **1975,** *48,* 1423.

6681. Shiro, Y., M. Ohsaku, et al., *Bull. Chem. Soc. Jpn.* **1970,** *43,* 609.

6682. Shishkina, M. V., *Neftekhimiya* **1961,** *1,* 255; *C.A.* **1962,** *57,* 6765.

6683. Shmulyakovskii, Ya. E., *Vestn. Leningrad Univ.* **1950,** No. 3, 65; *C.A.* **1954,** *48,* 12609.

6684. Shoitov, Yu. S., G. M. Pan'Kevich, et al., *Teploenergetika* **1968,** *15,* 76; *C.A.* **1969,** *70,* 51247.

6685. Shokal, E. C., and T. W. Evans, U.S. Patent 2,428,590, Oct. 7, 1947; *Official Gaz. U.S. Patent Office* **1947,** *603,* 124; British Patent 577,992, June 11, 1946.

6686. Shorygin, P. P., V. A. Petukhov, et al., *Zh. Fiz. Khim.* **1968,** *42,* 1057; *Russ. J. Phys. Chem.* **1968,** *42,* 555.

6687. Shorygin, P. P., T. N. Shkurina, et al., *Izv. Akad. Nauk SSSR, Otdel. Khim. Nauk* **1959,** 2208; *C.A.* **1960,** *54,* 10515.

6688. Shorygin, P. P., A. A. Simanovskaya, et al., *J. Appl. Chem.* (USSR) **1936,** *9,* 1442; *C.A.* **1937,** *31,* 2590.

6689. Shostakovskiĭ, M. F., *J. Appl. Chem.* (USSR) **1943,** *16,* 66; *C.A.* **1944,** *38,* 3020.

6690. Shostakovskiĭ, M. F., and N. A. Gershteĭn, *Akad. Nauk SSSR, Inst. Org. Khim., Sintezy Org. Saedineniĭ, Sb.* **1952,** *2,* 154; *C.A.* **1954,** *48,* 569.

6691. Shostakovskiĭ, M. F., and N. A. Gershteĭn, *J. Gen. Chem.* (USSR) **1946,** *16,* 937; *C.A.* **1947,** *41,* 1999.

6692. Shostakovskiĭ, M. F., and E. N. Prilezheava, *J. Gen. Chem.* (USSR) **1947,** *17,* 1129; *C.A.* **1948,** *42,* 3633.

6693. Shott-L'vova, E. A., and Ya. K. Syrkin, *J. Phys. Chem.* (USSR) **1938,** *12,* 479; *C.A.* **1939,** *33,* 4839.

6694. Shreve, O. D., M. R. Heether, et al., *Anal. Chem.* **1950,** *22,* 1261, 1498.

6695. Shreve, O. D., M. R. Heether, et al., *Anal. Chem.* **1951,** *23,* 277.

6696. Shreve, O. D., M. R. Heether, et al., *Anal. Chem.* **1951,** *23*, 282.
6697. Shubin, A. A., *Izv. Akad. Nauk SSSR, Ser. Fiz.* **1950,** *14*, 422; *C.A.* **1951,** *45*, 3245.
6698. Shugaev, V. A., *Gig. Sanit.* **1963,** *28*, 95; *C.A.* **1963,** *59*, 9230.
6699. Shugaev, V. A., *Kholod. Tekh.* **1971,** *48*, 33; *C.A.* **1971,** *75*, 107715.
6700. Shuikin, N. I., E. A. Viktorova, et al., *Vest. Moskov. Univ., Ser. Mat., Mekh., Astron., Fiz., Khim.* **1957,** *12*, No. 3, 141; *C.A.* **1958,** *52*, 6244.
6701. Shuler, W. E., and R. C. Axtmann, *U.S. At. Energy Comm.* **1960,** DP474; *C.A.* **1961,** *55*, 115.
6702. Shulmann, R. G., B. P. Dailey, et al., *Phys. Rev.* **1950,** *78*, 145.
6703. Shuman, R. L., *Paint, Oil and Chem. Rev.* **1935,** *97*, 34; *C.A.* **1935,** *29*, 6665.
6704. Shur, E. G., Chairman, "Comparative Evaporation Rate of Paint Solvents: II," Tech. Subcommittee 66, *ASTM Official Digest* **1956,** 1060.
6705. Shurvell, H. F., F. Cahill, et al., *Can. J. Chem.* **1976,** *54*, 2220.
6706. Siddiqi, K., *Current Sci.* **1943,** *12*, 253; *C.A.* **1944,** *38*, 1169.
6707. Sideri, C. N., and A. Osol, *J. Amer. Chem. Assoc., Sci. Ed.* **1953,** *42,* 586.
6708. Sidgwick, N. V., *Proc. Chem. Soc.* (London) *26*, 60; *Chem. Zentr.* **1910,** *I*, 1828.
6709. Sidgwick, N. V., and E. N. Allott, *J. Chem Soc.* **1923,** 2819.
6710. Sidgwick, N. V., and N. S. Bayliss, *J. Chem Soc.* **1930,** 2027.
6711. Sidgwick, N. V., and E. W. Ewbank, *J. Chem Soc.* **1924,** 2268.
6712. Sidgwick, N. V., and H. E. Rubie, *J. Chem Soc.* **1921,** 1013.
6713. Sidgwick, N. V., W. J. Spurrell, et al., *J. Chem Soc.* **1915,** 1202.
6714. Sidgwick, N. V., and L. E. Sutton, *J. Chem Soc.* **1930,** 1323.
6715. Sidorova, N. G., and I. A. Tuchinskaya, *Zh. Obsch. Khim.* **1957,** *27*, 1763; *C.A.* **1958,** *52*, 4568.
6716. Siedler, R., and H.-J. Bittrich, *J. Prakt. Chem.* **1969,** *311*, 721.
6717. Siegel, A. S., *Org. Mass. Spectrom.* **1970,** *3*, 1417.
6718. Siegel, S., *J. Amer. Chem. Soc.* **1953,** *75*, 1317.
6719. Siegel, S., and M. Dunkel, *Advances in Catalysis* **1957,** *9*, 15.
6720. Sieglitz, A., and O. Horn, *Chem. Ber.* **1951,** *84*, 607.
6721. Signaigo, F. K., and P. L. Cramer, *J. Amer. Chem. Soc.* **1933,** *55*, 3326.
6722. Silbermann, G. B., and S. T. Raschewskaja, *Z. Prikl. Chem.* **1936,** *9*, 1832; *Chem. Zentr.* **1937,** *I*, 4786.
6723. Silverberg, P. M., and L. A. Wenzel, *J. Chem. Eng. Data* **1965,** *10*, 363.
6724. Silvi, B., and J. P. Perchard, *Spectrochim. Acta, Part A* **1976,** *32A*, 11.
6725. Silvi, B., and J. P. Perchard, *Spectrochim. Acta, Part A* **1976,** *32A*, 23.
6726. Silvestro, G., and C. Lenchitz, *J. Phys. Chem.* **1961,** *65*, 694.
6727. Simanouti, T., and S.-I. Mizushima, *J. Chem. Phys.* **1949,** *17*, 1102.
6728. Simanouti, T., H. Turuta, et al., *Sci. Papers Inst. Phys. Chem. Res.* (Tokyo) **1946,** *42*, 51; *C.A.* **1948,** *42*, 459.
6729. Simeral, L., and R. L. Amey, *J. Phys. Chem.* **1970,** *74*, 1443.
6730. Simon, A., and G. Heintz, *Chem. Ber.* **1962,** *95*, 2333.
6731. Simon, A., and J. Huter, *Z. Electrochem.* **1935,** *41*, 28; *C.A.* **1935,** *29*, 5002.
6732. Simon, A., and G. Schulze, *Z. Anorg. Chem.* **1939,** *242*, 313.
6733. Simon, I., *Bull. Soc. Chim. Belges* **1929,** *38*, 47; *Chem. Zentr.* **1929,** *I*, 2519.
6734. Simon, M., *Annual Tables of Physical Constants*, Sec. 514(c), 1941.
6735. Simonetta, M., *Chimica e Industria* (Milan) **1947,** *29*, 37; *C.A.* **1947,** *41*, 7165.
6736. Simonetta, M., and E. Mugnaini, *Chim. Ind.* (Milan) **1948,** *30*, 73; *Brit. Chem. Abstr.* **1948,** *A*, 338.
6737. Simonov, V. D., V. E. Pogulyai, et al., *Zh. Fiz. Khim.* **1970,** *44*, 3087; *Russ. J. Phys. Chem.* **1970,** *44*, 1755.
6738. Simons, J. H., *Fluorine Chemistry, Vol. 1*, Academic Press, New York, 1950.
6739. Simons, J. H., and S. Archer, *J. Amer. Chem. Soc.* **1938,** *60*, 2952.
6740. Simons, J. H., and L. P. Block, *J. Amer. Chem. Soc.* **1939,** *61*, 2962.
6741. Simons, J. H., R. L. Bond, et al., *J. Amer. Chem. Soc.* **1940,** *62*, 3477.
6742. Simons, J. H., and R. D. Dresdner, *J. Electrochem. Soc.* **1949,** *95*, 64.
6743. Simons, J. H., and H. Hart, *J. Amer. Chem. Soc.* **1944,** *66*, 1309.
6744. Simons, J. H., and H. Hart, *J. Amer. Chem. Soc.* **1947,** *69*, 979.
6745. Simons, J. H., and C. J. Lewis, *J. Amer. Chem. Soc.* **1938,** *60*, 492.
6746. Simons, J. H., and K. E. Lorentzen, *J. Amer. Chem. Soc.* **1950,** *72*, 1426.

6747. Simons, J. H., and J. W. Mausteller, *J. Chem. Phys.* **1952**, *20*, 1516.

6748. Simons, W. W., ed., *Sadtler Handbook of Proton NMR Spectra*, Sadtler Research, Philadelphia, 1978.

6749. Simons, W. W., ed., *Sadtler Handbook of Ultraviolet Spectra*, Sadtler Research, Philadelphia, 1979.

6750. Simonsen, D. R., and E. R. Washburn, *J. Amer. Chem. Soc.* **1946**, *68*, 235.

6751. Simonsen, J. L., and L. N. Owen, *The Terpenes, Vol. 1*, Cambridge University Press, Cambridge, 1947.

6752. Simpson, D., and E. K. Plyler, *J. Res. Nat. Bur. Stand.* **1953**, *50*, 223.

6753. Sims, L. L., U.S. Patent 3,159,582, Dec. 1, 1964.

6754. Sinclair, R. G., A. F. McKay, et al., *J. Amer. Chem. Soc.* **1952**, *74*, 2578.

6755. Singh, B. K., and B. K. K. Nayar, *Proc. Indian Acad. Sci.* **1948**, *27A*, 61; *C.A.* **1949**, *43*, 610.

6756. Singh, R., and L. W. Shemil, *J. Chem. Phys.* **1955**, *23*, 1370.

6757. Singh, R. D., and R. S. Singh, *Indian J. Pure Appl. Phys.* **1970**, *8*, 348; *C.A.* **1970**, *73*, 114630.

6758. Singh, R. P., and A. P. Kudchadker, *J. Chem. Thermodyn.* **1979**, *11*, 205.

6759. Singh, R. P., and C. P. Sinha, *Indian J. Chem., Sect. A* **1983**, *22A*, 282.

6760. Singh, R. P., and C. P. Sinha, *J. Chem. Eng. Data* **1982**, *27*, 283.

6761. Singh, R. P., and C. P. Sinha, *Z. Phys. Chem.* (Leipzig) **1982**, *263*, 1075.

6762. Singh, S., B. S. Lark, et al., *Indian J. Chem., Sect. A* **1982**, *21A*, 1116.

6763. Singh, V. B., R. N. Singh, et al., *Spectrochim. Acta* **1966**, *22*, 927; *C.A.* **1966**, *65*, 1600.

6764. Sinke, G. C., *J. Chem. Thermodyn.* **1974**, *6*, 311.

6765. Sinke, G. C., *J. Phys. Chem.* **1959**, *63*, 2063.

6766. Sinke, G. C., and D. L. Hildenbrand, *J. Chem. Eng. Data* **1962**, *7*, 74.

6767. Sinke, G. C., and F. L. Oetting, *J. Phys. Chem.* **1964**, *68*, 1354.

6768. Sinke, G. C., and D. R. Stull, *J. Phys. Chem.* **1958**, *62*, 397.

6769. Sinkuviene, D., *Gig. Sanit.* **1970**, *35*, 6; *C.A.* **1970**, *72*, 136068.

6770. Sinsheimer, J. E., and A. M. Kevhnelian, *Anal. Chem.* **1974**, *46*, 89.

6771. Sinyakov, Yu. I., A. I. Gorbanev, et al., *Izv. Akad. Nauk SSSR, Otdel. Khim. Nauk* **1961**, 1514; *C.A.* **1962**, *56*, 952.

6772. Sirkar, S. C., *Indian J. Phys.* **1932**, *7*, 257; *C.A.* **1932**, *26*, 5846.

6773. Sirkar, S. C., and B. M. Bishui, *Indian J. Phys.* **1946**, *20*, 35; *C.A.* **1947**, *41*, 4718.

6774. Sirkar, S. C., and B. M. Bishui, *Indian J. Phys.* **1946**, 20, 111; *C.A.* **1947**, *41*, 4719.

6775. Sirkar, S. C., and B. M. Bishui, *Proc. Nat. Inst. Sci. India* **1943**, *9*, 287; *C.A.* **1948**, *42*, 8651.

6776. Sirkar, S. C., and B. M. Bishui, *Science and Culture* **1945**, *11*, 273; *C.A.* **1946**, *40*, 2073.

6777. Sirkar, S. C., D. K. Mukheyee, et al., *Indian J. Phys.* **1964**, *38*, 610; *C.A.* **1965**, *62*, 14045.

6778. Sirkar, S. C., and S. B. Sanyal, *Indian J. Phys.* **1944**, *17*, 309; *C.A.* **1945**, *39*, 868.

6779. Sisler, H. H., H. H. Batey, et al., *J. Amer. Chem. Soc.* **1948**, *70*, 3821.

6780. Sivokova, M., A. Matejicek, et al., *Chem. Prom.* **1967**, *17*, 213; *C.A.* **1967**, *67*, 47435.

6781. Sjogren, H., *Arkiv. Fysik.* **1966**, *31*, 159; *C.A.* **1966**, *65*, 3718.

6782. Skala, V., and J. Kuthan, *Collect. Czech. Chem. Commun.* **1970**, *35*, 2378; *C.A.* **1970**, *73*, 69998.

6783. Skau, E. L., *J. Phys. Chem.* **1933**, *37*, 609.

6784. Skau, E. L., and R. McCullough, *J. Amer. Chem. Soc.* **1935**, *57*, 2439.

6785. Skeeters, M. J., U.S. Patent 3,076,040, Apr. 4, 1960; *C.A.* **1963**, *58*, 13792.

6786. Skeeters, M. J., and W. J. Esselstyn, U.S. Patent 2,567,621, Sept. 11, 1951,

6787. Skerkak, T., and B. Ninkov, *Glasnik Drustva Hemicara Tehnol. SR Bosne Hercogovine* 1962, No. 11, 43; *C.A.* **1964**, *61*, 2496.

6788. Skinner, G. S., *J. Amer. Chem. Soc.* **1933**, *55*, 2036.

6789. Skinner, H. A., *Royal Inst. Chem. Monograph* **1958**, No. 3; *J. Chem. Soc.* **1962**, 4396.

6790. Skinner, H. A., and A. Snelson, *Trans. Faraday Soc.* **1960**, *56*, 1776.

6791. Skinner, J. F., E. L. Cussler, et al., *J. Phys. Chem.* **1968**, *72*, 1057.

6792. Skita, A., *Ber. Deut. Chem. Gesell.* **1920**, *53*, 1792.

6793. Skita, A., and W. Faust, *Ber. Deut. Chem. Gesell.* **1931**, *64*, 2878.

6794. Skita, A., and W. Faust, *Ber. Deut. Chem. Gesell.* **1939**, *72B*, 1127.

6795. Skita, A., H. Hauber, et al., *Ann. Chem.* **1923**, *431*, 1; *Chem. Zentr.* **1923**, *I*, 1320.

6796. Skita, A., and A. Schneck, *Ber. Deut. Chem. Gesell.* **1922**, *55B*, 144; *C.A.* **1922**, *16*, 2132.

6797. Sklyarenko, S. I., B. I. Markin, et al., *Zh. Fiz. Khim.* **1958**, *32*, 1916; *C.A.* **1959**, *53*, 4848.

6798. Skold, R., J. Suurkuusk, et al., *J. Chem. Thermodyn.* **1976**, *8*, 1075.

6799. Skulski, T., *Bull. Acad. Polon. Sci., Ser. Sci. Chim.* **1966**, *14*, 23; *C.A.* **1966**, *65*, 1605.

6800. Skuratov, S. M., and M. P. Kozina, *Dokl. Akad. Nauk. SSSR* **1958**, *122*, 109; *C.A.* **1960**, *54*, 23701.
6801. Skuratov, S. M., A. A. Strepikheev, et al., *Dokl. Akad. Nauk SSSR* **1957**, *117*, 452; *C.A.* **1958**, *52*, 14598.
6802. Sladkov, A. M., and L. K. Luneva, *Tr., Nauchn.-Issled. Inst. Sintetich. Spirtov i Organ. Produktov* **1960**, No. 2, 253.
6803. Smakman, R., and T. J. DeBoer, *Org. Mass Spectrom.* **1968**, *1*, 403.
6804. Small, P. A., K. W. Small, et al., *Trans. Faraday Soc.* **1948**, *44*, 810.
6805. Smeets, A., and E. Ruppol, *Bull. Soc. Chim. Biol.* **1934**, *16*, 865; *Chem. Zentr.* **1935**, *I*, 2619.
6806. Smeets, F., and J. Verhulst, *Bull. Soc. Chim. Belges* **1952**, *61*, 694; *C.A.* **1954**, *48*, 6989.
6807. Smets, G., *Trav. Lab. Chim. Gen., Univ. Louvain*, 1942-7, 69 pp.; *Acad. Roy. Belges, Classe Sci. Mem.* **1947**, *21*, 3; *C.A.* **1950**, *44*, 8315.
6808. Smiley, H. M., and P. G. Sears, *Trans. Kentucky Acad. Sci.* **1957**, *18*, 40.
6809. Smit, W. M., *Purity Control by Thermal Analysis*, Elsevier, New York, 1957.
6810. Smith, A. L., *J. Chem. Phys.* **1955**, *21*, 1997.
6811. Smith, A. W., and C. E. Boord, *J. Amer. Chem. Soc.* **1926**, *48*, 1512.
6812. Smith, A. W., C. E. Boord, et al., *J. Amer. Chem. Soc.* **1927**, *49*, 1335.
6813. Smith, C., *J. Chem. Soc.* **1914**, 1703.
6814. Smith, C. H., and R. H. Thompson, *J. Mol. Spectrosc.* **1972**, *42*, 227.
6815. Smith, D. C., G. M. Brown, et al., *J. Chem. Phys.* **1952**, *20*, 473.
6816. Smith, D. C., J. R. Nielson, et al., *J. Chem. Phys.* **1950**, *18*, 326.
6817. Smith, D. C., C. -Y. Pan, et al., *J. Chem. Phys.* **1950**, *18*, 706.
6818. Smith, D. C., R. A. Saunders, et al., *J. Chem. Phys.* **1952**, *20*, 847.
6819. Smith, D. M., and W. M. D. Bryant, *J. Amer. Chem. Soc.* **1935**, *57*, 61.
6820. Smith, D. M., and W. M. D. Bryant, *J. Amer. Chem. Soc.* **1935**, *57*, 841.
6821. Smith, E. R., *J. Res. Nat. Bur. Stand.* **1940**, *24*, 229.
6822. Smith, E. R., *J. Res. Nat. Bur. Stand.* **1941**, *26*, 129.
6823. Smith, E. R., and H. Matheson, *J. Res. Nat. Bur. Stand.* **1938**, *20*, 641.
6824. Smith, E. R., and M. Wojciechowski, *J. Res. Nat. Bur. Stand.* **1937**, *18*, 461.
6825. Smith, F., and L. M. Turton, *J. Chem. Soc.* **1951**, 1701.
6826. Smith, F. A., and E. C. Creitz, *J. Res. Nat. Bur. Stand.* **1951**, *46*, 145.
6827. Smith, G. E., and W. Hunter, British Patent 606,607, Aug. 17, 1948; *C.A.* **1949**, *43*, 3839.
6828. Smith, G. F., *J. Chem. Soc.* **1931**, 3257.
6829. Smith, G. F., *J. Chem. Soc.* **1940**, 869.
6830. Smith, G. W., and L. V. Sorg, *J. Phys. Chem.* **1941**, *45*, 671.
6831. Smith, J. C., *J. Chem. Soc.* **1931**, 802.
6832. Smith, J. C., *J. Chem. Soc.* **1939**, 974.
6833. Smith, J. C., N. J. Foecking, et al., *Ind. Eng. Chem.* **1949**, *41*, 2289.
6834. Smith, J. F. D., *Ind. Eng. Chem.* **1930**, 22, 1246.
6835. Smith, J. M., *Chem. Eng. Progress* **1948**, *44*, 521.
6836. Smith, J. M., *Trans. Amer. Inst. Chem. Engrs.* **1946**, *42*, 983.
6837. Smith, J. W., and S. M. Walshaw, *J. Chem. Soc.* **1957**, 3217.
6838. Smith, L., *Acta Chem. Scand.* **1956**, *10*, 884; *C.A.* **1958**, *52*, 19898.
6839. Smith, L., and L. Bjellerup, *Acta Chem. Scand.* **1947**, *1*, 556; *C.A.* **1948**, *42*, 3248.
6840. Smith, L., L. Bjellerup, et al., *Acta Chem. Scand.* **1953**, *7*, 65.
6841. Smith, L., and S. Sunner, *The Svedberg* (Mem. Vol.) **1944**, 352; *C.A.* **1945**, *39*, 1100.
6842. Smith, L. I., ed., *Organic Syntheses, Vol. 23*, Wiley, New York, 1943.
6843. Smith, L. T., and H. V. Claborn, *Ind. Eng. Chem.* **1940**, *32*, 692.
6844. Smith, M. S., and P. A. D. de Maine, *J. Miss. Acad. Sci.* **1962**, *8*, 244; *C.A.* **1963**, *58*, 2028.
6845. Smith, N. K., and W. D. Good, *J. Chem. Eng. Data* **1967**, *12*, 572.
6846. Smith, R. G., A. Vanterpool, et al., *Can. J. Chem.* **1969**, *47*, 2015.
6847. Smith, R. H., and D. H. Andrews, *J. Amer. Chem. Soc.* **1931**, *53*, 3644.
6848. Smith, S. G., A. H. Fainberg, et al., *J. Amer. Chem. Soc.* **1961**, *83*, 618.
6849. Smith, T. E., and R. F. Bonner, *Anal. Chem.* **1952**, *24*, 517.
6850. Smith, T. E., and R. F. Bonner, *Ind. Eng. Chem.* **1951**, *43*, 1169.
6851. Smith, W. B., and J. L. Roark, *J. Chem. Eng. Data* **1967**, *12*, 587.
6852. Smith, W. B., and J. L. Roark, *J. Phys. Chem.* **1969**, *73*, 1049.

6853. Smith, W. M., K. C. Eberly, et al., *J. Amer. Chem. Soc.* **1956**, *78*, 626.
6854. Smith, W. T., S. Greenbaum, et al., *J. Phys. Chem.* **1954**, *58*, 443.
6855. Smith, W. W., and W. F. von Oettingen, *J. Ind. Hyg. Toxicol.* **1947**, *29*, 47; *C.A.* **1947**, *41*, 2494.
6856. Smittenberg, J., H. Hoog, et al., *J. Amer. Chem. Soc.* **1938**, *60*, 17.
6857. Smolinski, S., M. Balazy, et al., *Bull. Chem. Soc. Jpn.* **1982**, *55*, 1106.
6858. Smolova, N. T., *Zh. Fiz. Khim.* **1978**, *52*, 2905; *Russ. J. Phys. Chem.* **1978**, *52*, 1668.
6859. Smyrl, W. H., and C. W. Tobias, *J. Electrochem. Soc.* **1968**, *115*, 33.
6860. Smyth, C. P., *Dielectric Constant and Molecular Structure*, Chemical Catalog Co., New York, 1931.
6861. Smyth, C. P., *J. Amer. Chem. Soc.* **1925**, *47*, 1894.
6862. Smyth, C. P., *J. Amer. Chem. Soc.* **1941**, *63*, 57.
6863. Smyth, C. P., R. W. Dornte, et al., *J. Amer. Chem. Soc.* **1931**, *53*, 4242.
6864. Smyth, C. P., and E. W. Engel, *J. Amer. Chem. Soc.* **1929**, *51*, 2646.
6865. Smyth, C. P., and S. E. Kamerling, *J. Amer. Chem. Soc.* **1931**, *53*, 2988.
6866. Smyth, C. P., and K. B. McAlpine, *J. Amer. Chem. Soc.* **1934**, *56*, 1697.
6867. Smyth, C. P., and K. B. McAlpine, *J. Chem. Phys.* **1933**, *1*, 190.
6868. Smyth, C. P., and K. B. McAlpine, *J. Chem. Phys.* **1934**, *2*, 499.
6869. Smyth, C. P., and K. B. McAlpine, *J. Chem. Phys.* **1935**, *3*, 347.
6870. Smyth, C. P., and S. A. McNeight, *J. Amer. Chem. Soc.* **1936**, *58*, 1597.
6871. Smyth, C. P., and S. O. Morgan, *J. Amer. Chem. Soc.* **1927**, *49*, 1030.
6872. Smyth, C. P., and S. O. Morgan, *J. Amer. Chem. Soc.* **1928**, *50*, 1547.
6873. Smyth, C. P., and H. E. Rogers, *J. Amer. Chem. Soc.* **1930**, *52*, 1824.
6874. Smyth, C. P., and H. E. Rogers, *J. Amer. Chem. Soc.* **1930**, *52*, 2227.
6875. Smyth, C. P., and W. N. Stoops, *J. Amer. Chem. Soc.* **1928**, *50*, 1883.
6876. Smyth, C. P., and W. N. Stoops, *J. Amer. Chem. Soc.* **1929**, *51*, 3312, 3330.
6877. Smyth, C. P., and W. S. Walls, *J. Amer. Chem. Soc.* **1931**, *53*, 527, 2115.
6878. Smyth, C. P., and W. S. Walls, *J. Amer. Chem. Soc.* **1932**, *54*, 1854.
6879. Smyth, C. P., and W. S. Walls, *J. Amer. Chem. Soc.* **1932**, *54*, 2261.
6880. Smyth, C. P., and W. S. Walls, *J. Amer. Chem. Soc.* **1932**, *54*, 3230.
6881. Smyth, C. P., and W. S. Walls, *J. Chem. Phys.* **1935**, *3*, 557.
6882. Smyth, H. F., C. P. Carpenter, et al., *Arch. Ind. Hyg. Occupational Med.* **1951**, *4*, 119; *C.A.* **1951**, *45*, 9710.
6883. Smyth, H. F., C. P. Carpenter, et al., *Arch. Ind. Hyg. Occupational Med.* **1954**, *10*, 61; *C.A.* **1954**, *48*, 13951.
6884. Smyth, H. F., C. P. Carpenter, et al., *J. Ind. Hyg. Toxicol.* **1949**, *31*, 60.
6885. Smyth, H. F., U. C. Pozzani, et al., *Toxicol. Appl. Pharmacol.* **1969**, 282; *C.A.* **1969**, *71*, 89660.
6886. Snead, C. C., and H. L. Clever, *J. Chem. Eng. Data* **1962**, *7*, 393.
6887. Sneddon, R., *Petrol. Engr.* **1945**, *16*, 148; *C.A.* **1946**, *40*, 4568.
6888. Snelson, A., and H. A. Skinner, *Trans. Faraday Soc.* **1961**, *57*, 2125.
6889. Snethlage, H. C. S., *Z. Phys. Chem.* (Leipzig) **1915**, *90*, 1.
6890. Snider, O. E., R. H. Belden, et al., U.S. Patent 3,021,326, Feb. 13, 1962.
6891. Snyder, E. I., *J. Amer. Chem. Soc.* **1963**, *85*, 2624.
6892. Snyder, L. R., and B. E. Buell, *J. Chem. Eng. Data* **1966**, *11*, 545.
6893. Snyder, R. G., and J. H. Schachtschneider, *Spectrochim. Acta* **1963**, *19*, 85.
6894. Snyder, R. G., and G. Zerbi, *Spectrochim. Acta* **1967**, *23A*, 391.
6895. Sobolev, E. V., and A. V. Bobrov, *Opt. Spektrosk.* **1964**, *17*, 135; *C.A.* **1964**, *61*, 14056.
6896. Sobotka, H., and J. Kahn, *J. Amer. Chem. Soc.* **1931**, *53*, 2935.
6897. Soc. Anon. Distilleries des Deux-Sevres, British Patent 274,488, July 14, 1926; *C.A.* **1928**, *22*, 2171.
6898. Socrates, G., and M. W. Adlard, *J. Chem. Soc. B* **1971**, 733.
6899. Sokolov, N. M., K. A. Andrianov, *Zh. Obshch. Khim.* **1955**, *25*, 675; *J. Gen. Chem.* (USSR) **1955**, *25*, 647; *C.A.* **1956**, *50*, 307.
6900. Sokolov, V. B., M. Kh. Karapet'yants, et al., *Tr. Mosk. Khim.-Tekhnol. Inst.* **1970**, No. 67, 37; *C.A.* **1971**, *75*, 122149.
6901. Sokolova, T. D., N. K. Prokof'eva, et al., *Russ. J. Phys. Chem.* (Eng. trans.) **1973**, *47*, 154.
6902. Solimo, H. N., R. Riggio, et al., *Can. J. Chem.* **1976**, *54*, 3125.
6903. Solvay and Cie, Belgian Patent 625,490, May 29, 1963; *C.A.* **1964**, *60*, 12243.
6904. Solvay and Cie, Netherlands Appl. 289,737, May 11, 1964; *C.A.* **1965**, *62*, 2707.

6905. Solvents and Chemical Companies, "Organic Solvents" and "Physical Properties of Organic Solvents," Chicago, 1972.

6906. Som, J., and G. S. Kastha, *Indian J. Phys.* **1973,** *47,* 494.

6907. Somerville, W. C., *J. Phys. Chem.* **1931,** *35,* 2412.

6908. Sommer, L. H., H. D. Blankman, et al., *J. Amer. Chem. Soc.* **1954,** *76,* 803.

6909. Sommer, S., and G. Tauberger, *Arzneim.-Forsch.* **1964,** *14,* 1050; *C.A.* **1965,** *62,* 989.

6910. Somsen, G., and J. Coops, *Rec. Trav. Chim. Pays-Bas* **1965,** *84,* 985; *C.A.* **1965,** *63,* 17222.

6911. Sordes, J., *Compt. Rend.* **1932,** *195,* 247; *C.A.* **1932,** *26,* 5008.

6912. Sorensen, P., *Chem. Ind.* (London) **1959,** 1593.

6913. Sorenson, P., *J. Oil. Col. Chem. Assoc.* **1967,** *50,* 226.

6914. Sorriso, S., A. Ricci, et al., *J. Organomet. Chem.* **1975,** *87,* 61.

6915. Soulie, M. A., D. Bares, et al., *C.R. Hebd. Seances Acad. Sci., Ser. C* **1975,** *281,* 341.

6919. Soundararajan, S., *Indian J. Chem.* **1963,** *1,* 503.

6917. Southard, J. C., and D. H. Andrews, *J. Franklin Inst.* **1930,** *209,* 349; *C.A.* **1930,** *24,* 2332.

6918. Spaght, M. E., S. B. Thomas, et al., *J. Phys. Chem.* **1932,** *36,* 882.

6919. Spasovski, M., *Khigiena* (Sofia) **1964,** *7,* 38; *C.A.* **1965,** *62,* 9684.

6920. Spassow, A., *Ber. Deut. Chem. Gesell.* **1937,** *70,* 1926; *C.A.* **1937,** *31,* 8503.

6921. Späth, E., *Monatsh. Chem.* **1913,** *34,* 1965; *C.A.* **1914,** *8,* 503.

6922. Späth, E., *Monatsh. Chem.* **1914,** *35,* 324.

6923. Späth, E., and H. Bretschneider, *Ber. Deut. Chem. Gesell.* **1928,** *61,* 327.

6924. Spell, H. L., and J. Laane, *Spectrochim. Acta, Part A* **1972,** *28,* 295.

6925. Spells, K. E., *Trans. Faraday Soc.* **1936,** *32,* 530.

6926. Spencer, H. M., *J. Amer. Chem. Soc.* **1945,** *67,* 1859.

6927. Spencer, H. M., and G. N. Flannagan, *J. Amer. Chem. Soc.* **1942,** *64,* 2511.

6928. Speros, D. M., and F. D. Rossini, *J. Phys. Chem.* **1960,** *64,* 1723.

6929. Spiers, C. W. F., and J. P. Wibaut, *Rec. Trav. Chim. Pays-Bas* **1937,** *56,* 573; *C.A.* **1937,** *31,* 5272.

6930. Spiesecke, H., and W. G. Schneider, *J. Chem. Phys.* **1961,** *35,* 722.

6931. Spiesecke, H., and W. G. Schneider, *J. Phys. Chem.* **1961,** *35,* 731.

6932. Spinner, E., and A. Burawoy, *Spectrochim. Acta* **1961,** *17,* 558; *C.A.* **1961,** *55,* 19473.

6933. Spitzer, R., and H. M. Huffman, *J. Amer. Chem. Soc.* **1947,** *69,* 211.

6934. Spitzer, R., and K. S. Pitzer, *J. Amer. Chem. Soc.* **1946,** *68,* 2537.

6935. Sponer, H., *Chem. Rev.* **1947,** *41,* 281.

6936. Sponer, H., *Rev. Mod. Phys.* **1942,** *14,* 224; *C.A.* **1943,** *37,* 1651.

6937. Sponer, H., and M. B. Hall, *Contrib. Etude Structure Mol.,* Vol. Commen. Victor Henri **1947/48,** 222; *C.A.* **1949,** *43,* 2864.

6938. Sponer, H., and J. S. Kirby-Smith, *J. Chem. Phys.* **1941,** *9,* 667.

6939. Sponer, H., and H. Stucklen, *J. Chem. Phys.* **1946,** *14,* 101.

6940. Sporn, A., I. Dine, et al., *Igiena* (Bucharest) **1967,** *16,* 23; *C.A.* **1968,** *68,* 20949.

6941. Sportouch, S., C. Lacoste, et al., *J. Mol. Struct.* **1971,** *9,* 119; *C.A.* **1972,** *76,* 19809.

6942. Sprengling, G. R., and C. W. Lewis, *J. Amer. Chem. Soc.* **1953,** *75,* 5709.

6943. Springer, C. S., and D. W. Meek, *J. Phys. Chem.* **1966,** *70,* 481.

6944. Springman, H., *Fette, Seifen, Anstrichm* **1971,** *73,* 396; *C.A.* **1971,** *75,* 129294.

6945. Sreenivasulu, M., and P. R. Naidu, *Aust. J. Chem.* **1979,** *32,* 471.

6946. Sreenivasulu, M., and P. R. Naidu, *Indian J. Chem., Sect. A* **1980,** *19A,* 470.

6947. Srinivasan, K. R., and R. L. Kay, *J. Solution Chem.* **1977,** *6,* 357.

6948. Srivastava, G. P. *Current Sci.* (India) **1955,** *24,* 1409; *C.A.* **1956,** *50,* 6857.

6949. Srivastava, G. P., and P. C. Mathur, *Can. J. Phys.* **1967,** *45,* 1617.

6950. Srivastava, G. P., M. C. Mathur, et. al., *Can. J. Phys.* **1972,** *50,* 1449.

6951. Srivastava, G. P., and Y. P. Varshni, *Z. Phys. Chem.* (Leipzig) **1960,** *213,* 30.

6952. Srivastava, N. K., and J. B. Lal, *J. Sci. Soc., Harcourt Butler Technol. Inst. and Indian Inst. Sugor Technol., Kanpur, India* **1955,** *4,* 9; *C.A.* **1960,** *54,* 12051.

6953. Staedl, W., *Ber. Deut. Chem. Gesell.* **1882,** *15,* 2559.

6954. Staff, *Chemical Processing,* Feb. 1962, p. 49.

6955. Stafford, O. F., *J. Amer. Chem. Soc.* **1933,** *55,* 3987.

6956. Stair, R., *J. Res. Nat. Bur. Stand.* **1949,** *42,* 587.

6957. Stair, R., and W. W. Coblentz, *J. Res. Nat. Bur. Stand.* **1935**, *15*, 295.

6958. Stairs, R. A., W. T. Rispin, et al., *Can. J. Chem.* **1970**, *48*, 2755.

6959. Stanford, S. C., and W. Gordy, *J. Amer. Chem. Soc.* **1940**, *62*, 1247.

6960. Stankevich, V. V., and V. I. Osetrov, *Gigiena i Fiziol. Truda, Proizv. Toksikol., Klinika Prof. Zabolevanii Sb.* **1963**, 96; *C.A.* **1964**, *61*, 11228.

6961. Stanley, J. W., Phillips Petroleum Co., Bartlesville, Oklahoma, private communication.

6962. Starling, K. E., B. E. Eakin, et al., *AIChEJ* **1960**, *6*, 438.

6963. Starostin, A. D., A. V. Nikolaev, et al., *Izv. Akad. Nauk SSSR, Ser. Khim.* **1966**, 1303; *C.A.* **1967**, *66*, 54923.

6964. Stasenkova, K. P., *Toksikol. Novykh Prom. Khim. Veshchestv.* **1965**, No. 7, 5; *C.A.* **1965**, *63*, 8946.

6965. Stasenkova, K. P., G. I. Bondarev, et al., *Kauch. Rezina* **1978**, *3*, 29; *C.A.* **1978**, *89*, 1216.

6966. Stasenkova, K. P., and T. A. Kochetkova, *Toksikol. Novykh Prom. Khim. Veshchestv.* **1962**, No. 4, 19, *C.A.* **1963**, *58*, 9543.

6967. Stasenkova, K. P., and T. A. Kochetkova, *Toksikol. Novykh Prom. Khim. Veshchestv.* **1962**, No. 4, 29; *C.A.* **1963**, *58*, 10655.

6968. Stasenkova, K. P., and T. A. Kochetkova, *Toksikol. Novykh Prom. Khim. Veshchestv.* **1965**, No. 7, 27; *C.A.* **1965**, *63*, 8947.

6969. Stasenkova, K. P., T. A. Kochetkova, et al., *Toksikol. Novykh Prom. Khim. Veshchestv.* **1967**, No. 9, 106; *C.A.* **1969**, *70*, 1951.

6970. Stasenkova, K. P., L. M. Samoilova, et al., *Kauch. Rezina* **1972**, *31*, 27; *C.A.* **1972**, *77*, 135875.

6971. Staub, M., *Mitt. Lebensm. Hyg.* **1950**, *41*, 37; *C.A.* **1950**, *44*, 7441.

6972. Staudhammer, P., and W. F. Sayer, *J. Amer. Chem. Soc.* **1958**, *80*, 6491.

6973. Staudinger, H., *Z. Angew. Chem.* **1922**, *35*, 658.

6974. Staudinger, H., and A. Schwalbach, *Ann. Chim.* **1931**, *488*, 8; *C.A.* **1931**, *25*, 5138.

6975. Stauffer, "Chlorinated Hydrocarbons," Stauffer Chemical Co., New York, 1965–1966.

6976. Staveley, L. A. K., and A. K. Gupta, *Trans. Faraday Soc.* **1949**, *45*, 50; *C.A.* **1949**, *43*, 5276.

6977. Staveley, L. A. K., W. I. Tupman, et al., *Trans. Faraday Soc.* **1955**, *51*, 323.

6978. Staveley, L. A. K., J. B. Warren, et al., *J. Chem. Soc.* **1954**, 1992.

6979. Staverman, A. J., *Rec. Trav. Chim. Pays-Bas* **1941**, *60*, 836.

6980. Stearn, A. E., and C. P. Smyth, *J. Amer. Chem. Soc.* **1934**, *56*, 1667.

6981. Stecher, P. G., ed., *The Merck Index*, 8th ed., Merck & Co., Rahway, New Jersey, 1968.

6982. Steele, D., and D. H. Whiffen, *Trans. Faraday Soc.* **1959**, *55*, 369.

6983. Steele, W. C., and F. G. A. Stone, *J. Amer. Chem. Soc.* **1962**, *84*, 3450.

6984. Steele, W. V., *J. Chem. Thermodyn.* **1979**, *11*, 1185.

6985. Steffen, M., *Ber. Bunsenges. Phys. Chem.* **1970**, *74*, 505.

6986. Stehling, F. C., *Anal. Chem.* **1963**, *35*, 773.

6987. Steiger, O., *Phys. Z.* **1931**, *32*, 425; *C.A.* **1931**, *25*, 4452.

6988. Stein, E., and O. Bayer, German Patent 888,689, Nov. 17, 1951; from [4718].

6989. Steiner, E. C., and J. M. Gilbert, *J. Amer. Chem. Soc.* **1965**, *87*, 382.

6990. Steiner, L. A., *Ind. Eng. Chem., Anal. Ed.* **1938**, *10*, 582.

6991. Steiner, L. E., and J. Johnston, *J. Phys. Chem.* **1928**, *32*, 912.

6992. Steinkopf, W., and G. Schwen, *J. Prakt. Chem.* **1921**, *102*, 363; *C.A.* **1922**, *16*, 1390.

6993. Stengle, T. R., and R. C. Taylor, *J. Mol. Spectrosc.* **1970**, *34*, 33.

6994. Stepanenko, N., *J. Exptl. Theoret. Phys.* (USSR) **1944**, *14*, 163; *C.A.* **1945**, *39*, 1337.

6995. Stepanenko, N. N., B. A. Agranat, et al., *J. Phys. Chem.* (USSR) **1947**, *21*, 893; *C.A.* **1948**, *42*, 2485.

6996. Stepanov, N. G., *Russ. J. Phys. Chem.* (Eng. trans.) **1972**, *46*, 464.

6997. Stepanov, N. G., and V. F. Nodzrev, *Russ. J. Phys. Chem.* (Eng. trans.) **1968**, *42*, 1300.

6998. Stepanova, I. P., A. V. Bondarenko, et al., *Neftekhimiya* **1976**, *16*, 471; *C.A.* **1976**, *85*, 93736.

6999. Stephen, H., and T. Stephen, *Solubilities of Inorganic and Organic Compounds, Vol. 1*, Binary Systems, Pts. 1 and 2, Macmillan, New York, 1963.

7000. Stephens, M. A., and W. S. Tamplin, *J. Chem. Eng. Data* **1979**, *24*, 81.

7001. Stephens, O. C., *J. Soc. Chem. Ind.* **1924**, *43*, 313T, 327T.

7002. Stephenson, C. V., and W. C. Coburn, *J. Chem. Phys.* **1965**, *42*, 35.

7003. Stern, A., and K. Thalmayer, *Z. Phys. Chem.* (Leipzig) **1936**, *31B*, 403; *Chem. Zentr.* **1936**, *II*, 1148.

7004. Stevels, J. M., *Trans. Faraday Soc.* **1937,** *33,* 1381.

7005. Stevens, D. R., *Ind. Eng. Chem.* **1943,** *35,* 655.

7006. Stevęns, T. E., *J. Org. Chem.* **1961,** *26,* 1627.

7007. Stevenson, D. P., and J. Y. Beach, *J. Chem. Phys.* **1938,** *6,* 25.

7008. Stevenson, D. P., and J. A. Hipple, *J. Amer. Chem. Soc.* **1942,** *64,* 1558.

7009. Stewart, A. W., *J. Chem. Soc.* **1908,** 1059.

7010. Stewart, J. E., *J. Chem. Phys.* **1955,** *23,* 986.

7011. Stewart, J. E., R. O. Brace, et al., *Nature* **1960,** *186,* 628.

7012. Stewart, P. B., and E. S. Starkman, *Chem. Eng. Progr.* **1957,** *53,* 41J.

7013. Stewart, R., and J. P. O'Donnell, *Can. J. Chem.* **1964,** *42,* 1681.

7014. Stewart, R., and K. Yates, *J. Amer. Chem. Soc.* **1958,** *80,* 6355.

7015. Stewart, T. D., and B. J. Fontana, *J. Amer. Chem. Soc.* **1940,** *62,* 3281.

7016. Stich, K., G. Rotzler, et al., *Helv. Chim. Acta* **1959,** *42,* 1480.

7017. Stock, A., and P. Seelig, *Ber. Deut. Chem. Gesell.* **1919,** *52,* 672.

7018. Stockhardt, J. S., and C. M. Hull, *Ind. Eng. Chem.* **1931,** *23,* 1438.

7019. Stokes, R. H., *J. Chem. Thermodyn.* **1973,** *5,* 379.

7020. Stokes, R. H., and R. P. Tomlins, *J. Chem. Thermodyn.* **1974,** *6,* 379.

7021. Stokes, S., *J. Chem. Phys.* **1963,** *39,* 2309.

7022. Stokinger, H. E., chairman, *Documentation of Threshold Limit Values,* revised ed., American Conference of Governmental Industrial Hygienists, Cincinnati, Ohio, 1966.

7023. Stone, H., and H. Shechter, *J. Org. Chem.* **1950,** *15,* 491.

7024. Storey, W. H., *J. Amer. Oil Chem. Soc.* **1960,** *37,* 676.

7025. Stork, G., and W. N. White, *J. Amer. Chem. Soc.* **1956,** *78,* 4609.

7026. Stothers, J. B., and P. C. Lauterbur, *Can. J. Chem.* **1964,** *42,* 1563.

7027. Stotz, E., W. W. Westerfield, et al., *J. Biol. Chem.* **1944,** *152,* 41.

7028. Stoughton, R. W., and P. D. Lamson, *J. Pharmacol.* **1936,** *58,* 74; *C.A.* **1937,** *31,* 465.

7029. Stout, A. W., and H. A. Schuette, *Ind. Eng. Chem., Anal. Ed.* **1933,** *5,* 100.

7030. Stout, J. W., and L. H. Fisher, *J. Chem. Phys.* **1941,** *9,* 163.

7031. Stranathan, J. D., *J. Chem. Phys.* **1937,** *5,* 828.

7032. Stranathan, J. D., *J. Chem. Phys.* **1938,** *6,* 395.

7033. Strecher, P. E., ed., *The Merck Index of Chemicals and Drugs,* 7th ed., Merck, Rahway, New Jersey, 1960.

7034. Strecker, W., and R. Spitaler, *Ber. Deut. Chem. Gesell.* **1926,** *59,* 1754.

7035. Streiff, A. J., A. R. Hulme, et al., *Ber. Deut. Chem. Gesell.* **1926,** *59,* 1754.

7036. Streiff, A. J., E. T. Murphy, et al., *J. Res. Nat. Bur. Stand.* **1946,** *37,* 331.

7037. Streiff, A. J., E. T. Murphy, et al., *J. Res. Nat. Bur. Stand.* **1947,** *38,* 53.

7038. Streiff, A. J., E. T. Murphy, et al., *J. Res. Nat. Bur. Stand.* **1947,** *39,* 321.

7039. Streiff, A. J., and F. D. Rossini, *J. Res. Nat. Bur. Stand.* **1944,** *32,* 185.

7040. Streiff, A. J., L. H. Schultz, et al., *Anal. Chem.* **1957,** *29,* 361.

7041. Streiff, A. J., L. F. Soule, et al., *J. Res. Nat. Bur. Stand.* **1950,** *45,* 173.

7042. Streiff, A. J., J. C. Zimmerman, et al., *J. Res. Nat. Bur. Stand.* **1948,** *41,* 323.

7043. Streim, H. G., E. A. Boyce, et al., *Anal. Chem.* **1961,** *33,* 85.

7044. Strepikheev, A. A., S. M. Skuratov, et al., *Dokl. Akad. Nauk SSSR* **1955,** *102,* 105; *C.A.* **1956,** *50,* 4903.

7045. Strey, R., and T. Schmeling, *Ber. Busen-Ges. Phys. Chem.* **1983,** *87,* 324.

7046. Stridh, G., *J. Chem. Thermodyn.* **1976,** *8,* 895.

7047. Stridh, G., and S. Sunner, *J. Chem. Thermodyn.* **1975,** *7,* 161.

7048. Stridh, G., S. Sunner, et al., *J. Chem. Thermodyn.* **1977,** *9,* 1005.

7049. Stroh, H., and G. Westphal, *Chem. Ber.* **1963,** *96,* 184.

7050. Strohmeier, W., and K. Miltenberger, *Z. Phys. Chem.* (Frankfurt) **1958,** *17,* 274.

7051. Stromsoe, E., H. G. Ronne, et al., *J. Chem. Eng. Data* **1970,** *15,* 286.

7052. Stross, F. H., C. M. Gable, et al., *J. Amer. Chem. Soc.* **1947,** *69,* 1629.

7053. Stross, F. H., J. M. Monger, et al., *J. Amer. Chem. Soc.* **1947,** *69,* 1627.

7054. Stuart, A. H., *Petroleum* (London) **1947,** *10,* 74; *C.A.* **1947,** *41,* 5293.

7055. Stuart, A. V., and G. B. B. M. Southerland, *J. Chem. Phys.* **1956,** *24,* 559.

7056. Stuchkov, G. S., D. S. Petrenko, et al., *Khim Tekhnol.* (Kiev) **1974,** 34; *C.A.* **1974,** *81,* 49194.

7057. Stuckey, J. M., and J. H. Saylor, *J. Amer. Chem. Soc.* **1940,** *62*, 2922.

7058. Stull, D. R., *Ind. Eng. Chem.* **1947,** *39*, 517.

7059. Stull, D. R., *J. Amer. Chem. Soc.* **1937,** *59*, 2796.

7060. Stull, D. R., E. F. Westrum, et al., *The Chemical Thermodynamics of Organic Compounds*, Wiley, New York, 1969.

7061. Su, Fu-Ti, and Yu-L. P'an, *Union Ind. Res. Inst. Rept.* **1957,** No. 24; *C.A.* **1960,** *54*, 22460.

7062. Subbotin, V. G., *Gig. Sanit.* **1967,** *32*, 9; *C.A.* **1967,** *67*, 120062.

7063. Sucharda, E., and H. Kuczyński, *Rocz. Chem.* **1934,** *14*, 1182; *C.A.* **1935,** *29*, 6214.

7064. Sudmeier, J. L., and C. N. Reilley, *Anal. Chem.* **1964,** *36*, 1698.

7065. Sue, P., J. Pauley, et al., *Bull. Chim. Soc. Fr.* **1958,** *5*, 593; *C.A.* **1958,** *52*, 15177; Ref. [8056].

7066. Suetaka, W., *Gazz. Chim. Ital.* **1956,** *86*, 783; *C.A.* **1958,** *52*, 5974.

7067. Suetaka, W., and M. Sanesi, *Ann. Chim.* (Rome) **1956,** *46*, 1133; *C.A.* **1957,** *51*, 8032.

7068. Suga, H., and S. Seki, *Bull. Chem. Soc. Jpn.* **1962,** *35*, 1905; *C.A.* **1963,** *58*, 3027.

7069. Sugarman, B., *Proc. Phys. Soc.* (London) **1943,** *55*, 429; *C.A.* **1944,** *38*, 1427.

7070. Sugden, S., *J. Chem. Soc.* **1924,** 1167.

7071. Sugden, S., *J. Chem. Soc.* **1929,** 316.

7072. Sugden, S., *J. Chem. Soc.* **1933,** 768.

7073. Sugden, S., J. B. Reed, et al., *J. Chem. Soc.* **1925,** *127*, 1525.

7074. Sugden, S., and H. Whittaker, *J. Chem. Soc.* **1925,** 1868.

7075. Sugden, S., and H. Wilkins, *J. Chem. Soc.* **1927,** 139.

7076. Suhr, H., *Chem. Ber.* **1963,** *96*, 1720.

7077. Suhrmann, R., and P. Klein, *Z. Phys. Chem.* (Leipzig) **1941,** *50B*, 23; *Chem. Zentr.* **1942,** *I*, 2251; *C.A.* **1943,** *37*, 3343.

7078. Sukhodol, V. F., and P. A. Chatskii, *Spirt. Prom.* **1962,** *28*, No. 3, 35; *C.A.* **1962,** *57*, 5124.

7079. Sullivan, M. V., J. K. Wolfe, et al., *Ind. Eng. Chem.* **1947,** *39*, 1607.

7080. Sultanov, G. A., and M. M. Musaev, *Izv. Akad. Nauk Azerb. SSR* **1957,** No. 1, 31; *C.A.* **1957,** *51*, 11072.

7081. Sumer, K. M., and A. R. Thompson, *J. Chem. Eng. Data* **1967,** *12*, 489.

7082. Sundara Rao, A. L., *J. Indian Chem. Soc.* **1945,** *22*, 260; *C.A.* **1946,** *40*, 4292.

7083. Sunier, A. A., *J. Chem. Ed.* **1972,** *49*, 805.

7084. Sunier, A. A., *J. Phys. Chem.* **1931,** *35*, 1756.

7085. Sunner, S., *Acta Chem. Scand.* **1955,** *9*, 837, 847; *C.A.* **1956,** *50*, 2269, 2270.

7086. Sunner, S., C. Svensson, et al., *J. Chem. Thermodyn.* **1979,** *11*, 491.

7087. Suri, S. K., and V. Ramakrishna, *Indian J. Chem.* **1969,** *7*, 243.

7088. Surinder, P. V., and K. A. Kobe, *J. Chem. Eng. Data* **1959,** *4*, 329.

7089. Sus, A. N., *Compt. Rend. Acad. Sci. URSS* **1941,** *33*, 310; *C.A.* **1943,** *37*, 1308.

7090. Susz, B., and E. Briner, *Helv. Chim. Acta* **1939,** *22*, 117.

7091. Sutton, C., and J. A. Calder, *Environ. Sci. Technol.* **1974,** *8*, 654.

7092. Sutton, C., and J. A. Calder, *J. Chem. Eng. Data* **1975,** *20*, 320.

7093. Sutton, L. E., *Proc. Roy. Soc.* (London) **1931,** *133A*, 668; *C.A.* **1932,** *26*, 886.

7094. Sutton, L. E., R. G. A. New, et al., *J. Chem. Soc.* **1933,** 652.

7095. Suyentoslavskii, V., *J. Russ. Phys. Chem. Soc.*, *41*, 920; *Chem. Zentr.* **1909,** *II*, 2144.

7096. Suyver, J. F., and J. P. Wibaut, *Rec. Trav. Chim. Pays-Bas* **1945,** *64*, 65; *C.A.* **1946,** *40*, 3428; *Beil.* **EIII5,** 1580.

7097. Suzuki, I., *Bull. Chem. Soc. Jpn.* **1962,** *35*, 540; from [4407].

7098. Suzuki, K., *Rev. Phys. Chem. Jpn.* **1953,** *23*, 57; *C.A.* **1955,** *49*, 3007.

7099. Suzuki, S., *Kagaku* **1953,** *23*, 535; *C.A.* **1953,** *47*, 11892.

7100. Svetlanov, E. B., S. M. Velichko, et al., *Zh. Fiz. Khim.* **1971,** *45*, 877; *Russ. J. Phys. Chem.* **1971,** *45*, 488.

7101. Svetlov, Y. M., and N. S. Vul'fson, *J. Appl. Chem.* (USSR) **1936,** *9* 1613; *C.A.* **1937,** *31*, 6670.

7102. Svirbely, J. L., B. Highman, et al., *J. Ind. Hyg. Toxicol.* **1947,** *29*, 382; *C.A.* **1948,** *42*, 987.

7103. Svirbely, W. J., J. E. Ablard, et al., *J. Amer. Chem. Soc.* **1935,** *57*, 652.

7104. Svirbely, W. J., and J. J. Lander, *J. Amer. Chem. Soc.* **1948,** *70*, 4121.

7105. Svirbely, W. J., W. M. Eareckson, et al., *J. Amer. Chem. Soc.* **1949,** *71*, 507.

7106. Svoboda, V., V. Charvatova, et al., *Collect. Czech. Chem. Commun.* **1981,** *46*, 2983.

7107. Svoboda, V., V. Charvatova, et al., *Collect. Czech. Chem. Commun.* **1982,** *47*, 543.

7108. Svoboda, V., V. Majer, et al., *Collect. Czech. Chem. Commun.* **1977,** *42,* 1755.
7109. Svoboda, V., V. Uchytilova, et al., *Collect. Czech. Chem. Commun.* **1980,** *45,* 3233.
7110. Svoboda, V., F. Vesely, et al., *Collect. Czech. Chem. Commun.* **1973,** *38,* 3539.
7111. Swalen, J. D., and D. R. Herschbach, *J. Chem. Phys.* **1957,** *27,* 100.
7112. Swallen, L. C., and C. E. Boord, *J. Amer. Chem. Soc.* **1930,** *52,* 651.
7113. Swallow, J. C., and R. O. Gibson, *J. Chem. Soc.* **1934,** 440.
7114. Swamy, H. N., *Indian J. Phys.* **1952,** *26,* 119; *C.A.* **1953,** *47,* 1489.
7115. Swamy, H. N., *Indian J. Phys.* **1952,** *26,* 445.
7116. Swamy, P. A., and M. Van Winkle, *J. Chem. Eng. Data* **1965,** *10,* 214
7117. Swarts, F., *Bull. Acad. Roy. Belg. Classe Sci.* **1909,** 43; *Chem. Zentr.* **1909,** *I,* 1989.
7118. Swarts, F., *Bull. Acad. Roy. Belg.* **1910,** 113; *C.A.* **1911,** *5,* 1086.
7119. Swarts, F., *Bull. Sci. Acad. Roy. Belg.* **1922,** *8,* 343; *C.A.* **1923,** *17,* 2873.
7120. Swarts, F., *Bull. Soc. Chim. Belges* **1922,** *31,* 375; *C.A.* **1923,** *17,* 2873.
7121. Swarts, F., *Bull. Soc. Chim. Belges* **1929,** *38,* 99.
7122. Swarts, F., *Bull. Soc. Chim. Belges* **1934,** *43,* 475.
7123. Swarts, F., *Chem. Zentr.* **1898,** *II,* 26.
7124. Swarts, F., *Chem. Zentr.* **1901,** *II,* 804.
7125. Swarts, F., *J. Chim. Phys.* **1923,** *20,* 30; *C.A.* **1923,** *17,* 2383.
7126. Swarts, F., *J. Chim. Phys.* **1931,** *28,* 622; *Beil.* **EIII5,** 676, 678.
7127. Swarts, F., *J. Pharm. Chim.* **1931,** *28,* 622.
7128. Swarts, F., *Rec. Trav. Chim. Pays-Bas* **1914,** *33,* 281; *Bull. Acad. Roy. Belg.* **1914,** 18; *C.A.* **1915,** *9,* 445.
7129. Swensen, R. F., and D. A. Keyworth, *Anal. Chem.* **1963,** *35,* 863.
7130. Swern, D., ed., *Bailey's Industrial Oil and Fat Products,* 3rd ed., Interscience, New York, 1964.
7131. Swietoslawski, W., *Ebulliometric Measurements,* Reinhold, New York, 1945.
7132. Swietoslawski, W., *J. Phys. Chem.* **1934,** *31,* 1169.
7133. Swietoslawski, W., *Przeglad Chem.* **1948,** *6,* 249; *C.A.* **1949,** *43,* 8977.
7134. Swietoslawski, W., *Z. Phys. Chem.* (Leipzig) **1909,** *65,* 513.
7135. Swietoslawski, W., *Z. Phys. Chem.* (Leipzig) **1910,** *72,* 49.
7136. Swietoslawski, W., and E. Bartoszewicz, *Rocz. Chem.* **1931,** *11,* 78; *Chem. Zentr.* **1931,** *I,* 2787.
7137. Swietoslawski, W., and M. Popow, *J. Chim. Phys.* **1925,** *22,* 395; *Chem. Zentr.* **1926,** *I,* 600.
7138. Swietoslawski, W., S. Rybicka, et al., *Rocz. Chem.* **1931,** *11,* 35.
7139. Swift, E., *J. Amer. Chem. Soc.* **1942,** *64,* 115.
7140. Swift, E., and C. R. Calkins, *J. Amer. Chem. Soc.* **1943,** *65,* 2415.
7141. Swift, G. W., J. Lohrenz, et al., *AIChEJ* **1960,** *6,* 415.
7142. Swindells, J. F., J. R. Coe, et al., *J. Res. Nat. Bur. Stand.* **1952,** *48,* 1.
7143. Sydow, E. v., *Acta Chem. Scand.* **1963,** *17,* 2504.
7144. Sydow, E. v., *Acta Chem. Scand.* **1964,** *18,* 1099; *C.A.* **1964,** *61,* 15504.
7145. Szasz, G. J., J. H. Morrison, et al., *J. Chem. Phys.* **1947,** *15,* 562.
7146. Szekely, A., *Acta Chim. Acad. Sci. Hung.* **1955,** *5,* 317; *C.A.* **1955,** *49,* 15426.
7147. Szobel, L., *Compt. Rend.* **1944,** *218,* 347; *C.A.* **1946,** *40,* 2740.
7148. Szyper, M., and P. Zuman, *Anal. Chim. Acta* **1976,** *85,* 357.

T

7149. Tabuteau, J., and H. Gautier, French Patent 860,484, Jan. 16, 1941; *C.A.* **1948,** *42,* 6841.
7150. Tadayon, J., A. H. Nissan, et al., *Anal. Chem.* **1949,** *21,* 1532.
7151. Tafel, J., *Z. Elektrochem.* **1911,** *17,* 1973.
7152. Tafel, J., and G. Frederichs, *Ber. Deut. Chem. Gesell.* **1904,** *37,* 3187.
7153. Tafel, J., and W. Jurgens, *Ber. Deut. Chem. Gesell.* **1909,** *42,* 2548; *C.A.* **1909,** *3,* 2567.
7154. Tafel, J., and O. Wassmuth, *Ber. Deut. Chem. Gesell.* **1907,** *40,* 2831.
7155. Takada, M., and H. S. Gutowsky, *J. Chem. Phys.* **1957,** *26,* 577.
7156. Takagi, T., *J. Chem. Thermodyn.* **1981,** *13,* 291.
7157. Takagi, T., and H. Teranishi, *J. Chem. Thermodyn.* **1982,** *14,* 577.
7158. Takagi, T., and H. Teranishi, *J. Chem. Thermodyn.* **1982,** *14,* 1167.
7159. Takamatsu, T., *Bull. Chem. Soc. Jpn.* **1974,** *47,* 1285.
7160. Takamitsu, N., and H. Sumida, *J. Chem. Soc. Jpn., Ind. Chem. Sect.* **1967,** *70,* 2138.

7161. Takao, M., and M. Kumamoto, Japanese Patent 6566, Oct. 15, 1954; *C.A.* **1956,** *50,* 4194.

7162. Takeda, M., *J. Chem. Soc. Jpn.* **1941,** *62,* 896; *C.A.* **1947,** *41,* 5390.

7163. Takhirov, M. T., *Gig. Sanit.* **1969,** *34,* 103; *C.A.* **1969,** *71,* 24536.

7164. Tallman, R. C., *J. Amer. Chem. Soc.* **1934,** *56,* 126.

7165. Talvari, A., S. Rang, et al., *Eesti NSV Tead. Akad. Toim., Keem., Geol.* **1974,** *23(4),* 307; *C.A.* **1975,** *82,* 85548.

7166. Tamas, J., *Acta Chim. Acad. Sci. Hung.* **1968,** 56; *C.A.* **1968,** *69,* 31402.

7167. Tammann, G., *Z. Phys. Chem.* (Leipzig) **1913,** *85,* 273.

7168. Tamris, M., S. Searles, et al., *J. Amer. Chem. Soc.* **1954,** *76,* 3983.

7169. Tamura, K., and S. Murakami, *J. Chem. Thermodyn.* **1984,** *16,* 33.

7170. Tamura, K., K. Ohomura, et al., *J. Chem. Thermodyn.* **1984,** *16,* 121.

7171. Tamura, K., M. K. Kumaran, et al., *J. Chem. Thermodyn.* **1984,** *16,* 145.

7172. Tamura, Z., and S. Kawai, *Bunseki Kagaku* **1966,** *15,* 64; *C.A.* **1966,** *65,* 6522.

7173. Tanabe, K., and S. Saeki, *Bull. Chem. Soc. Jpn.* **1974,** *47,* 2545.

7174. Tanaka, C., *Mem. Coll. Sci., Kyoto Imp. Univ.* **1930,** Ser. A., *13,* 239; *Chem. Zentr.* **1930,** *II,* 1523.

7175. Tanaka, C., *Nippon Kagaku Zasshi* **1962,** *83,* 521; *C.A.* **1963,** *58,* 10882.

7176. Tanaka, E., and M. Ichikawa, Japanese Patent 74 52,791, May 22, 1974; *C.A.* **1974,** *81,* 169084.

7177. Tanaka, N., *J. Chem. Soc. Jpn.* **1942,** *63,* 134; *C.A.* **1947,** *41,* 3063.

7178. Tanaka, R., *J. Chem. Thermodyn.* **1982,** *14,* 259.

7179. Tanaka, S., *Japan Analyst* **1955,** *4,* 109; *C.A.* **1956,** *50,* 5465.

7180. Tanaka, T., *Bull. Chem. Soc. Jpn.* **1965,** *38,* 1465; *C.A.* **1966,** *64,* 159.

7181. Tanaka, T., and M. Hashimoto, *Sci. Repts. Saitama Univ.* **1956,** Ser. A, 2, 117; *C.A.* **1957,** *51,* 6334.

7182. Tannenbaum, S., *J. Amer. Chem. Soc.* **1954,** *76,* 1027.

7183. Tannenbaum, S., S. Kaye, et al., *J. Amer. Chem. Soc.* **1953,** *75,* 3753.

7184. Tanner, H. C., and A. F. Benning, *Ind. Eng. Chem.* **1939,** *31,* 878.

7185. Tarakad, R. R., and W. A. Scheller, *J. Chem. Eng. Data* **1979,** *24,* 119.

7186. Tarasenkov, D. N., and E. N. Polozhintzwa, *Ber. Deut. Chem. Gesell.* **1932,** *65B,* 184.

7187. Tarte, P., and P. A. Laurent, *Bull. Soc. Chim. Belges* **1960,** *69,* 109; *C.A.* **1960,** *54,* 16179.

7188. Tatsuya, B., I. Kobayashi, et al., Japanese Patent 74 03,965, Jan. 29, 1974; *C.A.* **1975,** *82,* 111560.

7189. Taufen, H. J., M. J. Murray, et al., *J. Amer. Chem. Soc.* **1943,** *65,* 1130.

7190. Tavernier, P., and M. Lamouroux, *Mem. Poudres.* **1956,** *38,* 65; *C.A.* **1957,** *51,* 14403.

7191. Taylor, C. A., and W. H. Rinkenbach, *Ind. Eng. Chem.* **1926,** *18,* 676.

7192. Taylor, C. A., and W. H. Rinkenbach, *J. Amer. Chem. Soc.* **1926,** *48,* 1305.

7193. Taylor, G. B., and M. B. Hall, *Anal. Chem.* **1951,** *23,* 947.

7194. Taylor, H. A., and T. W. Davis, *J. Phys. Chem.* **1928,** *32,* 1467.

7195. Taylor, J. M., P. M. Jenner, et al., *Toxicol. and Appl. Pharmacol.* **1964,** *6,* 378.

7196. Taylor, M. D., and L. R. Grant, *J. Org. Chem.* **1960,** *25,* 678.

7197. Taylor, M. D., and M. B. Templeman, *J. Amer. Chem. Soc.* **1956,** *78,* 2950.

7189. Taylor, R. C., *J. Chem. Phys.* **1954,** *22,* 714.

7199. Taylor, R. C., R. A. Brown, et al., *Anal. Chem.* **1948,** *20,* 396.

7200. Taylor, R. C., and G. L. Vidale, *J. Chem. Phys.* **1957,** *26,* 122.

7201. Taylor, R. K., *J. Amer. Chem. Soc.* **1953,** *75,* 2521.

7202. Taylor, W. J., and F. D. Rossini, *J. Res. Nat. Bur. Stand.* **1944,** *32,* 197.

7203. Taylor, W. J., D. D. Wagman, et al., *J. Res. Nat. Bur. Stand.* **1946,** *37,* 95.

7204. Technical Committee, Natural Gas Association of America, *Ind. Eng. Chem.* **1942,** *34,* 1240.

7205. Tedder, J. M., *J. Chem. Soc.* **1954,** 2646.

7206. Teeple, J. E., *J. Amer. Chem. Soc.* **1908,** *30,* 412.

7207. Telang, M. S., *J. Phys. Chem.* **1946,** *50,* 373.

7208. Tel'noi, V. I., I. B. Rabinovich, et al., *Dokl. Akad. Nauk SSSR* **1964,** *159,* 1106.

7209. Tel'noi, V. I., and I. B. Rabinovich, *Russ. J. Phys. Chem.* (Eng. trans.) **1965,** *39,* 1108.

7210. Tel'noi, V. I., and I. B. Rabinovich, *Russ. J. Phys. Chem.* (Eng. trans.) **1966,** *40,* 842.

7211. Templeton, D. H., D. D. Davies, et al., *J. Amer. Chem. Soc.* **1944,** *66,* 2033.

7212. Teplyakova, E. V., S. N. Zyabbarova, et al., *Gig. Sanit.* **1970,** *35,* 103; *C.A.* **1970,** *73,* 6962.

7213. Teramoto, S., M. Ishikawa, et al., *J. Fermentation Technol.* (Japan) **1955,** *33,* 307; *C.A.* **1955,** *49,* 16323.

7214. Teramura, K., and R. Oda, *J. Chem. Soc. Jpn., Ind. Chem. Sect.* **1951,** *54,* 605; *C.A.* **1954,** *48,* 2567.

7215. Teranishi, R., R. A. Flath, et al., *J. Agric. Food Chem.* **1966,** *14(3),* 253.

7216. Terry, T. D., R. E. Kepner, et al., *J. Chem. Eng. Data* **1960,** *5,* 403.

7217. Tess, R. W., R. D. Harline, et al., *Ind. Eng. Chem.* **1957,** *49,* 374.

7218. Teter, J. W., and W. J. Merwin, U.S. Patent 2,411,346, Nov. 19, 1946; *C.A.* **1947,** *41,* 983.

7219. Tewari, Y. B., M. M. Miller, et al., *J. Chem. Eng. Data* **1982,** *27,* 451.

7220. Teze, M., and R. Schaal, *Bull. Soc. Chim. Fr.* **1962,** 1372; *Chem. Zentr.* **1963,** 7875.

7221. Thaker, K. A., and N. S. Dave, *J. Sci. Ind. Research* (India) **1961,** *20B,* 329; *C. A.* **1962,** *56,* 2319.

7222. Thatte, V. N., and A. S. Ganesan, *Phil. Mag.* **1931,** *12,* 823; *C.A.* **1932,** *26,* 655.

7223. Thatte, V. N., and M. S. Joglekar, *Phil. Mag.* **1935,** *19,* 1116; *C.A.* **1935,** *29,* 6505.

7224. Thau-Alexandrowicz, M., *J. Chem. Eng. Data* **1972,** *17,* 339.

7225–7374. Thermodynamics Research Center Data Project (formerly Manufacturing Chemists Association Research Project), Thermodynamics Research Center, College Station, Texas A & M University. Specific references:

7225. MCA Catalog of Infrared Spectra.

7226. Thermodynamics Research Center Data Project, Catalog of Mass Spectra.

7227. MCA Catalog of Nuclear Magnetic Resonance Spectra.

7228. MCA Catalog of Raman Spectra.

7229. MCA Catalog of Ultraviolet Spectra.

7230. Table 2-1-a, Dec. 31, 1962.

7231. Table 2-1-i, June 30, 1959.

7232. Table 2-1-(1.02)-fa, Dec. 31, 1966.

7233. Table 23-2-1-(1.1000)-a, Dec. 31, 1966.

7234. Table 23-2-1-(1.1000)-i, Dec. 31, 1969.

7235. Table 23-2-1-(1.1000)-k, June 30, 1965.

7236. Table 23-2-1-(1.1000)-m, Dec. 31, 1970.

7237. Table 23-2-1-(1.1000)-p, Dec. 31, 1965.

7238. Table 23-2-1-(1.1001)-i, June 30, 1971.

7239. Table 23-2-1-(1.1020)-a, Dec. 31, 1967.

7240. Table 23-2-1-(1.1020)-c, Dec. 31, 1971.

7241. Table 23-2-1-(1.1020)-d, June 30, 1966.

7242. Table 23-2-1-(1.1020)-fb, Dec. 31, 1970.

7243. Table 23-2-1-(1.1020)-i, Dec. 31, 1969.

7244. Table 23-2-1-(1.1020)-k, June 30, 1965.

7245. Table 23-2-1-(1.1020)-m, Dec. 31, 1970.

7246. Table 23-2-1-(1.1020)-p, June 30, 1965.

7247. Table 23-2-1-(1.1020)-v, Dec. 31, 1965.

7248. Table 23-2-1-(1.1021)-a, Dec. 31, 1967.

7249. Table 23-2-1-(1.1021)-d, June 30, 1968.

7250. Table 23-2-1-(1.1021)-k, June 30, 1965.

7251. Table 23-2-1-(1.1022)-a, Dec. 31, 1967.

7252. Table 23-2-1-(1.1022)-k, June 30, 1968.

7253. Table 23-2-1-(1.1023)-a, June 30, 1968.

7254. Table 23-2-1-(1.1023)-k, June 30, 1968.

7255. Table 23-2-1-(1.1036)-a, June 30, 1966.

7256. Table 23-2-1-(1.1036)-k, June 30, 1966.

7257. Table 23-2-1-(1.1038)-a, Dec. 31, 1968.

7258. Table 23-2-1-(1.1100)-k, Dec. 31, 1960.

7259. Table 23-2-1-(1.1105)-a, Dec. 31, 1964.

7260. Table 23-2-1-(1.1120)-a, Dec. 31, 1965.

7261. Table 23-2-1-(1.1120)-k, Dec. 31, 1965.

7262. Table 23-2-1-(1.1130)-a, Dec. 31, 1965.

7263. Table 23-2-1-(1.1130)-i, June 30, 1966.

7264. Table 23-2-1-(1.1130)-k, Dec. 31, 1965.

7265. Table 23-2-1-(1.1131)-a. Dec. 31, 1965.

7266. Table 23-2-1-(1.1131)-k, Dec. 31, 1965.

7267. Table 23-2-1-(1.1200)-a, June 30, 1961.
7268. Table 23-2-1-(1.1200)-k, Dec. 31, 1960.
7269. Table 23-2-1-(1.1210)-i, Dec. 31, 1973.
7270. Table 23-2-1-(1.1300)-a, Dec. 31, 1969.
7271. Table 23-2-1-(1.1300)-d, Dec. 31, 1969.
7272. Table 23-2-1-(1.1300)-i, June 30, 1973.
7273. Table 23-2-1-(1.1300)-k, Dec. 31, 1969.
7274. Table 23-2-1-(1.1310)-a, Dec. 31, 1969.
7275. Table 23-2-1-(1.1310)-d, Dec. 31, 1969.
7276. Table 23-2-1-(1.1310)-i, Dec. 31, 1970.
7277. Table 23-2-1-(1.1310)-k, Dec. 31, 1969.
7278. Table 23-2-1-(1.1330)-a, Dec. 31, 1969.
7279. Table 23-2-1-(1.1330)-d, Dec. 31, 1969.
7280. Table 23-2-1-(1.1330)-i, June 30, 1973.
7281. Table 23-2-1-(1.1330)-k, Dec. 31, 1969.
7282. Table 23-2-1-(1.1340)-i, Dec. 31, 1973.
7283. Table 23-2-1-(1.1350)-a, Dec. 31, 1969.
7284. Table 23-2-1-(1.1350)-k, Dec. 31, 1969.
7285. Table 23-2-1-(1.1360)-i, Dec. 31, 1973.
7286. Table 23-2-1-(1.1374)-i, Dec. 31, 1973.
7287. Table 23-2-1-(1.1404)-i, Dec. 31, 1973.
7288. Table 23-2-1-(1.2020)-i, June 30, 1973.
7289. Table 23-2-1-(1.2115)-a, Dec. 31, 1965.
7290. Table 23-2-1-(1.2115)-m, Dec. 31, 1965.
7291. Table 23-2-1-(1.2115)-p, Dec. 31, 1965.
7292. Table 23-2-1-(1.2120)-i, Dec. 31, 1973.
7293. Table 23-2-1-(1.2120)-k, June 30, 1973.
7294. Table 23-2-1-(1.2121)-a, June 30, 1963.
7295. Table 23-2-1-(1.2121)-m, Dec. 31, 1965.
7296. Table 23-2-1-(1.2121)-p, Dec. 31, 1965.
7297. Table 23-2-1-(33.1000)-a, June 30, 1962.
7298. Table 23-2-1-(33.1000)-d, Dec. 31, 1962.
7299. Table 23-2-1-(33.1000)-i, Dec. 31, 1963.
7300. Table 23-2-1-(33.1000)-k, June 30, 1962.
7301. Table 23-2-1-(33.7000)-m, Dec. 31, 1965.
7302. Table 23-2-1-(33.7000)-p, Dec. 31, 1965.
7303. Table 23-9-(12.011)-i, Dec. 31, 1968.
7304. Table 23-9-(12.023)-i, Dec. 31, 1966.
7305. Table 23-9-2-(2.011)-a, Dec. 31, 1964.
7306. Table 23-9-2-(2.0110)-i, Dec. 31, 1966.
7307. Table 23-9-2-(2.0110)-v, June 30, 1981.
7308. Table 23-9-2-(2.0310)-i, Dec. 31, 1966.
7309. Table 23-9-2-(3.011)-a, June 30, 1957.
7310. Table 23-9-2-(3.011)-i, Dec. 31, 1966.
7311. Table 23-9-2-(3.0110)-v, June 30, 1981.
7312. Table 23-9-2-(5.011)-a, June 30, 1963.
7313. Table 23-9-2-(10.011)-k, Dec. 31, 1973.
7314. Table 23-9-2-(10.011)-w, Dec. 31, 1974.
7315. Table 23-10-2-(1.011)-a, June 30, 1981.
7316. Table 23-10-2-(1.0110)-i, Dec. 31, 1966.
7317. Table 23-10-2-(1.0110)-v, June 30, 1981.
7318. Table 23-10-2-(1.0110)-w, June 30, 1981.
7319. Table 23-10-2-(1.0111)-k, June 30, 1956.
7320. Table 23-10-2-(1.0111)-k, June 30, 1981.
7321. Table 23-10-2-(1.013)-a, June 30, 1959.

7322. Table 23-10-2-(1.013)-i, Dec. 31, 1966.
7323. Table 23-10-2-(1.013)-k, June 30, 1981.
7324. Table 23-10-2-(1.0311)-a, June 30, 1956.
7325. Table 23-10-2-(2.011)-a, Dec. 31, 1964.
7326. Table 23-10-2-(2.0111)-k, Dec. 31, 1957.
7327. Table 23-10-2-(2.013)-a, Dec. 31, 1964.
7328. Table 23-10-2-(2.013)-i, Dec. 31, 1966.
7329. Table 23-10-2-(3.011)-i, Dec. 31, 1966.
7330. Table 23-10-2-(3.013)-a, June 30, 1960.
7331. Table 23-10-2-(4.011)-i, Dec. 31, 1966.
7332. Table 23-10-2-(10.011)-a, Dec. 31, 1974.
7333. Table 23-10-2-(10.011)-k, Dec. 31, 1973.
7334. Table 23-10-2-(10.011)-v, Dec. 31, 1981.
7335. Table 23-10-2-(10.011)-w, Dec. 31, 1974.
7336. Table 23-10-2-(10.011)-w, June 30, 1981.
7337. Table 23-10-2-(10.011)-d, June 30, 1980.
7338. Table 23-10-9-2-(10.011)-k, June 30, 1980.
7339. Table 23-10-9-2-(10.011)-v, Dec. 31, 1967.
7340. Table 23-10-9-2-(10.011)-v, June 30, 1975.
7341. Table 23-10-9-2-(10.011)-w, Dec. 31, 1967.
7342. Table 23-10-9-2-(10.011)-w, June 30, 1975.
7343. Table 23-11-2-(1.011)-a, Dec. 31, 1958.
7344. Table 23-11-2-(1.0110)-i, Dec. 31, 1966.
7345. Table 23-11-2-(1.0111)-k, June 30, 1956.
7346. Table 23-11-2-(1.013)-a, Dec. 31, 1958.
7347. Table 23-11-2-(1.0313)-a, June 30, 1956.
7348. Table 23-11-2-(2.013)-a, Dec. 31, 1964.
7349. Table 23-11-2-(2.013)-i, Dec. 31, 1966.
7350. Table 23-11-2-(10.011)-d, June 30, 1980.
7351. Table 23-11-2-(10.011)-k, June 30, 1980.
7352. Table 23-11-2-(10.011)-p, June 30, 1979.
7353. Table 23-11-9-2-(20.011)-p, June 30, 1979.
7354. Table 23-11-10-2-(10.011)-k, June 30, 1980.
7355. Table 23-11-10-2-(10.011)-p, Dec. 31, 1979.
7356. Table 23-11-10-9-2-(10.011)-d, June 30, 1980.
7357. Table 23-11-10-9-2-(10.011)-p, Dec. 31, 1979.
7358. Table 23-12-2-(1.0111)-k, June 30, 1956.
7359. Table 23-12-2-(1.013)-a, Dec. 31, 1964.
7360. Table 23-12-2-(2.011)-a, Dec. 31, 1964.
7361. Table 23-12-2-(2.011)-k, Dec. 31, 1963.
7362. Table 23-12-2-(10.011)-d, June 30, 1980.
7363. Table 23-12-2-(10.011)-p, June 30, 1979.
7364. Table 23-18-2-(1.0111)-a, Dec. 31, 1961.
7365. Table 23-18-2-(1.0111)-k, Dec. 31, 1956.
7366. Table 23-18-2-(1.0113)-a, Dec. 31, 1961.
7367. Table 23-18-2-(1.0113)-i, Dec. 31, 1968.
7368. Table 23-18-2-(1.0125)-a, Dec. 31, 1964.
7369. Table 23-18-2-(1.0125)-i, Dec. 31, 1968.
7370. Table 23-18-2-(1.0122)-a.
7371. Table 23-18-2-(1.0135)-a, Dec. 31, 1964.
7372. Table 23-18-2-(1.0135)-i, Dec. 31, 1968.
7373. Table 23-18-2-(1.0211)-a, June 30, 1960.
7374. Table 23-18-2-(1.0213)-a, June 30, 1960.
7375–7394 Thermodynamics Research Center Hydrocarbon Project.
7375. Table 23-2-(1.101)-v, April 30, 1981.

7376. Table 23-2-(1.101)-w, April 30, 1981.

7377. Table 23-2-(1.1100)-v, Oct. 31, 1981.

7378. Table 23-2-(1.1100)-w, April 30, 1981.

7379. Table 23-2-(1.200)-v, April 30, 1981.

7380. Table 23-2-(1.200)-w, Oct. 31, 1981.

7381. Table 23-2-(1.204)-p, April 30, 1981.

7382. Table 23-2-(1.205)-p, April 30, 1981.

7383. Table 23-2-(1.205)-vc, Oct. 31, 1979.

7384. Table 23-2-(1.207)-m, April 30, 1981.

7385. Table 23-2-(1.207)-n, April 30, 1979.

7386. Table 23-2-(1.207)-p, April 30, 1979.

7387. Table 23-2-(5.1101)-m, Oct. 31, 1981.

7388. Table 23-2-(5.1101)-n, Oct. 31, 1981.

7389. Table 23-2-(5.1101)-p, Oct. 31, 1981.

7390. Table 23-2-(33.1110)-i, April 30, 1979.

7391. Table 23-2-(33.1110)-vc, April 30, 1978.

7392. Table 23-2-(35.5214)-n, April 30, 1978.

7393. Table 23-2-(35.5214)-p, April 30, 1978.

7394. Table 23-2-(35.5214)-v, April 30, 1978.

7395. Thiessen, P. A., and C. Stuber, *Ber. Deut. Chem. Gesell.* **1938,** *71B,* 2103.

7396. Thinh, T. P., R. S. Ramalho, et al., *J. Chem. Eng. Data* **1974,** *19,* 193.

7397. Thirion, P., and E. C. Craven, *J. Appl. Chem.* (London) **1952,** *2,* 210.

7398. Thizy, A., *Compr. Rend. Congr. Intern. Chim.* (Liege) **1958,** 31e; *Ind. Chim. Belg. Suppl., 1,* 619; *C.A.* **1960,** *54,* 3274.

7399. Thodos, G. J., *AIChEJ* **1955,** *1,* 165.

7400. Thodos, G., *AIChEJ* **1955,** *1,* 168.

7401. Thole, F. B., *J. Chem. Soc.* **1910,** 2596.

7402. Thole, F. B., *J. Chem. Soc.* **1913,** 19.

7403. Thole, F. B., *J. Chem. Soc.* **1913,** 317.

7404. Thole, F. B., *J. Chem. Soc.* **1914,** 2004.

7405. Thole, F. B., *Z. Phys. Chem.* (Leipzig) **1910,** *74,* 683.

7406. Thomas, A. B., and E. G. Rochow, *J. Amer. Chem. Soc.* **1957,** *79,* 1843.

7407. Thomas, A. F., and B. Willhalm, *Helv. Chim. Acta* **1964,** *47,* 475.

7408. Thomas, B. W., and W. D. Seyfried, *Anal. Chem.* **1949,** *21,* 1949.

7409. Thomas, G. A., and J. E. Hawkins, *J. Amer. Chem. Soc.* **1954,** *76,* 4856.

7410. Thomas, G. L., and S. Young, *J. Chem. Soc.* **1895,** 1071.

7411. Thomas, J. R., and W. D. Gwinn, *J. Amer. Chem. Soc.* **1949,** *71,* 2785.

7412. Thomas, L., and E. Maxum, *Z. Phys. Chem.* (Leipzig) **1929,** *143A,* 191.

7413. Thomas, L. C., and R. A. Chittenden, *Spectrochim. Acta* **1964,** *20,* 467.

7414. Thomas, L. H., *J. Chem. Soc.* **1960,** 4960.

7415. Thomas, L. H., *Nature* **1946,** *158,* 622.

7416. Thomas, L. H., and R. Meatyard, *J. Chem. Soc.* **1963,** 1986.

7417. Thomas, L. H., and R. Meatyard, *J. Chem. Soc.* **1966,** 92.

7418. Thomas, L. H., R. Meatyard, et al., *J. Chem. Eng. Data* **1979,** *24,* 159.

7419. Thomas, L. H., R. Meatyard, et al., *J. Chem. Eng. Data* **1979,** *24,* 161.

7420. Thomas, R. H. P., and R. H. Harrison, *J. Chem. Eng. Data* **1982,** *27,* 1.

7421. Thomas, R. M., and G. K. Weisse, U.S. Patent 3,020,316, Feb. 6, 1962; *Official Gaz. U.S. Pat. Office* **1962,** *775,* 245.

7422. Thome, L. G., *Ber. Deut. Chem. Gesell.* **1903,** *36,* 582.

7423. Thompson, C. J., H. J. Coleman, et al., *J. Amer. Chem. Soc.* **1954,** *76,* 3445.

7424. Thompson, H. B., and L. A. LaPlanche, *J. Phys. Chem.* **1963,** *67,* 2230.

7425. Thompson, H. W., *Endeavor* **1945,** *4,* 154; *C.A.* **1946,** *40,* 2729.

7426. Thompson, H. W., *J. Amer. Chem. Soc.* **1939,** *61,* 1396.

7427. Thompson, H. W., *J. Chem. Phys.* **1939,** *7,* 441.

7428. Thompson, H. W., *J. Chem. Phys.* **1939,** *7,* 453.

7429. Thompson, H. W., and W. T. Cave, *Trans. Faraday Soc.* **1951,** *47,* 951; *C.A.* **1952,** *46,* 2908.
7430. Thompson, H. W., and J. J. Frewing, *Nature* **1935,** *135,* 507.
7431. Thompson, H. W., and G. P. Harris, *Trans. Faraday Soc.* **1942,** *38,* 37; *C.A.* **1942,** *36,* 5091.
7432. Thompson, H. W., and J. W. Linnett, *Trans. Faraday Soc.* **1936,** *32,* 681; *C.A.* **1936,** *30,* 4371.
7433. Thompson, H. W., and C. H. Purkis, *Trans. Faraday Soc.* **1936,** *32,* 674; *C.A.* **1936,** *30,* 4402.
7434. Thompson, H. W., and R. B. Temple, *J. Chem. Soc.* **1948,** 1428.
7435. Thompson, H. W., and R. B. Temple, *J. Chem. Soc.* **1948,** 1432.
7436. Thompson, H. W., and R. B. Temple, *Trans. Faraday Soc.* **1945,** *41,* 27; *C.A.* **1945,** *39,* 2454.
7437. Thompson, H. W., and P. Torkington, *J. Chem. Soc.* **1944,** 597.
7438. Thompson, H. W., and P. Torkington, *J. Chem. Soc.* **1945,** 640.
7439. Thompson, H. W., and P. Torkington, *Proc. Roy. Soc.* (London) **1945,** *184A,* 3; *C.A.* **1946,** *40,* 276.
7440. Thompson, H. W., and P. Torkington, *Proc. Roy. Soc.* (London) **1945,** *184A,* 21; *C.A.* **1946,** *40,* 277.
7441. Thompson, H. W., and P. Torkington, *Trans. Faraday Soc.* **1946,** *42,* 432; *C.A.* **1946,** *40,* 5641.
7442. Thompson, P. T., R. E. Taylor, et al., *J. Chem. Thermodyn.* **1975,** 7, 547.
7443. Thompson, R. J., *Ind. Eng. Chem.* **1932,** *24,* 620.
7444. Thompson, S. D., D. G. Carroll, et al., *J. Chem. Phys.* **1966,** *45,* 1367.
7445. Thompson, W. E., R. J. Warren, et al., *J. Pharm. Sci.* **1966,** *55,* 110; *C.A.* **1966,** *64,* 6454.
7446. Thomson, J., *Z. Phys. Chem.* (Leipzig) **1905,** *52,* 343.
7447. Thomson, G., *J. Chem. Soc.* **1937,** 1051.
7448. Thomson, G., *J. Chem. Soc.* **1939,** 1118.
7449. Thomson, G. W., *Chem. Rev.* **1946,** *38,* 1.
7450. Thornton, N. V., A. B. Burg, et al., *J. Amer. Chem. Soc.* **1933,** *55,* 3177.
7451. Thorpe, J. F., and G. A. R. Kon, *Organic Syntheses* **1925,** *5,* 37.
7452. Thorpe, T. E., and J. W. Rodger, *J. Chem. Soc.* **1897,** 360.
7453. Thorpe, T. E., and J. W. Rodger, *Phil. Trans.* **1894,** *185A,* 397.
7454. Thosar, B. V., *Z. Phys.* **1937,** *107,* 780; *C.A.* **1938,** *32,* 2027.
7455. Thosar, B. V., and B. K. Singh, *Proc. Indian Acad. Sci.* **1937,** *6A,* 105; *C.A.* **1938,** *32,* 48.
7456. Thronson, E. A., and A. I. Mendolia, U.S. Patent 2,421,441, June 3, 1947; *C.A.* **1947,** *41,* 5543.
7457. Thurber, F. H., and R. C. Thielke, *J. Amer. Chem. Soc.* **1931,** *53,* 1030.
7458. Thwing, C. B., *Z. Phys. Chem.* (Leipzig) **1894,** *14,* 286.
7459. Tickner, A. W., and F. P. Lossing, *J. Phys. Colloid. Chem.* **1951,** *55,* 733.
7460. Tiers, G. V. D., *J. Phys. Chem.* **1963,** *67,* 929.
7461. Tiganik, L., *Z. Phys. Chem.* (Leipzig) **1931,** *13B,* 425; *C.A.* **1931,** *25,* 5805.
7462. Tiganik, L., *Z. Phys. Chem.* (Leipzig) **1931,** *14B,* 135.
7463. Tikhomirova, E. N., I. F. Kovalev, et al., *Opt. Spektrosk.* **1969,** *27,* No. 4, 615; *C.A.* **1970,** *73,* 37301.
7464. Tilicheev, M. D., and Yu. M. Kachmarchik, *Zh. Obshch. Khim.* **1951,** *21,* 78; *C.A.* **1951,** *45,* 7513.
7465. Tilicheev, M. D., and N. V. Miloridora, *Zh. Prikl. Khim.* **1949,** *22,* 611; *C.A.* **1949,** *43,* 8124.
7466. Tilicheev, M. D., V. P. Peshkov, et al., *Zh. Obshch. Khim.* **1951,** *21,* 1229; *C.A.* **1952,** *46,* 1953.
7467. Tilitschejew, M., *J. Russ. Phys. Chem. Soc.* **1926,** *58,* 447; *Chem. Zentr.* **1927,** *I,* 440.
7468. Tilton, L. W., and J. K. Taylor, *J. Res. Nat. Bur. Stand.* **1938,** *20,* 419.
7469. Timm, B., and R. Mecke, *Z. Phys.* **1935,** *97,* 221; *C.A.* **1936,** *30,* 383.
7470. Timmermans, J., *Bull. Soc. Chim. Belges* **1910,** *24,* 244; *C.A.* **1910,** *4,* 2896.
7471. Timmermans, J., *Bull. Soc. Chim. Belges* **1911,** *25,* 300; *Chem. Zentr.* **1911,** *II,* 1015.
7472. Timmermans, J., *Bull. Soc. Chim. Belges* **1914,** *27,* 334; *Chem. Zentr.* **1914,** *I,* 618.
7473. Timmermans, J., *Bull. Soc. Chim. Belges* **1921,** *30,* 62; *Chem. Zentr.* **1921,** *III,* 287.
7474. Timmermans, J., *Bull. Soc. Chim. Belges* **1921,** *30,* 213.
7475. Timmermans, J., *Bull. Soc. Chim. Belges* **1923,** *31,* 389; *Chem. Zentr.* **1923,** *III,* 1137.
7476. Timmermans, J., *Bull. Soc. Chim. Belges* **1927,** *36,* 502; *C.A.* **1928,** *22,* 56.
7477. Timmermans, J., *Bull. Soc. Chim. Belges* **1934,** *43,* 626; *C.A.* **1935,** *29,* 1699.
7478. Timmermans, J., *Bull. Soc. Chim. Belges* **1935,** *44,* 17; *C.A.* **1935,** *29,* 2433.
7479. Timmermans, J., *Bull. Soc. Chim. Belges* **1952,** *61,* 393; *C.A.* **1954,** *48,* 540.
7480. Timmermans, J., *Comm. Phys. Lab. Univ. Leiden,* Suppl. No. *64,* **1929,** 3; *C.A.* **1929,** *23,* 4390.
7481. Timmermans, J., *Compt. Rend.* **1914,** *158,* 790.

7482. Timmermans, J., *Physico-Chemical Constant of Pure Organic Compounds*, Elsevier, New York, 1950.

7483. Timmermans, J., *Physico-Chemical Constants of Pure Organic Compounds, Vol. 2*, Elsevier, New York, 1965.

7484. Timmermans, J., previous publication or unpublished "Annual Tables of Physical Constants," 1941, Sec. 505(c).

7485. Timmermans, J., and Y. Delcourt, *J. Chim. Phys.* **1934**, *31*, 85; *C.A.* **1934**, *28*, 3958.

7486. Timmermans, J., and L. Gillo, *Rocz. Chem.* **1938**, *18*, 812; *C.A.* **1939**, *33*, 4582.

7487. Timmermans, J., and Mme. Hennaut-Roland, *Anales Soc. Espan. Fis. Quim.* **1929**, *27*, 460; *Chem. Zentr.* **1930**, *I*, 1612.

7488. Timmermans, J., and Mme. Hennaut-Roland, *J. Chim. Phys.* **1930**, *27*, 401; *C.A.* **1931**, *25*, 2038; *Chem. Zentr.* **1931**, *I*, 236.

7489. Timmermans, J., and Mme. Hennaut-Roland, *J. Chim. Phys.* **1932**, *29*, 529; *C.A.* **1933**, *27*, 2079; *Beil.* **EIII5**, 571.

7490. Timmermans, J., and Mme. Hennaut-Roland, *J. Chim. Phys.* **1935**, *32*, 501, 589; *C.A.* **1936**, *30*, 2072.

7491. Timmermans, J., and Mme. Hennaut-Roland, *J. Chim. Phys.* **1937**, *34*, 693.

7492. Timmermans, J., and Mme. Hennaut-Roland, *J. Chim. Phys.* **1955**, *52*, 223; *C.A.* **1955**, *49*, 11333.

7493. Timmermans, J., and Mme. Hennaut-Roland, *J. Chim. Phys.* **1959**, *56*, 984; from [7482].

7494. Timmermans, J., and F. Martin, *J. Chim. Phys.* **1926**, *23*, 747; *Chem. Zentr.* **1927**, *I*, 836.

7495. Timmermans, J., and F. Martin, *J. Chim. Phys.* **1928**, *25*, 411; *C.A.* **1928**, *22*, 4024.

7496. Timmermans, J., and R. E. Oesper, *Chemical Species*, Chemical Publishing Co., New York, 1940.

7497. Timmermans, J., A. M. Piette, et al., *Bull. Soc. Chim. Belges* **1955**, *64*, 5; *C.A.* **1955**, *49*, 13711.

7498. Timmermans, J., H. Van der Horst, et al., *Arch. Neerland* **1923**, *6*, No. IIIA, 180.

7499. Timmermans, J., H. Van der Horst, et al., *Compt. Rend.* **1922**, *174*, 365.

7500. Timofejew, G., *Isv. Kieff Polytechn. Inst.* **1905**, 1.

7501. Tindall, J. B., Commercial Solvents Corp., private communication.

7502. Tintea, H., *Bull. Sect. Sci. Acad. Roumaine* **1939**, *22*, 16; *C.A.* **1940**, *34*, 5753.

7503. Tintea, H., *Bull. Soc. Roumaine Phys.* **1942**, *43*, 9; *Chem. Zentr.* **1943**, *I*, 2188; *C.A.* **1944**, *38*, 4512.

7504. Tissier, *Ann. Chim. Phys.* 29, No. 6, 340; *Beil.* **H1**, 406.

7505. Titani, T., *Bull. Inst. Res. Jpn.* **1927**, 671; *Bull. Chem. Soc. Jpn.* **1927**, *2*, 95, 161, 196, 225.

7506. Tjebbes, J., *Acta Chem. Scand.* **1962**, *16*, 953; *C.A.* **1962**, *57*, 9298.

7507. Tjebbes, J., *Pure Appl. Chem.* **1961**, *2*, 129.

7508. Tkachev, P. G., *Gig. Sanit.* **1963**, *28*, No. 4, 3; *C.A.* **1963**, *59*, 4471.

7509. Tkachev, P. G., *Gig. Sanit.* **1971**, *36*, 8; *C.A.* **1972**, *76*, 10848.

7510. Tobin, M. C., *J. Phys. Chem.* **1957**, *61*, 1392.

7511. Toczylkin, L. S., and C. L. Young, *Aust. J. Chem.* **1977**, *30*, 1591.

7512. Todd, S. S., I. A. Hossenlopp, et al., *J. Chem. Thermodyn.* **1978**, *10*, 641.

7513. Todd, S. S., G. D. Oliver, et al., *J. Amer. Chem. Soc.* **1947**, *69*, 1519.

7514. Tokuhiro, T., *J. Chem. Phys.* **1964**, *41*, 438.

7515. Tolkmith, H., *J. Amer. Chem. Soc.* **1953**, *75*, 5273.

7516. Tollens, B., *Ber. Deut. Chem. Gesell.* **1872**, *5*, 68.

7517. Tomashevskaya, L. A., and Z. I. Zholdakova, *Vrach Delo* **1979**, 105; *C.A.* **1979**, *91*, 118197.

7518. Tomecko, J. W., and W. H. Hatcher, *Trans. Roy. Soc. Can.* **1951**, *45*, 39; *C.A.* **1952**, *46*, 9362.

7519. Tong, L. K. J., and W. D. Kenyon, *J. Amer. Chem. Soc.* **1949**, *71*, 1925.

7520. Tongberg, C. O., and M. R. Fenske, *Ind. Eng. Chem.* **1932**, *24*, 814.

7521. Tongberg, C. O., and F. Johnston, *Ind. Eng. Chem.* **1933**, *25*, 733.

7522. Tonomura, T., *Sci. Repts. Tohoku Imp. Univ.* 1st series **1933**, *22*, 104.

7523. Tooke, J. W., U.S. Patent 2,368,050, Jan. 23, 1945; *C.A.* **1945**, *39*, 4091.

7524. Toops, E. E., *J. Phys. Chem.* **1956**, *60*, 304.

7525. Toops, E. E., previously unpublished data.

7526. Toriyama, K., and T. Shimanouchi, *Proc. Intern. Symp. Mol. Struct. Spectry. Tokyo* **1962**, (A103); *C.A.* **1964**, *61*, 1397.

7527. Torkelson, T. R., F. Oyen, et al., *Amer. Ind. Hyg. Assoc. J.* **1960**, *21*, 275; *C.A.* **1960**, *54*, 23047.

7528. Torkelson, T. R., M. A. Wolf, et al., *Amer. Ind. Hyg. Assoc. J.* **1959**, *20*, 224; *C.A.* **1960**, *54*, 6967.

7529. Torkington, P., *Trans. Faraday Soc.* **1949**, *45*, 445; *C.A.* **1949**, *43*, 7342.

7530. Torkington, P., *Trans. Faraday Soc.* **1945,** *41,* 236; *C.A.* **1945,** *39,* 4548.
7531. Tornoe, H., *Ber. Deut. Chem. Gesell.* **1891,** *24,* 2670.
7532. Toropov, A. P., and A. M. Yakubov, *Zh. Fiz. Khim.* **1956,** *30,* 1702; *C.A.* **1957,** *51,* 7193.
7533. Tortorello, A., M. A. Kinsella, et al., *Macromol. Solutions: Solvent-Prop. Relat. Polym.* (Pap. Symph) **1981,** (Pub. 1982), 26; *C.A.* **1982,** *97,* 146.
7534. Toshiyasu, Y., K. Kimura, et al., *Bull. Chem. Soc. Jpn.* **1970,** *43,* 2676.
7535. Tourky, A. R., H. A. Rizk, et al., *Z. Phys. Chem.* (Frankfurt) **1962,** *31,* 161; *C.A.* **1962,** *57,* 201.
7536. Toussaint, W. J., U.S. Patent 2,060,267, Nov. 10, 1936; *C.A.* **1937,** *31,* 420.
7537. Trabert, E., and K. Schaum, *Z. Wiss. Phot.* **1936,** *35,* 153; *C.A.* **1937,** *31,* 319.
7538. Trachuk, S. V., *Khim. Prom.* **1963,** (2), 14; *C.A.* **1963,** *59,* 8374.
7539. Tranchant, J., *Bull. Soc. Chem. Fr.* **1968,** 2216.
7540. Transue, L. F., E. R. Washburn, et al., *J. Amer. Chem. Soc.* **1942,** *64,* 274.
7541. Traube, J., *Ber. Deut. Chem. Gesell.* **1896,** *29,* 1715.
7542. Traynard, P., *Compt. Rend.* **1946,** *223,* 991; *C.A.* **1947,** *41,* 5021.
7543. Traynham, J. G., and G. A. Knesel, *J. Amer. Chem. Soc.* **1965,** *87,* 4220.
7544. Treibs, W., and H. Rohnert, *Chem. Ber.* **1951,** *84,* 433.
7545. Treon, J. F., *Ind. Hyg. Toxicol.* **1943,** *25,* 199.
7546. Treon, J. F., W. E. Crutchfield, et al., *J. Ind. Hyg. Toxicol.* **1943,** *25,* 323.
7547. Treszczanowiez, E., and S. Bakowski, *Przemysl Chem.* **1947,** *3,* 429; *C.A.* **1948,** *42,* 2920.
7548. Trevoy, D. J., and W. A. Torpey, *Anal. Chem.* **1954,** *26,* 492.
7549. Trew, V. C. G., and G. M. C. Watkins, *Trans. Faraday Soc.* **1933,** *29,* 1310; *C.A.* **1934,** *28,* 3954.
7550. Trickey, J. P., *Ind. Eng. Chem.* **1927,** *19,* 643.
7551. Trieschmann, H. G., *Z. Phys. Chem.* (Leipzig) **1935,** *29B,* 328.
7552. Trimble, F., *Ind. Eng. Chem.* **1941,** *33,* 660.
7553. Trimble, H. M., and W. Potts, *Ind. Eng. Chem.* **1935,** *27,* 66.
7554. Trinh, N. Q., *Compt. Rend.* **1948,** *227,* 394; *Beil.* **EIII1,** 2256.
7555. Tromp, S. T. J., *Rec. Trav. Chem. Pays-Bas* **1922,** *41,* 278; *Chem. Zentr.* **1922,** *III,* 755.
7556. Troshina, M. M., *Toksikol. Novykh Prom. Khim. Veshchestv.* **1964,** No. 6, 45; *C.A.* **1965,** *63,* 17017.
7557. Trotter, I. F., and H. W. Thompson, *J. Chem. Soc.* **1946,** 481.
7558. Trunel, P., *Ann. Chim.* **1939,** *12,* 93; *C.A.* **1939,** *33,* 8066.
7559. Trusell, F., and H. Diehl, *Anal. Chem.* **1963,** *35,* 674.
7560. Tschamler, H., *Monatsh. Chem.* **1948,** *79,* 162; *C.A.* **1948,** *42,* 4609; *Beil.* **EIII1,** 1349.
7561. Tschamler, H., and H. Voetter, *Monatsh. Chem.* **1952,** *83,* 302; *C.A.* **1952,** *46,* 6935.
7562. Tschamler, H., F. Wettig, et al., *Monatsh. Chem.* **1949,** *80,* 572; *C.A.* **1950,** *44,* 421; *Beil.* **EIII1,** 1350.
7563. Tschelinzew, W., and B. Pawlow, *J. Russ. Phys. Chem. Soc.* **1913,** *45,* 298.
7564. Tschelinzew, W., and B. Tronow, *J. Russ. Phys. Chem. Gesell.* **1876,** 46; *Chem. Zentr.* **1915,** *II,* 470; *Beil.* **EI20,** 36.
7565. Tschugaeff, L., *Ber. Deut. Chem. Gesell.* **1899,** *32,* 3332.
7566. Tschugaeff, L., *Ber. Deut. Chem. Gesell.* **1900,** *33,* 735.
7567. Tseng, C. L., and C. S. Hou, *J. Chinese Chem. Soc.* **1934,** *2,* 57; *C.A.* **1934,** *28,* 3711.
7568. Tseng, Y-M., and A. R. Thompson, *J. Chem. Eng. Data* **1962,** *7,* 483.
7569. Tseng, Y-M., and A. R. Thompson, *J. Chem. Eng. Data* **1964,** *9,* 264.
7570. Tsirlin, Y. A., and G. S. Muromtseva, *Sb. Tr., Gos. Nauch.-Issled. Inst. Gidrolizn. Sul'fitno-Spirt. Prom.* **1961,** *9,* 205; *C.A.* **1964,** *60,* 11965.
7571. Tsubomura, H., K. Kimura, et al., *Bull. Chem. Soc. Jpn.* **1964,** *37,* 417; *C.A.* **1964,** *61,* 184.
7572. Tsuchiya, M., S. Matsuhira, et al., *Buseki Kagaku* **1965,** *14,* 465; *C.A.* **1965,** *63,* 5499.
7573. Tsuda, K., R. Hayatsu, et al., *Chem. Ind.* (London) **1959,** 1411.
7574. Tsuji, M., Y. Fujisaki, et al., *Oyo Yukuri* **1975,** *9,* 387; *C.A.* **1975,** *83,* 90979.
7575. Tsuzuki, Y., K. Tanabe, et al., *Bull. Chem. Soc. Jpn.* **1966,** *39,* 1389; *C.A.* **1966,** *65,* 13508.
7576. Tuck, D. G., *Trans. Faraday Soc.* **1961,** *57,* 1297.
7577. Tucker, S. W., and S. Walker, *J. Chem. Phys.* **1966,** *45,* 1302.
7578. Tucker, W. C., and J. E. Hawkins, *Ind. Eng. Chem.* **1954,** *46,* 2387.
7579. Tuot, M., and J. Lecomte, *Bull. Soc. Chim.* **1943,** *10,* 542; *C.A.* **1944,** *38,* 5146.
7580. Turk, A., U.S. Patent 2,513,534, July 4, 1950; *C.A.* **1950,** *44,* 8948.

7581. Turkevich, A., and C. P. Smyth, *J. Amer. Chem. Soc.* **1940,** *62,* 2468.
7582. Turkevich, A., and C. P. Smyth, *J. Amer. Chem. Soc.* **1942, *64,* 737.
7583. Turkevich, J., and P. C. Stevenson, *J. Chem. Phys.* **1943,** *11,* 328.
7584. Turnbull, D., and S. H. Maron, *J. Amer. Chem. Soc.* **1943,** *65,* 212.
7585. Turner, D. W., *J. Chem. Soc.* **1959,** 30.
7586. Turner, D. W., *J. Chem. Soc.* **1962,** 847.
7587. Turner, R. B., W. R. Meador, et al., *J. Amer. Chem. Soc.* **1957, *79,* 4127.
7588. Turner, W. E. S., *J. Chem. Soc.* **1910,** 1184.
7589. Turner, W. E. S., and E. W. Merry, *J. Chem. Soc.* **1910,** 2069.
7590. Turner, W. E. S., and C. T. Pollard, *J. Chem. Soc.* **1914,** 1751.
7591. Turova-Pollack, M. B., and M. A. Maslova, *Zh. Obshch. Khim.* **1956,** *26,* 2185; *C.A.* **1957,** *51,* 4972.
7592. Tyrer, D., *J. Chem. Soc.* **1911,** 1633.
7593. Tyrer, D., *J. Chem. Soc.* **1912,** 81.

U

7594. Udovenko, V. V., and R. P. Airapetova, *J. Gen. Chem.* (USSR) **1947,** *17,* 655; *C.A.* **1948,** *42,* 3650.
7595. Udovenko, V. V., and L. P. Aleksanorova, *Zh. Fiz. Khim.* **1960,** *34,* 1366; *Anal. Chem.* **1964,** *36,* 56R, ref. 70F.
7596. Udovenko, V. V., and T. B. Frid, *Zh. Fiz. Khim.* **1948,** *22,* 1126; *C.A.* **1949,** *43,* 476.
7597. Udovenko, V. V., and R. I. Khomenko, *Zh. Obshch. Khim.* **1956,** *26,* 3270; *Beil.* **EIV1,** 3663.
7598. Udránsky, L. V., *Z. Phys. Chem.* (Leipzig) **1888,** *12,* 385; *11,* 537; *12,* 33.
7599. Udránsky, L. V., *Z. Phys. Chem.* (Leipzig) **1889,** *13,* 251.
7600. Uenaka, F., and T. Shono, *Kagaku To Kogyo* (Osaka) **1961,** *35,* 175; *C.A.* **1961,** *55,* 25223.
7601. Ueno, S., and S. Komori, *J. Soc. Chem. Ind. Jpn.* **1946,** *49,* 161; *C.A.* **1948,** *42,* 6738.
7602. Ulich, H., and W. Nespital, *Z. Phys. Chem.* (Leipzig) **1932,** *16B,* 221.
7603. *Ullmanns Encyklopadie der technischen Chemie,* W. Foerst, ed., Urban & Schwarzenberger, Munchen-Berlin.
7604. Ul'yanova, O. D., and V. M. Tatevskii, *Vestn. Moskov Univ. 6, No. 5, Ser. Fiz.-Mat., Estest. Nauk* **1951,** No. 3, 87; *C.A.* **1952,** *46,* 343.
7605. Underwood, H. W., and G. C. Toone, *J. Amer. Chem. Soc.* **1930,** *52,* 391.
7606. Ungnade, H. E., *J. Amer. Chem. Soc.* **1953,** *75,* 432.
7607. Ungnade, H. E., V. Kerr, et al., *Science* **1951,** *113,* 601.
7608. Ungnade, H. E., and D. V. Nightingale, *J. Amer. Chem. Soc.* **1944,** *66,* 1218.
7609. Union Camp Corp., "Capryl Alcohol," Jacksonville, Florida, 1980.
7610. Union Carbide Chemicals Co., "Acids and Anhydrides," New York, 1960.
7611. Union Carbide Chemicals Co., "Acrylonitrile," New York, 1960.
7612. Union Carbide Chemicals Co., "Alkanolamines and Morpholines," New York, 1960.
7613. Union Carbide Chemicals Co., "Alkylene Oxides," New York, 1961.
7614. Union Carbide Chemicals Co., "Esters," New York, 1962.
7615. Union Carbide Chemicals Co., "Ethers and Oxides," New York, 1957.
7616. Union Carbide Chemicals Co., "Organic Chlorine Compounds," New York, 1960.
7617. Union Carbide Corp., "Acetic Acid and Acetic Anhydride," New York, 1971.
7618. Union Carbide Corp., "Alcohols," New York, 1969.
7619. Union Carbide Corp., "Aldehydes," New York, 1960.
7620. Union Carbide Corp., "Alkyl Amines," New York, 1969.
7621. Union Carbide Corp., "Alkyl Amines," New York, 1971.
7622. Union Carbide Corp., "Butyric Acid," New York, 1967.
7623. Union Carbide Corp., "Chemicals and Plastics, Physical Properties," New York, 1971.
7624. Union Carbide Corp., "1,4-Dioxane," New York, 1970.
7625. Union Carbide Corp., "Esters," New York, 1971.
7626. Union Carbide Corp., "Ethanolamines," New York, 1970.
7627. Union Carbide Corp., "Ethylene Amines," New York, 1969.
7628. Union Carbide Corp., "Flexol Plasticizer DOP," Technical Information, New York, 1971.
7629. Union Carbide Corp., "Formic Acid," Product Information, New York, 1969.

7630. Union Carbide Corp., "Glycols," New York, 1971.

7631. Union Carbide Corp., "Isopropanol," Product Bulletin, New York, 1965.

7632. Union Carbide Corp., "Morpholine," Product Information, New York, 1969.

7633. Union Carbide Corp., "Organic Acids," New York, 1971.

7634. Union Carbide Corp., "Propionic Acid and Propionic Anhydride," Product Information, New York, 1968.

7635. Union Carbide Corp., "Solvent Selector," New York, 1968.

7636. Union Carbide Corp., "Valeric Acid," New York, 1967.

7637. Union Carbide Corp., Chemicals Division, "Physical Properties Synthetic Organic Chemicals," New York, 1965.

7638. Unni, A. K. R., L. Elias, et al., *J. Phys. Chem.* **1963,** *67,* 1216.

7639. Uno, T., K. Machida, et al., *Bull. Chem. Soc. Jpn.* **1969,** *42,* 897.

7640. Upadhya, K. N., *J. Sci. Res. Banaras Hindu Univ.* **1963–64,** *14,* 173; *C.A.* **1964,** *61,* 11485.

7641. Upadhyay, R. K., and B. Krishna, *J. Chem. Soc. A* **1970,** 3144.

7642. Urazovskii, S. S., and P. A. Chernyavskii, *Trudy Khar'kov. Politekh. Inst. im. V.I. Lenina 26, Ser. Khim.-Tekhnol.* **1959,** No. 6, 29; *C.A.* **1962,** *56,* 10930.

7643. Urazovskii, S. S., and I. I. Ezhik, *Zh. Fiz. Khim.* **1962,** *36,* 951; *C.A.* **1963,** *58,* 5123.

7644. Urbancova, L., *Chem. Zvesti.* **1959,** *13,* 224; *C.A.* **1959,** *53,* 19500.

7645. Urbancova, L., and F. Gregor, *Plaste Kautschuk* **1963,** *10,* 532; *C.A.* **1964,** *60,* 1528.

7646. Urbani, H., R. Gigli, et al., *J. Chem. Eng. Data* **1980,** *25,* 97.

7647. Urbanski, T., *Tetrahedron* **1959,** *6,* 1.

7648. Urone, P., J. E. Smith, et al., *Anal. Chem.* **1962,** *34,* 476.

7649. Ushakov, S., and J. Feinstein, *Ind. Eng. Chem.* **1934,** *26,* 561.

7650. Usines de Melle, British Patent 586,470, Mar. 20, 1947; *C.A.* **1947,** *41,* 6894.

7651. Utermark, W., and W. Schicke, *Melting Point Tables of Organic Compounds,* 2nd revised and suppl. ed., Interscience-Wiley, New York, 1963.

7652. Uzhdovini, E. R., I. K. Astaf'eva, et al., *Gig. Tr. Prof. Zabol.* **1974,** 58; *C.A.* **1974,** *81,* 418.

V

7653. Vago, E. E., E. M. Tanner, et al., *J. Inst. Petroleum* **1949,** *35,* 293; *C.A.* **1949,** *43,* 6512.

7654. Vahrenkamp, H., and H. Noth, *J. Organometal. Chem.* **1968,** *12,* 281.

7655. Vahrman, M., *J. Appl. Chem.* **1958,** *8,* 485.

7656. Vakhlyueva, V. I., N. A. Kuznetsov, et al., *Opt. Spektrosk.* **1974,** *36,* 481; *Opt. Spectrosc.* **1974,** *36,* 277.

7657. Valette, A., *Ann. Chim.* **1948,** *3,* 644; *C.A.* **1949,** *43,* 2577.

7658. Valovi, V. A., and A. A. Polyakova, *Zh. Org. Khim.* **1967,** *3,* 842; *C.A.* **1967,** *67,* 43241.

7659. Valyashko, N. A., and Y. S. Rozum, *J. Gen. Chem.* (USSR) **1946,** *16,* 593; *C.A.* **1947,** *41,* 1152.

7660. Van Arkel, A. E., P. Meerburg, et al., *Rec. Trav. Chim. Pays-Bas* **1942,** *61,* 767; *C.A.* **1944,** *38,* 3174.

7661. Van Arkel, A. E., and J. L. Snoek, *Z. Phys. Chem.* (Leipzig) **1932,** *18B,* 159.

7662. Van Arkel, A. E., and S. E.Vles, *Rec. Trav. Chim. Pays-Bas* **1936,** *55,* 407; *Chem. Zentr.* **1936,** *II,* 1329; *Beil.* **EIII1,** 154.

7663. Van Dalen, M. J., and P. J. Van den Berg, *J. Organometal. Chem.* **1970,** *24,* 277.

7664. Van Dalen, M. J., and P. J. Van den Berg, *J. Organometal. Chem.* **1969,** *16,* 381.

7665. van den Hande, A., *Bull. Soc. Chim. Belges* **1945,** *54,* 88; *C.A.* **1946,** *40,* 6271.

7666. van den Hande, A., *Bull. Soc. Chim. Belges* **1947,** *56,* 328; *C.A.* **1948,** *42,* 6667.

7667. van der Kelen, G. P., *Bull. Soc. Chim. Belges* **1962,** *71,* 421; *C.A.* **1963,** *58,* 2934.

7668. van der Kelen, O. Volders, et al., *Z. Anorg. Allg. Chem.* **1965,** *338,* 106; *C.A.* **1965,** *63,* 6446.

7669. Van De Rostyne, C., and J. M. Prausnitz, *J. Chem. Eng. Data* **1980,** *25,* 1.

7670. van der Vlies, C., *Rec. Trav. Chim. Pays-Bas* **1965,** *84,* 1289; *C.A.* **1966,** *64,* 4474.

7671. van de Vloed, A., *Bull. Soc. Chim. Belges* **1939,** *48,* 229; *Chem. Zentr.* **1939,** *II,* 3973.

7672. Van Loon, R., S. Fuks, et al., *Bull. Soc. Chim. Belges* **1967,** *76,* 202.

7673. van Os, D., *Chem. Weekblad* **1925,** *22,* 18; *Chem. Zentr.* **1925,** *I,* 1076.

7674. van Pelt, A. J., and J. P. Wibaut, *Rec. Trav. Chim. Pays-Bas* **1941,** *60,* 55; *C.A.* **1941,** *35,* 5089.

7675. Van Reis, M. A., *Ann. Phys.* (Leipzig) **1881,** *13,* 447.

7676. van Risseghem, H., *Bull. Soc. Chim. Belges* **1923**, *32*, 144; *C.A.* **1923**, *17*, 2262.

7677. van Risseghem, H., *Bull. Soc. Chim. Belges* **1926**, *35*, 328; *Chem. Zentr.* **1927**, *I*, 53.

7678. Van Tussenbroeck, M. J., thesis, Delft, 1929; "Annual Tables of Physical Constants," Sec. 921(c), 1941.

7679. Van Wazer, J. R., and C. F. Callis, *J. Amer. Chem. Soc.* **1956**, *78*, 5715.

7680. Van Winkle, W., *J. Pharmacol.* **1941**, *72*, 227; *C.A.* **1941**, *35*, 5989.

7681. Varma, P. S., and C. Sreenivasmurthyachar, *J. Indian Chem. Soc.* **1936**, *13*, 187; *C.A.* **1936**, *30*, 5949.

7682. Varoanyi, G., T. Farago, et al., *Acta Chim. Acad. Sci. Hung.* **1965**, *43*, 205.

7683. Varshni, Y. P., *Z. Phys. Chem.* (Leipzig) **1954**, *203*, 247.

7684. Varshni, Y. P., and S. S. Mitra, *Z. Phys. Chem.* (Leipzig) **1954**, *203*, 380.

7685. Varushchenko, R. M., and L. L. Bulgakova, *Tr. Khim. Khim. Tekhnol.* **1974**, 69; *C.A.* **1975**, *82*, 65231.

7686. Vasil'ev, I. A., and A. D. Korkhov, *Tr. Khim. Khim. Tekhnol.* **1974**, 103; *C.A.* **1975**, *82*, 65233.

7687. Vasil'ev, I. A., and A. D. Korkhov, *Zh. Fiz. Khim.* **1973**, *47*, 2710; *Russ. J. Phys. Chem.* **1973**, *47*, 1527.

7688. Vasil'eva, T. F., E. N. Zhil'tsova, et al., *Zh. Fiz. Khim.* **1972**, *46*, 541; *Russ. J. Phys. Chem.* **1972**, *46*, 315.

7689. Vasil'eva, T. F., E. N. Zhil'tsova, et al., *Zh. Fiz. Khim.* **1972**, *46*, 541; *Russ. J. Phys. Chem.* **1972**, *46*, 315.

7690. Vasil'eva, V. N., K. A. Kocheshkov, et al., *Dokl. Akad. Nauk SSSR* **1962**, *143*, 844; *C.A.* **1962**, *57*, 3328.

7691. Vaughan, M. F., *Natl. Phys. Lab. (U.K.), Div. Chem. Stand., Rep.* **1978**, *86*, 42 pp; *C.A.* **1978**, *89*, 199507.

7692. Vaughan, W. C., *Phil. Mag.* **1939**, *27*, 669; *C.A.* **1939**, *33*, 7162.

7693. Vaughan, W. E., F. F. Rust, et al., *J. Org. Chem.* **1942**, *7*, 477.

7694. Vaughn, J. W., and C. F. Hawkins, *J. Chem. Eng. Data* **1964**, *9*, 140.

7695. Vaughn, W. E., S. B. W. Roeder, et al., *J. Chem. Phys.* **1963**, *39*, 701.

7696. Vavon, G., and P. Mottez, *Compt. Rend.* **1944**, *218*, 557.

7697. Vay, P. M., *J. Chim. Phys. Physicochim. Biol.* **1971**, *68*, 1757.

7698. Vdovenko, S. E., V. Ya. Semenii, et al., *Zh. Obshch. Khim.* **1976**, *46*, 2613; *J. Gen. Chem.* (USSR) **1976**, *46*, 2491.

7699. Vdovin, G. P., Yu. P. Egorov, et al., *Zh. Org. Khim.* **1974**, *10*, 1355; *C.A.* **1974**, *81*, 90725.

7700. Veatch, F., J. D. Idol, et al., *Hydrocarbon Process. Petrol. Refiner* **1964**, *43*, No. 4, 177.

7701. Vecerek, B., G. I. Kondraskin, et al., *Bratisl. Lek. Listy* **1976**, *65*, 9; *C.A.* **1977**, *86*, 51207.

7702. Vedal, D., O. H. Ellestad, et al., *Spectrochim. Acta, Part A.* **1975**, *31A*, 339.

7703. Vedal, D., O. H. Ellestad, et al., *Spectrochim. Acta, Part A.* **1976**, *32A*, 877.

7704. Veibel, S., F. Lundqvist, et al., *Bull. Soc. Chim. Pays-Bas* **1939**, *6*, 990; *C.A.* **1939**, *33*, 7273.

7705. Veksler, M. A., *J. Exptl. Theoret. Phys.* (USSR) **1939**, *9*, 616; *C.A.* **1939**, *33*, 9072.

7706. Veksler, R. I., *Zh. Anal. Khim.* **1949**, *4*, 14; *C.A.* **1950**, *44*, 484.

7707. Velasco-Durantez, M., *An. Real Soc. Espan. Fis. Quim.* **1927**, *25*, 252; *Chem. Zentr.* **1927**, *II*, 2649.

7708. Velsicol Chemical Corp., "Technical Data Bulletin," Sales Service Bulletin No. 8163, Chattanooga, 1965.

7709. Vencov, St., *Bull. Sect. Sci. Acad. Roumaine* **1943**, *26*, 89; *C.A.* **1945**, *39*, 2696.

7710. Veneraki, I. E., and N. K. Bolotin, *Gaz. Prom.* **1964**, *9*(3), 30; *C.A.* **1964**, *60*, 15457.

7711. Venkareswarlu, K., and G. Thyagarajan, *Z. Phys.* **1959**, *154*, 70; *C.A.* **1959**, *53*, 5871.

7712. Venkatasetty, H. V., and G. H. Brown, *J. Phys. Chem.* **1962**, *66*, 2075.

7713. Venkateswaran, C. S., and N. S. Pandya, *Proc. Indian Acad. Sci.* **1942**, *15A*, 390; *C.A.* **1942**, *36*, 6904.

7714. Venkateswaran, S., *Nature* **1928**, *122*, 506; *Chem. Zentr.* **1928**, *II*, 2705.

7715. Venkateswaran, S., *Phil Mag.* **1929**, *7*, 597; *C.A.* **1929**, *23*, 5108.

7716. Venkateswaran, S., and S. Bhagavantam, *Indian J. Physics* **1933**, *7*, 585; *C.A.* **1933**, *27*, 4481.

7717. Venkateswaran, S., and S. Bhagavantam, *Proc. Roy. Soc.* (London) **1930**, *128A*, 252; *C.A.* **1930**, *24*, 5627.

7718. Venkateswarlu, K., *Curr. Sci.* **1947**, *16*, 21; *C.A.* **1947**, *41*, 3697.

7719. Venkateswarlu, K., *Proc. Indian Acad. Sci.* **1944**, *19A*, 111; *C.A.* **1944**, *38*, 4867.
7720. Venkateswarlu, K., and S. Mariam, *Z. Phys.* **1962**, *168*, 195; *C.A.* **1962**, *57*, 4202.
7721. Venkateswarlu, K., M. G. Pillai, et al., *Proc. Indian Acad. Sci.* **1961**, *53A*, 195; *C.A.* **1961**, *55*, 24232.
7722. Verdonck, L., and G. P. Van der Kelen, *Spectrochim. Acta, Part A* **1972**, *28*, 55.
7723. Vereinigte Glanzstoff-Fabriken Akt.-Ges., British Patent 803,375, Oct. 22, 1958; C.A. **1959**, *53*, 4822.
7724. Vergnoux, A. M., and R. Dadillon, *Compt. Rend.* **1941**, *213*, 166; *C.A.* **1942**, *36*, 5705.
7725. Verhoek, F. H., *J. Amer. Chem. Soc.* **1936**, *58*, 2577.
7726. Verhoek, F. H., and A. L. Marshall, *J. Amer. Chem. Soc.* **1939**, *61*, 2737.
7727. Verhulst, G., and C. Glorieux, *Bull. Soc. Chim. Belges* **1932**, *41*, 501; *C.A.* **1933**, *27*, 702.
7728. Verkade, P. E., and J. Coops, *Rec. Trav. Chim. Pays-Bas* **1927**, *46*, 903; *C.A.* **1928**, *22*, 571.
7729. Verley, A., *Bull. Soc. Chim. Fr.* **1899**, *21*, 275.
7730. Vermillion, G., and M. A. Hill, *J. Amer. Chem. Soc.* **1945**, *67*, 2209.
7731. Vermillion, H. E., B. Werbel, et al., *J. Amer. Chem. Soc.* **1941**, *63*, 1346.
7732. Vernon, A. A., and E. V. Kring, *J. Amer. Chem. Soc.* **1949**, *71*, 1888.
7733. Vershinin, P. V., E. V. Varshaver, et al., USSR Patent 206,576, Dec. 8, 1967; *C.A.* **1968**, *69*, 51533.
7734. Vertut, M. C., J. P. Fayet, et al., *Bull. Soc. Chim. Fr.* **1972**, 166.
7735. Veselinovic, D. S., and D. L. Malesev, *Glas. Hem. Drus. Beograd* **1982**, *47*, 347; *C.A.* **1982**, *97*, 189010.
7736. Veselov, V. G., *Nauch. Tr. Omsk. Med. Inst.* **1969**, No. 88, 314; *C.A.* **1971**, *75*, 116956.
7737. Veselov, V. G., *Mater. Nauch.-Prakt. Konf. Molodykh. Gig. Sanit. Vrachei 11th* **1967**, 121; *C.A.* **1970**, *72*, 41134.
7738. Vesely, F., M. Zabransky, et al., *Collect. Czech. Chem. Commun.* **1979**, *44*, 3529.
7739. Vespignani, G. B., *Gazz. Chim. Ital.* **1903**, *33*, 73; *J. Chem. Soc.* **1903Ai**, 545.
7740. Vidulich, G. A., D. F. Evans, et al., *J. Phys. Chem.* **1967**, *71*, 656.
7741. Vig, O. P., J. C. Kapur, et al., *J. Indian Chem. Soc.* **1968**, *45*, 1026.
7742. Vilallonga, F., *Pubs. Centro Invest. Tisiol.* (Buenos Aires) **1946**, *10*, 59; *C.A.* **1947**, *41*, 5020.
7743. Vilcu, R., and St. Perisanu, *Rev. Roum. Chim.* **1979**, *24*, 237.
7744. Vilcu, R., and St. Perisanu, *Rev. Roum. Chim.* **1980**, *25*, 619.
7745. Villamanan, M. A., C. Casanova, et al., *Thermochim. Acta* **1982**, *52*, 279.
7746. Vinokurov, D. M., *Izv. Vyssh. Ucheb. Zaved., Lesnai Zhur.* **1959**, 2, No. 2, 155; *C.A.* **1960**, *54*, 266.
7747. Vinokurov, D. M., *Nauch. Dokl. Vysshei Shkoly. Lessinzhener Delo* **1958**, No. 4, 193. *C.A.* **1959**, *53*, 21628.
7748. Vinson, C. G., and J. J. Martin, *J. Chem. Eng. Data* **1963**, *8*, 74.
7749. Virginia Chemicals, Inc., "Virginia Amines," West Norfolk, Virginia, 1967.
7750. Virin, L. I., Ya. A. Safin, et al., *Zh. Fiz. Khim.* **1965**, *39*, 2824; *C.A.* **1966**, *64*, 9054.
7751. Virobyants, R. A., M. A. Nechaeva, et al., *Khim. Sera-organ. Soedin.*, *Soderzhashch. Neft. Nefte-prod.*, *Akad. Nauk SSSR, Bashkirsk. Filial* **1964**, *6*, 193; *C.A.* **1964**, *61*, 5592.
7752. Virtanen, P. O. I., and T. Sodervall, *Kemistilehti* **1967**, *B40*, 337.
7753. Vistron Corp., "Methacrylonitrile," Technical Bulletin, Cleveland, Ohio, 1967.
7754. Viswanath, D. S., and N. R. Kuloor, *J. Chem. Eng. Data* **1966**, *11*, 69.
7755. Viswanath, D. S., and N. R. Kuloor, *J. Chem. Eng. Data* **1966**, *11*, 544.
7756. Vitali, G., M. A. Berchiesi, et al., *J. Chem. Eng. Data* **1979**, *24*, 169.
7757. Vogel, A. I., *J. Chem. Soc.* **1934**, 333.
7758. Vogel, A. I., *J. Chem. Soc.* **1938**, 1323.
7759. Vogel, A. I., *J. Chem. Soc.* **1943**, 636.
7760. Vogel, A. I., *J. Chem. Soc.* **1946**, 133.
7761. Vogel, A. I., *J. Chem. Soc.* **1948**, 607.
7762. Vogel, A. I., *J. Chem. Soc.* **1948**, 610.
7763. Vogel, A. I., *J. Chem. Soc.* **1948**, 616.
7764. Vogel, A. I., *J. Chem. Soc.* **1948**, 624.
7765. Vogel, A. I., *J. Chem. Soc.* **1948**, 644.
7766. Vogel, A. I., *J. Chem. Soc.* **1948**, 654.

7767. Vogel, A. I., *J. Chem. Soc.* **1948,** 1809.

7768. Vogel, A. I., *J. Chem. Soc.* **1948,** 1814.

7769. Vogel, A. I., *J. Chem. Soc.* **1948,** 1820.

7770. Vogel, A. I., *J. Chem. Soc.* **1948,** 1825.

7771. Vogel, A. I., *J. Chem. Soc.* **1948,** 1833.

7772. Vogel, A. I., and D. M. Cowan, *J. Chem. Soc.* **1943,** 16.

7773. Vogel, A. I., W. T. Cresswell, et al., *J. Chem. Soc.* **1952,** 514.

7774. Vogel, I., *J. Chem. Soc.* **1928,** 2010.

7775. Vogel-Hogler, R., *Acta Phys. Austriaca* **1948,** *1,* 311; *C.A.* **1948,** *42,* 6663.

7776. Vohra, S. P., and K. A. Kobe, *J. Chem. Eng. Data* **1959,** *4,* 329.

7777. Voigt, D., *Anal. Chem.* **1949,** *4,* 429; *Annales de Chimie* **1949,** *4,* 393.

7778. Volarovich, M. P., and N. N. Stepanenko, *J. Exptl. Theoret. Phys.* (USSR) **1940,** *10,* 817; *C.A.* **1941,** *35,* 5365.

7779. Volarovich, M., and N. Stepanenko, *J. Exptl. Theoret. Phys.* (USSR) **1944,** *14,* 313; *C.A.* **1944,** *39,* 1338.

7780. Vold, R. D., *J. Amer. Chem. Soc.* **1937,** *59,* 1515.

7781. Vol'kenshtein, M. V., *Acta Physicochim. URSS* **1942,** *15,* 120; *C.A.* **1943,** *37,* 2267.

7782. Vol'kenshtein, M. V., and M. A. Eliashevich, *Compt. Rend. Acad. Sci. URSS* **1944,** *43,* 51; *C.A.* **1945,** *39,* 2030.

7783. Volmer, M., and M. Marder, *Z. Phys. Chem.* (Leipzig) **1931,** *154A,* 971.

7784. Volyak, L. D., *Zh. Fiz. Khim.* **1956,** *30,* 2244; *C.A.* **1957,** *51,* 9284.

7785. Von Buenau, G., G. Schade, et al., *Fresenius' Z. Anal. Chem.* **1969,** *244,* 7; *C.A.* **1969,** *70,* 106671.

7786. Vondráček, R., and J. Dostál, *Chem. Listy* **1937,** *31,* 15; *C.A.* **1937,** *31,* 3675.

7787. von Euler, H., and A. Olander, *Z. Phys. Chem.* (Leipzig) **1927,** *131,* 107.

7788. von Mikusch, J. E., *Farbe Lack* **1949,** *55,* 241; *C.A.* **1949,** *43,* 8700.

7789. Vorländer, D., *Ann. Chim.* (Paris) **1894,** *280,* 167; *J. Chem. Soc. A* **1895,** 17.

7790. Vorländer, D., *Justus Liebigs Ann. Chem.* **1894,** *280,* 167.

7791. Vorländer, D., and R. Walter, *Z. Phys. Chem.* (Leipzig) **1925,** *118,* 1.

7792. Vorob'ev, A. F., and P. N. Yakovlev, *Zh. Fiz. Khim.* **1982,** *56,* 1933; *Russ. J. Phys. Chem.* **1982,** *56,* 1181.

7793. Voronkov, M. G., *Zh. Fiz. Khim.* **1949,** *23,* 1311; *C.A.* **1950,** *44,* 3759.

7794. Voronkov, M. G., *Zh. Fiz. Khim.* **1952,** *26,* 813; *C.A.* **1952,** *46,* 11125.

7795. Voronkov, M. G., *Zh. Obshch. Khim.* **1950,** *20,* 2060; *C.A.* **1951,** *45,* 5607; *Chem. Zentr.* **1951,** *II,* 661.

7796. Voroshin, E. M., *Izv. Akad. Nauk SSSR, Ser. Fiz.* **1953,** *17,* 717; *C.A.* **1954,** *48,* 6826.

7797. Voronkov, M. G., V. Yu. Emel'yanov, et al., *Zh. Obshch. Khim.* **1978,** *48,* 429; *J. Gen. Chem.* (USSR) **1978,** *48,* 384.

7798. Voronkov, M. G., I. T. Yushmanova, et al., *Zh. Org. Khim.* **1975,** *11,* 902; *C.A.* **1975,** *82,* 170243.

7799. Vorozhtsov, N. N., V. E. Platonov, et al., *Izv. Akad. Nauk SSSR, Ser. Khim.* **1963,** 1524; *C.A.* **1963,** *59,* 13846.

7800. Vostrikov, A. A., E. P. Sheludyakov, et al., *Izv. Sib. Otd. Akad. Nauk SSSR, Ser. Tekh. Nauk* **1969,** 105; *C.A.* **1970,** *72,* 47683.

7801. Vriens, G. N., and A. G. Hill, *Ind. Eng. Chem.* **1952,** *44,* 2732.

7802. Vuks, M., *Acta Physico-Chim.* (URSS) **1945,** *20,* 851; *C.A.* **1946,** *40,* 4293.

7803. Vul'fson, S. G., "Vest. Stud. Nauch. Obshchest., Kazan. Gos. Univ." **1966,** No. 3, 94; *C.A.* **1968,** *69,* 55330.

7804. Vul'fson, N. S., L. S. Golovkina, et al., *Izv. Akad. Nauk SSSR, Ser. Khim.* **1967,** 2415; *C.A.* **1968,** *68,* 57291.

7805. Vvedenskii, A. A., P. Y. Ivannikov, et al., *Zh. Obshch. Khim.* **1949,** *19,* 1094; *C.A.* **1949,** *43,* 8247.

7806. Vvedenskii, A. A., and D. M. Maiorov, *Zh. Obshch. Khim.* **1957,** *27,* 2052; *C.A.* **1958,** *52,* 4968.

7807. Vvedenskii, A. A., and Y. A. Timin, *Novosti Neft. Gaz. Tekhn. Neftepererabolka Neftekhim.* **1961,** No. 11, 31; *C.A.* **1963,** *58,* 3962.

7808. Vvedenskii, A. A., M. I. Yakushkin, et al., *Khim. Prom.* **1962,** 11; *C.A.* **1963,** *58,* 2366.

7809. Vyakhirev, D. A., L. I. Galkin, et al., *Khim. Prom.-st.* (Moscow) **1976,***(10),* 739; *C.A.* **1977,** *86,* 189052.

7810. Vyas, A., and H. N. Srivastava, *J. Sci. and Ind. Research* (India) **1959**, *18B*, 399; *Beil.* **EIV1**, 3271.
7811. Vymetal, J., and L. Kulhanek, *Chem. Prum.* **1970**, *20(12)*, 565; *C.A.* **1971**, *74*, 144311.

W

7812. Wacker-Chemie G.m.b.H., British Patent 842,731, July 27, 1960; *C.A.* **1961**, *55*, 3432.
7813. Wacker, P. F., R. K. Cheney, et al., *J. Res. Nat. Bur. Stand.* **1947**, *38*, 651.
7814. Wackher, R. C., C. B. Linn, et al., *Ind. Eng. Chem.* **1945**, *37*, 464.
7815. Waclawek, W., and W. Guzovski, *Rocz. Chem.* **1968**, *42*, 2183.
7816. Waddington, G., and D. R. Douslin, *J. Amer. Chem. Soc.* **1947**, *69*, 2275.
7817. Waddington, G., J. W. Knowlton, et al., *J. Amer. Chem. Soc.* **1949**, *71*, 797.
7818. Waddington, G., J. C. Smith, et al., *J. Phys. Chem.* **1962**, *66*, 1074.
7819. Waddington, G., and S. S. Todd, *J. Amer. Chem. Soc.* **1947**, *69*, 22.
7820. Wade, J., and H. Finnemore, *J. Chem. Soc.* **1909**, 1842.
7821. Wadso, I., *Acta Chem. Scand.* **1966**, *20*, 544.
7822. Wadso, I., *Acta Chem. Scand.* **1968**, *22*, 2438.
7823. Wadso, I., *Acta Chem. Scand.* **1969**, *23*, 2061.
7824. Waeschke, H., and R. Mitzner, *Z. Chem.* **1979**, *19*, 379.
7825. Wafa, O. A., A. Lentz, et al., *Z. Anorg. Allg. Chem.* **1971**, *380*, 128.
7826. Wagman, D. D., W. H. Evans, et al., "Selected Values of Chemical Thermodynamic Properties," NBS Technical Note 270-3, U.S. Govt. Printing Office, Washington, 1968.
7827. Wagman, D. D., and W. H. Evans, *J. Phys. Chem. Ref. Data* **1982**, *11*, Supplement 2.
7828. Wagman, D. D., K. E. Kilpatrick, et al., *J. Res. Nat. Bur. Stand.* **1945**, *34*, 143.
7829. Wagner, C. D., R. H. Smith, et al., *Anal. Chem.* **1947**, *19*, 976.
7830. Wagner, E. C., *J. Chem. Ed.* **1930**, *7*, 1135.
7831. Wagner, J., *Z. Phys. Chem.* (Leipzig) **1938**, *40B*, 36; *C.A.* **1938**, *32*, 6549.
7832. Wagner, J., *Z. Phys. Chem.* (Leipzig) **1938**, *40B*, 439; *C.A.* **1938**, *32*, 8937.
7833. Wagner, J., *Z. Phys. Chem.* (Leipzig) **1939**, *45B*, 69; *C.A.* **1940**, *34*, 904.
7834. Wahl, H., *Compt. Rend.* **1936**, *20*, 2161; *C.A.* **1936**, *30*, 6346.
7835. Waksmundzki, A., *Rocz. Chem.* **1938**, *18*, 865; *C.A.* **1939**, *33*, 6688.
7836. Walden, P., *Ber. Deut. Chem. Gesell.* **1907**, *40*, 3214.
7837. Walden, P., *Bull. Acad. St. Petersburg Acad. Sci.* **1911**, 1055; *Nature* **1912**, 387; *Pharm. J.*, *88*, 125; *C.A.* **1912**, *6*, 1394; *Chem. Zentr.* **1912**, *I*, 122.
7838. Walden, P., *Bull. Acad. St. Petersburg* **1913**, 559; *Chem. Zentr.* **1913**, *II*, 331.
7839. Walden, P., *Bull. Acad. St. Petersburg* **1915**, 1485; *Chem. Zentr.* **1925**, *I*, 1676.
7840. Walden, P., *Z. Anorg. Chem.* (Leipzig) **1910**, *68*, 307.
7841. Walden, P., *Z. Phys. Chem.* (Leipzig) **1903**, *43*, 385.
7842. Walden, P., *Z. Phys. Chem.* (Leipzig) **1903**, *46*, 103.
7843. Walden, P., *Z. Phys. Chem.* (Leipzig) **1906**, *54*, 129.
7844. Walden, P., *Z. Phys. Chem.* (Leipzig) **1906**, *55*, 207.
7845. Walden, P., *Z. Phys. Chem.* (Leipzig) **1906**, *55*, 281.
7846. Walden, P., *Z. Phys. Chem.* (Leipzig) **1907**, *58*, 479.
7847. Walden, P., *Z. Phys. Chem.* (Leipzig) **1907**, *59*, 385.
7848. Walden, P., *Z. Phys. Chem.* (Leipzig) **1909**, *65*, 129.
7849. Walden, P., *Z. Phys. Chem.* (Leipzig) **1909**, *66*, 385.
7850. Walden, P., *Z. Phys. Chem.* (Leipzig) **1910**, *70*, 569.
7851. Walden, P., *Z. Phys. Chem.* (Leipzig) **1910**, *73*, 257.
7852. Walden, P., *Z. Phys. Chem.* (Leipzig) **1912**, *78*, 257.
7853. Walden, P., *Z. Phys. Chem.* (Leipzig) **1930**, *147A*, 1.
7854. Walden, P., *Z. Phys. Chem.* (Leipzig) **1932**, *162A*, 1; *Chem. Zentr.* **1933**, *I*, 1252.
7855. Walden, P., L. F. Audrieth, et al., *Z. Phys. Chem.* (Leipzig) **1932**, *160A*, 337.
7856. Walden, P., and E. J. Birr, *Z. Phys. Chem.* (Leipzig) **1929**, *144A*, 269.
7857. Walden, P., and E. J. Birr, *Z. Phys. Chem.* (Leipzig) **1931**, *153A*, 1.
7858. Walden, P., and E. J. Birr, *Z. Phys. Chem.* (Leipzig) **1933**, *163A*, 281.
7859. Walden, P., and E. J. Birr, *Z. Phys. Chem.* (Leipzig) **1933**, *165A*, 26.
7860. Walden, P., and E. J. Birr, *Z. Phys. Chem.* (Leipzig) **1934**, *168A*, 107.
7861. Walden, P., and G. Busch, *Z. Phys. Chem.* (Leipzig) **1929**, *140A*, 89.

7862. Walden, P., and M. Centnerszwer, Z. Phys. Chem. (Leipzig) **1906**, *55*, 321.

7863. Walden, P., and H. Gloy, Z. Phys. Chem. (Leipzig) **1929**, *144A*, 395.

7864. Walden, P., and R. Swinne, Z. Phys. Chem. (Leipzig) **1912**, *79*, 700.

7865. Walden, P., H. Ulich, et al., Z. Phys. Chem. (Leipzig) **1925**, *114*, 275.

7866. Walden, P., H. Ulich, et al., Z. Phys. Chem. (Leipzig) **1925**, *116*, 261.

7867. Walden, P., and O. Werner, Z. Phys. Chem. (Leipzig) **1924**, *111*, 465.

7868. Walden, P., and O. Werner, Z. Phys. Chem. (Leipzig) **1926**, *124*, 405.

7869. Walden, P., S. Zastrow, et al., Ann. Acad. Sci. Fennical **1929**, *29A*, No. 23.

7870. Waldron, J. W., U.S. Patent 2,454,019, Nov. 16, 1948; C.A. **1949**, *43*, 1810.

7871. Walisch, W., and H. G. Hahn, Chem. Ber. **1961**, *94*, 790.

7872. Walisch, W., and H. P. Ruppersberg, Chem. Ber. **1959**, *92*, 2622.

7873. Walker, C. A., Ind. Eng. Chem. **1949**, *41*, 2640.

7874. Walker, J., Z. Phys. Chem. (Leipzig) **1889**, *4*, 319.

7875. Walker, J. W., and F. M. G. Johnson, J. Chem. Soc. **1905**, 1597.

7876. Wall, F. T., and W. F. Claussen, J. Amer. Chem. Soc. **1939**, *61*, 2679.

7877. Wall, F. T., and W. F. Claussen, J. Amer. Chem. Soc. **1939**, *61*, 2812.

7878. Wallace, T. A., and W. R. G. Atkins, J. Chem. Soc. **1912**, 1179.

7879. Wallace, T. A., and W. R. G. Atkins, J. Chem. Soc. **1912**, 1958.

7880. Wallace, W. J., and A. L. Mathews, J. Chem. Eng. Data **1964**, *9*, 267.

7881. Wallace, W. J., and A. L. Mathews, J. Chem. Eng. Data **1963**, *8*, 496.

7882. Wallach, O., Justus Liebigs Ann. Chem. **1893**, *275*, 104.

7883. Wallach, O., Justus Liebigs Ann. Chem. **1888**, *275*, 195.

7884. Wallach, O., Justus Liebigs Ann. Chem. **1906**, *350*, 141; C.A. **1907**, *1*, 848.

7885. Wallach, O., Justus Liebigs Ann. Chem. **1908**, *359*, 287; C.A. **1908**, *2*, 1831.

7886. Wallach, O., Justus Liebigs Ann. Chem. **1908**, *360*, 82; C.A. **1908**, *2*, 2085.

7887. Wallach, O., Justus Liebigs Ann. Chem. **1918**, *414*, 312.

7888. Wallach, O., Ber. Deut. Chem. Gesell. **1907**, *40*, 600; C.A. **1907**, *1*, 1284.

7889. Wallerstein, R., T. Alba, et al., U.S. Patent 2,366,990, Jan. 9, 1945; C.A. **1945**, *39*, 2378.

7890. Walls, W. S., and C. P. Smyth, J. Chem. Phys. **1933**, *1*, 337.

7891. Walsh, A. D., Trans. Faraday Soc. **1946**, *42*, 62; C.A. **1946**, *40*, 4957.

7892. Walsh, A. D., and P. A. Warsop, Trans. Faraday Soc. **1967**, *63*, 524.

7893. Walsh, A. D., and P. A. Warsop, Trans. Faraday Soc. **1968**, *64*, 1418.

7894. Walsh, A. D., and P. A. Warsop, Trans. Faraday Soc. **1968**, *64*, 1425.

7895. Walsh, A. D., P. A. Warsop, et al., Trans. Faraday Soc. **1968**, *64*, 1432.

7896. Walsh, P. N., and N. O. Smith, Chem. Eng. Data **1961**, *6*, 33.

7897. Walter, M., and L. Ramaley, Anal. Chem. **1973**, *45*, 165.

7898. Walther, R., Arch. Gewerbepath. Gewerbehyg. **1942**, *11*, 326; C.A. **1943**, *37*, 5492.

7899. Walton, J. H., and L. L. Withrow, J. Amer. Chem. Soc. **1923**, *45*, 2689.

7900. Wan, S. W., J. Phys. Chem. **1941**, *45*, 903.

7901. Wang, T. S., J. Org. Chem. **1963**, *28*, 245.

7902. Wang, W., and H. Shih, Chemistry (Taiwan) **1955**, 325; C.A. **1956**, *50*, 11226.

7903. Wang, Y. L., Z. Phys. Chem. (Leipzig) **1940**, *B45*, 323; C.A. **1940**, *34*, 4626.

7904. Ward, A. L., and S. S. Kurtz, Ind. Eng. Chem., Anal. Ed. **1938**, *10*, 559.

7905. Ward, G. E., O. G. Pettijohn, et al., J. Amer. Chem. Soc. **1944**, *66*, 541.

7906. Ward, H. L., J. Phys. Chem. **1934**, *38*, 761.

7907. Ware, D. R., and B. A. Dunell, Trans. Faraday Soc. **1965**, *61*, 834.

7908. Waring, W., Chem. Rev. **1952**, *51*, 171.

7909. Warshowsky, B., and P. J. Elving, Ind. Eng. Chem., Anal. Ed. **1946**, *18*, 253.

7910. Wartenberg, H. v., and J. Schiefer, Z. Anorg. Allg. Chem. **1955**, *278*, 326; C.A. **1955**, *49*, 15436.

7911. Washburn, E. R., and C. H. Shildneck, J. Amer. Chem. Soc. **1933**, *55*, 2354.

7912. Washburn, E. W., and J. W. Read, J. Amer. Chem. Soc. **1919**, *41*, 729.

7913. Washburn, H. W., H. F. Wiley, et al., Ind. Eng. Chem., Anal. Ed. **1943**, *15*, 541.

7914. Washburn, H. W., H. F. Wiley, et al., Ind. Eng. Chem., Anal. Ed. **1945**, *17*, 74.

7915. Washburn, R. M., E. Levens, et al., Metallic-Organic Compounds, pp. 129 ff. (Advances in Chemistry Series, Vol 23), American Chemical Society, Washington, 1959.

7916. Washburn, W. H., Anal. Chem. **1958**, *30*, 156.

7917. Wasserman, A., Z. Phys. Chem. (Leipzig) **1940**, *146*, 418.

7918. Wasylishen, R., and T. Schaefer, Can. J. Chem. **1971**, *49*, 94.

7919. Wasylishen, R., and T. Schaefer, Can. J. Chem. **1971**, *49*, 3627.

7920. Watanabe, M., and R. M. Fuoss, J. Amer. Chem. Soc. **1956**, *78*, 527.

7921. Waterman, H. I., and S. H. Bertram, Rec. Trav. Chim. Pays-Bas **1927**, *46*, 699; C.A. **1928**, *22*, 218.

7922. Waterman, H. I., and W. J. C. de Kok, Rec. Trav. Chim. Pays-Bas **1933**, *52*, 251; C.A. **1933**, *27*, 2931.

7923. Waterman, H. I., and W. J. C. de Kok, Rec. Trav. Chim. Pays-Bas **1933**, *52*, 298; C.A. **1933**, *27*, 2932.

7924. Waterman, H. I., and W. J. C. de Kok, Rec. Trav. Chim. Pays-Bas **1934**, *53*, 725; C.A. **1934**, *28*, 5397.

7925. Waterman, H. I., P. van't Spijker, et al., Rec. Trav. Chim. Pays-Bas **1929**, *48*, 1191; C.A. **1930**, *24*, 824.

7926. Waterman, H. I., and H. A. Van Westen, Rec. Trav. Chim. Pays-Bas **1929**, *48*, 637; C.A. **1929**, *23*, 4455.

7927. Waters Associates, Inc., "Know More About Polymers," Milford, Massachusetts, 1975.

7928. Walters, D. N., and L. A. Woodward, Proc. Roy. Soc. (London) **1958**, *246A*, 119; C.A. **1958**, *52*, 19514.

7929. Watson, H. E., G. P. Kane, et al., Proc. Roy. Soc. (London) **1936**, *156A*, 137.

7930. Watson, H. E., and K. L. Ramaswamy, Proc. Roy. Soc (London) **1936**, *156A*, 130.

7931. Watson, R. W., J. A. R. Coope, et al., Can. J. Chem. **1951**, *29*, 885.

7932. Watt, G. W., and J. B. Otto, J. Amer. Chem. Soc. **1947**, *69*, 836.

7933. Wauchope, R. D., and F. W. Getzen, J. Chem. Eng. Data **1972**, *17*, 38.

7934. Wawzonek, S., and M. E. Runner, J. Electrochem. Soc. **1952**, *99*, 457.

7935. Way, K. R., and M. E. Russell, J. Phys. Chem. **1965**, *69*, 4420.

7936. Wayland, B. B., and R. S. Drago, J. Amer. Chem. Soc. **1965**, *87*, 2372.

7937. Wear, R. L., J. Amer. Chem. Soc. **1951**, *73*, 2390.

7938. Wearn, R. B., W. M. Murray, et al., Anal. Chem. **1948**, *20*, 922.

7939. Weast, R. C. ed., Handbook of Chemistry and Physics, 50th ed., The Chemical Rubber Co., Cleveland, 1970.

7940. Weatherall, I. L., and I. C. Bathurst, Lab. Pract. **1970**, *19*, 606; C.A. **1970**, *73*, 109603.

7941. Webb, J. L. A., and J. D. Webb, J. Amer. Chem. Soc. **1949**, *71*, 2285.

7942. Webb, T. J., and C. H. Lindsley, J. Amer. Chem. Soc. **1934**, *56*, 874.

7943. Webber, H., Res. Develop. Ind. **1962**, No. 10, 47; C.A. **1963**, *58*, 3608.

7944. Weber, A., A. G. Meister, et al., J. Chem. Phys. **1953**, *21*, 930.

7945. Weber, L. A., and J. E. Kilpatrick, J. Chem. Phys. **1962**, *36*, 829.

7946. Wedekind, E., and T. Goost, Ber. Deut. Chem. Gesell. **1916**, *49*, 942.

7947. Weeks, R. H., U.S. Patent 3,696,156, Oct. 3, 1972; C.A. **1973**, *78*, 29313.

7948. Weger, F., Liebigs Annalen der Chemie **1883**, *221*, 61.

7949. Wehle, H., Z. Anal. Chem. **1959**, *169*, 241; Chem. Zentr. **1960**, *1279*.

7950. Wehman, T. C., private communication.

7951. Weidmann, H., and H. K. Zimmerman, Texas J. Sci. **1959**, *11*, 212; C.A. **1960**, *54*, 1987.

7952. Weigert, F., Optische Methoden der Chemie, Akademische Verlagsgesellschaft, Leipzig, 1927.

7953. Weiler-Feilchenfeld, H., A. Pullman, et al., J. Mol. Struct. **1970**, *6*, 297.

7954. Weinberg, D. S., and C. Djerassi, J. Org. Chem. **1966**, *31*, 115.

7955. Weinberg, D. S., C. Stafford, et al., Tetrahedron **1968**, *24*, 5409.

7956. Weisburger, E. K., B. M. Ulland, et al., J. Natl. Cancer Inst. **1981**, *67*, 75.

7957. Weiss, S., and G. E. Leroi, Spectrochim. Acta, Part A **1969**, *25*, 1759.

7958. Weissberger, A., Physical Methods of Organic Chemistry, 3rd ed., Vol. I, Pt. I, Interscience, New York, 1959.

7959. Weissberger, A., Physical Methods of Organic Chemistry, 2nd ed., Vol. I, Pt. II, Interscience, New York, 1950.

7960. Weissberger, A., Physical Methods of Organic Chemistry, 3rd ed., Vol. I, Pt. IV, Interscience, New York, 1960.

7961. Weissberger, A., and M. R. J. Dack, eds., Techniques of Chemistry, Vol. VIII, Part I, Wiley-Interscience, New York, 1975.

7962. Weissberger, A., and K. Fasold, *Z. Phys. Chem.* (Leipzig) **1931**, *157A*, 65.
7963. Weissberger, A., and B. W. Rossiter, eds., *Techniques of Chemistry, Vol. I, Physical Methods of Chemistry*, Part V, Wiley-Interscience, New York, 1971.
7964. Weissberger, A., and R. Sangerwald, *Ber. Deut. Chem. Gesell.* **1932**, *65*, 701.
7965. Weissberger, A., and R. Sangerwald, *Z. Phys. Chem.* (Leipzig) **1933**, *20B*, 145; *C.A.* **1933**, *27*, 2139.
7966. Weissberger, A., and E. Strasser, *J. Prakt. Chem.* **1933**, *135*, 209.
7967. Weissenberger, G., R. Henke, et al., *Monatsh. Chem.* **1925**, *46*, 483; *C.A.* **1926**, *20*, 2851.
7968. Weissenberger, G., F. Schuster, et al., *Monatsh. Chem.* **1926**, *46*, 281; *C.A.* **1926**, *20*, 1785; *Chem. Zentr.* **1926**, *I*, 2454.
7969. Weissgerber, R., and O. Kruber, *Ber. Deut. Chem. Gesell.* **1919**, *52B*, 346; *C.A.* **1919**, *13*, 2357.
7970. Weissgerber, R., and O. Kruber, *Ber. Deut. Chem. Gesell.* **1920**, *53*, 1551.
7971. Weissler, A., *J. Amer. Chem. Soc.* **1948**, *70*, 1634.
7972. Weissman, H. B., R. B. Bernstein, et al., *J. Chem. Phys.* **1955**, *23*, 544.
7973. Weitz, E., A. Roth, et al., *Justus Liebigs Ann. Chem.* **1921**, *425*, 176.
7974. Weizmann, C., U.S. Patent 2,456,549, Dec. 14, 1948; *C.A.* **1949**, *43*, 2632.
7975. Weizmann, C., E. Bergmann, et al., *Chemistry and Industry* **1937**, 587; *C.A.* **1937**, *31*, 7408.
7976. Welch, S. C., and M. E. Walters, *J. Org. Chem.* **1978**, *43*, 2715.
7977. Wells, A. J., and E. B. Wilson, *J. Chem. Phys.* **1941**, *9*, 314.
7978. Weltner, W., and K. S. Pitzer, *J. Amer. Chem. Soc.* **1951**, *73*, 2606.
7979. Weniger, W., *Phys. Rev.* **1910**, *31*, 388.
7980. Wentworth, T. O., British Patent 567,288, Feb. 7, 1945; *C.A.* **1947**, *41*, 1805.
7981. Wepster, B. M., *Rec. Trav. Chim.* **1957**, *76*, 357.
7982. Werner, A., and P. Ferchland, *Z. Anorg. Chem.* **1897**, *15*, 1; *Chem. Zentr.* **1897**, *II*, 461.
7983. Werner, E. A., *Analyst* **1933**, *58*, 335.
7984. Werner, M., *Mitt. Ver. Grosskesselbesitzer* **1956**, *41*, 119; *C.A.* **1956**, *50*, 12606.
7985. Werner, R. L., and K. G. O'Brien, *Aust. J. Chem.* **1955**, *8*, 355; *C.A.* **1955**, *49*, 15486.
7986. Wertyporoch, E., and B. Adamus, *Z. Phys. Chem.* (Leipzig) **1934**, *168A*, 31.
7987. Wertyporoch, E., and W. Firla, *Z. Phys. Chem.* (Leipzig) **1932**, *162A*, 398.
7988. West, W., and M. Farnsworth, *Trans. Faraday Soc.* **1931**, *27*, 145; *C.A.* **1931**, *25*, 3916.
7989. Westermark, H., *Acta Chem. Scand.* **1955**, *9*, 54; *C.A.* **1956**, *50*, 3084.
7990. Wettaw, J. F., E. E. McEnary, et al., *J. Chem. Eng. Data* **1969**, *14*, 181.
7991. Whalley, W. B., *J. Chem. Soc. Ind.* (London) **1947**, *66*, 427; *C.A.* **1948**, *42*, 3310.
7992. Whanger, J. G., and H. H. Sisler, *J. Amer. Chem. Soc.* **1953**, *75*, 5188.
7993. Wheeler, D. H., and R. W. Riemenschneider, *Oil & Soap* **1939**, *16*, 207; *C.A.* **1940**, *34*, 278.
7994. Wheeler, O. H., and J. L. Mateos, *Can. J. Chem.* **1958**, *36*, 1431.
7995. Wheeler, O. H., and J. L. Mateos, *J. Org. Chem.* **1956**, *21*, 1110.
7996. Wheland, G. W., and J. Farr, *J. Amer. Chem. Soc.* **1943**, *65*, 1433.
7997. Whetsel, K., W. E. Roberson, et al., *Anal. Chem.* **1958**, *30*, 1594.
7998. Whetstone, R. R., and A. B. Seaver (to Shell Development), U.S. Patent 2,639,297, May 19, 1953; *C.A.* **1954**, *48*, 3996.
7999. Whiffen, D. H., *J. Chem. Soc.* **1956**, 1350.
8000. Whiffen, D. H., and H. W. Thompson, *J. Chem. Soc.* **1945**, 268.
8001. Whipple, E. B., J. H. Goldstein, et al., *J. Amer. Chem. Soc.* **1960**, *82*, 3010.
8002. Whipple, E. B., W. E. Stewart, et al., *J. Chem. Phys.* **1961**, *34*, 2136.
8003. Whitby, G. S., *J. Chem. Soc.* **1926**, 1458.
8004. White, A. H., and S. O. Morgan, *J. Chem. Phys.* **1937**, *5*, 655.
8005. White, A. H., and S. O. Morgan, *Physics* **1932**, *2*, 313.
8006. White, G. F., and H. C. Jones, *Amer. Chem. J.* **1910**, *44*, 157.
8007. White, G. F., A. B. Morrison, et al., *J. Amer. Chem. Soc.* **1924**, *46*, 961.
8008. White, H. F., *Anal. Chem.* **1964**, *36*, 1291.
8009. White, J. D., and F. W. Rose, *J. Res. Nat. Bur. Stand.* **1931**, *7*, 907.
8010. White, J. D., and F. W. Rose, *J. Res. Nat. Bur. Stand.* **1932**, *9*, 711.
8011. White, J. L., and P. P. Somers, *J. Ind. Hyg.* **1931**, *13*, 273; *C.A.* **1932**, *26*, 3294.
8012. White, J. R., *J. Amer. Chem. Soc.* **1950**, *72*, 1859.
8013. White, P. T., D. G. Barnard-Smith, et al., *Ind. Eng. Chem.* **1952**, *44*, 1430.
8014. White, T., *J. Chem. Soc.* **1943**, 238.

8015. White, W. P., *J. Phys. Chem.* **1920,** *24,* 393.

8016. Whitehead, E. V., R. A. Dean, et al., *J. Amer. Chem. Soc.* **1951,** *73,* 3632.

8017. Whitford, E. L., *J. Amer. Chem. Soc.* **1925,** *47,* 2934.

8018. Whitman, G. M., U.S. Patent 2,418,441, Apr. 1, 1947; *C.A.* **1947,** *41,* 4169.

8019. Whitman, J. L., and D. M. Hurt, *J. Amer. Chem. Soc.* **1930,** *52,* 4762.

8020. Whitmore, F. C., *Organic Chemistry,* Van Nostrand, Princeton, New Jersey, 1937.

8021. Whitmore, F. C., *Organic Syntheses, Vol. 7,* Wiley, New York, 1927.

8022. Whitmore, F. C., *Organic Syntheses, Vol. 12,* Wiley, New York, 1932.

8023. Whitmore, F. C., and G. H. Fleming, *J. Amer. Chem. Soc.* **1933,** *55,* 3803.

8024. Whitmore, F. C., and J. M. Herndon, *J. Amer. Chem. Soc.* **1933,** *55,* 3428.

8025. Whitmore, F. C., and F. A. Karnatz, *J. Amer. Chem. Soc.* **1938,** *60,* 2533.

8026. Whitmore, F. C., and J. H. Olewine, *J. Amer. Chem. Soc.* **1938,** *60,* 2569.

8027. Whitmore, F. C., and H. S. Rothrock, *J. Amer. Chem. Soc.* **1932,** *54,* 3431.

8028. Whitmore, F. C., and C. T. Simpson, *J. Amer. Chem. Soc.* **1933,** *55,* 3809.

8029. Whitmore, F. C., L. H. Sommer, et al., *J. Amer. Chem. Soc.* **1946,** *68,* 475.

8030. Whitmore, F. C., and L. H. Sommer, *J. Amer. Chem. Soc.* **1946,** *68,* 481.

8031. Whitmore, F. C., C. J. Stehman, et al., *J. Amer. Chem. Soc.* **1933,** *55,* 3807.

8032. Whitmore, F. C., and J. D. Surmatis, *J. Amer. Chem. Soc.* **1940,** *62,* 995.

8033. Whitmore, F. C., L. H. Sutherland, et al., *J. Amer. Chem. Soc.* **1942,** *64,* 1360.

8034. Whitmore, F. C., J. S. Whitaker, et al., *J. Amer. Chem. Soc.* **1941,** *63,* 643.

8035. Whitmore, W. F., and E. Lieber, *Ind. Eng. Chem., Anal. Ed.* **1935,** *7,* 127.

8036. Whitnack, G. C., and R. W. Moshier, *Ind. Eng. Chem., Anal. Ed.* **1944,** *16,* 496.

8037. Whittaker, A. G., and S. Siegel, *J. Chem. Phys.* **1965,** *42,* 3320.

8038. Wibaut, J. P., and H. Geldof, *Rec. Trav. Chim. Pays-Bas* **1946,** *65,* 125; *C.A.* **1946,** *40,* 3952.

8039. Wibaut, J. P., H. Hoog, et al., *Rec. Trav. Chim. Pays-Bas* **1939,** *58,* 329; *C.A.* **1939,** *33,* 6231.

8040. Wibaut, J. P., and S. L. Langedijk, *Rec. Trav. Chim. Pays-Bas* **1940,** *59,* 1220; *C.A.* **1941,** *35,* 5089.

8041. Wibaut, J. P., and A. J. van Pelt, *Rec. Trav. Chim. Pays-Bas* **1938,** *57,* 1055; *C.A.* **1939,** *33,* 1263.

8042. Wicker, R. J., *J. Chem. Soc.* **1956,** 2165.

8043. Wicker, R. J., *J. Chem. Soc.* **1957,** 3299.

8044. Widiger, A. H., U.S. Patent 3,087,969, Apr. 30, 1963; *Official Gaz. U.S. Patent Office* **1963,** 789.

8045. Widmark, G., *Acta Chem. Scand.* **1955,** *9,* 925; *C.A.* **1956,** *50,* 4503.

8046. Widmark, G., *Acta Chem. Scand.* **1955,** *9,* 938; *C.A.* **1956,** *50,* 4077.

8047. Wieczorek, S. A., *J. Chem. Thermodyn.* **1984,** *16,* 115.

8048. Wieczorek, S. A., and R. Kobayashi, *J. Chem. Eng. Data* **1980,** *25,* 302.

8049. Wieczorek, S. A., and R. Kobayashi, *J. Chem. Eng. Data* **1981,** *26,* 11.

8050. Wiedemann, H. G., *Thermochimia Acta* **1972,** *3,* 355.

8051. Wiemann, J., O. Convert, et al., *Bull. Soc. Chim. Fr.* **1966,** 1760; *C.A.* **1966,** *65,* 12087.

8052. Wienhaus, H., *Ber. Deut. Chem. Gesell.* **1920,** *53,* 1656.

8053. Wiezevich, P. J., and P. K. Frolich, *Ind. Eng. Chem.* **1934,** *26,* 267.

8054. Wightman, E. P., and H. C. Jones, *Amer. Chem. J.* **1911,** *46,* 56.

8055. Wikoff, H. L., B. R. Cohen, et al., *Ind. Eng. Chem., Anal. Ed.* **1940,** *12,* 92.

8056. Wilcox, W. R., R. Friedenberg, et al., *Chem. Rev.* **1964,** *64,* 187.

8057. Wilde, J. H. de, *Z. Anorg. Allg. Chem.* **1937,** *233,* 411; *C.A.* **1938,** *32,* 13.

8058. Wiley, G. A., R. L. Hershkowitz, et al., *J. Amer. Chem. Soc.* **1964,** *86,* 964.

8059. Wiley, P. F., and V. Hsiung, *Spectrochim. Acta, Part A* **1970,** *26,* 2229.

8060. Wiley, R. H., *J. Polym. Sci., Part A-1* **1971,** *9,* 129.

8061. Wiley, R. H., and R. R. Garrett, *J. Polym. Sci.* **1951,** *7,* 121; *C.A.* **1951,** *45,* 9998.

8062. Wilf, J., and A. Ben-Naim, *J. Chem. Phys.* **1979,** *70,* 3079.

8063. Wilhelm, E., A. Faradjzadeh, et al., *J. Chem. Thermodyn.* **1979,** *11,* 979.

8064. Wilhelm, E., R. Schano, et al., *Trans. Faraday Soc.* **1969,** *65,* 1443.

8065. Wilhoit, R. C., and B. J. Zwolinski, *J. Phys. Chem. Ref. Data* **1973,** *2* (Suppl. 1).

8066. Wilkie, A. L., and B. D. Shaw, *J. Chem. Soc. Ind.* **1927,** *46,* 469T.

8067. Wilkinson, R., *J. Chem. Soc.* **1931,** 3057.

8068. Wille, H., R. Oberkobusch, et al., German Patent 1,245,963, Aug. 3, 1967; *C.A.* **1968,** *68,* 87192.

8069. Wille, H., and L. Rappen, German Patent 1,099,546, Feb. 16, 1961; *C.A.* **1961,** *55,* 27860.

8070. Wille, H., and L. Rappen, German Patent 1,124,046, Feb. 22, 1962; *C.A.* **1962,** *57,* 7174.

8071. Willhite, C. C., *Clin. Toxicol.* **1981**, *18*, 991; *C.A.* **1981**, *95*, 109705.
8072. Williams, D., *Phys. Rev.* **1938**, *54*, 504.
8073. Williams, D., T. Gatica, et al., *J. Phys. Chem.* **1937**, *41*, 645.
8074. Williams, F. T., P. W. K. Flanagan, et al., *J. Org. Chem.* **1965**, *30*, 2674.
8075. Williams, J. W., *J. Amer. Chem. Soc.* **1930**, *52*, 1831.
8076. Williams, J. W., *Phys. Z.* **1928**, *29*, 174; *C.A.* **1928**, *22*, 1897.
8077. Williams, J. W., *Phys. Z.* **1928**, *29*, 204.
8078. Williams, J. W., *Phys. Z.* **1928**, *29*, 271; *C.A.* **1928**, *22*, 3089.
8079. Williams, J. W., *Z. Phys. Chem.* (Leipzig) **1928**, *138A*, 75; *C.A.* **1929**, *23*, 1027.
8080. Williams, J. W., and F. Daniels, *J. Amer. Chem. Soc.* **1924**, *46*, 903.
8081. Williams, J. W., and F. Daniels, *J. Amer. Chem. Soc.* **1925**, *47*, 1490.
8082. Williams, J. W., and J. M. Fogelberg, *J. Amer. Chem. Soc.* **1931**, *53*, 2096.
8083. Williams, J. W., and I. J. Krchma, *J. Amer. Chem. Soc.* **1926**, *48*, 1888.
8084. Williams, J. W., and E. F. Ogg, *J. Amer. Chem. Soc.* **1928**, *50*, 94.
8085. Williams, R. B., S. H. Hastings, et al., *Anal. Chem.* **1952**, *24*, 1911.
8086. Williams, T., and J. J. Sudborough, *J. Chem. Soc.* **1912**, 414.
8087. Williamson, A. G., and N. F. Catherall, *J. Chem. Eng. Data* **1971**, *16*, 335.
8088. Williamson, I., *Nature* **1951**, *167*, 316.
8089. Williamson, K. D., and R. H. Harrison, *J. Chem. Phys.* **1957**, *26*, 1409.
8090. Willingham, C. B., and F. D. Rossini, *J. Res. Nat. Bur. Stand.* **1946**, *37*, 15.
8091. Willingham, C. B., W. J. Taylor, et al., *J. Res. Nat. Bur. Stand.* **1945**, *35*, 219.
8092. Willis, W. L., *Cryogenics* **1966**, *6*, 279.
8093. Willson, J. E., D. E. Brown, et al., *Toxicol. Appl. Pharmacol.* **1965**, *7*, 104.
8094. Willstätter, R., and V. L. King, *Ber. Deut. Chem. Gesell.* **1913**, *46*, 527.
8095. Willstätter, R., and F. Seitz, *Ber. Deut. Chem. Gesell.* **1923**, *56*, 1388.
8096. Willstätter, R., and F. Seitz, *Ber. Deut. Chem. Gesell.* **1924**, *57*, 683.
8097. Wilmshurst, J. K., *Can. J. Chem.* **1958**, *36*, 285.
8098. Wilmshurst, J. K., and H. J. Bernstein, *Can. J. Chem.* **1957**, *35*, 734.
8099. Wilmshurst, J. K., and H. J. Bernstein, *Can. J. Chem.* **1957**, *35*, 911.
8100. Wilson, A. L., *Ind. Eng. Chem.* **1935**, *27*, 867.
8101. Wilson, C. E., and H. J. Lucas, *J. Amer. Chem. Soc.* **1936**, *58*, 2396.
8102. Wilson, C. L., British Patent 569,625, July 1, 1945; *C.A.* **1947**, *41*, 6275.
8103. Wilson, H. N., and A. E. Heron, *J. Chem. Soc. Ind.* **1941**, *60*, 168.
8104. Wilson, H. N., and W. C. Hughes, *J. Soc. Chem. Ind.* **1939**, *58*, 74.
8105. Wilson, H. W., *Appl. Spectrosc.* **1976**, *30*, 209.
8106. Wilson, H. W., R. W. MacNamee, et al., *J. Raman Spectrosc.* **1981**, *11*, 252.
8107. Wilson, J. M., *Experimentia* **1960**, *16*, 403; *C.A.* **1961**, *55*, 7043.
8108. Wimer, W. E., B. O. Bowers, et al., British Patent 1,103,230, Feb. 14, 1968; *C.A.* **1968**, *68*, 114044.
8109. Wincel, H., and Z. Kecki, *Nukleonika* **1965**, *10*, 567; *C.A.* **1966**, *64*, 17390.
8110. Windwer, S., and B. R. Sundheim, *J. Phys. Chem.* **1962**, *66*, 1254.
8111. Wing, J., and W. H. Johnston, *J. Amer. Chem. Soc.* **1957**, *79*, 864.
8112. Wingefors, S., *Acta Chim. Scand., Ser. A* **1980**, *34*, 297.
8113. Winkler, L. W., *Ber. Deut. Chem. Gesell.* **1905**, *38*, 3612.
8114. Winters, G., and J. Shelmerdine, British Patent 777,961, July 3, 1957; *C.A.* **1957**, *51*, 15098.
8115. Wirth, H. E., and P. I. Slick, *J. Phys. Chem.* **1962**, *66*, 2277.
8116. Wislicenus, J., *Justus Liebigs Ann. Chem.* **1878**, *92*, 106; *J. Chem. Soc.* **1878A**, 776.
8117. Wislicenus, J., *Ber. Deut. Chem. Gesell.* **1874**, *7*, 893.
8118. Wislicenus, J., and L. Kaufmann, *Ber. Deut. Chem. Gesell.* **1895**, *28*, 1323; *J. Prakt. Chem.* **1896**, *54*, 54.
8119. Wiswall, R. H., and C. P. Smyth, *J. Chem. Phys.* **1941**, *9*, 356.
8120. Witanowski, M., Y. Urbanski, et al., *J. Amer. Chem. Soc.* **1964**, *86*, 2569.
8121. Witschonke, C. R., *Anal. Chem.* **1954**, *26*, 562.
8122. Wittek, H., *Monatsh. Chem.* **1941**, *73*, 231; *C.A.* **1942**, *36*, 5092.
8123. Wittek, H., *Z. Phys. Chem.* (Leipzig) **1940**, *48B*, 1; *C.A.* **1941**, *35*, 3172.
8124. Wittek, H., *Z. Phys. Chem.* (Leipzig) **1942**, *51B*, 187; *C.A.* **1943**, *37*, 4013.
8125. Wittek, H., *Z. Phys. Chem.* (Leipzig) **1942**, *52B*, 315, 330; *Beil.* **EIII6**, 799.

8126. Wladislaw, B., A. Giora, et al., *J. Chem. Soc. B* **1966,** 586.
8127. Wohler, F., *Justus Liebigs Ann. Chem.* **1852,** *81*, 376.
8128. Wojciechowski, M., *J. Res. Nat. Bur. Stand.* **1936,** *17*, 453.
8129. Wojciechowski, M., *J. Res. Nat. Bur. Stand.* **1936,** *17*, 459.
8130. Wojciechowski, M., *J. Res. Nat. Bur. Stand.* **1936,** *17*, 721.
8131. Wojciechowski, M., *J. Res. Nat. Bur. Stand.* **1937,** *19*, 347.
8132. Wojciechowski, M., *Ind. Chim.* (Warsaw) **1939,** *5*, 129.
8133. Wojciechowski, M., and E. R. Smith, *J. Res. Nat. Bur. Stand.* **1937,** *18*, 499.
8134. Wojciechowski, M., and E. R. Smith, *Rocz. Chem.* **1937,** *17*, 118; *Chem. Zentr.* **1937,** *II*, 1554.
8135. Wojcik, B., and H. Adkins, *J. Amer. Chem. Soc.* **1933,** *55*, 1293.
8136. Wolf, G., *Helv. Chim. Acta* **1972,** *55*, 1446.
8137. Wolf, H. C., *Z. Naturforsch.* **1955,** *10a*, 270; *C.A.* **1957,** *51*, 3289.
8138. Wolf, H. C., and H. P. Deutsch, *Naturwissenschaften* **1954,** *41*, 425; *Chem. Zentr.* **1955,** 10708.
8139. Wolf, K. L., *Phys. Z.* **1930,** *31*, 227; *Chem. Zentr.* **1930,** *I*, 2521.
8140. Wolf, K. L., *Z. Phys. Chem.* (Leipzig) **1929,** *2B*, 39; *Beil.* **EII1,** 758.
8141. Wolf, K. L., and W. J. Gross, *Z. Phys. Chem.* (Leipzig) **1931,** *14B*, 305; *Chem. Zentr.* **1931,** *II*, 3581.
8142. Wolf, K. L., and W. Herold, *Z. Phys. Chem.* (Leipzig) **1931,** *13B*, 201.
8143. Wolf, K. L., H. Pahlke, et al., *Z. Phys. Chem.* (Leipzig) **1935,** *28B*, 1.
8144. Wolf, K. L., and O. Strasser, *Z. Phys. Chem.* (Leipzig) **1933,** *21B*, 389.
8145. Wolf, K. L., and H. Weghofer, *Z. Phys. Chem.* (Leipzig) **1938,** *39B*, 194; *Chem. Zentr.* **1938,** *II*, 2722.
8146. Wolfe, W. C., *Anal. Chem.* **1962,** *34*, 1328.
8147. Wolff, H., and H. Ludwig, *Ber. Bunsenges. Phys. Chem.* **1964,** *68*, 143.
8148. Wolff, H., and U. Schmidt, *Ber. Bunsenges. Phys. Chem.* **1964,** *68*, 579.
8149. Wolff, H., and D. Staschewski, *Ber. Bunsenges. Phys. Chem.* **1964,** *68*, 135.
8150. Wolff, H., and D. Staschewski, *Z. Elektrochem.* **1961,** 65, 840; *C.A.* **1962,** *57*, 6758.
8151. Wolfhagen, J. L., and E. J. Soltes, *J. Chem. Ed.* **1977,** *54*, 619.
8152. Wollman, S. H., *J. Chem. Phys.* **1946,** *14*, 123.
8153. Wood, C. E., J. E. Such, et al., *J. Chem. Soc.* **1923,** 600.
8154. Wood, S. E., *J. Amer. Chem. Soc.* **1937,** *59*, 1510.
8155. Wood, S. E., and J. P. Brusie, *J. Amer. Chem. Soc.* **1943,** *65*, 1891.
8156. Wood, S. E., and J. A. Gray, *J. Amer. Chem. Soc.* **1952,** *74*, 3729.
8157. Woodman, A. L., W. J. Murbach, et al., *J. Phys. Chem.* **1960,** *64*, 658.
8158. Woolley, E. M., and R. E. George, *J. Soln. Chem.* **1974,** *3*, 119.
8159. Worden, E. F., *U.S. Atomic Energy Comm.* UCRL-8508; *C.A.* **1959,** *53*, 13772.
8160. Woringer, B., *Z. Phys. Chem.* (Leipzig) **1900,** *34*, 257.
8161. Woronkow, M. G., *J. Phys. Chem.* (Leningrad) **1948,** *22*, 975; *Chem. Zentr.* **1948,** *I*, E193.
8162. Worsham, J. E., and M. E. Hobbs, *J. Amer. Chem. Soc.* **1954,** *76*, 206.
8163. Wotiz, J. H., and F. A. Miller, *J. Amer. Chem. Soc.* **1949,** *71*, 3441.
8164. Wotiz, J. H., F. A. Miller, et al., *J. Amer. Chem. Soc.* **1950,** *72*, 5055.
8165. Wowzonek, S., and M. E. Runner, *J. Electrochem. Soc.* **1952,** *99*, 457.
8166. Wrewsky, M. S., and A. A. Glagoleva, *Z. Phys. Chem.* (Leipzig) **1928,** *133A*, 370.
8167. Wrewsky, M. S., K. P. Miscenko, et al., *Z. Phys. Chem.* (Leipzig) **1928,** *133A*, 362.
8168. Wright, C. P., M. Murray-Rust, et al., *J. Chem. Soc.* **1931,** 199.
8169. Wright, F. J., *J. Chem. Eng. Data* **1961,** *6*, 454.
8170. Wright, O. L., and L. E. Mura, *J. Chem. Ed.* **1966,** *43*, 150.
8171. Wright, R. H., and R. S. E. Serenius, *J. Appl. Chem.* (London) **1954,** *4*, 615.
8172. Wrobel, J., and K. Galuszko, *Tetrahedron Lett.* **1965,** 4381; *C.A.* **1966,** *64*, 9550.
8173. Wu, T., *Phys. Rev.* **1934,** *46*, 465.
8174. Wu, T. K., and B. P. Dailey, *J. Chem. Phys.* **1964,** *41*, 3307.
8175. Wu, T. Y., *J. Chem. Phys.* **1939,** *7*, 965.
8176. Wu, Y., and H. L. Friedman, *J. Phys. Chem.* **1966,** *70*, 1501.
8177. Wuerflinger, A., *Ber. Bunsenges. Phys. Chem.* **1980,** *84*, 653.
8178. Wulf, O. R., E. J. Jones, et al., *J. Chem. Phys.* **1940,** *8*, 753.
8179. Wulff, C. A., and E. F. Westrum, *J. Phys. Chem.* **1963,** *67*, 2376.

8180. Wurrey, C. J., W. E. Bucy, et al., *J. Phys. Chem.* **1976,** *80,* 1129.
8181. Wurtz, C. A., *Justus Liebigs Ann. Chem.* **1855,** *93,* 116.
8182. Wuyfs, H., and R. Billeux, *Bull. Acad. Roy. Belges* **1920,** *29,* 55; *Chem. Zentr.* **1920,** *I,* 817; *C.A.* **1922,** *16,* 2111.
8183. Wuyts, H., and A. Lacourt, *Bull. Soc. Chim. Belges* **1930,** *39,* 157.
8184. Wyman, J., *Phys. Rev.* **1930,** *35,* 623.
8185. Wymore, C. E., *Ind. Eng. Chem., Prod. Res. Devel.* **1962,** *1,* 173.
8186. Wynn, E. A., *Microchem. J.* **1961,** *5,* 175; *Chem. Zentr.* **1962,** 16789.
8187. Wynne, W. P., *J. Chem. Soc.* **1892,** 1042.
8188. Wyss, R., R. D. Werder, et al., *Spectrochim. Acta* **1964,** *20,* 573; *C.A.* **1964,** *61,* 5095.

Y

8189. Yabroff, D. L., *Ind. Eng. Chem.* **1940,** *32,* 257.
8190. Yada, S., T. Yamauchi, et al., *Shokubai* (Tokyo) **1963,** *5*(1), 2; *C.A.* **1963,** *59,* 8582.
8191. Yagudaev, M. R., and E. M. Popov, *Izv. Akad. Nauk SSSR, Ser. Khim.* **1964,** 1189; *C.A.* **1966,** *64,* 18705.
8192. Yakerson, V. I., L. I. Lafer, et al., *J. Chromatog.* **1966,** *23,* 67.
8193. Yale, H. L., U.S. Patent 3,381,042, Apr. 30, **1968;** *C.A.* **1968,** *69,* 26746.
8194. Yamaguchi, I., *Bull. Chem. Soc. Jpn.* **1961,** *34,* 353; *C.A.* **1961,** *55,* 22210.
8195. Yamaguchi, I., *Bull. Chem. Soc. Jpn.* **1961,** *34,* 1606.
8196. Yamaguchi, I., and N. Hayakawa, *Bull. Chem. Soc. Jpan.* **1960,** *33,* 1128.
8197. Yamamoto, A., and S. Kambara, *J. Amer. Chem. Soc.* **1959,** *81,* 2663.
8198. Yamamoto, K., Y. Hirosawa, et al., *Bull. Chem. Soc. Jpn.* **1954,** *27,* 386; *C.A.* **1956,** *50,* 159.
8199. Yamamoto, M., and S. Ohara, *J. Pharm. Soc. Jpn.* **1940,** *60,* 535; *C.A.* **1941,** *35,* 1773.
8200. Yamamoto, O., K. Hayamizu, et al., *Anal. Chem.* **1972,** *44,* 1794.
8201. Yamamoto, O., and K. Hayamizu, *Bull. Chem. Soc. Jpn.* **1965,** *38,* 537.
8202. Yamamoto, T., *Synth. Commun.* **1979,** *9,* 219.
8203. Yamamoto, Y., H. Toi, et al., *J. Amer. Chem. Soc.* **1976,** *98,* 1965.
8204. Yamashita, M., and T. Matsumura, *J. Chem. Soc. Jpn.* **1943,** *64,* 506; *C.A.* **1947,** *41,* 3753.
8205. Yarborough, V. A., *Anal. Chem.* **1953,** *25,* 1914.
8206. Yarborough, V. A., J. F. Haskin, et al., *Anal. Chem.* **1954,** *26,* 1576.
8207. Yaroslavskii, N. G., *J. Phys. Chem.* (USSR) **1948,** *22,* 265; *C.A.* **1948,** *42,* 5772.
8208. Yaroslavskii, N. G., and A. N. Terenin, *Doklady Akad. Nauk SSSR* **1949,** *66,* 885; *C.A.* **1949,** *43,* 7343.
8209. Yarym-Agaev, N. L., L. D. Afanasenko, et al., *Ukr. Khi. Zh.* (Russ. ed.) **1980,** *46,* 1331; *C.A.* **1981,** *94,* 53183.
8210. Yates, K., and R. Stewart, *Can. J. Chem.* **1959,** *37,* 664.
8211. Yates, W. R., and J. L. Ihrig, *J. Amer. Chem. Soc.* **1965,** *87,* 710.
8212. Yeh, P.-Y., C.-T. Chen, et al., *Formosan Sci.* **1953,** *7,* 27; *C.A.* **1955,** *49,* 3033.
8213. Yeo, A. N. H., and D. H. Williams, *J. Chem. Soc. D* **1970,** 886.
8214. Yerger, E. A., thesis, Northwestern University, Evanston, Illinois, 1956.
8215. Yew, F. F., R. J. Kurland, et al., *Anal. Chem.* **1964,** *36,* 843.
8216. Yoke, G. R., H. U. Louder, et al., *J. Chem. Ed.* **1933,** *10,* 374.
8217. Yokota, H., H. Murata, et al., *Chem. High Polymers* (Japan) **1951,** *8,* 227; *C.A.* **1953,** *47,* 1424.
8218. Yokoto, K., and Y. Ishii, *Kogyo Kagaku Zasshi* **1966,** *69,* 1083; *C.A.* **1966,** *65,* 12303.
8219. Yonemoto, T., W. F. Reynolds, et al., *Can. J. Chem.* **1965,** *43,* 2668.
8220. Yonezawa, T., and I. Morishima, *Bull. Chem. Soc. Jpn.* **1966,** *39,* 2346.
8221. Yonezawa, T., I. Morishima, et al., *Bull. Chem. Soc. Jpn.* **1966,** *39,* 1398.
8222. Yonezawa, T., and H. Saito, *Bull. Chem. Soc. Jpn.* **1965,** *38,* 1431.
8223. Yoshida, Z., E. Osawa, et al., *J. Phys. Chem.* **1964,** *68,* 2895.
8224. Yoshino, T., and H. J. Bernstein, *Can. J. Chem.* **1957,** *35,* 1957.
8225. Young, C. W., J. S. Koehler, et al., *J. Amer. Chem. Soc.* **1947,** *69,* 1410.
8226. Young, S., *J. Chem. Soc.* **1889,** 483.
8227. Young, S., *J. Chem. Soc.* **1897,** 446.
8228. Young, S., *J. Chem. Soc.* **1898,** 675.
8229. Young, S., *J. Chem. Soc.* **1899,** 172.

8230. Young, S., *J. Chem. Soc.* **1902**, 707.
8231. Young, S., *Sci. Proc. Roy. Dublin Soc.* **1909–1910**, *12*, 274; *C.A.* **1911**, *5*, 406.
8232. Young, S., *Z. Phys. Chem.* (Leipzig) **1899**, *29*, 193.
8233. Young, S., *Z. Phys. Chem.* (Leipzig) **1910**, *70*, 620.
8234. Young, S., and E. C. Fortey, *J. Chem. Soc.* **1899**, 873.
8235. Young, S., and E. Fortney, *J. Chem. Soc.* **1903**, 45.
8236. Young, S., and G. L. Thomas, *J. Chem. Soc.* **1893**, 1191.
8237. Young, W., *J. Chem. Soc. Ind.* **1933**, *52*, 449.
8238. Young, W. G., and L. J. Andrews, *J. Amer. Chem. Soc.* **1944**, *66*, 421.
8239. Young, W. G., W. H. Hartung, et al., *J. Amer. Chem. Soc.* **1936**, *58*, 100.
8240. Ypenburg, J. W., and H. Gerding, *Rec. Trav. Chim. Pays-Bas* **1971**, *90*, 885.
8241. Yuan, H.-S., and Tseng, C.-Y., *Chlorine and Alkali News* (Formosa) **1953**, No. 12, 42; *C.A.* **1955**, *49*, 6819.
8242. Yukawa, Y., *Handbook of Organic Structural Analysis*, W. A. Benjamin, New York, 1965.
8243. Yuki, H., K. Hatada, et al., *Bull. Chem. Soc. Jpn.* **1969**, *42*, 3546.
8244. Yura, S., and R. Oda, *J. Soc. Chem. Ind. Jpn.* **1943**, *46*, 82; *C.A.* **1948**, *42*, 6339.
8245. Yur'ev, Y. K., and L. I. Khmal'nitskii, *Zh. Obshch. Khim.* **1953**, *23*, 1725; *C.A.* **1954**, *58*, 13680.
8246. Yvernault, T., *Aleagineux* **1946**, *1*, 189; *C.A.* **1947**, *41*, 3697.

Z

8247. Zabetakis, M. G., *Flammable Characteristics of Combustible Gases and Vapors*, Bulletin 627, Bureau of Mines, Washington, 1965.
8248. Zabetakis, M. G., G. S. Scott, et al., *Ind. Eng. Chem.* **1951**, *43*, 2120.
8249. Zaeva, G. N., K. L. Vinogradova, et al., *Toksikol. Novykh Prom. Khim. Veshchesty* **1967**, *9*, 163; *C.A.* **1969**, *70*, 18516.
8250. Zafiriadis, Z., and P. Mastagli, *Compt. Rend.* **1954**, *238*, 821; *C.A.* **1955**, *49*, 3779.
8251. Zagradnik, R., M. Chvapil, et al., *Farmakol. i Toksikol.* **1962**, *25*, 618; *C.A.* **1963**, *58*, 11886.
8252. Zahn, C. T., *Phys. Rev.* **1931**, *37*, 1516; *C.A.* **1931**, *25*, 5320.
8253. Zahn, C. T., *Phys. Rev.* **1932**, *40*, 291.
8254. Zahn, C. T., *Physik. Z.* **1932**, *33*, 525; *C.A.* **1932**, *26*, 5803.
8255. Zahn, C. T., *Physik. Z.* **1933**, *34*, 570; *Chem. Zentr.* **1933**, *II*, 2647.
8256. Zakharkin, L. I., and I. M. Khorlina, *Dokl. Akad. Nauk SSSR* **1957**, *116*, 422; *C.A.* **1958**, *52*, 8040.
8257. Zakharkin, L. I., O. Y. Okhlobystin, et al., *Tetrahedron* **1965**, *21*, 881.
8258. Zakharova, A. I., *J. Gen. Chem.* (USSR) **1945**, *15*, 429; *C.A.* **1946**, *40*, 4655.
8259. Zal'kind, Y. S., and I. F. Markov, *J. Appl. Chem.* (USSR) **1939**, *12*, 437; *C.A.* **1939**, *33*, 6809.
8260. Zaoral, M., *Chem. Listy* **1953**, *47*, 1872; *C.A.* **1955**, *49*, 966.
8261. Zapletálek, A., *Collect. Czech. Chem. Commun.* **1939**, *11*, 28; *C.A.* **1939**, *33*, 4550.
8262. Zappi, E. V., and H. Degiorgi, *Anal. Asoc. Quim. Argentina* **1930**, *18*, 210; *Chem. Zentr.* **1931**, *II*, 1121.
8263. Zaugg, H. E., B. W. Horrom, et al., *J. Amer. Chem. Soc.* **1960**, *82*, 2895.
8264. Zaugg, H. E., and A. D. Schaefer, *Anal. Chem.* **1964**, *36*, 2121.
8265. Zaugg, H. E., and A. D. Schaefer, *J. Amer. Chem. Soc.* **1965**, *87*, 1857.
8266. Zavitsas, A. A., *J. Chem. Eng. Data* **1967**, *12*, 94.
8267. Zawisza, A., and J. Vejrosta, *J. Chem. Thermodyn.* **1982**, *14*, 239.
8268. Zecherl, M. K., *Mikrochim. Acta* **1948**, *33*, 387; *C.A.* **1948**, *42*, 6537.
8269. Zegers, H. C., and G. Somsen, *J. Chem. Thermodyn.* **1984**, *16*, 225.
8270. Zeiss, H. H., and M. Tsutsui, *J. Amer. Chem. Soc.* **1953**, *75*, 897.
8271. Zelinski, R., and H. J. Eichel, *J. Org. Chem.* **1958**, *23*, 462.
8272. Zelinskii, N. D., and A. Chuksanov, *Ber. Deut. Chem. Gesell.* **1924**, *57B*, 42; *C.A.* **1924**, *18*, 2148.
8273. Zelinsky, N., *Ber. Deut. Chem. Gesell.* **1901**, *34*, 2877.
8274. Zelinsky, N., and N. Rosanoff, *Z. Phys. Chem.* (Leipzig) **1912**, *78*, 629.
8275. Zelinsky, N. D., and M. B. Turowa-Pollak, *Ber. Deut. Chem. Gesell.* **1932**, *65*, 1299.
8276. Zellhoefer, G. F., M. J. Copley, et al., *J. Amer. Chem. Soc.* **1938**, *60*, 1337.
8277. Zellner, R. J., and C. A. R. Johnson, U.S. Patent 3,031,511, Apr. 24, 1962; *C.A.* **1962**, *57*, 13728.
8278. Zemany, P. D., and F. P. Price, *J. Amer. Chem. Soc.* **1948**, *70*, 4222.

8279. Zepalova-Mikhailova, L. A., *Trans. Int. Pure Chem. Reagents* (USSR) **1939**, No. 16, 51; *Khim. Referat. Zhur.* **1939**, No. 6, 70; *C.A.* **1938**, *32*, 5288; *C.A.* **1940**, *34*, 3960.

8280. Zerewitinoff, T., *Z. Anal. Chem.* **1911**, *50*, 683; *Ber. Deut. Chem. Gesell.* **1914**, *47*, 2417.

8281. Zerweck, W., H. Ritter, et al., German Patent 855,107, Oct. 11, 1952; *Chem. Zentr.* **1953**, *I*, 2353.

8282. Zetta, L., and G. Gatti, *Org. Magn. Resonance* **1972**, *4*, 585.

8283. Zetzsche, F., and P. Zala, *Helv. Chim. Acta* **1926**, *9*, 288; *C.A.* **1926**, *20*, 2996.

8284. Zhdanov, A. K., *J. Gen. Chem.* (USSR) **1941**, *11*, No. 7, 471; *C.A.* **1941**, *35*, 7275.

8285. Zhizhin, G. N., Z. B. Barinova, et al., *Izv. Akad. Nauk SSSR, Ser. Fiz* **1962**, *26*, 1263; *C.A.* **1963**, *58*, 7518.

8286. Zhumaev, T., and M. D. Adamenkova, *Zh. Fiz. Khim.* **1983**, *57*, 773; *Russ. J. Phys. Chem.* **1983**, *57*, 474.

8287. Zhumaev, T., and M. D. Adamenkova, *Zh. Fiz. Khim.* **1983**, *57*, 775; *Russ. J. Phys. Chem.* **1983**, *57*, 475.

8288. Zhuravlev, D. I., *J. Phys. Chem.* (USSR) **1937**, *9*, 875; *C.A.* **1937**, *31*, 8287.

8289. Ziebell, G., German (East) Patent 26,288, Oct. 26, 1963; *C.A.* **1964**, *60*, 15734.

8290. Zief, M., and W. R. Wilcox, eds., *Fractional Solidification, Vol. 1*, Dekker, New York, 1967.

8291. Ziegler, R., *Phys. Z.* **1940**, *116*, 716; *C.A.* **1941**, *35*, 4682.

8292. Ziemecki, S., *Phys. Z.* **1932**, *78*, 123; *C.A.* **1933**, *27*, 902.

8293. Zietlow, J. P., F. F. Cleveland, et al., *J. Chem. Phys.* **1950**, *18*, 1076.

8294. Zikalov, P., A. Astrug, et al., *Talanta* **1975**, *22*, 511.

8295. Zilberman, E. N., *Zh. Prikl. Khim.* **1951**, *24*, 776; *C.A.* **1953**, *47*, 4172.

8296. Zil'berman-Granovskaya, A. A., *J. Phys. Chem.* (USSR) **1940**, *14*, 759; *C.A.* **1941**, *35*, 3867.

8297. Zil'berman-Granovskaya, A. A., *J. Phys. Chem.* (USSR) **1940**, *14*, 768; *C.A.* **1941**, *35*, 3867.

8298. Zil'berman-Granovskaya, A. A., *J. Phys. Chem.* (USSR) **1940**, *14*, 1004; *C.A.* **1941**, *35*, 3867.

8299. Zimakov, P. V., and V. A. Sokolova, *Zh. Fiz. Khim.* **1953**, *27*, 1079; *C.A.* **1955**, *49*, 3597.

8300. Zimm, B. H., *J. Phys. Colloid. Chem.* **1950**, *54*, 1306.

8301. Zincke, T., *Liebigs Annalen Chemie* **1869**, *152*, 5.

8302. Zirnitis, V. A., and M. M. Suschinskii, *Opt. Spektrosk. Akad. Nauk SSSR, Otd. Fiz.-Mat. Nauk, Sb. Statei* **1963**, *2*, 153; *C.A.* **1963**, *59*, 13500.

8303. Zirnitis, V. A., and M. M. Suschinskii, *Optika i Spektroskopiya* **1963**, *15*, 190; *C.A.* **1963**, *59*, 13502.

8304. Zmaczynski, A., *J. Chim. Phys.* **1930**, *27*, 503; *Brit. Abstr.* **1931A**, 156.

8305. Zmaczynski, M. A., *Svensk Kem. Tidskr.* **1936**, *48*, 268; *Chem. Zentr.* **1937**, *I*, 2762, 4994; *Rocz. Chem.* **1936**, *16*, 486.

8306. Zobel, R., and A. B. F. Duncan, *J. Amer. Chem. Soc.* **1955**, *77*, 2611.

8307. Zolotov, P. A., and O. A. Svintukhovskii, *Gig. Sanit.* **1972**, *37*, 104; *C.A.* **1972**, *77*, 71003.

8308. Zook, M. D., W. L. Kelly, et al., *J. Org. Chem.* **1968**, *33*, 3477.

8309. Zordan, T. A., D. G. Hurkot, et al., *Thermochimica Acta* **1972**, *5*, 21; *C.A.* **1973**, *78*, 34606.

8310. Zozulya, A. P., and E. V. Novikova, *Zh. Analit. Khim.* **1963**, *18*, 1105; *C.A.* **1964**, *60*, 1119.

8311. Zul'fugarzade, K. E., Kh. T. Turakulov, et al., *Zh. Fiz. Khim.* **1974**, *48*, 2503; *Russ. J. Phys. Chem.* **1974**, *48*, 1479.

8312. Zumwalt, L. R., and R. M. Badger, *J. Amer. Chem. Soc.* **1940**, *62*, 305.

8313. Zweig, A., J. E. Lehnsen, et al., *J. Amer. Chem. Soc.* **1963**, *85*, 3933.

8314. Zweig, A., J. E. Lehnsen, et al., *J. Amer. Chem. Soc.* **1963**, *85*, 3940.

8315. Zwolenik, J. J., and R. M. Fuoss, *J. Phys. Chem.* **1964**, *68*, 903.

8316. Zwolinski, B. J., and R. C. Wilhoit, *Heats of Formation and Heats of Combustion*, Thermodynamics Research Center, College Station, Texas, 1968.

8317. Zwolinski, B. J., and R. C. Wilhoit, *Handbook of Vapor Pressures and Heats of Vaporization of Hydrocarbons and Related Compounds*, Thermodynamics Research Center, College Station, Texas, 1971.

INDEX

The number immediately following the name of the solvent is its code number and is in (parentheses). The code number is followed by a colon. The first number following the colon refers to the page number of the table of physical properties of the subject solvent in Chapter III. The second number refers to the methods of preparation, purification, criteria of purity, and safety in Chapter V. Other pertinent information is listed as subentries.
Example:

2-Methylpentane, (11): 94, 826
    special grades available, 809
    viscosity equation, 32, 33

Synonyms listed in Chapter I for the solvents are included in the index to facilitate the use of the book. They are followed, in order, by the abbreviation "syn," the name of the solvent, the page number of the table of physical properties, and the page number of preparation, purification, etc. Example: Allyl alcohol, syn., 2-propen-1-ol, 255, 908. Complete indexing will be found under the primary name of the solvent.

1296   INDEX